PHYSICS
SEVENTH EDITION

PRINCIPLES WITH APPLICATIONS

DOUGLAS C. GIANCOLI

PEARSON

Boston   Columbus   Indianapolis   New York   San Francisco   Upper Saddle River
Amsterdam   Cape Town   Dubai   London   Madrid   Milan   Munich   Paris   Montréal   Toronto
Delhi   Mexico City   São Paulo   Sydney   Hong Kong   Seoul   Singapore   Taipei   Tokyo

President, Science, Business and Technology: Paul Corey
Publisher: Jim Smith
Executive Development Editor: Karen Karlin
Production Project Manager: Elisa Mandelbaum / Laura Ross
Marketing Manager: Will Moore
Senior Managing Editor: Corinne Benson
Managing Development Editor: Cathy Murphy
Copyeditor: Joanna Dinsmore
Proofreaders: Susan Fisher, Donna Young
Interior Designer: Mark Ong
Cover Designer: Derek Bacchus
Photo Permissions Management: Maya Melenchuk
Photo Research Manager: Eric Schrader
Photo Researcher: Mary Teresa Giancoli
Senior Administrative Assistant: Cathy Glenn
Senior Administrative Coordinator: Trisha Tarricone
Text Permissions Project Manager: Joseph Croscup
Editorial Media Producer: Kelly Reed
Manufacturing Buyer: Jeffrey Sargent
Indexer: Carol Reitz
Compositor: Preparé, Inc.
Illustrations: Precision Graphics

Cover Photo Credit: North Peak, California (D. Giancoli); Insets: left, analog to digital (page 488); right, electron microscope image—retina of human eye with cones artificially colored green, rods beige (page 785).
Back Cover Photo Credit: D. Giancoli

Credits and acknowledgments for materials borrowed from other sources and reproduced, with permission, in this textbook appear on page A-69.

Copyright © 2014, 2005, 1998, 1995, 1991, 1985, 1980 by Douglas C. Giancoli
Published by Pearson Education, Inc. All rights reserved. Manufactured in the United States of America. This publication is protected by Copyright, and permission should be obtained from the publisher prior to any prohibited reproduction, storage in a retrieval system, or transmission in any form or by any means, electronic, mechanical, photocopying, recording, or likewise. To obtain permission(s) to use material from this work, please submit a written request to Pearson Education, Inc., Permissions Department, 1900 E. Lake Ave., Glenview, IL 60025. For information regarding permissions, call (847) 486-2635.

Pearson Prentice Hall is a trademark, in the U.S. and/or other countries, of Pearson Education, Inc. or its affiliates.

Library of Congress Cataloging-in-Publication Data on file

ISBN-10: 0-321-62592-7
ISBN-13: 978-0-321-62592-2
ISBN-10: 0-321-86911-7: ISBN-13: 978-0-321-86911-1 (Books a la Carte editon)
ISBN-10: 0-321-76791-8: ISBN-13: 978-0-321-76791-2 (Instructor Review Copy)

10 2019

www.pearsonhighered.com

Contents

Applications List — x
Preface — xiii
To Students — xviii
Use of Color — xix

1 INTRODUCTION, MEASUREMENT, ESTIMATING — 1

1–1 The Nature of Science — 2
1–2 Physics and its Relation to Other Fields — 4
1–3 Models, Theories, and Laws — 5
1–4 Measurement and Uncertainty; Significant Figures — 5
1–5 Units, Standards, and the SI System — 8
1–6 Converting Units — 11
1–7 Order of Magnitude: Rapid Estimating — 13
*1–8 Dimensions and Dimensional Analysis — 16
Questions, MisConceptual Questions 17
Problems, Search and Learn 18–20

2 DESCRIBING MOTION: KINEMATICS IN ONE DIMENSION — 21

2–1 Reference Frames and Displacement — 22
2–2 Average Velocity — 23
2–3 Instantaneous Velocity — 25
2–4 Acceleration — 26
2–5 Motion at Constant Acceleration — 28
2–6 Solving Problems — 30
2–7 Freely Falling Objects — 33
2–8 Graphical Analysis of Linear Motion — 39
Questions, MisConceptual Questions 41–42
Problems, Search and Learn 43–48

3 KINEMATICS IN TWO DIMENSIONS; VECTORS — 49

3–1 Vectors and Scalars — 50
3–2 Addition of Vectors—Graphical Methods — 50
3–3 Subtraction of Vectors, and Multiplication of a Vector by a Scalar — 52
3–4 Adding Vectors by Components — 53
3–5 Projectile Motion — 58
3–6 Solving Projectile Motion Problems — 60
*3–7 Projectile Motion Is Parabolic — 64
3–8 Relative Velocity — 65
Questions, MisConceptual Questions 67–68
Problems, Search and Learn 68–74

4 DYNAMICS: NEWTON'S LAWS OF MOTION — 75

4–1 Force — 76
4–2 Newton's First Law of Motion — 76
4–3 Mass — 78
4–4 Newton's Second Law of Motion — 78
4–5 Newton's Third Law of Motion — 81
4–6 Weight—the Force of Gravity; and the Normal Force — 84
4–7 Solving Problems with Newton's Laws: Free-Body Diagrams — 87
4–8 Problems Involving Friction, Inclines — 93
Questions, MisConceptual Questions 98–100
Problems, Search and Learn 101–8

5 CIRCULAR MOTION; GRAVITATION — 109

5–1 Kinematics of Uniform Circular Motion — 110
5–2 Dynamics of Uniform Circular Motion — 112
5–3 Highway Curves: Banked and Unbanked — 115
*5–4 Nonuniform Circular Motion — 118
5–5 Newton's Law of Universal Gravitation — 119
5–6 Gravity Near the Earth's Surface — 121
5–7 Satellites and "Weightlessness" — 122
5–8 Planets, Kepler's Laws, and Newton's Synthesis — 125
5–9 Moon Rises an Hour Later Each Day — 129
5–10 Types of Forces in Nature — 129
Questions, MisConceptual Questions 130–32
Problems, Search and Learn 132–37

6 Work and Energy — 138

6–1 Work Done by a Constant Force — 139
*6–2 Work Done by a Varying Force — 142
6–3 Kinetic Energy, and the Work-Energy Principle — 142
6–4 Potential Energy — 145
6–5 Conservative and Nonconservative Forces — 149
6–6 Mechanical Energy and Its Conservation — 150
6–7 Problem Solving Using Conservation of Mechanical Energy — 151
6–8 Other Forms of Energy and Energy Transformations; The Law of Conservation of Energy — 155
6–9 Energy Conservation with Dissipative Forces: Solving Problems — 156
6–10 Power — 159
Questions, MisConceptual Questions 161–63
Problems, Search and Learn 164–69

7 Linear Momentum — 170

7–1 Momentum and Its Relation to Force — 171
7–2 Conservation of Momentum — 173
7–3 Collisions and Impulse — 176
7–4 Conservation of Energy and Momentum in Collisions — 177
7–5 Elastic Collisions in One Dimension — 178
7–6 Inelastic Collisions — 180
*7–7 Collisions in Two Dimensions — 182
7–8 Center of Mass (CM) — 184
*7–9 CM for the Human Body — 186
*7–10 CM and Translational Motion — 187
Questions, MisConceptual Questions 190–91
Problems, Search and Learn 192–97

8 Rotational Motion — 198

8–1 Angular Quantities — 199
8–2 Constant Angular Acceleration — 203
8–3 Rolling Motion (Without Slipping) — 204
8–4 Torque — 206
8–5 Rotational Dynamics; Torque and Rotational Inertia — 208
8–6 Solving Problems in Rotational Dynamics — 210
8–7 Rotational Kinetic Energy — 212
8–8 Angular Momentum and Its Conservation — 215
*8–9 Vector Nature of Angular Quantities — 217
Questions, MisConceptual Questions 220–21
Problems, Search and Learn 222–29

9 Static Equilibrium; Elasticity and Fracture — 230

9–1 The Conditions for Equilibrium — 231
9–2 Solving Statics Problems — 233
9–3 Applications to Muscles and Joints — 238
9–4 Stability and Balance — 240
9–5 Elasticity; Stress and Strain — 241
9–6 Fracture — 245
*9–7 Spanning a Space: Arches and Domes — 246
Questions, MisConceptual Questions 250–51
Problems, Search and Learn 252–59

10 Fluids — 260

10–1 Phases of Matter — 261
10–2 Density and Specific Gravity — 261
10–3 Pressure in Fluids — 262
10–4 Atmospheric Pressure and Gauge Pressure — 264
10–5 Pascal's Principle — 265
10–6 Measurement of Pressure; Gauges and the Barometer — 266
10–7 Buoyancy and Archimedes' Principle — 268
10–8 Fluids in Motion; Flow Rate and the Equation of Continuity — 272
10–9 Bernoulli's Equation — 274
10–10 Applications of Bernoulli's Principle: Torricelli, Airplanes, Baseballs, Blood Flow — 276
*10–11 Viscosity — 279
*10–12 Flow in Tubes: Poiseuille's Equation, Blood Flow — 279
*10–13 Surface Tension and Capillarity — 280
*10–14 Pumps, and the Heart — 282
Questions, MisConceptual Questions 283–85
Problems, Search and Learn 285–91

11 Oscillations and Waves 292

- 11–1 Simple Harmonic Motion—Spring Oscillations 293
- 11–2 Energy in Simple Harmonic Motion 295
- 11–3 The Period and Sinusoidal Nature of SHM 298
- 11–4 The Simple Pendulum 301
- 11–5 Damped Harmonic Motion 303
- 11–6 Forced Oscillations; Resonance 304
- 11–7 Wave Motion 305
- 11–8 Types of Waves and Their Speeds: Transverse and Longitudinal 307
- 11–9 Energy Transported by Waves 310
- 11–10 Reflection and Transmission of Waves 312
- 11–11 Interference; Principle of Superposition 313
- 11–12 Standing Waves; Resonance 315
- *11–13 Refraction 317
- *11–14 Diffraction 318
- *11–15 Mathematical Representation of a Traveling Wave 319

Questions, MisConceptual Questions 320–22
Problems, Search and Learn 322–27

12 Sound 328

- 12–1 Characteristics of Sound 329
- 12–2 Intensity of Sound: Decibels 331
- *12–3 The Ear and Its Response; Loudness 334
- 12–4 Sources of Sound: Vibrating Strings and Air Columns 335
- *12–5 Quality of Sound, and Noise; Superposition 340
- 12–6 Interference of Sound Waves; Beats 341
- 12–7 Doppler Effect 344
- *12–8 Shock Waves and the Sonic Boom 348
- *12–9 Applications: Sonar, Ultrasound, and Medical Imaging 349

Questions, MisConceptual Questions 352–53
Problems, Search and Learn 354–58

13 Temperature and Kinetic Theory 359

- 13–1 Atomic Theory of Matter 359
- 13–2 Temperature and Thermometers 361
- 13–3 Thermal Equilibrium and the Zeroth Law of Thermodynamics 363
- 13–4 Thermal Expansion 364
- 13–5 The Gas Laws and Absolute Temperature 367
- 13–6 The Ideal Gas Law 369
- 13–7 Problem Solving with the Ideal Gas Law 370
- 13–8 Ideal Gas Law in Terms of Molecules: Avogadro's Number 372
- 13–9 Kinetic Theory and the Molecular Interpretation of Temperature 373
- 13–10 Distribution of Molecular Speeds 376
- 13–11 Real Gases and Changes of Phase 377
- 13–12 Vapor Pressure and Humidity 379
- *13–13 Diffusion 381

Questions, MisConceptual Questions 384–85
Problems, Search and Learn 385–89

14 Heat 390

- 14–1 Heat as Energy Transfer 391
- 14–2 Internal Energy 392
- 14–3 Specific Heat 393
- 14–4 Calorimetry—Solving Problems 394
- 14–5 Latent Heat 397
- 14–6 Heat Transfer: Conduction 400
- 14–7 Heat Transfer: Convection 402
- 14–8 Heat Transfer: Radiation 403

Questions, MisConceptual Questions 406–8
Problems, Search and Learn 408–11

15 The Laws of Thermodynamics 412

- 15–1 The First Law of Thermodynamics 413
- 15–2 Thermodynamic Processes and the First Law 414
- *15–3 Human Metabolism and the First Law 418
- 15–4 The Second Law of Thermodynamics—Introduction 419
- 15–5 Heat Engines 420
- 15–6 Refrigerators, Air Conditioners, and Heat Pumps 425
- 15–7 Entropy and the Second Law of Thermodynamics 428
- 15–8 Order to Disorder 430
- 15–9 Unavailability of Energy; Heat Death 431
- *15–10 Statistical Interpretation of Entropy and the Second Law 432
- *15–11 Thermal Pollution, Global Warming, and Energy Resources 434

Questions, MisConceptual Questions 437–38
Problems, Search and Learn 438–42

16 Electric Charge and Electric Field 443

- 16–1 Static Electricity; Electric Charge and Its Conservation — 444
- 16–2 Electric Charge in the Atom — 445
- 16–3 Insulators and Conductors — 445
- 16–4 Induced Charge; the Electroscope — 446
- 16–5 Coulomb's Law — 447
- 16–6 Solving Problems Involving Coulomb's Law and Vectors — 450
- 16–7 The Electric Field — 453
- 16–8 Electric Field Lines — 457
- 16–9 Electric Fields and Conductors — 459
- *16–10 Electric Forces in Molecular Biology: DNA Structure and Replication — 460
- *16–11 Photocopy Machines and Computer Printers Use Electrostatics — 462
- *16–12 Gauss's Law — 463

Questions, MisConceptual Questions 467–68
Problems, Search and Learn 469–72

17 Electric Potential 473

- 17–1 Electric Potential Energy and Potential Difference — 474
- 17–2 Relation between Electric Potential and Electric Field — 477
- 17–3 Equipotential Lines and Surfaces — 478
- 17–4 The Electron Volt, a Unit of Energy — 478
- 17–5 Electric Potential Due to Point Charges — 479
- *17–6 Potential Due to Electric Dipole; Dipole Moment — 482
- 17–7 Capacitance — 482
- 17–8 Dielectrics — 485
- 17–9 Storage of Electric Energy — 486
- 17–10 Digital; Binary Numbers; Signal Voltage — 488
- *17–11 TV and Computer Monitors: CRTs, Flat Screens — 490
- *17–12 Electrocardiogram (ECG or EKG) — 493

Questions, MisConceptual Questions 494–95
Problems, Search and Learn 496–500

18 Electric Currents 501

- 18–1 The Electric Battery — 502
- 18–2 Electric Current — 504
- 18–3 Ohm's Law: Resistance and Resistors — 505
- 18–4 Resistivity — 508
- 18–5 Electric Power — 510
- 18–6 Power in Household Circuits — 512
- 18–7 Alternating Current — 514
- *18–8 Microscopic View of Electric Current — 516
- *18–9 Superconductivity — 517
- *18–10 Electrical Conduction in the Human Nervous System — 517

Questions, MisConceptual Questions 520–21
Problems, Search and Learn 521–25

19 DC Circuits 526

- 19–1 EMF and Terminal Voltage — 527
- 19–2 Resistors in Series and in Parallel — 528
- 19–3 Kirchhoff's Rules — 532
- 19–4 EMFs in Series and in Parallel; Charging a Battery — 536
- 19–5 Circuits Containing Capacitors in Series and in Parallel — 538
- 19–6 RC Circuits—Resistor and Capacitor in Series — 539
- 19–7 Electric Hazards — 543
- 19–8 Ammeters and Voltmeters—Measurement Affects the Quantity Being Measured — 546

Questions, MisConceptual Questions 549–51
Problems, Search and Learn 552–59

20 Magnetism 560

- 20–1 Magnets and Magnetic Fields — 560
- 20–2 Electric Currents Produce Magnetic Fields — 563
- 20–3 Force on an Electric Current in a Magnetic Field; Definition of $\vec{B}$ — 564
- 20–4 Force on an Electric Charge Moving in a Magnetic Field — 566
- 20–5 Magnetic Field Due to a Long Straight Wire — 570
- 20–6 Force between Two Parallel Wires — 571
- 20–7 Solenoids and Electromagnets — 572
- 20–8 Ampère's Law — 573
- 20–9 Torque on a Current Loop; Magnetic Moment — 575
- 20–10 Applications: Motors, Loudspeakers, Galvanometers — 576
- *20–11 Mass Spectrometer — 578
- *20–12 Ferromagnetism: Domains and Hysteresis — 579

Questions, MisConceptual Questions 581–83
Problems, Search and Learn 583–89

21 ELECTROMAGNETIC INDUCTION AND FARADAY'S LAW 590

- 21–1 Induced EMF 591
- 21–2 Faraday's Law of Induction; Lenz's Law 592
- 21–3 EMF Induced in a Moving Conductor 596
- 21–4 Changing Magnetic Flux Produces an Electric Field 597
- 21–5 Electric Generators 597
- 21–6 Back EMF and Counter Torque; Eddy Currents 599
- 21–7 Transformers and Transmission of Power 601
- *21–8 Information Storage: Magnetic and Semiconductor; Tape, Hard Drive, RAM 604
- *21–9 Applications of Induction: Microphone, Seismograph, GFCI 606
- *21–10 Inductance 608
- *21–11 Energy Stored in a Magnetic Field 610
- *21–12 *LR* Circuit 610
- *21–13 AC Circuits and Reactance 611
- *21–14 *LRC* Series AC Circuit 614
- *21–15 Resonance in AC Circuits 616

Questions, MisConceptual Questions 617–19
Problems, Search and Learn 620–24

22 ELECTROMAGNETIC WAVES 625

- 22–1 Changing Electric Fields Produce Magnetic Fields; Maxwell's Equations 626
- 22–2 Production of Electromagnetic Waves 627
- 22–3 Light as an Electromagnetic Wave and the Electromagnetic Spectrum 629
- 22–4 Measuring the Speed of Light 632
- 22–5 Energy in EM Waves 633
- 22–6 Momentum Transfer and Radiation Pressure 635
- 22–7 Radio and Television; Wireless Communication 636

Questions, MisConceptual Questions 640
Problems, Search and Learn 641–43

23 LIGHT: GEOMETRIC OPTICS 644

- 23–1 The Ray Model of Light 645
- 23–2 Reflection; Image Formation by a Plane Mirror 645
- 23–3 Formation of Images by Spherical Mirrors 649
- 23–4 Index of Refraction 656
- 23–5 Refraction: Snell's Law 657
- 23–6 Total Internal Reflection; Fiber Optics 659
- 23–7 Thin Lenses; Ray Tracing 661
- 23–8 The Thin Lens Equation 664
- *23–9 Combinations of Lenses 668
- *23–10 Lensmaker's Equation 670

Questions, MisConceptual Questions 671–73
Problems, Search and Learn 673–78

24 THE WAVE NATURE OF LIGHT 679

- 24–1 Waves vs. Particles; Huygens' Principle and Diffraction 680
- *24–2 Huygens' Principle and the Law of Refraction 681
- 24–3 Interference—Young's Double-Slit Experiment 682
- 24–4 The Visible Spectrum and Dispersion 685
- 24–5 Diffraction by a Single Slit or Disk 687
- 24–6 Diffraction Grating 690
- 24–7 The Spectrometer and Spectroscopy 692
- 24–8 Interference in Thin Films 693
- *24–9 Michelson Interferometer 698
- 24–10 Polarization 699
- *24–11 Liquid Crystal Displays (LCD) 703
- *24–12 Scattering of Light by the Atmosphere 704

Questions, MisConceptual Questions 705–7
Problems, Search and Learn 707–12

25 OPTICAL INSTRUMENTS 713

- 25–1 Cameras: Film and Digital 713
- 25–2 The Human Eye; Corrective Lenses 719
- 25–3 Magnifying Glass 722
- 25–4 Telescopes 723
- 25–5 Compound Microscope 726
- 25–6 Aberrations of Lenses and Mirrors 727
- 25–7 Limits of Resolution; Circular Apertures 728
- 25–8 Resolution of Telescopes and Microscopes; the λ Limit 730
- 25–9 Resolution of the Human Eye and Useful Magnification 732
- *25–10 Specialty Microscopes and Contrast 733
- 25–11 X-Rays and X-Ray Diffraction 733
- *25–12 X-Ray Imaging and Computed Tomography (CT Scan) 735

Questions, MisConceptual Questions 738–39
Problems, Search and Learn 740–43

26 THE SPECIAL THEORY OF RELATIVITY 744

- 26–1 Galilean–Newtonian Relativity 745
- 26–2 Postulates of the Special Theory of Relativity 748
- 26–3 Simultaneity 749
- 26–4 Time Dilation and the Twin Paradox 750
- 26–5 Length Contraction 756
- 26–6 Four-Dimensional Space–Time 758
- 26–7 Relativistic Momentum 759
- 26–8 The Ultimate Speed 760
- 26–9 $E = mc^2$; Mass and Energy 760
- 26–10 Relativistic Addition of Velocities 764
- 26–11 The Impact of Special Relativity 765
  Questions, MisConceptual Questions 766–67
  Problems, Search and Learn 767–70

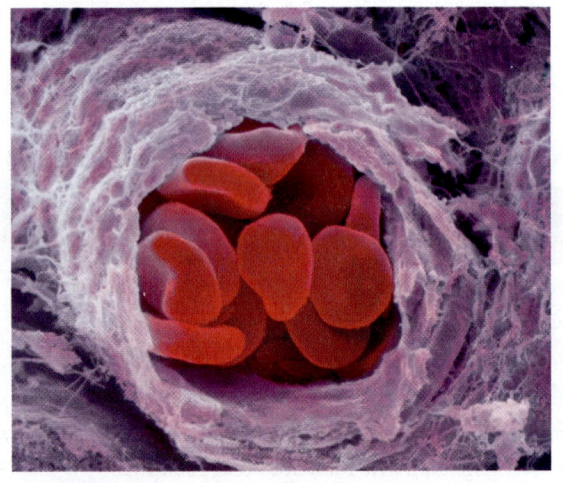

27 EARLY QUANTUM THEORY AND MODELS OF THE ATOM 771

- 27–1 Discovery and Properties of the Electron 772
- 27–2 Blackbody Radiation; Planck's Quantum Hypothesis 774
- 27–3 Photon Theory of Light and the Photoelectric Effect 775
- 27–4 Energy, Mass, and Momentum of a Photon 779
- *27–5 Compton Effect 780
- 27–6 Photon Interactions; Pair Production 781
- 27–7 Wave–Particle Duality; the Principle of Complementarity 782
- 27–8 Wave Nature of Matter 782
- 27–9 Electron Microscopes 785
- 27–10 Early Models of the Atom 786
- 27–11 Atomic Spectra: Key to the Structure of the Atom 787
- 27–12 The Bohr Model 789
- 27–13 de Broglie's Hypothesis Applied to Atoms 795
  Questions, MisConceptual Questions 797–98
  Problems, Search and Learn 799–802

28 QUANTUM MECHANICS OF ATOMS 803

- 28–1 Quantum Mechanics—A New Theory 804
- 28–2 The Wave Function and Its Interpretation; the Double-Slit Experiment 804
- 28–3 The Heisenberg Uncertainty Principle 806
- 28–4 Philosophic Implications; Probability versus Determinism 810
- 28–5 Quantum-Mechanical View of Atoms 811
- 28–6 Quantum Mechanics of the Hydrogen Atom; Quantum Numbers 812
- 28–7 Multielectron Atoms; the Exclusion Principle 815
- 28–8 The Periodic Table of Elements 816
- *28–9 X-Ray Spectra and Atomic Number 817
- *28–10 Fluorescence and Phosphorescence 820
- 28–11 Lasers 820
- *28–12 Holography 823
  Questions, MisConceptual Questions 825–26
  Problems, Search and Learn 826–28

29 MOLECULES AND SOLIDS 829

- *29–1 Bonding in Molecules 829
- *29–2 Potential-Energy Diagrams for Molecules 832
- *29–3 Weak (van der Waals) Bonds 834
- *29–4 Molecular Spectra 837
- *29–5 Bonding in Solids 840
- *29–6 Free-Electron Theory of Metals; Fermi Energy 841
- *29–7 Band Theory of Solids 842
- *29–8 Semiconductors and Doping 844
- *29–9 Semiconductor Diodes, LEDs, OLEDs 845
- *29–10 Transistors: Bipolar and MOSFETs 850
- *29–11 Integrated Circuits, 22-nm Technology 851
  Questions, MisConceptual Questions 852–53
  Problems, Search and Learn 854–56

30 NUCLEAR PHYSICS AND RADIOACTIVITY 857

- 30–1 Structure and Properties of the Nucleus 858
- 30–2 Binding Energy and Nuclear Forces 860
- 30–3 Radioactivity 863
- 30–4 Alpha Decay 864
- 30–5 Beta Decay 866
- 30–6 Gamma Decay 868
- 30–7 Conservation of Nucleon Number and Other Conservation Laws 869
- 30–8 Half-Life and Rate of Decay 869
- 30–9 Calculations Involving Decay Rates and Half-Life 872
- 30–10 Decay Series 873
- 30–11 Radioactive Dating 874
- *30–12 Stability and Tunneling 876
- 30–13 Detection of Particles 877
  Questions, MisConceptual Questions 879–81
  Problems, Search and Learn 881–84

31 NUCLEAR ENERGY; EFFECTS AND USES OF RADIATION 885

- 31–1 Nuclear Reactions and the Transmutation of Elements 885
- 31–2 Nuclear Fission; Nuclear Reactors 889
- 31–3 Nuclear Fusion 894
- 31–4 Passage of Radiation Through Matter; Biological Damage 898
- 31–5 Measurement of Radiation—Dosimetry 899
- *31–6 Radiation Therapy 903
- *31–7 Tracers in Research and Medicine 904
- *31–8 Emission Tomography: PET and SPECT 905
- 31–9 Nuclear Magnetic Resonance (NMR) and Magnetic Resonance Imaging (MRI) 906

Questions, MisConceptual Questions 909–10
Problems, Search and Learn 911–14

32 ELEMENTARY PARTICLES 915

- 32–1 High-Energy Particles and Accelerators 916
- 32–2 Beginnings of Elementary Particle Physics—Particle Exchange 922
- 32–3 Particles and Antiparticles 924
- 32–4 Particle Interactions and Conservation Laws 926
- 32–5 Neutrinos 928
- 32–6 Particle Classification 930
- 32–7 Particle Stability and Resonances 932
- 32–8 Strangeness? Charm? Towards a New Model 932
- 32–9 Quarks 933
- 32–10 The Standard Model: QCD and Electroweak Theory 936
- 32–11 Grand Unified Theories 939
- 32–12 Strings and Supersymmetry 942

Questions, MisConceptual Questions 943–44
Problems, Search and Learn 944–46

33 ASTROPHYSICS AND COSMOLOGY 947

- 33–1 Stars and Galaxies 948
- 33–2 Stellar Evolution: Birth and Death of Stars, Nucleosynthesis 951
- 33–3 Distance Measurements 957
- 33–4 General Relativity: Gravity and the Curvature of Space 959
- 33–5 The Expanding Universe: Redshift and Hubble's Law 964
- 33–6 The Big Bang and the Cosmic Microwave Background 967
- 33–7 The Standard Cosmological Model: Early History of the Universe 970
- 33–8 Inflation: Explaining Flatness, Uniformity, and Structure 973
- 33–9 Dark Matter and Dark Energy 975
- 33–10 Large-Scale Structure of the Universe 977
- 33–11 Finally . . . 978

Questions, MisConceptual Questions 980–81
Problems, Search and Learn 981–83

APPENDICES

A	**Mathematical Review**	A-1
A-1	Relationships, Proportionality, and Equations	A-1
A-2	Exponents	A-2
A-3	Powers of 10, or Exponential Notation	A-3
A-4	Algebra	A-3
A-5	The Binomial Expansion	A-6
A-6	Plane Geometry	A-7
A-7	Trigonometric Functions and Identities	A-8
A-8	Logarithms	A-10
B	**Selected Isotopes**	A-12
C	**Rotating Frames of Reference; Inertial Forces; Coriolis Effect**	A-16
D	**Molar Specific Heats for Gases, and the Equipartition of Energy**	A-19
E	**Galilean and Lorentz Transformations**	A-22

Answers to Odd-Numbered Problems A-27

Index A-43

Photo Credits A-69

Applications to Biology and Medicine (Selected)

Chapter 4
How we walk 82
Chapter 5
Weightlessness 124–25
Chapter 6
Cardiac treadmill 168
Chapter 7
Body parts, center of mass 186–87
Impulse, don't break a leg 193
Chapter 8
Bird of prey 200
Centrifuge 204, 222
Torque with muscles 207, 223
Chapter 9
Teeth straightening 231
Forces in muscles and joints 238–39, 255
Human body stability 240
Leg stress in fall 259
Chapter 10
Pressure in cells 264
Blood flow 274, 278, 280
Blood loss to brain, TIA 278
Underground animals, air circulation 278
Blood flow and heart disease 280
Walking on water (insect) 281
Heart as a pump 282
Blood pressure 283
Blood transfusion 288
Chapter 11
Spider web 298
Echolocation by animals 309
Chapter 12
Ear and hearing range 331, 334–35
Doppler, blood speed; bat position 347, 358
Ultrasound medical imaging 350–51
Chapter 13
Life under ice 366–67
Molecules in a breath 373
Evaporation cools 379, 400

Humidity and comfort 380
Diffusion in living organisms 383
Chapter 14
Working off Calories 392
Convection by blood 402
Human radiative heat loss 404
Room comfort and metabolism 404
Medical thermography 405
Chapter 15
Energy in the human body 418–19
Biological evolution, development 430–31
Trees offset CO_2 emission 442
Chapter 16
Cells: electric forces, kinetic theory 460–62
DNA structure, replication 460–61
Chapter 17
Heart-beat scan (ECG or EKG) 473
Dipoles in molecular biology 482
Capacitor burn or shock 487
Heart defibrillator 487, 559
Electrocardiogram (ECG) 493
Chapter 18
Electrical conduction in the human nervous system 517–19
Chapter 19
Blood sugar phone app 526
Pacemaker, ventricular fibrillation 543
Electric shock, grounding 544–45
Chapter 20
Blood flow rate 584
Electromagnetic pump 589
Chapter 21
EM blood-flow measurement 596
Ground fault interrupter (GFCI) 607
Pacemaker 608
Chapter 22
Optical tweezers 636
Chapter 23
Medical endoscopes 660

Chapter 24
Spectroscopic analysis 693
Chapter 25
Human eye 719
Corrective lenses 719–21
Contact lenses 721
Seeing under water 721
Light microscopes 726
Resolution of eye 730, 732
X-ray diffraction in biology 735
Medical imaging: X-rays, CT 735–37
Cones in fovea 740
Chapter 27
Electron microscope images: blood vessel, blood clot, retina, viruses 771, 785–86
Photosynthesis 779
Measuring bone density 780
Chapter 28
Laser surgery 823
Chapter 29
Cell energy—ATP 833–34
Weak bonds in cells, DNA 834–35
Protein synthesis 836–37
Pulse oximeter 848
Chapter 31
Biological radiation damage 899
Radiation dosimetry 899–903
Radon 901
Radiation exposure; film badge 901
Radiation sickness 901
Radon exposure calculation 902–3
Radiation therapy 903
Proton therapy 904
Tracers in medicine and biology 904–5
Medical imaging: PET, SPECT 905–6
NMR and MRI 906–8
Radiation and thyroid 912
Chapter 32
Linacs and tumor irradiation 920

Applications to Other Fields and Everyday Life (Selected)

Chapter 1
The 8000-m peaks 11
Estimating volume of a lake 13
Height by triangulation 14
Measuring Earth's radius 15
Chapter 2
Braking distances 32
Rapid transit 47
Chapter 3
Sports 49, 58, 67, 68, 69, 73, 74
Kicked football 62, 64
Chapter 4
Rocket acceleration 82
What force accelerates car? 82
Elevator and counterweight 91
Mechanical advantage of pulley 92
Skiing 97, 100, 138
Bear sling 100, 252
City planning, cars on hills 105
Chapter 5
Not skidding on a curve 116
Antilock brakes 116
Banked highways 117
Artificial Earth satellites 122–23, 134
Free fall in athletics 125
Planets 125–28, 134, 137, 189, 197, 228

Determining the Sun's mass 127
Moon's orbit, phases, periods, diagram 129
Simulated gravity 130, 132
Near-Earth orbit 134
Comets 135
Asteroids, moons 135, 136, 196, 228
Rings of Saturn, galaxy 136
GPS, Milky Way 136
Chapter 6
Work done on a baseball, skiing 138
Car stopping distance $\propto v^2$ 145
Roller coaster 152, 158
Pole vault, high jump 153, 165
Stair-climbing power output 159
Horsepower, car needs 159–61
Lever 164
Spiderman 167
Chapter 7
Billiards 170, 179, 183
Tennis serve 172, 176
Rocket propulsion 175, 188–89
Rifle recoil 176
Nuclear collisions 180, 182
Ballistic pendulum 181
High jump 187
Distant planets discovered 189

Chapter 8
Rotating carnival rides 198, 201, 202
Bicycle 205, 227, 229
Rotating skaters, divers 216
Neutron star collapse 217
Strange spinning bike wheel 218
Tightrope walker 220
Hard drive 222
Total solar eclipses 229
Chapter 9
Tragic collapse 231, 246
Lever's mechanical advantage 233
Cantilever 235
Architecture: columns, arches, domes 243, 246–49
Fracture 245–46
Concrete, prestressed 246
Tower crane 252
Chapter 10
Glaciers 260
Hydraulic lift, brakes, press 265, 286
Hydrometer 271
Continental drift, plate tectonics 272
Helium balloon lift 272
Airplane wings, dynamic lift 277
Sailing against the wind 277
Baseball curve 278

Smoke up a chimney	278
Surface tension, capillarity	280–82
Pumps	282
Siphon	284, 290
Hurricane	287
Reynolds number	288

Chapter 11

Car springs	295
Unwanted floor vibrations	299
Pendulum clock	302
Car shock absorbers, building dampers	303
Child on a swing	304
Shattering glass via resonance	304
Resonant bridge collapse	304
Tsunami	306, 327
Earthquake waves	309, 311, 318, 324

Chapter 12

Count distance from lightning	329
Autofocus camera	330
Loudspeaker response	332
Musical scale	335
Stringed instruments	336–37
Wind instruments	337–40
Tuning with beats	343
Doppler: speed, weather forecasting	347–48
Sonic boom, sound barrier	349
Sonar: depth finding, Earth soundings	349

Chapter 13

Hot-air balloon	359
Expansion joints	361, 365, 367
Opening a tight lid	365
Gas tank overflow	366
Mass (and weight) of air in a room	371
Cold and hot tire pressure	372
Temperature dependent chemistry	377
Humidity and weather	381
Thermostat	384
Pressure cooker	388

Chapter 14

Effects of water's high specific heat	393
Thermal windows	401
How clothes insulate	401, 403
R-values of thermal insulation	402
Convective home heating	402
Astronomy—size of a star	406
Loft of goose down	407

Chapter 15

Steam engine	420–21
Internal combustion engine	421
Refrigerators	425–26
Air conditioners, heat pump	426–27
SEER rating	427
Thermal pollution, global warming	434
Energy resources	435

Chapter 16

Static electricity	443, 444
Photocopy machines	454, 462
Electrical shielding, safety	459
Laser printers and inkjet printers	463

Chapter 17

Capacitor uses in backups, surge protectors, memory	482, 484
Very high capacitance	484
Condenser microphone	484
Computer key	484
Camera flash	486–87
Signal and supply voltages	488
Digital, analog, bits, bytes	488–89
Digital coding	488–89
Analog-to-digital converter	489, 559
Sampling rate	488–89

Digital compression	489
CRT, TV and computer monitors	490
Flat screens, addressing pixels	491–92
Digital TV, matrix, refresh rate	491–92
Oscilloscope	492
Photocell	499
Lightning bolt (Pr90, S&L3)	499, 500

Chapter 18

Electric cars	504
Resistance thermometer	510
Heating element	510
Why bulbs burn out at turn on	511
Lightning bolt	512
Household circuits	512–13
Fuses, circuit breakers, shorts	512–13
Extension cord danger	513
Hair dryer	515
Superconductors	517
Halogen incandescent lamp	525
Strain gauge	525

Chapter 19

Car battery charging	536–37
Jump start safety	537
RC applications: flashers, wipers	542–43
Electric safety	543–45
Proper grounding, plugs	544–45
Leakage current	545
Downed power lines	545
Meters, analog and digital	546–48
Meter connection, corrections	547–48
Potentiometers and bridges	556, 559
Car battery corrosion	558
Digital-to-analog converter	559

Chapter 20

Declination, compass	562
Aurora borealis	569
Solenoids and electromagnets	572–73
Solenoid switch: car starter, doorbell	573
Magnetic circuit breaker	573
Motors, loudspeakers	576–77
Mass spectrometer	578
Relay	582

Chapter 21

Generators, alternators	597–99
Motor overload	599–600
Magnetic damping	600, 618
Airport metal detector	601
Transformers, power transmission	601–4
Cell phone charger	602
Car ignition	602
Electric power transmission	603–4
Power transfer by induction	604
Information storage	604–6
Hard drives, tape, DVD	604–5
Computer DRAM, flash	605–6
Microphone, credit card swipe	606
Seismograph	607
Ground fault interrupter (GFCI)	607
Capacitors as filters	613
Loudspeaker cross-over	613
Shielded cable	617
Sort recycled waste	618

Chapter 22

TV from the Moon	625, 639
Coaxial cable	631
Phone call time lag	632
Solar sail	636
Wireless: TV and radio	636–38
Satellite dish	638
Cell phones, remotes	639

Chapter 23

How tall a mirror do you need	648

Magnifying and wide-view mirrors	649, 655, 656
Where you can see *yourself* in a concave mirror	654
Optical illusions	657
Apparent depth in water	658
Fiber optics in telecommunications	660
Where you can *see* a lens image	663

Chapter 24

Soap bubbles and oil films	679, 693, 696–97
Mirages	682
Rainbows and diamonds	686
Colors underwater	687
Spectroscopy	692–93
Colors in thin soap film, details	696–97
Lens coatings	697–98
Polaroids, sunglasses	699–700
LCDs—liquid crystal displays	703–4
Sky color, cloud color, sunsets	704

Chapter 25

Cameras, digital and film; lenses	713–18
Pixel arrays, digital artifacts	714
Pixels, resolution, sharpness	717–18
Magnifying glass	713, 722–23
Telescopes	723–25, 730, 731
Microscopes	726–27, 730, 731
Telescope and microscope resolution, the λ rule	730–32
Radiotelescopes	731
Specialty microscopes	733
X-ray diffraction	733–35

Chapter 26

Space travel	754
Global positioning system (GPS)	755

Chapter 27

Photocells, photodiodes	776, 778
Electron microscopes	785–86

Chapter 28

Neon tubes	803
Fluorescence and phosphorescence	820
Lasers and their uses	820–23
DVD, CD, bar codes	822–23
Holography	823–24

Chapter 29

Integrated circuits (chips), 22-nm technology	829, 851
Semiconductor diodes, transistors	845–50
Solar cells	847
LEDs	847–48
Diode lasers	848
OLEDs	849–50
Transistors	850–51

Chapter 30

Smoke detectors	866
Carbon-14 dating	874–75
Archeological, geological dating	875, 876, 882, 883
Oldest Earth rocks and earliest life	876

Chapter 31

Nuclear reactors and power	891–93
Manhattan Project	893–94
Fusion energy reactors	896–98
Radon gas pollution	901

Chapter 32

Antimatter	925–26, 941

Chapter 33

Stars and galaxies	947, 948–51
Black holes	956, 962–63
Big Bang	966, 967–70
Evolution of universe	970–73
Dark matter and dark energy	975–77

Student Supplements

- **MasteringPhysics**™ (www.masteringphysics.com) is a homework, tutorial, and assessment system based on years of research into how students work physics problems and precisely where they need help. Studies show that students who use MasteringPhysics significantly increase their final scores compared to hand-written homework. MasteringPhysics achieves this improvement by providing students with instantaneous feedback specific to their wrong answers, simpler sub-problems upon request when they get stuck, and partial credit for their method(s) used. This individualized, 24/7 Socratic tutoring is recommended by nine out of ten students to their peers as the most effective and time-efficient way to study.

- The **Student Study Guide with Selected Solutions**, **Volume I** (Chapters 1–15, ISBN 978-0-321-76240-5) and **Volume II** (Chapters 16–33, ISBN 978-0-321-76808-7), written by Joseph Boyle (Miami-Dade Community College), contains overviews, key terms and phrases, key equations, self-study exams, problems for review, problem solving skills, and answers and solutions to selected end-of-chapter questions and problems for each chapter of this textbook.

- **Pearson eText** is available through MasteringPhysics, either automatically when MasteringPhysics is packaged with new books, or available as a purchased upgrade online. Allowing students access to the text wherever they have access to the Internet, Pearson eText comprises the full text, including figures that can be enlarged for better viewing. Within eText, students are also able to pop up definitions and terms to help with vocabulary and the reading of the material. Students can also take notes in eText using the annotation feature at the top of each page.

- **Pearson Tutor Services** (www.pearsontutorservices.com): Each student's subscription to MasteringPhysics also contains complimentary access to Pearson Tutor Services, powered by Smarthinking, Inc. By logging in with their MasteringPhysics ID and password, they will be connected to highly qualified e-instructors™ who provide additional, interactive online tutoring on the major concepts of physics.

- **ActivPhysics OnLine**™ (accessed through the Self Study area within www.masteringphysics.com) provides students with a group of highly regarded applet-based tutorials (see above). The following workbooks help students work though complex concepts and understand them more clearly.

- **ActivPhysics OnLine Workbook Volume 1: Mechanics • Thermal Physics • Oscillations & Waves**
  (ISBN 978-0-805-39060-5)

- **ActivPhysics OnLine Workbook Volume 2: Electricity & Magnetism • Optics • Modern Physics**
  (ISBN 978-0-805-39061-2)

Preface

What's New?

Lots! Much is new and unseen before. Here are the big four:

1. Multiple-choice Questions added to the end of each Chapter. They are not the usual type. These are called **MisConceptual Questions** because the responses (*a*, *b*, *c*, *d*, etc.) are intended to include common student misconceptions. Thus they are as much, or more, a learning experience than simply a testing experience.

2. **Search and Learn Problems** at the very end of each Chapter, after the other Problems. Some are pretty hard, others are fairly easy. They are intended to encourage students to go back and reread some part or parts of the text, and in this search for an answer they will hopefully learn more—if only because they have to read some material again.

3. **Chapter-Opening Questions** (COQ) that start each Chapter, a sort of "stimulant." Each is multiple choice, with responses including common misconceptions—to get preconceived notions out on the table right at the start. Where the relevant material is covered in the text, students find an Exercise asking them to return to the COQ to rethink and answer again.

4. **Digital.** Biggest of all. Crucial new applications. Today we are surrounded by digital electronics. How does it work? If you try to find out, say on the Internet, you won't find much physics: you may find shallow hand-waving with no real content, or some heavy jargon whose basis might take months or years to understand. So, for the first time, I have tried to explain

    - The basis of digital in bits and bytes, how analog gets transformed into digital, sampling rate, bit depth, quantization error, compression, noise (Section 17–10).
    - How digital TV works, including how each pixel is addressed for each frame, data stream, refresh rate (Section 17–11).
    - Semiconductor computer memory, DRAM, and flash (Section 21–8).
    - Digital cameras and sensors—revised and expanded Section 25–1.
    - New semiconductor physics, some of which is used in digital devices, including LED and OLED—how they work and what their uses are—plus more on transistors (MOSFET), chips, and technology generation as in 22-nm technology (Sections 29–9, 10, 11).

Besides those above, this new seventh edition includes

5. *New topics, new applications, principal revisions.*
    - *You* can measure the Earth's radius (Section 1–7).
    - Improved graphical analysis of linear motion (Section 2–8).
    - Planets (how first seen), heliocentric, geocentric (Section 5–8).
    - The Moon's orbit around the Earth: its phases and periods with diagram (Section 5–9).
    - Explanation of lake level change when large rock thrown from boat (Example 10–11).

- Biology and medicine, including:
  - Blood measurements (flow, sugar)—Chapters 10, 12, 14, 19, 20, 21;
  - Trees help offset CO_2 buildup—Chapter 15;
  - Pulse oximeter—Chapter 29;
  - Proton therapy—Chapter 31;
  - Radon exposure calculation—Chapter 31;
  - Cell phone use and brain—Chapter 31.
- Colors as seen underwater (Section 24–4).
- Soap film sequence of colors explained (Section 24–8).
- Solar sails (Section 22–6).
- Lots on sports.
- Symmetry—more emphasis and using italics or boldface to make visible.
- Flat screens (Sections 17–11, 24–11).
- Free-electron theory of metals, Fermi gas, Fermi level. New Section 29–6.
- Semiconductor devices—new details on diodes, LEDs, OLEDs, solar cells, compound semiconductors, diode lasers, MOSFET transistors, chips, 22-nm technology (Sections 29–9, 10, 11).
- Cross section (Chapter 31).
- Length of an object is a script ℓ rather than normal l, which looks like 1 or I (moment of inertia, current), as in $F = I\ell B$. Capital L is for angular momentum, latent heat, inductance, dimensions of length $[L]$.

6. *New photographs* taken by students and instructors (we asked).

7. *Page layout*: More than in previous editions, serious attention to how each page is formatted. Important derivations and Examples are on facing pages: no turning a page back in the middle of a derivation or Example. Throughout, readers see, on two facing pages, an important slice of physics.

8. *Greater clarity*: No topic, no paragraph in this book was overlooked in the search to improve the clarity and conciseness of the presentation. Phrases and sentences that may slow down the principal argument have been eliminated: keep to the essentials at first, give the elaborations later.

9. Much use has been made of physics education research. See the new powerful pedagogic features listed first.

10. *Examples modified*: More math steps are spelled out, and many new Examples added. About 10% of all Examples are Estimation Examples.

11. *This Book is Shorter* than other complete full-service books at this level. Shorter explanations are easier to understand and more likely to be read.

12. *Cosmological Revolution*: With generous help from top experts in the field, readers have the latest results.

See the World through Eyes that Know Physics

I was motivated from the beginning to write a textbook different from the others which present physics as a sequence of facts, like a catalog: "Here are the facts and you better learn them." Instead of beginning formally and dogmatically, I have sought to begin each topic with concrete observations and experiences students can relate to: start with specifics, and after go to the great generalizations and the more formal aspects of a topic, showing *why* we believe what we believe. This approach reflects how science is actually practiced.

The ultimate aim is to give students a thorough understanding of the basic concepts of physics in all its aspects, from mechanics to modern physics. A second objective is to show students how useful physics is in their own everyday lives and in their future professions by means of interesting applications to biology, medicine, architecture, and more.

Also, much effort has gone into techniques and approaches for solving problems: worked-out Examples, Problem Solving sections (Sections 2–6, 3–6, 4–7, 4–8, 6–7, 6–9, 8–6, 9–2, 13–7, 14–4, and 16–6), and Problem Solving Strategies (pages 30, 57, 60, 88, 115, 141, 158, 184, 211, 234, 399, 436, 456, 534, 568, 594, 655, 666, and 697).

This textbook is especially suited for students taking a one-year introductory course in physics that uses algebra and trigonometry but not calculus.[†] Many of these students are majoring in biology or premed, as well as architecture, technology, and the earth and environmental sciences. Many applications to these fields are intended to answer that common student query: "Why must I study physics?" The answer is that physics is fundamental to a full understanding of these fields, and here they can see how. Physics is everywhere around us in the everyday world. It is the goal of this book to help students "see the world through eyes that know physics."

A major effort has been made to not throw too much material at students reading the first few chapters. The basics have to be learned first. Many aspects can come later, when students are less overloaded and more prepared. If we don't overwhelm students with too much detail, especially at the start, maybe they can find physics interesting, fun, and helpful—and those who were afraid may lose their fear.

Chapter 1 is *not* a throwaway. It is fundamental to physics to realize that every measurement has an *uncertainty*, and how significant figures are used. Converting units and being able to make rapid *estimates* are also basic.

Mathematics can be an obstacle to students. I have aimed at including all steps in a derivation. Important mathematical tools, such as addition of vectors and trigonometry, are incorporated in the text where first needed, so they come with a context rather than in a scary introductory Chapter. Appendices contain a review of algebra and geometry (plus a few advanced topics).

Color is used pedagogically to bring out the physics. Different types of vectors are given different colors (see the chart on page xix).

Sections marked with a star * are considered optional. These contain slightly more advanced physics material, or material not usually covered in typical courses and/or interesting applications; they contain no material needed in later Chapters (except perhaps in later optional Sections).

For a brief course, all optional material could be dropped as well as significant parts of Chapters 1, 10, 12, 22, 28, 29, 32, and selected parts of Chapters 7, 8, 9, 15, 21, 24, 25, 31. Topics not covered in class can be a valuable resource for later study by students. Indeed, this text can serve as a useful reference for years because of its wide range of coverage.

[†]It is fine to take a calculus course. But mixing calculus with physics for these students may often mean not learning the physics because of stumbling over the calculus.

Thanks

Many physics professors provided input or direct feedback on every aspect of this textbook. They are listed below, and I owe each a debt of gratitude.

Edward Adelson, The Ohio State University
Lorraine Allen, United States Coast Guard Academy
Zaven Altounian, McGill University
Leon Amstutz, Taylor University
David T. Bannon, Oregon State University
Bruce Barnett, Johns Hopkins University
Michael Barnett, Lawrence Berkeley Lab
Anand Batra, Howard University
Cornelius Bennhold, George Washington University
Bruce Birkett, University of California Berkeley
Steven Boggs, University of California Berkeley
Robert Boivin, Auburn University
Subir Bose, University of Central Florida
David Branning, Trinity College
Meade Brooks, Collin County Community College
Bruce Bunker, University of Notre Dame
Grant Bunker, Illinois Institute of Technology
Wayne Carr, Stevens Institute of Technology
Charles Chiu, University of Texas Austin
Roger N. Clark, U. S. Geological Survey
Russell Clark, University of Pittsburgh
Robert Coakley, University of Southern Maine
David Curott, University of North Alabama
Biman Das, SUNY Potsdam
Bob Davis, Taylor University
Kaushik De, University of Texas Arlington
Michael Dennin, University of California Irvine
Karim Diff, Santa Fe College
Kathy Dimiduk, Cornell University
John DiNardo, Drexel University
Scott Dudley, United States Air Force Academy
Paul Dyke
John Essick, Reed College
Kim Farah, Lasell College
Cassandra Fesen, Dartmouth College
Leonard Finegold, Drexel University
Alex Filippenko, University of California Berkeley
Richard Firestone, Lawrence Berkeley Lab
Allen Flora, Hood College
Mike Fortner, Northern Illinois University
Tom Furtak, Colorado School of Mines
Edward Gibson, California State University Sacramento
John Hardy, Texas A&M
Thomas Hemmick, State University of New York Stonybrook
J. Erik Hendrickson, University of Wisconsin Eau Claire
Laurent Hodges, Iowa State University
David Hogg, New York University
Mark Hollabaugh, Normandale Community College
Andy Hollerman, University of Louisiana at Lafayette
Russell Holmes, University of Minnesota Twin Cities
William Holzapfel, University of California Berkeley
Chenming Hu, University of California Berkeley
Bob Jacobsen, University of California Berkeley
Arthur W. John, Northeastern University
Teruki Kamon, Texas A&M
Daryao Khatri, University of the District of Columbia
Tsu-Jae King Liu, University of California Berkeley
Richard Kronenfeld, South Mountain Community College
Jay Kunze, Idaho State University
Jim LaBelle, Dartmouth College
Amer Lahamer, Berea College
David Lamp, Texas Tech University
Kevin Lear, SpatialGraphics.com
Ran Li, Kent State University
Andreí Linde, Stanford University
M.A.K. Lodhi, Texas Tech
Lisa Madewell, University of Wisconsin

Bruce Mason, University of Oklahoma
Mark Mattson, James Madison University
Dan Mazilu, Washington and Lee University
Linda McDonald, North Park College
Bill McNairy, Duke University
Jo Ann Merrell, Saddleback College
Raj Mohanty, Boston University
Giuseppe Molesini, Istituto Nazionale di Ottica Florence
Wouter Montfrooij, University of Missouri
Eric Moore, Frostburg State University
Lisa K. Morris, Washington State University
Richard Muller, University of California Berkeley
Blaine Norum, University of Virginia
Lauren Novatne, Reedley College
Alexandria Oakes, Eastern Michigan University
Ralph Oberly, Marshall University
Michael Ottinger, Missouri Western State University
Lyman Page, Princeton and WMAP
Laurence Palmer, University of Maryland
Bruce Partridge, Haverford College
R. Daryl Pedigo, University of Washington
Robert Pelcovitz, Brown University
Saul Perlmutter, University of California Berkeley
Vahe Peroomian, UCLA
Harvey Picker, Trinity College
Amy Pope, Clemson University
James Rabchuk, Western Illinois University
Michele Rallis, Ohio State University
Paul Richards, University of California Berkeley
Peter Riley, University of Texas Austin
Dennis Rioux, University of Wisconsin Oshkosh
John Rollino, Rutgers University
Larry Rowan, University of North Carolina Chapel Hill
Arthur Schmidt, Northwestern University
Cindy Schwarz-Rachmilowitz, Vassar College
Peter Sheldon, Randolph-Macon Woman's College
Natalia A. Sidorovskaia, University of Louisiana at Lafayette
James Siegrist, University of California Berkeley
Christopher Sirola, University of Southern Mississippi
Earl Skelton, Georgetown University
George Smoot, University of California Berkeley
David Snoke, University of Pittsburgh
Stanley Sobolewski, Indiana University of Pennsylvania
Mark Sprague, East Carolina University
Michael Strauss, University of Oklahoma
Laszlo Takac, University of Maryland Baltimore Co.
Leo Takahashi, Pennsylvania State University
Richard Taylor, University of Oregon
Oswald Tekyi-Mensah, Alabama State University
Franklin D. Trumpy, Des Moines Area Community College
Ray Turner, Clemson University
Som Tyagi, Drexel University
David Vakil, El Camino College
Trina VanAusdal, Salt Lake Community College
John Vasut, Baylor University
Robert Webb, Texas A&M
Robert Weidman, Michigan Technological University
Edward A. Whittaker, Stevens Institute of Technology
Lisa M. Will, San Diego City College
Suzanne Willis, Northern Illinois University
John Wolbeck, Orange County Community College
Stanley George Wojcicki, Stanford University
Mark Worthy, Mississippi State University
Edward Wright, UCLA and WMAP
Todd Young, Wayne State College
William Younger, College of the Albemarle
Hsiao-Ling Zhou, Georgia State University
Michael Ziegler, The Ohio State University
Ulrich Zurcher, Cleveland State University

New photographs were offered by Professors Vickie Frohne (Holy Cross Coll.), Guillermo Gonzales (Grove City Coll.), Martin Hackworth (Idaho State U.), Walter H. G. Lewin (MIT), Nicholas Murgo (NEIT), Melissa Vigil (Marquette U.), Brian Woodahl (Indiana U. at Indianapolis), and Gary Wysin (Kansas State U.). New photographs shot by students are from the AAPT photo contest: Matt Buck, (John Burroughs School), Matthew Claspill (Helias H. S.), Greg Gentile (West Forsyth H. S.), Shilpa Hampole (Notre Dame H. S.), Sarah Lampen (John Burroughs School), Mrinalini Modak (Fayetteville–Manlius H. S.), Joey Moro (Ithaca H. S.), and Anna Russell and Annacy Wilson (both Tamalpais H. S.).

I owe special thanks to Prof. Bob Davis for much valuable input, and especially for working out all the Problems and producing the Solutions Manual for all Problems, as well as for providing the answers to odd-numbered Problems at the back of the book. Many thanks also to J. Erik Hendrickson who collaborated with Bob Davis on the solutions, and to the team they managed (Profs. Karim Diff, Thomas Hemmick, Lauren Novatne, Michael Ottinger, and Trina VanAusdal).

I am grateful to Profs. Lorraine Allen, David Bannon, Robert Coakley, Kathy Dimiduk, John Essick, Dan Mazilu, John Rollino, Cindy Schwarz, Earl Skelton, Michael Strauss, Ray Turner, Suzanne Willis, and Todd Young, who helped with developing the new MisConceptual Questions and Search and Learn Problems, and offered other significant clarifications.

Crucial for rooting out errors, as well as providing excellent suggestions, were Profs. Lorraine Allen, Kathy Dimiduk, Michael Strauss, Ray Turner, and David Vakil. A huge thank you to them and to Prof. Giuseppe Molesini for his suggestions and his exceptional photographs for optics.

For Chapters 32 and 33 on Particle Physics and Cosmology and Astrophysics, I was fortunate to receive generous input from some of the top experts in the field, to whom I owe a debt of gratitude: Saul Perlmutter, George Smoot, Richard Muller, Steven Boggs, Alex Filippenko, Paul Richards, James Siegrist, and William Holzapfel (UC Berkeley), Andreí Linde (Stanford U.), Lyman Page (Princeton and WMAP), Edward Wright (UCLA and WMAP), Michael Strauss (University of Oklahoma), Michael Barnett (LBNL), and Bob Jacobsen (UC Berkeley; so helpful in many areas, including digital and pedagogy).

I also wish to thank Profs. Howard Shugart, Chair Frances Hellman, and many others at the University of California, Berkeley, Physics Department for helpful discussions, and for hospitality. Thanks also to Profs. Tito Arecchi, Giuseppe Molesini, and Riccardo Meucci at the Istituto Nazionale di Ottica, Florence, Italy.

Finally, I am grateful to the many people at Pearson Education with whom I worked on this project, especially Paul Corey and the ever-perspicacious Karen Karlin.

The final responsibility for all errors lies with me. I welcome comments, corrections, and suggestions as soon as possible to benefit students for the next reprint.

D.C.G.

email: Jim.Smith@Pearson.com
Post: Jim Smith
      1301 Sansome Street
      San Francisco, CA 94111

About the Author

Douglas C. Giancoli obtained his BA in physics (summa cum laude) from UC Berkeley, his MS in physics at MIT, and his PhD in elementary particle physics back at UC Berkeley. He spent 2 years as a post-doctoral fellow at UC Berkeley's Virus lab developing skills in molecular biology and biophysics. His mentors include Nobel winners Emilio Segrè and Donald Glaser.

He has taught a wide range of undergraduate courses, traditional as well as innovative ones, and continues to update his textbooks meticulously, seeking ways to better provide an understanding of physics for students.

Doug's favorite spare-time activity is the outdoors, especially climbing peaks. He says climbing peaks is like learning physics: it takes effort and the rewards are great.

To Students

HOW TO STUDY

1. Read the Chapter. Learn new vocabulary and notation. Try to respond to questions and exercises as they occur.
2. Attend all class meetings. Listen. Take notes, especially about aspects you do not remember seeing in the book. Ask questions (everyone wants to, but maybe you will have the courage). You will get more out of class if you read the Chapter first.
3. Read the Chapter again, paying attention to details. Follow derivations and worked-out Examples. Absorb their logic. Answer Exercises and as many of the end-of-Chapter Questions as you can, and all MisConceptual Questions.
4. Solve at least 10 to 20 end of Chapter Problems, especially those assigned. In doing Problems you find out what you learned and what you didn't. Discuss them with other students. Problem solving is one of the great learning tools. Don't just look for a formula—it might be the wrong one.

NOTES ON THE FORMAT AND PROBLEM SOLVING

1. Sections marked with a star (*) are considered **optional**. They can be omitted without interrupting the main flow of topics. No later material depends on them except possibly later starred Sections. They may be fun to read, though.
2. The customary **conventions** are used: symbols for quantities (such as m for mass) are italicized, whereas units (such as m for meter) are not italicized. Symbols for vectors are shown in boldface with a small arrow above: $\vec{\mathbf{F}}$.
3. Few equations are valid in all situations. Where practical, the **limitations** of important equations are stated in square brackets next to the equation. The equations that represent the great laws of physics are displayed with a tan background, as are a few other indispensable equations.
4. At the end of each Chapter is a set of **Questions** you should try to answer. Attempt all the multiple-choice **MisConceptual Questions**. Most important are **Problems** which are ranked as Level I, II, or III, according to estimated difficulty. Level I Problems are easiest, Level II are standard Problems, and Level III are "challenge problems." These ranked Problems are arranged by Section, but Problems for a given Section may depend on earlier material too. There follows a group of **General Problems**, not arranged by Section or ranked. Problems that relate to optional Sections are starred (*). Answers to odd-numbered Problems are given at the end of the book. **Search and Learn Problems** at the end are meant to encourage you to return to parts of the text to find needed detail, and at the same time help you to learn.
5. Being able to solve **Problems** is a crucial part of learning physics, and provides a powerful means for understanding the concepts and principles. This book contains many aids to problem solving: (a) worked-out **Examples**, including an Approach and Solution, which should be studied as an integral part of the text; (b) some of the worked-out Examples are **Estimation Examples**, which show how rough or approximate results can be obtained even if the given data are sparse (see Section 1–7); (c) **Problem Solving Strategies** placed throughout the text to suggest a step-by-step approach to problem solving for a particular topic—but remember that the basics remain the same; most of these "Strategies" are followed by an Example that is solved by explicitly following the suggested steps; (d) special problem-solving Sections; (e) "Problem Solving" marginal notes which refer to hints within the text for solving Problems; (f) **Exercises** within the text that you should work out immediately, and then check your response against the answer given at the bottom of the last page of that Chapter; (g) the Problems themselves at the end of each Chapter (point 4 above).
6. **Conceptual Examples** pose a question which hopefully starts you to think and come up with a response. Give yourself a little time to come up with your own response before reading the Response given.
7. **Math** review, plus additional topics, are found in Appendices. Useful **data**, **conversion factors**, and math **formulas** are found inside the front and back covers.

USE OF COLOR

Vectors

A general vector	→
resultant vector (sum) is slightly thicker	⇒
components of any vector are dashed	‐‐‐→
Displacement ($\vec{D}, \vec{r}$)	→
Velocity ($\vec{v}$)	→
Acceleration ($\vec{a}$)	→
Force ($\vec{F}$)	→
Force on second object	→
or third object in same figure	→
Momentum ($\vec{p}$ or $m\vec{v}$)	→
Angular momentum ($\vec{L}$)	→
Angular velocity ($\vec{\omega}$)	→
Torque ($\vec{\tau}$)	→
Electric field ($\vec{E}$)	→
Magnetic field ($\vec{B}$)	→

Electricity and magnetism

Electric field lines	⌒
Equipotential lines	‐ ‐ ‐ ‐
Magnetic field lines	⌒
Electric charge (+)	⊕ or • +
Electric charge (−)	⊖ or • −

Electric circuit symbols

Wire, with switch S	─/S─
Resistor	─/\/\/\─
Capacitor	─||─
Inductor	─∿∿∿─
Battery	─||─
Ground	⏚

Optics

Light rays	→
Object	↑
Real image (dashed)	↑
Virtual image (dashed and paler)	↑

Other

Energy level (atom, etc.)	──
Measurement lines	⊢—1.0 m—⊣
Path of a moving object	‐ ‐ ‐→
Direction of motion or current	→

Image of the Earth from a NASA satellite. The sky appears black from out in space because there are so few molecules to reflect light. (Why the sky appears blue to us on Earth has to do with scattering of light by molecules of the atmosphere, as discussed in Chapter 24.) Note the storm off the coast of Mexico.

Introduction, Measurement, Estimating

CHAPTER 1

CHAPTER-OPENING QUESTIONS—Guess now!

1. How many cm³ are in 1.0 m³?
   (a) 10. (b) 100. (c) 1000. (d) 10,000. (e) 100,000. (f) 1,000,000.

2. Suppose you wanted to actually measure the radius of the Earth, at least roughly, rather than taking other people's word for what it is. Which response below describes the best approach?
   (a) Use an extremely long measuring tape.
   (b) It is only possible by flying high enough to see the actual curvature of the Earth.
   (c) Use a standard measuring tape, a step ladder, and a large smooth lake.
   (d) Use a laser and a mirror on the Moon or on a satellite.
   (e) Give up; it is impossible using ordinary means.

[*We start each Chapter with a Question—sometimes two. Try to answer right away. Don't worry about getting the right answer now—the idea is to get your preconceived notions out on the table. If they are misconceptions, we expect them to be cleared up as you read the Chapter. You will usually get another chance at the Question(s) later in the Chapter when the appropriate material has been covered. These Chapter-Opening Questions will also help you see the power and usefulness of physics.*]

CONTENTS

- 1–1 The Nature of Science
- 1–2 Physics and its Relation to Other Fields
- 1–3 Models, Theories, and Laws
- 1–4 Measurement and Uncertainty; Significant Figures
- 1–5 Units, Standards, and the SI System
- 1–6 Converting Units
- 1–7 Order of Magnitude: Rapid Estimating
- *1–8 Dimensions and Dimensional Analysis

Physics is the most basic of the sciences. It deals with the behavior and structure of matter. The field of physics is usually divided into *classical physics* which includes motion, fluids, heat, sound, light, electricity, and magnetism; and *modern physics* which includes the topics of relativity, atomic structure, quantum theory, condensed matter, nuclear physics, elementary particles, and cosmology and astrophysics. We will cover all these topics in this book, beginning with motion (or mechanics, as it is often called) and ending with the most recent results in fundamental particles and the cosmos. But before we begin on the physics itself, we take a brief look at how this overall activity called "science," including physics, is actually practiced.

1–1 The Nature of Science

The principal aim of all sciences, including physics, is generally considered to be the search for order in our observations of the world around us. Many people think that science is a mechanical process of collecting facts and devising theories. But it is not so simple. Science is a creative activity that in many respects resembles other creative activities of the human mind.

One important aspect of science is **observation** of events, which includes the design and carrying out of experiments. But observation and experiments require imagination, because scientists can never include everything in a description of what they observe. Hence, scientists must make judgments about what is relevant in their observations and experiments.

Consider, for example, how two great minds, Aristotle (384–322 B.C.; Fig. 1–1) and Galileo (1564–1642; Fig. 2–18), interpreted motion along a horizontal surface. Aristotle noted that objects given an initial push along the ground (or on a tabletop) always slow down and stop. Consequently, Aristotle argued, the natural state of an object is to be at rest. Galileo, the first true experimentalist, reexamined horizontal motion in the 1600s. He imagined that if friction could be eliminated, an object given an initial push along a horizontal surface would continue to move indefinitely without stopping. He concluded that for an object to be in motion was just as natural as for it to be at rest. By inventing a new way of thinking about the same data, Galileo founded our modern view of motion (Chapters 2, 3, and 4), and he did so with a leap of the imagination. Galileo made this leap conceptually, without actually eliminating friction.

FIGURE 1–1 Aristotle is the central figure (dressed in blue) at the top of the stairs (the figure next to him is Plato) in this famous Renaissance portrayal of *The School of Athens*, painted by Raphael around 1510. Also in this painting, considered one of the great masterpieces in art, are Euclid (drawing a circle at the lower right), Ptolemy (extreme right with globe), Pythagoras, Socrates, and Diogenes.

Observation, with careful experimentation and measurement, is one side of the scientific process. The other side is the invention or creation of **theories** to explain and order the observations. Theories are never derived directly from observations. Observations may help inspire a theory, and theories are accepted or rejected based on the results of observation and experiment.

Theories are inspirations that come from the minds of human beings. For example, the idea that matter is made up of atoms (the atomic theory) was not arrived at by direct observation of atoms—we can't see atoms directly. Rather, the idea sprang from creative minds. The theory of relativity, the electromagnetic theory of light, and Newton's law of universal gravitation were likewise the result of human imagination.

The great theories of science may be compared, as creative achievements, with great works of art or literature. But how does science differ from these other creative activities? One important difference is that science requires **testing** of its ideas or theories to see if their predictions are borne out by experiment. But theories are not "proved" by testing. First of all, no measuring instrument is perfect, so exact confirmation is not possible. Furthermore, it is not possible to test a theory for every possible set of circumstances. Hence a theory cannot be absolutely verified. Indeed, the history of science tells us that long-held theories can sometimes be replaced by new ones, particularly when new experimental techniques provide new or contradictory data.

A new theory is accepted by scientists in some cases because its predictions are quantitatively in better agreement with experiment than those of the older theory. But in many cases, a new theory is accepted only if it explains a greater *range* of phenomena than does the older one. Copernicus's Sun-centered theory of the universe (Fig. 1–2b), for example, was originally no more accurate than Ptolemy's Earth-centered theory (Fig. 1–2a) for predicting the motion of heavenly bodies (Sun, Moon, planets). But Copernicus's theory had consequences that Ptolemy's did not, such as predicting the moonlike phases of Venus. A simpler and richer theory, one which unifies and explains a greater variety of phenomena, is more useful and beautiful to a scientist. And this aspect, as well as quantitative agreement, plays a major role in the acceptance of a theory.

FIGURE 1–2 (a) Ptolemy's geocentric view of the universe. Note at the center the four elements of the ancients: Earth, water, air (clouds around the Earth), and fire; then the circles, with symbols, for the Moon, Mercury, Venus, Sun, Mars, Jupiter, Saturn, the fixed stars, and the signs of the zodiac. (b) An early representation of Copernicus's heliocentric view of the universe with the Sun at the center. (See Chapter 5.)

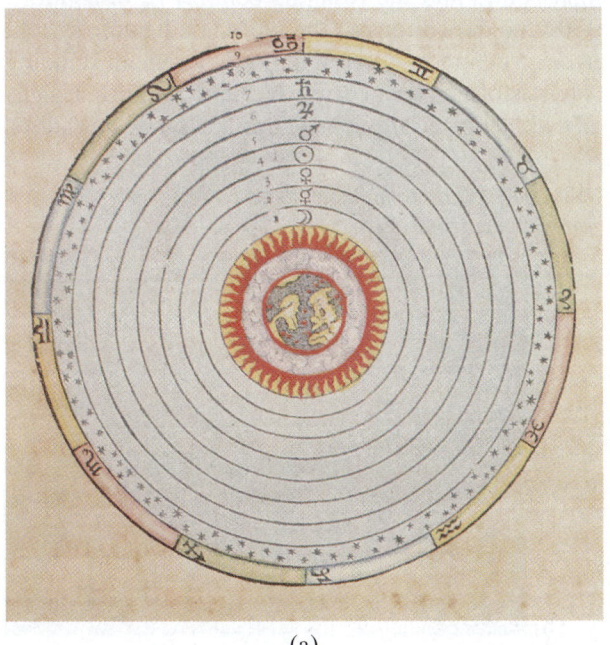
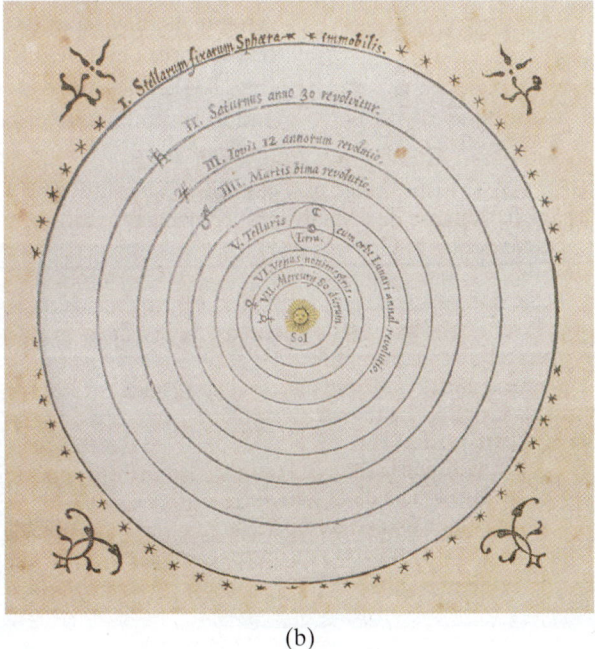

(a)   (b)

An important aspect of any theory is how well it can quantitatively predict phenomena, and from this point of view a new theory may often seem to be only a minor advance over the old one. For example, Einstein's theory of relativity gives predictions that differ very little from the older theories of Galileo and Newton in nearly all everyday situations. Its predictions are better mainly in the extreme case of very high speeds close to the speed of light. But quantitative prediction is not the only important outcome of a theory. Our view of the world is affected as well. As a result of Einstein's theory of relativity, for example, our concepts of space and time have been completely altered, and we have come to see mass and energy as a single entity (via the famous equation $E = mc^2$).

1–2 Physics and its Relation to Other Fields

For a long time science was more or less a united whole known as natural philosophy. Not until a century or two ago did the distinctions between physics and chemistry and even the life sciences become prominent. Indeed, the sharp distinction we now see between the arts and the sciences is itself only a few centuries old. It is no wonder then that the development of physics has both influenced and been influenced by other fields. For example, the notebooks (Fig. 1–3) of Leonardo da Vinci, the great Renaissance artist, researcher, and engineer, contain the first references to the forces acting within a structure, a subject we consider as physics today; but then, as now, it has great relevance to architecture and building.

Early work in electricity that led to the discovery of the electric battery and electric current was done by an eighteenth-century physiologist, Luigi Galvani (1737–1798). He noticed the twitching of frogs' legs in response to an electric spark and later that the muscles twitched when in contact with two dissimilar metals (Chapter 18). At first this phenomenon was known as "animal electricity," but it shortly became clear that electric current itself could exist in the absence of an animal.

Physics is used in many fields. A zoologist, for example, may find physics useful in understanding how prairie dogs and other animals can live underground without suffocating. A physical therapist will be more effective if aware of the principles of center of gravity and the action of forces within the human body. A knowledge of the operating principles of optical and electronic equipment is helpful in a variety of fields. Life scientists and architects alike will be interested in the nature of heat loss and gain in human beings and the resulting comfort or discomfort. Architects may have to calculate the dimensions of the pipes in a heating system or the forces involved in a given structure to determine if it will remain standing (Fig. 1–4). They must know physics principles in order to make realistic designs and to communicate effectively with engineering consultants and other specialists.

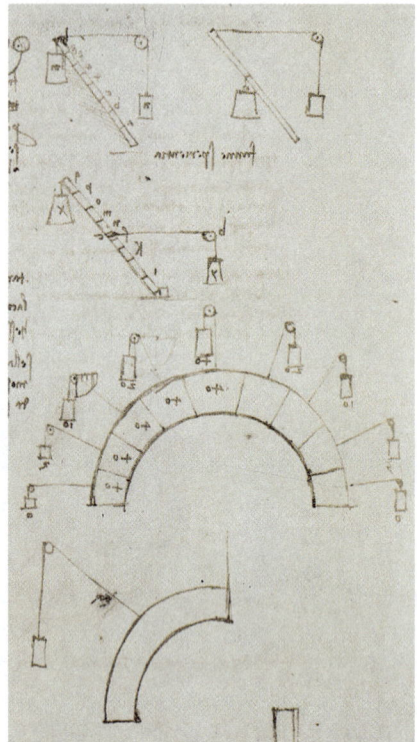

FIGURE 1–3 Studies on the forces in structures by Leonardo da Vinci (1452–1519).

FIGURE 1–4 (a) This bridge over the River Tiber in Rome was built 2000 years ago and still stands. (b) The 2007 collapse of a Mississippi River highway bridge built only 40 years before.

(a)

(b)

From the aesthetic or psychological point of view, too, architects must be aware of the forces involved in a structure—for example instability, even if only illusory, can be discomforting to those who must live or work in the structure.

The list of ways in which physics relates to other fields is extensive. In the Chapters that follow we will discuss many such applications as we carry out our principal aim of explaining basic physics.

1–3 Models, Theories, and Laws

When scientists are trying to understand a particular set of phenomena, they often make use of a **model**. A model, in the scientific sense, is a kind of analogy or mental image of the phenomena in terms of something else we are already familiar with. One example is the wave model of light. We cannot see waves of light as we can water waves. But it is valuable to think of light as made up of waves, because experiments indicate that light behaves in many respects as water waves do.

The purpose of a model is to give us an approximate mental or visual picture—something to hold on to—when we cannot see what actually is happening. Models often give us a deeper understanding: the analogy to a known system (for instance, the water waves above) can suggest new experiments to perform and can provide ideas about what other related phenomena might occur.

You may wonder what the difference is between a theory and a model. Usually a model is relatively simple and provides a structural similarity to the phenomena being studied. A **theory** is broader, more detailed, and can give quantitatively testable predictions, often with great precision. It is important, however, not to confuse a model or a theory with the real system or the phenomena themselves.

Scientists have given the title **law** to certain concise but general statements about how nature behaves (that electric charge is conserved, for example). Often the statement takes the form of a relationship or equation between quantities (such as Newton's second law, $F = ma$).

Statements that we call laws are usually experimentally valid over a wide range of observed phenomena. For less general statements, the term **principle** is often used (such as Archimedes' principle). We use "theory" for a more general picture of the phenomena dealt with.

Scientific laws are different from political laws in that the latter are *prescriptive*: they tell us how we ought to behave. Scientific laws are *descriptive*: they do not say how nature *should* behave, but rather are meant to describe how nature *does* behave. As with theories, laws cannot be tested in the infinite variety of cases possible. So we cannot be sure that any law is absolutely true. We use the term "law" when its validity has been tested over a wide range of cases, and when any limitations and the range of validity are clearly understood.

Scientists normally do their research as if the accepted laws and theories were true. But they are obliged to keep an open mind in case new information should alter the validity of any given law or theory.

1–4 Measurement and Uncertainty; Significant Figures

In the quest to understand the world around us, scientists seek to find relationships among physical quantities that can be measured.

Uncertainty

Reliable measurements are an important part of physics. But no measurement is absolutely precise. There is an uncertainty associated with every measurement.

Among the most important sources of uncertainty, other than blunders, are the limited accuracy of every measuring instrument and the inability to read an instrument beyond some fraction of the smallest division shown. For example, if you were to use a centimeter ruler to measure the width of a board (Fig. 1–5), the result could be claimed to be precise to about 0.1 cm (1 mm), the smallest division on the ruler, although half of this value might be a valid claim as well. The reason is that it is difficult for the observer to estimate (or "interpolate") between the smallest divisions. Furthermore, the ruler itself may not have been manufactured to an accuracy very much better than this.

When giving the result of a measurement, it is important to state the **estimated uncertainty** in the measurement. For example, the width of a board might be written as 8.8 ± 0.1 cm. The ± 0.1 cm ("plus or minus 0.1 cm") represents the estimated uncertainty in the measurement, so that the actual width most likely lies between 8.7 and 8.9 cm. The **percent uncertainty** is the ratio of the uncertainty to the measured value, multiplied by 100. For example, if the measurement is 8.8 cm and the uncertainty about 0.1 cm, the percent uncertainty is

$$\frac{0.1}{8.8} \times 100\% \approx 1\%,$$

where $\approx$ means "is approximately equal to."

Often the uncertainty in a measured value is not specified explicitly. In such cases, the

> uncertainty in a numerical value is assumed to be *one or a few units* in the last digit specified.

For example, if a length is given as 8.8 cm, the uncertainty is assumed to be about 0.1 cm or 0.2 cm, or possibly even 0.3 cm. It is important in this case that you do not write 8.80 cm, because this implies an uncertainty on the order of 0.01 cm; it assumes that the length is probably between about 8.79 cm and 8.81 cm, when actually you believe it is between about 8.7 and 8.9 cm.

FIGURE 1–5 Measuring the width of a board with a centimeter ruler. Accuracy is about ± 1 mm.

CONCEPTUAL EXAMPLE 1–1 Is the diamond yours? A friend asks to borrow your precious diamond for a day to show her family. You are a bit worried, so you carefully have your diamond weighed on a scale which reads 8.17 grams. The scale's accuracy is claimed to be ± 0.05 gram. The next day you weigh the returned diamond again, getting 8.09 grams. Is this your diamond?

RESPONSE The scale readings are measurements and are not perfect. They do not necessarily give the "true" value of the mass. Each measurement could have been high or low by up to 0.05 gram or so. The actual mass of your diamond lies most likely between 8.12 grams and 8.22 grams. The actual mass of the returned diamond is most likely between 8.04 grams and 8.14 grams. These two ranges overlap, so the data do not give you a strong reason to doubt that the returned diamond is yours.

Significant Figures

The number of reliably known digits in a number is called the number of **significant figures**. Thus there are four significant figures in the number 23.21 cm and two in the number 0.062 cm (the zeros in the latter are merely place holders that show where the decimal point goes). The number of significant figures may not always be clear. Take, for example, the number 80. Are there one or two significant figures? We need words here: If we say it is *roughly* 80 km between two cities, there is only one significant figure (the 8) since the zero is merely a place holder. If there is no suggestion that the 80 is a rough approximation, then we can often assume (as we will in this book) that it has 2 significant figures: so it is 80 km within an accuracy of about 1 or 2 km. If it is precisely 80 km, to within ± 0.1 or ± 0.2 km, then we write 80.0 km (three significant figures).

When specifying numerical results, you should avoid the temptation to keep more digits in the final answer than is justified: see boldface statement on previous page. For example, to calculate the area of a rectangle 11.3 cm by 6.8 cm, the result of multiplication would be 76.84 cm^2. But this answer can not be accurate to the implied 0.01 cm^2 uncertainty, because (using the outer limits of the assumed uncertainty for each measurement) the result could be between 11.2 cm × 6.7 cm = 75.04 cm^2 and 11.4 cm × 6.9 cm = 78.66 cm^2. At best, we can quote the answer as 77 cm^2, which implies an uncertainty of about 1 or 2 cm^2. The other two digits (in the number 76.84 cm^2) must be dropped (rounded off) because they are not significant. As a rough general "significant figure" rule

the final result of a multiplication or division should have no more digits than the numerical value with the fewest significant figures.

In our example, 6.8 cm has the least number of significant figures, namely two. Thus the result 76.84 cm^2 needs to be rounded off to 77 cm^2.

EXERCISE A The area of a rectangle 4.5 cm by 3.25 cm is correctly given by (*a*) 14.625 cm^2; (*b*) 14.63 cm^2; (*c*) 14.6 cm^2; (*d*) 15 cm^2.

When adding or subtracting numbers, the final result should contain no more decimal places than the number with the fewest decimal places. For example, the result of subtracting 0.57 from 3.6 is 3.0 (not 3.03). Similarly 36 + 8.2 = 44, not 44.2.

Be careful not to confuse significant figures with the number of decimal places.

EXERCISE B For each of the following numbers, state the number of significant figures and the number of decimal places: (*a*) 1.23; (*b*) 0.123; (*c*) 0.0123.

Keep in mind when you use a calculator that all the digits it produces may not be significant. When you divide 2.0 by 3.0, the proper answer is 0.67, and not 0.666666666 as calculators give (Fig. 1–6a). Digits should not be quoted in a result unless they are truly significant figures. However, to obtain the most accurate result, you should normally *keep one or more extra significant figures throughout a calculation, and round off only in the final result.* (With a calculator, you can keep all its digits in intermediate results.) Note also that calculators sometimes give too few significant figures. For example, when you multiply 2.5 × 3.2, a calculator may give the answer as simply 8. But the answer is accurate to two significant figures, so the proper answer is 8.0. See Fig. 1–6b.

CONCEPTUAL EXAMPLE 1–2 Significant figures. Using a protractor (Fig. 1–7), you measure an angle to be 30°. (*a*) How many significant figures should you quote in this measurement? (*b*) Use a calculator to find the cosine of the angle you measured.

RESPONSE (*a*) If you look at a protractor, you will see that the precision with which you can measure an angle is about one degree (certainly not 0.1°). So you can quote two significant figures, namely 30° (not 30.0°). (*b*) If you enter cos 30° in your calculator, you will get a number like 0.866025403. But the angle you entered is known only to two significant figures, so its cosine is correctly given by 0.87; you must round your answer to two significant figures.

NOTE Trigonometric functions, like cosine, are reviewed in Chapter 3 and Appendix A.

(a)

(b)

FIGURE 1–6 These two calculators show the wrong number of significant figures. In (a), 2.0 was divided by 3.0. The correct final result should be 0.67. In (b), 2.5 was multiplied by 3.2. The correct result is 8.0.

PROBLEM SOLVING
Report only the proper number of significant figures in the final result. But keep extra digits during the calculation

FIGURE 1–7 Example 1–2. A protractor used to measure an angle.

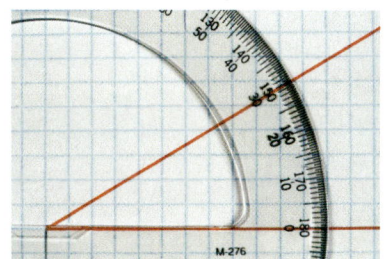

Scientific Notation

We commonly write numbers in "powers of ten," or "scientific" notation—for instance 36,900 as 3.69×10^4, or 0.0021 as 2.1×10^{-3}. One advantage of scientific notation (reviewed in Appendix A) is that it allows the number of significant figures to be clearly expressed. For example, it is not clear whether 36,900 has three, four, or five significant figures. With powers of 10 notation the ambiguity can be avoided: if the number is known to three significant figures, we write 3.69×10^4, but if it is known to four, we write 3.690×10^4.

EXERCISE C Write each of the following in scientific notation and state the number of significant figures for each: (*a*) 0.0258; (*b*) 42,300; (*c*) 344.50.

*Percent Uncertainty vs. Significant Figures

The significant figures rule is only approximate, and in some cases may underestimate the accuracy (or uncertainty) of the answer. Suppose for example we divide 97 by 92:

$$\frac{97}{92} = 1.05 \approx 1.1.$$

Both 97 and 92 have two significant figures, so the rule says to give the answer as 1.1. Yet the numbers 97 and 92 both imply an uncertainty of ± 1 if no other uncertainty is stated. Both 92 ± 1 and 97 ± 1 imply an uncertainty of about 1% ($1/92 \approx 0.01 = 1\%$). But the final result to two significant figures is 1.1, with an implied uncertainty of ± 0.1, which is an uncertainty of about 10% ($0.1/1.1 \approx 0.1 \approx 10\%$). It is better in this case to give the answer as 1.05 (which is three significant figures). Why? Because 1.05 implies an uncertainty of ± 0.01 which is $0.01/1.05 \approx 0.01 \approx 1\%$, just like the uncertainty in the original numbers 92 and 97.

SUGGESTION: Use the significant figures rule, but consider the % uncertainty too, and add an extra digit if it gives a more realistic estimate of uncertainty.

Approximations

Much of physics involves approximations, often because we do not have the means to solve a problem precisely. For example, we may choose to ignore air resistance or friction in doing a Problem even though they are present in the real world, and then our calculation is only an approximation. In doing Problems, we should be aware of what approximations we are making, and be aware that the precision of our answer may not be nearly as good as the number of significant figures given in the result.

Accuracy vs. Precision

There is a technical difference between "precision" and "accuracy." **Precision** in a strict sense refers to the repeatability of the measurement using a given instrument. For example, if you measure the width of a board many times, getting results like 8.81 cm, 8.85 cm, 8.78 cm, 8.82 cm (interpolating between the 0.1 cm marks as best as possible each time), you could say the measurements give a *precision* a bit better than 0.1 cm. **Accuracy** refers to how close a measurement is to the true value. For example, if the ruler shown in Fig. 1–5 was manufactured with a 2% error, the accuracy of its measurement of the board's width (about 8.8 cm) would be about 2% of 8.8 cm or about ± 0.2 cm. Estimated uncertainty is meant to take both accuracy and precision into account.

1–5 Units, Standards, and the SI System

The measurement of any quantity is made relative to a particular standard or **unit**, and this unit must be specified along with the numerical value of the quantity. For example, we can measure length in British units such as inches, feet, or miles, or in the metric system in centimeters, meters, or kilometers. To specify that the length of a particular object is 18.6 is insufficient. The unit *must* be given, because 18.6 meters is very different from 18.6 inches or 18.6 millimeters.

For any unit we use, such as the meter for distance or the second for time, we need to define a **standard** which defines exactly how long one meter or one second is. It is important that standards be chosen that are readily reproducible so that anyone needing to make a very accurate measurement can refer to the standard in the laboratory and communicate with other people.

Length

The first truly international standard was the **meter** (abbreviated m) established as the standard of **length** by the French Academy of Sciences in the 1790s. The standard meter was originally chosen to be one ten-millionth of the distance from the Earth's equator to either pole,[†] and a platinum rod to represent this length was made. (One meter is, very roughly, the distance from the tip of your nose to the tip of your finger, with arm and hand stretched out horizontally.) In 1889, the meter was defined more precisely as the distance between two finely engraved marks on a particular bar of platinum–iridium alloy. In 1960, to provide even greater precision and reproducibility, the meter was redefined as 1,650,763.73 wavelengths of a particular orange light emitted by the gas krypton-86. In 1983 the meter was again redefined, this time in terms of the speed of light (whose best measured value in terms of the older definition of the meter was 299,792,458 m/s, with an uncertainty of 1 m/s). The new definition reads: "The meter is the length of path traveled by light in vacuum during a time interval of 1/299,792,458 of a second."[‡]

British units of length (inch, foot, mile) are now defined in terms of the meter. The inch (in.) is defined as exactly 2.54 centimeters (cm; 1 cm = 0.01 m). Other conversion factors are given in the Table on the inside of the front cover of this book. Table 1–1 presents some typical lengths, from very small to very large, rounded off to the nearest power of 10. See also Fig. 1–8. [Note that the abbreviation for inches (in.) is the only one with a period, to distinguish it from the word "in".]

FIGURE 1–8 Some lengths: (a) viruses (about 10^{-7} m long) attacking a cell; (b) Mt. Everest's height is on the order of 10^4 m (8850 m above sea level, to be precise).

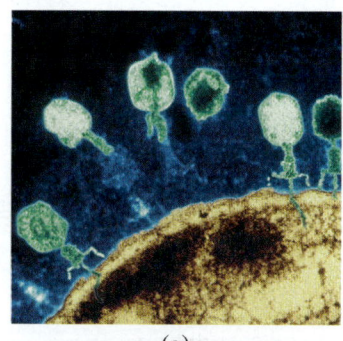

(a)

Time

The standard unit of **time** is the **second** (s). For many years, the second was defined as 1/86,400 of a mean solar day (24 h/day × 60 min/h × 60 s/min = 86,400 s/day). The standard second is now defined more precisely in terms of the frequency of radiation emitted by cesium atoms when they pass between two particular states. [Specifically, one second is defined as the time required for 9,192,631,770 oscillations of this radiation.] There are, by definition, 60 s in one minute (min) and 60 minutes in one hour (h). Table 1–2 presents a range of measured time intervals, rounded off to the nearest power of 10.

(b)

[†]Modern measurements of the Earth's circumference reveal that the intended length is off by about one-fiftieth of 1%. Not bad!

[‡]The new definition of the meter has the effect of giving the speed of light the exact value of 299,792,458 m/s.

TABLE 1–1 Some Typical Lengths or Distances (order of magnitude)	
Length (or Distance)	**Meters (approximate)**
Neutron or proton (diameter)	10^{-15} m
Atom (diameter)	10^{-10} m
Virus [see Fig. 1–8a]	10^{-7} m
Sheet of paper (thickness)	10^{-4} m
Finger width	10^{-2} m
Football field length	10^2 m
Height of Mt. Everest [see Fig. 1–8b]	10^4 m
Earth diameter	10^7 m
Earth to Sun	10^{11} m
Earth to nearest star	10^{16} m
Earth to nearest galaxy	10^{22} m
Earth to farthest galaxy visible	10^{26} m

TABLE 1–2 Some Typical Time Intervals (order of magnitude)	
Time Interval	**Seconds (approximate)**
Lifetime of very unstable subatomic particle	10^{-23} s
Lifetime of radioactive elements	10^{-22} s to 10^{28} s
Lifetime of muon	10^{-6} s
Time between human heartbeats	10^0 s (= 1 s)
One day	10^5 s
One year	3×10^7 s
Human life span	2×10^9 s
Length of recorded history	10^{11} s
Humans on Earth	10^{13} s
Age of Earth	10^{17} s
Age of Universe	4×10^{17} s

TABLE 1–3 Some Masses

Object	Kilograms (approximate)
Electron	10^{-30} kg
Proton, neutron	10^{-27} kg
DNA molecule	10^{-17} kg
Bacterium	10^{-15} kg
Mosquito	10^{-5} kg
Plum	10^{-1} kg
Human	10^{2} kg
Ship	10^{8} kg
Earth	6×10^{24} kg
Sun	2×10^{30} kg
Galaxy	10^{41} kg

Mass

The standard unit of **mass** is the **kilogram** (kg). The standard mass is a particular platinum–iridium cylinder, kept at the International Bureau of Weights and Measures near Paris, France, whose mass is defined as exactly 1 kg. A range of masses is presented in Table 1–3. [For practical purposes, 1 kg weighs about 2.2 pounds on Earth.]

When dealing with atoms and molecules, we usually use the **unified atomic mass unit** (u or amu). In terms of the kilogram,

$$1 \text{ u} = 1.6605 \times 10^{-27} \text{ kg}.$$

Precise values of this and other useful numbers are given inside the front cover.

The definitions of other standard units for other quantities will be given as we encounter them in later Chapters.

Unit Prefixes

In the metric system, the larger and smaller units are defined in multiples of 10 from the standard unit, and this makes calculation particularly easy. Thus 1 kilometer (km) is 1000 m, 1 centimeter is $\frac{1}{100}$ m, 1 millimeter (mm) is $\frac{1}{1000}$ m or $\frac{1}{10}$ cm, and so on. The prefixes "centi-," "kilo-," and others are listed in Table 1–4 and can be applied not only to units of length but to units of volume, mass, or any other unit. For example, a centiliter (cL) is $\frac{1}{100}$ liter (L), and a kilogram (kg) is 1000 grams (g). An 8.2-megapixel camera has a detector with 8,200,000 pixels (individual "picture elements").

In common usage, $1 \mu m \ (= 10^{-6} \text{ m})$ is called 1 **micron**.

Systems of Units

When dealing with the laws and equations of physics it is very important to use a consistent set of units. Several systems of units have been in use over the years. Today the most important is the **Système International** (French for International System), which is abbreviated SI. In SI units, the standard of length is the meter, the standard for time is the second, and the standard for mass is the kilogram. This system used to be called the MKS (meter-kilogram-second) system.

A second metric system is the **cgs system**, in which the centimeter, gram, and second are the standard units of length, mass, and time, as abbreviated in the title. The **British engineering system** (although more used in the U.S. than Britain) has as its standards the foot for length, the pound for force, and the second for time.

We use SI units almost exclusively in this book.

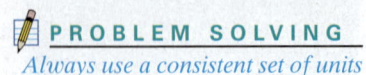
PROBLEM SOLVING
Always use a consistent set of units

*Base vs. Derived Quantities

Physical quantities can be divided into two categories: *base quantities* and *derived quantities*. The corresponding units for these quantities are called *base units* and *derived units*. A **base quantity** must be defined in terms of a standard. Scientists, in the interest of simplicity, want the smallest number of base quantities possible consistent with a full description of the physical world. This number turns out to be seven, and those used in the SI are given in Table 1–5.

TABLE 1–4 Metric (SI) Prefixes

Prefix	Abbreviation	Value
yotta	Y	10^{24}
zetta	Z	10^{21}
exa	E	10^{18}
peta	P	10^{15}
tera	T	10^{12}
giga	G	10^{9}
mega	M	10^{6}
kilo	k	10^{3}
hecto	h	10^{2}
deka	da	10^{1}
deci	d	10^{-1}
centi	c	10^{-2}
milli	m	10^{-3}
micro[†]	μ	10^{-6}
nano	n	10^{-9}
pico	p	10^{-12}
femto	f	10^{-15}
atto	a	10^{-18}
zepto	z	10^{-21}
yocto	y	10^{-24}

[†]μ is the Greek letter "mu."

TABLE 1–5 SI Base Quantities and Units

Quantity	Unit	Unit Abbreviation
Length	meter	m
Time	second	s
Mass	kilogram	kg
Electric current	ampere	A
Temperature	kelvin	K
Amount of substance	mole	mol
Luminous intensity	candela	cd

All other quantities can be defined in terms of these seven base quantities,[†] and hence are referred to as **derived quantities**. An example of a derived quantity is speed, which is defined as distance divided by the time it takes to travel that distance. A Table inside the front cover lists many derived quantities and their units in terms of base units. To define any quantity, whether base or derived, we can specify a rule or procedure, and this is called an **operational definition**.

1–6 Converting Units

Any quantity we measure, such as a length, a speed, or an electric current, consists of a number *and* a unit. Often we are given a quantity in one set of units, but we want it expressed in another set of units. For example, suppose we measure that a shelf is 21.5 inches wide, and we want to express this in centimeters. We must use a **conversion factor**, which in this case is, *by definition*, exactly

$$1 \text{ in.} = 2.54 \text{ cm}$$

or, written another way,

$$1 = 2.54 \text{ cm/in.}$$

Since multiplying by the number one does not change anything, the width of our shelf, in cm, is

$$21.5 \text{ inches} = (21.5 \text{ in.}) \times \left(2.54 \frac{\text{cm}}{\text{in.}}\right) = 54.6 \text{ cm}.$$

Note how the units (inches in this case) cancelled out (thin red lines). A Table containing many unit conversions is found inside the front cover of this book. Let's consider some Examples.

EXAMPLE 1–3 The 8000-m peaks. There are only 14 peaks whose summits are over 8000 m above sea level. They are the tallest peaks in the world (Fig. 1–9 and Table 1–6) and are referred to as "eight-thousanders." What is the elevation, in feet, of an elevation of 8000 m?

APPROACH We need to convert meters to feet, and we can start with the conversion factor 1 in. = 2.54 cm, which is exact. That is, 1 in. = 2.5400 cm to any number of significant figures, because it is *defined* to be.

SOLUTION One foot is 12 in., so we can write

$$1 \text{ ft} = (12 \text{ in.})\left(2.54 \frac{\text{cm}}{\text{in.}}\right) = 30.48 \text{ cm} = 0.3048 \text{ m},$$

which is exact. Note how the units cancel (colored slashes). We can rewrite this equation to find the number of feet in 1 meter:

$$1 \text{ m} = \frac{1 \text{ ft}}{0.3048} = 3.28084 \text{ ft}.$$

(We could carry the result to 6 significant figures because 0.3048 is exact, 0.304800···.) We multiply this equation by 8000.0 (to have five significant figures):

$$8000.0 \text{ m} = (8000.0 \text{ m})\left(3.28084 \frac{\text{ft}}{\text{m}}\right) = 26{,}247 \text{ ft}.$$

An elevation of 8000 m is 26,247 ft above sea level.

NOTE We could have done the unit conversions all in one line:

$$8000.0 \text{ m} = (8000.0 \text{ m})\left(\frac{100 \text{ cm}}{1 \text{ m}}\right)\left(\frac{1 \text{ in.}}{2.54 \text{ cm}}\right)\left(\frac{1 \text{ ft}}{12 \text{ in.}}\right) = 26{,}247 \text{ ft}.$$

The key is to multiply conversion factors, each equal to one (= 1.0000), and to make sure which units cancel.

[†]Some exceptions are for angle (radians—see Chapter 8), solid angle (steradian), and sound level (bel or decibel, Chapter 12). No general agreement has been reached as to whether these are base or derived quantities.

FIGURE 1–9 The world's second highest peak, K2, whose summit is considered the most difficult of the "8000-ers." K2 is seen here from the south (Pakistan). Example 1–3.

PHYSICS APPLIED
The world's tallest peaks

TABLE 1–6 The 8000-m Peaks	
Peak	**Height (m)**
Mt. Everest	8850
K2	8611
Kangchenjunga	8586
Lhotse	8516
Makalu	8462
Cho Oyu	8201
Dhaulagiri	8167
Manaslu	8156
Nanga Parbat	8125
Annapurna	8091
Gasherbrum I	8068
Broad Peak	8047
Gasherbrum II	8035
Shisha Pangma	8013

EXAMPLE 1-4 Apartment area. You have seen a nice apartment whose floor area is 880 square feet (ft^2). What is its area in square meters?

APPROACH We use the same conversion factor, 1 in. = 2.54 cm, but this time we have to use it twice.

SOLUTION Because 1 in. = 2.54 cm = 0.0254 m, then

$$1 \, ft^2 = (12 \, in.)^2 (0.0254 \, m/in.)^2 = 0.0929 \, m^2.$$

So

$$880 \, ft^2 = (880 \, ft^2)(0.0929 \, m^2/ft^2) \approx 82 \, m^2.$$

NOTE As a rule of thumb, an area given in ft^2 is roughly 10 times the number of square meters (more precisely, about 10.8×).

EXAMPLE 1-5 Speeds. Where the posted speed limit is 55 miles per hour (mi/h or mph), what is this speed (a) in meters per second (m/s) and (b) in kilometers per hour (km/h)?

APPROACH We again use the conversion factor 1 in. = 2.54 cm, and we recall that there are 5280 ft in a mile and 12 inches in a foot; also, one hour contains (60 min/h) × (60 s/min) = 3600 s/h.

SOLUTION (a) We can write 1 mile as

$$1 \, mi = (5280 \, \cancel{ft})\left(12 \, \frac{\cancel{in.}}{\cancel{ft}}\right)\left(2.54 \, \frac{\cancel{cm}}{\cancel{in.}}\right)\left(\frac{1 \, m}{100 \, \cancel{cm}}\right)$$

$$= 1609 \, m.$$

We also know that 1 hour contains 3600 s, so

$$55 \, \frac{mi}{h} = \left(55 \, \frac{\cancel{mi}}{\cancel{h}}\right)\left(1609 \, \frac{m}{\cancel{mi}}\right)\left(\frac{1 \, \cancel{h}}{3600 \, s}\right)$$

$$= 25 \, \frac{m}{s},$$

where we rounded off to two significant figures.

(b) Now we use 1 mi = 1609 m = 1.609 km; then

$$55 \, \frac{mi}{h} = \left(55 \, \frac{\cancel{mi}}{h}\right)\left(1.609 \, \frac{km}{\cancel{mi}}\right)$$

$$= 88 \, \frac{km}{h}.$$

NOTE Each conversion factor is equal to one. You can look up most conversion factors in the Table inside the front cover.

PROBLEM SOLVING
Conversion factors = 1

EXERCISE D Return to the first Chapter-Opening Question, page 1, and answer it again now. Try to explain why you may have answered differently the first time.

EXERCISE E Would a driver traveling at 15 m/s in a 35 mi/h zone be exceeding the speed limit? Why or why not?

PROBLEM SOLVING
Unit conversion is wrong if units do not cancel

When changing units, you can avoid making an error in the use of conversion factors by checking that units cancel out properly. For example, in our conversion of 1 mi to 1609 m in Example 1-5(a), if we had incorrectly used the factor  instead of $\left(\frac{1 \, m}{100 \, cm}\right)$, the centimeter units would not have cancelled out; we would not have ended up with meters.

1–7 Order of Magnitude: Rapid Estimating

We are sometimes interested only in an approximate value for a quantity. This might be because an accurate calculation would take more time than it is worth or would require additional data that are not available. In other cases, we may want to make a rough estimate in order to check a calculation made on a calculator, to make sure that no blunders were made when the numbers were entered.

A rough estimate can be made by rounding off all numbers to one significant figure and its power of 10, and after the calculation is made, again keeping only one significant figure. Such an estimate is called an **order-of-magnitude estimate** and can be accurate within a factor of 10, and often better. In fact, the phrase "order of magnitude" is sometimes used to refer simply to the power of 10.

Let's do some Examples.

PROBLEM SOLVING
How to make a rough estimate

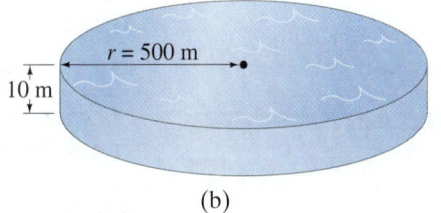

(b)

FIGURE 1–10 Example 1–6. (a) How much water is in this lake? (Photo is one of the Rae Lakes in the Sierra Nevada of California.) (b) Model of the lake as a cylinder. [We could go one step further and estimate the mass or weight of this lake. We will see later that water has a density of 1000 kg/m^3, so this lake has a mass of about $(10^3 \text{ kg/m}^3)(10^7 \text{ m}^3) \approx 10^{10} \text{ kg}$, which is about 10 billion kg or 10 million metric tons. (A metric ton is 1000 kg, about 2200 lb, slightly larger than a British ton, 2000 lb.)]

(a)

EXAMPLE 1–6 ESTIMATE **Volume of a lake.** Estimate how much water there is in a particular lake, Fig. 1–10a, which is roughly circular, about 1 km across, and you guess it has an average depth of about 10 m.

APPROACH No lake is a perfect circle, nor can lakes be expected to have a perfectly flat bottom. We are only estimating here. To estimate the volume, we can use a simple model of the lake as a cylinder: we multiply the average depth of the lake times its roughly circular surface area, as if the lake were a cylinder (Fig. 1–10b).

SOLUTION The volume V of a cylinder is the product of its height h times the area of its base: $V = h\pi r^2$, where r is the radius of the circular base.† The radius r is $\frac{1}{2}$ km = 500 m, so the volume is approximately

$$V = h\pi r^2 \approx (10 \text{ m}) \times (3) \times (5 \times 10^2 \text{ m})^2 \approx 8 \times 10^6 \text{ m}^3 \approx 10^7 \text{ m}^3,$$

where π was rounded off to 3. So the volume is on the order of 10^7 m^3, ten million cubic meters. Because of all the estimates that went into this calculation, the order-of-magnitude estimate (10^7 m^3) is probably better to quote than the $8 \times 10^6 \text{ m}^3$ figure.

NOTE To express our result in U.S. gallons, we see in the Table on the inside front cover that 1 liter = $10^{-3} \text{ m}^3 \approx \frac{1}{4}$ gallon. Hence, the lake contains $(8 \times 10^6 \text{ m}^3)(1 \text{ gallon}/4 \times 10^{-3} \text{ m}^3) \approx 2 \times 10^9$ gallons of water.

PHYSICS APPLIED
Estimating the volume (or mass) of a lake; see also Fig. 1–10

---
†Formulas like this for volume, area, etc., are found inside the back cover of this book.

FIGURE 1–11 Example 1–7. Micrometer used for measuring small thicknesses.

FIGURE 1–12 Example 1–8. Diagrams are really useful!

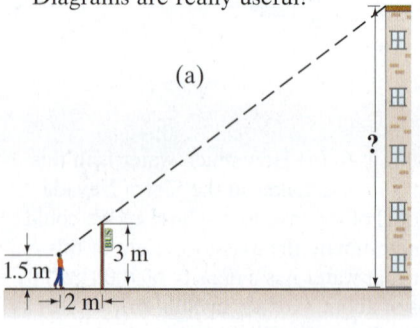

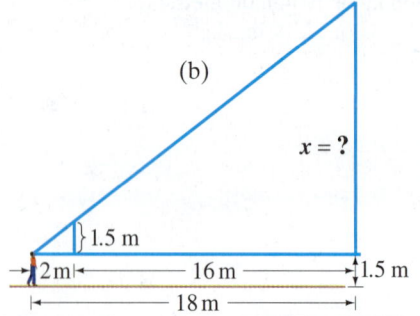

FIGURE 1–13 Enrico Fermi. Fermi contributed significantly to both theoretical and experimental physics, a feat almost unique in modern times.

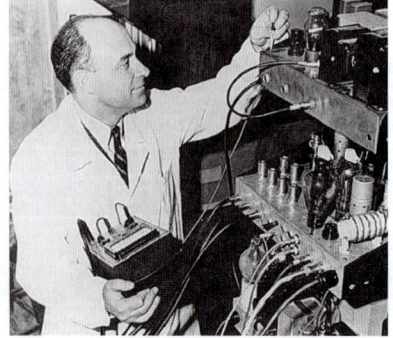

EXAMPLE 1–7 ESTIMATE **Thickness of a sheet of paper.** Estimate the thickness of a page of this book.

APPROACH At first you might think that a special measuring device, a micrometer (Fig. 1–11), is needed to measure the thickness of one page since an ordinary ruler can not be read so finely. But we can use a trick or, to put it in physics terms, make use of a *symmetry*: we can make the reasonable assumption that all the pages of this book are equal in thickness.

SOLUTION We can use a ruler to measure hundreds of pages at once. If you measure the thickness of the first 500 pages of this book (page 1 to page 500), you might get something like 1.5 cm. Note that 500 numbered pages, counted front and back, is 250 separate pieces of paper. So one sheet must have a thickness of about

$$\frac{1.5 \text{ cm}}{250 \text{ sheets}} \approx 6 \times 10^{-3} \text{ cm} = 6 \times 10^{-2} \text{ mm},$$

or less than a tenth of a millimeter (0.1 mm).

It cannot be emphasized enough how important it is to draw a diagram when solving a physics Problem, as the next Example shows.

EXAMPLE 1–8 ESTIMATE **Height by triangulation.** Estimate the height of the building shown in Fig. 1–12, by "triangulation," with the help of a bus-stop pole and a friend.

APPROACH By standing your friend next to the pole, you estimate the height of the pole to be 3 m. You next step away from the pole until the top of the pole is in line with the top of the building, Fig. 1–12a. You are 5 ft 6 in. tall, so your eyes are about 1.5 m above the ground. Your friend is taller, and when she stretches out her arms, one hand touches you, and the other touches the pole, so you estimate that distance as 2 m (Fig. 1–12a). You then pace off the distance from the pole to the base of the building with big, 1-m-long steps, and you get a total of 16 steps or 16 m.

SOLUTION Now you draw, to scale, the diagram shown in Fig. 1–12b using these measurements. You can measure, right on the diagram, the last side of the triangle to be about $x = 13$ m. Alternatively, you can use similar triangles to obtain the height x:

$$\frac{1.5 \text{ m}}{2 \text{ m}} = \frac{x}{18 \text{ m}},$$

so

$$x \approx 13\tfrac{1}{2} \text{ m}.$$

Finally you add in your eye height of 1.5 m above the ground to get your final result: the building is about 15 m tall.

Another approach, this one made famous by Enrico Fermi (1901–1954, Fig. 1–13), was to show his students how to estimate the number of piano tuners in a city, say, Chicago or San Francisco. To get a rough order-of-magnitude estimate of the number of piano tuners today in San Francisco, a city of about 800,000 inhabitants, we can proceed by estimating the number of functioning pianos, how often each piano is tuned, and how many pianos each tuner can tune. To estimate the number of pianos in San Francisco, we note that certainly not everyone has a piano. A guess of 1 family in 3 having a piano would correspond to 1 piano per 12 persons, assuming an average family of 4 persons.

As an order of magnitude, let's say 1 piano per 10 people. This is certainly more reasonable than 1 per 100 people, or 1 per every person, so let's proceed with the estimate that 1 person in 10 has a piano, or about 80,000 pianos in San Francisco. Now a piano tuner needs an hour or two to tune a piano. So let's estimate that a tuner can tune 4 or 5 pianos a day. A piano ought to be tuned every 6 months or a year—let's say once each year. A piano tuner tuning 4 pianos a day, 5 days a week, 50 weeks a year can tune about 1000 pianos a year. So San Francisco, with its (very) roughly 80,000 pianos, needs about 80 piano tuners. This is, of course, only a rough estimate.[†] It tells us that there must be many more than 10 piano tuners, and surely not as many as 1000.

PROBLEM SOLVING
Estimating how many piano tuners there are in a city

A Harder Example—But Powerful

EXAMPLE 1–9 ESTIMATE **Estimating the radius of Earth.** Believe it or not, you can estimate the radius of the Earth without having to go into space (see the photograph on page 1). If you have ever been on the shore of a large lake, you may have noticed that you cannot see the beaches, piers, or rocks at water level across the lake on the opposite shore. The lake seems to bulge out between you and the opposite shore—a good clue that the Earth is round. Suppose you climb a stepladder and discover that when your eyes are 10 ft (3.0 m) above the water, you can just see the rocks at water level on the opposite shore. From a map, you estimate the distance to the opposite shore as $d \approx 6.1$ km. Use Fig. 1–14 with $h = 3.0$ m to estimate the radius R of the Earth.

APPROACH We use simple geometry, including the theorem of Pythagoras,

$$c^2 = a^2 + b^2,$$

where c is the length of the hypotenuse of any right triangle, and a and b are the lengths of the other two sides.

SOLUTION For the right triangle of Fig. 1–14, the two sides are the radius of the Earth R and the distance $d = 6.1$ km $= 6100$ m. The hypotenuse is approximately the length $R + h$, where $h = 3.0$ m. By the Pythagorean theorem,

$$R^2 + d^2 \approx (R + h)^2$$
$$\approx R^2 + 2hR + h^2.$$

We solve algebraically for R, after cancelling R^2 on both sides:

$$R \approx \frac{d^2 - h^2}{2h} = \frac{(6100\,\text{m})^2 - (3.0\,\text{m})^2}{6.0\,\text{m}}$$
$$= 6.2 \times 10^6\,\text{m}$$
$$= 6200\,\text{km}.$$

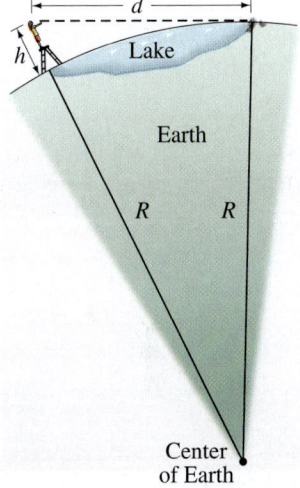

FIGURE 1–14 Example 1–9, but not to scale. You can just barely see rocks at water level on the opposite shore of a lake 6.1 km wide if you stand on a stepladder.

NOTE Precise measurements give 6380 km. But look at your achievement! With a few simple rough measurements and simple geometry, you made a good estimate of the Earth's radius. You did not need to go out in space, nor did you need a very long measuring tape.

EXERCISE F Return to the second Chapter-Opening Question, page 1, and answer it again now. Try to explain why you may have answered differently the first time.

[†]A check of the San Francisco Yellow Pages (done after this calculation) reveals about 60 listings. Each of these listings may employ more than one tuner, but on the other hand, each may also do repairs as well as tuning. In any case, our estimate is reasonable.

*1–8 Dimensions and Dimensional Analysis

When we speak of the **dimensions** of a quantity, we are referring to the type of base units or base quantities that make it up. The dimensions of area, for example, are always length squared, abbreviated $[L^2]$, using square brackets; the units can be square meters, square feet, cm^2, and so on. Velocity, on the other hand, can be measured in units of km/h, m/s, or mi/h, but the dimensions are always a length $[L]$ divided by a time $[T]$: that is, $[L/T]$.

The formula for a quantity may be different in different cases, but the dimensions remain the same. For example, the area of a triangle of base b and height h is $A = \frac{1}{2}bh$, whereas the area of a circle of radius r is $A = \pi r^2$. The formulas are different in the two cases, but the dimensions of area are always $[L^2]$.

Dimensions can be used as a help in working out relationships, a procedure referred to as **dimensional analysis**. One useful technique is the use of dimensions to check if a relationship is *incorrect*. Note that we add or subtract quantities only if they have the same dimensions (we don't add centimeters and hours); and the quantities on each side of an equals sign must have the same dimensions. (In numerical calculations, the units must also be the same on both sides of an equation.)

For example, suppose you derived the equation $v = v_0 + \frac{1}{2}at^2$, where v is the speed of an object after a time t, v_0 is the object's initial speed, and the object undergoes an acceleration a. Let's do a dimensional check to see if this equation could be correct or is surely incorrect. Note that numerical factors, like the $\frac{1}{2}$ here, do not affect dimensional checks. We write a dimensional equation as follows, remembering that the dimensions of speed are $[L/T]$ and (as we shall see in Chapter 2) the dimensions of acceleration are $[L/T^2]$:

$$\left[\frac{L}{T}\right] \stackrel{?}{=} \left[\frac{L}{T}\right] + \left[\frac{L}{T^2}\right][T^2]$$

$$\stackrel{?}{=} \left[\frac{L}{T}\right] + [L].$$

The dimensions are incorrect: on the right side, we have the sum of quantities whose dimensions are not the same. Thus we conclude that an error was made in the derivation of the original equation.

A dimensional check can only tell you when a relationship is wrong. It can't tell you if it is completely right. For example, a dimensionless numerical factor (such as $\frac{1}{2}$ or 2π) could be missing.

Dimensional analysis can also be used as a quick check on an equation you are not sure about. For example, consider a simple pendulum of length ℓ. Suppose that you can't remember whether the equation for the period T (the time to make one back-and-forth swing) is $T = 2\pi\sqrt{\ell/g}$ or $T = 2\pi\sqrt{g/\ell}$, where g is the acceleration due to gravity and, like all accelerations, has dimensions $[L/T^2]$. (Do not worry about these formulas—the correct one will be derived in Chapter 11; what we are concerned about here is a person's recalling whether it contains ℓ/g or g/ℓ.) A dimensional check shows that the former (ℓ/g) is correct:

$$[T] = \sqrt{\frac{[L]}{[L/T^2]}} = \sqrt{[T^2]} = [T],$$

whereas the latter (g/ℓ) is not:

$$[T] \neq \sqrt{\frac{[L/T^2]}{[L]}} = \sqrt{\frac{1}{[T^2]}} = \frac{1}{[T]}.$$

The constant 2π has no dimensions and so can't be checked using dimensions.

---

*Some Sections of this book, such as this one, may be considered *optional* at the discretion of the instructor, and they are marked with an asterisk (*). See the Preface for more details.

Summary

[*The Summary that appears at the end of each Chapter in this book gives a brief overview of the main ideas of the Chapter. The Summary cannot serve to give an understanding of the material, which can be accomplished only by a detailed reading of the Chapter.*]

Physics, like other sciences, is a creative endeavor. It is not simply a collection of facts. Important **theories** are created with the idea of explaining **observations**. To be accepted, theories are "tested" by comparing their predictions with the results of actual experiments. Note that, in general, a theory cannot be "proved" in an absolute sense.

Scientists often devise models of physical phenomena. A **model** is a kind of picture or analogy that helps to describe the phenomena in terms of something we already know. A **theory**, often developed from a model, is usually deeper and more complex than a simple model.

A scientific **law** is a concise statement, often expressed in the form of an equation, which quantitatively describes a wide range of phenomena.

Measurements play a crucial role in physics, but can never be perfectly precise. It is important to specify the **uncertainty** of a measurement either by stating it directly using the ± notation, and/or by keeping only the correct number of **significant figures**.

Physical quantities are always specified relative to a particular standard or **unit**, and the unit used should always be stated. The commonly accepted set of units today is the **Système International** (SI), in which the standard units of length, mass, and time are the **meter**, **kilogram**, and **second**.

When converting units, check all **conversion factors** for correct cancellation of units.

Making rough, **order-of-magnitude estimates** is a very useful technique in science as well as in everyday life.

[*The **dimensions** of a quantity refer to the combination of base quantities that comprise it. Velocity, for example, has dimensions of [length/time] or $[L/T]$. Working with only the dimensions of the various quantities in a given relationship (this technique is called **dimensional analysis**) makes it possible to check a relationship for correct form.]

Questions

1. What are the merits and drawbacks of using a person's foot as a standard? Consider both (*a*) a particular person's foot, and (*b*) any person's foot. Keep in mind that it is advantageous that fundamental standards be accessible (easy to compare to), invariable (do not change), indestructible, and reproducible.

2. What is wrong with this road sign:

    Memphis 7 mi (11.263 km)?

3. Why is it incorrect to think that the more digits you include in your answer, the more accurate it is?

4. For an answer to be complete, the units need to be specified. Why?

5. You measure the radius of a wheel to be 4.16 cm. If you multiply by 2 to get the diameter, should you write the result as 8 cm or as 8.32 cm? Justify your answer.

6. Express the sine of 30.0° with the correct number of significant figures.

7. List assumptions useful to estimate the number of car mechanics in (*a*) San Francisco, (*b*) your hometown, and then make the estimates.

MisConceptual Questions

[List all answers that are valid.]

1. A student weighs herself on a digital bathroom scale as 117.4 lb. If all the digits displayed reflect the true precision of the scale, then probably her weight is
   (*a*) within 1% of 117.4 lb.
   (*b*) exactly 117.4 lb.
   (*c*) somewhere between 117.38 and 117.42 lb.
   (*d*) roughly between 117.2 and 117.6 lb.

2. Four students use different instruments to measure the length of the same pen. Which measurement implies the greatest precision?
   (*a*) 160.0 mm. (*b*) 16.0 cm. (*c*) 0.160 m. (*d*) 0.00016 km.
   (*e*) Need more information.

3. The number 0.0078 has how many significant figures?
   (*a*) 1.   (*b*) 2.   (*c*) 3.   (*d*) 4.

4. How many significant figures does 1.362 + 25.2 have?
   (*a*) 2.   (*b*) 3.   (*c*) 4.   (*d*) 5.

5. Accuracy represents
   (*a*) repeatability of a measurement, using a given instrument.
   (*b*) how close a measurement is to the true value.
   (*c*) an ideal number of measurements to make.
   (*d*) how poorly an instrument is operating.

6. To convert from ft^2 to yd^2, you should
   (*a*) multiply by 3.
   (*b*) multiply by 1/3.
   (*c*) multiply by 9.
   (*d*) multiply by 1/9.
   (*e*) multiply by 6.
   (*f*) multiply by 1/6.

7. Which is *not* true about an order-of-magnitude estimation?
   (*a*) It gives you a rough idea of the answer.
   (*b*) It can be done by keeping only one significant figure.
   (*c*) It can be used to check if an exact calculation is reasonable.
   (*d*) It may require making some reasonable assumptions in order to calculate the answer.
   (*e*) It will always be accurate to at least two significant figures.

*8. $[L^2]$ represents the dimensions for which of the following?
   (*a*) cm^2.
   (*b*) square feet.
   (*c*) m^2.
   (*d*) All of the above.

For assigned homework and other learning materials, go to the MasteringPhysics website.

Problems

[*The Problems at the end of each Chapter are ranked* I, II, *or* III *according to estimated difficulty, with* (I) *Problems being easiest. Level* III *are meant as challenges for the best students. The Problems are arranged by Section, meaning that the reader should have read up to and including that Section, but not only that Section—Problems often depend on earlier material. Next is a set of "General Problems" not arranged by Section and not ranked. Finally, there are "Search and Learn" Problems that require rereading parts of the Chapter.*]

1–4 Measurement, Uncertainty, Significant Figures

(*Note: In Problems, assume a number like 6.4 is accurate to ± 0.1; and 950 is ± 10 unless 950 is said to be "precisely" or "very nearly" 950, in which case assume 950 ± 1.*)

1. (I) How many significant figures do each of the following numbers have: (*a*) 214, (*b*) 81.60, (*c*) 7.03, (*d*) 0.03, (*e*) 0.0086, (*f*) 3236, and (*g*) 8700?
2. (I) Write the following numbers in powers of 10 notation: (*a*) 1.156, (*b*) 21.8, (*c*) 0.0068, (*d*) 328.65, (*e*) 0.219, and (*f*) 444.
3. (I) Write out the following numbers in full with the correct number of zeros: (*a*) 8.69×10^4, (*b*) 9.1×10^3, (*c*) 8.8×10^{-1}, (*d*) 4.76×10^2, and (*e*) 3.62×10^{-5}.
4. (II) The age of the universe is thought to be about 14 billion years. Assuming two significant figures, write this in powers of 10 in (*a*) years, (*b*) seconds.
5. (II) What is the percent uncertainty in the measurement 5.48 ± 0.25 m?
6. (II) Time intervals measured with a stopwatch typically have an uncertainty of about 0.2 s, due to human reaction time at the start and stop moments. What is the percent uncertainty of a hand-timed measurement of (*a*) 5.5 s, (*b*) 55 s, (*c*) 5.5 min?
7. (II) Add $(9.2 \times 10^3 \text{ s}) + (8.3 \times 10^4 \text{ s}) + (0.008 \times 10^6 \text{ s})$.
8. (II) Multiply 3.079×10^2 m by 0.068×10^{-1} m, taking into account significant figures.
9. (II) What, approximately, is the percent uncertainty for a measurement given as 1.57 m^2?
10. (III) What, roughly, is the percent uncertainty in the volume of a spherical beach ball of radius $r = 0.84 \pm 0.04$ m?
11. (III) What is the area, and its approximate uncertainty, of a circle of radius 3.1×10^4 cm?

1–5 and 1–6 Units, Standards, SI, Converting Units

12. (I) Write the following as full (decimal) numbers without prefixes on the units: (*a*) 286.6 mm, (*b*) 85 μV, (*c*) 760 mg, (*d*) 62.1 ps, (*e*) 22.5 nm, (*f*) 2.50 gigavolts.
13. (I) Express the following using the prefixes of Table 1–4: (*a*) 1×10^6 volts, (*b*) 2×10^{-6} meters, (*c*) 6×10^3 days, (*d*) 18×10^2 bucks, and (*e*) 7×10^{-7} seconds.
14. (I) One hectare is defined as $1.000 \times 10^4 \text{ m}^2$. One acre is $4.356 \times 10^4 \text{ ft}^2$. How many acres are in one hectare?
15. (II) The Sun, on average, is 93 million miles from Earth. How many meters is this? Express (*a*) using powers of 10, and (*b*) using a metric prefix (km).
16. (II) Express the following sum with the correct number of significant figures: $1.80 \text{ m} + 142.5 \text{ cm} + 5.34 \times 10^5 \mu\text{m}$.
17. (II) A typical atom has a diameter of about 1.0×10^{-10} m. (*a*) What is this in inches? (*b*) Approximately how many atoms are along a 1.0-cm line, assuming they just touch?
18. (II) Determine the conversion factor between (*a*) km/h and mi/h, (*b*) m/s and ft/s, and (*c*) km/h and m/s.
19. (II) A **light-year** is the distance light travels in one year (at speed $= 2.998 \times 10^8$ m/s). (*a*) How many meters are there in 1.00 light-year? (*b*) An astronomical unit (AU) is the average distance from the Sun to Earth, 1.50×10^8 km. How many AU are there in 1.00 light-year?
20. (II) How much longer (percentage) is a one-mile race than a 1500-m race ("the metric mile")?
21. (II) American football uses a field that is 100.0 yd long, whereas a soccer field is 100.0 m long. Which field is longer, and by how much (give yards, meters, and percent)?
22. (II) (*a*) How many seconds are there in 1.00 year? (*b*) How many nanoseconds are there in 1.00 year? (*c*) How many years are there in 1.00 second?
23. (II) Use Table 1–3 to estimate the total number of protons or neutrons in (*a*) a bacterium, (*b*) a DNA molecule, (*c*) the human body, (*d*) our Galaxy.
24. (III) A standard baseball has a circumference of approximately 23 cm. If a baseball had the same mass per unit volume (see Tables in Section 1–5) as a neutron or a proton, about what would its mass be?

1–7 Order-of-Magnitude Estimating

(*Note: Remember that for rough estimates, only round numbers are needed both as input to calculations and as final results.*)

25. (I) Estimate the order of magnitude (power of 10) of: (*a*) 2800, (*b*) 86.30×10^3, (*c*) 0.0076, and (*d*) 15.0×10^8.
26. (II) Estimate how many books can be shelved in a college library with 3500 m² of floor space. Assume 8 shelves high, having books on both sides, with corridors 1.5 m wide. Assume books are about the size of this one, on average.
27. (II) Estimate how many hours it would take to run (at 10 km/h) across the U.S. from New York to California.
28. (II) Estimate the number of liters of water a human drinks in a lifetime.
29. (II) Estimate how long it would take one person to mow a football field using an ordinary home lawn mower (Fig. 1–15). (State your assumption, such as the mower moves with a 1-km/h speed, and has a 0.5-m width.)

FIGURE 1–15 Problem 29.

30. (II) Estimate the number of gallons of gasoline consumed by the total of all automobile drivers in the U.S., per year.
31. (II) Estimate the number of dentists (*a*) in San Francisco and (*b*) in your town or city.
32. (III) You are in a hot air balloon, 200 m above the flat Texas plains. You look out toward the horizon. How far out can you see—that is, how far is your horizon? The Earth's radius is about 6400 km.

33. (III) I agree to hire you for 30 days. You can decide between two methods of payment: either (1) $1000 a day, or (2) one penny on the first day, two pennies on the second day and continue to double your daily pay each day up to day 30. Use quick estimation to make your decision, and justify it.

34. (III) Many sailboats are docked at a marina 4.4 km away on the opposite side of a lake. You stare at one of the sailboats because, when you are lying flat at the water's edge, you can just see its deck but none of the side of the sailboat. You then go to that sailboat on the other side of the lake and measure that the deck is 1.5 m above the level of the water. Using Fig. 1–16, where $h = 1.5$ m, estimate the radius R of the Earth.

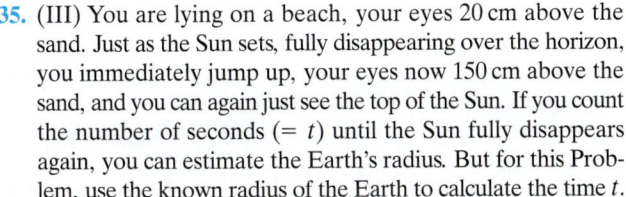

FIGURE 1–16 Problem 34. You see a sailboat across a lake (not to scale). R is the radius of the Earth. Because of the curvature of the Earth, the water "bulges out" between you and the boat.

35. (III) You are lying on a beach, your eyes 20 cm above the sand. Just as the Sun sets, fully disappearing over the horizon, you immediately jump up, your eyes now 150 cm above the sand, and you can again just see the top of the Sun. If you count the number of seconds ($= t$) until the Sun fully disappears again, you can estimate the Earth's radius. But for this Problem, use the known radius of the Earth to calculate the time t.

*1–8 Dimensions

*36. (I) What are the dimensions of density, which is mass per volume?

*37. (II) The speed v of an object is given by the equation $v = At^3 - Bt$, where t refers to time. (a) What are the dimensions of A and B? (b) What are the SI units for the constants A and B?

*38. (II) Three students derive the following equations in which x refers to distance traveled, v the speed, a the acceleration (m/s^2), t the time, and the subscript zero ($_0$) means a quantity at time $t = 0$. Here are their equations: (a) $x = vt^2 + 2at$, (b) $x = v_0 t + \frac{1}{2}at^2$, and (c) $x = v_0 t + 2at^2$. Which of these could possibly be correct according to a dimensional check, and why?

*39. (III) The smallest meaningful measure of length is called the **Planck length**, and is defined in terms of three fundamental constants in nature: the speed of light $c = 3.00 \times 10^8$ m/s, the gravitational constant $G = 6.67 \times 10^{-11}$ m^3/kg·s^2, and Planck's constant $h = 6.63 \times 10^{-34}$ kg·m^2/s. The Planck length ℓ_P is given by the following combination of these three constants:

$$\ell_P = \sqrt{\frac{Gh}{c^3}}.$$

Show that the dimensions of ℓ_P are length $[L]$, and find the order of magnitude of ℓ_P. [Recent theories (Chapters 32 and 33) suggest that the smallest particles (quarks, leptons) are "strings" with lengths on the order of the Planck length, 10^{-35} m. These theories also suggest that the "Big Bang," with which the universe is believed to have begun, started from an initial size on the order of the Planck length.]

General Problems

40. **Global positioning satellites (GPS)** can be used to determine your position with great accuracy. If one of the satellites is 20,000 km from you, and you want to know your position to ± 2 m, what percent uncertainty in the distance is required? How many significant figures are needed in the distance?

41. **Computer chips** (Fig. 1–17) are etched on circular silicon wafers of thickness 0.300 mm that are sliced from a solid cylindrical silicon crystal of length 25 cm. If each wafer can hold 400 chips, what is the maximum number of chips that can be produced from one entire cylinder?

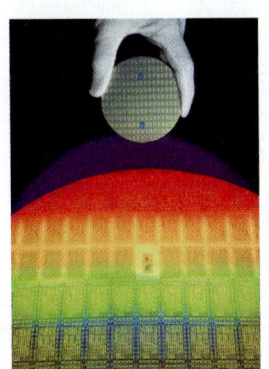

FIGURE 1–17 Problem 41. The wafer held by the hand is shown below, enlarged and illuminated by colored light. Visible are rows of integrated circuits (chips).

42. A typical adult human lung contains about 300 million tiny cavities called alveoli. Estimate the average diameter of a single alveolus.

43. If you used only a keyboard to enter data, how many years would it take to fill up the hard drive in a computer that can store 1.0 terabytes (1.0×10^{12} bytes) of data? Assume 40-hour work weeks, and that you can type 180 characters per minute, and that one byte is one keyboard character.

44. An average family of four uses roughly 1200 L (about 300 gallons) of water per day (1 L = 1000 cm^3). How much depth would a lake lose per year if it covered an area of 50 km^2 with uniform depth and supplied a local town with a population of 40,000 people? Consider only population uses, and neglect evaporation, rain, creeks and rivers.

45. Estimate the number of jelly beans in the jar of Fig. 1–18.

FIGURE 1–18 Problem 45. Estimate the number of jelly beans in the jar.

46. How big is a ton? That is, what is the volume of something that weighs a ton? To be specific, estimate the diameter of a 1-ton rock, but first make a wild guess: will it be 1 ft across, 3 ft, or the size of a car? [*Hint*: Rock has mass per volume about 3 times that of water, which is 1 kg per liter (10^3 cm^3) or 62 lb per cubic foot.]
47. A certain compact disc (CD) contains 783.216 megabytes of digital information. Each byte consists of exactly 8 bits. When played, a CD player reads the CD's information at a constant rate of 1.4 megabits per second. How many minutes does it take the player to read the entire CD?
48. Hold a pencil in front of your eye at a position where its blunt end just blocks out the Moon (Fig. 1–19). Make appropriate measurements to estimate the diameter of the Moon, given that the Earth–Moon distance is 3.8×10^5 km.

FIGURE 1–19 Problem 48. How big is the Moon?

49. A storm dumps 1.0 cm of rain on a city 6 km wide and 8 km long in a 2-h period. How many metric tons (1 metric ton = 10^3 kg) of water fell on the city? (1 cm^3 of water has a mass of 1 g = 10^{-3} kg.) How many gallons of water was this?
50. Estimate how many days it would take to walk around the Earth, assuming 12 h walking per day at 4 km/h.
51. One liter (1000 cm^3) of oil is spilled onto a smooth lake. If the oil spreads out uniformly until it makes an oil slick just one molecule thick, with adjacent molecules just touching, estimate the diameter of the oil slick. Assume the oil molecules have a diameter of 2×10^{-10} m.
52. A watch manufacturer claims that its watches gain or lose no more than 8 seconds in a year. How accurate are these watches, expressed as a percentage?
53. An angstrom (symbol Å) is a unit of length, defined as 10^{-10} m, which is on the order of the diameter of an atom. (a) How many nanometers are in 1.0 angstrom? (b) How many femtometers or fermis (the common unit of length in nuclear physics) are in 1.0 angstrom? (c) How many angstroms are in 1.0 m? (d) How many angstroms are in 1.0 light-year (see Problem 19)?
54. Jim stands beside a wide river and wonders how wide it is. He spots a large rock on the bank directly across from him. He then walks upstream 65 strides and judges that the angle between him and the rock, which he can still see, is now at an angle of 30° downstream (Fig. 1–20). Jim measures his stride to be about 0.8 m long. Estimate the width of the river.

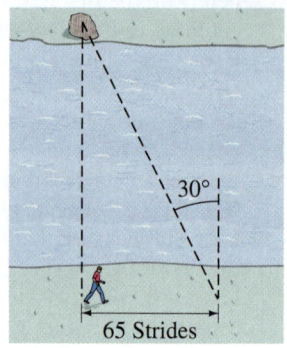

FIGURE 1–20 Problem 54.

55. Determine the percent uncertainty in θ, and in sin θ, when (a) $\theta = 15.0° \pm 0.5°$, (b) $\theta = 75.0° \pm 0.5°$.
56. If you walked north along one of Earth's lines of longitude until you had changed latitude by 1 minute of arc (there are 60 minutes per degree), how far would you have walked (in miles)? This distance is a **nautical mile**.
57. Make a rough estimate of the volume of your body (in m^3).
58. The following formula estimates an average person's lung capacity V (in liters, where 1 L = 10^3 cm^3):
$$V = 4.1H - 0.018A - 2.7,$$
where H and A are the person's height (in meters) and age (in years), respectively. In this formula, what are the units of the numbers 4.1, 0.018, and 2.7?
59. One mole of atoms consists of 6.02×10^{23} individual atoms. If a mole of atoms were spread uniformly over the Earth's surface, how many atoms would there be per square meter?
60. The density of an object is defined as its mass divided by its volume. Suppose a rock's mass and volume are measured to be 6 g and 2.8325 cm^3. To the correct number of significant figures, determine the rock's density (mass/volume).
61. Recent findings in astrophysics suggest that the observable universe can be modeled as a sphere of radius $R = 13.7 \times 10^9$ light-years = 13.0×10^{25} m with an average total mass density of about 1×10^{-26} kg/m^3. Only about 4% of total mass is due to "ordinary" matter (such as protons, neutrons, and electrons). Estimate how much ordinary matter (in kg) there is in the observable universe. (For the light-year, see Problem 19.)

Search and Learn

1. Galileo is to Aristotle as Copernicus is to Ptolemy. See Section 1–1 and explain this analogy.
2. How many wavelengths of orange krypton-86 light (Section 1–5) would fit into the thickness of one page of this book?
3. Using the French Academy of Sciences' original definition of the meter, determine Earth's circumference and radius in those meters.
4. Estimate the ratio (order of magnitude) of the mass of a human to the mass of a DNA molecule.
5. To the correct number of significant figures, use the information inside the front cover of this book to determine the ratio of (a) the surface area of Earth compared to the surface area of the Moon; (b) the volume of Earth compared to the volume of the Moon.

ANSWERS TO EXERCISES

A: (d).
B: All three have three significant figures; the number of decimal places is (a) 2, (b) 3, (c) 4.
C: (a) 2.58×10^{-2}, 3; (b) 4.23×10^4, 3 (probably); (c) 3.4450×10^2, 5.
D: (f).
E: No: 15 m/s ≈ 34 mi/h.
F: (c).

The space shuttle has released a parachute to reduce its speed quickly. The directions of the shuttle's velocity and acceleration are shown by the green ($\vec{v}$) and gold ($\vec{a}$) arrows.

Motion is described using the concepts of velocity and acceleration. In the case shown here, the velocity $\vec{v}$ is to the right, in the direction of motion. The acceleration $\vec{a}$ is in the opposite direction from the velocity $\vec{v}$, which means the object is slowing down.

We examine in detail motion with constant acceleration, including the vertical motion of objects falling under gravity.

CHAPTER 2

Describing Motion: Kinematics in One Dimension

CHAPTER-OPENING QUESTION—Guess now!

[*Don't worry about getting the right answer now—you will get another chance later in the Chapter. See also p. 1 of Chapter 1 for more explanation.*]

Two small heavy balls have the same diameter but one weighs twice as much as the other. The balls are dropped from a second-story balcony at the exact same time. The time to reach the ground below will be:

(a) twice as long for the lighter ball as for the heavier one.
(b) longer for the lighter ball, but not twice as long.
(c) twice as long for the heavier ball as for the lighter one.
(d) longer for the heavier ball, but not twice as long.
(e) nearly the same for both balls.

CONTENTS

2–1 Reference Frames and Displacement
2–2 Average Velocity
2–3 Instantaneous Velocity
2–4 Acceleration
2–5 Motion at Constant Acceleration
2–6 Solving Problems
2–7 Freely Falling Objects
2–8 Graphical Analysis of Linear Motion

The motion of objects—baseballs, automobiles, joggers, and even the Sun and Moon—is an obvious part of everyday life. It was not until the sixteenth and seventeenth centuries that our modern understanding of motion was established. Many individuals contributed to this understanding, particularly Galileo Galilei (1564–1642) and Isaac Newton (1642–1727).

The study of the motion of objects, and the related concepts of force and energy, form the field called **mechanics**. Mechanics is customarily divided into two parts: **kinematics**, which is the description of how objects move, and **dynamics**, which deals with force and why objects move as they do. This Chapter and the next deal with kinematics.

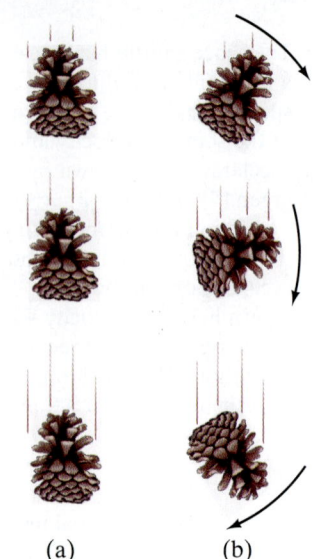

FIGURE 2–1 A falling pinecone undergoes (a) pure translation; (b) it is rotating as well as translating.

For now we only discuss objects that move without rotating (Fig. 2–1a). Such motion is called **translational motion**. In this Chapter we will be concerned with describing an object that moves along a straight-line path, which is one-dimensional translational motion. In Chapter 3 we will describe translational motion in two (or three) dimensions along paths that are not straight. (Rotation, shown in Fig. 2–1b, is discussed in Chapter 8.)

We will often use the concept, or *model*, of an idealized **particle** which is considered to be a mathematical **point** with no spatial extent (no size). A point particle can undergo only translational motion. The particle model is useful in many real situations where we are interested only in translational motion and the object's size is not significant. For example, we might consider a billiard ball, or even a spacecraft traveling toward the Moon, as a particle for many purposes.

2–1 Reference Frames and Displacement

Any measurement of position, distance, or speed must be made with respect to a **reference frame**, or **frame of reference**. For example, while you are on a train traveling at 80 km/h, suppose a person walks past you toward the front of the train at a speed of, say, 5 km/h (Fig. 2–2). This 5 km/h is the person's speed with respect to the train as frame of reference. With respect to the ground, that person is moving at a speed of 80 km/h + 5 km/h = 85 km/h. It is always important to specify the frame of reference when stating a speed. In everyday life, we usually mean "with respect to the Earth" without even thinking about it, but the reference frame must be specified whenever there might be confusion.

FIGURE 2–2 A person walks toward the front of a train at 5 km/h. The train is moving 80 km/h with respect to the ground, so the walking person's speed, relative to the ground, is 85 km/h.

When specifying the motion of an object, it is important to specify not only the speed but also the direction of motion. Often we can specify a direction by using north, east, south, and west, and by "up" and "down." In physics, we often draw a set of **coordinate axes**, as shown in Fig. 2–3, to represent a frame of reference. We can always place the origin 0, and the directions of the x and y axes, as we like for convenience. The x and y axes are always perpendicular to each other. The **origin** is where $x = 0$, $y = 0$. Objects positioned to the right of the origin of coordinates (0) on the x axis have an x coordinate which we almost always choose to be positive; then points to the left of 0 have a negative x coordinate. The position along the y axis is usually considered positive when above 0, and negative when below 0, although the reverse convention can be used if convenient. Any point on the plane can be specified by giving its x and y coordinates. In three dimensions, a z axis perpendicular to the x and y axes is added.

For one-dimensional motion, we often choose the x axis as the line along which the motion takes place. Then the **position** of an object at any moment is given by its x coordinate. If the motion is vertical, as for a dropped object, we usually use the y axis.

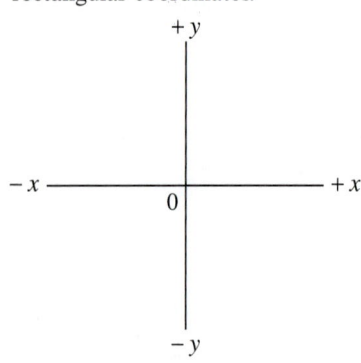

FIGURE 2–3 Standard set of xy coordinate axes, sometimes called "rectangular coordinates."

We need to make a distinction between the *distance* an object has traveled and its **displacement**, which is defined as the *change in position* of the object. That is, *displacement is how far the object is from its starting point.* To see the distinction between total distance and displacement, imagine a person walking 70 m to the east and then turning around and walking back (west) a distance of 30 m (see Fig. 2–4). The total *distance* traveled is 100 m, but the *displacement* is only 40 m since the person is now only 40 m from the starting point.

Displacement is a quantity that has both magnitude and direction. Such quantities are called **vectors**, and are represented by arrows in diagrams. For example, in Fig. 2–4, the blue arrow represents the displacement whose magnitude is 40 m and whose direction is to the right (east).

We will deal with vectors more fully in Chapter 3. For now, we deal only with motion in one dimension, along a line. In this case, vectors which point in one direction will be positive (typically to the right along the x axis). Vectors that point in the opposite direction will have a negative sign in front of their magnitude.

Consider the motion of an object over a particular time interval. Suppose that at some initial time, call it t_1, the object is on the x axis at the position x_1 in the coordinate system shown in Fig. 2–5. At some later time, t_2, suppose the object has moved to position x_2. The displacement of our object is $x_2 - x_1$, and is represented by the arrow pointing to the right in Fig. 2–5. It is convenient to write

$$\Delta x = x_2 - x_1,$$

where the symbol Δ (Greek letter delta) means "change in." Then Δx means "the change in x," or "change in position," which is the displacement. The **change in** any quantity means *the final value of that quantity, minus the initial value.* Suppose $x_1 = 10.0$ m and $x_2 = 30.0$ m, as in Fig. 2–5. Then

$$\Delta x = x_2 - x_1 = 30.0 \text{ m} - 10.0 \text{ m} = 20.0 \text{ m},$$

so the displacement is 20.0 m in the positive direction, Fig. 2–5.

Now consider an object moving to the left as shown in Fig. 2–6. Here the object, a person, starts at $x_1 = 30.0$ m and walks to the left to the point $x_2 = 10.0$ m. In this case her displacement is

$$\Delta x = x_2 - x_1 = 10.0 \text{ m} - 30.0 \text{ m} = -20.0 \text{ m},$$

and the blue arrow representing the vector displacement points to the left. For one-dimensional motion along the x axis, a vector pointing to the right is positive, whereas a vector pointing to the left has a negative sign.

EXERCISE A An ant starts at $x = 20$ cm on a piece of graph paper and walks along the x axis to $x = -20$ cm. It then turns around and walks back to $x = -10$ cm. Determine (a) the ant's displacement and (b) the total distance traveled.

2–2 Average Velocity

An important aspect of the motion of a moving object is how *fast* it is moving—its speed or velocity.

The term "speed" refers to how far an object travels in a given time interval, regardless of direction. If a car travels 240 kilometers (km) in 3 hours (h), we say its average speed was 80 km/h. In general, the **average speed** of an object is defined as *the total distance traveled along its path divided by the time it takes to travel this distance:*

$$\text{average speed} = \frac{\text{distance traveled}}{\text{time elapsed}}. \quad (2\text{–}1)$$

The terms "velocity" and "speed" are often used interchangeably in ordinary language. But in physics we make a distinction between the two. Speed is simply a positive number, with units. **Velocity**, on the other hand, is used to signify both the *magnitude* (numerical value) of how fast an object is moving and also the *direction* in which it is moving. Velocity is therefore a *vector*.

CAUTION
The displacement may not equal the total distance traveled

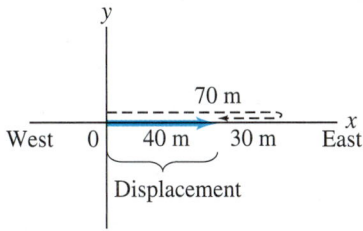

FIGURE 2–4 A person walks 70 m east, then 30 m west. The total distance traveled is 100 m (path is shown dashed in black); but the displacement, shown as a solid blue arrow, is 40 m to the east.

FIGURE 2–5 The arrow represents the displacement $x_2 - x_1$. Distances are in meters.

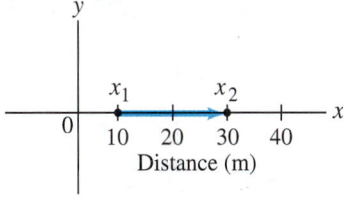

FIGURE 2–6 For the displacement $\Delta x = x_2 - x_1 = 10.0 \text{ m} - 30.0 \text{ m}$, the displacement vector points left.

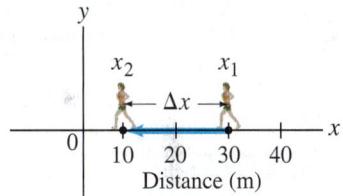

There is a second difference between speed and velocity: namely, the *average velocity* is defined in terms of *displacement*, rather than total distance traveled:

$$\text{average velocity} = \frac{\text{displacement}}{\text{time elapsed}} = \frac{\text{final position} - \text{initial position}}{\text{time elapsed}}.$$

> **CAUTION**
> *Average speed is not necessarily equal to the magnitude of the average velocity*

Average speed and average velocity have the same magnitude when the motion is all in one direction. In other cases, they may differ: recall the walk we described earlier, in Fig. 2–4, where a person walked 70 m east and then 30 m west. The total distance traveled was 70 m + 30 m = 100 m, but the displacement was 40 m. Suppose this walk took 70 s to complete. Then the average speed was:

$$\frac{\text{distance}}{\text{time elapsed}} = \frac{100 \text{ m}}{70 \text{ s}} = 1.4 \text{ m/s}.$$

The magnitude of the average velocity, on the other hand, was:

$$\frac{\text{displacement}}{\text{time elapsed}} = \frac{40 \text{ m}}{70 \text{ s}} = 0.57 \text{ m/s}.$$

To discuss one-dimensional motion of an object in general, suppose that at some moment in time, call it t_1, the object is on the x axis at position x_1 in a coordinate system, and at some later time, t_2, suppose it is at position x_2. The **elapsed time** (= change in time) is $\Delta t = t_2 - t_1$; during this time interval the displacement of our object is $\Delta x = x_2 - x_1$. Then the **average velocity**, defined as *the displacement divided by the elapsed time*, can be written

$$\bar{v} = \frac{x_2 - x_1}{t_2 - t_1} = \frac{\Delta x}{\Delta t}, \qquad \text{[average velocity]} \quad (2\text{–}2)$$

where v stands for velocity and the bar (¯) over the v is a standard symbol meaning "average."

For one-dimensional motion in the usual case of the $+x$ axis to the right, note that if x_2 is less than x_1, the object is moving to the left, and then $\Delta x = x_2 - x_1$ is less than zero. The sign of the displacement, and thus of the average velocity, indicates the direction: the average velocity is positive for an object moving to the right along the x axis and negative when the object moves to the left. The direction of the average velocity is always the same as the direction of the displacement.

> **PROBLEM SOLVING**
> *+ or − sign can signify the direction for linear motion*

> **CAUTION**
> *Time interval = elapsed time*

It is always important to choose (and state) the *elapsed time*, or **time interval**, $t_2 - t_1$, the time that passes during our chosen period of observation.

EXAMPLE 2–1 Runner's average velocity. The position of a runner is plotted as moving along the x axis of a coordinate system. During a 3.00-s time interval, the runner's position changes from $x_1 = 50.0$ m to $x_2 = 30.5$ m, as shown in Fig. 2–7. What is the runner's average velocity?

APPROACH We want to find the average velocity, which is the displacement divided by the elapsed time.

SOLUTION The displacement is

$$\Delta x = x_2 - x_1$$
$$= 30.5 \text{ m} - 50.0 \text{ m} = -19.5 \text{ m}.$$

The elapsed time, or time interval, is given as $\Delta t = 3.00$ s. The average velocity (Eq. 2–2) is

$$\bar{v} = \frac{\Delta x}{\Delta t} = \frac{-19.5 \text{ m}}{3.00 \text{ s}} = -6.50 \text{ m/s}.$$

The displacement and average velocity are negative, which tells us that the runner is moving to the left along the x axis, as indicated by the arrow in Fig. 2–7. The runner's average velocity is 6.50 m/s to the left.

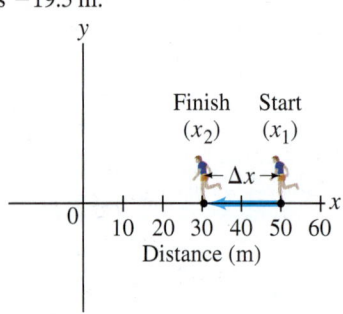

FIGURE 2–7 Example 2–1. A person runs from $x_1 = 50.0$ m to $x_2 = 30.5$ m. The displacement is −19.5 m.

EXAMPLE 2–2 Distance a cyclist travels. How far can a cyclist travel in 2.5 h along a straight road if her average velocity is 18 km/h?

APPROACH We want to find the distance traveled, so we solve Eq. 2–2 for Δx.

SOLUTION In Eq. 2–2, $\bar{v} = \Delta x/\Delta t$, we multiply both sides by Δt and obtain
$$\Delta x = \bar{v}\,\Delta t = (18\,\text{km/h})(2.5\,\text{h}) = 45\,\text{km}.$$

EXAMPLE 2–3 Car changes speed. A car travels at a constant 50 km/h for 100 km. It then speeds up to 100 km/h and is driven another 100 km. What is the car's average speed for the 200-km trip?

APPROACH At 50 km/h, the car takes 2.0 h to travel 100 km. At 100 km/h it takes only 1.0 h to travel 100 km. We use the defintion of average velocity, Eq. 2–2.

SOLUTION Average velocity (Eq. 2–2) is
$$\bar{v} = \frac{\Delta x}{\Delta t} = \frac{100\,\text{km} + 100\,\text{km}}{2.0\,\text{h} + 1.0\,\text{h}} = 67\,\text{km/h}.$$

NOTE Averaging the two speeds, (50 km/h + 100 km/h)/2 = 75 km/h, gives a wrong answer. Can you see why? You must use the definition of $\bar{v}$, Eq. 2–2.

2–3 Instantaneous Velocity

If you drive a car along a straight road for 150 km in 2.0 h, the magnitude of your average velocity is 75 km/h. It is unlikely, though, that you were moving at precisely 75 km/h at every instant. To describe this situation we need the concept of *instantaneous velocity*, which is the velocity at any instant of time. (Its magnitude is the number, with units, indicated by a speedometer, Fig. 2–8.) More precisely, the **instantaneous velocity** at any moment is defined as *the average velocity over an infinitesimally short time interval*. That is, Eq. 2–2 is to be evaluated in the limit of Δt becoming extremely small, approaching zero. We can write the definition of instantaneous velocity, v, for one-dimensional motion as

$$v = \lim_{\Delta t \to 0} \frac{\Delta x}{\Delta t}. \quad \text{[instantaneous velocity]} \quad (2\text{–}3)$$

The notation $\lim_{\Delta t \to 0}$ means the ratio $\Delta x/\Delta t$ is to be evaluated in the limit of Δt approaching zero.[†]

For instantaneous velocity we use the symbol v, whereas for average velocity we use $\bar{v}$, with a bar above. In the rest of this book, when we use the term "velocity" it will refer to instantaneous velocity. When we want to speak of the average velocity, we will make this clear by including the word "average."

Note that the *instantaneous speed* always equals the magnitude of the instantaneous velocity. Why? Because distance traveled and the magnitude of the displacement become the same when they become infinitesimally small.

If an object moves at a uniform (that is, constant) velocity during a particular time interval, then its instantaneous velocity at any instant is the same as its average velocity (see Fig. 2–9a). But in many situations this is not the case. For example, a car may start from rest, speed up to 50 km/h, remain at that velocity for a time, then slow down to 20 km/h in a traffic jam, and finally stop at its destination after traveling a total of 15 km in 30 min. This trip is plotted on the graph of Fig. 2–9b. Also shown on the graph is the average velocity (dashed line), which is $\bar{v} = \Delta x/\Delta t = 15\,\text{km}/0.50\,\text{h} = 30\,\text{km/h}$.

Graphs are often useful for analysis of motion; we discuss additional insights graphs can provide as we go along, especially in Section 2–8.

EXERCISE B What is your instantaneous speed at the instant you turn around to move in the opposite direction? (*a*) Depends on how quickly you turn around; (*b*) always zero; (*c*) always negative; (*d*) none of the above.

FIGURE 2–8 Car speedometer showing mi/h in white, and km/h in orange.

FIGURE 2–9 Velocity of a car as a function of time: (a) at constant velocity; (b) with velocity varying in time.

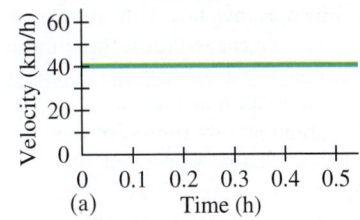

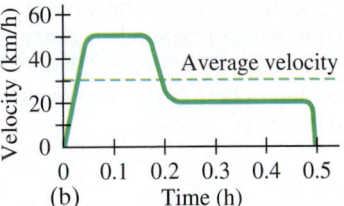

[†]We do not simply set $\Delta t = 0$ in this definition, for then Δx would also be zero, and we would have an undetermined number. Rather, we consider the *ratio* $\Delta x/\Delta t$, as a whole. As we let Δt approach zero, Δx approaches zero as well. But the ratio $\Delta x/\Delta t$ approaches some definite value, which is the instantaneous velocity at a given instant.

2-4 Acceleration

An object whose velocity is changing is said to be accelerating. For instance, a car whose velocity increases in magnitude from zero to 80 km/h is accelerating. Acceleration specifies how *rapidly* the velocity of an object is changing.

Average acceleration is defined as the change in velocity divided by the time taken to make this change:

$$\text{average acceleration} = \frac{\text{change of velocity}}{\text{time elapsed}}.$$

In symbols, the **average acceleration**, $\bar{a}$, over a time interval $\Delta t = t_2 - t_1$, during which the velocity changes by $\Delta v = v_2 - v_1$, is defined as

$$\bar{a} = \frac{v_2 - v_1}{t_2 - t_1} = \frac{\Delta v}{\Delta t}. \qquad \text{[average acceleration]} \quad \textbf{(2-4)}$$

We saw that velocity is a vector (it has magnitude and direction), so acceleration is a vector too. But for one dimensional motion, we need only use a plus or minus sign to indicate acceleration direction relative to a chosen coordinate axis. (Usually, right is $+$, left is $-$.)

The **instantaneous acceleration**, a, can be defined in analogy to instantaneous velocity as the average acceleration over an infinitesimally short time interval at a given instant:

$$a = \lim_{\Delta t \to 0} \frac{\Delta v}{\Delta t}. \qquad \text{[instantaneous acceleration]} \quad \textbf{(2-5)}$$

Here Δv is the very small change in velocity during the very short time interval Δt.

EXAMPLE 2-4 **Average acceleration.** A car accelerates on a straight road from rest to 75 km/h in 5.0 s, Fig. 2-10. What is the magnitude of its average acceleration?

APPROACH Average acceleration is the change in velocity divided by the elapsed time, 5.0 s. The car starts from rest, so $v_1 = 0$. The final velocity is $v_2 = 75$ km/h.

SOLUTION From Eq. 2-4, the average acceleration is

$$\bar{a} = \frac{v_2 - v_1}{t_2 - t_1} = \frac{75 \text{ km/h} - 0 \text{ km/h}}{5.0 \text{ s}} = 15 \frac{\text{km/h}}{\text{s}}.$$

This is read as "fifteen kilometers per hour per second" and means that, on average, the velocity changed by 15 km/h during each second. That is, assuming the acceleration was constant, during the first second the car's velocity increased from zero to 15 km/h. During the next second its velocity increased by another 15 km/h, reaching a velocity of 30 km/h at $t = 2.0$ s, and so on. See Fig. 2-10.

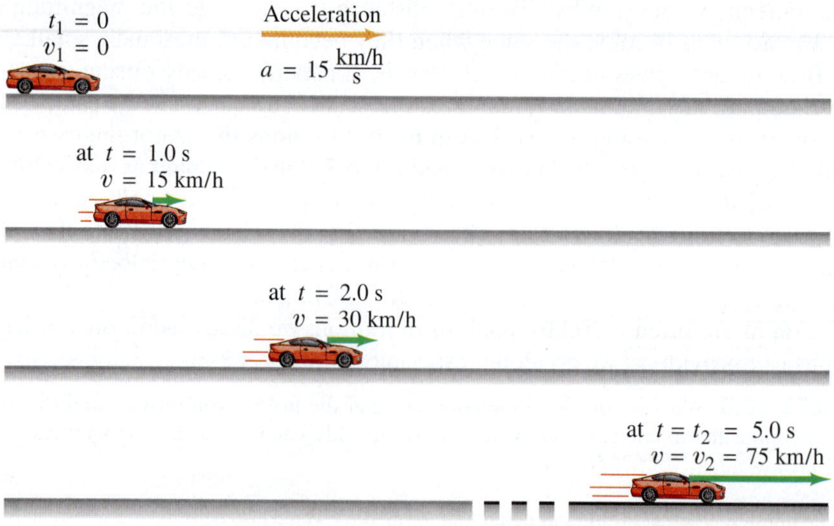

FIGURE 2-10 Example 2-4. The car is shown at the start with $v_1 = 0$ at $t_1 = 0$. The car is shown three more times, at $t = 1.0$ s, $t = 2.0$ s, and at the end of our time interval, $t_2 = 5.0$ s. The green arrows represent the velocity vectors, whose length represents the magnitude of the velocity at that moment. The acceleration vector is the orange arrow, whose magnitude is constant and equals 15 km/h/s or 4.2 m/s² (see top of next page). Distances are not to scale.

26 CHAPTER 2 Describing Motion: Kinematics in One Dimension

Our result in Example 2–4 contains two different time units: hours and seconds. We usually prefer to use only seconds. To do so we can change km/h to m/s (see Section 1–6, and Example 1–5):

$$75 \text{ km/h} = \left(75 \frac{\text{km}}{\text{h}}\right)\left(\frac{1000 \text{ m}}{1 \text{ km}}\right)\left(\frac{1 \text{ h}}{3600 \text{ s}}\right) = 21 \text{ m/s}.$$

Then

$$\bar{a} = \frac{21 \text{ m/s} - 0.0 \text{ m/s}}{5.0 \text{ s}} = 4.2 \frac{\text{m/s}}{\text{s}} = 4.2 \frac{\text{m}}{\text{s}^2}.$$

We almost always write the units for acceleration as m/s² (meters per second squared) instead of m/s/s. This is possible because:

$$\frac{\text{m/s}}{\text{s}} = \frac{\text{m}}{\text{s} \cdot \text{s}} = \frac{\text{m}}{\text{s}^2}.$$

Note that *acceleration tells us how quickly the velocity changes*, whereas *velocity tells us how quickly the position changes*.

> **CAUTION**
> *Distinguish velocity from acceleration*

CONCEPTUAL EXAMPLE 2–5 **Velocity and acceleration.** (*a*) If the velocity of an object is zero, does it mean that the acceleration is zero? (*b*) If the acceleration is zero, does it mean that the velocity is zero? Think of some examples.

RESPONSE A zero velocity does not necessarily mean that the acceleration is zero, nor does a zero acceleration mean that the velocity is zero. (*a*) For example, when you put your foot on the gas pedal of your car which is at rest, the velocity starts from zero but the acceleration is not zero since the velocity of the car changes. (How else could your car start forward if its velocity weren't changing—that is, accelerating?) (*b*) As you cruise along a straight highway at a constant velocity of 100 km/h, your acceleration is zero: $a = 0$, $v \neq 0$.

> **CAUTION**
> *If v or a is zero, is the other zero too?*

EXAMPLE 2–6 **Car slowing down.** An automobile is moving to the right along a straight highway, which we choose to be the positive x axis (Fig. 2–11). Then the driver steps on the brakes. If the initial velocity (when the driver hits the brakes) is $v_1 = 15.0 \text{ m/s}$, and it takes 5.0 s to slow down to $v_2 = 5.0 \text{ m/s}$, what was the car's average acceleration?

APPROACH We put the given initial and final velocities, and the elapsed time, into Eq. 2–4 for $\bar{a}$.

SOLUTION In Eq. 2–4, we call the initial time $t_1 = 0$, and set $t_2 = 5.0 \text{ s}$:

$$\bar{a} = \frac{5.0 \text{ m/s} - 15.0 \text{ m/s}}{5.0 \text{ s}} = -2.0 \text{ m/s}^2.$$

The negative sign appears because the final velocity is less than the initial velocity. In this case the direction of the acceleration is to the left (in the negative x direction)—even though the velocity is always pointing to the right. We say that the acceleration is 2.0 m/s² to the left, and it is shown in Fig. 2–11 as an orange arrow.

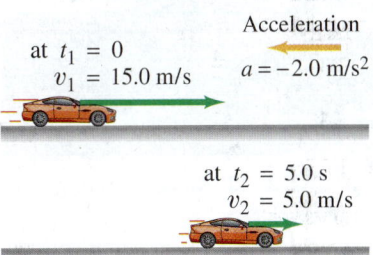

FIGURE 2–11 Example 2–6, showing the position of the car at times t_1 and t_2, as well as the car's velocity represented by the green arrows. The acceleration vector (orange) points to the left because the car slows down as it moves to the right.

Deceleration

When an object is slowing down, we can say it is **decelerating**. But be careful: deceleration does *not* mean that the acceleration is necessarily negative. The velocity of an object moving to the right along the positive x axis is positive; if the object is slowing down (as in Fig. 2–11), the acceleration is negative. But the same car moving to the left (decreasing x), and slowing down, has positive acceleration that points to the right, as shown in Fig. 2–12. We have a deceleration whenever the magnitude of the velocity is decreasing; thus the *velocity and acceleration point in opposite directions* when there is deceleration.

FIGURE 2–12 The car of Example 2–6, now moving to the *left* and decelerating. The acceleration is $a = (v_2 - v_1)/\Delta t$, or

$$a = \frac{(-5.0 \text{ m/s}) - (-15.0 \text{ m/s})}{5.0 \text{ s}}$$

$$= \frac{-5.0 \text{ m/s} + 15.0 \text{ m/s}}{5.0 \text{ s}} = +2.0 \text{ m/s}^2.$$

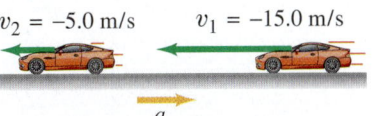

EXERCISE C A car moves along the x axis. What is the sign of the car's acceleration if it is moving in the positive x direction with (*a*) increasing speed or (*b*) decreasing speed? What is the sign of the acceleration if the car moves in the negative x direction with (*c*) increasing speed or (*d*) decreasing speed?

2–5 Motion at Constant Acceleration

We now examine motion in a straight line when the magnitude of the acceleration is constant. In this case, the instantaneous and average accelerations are equal. We use the definitions of average velocity and acceleration to derive a set of valuable equations that relate x, v, a, and t when a is constant, allowing us to determine any one of these variables if we know the others. We can then solve many interesting Problems.

Notation in physics varies from book to book; and different instructors use different notation. We are now going to change our notation, to simplify it a bit for our discussion here of **constant acceleration**. First we choose the initial time in any discussion to be zero, and we call it t_0. That is, $t_1 = t_0 = 0$. (This is effectively starting a stopwatch at t_0.) We can then let $t_2 = t$ be the elapsed time. The initial position (x_1) and the initial velocity (v_1) of an object will now be represented by x_0 and v_0, since they represent x and v at $t = 0$. At time t the position and velocity will be called x and v (rather than x_2 and v_2). The average velocity during the time interval $t - t_0$ will be (Eq. 2–2)

$$\bar{v} = \frac{\Delta x}{\Delta t} = \frac{x - x_0}{t - t_0} = \frac{x - x_0}{t}$$

since we chose $t_0 = 0$. The acceleration, assumed constant in time, is $a = \Delta v / \Delta t$ (Eq. 2–4), so

$$a = \frac{v - v_0}{t}.$$

A common problem is to determine the velocity of an object after any elapsed time t, when we are given the object's constant acceleration. We can solve such problems[†] by solving for v in the last equation: first we multiply both sides by t,

$$at = v - v_0 \quad \text{or} \quad v - v_0 = at.$$

Then, adding v_0 to both sides, we obtain

$$v = v_0 + at. \qquad \text{[constant acceleration]} \quad (2\text{–}6)$$

FIGURE 2–13 An accelerating motorcycle.

If an object, such as a motorcycle (Fig. 2–13), starts from rest $(v_0 = 0)$ and accelerates at 4.0 m/s^2, after an elapsed time $t = 6.0 \text{ s}$ its velocity will be $v = 0 + at = (4.0 \text{ m/s}^2)(6.0 \text{ s}) = 24 \text{ m/s}$.

Next, let us see how to calculate the position x of an object after a time t when it undergoes constant acceleration. The definition of average velocity (Eq. 2–2) is $\bar{v} = (x - x_0)/t$, which we can rewrite by multiplying both sides by t:

$$x = x_0 + \bar{v}t. \qquad (2\text{–}7)$$

Because the velocity increases at a uniform rate, the average velocity, $\bar{v}$, will be midway between the initial and final velocities:

$$\bar{v} = \frac{v_0 + v}{2}. \qquad \text{[constant acceleration]} \quad (2\text{–}8)$$

CAUTION
Average velocity, but only if $a = $ constant

(Careful: Equation 2–8 is not necessarily valid if the acceleration is not constant.) We combine the last two Equations with Eq. 2–6 and find, starting with Eq. 2–7,

$$x = x_0 + \bar{v}t$$
$$= x_0 + \left(\frac{v_0 + v}{2}\right)t$$
$$= x_0 + \left(\frac{v_0 + v_0 + at}{2}\right)t$$

or

$$x = x_0 + v_0 t + \tfrac{1}{2}at^2. \qquad \text{[constant acceleration]} \quad (2\text{–}9)$$

Equations 2–6, 2–8, and 2–9 are three of the four most useful equations for motion at constant acceleration. We now derive the fourth equation, which is useful

[†]Appendix A–4 summarizes simple algebraic manipulations.

in situations where the time t is not known. We substitute Eq. 2–8 into Eq. 2–7:

$$x = x_0 + \bar{v}t = x_0 + \left(\frac{v + v_0}{2}\right)t.$$

Next we solve Eq. 2–6 for t, obtaining (see Appendix A–4 for a quick review)

$$t = \frac{v - v_0}{a},$$

and substituting this into the previous equation we have

$$x = x_0 + \left(\frac{v + v_0}{2}\right)\left(\frac{v - v_0}{a}\right) = x_0 + \frac{v^2 - v_0^2}{2a}.$$

We solve this for v^2 and obtain

$$v^2 = v_0^2 + 2a(x - x_0), \qquad \text{[constant acceleration]} \quad (2\text{–}10)$$

which is the other useful equation we sought.

We now have four equations relating position, velocity, acceleration, and time, when the acceleration a is constant. We collect these *kinematic equations for constant acceleration* here in one place for future reference (the tan background screen emphasizes their usefulness):

$$v = v_0 + at \qquad [a = \text{constant}] \quad (2\text{–}11\text{a})$$
$$x = x_0 + v_0 t + \tfrac{1}{2}at^2 \qquad [a = \text{constant}] \quad (2\text{–}11\text{b})$$
$$v^2 = v_0^2 + 2a(x - x_0) \qquad [a = \text{constant}] \quad (2\text{–}11\text{c})$$
$$\bar{v} = \frac{v + v_0}{2}. \qquad [a = \text{constant}] \quad (2\text{–}11\text{d})$$

Kinematic equations for constant acceleration (we'll use them a lot)

These useful equations are not valid unless a is a constant. In many cases we can set $x_0 = 0$, and this simplifies the above equations a bit. Note that x represents position (not distance), also that $x - x_0$ is the displacement, and that t is the elapsed time. Equations 2–11 are useful also when a is approximately constant to obtain reasonable estimates.

EXAMPLE 2–7 **Runway design.** You are designing an airport for small planes. One kind of airplane that might use this airfield must reach a speed before takeoff of at least 27.8 m/s (100 km/h), and can accelerate at 2.00 m/s². (*a*) If the runway is 150 m long, can this airplane reach the required speed for takeoff? (*b*) If not, what minimum length must the runway have?

APPROACH Assuming the plane's acceleration is constant, we use the kinematic equations for constant acceleration. In (*a*), we want to find v, and what we are given is shown in the Table in the margin.

SOLUTION (*a*) Of the above four equations, Eq. 2–11c will give us v when we know v_0, a, x, and x_0:

$$v^2 = v_0^2 + 2a(x - x_0)$$
$$= 0 + 2(2.00 \text{ m/s}^2)(150 \text{ m}) = 600 \text{ m}^2/\text{s}^2$$
$$v = \sqrt{600 \text{ m}^2/\text{s}^2} = 24.5 \text{ m/s}.$$

This runway length is *not* sufficient, because the minimum speed is not reached. (*b*) Now we want to find the minimum runway length, $x - x_0$, for a plane to reach $v = 27.8$ m/s, given $a = 2.00$ m/s². We again use Eq. 2–11c, but rewritten as

$$(x - x_0) = \frac{v^2 - v_0^2}{2a} = \frac{(27.8 \text{ m/s})^2 - 0}{2(2.00 \text{ m/s}^2)} = 193 \text{ m}.$$

A 200-m runway is more appropriate for this plane.

NOTE We did this Example as if the plane were a particle, so we round off our answer to 200 m.

PHYSICS APPLIED
Airport design

Known	Wanted
$x_0 = 0$	v
$v_0 = 0$	
$x = 150$ m	
$a = 2.00$ m/s²	

PROBLEM SOLVING
Equations 2–11 are valid only when the acceleration is constant, which we assume in this Example

EXERCISE D A car starts from rest and accelerates at a constant 10 m/s^2 during a $\frac{1}{4}$-mile (402 m) race. How fast is the car going at the finish line? (a) 8040 m/s; (b) 90 m/s; (c) 81 m/s; (d) 804 m/s.

2–6 Solving Problems

Before doing more worked-out Examples, let us look at how to approach problem solving. First, it is important to note that physics is *not* a collection of equations to be memorized. Simply searching for an equation that might work can lead you to a wrong result and will not help you understand physics (Fig. 2–14). A better approach is to use the following (rough) procedure, which we present as a special "Problem Solving Strategy." (Other such Problem Solving Strategies will be found throughout the book.)

FIGURE 2–14 Read the book, study carefully, and work the Problems using your reasoning abilities.

PROBLEM SOLVING

1. Read and **reread** the whole problem carefully before trying to solve it.

2. Decide what **object** (or objects) you are going to study, and for what **time interval**. You can often choose the initial time to be $t = 0$.

3. **Draw** a **diagram** or picture of the situation, with coordinate axes wherever applicable. [You can place the origin of coordinates and the axes wherever you like to make your calculations easier. You also choose which direction is positive and which is negative. Usually we choose the x axis to the right as positive.]

4. Write down what quantities are "**known**" or "given," and then what you *want* to know. Consider quantities both at the beginning and at the end of the chosen time interval. You may need to "translate" language into physical terms, such as "starts from rest" means $v_0 = 0$.

5. Think about which **principles of physics** apply in this problem. Use common sense and your own experiences. Then plan an approach.

6. Consider which **equations** (and/or definitions) relate the quantities involved. Before using them, be sure their **range of validity** includes your problem (for example, Eqs. 2–11 are valid only when the acceleration is constant). If you find an applicable equation that involves only known quantities and one desired unknown, **solve** the equation algebraically for the unknown. Sometimes several sequential calculations, or a combination of equations, may be needed. It is often preferable to solve algebraically for the desired unknown before putting in numerical values.

7. Carry out the **calculation** if it is a numerical problem. Keep one or two extra digits during the calculations, but round off the final answer(s) to the correct number of significant figures (Section 1–4).

8. Think carefully about the result you obtain: Is it **reasonable**? Does it make sense according to your own intuition and experience? A good check is to do a rough **estimate** using only powers of 10, as discussed in Section 1–7. Often it is preferable to do a rough estimate at the *start* of a numerical problem because it can help you focus your attention on finding a path toward a solution.

9. A very important aspect of doing problems is keeping track of **units**. An equals sign implies the units on each side must be the same, just as the numbers must. If the units do not balance, a mistake has been made. This can serve as a **check** on your solution (but it only tells you if you're wrong, not if you're right). Always use a consistent set of units.

FIGURE 2–18 Painting of Galileo demonstrating to the Grand Duke of Tuscany his argument for the action of gravity being uniform acceleration. He used an inclined plane to slow down the action. A ball rolling down the plane still accelerates. Tiny bells placed at equal distances along the inclined plane would ring at shorter time intervals as the ball "fell," indicating that the speed was increasing.

2–7 Freely Falling Objects

One of the most common examples of uniformly accelerated motion is that of an object allowed to fall freely near the Earth's surface. That a falling object is accelerating may not be obvious at first. And beware of thinking, as was widely believed before the time of Galileo (Fig. 2–18), that heavier objects fall faster than lighter objects and that the speed of fall is proportional to how heavy the object is. *The speed of a falling object is* not *proportional to its mass.*

Galileo made use of his new technique of imagining what would happen in idealized (simplified) cases. For free fall, he postulated that *all objects would fall with the same constant acceleration in the absence of air or other resistance.* He showed that this postulate predicts that for an object falling from rest, the distance traveled will be proportional to the square of the time (Fig. 2–19); that is, $d \propto t^2$. We can see this from Eq. 2–11b for constant acceleration; but Galileo was the first to derive this mathematical relation.

To support his claim that falling objects increase in speed as they fall, Galileo made use of a clever argument: a heavy stone dropped from a height of 2 m will drive a stake into the ground much further than will the same stone dropped from a height of only 0.2 m. Clearly, the stone must be moving faster in the former case.

Galileo claimed that *all* objects, light or heavy, fall with the *same* acceleration, at least in the absence of air. If you hold a piece of paper flat and horizontal in one hand, and a heavier object like a baseball in the other, and release them at the same time as in Fig. 2–20a, the heavier object will reach the ground first. But if you repeat the experiment, this time crumpling the paper into a small wad, you will find (see Fig. 2–20b) that the two objects reach the floor at nearly the same time.

Galileo was sure that air acts as a resistance to very light objects that have a large surface area. But in many ordinary circumstances this air resistance is negligible. In a chamber from which the air has been removed, even light objects like a feather or a horizontally held piece of paper will fall with the same acceleration as any other object (see Fig. 2–21). Such a demonstration in vacuum was not possible in Galileo's time, which makes Galileo's achievement all the greater. Galileo is often called the "father of modern science," not only for the *content* of his science (astronomical discoveries, inertia, free fall) but also for his new methods of *doing* science (idealization and simplification, mathematization of theory, theories that have testable consequences, experiments to test theoretical predictions).

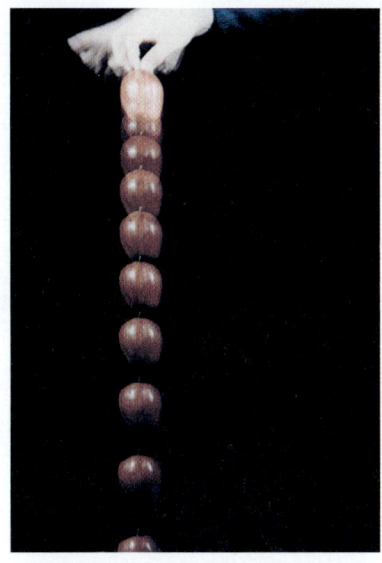

FIGURE 2–19 Multiflash photograph of a falling apple, at equal time intervals. The apple falls farther during each successive interval, which means it is accelerating.

FIGURE 2–20 (a) A ball and a light piece of paper are dropped at the same time. (b) Repeated, with the paper wadded up.

FIGURE 2–21 A rock and a feather are dropped simultaneously (a) in air, (b) in a vacuum.

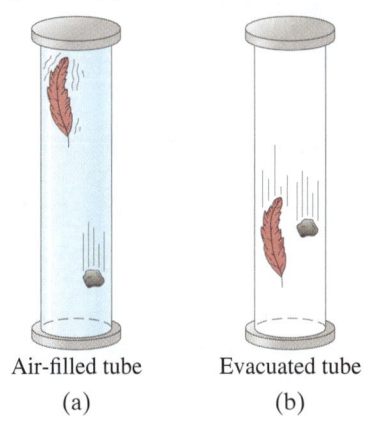

Galileo's specific contribution to our understanding of the motion of falling objects can be summarized as follows:

at a given location on the Earth and in the absence of air resistance, all objects fall with the same constant acceleration.

We call this acceleration the **acceleration due to gravity** at the surface of the Earth, and we give it the symbol g. Its magnitude is approximately

$$g = 9.80 \text{ m/s}^2. \qquad \left[\begin{array}{l}\text{acceleration due to gravity}\\\text{at surface of Earth}\end{array}\right]$$

In British units g is about 32 ft/s^2. Actually, g varies slightly according to latitude and elevation on the Earth's surface, but these variations are so small that we will ignore them for most purposes. (Acceleration of gravity in space beyond the Earth's surface is treated in Chapter 5.) The effects of air resistance are often small, and we will neglect them for the most part. However, air resistance will be noticeable even on a reasonably heavy object if the velocity becomes large.[†] Acceleration due to gravity is a vector, as is any acceleration, and its direction is downward toward the center of the Earth.

When dealing with freely falling objects we can make use of Eqs. 2–11, where for a we use the value of g given above. Also, since the motion is vertical we will substitute y in place of x, and y_0 in place of x_0. We take $y_0 = 0$ unless otherwise specified. *It is arbitrary whether we choose y to be positive in the upward direction or in the downward direction; but we must be consistent about it throughout a problem's solution.*

📝 **PROBLEM SOLVING**
You can choose y to be positive either up or down

EXERCISE E Return to the Chapter-Opening Question, page 21, and answer it again now, assuming minimal air resistance. Try to explain why you may have answered differently the first time.

FIGURE 2–22 Example 2–10. (a) An object dropped from a tower falls with progressively greater speed and covers greater distance with each successive second. (See also Fig. 2–19.) (b) Graph of y vs. t.

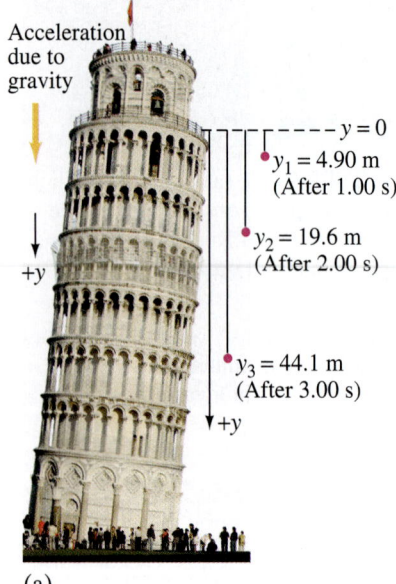

(a)

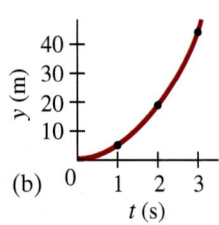

(b)

EXAMPLE 2–10 **Falling from a tower.** Suppose that a ball is dropped ($v_0 = 0$) from a tower. How far will it have fallen after a time $t_1 = 1.00$ s, $t_2 = 2.00$ s, and $t_3 = 3.00$ s? Ignore air resistance.

APPROACH Let us take y as positive downward, so the acceleration is $a = g = +9.80 \text{ m/s}^2$. We set $v_0 = 0$ and $y_0 = 0$. We want to find the position y of the ball after three different time intervals. Equation 2–11b, with x replaced by y, relates the given quantities (t, a, and v_0) to the unknown y.

SOLUTION We set $t = t_1 = 1.00$ s in Eq. 2–11b:

$$y_1 = v_0 t_1 + \tfrac{1}{2}at_1^2$$
$$= 0 + \tfrac{1}{2}at_1^2 = \tfrac{1}{2}(9.80 \text{ m/s}^2)(1.00 \text{ s})^2 = 4.90 \text{ m}.$$

The ball has fallen a distance of 4.90 m during the time interval $t = 0$ to $t_1 = 1.00$ s. Similarly, after 2.00 s ($= t_2$), the ball's position is

$$y_2 = \tfrac{1}{2}at_2^2 = \tfrac{1}{2}(9.80 \text{ m/s}^2)(2.00 \text{ s})^2 = 19.6 \text{ m}.$$

Finally, after 3.00 s ($= t_3$), the ball's position is (see Fig. 2–22)

$$y_3 = \tfrac{1}{2}at_3^2 = \tfrac{1}{2}(9.80 \text{ m/s}^2)(3.00 \text{ s})^2 = 44.1 \text{ m}.$$

NOTE Whenever we say "dropped," it means $v_0 = 0$. Note also the graph of y vs. t (Fig. 2–22b): the curve is not straight but bends upward because y is proportional to t^2.

[†]The speed of an object falling in air (or other fluid) does not increase indefinitely. If the object falls far enough, it will reach a maximum velocity called the **terminal velocity** due to air resistance.

EXAMPLE 2–11 Thrown down from a tower. Suppose the ball in Example 2–10 is *thrown* downward with an initial velocity of 3.00 m/s, instead of being dropped. (*a*) What then would be its position after 1.00 s and 2.00 s? (*b*) What would its speed be after 1.00 s and 2.00 s? Compare with the speeds of a dropped ball.

APPROACH Again we use Eq. 2–11b, but now v_0 is not zero, it is $v_0 = 3.00 \text{ m/s}$.

SOLUTION (*a*) At $t_1 = 1.00 \text{ s}$, the position of the ball as given by Eq. 2–11b is

$$y = v_0 t + \tfrac{1}{2} a t^2 = (3.00 \text{ m/s})(1.00 \text{ s}) + \tfrac{1}{2}(9.80 \text{ m/s}^2)(1.00 \text{ s})^2 = 7.90 \text{ m}.$$

At $t_2 = 2.00 \text{ s}$ (time interval $t = 0$ to $t = 2.00 \text{ s}$), the position is

$$y = v_0 t + \tfrac{1}{2} a t^2 = (3.00 \text{ m/s})(2.00 \text{ s}) + \tfrac{1}{2}(9.80 \text{ m/s}^2)(2.00 \text{ s})^2 = 25.6 \text{ m}.$$

As expected, the ball falls farther each second than if it were dropped with $v_0 = 0$.

(*b*) The velocity is obtained from Eq. 2–11a:

$$\begin{aligned} v &= v_0 + at \\ &= 3.00 \text{ m/s} + (9.80 \text{ m/s}^2)(1.00 \text{ s}) = 12.8 \text{ m/s} \quad [\text{at } t_1 = 1.00 \text{ s}] \\ &= 3.00 \text{ m/s} + (9.80 \text{ m/s}^2)(2.00 \text{ s}) = 22.6 \text{ m/s}. \quad [\text{at } t_2 = 2.00 \text{ s}] \end{aligned}$$

In Example 2–10, when the ball was dropped ($v_0 = 0$), the first term (v_0) in these equations was zero, so

$$\begin{aligned} v &= 0 + at \\ &= (9.80 \text{ m/s}^2)(1.00 \text{ s}) = 9.80 \text{ m/s} \quad [\text{at } t_1 = 1.00 \text{ s}] \\ &= (9.80 \text{ m/s}^2)(2.00 \text{ s}) = 19.6 \text{ m/s}. \quad [\text{at } t_2 = 2.00 \text{ s}] \end{aligned}$$

NOTE For both Examples 2–10 and 2–11, the speed increases linearly in time by 9.80 m/s during each second. But the speed of the downwardly thrown ball at any instant is always 3.00 m/s (its initial speed) higher than that of a dropped ball.

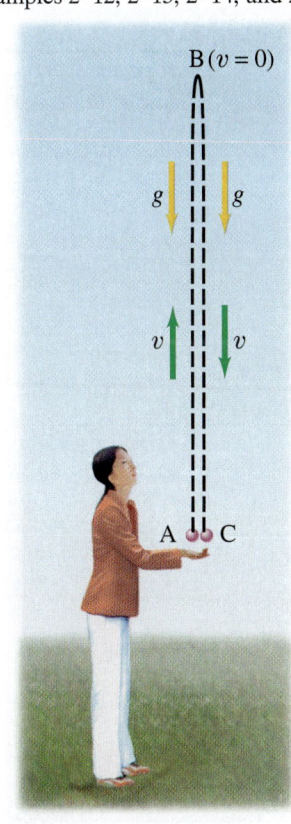

FIGURE 2–23 An object thrown into the air leaves the thrower's hand at A, reaches its maximum height at B, and returns to the original position at C. Examples 2–12, 2–13, 2–14, and 2–15.

EXAMPLE 2–12 Ball thrown upward. A person throws a ball *upward* into the air with an initial velocity of 15.0 m/s. Calculate how high it goes. Ignore air resistance.

APPROACH We are not concerned here with the throwing action, but only with the motion of the ball *after* it leaves the thrower's hand (Fig. 2–23) and until it comes back to the hand again. Let us choose y to be positive in the upward direction and negative in the downward direction. (This is a different convention from that used in Examples 2–10 and 2–11, and so illustrates our options.) The acceleration due to gravity is downward and so will have a negative sign, $a = -g = -9.80 \text{ m/s}^2$. As the ball rises, its speed decreases until it reaches the highest point (B in Fig. 2–23), where its speed is zero for an instant; then it descends, with increasing speed.

SOLUTION We consider the time interval from when the ball leaves the thrower's hand until the ball reaches the highest point. To determine the maximum height, we calculate the position of the ball when its velocity equals zero ($v = 0$ at the highest point). At $t = 0$ (point A in Fig. 2–23) we have $y_0 = 0$, $v_0 = 15.0 \text{ m/s}$, and $a = -9.80 \text{ m/s}^2$. At time t (maximum height), $v = 0$, $a = -9.80 \text{ m/s}^2$, and we wish to find y. We use Eq. 2–11c, replacing x with y: $v^2 = v_0^2 + 2ay$. We solve this equation for y:

$$y = \frac{v^2 - v_0^2}{2a} = \frac{0 - (15.0 \text{ m/s})^2}{2(-9.80 \text{ m/s}^2)} = 11.5 \text{ m}.$$

The ball reaches a height of 11.5 m above the hand.

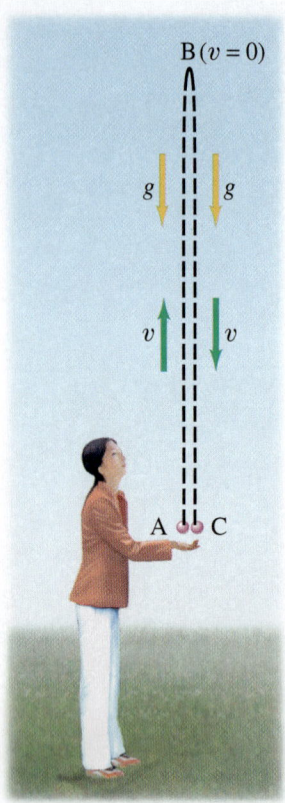

FIGURE 2–23 (Repeated.) An object thrown into the air leaves the thrower's hand at A, reaches its maximum height at B, and returns to the original position at C. Examples 2–12, 2–13, 2–14, and 2–15.

⚠ **CAUTION**
Quadratic equations have two solutions. Sometimes only one corresponds to reality, sometimes both

⚠ **CAUTION**
(1) Velocity and acceleration are not always in the same direction; the acceleration (of gravity) always points down
(2) $a \neq 0$ even at the highest point of a trajectory

EXAMPLE 2–13 **Ball thrown upward, II.** In Fig. 2–23, Example 2–12, how long is the ball in the air before it comes back to the hand?

APPROACH We need to choose a time interval to calculate how long the ball is in the air before it returns to the hand. We could do this calculation in two parts by first determining the time required for the ball to reach its highest point, and then determining the time it takes to fall back down. However, it is simpler to consider the time interval for the entire motion from A to B to C (Fig. 2–23) in one step and use Eq. 2–11b. We can do this because y is position or displacement, and not the total distance traveled. Thus, at both points A and C, $y = 0$.

SOLUTION We use Eq. 2–11b with $a = -9.80 \text{ m/s}^2$ and find
$$y = y_0 + v_0 t + \tfrac{1}{2}at^2$$
$$0 = 0 + (15.0 \text{ m/s})t + \tfrac{1}{2}(-9.80 \text{ m/s}^2)t^2.$$

This equation can be factored (we factor out one t):
$$(15.0 \text{ m/s} - 4.90 \text{ m/s}^2 \, t)t = 0.$$

There are two solutions:
$$t = 0 \quad \text{and} \quad t = \frac{15.0 \text{ m/s}}{4.90 \text{ m/s}^2} = 3.06 \text{ s}.$$

The first solution ($t = 0$) corresponds to the initial point (A) in Fig. 2–23, when the ball was first thrown from $y = 0$. The second solution, $t = 3.06$ s, corresponds to point C, when the ball has returned to $y = 0$. Thus the ball is in the air 3.06 s.

NOTE We have ignored air resistance in these last two Examples, which could be significant, so our result is only an approximation to a real, practical situation.

We did not consider the throwing action in these Examples. Why? Because during the throw, the thrower's hand is touching the ball and accelerating the ball at a rate unknown to us—the acceleration is *not* g. We consider only the time when the ball is in the air and the acceleration is equal to g.

Every quadratic equation (where the variable is squared) mathematically produces two solutions. In physics, sometimes only one solution corresponds to the real situation, as in Example 2–8, in which case we ignore the "unphysical" solution. But in Example 2–13, both solutions to our equation in t^2 are physically meaningful: $t = 0$ and $t = 3.06$ s.

CONCEPTUAL EXAMPLE 2–14 **Two possible misconceptions.** Give examples to show the error in these two common misconceptions: (1) that acceleration and velocity are always in the same direction, and (2) that an object thrown upward has zero acceleration at the highest point (B in Fig. 2–23).

RESPONSE Both are wrong. (1) Velocity and acceleration are *not* necessarily in the same direction. When the ball in Fig. 2–23 is moving upward, its velocity is positive (upward), whereas the acceleration is negative (downward). (2) At the highest point (B in Fig. 2–23), the ball has zero velocity for an instant. Is the acceleration also zero at this point? No. The velocity near the top of the arc points upward, then becomes zero for an instant (zero time) at the highest point, and then points downward. Gravity does not stop acting, so $a = -g = -9.80 \text{ m/s}^2$ even there. Thinking that $a = 0$ at point B would lead to the conclusion that upon reaching point B, the ball would stay there: if the acceleration (= rate of change of velocity) were zero, the velocity would stay zero at the highest point, and the ball would stay up there without falling. Remember: the acceleration of gravity always points down toward the Earth, even when the object is moving up.

EXAMPLE 2–15 Ball thrown upward, III. Let us consider again the ball thrown upward of Examples 2–12 and 2–13, and make more calculations. Calculate (a) how much time it takes for the ball to reach the maximum height (point B in Fig. 2–23), and (b) the velocity of the ball when it returns to the thrower's hand (point C).

APPROACH Again we assume the acceleration is constant, so we can use Eqs. 2–11. We have the maximum height of 11.5 m and initial speed of 15.0 m/s from Example 2–12. Again we take y as positive upward.

SOLUTION (a) We consider the time interval between the throw ($t = 0$, $v_0 = 15.0$ m/s) and the top of the path ($y = +11.5$ m, $v = 0$), and we want to find t. The acceleration is constant at $a = -g = -9.80$ m/s². Both Eqs. 2–11a and 2–11b contain the time t with other quantities known. Let us use Eq. 2–11a with $a = -9.80$ m/s², $v_0 = 15.0$ m/s, and $v = 0$:

$$v = v_0 + at;$$

setting $v = 0$ gives $0 = v_0 + at$, which we rearrange to solve for t: $at = -v_0$ or

$$t = -\frac{v_0}{a}$$
$$= -\frac{15.0 \text{ m/s}}{-9.80 \text{ m/s}^2} = 1.53 \text{ s}.$$

This is just half the time it takes the ball to go up and fall back to its original position [3.06 s, calculated in Example 2–13]. Thus it takes the same time to reach the maximum height as to fall back to the starting point.

(b) Now we consider the time interval from the throw ($t = 0$, $v_0 = 15.0$ m/s) until the ball's return to the hand, which occurs at $t = 3.06$ s (as calculated in Example 2–13), and we want to find v when $t = 3.06$ s:

$$v = v_0 + at$$
$$= 15.0 \text{ m/s} - (9.80 \text{ m/s}^2)(3.06 \text{ s}) = -15.0 \text{ m/s}.$$

NOTE The ball has the same speed (magnitude of velocity) when it returns to the starting point as it did initially, but in the opposite direction (this is the meaning of the negative sign). And, as we saw in part (a), the time is the same up as down. Thus the motion is *symmetrical* about the maximum height.

The acceleration of objects such as rockets and fast airplanes is often given as a multiple of $g = 9.80$ m/s². For example, a plane pulling out of a dive (see Fig. 2–24) and undergoing 3.00 g's would have an acceleration of $(3.00)(9.80 \text{ m/s}^2) = 29.4$ m/s².

PROBLEM SOLVING
Acceleration in g's

FIGURE 2–24 Several planes, in formation, are just coming out of a downward dive.

EXERCISE F Two balls are thrown from a cliff. One is thrown directly up, the other directly down. Both balls have the same initial speed, and both hit the ground below the cliff but at different times. Which ball hits the ground at the greater speed: (a) the ball thrown upward, (b) the ball thrown downward, or (c) both the same? Ignore air resistance.

Additional Example—Using the Quadratic Formula

EXAMPLE 2–16 Ball thrown upward at edge of cliff. Suppose that the person of Examples 2–12, 2–13, and 2–15 throws the ball upward at 15.0 m/s while standing on the edge of a cliff, so that the ball can fall to the base of the cliff 50.0 m below, as shown in Fig. 2–25a. (a) How long does it take the ball to reach the base of the cliff? (b) What is the total distance traveled by the ball? Ignore air resistance (likely to be significant, so our result is an approximation).

APPROACH We again use Eq. 2–11b, with y as + upward, but this time we set $y = -50.0 \text{ m}$, the bottom of the cliff, which is 50.0 m below the initial position ($y_0 = 0$); hence the minus sign.

SOLUTION (a) We use Eq. 2–11b with $a = -9.80 \text{ m/s}^2$, $v_0 = 15.0 \text{ m/s}$, $y_0 = 0$, and $y = -50.0 \text{ m}$:

$$y = y_0 + v_0 t + \tfrac{1}{2}at^2$$
$$-50.0 \text{ m} = 0 + (15.0 \text{ m/s})t - \tfrac{1}{2}(9.80 \text{ m/s}^2)t^2.$$

To solve any quadratic equation of the form

$$at^2 + bt + c = 0,$$

where a, b, and c are constants (a is *not* acceleration here), we use the **quadratic formula** (see Appendix A–4):

$$t = \frac{-b \pm \sqrt{b^2 - 4ac}}{2a}.$$

We rewrite our y equation just above in standard form, $at^2 + bt + c = 0$:

$$(4.90 \text{ m/s}^2)t^2 - (15.0 \text{ m/s})t - (50.0 \text{ m}) = 0.$$

Using the quadratic formula, we find as solutions

$$t = 5.07 \text{ s}$$

and

$$t = -2.01 \text{ s}.$$

The first solution, $t = 5.07 \text{ s}$, is the answer we are seeking: the time it takes the ball to rise to its highest point and then fall to the base of the cliff. To rise and fall back to the top of the cliff took 3.06 s (Example 2–13); so it took an additional 2.01 s to fall to the base. But what is the meaning of the other solution, $t = -2.01 \text{ s}$? This is a time before the throw, when our calculation begins, so it isn't relevant here. It is outside our chosen time interval, and so is an *unphysical* solution (also in Example 2–8).

(b) From Example 2–12, the ball moves up 11.5 m, falls 11.5 m back down to the top of the cliff, and then down another 50.0 m to the base of the cliff, for a total distance traveled of 73.0 m. [Note that the *displacement*, however, was −50.0 m.] Figure 2–25b shows the y vs. t graph for this situation.

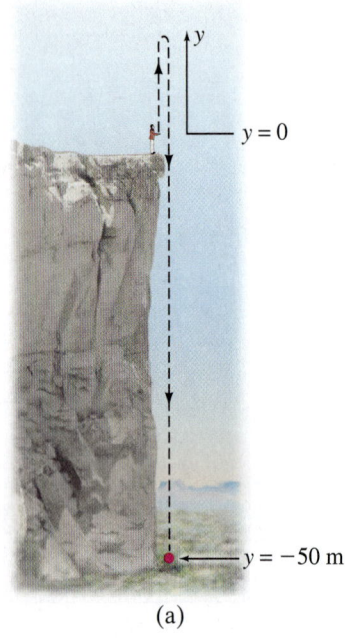

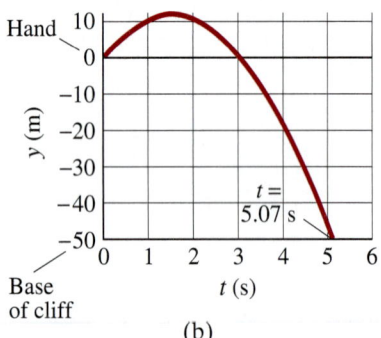

FIGURE 2–25 Example 2–16. (a) A person stands on the edge of a cliff. A ball is thrown upward, then falls back down past the thrower to the base of the cliff, 50.0 m below. (b) The y vs. t graph.

CAUTION
Sometimes a solution to a quadratic equation does not apply to the actual physical conditions of the Problem

2–8 Graphical Analysis of Linear Motion

Velocity as Slope

Analysis of motion using graphs can give us additional insight into kinematics. Let us draw a graph of x vs. t, making the choice that at $t = 0$, the position of an object is $x = 0$, and the object is moving at a constant velocity, $v = \bar{v} = 11$ m/s (40 km/h). Our graph starts at $x = 0$, $t = 0$ (the origin). The graph of the position increases linearly in time because, by Eq. 2–2, $\Delta x = \bar{v}\,\Delta t$ and $\bar{v}$ is a constant. So the graph of x vs. t is a straight line, as shown in Fig. 2–26. The small (shaded) triangle on the graph indicates the **slope** of the straight line:

$$\text{slope} = \frac{\Delta x}{\Delta t}.$$

We see, using the definition of average velocity (Eq. 2–2), that the *slope of the x vs. t graph is equal to the velocity*. And, as can be seen from the small triangle on the graph, $\Delta x / \Delta t = (11\text{ m})/(1.0\text{ s}) = 11$ m/s, which is the given velocity.

If the object's velocity changes in time, we might have an x vs. t graph like that shown in Fig. 2–27. (Note that this graph is different from showing the "path" of an object on an x vs. y plot.) Suppose the object is at position x_1 at time t_1, and at position x_2 at time t_2. P_1 and P_2 represent these two points on the graph. A straight line drawn from point $P_1(x_1, t_1)$ to point $P_2(x_2, t_2)$ forms the hypotenuse of a right triangle whose sides are Δx and Δt. The ratio $\Delta x / \Delta t$ is the **slope** of the straight line $P_1 P_2$. But $\Delta x / \Delta t$ is also the average velocity of the object during the time interval $\Delta t = t_2 - t_1$. Therefore, we conclude that the *average velocity of an object during any time interval $\Delta t = t_2 - t_1$ is equal to the slope of the straight line* (or **chord**) connecting the two points (x_1, t_1) and (x_2, t_2) on an x vs. t graph.

Consider now a time intermediate between t_1 and t_2, call it t_3, at which moment the object is at x_3 (Fig. 2–28). The slope of the straight line $P_1 P_3$ is less than the slope of $P_1 P_2$. Thus the average velocity during the time interval $t_3 - t_1$ is less than during the time interval $t_2 - t_1$.

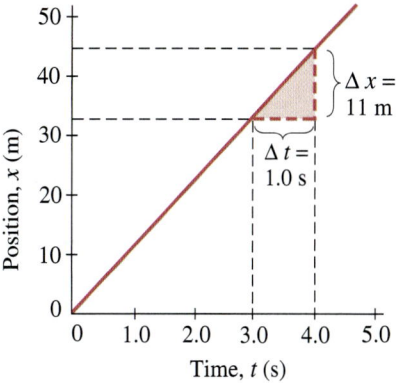

FIGURE 2–26 Graph of position vs. time for an object moving at a constant velocity of 11 m/s.

FIGURE 2–27 Graph of an object's position x vs. time t. The slope of the straight line $P_1 P_2$ represents the average velocity of the object during the time interval $\Delta t = t_2 - t_1$.

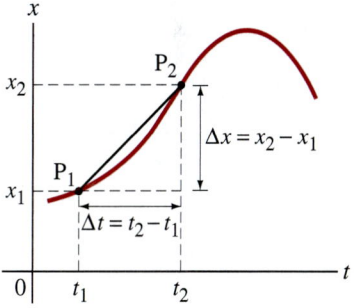

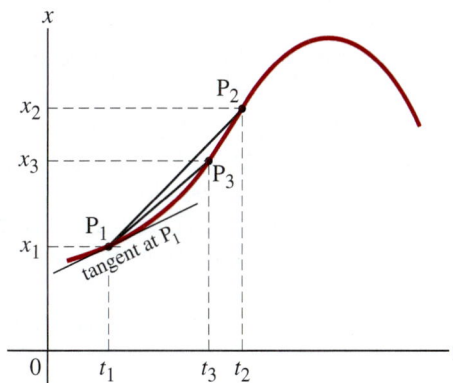

FIGURE 2–28 Same position vs. time curve as in Fig. 2–27. Note that the average velocity over the time interval $t_3 - t_1$ (which is the slope of $P_1 P_3$) is less than the average velocity over the time interval $t_2 - t_1$. The slope of the line tangent to the curve at point P_1 equals the *instantaneous* velocity at time t_1.

Next let us take point P_3 in Fig. 2–28 to be closer and closer to point P_1. That is, we let the interval $t_3 - t_1$, which we now call Δt, to become smaller and smaller. The slope of the line connecting the two points becomes closer and closer to the slope of a line **tangent**† to the curve at point P_1. The average velocity (equal to the slope of the chord) thus approaches the slope of the tangent at point P_1. The definition of the instantaneous velocity (Eq. 2–3) is the limiting value of the average velocity as Δt approaches zero. Thus *instantaneous velocity equals the slope of the tangent to the curve of x vs. t at any chosen point* (which we can simply call "the slope of the curve" at that point).

PROBLEM SOLVING
Velocity equals slope of x vs. t graph at any instant

---
†The tangent is a straight line that touches the curve only at the one chosen point, without passing across or through the curve at that point.

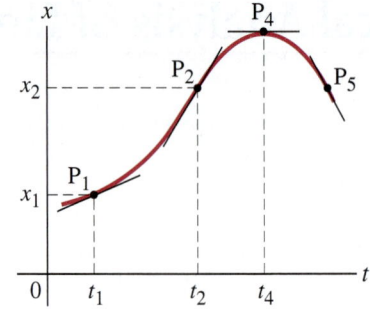

FIGURE 2–29 Same x vs. t curve as in Figs. 2–27 and 2–28, but here showing the slope at four different points: At P_4, the slope is zero, so $v = 0$. At P_5 the slope is negative, so $v < 0$.

We can obtain the velocity of an object at any instant from its graph of x vs. t. For example, in Fig. 2–29 (which shows the same graph as in Figs. 2–27 and 2–28), as our object moves from x_1 to x_2, the slope continually increases, so the velocity is increasing. For times after t_2, the slope begins to decrease and reaches zero ($v = 0$) where x has its maximum value, at point P_4 in Fig. 2–29. Beyond point P_4, the slope is negative, as for point P_5. The velocity is therefore negative, which makes sense since x is now decreasing—the particle is moving toward decreasing values of x, to the left on a standard xy plot.

Slope and Acceleration

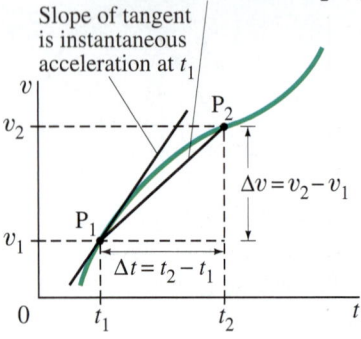

FIGURE 2–30 A graph of velocity v vs. time t. The average acceleration over a time interval $\Delta t = t_2 - t_1$ is the slope of the straight line $P_1 P_2$: $\bar{a} = \Delta v/\Delta t$. The instantaneous acceleration at time t_1 is the slope of the v vs. t curve at that instant.

We can also draw a graph of the *velocity*, v, vs. time, t, as shown in Fig. 2–30. Then the average acceleration over a time interval $\Delta t = t_2 - t_1$ is represented by the slope of the straight line connecting the two points P_1 and P_2 as shown. [Compare this to the position vs. time graph of Fig. 2–27 for which the slope of the straight line represents the average velocity.] The instantaneous acceleration at any time, say t_1, is the slope of the tangent to the v vs. t curve at that time, which is also shown in Fig. 2–30. Using this fact for the situation graphed in Fig. 2–30, as we go from time t_1 to time t_2 the velocity continually increases, but the acceleration (the rate at which the velocity changes) is decreasing since the slope of the curve is decreasing.

> **CONCEPTUAL EXAMPLE 2–17 Analyzing with graphs.** Figure 2–31 shows the velocity as a function of time for two cars accelerating from 0 to 100 km/h in a time of 10.0 s. Compare (a) the average acceleration; (b) the instantaneous acceleration; and (c) the total distance traveled for the two cars.
>
> **RESPONSE** (a) Average acceleration is $\Delta v/\Delta t$. Both cars have the same Δv (100 km/h) over the same time interval $\Delta t = 10.0$ s, so the average acceleration is the same for both cars. (b) Instantaneous acceleration is the slope of the tangent to the v vs. t curve. For the first 4 s or so, the top curve (car A) is steeper than the bottom curve, so car A has a greater acceleration during this interval. The bottom curve is steeper during the last 6 s, so car B has the larger acceleration for this period. (c) Except at $t = 0$ and $t = 10.0$ s, car A is always going faster than car B. Since it is going faster, it will go farther in the same time.

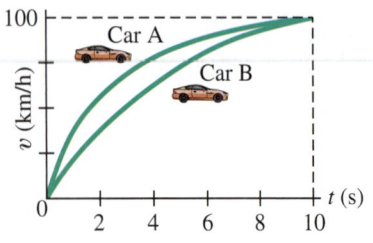

FIGURE 2–31 (below) Example 2–17.

Summary

[The Summary that appears at the end of each Chapter in this book gives a brief overview of the main ideas of the Chapter. The Summary cannot serve to give an understanding of the material, which can be accomplished only by a detailed reading of the Chapter.]

Kinematics deals with the description of how objects move. The description of the motion of any object must always be given relative to some particular **reference frame**.

The **displacement** of an object is the change in position of the object.

Average speed is the distance traveled divided by the **elapsed time** or **time interval**, Δt (the time period over which we choose to make our observations). An object's **average velocity** over a particular time interval is

$$\bar{v} = \frac{\Delta x}{\Delta t}, \qquad (2\text{–}2)$$

where Δx is the displacement during the time interval Δt.

The **instantaneous velocity**, whose magnitude is the same as the *instantaneous speed*, is defined as the average velocity taken over an infinitesimally short time interval.

Acceleration is the change of velocity per unit time. An object's **average acceleration** over a time interval Δt is

$$\bar{a} = \frac{\Delta v}{\Delta t}, \qquad (2\text{-}4)$$

where Δv is the change of velocity during the time interval Δt. **Instantaneous acceleration** is the average acceleration taken over an infinitesimally short time interval.

If an object has position x_0 and velocity v_0 at time $t = 0$ and moves in a straight line with **constant acceleration**, the velocity v and position x at a later time t are related to the acceleration a, the initial position x_0, and the initial velocity v_0 by Eqs. 2–11:

$$\begin{aligned} v &= v_0 + at, \\ x &= x_0 + v_0 t + \tfrac{1}{2} a t^2, \\ v^2 &= v_0^2 + 2a(x - x_0), \\ \bar{v} &= \frac{v + v_0}{2}. \end{aligned} \qquad (2\text{-}11)$$

Objects that move vertically near the surface of the Earth, either falling or having been projected vertically up or down, move with the constant downward **acceleration due to gravity**, whose magnitude is $g = 9.80 \text{ m/s}^2$ if air resistance can be ignored. We can apply Eqs. 2–11 for constant acceleration to objects that move up or down freely near the Earth's surface.

The slope of a curve at any point on a graph is the slope of the tangent to the curve at that point. On a graph of position vs. time, the **slope** is equal to the instantaneous velocity. On a graph of velocity vs. time, the slope is the acceleration.

Questions

1. Does a car speedometer measure speed, velocity, or both? Explain.
2. When an object moves with constant velocity, does its average velocity during any time interval differ from its instantaneous velocity at any instant? Explain.
3. If one object has a greater speed than a second object, does the first necessarily have a greater acceleration? Explain, using examples.
4. Compare the acceleration of a motorcycle that accelerates from 80 km/h to 90 km/h with the acceleration of a bicycle that accelerates from rest to 10 km/h in the same time.
5. Can an object have a northward velocity and a southward acceleration? Explain.
6. Can the velocity of an object be negative when its acceleration is positive? What about vice versa? If yes, give examples in each case.
7. Give an example where both the velocity and acceleration are negative.
8. Can an object be increasing in speed as its acceleration decreases? If so, give an example. If not, explain.
9. Two cars emerge side by side from a tunnel. Car A is traveling with a speed of 60 km/h and has an acceleration of 40 km/h/min. Car B has a speed of 40 km/h and has an acceleration of 60 km/h/min. Which car is passing the other as they come out of the tunnel? Explain your reasoning.
10. A baseball player hits a ball straight up into the air. It leaves the bat with a speed of 120 km/h. In the absence of air resistance, how fast would the ball be traveling when it is caught at the same height above the ground as it left the bat? Explain.
11. As a freely falling object speeds up, what is happening to its acceleration—does it increase, decrease, or stay the same? (a) Ignore air resistance. (b) Consider air resistance.
12. You travel from point A to point B in a car moving at a constant speed of 70 km/h. Then you travel the same distance from point B to another point C, moving at a constant speed of 90 km/h. Is your average speed for the entire trip from A to C equal to 80 km/h? Explain why or why not.
13. Can an object have zero velocity and nonzero acceleration at the same time? Give examples.
14. Can an object have zero acceleration and nonzero velocity at the same time? Give examples.
15. Which of these motions is *not* at constant acceleration: a rock falling from a cliff, an elevator moving from the second floor to the fifth floor making stops along the way, a dish resting on a table? Explain your answers.
16. Describe in words the motion plotted in Fig. 2–32 in terms of velocity, acceleration, etc. [*Hint*: First try to duplicate the motion plotted by walking or moving your hand.]

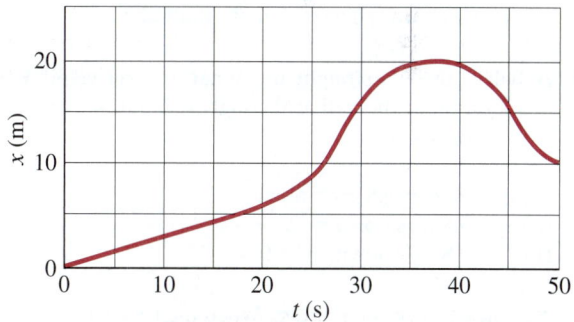

FIGURE 2–32 Question 16.

17. Describe in words the motion of the object graphed in Fig. 2–33.

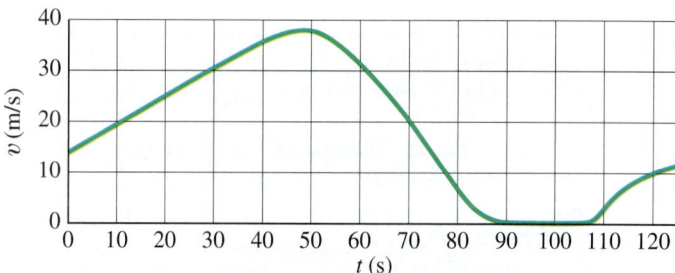

FIGURE 2–33 Question 17.

MisConceptual Questions

[List all answers that are valid.]

1. Which of the following should be part of solving any problem in physics? Select all that apply:
   (a) Read the problem carefully.
   (b) Draw a picture of the situation.
   (c) Write down the variables that are given.
   (d) Think about which physics principles to apply.
   (e) Determine which equations can be used to apply the correct physics principles.
   (f) Check the units when you have completed your calculation.
   (g) Consider whether your answer is reasonable.

2. In which of the following cases does a car have a negative velocity and a positive acceleration? A car that is traveling in the
   (a) $-x$ direction at a constant 20 m/s.
   (b) $-x$ direction increasing in speed.
   (c) $+x$ direction increasing in speed.
   (d) $-x$ direction decreasing in speed.
   (e) $+x$ direction decreasing in speed.

3. At time $t = 0$ an object is traveling to the right along the $+x$ axis at a speed of 10.0 m/s with acceleration -2.0 m/s^2. Which statement is true?
   (a) The object will slow down, eventually coming to a complete stop.
   (b) The object cannot have a negative acceleration and be moving to the right.
   (c) The object will continue to move to the right, slowing down but never coming to a complete stop.
   (d) The object will slow down, momentarily stopping, then pick up speed moving to the left.

4. A ball is thrown straight up. What are the velocity and acceleration of the ball at the highest point in its path?
   (a) $v = 0$, $a = 0$.
   (b) $v = 0$, $a = 9.8 \text{ m/s}^2$ up.
   (c) $v = 0$, $a = 9.8 \text{ m/s}^2$ down.
   (d) $v = 9.8$ m/s up, $a = 0$.
   (e) $v = 9.8$ m/s down, $a = 0$.

5. You drop a rock off a bridge. When the rock has fallen 4 m, you drop a second rock. As the two rocks continue to fall, what happens to their velocities?
   (a) Both increase at the same rate.
   (b) The velocity of the first rock increases faster than the velocity of the second.
   (c) The velocity of the second rock increases faster than the velocity of the first.
   (d) Both velocities stay constant.

6. You drive 4 km at 30 km/h and then another 4 km at 50 km/h. What is your average speed for the whole 8-km trip?
   (a) More than 40 km/h.
   (b) Equal to 40 km/h.
   (c) Less than 40 km/h.
   (d) Not enough information.

7. A ball is dropped from the top of a tall building. At the same instant, a second ball is thrown upward from ground level. When the two balls pass one another, one on the way up, the other on the way down, compare the magnitudes of their acceleration:
   (a) The acceleration of the dropped ball is greater.
   (b) The acceleration of the ball thrown upward is greater.
   (c) The acceleration of both balls is the same.
   (d) The acceleration changes during the motion, so you cannot predict the exact value when the two balls pass each other.
   (e) The accelerations are in opposite directions.

8. A ball is thrown downward at a speed of 20 m/s. Choosing the $+y$ axis pointing up and neglecting air resistance, which equation(s) could be used to solve for other variables? The acceleration due to gravity is $g = 9.8 \text{ m/s}^2$ downward.
   (a) $v = (20 \text{ m/s}) - gt$.
   (b) $y = y_0 + (-20 \text{ m/s})t - (1/2)gt^2$.
   (c) $v^2 = (20 \text{ m/s})^2 - 2g(y - y_0)$.
   (d) $(20 \text{ m/s}) = (v + v_0)/2$.
   (e) All of the above.

9. A car travels along the x axis with increasing speed. We don't know if to the left or the right. Which of the graphs in Fig. 2–34 most closely represents the motion of the car?

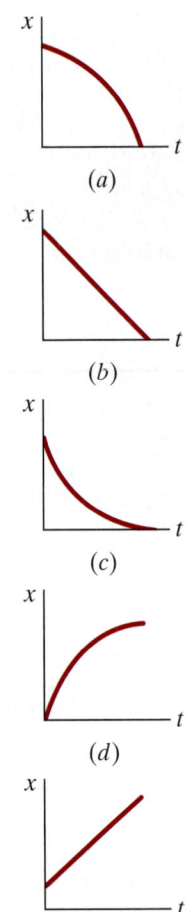

FIGURE 2–34 MisConceptual Question 9.

For assigned homework and other learning materials, go to the MasteringPhysics website.

Problems

[*The Problems at the end of each Chapter are ranked* I, II, *or* III *according to estimated difficulty, with level* I *Problems being easiest. Level* III *are meant as challenges for the best students. The Problems are arranged by Section, meaning that the reader should have read up to and including that Section, but not only that Section—Problems often depend on earlier material. Next is a set of "General Problems" not arranged by Section and not ranked. Finally, there are "Search and Learn" Problems that require rereading parts of the Chapter and sometimes earlier Chapters.*]

(Note: *In Problems, assume a number like* 6.4 *is accurate to* ± 0.1; *and* 950 *is* ± 10 *unless* 950 *is said to be "precisely" or "very nearly"* 950, *in which case assume* 950 ± 1. *See Section 1–4.*)

2–1 to 2–3 Speed and Velocity

1. (I) If you are driving 95 km/h along a straight road and you look to the side for 2.0 s, how far do you travel during this inattentive period?
2. (I) What must your car's average speed be in order to travel 235 km in 2.75 h?
3. (I) A particle at $t_1 = -2.0$ s is at $x_1 = 4.8$ cm and at $t_2 = 4.5$ s is at $x_2 = 8.5$ cm. What is its average velocity over this time interval? Can you calculate its average speed from these data? Why or why not?
4. (I) A rolling ball moves from $x_1 = 8.4$ cm to $x_2 = -4.2$ cm during the time from $t_1 = 3.0$ s to $t_2 = 6.1$ s. What is its average velocity over this time interval?
5. (I) A bird can fly 25 km/h. How long does it take to fly 3.5 km?
6. (II) According to a rule-of-thumb, each five seconds between a lightning flash and the following thunder gives the distance to the flash in miles. (*a*) Assuming that the flash of light arrives in essentially no time at all, estimate the speed of sound in m/s from this rule. (*b*) What would be the rule for kilometers?
7. (II) You are driving home from school steadily at 95 km/h for 180 km. It then begins to rain and you slow to 65 km/h. You arrive home after driving 4.5 h. (*a*) How far is your hometown from school? (*b*) What was your average speed?
8. (II) A horse trots away from its trainer in a straight line, moving 38 m away in 9.0 s. It then turns abruptly and gallops halfway back in 1.8 s. Calculate (*a*) its average speed and (*b*) its average velocity for the entire trip, using "away from the trainer" as the positive direction.
9. (II) A person jogs eight complete laps around a 400-m track in a total time of 14.5 min. Calculate (*a*) the average speed and (*b*) the average velocity, in m/s.
10. (II) Every year the Earth travels about 10^9 km as it orbits the Sun. What is Earth's average speed in km/h?
11. (II) A car traveling 95 km/h is 210 m behind a truck traveling 75 km/h. How long will it take the car to reach the truck?
12. (II) Calculate the average speed and average velocity of a complete round trip in which the outgoing 250 km is covered at 95 km/h, followed by a 1.0-h lunch break, and the return 250 km is covered at 55 km/h.
13. (II) Two locomotives approach each other on parallel tracks. Each has a speed of 155 km/h with respect to the ground. If they are initially 8.5 km apart, how long will it be before they reach each other? (See Fig. 2–35.)

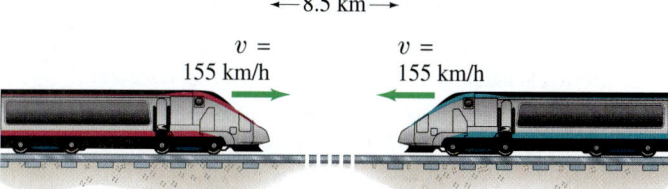

FIGURE 2–35 Problem 13.

14. (II) Digital bits on a 12.0-cm diameter audio CD are encoded along an outward spiraling path that starts at radius $R_1 = 2.5$ cm and finishes at radius $R_2 = 5.8$ cm. The distance between the centers of neighboring spiral-windings is 1.6 μm ($= 1.6 \times 10^{-6}$ m). (*a*) Determine the total length of the spiraling path. [*Hint*: Imagine "unwinding" the spiral into a straight path of width 1.6 μm, and note that the original spiral and the straight path both occupy the same area.] (*b*) To read information, a CD player adjusts the rotation of the CD so that the player's readout laser moves along the spiral path at a constant speed of about 1.2 m/s. Estimate the maximum playing time of such a CD.
15. (III) A bowling ball traveling with constant speed hits the pins at the end of a bowling lane 16.5 m long. The bowler hears the sound of the ball hitting the pins 2.80 s after the ball is released from his hands. What is the speed of the ball, assuming the speed of sound is 340 m/s?
16. (III) An automobile traveling 95 km/h overtakes a 1.30-km-long train traveling in the same direction on a track parallel to the road. If the train's speed is 75 km/h, how long does it take the car to pass it, and how far will the car have traveled in this time? See Fig. 2–36. What are the results if the car and train are traveling in opposite directions?

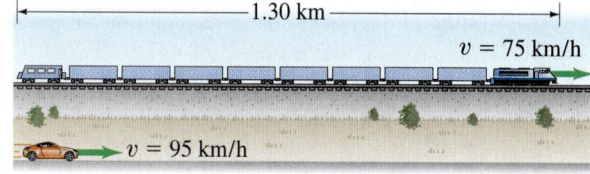

FIGURE 2–36 Problem 16.

2–4 Acceleration

17. (I) A sports car accelerates from rest to 95 km/h in 4.3 s. What is its average acceleration in m/s^2?
18. (I) A sprinter accelerates from rest to 9.00 m/s in 1.38 s. What is her acceleration in (*a*) m/s^2; (*b*) km/h^2?
19. (II) A sports car moving at constant velocity travels 120 m in 5.0 s. If it then brakes and comes to a stop in 4.0 s, what is the magnitude of its acceleration (assumed constant) in m/s^2, and in *g*'s ($g = 9.80$ m/s^2)?

20. (II) At highway speeds, a particular automobile is capable of an acceleration of about 1.8 m/s^2. At this rate, how long does it take to accelerate from 65 km/h to 120 km/h?

21. (II) A car moving in a straight line starts at $x = 0$ at $t = 0$. It passes the point $x = 25.0 \text{ m}$ with a speed of 11.0 m/s at $t = 3.00 \text{ s}$. It passes the point $x = 385 \text{ m}$ with a speed of 45.0 m/s at $t = 20.0 \text{ s}$. Find (a) the average velocity, and (b) the average acceleration, between $t = 3.00 \text{ s}$ and $t = 20.0 \text{ s}$.

2–5 and 2–6 Motion at Constant Acceleration

22. (I) A car slows down from 28 m/s to rest in a distance of 88 m. What was its acceleration, assumed constant?

23. (I) A car accelerates from 14 m/s to 21 m/s in 6.0 s. What was its acceleration? How far did it travel in this time? Assume constant acceleration.

24. (I) A light plane must reach a speed of 35 m/s for takeoff. How long a runway is needed if the (constant) acceleration is 3.0 m/s^2?

25. (II) A baseball pitcher throws a baseball with a speed of 43 m/s. Estimate the average acceleration of the ball during the throwing motion. In throwing the baseball, the pitcher accelerates it through a displacement of about 3.5 m, from behind the body to the point where it is released (Fig. 2–37).

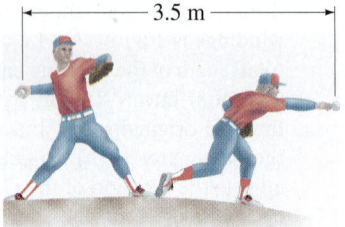

FIGURE 2–37 Problem 25.

26. (II) A world-class sprinter can reach a top speed (of about 11.5 m/s) in the first 18.0 m of a race. What is the average acceleration of this sprinter and how long does it take her to reach that speed?

27. (II) A car slows down uniformly from a speed of 28.0 m/s to rest in 8.00 s. How far did it travel in that time?

28. (II) In coming to a stop, a car leaves skid marks 65 m long on the highway. Assuming a deceleration of 4.00 m/s^2, estimate the speed of the car just before braking.

29. (II) A car traveling at 95 km/h strikes a tree. The front end of the car compresses and the driver comes to rest after traveling 0.80 m. What was the magnitude of the average acceleration of the driver during the collision? Express the answer in terms of "g's," where $1.00 g = 9.80 \text{ m/s}^2$.

30. (II) A car traveling 75 km/h slows down at a constant 0.50 m/s^2 just by "letting up on the gas." Calculate (a) the distance the car coasts before it stops, (b) the time it takes to stop, and (c) the distance it travels during the first and fifth seconds.

31. (II) Determine the stopping distances for an automobile going a constant initial speed of 95 km/h and human reaction time of 0.40 s: (a) for an acceleration $a = -3.0 \text{ m/s}^2$; (b) for $a = -6.0 \text{ m/s}^2$.

32. (II) A driver is traveling 18.0 m/s when she sees a red light ahead. Her car is capable of decelerating at a rate of 3.65 m/s^2. If it takes her 0.350 s to get the brakes on and she is 20.0 m from the intersection when she sees the light, will she be able to stop in time? How far from the beginning of the intersection will she be, and in what direction?

33. (II) A 75-m-long train begins uniform acceleration from rest. The front of the train has a speed of 18 m/s when it passes a railway worker who is standing 180 m from where the front of the train started. What will be the speed of the last car as it passes the worker? (See Fig. 2–38.)

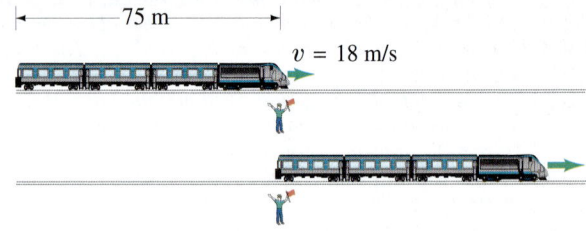

FIGURE 2–38 Problem 33.

34. (II) A space vehicle accelerates uniformly from 85 m/s at $t = 0$ to 162 m/s at $t = 10.0 \text{ s}$. How far did it move between $t = 2.0 \text{ s}$ and $t = 6.0 \text{ s}$?

35. (II) A runner hopes to complete the 10,000-m run in less than 30.0 min. After running at constant speed for exactly 27.0 min, there are still 1200 m to go. The runner must then accelerate at 0.20 m/s^2 for how many seconds in order to achieve the desired time?

36. (III) A fugitive tries to hop on a freight train traveling at a constant speed of 5.0 m/s. Just as an empty box car passes him, the fugitive starts from rest and accelerates at $a = 1.4 \text{ m/s}^2$ to his maximum speed of 6.0 m/s, which he then maintains. (a) How long does it take him to catch up to the empty box car? (b) What is the distance traveled to reach the box car?

37. (III) Mary and Sally are in a foot race (Fig. 2–39). When Mary is 22 m from the finish line, she has a speed of 4.0 m/s and is 5.0 m behind Sally, who has a speed of 5.0 m/s. Sally thinks she has an easy win and so, during the remaining portion of the race, decelerates at a constant rate of 0.40 m/s^2 to the finish line. What constant acceleration does Mary now need during the remaining portion of the race, if she wishes to cross the finish line side-by-side with Sally?

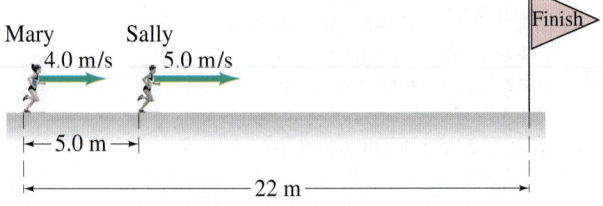

FIGURE 2–39 Problem 37.

38. (III) An unmarked police car traveling a constant 95 km/h is passed by a speeder traveling 135 km/h. Precisely 1.00 s after the speeder passes, the police officer steps on the accelerator; if the police car's acceleration is 2.60 m/s^2, how much time passes before the police car overtakes the speeder (assumed moving at constant speed)?

2–7 Freely Falling Objects (neglect air resistance)

39. (I) A stone is dropped from the top of a cliff. It is seen to hit the ground below after 3.55 s. How high is the cliff?

40. (I) Estimate (a) how long it took King Kong to fall straight down from the top of the Empire State Building (380 m high), and (b) his velocity just before "landing."

41. (II) A ball player catches a ball 3.4 s after throwing it vertically upward. With what speed did he throw it, and what height did it reach?

42. (II) A baseball is hit almost straight up into the air with a speed of 25 m/s. Estimate (a) how high it goes, (b) how long it is in the air. (c) What factors make this an estimate?

43. (II) A kangaroo jumps straight up to a vertical height of 1.45 m. How long was it in the air before returning to Earth?

44. (II) The best rebounders in basketball have a vertical leap (that is, the vertical movement of a fixed point on their body) of about 120 cm. (a) What is their initial "launch" speed off the ground? (b) How long are they in the air?

45. (II) An object starts from rest and falls under the influence of gravity. Draw graphs of (a) its speed and (b) the distance it has fallen, as a function of time from $t = 0$ to $t = 5.00$ s. Ignore air resistance.

46. (II) A stone is thrown vertically upward with a speed of 24.0 m/s. (a) How fast is it moving when it is at a height of 13.0 m? (b) How much time is required to reach this height? (c) Why are there two answers to (b)?

47. (II) For an object falling freely from rest, show that the distance traveled *during* each successive second increases in the ratio of successive odd integers (1, 3, 5, etc.). (This was first shown by Galileo.) See Figs. 2–19 and 2–22.

48. (II) A rocket rises vertically, from rest, with an acceleration of 3.2 m/s² until it runs out of fuel at an altitude of 775 m. After this point, its acceleration is that of gravity, downward. (a) What is the velocity of the rocket when it runs out of fuel? (b) How long does it take to reach this point? (c) What maximum altitude does the rocket reach? (d) How much time (total) does it take to reach maximum altitude? (e) With what velocity does it strike the Earth? (f) How long (total) is it in the air?

49. (II) A helicopter is ascending vertically with a speed of 5.40 m/s. At a height of 105 m above the Earth, a package is dropped from the helicopter. How much time does it take for the package to reach the ground? [*Hint*: What is v_0 for the package?]

50. (II) Roger sees water balloons fall past his window. He notices that each balloon strikes the sidewalk 0.83 s after passing his window. Roger's room is on the third floor, 15 m above the sidewalk. (a) How fast are the balloons traveling when they pass Roger's window? (b) Assuming the balloons are being released from rest, from what floor are they being released? Each floor of the dorm is 5.0 m high.

51. (II) Suppose you adjust your garden hose nozzle for a fast stream of water. You point the nozzle vertically upward at a height of 1.8 m above the ground (Fig. 2–40). When you quickly turn off the nozzle, you hear the water striking the ground next to you for another 2.5 s. What is the water speed as it leaves the nozzle?

FIGURE 2–40 Problem 51.

52. (III) A baseball is seen to pass upward by a window with a vertical speed of 14 m/s. If the ball was thrown by a person 18 m below on the street, (a) what was its initial speed, (b) what altitude does it reach, (c) when was it thrown, and (d) when does it reach the street again?

53. (III) A falling stone takes 0.31 s to travel past a window 2.2 m tall (Fig. 2–41). From what height above the top of the window did the stone fall?

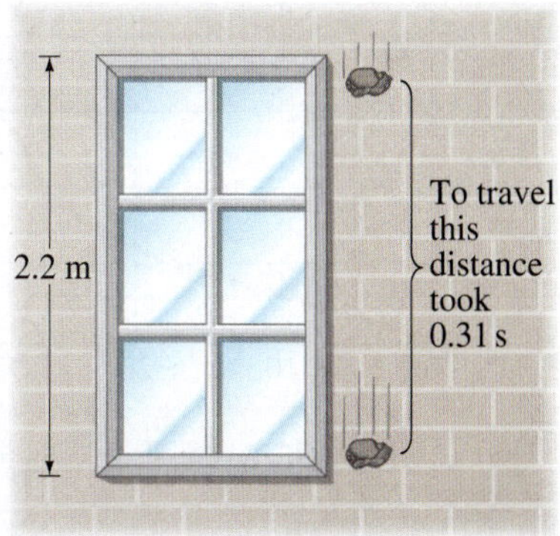

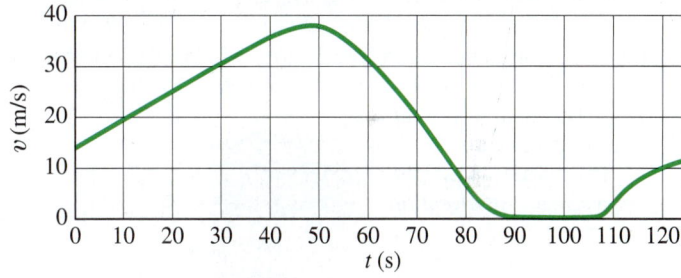

FIGURE 2–41 Problem 53.

54. (III) A rock is dropped from a sea cliff, and the sound of it striking the ocean is heard 3.4 s later. If the speed of sound is 340 m/s, how high is the cliff?

2–8 Graphical Analysis

55. (II) Figure 2–42 shows the velocity of a train as a function of time. (a) At what time was its velocity greatest? (b) During what periods, if any, was the velocity constant? (c) During what periods, if any, was the acceleration constant? (d) When was the magnitude of the acceleration greatest?

FIGURE 2–42 Problem 55.

56. (II) A sports car accelerates approximately as shown in the velocity–time graph of Fig. 2–43. (The short flat spots in the curve represent manual shifting of the gears.) Estimate the car's average acceleration in (a) second gear and (b) fourth gear.

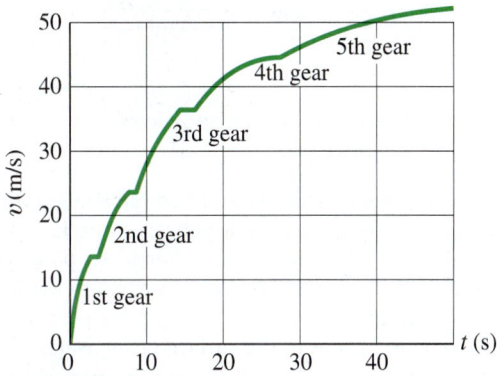

FIGURE 2–43 Problem 56. The velocity of a car as a function of time, starting from a dead stop. The flat spots in the curve represent gear shifts.

57. (II) The position of a rabbit along a straight tunnel as a function of time is plotted in Fig. 2–44. What is its instantaneous velocity (a) at $t = 10.0\,\text{s}$ and (b) at $t = 30.0\,\text{s}$? What is its average velocity (c) between $t = 0$ and $t = 5.0\,\text{s}$, (d) between $t = 25.0\,\text{s}$ and $t = 30.0\,\text{s}$, and (e) between $t = 40.0\,\text{s}$ and $t = 50.0\,\text{s}$?

58. (II) In Fig. 2–44, (a) during what time periods, if any, is the velocity constant? (b) At what time is the velocity greatest? (c) At what time, if any, is the velocity zero? (d) Does the object move in one direction or in both directions during the time shown?

59. (III) Sketch the v vs. t graph for the object whose displacement as a function of time is given by Fig. 2–44.

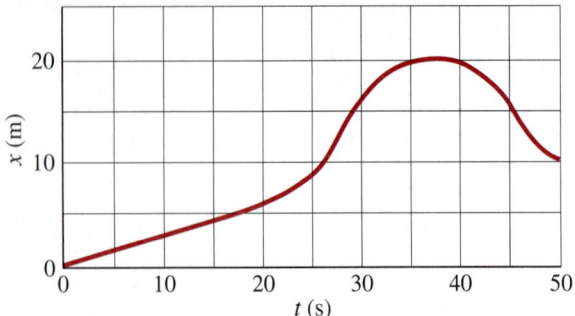

FIGURE 2–44 Problems 57, 58, and 59.

General Problems

60. The acceleration due to gravity on the Moon is about one-sixth what it is on Earth. If an object is thrown vertically upward on the Moon, how many times higher will it go than it would on Earth, assuming the same initial velocity?

61. A person who is properly restrained by an over-the-shoulder seat belt has a good chance of surviving a car collision if the deceleration does not exceed 30 "g's" $(1.00\,g = 9.80\,\text{m/s}^2)$. Assuming uniform deceleration at 30 g's, calculate the distance over which the front end of the car must be designed to collapse if a crash brings the car to rest from 95 km/h.

62. A person jumps out a fourth-story window 18.0 m above a firefighter's safety net. The survivor stretches the net 1.0 m before coming to rest, Fig. 2–45. (a) What was the average deceleration experienced by the survivor when she was slowed to rest by the net? (b) What would you do to make it "safer" (that is, to generate a smaller deceleration): would you stiffen or loosen the net? Explain.

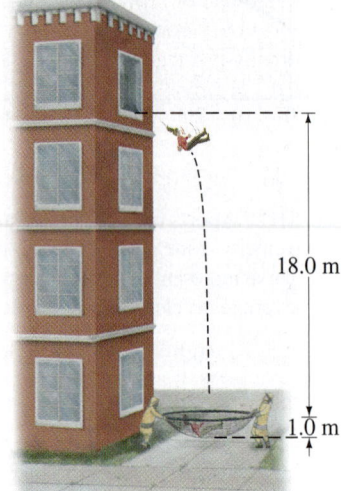

FIGURE 2–45 Problem 62.

63. Pelicans tuck their wings and free-fall straight down when diving for fish. Suppose a pelican starts its dive from a height of 14.0 m and cannot change its path once committed. If it takes a fish 0.20 s to perform evasive action, at what minimum height must it spot the pelican to escape? Assume the fish is at the surface of the water.

64. A bicyclist in the Tour de France crests a mountain pass as he moves at 15 km/h. At the bottom, 4.0 km farther, his speed is 65 km/h. Estimate his average acceleration (in m/s²) while riding down the mountain.

65. Consider the street pattern shown in Fig. 2–46. Each intersection has a traffic signal, and the speed limit is 40 km/h. Suppose you are driving from the west at the speed limit. When you are 10.0 m from the first intersection, all the lights turn green. The lights are green for 13.0 s each. (a) Calculate the time needed to reach the third stoplight. Can you make it through all three lights without stopping? (b) Another car was stopped at the first light when all the lights turned green. It can accelerate at the rate of $2.00\,\text{m/s}^2$ to the speed limit. Can the second car make it through all three lights without stopping? By how many seconds would it make it, or not make it?

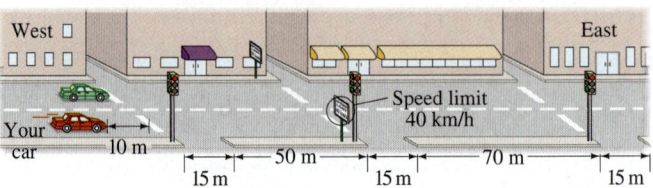

FIGURE 2–46 Problem 65.

66. An airplane travels 2100 km at a speed of 720 km/h, and then encounters a tailwind that boosts its speed to 990 km/h for the next 2800 km. What was the total time for the trip? What was the average speed of the plane for this trip? [*Hint*: Does Eq. 2–11d apply?]

67. Suppose a car manufacturer tested its cars for front-end collisions by hauling them up on a crane and dropping them from a certain height. (a) Show that the speed just before a car hits the ground, after falling from rest a vertical distance H, is given by $\sqrt{2gH}$. What height corresponds to a collision at (b) 35 km/h? (c) 95 km/h?

68. A stone is dropped from the roof of a high building. A second stone is dropped 1.30 s later. How far apart are the stones when the second one has reached a speed of 12.0 m/s?

69. A person jumps off a diving board 4.0 m above the water's surface into a deep pool. The person's downward motion stops 2.0 m below the surface of the water. Estimate the average deceleration of the person while under the water.

70. In putting, the force with which a golfer strikes a ball is planned so that the ball will stop within some small distance of the cup, say 1.0 m long or short, in case the putt is missed. Accomplishing this from an uphill lie (that is, putting the ball downhill, see Fig. 2–47) is more difficult than from a downhill lie. To see why, assume that on a particular green the ball decelerates constantly at 1.8 m/s^2 going downhill, and constantly at 2.6 m/s^2 going uphill. Suppose we have an uphill lie 7.0 m from the cup. Calculate the allowable range of initial velocities we may impart to the ball so that it stops in the range 1.0 m short to 1.0 m long of the cup. Do the same for a downhill lie 7.0 m from the cup. What in your results suggests that the downhill putt is more difficult?

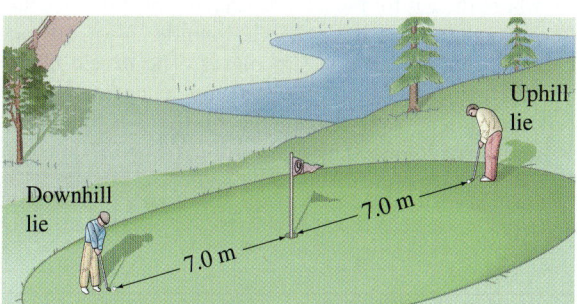

FIGURE 2–47 Problem 70.

71. A stone is thrown vertically upward with a speed of 15.5 m/s from the edge of a cliff 75.0 m high (Fig. 2–48).
(a) How much later does it reach the bottom of the cliff?
(b) What is its speed just before hitting?
(c) What total distance did it travel?

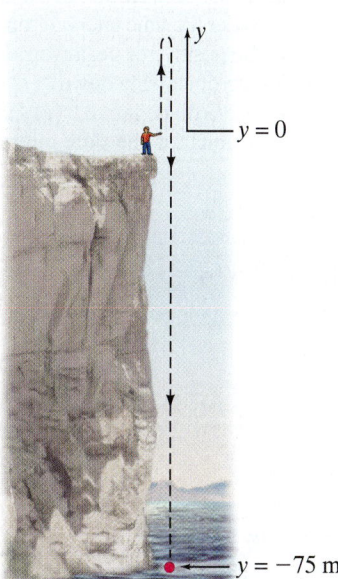

FIGURE 2–48 Problem 71.

72. In the design of a **rapid transit system**, it is necessary to balance the average speed of a train against the distance between station stops. The more stops there are, the slower the train's average speed. To get an idea of this problem, calculate the time it takes a train to make a 15.0-km trip in two situations: (a) the stations at which the trains must stop are 3.0 km apart (a total of 6 stations, including those at the ends); and (b) the stations are 5.0 km apart (4 stations total). Assume that at each station the train accelerates at a rate of 1.1 m/s^2 until it reaches 95 km/h, then stays at this speed until its brakes are applied for arrival at the next station, at which time it decelerates at -2.0 m/s^2. Assume it stops at each intermediate station for 22 s.

73. A person driving her car at 35 km/h approaches an intersection just as the traffic light turns yellow. She knows that the yellow light lasts only 2.0 s before turning to red, and she is 28 m away from the near side of the intersection (Fig. 2–49). Should she try to stop, or should she speed up to cross the intersection before the light turns red? The intersection is 15 m wide. Her car's maximum deceleration is -5.8 m/s^2, whereas it can accelerate from 45 km/h to 65 km/h in 6.0 s. Ignore the length of her car and her reaction time.

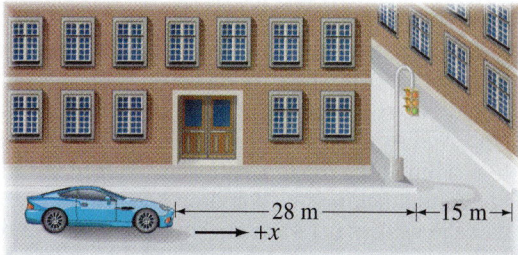

FIGURE 2–49 Problem 73.

74. A car is behind a truck going 18 m/s on the highway. The car's driver looks for an opportunity to pass, guessing that his car can accelerate at 0.60 m/s^2 and that he has to cover the 20-m length of the truck, plus 10 m extra space at the rear of the truck and 10 m more at the front of it. In the oncoming lane, he sees a car approaching, probably at the speed limit, 25 m/s (55 mph). He estimates that the car is about 500 m away. Should he attempt the pass? Give details.

75. Agent Bond is standing on a bridge, 15 m above the road below, and his pursuers are getting too close for comfort. He spots a flatbed truck approaching at 25 m/s, which he measures by knowing that the telephone poles the truck is passing are 25 m apart in this region. The roof of the truck is 3.5 m above the road, and Bond quickly calculates how many poles away the truck should be when he drops down from the bridge onto the truck, making his getaway. How many poles is it?

76. A conveyor belt is used to send burgers through a grilling machine. If the grilling machine is 1.2 m long and the burgers require 2.8 min to cook, how fast must the conveyor belt travel? If the burgers are spaced 25 cm apart, what is the rate of burger production (in burgers/min)?

77. Two students are asked to find the height of a particular building using a barometer. Instead of using the barometer as an altitude measuring device, they take it to the roof of the building and drop it off, timing its fall. One student reports a fall time of 2.0 s, and the other, 2.3 s. What % difference does the 0.3 s make for the estimates of the building's height?

78. Figure 2–50 shows the position vs. time graph for two bicycles, A and B. (a) Identify any instant at which the two bicycles have the same velocity. (b) Which bicycle has the larger acceleration? (c) At which instant(s) are the bicycles passing each other? Which bicycle is passing the other? (d) Which bicycle has the larger instantaneous velocity? (e) Which bicycle has the larger average velocity?

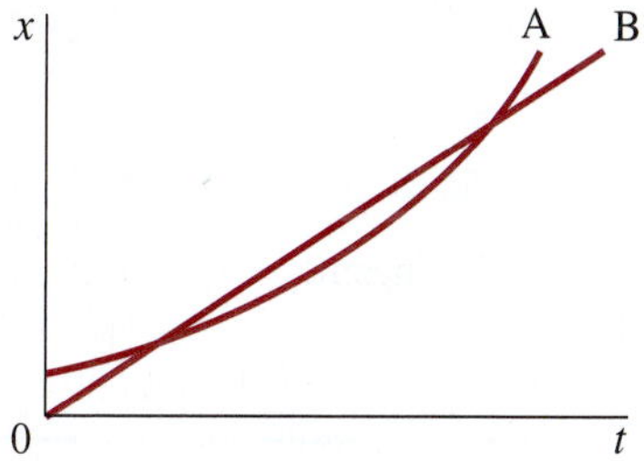

FIGURE 2–50 Problem 78.

79. A race car driver must average 200.0 km/h over the course of a time trial lasting ten laps. If the first nine laps were done at an average speed of 196.0 km/h, what average speed must be maintained for the last lap?

80. Two children are playing on two trampolines. The first child bounces up one-and-a-half times higher than the second child. The initial speed up of the second child is 4.0 m/s. (a) Find the maximum height the second child reaches. (b) What is the initial speed of the first child? (c) How long was the first child in the air?

81. If there were no air resistance, how long would it take a free-falling skydiver to fall from a plane at 3200 m to an altitude of 450 m, where she will open her parachute? What would her speed be at 450 m? (In reality, the air resistance will restrict her speed to perhaps 150 km/h.)

82. You stand at the top of a cliff while your friend stands on the ground below you. You drop a ball from rest and see that she catches it 1.4 s later. Your friend then throws the ball up to you, such that it just comes to rest in your hand. What is the speed with which your friend threw the ball?

83. On an audio compact disc (CD), digital bits of information are encoded sequentially along a spiral path. Each bit occupies about $0.28\,\mu\text{m}$. A CD player's readout laser scans along the spiral's sequence of bits at a constant speed of about 1.2 m/s as the CD spins. (a) Determine the number N of digital bits that a CD player reads every second. (b) The audio information is sent to each of the two loudspeakers 44,100 times per second. Each of these samplings requires 16 bits, and so you might expect the required bit rate for a CD player to be

$$N_0 = 2\left(44{,}100\,\frac{\text{samplings}}{\text{s}}\right)\left(16\,\frac{\text{bits}}{\text{sampling}}\right) = 1.4 \times 10^6\,\frac{\text{bits}}{\text{s}},$$

where the 2 is for the 2 loudspeakers (the 2 stereo channels). Note that N_0 is less than the number N of bits actually read per second by a CD player. The excess number of bits $(= N - N_0)$ is needed for encoding and error-correction. What percentage of the bits on a CD are dedicated to encoding and error-correction?

Search and Learn

1. Discuss two conditions given in Section 2–7 for being able to use a constant acceleration of magnitude $g = 9.8\,\text{m/s}^2$. Give an example in which one of these conditions would not be met and would not even be a reasonable approximation of motion.

2. In a lecture demonstration, a 3.0-m-long vertical string with ten bolts tied to it at equal intervals is dropped from the ceiling of the lecture hall. The string falls on a tin plate, and the class hears the clink of each bolt as it hits the plate. (a) The sounds will not occur at equal time intervals. Why? (b) Will the time between clinks increase or decrease as the string falls? (c) How could the bolts be tied so that the clinks occur at equal intervals? (Assume the string is vertical with the bottom bolt touching the tin plate when the string is released.)

3. A police car at rest is passed by a speeder traveling at a constant 140 km/h. The police officer takes off in hot pursuit and catches up to the speeder in 850 m, maintaining a constant acceleration. (a) Qualitatively plot the position vs. time graph for both cars from the police car's start to the catch-up point. Calculate (b) how long it took the police officer to overtake the speeder, (c) the required police car acceleration, and (d) the speed of the police car at the overtaking point.

4. Figure 2–51 is a position versus time graph for the motion of an object along the x axis. Consider the time interval from A to B. (a) Is the object moving in the positive or negative x direction? (b) Is the object speeding up or slowing down? (c) Is the acceleration of the object positive or negative? Now consider the time interval from D to E. (d) Is the object moving in the positive or negative x direction? (e) Is the object speeding up or slowing down? (f) Is the acceleration of the object positive or negative? (g) Finally, answer these same three questions for the time interval from C to D.

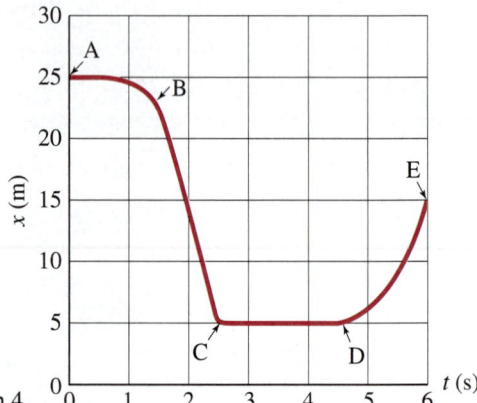

FIGURE 2–51
Search and Learn 4.

5. The position of a ball rolling in a straight line is given by $x = 2.0 - 3.6t + 1.7t^2$, where x is in meters and t in seconds. (a) What do the numbers 2.0, 3.6, and 1.7 refer to? (b) What are the units of each of these numbers? (c) Determine the position of the ball at $t = 1.0\,\text{s}$, 2.0 s, and 3.0 s. (d) What is the average velocity over the interval $t = 1.0\,\text{s}$ to $t = 3.0\,\text{s}$?

ANSWERS TO EXERCISES

A: (a) displacement = -30 cm; (b) total distance = 50 cm.
B: (b).
C: (a) +; (b) −; (c) −; (d) +.
D: (b).
E: (e).
F: (c).

This snowboarder flying through the air shows an example of motion in two dimensions. In the absence of air resistance, the path would be a perfect parabola. The gold arrow represents the downward acceleration of gravity, $\vec{g}$. Galileo analyzed the motion of objects in 2 dimensions under the action of gravity near the Earth's surface (now called "projectile motion") into its horizontal and vertical components.

We will discuss vectors and how to add them. Besides analyzing projectile motion, we will also see how to work with relative velocity.

Kinematics in Two Dimensions; Vectors

CHAPTER 3

CHAPTER-OPENING QUESTION—Guess now!

[*Don't worry about getting the right answer now—you will get another chance later in the Chapter. See also p. 1 of Chapter 1 for more explanation.*]

A small heavy box of emergency supplies is dropped from a moving helicopter at point A as it flies at constant speed in a horizontal direction. Which path in the drawing below best describes the path of the box (neglecting air resistance) as seen by a person standing on the ground?

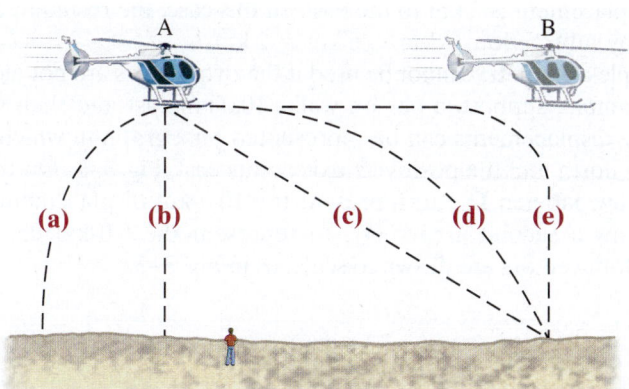

CONTENTS

- 3–1 Vectors and Scalars
- 3–2 Addition of Vectors— Graphical Methods
- 3–3 Subtraction of Vectors, and Multiplication of a Vector by a Scalar
- 3–4 Adding Vectors by Components
- 3–5 Projectile Motion
- 3–6 Solving Projectile Motion Problems
- *3–7 Projectile Motion Is Parabolic
- 3–8 Relative Velocity

In Chapter 2 we dealt with motion along a straight line. We now consider the motion of objects that move in paths in two (or three) dimensions. In particular, we discuss an important type of motion known as *projectile motion*: objects projected outward near the Earth's surface, such as struck baseballs and golf balls, kicked footballs, and other projectiles. Before beginning our discussion of motion in two dimensions, we will need a new tool, vectors, and how to add them.

49

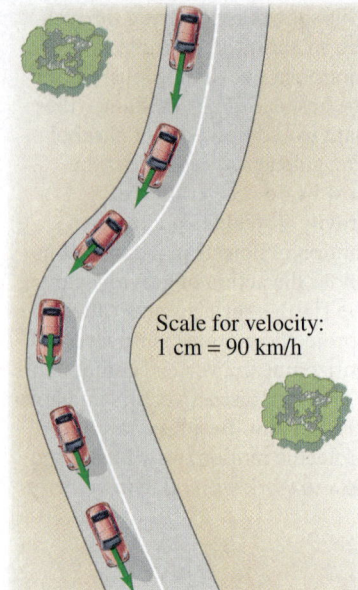

FIGURE 3–1 Car traveling on a road, slowing down to round the curve. The green arrows represent the velocity vector at each position.

3–1 Vectors and Scalars

We mentioned in Chapter 2 that the term *velocity* refers not only to how fast an object is moving but also to its direction. A quantity such as velocity, which has *direction* as well as *magnitude*, is a **vector** quantity. Other quantities that are also vectors are displacement, force, and momentum. However, many quantities have no direction associated with them, such as mass, time, and temperature. They are specified completely by a number and units. Such quantities are called **scalar** quantities.

Drawing a diagram of a particular physical situation is always helpful in physics, and this is especially true when dealing with vectors. On a diagram, each vector is represented by an arrow. The arrow is always drawn so that it points in the direction of the vector quantity it represents. The length of the arrow is drawn proportional to the magnitude of the vector quantity. For example, in Fig. 3–1, green arrows have been drawn representing the velocity of a car at various places as it rounds a curve. The magnitude of the velocity at each point can be read off Fig. 3–1 by measuring the length of the corresponding arrow and using the scale shown (1 cm = 90 km/h).

When we write the symbol for a vector, we will always use boldface type, with a tiny arrow over the symbol. Thus for velocity we write $\vec{v}$. If we are concerned only with the magnitude of the vector, we will write simply v, in italics, as we do for other symbols.

3–2 Addition of Vectors—Graphical Methods

Because vectors are quantities that have direction as well as magnitude, they must be added in a special way. In this Chapter, we will deal mainly with displacement vectors, for which we now use the symbol $\vec{D}$, and velocity vectors, $\vec{v}$. But the results will apply for other vectors we encounter later.

We use simple arithmetic for adding scalars. Simple arithmetic can also be used for adding vectors if they are in the same direction. For example, if a person walks 8 km east one day, and 6 km east the next day, the person will be 8 km + 6 km = 14 km east of the point of origin. We say that the *net* or *resultant* displacement is 14 km to the east (Fig. 3–2a). If, on the other hand, the person walks 8 km east on the first day, and 6 km west (in the reverse direction) on the second day, then the person will end up 2 km from the origin (Fig. 3–2b), so the resultant displacement is 2 km to the east. In this case, the resultant displacement is obtained by subtraction: 8 km − 6 km = 2 km.

But simple arithmetic cannot be used if the two vectors are not along the same line. For example, suppose a person walks 10.0 km east and then walks 5.0 km north. These displacements can be represented on a graph in which the positive y axis points north and the positive x axis points east, Fig. 3–3. On this graph, we draw an arrow, labeled $\vec{D}_1$, to represent the 10.0-km displacement to the east. Then we draw a second arrow, $\vec{D}_2$, to represent the 5.0-km displacement to the north. Both vectors are drawn to scale, as in Fig. 3–3.

FIGURE 3–2 Combining vectors in one dimension.

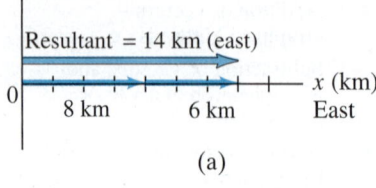

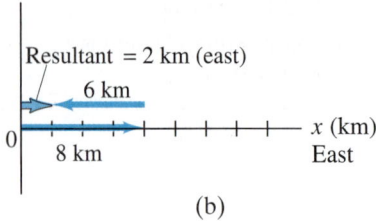

FIGURE 3–3 A person walks 10.0 km east and then 5.0 km north. These two displacements are represented by the vectors $\vec{D}_1$ and $\vec{D}_2$, which are shown as arrows. Also shown is the resultant displacement vector, $\vec{D}_R$, which is the vector sum of $\vec{D}_1$ and $\vec{D}_2$. Measurement on the graph with ruler and protractor shows that $\vec{D}_R$ has a magnitude of 11.2 km and points at an angle $\theta = 27°$ north of east.

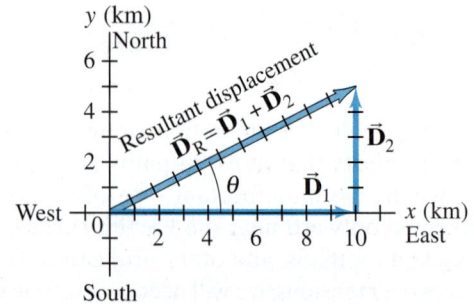

After taking this walk, the person is now 10.0 km east and 5.0 km north of the point of origin. The **resultant displacement** is represented by the arrow labeled $\vec{D}_R$ in Fig. 3–3. (The subscript R stands for resultant.) Using a ruler and a protractor, you can measure on this diagram that the person is 11.2 km from the origin at an angle $\theta = 27°$ north of east. In other words, the resultant displacement vector has a magnitude of 11.2 km and makes an angle $\theta = 27°$ with the positive x axis. The magnitude (length) of $\vec{D}_R$ can also be obtained using the theorem of Pythagoras in this case, because D_1, D_2, and D_R form a right triangle with D_R as the hypotenuse. Thus

$$D_R = \sqrt{D_1^2 + D_2^2} = \sqrt{(10.0\,\text{km})^2 + (5.0\,\text{km})^2}$$
$$= \sqrt{125\,\text{km}^2} = 11.2\,\text{km}.$$

You can use the Pythagorean theorem only when the vectors are *perpendicular* to each other.

The resultant displacement vector, $\vec{D}_R$, is the sum of the vectors $\vec{D}_1$ and $\vec{D}_2$. That is,

$$\vec{D}_R = \vec{D}_1 + \vec{D}_2.$$

This is a *vector* equation. An important feature of adding two vectors that are not along the same line is that the magnitude of the resultant vector is not equal to the sum of the magnitudes of the two separate vectors, but is smaller than their sum. That is,

$$D_R \leq (D_1 + D_2),$$

where the equals sign applies only if the two vectors point in the same direction. In our example (Fig. 3–3), $D_R = 11.2\,\text{km}$, whereas $D_1 + D_2$ equals 15 km, which is the total distance traveled. Note also that we cannot set $\vec{D}_R$ equal to 11.2 km, because we have a vector equation and 11.2 km is only a part of the resultant vector, its magnitude. We could write something like this, though: $\vec{D}_R = \vec{D}_1 + \vec{D}_2 = (11.2\,\text{km},\, 27°\,\text{N of E})$.

Figure 3–3 illustrates the general rules for graphically adding two vectors together, no matter what angles they make, to get their sum. The rules are as follows:

1. On a diagram, draw one of the vectors—call it $\vec{D}_1$—to scale.
2. Next draw the second vector, $\vec{D}_2$, to scale, placing its tail at the tip of the first vector and being sure its direction is correct.
3. The arrow drawn from the tail of the first vector to the tip of the second vector represents the *sum*, or **resultant**, of the two vectors.

The length of the resultant vector represents its magnitude. Note that vectors can be moved parallel to themselves on paper (maintaining the same length and angle) to accomplish these manipulations. The length of the resultant can be measured with a ruler and compared to the scale. Angles can be measured with a protractor. This method is known as the **tail-to-tip method of adding vectors**.

The resultant is not affected by the order in which the vectors are added. For example, a displacement of 5.0 km north, to which is added a displacement of 10.0 km east, yields a resultant of 11.2 km and angle $\theta = 27°$ (see Fig. 3–4), the same as when they were added in reverse order (Fig. 3–3). That is, now using $\vec{V}$ to represent any type of vector,

$$\vec{V}_1 + \vec{V}_2 = \vec{V}_2 + \vec{V}_1.$$

[Mathematicians call this equation the *commutative* property of vector addition.]

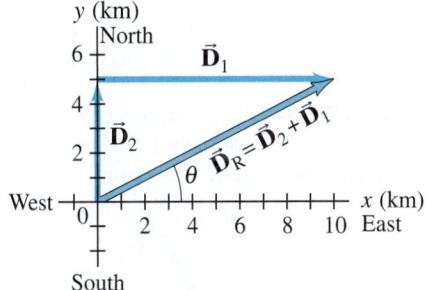

FIGURE 3–4 If the vectors are added in reverse order, the resultant is the same. (Compare to Fig. 3–3.)

The tail-to-tip method of adding vectors can be extended to three or more vectors. The resultant is drawn from the tail of the first vector to the tip of the last one added. An example is shown in Fig. 3–5; the three vectors could represent displacements (northeast, south, west) or perhaps three forces. Check for yourself that you get the same resultant no matter in which order you add the three vectors.

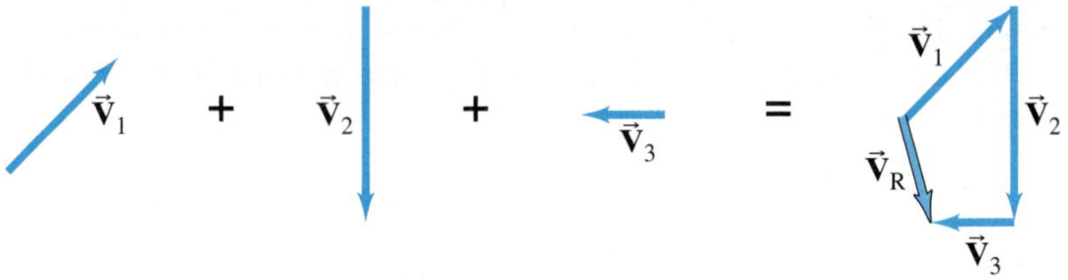

FIGURE 3–5 The resultant of three vectors: $\vec{V}_R = \vec{V}_1 + \vec{V}_2 + \vec{V}_3$.

A second way to add two vectors is the **parallelogram method**. It is fully equivalent to the tail-to-tip method. In this method, the two vectors are drawn starting from a common origin, and a parallelogram is constructed using these two vectors as adjacent sides as shown in Fig. 3–6b. The resultant is the diagonal drawn from the common origin. In Fig. 3–6a, the tail-to-tip method is shown, and we can see that both methods yield the same result.

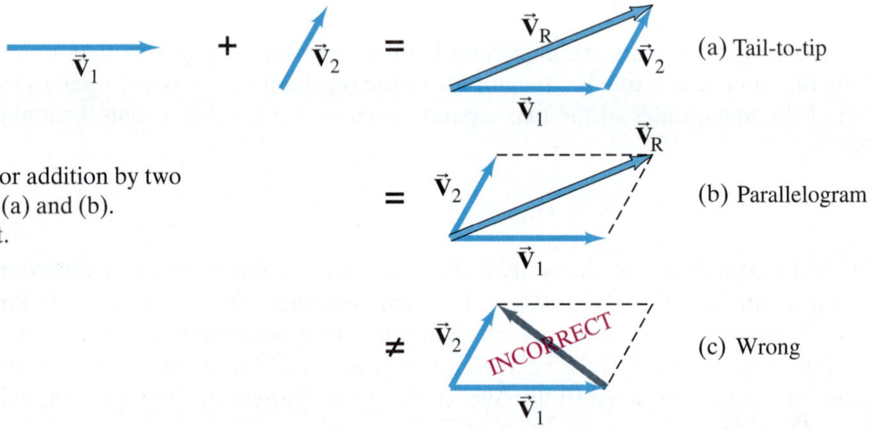

FIGURE 3–6 Vector addition by two different methods, (a) and (b). Part (c) is incorrect.

CAUTION
Be sure to use the correct diagonal on the parallelogram to get the resultant

It is a common error to draw the sum vector as the diagonal running between the tips of the two vectors, as in Fig. 3–6c. *This is incorrect*: it does not represent the sum of the two vectors. (In fact, it represents their difference, $\vec{V}_2 - \vec{V}_1$, as we will see in the next Section.)

CONCEPTUAL EXAMPLE 3–1 **Range of vector lengths.** Suppose two vectors each have length 3.0 units. What is the range of possible lengths for the vector representing the sum of the two?

RESPONSE The sum can take on any value from 6.0 $(= 3.0 + 3.0)$ where the vectors point in the same direction, to 0 $(= 3.0 - 3.0)$ when the vectors are antiparallel. Magnitudes between 0 and 6.0 occur when the two vectors are at an angle other than 0° and 180°.

EXERCISE A If the two vectors of Example 3–1 are perpendicular to each other, what is the resultant vector length?

3–3 Subtraction of Vectors, and Multiplication of a Vector by a Scalar

Given a vector $\vec{V}$, we define the *negative* of this vector $(-\vec{V})$ to be a vector with the same magnitude as $\vec{V}$ but opposite in direction, Fig. 3–7. Note, however, that no vector is ever negative in the sense of its magnitude: the magnitude of every vector is positive. Rather, a minus sign tells us about its direction.

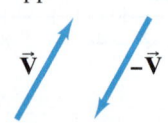

FIGURE 3–7 The negative of a vector is a vector having the same length but opposite direction.

52 CHAPTER 3 Kinematics in Two Dimensions; Vectors

We can now define the subtraction of one vector from another: the difference between two vectors $\vec{V}_2 - \vec{V}_1$ is defined as

$$\vec{V}_2 - \vec{V}_1 = \vec{V}_2 + (-\vec{V}_1).$$

That is, the difference between two vectors is equal to the sum of the first plus the negative of the second. Thus our rules for addition of vectors can be applied as shown in Fig. 3–8 using the tail-to-tip method.

FIGURE 3–8 Subtracting two vectors: $\vec{V}_2 - \vec{V}_1$.

A vector $\vec{V}$ can be multiplied by a scalar c. We define their product so that $c\vec{V}$ has the same direction as $\vec{V}$ and has magnitude cV. That is, multiplication of a vector by a positive scalar c changes the magnitude of the vector by a factor c but doesn't alter the direction. If c is a negative scalar (such as -2.0), the magnitude of the product $c\vec{V}$ is changed by the factor $|c|$ (where $|c|$ means the magnitude of c), but the direction is precisely opposite to that of $\vec{V}$. See Fig. 3–9.

FIGURE 3–9 Multiplying a vector $\vec{V}$ by a scalar c gives a vector whose magnitude is c times greater and in the same direction as $\vec{V}$ (or opposite direction if c is negative).

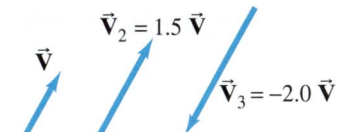

EXERCISE B What does the "incorrect" vector in Fig. 3–6c represent? (a) $\vec{V}_2 - \vec{V}_1$; (b) $\vec{V}_1 - \vec{V}_2$; (c) something else (specify).

3–4 Adding Vectors by Components

Adding vectors graphically using a ruler and protractor is often not sufficiently accurate and is not useful for vectors in three dimensions. We discuss now a more powerful and precise method for adding vectors. But do not forget graphical methods—they are useful for visualizing, for checking your math, and thus for getting the correct result.

Components

Consider first a vector $\vec{V}$ that lies in a particular plane. It can be expressed as the sum of two other vectors, called the **components** of the original vector. The components are usually chosen to be along two perpendicular directions, such as the x and y axes. The process of finding the components is known as **resolving the vector into its components**. An example is shown in Fig. 3–10; the vector $\vec{V}$ could be a displacement vector that points at an angle $\theta = 30°$ north of east, where we have chosen the positive x axis to be to the east and the positive y axis north. This vector $\vec{V}$ is resolved into its x and y components by drawing dashed lines (AB and AC) out from the tip (A) of the vector, making them perpendicular to the x and y axes. Then the lines 0B and 0C represent the x and y components of $\vec{V}$, respectively, as shown in Fig. 3–10b. These *vector components* are written $\vec{V}_x$ and $\vec{V}_y$. In this book we usually show vector components as arrows, like vectors, but dashed. The *scalar components*, V_x and V_y, are the magnitudes of the vector components, with units, accompanied by a positive or negative sign depending on whether they point along the positive or negative x or y axis. As can be seen in Fig. 3–10, $\vec{V}_x + \vec{V}_y = \vec{V}$ by the parallelogram method of adding vectors.

Space is made up of three dimensions, and sometimes it is necessary to resolve a vector into components along three mutually perpendicular directions. In rectangular coordinates the components are $\vec{V}_x$, $\vec{V}_y$, and $\vec{V}_z$.

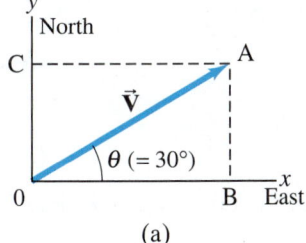
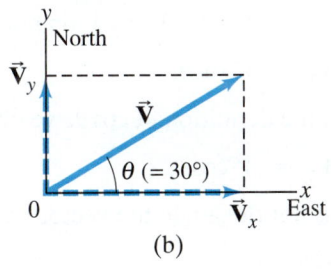

FIGURE 3–10 Resolving a vector $\vec{V}$ into its components along a chosen set of x and y axes. The components, once found, themselves represent the vector. That is, the components contain as much information as the vector itself.

To add vectors using the method of components, we need to use the trigonometric functions sine, cosine, and tangent, which we now review.

Given any angle θ, as in Fig. 3–11a, a right triangle can be constructed by drawing a line perpendicular to one of its sides, as in Fig. 3–11b. The longest side of a right triangle, opposite the right angle, is called the hypotenuse, which we label h. The side opposite the angle θ is labeled o, and the side adjacent is labeled a. We let h, o, and a represent the lengths of these sides, respectively.

FIGURE 3–11 Starting with an angle θ as in (a), we can construct right triangles of different sizes, (b) and (c), but the ratio of the lengths of the sides does not depend on the size of the triangle.

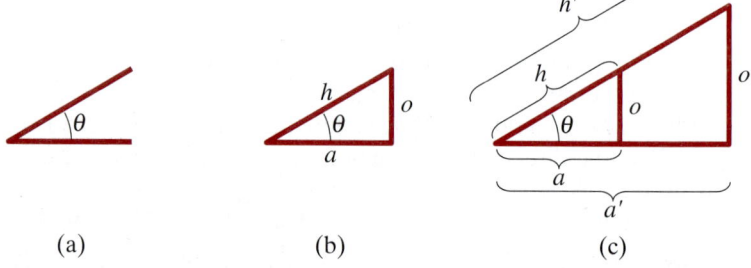

We now define the three trigonometric functions, sine, cosine, and tangent (abbreviated sin, cos, tan), in terms of the right triangle, as follows:

$$\sin \theta = \frac{\text{side opposite}}{\text{hypotenuse}} = \frac{o}{h}$$

$$\cos \theta = \frac{\text{side adjacent}}{\text{hypotenuse}} = \frac{a}{h} \quad \quad (3\text{–}1)$$

$$\tan \theta = \frac{\text{side opposite}}{\text{side adjacent}} = \frac{o}{a}.$$

If we make the triangle bigger, but keep the same angles, then the ratio of the length of one side to the other, or of one side to the hypotenuse, remains the same. That is, in Fig. 3–11c we have: $a/h = a'/h'$; $o/h = o'/h'$; and $o/a = o'/a'$. Thus the values of sine, cosine, and tangent do not depend on how big the triangle is. They depend only on the size of the angle. The values of sine, cosine, and tangent for different angles can be found using a scientific calculator, or from the Table in Appendix A.

A useful trigonometric identity is

$$\sin^2 \theta + \cos^2 \theta = 1 \quad \quad (3\text{–}2)$$

which follows from the Pythagorean theorem ($o^2 + a^2 = h^2$ in Fig. 3–11). That is:

$$\sin^2 \theta + \cos^2 \theta = \frac{o^2}{h^2} + \frac{a^2}{h^2} = \frac{o^2 + a^2}{h^2} = \frac{h^2}{h^2} = 1.$$

(See Appendix A and inside the rear cover for other details on trigonometric functions and identities.)

The use of trigonometric functions for finding the components of a vector is illustrated in Fig. 3–12, where a vector and its two components are thought of as making up a right triangle. We then see that the sine, cosine, and tangent are as given in Fig. 3–12, where θ is the angle $\vec{V}$ makes with the $+x$ axis. If we multiply the definition of $\sin \theta = V_y/V$ by V on both sides, we get

$$V_y = V \sin \theta. \quad \quad (3\text{–}3\text{a})$$

Similarly, from the definition of $\cos \theta$, we obtain

$$V_x = V \cos \theta. \quad \quad (3\text{–}3\text{b})$$

Note that if θ is not the angle the vector makes with the positive x axis, Eqs. 3–3 are not valid.

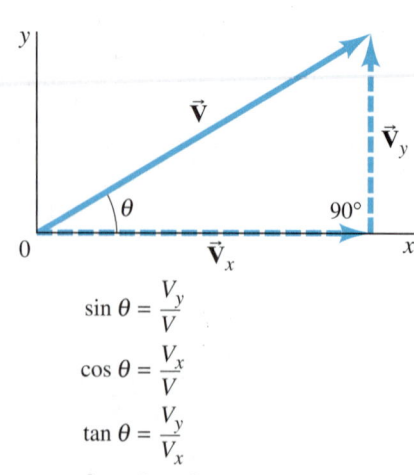

FIGURE 3–12 Finding the components of a vector using trigonometric functions. The equations are valid only if θ is the angle $\vec{V}$ makes with the positive x axis.

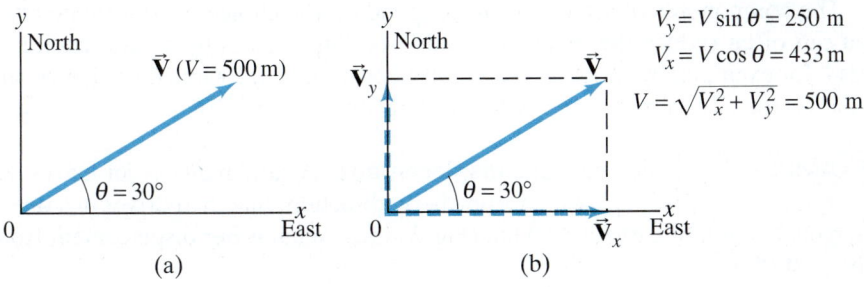

FIGURE 3–13 (a) Vector $\vec{V}$ represents a displacement of 500 m at a 30° angle north of east. (b) The components of $\vec{V}$ are $\vec{V}_x$ and $\vec{V}_y$, whose magnitudes are given on the right in the diagram.

Using Eqs. 3–3, we can calculate V_x and V_y for any vector, such as that illustrated in Fig. 3–10 or Fig. 3–12. Suppose $\vec{V}$ represents a displacement of 500 m in a direction 30° north of east, as shown in Fig. 3–13. Then $V = 500$ m. From a calculator or Tables, $\sin 30° = 0.500$ and $\cos 30° = 0.866$. Then

$$V_x = V \cos\theta = (500 \text{ m})(0.866) = 433 \text{ m (east)},$$
$$V_y = V \sin\theta = (500 \text{ m})(0.500) = 250 \text{ m (north)}.$$

There are two ways to specify a vector in a given coordinate system:

1. We can give its components, V_x and V_y.
2. We can give its magnitude V and the angle θ it makes with the positive x axis.

We can shift from one description to the other using Eqs. 3–3, and, for the reverse, by using the theorem of Pythagoras† and the definition of tangent:

$$V = \sqrt{V_x^2 + V_y^2} \qquad (3\text{–}4\text{a})$$

$$\tan\theta = \frac{V_y}{V_x} \qquad (3\text{–}4\text{b})$$

as can be seen in Fig. 3–12.

Adding Vectors

We can now discuss how to add vectors using components. The first step is to resolve each vector into its components. Next we can see, using Fig. 3–14, that the addition of any two vectors $\vec{V}_1$ and $\vec{V}_2$ to give a resultant, $\vec{V}_R = \vec{V}_1 + \vec{V}_2$, implies that

$$\begin{aligned} V_{Rx} &= V_{1x} + V_{2x} \\ V_{Ry} &= V_{1y} + V_{2y}. \end{aligned} \qquad (3\text{–}5)$$

That is, the sum of the x components equals the x component of the resultant vector, and the sum of the y components equals the y component of the resultant, as can be verified by a careful examination of Fig. 3–14. Note that we do *not* add x components to y components.

If the magnitude and direction of the resultant vector are desired, they can be obtained using Eqs. 3–4.

†In three dimensions, the theorem of Pythagoras becomes $V = \sqrt{V_x^2 + V_y^2 + V_z^2}$, where V_z is the component along the third, or z, axis.

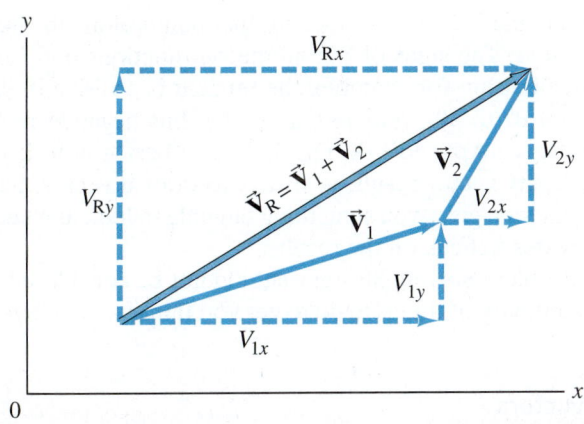

FIGURE 3–14 The components of $\vec{V}_R = \vec{V}_1 + \vec{V}_2$ are $V_{Rx} = V_{1x} + V_{2x}$ and $V_{Ry} = V_{1y} + V_{2y}$.

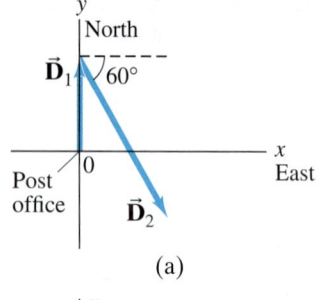
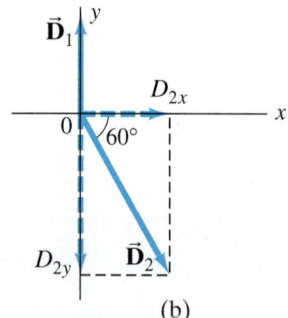
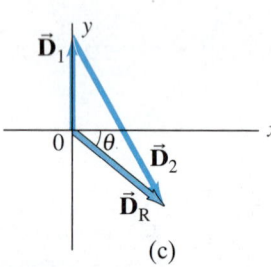

FIGURE 3–15 Example 3–2.
(a) The two displacement vectors, $\vec{D}_1$ and $\vec{D}_2$. (b) $\vec{D}_2$ is resolved into its components. (c) $\vec{D}_1$ and $\vec{D}_2$ are added to obtain the resultant $\vec{D}_R$. The component method of adding the vectors is explained in the Example.

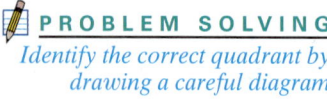

PROBLEM SOLVING
Identify the correct quadrant by drawing a careful diagram

The components of a given vector depend on the choice of coordinate axes. You can often reduce the work involved in adding vectors by a good choice of axes—for example, by choosing one of the axes to be in the same direction as one of the vectors. Then that vector will have only one nonzero component.

EXAMPLE 3–2 Mail carrier's displacement. A rural mail carrier leaves the post office and drives 22.0 km in a northerly direction. She then drives in a direction 60.0° south of east for 47.0 km (Fig. 3–15a). What is her displacement from the post office?

APPROACH We choose the positive x axis to be east and the positive y axis to be north, since those are the compass directions used on most maps. The origin of the xy coordinate system is at the post office. We resolve each vector into its x and y components. We add the x components together, and then the y components together, giving us the x and y components of the resultant.

SOLUTION Resolve each displacement vector into its components, as shown in Fig. 3–15b. Since $\vec{D}_1$ has magnitude 22.0 km and points north, it has only a y component:

$$D_{1x} = 0, \quad D_{1y} = 22.0 \text{ km}.$$

$\vec{D}_2$ has both x and y components:

$$D_{2x} = +(47.0 \text{ km})(\cos 60°) = +(47.0 \text{ km})(0.500) = +23.5 \text{ km}$$
$$D_{2y} = -(47.0 \text{ km})(\sin 60°) = -(47.0 \text{ km})(0.866) = -40.7 \text{ km}.$$

Notice that D_{2y} is negative because this vector component points along the negative y axis. The resultant vector, $\vec{D}_R$, has components:

$$D_{Rx} = D_{1x} + D_{2x} = 0 \text{ km} + 23.5 \text{ km} = +23.5 \text{ km}$$
$$D_{Ry} = D_{1y} + D_{2y} = 22.0 \text{ km} + (-40.7 \text{ km}) = -18.7 \text{ km}.$$

This specifies the resultant vector completely:

$$D_{Rx} = 23.5 \text{ km}, \quad D_{Ry} = -18.7 \text{ km}.$$

We can also specify the resultant vector by giving its magnitude and angle using Eqs. 3–4:

$$D_R = \sqrt{D_{Rx}^2 + D_{Ry}^2} = \sqrt{(23.5 \text{ km})^2 + (-18.7 \text{ km})^2} = 30.0 \text{ km}$$

$$\tan \theta = \frac{D_{Ry}}{D_{Rx}} = \frac{-18.7 \text{ km}}{23.5 \text{ km}} = -0.796.$$

A calculator with a key labeled INV TAN, or ARC TAN, or TAN^{-1} gives $\theta = \tan^{-1}(-0.796) = -38.5°$. The negative sign means $\theta = 38.5°$ below the x axis, Fig. 3–15c. So, the resultant displacement is 30.0 km directed at 38.5° in a southeasterly direction.

NOTE Always be attentive about the quadrant in which the resultant vector lies. An electronic calculator does not fully give this information, but a good diagram does.

As we saw in Example 3–2, any component that points along the negative x or y axis gets a minus sign. The signs of trigonometric functions depend on which "quadrant" the angle falls in: for example, the tangent is positive in the first and third quadrants (from 0° to 90°, and 180° to 270°), but negative in the second and fourth quadrants; see Appendix A, Fig. A–7. The best way to keep track of angles, and to check any vector result, is always to draw a vector diagram, like Fig. 3–15. A vector diagram gives you something tangible to look at when analyzing a problem, and provides a check on the results.

The following Problem Solving Strategy should not be considered a prescription. Rather it is a summary of things to do to get you thinking and involved in the problem at hand.

PROBLEM SOLVING

Adding Vectors

Here is a brief summary of how to add two or more vectors using components:

1. **Draw a diagram**, adding the vectors graphically by either the parallelogram or tail-to-tip method.
2. **Choose x and y axes.** Choose them in a way, if possible, that will make your work easier. (For example, choose one axis along the direction of one of the vectors, which then will have only one component.)
3. **Resolve** each vector into its x and y **components**, showing each component along its appropriate (x or y) axis as a (dashed) arrow.
4. **Calculate each component** (when not given) using sines and cosines. If θ_1 is the angle that vector $\vec{V}_1$ makes with the positive x axis, then:
$$V_{1x} = V_1 \cos\theta_1, \quad V_{1y} = V_1 \sin\theta_1.$$
Pay careful attention to **signs**: any component that points along the negative x or y axis gets a minus sign.

5. **Add** the x components together to get the x component of the resultant. Similarly for y:
$$V_{Rx} = V_{1x} + V_{2x} + \text{any others}$$
$$V_{Ry} = V_{1y} + V_{2y} + \text{any others}.$$
This is the answer: the components of the resultant vector. Check signs to see if they fit the quadrant shown in your diagram (point 1 above).

6. If you want to know the **magnitude and direction** of the resultant vector, use Eqs. 3–4:
$$V_R = \sqrt{V_{Rx}^2 + V_{Ry}^2}, \quad \tan\theta = \frac{V_{Ry}}{V_{Rx}}.$$
The vector diagram you already drew helps to obtain the correct position (quadrant) of the angle θ.

EXAMPLE 3–3 **Three short trips.** An airplane trip involves three legs, with two stopovers, as shown in Fig. 3–16a. The first leg is due east for 620 km; the second leg is southeast (45°) for 440 km; and the third leg is at 53° south of west, for 550 km, as shown. What is the plane's total displacement?

APPROACH We follow the steps in the Problem Solving Strategy above.

SOLUTION

1. **Draw a diagram** such as Fig. 3–16a, where $\vec{D}_1$, $\vec{D}_2$, and $\vec{D}_3$ represent the three legs of the trip, and $\vec{D}_R$ is the plane's total displacement.
2. **Choose axes**: Axes are also shown in Fig. 3–16a: x is east, y north.
3. **Resolve components**: It is imperative to draw a good diagram. The components are drawn in Fig. 3–16b. Instead of drawing all the vectors starting from a common origin, as we did in Fig. 3–15b, here we draw them "tail-to-tip" style, which is just as valid and may make it easier to see.
4. **Calculate the components**:

$$\vec{D}_1: D_{1x} = +D_1 \cos 0° = D_1 = 620 \text{ km}$$
$$D_{1y} = +D_1 \sin 0° = 0 \text{ km}$$
$$\vec{D}_2: D_{2x} = +D_2 \cos 45° = +(440 \text{ km})(0.707) = +311 \text{ km}$$
$$D_{2y} = -D_2 \sin 45° = -(440 \text{ km})(0.707) = -311 \text{ km}$$
$$\vec{D}_3: D_{3x} = -D_3 \cos 53° = -(550 \text{ km})(0.602) = -331 \text{ km}$$
$$D_{3y} = -D_3 \sin 53° = -(550 \text{ km})(0.799) = -439 \text{ km}.$$

We have given a minus sign to each component that in Fig. 3–16b points in the $-x$ or $-y$ direction. The components are shown in the Table in the margin.

5. **Add the components**: We add the x components together, and we add the y components together to obtain the x and y components of the resultant:

$$D_{Rx} = D_{1x} + D_{2x} + D_{3x} = 620 \text{ km} + 311 \text{ km} - 331 \text{ km} = 600 \text{ km}$$
$$D_{Ry} = D_{1y} + D_{2y} + D_{3y} = 0 \text{ km} - 311 \text{ km} - 439 \text{ km} = -750 \text{ km}.$$

The x and y components of the resultant are 600 km and −750 km, and point respectively to the east and south. This is one way to give the answer.

6. **Magnitude and direction**: We can also give the answer as
$$D_R = \sqrt{D_{Rx}^2 + D_{Ry}^2} = \sqrt{(600)^2 + (-750)^2} \text{ km} = 960 \text{ km}$$
$$\tan\theta = \frac{D_{Ry}}{D_{Rx}} = \frac{-750 \text{ km}}{600 \text{ km}} = -1.25, \quad \text{so } \theta = -51°.$$

Thus, the total displacement has magnitude 960 km and points 51° below the x axis (south of east), as was shown in our original sketch, Fig. 3–16a.

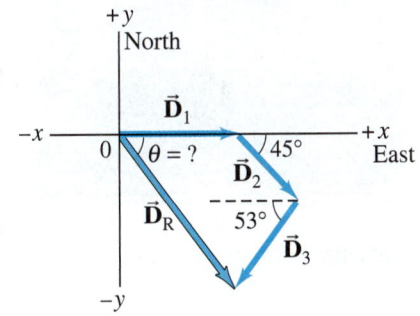
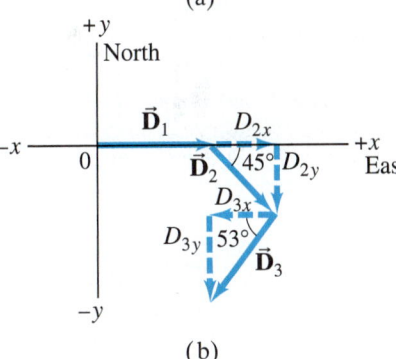

FIGURE 3–16 Example 3–3.

Vector	Components x (km)	y (km)
$\vec{D}_1$	620	0
$\vec{D}_2$	311	−311
$\vec{D}_3$	−331	−439
$\vec{D}_R$	600	−750

3–5 Projectile Motion

In Chapter 2, we studied the one-dimensional motion of an object in terms of displacement, velocity, and acceleration, including purely vertical motion of a falling object undergoing acceleration due to gravity. Now we examine the more general translational motion of objects moving through the air in two dimensions near the Earth's surface, such as a golf ball, a thrown or batted baseball, kicked footballs, and speeding bullets. These are all examples of **projectile motion** (see Fig. 3–17), which we can describe as taking place in two dimensions if there is no wind.

Although air resistance is often important, in many cases its effect can be ignored, and we will ignore it in the following analysis. We will not be concerned now with the process by which the object is thrown or projected. We consider only its motion *after* it has been projected, and *before* it lands or is caught—that is, we analyze our projected object only when it is moving freely through the air under the action of gravity alone. Then the acceleration of the object is that due to gravity, which acts downward with magnitude $g = 9.80 \text{ m/s}^2$, and we assume it is constant.†

Galileo was the first to describe projectile motion accurately. He showed that it could be understood by analyzing the horizontal and vertical components of the motion separately. For convenience, we assume that the motion begins at time $t = 0$ at the origin of an xy coordinate system (so $x_0 = y_0 = 0$).

Let us look at a (tiny) ball rolling off the end of a horizontal table with an initial velocity in the horizontal (x) direction, $\vec{v}_{x0}$. See Fig. 3–18, where an object falling vertically is also shown for comparison. The velocity vector $\vec{v}$ at each instant points in the direction of the ball's motion at that instant and is thus always tangent to the path. Following Galileo's ideas, we treat the horizontal and vertical components of velocity and acceleration separately, and we can apply the kinematic equations (Eqs. 2–11a through 2–11c) to the x and y components of the motion.

First we examine the vertical (y) component of the motion. At the instant the ball leaves the table's top ($t = 0$), it has only an x component of velocity. Once the ball leaves the table (at $t = 0$), it experiences a vertically downward acceleration g, the acceleration due to gravity. Thus v_y is initially zero ($v_{y0} = 0$) but increases continually in the downward direction (until the ball hits the ground). Let us take y to be positive upward. Then the acceleration due to gravity is in the $-y$ direction, so $a_y = -g$. From Eq. 2–11a (using y in place of x) we can write $v_y = v_{y0} + a_y t = -gt$ since we set $v_{y0} = 0$. The vertical displacement is given by Eq. 2–11b written in terms of y: $y = y_0 + v_{y0}t + \frac{1}{2}a_y t^2$. Given $y_0 = 0$, $v_{y0} = 0$, and $a_y = -g$, then $y = -\frac{1}{2}gt^2$.

†This restricts us to objects whose distance traveled and maximum height above the Earth are small compared to the Earth's radius (6400 km).

FIGURE 3–17 Photographs of (a) a bouncing ball and (b) a thrown basketball, each showing the characteristic "parabolic" path of projectile motion.

FIGURE 3–18 Projectile motion of a small ball projected horizontally with initial velocity $\vec{v} = \vec{v}_{x0}$. The dashed black line represents the path of the object. The velocity vector $\vec{v}$ is in the direction of motion at each point, and thus is tangent to the path. The velocity vectors are green arrows, and velocity components are dashed. (A vertically falling object starting from rest at the same place and time is shown at the left for comparison; v_y is the same at each instant for the falling object and the projectile.)

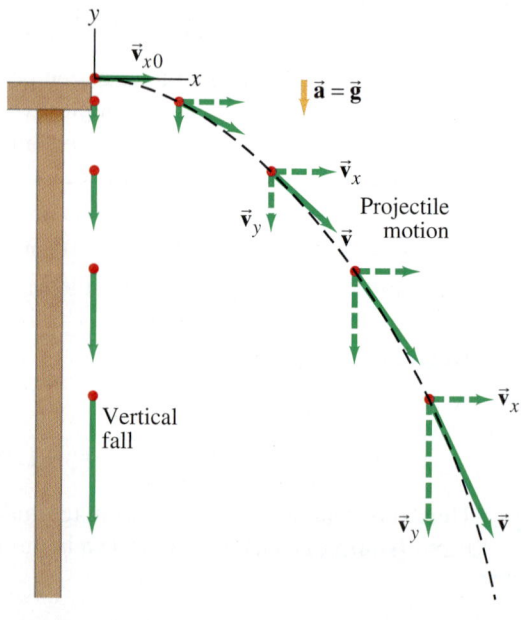

In the horizontal direction, on the other hand, there is no acceleration (we are ignoring air resistance). With $a_x = 0$, the horizontal component of velocity, v_x, remains constant, equal to its initial value, v_{x0}, and thus has the same magnitude at each point on the path. The horizontal displacement (with $a_x = 0$) is given by $x = v_{x0}t + \frac{1}{2}a_x t^2 = v_{x0}t$. The two vector components, $\vec{v}_x$ and $\vec{v}_y$, can be added vectorially at any instant to obtain the velocity $\vec{v}$ at that time (that is, for each point on the path), as shown in Fig. 3–18.

One result of this analysis, which Galileo himself predicted, is that *an object projected horizontally will reach the ground in the same time as an object dropped vertically*. This is because the vertical motions are the same in both cases, as shown in Fig. 3–18. Figure 3–19 is a multiple-exposure photograph of an experiment that confirms this.

| **EXERCISE C** Two balls having different speeds roll off the edge of a horizontal table at the same time. Which hits the floor sooner, the faster ball or the slower one?

If an object is projected at an upward angle, as in Fig. 3–20, the analysis is similar, except that now there is an initial vertical component of velocity, v_{y0}. Because of the downward acceleration of gravity, the upward component of velocity v_y gradually decreases with time until the object reaches the highest point on its path, at which point $v_y = 0$. Subsequently the object moves downward (Fig. 3–20) and v_y increases in the downward direction, as shown (that is, becoming more negative). As before, v_x remains constant.

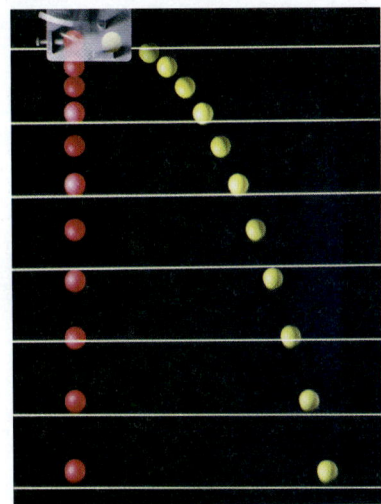

FIGURE 3–19 Multiple-exposure photograph showing positions of two balls at equal time intervals. One ball was dropped from rest at the same time the other ball was projected horizontally outward. The vertical position of each ball is seen to be the same at each instant.

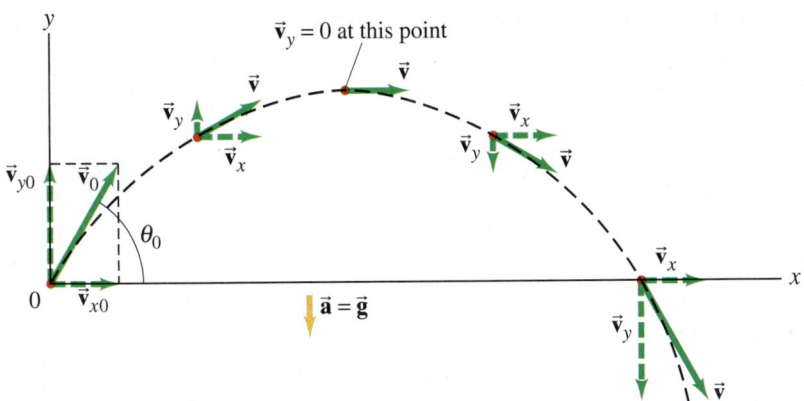

FIGURE 3–20 Path of a projectile launched with initial velocity $\vec{v}_0$ at angle θ_0 to the horizontal. Path is shown dashed in black, the velocity vectors are green arrows, and velocity components are dashed. The figure does not show where the projectile hits the ground (at that point, projectile motion ceases).

| **EXERCISE D** Where in Fig. 3–20 is (i) $\vec{v} = 0$, (ii) $v_y = 0$, and (iii) $v_x = 0$?

CONCEPTUAL EXAMPLE 3–4 **Where does the apple land?** A child sits upright in a wagon which is moving to the right at constant speed as shown in Fig. 3–21. The child extends her hand and throws an apple straight upward (from her own point of view, Fig. 3–21a), while the wagon continues to travel forward at constant speed. If air resistance is neglected, will the apple land (*a*) behind the wagon, (*b*) in the wagon, or (*c*) in front of the wagon?

RESPONSE The child throws the apple straight up from her own reference frame with initial velocity $\vec{v}_{y0}$ (Fig. 3–21a). But when viewed by someone on the ground, the apple also has an initial horizontal component of velocity equal to the speed of the wagon, $\vec{v}_{x0}$. Thus, to a person on the ground, the apple will follow the path of a projectile as shown in Fig. 3–21b. The apple experiences no horizontal acceleration, so $\vec{v}_{x0}$ will stay constant and equal to the speed of the wagon. As the apple follows its arc, the wagon will be directly under the apple at all times because they have the same horizontal velocity. When the apple comes down, it will drop right into the outstretched hand of the child. The answer is (*b*).

FIGURE 3–21 Example 3–4.

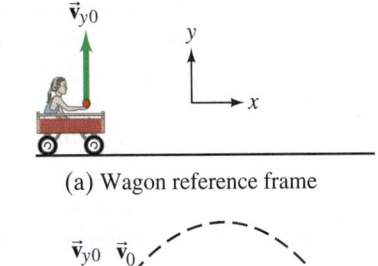

(a) Wagon reference frame

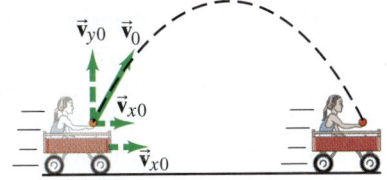

(b) Ground reference frame

| **EXERCISE E** Return to the Chapter-Opening Question, page 49, and answer it again now. Try to explain why you may have answered differently the first time. Describe the role of the helicopter in this example of projectile motion.

3–6 Solving Projectile Motion Problems

We now work through several Examples of projectile motion quantitatively. We use the kinematic equations (2–11a through 2–11c) separately for the vertical and horizontal components of the motion. These equations are shown separately for the x and y components of the motion in Table 3–1, for the general case of two-dimensional motion at constant acceleration. Note that x and y are the respective displacements, that v_x and v_y are the components of the velocity, and that a_x and a_y are the components of the acceleration, each of which is constant. The subscript 0 means "at $t = 0$."

TABLE 3–1 General Kinematic Equations for Constant Acceleration in Two Dimensions

x component (horizontal)		y component (vertical)
$v_x = v_{x0} + a_x t$	(Eq. 2–11a)	$v_y = v_{y0} + a_y t$
$x = x_0 + v_{x0}t + \frac{1}{2}a_x t^2$	(Eq. 2–11b)	$y = y_0 + v_{y0}t + \frac{1}{2}a_y t^2$
$v_x^2 = v_{x0}^2 + 2a_x(x - x_0)$	(Eq. 2–11c)	$v_y^2 = v_{y0}^2 + 2a_y(y - y_0)$

We can simplify Eqs. 2–11 to use for projectile motion because we can set $a_x = 0$. See Table 3–2, which assumes y is positive upward, so $a_y = -g = -9.80 \text{ m/s}^2$.

TABLE 3–2 Kinematic Equations for Projectile Motion
(y positive upward; $a_x = 0$, $a_y = -g = -9.80 \text{ m/s}^2$)

Horizontal Motion ($a_x = 0$, $v_x =$ constant)		Vertical Motion[†] ($a_y = -g =$ constant)
$v_x = v_{x0}$	(Eq. 2–11a)	$v_y = v_{y0} - gt$
$x = x_0 + v_{x0}t$	(Eq. 2–11b)	$y = y_0 + v_{y0}t - \frac{1}{2}gt^2$
	(Eq. 2–11c)	$v_y^2 = v_{y0}^2 - 2g(y - y_0)$

[†] If y is taken positive downward, the minus (−) signs in front of g become + signs.

If the projection angle θ_0 is chosen relative to the $+x$ axis (Fig. 3–20), then

$$v_{x0} = v_0 \cos \theta_0, \quad \text{and} \quad v_{y0} = v_0 \sin \theta_0.$$

PROBLEM SOLVING
Choice of time interval

In doing Problems involving projectile motion, we must consider a time interval for which our chosen object is in the air, influenced only by gravity. We do not consider the throwing (or projecting) process, nor the time after the object lands or is caught, because then other influences act on the object, and we can no longer set $\vec{a} = \vec{g}$.

Projectile Motion

Our approach to solving Problems in Section 2–6 also applies here. Solving Problems involving projectile motion can require creativity, and cannot be done just by following some rules. Certainly you must avoid just plugging numbers into equations that seem to "work."

1. As always, **read** carefully; **choose** the **object** (or objects) you are going to analyze.
2. **Draw** a careful **diagram** showing what is happening to the object.
3. **Choose** an origin and an **xy coordinate system**.
4. **Decide** on the **time interval**, which for projectile motion can only include motion under the effect of gravity alone, not throwing or landing. The time interval must be the same for the x and y analyses. The x and y motions are connected by the common time, t.
5. **Examine** the horizontal (x) and vertical (y) **motions** separately. If you are given the initial velocity, you may want to resolve it into its x and y components.
6. List the **known** and **unknown** quantities, choosing $a_x = 0$ and $a_y = -g$ or $+g$, where $g = 9.80 \text{ m/s}^2$, and using the − or + sign, depending on whether you choose y positive up or down. Remember that v_x never changes throughout the trajectory, and that $v_y = 0$ at the highest point of any trajectory that returns downward. The velocity just before landing is generally not zero.
7. Think for a minute before jumping into the equations. A little planning goes a long way. **Apply** the **relevant equations** (Table 3–2), combining equations if necessary. You may need to combine components of a vector to get magnitude and direction (Eqs. 3–4).

EXAMPLE 3–5 Driving off a cliff. A movie stunt driver on a motorcycle speeds horizontally off a 50.0-m-high cliff. How fast must the motorcycle leave the cliff top to land on level ground below, 90.0 m from the base of the cliff where the cameras are? Ignore air resistance.

APPROACH We explicitly follow the steps of the Problem Solving Strategy on the previous page.

SOLUTION

1. and 2. **Read, choose the object, and draw a diagram**. Our object is the motorcycle and driver, taken as a single unit. The diagram is shown in Fig. 3–22.

3. **Choose a coordinate system**. We choose the y direction to be positive upward, with the top of the cliff as $y_0 = 0$. The x direction is horizontal with $x_0 = 0$ at the point where the motorcycle leaves the cliff.

4. **Choose a time interval**. We choose our time interval to begin ($t = 0$) just as the motorcycle leaves the cliff top at position $x_0 = 0$, $y_0 = 0$. Our time interval ends just before the motorcycle touches the ground below.

5. **Examine x and y motions**. In the horizontal (x) direction, the acceleration $a_x = 0$, so the velocity is constant. The value of x when the motorcycle reaches the ground is $x = +90.0$ m. In the vertical direction, the acceleration is the acceleration due to gravity, $a_y = -g = -9.80 \text{ m/s}^2$. The value of y when the motorcycle reaches the ground is $y = -50.0$ m. The initial velocity is horizontal and is our unknown, v_{x0}; the initial vertical velocity is zero, $v_{y0} = 0$.

6. **List knowns and unknowns**. See the Table in the margin. Note that in addition to not knowing the initial horizontal velocity v_{x0} (which stays constant until landing), we also do not know the time t when the motorcycle reaches the ground.

7. **Apply relevant equations**. The motorcycle maintains constant v_x as long as it is in the air. The time it stays in the air is determined by the y motion—when it reaches the ground. So we first find the time using the y motion, and then use this time value in the x equations. To find out how long it takes the motorcycle to reach the ground below, we use Eq. 2–11b (Tables 3–1 and 3–2) for the vertical (y) direction with $y_0 = 0$ and $v_{y0} = 0$:

$$y = y_0 + v_{y0}t + \tfrac{1}{2}a_y t^2$$
$$= 0 + 0 + \tfrac{1}{2}(-g)t^2$$

or

$$y = -\tfrac{1}{2}gt^2.$$

We solve for t and set $y = -50.0$ m:

$$t = \sqrt{\frac{2y}{-g}} = \sqrt{\frac{2(-50.0 \text{ m})}{-9.80 \text{ m/s}^2}} = 3.19 \text{ s}.$$

To calculate the initial velocity, v_{x0}, we again use Eq. 2–11b, but this time for the horizontal (x) direction, with $a_x = 0$ and $x_0 = 0$:

$$x = x_0 + v_{x0}t + \tfrac{1}{2}a_x t^2$$
$$= 0 + v_{x0}t + 0$$

or

$$x = v_{x0}t.$$

Then

$$v_{x0} = \frac{x}{t} = \frac{90.0 \text{ m}}{3.19 \text{ s}} = 28.2 \text{ m/s},$$

which is about 100 km/h (roughly 60 mi/h).

NOTE In the time interval of the projectile motion, the only acceleration is g in the negative y direction. The acceleration in the x direction is zero.

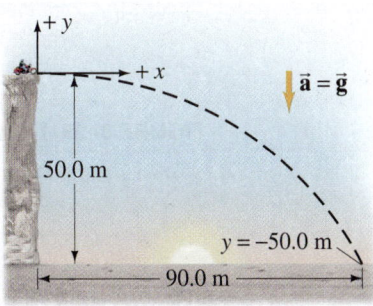

FIGURE 3–22 Example 3–5.

Known	Unknown
$x_0 = y_0 = 0$	v_{x0}
$x = 90.0$ m	t
$y = -50.0$ m	
$a_x = 0$	
$a_y = -g = -9.80 \text{ m/s}^2$	
$v_{y0} = 0$	

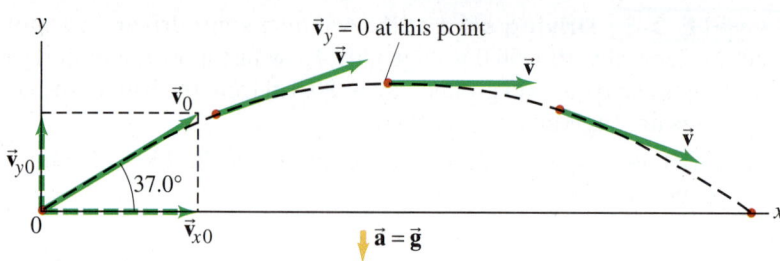

FIGURE 3–23 Example 3–6.

PHYSICS APPLIED
Sports

EXAMPLE 3–6 **A kicked football.** A kicked football leaves the ground at an angle $\theta_0 = 37.0°$ with a velocity of 20.0 m/s, as shown in Fig. 3–23. Calculate (a) the maximum height, (b) the time of travel before the football hits the ground, and (c) how far away it hits the ground. Assume the ball leaves the foot at ground level, and ignore air resistance and rotation of the ball.

APPROACH This may seem difficult at first because there are so many questions. But we can deal with them one at a time. We take the y direction as positive upward, and treat the x and y motions separately. The total time in the air is again determined by the y motion. The x motion occurs at constant velocity. The y component of velocity varies, being positive (upward) initially, decreasing to zero at the highest point, and then becoming negative as the football falls.

SOLUTION We resolve the initial velocity into its components (Fig. 3–23):

$$v_{x0} = v_0 \cos 37.0° = (20.0 \text{ m/s})(0.799) = 16.0 \text{ m/s}$$
$$v_{y0} = v_0 \sin 37.0° = (20.0 \text{ m/s})(0.602) = 12.0 \text{ m/s}.$$

(a) To find the maximum height, we consider a time interval that begins just after the football loses contact with the foot until the ball reaches its maximum height. During this time interval, the acceleration is g downward. At the maximum height, the velocity is horizontal (Fig. 3–23), so $v_y = 0$. This occurs at a time given by $v_y = v_{y0} - gt$ with $v_y = 0$ (see Eq. 2–11a in Table 3–2), so $v_{y0} = gt$ and

$$t = \frac{v_{y0}}{g} = \frac{(12.0 \text{ m/s})}{(9.80 \text{ m/s}^2)} = 1.224 \text{ s} \approx 1.22 \text{ s}.$$

From Eq. 2–11b, with $y_0 = 0$, we can solve for y at this time ($t = v_{y0}/g$):

$$y = v_{y0}t - \tfrac{1}{2}gt^2 = \frac{v_{y0}^2}{g} - \frac{1}{2}\frac{v_{y0}^2}{g} = \frac{v_{y0}^2}{2g} = \frac{(12.0 \text{ m/s})^2}{2(9.80 \text{ m/s}^2)} = 7.35 \text{ m}.$$

The maximum height is 7.35 m. [Solving Eq. 2–11c for y gives the same result.]

(b) To find the time it takes for the ball to return to the ground, we consider a different time interval, starting at the moment the ball leaves the foot ($t = 0$, $y_0 = 0$) and ending just before the ball touches the ground ($y = 0$ again). We can use Eq. 2–11b with $y_0 = 0$ and also set $y = 0$ (ground level):

$$y = y_0 + v_{y0}t - \tfrac{1}{2}gt^2$$
$$0 = 0 + v_{y0}t - \tfrac{1}{2}gt^2.$$

This equation can be factored:

$$t\left(\tfrac{1}{2}gt - v_{y0}\right) = 0.$$

There are two solutions, $t = 0$ (which corresponds to the initial point, y_0), and

$$t = \frac{2v_{y0}}{g} = \frac{2(12.0 \text{ m/s})}{(9.80 \text{ m/s}^2)} = 2.45 \text{ s},$$

which is the total travel time of the football.

(c) The total distance traveled in the x direction is found by applying Eq. 2–11b with $x_0 = 0$, $a_x = 0$, $v_{x0} = 16.0$ m/s, and $t = 2.45$ s:

$$x = v_{x0}t = (16.0 \text{ m/s})(2.45 \text{ s}) = 39.2 \text{ m}.$$

NOTE In (b), the time needed for the whole trip, $t = 2v_{y0}/g = 2.45$ s, is double the time to reach the highest point, calculated in (a). That is, the time to go up equals the time to come back down to the same level (ignoring air resistance).

PROBLEM SOLVING
Symmetry

EXERCISE F In Example 3–6, what is (*a*) the velocity vector at the maximum height, and (*b*) the acceleration vector at maximum height?

In Example 3–6, we treated the football as if it were a particle, ignoring its rotation. We also ignored air resistance. Because air resistance is significant on a football, our results are only estimates (mainly overestimates).

CONCEPTUAL EXAMPLE 3–7 The wrong strategy. A boy on a small hill aims his water-balloon slingshot horizontally, straight at a second boy hanging from a tree branch a distance *d* away, Fig. 3–24. At the instant the water balloon is released, the second boy lets go and falls from the tree, hoping to avoid being hit. Show that he made the wrong move. (He hadn't studied physics yet.) Ignore air resistance.

RESPONSE Both the water balloon and the boy in the tree start falling at the same instant, and in a time t they each fall the same vertical distance $y = \frac{1}{2}gt^2$, much like Fig. 3–19. In the time it takes the water balloon to travel the horizontal distance d, the balloon will have the same y position as the falling boy. Splat. If the boy had stayed in the tree, he would have avoided the humiliation.

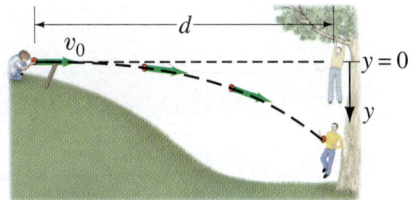

FIGURE 3–24 Example 3–7.

Level Horizontal Range

The total distance the football traveled in Example 3–6 is called the horizontal **range** R. We now derive a formula for the range, which applies to a projectile that lands at the same level it started ($= y_0$): that is, y (final) $= y_0$ (see Fig. 3–25a). Looking back at Example 3–6 part (*c*), we see that $x = R = v_{x0}t$ where (from part *b*) $t = 2v_{y0}/g$. Thus

$$R = v_{x0}t = v_{x0}\left(\frac{2v_{y0}}{g}\right) = \frac{2v_{x0}v_{y0}}{g} = \frac{2v_0^2 \sin\theta_0 \cos\theta_0}{g}, \quad [y = y_0]$$

where $v_{x0} = v_0 \cos\theta_0$ and $v_{y0} = v_0 \sin\theta_0$. This can be rewritten, using the trigonometric identity $2\sin\theta\cos\theta = \sin 2\theta$ (Appendix A or inside the rear cover):

$$R = \frac{v_0^2 \sin 2\theta_0}{g}. \quad \text{[only if } y \text{ (final)} = y_0]$$

Note that the *maximum* range, for a given initial velocity v_0, is obtained when $\sin 2\theta$ takes on its maximum value of 1.0, which occurs for $2\theta_0 = 90°$; so

$$\theta_0 = 45° \text{ for maximum range, and } R_{max} = v_0^2/g.$$

The maximum range increases by the square of v_0, so doubling the muzzle velocity of a cannon increases its maximum range by a factor of 4.

When air resistance is important, the range is less for a given v_0, and the maximum range is obtained at an angle smaller than 45°.

FIGURE 3–25 (a) The range R of a projectile. (b) There are generally two angles θ_0 that will give the same range. If one angle is θ_{01}, the other is $\theta_{02} = 90° - \theta_{01}$. Example 3–8.

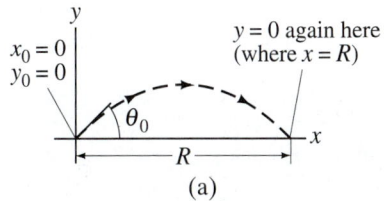

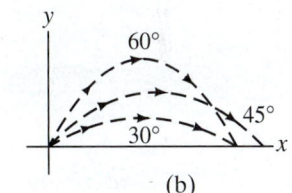

EXAMPLE 3–8 Range of a cannon ball. Suppose one of Napoleon's cannons had a muzzle speed, v_0, of 60.0 m/s. At what angle should it have been aimed (ignore air resistance) to strike a target 320 m away?

APPROACH We use the equation just derived for the range, $R = v_0^2 \sin 2\theta_0 / g$, with $R = 320$ m.

SOLUTION We solve for $\sin 2\theta_0$ in the range formula:

$$\sin 2\theta_0 = \frac{Rg}{v_0^2} = \frac{(320 \text{ m})(9.80 \text{ m/s}^2)}{(60.0 \text{ m/s})^2} = 0.871.$$

We want to solve for an angle θ_0 that is between 0° and 90°, which means $2\theta_0$ in this equation can be as large as 180°. Thus, $2\theta_0 = 60.6°$ is a solution, so $\theta_0 = 30.3°$. But $2\theta_0 = 180° - 60.6° = 119.4°$ is also a solution (see Appendix A–7), so θ_0 can also be $\theta_0 = 59.7°$. In general we have two solutions (see Fig. 3–25b), which in the present case are given by

$$\theta_0 = 30.3° \quad \text{or} \quad 59.7°.$$

Either angle gives the same range. Only when $\sin 2\theta_0 = 1$ (so $\theta_0 = 45°$) is there a single solution (that is, both solutions are the same).

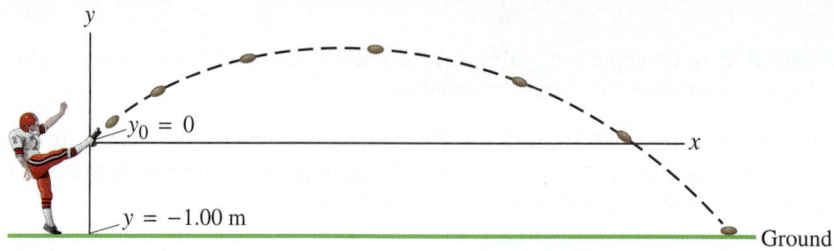

FIGURE 3–26 Example 3–9: the football leaves the punter's foot at $y = 0$, and reaches the ground where $y = -1.00$ m.

PHYSICS APPLIED
Sports

PROBLEM SOLVING
Do not use any formula unless you are sure its range of validity fits the problem; the range formula does not apply here because $y \neq y_0$

EXAMPLE 3–9 A punt. Suppose the football in Example 3–6 was punted, and left the punter's foot at a height of 1.00 m above the ground. How far did the football travel before hitting the ground? Set $x_0 = 0$, $y_0 = 0$.

APPROACH The only difference here from Example 3–6 is that the football hits the ground *below* its starting point of $y_0 = 0$. That is, the ball hits the ground at $y = -1.00$ m. See Fig. 3–26. Thus we cannot use the range formula which is valid only if y (final) $= y_0$. As in Example 3–6, $v_0 = 20.0$ m/s, $\theta_0 = 37.0°$.

SOLUTION With $y = -1.00$ m and $v_{y0} = 12.0$ m/s (see Example 3–6), we use the y version of Eq. 2–11b with $a_y = -g$,

$$y = y_0 + v_{y0}t - \tfrac{1}{2}gt^2,$$

and obtain

$$-1.00 \text{ m} = 0 + (12.0 \text{ m/s})t - (4.90 \text{ m/s}^2)t^2.$$

We rearrange this equation into standard form $(ax^2 + bx + c = 0)$ so we can use the quadratic formula:

$$(4.90 \text{ m/s}^2)t^2 - (12.0 \text{ m/s})t - (1.00 \text{ m}) = 0.$$

The quadratic formula (Appendix A–4) gives

$$t = \frac{12.0 \text{ m/s} \pm \sqrt{(-12.0 \text{ m/s})^2 - 4(4.90 \text{ m/s}^2)(-1.00 \text{ m})}}{2(4.90 \text{ m/s}^2)}$$

$$= 2.53 \text{ s} \quad \text{or} \quad -0.081 \text{ s}.$$

The second solution would correspond to a time prior to the kick, so it doesn't apply. With $t = 2.53$ s for the time at which the ball touches the ground, the horizontal distance the ball traveled is (using $v_{x0} = 16.0$ m/s from Example 3–6):

$$x = v_{x0}t = (16.0 \text{ m/s})(2.53 \text{ s}) = 40.5 \text{ m}.$$

Our assumption in Example 3–6 that the ball leaves the foot at ground level would result in an underestimate of about 1.3 m in the distance our punt traveled.

*3–7 Projectile Motion Is Parabolic

We now show that the path followed by any projectile is a *parabola*, if we can ignore air resistance and can assume that $\vec{g}$ is constant. To do so, we need to find y as a function of x by eliminating t between the two equations for horizontal and vertical motion (Eq. 2–11b in Table 3–2), and for simplicity we set $x_0 = y_0 = 0$:

$$x = v_{x0}t$$
$$y = v_{y0}t - \tfrac{1}{2}gt^2.$$

From the first equation, we have $t = x/v_{x0}$, and we substitute this into the second one to obtain

$$y = \left(\frac{v_{y0}}{v_{x0}}\right)x - \left(\frac{g}{2v_{x0}^2}\right)x^2. \tag{3–6}$$

We see that y as a function of x has the form

$$y = Ax - Bx^2,$$

where A and B are constants for any specific projectile motion. This is the standard equation for a parabola. See Figs. 3–17 and 3–27.

The idea that projectile motion is parabolic was, in Galileo's day, at the forefront of physics research. Today we discuss it in Chapter 3 of introductory physics!

FIGURE 3–27 Examples of projectile motion: a boy jumping, and glowing lava from the volcano Stromboli.

*Some Sections of this book, such as this one, may be considered *optional* at the discretion of the instructor. See the Preface for more details.

3–8 Relative Velocity

We now consider how observations made in different frames of reference are related to each other. For example, consider two trains approaching one another, each with a speed of 80 km/h with respect to the Earth. Observers on the Earth beside the train tracks will measure 80 km/h for the speed of each of the trains. Observers on either one of the trains (a different frame of reference) will measure a speed of 160 km/h for the train approaching them.

Similarly, when one car traveling 90 km/h passes a second car traveling in the same direction at 75 km/h, the first car has a speed relative to the second car of 90 km/h − 75 km/h = 15 km/h.

When the velocities are along the same line, simple addition or subtraction is sufficient to obtain the **relative velocity**. But if they are not along the same line, we must make use of vector addition. We emphasize, as mentioned in Section 2–1, that when specifying a velocity, it is important to specify what the reference frame is.

When determining relative velocity, it is easy to make a mistake by adding or subtracting the wrong velocities. It is important, therefore, to draw a diagram and use a careful labeling process. Each velocity is labeled by *two subscripts: the first refers to the object, the second to the reference frame in which it has this velocity.* For example, suppose a boat heads directly across a river, as shown in Fig. 3–28. We let $\vec{v}_{BW}$ be the velocity of the **B**oat with respect to the **W**ater. (This is also what the boat's velocity would be relative to the shore if the water were still.) Similarly, $\vec{v}_{BS}$ is the velocity of the **B**oat with respect to the **S**hore, and $\vec{v}_{WS}$ is the velocity of the **W**ater with respect to the **S**hore (this is the river current). Note that $\vec{v}_{BW}$ is what the boat's motor produces (against the water), whereas $\vec{v}_{BS}$ is equal to $\vec{v}_{BW}$ plus the effect of the current, $\vec{v}_{WS}$. Therefore, the velocity of the boat relative to the shore is (see vector diagram, Fig. 3–28)

$$\vec{v}_{BS} = \vec{v}_{BW} + \vec{v}_{WS}. \tag{3-7}$$

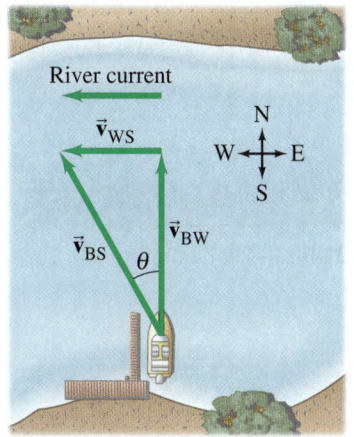

FIGURE 3–28 A boat heads north directly across a river which flows west. Velocity vectors are shown as green arrows:

$\vec{v}_{BS}$ = velocity of **B**oat with respect to the **S**hore,

$\vec{v}_{BW}$ = velocity of **B**oat with respect to the **W**ater,

$\vec{v}_{WS}$ = velocity of **W**ater with respect to the **S**hore (river current).

As it crosses the river, the boat is dragged downstream by the current.

By writing the subscripts using this convention, we see that the inner subscripts (the two W's) on the right-hand side of Eq. 3–7 are the same; also, the outer subscripts on the right of Eq. 3–7 (the B and the S) are the same as the two subscripts for the sum vector on the left, $\vec{v}_{BS}$. By following this convention (first subscript for the object, second for the reference frame), you can write down the correct equation relating velocities in different reference frames.[†]

Equation 3–7 is valid in general and can be extended to three or more velocities. For example, if a fisherman on the boat walks with a velocity $\vec{v}_{FB}$ relative to the boat, his velocity relative to the shore is $\vec{v}_{FS} = \vec{v}_{FB} + \vec{v}_{BW} + \vec{v}_{WS}$. The equations involving relative velocity will be correct when adjacent inner subscripts are identical and when the outermost ones correspond exactly to the two on the velocity on the left of the equation. But this works only with plus signs (on the right), not minus signs.

It is often useful to remember that for any two objects or reference frames, A and B, the velocity of A relative to B has the same magnitude, but opposite direction, as the velocity of B relative to A:

$$\vec{v}_{BA} = -\vec{v}_{AB}. \tag{3-8}$$

For example, if a train is traveling 100 km/h relative to the Earth in a certain direction, objects on the Earth (such as trees) appear to an observer on the train to be traveling 100 km/h in the opposite direction.

[†]We thus can see, for example, that the equation $\vec{v}_{BW} = \vec{v}_{BS} + \vec{v}_{WS}$ is wrong: the inner subscripts are not the same, and the outer ones on the right do not correspond to the subscripts on the left.

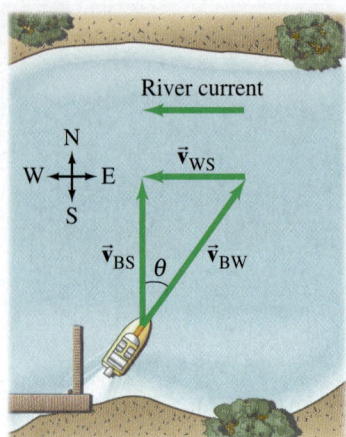

FIGURE 3–29 Example 3–10.

EXAMPLE 3–10 **Heading upstream.** A boat's speed in still water is v_{BW} = 1.85 m/s. If the boat is to travel north directly across a river whose westward current has speed v_{WS} = 1.20 m/s, at what upstream angle must the boat head? (See Fig. 3–29.)

APPROACH If the boat heads straight across the river, the current will drag the boat downstream (westward). To overcome the river's current, the boat must have an upstream (eastward) component of velocity as well as a cross-stream (northward) component. Figure 3–29 has been drawn with $\vec{v}_{BS}$, the velocity of the **B**oat relative to the **S**hore, pointing directly across the river because this is where the boat is supposed to go. (Note that $\vec{v}_{BS} = \vec{v}_{BW} + \vec{v}_{WS}$.)

SOLUTION Vector $\vec{v}_{BW}$ points upstream at angle θ as shown. From the diagram,

$$\sin\theta = \frac{v_{WS}}{v_{BW}} = \frac{1.20 \text{ m/s}}{1.85 \text{ m/s}} = 0.6486.$$

Thus $\theta = 40.4°$, so the boat must head upstream at a 40.4° angle.

EXAMPLE 3–11 **Heading across the river.** The same boat (v_{BW} = 1.85 m/s) now heads directly across the river whose current is still 1.20 m/s. (*a*) What is the velocity (magnitude and direction) of the boat relative to the shore? (*b*) If the river is 110 m wide, how long will it take to cross and how far downstream will the boat be then?

APPROACH The boat now heads directly across the river and is pulled downstream by the current, as shown in Fig. 3–30. The boat's velocity with respect to the shore, $\vec{v}_{BS}$, is the sum of its velocity with respect to the water, $\vec{v}_{BW}$, plus the velocity of the water with respect to the shore, $\vec{v}_{WS}$: just as before,

$$\vec{v}_{BS} = \vec{v}_{BW} + \vec{v}_{WS}.$$

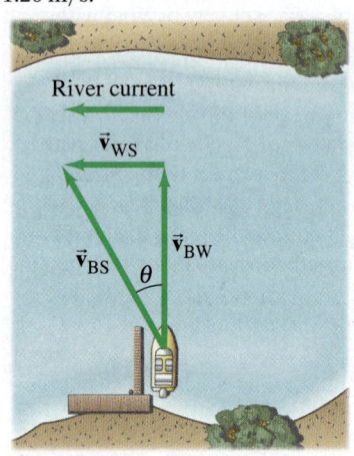

FIGURE 3–30 Example 3–11. A boat heading directly across a river whose current moves at 1.20 m/s.

SOLUTION (*a*) Since $\vec{v}_{BW}$ is perpendicular to $\vec{v}_{WS}$, we can get v_{BS} using the theorem of Pythagoras:

$$v_{BS} = \sqrt{v_{BW}^2 + v_{WS}^2} = \sqrt{(1.85 \text{ m/s})^2 + (1.20 \text{ m/s})^2} = 2.21 \text{ m/s}.$$

We can obtain the angle (note how θ is defined in Fig. 3–30) from:

$$\tan\theta = v_{WS}/v_{BW} = (1.20 \text{ m/s})/(1.85 \text{ m/s}) = 0.6486.$$

A calculator with a key INV TAN or ARC TAN or TAN^{-1} gives $\theta = \tan^{-1}(0.6486) = 33.0°$. Note that this angle is not equal to the angle calculated in Example 3–10.
(*b*) The travel time for the boat is determined by the time it takes to cross the river. Given the river's width D = 110 m, we can use the velocity component in the direction of D, $v_{BW} = D/t$. Solving for t, we get t = 110 m/1.85 m/s = 59.5 s. The boat will have been carried downstream, in this time, a distance

$$d = v_{WS}t = (1.20 \text{ m/s})(59.5 \text{ s}) = 71.4 \text{ m} \approx 71 \text{ m}.$$

NOTE There is no acceleration in this Example, so the motion involves only constant velocities (of the boat or of the river).

Summary

A quantity such as velocity, that has both a magnitude and a direction, is called a **vector**. A quantity such as mass, that has only a magnitude, is called a **scalar**. On diagrams, vectors are represented by arrows.

Addition of vectors can be done graphically by placing the tail of each successive arrow at the tip of the previous one. The sum, or **resultant vector**, is the arrow drawn from the tail of the first vector to the tip of the last vector. Two vectors can also be added using the parallelogram method.

Vectors can be added more accurately by adding their **components** along chosen axes with the aid of trigonometric functions. A vector of magnitude V making an angle θ with the $+x$ axis has components

$$V_x = V\cos\theta, \quad V_y = V\sin\theta. \quad \text{(3–3)}$$

Given the components, we can find a vector's magnitude and direction from

$$V = \sqrt{V_x^2 + V_y^2}, \quad \tan\theta = \frac{V_y}{V_x}. \quad \text{(3–4)}$$

Projectile motion is the motion of an object in the air near the Earth's surface under the effect of gravity alone. It can be analyzed as two separate motions if air resistance can be ignored. The horizontal component of motion is at constant velocity, whereas the vertical component is at constant acceleration, $\vec{g}$, just as for an object falling vertically under the action of gravity.

The velocity of an object relative to one frame of reference can be found by vector addition if its velocity relative to a second frame of reference, and the **relative velocity** of the two reference frames, are known.

Questions

1. One car travels due east at 40 km/h, and a second car travels north at 40 km/h. Are their velocities equal? Explain.
2. Can you conclude that a car is not accelerating if its speedometer indicates a steady 60 km/h? Explain.
3. Give several examples of an object's motion in which a great distance is traveled but the displacement is zero.
4. Can the displacement vector for a particle moving in two dimensions be longer than the length of path traveled by the particle over the same time interval? Can it be less? Discuss.
5. During baseball practice, a player hits a very high fly ball and then runs in a straight line and catches it. Which had the greater displacement, the player or the ball? Explain.
6. If $\vec{V} = \vec{V}_1 + \vec{V}_2$, is V necessarily greater than V_1 and/or V_2? Discuss.
7. Two vectors have length $V_1 = 3.5$ km and $V_2 = 4.0$ km. What are the maximum and minimum magnitudes of their vector sum?
8. Can two vectors, of unequal magnitude, add up to give the zero vector? Can *three* unequal vectors? Under what conditions?
9. Can the magnitude of a vector ever (a) equal, or (b) be less than, one of its components?
10. Does the odometer of a car measure a scalar or a vector quantity? What about the speedometer?
11. How could you determine the speed a slingshot imparts to a rock, using only a meter stick, a rock, and the slingshot?
12. In archery, should the arrow be aimed directly at the target? How should your angle of aim depend on the distance to the target?
13. It was reported in World War I that a pilot flying at an altitude of 2 km caught in his bare hands a bullet fired at the plane! Using the fact that a bullet slows down considerably due to air resistance, explain how this incident occurred.
14. You are on the street trying to hit a friend in his dorm window with a water balloon. He has a similar idea and is aiming at you with *his* water balloon. You aim straight at each other and throw at the same instant. Do the water balloons hit each other? Explain why or why not.
15. A projectile is launched at an upward angle of 30° to the horizontal with a speed of 30 m/s. How does the horizontal component of its velocity 1.0 s after launch compare with its horizontal component of velocity 2.0 s after launch, ignoring air resistance? Explain.
16. A projectile has the least speed at what point in its path?
17. Two cannonballs, A and B, are fired from the ground with identical initial speeds, but with θ_A larger than θ_B. (a) Which cannonball reaches a higher elevation? (b) Which stays longer in the air? (c) Which travels farther? Explain.
18. A person sitting in an enclosed train car, moving at constant velocity, throws a ball straight up into the air in her reference frame. (a) Where does the ball land? What is your answer if the car (b) accelerates, (c) decelerates, (d) rounds a curve, (e) moves with constant velocity but is open to the air?
19. If you are riding on a train that speeds past another train moving in the same direction on an adjacent track, it appears that the other train is moving backward. Why?
20. Two rowers, who can row at the same speed in still water, set off across a river at the same time. One heads straight across and is pulled downstream somewhat by the current. The other one heads upstream at an angle so as to arrive at a point opposite the starting point. Which rower reaches the opposite side first? Explain.
21. If you stand motionless under an umbrella in a rainstorm where the drops fall vertically, you remain relatively dry. However, if you start running, the rain begins to hit your legs even if they remain under the umbrella. Why?

MisConceptual Questions

1. You are adding vectors of length 20 and 40 units. Which of the following choices is a possible resultant magnitude?
   (a) 0.
   (b) 18.
   (c) 37.
   (d) 64.
   (e) 100.
2. The magnitude of a component of a vector must be
   (a) less than or equal to the magnitude of the vector.
   (b) equal to the magnitude of the vector.
   (c) greater than or equal to the magnitude of the vector.
   (d) less than, equal to, or greater than the magnitude of the vector.
3. You are in the middle of a large field. You walk in a straight line for 100 m, then turn left and walk 100 m more in a straight line before stopping. When you stop, you are 100 m from your starting point. By how many degrees did you turn?
   (a) 90°.
   (b) 120°.
   (c) 30°.
   (d) 180°.
   (e) This is impossible. You cannot walk 200 m and be only 100 m away from where you started.
4. A bullet fired from a rifle begins to fall
   (a) as soon as it leaves the barrel.
   (b) after air friction reduces its speed.
   (c) not at all if air resistance is ignored.
5. A baseball player hits a ball that soars high into the air. After the ball has left the bat, and while it is traveling upward (at point P in Fig. 3–31), what is the direction of acceleration? Ignore air resistance.

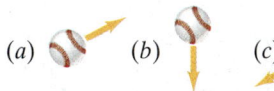

FIGURE 3–31 MisConceptual Question 5.

6. One ball is dropped vertically from a window. At the same instant, a second ball is thrown horizontally from the same window. Which ball has the greater speed at ground level?
   (a) The dropped ball.
   (b) The thrown ball.
   (c) Neither—they both have the same speed on impact.
   (d) It depends on how hard the ball was thrown.

7. You are riding in an enclosed train car moving at 90 km/h. If you throw a baseball straight up, where will the baseball land?
   (a) In front of you.
   (b) Behind you.
   (c) In your hand.
   (d) Can't decide from the given information.

8. Which of the three kicks in Fig. 3–32 is in the air for the longest time? They all reach the same maximum height h. Ignore air resistance.
   (a), (b), (c), or (d) all the same time.

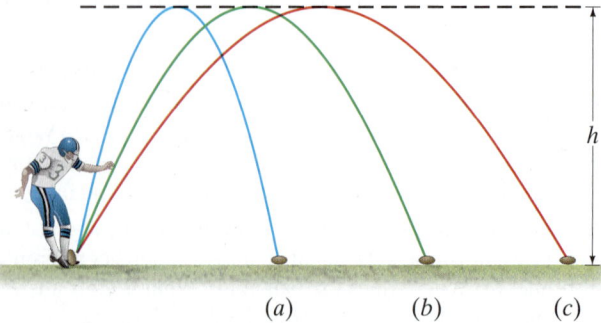

FIGURE 3–32 MisConceptual Question 8.

9. A baseball is hit high and far. Which of the following statements is true? At the highest point,
   (a) the magnitude of the acceleration is zero.
   (b) the magnitude of the velocity is zero.
   (c) the magnitude of the velocity is the slowest.
   (d) more than one of the above is true.
   (e) none of the above are true.

10. A hunter is aiming horizontally at a monkey who is sitting in a tree. The monkey is so terrified when it sees the gun that it falls off the tree. At that very instant, the hunter pulls the trigger. What will happen?
    (a) The bullet will miss the monkey because the monkey falls down while the bullet speeds straight forward.
    (b) The bullet will hit the monkey because both the monkey and the bullet are falling downward at the same rate due to gravity.
    (c) The bullet will miss the monkey because although both the monkey and the bullet are falling downward due to gravity, the monkey is falling faster.
    (d) It depends on how far the hunter is from the monkey.

11. Which statements are *not* valid for a projectile? Take up as positive.
    (a) The projectile has the same x velocity at any point on its path.
    (b) The acceleration of the projectile is positive and decreasing when the projectile is moving upwards, zero at the top, and increasingly negative as the projectile descends.
    (c) The acceleration of the projectile is a constant negative value.
    (d) The y component of the velocity of the projectile is zero at the highest point of the projectile's path.
    (e) The velocity at the highest point is zero.

12. A car travels 10 m/s east. Another car travels 10 m/s north. The relative speed of the first car with respect to the second is
    (a) less than 20 m/s.
    (b) exactly 20 m/s.
    (c) more than 20 m/s.

For assigned homework and other learning materials, go to the MasteringPhysics website.

Problems

3–2 to 3–4 Vector Addition

1. (I) A car is driven 225 km west and then 98 km southwest (45°). What is the displacement of the car from the point of origin (magnitude and direction)? Draw a diagram.

2. (I) A delivery truck travels 21 blocks north, 16 blocks east, and 26 blocks south. What is its final displacement from the origin? Assume the blocks are equal length.

3. (I) If $V_x = 9.80$ units and $V_y = -6.40$ units, determine the magnitude and direction of $\vec{V}$.

4. (II) Graphically determine the resultant of the following three vector displacements: (1) 24 m, 36° north of east; (2) 18 m, 37° east of north; and (3) 26 m, 33° west of south.

5. (II) $\vec{V}$ is a vector 24.8 units in magnitude and points at an angle of 23.4° above the negative x axis. (a) Sketch this vector. (b) Calculate V_x and V_y. (c) Use V_x and V_y to obtain (again) the magnitude and direction of $\vec{V}$. [*Note*: Part (c) is a good way to check if you've resolved your vector correctly.]

6. (II) Vector $\vec{V}_1$ is 6.6 units long and points along the negative x axis. Vector $\vec{V}_2$ is 8.5 units long and points at +55° to the positive x axis. (a) What are the x and y components of each vector? (b) Determine the sum $\vec{V}_1 + \vec{V}_2$ (magnitude and angle).

7. (II) Figure 3–33 shows two vectors, $\vec{A}$ and $\vec{B}$, whose magnitudes are $A = 6.8$ units and $B = 5.5$ units. Determine $\vec{C}$ if (a) $\vec{C} = \vec{A} + \vec{B}$, (b) $\vec{C} = \vec{A} - \vec{B}$, (c) $\vec{C} = \vec{B} - \vec{A}$. Give the magnitude and direction for each.

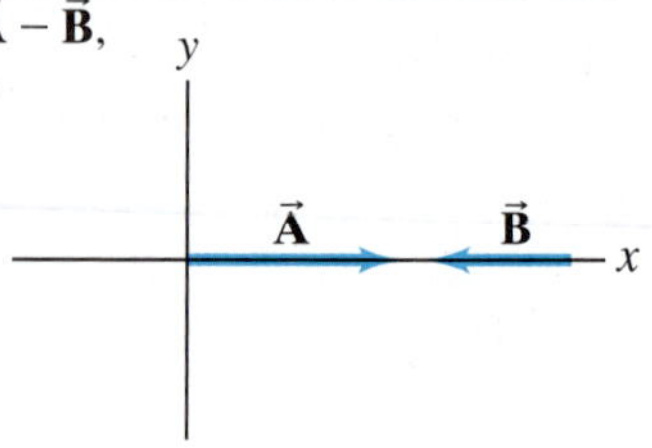

FIGURE 3–33 Problem 7.

8. (II) An airplane is traveling 835 km/h in a direction 41.5° west of north (Fig. 3–34).
   (a) Find the components of the velocity vector in the northerly and westerly directions. (b) How far north and how far west has the plane traveled after 1.75 h?

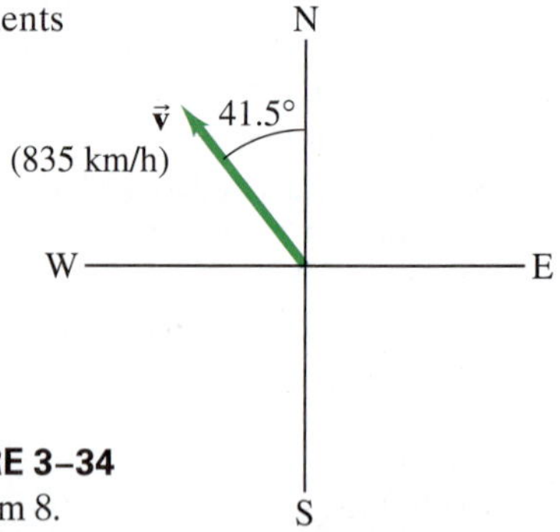

FIGURE 3–34 Problem 8.

9. (II) Three vectors are shown in Fig. 3–35. Their magnitudes are given in arbitrary units. Determine the sum of the three vectors. Give the resultant in terms of (a) components, (b) magnitude and angle with the +x axis.

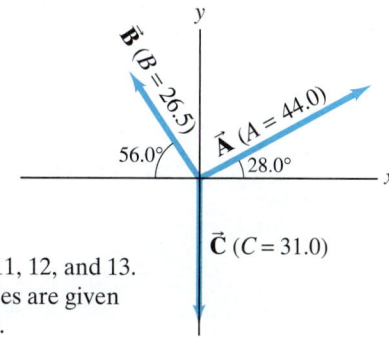

FIGURE 3–35
Problems 9, 10, 11, 12, and 13. Vector magnitudes are given in arbitrary units.

10. (II) (a) Given the vectors $\vec{A}$ and $\vec{B}$ shown in Fig. 3–35, determine $\vec{B} - \vec{A}$. (b) Determine $\vec{A} - \vec{B}$ without using your answer in (a). Then compare your results and see if they are opposite.

11. (II) Determine the vector $\vec{A} - \vec{C}$, given the vectors $\vec{A}$ and $\vec{C}$ in Fig. 3–35.

12. (II) For the vectors shown in Fig. 3–35, determine (a) $\vec{B} - 3\vec{A}$, (b) $2\vec{A} - 3\vec{B} + 2\vec{C}$.

13. (II) For the vectors given in Fig. 3–35, determine (a) $\vec{A} - \vec{B} + \vec{C}$, (b) $\vec{A} + \vec{B} - \vec{C}$, and (c) $\vec{C} - \vec{A} - \vec{B}$.

14. (II) Suppose a vector $\vec{V}$ makes an angle ϕ with respect to the y axis. What could be the x and y components of the vector $\vec{V}$?

15. (II) The summit of a mountain, 2450 m above base camp, is measured on a map to be 4580 m horizontally from the camp in a direction 38.4° west of north. What are the components of the displacement vector from camp to summit? What is its magnitude? Choose the x axis east, y axis north, and z axis up.

16. (III) You are given a vector in the xy plane that has a magnitude of 90.0 units and a y component of −65.0 units. (a) What are the two possibilities for its x component? (b) Assuming the x component is known to be positive, specify the vector which, if you add it to the original one, would give a resultant vector that is 80.0 units long and points entirely in the −x direction.

3–5 and 3–6 Projectile Motion (neglect air resistance)

17. (I) A tiger leaps horizontally from a 7.5-m-high rock with a speed of 3.0 m/s. How far from the base of the rock will she land?

18. (I) A diver running 2.5 m/s dives out horizontally from the edge of a vertical cliff and 3.0 s later reaches the water below. How high was the cliff and how far from its base did the diver hit the water?

19. (II) Estimate by what factor a person can jump farther on the Moon as compared to the Earth if the takeoff speed and angle are the same. The acceleration due to gravity on the Moon is one-sixth what it is on Earth.

20. (II) A ball is thrown horizontally from the roof of a building 7.5 m tall and lands 9.5 m from the base. What was the ball's initial speed?

21. (II) A ball thrown horizontally at 12.2 m/s from the roof of a building lands 21.0 m from the base of the building. How high is the building?

22. (II) A football is kicked at ground level with a speed of 18.0 m/s at an angle of 31.0° to the horizontal. How much later does it hit the ground?

23. (II) A fire hose held near the ground shoots water at a speed of 6.5 m/s. At what angle(s) should the nozzle point in order that the water land 2.5 m away (Fig. 3–36)? Why are there two different angles? Sketch the two trajectories.

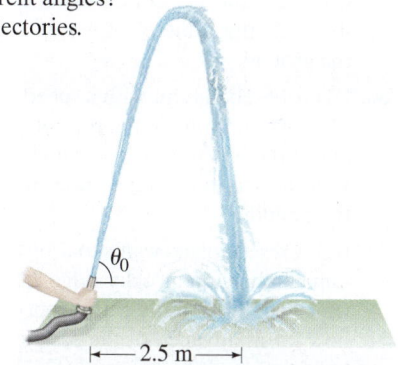

FIGURE 3–36
Problem 23.

24. (II) You buy a plastic dart gun, and being a clever physics student you decide to do a quick calculation to find its maximum horizontal range. You shoot the gun straight up, and it takes 4.0 s for the dart to land back at the barrel. What is the maximum horizontal range of your gun?

25. (II) A grasshopper hops along a level road. On each hop, the grasshopper launches itself at angle $\theta_0 = 45°$ and achieves a range $R = 0.80$ m. What is the average horizontal speed of the grasshopper as it hops along the road? Assume that the time spent on the ground between hops is negligible.

26. (II) Extreme-sports enthusiasts have been known to jump off the top of El Capitan, a sheer granite cliff of height 910 m in Yosemite National Park. Assume a jumper runs horizontally off the top of El Capitan with speed 4.0 m/s and enjoys a free fall until she is 150 m above the valley floor, at which time she opens her parachute (Fig. 3–37). (a) How long is the jumper in free fall? Ignore air resistance. (b) It is important to be as far away from the cliff as possible before opening the parachute. How far from the cliff is this jumper when she opens her chute?

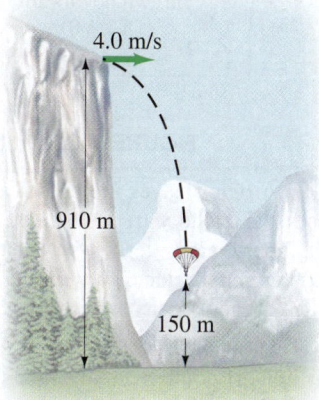

FIGURE 3–37
Problem 26.

27. (II) A projectile is fired with an initial speed of 36.6 m/s at an angle of 42.2° above the horizontal on a long flat firing range. Determine (a) the maximum height reached by the projectile, (b) the total time in the air, (c) the total horizontal distance covered (that is, the range), and (d) the speed of the projectile 1.50 s after firing.

28. (II) An athlete performing a long jump leaves the ground at a 27.0° angle and lands 7.80 m away. (a) What was the takeoff speed? (b) If this speed were increased by just 5.0%, how much longer would the jump be?

29. (II) A shot-putter throws the "shot" (mass = 7.3 kg) with an initial speed of 14.4 m/s at a 34.0° angle to the horizontal. Calculate the horizontal distance traveled by the shot if it leaves the athlete's hand at a height of 2.10 m above the ground.

30. (II) A baseball is hit with a speed of 27.0 m/s at an angle of 45.0°. It lands on the flat roof of a 13.0-m-tall nearby building. If the ball was hit when it was 1.0 m above the ground, what horizontal distance does it travel before it lands on the building?

31. (II) A rescue plane wants to drop supplies to isolated mountain climbers on a rocky ridge 235 m below. If the plane is traveling horizontally with a speed of 250 km/h (69.4 m/s), how far in advance of the recipients (horizontal distance) must the goods be dropped (Fig. 3–38)?

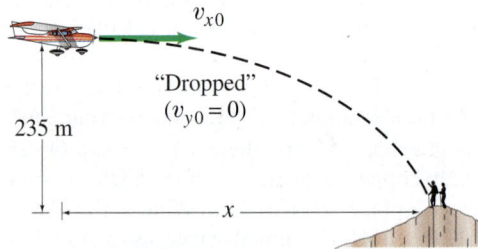

FIGURE 3–38 Problem 31.

32. (III) Suppose the rescue plane of Problem 31 releases the supplies a horizontal distance of 425 m in advance of the mountain climbers. What vertical velocity (up or down) should the supplies be given so that they arrive precisely at the climbers' position (Fig. 3–39)? With what speed do the supplies land?

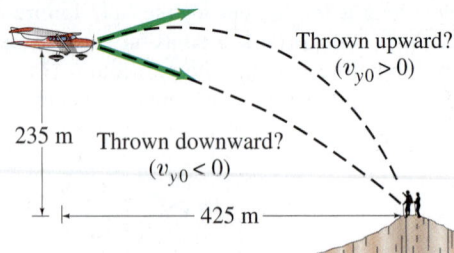

FIGURE 3–39 Problem 32.

33. (III) A diver leaves the end of a 4.0-m-high diving board and strikes the water 1.3 s later, 3.0 m beyond the end of the board. Considering the diver as a particle, determine: (a) her initial velocity, $\vec{v}_0$; (b) the maximum height reached; and (c) the velocity $\vec{v}_f$ with which she enters the water.

34. (III) Show that the time required for a projectile to reach its highest point is equal to the time for it to return to its original height if air resistance is neglible.

35. (III) Suppose the kick in Example 3–6 is attempted 36.0 m from the goalposts, whose crossbar is 3.05 m above the ground. If the football is directed perfectly between the goalposts, will it pass over the bar and be a field goal? Show why or why not. If not, from what horizontal distance must this kick be made if it is to score?

36. (III) Revisit Example 3–7, and assume that the boy with the slingshot is *below* the boy in the tree (Fig. 3–40) and so aims *upward*, directly at the boy in the tree. Show that again the boy in the tree makes the wrong move by letting go at the moment the water balloon is shot.

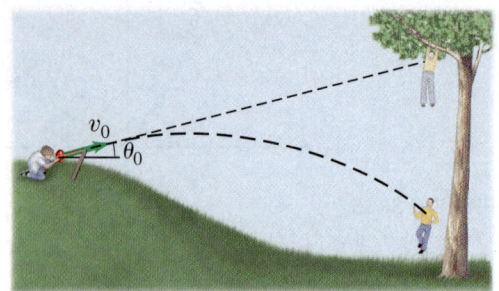

FIGURE 3–40 Problem 36.

37. (III) A stunt driver wants to make his car jump over 8 cars parked side by side below a horizontal ramp (Fig. 3–41). (a) With what minimum speed must he drive off the horizontal ramp? The vertical height of the ramp is 1.5 m above the cars and the horizontal distance he must clear is 22 m. (b) If the ramp is now tilted upward, so that "takeoff angle" is 7.0° above the horizontal, what is the new minimum speed?

FIGURE 3–41 Problem 37.

3–8 Relative Velocity

38. (I) A person going for a morning jog on the deck of a cruise ship is running toward the bow (front) of the ship at 2.0 m/s while the ship is moving ahead at 8.5 m/s. What is the velocity of the jogger relative to the water? Later, the jogger is moving toward the stern (rear) of the ship. What is the jogger's velocity relative to the water now?

39. (I) Huck Finn walks at a speed of 0.70 m/s across his raft (that is, he walks perpendicular to the raft's motion relative to the shore). The heavy raft is traveling down the Mississippi River at a speed of 1.50 m/s relative to the river bank (Fig. 3–42). What is Huck's velocity (speed and direction) relative to the river bank?

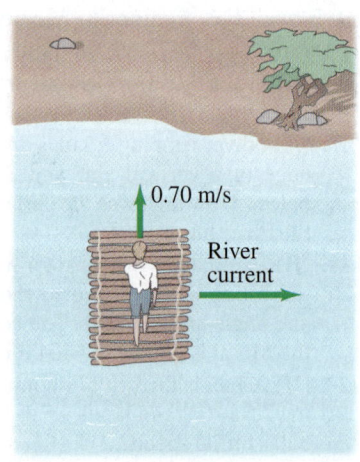

FIGURE 3–42 Problem 39.

40. (II) Determine the speed of the boat with respect to the shore in Example 3–10.

41. (II) Two planes approach each other head-on. Each has a speed of 780 km/h, and they spot each other when they are initially 10.0 km apart. How much time do the pilots have to take evasive action?

42. (II) A passenger on a boat moving at 1.70 m/s on a still lake walks up a flight of stairs at a speed of 0.60 m/s, Fig. 3–43. The stairs are angled at 45° pointing in the direction of motion as shown. What is the velocity of the passenger relative to the water?

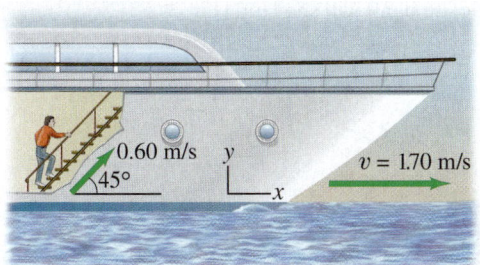

FIGURE 3–43 Problem 42.

43. (II) A person in the passenger basket of a hot-air balloon throws a ball horizontally outward from the basket with speed 10.0 m/s (Fig. 3–44). What initial velocity (magnitude and direction) does the ball have relative to a person standing on the ground (*a*) if the hot-air balloon is rising at 3.0 m/s relative to the ground during this throw, (*b*) if the hot-air balloon is descending at 3.0 m/s relative to the ground?

FIGURE 3–44 Problem 43.

44. (II) An airplane is heading due south at a speed of 688 km/h. If a wind begins blowing from the southwest at a speed of 90.0 km/h (average), calculate (*a*) the velocity (magnitude and direction) of the plane, relative to the ground, and (*b*) how far from its intended position it will be after 11.0 min if the pilot takes no corrective action. [*Hint*: First draw a diagram.]

45. (II) In what direction should the pilot aim the plane in Problem 44 so that it will fly due south?

46. (II) A swimmer is capable of swimming 0.60 m/s in still water. (*a*) If she aims her body directly across a 45-m-wide river whose current is 0.50 m/s, how far downstream (from a point opposite her starting point) will she land? (*b*) How long will it take her to reach the other side?

47. (II) (*a*) At what upstream angle must the swimmer in Problem 46 aim, if she is to arrive at a point directly across the stream? (*b*) How long will it take her?

48. (II) A boat, whose speed in still water is 2.50 m/s, must cross a 285-m-wide river and arrive at a point 118 m upstream from where it starts (Fig. 3–45). To do so, the pilot must head the boat at a 45.0° upstream angle. What is the speed of the river's current?

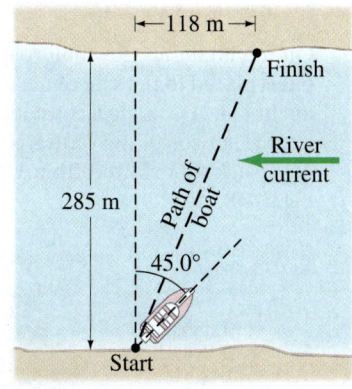

FIGURE 3–45 Problem 48.

49. (II) A child, who is 45 m from the bank of a river, is being carried helplessly downstream by the river's swift current of 1.0 m/s. As the child passes a lifeguard on the river's bank, the lifeguard starts swimming in a straight line (Fig. 3–46) until she reaches the child at a point downstream. If the lifeguard can swim at a speed of 2.0 m/s relative to the water, how long does it take her to reach the child? How far downstream does the lifeguard intercept the child?

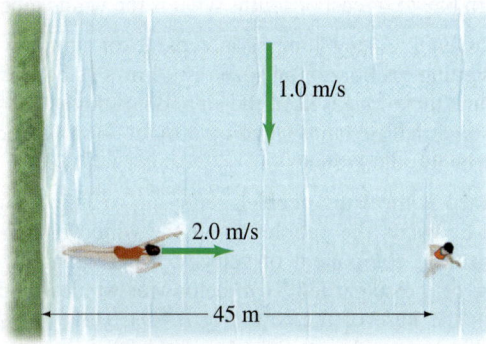

FIGURE 3–46 Problem 49.

50. (III) An airplane, whose air speed is 580 km/h, is supposed to fly in a straight path 38.0° N of E. But a steady 82 km/h wind is blowing from the north. In what direction should the plane head? [*Hint*: Use the law of sines, Appendix A–7.]

51. (III) Two cars approach a street corner at right angles to each other (Fig. 3–47). Car 1 travels at a speed relative to Earth $v_{1E} = 35$ km/h, and car 2 at $v_{2E} = 55$ km/h. What is the relative velocity of car 1 as seen by car 2? What is the velocity of car 2 relative to car 1?

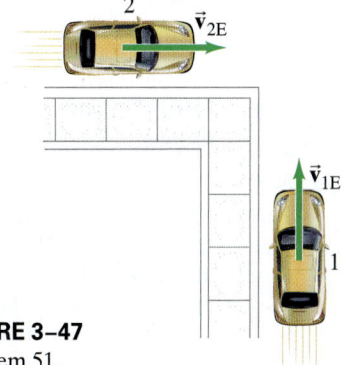

FIGURE 3–47 Problem 51.

General Problems

52. Two vectors, $\vec{V}_1$ and $\vec{V}_2$, add to a resultant $\vec{V}_R = \vec{V}_1 + \vec{V}_2$. Describe $\vec{V}_1$ and $\vec{V}_2$ if (a) $V_R = V_1 + V_2$, (b) $V_R^2 = V_1^2 + V_2^2$, (c) $V_1 + V_2 = V_1 - V_2$.

53. On mountainous downhill roads, escape routes are sometimes placed to the side of the road for trucks whose brakes might fail. Assuming a constant upward slope of 26°, calculate the horizontal and vertical components of the acceleration of a truck that slowed from 110 km/h to rest in 7.0 s. See Fig. 3–48.

FIGURE 3–48 Problem 53.

54. A light plane is headed due south with a speed relative to still air of 185 km/h. After 1.00 h, the pilot notices that they have covered only 135 km and their direction is not south but 15.0° east of south. What is the wind velocity?

55. An Olympic long jumper is capable of jumping 8.0 m. Assuming his horizontal speed is 9.1 m/s as he leaves the ground, how long is he in the air and how high does he go? Assume that he lands standing upright—that is, the same way he left the ground.

56. Romeo is throwing pebbles gently up to Juliet's window, and he wants the pebbles to hit the window with only a horizontal component of velocity. He is standing at the edge of a rose garden 8.0 m below her window and 8.5 m from the base of the wall (Fig. 3–49). How fast are the pebbles going when they hit her window?

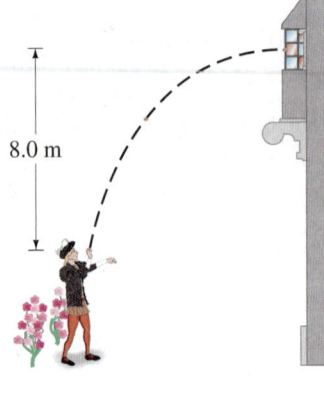

FIGURE 3–49 Problem 56.

57. *Apollo* astronauts took a "nine iron" to the Moon and hit a golf ball about 180 m. Assuming that the swing, launch angle, and so on, were the same as on Earth where the same astronaut could hit it only 32 m, estimate the acceleration due to gravity on the surface of the Moon. (We neglect air resistance in both cases, but on the Moon there is none.)

58. (a) A long jumper leaves the ground at 45° above the horizontal and lands 8.0 m away. What is her "takeoff" speed v_0? (b) Now she is out on a hike and comes to the left bank of a river. There is no bridge and the right bank is 10.0 m away horizontally and 2.5 m vertically below. If she long jumps from the edge of the left bank at 45° with the speed calculated in (a), how long, or short, of the opposite bank will she land (Fig. 3–50)?

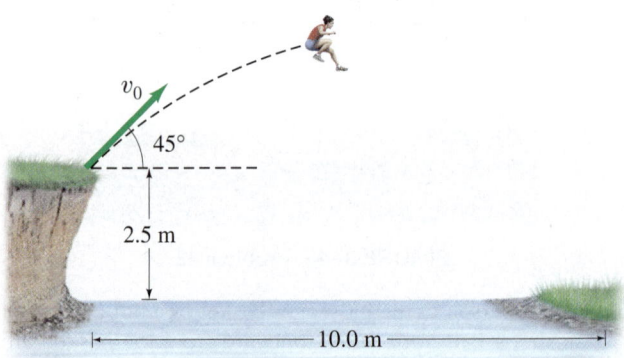

FIGURE 3–50 Problem 58.

59. A projectile is shot from the edge of a cliff 115 m above ground level with an initial speed of 65.0 m/s at an angle of 35.0° with the horizontal, as shown in Fig. 3–51. (a) Determine the time taken by the projectile to hit point P at ground level. (b) Determine the distance X of point P from the base of the vertical cliff. At the instant just before the projectile hits point P, find (c) the horizontal and the vertical components of its velocity, (d) the magnitude of the velocity, and (e) the angle made by the velocity vector with the horizontal. (f) Find the maximum height above the cliff top reached by the projectile.

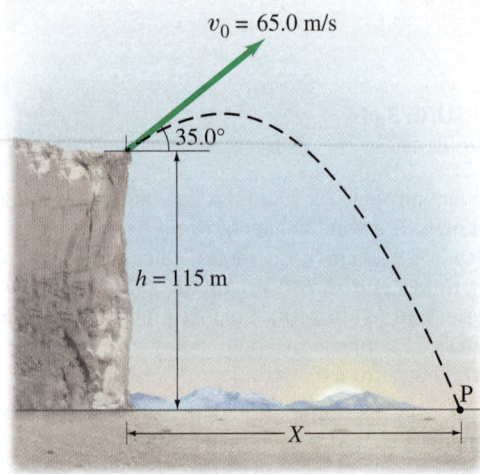

FIGURE 3–51 Problem 59.

60. William Tell must split the apple on top of his son's head from a distance of 27 m. When William aims directly at the apple, the arrow is horizontal. At what angle should he aim the arrow to hit the apple if the arrow travels at a speed of 35 m/s?

61. Raindrops make an angle θ with the vertical when viewed through a moving train window (Fig. 3–52). If the speed of the train is v_T, what is the speed of the raindrops in the reference frame of the Earth in which they are assumed to fall vertically?

FIGURE 3–52 Problem 61.

62. A car moving at 95 km/h passes a 1.00-km-long train traveling in the same direction on a track that is parallel to the road. If the speed of the train is 75 km/h, how long does it take the car to pass the train, and how far will the car have traveled in this time? What are the results if the car and train are instead traveling in opposite directions?

63. A hunter aims directly at a target (on the same level) 38.0 m away. (a) If the arrow leaves the bow at a speed of 23.1 m/s, by how much will it miss the target? (b) At what angle should the bow be aimed so the target will be hit?

64. The cliff divers of Acapulco push off horizontally from rock platforms about 35 m above the water, but they must clear rocky outcrops at water level that extend out into the water 5.0 m from the base of the cliff directly under their launch point. See Fig. 3–53. What minimum pushoff speed is necessary to clear the rocks? How long are they in the air?

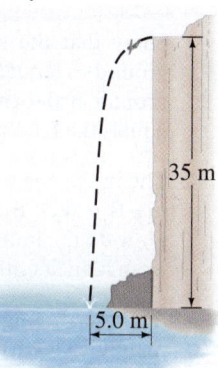

FIGURE 3–53 Problem 64.

65. When Babe Ruth hit a homer over the 8.0-m-high right-field fence 98 m from home plate, roughly what was the minimum speed of the ball when it left the bat? Assume the ball was hit 1.0 m above the ground and its path initially made a 36° angle with the ground.

66. At serve, a tennis player aims to hit the ball horizontally. What minimum speed is required for the ball to clear the 0.90-m-high net about 15.0 m from the server if the ball is "launched" from a height of 2.50 m? Where will the ball land if it just clears the net (and will it be "good" in the sense that it lands within 7.0 m of the net)? How long will it be in the air? See Fig. 3–54.

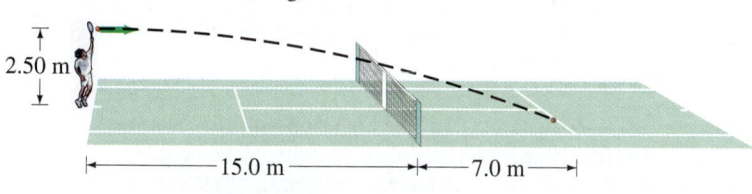

FIGURE 3–54 Problem 66.

67. Spymaster Chris, flying a constant 208 km/h horizontally in a low-flying helicopter, wants to drop secret documents into her contact's open car which is traveling 156 km/h on a level highway 78.0 m below. At what angle (with the horizontal) should the car be in her sights when the packet is released (Fig. 3–55)?

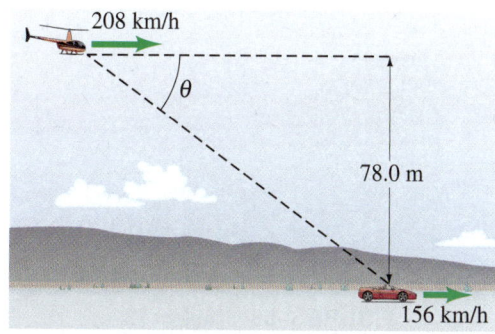

FIGURE 3–55 Problem 67.

68. A basketball leaves a player's hands at a height of 2.10 m above the floor. The basket is 3.05 m above the floor. The player likes to shoot the ball at a 38.0° angle. If the shot is made from a horizontal distance of 11.00 m and must be accurate to ±0.22 m (horizontally), what is the range of initial speeds allowed to make the basket?

69. A boat can travel 2.20 m/s in still water. (a) If the boat points directly across a stream whose current is 1.20 m/s, what is the velocity (magnitude and direction) of the boat relative to the shore? (b) What will be the position of the boat, relative to its point of origin, after 3.00 s?

70. A projectile is launched from ground level to the top of a cliff which is 195 m away and 135 m high (see Fig. 3–56). If the projectile lands on top of the cliff 6.6 s after it is fired, find the initial velocity of the projectile (magnitude and direction). Neglect air resistance.

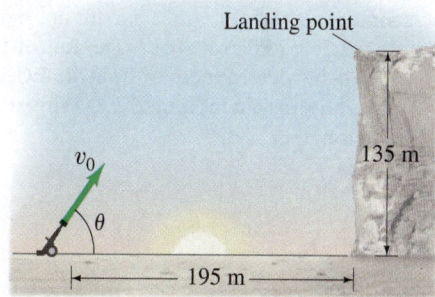

FIGURE 3–56 Problem 70.

71. A basketball is shot from an initial height of 2.40 m (Fig. 3–57) with an initial speed $v_0 = 12$ m/s directed at an angle $\theta_0 = 35°$ above the horizontal. (a) How far from the basket was the player if he made a basket? (b) At what angle to the horizontal did the ball enter the basket?

FIGURE 3–57 Problem 71.

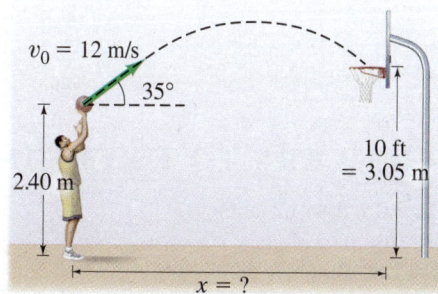

General Problems 73

72. A rock is kicked horizontally at 15 m/s from a hill with a 45° slope (Fig. 3–58). How long does it take for the rock to hit the ground?

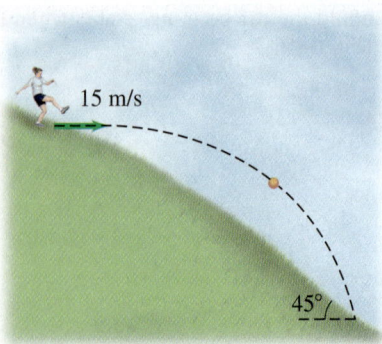

FIGURE 3–58 Problem 72.

73. A batter hits a fly ball which leaves the bat 0.90 m above the ground at an angle of 61° with an initial speed of 28 m/s heading toward centerfield. Ignore air resistance. (a) How far from home plate would the ball land if not caught? (b) The ball is caught by the centerfielder who, starting at a distance of 105 m from home plate just as the ball was hit, runs straight toward home plate at a constant speed and makes the catch at ground level. Find his speed.

74. A ball is shot from the top of a building with an initial velocity of 18 m/s at an angle $\theta = 42°$ above the horizontal. (a) What are the horizontal and vertical components of the initial velocity? (b) If a nearby building is the same height and 55 m away, how far below the top of the building will the ball strike the nearby building?

75. If a baseball pitch leaves the pitcher's hand horizontally at a velocity of 150 km/h, by what % will the pull of gravity change the magnitude of the velocity when the ball reaches the batter, 18 m away? For this estimate, ignore air resistance and spin on the ball.

Search and Learn

1. Here is something to try at a sporting event. Show that the maximum height h attained by an object projected into the air, such as a baseball, football, or soccer ball, is approximately given by

    $$h \approx 1.2 t^2 \text{ m},$$

    where t is the total time of flight for the object in seconds. Assume that the object returns to the same level as that from which it was launched, as in Fig. 3–59. For example, if you count to find that a baseball was in the air for $t = 5.0$ s, the maximum height attained was $h = 1.2 \times (5.0)^2 = 30$ m. The fun of this relation is that h can be determined without knowledge of the launch speed v_0 or launch angle θ_0. Why is that exactly? See Section 3–6.

2. Two balls are thrown in the air at different angles, but each reaches the same height. Which ball remains in the air longer? Explain, using equations.

3. Show that the speed with which a projectile leaves the ground is equal to its speed just before it strikes the ground at the end of its journey, assuming the firing level equals the landing level.

4. The initial angle of projectile A is 30°, while that of projectile B is 60°. Both have the same level horizontal range. How do the initial velocities and flight times (elapsed time from launch until landing) compare for A and B?

5. You are driving south on a highway at 12 m/s (approximately 25 mi/h) in a snowstorm. When you last stopped, you noticed that the snow was coming down vertically, but it is passing the windows of the moving car at an angle of 7.0° to the horizontal. Estimate the speed of the vertically falling snowflakes relative to the ground. [*Hint*: Construct a relative velocity diagram similar to Fig. 3–29 or 3–30. Be careful about which angle is the angle given.]

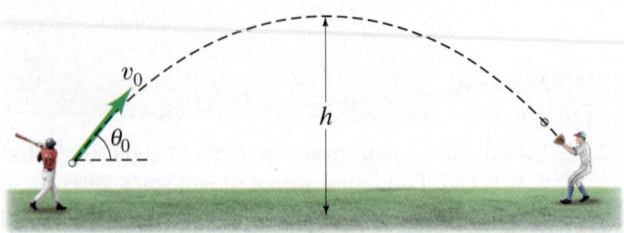

FIGURE 3–59 Search and Learn 1.

ANSWERS TO EXERCISES

A: $3.0\sqrt{2} \approx 4.2$ units.
B: (a).
C: They hit at the same time.
D: (i) Nowhere; (ii) at the highest point; (iii) nowhere.
E: (d). It provides the initial velocity of the box.
F: (a) $v = v_{x0} = 16.0$ m/s, horizontal; (b) 9.80 m/s^2 down.

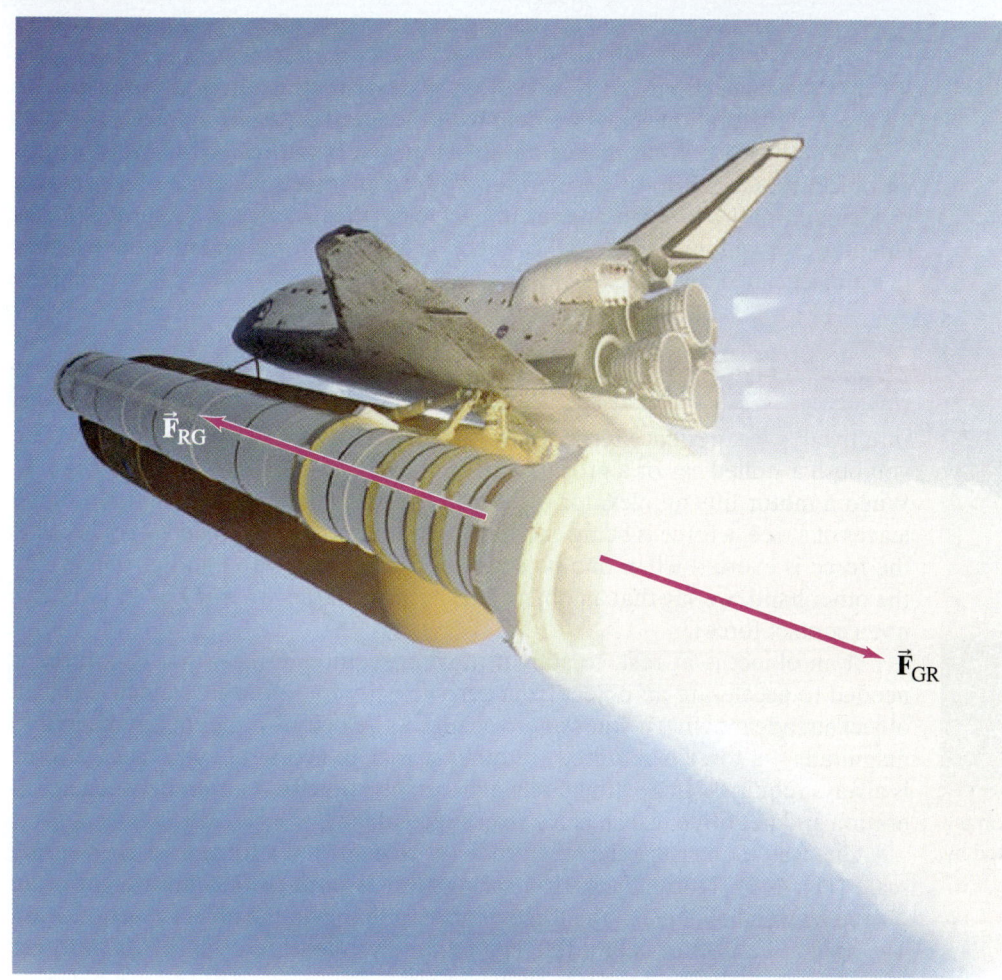

A space shuttle is carried out into space by powerful rockets. They are accelerating, increasing in speed rapidly. To do so, a force must be exerted on them according to Newton's second law, $\Sigma \vec{F} = m\vec{a}$. What exerts this force? The **r**ocket engines exert a force on the **g**ases they push out (expel) from the rear of the rockets (labeled $\vec{F}_{GR}$). According to Newton's third law, these ejected gases exert an equal and opposite force on the rockets in the forward direction. It is this "reaction" force exerted *on* the **r**ockets *by* the **g**ases, labeled $\vec{F}_{RG}$, that accelerates the rockets forward.

Dynamics: Newton's Laws of Motion

CHAPTER 4

CHAPTER-OPENING QUESTIONS—Guess now!

1. A 150-kg football player collides head-on with a 75-kg running back. During the collision, the heavier player exerts a force of magnitude F_A on the smaller player. If the smaller player exerts a force F_B back on the heavier player, which response is most accurate?
  (a) $F_B = F_A$.
  (b) $F_B < F_A$.
  (c) $F_B > F_A$.
  (d) $F_B = 0$.
  (e) We need more information.

2. A line by the poet T. S. Eliot (from *Murder in the Cathedral*) has the women of Canterbury say "the earth presses up against our feet." What force is this?
  (a) Gravity.
  (b) The normal force.
  (c) A friction force.
  (d) Centrifugal force.
  (e) No force—they are being poetic.

CONTENTS

4–1 Force
4–2 Newton's First Law of Motion
4–3 Mass
4–4 Newton's Second Law of Motion
4–5 Newton's Third Law of Motion
4–6 Weight—the Force of Gravity; and the Normal Force
4–7 Solving Problems with Newton's Laws: Free-Body Diagrams
4–8 Problems Involving Friction, Inclines

We have discussed how motion is described in terms of velocity and acceleration. Now we deal with the question of *why* objects move as they do: What makes an object at rest begin to move? What causes an object to accelerate or decelerate? What is involved when an object moves in a curved path? We can answer in each case that a force is required. In this Chapter[†], we will investigate the connection between force and motion, which is the subject called **dynamics**.

4–1 Force

FIGURE 4–1 A force exerted on a grocery cart—in this case exerted by a person.

Intuitively, we experience **force** as any kind of a push or a pull on an object. When you push a stalled car or a grocery cart (Fig. 4–1), you are exerting a force on it. When a motor lifts an elevator, or a hammer hits a nail, or the wind blows the leaves of a tree, a force is being exerted. We often call these *contact forces* because the force is exerted when one object comes in contact with another object. On the other hand, we say that an object falls because of the *force of gravity* (which is not a contact force).

If an object is at rest, to start it moving requires force—that is, a force is needed to accelerate an object from zero velocity to a nonzero velocity. For an object already moving, if you want to change its velocity—either in direction or in magnitude—a force is required. In other words, to accelerate an object, a force is always required. In Section 4–4 we discuss the precise relation between acceleration and net force, which is Newton's second law.

One way to measure the magnitude (or strength) of a force is to use a spring scale (Fig. 4–2). Normally, such a spring scale is used to find the weight of an object; by weight we mean the force of gravity acting on the object (Section 4–6). The spring scale, once calibrated, can be used to measure other kinds of forces as well, such as the pulling force shown in Fig. 4–2.

A force exerted in a different direction has a different effect. Force has direction as well as magnitude, and is indeed a vector that follows the rules of vector addition discussed in Chapter 3. We can represent any force on a diagram by an arrow, just as we did with velocity. The direction of the arrow is the direction of the push or pull, and its length is drawn proportional to the magnitude of the force.

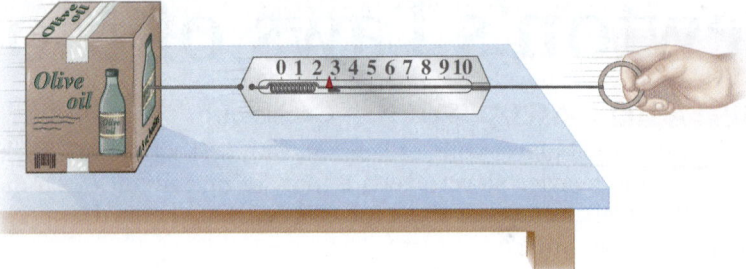

FIGURE 4–2 A spring scale used to measure a force.

4–2 Newton's First Law of Motion

What is the relationship between force and motion? Aristotle (384–322 B.C.) believed that a force was required to keep an object moving along a horizontal plane. To Aristotle, the natural state of an object was at rest, and a force was believed necessary to keep an object in motion. Furthermore, Aristotle argued, the greater the force on the object, the greater its speed.

Some 2000 years later, Galileo disagreed: he maintained that it is just as natural for an object to be in motion with a constant velocity as it is for it to be at rest.

[†]We treat everyday objects in motion here. When velocities are extremely high, close to the speed of light $(3.0 \times 10^8 \, \text{m/s})$, we use the theory of relativity (Chapter 26), and in the submicroscopic world of atoms and molecules we use quantum theory (Chapter 27 *ff*).

To understand Galileo's idea, consider the following observations involving motion along a horizontal plane. To push an object with a rough surface along a tabletop at constant speed requires a certain amount of force. To push an equally heavy object with a very smooth surface across the table at the same speed will require less force. If a layer of oil or other lubricant is placed between the surface of the object and the table, then almost no force is required to keep the object moving. Notice that in each successive step, less force is required. As the next step, we imagine there is no friction at all, that the object does not rub against the table—or there is a perfect lubricant between the object and the table—and theorize that once started, the object would move across the table at constant speed with *no* force applied. A steel ball bearing rolling on a hard horizontal surface approaches this situation. So does a puck on an air table, in which a thin layer of air reduces friction almost to zero.

It was Galileo's genius to imagine such an idealized world—in this case, one where there is no friction—and to see that it could lead to a more accurate and richer understanding of the real world. This idealization led him to his remarkable conclusion that if no force is applied to a moving object, it will continue to move with constant speed in a straight line. An object slows down only if a force is exerted on it. Galileo thus interpreted friction as a force akin to ordinary pushes and pulls.

To push an object across a table at constant speed requires a force from your hand that can balance the force of friction (Fig. 4–3). When the object moves at constant speed, your pushing force is equal in magnitude to the friction force; but these two forces are in opposite directions, so the *net* force on the object (the vector sum of the two forces) is zero. This is consistent with Galileo's viewpoint, for the object moves with constant velocity when no *net* force is exerted on it.

FIGURE 4–3 $\vec{F}$ represents the force applied by the person and $\vec{F}_{fr}$ represents the force of friction.

Upon this foundation laid by Galileo, Isaac Newton (Fig. 4–4) built his great theory of motion. Newton's analysis of motion is summarized in his famous "three laws of motion." In his great work, the *Principia* (published in 1687), Newton readily acknowledged his debt to Galileo. In fact, **Newton's first law of motion** is close to Galileo's conclusions. It states that

> **Every object continues in its state of rest, or of uniform velocity in a straight line, as long as no net force acts on it.**

NEWTON'S FIRST LAW OF MOTION

The tendency of an object to maintain its state of rest or of uniform velocity in a straight line is called **inertia**. As a result, Newton's first law is often called the **law of inertia**.

CONCEPTUAL EXAMPLE 4–1 **Newton's first law.** A school bus comes to a sudden stop, and all of the backpacks on the floor start to slide forward. What force causes them to do that?

RESPONSE It isn't "force" that does it. By Newton's first law, the backpacks continue their state of motion, maintaining their velocity. The backpacks slow down if a force is applied, such as friction with the floor.

FIGURE 4–4 Isaac Newton (1642–1727). Besides developing mechanics, including his three great laws of motion and the law of universal gravitation, he also tried to understand the nature of light.

Inertial Reference Frames

Newton's first law does not hold in every reference frame. For example, if your reference frame is an accelerating car, an object such as a cup resting on the dashboard may begin to move toward you (it stayed at rest as long as the car's velocity remained constant). The cup accelerated toward you, but neither you nor anything else exerted a force on it in that direction. Similarly, in the reference frame of the decelerating bus in Example 4–1, there was no force pushing the backpacks forward. In accelerating reference frames, Newton's first law does not hold. Physics is easier in reference frames in which Newton's first law *does* hold, and they are called **inertial reference frames** (the law of inertia is valid in them). For most purposes, we usually make the approximation that a reference frame fixed on the Earth is an inertial frame. This is not precisely true, due to the Earth's rotation, but usually it is close enough.

Any reference frame that moves with constant velocity (say, a car or an airplane) relative to an inertial frame is also an inertial reference frame. Reference frames where the law of inertia does *not* hold, such as the accelerating reference frames discussed above, are called **noninertial** reference frames. How can we be sure a reference frame is inertial or not? By checking to see if Newton's first law holds. Thus Newton's first law serves as the definition of inertial reference frames.

4–3 Mass

Newton's second law, which we come to in the next Section, makes use of the concept of mass. Newton used the term *mass* as a synonym for "quantity of matter." This intuitive notion of the mass of an object is not very precise because the concept "quantity of matter" is not very well defined. More precisely, we can say that **mass** is a *measure of the inertia* of an object. The more mass an object has, the greater the force needed to give it a particular acceleration. It is harder to start it moving from rest, or to stop it when it is moving, or to change its velocity sideways out of a straight-line path. A truck has much more inertia than a baseball moving at the same speed, and a much greater force is needed to change the truck's velocity at the same rate as the ball's. The truck therefore has much more mass.

To quantify the concept of mass, we must define a standard. In SI units, the unit of mass is the **kilogram** (kg) as we discussed in Chapter 1, Section 1–5.

CAUTION
Distinguish mass from weight

The terms *mass* and *weight* are often confused with one another, but it is important to distinguish between them. Mass is a property of an object itself (a measure of an object's inertia, or its "quantity of matter"). Weight, on the other hand, is a force, the pull of gravity acting on an object. To see the difference, suppose we take an object to the Moon. The object will weigh only about one-sixth as much as it did on Earth, since the force of gravity is weaker. But its mass will be the same. It will have the same amount of matter as on Earth, and will have just as much inertia—in the absence of friction, it will be just as hard to start it moving on the Moon as on Earth, or to stop it once it is moving. (More on weight in Section 4–6.)

4–4 Newton's Second Law of Motion

Newton's first law states that if no net force is acting on an object at rest, the object remains at rest; or if the object is moving, it continues moving with constant speed in a straight line. But what happens if a net force *is* exerted on an object? Newton perceived that the object's velocity will change (Fig. 4–5). A net force exerted on an object may make its velocity increase. Or, if the net force is in a direction opposite to the motion, that force will reduce the object's velocity. If the net force acts sideways on a moving object, the *direction* of the object's velocity changes. That change in the *direction* of the velocity is also an acceleration. So a sideways net force on an object also causes acceleration. In general, we can say that *a net force causes acceleration*.

FIGURE 4–5 The bobsled accelerates because the team exerts a force.

What precisely is the relationship between acceleration and force? Everyday experience can suggest an answer. Consider the force required to push a cart when friction is small enough to ignore. (If there is friction, consider the *net* force, which is the force you exert minus the force of friction.) If you push the cart horizontally with a gentle but constant force for a certain period of time, you will make the cart accelerate from rest up to some speed, say 3 km/h. If you push with twice the force, the cart will reach 3 km/h in half the time. The acceleration will be twice as great. If you triple the force, the acceleration is tripled, and so on. Thus, the acceleration of an object is directly proportional[†] to the net applied force. But the acceleration depends on the mass of the object as well. If you push an empty grocery cart with the same force as you push one that is filled with groceries, you will find that the full cart accelerates more slowly.

[†]A review of proportionality is given in Appendix A.

The greater the mass, the less the acceleration for the same net force. The mathematical relation, as Newton argued, is that the acceleration of an object is inversely proportional to its mass. These relationships are found to hold in general and can be summarized as follows:

> **The acceleration of an object is directly proportional to the net force acting on it, and is inversely proportional to the object's mass. The direction of the acceleration is in the direction of the net force acting on the object.**

This is **Newton's second law of motion**.

Newton's second law can be written as an equation:

$$\vec{a} = \frac{\Sigma \vec{F}}{m},$$

where $\vec{a}$ stands for acceleration, m for the mass, and $\Sigma \vec{F}$ for the *net force* on the object. The symbol Σ (Greek "sigma") stands for "sum of"; $\vec{F}$ stands for force, so $\Sigma \vec{F}$ means the *vector sum of all forces* acting on the object, which we define as the **net force**.

We rearrange this equation to obtain the familiar statement of Newton's second law:

$$\Sigma \vec{F} = m\vec{a}. \quad (4\text{--}1)$$

> NEWTON'S SECOND LAW OF MOTION

> NEWTON'S SECOND LAW OF MOTION

Newton's second law relates the description of motion to the cause of motion, force. It is one of the most fundamental relationships in physics. From Newton's second law we can make a more precise definition of **force** as *an action capable of accelerating an object*.

Every force $\vec{F}$ is a vector, with magnitude and direction. Equation 4–1 is a vector equation valid in any inertial reference frame. It can be written in component form in rectangular coordinates as

$$\Sigma F_x = ma_x, \quad \Sigma F_y = ma_y, \quad \Sigma F_z = ma_z.$$

If the motion is all along a line (one-dimensional), we can leave out the subscripts and simply write $\Sigma F = ma$. Again, a is the acceleration of an object of mass m, and ΣF includes all the forces acting on that object, and *only* forces acting on that object. (Sometimes the net force ΣF is written as F_{net}, so $F_{net} = ma$.)

In SI units, with the mass in kilograms, the unit of force is called the **newton** (N). One newton is the force required to impart an acceleration of 1 m/s^2 to a mass of 1 kg. Thus $1 \text{ N} = 1 \text{ kg} \cdot \text{m/s}^2$.

In cgs units, the unit of mass is the gram† (g). The unit of force is the *dyne*, which is defined as the net force needed to impart an acceleration of 1 cm/s^2 to a mass of 1 g. Thus $1 \text{ dyne} = 1 \text{ g} \cdot \text{cm/s}^2$. Because $1 \text{ g} = 10^{-3} \text{ kg}$ and $1 \text{ cm} = 10^{-2} \text{ m}$, then $1 \text{ dyne} = 10^{-5} \text{ N}$.

In the British system, which we rarely use, the unit of force is the *pound* (abbreviated lb), where $1 \text{ lb} = 4.44822 \text{ N} \approx 4.45 \text{ N}$. The unit of mass is the *slug*, which is defined as that mass which will undergo an acceleration of 1 ft/s^2 when a force of 1 lb is applied to it. Thus $1 \text{ lb} = 1 \text{ slug} \cdot \text{ft/s}^2$. Table 4–1 summarizes the units in the different systems.

TABLE 4–1 Units for Mass and Force

System	Mass	Force
SI	kilogram (kg)	newton (N) (= kg·m/s²)
cgs	gram (g)	dyne (= g·cm/s²)
British	slug	pound (lb)

Conversion factors: $1 \text{ dyne} = 10^{-5} \text{ N}$; $1 \text{ lb} \approx 4.45 \text{ N}$; $1 \text{ slug} \approx 14.6 \text{ kg}$.

It is very important that only one set of units be used in a given calculation or Problem, with the SI being what we almost always use. If the force is given in, say, newtons, and the mass in grams, then before attempting to solve for the acceleration in SI units, we must change the mass to kilograms. For example, if the force is given as 2.0 N along the x axis and the mass is 500 g, we change the latter to 0.50 kg, and the acceleration will then automatically come out in m/s² when Newton's second law is used:

$$a_x = \frac{\Sigma F_x}{m} = \frac{2.0 \text{ N}}{0.50 \text{ kg}} = \frac{2.0 \text{ kg} \cdot \text{m/s}^2}{0.50 \text{ kg}} = 4.0 \text{ m/s}^2,$$

where we set $1 \text{ N} = 1 \text{ kg} \cdot \text{m/s}^2$.

Use a consistent set of units

---
†Be careful not to confuse g for gram with g for the acceleration due to gravity. The latter is always italicized (or boldface when shown as a vector).

EXAMPLE 4-2 ESTIMATE **Force to accelerate a fast car.** Estimate the net force needed to accelerate (a) a 1000-kg car at $\frac{1}{2}g$; (b) a 200-gram apple at the same rate.

APPROACH We use Newton's second law to find the net force needed for each object; we are given the mass and the acceleration. This is an estimate (the $\frac{1}{2}$ is not said to be precise) so we round off to one significant figure.

SOLUTION (a) The car's acceleration is $a = \frac{1}{2}g = \frac{1}{2}(9.8 \text{ m/s}^2) \approx 5 \text{ m/s}^2$. We use Newton's second law to get the net force needed to achieve this acceleration:

$$\Sigma F = ma \approx (1000 \text{ kg})(5 \text{ m/s}^2) = 5000 \text{ N}.$$

(If you are used to British units, to get an idea of what a 5000-N force is, you can divide by 4.45 N/lb and get a force of about 1000 lb.)

(b) For the apple, $m = 200 \text{ g} = 0.2 \text{ kg}$, so

$$\Sigma F = ma \approx (0.2 \text{ kg})(5 \text{ m/s}^2) = 1 \text{ N}.$$

EXAMPLE 4-3 **Force to stop a car.** What average net force is required to bring a 1500-kg car to rest from a speed of 100 km/h within a distance of 55 m?

APPROACH We use Newton's second law, $\Sigma F = ma$, to determine the force, but first we need to calculate the acceleration a. We assume the acceleration is constant so that we can use the kinematic equations, Eqs. 2–11, to calculate it.

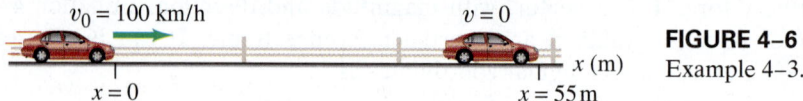

FIGURE 4-6 Example 4-3.

SOLUTION We assume the motion is along the $+x$ axis (Fig. 4–6). We are given the initial velocity $v_0 = 100 \text{ km/h} = 27.8 \text{ m/s}$ (Section 1–6), the final velocity $v = 0$, and the distance traveled $x - x_0 = 55 \text{ m}$. From Eq. 2–11c, we have

$$v^2 = v_0^2 + 2a(x - x_0),$$

so

$$a = \frac{v^2 - v_0^2}{2(x - x_0)} = \frac{0 - (27.8 \text{ m/s})^2}{2(55 \text{ m})} = -7.0 \text{ m/s}^2.$$

The net force required is then

$$\Sigma F = ma = (1500 \text{ kg})(-7.0 \text{ m/s}^2) = -1.1 \times 10^4 \text{ N},$$

or 11,000 N. The force must be exerted in the direction *opposite* to the initial velocity, which is what the negative sign means.

NOTE If the acceleration is not precisely constant, then we are determining an "average" acceleration and we obtain an "average" net force.

Newton's second law, like the first law, is valid only in inertial reference frames (Section 4–2). In the noninertial reference frame of a car that begins accelerating, a cup on the dashboard starts sliding—it accelerates—even though the net force on it is zero. Thus $\Sigma \vec{F} = m\vec{a}$ does not work in such an accelerating reference frame ($\Sigma \vec{F} = 0$, but $\vec{a} \neq 0$ in this noninertial frame).

EXERCISE A Suppose you watch a cup slide on the (smooth) dashboard of an accelerating car as we just discussed, but this time from an inertial reference frame outside the car, on the street. From your inertial frame, Newton's laws are valid. What force pushes the cup off the dashboard?

4–5 Newton's Third Law of Motion

Newton's second law of motion describes quantitatively how forces affect motion. But where, we may ask, do forces come from? Observations suggest that a force exerted on any object is always exerted *by another object*. A horse pulls a wagon, a person pushes a grocery cart, a hammer pushes on a nail, a magnet attracts a paper clip. In each of these examples, a force is exerted *on* one object, and that force is exerted *by* another object. For example, the force exerted *on* the nail is exerted *by* the hammer.

But Newton realized that things are not so one-sided. True, the hammer exerts a force on the nail (Fig. 4–7). But the nail evidently exerts a force back on the hammer as well, for the hammer's speed is rapidly reduced to zero upon contact. Only a strong force could cause such a rapid deceleration of the hammer. Thus, said Newton, the two objects must be treated on an equal basis. The hammer exerts a force on the nail, and the nail exerts a force back on the hammer. This is the essence of **Newton's third law of motion**:

> **Whenever one object exerts a force on a second object, the second object exerts an equal force in the opposite direction on the first.**

This law is sometimes paraphrased as "to every action there is an equal and opposite reaction." This is perfectly valid. But to avoid confusion, it is very important to remember that the "action" force and the "reaction" force are acting on *different* objects.

As evidence for the validity of Newton's third law, look at your hand when you push against the edge of a desk, Fig. 4–8. Your hand's shape is distorted, clear evidence that a force is being exerted on it. You can *see* the edge of the desk pressing into your hand. You can even *feel* the desk exerting a force on your hand; it hurts! The harder you push against the desk, the harder the desk pushes back on your hand. (You only feel forces exerted *on* you; when you exert a force on another object, what you feel is that object pushing back on you.)

FIGURE 4–7 A hammer striking a nail. The hammer exerts a force on the nail and the nail exerts a force back on the hammer. The latter force decelerates the hammer and brings it to rest.

> NEWTON'S THIRD LAW OF MOTION

> ⚠ **CAUTION**
> *Action and reaction forces act on different objects*

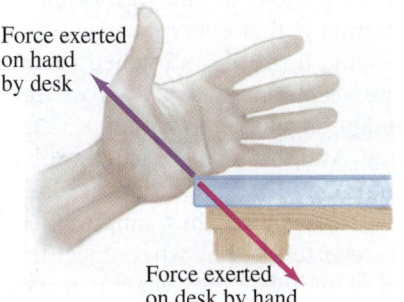

FIGURE 4–8 If your hand pushes against the edge of a desk (the force vector is shown in red), the desk pushes back against your hand (this force vector is shown in a different color, violet, to remind us that this force acts on a different object).

The force the desk exerts on your hand has the same magnitude as the force your hand exerts on the desk. This is true not only if the desk is at rest but is true even if the desk is accelerating due to the force your hand exerts.

As another demonstration of Newton's third law, consider the ice skater in Fig. 4–9. There is very little friction between her skates and the ice, so she will move freely if a force is exerted on her. She pushes against the wall; and then *she* starts moving backward. The force she exerts on the wall cannot make *her* start moving, because that force acts on the wall. Something had to exert a force *on her* to start her moving, and that force could only have been exerted by the wall. The force with which the wall pushes on her is, by Newton's third law, equal and opposite to the force she exerts on the wall.

When a person throws a package out of a small boat (initially at rest), the boat starts moving in the opposite direction. The person exerts a force on the package. The package exerts an equal and opposite force back on the person, and this force propels the person (and the boat) backward slightly.

FIGURE 4–9 An example of Newton's third law: when an ice skater pushes against the wall, the wall pushes back and this force causes her to accelerate away.

FIGURE 4–10 Another example of Newton's third law: the launch of a rocket. The rocket engine pushes the gases downward, and the gases exert an equal and opposite force upward on the rocket, accelerating it upward. (A rocket does *not* accelerate as a result of its expelled gases pushing against the ground.)

FIGURE 4–11 We can walk forward because, when one foot pushes backward against the ground, the ground pushes forward on that foot (Newton's third law). The two forces shown *act on different objects*.

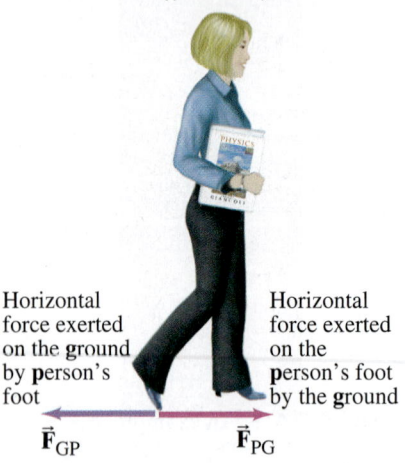

Horizontal force exerted on the ground by **p**erson's foot
$\vec{F}_{GP}$

Horizontal force exerted on the **p**erson's foot by the **g**round
$\vec{F}_{PG}$

NEWTON'S THIRD LAW OF MOTION

Rocket propulsion also is explained using Newton's third law (Fig. 4–10). A common misconception is that rockets accelerate because the gases rushing out the back of the engine push against the ground or the atmosphere. Not true. What happens, instead, is that a rocket exerts a strong force on the gases, expelling them; and the gases exert an equal and opposite force *on the rocket*. It is this latter force that propels the rocket forward—the force exerted *on* the rocket *by* the gases (see Chapter-Opening Photo, page 75). Thus, a space vehicle is maneuvered in empty space by firing its rockets in the direction opposite to that in which it needs to accelerate. When the rocket pushes on the gases in one direction, the gases push back on the rocket in the opposite direction. Jet aircraft too accelerate because the gases they thrust out backwards exert a forward force on the engines (Newton's third law).

Consider how we walk. A person begins walking by pushing with the foot backward against the ground. The ground then exerts an equal and opposite force forward on the person (Fig. 4–11), and it is this force, *on the person*, that moves the person forward. (If you doubt this, try walking normally where there is no friction, such as on very smooth slippery ice.) In a similar way, a bird flies forward by exerting a backward force on the air, but it is the air pushing forward (Newton's third law) on the bird's wings that propels the bird forward.

CONCEPTUAL EXAMPLE 4–4 | **What exerts the force to move a car?** What makes a car go forward?

RESPONSE A common answer is that the engine makes the car move forward. But it is not so simple. The engine makes the wheels go around. But if the tires are on slick ice or wet mud, they just spin. Friction is needed. On firm ground, the tires push backward against the ground because of friction. By Newton's third law, the ground pushes on the tires in the opposite direction, accelerating the car forward.

We tend to associate forces with active objects such as humans, animals, engines, or a moving object like a hammer. It is often difficult to see how an inanimate object at rest, such as a wall or a desk, or the wall of an ice rink (Fig. 4–9), can exert a force. The explanation is that every material, no matter how hard, is elastic (springy) at least to some degree. A stretched rubber band can exert a force on a wad of paper and accelerate it to fly across the room. Other materials may not stretch as readily as rubber, but they do stretch or compress when a force is applied to them. And just as a stretched rubber band exerts a force, so does a stretched (or compressed) wall, desk, or car fender.

From the examples discussed above, we can see how important it is to remember *on* what object a given force is exerted and *by* what object that force is exerted. A force influences the motion of an object only when it is applied *on* that object. A force exerted *by* an object does not influence that same object; it only influences the other object *on* which it is exerted. Thus, to avoid confusion, the two prepositions *on* and *by* must always be used—and used with care.

One way to keep clear which force acts on which object is to use double subscripts. For example, the force exerted *on* the **P**erson *by* the **G**round as the person walks in Fig. 4–11 can be labeled $\vec{F}_{PG}$. And the force exerted on the ground by the person is $\vec{F}_{GP}$. By Newton's third law

$$\vec{F}_{GP} = -\vec{F}_{PG}. \quad (4\text{–}2)$$

$\vec{F}_{GP}$ and $\vec{F}_{PG}$ have the same magnitude (Newton's third law), and the minus sign reminds us that these two forces are in opposite directions.

Note carefully that the two forces shown in Fig. 4–11 act on different objects—to emphasize this we used slightly different colors for the vector arrows representing these forces. These two forces would never appear together in a sum of forces in Newton's second law, $\Sigma\vec{F} = m\vec{a}$. Why not? Because they act on different objects: $\vec{a}$ is the acceleration of one particular object, and $\Sigma\vec{F}$ must include *only* the forces on that *one* object.

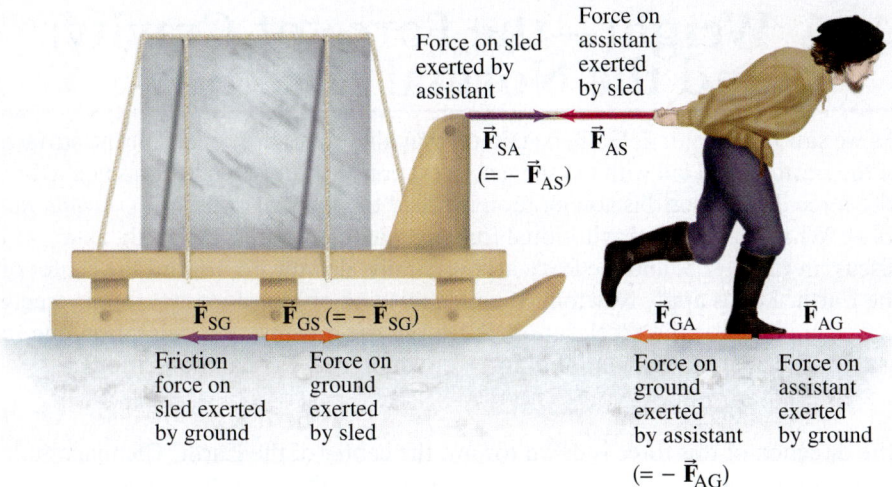

FIGURE 4–12 Example 4–5, showing only horizontal forces. Michelangelo has selected a fine block of marble for his next sculpture. Shown here is his assistant pulling it on a sled away from the quarry. Forces on the assistant are shown as red (magenta) arrows. Forces on the sled are purple arrows. Forces acting on the ground are orange arrows. Action–reaction forces that are equal and opposite are labeled by the same subscripts but reversed (such as $\vec{F}_{GA}$ and $\vec{F}_{AG}$) and are of different colors because they act on different objects.

CONCEPTUAL EXAMPLE 4–5 **Third law clarification.** Michelangelo's assistant has been assigned the task of moving a block of marble using a sled (Fig. 4–12). He says to his boss, "When I exert a forward force on the sled, the sled exerts an equal and opposite force backward. So how can I ever start it moving? No matter how hard I pull, the backward reaction force always equals my forward force, so the net force must be zero. I'll never be able to move this load." Is he correct?

RESPONSE No. Although it is true that the action and reaction forces are equal in magnitude, the assistant has forgotten that they are exerted on different objects. The forward ("action") force is exerted by the assistant on the sled (Fig. 4–12), whereas the backward "reaction" force is exerted by the sled on the assistant. To determine if the *assistant* moves or not, we must consider only the forces *on the assistant* and then apply $\Sigma \vec{F} = m\vec{a}$, where $\Sigma \vec{F}$ is the net force *on the assistant*, $\vec{a}$ is the acceleration of the assistant, and m is the assistant's mass. There are two forces on the assistant that affect his forward motion; they are shown as bright red (magenta) arrows in Figs. 4–12 and 4–13: they are (1) the horizontal force $\vec{F}_{AG}$ exerted on the assistant by the ground (the harder he pushes backward against the ground, the harder the ground pushes forward on him—Newton's third law), and (2) the force $\vec{F}_{AS}$ exerted on the assistant by the sled, pulling backward on him; see Fig. 4–13. If he pushes hard enough on the ground, the force on him exerted by the ground, $\vec{F}_{AG}$, will be larger than the sled pulling back, $\vec{F}_{AS}$, and the assistant accelerates forward (Newton's second law). The sled, on the other hand, accelerates forward when the force on *it* exerted by the assistant is greater than the frictional force exerted backward on it by the ground (that is, when $\vec{F}_{SA}$ has greater magnitude than $\vec{F}_{SG}$ in Fig. 4–12).

PROBLEM SOLVING
A study of Newton's second and third laws

FIGURE 4–13 Example 4–5. The horizontal forces on the assistant.

Using double subscripts to clarify Newton's third law can become cumbersome, and we won't usually use them in this way. We will usually use a single subscript referring to what exerts the force on the object being discussed. Nevertheless, if there is any confusion in your mind about a given force, go ahead and use two subscripts to identify *on* what object and *by* what object the force is exerted.

EXERCISE B Return to the first Chapter-Opening Question, page 75, and answer it again now. Try to explain why you may have answered differently the first time.

EXERCISE C A tennis ball collides head-on with a more massive baseball. (i) Which ball experiences the greater force of impact? (ii) Which experiences the greater acceleration during the impact? (iii) Which of Newton's laws are useful to obtain the correct answers?

EXERCISE D If you push on a heavy desk, does it always push back on you? (*a*) No. (*b*) Yes. (*c*) Not unless someone else also pushes on it. (*d*) Yes, if it is out in space. (*e*) A desk never pushes to start with.

SECTION 4–5 Newton's Third Law of Motion

4–6 Weight—the Force of Gravity; and the Normal Force

As we saw in Chapter 2, Galileo claimed that all objects dropped near the surface of the Earth would fall with the same acceleration, $\vec{g}$, if air resistance was negligible. The force that causes this acceleration is called the *force of gravity* or *gravitational force*. What exerts the gravitational force on an object? It is the Earth, as we will discuss in Chapter 5, and the force acts vertically[†] downward, toward the center of the Earth. Let us apply Newton's second law to an object of mass m falling freely due to gravity. For the acceleration, $\vec{a}$, we use the downward acceleration due to gravity, $\vec{g}$. Thus, the **gravitational force** on an object, $\vec{F}_G$, can be written as

$$\vec{F}_G = m\vec{g}. \qquad (4\text{–}3)$$

The direction of this force is down toward the center of the Earth. The magnitude of the force of gravity on an object, mg, is commonly called the object's **weight**.

In SI units, $g = 9.80 \text{ m/s}^2 = 9.80 \text{ N/kg}$,[‡] so the weight of a 1.00-kg mass on Earth is $1.00 \text{ kg} \times 9.80 \text{ m/s}^2 = 9.80 \text{ N}$. We will mainly be concerned with the weight of objects on Earth, but we note that on the Moon, on other planets, or in space, the weight of a given mass will be different than it is on Earth. For example, on the Moon the acceleration due to gravity is about one-sixth what it is on Earth, and a 1.0-kg mass weighs only 1.6 N. Although we will not use British units, we note that for practical purposes on the Earth, a mass of 1.0 kg weighs about 2.2 lb. (On the Moon, 1 kg weighs only about 0.4 lb.)

The force of gravity acts on an object when it is falling. When an object is at rest on the Earth, the gravitational force on it does not disappear, as we know if we weigh it on a spring scale. The same force, given by Eq. 4–3, continues to act. Why, then, doesn't the object move? From Newton's second law, the net force on an object that remains at rest is zero. There must be another force on the object to balance the gravitational force. For an object resting on a table, the table exerts this upward force; see Fig. 4–14a. The table is compressed slightly beneath the object, and due to its elasticity, it pushes up on the object as shown. The force exerted by the table is often called a **contact force**, since it occurs when two objects are in contact. (The force of your hand pushing on a cart is also a contact force.) When a contact force acts *perpendicular* to the common surface of contact, it is referred to as the **normal force** ("normal" means perpendicular); hence it is labeled $\vec{F}_N$ in Fig. 4–14a.

The two forces shown in Fig. 4–14a are both acting on the statue, which remains at rest, so the vector sum of these two forces must be zero (Newton's second law). Hence $\vec{F}_G$ and $\vec{F}_N$ must be of equal magnitude and in opposite directions. But they are *not* the equal and opposite forces spoken of in Newton's third law. The action and reaction forces of Newton's third law act on *different objects*, whereas the two forces shown in Fig. 4–14a act on the *same* object. For each of the forces shown in Fig. 4–14a, we can ask, "What is the reaction force?" The upward force $\vec{F}_N$ on the statue is exerted by the table. The reaction to this force is a force exerted by the statue downward on the table. It is shown in Fig. 4–14b, where it is labeled $\vec{F}'_N$. This force, $\vec{F}'_N$, exerted on the table by the statue, is the reaction force to $\vec{F}_N$ in accord with Newton's third law. What about the other force on the statue, the force of gravity $\vec{F}_G$ exerted by the Earth? Can you guess what the reaction is to this force? We will see in Chapter 5 that the reaction force is also a gravitational force, exerted on the Earth by the statue.

> **EXERCISE E** Return to the second Chapter-Opening Question, page 75, and answer it again now. Try to explain why you may have answered differently the first time.

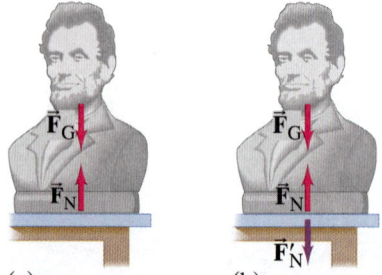

FIGURE 4–14 (a) The net force on an object at rest is zero according to Newton's second law. Therefore the downward force of gravity ($\vec{F}_G$) on an object at rest must be balanced by an upward force (the normal force $\vec{F}_N$) exerted by the table in this case. (b) $\vec{F}'_N$ is the force exerted on the table by the statue and is the reaction force to $\vec{F}_N$ by Newton's third law. ($\vec{F}'_N$ is shown in a different color to remind us it acts on a different object.) The reaction force to $\vec{F}_G$ is not shown.

CAUTION
*Weight and normal force are **not** action–reaction pairs*

[†]The concept of "vertical" is tied to gravity. The best definition of *vertical* is that it is the direction in which objects fall. A surface that is "horizontal," on the other hand, is a surface on which a round object won't start rolling: gravity has no effect. Horizontal is perpendicular to vertical.

[‡]Since $1 \text{ N} = 1 \text{ kg} \cdot \text{m/s}^2$ (Section 4–4), then $1 \text{ m/s}^2 = 1 \text{ N/kg}$.

EXAMPLE 4–6 **Weight, normal force, and a box.** A friend has given you a special gift, a box of mass 10.0 kg with a mystery surprise inside. The box is resting on the smooth (frictionless) horizontal surface of a table (Fig. 4–15a). (*a*) Determine the weight of the box and the normal force exerted on it by the table. (*b*) Now your friend pushes down on the box with a force of 40.0 N, as in Fig. 4–15b. Again determine the normal force exerted on the box by the table. (*c*) If your friend pulls upward on the box with a force of 40.0 N (Fig. 4–15c), what now is the normal force exerted on the box by the table?

APPROACH The box is at rest on the table, so the net force on the box in each case is zero (Newton's first or second law). The weight of the box has magnitude mg in all three cases.

SOLUTION (*a*) The weight of the box is $mg = (10.0\,\text{kg})(9.80\,\text{m/s}^2) = 98.0\,\text{N}$, and this force acts downward. The only other force on the box is the normal force exerted upward on it by the table, as shown in Fig. 4–15a. We choose the upward direction as the positive y direction; then the net force ΣF_y on the box is $\Sigma F_y = F_N - mg$; the minus sign means mg acts in the negative y direction (m and g are magnitudes). The box is at rest, so the net force on it must be zero (Newton's second law, $\Sigma F_y = ma_y$, and $a_y = 0$). Thus

$$\Sigma F_y = ma_y$$
$$F_N - mg = 0,$$

so we have

$$F_N = mg.$$

The normal force on the box, exerted by the table, is 98.0 N upward, and has magnitude equal to the box's weight.

(*b*) Your friend is pushing down on the box with a force of 40.0 N. So instead of only two forces acting on the box, now there are three forces acting on the box, as shown in Fig. 4–15b. The weight of the box is still $mg = 98.0\,\text{N}$. The net force is $\Sigma F_y = F_N - mg - 40.0\,\text{N}$, and is equal to zero because the box remains at rest ($a = 0$). Newton's second law gives

$$\Sigma F_y = F_N - mg - 40.0\,\text{N} = 0.$$

We solve this equation for the normal force:

$$F_N = mg + 40.0\,\text{N} = 98.0\,\text{N} + 40.0\,\text{N} = 138.0\,\text{N},$$

which is greater than in (*a*). The table pushes back with more force when a person pushes down on the box. The normal force is not always equal to the weight!

(*c*) The box's weight is still 98.0 N and acts downward. The force exerted by your friend and the normal force both act upward (positive direction), as shown in Fig. 4–15c. The box doesn't move since your friend's upward force is less than the weight. The net force, again set to zero in Newton's second law because $a = 0$, is

$$\Sigma F_y = F_N - mg + 40.0\,\text{N} = 0,$$

so

$$F_N = mg - 40.0\,\text{N} = 98.0\,\text{N} - 40.0\,\text{N} = 58.0\,\text{N}.$$

The table does not push against the full weight of the box because of the upward force exerted by your friend.

NOTE The weight of the box ($= mg$) does not change as a result of your friend's push or pull. Only the normal force is affected.

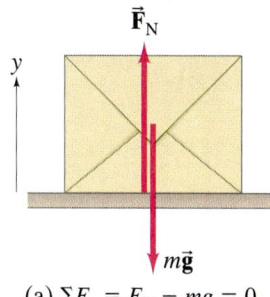

(a) $\Sigma F_y = F_N - mg = 0$

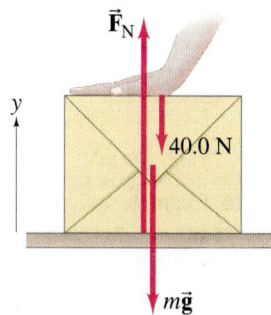

(b) $\Sigma F_y = F_N - mg - 40.0\,\text{N} = 0$

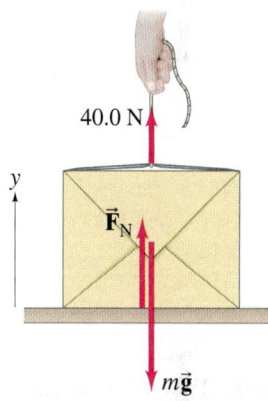

(c) $\Sigma F_y = F_N - mg + 40.0\,\text{N} = 0$

FIGURE 4–15 Example 4–6. (a) A 10-kg gift box is at rest on a table. (b) A person pushes down on the box with a force of 40.0 N. (c) A person pulls upward on the box with a force of 40.0 N. The forces are all assumed to act along a line; they are shown slightly displaced in order to be distinguishable. Only forces acting on the box are shown.

⚠ **CAUTION**
The normal force is not always equal to the weight

Recall that the normal force is elastic in origin (the table in Fig. 4–15 sags slightly under the weight of the box). The normal force in Example 4–6 is vertical, perpendicular to the horizontal table. The normal force is not always vertical, however. When you push against a wall, for example, the normal force with which the wall pushes back on you is horizontal (Fig. 4–9). For an object on a plane inclined at an angle to the horizontal, such as a skier or car on a hill, the normal force acts perpendicular to the plane and so is not vertical.

⚠ **CAUTION**
The normal force, $\vec{F}_N$, is not necessarily vertical

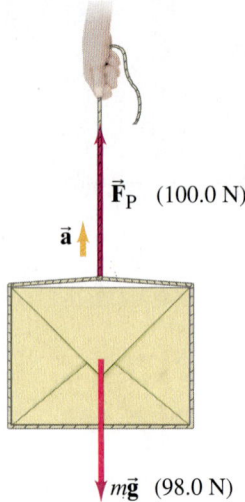

FIGURE 4–16 Example 4–7. The box accelerates upward because $F_P > mg$.

EXAMPLE 4–7 Accelerating the box. What happens when a person pulls upward on the box in Example 4–6c with a force equal to, or greater than, the box's weight? For example, let $F_P = 100.0$ N (Fig. 4–16) rather than the 40.0 N shown in Fig. 4–15c.

APPROACH We can start just as in Example 4–6, but be ready for a surprise.

SOLUTION The net force on the box is

$$\Sigma F_y = F_N - mg + F_P$$
$$= F_N - 98.0\text{ N} + 100.0\text{ N},$$

and if we set this equal to zero (thinking the acceleration might be zero), we would get $F_N = -2.0$ N. This is nonsense, since the negative sign implies F_N points downward, and the table surely cannot *pull* down on the box (unless there's glue on the table). The least F_N can be is zero, which it will be in this case. What really happens here is that the box accelerates upward ($a \neq 0$) because the net force is not zero. The net force (setting the normal force $F_N = 0$) is

$$\Sigma F_y = F_P - mg = 100.0\text{ N} - 98.0\text{ N}$$
$$= 2.0\text{ N}$$

upward. See Fig. 4–16. We apply Newton's second law and see that the box moves upward with an acceleration

$$a_y = \frac{\Sigma F_y}{m} = \frac{2.0\text{ N}}{10.0\text{ kg}}$$
$$= 0.20\text{ m/s}^2.$$

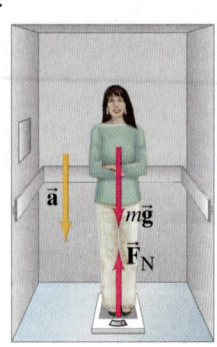

FIGURE 4–17 Example 4–8. The acceleration vector is shown in gold to distinguish it from the red force vectors.

EXAMPLE 4–8 Apparent weight loss. A 65-kg woman descends in an elevator that briefly accelerates at $0.20g$ downward. She stands on a scale that reads in kg. (*a*) During this acceleration, what is her weight and what does the scale read? (*b*) What does the scale read when the elevator descends at a constant speed of 2.0 m/s?

APPROACH Figure 4–17 shows all the forces that act on the woman (and *only* those that act on her). The direction of the acceleration is downward, so we choose the positive direction as down (this is the opposite choice from Examples 4–6 and 4–7).

SOLUTION (*a*) From Newton's second law,

$$\Sigma F = ma$$
$$mg - F_N = m(0.20g).$$

We solve for F_N:

$$F_N = mg - 0.20mg$$
$$= 0.80mg,$$

and it acts upward. The normal force $\vec{F}_N$ is the force the scale exerts on the person, and is equal and opposite to the force she exerts on the scale: $F'_N = 0.80mg$ downward. Her weight (force of gravity on her) is still $mg = (65\text{ kg})(9.8\text{ m/s}^2) = 640$ N. But the scale, needing to exert a force of only $0.80mg$, will give a reading of $0.80m = 52$ kg.

(*b*) Now there is no acceleration, $a = 0$, so by Newton's second law, $mg - F_N = 0$ and $F_N = mg$. The scale reads her true mass of 65 kg.

NOTE The scale in (*a*) gives a reading of 52 kg (as an "apparent mass"), but her mass doesn't change as a result of the acceleration: it stays at 65 kg.

86 CHAPTER 4 Dynamics: Newton's Laws of Motion

4–7 Solving Problems with Newton's Laws: Free-Body Diagrams

Newton's second law tells us that the acceleration of an object is proportional to the *net force* acting on the object. The **net force**, as mentioned earlier, is the *vector sum* of all forces acting on the object. Indeed, extensive experiments have shown that forces do add together as vectors precisely according to the rules we developed in Chapter 3. For example, in Fig. 4–18, two forces of equal magnitude (100 N each) are shown acting on an object at right angles to each other. Intuitively, we can see that the object will start moving at a 45° angle and thus the net force acts at a 45° angle. This is just what the rules of vector addition give. From the theorem of Pythagoras, the magnitude of the resultant force is $F_R = \sqrt{(100\,\text{N})^2 + (100\,\text{N})^2} = 141\,\text{N}$.

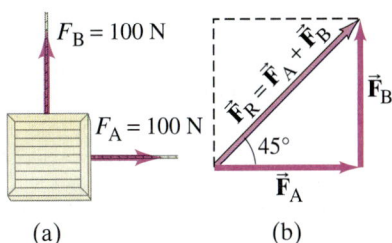

FIGURE 4–18 (a) Two horizontal forces, $\vec{F}_A$ and $\vec{F}_B$, exerted by workers A and B, act on a crate (we are looking down from above). (b) The sum, or resultant, of $\vec{F}_A$ and $\vec{F}_B$ is $\vec{F}_R$.

EXAMPLE 4–9 Adding force vectors. Calculate the sum of the two forces exerted on the boat by workers A and B in Fig. 4–19a.

APPROACH We add force vectors like any other vectors as described in Chapter 3. The first step is to choose an *xy* coordinate system (see Fig. 4–19a), and then resolve vectors into their components.

SOLUTION The two force vectors are shown resolved into components in Fig. 4–19b. We add the forces using the method of components. The components of $\vec{F}_A$ are

$$F_{Ax} = F_A \cos 45.0° = (40.0\,\text{N})(0.707) = 28.3\,\text{N},$$
$$F_{Ay} = F_A \sin 45.0° = (40.0\,\text{N})(0.707) = 28.3\,\text{N}.$$

The components of $\vec{F}_B$ are

$$F_{Bx} = +F_B \cos 37.0° = +(30.0\,\text{N})(0.799) = +24.0\,\text{N},$$
$$F_{By} = -F_B \sin 37.0° = -(30.0\,\text{N})(0.602) = -18.1\,\text{N}.$$

F_{By} is negative because it points along the negative y axis. The components of the resultant force are (see Fig. 4–19c):

$$F_{Rx} = F_{Ax} + F_{Bx} = 28.3\,\text{N} + 24.0\,\text{N} = 52.3\,\text{N},$$
$$F_{Ry} = F_{Ay} + F_{By} = 28.3\,\text{N} - 18.1\,\text{N} = 10.2\,\text{N}.$$

To find the magnitude of the resultant force, we use the Pythagorean theorem,

$$F_R = \sqrt{F_{Rx}^2 + F_{Ry}^2} = \sqrt{(52.3)^2 + (10.2)^2}\,\text{N} = 53.3\,\text{N}.$$

The only remaining question is the angle θ that the net force $\vec{F}_R$ makes with the x axis. We use:

$$\tan\theta = \frac{F_{Ry}}{F_{Rx}} = \frac{10.2\,\text{N}}{52.3\,\text{N}} = 0.195,$$

and $\tan^{-1}(0.195) = 11.0°$. The net force on the boat has magnitude 53.3 N and acts at an 11.0° angle to the x axis.

FIGURE 4–19 Example 4–9: Two force vectors act on a boat.

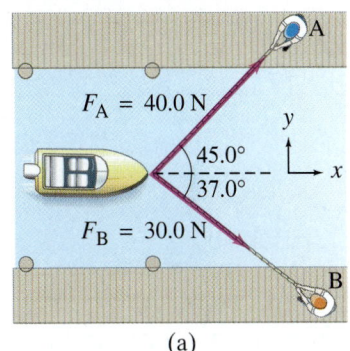

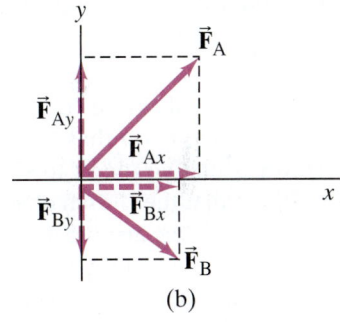

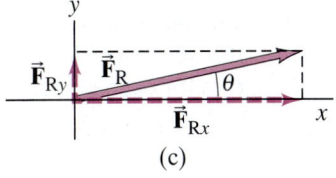

When solving problems involving Newton's laws and force, it is very important to draw a diagram showing all the forces acting *on* each object involved. Such a diagram is called a **free-body diagram**, or **force diagram**: choose one object, and draw an arrow to represent each force acting on it. Include *every* force acting on that object. Do not show forces that the chosen object exerts on *other* objects. To help you identify each and every force that is exerted on your chosen object, ask yourself what other objects could exert a force on it. If your problem involves more than one object, a separate free-body diagram is needed for each object. For now, the likely forces that could be acting are *gravity* and *contact forces* (one object pushing or pulling another, normal force, friction). Later we will consider other types of force such as buoyancy, fluid pressure, and electric and magnetic forces.

PROBLEM SOLVING
Free-body diagram

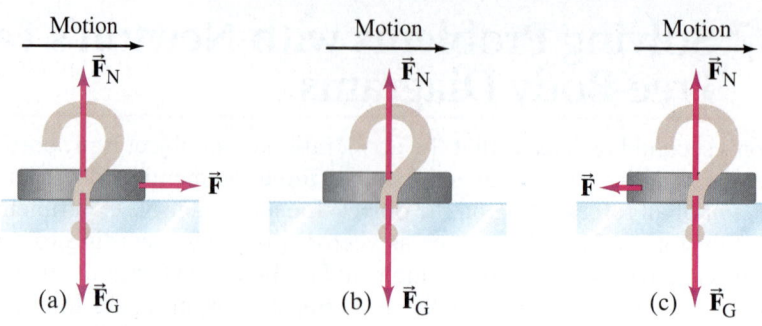

FIGURE 4–20 Example 4–10. Which is the correct free-body diagram for a hockey puck sliding across frictionless ice?

CONCEPTUAL EXAMPLE 4–10 **The hockey puck.** A hockey puck is sliding at constant velocity across a flat horizontal ice surface that is assumed to be frictionless. Which of the sketches in Fig. 4–20 is the correct free-body diagram for this puck? What would your answer be if the puck slowed down?

RESPONSE Did you choose (*a*)? If so, can you answer the question: what exerts the horizontal force labeled $\vec{F}$ on the puck? If you say that it is the force needed to maintain the motion, ask yourself: what exerts this force? Remember that another object must exert any force—and there simply isn't any possibility here. Therefore, (*a*) is wrong. Besides, the force $\vec{F}$ in Fig. 4–20a would give rise to an acceleration by Newton's second law. It is (*b*) that is correct. No net force acts on the puck, and the puck slides at constant velocity across the ice.

In the real world, where even smooth ice exerts at least a tiny friction force, then (*c*) is the correct answer. The tiny friction force is in the direction opposite to the motion, and the puck's velocity decreases, even if very slowly.

PROBLEM SOLVING

Newton's Laws; Free-Body Diagrams

1. **Draw a sketch** of the situation, after carefully reading the Problem at least twice.

2. Consider only one object (at a time), and draw a **free-body diagram** for that object, showing *all* the forces acting *on* that object. Include any unknown forces that you have to solve for. Do not show any forces that the chosen object exerts on other objects.

   Draw the arrow for each force vector reasonably accurately for direction and magnitude. Label each force acting on the object, including forces you must solve for, according to its source (gravity, person, friction, and so on).

   If several objects are involved, draw a free-body diagram for each object *separately*. For each object, show all the forces acting *on that object* (and *only* forces acting on that object). For each (and every) force, you must be clear about: *on* what object that force acts, and *by* what object that force is exerted. Only forces acting *on* a given object can be included in $\Sigma \vec{F} = m\vec{a}$ for that object.

3. Newton's second law involves vectors, and it is usually important to **resolve vectors** into components. **Choose** x and y **axes** in a way that simplifies the calculation. For example, it often saves work if you choose one coordinate axis to be in the direction of the acceleration (if known).

4. For each object, **apply Newton's second law** to the x and y components separately. That is, the x component of the net force on that object is related to the x component of that object's acceleration: $\Sigma F_x = ma_x$, and similarly for the y direction.

5. **Solve** the equation or equations for the unknown(s). Put in numerical values only at the end, and keep track of units.

This Problem Solving Strategy should not be considered a prescription. Rather it is a summary of things to do that will start you thinking and getting involved in the problem at hand.

When we are concerned only about translational motion, all the forces on a given object can be drawn as acting at the center of the object, thus treating the object as a *point particle*. However, for problems involving rotation or statics, the place *where* each force acts is also important, as we shall see in Chapters 8 and 9.

In the Examples in this Section, we assume that all surfaces are very smooth so that friction can be ignored. (Friction, and Examples using it, are discussed in Section 4–8.)

⚠ **CAUTION**
Treating an object as a particle

EXAMPLE 4–11 **Pulling the mystery box.** Suppose a friend asks to examine the 10.0-kg box you were given (Example 4–6, Fig. 4–15), hoping to guess what is inside; and you respond, "Sure, pull the box over to you." She then pulls the box by the attached cord, as shown in Fig. 4–21a, along the smooth surface of the table. The magnitude of the force exerted by the person is $F_P = 40.0$ N, and it is exerted at a 30.0° angle as shown. Calculate (a) the acceleration of the box, and (b) the magnitude of the upward force F_N exerted by the table on the box. Assume that friction can be neglected.

APPROACH We follow the Problem Solving Strategy on the previous page.

SOLUTION

1. **Draw a sketch**: The situation is shown in Fig. 4–21a; it shows the box and the force applied by the person, F_P.
2. **Free-body diagram**: Figure 4–21b shows the free-body diagram of the box. To draw it correctly, we show *all* the forces acting on the box and *only* the forces acting on the box. They are: the force of gravity $m\vec{g}$; the normal force exerted by the table $\vec{F}_N$; and the force exerted by the person $\vec{F}_P$. We are interested only in translational motion, so we can show the three forces acting at a point, Fig. 4–21c.
3. **Choose axes and resolve vectors**: We expect the motion to be horizontal, so we choose the x axis horizontal and the y axis vertical. The pull of 40.0 N has components

$$F_{Px} = (40.0\,\text{N})(\cos 30.0°) = (40.0\,\text{N})(0.866) = 34.6\,\text{N},$$
$$F_{Py} = (40.0\,\text{N})(\sin 30.0°) = (40.0\,\text{N})(0.500) = 20.0\,\text{N}.$$

In the horizontal (x) direction, $\vec{F}_N$ and $m\vec{g}$ have zero components. Thus the horizontal component of the net force is F_{Px}.

4. (a) **Apply Newton's second law** to get the x component of the acceleration:

$$F_{Px} = ma_x.$$

5. (a) **Solve**:

$$a_x = \frac{F_{Px}}{m} = \frac{(34.6\,\text{N})}{(10.0\,\text{kg})} = 3.46\,\text{m/s}^2.$$

The acceleration of the box is 3.46 m/s² to the right.

(b) Next we want to find F_N.

4′. (b) **Apply Newton's second law** to the vertical (y) direction, with upward as positive:

$$\Sigma F_y = ma_y$$
$$F_N - mg + F_{Py} = ma_y.$$

5′. (b) **Solve**: We have $mg = (10.0\,\text{kg})(9.80\,\text{m/s}^2) = 98.0\,\text{N}$ and, from point 3 above, $F_{Py} = 20.0\,\text{N}$. Furthermore, since $F_{Py} < mg$, the box does not move vertically, so $a_y = 0$. Thus

$$F_N - 98.0\,\text{N} + 20.0\,\text{N} = 0,$$

so

$$F_N = 78.0\,\text{N}.$$

NOTE F_N is less than mg: the table does not push against the full weight of the box because part of the pull exerted by the person is in the upward direction.

EXERCISE F A 10.0-kg box is dragged on a horizontal frictionless surface by a horizontal force of 10.0 N. If the applied force is doubled, the normal force on the box will (*a*) increase; (*b*) remain the same; (*c*) decrease.

Tension in a Flexible Cord

When a flexible cord pulls on an object, the cord is said to be under **tension**, and the force it exerts on the object is the tension F_T. If the cord has negligible mass, the force exerted at one end is transmitted undiminished to each adjacent piece of cord along the entire length to the other end. Why? Because $\Sigma \vec{F} = m\vec{a} = 0$ for the cord if the cord's mass m is zero (or negligible) no matter what $\vec{a}$ is. Hence the forces pulling on the cord at its two ends must add up to zero (F_T and $-F_T$). Note that flexible cords and strings can only pull. They can't push because they bend.

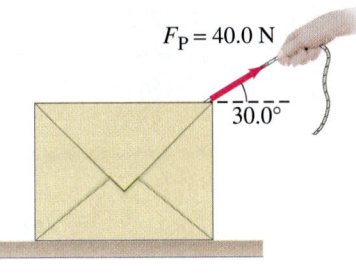

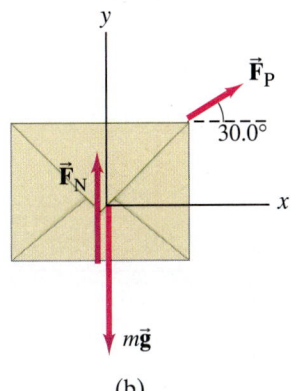

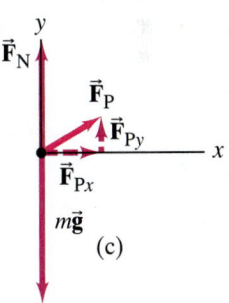

FIGURE 4–21 (a) Pulling the box, Example 4–11; (b) is the free-body diagram for the box, and (c) is the free-body diagram considering all the forces to act at a point (translational motion only, which is what we have here).

PROBLEM SOLVING
Cords can pull but can't push; tension exists throughout a taut cord

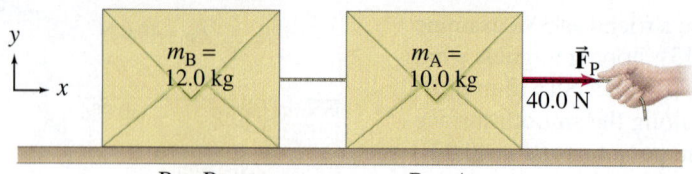

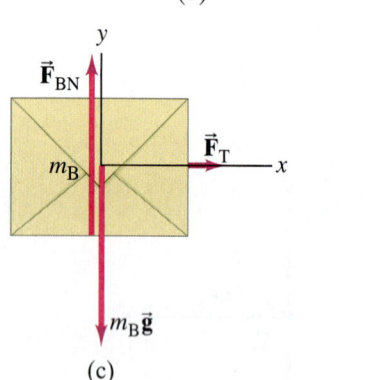

FIGURE 4–22 Example 4–12. (a) Two boxes, A and B, are connected by a cord. A person pulls horizontally on box A with force $F_P = 40.0$ N. (b) Free-body diagram for box A. (c) Free-body diagram for box B.

Our next Example involves two boxes connected by a cord. We can refer to this group of objects as a system. A **system** is any group of one or more objects we choose to consider and study.

EXAMPLE 4–12 Two boxes connected by a cord. Two boxes, A and B, are connected by a lightweight cord and are resting on a smooth (frictionless) table. The boxes have masses of 12.0 kg and 10.0 kg. A horizontal force F_P of 40.0 N is applied to the 10.0-kg box, as shown in Fig. 4–22a. Find (a) the acceleration of each box, and (b) the tension in the cord connecting the boxes.

APPROACH We streamline our approach by not listing each step. We have two boxes so we draw a free-body diagram for each. To draw them correctly, we must consider the forces on *each* box by itself, so that Newton's second law can be applied to each. The person exerts a force F_P on box A. Box A exerts a force F_T on the connecting cord, and the cord exerts an opposite but equal magnitude force F_T back on box A (Newton's third law). The two horizontal forces on box A are shown in Fig. 4–22b, along with the force of gravity $m_A \vec{g}$ downward and the normal force $\vec{F}_{AN}$ exerted upward by the table. The cord is light, so we neglect its mass. The tension at each end of the cord is thus the same. Hence the cord exerts a force F_T on the second box. Figure 4–22c shows the forces on box B, which are $\vec{F}_T$, $m_B \vec{g}$, and the normal force $\vec{F}_{BN}$. There will be only horizontal motion. We take the positive x axis to the right.

SOLUTION (a) We apply $\Sigma F_x = ma_x$ to box A:

$$\Sigma F_x = F_P - F_T = m_A a_A. \qquad \text{[box A]}$$

For box B, the only horizontal force is F_T, so

$$\Sigma F_x = F_T = m_B a_B. \qquad \text{[box B]}$$

The boxes are connected, and if the cord remains taut and doesn't stretch, then the two boxes will have the same acceleration a. Thus $a_A = a_B = a$. We are given $m_A = 10.0$ kg and $m_B = 12.0$ kg. We can add the two equations above to eliminate an unknown (F_T) and obtain

$$(m_A + m_B)a = F_P - F_T + F_T = F_P$$

or

$$a = \frac{F_P}{m_A + m_B} = \frac{40.0 \text{ N}}{22.0 \text{ kg}} = 1.82 \text{ m/s}^2.$$

This is what we sought.

(b) From the equation for box B above $(F_T = m_B a_B)$, the tension in the cord is

$$F_T = m_B a = (12.0 \text{ kg})(1.82 \text{ m/s}^2) = 21.8 \text{ N}.$$

Thus, $F_T < F_P (= 40.0 \text{ N})$, as we expect, since F_T acts to accelerate only m_B.

Alternate Solution to (a) We would have obtained the same result had we considered a single system, of mass $m_A + m_B$, acted on by a net horizontal force equal to F_P. (The tension forces F_T would then be considered internal to the system as a whole, and summed together would make zero contribution to the net force on the *whole* system.)

NOTE It might be tempting to say that the force the person exerts, F_P, acts not only on box A but also on box B. It doesn't. F_P acts only on box A. It affects box B via the tension in the cord, F_T, which acts on box B and accelerates it. (You could look at it this way: $F_T < F_P$ because F_P accelerates *both* boxes whereas F_T only accelerates box B.)

> ⚠ **CAUTION**
> *For any object, use only the forces on that object in calculating $\Sigma F = ma$*

EXAMPLE 4–13 **Elevator and counterweight (Atwood machine).** A system of two objects suspended over a pulley by a flexible cable, as shown in Fig. 4–23a, is sometimes referred to as an *Atwood machine*. Consider the real-life application of an elevator (m_E) and its counterweight (m_C). To minimize the work done by the motor to raise and lower the elevator safely, m_E and m_C are made similar in mass. We leave the motor out of the system for this calculation, and assume that the cable's mass is negligible and that the mass of the pulley, as well as any friction, is small and ignorable. These assumptions ensure that the tension F_T in the cable has the same magnitude on both sides of the pulley. Let the mass of the counterweight be $m_C = 1000$ kg. Assume the mass of the empty elevator is 850 kg, and its mass when carrying four passengers is $m_E = 1150$ kg. For the latter case ($m_E = 1150$ kg), calculate (a) the acceleration of the elevator and (b) the tension in the cable.

PHYSICS APPLIED
Elevator (as Atwood machine)

APPROACH Again we have two objects, and we will need to apply Newton's second law to each of them separately. Each mass has two forces acting on it: gravity downward and the cable tension pulling upward, $\vec{F}_T$. Figures 4–23b and c show the free-body diagrams for the elevator (m_E) and for the counterweight (m_C). The elevator, being the heavier, will accelerate downward, whereas the counterweight will accelerate upward. The magnitudes of their accelerations will be equal (we assume the cable is massless and doesn't stretch). For the counterweight, $m_C g = (1000 \text{ kg})(9.80 \text{ m/s}^2) = 9800$ N, so F_T must be greater than 9800 N (in order that m_C will accelerate upward). For the elevator, $m_E g = (1150 \text{ kg})(9.80 \text{ m/s}^2) = 11{,}300$ N, which must have greater magnitude than F_T so that m_E accelerates downward. Thus our calculation must give F_T between 9800 N and 11,300 N.

SOLUTION (a) To find F_T as well as the acceleration a, we apply Newton's second law, $\Sigma F = ma$, to each object. We take upward as the positive y direction for both objects. With this choice of axes, $a_C = a$ because m_C accelerates upward, and $a_E = -a$ because m_E accelerates downward. Thus

$$F_T - m_E g = m_E a_E = -m_E a$$
$$F_T - m_C g = m_C a_C = +m_C a.$$

We can subtract the first equation from the second to get

$$(m_E - m_C)g = (m_E + m_C)a,$$

where a is now the only unknown. We solve this for a:

$$a = \frac{m_E - m_C}{m_E + m_C}g = \frac{1150 \text{ kg} - 1000 \text{ kg}}{1150 \text{ kg} + 1000 \text{ kg}}g = 0.070g = 0.68 \text{ m/s}^2.$$

The elevator (m_E) accelerates downward (and the counterweight m_C upward) at $a = 0.070g = 0.68 \text{ m/s}^2$.

(b) The tension in the cable F_T can be obtained from either of the two $\Sigma F = ma$ equations at the start of our solution, setting $a = 0.070g = 0.68 \text{ m/s}^2$:

$$F_T = m_E g - m_E a = m_E(g - a)$$
$$= 1150 \text{ kg } (9.80 \text{ m/s}^2 - 0.68 \text{ m/s}^2) = 10{,}500 \text{ N},$$

or

$$F_T = m_C g + m_C a = m_C(g + a)$$
$$= 1000 \text{ kg } (9.80 \text{ m/s}^2 + 0.68 \text{ m/s}^2) = 10{,}500 \text{ N},$$

which are consistent. As predicted, our result lies between 9800 N and 11,300 N.

NOTE We can check our equation for the acceleration a in this Example by noting that if the masses were equal ($m_E = m_C$), then our equation above for a would give $a = 0$, as we should expect. Also, if one of the masses is zero (say, $m_C = 0$), then the other mass ($m_E \neq 0$) would be predicted by our equation to accelerate at $a = g$, again as expected.

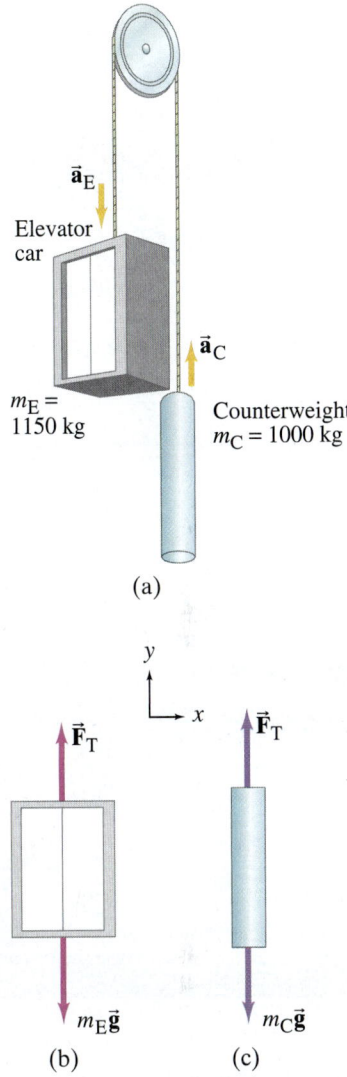

FIGURE 4–23 Example 4–13. (a) Atwood machine in the form of an elevator–counterweight system. (b) and (c) Free-body diagrams for the two objects.

PROBLEM SOLVING
Check your result by seeing if it works in situations where the answer is easily guessed

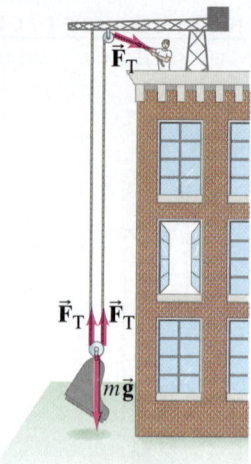

FIGURE 4–24 Example 4–14.

CONCEPTUAL EXAMPLE 4–14 | **The advantage of a pulley.** A mover is trying to lift a piano (slowly) up to a second-story apartment (Fig. 4–24). He is using a rope looped over two pulleys as shown. What force must he exert on the rope to slowly lift the piano's 1600-N weight?

RESPONSE The magnitude of the tension force F_T within the rope is the same at any point along the rope if we assume we can ignore its mass. First notice the forces acting on the lower pulley at the piano. The weight of the piano ($= mg$) pulls down on the pulley. The tension in the rope, looped through this pulley, pulls up *twice*, once on each side of the pulley. Let us apply Newton's second law to the pulley–piano combination (of mass m), choosing the upward direction as positive:

$$2F_T - mg = ma.$$

To move the piano with constant speed (set $a = 0$ in this equation) thus requires a tension in the rope, and hence a pull on the rope, of $F_T = mg/2$. The piano mover can exert a force equal to half the piano's weight.

NOTE We say the pulley has given a **mechanical advantage** of 2, since without the pulley the mover would have to exert twice the force.

PHYSICS APPLIED
Accelerometer

FIGURE 4–25 Example 4–15.

(a)

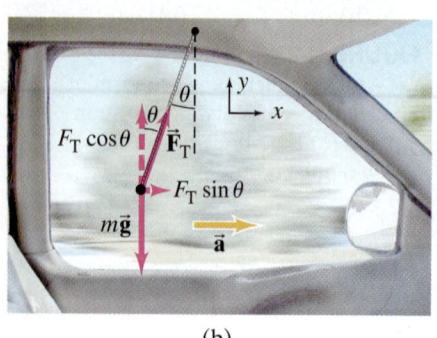

(b)

EXAMPLE 4–15 **Accelerometer.** A small mass m hangs from a thin string and can swing like a pendulum. You attach it above the window of your car as shown in Fig. 4–25a. When the car is at rest, the string hangs vertically. What angle θ does the string make (a) when the car accelerates at a constant $a = 1.20 \text{ m/s}^2$, and (b) when the car moves at constant velocity, $v = 90$ km/h?

APPROACH The free-body diagram of Fig. 4–25b shows the pendulum at some angle θ relative to the vertical, and the forces on it: $m\vec{g}$ downward, and the tension $\vec{F}_T$ in the cord (including its components). These forces do not add up to zero if $\theta \neq 0$; and since we have an acceleration a, we expect $\theta \neq 0$.

SOLUTION (a) The acceleration $a = 1.20 \text{ m/s}^2$ is horizontal ($= a_x$), and the only horizontal force is the x component of $\vec{F}_T$, $F_T \sin\theta$ (Fig. 4–25b). Then from Newton's second law,

$$ma = F_T \sin\theta.$$

The vertical component of Newton's second law gives, since $a_y = 0$,

$$0 = F_T \cos\theta - mg.$$

So

$$mg = F_T \cos\theta.$$

Dividing these two equations, we obtain

$$\tan\theta = \frac{F_T \sin\theta}{F_T \cos\theta} = \frac{ma}{mg} = \frac{a}{g}$$

or

$$\tan\theta = \frac{1.20 \text{ m/s}^2}{9.80 \text{ m/s}^2}$$

$$= 0.122,$$

so

$$\theta = 7.0°.$$

(b) The velocity is constant, so $a = 0$ and $\tan\theta = 0$. Hence the pendulum hangs vertically ($\theta = 0°$).

NOTE This simple device is an **accelerometer**—it can be used to determine acceleration, by measuring the angle θ.

4–8 Problems Involving Friction, Inclines

Friction

Until now we have ignored friction, but it must be taken into account in most practical situations. Friction exists between two solid surfaces because even the smoothest looking surface is quite rough on a microscopic scale, Fig. 4–26. When we try to slide an object across a surface, these microscopic bumps impede the motion. Exactly what is happening at the microscopic level is not yet fully understood. It is thought that the atoms on a bump of one surface may come so close to the atoms of the other surface that attractive electric forces between the atoms could "bond" as a tiny weld between the two surfaces. Sliding an object across a surface is often jerky, perhaps due to the making and breaking of these bonds. Even when a round object rolls across a surface, there is still some friction, called *rolling friction*, although it is generally much less than when an object slides across a surface. We focus now on sliding friction, which is usually called **kinetic friction** (*kinetic* is from the Greek for "moving").

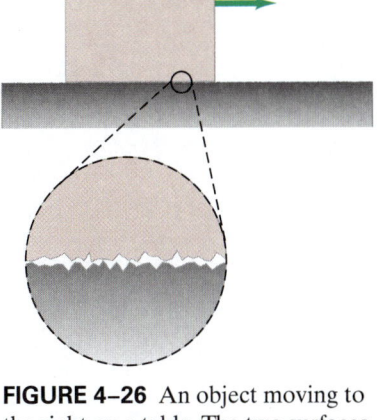

FIGURE 4–26 An object moving to the right on a table. The two surfaces in contact are assumed smooth, but are rough on a microscopic scale.

When an object slides along a rough surface, the force of kinetic friction acts opposite to the direction of the object's velocity. The magnitude of the force of kinetic friction depends on the nature of the two sliding surfaces. For given surfaces, experiment shows that the friction force is approximately proportional to the *normal force* between the two surfaces, which is the force that either object exerts on the other and is perpendicular to their common surface of contact (see Fig. 4–27). The force of friction between hard surfaces in many cases depends very little on the total surface area of contact; that is, the friction force on this book is roughly the same whether it is being slid across a table on its wide face or on its spine, assuming the surfaces have the same smoothness. We consider a simple model of friction in which we make this assumption that the friction force is independent of area. Then we write the proportionality between the magnitudes of the friction force F_{fr} and the normal force F_N as an equation by inserting a constant of proportionality, μ_k:

$$F_{fr} = \mu_k F_N. \qquad \text{[kinetic friction]}$$

FIGURE 4–27 When an object is pulled along a surface by an applied force ($\vec{F}_A$), the force of friction $\vec{F}_{fr}$ opposes the motion. The magnitude of $\vec{F}_{fr}$ is proportional to the magnitude of the normal force (F_N).

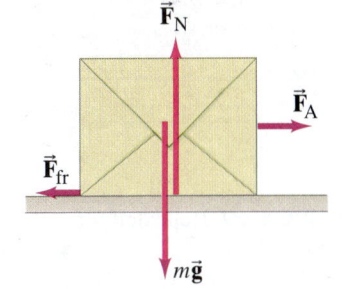

This relation is not a fundamental law; it is an experimental relation between the magnitude of the friction force F_{fr}, which acts parallel to the two surfaces, and the magnitude of the normal force F_N, which acts perpendicular to the surfaces. It is *not* a vector equation since the two forces have different directions, perpendicular to one another. The term μ_k is called the *coefficient of kinetic friction*, and its value depends on the nature of the two surfaces. Measured values for a variety of surfaces are given in Table 4–2. These are only approximate, however, since μ depends on whether the surfaces are wet or dry, on how much they have been sanded or rubbed, if any burrs remain, and other such factors. But μ_k (which has no units) is roughly independent of the sliding speed, as well as the area in contact.

⚠️ **CAUTION**

TABLE 4–2 Coefficients of Friction[†]

Surfaces	Coefficient of Static Friction, μ_s	Coefficient of Kinetic Friction, μ_k
Wood on wood	0.4	0.2
Ice on ice	0.1	0.03
Metal on metal (lubricated)	0.15	0.07
Steel on steel (unlubricated)	0.7	0.6
Rubber on dry concrete	1.0	0.8
Rubber on wet concrete	0.7	0.5
Rubber on other solid surfaces	1–4	1
Teflon® on Teflon in air	0.04	0.04
Teflon on steel in air	0.04	0.04
Lubricated ball bearings	<0.01	<0.01
Synovial joints (in human limbs)	0.01	0.01

[†] Values are approximate and intended only as a guide.

What we have been discussing up to now is *kinetic friction*, when one object slides over another. There is also **static friction**, which refers to a force parallel to the two surfaces that can arise even when they are not sliding. Suppose an object such as a desk is resting on a horizontal floor. If no horizontal force is exerted on the desk, there also is no friction force. But now suppose you try to push the desk, and it doesn't move. You are exerting a horizontal force, but the desk isn't moving, so there must be another force on the desk keeping it from moving (the net force is zero on an object at rest). This is the force of *static friction* exerted by the floor on the desk. If you push with a greater force without moving the desk, the force of static friction also has increased. If you push hard enough, the desk will eventually start to move, and kinetic friction takes over. At this point, you have exceeded the maximum force of static friction, which is given by $(F_{fr})_{max} = \mu_s F_N$, where μ_s is the *coefficient of static friction* (Table 4–2). Because the force of static friction can vary from zero to this maximum value, we write

$$F_{fr} \leq \mu_s F_N. \qquad \text{[static friction]}$$

You may have noticed that it is often easier to keep a heavy object sliding than it is to start it sliding in the first place. This is consistent with μ_s generally being greater than μ_k (see Table 4–2).

EXAMPLE 4–16 Friction: static and kinetic. Our 10.0-kg mystery box rests on a horizontal floor. The coefficient of static friction is $\mu_s = 0.40$ and the coefficient of kinetic friction is $\mu_k = 0.30$. Determine the force of friction, F_{fr}, acting on the box if a horizontal applied force F_A is exerted on it of magnitude: (a) 0, (b) 10 N, (c) 20 N, (d) 38 N, and (e) 40 N.

APPROACH We don't know, right off, if we are dealing with static friction or kinetic friction, nor if the box remains at rest or accelerates. We need to draw a free-body diagram, and then determine in each case whether or not the box will move: the box starts moving if F_A is greater than the maximum static friction force (Newton's second law). The forces on the box are gravity $m\vec{g}$, the normal force exerted by the floor $\vec{F}_N$, the horizontal applied force $\vec{F}_A$, and the friction force $\vec{F}_{fr}$, as shown in Fig. 4–27.

SOLUTION The free-body diagram of the box is shown in Fig. 4–27. In the vertical direction there is no motion, so Newton's second law in the vertical direction gives $\Sigma F_y = ma_y = 0$, which tells us $F_N - mg = 0$. Hence the normal force is

$$F_N = mg = (10.0\ \text{kg})(9.80\ \text{m/s}^2) = 98.0\ \text{N}.$$

(a) Because $F_A = 0$ in this first case, the box doesn't move, and $F_{fr} = 0$.
(b) The force of static friction will oppose any applied force up to a maximum of

$$\mu_s F_N = (0.40)(98.0\ \text{N}) = 39\ \text{N}.$$

When the applied force is $F_A = 10\ \text{N}$, the box will not move. Newton's second law gives $\Sigma F_x = F_A - F_{fr} = 0$, so $F_{fr} = 10\ \text{N}$.
(c) An applied force of 20 N is also not sufficient to move the box. Thus $F_{fr} = 20\ \text{N}$ to balance the applied force.
(d) The applied force of 38 N is still not quite large enough to move the box; so the friction force has now increased to 38 N to keep the box at rest.
(e) A force of 40 N will start the box moving since it exceeds the maximum force of static friction, $\mu_s F_N = (0.40)(98\ \text{N}) = 39\ \text{N}$. Instead of static friction, we now have kinetic friction, and its magnitude is

$$F_{fr} = \mu_k F_N = (0.30)(98.0\ \text{N}) = 29\ \text{N}.$$

There is now a net (horizontal) force on the box of magnitude $F = 40\ \text{N} - 29\ \text{N} = 11\ \text{N}$, so the box will accelerate at a rate

$$a_x = \frac{\Sigma F}{m} = \frac{11\ \text{N}}{10.0\ \text{kg}} = 1.1\ \text{m/s}^2$$

as long as the applied force is 40 N. Figure 4–28 shows a graph that summarizes this Example.

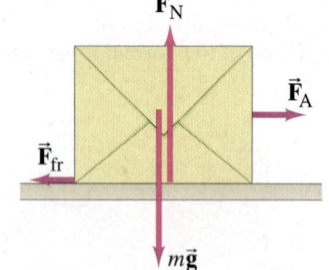

FIGURE 4–27 Repeated for Example 4–16.

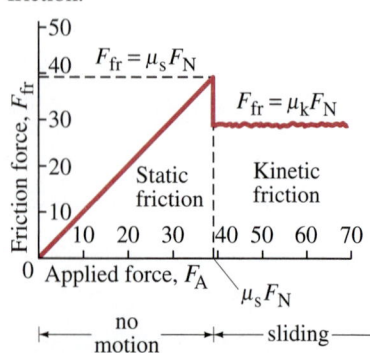

FIGURE 4–28 Example 4–16. Magnitude of the force of friction as a function of the external force applied to an object initially at rest. As the applied force is increased in magnitude, the force of static friction increases in proportion until the applied force equals $\mu_s F_N$. If the applied force increases further, the object will begin to move, and the friction force drops to a roughly constant value characteristic of kinetic friction.

Friction can be a hindrance. It slows down moving objects and causes heating and binding of moving parts in machinery. Friction can be reduced by using lubricants such as oil. More effective in reducing friction between two surfaces is to maintain a layer of air or other gas between them. Devices using this concept, which is not practical for most situations, include air tracks and air tables in which the layer of air is maintained by forcing air through many tiny holes. Another technique to maintain the air layer is to suspend objects in air using magnetic fields ("magnetic levitation").

On the other hand, friction can be helpful. Our ability to walk depends on friction between the soles of our shoes (or feet) and the ground. (Walking involves static friction, not kinetic friction. Why?) The movement of a car, and also its stability, depend on friction. When friction is low, such as on ice, safe walking or driving becomes difficult.

CONCEPTUAL EXAMPLE 4–17 **A box against a wall.** You can hold a box against a rough wall (Fig. 4–29) and prevent it from slipping down by pressing hard horizontally. How does the application of a horizontal force keep an object from moving vertically?

RESPONSE This won't work well if the wall is slippery. You need friction. Even then, if you don't press hard enough, the box will slip. The horizontal force you apply produces a normal force on the box exerted by the wall (the net force horizontally is zero since the box doesn't move horizontally). The force of gravity mg, acting downward on the box, can now be balanced by an upward static friction force whose maximum magnitude is proportional to the normal force. The harder you push, the greater F_N is and the greater F_{fr} can be. If you don't press hard enough, then $mg > \mu_s F_N$ and the box begins to slide down.

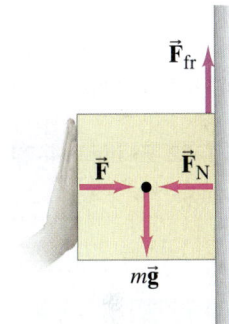

FIGURE 4–29 Example 4–17.

EXERCISE G If $\mu_s = 0.40$ and $mg = 20$ N, what minimum force F will keep the box from falling: (a) 100 N; (b) 80 N; (c) 50 N; (d) 20 N; (e) 8 N?

EXAMPLE 4–18 **Pulling against friction.** A 10.0-kg box is pulled along a horizontal surface by a force F_P of 40.0 N applied at a 30.0° angle above horizontal. This is like Example 4–11 except now there is friction, and we assume a coefficient of kinetic friction of 0.30. Calculate the acceleration.

APPROACH The free-body diagram is shown in Fig. 4–30. It is much like that in Fig. 4–21b, but with one more force, friction.

FIGURE 4–30 Example 4–18.

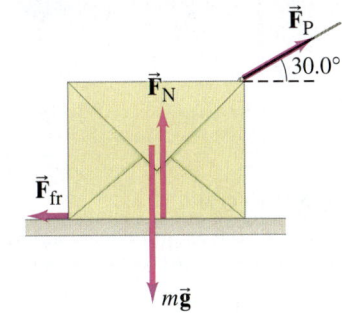

SOLUTION The calculation for the vertical (y) direction is just the same as in Example 4–11b, $mg = (10.0 \text{ kg})(9.80 \text{ m/s}^2) = 98.0$ N and $F_{Py} = (40.0 \text{ N})(\sin 30.0°) = 20.0$ N. With y positive upward and $a_y = 0$, we have

$$F_N - mg + F_{Py} = ma_y$$
$$F_N - 98.0 \text{ N} + 20.0 \text{ N} = 0,$$

so the normal force is $F_N = 78.0$ N. Now we apply Newton's second law for the horizontal (x) direction (positive to the right), and include the friction force:

$$F_{Px} - F_{fr} = ma_x.$$

The friction force is kinetic friction as long as $F_{fr} = \mu_k F_N$ is less than $F_{Px} = (40.0 \text{ N}) \cos 30.0° = 34.6$ N, which it is:

$$F_{fr} = \mu_k F_N = (0.30)(78.0 \text{ N}) = 23.4 \text{ N}.$$

Hence the box does accelerate:

$$a_x = \frac{F_{Px} - F_{fr}}{m} = \frac{34.6 \text{ N} - 23.4 \text{ N}}{10.0 \text{ kg}} = 1.1 \text{ m/s}^2.$$

In the absence of friction, as we saw in Example 4–11, the acceleration would be much greater than this.

NOTE Our final answer has only two significant figures because our least significant input value ($\mu_k = 0.30$) has two.

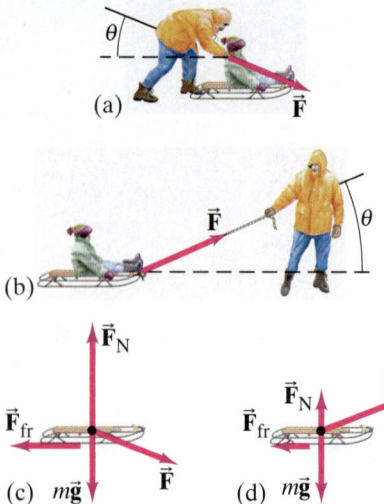

FIGURE 4–31 Example 4–19.

CONCEPTUAL EXAMPLE 4–19 **To push or to pull a sled?** Your little sister wants a ride on her sled. If you are on flat ground, will you exert less force if you push her or pull her? See Figs. 4–31a and b. Assume the same angle θ in each case.

RESPONSE Let us draw free-body diagrams for the sled–sister combination, as shown in Figs. 4–31c and d. They show, for the two cases, the forces exerted by you, $\vec{F}$ (an unknown), by the snow, $\vec{F}_N$ and $\vec{F}_{fr}$, and gravity $m\vec{g}$. (a) If you push her, and $\theta > 0$, there is a vertically downward component to your force. Hence the normal force upward exerted by the ground (Fig. 4–31c) will be larger than mg (where m is the mass of sister plus sled). (b) If you pull her, your force has a vertically upward component, so the normal force F_N will be less than mg, Fig. 4–31d. Because the friction force is proportional to the normal force, F_{fr} will be less if you pull her. So you exert less force if you pull her.

EXAMPLE 4–20 **Two boxes and a pulley.** In Fig. 4–32a, two boxes are connected by a cord running over a pulley. The coefficient of kinetic friction between box A and the table is 0.20. We ignore the mass of the cord and pulley and any friction in the pulley, which means we can assume that a force applied to one end of the cord will have the same magnitude at the other end. We wish to find the acceleration, a, of the system, which will have the same magnitude for both boxes assuming the cord doesn't stretch. As box B moves down, box A moves to the right.

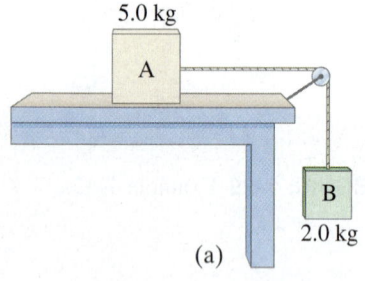

FIGURE 4–32 Example 4–20.

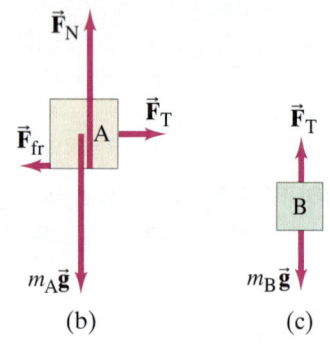

⚠ **CAUTION**
Tension in a cord supporting a falling object may not equal object's weight

APPROACH The free-body diagrams for each box are shown in Figs. 4–32b and c. The forces on box A are the pulling force of the cord F_T, gravity $m_A g$, the normal force exerted by the table F_N, and a friction force exerted by the table F_{fr}; the forces on box B are gravity $m_B g$, and the cord pulling up, F_T.

SOLUTION Box A does not move vertically, so Newton's second law tells us the normal force just balances the weight,

$$F_N = m_A g = (5.0\,\text{kg})(9.8\,\text{m/s}^2) = 49\,\text{N}.$$

In the horizontal direction, there are two forces on box A (Fig. 4–32b): F_T, the tension in the cord (whose value we don't know), and the force of friction

$$F_{fr} = \mu_k F_N = (0.20)(49\,\text{N}) = 9.8\,\text{N}.$$

The horizontal acceleration (box A) is what we wish to find; we use Newton's second law in the x direction, $\Sigma F_{Ax} = m_A a_x$, which becomes (taking the positive direction to the right and setting $a_{Ax} = a$):

$$\Sigma F_{Ax} = F_T - F_{fr} = m_A a. \qquad [\text{box A}]$$

Next consider box B. The force of gravity $m_B g = (2.0\,\text{kg})(9.8\,\text{m/s}^2) = 19.6\,\text{N}$ pulls downward; and the cord pulls upward with a force F_T. So we can write Newton's second law for box B (taking the downward direction as positive):

$$\Sigma F_{By} = m_B g - F_T = m_B a. \qquad [\text{box B}]$$

[Notice that if $a \neq 0$, then F_T is not equal to $m_B g$.]

We have two unknowns, a and F_T, and we also have two equations. We solve the box A equation for F_T:

$$F_T = F_{fr} + m_A a,$$

and substitute this into the box B equation:

$$m_B g - F_{fr} - m_A a = m_B a.$$

Now we solve for a and put in numerical values:

$$a = \frac{m_B g - F_{fr}}{m_A + m_B} = \frac{19.6\,\text{N} - 9.8\,\text{N}}{5.0\,\text{kg} + 2.0\,\text{kg}} = 1.4\,\text{m/s}^2,$$

which is the acceleration of box A to the right, and of box B down.

If we wish, we can calculate F_T using the third equation up from here:

$$F_T = F_{fr} + m_A a = 9.8\,\text{N} + (5.0\,\text{kg})(1.4\,\text{m/s}^2) = 17\,\text{N}.$$

NOTE Box B is not in free fall. It does not fall at $a = g$ because an additional force, F_T, is acting upward on it.

Inclines

Now we consider what happens when an object slides down an incline, such as a hill or ramp. Such problems are interesting because gravity is the accelerating force, yet the acceleration is not vertical. Solving problems is usually easier if we choose the xy coordinate system so the x axis points along the incline (the direction of motion) and the y axis is perpendicular to the incline, as shown in Fig. 4–33. Note also that the normal force is not vertical, but is perpendicular to the sloping surface of the plane, along the y axis in Fig. 4–33.

EXERCISE H Is the normal force always perpendicular to an inclined plane? Is it always vertical?

EXAMPLE 4–21 The skier. The skier in Fig. 4–34a has begun descending the 30° slope. If the coefficient of kinetic friction is 0.10, what is her acceleration?

APPROACH We choose the x axis along the slope, positive downslope in the direction of the skier's motion. The y axis is perpendicular to the surface. The forces acting on the skier are gravity, $\vec{F}_G = m\vec{g}$, which points vertically downward (*not* perpendicular to the slope), and the two forces exerted on her skis by the snow—the normal force perpendicular to the snowy slope (*not* vertical), and the friction force parallel to the surface. These three forces are shown acting at one point in Fig. 4–34b, which is our free-body diagram for the skier.

SOLUTION We have to resolve only one vector into components, the weight $\vec{F}_G$, and its components are shown as dashed lines in Fig. 4–34c. To be general, we use θ rather than 30° for now. We use the definitions of sine ("side opposite") and cosine ("side adjacent") to obtain the components:

$$F_{Gx} = mg \sin \theta,$$
$$F_{Gy} = -mg \cos \theta$$

where F_{Gy} is in the negative y direction. To calculate the skier's acceleration down the hill, a_x, we apply Newton's second law to the x direction:

$$\Sigma F_x = ma_x$$
$$mg \sin \theta - \mu_k F_N = ma_x$$

where the two forces are the x component of the gravity force ($+x$ direction) and the friction force ($-x$ direction). We want to find the value of a_x, but we don't yet know F_N in the last equation. Let's see if we can get F_N from the y component of Newton's second law:

$$\Sigma F_y = ma_y$$
$$F_N - mg \cos \theta = ma_y = 0$$

where we set $a_y = 0$ because there is no motion in the y direction (perpendicular to the slope). Thus we can solve for F_N:

$$F_N = mg \cos \theta$$

and we can substitute this into our equation above for ma_x:

$$mg \sin \theta - \mu_k (mg \cos \theta) = ma_x.$$

There is an m in each term which can be canceled out. Thus (setting $\theta = 30°$ and $\mu_k = 0.10$):

$$a_x = g \sin 30° - \mu_k g \cos 30°$$
$$= 0.50g - (0.10)(0.866)g = 0.41g.$$

The skier's acceleration is 0.41 times the acceleration of gravity, which in numbers[†] is $a = (0.41)(9.8 \text{ m/s}^2) = 4.0 \text{ m/s}^2$.

NOTE The mass canceled out, so we have the useful conclusion that *the acceleration doesn't depend on the mass*. That such a cancellation sometimes occurs, and thus may give a useful conclusion as well as saving calculation, is a big advantage of working with the algebraic equations and putting in the numbers only at the end.

[†] We used values rounded off to 2 significant figures to obtain $a = 4.0 \text{ m/s}^2$. If we kept all the extra digits in our calculator, we would find $a = 0.4134g \approx 4.1 \text{ m/s}^2$. This difference is within the expected precision (number of significant figures, Section 1–4).

PROBLEM SOLVING
Good choice of coordinate system simplifies the calculation

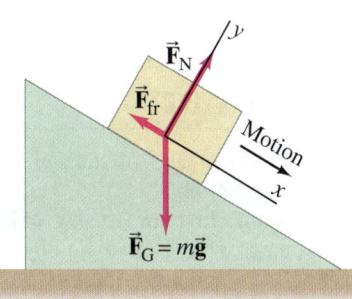
FIGURE 4–33 Forces on an object sliding down an incline.

PHYSICS APPLIED
Skiing

FIGURE 4–34 Example 4–21. Skier descending a slope; $\vec{F}_G = m\vec{g}$ is the force of gravity (weight) on the skier.

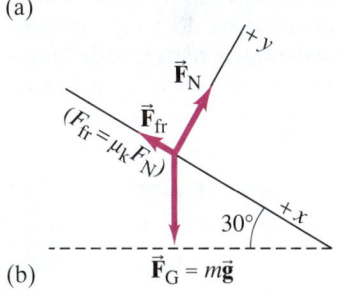

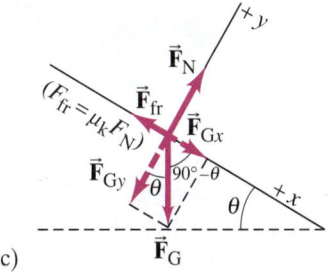

PROBLEM SOLVING
It is often helpful to put in numbers only at the end

CAUTION
Directions of gravity and the normal force

In Problems involving a slope or an "inclined plane," avoid making errors in the directions of the normal force and gravity. The normal force on an incline is *not* vertical: it is perpendicular to the slope or plane. And gravity is *not* perpendicular to the slope—gravity acts vertically downward toward the center of the Earth.

Summary

Newton's three laws of motion are the basic classical laws describing motion.

Newton's first law (the **law of inertia**) states that if the net force on an object is zero, an object originally at rest remains at rest, and an object in motion remains in motion in a straight line with constant velocity.

Newton's second law states that the acceleration of an object is directly proportional to the net force acting on it, and inversely proportional to its mass:

$$\Sigma \vec{F} = m\vec{a}. \qquad (4\text{--}1)$$

Newton's second law is one of the most important and fundamental laws in classical physics.

Newton's third law states that whenever one object exerts a force on a second object, the second object always exerts a force on the first object which is equal in magnitude but opposite in direction:

$$\vec{F}_{AB} = -\vec{F}_{BA} \qquad (4\text{--}2)$$

where $\vec{F}_{BA}$ is the force on object B exerted by object A.

The tendency of an object to resist a change in its motion is called **inertia**. **Mass** is a measure of the inertia of an object.

Weight refers to the **gravitational force** on an object, and is equal to the product of the object's mass m and the acceleration of gravity $\vec{g}$:

$$\vec{F}_G = m\vec{g}. \qquad (4\text{--}3)$$

Force, which is a vector, can be considered as a push or pull; or, from Newton's second law, force can be defined as an action capable of giving rise to acceleration. The **net force** on an object is the vector sum of all forces acting on that object.

When two objects slide over one another, the force of friction that each object exerts on the other can be written approximately as $F_{fr} = \mu_k F_N$, where F_N is the **normal force** (the force each object exerts on the other perpendicular to their contact surfaces), and μ_k is the coefficient of **kinetic friction**. If the objects are at rest relative to each other, then F_{fr} is just large enough to hold them at rest and satisfies the inequality $F_{fr} < \mu_s F_N$, where μ_s is the coefficient of **static friction**.

For solving problems involving the forces on one or more objects, it is essential to draw a **free-body diagram** for each object, showing all the forces acting on only that object. Newton's second law can be applied to the vector components for each object.

Questions

1. Why does a child in a wagon seem to fall backward when you give the wagon a sharp pull forward?

2. A box rests on the (frictionless) bed of a truck. The truck driver starts the truck and accelerates forward. The box immediately starts to slide toward the rear of the truck bed. Discuss the motion of the box, in terms of Newton's laws, as seen (*a*) by Mary standing on the ground beside the truck, and (*b*) by Chris who is riding on the truck (Fig. 4–35).

FIGURE 4–35 Question 2.

3. If an object is moving, is it possible for the net force acting on it to be zero? Explain.

4. If the acceleration of an object is zero, are no forces acting on it? Explain.

5. Only one force acts on an object. Can the object have zero acceleration? Can it have zero velocity? Explain.

6. When a golf ball is dropped to the pavement, it bounces back up. (*a*) Is a force needed to make it bounce back up? (*b*) If so, what exerts the force?

7. If you walk along a log floating on a lake, why does the log move in the opposite direction?

8. (*a*) Why do you push down harder on the pedals of a bicycle when first starting out than when moving at constant speed? (*b*) Why do you need to pedal at all when cycling at constant speed?

9. A stone hangs by a fine thread from the ceiling, and a section of the same thread dangles from the bottom of the stone (Fig. 4–36). If a person gives a sharp pull on the dangling thread, where is the thread likely to break: below the stone or above it? What if the person gives a slow and steady pull? Explain your answers.

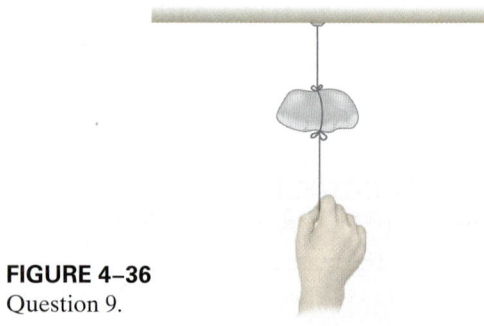

FIGURE 4–36 Question 9.

10. The force of gravity on a 2-kg rock is twice as great as that on a 1-kg rock. Why then doesn't the heavier rock fall faster?

11. (a) You pull a box with a constant force across a frictionless table using an attached rope held horizontally. If you now pull the rope with the same force at an angle to the horizontal (with the box remaining flat on the table), does the acceleration of the box increase, decrease, or remain the same? Explain. (b) What if there is friction?

12. When an object falls freely under the influence of gravity there is a net force mg exerted on it by the Earth. Yet by Newton's third law the object exerts an equal and opposite force on the Earth. Does the Earth move? Explain.

13. Compare the effort (or force) needed to lift a 10-kg object when you are on the Moon with the force needed to lift it on Earth. Compare the force needed to throw a 2-kg object horizontally with a given speed on the Moon and on Earth.

14. According to Newton's third law, each team in a tug of war (Fig. 4–37) pulls with equal force on the other team. What, then, determines which team will win?

FIGURE 4–37 Question 14. A tug of war. Describe the forces on each of the teams and on the rope.

15. When you stand still on the ground, how large a force does the ground exert on you? Why doesn't this force make you rise up into the air?

16. Whiplash sometimes results from an automobile accident when the victim's car is struck violently from the rear. Explain why the head of the victim seems to be thrown backward in this situation. Is it really?

17. Mary exerts an upward force of 40 N to hold a bag of groceries. Describe the "reaction" force (Newton's third law) by stating (a) its magnitude, (b) its direction, (c) on what object it is exerted, and (d) by what object it is exerted.

18. A father and his young daughter are ice skating. They face each other at rest and push each other, moving in opposite directions. Which one has the greater final speed? Explain.

19. A heavy crate rests on the bed of a flatbed truck. When the truck accelerates, the crate stays fixed on the truck, so it, too, accelerates. What force causes the crate to accelerate?

20. A block is given a brief push so that it slides up a ramp. After the block reaches its highest point, it slides back down, but the magnitude of its acceleration is less on the descent than on the ascent. Why?

21. Why is the stopping distance of a truck much shorter than for a train going the same speed?

22. What would your bathroom scale read if you weighed yourself on an inclined plane? Assume the mechanism functions properly, even at an angle.

MisConceptual Questions

1. A truck is traveling horizontally to the right (Fig. 4–38). When the truck starts to slow down, the crate on the (frictionless) truck bed starts to slide. In what direction could the net force be on the crate?
   (a) No direction. The net force is zero.
   (b) Straight down (because of gravity).
   (c) Straight up (the normal force).
   (d) Horizontal and to the right.
   (e) Horizontal and to the left.

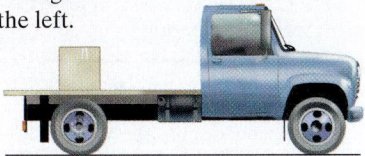

FIGURE 4–38 MisConceptual Question 1.

2. You are trying to push your stalled car. Although you apply a horizontal force of 400 N to the car, it doesn't budge, and neither do you. Which force(s) must also have a magnitude of 400 N?
   (a) The force exerted by the car on you.
   (b) The friction force exerted by the car on the road.
   (c) The normal force exerted by the road on you.
   (d) The friction force exerted by the road on you.

3. Matt, in the foreground of Fig. 4–39, is able to move the large truck because
   (a) he is stronger than the truck.
   (b) he is heavier in some respects than the truck.
   (c) he exerts a greater force on the truck than the truck exerts back on him.
   (d) the ground exerts a greater friction force on Matt than it does on the truck.
   (e) the truck offers no resistance because its brakes are off.

FIGURE 4–39 MisConceptual Question 3.

4. A bear sling, Fig. 4–40, is used in some national parks for placing backpackers' food out of the reach of bears. As the backpacker raises the pack by pulling down on the rope, the force F needed:
   (a) decreases as the pack rises until the rope is straight across.
   (b) doesn't change.
   (c) increases until the rope is straight.
   (d) increases but the rope always sags where the pack hangs.

FIGURE 4–40 MisConceptual Question 4.

5. What causes the boat in Fig. 4–41 to move forward?
   (a) The force the man exerts on the paddle.
   (b) The force the paddle exerts on the water.
   (c) The force the water exerts on the paddle.
   (d) The motion of the water itself.

FIGURE 4–41 MisConceptual Question 5.

6. A person stands on a scale in an elevator. His apparent weight will be the greatest when the elevator
   (a) is standing still.
   (b) is moving upward at constant velocity.
   (c) is accelerating upward.
   (d) is moving downward at constant velocity.
   (e) is accelerating downward.

7. When a skier skis down a hill, the normal force exerted on the skier by the hill is
   (a) equal to the weight of the skier.
   (b) greater than the weight of the skier.
   (c) less than the weight of the skier.

8. A golf ball is hit with a golf club. While the ball flies through the air, which forces act on the ball? Neglect air resistance.
   (a) The force of the golf club acting on the ball.
   (b) The force of gravity acting on the ball.
   (c) The force of the ball moving forward through the air.
   (d) All of the above.
   (e) Both (a) and (c).

9. Suppose an object is accelerated by a force of 100 N. Suddenly a second force of 100 N in the opposite direction is exerted on the object, so that the forces cancel. The object
   (a) is brought to rest rapidly.
   (b) decelerates gradually to rest.
   (c) continues at the velocity it had before the second force was applied.
   (d) is brought to rest and then accelerates in the direction of the second force.

10. You are pushing a heavy box across a rough floor. When you are initially pushing the box and it is accelerating,
    (a) you exert a force on the box, but the box does not exert a force on you.
    (b) the box is so heavy it exerts a force on you, but you do not exert a force on the box.
    (c) the force you exert on the box is greater than the force of the box pushing back on you.
    (d) the force you exert on the box is equal to the force of the box pushing back on you.
    (e) the force that the box exerts on you is greater than the force you exert on the box.

11. A 50-N crate sits on a horizontal floor where the coefficient of static friction between the crate and the floor is 0.50. A 20-N force is applied to the crate acting to the right. What is the resulting static friction force acting on the crate?
    (a) 20 N to the right.
    (b) 20 N to the left.
    (c) 25 N to the right.
    (d) 25 N to the left.
    (e) None of the above; the crate starts to move.

12. The normal force on an extreme skier descending a very steep slope (Fig. 4–42) can be zero if
    (a) his speed is great enough.
    (b) he leaves the slope (no longer touches the snow).
    (c) the slope is greater than 75°.
    (d) the slope is vertical (90°).

FIGURE 4–42 MisConceptual Question 12.

13. To pull an old stump out of the ground, you and a friend tie two ropes to the stump. You pull on it with a force of 500 N to the north while your friend pulls with a force of 450 N to the northwest. The total force from the two ropes is
    (a) less than 950 N.
    (b) exactly 950 N.
    (c) more than 950 N.

For assigned homework and other learning materials, go to the MasteringPhysics website.

Problems

[It would be wise, before starting the Problems, to reread the Problem Solving Strategies on pages 30, 60, and 88.]

4–4 to 4–6 Newton's Laws, Gravitational Force, Normal Force [Assume no friction.]

1. (I) What force is needed to accelerate a sled (mass = 55 kg) at 1.4 m/s^2 on horizontal frictionless ice?

2. (I) What is the weight of a 68-kg astronaut (*a*) on Earth, (*b*) on the Moon ($g = 1.7 \text{ m/s}^2$), (*c*) on Mars ($g = 3.7 \text{ m/s}^2$), (*d*) in outer space traveling with constant velocity?

3. (I) How much tension must a rope withstand if it is used to accelerate a 1210-kg car horizontally along a frictionless surface at 1.20 m/s^2?

4. (II) According to a simplified model of a mammalian heart, at each pulse approximately 20 g of blood is accelerated from 0.25 m/s to 0.35 m/s during a period of 0.10 s. What is the magnitude of the force exerted by the heart muscle?

5. (II) Superman must stop a 120-km/h train in 150 m to keep it from hitting a stalled car on the tracks. If the train's mass is 3.6×10^5 kg, how much force must he exert? Compare to the weight of the train (give as %). How much force does the train exert on Superman?

6. (II) A person has a reasonable chance of surviving an automobile crash if the deceleration is no more than 30 *g*'s. Calculate the force on a 65-kg person accelerating at this rate. What distance is traveled if brought to rest at this rate from 95 km/h?

7. (II) What average force is required to stop a 950-kg car in 8.0 s if the car is traveling at 95 km/h?

8. (II) Estimate the average force exerted by a shot-putter on a 7.0-kg shot if the shot is moved through a distance of 2.8 m and is released with a speed of 13 m/s.

9. (II) A 0.140-kg baseball traveling 35.0 m/s strikes the catcher's mitt, which, in bringing the ball to rest, recoils backward 11.0 cm. What was the average force applied by the ball on the glove?

10. (II) How much tension must a cable withstand if it is used to accelerate a 1200-kg car vertically upward at 0.70 m/s^2?

11. (II) A 20.0-kg box rests on a table. (*a*) What is the weight of the box and the normal force acting on it? (*b*) A 10.0-kg box is placed on top of the 20.0-kg box, as shown in Fig. 4–43. Determine the normal force that the table exerts on the 20.0-kg box and the normal force that the 20.0-kg box exerts on the 10.0-kg box.

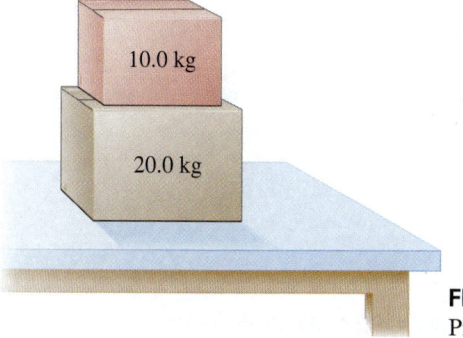

FIGURE 4–43 Problem 11.

12. (II) A 14.0-kg bucket is lowered vertically by a rope in which there is 163 N of tension at a given instant. What is the acceleration of the bucket? Is it up or down?

13. (II) A 75-kg petty thief wants to escape from a third-story jail window. Unfortunately, a makeshift rope made of sheets tied together can support a mass of only 58 kg. How might the thief use this "rope" to escape? Give a quantitative answer.

14. (II) An elevator (mass 4850 kg) is to be designed so that the maximum acceleration is 0.0680*g*. What are the maximum and minimum forces the motor should exert on the supporting cable?

15. (II) Can cars "stop on a dime"? Calculate the acceleration of a 1400-kg car if it can stop from 35 km/h on a dime (diameter = 1.7 cm). How many *g*'s is this? What is the force felt by the 68-kg occupant of the car?

16. (II) A woman stands on a bathroom scale in a motionless elevator. When the elevator begins to move, the scale briefly reads only 0.75 of her regular weight. Calculate the acceleration of the elevator, and find the direction of acceleration.

17. (II) (*a*) What is the acceleration of two falling sky divers (total mass = 132 kg including parachute) when the upward force of air resistance is equal to one-fourth of their weight? (*b*) After opening the parachute, the divers descend leisurely to the ground at constant speed. What now is the force of air resistance on the sky divers and their parachute? See Fig. 4–44.

FIGURE 4–44 Problem 17.

18. (II) The cable supporting a 2125-kg elevator has a maximum strength of 21,750 N. What maximum upward acceleration can it give the elevator without breaking?

19. (III) A person jumps from the roof of a house 2.8 m high. When he strikes the ground below, he bends his knees so that his torso decelerates over an approximate distance of 0.70 m. If the mass of his torso (excluding legs) is 42 kg, find (*a*) his velocity just before his feet strike the ground, and (*b*) the average force exerted on his torso by his legs during deceleration.

4–7 Newton's Laws and Vectors [Ignore friction.]

20. (I) A box weighing 77.0 N rests on a table. A rope tied to the box runs vertically upward over a pulley and a weight is hung from the other end (Fig. 4–45). Determine the force that the table exerts on the box if the weight hanging on the other side of the pulley weighs (a) 30.0 N, (b) 60.0 N, and (c) 90.0 N.

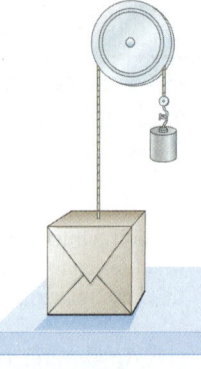

FIGURE 4–45 Problem 20.

21. (I) Draw the free-body diagram for a basketball player (a) just before leaving the ground on a jump, and (b) while in the air. See Fig. 4–46.

FIGURE 4–46 Problem 21.

22. (I) Sketch the free-body diagram of a baseball (a) at the moment it is hit by the bat, and again (b) after it has left the bat and is flying toward the outfield. Ignore air resistance.

23. (II) Arlene is to walk across a "high wire" strung horizontally between two buildings 10.0 m apart. The sag in the rope when she is at the midpoint is 10.0°, as shown in Fig. 4–47. If her mass is 50.0 kg, what is the tension in the rope at this point?

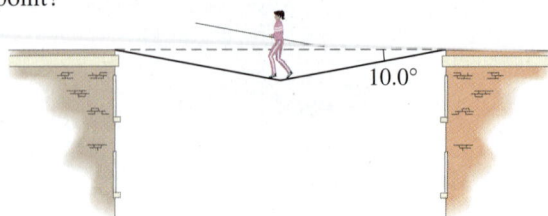

FIGURE 4–47 Problem 23.

24. (II) A window washer pulls herself upward using the bucket–pulley apparatus shown in Fig. 4–48. (a) How hard must she pull downward to raise herself slowly at constant speed? (b) If she increases this force by 15%, what will her acceleration be? The mass of the person plus the bucket is 72 kg.

FIGURE 4–48 Problem 24.

25. (II) One 3.2-kg paint bucket is hanging by a massless cord from another 3.2-kg paint bucket, also hanging by a massless cord, as shown in Fig. 4–49. (a) If the buckets are at rest, what is the tension in each cord? (b) If the two buckets are pulled upward with an acceleration of 1.25 m/s² by the upper cord, calculate the tension in each cord.

FIGURE 4–49 Problem 25.

26. (II) Two snowcats in Antarctica are towing a housing unit north, as shown in Fig. 4–50. The sum of the forces $\vec{F}_A$ and $\vec{F}_B$ exerted on the unit by the horizontal cables is north, parallel to the line L, and $F_A = 4500$ N. Determine F_B and the magnitude of $\vec{F}_A + \vec{F}_B$.

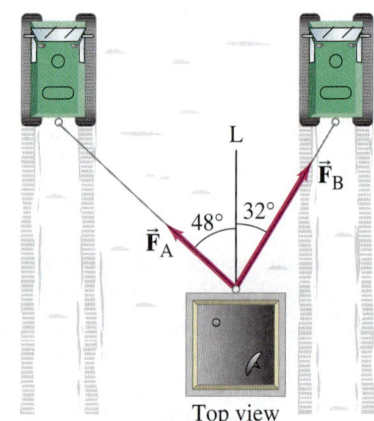

FIGURE 4–50 Problem 26. Top view

27. (II) A train locomotive is pulling two cars of the same mass behind it, Fig. 4–51. Determine the ratio of the tension in the coupling (think of it as a cord) between the locomotive and the first car (F_{T1}), to that between the first car and the second car (F_{T2}), for any nonzero acceleration of the train.

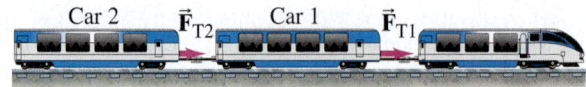

FIGURE 4–51 Problem 27.

28. (II) The two forces $\vec{F}_1$ and $\vec{F}_2$ shown in Fig. 4–52a and b (looking down) act on an 18.5-kg object on a frictionless tabletop. If $F_1 = 10.2$ N and $F_2 = 16.0$ N, find the net force on the object and its acceleration for (a) and (b).

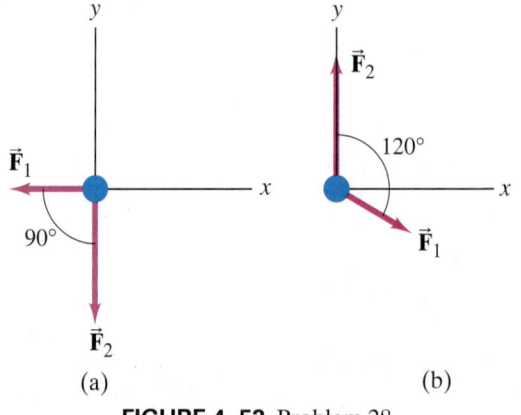

FIGURE 4–52 Problem 28.

102 CHAPTER 4 Dynamics: Newton's Laws of Motion

29. (II) At the instant a race began, a 65-kg sprinter exerted a force of 720 N on the starting block at a 22° angle with respect to the ground. (a) What was the horizontal acceleration of the sprinter? (b) If the force was exerted for 0.32 s, with what speed did the sprinter leave the starting block?

30. (II) A 27-kg chandelier hangs from a ceiling on a vertical 4.0-m-long wire. (a) What horizontal force would be necessary to displace its position 0.15 m to one side? (b) What will be the tension in the wire?

31. (II) An object is hanging by a string from your rearview mirror. While you are decelerating at a constant rate from 25 m/s to rest in 6.0 s, (a) what angle does the string make with the vertical, and (b) is it toward the windshield or away from it? [*Hint*: See Example 4–15.]

32. (II) Figure 4–53 shows a block (mass m_A) on a smooth horizontal surface, connected by a thin cord that passes over a pulley to a second block (m_B), which hangs vertically. (a) Draw a free-body diagram for each block, showing the force of gravity on each, the force (tension) exerted by the cord, and any normal force. (b) Apply Newton's second law to find formulas for the acceleration of the system and for the tension in the cord. Ignore friction and the masses of the pulley and cord.

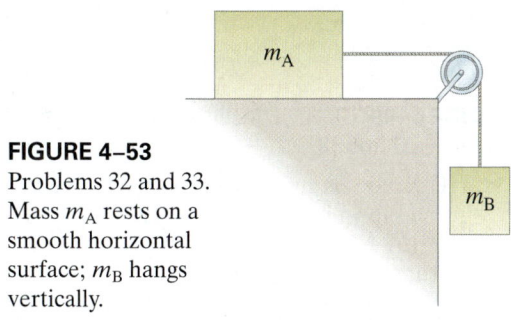

FIGURE 4–53
Problems 32 and 33.
Mass m_A rests on a smooth horizontal surface; m_B hangs vertically.

33. (II) (a) If $m_A = 13.0$ kg and $m_B = 5.0$ kg in Fig. 4–53, determine the acceleration of each block. (b) If initially m_A is at rest 1.250 m from the edge of the table, how long does it take to reach the edge of the table if the system is allowed to move freely? (c) If $m_B = 1.0$ kg, how large must m_A be if the acceleration of the system is to be kept at $\frac{1}{100}g$?

34. (III) Three blocks on a frictionless horizontal surface are in contact with each other as shown in Fig. 4–54. A force $\vec{F}$ is applied to block A (mass m_A). (a) Draw a free-body diagram for each block. Determine (b) the acceleration of the system (in terms of m_A, m_B, and m_C), (c) the net force on each block, and (d) the force of contact that each block exerts on its neighbor. (e) If $m_A = m_B = m_C = 10.0$ kg and $F = 96.0$ N, give numerical answers to (b), (c), and (d). Explain how your answers make sense intuitively.

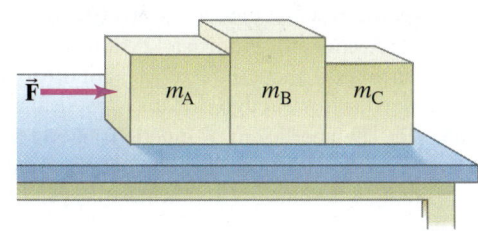

FIGURE 4–54
Problem 34.

35. (III) Suppose the pulley in Fig. 4–55 is suspended by a cord C. Determine the tension in this cord after the masses are released and before one hits the ground. Ignore the mass of the pulley and cords.

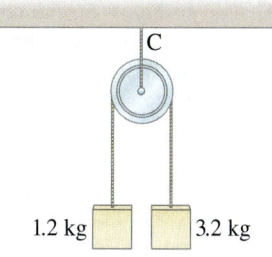

FIGURE 4–55
Problem 35.

4–8 Newton's Laws with Friction, Inclines

36. (I) If the coefficient of kinetic friction between a 22-kg crate and the floor is 0.30, what horizontal force is required to move the crate at a steady speed across the floor? What horizontal force is required if μ_k is zero?

37. (I) A force of 35.0 N is required to start a 6.0-kg box moving across a horizontal concrete floor. (a) What is the coefficient of static friction between the box and the floor? (b) If the 35.0-N force continues, the box accelerates at 0.60 m/s^2. What is the coefficient of kinetic friction?

38. (I) Suppose you are standing on a train accelerating at $0.20\,g$. What minimum coefficient of static friction must exist between your feet and the floor if you are not to slide?

39. (II) The coefficient of static friction between hard rubber and normal street pavement is about 0.90. On how steep a hill (maximum angle) can you leave a car parked?

40. (II) A flatbed truck is carrying a heavy crate. The coefficient of static friction between the crate and the bed of the truck is 0.75. What is the maximum rate at which the driver can decelerate and still avoid having the crate slide against the cab of the truck?

41. (II) A 2.0-kg silverware drawer does not slide readily. The owner gradually pulls with more and more force, and when the applied force reaches 9.0 N, the drawer suddenly opens, throwing all the utensils to the floor. What is the coefficient of static friction between the drawer and the cabinet?

42. (II) A box is given a push so that it slides across the floor. How far will it go, given that the coefficient of kinetic friction is 0.15 and the push imparts an initial speed of 3.5 m/s?

43. (II) A 1280-kg car pulls a 350-kg trailer. The car exerts a horizontal force of 3.6×10^3 N against the ground in order to accelerate. What force does the car exert on the trailer? Assume an effective friction coefficient of 0.15 for the trailer.

44. (II) Police investigators, examining the scene of an accident involving two cars, measure 72-m-long skid marks of one of the cars, which nearly came to a stop before colliding. The coefficient of kinetic friction between rubber and the pavement is about 0.80. Estimate the initial speed of that car assuming a level road.

45. (II) Drag-race tires in contact with an asphalt surface have a very high coefficient of static friction. Assuming a constant acceleration and no slipping of tires, estimate the coefficient of static friction needed for a drag racer to cover 1.0 km in 12 s, starting from rest.

46. (II) For the system of Fig. 4–32 (Example 4–20), how large a mass would box A have to have to prevent any motion from occurring? Assume $\mu_s = 0.30$.

47. (II) In Fig. 4–56 the coefficient of static friction between mass m_A and the table is 0.40, whereas the coefficient of kinetic friction is 0.20. (a) What minimum value of m_A will keep the system from starting to move? (b) What value(s) of m_A will keep the system moving at constant speed? [Ignore masses of the cord and the (frictionless) pulley.]

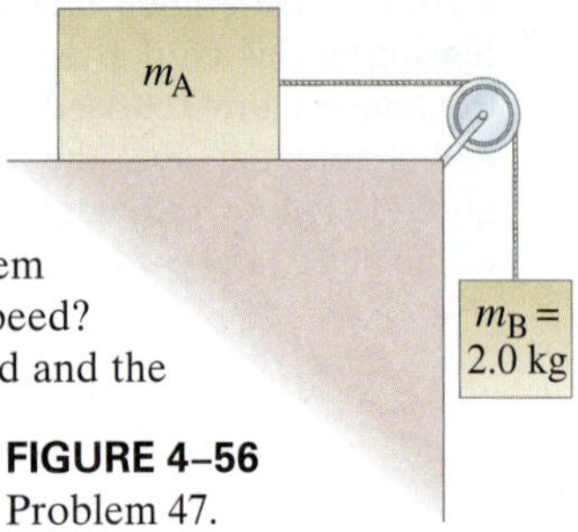

FIGURE 4–56
Problem 47.

48. (II) A small box is held in place against a rough vertical wall by someone pushing on it with a force directed upward at 28° above the horizontal. The coefficients of static and kinetic friction between the box and wall are 0.40 and 0.30, respectively. The box slides down unless the applied force has magnitude 23 N. What is the mass of the box?

49. (II) Two crates, of mass 65 kg and 125 kg, are in contact and at rest on a horizontal surface (Fig. 4–57). A 650-N force is exerted on the 65-kg crate. If the coefficient of kinetic friction is 0.18, calculate (a) the acceleration of the system, and (b) the force that each crate exerts on the other. (c) Repeat with the crates reversed.

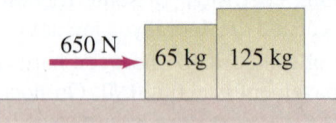

FIGURE 4–57
Problem 49.

50. (II) A person pushes a 14.0-kg lawn mower at constant speed with a force of $F = 88.0$ N directed along the handle, which is at an angle of 45.0° to the horizontal (Fig. 4–58). (a) Draw the free-body diagram showing all forces acting on the mower. Calculate (b) the horizontal friction force on the mower, then (c) the normal force exerted vertically upward on the mower by the ground. (d) What force must the person exert on the lawn mower to accelerate it from rest to 1.5 m/s in 2.5 seconds, assuming the same friction force?

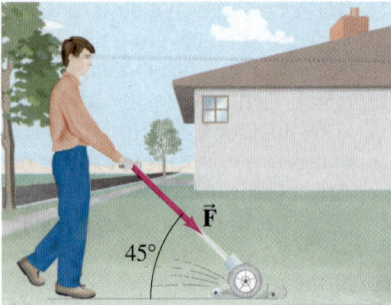

FIGURE 4–58
Problem 50.

51. (II) A child on a sled reaches the bottom of a hill with a velocity of 10.0 m/s and travels 25.0 m along a horizontal straightaway to a stop. If the child and sled together have a mass of 60.0 kg, what is the average retarding force on the sled on the horizontal straightaway?

52. (II) (a) A box sits at rest on a rough 33° inclined plane. Draw the free-body diagram, showing all the forces acting on the box. (b) How would the diagram change if the box were sliding down the plane? (c) How would it change if the box were sliding up the plane after an initial shove?

53. (II) A wet bar of soap slides down a ramp 9.0 m long inclined at 8.0°. How long does it take to reach the bottom? Assume $\mu_k = 0.060$.

54. (II) A skateboarder, with an initial speed of 2.0 m/s, rolls virtually friction free down a straight incline of length 18 m in 3.3 s. At what angle θ is the incline oriented above the horizontal?

55. (II) Uphill escape ramps are sometimes provided to the side of steep downhill highways for trucks with overheated brakes. For a simple 11° upward ramp, what minimum length would be needed for a runaway truck traveling 140 km/h? Note the large size of your calculated length. (If sand is used for the bed of the ramp, its length can be reduced by a factor of about 2.)

56. (II) A 25.0-kg box is released on a 27° incline and accelerates down the incline at 0.30 m/s². Find the friction force impeding its motion. What is the coefficient of kinetic friction?

57. (II) The block shown in Fig. 4–59 has mass $m = 7.0$ kg and lies on a fixed smooth frictionless plane tilted at an angle $\theta = 22.0°$ to the horizontal. (a) Determine the acceleration of the block as it slides down the plane. (b) If the block starts from rest 12.0 m up the plane from its base, what will be the block's speed when it reaches the bottom of the incline?

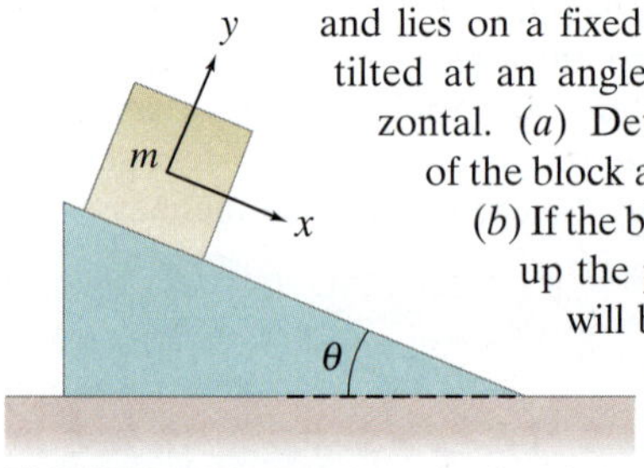

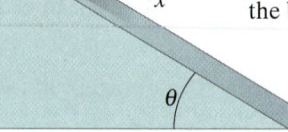

FIGURE 4–59 Block on inclined plane.
Problems 57 and 58.

58. (II) A block is given an initial speed of 4.5 m/s up the 22.0° plane shown in Fig. 4–59. (a) How far up the plane will it go? (b) How much time elapses before it returns to its starting point? Ignore friction.

59. (II) The crate shown in Fig. 4–60 lies on a plane tilted at an angle $\theta = 25.0°$ to the horizontal, with $\mu_k = 0.19$. (a) Determine the acceleration of the crate as it slides down the plane. (b) If the crate starts from rest 8.15 m up along the plane from its base, what will be the crate's speed when it reaches the bottom of the incline?

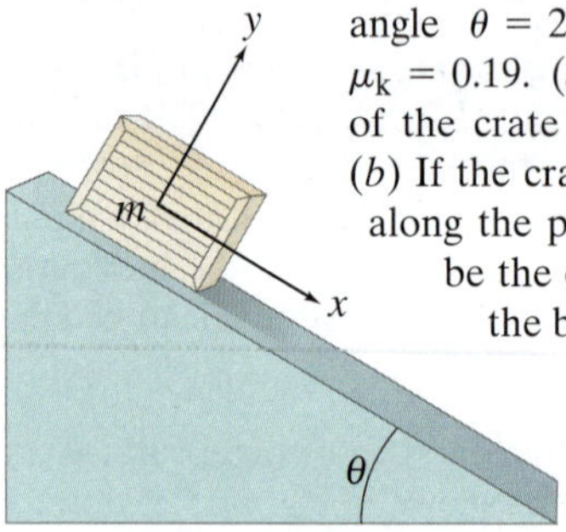

FIGURE 4–60
Crate on inclined plane.
Problems 59 and 60.

60. (II) A crate is given an initial speed of 3.0 m/s up the 25.0° plane shown in Fig. 4–60. (a) How far up the plane will it go? (b) How much time elapses before it returns to its starting point? Assume $\mu_k = 0.12$.

61. (II) A car can decelerate at -3.80 m/s² without skidding when coming to rest on a level road. What would its deceleration be if the road is inclined at 9.3° and the car moves uphill? Assume the same static friction coefficient.

62. (II) A skier moves down a 12° slope at constant speed. What can you say about the coefficient of friction, μ_k? Assume the speed is low enough that air resistance can be ignored.

63. (II) The coefficient of kinetic friction for a 22-kg bobsled on a track is 0.10. What force is required to push it down along a 6.0° incline and achieve a speed of 60 km/h at the end of 75 m?

64. (II) On an icy day, you worry about parking your car in your driveway, which has an incline of 12°. Your neighbor's driveway has an incline of 9.0°, and the driveway across the street is at 6.0°. The coefficient of static friction between tire rubber and ice is 0.15. Which driveway(s) will be safe to park in?

65. (III) Two masses $m_A = 2.0$ kg and $m_B = 5.0$ kg are on inclines and are connected together by a string as shown in Fig. 4–61. The coefficient of kinetic friction between each mass and its incline is $\mu_k = 0.30$. If m_A moves up, and m_B moves down, determine their acceleration. [Ignore masses of the (frictionless) pulley and the cord.]

66. (III) A child slides down a slide with a 34° incline, and at the bottom her speed is precisely half what it would have been if the slide had been frictionless. Calculate the coefficient of kinetic friction between the slide and the child.

67. (III) (a) Suppose the coefficient of kinetic friction between m_A and the plane in Fig. 4–62 is $\mu_k = 0.15$, and that $m_A = m_B = 2.7$ kg. As m_B moves down, determine the magnitude of the acceleration of m_A and m_B, given $\theta = 34°$. (b) What smallest value of μ_k will keep the system from accelerating? [Ignore masses of the (frictionless) pulley and the cord.]

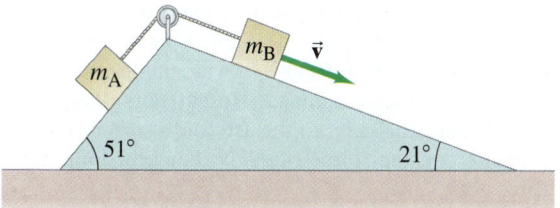

FIGURE 4–61 Problem 65.

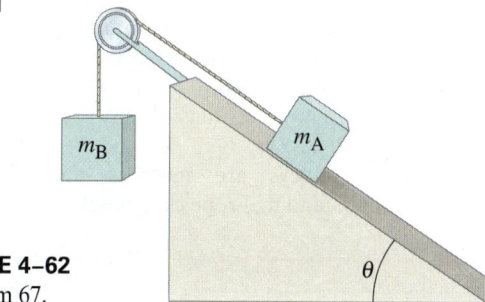

FIGURE 4–62 Problem 67.

General Problems

68. A 2.0-kg purse is dropped from the top of the Leaning Tower of Pisa and falls 55 m before reaching the ground with a speed of 27 m/s. What was the average force of air resistance?

69. A crane's trolley at point P in Fig. 4–63 moves for a few seconds to the right with constant acceleration, and the 870-kg load hangs on a light cable at a 5.0° angle to the vertical as shown. What is the acceleration of the trolley and load?

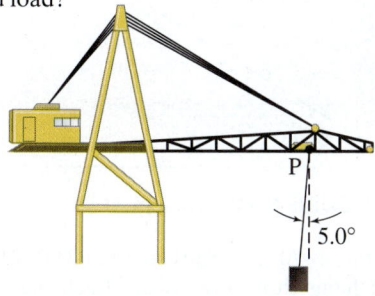

FIGURE 4–63 Problem 69.

70. A 75.0-kg person stands on a scale in an elevator. What does the scale read (in N and in kg) when (a) the elevator is at rest, (b) the elevator is climbing at a constant speed of 3.0 m/s, (c) the elevator is descending at 3.0 m/s, (d) the elevator is accelerating upward at 3.0 m/s², (e) the elevator is accelerating downward at 3.0 m/s²?

71. A city planner is working on the redesign of a hilly portion of a city. An important consideration is how steep the roads can be so that even low-powered cars can get up the hills without slowing down. A particular small car, with a mass of 920 kg, can accelerate on a level road from rest to 21 m/s (75 km/h) in 12.5 s. Using these data, calculate the maximum steepness of a hill.

72. If a bicyclist of mass 65 kg (including the bicycle) can coast down a 6.5° hill at a steady speed of 6.0 km/h because of air resistance, how much force must be applied to climb the hill at the same speed (and the same air resistance)?

73. Francesca dangles her watch from a thin piece of string while the jetliner she is in accelerates for takeoff, which takes about 16 s. Estimate the takeoff speed of the aircraft if the string makes an angle of 25° with respect to the vertical, Fig. 4–64.

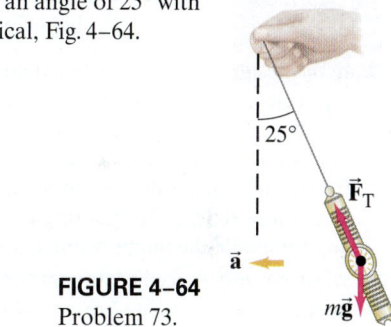

FIGURE 4–64 Problem 73.

74. Bob traverses a chasm by stringing a rope between a tree on one side of the chasm and a tree on the opposite side, 25 m away, Fig. 4–65. Assume the rope can provide a tension force of up to 29 kN before breaking, and use a "safety factor" of 10 (that is, the rope should only be required to undergo a tension force of 2.9 kN). (a) If Bob's mass is 72.0 kg, determine the distance x that the rope must sag at a point halfway across if it is to be within its recommended safety range. (b) If the rope sags by only one-fourth the distance found in (a), determine the tension force in the rope. Will the rope break?

FIGURE 4–65 Problem 74.

75. Piles of snow on slippery roofs can become dangerous projectiles as they melt. Consider a chunk of snow at the ridge of a roof with a slope of 34°. (a) What is the minimum value of the coefficient of static friction that will keep the snow from sliding down? (b) As the snow begins to melt, the coefficient of static friction decreases and the snow finally slips. Assuming that the distance from the chunk to the edge of the roof is 4.0 m and the coefficient of kinetic friction is 0.10, calculate the speed of the snow chunk when it slides off the roof. (c) If the roof edge is 10.0 m above ground, estimate the speed of the snow when it hits the ground.

76. (a) What minimum force F is needed to lift the piano (mass M) using the pulley apparatus shown in Fig. 4–66? (b) Determine the tension in each section of rope: F_{T1}, F_{T2}, F_{T3}, and F_{T4}. Assume pulleys are massless and frictionless, and that ropes are massless.

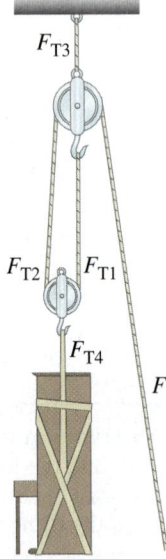

FIGURE 4–66
Problem 76.

77. In the design of a supermarket, there are to be several ramps connecting different parts of the store. Customers will have to push grocery carts up the ramps and it is desirable that this not be too difficult. The engineer has done a survey and found that almost no one complains if the force required is no more than 18 N. Ignoring friction, at what maximum angle θ should the ramps be built, assuming a full 25-kg cart?

78. A jet aircraft is accelerating at 3.8 m/s² as it climbs at an angle of 18° above the horizontal (Fig. 4–67). What is the total force that the cockpit seat exerts on the 75-kg pilot?

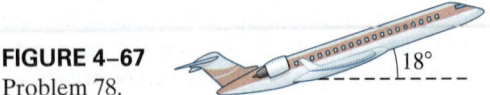

FIGURE 4–67
Problem 78.

79. A 7180-kg helicopter accelerates upward at 0.80 m/s² while lifting a 1080-kg frame at a construction site, Fig. 4–68. (a) What is the lift force exerted by the air on the helicopter rotors? (b) What is the tension in the cable (ignore its mass) which connects the frame to the helicopter? (c) What force does the cable exert on the helicopter?

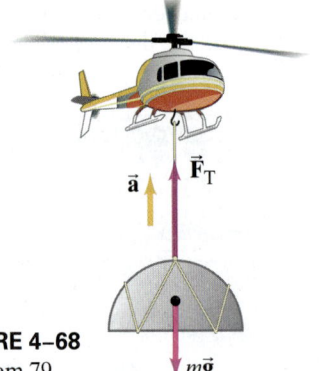

FIGURE 4–68
Problem 79.

80. An elevator in a tall building is allowed to reach a maximum speed of 3.5 m/s going down. What must the tension be in the cable to stop this elevator over a distance of 2.6 m if the elevator has a mass of 1450 kg including occupants?

81. A fisherman in a boat is using a "10-lb test" fishing line. This means that the line can exert a force of 45 N without breaking (1 lb = 4.45 N). (a) How heavy a fish can the fisherman land if he pulls the fish up vertically at constant speed? (b) If he accelerates the fish upward at 2.0 m/s², what maximum weight fish can he land? (c) Is it possible to land a 15-lb trout on 10-lb test line? Why or why not?

82. A "doomsday" asteroid with a mass of 1.0×10^{10} kg is hurtling through space. Unless the asteroid's speed is changed by about 0.20 cm/s, it will collide with Earth and cause tremendous damage. Researchers suggest that a small "space tug" sent to the asteroid's surface could exert a gentle constant force of 2.5 N. For how long must this force act?

83. Three mountain climbers who are roped together in a line are ascending an icefield inclined at 31.0° to the horizontal (Fig. 4–69). The last climber slips, pulling the second climber off his feet. The first climber is able to hold them both. If each climber has a mass of 75 kg, calculate the tension in each of the two sections of rope between the three climbers. Ignore friction between the ice and the fallen climbers.

FIGURE 4–69 Problem 83.

84. As shown in Fig. 4–70, five balls (masses 2.00, 2.05, 2.10, 2.15, 2.20 kg) hang from a crossbar. Each mass is supported by "5-lb test" fishing line which will break when its tension force exceeds 22.2 N (= 5.00 lb). When this device is placed in an elevator, which accelerates upward, only the lines attached to the 2.05 and 2.00 kg masses do not break. Within what range is the elevator's acceleration?

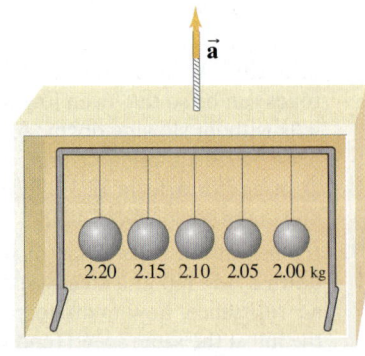

FIGURE 4–70
Problem 84.

85. Two rock climbers, Jim and Karen, use safety ropes of similar length. Karen's rope is more elastic, called a *dynamic rope* by climbers. Jim has a *static rope*, not recommended for safety purposes in pro climbing. (*a*) Karen (Fig. 4–71) falls freely about 2.0 m and then the rope stops her over a distance of 1.0 m. Estimate how large a force (assume constant) she will feel from the rope. (Express the result in multiples of her weight.) (*b*) In a similar fall, Jim's rope stretches by only 30 cm. How many times his weight will the rope pull on him? Which climber is more likely to be hurt?

FIGURE 4–71 Problem 85.

86. A coffee cup on the horizontal dashboard of a car slides forward when the driver decelerates from 45 km/h to rest in 3.5 s or less, but not if she decelerates in a longer time. What is the coefficient of static friction between the cup and the dash? Assume the road and the dashboard are level (horizontal).

87. A roller coaster reaches the top of the steepest hill with a speed of 6.0 km/h. It then descends the hill, which is at an average angle of 45° and is 45.0 m long. What will its speed be when it reaches the bottom? Assume $\mu_k = 0.12$.

88. A motorcyclist is coasting with the engine off at a steady speed of 20.0 m/s but enters a sandy stretch where the coefficient of kinetic friction is 0.70. Will the cyclist emerge from the sandy stretch without having to start the engine if the sand lasts for 15 m? If so, what will be the speed upon emerging?

89. The 70.0-kg climber in Fig. 4–72 is supported in the "chimney" by the friction forces exerted on his shoes and back. The static coefficients of friction between his shoes and the wall, and between his back and the wall, are 0.80 and 0.60, respectively. What is the minimum normal force he must exert? Assume the walls are vertical and that the static friction forces are both at their maximum. Ignore his grip on the rope.

FIGURE 4–72 Problem 89.

90. A 28.0-kg block is connected to an empty 2.00-kg bucket by a cord running over a frictionless pulley (Fig. 4–73). The coefficient of static friction between the table and the block is 0.45 and the coefficient of kinetic friction between the table and the block is 0.32. Sand is gradually added to the bucket until the system just begins to move. (*a*) Calculate the mass of sand added to the bucket. (*b*) Calculate the acceleration of the system. Ignore mass of cord.

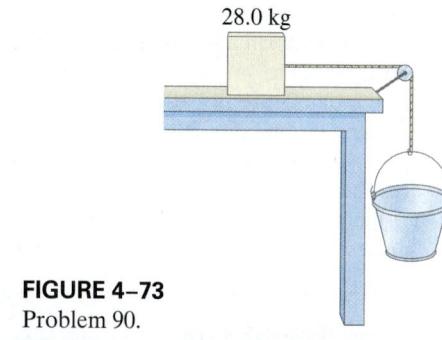

FIGURE 4–73 Problem 90.

91. A 72-kg water skier is being accelerated by a ski boat on a flat ("glassy") lake. The coefficient of kinetic friction between the skier's skis and the water surface is $\mu_k = 0.25$ (Fig. 4–74). (*a*) What is the skier's acceleration if the rope pulling the skier behind the boat applies a horizontal tension force of magnitude $F_T = 240$ N to the skier ($\theta = 0°$)? (*b*) What is the skier's horizontal acceleration if the rope pulling the skier exerts a force of $F_T = 240$ N on the skier at an upward angle $\theta = 12°$? (*c*) Explain why the skier's acceleration in part (*b*) is greater than that in part (*a*).

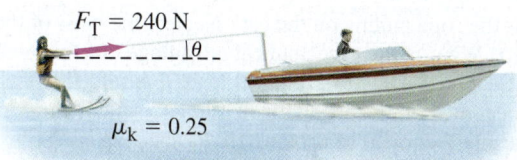

FIGURE 4–74 Problem 91.

92. A 75-kg snowboarder has an initial velocity of 5.0 m/s at the top of a 28° incline (Fig. 4–75). After sliding down the 110-m-long incline (on which the coefficient of kinetic friction is $\mu_k = 0.18$), the snowboarder has attained a velocity v. The snowboarder then slides along a flat surface (on which $\mu_k = 0.15$) and comes to rest after a distance x. Use Newton's second law to find the snowboarder's acceleration while on the incline and while on the flat surface. Then use these accelerations to determine x.

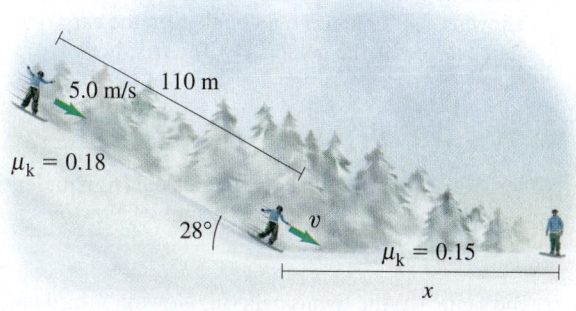

FIGURE 4–75 Problem 92.

93. (a) If the horizontal acceleration produced briefly by an earthquake is a, and if an object is going to "hold its place" on the ground, show that the coefficient of static friction with the ground must be at least $\mu_s = a/g$. (b) The famous Loma Prieta earthquake that stopped the 1989 World Series produced ground accelerations of up to 4.0 m/s^2 in the San Francisco Bay Area. Would a chair have started to slide on a floor with coefficient of static friction 0.25?

94. Two blocks made of different materials, connected by a thin cord, slide down a plane ramp inclined at an angle θ to the horizontal, Fig. 4–76 (block B is above block A). The masses of the blocks are m_A and m_B, and the coefficients of friction are μ_A and μ_B. If $m_A = m_B = 5.0 \text{ kg}$, and $\mu_A = 0.20$ and $\mu_B = 0.30$, determine (a) the acceleration of the blocks and (b) the tension in the cord, for an angle $\theta = 32°$.

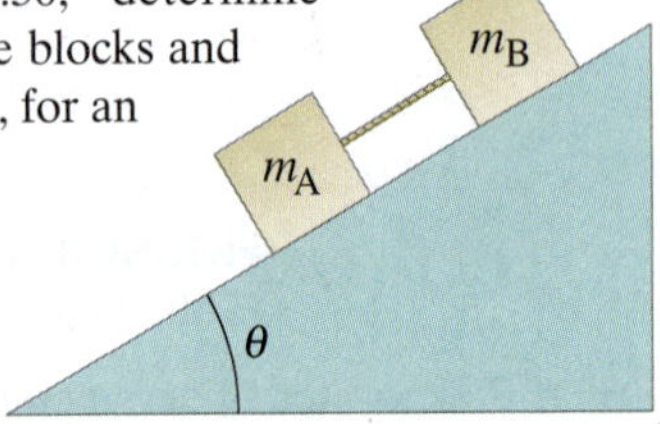

FIGURE 4–76 Problem 94.

95. A car starts rolling down a 1-in-4 hill (1-in-4 means that for each 4 m traveled along the sloping road, the elevation change is 1 m). How fast is it going when it reaches the bottom after traveling 55 m? (a) Ignore friction. (b) Assume an effective coefficient of friction equal to 0.10.

96. A 65-kg ice skater coasts with no effort for 75 m until she stops. If the coefficient of kinetic friction between her skates and the ice is $\mu_k = 0.10$, how fast was she moving at the start of her coast?

97. An 18-kg child is riding in a child-restraint chair, securely fastened to the seat of a car (Fig. 4–77). Assume the car has speed 45 km/h when it hits a tree and is brought to rest in 0.20 s. Assuming constant deceleration during the collision, estimate the net horizontal force F that the straps of the restraint chair exert on the child to hold her in the chair.

FIGURE 4–77 Problem 97.

Search and Learn

1. (a) Finding her car stuck in the mud, a bright graduate of a good physics course ties a strong rope to the back bumper of the car, and the other end to a boulder, as shown in Fig. 4–78a. She pushes at the midpoint of the rope with her maximum effort, which she estimates to be a force $F_P \approx 300 \text{ N}$. The car just begins to budge with the rope at an angle θ, which she estimates to be 5°. With what force is the rope pulling on the car? Neglect the mass of the rope. (b) What is the "mechanical advantage" of this technique [Section 4–7]? (c) At what angle θ would this technique become counterproductive? [*Hint*: Consider the forces on a small segment of rope where $\vec{F}_P$ acts, Fig. 4–78b.]

2. (a) Show that the minimum stopping distance for an automobile traveling on a level road at speed v is equal to $v^2/(2\mu_s g)$, where μ_s is the coefficient of static friction between the tires and the road, and g is the acceleration of gravity. (b) What is this distance for a 1200-kg car traveling 95 km/h if $\mu_s = 0.65$? (c) What would it be if the car were on the Moon (the acceleration of gravity on the Moon is about $g/6$) but all else stayed the same?

3. In the equation for static friction in Section 4–8, what is the significance of the $<$ sign? When should you use the equals sign in the static friction equation?

4. Referring to Example 4–21, show that if a skier moves at constant speed straight down a slope of angle θ, then the coefficient of kinetic friction between skis and snow is $\mu_k = \tan \theta$.

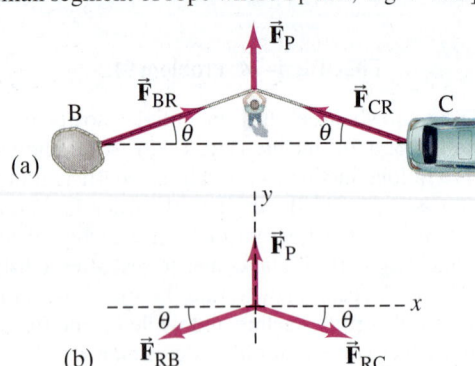

FIGURE 4–78 (a) Getting a car out of the mud, showing the forces on the boulder, on the car, and exerted by the person. (b) The free-body diagram: forces on a small segment of rope.

ANSWERS TO EXERCISES

A: No force is needed. The car accelerates out from under the cup, which tends to remain at rest. Think of Newton's first law (see Example 4–1).
B: (a).
C: (i) The same; (ii) the tennis ball; (iii) Newton's third law for part (i), second law for part (ii).
D: (b).
E: (b).
F: (b).
G: (c).
H: Yes; no.

The astronauts in the upper left of this photo are working on a space shuttle. As they orbit the Earth—at a rather high speed—they experience apparent weightlessness. The Moon, in the background seen against the blackness of space, also is orbiting the Earth at high speed. Both the Moon and the space shuttle move in nearly circular orbits, and each undergoes a centripetal acceleration. What keeps the Moon and the space shuttle (and its astronauts) from moving off in a straight line away from Earth? It is the force of gravity. Newton's law of universal gravitation states that all objects attract all other objects with a force that depends on their masses and the square of the distance between them.

Circular Motion; Gravitation

CHAPTER 5

CHAPTER-OPENING QUESTIONS—Guess now!

1. You revolve a ball around you in a horizontal circle at constant speed on a string, as shown here from above. Which path will the ball follow if you let go of the string when the ball is at point P?

2. A space station revolves around the Earth as a satellite, 100 km above Earth's surface. What is the net force on an astronaut at rest inside the space station?

(a) Equal to her weight on Earth.
(b) A little less than her weight on Earth.
(c) Less than half her weight on Earth.
(d) Zero (she is weightless).
(e) Somewhat larger than her weight on Earth.

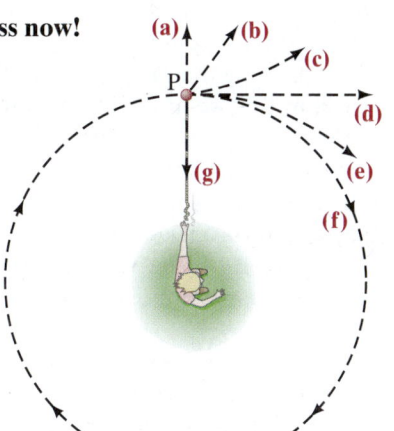

CONTENTS

5–1 Kinematics of Uniform Circular Motion
5–2 Dynamics of Uniform Circular Motion
5–3 Highway Curves: Banked and Unbanked
*5–4 Nonuniform Circular Motion
5–5 Newton's Law of Universal Gravitation
5–6 Gravity Near the Earth's Surface
5–7 Satellites and "Weightlessness"
5–8 Planets, Kepler's Laws, and Newton's Synthesis
5–9 Moon Rises an Hour Later Each Day
5–10 Types of Forces in Nature

An object moves in a straight line if the net force on it acts along the direction of motion, or the net force is zero. If the net force acts at an angle to the direction of motion at any moment, then the object moves in a curved path. An example of the latter is projectile motion, which we discussed in Chapter 3. Another important case is that of an object moving in a circle, such as a ball at the end of a string being swung in a circle above one's head, or the nearly circular motion of the Moon about the Earth.

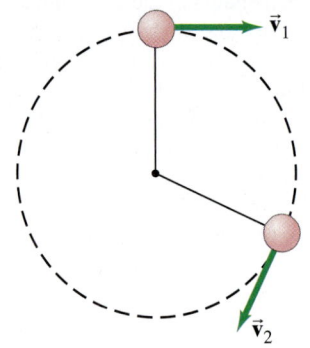

FIGURE 5–1 A small object moving in a circle, showing how the velocity changes. At each point, the instantaneous velocity is in a direction tangent to the circular path.

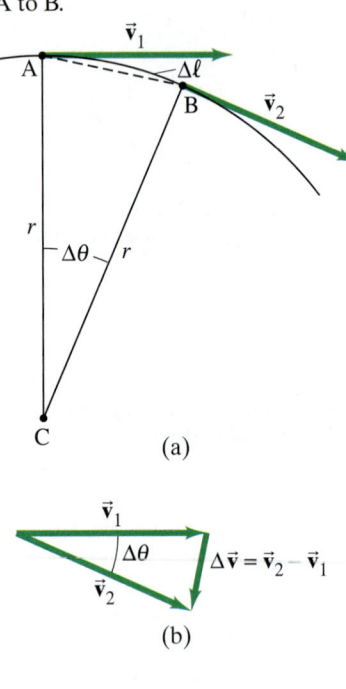

FIGURE 5–2 Determining the change in velocity, $\Delta\vec{v}$, for a particle moving in a circle. The length $\Delta\ell$ is the distance along the arc, from A to B.

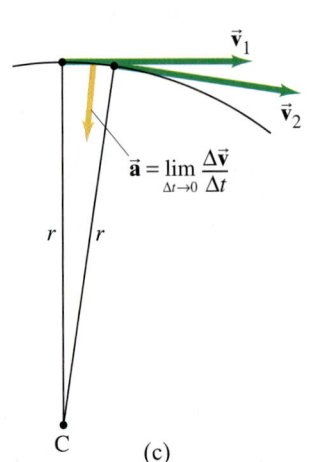

In this Chapter, we study the circular motion of objects, and how Newton's laws of motion apply. We also discuss how Newton conceived of another great law by applying the concepts of circular motion to the motion of the Moon and the planets. This is the law of universal gravitation, which was the capstone of Newton's analysis of the physical world.

5–1 Kinematics of Uniform Circular Motion

An object that moves in a circle at constant speed v is said to experience **uniform circular motion**. The *magnitude* of the velocity remains constant in this case, but the *direction* of the velocity continuously changes as the object moves around the circle (Fig. 5–1). Because acceleration is defined as the rate of change of velocity, a change in direction of velocity is an acceleration, just as a change in its magnitude is. Thus, an object revolving in a circle is continuously accelerating, even when the speed remains constant ($v_1 = v_2 = v$ in Fig. 5–1). We now investigate this acceleration quantitatively.

Acceleration is defined as

$$\vec{a} = \frac{\vec{v}_2 - \vec{v}_1}{\Delta t} = \frac{\Delta\vec{v}}{\Delta t},$$

where $\Delta\vec{v}$ is the change in velocity during the short time interval Δt. We will eventually consider the situation in which Δt approaches zero and thus obtain the instantaneous acceleration. But for purposes of making a clear drawing, Fig. 5–2, we consider a nonzero time interval. During the time interval Δt, the particle in Fig. 5–2a moves from point A to point B, covering a distance $\Delta\ell$ *along the arc* which subtends an angle $\Delta\theta$. The change in the velocity vector is $\vec{v}_2 - \vec{v}_1 = \Delta\vec{v}$, and is shown in Fig. 5–2b (note that $\vec{v}_2 = \vec{v}_1 + \Delta\vec{v}$).

Now we let Δt be very small, approaching zero. Then $\Delta\ell$ and $\Delta\theta$ are also very small, and $\vec{v}_2$ will be almost parallel to $\vec{v}_1$, Fig. 5–2c; $\Delta\vec{v}$ will be essentially perpendicular to them. Thus $\Delta\vec{v}$ points toward the center of the circle. Since $\vec{a}$, by definition, is in the same direction as $\Delta\vec{v}$ (equation above), it too must point toward the center of the circle. Therefore, this acceleration is called **centripetal acceleration** ("center-pointing" acceleration) or **radial acceleration** (since it is directed along the radius, toward the center of the circle), and we denote it by $\vec{a}_R$.

Now that we have determined the direction, next we find the magnitude of the radial (centripetal) acceleration, a_R. Because the line CA in Fig. 5–2a is perpendicular to $\vec{v}_1$, and line CB is perpendicular to $\vec{v}_2$, then the angle $\Delta\theta$ between CA and CB is also the angle between $\vec{v}_1$ and $\vec{v}_2$. Hence the vectors $\vec{v}_1$, $\vec{v}_2$, and $\Delta\vec{v}$ in Fig. 5–2b form a triangle that is geometrically similar[†] to triangle ACB in Fig. 5–2a. If we take $\Delta\theta$ to be very small (letting Δt be very small) and set $v = v_1 = v_2$ because the magnitude of the velocity is assumed not to change, we can write

$$\frac{\Delta v}{v} \approx \frac{\Delta\ell}{r}.$$

This is an exact equality when Δt approaches zero, for then the arc length $\Delta\ell$ equals the chord length AB. We want to find the instantaneous acceleration, so we let Δt approach zero, write the above expression as an equality, and then solve for Δv:

$$\Delta v = \frac{v}{r}\Delta\ell. \qquad [\Delta t \to 0]$$

To get the centripetal acceleration, a_R, we divide Δv by Δt:

$$a_R = \frac{\Delta v}{\Delta t} = \frac{v}{r}\frac{\Delta\ell}{\Delta t}. \qquad [\Delta t \to 0]$$

But $\Delta\ell/\Delta t$ is the linear speed, v, of the object, so the radial (centripetal)

[†]Appendix A contains a review of geometry.

acceleration is

$$a_R = \frac{v^2}{r}.$$ [radial (centripetal) acceleration]   (5-1)

[Equation 5-1 is valid at any instant in circular motion, and even when v is not constant.]

To summarize, *an object moving in a circle of radius r at constant speed v has an acceleration whose direction is toward the center of the circle and whose magnitude is* $a_R = v^2/r$. It is not surprising that this acceleration depends on v and r. The greater the speed v, the faster the velocity changes direction; and the larger the radius, the less rapidly the velocity changes direction.

The acceleration vector points toward the center of the circle when v is constant. But the velocity vector always points in the direction of motion, which is tangential to the circle. Thus the velocity and acceleration vectors are perpendicular to each other at every point in the path for uniform circular motion (Fig. 5-3). This is another example that illustrates the error in thinking that acceleration and velocity are always in the same direction. For an object falling in a vertical path, $\vec{a}$ and $\vec{v}$ are indeed parallel. But in uniform circular motion, $\vec{a}$ and $\vec{v}$ are perpendicular, not parallel (nor were they parallel in projectile motion, Section 3-5).

Circular motion is often described in terms of the **frequency** f, the number of revolutions per second. The **period** T of an object revolving in a circle is the time required for one complete revolution. Period and frequency are related by

$$T = \frac{1}{f}.$$ (5-2)

For example, if an object revolves at a frequency of 3 rev/s, then each revolution (= rev) takes $\frac{1}{3}$s. An object revolving in a circle (of circumference $2\pi r$) at constant speed v travels a distance $2\pi r$ in one revolution which takes a time T. Thus

$$v = \frac{\text{distance}}{\text{time}} = \frac{2\pi r}{T}.$$

> **CAUTION**
> *In uniform circular motion, the speed is constant, but the acceleration is not zero*

> **CAUTION**
> *The direction of motion ($\vec{v}$) and the acceleration ($\vec{a}$) are not in the same direction; instead, $\vec{a} \perp \vec{v}$*

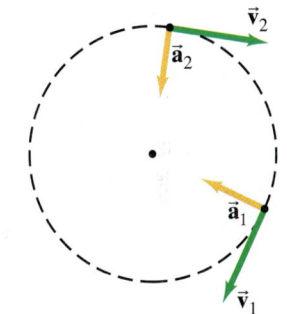

FIGURE 5-3 For uniform circular motion, $\vec{a}$ is always perpendicular to $\vec{v}$.

EXAMPLE 5-1 **Acceleration of a revolving ball.** A 150-g ball at the end of a string is revolving uniformly in a horizontal circle of radius 0.600 m, as in Fig. 5-1 or 5-3. The ball makes 2.00 revolutions in a second. What is its centripetal acceleration?

APPROACH The centripetal acceleration is $a_R = v^2/r$. We are given r, and we can find the speed of the ball, v, from the given radius and frequency.

SOLUTION If the ball makes 2.00 complete revolutions per second, then the ball travels in a complete circle in a time interval equal to 0.500 s, which is its period T. The distance traveled in this time is the circumference of the circle, $2\pi r$, where r is the radius of the circle. Therefore, the ball has speed

$$v = \frac{2\pi r}{T} = \frac{2\pi(0.600 \text{ m})}{(0.500 \text{ s})} = 7.54 \text{ m/s}.$$

The centripetal acceleration[†] is

$$a_R = \frac{v^2}{r} = \frac{(7.54 \text{ m/s})^2}{(0.600 \text{ m})} = 94.7 \text{ m/s}^2.$$

EXERCISE A In Example 5-1, if the radius is doubled to 1.20 m, but the period stays the same, the centripetal acceleration will change by a factor of:
(a) 2; (b) 4; (c) $\frac{1}{2}$; (d) $\frac{1}{4}$; (e) none of these.

[†]Differences in the final digit can depend on whether you keep all digits in your calculator for v (which gives $a_R = 94.7 \text{ m/s}^2$), or if you use $v = 7.54 \text{ m/s}$ (which gives $a_R = 94.8 \text{ m/s}^2$). Both results are valid since our assumed accuracy is about $\pm 0.1 \text{ m/s}^2$ (see Section 1-4).

EXAMPLE 5–2 **Moon's centripetal acceleration.** The Moon's nearly circular orbit around the Earth has a radius of about 384,000 km and a period T of 27.3 days. Determine the acceleration of the Moon toward the Earth.

APPROACH Again we need to find the velocity v in order to find a_R.

SOLUTION In one orbit around the Earth, the Moon travels a distance $2\pi r$, where $r = 3.84 \times 10^8$ m is the radius of its circular path. The time required for one complete orbit is the Moon's period of 27.3 d. The speed of the Moon in its orbit about the Earth is $v = 2\pi r/T$. The period T in seconds is $T = (27.3 \text{ d})(24.0 \text{ h/d})(3600 \text{ s/h}) = 2.36 \times 10^6$ s. Therefore,

$$a_R = \frac{v^2}{r} = \frac{(2\pi r)^2}{T^2 r} = \frac{4\pi^2 r}{T^2} = \frac{4\pi^2 (3.84 \times 10^8 \text{ m})}{(2.36 \times 10^6 \text{ s})^2}$$

$$= 0.00272 \text{ m/s}^2 = 2.72 \times 10^{-3} \text{ m/s}^2.$$

We can write this acceleration in terms of $g = 9.80$ m/s^2 (the acceleration of gravity at the Earth's surface) as

$$a_R = 2.72 \times 10^{-3} \text{ m/s}^2 \left(\frac{g}{9.80 \text{ m/s}^2}\right) = 2.78 \times 10^{-4} g$$

$$\approx 0.0003 \, g.$$

> **CAUTION**
> *Distinguish the Moon's gravity on objects at its surface from the Earth's gravity acting on the Moon (this Example)*

NOTE The centripetal acceleration of the Moon, $a_R = 2.78 \times 10^{-4} g$, is *not* the acceleration of gravity for objects at the Moon's surface due to the Moon's gravity. Rather, it is the acceleration due to the *Earth's* gravity for any object (such as the Moon) that is 384,000 km from the Earth. Notice how small this acceleration is compared to the acceleration of objects near the Earth's surface.

5–2 Dynamics of Uniform Circular Motion

According to Newton's second law $(\Sigma \vec{F} = m\vec{a})$, an object that is accelerating must have a net force acting on it. An object moving in a circle, such as a ball on the end of a string, must therefore have a force applied to it to keep it moving in that circle. That is, a net force is necessary to give it centripetal acceleration. The magnitude of the required force can be calculated using Newton's second law for the radial component, $\Sigma F_R = ma_R$, where a_R is the centripetal acceleration, $a_R = v^2/r$, and ΣF_R is the total (or net) force in the radial direction:

$$\Sigma F_R = ma_R = m\frac{v^2}{r}. \qquad \text{[circular motion]} \quad (5\text{–}3)$$

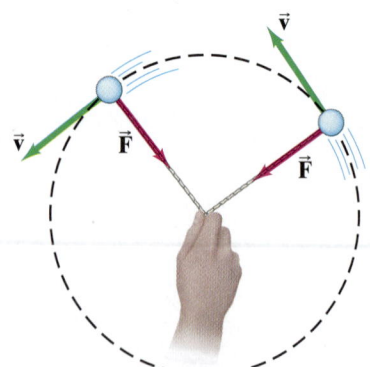

FIGURE 5–4 A force is required to keep an object moving in a circle. If the speed is constant, the force is directed toward the circle's center.

> **CAUTION**
> *Centripetal force is not a new kind of force (Every force must be exerted by an object)*

For uniform circular motion (v = constant), the acceleration is a_R, which is directed toward the center of the circle at all times. Thus the *net force too must be directed toward the center of the circle* (Fig. 5–4). A net force is necessary because if no net force were exerted on the object, it would not move in a circle but in a straight line, as Newton's first law tells us. The direction of the net force is continually changing so that it is always directed toward the center of the circle. This force is sometimes called a centripetal ("pointing toward the center") force. But be aware that "centripetal force" does not indicate some new kind of force. The term "centripetal force" merely describes the *direction* of the net force needed to provide a circular path: the net force is directed toward the circle's center. The force must be applied by other objects. For example, to swing a ball in a circle on the end of a string, you pull on the string and the string exerts the force on the ball. (Try it.) Here, the "centripetal force" that provides the centripetal acceleration is tension in the string. In other cases it can be gravity (on the Moon, for example), a normal force, or even an electric force.

There is a common misconception that an object moving in a circle has an outward force acting on it, a so-called centrifugal ("center-fleeing") force. This is incorrect: *there is no outward force* on the revolving object. Consider, for example, a person swinging a ball on the end of a string around her head (Fig. 5–5). If you have ever done this yourself, you know that you feel a force pulling outward on your hand. The misconception arises when this pull is interpreted as an outward "centrifugal" force pulling on the ball that is transmitted along the string to your hand. This is not what is happening at all. To keep the ball moving in a circle, you pull *inwardly* on the string, and the string exerts this inward force on the ball. The ball exerts an equal and opposite force on the string (Newton's third law), and *this* is the outward force your hand feels (see Fig. 5–5).

The force *on the ball* in Fig. 5–5 is the one exerted *inwardly* on it by you, via the string. To see even more convincing evidence that a "centrifugal force" does not act on the ball, consider what happens when you let go of the string. If a centrifugal force were acting, the ball would fly outward, as shown in Fig. 5–6a. But it doesn't; the ball flies off tangentially (Fig. 5–6b), in the direction of the velocity it had at the moment it was released, because the inward force no longer acts. Try it and see!

EXERCISE B Return to Chapter-Opening Question 1, page 109, and answer it again now. Try to explain why you may have answered differently the first time.

> ⚠ **CAUTION**
> *There is no real "centrifugal force"*

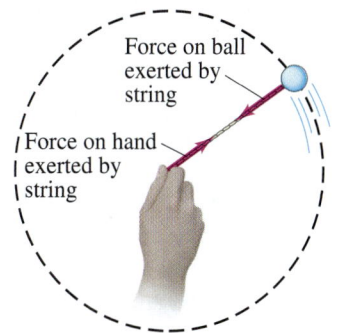

FIGURE 5–5 Swinging a ball on the end of a string (looking down from above).

FIGURE 5–6 If centrifugal force existed, the revolving ball would fly outward as in (a) when released. In fact, it flies off tangentially as in (b). In (c) sparks fly in straight lines tangentially from the edge of a rotating grinding wheel.

EXAMPLE 5–3 ESTIMATE Force on revolving ball (horizontal). Estimate the force a person must exert on a string attached to a 0.150-kg ball to make the ball revolve in a horizontal circle of radius 0.600 m. The ball makes 2.00 revolutions per second ($T = 0.500$ s), as in Example 5–1. Ignore the string's mass.

APPROACH First we need to draw the free-body diagram for the ball. The forces acting on the ball are the force of gravity, $m\vec{g}$ downward, and the tension force $\vec{F}_T$ that the string exerts toward the hand at the center (which occurs because the person exerts that same force on the string). The free-body diagram for the ball is shown in Fig. 5–7. The ball's weight complicates matters and makes it impossible to revolve a ball with the cord perfectly horizontal. We estimate the force assuming the weight is small, and letting $\phi \approx 0$ in Fig. 5–7. Then $\vec{F}_T$ will act nearly horizontally and, in any case, provides the force necessary to give the ball its centripetal acceleration.

SOLUTION We apply Newton's second law to the radial direction, which we assume is horizontal:

$$(\Sigma F)_R = ma_R,$$

where $a_R = v^2/r$ and $v = 2\pi r/T = 2\pi(0.600 \text{ m})/(0.500 \text{ s}) = 7.54 \text{ m/s}$. Thus

$$F_T = m\frac{v^2}{r}$$

$$= (0.150 \text{ kg})\frac{(7.54 \text{ m/s})^2}{(0.600 \text{ m})} \approx 14 \text{ N}.$$

NOTE We keep only two significant figures in the answer because we ignored the ball's weight; it is $mg = (0.150 \text{ kg})(9.80 \text{ m/s}^2) = 1.5$ N, about $\frac{1}{10}$ of our result, which is small but maybe not so small as to justify stating a more precise answer for F_T.

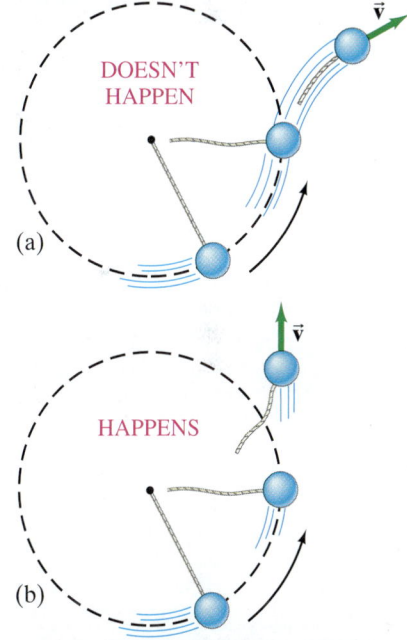

(c)

FIGURE 5–7 Example 5–3.

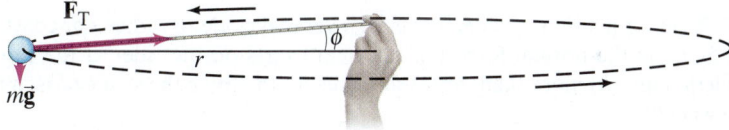

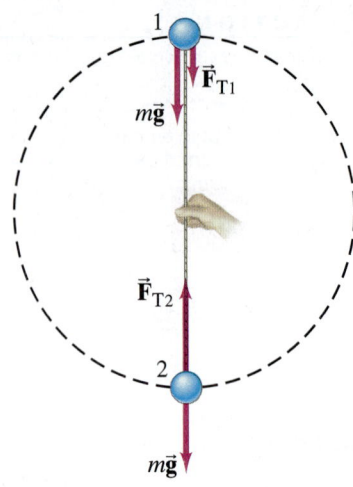

FIGURE 5–8 Example 5–4. Free-body diagrams for positions 1 and 2.

⚠ **CAUTION**
Circular motion only if cord is under tension

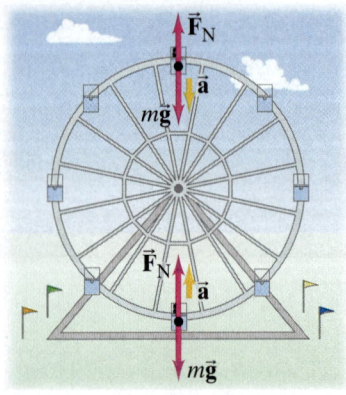

FIGURE 5–9 Exercise C.

EXAMPLE 5–4 Revolving ball (vertical circle). A 0.150-kg ball on the end of a 1.10-m-long cord (negligible mass) is swung in a *vertical* circle. (*a*) Determine the minimum speed the ball must have at the top of its arc so that the ball continues moving in a circle. (*b*) Calculate the tension in the cord at the bottom of the arc, assuming the ball is moving at twice the speed of part (*a*).

APPROACH The ball moves in a vertical circle and is *not* undergoing uniform circular motion. The radius is assumed constant, but the speed v changes because of gravity. Nonetheless, Eq. 5–1 ($a_R = v^2/r$) is valid at each point along the circle, and we use it at the top and bottom points. The free-body diagram is shown in Fig. 5–8 for both positions.

SOLUTION (*a*) At the top (point 1), two forces act on the ball: $m\vec{g}$, the force of gravity, and $\vec{F}_{T1}$, the tension force the cord exerts at point 1. Both act downward, and their vector sum acts to give the ball its centripetal acceleration a_R. We apply Newton's second law, for the vertical direction, choosing downward as positive since the acceleration is downward (toward the center):

$$(\Sigma F)_R = ma_R$$
$$F_{T1} + mg = m\frac{v_1^2}{r}. \qquad \text{[at top]}$$

From this equation we can see that the tension force F_{T1} at point 1 will get larger if v_1 (ball's speed at top of circle) is made larger, as expected. But we are asked for the *minimum* speed to keep the ball moving in a circle. The cord will remain taut as long as there is tension in it. But if the tension disappears (because v_1 is too small) the cord can go limp, and the ball will fall out of its circular path. Thus, the minimum speed will occur if $F_{T1} = 0$ (the ball at the topmost point), for which the equation above becomes

$$mg = m\frac{v_1^2}{r}. \qquad \text{[minimum speed at top]}$$

We solve for v_1, keeping an extra digit for use in (*b*):

$$v_1 = \sqrt{gr} = \sqrt{(9.80 \text{ m/s}^2)(1.10 \text{ m})}$$
$$= 3.283 \text{ m/s} \approx 3.28 \text{ m/s}.$$

This is the minimum speed at the top of the circle if the ball is to continue moving in a circular path.

(*b*) When the ball is at the bottom of the circle (point 2 in Fig. 5–8), the cord exerts its tension force F_{T2} upward, whereas the force of gravity, $m\vec{g}$, still acts downward. Choosing *upward* as positive, Newton's second law gives:

$$(\Sigma F)_R = ma_R$$
$$F_{T2} - mg = m\frac{v_2^2}{r}. \qquad \text{[at bottom]}$$

The speed v_2 is given as twice that in (*a*), namely 6.566 m/s. We solve for F_{T2}:

$$F_{T2} = m\frac{v_2^2}{r} + mg$$
$$= (0.150 \text{ kg})\frac{(6.566 \text{ m/s})^2}{(1.10 \text{ m})} + (0.150 \text{ kg})(9.80 \text{ m/s}^2) = 7.35 \text{ N}.$$

EXERCISE C A rider on a Ferris wheel moves in a vertical circle of radius r at constant speed v (Fig. 5–9). Is the normal force that the seat exerts on the rider at the top of the wheel (*a*) less than, (*b*) more than, or (*c*) the same as, the force the seat exerts at the bottom of the wheel?

CONCEPTUAL EXAMPLE 5–5 **Tetherball.** The game of tetherball is played with a ball tied to a pole with a cord. After the ball is struck, it revolves around the pole as shown in Fig. 5–10. In what direction is the acceleration of the ball, and what force causes the acceleration, assuming constant speed?

RESPONSE If the ball revolves in a horizontal plane as shown, then the acceleration points horizontally toward the center of the ball's circular path (not toward the top of the pole). The force responsible for the acceleration may not be obvious at first, since there seems to be no force pointing directly horizontally. But it is the *net* force (the sum of $m\vec{g}$ and $\vec{F}_T$ here) that must point in the direction of the acceleration. The vertical component of the cord tension, F_{Ty}, balances the ball's weight, $m\vec{g}$. The horizontal component of the cord tension, F_{Tx}, is the force that produces the centripetal acceleration toward the center.

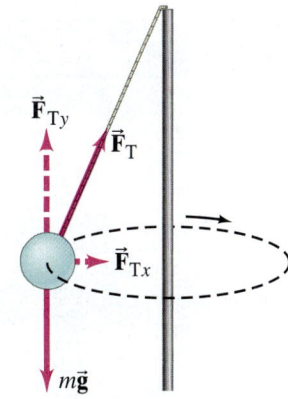

FIGURE 5–10 Example 5–5.

Uniform Circular Motion

1. **Draw a free-body diagram**, showing all the forces acting on each object under consideration. Be sure you can identify the source of each force (tension in a cord, Earth's gravity, friction, normal force, and so on). Don't put in something that doesn't belong (like a centrifugal force).

2. **Determine** which of the forces, or which of their components, act to provide the centripetal acceleration—that is, all the **forces or components that act radially**, toward or away from the center of the circular path. The sum of these forces (or components) provides the centripetal acceleration, $a_R = v^2/r$.

3. **Choose a convenient coordinate system**, preferably with one axis along the acceleration direction.

4. **Apply Newton's second law** to the radial component:

$$\Sigma F_R = ma_R = m\frac{v^2}{r}. \quad \text{[radial direction]}$$

5–3 Highway Curves: Banked and Unbanked

An example of circular dynamics occurs when an automobile rounds a curve, say to the left. In such a situation, you may feel that you are thrust outward toward the right side door. But there is no mysterious centrifugal force pulling on you. What is happening is that you tend to move in a straight line, whereas the car has begun to follow a curved path. To make you go in the curved path, the seat (friction) or the door of the car (direct contact) exerts a force on you (Fig. 5–11). The car also must have a force exerted on it toward the center of the curve if it is to move in that curve. On a flat road, this force is supplied by friction between the tires and the pavement.

PHYSICS APPLIED
Driving around a curve

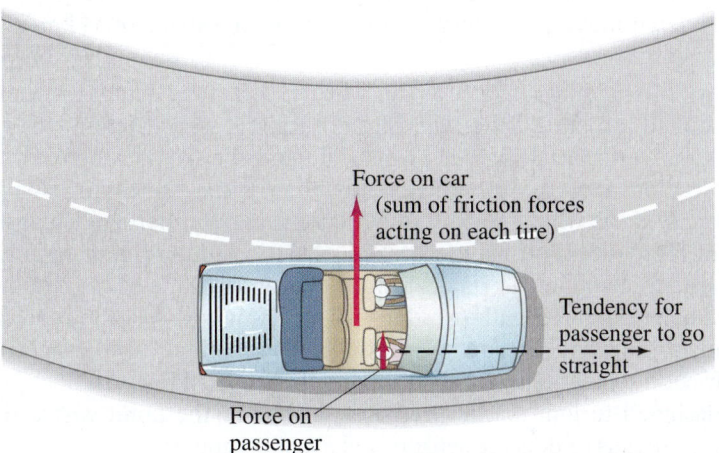

FIGURE 5–11 The road exerts an inward force (friction against the tires) on a car to make it move in a circle. The car exerts an inward force on the passenger.

FIGURE 5–12 Race car heading into a curve. From the tire marks we see that most cars experienced a sufficient friction force to give them the needed centripetal acceleration for rounding the curve safely. But, we also see tire tracks of cars on which there was not sufficient force—and which unfortunately followed more nearly straight-line paths.

FIGURE 5–13 Example 5–6. Forces on a car rounding a curve on a flat road. (a) Front view, (b) top view.

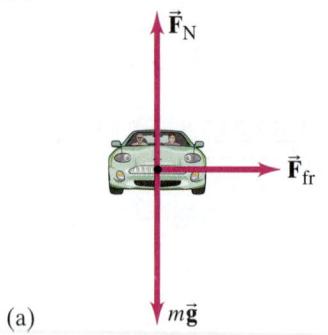

(a)

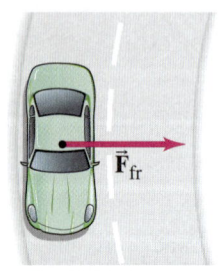

(b)

PHYSICS APPLIED
Antilock brakes

If the wheels and tires of the car are rolling normally without slipping or sliding, the bottom of the tire is at rest against the road at each instant. So the friction force the road exerts on the tires is static friction. But if static friction is not great enough, as under icy conditions or high speed, the static friction force is less than mv^2/r and the car will skid out of a circular path into a more nearly straight path. See Fig. 5–12. Once a car skids or slides, the friction force becomes kinetic friction, which is smaller than static friction.

EXAMPLE 5–6 Skidding on a curve. A 1000-kg car rounds a curve on a flat road of radius 50 m at a speed of 15 m/s (54 km/h). Will the car follow the curve, or will it skid? Assume: (a) the pavement is dry and the coefficient of static friction is $\mu_s = 0.60$; (b) the pavement is icy and $\mu_s = 0.25$.

APPROACH The forces on the car are gravity mg downward, the normal force F_N exerted upward by the road, and a horizontal friction force due to the road. They are shown in Fig. 5–13, which is the free-body diagram for the car. The car will follow the curve if the maximum static friction force is greater than the mass times the centripetal acceleration.

SOLUTION In the vertical direction (y) there is no acceleration. Newton's second law tells us that the normal force F_N on the car is equal to the weight mg since the road is flat:

$$0 = \Sigma F_y = F_N - mg$$

so

$$F_N = mg = (1000 \text{ kg})(9.80 \text{ m/s}^2) = 9800 \text{ N}.$$

In the horizontal direction the only force is friction, and we must compare it to the force needed to produce the centripetal acceleration to see if it is sufficient. The net horizontal force required to keep the car moving in a circle around the curve is

$$(\Sigma F)_R = ma_R = m\frac{v^2}{r} = (1000 \text{ kg})\frac{(15 \text{ m/s})^2}{(50 \text{ m})} = 4500 \text{ N}.$$

Now we compute the maximum total static friction force (the sum of the friction forces acting on each of the four tires) to see if it can be large enough to provide a safe centripetal acceleration. For (a), $\mu_s = 0.60$, and the maximum friction force attainable (recall from Section 4–8 that $F_{fr} \leq \mu_s F_N$) is

$$(F_{fr})_{max} = \mu_s F_N = (0.60)(9800 \text{ N}) = 5880 \text{ N}.$$

Since a force of only 4500 N is needed, and that is, in fact, how much will be exerted by the road as a static friction force, the car can follow the curve. But in (b) the maximum static friction force possible is

$$(F_{fr})_{max} = \mu_s F_N = (0.25)(9800 \text{ N}) = 2450 \text{ N}.$$

The car will skid because the ground cannot exert sufficient force (4500 N is needed) to keep it moving in a curve of radius 50 m at a speed of 54 km/h.

The possibility of skidding is worse if the wheels lock (stop rotating) when the brakes are applied too hard. When the tires are rolling, static friction exists. But if the wheels lock (stop rotating), the tires slide and the friction force, which is now kinetic friction, is less. More importantly, the *direction* of the friction force changes suddenly if the wheels lock. Static friction can point perpendicular to the velocity, as in Fig. 5–13b; but if the car slides, kinetic friction points *opposite* to the velocity. The force no longer points toward the center of the circle, and the car cannot continue in a curved path (see Fig. 5–12). Even worse, if the road is wet or icy, locking of the wheels occurs with less force on the brake pedal since there is less road friction to keep the wheels turning rather than sliding. Antilock brakes (ABS) are designed to limit brake pressure just before the point where sliding would occur, by means of delicate sensors and a fast computer.

EXERCISE D To negotiate a flat (unbanked) curve at a *faster* speed, a driver puts a couple of sand bags in his van aiming to increase the force of friction between the tires and the road. Will the sand bags help?

The banking of curves can reduce the chance of skidding. The normal force exerted by a banked road, acting perpendicular to the road, will have a component toward the center of the circle (Fig. 5–14), thus reducing the reliance on friction. For a given banking angle θ, there will be one speed for which no friction at all is required. This will be the case when the horizontal component of the normal force toward the center of the curve, $F_N \sin\theta$ (see Fig. 5–14), is just equal to the force required to give a vehicle its centripetal acceleration—that is, when

$$F_N \sin\theta = m\frac{v^2}{r}.\qquad \text{[no friction required]}$$

The banking angle of a road, θ, is chosen so that this condition holds for a particular speed, called the "design speed."

PHYSICS APPLIED
Banked curves

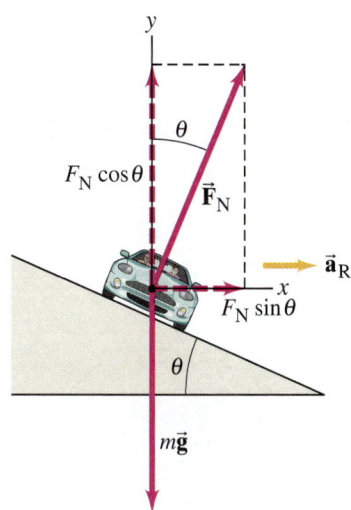

EXAMPLE 5–7 **Banking angle.** (*a*) For a car traveling with speed v around a curve of radius r, determine a formula for the angle at which a road should be banked so that no friction is required. (*b*) What is this angle for a road which has a curve of radius 50 m with a design speed of 50 km/h?

APPROACH Even though the road is banked, the car is still moving along a horizontal circle, so the centripetal acceleration needs to be horizontal. We choose our x and y axes as horizontal and vertical so that a_R, which is horizontal, is along the x axis. The forces on the car are the Earth's gravity mg downward, and the normal force F_N exerted by the road perpendicular to its surface. See Fig. 5–14, where the components of F_N are also shown. We don't need to consider the friction of the road because we are designing a road to be banked so as to eliminate dependence on friction.

SOLUTION (*a*) Since there is no vertical motion, $a_y = 0$ and $\Sigma F_y = ma_y$ gives

$$F_N \cos\theta - mg = 0$$

or

$$F_N = \frac{mg}{\cos\theta}.$$

FIGURE 5–14 Normal force on a car rounding a banked curve, resolved into its horizontal and vertical components. The centripetal acceleration is horizontal (*not* parallel to the sloping road). The friction force on the tires, not shown, could point up or down along the slope, depending on the car's speed. The friction force will be zero for one particular speed.

CAUTION
F_N is not always equal to mg

[Note in this case that $F_N \geq mg$ because $\cos\theta \leq 1$.]
We substitute this relation for F_N into the equation for the horizontal motion,

$$F_N \sin\theta = m\frac{v^2}{r},$$

which becomes

$$\frac{mg}{\cos\theta}\sin\theta = m\frac{v^2}{r}$$

or

$$\tan\theta = \frac{v^2}{rg}.$$

This is the formula for the banking angle θ: no friction needed at this speed v.
(*b*) For $r = 50$ m and $v = 50$ km/h ($= 14$ m/s),

$$\tan\theta = \frac{(14\text{ m/s})^2}{(50\text{ m})(9.8\text{ m/s}^2)} = 0.40,$$

so $\theta = \tan^{-1}(0.40) = 22°$.

We have been using the centripetal acceleration $a = v^2/r$ where r is the radius of a circle. For a road, and in many other situations, we don't have a full circle, but only a portion of a circle: $a = v^2/r$ still works and we often call r the **radius of curvature** of that portion of a circle we are dealing with.

SECTION 5–3 Highway Curves: Banked and Unbanked **117**

*5–4 Nonuniform Circular Motion

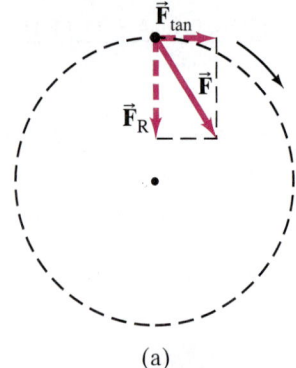

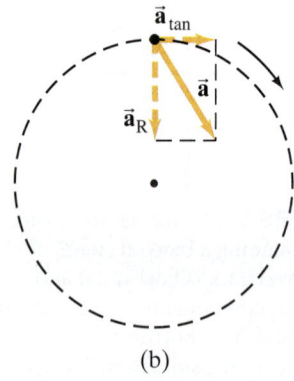

FIGURE 5–15 The speed of an object moving in a circle changes if the force on it has a tangential component, $F_{\tan}$. Part (a) shows the force $\vec{F}$ and its vector components; part (b) shows the acceleration vector and its vector components.

Circular motion at constant speed occurs when the net force on an object is exerted toward the center of the circle. If the net force is not directed toward the center but is at an angle, as shown in Fig. 5–15a, the force has two components. The component directed toward the center of the circle, $\vec{F}_R$, gives rise to the centripetal acceleration, $\vec{a}_R$, and keeps the object moving in a circle. The component tangent to the circle, $\vec{F}_{\tan}$, acts to increase (or decrease) the speed, and thus gives rise to a component of the acceleration tangent to the circle, $\vec{a}_{\tan}$. When the speed of the object is changing, a tangential component of force is acting.

When you first start revolving a ball on the end of a string around your head, you must give it tangential acceleration. You do this by pulling on the string with your hand displaced from the center of the circle. In athletics, a hammer thrower accelerates the hammer tangentially in a similar way so that it reaches a high speed before release.

The tangential component of the acceleration, $a_{\tan}$, has magnitude equal to the rate of change of the *magnitude* of the object's velocity:

$$a_{\tan} = \frac{\Delta v}{\Delta t}.$$

The radial (centripetal) acceleration arises from the change in *direction* of the velocity and, as we have seen (Eq. 5–1), has magnitude

$$a_R = \frac{v^2}{r}.$$

The tangential acceleration always points in a direction tangent to the circle, and is in the direction of motion (parallel to $\vec{v}$, which is always tangent to the circle) if the speed is increasing, as shown in Fig. 5–15b. If the speed is decreasing, $\vec{a}_{\tan}$ points antiparallel to $\vec{v}$. In either case, $\vec{a}_{\tan}$ and $\vec{a}_R$ are always perpendicular to each other; and *their directions change* continually as the object moves along its circular path. The total vector acceleration $\vec{a}$ is the sum of the two components:

$$\vec{a} = \vec{a}_{\tan} + \vec{a}_R.$$

Since $\vec{a}_R$ and $\vec{a}_{\tan}$ are always perpendicular to each other, the magnitude of $\vec{a}$ at any moment is

$$a = \sqrt{a_{\tan}^2 + a_R^2}.$$

EXAMPLE 5–8 Two components of acceleration. A race car starts from rest in the pit area and accelerates at a uniform rate to a speed of 35 m/s in 11 s, moving on a circular track of radius 500 m. Assuming constant tangential acceleration, find (*a*) the tangential acceleration, and (*b*) the radial acceleration, at the instant when the speed is $v = 15$ m/s.

APPROACH The tangential acceleration relates to the change in speed of the car, and can be calculated as $a_{\tan} = \Delta v/\Delta t$. The centripetal acceleration relates to the change in the *direction* of the velocity vector and is calculated using $a_R = v^2/r$.

SOLUTION (*a*) During the 11-s time interval, we assume the tangential acceleration $a_{\tan}$ is constant. Its magnitude is

$$a_{\tan} = \frac{\Delta v}{\Delta t} = \frac{(35 \text{ m/s} - 0 \text{ m/s})}{11 \text{ s}} = 3.2 \text{ m/s}^2.$$

(*b*) When $v = 15$ m/s, the centripetal acceleration is

$$a_R = \frac{v^2}{r} = \frac{(15 \text{ m/s})^2}{(500 \text{ m})} = 0.45 \text{ m/s}^2.$$

NOTE The radial (centripetal) acceleration increases continually, whereas the tangential acceleration stays constant.

5–5 Newton's Law of Universal Gravitation

Besides developing the three laws of motion, Isaac Newton also examined the motion of the planets and the Moon. In particular, he wondered about the nature of the force that must act to keep the Moon in its nearly circular orbit around the Earth.

Newton was also thinking about the problem of gravity. Since falling objects accelerate, Newton had concluded that they must have a force exerted on them, a force we call the force of gravity. Whenever an object has a force exerted *on* it, that force is exerted *by* some other object. But what *exerts* the force of gravity? Every object on the surface of the Earth feels the force of gravity F_G, and no matter where the object is, the force is directed toward the center of the Earth (Fig. 5–16). Newton concluded that it must be the Earth itself that exerts the gravitational force on objects at its surface.

According to legend, Newton noticed an apple drop from a tree. He is said to have been struck with a sudden inspiration: If gravity acts at the tops of trees, and even at the tops of mountains, then perhaps it acts all the way to the Moon! With this idea that it is the Earth's gravity that holds the Moon in its orbit, Newton developed his great theory of gravitation. But there was controversy at the time. Many thinkers had trouble accepting the idea of a force "acting at a distance." Typical forces act through contact—your hand pushes a cart and pulls a wagon, a bat hits a ball, and so on. But gravity acts without contact, said Newton: the Earth exerts a force on a falling apple and on the Moon, even though there is no contact, and the two objects may even be very far apart.[†]

Newton set about determining the magnitude of the gravitational force that the Earth exerts on the Moon as compared to the gravitational force on objects at the Earth's surface. The centripetal acceleration of the Moon, as we calculated in Example 5–2, is $a_R = 0.00272$ m/s². In terms of the acceleration of gravity at the Earth's surface, $g = 9.80$ m/s²,

$$a_R = \frac{0.00272 \text{ m/s}^2}{9.80 \text{ m/s}^2} g \approx \frac{1}{3600} g.$$

That is, the acceleration of the Moon toward the Earth is about $\frac{1}{3600}$ as great as the acceleration of objects at the Earth's surface. The Moon is 384,000 km from the Earth, which is about 60 times the Earth's radius of 6380 km. That is, the Moon is 60 times farther from the Earth's center than are objects at the Earth's surface. But $60 \times 60 = 60^2 = 3600$. Again that number 3600! Newton concluded that the gravitational force F_{grav} or F_G exerted by the Earth on any object decreases with the square of its distance r from the Earth's center:

$$F_G \propto \frac{1}{r^2}.$$

The Moon is 60 Earth radii away, so it feels a gravitational force only $\frac{1}{60^2} = \frac{1}{3600}$ times as strong as it would if it were at a point on the Earth's surface.

Newton realized that the force of gravity on an object depends not only on distance but also on the object's mass. In fact, it is directly proportional to its mass, as we have seen (Eq. 4–3). According to Newton's third law, when the Earth exerts its gravitational force on any object, such as the Moon, that object exerts an equal and opposite force on the Earth (Fig. 5–17). Because of this *symmetry*, Newton reasoned, the magnitude of the force of gravity must be proportional to *both* masses:

$$F_G \propto \frac{m_E m_{Obj}}{r^2},$$

where m_E and m_{Obj} are the masses of the Earth and the other object, respectively, and r is the distance from the Earth's center to the center of the other object.

[†]To deal with the conceptual difficulty of "action at a distance," the idea of a *gravitational field* was introduced many years later: every object that has mass produces a gravitational field in space. The force one object exerts on a second object is then due to the gravitational field produced by the first object at the position of the second object. We discuss fields in Section 16–7.

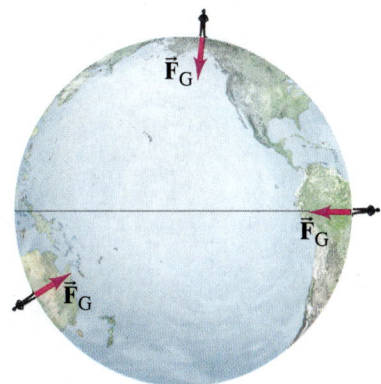

FIGURE 5–16 Anywhere on Earth, whether in Alaska, Peru, or Australia, the force of gravity acts downward toward the Earth's center.

FIGURE 5–17 The gravitational force one object exerts on a second object is directed toward the first object; and, by Newton's third law, is equal and opposite to the force exerted by the second object on the first. In the case shown, the gravitational force on the Moon due to Earth, $\vec{F}_{ME}$, is equal and opposite to the gravitational force on Earth due to the Moon, $\vec{F}_{EM}$. That is, $\vec{F}_{ME} = -\vec{F}_{EM}$.

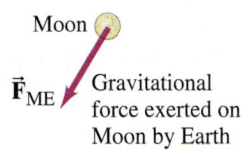

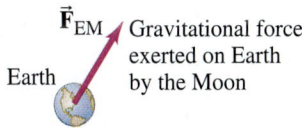

Newton went a step further in his analysis of gravity. In his examination of the orbits of the planets, he concluded that the force required to hold the different planets in their orbits around the Sun seems to diminish as the inverse square of their distance from the Sun. This led him to believe that it is also the gravitational force that acts between the Sun and each of the planets to keep them in their orbits. And if gravity acts between these objects, why not between all objects? Thus he proposed his **law of universal gravitation**, which we can state as follows:

NEWTON'S LAW OF UNIVERSAL GRAVITATION

> Every particle in the universe attracts every other particle with a force that is proportional to the product of their masses and inversely proportional to the square of the distance between them. This force acts along the line joining the two particles.

The magnitude of the gravitational force can be written as

$$F_G = G \frac{m_1 m_2}{r^2}, \qquad (5\text{--}4)$$

where m_1 and m_2 are the masses of the two particles, r is the distance between them, and G is a universal constant which must be measured experimentally.

The value of G must be very small, since we are not aware of any force of attraction between ordinary-sized objects, such as between two baseballs. The force between two ordinary objects was first measured by Henry Cavendish in 1798, over 100 years after Newton published his law. To detect and measure the incredibly small force between ordinary objects, he used an apparatus like that shown in Fig. 5–18. Cavendish confirmed Newton's hypothesis that two objects attract one another and that Eq. 5–4 accurately describes this force. In addition, because Cavendish could measure F_G, m_1, m_2, and r accurately, he was able to determine the value of the constant G as well. The accepted value today is

$$G = 6.67 \times 10^{-11} \, \text{N} \cdot \text{m}^2/\text{kg}^2.$$

(See Table inside front cover for values of all constants to highest known precision.) Equation 5–4 is called an **inverse square law** because the force is inversely proportional to r^2.

[Strictly speaking, Eq. 5–4 gives the magnitude of the gravitational force that one particle exerts on a second particle that is a distance r away. For an extended object (that is, not a point), we must consider how to measure the distance r. A correct calculation treats each extended body as a collection of particles, and the total force is the sum of the forces due to all the particles. The sum over all these particles is often done using integral calculus, which Newton himself invented. When extended bodies are small compared to the distance between them (as for the Earth–Sun system), little inaccuracy results from considering them as point particles. Newton was able to show that the *gravitational force exerted on a particle outside a uniform sphere is the same as if the entire mass of the sphere was concentrated at its center*.† Thus Eq. 5–4 gives the correct force between two uniform spheres where r is the distance between their centers.]

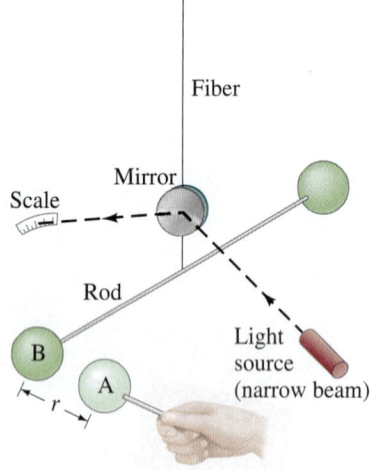

FIGURE 5–18 Schematic diagram of Cavendish's apparatus. Two spheres are attached to a light horizontal rod, which is suspended at its center by a thin fiber. When a third sphere (labeled A) is brought close to one of the suspended spheres (labeled B), the gravitational force causes the latter to move, and this twists the fiber slightly. The tiny movement is magnified by the use of a narrow light beam directed at a mirror mounted on the fiber. The beam reflects onto a scale. Previous determination of how large a force will twist the fiber a given amount then allows the experimenter to determine the magnitude of the gravitational force between the two objects, A and B.

EXAMPLE 5–9 ESTIMATE Can you attract another person gravitationally? A 50-kg person and a 70-kg person are sitting on a bench close to each other. Estimate the magnitude of the gravitational force each exerts on the other.

APPROACH This is an estimate: we let the distance between the centers of the two people be $\frac{1}{2}$ m (about as close as you can get).

SOLUTION We use Eq. 5–4, which gives

$$F_G = G \frac{m_1 m_2}{r^2} \approx \frac{(6.67 \times 10^{-11} \, \text{N} \cdot \text{m}^2/\text{kg}^2)(50 \, \text{kg})(70 \, \text{kg})}{(0.5 \, \text{m})^2} \approx 10^{-6} \, \text{N},$$

rounded off to an order of magnitude. Such a force is unnoticeably small unless extremely sensitive instruments are used ($< 1/100,000$ of a pound).

---

†We demonstrate this result in Section 16–12.

EXAMPLE 5–10 **Spacecraft at $2r_E$.** What is the force of gravity acting on a 2000-kg spacecraft when it orbits two Earth radii from the Earth's center (that is, a distance $r_E = 6380$ km above the Earth's surface, Fig. 5–19)? The mass of the Earth is $m_E = 5.98 \times 10^{24}$ kg.

APPROACH We could plug all the numbers into Eq. 5–4, but there is a simpler approach. The spacecraft is twice as far from the Earth's center as when it is at the surface of the Earth. Therefore, since the force of gravity F_G decreases as the square of the distance (and $\frac{1}{2^2} = \frac{1}{4}$), the force of gravity on the satellite will be only one-fourth its weight at the Earth's surface.

SOLUTION At the surface of the Earth, $F_G = mg$. At a distance from the Earth's center of $2r_E$, F_G is $\frac{1}{4}$ as great:

$$F_G = \tfrac{1}{4}mg = \tfrac{1}{4}(2000 \text{ kg})(9.80 \text{ m/s}^2) = 4900 \text{ N}.$$

Note carefully that the law of universal gravitation describes a *particular* force (gravity), whereas Newton's second law of motion ($F = ma$) tells how an object accelerates due to *any* type of force.

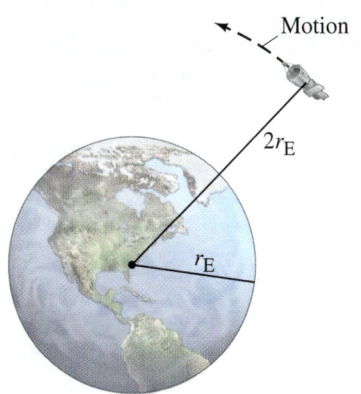

FIGURE 5–19 Example 5–10; a spacecraft in orbit at $r = 2r_E$.

⚠️ **CAUTION**
Distinguish Newton's second law from the law of universal gravitation

5–6 Gravity Near the Earth's Surface

When Eq. 5–4 is applied to the gravitational force between the Earth and an object at its surface, m_1 becomes the mass of the Earth m_E, m_2 becomes the mass of the object m, and r becomes the distance of the object from the Earth's center, which is the radius of the Earth r_E. This force of gravity due to the Earth is the weight of the object on Earth, which we have been writing as mg. Thus,

$$mg = G\frac{mm_E}{r_E^2}.$$

We can solve this for g, the acceleration of gravity at the Earth's surface:

$$g = G\frac{m_E}{r_E^2}. \qquad (5\text{–}5)$$

Thus, the acceleration of gravity at the surface of the Earth, g, is determined by m_E and r_E. (Don't confuse G with g; they are very different quantities, but are related by Eq. 5–5.)

Until G was measured, the mass of the Earth was not known. But once G was measured, Eq. 5–5 could be used to calculate the Earth's mass, and Cavendish was the first to do so. Since $g = 9.80 \text{ m/s}^2$ and the radius of the Earth is $r_E = 6.38 \times 10^6$ m, then, from Eq. 5–5, we obtain the mass of the Earth to be

$$m_E = \frac{gr_E^2}{G} = \frac{(9.80 \text{ m/s}^2)(6.38 \times 10^6 \text{ m})^2}{6.67 \times 10^{-11} \text{ N}\cdot\text{m}^2/\text{kg}^2} = 5.98 \times 10^{24} \text{ kg}.$$

Equation 5–5 can be applied to other planets, where g, m, and r would refer to that planet.

⚠️ **CAUTION**
Distinguish G from g

FIGURE 5–20 Example 5–11. Mount Everest, 8850 m (29,035 ft) above sea level; in the foreground, the author with sherpas at 5500 m (18,000 ft).

EXAMPLE 5–11 **ESTIMATE** **Gravity on Everest.** Estimate the effective value of g on the top of Mt. Everest, 8850 m (29,035 ft) above sea level (Fig. 5–20). That is, what is the acceleration due to gravity of objects allowed to fall freely at this altitude? Ignore the mass of the mountain itself.

APPROACH The force of gravity (and the acceleration due to gravity g) depends on the distance from the center of the Earth, so there will be an effective value g' on top of Mt. Everest which will be smaller than g at sea level. We assume the Earth is a uniform sphere (a reasonable "estimate").

SOLUTION We use Eq. 5–5, with r_E replaced by $r = 6380$ km $+$ 8.9 km $=$ 6389 km $= 6.389 \times 10^6$ m:

$$g' = G\frac{m_E}{r^2} = \frac{(6.67 \times 10^{-11} \text{ N}\cdot\text{m}^2/\text{kg}^2)(5.98 \times 10^{24} \text{ kg})}{(6.389 \times 10^6 \text{ m})^2} = 9.77 \text{ m/s}^2,$$

which is a reduction of about 3 parts in a thousand (0.3%).

TABLE 5–1 Acceleration Due to Gravity at Various Locations		
Location	Elevation (m)	g (m/s^2)
New York	0	9.803
San Francisco	0	9.800
Denver	1650	9.796
Pikes Peak	4300	9.789
Sydney, Australia	0	9.798
Equator	0	9.780
North Pole (calculated)	0	9.832

Note that Eq. 5–5 does not give precise values for g at different locations because the Earth is not a perfect sphere. The Earth not only has mountains and valleys, and it bulges at the equator, but also its mass is not distributed precisely uniformly. (See Table 5–1.) The Earth's rotation also affects the value of g. However, for most practical purposes, when an object is near the Earth's surface, we will simply use $g = 9.80 \text{ m/s}^2$ and write the weight of an object as mg.

EXERCISE E Suppose you could double the mass of a planet but keep its volume the same. How would the acceleration of gravity, g, at the surface change?

5–7 Satellites and "Weightlessness"

Satellite Motion

PHYSICS APPLIED
Artificial Earth satellites

Artificial satellites circling the Earth are now commonplace (Fig. 5–21). A satellite is put into orbit by accelerating it to a sufficiently high tangential speed with the use of rockets, as shown in Fig. 5–22. If the speed is too high, the spacecraft will not be confined by the Earth's gravity and will escape, never to return. If the speed is too low, it will return to Earth. Satellites are typically put into circular (or nearly circular) orbits, because such orbits require the least takeoff speed.

FIGURE 5–21 A satellite, the International Space Station, circling the Earth.

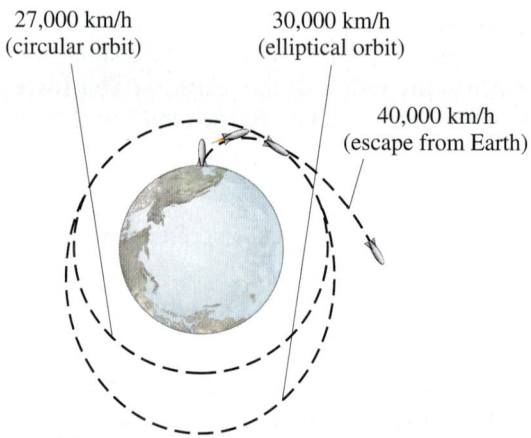

FIGURE 5–22 Artificial satellites launched at different speeds. Escape velocity from Earth is about 40,000 km/h.

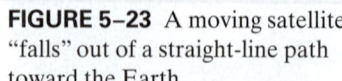

FIGURE 5–23 A moving satellite "falls" out of a straight-line path toward the Earth.

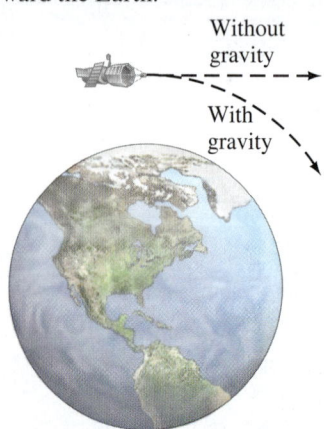

It is sometimes asked: "What keeps a satellite up?" The answer is: its high speed. If a satellite in orbit stopped moving, it would fall directly to Earth. But at the very high speed a satellite has, it would quickly fly out into space (Fig. 5–23) if it weren't for the gravitational force of the Earth pulling it into orbit. In fact, a satellite in orbit *is* falling (accelerating) toward Earth, but its high tangential speed keeps it from hitting Earth.

For satellites that move in a circle (at least approximately), the needed acceleration is centripetal and equals v^2/r. The force that gives a satellite this acceleration is the force of gravity exerted by the Earth, and since a satellite may be at a considerable distance from the Earth, we must use Newton's law of universal gravitation (Eq. 5–4) for the force acting on it. When we apply Newton's second law, $\Sigma F_R = ma_R$ in the radial direction, we find

$$G\frac{mm_E}{r^2} = m\frac{v^2}{r}, \qquad (5\text{–}6)$$

where m is the mass of the satellite. This equation relates the distance of the satellite from the Earth's center, r, to its speed, v, in a circular orbit. Note that only one force—gravity—is acting on the satellite, and that r is the sum of the Earth's radius r_E plus the satellite's height h above the Earth: $r = r_E + h$.

If we solve Eq. 5–6 for v, we find $v = \sqrt{Gm_E/r}$ and we see that a satellite's speed does not depend on its own mass. Satellites of different mass orbiting at the same distance above Earth have the same speed and period.

EXAMPLE 5–12 **Geosynchronous satellite.** A *geosynchronous* satellite is one that stays above the same point on the Earth, which is possible only if it is above a point on the equator. Why? Because the center of a satellite orbit is always at the center of the Earth; so it is not possible to have a satellite orbiting above a fixed point on the Earth at any latitude other than 0°. Geosynchronous satellites are commonly used for TV and radio transmission, for weather forecasting, and as communication relays.[†] Determine (*a*) the height above the Earth's surface such a satellite must orbit, and (*b*) such a satellite's speed. (*c*) Compare to the speed of a satellite orbiting 200 km above Earth's surface.

PHYSICS APPLIED
Geosynchronous satellites

APPROACH To remain above the same point on Earth as the Earth rotates, the satellite must have a period of 24 hours. We can apply Newton's second law, $F = ma$, where $a = v^2/r$ if we assume the orbit is circular.

SOLUTION (*a*) The only force on the satellite is the gravitational force due to the Earth. (We can ignore the gravitational force exerted by the Sun. Why?) We apply Eq. 5–6 assuming the satellite moves in a circle:

$$G\frac{m_{\text{Sat}} m_E}{r^2} = m_{\text{Sat}}\frac{v^2}{r}.$$

This equation has two unknowns, r and v. So we need a second equation. The satellite revolves around the Earth with the same period that the Earth rotates on its axis, namely once in 24 hours. Thus the speed of the satellite must be

$$v = \frac{2\pi r}{T},$$

where $T = 1$ day $= (24\text{ h})(3600\text{ s/h}) = 86{,}400$ s. We substitute this into the "satellite equation" above and obtain (after cancelling m_{Sat} on both sides):

$$G\frac{m_E}{r^2} = \frac{(2\pi r)^2}{rT^2}.$$

After cancelling an r, we can solve for r^3:

$$r^3 = \frac{Gm_E T^2}{4\pi^2} = \frac{(6.67 \times 10^{-11}\text{ N}\cdot\text{m}^2/\text{kg}^2)(5.98 \times 10^{24}\text{ kg})(86{,}400\text{ s})^2}{4\pi^2}$$
$$= 7.54 \times 10^{22}\text{ m}^3.$$

We take the cube root and find

$$r = 4.22 \times 10^7\text{ m},$$

or 42,200 km from the Earth's center. We subtract the Earth's radius of 6380 km to find that a geosynchronous satellite must orbit about 36,000 km (about $6r_E$) above the Earth's surface.

(*b*) We solve for v in the satellite equation, Eq. 5–6:

$$v = \sqrt{\frac{Gm_E}{r}} = \sqrt{\frac{(6.67 \times 10^{-11}\text{ N}\cdot\text{m}^2/\text{kg}^2)(5.98 \times 10^{24}\text{ kg})}{(4.22 \times 10^7\text{ m})}} = 3070\text{ m/s},$$

or about 11,000 km/h ($\approx$ 7000 mi/h). We get the same result if we use $v = 2\pi r/T$.

(*c*) The equation in part (*b*) for v shows $v \propto \sqrt{1/r}$. So for $r = r_E + h = 6380$ km $+$ 200 km $=$ 6580 km, we get

$$v' = v\sqrt{\frac{r}{r'}} = (3070\text{ m/s})\sqrt{\frac{(42{,}200\text{ km})}{(6580\text{ km})}} = 7770\text{ m/s},$$

or about 28,000 km/h ($\approx$ 17,000 mi/h).

[†]Geosynchronous satellites are useful because receiving and transmitting antennas at a given place on Earth can stay fixed on such a satellite (no tracking and no switching satellites is needed).

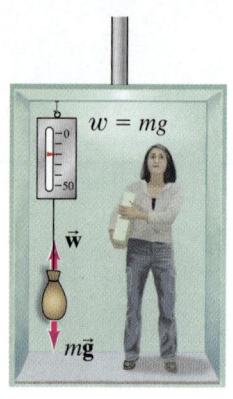

(a) $a = 0$

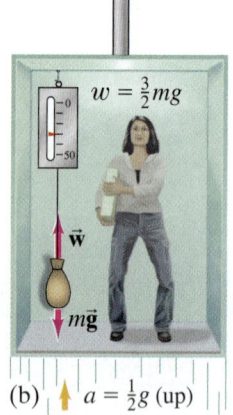

(b) $a = \tfrac{1}{2}g$ (up)

(c) $a = g$ (down)

FIGURE 5–24 (a) A bag in an elevator at rest exerts a force on a spring scale equal to its weight. (b) In an elevator accelerating upward at $\tfrac{1}{2}g$, the bag's apparent weight is $1\tfrac{1}{2}$ times larger than its true weight. (c) In a freely falling elevator, the bag experiences "weightlessness": the scale reads zero.

Weightlessness

People and other objects in a satellite circling the Earth are said to experience apparent weightlessness. Let us first look at a simpler case: a falling elevator. In Fig. 5–24a, an elevator is at rest with a bag hanging from a spring scale. The scale reading indicates the downward force exerted on it by the bag. This force, exerted *on* the scale, is equal and opposite to the force exerted *by* the scale upward on the bag, and we call its magnitude w (for "weight"). Two forces act on the bag: the downward gravitational force and the upward force exerted by the scale equal to w. Because the bag is not accelerating ($a = 0$), when we apply $\Sigma F = ma$ to the bag in Fig. 5–24a we obtain

$$w - mg = 0,$$

where mg is the weight of the bag. Thus, $w = mg$, and since the scale indicates the force w exerted on it by the bag, it registers a force equal to the weight of the bag, as we expect.

Now let the elevator have an acceleration, a. Applying Newton's second law, $\Sigma F = ma$, to the bag as seen from an inertial reference frame (the elevator itself is not now an inertial frame) we have

$$w - mg = ma.$$

Solving for w, we have

$$w = mg + ma. \qquad [a \text{ is } + \text{ upward}]$$

We have chosen the positive direction up. Thus, if the acceleration a is up, a is positive; and the scale, which measures w, will read more than mg. We call w the *apparent weight* of the bag, which in this case would be greater than its actual weight (mg). If the elevator accelerates downward, a will be negative and w, the apparent weight, will be less than mg. The direction of the velocity $\vec{v}$ doesn't matter. Only the direction of the acceleration $\vec{a}$ (and its magnitude) influences the scale reading.

Suppose the elevator's acceleration is $\tfrac{1}{2}g$ upward; then we find

$$w = mg + m(\tfrac{1}{2}g)$$
$$= \tfrac{3}{2}mg.$$

That is, the scale reads $1\tfrac{1}{2}$ times the actual weight of the bag (Fig. 5–24b). The apparent weight of the bag is $1\tfrac{1}{2}$ times its real weight. The same is true of the person: her apparent weight (equal to the normal force exerted on her by the elevator floor) is $1\tfrac{1}{2}$ times her real weight. We can say that she is experiencing $1\tfrac{1}{2}$ g's, just as astronauts experience so many g's at a rocket's launch.

If, instead, the elevator's acceleration is $a = -\tfrac{1}{2}g$ (downward), then $w = mg - \tfrac{1}{2}mg = \tfrac{1}{2}mg$. That is, the scale reads half the actual weight. If the elevator is in *free fall* (for example, if the cables break), then $a = -g$ and $w = mg - mg = 0$. The scale reads zero. See Fig. 5–24c. The bag appears weightless. If the person in the elevator accelerating at $-g$ let go of a box, it would not fall to the floor. True, the box would be falling with acceleration g. But so would the floor of the elevator and the person. The box would hover right in front of the person. This phenomenon is called **apparent weightlessness** because in the reference frame of the person, objects don't fall or seem to have weight—yet gravity does not disappear. Gravity is still acting on each object, whose weight is still mg.

The "weightlessness" experienced by people in a satellite orbit close to the Earth (Fig. 5–25) is the same apparent weightlessness experienced in a freely falling elevator. It may seem strange, at first, to think of a satellite as freely falling. But a satellite is indeed falling toward the Earth, as was shown in Fig. 5–23. The force of gravity causes it to "fall" out of its natural straight-line path. The acceleration of the satellite must be the acceleration due to gravity at that point, because the only force acting on it is gravity. Thus, although the force of gravity acts on objects within the satellite, the objects experience an apparent weightlessness because they, and the satellite, are accelerating together as in free fall.

Figure 5–26 shows some examples of "free fall," or apparent weightlessness, experienced by people on Earth for brief moments.

A completely different situation occurs if a spacecraft is out in space far from the Earth, the Moon, and other attracting bodies. The force of gravity due to the Earth and other celestial bodies will then be quite small because of the distances involved, and persons in such a spacecraft would experience real weightlessness.

| **EXERCISE F** Return to Chapter-Opening Question 2, page 109, and answer it again now. Try to explain why you may have answered differently the first time.

(a) (b) (c)

FIGURE 5–25 This astronaut is outside the International Space Station. He must feel very free because he is experiencing apparent weightlessness.

FIGURE 5–26 Experiencing "weightlessness" on Earth.

5–8 Planets, Kepler's Laws, and Newton's Synthesis

Where did we first get the idea of planets? Have you ever escaped the lights of the city to gaze late at night at the multitude of stars in the night sky? It is a moving experience. Thousands of years ago, the ancients saw this sight every cloudless night, and were fascinated. They noted that the vast majority of stars, bright or dim, seemed to maintain fixed positions relative to each other. The ancients imagined these **fixed stars** as being attached to a huge inverted bowl, or sphere. This **celestial sphere** revolved around the Earth almost exactly once a day (Fig. 5–27), from east to west. Among all the stars that were visible to the naked eye (there were no telescopes until much later, about 1600), the ancients saw five stars that changed position relative to the fixed stars over weeks and months. These five wandering stars were called **planets** (Greek for wandering). Planets were thus visible at night as tiny points of light like other stars.

The ancient idea that the Sun, Moon, and planets revolve around the Earth is called the **geocentric** view (geo = Earth in Greek). It was developed into a fine theoretical system by Ptolemy in the second century B.C. Today we believe in a **heliocentric** system (helios = Sun in Greek), where the Earth is just another planet, between Venus and Mars, orbiting around the Sun. Although a heliocentric view was proposed in ancient times, it was largely ignored until Renaissance Italy of the fifteenth century. The real theory change (see Section 1–1 and Fig. 1–2) began with the heliocentric theory of Nicolaus Copernicus (1473–1543) and then was greatly advanced by the experimental observations of Galileo around 1610 using his newly developed 30× telescope. Galileo observed that the planet Jupiter has moons (like a miniature solar system) and that Venus has phases like our Moon, not explainable by Ptolemy's geocentric system. [Galileo's famous encounter with the Church had little to do with religious faith, but rather with politics, personality conflict, and authority. Today it is generally understood that science and faith are different approaches that are not in conflict.]

FIGURE 5–27 Time exposure showing movement of stars over a period of several hours.

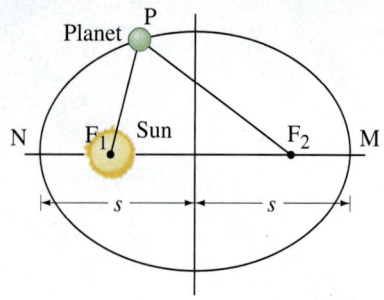

FIGURE 5–28 *Kepler's first law.* An ellipse is a closed curve such that the sum of the distances from any point P on the curve to two fixed points (called the foci, F_1 and F_2) remains constant. That is, the sum of the distances, $F_1P + F_2P$, is the same for all points on the curve. A circle is a special case of an ellipse in which the two foci coincide, at the center of the circle.

FIGURE 5–29 *Kepler's second law.* The two shaded regions have equal areas. The planet moves from point 1 to point 2 in the same time it takes to move from point 3 to point 4. Planets move fastest when closest to the Sun.

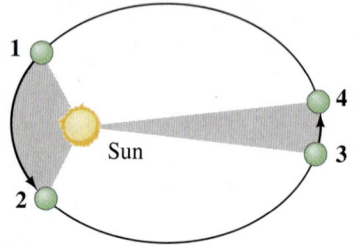

TABLE 5–2 Planetary Data Applied to Kepler's Third Law

Planet	Mean Distance to Sun, s (10^6 km)	Period, T (Earth yr)	s^3/T^2 ($10^{24} \frac{km^3}{yr^2}$)
Mercury	57.9	0.241	3.34
Venus	108.2	0.615	3.35
Earth	149.6	1.000	3.35
Mars	227.9	1.88	3.35
Jupiter	778.3	11.86	3.35
Saturn	1427	29.5	3.34
Uranus	2870	84.0	3.35
Neptune	4497	165	3.34
(Pluto)†	5900	248	3.34

†Pluto, since its discovery in 1930, was considered a ninth planet. But its small mass and the recent discovery of other objects beyond Neptune with similar masses has led to calling these smaller objects, including Pluto, "dwarf planets." We keep it in the Table to indicate its great distance, and its consistency with Kepler's third law.

Kepler's Laws

Also about 1600, more than a half century before Newton proposed his three laws of motion and his law of universal gravitation, the German astronomer Johannes Kepler (1571–1630) had worked out a detailed description of the motion of the planets around the Sun. Kepler's work resulted in part from the many years he spent examining data collected (without a telescope) by Tycho Brahe (1546–1601) on the positions of the planets in their motion through the night sky.

Among Kepler's writings were three empirical findings that we now refer to as **Kepler's laws of planetary motion**. These are summarized as follows, with additional explanation in Figs. 5–28 and 5–29.

Kepler's first law: The path of each planet around the Sun is an ellipse with the Sun at one focus (Fig. 5–28).

Kepler's second law: Each planet moves so that an imaginary line drawn from the Sun to the planet sweeps out equal areas in equal periods of time (Fig. 5–29).

Kepler's third law: The ratio of the squares of the periods T of any two planets revolving around the Sun is equal to the ratio of the cubes of their mean distances from the Sun. [The mean distance equals the semimajor axis s (= half the distance from the planet's near point N and far point M from the Sun, Fig. 5–28).] That is, if T_1 and T_2 represent the periods (the time needed for one revolution about the Sun) for any two planets, and s_1 and s_2 represent their mean distances from the Sun, then

$$\left(\frac{T_1}{T_2}\right)^2 = \left(\frac{s_1}{s_2}\right)^3.$$

We can rewrite Kepler's third law as

$$\frac{s_1^3}{T_1^2} = \frac{s_2^3}{T_2^2},$$

meaning that s^3/T^2 should be the same for each planet. Present-day data are given in Table 5–2; see the last column.

In Examples and Problems we usually will assume the orbits are circles, although it is not quite true in general.

EXAMPLE 5–13 Where is Mars? Mars' period (its "year") was noted by Kepler to be about 687 days (Earth days), which is (687 d/365 d) = 1.88 yr (Earth years). Determine the mean distance of Mars from the Sun using the Earth as a reference.

APPROACH We are given the ratio of the periods of Mars and Earth. We can find the distance from Mars to the Sun using Kepler's third law, given the Earth–Sun distance as 1.50×10^{11} m (Table 5–2; also Table inside front cover).

SOLUTION Let the distance of Mars from the Sun be s_{MS}, and the Earth–Sun distance be $s_{ES} = 1.50 \times 10^{11}$ m. From Kepler's third law:

$$\frac{s_{MS}}{s_{ES}} = \left(\frac{T_M}{T_E}\right)^{\frac{2}{3}} = \left(\frac{1.88 \text{ yr}}{1 \text{ yr}}\right)^{\frac{2}{3}} = 1.52.$$

So Mars is 1.52 times the Earth's distance from the Sun, or 2.28×10^{11} m.

Kepler's Third Law Derived, Sun's Mass, Perturbations

We will derive Kepler's third law for the special case of a circular orbit, in which case the mean distance s is the radius r of the circle. (Most planetary orbits are close to a circle.) First, we write Newton's second law of motion, $\Sigma F = ma$. For F we use the law of universal gravitation (Eq. 5–4) for the force between the Sun and a planet of mass m_1, and for a the centripetal acceleration, v^2/r. We assume the mass of the Sun M_S is much greater than the mass of its planets, so we ignore the effects of the planets on each other. Then

$$\Sigma F = ma$$
$$G\frac{m_1 M_S}{r_1^2} = m_1 \frac{v_1^2}{r_1}.$$

Here m_1 is the mass of a particular planet, r_1 its distance from the Sun, and v_1 its speed in orbit; M_S is the mass of the Sun, since it is the gravitational attraction of the Sun that keeps each planet in its orbit. The period T_1 of the planet is the time required for one complete orbit, which is a distance equal to $2\pi r_1$, the circumference of a circle. Thus

$$v_1 = \frac{2\pi r_1}{T_1}.$$

We substitute this formula for v_1 into the previous equation:

$$G\frac{m_1 M_S}{r_1^2} = m_1 \frac{4\pi^2 r_1}{T_1^2}.$$

We rearrange this to get

$$\frac{T_1^2}{r_1^3} = \frac{4\pi^2}{GM_S}. \tag{5-7a}$$

We derived this for planet 1 (say, Mars). The same derivation would apply for a second planet (say, Saturn) orbiting the Sun,

$$\frac{T_2^2}{r_2^3} = \frac{4\pi^2}{GM_S},$$

where T_2 and r_2 are the period and orbit radius, respectively, for the second planet. Since the right sides of the two previous equations are equal, we have $T_1^2/r_1^3 = T_2^2/r_2^3$ or, rearranging,

$$\left(\frac{T_1}{T_2}\right)^2 = \left(\frac{r_1}{r_2}\right)^3, \tag{5-7b}$$

Kepler's third law

which is Kepler's third law. Equations 5–7a and 5–7b are valid also for elliptical orbits if we replace r with the semimajor axis s.

EXAMPLE 5–14 **The Sun's mass determined.** Determine the mass of the Sun given the Earth's distance from the Sun as $r_{ES} = 1.5 \times 10^{11}$ m.

APPROACH Equation 5–7a relates the mass of the Sun M_S to the period and distance of any planet. We use the Earth.

SOLUTION The Earth's period is $T_E = 1$ yr $= (365\frac{1}{4}\text{d})(24\text{ h/d})(3600\text{ s/h}) = 3.16 \times 10^7$ s. We solve Eq. 5–7a for M_S:

$$M_S = \frac{4\pi^2 r_{ES}^3}{GT_E^2} = \frac{4\pi^2 (1.5 \times 10^{11}\text{ m})^3}{(6.67 \times 10^{-11}\text{ N}\cdot\text{m}^2/\text{kg}^2)(3.16 \times 10^7\text{ s})^2} = 2.0 \times 10^{30}\text{ kg}.$$

Determining the Sun's mass

Accurate measurements on the orbits of the planets indicated that they did not precisely follow Kepler's laws. For example, slight deviations from perfectly elliptical orbits were observed. Newton was aware that this was to be expected because any planet would be attracted gravitationally not only by the Sun but also (to a much lesser extent) by the other planets. Such deviations, or **perturbations**, in the orbit of Saturn were a hint that helped Newton formulate the law of universal gravitation, that all objects attract each other gravitationally. Observation of other perturbations later led to the discovery of Neptune. Deviations in the orbit of Uranus could not all be accounted for by perturbations due to the other known planets. Careful calculation in the nineteenth century indicated that these deviations could be accounted for if another planet existed farther out in the solar system. The position of this planet was predicted from the deviations in the orbit of Uranus, and telescopes focused on that region of the sky quickly found it; the new planet was called Neptune. Similar but much smaller perturbations of Neptune's orbit led to the discovery of Pluto in 1930.

Perturbations and discovery of planets

Other Centers for Kepler's Laws

The derivation of Eq. 5–7b, Kepler's third law, compared two planets revolving around the Sun. But the derivation is general enough to be applied to other systems. For example, we could apply Eq. 5–7b to compare an artificial satellite and our Moon, both revolving around Earth (then M_S would be replaced by M_E, the mass of the Earth). Or we could apply Eq. 5–7b to compare two moons revolving around Jupiter. But Kepler's third law, Eq. 5–7b, applies only to objects orbiting the same attracting center. Do not use Eq. 5–7b to compare, say, the Moon's orbit around Earth to the orbit of Mars around the Sun: they depend on different attracting centers.

CAUTION
Compare orbits of objects only around the same center

PHYSICS APPLIED
Planets around other stars

Distant Planetary Systems

Starting in the mid-1990s, planets revolving around distant stars (Fig. 5–30) were inferred from the regular "wobble" in position of each star due to the gravitational attraction of the revolving planet(s). Many such "extrasolar" planets are now known.

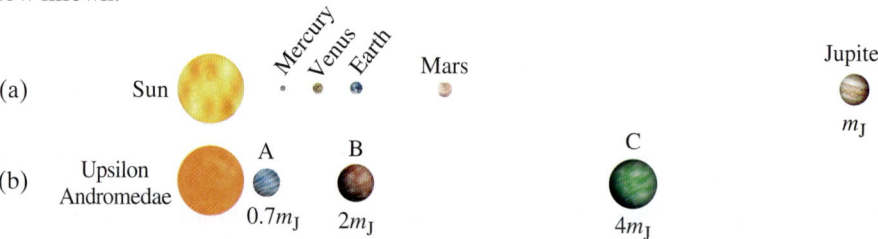

FIGURE 5–30 Our solar system (a) is compared to recently discovered planets orbiting (b) the star Upsilon Andromedae with at least three planets. m_J is the mass of Jupiter. (Sizes are not to scale.)

Newton's Synthesis

Kepler arrived at his laws through careful analysis of experimental data. Fifty years later, Newton was able to show that Kepler's laws could be derived mathematically from the law of universal gravitation and the laws of motion. Newton also showed that for any reasonable form for the gravitational force law, only one that depends on the inverse square of the distance is fully consistent with Kepler's laws. He thus used Kepler's laws as evidence in favor of his law of universal gravitation, Eq. 5–4.

The development by Newton of the law of universal gravitation and the three laws of motion was a major intellectual achievement. With these laws, he was able to describe the motion of objects on Earth and of the far-away planets seen in the night sky. The motions of the planets through the heavens and of objects on Earth were seen to follow the same laws (not recognized previously). For this reason, and also because Newton integrated the results of earlier scientists into his system, we sometimes speak of **Newton's synthesis**.

The laws formulated by Newton are referred to as **causal laws**. By **causality** we mean that one occurrence can cause another. When a rock strikes a window, we infer the rock *causes* the window to break. This idea of "cause and effect" relates to Newton's laws: the acceleration of an object was seen to be *caused* by the net force acting on it.

As a result of Newton's theories, the universe came to be viewed by many as a machine whose parts move in a **deterministic** way. This deterministic view of the universe had to be modified in the twentieth century (Chapter 28).

Sun/Earth Reference Frames

The geocentric–heliocentric controversy (page 125) may be seen today as a matter of frame of reference. From the reference frame of Earth, we see the Sun and Moon as revolving around us with average periods of 24 h (= definition of 1 day) and almost 25 h, respectively, roughly in circles. The orbits of the planets as seen from Earth are very complicated, however.

In the Sun's reference frame, Earth makes one revolution (= definition of the year) in 365.256 days, in an ellipse that is nearly a circle. The Sun's reference frame has the advantage that the other planets also have simple elliptical orbits. (Or nearly so—each planet's gravity pulls on the others, causing small perturbations.) The Sun's vastly greater mass ($>10^5 \times$ Earth's) allows it to be an easier reference frame to use.

The Sun itself (and the Earth with it) revolves around the center of our Galaxy (see Fig. 33–2 or 5–49) which itself moves relative to other galaxies. Indeed, there is no one reference frame that we can consider as preferred or central.

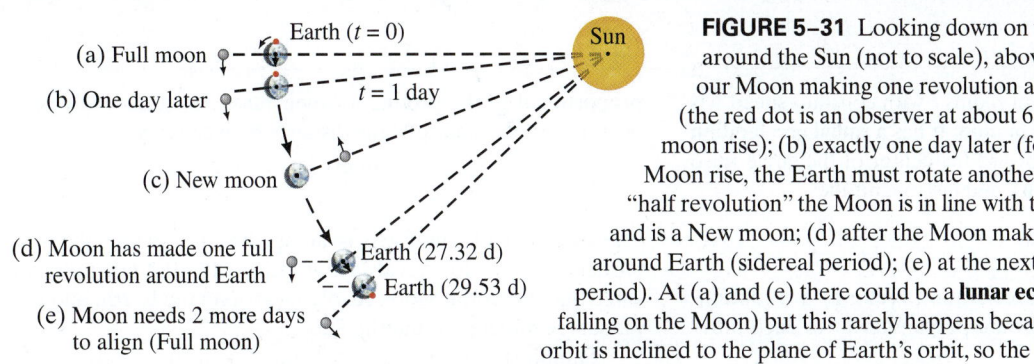

FIGURE 5–31 Looking down on the plane of Earth's orbit around the Sun (not to scale), above Earth's north pole, showing our Moon making one revolution about Earth: (a) at a Full moon (the red dot is an observer at about 6 PM who can just see the Full moon rise); (b) exactly one day later (for the red dot to see the Moon rise, the Earth must rotate another 50 min); (c) after making a "half revolution" the Moon is in line with the Sun, on the Sun's side, and is a New moon; (d) after the Moon makes one complete revolution around Earth (sidereal period); (e) at the next Full moon (synodic period). At (a) and (e) there could be a **lunar eclipse** (Earth's shadow falling on the Moon) but this rarely happens because the plane of Moon's orbit is inclined to the plane of Earth's orbit, so the Moon is usually above or below the Earth's orbital plane. At (c) there could be a **solar eclipse**, also rare.

5–9 Moon Rises an Hour Later Each Day

From the Earth's reference frame, our Moon revolves on average in 24 h, 50 min, which means the Moon rises nearly an hour later each day; and it is at its highest point in the sky about an hour later each day. When the Moon is on the direct opposite side of Earth from the Sun, the Sun's light fully illuminates the Moon and we call it a **Full moon** (Fig. 5–31a). When the Moon is on the same side of the Earth as the Sun, and nearly aligned with both, we see the Moon as a thin sliver—most or all of it is in shadow (= a **New moon**). The phases of the Moon (new, first quarter, full, third quarter) take it from one Full moon to the next Full moon in 29.53 days (= **synodic period**) on average, as seen from the Earth as reference frame (Fig. 5–31e). In the Sun's frame of reference, the Moon revolves around the Earth in 27.32 days (**sidereal period**, Fig. 5–31d). This small difference arises because, when the Moon has made one complete revolution around the Earth, the Earth itself has moved in its orbit relative to the Sun. So the Moon needs more time ($\approx$ 2 days) to be fully aligned with the Sun and Earth and be a Full moon, Fig. 5–31e. The red dot in Figs. 5–31a, b, and e represents an observer at the same location on Earth, which in (a) is when the Full moon is rising and the Sun is just setting.

5–10 Types of Forces in Nature

We have already discussed that Newton's law of universal gravitation, Eq. 5–4, describes how a particular type of force—gravity—depends on the masses of the objects involved and the distance between them. Newton's second law, $\Sigma \vec{F} = m\vec{a}$, on the other hand, tells how an object will accelerate due to *any* type of force. But what are the types of forces that occur in nature besides gravity?

In the twentieth century, physicists came to recognize four fundamental forces in nature: (1) the gravitational force; (2) the electromagnetic force (we shall see later that electric and magnetic forces are intimately related); (3) the strong nuclear force (which holds protons and neutrons together to form atomic nuclei); and (4) the weak nuclear force (involved in radioactivity). In this Chapter, we discussed the gravitational force in detail. The nature of the electromagnetic force will be discussed in Chapters 16 to 22. The strong and weak nuclear forces, which are discussed in Chapters 30 to 32, operate at the level of the atomic nucleus and are much less obvious in our daily lives.

Physicists have been working on theories that would unify these four forces—that is, to consider some or all of these forces as different manifestations of the same basic force. So far, the electromagnetic and weak nuclear forces have been theoretically united to form *electroweak* theory, in which the electromagnetic and weak forces are seen as two aspects of a single *electroweak force*. Attempts to further unify the forces, such as in *grand unified theories* (GUT), are hot research topics today.

But where do everyday forces fit? Ordinary forces, other than gravity, such as pushes, pulls, and other contact forces like the normal force and friction, are today considered to be due to the electromagnetic force acting at the atomic level. For example, the force your fingers exert on a pencil is the result of electrical repulsion between the outer electrons of the atoms of your finger and those of the pencil.

Summary

An object moving in a circle of radius r with constant speed v is said to be in **uniform circular motion**. It has a **radial acceleration** a_R that is directed radially toward the center of the circle (also called **centripetal acceleration**), and has magnitude

$$a_R = \frac{v^2}{r}. \quad (5\text{--}1)$$

The velocity vector and the acceleration vector $\vec{a}_R$ are continually changing in direction, but are perpendicular to each other at each moment.

A force is needed to keep an object revolving in a circle, and the direction of this force is toward the center of the circle. This force could be due to gravity (as for the Moon), to tension in a cord, to a component of the normal force, or to another type of force or combination of forces.

[*When the speed of circular motion is not constant, the acceleration has two components, tangential as well as centripetal.]

Newton's **law of universal gravitation** states that every particle in the universe attracts every other particle with a force proportional to the product of their masses and inversely proportional to the square of the distance between them:

$$F_G = G\frac{m_1 m_2}{r^2}. \quad (5\text{--}4)$$

The direction of this force is along the line joining the two particles, and the force is always attractive. It is this gravitational force that keeps the Moon revolving around the Earth, and the planets revolving around the Sun.

Satellites revolving around the Earth are acted on by gravity, but "stay up" because of their high tangential speed.

Newton's three laws of motion, plus his law of universal gravitation, constituted a wide-ranging theory of the universe. With them, motion of objects on Earth and in space could be accurately described. And they provided a theoretical base for **Kepler's laws** of planetary motion.

The four fundamental forces in nature are (1) the gravitational force, (2) the electromagnetic force, (3) the strong nuclear force, and (4) the weak nuclear force. The first two fundamental forces are responsible for nearly all "everyday" forces.

Questions

1. How many "accelerators" do you have in your car? There are at least three controls in the car which can be used to cause the car to accelerate. What are they? What accelerations do they produce?

2. A car rounds a curve at a steady 50 km/h. If it rounds the same curve at a steady 70 km/h, will its acceleration be any different? Explain.

3. Will the acceleration of a car be the same when a car travels around a sharp curve at a constant 60 km/h as when it travels around a gentle curve at the same speed? Explain.

4. Describe all the forces acting on a child riding a horse on a merry-go-round. Which of these forces provides the centripetal acceleration of the child?

5. A child on a sled comes flying over the crest of a small hill, as shown in Fig. 5–32. His sled does not leave the ground, but he feels the normal force between his chest and the sled decrease as he goes over the hill. Explain this decrease using Newton's second law.

FIGURE 5–32 Question 5.

6. Sometimes it is said that water is removed from clothes in the spin dryer by centrifugal force throwing the water outward. Is this correct? Discuss.

7. A girl is whirling a ball on a string around her head in a horizontal plane. She wants to let go at precisely the right time so that the ball will hit a target on the other side of the yard. When should she let go of the string?

8. A bucket of water can be whirled in a vertical circle without the water spilling out, even at the top of the circle when the bucket is upside down. Explain.

9. Astronauts who spend long periods in outer space could be adversely affected by weightlessness. One way to simulate gravity is to shape the spaceship like a cylindrical shell that rotates, with the astronauts walking on the inside surface (Fig. 5–33). Explain how this simulates gravity. Consider (a) how objects fall, (b) the force we feel on our feet, and (c) any other aspects of gravity you can think of.

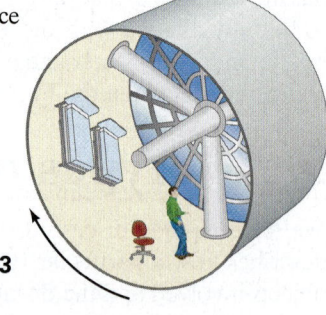

FIGURE 5–33 Question 9.

10. A car maintains a constant speed v as it traverses the hill and valley shown in Fig. 5–34. Both the hill and valley have a radius of curvature R. At which point, A, B, or C, is the normal force acting on the car (a) the largest, (b) the smallest? Explain. (c) Where would the driver feel heaviest and (d) lightest? Explain. (e) How fast can the car go without losing contact with the road at A?

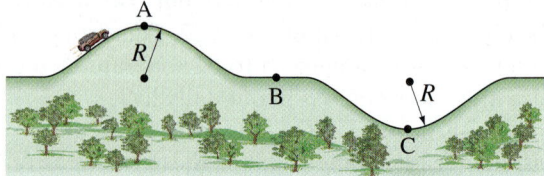

FIGURE 5–34 Question 10.

11. Can a particle with constant speed be accelerating? What if it has constant velocity? Explain.

12. Why do airplanes bank when they turn? How would you compute the banking angle given the airspeed and radius of the turn? [*Hint*: Assume an aerodynamic "lift" force acts perpendicular to the wings. See also Example 5–7.]

13. Does an apple exert a gravitational force on the Earth? If so, how large a force? Consider an apple (*a*) attached to a tree and (*b*) falling.
14. Why is more fuel required for a spacecraft to travel from the Earth to the Moon than to return from the Moon to the Earth?
15. Would it require less speed to launch a satellite (*a*) toward the east or (*b*) toward the west? Consider the Earth's rotation direction and explain your choice.
16. An antenna loosens and becomes detached from a satellite in a circular orbit around the Earth. Describe the antenna's subsequent motion. If it will land on the Earth, describe where; if not, describe how it could be made to land on the Earth.
17. The Sun is below us at midnight, nearly in line with the Earth's center. Are we then heavier at midnight, due to the Sun's gravitational force on us, than we are at noon? Explain.
18. (*a*) When will your apparent weight be the greatest, as measured by a scale in a moving elevator: when the elevator (i) accelerates downward, (ii) accelerates upward, (iii) is in free fall, or (iv) moves upward at constant speed? (*b*) In which case would your apparent weight be the least? (*c*) When would it be the same as when you are on the ground? Explain.
19. The source of the Mississippi River is closer to the center of the Earth than is its outlet in Louisiana (because the Earth is fatter at the equator than at the poles). Explain how the Mississippi can flow "uphill."
20. People sometimes ask, "What keeps a satellite up in its orbit around the Earth?" How would you respond?
21. Is the centripetal acceleration of Mars in its orbit around the Sun larger or smaller than the centripetal acceleration of the Earth? Explain.
22. The mass of the "planet" Pluto was not known until it was discovered to have a moon. Explain how this enabled an estimate of Pluto's mass.
23. The Earth moves faster in its orbit around the Sun in January than in July. Is the Earth closer to the Sun in January, or in July? Explain. [*Note*: This is not much of a factor in producing the seasons—the main factor is the tilt of the Earth's axis relative to the plane of its orbit.]

MisConceptual Questions

1. While driving fast around a sharp right turn, you find yourself pressing against the car door. What is happening?
   (*a*) Centrifugal force is pushing you into the door.
   (*b*) The door is exerting a rightward force on you.
   (*c*) Both of the above.
   (*d*) Neither of the above.
2. Which of the following point towards the center of the circle in uniform circular motion?
   (*a*) Acceleration.
   (*b*) Velocity, acceleration, net force.
   (*c*) Velocity, acceleration.
   (*d*) Velocity, net force.
   (*e*) Acceleration, net force.
3. A Ping-Pong ball is shot into a circular tube that is lying flat (horizontal) on a tabletop. When the Ping-Pong ball exits the tube, which path will it follow in Fig. 5–35?

   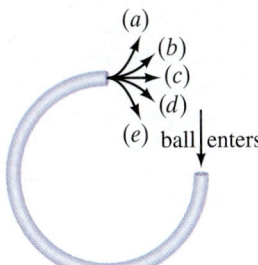

   FIGURE 5–35 MisConceptual Question 3.

4. A car drives at steady speed around a perfectly circular track.
   (*a*) The car's acceleration is zero.
   (*b*) The net force on the car is zero.
   (*c*) Both the acceleration and net force on the car point outward.
   (*d*) Both the acceleration and net force on the car point inward.
   (*e*) If there is no friction, the acceleration is outward.

5. A child whirls a ball in a vertical circle. Assuming the speed of the ball is constant (an approximation), when would the tension in the cord connected to the ball be greatest?
   (*a*) At the top of the circle.
   (*b*) At the bottom of the circle.
   (*c*) A little after the bottom of the circle when the ball is climbing.
   (*d*) A little before the bottom of the circle when the ball is descending quickly.
   (*e*) Nowhere; the cord is stretched the same amount at all points.

6. In a rotating vertical cylinder (Rotor-ride) a rider finds herself pressed with her back to the rotating wall. Which is the correct free-body diagram for her (Fig. 5–36)?

   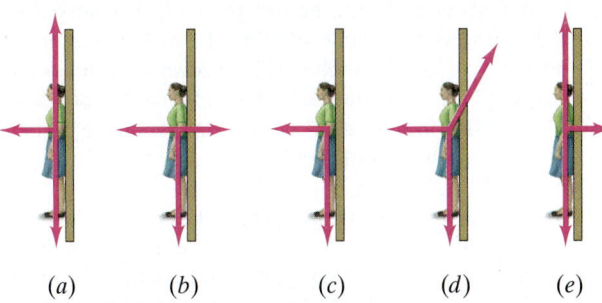

   FIGURE 5–36 MisConceptual Question 6.

7. The Moon does not crash into the Earth because:
   (*a*) the net force on it is zero.
   (*b*) it is beyond the main pull of the Earth's gravity.
   (*c*) it is being pulled by the Sun as well as by the Earth.
   (*d*) it is freely falling but it has a high tangential velocity.

8. Which pulls harder gravitationally, the Earth on the Moon, or the Moon on the Earth? Which accelerates more?
   (a) The Earth on the Moon; the Earth.
   (b) The Earth on the Moon; the Moon.
   (c) The Moon on the Earth; the Earth.
   (d) The Moon on the Earth; the Moon.
   (e) Both the same; the Earth.
   (f) Both the same; the Moon.
9. In the International Space Station which orbits Earth, astronauts experience apparent weightlessness because
   (a) the station is so far away from the center of the Earth.
   (b) the station is kept in orbit by a centrifugal force that counteracts the Earth's gravity.
   (c) the astronauts and the station are in free fall towards the center of the Earth.
   (d) there is no gravity in space.
   (e) the station's high speed nullifies the effects of gravity.
10. Two satellites orbit the Earth in circular orbits of the same radius. One satellite is twice as massive as the other. Which statement is true about the speeds of these satellites?
    (a) The heavier satellite moves twice as fast as the lighter one.
    (b) The two satellites have the same speed.
    (c) The lighter satellite moves twice as fast as the heavier one.
    (d) The ratio of their speeds depends on the orbital radius.

11. A space shuttle in orbit around the Earth carries its payload with its mechanical arm. Suddenly, the arm malfunctions and releases the payload. What will happen to the payload?
    (a) It will fall straight down and hit the Earth.
    (b) It will follow a curved path and eventually hit the Earth.
    (c) It will remain in the same orbit with the shuttle.
    (d) It will drift out into deep space.

*12. A penny is placed on a turntable which is spinning clockwise as shown in Fig. 5–37. If the power to the turntable is turned off, which arrow best represents the direction of the acceleration of the penny at point P while the turntable is still spinning but slowing down?

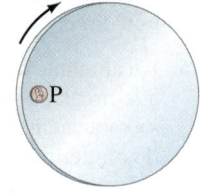

FIGURE 5–37 MisConceptual Question 12.

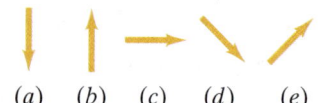

(a)   (b)   (c)   (d)   (e)

For assigned homework and other learning materials, go to the MasteringPhysics website.

Problems

5–1 to 5–3 Uniform Circular Motion

1. (I) A child sitting 1.20 m from the center of a merry-go-round moves with a speed of 1.10 m/s. Calculate (a) the centripetal acceleration of the child and (b) the net horizontal force exerted on the child (mass = 22.5 kg).
2. (I) A jet plane traveling 1890 km/h (525 m/s) pulls out of a dive by moving in an arc of radius 5.20 km. What is the plane's acceleration in g's?
3. (I) A horizontal force of 310 N is exerted on a 2.0-kg ball as it rotates (at arm's length) uniformly in a horizontal circle of radius 0.90 m. Calculate the speed of the ball.
4. (II) What is the magnitude of the acceleration of a speck of clay on the edge of a potter's wheel turning at 45 rpm (revolutions per minute) if the wheel's diameter is 35 cm?
5. (II) A 0.55-kg ball, attached to the end of a horizontal cord, is revolved in a circle of radius 1.3 m on a frictionless horizontal surface. If the cord will break when the tension in it exceeds 75 N, what is the maximum speed the ball can have?
6. (II) How fast (in rpm) must a centrifuge rotate if a particle 7.00 cm from the axis of rotation is to experience an acceleration of 125,000 g's?
7. (II) A car drives straight down toward the bottom of a valley and up the other side on a road whose bottom has a radius of curvature of 115 m. At the very bottom, the normal force on the driver is twice his weight. At what speed was the car traveling?
8. (II) How large must the coefficient of static friction be between the tires and the road if a car is to round a level curve of radius 125 m at a speed of 95 km/h?
9. (II) What is the maximum speed with which a 1200-kg car can round a turn of radius 90.0 m on a flat road if the coefficient of friction between tires and road is 0.65? Is this result independent of the mass of the car?
10. (II) A bucket of mass 2.00 kg is whirled in a vertical circle of radius 1.20 m. At the lowest point of its motion the tension in the rope supporting the bucket is 25.0 N. (a) Find the speed of the bucket. (b) How fast must the bucket move at the top of the circle so that the rope does not go slack?
11. (II) How many revolutions per minute would a 25-m-diameter Ferris wheel need to make for the passengers to feel "weightless" at the topmost point?
12. (II) A jet pilot takes his aircraft in a vertical loop (Fig. 5–38).
    (a) If the jet is moving at a speed of 840 km/h at the lowest point of the loop, determine the minimum radius of the circle so that the centripetal acceleration at the lowest point does not exceed 6.0 g's. (b) Calculate the 78-kg pilot's effective weight (the force with which the seat pushes up on him) at the bottom of the circle, and (c) at the top of the circle (assume the same speed).

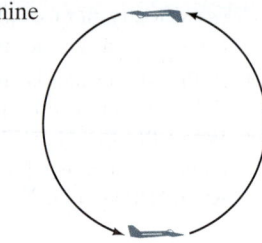

FIGURE 5–38 Problem 12.

13. (II) A proposed space station consists of a circular tube that will rotate about its center (like a tubular bicycle tire), Fig. 5–39. The circle formed by the tube has a diameter of 1.1 km. What must be the rotation speed (revolutions per day) if an effect nearly equal to gravity at the surface of the Earth (say, 0.90 g) is to be felt?

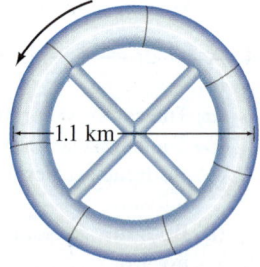

FIGURE 5–39 Problem 13.

132 CHAPTER 5 Circular Motion; Gravitation

14. (II) On an ice rink two skaters of equal mass grab hands and spin in a mutual circle once every 2.5 s. If we assume their arms are each 0.80 m long and their individual masses are 55.0 kg, how hard are they pulling on one another?

15. (II) A coin is placed 13.0 cm from the axis of a rotating turntable of variable speed. When the speed of the turntable is slowly increased, the coin remains fixed on the turntable until a rate of 38.0 rpm (revolutions per minute) is reached, at which point the coin slides off. What is the coefficient of static friction between the coin and the turntable?

16. (II) The design of a new road includes a straight stretch that is horizontal and flat but that suddenly dips down a steep hill at 18°. The transition should be rounded with what minimum radius so that cars traveling 95 km/h will not leave the road (Fig. 5–40)?

FIGURE 5–40 Problem 16.

17. (II) Two blocks, with masses m_A and m_B, are connected to each other and to a central post by thin rods as shown in Fig. 5–41. The blocks revolve about the post at the same frequency f (revolutions per second) on a frictionless horizontal surface at distances r_A and r_B from the post. Derive an algebraic expression for the tension in each rod.

FIGURE 5–41 Problem 17.

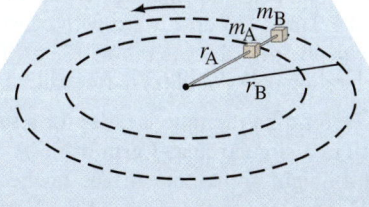

18. (II) Tarzan plans to cross a gorge by swinging in an arc from a hanging vine (Fig. 5–42). If his arms are capable of exerting a force of 1150 N on the vine, what is the maximum speed he can tolerate at the lowest point of his swing? His mass is 78 kg and the vine is 4.7 m long.

FIGURE 5–42 Problem 18.

19. (II) A 975-kg sports car (including driver) crosses the rounded top of a hill (radius = 88.0 m) at 18.0 m/s. Determine (a) the normal force exerted by the road on the car, (b) the normal force exerted by the car on the 62.0-kg driver, and (c) the car speed at which the normal force on the driver equals zero.

20. (II) Highway curves are marked with a suggested speed. If this speed is based on what would be safe in wet weather, estimate the radius of curvature for an unbanked curve marked 50 km/h. Use Table 4–2 (coefficients of friction).

21. (III) A pilot performs an evasive maneuver by diving vertically at 270 m/s. If he can withstand an acceleration of 8.0 g's without blacking out, at what altitude must he begin to pull his plane out of the dive to avoid crashing into the sea?

22. (III) If a curve with a radius of 95 m is properly banked for a car traveling 65 km/h, what must be the coefficient of static friction for a car not to skid when traveling at 95 km/h?

23. (III) A curve of radius 78 m is banked for a design speed of 85 km/h. If the coefficient of static friction is 0.30 (wet pavement), at what range of speeds can a car safely make the curve? [*Hint*: Consider the direction of the friction force when the car goes too slow or too fast.]

*5–4 Nonuniform Circular Motion

*24. (I) Determine the tangential and centripetal components of the net force exerted on the car (by the ground) in Example 5–8 when its speed is 15 m/s. The car's mass is 950 kg.

*25. (II) A car at the Indianapolis 500 accelerates uniformly from the pit area, going from rest to 270 km/h in a semicircular arc with a radius of 220 m. Determine the tangential and radial acceleration of the car when it is halfway through the arc, assuming constant tangential acceleration. If the curve were flat, what coefficient of static friction would be necessary between the tires and the road to provide this acceleration with no slipping or skidding?

*26. (II) For each of the cases described below, sketch and label the total acceleration vector, the radial acceleration vector, and the tangential acceleration vector. (a) A car is accelerating from 55 km/h to 70 km/h as it rounds a curve of constant radius. (b) A car is going a constant 65 km/h as it rounds a curve of constant radius. (c) A car slows down while rounding a curve of constant radius.

*27. (III) A particle revolves in a horizontal circle of radius 1.95 m. At a particular instant, its acceleration is 1.05 m/s^2, in a direction that makes an angle of 25.0° to its direction of motion. Determine its speed (a) at this moment, and (b) 2.00 s later, assuming constant tangential acceleration.

5–5 and 5–6 Law of Universal Gravitation

28. (I) Calculate the force of Earth's gravity on a spacecraft 2.00 Earth radii above the Earth's surface if its mass is 1850 kg.

29. (I) At the surface of a certain planet, the gravitational acceleration g has a magnitude of 12.0 m/s^2. A 24.0-kg brass ball is transported to this planet. What is (a) the mass of the brass ball on the Earth and on the planet, and (b) the weight of the brass ball on the Earth and on the planet?

30. (II) At what distance from the Earth will a spacecraft traveling directly from the Earth to the Moon experience zero net force because the Earth and Moon pull in opposite directions with equal force?

31. (II) Two objects attract each other gravitationally with a force of 2.5×10^{-10} N when they are 0.25 m apart. Their total mass is 4.00 kg. Find their individual masses.

32. (II) A hypothetical planet has a radius 2.0 times that of Earth, but has the same mass. What is the acceleration due to gravity near its surface?

33. (II) Calculate the acceleration due to gravity on the Moon, which has radius 1.74×10^6 m and mass 7.35×10^{22} kg.

34. (II) Estimate the acceleration due to gravity at the surface of Europa (one of the moons of Jupiter) given that its mass is 4.9×10^{22} kg and making the assumption that its mass per unit volume is the same as Earth's.

35. (II) Given that the acceleration of gravity at the surface of Mars is 0.38 of what it is on Earth, and that Mars' radius is 3400 km, determine the mass of Mars.

36. (II) Find the net force on the Moon ($m_M = 7.35 \times 10^{22}$ kg) due to the gravitational attraction of both the Earth ($m_E = 5.98 \times 10^{24}$ kg) and the Sun ($m_S = 1.99 \times 10^{30}$ kg), assuming they are at right angles to each other, Fig. 5–43.

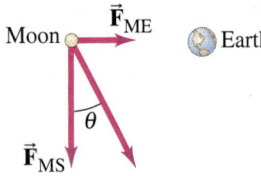

FIGURE 5–43 Problem 36. Orientation of Sun (S), Earth (E), and Moon (M) at right angles to each other (not to scale).

37. (II) A hypothetical planet has a mass 2.80 times that of Earth, but has the same radius. What is g near its surface?

38. (II) If you doubled the mass and tripled the radius of a planet, by what factor would g at its surface change?

39. (II) Calculate the effective value of g, the acceleration of gravity, at (a) 6400 m, and (b) 6400 km, above the Earth's surface.

40. (II) You are explaining to friends why an astronaut feels weightless orbiting in the space shuttle, and they respond that they thought gravity was just a lot weaker up there. Convince them that it isn't so by calculating how much weaker (in %) gravity is 380 km above the Earth's surface.

41. (II) Every few hundred years most of the planets line up on the same side of the Sun. Calculate the total force on the Earth due to Venus, Jupiter, and Saturn, assuming all four planets are in a line, Fig. 5–44. The masses are $m_V = 0.815\, m_E$, $m_J = 318\, m_E$, $m_{Sat} = 95.1\, m_E$, and the mean distances of the four planets from the Sun are 108, 150, 778, and 1430 million km. What fraction of the Sun's force on the Earth is this?

FIGURE 5–44 Problem 41 (not to scale).

42. (II) Four 7.5-kg spheres are located at the corners of a square of side 0.80 m. Calculate the magnitude and direction of the gravitational force exerted on one sphere by the other three.

43. (II) Determine the distance from the Earth's center to a point outside the Earth where the gravitational acceleration due to the Earth is $\frac{1}{10}$ of its value at the Earth's surface.

44. (II) A certain neutron star has five times the mass of our Sun packed into a sphere about 10 km in radius. Estimate the surface gravity on this monster.

5–7 Satellites and Weightlessness

45. (I) A space shuttle releases a satellite into a circular orbit 780 km above the Earth. How fast must the shuttle be moving (relative to Earth's center) when the release occurs?

46. (I) Calculate the speed of a satellite moving in a stable circular orbit about the Earth at a height of 4800 km.

47. (II) You know your mass is 62 kg, but when you stand on a bathroom scale in an elevator, it says your mass is 77 kg. What is the acceleration of the elevator, and in which direction?

48. (II) A 12.0-kg monkey hangs from a cord suspended from the ceiling of an elevator. The cord can withstand a tension of 185 N and breaks as the elevator accelerates. What was the elevator's minimum acceleration (magnitude and direction)?

49. (II) Calculate the period of a satellite orbiting the Moon, 95 km above the Moon's surface. Ignore effects of the Earth. The radius of the Moon is 1740 km.

50. (II) Two satellites orbit Earth at altitudes of 7500 km and 15,000 km above the Earth's surface. Which satellite is faster, and by what factor?

51. (II) What will a spring scale read for the weight of a 58.0-kg woman in an elevator that moves (a) upward with constant speed 5.0 m/s, (b) downward with constant speed 5.0 m/s, (c) with an upward acceleration 0.23 g, (d) with a downward acceleration 0.23 g, and (e) in free fall?

52. (II) Determine the time it takes for a satellite to orbit the Earth in a circular **near-Earth orbit**. A "near-Earth" orbit is at a height above the surface of the Earth that is very small compared to the radius of the Earth. [*Hint:* You may take the acceleration due to gravity as essentially the same as that on the surface.] Does your result depend on the mass of the satellite?

53. (II) What is the apparent weight of a 75-kg astronaut 2500 km from the center of the Moon in a space vehicle (a) moving at constant velocity and (b) accelerating toward the Moon at 1.8 m/s²? State "direction" in each case.

54. (II) A Ferris wheel 22.0 m in diameter rotates once every 12.5 s (see Fig. 5–9). What is the ratio of a person's apparent weight to her real weight at (a) the top, and (b) the bottom?

55. (II) At what rate must a cylindrical spaceship rotate if occupants are to experience simulated gravity of 0.70 g? Assume the spaceship's diameter is 32 m, and give your answer as the time needed for one revolution. (See Question 9, Fig 5–33.)

56. (III) (a) Show that if a satellite orbits very near the surface of a planet with period T, the density (= mass per unit volume) of the planet is $\rho = m/V = 3\pi/GT^2$. (b) Estimate the density of the Earth, given that a satellite near the surface orbits with a period of 85 min. Approximate the Earth as a uniform sphere.

5–8 Kepler's Laws

57. (I) Neptune is an average distance of 4.5×10^9 km from the Sun. Estimate the length of the Neptunian year using the fact that the Earth is 1.50×10^8 km from the Sun on average.

58. (I) The *asteroid* Icarus, though only a few hundred meters across, orbits the Sun like the planets. Its period is 410 d. What is its mean distance from the Sun?

59. (I) Use Kepler's laws and the period of the Moon (27.4 d) to determine the period of an artificial satellite orbiting very near the Earth's surface.

60. (II) Determine the mass of the Earth from the known period and distance of the Moon.

61. (II) Our Sun revolves about the center of our Galaxy ($m_G \approx 4 \times 10^{41}$ kg) at a distance of about 3×10^4 light-years $[1\,\text{ly} = (3.00 \times 10^8\,\text{m/s}) \cdot (3.16 \times 10^7\,\text{s/yr}) \cdot (1.00\,\text{yr})]$. What is the period of the Sun's orbital motion about the center of the Galaxy?

62. (II) Table 5–3 gives the mean distance, period, and mass for the four largest moons of Jupiter (those discovered by Galileo in 1609). Determine the mass of Jupiter: (*a*) using the data for Io; (*b*) using data for each of the other three moons. Are the results consistent?

TABLE 5–3 Principal Moons of Jupiter
(Problems 62 and 63)

Moon	Mass (kg)	Period (Earth days)	Mean distance from Jupiter (km)
Io	8.9×10^{22}	1.77	422×10^3
Europa	4.9×10^{22}	3.55	671×10^3
Ganymede	15×10^{22}	7.16	1070×10^3
Callisto	11×10^{22}	16.7	1883×10^3

63. (II) Determine the mean distance from Jupiter for each of Jupiter's principal moons, using Kepler's third law. Use the distance of Io and the periods given in Table 5–3. Compare your results to the values in Table 5–3.

64. (II) Planet A and planet B are in circular orbits around a distant star. Planet A is 7.0 times farther from the star than is planet B. What is the ratio of their speeds v_A/v_B?

65. (II) *Halley's comet* orbits the Sun roughly once every 76 years. It comes very close to the surface of the Sun on its closest approach (Fig. 5–45). Estimate the greatest distance of the comet from the Sun. Is it still "in" the solar system? What planet's orbit is nearest when it is out there?

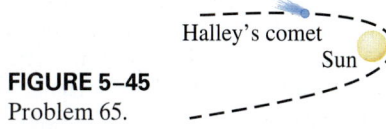

FIGURE 5–45 Problem 65.

66. (III) The *comet Hale–Bopp* has an orbital period of 2400 years. (*a*) What is its mean distance from the Sun? (*b*) At its closest approach, the comet is about 1.0 AU from the Sun (1 AU = distance from Earth to the Sun). What is the farthest distance? (*c*) What is the ratio of the speed at the closest point to the speed at the farthest point?

General Problems

67. Calculate the centripetal acceleration of the Earth in its orbit around the Sun, and the net force exerted on the Earth. What exerts this force on the Earth? Assume that the Earth's orbit is a circle of radius 1.50×10^{11} m.

68. A flat puck (mass M) is revolved in a circle on a frictionless air hockey table top, and is held in this orbit by a massless cord which is connected to a dangling mass (mass m) through a central hole as shown in Fig. 5–46. Show that the speed of the puck is given by $v = \sqrt{mgR/M}$.

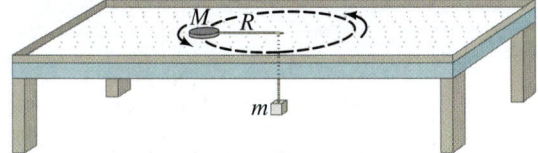

FIGURE 5–46 Problem 68.

69. A device for training astronauts and jet fighter pilots is designed to move the trainee in a horizontal circle of radius 11.0 m. If the force felt by the trainee is 7.45 times her own weight, how fast is she revolving? Express your answer in both m/s and rev/s.

70. A 1050-kg car rounds a curve of radius 72 m banked at an angle of 14°. If the car is traveling at 85 km/h, will a friction force be required? If so, how much and in what direction?

71. In a "Rotor-ride" at a carnival, people rotate in a vertical cylindrically walled "room." (See Fig. 5–47.) If the room radius is 5.5 m, and the rotation frequency 0.50 revolutions per second when the floor drops out, what minimum coefficient of static friction keeps the people from slipping down? People on this ride said they were "pressed against the wall." Is there really an outward force pressing them against the wall? If so, what is its source? If not, what is the proper description of their situation (besides nausea)? [*Hint:* Draw a free-body diagram for a person.]

FIGURE 5–47 Problem 71.

72. While fishing, you get bored and start to swing a sinker weight around in a circle below you on a 0.25-m piece of fishing line. The weight makes a complete circle every 0.75 s. What is the angle that the fishing line makes with the vertical? [*Hint:* See Fig. 5–10.]

73. At what minimum speed must a roller coaster be traveling so that passengers upside down at the top of the circle (Fig. 5–48) do not fall out? Assume a radius of curvature of 8.6 m.

FIGURE 5–48 Problem 73.

74. Consider a train that rounds a curve with a radius of 570 m at a speed of 160 km/h (approximately 100 mi/h). (a) Calculate the friction force needed on a train passenger of mass 55 kg if the track is not banked and the train does not tilt. (b) Calculate the friction force on the passenger if the train tilts at an angle of 8.0° toward the center of the curve.

75. Two equal-mass stars maintain a constant distance apart of 8.0×10^{11} m and revolve about a point midway between them at a rate of one revolution every 12.6 yr. (a) Why don't the two stars crash into one another due to the gravitational force between them? (b) What must be the mass of each star?

76. How far above the Earth's surface will the acceleration of gravity be half what it is at the surface?

77. Is it possible to whirl a bucket of water fast enough in a vertical circle so that the water won't fall out? If so, what is the minimum speed? Define all quantities needed.

78. How long would a day be if the Earth were rotating so fast that objects at the equator were apparently weightless?

79. The *rings of Saturn* are composed of chunks of ice that orbit the planet. The inner radius of the rings is 73,000 km, and the outer radius is 170,000 km. Find the period of an orbiting chunk of ice at the inner radius and the period of a chunk at the outer radius. Compare your numbers with Saturn's own rotation period of 10 hours and 39 minutes. The mass of Saturn is 5.7×10^{26} kg.

80. During an *Apollo* lunar landing mission, the command module continued to orbit the Moon at an altitude of about 100 km. How long did it take to go around the Moon once?

81. The **Navstar Global Positioning System** (GPS) utilizes a group of 24 satellites orbiting the Earth. Using "triangulation" and signals transmitted by these satellites, the position of a receiver on the Earth can be determined to within an accuracy of a few centimeters. The satellite orbits are distributed around the Earth, allowing continuous navigational "fixes." The satellites orbit at an altitude of approximately 11,000 nautical miles [1 nautical mile = 1.852 km = 6076 ft]. (a) Determine the speed of each satellite. (b) Determine the period of each satellite.

82. The *Near Earth Asteroid Rendezvous* (NEAR) spacecraft, after traveling 2.1 billion km, is meant to orbit the asteroid Eros with an orbital radius of about 20 km. Eros is roughly 40 km × 6 km × 6 km. Assume Eros has a density (mass/volume) of about 2.3×10^3 kg/m³. (a) If Eros were a sphere with the same mass and density, what would its radius be? (b) What would g be at the surface of a spherical Eros? (c) Estimate the orbital period of NEAR as it orbits Eros, as if Eros were a sphere.

83. A train traveling at a constant speed rounds a curve of radius 215 m. A lamp suspended from the ceiling swings out to an angle of 16.5° throughout the curve. What is the speed of the train?

84. The Sun revolves around the center of the *Milky Way Galaxy* (Fig. 5–49) at a distance of about 30,000 light-years from the center $(1 \text{ ly} = 9.5 \times 10^{15} \text{ m})$. If it takes about 200 million years to make one revolution, estimate the mass of our Galaxy. Assume that the mass distribution of our Galaxy is concentrated mostly in a central uniform sphere. If all the stars had about the mass of our Sun $(2 \times 10^{30}$ kg), how many stars would there be in our Galaxy?

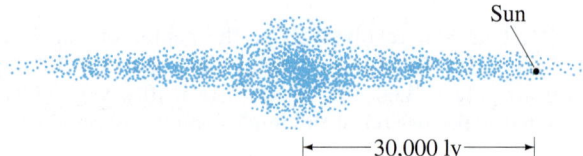

FIGURE 5–49 Edge-on view of our galaxy. Problem 84.

85. A satellite of mass 5500 kg orbits the Earth and has a period of 6600 s. Determine (a) the radius of its circular orbit, (b) the magnitude of the Earth's gravitational force on the satellite, and (c) the altitude of the satellite.

86. Astronomers using the Hubble Space Telescope deduced the presence of an extremely massive core in the distant *galaxy* M87, so dense that it could be a black hole (from which no light escapes). They did this by measuring the speed of gas clouds orbiting the core to be 780 km/s at a distance of 60 light-years $(= 5.7 \times 10^{17}$ m) from the core. Deduce the mass of the core, and compare it to the mass of our Sun.

87. Suppose all the mass of the Earth were compacted into a small spherical ball. What radius must the sphere have so that the acceleration due to gravity at the Earth's new surface would equal the acceleration due to gravity at the surface of the Sun?

88. A science-fiction tale describes an artificial "planet" in the form of a band completely encircling a sun (Fig. 5–50). The inhabitants live on the inside surface (where it is always noon). Imagine that this sun is exactly like our own, that the distance to the band is the same as the Earth–Sun distance (to make the climate livable), and that the ring rotates quickly enough to produce an apparent gravity of g as on Earth. What will be the period of revolution, this planet's year, in Earth days?

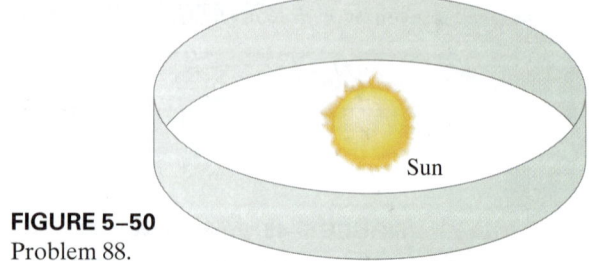

FIGURE 5–50 Problem 88.

89. An asteroid of mass m is in a circular orbit of radius r around the Sun with a speed v. It has an impact with another asteroid of mass M and is kicked into a new circular orbit with a speed of $1.5\,v$. What is the radius of the new orbit in terms of r?

*90. Use **dimensional analysis** (Section 1–8) to obtain the form for the centripetal acceleration, $a_R = v^2/r$.

Search and Learn

1. Reread each Example in this Chapter and identify (i) the object undergoing centripetal acceleration (if any), and (ii) the force, or force component, that causes the circular motion.

2. Redo Example 5–3, precisely this time, by not ignoring the weight of the ball which revolves on a string 0.600 m long. In particular, find the magnitude of $\vec{F}_T$, and the angle it makes with the horizontal. [*Hint*: Set the horizontal component of $\vec{F}_T$ equal to ma_R; also, since there is no vertical motion, what can you say about the vertical component of $\vec{F}_T$?]

3. A banked curve of radius R in a new highway is designed so that a car traveling at speed v_0 can negotiate the turn safely on glare ice (zero friction). If a car travels too slowly, then it will slip toward the center of the circle. If it travels too fast, it will slip away from the center of the circle. If the coefficient of static friction increases, it becomes possible for a car to stay on the road while traveling at a speed within a range from v_{min} to v_{max}. Derive formulas for v_{min} and v_{max} as functions of μ_s, v_0, and R.

4. *Earth is not quite an inertial frame.* We often make measurements in a reference frame fixed on the Earth, assuming Earth is an inertial reference frame [Section 4–2]. But the Earth rotates, so this assumption is not quite valid. Show that this assumption is off by 3 parts in 1000 by calculating the acceleration of an object at Earth's equator due to Earth's daily rotation, and compare to $g = 9.80 \text{ m/s}^2$, the acceleration due to gravity.

5. A certain white dwarf star was once an average star like our Sun. But now it is in the last stage of its evolution and is the size of our Moon but has the mass of our Sun. (*a*) Estimate the acceleration due to gravity on the surface of this star. (*b*) How much would a 65-kg person weigh on this star? Give as a percentage of the person's weight on Earth. (*c*) What would be the speed of a baseball dropped from a height of 1.0 m when it hit the surface?

6. Jupiter is about 320 times as massive as the Earth. Thus, it has been claimed that a person would be crushed by the force of gravity on a planet the size of Jupiter because people cannot survive more than a few g's. Calculate the number of g's a person would experience at Jupiter's equator, using the following data for Jupiter: mass = 1.9×10^{27} kg, equatorial radius = 7.1×10^4 km, rotation period = 9 hr 55 min. Take the centripetal acceleration into account. [See Sections 5–2, 5–6, and 5–7.]

7. A plumb bob (a mass m hanging on a string) is deflected from the vertical by an angle θ due to a massive mountain nearby (Fig. 5–51). (*a*) Find an approximate formula for θ in terms of the mass of the mountain, m_M, the distance to its center, D_M, and the radius and mass of the Earth. (*b*) Make a rough estimate of the mass of Mt. Everest, assuming it has the shape of a cone 4000 m high and base of diameter 4000 m. Assume its mass per unit volume is 3000 kg per m³. (*c*) Estimate the angle θ of the plumb bob if it is 5 km from the center of Mt. Everest.

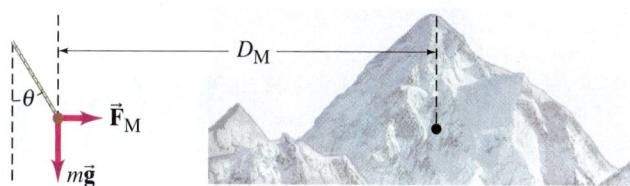

FIGURE 5–51 Search and Learn 7.

8. (*a*) Explain why a Full moon always rises at sunset. (*b*) Explain how the position of the Moon in Fig. 5–31b cannot be seen yet by the person at the red dot (shown at 6 PM). (*c*) Explain why the red dot is where it is in parts (b) and (e), and show where it should be in part (d). (*d*) PRETTY HARD. Determine the average period of the Moon around the Earth (sidereal period) starting with the synodic period of 29.53 days as observed from Earth. [*Hint*: First determine the angle of the Moon in Fig. 5–31e relative to "horizontal," as in part (a).]

ANSWERS TO EXERCISES

A: (*a*).
B: (*d*).
C: (*a*).
D: No.
E: *g* would double.
F: (*b*).

This baseball pitcher is about to accelerate the baseball to a high velocity by exerting a force on it. He will be doing work on the ball as he exerts the force over a displacement of perhaps several meters, from behind his head until he releases the ball with arm outstretched in front of him. The total work done on the ball will be equal to the kinetic energy ($\frac{1}{2}mv^2$) acquired by the ball, a result known as the work-energy principle.

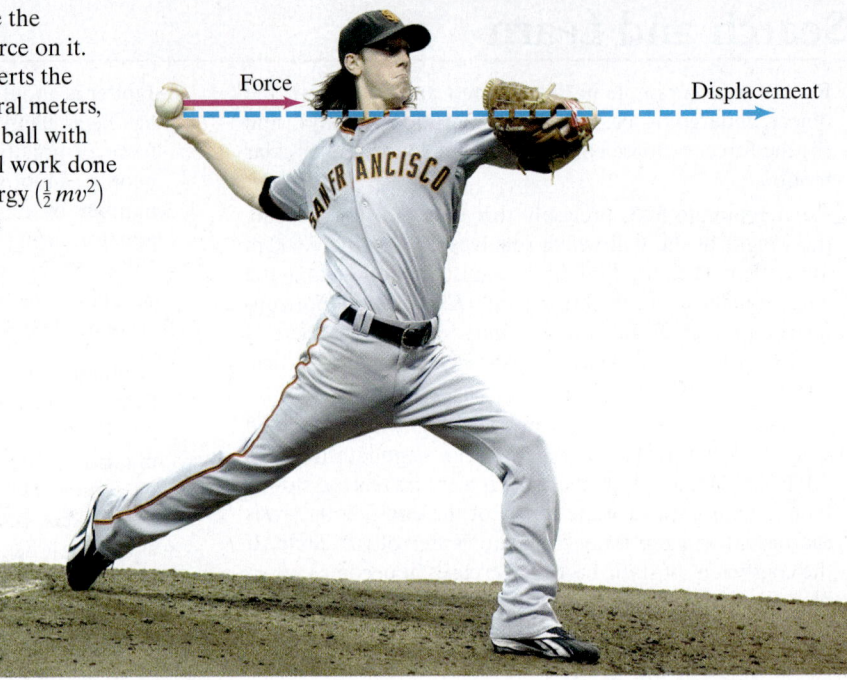

CHAPTER 6

Work and Energy

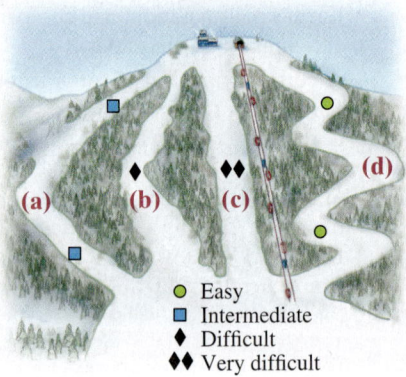

CHAPTER-OPENING QUESTION—Guess now!

A skier starts at the top of a hill. On which run does her gravitational potential energy change the most: **(a)**, **(b)**, **(c)**, or **(d)**; or are they **(e)** all the same? On which run would her speed at the bottom be the fastest if the runs are icy and we assume no friction or air resistance? Recognizing that there is always some friction, answer the above two questions again. List your four answers now.

CONTENTS

- 6–1 Work Done by a Constant Force
- *6–2 Work Done by a Varying Force
- 6–3 Kinetic Energy, and the Work-Energy Principle
- 6–4 Potential Energy
- 6–5 Conservative and Nonconservative Forces
- 6–6 Mechanical Energy and Its Conservation
- 6–7 Problem Solving Using Conservation of Mechanical Energy
- 6–8 Other Forms of Energy and Energy Transformations; The Law of Conservation of Energy
- 6–9 Energy Conservation with Dissipative Forces: Solving Problems
- 6–10 Power

Until now we have been studying the translational motion of an object in terms of Newton's three laws of motion. In that analysis, *force* has played a central role as the quantity determining the motion. In this Chapter and the next, we discuss an alternative analysis of the translational motion of objects in terms of the quantities *energy* and *momentum*. The significance of energy and momentum is that they are *conserved*. That is, in quite general circumstances they remain constant. That conserved quantities exist gives us not only a deeper insight into the nature of the world, but also gives us another way to approach solving practical problems.

The conservation laws of energy and momentum are especially valuable in dealing with systems of many objects, in which a detailed consideration of the forces involved would be difficult or impossible. These laws apply to a wide range of phenomena. They even apply in the atomic and subatomic worlds, where Newton's laws are not sufficient.

This Chapter is devoted to the very important concept of *energy* and the closely related concept of *work*. These two quantities are scalars and so have no direction associated with them, which often makes them easier to work with than vector quantities such as acceleration and force.

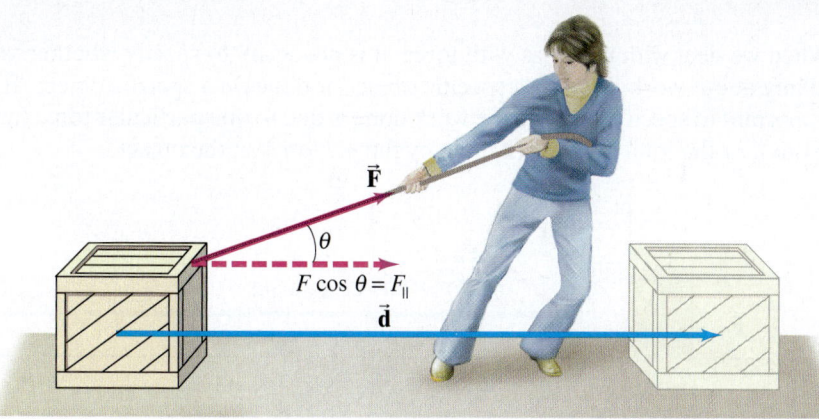

FIGURE 6–1 A person pulling a crate along the floor. The work done by the force $\vec{F}$ is $W = Fd\cos\theta$, where $\vec{d}$ is the displacement.

6–1 Work Done by a Constant Force

The word *work* has a variety of meanings in everyday language. But in physics, work is given a very specific meaning to describe what is accomplished when a force acts on an object, and the object moves through a distance. We consider only translational motion for now and, unless otherwise explained, objects are assumed to be rigid with no complicating internal motion, and can be treated like particles. Then the **work** done on an object by a constant force (constant in both magnitude and direction) is defined to be *the product of the magnitude of the displacement times the component of the force parallel to the displacement*. In equation form, we can write

$$W = F_\parallel d,$$

where $F_\parallel$ is the component of the constant force $\vec{F}$ parallel to the displacement $\vec{d}$. We can also write

$$W = Fd\cos\theta, \tag{6–1}$$

where F is the magnitude of the constant force, d is the magnitude of the displacement of the object, and θ is the angle between the directions of the force and the displacement (Fig. 6–1). The $\cos\theta$ factor appears in Eq. 6–1 because $F\cos\theta$ ($= F_\parallel$) is the component of $\vec{F}$ that is parallel to $\vec{d}$. Work is a scalar quantity—it has no direction, but only magnitude, which can be positive or negative.

Let us consider the case in which the motion and the force are in the same direction, so $\theta = 0$ and $\cos\theta = 1$; in this case, $W = Fd$. For example, if you push a loaded grocery cart a distance of 50 m by exerting a horizontal force of 30 N on the cart, you do $30\,\text{N} \times 50\,\text{m} = 1500\,\text{N}\cdot\text{m}$ of work on the cart.

As this example shows, in SI units work is measured in newton-meters ($\text{N}\cdot\text{m}$). A special name is given to this unit, the **joule** (J): $1\,\text{J} = 1\,\text{N}\cdot\text{m}$.

[In the cgs system, the unit of work is called the *erg* and is defined as $1\,\text{erg} = 1\,\text{dyne}\cdot\text{cm}$. In British units, work is measured in foot-pounds. Their equivalence is $1\,\text{J} = 10^7\,\text{erg} = 0.7376\,\text{ft}\cdot\text{lb}$.]

A force can be exerted on an object and yet do no work. If you hold a heavy bag of groceries in your hands at rest, you do no work on it. You do exert a force on the bag, but the displacement of the bag is zero, so the work done by you on the bag is $W = 0$. You need both a force and a displacement to do work. You also do no work on the bag of groceries if you carry it as you walk horizontally across the floor at constant velocity, as shown in Fig. 6–2. No horizontal force is required to move the bag at a constant velocity. The person shown in Fig. 6–2 exerts an upward force $\vec{F}_\text{P}$ on the bag equal to its weight. But this upward force is perpendicular to the horizontal displacement of the bag and thus is doing no work. This conclusion comes from our definition of work, Eq. 6–1: $W = 0$, because $\theta = 90°$ and $\cos 90° = 0$. Thus, when a particular force is perpendicular to the displacement, no work is done by that force. When you start or stop walking, there is a horizontal acceleration and you do briefly exert a horizontal force, and thus do work on the bag.

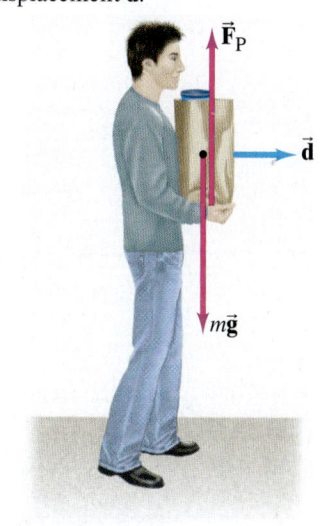

FIGURE 6–2 The person does no work on the bag of groceries because $\vec{F}_\text{P}$ is perpendicular to the displacement $\vec{d}$.

CAUTION
Force without work

CAUTION
State that work is done on or by an object

When we deal with work, as with force, it is necessary to specify whether you are talking about work done *by* a specific object or done *on* a specific object. It is also important to specify whether the work done is due to one particular force (and which one), or the total (net) work done by the *net force* on the object.

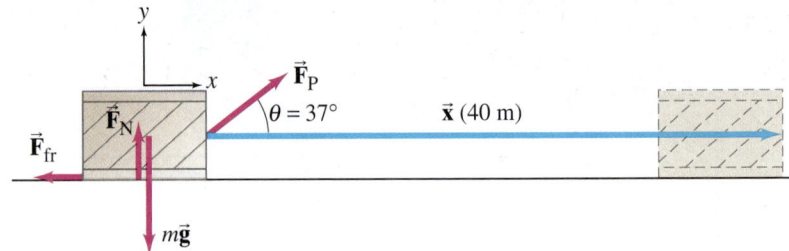

FIGURE 6–3 Example 6–1. A 50-kg crate is pulled along a floor.

EXAMPLE 6–1 Work done on a crate. A person pulls a 50-kg crate 40 m along a horizontal floor by a constant force $F_P = 100$ N, which acts at a 37° angle as shown in Fig. 6–3. The floor is rough and exerts a friction force $\vec{F}_{fr} = 50$ N. Determine (*a*) the work done by each force acting on the crate, and (*b*) the net work done on the crate.

APPROACH We choose our coordinate system so that the vector that represents the 40-m displacement is $\vec{x}$ (that is, along the *x* axis). Four forces act on the crate, as shown in the free-body diagram in Fig. 6–3: the force exerted by the person $\vec{F}_P$; the friction force $\vec{F}_{fr}$; the gravitational force exerted by the Earth, $\vec{F}_G = m\vec{g}$; and the normal force $\vec{F}_N$ exerted upward by the floor. The net force on the crate is the vector sum of these four forces.

SOLUTION (*a*) The work done by the gravitational force ($\vec{F}_G$) and by the normal force ($\vec{F}_N$) is zero, because they are perpendicular to the displacement $\vec{x}$ ($\theta = 90°$ in Eq. 6–1):

$$W_G = mgx \cos 90° = 0$$
$$W_N = F_N x \cos 90° = 0.$$

The work done by $\vec{F}_P$ is

$$W_P = F_P x \cos\theta = (100 \text{ N})(40 \text{ m}) \cos 37° = 3200 \text{ J}.$$

The work done by the friction force is

$$W_{fr} = F_{fr} x \cos 180° = (50 \text{ N})(40 \text{ m})(-1) = -2000 \text{ J}.$$

The angle between the displacement $\vec{x}$ and $\vec{F}_{fr}$ is 180° because they point in opposite directions. Since the force of friction is opposing the motion (and $\cos 180° = -1$), the work done by friction on the crate is *negative*.

(*b*) The net work can be calculated in two equivalent ways.
(1) The net work done on an object is the algebraic sum of the work done by each force, since work is a scalar:

$$W_{net} = W_G + W_N + W_P + W_{fr}$$
$$= 0 + 0 + 3200 \text{ J} - 2000 \text{ J} = 1200 \text{ J}.$$

(2) The net work can also be calculated by first determining the net force on the object and then taking the component of this net force along the displacement: $(F_{net})_x = F_P \cos\theta - F_{fr}$. Then the net work is

$$W_{net} = (F_{net})_x x = (F_P \cos\theta - F_{fr})x$$
$$= (100 \text{ N} \cos 37° - 50 \text{ N})(40 \text{ m}) = 1200 \text{ J}.$$

In the vertical (*y*) direction, there is no displacement and no work done.

CAUTION
Negative work

In Example 6–1 we saw that friction did negative work. In general, the work done by a force is negative whenever the force (or the component of the force, $F_\parallel$) acts in the direction opposite to the direction of motion.

EXERCISE A A box is dragged a distance *d* across a floor by a force $\vec{F}_P$ which makes an angle θ with the horizontal as in Fig. 6–1 or 6–3. If the magnitude of $\vec{F}_P$ is held constant but the angle θ is increased, the work done by $\vec{F}_P$ (*a*) remains the same; (*b*) increases; (*c*) decreases; (*d*) first increases, then decreases.

PROBLEM SOLVING

Work

1. **Draw a free-body diagram** showing all the forces acting on the object you choose to study.

2. **Choose** an *xy* **coordinate system**. If the object is in motion, it may be convenient to choose one of the coordinate directions as the direction of one of the forces, or as the direction of motion. [Thus, for an object on an incline, you might choose one coordinate axis to be parallel to the incline.]

3. **Apply Newton's laws** to determine unknown forces.

4. Find the **work done** *by a specific force* **on** the object by using $W = Fd \cos\theta$ for a constant force. The work done is negative when a force opposes the displacement.

5. To find the **net work** done on the object, either (*a*) find the work done by each force and add the results algebraically; or (*b*) find the net force on the object, F_{net}, and then use it to find the net work done, which for constant net force is:
$$W_{net} = F_{net} d \cos\theta.$$

EXAMPLE 6–2 **Work on a backpack.** (*a*) Determine the work a hiker must do on a 15.0-kg backpack to carry it up a hill of height $h = 10.0$ m, as shown in Fig. 6–4a. Determine also (*b*) the work done by gravity on the backpack, and (*c*) the net work done on the backpack. For simplicity, assume the motion is smooth and at constant velocity (i.e., acceleration is zero).

FIGURE 6–4 Example 6–2.

APPROACH We explicitly follow the steps of the Problem Solving Strategy above.

SOLUTION

1. **Draw a free-body diagram.** The forces on the backpack are shown in Fig. 6–4b: the force of gravity, $m\vec{g}$, acting downward; and $\vec{F}_H$, the force the hiker must exert upward to support the backpack. The acceleration is zero, so horizontal forces on the backpack are negligible.

2. **Choose a coordinate system.** We are interested in the vertical motion of the backpack, so we choose the *y* coordinate as positive vertically upward.

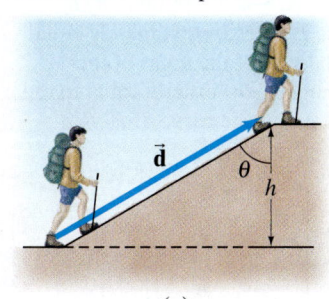

(a)

3. **Apply Newton's laws.** Newton's second law applied in the vertical direction to the backpack gives (with $a_y = 0$)
$$\Sigma F_y = ma_y$$
$$F_H - mg = 0.$$
So,
$$F_H = mg = (15.0 \text{ kg})(9.80 \text{ m/s}^2) = 147 \text{ N}.$$

4. **Work done by a specific force.** (*a*) To calculate the work done by the hiker on the backpack, we use Eq. 6–1, where θ is shown in Fig. 6–4c,
$$W_H = F_H(d \cos\theta),$$
and we note from Fig. 6–4a that $d \cos\theta = h$. So the work done by the hiker is
$$W_H = F_H(d \cos\theta) = F_H h = mgh = (147 \text{ N})(10.0 \text{ m}) = 1470 \text{ J}.$$

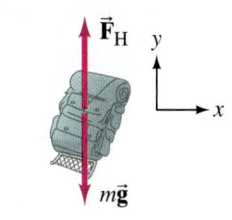

(b)

The work done depends only on the elevation change and not on the angle of the hill, θ. The hiker would do the same work to lift the pack vertically by height h.

(*b*) The work done by gravity on the backpack is (from Eq. 6–1 and Fig. 6–4c)
$$W_G = mg \, d \cos(180° - \theta).$$
Since $\cos(180° - \theta) = -\cos\theta$ (Appendix A–7), we have
$$W_G = mg(-d \cos\theta)$$
$$= -mgh$$
$$= -(15.0 \text{ kg})(9.80 \text{ m/s}^2)(10.0 \text{ m}) = -1470 \text{ J}.$$

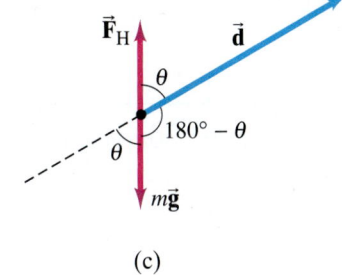

(c)

NOTE The work done by gravity (which is negative here) does not depend on the angle of the incline, only on the vertical height h of the hill.

5. **Net work done.** (*c*) The *net* work done on the backpack is $W_{net} = 0$, because the net force on the backpack is zero (it is assumed not to accelerate significantly). We can also get the net work done by adding the work done by each force:
$$W_{net} = W_G + W_H = -1470 \text{ J} + 1470 \text{ J} = 0.$$

NOTE Even though the *net* work done by all the forces on the backpack is zero, the hiker *does* do work on the backpack equal to 1470 J.

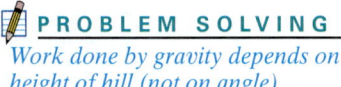

Work done by gravity depends on height of hill (not on angle)

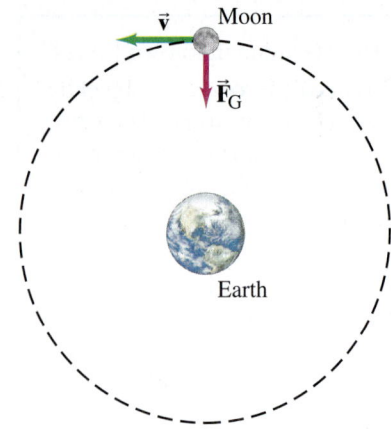

FIGURE 6–5 Example 6–3.

FIGURE 6–6 Work done by a force F is (a) approximately equal to the sum of the areas of the rectangles, (b) exactly equal to the area under the curve of $F_\parallel$ vs. d.

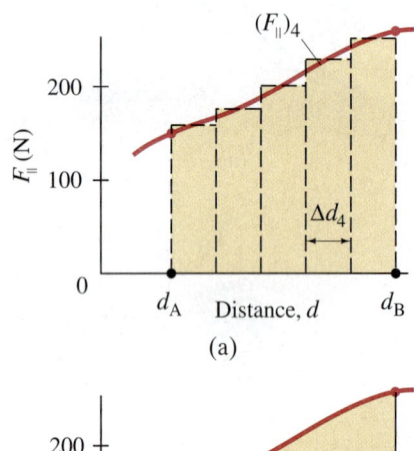

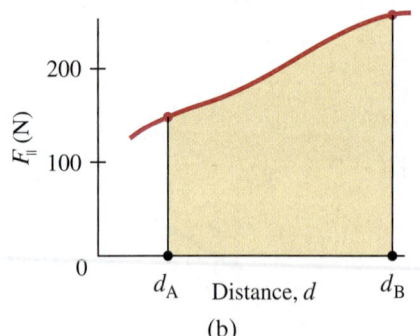

CONCEPTUAL EXAMPLE 6–3 **Does the Earth do work on the Moon?** The Moon revolves around the Earth in a nearly circular orbit, kept there by the gravitational force exerted by the Earth. Does gravity do (a) positive work, (b) negative work, or (c) no work on the Moon?

RESPONSE The gravitational force $\vec{F}_G$ exerted by the Earth on the Moon (Fig. 6–5) acts toward the Earth and provides its centripetal acceleration, inward along the radius of the Moon's orbit. The Moon's displacement at any moment is tangent to the circle, in the direction of its velocity, perpendicular to the radius and perpendicular to the force of gravity. Hence the angle θ between the force $\vec{F}_G$ and the instantaneous displacement of the Moon is 90°, and the work done by gravity is therefore zero (cos 90° = 0). This is why the Moon, as well as artificial satellites, can stay in orbit without expenditure of fuel: no work needs to be done against the force of gravity.

*6–2 Work Done by a Varying Force

If the force acting on an object is constant, the work done by that force can be calculated using Eq. 6–1. But in many cases, the force varies in magnitude or direction during a process. For example, as a rocket moves away from Earth, work is done to overcome the force of gravity, which varies as the inverse square of the distance from the Earth's center. Other examples are the force exerted by a spring, which increases with the amount of stretch, or the work done by a varying force that pulls a box or cart up an uneven hill.

The work done by a varying force can be determined graphically. To do so, we plot $F_\parallel$ ($= F \cos\theta$, the component of $\vec{F}$ parallel to the direction of motion at any point) as a function of distance d, as in Fig. 6–6a. We divide the distance into small segments Δd. For each segment, we indicate the average of $F_\parallel$ by a horizontal dashed line. Then the work done for each segment is $\Delta W = F_\parallel \Delta d$, which is the area of a rectangle Δd wide and $F_\parallel$ high. The total work done to move the object a total distance $d = d_B - d_A$ is the sum of the areas of the rectangles (five in the case shown in Fig. 6–6a). Usually, the average value of $F_\parallel$ for each segment must be estimated, and a reasonable approximation of the work done can then be made.

If we subdivide the distance into many more segments, Δd can be made smaller and our estimate of the work done would be more accurate. In the limit as Δd approaches zero, the total area of the many narrow rectangles approaches the area under the curve, Fig. 6–6b. That is, *the work done by a variable force in moving an object between two points is equal to the area under the $F_\parallel$ vs. d curve between those two points.*

6–3 Kinetic Energy, and the Work-Energy Principle

Energy is one of the most important concepts in science. Yet we cannot give a simple general definition of energy in only a few words. Nonetheless, each specific type of energy can be defined fairly simply. In this Chapter we define translational kinetic energy and some types of potential energy. In later Chapters, we will examine other types of energy, such as that related to heat and electricity. The crucial aspect of energy is that the sum of all types, the *total energy*, is the same after any process as it was before: that is, energy is a conserved quantity.

For the purposes of this Chapter, we can define energy in the traditional way as "the ability to do work." This simple definition is not always applicable,[†] but it is valid for mechanical energy which we discuss in this Chapter. We now define and discuss one of the basic types of energy, kinetic energy.

[†]Energy associated with heat is often not available to do work, as we will discuss in Chapter 15.

A moving object can do work on another object it strikes. A flying cannonball does work on a brick wall it knocks down; a moving hammer does work on a nail it drives into wood. In either case, a moving object exerts a force on a second object which undergoes a displacement. An object in motion has the ability to do work and thus can be said to have energy. The energy of motion is called **kinetic energy**, from the Greek word *kinetikos*, meaning "motion."

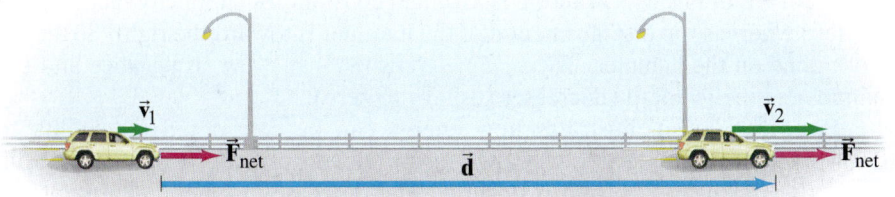

FIGURE 6–7 A constant net force F_{net} accelerates a car from speed v_1 to speed v_2 over a displacement d. The net work done is $W_{net} = F_{net}d$.

To obtain a quantitative definition for kinetic energy, let us consider a simple rigid object of mass m (treated as a particle) that is moving in a straight line with an initial speed v_1. To accelerate it uniformly to a speed v_2, a constant net force F_{net} is exerted on it parallel to its motion over a displacement d, Fig. 6–7. Then the net work done on the object is $W_{net} = F_{net}d$. We apply Newton's second law, $F_{net} = ma$, and use Eq. 2–11c $(v_2^2 = v_1^2 + 2ad)$, which we rewrite as

$$a = \frac{v_2^2 - v_1^2}{2d},$$

where v_1 is the initial speed and v_2 is the final speed. Substituting this into $F_{net} = ma$, we determine the work done:

$$W_{net} = F_{net}d = mad = m\left(\frac{v_2^2 - v_1^2}{2d}\right)d = m\left(\frac{v_2^2 - v_1^2}{2}\right)$$

or

$$W_{net} = \tfrac{1}{2}mv_2^2 - \tfrac{1}{2}mv_1^2. \quad (6\text{–}2)$$

We *define* the quantity $\tfrac{1}{2}mv^2$ to be the **translational kinetic energy (KE)** of the object:

$$\text{KE} = \tfrac{1}{2}mv^2. \quad (6\text{–}3)$$

Kinetic energy (defined)

(We call this "translational" kinetic energy to distinguish it from rotational kinetic energy, which we will discuss in Chapter 8.) Equation 6–2, derived here for one-dimensional motion with a constant force, is valid in general for translational motion of an object in three dimensions and even if the force varies.

We can rewrite Eq. 6–2 as:

$$W_{net} = \text{KE}_2 - \text{KE}_1$$

or

$$W_{net} = \Delta \text{KE} = \tfrac{1}{2}mv_2^2 - \tfrac{1}{2}mv_1^2. \quad (6\text{–}4)$$

WORK-ENERGY PRINCIPLE

Equation 6–4 is a useful result known as the **work-energy principle**. It can be stated in words:

> **The net work done on an object is equal to the change in the object's kinetic energy.**

WORK-ENERGY PRINCIPLE

Notice that we made use of Newton's second law, $F_{net} = ma$, where F_{net} is the *net force*—the sum of all forces acting on the object. Thus, the work-energy principle is valid only if W is the *net work* done on the object—that is, the work done by all forces acting on the object.

CAUTION
Work-energy valid only for net work

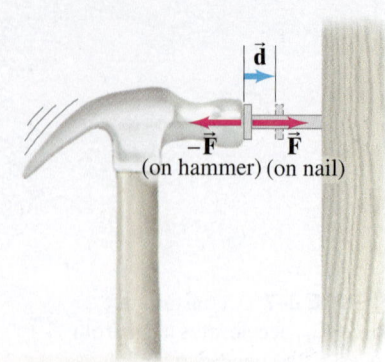

FIGURE 6–8 A moving hammer strikes a nail and comes to rest. The hammer exerts a force F on the nail; the nail exerts a force $-F$ on the hammer (Newton's third law). The work done on the nail by the hammer is positive ($W_n = Fd > 0$). The work done on the hammer by the nail is negative ($W_h = -Fd$).

The work-energy principle is a very useful reformulation of Newton's laws. It tells us that if (positive) net work W is done on an object, the object's kinetic energy increases by an amount W. The principle also holds true for the reverse situation: if the net work W done on an object is negative, the object's kinetic energy decreases by an amount W. That is, a net force exerted on an object opposite to the object's direction of motion decreases its speed and its kinetic energy. An example is a moving hammer (Fig. 6–8) striking a nail. The net force on the hammer ($-\vec{F}$ in Fig. 6–8, where $\vec{F}$ is assumed constant for simplicity) acts toward the left, whereas the displacement $\vec{d}$ of the hammer is toward the right. So the net work done on the hammer, $W_h = (F)(d)(\cos 180°) = -Fd$, is negative and the hammer's kinetic energy decreases (usually to zero).

Figure 6–8 also illustrates how energy can be considered the ability to do work. The hammer, as it slows down, does positive work on the nail: $W_n = (+F)(+d) = Fd$ and is positive. The decrease in kinetic energy of the hammer ($= Fd$ by Eq. 6–4) is equal to the work the hammer can do on another object, the nail in this case.

The translational kinetic energy ($= \frac{1}{2}mv^2$) is directly proportional to the mass of the object, and it is also proportional to the *square* of the speed. Thus, if the mass is doubled, the kinetic energy is doubled. But if the speed is doubled, the object has four times as much kinetic energy and is therefore capable of doing four times as much work.

Because of the direct connection between work and kinetic energy, energy is measured in the same units as work: joules in SI units. [The energy unit is ergs in the cgs, and foot-pounds in the British system.] Like work, kinetic energy is a scalar quantity. The kinetic energy of a group of objects is the sum of the kinetic energies of the individual objects.

The work-energy principle can be applied to a particle, and also to an object that can be approximated as a particle, such as an object that is rigid or whose internal motions are insignificant. It is very useful in simple situations, as we will see in the Examples below.

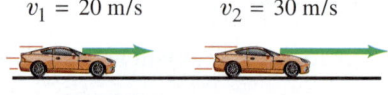

FIGURE 6–9 Example 6–4.

EXAMPLE 6–4 ESTIMATE **Work on a car, to increase its kinetic energy.** How much net work is required to accelerate a 1000-kg car from 20 m/s to 30 m/s (Fig. 6–9)?

APPROACH A car is a complex system. The engine turns the wheels and tires which push against the ground, and the ground pushes back (see Example 4–4). We aren't interested right now in those complications. Instead, we can get a useful result using the work-energy principle, but only if we model the car as a particle or simple rigid object.

SOLUTION The net work needed is equal to the increase in kinetic energy:

$$\begin{aligned} W &= \text{KE}_2 - \text{KE}_1 \\ &= \tfrac{1}{2}mv_2^2 - \tfrac{1}{2}mv_1^2 \\ &= \tfrac{1}{2}(1000 \text{ kg})(30 \text{ m/s})^2 - \tfrac{1}{2}(1000 \text{ kg})(20 \text{ m/s})^2 \\ &= 2.5 \times 10^5 \text{ J}. \end{aligned}$$

EXERCISE B (*a*) Make a guess: will the work needed to accelerate the car in Example 6–4 from rest to 20 m/s be more than, less than, or equal to the work already calculated to accelerate it from 20 m/s to 30 m/s? (*b*) Make the calculation.

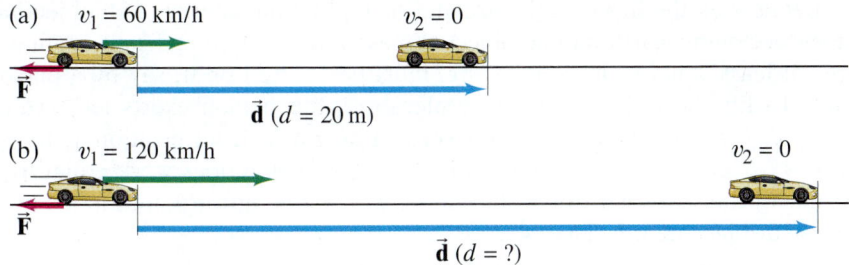

FIGURE 6–10 Example 6–5. A moving car comes to a stop. Initial velocity is (a) 60 km/h, (b) 120 km/h.

CONCEPTUAL EXAMPLE 6–5 **Work to stop a car.** A car traveling 60 km/h can brake to a stop in a distance d of 20 m (Fig. 6–10a). If the car is going twice as fast, 120 km/h, what is its stopping distance (Fig. 6–10b)? Assume the maximum braking force is approximately independent of speed.

RESPONSE Again we model the car as if it were a particle. Because the net stopping force F is approximately constant, the work needed to stop the car, Fd, is proportional to the distance traveled. We apply the work-energy principle, noting that $\vec{F}$ and $\vec{d}$ are in opposite directions and that the final speed of the car is zero:
$$W_{net} = Fd \cos 180° = -Fd.$$
Then
$$-Fd = \Delta KE = \tfrac{1}{2}mv_2^2 - \tfrac{1}{2}mv_1^2$$
$$= 0 - \tfrac{1}{2}mv_1^2.$$

Thus, since the force and mass are constant, we see that the stopping distance, d, increases with the square of the speed:
$$d \propto v^2.$$
If the car's initial speed is doubled, the stopping distance is $(2)^2 = 4$ times as great, or 80 m.

PHYSICS APPLIED
Car's stopping distance $\propto$ to initial speed squared

EXERCISE C Can kinetic energy ever be negative?

EXERCISE D (*a*) If the kinetic energy of a baseball is doubled, by what factor has its speed increased? (*b*) If its speed is doubled, by what factor does its kinetic energy increase?

6–4 Potential Energy

We have just discussed how an object is said to have energy by virtue of its motion, which we call kinetic energy. But it is also possible to have **potential energy**, which is the energy associated with forces that depend on the position or configuration of an object (or objects) relative to the surroundings. Various types of potential energy (PE) can be defined, and each type is associated with a particular force.

The spring of a wind-up toy is an example of an object with potential energy. The spring acquired its potential energy because work was done *on* it by the person winding the toy. As the spring unwinds, it exerts a force and does work to make the toy move.

Gravitational Potential Energy

Perhaps the most common example of potential energy is *gravitational potential energy*. A heavy brick held high above the ground has potential energy because of its position relative to the Earth. The raised brick has the ability to do work, for if it is released, it will fall to the ground due to the gravitational force, and can do work on, say, a stake, driving it into the ground.

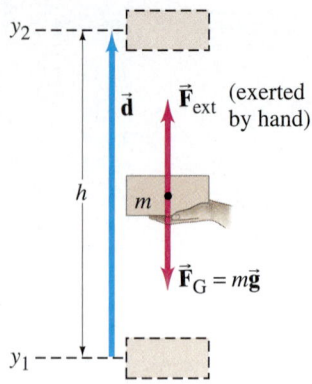

FIGURE 6–11 A person exerts an upward force $F_{ext} = mg$ to lift a brick from y_1 to y_2.

Let us seek the form for the gravitational potential energy of an object near the surface of the Earth. For an object of mass m to be lifted vertically, an upward force at least equal to its weight, mg, must be exerted on it, say by a person's hand. To lift the object without acceleration, the person exerts an "external force" $F_{ext} = mg$. If it is raised a vertical height h, from position y_1 to y_2 in Fig. 6–11 (upward direction chosen positive), a person does work equal to the product of the "external" force she exerts, $F_{ext} = mg$ upward, multiplied by the vertical displacement h. That is,

$$W_{ext} = F_{ext} d \cos 0° = mgh$$
$$= mg(y_2 - y_1). \quad (6\text{–}5a)$$

Gravity is also acting on the object as it moves from y_1 to y_2, and does work on the object equal to

$$W_G = F_G d \cos \theta = mgh \cos 180°,$$

where $\theta = 180°$ because $\vec{F}_G$ and $\vec{d}$ point in opposite directions. So

$$W_G = -mgh$$
$$= -mg(y_2 - y_1). \quad (6\text{–}5b)$$

Next, if we allow the object to start from rest at y_2 and fall freely under the action of gravity, it acquires a velocity given by $v^2 = 2gh$ (Eq. 2–11c) after falling a height h. It then has kinetic energy $\frac{1}{2}mv^2 = \frac{1}{2}m(2gh) = mgh$, and if it strikes a stake, it can do work on the stake equal to mgh (Section 6–3).

Thus, to raise an object of mass m to a height h *requires* an amount of work equal to mgh (Eq. 6–5a). And once at height h, the object has the *ability* to do an amount of work equal to mgh. We can say that the work done in lifting the object has been stored as gravitational potential energy.

We therefore define the **gravitational potential energy** of an object, due to Earth's gravity, as the product of the object's weight mg and its height y above some reference level (such as the ground):

$$\text{PE}_G = mgy. \quad (6\text{–}6)$$

The higher an object is above the ground, the more gravitational potential energy it has. We combine Eq. 6–5a with Eq. 6–6:

$$W_{ext} = mg(y_2 - y_1)$$
$$W_{ext} = \text{PE}_2 - \text{PE}_1 = \Delta \text{PE}_G. \quad (6\text{–}7a)$$

That is, the change in gravitational potential energy when an object is moved from height y_1 to height y_2 is equal to the work done by the net external force that accomplishes this without acceleration.

A more direct way to define the *change in gravitational potential energy*, ΔPE_G, is that it is equal to the *negative of the work done by gravity* itself (Eq. 6–5b):

$$W_G = -mg(y_2 - y_1)$$
$$W_G = -(\text{PE}_2 - \text{PE}_1) = -\Delta \text{PE}_G$$

or

$$\Delta \text{PE}_G = -W_G. \quad (6\text{–}7b)$$

Gravitational potential energy depends on the *vertical height* of the object *above some reference level* (Eq. 6–6). In some situations, you may wonder from what point to measure the height y. The gravitational potential energy of a book held high above a table, for example, depends on whether we measure y from the top of the table, from the floor, or from some other reference point. What is physically important in any situation is the *change* in potential energy, ΔPE, because that is what is related to the work done, Eqs. 6–7; and it is ΔPE that can be measured. We can thus choose to measure y from any reference level that is convenient, but we must choose the reference level at the start and be consistent throughout. The *change* in potential energy between any two points does not depend on this choice.

Change in PE is what is physically meaningful

An important result we discussed earlier (see Example 6–2 and Fig. 6–4) concerns the gravity force, which does work only in the vertical direction: the work done by gravity depends only on the vertical height h, and not on the path taken, whether it be purely vertical motion or, say, motion along an incline. Thus, from Eqs. 6–7 we see that changes in gravitational potential energy depend only on the change in vertical height and not on the path taken.

Potential energy belongs to a system, and not to a single object alone. Potential energy is associated with a force, and a force on one object is always exerted by some other object. Thus potential energy is a property of the system as a whole. For an object raised to a height y above the Earth's surface, the change in gravitational potential energy is mgy. The system here is the object plus the Earth, and properties of both are involved: object (m) and Earth (g).

> **CAUTION**
> *Potential energy belongs to a system, not to a single object*

EXAMPLE 6–6 Potential energy changes for a roller coaster. A 1000-kg roller-coaster car moves from point 1, Fig. 6–12, to point 2 and then to point 3. (a) What is the gravitational potential energy at points 2 and 3 relative to point 1? That is, take $y = 0$ at point 1. (b) What is the change in potential energy when the car goes from point 2 to point 3? (c) Repeat parts (a) and (b), but take the reference point ($y = 0$) to be at point 3.

APPROACH We are interested in the potential energy of the car–Earth system. We take upward as the positive y direction, and use the definition of gravitational potential energy to calculate the potential energy.

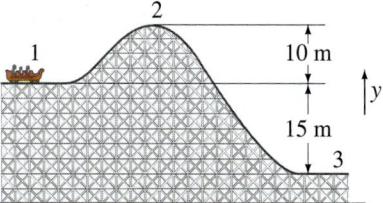

FIGURE 6–12 Example 6–6.

SOLUTION (a) We measure heights from point 1 ($y_1 = 0$), which means initially that the gravitational potential energy is zero. At point 2, where $y_2 = 10$ m,

$$PE_2 = mgy_2 = (1000 \text{ kg})(9.8 \text{ m/s}^2)(10 \text{ m}) = 9.8 \times 10^4 \text{ J}.$$

At point 3, $y_3 = -15$ m, since point 3 is below point 1. Therefore,

$$PE_3 = mgy_3 = (1000 \text{ kg})(9.8 \text{ m/s}^2)(-15 \text{ m}) = -1.5 \times 10^5 \text{ J}.$$

(b) In going from point 2 to point 3, the potential energy change ($PE_\text{final} - PE_\text{initial}$) is

$$PE_3 - PE_2 = (-1.5 \times 10^5 \text{ J}) - (9.8 \times 10^4 \text{ J}) = -2.5 \times 10^5 \text{ J}.$$

The gravitational potential energy decreases by 2.5×10^5 J.

(c) Now we set $y_3 = 0$. Then $y_1 = +15$ m at point 1, so the potential energy initially is

$$PE_1 = (1000 \text{ kg})(9.8 \text{ m/s}^2)(15 \text{ m}) = 1.5 \times 10^5 \text{ J}.$$

At point 2, $y_2 = 25$ m, so the potential energy is

$$PE_2 = 2.5 \times 10^5 \text{ J}.$$

At point 3, $y_3 = 0$, so the potential energy is zero. The change in potential energy going from point 2 to point 3 is

$$PE_3 - PE_2 = 0 - 2.5 \times 10^5 \text{ J} = -2.5 \times 10^5 \text{ J},$$

which is the same as in part (b).

NOTE Work done by gravity depends only on the vertical height, so changes in gravitational potential energy do not depend on the path taken.

Potential Energy Defined in General

There are other kinds of potential energy besides gravitational. Each form of potential energy is associated with a particular force, and can be defined analogously to gravitational potential energy. In general, the *change in potential energy associated with a particular force is equal to the negative of the work done by that force when the object is moved from one point to a second point* (as in Eq. 6–7b for gravity). An alternate and sometimes useful way to define the change in potential energy is: the work required of a net external force to move the object without acceleration between the two points, as in Eq. 6–7a.

Potential Energy defined

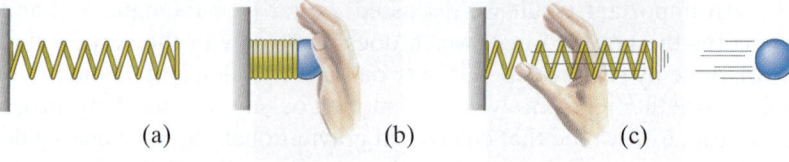

FIGURE 6–13 A spring (a) can store energy (elastic PE) when compressed as in (b) and can do work when released (c).

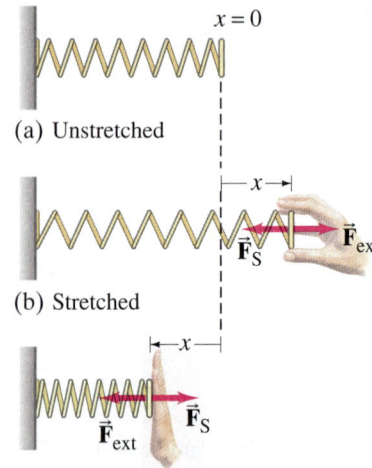

(a) Unstretched

(b) Stretched

(c) Compressed

FIGURE 6–14 (a) Spring in natural (unstretched) position. (b) Spring is stretched by a person exerting a force $\vec{F}_{ext}$ to the right (positive direction). The spring pulls back with a force $\vec{F}_S$, where $F_S = -kx$. (c) Person compresses the spring ($x < 0$) by exerting an external force $\vec{F}_{ext}$ to the left; the spring pushes back with a force $F_S = -kx$, where $F_S > 0$ because $x < 0$.

FIGURE 6–15 As a spring is stretched (or compressed), the magnitude of the force needed increases linearly as x increases: graph of $F = kx$ vs. x from $x = 0$ to $x = x_f$.

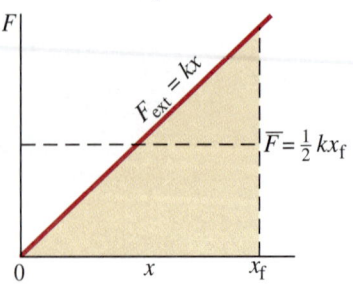

Potential Energy of Elastic Spring

We now consider potential energy associated with elastic materials, which includes a great variety of practical applications. Consider the simple coil spring shown in Fig. 6–13. The spring has potential energy when compressed (or stretched), because when it is released, it can do work on a ball as shown. To hold a spring either stretched or compressed an amount x from its natural (unstretched) length requires the hand to exert an external force on the spring of magnitude F_{ext} which is directly proportional to x. That is,

$$F_{ext} = kx,$$

where k is a constant, called the *spring stiffness constant* (or simply **spring constant**), and is a measure of the stiffness of the particular spring. The stretched or compressed spring itself exerts a force F_S in the opposite direction on the hand, as shown in Fig. 6–14:

$$F_S = -kx. \qquad \text{[spring force]} \quad (6\text{–}8)$$

This force is sometimes called a "restoring force" because the spring exerts its force in the direction opposite the displacement (hence the minus sign), acting to return it to its natural length. Equation 6–8 is known as the **spring equation** and also as **Hooke's law**, and is accurate for springs as long as x is not too great.

To calculate the potential energy of a stretched spring, let us calculate the work required to stretch it (Fig. 6–14b). We might hope to use Eq. 6–1 for the work done on it, $W = Fx$, where x is the amount it is stretched from its natural length. But this would be incorrect since the force F_{ext} ($= kx$) is not constant but varies over the distance x, becoming greater the more the spring is stretched, as shown graphically in Fig. 6–15. So let us use the average force, $\bar{F}$. Since F_{ext} varies linearly, from zero at the unstretched position to kx when stretched to x, the average force is $\bar{F} = \frac{1}{2}[0 + kx] = \frac{1}{2}kx$, where x here is the final amount stretched (shown as x_f in Fig. 6–15 for clarity). The work done is then

$$W_{ext} = \bar{F}x = \left(\tfrac{1}{2}kx\right)(x) = \tfrac{1}{2}kx^2.$$

Hence the **elastic potential energy**, PE_{el}, is proportional to the square of the amount stretched:

$$PE_{el} = \tfrac{1}{2}kx^2. \qquad \text{[elastic spring]} \quad (6\text{–}9)$$

If a spring is *compressed* a distance x from its natural ("equilibrium") length, the average force again has magnitude $\bar{F} = \frac{1}{2}kx$, and again the potential energy is given by Eq. 6–9. Thus x can be either the amount compressed or amount stretched from the spring's natural length.[†] Note that for a spring, we choose the reference point for zero PE at the spring's natural position.

Potential Energy as Stored Energy

In the above examples of potential energy—from a brick held at a height y, to a stretched or compressed spring—an object has the capacity or *potential* to do work even though it is not yet actually doing it. These examples show that energy can be *stored*, for later use, in the form of potential energy (as in Fig. 6–13, for a spring).

Note that there is a single universal formula for the translational kinetic energy of an object, $\frac{1}{2}mv^2$, but there is no single formula for potential energy. Instead, the mathematical form of the potential energy depends on the force involved.

[†]We can also obtain Eq. 6–9 using Section 6–2. The work done, and hence ΔPE, equals the area under the F vs. x graph of Fig. 6–15. This area is a triangle (colored in Fig. 6–15) of altitude kx and base x, and hence of area (for a triangle) equal to $\frac{1}{2}(kx)(x) = \frac{1}{2}kx^2$.

148 CHAPTER 6 Work and Energy

6–5 Conservative and Nonconservative Forces

The work done against gravity in moving an object from one point to another does not depend on the path taken. For example, it takes the same work ($= mgh$) to lift an object of mass m vertically a height h as to carry it up an incline of the same vertical height, as in Fig. 6–4 (see Example 6–2). Forces such as gravity, for which the work done does not depend on the path taken but only on the initial and final positions, are called **conservative forces**. The elastic force of a spring (or other elastic material), in which $F = -kx$, is also a conservative force. An object that starts at a given point and returns to that same point under the action of a conservative force has no net work done on it because the potential energy is the same at the start and the finish of such a round trip.

Many forces, such as friction and a push or pull exerted by a person, are **nonconservative forces** since any work they do depends on the path. For example, if you push a crate across a floor from one point to another, the work you do depends on whether the path taken is straight or is curved. As shown in Fig. 6–16, if a crate is pushed slowly from point 1 to point 2 along the longer semicircular path, you do more work against friction than if you push it along the straight path.

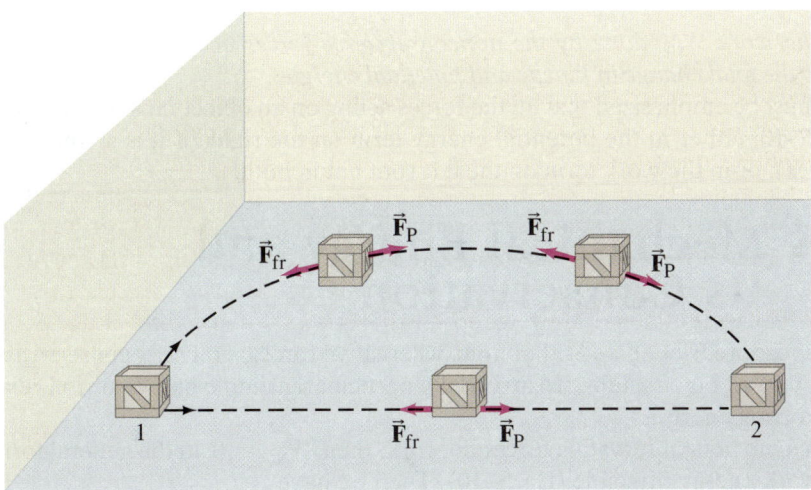

FIGURE 6–16 A crate is pushed slowly at constant speed across a rough floor from position 1 to position 2 via two paths, one straight and one curved. The pushing force $\vec{F}_P$ is in the direction of motion at each point. (The friction force opposes the motion.) Hence for a constant magnitude pushing force, the work it does is $W = F_P d$, so if the distance traveled d is greater (as for the curved path), then W is greater. The work done does not depend only on points 1 and 2; it also depends on the path taken.

You do more work on the curved path because the distance is greater and, unlike the gravitational force, the pushing force $\vec{F}_P$ is in the direction of motion at each point. Thus the work done by the person in Fig. 6–16 does not depend *only* on points 1 and 2; it depends also on the path taken. The force of kinetic friction, also shown in Fig. 6–16, always opposes the motion; it too is a nonconservative force, and we discuss how to treat it later in this Chapter (Section 6–9). Table 6–1 lists a few conservative and nonconservative forces.

Because potential energy is energy associated with the position or configuration of objects, potential energy can only make sense if it can be stated uniquely for a given point. This cannot be done with nonconservative forces because the work done depends on the path taken (as in Fig. 6–16). Hence, *potential energy can be defined only for a conservative force.* Thus, although potential energy is always associated with a force, not all forces have a potential energy. For example, there is no potential energy for friction.

TABLE 6–1 Conservative and Nonconservative Forces

Conservative Forces	Nonconservative Forces
Gravitational	Friction
Elastic	Air resistance
Electric	Tension in cord
	Motor or rocket propulsion
	Push or pull by a person

EXERCISE E An object acted on by a constant force F moves from point 1 to point 2 and back again. The work done by the force F in this round trip is 60 J. Can you determine from this information if F is a conservative or nonconservative force?

Work-Energy Extended

We can extend the **work-energy principle** (discussed in Section 6–3) to include potential energy. Suppose several forces act on an object which can undergo translational motion. And suppose only some of these forces are conservative. We write the total (net) work W_{net} as a sum of the work done by conservative forces, W_C, and the work done by nonconservative forces, W_{NC}:

$$W_{net} = W_C + W_{NC}.$$

Then, from the work-energy principle, Eq. 6–4, we have

$$W_{net} = \Delta KE$$
$$W_C + W_{NC} = \Delta KE$$

where $\Delta KE = KE_2 - KE_1$. Then

$$W_{NC} = \Delta KE - W_C.$$

Work done by a conservative force can be written in terms of potential energy, as we saw in Eq. 6–7b for gravitational potential energy:

$$W_C = -\Delta PE.$$

We combine these last two equations:

$$W_{NC} = \Delta KE + \Delta PE. \qquad (6\text{–}10)$$

Thus, *the work W_{NC} done by the nonconservative forces acting on an object is equal to the total change in kinetic and potential energies.*

It must be emphasized that *all* the forces acting on an object must be included in Eq. 6–10, either in the potential energy term on the right (if it is a conservative force), or in the work term on the left (but not in both!).

6–6 Mechanical Energy and Its Conservation

If we can ignore friction and other nonconservative forces, or if only conservative forces do work on a system, we arrive at a particularly simple and beautiful relation involving energy.

When no nonconservative forces do work, then $W_{NC} = 0$ in the general form of the work-energy principle (Eq. 6–10). Then we have

$$\Delta KE + \Delta PE = 0 \qquad \left[\begin{array}{c}\text{conservative}\\ \text{forces only}\end{array}\right] \qquad (6\text{–}11a)$$

or

$$(KE_2 - KE_1) + (PE_2 - PE_1) = 0. \qquad \left[\begin{array}{c}\text{conservative}\\ \text{forces only}\end{array}\right] \qquad (6\text{–}11b)$$

We now define a quantity E, called the **total mechanical energy** of our system, as the sum of the kinetic and potential energies at any moment:

$$E = KE + PE.$$

Now we can rewrite Eq. 6–11b as

$$KE_2 + PE_2 = KE_1 + PE_1 \qquad \left[\begin{array}{c}\text{conservative}\\ \text{forces only}\end{array}\right] \qquad (6\text{–}12a)$$

or

$$E_2 = E_1 = \text{constant}. \qquad \left[\begin{array}{c}\text{conservative}\\ \text{forces only}\end{array}\right] \qquad (6\text{–}12b)$$

CONSERVATION OF MECHANICAL ENERGY

Equations 6–12 express a useful and profound principle regarding the total mechanical energy of a system—namely, that it is a **conserved quantity**. The total mechanical energy E remains constant as long as no nonconservative forces do work: $KE + PE$ at some initial time 1 is equal to the $KE + PE$ at any later time 2.

To say it another way, consider Eq. 6–11a which tells us $\Delta \text{PE} = -\Delta \text{KE}$; that is, if the kinetic energy KE of a system increases, then the potential energy PE must decrease by an equivalent amount to compensate. Thus, the total, KE + PE, remains constant:

If only conservative forces do work, the total mechanical energy of a system neither increases nor decreases in any process. It stays constant—it is conserved.

> CONSERVATION OF MECHANICAL ENERGY

This is the **principle of conservation of mechanical energy** for conservative forces.

In the next Section we shall see the great usefulness of the conservation of mechanical energy principle in a variety of situations, and how it is often easier to use than the kinematic equations or Newton's laws. After that we will discuss how other forms of energy can be included in the general conservation of energy law, such as energy associated with friction.

6–7 Problem Solving Using Conservation of Mechanical Energy

A simple example of the conservation of mechanical energy (neglecting air resistance) is a rock allowed to fall due to Earth's gravity from a height h above the ground, as shown in Fig. 6–17. If the rock starts from rest, all of the initial energy is potential energy. As the rock falls, the potential energy mgy decreases (because the rock's height above the ground y decreases), but the rock's kinetic energy increases to compensate, so that the sum of the two remains constant. At any point along the path, the total mechanical energy is given by

$$E = \text{KE} + \text{PE} = \tfrac{1}{2}mv^2 + mgy$$

where v is its speed at that point. If we let the subscript 1 represent the rock at one point along its path (for example, the initial point), and the subscript 2 represent it at some other point, then we can write

total mechanical energy at point 1 = total mechanical energy at point 2

or (see also Eq. 6–12a)

$$\tfrac{1}{2}mv_1^2 + mgy_1 = \tfrac{1}{2}mv_2^2 + mgy_2. \qquad \text{[gravity only]} \quad (6\text{–}13)$$

Just before the rock hits the ground, where we chose $y = 0$, all of the initial potential energy will have been transformed into kinetic energy.

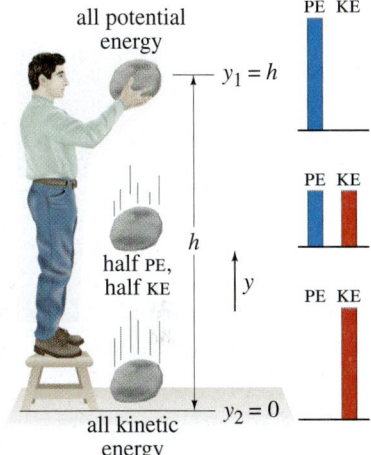

FIGURE 6–17 The rock's potential energy changes to kinetic energy as it falls. Note bar graphs representing potential energy PE and kinetic energy KE for the three different positions.

EXAMPLE 6–7 Falling rock. If the initial height of the rock in Fig. 6–17 is $y_1 = h = 3.0 \text{ m}$, calculate the rock's velocity when it has fallen to 1.0 m above the ground.

APPROACH We apply the principle of conservation of mechanical energy, Eq. 6–13, with only gravity acting on the rock. We choose the ground as our reference level ($y = 0$).

SOLUTION At the moment of release (point 1) the rock's position is $y_1 = 3.0 \text{ m}$ and it is at rest: $v_1 = 0$. We want to find v_2 when the rock is at position $y_2 = 1.0 \text{ m}$. Equation 6–13 gives

$$\tfrac{1}{2}mv_1^2 + mgy_1 = \tfrac{1}{2}mv_2^2 + mgy_2.$$

The m's cancel out and $v_1 = 0$, so

$$gy_1 = \tfrac{1}{2}v_2^2 + gy_2.$$

Solving for v_2 we find

$$v_2 = \sqrt{2g(y_1 - y_2)} = \sqrt{2(9.8 \text{ m/s}^2)[(3.0 \text{ m}) - (1.0 \text{ m})]} = 6.3 \text{ m/s}.$$

The rock's velocity 1.0 m above the ground is 6.3 m/s downward.

NOTE The velocity of the rock is independent of the rock's mass.

FIGURE 6–18 A roller-coaster car moving without friction illustrates the conservation of mechanical energy.

Equation 6–13 can be applied to any object moving without friction under the action of gravity. For example, Fig. 6–18 shows a roller-coaster car starting from rest at the top of a hill and coasting without friction to the bottom and up the hill on the other side. True, there is another force besides gravity acting on the car, the normal force exerted by the tracks. But the normal force acts perpendicular to the direction of motion at each point and so does zero work. We ignore rotational motion of the car's wheels and treat the car as a particle undergoing simple translation. Initially, the car has only potential energy. As it coasts down the hill, it loses potential energy and gains in kinetic energy, but the sum of the two remains constant. At the bottom of the hill it has its maximum kinetic energy, and as it climbs up the other side the kinetic energy changes back to potential energy. When the car comes to rest again at the same height from which it started, all of its energy will be potential energy. Given that the gravitational potential energy is proportional to the vertical height, energy conservation tells us that (in the absence of friction) the car comes to rest at a height equal to its original height. If the two hills are the same height, the car will just barely reach the top of the second hill when it stops. If the second hill is lower than the first, not all of the car's kinetic energy will be transformed to potential energy and the car can continue over the top and down the other side. If the second hill is higher, the car will reach a maximum height on it equal to its original height on the first hill. This is true (in the absence of friction) no matter how steep the hill is, since potential energy depends only on the vertical height (Eq. 6–6).

EXAMPLE 6–8 Roller-coaster car speed using energy conservation. Assuming the height of the hill in Fig. 6–18 is 40 m, and the roller-coaster car starts from rest at the top, calculate (a) the speed of the roller-coaster car at the bottom of the hill, and (b) at what height it will have half this speed. Take $y = 0$ at the bottom of the hill.

APPROACH We use conservation of mechanical energy. We choose point 1 to be where the car starts from rest $(v_1 = 0)$ at the top of the hill $(y_1 = 40 \text{ m})$. In part (a), point 2 is the bottom of the hill, which we choose as our reference level, so $y_2 = 0$. In part (b) we let y_2 be the unknown.

SOLUTION (a) We use Eq. 6–13 with $v_1 = 0$ and $y_2 = 0$, which gives

$$mgy_1 = \tfrac{1}{2}mv_2^2$$

or

$$v_2 = \sqrt{2gy_1}$$
$$= \sqrt{2(9.8 \text{ m/s}^2)(40 \text{ m})} = 28 \text{ m/s}.$$

(b) Now y_2 will be an unknown. We again use conservation of energy,

$$\tfrac{1}{2}mv_1^2 + mgy_1 = \tfrac{1}{2}mv_2^2 + mgy_2,$$

but now $v_2 = \tfrac{1}{2}(28 \text{ m/s}) = 14 \text{ m/s}$ and $v_1 = 0$. Solving for the unknown y_2 gives

$$y_2 = y_1 - \frac{v_2^2}{2g} = 40 \text{ m} - \frac{(14 \text{ m/s})^2}{2(9.8 \text{ m/s}^2)} = 30 \text{ m}.$$

That is, the car has a speed of 14 m/s when it is 30 *vertical* meters above the lowest point, both when descending the left-hand hill and when ascending the right-hand hill.

The mathematics of the roller-coaster Example 6–8 is almost the same as in Example 6–7. But there is an important difference between them. In Example 6–7 the motion is all vertical and could have been solved using force, acceleration, and the kinematic equations (Eqs. 2–11). For the roller coaster, where the motion is not vertical, we could *not* have used Eqs. 2–11 because a is not constant on the curved track of Example 6–8. But energy conservation readily gives us the answer.

CONCEPTUAL EXAMPLE 6–9 **Speeds on two water slides.** Two water slides at a pool are shaped differently, but start at the same height h (Fig. 6–19). Two riders start from rest at the same time on different slides. (a) Which rider, Paul or Corinne, is traveling faster at the bottom? (b) Which rider makes it to the bottom first? Ignore friction and assume both slides have the same path length.

RESPONSE (a) Each rider's initial potential energy mgh gets transformed to kinetic energy, so the speed v at the bottom is obtained from $\frac{1}{2}mv^2 = mgh$. The mass cancels and so the speed will be the same, regardless of the mass of the rider. Since they descend the same vertical height, they will finish with the same speed. (b) Note that Corinne is consistently at a lower elevation than Paul at any instant, until the end. This means she has converted her potential energy to kinetic energy earlier. Consequently, she is traveling faster than Paul for the whole trip, and because the distance is the same, Corinne gets to the bottom first.

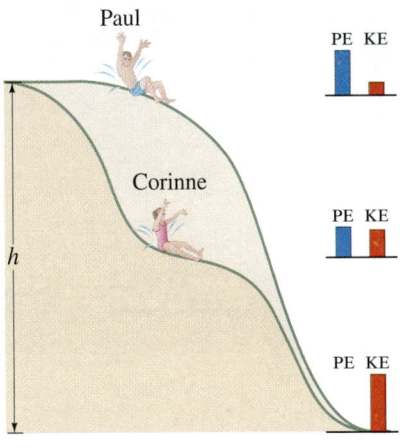

FIGURE 6–19 Example 6–9.

FIGURE 6–20 Transformation of energy during a pole vault: $KE \rightarrow PE_{el} \rightarrow PE_G$.

There are many interesting examples of the conservation of energy in sports, such as the pole vault illustrated in Fig. 6–20. We often have to make approximations, but the sequence of events in broad outline for the pole vault is as follows. The initial kinetic energy of the running athlete is transformed into elastic potential energy of the bending pole and, as the athlete leaves the ground, into gravitational potential energy. When the vaulter reaches the top and the pole has straightened out again, the energy has all been transformed into gravitational potential energy (if we ignore the vaulter's low horizontal speed over the bar). The pole does not supply any energy, but it acts as a device to *store* energy and thus aid in the transformation of kinetic energy into gravitational potential energy, which is the net result. The energy required to pass over the bar depends on how high the center of mass (CM) of the vaulter must be raised. By bending their bodies, pole vaulters keep their CM so low that it can actually pass slightly beneath the bar (Fig. 6–21), thus enabling them to cross over a higher bar than would otherwise be possible. (Center of mass is covered in Chapter 7.)

As another example of the conservation of mechanical energy, let us consider an object of mass m connected to a compressed horizontal spring (Fig. 6–13b) whose own mass can be neglected and whose spring stiffness constant is k. When the spring is released, the mass m has speed v at any moment. The potential energy of the system (object plus spring) is $\frac{1}{2}kx^2$, where x is the displacement of the spring from its unstretched length (Eq. 6–9). If neither friction nor any other force is acting, conservation of mechanical energy tells us that

$$\tfrac{1}{2}mv_1^2 + \tfrac{1}{2}kx_1^2 = \tfrac{1}{2}mv_2^2 + \tfrac{1}{2}kx_2^2, \qquad \text{[elastic PE only]} \quad (6\text{–}14)$$

where the subscripts 1 and 2 refer to the velocity and displacement at two different moments.

FIGURE 6–21 By bending her body, a pole vaulter can keep her center of mass so low that it may even pass below the bar.

SECTION 6–7 Problem Solving Using Conservation of Mechanical Energy

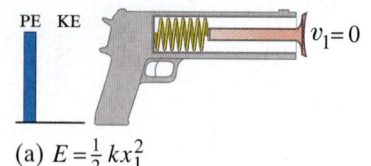

(a) $E = \tfrac{1}{2}kx_1^2$

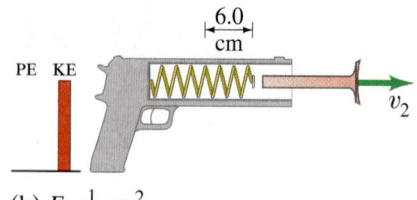

(b) $E = \tfrac{1}{2}mv_2^2$

FIGURE 6–22 Example 6–10. (a) A dart is pushed against a spring, compressing it 6.0 cm. The dart is then released, and in (b) it leaves the spring at velocity v_2.

FIGURE 6–23 Example 6–11. A falling ball compresses a spring.

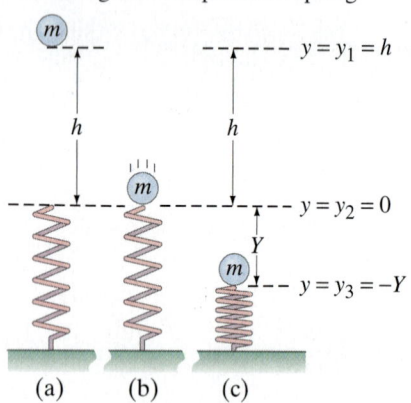

EXAMPLE 6–10 **Toy dart gun.** A dart of mass 0.100 kg is pressed against the spring of a toy dart gun as shown in Fig. 6–22a. The spring, with spring stiffness constant $k = 250$ N/m and ignorable mass, is compressed 6.0 cm and released. If the dart detaches from the spring when the spring reaches its natural length ($x = 0$), what speed does the dart acquire?

APPROACH The dart is initially at rest (point 1), so $KE_1 = 0$. We ignore friction and use conservation of mechanical energy; the only potential energy is elastic.

SOLUTION We use Eq. 6–14 with point 1 being at the maximum compression of the spring, so $v_1 = 0$ (dart not yet released) and $x_1 = -0.060$ m. Point 2 we choose to be the instant the dart flies off the end of the spring (Fig. 6–22b), so $x_2 = 0$ and we want to find v_2. Thus Eq. 6–14 can be written

$$0 + \tfrac{1}{2}kx_1^2 = \tfrac{1}{2}mv_2^2 + 0.$$

Then

$$v_2^2 = \frac{kx_1^2}{m} = \frac{(250\ \text{N/m})(-0.060\ \text{m})^2}{(0.100\ \text{kg})} = 9.0\ \text{m}^2/\text{s}^2,$$

and $v_2 = \sqrt{v_2^2} = 3.0$ m/s.

EXAMPLE 6–11 **Two kinds of potential energy.** A ball of mass $m = 2.60$ kg, starting from rest, falls a vertical distance $h = 55.0$ cm before striking a vertical coiled spring, which it compresses an amount $Y = 15.0$ cm (Fig. 6–23). Determine the spring stiffness constant k of the spring. Assume the spring has negligible mass, and ignore air resistance. Measure all distances from the point where the ball first touches the uncompressed spring ($y = 0$ at this point).

APPROACH The forces acting on the ball are the gravitational pull of the Earth and the elastic force exerted by the spring. Both forces are conservative, so we can use conservation of mechanical energy, including both types of potential energy. We must be careful, however: gravity acts throughout the fall (Fig. 6–23), whereas the elastic force does not act until the ball touches the spring (Fig. 6–23b). We choose y positive upward, and $y = 0$ at the end of the spring in its natural (uncompressed) state.

SOLUTION We divide this solution into two parts. (An alternate solution follows.)
Part 1: Let us first consider the energy changes as the ball falls from a height $y_1 = h = 0.550$ m, Fig. 6–23a, to $y_2 = 0$, just as it touches the spring, Fig. 6–23b. Our system is the ball acted on by gravity plus the spring (which up to this point doesn't do anything). Thus

$$\tfrac{1}{2}mv_1^2 + mgy_1 = \tfrac{1}{2}mv_2^2 + mgy_2$$
$$0 + mgh = \tfrac{1}{2}mv_2^2 + 0.$$

We solve for $v_2 = \sqrt{2gh} = \sqrt{2(9.80\ \text{m/s}^2)(0.550\ \text{m})} = 3.283$ m/s ≈ 3.28 m/s. This is the speed of the ball just as it touches the top of the spring, Fig. 6–23b.
Part 2: As the ball compresses the spring, Figs. 6–23b to c, there are two conservative forces on the ball—gravity and the spring force. So our conservation of energy equation is

$$E_2\ (\text{ball touches spring}) = E_3\ (\text{spring compressed})$$
$$\tfrac{1}{2}mv_2^2 + mgy_2 + \tfrac{1}{2}ky_2^2 = \tfrac{1}{2}mv_3^2 + mgy_3 + \tfrac{1}{2}ky_3^2.$$

Substituting $y_2 = 0$, $v_2 = 3.283$ m/s, $v_3 = 0$ (the ball comes to rest for an instant), and $y_3 = -Y = -0.150$ m, we have

$$\tfrac{1}{2}mv_2^2 + 0 + 0 = 0 - mgY + \tfrac{1}{2}k(-Y)^2.$$

We know m, v_2, and Y, so we can solve for k:

$$k = \frac{2}{Y^2}\left[\tfrac{1}{2}mv_2^2 + mgY\right] = \frac{m}{Y^2}\left[v_2^2 + 2gY\right]$$
$$= \frac{(2.60\ \text{kg})}{(0.150\ \text{m})^2}\left[(3.283\ \text{m/s})^2 + 2(9.80\ \text{m/s}^2)(0.150\ \text{m})\right] = 1590\ \text{N/m}.$$

Alternate Solution Instead of dividing the solution into two parts, we can do it all at once. After all, we get to choose what two points are used on the left and right of the energy equation. Let us write the energy equation for points 1 and 3 in Fig. 6–23. Point 1 is the initial point just before the ball starts to fall (Fig. 6–23a), so $v_1 = 0$, and $y_1 = h = 0.550$ m. Point 3 is when the spring is fully compressed (Fig. 6–23c), so $v_3 = 0$, $y_3 = -Y = -0.150$ m. The forces on the ball in this process are gravity and (at least part of the time) the spring. So conservation of energy tells us

PROBLEM SOLVING
Quicker Solution

$$\tfrac{1}{2}mv_1^2 + mgy_1 + \tfrac{1}{2}k(0)^2 = \tfrac{1}{2}mv_3^2 + mgy_3 + \tfrac{1}{2}ky_3^2$$
$$0 + mgh + 0 = 0 - mgY + \tfrac{1}{2}kY^2$$

where we have set $y = 0$ for the spring at point 1 because it is not acting and is not compressed or stretched. We solve for k:

$$k = \frac{2mg(h + Y)}{Y^2} = \frac{2(2.60 \text{ kg})(9.80 \text{ m/s}^2)(0.550 \text{ m} + 0.150 \text{ m})}{(0.150 \text{ m})^2} = 1590 \text{ N/m}$$

just as in our first method of solution.

6–8 Other Forms of Energy and Energy Transformations; The Law of Conservation of Energy

Besides the kinetic energy and potential energy of mechanical systems, other forms of energy can be defined as well. These include electric energy, nuclear energy, thermal energy, and the chemical energy stored in food and fuels. These other forms of energy are considered to be kinetic or potential energy at the atomic or molecular level. For example, according to atomic theory, thermal energy is the kinetic energy of rapidly moving molecules—when an object is heated, the molecules that make up the object move faster. On the other hand, the energy stored in food or in a fuel such as gasoline is regarded as potential energy stored by virtue of the relative positions of the atoms within a molecule due to electric forces between the atoms (chemical bonds). The energy in chemical bonds can be released through chemical reactions. This is analogous to a compressed spring which, when released, can do work. Electric, magnetic, and nuclear energies also can be considered examples of kinetic and potential (or stored) energies. We will deal with these other forms of energy in later Chapters.

Energy can be transformed from one form to another. For example, a rock held high in the air has potential energy; as it falls, it loses potential energy and gains in kinetic energy. Potential energy is being transformed into kinetic energy.

Often the transformation of energy involves a transfer of energy from one object to another. The potential energy stored in the spring of Fig. 6–13b is transformed into the kinetic energy of the ball, Fig. 6–13c. Water at the top of a waterfall (Fig. 6–24) or a dam has potential energy, which is transformed into kinetic energy as the water falls. At the base of a dam, the kinetic energy of the water can be transferred to turbine blades and further transformed into electric energy, as discussed later. The potential energy stored in a bent bow can be transformed into kinetic energy of the arrow (Fig. 6–25).

In each of these examples, the transfer of energy is accompanied by the performance of work. The spring of Fig. 6–13 does work on the ball. Water does work on turbine blades. A bow does work on an arrow. This observation gives us a further insight into the relation between work and energy: *work is done when energy is transferred from one object to another.*[†]

FIGURE 6–24 Gravitational potential energy of water at the top of Yosemite Falls gets transformed into kinetic energy as the water falls. (Some of the energy is transformed into heat by air resistance, and some into sound.)

FIGURE 6–25 Potential energy of a bent bow about to be transformed into kinetic energy of an arrow.

[†]If the objects are at different temperatures, heat can flow between them instead, or in addition. See Chapters 14 and 15.

One of the great results of physics is that whenever energy is transferred or transformed, it is found that no energy is gained or lost in the process.

This is the **law of conservation of energy**, one of the most important principles in physics; it can be stated as:

> **LAW OF CONSERVATION OF ENERGY**
>
> **The total energy is neither increased nor decreased in any process. Energy can be transformed from one form to another, and transferred from one object to another, but the total amount remains constant.**

We have already discussed the conservation of energy for mechanical systems involving conservative forces, and we saw how it could be derived from Newton's laws and thus is equivalent to them. But in its full generality, the validity of the law of conservation of energy, encompassing all forms of energy including those associated with nonconservative forces like friction, rests on experimental observation. Even though Newton's laws are found to fail in the submicroscopic world of the atom, the law of conservation of energy has been found to hold in every experimental situation so far tested.

6–9 Energy Conservation with Dissipative Forces: Solving Problems

In our applications of energy conservation in Section 6–7, we neglected friction and other nonconservative forces. But in many situations they cannot be ignored. In a real situation, the roller-coaster car in Fig. 6–18, for example, will not in fact reach the same height on the second hill as it had on the first hill because of friction. In this, and in other natural processes, the mechanical energy (sum of the kinetic and potential energies) does not remain constant but decreases. Because frictional forces reduce the mechanical energy (but *not* the total energy), they are called **dissipative forces**. Historically, the presence of dissipative forces hindered the formulation of a comprehensive conservation of energy law until well into the nineteenth century. It was only then that heat, which is always produced when there is friction (try rubbing your hands together), was interpreted in terms of energy. Quantitative studies by nineteenth-century scientists (discussed in Chapters 14 and 15) demonstrated that if heat is considered as a transfer of energy (thermal energy), then the total energy is conserved in any process. For example, if the roller-coaster car in Fig. 6–18 is subject to frictional forces, then the initial total energy of the car will be equal to the kinetic plus potential energy of the car at any subsequent point along its path plus the amount of thermal energy produced in the process (equal to the work done by friction).

Let us recall the general form of the work-energy principle, Eq. 6–10:

$$W_{NC} = \Delta KE + \Delta PE,$$

where W_{NC} is the work done by nonconservative forces such as friction. Consider an object, such as a roller-coaster car, as a particle moving under gravity with nonconservative forces like friction acting on it. When the object moves from some point 1 to another point 2, then

$$W_{NC} = KE_2 - KE_1 + PE_2 - PE_1.$$

We can rewrite this as

$$KE_1 + PE_1 + W_{NC} = KE_2 + PE_2. \quad (6\text{–}15)$$

For the case of friction, $W_{NC} = -F_{fr}d$, where d is the distance over which the friction (assumed constant) acts as the object moves from point 1 to point 2. ($\vec{F}$ and $\vec{d}$ are in opposite directions, hence the minus sign from $\cos 180° = -1$ in Eq. 6–1.)

With $\text{KE} = \frac{1}{2}mv^2$ and $\text{PE} = mgy$, Eq. 6–15 with $W_{\text{NC}} = -F_{\text{fr}}d$ becomes

$$\tfrac{1}{2}mv_1^2 + mgy_1 - F_{\text{fr}}d = \tfrac{1}{2}mv_2^2 + mgy_2. \quad \begin{bmatrix}\text{gravity and}\\\text{friction acting}\end{bmatrix} \quad (6\text{–}16\text{a})$$

That is, the initial mechanical energy is reduced by the amount $F_{\text{fr}}d$. We could also write this equation as

or
$$\tfrac{1}{2}mv_1^2 + mgy_1 = \tfrac{1}{2}mv_2^2 + mgy_2 + F_{\text{fr}}d$$
$$\text{KE}_1 + \text{PE}_1 = \text{KE}_2 + \text{PE}_2 + F_{\text{fr}}d, \quad \begin{bmatrix}\text{gravity and}\\\text{friction}\\\text{acting}\end{bmatrix} \quad (6\text{–}16\text{b})$$

and state equally well that the initial mechanical energy of the car (point 1) equals the (reduced) final mechanical energy of the car plus the energy transformed by friction into thermal energy.

Equations 6–16 can be seen to be Eq. 6–13 modified to include nonconservative forces such as friction. As such, they are statements of conservation of energy. When other forms of energy are involved, such as chemical or electrical energy, the total amount of energy is always found to be conserved. Hence the law of conservation of energy is believed to be universally valid.

EXERCISE F Return to the Chapter-Opening Question, page 138, and answer it again now. Try to explain why you may have answered differently the first time.

Work-Energy versus Energy Conservation

The law of conservation of energy is more general and more powerful than the work-energy principle. Indeed, the work-energy principle should not be viewed as a statement of conservation of energy. It is nonetheless useful for mechanical problems; and whether you use it, or use the more powerful conservation of energy, can depend on your *choice of the system* under study. If you choose as your system a particle or rigid object on which external forces do work, then you can use the work-energy principle: the work done by the external forces on your object equals the change in its kinetic energy.

On the other hand, if you choose a system on which no external forces do work, then you need to apply conservation of energy to that system directly.

Consider, for example, a spring connected to a block on a frictionless table (Fig. 6–26). If you choose the block as your system, then the work done on the block by the spring equals the change in kinetic energy of the block: the work-energy principle. (Energy conservation does not apply to this system—the block's energy changes.) If instead you choose the block plus the spring as your system, no external forces do work (since the spring is part of the chosen system). To this system you need to apply conservation of energy: if you compress the spring and then release it, the spring still exerts a force[†] on the block, but the subsequent motion can be discussed in terms of kinetic energy $(\tfrac{1}{2}mv^2)$ plus potential energy $(\tfrac{1}{2}kx^2)$, whose total remains constant.

You may also wonder sometimes whether to approach a problem using work and energy, or instead to use Newton's laws. As a rough guideline, if the force(s) involved are constant, either approach may succeed. If the forces are not constant, and/or the path is not simple, energy may be the better approach because it is a scalar.

Problem solving is not a process that can be done by simply following a set of rules. The Problem Solving Strategy on the next page, like all others, is thus *not* a prescription, but is a summary to help you get started solving problems involving energy.

Choosing the system

FIGURE 6–26 A spring connected to a block on a frictionless table. If you choose your system to be the block plus spring, then
$$E = \tfrac{1}{2}mv^2 + \tfrac{1}{2}kx^2$$
is conserved.

Use energy, or Newton's laws?

[†]The force the spring exerts on the block, and the force the block exerts back on the spring, are not "external" forces—they are within the system.

PROBLEM SOLVING

Conservation of Energy

1. **Draw a picture** of the physical situation.

2. Determine **the system** for which you will apply energy conservation: the object or objects and the forces acting.

3. Ask yourself what quantity you are looking for, and **choose initial** (point 1) **and final** (point 2) **positions**.

4. If the object under investigation changes its height during the problem, then **choose a reference frame** with a convenient $y = 0$ level for gravitational potential energy; the lowest point in the situation is often a good choice.

    If springs are involved, choose the unstretched spring position to be x (or y) $= 0$.

5. **Is mechanical energy conserved?** If no friction or other nonconservative forces act, then conservation of mechanical energy holds:
$$KE_1 + PE_1 = KE_2 + PE_2. \quad (6\text{–}12a)$$

6. **Apply conservation of energy.** If friction (or other nonconservative forces) are present, then an additional term (W_{NC}) will be needed:
$$W_{NC} = \Delta KE + \Delta PE. \quad (6\text{–}10)$$
For a constant friction force acting over a distance d
$$KE_1 + PE_1 = KE_2 + PE_2 + F_{fr}d. \quad (6\text{–}16b)$$
For other nonconservative forces use your intuition for the sign of W_{NC}: is the total mechanical energy increased or decreased in the process?

7. Use the equation(s) you develop to **solve** for the unknown quantity.

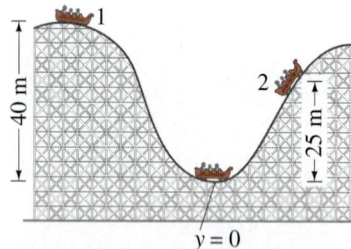

FIGURE 6–27 Example 6–12. Because of friction, a roller-coaster car does not reach the original height on the second hill. (Not to scale.)

EXAMPLE 6–12 ESTIMATE Friction on the roller-coaster car. The roller-coaster car in Example 6–8 reaches a vertical height of only 25 m on the second hill, where it slows to a momentary stop, Fig. 6–27. It traveled a total distance of 400 m. Determine the thermal energy produced and estimate the average friction force (assume it is roughly constant) on the car, whose mass is 1000 kg.

APPROACH We explicitly follow the Problem Solving Strategy above.

SOLUTION

1. **Draw a picture.** See Fig. 6–27.
2. **The system.** The system is the roller-coaster car and the Earth (which exerts the gravitational force). The forces acting on the car are gravity and friction. (The normal force also acts on the car, but does no work, so it does not affect the energy.) Gravity is accounted for as potential energy, and friction as a term $F_{fr}d$.
3. **Choose initial and final positions.** We take point 1 to be the instant when the car started coasting (at the top of the first hill), and point 2 to be the instant it stopped at a height of 25 m up the second hill.
4. **Choose a reference frame.** We choose the lowest point in the motion to be $y = 0$ for the gravitational potential energy.
5. **Is mechanical energy conserved?** No. Friction is present.
6. **Apply conservation of energy.** There is friction acting on the car, so we use conservation of energy in the form of Eq. 6–16b, with $v_1 = 0$, $y_1 = 40\,\text{m}$, $v_2 = 0$, $y_2 = 25\,\text{m}$, and $d = 400\,\text{m}$. Thus
$$0 + (1000\,\text{kg})(9.8\,\text{m/s}^2)(40\,\text{m}) = 0 + (1000\,\text{kg})(9.8\,\text{m/s}^2)(25\,\text{m}) + F_{fr}d.$$
7. **Solve.** We solve the above equation for $F_{fr}d$, the energy dissipated to thermal energy:
$$F_{fr}d = mg\,\Delta h = (1000\,\text{kg})(9.8\,\text{m/s}^2)(40\,\text{m} - 25\,\text{m}) = 147{,}000\,\text{J}.$$
The friction force, which acts over a distance of 400 m, averages out to be
$$F_{fr} = (1.47 \times 10^5\,\text{J})/400\,\text{m} = 370\,\text{N}.$$

NOTE This result is only a rough average: the friction force at various points depends on the normal force, which varies with slope.

6–10 Power

Power is defined as the *rate at which work is done*. Average power equals the work done divided by the time to do it. Power can also be defined as the *rate at which energy is transformed*. Thus

$$\overline{P} = \text{average power} = \frac{\text{work}}{\text{time}} = \frac{\text{energy transformed}}{\text{time}}. \quad (6\text{–}17)$$

The power rating of an engine refers to how much chemical or electrical energy can be transformed into mechanical energy per unit time. In SI units, power is measured in joules per second, and this unit is given a special name, the **watt** (W): 1 W = 1 J/s. We are most familiar with the watt for electrical devices, such as the rate at which an electric lightbulb or heater changes electric energy into light or thermal energy. But the watt is used for other types of energy transformations as well.

In the British system, the unit of power is the foot-pound per second (ft·lb/s). For practical purposes, a larger unit is often used, the **horsepower**. One horsepower[†] (hp) is defined as 550 ft·lb/s, which equals 746 W. An engine's power is usually specified in hp or in kW $\left(1 \text{ kW} \approx 1\tfrac{1}{3} \text{hp}\right)$[‡].

To see the distinction between energy and power, consider the following example. A person is limited in the work he or she can do, not only by the total energy required, but also by how fast this energy is transformed: that is, by power. For example, a person may be able to walk a long distance or climb many flights of stairs before having to stop because so much energy has been expended. On the other hand, a person who runs very quickly up stairs may feel exhausted after only a flight or two. He or she is limited in this case by power, the rate at which his or her body can transform chemical energy into mechanical energy.

CAUTION
Distinguish between power and energy

EXAMPLE 6–13 **Stair-climbing power.** A 60-kg jogger runs up a long flight of stairs in 4.0 s (Fig. 6–28). The vertical height of the stairs is 4.5 m. (*a*) Estimate the jogger's power output in watts and horsepower. (*b*) How much energy did this require?

APPROACH The work done by the jogger is against gravity, and equals $W = mgy$. To get her average power output, we divide W by the time it took.

SOLUTION (*a*) The average power output was

$$\overline{P} = \frac{W}{t} = \frac{mgy}{t} = \frac{(60 \text{ kg})(9.8 \text{ m/s}^2)(4.5 \text{ m})}{4.0 \text{ s}} = 660 \text{ W}.$$

Since there are 746 W in 1 hp, the jogger is doing work at a rate of just under 1 hp. A human cannot do work at this rate for very long.
(*b*) The energy required is $E = \overline{P}t = (660 \text{ J/s})(4.0 \text{ s}) = 2600 \text{ J}.$ This result equals $W = mgy.$

NOTE The person had to transform more energy than this 2600 J. The total energy transformed by a person or an engine always includes some thermal energy (recall how hot you get running up stairs).

FIGURE 6–28 Example 6–13.

Automobiles do work to overcome the force of friction and air resistance, to climb hills, and to accelerate. A car is limited by the rate at which it can do work, which is why automobile engines are rated in horsepower or kilowatts.

PHYSICS APPLIED
Power needs of a car

[†]The unit was chosen by James Watt (1736–1819), who needed a way to specify the power of his newly developed steam engines. He found by experiment that a good horse can work all day at an average rate of about 360 ft·lb/s. So as not to be accused of exaggeration in the sale of his steam engines, he multiplied this by $1\tfrac{1}{2}$ when he defined the hp.

[‡]1 kW = (1000 W)/(746 W/hp) ≈ $1\tfrac{1}{3}$ hp.

A car needs power most when climbing hills and when accelerating. In the next Example, we will calculate how much power is needed in these situations for a car of reasonable size. Even when a car travels on a level road at constant speed, it needs some power just to do work to overcome the retarding forces of internal friction and air resistance. These forces depend on the conditions and speed of the car, but are typically in the range 400 N to 1000 N.

It is often convenient to write power in terms of the net force F applied to an object and its speed v. This is readily done because $\overline{P} = W/t$ and $W = Fd$, where d is the distance traveled. Then

$$\overline{P} = \frac{W}{t} = \frac{Fd}{t} = F\overline{v}, \quad (6\text{-}18)$$

where $\overline{v} = d/t$ is the average speed of the object.

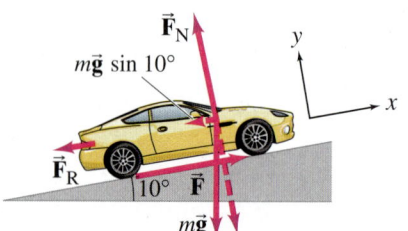

FIGURE 6–29 Example 6–14. Calculation of power needed for a car to climb a hill.

EXAMPLE 6–14 Power needs of a car. Calculate the power required of a 1400-kg car under the following circumstances: (a) the car climbs a 10° hill (a fairly steep hill) at a steady 80 km/h; and (b) the car accelerates along a level road from 90 to 110 km/h in 6.0 s to pass another car. Assume the average retarding force on the car is $F_R = 700$ N throughout. See Fig. 6–29.

APPROACH First we must be careful not to confuse $\vec{F}_R$, which is due to air resistance and friction that retards the motion, with the force $\vec{F}$ needed to accelerate the car, which is the frictional force exerted by the road on the tires—the reaction to the motor-driven tires pushing against the road. We must determine the magnitude of the force F before calculating the power.

SOLUTION (a) To move at a steady speed up the hill, the car must, by Newton's second law, exert a force F equal to the sum of the retarding force, 700 N, and the component of gravity parallel to the hill, $mg \sin 10°$, Fig. 6–29. Thus

$$F = 700 \text{ N} + mg \sin 10°$$
$$= 700 \text{ N} + (1400 \text{ kg})(9.80 \text{ m/s}^2)(0.174) = 3100 \text{ N}.$$

Since $\overline{v} = 80$ km/h $= 22$ m/s† and is parallel to $\vec{F}$, then (Eq. 6–18) the power is

$$\overline{P} = F\overline{v} = (3100 \text{ N})(22 \text{ m/s}) = 6.8 \times 10^4 \text{ W} = 68 \text{ kW} = 91 \text{ hp}.$$

(b) The car accelerates from 25.0 m/s to 30.6 m/s (90 to 110 km/h) on the flat. The car must exert a force that overcomes the 700-N retarding force plus that required to give it the acceleration

$$\overline{a}_x = \frac{(30.6 \text{ m/s} - 25.0 \text{ m/s})}{6.0 \text{ s}} = 0.93 \text{ m/s}^2.$$

We apply Newton's second law with x being the horizontal direction of motion (no component of gravity):

$$ma_x = \Sigma F_x = F - F_R.$$

We solve for the force required, F:

$$F = ma_x + F_R$$
$$= (1400 \text{ kg})(0.93 \text{ m/s}^2) + 700 \text{ N} = 1300 \text{ N} + 700 \text{ N} = 2000 \text{ N}.$$

Since $\overline{P} = F\overline{v}$, the required power increases with speed and the motor must be able to provide a maximum power output in this case of

$$\overline{P} = (2000 \text{ N})(30.6 \text{ m/s}) = 6.1 \times 10^4 \text{ W} = 61 \text{ kW} = 82 \text{ hp}.$$

NOTE Even taking into account the fact that only 60 to 80% of the engine's power output reaches the wheels, it is clear from these calculations that an engine of 75 to 100 kW (100 to 130 hp) is adequate from a practical point of view.

†Recall 1 km/h = 1000 m/3600 s = 0.278 m/s.

We mentioned in Example 6–14 that only part of the energy output of a car engine reaches the wheels. Not only is some energy wasted in getting from the engine to the wheels, in the engine itself most of the input energy (from the burning of gasoline or other fuel) does not do useful work. An important characteristic of all engines is their overall **efficiency** e, defined as the ratio of the useful power output of the engine, P_{out}, to the power input, P_{in} (provided by burning of gasoline, for example):

$$e = \frac{P_{\text{out}}}{P_{\text{in}}}.$$

The efficiency is always less than 1.0 because no engine can create energy, and no engine can even transform energy from one form to another without some energy going to friction, thermal energy, and other nonuseful forms of energy. For example, an automobile engine converts chemical energy released in the burning of gasoline into mechanical energy that moves the pistons and eventually the wheels. But nearly 85% of the input energy is "wasted" as thermal energy that goes into the cooling system or out the exhaust pipe, plus friction in the moving parts. Thus car engines are roughly only about 15% efficient. We will discuss efficiency in more detail in Chapter 15.

Summary

Work is done on an object by a force when the object moves through a distance d. If the direction of a constant force $\vec{F}$ makes an angle θ with the direction of motion, the work done by this force is

$$W = Fd\cos\theta. \qquad (6\text{–}1)$$

Energy can be defined as the ability to do work. In SI units, work and energy are measured in **joules** ($1\ \text{J} = 1\ \text{N}\cdot\text{m}$).

Kinetic energy (KE) is energy of motion. An object of mass m and speed v has **translational kinetic energy**

$$\text{KE} = \tfrac{1}{2}mv^2. \qquad (6\text{–}3)$$

The **work-energy principle** states that the *net* work done on an object (by the *net* force) equals the change in kinetic energy of that object:

$$W_{\text{net}} = \Delta\text{KE} = \tfrac{1}{2}mv_2^2 - \tfrac{1}{2}mv_1^2. \qquad (6\text{–}4)$$

Potential energy (PE) is energy associated with forces that depend on the position or configuration of objects. Gravitational potential energy is

$$\text{PE}_G = mgy, \qquad (6\text{–}6)$$

where y is the height of the object of mass m above an arbitrary reference point. Elastic potential energy is given by

$$\text{PE}_{\text{el}} = \tfrac{1}{2}kx^2 \qquad (6\text{–}9)$$

for a stretched or compressed spring, where x is the displacement from the unstretched position and k is the **spring stiffness constant**. Other potential energies include chemical, electrical, and nuclear energy. The *change in potential energy* when an object changes position *is equal to the external work* needed to take the object from one position to the other.

Potential energy is associated only with **conservative forces**, for which the work done by the force in moving an object from one position to another depends only on the two positions and not on the path taken. **Nonconservative forces** like friction are different—work done by them does depend on the path taken and potential energy cannot be defined for them.

The **law of conservation of energy** states that energy can be transformed from one type to another, but the total energy remains constant. It is valid even when friction is present, because the heat generated can be considered a form of energy transfer. When only *conservative forces* act, the **total mechanical energy** is conserved:

$$\text{KE} + \text{PE} = \text{constant}. \qquad (6\text{–}12)$$

When nonconservative forces such as friction act, then

$$W_{\text{NC}} = \Delta\text{KE} + \Delta\text{PE}, \qquad (6\text{–}10, 6\text{–}15)$$

where W_{NC} is the work done by nonconservative forces.

Power is defined as the rate at which work is done, or the rate at which energy is transformed. The SI unit of power is the **watt** ($1\ \text{W} = 1\ \text{J/s}$).

Questions

1. In what ways is the word "work" as used in everyday language the same as it is defined in physics? In what ways is it different? Give examples of both.
2. Can a centripetal force ever do work on an object? Explain.
3. Why is it tiring to push hard against a solid wall even though you are doing no work?
4. Can the normal force on an object ever do work? Explain.
5. You have two springs that are identical except that spring 1 is stiffer than spring 2 ($k_1 > k_2$). On which spring is more work done: (*a*) if they are stretched using the same force; (*b*) if they are stretched the same distance?
6. If the speed of a particle triples, by what factor does its kinetic energy increase?
7. List some everyday forces that are not conservative, and explain why they aren't.

8. A hand exerts a constant horizontal force on a block that is free to slide on a frictionless surface (Fig. 6–30). The block starts from rest at point A, and by the time it has traveled a distance d to point B it is traveling with speed v_B. When the block has traveled another distance d to point C, will its speed be greater than, less than, or equal to $2v_B$? Explain your reasoning.

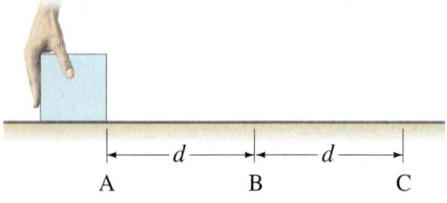

FIGURE 6–30 Question 8.

9. You lift a heavy book from a table to a high shelf. List the forces on the book during this process, and state whether each is conservative or nonconservative.

10. A hill has a height h. A child on a sled (total mass m) slides down starting from rest at the top. Does the speed at the bottom depend on the angle of the hill if (a) it is icy and there is no friction, and (b) there is friction (deep snow)? Explain your answers.

11. Analyze the motion of a simple swinging pendulum in terms of energy, (a) ignoring friction, and (b) taking friction into account. Explain why a grandfather clock has to be wound up.

12. In Fig. 6–31, water balloons are tossed from the roof of a building, all with the same speed but with different launch angles. Which one has the highest speed when it hits the ground? Ignore air resistance. Explain your answer.

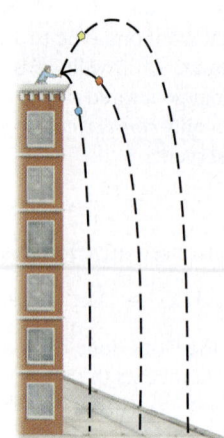

FIGURE 6–31 Question 12.

13. What happens to the gravitational potential energy when water at the top of a waterfall falls to the pool below?

14. Experienced hikers prefer to step over a fallen log in their path rather than stepping on top and stepping down on the other side. Explain.

15. The energy transformations in pole vaulting and archery are discussed in this Chapter. In a similar fashion, discuss the energy transformations related to: (a) hitting a golf ball; (b) serving a tennis ball; and (c) shooting a basket in basketball.

16. Describe precisely what is "wrong" physically in the famous Escher drawing shown in Fig. 6–32.

FIGURE 6–32 Question 16.

17. Two identical arrows, one with twice the speed of the other, are fired into a bale of hay. Assuming the hay exerts a constant "frictional" force on the arrows, the faster arrow will penetrate how much farther than the slower arrow? Explain.

18. A heavy ball is hung from the ceiling by a steel wire. The instructor pulls the ball back and stands against the wall with the ball against his chin. To avoid injury the instructor is supposed to release the ball without pushing it (Fig. 6–33). Why?

FIGURE 6–33 Question 18.

19. Describe the energy transformations when a child hops around on a pogo stick (there is a spring inside).

20. Describe the energy transformations that take place when a skier starts skiing down a hill, but after a time is brought to rest by striking a snowdrift.

21. Suppose you lift a suitcase from the floor to a table. The work you do on the suitcase depends on which of the following: (a) whether you lift it straight up or along a more complicated path, (b) the time the lifting takes, (c) the height of the table, and (d) the weight of the suitcase?

22. Repeat Question 21 for the *power* needed instead of the work.

23. Why is it easier to climb a mountain via a zigzag trail rather than to climb straight up?

MisConceptual Questions

1. You push very hard on a heavy desk, trying to move it. You do work on the desk:
   (a) whether or not it moves, as long as you are exerting a force.
   (b) only if it starts moving.
   (c) only if it doesn't move.
   (d) never—it does work on you.
   (e) None of the above.

2. A satellite in circular orbit around the Earth moves at constant speed. This orbit is maintained by the force of gravity between the Earth and the satellite, yet no work is done on the satellite. How is this possible?
   (a) No work is done if there is no contact between objects.
   (b) No work is done because there is no gravity in space.
   (c) No work is done if the direction of motion is perpendicular to the force.
   (d) No work is done if objects move in a circle.

3. When the speed of your car is doubled, by what factor does its kinetic energy increase?
   (a) $\sqrt{2}$. (b) 2. (c) 4. (d) 8.

4. A car traveling at a velocity v can stop in a minimum distance d. What would be the car's minimum stopping distance if it were traveling at a velocity of $2v$?
   (a) d. (b) $\sqrt{2}\,d$. (c) $2d$. (d) $4d$. (e) $8d$.

5. A bowling ball is dropped from a height h onto the center of a trampoline, which launches the ball back up into the air. How high will the ball rise?
   (a) Significantly less than h.
   (b) More than h. The exact amount depends on the mass of the ball and the springiness of the trampoline.
   (c) No more than h—probably a little less.
   (d) Cannot tell without knowing the characteristics of the trampoline.

6. A ball is thrown straight up. At what point does the ball have the most energy? Ignore air resistance.
   (a) At the highest point of its path.
   (b) When it is first thrown.
   (c) Just before it hits the ground.
   (d) When the ball is halfway to the highest point of its path.
   (e) Everywhere; the energy of the ball is the same at all of these points.

7. A car accelerates from rest to 30 km/h. Later, on a highway it accelerates from 30 km/h to 60 km/h. Which takes more energy, going from 0 to 30, or from 30 to 60?
   (a) 0 to 30 km/h.
   (b) 30 to 60 km/h.
   (c) Both are the same.

8. Engines, including car engines, are rated in horsepower. What is horsepower?
   (a) The force needed to start the engine.
   (b) The force needed to keep the engine running at a steady rate.
   (c) The energy the engine needs to obtain from gasoline or some other source.
   (d) The rate at which the engine can do work.
   (e) The amount of work the engine can perform.

9. Two balls are thrown off a building with the same speed, one straight up and one at a 45° angle. Which statement is true if air resistance can be ignored?
   (a) Both hit the ground at the same time.
   (b) Both hit the ground with the same speed.
   (c) The one thrown at an angle hits the ground with a lower speed.
   (d) The one thrown at an angle hits the ground with a higher speed.
   (e) Both (a) and (b).

10. A skier starts from rest at the top of each of the hills shown in Fig. 6–34. On which hill will the skier have the highest speed at the bottom if we ignore friction: (a), (b), (c), (d), or (e) c and d equally?

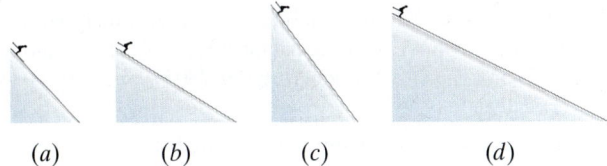

(a)   (b)   (c)   (d)

FIGURE 6–34 MisConceptual Questions 10 and 11.

11. Answer MisConceptual Question 10 assuming a small amount of friction.

12. A man pushes a block up an incline at a constant speed. As the block moves up the incline,
    (a) its kinetic energy and potential energy both increase.
    (b) its kinetic energy increases and its potential energy remains the same.
    (c) its potential energy increases and its kinetic energy remains the same.
    (d) its potential energy increases and its kinetic energy decreases by the same amount.

13. You push a heavy crate *down* a ramp at a constant velocity. Only four forces act on the crate. Which force does the greatest magnitude of work on the crate?
    (a) The force of friction.
    (b) The force of gravity.
    (c) The normal force.
    (d) The force of you pushing.
    (e) The net force.

14. A ball is thrown straight up. Neglecting air resistance, which statement is *not* true regarding the energy of the ball?
    (a) The potential energy decreases while the ball is going up.
    (b) The kinetic energy decreases while the ball is going up.
    (c) The sum of the kinetic energy and potential energy is constant.
    (d) The potential energy decreases when the ball is coming down.
    (e) The kinetic energy increases when the ball is coming down.

Problems

6–1 Work, Constant Force

1. (I) A 75.0-kg firefighter climbs a flight of stairs 28.0 m high. How much work does he do?

2. (I) The head of a hammer with a mass of 1.2 kg is allowed to fall onto a nail from a height of 0.50 m. What is the maximum amount of work it could do on the nail? Why do people not just "let it fall" but add their own force to the hammer as it falls?

3. (II) How much work did the movers do (horizontally) pushing a 46.0-kg crate 10.3 m across a rough floor without acceleration, if the effective coefficient of friction was 0.50?

4. (II) A 1200-N crate rests on the floor. How much work is required to move it at constant speed (a) 5.0 m along the floor against a friction force of 230 N, and (b) 5.0 m vertically?

5. (II) What is the minimum work needed to push a 950-kg car 710 m up along a 9.0° incline? Ignore friction.

6. (II) Estimate the work you do to mow a lawn 10 m by 20 m with a 50-cm-wide mower. Assume you push with a force of about 15 N.

7. (II) In a certain library the first shelf is 15.0 cm off the ground, and the remaining four shelves are each spaced 38.0 cm above the previous one. If the average book has a mass of 1.40 kg with a height of 22.0 cm, and an average shelf holds 28 books (standing vertically), how much work is required to fill all the shelves, assuming the books are all laying flat on the floor to start?

8. (II) A **lever** such as that shown in Fig. 6–35 can be used to lift objects we might not otherwise be able to lift. Show that the ratio of output force, F_O, to input force, F_I, is related to the lengths ℓ_I and ℓ_O from the pivot by $F_O/F_I = \ell_I/\ell_O$. Ignore friction and the mass of the lever, and assume the work output equals the work input.

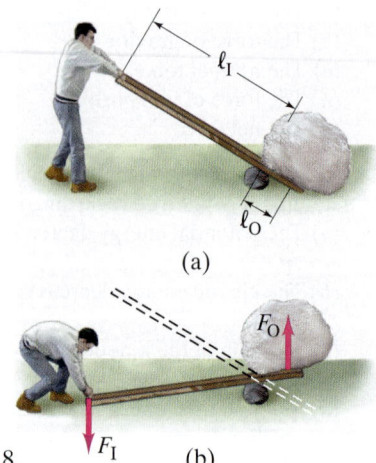

FIGURE 6–35 A lever. Problem 8.

9. (II) A box of mass 4.0 kg is accelerated from rest by a force across a floor at a rate of 2.0 m/s² for 7.0 s. Find the net work done on the box.

10. (II) A 380-kg piano slides 2.9 m down a 25° incline and is kept from accelerating by a man who is pushing back on it *parallel to the incline* (Fig. 6–36). Determine: (a) the force exerted by the man, (b) the work done on the piano by the man, (c) the work done on the piano by the force of gravity, and (d) the net work done on the piano. Ignore friction.

FIGURE 6–36 Problem 10.

11. (II) Recall from Chapter 4, Example 4–14, that you can use a pulley and ropes to decrease the force needed to raise a heavy load (see Fig. 6–37). But for every meter the load is raised, how much rope must be pulled up? Account for this, using energy concepts.

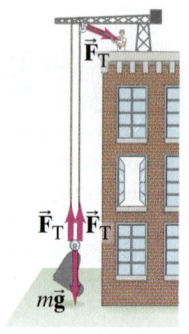

FIGURE 6–37 Problem 11.

12. (III) A grocery cart with mass of 16 kg is being pushed at constant speed up a 12° ramp by a force F_P which acts at an angle of 17° below the horizontal. Find the work done by each of the forces $(m\vec{g}, \vec{F}_N, \vec{F}_P)$ on the cart if the ramp is 7.5 m long.

*6–2 Work, Varying Force

*13. (II) The force on a particle, acting along the x axis, varies as shown in Fig. 6–38. Determine the work done by this force to move the particle along the x axis: (a) from $x = 0.0$ to $x = 10.0$ m; (b) from $x = 0.0$ to $x = 15.0$ m.

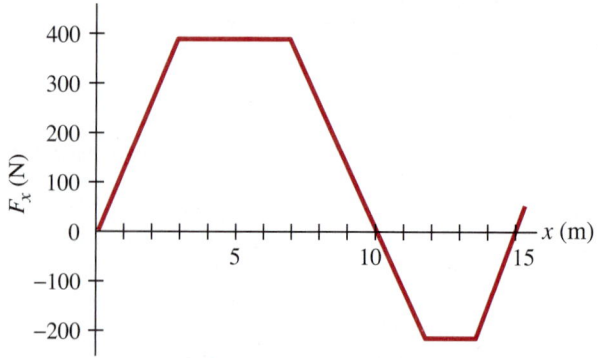

FIGURE 6–38 Problem 13.

*14. (III) A 17,000-kg jet takes off from an aircraft carrier via a catapult (Fig. 6–39a). The gases thrust out from the jet's engines exert a constant force of 130 kN on the jet; the force exerted on the jet by the catapult is plotted in Fig. 6–39b. Determine the work done on the jet: (a) by the gases expelled by its engines during launch of the jet; and (b) by the catapult during launch of the jet.

(a)

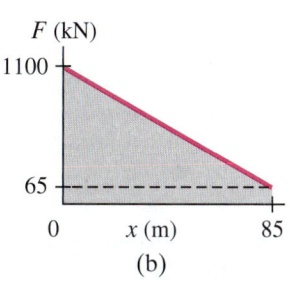
(b)

FIGURE 6–39 Problem 14.

6–3 Kinetic Energy; Work-Energy Principle

15. (I) At room temperature, an oxygen molecule, with mass of 5.31×10^{-26} kg, typically has a kinetic energy of about 6.21×10^{-21} J. How fast is it moving?

16. (I) (a) If the kinetic energy of a particle is tripled, by what factor has its speed increased? (b) If the speed of a particle is halved, by what factor does its kinetic energy change?

17. (I) How much work is required to stop an electron $(m = 9.11 \times 10^{-31}$ kg) which is moving with a speed of 1.10×10^6 m/s?

18. (I) How much work must be done to stop a 925-kg car traveling at 95 km/h?

19. (II) Two bullets are fired at the same time with the same kinetic energy. If one bullet has twice the mass of the other, which has the greater speed and by what factor? Which can do the most work?

20. (II) A baseball $(m = 145$ g) traveling 32 m/s moves a fielder's glove backward 25 cm when the ball is caught. What was the average force exerted by the ball on the glove?

21. (II) An 85-g arrow is fired from a bow whose string exerts an average force of 105 N on the arrow over a distance of 75 cm. What is the speed of the arrow as it leaves the bow?

22. (II) If the speed of a car is increased by 50%, by what factor will its minimum braking distance be increased, assuming all else is the same? Ignore the driver's reaction time.

23. (II) At an accident scene on a level road, investigators measure a car's skid mark to be 78 m long. It was a rainy day and the coefficient of friction was estimated to be 0.30. Use these data to determine the speed of the car when the driver slammed on (and locked) the brakes. (Why does the car's mass not matter?)

24. (III) One car has twice the mass of a second car, but only half as much kinetic energy. When both cars increase their speed by 8.0 m/s, they then have the same kinetic energy. What were the original speeds of the two cars?

25. (III) A 265-kg load is lifted 18.0 m vertically with an acceleration $a = 0.160 g$ by a single cable. Determine (a) the tension in the cable; (b) the net work done on the load; (c) the work done by the cable on the load; (d) the work done by gravity on the load; (e) the final speed of the load assuming it started from rest.

6–4 and 6–5 Potential Energy

26. (I) By how much does the gravitational potential energy of a 54-kg pole vaulter change if her center of mass rises about 4.0 m during the jump?

27. (I) A spring has a spring constant k of 88.0 N/m. How much must this spring be compressed to store 45.0 J of potential energy?

28. (II) If it requires 6.0 J of work to stretch a particular spring by 2.0 cm from its equilibrium length, how much more work will be required to stretch it an additional 4.0 cm?

29. (II) A 66.5-kg hiker starts at an elevation of 1270 m and climbs to the top of a peak 2660 m high. (a) What is the hiker's change in potential energy? (b) What is the minimum work required of the hiker? (c) Can the actual work done be greater than this? Explain.

30. (II) A 1.60-m-tall person lifts a 1.65-kg book off the ground so it is 2.20 m above the ground. What is the potential energy of the book relative to (a) the ground, and (b) the top of the person's head? (c) How is the work done by the person related to the answers in parts (a) and (b)?

6–6 and 6–7 Conservation of Mechanical Energy

31. (I) A novice skier, starting from rest, slides down an icy frictionless 8.0° incline whose vertical height is 105 m. How fast is she going when she reaches the bottom?

32. (I) Jane, looking for Tarzan, is running at top speed (5.0 m/s) and grabs a vine hanging vertically from a tall tree in the jungle. How high can she swing upward? Does the length of the vine affect your answer?

33. (II) A sled is initially given a shove up a frictionless 23.0° incline. It reaches a maximum vertical height 1.22 m higher than where it started at the bottom. What was its initial speed?

34. (II) In the *high jump*, the kinetic energy of an athlete is transformed into gravitational potential energy without the aid of a pole. With what minimum speed must the athlete leave the ground in order to lift his center of mass 2.10 m and cross the bar with a speed of 0.50 m/s?

35. (II) A spring with $k = 83$ N/m hangs vertically next to a ruler. The end of the spring is next to the 15-cm mark on the ruler. If a 2.5-kg mass is now attached to the end of the spring, and the mass is allowed to fall, where will the end of the spring line up with the ruler marks when the mass is at its lowest position?

36. (II) A 0.48-kg ball is thrown with a speed of 8.8 m/s at an upward angle of 36°. (a) What is its speed at its highest point, and (b) how high does it go? (Use conservation of energy.)

37. (II) A 1200-kg car moving on a horizontal surface has speed $v = 85$ km/h when it strikes a horizontal coiled spring and is brought to rest in a distance of 2.2 m. What is the spring stiffness constant of the spring?

38. (II) A 62-kg trampoline artist jumps upward from the top of a platform with a vertical speed of 4.5 m/s. (a) How fast is he going as he lands on the trampoline, 2.0 m below (Fig. 6–40)? (b) If the trampoline behaves like a spring of spring constant 5.8×10^4 N/m, how far down does he depress it?

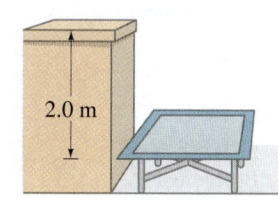

FIGURE 6–40 Problem 38.

39. (II) A vertical spring (ignore its mass), whose spring constant is 875 N/m, is attached to a table and is compressed down by 0.160 m. (a) What upward speed can it give to a 0.380-kg ball when released? (b) How high above its original position (spring compressed) will the ball fly?

40. (II) A roller-coaster car shown in Fig. 6–41 is pulled up to point 1 where it is released from rest. Assuming no friction, calculate the speed at points 2, 3, and 4.

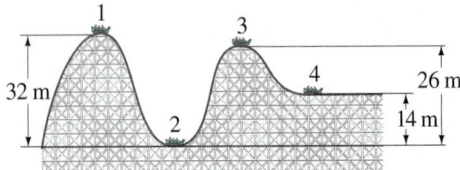

FIGURE 6–41 Problems 40 and 50.

41. (II) Chris jumps off a bridge with a bungee cord (a heavy stretchable cord) tied around his ankle, Fig. 6–42. He falls for 15 m before the bungee cord begins to stretch. Chris's mass is 75 kg and we assume the cord obeys Hooke's law, $F = -kx$, with $k = 55$ N/m. If we neglect air resistance, estimate what distance d below the bridge Chris's foot will be before coming to a stop. Ignore the mass of the cord (not realistic, however) and treat Chris as a particle.

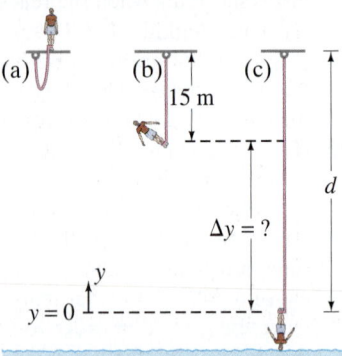

FIGURE 6–42
Problem 41. (a) Bungee jumper about to jump.
(b) Bungee cord at its unstretched length.
(c) Maximum stretch of cord.

42. (II) What should be the spring constant k of a spring designed to bring a 1200-kg car to rest from a speed of 95 km/h so that the occupants undergo a maximum acceleration of 4.0 g?

43. (III) An engineer is designing a spring to be placed at the bottom of an elevator shaft. If the elevator cable breaks when the elevator is at a height h above the top of the spring, calculate the value that the spring constant k should have so that passengers undergo an acceleration of no more than 5.0 g when brought to rest. Let M be the total mass of the elevator and passengers.

44. (III) A block of mass m is attached to the end of a spring (spring stiffness constant k), Fig. 6–43. The mass is given an initial displacement x_0 from equilibrium, and an initial speed v_0. Ignoring friction and the mass of the spring, use energy methods to find (a) its maximum speed, and (b) its maximum stretch from equilibrium, in terms of the given quantities.

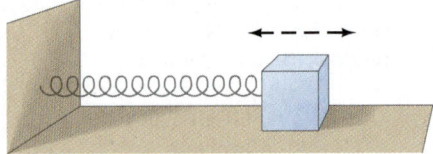

FIGURE 6–43
Problem 44.

45. (III) A cyclist intends to cycle up a 7.50° hill whose vertical height is 125 m. The pedals turn in a circle of diameter 36.0 cm. Assuming the mass of bicycle plus person is 75.0 kg, (a) calculate how much work must be done against gravity. (b) If each complete revolution of the pedals moves the bike 5.10 m along its path, calculate the average force that must be exerted on the pedals tangent to their circular path. Neglect work done by friction and other losses.

6–8 and 6–9 Law of Conservation of Energy

46. (I) Two railroad cars, each of mass 66,000 kg, are traveling 85 km/h toward each other. They collide head-on and come to rest. How much thermal energy is produced in this collision?

47. (I) A 16.0-kg child descends a slide 2.20 m high and, starting from rest, reaches the bottom with a speed of 1.25 m/s. How much thermal energy due to friction was generated in this process?

48. (II) A ski starts from rest and slides down a 28° incline 85 m long. (a) If the coefficient of friction is 0.090, what is the ski's speed at the base of the incline? (b) If the snow is level at the foot of the incline and has the same coefficient of friction, how far will the ski travel along the level? Use energy methods.

49. (II) A 145-g baseball is dropped from a tree 12.0 m above the ground. (a) With what speed would it hit the ground if air resistance could be ignored? (b) If it actually hits the ground with a speed of 8.00 m/s, what is the average force of air resistance exerted on it?

50. (II) Suppose the roller-coaster car in Fig. 6–41 passes point 1 with a speed of 1.30 m/s. If the average force of friction is equal to 0.23 of its weight, with what speed will it reach point 2? The distance traveled is 45.0 m.

51. (II) A skier traveling 11.0 m/s reaches the foot of a steady upward 19° incline and glides 15 m up along this slope before coming to rest. What was the average coefficient of friction?

52. (II) You drop a ball from a height of 2.0 m, and it bounces back to a height of 1.6 m. (a) What fraction of its initial energy is lost during the bounce? (b) What is the ball's speed just before and just after the bounce? (c) Where did the energy go?

53. (II) A 66-kg skier starts from rest at the top of a 1200-m-long trail which drops a total of 230 m from top to bottom. At the bottom, the skier is moving 11.0 m/s. How much energy was dissipated by friction?

54. (II) A projectile is fired at an upward angle of 38.0° from the top of a 135-m-high cliff with a speed of 165 m/s. What will be its speed when it strikes the ground below? (Use conservation of energy.)

55. (II) The Lunar Module could make a safe landing if its vertical velocity at impact is 3.0 m/s or less. Suppose that you want to determine the greatest height h at which the pilot could shut off the engine if the velocity of the lander relative to the surface at that moment is (a) zero; (b) 2.0 m/s downward; (c) 2.0 m/s upward. Use conservation of energy to determine h in each case. The acceleration due to gravity at the surface of the Moon is 1.62 m/s².

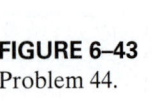

166 CHAPTER 6 Work and Energy

56. (III) Early test flights for the space shuttle used a "glider" (mass of 980 kg including pilot). After a horizontal launch at 480 km/h at a height of 3500 m, the glider eventually landed at a speed of 210 km/h. (a) What would its landing speed have been in the absence of air resistance? (b) What was the average force of air resistance exerted on it if it came in at a constant glide angle of 12° to the Earth's surface?

6–10 Power

57. (I) How long will it take a 2750-W motor to lift a 385-kg piano to a sixth-story window 16.0 m above?

58. (I) (a) Show that one British horsepower (550 ft·lb/s) is equal to 746 W. (b) What is the horsepower rating of a 75-W lightbulb?

59. (I) An 85-kg football player traveling 5.0 m/s is stopped in 1.0 s by a tackler. (a) What is the original kinetic energy of the player? (b) What average power is required to stop him?

60. (II) If a car generates 18 hp when traveling at a steady 95 km/h, what must be the average force exerted on the car due to friction and air resistance?

61. (II) An outboard motor for a boat is rated at 35 hp. If it can move a particular boat at a steady speed of 35 km/h, what is the total force resisting the motion of the boat?

62. (II) A shot-putter accelerates a 7.3-kg shot from rest to 14 m/s in 1.5 s. What average power was developed?

63. (II) A driver notices that her 1080-kg car, when in neutral, slows down from 95 km/h to 65 km/h in about 7.0 s on a flat horizontal road. Approximately what power (watts and hp) is needed to keep the car traveling at a constant 80 km/h?

64. (II) How much work can a 2.0-hp motor do in 1.0 h?

65. (II) A 975-kg sports car accelerates from rest to 95 km/h in 6.4 s. What is the average power delivered by the engine?

66. (II) During a workout, football players ran up the stadium stairs in 75 s. The distance along the stairs is 83 m and they are inclined at a 33° angle. If a player has a mass of 82 kg, estimate his average power output on the way up. Ignore friction and air resistance.

67. (II) A pump lifts 27.0 kg of water per minute through a height of 3.50 m. What minimum output rating (watts) must the pump motor have?

68. (II) A ski area claims that its lifts can move 47,000 people per hour. If the average lift carries people about 200 m (vertically) higher, estimate the maximum total power needed.

69. (II) A 65-kg skier grips a moving rope that is powered by an engine and is pulled at constant speed to the top of a 23° hill. The skier is pulled a distance $x = 320$ m along the incline and it takes 2.0 min to reach the top of the hill. If the coefficient of kinetic friction between the snow and skis is $\mu_k = 0.10$, what horsepower engine is required if 30 such skiers (max) are on the rope at one time?

70. (II) What minimum horsepower must a motor have to be able to drag a 370-kg box along a level floor at a speed of 1.20 m/s if the coefficient of friction is 0.45?

71. (III) A bicyclist coasts down a 6.0° hill at a steady speed of 4.0 m/s. Assuming a total mass of 75 kg (bicycle plus rider), what must be the cyclist's power output to climb the same hill at the same speed?

General Problems

72. Spiderman uses his spider webs to save a runaway train moving about 60 km/h, Fig. 6–44. His web stretches a few city blocks (500 m) before the 10^4-kg train comes to a stop. Assuming the web acts like a spring, estimate the effective spring constant.

FIGURE 6–44 Problem 72.

73. A 36.0-kg crate, starting from rest, is pulled across a floor with a constant horizontal force of 225 N. For the first 11.0 m the floor is frictionless, and for the next 10.0 m the coefficient of friction is 0.20. What is the final speed of the crate after being pulled these 21.0 m?

74. How high will a 1.85-kg rock go from the point of release if thrown straight up by someone who does 80.0 J of work on it? Neglect air resistance.

75. A mass m is attached to a spring which is held stretched a distance x by a force F, Fig. 6–45, and then released. The spring pulls the mass to the left, towards its natural equilibrium length. Assuming there is no friction, determine the speed of the mass m when the spring returns: (a) to its normal length ($x = 0$); (b) to half its original extension ($x/2$).

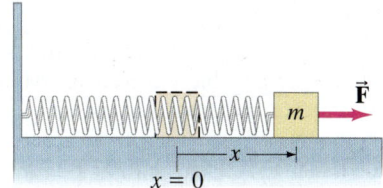

FIGURE 6–45 Problem 75.

76. An elevator cable breaks when a 925-kg elevator is 28.5 m above the top of a huge spring ($k = 8.00 \times 10^4$ N/m) at the bottom of the shaft. Calculate (a) the work done by gravity on the elevator before it hits the spring; (b) the speed of the elevator just before striking the spring; (c) the amount the spring compresses (note that here work is done by both the spring and gravity).

77. (a) A 3.0-g locust reaches a speed of 3.0 m/s during its jump. What is its kinetic energy at this speed? (b) If the locust transforms energy with 35% efficiency, how much energy is required for the jump?

78. In a common test for cardiac function (the "stress test"), the patient walks on an inclined treadmill (Fig. 6–46). Estimate the power required from a 75-kg patient when the treadmill is sloping at an angle of 12° and the velocity is 3.1 km/h. (How does this power compare to the power rating of a lightbulb?)

FIGURE 6–46 Problem 78.

79. An airplane pilot fell 370 m after jumping from an aircraft without his parachute opening. He landed in a snowbank, creating a crater 1.1 m deep, but survived with only minor injuries. Assuming the pilot's mass was 88 kg and his speed at impact was 45 m/s, estimate: (a) the work done by the snow in bringing him to rest; (b) the average force exerted on him by the snow to stop him; and (c) the work done on him by air resistance as he fell. Model him as a particle.

80. Many cars have "5 mi/h (8 km/h) bumpers" that are designed to compress and rebound elastically without any physical damage at speeds below 8 km/h. If the material of the bumpers permanently deforms after a compression of 1.5 cm, but remains like an elastic spring up to that point, what must be the effective spring constant of the bumper material, assuming the car has a mass of 1050 kg and is tested by ramming into a solid wall?

81. In climbing up a rope, a 62-kg athlete climbs a vertical distance of 5.0 m in 9.0 s. What minimum power output was used to accomplish this feat?

82. If a 1300-kg car can accelerate from 35 km/h to 65 km/h in 3.8 s, how long will it take to accelerate from 55 km/h to 95 km/h? Assume the power stays the same, and neglect frictional losses.

83. A cyclist starts from rest and coasts down a 4.0° hill. The mass of the cyclist plus bicycle is 85 kg. After the cyclist has traveled 180 m, (a) what was the net work done by gravity on the cyclist? (b) How fast is the cyclist going? Ignore air resistance and friction.

84. A film of Jesse Owens's famous long jump (Fig. 6–47) in the 1936 Olympics shows that his center of mass rose 1.1 m from launch point to the top of the arc. What minimum speed did he need at launch if he was traveling at 6.5 m/s at the top of the arc?

FIGURE 6–47 Problem 84.

85. Water flows over a dam at the rate of 680 kg/s and falls vertically 88 m before striking the turbine blades. Calculate (a) the speed of the water just before striking the turbine blades (neglect air resistance), and (b) the rate at which mechanical energy is transferred to the turbine blades, assuming 55% efficiency.

86. A 55-kg skier starts from rest at the top of a ski jump, point A in Fig. 6–48, and travels down the ramp. If friction and air resistance can be neglected, (a) determine her speed v_B when she reaches the horizontal end of the ramp at B. (b) Determine the distance s to where she strikes the ground at C.

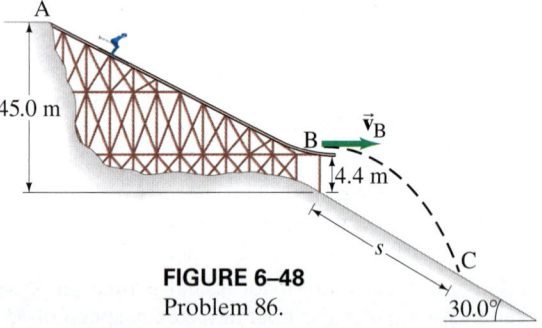

FIGURE 6–48 Problem 86.

87. Electric energy units are often expressed in "kilowatt-hours." (a) Show that one kilowatt-hour (kWh) is equal to 3.6×10^6 J. (b) If a typical family of four uses electric energy at an average rate of 580 W, how many kWh would their electric bill show for one month, and (c) how many joules would this be? (d) At a cost of $0.12 per kWh, what would their monthly bill be in dollars? Does the monthly bill depend on the *rate* at which they use the electric energy?

88. If you stand on a bathroom scale, the spring inside the scale compresses 0.60 mm, and it tells you your weight is 760 N. Now if you jump on the scale from a height of 1.0 m, what does the scale read at its peak?

89. A 65-kg hiker climbs to the top of a mountain 4200 m high. The climb is made in 4.6 h starting at an elevation of 2800 m. Calculate (a) the work done by the hiker against gravity, (b) the average power output in watts and in horsepower, and (c) assuming the body is 15% efficient, what rate of energy input was required.

90. A ball is attached to a horizontal cord of length ℓ whose other end is fixed, Fig. 6–49. (a) If the ball is released, what will be its speed at the lowest point of its path? (b) A peg is located a distance h directly below the point of attachment of the cord. If $h = 0.80\ell$, what will be the speed of the ball when it reaches the top of its circular path about the peg?

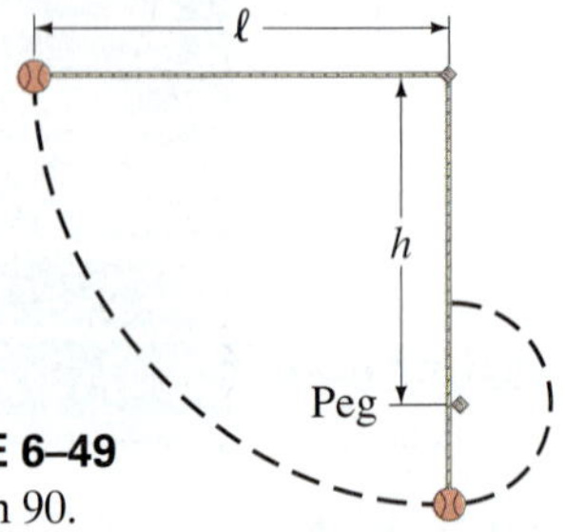

FIGURE 6–49 Problem 90.

91. An 18-kg sled starts up a 28° incline with a speed of 2.3 m/s. The coefficient of kinetic friction is $\mu_k = 0.25$. (a) How far up the incline does the sled travel? (b) What condition must you put on the coefficient of static friction if the sled is not to get stuck at the point determined in part (a)? (c) If the sled slides back down, what is its speed when it returns to its starting point?

92. A 56-kg student runs at 6.0 m/s, grabs a hanging 10.0-m-long rope, and swings out over a lake (Fig. 6–50). He releases the rope when his velocity is zero. (a) What is the angle θ when he releases the rope? (b) What is the tension in the rope just before he releases it? (c) What is the maximum tension in the rope during the swing?

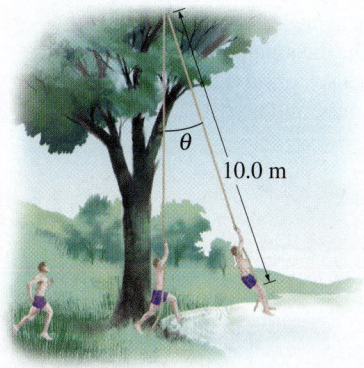

FIGURE 6–50 Problem 92.

93. Some electric power companies use water to store energy. Water is pumped from a low reservoir to a high reservoir. To store the energy produced in 1.0 hour by a 180-MW electric power plant, how many cubic meters of water will have to be pumped from the lower to the upper reservoir? Assume the upper reservoir is an average of 380 m above the lower one. Water has a mass of 1.00×10^3 kg for every 1.0 m^3.

94. A softball having a mass of 0.25 kg is pitched horizontally at 120 km/h. By the time it reaches the plate, it may have slowed by 10%. Neglecting gravity, estimate the average force of air resistance during a pitch. The distance between the plate and the pitcher is about 15 m.

Search and Learn

1. We studied forces earlier and used them to solve Problems. Now we are using energy to solve Problems, even some that could be solved with forces. (a) Give at least three advantages of using energy to solve a Problem. (b) When must you use energy to solve a Problem? (c) When must you use forces to solve a Problem? (d) What information is not available when solving Problems with energy? Look at the Examples in Chapters 6 and 4.

2. The brakes on a truck can overheat and catch on fire if the truck goes down a long steep hill without shifting into a lower gear. (a) Explain why this happens in terms of energy and power. (b) Would it matter if the same elevation change was made going down a steep hill or a gradual hill? Explain your reasoning. [*Hint*: Read Sections 6–4, 6–9, and 6–10 carefully.] (c) Why does shifting into a lower gear help? [*Hint*: Use your own experience, downshifting in a car.] (d) Calculate the thermal energy dissipated from the brakes in an 8000-kg truck that descends a 12° hill. The truck begins braking when its speed is 95 km/h and slows to a speed of 35 km/h in a distance of 0.36 km measured along the road.

3. (a) Only two conservative forces are discussed in this Chapter. What are they, and how are they accounted for when you are dealing with conservation of energy? (b) Not mentioned is the force of water on a swimmer. Is it conservative or nonconservative?

4. Give at least two examples of friction doing positive work. Reread parts of Chapters 4 and 6.

5. Show that on a roller coaster with a circular vertical loop (Fig. 6–51), the difference in your apparent weight at the top of the loop and the bottom of the loop is 6.0 times your weight. Ignore friction. Show also that as long as your speed is above the minimum needed (so the car holds the track), this answer doesn't depend on the size of the loop or how fast you go through it. [Reread Sections 6–6, 5–2, and 4–6.]

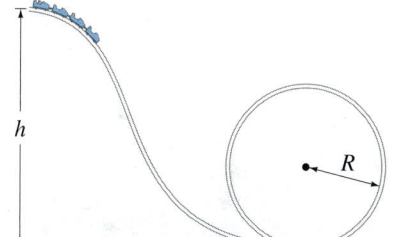

FIGURE 6–51 Search and Learn 5 and 6.

6. Suppose that the track in Fig. 6–51 is not frictionless and the values of h and R are given. (See Sections 6–9 and 6–1.) (a) If you measure the velocity of the roller coaster at the top of the hill (of height h) and at the top of the circle (of height $2R$), can you determine the work done by friction during the time the roller coaster moves between those two points? Why or why not? (b) Can you determine the average force of friction between those two points? Why or why not? If not, what additional information do you need?

ANSWERS TO EXERCISES

A: (c).
B: (a) Less, because $(20)^2 = 400 < (30)^2 - (20)^2 = 500$; (b) 2.0×10^5 J.
C: No, because the speed v would be the square root of a negative number, which is not real.
D: (a) $\sqrt{2}$; (b) 4.
E: Yes. It is nonconservative, because for a conservative force $W = 0$ in a round trip.
F: (e), (e); (e), (c).

Conservation of linear momentum is another great conservation law of physics. Collisions, such as between billiard or pool balls, illustrate this law very nicely: the total vector momentum just before the collision equals the total vector momentum just after the collision. In this photo, the moving cue ball makes a glancing collision with the 11 ball which is initially at rest. After the collision, both balls move at angles, but the sum of their vector momenta equals the initial vector momentum of the incoming cue ball.

We will consider both elastic collisions (where kinetic energy is also conserved) and inelastic collisions. We also examine the concept of center of mass, and how it helps us in the study of complex motion.

7 Linear Momentum

CONTENTS

7–1 Momentum and Its Relation to Force
7–2 Conservation of Momentum
7–3 Collisions and Impulse
7–4 Conservation of Energy and Momentum in Collisions
7–5 Elastic Collisions in One Dimension
7–6 Inelastic Collisions
*7–7 Collisions in Two Dimensions
7–8 Center of Mass (CM)
*7–9 CM for the Human Body
*7–10 CM and Translational Motion

CHAPTER-OPENING QUESTIONS—Guess now!

1. A railroad car loaded with rocks coasts on a level track without friction. A worker at the back of the car starts throwing the rocks horizontally backward from the car. Then what happens?
 (a) The car slows down.
 (b) The car speeds up.
 (c) First the car speeds up and then it slows down.
 (d) The car's speed remains constant.
 (e) None of these.

2. Which answer would you choose if the rocks fall out through a hole in the floor of the car, one at a time?

The law of conservation of energy, which we discussed in the previous Chapter, is one of several great conservation laws in physics. Among the other quantities found to be conserved are linear momentum, angular momentum, and electric charge. We will eventually discuss all of these because the conservation laws are among the most important ideas in science. In this Chapter we discuss linear momentum and its conservation. The law of conservation of momentum is essentially a reworking of Newton's laws that gives us tremendous physical insight and problem-solving power.

The law of conservation of momentum is particularly useful when dealing with a system of two or more objects that interact with each other, such as in collisions of ordinary objects or nuclear particles.

Our focus up to now has been mainly on the motion of a single object, often thought of as a "particle" in the sense that we have ignored any rotation or internal motion. In this Chapter we will deal with systems of two or more objects, and—toward the end of the Chapter—the concept of center of mass.

7–1 Momentum and Its Relation to Force

The **linear momentum** (or "momentum" for short) of an object is defined as the product of its mass and its velocity. Momentum (plural is *momenta*—from Latin) is represented by the symbol $\vec{\mathbf{p}}$. If we let m represent the mass of an object and $\vec{\mathbf{v}}$ represent its velocity, then its momentum $\vec{\mathbf{p}}$ is defined as

$$\vec{\mathbf{p}} = m\vec{\mathbf{v}}. \tag{7–1}$$

Velocity is a vector, so momentum too is a vector. The direction of the momentum is the direction of the velocity, and the magnitude of the momentum is $p = mv$. Because velocity depends on the reference frame, so does momentum; thus the reference frame must be specified. The unit of momentum is that of mass × velocity, which in SI units is kg·m/s. There is no special name for this unit.

Everyday usage of the term *momentum* is in accord with the definition above. According to Eq. 7–1, a fast-moving car has more momentum than a slow-moving car of the same mass; a heavy truck has more momentum than a small car moving with the same speed. The more momentum an object has, the harder it is to stop it, and the greater effect it will have on another object if it is brought to rest by striking that object. A football player is more likely to be stunned if tackled by a heavy opponent running at top speed than by a lighter or slower-moving tackler. A heavy, fast-moving truck can do more damage than a slow-moving motorcycle.

EXERCISE A Can a small sports car ever have the same momentum as a large sport-utility vehicle with three times the sports car's mass? Explain.

A force is required to change the momentum of an object, whether to increase the momentum, to decrease it, or to change its direction. Newton originally stated his second law in terms of momentum (although he called the product mv the "quantity of motion"). Newton's statement of the **second law of motion**, translated into modern language, is as follows:

> **The rate of change of momentum of an object is equal to the net force applied to it.**

NEWTON'S SECOND LAW

We can write this as an equation,

$$\Sigma\vec{\mathbf{F}} = \frac{\Delta\vec{\mathbf{p}}}{\Delta t}, \tag{7–2}$$

NEWTON'S SECOND LAW

where $\Sigma\vec{\mathbf{F}}$ is the net force applied to the object (the vector sum of all forces acting on it) and $\Delta\vec{\mathbf{p}}$ is the resulting momentum change that occurs during the time interval[†] Δt.

⚠ **CAUTION**
The change in the momentum vector is in the direction of the net force

We can readily derive the familiar form of the second law, $\Sigma\vec{\mathbf{F}} = m\vec{\mathbf{a}}$, from Eq. 7–2 for the case of constant mass. If $\vec{\mathbf{v}}_1$ is the initial velocity of an object and $\vec{\mathbf{v}}_2$ is its velocity after a time interval Δt has elapsed, then

$$\Sigma\vec{\mathbf{F}} = \frac{\Delta\vec{\mathbf{p}}}{\Delta t} = \frac{m\vec{\mathbf{v}}_2 - m\vec{\mathbf{v}}_1}{\Delta t} = \frac{m(\vec{\mathbf{v}}_2 - \vec{\mathbf{v}}_1)}{\Delta t} = m\frac{\Delta\vec{\mathbf{v}}}{\Delta t}.$$

By definition, $\vec{\mathbf{a}} = \Delta\vec{\mathbf{v}}/\Delta t$, so

$$\Sigma\vec{\mathbf{F}} = m\vec{\mathbf{a}}. \qquad \text{[constant mass]}$$

Equation 7–2 is a more general statement of Newton's second law than the more familiar version $(\Sigma\vec{\mathbf{F}} = m\vec{\mathbf{a}})$ because it includes the situation in which the mass may change. A change in mass occurs in certain circumstances, such as for rockets which lose mass as they expel burnt fuel.

[†]Normally we think of Δt as being a small time interval. If it is not small, then Eq. 7–2 is valid if $\Sigma\vec{\mathbf{F}}$ is constant during that time interval, or if $\Sigma\vec{\mathbf{F}}$ is the average net force during that time interval.

FIGURE 7–1 Example 7–1.

EXAMPLE 7–1 ESTIMATE **Force of a tennis serve.** For a top player, a tennis ball may leave the racket on the serve with a speed of 55 m/s (about 120 mi/h), Fig. 7–1. If the ball has a mass of 0.060 kg and is in contact with the racket for about 4 ms $(4 \times 10^{-3}\,\text{s})$, estimate the average force on the ball. Would this force be large enough to lift a 60-kg person?

APPROACH We write Newton's second law, Eq. 7–2, for the average force as

$$F_{\text{avg}} = \frac{\Delta p}{\Delta t} = \frac{mv_2 - mv_1}{\Delta t},$$

where mv_1 and mv_2 are the initial and final momenta. The tennis ball is hit when its initial velocity v_1 is very nearly zero at the top of the throw, so we set $v_1 = 0$, and we assume $v_2 = 55$ m/s is in the horizontal direction. We ignore all other forces on the ball during this brief time interval, such as gravity, in comparison to the force exerted by the tennis racket.

SOLUTION The force exerted on the ball by the racket is

$$F_{\text{avg}} = \frac{\Delta p}{\Delta t} = \frac{mv_2 - mv_1}{\Delta t} = \frac{(0.060\,\text{kg})(55\,\text{m/s}) - 0}{0.004\,\text{s}} \approx 800\,\text{N}.$$

This is a large force, larger than the weight of a 60-kg person, which would require a force $mg = (60\,\text{kg})(9.8\,\text{m/s}^2) \approx 600\,\text{N}$ to lift.

NOTE The force of gravity acting on the tennis ball is $mg = (0.060\,\text{kg})(9.8\,\text{m/s}^2) = 0.59\,\text{N}$, which justifies our ignoring it compared to the enormous force the racket exerts.

NOTE High-speed photography and radar can give us an estimate of the contact time and the velocity of the ball leaving the racket. But a direct measurement of the force is not practical. Our calculation shows a handy technique for determining an unknown force in the real world.

FIGURE 7–2 Example 7–2.

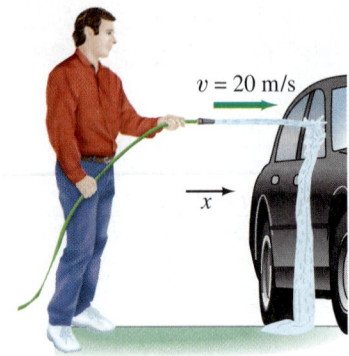

EXAMPLE 7–2 **Washing a car: momentum change and force.** Water leaves a hose at a rate of 1.5 kg/s with a speed of 20 m/s and is aimed at the side of a car, which stops it, Fig. 7–2. (That is, we ignore any splashing back.) What is the force exerted by the water on the car?

APPROACH The water leaving the hose has mass and velocity, so it has a momentum p_{initial} in the horizontal (x) direction, and we assume gravity doesn't pull the water down significantly. When the water hits the car, the water loses this momentum ($p_{\text{final}} = 0$). We use Newton's second law in the momentum form, Eq. 7–2, to find the force that the car exerts on the water to stop it. By Newton's third law, the force exerted by the water on the car is equal and opposite. We have a continuing process: 1.5 kg of water leaves the hose in each 1.0-s time interval. So let us write $F = \Delta p/\Delta t$ where $\Delta t = 1.0$ s, and $mv_{\text{initial}} = (1.5\,\text{kg})(20\,\text{m/s}) = 30\,\text{kg}\cdot\text{m/s}$.

SOLUTION The force (assumed constant) that the car must exert to change the momentum of the water is

$$F = \frac{\Delta p}{\Delta t} = \frac{p_{\text{final}} - p_{\text{initial}}}{\Delta t} = \frac{0 - 30\,\text{kg}\cdot\text{m/s}}{1.0\,\text{s}} = -30\,\text{N}.$$

The minus sign indicates that the force exerted by the car on the water is opposite to the water's original velocity. The car exerts a force of 30 N to the left to stop the water, so by Newton's third law, the water exerts a force of 30 N to the right on the car.

NOTE Keep track of signs, although common sense helps too. The water is moving to the right, so common sense tells us the force on the car must be to the right.

EXERCISE B If the water splashes back from the car in Example 7–2, would the force on the car be larger or smaller?

7–2 Conservation of Momentum

The concept of momentum is particularly important because, if no net external force acts on a system, the total momentum of the system is a conserved quantity. This was expressed in Eq. 7–2 for a single object, but it holds also for a system as we shall see.

Consider the head-on collision of two billiard balls, as shown in Fig. 7–3. We assume the net external force on this system of two balls is zero—that is, the only significant forces during the collision are the forces that each ball exerts on the other. Although the momentum of each of the two balls changes as a result of the collision, the *sum* of their momenta is found to be the same before as after the collision. If $m_A \vec{v}_A$ is the momentum of ball A and $m_B \vec{v}_B$ the momentum of ball B, both measured just before the collision, then the total momentum of the two balls before the collision is the vector sum $m_A \vec{v}_A + m_B \vec{v}_B$. Immediately after the collision, the balls each have a different velocity and momentum, which we designate by a "prime" on the velocity: $m_A \vec{v}'_A$ and $m_B \vec{v}'_B$. The total momentum after the collision is the vector sum $m_A \vec{v}'_A + m_B \vec{v}'_B$. No matter what the velocities and masses are, experiments show that the total momentum before the collision is the same as afterward, whether the collision is head-on or not, as long as no net external force acts:

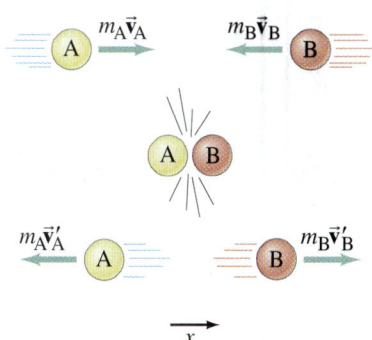

FIGURE 7–3 Momentum is conserved in a collision of two balls, labeled A and B.

momentum before = momentum after

$$m_A \vec{v}_A + m_B \vec{v}_B = m_A \vec{v}'_A + m_B \vec{v}'_B. \qquad [\Sigma \vec{F}_{ext} = 0] \quad (7\text{–}3)$$

CONSERVATION OF MOMENTUM
(two objects colliding)

That is, the total vector momentum of the system of two colliding balls is conserved: it stays constant. (We saw this result in this Chapter's opening photograph.)

Although the law of conservation of momentum was discovered experimentally, it can be derived from Newton's laws of motion, which we now show.

Let us consider two objects of mass m_A and m_B that have momenta $\vec{p}_A$ ($= m_A \vec{v}_A$) and $\vec{p}_B$ ($= m_B \vec{v}_B$) before they collide and $\vec{p}'_A$ and $\vec{p}'_B$ after they collide, as in Fig. 7–4. During the collision, suppose that the force exerted by object A on object B at any instant is $\vec{F}$. Then, by Newton's third law, the force exerted by object B on object A is $-\vec{F}$. During the brief collision time, we assume no other (external) forces are acting (or that $\vec{F}$ is much greater than any other external forces acting). Over a very short time interval Δt we have

$$\vec{F} = \frac{\Delta \vec{p}_B}{\Delta t} = \frac{\vec{p}'_B - \vec{p}_B}{\Delta t}$$

and

$$-\vec{F} = \frac{\Delta \vec{p}_A}{\Delta t} = \frac{\vec{p}'_A - \vec{p}_A}{\Delta t}.$$

We add these two equations together and find

$$0 = \frac{\Delta \vec{p}_B + \Delta \vec{p}_A}{\Delta t} = \frac{(\vec{p}'_B - \vec{p}_B) + (\vec{p}'_A - \vec{p}_A)}{\Delta t}.$$

This means

$$\vec{p}'_B - \vec{p}_B + \vec{p}'_A - \vec{p}_A = 0,$$

or

$$\vec{p}'_A + \vec{p}'_B = \vec{p}_A + \vec{p}_B.$$

This is Eq. 7–3. The total momentum is conserved.

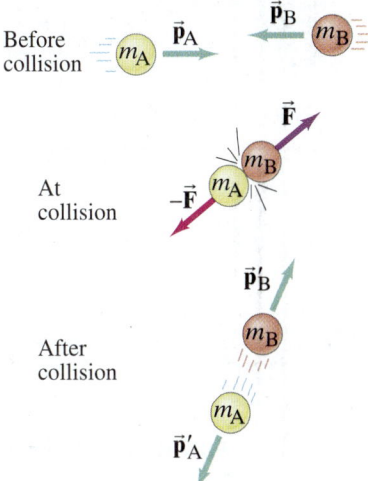

FIGURE 7–4 Collision of two objects. Their momenta before collision are $\vec{p}_A$ and $\vec{p}_B$, and after collision are $\vec{p}'_A$ and $\vec{p}'_B$. At any moment during the collision each exerts a force on the other of equal magnitude but opposite direction.

We have put this derivation in the context of a collision. As long as no external forces act, it is valid over any time interval, and conservation of momentum is always valid as long as no external forces act on the chosen system. In the real world, external forces do act: friction on billiard balls, gravity acting on a tennis ball, and so on. So we often want our "observation time" (before and after) to be small. When a racket hits a tennis ball or a bat hits a baseball, both before and after the "collision" the ball moves as a projectile under the action of gravity and air resistance.

However, when the bat or racket hits the ball, during the brief time of the collision those external forces are insignificant compared to the collision force the bat or racket exerts on the ball. Momentum is conserved (or very nearly so) as long as we measure $\vec{p}_A$ and $\vec{p}_B$ just before the collision and $\vec{p}'_A$ and $\vec{p}'_B$ immediately after the collision (Eq. 7–3). We can not wait for external forces to produce their effect before measuring $\vec{p}'_A$ and $\vec{p}'_B$.

The above derivation can be extended to include any number of interacting objects. To show this, we let $\vec{p}$ in Eq. 7–2 ($\Sigma \vec{F} = \Delta \vec{p}/\Delta t$) represent the total momentum of a system—that is, the vector sum of the momenta of all objects in the system. (For our two-object system above, $\vec{p} = m_A \vec{v}_A + m_B \vec{v}_B$.) If the net force $\Sigma \vec{F}$ on the system is zero [as it was above for our two-object system, $\vec{F} + (-\vec{F}) = 0$], then from Eq. 7–2, $\Delta \vec{p} = \Sigma \vec{F} \Delta t = 0$, so the total momentum doesn't change. The general statement of the **law of conservation of momentum** is

LAW OF CONSERVATION OF MOMENTUM

The total momentum of an isolated system of objects remains constant.

By a **system**, we simply mean a set of objects that we choose, and which may interact with each other. An **isolated system** is one in which the only (significant) forces are those between the objects in the system. The sum of all these "internal" forces within the system will be zero because of Newton's third law. If there are *external forces*—by which we mean forces exerted by objects *outside* the system— and they don't add up to zero, then the total momentum of the system won't be conserved. However, if the system can be redefined so as to include the other objects exerting these forces, then the conservation of momentum principle can apply. For example, if we take as our system a falling rock, it does not conserve momentum because an external force, the force of gravity exerted by the Earth, accelerates the rock and changes its momentum. However, if we include the Earth in the system, the total momentum of rock plus Earth is conserved. (This means that the Earth comes up to meet the rock. But the Earth's mass is so great, its upward velocity is very tiny.)

Although the law of conservation of momentum follows from Newton's second law, as we have seen, it is in fact more general than Newton's laws. In the tiny world of the atom, Newton's laws fail, but the great conservation laws— those of energy, momentum, angular momentum, and electric charge—have been found to hold in every experimental situation tested. It is for this reason that the conservation laws are considered more basic than Newton's laws.

EXAMPLE 7–3 Railroad cars collide: momentum conserved. A 10,000-kg railroad car, A, traveling at a speed of 24.0 m/s strikes an identical car, B, at rest. If the cars lock together as a result of the collision, what is their common speed just afterward? See Fig. 7–5.

APPROACH We choose our system to be the two railroad cars. We consider a very brief time interval, from just before the collision until just after, so that external forces such as friction can be ignored. Then we apply conservation of momentum.

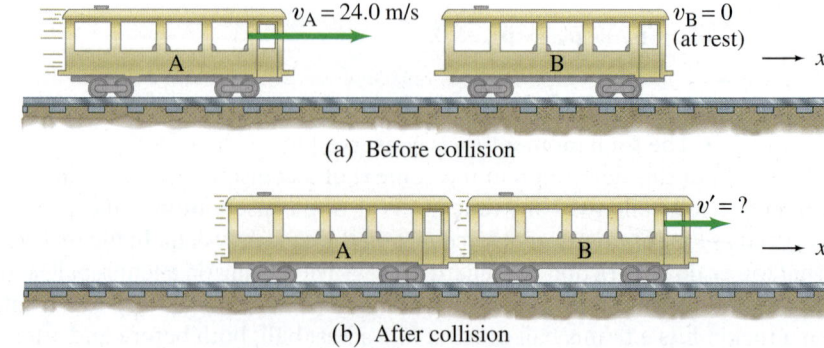

FIGURE 7–5 Example 7–3.

SOLUTION The initial total momentum is

$$p_{\text{initial}} = m_A v_A + m_B v_B = m_A v_A$$

because car B is at rest initially $(v_B = 0)$. The direction is to the right in the $+x$ direction. After the collision, the two cars become attached, so they will have the same speed, call it v'. Then the total momentum after the collision is

$$p_{\text{final}} = (m_A + m_B) v'.$$

We have assumed there are no external forces, so momentum is conserved:

$$p_{\text{initial}} = p_{\text{final}}$$
$$m_A v_A = (m_A + m_B) v'.$$

Solving for v', we obtain

$$v' = \frac{m_A}{m_A + m_B} v_A = \left(\frac{10{,}000 \text{ kg}}{10{,}000 \text{ kg} + 10{,}000 \text{ kg}}\right)(24.0 \text{ m/s}) = 12.0 \text{ m/s},$$

to the right. Their mutual speed after collision is half the initial speed of car A.

NOTE We kept symbols until the very end, so we have an equation we can use in other (related) situations.

NOTE We haven't included friction here. Why? Because we are examining speeds just before and just after the very brief time interval of the collision, and during that brief time friction can't do much—it is ignorable (but not for long: the cars will slow down because of friction).

EXERCISE C In Example 7–3, $m_A = m_B$, so in the last equation, $m_A/(m_A + m_B) = \frac{1}{2}$. Hence $v' = \frac{1}{2} v_A$. What result do you get if (a) $m_B = 3 m_A$, (b) m_B is much larger than m_A $(m_B \gg m_A)$, and (c) $m_B \ll m_A$?

EXERCISE D A 50-kg child runs off a dock at 2.0 m/s (horizontally) and lands in a waiting rowboat of mass 150 kg. At what speed does the rowboat move away from the dock?

The law of conservation of momentum is particularly useful when we are dealing with fairly simple systems such as colliding objects and certain types of "explosions." For example, *rocket propulsion*, which we saw in Chapter 4 can be understood on the basis of action and reaction, can also be explained on the basis of the conservation of momentum. We can consider the rocket plus its fuel as an isolated system if it is far out in space (no external forces). In the reference frame of the rocket before any fuel is ejected, the total momentum of rocket plus fuel is zero. When the fuel burns, the total momentum remains unchanged: the backward momentum of the expelled gases is just balanced by the forward momentum gained by the rocket itself (see Fig. 7–6). Thus, a rocket can accelerate in empty space. There is no need for the expelled gases to push against the Earth or the air (as is sometimes erroneously thought). Similar examples of (nearly) isolated systems where momentum is conserved are the recoil of a gun when a bullet is fired (Example 7–5), and the movement of a rowboat just after a package is thrown from it.

CONCEPTUAL EXAMPLE 7–4 Falling on or off a sled. (a) An empty sled is sliding on frictionless ice when Susan drops vertically from a tree down onto the sled. When she lands, does the sled speed up, slow down, or keep the same speed? (b) Later: Susan falls sideways off the sled. When she drops off, does the sled speed up, slow down, or keep the same speed?

RESPONSE (a) Because Susan falls vertically onto the sled, she has no initial horizontal momentum. Thus the total horizontal momentum afterward equals the momentum of the sled initially. Since the mass of the system (sled + person) has increased, the speed must decrease.

(b) At the instant Susan falls off, she is moving with the same horizontal speed as she was while on the sled. At the moment she leaves the sled, she has the same momentum she had an instant before. Because her momentum does not change, neither does the sled's (total momentum conserved); the sled keeps the same speed.

PHYSICS APPLIED
Rocket propulsion

CAUTION
A rocket does not push on the Earth; it is propelled by pushing out the gases it burned as fuel

FIGURE 7–6 (a) A rocket, containing fuel, at rest in some reference frame. (b) In the same reference frame, the rocket fires and gases are expelled at high speed out the rear. The total vector momentum, $\vec{p}_{\text{gas}} + \vec{p}_{\text{rocket}}$, remains zero.

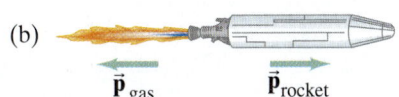

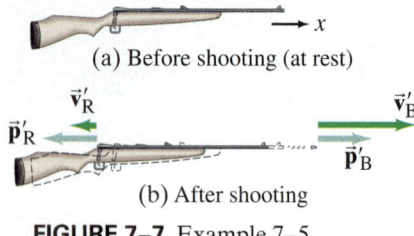

(a) Before shooting (at rest)

(b) After shooting

FIGURE 7-7 Example 7-5.

EXAMPLE 7-5 **Rifle recoil.** Calculate the recoil velocity of a 5.0-kg rifle that shoots a 0.020-kg bullet at a speed of 620 m/s, Fig. 7-7.

APPROACH Our system is the rifle and the bullet, both at rest initially, just before the trigger is pulled. The trigger is pulled, an explosion occurs inside the bullet's shell, and we look at the rifle and bullet just as the bullet leaves the barrel (Fig. 7-7b). The bullet moves to the right ($+x$), and the gun recoils to the left. During the very short time interval of the explosion, we can assume the external forces are small compared to the forces exerted by the exploding gunpowder. Thus we can apply conservation of momentum, at least approximately.

SOLUTION Let subscript B represent the bullet and R the rifle; the final velocities are indicated by primes. Then momentum conservation in the x direction gives

$$\text{momentum before} = \text{momentum after}$$
$$m_B v_B + m_R v_R = m_B v'_B + m_R v'_R$$
$$0 + 0 = m_B v'_B + m_R v'_R.$$

We solve for the unknown v'_R, and find

$$v'_R = -\frac{m_B v'_B}{m_R} = -\frac{(0.020 \text{ kg})(620 \text{ m/s})}{(5.0 \text{ kg})} = -2.5 \text{ m/s}.$$

Since the rifle has a much larger mass, its (recoil) velocity is much less than that of the bullet. The minus sign indicates that the velocity (and momentum) of the rifle is in the negative x direction, opposite to that of the bullet.

EXERCISE E Return to the Chapter-Opening Questions, page 170, and answer them again now. Try to explain why you may have answered differently the first time.

7-3 Collisions and Impulse

Collisions are a common occurrence in everyday life: a tennis racket or a baseball bat striking a ball, billiard balls colliding, a hammer hitting a nail. When a collision occurs, the interaction between the objects involved is usually far stronger than any external forces. We can then ignore the effects of any other forces during the brief time interval of the collision.

During a collision of two ordinary objects, both objects are deformed, often considerably, because of the large forces involved (Fig. 7-8). When the collision occurs, the force each exerts on the other usually jumps from zero at the moment of contact to a very large force within a very short time, and then rapidly returns to zero again. A graph of the magnitude of the force that one object exerts on the other during a collision, as a function of time, is something like the red curve in Fig. 7-9. The time interval Δt is usually very distinct and very small, typically milliseconds for a macroscopic collision.

FIGURE 7-8 Tennis racket striking a ball. Both the ball and the racket strings are deformed due to the large force each exerts on the other.

FIGURE 7-9 Force as a function of time during a typical collision. F can become very large; Δt is typically milliseconds for macroscopic collisions.

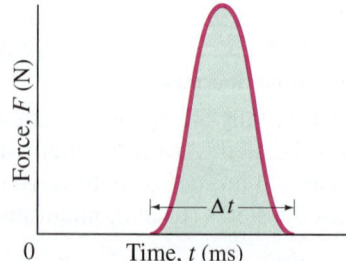

176 CHAPTER 7 Linear Momentum

From Newton's second law, Eq. 7–2, the *net* force on an object is equal to the rate of change of its momentum:

$$\vec{F} = \frac{\Delta \vec{p}}{\Delta t}.$$

(We have written $\vec{F}$ instead of $\Sigma\vec{F}$ for the net force, which we assume is entirely due to the brief but large average force that acts during the collision.) This equation applies to *each* of the two objects in a collision. We multiply both sides of this equation by the time interval Δt, and obtain

$$\vec{F}\,\Delta t = \Delta \vec{p}. \qquad (7\text{–}4)$$

The quantity on the left, the product of the force $\vec{F}$ times the time Δt over which the force acts, is called the **impulse**:

$$\text{Impulse} = \vec{F}\,\Delta t. \qquad (7\text{–}5)$$

We see that the total change in momentum is equal to the impulse. The concept of impulse is useful mainly when dealing with forces that act during a short time interval, as when a bat hits a baseball. The force is generally not constant, and often its variation in time is like that graphed in Figs. 7–9 and 7–10. We can often approximate such a varying force as an average force $\bar{F}$ acting during a time interval Δt, as indicated by the dashed line in Fig. 7–10. $\bar{F}$ is chosen so that the area shown shaded in Fig. 7–10 (equal to $\bar{F} \times \Delta t$) is equal to the area under the actual curve of F vs. t, Fig. 7–9 (which represents the actual impulse).

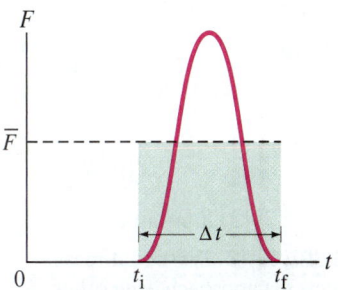

FIGURE 7–10 The average force $\bar{F}$ acting over a very brief time interval Δt gives the same impulse ($\bar{F}\,\Delta t$) as the actual force.

EXERCISE F Suppose Fig. 7–9 shows the force on a golf ball vs. time during the time interval when the ball hits a wall. How would the shape of this curve change if a softer rubber ball with the same mass and speed hit the same wall?

EXAMPLE 7–6 ESTIMATE Karate blow. Estimate the impulse and the average force delivered by a karate blow that breaks a board (Fig. 7–11). Assume the hand moves at roughly 10 m/s when it hits the board.

APPROACH We use the momentum-impulse relation, Eq. 7–4. The hand's speed changes from 10 m/s to zero over a distance of perhaps one cm (roughly how much your hand and the board compress before your hand comes to a stop, and the board begins to give way). The hand's mass should probably include part of the arm, and we take it to be roughly $m \approx 1$ kg.

SOLUTION The impulse $F\,\Delta t$ equals the change in momentum

$$\bar{F}\,\Delta t = \Delta p = m\,\Delta v \approx (1\text{ kg})(10\text{ m/s} - 0) = 10\text{ kg}\cdot\text{m/s}.$$

FIGURE 7–11 Example 7–6.

We can obtain the force if we know Δt. The hand is brought to rest over the distance of roughly a centimeter: $\Delta x \approx 1$ cm. The average speed during the impact is $\bar{v} = (10\text{ m/s} + 0)/2 = 5\text{ m/s}$ and equals $\Delta x/\Delta t$. Thus $\Delta t = \Delta x/\bar{v} \approx (10^{-2}\text{ m})/(5\text{ m/s}) = 2 \times 10^{-3}$ s or 2 ms. The average force is thus (Eq. 7–4) about

$$\bar{F} = \frac{\Delta p}{\Delta t} = \frac{10\text{ kg}\cdot\text{m/s}}{2 \times 10^{-3}\text{ s}} \approx 5000\text{ N} = 5\text{ kN}.$$

7–4 Conservation of Energy and Momentum in Collisions

During most collisions, we usually don't know how the collision force varies over time, and so analysis using Newton's second law becomes difficult or impossible. But by making use of the conservation laws for momentum and energy, we can still determine a lot about the motion after a collision, given the motion before the collision. We saw in Section 7–2 that in the collision of two objects such as billiard balls, the total momentum is conserved. If the two objects are very hard and no heat or other energy is produced in the collision, then the total kinetic energy of the two objects is the same after the collision as before. For the brief moment during which the two objects are in contact, some (or all) of the energy is stored momentarily in the form of elastic potential energy.

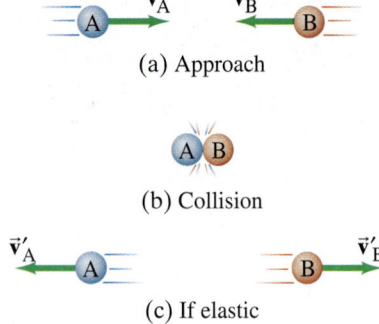

FIGURE 7–12 Two equal-mass objects (a) approach each other with equal speeds, (b) collide, and then (c) bounce off with equal speeds in the opposite directions if the collision is elastic, or (d) bounce back much less or not at all if the collision is inelastic (some of the KE is transformed to other forms of energy such as sound and heat).

But if we compare the total kinetic energy just before the collision with the total kinetic energy just after the collision, and they are found to be the same, then we say that the total kinetic energy is conserved. Such a collision is called an **elastic collision**. If we use the subscripts A and B to represent the two objects, we can write the equation for conservation of total kinetic energy as

total KE before = total KE after

$$\tfrac{1}{2}m_A v_A^2 + \tfrac{1}{2}m_B v_B^2 = \tfrac{1}{2}m_A v_A'^2 + \tfrac{1}{2}m_B v_B'^2. \qquad \text{[elastic collision]} \quad \textbf{(7–6)}$$

Primed quantities (') mean after the collision, and unprimed mean before the collision, just as in Eq. 7–3 for conservation of momentum.

At the atomic level the collisions of atoms and molecules are often elastic. But in the "macroscopic" world of ordinary objects, an elastic collision is an ideal that is never quite reached, since at least a little thermal energy is always produced during a collision (also perhaps sound and other forms of energy). The collision of two hard elastic balls, such as billiard balls, however, is very close to being perfectly elastic, and we often treat it as such.

We do need to remember that even when kinetic energy is not conserved, the *total* energy is always conserved.

Collisions in which kinetic energy is not conserved are said to be **inelastic collisions**. The kinetic energy that is lost is changed into other forms of energy, often thermal energy, so that the total energy (as always) is conserved. In this case,

$$KE_A + KE_B = KE_A' + KE_B' + \text{thermal and other forms of energy}.$$

See Fig. 7–12, and the details in its caption.

7–5 Elastic Collisions in One Dimension

We now apply the conservation laws for momentum and kinetic energy to an elastic collision between two small objects that collide head-on, so all the motion is along a line. To be general, we assume that the two objects are moving, and their velocities are v_A and v_B along the x axis before the collision, Fig. 7–13a. After the collision, their velocities are v_A' and v_B', Fig. 7–13b. For any $v > 0$, the object is moving to the right (increasing x), whereas for $v < 0$, the object is moving to the left (toward decreasing values of x).

From conservation of momentum, we have

$$m_A v_A + m_B v_B = m_A v_A' + m_B v_B'.$$

Because the collision is assumed to be elastic, kinetic energy is also conserved:

$$\tfrac{1}{2}m_A v_A^2 + \tfrac{1}{2}m_B v_B^2 = \tfrac{1}{2}m_A v_A'^2 + \tfrac{1}{2}m_B v_B'^2.$$

We have two equations, so we can solve for two unknowns. If we know the masses and velocities before the collision, then we can solve these two equations for the velocities after the collision, v_A' and v_B'. We derive a helpful result by rewriting the momentum equation as

$$m_A(v_A - v_A') = m_B(v_B' - v_B), \qquad \textbf{(i)}$$

and we rewrite the kinetic energy equation as

$$m_A(v_A^2 - v_A'^2) = m_B(v_B'^2 - v_B^2).$$

Noting that algebraically $(a^2 - b^2) = (a - b)(a + b)$, we write this last equation as

$$m_A(v_A - v_A')(v_A + v_A') = m_B(v_B' - v_B)(v_B' + v_B). \qquad \textbf{(ii)}$$

We divide Eq. (ii) by Eq. (i), and (assuming $v_A \neq v_A'$ and $v_B \neq v_B'$)[†] obtain

$$v_A + v_A' = v_B' + v_B.$$

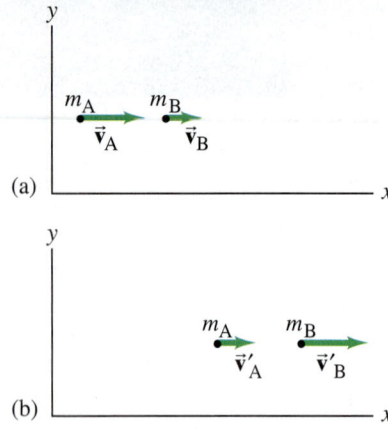

FIGURE 7–13 Two small objects of masses m_A and m_B, (a) before the collision and (b) after the collision.

[†]Note that Eqs. (i) and (ii), which are the conservation laws for momentum and kinetic energy, are both satisfied by the solution $v_A' = v_A$ and $v_B' = v_B$. This is a valid solution, but not very interesting. It corresponds to no collision at all—when the two objects miss each other.

We can rewrite this equation as

$$v_A - v_B = v'_B - v'_A$$

or

$$v_A - v_B = -(v'_A - v'_B). \quad \text{[head-on (1-D) elastic collision]} \quad (7\text{--}7)$$

> **CAUTION**
> *Relative speeds (one dimension only)*

This is an interesting result: it tells us that for any elastic head-on collision, the relative speed of the two objects after the collision $(v'_A - v'_B)$ has the same magnitude (but opposite direction) as before the collision, no matter what the masses are.

Equation 7–7 was derived from conservation of kinetic energy for elastic collisions, and can be used in place of it. Because the v's are not squared in Eq. 7–7, it is simpler to use in calculations than the conservation of kinetic energy equation (Eq. 7–6) directly.

EXAMPLE 7–7 **Equal masses.** Billiard ball A of mass m moving with speed v_A collides head-on with ball B of equal mass. What are the speeds of the two balls after the collision, assuming it is elastic? Assume (a) both balls are moving initially (v_A and v_B), (b) ball B is initially at rest ($v_B = 0$).

APPROACH There are two unknowns, v'_A and v'_B, so we need two independent equations. We focus on the time interval from just before the collision until just after. No net external force acts on our system of two balls (mg and the normal force cancel), so momentum is conserved. Conservation of kinetic energy applies as well because we are told the collision is elastic.

SOLUTION (a) The masses are equal $(m_A = m_B = m)$ so conservation of momentum gives

$$v_A + v_B = v'_A + v'_B.$$

We need a second equation, because there are two unknowns. We could use the conservation of kinetic energy equation, or the simpler Eq. 7–7 derived from it:

$$v_A - v_B = v'_B - v'_A.$$

We add these two equations and obtain

$$v'_B = v_A$$

and then subtract the two equations to obtain

$$v'_A = v_B.$$

That is, the balls exchange velocities as a result of the collision: ball B acquires the velocity that ball A had before the collision, and vice versa.

(b) If ball B is at rest initially, so that $v_B = 0$, we have

$$v'_B = v_A$$

and

$$v'_A = 0.$$

That is, ball A is brought to rest by the collision, whereas ball B acquires the original velocity of ball A. See Fig. 7–14.

NOTE Our result in part (b) is often observed by billiard and pool players, and is valid only if the two balls have equal masses (and no spin is given to the balls).

FIGURE 7–14 In this multiflash photo of a head-on collision between two balls of equal mass, the white cue ball is accelerated from rest by the cue stick and then strikes the red ball, initially at rest. The white ball stops in its tracks, and the (equal-mass) red ball moves off with the same speed as the white ball had before the collision. See Example 7–7, part (b).

SECTION 7–5 Elastic Collisions in One Dimension

EXAMPLE 7-8 **A nuclear collision.** A proton (p) of mass 1.01 u (unified atomic mass units) traveling with a speed of 3.60×10^4 m/s has an elastic head-on collision with a helium (He) nucleus $(m_{He} = 4.00 \text{ u})$ initially at rest. What are the velocities of the proton and helium nucleus after the collision? (As mentioned in Chapter 1, 1 u = 1.66×10^{-27} kg, but we won't need this fact.) Assume the collision takes place in nearly empty space.

APPROACH Like Example 7-7, this is an elastic head-on collision, but now the masses of our two particles are not equal. The only external force could be Earth's gravity, but it is insignificant compared to the powerful forces between the two particles at the moment of collision. So again we use the conservation laws of momentum and of kinetic energy, and apply them to our system of two particles.

SOLUTION We use the subscripts p for the proton and He for the helium nucleus. We are given $v_{He} = 0$ and $v_p = 3.60 \times 10^4$ m/s. We want to find the velocities v'_p and v'_{He} after the collision. From conservation of momentum,

$$m_p v_p + 0 = m_p v'_p + m_{He} v'_{He}.$$

Because the collision is elastic, the kinetic energy of our system of two particles is conserved and we can use Eq. 7-7, which becomes

$$v_p - 0 = v'_{He} - v'_p.$$

Thus

$$v'_p = v'_{He} - v_p,$$

and substituting this into our momentum equation displayed above, we get

$$m_p v_p = m_p v'_{He} - m_p v_p + m_{He} v'_{He}.$$

Solving for v'_{He}, we obtain

$$v'_{He} = \frac{2 m_p v_p}{m_p + m_{He}} = \frac{2(1.01 \text{ u})(3.60 \times 10^4 \text{ m/s})}{(4.00 \text{ u} + 1.01 \text{ u})} = 1.45 \times 10^4 \text{ m/s}.$$

The other unknown is v'_p, which we can now obtain from

$$v'_p = v'_{He} - v_p = (1.45 \times 10^4 \text{ m/s}) - (3.60 \times 10^4 \text{ m/s}) = -2.15 \times 10^4 \text{ m/s}.$$

The minus sign for v'_p tells us that the proton reverses direction upon collision, and we see that its speed is less than its initial speed (see Fig. 7-15).

NOTE This result makes sense: the lighter proton would be expected to "bounce back" from the more massive helium nucleus, but not with its full original velocity as from a rigid wall (which corresponds to extremely large, or infinite, mass).

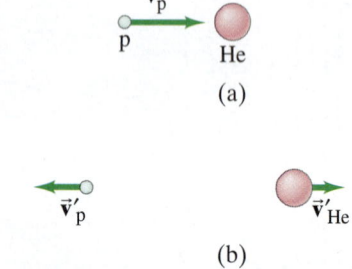

FIGURE 7-15 Example 7-8: (a) before collision, (b) after collision.

7-6 Inelastic Collisions

Collisions in which kinetic energy is not conserved are called **inelastic collisions**. Some of the initial kinetic energy is transformed into other types of energy, such as thermal or potential energy, so the total kinetic energy after the collision is less than the total kinetic energy before the collision. The inverse can also happen when potential energy (such as chemical or nuclear) is released, in which case the total kinetic energy after the interaction can be greater than the initial kinetic energy. Explosions are examples of this type.

Typical macroscopic collisions are inelastic, at least to some extent, and often to a large extent. If two objects stick together as a result of a collision, the collision is said to be **completely inelastic**. Two colliding balls of putty that stick together or two railroad cars that couple together when they collide are examples of completely inelastic collisions. The kinetic energy in some cases is all transformed to other forms of energy in an inelastic collision, but in other cases only part of it is. In Example 7-3, for instance, we saw that when a traveling railroad car collided with a stationary one, the coupled cars traveled off with some kinetic energy. In a completely inelastic collision, the maximum amount of kinetic energy is transformed to other forms consistent with conservation of momentum. Even though kinetic energy is not conserved in inelastic collisions, the total energy is always conserved, and the total vector momentum is also conserved.

EXAMPLE 7-9 **Ballistic pendulum.** The *ballistic pendulum* is a device used to measure the speed of a projectile, such as a bullet. The projectile, of mass m, is fired into a large block (of wood or other material) of mass M, which is suspended like a pendulum. (Usually, M is somewhat greater than m.) As a result of the collision, the pendulum and projectile together swing up to a maximum height h, Fig. 7–16. Determine the relationship between the initial horizontal speed of the projectile, v, and the maximum height h.

PHYSICS APPLIED
Ballistic pendulum

APPROACH We can analyze the process by dividing it into two parts or two time intervals: (1) the time interval from just before to just after the collision itself, and (2) the subsequent time interval in which the pendulum moves from the vertical hanging position to the maximum height h.

In part (1), Fig. 7–16a, we assume the collision time is very short, so that the projectile is embedded in the block before the block has moved significantly from its rest position directly below its support. Thus there is effectively no net external force, and we can apply conservation of momentum to this completely inelastic collision.

In part (2), Fig. 7–16b, the pendulum begins to move, subject to a net external force (gravity, tending to pull it back to the vertical position); so for part (2), we cannot use conservation of momentum. But we can use conservation of mechanical energy because gravity is a conservative force (Chapter 6). The kinetic energy immediately after the collision is changed entirely to gravitational potential energy when the pendulum reaches its maximum height, h.

SOLUTION In part (1) momentum is conserved:

$$\text{total } p \text{ before} = \text{total } p \text{ after}$$
$$mv = (m + M)v', \quad \text{(i)}$$

where v' is the speed of the block and embedded projectile just after the collision, before they have moved significantly.

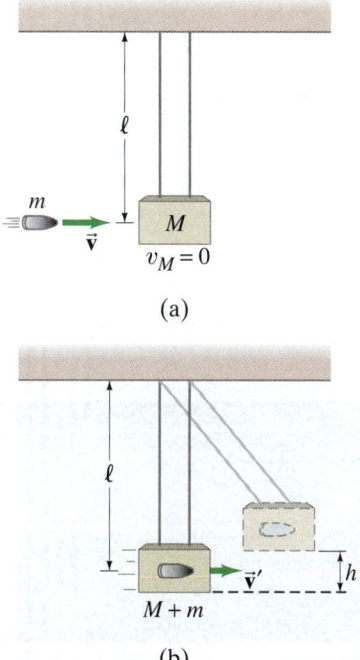

FIGURE 7–16 Ballistic pendulum. Example 7–9.

In part (2), mechanical energy is conserved. We choose $y = 0$ when the pendulum hangs vertically, and then $y = h$ when the pendulum–projectile system reaches its maximum height. Thus we write

(KE + PE) just after collision = (KE + PE) at pendulum's maximum height

or

$$\tfrac{1}{2}(m + M)v'^2 + 0 = 0 + (m + M)gh. \quad \text{(ii)}$$

We solve for v':

$$v' = \sqrt{2gh}.$$

Inserting this result for v' into Eq. (i) above, and solving for v, gives

$$v = \frac{m + M}{m} v' = \frac{m + M}{m} \sqrt{2gh},$$

which is our final result.

NOTE The separation of the process into two parts was crucial. Such an analysis is a powerful problem-solving tool. But how do you decide how to make such a division? Think about the conservation laws. They are your *tools*. Start a problem by asking yourself whether the conservation laws apply in the given situation. Here, we determined that momentum is conserved only during the brief collision, which we called part (1). But in part (1), because the collision is inelastic, the conservation of mechanical energy is not valid. Then in part (2), conservation of mechanical energy is valid, but not conservation of momentum.

Note, however, that if there had been significant motion of the pendulum during the deceleration of the projectile in the block, then there *would* have been an external force (gravity) during the collision, so conservation of momentum would not have been valid in part (1).

PROBLEM SOLVING
Use the conservation laws to analyze a problem

EXAMPLE 7–10 Railroad cars again. For the completely inelastic collision of the two railroad cars that we considered in Example 7–3, calculate how much of the initial kinetic energy is transformed to thermal or other forms of energy.

APPROACH The railroad cars stick together after the collision, so this is a completely inelastic collision. By subtracting the total kinetic energy after the collision from the total initial kinetic energy, we can find how much energy is transformed to other types of energy.

SOLUTION Before the collision, only car A is moving, so the total initial kinetic energy is $\frac{1}{2} m_A v_A^2 = \frac{1}{2}(10{,}000 \text{ kg})(24.0 \text{ m/s})^2 = 2.88 \times 10^6 \text{ J}$. After the collision, both cars are moving with half the speed, $v' = 12.0$ m/s, by conservation of momentum (Example 7–3). So the total kinetic energy afterward is $\text{KE}' = \frac{1}{2}(m_A + m_B)v'^2 = \frac{1}{2}(20{,}000 \text{ kg})(12.0 \text{ m/s})^2 = 1.44 \times 10^6 \text{ J}$. Hence the energy transformed to other forms is

$$(2.88 \times 10^6 \text{ J}) - (1.44 \times 10^6 \text{ J}) = 1.44 \times 10^6 \text{ J},$$

which is half the original kinetic energy.

*7–7 Collisions in Two Dimensions

Conservation of momentum and energy can also be applied to collisions in two or three dimensions, where the vector nature of momentum is especially important. One common type of non-head-on collision is that in which a moving object (called the "projectile") strikes a second object initially at rest (the "target"). This is the common situation in games such as billiards and pool, and for experiments in atomic and nuclear physics (the projectiles, from radioactive decay or a high-energy accelerator, strike a stationary target nucleus, Fig. 7–17).

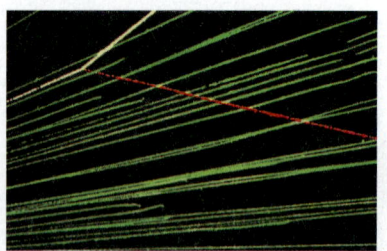

FIGURE 7–17 A recent color-enhanced version of a cloud-chamber photograph made in the early days (1920s) of nuclear physics. Green lines are paths of helium nuclei (He) coming from the left. One He, highlighted in yellow, strikes a proton of the hydrogen gas in the chamber, and both scatter at an angle; the scattered proton's path is shown in red.

Figure 7–18 shows the incoming projectile, m_A, heading along the x axis toward the target object, m_B, which is initially at rest. If these are billiard balls, m_A strikes m_B not quite head-on and they go off at the angles θ'_A and θ'_B, respectively, which are measured relative to m_A's initial direction (the x axis).[†]

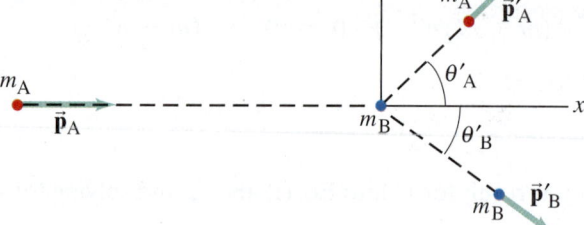

FIGURE 7–18 Object A, the projectile, collides with object B, the target. After the collision, they move off with momenta $\vec{p}'_A$ and $\vec{p}'_B$ at angles θ'_A and θ'_B.

Let us apply the law of conservation of momentum to a collision like that of Fig. 7–18. We choose the xy plane to be the plane in which the initial and final momenta lie. Momentum is a vector, and because the total momentum is conserved, its components in the x and y directions also are conserved. The x component of momentum conservation gives

$$p_{Ax} + p_{Bx} = p'_{Ax} + p'_{Bx}$$

or, with $p_{Bx} = m_B v_{Bx} = 0$,

$$m_A v_A = m_A v'_A \cos\theta'_A + m_B v'_B \cos\theta'_B, \qquad (7\text{–}8\text{a})$$

where primes (') refer to quantities *after* the collision. There is no motion in the y direction initially, so the y component of the total momentum is zero before the collision.

[†]The objects may begin to deflect even before they touch if electric, magnetic, or nuclear forces act between them. You might think, for example, of two magnets oriented so that they repel each other: when one moves toward the other, the second moves away before the first one touches it.

The y component equation of momentum conservation is then

$$p_{Ay} + p_{By} = p'_{Ay} + p'_{By}$$

or

$$0 = m_A v'_A \sin\theta'_A + m_B v'_B \sin\theta'_B. \qquad (7\text{-}8b)$$

When we have two independent equations, we can solve for two unknowns at most.

EXAMPLE 7–11 Billiard ball collision in 2-D. Billiard ball A moving with speed $v_A = 3.0$ m/s in the $+x$ direction (Fig. 7–19) strikes an equal-mass ball B initially at rest. The two balls are observed to move off at 45° to the x axis, ball A above the x axis and ball B below. That is, $\theta'_A = 45°$ and $\theta'_B = -45°$ in Fig. 7–19. What are the speeds of the two balls after the collision?

APPROACH There is no net external force on our system of two balls, assuming the table is level (the normal force balances gravity). Thus momentum conservation applies, and we apply it to both the x and y components using the xy coordinate system shown in Fig. 7–19. We get two equations, and we have two unknowns, v'_A and v'_B. From symmetry we might guess that the two balls have the same speed. But let us not assume that now. Even though we are not told whether the collision is elastic or inelastic, we can still use conservation of momentum.

SOLUTION We apply conservation of momentum for the x and y components, Eqs. 7–8a and b, and we solve for v'_A and v'_B. We are given $m_A = m_B (= m)$, so

(for x) $\qquad mv_A = mv'_A \cos(45°) + mv'_B \cos(-45°)$

and

(for y) $\qquad 0 = mv'_A \sin(45°) + mv'_B \sin(-45°).$

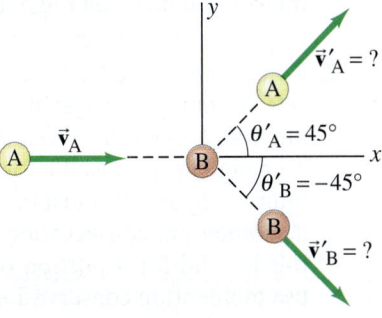

FIGURE 7–19 Example 7–11.

The m's cancel out in both equations (the masses are equal). The second equation yields [recall from trigonometry that $\sin(-\theta) = -\sin\theta$]:

$$v'_B = -v'_A \frac{\sin(45°)}{\sin(-45°)} = -v'_A \left(\frac{\sin 45°}{-\sin 45°}\right) = v'_A.$$

So they do have equal speeds as we guessed at first. The x component equation gives [recall that $\cos(-\theta) = \cos\theta$]:

$$v_A = v'_A \cos(45°) + v'_B \cos(45°) = 2v'_A \cos(45°);$$

solving for v'_A (which also equals v'_B) gives

$$v'_A = \frac{v_A}{2\cos(45°)} = \frac{3.0 \text{ m/s}}{2(0.707)} = 2.1 \text{ m/s}.$$

If we know that a collision is elastic, we can also apply conservation of kinetic energy and obtain a third equation in addition to Eqs. 7–8a and b:

$$KE_A + KE_B = KE'_A + KE'_B$$

or, for the collision shown in Fig. 7–18 or 7–19 (where $KE_B = 0$),

$$\tfrac{1}{2}m_A v_A^2 = \tfrac{1}{2}m_A v'^2_A + \tfrac{1}{2}m_B v'^2_B. \qquad \text{[elastic collision]} \quad (7\text{-}8c)$$

If the collision is elastic, we have three independent equations and can solve for three unknowns. If we are given m_A, m_B, v_A (and v_B, if it is not zero), we cannot, for example, predict the final variables, v'_A, v'_B, θ'_A, and θ'_B, because there are four of them. However, if we measure one of these variables, say θ'_A, then the other three variables (v'_A, v'_B, and θ'_B) are uniquely determined, and we can determine them using Eqs. 7–8a, b, c.

A note of caution: Eq. 7–7 (page 179) does *not* apply for two-dimensional collisions. It works only when a collision occurs along a line.

⚠ **CAUTION**
Equation 7–7 applies only in 1-D

PROBLEM SOLVING

Momentum Conservation and Collisions

1. Choose your **system**. If the situation is complex, think about how you might break it up into separate **parts** when one or more conservation laws apply.
2. Consider whether a significant **net external force** acts on your chosen system; if it does, be sure the time interval Δt is so short that the effect on momentum is negligible. That is, the forces that act between the interacting objects must be the only significant ones if momentum conservation is to be used. [Note: If this is valid for a portion of the problem, you can use momentum conservation only for that portion.]
3. Draw a **diagram** of the initial situation, just before the interaction (collision, explosion) takes place, and represent the momentum of each object with an arrow and a label. Do the same for the final situation, just after the interaction.
4. Choose a **coordinate system** and "+" and "−" directions. (For a head-on collision, you will need only an x axis.) It is often convenient to choose the $+x$ axis in the direction of one object's initial velocity.
5. Apply the **momentum conservation** equation(s):

    total initial momentum = total final momentum.

    You have one equation for each component (x, y, z): only one equation for a head-on collision. [Don't forget that it is the *total* momentum of the system that is conserved, not the momenta of individual objects.]
6. If the collision is elastic, you can also write down a **conservation of kinetic energy** equation:

    total initial KE = total final KE.

    [Alternatively, you could use Eq. 7–7:

    $$v_A - v_B = v'_B - v'_A,$$

    if the collision is one dimensional (head-on).]
7. Solve for the **unknown(s)**.
8. **Check** your work, check the units, and ask yourself whether the results are reasonable.

(a)

(b)

7–8 Center of Mass (CM)

Momentum is a powerful concept not only for analyzing collisions but also for analyzing the translational motion of real extended objects. Until now, whenever we have dealt with the motion of an extended object (that is, an object that has size), we have assumed that it could be approximated as a point particle or that it undergoes only translational motion. Real extended objects, however, can undergo rotational and other types of motion as well. For example, the diver in Fig. 7–20a undergoes only translational motion (all parts of the object follow the same path), whereas the diver in Fig. 7–20b undergoes both translational and rotational motion. We will refer to motion that is not pure translation as *general motion*.

Observations indicate that even if an object rotates, or several parts of a system of objects move relative to one another, there is one point that moves in the same path that a particle would move if subjected to the same net force. This point is called the **center of mass** (abbreviated CM). The general motion of an extended object (or system of objects) can be considered as *the sum of the translational motion of the CM, plus rotational, vibrational, or other types of motion about the CM*.

As an example, consider the motion of the center of mass of the diver in Fig. 7–20; the CM follows a parabolic path even when the diver rotates, as shown in Fig. 7–20b. This is the same parabolic path that a projected particle follows when acted on only by the force of gravity (projectile motion, Chapter 3). Other points in the rotating diver's body, such as her feet or head, follow more complicated paths.

FIGURE 7–20 The motion of the diver is pure translation in (a), but is translation plus rotation in (b). The black dot represents the diver's CM at each moment.

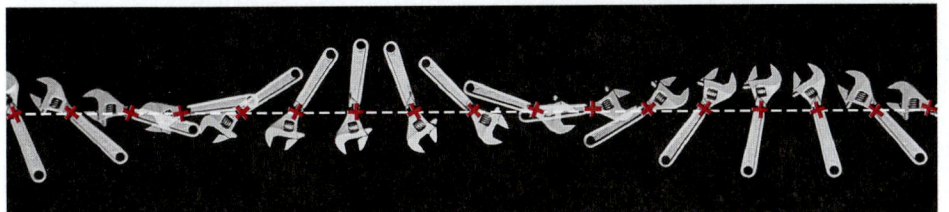

FIGURE 7–21 Translation plus rotation: a wrench moving over a smooth horizontal surface. The CM, marked with a red cross, moves in a straight line because no net force acts on the wrench.

Figure 7–21 shows a wrench acted on by zero net force, translating and rotating along a horizontal surface. Note that its CM, marked by a red cross, moves in a straight line, as shown by the dashed white line.

We will show in Section 7–10 that the important properties of the CM follow from Newton's laws if the CM is defined in the following way. We can consider any extended object as being made up of many tiny particles. But first we consider a system made up of only two particles (or small objects), of masses m_A and m_B. We choose a coordinate system so that both particles lie on the x axis at positions x_A and x_B, Fig. 7–22. The center of mass of this system is defined to be at the position x_{CM}, given by

$$x_{CM} = \frac{m_A x_A + m_B x_B}{m_A + m_B} = \frac{m_A x_A + m_B x_B}{M},$$

where $M = m_A + m_B$ is the total mass of the system. The center of mass lies on the line joining m_A and m_B. If the two masses are equal $(m_A = m_B = m)$, then x_{CM} is midway between them, because in this case

$$x_{CM} = \frac{m(x_A + x_B)}{2m} = \frac{(x_A + x_B)}{2}.$$

If one mass is greater than the other, then the CM is closer to the larger mass.

FIGURE 7–22 The center of mass of a two-particle system lies on the line joining the two masses. Here $m_A > m_B$, so the CM is closer to m_A than to m_B.

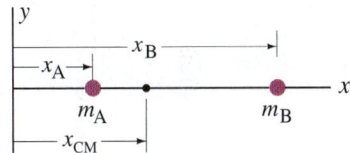

If there are more than two particles along a line, there will be additional terms:

$$x_{CM} = \frac{m_A x_A + m_B x_B + m_C x_C + \cdots}{m_A + m_B + m_C + \cdots} = \frac{m_A x_A + m_B x_B + m_C x_C + \cdots}{M}, \quad (7\text{–}9a)$$

where M is the total mass of all the particles.

EXAMPLE 7–12 CM of three guys on a raft. On a lightweight (air-filled) "banana boat," three people of roughly equal mass m sit along the x axis at positions $x_A = 1.0$ m, $x_B = 5.0$ m, and $x_C = 6.0$ m, measured from the left-hand end as shown in Fig. 7–23. Find the position of the CM. Ignore the mass of the boat.

APPROACH We are given the mass and location of the three people, so we use three terms in Eq. 7–9a. We approximate each person as a point particle. Equivalently, the location of each person is the position of that person's own CM.

SOLUTION We use Eq. 7–9a with three terms:

$$x_{CM} = \frac{mx_A + mx_B + mx_C}{m + m + m} = \frac{m(x_A + x_B + x_C)}{3m}$$
$$= \frac{(1.0\text{ m} + 5.0\text{ m} + 6.0\text{ m})}{3} = \frac{12.0\text{ m}}{3} = 4.0\text{ m}.$$

The CM is 4.0 m from the left-hand end of the boat.

FIGURE 7–23 Example 7–12.

EXERCISE G Calculate the CM of the three people in Example 7–12, taking the origin at the driver $(x_C = 0)$ on the right. Is the physical location of the CM the same?

Note that the coordinates of the CM depend on the reference frame or coordinate system chosen. But the physical location of the CM is independent of that choice.

If the particles are spread out in two or three dimensions, then we must specify not only the x coordinate of the CM (x_{CM}), but also the y and z coordinates, which will be given by formulas like Eq. 7–9a. For example, the y coordinate of the CM will be

$$y_{CM} = \frac{m_A y_A + m_B y_B + \cdots}{m_A + m_B + \cdots} = \frac{m_A y_A + m_B y_B + \cdots}{M} \quad (7\text{–}9b)$$

where M is the total mass of all the particles.

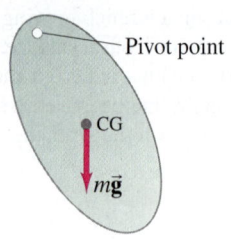

FIGURE 7–24 The force of gravity, considered to act at the CG, causes this object to rotate about the pivot point; if the CG were on a vertical line directly below the pivot, the object would remain at rest.

FIGURE 7–25 Finding the CG.

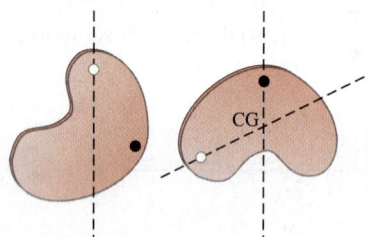

A concept similar to *center of mass* is **center of gravity** (CG). An object's CG is that point at which the force of gravity can be considered to act. The force of gravity actually acts on *all* the different parts or particles of an object, but for purposes of determining the translational motion of an object as a whole, we can assume that the entire weight of the object (which is the sum of the weights of all its parts) acts at the CG. There is a conceptual difference between the center of gravity and the center of mass, but for nearly all practical purposes, they are at the same point.[†]

It is often easier to determine the CM or CG of an extended object experimentally rather than analytically. If an object is suspended from any point, it will swing (Fig. 7–24) due to the force of gravity on it, unless it is placed so its CG lies on a vertical line directly below the point from which it is suspended. If the object is two dimensional, or has a plane of symmetry, it need only be hung from two different pivot points and the respective vertical (plumb) lines drawn. Then the center of gravity will be at the intersection of the two lines, as in Fig. 7–25. If the object doesn't have a plane of symmetry, the CG with respect to the third dimension is found by suspending the object from at least three points whose plumb lines do not lie in the same plane.

For symmetrically shaped objects such as uniform cylinders (wheels), spheres, and rectangular solids, the CM is located at the geometric center of the object.

To locate the center of mass of a group of extended objects, we can use Eqs. 7–9, where the m's are the masses of these objects and the x's, y's, and z's are the coordinates of the CM of each of the objects.

*7–9 CM for the Human Body

For a group of extended objects, each of whose CM is known, we can find the CM of the group using Eqs. 7–9a and b. As an example, we consider the human body. Table 7–1 indicates the CM and hinge points (joints) for the different components of a "representative" person. Of course, there are wide variations among people, so these data represent only a very rough average. The numbers represent a *percentage* of the total height, which is regarded as 100 units; similarly, the total mass is 100 units. For example, if a person is 1.70 m tall, his or her shoulder joint would be $(1.70\,\text{m})(81.2/100) = 1.38\,\text{m}$ above the floor.

TABLE 7–1 Center of Mass of Parts of Typical Human Body, given as %
(full height and mass = 100 units)

Distance of Hinge Points from Floor (%)	Hinge Points (•) (Joints)	Center of Mass (×) (% Height Above Floor)		Percent Mass
91.2%	Base of skull on spine	Head	93.5%	6.9%
81.2%	Shoulder joint	Trunk and neck	71.1%	46.1%
	elbow 62.2%[‡]	Upper arms	71.7%	6.6%
	wrist 46.2%[‡]	Lower arms	55.3%	4.2%
52.1%	Hip joint	Hands	43.1%	1.7%
		Upper legs (thighs)	42.5%	21.5%
28.5%	Knee joint			
		Lower legs	18.2%	9.6%
4.0%	Ankle joint	Feet	1.8%	3.4%
		Body CM =	58.0%	100.0%

[‡] For arm hanging vertically.

[†]There would be a difference between the CM and CG only in the unusual case of an object so large that the acceleration due to gravity, g, was different at different parts of the object.

EXAMPLE 7–13 **A leg's CM.** Determine the position of the CM of a whole leg (a) when stretched out, and (b) when bent at 90°. See Fig. 7–26. Assume the person is 1.70 m tall.

APPROACH Our system consists of three objects: upper leg, lower leg, and foot. The location of the CM of each object, as well as the mass of each, is given in Table 7–1, where they are expressed in percentage units. To express the results in meters, these percentage values need to be multiplied by (1.70 m/100). When the leg is stretched out, the problem is one dimensional and we can solve for the x coordinate of the CM. When the leg is bent, the problem is two dimensional and we need to find both the x and y coordinates.

SOLUTION (a) We determine the distances from the hip joint using Table 7–1 and obtain the numbers (%) shown in Fig. 7–26a. Using Eq. 7–9a, we obtain ($u\ell$ = upper leg, etc.)

$$x_{CM} = \frac{m_{u\ell}x_{u\ell} + m_{\ell\ell}x_{\ell\ell} + m_f x_f}{m_{u\ell} + m_{\ell\ell} + m_f}$$

$$= \frac{(21.5)(9.6) + (9.6)(33.9) + (3.4)(50.3)}{21.5 + 9.6 + 3.4} = 20.4 \text{ units.}$$

Thus, the center of mass of the leg and foot is 20.4 units from the hip joint, or $52.1 - 20.4 = 31.7$ units from the base of the foot. Since the person is 1.70 m tall, this is $(1.70 \text{ m})(31.7/100) = 0.54 \text{ m}$ above the bottom of the foot.

(b) We use an xy coordinate system, as shown in Fig. 7–26b. First, we calculate how far to the right of the hip joint the CM lies, accounting for all three parts:

$$x_{CM} = \frac{(21.5)(9.6) + (9.6)(23.6) + (3.4)(23.6)}{21.5 + 9.6 + 3.4} = 14.9 \text{ units.}$$

For our 1.70-m-tall person, this is $(1.70 \text{ m})(14.9/100) = 0.25 \text{ m}$ from the hip joint. Next, we calculate the distance, y_{CM}, of the CM above the floor:

$$y_{CM} = \frac{(3.4)(1.8) + (9.6)(18.2) + (21.5)(28.5)}{3.4 + 9.6 + 21.5} = 23.0 \text{ units,}$$

or $(1.70 \text{ m})(23.0/100) = 0.39 \text{ m}$. Thus, the CM is located 39 cm above the floor and 25 cm to the right of the hip joint.

NOTE The CM lies outside the body in (b).

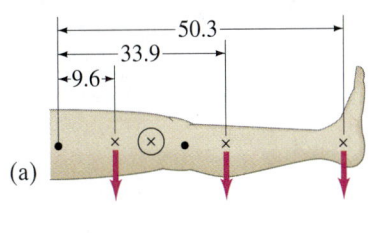
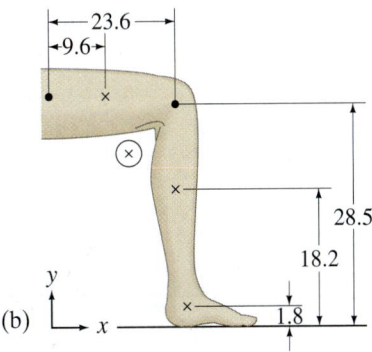

FIGURE 7–26 Example 7–13: finding the CM of a leg in two different positions using percentages from Table 7–1. (⊗ represents the calculated CM.)

Knowing the CM of the body when it is in various positions is of great use in studying body mechanics. One simple example from athletics is shown in Fig. 7–27. If high jumpers can get into the position shown, their CM can pass below the bar which their bodies go over, meaning that for a particular takeoff speed, they can clear a higher bar. This is indeed what they try to do.

FIGURE 7–27 A high jumper's CM may actually pass beneath the bar.

PHYSICS APPLIED
The high jump

*7–10 CM and Translational Motion

As mentioned in Section 7–8, a major reason for the importance of the concept of center of mass is that the motion of the CM for a system of particles (or an extended object) is directly related to the net force acting on the system as a whole. We now show this, taking the simple case of one-dimensional motion (x direction) and only three particles, but the extension to more objects and to three dimensions follows the same reasoning.

Suppose the three particles lie on the x axis and have masses m_A, m_B, m_C, and positions x_A, x_B, x_C. From Eq. 7–9a for the center of mass, we can write

$$Mx_{CM} = m_A x_A + m_B x_B + m_C x_C,$$

where $M = m_A + m_B + m_C$ is the total mass of the system. If these particles are in motion (say, along the x axis with velocities v_A, v_B, and v_C, respectively), then in a short time interval Δt each particle and the CM will have traveled a distance $\Delta x = v\Delta t$, so that

$$Mv_{CM}\Delta t = m_A v_A \Delta t + m_B v_B \Delta t + m_C v_C \Delta t.$$

We cancel Δt and get

$$Mv_{CM} = m_A v_A + m_B v_B + m_C v_C. \qquad (7\text{–}10)$$

Since $m_A v_A + m_B v_B + m_C v_C$ is the sum of the momenta of the particles of the system, it represents the *total momentum* of the system. Thus we see from Eq. 7–10 that *the total (linear) momentum of a system of particles is equal to the product of the total mass M and the velocity of the center of mass of the system.* Or, *the linear momentum of an extended object is the product of the object's mass and the velocity of its CM.*

If forces are acting on the particles, then the particles may be accelerating. In a short time interval Δt, each particle's velocity will change by an amount $\Delta v = a\,\Delta t$. If we use the same reasoning as we did to obtain Eq. 7–10, we find

$$Ma_{CM} = m_A a_A + m_B a_B + m_C a_C.$$

According to Newton's second law, $m_A a_A = F_A$, $m_B a_B = F_B$, and $m_C a_C = F_C$, where F_A, F_B, and F_C are the net forces on the three particles, respectively. Thus we get for the system as a whole $Ma_{CM} = F_A + F_B + F_C$, or

NEWTON'S SECOND LAW *(for a system)*

$$Ma_{CM} = F_{net}. \qquad (7\text{–}11)$$

That is, *the sum of all the forces acting on the system is equal to the total mass of the system times the acceleration of its center of mass.* This is **Newton's second law** for a system of particles. It also applies to an extended object (which can be thought of as a collection of particles). Thus the center of mass of a system of particles (or of an object) with total mass M moves as if all its mass were concentrated at the center of mass and all the external forces acted at that point. We can thus treat the *translational motion* of any object or system of objects as the motion of a particle (see Figs. 7–20 and 7–21). This result simplifies our analysis of the motion of complex systems and extended objects. Although the motion of various parts of the system may be complicated, we may often be satisfied with knowing the motion of the center of mass. This result also allows us to solve certain types of problems very easily, as illustrated by the following Example.

CONCEPTUAL EXAMPLE 7–14 **A two-stage rocket.** A rocket is shot into the air as shown in Fig. 7–28. At the moment the rocket reaches its highest point, a horizontal distance d from its starting point, a prearranged explosion separates it into two parts of equal mass. Part I is stopped in midair by the explosion, and it falls vertically to Earth. Where does part II land? Assume $\vec{g}$ = constant.

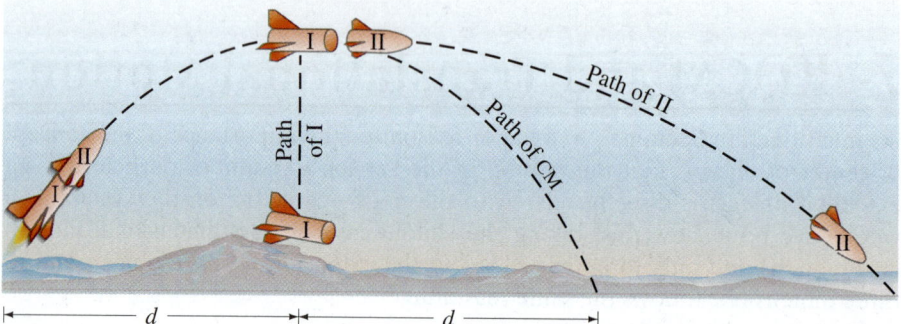

FIGURE 7–28 Example 7–14.

RESPONSE After the rocket is fired, the path of the CM of the system continues to follow the parabolic trajectory of a projectile acted on by only a constant gravitational force. The CM will thus land at a point $2d$ from the starting point. Since the masses of I and II are equal, the CM must be midway between them at any time. Therefore, part II lands a distance $3d$ from the starting point.

NOTE If part I had been given a kick up or down, instead of merely falling, the solution would have been more complicated.

EXERCISE H A woman stands up in a rowboat and walks from one end of the boat to the other. How does the boat move, as seen from the shore?

PHYSICS APPLIED
Distant planets discovered

An interesting application is the discovery of nearby stars (see Section 5–8) that seem to "wobble." What could cause such a wobble? It could be that a planet orbits the star, and each exerts a gravitational force on the other. The planets are too small and too far away to be observed directly by telescopes. But the slight wobble in the motion of the star suggests that both the planet and the star (its sun) orbit about their mutual center of mass, and hence the star appears to have a wobble. Irregularities in the star's motion can be measured to high accuracy, yielding information on the size of the planets' orbits and their masses. See Fig. 5–30 in Chapter 5.

Summary

The **linear momentum**, $\vec{p}$, of an object is defined as the product of its mass times its velocity,

$$\vec{p} = m\vec{v}. \tag{7-1}$$

In terms of momentum, **Newton's second law** can be written as

$$\Sigma \vec{F} = \frac{\Delta \vec{p}}{\Delta t}. \tag{7-2}$$

That is, the rate of change of momentum of an object equals the net force exerted on it.

When the net external force on a system of objects is zero, the total momentum remains constant. This is the **law of conservation of momentum**. Stated another way, the total momentum of an isolated system of objects remains constant.

The law of conservation of momentum is very useful in dealing with **collisions**. In a collision, two (or more) objects interact with each other over a very short time interval, and the force each exerts on the other during this time interval is very large compared to any other forces acting.

The **impulse** delivered by a force on an object is defined as

$$\text{Impulse} = \vec{F}\,\Delta t, \tag{7-5}$$

where $\vec{F}$ is the average force acting during the (usually very short) time interval Δt. The impulse is equal to the change in momentum of the object:

$$\text{Impulse} = \vec{F}\,\Delta t = \Delta \vec{p}. \tag{7-4}$$

Total momentum is conserved in *any* collision as long as any net external force is zero or negligible. If $m_A \vec{v}_A$ and $m_B \vec{v}_B$ are the momenta of two objects before the collision and $m_A \vec{v}'_A$ and $m_B \vec{v}'_B$ are their momenta after, then momentum conservation tells us that

$$m_A \vec{v}_A + m_B \vec{v}_B = m_A \vec{v}'_A + m_B \vec{v}'_B \tag{7-3}$$

for this two-object system.

Total energy is also conserved. But this may not be helpful unless kinetic energy is conserved, in which case the collision is called an **elastic collision** and we can write

$$\tfrac{1}{2} m_A v_A^2 + \tfrac{1}{2} m_B v_B^2 = \tfrac{1}{2} m_A v'^2_A + \tfrac{1}{2} m_B v'^2_B. \tag{7-6}$$

If kinetic energy is not conserved, the collision is called **inelastic**. Macroscopic collisions are generally inelastic. A **completely inelastic** collision is one in which the colliding objects stick together after the collision.

The **center of mass** (CM) of an extended object (or group of objects) is that point at which the net force can be considered to act, for purposes of determining the translational motion of the object as a whole. The x component of the CM for objects with mass m_A, m_B, …, is given by

$$x_{CM} = \frac{m_A x_A + m_B x_B + \cdots}{m_A + m_B + \cdots}. \tag{7-9a}$$

[*The center of mass of a system of total mass M moves in the same path that a particle of mass M would move if subjected to the same net external force. In equation form, this is Newton's second law for a system of particles (or extended objects):

$$M a_{CM} = F_{net} \tag{7-11}$$

where M is the total mass of the system, a_{CM} is the acceleration of the CM of the system, and F_{net} is the total (net) external force acting on all parts of the system.]

Questions

1. We claim that momentum is conserved. Yet most moving objects eventually slow down and stop. Explain.

2. A light object and a heavy object have the same kinetic energy. Which has the greater momentum? Explain.

3. When a person jumps from a tree to the ground, what happens to the momentum of the person upon striking the ground?

4. When you release an inflated but untied balloon, why does it fly across the room?

5. Explain, on the basis of conservation of momentum, how a fish propels itself forward by swishing its tail back and forth.

6. Two children float motionlessly in a space station. The 20-kg girl pushes on the 40-kg boy and he sails away at 1.0 m/s. The girl (*a*) remains motionless; (*b*) moves in the same direction at 1.0 m/s; (*c*) moves in the opposite direction at 1.0 m/s; (*d*) moves in the opposite direction at 2.0 m/s; (*e*) none of these.

7. According to Eq. 7–4, the longer the impact time of an impulse, the smaller the force can be for the same momentum change, and hence the smaller the deformation of the object on which the force acts. On this basis, explain the value of air bags, which are intended to inflate during an automobile collision and reduce the possibility of fracture or death.

8. If a falling ball were to make a perfectly elastic collision with the floor, would it rebound to its original height? Explain.

9. A boy stands on the back of a rowboat and dives into the water. What happens to the boat as he leaves it? Explain.

10. It is said that in ancient times a rich man with a bag of gold coins was stranded on the surface of a frozen lake. Because the ice was frictionless, he could not push himself to shore and froze to death. What could he have done to save himself had he not been so miserly?

11. The speed of a tennis ball on the return of a serve can be just as fast as the serve, even though the racket isn't swung very fast. How can this be?

12. Is it possible for an object to receive a larger impulse from a small force than from a large force? Explain.

13. In a collision between two cars, which would you expect to be more damaging to the occupants: if the cars collide and remain together, or if the two cars collide and rebound backward? Explain.

14. A very elastic "superball" is dropped from a height *h* onto a hard steel plate (fixed to the Earth), from which it rebounds at very nearly its original speed. (*a*) Is the momentum of the ball conserved during any part of this process? (*b*) If we consider the ball and the Earth as our system, during what parts of the process is momentum conserved? (*c*) Answer part (*b*) for a piece of putty that falls and sticks to the steel plate.

15. Cars used to be built as rigid as possible to withstand collisions. Today, though, cars are designed to have "crumple zones" that collapse upon impact. What is the advantage of this new design?

16. At a hydroelectric power plant, water is directed at high speed against turbine blades on an axle that turns an electric generator. For maximum power generation, should the turbine blades be designed so that the water is brought to a dead stop, or so that the water rebounds?

17. A squash ball hits a wall at a 45° angle as shown in Fig. 7–29. What is the direction (*a*) of the change in momentum of the ball, (*b*) of the force on the wall?

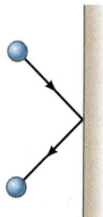

FIGURE 7–29 Question 17.

18. Why can a batter hit a pitched baseball farther than a ball he himself has tossed up in the air?

19. Describe a collision in which all kinetic energy is lost.

20. If a 20-passenger plane is not full, sometimes passengers are told they must sit in certain seats and may not move to empty seats. Why might this be?

21. Why do you tend to lean backward when carrying a heavy load in your arms?

22. Why is the CM of a 1-m length of pipe at its midpoint, whereas this is not true for your arm or leg?

23. How can a rocket change direction when it is far out in space and essentially in a vacuum?

24. Bob and Jim decide to play tug-of-war on a frictionless (icy) surface. Jim is considerably stronger than Bob, but Bob weighs 160 lb whereas Jim weighs 145 lb. Who loses by crossing over the midline first? Explain.

*25. In one type of nuclear radioactive decay, an electron and a recoil nucleus are emitted but often do not separate along the same line. Use conservation of momentum in two dimensions to explain why this implies the emission of at least one other particle (it came to be called a "neutrino").

*26. Show on a diagram how your CM shifts when you move from a lying position to a sitting position.

*27. If only an external force can change the momentum of the center of mass of an object, how can the internal force of the engine accelerate a car?

*28. A rocket following a parabolic path through the air suddenly explodes into many pieces. What can you say about the motion of this system of pieces?

MisConceptual Questions

1. A truck going 15 km/h has a head-on collision with a small car going 30 km/h. Which statement best describes the situation?
   (a) The truck has the greater change of momentum because it has the greater mass.
   (b) The car has the greater change of momentum because it has the greater speed.
   (c) Neither the car nor the truck changes its momentum in the collision because momentum is conserved.
   (d) They both have the same change in magnitude of momentum because momentum is conserved.
   (e) None of the above is necessarily true.

2. A small boat coasts at constant speed under a bridge. A heavy sack of sand is dropped from the bridge onto the boat. The speed of the boat
   (a) increases.
   (b) decreases.
   (c) does not change.
   (d) Without knowing the mass of the boat and the sand, we can't tell.

3. Two identical billiard balls traveling at the same speed have a head-on collision and rebound. If the balls had twice the mass, but maintained the same size and speed, how would the rebound be different?
   (a) At a higher speed.
   (b) At slower speed.
   (c) No difference.

4. An astronaut is a short distance away from her space station without a tether rope. She has a large wrench. What should she do with the wrench to move toward the space station?
   (a) Throw it directly away from the space station.
   (b) Throw it directly toward the space station.
   (c) Throw it toward the station without letting go of it.
   (d) Throw it parallel to the direction of the station's orbit.
   (e) Throw it opposite to the direction of the station's orbit.

5. A space vehicle, in circular orbit around the Earth, collides with a small asteroid which ends up in the vehicle's storage bay. For this collision,
   (a) only momentum is conserved.
   (b) only kinetic energy is conserved.
   (c) both momentum and kinetic energy are conserved.
   (d) neither momentum nor kinetic energy is conserved.

6. A golf ball and an equal-mass bean bag are dropped from the same height and hit the ground. The bean bag stays on the ground while the golf ball rebounds. Which experiences the greater impulse from the ground?
   (a) The golf ball.
   (b) The bean bag.
   (c) Both the same.
   (d) Not enough information.

7. You are lying in bed and want to shut your bedroom door. You have a bouncy ball and a blob of clay, both with the same mass. Which one would be more effective to throw at your door to close it?
   (a) The bouncy ball.
   (b) The blob of clay.
   (c) Both the same.
   (d) Neither will work.

8. A baseball is pitched horizontally toward home plate with a velocity of 110 km/h. In which of the following scenarios does the change in momentum of the baseball have the largest magnitude?
   (a) The catcher catches the ball.
   (b) The ball is popped straight up at a speed of 110 km/h.
   (c) The baseball is hit straight back to the pitcher at a speed of 110 km/h.
   (d) Scenarios (a) and (b) have the same change in momentum.
   (e) Scenarios (a), (b), and (c) have the same change in momentum.

9. A small car and a heavy pickup truck are both out of gas. The truck has twice the mass of the car. After you push first the car and then the truck for the *same amount of time* with the same force, what can you say about the momentum and kinetic energy (KE) of the car and the truck? Ignore friction.
   (a) They have the same momentum and the same KE.
   (b) The car has more momentum and more KE than the truck.
   (c) The truck has more momentum and more KE than the car.
   (d) They have the same momentum, but the car has more kinetic energy than the truck.
   (e) They have the same kinetic energy, but the truck has more momentum than the car.

10. Answer the previous Question (# 9) but now assume that you push both the car and the truck for the *same distance* with the same force. [*Hint*: See also Chapter 6.]

11. A railroad tank car contains milk and rolls at a constant speed along a level track. The milk begins to leak out the bottom. The car then
    (a) slows down.
    (b) speeds up.
    (c) maintains a constant speed.
    (d) Need more information about the rate of the leak.

12. A bowling ball hangs from a 1.0-m-long cord, Fig. 7–30: (i) A 200-gram putty ball moving 5.0 m/s hits the bowling ball and sticks to it, causing the bowling ball to swing up; (ii) a 200-gram rubber ball moving 5.0 m/s hits the bowling ball and bounces straight back at nearly 5.0 m/s, causing the bowling ball to swing up. Describe what happens.
    (a) The bowling ball swings up by the same amount in both (i) and (ii).
    (b) The ball swings up farther in (i) than in (ii).
    (c) The ball swings up farther in (ii) than in (i).
    (d) Not enough information is given; we need the contact time between the rubber ball and the bowling ball.

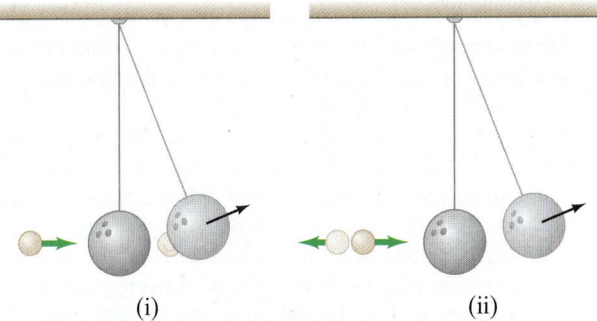

FIGURE 7–30 MisConceptual Question 12.

For assigned homework and other learning materials, go to the MasteringPhysics website.

Problems

7–1 and 7–2 Momentum and Its Conservation

1. (I) What is the magnitude of the momentum of a 28-g sparrow flying with a speed of 8.4 m/s?

2. (I) A constant friction force of 25 N acts on a 65-kg skier for 15 s on level snow. What is the skier's change in velocity?

3. (I) A 7150-kg railroad car travels alone on a level frictionless track with a constant speed of 15.0 m/s. A 3350-kg load, initially at rest, is dropped onto the car. What will be the car's new speed?

4. (I) A 110-kg tackler moving at 2.5 m/s meets head-on (and holds on to) an 82-kg halfback moving at 5.0 m/s. What will be their mutual speed immediately after the collision?

5. (II) Calculate the force exerted on a rocket when the propelling gases are being expelled at a rate of 1300 kg/s with a speed of 4.5×10^4 m/s.

6. (II) A 7700-kg boxcar traveling 14 m/s strikes a second car at rest. The two stick together and move off with a speed of 5.0 m/s. What is the mass of the second car?

7. (II) A child in a boat throws a 5.30-kg package out horizontally with a speed of 10.0 m/s, Fig. 7–31. Calculate the velocity of the boat immediately after, assuming it was initially at rest. The mass of the child is 24.0 kg and the mass of the boat is 35.0 kg.

FIGURE 7–31 Problem 7.

8. (II) An atomic nucleus at rest decays radioactively into an alpha particle and a different nucleus. What will be the speed of this recoiling nucleus if the speed of the alpha particle is 2.8×10^5 m/s? Assume the recoiling nucleus has a mass 57 times greater than that of the alpha particle.

9. (II) An atomic nucleus initially moving at 320 m/s emits an alpha particle in the direction of its velocity, and the remaining nucleus slows to 280 m/s. If the alpha particle has a mass of 4.0 u and the original nucleus has a mass of 222 u, what speed does the alpha particle have when it is emitted?

10. (II) An object at rest is suddenly broken apart into two fragments by an explosion. One fragment acquires twice the kinetic energy of the other. What is the ratio of their masses?

11. (II) A 22-g bullet traveling 240 m/s penetrates a 2.0-kg block of wood and emerges going 150 m/s. If the block is stationary on a frictionless surface when hit, how fast does it move after the bullet emerges?

12. (III) A 0.145-kg baseball pitched horizontally at 27.0 m/s strikes a bat and pops straight up to a height of 31.5 m. If the contact time between bat and ball is 2.5 ms, calculate the average force between the ball and bat during contact.

13. (III) Air in a 120-km/h wind strikes head-on the face of a building 45 m wide by 75 m high and is brought to rest. If air has a mass of 1.3 kg per cubic meter, determine the average force of the wind on the building.

14. (III) A 725-kg two-stage rocket is traveling at a speed of 6.60×10^3 m/s away from Earth when a predesigned explosion separates the rocket into two sections of equal mass that then move with a speed of 2.80×10^3 m/s relative to each other along the original line of motion. (a) What is the speed and direction of each section (relative to Earth) after the explosion? (b) How much energy was supplied by the explosion? [*Hint*: What is the change in kinetic energy as a result of the explosion?]

7–3 Collisions and Impulse

15. (I) A 0.145-kg baseball pitched at 31.0 m/s is hit on a horizontal line drive straight back at the pitcher at 46.0 m/s. If the contact time between bat and ball is 5.00×10^{-3} s, calculate the force (assumed to be constant) between the ball and bat.

16. (II) A golf ball of mass 0.045 kg is hit off the tee at a speed of 38 m/s. The golf club was in contact with the ball for 3.5×10^{-3} s. Find (*a*) the impulse imparted to the golf ball, and (*b*) the average force exerted on the ball by the golf club.

17. (II) A 12-kg hammer strikes a nail at a velocity of 7.5 m/s and comes to rest in a time interval of 8.0 ms. (*a*) What is the impulse given to the nail? (*b*) What is the average force acting on the nail?

18. (II) A tennis ball of mass $m = 0.060$ kg and speed $v = 28$ m/s strikes a wall at a 45° angle and rebounds with the same speed at 45° (Fig. 7–32). What is the impulse (magnitude and direction) given to the ball?

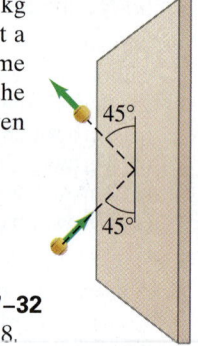

FIGURE 7–32 Problem 18.

19. (II) A 125-kg astronaut (including space suit) acquires a speed of 2.50 m/s by pushing off with her legs from a 1900-kg space capsule. (*a*) What is the change in speed of the space capsule? (*b*) If the push lasts 0.600 s, what is the average force exerted by each on the other? As the reference frame, use the position of the capsule before the push. (*c*) What is the kinetic energy of each after the push?

20. (II) Rain is falling at the rate of 2.5 cm/h and accumulates in a pan. If the raindrops hit at 8.0 m/s, estimate the force on the bottom of a 1.0-m² pan due to the impacting rain which we assume does not rebound. Water has a mass of 1.00×10^3 kg per m³.

21. (II) A 95-kg fullback is running at 3.0 m/s to the east and is stopped in 0.85 s by a head-on tackle by a tackler running due west. Calculate (*a*) the original momentum of the fullback, (*b*) the impulse exerted on the fullback, (*c*) the impulse exerted on the tackler, and (*d*) the average force exerted on the tackler.

22. (II) With what impulse does a 0.50-kg newspaper have to be thrown to give it a velocity of 3.0 m/s?

*23. (III) Suppose the force acting on a tennis ball (mass 0.060 kg) points in the $+x$ direction and is given by the graph of Fig. 7–33 as a function of time. (a) Use graphical methods (count squares) to estimate the total impulse given the ball. (b) Estimate the velocity of the ball after being struck, assuming the ball is being served so it is nearly at rest initially. [Hint: See Section 6–2.]

FIGURE 7–33 Problem 23.

24. (III) (a) Calculate the impulse experienced when a 55-kg person lands on firm ground after jumping from a height of 2.8 m. (b) Estimate the average force exerted on the person's feet by the ground if the landing is stiff-legged, and again (c) with bent legs. With stiff legs, assume the body moves 1.0 cm during impact, and when the legs are bent, about 50 cm. [Hint: The average net force on him, which is related to impulse, is the vector sum of gravity and the force exerted by the ground. See Fig. 7–34.] We will see in Chapter 9 that the force in (b) exceeds the ultimate strength of bone (Table 9–2).

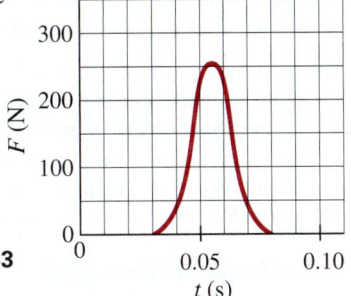

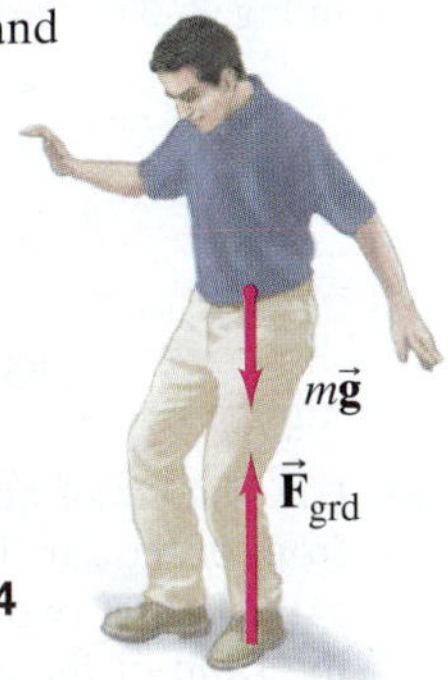

FIGURE 7–34 Problem 24.

7–4 and 7–5 Elastic Collisions

25. (II) A ball of mass 0.440 kg moving east ($+x$ direction) with a speed of 3.80 m/s collides head-on with a 0.220-kg ball at rest. If the collision is perfectly elastic, what will be the speed and direction of each ball after the collision?

26. (II) A 0.450-kg hockey puck, moving east with a speed of 5.80 m/s, has a head-on collision with a 0.900-kg puck initially at rest. Assuming a perfectly elastic collision, what will be the speed and direction of each puck after the collision?

27. (II) A 0.060-kg tennis ball, moving with a speed of 5.50 m/s, has a head-on collision with a 0.090-kg ball initially moving in the same direction at a speed of 3.00 m/s. Assuming a perfectly elastic collision, determine the speed and direction of each ball after the collision.

28. (II) Two billiard balls of equal mass undergo a perfectly elastic head-on collision. If one ball's initial speed was 2.00 m/s, and the other's was 3.60 m/s in the opposite direction, what will be their speeds and directions after the collision?

29. (II) A 0.280-kg croquet ball makes an elastic head-on collision with a second ball initially at rest. The second ball moves off with half the original speed of the first ball. (a) What is the mass of the second ball? (b) What fraction of the original kinetic energy ($\Delta KE/KE$) gets transferred to the second ball?

30. (II) A ball of mass m makes a head-on elastic collision with a second ball (at rest) and rebounds with a speed equal to 0.450 its original speed. What is the mass of the second ball?

31. (II) A ball of mass 0.220 kg that is moving with a speed of 5.5 m/s collides head-on and elastically with another ball initially at rest. Immediately after the collision, the incoming ball bounces backward with a speed of 3.8 m/s. Calculate (a) the velocity of the target ball after the collision, and (b) the mass of the target ball.

32. (II) Determine the fraction of kinetic energy lost by a neutron ($m_1 = 1.01$ u) when it collides head-on and elastically with a target particle at rest which is (a) $^1_1$H ($m = 1.01$ u); (b) $^2_1$H (heavy hydrogen, $m = 2.01$ u); (c) $^{12}_6$C ($m = 12.00$ u); (d) $^{208}_{82}$Pb (lead, $m = 208$ u).

7–6 Inelastic Collisions

33. (I) In a ballistic pendulum experiment, projectile 1 results in a maximum height h of the pendulum equal to 2.6 cm. A second projectile (of the same mass) causes the pendulum to swing twice as high, $h_2 = 5.2$ cm. The second projectile was how many times faster than the first?

34. (II) (a) Derive a formula for the fraction of kinetic energy lost, $\Delta KE/KE$, in terms of m and M for the ballistic pendulum collision of Example 7–9. (b) Evaluate for $m = 18.0$ g and $M = 380$ g.

35. (II) A 28-g rifle bullet traveling 190 m/s embeds itself in a 3.1-kg pendulum hanging on a 2.8-m-long string, which makes the pendulum swing upward in an arc. Determine the vertical and horizontal components of the pendulum's maximum displacement.

36. (II) An internal explosion breaks an object, initially at rest, into two pieces, one of which has 1.5 times the mass of the other. If 5500 J is released in the explosion, how much kinetic energy does each piece acquire?

37. (II) A 980-kg sports car collides into the rear end of a 2300-kg SUV stopped at a red light. The bumpers lock, the brakes are locked, and the two cars skid forward 2.6 m before stopping. The police officer, estimating the coefficient of kinetic friction between tires and road to be 0.80, calculates the speed of the sports car at impact. What was that speed?

38. (II) You drop a 14-g ball from a height of 1.5 m and it only bounces back to a height of 0.85 m. What was the total impulse on the ball when it hit the floor? (Ignore air resistance.)

39. Croquet ball A moving at 4.3 m/s makes a head on collision with ball B of equal mass and initially at rest. Immediately after the collision ball B moves forward at 3.0 m/s. What fraction of the initial kinetic energy is lost in the collision?

40. (II) A wooden block is cut into two pieces, one with three times the mass of the other. A depression is made in both faces of the cut, so that a firecracker can be placed in it with the block reassembled. The reassembled block is set on a rough-surfaced table, and the fuse is lit. When the firecracker explodes inside, the two blocks separate and slide apart. What is the ratio of distances each block travels?

41. (II) A 144-g baseball moving 28.0 m/s strikes a stationary 5.25-kg brick resting on small rollers so it moves without significant friction. After hitting the brick, the baseball bounces straight back, and the brick moves forward at 1.10 m/s. (a) What is the baseball's speed after the collision? (b) Find the total kinetic energy before and after the collision.

42. (III) A pendulum consists of a mass M hanging at the bottom end of a massless rod of length ℓ, which has a frictionless pivot at its top end. A mass m, moving as shown in Fig. 7–35 with velocity v, impacts M and becomes embedded. What is the smallest value of v sufficient to cause the pendulum (with embedded mass m) to swing clear over the top of its arc?

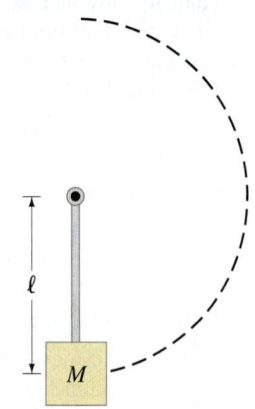

FIGURE 7–35 Problem 42.

43. (III) A bullet of mass $m = 0.0010$ kg embeds itself in a wooden block with mass $M = 0.999$ kg, which then compresses a spring ($k = 140$ N/m) by a distance $x = 0.050$ m before coming to rest. The coefficient of kinetic friction between the block and table is $\mu = 0.50$. (a) What is the initial velocity (assumed horizontal) of the bullet? (b) What fraction of the bullet's initial kinetic energy is dissipated (in damage to the wooden block, rising temperature, etc.) in the collision between the bullet and the block?

*7–7 Collisions in Two Dimensions

*44. (II) Billiard ball A of mass $m_A = 0.120$ kg moving with speed $v_A = 2.80$ m/s strikes ball B, initially at rest, of mass $m_B = 0.140$ kg. As a result of the collision, ball A is deflected off at an angle of 30.0° with a speed $v'_A = 2.10$ m/s. (a) Taking the x axis to be the original direction of motion of ball A, write down the equations expressing the conservation of momentum for the components in the x and y directions separately. (b) Solve these equations for the speed, v'_B, and angle, θ'_B, of ball B after the collision. Do not assume the collision is elastic.

*45. (II) A radioactive nucleus at rest decays into a second nucleus, an electron, and a neutrino. The electron and neutrino are emitted at right angles and have momenta of 9.6×10^{-23} kg·m/s and 6.2×10^{-23} kg·m/s, respectively. Determine the magnitude and the direction of the momentum of the second (recoiling) nucleus.

*46. (III) Billiard balls A and B, of equal mass, move at right angles and meet at the origin of an xy coordinate system as shown in Fig. 7–36. Initially ball A is moving along the y axis at $+2.0$ m/s, and ball B is moving to the right along the x axis with speed $+3.7$ m/s. After the collision (assumed elastic), ball B is moving along the positive y axis (Fig. 7–36) with velocity v'_B. What is the final direction of ball A, and what are the speeds of the two balls?

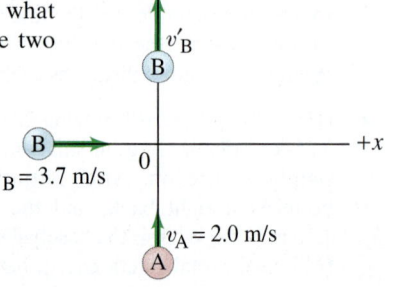

FIGURE 7–36 Problem 46. (Ball A after the collision is not shown.)

*47. (III) An atomic nucleus of mass m traveling with speed v collides elastically with a target particle of mass $2m$ (initially at rest) and is scattered at 90°. (a) At what angle does the target particle move after the collision? (b) What are the final speeds of the two particles? (c) What fraction of the initial kinetic energy is transferred to the target particle?

*48. (III) A neon atom ($m = 20.0$ u) makes a perfectly elastic collision with another atom at rest. After the impact, the neon atom travels away at a 55.6° angle from its original direction and the unknown atom travels away at a $-50.0°$ angle. What is the mass (in u) of the unknown atom? [*Hint*: You could use the law of sines.]

7–8 Center of Mass (CM)

49. (I) The distance between a carbon atom ($m = 12$ u) and an oxygen atom ($m = 16$ u) in the CO molecule is 1.13×10^{-10} m. How far from the carbon atom is the center of mass of the molecule?

50. (I) Find the center of mass of the three-mass system shown in Fig. 7–37 relative to the 1.00-kg mass.

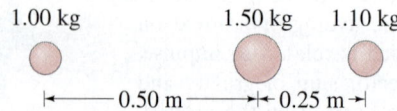

FIGURE 7–37 Problem 50.

51. (II) The CM of an empty 1250-kg car is 2.40 m behind the front of the car. How far from the front of the car will the CM be when two people sit in the front seat 2.80 m from the front of the car, and three people sit in the back seat 3.90 m from the front? Assume that each person has a mass of 65.0 kg.

52. (II) Three cubes, of side ℓ_0, $2\ell_0$, and $3\ell_0$, are placed next to one another (in contact) with their centers along a straight line as shown in Fig. 7–38. What is the position, along this line, of the CM of this system? Assume the cubes are made of the same uniform material.

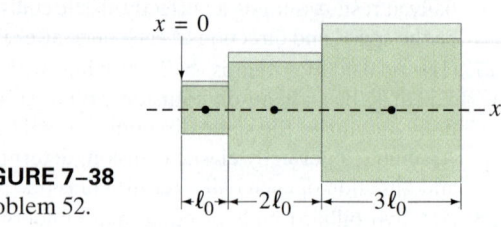

FIGURE 7–38 Problem 52.

53. (II) A (lightweight) pallet has a load of ten identical cases of tomato paste (see Fig. 7–39), each of which is a cube of length ℓ. Find the center of gravity in the horizontal plane, so that the crane operator can pick up the load without tipping it.

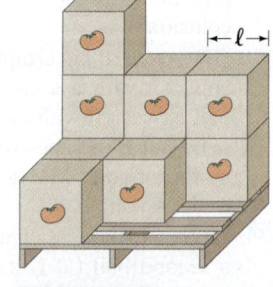

FIGURE 7–39 Problem 53.

194 CHAPTER 7 Linear Momentum

54. (III) Determine the CM of the uniform thin L-shaped construction brace shown in Fig. 7–40.

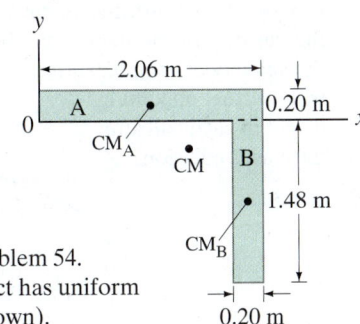

FIGURE 7–40 Problem 54. This L-shaped object has uniform thickness d (not shown).

55. (III) A uniform circular plate of radius $2R$ has a circular hole of radius R cut out of it. The center C' of the smaller circle is a distance $0.80R$ from the center C of the larger circle, Fig. 7–41. What is the position of the center of mass of the plate? [*Hint*: Try subtraction.]

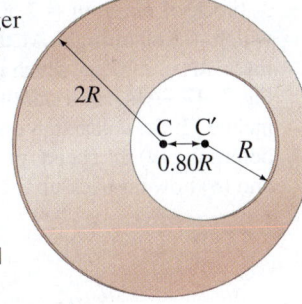

FIGURE 7–41 Problem 55.

*7–9 CM for the Human Body

*56. (I) Assume that your proportions are the same as those in Table 7–1, and calculate the mass of one of your legs.

*57. (I) Determine the CM of an outstretched arm using Table 7–1.

*58. (II) Use Table 7–1 to calculate the position of the CM of an arm bent at a right angle. Assume that the person is 155 cm tall.

*59. (II) When a high jumper is in a position such that his arms and lower legs are hanging vertically, and his thighs, trunk, and head are horizontal just above the bar, estimate how far below the torso's median line the CM will be. Will this CM be outside the body? Use Table 7–1.

*60. (III) Repeat Problem 59 assuming the body bends at the hip joint by about 15°. Estimate, using Fig. 7–27 as a model.

*7–10 CM and Translational Motion

*61. (II) The masses of the Earth and Moon are 5.98×10^{24} kg and 7.35×10^{22} kg, respectively, and their centers are separated by 3.84×10^{8} m. (*a*) Where is the CM of the Earth–Moon system located? (*b*) What can you say about the motion of the Earth–Moon system about the Sun, and of the Earth and Moon separately about the Sun?

*62. (II) A mallet consists of a uniform cylindrical head of mass 2.30 kg and a diameter 0.0800 m mounted on a uniform cylindrical handle of mass 0.500 kg and length 0.240 m, as shown in Fig. 7–42. If this mallet is tossed, spinning, into the air, how far above the bottom of the handle is the point that will follow a parabolic trajectory?

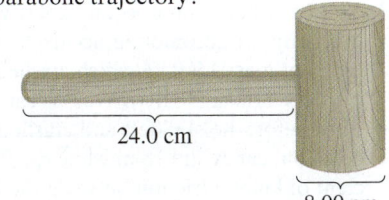

FIGURE 7–42 Problem 62.

*63. (II) A 52-kg woman and a 72-kg man stand 10.0 m apart on nearly frictionless ice. (*a*) How far from the woman is their CM? (*b*) If each holds one end of a rope, and the man pulls on the rope so that he moves 2.5 m, how far from the woman will he be now? (*c*) How far will the man have moved when he collides with the woman?

*64. (II) Suppose that in Example 7–14 (Fig. 7–28), $m_{II} = 3m_I$. (*a*) Where then would m_{II} land? (*b*) What if $m_I = 3m_{II}$?

*65. (II) Two people, one of mass 85 kg and the other of mass 55 kg, sit in a rowboat of mass 58 kg. With the boat initially at rest, the two people, who have been sitting at opposite ends of the boat, 3.0 m apart from each other, now exchange seats. How far and in what direction will the boat move?

*66. (III) A huge balloon and its gondola, of mass M, are in the air and stationary with respect to the ground. A passenger, of mass m, then climbs out and slides down a rope with speed v, measured with respect to the balloon. With what speed and direction (relative to Earth) does the balloon then move? What happens if the passenger stops?

General Problems

67. Two astronauts, one of mass 55 kg and the other 85 kg, are initially at rest together in outer space. They then push each other apart. How far apart are they when the lighter astronaut has moved 12 m?

68. Two asteroids strike head-on: before the collision, asteroid A ($m_A = 7.5 \times 10^{12}$ kg) has velocity 3.3 km/s and asteroid B ($m_B = 1.45 \times 10^{13}$ kg) has velocity 1.4 km/s in the opposite direction. If the asteroids stick together, what is the velocity (magnitude and direction) of the new asteroid after the collision?

69. A ball is dropped from a height of 1.60 m and rebounds to a height of 1.20 m. Approximately how many rebounds will the ball make before losing 90% of its energy?

70. A 4800-kg open railroad car coasts at a constant speed of 7.60 m/s on a level track. Snow begins to fall vertically and fills the car at a rate of 3.80 kg/min. Ignoring friction with the tracks, what is the car's speed after 60.0 min? (See Section 7–2.)

71. Two bumper cars in an amusement park ride collide elastically as one approaches the other directly from the rear (Fig. 7–43). Car A has a mass of 435 kg and car B 495 kg, owing to differences in passenger mass. If car A approaches at 4.50 m/s and car B is moving at 3.70 m/s, calculate (*a*) their velocities after the collision, and (*b*) the change in momentum of each.

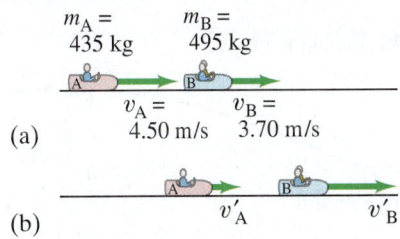

FIGURE 7–43 Problem 71: (a) before collision, (b) after collision.

72. A gun fires a bullet vertically into a 1.40-kg block of wood at rest on a thin horizontal sheet, Fig. 7–44. If the bullet has a mass of 25.0 g and a speed of 230 m/s, how high will the block rise into the air after the bullet becomes embedded in it?

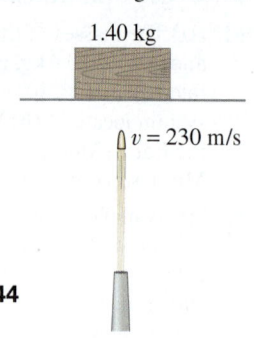

FIGURE 7–44
Problem 72.

73. You have been hired as an expert witness in a court case involving an automobile accident. The accident involved car A of mass 1500 kg which crashed into stationary car B of mass 1100 kg. The driver of car A applied his brakes 15 m before he skidded and crashed into car B. After the collision, car A slid 18 m while car B slid 30 m. The coefficient of kinetic friction between the locked wheels and the road was measured to be 0.60. Show that the driver of car A was exceeding the 55-mi/h (90-km/h) speed limit before applying the brakes.

74. A meteor whose mass was about 1.5×10^8 kg struck the Earth ($m_E = 6.0 \times 10^{24}$ kg) with a speed of about 25 km/s and came to rest in the Earth. (a) What was the Earth's recoil speed (relative to Earth at rest before the collision)? (b) What fraction of the meteor's kinetic energy was transformed to kinetic energy of the Earth? (c) By how much did the Earth's kinetic energy change as a result of this collision?

75. A 28-g bullet strikes and becomes embedded in a 1.35-kg block of wood placed on a horizontal surface just in front of the gun. If the coefficient of kinetic friction between the block and the surface is 0.28, and the impact drives the block a distance of 8.5 m before it comes to rest, what was the muzzle speed of the bullet?

76. You are the design engineer in charge of the crashworthiness of new automobile models. Cars are tested by smashing them into fixed, massive barriers at 45 km/h. A new model of mass 1500 kg takes 0.15 s from the time of impact until it is brought to rest. (a) Calculate the average force exerted on the car by the barrier. (b) Calculate the average deceleration of the car in g's.

77. A 0.25-kg skeet (clay target) is fired at an angle of 28° to the horizontal with a speed of 25 m/s (Fig. 7–45). When it reaches the maximum height, h, it is hit from below by a 15-g pellet traveling vertically upward at a speed of 230 m/s. The pellet is embedded in the skeet. (a) How much higher, h', does the skeet go up? (b) How much extra distance, Δx, does the skeet travel because of the collision?

78. Two balls, of masses $m_A = 45$ g and $m_B = 65$ g, are suspended as shown in Fig. 7–46. The lighter ball is pulled away to a 66° angle with the vertical and released. (a) What is the velocity of the lighter ball before impact? (b) What is the velocity of each ball after the elastic collision? (c) What will be the maximum height of each ball after the elastic collision?

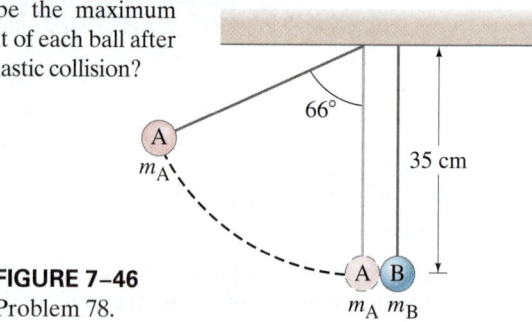

FIGURE 7–46
Problem 78.

79. A block of mass $m = 2.50$ kg slides down a 30.0° incline which is 3.60 m high. At the bottom, it strikes a block of mass $M = 7.00$ kg which is at rest on a horizontal surface, Fig. 7–47. (Assume a smooth transition at the bottom of the incline.) If the collision is elastic, and friction can be ignored, determine (a) the speeds of the two blocks after the collision, and (b) how far back up the incline the smaller mass will go.

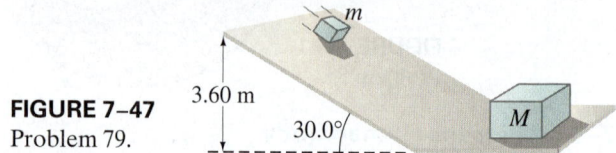

FIGURE 7–47
Problem 79.

80. The space shuttle launches an 850-kg satellite by ejecting it from the cargo bay. The ejection mechanism is activated and is in contact with the satellite for 4.8 s to give it a velocity of 0.30 m/s in the x direction relative to the shuttle. The mass of the shuttle is 92,000 kg. (a) Determine the component of velocity v_f of the shuttle in the minus x direction resulting from the ejection. (b) Find the average force that the shuttle exerts on the satellite during the ejection.

81. Astronomers estimate that a 2.0-km-diameter asteroid collides with the Earth once every million years. The collision could pose a threat to life on Earth. (a) Assume a spherical asteroid has a mass of 3200 kg for each cubic meter of volume and moves toward the Earth at 15 km/s. How much destructive energy could be released when it embeds itself in the Earth? (b) For comparison, a nuclear bomb could release about 4.0×10^{16} J. How many such bombs would have to explode simultaneously to release the destructive energy of the asteroid collision with the Earth?

82. An astronaut of mass 210 kg including his suit and jet pack wants to acquire a velocity of 2.0 m/s to move back toward his space shuttle. Assuming the jet pack can eject gas with a velocity of 35 m/s, what mass of gas will need to be ejected?

FIGURE 7–45 Problem 77.

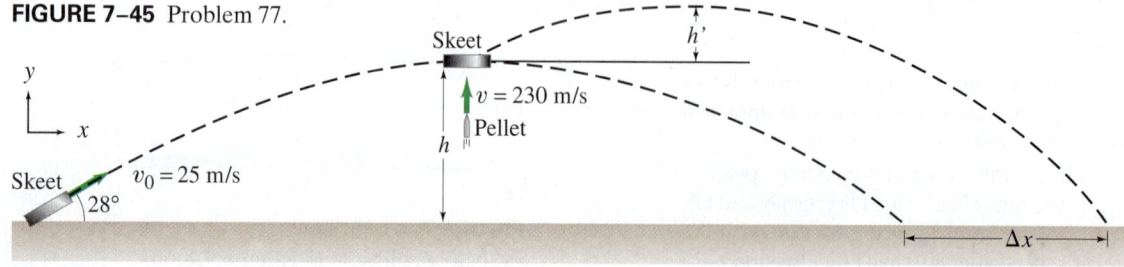

196   CHAPTER 7   Linear Momentum

83. Two blocks of mass m_A and m_B, resting on a frictionless table, are connected by a stretched spring and then released (Fig. 7–48). (a) Is there a net external force on the system before release? (b) Determine the ratio of their speeds, v_A/v_B. (c) What is the ratio of their kinetic energies? (d) Describe the motion of the CM of this system. Ignore mass of spring.

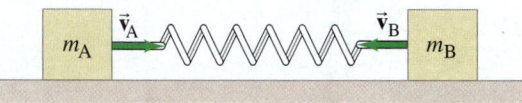

FIGURE 7–48 Problem 83.

84. A golf ball rolls off the top of a flight of concrete steps of total vertical height 4.00 m. The ball hits four times on the way down, each time striking the horizontal part of a different step 1.00 m lower. If all collisions are perfectly elastic, what is the bounce height on the fourth bounce when the ball reaches the bottom of the stairs?

85. A massless spring with spring constant k is placed between a block of mass m and a block of mass $3m$. Initially the blocks are at rest on a frictionless surface and they are held together so that the spring between them is compressed by an amount D from its equilibrium length. The blocks are then released and the spring pushes them off in opposite directions. Find the speeds of the two blocks when they detach from the spring.

*86. A novice pool player is faced with the corner pocket shot shown in Fig. 7–49. Relative dimensions are also shown. Should the player worry that this might be a "scratch shot," in which the cue ball will also fall into a pocket? Give details. Assume equal-mass balls and an elastic collision. Ignore spin.

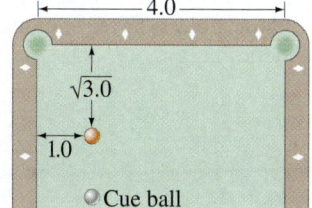

FIGURE 7–49 Problem 86.

Search and Learn

1. Consider the Examples in this Chapter involving $\Sigma \vec{F}_{ext} = \Delta \vec{p}/\Delta t$. Provide some general guidelines as to when it is best to solve the problem using $\Sigma \vec{F}_{ext} = 0$ so $\Sigma \vec{p}_i = \Sigma \vec{p}_f$, and when to use the principle of impulse instead so that $\Sigma \vec{F}_{ext} \Delta t = \Delta \vec{p}$.

2. A 6.0-kg object moving in the $+x$ direction at 6.5 m/s collides head-on with an 8.0-kg object moving in the $-x$ direction at 4.0 m/s. Determine the final velocity of each object if: (a) the objects stick together; (b) the collision is elastic; (c) the 6.0-kg object is at rest after the collision; (d) the 8.0-kg object is at rest after the collision; (e) the 6.0-kg object has a velocity of 4.0 m/s in the $-x$ direction after the collision. Finally, (f) are the results in (c), (d), and (e) "reasonable"? Explain.

3. In a physics lab, a cube slides down a frictionless incline as shown in Fig. 7–50 and elastically strikes another cube at the bottom that is only one-half its mass. If the incline is 35 cm high and the table is 95 cm off the floor, where does each cube land? [Hint: Both leave the incline moving horizontally.]

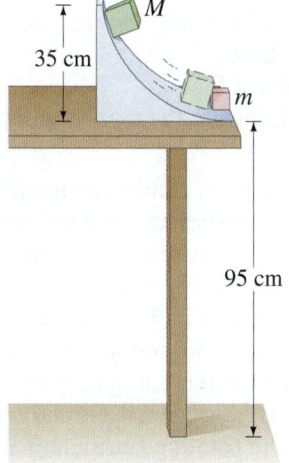

FIGURE 7–50 Search and Learn 3.

4. **The gravitational slingshot effect.** Figure 7–51 shows the planet Saturn moving in the negative x direction at its orbital speed (with respect to the Sun) of 9.6 km/s. The mass of Saturn is 5.69×10^{26} kg. A spacecraft with mass 825 kg approaches Saturn. When far from Saturn, it moves in the $+x$ direction at 10.4 km/s. The gravitational attraction of Saturn (a conservative force) acting on the spacecraft causes it to swing around the planet (orbit shown as dashed line) and head off in the opposite direction. Estimate the final speed of the spacecraft after it is far enough away to be considered free of Saturn's gravitational pull.

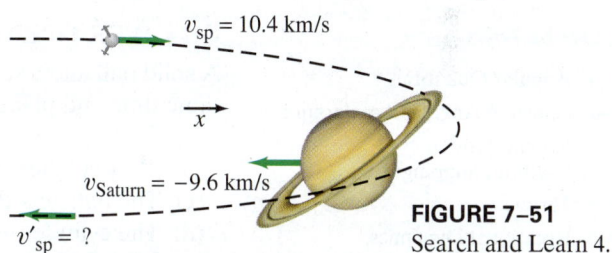

FIGURE 7–51 Search and Learn 4.

5. Take the general case of an object of mass m_A and velocity v_A elastically striking a stationary $(v_B = 0)$ object of mass m_B head-on. (a) Show that the final velocities v'_A and v'_B are given by

$$v'_A = \left(\frac{m_A - m_B}{m_A + m_B}\right)v_A, \qquad v'_B = \left(\frac{2m_A}{m_A + m_B}\right)v_A.$$

(b) What happens in the extreme case when m_A is much smaller than m_B? Cite a common example of this. (c) What happens in the extreme case when m_A is much larger than m_B? Cite a common example of this. (d) What happens in the case when $m_A = m_B$? Cite a common example.

ANSWERS TO EXERCISES

A: Yes, if the sports car's speed is three times greater.
B: Larger (Δp is greater).
C: (a) 6.0 m/s; (b) almost zero; (c) almost 24.0 m/s.
D: 0.50 m/s.
E: (b); (d).
F: The curve would be wider and less high.
G: $x_{CM} = -2.0$ m; yes.
H: The boat moves in the opposite direction.

You too can experience rapid rotation—if your stomach can take the high angular velocity and centripetal acceleration of some of the faster amusement park rides. If not, try the slower merry-go-round or Ferris wheel. Rotating carnival rides have rotational kinetic energy as well as angular momentum. Angular acceleration is produced by a net torque, and rotating objects have rotational kinetic energy.

CHAPTER 8

Rotational Motion

CONTENTS

8–1 Angular Quantities
8–2 Constant Angular Acceleration
8–3 Rolling Motion (Without Slipping)
8–4 Torque
8–5 Rotational Dynamics; Torque and Rotational Inertia
8–6 Solving Problems in Rotational Dynamics
8–7 Rotational Kinetic Energy
8–8 Angular Momentum and Its Conservation
*8–9 Vector Nature of Angular Quantities

CHAPTER-OPENING QUESTION—Guess now!

A solid ball and a solid cylinder roll down a ramp. They both start from rest at the same time and place. Which gets to the bottom first?

(a) They get there at the same time.
(b) They get there at almost exactly the same time except for frictional differences.
(c) The ball gets there first.
(d) The cylinder gets there first.
(e) Can't tell without knowing the mass and radius of each.

Until now, we have been concerned mainly with translational motion. We discussed the kinematics and dynamics of translational motion (the role of force). We also discussed the energy and momentum for translational motion. In this Chapter we will deal with rotational motion. We will discuss the kinematics of rotational motion and then its dynamics (involving torque), as well as rotational kinetic energy and angular momentum (the rotational analog of linear momentum). Our understanding of the world around us will be increased significantly—from rotating bicycle wheels and compact discs to amusement park rides, a spinning skater, the rotating Earth, and a centrifuge—and there may be a few surprises.

We will consider mainly the rotation of rigid objects about a fixed axis. A **rigid object** is an object with a definite shape that doesn't change, so that the particles composing it stay in fixed positions relative to one another. Any real object is capable of vibrating or deforming when a force is exerted on it. But these effects are often very small, so the concept of an ideal rigid object is very useful as a good approximation.

8–1 Angular Quantities

The motion of a rigid object can be analyzed as the translational motion of the object's center of mass, plus rotational motion *about* its center of mass (Section 7–8). We have already discussed translational motion in detail, so now we focus on purely rotational motion. By *purely rotational motion* we mean that all points in the object move in circles, such as the point P in the rotating wheel of Fig. 8–1, and that the centers of these circles all lie on one line called the **axis of rotation**. In Fig. 8–1 the axis of rotation is perpendicular to the page and passes through point O.

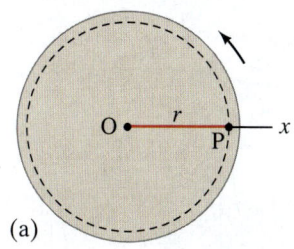

(a)

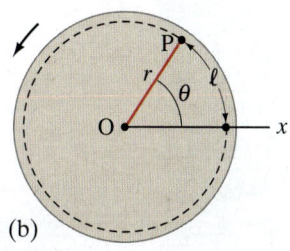

(b)

FIGURE 8–1 Looking at a wheel that is rotating counterclockwise about an axis through the wheel's center at O (axis perpendicular to the page). Each point, such as point P, moves in a circular path; ℓ is the distance P travels as the wheel rotates through the angle θ.

Every point in an object rotating about a fixed axis moves in a circle (shown dashed in Fig. 8–1 for point P) whose center is on the axis of rotation and whose radius is r, the distance of that point from the axis of rotation. A straight line drawn from the axis to any point in the object sweeps out the same angle θ in the same time interval.

To indicate the **angular position** of a rotating object, or how far it has rotated, we specify the angle θ of some particular line in the object (red in Fig. 8–1) with respect to a reference line, such as the x axis in Fig. 8–1. A point in the object, such as P in Fig. 8–1, moves through an angle θ when it travels the distance ℓ measured along the circumference of its circular path. Angles are commonly measured in degrees, but the mathematics of circular motion is much simpler if we use the *radian* for angular measure. One **radian** (abbreviated rad) is defined as the angle subtended by an arc whose length is equal to the radius. For example, in Fig. 8–1b, point P is a distance r from the axis of rotation, and it has moved a distance ℓ along the arc of a circle. The arc length ℓ is said to "subtend" the angle θ. In radians, any angle θ is given by

$$\theta = \frac{\ell}{r}, \qquad [\theta \text{ in radians}] \quad \textbf{(8–1a)}$$

where r is the radius of the circle, and ℓ is the arc length subtended by the angle θ specified in radians. If $\ell = r$, then $\theta = 1$ rad.

The radian is dimensionless since it is the ratio of two lengths. Nonetheless when giving an angle in radians, we always mention rad to remind us it is not degrees. It is often useful to rewrite Eq. 8–1a in terms of arc length ℓ:

$$\ell = r\theta. \qquad \textbf{(8–1b)}$$

**CAUTION**
Use radians in calculating, not degrees

Radians can be related to degrees in the following way. In a complete circle there are $360°$, which must correspond to an arc length equal to the circumference of the circle, $\ell = 2\pi r$. For a full circle, $\theta = \ell/r = 2\pi r/r = 2\pi$ rad. Thus

$$360° = 2\pi \text{ rad}.$$

One radian is then $360°/2\pi \approx 360°/6.28 \approx 57.3°$. An object that makes one complete revolution (rev) has rotated through $360°$, or 2π radians:

$$1 \text{ rev} = 360° = 2\pi \text{ rad}.$$

EXAMPLE 8–1 **Bike wheel.** A bike wheel rotates 4.50 revolutions. How many radians has it rotated?

APPROACH All we need is a conversion of units using

$$1 \text{ revolution} = 360° = 2\pi \text{ rad} = 6.28 \text{ rad}.$$

SOLUTION

$$4.50 \text{ revolutions} = (4.50 \text{ rev})\left(2\pi \frac{\text{rad}}{\text{rev}}\right) = 9.00\pi \text{ rad} = 28.3 \text{ rad}.$$

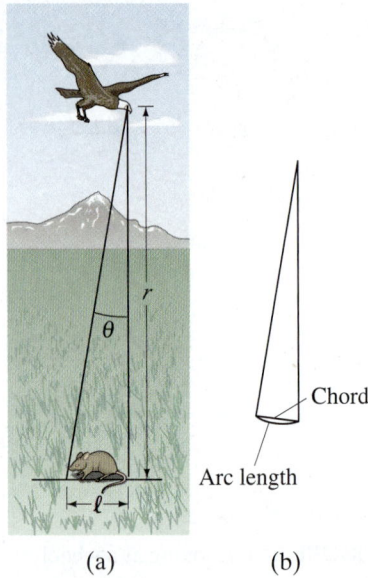

FIGURE 8–2 (a) Example 8–2. (b) For small angles, arc length and the chord length (straight line) are nearly equal.

EXAMPLE 8–2 Birds of prey—in radians. A particular bird's eye can just distinguish objects that subtend an angle no smaller than about 3×10^{-4} rad. (a) How many degrees is this? (b) How small an object can the bird just distinguish when flying at a height of 100 m (Fig. 8–2a)?

APPROACH For (a) we use the relation $360° = 2\pi$ rad. For (b) we use Eq. 8–1b, $\ell = r\theta$, to find the arc length.

SOLUTION (a) We convert 3×10^{-4} rad to degrees:

$$(3 \times 10^{-4} \text{ rad})\left(\frac{360°}{2\pi \text{ rad}}\right) = 0.017°.$$

(b) We use Eq. 8–1b, $\ell = r\theta$. For small angles, the arc length ℓ and the chord length are approximately[†] the same (Fig. 8–2b). Since $r = 100$ m and $\theta = 3 \times 10^{-4}$ rad, we find

$$\ell = r\theta = (100 \text{ m})(3 \times 10^{-4} \text{ rad}) = 3 \times 10^{-2} \text{ m} = 3 \text{ cm}.$$

A bird can distinguish a small mouse (about 3 cm long) from a height of 100 m. That is good eyesight.

NOTE Had the angle been given in degrees, we would first have had to convert it to radians to make this calculation. Equations 8–1 are valid *only* if the angle is specified in radians. Degrees (or revolutions) won't work.

To describe rotational motion, we make use of angular quantities, such as angular velocity and angular acceleration. These are defined in analogy to the corresponding quantities in linear motion, and are chosen to describe the rotating object as a whole, so they are the same for each point in the rotating object. Each point in a rotating object may also have translational velocity and acceleration, but they have different values for different points in the object.

When an object such as the bicycle wheel in Fig. 8–3 rotates from some initial position, specified by θ_1, to some final position, θ_2, its *angular displacement* is

$$\Delta\theta = \theta_2 - \theta_1.$$

The *angular velocity* (denoted by ω, the Greek lowercase letter omega) is defined in analogy with linear (translational) velocity that was discussed in Chapter 2. Instead of linear displacement, we use the angular displacement. Thus the **average angular velocity** of an object rotating about a fixed axis is defined as

$$\bar{\omega} = \frac{\Delta\theta}{\Delta t}, \quad (8\text{–}2a)$$

where $\Delta\theta$ is the angle through which the object has rotated in the time interval Δt. The **instantaneous angular velocity** is the limit of this ratio as Δt approaches zero:

$$\omega = \lim_{\Delta t \to 0} \frac{\Delta\theta}{\Delta t}. \quad (8\text{–}2b)$$

Angular velocity is generally specified in radians per second (rad/s). Note that *all points in a rigid object rotate with the same angular velocity*, since every position in the object moves through the same angle in the same time interval.

An object such as the wheel in Fig. 8–3 can rotate about a fixed axis either clockwise or counterclockwise. The direction can be specified with a + or − sign. The usual convention is to choose the angular displacement $\Delta\theta$ and angular velocity ω as positive when the wheel rotates counterclockwise. If the rotation is clockwise, then θ would decrease, so $\Delta\theta$ and ω would be negative.

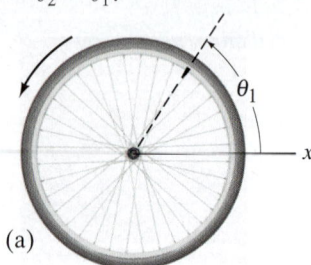
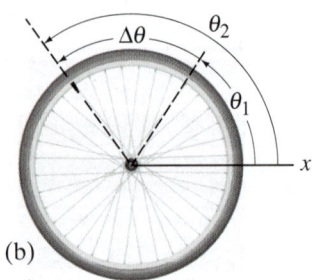

FIGURE 8–3 A wheel rotates about its axle from (a) initial position θ_1 to (b) final position θ_2. The angular displacement is $\Delta\theta = \theta_2 - \theta_1$.

[†]Even for an angle as large as 15°, the error in making this estimate is only 1%, but for larger angles the error increases rapidly. (The **chord** is the straight-line distance between the ends of the arc.)

Angular acceleration (denoted by α, the Greek lowercase letter alpha), in analogy to linear acceleration, is defined as the change in angular velocity divided by the time required to make this change. The **average angular acceleration** is defined as

$$\bar{\alpha} = \frac{\omega_2 - \omega_1}{\Delta t} = \frac{\Delta \omega}{\Delta t}, \quad (8\text{-}3\text{a})$$

where ω_1 is the angular velocity initially, and ω_2 is the angular velocity after a time interval Δt. **Instantaneous angular acceleration** is defined as the limit of this ratio as Δt approaches zero:

$$\alpha = \lim_{\Delta t \to 0} \frac{\Delta \omega}{\Delta t}. \quad (8\text{-}3\text{b})$$

Since ω is the same for all points of a rotating object, Eq. 8–3 tells us that α also will be the same for all points. Thus, ω and α are properties of the rotating object as a whole. With ω measured in radians per second and t in seconds, α has units of radians per second squared (rad/s^2).

Each point or particle of a rotating object has, at any moment, a linear velocity v and a linear acceleration a. We can now relate the linear quantities at each point, v and a, to the angular quantities, ω and α, for a rigid object rotating about a fixed axis. Consider a point P located a distance r from the axis of rotation, as in Fig. 8–4. If the object rotates with angular velocity ω, any point will have a linear velocity whose direction is tangent to its circular path. The magnitude of that point's linear velocity is $v = \Delta \ell / \Delta t$. From Eq. 8–1b, a change in rotation angle $\Delta \theta$ (in radians) is related to the linear distance traveled by $\Delta \ell = r \Delta \theta$. Hence

$$v = \frac{\Delta \ell}{\Delta t} = r \frac{\Delta \theta}{\Delta t}$$

or (since $\Delta \theta / \Delta t = \omega$)

$$v = r\omega. \quad (8\text{-}4)$$

In this very useful Eq. 8–4, r is the distance of a point from the rotation axis and ω is given in rad/s. Thus, although ω is the same for every point in the rotating object at any instant, the linear velocity v is greater for points farther from the axis (Fig. 8–5). Note that Eq. 8–4 is valid both instantaneously and on average.

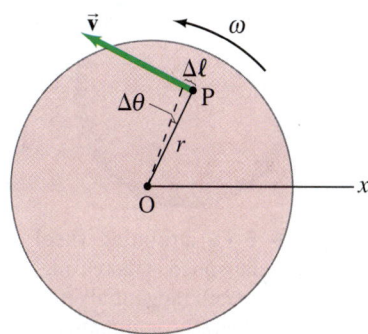

FIGURE 8–4 A point P on a rotating wheel has a linear velocity $\vec{v}$ at any moment.

FIGURE 8–5 A wheel rotating uniformly counterclockwise. Two points on the wheel, at distances r_A and r_B from the center, have the same angular velocity ω because they travel through the same angle θ in the same time interval. But the two points have different linear velocities because they travel different distances in the same time interval. Since $v = r\omega$ and $r_B > r_A$, then $v_B > v_A$.

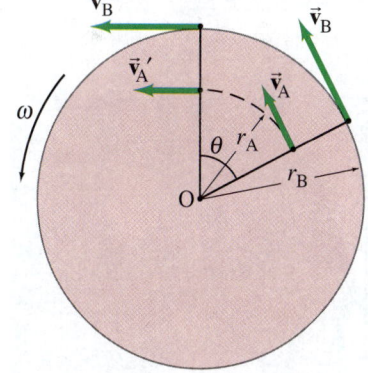

CONCEPTUAL EXAMPLE 8–3 **Is the lion faster than the horse?** On a rotating carousel or merry-go-round, one child sits on a horse near the outer edge and another child sits on a lion halfway out from the center. (*a*) Which child has the greater linear velocity? (*b*) Which child has the greater angular velocity?

RESPONSE (*a*) The *linear* velocity is the distance traveled divided by the time interval. In one rotation the child on the outer edge travels a longer distance than the child near the center, but the time interval is the same for both. Thus the child at the outer edge, on the horse, has the greater linear velocity.

(*b*) The *angular* velocity is the angle of rotation of the carousel as a whole divided by the time interval. For example, in one rotation both children rotate through the same angle (360° or 2π radians). The two children have the same angular velocity.

If the angular velocity of a rotating object changes, the object as a whole—and each point in it—has an angular acceleration. Each point also has a linear acceleration whose direction is tangent to that point's circular path. We use Eq. 8–4 ($v = r\omega$) to see that the angular acceleration α is related to the tangential linear acceleration a_{tan} of a point in the rotating object by

$$a_{\text{tan}} = \frac{\Delta v}{\Delta t} = r \frac{\Delta \omega}{\Delta t}$$

or (using Eq. 8–3)

$$a_{\text{tan}} = r\alpha. \quad (8\text{-}5)$$

In this equation, r is the radius of the circle in which the particle is moving, and the subscript "tan" in a_{tan} stands for "tangential."

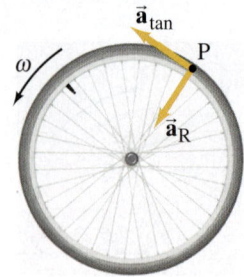

FIGURE 8–6 On a rotating wheel whose angular speed is increasing, a point P has both tangential and radial (centripetal) components of linear acceleration. (See also Chapter 5.)

The total linear acceleration of a point in the rotating object is the vector sum of two components:
$$\vec{a} = \vec{a}_{tan} + \vec{a}_R,$$
where the radial component, $\vec{a}_R$, is the radial or "centripetal" acceleration and its direction is toward the center of the point's circular path; see Fig. 8–6. We saw in Chapter 5 (Eq. 5–1) that a particle moving in a circle of radius r with linear speed v has a radial acceleration $a_R = v^2/r$. We can rewrite this in terms of ω using Eq. 8–4:

$$a_R = \frac{v^2}{r} = \frac{(r\omega)^2}{r} = \omega^2 r. \qquad (8\text{–}6)$$

Thus the centripetal acceleration is greater the farther you are from the axis of rotation: the children farthest out on a carousel feel the greatest acceleration.

Equations 8–1, 8–4, 8–5, and 8–6 relate the angular quantities describing the rotation of an object to the linear quantities for each point of a rotating object. Table 8–1 summarizes these relationships.

TABLE 8–1 Linear and Rotational Quantities

Linear	Type	Rotational	Relation[‡]
x	displacement	θ	$x = r\theta$
v	velocity	ω	$v = r\omega$
a_{tan}	acceleration	α	$a_{tan} = r\alpha$

[‡] You must use radians.

FIGURE 8–7 Examples 8–4 and 8–5. The total acceleration vector is $\vec{a} = \vec{a}_{tan} + \vec{a}_R$, at $t = 8.0$ s.

EXAMPLE 8–4 Angular and linear velocities. A carousel is initially at rest. At $t = 0$ it is given a constant angular acceleration $\alpha = 0.060 \text{ rad/s}^2$, which increases its angular velocity for 8.0 s. At $t = 8.0$ s, determine (a) the angular velocity of the carousel, and (b) the linear velocity of a child (Fig. 8–7a) located 2.5 m from the center, point P in Fig. 8–7b.

APPROACH The angular acceleration α is constant, so we can use $\alpha = \Delta\omega/\Delta t$ (Eq. 8–3a) to solve for ω after a time $t = 8.0$ s. With this ω, we determine the linear velocity using Eq. 8–4, $v = r\omega$.

SOLUTION (a) In Eq. 8–3a, $\bar{\alpha} = (\omega_2 - \omega_1)/\Delta t$, we put $\Delta t = 8.0$ s, $\bar{\alpha} = 0.060 \text{ rad/s}^2$, and $\omega_1 = 0$. Solving for ω_2, we get
$$\omega_2 = \omega_1 + \bar{\alpha}\,\Delta t = 0 + (0.060 \text{ rad/s}^2)(8.0 \text{ s}) = 0.48 \text{ rad/s}.$$
During the 8.0-s time interval, the carousel accelerates from $\omega_1 = 0$ to $\omega_2 = 0.48$ rad/s.

(b) The linear velocity of the child with $r = 2.5$ m at time $t = 8.0$ s is found using Eq. 8–4:
$$v = r\omega = (2.5 \text{ m})(0.48 \text{ rad/s}) = 1.2 \text{ m/s}.$$
Note that the "rad" has been omitted in the final result because it is dimensionless (and only a reminder)—it is a ratio of two distances, Eq. 8–1a.

EXAMPLE 8–5 Angular and linear accelerations. For the child on the rotating carousel of Example 8–4, determine that child's (a) tangential (linear) acceleration, (b) centripetal acceleration, (c) total acceleration.

APPROACH We use the relations discussed above, Eqs. 8–5 and 8–6.

SOLUTION (a) The child's tangential acceleration is given by Eq. 8–5:
$$a_{tan} = r\alpha = (2.5 \text{ m})(0.060 \text{ rad/s}^2) = 0.15 \text{ m/s}^2,$$
and it is the same throughout the 8.0-s acceleration period.

(b) The child's centripetal acceleration at $t = 8.0$ s is given by Eq. 8–6:
$$a_R = \frac{v^2}{r} = \frac{(1.2 \text{ m/s})^2}{(2.5 \text{ m})} = 0.58 \text{ m/s}^2.$$

(*c*) The two components of linear acceleration calculated in parts (*a*) and (*b*) are perpendicular to each other. Thus the total linear acceleration at $t = 8.0\,\text{s}$ has magnitude

$$a = \sqrt{a_{\text{tan}}^2 + a_R^2} = \sqrt{(0.15\,\text{m/s}^2)^2 + (0.58\,\text{m/s}^2)^2} = 0.60\,\text{m/s}^2.$$

NOTE The linear acceleration at this chosen instant is mostly centripetal, and keeps the child moving in a circle with the carousel. The tangential component that speeds up the circular motion is smaller.

NOTE The direction of the linear acceleration (magnitude calculated above as $0.60\,\text{m/s}^2$) is at the angle θ shown in Fig. 8–7b:

$$\theta = \tan^{-1}\left(\frac{a_{\text{tan}}}{a_R}\right) = \tan^{-1}\left(\frac{0.15\,\text{m/s}^2}{0.58\,\text{m/s}^2}\right) = 0.25\,\text{rad},$$

so $\theta \approx 15°$.

We can relate the angular velocity ω to the frequency of rotation, f. The **frequency** is the number of complete revolutions (rev) per second, as we saw in Chapter 5. One revolution (of a wheel, say) corresponds to an angle of 2π radians, and thus $1\,\text{rev/s} = 2\pi\,\text{rad/s}$. Hence, in general, the frequency f is related to the angular velocity ω by

$$f = \frac{\omega}{2\pi}$$

or

$$\omega = 2\pi f. \qquad (8\text{–}7)$$

The unit for frequency, revolutions per second (rev/s), is given the special name the hertz (Hz). That is,

$$1\,\text{Hz} = 1\,\text{rev/s}.$$

Note that "revolution" is not really a unit, so we can also write $1\,\text{Hz} = 1\,\text{s}^{-1}$.

The time required for one complete revolution is called the **period** T, and it is related to the frequency by

$$T = \frac{1}{f}. \qquad (8\text{–}8)$$

If a particle rotates at a frequency of three revolutions per second, then the period of each revolution is $\tfrac{1}{3}\,\text{s}$.

EXERCISE A In Example 8–4 we found that the carousel, after 8.0 s, rotates at an angular velocity $\omega = 0.48\,\text{rad/s}$, and continues to do so after $t = 8.0\,\text{s}$ because the acceleration ceased. What are the frequency and period of the carousel when rotating at this constant angular velocity $\omega = 0.48\,\text{rad/s}$?

8–2 Constant Angular Acceleration

In Chapter 2, we derived the useful kinematic equations (Eqs. 2–11) that relate acceleration, velocity, distance, and time for the special case of uniform linear acceleration. Those equations were derived from the definitions of linear velocity and acceleration, assuming constant acceleration. The definitions of angular velocity and angular acceleration (Eqs. 8–2 and 8–3) are just like those for their linear counterparts, except that θ replaces the linear displacement x, ω replaces v, and α replaces a. Therefore, the angular equations for **constant angular acceleration** will be analogous to Eqs. 2–11 with x replaced by θ, v by ω, and a by α, and they can be derived in exactly the same way.

We summarize these angular equations here, opposite their linear equivalents, Eqs. 2–11 (for simplicity we choose $\theta_0 = 0$ and $x_0 = 0$ at the initial time $t_0 = 0$):

Kinematic equations for constant angular acceleration
$[x_0 = 0, \theta_0 = 0]$

Angular	Linear		
$\omega = \omega_0 + \alpha t$	$v = v_0 + at$	[constant α, a]	(8–9a)
$\theta = \omega_0 t + \frac{1}{2}\alpha t^2$	$x = v_0 t + \frac{1}{2}at^2$	[constant α, a]	(8–9b)
$\omega^2 = \omega_0^2 + 2\alpha\theta$	$v^2 = v_0^2 + 2ax$	[constant α, a]	(8–9c)
$\bar{\omega} = \dfrac{\omega + \omega_0}{2}$	$\bar{v} = \dfrac{v + v_0}{2}$	[constant α, a]	(8–9d)

Note that ω_0 represents the angular velocity at $t_0 = 0$, whereas θ and ω represent the angular position and velocity, respectively, at time t. Since the angular acceleration is constant, $\alpha = \bar{\alpha}$.

PHYSICS APPLIED
Centrifuge

EXAMPLE 8–6 Centrifuge acceleration. A centrifuge rotor is accelerated for 30 s from rest to 20,000 rpm (revolutions per minute). (*a*) What is its average angular acceleration? (*b*) Through how many revolutions has the centrifuge rotor turned during its acceleration period, assuming constant angular acceleration?

APPROACH To determine $\bar{\alpha} = \Delta\omega/\Delta t$, we need the initial and final angular velocities. For (*b*), we use Eqs. 8–9 (recall that one revolution corresponds to $\theta = 2\pi$ rad).

SOLUTION (*a*) The initial angular velocity is $\omega_0 = 0$. The final angular velocity is

$$\omega = 2\pi f = (2\pi \text{ rad/rev}) \frac{(20{,}000 \text{ rev/min})}{(60 \text{ s/min})} = 2100 \text{ rad/s}.$$

Then, since $\bar{\alpha} = \Delta\omega/\Delta t$ and $\Delta t = 30$ s, we have

$$\bar{\alpha} = \frac{\omega - \omega_0}{\Delta t} = \frac{2100 \text{ rad/s} - 0}{30 \text{ s}} = 70 \text{ rad/s}^2.$$

That is, every second the rotor's angular velocity increases by 70 rad/s, or by $(70 \text{ rad/s})(1 \text{ rev}/2\pi \text{ rad}) = 11$ revolutions per second.

(*b*) To find θ we could use either Eq. 8–9b or 8–9c (or both to check our answer). The former gives

$$\theta = \omega_0 t + \tfrac{1}{2}\alpha t^2 = 0 + \tfrac{1}{2}(70 \text{ rad/s}^2)(30 \text{ s})^2 = 3.15 \times 10^4 \text{ rad},$$

where we have kept an extra digit because this is an intermediate result. To find the total number of revolutions, we divide by 2π rad/rev and obtain

$$\frac{3.15 \times 10^4 \text{ rad}}{2\pi \text{ rad/rev}} = 5.0 \times 10^3 \text{ rev}.$$

NOTE Let us calculate θ using Eq. 8–9c:

$$\theta = \frac{\omega^2 - \omega_0^2}{2\alpha} = \frac{(2100 \text{ rad/s})^2 - 0}{2(70 \text{ rad/s}^2)} = 3.15 \times 10^4 \text{ rad}$$

which checks our answer above from Eq. 8–9b perfectly.

8–3 Rolling Motion (Without Slipping)

The rolling motion of a ball or wheel is familiar in everyday life: a ball rolling across the floor, or the wheels and tires of a car or bicycle rolling along the pavement. Rolling *without slipping* depends on static friction between the rolling object and the ground. The friction is static because the rolling object's point of contact with the ground is at rest at each moment.

Rolling without slipping involves both rotation and translation. There is a simple relation between the linear speed v of the axle and the angular velocity ω of the rotating wheel or sphere: namely, $v = r\omega$ (where r is the radius) as we now show.

Figure 8–8a shows a wheel rolling to the right without slipping. At the instant shown, point P on the wheel is in contact with the ground and is momentarily at rest. (If P was not at rest, the wheel would be slipping.) The velocity of the axle at the wheel's center C is $\vec{v}$. In Fig. 8–8b we have put ourselves in the reference frame of the wheel—that is, we are moving to the right with velocity $\vec{v}$ relative to the ground. In this reference frame the axle C is at rest, whereas the ground and point P are moving to the left with velocity $-\vec{v}$ as shown. In Fig. 8–8b we are seeing pure rotation. So we can use Eq. 8–4 to obtain $v = r\omega$, where r is the radius of the wheel. This is the same v as in Fig. 8–8a, so we see that the linear speed v of the axle relative to the ground is related to the angular velocity ω of the wheel by

$$v = r\omega. \qquad \text{[rolling without slipping]}$$

This relationship is valid only if there is no slipping.

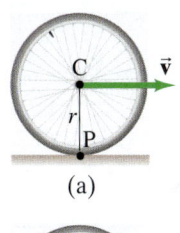

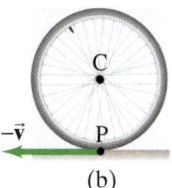

(a)

(b)

FIGURE 8–8 (a) A wheel rolling to the right. Its center C moves with velocity $\vec{v}$. Point P is at rest at the instant shown. (b) The same wheel as seen from a reference frame in which the axle of the wheel C is at rest—that is, we are moving to the right with velocity $\vec{v}$ relative to the ground. Point P, which was at rest in (a), here in (b) is moving to the left with velocity $-\vec{v}$ as shown. (See also Section 3–8 on relative velocity.) Thus $v = r\omega$.

EXAMPLE 8–7 Bicycle. A bicycle slows down uniformly from $v_0 = 8.40$ m/s to rest over a distance of 115 m, Fig. 8–9. Each wheel and tire has an overall diameter of 68.0 cm. Determine (a) the angular velocity of the wheels at the initial instant ($t = 0$); (b) the total number of revolutions each wheel rotates before coming to rest; (c) the angular acceleration of the wheel; and (d) the time it took to come to a stop.

APPROACH We assume the bicycle wheels roll without slipping and the tire is in firm contact with the ground. The speed of the bike v and the angular velocity of the wheels ω are related by $v = r\omega$. The bike slows down uniformly, so the angular acceleration is constant and we can use Eqs. 8–9.

SOLUTION (a) The initial angular velocity of the wheel, whose radius is 34.0 cm, is

$$\omega_0 = \frac{v_0}{r} = \frac{8.40 \text{ m/s}}{0.340 \text{ m}} = 24.7 \text{ rad/s}.$$

(b) In coming to a stop, the bike passes over 115 m of ground. The circumference of the wheel is $2\pi r$, so each revolution of the wheel corresponds to a distance traveled of $2\pi r = (2\pi)(0.340 \text{ m})$. Thus the number of revolutions the wheel makes in coming to a stop is

$$\frac{115 \text{ m}}{2\pi r} = \frac{115 \text{ m}}{(2\pi)(0.340 \text{ m})} = 53.8 \text{ rev}.$$

(c) The angular acceleration of the wheel can be obtained from Eq. 8–9c, for which we set $\omega = 0$ and $\omega_0 = 24.7$ rad/s. Because each revolution corresponds to 2π radians of angle, then $\theta = 2\pi$ rad/rev × 53.8 rev ($= 338$ rad) and

$$\alpha = \frac{\omega^2 - \omega_0^2}{2\theta} = \frac{0 - (24.7 \text{ rad/s})^2}{2(2\pi \text{ rad/rev})(53.8 \text{ rev})} = -0.902 \text{ rad/s}^2.$$

(d) Equation 8–9a or b allows us to solve for the time. The first is easier:

$$t = \frac{\omega - \omega_0}{\alpha} = \frac{0 - 24.7 \text{ rad/s}}{-0.902 \text{ rad/s}^2} = 27.4 \text{ s}.$$

NOTE When the bike tire completes one revolution, the bike advances linearly a distance equal to the outer circumference ($2\pi r$) of the tire, as long as there is no slipping or sliding.

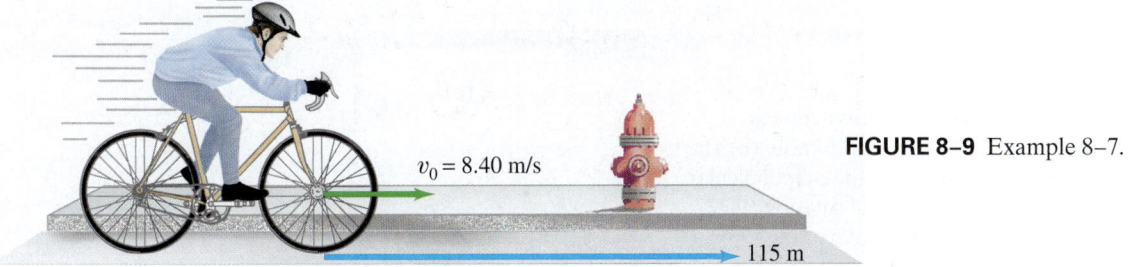

$v_0 = 8.40$ m/s

115 m

Bike as seen from the ground at $t = 0$

FIGURE 8–9 Example 8–7.

8–4 Torque

We have so far discussed rotational kinematics—the description of rotational motion in terms of angular position, angular velocity, and angular acceleration. Now we discuss the dynamics, or causes, of rotational motion. Just as we found analogies between linear and rotational motion for the description of motion, so rotational equivalents for dynamics exist as well.

To make an object start rotating about an axis clearly requires a force. But the direction of this force, and where it is applied, are also important. Take, for example, an ordinary situation such as the overhead view of the door in Fig. 8–10. If you apply a force $\vec{F}_A$ perpendicular to the door as shown, you will find that the greater the magnitude, F_A, the more quickly the door opens. But now if you apply the same force at a point closer to the hinge—say, $\vec{F}_B$ in Fig. 8–10—the door will not open so quickly. The effect of the force is less: *where* the force acts, as well as its magnitude and direction, affects how quickly the door opens. Indeed, if only this one force acts, the angular acceleration of the door is proportional not only to the magnitude of the force, but is also directly proportional to *the perpendicular distance from the axis of rotation to the line along which the force acts*. This distance is called the **lever arm**, or **moment arm**, of the force, and is labeled r_A and r_B for the two forces in Fig. 8–10. Thus, if r_A in Fig. 8–10 is three times larger than r_B, then the angular acceleration of the door will be three times as great, assuming that the magnitudes of the forces are the same. To say it another way, if $r_A = 3r_B$, then F_B must be three times as large as F_A to give the same angular acceleration. (Figure 8–11 shows two examples of tools whose long lever arms are very effective.)

The angular acceleration, then, is proportional to the product of the *force times the lever arm*. This product is called the *moment of the force* about the axis, or, more commonly, it is called the **torque**, and is represented by τ (Greek lowercase letter tau). Thus, the angular acceleration α of an object is directly proportional to the net applied torque τ:

$$\alpha \propto \tau,$$

and we see that it is torque that gives rise to angular acceleration. This is the rotational analog of Newton's second law for linear motion, $a \propto F$.

We defined the lever arm as the *perpendicular* distance from the axis of rotation to the line of action of the force—that is, the distance which is perpendicular both to the axis of rotation and to an imaginary line drawn along the direction of the force. We do this to take into account the effect of forces acting at an angle. It is clear that a force applied at an angle, such as $\vec{F}_C$ in Fig. 8–12, will be less effective than the same magnitude force applied perpendicular to the door, such as $\vec{F}_A$ (Fig. 8–12a). And if you push on the end of the door so that the force is directed at the hinge (the axis of rotation), as indicated by $\vec{F}_D$, the door will not rotate at all.

The lever arm for a force such as $\vec{F}_C$ is found by drawing a line along the direction of $\vec{F}_C$ (this is the "line of action" of $\vec{F}_C$). Then we draw another line, perpendicular to this line of action, that goes to the axis of rotation and is perpendicular also to it. The length of this second line is the lever arm for $\vec{F}_C$ and is labeled r_C in Fig. 8–12b. The lever arm for $\vec{F}_A$ is the full distance from the hinge to the doorknob, r_A (just as in Fig. 8–10). Thus r_C is much smaller than r_A.

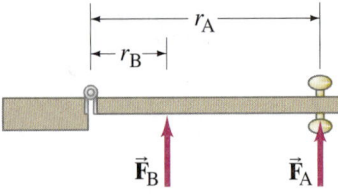

FIGURE 8–10 Top view of a door. Applying the same force with different lever arms, r_A and r_B. If $r_A = 3r_B$, then to create the same effect (angular acceleration), F_B needs to be three times F_A.

FIGURE 8–11 (a) A plumber can exert greater torque using a wrench with a long lever arm. (b) A tire iron too can have a long lever arm.

(a)   (b)

FIGURE 8–12 (a) Forces acting at different angles at the doorknob. (b) The lever arm is defined as the perpendicular distance from the axis of rotation (the hinge) to the line of action of the force (r_C for the force $\vec{F}_C$).

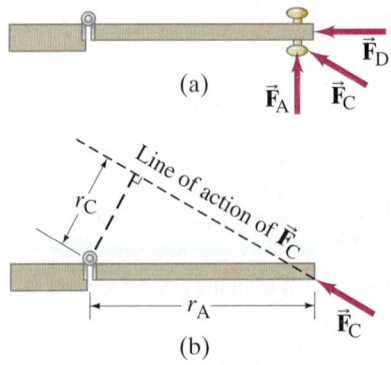

The magnitude of the torque associated with $\vec{F}_C$ is then $r_C F_C$. This short lever arm r_C and the corresponding smaller torque associated with $\vec{F}_C$ are consistent with the observation that $\vec{F}_C$ is less effective in accelerating the door than is $\vec{F}_A$ with its larger lever arm. When the lever arm is defined in this way, experiment shows that the relation $\alpha \propto \tau$ is valid in general. Notice in Fig. 8–12 that the line of action of the force $\vec{F}_D$ passes through the hinge, and hence its lever arm is zero. Consequently, zero torque is associated with $\vec{F}_D$ and it gives rise to no angular acceleration, in accord with everyday experience (you can't get a door to start moving by pushing directly at the hinge).

In general, then, we can write the magnitude of the torque about a given axis as

$$\tau = r_\perp F, \tag{8-10a}$$

where $r_\perp$ is the lever arm, and the perpendicular symbol ($\perp$) reminds us that we must use the distance from the axis of rotation that is perpendicular to the line of action of the force (Fig. 8–13a).

An equivalent way of determining the torque associated with a force is to resolve the force into components parallel and perpendicular to the line that connects the axis to the point of application of the force, as shown in Fig. 8–13b. The component $F_\parallel$ exerts no torque since it is directed at the rotation axis (its lever arm is zero). Hence the torque will be equal to $F_\perp$ times the distance r from the axis to the point of application of the force:

$$\tau = rF_\perp. \tag{8-10b}$$

This gives the same result as Eq. 8–10a because $F_\perp = F \sin\theta$ and $r_\perp = r \sin\theta$. Thus

$$\tau = rF \sin\theta \tag{8-10c}$$

in either case. [Note that θ is the angle between the directions of $\vec{F}$ and r (radial line from the axis to the point where $\vec{F}$ acts).] We can use any of Eqs. 8–10 to calculate the torque, whichever is easiest.

Because torque is a distance times a force, it is measured in units of $m \cdot N$ in SI units,[†] $cm \cdot dyne$ in the cgs system, and $ft \cdot lb$ in the English system.

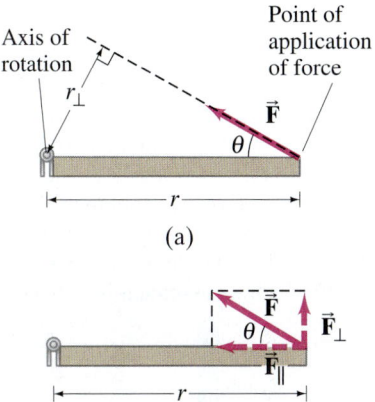

FIGURE 8–13 Torque $= r_\perp F = rF_\perp$.

EXAMPLE 8–8 Biceps torque. The biceps muscle exerts a vertical force on the lower arm, bent as shown in Figs. 8–14a and b. For each case, calculate the torque about the axis of rotation through the elbow joint, assuming the muscle is attached 5.0 cm from the elbow as shown.

APPROACH The force is given, and the lever arm in (a) is given. In (b) we have to take into account the angle to get the lever arm.

SOLUTION (a) $F = 700 \text{ N}$ and $r_\perp = 0.050 \text{ m}$, so

$$\tau = r_\perp F = (0.050 \text{ m})(700 \text{ N}) = 35 \text{ m} \cdot \text{N}.$$

(b) Because the arm is at an angle below the horizontal, the lever arm is shorter (Fig. 8–14c) than in part (a): $r_\perp = (0.050 \text{ m})(\sin 60°)$, where $\theta = 60°$ is the angle between $\vec{F}$ and r. F is still 700 N, so

$$\tau = (0.050 \text{ m})(0.866)(700 \text{ N}) = 30 \text{ m} \cdot \text{N}.$$

The arm can exert less torque at this angle than when it is at 90°. Weight machines at gyms are often designed to take this variation with angle into account.

NOTE In (b), we could instead have used $\tau = rF_\perp$. As shown in Fig. 8–14d, $F_\perp = F \sin 60°$. Then $\tau = rF_\perp = rF \sin\theta = (0.050 \text{ m})(700 \text{ N})(0.866)$ gives the same result.

FIGURE 8–14 Example 8–8.

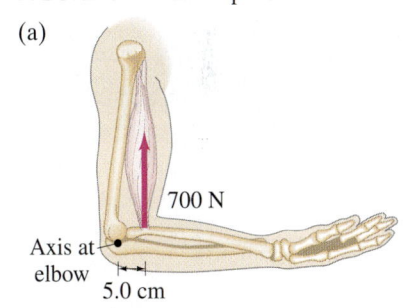

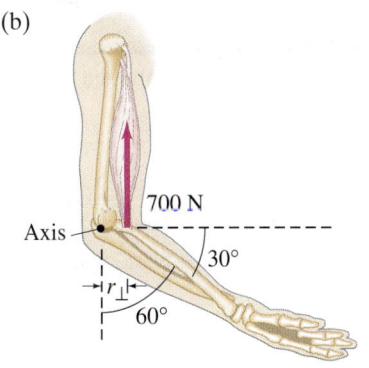

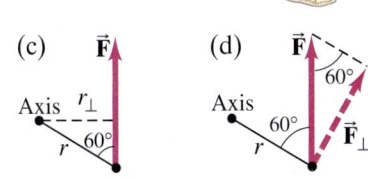

[†]Note that the units for torque are the same as those for energy. We write the unit for torque here as $m \cdot N$ (in SI) to distinguish it from energy ($N \cdot m$) because the two quantities are very different. The special name *joule* (1 J = 1 N·m) is used only for energy (and for work), *never* for torque.

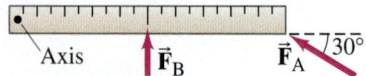

FIGURE 8–15 Exercise B.

EXERCISE B Two forces ($F_B = 20$ N and $F_A = 30$ N) are applied to a meter stick which can rotate about its left end, Fig. 8–15. Force $\vec{F}_B$ is applied perpendicularly at the midpoint. Which force exerts the greater torque: F_A, F_B, or both the same?

When more than one torque acts on an object, the angular acceleration α is found to be proportional to the *net* torque. If all the torques acting on an object tend to rotate it in the same direction about a fixed axis of rotation, the net torque is the sum of the torques. But if, say, one torque acts to rotate an object in one direction, and a second torque acts to rotate the object in the opposite direction, the net torque is the difference of the two torques. We normally assign a positive sign to torques that act to rotate the object counterclockwise (just as θ is usually positive counterclockwise), and a negative sign to torques that act to rotate the object clockwise.

FIGURE 8–16 Only the component of $\vec{F}$ that acts in the plane perpendicular to the rotation axis, $\vec{F}_\perp$, acts to accelerate the wheel about the axis. The component parallel to the axis, $\vec{F}_\parallel$, would tend to move the axis itself, which we assume is held fixed.

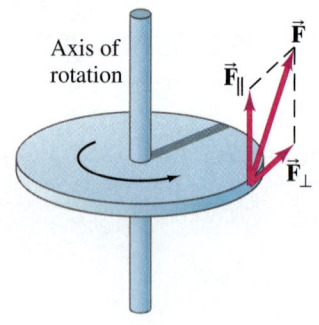

*Forces that Act to Tilt the Axis

We have been considering only rotation about a fixed axis, and so we considered only forces that act in a plane perpendicular to the axis of rotation. If there is a force (or component of a force) acting parallel to the axis of rotation, it will tend to tilt the axis of rotation—the component $\vec{F}_\parallel$ in Fig. 8–16 is an example. Since we are assuming the axis remains fixed in direction, either there can be no such forces or else the axis must be mounted in bearings or hinges that hold the axis fixed. Thus, only a force, or component of a force ($\vec{F}_\perp$ in Fig. 8–16), in a plane perpendicular to the axis will give rise to rotational acceleration about the axis.

8–5 Rotational Dynamics; Torque and Rotational Inertia

We discussed in Section 8–4 that the angular acceleration α of a rotating object is proportional to the net torque τ applied to it:

$$\alpha \propto \Sigma \tau.$$

We write $\Sigma \tau$ to remind us that it is the *net* torque (sum of all torques acting on the object) that is proportional to α. This corresponds to Newton's second law for translational motion, $a \propto \Sigma F$. In the translational case, the acceleration is not only proportional to the net force, but it is also inversely proportional to the inertia of the object, which we call its mass, m. Thus we wrote $a = \Sigma F/m$. But what plays the role of mass for the rotational case? That is what we now set out to determine. At the same time, we will see that the relation $\alpha \propto \Sigma \tau$ follows directly from Newton's second law, $\Sigma F = ma$.

We first examine a very simple case: a particle of mass m revolving in a circle of radius r at the end of a string or rod whose mass we can ignore compared to m (Fig. 8–17). Consider a force F that acts on the mass m tangent to the circle as shown. The torque that gives rise to an angular acceleration is $\tau = rF$. If we use Newton's second law for linear quantities, $\Sigma F = ma$, and Eq. 8–5 relating the angular acceleration to the tangential linear acceleration, $a_{\text{tan}} = r\alpha$, then we have

$$F = ma$$
$$= mr\alpha.$$

When we multiply both sides of this equation by r, we find that the torque

$$\tau = rF = r(mr\alpha),$$

or

$$\tau = mr^2\alpha. \qquad \text{[single particle]} \quad (8\text{–}11)$$

FIGURE 8–17 A mass m revolving in a circle of radius r about a fixed point C.

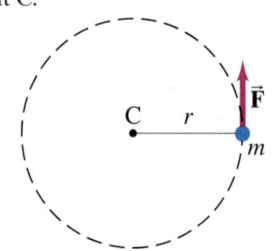

Here at last we have a direct relation between the angular acceleration and the applied torque τ. The quantity mr^2 represents the *rotational inertia* of the particle and is called its **moment of inertia**.

Now let us consider a rotating rigid object, such as a wheel rotating about a fixed axis (an axle) through its center. We can think of the wheel as consisting of many particles located at various distances from the axis of rotation. We can apply Eq. 8–11 to each particle of the object, and then sum over all the particles. The sum of the various torques is the net torque, $\Sigma \tau$, so we obtain:

$$\Sigma \tau = (\Sigma m r^2) \alpha \tag{8-12}$$

where we factored out α because it is the same for all the particles of a rigid object. The sum $\Sigma m r^2$ represents the sum of the masses of each particle in the object multiplied by the square of the distance of that particle from the axis of rotation. If we assign each particle a number (1, 2, 3, ...), then $\Sigma m r^2 = m_1 r_1^2 + m_2 r_2^2 + m_3 r_3^2 + \cdots$. This sum is called the **moment of inertia** (or *rotational inertia*) I of the object:

$$I = \Sigma m r^2 = m_1 r_1^2 + m_2 r_2^2 + \cdots. \tag{8-13}$$

Combining Eqs. 8–12 and 8–13, we can write

$$\boxed{\Sigma \tau = I \alpha.} \tag{8-14}$$

NEWTON'S SECOND LAW FOR ROTATION

This is the rotational equivalent of Newton's second law. It is valid for the rotation of a rigid object about a fixed axis. [It is also valid when the object is rotating while translating with acceleration, as long as I and α are calculated about the center of mass of the object, and the rotation axis through the CM doesn't change direction. A ball rolling down a ramp is an example.]

We see that the moment of inertia, I, which is a measure of the rotational inertia of an object, plays the same role for rotational motion that mass does for translational motion. As can be seen from Eq. 8–13, the rotational inertia of a rigid object depends not only on its mass, but also on how that mass is distributed with respect to the axis. For example, a large-diameter cylinder will have greater rotational inertia than one of equal mass but smaller diameter, Fig. 8–18. The former will be harder to start rotating, and harder to stop. When the mass is concentrated farther from the axis of rotation, the rotational inertia is greater. For rotational motion, the mass of an object can *not* be considered as concentrated at its center of mass.

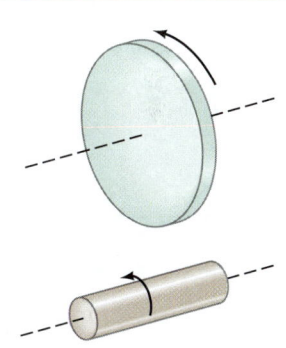

FIGURE 8–18 A large-diameter cylinder has greater rotational inertia than one of smaller diameter but equal mass.

EXAMPLE 8–9 Two weights on a bar: different axis, different I. Two small "weights," of mass 5.0 kg and 7.0 kg, are mounted 4.0 m apart on a light rod (whose mass can be ignored), as shown in Fig. 8–19. Calculate the moment of inertia of the system (a) when rotated about an axis halfway between the weights, Fig. 8–19a, and (b) when rotated about an axis 0.50 m to the left of the 5.0-kg mass (Fig. 8–19b).

APPROACH In each case, the moment of inertia of the system is found by summing over the two parts using Eq. 8–13.

SOLUTION (a) Both weights are the same distance, 2.0 m, from the axis of rotation. Thus

$$I = \Sigma m r^2 = (5.0\,\text{kg})(2.0\,\text{m})^2 + (7.0\,\text{kg})(2.0\,\text{m})^2$$
$$= 20\,\text{kg}\cdot\text{m}^2 + 28\,\text{kg}\cdot\text{m}^2 = 48\,\text{kg}\cdot\text{m}^2.$$

(b) The 5.0-kg mass is now 0.50 m from the axis, and the 7.0-kg mass is 4.50 m from the axis. Then

$$I = \Sigma m r^2 = (5.0\,\text{kg})(0.50\,\text{m})^2 + (7.0\,\text{kg})(4.5\,\text{m})^2$$
$$= 1.3\,\text{kg}\cdot\text{m}^2 + 142\,\text{kg}\cdot\text{m}^2 = 143\,\text{kg}\cdot\text{m}^2.$$

NOTE This Example illustrates two important points. First, the moment of inertia of a given system is different for different axes of rotation. Second, we see in part (b) that mass close to the axis of rotation contributes little to the total moment of inertia; here, the 5.0-kg object contributed less than 1% to the total.

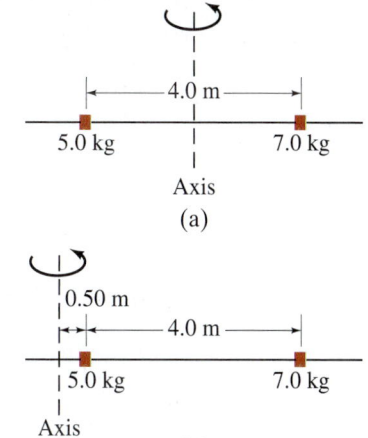

FIGURE 8–19 Example 8–9: calculating the moment of inertia.

CAUTION
I depends on axis of rotation and on distribution of mass

Object	Location of axis		Moment of inertia
(a) **Thin hoop**, radius R	Through center		MR^2
(b) **Thin hoop**, radius R width w	Through central diameter		$\frac{1}{2}MR^2 + \frac{1}{12}Mw^2$
(c) **Solid cylinder**, radius R	Through center		$\frac{1}{2}MR^2$
(d) **Hollow cylinder**, inner radius R_1 outer radius R_2	Through center		$\frac{1}{2}M(R_1^2 + R_2^2)$
(e) **Uniform sphere**, radius R	Through center		$\frac{2}{5}MR^2$
(f) **Long uniform rod**, length ℓ	Through center		$\frac{1}{12}M\ell^2$
(g) **Long uniform rod**, length ℓ	Through end		$\frac{1}{3}M\ell^2$
(h) **Rectangular thin plate**, length ℓ, width w	Through center		$\frac{1}{12}M(\ell^2 + w^2)$

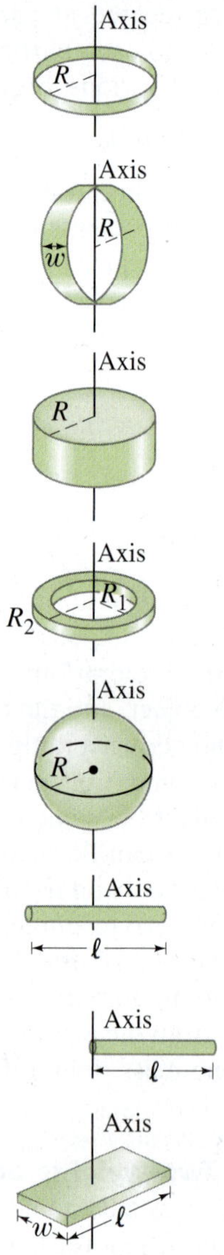

FIGURE 8–20 Moments of inertia for various objects of uniform composition, each with mass M.

For most ordinary objects, the mass is distributed continuously, and the calculation of the moment of inertia, Σmr^2, can be difficult. Expressions can, however, be worked out (using calculus) for the moments of inertia of regularly shaped objects in terms of the dimensions of the objects. Figure 8–20 gives these expressions for a number of solids rotated about the axes specified. The only one for which the result is obvious is that for the thin hoop or ring rotated about an axis passing through its center perpendicular to the plane of the hoop (Fig. 8–20a). For a hoop, all the mass is concentrated at the same distance from the axis, R. Thus $\Sigma mr^2 = (\Sigma m)R^2 = MR^2$, where M is the total mass of the hoop. In Fig. 8–20, we use capital R to refer to the outer radius of an object (in (d) also the inner radius).

When calculation is difficult, I can be determined experimentally by measuring the angular acceleration α about a fixed axis due to a known net torque, $\Sigma \tau$, and applying Newton's second law, $I = \Sigma \tau / \alpha$, Eq. 8–14.

8–6 Solving Problems in Rotational Dynamics

When working with torque and angular acceleration (Eq. 8–14), it is important to use a consistent set of units, which in SI is: α in rad/s²; τ in m·N; and the moment of inertia, I, in kg·m².

PROBLEM SOLVING

Rotational Motion

1. As always, draw a clear and complete **diagram**.
2. Choose the object or objects that will be the **system** to be studied.
3. Draw a **free-body diagram** for the object under consideration (or for each object, if more than one), showing all (and only) the forces acting on that object and exactly where they act, so you can determine the torque due to each. Gravity acts at the CM of the object (Section 7–8).
4. Identify the axis of rotation and determine the **torques** about it. Choose positive and negative directions of rotation (counterclockwise and clockwise), and assign the correct sign to each torque.
5. Apply **Newton's second law for rotation**, $\Sigma\tau = I\alpha$. If the moment of inertia is not given, and it is not the unknown sought, you need to determine it first. Use consistent units, which in SI are: α in rad/s^2; τ in m·N; and I in kg·m^2.
6. Also apply **Newton's second law for translation**, $\Sigma\vec{F} = m\vec{a}$, and **other** laws or principles as needed.
7. **Solve** the resulting equation(s) for the unknown(s).
8. Do a rough **estimate** to determine if your answer is reasonable.

EXAMPLE 8–10 **A heavy pulley.** A 15.0-N force (represented by $\vec{F}_T$) is applied to a cord wrapped around a pulley of mass $M = 4.00$ kg and radius $R = 33.0$ cm, Fig. 8–21. The pulley accelerates uniformly from rest to an angular speed of 30.0 rad/s in 3.00 s. If there is a frictional torque $\tau_{fr} = 1.10$ m·N at the axle, determine the moment of inertia of the pulley. The pulley rotates about its center.

APPROACH We follow the steps of the Problem Solving Strategy above.

SOLUTION
1. **Draw a diagram.** The pulley and the attached cord are shown in Fig. 8–21.
2. **Choose the system**: the pulley.
3. **Draw a free-body diagram.** The force that the cord exerts on the pulley is shown as $\vec{F}_T$ in Fig. 8–21. The friction force acts all around the axle, retarding the motion, as suggested by $\vec{F}_{fr}$ in Fig. 8–21. We are given only its torque, which is what we need. Two other forces could be included in the diagram: the force of gravity mg down and whatever force keeps the axle in place (they balance each other). They do not contribute to the torque (their lever arms are zero) and so we omit them to keep our diagram simple.
4. **Determine the torques.** The cord exerts a force $\vec{F}_T$ that acts at the edge of the pulley, so its lever arm is R. The torque exerted by the cord equals RF_T and is counterclockwise, which we choose to be positive. The frictional torque is given as $\tau_{fr} = 1.10$ m·N; it opposes the motion and is negative.
5. **Apply Newton's second law for rotation.** The net torque is
$$\Sigma\tau = RF_T - \tau_{fr} = (0.330\text{ m})(15.0\text{ N}) - 1.10\text{ m·N} = 3.85\text{ m·N}.$$
The angular acceleration α is found from the given data that it takes 3.00 s to accelerate the pulley from rest to $\omega = 30.0$ rad/s:
$$\alpha = \frac{\Delta\omega}{\Delta t} = \frac{30.0\text{ rad/s} - 0}{3.00\text{ s}} = 10.0\text{ rad/s}^2.$$
Newton's second law, $\Sigma\tau = I\alpha$, can be solved for I which is the unknown: $I = \Sigma\tau/\alpha$.
6. **Other calculations**: None needed.
7. **Solve for unknowns.** From Newton's second law,
$$I = \frac{\Sigma\tau}{\alpha} = \frac{3.85\text{ m·N}}{10.0\text{ rad/s}^2} = 0.385\text{ kg·m}^2.$$
8. **Do a rough estimate.** We can do a rough estimate of the moment of inertia by assuming the pulley is a uniform cylinder and using Fig. 8–20c:
$$I \approx \tfrac{1}{2}MR^2 = \tfrac{1}{2}(4.00\text{ kg})(0.330\text{ m})^2 = 0.218\text{ kg·m}^2.$$
This is the same order of magnitude as our result, but numerically somewhat less. This makes sense, though, because a pulley is not usually a uniform cylinder but instead has more of its mass concentrated toward the outside edge. Such a pulley would be expected to have a greater moment of inertia than a solid cylinder of equal mass. A thin hoop, Fig. 8–20a, ought to have a greater I than our pulley, and indeed it does: $I = MR^2 = 0.436\text{ kg·m}^2$.

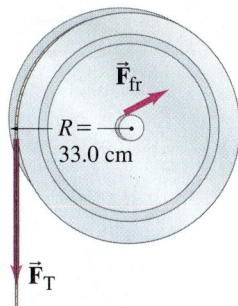

FIGURE 8–21 Example 8–10.

PROBLEM SOLVING
Usefulness and power of rough estimates

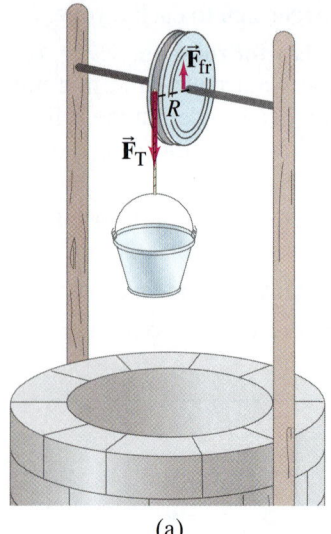

Additional Example—a bit more challenging

EXAMPLE 8–11 Pulley and bucket. Consider again the pulley in Example 8–10. But instead of a constant 15.0-N force being exerted on the cord, we now have a bucket of weight $w = 15.0$ N (mass $m = w/g = 1.53$ kg) hanging from the cord. See Fig. 8–22a. We assume the cord has negligible mass and does not stretch or slip on the pulley. Calculate the angular acceleration α of the pulley and the linear acceleration a of the bucket. Assume the same frictional torque $\tau_{\text{fr}} = 1.10$ m·N acts.

APPROACH This situation looks a lot like Example 8–10, Fig. 8–21. But there is a big difference: the tension in the cord is now an unknown, and it is no longer equal to the weight of the bucket if the bucket accelerates. Our system has two parts: the bucket, which can undergo translational motion (Fig. 8–22b is its free-body diagram); and the pulley. The pulley does not translate, but it can rotate. We apply the rotational version of Newton's second law to the pulley, $\Sigma \tau = I\alpha$, and the linear version to the bucket, $\Sigma F = ma$.

SOLUTION Let F_T be the tension in the cord. Then a force F_T acts at the edge of the pulley, and we apply Newton's second law, Eq. 8–14, for the rotation of the pulley:

$$I\alpha = \Sigma\tau = RF_T - \tau_{\text{fr}}. \quad \text{[pulley]}$$

Next we look at the (linear) motion of the bucket of mass m. Figure 8–22b, the free-body diagram for the bucket, shows that two forces act on the bucket: the force of gravity mg acts downward, and the tension of the cord F_T pulls upward. Applying Newton's second law, $\Sigma F = ma$, for the bucket, we have (taking downward as positive):

$$mg - F_T = ma. \quad \text{[bucket]}$$

Note that the tension F_T, which is the force exerted on the edge of the pulley, is *not* equal to the weight of the bucket ($= mg = 15.0$ N). There must be a net force on the bucket if it is accelerating, so $F_T < mg$. We can also see this from the last equation above, $F_T = mg - ma$.

To obtain α, we note that the tangential acceleration of a point on the edge of the pulley is the same as the acceleration of the bucket if the cord doesn't stretch or slip. Hence we can use Eq. 8–5, $a_{\text{tan}} = a = R\alpha$. Substituting $F_T = mg - ma = mg - mR\alpha$ into the first equation above (Newton's second law for rotation of the pulley), we obtain

$$I\alpha = \Sigma\tau = RF_T - \tau_{\text{fr}} = R(mg - mR\alpha) - \tau_{\text{fr}} = mgR - mR^2\alpha - \tau_{\text{fr}}.$$

The unknown α appears on the left and in the second term on the far right, so we bring that term to the left side and solve for α:

$$\alpha = \frac{mgR - \tau_{\text{fr}}}{I + mR^2}.$$

The numerator $(mgR - \tau_{\text{fr}})$ is the net torque, and the denominator $(I + mR^2)$ is the total rotational inertia of the system. With $mg = 15.0$ N ($m = 1.53$ kg) and, from Example 8–10, $I = 0.385$ kg·m² and $\tau_{\text{fr}} = 1.10$ m·N, then

$$\alpha = \frac{(15.0 \text{ N})(0.330 \text{ m}) - 1.10 \text{ m·N}}{0.385 \text{ kg·m}^2 + (1.53 \text{ kg})(0.330 \text{ m})^2} = 6.98 \text{ rad/s}^2.$$

The angular acceleration is somewhat less in this case than the 10.0 rad/s² of Example 8–10. Why? Because F_T ($= mg - ma = 15.0$ N $- ma$) is less than the 15.0-N force in Example 8–10. The linear acceleration of the bucket is

$$a = R\alpha = (0.330 \text{ m})(6.98 \text{ rad/s}^2) = 2.30 \text{ m/s}^2.$$

NOTE The tension in the cord F_T is less than mg because the bucket accelerates.

FIGURE 8–22 Example 8–11. (a) Pulley and falling bucket of mass m. This is also the free-body diagram for the pulley. (b) Free-body diagram for the bucket.

8–7 Rotational Kinetic Energy

The quantity $\frac{1}{2}mv^2$ is the kinetic energy of an object undergoing translational motion. An object rotating about an axis is said to have **rotational kinetic energy**. By analogy with translational kinetic energy, we might expect this to be given by the expression $\frac{1}{2}I\omega^2$, where I is the moment of inertia of the object and ω is its angular velocity. We can indeed show that this is true.

Consider any rigid rotating object as made up of many tiny particles, each of mass m. If we let r represent the distance of any one particle from the axis of rotation, then its linear velocity is $v = r\omega$. The total kinetic energy of the whole object will be the sum of the kinetic energies of all its particles:

$$KE = \Sigma(\tfrac{1}{2}mv^2) = \Sigma(\tfrac{1}{2}mr^2\omega^2)$$
$$= \tfrac{1}{2}(\Sigma mr^2)\omega^2.$$

We have factored out the $\tfrac{1}{2}$ and the ω^2 since they are the same for every particle of a rigid object. Since $\Sigma mr^2 = I$, the moment of inertia, we see that the kinetic energy of a rigid rotating object is

$$\text{rotational KE} = \tfrac{1}{2}I\omega^2. \tag{8–15}$$

The units are joules, as with all other forms of energy.

An object that rotates while its center of mass (CM) undergoes translational motion will have both translational and rotational kinetic energy. Equation 8–15 gives the rotational kinetic energy if the rotation axis is fixed. If the object is moving, such as a wheel rolling down a hill, this equation is still valid as long as the rotation axis is fixed in direction. Then the total kinetic energy is

$$KE = \tfrac{1}{2}Mv_{CM}^2 + \tfrac{1}{2}I_{CM}\omega^2, \tag{8–16}$$

where v_{CM} is the linear velocity of the center of mass, I_{CM} is the moment of inertia about an axis through the center of mass, ω is the angular velocity about this axis, and M is the total mass of the object.

EXAMPLE 8–12 Sphere rolling down an incline. What will be the speed of a solid sphere of mass M and radius R when it reaches the bottom of an incline if it starts from rest at a vertical height H and rolls without slipping? See Fig. 8–23. (Assume sufficient static friction so no slipping occurs: we will see shortly that static friction does no work.) Compare your result to that for an object *sliding* down a frictionless incline.

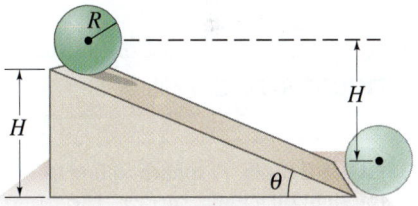

APPROACH We use the law of conservation of energy with gravitational potential energy, now including rotational kinetic energy as well as translational KE.

SOLUTION The total energy at any point a vertical distance y above the base of the incline is

FIGURE 8–23 A sphere rolling down a hill has both translational and rotational kinetic energy. Example 8–12.

$$E = \tfrac{1}{2}Mv^2 + \tfrac{1}{2}I_{CM}\omega^2 + Mgy,$$

where v is the speed of the center of mass, and Mgy is the gravitational potential energy. Applying conservation of energy, we equate the total energy at the top ($y = H$, $v = 0$, $\omega = 0$) to the total energy at the bottom ($y = 0$):

Rotational energy adds to other forms of energy to get the total energy which is conserved

$$E_{top} = E_{bottom}$$
$$0 + 0 + MgH = \tfrac{1}{2}Mv^2 + \tfrac{1}{2}I_{CM}\omega^2 + 0. \quad \text{[energy conservation]}$$

The moment of inertia of a solid sphere about an axis through its center of mass is $I_{CM} = \tfrac{2}{5}MR^2$, Fig. 8–20e. Since the sphere rolls without slipping, we have $\omega = v/R$ (recall Fig. 8–8). Hence

$$MgH = \tfrac{1}{2}Mv^2 + \tfrac{1}{2}(\tfrac{2}{5}MR^2)\left(\frac{v^2}{R^2}\right).$$

Canceling the M's and R's, we obtain

$$(\tfrac{1}{2} + \tfrac{1}{5})v^2 = gH$$

or

$$v = \sqrt{\tfrac{10}{7}gH}. \quad \text{[rolling sphere]}$$

We can compare this result for the speed of a rolling sphere to that for an object sliding down a plane without rotating and without friction, $\tfrac{1}{2}mv^2 = mgH$ (see our energy conservation equation above, removing the rotational term). For the sliding object, $v = \sqrt{2gH}$, which is greater than our result for a rolling sphere ($2 > 10/7$). An object sliding without friction or rotation transforms its initial potential energy entirely into translational kinetic energy (none into rotational kinetic energy), so the speed of its center of mass is greater.

NOTE Our result for the rolling sphere shows (perhaps surprisingly) that v is independent of both the mass M and the radius R of the sphere.

CONCEPTUAL EXAMPLE 8–13 **Which is fastest?** Several objects roll without slipping down an incline of vertical height H, all starting from rest at the same moment. The objects are a thin hoop (or a plain wedding band), a spherical marble, a solid cylinder (a D-cell battery), and an empty soup can. In addition, a greased box slides down without friction. In what order do they reach the bottom of the incline?

RESPONSE We use conservation of energy with gravitational potential energy plus rotational and translational kinetic energy. The sliding box would be fastest because the potential energy loss (MgH) is transformed completely into translational kinetic energy of the box, whereas for rolling objects the initial potential energy is shared between translational and rotational kinetic energies, and so the speed of the CM is less. For each of the rolling objects we can state that the decrease in potential energy equals the increase in translational plus rotational kinetic energy:

$$MgH = \tfrac{1}{2}Mv^2 + \tfrac{1}{2}I_{CM}\omega^2.$$

For all our rolling objects, the moment of inertia I_{CM} is a numerical factor times the mass M and the radius R^2 (Fig. 8–20). The mass M is in each term, so the translational speed v doesn't depend on M; nor does it depend on the radius R since $\omega = v/R$, so R^2 cancels out for all the rolling objects. Thus the speed v at the bottom of the incline depends only on that numerical factor in I_{CM} which expresses how the mass is distributed. The hoop, with all its mass concentrated at radius R ($I_{CM} = MR^2$), has the largest moment of inertia; hence it will have the lowest speed and will arrive at the bottom behind the D-cell ($I_{CM} = \tfrac{1}{2}MR^2$), which in turn will be behind the marble ($I_{CM} = \tfrac{2}{5}MR^2$). The empty can, which is mainly a hoop plus a thin disk, has most of its mass concentrated at R; so it will be a bit faster than the pure hoop but slower than the D-cell. See Fig. 8–24.

NOTE The rolling objects do not even have to have the same radius: the speed at the bottom does not depend on the object's mass M or radius R, but only on the shape (and the height of the incline H).

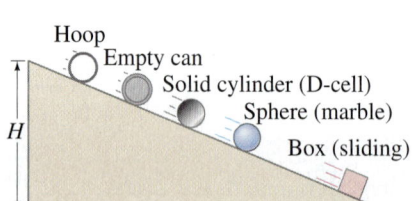

FIGURE 8–24 Example 8–13.

FIGURE 8–25 A sphere rolling to the right on a plane surface. The point in contact with the ground at any moment, point P, is momentarily at rest. Point A to the left of P is moving nearly vertically upward at the instant shown, and point B to the right is moving nearly vertically downward. An instant later, point B will touch the plane and be at rest momentarily. Thus no work is done by the force of static friction.

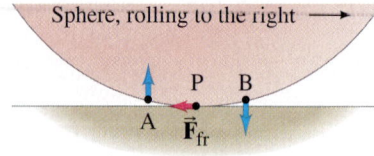

If there had been little or no static friction between the rolling objects and the plane in these Examples, the round objects would have slid rather than rolled, or a combination of both. Static friction must be present to make a round object roll. We did not need to take friction into account in the energy equation for the rolling objects because it is *static* friction and does no work—the point of contact of a sphere at each instant does not slide, but moves perpendicular to the plane (first down and then up as shown in Fig. 8–25) as it rolls. Thus, no work is done by the static friction force because the force and the motion (displacement) are perpendicular. The reason the rolling objects in Examples 8–12 and 8–13 move down the slope more slowly than if they were sliding is *not* because friction slows them down. Rather, it is because some of the gravitational potential energy is converted to rotational kinetic energy, leaving less for the translational kinetic energy.

EXERCISE C Return to the Chapter-Opening Question, page 198, and answer it again now. Try to explain why you may have answered differently the first time.

Work Done by Torque

The work done on an object rotating about a fixed axis, such as the pulleys in Figs. 8–21 and 8–22, can be written using angular quantities. As shown in Fig. 8–26, a force F exerting a torque $\tau = rF$ on a wheel does work $W = F\,\Delta\ell$ in rotating the wheel a small distance $\Delta\ell$ at the point of application of $\vec{F}$. The wheel has rotated through a small angle $\Delta\theta = \Delta\ell/r$ (Eq. 8–1). Hence

$$W = F\,\Delta\ell = Fr\,\Delta\theta.$$

Because $\tau = rF$, then

$$W = \tau\,\Delta\theta \tag{8-17}$$

is the work done by the torque τ when rotating the wheel through an angle $\Delta\theta$. Finally, power P is the rate work is done:

$$P = W/\Delta t = \tau\,\Delta\theta/\Delta t = \tau\omega,$$

which is analogous to the translational version, $P = Fv$ (see Eq. 6–18).

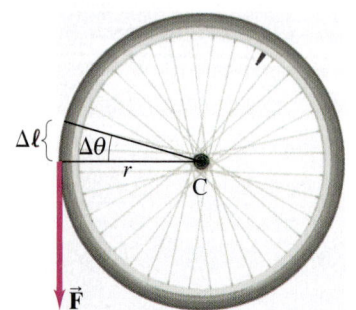

FIGURE 8–26 Torque $\tau = rF$ does work when rotating a wheel equal to $W = F\,\Delta\ell = Fr\,\Delta\theta = \tau\,\Delta\theta$.

8–8 Angular Momentum and Its Conservation

Throughout this Chapter we have seen that if we use the appropriate angular variables, the kinematic and dynamic equations for rotational motion are analogous to those for ordinary linear motion. We saw in the previous Section, for example, that rotational kinetic energy can be written as $\frac{1}{2}I\omega^2$, which is analogous to the translational kinetic energy, $\frac{1}{2}mv^2$. In like manner, the linear momentum, $p = mv$, has a rotational analog. It is called **angular momentum**, L. For a symmetrical object rotating about a fixed axis through the CM, the angular momentum is

$$L = I\omega, \tag{8-18}$$

where I is the moment of inertia and ω is the angular velocity about the axis of rotation. The SI units for L are $\text{kg} \cdot \text{m}^2/\text{s}$, which has no special name.

We saw in Chapter 7 (Section 7–1) that Newton's second law can be written not only as $\Sigma F = ma$ but also more generally in terms of momentum (Eq. 7–2), $\Sigma F = \Delta p/\Delta t$. In a similar way, the rotational equivalent of Newton's second law, which we saw in Eq. 8–14 can be written as $\Sigma \tau = I\alpha$, can also be written in terms of angular momentum:

$$\Sigma \tau = \frac{\Delta L}{\Delta t}, \tag{8-19}$$

> NEWTON'S SECOND LAW FOR ROTATION

where $\Sigma \tau$ is the net torque acting to rotate the object, and ΔL is the change in angular momentum in a time interval Δt. Equation 8–14, $\Sigma \tau = I\alpha$, is a special case of Eq. 8–19 when the moment of inertia is constant. This can be seen as follows. If an object has angular velocity ω_0 at time $t = 0$, and angular velocity ω after a time interval Δt, then its angular acceleration (Eq. 8–3) is

$$\alpha = \frac{\Delta \omega}{\Delta t} = \frac{\omega - \omega_0}{\Delta t}.$$

Then from Eq. 8–19, we have

$$\Sigma \tau = \frac{\Delta L}{\Delta t} = \frac{I\omega - I\omega_0}{\Delta t} = \frac{I(\omega - \omega_0)}{\Delta t} = I\frac{\Delta \omega}{\Delta t} = I\alpha,$$

which is Eq. 8–14.

Angular momentum is an important concept in physics because, under certain conditions, it is a conserved quantity. We can see from Eq. 8–19 that if the net torque $\Sigma \tau$ on an object is zero, then $\Delta L/\Delta t$ equals zero. That is, $\Delta L = 0$, so L does not change. This is the **law of conservation of angular momentum** for a rotating object:

> **The total angular momentum of a rotating object remains constant if the net torque acting on it is zero.**

> CONSERVATION OF ANGULAR MOMENTUM

The law of conservation of angular momentum is one of the great conservation laws of physics, along with those for energy and linear momentum.

When there is zero net torque acting on an object, and the object is rotating about a fixed axis or about an axis through its center of mass whose direction doesn't change, we can write

$$I\omega = I_0\omega_0 = \text{constant}. \tag{8-20}$$

I_0 and ω_0 are the moment of inertia and angular velocity, respectively, about that axis at some initial time $(t = 0)$, and I and ω are their values at some other time. The parts of the object may alter their positions relative to one another, so that I changes. But then ω changes as well, so that the product $I\omega$ remains constant.

FIGURE 8–27 A skater spinning on ice, illustrating conservation of angular momentum: (a) I is large and ω is small; (b) I is smaller so ω is larger.

FIGURE 8–28 A diver rotates faster when arms and legs are tucked in than when they are outstretched. Angular momentum is conserved.

FIGURE 8–29 Example 8–14.

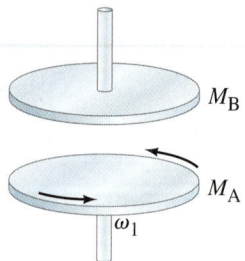

Many interesting phenomena can be understood on the basis of conservation of angular momentum. Consider a skater doing a spin on the tips of her skates, Fig. 8–27. She rotates at a relatively low speed when her arms are outstretched; when she brings her arms in close to her body, she suddenly spins much faster. From the definition of moment of inertia, $I = \Sigma mr^2$, it is clear that when she pulls her arms in closer to the axis of rotation, r is reduced for the arms so her moment of inertia is reduced. Since the angular momentum $I\omega$ remains constant (we ignore the small torque due to friction), if I decreases, then the angular velocity ω must increase. If the skater reduces her moment of inertia by a factor of 2, she will then rotate with twice the angular velocity.

EXERCISE D When a spinning figure skater pulls in her arms, her moment of inertia decreases; to conserve angular momentum, her angular velocity increases. Does her rotational kinetic energy also increase? If so, where does the energy come from?

A similar example is the diver shown in Fig. 8–28. The push as she leaves the board gives her an initial angular momentum about her center of mass. When she curls herself into the tuck position, she rotates quickly one or more times. She then stretches out again, increasing her moment of inertia which reduces the angular velocity to a small value, and then she enters the water. The change in moment of inertia from the straight position to the tuck position can be a factor of as much as $3\tfrac{1}{2}$.

Note that for angular momentum to be conserved, the net torque must be zero; but the net force does not necessarily have to be zero. The net force on the diver in Fig. 8–28, for example, is not zero (gravity is acting), but the net torque about her CM is zero because the force of gravity acts at her center of mass.

EXAMPLE 8–14 Clutch. A simple clutch consists of two cylindrical plates that can be pressed together to connect two sections of an axle, as needed, in a piece of machinery. The two plates have masses $M_A = 6.0\,\text{kg}$ and $M_B = 9.0\,\text{kg}$, with equal radii $R = 0.60\,\text{m}$. They are initially separated (Fig. 8–29). Plate M_A is accelerated from rest to an angular velocity $\omega_1 = 7.2\,\text{rad/s}$ in time $\Delta t = 2.0\,\text{s}$. Calculate (a) the angular momentum of M_A, and (b) the torque required to accelerate M_A from rest to ω_1. (c) Next, plate M_B, initially at rest but free to rotate without friction, is placed in firm contact with freely rotating plate M_A, and the two plates then both rotate at a constant angular velocity ω_2, which is considerably less than ω_1. Why does this happen, and what is ω_2?

APPROACH We use angular momentum, $L = I\omega$ (Eq. 8–18), plus Newton's second law for rotation, Eq. 8–19.

SOLUTION (a) The angular momentum of M_A, a cylinder, is

$$L_A = I_A \omega_1 = \tfrac{1}{2} M_A R^2 \omega_1 = \tfrac{1}{2}(6.0\,\text{kg})(0.60\,\text{m})^2(7.2\,\text{rad/s}) = 7.8\,\text{kg}\cdot\text{m}^2/\text{s}.$$

(b) The plate started from rest so the torque, assumed constant, was

$$\tau = \frac{\Delta L}{\Delta t} = \frac{7.8\,\text{kg}\cdot\text{m}^2/\text{s} - 0}{2.0\,\text{s}} = 3.9\,\text{m}\cdot\text{N}.$$

(c) Initially, before contact, M_A is rotating at constant ω_1 (we ignore friction). When plate B comes in contact, why is their joint rotation speed less? You might think in terms of the torque each exerts on the other upon contact. But quantitatively, it's easier to use conservation of angular momentum, Eq. 8–20, since no external torques are assumed to act. Thus

angular momentum before = angular momentum after

$$I_A \omega_1 = (I_A + I_B)\omega_2.$$

Solving for ω_2 we find (after cancelling factors of R^2)

$$\omega_2 = \left(\frac{I_A}{I_A + I_B}\right)\omega_1 = \left(\frac{M_A}{M_A + M_B}\right)\omega_1 = \left(\frac{6.0\,\text{kg}}{15.0\,\text{kg}}\right)(7.2\,\text{rad/s}) = 2.9\,\text{rad/s}.$$

EXAMPLE 8–15 ESTIMATE **Neutron star.** Astronomers detect stars that are rotating extremely rapidly, known as neutron stars. A neutron star is believed to form from the inner core of a larger star that collapsed, under its own gravitation, to a star of very small radius and very high density. Before collapse, suppose the core of such a star is the size of our Sun $(R \approx 7 \times 10^5 \text{ km})$ with mass 2.0 times as great as the Sun, and is rotating at a frequency of 1.0 revolution every 100 days. If it were to undergo gravitational collapse to a neutron star of radius 10 km, what would its rotation frequency be? Assume the star is a uniform sphere at all times, and loses no mass.

APPROACH We assume the star is isolated (no external forces), so we can use conservation of angular momentum for this process.

SOLUTION From conservation of angular momentum, Eq. 8–20,

$$I_1 \omega_1 = I_2 \omega_2,$$

where the subscripts 1 and 2 refer to initial (normal star) and final (neutron star), respectively. Then, assuming no mass is lost in the process $(M_1 = M_2)$,

$$\omega_2 = \left(\frac{I_1}{I_2}\right)\omega_1 = \left(\frac{\frac{2}{5} M_1 R_1^2}{\frac{2}{5} M_2 R_2^2}\right)\omega_1 = \frac{R_1^2}{R_2^2}\omega_1.$$

The frequency $f = \omega/2\pi$, so

$$f_2 = \frac{\omega_2}{2\pi} = \frac{R_1^2}{R_2^2} f_1$$

$$= \left(\frac{7 \times 10^5 \text{ km}}{10 \text{ km}}\right)^2 \left(\frac{1.0 \text{ rev}}{100 \text{ d (24 h/d)(3600 s/h)}}\right) \approx 6 \times 10^2 \text{ rev/s},$$

which is 600 Hz or $(600 \text{ rev/s})(60 \text{ s/min}) = 36{,}000 \text{ rpm}$.

PHYSICS APPLIED
Neutron star

*8–9 Vector Nature of Angular Quantities

Up to now we have considered only the magnitudes of angular quantities such as ω, α, and L. But they have a vector aspect too, and now we consider the directions. In fact, we have to *define* the directions for rotational quantities. We consider first the angular velocity, $\vec{\omega}$.

Consider the rotating wheel shown in Fig. 8–30a. The linear velocities of different particles of the wheel point in all different directions. The only unique direction in space associated with the rotation is along the axis of rotation, perpendicular to the actual motion. We therefore choose the axis of rotation to be the direction of the angular velocity vector, $\vec{\omega}$. Actually, there is still an ambiguity since $\vec{\omega}$ could point in either direction along the axis of rotation (up or down in Fig. 8–30a). The convention we use, called **the right-hand rule**, is this: when the fingers of the right hand are curled around the rotation axis and point in the direction of the rotation, then the thumb points in the direction of $\vec{\omega}$. This is shown in Fig. 8–30b. Note that $\vec{\omega}$ points in the direction a right-handed screw would move when turned in the direction of rotation. Thus, if the rotation of the wheel in Fig. 8–30a is counterclockwise, the direction of $\vec{\omega}$ is upward as shown in Fig. 8–30b. If the wheel rotates clockwise, then $\vec{\omega}$ points in the opposite direction, downward. Note that no part of the rotating object moves in the direction of $\vec{\omega}$.

If the axis of rotation is fixed, then $\vec{\omega}$ can change only in magnitude. Thus $\vec{\alpha} = \Delta\vec{\omega}/\Delta t$ must also point along the axis of rotation. If the rotation is counterclockwise as in Fig. 8–30a and the magnitude of ω is increasing, then $\vec{\alpha}$ points upward; but if ω is decreasing (the wheel is slowing down), $\vec{\alpha}$ points downward. If the rotation is clockwise, $\vec{\alpha}$ points downward if ω is increasing, and $\vec{\alpha}$ points upward if ω is decreasing.

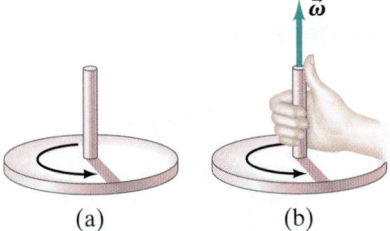

FIGURE 8–30 (a) Rotating wheel. (b) Right-hand rule for obtaining the direction of $\vec{\omega}$.

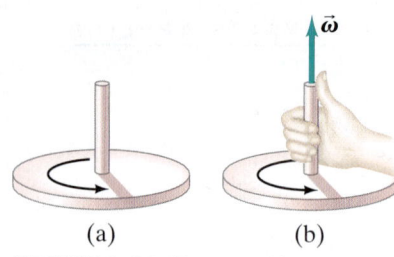

FIGURE 8–30 (Repeated.)
(a) Rotating wheel. (b) Right-hand rule for obtaining the direction of $\vec{\omega}$.

FIGURE 8–31 (a) A person standing on a circular platform, initially at rest, begins walking along the edge at speed v. The platform, mounted on nearly friction-free bearings, begins rotating in the opposite direction, so that (b) the total angular momentum remains zero ($\vec{L}_{\text{platform}} = -\vec{L}_{\text{person}}$).

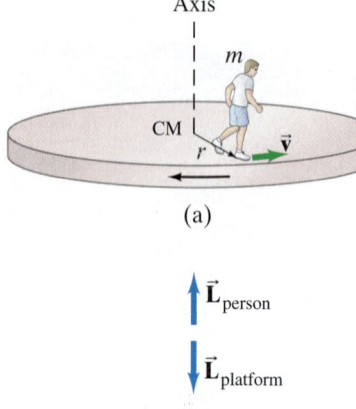

FIGURE 8–32 Example 8–16.

Angular momentum, like linear momentum, is a vector quantity. For a symmetrical object rotating about a symmetry axis (such as a wheel, cylinder, hoop, or sphere), we can write the vector angular momentum as

$$\vec{L} = I\vec{\omega}. \tag{8-21}$$

The angular velocity vector $\vec{\omega}$ (and therefore also $\vec{L}$) points along the axis of rotation in the direction given by the right-hand rule (Fig. 8–30b).

The vector nature of angular momentum can be used to explain a number of interesting (and sometimes surprising) phenomena. For example, consider a person standing at rest on a circular platform capable of rotating without friction about an axis through its center (that is, a simplified merry-go-round). If the person now starts to walk along the edge of the platform, Fig. 8–31a, the platform starts rotating in the opposite direction. Why? One explanation is that the person's foot exerts a force on the platform. Another explanation (and this is the most useful analysis here) is that this is an example of the conservation of angular momentum. If the person starts walking counterclockwise, the person's angular momentum will point upward along the axis of rotation (remember how we defined the direction of $\vec{\omega}$ using the right-hand rule). The magnitude of the person's angular momentum will be $L = I\omega = (mr^2)(v/r)$, where v is the person's speed (relative to the Earth, not to the platform), r is his distance from the rotation axis, m is his mass, and mr^2 is his moment of inertia if we consider him a particle (mass concentrated at one point, Eq. 8–11). The platform rotates in the opposite direction, so its angular momentum points downward. If the total angular momentum of the system is initially zero (person and platform at rest), it will remain zero after the person starts walking. That is, the upward angular momentum of the person just balances the oppositely directed downward angular momentum of the platform (Fig. 8–31b), so the total vector angular momentum remains zero. Even though the person exerts a force (and torque) on the platform, the platform exerts an equal and opposite torque on the person. So the net torque on the *system* of person plus platform is zero (ignoring friction), and the total angular momentum remains constant.

CONCEPTUAL EXAMPLE 8–16 **Spinning bicycle wheel.** Your physics teacher is holding a spinning bicycle wheel while he stands with feet fixed on a stationary frictionless turntable (Fig. 8–32). What will happen if the teacher suddenly flips the bicycle wheel over so that it is spinning in the opposite direction?

RESPONSE We consider the system of turntable, teacher, and bicycle wheel. The total angular momentum initially is $\vec{L}$ vertically upward. That is also what the system's angular momentum must be afterward, since $\vec{L}$ is conserved when there is no net torque. Thus, if the wheel's angular momentum after being flipped over is $-\vec{L}$ downward, then the angular momentum of teacher plus turntable will have to be $+2\vec{L}$ upward. We can safely predict that the turntable (and teacher) will begin revolving in the same direction that the bicycle wheel was spinning originally.

EXERCISE E In Example 8–16, what if he moves the axis only 90° so it is horizontal? (*a*) The same direction and speed as above; (*b*) the same as above, but slower; (*c*) the opposite result.

EXERCISE F Suppose you are standing on the edge of a large freely rotating turntable. If you walk toward the center, (*a*) the turntable slows down; (*b*) the turntable speeds up; (*c*) its rotation speed is unchanged; (*d*) the turntable stops; (*e*) you need to know the walking speed to answer.

One final note: the motion of particles and objects in **rotating frames of reference** is extremely interesting, though a bit advanced and so is treated at the end of the book in Appendix C.

Summary

When a rigid object rotates about a fixed axis, each point of the object moves in a circular path. Lines drawn perpendicularly from the rotation axis to various points in the object all sweep out the same angle θ in any given time interval.

Angles are conventionally measured in **radians**, where one radian is the angle subtended by an arc whose length is equal to the radius, or

$$2\pi \text{ rad} = 360°$$
$$1 \text{ rad} \approx 57.3°.$$

Angular velocity, ω, is defined as the rate of change of angular position:

$$\omega = \frac{\Delta\theta}{\Delta t}. \quad (8\text{–}2)$$

All parts of a rigid object rotating about a fixed axis have the same angular velocity at any instant.

Angular acceleration, α, is defined as the rate of change of angular velocity:

$$\alpha = \frac{\Delta\omega}{\Delta t}. \quad (8\text{–}3)$$

The linear velocity v and acceleration a of a point located a distance r from the axis of rotation are related to ω and α by

$$v = r\omega, \quad (8\text{–}4)$$
$$a_{\tan} = r\alpha, \quad (8\text{–}5)$$
$$a_R = \omega^2 r, \quad (8\text{–}6)$$

where $a_{\tan}$ and a_R are the tangential and radial (centripetal) components of the linear acceleration, respectively.

The frequency f is related to ω by

$$\omega = 2\pi f, \quad (8\text{–}7)$$

and to the period T by

$$T = 1/f. \quad (8\text{–}8)$$

If a rigid object undergoes uniformly accelerated rotational motion (α = constant), equations analogous to those for linear motion are valid:

$$\omega = \omega_0 + \alpha t, \quad \theta = \omega_0 t + \tfrac{1}{2}\alpha t^2,$$
$$\omega^2 = \omega_0^2 + 2\alpha\theta, \quad \bar{\omega} = \frac{\omega + \omega_0}{2}. \quad (8\text{–}9)$$

The **torque** due to a force $\vec{F}$ exerted on a rigid object is equal to

$$\tau = r_\perp F = r F_\perp = rF\sin\theta, \quad (8\text{–}10)$$

where $r_\perp$, called the **lever arm**, is the perpendicular distance from the axis of rotation to the line along which the force acts, and θ is the angle between $\vec{F}$ and r.

The rotational equivalent of Newton's second law is

$$\Sigma\tau = I\alpha, \quad (8\text{–}14)$$

where $I = \Sigma mr^2$ is the **moment of inertia** of the object about the axis of rotation. I depends not only on the mass of the object but also on how the mass is distributed relative to the axis of rotation. For a uniform solid cylinder or sphere of radius R and mass M, I has the form $I = \tfrac{1}{2}MR^2$ or $\tfrac{2}{5}MR^2$, respectively (see Fig. 8–20).

The **rotational kinetic energy** of an object rotating about a fixed axis with angular velocity ω is

$$\text{KE} = \tfrac{1}{2}I\omega^2. \quad (8\text{–}15)$$

For an object both translating and rotating, the total kinetic energy is the sum of the translational kinetic energy of the object's center of mass plus the rotational kinetic energy of the object about its center of mass:

$$\text{KE} = \tfrac{1}{2}Mv_{\text{CM}}^2 + \tfrac{1}{2}I_{\text{CM}}\omega^2 \quad (8\text{–}16)$$

as long as the rotation axis is fixed in direction.

The **angular momentum** L of an object rotating about a fixed rotation axis is given by

$$L = I\omega. \quad (8\text{–}18)$$

Newton's second law, in terms of angular momentum, is

$$\Sigma\tau = \frac{\Delta L}{\Delta t}. \quad (8\text{–}19)$$

If the net torque on an object is zero, $\Delta L/\Delta t = 0$, so L = constant. This is the **law of conservation of angular momentum** for a rotating object.

The following Table summarizes angular (or rotational) quantities, comparing them to their translational analogs.

Translation	Rotation	Connection
x	θ	$x = r\theta$
v	ω	$v = r\omega$
a	α	$a_{\tan} = r\alpha$
m	I	$I = \Sigma mr^2$
F	τ	$\tau = rF\sin\theta$
$\text{KE} = \tfrac{1}{2}mv^2$	$\tfrac{1}{2}I\omega^2$	
$p = mv$	$L = I\omega$	
$W = Fd$	$W = \tau\theta$	
$\Sigma F = ma$	$\Sigma\tau = I\alpha$	
$\Sigma F = \dfrac{\Delta p}{\Delta t}$	$\Sigma\tau = \dfrac{\Delta L}{\Delta t}$	

[*Angular velocity, angular acceleration, and angular momentum are vectors. For a rigid object rotating about a fixed axis, the vectors $\vec{\omega}$, $\vec{\alpha}$, and $\vec{L}$ point along the rotation axis. The direction of $\vec{\omega}$ or $\vec{L}$ is given by the **right-hand rule**.]

Questions

1. A bicycle odometer (which counts revolutions and is calibrated to report distance traveled) is attached near the wheel axle and is calibrated for 27-inch wheels. What happens if you use it on a bicycle with 24-inch wheels?
2. Suppose a disk rotates at constant angular velocity. (a) Does a point on the rim have radial and/or tangential acceleration? (b) If the disk's angular velocity increases uniformly, does the point have radial and/or tangential acceleration? (c) For which cases would the magnitude of either component of linear acceleration change?
3. Can a small force ever exert a greater torque than a larger force? Explain.
4. Why is it more difficult to do a sit-up with your hands behind your head than when your arms are stretched out in front of you? A diagram may help you to answer this.
5. If the net force on a system is zero, is the net torque also zero? If the net torque on a system is zero, is the net force zero? Explain and give examples.
6. Mammals that depend on being able to run fast have slender lower legs with flesh and muscle concentrated high, close to the body (Fig. 8–33). On the basis of rotational dynamics, explain why this distribution of mass is advantageous.

FIGURE 8–33 Question 6. A gazelle.

7. This book has three symmetry axes through its center, all mutually perpendicular. The book's moment of inertia would be smallest about which of the three? Explain.
8. Can the mass of a rigid object be considered concentrated at its CM for rotational motion? Explain.
9. The moment of inertia of a rotating solid disk about an axis through its CM is $\frac{1}{2}MR^2$ (Fig. 8–20c). Suppose instead that a parallel axis of rotation passes through a point on the edge of the disk. Will the moment of inertia be the same, larger, or smaller? Explain why.
10. Two inclines have the same height but make different angles with the horizontal. The same steel ball rolls without slipping down each incline. On which incline will the speed of the ball at the bottom be greater? Explain.
11. Two spheres look identical and have the same mass. However, one is hollow and the other is solid. Describe an experiment to determine which is which.
12. A sphere and a cylinder have the same radius and the same mass. They start from rest at the top of an incline. (a) Which reaches the bottom first? (b) Which has the greater speed at the bottom? (c) Which has the greater total kinetic energy at the bottom? (d) Which has the greater rotational kinetic energy? Explain your answers.

13. Why do tightrope walkers (Fig. 8–34) carry a long, narrow rod?

FIGURE 8–34 Question 13.

14. We claim that momentum and angular momentum are conserved. Yet most moving or rotating objects eventually slow down and stop. Explain.
15. Can the diver of Fig. 8–28 do a somersault without having any initial rotation when she leaves the board? Explain.
16. When a motorcyclist leaves the ground on a jump and leaves the throttle on (so the rear wheel spins), why does the front of the cycle rise up?
17. A shortstop may leap into the air to catch a ball and throw it quickly. As he throws the ball, the upper part of his body rotates. If you look quickly you will notice that his hips and legs rotate in the opposite direction (Fig. 8–35). Explain.

FIGURE 8–35 Question 17. A shortstop in the air, throwing the ball.

*18. The angular velocity of a wheel rotating on a horizontal axle points west. In what direction is the linear velocity of a point on the top of the wheel? If the angular acceleration points east, describe the tangential linear acceleration of this point at the top of the wheel. Is the angular speed increasing or decreasing?

*19. In what direction is the Earth's angular velocity vector as it rotates daily about its axis, north or south?

*20. On the basis of the law of conservation of angular momentum, discuss why a helicopter must have more than one rotor (or propeller). Discuss one or more ways the second propeller can operate in order to keep the helicopter stable.

MisConceptual Questions

1. Bonnie sits on the outer rim of a merry-go-round, and Jill sits midway between the center and the rim. The merry-go-round makes one complete revolution every 2 seconds. Jill's linear velocity is:
   (a) the same as Bonnie's.
   (b) twice Bonnie's.
   (c) half of Bonnie's.
   (d) one-quarter of Bonnie's.
   (e) four times Bonnie's.

2. An object at rest begins to rotate with a constant angular acceleration. If this object rotates through an angle θ in time t, through what angle did it rotate in the time $\frac{1}{2}t$?
   (a) $\frac{1}{2}\theta$. (b) $\frac{1}{4}\theta$. (c) θ. (d) 2θ. (e) 4θ.

3. A car speedometer that is supposed to read the linear speed of the car uses a device that actually measures the angular speed of the tires. If larger-diameter tires are mounted on the car instead, how will that affect the speedometer reading? The speedometer
   (a) will still read the speed accurately.
   (b) will read low.
   (c) will read high.

4. The solid dot shown in Fig. 8–36 is a pivot point. The board can rotate about the pivot. Which force shown exerts the largest magnitude torque on the board?

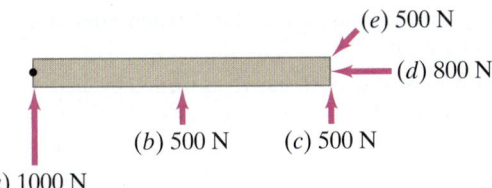

   FIGURE 8–36 MisConceptual Question 4.

5. Consider a force $F = 80$ N applied to a beam as shown in Fig. 8–37. The length of the beam is $\ell = 5.0$ m, and $\theta = 37°$, so that $x = 3.0$ m and $y = 4.0$ m. Of the following expressions, which ones give the correct torque produced by the force $\vec{F}$ around point P?
   (a) 80 N.
   (b) (80 N)(5.0 m).
   (c) (80 N)(5.0 m)(sin 37°).
   (d) (80 N)(4.0 m).
   (e) (80 N)(3.0 m).
   (f) (48 N)(5.0 m).
   (g) (48 N)(4.0 m)(sin 37°).

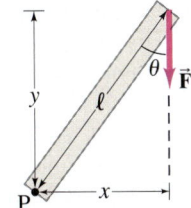

   FIGURE 8–37 MisConceptual Question 5.

6. Two spheres have the same radius and equal mass. One sphere is solid, and the other is hollow and made of a denser material. Which one has the bigger moment of inertia about an axis through its center?
   (a) The solid one.
   (b) The hollow one.
   (c) Both the same.

7. Two wheels having the same radius and mass rotate at the same angular velocity (Fig. 8–38). One wheel is made with spokes so nearly all the mass is at the rim. The other is a solid disk. How do their rotational kinetic energies compare?
   (a) They are nearly the same.
   (b) The wheel with spokes has about twice the KE.
   (c) The wheel with spokes has higher KE, but not twice as high.
   (d) The solid wheel has about twice the KE.
   (e) The solid wheel has higher KE, but not twice as high.

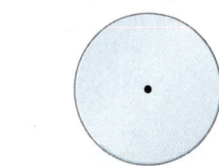

   FIGURE 8–38 MisConceptual Question 7.

8. If you used 1000 J of energy to throw a ball, would it travel faster if you threw the ball (ignoring air resistance)
   (a) so that it was also rotating?
   (b) so that it wasn't rotating?
   (c) It makes no difference.

9. A small solid sphere and a small thin hoop are rolling along a horizontal surface with the same translational speed when they encounter a 20° rising slope. If these two objects roll up the slope without slipping, which will rise farther up the slope?
   (a) The sphere.
   (b) The hoop.
   (c) Both the same.
   (d) More information about the objects' mass and diameter is needed.

10. A small mass m on a string is rotating without friction in a circle. The string is shortened by pulling it through the axis of rotation without any external torque, Fig. 8–39. What happens to the angular velocity of the object?
    (a) It increases.
    (b) It decreases.
    (c) It remains the same.

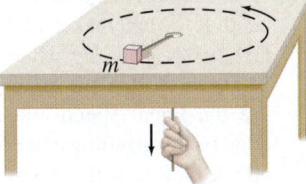

    FIGURE 8–39 MisConceptual Questions 10 and 11.

11. A small mass m on a string is rotating without friction in a circle. The string is shortened by pulling it through the axis of rotation without any external torque, Fig. 8–39. What happens to the tangential velocity of the object?
    (a) It increases.
    (b) It decreases.
    (c) It remains the same.

12. If there were a great migration of people toward the Earth's equator, the length of the day would
    (a) increase because of conservation of angular momentum.
    (b) decrease because of conservation of angular momentum.
    (c) decrease because of conservation of energy.
    (d) increase because of conservation of energy.
    (e) remain unaffected.

13. Suppose you are sitting on a rotating stool holding a 2-kg mass in each outstretched hand. If you suddenly drop the masses, your angular velocity will
    (a) increase. (b) decrease. (c) stay the same.

For assigned homework and other learning materials, go to the MasteringPhysics website.

Problems

8–1 Angular Quantities

1. (I) Express the following angles in radians: (a) 45.0°, (b) 60.0°, (c) 90.0°, (d) 360.0°, and (e) 445°. Give as numerical values and as fractions of π.

2. (I) The Sun subtends an angle of about 0.5° to us on Earth, 150 million km away. Estimate the radius of the Sun.

3. (I) A laser beam is directed at the Moon, 380,000 km from Earth. The beam diverges at an angle θ (Fig. 8–40) of 1.4×10^{-5} rad. What diameter spot will it make on the Moon?

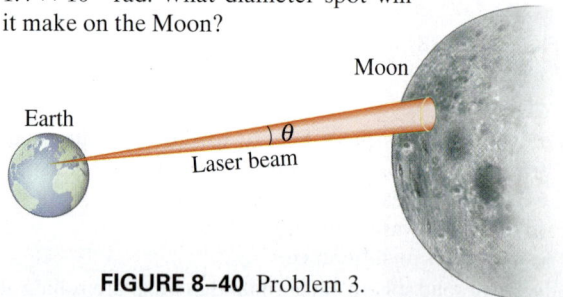

FIGURE 8–40 Problem 3.

4. (I) The blades in a blender rotate at a rate of 6500 rpm. When the motor is turned off during operation, the blades slow to rest in 4.0 s. What is the angular acceleration as the blades slow down?

5. (II) The platter of the **hard drive** of a computer rotates at 7200 rpm (rpm = revolutions per minute = rev/min). (a) What is the angular velocity (rad/s) of the platter? (b) If the reading head of the drive is located 3.00 cm from the rotation axis, what is the linear speed of the point on the platter just below it? (c) If a single bit requires 0.50 μm of length along the direction of motion, how many bits per second can the writing head write when it is 3.00 cm from the axis?

6. (II) A child rolls a ball on a level floor 3.5 m to another child. If the ball makes 12.0 revolutions, what is its diameter?

7. (II) (a) A grinding wheel 0.35 m in diameter rotates at 2200 rpm. Calculate its angular velocity in rad/s. (b) What are the linear speed and acceleration of a point on the edge of the grinding wheel?

8. (II) A bicycle with tires 68 cm in diameter travels 9.2 km. How many revolutions do the wheels make?

9. (II) Calculate the angular velocity (a) of a clock's second hand, (b) its minute hand, and (c) its hour hand. State in rad/s. (d) What is the angular acceleration in each case?

10. (II) A rotating merry-go-round makes one complete revolution in 4.0 s (Fig. 8–41). (a) What is the linear speed of a child seated 1.2 m from the center? (b) What is her acceleration (give components)?

FIGURE 8–41 Problem 10.

11. (II) What is the linear speed, due to the Earth's rotation, of a point (a) on the equator, (b) on the Arctic Circle (latitude 66.5° N), and (c) at a latitude of 42.0° N?

12. (II) Calculate the angular velocity of the Earth (a) in its orbit around the Sun, and (b) about its axis.

13. (II) How fast (in rpm) must a centrifuge rotate if a particle 8.0 cm from the axis of rotation is to experience an acceleration of 100,000 g's?

14. (II) A 61-cm-diameter wheel accelerates uniformly about its center from 120 rpm to 280 rpm in 4.0 s. Determine (a) its angular acceleration, and (b) the radial and tangential components of the linear acceleration of a point on the edge of the wheel 2.0 s after it has started accelerating.

15. (II) In traveling to the Moon, astronauts aboard the *Apollo* spacecraft put the spacecraft into a slow rotation to distribute the Sun's energy evenly (so one side would not become too hot). At the start of their trip, they accelerated from no rotation to 1.0 revolution every minute during a 12-min time interval. Think of the spacecraft as a cylinder with a diameter of 8.5 m rotating about its cylindrical axis. Determine (a) the angular acceleration, and (b) the radial and tangential components of the linear acceleration of a point on the skin of the ship 6.0 min after it started this acceleration.

16. (II) A turntable of radius R_1 is turned by a circular rubber roller of radius R_2 in contact with it at their outer edges. What is the ratio of their angular velocities, ω_1/ω_2?

8–2 and 8–3 Constant Angular Acceleration; Rolling

17. (I) An automobile engine slows down from 3500 rpm to 1200 rpm in 2.5 s. Calculate (a) its angular acceleration, assumed constant, and (b) the total number of revolutions the engine makes in this time.

18. (I) A centrifuge accelerates uniformly from rest to 15,000 rpm in 240 s. Through how many revolutions did it turn in this time?

19. (I) Pilots can be tested for the stresses of flying high-speed jets in a whirling "human centrifuge," which takes 1.0 min to turn through 23 complete revolutions before reaching its final speed. (a) What was its angular acceleration (assumed constant), and (b) what was its final angular speed in rpm?

20. (II) A cooling fan is turned off when it is running at 850 rev/min. It turns 1250 revolutions before it comes to a stop. (a) What was the fan's angular acceleration, assumed constant? (b) How long did it take the fan to come to a complete stop?

21. (II) A wheel 31 cm in diameter accelerates uniformly from 240 rpm to 360 rpm in 6.8 s. How far will a point on the edge of the wheel have traveled in this time?

22. (II) The tires of a car make 75 revolutions as the car reduces its speed uniformly from 95 km/h to 55 km/h. The tires have a diameter of 0.80 m. (a) What was the angular acceleration of the tires? If the car continues to decelerate at this rate, (b) how much more time is required for it to stop, and (c) how far does it go?

23. (II) A small rubber wheel is used to drive a large pottery wheel. The two wheels are mounted so that their circular edges touch. The small wheel has a radius of 2.0 cm and accelerates at the rate of 7.2 rad/s^2, and it is in contact with the pottery wheel (radius 27.0 cm) without slipping. Calculate (a) the angular acceleration of the pottery wheel, and (b) the time it takes the pottery wheel to reach its required speed of 65 rpm.

8–4 Torque

24. (I) A 52-kg person riding a bike puts all her weight on each pedal when climbing a hill. The pedals rotate in a circle of radius 17 cm. (a) What is the maximum torque she exerts? (b) How could she exert more torque?

25. (II) Calculate the net torque about the axle of the wheel shown in Fig. 8–42. Assume that a friction torque of 0.60 m·N opposes the motion.

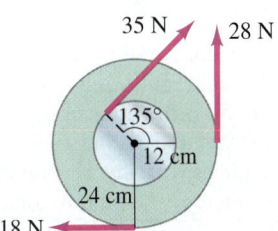

FIGURE 8–42 Problem 25.

26. (II) A person exerts a horizontal force of 42 N on the end of a door 96 cm wide. What is the magnitude of the torque if the force is exerted (a) perpendicular to the door and (b) at a 60.0° angle to the face of the door?

27. (II) Two blocks, each of mass m, are attached to the ends of a massless rod which pivots as shown in Fig. 8–43. Initially the rod is held in the horizontal position and then released. Calculate the magnitude and direction of the net torque on this system when it is first released.

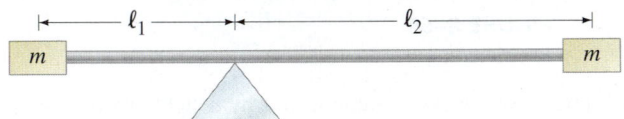

FIGURE 8–43 Problem 27.

28. (II) The bolts on the cylinder head of an engine require tightening to a torque of 95 m·N. If a wrench is 28 cm long, what force perpendicular to the wrench must the mechanic exert at its end? If the six-sided bolt head is 15 mm across (Fig. 8–44), estimate the force applied near each of the six points by a wrench.

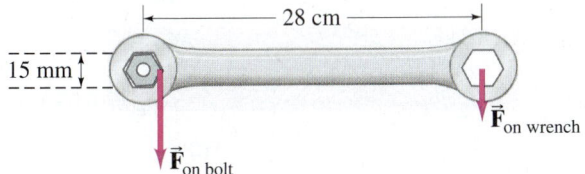

FIGURE 8–44 Problem 28.

29. (II) Determine the net torque on the 2.0-m-long uniform beam shown in Fig. 8–45. All forces are shown. Calculate about (a) point C, the CM, and (b) point P at one end.

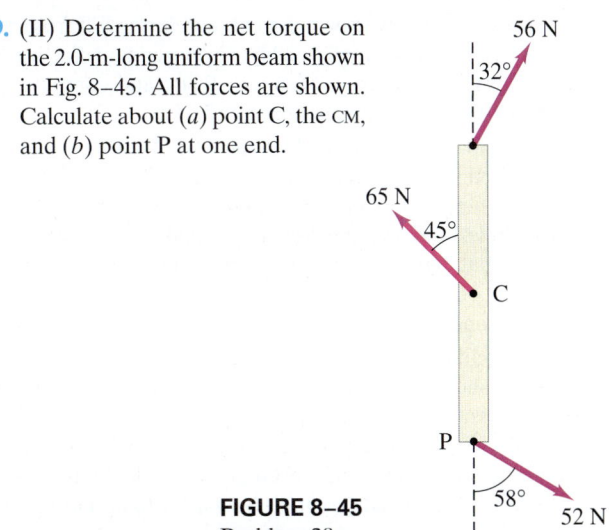

FIGURE 8–45 Problem 29.

8–5 and 8–6 Rotational Dynamics

30. (I) Determine the moment of inertia of a 10.8-kg sphere of radius 0.648 m when the axis of rotation is through its center.

31. (I) Estimate the moment of inertia of a bicycle wheel 67 cm in diameter. The rim and tire have a combined mass of 1.1 kg. The mass of the hub (at the center) can be ignored (why?).

32. (II) A merry-go-round accelerates from rest to 0.68 rad/s in 34 s. Assuming the merry-go-round is a uniform disk of radius 7.0 m and mass 31,000 kg, calculate the net torque required to accelerate it.

33. (II) An oxygen molecule consists of two oxygen atoms whose total mass is 5.3×10^{-26} kg and whose moment of inertia about an axis perpendicular to the line joining the two atoms, midway between them, is 1.9×10^{-46} kg·m^2. From these data, estimate the effective distance between the atoms.

34. (II) A grinding wheel is a uniform cylinder with a radius of 8.50 cm and a mass of 0.380 kg. Calculate (a) its moment of inertia about its center, and (b) the applied torque needed to accelerate it from rest to 1750 rpm in 5.00 s. Take into account a frictional torque that has been measured to slow down the wheel from 1500 rpm to rest in 55.0 s.

35. (II) The forearm in Fig. 8–46 accelerates a 3.6-kg ball at 7.0 m/s^2 by means of the triceps muscle, as shown. Calculate (a) the torque needed, and (b) the force that must be exerted by the triceps muscle. Ignore the mass of the arm.

36. (II) Assume that a 1.00-kg ball is thrown solely by the action of the forearm, which rotates about the elbow joint under the action of the triceps muscle, Fig. 8–46. The ball is accelerated uniformly from rest to 8.5 m/s in 0.38 s, at which point it is released. Calculate (a) the angular acceleration of the arm, and (b) the force required of the triceps muscle. Assume that the forearm has a mass of 3.7 kg and rotates like a uniform rod about an axis at its end.

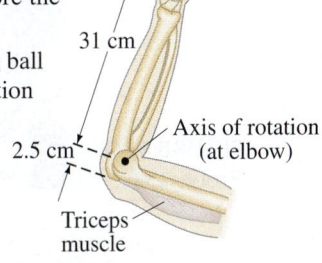

FIGURE 8–46 Problems 35 and 36.

37. (II) A softball player swings a bat, accelerating it from rest to 2.6 rev/s in a time of 0.20 s. Approximate the bat as a 0.90-kg uniform rod of length 0.95 m, and compute the torque the player applies to one end of it.

38. (II) A small 350-gram ball on the end of a thin, light rod is rotated in a horizontal circle of radius 1.2 m. Calculate (a) the moment of inertia of the ball about the center of the circle, and (b) the torque needed to keep the ball rotating at constant angular velocity if air resistance exerts a force of 0.020 N on the ball. Ignore air resistance on the rod and its moment of inertia.

39. (II) Calculate the moment of inertia of the array of point objects shown in Fig. 8–47 about (a) the y axis, and (b) the x axis. Assume $m = 2.2$ kg, $M = 3.4$ kg, and the objects are wired together by very light, rigid pieces of wire. The array is rectangular and is split through the middle by the x axis. (c) About which axis would it be harder to accelerate this array?

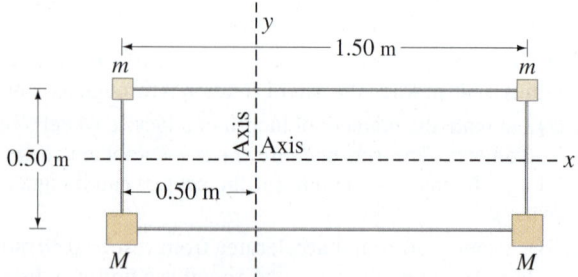

FIGURE 8–47 Problem 39.

40. (II) A potter is shaping a bowl on a potter's wheel rotating at constant angular velocity of 1.6 rev/s (Fig. 8–48). The friction force between her hands and the clay is 1.5 N total. (a) How large is her torque on the wheel, if the diameter of the bowl is 9.0 cm? (b) How long would it take for the potter's wheel to stop if the only torque acting on it is due to the potter's hands? The moment of inertia of the wheel and the bowl is 0.11 kg·m².

FIGURE 8–48 Problem 40.

41. (II) A dad pushes tangentially on a small hand-driven merry-go-round and is able to accelerate it from rest to a frequency of 15 rpm in 10.0 s. Assume the merry-go-round is a uniform disk of radius 2.5 m and has a mass of 560 kg, and two children (each with a mass of 25 kg) sit opposite each other on the edge. Calculate the torque required to produce the acceleration, neglecting frictional torque. What force is required at the edge?

42. (II) A 0.72-m-diameter solid sphere can be rotated about an axis through its center by a torque of 10.8 m·N which accelerates it uniformly from rest through a total of 160 revolutions in 15.0 s. What is the mass of the sphere?

43. (II) Let us treat a helicopter rotor blade as a long thin rod, as shown in Fig. 8–49. (a) If each of the three rotor helicopter blades is 3.75 m long and has a mass of 135 kg, calculate the moment of inertia of the three rotor blades about the axis of rotation. (b) How much torque must the motor apply to bring the blades from rest up to a speed of 6.0 rev/s in 8.0 s?

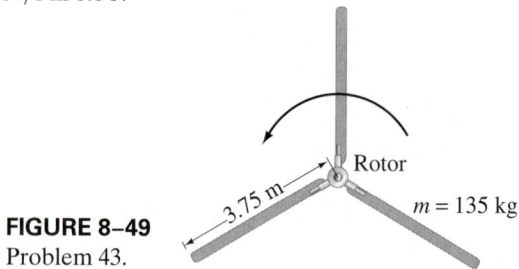

FIGURE 8–49 Problem 43.

44. (II) A centrifuge rotor rotating at 9200 rpm is shut off and is eventually brought uniformly to rest by a frictional torque of 1.20 m·N. If the mass of the rotor is 3.10 kg and it can be approximated as a solid cylinder of radius 0.0710 m, through how many revolutions will the rotor turn before coming to rest, and how long will it take?

45. (II) To get a flat, uniform cylindrical satellite spinning at the correct rate, engineers fire four tangential rockets as shown in Fig. 8–50. Suppose that the satellite has a mass of 3600 kg and a radius of 4.0 m, and that the rockets each add a mass of 250 kg. What is the steady force required of each rocket if the satellite is to reach 32 rpm in 5.0 min, starting from rest?

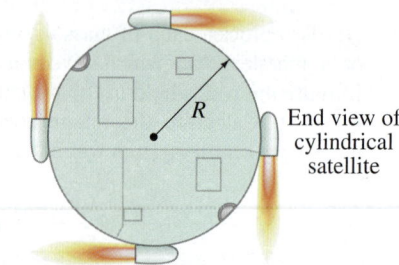

FIGURE 8–50 Problem 45.

46. (III) Two blocks are connected by a light string passing over a pulley of radius 0.15 m and moment of inertia I. The blocks move (towards the right) with an acceleration of 1.00 m/s² along their frictionless inclines (see Fig. 8–51). (a) Draw free-body diagrams for each of the two blocks and the pulley. (b) Determine F_{TA} and F_{TB}, the tensions in the two parts of the string. (c) Find the net torque acting on the pulley, and determine its moment of inertia, I.

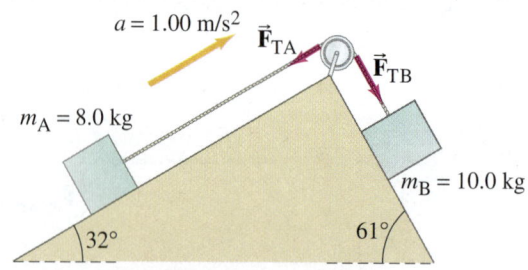

FIGURE 8–51 Problem 46.

224 CHAPTER 8 Rotational Motion

47. (III) An *Atwood machine* consists of two masses, $m_A = 65$ kg and $m_B = 75$ kg, connected by a massless inelastic cord that passes over a pulley free to rotate, Fig. 8–52. The pulley is a solid cylinder of radius $R = 0.45$ m and mass 6.0 kg. (*a*) Determine the acceleration of each mass. (*b*) What % error would be made if the moment of inertia of the pulley is ignored? [*Hint*: The tensions F_{TA} and F_{TB} are not equal. We discussed the Atwood machine in Example 4–13, assuming $I = 0$ for the pulley.]

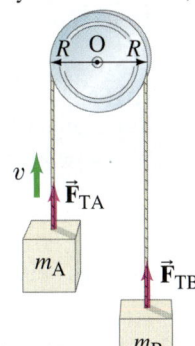

FIGURE 8–52 Problem 47. Atwood machine.

48. (III) A hammer thrower accelerates the hammer (mass = 7.30 kg) from rest within four full turns (revolutions) and releases it at a speed of 26.5 m/s. Assuming a uniform rate of increase in angular velocity and a horizontal circular path of radius 1.20 m, calculate (*a*) the angular acceleration, (*b*) the (linear) tangential acceleration, (*c*) the centripetal acceleration just before release, (*d*) the net force being exerted on the hammer by the athlete just before release, and (*e*) the angle of this force with respect to the radius of the circular motion. Ignore gravity.

8–7 Rotational Kinetic Energy

49. (I) An automobile engine develops a torque of 265 m·N at 3350 rpm. What is the horsepower of the engine?

50. (I) A centrifuge rotor has a moment of inertia of 3.25×10^{-2} kg·m². How much energy is required to bring it from rest to 8750 rpm?

51. (I) Calculate the translational speed of a cylinder when it reaches the foot of an incline 7.20 m high. Assume it starts from rest and rolls without slipping.

52. (II) A bowling ball of mass 7.25 kg and radius 10.8 cm rolls without slipping down a lane at 3.10 m/s. Calculate its total kinetic energy.

53. (II) Estimate the kinetic energy of the Earth with respect to the Sun as the sum of two terms, (*a*) that due to its daily rotation about its axis, and (*b*) that due to its yearly revolution about the Sun. [Assume the Earth is a uniform sphere with mass = 6.0×10^{24} kg, radius = 6.4×10^{6} m, and is 1.5×10^{8} km from the Sun.]

54. (II) A rotating uniform cylindrical platform of mass 220 kg and radius 5.5 m slows down from 3.8 rev/s to rest in 16 s when the driving motor is disconnected. Estimate the power output of the motor (hp) required to maintain a steady speed of 3.8 rev/s.

55. (II) A merry-go-round has a mass of 1440 kg and a radius of 7.50 m. How much net work is required to accelerate it from rest to a rotation rate of 1.00 revolution per 7.00 s? Assume it is a solid cylinder.

56. (II) A sphere of radius $r = 34.5$ cm and mass $m = 1.80$ kg starts from rest and rolls without slipping down a 30.0° incline that is 10.0 m long. (*a*) Calculate its translational and rotational speeds when it reaches the bottom. (*b*) What is the ratio of translational to rotational kinetic energy at the bottom? Avoid putting in numbers until the end so you can answer: (*c*) do your answers in (*a*) and (*b*) depend on the radius of the sphere or its mass?

57. (II) A ball of radius r rolls on the inside of a track of radius R (see Fig. 8–53). If the ball starts from rest at the vertical edge of the track, what will be its speed when it reaches the lowest point of the track, rolling without slipping?

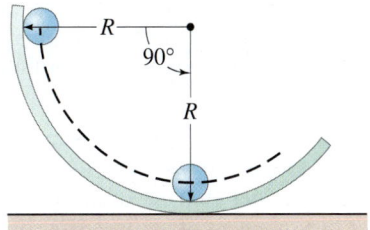

FIGURE 8–53 Problem 57.

58. (II) Two masses, $m_A = 32.0$ kg and $m_B = 38.0$ kg, are connected by a rope that hangs over a pulley (as in Fig. 8–54). The pulley is a uniform cylinder of radius $R = 0.311$ m and mass 3.1 kg. Initially m_A is on the ground and m_B rests 2.5 m above the ground. If the system is released, use conservation of energy to determine the speed of m_B just before it strikes the ground. Assume the pulley bearing is frictionless.

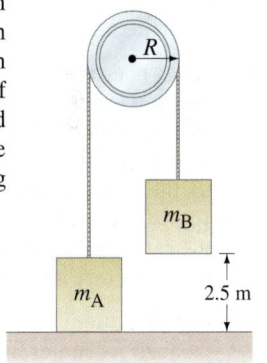

FIGURE 8–54 Problem 58.

59. (III) A 1.80-m-long pole is balanced vertically with its tip on the ground. It starts to fall and its lower end does not slip. What will be the speed of the upper end of the pole just before it hits the ground? [*Hint*: Use conservation of energy.]

8–8 Angular Momentum

60. (I) What is the angular momentum of a 0.270-kg ball revolving on the end of a thin string in a circle of radius 1.35 m at an angular speed of 10.4 rad/s?

61. (I) (*a*) What is the angular momentum of a 2.8-kg uniform cylindrical grinding wheel of radius 28 cm when rotating at 1300 rpm? (*b*) How much torque is required to stop it in 6.0 s?

62. (II) A person stands, hands at his side, on a platform that is rotating at a rate of 0.90 rev/s. If he raises his arms to a horizontal position, Fig. 8–55, the speed of rotation decreases to 0.60 rev/s. (*a*) Why? (*b*) By what factor has his moment of inertia changed?

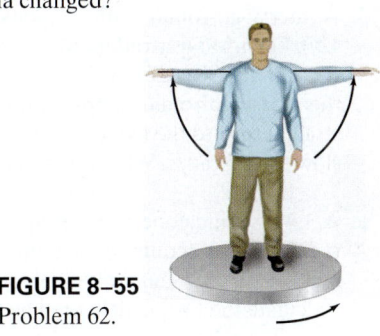

FIGURE 8–55 Problem 62.

63. (II) A nonrotating cylindrical disk of moment of inertia I is dropped onto an identical disk rotating at angular speed ω. Assuming no external torques, what is the final common angular speed of the two disks?

64. (II) A diver (such as the one shown in Fig. 8–28) can reduce her moment of inertia by a factor of about 3.5 when changing from the straight position to the tuck position. If she makes 2.0 rotations in 1.5 s when in the tuck position, what is her angular speed (rev/s) when in the straight position?

65. (II) A figure skater can increase her spin rotation rate from an initial rate of 1.0 rev every 1.5 s to a final rate of 2.5 rev/s. If her initial moment of inertia was $4.6 \text{ kg} \cdot \text{m}^2$, what is her final moment of inertia? How does she physically accomplish this change?

66. (II) (a) What is the angular momentum of a figure skater spinning at 3.0 rev/s with arms in close to her body, assuming her to be a uniform cylinder with a height of 1.5 m, a radius of 15 cm, and a mass of 48 kg? (b) How much torque is required to slow her to a stop in 4.0 s, assuming she does *not* move her arms?

67. (II) A person of mass 75 kg stands at the center of a rotating merry-go-round platform of radius 3.0 m and moment of inertia $820 \text{ kg} \cdot \text{m}^2$. The platform rotates without friction with angular velocity 0.95 rad/s. The person walks radially to the edge of the platform. (a) Calculate the angular velocity when the person reaches the edge. (b) Calculate the rotational kinetic energy of the system of platform plus person before and after the person's walk.

68. (II) A potter's wheel is rotating around a vertical axis through its center at a frequency of 1.5 rev/s. The wheel can be considered a uniform disk of mass 5.0 kg and diameter 0.40 m. The potter then throws a 2.6-kg chunk of clay, approximately shaped as a flat disk of radius 7.0 cm, onto the center of the rotating wheel. What is the frequency of the wheel after the clay sticks to it? Ignore friction.

69. (II) A 4.2-m-diameter merry-go-round is rotating freely with an angular velocity of 0.80 rad/s. Its total moment of inertia is $1360 \text{ kg} \cdot \text{m}^2$. Four people standing on the ground, each of mass 65 kg, suddenly step onto the edge of the merry-go-round. (a) What is the angular velocity of the merry-go-round now? (b) What if the people were on it initially and then jumped off in a radial direction (relative to the merry-go-round)?

70. (II) A uniform horizontal rod of mass M and length ℓ rotates with angular velocity ω about a vertical axis through its center. Attached to each end of the rod is a small mass m. Determine the angular momentum of the system about the axis.

71. (II) Suppose our Sun eventually collapses into a white dwarf, losing about half its mass in the process, and winding up with a radius 1.0% of its existing radius. Assuming the lost mass carries away no angular momentum, (a) what would the Sun's new rotation rate be? Take the Sun's current period to be about 30 days. (b) What would be its final kinetic energy in terms of its initial kinetic energy of today?

72. (II) A uniform disk turns at 3.3 rev/s around a frictionless central axis. A nonrotating rod, of the same mass as the disk and length equal to the disk's diameter, is dropped onto the freely spinning disk, Fig. 8–56. They then turn together around the axis with their centers superposed. What is the angular frequency in rev/s of the combination?

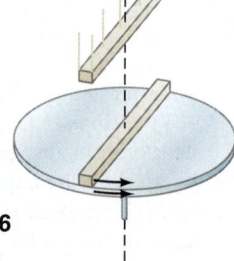

FIGURE 8–56
Problem 72.

73. (III) An asteroid of mass 1.0×10^5 kg, traveling at a speed of 35 km/s relative to the Earth, hits the Earth at the equator tangentially, in the direction of Earth's rotation, and is embedded there. Use angular momentum to estimate the percent change in the angular speed of the Earth as a result of the collision.

*8–9 Angular Quantities as Vectors

74. (III) Suppose a 65-kg person stands at the edge of a 5.5-m diameter merry-go-round turntable that is mounted on frictionless bearings and has a moment of inertia of $1850 \text{ kg} \cdot \text{m}^2$. The turntable is at rest initially, but when the person begins running at a speed of 4.0 m/s (with respect to the turntable) around its edge, the turntable begins to rotate in the opposite direction. Calculate the angular velocity of the turntable.

General Problems

75. A merry-go-round with a moment of inertia equal to $1260 \text{ kg} \cdot \text{m}^2$ and a radius of 2.5 m rotates with negligible friction at 1.70 rad/s. A child initially standing still next to the merry-go-round jumps onto the edge of the platform straight toward the axis of rotation, causing the platform to slow to 1.35 rad/s. What is her mass?

76. A 1.6-kg grindstone in the shape of a uniform cylinder of radius 0.20 m acquires a rotational rate of 24 rev/s from rest over a 6.0-s interval at constant angular acceleration. Calculate the torque delivered by the motor.

77. On a 12.0-cm-diameter audio compact disc (CD), digital bits of information are encoded sequentially along an outward spiraling path. The spiral starts at radius $R_1 = 2.5$ cm and winds its way out to radius $R_2 = 5.8$ cm. To read the digital information, a CD player rotates the CD so that the player's readout laser scans along the spiral's sequence of bits at a constant linear speed of 1.25 m/s. Thus the player must accurately adjust the rotational frequency f of the CD as the laser moves outward. Determine the values for f (in units of rpm) when the laser is located at R_1 and when it is at R_2.

78. (a) A yo-yo is made of two solid cylindrical disks, each of mass 0.050 kg and diameter 0.075 m, joined by a (concentric) thin solid cylindrical hub of mass 0.0050 kg and diameter 0.013 m. Use conservation of energy to calculate the linear speed of the yo-yo just before it reaches the end of its 1.0-m-long string, if it is released from rest. (b) What fraction of its kinetic energy is rotational?

79. A cyclist accelerates from rest at a rate of 1.00 m/s^2. How fast will a point at the top of the rim of the tire (diameter = 68.0 cm) be moving after 2.25 s? [*Hint*: At any moment, the lowest point on the tire is in contact with the ground and is at rest—see Fig. 8–57.]

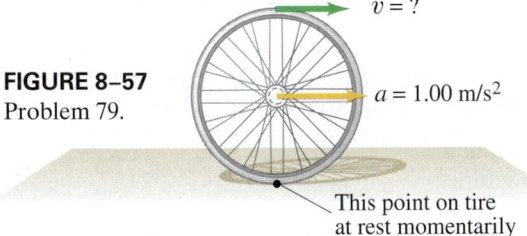

FIGURE 8–57 Problem 79.

80. Suppose David puts a 0.60-kg rock into a sling of length 1.5 m and begins whirling the rock in a nearly horizontal circle, accelerating it from rest to a rate of 75 rpm after 5.0 s. What is the torque required to achieve this feat, and where does the torque come from?

81. **Bicycle gears**: (a) How is the angular velocity ω_R of the rear wheel of a bicycle related to the angular velocity ω_F of the front sprocket and pedals? Let N_F and N_R be the number of teeth on the front and rear sprockets, respectively, Fig. 8–58. The teeth are spaced the same on both sprockets and the rear sprocket is firmly attached to the rear wheel. (b) Evaluate the ratio ω_R/ω_F when the front and rear sprockets have 52 and 13 teeth, respectively, and (c) when they have 42 and 28 teeth.

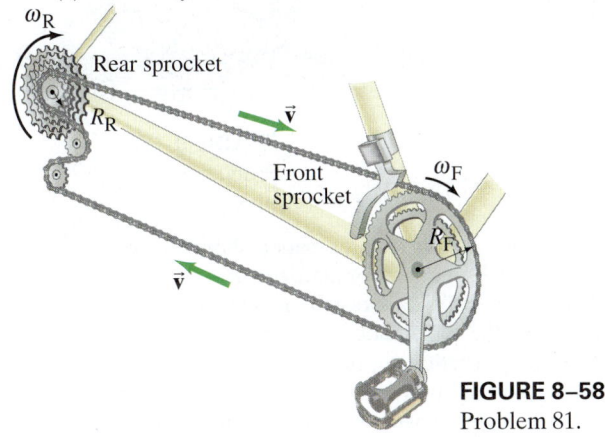

FIGURE 8–58 Problem 81.

82. Figure 8–59 illustrates an H_2O molecule. The O—H bond length is 0.096 nm and the H—O—H bonds make an angle of 104°. Calculate the moment of inertia of the H_2O molecule (assume the atoms are points) about an axis passing through the center of the oxygen atom (a) perpendicular to the plane of the molecule, and (b) in the plane of the molecule, bisecting the H—O—H bonds.

FIGURE 8–59 Problem 82.

83. A hollow cylinder (hoop) is rolling on a horizontal surface at speed $v = 3.0 \text{ m/s}$ when it reaches a 15° incline. (a) How far up the incline will it go? (b) How long will it be on the incline before it arrives back at the bottom?

84. Determine the angular momentum of the Earth (a) about its rotation axis (assume the Earth is a uniform sphere), and (b) in its orbit around the Sun (treat the Earth as a particle orbiting the Sun).

85. A wheel of mass M has radius R. It is standing vertically on the floor, and we want to exert a horizontal force F at its axle so that it will climb a step against which it rests (Fig. 8–60). The step has height h, where $h < R$. What minimum force F is needed?

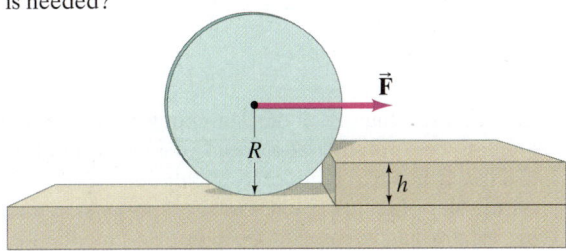

FIGURE 8–60 Problem 85.

86. If the coefficient of static friction between a car's tires and the pavement is 0.65, calculate the minimum torque that must be applied to the 66-cm-diameter tire of a 1080-kg automobile in order to "lay rubber" (make the wheels spin, slipping as the car accelerates). Assume each wheel supports an equal share of the weight.

87. A 4.00-kg mass and a 3.00-kg mass are attached to opposite ends of a very light 42.0-cm-long horizontal rod (Fig. 8–61). The system is rotating at angular speed $\omega = 5.60 \text{ rad/s}$ about a vertical axle at the center of the rod. Determine (a) the kinetic energy KE of the system, and (b) the net force on each mass.

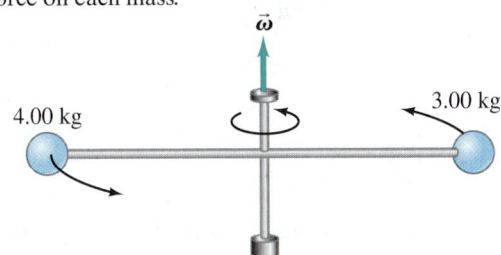

FIGURE 8–61 Problem 87.

88. A small mass m attached to the end of a string revolves in a circle on a frictionless tabletop. The other end of the string passes through a hole in the table (Fig. 8–62). Initially, the mass revolves with a speed $v_1 = 2.4 \text{ m/s}$ in a circle of radius $r_1 = 0.80 \text{ m}$. The string is then pulled slowly through the hole so that the radius is reduced to $r_2 = 0.48 \text{ m}$. What is the speed, v_2, of the mass now?

FIGURE 8–62 Problem 88.

General Problems

89. A uniform rod of mass M and length ℓ can pivot freely (i.e., we ignore friction) about a hinge attached to a wall, as in Fig. 8–63. The rod is held horizontally and then released. At the moment of release, determine (a) the angular acceleration of the rod, and (b) the linear acceleration of the tip of the rod. Assume that the force of gravity acts at the center of mass of the rod, as shown. [*Hint*: See Fig. 8–20g.]

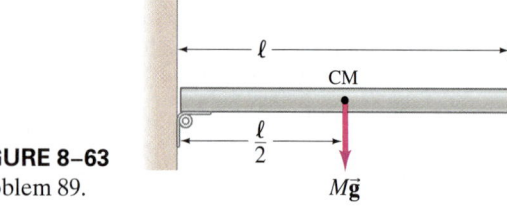

FIGURE 8–63 Problem 89.

90. Suppose a star the size of our Sun, but with mass 8.0 times as great, were rotating at a speed of 1.0 revolution every 9.0 days. If it were to undergo gravitational collapse to a neutron star of radius 12 km, losing $\frac{3}{4}$ of its mass in the process, what would its rotation speed be? Assume the star is a uniform sphere at all times. Assume also that the thrown-off mass carries off either (a) no angular momentum, or (b) its proportional share ($\frac{3}{4}$) of the initial angular momentum.

91. A large spool of rope rolls on the ground with the end of the rope lying on the top edge of the spool. A person grabs the end of the rope and walks a distance ℓ, holding onto it, Fig. 8–64. The spool rolls behind the person without slipping. What length of rope unwinds from the spool? How far does the spool's center of mass move?

FIGURE 8–64 Problem 91.

92. The Moon orbits the Earth such that the same side always faces the Earth. Determine the ratio of the Moon's spin angular momentum (about its own axis) to its orbital angular momentum. (In the latter case, treat the Moon as a particle orbiting the Earth.)

93. A spherical asteroid with radius $r = 123$ m and mass $M = 2.25 \times 10^{10}$ kg rotates about an axis at four revolutions per day. A "tug" spaceship attaches itself to the asteroid's south pole (as defined by the axis of rotation) and fires its engine, applying a force F tangentially to the asteroid's surface as shown in Fig. 8–65. If $F = 285$ N, how long will it take the tug to rotate the asteroid's axis of rotation through an angle of 5.0° by this method?

FIGURE 8–65 Problem 93.

94. Most of our Solar System's mass is contained in the Sun, and the planets possess almost all of the Solar System's angular momentum. This observation plays a key role in theories attempting to explain the formation of our Solar System. Estimate the fraction of the Solar System's total angular momentum that is possessed by planets using a simplified model which includes only the large outer planets with the most angular momentum. The central Sun (mass 1.99×10^{30} kg, radius 6.96×10^8 m) spins about its axis once every 25 days and the planets Jupiter, Saturn, Uranus, and Neptune move in nearly circular orbits around the Sun with orbital data given in the Table below. Ignore each planet's spin about its own axis.

Planet	Mean Distance from Sun ($\times 10^6$ km)	Orbital Period (Earth Years)	Mass ($\times 10^{25}$ kg)
Jupiter	778	11.9	190
Saturn	1427	29.5	56.8
Uranus	2870	84.0	8.68
Neptune	4500	165	10.2

95. Water drives a waterwheel (or turbine) of radius $R = 3.0$ m as shown in Fig. 8–66. The water enters at a speed $v_1 = 7.0$ m/s and exits from the waterwheel at a speed $v_2 = 3.8$ m/s. (a) If 85 kg of water passes through per second, what is the rate at which the water delivers angular momentum to the waterwheel? (b) What is the torque the water applies to the waterwheel? (c) If the water causes the waterwheel to make one revolution every 5.5 s, how much power is delivered to the wheel?

FIGURE 8–66 Problem 95.

96. The radius of the roll of paper shown in Fig. 8–67 is 7.6 cm and its moment of inertia is $I = 3.3 \times 10^{-3}$ kg·m². A force of 3.5 N is exerted on the end of the roll for 1.3 s, but the paper does not tear so it begins to unroll. A constant friction torque of 0.11 m·N is exerted on the roll which gradually brings it to a stop. Assuming that the paper's thickness is negligible, calculate (a) the length of paper that unrolls during the time that the force is applied (1.3 s) and (b) the length of paper that unrolls from the time the force ends to the time when the roll has stopped moving.

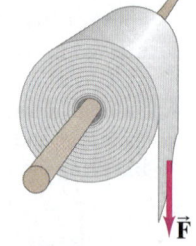

FIGURE 8–67 Problem 96.

78. (a) A yo-yo is made of two solid cylindrical disks, each of mass 0.050 kg and diameter 0.075 m, joined by a (concentric) thin solid cylindrical hub of mass 0.0050 kg and diameter 0.013 m. Use conservation of energy to calculate the linear speed of the yo-yo just before it reaches the end of its 1.0-m-long string, if it is released from rest. (b) What fraction of its kinetic energy is rotational?

79. A cyclist accelerates from rest at a rate of 1.00 m/s^2. How fast will a point at the top of the rim of the tire (diameter = 68.0 cm) be moving after 2.25 s? [*Hint*: At any moment, the lowest point on the tire is in contact with the ground and is at rest—see Fig. 8–57.]

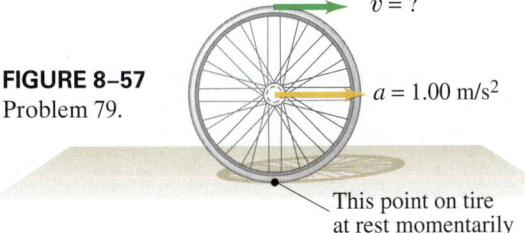

FIGURE 8–57 Problem 79.

80. Suppose David puts a 0.60-kg rock into a sling of length 1.5 m and begins whirling the rock in a nearly horizontal circle, accelerating it from rest to a rate of 75 rpm after 5.0 s. What is the torque required to achieve this feat, and where does the torque come from?

81. **Bicycle gears**: (a) How is the angular velocity ω_R of the rear wheel of a bicycle related to the angular velocity ω_F of the front sprocket and pedals? Let N_F and N_R be the number of teeth on the front and rear sprockets, respectively, Fig. 8–58. The teeth are spaced the same on both sprockets and the rear sprocket is firmly attached to the rear wheel. (b) Evaluate the ratio ω_R/ω_F when the front and rear sprockets have 52 and 13 teeth, respectively, and (c) when they have 42 and 28 teeth.

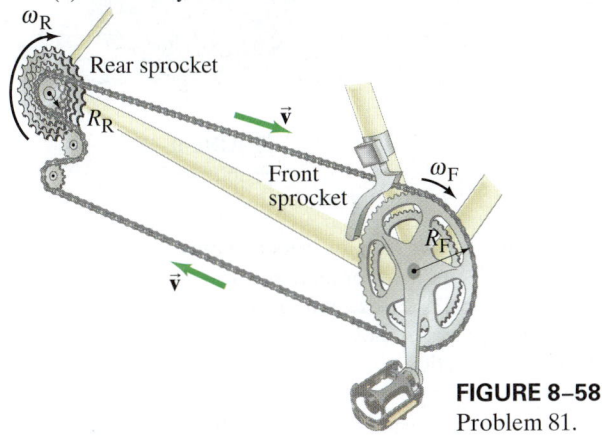

FIGURE 8–58 Problem 81.

82. Figure 8–59 illustrates an H_2O molecule. The O—H bond length is 0.096 nm and the H—O—H bonds make an angle of 104°. Calculate the moment of inertia of the H_2O molecule (assume the atoms are points) about an axis passing through the center of the oxygen atom (a) perpendicular to the plane of the molecule, and (b) in the plane of the molecule, bisecting the H—O—H bonds.

FIGURE 8–59 Problem 82.

83. A hollow cylinder (hoop) is rolling on a horizontal surface at speed $v = 3.0 \text{ m/s}$ when it reaches a 15° incline. (a) How far up the incline will it go? (b) How long will it be on the incline before it arrives back at the bottom?

84. Determine the angular momentum of the Earth (a) about its rotation axis (assume the Earth is a uniform sphere), and (b) in its orbit around the Sun (treat the Earth as a particle orbiting the Sun).

85. A wheel of mass M has radius R. It is standing vertically on the floor, and we want to exert a horizontal force F at its axle so that it will climb a step against which it rests (Fig. 8–60). The step has height h, where $h < R$. What minimum force F is needed?

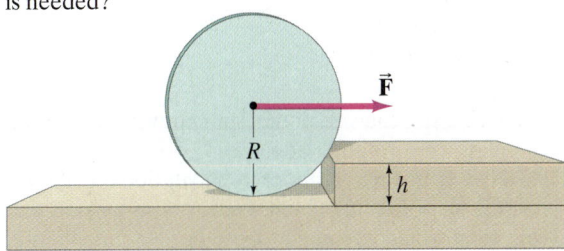

FIGURE 8–60 Problem 85.

86. If the coefficient of static friction between a car's tires and the pavement is 0.65, calculate the minimum torque that must be applied to the 66-cm-diameter tire of a 1080-kg automobile in order to "lay rubber" (make the wheels spin, slipping as the car accelerates). Assume each wheel supports an equal share of the weight.

87. A 4.00-kg mass and a 3.00-kg mass are attached to opposite ends of a very light 42.0-cm-long horizontal rod (Fig. 8–61). The system is rotating at angular speed $\omega = 5.60 \text{ rad/s}$ about a vertical axle at the center of the rod. Determine (a) the kinetic energy KE of the system, and (b) the net force on each mass.

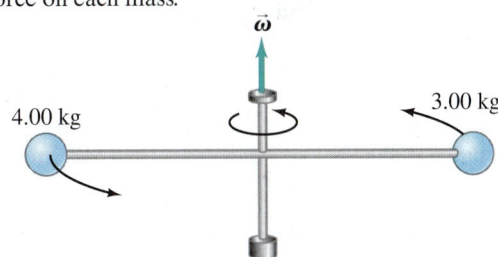

FIGURE 8–61 Problem 87.

88. A small mass m attached to the end of a string revolves in a circle on a frictionless tabletop. The other end of the string passes through a hole in the table (Fig. 8–62). Initially, the mass revolves with a speed $v_1 = 2.4 \text{ m/s}$ in a circle of radius $r_1 = 0.80 \text{ m}$. The string is then pulled slowly through the hole so that the radius is reduced to $r_2 = 0.48 \text{ m}$. What is the speed, v_2, of the mass now?

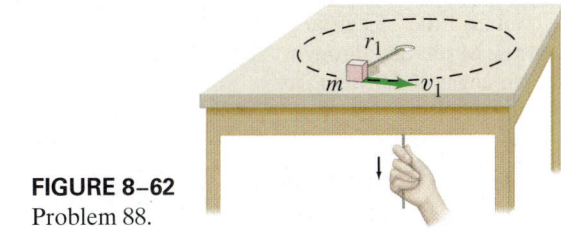

FIGURE 8–62 Problem 88.

General Problems 227

89. A uniform rod of mass M and length ℓ can pivot freely (i.e., we ignore friction) about a hinge attached to a wall, as in Fig. 8-63. The rod is held horizontally and then released. At the moment of release, determine (a) the angular acceleration of the rod, and (b) the linear acceleration of the tip of the rod. Assume that the force of gravity acts at the center of mass of the rod, as shown. [*Hint*: See Fig. 8-20g.]

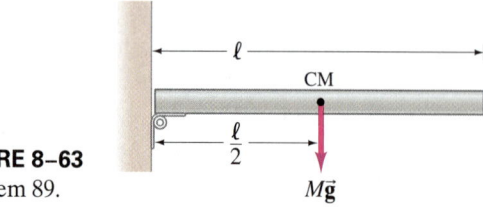

FIGURE 8-63
Problem 89.

90. Suppose a star the size of our Sun, but with mass 8.0 times as great, were rotating at a speed of 1.0 revolution every 9.0 days. If it were to undergo gravitational collapse to a neutron star of radius 12 km, losing $\frac{3}{4}$ of its mass in the process, what would its rotation speed be? Assume the star is a uniform sphere at all times. Assume also that the thrown-off mass carries off either (a) no angular momentum, or (b) its proportional share ($\frac{3}{4}$) of the initial angular momentum.

91. A large spool of rope rolls on the ground with the end of the rope lying on the top edge of the spool. A person grabs the end of the rope and walks a distance ℓ, holding onto it, Fig. 8-64. The spool rolls behind the person without slipping. What length of rope unwinds from the spool? How far does the spool's center of mass move?

FIGURE 8-64
Problem 91.

92. The Moon orbits the Earth such that the same side always faces the Earth. Determine the ratio of the Moon's spin angular momentum (about its own axis) to its orbital angular momentum. (In the latter case, treat the Moon as a particle orbiting the Earth.)

93. A spherical asteroid with radius $r = 123$ m and mass $M = 2.25 \times 10^{10}$ kg rotates about an axis at four revolutions per day. A "tug" spaceship attaches itself to the asteroid's south pole (as defined by the axis of rotation) and fires its engine, applying a force F tangentially to the asteroid's surface as shown in Fig. 8-65. If $F = 285$ N, how long will it take the tug to rotate the asteroid's axis of rotation through an angle of $5.0°$ by this method?

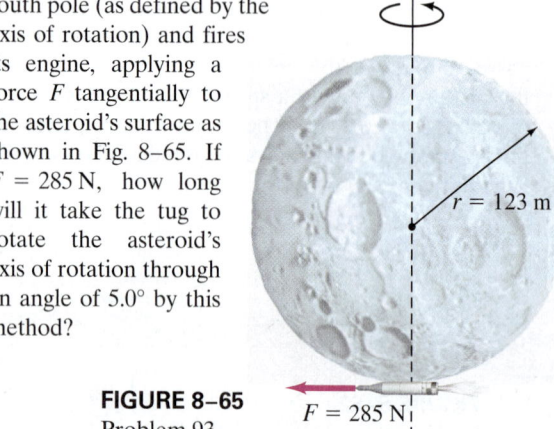

FIGURE 8-65
Problem 93.

94. Most of our Solar System's mass is contained in the Sun, and the planets possess almost all of the Solar System's angular momentum. This observation plays a key role in theories attempting to explain the formation of our Solar System. Estimate the fraction of the Solar System's total angular momentum that is possessed by planets using a simplified model which includes only the large outer planets with the most angular momentum. The central Sun (mass 1.99×10^{30} kg, radius 6.96×10^8 m) spins about its axis once every 25 days and the planets Jupiter, Saturn, Uranus, and Neptune move in nearly circular orbits around the Sun with orbital data given in the Table below. Ignore each planet's spin about its own axis.

Planet	Mean Distance from Sun ($\times 10^6$ km)	Orbital Period (Earth Years)	Mass ($\times 10^{25}$ kg)
Jupiter	778	11.9	190
Saturn	1427	29.5	56.8
Uranus	2870	84.0	8.68
Neptune	4500	165	10.2

95. Water drives a waterwheel (or turbine) of radius $R = 3.0$ m as shown in Fig. 8-66. The water enters at a speed $v_1 = 7.0$ m/s and exits from the waterwheel at a speed $v_2 = 3.8$ m/s. (a) If 85 kg of water passes through per second, what is the rate at which the water delivers angular momentum to the waterwheel? (b) What is the torque the water applies to the waterwheel? (c) If the water causes the waterwheel to make one revolution every 5.5 s, how much power is delivered to the wheel?

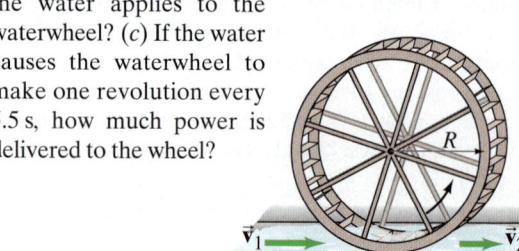

FIGURE 8-66
Problem 95.

96. The radius of the roll of paper shown in Fig. 8-67 is 7.6 cm and its moment of inertia is $I = 3.3 \times 10^{-3}$ kg·m². A force of 3.5 N is exerted on the end of the roll for 1.3 s, but the paper does not tear so it begins to unroll. A constant friction torque of 0.11 m·N is exerted on the roll which gradually brings it to a stop. Assuming that the paper's thickness is negligible, calculate (a) the length of paper that unrolls during the time that the force is applied (1.3 s) and (b) the length of paper that unrolls from the time the force ends to the time when the roll has stopped moving.

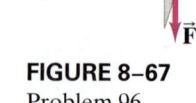

FIGURE 8-67
Problem 96.

Search and Learn

1. Why are Eqs. 8–4 and 8–5 valid for radians but not for revolutions or degrees? Read Section 8–1 and follow the derivations carefully to find the answer.

2. **Total solar eclipses** can happen on Earth because of amazing coincidences: for one, the sometimes near-perfect alignment of Earth, Moon, and Sun. Secondly, using the information inside the front cover, calculate the angular diameters (in radians) of the Sun and the Moon, as seen from Earth, and then comment.

3. Two uniform spheres simultaneously start rolling (from rest) down an incline. One sphere has twice the radius and twice the mass of the other. (a) Which reaches the bottom of the incline first? (b) Which has the greater speed there? (c) Which has the greater total kinetic energy at the bottom? Explain your answers.

4. A bicyclist traveling with speed $v = 8.2$ m/s on a flat road is making a turn with a radius $r = 13$ m. There are three forces acting on the cyclist and cycle: the normal force ($\vec{F}_N$) and friction force ($\vec{F}_{fr}$) exerted by the road on the tires; and $m\vec{g}$, the total weight of the cyclist and cycle. Ignore the small mass of the wheels. (a) Explain carefully why the angle θ the bicycle makes with the vertical (Fig. 8–68) must be given by $\tan\theta = F_{fr}/F_N$ if the cyclist is to maintain balance. (b) Calculate θ for the values given. [*Hint*: Consider the "circular" translational motion of the bicycle and rider.] (c) If the coefficient of static friction between tires and road is $\mu_s = 0.65$, what is the minimum turning radius?

5. Model a figure skater's body as a solid cylinder and her arms as thin rods, making reasonable estimates for the dimensions. Then calculate the ratio of the angular speeds for a spinning skater with outstretched arms, and with arms held tightly against her body. Check Sections 8–5 and 8–8.

6. One possibility for a low-pollution automobile is for it to use energy stored in a heavy rotating **flywheel**. Suppose such a car has a total mass of 1100 kg, uses a uniform cylindrical flywheel of diameter 1.50 m and mass 270 kg, and should be able to travel 350 km without needing a flywheel "spinup." (a) Make reasonable assumptions (average frictional retarding force on car = 450 N, thirty acceleration periods from rest to 95 km/h, equal uphill and downhill, and that energy can be put back into the flywheel as the car goes downhill), and estimate what total energy needs to be stored in the flywheel. (b) What is the angular velocity of the flywheel when it has a full "energy charge"? (c) About how long would it take a 150-hp motor to give the flywheel a full energy charge before a trip?

*7. A person stands on a platform, initially at rest, that can rotate freely without friction. The moment of inertia of the person plus the platform is I_P. The person holds a spinning bicycle wheel with its axis horizontal. The wheel has moment of inertia I_W and angular velocity ω_W. What will be the angular velocity ω_P of the platform if the person moves the axis of the wheel so that it points (a) vertically upward, (b) at a 60° angle to the vertical, (c) vertically downward? (d) What will ω_P be if the person reaches up and stops the wheel in part (a)? See Sections 8–8 and 8–9.

(a)

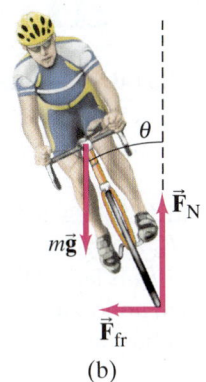
(b)

FIGURE 8–68 Search and Learn 4.

ANSWERS TO EXERCISES

A: $f = 0.076$ Hz; $T = 13$ s.
B: $\vec{F}_A$.
C: (c).
D: Yes; she does work to pull in her arms.
E: (b).
F: (b).

Our whole built environment, from modern bridges to skyscrapers, has required architects and engineers to determine the forces and stresses within these structures. The object is to keep these structures standing, or "static"—that is, not in motion, especially not falling down.

The study of statics applies equally well to the human body, including balance, the forces in muscles, joints, and bones, and ultimately the possibility of fracture.

Static Equilibrium; Elasticity and Fracture

CONTENTS

9–1 The Conditions for Equilibrium
9–2 Solving Statics Problems
9–3 Applications to Muscles and Joints
9–4 Stability and Balance
9–5 Elasticity; Stress and Strain
9–6 Fracture
*9–7 Spanning a Space: Arches and Domes

CHAPTER-OPENING QUESTION—Guess now!

The diving board shown here is held by two supports at A and B. Which statement is true about the forces exerted *on* the diving board at A and B?

(a) $\vec{F}_A$ is down, $\vec{F}_B$ is up, and F_B is larger than F_A.
(b) Both forces are up and F_B is larger than F_A.
(c) $\vec{F}_A$ is down, $\vec{F}_B$ is up, and F_A is larger than F_B.
(d) Both forces are down and approximately equal.
(e) $\vec{F}_B$ is down, $\vec{F}_A$ is up, and they are equal.

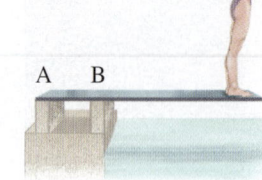

In this Chapter, we will study a special case in mechanics—when the net force and the net torque on an object, or system of objects, are both zero. In this case both the linear acceleration and the angular acceleration of the object or system are zero. The object is either at rest, or its center of mass is moving at constant velocity. We will be concerned mainly with the first situation, in which the object or objects are all at rest, or *static* (= not moving).

The net force and the net torque can be zero, but this does not imply that no forces at all act on the objects. In fact it is virtually impossible to find an object on which no forces act. Just how and where these forces act can be very important, both for buildings and other structures, and in the human body.

Sometimes, as we shall see in this Chapter, the forces may be so great that the object is seriously *deformed*, or it may even *fracture* (break)—and avoiding such problems gives this field of *statics* even greater importance.

Statics is concerned with the calculation of the forces acting on and within structures that are in *equilibrium*. Determination of these forces, which occupies us in the first part of this Chapter, then allows a determination of whether the structures can sustain the forces without significant deformation or fracture, subjects we discuss later in this Chapter. These techniques can be applied in a wide range of fields. Architects and engineers must be able to calculate the forces on the structural components of buildings, bridges, machines, vehicles, and other structures, since any material will buckle or break if too much force is applied (Fig. 9–1). In the human body a knowledge of the forces in muscles and joints is of great value for doctors, physical therapists, and athletes.

9–1 The Conditions for Equilibrium

Objects in daily life have at least one force acting on them (gravity). If they are at rest, then there must be other forces acting on them as well so that the net force is zero. A book at rest on a table, for example, has two forces acting on it, the downward force of gravity and the normal force the table exerts upward on it (Fig. 9–2). Because the book is at rest, Newton's second law tells us that the net force on it is zero. Thus the upward force exerted by the table on the book must be equal in magnitude to the force of gravity acting downward on the book. Such an object is said to be in **equilibrium** (Latin for "equal forces" or "balance") under the action of these two forces.

Do not confuse the two forces in Fig. 9–2 with the equal and opposite forces of Newton's third law, which act on different objects. In Fig. 9–2, both forces act on the same object; and they happen to add up to zero.

FIGURE 9–1 Elevated walkway collapse in a Kansas City hotel in 1981. How a simple physics calculation could have prevented the tragic loss of over 100 lives is considered in Example 9–12.

FIGURE 9–2 The book is in equilibrium; the net force on it is zero.

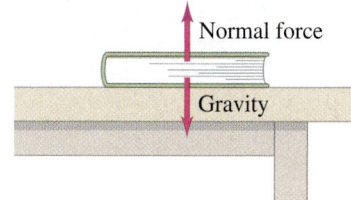

EXAMPLE 9–1 Straightening teeth. The wire band shown in Fig. 9–3a has a tension F_T of 2.0 N along it. It therefore exerts forces of 2.0 N on the highlighted tooth (to which it is attached) in the two directions shown. Calculate the resultant force on the tooth due to the wire, F_R.

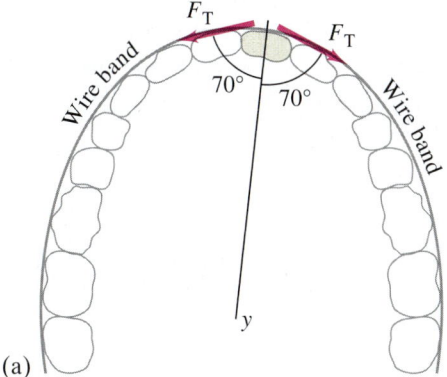
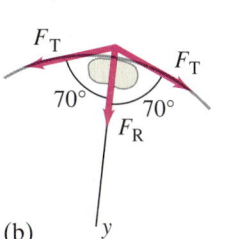

FIGURE 9–3 Forces on a tooth. Example 9–1.

APPROACH Since the two forces F_T are equal, their sum will be directed along the line that bisects the angle between them, which we have chosen to be the y axis. The x components of the two forces add up to zero.

SOLUTION The y component of each force is $(2.0\,\text{N})(\cos 70°) = 0.68\,\text{N}$: adding the two together, we get a resultant force $F_R = 1.4\,\text{N}$ as shown in Fig. 9–3b. We assume that the tooth is in equilibrium because the gums exert a nearly equal magnitude force in the opposite direction. Actually that is not quite so since the objective is to move the tooth ever so slowly.

NOTE If the wire is firmly attached to the tooth, the tension to the right, say, can be made larger than that to the left, and the resultant force would correspondingly be directed more toward the right.

PHYSICS APPLIED
Braces for teeth

The First Condition for Equilibrium

For an object to be at rest, Newton's second law tells us that the sum of the forces acting on it must add up to zero. Since force is a vector, the components of the net force must each be zero. Hence, a condition for equilibrium is that

$$\Sigma F_x = 0, \qquad \Sigma F_y = 0, \qquad \Sigma F_z = 0. \qquad (9\text{–}1)$$

We will mainly be dealing with forces that act in a plane, so we usually need only the x and y components. We must remember that if a particular force component points along the negative x or y axis, it must have a negative sign. Equations 9–1 represent the **first condition for equilibrium.**

We saw in Chapter 4 that to solve Problems involving forces, we need to draw a *free-body diagram*, indicating *all* the forces on a given object (see Section 4–7).

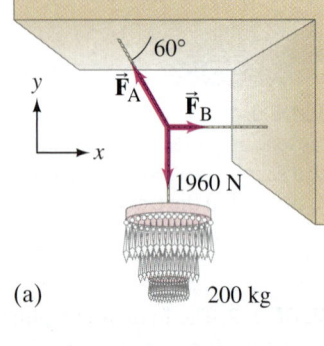

EXAMPLE 9–2 **Chandelier cord tension.** Calculate the tensions $\vec{F}_A$ and $\vec{F}_B$ in the two cords that are connected to the vertical cord supporting the 200-kg chandelier in Fig. 9–4. Ignore the mass of the cords.

APPROACH We need a free-body diagram, but for which object? If we choose the chandelier, the cord supporting it must exert a force equal to the chandelier's weight $mg = (200\ \text{kg})(9.80\ \text{m/s}^2) = 1960\ \text{N}$. But the forces $\vec{F}_A$ and $\vec{F}_B$ don't get involved. Instead, let us choose as our object the point where the three cords join (it could be a knot). The free-body diagram is then as shown in Fig. 9–4a. The three forces—$\vec{F}_A$, $\vec{F}_B$, and the tension in the vertical cord equal to the weight of the 200-kg chandelier—act at this point where the three cords join. For this junction point we write $\Sigma F_x = 0$ and $\Sigma F_y = 0$, since the problem is laid out in two dimensions. The directions of $\vec{F}_A$ and $\vec{F}_B$ are known, since tension in a cord can only be along the cord—any other direction would cause the cord to bend, as already pointed out in Chapter 4. Thus, our unknowns are the magnitudes F_A and F_B.

FIGURE 9–4 Example 9–2.

SOLUTION We first resolve $\vec{F}_A$ into its horizontal (x) and vertical (y) components. Although we don't know the value of F_A, we can write (see Fig. 9–4b) $F_{Ax} = -F_A \cos 60°$ and $F_{Ay} = F_A \sin 60°$. $\vec{F}_B$ has only an x component. In the vertical direction, we have the downward force exerted by the vertical cord equal to the weight of the chandelier $mg = (200\ \text{kg})(g)$, and the vertical component of $\vec{F}_A$ upward:

$$\Sigma F_y = 0$$
$$F_A \sin 60° - (200\ \text{kg})(g) = 0$$

so

$$F_A = \frac{(200\ \text{kg})g}{\sin 60°} = (231\ \text{kg})g = (231\ \text{kg})(9.80\ \text{m/s}^2) = 2260\ \text{N}.$$

In the horizontal direction, with $\Sigma F_x = 0$,

$$\Sigma F_x = F_B - F_A \cos 60° = 0.$$

Thus

$$F_B = F_A \cos 60° = (231\ \text{kg})(g)(0.500) = (115\ \text{kg})g = 1130\ \text{N}.$$

The magnitudes of $\vec{F}_A$ and $\vec{F}_B$ determine the strength of cord or wire that must be used. In this case, the cord must be able to support a mass of more than 230 kg.

NOTE We didn't insert the value of g, the acceleration due to gravity, until the end. In this way we found the magnitude of the force in terms of g times the number of kilograms (which may be a more familiar quantity than newtons).

EXERCISE A In Example 9–2, F_A has to be greater than the chandelier's weight, mg. Why?

The Second Condition for Equilibrium

Although Eqs. 9–1 are a necessary condition for an object to be in equilibrium, they are not always a sufficient condition. Figure 9–5 shows an object on which the net force is zero. Although the two forces labeled $\vec{F}$ add up to give zero net force on the object, they do give rise to a net torque that will rotate the object.

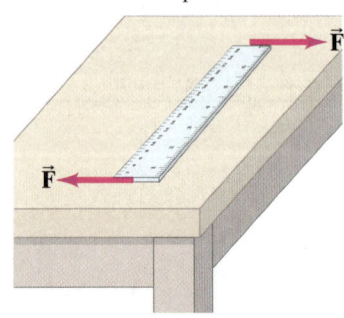

FIGURE 9–5 Although the net force on it is zero, the ruler will move (rotate). A pair of equal forces acting in opposite directions but at different points on an object (as shown here) is referred to as a *couple.*

Referring to Eq. 8–14, $\Sigma\tau = I\alpha$, we see that if an object is to remain at rest, the net torque applied to it (calculated about *any* axis) must be zero. Thus we have the **second condition for equilibrium**: that the sum of the torques acting on an object, as calculated about any axis, must be zero:

$$\Sigma\tau = 0. \qquad (9\text{–}2)$$

This condition will ensure that the angular acceleration, α, about any axis will be zero. If the object is not rotating initially ($\omega = 0$), it will not start rotating. Equations 9–1 and 9–2 are the only requirements for an object to be in equilibrium.

We will mainly consider cases in which the forces all act in a plane (we call it the *xy* plane). In such cases the torque is calculated about an axis that is perpendicular to the *xy* plane. *The choice of this axis is arbitrary.* If the object is at rest, then $\Sigma\tau = 0$ is valid about any axis. Therefore we can choose any axis that makes our calculation easier. Once the axis is chosen, all torques must be calculated about that axis.

CAUTION
Axis choice for $\Sigma\tau = 0$ is arbitrary. All torques must be calculated about the same axis.

CONCEPTUAL EXAMPLE 9–3 **A lever.** The bar in Fig. 9–6 is being used as a lever to pry up a large rock. The small rock acts as a fulcrum (pivot point). The force F_P required at the long end of the bar can be quite a bit smaller than the rock's weight mg, since it is the *torques* that balance in the rotation about the fulcrum. If, however, the leverage isn't sufficient, and the large rock isn't budged, what are two ways to increase the lever arm?

RESPONSE One way is to increase the lever arm of the force F_P by slipping a pipe over the end of the bar and thereby pushing with a longer lever arm. A second way is to move the fulcrum closer to the large rock. This may change the long lever arm R only a little, but it changes the short lever arm r by a substantial fraction and therefore changes the ratio of R/r dramatically. In order to pry the rock, the torque due to F_P must at least balance the torque due to mg; that is, $mgr = F_P R$ and

$$\frac{r}{R} = \frac{F_P}{mg}.$$

With r smaller, the weight mg can be balanced with less force F_P. The ratio R/r is the **mechanical advantage** of the system. A lever is a "simple machine." We discussed another simple machine, the pulley, in Chapter 4, Example 4–14.

PHYSICS APPLIED
The lever

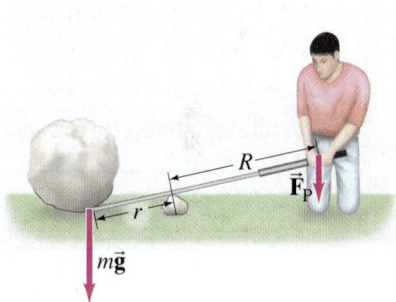

FIGURE 9–6 Example 9–3. A lever can "multiply" your force.

EXERCISE B For simplicity, we wrote the equation in Example 9–3 as if the lever were perpendicular to the forces. Would the equation be valid even for a lever at an angle as shown in Fig. 9–6?

9–2 Solving Statics Problems

The subject of statics is important because it allows us to calculate certain forces on (or within) a structure when some of the forces on it are already known. We will mainly consider situations in which all the forces act in a plane, so we can have two force equations (*x* and *y* components) and one torque equation, for a total of three equations. Of course, you do not have to use all three equations if they are not needed. When using a torque equation, a torque that tends to rotate the object counterclockwise is usually considered positive, whereas a torque that tends to rotate it clockwise is considered negative. (But the opposite convention would be OK too.)

One of the forces that acts on objects is the force of gravity. As we discussed in Section 7–8, we can consider the force of gravity on an object as acting at its center of gravity (CG) or center of mass (CM), which for practical purposes are the same point. For uniform symmetrically shaped objects, the CG is at the geometric center. For more complicated objects, the CG can be determined as discussed in Section 7–8.

PROBLEM SOLVING
$\tau > 0$ counterclockwise
$\tau < 0$ clockwise

PROBLEM SOLVING

There is no single technique for attacking statics problems, but the following procedure may be helpful.

Statics

1. Choose one object at a time for consideration. Make a careful **free-body diagram** by showing all the forces acting on that object, including gravity, and the points at which these forces act. If you aren't sure of the direction of a force, choose a direction; if the actual direction of the force (or component of a force) is opposite, your eventual calculation will give a result with a minus sign.
2. Choose a convenient **coordinate system**, and resolve the forces into their components.
3. Using letters to represent unknowns, write down the **equilibrium equations** for the **forces**:

$$\Sigma F_x = 0$$

and

$$\Sigma F_y = 0,$$

assuming all the forces act in a plane.

4. For the **torque equation**,

$$\Sigma \tau = 0,$$

choose any axis perpendicular to the xy plane that might make the calculation easier. (For example, you can reduce the number of unknowns in the resulting equation by choosing the axis so that one of the unknown forces acts through that axis; then this force will have zero lever arm and produce zero torque, and so won't appear in the torque equation.) Pay careful attention to determining the lever arm for each force correctly. Give each torque a + or − sign to indicate torque direction. For example, if torques tending to rotate the object counterclockwise are positive, then those tending to rotate it clockwise are negative.

5. **Solve** these equations for the unknowns. Three equations allow a maximum of three unknowns to be solved for. They can be forces, distances, or even angles.

PHYSICS APPLIED
Balancing a seesaw

EXAMPLE 9–4 Balancing a seesaw. A board of mass $M = 4.0$ kg serves as a seesaw for two children, as shown in Fig. 9–7a. Child A has a mass of 30 kg and sits 2.5 m from the pivot point, P (his center of gravity is 2.5 m from the pivot). At what distance x from the pivot must child B, of mass 25 kg, place herself to balance the seesaw? Assume the board is uniform and centered over the pivot.

APPROACH We follow the steps of the Problem Solving Strategy above.

SOLUTION

1. **Free-body diagram.** We choose the board as our object, and assume it is horizontal. Its free-body diagram is shown in Fig. 9–7b. The forces acting on the board are the forces exerted downward on it by each child, $\vec{F}_A$ and $\vec{F}_B$, the upward force exerted by the pivot $\vec{F}_N$, and the force of gravity on the board ($= M\vec{g}$) which acts at the center of the uniform board.
2. **Coordinate system.** We choose y to be vertical, with positive upward, and x horizontal to the right, with origin at the pivot.
3. **Force equation.** All the forces are in the y (vertical) direction, so

$$\Sigma F_y = 0$$
$$F_N - m_A g - m_B g - Mg = 0,$$

where $F_A = m_A g$ and $F_B = m_B g$.

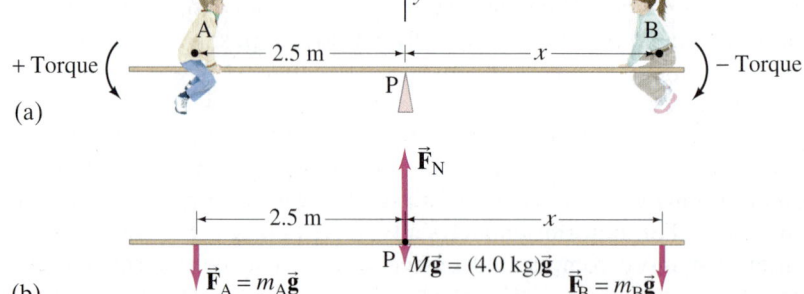

FIGURE 9–7 (a) Two children on a seesaw, Example 9–4. (b) Free-body diagram of the board.

4. **Torque equation.** Let us calculate the torque about an axis through the board at the pivot point, P. Then the lever arms for F_N and for the weight of the board are zero, and they will contribute zero torque about point P. Thus the torque equation will involve only the forces $\vec{F}_A$ and $\vec{F}_B$, which are equal to the weights of the children. The torque exerted by each child will be mg times the appropriate lever arm, which here is the distance of each child from the pivot point. $\vec{F}_A$ tends to rotate the board counterclockwise (+) and $\vec{F}_B$ clockwise (−), so the torque equation is

$$\Sigma\tau = 0$$
$$m_A g(2.5\text{ m}) - m_B g x + Mg(0\text{ m}) + F_N(0\text{ m}) = 0$$

or

$$m_A g(2.5\text{ m}) - m_B g x = 0,$$

where two terms were dropped because their lever arms were zero.

5. **Solve.** We solve the torque equation for x and find

$$x = \frac{m_A}{m_B}(2.5\text{ m}) = \frac{30\text{ kg}}{25\text{ kg}}(2.5\text{ m}) = 3.0\text{ m}.$$

To balance the seesaw, child B must sit so that her CG is 3.0 m from the pivot point. This makes sense: since she is lighter, she must sit farther from the pivot than the heavier child in order to provide torques of equal magnitude.

EXERCISE C We did not need to use the force equation to solve Example 9–4 because of our choice of the axis. Use the force equation to find the force exerted by the pivot.

EXAMPLE 9–5 **Forces on a beam and supports.** A uniform 1500-kg beam, 20.0 m long, supports a 15,000-kg printing press 5.0 m from the right support column (Fig. 9–8). Calculate the force on each of the vertical support columns.

APPROACH We analyze the forces on the beam (the force the beam exerts on each column is equal and opposite to the force exerted by the column on the beam). We label these forces $\vec{F}_A$ and $\vec{F}_B$ in Fig. 9–8. The weight of the beam itself acts at its center of gravity, 10.0 m from either end. We choose a convenient axis for writing the torque equation: the point of application of $\vec{F}_A$ (labeled P), so $\vec{F}_A$ will not enter the equation (its lever arm will be zero) and we will have an equation in only one unknown, F_B.

SOLUTION The torque equation, $\Sigma\tau = 0$, with the counterclockwise direction as positive, gives

$$\Sigma\tau = -(10.0\text{ m})(1500\text{ kg})g - (15.0\text{ m})(15,000\text{ kg})g + (20.0\text{ m})F_B = 0.$$

Solving for F_B, we find $F_B = (12,000\text{ kg})g = 118,000\text{ N}$. To find F_A, we use $\Sigma F_y = 0$, with $+y$ upward:

$$\Sigma F_y = F_A - (1500\text{ kg})g - (15,000\text{ kg})g + F_B = 0.$$

Putting in $F_B = (12,000\text{ kg})g$, we find that $F_A = (4500\text{ kg})g = 44,100\text{ N}$.

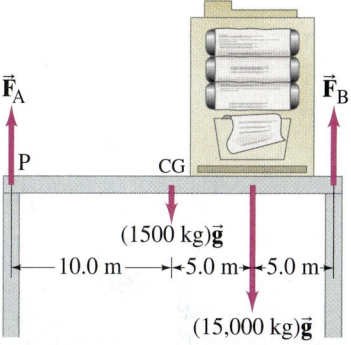

FIGURE 9–8 A 1500-kg beam supports a 15,000-kg machine. Example 9–5.

Figure 9–9 shows a uniform beam that extends beyond its support like a diving board. Such a beam is called a **cantilever**. The forces acting on the beam in Fig. 9–9 are those due to the supports, $\vec{F}_A$ and $\vec{F}_B$, and the force of gravity which acts at the CG, 5.0 m to the right of the right-hand support. If you follow the procedure of the last Example and calculate F_A and F_B, assuming they point upward as shown in Fig. 9–9, you will find that F_A comes out negative. If the beam has a mass of 1200 kg and a weight $mg = 12,000\text{ N}$, then $F_B = 15,000\text{ N}$ and $F_A = -3000\text{ N}$ (see Problem 10). Whenever an unknown force comes out negative, it merely means that the force actually points in the opposite direction from what you assumed. Thus in Fig. 9–9, $\vec{F}_A$ actually must pull downward (by means of bolts, screws, fasteners, and/or glue). To see why $\vec{F}_A$ has to act downward, note that the board's weight acting at the CG would otherwise rotate the board clockwise about support B.

PHYSICS APPLIED
Cantilever

PROBLEM SOLVING
If a force comes out negative

FIGURE 9–9 A cantilever.

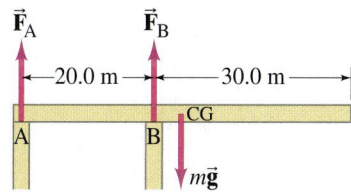

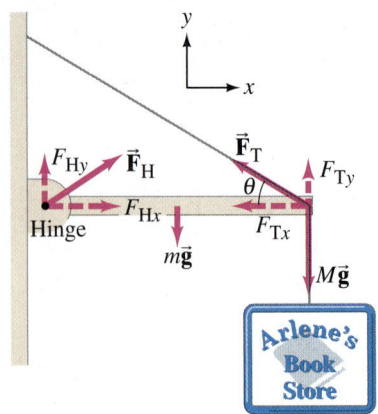

FIGURE 9–10 Example 9–6.

EXERCISE D Return to the Chapter-Opening Question, page 230, and answer it again now. Try to explain why you may have answered differently the first time.

Our next Example involves a beam that is attached to a wall by a hinge and is supported by a cable or cord (Fig. 9–10). It is important to remember that a flexible cable can support a force only along its length. (If there were a component of force perpendicular to the cable, it would bend because it is flexible.) But for a rigid device, such as the hinge in Fig. 9–10, the force can be in any direction and we can know the direction only after solving the equations. (The hinge is assumed small and smooth, so it can exert no internal torque on the beam.)

EXAMPLE 9–6 **Hinged beam and cable.** A uniform beam, 2.20 m long with mass $m = 25.0$ kg, is mounted by a small hinge on a wall as shown in Fig. 9–10. The beam is held in a horizontal position by a cable that makes an angle $\theta = 30.0°$. The beam supports a sign of mass $M = 28.0$ kg suspended from its end. Determine the components of the force $\vec{F}_H$ that the (smooth) hinge exerts on the beam, and the tension F_T in the supporting cable.

APPROACH Figure 9–10 is the free-body diagram for the beam, showing all the forces acting on the beam. It also shows the components of $\vec{F}_T$ and a guess for the direction of $\vec{F}_H$. We have three unknowns, F_{Hx}, F_{Hy}, and F_T (we are given θ), so we will need all three equations, $\Sigma F_x = 0$, $\Sigma F_y = 0$, $\Sigma \tau = 0$.

SOLUTION The sum of the forces in the vertical (y) direction is

$$\Sigma F_y = 0$$
$$F_{Hy} + F_{Ty} - mg - Mg = 0. \quad \text{(i)}$$

In the horizontal (x) direction, the sum of the forces is

$$\Sigma F_x = 0$$
$$F_{Hx} - F_{Tx} = 0. \quad \text{(ii)}$$

For the torque equation, we choose the axis at the point where $\vec{F}_T$ and $M\vec{g}$ act. Then our torque equation will contain only one unknown, F_{Hy}, because the lever arms for $\vec{F}_T$, $M\vec{g}$, and F_{Hx} are zero. We choose torques that tend to rotate the beam counterclockwise as positive. The weight mg of the (uniform) beam acts at its center, so we have

$$\Sigma \tau = 0$$
$$-(F_{Hy})(2.20 \text{ m}) + mg(1.10 \text{ m}) = 0.$$

We solve for F_{Hy}:

$$F_{Hy} = \left(\frac{1.10 \text{ m}}{2.20 \text{ m}}\right)mg = (0.500)(25.0 \text{ kg})(9.80 \text{ m/s}^2) = 123 \text{ N}. \quad \text{(iii)}$$

Next, since the tension $\vec{F}_T$ in the cable acts along the cable ($\theta = 30.0°$), we see from Fig. 9–10 that $\tan \theta = F_{Ty}/F_{Tx}$, or

$$F_{Ty} = F_{Tx} \tan \theta = F_{Tx} (\tan 30.0°). \quad \text{(iv)}$$

Equation (i) above gives

$$F_{Ty} = (m + M)g - F_{Hy} = (53.0 \text{ kg})(9.80 \text{ m/s}^2) - 123 \text{ N} = 396 \text{ N}.$$

Equations (iv) and (ii) give

$$F_{Tx} = F_{Ty}/\tan 30.0° = 396 \text{ N}/\tan 30.0° = 686 \text{ N};$$
$$F_{Hx} = F_{Tx} = 686 \text{ N}.$$

The components of $\vec{F}_H$ are $F_{Hy} = 123$ N and $F_{Hx} = 686$ N. The tension in the wire is $F_T = \sqrt{F_{Tx}^2 + F_{Ty}^2} = \sqrt{(686 \text{ N})^2 + (396 \text{ N})^2} = 792 \text{ N}.$[†]

[†]Our calculation used numbers rounded off to 3 significant figures. If you keep an extra digit, or leave the numbers in your calculator, you get $F_{Ty} = 396.5$ N, $F_{Tx} = 686.8$ N, and $F_T = 793$ N, all within the expected precision of 3 significant figures (Section 1–4).

Alternate Solution Let us see the effect of choosing a different axis for calculating torques, such as an axis through the hinge. Then the lever arm for F_H is zero, and the torque equation ($\Sigma \tau = 0$) becomes

$$-mg(1.10 \text{ m}) - Mg(2.20 \text{ m}) + F_{Ty}(2.20 \text{ m}) = 0.$$

We solve this for F_{Ty} and find

$$F_{Ty} = \frac{m}{2}g + Mg = (12.5 \text{ kg} + 28.0 \text{ kg})(9.80 \text{ m/s}^2) = 397 \text{ N}.$$

We get the same result, within the precision of our significant figures.

NOTE It doesn't matter which axis we choose for $\Sigma \tau = 0$. Using a second axis can serve as a check.

*A More Difficult Example—The Ladder

EXAMPLE 9–7 **Ladder.** A 5.0-m-long ladder leans against a wall at a point 4.0 m above a cement floor as shown in Fig. 9–11. The ladder is uniform and has mass $m = 12.0$ kg. Assuming the wall is frictionless, but the floor is not, determine the forces exerted on the ladder by the floor and by the wall.

APPROACH Figure 9–11 is the free-body diagram for the ladder, showing all the forces acting on the ladder. The wall, since it is frictionless, can exert a force only perpendicular to the wall, and we label that force $\vec{F}_W$. The cement floor exerts a force $\vec{F}_C$ which has both horizontal and vertical force components: F_{Cx} is frictional and F_{Cy} is the normal force. Finally, gravity exerts a force $mg = (12.0 \text{ kg})(9.80 \text{ m/s}^2) = 118 \text{ N}$ on the ladder at its midpoint, since the ladder is uniform.

SOLUTION Again we use the equilibrium conditions, $\Sigma F_x = 0$, $\Sigma F_y = 0$, $\Sigma \tau = 0$. We will need all three since there are three unknowns: F_W, F_{Cx}, and F_{Cy}. The y component of the force equation is

$$\Sigma F_y = F_{Cy} - mg = 0,$$

so immediately we have

$$F_{Cy} = mg = 118 \text{ N}.$$

The x component of the force equation is

$$\Sigma F_x = F_{Cx} - F_W = 0.$$

To determine both F_{Cx} and F_W, we need a torque equation. If we choose to calculate torques about an axis through the point where the ladder touches the cement floor, then $\vec{F}_C$, which acts at this point, will have a lever arm of zero and so won't enter the equation. The ladder touches the floor a distance $x_0 = \sqrt{(5.0 \text{ m})^2 - (4.0 \text{ m})^2} = 3.0$ m from the wall (right triangle, $c^2 = a^2 + b^2$). The lever arm for mg is half this, or 1.5 m, and the lever arm for F_W is 4.0 m, Fig. 9–11. The torque equation about the ladder's contact point on the cement is

$$\Sigma \tau = (4.0 \text{ m})F_W - (1.5 \text{ m})mg = 0.$$

Thus

$$F_W = \frac{(1.5 \text{ m})(12.0 \text{ kg})(9.8 \text{ m/s}^2)}{4.0 \text{ m}} = 44 \text{ N}.$$

Then, from the x component of the force equation,

$$F_{Cx} = F_W = 44 \text{ N}.$$

Since the components of $\vec{F}_C$ are $F_{Cx} = 44$ N and $F_{Cy} = 118$ N, then

$$F_C = \sqrt{(44 \text{ N})^2 + (118 \text{ N})^2} = 126 \text{ N} \approx 130 \text{ N}$$

(rounded off to two significant figures), and it acts at an angle to the floor of

$$\theta = \tan^{-1}(118 \text{ N}/44 \text{ N}) = 70°.$$

NOTE The force $\vec{F}_C$ does *not* have to act along the ladder's direction because the ladder is rigid and not flexible like a cord or cable (see page 89, Chapter 4).

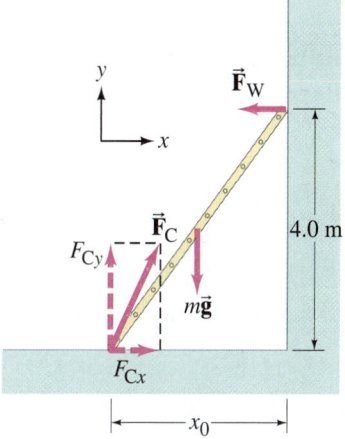

FIGURE 9–11 A ladder leaning against a wall. Example 9–7. The force $\vec{F}_C$ that the cement floor exerts on the ladder need not be along the ladder which (unlike a cord) is rigid.

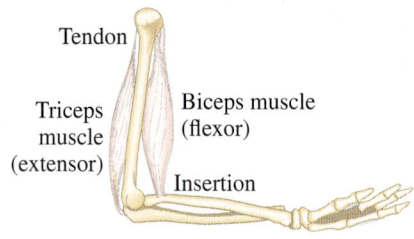

FIGURE 9–12 The biceps (flexor) and triceps (extensor) muscles in the human arm.

Forces in muscles and joints

FIGURE 9–13 Example 9–8, forces on forearm.

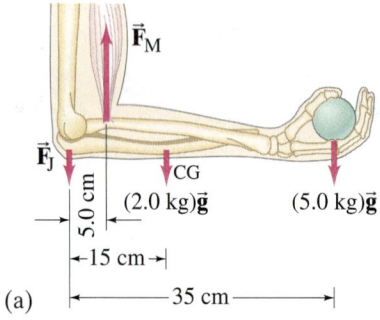

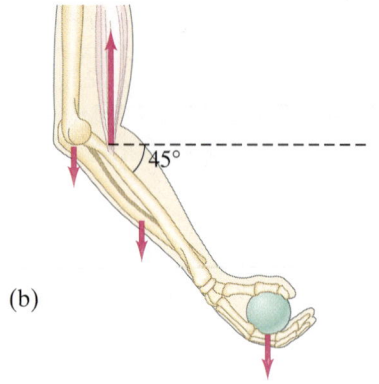

9–3 Applications to Muscles and Joints

The techniques we have been discussing for calculating forces on objects in equilibrium can be applied to the human body to study forces on muscles, bones, joints. Generally a muscle is attached, via tendons, to two different bones, as in Fig. 9–12. The points of attachment are called *insertions*. Two bones are flexibly connected at a *joint*, such as at the elbow, knee, hip, and shoulder. A muscle exerts a pull when its fibers contract under stimulation by a nerve, but a muscle can *not* exert a push. Muscles that tend to bring two limbs closer together, such as the biceps muscle in the upper arm (Fig. 9–12), are called *flexors*; those that act to extend a limb outward, such as the triceps muscle in Fig. 9–12, are called *extensors*. You use the flexor muscle in the upper arm when lifting an object in your hand; you use the extensor muscle when throwing a ball.

EXAMPLE 9–8 Force exerted by biceps muscle. How much force must the biceps muscle exert when a 5.0-kg ball is held in the hand (*a*) with the arm horizontal as in Fig. 9–13a, and (*b*) when the arm is at a 45° angle as in Fig. 9–13b? The biceps muscle is connected to the forearm by a tendon attached 5.0 cm from the elbow joint. Assume that the mass of forearm and hand together is 2.0 kg and their CG is as shown.

APPROACH The free-body diagram for the forearm is shown in Fig. 9–13; the forces are the weights of the arm and ball, the upward force $\vec{F}_M$ exerted by the muscle, and a force $\vec{F}_J$ exerted at the joint by the bone in the upper arm (all assumed to act vertically). We wish to find the magnitude of $\vec{F}_M$, which can be done using the torque equation and by choosing our axis through the joint so that $\vec{F}_J$ contributes zero torque.

SOLUTION (*a*) We calculate torques about the point where $\vec{F}_J$ acts in Fig. 9–13a. The $\Sigma\tau = 0$ equation gives

$$(0.050 \text{ m})F_M - (0.15 \text{ m})(2.0 \text{ kg})g - (0.35 \text{ m})(5.0 \text{ kg})g = 0.$$

We solve for F_M:

$$F_M = \frac{(0.15 \text{ m})(2.0 \text{ kg})g + (0.35 \text{ m})(5.0 \text{ kg})g}{0.050 \text{ m}} = (41 \text{ kg})g = 400 \text{ N}.$$

(*b*) The lever arm, as calculated about the joint, is reduced by the factor cos 45° for all three forces. Our torque equation will look like the one just above, except that each term will have its lever arm reduced by the same factor, which will cancel out. The same result is obtained, $F_M = 400$ N.

NOTE The force required of the muscle (400 N) is quite large compared to the weight of the object lifted ($= mg = 49$ N). Indeed, the muscles and joints of the body are generally subjected to quite large forces.

NOTE Forces exerted on joints can be large and even painful or injurious. Using $\Sigma F_y = 0$ we calculate for this case $F_J = F_M - (2.0 \text{ kg})g - (5.0 \text{ kg})g = 330$ N.

Muscle insertion and lever arm

Forces on the spine, and back pain

The point of insertion of a muscle varies from person to person. A slight increase in the distance of the joint to the point of insertion of the biceps muscle from 5.0 cm to 5.5 cm can be a considerable advantage for lifting and throwing. Champion athletes are often found to have muscle insertions farther from the joint than the average person, and if this applies to one muscle, it usually applies to all.

As another example of the large forces acting within the human body, we consider the muscles used to support the body (trunk) when you bend forward (Fig. 9–14a). The lowest vertebra on the spinal column (fifth lumbar vertebra) acts as a fulcrum for this bending position. The "erector spinae" muscles in the back that support the trunk act at an effective angle of about 12° to the axis of the spine. Let us assume the trunk makes an angle of 30° with the horizontal.

Figure 9–14b is a simplified schematic drawing showing the forces on the upper body. The force exerted by the back muscles is represented by $\vec{F}_M$, the force exerted on the base of the spine at the lowest vertebra is $\vec{F}_V$, and $\vec{w}_H$, $\vec{w}_A$, and $\vec{w}_T$ represent the weights of the **h**ead, freely hanging **a**rms, and **t**runk, respectively. The values shown are approximations. The distances (in cm) refer to a person 180 cm tall, but are approximately in the same ratio of 1:2:3 for an average person of any height, and the result in the following Example is then independent of the height of the person.

EXAMPLE 9–9 **Forces on your back.** Calculate the magnitude and direction of the force $\vec{F}_V$ acting on the fifth lumbar vertebra as represented in Fig. 9–14b.

APPROACH We use the model of the upper body described above and shown in Fig. 9–14b. We can calculate F_M using the torque equation if we take the axis at the base of the spine (point S); with this choice, the other unknown, F_V, doesn't appear in the equation because its lever arm is zero. To figure the lever arms, we need to use trigonometric functions.

SOLUTION For $\vec{F}_M$, the lever arm (perpendicular distance from axis to line of action of the force) will be the real distance to where the force acts (48 cm) multiplied by sin 12°, as shown in Fig. 9–14c. The lever arms for $\vec{w}_H$, $\vec{w}_A$, and $\vec{w}_T$ can be seen from Fig. 9–14b to be their respective distances from S times sin 60°. F_M tends to rotate the trunk counterclockwise, which we take to be positive. Then $\vec{w}_H$, $\vec{w}_A$, $\vec{w}_T$ will contribute negative torques. Thus $\Sigma\tau = 0$ gives

$$(0.48 \text{ m})(\sin 12°)(F_M) - (0.72 \text{ m})(\sin 60°)(w_H)$$
$$- (0.48 \text{ m})(\sin 60°)(w_A) - (0.36 \text{ m})(\sin 60°)(w_T) = 0.$$

Solving for F_M and putting in the values for w_H, w_A, w_T given in Fig. 9–14b, we find

$$F_M = \frac{(0.72 \text{ m})(0.07w) + (0.48 \text{ m})(0.12w) + (0.36 \text{ m})(0.46w)}{(0.48 \text{ m})(\sin 12°)} (\sin 60°)$$
$$= 2.37w \approx 2.4w,$$

where w is the total weight of the body. To get the components of $\vec{F}_V$ we use the x and y components of the force equation (noting that 30° − 12° = 18°):

$$\Sigma F_y = F_{Vy} - F_M \sin 18° - w_H - w_A - w_T = 0$$

so

$$F_{Vy} = 1.38w \approx 1.4w,$$

and

$$\Sigma F_x = F_{Vx} - F_M \cos 18° = 0$$

so

$$F_{Vx} = 2.25w \approx 2.3w,$$

where we keep 3 significant figures for calculating, but round off to 2 for giving the answer. Then

$$F_V = \sqrt{F_{Vx}^2 + F_{Vy}^2} = 2.6w.$$

The angle θ that F_V makes with the horizontal is given by $\tan\theta = F_{Vy}/F_{Vx} = 0.61$, so $\theta = 32°$.

NOTE The force on the lowest vertebra is over $2\frac{1}{2}$ times the total body weight! This force is exerted by the "sacral" bone at the base of the spine, through the somewhat flexible *intervertebral disk*. The disks at the base of the spine are clearly being compressed under very large forces. [If the body was less bent over (say, the 30° angle in Fig. 9–14b becomes 40° or 50°), then the stress on the lower back will be less (see Problem 33).]

If the person in Fig. 9–14 has a mass of 90 kg and is holding 20 kg in his hands (this increases w_A to $0.34w$), then F_V is increased to almost four times the person's weight ($3.7w$). For this 200-lb person, the force on the disk would be over 700 lb! With such strong forces acting, it is little wonder that so many people suffer from low back pain at one time or another.

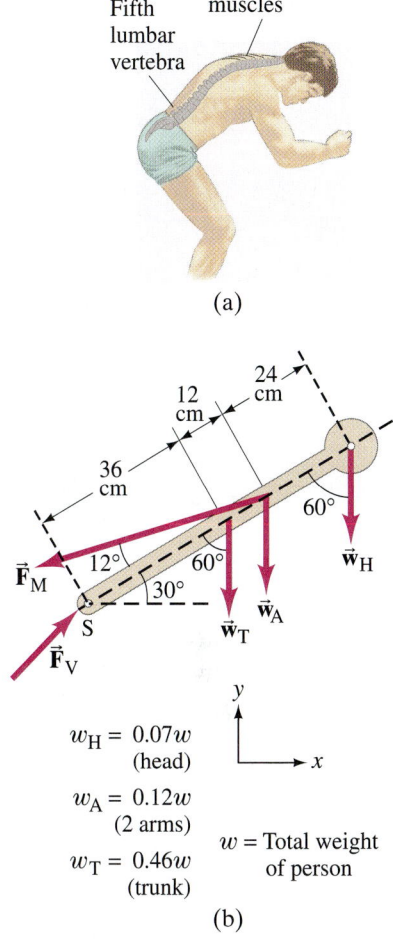

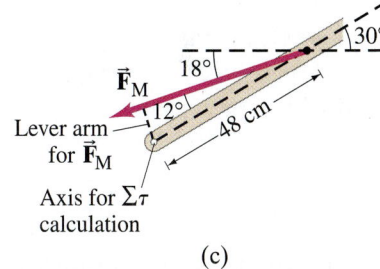

FIGURE 9–14 (a) A person bending over. (b) Forces on the back exerted by the back muscles ($\vec{F}_M$) and by the vertebrae ($\vec{F}_V$) when a person bends over. (c) Finding the lever arm for $\vec{F}_M$.

9–4 Stability and Balance

An object in static equilibrium, if left undisturbed, will undergo no translational or rotational acceleration since the sum of all the forces and the sum of all the torques acting on it are zero. However, if the object is displaced slightly, three outcomes are possible: (1) the object returns to its original position, in which case it is said to be in **stable equilibrium**; (2) the object moves even farther from its original position, and it is said to be in **unstable equilibrium**; or (3) the object remains in its new position, and it is said to be in **neutral equilibrium**.

Consider the following examples. A ball suspended freely from a string is in stable equilibrium, for if it is displaced to one side, it will return to its original position (Fig. 9–15a) due to the net force and torque exerted on it. On the other hand, a pencil standing on its point is in unstable equilibrium. If its center of gravity is directly over its tip (Fig. 9–15b), the net force and net torque on it will be zero. But if it is displaced ever so slightly as shown—say, by a slight vibration or tiny air current—there will be a torque on it, and this torque acts to make the pencil continue to fall in the direction of the original displacement. Finally, an example of an object in neutral equilibrium is a sphere resting on a horizontal tabletop. If it is moved slightly to one side, it will remain in its new position—no net torque acts on it.

FIGURE 9–15 (a) Stable equilibrium, and (b) unstable equilibrium.

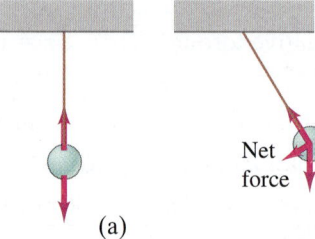
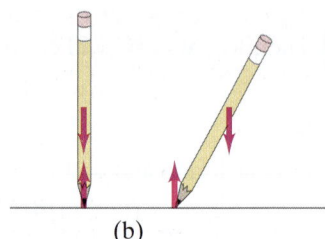

(a)   (b)

In most situations, such as in the design of structures and in working with the human body, we are interested in maintaining stable equilibrium, or *balance*, as we sometimes say. In general, an object whose center of gravity (CG) is below its point of support, such as a ball on a string, will be in stable equilibrium. If the CG is above the base of support, we have a more complicated situation. Consider a standing refrigerator (Fig. 9–16a). If it is tipped slightly, it will return to its original position due to the torque on it as shown in Fig. 9–16b. But if it is tipped too far, Fig. 9–16c, it will fall over. The critical point is reached when the CG shifts from one side of the pivot point to the other. When the CG is on one side, the torque pulls the object back onto its original base of support, Fig. 9–16b. If the object is tipped further, the CG goes past the pivot point and the torque causes the object to topple, Fig. 9–16c. In general, *an object whose center of gravity is above its base of support will be stable if a vertical line projected downward from the CG falls within the base of support*. This is because the normal force upward on the object (which balances out gravity) can be exerted only within the area of contact, so if the force of gravity acts beyond this area, a net torque will act to topple the object.

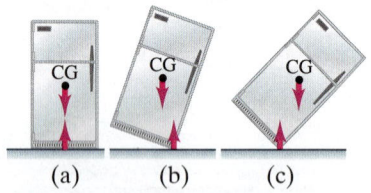

FIGURE 9–16 Equilibrium of a refrigerator resting on a flat floor.

Stability, then, can be relative. A brick lying on its widest face is more stable than a brick standing on its end, for it will take more of an effort to tip it over. In the extreme case of the pencil in Fig. 9–15b, the base is practically a point and the slightest disturbance will topple it. In general, the larger the base and the lower the CG, the more stable the object.

In this sense, humans are less stable than four-legged mammals, which have a larger base of support because of their four legs, and most also have a lower center of gravity. When walking and performing other kinds of movement, a person continually shifts the body so that its CG is over the feet, although in the normal adult this requires no conscious thought. Even as simple a movement as bending over requires moving the hips backward so that the CG remains over the feet, and you do this repositioning without thinking about it. To see this, position yourself with your heels and back to a wall and try to touch your toes. You won't be able to do it without falling. Persons carrying heavy loads automatically adjust their posture so that the CG of the total mass is over their feet, Fig. 9–17.

FIGURE 9–17 Humans adjust their posture to achieve stability when carrying loads.

PHYSICS APPLIED
Humans and balance

9–5 Elasticity; Stress and Strain

In the first part of this Chapter we studied how to calculate the forces on objects in equilibrium. In this Section we study the effects of these forces: any object changes shape under the action of applied forces. If the forces are great enough, the object will break, or *fracture*, as we will discuss in Section 9–6.

Elasticity and Hooke's Law

If a force is exerted on an object, such as the vertically suspended metal rod shown in Fig. 9–18, the length of the object changes. If the amount of elongation, $\Delta \ell$, is small compared to the length of the object, experiment shows that $\Delta \ell$ is proportional to the force exerted on the object. This proportionality can be written as an equation:

$$F = k \Delta \ell. \tag{9-3}$$

Here F represents the force pulling on the object, $\Delta \ell$ is the change in length, and k is a proportionality constant. Equation 9–3, which is sometimes called **Hooke's law**[†] after Robert Hooke (1635–1703), who first noted it, is found to be valid for almost any solid material from iron to bone—but it is valid only up to a point. For if the force is too great, the object stretches excessively and eventually breaks.

Figure 9–19 shows a typical graph of applied force versus elongation. Up to a point called the **proportional limit**, Eq. 9–3 is a good approximation for many common materials, and the curve is a straight line. Beyond this point, the graph deviates from a straight line, and no simple relationship exists between F and $\Delta \ell$. Nonetheless, up to a point farther along the curve called the **elastic limit**, the object will return to its original length if the applied force is removed. The region from the origin to the elastic limit is called the *elastic region*. If the object is stretched beyond the elastic limit, it enters the *plastic region*: it does not return to the original length upon removal of the external force, but remains permanently deformed (such as a bent paper clip). The maximum elongation is reached at the *breaking point*. The maximum force that can be applied without breaking is called the **ultimate strength** of the material (actually, force per unit area, as we discuss in Section 9–6).

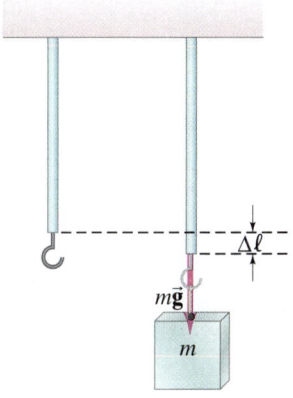

FIGURE 9–18 Hooke's law: $\Delta \ell \propto$ applied force.

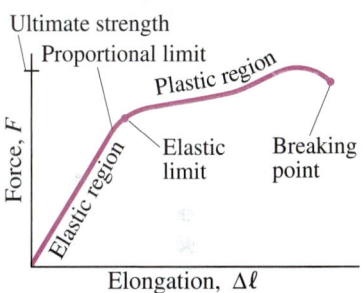

FIGURE 9–19 Applied force vs. elongation for a typical metal under tension.

Young's Modulus

The amount of elongation of an object, such as the rod shown in Fig. 9–18, depends not only on the force applied to it, but also on the material of which it is made and on its dimensions. That is, the constant k in Eq. 9–3 can be written in terms of these factors.

If we compare rods made of the same material but of different lengths and cross-sectional areas, it is found that for the same applied force, the amount of stretch (again assumed small compared to the total length) is proportional to the original length and inversely proportional to the cross-sectional area. That is, the longer the object, the more it elongates for a given force; and the thicker it is, the less it elongates. These findings can be combined with Eq. 9–3 to yield

$$\Delta \ell = \frac{1}{E} \frac{F}{A} \ell_0, \tag{9-4}$$

where ℓ_0 is the original length of the object, A is the cross-sectional area, and $\Delta \ell$ is the change in length due to the applied force F. E is a constant of proportionality[‡] known as the **elastic modulus**, or **Young's modulus**; its value depends only on the material.

[†] The term "law" applied to this relation is historical, but today it is not really appropriate. First of all, it is only an approximation, and second, it refers only to a limited set of phenomena. Most physicists today prefer to reserve the word "law" for those relations that are deeper and more encompassing and precise, such as Newton's laws of motion or the law of conservation of energy.

[‡] The fact that E is in the denominator, so $1/E$ is the actual proportionality constant, is merely a convention. When we rewrite Eq. 9–4 to get Eq. 9–5, E is found in the numerator.

The value of Young's modulus for various materials is given in Table 9–1 (the shear modulus and bulk modulus in this Table are discussed later in this Section). Because E is a property only of the material and is independent of the object's size or shape, Eq. 9–4 is far more useful for practical calculation than Eq. 9–3.

TABLE 9–1 Elastic Moduli

Material	Young's Modulus, E (N/m^2)	Shear Modulus, G (N/m^2)	Bulk Modulus, B (N/m^2)
Solids			
Iron, cast	100×10^9	40×10^9	90×10^9
Steel	200×10^9	80×10^9	140×10^9
Brass	100×10^9	35×10^9	80×10^9
Aluminum	70×10^9	25×10^9	70×10^9
Concrete	20×10^9		
Brick	14×10^9		
Marble	50×10^9		70×10^9
Granite	45×10^9		45×10^9
Wood (pine) (parallel to grain)	10×10^9		
(perpendicular to grain)	1×10^9		
Nylon	$\approx 3 \times 10^9$		
Bone (limb)	15×10^9	80×10^9	
Liquids			
Water			2.0×10^9
Alcohol (ethyl)			1.0×10^9
Mercury			2.5×10^9
Gases[†]			
Air, H$_2$, He, CO$_2$			1.01×10^5

[†]At normal atmospheric pressure; no variation in temperature during process.

EXAMPLE 9–10 Tension in piano wire. A 1.60-m-long steel piano wire has a diameter of 0.20 cm. How great is the tension in the wire if it stretches 0.25 cm when tightened?

APPROACH We assume Hooke's law holds, and use it in the form of Eq. 9–4, finding E for steel in Table 9–1.

SOLUTION We solve for F in Eq. 9–4 and note that the area of the wire is $A = \pi r^2 = (3.14)(0.0010 \text{ m})^2 = 3.14 \times 10^{-6} \text{ m}^2$. Then

$$F = E \frac{\Delta \ell}{\ell_0} A$$

$$= (2.0 \times 10^{11} \text{ N/m}^2)\left(\frac{0.0025 \text{ m}}{1.60 \text{ m}}\right)(3.14 \times 10^{-6} \text{ m}^2)$$

$$= 980 \text{ N}.$$

NOTE The large tension in all the wires in a piano must be supported by a strong frame.

EXERCISE E Two steel wires have the same length and are under the same tension. But wire A has twice the diameter of wire B. Which of the following is true? (*a*) Wire B stretches twice as much as wire A. (*b*) Wire B stretches four times as much as wire A. (*c*) Wire A stretches twice as much as wire B. (*d*) Wire A stretches four times as much as wire B. (*e*) Both wires stretch the same amount.

Stress and Strain

From Eq. 9–4, we see that the change in length of an object is directly proportional to the product of the object's length ℓ_0 and the force per unit area F/A applied to it. It is general practice to define the force per unit area as the **stress**:

$$\text{stress} = \frac{\text{force}}{\text{area}} = \frac{F}{A},$$

which has SI units of N/m^2. Also, the **strain** is defined to be the ratio of the change in length to the original length:

$$\text{strain} = \frac{\text{change in length}}{\text{original length}} = \frac{\Delta \ell}{\ell_0},$$

and is dimensionless (no units). Strain is thus the fractional change in length of the object, and is a measure of how much the object has been deformed. Stress is applied to the material by external agents, whereas strain is the material's response to the stress. Equation 9–4 can be rewritten as

$$\frac{F}{A} = E \frac{\Delta \ell}{\ell_0} \tag{9-5}$$

or

$$E = \frac{F/A}{\Delta \ell / \ell_0} = \frac{\text{stress}}{\text{strain}}.$$

Thus we see that the strain is directly proportional to the stress, in the linear (elastic) region of Fig. 9–19.

Tension, Compression, and Shear Stress

The rod shown in Fig. 9–20a is said to be under *tension* or **tensile stress**. Not only is there a force pulling down on the rod at its lower end, but since the rod is in equilibrium we know that the support at the top is exerting an equal[†] upward force on the rod at its upper end, Fig. 9–20a. In fact, this tensile stress exists throughout the material. Consider, for example, the lower half of a suspended rod as shown in Fig. 9–20b. This lower half is in equilibrium, so there must be an upward force on it to balance the downward force at its lower end. What exerts this upward force? It must be the upper part of the rod. Thus we see that external forces applied to an object give rise to internal forces, or stress, within the material itself.

Strain or deformation due to tensile stress is but one type of stress to which materials can be subjected. There are two other common types of stress: compressive and shear. **Compressive stress** is the exact opposite of tensile stress. Instead of being stretched, the material is compressed: the forces act inwardly on the object. Columns that support a weight, such as the columns of a Greek temple (Fig. 9–21), are subjected to compressive stress. Equations 9–4 and 9–5 apply equally well to compression and tension, and the values for the modulus E are usually the same.

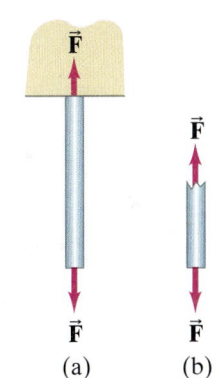

FIGURE 9–20 Stress exists *within* the material.

[†]Or a greater force if the weight of the rod cannot be ignored compared to F.

FIGURE 9–21 This Greek temple, in Agrigento, Sicily, built 2500 years ago, shows the post-and-beam construction. The columns are under compression.

FIGURE 9–22 The three types of stress for rigid objects.

Tension
(a)

Compression
(b)

Shear
(c)

Figure 9–22 compares tensile and compressive stresses as well as the third type, shear stress. An object under **shear stress** has equal and opposite forces applied *across* its opposite faces. A simple example is a book or brick firmly attached to a tabletop, on which a force is exerted parallel to the top surface. The table exerts an equal and opposite force along the bottom surface. Although the dimensions of the object do not change significantly, the shape of the object does change, Fig. 9–22c. An equation similar to Eq. 9–4 can be applied to calculate shear strain:

$$\Delta\ell = \frac{1}{G}\frac{F}{A}\ell_0 \qquad (9\text{–}6)$$

but $\Delta\ell$, ℓ_0, and A must be reinterpreted as indicated in Fig. 9–22c. Note that A is the area of the surface *parallel* to the applied force (and not perpendicular as for tension and compression), and $\Delta\ell$ is *perpendicular* to ℓ_0. The constant of proportionality G is called the **shear modulus** and is generally one-half to one-third the value of Young's modulus E (see Table 9–1). Figure 9–23 suggests why $\Delta\ell \propto \ell_0$: the fatter book shifts more for the same shearing force.

FIGURE 9–23 The fatter book (a) shifts more than the thinner book (b) with the same applied shear force.

(a) (b)

Volume Change—Bulk Modulus

If an object is subjected to inward forces from all sides, its volume will decrease. A common situation is an object submerged in a fluid. In this case, the fluid exerts a pressure on the object in all directions, as we shall see in Chapter 10. *Pressure* is defined as force per unit area, and thus is the equivalent of stress. For this situation the change in volume, ΔV, is proportional to the original volume, V_0, and to the change in the pressure, ΔP. We thus obtain a relation of the same form as Eq. 9–4 but with a proportionality constant called the **bulk modulus** B:

$$\frac{\Delta V}{V_0} = -\frac{1}{B}\Delta P \qquad (9\text{–}7)$$

or

$$B = -\frac{\Delta P}{\Delta V/V_0}.$$

The minus sign means the volume *decreases* with an increase in pressure.

Values for the bulk modulus are given in Table 9–1. Since liquids and gases do not have a fixed shape, only the bulk modulus (not the Young's or shear moduli) applies to them.

9–6 Fracture

If the stress on a solid object is too great, the object fractures, or breaks (Fig. 9–24). Table 9–2 lists the ultimate strengths for tension, compression, and shear for a variety of materials. These values give the maximum force per unit area, or stress, that an object can withstand under each of these three types of stress for various types of material. They are, however, representative values only, and the actual value for a given specimen can differ considerably. It is therefore necessary to maintain a **safety factor** of from 3 to perhaps 10 or more—that is, the actual stresses on a structure should not exceed one-tenth to one-third of the values given in the Table. You may encounter tables of "allowable stresses" in which appropriate safety factors have already been included.

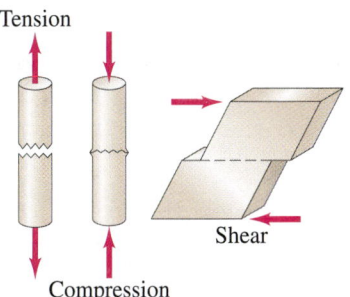

FIGURE 9–24 Fracture as a result of the three types of stress.

TABLE 9–2 Ultimate Strengths of Materials (force/area)

Material	Tensile Strength (N/m²)	Compressive Strength (N/m²)	Shear Strength (N/m²)
Iron, cast	170×10^6	550×10^6	170×10^6
Steel	$500\text{–}2500 \times 10^6$	500×10^6	250×10^6
Brass	250×10^6	250×10^6	200×10^6
Aluminum	200×10^6	200×10^6	200×10^6
Concrete	2×10^6	20×10^6	2×10^6
Brick		35×10^6	
Marble		80×10^6	
Granite		170×10^6	
Wood (pine) (parallel to grain)	40×10^6	35×10^6	5×10^6
(perpendicular to grain)		10×10^6	
Nylon	500×10^6		
Bone (limb)	130×10^6	170×10^6	

EXAMPLE 9–11 ESTIMATE **Breaking the piano wire.** The steel piano wire we discussed in Example 9–10 was 1.60 m long with a diameter of 0.20 cm. Approximately what tension force would break it?

APPROACH We set the tensile stress F/A equal to the ultimate tensile strength of steel given in Table 9–2, and we choose the highest value which represents high-carbon steel.

SOLUTION The wire's area is $A = \pi r^2$, where $r = 0.10 \text{ cm} = 1.0 \times 10^{-3}$ m. Table 9–2 tells us

$$\frac{F}{A} = 2500 \times 10^6 \text{ N/m}^2,$$

so the wire would likely break if the force exceeded

$$F = (2500 \times 10^6 \text{ N/m}^2)(\pi)(1.0 \times 10^{-3} \text{ m})^2 = 8000 \text{ N}.$$

As can be seen in Table 9–2, concrete (like stone and brick) is reasonably strong under compression but extremely weak under tension. Thus concrete can be used as vertical columns placed under compression, but is of little value as a beam because it cannot withstand the tensile forces that result from the inevitable sagging of the lower edge of a beam (see Fig. 9–25).

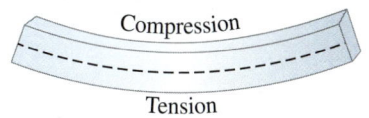

FIGURE 9–25 A beam sags, at least a little (but is exaggerated here), even under its own weight. The beam thus changes shape: the upper edge is compressed, and the lower edge is under tension (elongated). Shearing stress also occurs within the beam.

FIGURE 9–26 Steel rods around which concrete is poured for strength.

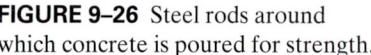

PHYSICS APPLIED
Reinforced concrete and prestressed concrete

Reinforced concrete, in which iron rods are embedded in the concrete (Fig. 9–26), is much stronger. But the concrete on the lower edge of a loaded beam still tends to crack because it is weak under tension. This problem is solved with *prestressed concrete*, which also contains iron rods or a wire mesh, but during the pouring of the concrete, the rods or wire are held under tension. After the concrete dries, the tension on the iron is released, putting the concrete under compression. The amount of compressive stress is carefully predetermined so that when loads are applied to the beam, the compression on the lower edge is never allowed to be reduced so far as to put the concrete into tension.

PHYSICS APPLIED
A tragic collapse

FIGURE 9–27 Example 9–12.

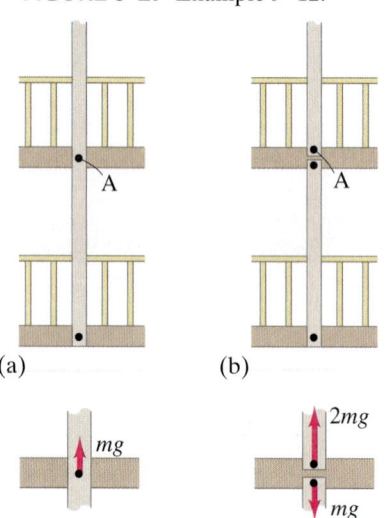

(a) (b)

(c) Force on pin A exerted by vertical rod

(d) Forces on pins at A exerted by vertical rods

CONCEPTUAL EXAMPLE 9–12 **A tragic substitution.** Two walkways, one above the other, are suspended from vertical rods attached to the ceiling of a high hotel lobby, Fig. 9–27a. The original design called for single rods 14 m long, but when such long rods proved to be unwieldy to install, it was decided to replace each long rod with two shorter ones as shown schematically in Fig. 9–27b. Determine the net force exerted by the rods on the supporting pin A (assumed to be the same size) for each design. Assume each vertical rod supports a mass m of each bridge.

RESPONSE The single long vertical rod in Fig. 9–27a exerts an upward force equal to mg on pin A to support the mass m of the upper bridge. Why? Because the pin is in equilibrium, and the other force that balances this is the downward force mg exerted on it by the upper bridge (Fig. 9–27c). There is thus a shear stress on the pin because the rod pulls up on one side of the pin, and the bridge pulls down on the other side. The situation when two shorter rods support the bridges (Fig. 9–27b) is shown in Fig. 9–27d, in which only the connections at the upper bridge are shown. The lower rod exerts a force of mg downward on the lower of the two pins because it supports the lower bridge. The upper rod exerts a force of $2mg$ on the upper pin (labelled A) because the upper rod supports both bridges. Thus we see that when the builders substituted two shorter rods for each single long one, the stress in the supporting pin A was *doubled*. What perhaps seemed like a simple substitution did, in fact, lead to a tragic collapse in 1981 with a loss of life of over 100 people (see Fig. 9–1). Having a feel for physics, and being able to make simple calculations based on physics, can have a great effect, literally, on people's lives.

*9–7 Spanning a Space: Arches and Domes

There are a great many areas where the arts and humanities overlap the sciences, and this is especially clear in architecture, where the forces in the materials that make up a structure need to be understood to avoid excessive deformation and collapse. Many of the features we admire in the architecture of the past were introduced not simply for their decorative effect, but for technical reasons. One example is the development of methods to span a space, from the simple beam supported by columns, to arches and domes.

The first important architectural invention was the **post-and-beam** (or post-and-lintel) construction, in which two upright posts or columns support a horizontal beam. Before steel was introduced in the nineteenth century, the length of a beam was quite limited because the strongest building materials were then stone and brick. Hence the width of a span was limited by the size of available stones. Equally important, stone and brick, though strong under compression—are very weak under tension and shear; all three types of stress occur in a beam (see Fig. 9–25). The minimal space that could be spanned using stone is shown by the closely spaced columns of the great Greek temples (Fig. 9–21).

The semicircular **arch** (Figs. 9–28a and b) was introduced by the ancient Romans 2000 years ago. Aside from its aesthetic appeal, it was a tremendous technological innovation. The advantage of the "true" or semicircular arch is that, if well designed, its wedge-shaped stones experience stress which is mainly compressive even when supporting a large load such as the wall and roof of a cathedral.

PHYSICS APPLIED
Architecture: beams, arches, and domes

(a)

(b)

FIGURE 9–28 (a) Round arches in the Roman Forum, 2000 years old. The one in the background is the Arch of Titus. (b) An arch is used here to good effect for a bridge over a chasm on the California coast.

Because the stones are forced to squeeze against one another, they are mainly under compression (see Fig. 9–29). A round arch consisting of many well-shaped stones could span a very wide space. However, because the arch transfers horizontal as well as vertical forces to the supports, considerable buttressing on the sides is needed, as we discuss shortly.

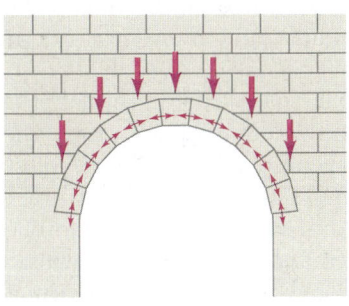

FIGURE 9–29 Stones in a round arch (see Fig. 9–28a) are mainly under compression.

The pointed arch came into use about A.D. 1100 and became the hallmark of the great Gothic cathedrals. It too was an important technical innovation, and was first used to support heavy loads such as the tower and central arch of a cathedral. Apparently the builders realized that, because of the steepness of the pointed arch, the forces due to the weight above could be brought down more nearly vertically, so less horizontal buttressing would be needed. The pointed arch reduced the load on the walls, so there could be more window openings and light. The smaller buttressing needed was provided on the outside by graceful flying buttresses (Fig. 9–30).

The technical innovation of the pointed arch was achieved not through calculation but through experience and intuition; it was not until much later that detailed calculations, such as those presented earlier in this Chapter, came into use.

FIGURE 9–30 Flying buttresses (on the cathedral of Notre Dame, in Paris).

*SECTION 9–7 Spanning a Space: Arches and Domes

FIGURE 9–31 (a) Forces in a round arch, compared (b) with those in a pointed arch.

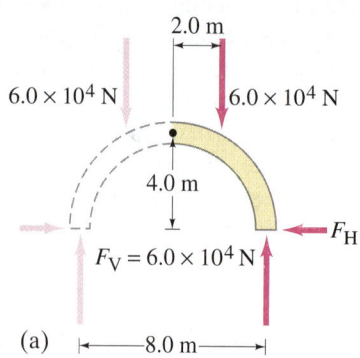
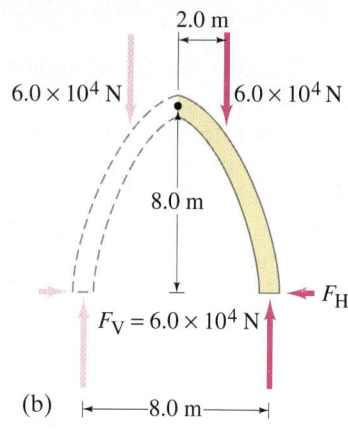

FIGURE 9–32 Interior of the Pantheon in Rome, built almost 2000 years ago. This view, showing the great dome and its central opening for light, was painted about 1740 by Panini. Photographs do not capture its grandeur as well as this painting does.

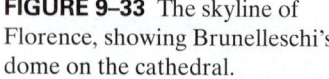

FIGURE 9–33 The skyline of Florence, showing Brunelleschi's dome on the cathedral.

To make an accurate analysis of a stone arch is quite difficult in practice. But if we make some simplifying assumptions, we can show why the horizontal component of the force at the base is less for a pointed arch than for a round one. Figure 9–31 shows a round arch and a pointed arch, each with an 8.0-m span. The height of the round arch is thus 4.0 m, whereas that of the pointed arch is larger and has been chosen to be 8.0 m. Each arch supports a weight of 12.0×10^4 N ($= 12{,}000$ kg $\times g$) which, for simplicity, we have divided into two parts (each 6.0×10^4 N) acting on the two halves of each arch as shown. To be in equilibrium, each of the supports must exert an upward force of 6.0×10^4 N. For rotational equilibrium, each support also exerts a horizontal force, F_H, at the base of the arch, and it is this we want to calculate. We focus only on the right half of each arch. We set equal to zero the total torque calculated about the apex of the arch due to the three forces exerted on that half arch. The torque equation ($\Sigma \tau = 0$) contains three terms: the weight above, the support F_V below, and the horizontal force F_H, which for the round arch (see Fig. 9–31a) is

$$-(2.0 \text{ m})(6.0 \times 10^4 \text{ N}) + (4.0 \text{ m})(6.0 \times 10^4 \text{ N}) - (4.0 \text{ m})(F_H) = 0.$$

Thus $F_H = 3.0 \times 10^4$ N for the round arch. For the pointed arch, the torque equation is (see Fig. 9–31b)

$$-(2.0 \text{ m})(6.0 \times 10^4 \text{ N}) + (4.0 \text{ m})(6.0 \times 10^4 \text{ N}) - (8.0 \text{ m})(F_H) = 0.$$

Solving, we find that $F_H = 1.5 \times 10^4$ N—only half as much as for the round arch! From this calculation we can see that the horizontal buttressing force required for a pointed arch is less because the arch is higher, and there is therefore a longer lever arm for this force. Indeed, the steeper the arch, the less the horizontal component of the force needs to be, and hence the more nearly vertical is the force exerted at the base of the arch.

Whereas an arch spans a two-dimensional space, a **dome**—which is basically an arch rotated about a vertical axis—spans a three-dimensional space. The Romans built the first large domes. Their shape was hemispherical and some still stand, such as that of the Pantheon in Rome (Fig. 9–32), built nearly 2000 years ago.

Fourteen centuries later, a new cathedral was being built in Florence.[†] It was to have a dome 43 m in diameter to rival that of the Pantheon, whose construction has remained a mystery. The new dome was to rest on a "drum" with no external abutments. Filippo Brunelleschi (1377–1446) designed a pointed dome (Fig. 9–33), since a pointed dome, like a pointed arch, exerts a smaller side thrust against its base. A dome, like an arch, is not stable until all the stones are in place. To support smaller domes during construction, wooden frameworks were used. But no trees big enough or strong enough could be found to span the 43-m space required. Brunelleschi built the dome in horizontal layers, each bonded to the previous one, holding it in place until the last stone of the circle was placed. Each closed ring was then strong enough to support the next layer. An amazing feat. Only in the twentieth century were larger domes built.

[†]Supervised by Arnolfo di Cambio, who wanted a dome but was unable to design it himself. But he was confident that Florentine genius would find a way, given that the ancient Romans had done it with the Pantheon, which still stands.

EXAMPLE 9-13 A modern dome. The 1.2×10^6 kg dome of the Small Sports Palace in Rome (Fig. 9–34a) is supported by 36 buttresses positioned at a 38° angle so that they connect smoothly with the dome. Calculate the components of the force, F_H and F_V, that each buttress exerts on the dome so that the force acts purely in compression—that is, at a 38° angle (Fig. 9–34b).

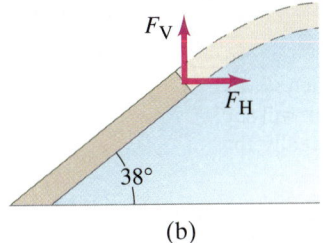

FIGURE 9–34 Example 9–13. (a) The dome of the Small Sports Palace in Rome, built by Pier Luigi Nervi for the 1960 Olympics. (b) The force components each buttress exerts on the dome.

APPROACH We can find the vertical component F_V exerted upward by each buttress because each supports $\frac{1}{36}$ of the dome's weight. We find F_H knowing that the buttress needs to be under compression so $\vec{F} = \vec{F}_V + \vec{F}_H$ acts at a 38° angle.

SOLUTION The vertical load on *each* buttress is $\frac{1}{36}$ of the total weight. Thus

$$F_V = \frac{mg}{36} = \frac{(1.2 \times 10^6 \text{ kg})(9.8 \text{ m/s}^2)}{36} = 330,000 \text{ N}.$$

The force must act at a 38° angle at the base of the dome in order to be purely compressive. Thus

$$\tan 38° = \frac{F_V}{F_H};$$

$$F_H = \frac{F_V}{\tan 38°} = \frac{330,000 \text{ N}}{\tan 38°} = 420,000 \text{ N}.$$

NOTE For each buttress to exert this 420,000-N horizontal force, a prestressed-concrete tension ring surrounds the base of the buttresses beneath the ground (see Problem 58 and Fig. 9–77).

Summary

An object at rest is said to be in **equilibrium**. The subject concerned with the determination of the forces within a structure at rest is called **statics**.

The two necessary conditions for an object to be in equilibrium are (1) the vector sum of all the forces on it must be zero, and (2) the sum of all the torques (calculated about any arbitrary axis) must also be zero. For a two-dimensional problem we can write

$$\Sigma F_x = 0, \quad \Sigma F_y = 0, \quad \Sigma \tau = 0. \quad \textbf{(9-1, 9-2)}$$

It is important when doing statics problems to apply the equilibrium conditions to only one object at a time.

An object in static equilibrium is said to be in (*a*) **stable**, (*b*) **unstable**, or (*c*) **neutral equilibrium**, depending on whether a slight displacement leads to (*a*) a return to the original position, (*b*) further movement away from the original position, or (*c*) rest in the new position. An object in stable equilibrium is also said to be in *balance*.

Hooke's law applies to many elastic solids, and states that the change in length of an object is proportional to the applied force:

$$F = k \Delta \ell. \quad \textbf{(9-3)}$$

If the force is too great, the object will exceed its **elastic limit**, which means it will no longer return to its original shape when the distorting force is removed. If the force is even greater, the **ultimate strength** of the material can be exceeded, and the object will **fracture**. The force per unit area acting on an object is the **stress**, and the resulting fractional change in length is the **strain**.

The stress on an object is present within the object and can be of three types: **compression**, **tension**, or **shear**. The ratio of stress to strain is called the **elastic modulus** of the material. **Young's modulus** applies for compression and tension, and the **shear modulus** for shear. **Bulk modulus** applies to an object whose volume changes as a result of pressure on all sides. All three moduli are constants for a given material when distorted within the elastic region.

[*Arches and domes are special ways to span a space that allow the stresses to be managed well.*]

Questions

1. Describe several situations in which an object is not in equilibrium, even though the net force on it is zero.

2. A bungee jumper momentarily comes to rest at the bottom of the dive before he springs back upward. At that moment, is the bungee jumper in equilibrium? Explain.

3. You can find the center of gravity of a meter stick by resting it horizontally on your two index fingers, and then slowly drawing your fingers together. First the meter stick will slip on one finger, and then on the other, but eventually the fingers meet at the CG. Why does this work?

4. Your doctor's scale has arms on which weights slide to counter your weight, Fig. 9–35. These weights are much lighter than you are. How does this work?

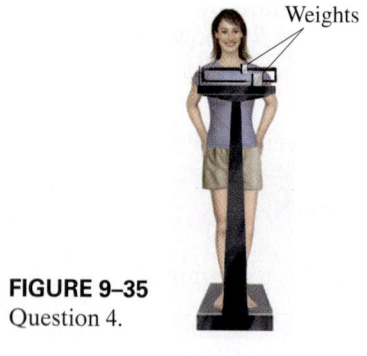

FIGURE 9–35 Question 4.

5. A ground retaining wall is shown in Fig. 9–36a. The ground, particularly when wet, can exert a significant force F on the wall. (*a*) What force produces the torque to keep the wall upright? (*b*) Explain why the retaining wall in Fig. 9–36b would be much less likely to overturn than that in Fig. 9–36a.

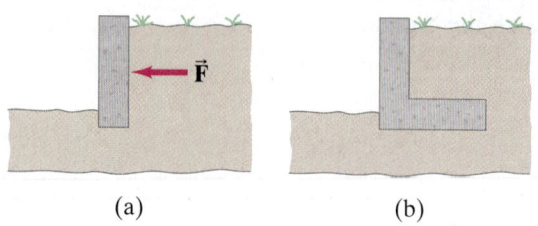

FIGURE 9–36 Question 5.

6. Can the sum of the torques on an object be zero while the net force on the object is nonzero? Explain.

7. A ladder, leaning against a wall, makes a 60° angle with the ground. When is it more likely to slip: when a person stands on the ladder near the top or near the bottom? Explain.

8. A uniform meter stick supported at the 25-cm mark is in equilibrium when a 1-kg rock is suspended at the 0-cm end (as shown in Fig. 9–37). Is the mass of the meter stick greater than, equal to, or less than the mass of the rock? Explain your reasoning.

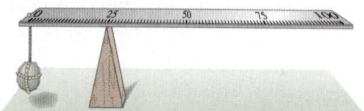

FIGURE 9–37 Question 8.

9. Why do you tend to lean backward when carrying a heavy load in your arms?

10. Figure 9–38 shows a cone. Explain how to lay it on a flat table so that it is in (*a*) stable equilibrium, (*b*) unstable equilibrium, (*c*) neutral equilibrium.

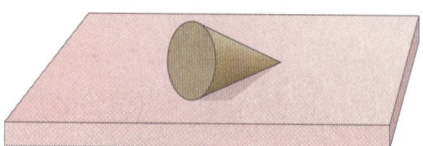

FIGURE 9–38 Question 10.

11. Place yourself facing the edge of an open door. Position your feet astride the door with your nose and abdomen touching the door's edge. Try to rise on your tiptoes. Why can't this be done?

12. Why is it not possible to sit upright in a chair and rise to your feet without first leaning forward?

13. Why is it more difficult to do sit-ups when your knees are bent than when your legs are stretched out?

14. Explain why touching your toes while you are seated on the floor with outstretched legs produces less stress on the lower spinal column than when touching your toes from a standing position. Use a diagram.

15. Which configuration of bricks, Fig. 9–39a or Fig. 9–39b, is the more likely to be stable? Why?

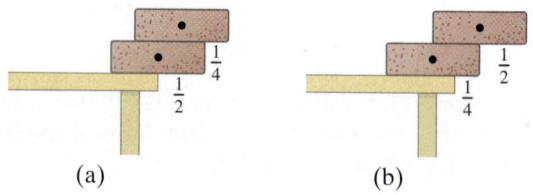

FIGURE 9–39 Question 15. The dots indicate the CG of each brick (assumed uniform). The fractions $\frac{1}{4}$ and $\frac{1}{2}$ indicate what portion of each brick is hanging beyond its support.

16. Name the type of equilibrium for each position of the ball in Fig. 9–40.

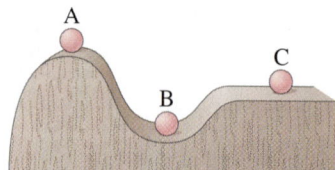

FIGURE 9–40 Question 16.

17. Is the Young's modulus for a bungee cord smaller or larger than that for an ordinary rope?

18. Examine how a pair of scissors or shears cuts through a piece of cardboard. Is the name "shears" justified? Explain.

19. Materials such as ordinary concrete and stone are very weak under tension or shear. Would it be wise to use such a material for either of the supports of the cantilever shown in Fig. 9–9? If so, which one(s)? Explain.

MisConceptual Questions

1. A 60-kg woman stands on the very end of a uniform board, of length ℓ, which is supported one-quarter of the way from one end and is balanced (Fig. 9–41). What is the mass of the board?
   (a) 15 kg. (b) 20 kg. (c) 30 kg. (d) 60 kg. (e) 120 kg.

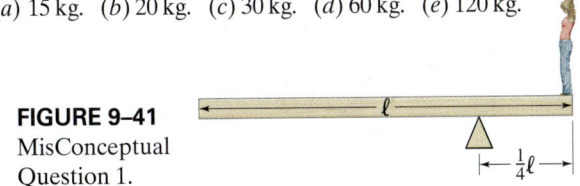

FIGURE 9–41 MisConceptual Question 1.

2. When you apply the torque equation $\Sigma\tau = 0$ to an object in equilibrium, the axis about which torques are calculated
   (a) must be located at a pivot.
   (b) must be located at the object's center of gravity.
   (c) should be located at the edge of the object.
   (d) can be located anywhere.

3. A uniform beam is hinged at one end and held in a horizontal position by a cable, as shown in Fig. 9–42. The tension in the cable
   (a) must be at least half the weight of the beam, no matter what the angle of the cable.
   (b) could be less than half the beam's weight for some angles.
   (c) will be half the beam's weight for all angles.
   (d) will equal the beam's weight for all angles.

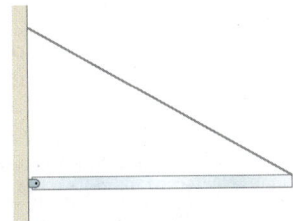

FIGURE 9–42 MisConceptual Question 3: beam and cable.

4. A heavy ball suspended by a cable is pulled to the side by a horizontal force $\vec{F}$ as shown in Fig. 9–43. If angle θ is small, the magnitude of the force F can be less than the weight of the ball because:
   (a) the force holds up only part of the ball's weight.
   (b) even though the ball is stationary, it is not really in equilibrium.
   (c) $\vec{F}$ is equal to only the x component of the tension in the cable.
   (d) the original statement is not true. To move the ball, $\vec{F}$ must be at least equal to the ball's weight.

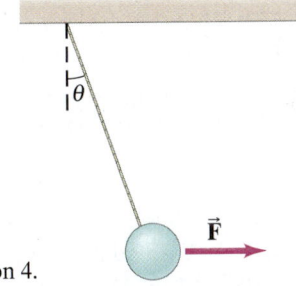

FIGURE 9–43 MisConceptual Question 4.

5. Two children are balanced on opposite sides of a seesaw. If one child leans inward toward the pivot point, her side will
   (a) rise. (b) fall. (c) neither rise nor fall.

6. A 10.0-N weight is suspended by two cords as shown in Fig. 9–44. What can you say about the tension in the two cords?
   (a) The tension in both cords is 5.0 N.
   (b) The tension in both cords is equal but not 5.0 N.
   (c) The tension in cord A is greater than that in cord B.
   (d) The tension in cord B is greater than that in cord A.

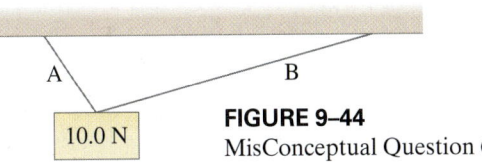

FIGURE 9–44 MisConceptual Question 6.

7. As you increase the force that you apply while pulling on a rope, which of the following is NOT affected?
   (a) The stress on the rope.
   (b) The strain on the rope.
   (c) The Young's modulus of the rope.
   (d) All of the above.
   (e) None of the above.

8. A woman is balancing on a high wire which is tightly strung, as shown in Fig. 9–45. The tension in the wire is
   (a) about half the woman's weight.
   (b) about twice the woman's weight.
   (c) about equal to the woman's weight.
   (d) much less than the woman's weight.
   (e) much more than the woman's weight.

FIGURE 9–45 MisConceptual Question 8.

9. A parking garage is designed for two levels of cars. To make more money, the owner decides to double the size of the garage in each dimension (length, width, and number of levels). For the support columns to hold up four floors instead of two, how should he change the columns' diameter?
   (a) Double the area of the columns by increasing their diameter by a factor of 2.
   (b) Double the area of the columns by increasing their diameter by a factor of $\sqrt{2}$.
   (c) Quadruple the area of the columns by increasing their diameter by a factor of 2.
   (d) Increase the area of the columns by a factor of 8 by increasing their diameter by a factor of $2\sqrt{2}$.
   (e) He doesn't need to increase the diameter of the columns.

10. A rubber band is stretched by 1.0 cm when a force of 0.35 N is applied to each end. If instead a force of 0.70 N is applied to each end, estimate how far the rubber band will stretch from its unstretched length:
    (a) 0.25 cm. (b) 0.5 cm. (c) 1.0 cm. (d) 2.0 cm. (e) 4.0 cm.

For assigned homework and other learning materials, go to the MasteringPhysics website.

Problems

9–1 and 9–2 Equilibrium

1. (I) Three forces are applied to a tree sapling, as shown in Fig. 9–46, to stabilize it. If $\vec{F}_A = 385$ N and $\vec{F}_B = 475$ N, find $\vec{F}_C$ in magnitude and direction.

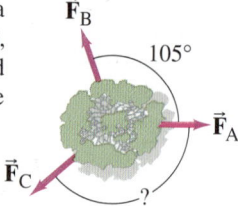

FIGURE 9–46 Problem 1.

2. (I) Calculate the mass m needed in order to suspend the leg shown in Fig. 9–47. Assume the leg (with cast) has a mass of 15.0 kg, and its CG is 35.0 cm from the hip joint; the cord holding the sling is 78.0 cm from the hip joint.

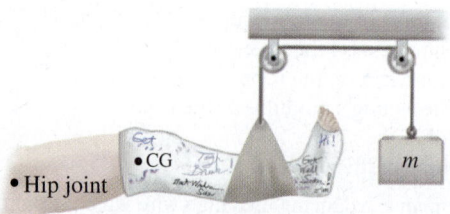

FIGURE 9–47 Problem 2.

3. (I) A tower crane (Fig. 9–48a) must always be carefully balanced so that there is no net torque tending to tip it. A particular crane at a building site is about to lift a 2800-kg air-conditioning unit. The crane's dimensions are shown in Fig. 9–48b. (a) Where must the crane's 9500-kg counterweight be placed when the load is lifted from the ground? (The counterweight is usually moved automatically via sensors and motors to precisely compensate for the load.) (b) Determine the maximum load that can be lifted with this counterweight when it is placed at its full extent. Ignore the mass of the beam.

(a)

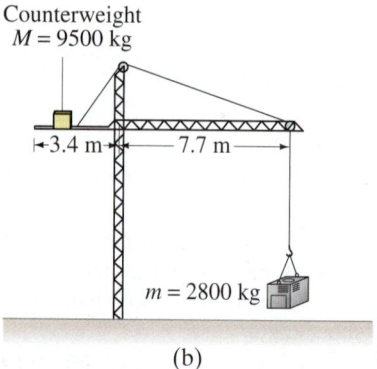

(b)

FIGURE 9–48 Problem 3.

4. (I) What is the mass of the diver in Fig. 9–49 if she exerts a torque of 1800 m·N on the board, relative to the left (A) support post?

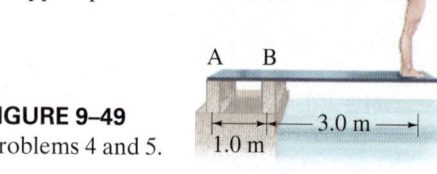

FIGURE 9–49 Problems 4 and 5.

5. (II) Calculate the forces F_A and F_B that the supports exert on the diving board of Fig. 9–49 when a 52-kg person stands at its tip. (a) Ignore the weight of the board. (b) Take into account the board's mass of 28 kg. Assume the board's CG is at its center.

6. (II) Figure 9–50 shows a pair of forceps used to hold a thin plastic rod firmly. If the thumb and finger each squeeze with a force $F_T = F_F = 11.0$ N, what force do the forceps jaws exert on the plastic rod?

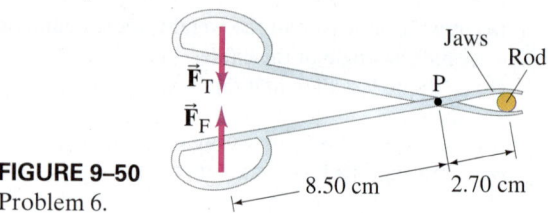

FIGURE 9–50 Problem 6.

7. (II) Two cords support a chandelier in the manner shown in Fig. 9–4 except that the upper cord makes an angle of 45° with the ceiling. If the cords can sustain a force of 1660 N without breaking, what is the maximum chandelier weight that can be supported?

8. (II) The two trees in Fig. 9–51 are 6.6 m apart. A backpacker is trying to lift his pack out of the reach of bears. Calculate the magnitude of the force $\vec{F}$ that he must exert downward to hold a 19-kg backpack so that the rope sags at its midpoint by (a) 1.5 m, (b) 0.15 m.

FIGURE 9–51 Problems 8 and 70.

9. (II) A 110-kg horizontal beam is supported at each end. A 320-kg piano rests a quarter of the way from one end. What is the vertical force on each of the supports?

10. (II) Calculate F_A and F_B for the uniform cantilever shown in Fig. 9–9 whose mass is 1200 kg.

11. (II) A 75-kg adult sits at one end of a 9.0-m-long board. His 25-kg child sits on the other end. (a) Where should the pivot be placed so that the board is balanced, ignoring the board's mass? (b) Find the pivot point if the board is uniform and has a mass of 15 kg.

252 CHAPTER 9 Static Equilibrium; Elasticity and Fracture

12. (II) Find the tension in the two cords shown in Fig. 9–52. Neglect the mass of the cords, and assume that the angle θ is 33° and the mass m is 190 kg.

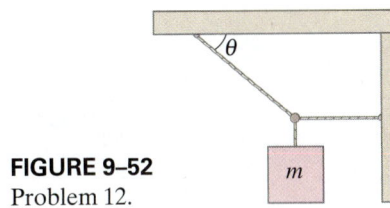

FIGURE 9–52 Problem 12.

13. (II) Find the tension in the two wires supporting the traffic light shown in Fig. 9–53.

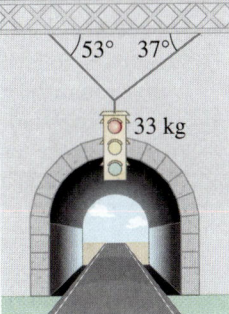

FIGURE 9–53 Problem 13.

14. (II) How close to the edge of the 24.0-kg table shown in Fig. 9–54 can a 66.0-kg person sit without tipping it over?

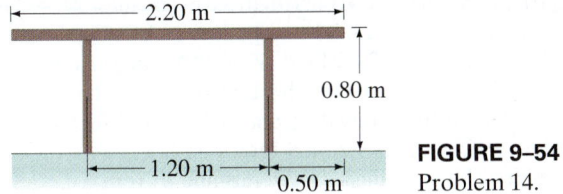

FIGURE 9–54 Problem 14.

15. (II) The force required to pull the cork out of the top of a wine bottle is in the range of 200 to 400 N. What range of forces F is required to open a wine bottle with the bottle opener shown in Fig. 9–55?

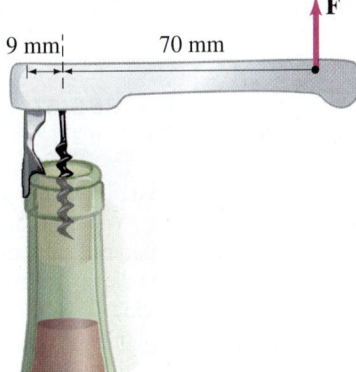

FIGURE 9–55 Problem 15.

16. (II) Calculate F_A and F_B for the beam shown in Fig. 9–56. The downward forces represent the weights of machinery on the beam. Assume the beam is uniform and has a mass of 280 kg.

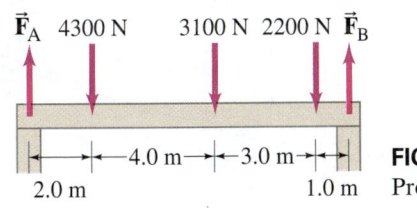

FIGURE 9–56 Problem 16.

17. (II) Three children are trying to balance on a seesaw, which includes a fulcrum rock acting as a pivot at the center, and a very light board 3.2 m long (Fig. 9–57). Two playmates are already on either end. Boy A has a mass of 45 kg, and boy B a mass of 35 kg. Where should girl C, whose mass is 25 kg, place herself so as to balance the seesaw?

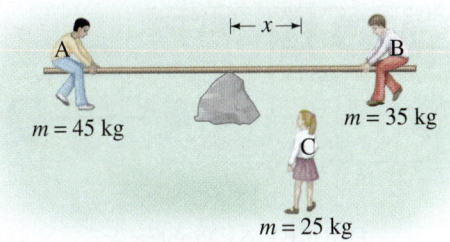

FIGURE 9–57 Problem 17.

18. (II) A shop sign weighing 215 N hangs from the end of a uniform 155-N beam as shown in Fig. 9–58. Find the tension in the supporting wire (at 35.0°), and the horizontal and vertical forces exerted by the hinge on the beam at the wall. [*Hint*: First draw a free-body diagram.]

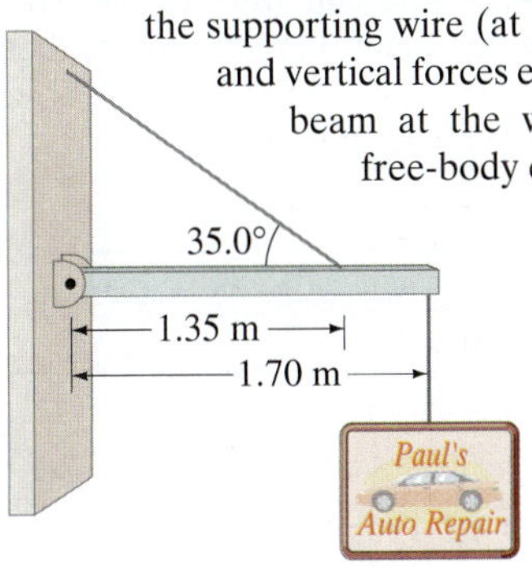

FIGURE 9–58 Problem 18.

19. (II) A traffic light hangs from a pole as shown in Fig. 9–59. The uniform aluminum pole AB is 7.20 m long and has a mass of 12.0 kg. The mass of the traffic light is 21.5 kg. Determine (*a*) the tension in the horizontal massless cable CD, and (*b*) the vertical and horizontal components of the force exerted by the pivot A on the aluminum pole.

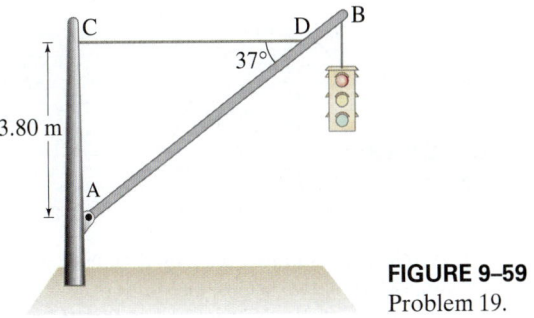

FIGURE 9–59 Problem 19.

20. (II) A uniform steel beam has a mass of 940 kg. On it is resting half of an identical beam, as shown in Fig. 9–60. What is the vertical support force at each end?

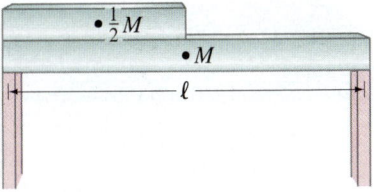

FIGURE 9–60 Problem 20.

21. (II) A 2500-kg trailer is attached to a stationary truck at point B, Fig. 9–61. Determine the normal force exerted by the road on the rear tires at A, and the vertical force exerted on the trailer by the support B.

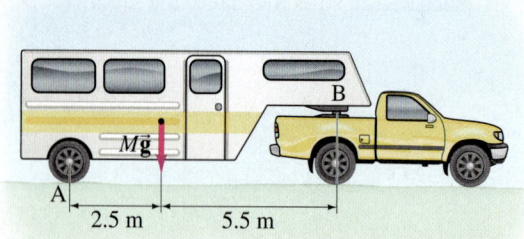

FIGURE 9–61 Problem 21.

22. (II) A 20.0-m-long uniform beam weighing 650 N rests on walls A and B, as shown in Fig. 9–62. (*a*) Find the maximum weight of a person who can walk to the extreme end D without tipping the beam. Find the forces that the walls A and B exert on the beam when the person is standing: (*b*) at D; (*c*) 2.0 m to the right of A.

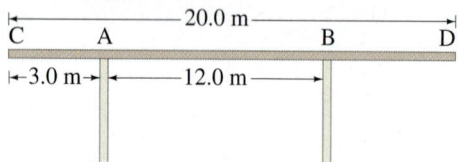

FIGURE 9–62 Problem 22.

23. (II) A 0.75-kg sheet is centered on a clothesline as shown in Fig. 9–63. The clothesline on either side of the hanging sheet makes an angle of 3.5° with the horizontal. Calculate the tension in the clothesline (ignore its mass) on either side of the sheet. Why is the tension so much greater than the weight of the sheet?

FIGURE 9–63
Problem 23.

24. (II) A 172-cm-tall person lies on a light (massless) board which is supported by two scales, one under the top of her head and one beneath the bottom of her feet (Fig. 9–64). The two scales read, respectively, 35.1 and 31.6 kg. What distance is the center of gravity of this person from the bottom of her feet?

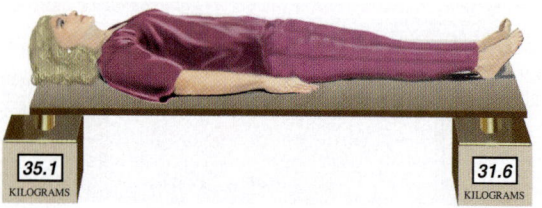

FIGURE 9–64 Problem 24.

25. (II) A man doing push-ups pauses in the position shown in Fig. 9–65. His mass $m = 68$ kg. Determine the normal force exerted by the floor (*a*) on each hand; (*b*) on each foot.

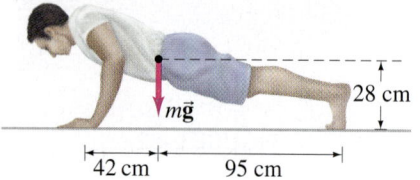

FIGURE 9–65 Problem 25.

26. (III) Two wires run from the top of a pole 2.6 m tall that supports a volleyball net. The two wires are anchored to the ground 2.0 m apart, and each is 2.0 m from the pole (Fig. 9–66). The tension in each wire is 115 N. What is the tension in the net, assumed horizontal and attached at the top of the pole?

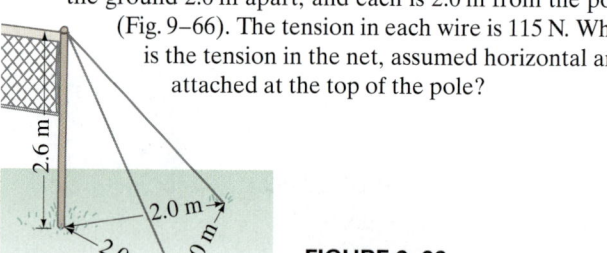

FIGURE 9–66
Problem 26.

27. (III) A uniform rod AB of length 5.0 m and mass $M = 3.8$ kg is hinged at A and held in equilibrium by a light cord, as shown in Fig. 9–67. A load $W = 22$ N hangs from the rod at a distance d so that the tension in the cord is 85 N. (*a*) Draw a free-body diagram for the rod. (*b*) Determine the vertical and horizontal forces on the rod exerted by the hinge. (*c*) Determine d from the appropriate torque equation.

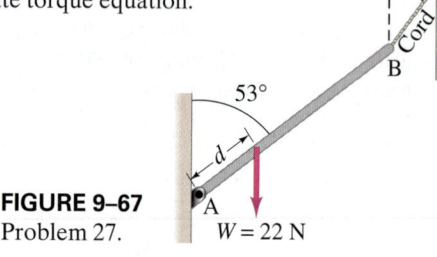

FIGURE 9–67
Problem 27.

28. (III) You are on a pirate ship and being forced to walk the plank (Fig. 9–68). You are standing at the point marked C. The plank is nailed onto the deck at point A, and rests on the support 0.75 m away from A. The center of mass of the uniform plank is located at point B. Your mass is 65 kg and the mass of the plank is 45 kg. What is the minimum downward force the nails must exert on the plank to hold it in place?

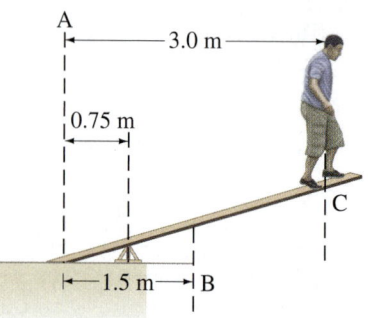

FIGURE 9–68
Problem 28.

254 CHAPTER 9 Static Equilibrium; Elasticity and Fracture

29. (III) A door 2.30 m high and 1.30 m wide has a mass of 13.0 kg. A hinge 0.40 m from the top and another hinge 0.40 m from the bottom each support half the door's weight (Fig. 9–69). Assume that the center of gravity is at the geometrical center of the door, and determine the horizontal and vertical force components exerted by each hinge on the door.

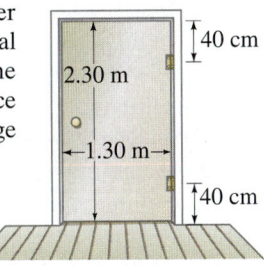

FIGURE 9–69 Problem 29.

*30. (III) A uniform ladder of mass m and length ℓ leans at an angle θ against a frictionless wall, Fig. 9–70. If the coefficient of static friction between the ladder and the ground is μ_s, determine a formula for the minimum angle at which the ladder will not slip.

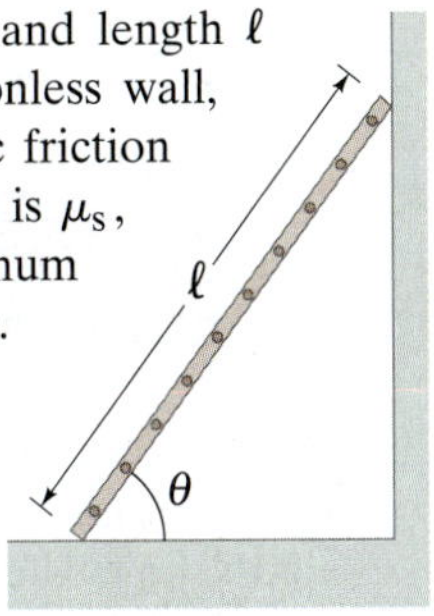

FIGURE 9–70 Problem 30.

9–3 Muscles and Joints

31. (I) Suppose the point of insertion of the biceps muscle into the lower arm shown in Fig. 9–13a (Example 9–8) is 6.0 cm instead of 5.0 cm; how much mass could the person hold with a muscle exertion of 450 N?

32. (I) Approximately what magnitude force, F_M, must the extensor muscle in the upper arm exert on the lower arm to hold a 7.3-kg shot put (Fig. 9–71)? Assume the lower arm has a mass of 2.3 kg and its CG is 12.0 cm from the elbow-joint pivot.

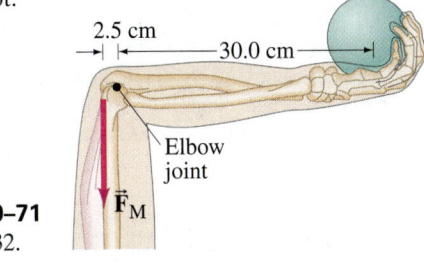

FIGURE 9–71 Problem 32.

33. (II) Redo Example 9–9, assuming now that the person is less bent over so that the 30° in Fig. 9–14b is instead 45°. What will be the magnitude of F_V on the vertebra?

34. (II) (a) Calculate the magnitude of the force, F_M, required of the "deltoid" muscle to hold up the outstretched arm shown in Fig. 9–72. The total mass of the arm is 3.3 kg. (b) Calculate the magnitude of the force F_J exerted by the shoulder joint on the upper arm and the angle (to the horizontal) at which it acts.

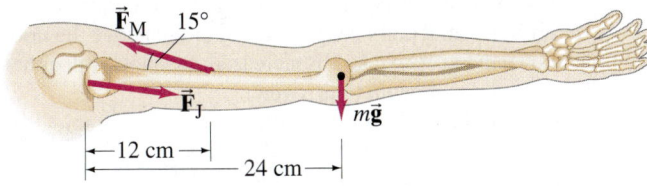

FIGURE 9–72 Problems 34 and 35.

35. (II) Suppose the hand in Problem 34 holds an 8.5-kg mass. What force, F_M, is required of the deltoid muscle, assuming the mass is 52 cm from the shoulder joint?

36. (II) The Achilles tendon is attached to the rear of the foot as shown in Fig. 9–73. When a person elevates himself just barely off the floor on the "ball of one foot," estimate the tension F_T in the Achilles tendon (pulling upward), and the (downward) force F_B exerted by the lower leg bone on the foot. Assume the person has a mass of 72 kg and D is twice as long as d.

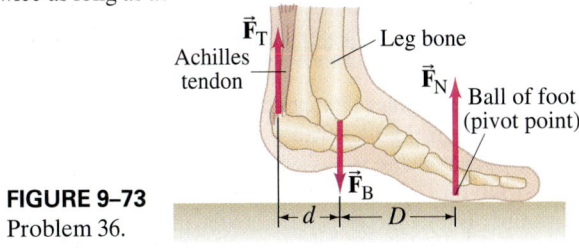

FIGURE 9–73 Problem 36.

37. (II) If 25 kg is the maximum mass m that a person can hold in a hand when the arm is positioned with a 105° angle at the elbow as shown in Fig. 9–74, what is the maximum force F_{max} that the biceps muscle exerts on the forearm? Assume the forearm and hand have a total mass of 2.0 kg with a CG that is 15 cm from the elbow, and that the biceps muscle attaches 5.0 cm from the elbow.

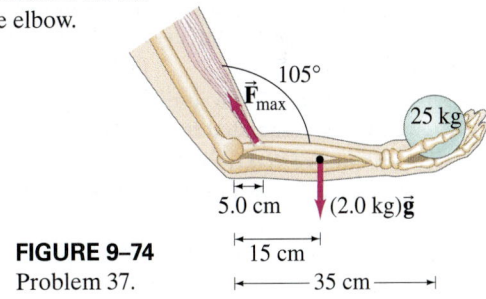

FIGURE 9–74 Problem 37.

9–4 Stability and Balance

38. (II) The Leaning Tower of Pisa is 55 m tall and about 7.7 m in radius. The top is 4.5 m off center. Is the tower in stable equilibrium? If so, how much farther can it lean before it becomes unstable? Assume the tower is of uniform composition.

39. (III) Four bricks are to be stacked at the edge of a table, each brick overhanging the one below it, so that the top brick extends as far as possible beyond the edge of the table. (a) To achieve this, show that successive bricks must extend no more than (starting at the top) $\frac{1}{2}$, $\frac{1}{4}$, $\frac{1}{6}$, and $\frac{1}{8}$ of their length beyond the one below (Fig. 9–75a). (b) Is the top brick completely beyond the base? (c) Determine a general formula for the maximum total distance spanned by n bricks if they are to remain stable. (d) A builder wants to construct a corbeled arch (Fig. 9–75b) based on the principle of stability discussed in (a) and (c) above. What minimum number of bricks, each 0.30 m long and uniform, is needed if the arch is to span 1.0 m?

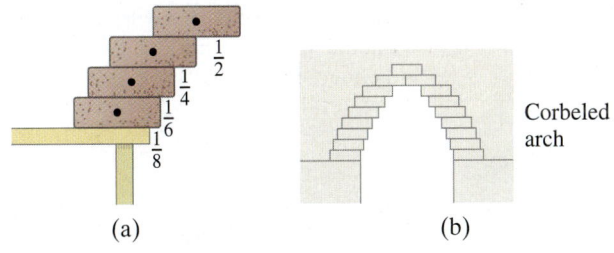

FIGURE 9–75 Problem 39.

9–5 Elasticity; Stress and Strain

40. (I) A nylon string on a tennis racket is under a tension of 275 N. If its diameter is 1.00 mm, by how much is it lengthened from its untensioned length of 30.0 cm?

41. (I) A marble column of cross-sectional area 1.4 m² supports a mass of 25,000 kg. (a) What is the stress within the column? (b) What is the strain?

42. (I) By how much is the column in Problem 41 shortened if it is 8.6 m high?

43. (I) A sign (mass 1700 kg) hangs from the bottom end of a vertical steel girder with a cross-sectional area of 0.012 m². (a) What is the stress within the girder? (b) What is the strain on the girder? (c) If the girder is 9.50 m long, how much is it lengthened? (Ignore the mass of the girder itself.)

44. (II) One liter of alcohol (1000 cm³) in a flexible container is carried to the bottom of the sea, where the pressure is $2.6 \times 10^6 \text{ N/m}^2$. What will be its volume there?

45. (II) How much pressure is needed to compress the volume of an iron block by 0.10%? Express your answer in N/m², and compare it to atmospheric pressure ($1.0 \times 10^5 \text{ N/m}^2$).

46. (II) A 15-cm-long tendon was found to stretch 3.7 mm by a force of 13.4 N. The tendon was approximately round with an average diameter of 8.5 mm. Calculate Young's modulus of this tendon.

47. (II) A steel wire 2.3 mm in diameter stretches by 0.030% when a mass is suspended from it. How large is the mass?

48. (II) At depths of 2000 m in the sea, the pressure is about 200 times atmospheric pressure (1 atm = $1.0 \times 10^5 \text{ N/m}^2$). By what percentage does the interior space of an iron bathysphere's volume change at this depth?

49. (III) A scallop forces open its shell with an elastic material called abductin, whose Young's modulus is about $2.0 \times 10^6 \text{ N/m}^2$. If this piece of abductin is 3.0 mm thick and has a cross-sectional area of 0.50 cm², how much potential energy does it store when compressed 1.0 mm?

9–6 Fracture

50. (I) The femur bone in the human leg has a minimum effective cross section of about 3.0 cm² ($= 3.0 \times 10^{-4} \text{ m}^2$). How much compressive force can it withstand before breaking?

51. (II) (a) What is the maximum tension possible in a 1.00-mm-diameter nylon tennis racket string? (b) If you want tighter strings, what do you do to prevent breakage: use thinner or thicker strings? Why? What causes strings to break when they are hit by the ball?

52. (II) If a compressive force of 3.3×10^4 N is exerted on the end of a 22-cm-long bone of cross-sectional area 3.6 cm², (a) will the bone break, and (b) if not, by how much does it shorten?

53. (II) (a) What is the minimum cross-sectional area required of a vertical steel cable from which is suspended a 270-kg chandelier? Assume a safety factor of 7.0. (b) If the cable is 7.5 m long, how much does it elongate?

54. (II) Assume the supports of the uniform cantilever shown in Fig. 9–76 ($m = 2900$ kg) are made of wood. Calculate the minimum cross-sectional area required of each, assuming a safety factor of 9.0.

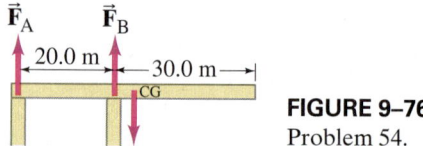

FIGURE 9–76
Problem 54.

55. (II) An iron bolt is used to connect two iron plates together. The bolt must withstand shear forces up to about 3300 N. Calculate the minimum diameter for the bolt, based on a safety factor of 7.0.

56. (III) A steel cable is to support an elevator whose total (loaded) mass is not to exceed 3100 kg. If the maximum acceleration of the elevator is 1.8 m/s², calculate the diameter of cable required. Assume a safety factor of 8.0.

*9–7 Arches and Domes

*57. (II) How high must a pointed arch be if it is to span a space 8.0 m wide and exert one-third the horizontal force at its base that a round arch would?

*58. (II) The subterranean tension ring that exerts the balancing horizontal force on the abutments for the dome in Fig. 9–34 is 36-sided, so each segment makes a 10° angle with the adjacent one (Fig. 9–77). Calculate the tension F that must exist in each segment so that the required force of 4.2×10^5 N can be exerted at each corner (Example 9–13).

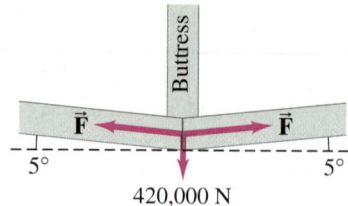

FIGURE 9–77
Problem 58.

General Problems

59. A woman holds a 2.0-m-long uniform 10.0-kg pole as shown in Fig. 9–78. (a) Determine the forces she must exert with each hand (magnitude and direction). To what position should she move her left hand so that neither hand has to exert a force greater than (b) 150 N? (c) 85 N?

FIGURE 9–78
Problem 59.

60. A cube of side ℓ rests on a rough floor. It is subjected to a steady horizontal pull F, exerted a distance h above the floor as shown in Fig. 9–79. As F is increased, the block will either begin to slide, or begin to tip over. Determine the coefficient of static friction μ_s so that (a) the block begins to slide rather than tip; (b) the block begins to tip. [*Hint*: Where will the normal force on the block act if it tips?]

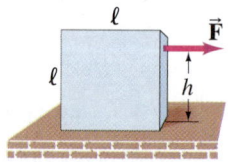

FIGURE 9–79
Problem 60.

61. A 50-story building is being planned. It is to be 180.0 m high with a base 46.0 m by 76.0 m. Its total mass will be about 1.8×10^7 kg, and its weight therefore about 1.8×10^8 N. Suppose a 200-km/h wind exerts a force of 950 N/m² over the 76.0-m-wide face (Fig. 9–80). Calculate the torque about the potential pivot point, the rear edge of the building (where $\vec{F}_E$ acts in Fig. 9–80), and determine whether the building will topple. Assume the total force of the wind acts at the midpoint of the building's face, and that the building is not anchored in bedrock. [Hint: $\vec{F}_E$ in Fig. 9–80 represents the force that the Earth would exert on the building in the case where the building would just begin to tip.]

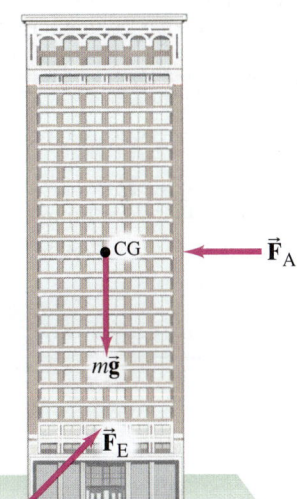

FIGURE 9–80 Forces on a building subjected to wind ($\vec{F}_A$), gravity ($m\vec{g}$), and the force $\vec{F}_E$ on the building due to the Earth if the building were just about to tip. Problem 61.

62. The center of gravity of a loaded truck depends on how the truck is packed. If it is 4.0 m high and 2.4 m wide, and its CG is 2.2 m above the ground, how steep a slope can the truck be parked on without tipping over (Fig. 9–81)?

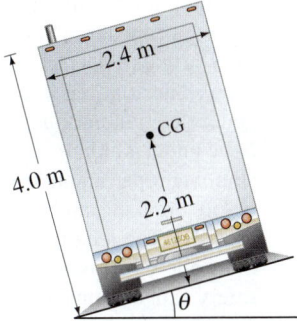

FIGURE 9–81 Problem 62.

63. A uniform meter stick with a mass of 180 g is supported horizontally by two vertical strings, one at the 0-cm mark and the other at the 90-cm mark (Fig. 9–82). What is the tension in the string (a) at 0 cm? (b) at 90 cm?

FIGURE 9–82 Problem 63.

64. There is a maximum height of a uniform vertical column made of any material that can support itself without buckling, and it is independent of the cross-sectional area (why?). Calculate this height for (a) steel (density 7.8×10^3 kg/m³), and (b) granite (density 2.7×10^3 kg/m³).

65. When a mass of 25 kg is hung from the middle of a fixed straight aluminum wire, the wire sags to make an angle of 12° with the horizontal as shown in Fig. 9–83. Determine the radius of the wire.

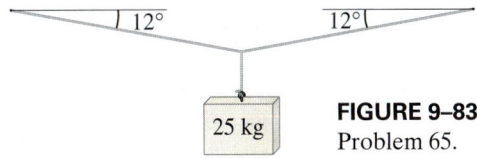

FIGURE 9–83 Problem 65.

66. A 65.0-kg painter is on a uniform 25-kg scaffold supported from above by ropes (Fig. 9–84). There is a 4.0-kg pail of paint to one side, as shown. Can the painter walk safely to both ends of the scaffold? If not, which end(s) is dangerous, and how close to the end can he approach safely?

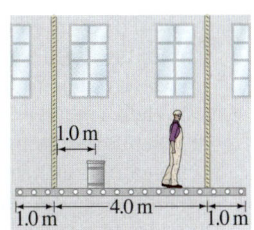

FIGURE 9–84 Problem 66.

67. A 15.0-kg ball is supported from the ceiling by rope A. Rope B pulls downward and to the side on the ball. If the angle of A to the vertical is 22° and if B makes an angle of 53° to the vertical (Fig. 9–85), find the tensions in ropes A and B.

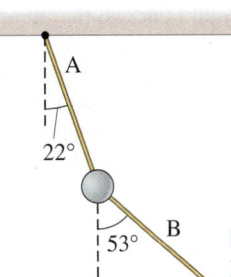

FIGURE 9–85 Problem 67.

68. The roof over a 9.0-m $\times$ 10.0-m room in a school has a total mass of 13,600 kg. The roof is to be supported by vertical wooden "2 $\times$ 4s" (actually about 4.0 cm $\times$ 9.0 cm) equally spaced along the 10.0-m sides. How many supports are required on each side, and how far apart must they be? Consider only compression, and assume a safety factor of 12.

69. A 25-kg object is being lifted by two people pulling on the ends of a 1.15-mm-diameter nylon cord that goes over two 3.00-m-high poles 4.0 m apart, as shown in Fig. 9–86. How high above the floor will the object be when the cord breaks?

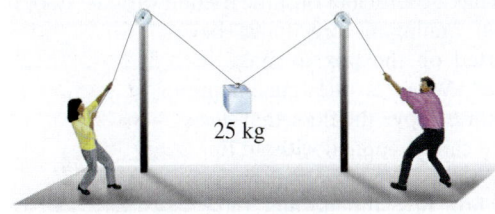

FIGURE 9–86 Problem 69.

70. A 23.0-kg backpack is suspended midway between two trees by a light cord as in Fig. 9–51. A bear grabs the backpack and pulls vertically downward with a constant force, so that each section of cord makes an angle of 27° below the horizontal. Initially, without the bear pulling, the angle was 15°; the tension in the cord with the bear pulling is double what it was when he was not. Calculate the force the bear is exerting on the backpack.

General Problems 257

71. Two identical, uniform beams are symmetrically set up against each other (Fig. 9–87) on a floor with which they have a coefficient of friction $\mu_s = 0.50$. What is the minimum angle the beams can make with the floor and still not fall?

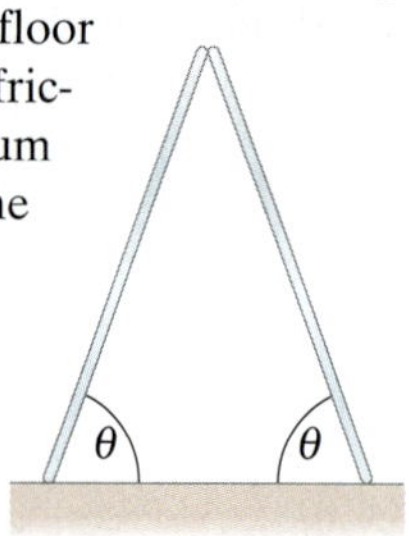

FIGURE 9–87
Problem 71.

72. A steel rod of radius $R = 15$ cm and length ℓ_0 stands upright on a firm surface. A 65-kg man climbs atop the rod. (a) Determine the percent decrease in the rod's length. (b) When a metal is compressed, each atom moves closer to its neighboring atom by exactly the same fractional amount. If iron atoms in steel are normally 2.0×10^{-10} m apart, by what distance did this interatomic spacing have to change in order to produce the normal force required to support the man? [*Note*: Neighboring atoms repel each other, and this repulsion accounts for the observed normal force.]

73. A home mechanic wants to raise the 280-kg engine out of a car. The plan is to stretch a rope vertically from the engine to a branch of a tree 6.0 m above, and back to the bumper (Fig. 9–88). When the mechanic climbs up a stepladder and pulls horizontally on the rope at its midpoint, the engine rises out of the car. (a) How much force must the mechanic exert to hold the engine 0.50 m above its normal position? (b) What is the system's mechanical advantage?

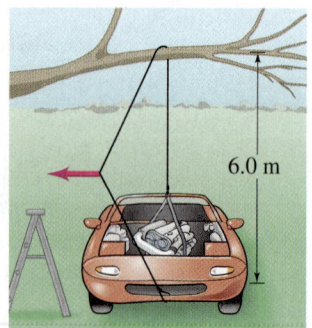

FIGURE 9–88
Problem 73.

74. A 2.0-m-high box with a 1.0-m-square base is moved across a rough floor as in Fig. 9–89. The uniform box weighs 250 N and has a coefficient of static friction with the floor of 0.60. What minimum force must be exerted on the box to make it slide? What is the maximum height h above the floor that this force can be applied without tipping the box over? Note that as the box tips, the normal force and the friction force will act at the lowest corner.

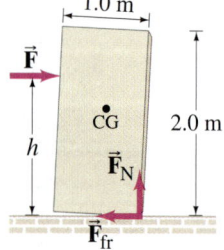

FIGURE 9–89
Problem 74.

75. A tightly stretched horizontal "high wire" is 36 m long. It sags vertically 2.1 m when a 60.0-kg tightrope walker stands at its center. What is the tension in the wire? Is it possible to increase the tension in the wire so that there is no sag?

76. Parachutists whose chutes have failed to open have been known to survive if they land in deep snow. Assume that a 75-kg parachutist hits the ground with an area of impact of 0.30 m^2 at a velocity of 55 m/s, and that the ultimate strength of body tissue is 5×10^5 N/m^2. Assume that the person is brought to rest in 1.0 m of snow. Show that the person may escape serious injury.

77. If the left vertical support column in Example 9–5 is made of steel, what is its cross-sectional area? Assume that a safety factor of 3 was used in its design to avoid fracture.

78. The mobile in Fig. 9–90 is in equilibrium. Object B has mass of 0.748 kg. Determine the masses of objects A, C, and D. (Neglect the weights of the crossbars.)

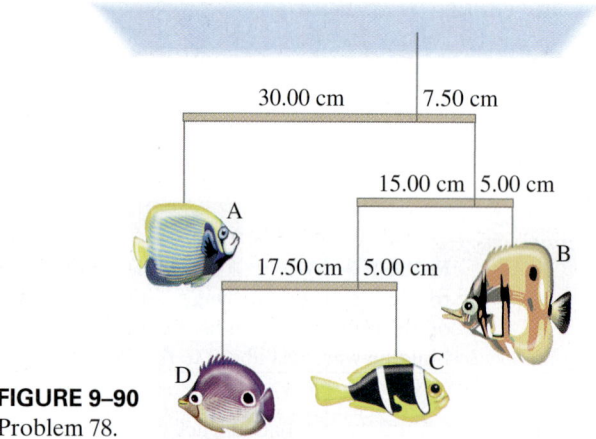

FIGURE 9–90
Problem 78.

79. In a mountain-climbing technique called the "Tyrolean traverse," a rope is anchored on both ends (to rocks or strong trees) across a deep chasm, and then a climber traverses the rope while attached by a sling as in Fig. 9–91. This technique generates tremendous forces in the rope and anchors, so a basic understanding of physics is crucial for safety. A typical climbing rope can undergo a tension force of perhaps 29 kN before breaking, and a "safety factor" of 10 is usually recommended. The length of rope used in the Tyrolean traverse must allow for some "sag" to remain in the recommended safety range. Consider a 75-kg climber at the center of a Tyrolean traverse, spanning a 25-m chasm. (a) To be within its recommended safety range, what minimum distance x must the rope sag? (b) If the Tyrolean traverse is set up incorrectly so that the rope sags by only one-fourth the distance found in (a), determine the tension in the rope. Ignore stretching of the rope. Will the rope break?

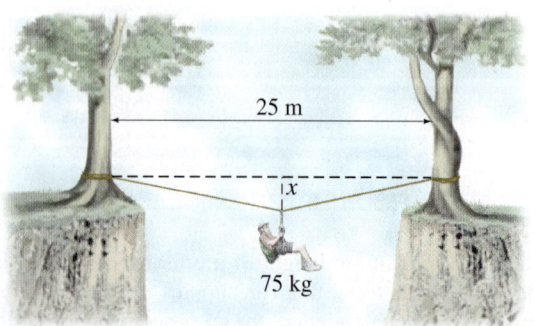

FIGURE 9–91
Problem 79.

80. When a wood shelf of mass 6.6 kg is fastened inside a slot in a vertical support as shown in Fig. 9–92, the support exerts a torque on the shelf. (*a*) Draw a free-body diagram for the shelf, assuming three vertical forces (two exerted by the support slot—explain why). Then calculate (*b*) the magnitudes of the three forces and (*c*) the torque exerted by the support (about the left end of the shelf).

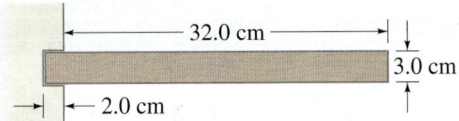

FIGURE 9–92 Problem 80.

81. A cubic crate of side $s = 2.0$ m is top-heavy: its CG is 18 cm above its true center. How steep an incline can the crate rest on without tipping over? [*Hint*: The normal force would act at the lowest corner.]

Search and Learn

1. Stand facing a wall with your toes tight against the wall, and go up on your tiptoes. Then turn around and push your back against the wall with your heels tight against the wall. Using the ideas of Section 9–4, explain why you can or cannot perform these motions.

2. From what minimum height must a 1.2-kg rectangular brick 15.0 cm × 6.0 cm × 4.0 cm be dropped above a rigid steel floor in order to break the brick? Assume the brick strikes the floor directly on its largest face, and that the compression of the brick is much greater than that of the steel (that is, ignore compression of the steel). State other simplifying assumptions that may be necessary.

3. Suppose a 65-kg person jumps from a height of 3.0 m down to the ground. (*a*) What is the speed of the person just before landing (Chapter 2)? (*b*) Estimate the average force on the person's feet exerted by the ground to bring the person to rest, if the knees are bent so the person's CG moves a distance $d = 50$ cm during the deceleration period (Fig. 9–93). [*Hint*: This force exerted by the ground ≠ net force. You may want to consult Chapters 2, 4, and 7, and be sure to draw a careful free-body diagram of the person.] (*c*) Estimate the decelerating force if the person lands stiff-legged so $d \approx 1.0$ cm. (*d*) Estimate the stress in the tibia (a lower leg bone of area $= 3.0 \times 10^{-4}$ m^2), and determine whether or not the bone will break if the landing is made with bent legs ($d = 50$ cm). (*e*) Estimate the stress and determine if the tibia will break in a stiff-legged landing ($d = 1.0$ cm).

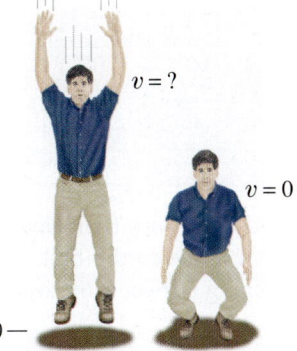

FIGURE 9–93 Search and Learn 3.

4. In Example 9–6, the torque is calculated around the axis where the cable is attached to the beam. (*a*) By using the $\Sigma\tau = 0$ equation with this axis of rotation, how do you know that the vertical force on the hinge points up and not down? (*b*) What advantage would be gained in solving part (*a*) if the axis of rotation were chosen around the hinge instead of around the point where the cable is attached? (*c*) Show that you get the same answer as in Example 9–6 if you solve the problem as in part (*b*). (*d*) In general, do you see any patterns in Sections 9–2 and 9–3 for choosing the axis of rotation to solve Problems in this Chapter?

5. Consider a ladder with a painter climbing up it (Fig. 9–94). The mass of the uniform ladder is 12.0 kg, and the mass of the painter is 55.0 kg. If the ladder begins to slip at its base when the painter's feet are 70% of the way up the length of the ladder, what is the coefficient of static friction between the ladder and the floor? Assume the wall is frictionless.

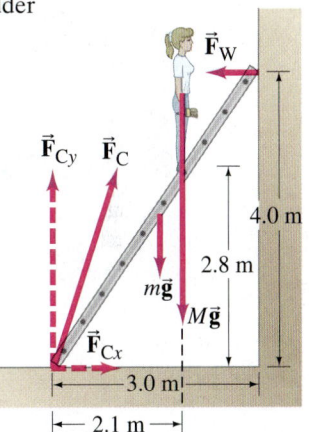

FIGURE 9–94 Search and Learn 5.

ANSWERS TO EXERCISES

A: F_A also has a component to balance the sideways force F_B.

B: Yes: $\cos\theta$ (angle of bar with ground) appears on both sides and cancels out.

C: $F_N = m_A g + m_B g + Mg$
$= (30\text{ kg} + 25\text{ kg} + 4.0\text{ kg}) g = 560$ N.

D: (*a*).

E: (*b*).

We start our study with fluids at rest, such as water in a glass or a lake. Pressure in a fluid increases with depth, a fact that allows less dense objects to float—the pressure underneath is higher than on top. When fluids flow, such as water or air, interesting effects occur because the pressure in the fluid is lower where the fluid velocity is higher (Bernoulli's principle).

The great mass of a glacier's ice (photos here) moves slowly, like a viscous liquid. The dark lines are "moraines," made up of rock broken off mountain walls by the moving ice, and represent streamlines. The two photos, taken in 1929 and 2009 by Italian expeditions to the mountain K2 (on the right in the distance), show the same glacier has become less thick, presumably due to global warming.

Fluids

CONTENTS

- 10–1 Phases of Matter
- 10–2 Density and Specific Gravity
- 10–3 Pressure in Fluids
- 10–4 Atmospheric Pressure and Gauge Pressure
- 10–5 Pascal's Principle
- 10–6 Measurement of Pressure; Gauges and the Barometer
- 10–7 Buoyancy and Archimedes' Principle
- 10–8 Fluids in Motion; Flow Rate and the Equation of Continuity
- 10–9 Bernoulli's Equation
- 10–10 Applications of Bernoulli's Principle: Torricelli, Airplanes, Baseballs, Blood Flow
- *10–11 Viscosity
- *10–12 Flow in Tubes: Poiseuille's Equation, Blood Flow
- *10–13 Surface Tension and Capillarity
- *10–14 Pumps, and the Heart

CHAPTER-OPENING QUESTIONS—Guess now!

1. Which container has the largest pressure at the bottom? Assume each container holds the same volume of water.

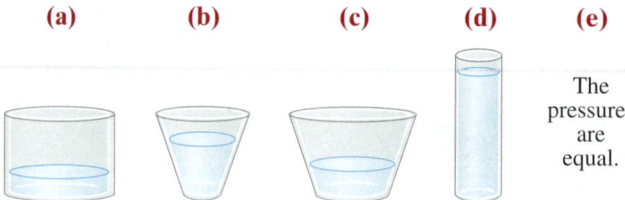

(a) (b) (c) (d) (e) The pressures are equal.

2. Two balloons are tied and hang with their nearest edges about 3 cm apart. If you blow between the balloons (not *at* the balloons, but at the opening between them), what will happen?

(a) Nothing.
(b) The balloons will move closer together.
(c) The balloons will move farther apart.

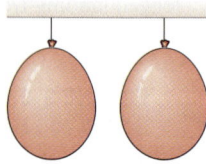

In previous Chapters we considered objects that were solid and assumed to maintain their shape except for a small amount of elastic deformation. We sometimes treated objects as point particles. Now we are going to shift our attention to materials that are very deformable and can flow. Such "fluids" include liquids and gases. We will examine fluids both at rest (fluid statics) and in motion (fluid dynamics).

10–1 Phases of Matter

The three common **phases**, or **states**, of matter are solid, liquid, and gas. A simple way to distinguish these three phases is as follows. A **solid** maintains a generally fixed size and shape; usually it requires a large force to change the volume or shape of a solid[†] (although a thin object might bend). A **liquid** does not maintain a fixed shape—it takes on the shape of its container, and it can flow; but like a solid it is not readily compressible, and its volume can be changed significantly only by a very large force. A **gas** has neither a fixed shape nor a fixed volume—it will expand to fill its container. For example, when air is pumped into an automobile tire, the air does not all run to the bottom of the tire as a liquid would; it spreads out to fill the whole volume of the tire.

Because liquids and gases do not maintain a fixed shape, they both have the ability to flow. They are thus referred to collectively as **fluids**.

The division of matter into three phases is not always simple. How, for example, should butter be classified? Furthermore, a fourth phase of matter can be distinguished, the **plasma** phase, which occurs only at very high temperatures and consists of ionized atoms (electrons separated from the nuclei). Some scientists believe that *colloids* (suspensions of tiny particles in a liquid) should also be considered a separate phase of matter. **Liquid crystals**, used in TV, cell phone, and computer screens, can be considered a phase of matter in between solids and liquids. For now, we will be interested in the three ordinary phases of matter.

10–2 Density and Specific Gravity

It is sometimes said that iron is "heavier" than wood. This cannot really be true since a large log clearly weighs more than an iron nail. What we should say is that iron is more *dense* than wood.

The **density**, ρ, of a substance (ρ is the lowercase Greek letter rho) is defined as its mass per unit volume:

$$\rho = \frac{m}{V}, \qquad (10\text{–}1)$$

where m is the mass of a sample of the substance and V its volume. Density is a characteristic property of any pure substance. Objects made of a particular pure substance, such as pure gold, can have any size or mass, but the density will be the same for each.

We can use the concept of density, Eq. 10–1, to write the mass of an object as

$$m = \rho V,$$

and the weight of an object as

$$mg = \rho V g.$$

The SI unit for density is kg/m^3. Sometimes densities are given in g/cm^3. Note that a density given in g/cm^3 must be multiplied by 1000 to give the result in kg/m^3 [$1 \text{ kg/m}^3 = 1000 \text{ g}/(100 \text{ cm})^3 = 10^3 \text{ g}/10^6 \text{ cm}^3 = 10^{-3} \text{ g/cm}^3$]. For example, the density of aluminum is $\rho = 2.70 \text{ g/cm}^3$, which equals 2700 kg/m^3. The densities of various substances are given in Table 10–1. The Table specifies temperature and atmospheric pressure because they affect density (the effect is slight for liquids and solids). Note that air is about 1000 times less dense than water.

EXAMPLE 10–1 Mass, given volume and density. What is the mass of a solid iron wrecking ball of radius 18 cm?

APPROACH First we use the standard formula $V = \frac{4}{3}\pi r^3$ (see inside rear cover) to obtain the sphere's volume. Then Eq. 10–1 and Table 10–1 give us the mass m.

SOLUTION The volume of the sphere is

$$V = \tfrac{4}{3}\pi r^3 = \tfrac{4}{3}(3.14)(0.18 \text{ m})^3 = 0.024 \text{ m}^3.$$

From Table 10–1, the density of iron is $\rho = 7800 \text{ kg/m}^3$, so Eq. 10–1 gives

$$m = \rho V = (7800 \text{ kg/m}^3)(0.024 \text{ m}^3) = 190 \text{ kg}.$$

[†]Section 9–5.

TABLE 10–1 Densities of Substances[‡]

Substance	Density, ρ (kg/m³)
Solids	
Aluminum	2.70×10^3
Iron and steel	7.8×10^3
Copper	8.9×10^3
Lead	11.3×10^3
Gold	19.3×10^3
Concrete	2.3×10^3
Granite	2.7×10^3
Wood (typical)	$0.3 – 0.9 \times 10^3$
Glass, common	$2.4 – 2.8 \times 10^3$
Ice (H_2O)	0.917×10^3
Bone	$1.7 – 2.0 \times 10^3$
Liquids	
Water (4°C)	1.000×10^3
Sea water	1.025×10^3
Blood, plasma	1.03×10^3
Blood, whole	1.05×10^3
Mercury	13.6×10^3
Alcohol, ethyl	0.79×10^3
Gasoline	$0.7 – 0.8 \times 10^3$
Gases	
Air	1.29
Helium	0.179
Carbon dioxide	1.98
Water (steam) (100°C)	0.598

[‡]Densities are given at 0°C and 1 atm pressure unless otherwise specified.

The **specific gravity** of a substance is defined as the ratio of the density of that substance to the density of water at 4.0°C. Because specific gravity (abbreviated SG) is a ratio, it is a simple number without dimensions or units. For example (see Table 10–1), the specific gravity of lead is 11.3 [$(11.3 \times 10^3 \, \text{kg/m}^3)/(1.00 \times 10^3 \, \text{kg/m}^3)$]. The SG of alcohol is 0.79.

The concepts of density and specific gravity are especially helpful in the study of fluids because we are not always dealing with a fixed volume or mass.

10–3 Pressure in Fluids

Pressure and force are related, but they are not the same thing. **Pressure** is defined as force per unit area, where the force F is understood to be the magnitude of the force acting perpendicular to the surface area A:

$$\text{pressure} = P = \frac{F}{A}. \quad (10\text{–}2)$$

CAUTION
Pressure is a scalar, not a vector

Although force is a vector, pressure is a scalar. Pressure has magnitude only. The SI unit of pressure is N/m^2. This unit has the official name **pascal** (Pa), in honor of Blaise Pascal (see Section 10–5); that is, $1 \, \text{Pa} = 1 \, \text{N/m}^2$. However, for simplicity, we will often use N/m^2. Other units sometimes used are dynes/cm^2, and lb/in.^2 (pounds per square inch, abbreviated "psi"). Several other units for pressure are discussed in Sections 10–4 and 10–6, along with conversions between them (see also the Table inside the front cover).

EXAMPLE 10–2 Calculating pressure. A 60-kg person's two feet cover an area of $500 \, \text{cm}^2$. (*a*) Determine the pressure exerted by the two feet on the ground. (*b*) If the person stands on one foot, what will be the pressure under that foot?

APPROACH Assume the person is at rest. Then the ground pushes up on her with a force equal to her weight mg, and she exerts a force mg on the ground where her feet (or foot) contact it. Because $1 \, \text{cm}^2 = (10^{-2} \, \text{m})^2 = 10^{-4} \, \text{m}^2$, then $500 \, \text{cm}^2 = 0.050 \, \text{m}^2$.

SOLUTION (*a*) The pressure on the ground exerted by the two feet is

$$P = \frac{F}{A} = \frac{mg}{A} = \frac{(60 \, \text{kg})(9.8 \, \text{m/s}^2)}{(0.050 \, \text{m}^2)} = 12 \times 10^3 \, \text{N/m}^2.$$

(*b*) If the person stands on one foot, the force is still equal to the person's weight, but the area will be half as much, so the pressure will be twice as much: $24 \times 10^3 \, \text{N/m}^2$.

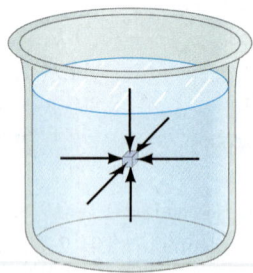

FIGURE 10–1 Pressure is the same in every direction in a nonmoving fluid at a given depth. If this weren't true, the fluid would be in motion.

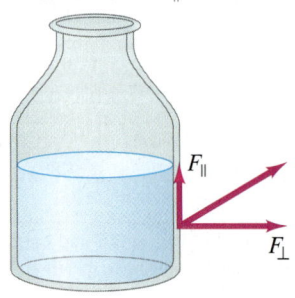

FIGURE 10–2 If there were a component of force parallel to the solid surface of the container, the liquid would move in response to it. For a liquid at rest, $F_\parallel = 0$.

Pressure is particularly useful for dealing with fluids. It is an experimental observation that *a fluid exerts pressure in every direction*. This is well known to swimmers and divers who feel the water pressure on all parts of their bodies. At any depth in a fluid at rest, the pressure is the same in all directions at that given depth. To see why, consider a tiny cube of the fluid (Fig. 10–1) which is so small that we can consider it a point and can ignore the force of gravity on it. The pressure on one side of it must equal the pressure on the opposite side. If this weren't true, there would be a net force on the cube and it would start moving. If the fluid is not flowing, then the pressures must be equal.

For a fluid at rest, the force due to fluid pressure always acts *perpendicular* to any solid surface it touches. If there were a component of the force parallel to the surface, as shown in Fig. 10–2, then according to Newton's third law the solid surface would exert a force back on the fluid, which would cause the fluid to flow—in contradiction to our assumption that the fluid is at rest. Thus the force due to the pressure in a fluid at rest is always perpendicular to the surface.

We now calculate quantitatively how the pressure in a liquid of uniform density varies with depth. Let us look at a depth h below the surface of the liquid as shown in Fig. 10–3 (that is, the liquid's top surface is a height h above this level). The pressure due to the liquid at this depth h is due to the weight of the column of liquid above it. Thus the force due to the weight of liquid acting on the area A is $F = mg = (\rho V)g = \rho A h g$, where Ah is the volume of the column of liquid, ρ is the density of the liquid (assumed to be constant), and g is the acceleration of gravity. The pressure P due to the weight of liquid is then

$$P = \frac{F}{A} = \frac{\rho A h g}{A}$$

$$P = \rho g h. \qquad \text{[liquid]} \quad (10\text{–}3\text{a})$$

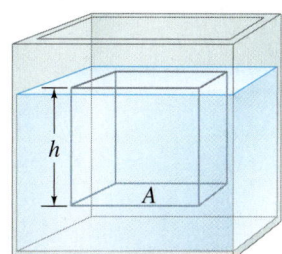

FIGURE 10–3 Calculating the pressure at a depth h in a liquid, due to the weight of the liquid above.

Note that the area A doesn't affect the pressure at a given depth. The fluid pressure is directly proportional to the density of the liquid and to the depth within the liquid. In general, *the pressure at equal depths within a uniform liquid is the same.*

EXERCISE A Return to Chapter-Opening Question 1, page 260, and answer it again now. Try to explain why you may have answered differently the first time.

Equation 10–3a is extremely useful. It is valid for fluids whose density is constant and does not change with depth—that is, if the fluid is *incompressible*. This is usually a good approximation for liquids (although at great depths in the ocean, the density of water is increased some by compression due to the great weight of water above).

If the density of a fluid does vary, a useful relation can be found by considering a thin horizontal slab of the fluid of thickness $\Delta h = h_2 - h_1$. The pressure on the top of the slab, at depth h_1, is $P_1 = \rho g h_1$. The pressure on the bottom of the slab (pushing upward), at depth h_2, is $P_2 = \rho g h_2$. The difference in pressure is

$$\Delta P = P_2 - P_1 = \rho g (h_2 - h_1)$$

or

$$\Delta P = \rho g \, \Delta h. \qquad [\rho \approx \text{constant over } \Delta h] \quad (10\text{–}3\text{b})$$

Equation 10–3b tells us how the pressure changes over a small change in depth (Δh) within a fluid, even if compressible.

Gases are very compressible, and density can vary significantly with depth. For this more general case, in which ρ may vary, we need to use Eq. 10–3b where Δh should be small if ρ varies significantly with depth (or height).

EXAMPLE 10–3 Pressure at a faucet. The surface of the water in a storage tank is 30 m above a water faucet in the kitchen of a house, Fig. 10–4. Calculate the difference in water pressure between the faucet and the surface of the water in the tank.

APPROACH Water is practically incompressible, so ρ is constant even for a $\Delta h = 30$ m when used in Eq. 10–3b. Only Δh matters; we can ignore the "route" of the pipe and its bends.

SOLUTION We assume the atmospheric pressure at the surface of the water in the storage tank is the same as at the faucet. So, the water pressure difference between the faucet and the surface of the water in the tank is

$$\Delta P = \rho g \, \Delta h = (1.0 \times 10^3 \text{ kg/m}^3)(9.8 \text{ m/s}^2)(30 \text{ m}) = 2.9 \times 10^5 \text{ N/m}^2.$$

NOTE The height Δh is sometimes called the **pressure head**. In this Example, the head of water is 30 m at the faucet. The very different diameters of the tank and faucet don't affect the result—only height does.

Water supply

FIGURE 10–4 Example 10–3.

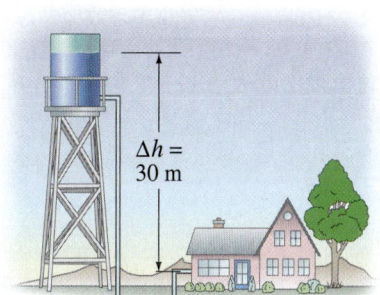

EXERCISE B A dam holds back a lake that is 85 m deep at the dam. If the lake is 20 km long, how much thicker should the dam be than if the lake were smaller, only 1.0 km long?

10–4 Atmospheric Pressure and Gauge Pressure

Atmospheric Pressure

The pressure of the Earth's atmosphere, as in any fluid, changes with depth. But the Earth's atmosphere is somewhat complicated: not only does the density of air vary greatly with altitude but there is no distinct top surface to the atmosphere from which h (in Eq. 10–3a) could be measured. We can, however, calculate the approximate difference in pressure between two altitudes above Earth's surface using Eq. 10–3b.

The pressure of the air at a given place varies slightly according to the weather. At sea level, the pressure of the atmosphere on average is $1.013 \times 10^5 \, \text{N/m}^2$ (or $14.7 \, \text{lb/in.}^2$). This value lets us define a commonly used unit of pressure, the **atmosphere** (abbreviated atm):

$$1 \, \text{atm} = 1.013 \times 10^5 \, \text{N/m}^2 = 101.3 \, \text{kPa}.$$

Another unit of pressure sometimes used (in meteorology and on weather maps) is the **bar**, which is defined as

$$1 \, \text{bar} = 1.000 \times 10^5 \, \text{N/m}^2.$$

Thus standard atmospheric pressure is slightly more than 1 bar.

The pressure due to the weight of the atmosphere is exerted on all objects immersed in this great sea of air, including our bodies. How does a human body withstand the enormous pressure on its surface? The answer is that living cells maintain an internal pressure that closely equals the external pressure, just as the pressure inside a balloon closely matches the outside pressure of the atmosphere. An automobile tire, because of its rigidity, can maintain internal pressures much greater than the external pressure.

> **PHYSICS APPLIED**
> *Pressure on living cells*

CONCEPTUAL EXAMPLE 10–4 | **Finger holds water in a straw.** You insert a straw of length ℓ into a tall glass of water. You place your finger over the top of the straw, capturing some air above the water but preventing any additional air from getting in or out, and then you lift the straw from the water. You find that the straw retains most of the water (Fig. 10–5a). Does the air in the space between your finger and the top of the water have a pressure P that is greater than, equal to, or less than, the atmospheric pressure P_0 outside the straw?

RESPONSE Consider the forces on the column of water (Fig. 10–5b). Atmospheric pressure outside the straw pushes upward on the water at the bottom of the straw, gravity pulls the water downward, and the air pressure inside the top of the straw pushes downward on the water. Since the water is in equilibrium, the upward force due to atmospheric pressure P_0 must balance the two downward forces. The only way this is possible is for the air pressure P inside the straw at the top to be *less than* the atmosphere pressure outside the straw. (When you initially remove the straw from the water glass, a little water may leave the bottom of the straw, thus increasing the volume of trapped air and reducing its density and pressure.)

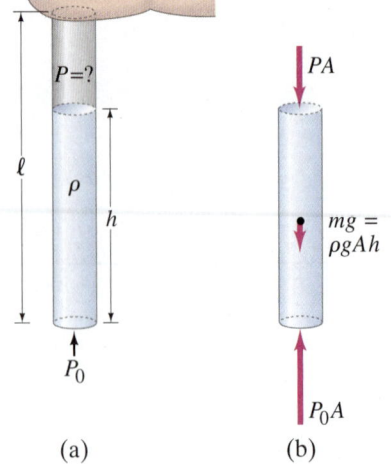

FIGURE 10–5 Example 10–4.

Gauge Pressure

It is important to note that tire gauges, and most other pressure gauges, register the pressure above and beyond atmospheric pressure. This is called **gauge pressure**. Thus, to get the **absolute pressure**, P, we must add the atmospheric pressure, P_0, to the gauge pressure, P_G:

$$P = P_G + P_0.$$

If a tire gauge registers 220 kPa, the absolute pressure within the tire is 220 kPa + 101 kPa = 321 kPa, equivalent to about 3.2 atm (2.2 atm gauge pressure).

10–5 Pascal's Principle

The Earth's atmosphere exerts a pressure on all objects with which it is in contact, including other fluids. External pressure acting on a fluid is transmitted throughout that fluid. For instance, according to Eq. 10–3a, the pressure due to the water at a depth of 100 m below the surface of a lake is $P = \rho g \Delta h = (1000 \text{ kg/m}^3)(9.8 \text{ m/s}^2)(100 \text{ m}) = 9.8 \times 10^5 \text{ N/m}^2$, or 9.7 atm. However, the total pressure at this point is due to the pressure of water plus the pressure of the air above it. Hence the total pressure (if the lake is near sea level) is 9.7 atm + 1.0 atm = 10.7 atm. This is just one example of a general principle attributed to the French philosopher and scientist Blaise Pascal (1623–1662). **Pascal's principle** states that *if an external pressure is applied to a confined fluid, the pressure at every point within the fluid increases by that amount.*

A number of practical devices make use of Pascal's principle. One example is the hydraulic lift, illustrated in Fig. 10–6a, in which a small input force is used to exert a large output force by making the area of the output piston larger than the area of the input piston. To see how this works, we assume the input and output pistons are at the same height (at least approximately). Then the external input force F_{in}, by Pascal's principle, increases the pressure equally throughout. Therefore, at the same level (see Fig. 10–6a),

$$P_{out} = P_{in}$$

where the input quantities are represented by the subscript "in" and the output by "out." Since $P = F/A$, we write the above equality as

$$\frac{F_{out}}{A_{out}} = \frac{F_{in}}{A_{in}},$$

or

$$\frac{F_{out}}{F_{in}} = \frac{A_{out}}{A_{in}}.$$

The quantity F_{out}/F_{in} is called the **mechanical advantage** of the hydraulic lift, and it is equal to the ratio of the areas. For example, if the area of the output piston is 20 times that of the input cylinder, the force is multiplied by a factor of 20. Thus a force of 200 lb could lift a 4000-lb car.

PHYSICS APPLIED
Hydraulic lift

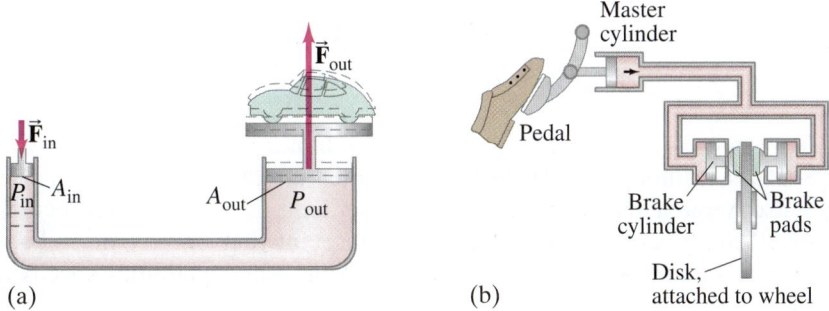

FIGURE 10–6 Applications of Pascal's principle: (a) hydraulic lift; (b) hydraulic brakes in a car.

Figure 10–6b illustrates the brake system of a car. When the driver presses the brake pedal, the pressure in the master cylinder increases. This pressure increase occurs throughout the brake fluid, thus pushing the brake pads against the disk attached to the car's wheel.

PHYSICS APPLIED
Hydraulic brakes

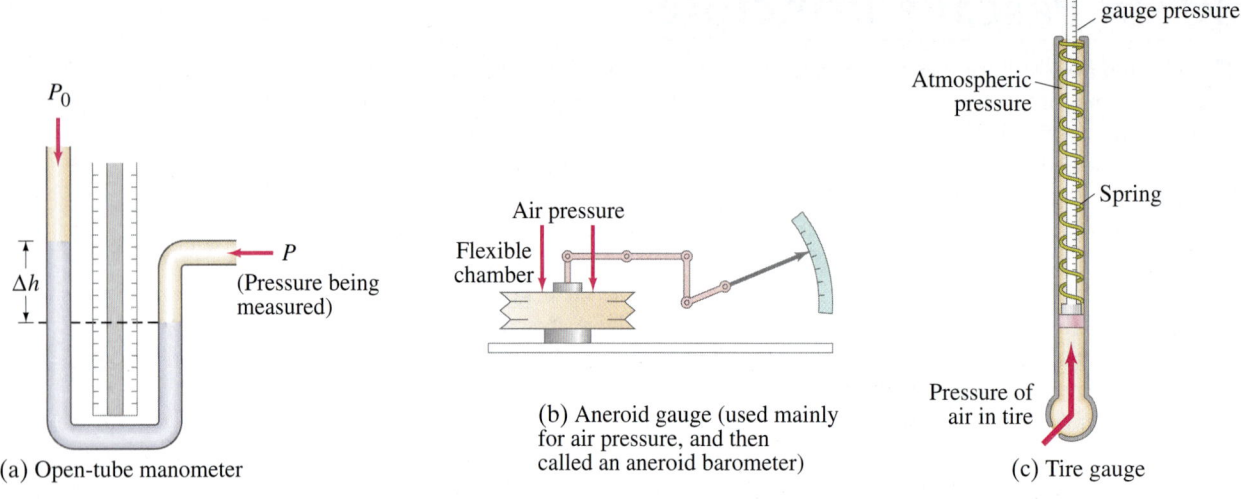

FIGURE 10–7 Pressure gauges: (a) open-tube manometer, (b) aneroid gauge, and (c) common tire pressure gauge.

10–6 Measurement of Pressure; Gauges and the Barometer

Many devices have been invented to measure pressure, some of which are shown in Fig. 10–7. The simplest is the *open-tube* **manometer** (Fig. 10–7a) which is a U-shaped tube partially filled with a liquid, usually mercury or water. The pressure P being measured is related (by Eq. 10–3b) to the difference in height Δh of the two levels of the liquid by the relation

$$P = P_0 + \rho g\, \Delta h, \qquad \text{[manometer]} \quad (10\text{–}3\text{c})$$

where P_0 is atmospheric pressure (acting on the top of the liquid in the left-hand tube), and ρ is the density of the liquid. Note that the quantity $\rho g\, \Delta h$ is the gauge pressure—the amount by which P exceeds atmospheric pressure P_0. If the liquid in the left-hand column were lower than that in the right-hand column, P would have to be less than atmospheric pressure (and Δh would be negative).

Instead of calculating the product $\rho g\, \Delta h$, sometimes only the change in height Δh is specified. In fact, pressures are sometimes specified as so many "millimeters of mercury" (mm-Hg) or "mm of water" (mm-H$_2$O). The unit mm-Hg is equivalent to a pressure of 133 N/m^2, because $\rho g\, \Delta h$ for 1 mm ($= 1.0 \times 10^{-3}$ m) of mercury gives

$$\rho g\, \Delta h = (13.6 \times 10^3 \text{ kg/m}^3)(9.80 \text{ m/s}^2)(1.00 \times 10^{-3} \text{ m})$$
$$= 1.33 \times 10^2 \text{ N/m}^2.$$

PROBLEM SOLVING
Use SI unit in calculations:
$1 \text{ Pa} = 1 \text{ N/m}^2$

The unit mm-Hg is also called the **torr** in honor of Evangelista Torricelli (1608–1647), a student of Galileo's who invented the barometer (see top of next page). Conversion factors among the various units of pressure (an incredible nuisance!) are given in Table 10–2. It is important that only N/m^2 = Pa, the proper SI unit, be used in calculations involving other quantities specified in SI units.

Another type of pressure gauge is the **aneroid gauge** (Fig. 10–7b) in which the pointer is linked to the flexible ends of an evacuated thin metal chamber. In electronic gauges, the pressure may be applied to a thin metal diaphragm whose resulting deformation is translated into an electrical signal by a transducer. A common tire gauge uses a spring, as shown in Fig. 10–7c.

TABLE 10–2 Conversion Factors Between Different Units of Pressure

In Terms of 1 Pa = 1 N/m²	1 atm in Different Units
1 atm = 1.013×10^5 N/m²	1 atm = 1.013×10^5 N/m²
= 1.013×10^5 Pa = 101.3 kPa	
1 bar = 1.000×10^5 N/m²	1 atm = 1.013 bar
1 dyne/cm² = 0.1 N/m²	1 atm = 1.013×10^6 dyne/cm²
1 lb/in.² = 6.90×10^3 N/m²	1 atm = 14.7 lb/in.²
1 lb/ft² = 47.9 N/m²	1 atm = 2.12×10^3 lb/ft²
1 cm-Hg = 1.33×10^3 N/m²	1 atm = 76.0 cm-Hg
1 mm-Hg = 133 N/m²	1 atm = 760 mm-Hg
1 torr = 133 N/m²	1 atm = 760 torr
1 mm-H₂O (4°C) = 9.80 N/m²	1 atm = 1.03×10^4 mm-H₂O (4°C)
	≈ 10 m of water

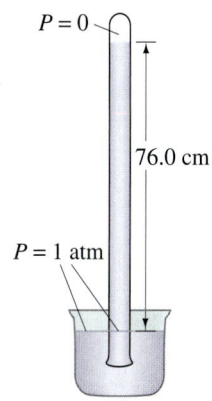

FIGURE 10–8 A mercury barometer, invented by Torricelli, is shown here when the air pressure is standard atmospheric, 76.0 cm-Hg.

Atmospheric pressure can be measured by a modified kind of mercury manometer with one end closed, called a mercury **barometer** (Fig. 10–8). The glass tube is completely filled with mercury and then inverted into the bowl of mercury. If the tube is long enough, the level of the mercury will drop, leaving a vacuum at the top of the tube, since atmospheric pressure can support a column of mercury only about 76 cm high (exactly 76.0 cm at standard atmospheric pressure). That is, a column of mercury 76 cm high exerts the same pressure as the atmosphere[†]:

$$P = \rho g \, \Delta h$$
$$= (13.6 \times 10^3 \text{ kg/m}^3)(9.80 \text{ m/s}^2)(0.760 \text{ m}) = 1.013 \times 10^5 \text{ N/m}^2 = 1.00 \text{ atm}.$$

Household barometers are usually of the aneroid type (Fig. 10–7b), either mechanical (with dial) or electronic.

A calculation similar to that just done will show that atmospheric pressure can maintain a column of water 10.3 m high in a tube whose top is under vacuum (Fig. 10–9). No matter how good a vacuum pump is, water cannot be made to rise more than about 10 m under normal atmospheric pressure. To pump water out of deep mine shafts with a vacuum pump requires multiple stages for depths greater than 10 m. Galileo studied this problem, and his student Torricelli was the first to explain it. The point is that a pump does not really suck water up a tube—it merely reduces the pressure at the top of the tube. Atmospheric air pressure *pushes* the water up the tube if the top end is at low pressure (under a vacuum), just as it is air pressure that pushes (or maintains) the mercury 76 cm high in a barometer. [Force pumps, Section 10–14, can push higher.]

FIGURE 10–9 A water barometer: a full tube of water (longer than 10 m), closed at the top, is inserted into a tub of water. When the submerged bottom end of the tube is unplugged, some water flows out of the tube into the tub, leaving a vacuum at the top of the tube above the water's upper surface. Why? Because air pressure can support a column of water only 10 m high.

CONCEPTUAL EXAMPLE 10–5 Suction. A novice engineer proposes suction cup shoes for space shuttle astronauts working on the exterior of a spacecraft. Having just studied this Chapter, you gently remind him of the fallacy of this plan. What is it?

RESPONSE Suction cups work by pushing out the air underneath the cup. What holds the suction cup in place is the air pressure outside it. (This can be a substantial force when on Earth. For example, a 10-cm-diameter suction cup has an area of 7.9×10^{-3} m². The force of the atmosphere on it is $(7.9 \times 10^{-3} \text{ m}^2)(1.0 \times 10^5 \text{ N/m}^2) \approx 800$ N, about 180 lbs!) But in outer space, there is no air pressure to push the suction cup onto the spacecraft.

We sometimes mistakenly think of suction as something we actively do. For example, we intuitively think that we pull the soda up through a straw. Instead, what we do is lower the pressure at the top of the straw, and the atmosphere *pushes* the soda up the straw.

[†]This calculation confirms the entry in Table 10–2, 1 atm = 76.0 cm-Hg.

10–7 Buoyancy and Archimedes' Principle

Objects submerged in a fluid appear to weigh less than they do when outside the fluid. For example, a large rock that you would have difficulty lifting off the ground can often be easily lifted from the bottom of a stream. When you lift the rock through the surface of the water, it suddenly seems to be much heavier. Many objects, such as wood, float on the surface of water. These are two examples of **buoyancy**. In each example, the force of gravity is acting downward. But in addition, an upward *buoyant force* is exerted by the liquid. The buoyant force on fish and underwater divers almost exactly balances the force of gravity downward, and allows them to "hover" in equilibrium.

The buoyant force occurs because the pressure in a fluid increases with depth. Thus the upward pressure on the bottom surface of a submerged object is greater than the downward pressure on its top surface. To see this effect, consider a cylinder of height Δh whose top and bottom ends have an area A and which is completely submerged in a fluid of density ρ_F, as shown in Fig. 10–10. The fluid exerts a pressure $P_1 = \rho_F g h_1$ at the top surface of the cylinder (Eq. 10–3a).

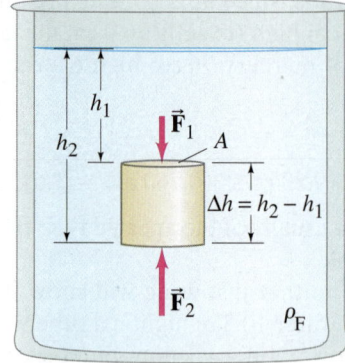

FIGURE 10–10 Determination of the buoyant force.

The force due to this pressure on top of the cylinder is $F_1 = P_1 A = \rho_F g h_1 A$, and it is directed downward. Similarly, the fluid exerts an upward force on the bottom of the cylinder equal to $F_2 = P_2 A = \rho_F g h_2 A$. The net force on the cylinder exerted by the fluid pressure, which is the **buoyant force**, $\vec{F}_B$, acts upward and has the magnitude

$$F_B = F_2 - F_1 = \rho_F g A (h_2 - h_1)$$
$$= \rho_F g A \, \Delta h$$
$$= \rho_F V g$$
$$= m_F g,$$

where $V = A \, \Delta h$ is the volume of the cylinder; the product $\rho_F V$ is the mass of the fluid displaced, and $\rho_F V g = m_F g$ is the weight of fluid which takes up a volume equal to the volume of the cylinder. Thus the buoyant force on the cylinder is equal to the weight of fluid displaced by the cylinder.

This result is valid no matter what the shape of the object. Its discovery is credited to Archimedes (287?–212 B.C.), and it is called **Archimedes' principle**:

> the buoyant force on an object immersed in a fluid is equal to the weight of the fluid displaced by that object.

By "fluid displaced," we mean a volume of fluid equal to the submerged volume of the object (or that part of the object that is submerged). If the object is placed in a glass or tub initially filled to the brim with water, the water that flows over the top represents the water displaced by the object.

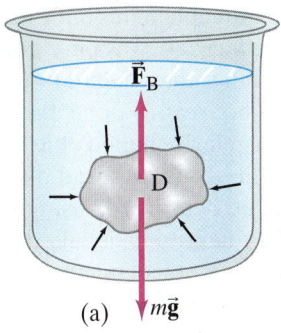
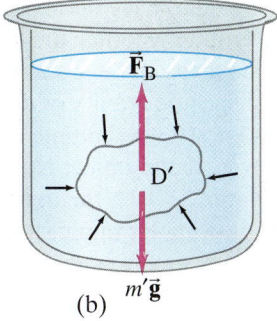

FIGURE 10–11 Archimedes' principle.

We can derive Archimedes' principle in general by the following simple but elegant argument. The irregularly shaped object D shown in Fig. 10–11a is acted on by the force of gravity (its weight, $m\vec{g}$, downward) and the buoyant force, $\vec{F}_B$, upward. We wish to determine F_B. To do so, we next consider a body (D′ in Fig. 10–11b), this time made of the fluid itself, with the same shape and size as the original object, and located at the same depth. You might think of this body of fluid as being separated from the rest of the fluid by an imaginary membrane. The buoyant force F_B on this body of fluid will be exactly the same as that on the original object since the surrounding fluid, which exerts F_B, is in exactly the same configuration. This body of fluid D′ is in equilibrium (the fluid as a whole is at rest). Therefore, $F_B = m'g$, where $m'g$ is the weight of the body of fluid D′. Hence the buoyant force F_B is equal to the weight of the body of fluid whose volume equals the volume of the original submerged object, which is Archimedes' principle.

Archimedes' discovery was made by experiment. What we have done is show that Archimedes' principle can be derived from Newton's laws.

CONCEPTUAL EXAMPLE 10–6 **Two pails of water.** Consider two identical pails of water filled to the brim. One pail contains only water, the other has a piece of wood floating in it. Which pail has the greater weight?

RESPONSE Both pails weigh the same. Recall Archimedes' principle: the wood displaces a volume of water with weight equal to the weight of the wood. Some water will overflow the pail, but Archimedes' principle tells us the spilled water has weight equal to the weight of the wood; so the pails have the same weight.

EXAMPLE 10–7 **Recovering a submerged statue.** A 70-kg ancient statue lies at the bottom of the sea. Its volume is $3.0 \times 10^4 \text{ cm}^3$. How much force is needed to lift it (without acceleration)?

APPROACH The force F needed to lift the statue is equal to the statue's weight mg minus the buoyant force F_B. Figure 10–12 is the free-body diagram.

SOLUTION We apply Newton's second law, $\Sigma F = ma = 0$, which gives $F + F_B - mg = 0$ or
$$F = mg - F_B.$$
The buoyant force on the statue due to the water is equal to the weight of $3.0 \times 10^4 \text{ cm}^3 = 3.0 \times 10^{-2} \text{ m}^3$ of water (for seawater, $\rho = 1.025 \times 10^3 \text{ kg/m}^3$):
$$F_B = m_{H_2O}\, g = \rho_{H_2O} V g = (1.025 \times 10^3 \text{ kg/m}^3)(3.0 \times 10^{-2} \text{ m}^3)(9.8 \text{ m/s}^2)$$
$$= 3.0 \times 10^2 \text{ N},$$
where we use the chemical symbol for water, H_2O, as a subscript. The weight of the statue is $mg = (70 \text{ kg})(9.8 \text{ m/s}^2) = 6.9 \times 10^2 \text{ N}$. Hence the force F needed to lift it is $690 \text{ N} - 300 \text{ N} = 390 \text{ N}$. It is as if the statue had a mass of only $(390 \text{ N})/(9.8 \text{ m/s}^2) = 40 \text{ kg}$.

NOTE Here $F = 390 \text{ N}$ is the force needed to lift the statue without acceleration when it is under water. As the statue comes *out* of the water, the force F increases, reaching 690 N when the statue is fully out of the water.

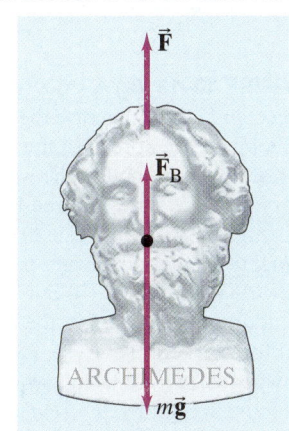

FIGURE 10–12 Example 10–7. The force needed to lift the statue is $\vec{F}$.

Archimedes is said to have discovered his principle in his bath while thinking how he might determine whether the king's new crown was pure gold or a fake. Gold has a specific gravity of 19.3, somewhat higher than that of most metals, but a determination of specific gravity or density is not readily done directly because, even if the mass is known, the volume of an irregularly shaped object is not easily calculated. However, if the object is weighed in air $(= w)$ and also "weighed" while it is under water $(= w')$, the density can be determined using Archimedes' principle, as the following Example shows. The quantity w' is called the **apparent weight** in water, and is what a scale reads when the object is submerged in water (see Fig. 10–13); w' equals the true weight $(w = mg)$ minus the buoyant force.

EXAMPLE 10–8 Archimedes: Is the crown gold? When a crown of mass 14.7 kg is submerged in water, an accurate scale reads only 13.4 kg. Is the crown made of gold?

APPROACH If the crown is gold, its density and specific gravity must be very high, SG = 19.3 (see Section 10–2 and Table 10–1). We determine the specific gravity using Archimedes' principle and the two free-body diagrams shown in Fig. 10–13.

SOLUTION The *apparent weight* of the submerged object (the crown) is w' (what the scale reads), and is the force pulling down on the scale hook. By Newton's third law, w' equals the force F'_T that the scale exerts on the crown in Fig. 10–13b. The sum of the forces on the crown is zero, so w' equals the actual weight w $(= mg)$ minus the buoyant force F_B:

$$w' = F'_T = w - F_B$$

so

$$w - w' = F_B.$$

Let V be the volume of the completely submerged object and ρ_O the object's density (so $\rho_O V$ is its mass), and let ρ_F be the density of the fluid (water). Then $(\rho_F V)g$ is the weight of fluid displaced $(= F_B)$. Now we can write

$$w = mg = \rho_O V g$$
$$w - w' = F_B = \rho_F V g.$$

We divide these two equations and obtain

$$\frac{w}{w - w'} = \frac{\rho_O V g}{\rho_F V g} = \frac{\rho_O}{\rho_F}.$$

We see that $w/(w - w')$ is equal to the specific gravity of the object (the crown) if the fluid in which it is submerged is water $(\rho_F = 1.00 \times 10^3 \text{ kg/m}^3)$. Thus

$$\frac{\rho_O}{\rho_{H_2O}} = \frac{w}{w - w'} = \frac{(14.7 \text{ kg})g}{(14.7 \text{ kg} - 13.4 \text{ kg})g} = \frac{14.7 \text{ kg}}{1.3 \text{ kg}} = 11.3.$$

This corresponds to a density of 11,300 kg/m³. The crown is not gold, but seems to be made of lead (see Table 10–1).

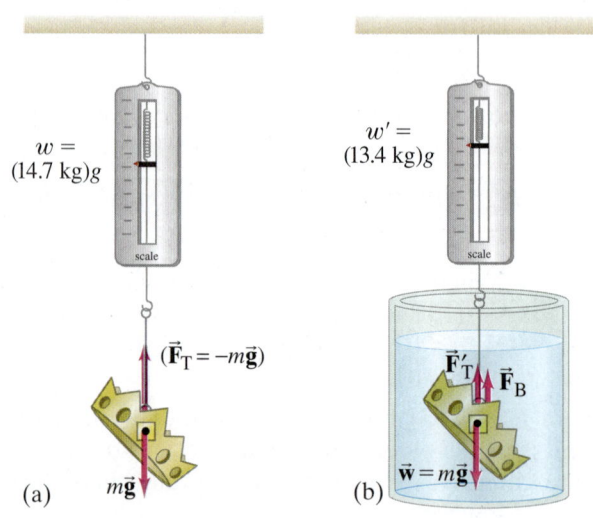

FIGURE 10–13 (a) A scale reads the mass of an object in air—in this case the crown of Example 10–8. All objects are at rest, so the tension F_T in the connecting cord equals the weight w of the object: $F_T = mg$. We show the free-body diagram of the crown, and F_T is what causes the scale reading (it is equal to the net downward force on the scale, by Newton's third law). (b) Submerged, the crown has an additional force on it, the buoyant force F_B. The net force is zero, so $F'_T + F_B = mg$ $(= w)$. The scale now reads $m' = 13.4$ kg, where m' is related to the effective weight by $w' = m'g$. Thus $F'_T = w' = w - F_B$.

Archimedes' principle applies equally well to objects that float, such as wood. In general, *an object floats on a fluid if its density* (ρ_O) *is less than that of the fluid* (ρ_F). This is readily seen from Fig. 10–14a, where a submerged log of mass m_O will experience a net upward force and float to the surface if $F_B > m_O g$; that is, if $\rho_F V g > \rho_O V g$ or $\rho_F > \rho_O$. At equilibrium—that is, when floating—the buoyant force on an object has magnitude equal to the weight of the object. For example, a log whose specific gravity is 0.60 and whose volume is 2.0 m³ has a mass

$$m_O = \rho_O V = (0.60 \times 10^3 \text{ kg/m}^3)(2.0 \text{ m}^3) = 1200 \text{ kg}.$$

If the log is fully submerged, it will displace a mass of water

$$m_F = \rho_F V = (1000 \text{ kg/m}^3)(2.0 \text{ m}^3) = 2000 \text{ kg}.$$

Hence the buoyant force on the log will be greater than its weight, and it will float upward to the surface (Fig. 10–14). The log will come to equilibrium when it displaces 1200 kg of water, which means that 1.2 m³ of its volume will be submerged. This 1.2 m³ corresponds to 60% of the volume of the log (= 1.2/2.0 = 0.60), so 60% of the log is submerged.

In general when an object floats, we have $F_B = m_O g$, which we can write as (see Fig. 10–15)

$$F_B = m_O g$$
$$\rho_F V_{\text{displ}} g = \rho_O V_O g,$$

where V_O is the full volume of the object and V_{displ} is the volume of fluid it displaces (= volume submerged). Thus

$$\frac{V_{\text{displ}}}{V_O} = \frac{\rho_O}{\rho_F}.$$

That is, the fraction of the object submerged is given by the ratio of the object's density to that of the fluid. If the fluid is water, this fraction equals the specific gravity of the object.

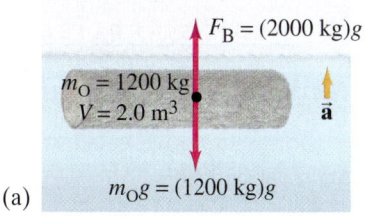

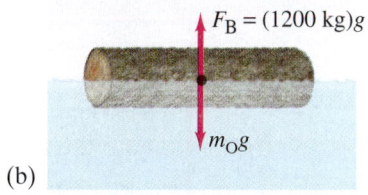

FIGURE 10–14 (a) The fully submerged log accelerates upward because $F_B > m_O g$. It comes to equilibrium (b) when $\Sigma F = 0$, so $F_B = m_O g = (1200 \text{ kg})g$. Then 1200 kg, or 1.2 m³, of water is displaced.

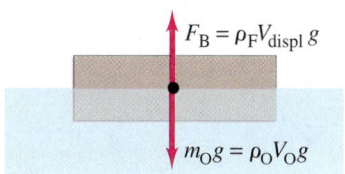

FIGURE 10–15 An object floating in equilibrium: $F_B = m_O g$.

EXAMPLE 10–9 **Hydrometer calibration.** A **hydrometer** is a simple instrument used to measure the specific gravity of a liquid by indicating how deeply the instrument sinks in the liquid. A particular hydrometer (Fig. 10–16) consists of a glass tube, weighted at the bottom, which is 25.0 cm long and 2.00 cm² in cross-sectional area, and has a mass of 45.0 g. How far from the weighted end should the 1.000 mark be placed?

APPROACH The hydrometer will float in water if its density ρ is less than $\rho_{H_2O} = 1.000 \text{ g/cm}^3$, the density of water. The fraction of the hydrometer submerged ($V_{\text{displaced}}/V_{\text{total}}$) is equal to the density ratio ρ/ρ_{H_2O}.

SOLUTION The hydrometer has an overall density

$$\rho = \frac{m}{V} = \frac{45.0 \text{ g}}{(2.00 \text{ cm}^2)(25.0 \text{ cm})} = 0.900 \text{ g/cm}^3.$$

Thus, when placed in water, it will come to equilibrium when 0.900 of its volume is submerged. Since it is of uniform cross section, $(0.900)(25.0 \text{ cm}) = 22.5$ cm of its length will be submerged. The specific gravity of water is defined to be 1.000, so the mark should be placed 22.5 cm from the weighted end.

NOTE Hydrometers can be used to measure the density of liquids like car antifreeze coolant, car battery acid (a measure of its charge), wine fermenting in casks, and many others.

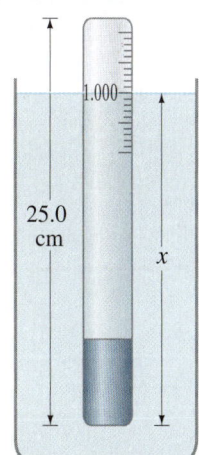

FIGURE 10–16 A hydrometer. Example 10–9.

EXERCISE C Which of the following objects, submerged in water, experiences the largest magnitude of the buoyant force? (*a*) A 1-kg helium balloon; (*b*) 1 kg of wood; (*c*) 1 kg of ice; (*d*) 1 kg of iron; (*e*) all the same.

EXERCISE D Which of the following objects, submerged in water, experiences the largest magnitude of the buoyant force? (*a*) A 1-m³ helium balloon; (*b*) 1 m³ of wood; (*c*) 1 m³ of ice; (*d*) 1 m³ of iron; (*e*) all the same.

PHYSICS APPLIED
Continental drift—plate tectonics

Archimedes' principle is also useful in geology. According to the theories of plate tectonics and continental drift, the continents float on a fluid "sea" of slightly deformable rock (mantle rock). Some interesting calculations can be done using very simple models, which we consider in the Problems at the end of the Chapter.

Air is a fluid, and it too exerts a buoyant force. Ordinary objects weigh less in air than they do in a vacuum. Because the density of air is so small, the effect for ordinary solids is slight. There are objects, however, that *float* in air—helium-filled balloons, for example, because the density of helium is less than the density of air.

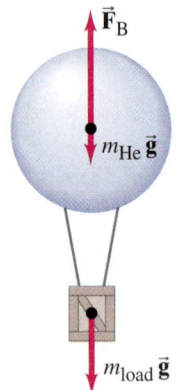

FIGURE 10–17 Example 10–10.

EXAMPLE 10–10 **Helium balloon.** What volume V of helium is needed if a balloon is to lift a load of 180 kg (including the weight of the empty balloon)?

APPROACH The buoyant force on the helium balloon, F_B, which is equal to the weight of displaced air, must be at least equal to the weight of the helium plus the weight of the balloon and load (Fig. 10–17). Table 10–1 gives the density of helium as 0.179 kg/m^3.

SOLUTION The buoyant force must have a minimum value of
$$F_B = (m_{He} + 180 \text{ kg})g.$$
This equation can be written in terms of density using Archimedes' principle:
$$\rho_{air} V g = (\rho_{He} V + 180 \text{ kg})g.$$
Solving now for V, we find
$$V = \frac{180 \text{ kg}}{\rho_{air} - \rho_{He}} = \frac{180 \text{ kg}}{(1.29 \text{ kg/m}^3 - 0.179 \text{ kg/m}^3)} = 160 \text{ m}^3.$$

NOTE This is the minimum volume needed near the Earth's surface, where $\rho_{air} = 1.29 \text{ kg/m}^3$. To reach a high altitude, a greater volume would be needed since the density of air decreases with altitude.

CONCEPTUAL EXAMPLE 10–11 **Throwing a rock overboard.** A rowboat carrying a large granite rock floats in a small lake. If the rock (SG ≈ 3, Table 10–1) is thrown overboard and sinks, does the lake level drop, rise, or stay the same?

RESPONSE Together the boat and rock float, so the buoyant force on them equals their total weight. The boat and rock displace a mass of water whose weight is equal to the weight of boat plus rock. When the rock is thrown into the lake, it displaces only its own volume, which is smaller than the volume of water the rock displaced when in the boat ($\approx \frac{1}{3}$ as much because the rock's density is ≈3 times greater than water). So less lake water is displaced and the water level of the lake *drops* when the rock is in the lake.

Maybe numbers can help. Suppose the boat and the rock each has a mass of 60 kg. Then the boat carrying the rock displaces 120 kg of water, which is a volume of 0.12 m^3 ($\rho = 1000 \text{ kg/m}^3$ for water, Table 10–1). When the rock is thrown into the lake, the boat alone now displaces 0.06 m^3. The rock displaces only its own volume of 0.02 m^3 ($\rho = m/V \approx 3$ so $V \approx 0.06 \text{ m}^3/3$). Thus a total of 0.08 m^3 of water is displaced. Less water is displaced so the water level of the lake drops.

EXERCISE E If you throw a flat 60-kg aluminum plate into water, the plate sinks. But if that aluminum is shaped into a rowboat, it floats. Explain.

10–8 Fluids in Motion; Flow Rate and the Equation of Continuity

We now turn to the subject of fluids in motion, which is called **fluid dynamics**, or (especially if the fluid is water) **hydrodynamics**.

We can distinguish two main types of fluid flow. If the flow is smooth, such that neighboring layers of the fluid slide by each other smoothly, the flow is said to be **streamline** or **laminar flow**.[†] In streamline flow, each particle of the fluid follows a smooth path, called a **streamline**, and these paths do not cross one another (Fig. 10–18a).

[†]The word laminar means "in layers."

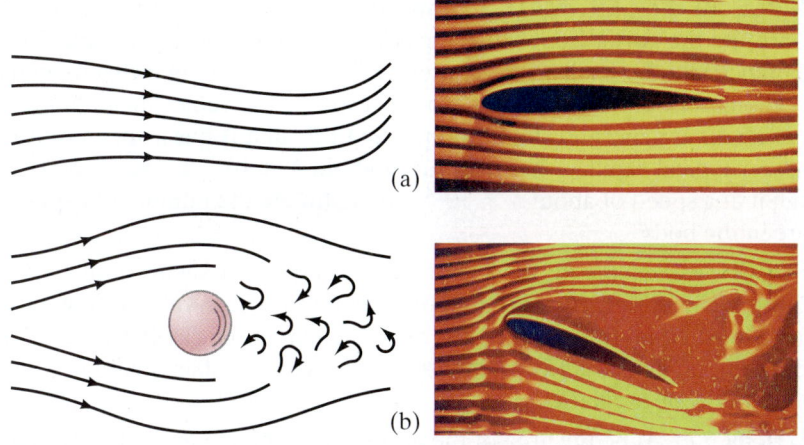

FIGURE 10–18 (a) Streamline, or laminar, flow; (b) turbulent flow. The photos show airflow around an airfoil or airplane wing (more in Section 10–10).

Above a certain speed, the flow becomes turbulent. **Turbulent flow** is characterized by erratic, small, whirlpool-like circles called *eddy currents* or *eddies* (Fig. 10–18b). Eddies absorb a great deal of energy, and although a certain amount of internal friction called **viscosity** is present even during streamline flow, it is much greater when the flow is turbulent. A few tiny drops of ink or food coloring dropped into a moving liquid can quickly reveal whether the flow is streamline or turbulent.

Let us consider the steady laminar flow of a fluid through an enclosed tube or pipe as shown in Fig. 10–19. First we determine how the speed of the fluid changes when the diameter of the tube changes. The mass **flow rate** is defined as the mass Δm of fluid that passes a given point per unit time Δt:

$$\text{mass flow rate} = \frac{\Delta m}{\Delta t}.$$

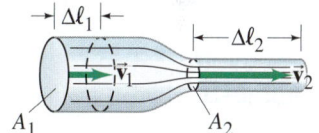

FIGURE 10–19 Fluid flow through a pipe of varying diameter.

In Fig. 10–19, the volume of fluid passing point 1 (through area A_1) in a time Δt is $A_1 \Delta \ell_1$, where $\Delta \ell_1$ is the distance the fluid moves in time Δt. The velocity† of fluid (density ρ_1) passing point 1 is $v_1 = \Delta \ell_1 / \Delta t$. Then the mass flow rate $\Delta m_1 / \Delta t$ through area A_1 is

$$\frac{\Delta m_1}{\Delta t} = \frac{\rho_1 \Delta V_1}{\Delta t} = \frac{\rho_1 A_1 \Delta \ell_1}{\Delta t} = \rho_1 A_1 v_1,$$

where $\Delta V_1 = A_1 \Delta \ell_1$ is the volume of mass Δm_1. Similarly, at point 2 (through area A_2), the flow rate is $\rho_2 A_2 v_2$. Since no fluid flows in or out the sides of the tube, the flow rates through A_1 and A_2 must be equal. Thus

$$\frac{\Delta m_1}{\Delta t} = \frac{\Delta m_2}{\Delta t},$$

and

$$\rho_1 A_1 v_1 = \rho_2 A_2 v_2. \quad (10\text{–}4\text{a})$$

This is called the **equation of continuity**.

If the fluid is incompressible (ρ doesn't change with pressure), which is an excellent approximation for liquids under most circumstances (and sometimes for gases as well), then $\rho_1 = \rho_2$, and the equation of continuity becomes

$$A_1 v_1 = A_2 v_2. \quad [\rho = \text{constant}] \quad (10\text{–}4\text{b})$$

The product Av represents the *volume rate of flow* (volume of fluid passing a given point per second), since $\Delta V / \Delta t = A \Delta \ell / \Delta t = Av$, which in SI units is m³/s. Equation 10–4b tells us that where the cross-sectional area is large, the velocity is small, and where the area is small, the velocity is large. That this is reasonable can be seen by looking at a river. A river flows slowly through a meadow where it is broad, but speeds up to torrential speed when passing through a narrow gorge.

---

†If there were no viscosity, the velocity would be the same across a cross section of the tube. Real fluids have viscosity, and this internal friction causes different layers of the fluid to flow at different speeds. In this case v_1 and v_2 represent the average speeds at each cross section.

PHYSICS APPLIED
Blood flow

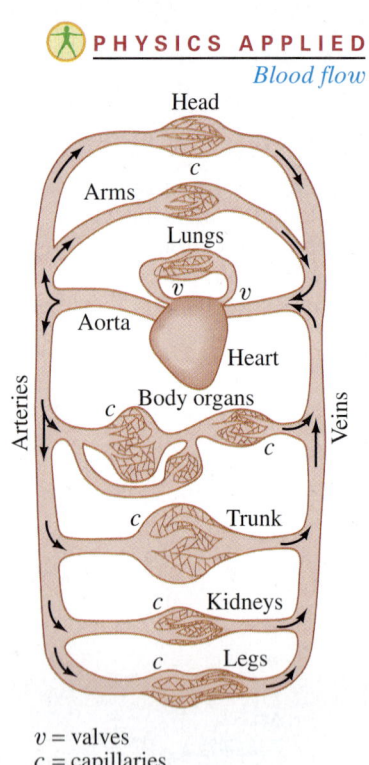

v = valves
c = capillaries

FIGURE 10–20 Human circulatory system.

PHYSICS APPLIED
Heating duct

FIGURE 10–21 Example 10–13.

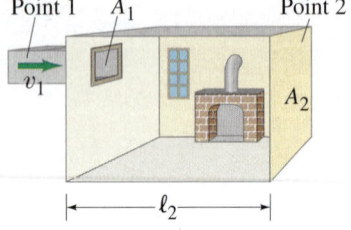

FIGURE 10–19 (Repeated.) Fluid flow through a pipe of varying diameter.

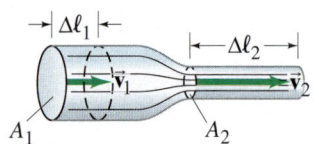

EXAMPLE 10–12 ESTIMATE Blood flow. In humans, blood flows from the heart into the aorta, from which it passes into the major arteries, Fig. 10–20. These branch into the small arteries (arterioles), which in turn branch into myriads of tiny capillaries. The blood returns to the heart via the veins. The radius of the aorta is about 1.2 cm, and the blood passing through it has a speed of about 40 cm/s. A typical capillary has a radius of about 4×10^{-4} cm, and blood flows through it at a speed of about 5×10^{-4} m/s. Estimate the number of capillaries that are in the body.

APPROACH We assume the density of blood doesn't vary significantly from the aorta to the capillaries. By the equation of continuity, the volume flow rate in the aorta must equal the volume flow rate through *all* the capillaries. The total area of all the capillaries is given by the area of a typical capillary multiplied by the total number N of capillaries.

SOLUTION Let A_1 be the area of the aorta and A_2 be the area of *all* the capillaries through which blood flows. Then $A_2 = N\pi r_{cap}^2$, where $r_{cap} \approx 4 \times 10^{-4}$ cm is the estimated average radius of one capillary. From the equation of continuity (Eq. 10–4b), we have

$$v_2 A_2 = v_1 A_1$$
$$v_2 N\pi r_{cap}^2 = v_1 \pi r_{aorta}^2$$

so

$$N = \frac{v_1}{v_2} \frac{r_{aorta}^2}{r_{cap}^2} = \left(\frac{0.40 \text{ m/s}}{5 \times 10^{-4} \text{ m/s}}\right)\left(\frac{1.2 \times 10^{-2} \text{ m}}{4 \times 10^{-6} \text{ m}}\right)^2 \approx 7 \times 10^9,$$

or on the order of 10 billion capillaries.

EXAMPLE 10–13 Heating duct to a room. What area must a heating duct have if air moving 3.0 m/s along it can replenish the air every 15 minutes in a room of volume 300 m³? Assume the air's density remains constant.

APPROACH We apply the equation of continuity at constant density, Eq. 10–4b, to the air that flows through the duct (point 1 in Fig. 10–21) and then into the room (point 2). The volume flow rate in the room equals the volume of the room divided by the 15-min replenishing time.

SOLUTION Consider the room as a large section of the duct, Fig. 10–21, and think of air equal to the volume of the room as passing by point 2 in $t = 15$ min $= 900$ s. Reasoning in the same way we did to obtain Eq. 10–4a (changing Δt to t), we write $v_2 = \ell_2/t$ so $A_2 v_2 = A_2 \ell_2/t = V_2/t$, where V_2 is the volume of the room. Then the equation of continuity becomes $A_1 v_1 = A_2 v_2 = V_2/t$ and

$$A_1 = \frac{V_2}{v_1 t} = \frac{300 \text{ m}^3}{(3.0 \text{ m/s})(900 \text{ s})} = 0.11 \text{ m}^2.$$

NOTE If the duct is square, then each side has length $\ell = \sqrt{A} = 0.33$ m, or 33 cm. A rectangular duct 20 cm $\times$ 55 cm will also do.

10–9 Bernoulli's Equation

Have you ever wondered why an airplane can fly, or how a sailboat can move against the wind? These are examples of a principle worked out by Daniel Bernoulli (1700–1782) concerning fluids in motion. In essence, **Bernoulli's principle** states that *where the velocity of a fluid is high, the pressure is low, and where the velocity is low, the pressure is high.* For example, if the pressure in the fluid is measured at points 1 and 2 of Fig. 10–19, it will be found that the pressure is lower at point 2, where the velocity is greater, than it is at point 1, where the velocity is smaller. At first glance, this might seem strange; you might expect that the greater speed at point 2 would imply a higher pressure. But this cannot be the case:

if the pressure in the fluid at point 2 were higher than at point 1, this higher pressure would slow the fluid down, whereas in fact it has sped up in going from point 1 to point 2. Thus the pressure at point 2 must be less than at point 1, to be consistent with the fact that the fluid accelerates.

To help clarify any misconceptions, a faster fluid might indeed exert a greater force bouncing off an obstacle placed in its path. But that is not what we mean by the pressure in a fluid. We are examining smooth streamline flow, with no obstacles that interrupt the flow. The fluid pressure is exerted on the walls of a tube or pipe, or on the surface of a material the fluid passes over.

Bernoulli developed an equation that expresses this principle quantitatively. To derive Bernoulli's equation, we assume the flow is steady and laminar, the fluid is incompressible, and the viscosity is small enough to be ignored. To be general, we assume the fluid is flowing in a tube of nonuniform cross section that varies in height above some reference level, Fig. 10–22. We will consider the volume of fluid shown in color and calculate the work done to move it from the position shown in Fig. 10–22a to that shown in Fig. 10–22b. In this process, fluid entering area A_1 flows a distance $\Delta\ell_1$ and forces the fluid at area A_2 to move a distance $\Delta\ell_2$. The fluid to the left of area A_1 exerts a pressure P_1 on our section of fluid and does an amount of work

$$W_1 = F_1 \Delta\ell_1 = P_1 A_1 \Delta\ell_1,$$

(since $P = F/A$). At point 2, the work done on our section of fluid is

$$W_2 = -P_2 A_2 \Delta\ell_2.$$

The negative sign is present because the force exerted on the fluid is opposite to the displacement. Work is also done on the fluid by the force of gravity. The net effect of the process shown in Fig. 10–22 is to move a mass m of volume $A_1 \Delta\ell_1$ ($= A_2 \Delta\ell_2$, since the fluid is incompressible) from point 1 to point 2, so the work done by gravity is

$$W_3 = -mg(y_2 - y_1),$$

where y_1 and y_2 are heights of the center of the tube above some (arbitrary) reference level. In the case shown in Fig. 10–22, this term is negative since the motion is uphill against the force of gravity. The net work W done on the fluid is thus

$$W = W_1 + W_2 + W_3$$
$$W = P_1 A_1 \Delta\ell_1 - P_2 A_2 \Delta\ell_2 - mgy_2 + mgy_1.$$

According to the work-energy principle (Section 6–3), the net work done on a system is equal to its change in kinetic energy. Hence

$$\tfrac{1}{2}mv_2^2 - \tfrac{1}{2}mv_1^2 = P_1 A_1 \Delta\ell_1 - P_2 A_2 \Delta\ell_2 - mgy_2 + mgy_1.$$

The mass m has volume $A_1 \Delta\ell_1 = A_2 \Delta\ell_2$ for an incompressible fluid. Thus we can substitute $m = \rho A_1 \Delta\ell_1 = \rho A_2 \Delta\ell_2$, and then divide through by $A_1 \Delta\ell_1 = A_2 \Delta\ell_2$, to obtain

$$\tfrac{1}{2}\rho v_2^2 - \tfrac{1}{2}\rho v_1^2 = P_1 - P_2 - \rho g y_2 + \rho g y_1,$$

which we rearrange to get

$$P_2 + \tfrac{1}{2}\rho v_2^2 + \rho g y_2 = P_1 + \tfrac{1}{2}\rho v_1^2 + \rho g y_1. \qquad (10\text{--}5) \quad \textit{Bernoulli's equation}$$

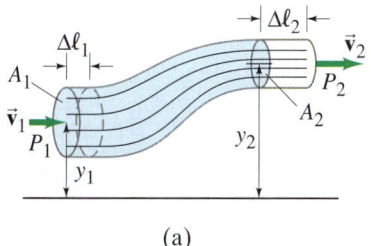
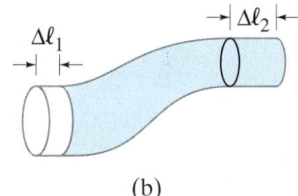

FIGURE 10–22 Fluid flow: for derivation of Bernoulli's equation.

This is **Bernoulli's equation**. Since points 1 and 2 can be any two points along a tube of flow, Bernoulli's equation can be written as

$$P + \tfrac{1}{2}\rho v^2 + \rho g y = \text{constant}$$

at every point in the fluid, where y is the height of the center of the tube above a fixed reference level. [Note that if there is no flow ($v_1 = v_2 = 0$), then Eq. 10–5 reduces to the hydrostatic equation, Eq. 10–3b or c: $P_1 - P_2 = \rho g(y_2 - y_1)$.]

Bernoulli's equation is an expression of the law of energy conservation, since we derived it from the work-energy principle.

> **EXERCISE F** As water in a level pipe passes from a narrow cross section of pipe to a wider cross section, how does the pressure against the walls change?

PHYSICS APPLIED
Hot-water heating system

EXAMPLE 10–14 Flow and pressure in a hot-water heating system. Water circulates throughout a house in a hot-water heating system. If the water is pumped at a speed of 0.50 m/s through a 4.0-cm-diameter pipe in the basement under a pressure of 3.0 atm, what will be the flow speed and pressure in a 2.6-cm-diameter pipe on the second floor 5.0 m above? Assume the pipes do not divide into branches.

APPROACH We use the equation of continuity at constant density to determine the flow speed on the second floor, and then Bernoulli's equation to find the pressure.

SOLUTION We take v_2 in the equation of continuity, Eq. 10–4, as the flow speed on the second floor, and v_1 as the flow speed in the basement. Noting that the areas are proportional to the radii squared $(A = \pi r^2)$, we obtain

$$v_2 = \frac{v_1 A_1}{A_2} = \frac{v_1 \pi r_1^2}{\pi r_2^2} = (0.50 \text{ m/s}) \frac{(0.020 \text{ m})^2}{(0.013 \text{ m})^2} = 1.2 \text{ m/s}.$$

To find the pressure on the second floor, we use Bernoulli's equation (Eq. 10–5):

$$\begin{aligned} P_2 &= P_1 + \rho g(y_1 - y_2) + \tfrac{1}{2}\rho(v_1^2 - v_2^2) \\ &= (3.0 \times 10^5 \text{ N/m}^2) + (1.0 \times 10^3 \text{ kg/m}^3)(9.8 \text{ m/s}^2)(-5.0 \text{ m}) \\ &\quad + \tfrac{1}{2}(1.0 \times 10^3 \text{ kg/m}^3)[(0.50 \text{ m/s})^2 - (1.2 \text{ m/s})^2] \\ &= (3.0 \times 10^5 \text{ N/m}^2) - (4.9 \times 10^4 \text{ N/m}^2) - (6.0 \times 10^2 \text{ N/m}^2) \\ &= 2.5 \times 10^5 \text{ N/m}^2 = 2.5 \text{ atm}. \end{aligned}$$

NOTE The velocity term contributes very little in this case.

10–10 Applications of Bernoulli's Principle: Torricelli, Airplanes, Baseballs, Blood Flow

Bernoulli's equation can be applied to many situations. One example is to calculate the velocity, v_1, of a liquid flowing out of a spigot at the bottom of a reservoir, Fig. 10–23. We choose point 2 in Eq. 10–5 to be the top surface of the liquid. Assuming the diameter of the reservoir is large compared to that of the spigot, v_2 will be almost zero. Points 1 (the spigot) and 2 (top surface) are open to the atmosphere, so the pressure at both points is equal to atmospheric pressure: $P_1 = P_2$. Then Bernoulli's equation becomes

$$\tfrac{1}{2}\rho v_1^2 + \rho g y_1 = \rho g y_2$$

or

$$v_1 = \sqrt{2g(y_2 - y_1)}. \quad (10\text{–}6)$$

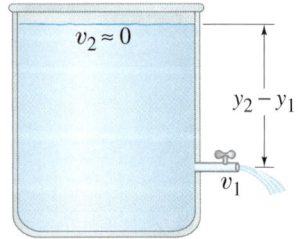

FIGURE 10–23 Torricelli's theorem: $v_1 = \sqrt{2g(y_2 - y_1)}$.

This result is called **Torricelli's theorem**. Although it is seen to be a special case of Bernoulli's equation, it was discovered a century earlier by Evangelista Torricelli. Equation 10–6 tells us that the liquid leaves the spigot with the same speed that a freely falling object would attain if falling from the same height. This should not be too surprising since the derivation of Bernoulli's equation relies on the conservation of energy.

Another special case of Bernoulli's equation arises when a fluid is flowing horizontally with no appreciable change in height; that is, $y_1 = y_2$. Then Eq. 10–5 becomes

$$P_1 + \tfrac{1}{2}\rho v_1^2 = P_2 + \tfrac{1}{2}\rho v_2^2, \quad (10\text{–}7)$$

which tells us quantitatively that the speed is high where the pressure is low, and vice versa. It explains many common phenomena, some of which are illustrated in Figs. 10–24 to 10–30. The pressure in the air blown at high speed across the top of the vertical tube of a perfume atomizer (Fig. 10–24a) is less than the normal air pressure acting on the surface of the liquid in the bowl. Thus atmospheric pressure in the bowl pushes the perfume up the tube because of the lower pressure at the top. A Ping-Pong ball can be made to float above a blowing jet of air (a hair dryer or a vacuum cleaner that can also blow air), Fig. 10–24b; if the ball begins to leave the jet of air, the higher pressure in the still air outside the jet pushes the ball back in.

EXERCISE G Return to Chapter-Opening Question 2, page 260, and answer it again now. Try to explain why you may have answered differently the first time. Try it and see.

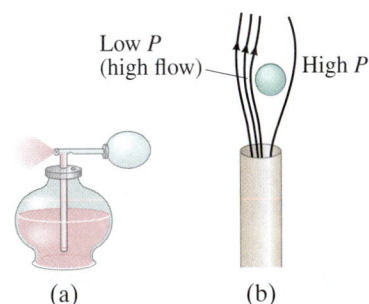

FIGURE 10–24 Examples of Bernoulli's principle: (a) atomizer, (b) Ping-Pong ball in jet of air.

Airplane Wings and Dynamic Lift

Airplanes experience a "lift" force on their wings, keeping them up in the air, if they are moving at a sufficiently high speed relative to the air and the wing is tilted upward at a small angle (the "attack angle"). See Fig. 10–25, where streamlines of air are shown rushing by the wing (we are in the reference frame of the wing, as if sitting on the wing). The upward tilt, as well as the rounded upper surface of the wing, causes the streamlines to be forced upward and to be crowded together above the wing. The area of air flowing between any two streamlines is smaller as the streamlines get closer together, so from the equation of continuity $(A_1 v_1 = A_2 v_2)$, the air speed increases above the wing where the streamlines are squished together. (Recall also how the crowded streamlines in a pipe constriction, Fig. 10–19, indicate the velocity is higher in the constriction.) Thus the air speed is greater above the wing than below it, so the pressure above the wing is less than the pressure below the wing (Bernoulli's principle). Hence there is a net upward force on the wing called **dynamic lift**. Experiments show that the speed of air above the wing can even be double the speed of the air below it. (Friction between the air and wing exerts a *drag force*, toward the rear, which must be overcome by the plane's engines.)

A flat wing, or one with symmetric cross section, will experience lift as long as the front of the wing is tilted upward (attack angle). The wing shown in Fig. 10–25 can experience lift even if the attack angle is zero, because the rounded upper surface deflects air up, squeezing the streamlines together. Airplanes can fly upside down, experiencing lift, if the attack angle is sufficient to deflect streamlines up and closer together.

Our picture considers streamlines; but if the attack angle is larger than about 15°, turbulence sets in (Fig. 10–18b) leading to greater drag and less lift, causing the plane to "stall" and then to drop.

From another point of view, the upward tilt of a wing means the air moving horizontally in front of the wing is deflected downward; the change in momentum of the rebounding air molecules results in an upward force on the wing (Newton's third law).

FIGURE 10–25 Lift on an airplane wing. We are in the reference frame of the wing, seeing the air flow by.

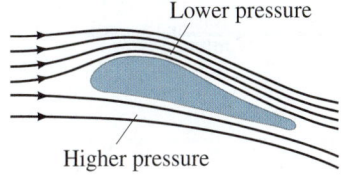

PHYSICS APPLIED
Airplanes and dynamic lift

FIGURE 10–26 Sailboat (a) sailing against the wind with (b) analysis.

Sailboats

A sailboat can move "against" the wind, with the aid of the Bernoulli effect, by setting the sails at an angle, as shown in Fig. 10–26. The air traveling rapidly over the bulging front surface of the mainsail exerts a smaller pressure than the relatively still air behind the sail. The result is a net force on the sail, $\vec{\mathbf{F}}_{\text{wind}}$, as shown in Fig. 10–26b. This force would tend to make the boat move sideways if it weren't for the keel that extends vertically downward beneath the water: the water exerts a force ($\vec{\mathbf{F}}_{\text{water}}$) on the keel nearly perpendicular to the keel. The resultant of these two forces ($\vec{\mathbf{F}}_R$) is almost directly forward as shown.

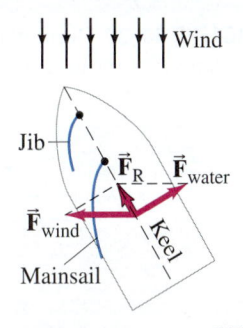

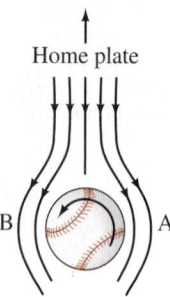

FIGURE 10–27 Looking down on a pitched baseball heading toward home plate. We are in the reference frame of the baseball, with the air flowing by.

FIGURE 10–28 Rear of the head and shoulders showing arteries leading to the brain and to the arms. High blood velocity past the constriction in the left subclavian artery causes low pressure in the left vertebral artery, in which a reverse (downward) blood flow can then occur, resulting in a TIA, a loss of blood to the brain.

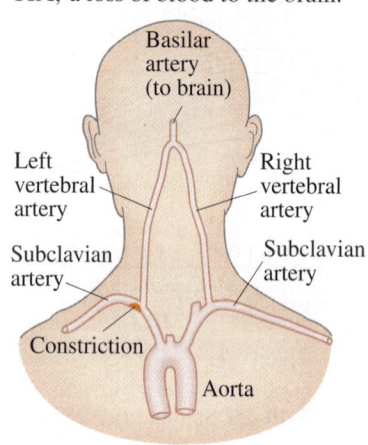

Baseball Curve

Why a spinning pitched baseball (or tennis ball) curves can also be explained using Bernoulli's principle. It is simplest if we put ourselves in the reference frame of the ball, with the air rushing by, just as we did for the airplane wing. Suppose the ball is rotating counterclockwise as seen from above, Fig. 10–27. A thin layer of air ("boundary layer") is being dragged around by the ball. We are looking down on the ball, and at point A in Fig. 10–27, this boundary layer tends to slow down the oncoming air. At point B, the air rotating with the ball adds its speed to that of the oncoming air, so the air speed is higher at B than at A. The higher speed at B means the pressure is lower at B than at A, resulting in a net force toward B. The ball's path curves toward the left (as seen by the pitcher).

Lack of Blood to the Brain—TIA

In medicine, one of many applications of Bernoulli's principle is to explain a TIA, a *transient ischemic attack* (meaning a temporary lack of blood supply to the brain). A person suffering a TIA may experience symptoms such as dizziness, double vision, headache, and weakness of the limbs. A TIA can occur as follows. Blood normally flows up to the brain at the back of the head via the two vertebral arteries—one going up each side of the neck—which meet to form the basilar artery just below the brain, as shown in Fig. 10–28. Each vertebral artery connects to the subclavian artery, as shown, before the blood passes to the arms. When an arm is exercised vigorously, blood flow increases to meet the needs of the arm's muscles. If the subclavian artery on one side of the body is partially blocked, however, as in arteriosclerosis (hardening of the arteries), the blood velocity will have to be higher on that side to supply the needed blood. (Recall the equation of continuity: smaller area means larger velocity for the same flow rate, Eqs. 10–4.) The increased blood velocity past the opening to the vertebral artery results in lower pressure (Bernoulli's principle). Thus, blood rising in the vertebral artery on the "good" side at normal pressure can be *diverted down* into the other vertebral artery because of the low pressure on that side, instead of passing upward to the brain. Hence the blood supply to the brain is reduced.

Other Applications

A **venturi tube** is essentially a pipe with a narrow constriction (the throat). The flowing fluid speeds up as it passes through this constriction, so the pressure is lower in the throat. A **venturi meter**, Fig. 10–29, is used to measure the flow speed of gases and liquids, including blood velocity in arteries. The velocity v_1 can be determined by measuring the pressure P_1 and P_2, the areas A_1 and A_2, as well as the density of the fluid. (The formula is given in Problem 56.)

FIGURE 10–29 Venturi meter.

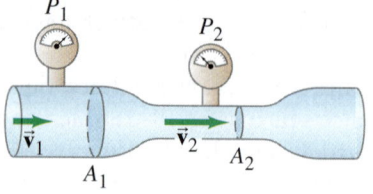

PHYSICS APPLIED
Smoke up a chimney
Underground air circulation

FIGURE 10–30 Bernoulli's principle explains air flow in underground burrows.

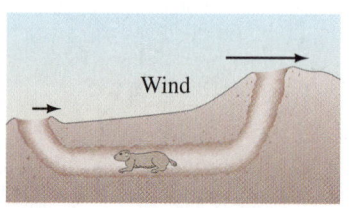

Why does smoke go up a chimney? It's partly because hot air rises (it's less dense and therefore buoyant). But Bernoulli's principle also plays a role. When wind blows across the top of a chimney, the pressure is less there than inside the house. Hence, air and smoke are pushed up the chimney by the higher indoor pressure. Even on an apparently still night there is usually enough ambient air flow at the top of a chimney to assist upward flow of smoke.

If gophers, prairie dogs, rabbits, and other animals that live underground are to avoid suffocation, the air must circulate in their burrows. The burrows always have at least two entrances (Fig. 10–30). The speed of air flow across different holes will usually be slightly different. This results in a slight pressure difference, which forces a flow of air through the burrow via Bernoulli's principle. The flow of air is enhanced if one hole is higher than the other (animals often build mounds) since wind speed tends to increase with height.

Bernoulli's equation ignores the effects of friction (viscosity) and the compressibility of the fluid. The energy that is transformed to internal (or potential) energy due to compression and to thermal energy by friction can be taken into account by adding terms to Eq. 10–5. These terms are difficult to calculate theoretically and are normally determined empirically for given situations. They do not significantly alter the explanations for the phenomena described above.

*10–11 Viscosity

Real fluids have a certain amount of internal friction called **viscosity**, as mentioned in Section 10–8. Viscosity exists in both liquids and gases, and is essentially a frictional force between adjacent layers of fluid as the layers move past one another. In liquids, viscosity is due to the electrical cohesive forces between the molecules. In gases, it arises from collisions between the molecules.

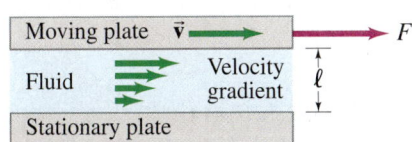

FIGURE 10–31 Determination of viscosity.

The viscosity of different fluids can be expressed quantitatively by a *coefficient of viscosity*, η (the Greek lowercase letter eta), which is defined in the following way. A thin layer of fluid is placed between two flat plates. One plate is stationary and the other is made to move, Fig. 10–31. The fluid directly in contact with each plate is held to the surface by the adhesive force between the molecules of the liquid and those of the plate. Thus the upper surface of the fluid moves with the same speed v as the upper plate, whereas the fluid in contact with the stationary plate remains stationary. The stationary layer of fluid retards the flow of the layer just above it, which in turn retards the flow of the next layer, and so on. Thus the velocity varies continuously from 0 to v, as shown. The increase in velocity divided by the distance over which this change is made—equal to v/ℓ—is called the *velocity gradient*. To move the upper plate requires a force, which you can verify by moving a flat plate across a puddle of syrup on a table. For a given fluid, it is found that the force required, F, is proportional to the area of fluid in contact with each plate, A, and to the speed, v, and is inversely proportional to the separation, ℓ, of the plates: $F \propto vA/\ell$. For different fluids, the more viscous the fluid, the greater is the required force. The proportionality constant for this equation is defined as the coefficient of viscosity, η:

$$F = \eta A \frac{v}{\ell}. \quad (10\text{–}8)$$

TABLE 10–3 Coefficients of Viscosity

Fluid (temperature in °C)	Coefficient of Viscosity, η (Pa·s)[†]
Water (0°)	1.8×10^{-3}
(20°)	1.0×10^{-3}
(100°)	0.3×10^{-3}
Whole blood (37°)	$\approx 4 \times 10^{-3}$
Blood plasma (37°)	$\approx 1.5 \times 10^{-3}$
Ethyl alcohol (20°)	1.2×10^{-3}
Engine oil (30°) (SAE 10)	200×10^{-3}
Glycerine (20°)	1500×10^{-3}
Air (20°)	0.018×10^{-3}
Hydrogen (0°)	0.009×10^{-3}
Water vapor (100°)	0.013×10^{-3}

[†] 1 Pa·s = 10 poise (P) = 1000 cP.

Solving for η, we find $\eta = F\ell/vA$. The SI unit for η is N·s/m² = Pa·s (pascal·second). In the cgs system, the unit is dyne·s/cm², which is called a *poise* (P). Viscosities are often given in centipoise ($1 \text{ cP} = 10^{-2} \text{ P} = 10^{-3} \text{ Pa·s}$). Table 10–3 lists the coefficient of viscosity for various fluids. The temperature is also specified, since it has a strong effect; the viscosity of liquids such as motor oil, for example, decreases rapidly as temperature increases.[‡]

*10–12 Flow in Tubes: Poiseuille's Equation, Blood Flow

If a fluid had no viscosity, it could flow through a level tube or pipe without a force being applied. Viscosity acts like a sort of friction (between fluid layers moving at slightly different speeds), so a pressure difference between the ends of a level tube is necessary for the steady flow of any real fluid, be it water or oil in a pipe, or blood in the circulatory system of a human.

[‡]The Society of Automotive Engineers assigns numbers to represent the viscosity of oils: 30-weight (SAE 30) is more viscous than 10-weight. Multigrade oils, such as 20–50, are designed to maintain viscosity as temperature increases; 20–50 means the oil acts like 20-weight when cool and is like 50-weight when it is hot (engine running temperature). In other words, the viscosity does not drop precipitously as the oil warms up, as a simple 20-weight oil would.

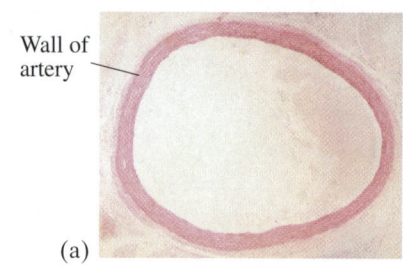

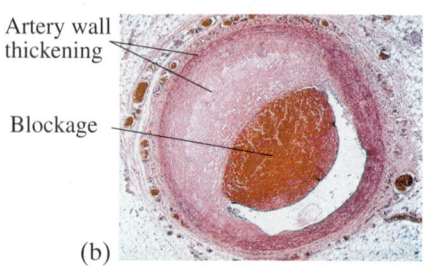

FIGURE 10–32 A cross section of a human artery that (a) is healthy, (b) is partly blocked as a result of arteriosclerosis.

PHYSICS APPLIED
Medicine—blood flow and heart disease

FIGURE 10–33 Spherical water droplets, dew on a blade of grass.

FIGURE 10–34 U-shaped wire apparatus holding a film of liquid to measure surface tension ($\gamma = F/2\ell$).

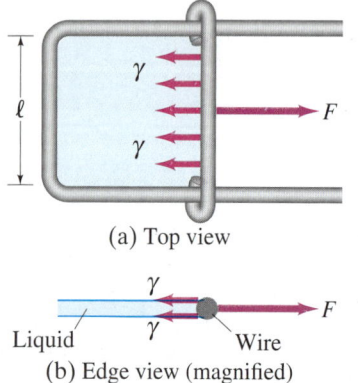

The French scientist J. L. Poiseuille (1799–1869), who was interested in the physics of blood circulation (and after whom the "poise" is named), determined how the variables affect the flow rate of an incompressible fluid undergoing laminar flow in a cylindrical tube. His result, known as **Poiseuille's equation**, is:

$$Q = \frac{\pi R^4 (P_1 - P_2)}{8\eta \ell}, \quad (10\text{–}9)$$

where R is the inside radius of the tube, ℓ is the tube length, $P_1 - P_2$ is the pressure difference between the ends, η is the coefficient of viscosity, and Q is the volume rate of flow (volume of fluid flowing past a given point per unit time which in SI has units of m³/s). Equation 10–9 applies only to laminar (streamline) flow.

Poiseuille's equation tells us that the flow rate Q is directly proportional to the "pressure gradient," $(P_1 - P_2)/\ell$, and it is inversely proportional to the viscosity of the fluid. This is just what we might expect. It may be surprising, however, that Q also depends on the *fourth* power of the tube's radius. This means that for the same pressure gradient, if the tube radius is halved, the flow rate is decreased by a factor of 16! Thus the rate of flow, or alternately the pressure required to maintain a given flow rate, is greatly affected by only a small change in tube radius.

An interesting example of this R^4 dependence is *blood flow* in the human body. Poiseuille's equation is valid only for the streamline flow of an incompressible fluid. So it cannot be precisely accurate for blood whose flow is not without turbulence and that contains blood cells (whose diameter is almost equal to that of a capillary). Nonetheless, Poiseuille's equation does give a reasonable first approximation. Because the radius of arteries is reduced as a result of arteriosclerosis (thickening and hardening of artery walls, Fig. 10–32) and by cholesterol buildup, the pressure gradient must be increased to maintain the same flow rate. If the radius is reduced by half, the heart would have to increase the pressure by a factor of about $2^4 = 16$ in order to maintain the same blood-flow rate. The heart must work much harder under these conditions, but usually cannot maintain the original flow rate. Thus, high blood pressure is an indication both that the heart is working harder and that the blood-flow rate is reduced.

*10–13 Surface Tension and Capillarity

The *surface* of a liquid at rest behaves in an interesting way, almost as if it were a stretched membrane under tension. For example, a drop of water on the end of a dripping faucet, or hanging from a thin branch in the early morning dew (Fig. 10–33), forms into a nearly spherical shape as if it were a tiny balloon filled with water. A steel needle can be made to float on the surface of water even though it is denser than the water. The surface of a liquid acts like it is under tension, and this tension, acting along the surface, arises from the attractive forces between the molecules. This effect is called **surface tension**. More specifically, a quantity called the *surface tension*, γ (the Greek letter gamma), is defined as the force F per unit length ℓ that acts perpendicular to any line or cut in a liquid surface, tending to pull the surface closed:

$$\gamma = \frac{F}{\ell}. \quad (10\text{–}10)$$

To understand this, consider the U-shaped apparatus shown in Fig. 10–34 which encloses a thin film of liquid (such as a liquid soap film). Because of surface tension, a force F is required to pull the movable wire and thus increase the surface area of the liquid. The liquid contained by the wire apparatus is a thin film having both a top and a bottom surface. Hence the total length of the surface being increased is 2ℓ, and the surface tension is $\gamma = F/2\ell$. A delicate apparatus of this type can be used to measure the surface tension of various liquids. The surface tension of water is 0.072 N/m at 20°C. Table 10–4 (next page) gives the values for several substances. Note that temperature has a considerable effect on the surface tension.

280 CHAPTER 10 Fluids

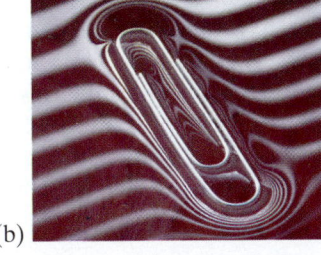

(a) (b)

FIGURE 10–35 (a) Water strider. (b) Paper clip (light coming through window blinds).

Because of surface tension, some insects (Fig. 10–35a) can walk on water, and objects more dense than water, such as a paper clip (Fig. 10–35b), can float on the surface. Figure 10–36a shows how the surface tension can support the weight w of an object. Actually, the object sinks slightly into the fluid, so w is the "effective weight" of that object—its true weight less the buoyant force.

EXAMPLE 10–15 ESTIMATE Insect walks on water. The base of an insect's leg is approximately spherical in shape, with a radius of about 2.0×10^{-5} m. The 0.0030-g mass of the insect is supported equally by its six legs. Estimate the angle θ at which the surface tension force acts (see Fig. 10–36) for an insect on the surface of water. Assume the water temperature is 20°C.

APPROACH Since the insect is in equilibrium, the upward surface tension force is equal to the pull of gravity downward on each leg. We ignore buoyant forces for this estimate.

SOLUTION For each leg, we assume the surface tension force acts all around a circle of radius r, at an angle θ, as shown in Fig. 10–36a. Only the vertical component, $\gamma \cos \theta$, acts to balance the weight mg. We set the length ℓ in Eq. 10–10 equal to the circumference of the circle, $\ell \approx 2\pi r$. Then the net upward force due to surface tension is $F_y \approx (\gamma \cos \theta) \ell \approx 2\pi r \gamma \cos \theta$. We set this surface tension force equal to one-sixth the weight of the insect since it has six legs:

$$2\pi r \gamma \cos \theta \approx \tfrac{1}{6} mg$$
$$(6.28)(2.0 \times 10^{-5}\,\text{m})(0.072\,\text{N/m}) \cos \theta \approx \tfrac{1}{6}(3.0 \times 10^{-6}\,\text{kg})(9.8\,\text{m/s}^2)$$
$$\cos \theta \approx 0.54.$$

So $\theta \approx 57°$.

NOTE If $\cos \theta$ had come out greater than 1, the surface tension would not have been great enough to support the insect's weight. If the insect is very light, it will sink less into the water and θ (Fig. 10–36a) will be larger than calculated above.

NOTE Our estimate ignored the buoyant force and ignored any difference between the radius of the insect's "foot" and the radius of the surface depression.

Soaps and detergents lower the surface tension of water. This is desirable for washing and cleaning since the high surface tension of pure water prevents it from penetrating easily between the fibers of material and into tiny crevices. Substances that reduce the surface tension of a liquid are called *surfactants*.

*Capillarity

Surface tension plays a role in another interesting phenomenon, *capillarity*. It is a common observation that water in a glass container rises up slightly where it touches the glass, Fig. 10–37a. The water is said to "wet" the glass. Mercury, on the other hand, is depressed when it touches the glass, Fig. 10–37b; the mercury does not wet the glass. Whether a liquid wets a solid surface is determined by the relative strength of the cohesive forces between the molecules of the liquid compared to the adhesive forces between the molecules of the liquid and those of the container. **Cohesion** refers to the force between molecules of the same type, whereas **adhesion** refers to the force between molecules of different types. Water wets glass because the water molecules are more strongly attracted to the glass molecules than they are to other water molecules. The opposite is true for mercury: the cohesive forces are stronger than the adhesive forces.

TABLE 10–4 Surface Tension of Some Substances

Substance	Surface Tension (N/m)
Mercury (20°C)	0.44
Blood, whole (37°C)	0.058
Blood, plasma (37°C)	0.073
Alcohol, ethyl (20°C)	0.023
Water (0°C)	0.076
(20°C)	0.072
(100°C)	0.059
Benzene (20°C)	0.029
Soap solution (20°C)	≈ 0.025
Oxygen (−193°C)	0.016

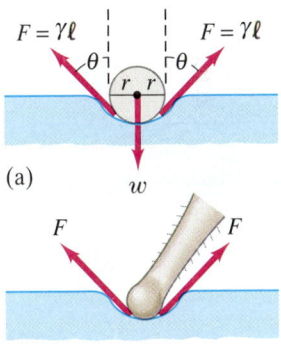

FIGURE 10–36 Surface tension acting on (a) a sphere, and (b) an insect leg. Example 10–15.

PHYSICS APPLIED
Soaps and detergents

FIGURE 10–37 (a) Water "wets" the surface of glass, whereas (b) mercury does not "wet" the glass.

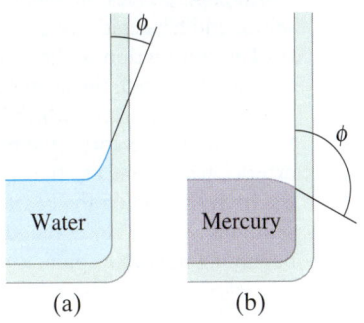

In tubes having very small diameters, liquids are observed to rise or fall relative to the level of the surrounding liquid. This phenomenon is called **capillarity**, and such thin tubes are called **capillaries**. Whether the liquid rises or falls (Fig. 10–38) depends on the relative strengths of the adhesive and cohesive forces. Thus water rises in a glass tube, whereas mercury falls. The actual amount of rise (or fall) depends on the surface tension—which is what keeps the liquid surface from breaking apart.

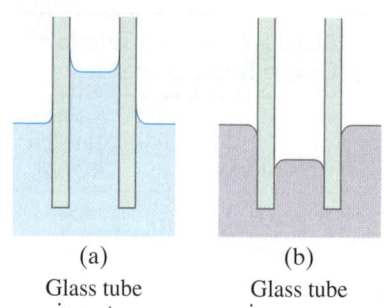

(a) Glass tube in water   (b) Glass tube in mercury

FIGURE 10–38 Capillarity.

*10–14 Pumps, and the Heart

We conclude this Chapter with a brief discussion of pumps, including the heart. Pumps can be classified into categories according to their function. A *vacuum pump* is designed to reduce the pressure (usually of air) in a given vessel. A *force pump*, on the other hand, is a pump that is intended to increase the pressure—for example, to lift a liquid (such as water from a well) or to push a fluid through a pipe. Figure 10–39 illustrates the principle behind a simple reciprocating pump. It could be a vacuum pump, in which case the intake is connected to the vessel to be evacuated. A similar mechanism is used in some force pumps, and in this case the fluid is forced under increased pressure through the outlet.

Another type of pump is the centrifugal pump, shown in Fig. 10–40. It, or any force pump, can be used as a *circulating pump*—that is, to circulate a fluid around a closed path, such as the cooling water or lubricating oil in an automobile.

FIGURE 10–39 One kind of pump (reciprocating type): the intake valve opens and air (or fluid that is being pumped) fills the empty space when the piston moves to the left. When the piston moves to the right (not shown), the outlet valve opens and fluid is forced out.

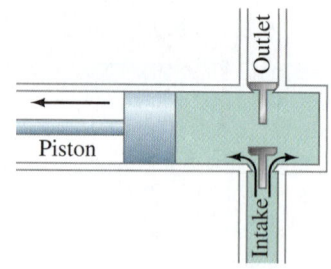

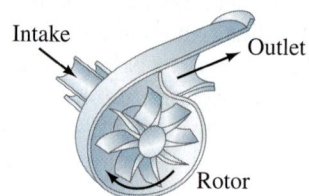

FIGURE 10–40 Centrifugal pump: the rotating blades force fluid through the outlet pipe; this kind of pump is used in vacuum cleaners and as a water pump in automobiles.

PHYSICS APPLIED
Heart as a pump

The heart of a human (and of other animals as well) is essentially a circulating pump. The action of a human heart is shown in Fig. 10–41. There are actually two separate paths for blood flow. The longer path takes blood to the parts of the body, via the arteries, bringing oxygen to body tissues and picking up carbon dioxide, which it carries back to the heart via veins. This blood is then pumped to the lungs (the second path), where the carbon dioxide is released and oxygen is taken up. The oxygen-laden blood is returned to the heart, where it is again pumped to the tissues of the body.

FIGURE 10–41 Pumping human heart. (a) In the diastole phase, the heart relaxes between beats. Blood moves into the heart; both atria fill rapidly. (b) When the atria contract, the systole or pumping phase begins. The contraction pushes the blood through the mitral and tricuspid valves into the ventricles. (c) The contraction of the ventricles forces the blood through the semilunar valves into the pulmonary artery, which leads to the lungs, and to the aorta (the body's largest artery), which leads to the arteries serving all the body. (d) When the heart relaxes, the semilunar valves close; blood fills the atria, beginning the cycle again.

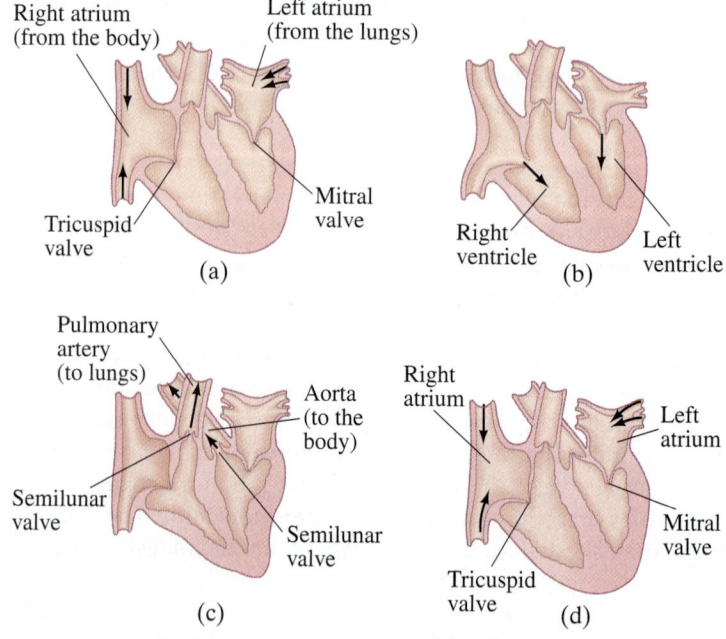

Blood pressure is measured using one of the types of gauge mentioned earlier (Section 10–6), and it is usually calibrated in mm-Hg. The gauge is attached to a closed, air-filled cuff that is wrapped around the upper arm at the level of the heart, Fig. 10–42. Two values of blood pressure are measured: the maximum pressure when the heart is pumping, called *systolic pressure*; and the pressure when the heart is in the resting part of the cycle, called *diastolic pressure*. Initially, the air pressure in the cuff is increased high above the systolic pressure by a pump, compressing the main (brachial) artery in the arm and briefly cutting off the flow of blood. The air pressure is then reduced slowly until blood again begins to flow into the arm; it can be detected by listening with a stethoscope to the characteristic tapping sound[†] of the blood returning to the forearm. At this point, systolic pressure is just equal to the air pressure in the arm cuff which can be read off the gauge. The air pressure is subsequently reduced further, and the tapping sound disappears when blood at low pressure can enter the artery. At this point, the gauge indicates the diastolic pressure. Normal systolic pressure is around 120 mm-Hg, whereas normal diastolic pressure is around 70 or 80 mm-Hg. Blood pressure is reported in the form 120/70.

PHYSICS APPLIED
Blood pressure

FIGURE 10–42 Device for measuring blood pressure.

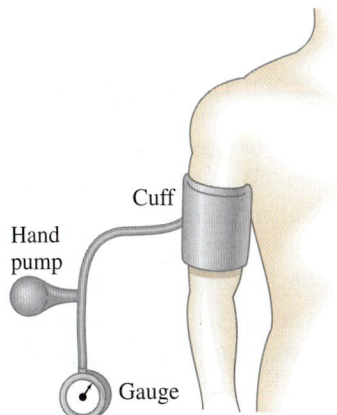

[†]When the blood starts flowing through the constriction caused by the tight cuff, its velocity is high and the flow is turbulent. It is the turbulence that causes the tapping sound.

Summary

The three common phases of matter are **solid**, **liquid**, and **gas**. Liquids and gases are collectively called **fluids**, meaning they have the ability to flow. The **density** of a material is defined as its mass per unit volume:

$$\rho = \frac{m}{V}. \quad (10\text{–}1)$$

Specific gravity (SG) is the ratio of the density of the material to the density of water (at 4°C).

Pressure is defined as force per unit area:

$$P = \frac{F}{A}. \quad (10\text{–}2)$$

The pressure P at a depth h in a liquid of constant density ρ, due to the weight of the liquid, is given by

$$P = \rho g h, \quad (10\text{–}3a)$$

where g is the acceleration due to gravity.

Pascal's principle says that an external pressure applied to a confined fluid is transmitted throughout the fluid.

Pressure is measured using a **manometer** or other type of gauge. A **barometer** is used to measure atmospheric pressure. Standard **atmospheric pressure** (average at sea level) is $1.013 \times 10^5 \text{ N/m}^2$. **Gauge pressure** is the total (absolute) pressure minus atmospheric pressure.

Archimedes' principle states that an object submerged wholly or partially in a fluid is buoyed up by a force equal to the weight of fluid it displaces ($F_B = m_F g = \rho_F V_{\text{displ}} g$).

Fluid flow can be characterized either as **streamline** (also called **laminar**), in which the layers of fluid move smoothly and regularly along paths called **streamlines**, or as **turbulent**, in which case the flow is not smooth and regular but is characterized by irregularly shaped whirlpools.

Fluid flow rate is the mass or volume of fluid that passes a given point per unit time. The **equation of continuity** states that for an incompressible fluid flowing in an enclosed tube, the product of the velocity of flow and the cross-sectional area of the tube remains constant:

$$Av = \text{constant}. \quad (10\text{–}4)$$

Bernoulli's principle tells us that where the velocity of a fluid is high, the pressure in it is low, and where the velocity is low, the pressure is high. For steady laminar flow of an incompressible and nonviscous fluid, **Bernoulli's equation**, which is based on the law of conservation of energy, is

$$P_2 + \tfrac{1}{2}\rho v_2^2 + \rho g y_2 = P_1 + \tfrac{1}{2}\rho v_1^2 + \rho g y_1, \quad (10\text{–}5)$$

for two points along the flow.

[***Viscosity** refers to friction within a fluid and is essentially a frictional force between adjacent layers of fluid as they move past one another.]

[*Liquid surfaces hold together as if under tension (**surface tension**), allowing drops to form and objects like needles and insects to stay on the surface.]

Questions

1. If one material has a higher density than another, must the molecules of the first be heavier than those of the second? Explain.
2. Consider what happens when you push both a pin and the blunt end of a pen against your skin with the same force. Decide what determines whether your skin is cut—the net force applied to it or the pressure.
3. A small amount of water is boiled in a 1-gallon metal can. The can is removed from the heat and the lid put on. As the can cools, it collapses and looks crushed. Explain.
4. An ice cube floats in a glass of water filled to the brim. What can you say about the density of ice? As the ice melts, will the water overflow? Explain.
5. Will an ice cube float in a glass of alcohol? Why or why not?

6. A submerged can of Coke® will sink, but a can of Diet Coke® will float. (Try it!) Explain.

7. Why don't ships made of iron sink?

8. A barge filled high with sand approaches a low bridge over the river and cannot quite pass under it. Should sand be added to, or removed from, the barge? [*Hint*: Consider Archimedes' principle.]

9. Explain why helium weather balloons, which are used to measure atmospheric conditions at high altitudes, are normally released while filled to only 10–20% of their maximum volume.

10. Will an empty balloon have precisely the same apparent weight on a scale as a balloon filled with air? Explain.

11. Why do you float higher in salt water than in fresh water?

12. Why does the stream of water from a faucet become narrower as it falls (Fig. 10–43)?

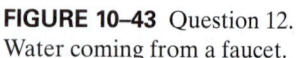

FIGURE 10–43 Question 12. Water coming from a faucet.

13. Children are told to avoid standing too close to a rapidly moving train because they might get sucked under it. Is this possible? Explain.

14. A tall Styrofoam cup is filled with water. Two holes are punched in the cup near the bottom, and water begins rushing out. If the cup is dropped so it falls freely, will the water continue to flow from the holes? Explain.

15. Why do airplanes normally take off into the wind?

16. Two ships moving in parallel paths close to one another risk colliding. Why?

17. If you dangle two pieces of paper vertically, a few inches apart (Fig. 10–44), and blow between them, how do you think the papers will move? Try it and see. Explain.

FIGURE 10–44 Question 17.

18. Why does the canvas top of a convertible bulge out when the car is traveling at high speed? [*Hint*: The windshield deflects air upward, pushing streamlines closer together.]

19. Roofs of houses are sometimes "blown" off (or are they pushed off?) during a tornado or hurricane. Explain using Bernoulli's principle.

20. Explain how the tube in Fig. 10–45, known as a **siphon**, can transfer liquid from one container to a lower one even though the liquid must flow uphill for part of its journey. (Note that the tube must be filled with liquid to start with.)

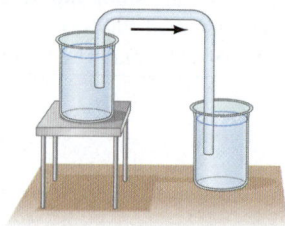

FIGURE 10–45 Question 20. A siphon.

*21. When blood pressure is measured, why must the arm cuff be held at the level of the heart?

MisConceptual Questions

1. You hold a piece of wood in one hand and a piece of iron in the other. Both pieces have the same volume, and you hold them fully under water at the same depth. At the moment you let go of them, which one experiences the greater buoyancy force?
   (a) The piece of wood.
   (b) The piece of iron.
   (c) They experience the same buoyancy force.
   (d) More information is needed.

2. Three containers are filled with water to the same height and have the same surface area at the base, but the total weight of water is different for each (Fig. 10–46). In which container does the water exert the greatest force on the bottom of the container?
   (a) Container A.
   (b) Container B.
   (c) Container C.
   (d) All three are equal.

FIGURE 10–46 MisConceptual Question 2.

3. Beaker A is filled to the brim with water. Beaker B is the same size and contains a small block of wood which floats when the beaker is filled with water to the brim. Which beaker weighs more?
   (a) Beaker A.
   (b) Beaker B.
   (c) The same for both.

4. Why does an ocean liner float?
   (a) It is made of steel, which floats.
   (b) Its very big size changes the way water supports it.
   (c) It is held up in the water by large Styrofoam compartments.
   (d) The average density of the ocean liner is less than that of seawater.
   (e) Remember the *Titanic*—ocean liners do not float.

5. A rowboat floats in a swimming pool, and the level of the water at the edge of the pool is marked. Consider the following situations. (i) The boat is removed from the water. (ii) The boat in the water holds an iron anchor which is removed from the boat and placed on the shore. For each situation, the level of the water will
   (a) rise.   (b) fall.   (c) stay the same.

6. You put two ice cubes in a glass and fill the glass to the rim with water. As the ice melts, the water level
   (a) drops below the rim.
   (b) rises and water spills out of the glass.
   (c) remains the same.
   (d) drops at first, then rises until a little water spills out.

7. Hot air is less dense than cold air. Could a hot-air balloon be flown on the Moon, where there is no atmosphere?
   (a) No, there is no cold air to displace, so no buoyancy force would exist.
   (b) Yes, warm air always rises, especially in a weak gravitational field like that of the Moon.
   (c) Yes, but the balloon would have to be filled with helium instead of hot air.

8. An object that can float in both water and in oil (whose density is less than that of water) experiences a buoyant force that is
   (a) greater when it is floating in oil than when floating in water.
   (b) greater when it is floating in water than when floating in oil.
   (c) the same when it is floating in water or in oil.

9. As water flows from a low elevation to a higher elevation through a pipe that changes in diameter,
   (a) the water pressure will increase.
   (b) the water pressure will decrease.
   (c) the water pressure will stay the same.
   (d) Need more information to determine how the water pressure changes.

10. Water flows in a horizontal pipe that is narrow but then widens and the speed of the water becomes less. The pressure in the water moving in the pipe is
    (a) greater in the wide part.
    (b) greater in the narrow part.
    (c) the same in both parts.
    (d) greater where the speed is higher.
    (e) greater where the speed is lower.

11. When a baseball curves to the right (a curveball), air is flowing
    (a) faster over the left side than over the right side.
    (b) faster over the right side than over the left side.
    (c) faster over the top than underneath.
    (d) at the same speed all around the baseball, but the ball curves as a result of the way the wind is blowing on the field.

12. How is the smoke drawn up a chimney affected when a wind is blowing outside?
    (a) Smoke rises more rapidly in the chimney.
    (b) Smoke rises more slowly in the chimney.
    (c) Smoke is forced back down the chimney.
    (d) Smoke is unaffected.

For assigned homework and other learning materials, go to the MasteringPhysics website.

Problems

10–2 Density and Specific Gravity

1. (I) The approximate volume of the granite monolith known as El Capitan in Yosemite National Park (Fig. 10–47) is about 10^8 m^3. What is its approximate mass?

FIGURE 10–47 Problem 1.

2. (I) What is the approximate mass of air in a living room 5.6 m × 3.6 m × 2.4 m?

3. (I) If you tried to smuggle gold bricks by filling your backpack, whose dimensions are 54 cm × 31 cm × 22 cm, what would its mass be?

4. (I) State your mass and then estimate your volume. [*Hint*: Because you can swim on or just under the surface of the water in a swimming pool, you have a pretty good idea of your density.]

5. (II) A bottle has a mass of 35.00 g when empty and 98.44 g when filled with water. When filled with another fluid, the mass is 89.22 g. What is the specific gravity of this other fluid?

6. (II) If 4.0 L of antifreeze solution (specific gravity = 0.80) is added to 5.0 L of water to make a 9.0-L mixture, what is the specific gravity of the mixture?

7. (III) The Earth is not a uniform sphere, but has regions of varying density. Consider a simple model of the Earth divided into three regions—inner core, outer core, and mantle. Each region is taken to have a unique constant density (the average density of that region in the real Earth):

Region	Radius (km)	Density (kg/m^3)
Inner Core	0–1220	13,000
Outer Core	1220–3480	11,100
Mantle	3480–6380	4400

(a) Use this model to predict the average density of the entire Earth. (b) If the radius of the Earth is 6380 km and its mass is 5.98×10^{24} kg, determine the actual average density of the Earth and compare it (as a percent difference) with the one you determined in (a).

Problems **285**

10–3 to 10–6 Pressure; Pascal's Principle

8. (I) Estimate the pressure needed to raise a column of water to the same height as a 46-m-tall pine tree.

9. (I) Estimate the pressure exerted on a floor by (a) one pointed heel of area = 0.45 cm^2, and (b) one wide heel of area 16 cm^2, Fig. 10–48. The person wearing the shoes has a mass of 56 kg.

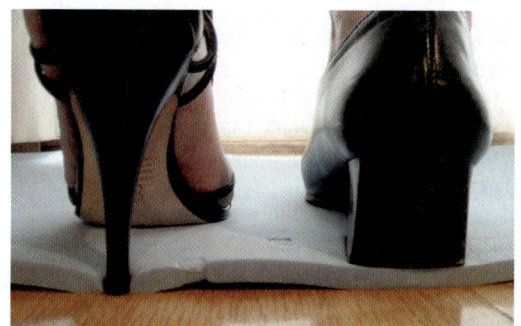

FIGURE 10–48 Problem 9.

10. (I) What is the difference in blood pressure (mm-Hg) between the top of the head and bottom of the feet of a 1.75-m-tall person standing vertically?

11. (I) (a) Calculate the total force of the atmosphere acting on the top of a table that measures $1.7 \text{ m} \times 2.6 \text{ m}$. (b) What is the total force acting upward on the underside of the table?

12. (II) How high would the level be in an alcohol barometer at normal atmospheric pressure?

13. (II) In a movie, Tarzan evades his captors by hiding under water for many minutes while breathing through a long, thin reed. Assuming the maximum pressure difference his lungs can manage and still breathe is -85 mm-Hg, calculate the deepest he could have been.

14. (II) The maximum gauge pressure in a hydraulic lift is 17.0 atm. What is the largest-size vehicle (kg) it can lift if the diameter of the output line is 25.5 cm?

15. (II) The gauge pressure in each of the four tires of an automobile is 240 kPa. If each tire has a "footprint" of 190 cm^2 (area touching the ground), estimate the mass of the car.

16. (II) (a) Determine the total force and the absolute pressure on the bottom of a swimming pool 28.0 m by 8.5 m whose uniform depth is 1.8 m. (b) What will be the pressure against the *side* of the pool near the bottom?

17. (II) A house at the bottom of a hill is fed by a full tank of water 6.0 m deep and connected to the house by a pipe that is 75 m long at an angle of 61° from the horizontal (Fig. 10–49). (a) Determine the water gauge pressure at the house. (b) How high could the water shoot if it came vertically out of a broken pipe in front of the house?

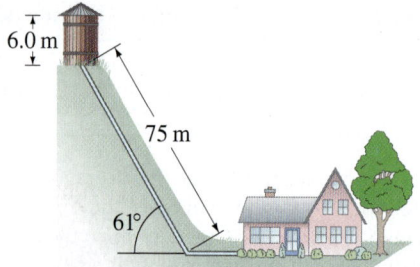

FIGURE 10–49 Problem 17.

18. (II) Water and then oil (which don't mix) are poured into a U-shaped tube, open at both ends. They come to equilibrium as shown in Fig. 10–50. What is the density of the oil? [*Hint*: Pressures at points a and b are equal. Why?]

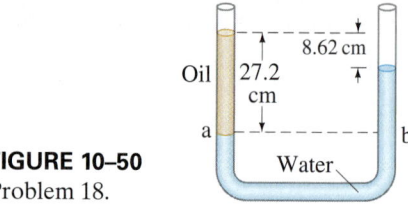

FIGURE 10–50 Problem 18.

19. (II) How high would the atmosphere extend if it were of uniform density throughout, equal to half the present density at sea level?

20. (II) Determine the minimum gauge pressure needed in the water pipe leading into a building if water is to come out of a faucet on the fourteenth floor, 44 m above that pipe.

21. (II) A **hydraulic press** for compacting powdered samples has a large cylinder which is 10.0 cm in diameter, and a small cylinder with a diameter of 2.0 cm (Fig. 10–51). A lever is attached to the small cylinder as shown. The sample, which is placed on the large cylinder, has an area of 4.0 cm^2. What is the pressure on the sample if 320 N is applied to the lever?

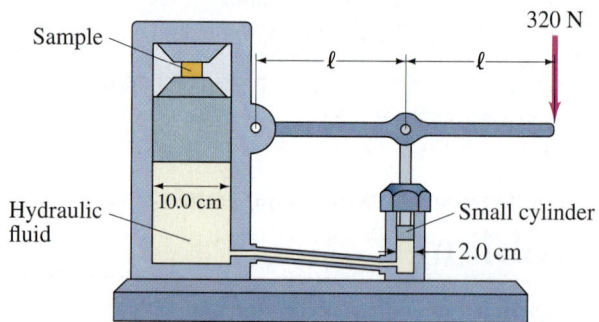

FIGURE 10–51 Problem 21.

22. (II) An open-tube mercury manometer is used to measure the pressure in an oxygen tank. When the atmospheric pressure is 1040 mbar, what is the absolute pressure (in Pa) in the tank if the height of the mercury in the open tube is (a) 18.5 cm higher, (b) 5.6 cm lower, than the mercury in the tube connected to the tank? See Fig. 10–7a.

10–7 Buoyancy and Archimedes' Principle

23. (II) What fraction of a piece of iron will be submerged when it floats in mercury?

24. (II) A geologist finds that a Moon rock whose mass is 9.28 kg has an apparent mass of 6.18 kg when submerged in water. What is the density of the rock?

25. (II) A crane lifts the 18,000-kg steel hull of a sunken ship out of the water. Determine (a) the tension in the crane's cable when the hull is fully submerged in the water, and (b) the tension when the hull is completely out of the water.

26. (II) A spherical balloon has a radius of 7.15 m and is filled with helium. How large a cargo can it lift, assuming that the skin and structure of the balloon have a mass of 930 kg? Neglect the buoyant force on the cargo volume itself.

27. (II) What is the likely identity of a metal (see Table 10–1) if a sample has a mass of 63.5 g when measured in air and an apparent mass of 55.4 g when submerged in water?

28. (II) Calculate the true mass (in vacuum) of a piece of aluminum whose apparent mass is 4.0000 kg when weighed in air.

29. (II) Because gasoline is less dense than water, drums containing gasoline will float in water. Suppose a 210-L steel drum is completely full of gasoline. What total volume of steel can be used in making the drum if the gasoline-filled drum is to float in fresh water?

30. (II) A scuba diver and her gear displace a volume of 69.6 L and have a total mass of 72.8 kg. (a) What is the buoyant force on the diver in seawater? (b) Will the diver sink or float?

31. (II) The specific gravity of ice is 0.917, whereas that of seawater is 1.025. What percent of an iceberg is above the surface of the water?

32. (II) Archimedes' principle can be used to determine the specific gravity of a solid using a known liquid (Example 10–8). The reverse can be done as well. (a) As an example, a 3.80-kg aluminum ball has an apparent mass of 2.10 kg when submerged in a particular liquid: calculate the density of the liquid. (b) Determine a formula for finding the density of a liquid using this procedure.

33. (II) A 32-kg child decides to make a raft out of empty 1.0-L soda bottles and duct tape. Neglecting the mass of the duct tape and plastic in the bottles, what minimum number of soda bottles will the child need to be able stay dry on the raft?

34. (II) An undersea research chamber is spherical with an external diameter of 5.20 m. The mass of the chamber, when occupied, is 74,400 kg. It is anchored to the sea bottom by a cable. What is (a) the buoyant force on the chamber, and (b) the tension in the cable?

35. (II) A 0.48-kg piece of wood floats in water but is found to sink in alcohol (SG = 0.79), in which it has an apparent mass of 0.047 kg. What is the SG of the wood?

36. (II) A two-component model used to determine percent body fat in a human body assumes that a fraction f (< 1) of the body's total mass m is composed of fat with a density of 0.90 g/cm^3, and that the remaining mass of the body is composed of fat-free tissue with a density of 1.10 g/cm^3. If the specific gravity of the entire body's density is X, show that the percent body fat ($= f \times 100$) is given by

$$\% \text{ Body fat} = \frac{495}{X} - 450.$$

37. (II) On dry land, an athlete weighs 70.2 kg. The same athlete, when submerged in a swimming pool and hanging from a scale, has an "apparent weight" of 3.4 kg. Using Example 10–8 as a guide, (a) find the total volume V of the submerged athlete. (b) Assume that when submerged, the athlete's body contains a residual volume $V_R = 1.3 \times 10^{-3} \text{ m}^3$ of air (mainly in the lungs). Taking $V - V_R$ to be the actual volume of the athlete's body, find the body's specific gravity, SG. (c) What is the athlete's percent body fat assuming it is given by the formula $(495/\text{SG}) - 450$?

38. (II) How many helium-filled balloons would it take to lift a person? Assume the person has a mass of 72 kg and that each helium-filled balloon is spherical with a diameter of 33 cm.

39. (III) A scuba tank, when fully submerged, displaces 15.7 L of seawater. The tank itself has a mass of 14.0 kg and, when "full," contains 3.00 kg of air. Assuming only its weight and the buoyant force act on the tank, determine the net force (magnitude and direction) on the fully submerged tank at the beginning of a dive (when it is full of air) and at the end of a dive (when it no longer contains any air).

40. (III) A 3.65-kg block of wood (SG = 0.50) floats on water. What minimum mass of lead, hung from the wood by a string, will cause the block to sink?

10–8 to 10–10 Fluid Flow, Bernoulli's Equation

41. (I) A 12-cm-radius air duct is used to replenish the air of a room 8.2 m × 5.0 m × 3.5 m every 12 min. How fast does the air flow in the duct?

42. (I) Calculate the average speed of blood flow in the major arteries of the body, which have a total cross-sectional area of about 2.0 cm^2. Use the data of Example 10–12.

43. (I) How fast does water flow from a hole at the bottom of a very wide, 4.7-m-deep storage tank filled with water? Ignore viscosity.

44. (I) Show that Bernoulli's equation reduces to the hydrostatic variation of pressure with depth (Eq. 10–3b) when there is no flow ($v_1 = v_2 = 0$).

45. (II) What is the volume rate of flow of water from a 1.85-cm-diameter faucet if the pressure head is 12.0 m?

46. (II) A fish tank has dimensions 36 cm wide by 1.0 m long by 0.60 m high. If the filter should process all the water in the tank once every 3.0 h, what should the flow speed be in the 3.0-cm-diameter input tube for the filter?

47. (II) What gauge pressure in the water pipes is necessary if a fire hose is to spray water to a height of 16 m?

48. (II) A $\frac{5}{8}$-in. (inside) diameter garden hose is used to fill a round swimming pool 6.1 m in diameter. How long will it take to fill the pool to a depth of 1.4 m if water flows from the hose at a speed of 0.40 m/s?

49. (II) A 180-km/h wind blowing over the flat roof of a house causes the roof to lift off the house. If the house is 6.2 m × 12.4 m in size, estimate the weight of the roof. Assume the roof is not nailed down.

50. (II) A 6.0-cm-diameter horizontal pipe gradually narrows to 4.5 cm. When water flows through this pipe at a certain rate, the gauge pressure in these two sections is 33.5 kPa and 22.6 kPa, respectively. What is the volume rate of flow?

51. (II) Estimate the air pressure inside a category 5 hurricane, where the wind speed is 300 km/h (Fig. 10–52).

FIGURE 10–52 Problem 51.

52. (II) What is the lift (in newtons) due to Bernoulli's principle on a wing of area 88 m² if the air passes over the top and bottom surfaces at speeds of 280 m/s and 150 m/s, respectively?

53. (II) Water at a gauge pressure of 3.8 atm at street level flows into an office building at a speed of 0.78 m/s through a pipe 5.0 cm in diameter. The pipe tapers down to 2.8 cm in diameter by the top floor, 16 m above (Fig. 10–53), where the faucet has been left open. Calculate the flow velocity and the gauge pressure in the pipe on the top floor. Assume no branch pipes and ignore viscosity.

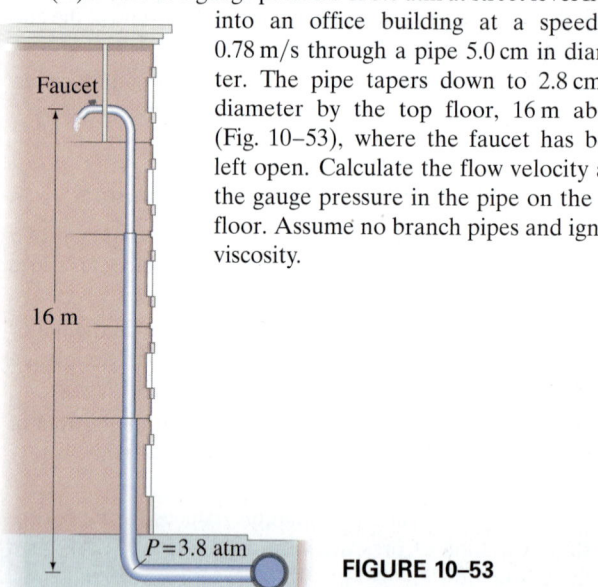

FIGURE 10–53
Problem 53.

54. (II) Show that the power needed to drive a fluid through a pipe with uniform cross-section is equal to the volume rate of flow, Q, times the pressure difference, $P_1 - P_2$. Ignore viscosity.

55. (III) In Fig. 10–54, take into account the speed of the top surface of the tank and show that the speed of fluid leaving an opening near the bottom is

$$v_1 = \sqrt{\frac{2gh}{(1 - A_1^2/A_2^2)}},$$

where $h = y_2 - y_1$, and A_1 and A_2 are the areas of the opening and of the top surface, respectively. Assume $A_1 \ll A_2$ so that the flow remains nearly steady and laminar.

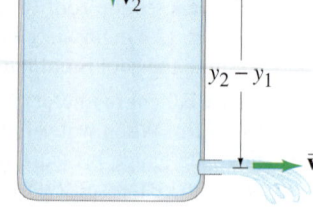

FIGURE 10–54
Problem 55.

56. (III) (a) Show that the flow speed measured by a venturi meter (see Fig. 10–29) is given by the relation

$$v_1 = A_2 \sqrt{\frac{2(P_1 - P_2)}{\rho(A_1^2 - A_2^2)}}.$$

(b) A venturi meter is measuring the flow of water; it has a main diameter of 3.5 cm tapering down to a throat diameter of 1.0 cm. If the pressure difference is measured to be 18 mm-Hg, what is the speed of the water entering the venturi throat?

57. (III) A fire hose exerts a force on the person holding it. This is because the water accelerates as it goes from the hose through the nozzle. How much force is required to hold a 7.0-cm-diameter hose delivering 420 L/min through a 0.75-cm-diameter nozzle?

*10–11 Viscosity

*__58.__ (II) A viscometer consists of two concentric cylinders, 10.20 cm and 10.60 cm in diameter. A liquid fills the space between them to a depth of 12.0 cm. The outer cylinder is fixed, and a torque of 0.024 m·N keeps the inner cylinder turning at a steady rotational speed of 57 rev/min. What is the viscosity of the liquid?

*10–12 Flow in Tubes; Poiseuille's Equation

*__59.__ (I) Engine oil (assume SAE 10, Table 10–3) passes through a fine 1.80-mm-diameter tube that is 10.2 cm long. What pressure difference is needed to maintain a flow rate of 6.2 mL/min?

*__60.__ (I) A gardener feels it is taking too long to water a garden with a $\frac{3}{8}$-in.-diameter hose. By what factor will the time be cut using a $\frac{5}{8}$-in.-diameter hose instead? Assume nothing else is changed.

*__61.__ (II) What diameter must a 15.5-m-long air duct have if the ventilation and heating system is to replenish the air in a room 8.0 m × 14.0 m × 4.0 m every 15.0 min? Assume the pump can exert a gauge pressure of 0.710×10^{-3} atm.

*__62.__ (II) What must be the pressure difference between the two ends of a 1.6-km section of pipe, 29 cm in diameter, if it is to transport oil ($\rho = 950$ kg/m³, $\eta = 0.20$ Pa·s) at a rate of 650 cm³/s?

*__63.__ (II) Poiseuille's equation does not hold if the flow velocity is high enough that turbulence sets in. The onset of turbulence occurs when the **Reynolds number**, Re, exceeds approximately 2000. Re is defined as

$$Re = \frac{2\bar{v}r\rho}{\eta},$$

where $\bar{v}$ is the average speed of the fluid, ρ is its density, η is its viscosity, and r is the radius of the tube in which the fluid is flowing. (a) Determine if blood flow through the aorta is laminar or turbulent when the average speed of blood in the aorta ($r = 0.80$ cm) during the resting part of the heart's cycle is about 35 cm/s. (b) During exercise, the blood-flow speed approximately doubles. Calculate the Reynolds number in this case, and determine if the flow is laminar or turbulent.

*__64.__ (II) Assuming a constant pressure gradient, if blood flow is reduced by 65%, by what factor is the radius of a blood vessel decreased?

*__65.__ (II) Calculate the pressure drop per cm along the aorta using the data of Example 10–12 and Table 10–3.

*__66.__ (III) A patient is to be given a blood transfusion. The blood is to flow through a tube from a raised bottle to a needle inserted in the vein (Fig. 10–55). The inside diameter of the 25-mm-long needle is 0.80 mm, and the required flow rate is 2.0 cm³ of blood per minute. How high h should the bottle be placed above the needle? Obtain ρ and η from the Tables. Assume the blood pressure is 78 torr above atmospheric pressure.

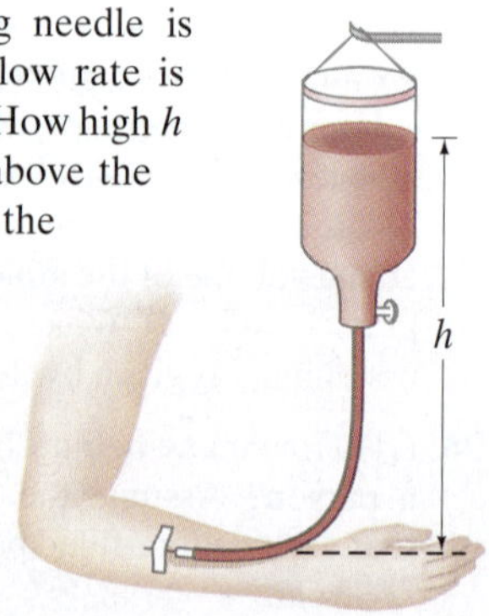

FIGURE 10–55
Problems 66 and 74.

*10–13 Surface Tension and Capillarity

*67. (I) If the force F needed to move the wire in Fig. 10–34 is 3.4×10^{-3} N, calculate the surface tension γ of the enclosed fluid. Assume $\ell = 0.070$ m.

*68. (I) Calculate the force needed to move the wire in Fig. 10–34 if it holds a soapy solution (Table 10–4) and the wire is 21.5 cm long.

*69. (II) The surface tension of a liquid can be determined by measuring the force F needed to just lift a circular platinum ring of radius r from the surface of the liquid. (a) Find a formula for γ in terms of F and r. (b) At 30°C, if $F = 6.20 \times 10^{-3}$ N and $r = 2.9$ cm, calculate γ for the tested liquid.

*70. (II) If the base of an insect's leg has a radius of about 3.0×10^{-5} m and the insect's mass is 0.016 g, would you expect the six-legged insect to remain on top of the water? Why or why not?

*71. (III) Estimate the diameter of a steel needle that can just barely remain on top of water due to surface tension.

*10–14 Pumps; the Heart

*72. (II) A physician judges the health of a heart by measuring the pressure with which it pumps blood. If the physician mistakenly attaches the pressurized cuff around a standing patient's calf (about 1 m below the heart) instead of the arm (Fig. 10–42), what error (in Pa) would be introduced in the heart's blood pressure measurement?

General Problems

73. A 3.2-N force is applied to the plunger of a hypodermic needle. If the diameter of the plunger is 1.3 cm and that of the needle is 0.20 mm, (a) with what force does the fluid leave the needle? (b) What force on the plunger would be needed to push fluid into a vein where the gauge pressure is 75 mm-Hg? Answer for the instant just before the fluid starts to move.

74. Intravenous transfusions are often made under gravity, as shown in Fig. 10–55. Assuming the fluid has a density of 1.00 g/cm³, at what height h should the bottle be placed so the liquid pressure is (a) 52 mm-Hg, and (b) 680 mm-H$_2$O? (c) If the blood pressure is 75 mm-Hg above atmospheric pressure, how high should the bottle be placed so that the fluid just barely enters the vein?

75. A beaker of water rests on an electronic balance that reads 975.0 g. A 2.6-cm-diameter solid copper ball attached to a string is submerged in the water, but does not touch the bottom. What are the tension in the string and the new balance reading?

76. Estimate the difference in air pressure between the top and the bottom of the Empire State Building in New York City. It is 380 m tall and is located at sea level. Express as a fraction of atmospheric pressure at sea level.

77. A hydraulic lift is used to jack a 960-kg car 42 cm off the floor. The diameter of the output piston is 18 cm, and the input force is 380 N. (a) What is the area of the input piston? (b) What is the work done in lifting the car 42 cm? (c) If the input piston moves 13 cm in each stroke, how high does the car move up for each stroke? (d) How many strokes are required to jack the car up 42 cm? (e) Show that energy is conserved.

78. When you ascend or descend a great deal when driving in a car, your ears "pop," which means that the pressure behind the eardrum is being equalized to that outside. If this did not happen, what would be the approximate force on an eardrum of area 0.20 cm² if a change in altitude of 1250 m takes place?

79. Giraffes are a wonder of cardiovascular engineering. Calculate the difference in pressure (in atmospheres) that the blood vessels in a giraffe's head must accommodate as the head is lowered from a full upright position to ground level for a drink. The height of an average giraffe is about 6 m.

80. How high should the pressure head be if water is to come from a faucet at a speed of 9.2 m/s? Ignore viscosity.

81. Suppose a person can reduce the pressure in his lungs to −75 mm-Hg gauge pressure. How high can water then be "sucked" up a straw?

82. A bicycle pump is used to inflate a tire. The initial tire (gauge) pressure is 210 kPa (30 psi). At the end of the pumping process, the final pressure is 310 kPa (45 psi). If the diameter of the plunger in the cylinder of the pump is 2.5 cm, what is the range of the force that needs to be applied to the pump handle from beginning to end?

83. Estimate the pressure on the mountains underneath the Antarctic ice sheet, which is typically 2 km thick.

84. A simple model (Fig. 10–56) considers a continent as a block (density ≈ 2800 kg/m³) floating in the mantle rock around it (density ≈ 3300 kg/m³). Assuming the continent is 35 km thick (the average thickness of the Earth's continental crust), estimate the height of the continent above the surrounding mantle rock.

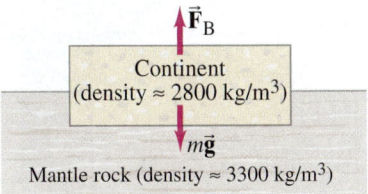

FIGURE 10–56 Problem 84.

85. A ship, carrying fresh water to a desert island in the Caribbean, has a horizontal cross-sectional area of 2240 m² at the waterline. When unloaded, the ship rises 8.25 m higher in the sea. How much water (m³) was delivered?

86. A raft is made of 12 logs lashed together. Each is 45 cm in diameter and has a length of 6.5 m. How many people can the raft hold before they start getting their feet wet, assuming the average person has a mass of 68 kg? Do *not* neglect the weight of the logs. Assume the specific gravity of wood is 0.60.

87. Estimate the total mass of the Earth's atmosphere, using the known value of atmospheric pressure at sea level.

88. During each heartbeat, approximately 70 cm³ of blood is pushed from the heart at an average pressure of 105 mm-Hg. Calculate the power output of the heart, in watts, assuming 70 beats per minute.

89. Four lawn sprinkler heads are fed by a 1.9-cm-diameter pipe. The water comes out of the heads at an angle of 35° above the horizontal and covers a radius of 6.0 m. (*a*) What is the velocity of the water coming out of each sprinkler head? (Assume zero air resistance.) (*b*) If the output diameter of each head is 3.0 mm, how many liters of water do the four heads deliver per second? (*c*) How fast is the water flowing inside the 1.9-cm-diameter pipe?

90. One arm of a U-shaped tube (open at both ends) contains water, and the other alcohol. If the two fluids meet at exactly the bottom of the U, and the alcohol is at a height of 16.0 cm, at what height will the water be?

91. The contraction of the left ventricle (chamber) of the heart pumps blood to the body. Assuming that the inner surface of the left ventricle has an area of 82 cm² and the maximum pressure in the blood is 120 mm-Hg, estimate the force exerted by that ventricle at maximum pressure.

92. An airplane has a mass of 1.7×10^6 kg, and the air flows past the lower surface of the wings at 95 m/s. If the wings have a surface area of 1200 m², how fast must the air flow over the upper surface of the wing if the plane is to stay in the air?

93. A drinking fountain shoots water about 12 cm up in the air from a nozzle of diameter 0.60 cm (Fig. 10–57). The pump at the base of the unit (1.1 m below the nozzle) pushes water into a 1.2-cm-diameter supply pipe that goes up to the nozzle. What gauge pressure does the pump have to provide? Ignore the viscosity; your answer will therefore be an underestimate.

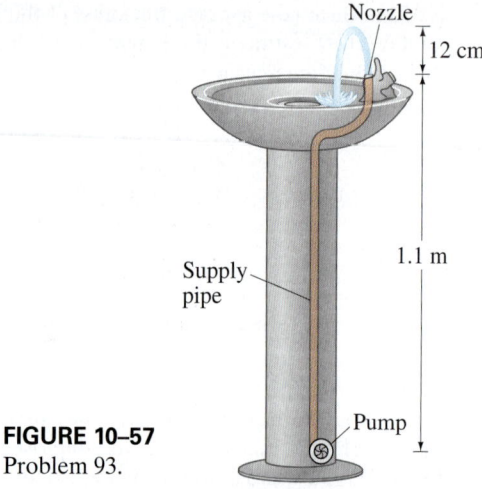

FIGURE 10–57 Problem 93.

94. A hurricane-force wind of 180 km/h blows across the face of a storefront window. Estimate the force on the 2.0 m × 3.0 m window due to the difference in air pressure inside and outside the window. Assume the store is airtight so the inside pressure remains at 1.0 atm. (This is why you should not tightly seal a building in preparation for a hurricane.)

*95. Blood is placed in a bottle 1.40 m above a 3.8-cm-long needle, of inside diameter 0.40 mm, from which it flows at a rate of 4.1 cm³/min. What is the viscosity of this blood?

96. You are watering your lawn with a hose when you put your finger over the hose opening to increase the distance the water reaches. If you are holding the hose horizontally, and the distance the water reaches increases by a factor of 4, what fraction of the hose opening did you block?

97. A copper (Cu) weight is placed on top of a 0.40-kg block of wood (density = 0.60×10^3 kg/m³) floating in water, as shown in Fig. 10–58. What is the mass of the copper if the top of the wood block is exactly at the water's surface?

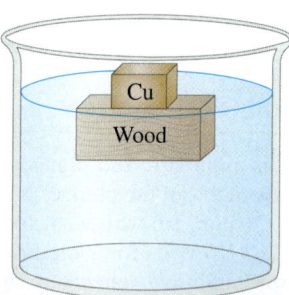

FIGURE 10–58 Problem 97.

98. You need to siphon water from a clogged sink. The sink has an area of 0.38 m² and is filled to a height of 4.0 cm. Your siphon tube rises 45 cm above the bottom of the sink and then descends 85 cm to a pail as shown in Fig. 10–59. The siphon tube has a diameter of 2.3 cm. (*a*) Assuming that the water level in the sink has almost zero velocity, use Bernoulli's equation to estimate the water velocity when it enters the pail. (*b*) Estimate how long it will take to empty the sink. Ignore viscosity.

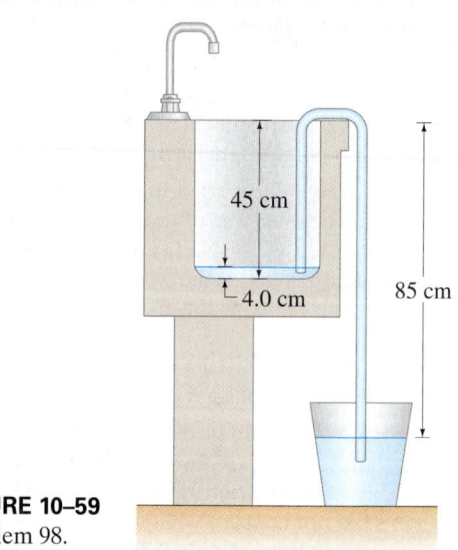

FIGURE 10–59 Problem 98.

*99. If cholesterol buildup reduces the diameter of an artery by 25%, by what % will the blood flow rate be reduced, assuming the same pressure difference?

Search and Learn

1. A 5.0-kg block and 4.0 kg of water in a 0.50-kg container are placed symmetrically on a board that can balance at the center (Fig. 10–60). A solid aluminum cube of sides 10.0 cm is lowered into the water. How much of the aluminum must be under water to make this system balance? How would your answer change for a lead cube of the same size? Explain. (See Sections 10–7 and 9–1.)

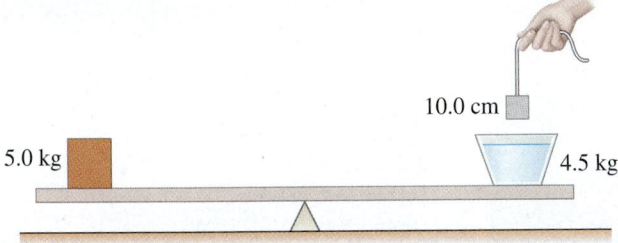

FIGURE 10–60 Search and Learn 1.

2. (a) Show that the buoyant force F_B on a partially submerged object such as a ship acts at the center of gravity of the fluid before it is displaced, Fig. 10–61. This point is called the **center of buoyancy**. (b) To ensure that a ship is in stable equilibrium, would it be better if its center of buoyancy was above, below, or at the same point as its center of gravity? Explain. (See Section 10–7 and Chapter 9.)

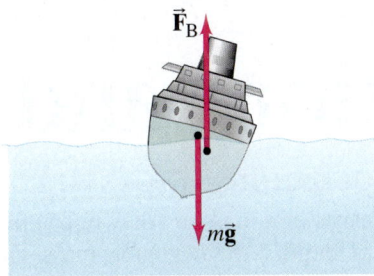

FIGURE 10–61 Search and Learn 2.

3. In working out his principle, Pascal showed dramatically how force can be multiplied with fluid pressure. He placed a long, thin tube of radius $r = 0.30$ cm vertically into a wine barrel of radius $R = 21$ cm, Fig. 10–62. He found that when the barrel was filled with water and the tube filled to a height of 12 m, the barrel burst. Calculate (a) the mass of water in the tube, and (b) the net force exerted by the water in the barrel on the lid just before rupture.

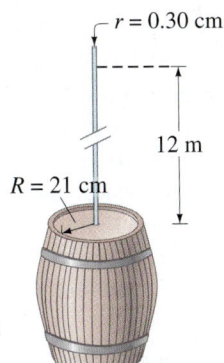

FIGURE 10–62 Search and Learn 3 (not to scale).

4. (a) When submerged in water, two objects with different volumes have the same *apparent* weight. When taken out of water, compare their weights in air. (b) Which object has the greater density?

5. A tub of water rests on a scale as shown in Fig. 10–63. The weight of the tub plus water is 100 N. A 50-N concrete brick is then lowered down from a fixed arm into the water but does not touch the tub. What does the scale read now? [*Hint*: Draw two free-body diagrams, one for the brick and a second one for the tub + water + brick.]

FIGURE 10–63 Search and Learn 5.

6. What approximations are made in the derivation of Bernoulli's equation? Qualitatively, how do you think Bernoulli's equation would change if each of these approximations was not made? (See Sections 10–8, 10–9, 10–11, and 10–12.)

*7. Estimate the density of the water 5.4 km deep in the sea. (See Table 9–1 and Section 9–5 regarding bulk modulus.) By what fraction does it differ from the density at the surface?

ANSWERS TO EXERCISES

A: (d).
B: The same. Pressure depends on depth, not on length.
C: (a).
D: (e).
E: The rowboat is shaped to have a lot of empty, air-filled space, so its "average" density is much lower than that of water (unless the boat becomes full of water, in which case it sinks). Steel ships float for the same reason.
F: Increases.
G: (b).

An object attached to a coil spring can exhibit oscillatory motion. Many kinds of oscillatory motion are sinusoidal in time, or nearly so, and are referred to as simple harmonic motion. Real systems generally have at least some friction, causing the motion to be damped. The automobile spring shown here has a shock absorber (yellow) that purposefully dampens the oscillation to make for a smooth ride. When an external sinusoidal force is exerted on a system able to oscillate, resonance occurs if the driving force is at or near the natural frequency of oscillation.

Vibrations can give rise to waves—such as water waves or waves traveling along a cord—which travel outward from their source.

CHAPTER 11

Oscillations and Waves

CONTENTS

- 11–1 Simple Harmonic Motion— Spring Oscillations
- 11–2 Energy in Simple Harmonic Motion
- 11–3 The Period and Sinusoidal Nature of SHM
- 11–4 The Simple Pendulum
- 11–5 Damped Harmonic Motion
- 11–6 Forced Oscillations; Resonance
- 11–7 Wave Motion
- 11–8 Types of Waves and Their Speeds: Transverse and Longitudinal
- 11–9 Energy Transported by Waves
- 11–10 Reflection and Transmission of Waves
- 11–11 Interference; Principle of Superposition
- 11–12 Standing Waves; Resonance
- *11–13 Refraction
- *11–14 Diffraction
- *11–15 Mathematical Representation of a Traveling Wave

CHAPTER-OPENING QUESTIONS—Guess now!

1. A simple pendulum consists of a mass m (the "bob") hanging on the end of a thin string of length ℓ and negligible mass. The bob is pulled sideways so the string makes a 5.0° angle to the vertical; when released, it oscillates back and forth at a frequency f. If the pendulum is started at a 10.0° angle instead, its frequency would be

(a) twice as great.
(b) half as great.
(c) the same, or very close to it.
(d) not quite twice as great.
(e) a bit more than half as great.

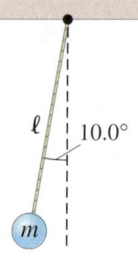

2. You drop a rock into a pond, and water waves spread out in circles.
(a) The waves carry water outward, away from where the rock hit. That moving water carries energy outward.
(b) The waves only make the water move up and down. No energy is carried outward from where the rock hit.
(c) The waves only make the water move up and down, but the waves do carry energy outward, away from where the rock hit.

Many objects vibrate or oscillate—an object on the end of a spring, a tuning fork, the balance wheel of an old watch, a pendulum, a plastic ruler held firmly over the edge of a table and gently struck, the strings of a guitar or piano. Spiders detect prey by the vibrations of their webs; cars oscillate up and down when they hit a bump; buildings and bridges vibrate when heavy trucks pass or the wind is fierce. Indeed, because most solids are elastic (see Section 9–5), they vibrate (at least briefly) when given an impulse. Electrical oscillations occur in radio and television sets. At the atomic level, atoms oscillate within a molecule, and the atoms of a solid oscillate about their relatively fixed positions.

Because it is so common in everyday life and occurs in so many areas of physics, oscillatory (or vibrational) motion is of great importance. Mechanical oscillations or vibrations are fully described on the basis of Newtonian mechanics.

Vibrations and wave motion are intimately related. Waves—whether ocean waves, waves on a string, earthquake waves, or sound waves in air—have as their source a vibration. In the case of sound, not only is the source a vibrating object, but so is the detector—the eardrum or the membrane of a microphone. Indeed, when a wave travels through a medium, the medium oscillates (such as air for sound waves). In the second half of this Chapter, after we discuss oscillations, we will discuss simple waves such as those on water or on a string. In Chapter 12 we will study sound waves, and in later Chapters we will encounter other forms of wave motion, including electromagnetic waves and light.

11–1 Simple Harmonic Motion— Spring Oscillations

When an object **vibrates** or **oscillates** back and forth, over the same path, each oscillation taking the same amount of time, the motion is **periodic**. The simplest form of periodic motion is represented by an object oscillating on the end of a uniform coil spring. Because many other types of oscillatory motion closely resemble this system, we will look at it in detail. We assume that the mass of the spring can be ignored, and that the spring is mounted horizontally, as shown in Fig. 11–1a, so that the object of mass m slides without friction on the horizontal surface. Any spring has a natural length at which it exerts no force on the mass m. The position of the mass at this point is called the **equilibrium position**. If the mass is moved either to the left, which compresses the spring, or to the right, which stretches it, the spring exerts a force on the mass that acts in the direction of returning the mass to the equilibrium position; hence it is called a *restoring force*. We consider the common situation where we can assume the restoring force F is directly proportional to the displacement x the spring has been stretched (Fig. 11–1b) or compressed (Fig. 11–1c) from the equilibrium position:

$$F = -kx. \qquad \text{[force exerted by spring]} \quad \textbf{(11–1)}$$

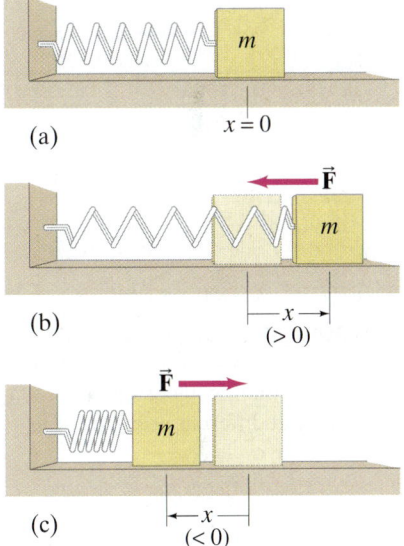

FIGURE 11–1 An object of mass m oscillating at the end of a uniform spring. The force $\vec{F}$ on the object at the different positions is shown *above* the object.

Note that the equilibrium position has been chosen at $x = 0$ and the minus sign in Eq. 11–1 indicates that the restoring force is always in the direction opposite to the displacement x. For example, if we choose the positive direction to the right in Fig. 11–1, x is positive when the spring is stretched (Fig. 11–1b), but the direction of the restoring force is to the left (negative direction). If the spring is compressed, x is negative (to the left) but the force F acts toward the right (Fig. 11–1c).

Equation 11–1 is often referred to as Hooke's law (Sections 6–4 and 9–5), and is accurate only if the spring is not compressed to where the coils are close to touching, or stretched beyond the elastic region (see Fig. 9–19). Hooke's law works not only for springs but for other oscillating solids as well; it thus has wide applicability, even though it is valid only over a certain range of F and x values.

The proportionality constant k in Eq. 11–1 is called the *spring constant* for that particular spring, or its *spring stiffness constant* (units = N/m). To stretch the spring a distance x, an (external) force must be exerted on the free end of the spring with a magnitude at least equal to

$$F_{\text{ext}} = +kx. \qquad \text{[external force on spring]}$$

The greater the value of k, the greater the force needed to stretch a spring a given distance. That is, the stiffer the spring, the greater the spring constant k.

Note that the force F in Eq. 11–1 is *not* a constant, but varies with position. Therefore the acceleration of the mass m is not constant, so we *cannot* use the equations for constant acceleration developed in Chapter 2.

Eqs. 2–11 for constant acceleration do not apply to a spring

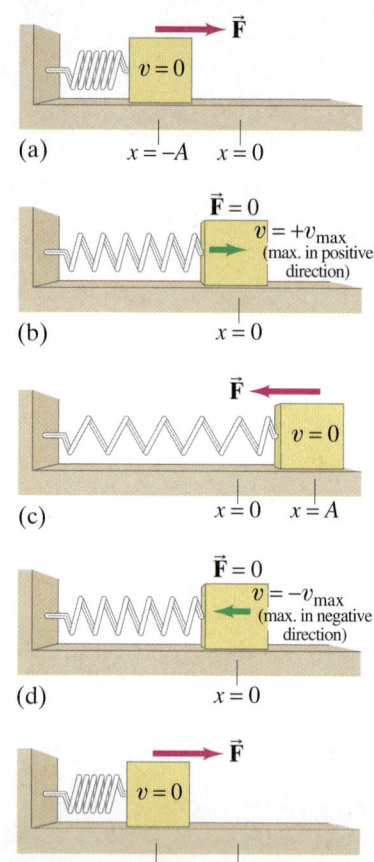

FIGURE 11–2 An object oscillating on a frictionless surface, indicating the force on the object and its velocity at different positions of its oscillation cycle.

Let us examine what happens when our uniform spring is initially compressed a distance $x = -A$, as shown in Fig. 11–2a, and then our object of mass m is released on the frictionless surface. The spring exerts a force on the mass that accelerates it toward the equilibrium position. Because the mass has inertia, it passes the equilibrium position with considerable speed. Indeed, as the mass reaches the equilibrium position, the force on it decreases to zero, but its speed at this point is a maximum, v_{max} (Fig. 11–2b). As the mass moves farther to the right, the force on it acts to slow it down, and it stops for an instant at $x = A$ (Fig. 11–2c). It then begins moving back in the opposite direction, accelerating until it passes the equilibrium point (Fig. 11–2d), and then slows down until it reaches zero speed at the original starting point, $x = -A$ (Fig. 11–2e). It then repeats the motion, moving back and forth symmetrically between $x = A$ and $x = -A$.

EXERCISE A A mass is oscillating on a frictionless surface at the end of a horizontal spring. Where, if anywhere, is the acceleration of the mass zero (see Fig. 11–2)? (*a*) At $x = -A$; (*b*) at $x = 0$; (*c*) at $x = +A$; (*d*) at both $x = -A$ and $x = +A$; (*e*) nowhere.

To discuss oscillatory motion, we need to define a few terms. The distance x of the mass from the equilibrium point at any moment is the **displacement** (with a + or − sign). The maximum displacement—the greatest distance from the equilibrium point—is called the **amplitude**, A. One **cycle** refers to the complete to-and-fro motion from some initial point back to that same point—say, from $x = -A$ to $x = +A$ and back to $x = -A$. The **period**, T, is defined as the time required to complete one cycle. Finally, the **frequency**, f, is the number of complete cycles per second. Frequency is generally specified in hertz (Hz), where $1\,\text{Hz} = 1$ cycle per second (s^{-1}). Given their definitions, frequency and period are inversely related, as we saw earlier (Eqs. 5–2 and 8–8):

$$f = \frac{1}{T} \quad \text{and} \quad T = \frac{1}{f}. \tag{11–2}$$

For example, if the frequency is 2 cycles per second, then each cycle takes $\tfrac{1}{2}$ s.

EXERCISE B If an oscillating mass has a frequency of 1.25 Hz, it makes 100 oscillations in (*a*) 12.5 s, (*b*) 125 s, (*c*) 80 s, (*d*) 8.0 s.

⚠️ **CAUTION**
For vertical spring, measure displacement (x or y) from the vertical equilibrium position

The oscillation of a spring hung vertically is similar to that of a horizontal spring; but because of gravity, the length of a vertical spring with a mass m on the end will be longer at equilibrium than when that same spring is horizontal. See Fig. 11–3. The spring is in equilibrium when $\Sigma F = 0 = mg - kx_0$, so the spring stretches an extra amount $x_0 = mg/k$ to be in equilibrium. If x is measured from this new equilibrium position, Eq. 11–1 can be used directly with the same value of k.

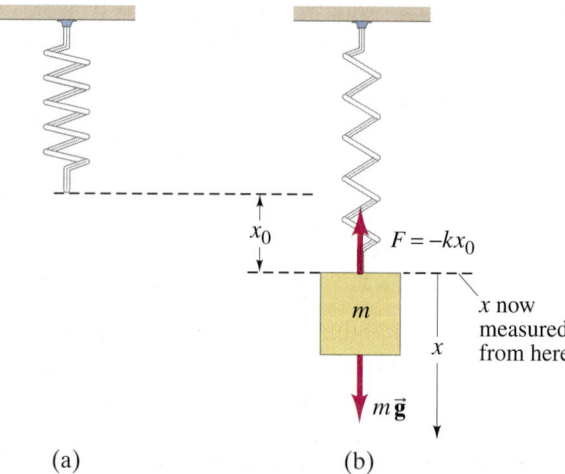

FIGURE 11–3
(a) Free spring, hung vertically.
(b) Mass m attached to spring in new equilibrium position, which occurs when $\Sigma F = 0 = mg - kx_0$.

EXAMPLE 11–1 Car springs. When a family of four with a total mass of 200 kg step into their 1200-kg car, the car's springs compress 3.0 cm. (a) What is the spring constant of the car's springs (Fig. 11–4), assuming they act as a single spring? (b) How far will the car lower if loaded with 300 kg rather than 200 kg?

APPROACH We use Hooke's law: the weight of the people, mg, causes a 3.0-cm displacement.

SOLUTION (a) The added force of $(200 \text{ kg})(9.8 \text{ m/s}^2) = 1960 \text{ N}$ causes the springs to compress 3.0×10^{-2} m. Therefore (Eq. 11–1), the spring constant is

$$k = \frac{F}{x} = \frac{1960 \text{ N}}{3.0 \times 10^{-2} \text{ m}} = 6.5 \times 10^4 \text{ N/m}.$$

(b) If the car is loaded with 300 kg, Hooke's law gives

$$x = \frac{F}{k} = \frac{(300 \text{ kg})(9.8 \text{ m/s}^2)}{(6.5 \times 10^4 \text{ N/m})} = 4.5 \times 10^{-2} \text{ m},$$

or 4.5 cm.

NOTE In (b), we could have obtained x without solving for k: since x is proportional to F, if 200 kg compresses the spring 3.0 cm, then 1.5 times the force will compress the spring 1.5 times as much, or 4.5 cm.

FIGURE 11–4 Photo of a car's spring. (Also visible is the shock absorber, in blue—see Section 11–5.)

Any oscillating system for which the net restoring force is directly proportional to the negative of the displacement (as in Eq. 11–1, $F = -kx$) is said to exhibit **simple harmonic motion** (SHM).[†] Such a system is often called a **simple harmonic oscillator** (SHO). We saw in Section 9–5 that most solid materials stretch or compress according to Eq. 11–1 as long as the displacement is not too great. Because of this, many natural oscillations are simple harmonic, or sufficiently close to it that they can be treated using this SHM model.

CONCEPTUAL EXAMPLE 11–2 Is the motion simple harmonic? Which of the following forces would cause an object to move in simple harmonic motion? (a) $F = -0.5x^2$, (b) $F = -2.3y$, (c) $F = 8.6x$, (d) $F = -4\theta$?

RESPONSE Both (b) and (d) will give simple harmonic motion because they give the force as minus a constant times a displacement. The displacement need not be x, but the minus sign is required to restore the system to equilibrium, which is why (c) does not produce SHM.

11–2 Energy in Simple Harmonic Motion

With forces that are not constant, such as here with simple harmonic motion, it is often convenient and useful to use the energy approach, as we saw in Chapter 6.

To stretch or compress a spring, work has to be done. Hence potential energy is stored in a stretched or compressed spring. We have already seen in Section 6–4 that elastic potential energy is given by

$$\text{PE} = \tfrac{1}{2}kx^2.$$

The total mechanical energy E is the sum of the kinetic and potential energies,

$$E = \tfrac{1}{2}mv^2 + \tfrac{1}{2}kx^2, \qquad (11\text{–}3)$$

where v is the speed of the mass m at a distance x from the equilibrium position.

[†]The word "harmonic" refers to the motion being sinusoidal, which we discuss in Section 11–3. It is "simple" when the motion is sinusoidal of a single frequency. This can happen only if friction or other forces are not acting.

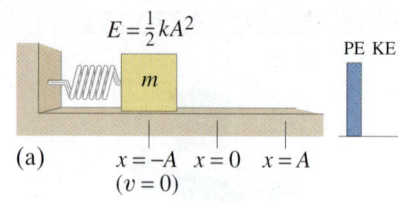
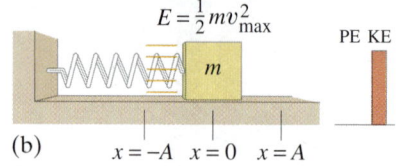
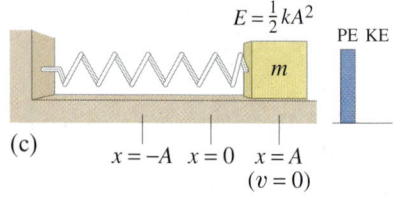
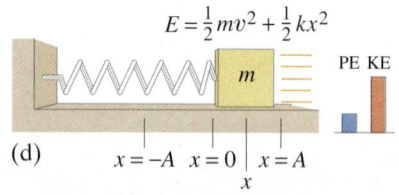

FIGURE 11–5 Energy changes from potential energy to kinetic energy and back again as the spring oscillates. Energy bar graphs (on the right) were used in Section 6–7.

SHM can occur only if friction is negligible so that the total mechanical energy E remains constant. As the mass oscillates back and forth, the energy continuously changes from potential energy to kinetic energy, and back again (Fig. 11–5). At the extreme points, $x = -A$ and $x = A$ (Fig. 11–5a, c), all the energy is stored in the spring as potential energy (and is the same whether the spring is compressed or stretched to the full amplitude). At these extreme points, the mass stops for an instant as it changes direction, so $v = 0$ and

$$E = \tfrac{1}{2}m(0)^2 + \tfrac{1}{2}kA^2 = \tfrac{1}{2}kA^2. \qquad (11\text{–}4\text{a})$$

Thus, the **total mechanical energy of a simple harmonic oscillator is proportional to the square of the amplitude.** At the equilibrium point, $x = 0$ (Fig. 11–5b), all the energy is kinetic:

$$E = \tfrac{1}{2}mv_{\max}^2 + \tfrac{1}{2}k(0)^2 = \tfrac{1}{2}mv_{\max}^2, \qquad (11\text{–}4\text{b})$$

where $v_{\max}$ is the maximum speed during the motion (which occurs at $x = 0$). At intermediate points (Fig. 11–5d), the energy is part kinetic and part potential; because energy is conserved (we use Eqs. 11–3 and 11–4a),

$$\tfrac{1}{2}mv^2 + \tfrac{1}{2}kx^2 = \tfrac{1}{2}kA^2. \qquad (11\text{–}4\text{c})$$

From this conservation of energy equation, we can obtain the velocity as a function of position. Solving for v^2, we have

$$v^2 = \frac{k}{m}(A^2 - x^2) = \frac{k}{m}A^2\left(1 - \frac{x^2}{A^2}\right).$$

From Eqs. 11–4a and 11–4b, we have $\tfrac{1}{2}mv_{\max}^2 = \tfrac{1}{2}kA^2$, so $v_{\max}^2 = (k/m)A^2$ or

$$v_{\max} = \sqrt{\frac{k}{m}}\,A. \qquad (11\text{–}5\text{a})$$

Inserting this equation into the equation just above it and taking the square root, we have

$$v = \pm v_{\max}\sqrt{1 - \frac{x^2}{A^2}}. \qquad (11\text{–}5\text{b})$$

This gives the velocity of the object at any position x. The object moves back and forth, so its velocity can be either in the $+$ or $-$ direction, but its magnitude depends only on its position x.

CONCEPTUAL EXAMPLE 11–3 Doubling the amplitude. Suppose the spring in Fig. 11–5 is stretched twice as far (to $x = 2A$). What happens to (a) the energy of the system, (b) the maximum velocity of the oscillating mass, (c) the maximum acceleration of the mass?

RESPONSE (a) From Eq. 11–4a, the total energy is proportional to the square of the amplitude A, so stretching it twice as far quadruples the energy ($2^2 = 4$). You may protest, "I did work stretching the spring from $x = 0$ to $x = A$. Don't I do the same work stretching it from A to $2A$?" No. The force you exert is proportional to the displacement x, so for the second displacement, from $x = A$ to $2A$, you do more work than for the first displacement ($x = 0$ to A). (b) From Eq. 11–5a, we can see that when the amplitude is doubled, the maximum velocity must be doubled. (c) Since the force is twice as great when we stretch the spring twice as far ($F = kx$), the acceleration is also twice as great: $a \propto F \propto x$.

EXERCISE C Suppose the spring in Fig. 11–5 is compressed to $x = -A$, but is given a push to the right so that the initial speed of the mass m is v_0. What effect does this push have on (a) the energy of the system, (b) the maximum velocity, (c) the maximum acceleration?

EXAMPLE 11-4 Spring calculations. A spring stretches 0.150 m when a 0.300-kg mass is gently suspended from it as in Fig. 11-3b. The spring is then set up horizontally with the 0.300-kg mass resting on a frictionless table as in Fig. 11-5. The mass is pulled so that the spring is stretched 0.100 m from the equilibrium point, and released from rest. Determine: (*a*) the spring stiffness constant k; (*b*) the amplitude of the horizontal oscillation A; (*c*) the magnitude of the maximum velocity v_{max}; (*d*) the magnitude of the velocity v when the mass is 0.050 m from equilibrium; and (*e*) the magnitude of the maximum acceleration a_{max} of the mass.

APPROACH Wow, a lot of questions, but we can take them one by one. When the 0.300-kg mass hangs at rest from the spring as in Fig. 11-3b, we apply Newton's second law for the vertical forces: $\Sigma F = 0 = mg - kx_0$, so $k = mg/x_0$. For the horizontal oscillations, the amplitude is given, the velocities are found using conservation of energy, and the acceleration is found from $F = ma$.

SOLUTION (*a*) The spring stretches 0.150 m due to the 0.300-kg load, so

$$k = \frac{F}{x_0} = \frac{mg}{x_0} = \frac{(0.300 \text{ kg})(9.80 \text{ m/s}^2)}{0.150 \text{ m}} = 19.6 \text{ N/m}.$$

(*b*) The spring is now horizontal (on a table). It is stretched 0.100 m from equilibrium and is given no initial speed, so $A = 0.100$ m.

(*c*) The maximum velocity v_{max} is attained as the mass passes through the equilibrium point where all the energy is kinetic. By comparing the total energy (see Eq. 11-3) at equilibrium with that at full extension, conservation of energy tells us that

$$\tfrac{1}{2}mv_{max}^2 + 0 = 0 + \tfrac{1}{2}kA^2,$$

where $A = 0.100$ m. Solving for v_{max} (or using Eq. 11-5a), we have

$$v_{max} = A\sqrt{\frac{k}{m}} = (0.100 \text{ m})\sqrt{\frac{19.6 \text{ N/m}}{0.300 \text{ kg}}} = 0.808 \text{ m/s}.$$

(*d*) We use conservation of energy, or Eq. 11-5b derived from it, and find that

$$v = v_{max}\sqrt{1 - \frac{x^2}{A^2}} = (0.808 \text{ m/s})\sqrt{1 - \frac{(0.050 \text{ m})^2}{(0.100 \text{ m})^2}} = 0.700 \text{ m/s}.$$

(*e*) By Newton's second law, $F = ma$. So the maximum acceleration occurs where the force is greatest—that is, when $x = A = 0.100$ m. Thus

$$a_{max} = \frac{F_{max}}{m} = \frac{kA}{m} = \frac{(19.6 \text{ N/m})(0.100 \text{ m})}{0.300 \text{ kg}} = 6.53 \text{ m/s}^2.$$

NOTE We cannot use the kinematic equations, Eqs. 2-11, because the acceleration is not constant in SHM.

EXAMPLE 11-5 Energy calculations. For the simple harmonic oscillator of Example 11-4, determine (*a*) the total energy, and (*b*) the kinetic and potential energies at half amplitude ($x = \pm A/2$).

APPROACH We use conservation of energy for a mass–spring system, Eqs. 11-3 and 11-4.

SOLUTION (*a*) With $k = 19.6$ N/m and $A = 0.100$ m, the total energy E from Eq. 11-4a is

$$E = \tfrac{1}{2}kA^2 = \tfrac{1}{2}(19.6 \text{ N/m})(0.100 \text{ m})^2 = 9.80 \times 10^{-2} \text{ J}.$$

(*b*) At $x = A/2 = 0.050$ m, we have

$$\text{PE} = \tfrac{1}{2}kx^2 = \tfrac{1}{2}(19.6 \text{ N/m})(0.050 \text{ m})^2 = 2.45 \times 10^{-2} \text{ J}.$$

By conservation of energy, the kinetic energy must be

$$\text{KE} = E - \text{PE} = 7.35 \times 10^{-2} \text{ J}.$$

11–3 The Period and Sinusoidal Nature of SHM

The period of a simple harmonic oscillator is found to depend on the stiffness of the spring and also on the mass m that is oscillating. But—strange as it may seem—the *period does not depend on the amplitude*. You can find this out for yourself by using a watch and timing 10 or 20 cycles of an oscillating spring for a small amplitude and then for a large amplitude.

The period T is given by (see derivation on next page):

$$T = 2\pi\sqrt{\frac{m}{k}}. \qquad (11\text{–}6\text{a})$$

We see that the larger the mass, the longer the period; and the stiffer the spring (larger k), the shorter the period. This makes sense since a larger mass means more inertia and therefore slower response (smaller acceleration). And larger k means greater force and therefore quicker response (larger acceleration). Notice that Eq. 11–6a is not a direct proportion: the period varies as the *square root* of m/k. For example, the mass must be quadrupled to double the period. Equation 11–6a is fully in accord with experiment and is valid not only for a spring, but for all kinds of simple harmonic motion—that is, for motion subject to a restoring force proportional to displacement, Eq. 11–1.

We can write the frequency using $f = 1/T$ (Eq. 11–2):

$$f = \frac{1}{T} = \frac{1}{2\pi}\sqrt{\frac{k}{m}}. \qquad (11\text{–}6\text{b})$$

EXERCISE D By how much should the mass on the end of a spring be changed to halve the frequency of its oscillations? (*a*) No change; (*b*) doubled; (*c*) quadrupled; (*d*) halved; (*e*) quartered.

FIGURE 11–6 Example 11–6. A spider waits for its prey (on the left).

EXAMPLE 11–6 ESTIMATE **Spider web.** A spider of mass 0.30 g waits in its web of negligible mass (Fig. 11–6). A slight movement causes the web to vibrate with a frequency of about 15 Hz. (*a*) Estimate the value of the spring stiffness constant k for the web. (*b*) At what frequency would you expect the web to vibrate if an insect of mass 0.10 g were trapped in addition to the spider?

APPROACH We can only make a rough estimate because a spider's web is fairly complicated and may vibrate with a mixture of frequencies. We use SHM as an approximate model.

SOLUTION (*a*) The frequency of SHM is given by Eq. 11–6b,

$$f = \frac{1}{2\pi}\sqrt{\frac{k}{m}}.$$

We solve for k:

$$k = (2\pi f)^2 m$$
$$= (2\pi)^2 (15\text{ s}^{-1})^2 (3.0 \times 10^{-4}\text{ kg}) = 2.7\text{ N/m}.$$

(*b*) The total mass is now $0.10\text{ g} + 0.30\text{ g} = 4.0 \times 10^{-4}\text{ kg}$. We could substitute $m = 4.0 \times 10^{-4}\text{ kg}$ into Eq. 11–6b. Instead, we notice that the frequency decreases with the square root of the mass. Since the new mass is 4/3 times the first mass, the frequency changes by a factor of $1/\sqrt{4/3} = \sqrt{3/4}$. Thus $f = (15\text{ Hz})(\sqrt{3/4}) = 13\text{ Hz}$.

NOTE Check this result by direct substitution of k, found in part (*a*), and the new mass m into Eq. 11–6b.

EXAMPLE 11–7 ESTIMATE **A vibrating floor.** A large motor in a factory causes the floor to vibrate up and down at a frequency of 10 Hz. The amplitude of the floor's motion near the motor is about 3.0 mm. Estimate the maximum acceleration of the floor near the motor.

APPROACH Assuming the motion of the floor is roughly SHM, we can make an estimate for the maximum acceleration using $F = ma$ and Eq. 11–6b.

SOLUTION The maximum acceleration occurs when the force $(F = kx)$ is largest, which is when $x = A$. Thus, $a_{max} = F_{max}/m = kA/m = (k/m)A$. From Eq. 11–6b, $(k/m) = (2\pi f)^2$, so

$$a_{max} = \frac{F_{max}}{m} = \left(\frac{k}{m}\right)A = (2\pi f)^2 A = (2\pi)^2 (10\text{ s}^{-1})^2 (3.0 \times 10^{-3}\text{ m}) = 12\text{ m/s}^2.$$

NOTE The maximum acceleration is a little over g, so when the floor accelerates down, objects sitting on the floor will actually lose contact with the floor momentarily, which will cause noise and serious wear.

PHYSICS APPLIED
Unwanted floor vibrations

Period and Frequency—Derivation

We can derive a formula for the period of simple harmonic motion (SHM) by comparing SHM to an object rotating uniformly in a circle. From this same "reference circle" we can obtain a second useful result—a formula for the position of an oscillating mass as a function of time. There is nothing actually rotating in a circle when a spring oscillates linearly, but it is the mathematical similarity that we find useful.

Consider a small object of mass m revolving counterclockwise in a circle of radius A, with constant speed v_{max}, on top of a table as shown in Fig. 11–7. As viewed from above, the motion is a circle in the xy plane. But a person who looks at the motion from the edge of the table sees an oscillatory motion back and forth, and this one-dimensional motion corresponds precisely to simple harmonic motion, as we shall now see.

What the person sees, and what we are interested in, is the projection of the circular motion onto the x axis (Fig. 11–7b). To see that this x motion is analogous to SHM, let us calculate the magnitude of the x component of the velocity v_{max}, which is labeled v in Fig. 11–7. The two triangles involving θ in Fig. 11–7a are similar, so

$$\frac{v}{v_{max}} = \frac{\sqrt{A^2 - x^2}}{A}$$

or

$$v = v_{max}\sqrt{1 - \frac{x^2}{A^2}}.$$

This is exactly the equation for the speed of a mass oscillating with SHM, as we saw in Eq. 11–5b. Thus the projection on the x axis of an object revolving in a circle has the same motion as a mass undergoing SHM.

We can now determine the period of SHM because it is equal to the time for our object revolving in a circle to make one complete revolution. First we note that the velocity v_{max} is equal to the circumference of the circle (distance) divided by the period T:

$$v_{max} = \frac{2\pi A}{T} = 2\pi A f. \qquad (11\text{–}7)$$

We solve for the period T in terms of A:

$$T = \frac{2\pi A}{v_{max}}.$$

From Eq. 11–5a, $A/v_{max} = \sqrt{m/k}$. Thus

$$T = 2\pi\sqrt{\frac{m}{k}},$$

which is Eq. 11–6a, the formula we were looking for. The period depends on the mass m and the spring stiffness constant k, but not on the amplitude A.

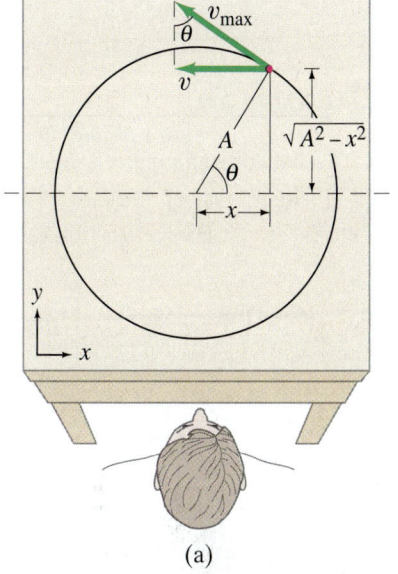
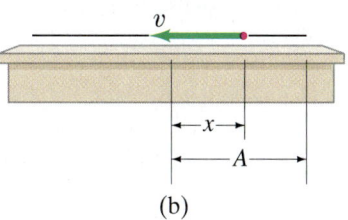

FIGURE 11–7 (a) Circular motion of a small (red) object. (b) Side view of circular motion (x component) is simple harmonic motion.

Position as a Function of Time

We now use the reference circle to find the position of a mass undergoing simple harmonic motion as a function of time. From Fig. 11–7, we see that $\cos\theta = x/A$, so the projection of the object's position on the x axis is

$$x = A\cos\theta.$$

The mass in the reference circle (Fig. 11–7) is rotating with uniform angular velocity ω. We then can write $\theta = \omega t$, where θ is in radians (Section 8–1). Thus

$$x = A\cos\omega t. \tag{11–8a}$$

Furthermore, since the angular velocity ω (specified in radians per second) can be written as $\omega = 2\pi f$, where f is the frequency (Eq. 8–7), we then write

$$x = A\cos(2\pi ft), \tag{11–8b}$$

or in terms of the period T,

$$x = A\cos(2\pi t/T). \tag{11–8c}$$

Notice in Eq. 11–8c that when $t = T$ (that is, after a time equal to one period), we have the cosine of 2π (or 360°), which is the same as the cosine of zero. This makes sense since the motion repeats itself after a time $t = T$.

Because the cosine function varies between 1 and -1, Eqs. 11–8 tell us that x varies between A and $-A$, as it must. If a pen is attached to a vibrating mass as a sheet of paper is moved at a steady rate beneath it (Fig. 11–8), a sinusoidal curve will be drawn that accurately follows Eqs. 11–8.

> **CAUTION**
> *t is a variable (time); T is a constant for a given situation*

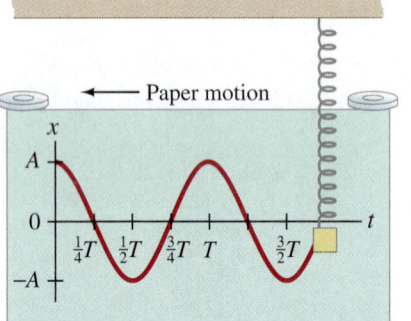

FIGURE 11–8 Position as a function of time for a simple harmonic oscillator, $x = A\cos(2\pi t/T)$.

EXAMPLE 11–8 Starting with $x = A\cos\omega t$. The displacement of an object is described by the following equation, where x is in meters and t is in seconds:

$$x = (0.30\,\text{m})\cos(8.0\,t).$$

Determine the oscillating object's (*a*) amplitude, (*b*) frequency, (*c*) period, (*d*) maximum speed, and (*e*) maximum acceleration.

APPROACH We start by comparing the given equation for x with Eq. 11–8b, $x = A\cos(2\pi ft)$.

SOLUTION From $x = A\cos(2\pi ft)$, we see by inspection that (*a*) the amplitude $A = 0.30\,\text{m}$, and (*b*) $2\pi f = 8.0\,\text{s}^{-1}$; so $f = (8.0\,\text{s}^{-1}/2\pi) = 1.27\,\text{Hz}$. (*c*) Then $T = 1/f = 0.79\,\text{s}$. (*d*) The maximum speed (see Eq. 11–7) is

$$v_{\max} = 2\pi Af$$
$$= (2\pi)(0.30\,\text{m})(1.27\,\text{s}^{-1}) = 2.4\,\text{m/s}.$$

(*e*) The maximum acceleration, by Newton's second law, is $a_{\max} = F_{\max}/m = kA/m$, because $F\,(= kx)$ is greatest when x is greatest. From Eq. 11–6b we see that $k/m = (2\pi f)^2$. Hence

$$a_{\max} = \frac{k}{m}A = (2\pi f)^2 A$$
$$= (2\pi)^2 (1.27\,\text{s}^{-1})^2 (0.30\,\text{m}) = 19\,\text{m/s}^2.$$

Sinusoidal Motion

Equation 11–8a, $x = A\cos\omega t$, assumes that the oscillating object starts from rest ($v = 0$) at its maximum displacement ($x = A$) at $t = 0$. Other equations for SHM are also possible, depending on the initial conditions (when you choose t to be zero).

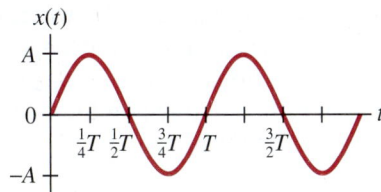

FIGURE 11–9 Sinusoidal nature of SHM, position as a function of time. In this case, $x = A \sin(2\pi t/T)$ because at $t = 0$ the mass is at the equilibrium position $x = 0$ and has (or is given) an initial speed at $t = 0$ that carries it to $x = A$ at $t = \frac{1}{4}T$.

For example, if at $t = 0$ the object is at the equilibrium position and the oscillations are begun by giving the object a push to the right $(+x)$, the equation would be

$$x = A \sin \omega t = A \sin(2\pi t/T).$$

This curve, shown in Fig. 11–9, has the same shape as the cosine curve shown in Fig. 11–8, except it is shifted to the right by a quarter cycle. Hence at $t = 0$ it starts out at $x = 0$ instead of at $x = A$.

Both sine and cosine curves are referred to as being **sinusoidal** (having the shape of a sine function). Thus simple harmonic motion[†] is said to be sinusoidal because the position varies as a sinusoidal function of time.

*Velocity and Acceleration as Functions of Time

Figure 11–10a, like Fig. 11–8, shows a graph of displacement x vs. time t, as given by Eqs. 11–8. We can also find the velocity v as a function of time from Fig. 11–7a. For the position shown (red dot in Fig. 11–7a), the magnitude of v is $v_{max} \sin \theta$, but $\vec{v}$ points to the left, so $v = -v_{max} \sin \theta$. Again setting $\theta = \omega t = 2\pi f t = 2\pi t/T$, we have

$$v = -v_{max} \sin \omega t = -v_{max} \sin(2\pi f t) = -v_{max} \sin(2\pi t/T). \quad (11\text{–}9)$$

Just after $t = 0$, the velocity is negative (points to the left) and remains so until $t = \frac{1}{2}T$ (corresponding to $\theta = 180° = \pi$ radians). After $t = \frac{1}{2}T$ until $t = T$ the velocity is positive. The velocity as a function of time (Eq. 11–9) is plotted in Fig. 11–10b. From Eqs. 11–6b and 11–7,

$$v_{max} = 2\pi A f = A\sqrt{\frac{k}{m}}.$$

For a given spring–mass system, the maximum speed v_{max} is higher if the amplitude is larger, and always occurs as the mass passes the equilibrium point.

Newton's second law and Eqs. 11–8 give us the acceleration as a function of time:

$$a = \frac{F}{m} = \frac{-kx}{m} = -\left(\frac{kA}{m}\right)\cos \omega t = -a_{max}\cos(2\pi t/T) \quad (11\text{–}10)$$

where the maximum acceleration is

$$a_{max} = kA/m.$$

Equation 11–10 is plotted in Fig. 11–10c. Because the acceleration of a SHO is *not* constant, the equations for uniformly accelerated motion do *not* apply to SHM.

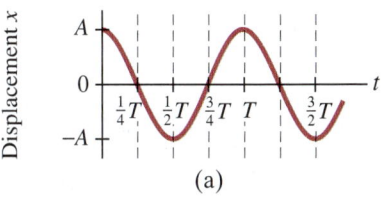
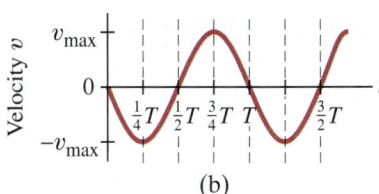
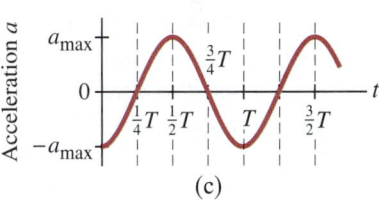

FIGURE 11–10 Graphs showing (a) displacement x as a function of time t: $x = A\cos(2\pi t/T)$; (b) velocity as a function of time: $v = -v_{max}\sin(2\pi t/T)$, where $v_{max} = A\sqrt{k/m}$; (c) acceleration as a function of time: $a = -a_{max}\cos(2\pi t/T)$, where $a_{max} = Ak/m$.

FIGURE 11–11 Strobe-light photo of an oscillating pendulum, at equal time intervals.

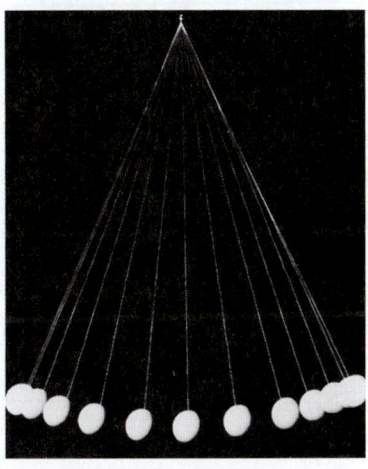

11–4 The Simple Pendulum

A **simple pendulum** consists of a small object (the pendulum bob) suspended from the end of a lightweight cord, Fig. 11–11. We assume that the cord does not stretch and that its mass can be ignored relative to that of the bob. The motion of a simple pendulum moving back and forth with negligible friction resembles simple harmonic motion: the pendulum bob oscillates along the arc of a circle with equal amplitude on either side of its equilibrium point, and as it passes through the equilibrium point (where it would hang vertically) it has its maximum speed. But is it really undergoing SHM? That is, is the restoring force proportional to its displacement? Let us find out.

[†]Simple harmonic motion can be *defined* as motion that is sinusoidal. This definition is fully consistent with our earlier definition in Section 11–1.

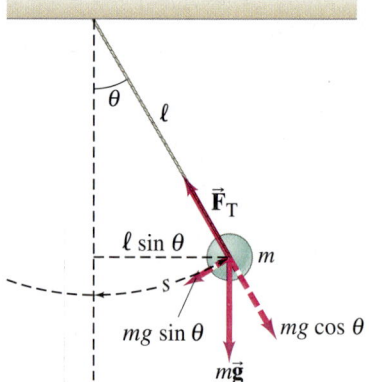

FIGURE 11–12 Simple pendulum, and a free-body diagram.

The displacement s of the pendulum along the arc is given by $s = \ell\theta$, where θ is the angle (in radians) that the cord makes with the vertical and ℓ is the length of the cord (Fig. 11–12). If the restoring force is proportional to s or to θ, the motion will be simple harmonic. The restoring force is the net force on the bob, which equals the component of the weight (mg) tangent to the arc:

$$F = -mg \sin \theta,$$

where g is the acceleration due to gravity. The minus sign here, as in Eq. 11–1, means the force is in the direction opposite to the angular displacement θ. Since F is proportional to the sine of θ and not to θ itself, the motion is *not* SHM. However, if θ is small, then $\sin \theta$ is very nearly equal to θ when the angle is specified in radians. This can be seen by noting in Fig. 11–12 that the arc length s ($= \ell\theta$) is nearly the same length as the chord ($= \ell \sin \theta$) indicated by the horizontal straight dashed line, *if θ is small*. For angles less than 15°, the difference between θ (in radians) and $\sin \theta$ is less than 1%—see Table 11–1. Thus, to a very good approximation for small angles,

$$F = -mg \sin \theta \approx -mg\theta.$$

Substituting $s = \ell\theta$, or $\theta = s/\ell$, we have

$$F \approx -\frac{mg}{\ell} s.$$

TABLE 11–1 Sin θ at Small Angles

θ (degrees)	θ (radians)	sin θ	% Difference
0	0	0	0
1°	0.01745	0.01745	0.005%
5°	0.08727	0.08716	0.1%
10°	0.17453	0.17365	0.5%
15°	0.26180	0.25882	1.1%
20°	0.34907	0.34202	2.0%
30°	0.52360	0.50000	4.5%

Thus, for small displacements, the motion can be modeled as being approximately simple harmonic, because this approximate equation fits Hooke's law, $F = -kx$, where in place of x we have arc length s. The effective force constant is $k = mg/\ell$. If we substitute $k = mg/\ell$ into Eq. 11–6a, we obtain the period of a simple pendulum:

$$T = 2\pi \sqrt{\frac{m}{k}} = 2\pi \sqrt{\frac{m}{mg/\ell}}$$

or

$$T = 2\pi \sqrt{\frac{\ell}{g}}. \qquad [\theta \text{ small}] \quad \textbf{(11–11a)}$$

The frequency is $f = 1/T$, so

$$f = \frac{1}{2\pi} \sqrt{\frac{g}{\ell}}. \qquad [\theta \text{ small}] \quad \textbf{(11–11b)}$$

The mass m of the pendulum bob does not appear in these formulas for T and f. Thus we have the surprising result that the period and frequency of a simple pendulum do not depend on the mass of the pendulum bob. You may have noticed this if you pushed a small child and then a large one on the same swing.

We also see from Eq. 11–11a that the period of a pendulum does not depend on the amplitude (like any SHM, Section 11–3), as long as the amplitude θ is small. Galileo is said to have first noted this fact while watching a swinging lamp in the cathedral at Pisa (Fig. 11–13). This discovery led to the invention of the pendulum clock, the first really precise timepiece, which became the standard for centuries.

FIGURE 11–13 The swinging motion of this elaborate lamp, hanging by a very long cord from the ceiling of the cathedral at Pisa, is said to have been observed by Galileo and to have inspired him to the conclusion that the period of a pendulum does not depend on amplitude.

PHYSICS APPLIED
Pendulum clock

| **EXERCISE E** Return to Chapter-Opening Question 1, page 292, and answer it again now. Try to explain why you may have answered differently the first time.

| **EXERCISE F** If a simple pendulum is taken from sea level to the top of a high mountain and started at the same angle of 5°, it would oscillate at the top of the mountain (*a*) slightly slower; (*b*) slightly faster; (*c*) at exactly the same frequency; (*d*) not at all—it would stop; (*e*) none of these.

Because a pendulum does not undergo *precisely* SHM, the period does depend slightly on the amplitude—the more so for large amplitudes. The accuracy of a pendulum clock would be affected, after many swings, by the decrease in amplitude due to friction. But the mainspring in a pendulum clock (or the falling weight in a grandfather clock) supplies energy to compensate for the friction and to maintain the amplitude constant, so that the timing remains precise.

EXAMPLE 11–9 Measuring g. A geologist uses a simple pendulum that has a length of 37.10 cm and a frequency of 0.8190 Hz at a particular location on the Earth. What is the acceleration due to gravity at this location?

APPROACH We can use the length ℓ and frequency f of the pendulum in Eq. 11–11b, which contains our unknown, g.

SOLUTION We solve Eq. 11–11b for g and obtain
$$g = (2\pi f)^2 \ell = (2\pi)^2 (0.8190 \text{ s}^{-1})^2 (0.3710 \text{ m}) = 9.824 \text{ m/s}^2.$$

11–5 Damped Harmonic Motion

The amplitude of any real oscillating spring or swinging pendulum slowly decreases in time until the oscillations stop altogether. Figure 11–14 shows a typical graph of the displacement as a function of time. This is called **damped harmonic motion**. The damping† is generally due to the resistance of air and to internal friction within the oscillating system. The energy that is dissipated to thermal energy results in a decreased amplitude of oscillation.

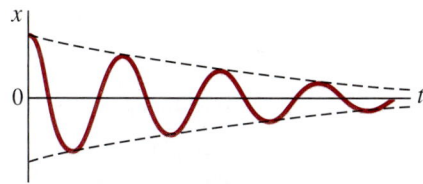

FIGURE 11–14 Damped harmonic motion.

Since natural oscillating systems are damped in general, why do we even talk about (undamped) simple harmonic motion? The answer is that SHM is much easier to deal with mathematically. And if the damping is not large, the oscillations can be thought of as simple harmonic motion on which the damping is superposed, as represented by the dashed curves in Fig. 11–14. Although damping does alter the frequency of vibration, the effect can be small if the damping is small; then Eqs. 11–6 can still be useful approximations.

Sometimes the damping is so large, however, that the motion no longer resembles simple harmonic motion. Three common cases of *heavily damped systems* are shown in Fig. 11–15. Curve A represents an **underdamped** situation, in which the system makes several oscillations before coming to rest; it corresponds to a more heavily damped version of Fig. 11–14. Curve C represents the **overdamped** situation, when the damping is so large that there is no oscillation and the system takes a long time to come to rest (equilibrium). Curve B represents **critical damping**: in this case the displacement reaches zero in the shortest time. These terms all derive from the use of practical damped systems such as door-closing mechanisms and **shock absorbers** in a car (Fig. 11–16), which are usually designed to give critical damping. But as they wear out, underdamping occurs: the door of a room slams and a car bounces up and down several times when it hits a bump.

FIGURE 11–15 Graphs that represent (A) underdamped, (B) critically damped, and (C) overdamped oscillatory motion.

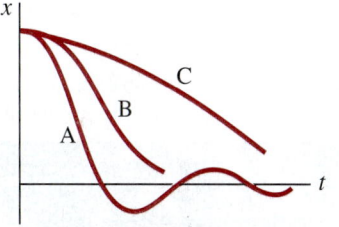

In many systems, the oscillatory motion is what counts, as in clocks and musical instruments, and damping may need to be minimized. In other systems, oscillations are the problem, such as a car's springs, so a proper amount of damping (i.e., critical) is desired. Well-designed damping is needed for all kinds of applications. Large buildings, especially in California, are now built (or retrofitted) with huge dampers to reduce possible earthquake damage (Fig. 11–17).

PHYSICS APPLIED
Shock absorbers and building dampers

†To "damp" means to diminish, restrain, or extinguish, as to "dampen one's spirits."

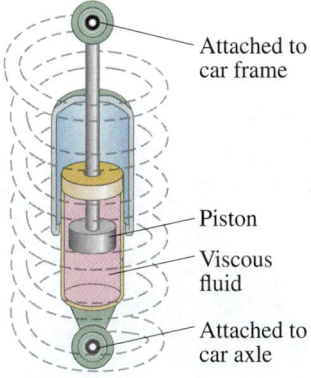

FIGURE 11–16 Automobile spring and shock absorber provide damping so that a car won't bounce up and down so much.

FIGURE 11–17 These huge dampers placed in a building look a lot like huge automobile shock absorbers, and they serve a similar purpose—to reduce the amplitude and the acceleration of movement when the shock of an earthquake hits.

11–6 Forced Oscillations; Resonance

When an oscillating system is set into motion, it oscillates at its natural frequency (Eqs. 11–6b and 11–11b). However, a system may have an external force applied to it that has its own particular frequency. Then we have a **forced oscillation**.

For example, we might pull the mass on the spring of Fig. 11–1 back and forth at an externally applied frequency f. The mass then oscillates at the external frequency f of the external force, even if this frequency is different from the **natural frequency** of the spring, which we will now denote by f_0, where (see Eq. 11–6b)

$$f_0 = \frac{1}{2\pi}\sqrt{\frac{k}{m}}.$$

For a forced oscillation with only light damping, the amplitude of oscillation is found to depend on the difference between f and f_0, and is a maximum when the frequency of the external force equals the natural frequency of the system—that is, when $f = f_0$. The amplitude is plotted in Fig. 11–18 as a function of the external frequency f. Curve A represents light damping and curve B heavy damping. When the external driving frequency f is near the natural frequency, $f \approx f_0$, the amplitude can become large if the damping is small. This effect of increased amplitude at $f = f_0$ is known as **resonance**. The natural oscillation frequency f_0 of a system is also called its **resonant frequency**.

A simple illustration of resonance is pushing a child on a swing. A swing, like any pendulum, has a natural frequency of oscillation. If you push on the swing at a random frequency, the swing bounces around and reaches no great amplitude. But if you push with a frequency equal to the natural frequency of the swing, the amplitude increases greatly. At resonance, relatively little effort is required to obtain and maintain a large amplitude.

The great tenor Enrico Caruso was said to be able to shatter a crystal goblet by singing a note of just the right frequency at full voice. This is an example of resonance, for the sound waves emitted by the voice act as a forced oscillation on the glass. At resonance, the resulting oscillation of the goblet may be large enough in amplitude that the glass exceeds its elastic limit and breaks (Fig. 11–19).

Since material objects are, in general, elastic, resonance is an important phenomenon in a variety of situations. It is particularly important in construction, although the effects are not always foreseen. For example, it has been reported that a railway bridge collapsed because a nick in one of the wheels of a crossing train set up a resonant oscillation in the bridge. Marching soldiers break step when crossing a bridge to avoid the possibility that their rhythmic march might match a resonant frequency of the bridge. The famous collapse of the Tacoma Narrows Bridge (Fig. 11–20a) in 1940 occurred as a result of strong gusting winds driving the span into large-amplitude oscillatory motion. Bridges and tall buildings are now designed with more inherent damping. The Oakland freeway collapse in the 1989 California earthquake (Fig. 11–20b) involved resonant oscillation of a section built on mudfill that readily transmitted that frequency.

Resonance can be very useful, too, and we will meet important examples later, such as in musical instruments and tuning a radio. We will also see that vibrating objects often have not one, but many resonant frequencies.

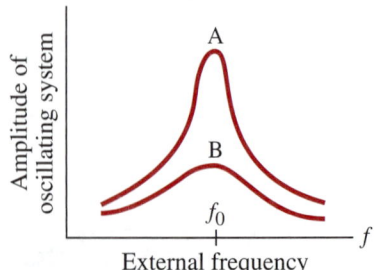

FIGURE 11–18 Amplitude as a function of driving frequency f, showing resonance for lightly damped (A) and heavily damped (B) systems.

PHYSICS APPLIED
Child on a swing

PHYSICS APPLIED
Shattering glass via resonance

FIGURE 11–19 This goblet breaks as it vibrates in resonance to a trumpet call.

PHYSICS APPLIED
Resonant collapse

FIGURE 11–20 (a) Large-amplitude oscillations of the Tacoma Narrows Bridge, due to gusty winds, led to its collapse (November 7, 1940). (b) Collapse of a freeway in California, due to the 1989 earthquake.

(a)

(b)

11–7 Wave Motion

When you throw a stone into a lake or pool of water, circular waves form and move outward, Fig. 11–21. Waves will also travel along a rope that is stretched out straight on a table if you vibrate one end back and forth as shown in Fig. 11–22. Water waves and waves on a rope or cord are two common examples of **mechanical waves**, which propagate as oscillations of matter. We will discuss other kinds of waves in later Chapters, including electromagnetic waves and light.

FIGURE 11–21 Water waves spreading outward from a source. In this case the source is a small spot of water oscillating up and down briefly where a rock hit (left photo).

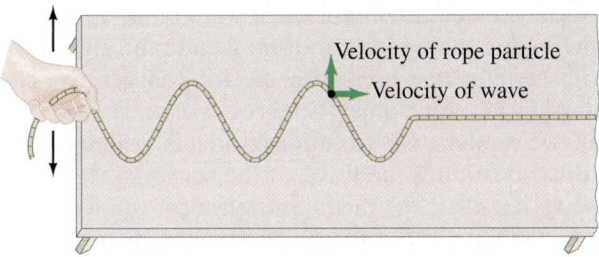

FIGURE 11–22 Wave traveling on a rope or cord. The wave travels to the right along the rope. Particles of the rope oscillate back and forth on the tabletop.

If you have ever watched ocean waves moving toward shore before they break, you may have wondered if the waves were carrying water from far out at sea onto the beach. They don't.[†] Water waves move with a recognizable velocity. But each particle (or molecule) of the water itself merely oscillates about an equilibrium point. This is clearly demonstrated by observing leaves on a pond as waves move by. The leaves (or a cork) are not carried forward by the waves, but oscillate more or less up and down about an equilibrium point because this is the motion of the water itself.

CONCEPTUAL EXAMPLE 11–10 **Wave vs. particle velocity.** Is the velocity of a wave moving along a rope the same as the velocity of a particle of the rope? See Fig. 11–22.

RESPONSE No. The two velocities are different, both in magnitude and direction. The wave on the rope of Fig. 11–22 moves to the right along the tabletop, but each piece of the rope only vibrates to and fro, perpendicular to the traveling wave. (The rope clearly does not travel in the direction that the wave on it does.)

Waves can move over large distances, but the medium (the water or the rope) itself has only a limited movement, oscillating about an equilibrium point as in simple harmonic motion. Thus, although a wave is not itself matter, the wave pattern can travel in matter. A wave consists of oscillations that move without carrying matter with them.

[†]Do not be confused by the "breaking" of ocean waves, which occurs when a wave interacts with the ground in shallow water and hence is no longer a simple wave.

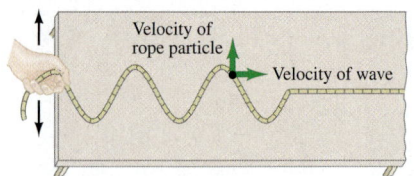

FIGURE 11–22 (Repeated.) Wave traveling on a rope or cord. The wave travels to the right along the rope. Particles of the rope oscillate back and forth on the tabletop.

FIGURE 11–23 A wave pulse is generated by a hand holding the end of a cord and moving up and down once. Motion of the wave pulse is to the right. Arrows indicate velocity of cord particles.

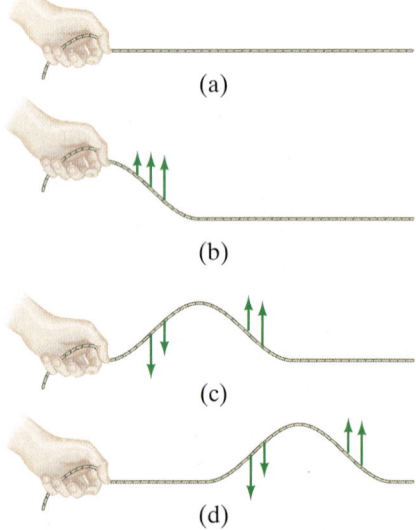

Waves carry energy from one place to another. Energy is given to a water wave, for example, by a rock thrown into the water, or by wind far out at sea. The energy is transported by waves to the shore. The oscillating hand in Fig. 11–22 transfers energy to the rope, and that energy is transported down the rope and can be transferred to an object at the other end. All forms of traveling waves transport energy.

EXERCISE G Return to Chapter-Opening Question 2, page 292, and answer it again now. Try to explain why you may have answered differently the first time.

Let us look more closely at how a wave is formed and how it comes to "travel." We first look at a single wave bump, or **pulse**. A single pulse can be formed on a cord by a quick up-and-down motion of the hand, Fig. 11–23. The hand pulls up on one end of the cord. Because the end section is attached to adjacent sections, these also feel an upward force and they too begin to move upward. As each succeeding section of cord moves upward, the wave crest moves outward along the cord. Meanwhile, the end section of cord has been returned to its original position by the hand. As each succeeding section of cord reaches its peak position, it too is pulled back down again by tension from the adjacent section of cord. Thus the source of a traveling wave pulse is a disturbance (or vibration), and cohesive forces between adjacent sections of cord cause the pulse to travel. Waves in other media are created and propagate outward in a similar fashion. A dramatic example of a wave pulse is a tsunami or tidal wave that is created by an earthquake in the Earth's crust under the ocean. The bang you hear when a door slams is a sound wave pulse.

A **continuous** or **periodic wave**, such as that shown in Fig. 11–22, has as its source a disturbance that is continuous and oscillating; that is, the source is a *vibration* or *oscillation*. In Fig. 11–22, a hand oscillates one end of the rope. Water waves may be produced by any vibrating object at the surface, such as your hand; or the water itself is made to vibrate when wind blows across it or a rock is thrown into it. A vibrating tuning fork or drum membrane gives rise to sound waves in air. We will see later that oscillating electric charges give rise to light waves. Indeed, almost any vibrating object sends out waves.

The source of any wave, then, is a vibration. And it is a *vibration* that propagates outward and thus constitutes the wave. If the source vibrates sinusoidally in SHM, then the wave itself—if the medium is elastic—will have a sinusoidal shape both in space and in time. (1) In space: if you take a picture of the wave in space at a given instant of time, the wave will have the shape of a sine or cosine as a function of position. (2) In time: if you look at the motion of the medium at one place over a long period of time—for example, if you look between two closely spaced posts of a pier or out of a ship's porthole as water waves pass by—the up-and-down motion of that small segment of water will be simple harmonic motion. The water moves up and down sinusoidally in time.

Some of the important quantities used to describe a periodic sinusoidal wave are shown in Fig. 11–24. The high points on a wave are called *crests*; the low points, *troughs*. The **amplitude**, A, is the maximum height of a crest, or depth of a trough, relative to the normal (or equilibrium) level. The total swing from a crest to a trough is $2A$ (twice the amplitude). The distance between two successive crests is the **wavelength**, λ (the Greek letter lambda). The wavelength is also equal to the distance between *any* two successive identical points on the wave. The **frequency**, f, is the number of crests—or complete cycles—that pass a given point per unit time. The **period**, T, equals $1/f$ and is the time elapsed between two successive crests passing by the same point in space.

FIGURE 11–24 Characteristics of a single-frequency continuous wave moving through space.

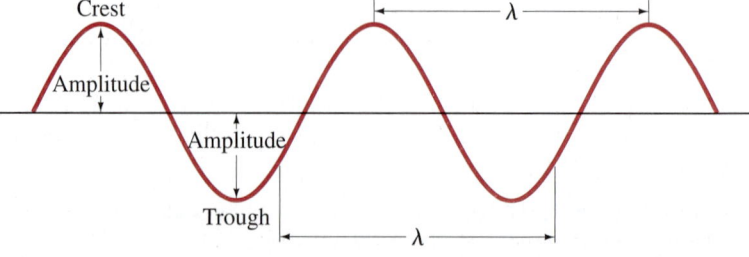

The **wave speed**, v, is the speed at which wave crests (or any other fixed point on the wave shape) move forward. The wave speed must be distinguished from the speed of a particle of the medium itself as we saw in Example 11–10.

A wave crest travels a distance of one wavelength, λ, in a time equal to one period, T. Thus the wave speed is $v = \lambda/T$. Then, since $1/T = f$,

$$v = \lambda f. \qquad (11\text{–}12)$$

For example, suppose a wave has a wavelength of 5 m and a frequency of 3 Hz. Since three crests pass a given point per second, and the crests are 5 m apart, the first crest (or any other part of the wave) must travel a distance of 15 m during the 1 s. So the wave speed is 15 m/s.

EXERCISE H You notice a water wave pass by the end of a pier, with about 0.5 s between crests. Therefore (*a*) the frequency is 0.5 Hz; (*b*) the velocity is 0.5 m/s; (*c*) the wavelength is 0.5 m; (*d*) the period is 0.5 s.

11–8 Types of Waves and Their Speeds: Transverse and Longitudinal

When a wave travels down a cord—say, from left to right as in Fig. 11–22—the particles of the cord vibrate back and forth in a direction transverse (that is, perpendicular) to the motion of the wave itself. Such a wave is called a **transverse wave** (Fig. 11–25a). There exists another type of wave known as a **longitudinal wave**. In a longitudinal wave, the vibration of the particles of the medium is *along* the direction of the wave's motion. Longitudinal waves are readily formed on a stretched spring or Slinky by alternately compressing and expanding one end. This is shown in Fig. 11–25b, and can be compared to the transverse wave in Fig. 11–25a. A series of compressions and expansions travel along the spring. The *compressions* are those areas where the coils are momentarily close together. *Expansions* (sometimes called *rarefactions*) are regions where the coils are momentarily far apart. Compressions and expansions correspond to the crests and troughs of a transverse wave.

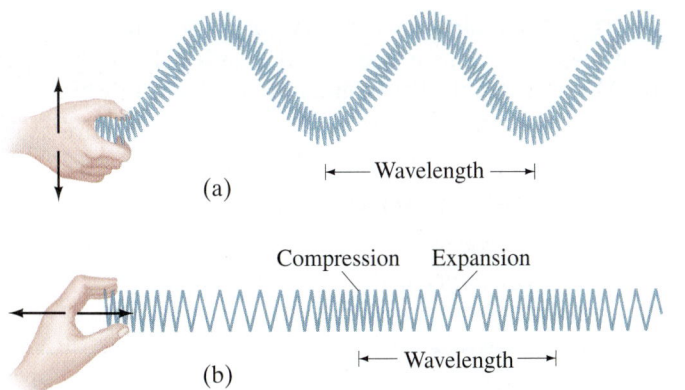

FIGURE 11–25
(a) Transverse wave;
(b) longitudinal wave.

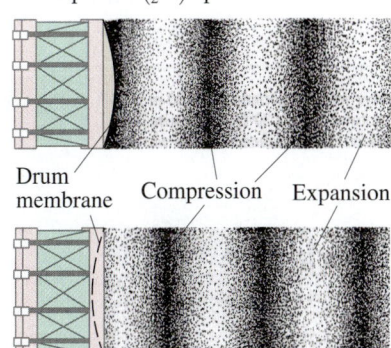

FIGURE 11–26 Production of a sound wave, which is longitudinal, shown at two moments in time about a half period ($\frac{1}{2}T$) apart.

An important example of a longitudinal wave is a sound wave in air. A vibrating drumhead, for instance, alternately compresses and expands the air in contact with it, producing a longitudinal wave that travels outward in the air, as shown in Fig. 11–26.

As in the case of transverse waves, each section of the medium in which a longitudinal wave passes oscillates over a very small distance, whereas the wave itself can travel large distances. Wavelength, frequency, and wave speed all have meaning for a longitudinal wave. The wavelength is the distance between successive compressions (or between successive expansions), and frequency is the number of compressions that pass a given point per second. The wave speed is the speed with which each compression appears to move; it is equal to the product of wavelength and frequency, $v = \lambda f$ (Eq. 11–12).

A longitudinal wave can be represented graphically by plotting the density of air molecules (or coils of a Slinky) versus position at a given instant, as shown in Fig. 11–27. Such a graphical representation makes it easy to illustrate what is happening. Note that the graph looks much like a transverse wave.

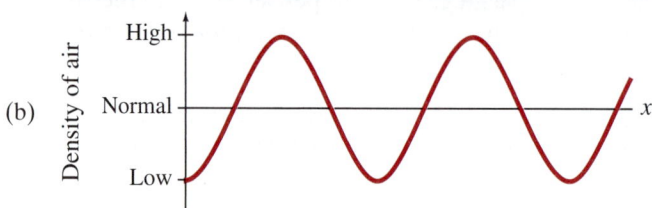

FIGURE 11–27 (a) A longitudinal wave in air, with (b) its graphical representation at a particular instant in time.

Speed of Transverse Waves

The speed of a wave depends on the properties of the medium in which it travels. The speed of a transverse wave on a stretched string or cord, for example, depends on the tension in the cord, F_T, and on the mass per unit length of the cord, μ (the Greek letter mu). If m is the mass of a length ℓ of wire, $\mu = m/\ell$. For waves of small amplitude, the wave speed is

$$v = \sqrt{\frac{F_T}{\mu}}. \qquad \left[\begin{array}{l}\text{transverse wave} \\ \text{on a cord}\end{array}\right] \quad (11\text{–}13)$$

This formula makes sense qualitatively on the basis of Newtonian mechanics. That is, we do expect the tension to be in the numerator and the mass per unit length in the denominator. Why? Because when the tension is greater, we expect the speed to be greater since each segment of cord is in tighter contact with its neighbor. Also, the greater the mass per unit length, the more inertia the cord has and the more slowly the wave would be expected to propagate.

EXAMPLE 11–11 Wave along a wire. A wave whose wavelength is 0.30 m is traveling down a 300-m-long wire whose total mass is 15 kg. If the wire is under a tension of 1000 N, what are the speed and frequency of this wave?

APPROACH We assume the velocity of this wave on a wire is given by Eq. 11–13. We get the frequency from Eq. 11–12, $f = v/\lambda$.

SOLUTION From Eq. 11–13, the velocity is

$$v = \sqrt{\frac{1000 \text{ N}}{(15 \text{ kg})/(300 \text{ m})}} = \sqrt{\frac{1000 \text{ N}}{(0.050 \text{ kg/m})}} = 140 \text{ m/s}.$$

The frequency is

$$f = \frac{v}{\lambda} = \frac{140 \text{ m/s}}{0.30 \text{ m}} = 470 \text{ Hz}.$$

NOTE A higher tension would increase both v and f, whereas a thicker, denser wire would reduce v and f.

Speed of Longitudinal Waves

The speed of a longitudinal wave has a form similar to that for a transverse wave on a cord (Eq. 11–13); that is,

$$v = \sqrt{\frac{\text{elastic force factor}}{\text{inertia factor}}}.$$

In particular, for a longitudinal wave traveling down a long solid rod,

$$v = \sqrt{\frac{E}{\rho}}, \qquad \left[\begin{array}{l}\text{longitudinal wave} \\ \text{in a long rod}\end{array}\right] \quad (11\text{–}14a)$$

where E is the elastic modulus (Section 9–5) of the material and ρ is its density.

For a longitudinal wave traveling in a liquid or gas,

$$v = \sqrt{\frac{B}{\rho}}, \quad \begin{bmatrix}\text{longitudinal wave}\\ \text{in a fluid}\end{bmatrix} \quad (11\text{-}14\text{b})$$

where B is the bulk modulus (Section 9–5) and ρ again is the density.

EXAMPLE 11–12 **Echolocation.** Echolocation is a form of sensory perception used by animals such as bats, dolphins, and toothed whales (Fig. 11–28). The animal emits a pulse of sound (a longitudinal wave) which, after reflection from objects, returns and is detected by the animal. Echolocation waves can have frequencies of about 100,000 Hz. (*a*) Estimate the wavelength of a sea animal's echolocation wave. (*b*) If an obstacle is 100 m from the animal, how long after the animal emits a wave is its reflection detected?

APPROACH We first compute the speed of longitudinal (sound) waves in sea water, using Eq. 11–14b and Tables 9–1 and 10–1. The wavelength is $\lambda = v/f$.

SOLUTION (*a*) The speed of longitudinal waves in sea water, which is slightly more dense than pure water, is (Tables 9–1 and 10–1)

$$v = \sqrt{\frac{B}{\rho}} = \sqrt{\frac{2.0 \times 10^9 \,\text{N/m}^2}{1.025 \times 10^3 \,\text{kg/m}^3}} = 1.4 \times 10^3 \,\text{m/s}.$$

Then, using Eq. 11–12, we find

$$\lambda = \frac{v}{f} = \frac{(1.4 \times 10^3 \,\text{m/s})}{(1.0 \times 10^5 \,\text{Hz})} = 14 \,\text{mm}.$$

(*b*) The time required for the round trip between the animal and the object is

$$t = \frac{\text{distance}}{\text{speed}} = \frac{2(100 \,\text{m})}{1.4 \times 10^3 \,\text{m/s}} = 0.14 \,\text{s}.$$

NOTE We shall see later that waves can be used to "resolve" (or detect) objects whose size is comparable to or larger than the wavelength. Thus, a dolphin can resolve objects on the order of a centimeter or larger in size.

Space perception by animals using sound waves

FIGURE 11–28 A toothed whale (Example 11–12).

Other Waves

Both transverse and longitudinal waves are produced when an **earthquake** occurs. The transverse waves that travel through the body of the Earth are called S waves (S for shear), and the longitudinal waves are called P waves (P for pressure) or *compression* waves. Both longitudinal and transverse waves can travel through a solid since the atoms or molecules can vibrate about their relatively fixed positions in any direction. But only longitudinal waves can propagate through a fluid, because any transverse motion would not experience any restoring force since a fluid is readily deformable. This fact was used by geophysicists to infer that a portion of the Earth's core must be liquid: after an earthquake, longitudinal waves are detected diametrically across the Earth, but not transverse waves.

Besides these two types of waves that can pass through the body of the Earth (or other substance), there can also be *surface waves* that travel along the boundary between two materials. A wave on water is actually a surface wave that moves on the boundary between water and air. The motion of each particle of water at the surface is circular or elliptical (Fig. 11–29), so it is a combination of horizontal and vertical motions. Below the surface, there is also horizontal plus vertical motion, as shown. At the bottom, the motion is only horizontal. (When a wave approaches shore, the water drags at the bottom and is slowed down, while the crests move ahead at higher speed (Fig. 11–30) and "spill" over the top.)

Surface waves are also set up on the Earth when an earthquake occurs. The waves that travel along the surface are mainly responsible for the damage caused by earthquakes.

Earthquake waves

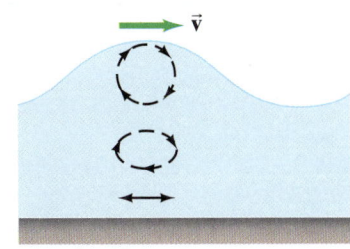

FIGURE 11–29 A shallow water wave is an example of a *surface wave*, which is a combination of transverse and longitudinal wave motions.

FIGURE 11–30 How a water wave breaks. The green arrows represent the local velocity of water molecules.

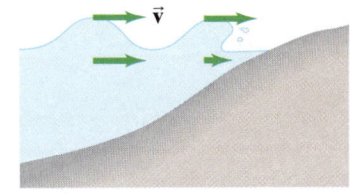

Waves which travel along a line in one dimension, such as transverse waves on a stretched string, or longitudinal waves in a rod or fluid-filled tube, are *linear* or *one-dimensional waves*. Surface waves, such as water waves (Fig. 11–21), are *two-dimensional waves*. Finally, waves that move out from a source in all directions, such as sound from a loudspeaker or earthquake waves through the Earth, are *three-dimensional waves*.

11–9 Energy Transported by Waves

Waves transport energy from one place to another. As waves travel through a medium, the energy is transferred as vibrational energy from particle to particle of the medium. For a sinusoidal wave of frequency f, the particles move in SHM as a wave passes, so each particle has an energy $E = \frac{1}{2}kA^2$, where A is the amplitude of its motion, either transversely or longitudinally. See Eq. 11–4a.

Thus, we have the important result that the **energy transported by a wave is proportional to the square of the amplitude**. The **intensity** I of a wave is defined as the power (energy per unit time) transported across unit area perpendicular to the direction of energy flow:

$$I = \frac{\text{energy/time}}{\text{area}} = \frac{\text{power}}{\text{area}}.$$

The SI unit of intensity is watts per square meter (W/m^2). Since the energy is proportional to the wave amplitude squared, so too is the intensity:

$$I \propto A^2. \qquad (11\text{--}15)$$

If a wave flows out from the source in all directions, it is a three-dimensional wave. Examples are sound traveling in open air, earthquake waves, and light waves. If the medium is isotropic (same in all directions), the wave is a *spherical wave* (Fig. 11–31). As the wave moves outward, the energy it carries is spread over a larger and larger area since the surface area of a sphere of radius r is $4\pi r^2$. Thus the intensity of a spherical wave is

$$I = \frac{\text{power}}{\text{area}} = \frac{P}{4\pi r^2}. \qquad \text{[spherical wave]} \quad (11\text{--}16\text{a})$$

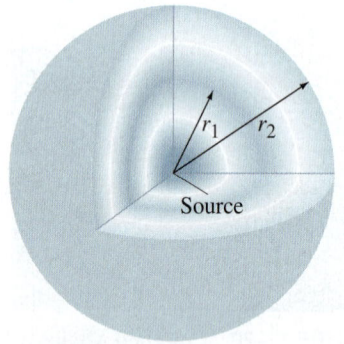

FIGURE 11–31 A wave traveling uniformly outward in three dimensions from a source is spherical. Two crests (or compressions) are shown, of radii r_1 and r_2.

If the power output P of the source is constant, then the intensity decreases as the inverse square of the distance from the source:

$$I \propto \frac{1}{r^2}. \qquad \text{[spherical wave]} \quad (11\text{--}16\text{b})$$

PROBLEM SOLVING
The $1/r^2$ law

This is often called the **inverse square law**, or the "one over r^2 law." If we consider two points at distances r_1 and r_2 from the source, as in Fig. 11–31, then $I_1 = P/4\pi r_1^2$ and $I_2 = P/4\pi r_2^2$, so

$$\frac{I_2}{I_1} = \frac{r_1^2}{r_2^2}. \qquad \text{[spherical wave]} \quad (11\text{--}16\text{c})$$

Thus, for example, when the distance doubles $(r_2/r_1 = 2)$, the intensity is reduced to $\frac{1}{4}$ its earlier value: $I_2/I_1 = \left(\frac{1}{2}\right)^2 = \frac{1}{4}$.

The amplitude of a wave also decreases with distance. Since the intensity is proportional to the square of the amplitude (Eq. 11–15), the amplitude A must decrease as $1/r$ so that $I \propto A^2$ will be proportional to $1/r^2$ (as in Eq. 11–16b). Hence

$$A \propto \frac{1}{r}.$$

If we consider again two distances from the source, r_1 and r_2, then

$$\frac{A_2}{A_1} = \frac{r_1}{r_2}. \qquad \text{[spherical wave]}$$

When the wave is twice as far from the source, the amplitude is half as large, and so on (ignoring damping due to friction).

EXAMPLE 11-13 Earthquake intensity. The intensity of an earthquake P wave traveling through the Earth and detected 100 km from the source is $1.0 \times 10^6 \text{ W/m}^2$. What is the intensity of that wave if detected 400 km from the source?

APPROACH We assume the wave is spherical, so the intensity decreases as the square of the distance from the source.

SOLUTION At 400 km the distance is 4 times greater than at 100 km, so the intensity will be $\left(\frac{1}{4}\right)^2 = \frac{1}{16}$ of its value at 100 km, or $(1.0 \times 10^6 \text{ W/m}^2)/16 = 6.3 \times 10^4 \text{ W/m}^2$.

NOTE Using Eq. 11–16c directly gives:
$$I_2 = I_1 r_1^2/r_2^2 = (1.0 \times 10^6 \text{ W/m}^2)(100 \text{ km})^2/(400 \text{ km})^2 = 6.3 \times 10^4 \text{ W/m}^2.$$

The situation is different for a one-dimensional wave, such as a transverse wave on a string or a longitudinal wave pulse traveling down a thin uniform metal rod. The area remains constant, so the amplitude A also remains constant (ignoring friction). Thus the amplitude and the intensity do not decrease with distance.

In practice, frictional damping is generally present, and some of the energy is transformed into thermal energy. Thus the amplitude and intensity of a one-dimensional wave will decrease with distance from the source. For a three-dimensional wave, the decrease will be greater than that discussed above, more than $1/r^2$, although the effect may often be small.

Intensity Related to Amplitude and Frequency

For a sinusoidal wave of frequency f, the particles move in SHM as a wave passes, so each particle has an energy $E = \frac{1}{2}kA^2$, where A is the amplitude of its motion. Using Eq. 11–6b, we can write k in terms of the frequency: $k = 4\pi^2 mf^2$, where m is the mass of a particle (or small volume) of the medium. Then

$$E = \tfrac{1}{2}kA^2 = 2\pi^2 mf^2 A^2.$$

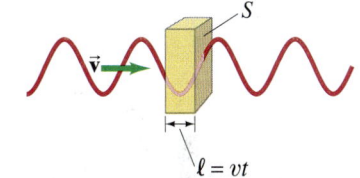

FIGURE 11–32 Calculating the energy carried by a wave moving with velocity v.

The mass $m = \rho V$, where ρ is the density of the medium and V is the volume of a small slice of the medium as shown in Fig. 11–32. The volume $V = S\ell$, where S is the cross-sectional surface area through which the wave travels. (We use S instead of A for area because we are using A for amplitude.) We can write ℓ as the distance the wave travels in a time t as $\ell = vt$, where v is the speed of the wave. Thus $m = \rho V = \rho S\ell = \rho Svt$, and

$$E = 2\pi^2 \rho Svt f^2 A^2. \quad (11\text{–}17\text{a})$$

From this equation, we see again the important result that the energy transported by a wave is proportional to the square of the amplitude. The average power transported, $\overline{P} = E/t$, is

$$\overline{P} = \frac{E}{t} = 2\pi^2 \rho S v f^2 A^2. \quad (11\text{–}17\text{b})$$

Finally, the **intensity** I of a wave is the average power transported across unit area perpendicular to the direction of energy flow:

$$I = \frac{\overline{P}}{S} = 2\pi^2 \rho v f^2 A^2. \quad (11\text{–}18)$$

This relation shows explicitly that the intensity of a wave is proportional both to the square of the wave amplitude A at any point and to the square of the frequency f.

11–10 Reflection and Transmission of Waves

When a wave strikes an obstacle, or comes to the end of the medium in which it is traveling, at least a part of the wave is reflected. You have probably seen water waves reflect off a rock or the side of a swimming pool. And you may have heard a shout reflected from a distant cliff—which we call an "echo."

A wave pulse traveling along a cord is reflected as shown in Fig. 11–33 (time increases going downward in both a and b). The reflected pulse returns inverted as in Fig. 11–33a if the end of the cord is fixed; it returns right side up if the end is free as in Fig. 11–33b. When the end is fixed to a support, as in Fig. 11–33a, the pulse reaching that fixed end exerts a force (upward) on the support. The support exerts an equal but opposite force downward on the cord (Newton's third law). This downward force on the cord is what "generates" the inverted reflected pulse.

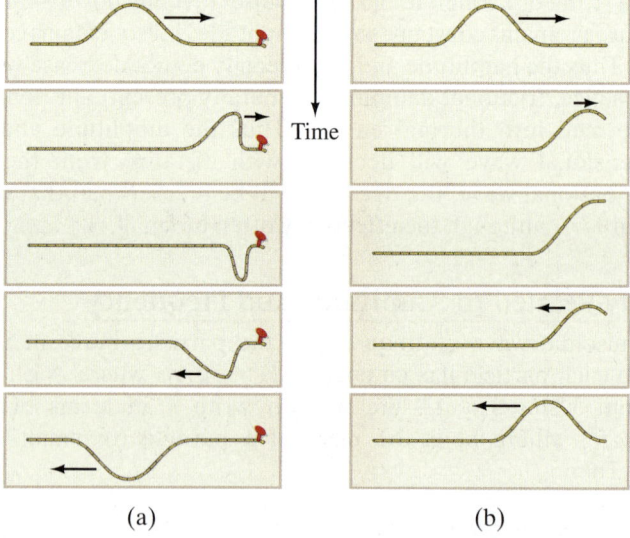

FIGURE 11–33 Reflection of a wave pulse traveling along a cord lying on a table. (Time increases going down.) (a) The end of the cord is fixed to a peg. (b) The end of the cord is free to move.

Consider next a pulse that travels along a cord which consists of a light section and a heavy section, as shown in Fig. 11–34. When the wave pulse reaches the boundary between the two sections, part of the pulse is reflected and part is transmitted, as shown. The heavier the second section of the cord, the less the energy that is transmitted. (When the second section is a wall or rigid support, very little is transmitted and most is reflected, as in Fig. 11–33a.) For a sinusoidal wave, the frequency of the transmitted wave does not change across the boundary because the boundary point oscillates at that frequency. Thus if the transmitted wave has a lower speed, its wavelength is also less ($\lambda = v/f$).

For a two or three dimensional wave, such as a water wave, we are concerned with **wave fronts**, by which we mean all the points along the wave forming the wave crest (what we usually refer to simply as a "wave" at the seashore). A line drawn in the direction of wave motion, perpendicular to the wave front, is called a **ray**, as shown in Fig. 11–35. Wave fronts far from the source have lost almost all their curvature (Fig. 11–35b) and are nearly straight, as ocean waves often are. They are then called **plane waves**.

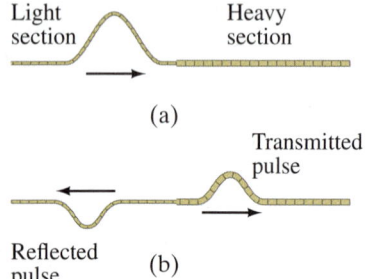

FIGURE 11–34 When a wave pulse traveling to the right along a thin cord (a) reaches a discontinuity where the cord becomes thicker and heavier, then part is reflected and part is transmitted (b).

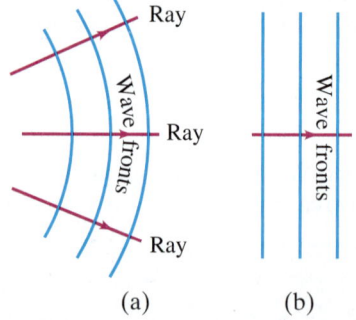

FIGURE 11–35 Rays, signifying the direction of wave motion, are always perpendicular to the wave fronts (wave crests). (a) Circular or spherical waves near the source. (b) Far from the source, the wave fronts are nearly straight or flat, and are called plane waves.

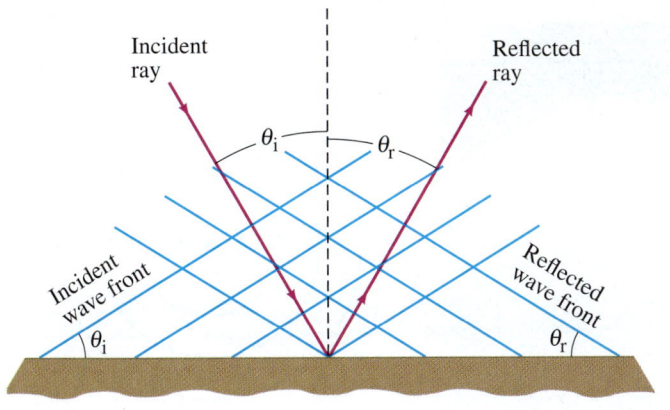

FIGURE 11–36 Law of reflection: $\theta_r = \theta_i$.

For reflection of a two or three dimensional plane wave, as shown in Fig. 11–36, the angle that the incoming or *incident wave* makes with the reflecting surface is equal to the angle made by the reflected wave. This is the **law of reflection:**

the angle of reflection equals the angle of incidence.

The **angle of incidence** is defined as the angle (θ_i) the incident ray makes with the perpendicular to the reflecting surface (or the wave front makes with the surface). The **angle of reflection** is the corresponding angle (θ_r) for the reflected wave.

11–11 Interference; Principle of Superposition

Interference refers to what happens when two waves pass through the same region of space at the same time. Consider, for example, the two wave pulses on a cord traveling toward each other as shown in Fig. 11–37 (time increases downward in both a and b). In Fig. 11–37a the two pulses have the same amplitude, but one is a crest and the other a trough; in Fig. 11–37b they are both crests. In both cases, the waves meet and pass right by each other. However, in the region where they overlap, the resultant displacement is the *algebraic sum of their separate displacements* (a crest is considered positive and a trough negative). This is the **principle of superposition**. In Fig. 11–37a, the two waves have opposite displacements at the instant they pass one another, and they add to zero. The result is called **destructive interference**. In Fig. 11–37b, at the instant the two pulses overlap, they produce a resultant displacement that is greater than the displacement of either separate pulse, and the result is **constructive interference**.

You may wonder where the energy is at the moment of destructive interference in Fig. 11–37a; the cord may be straight at this instant, but the central parts of it are still moving up or down (kinetic energy).

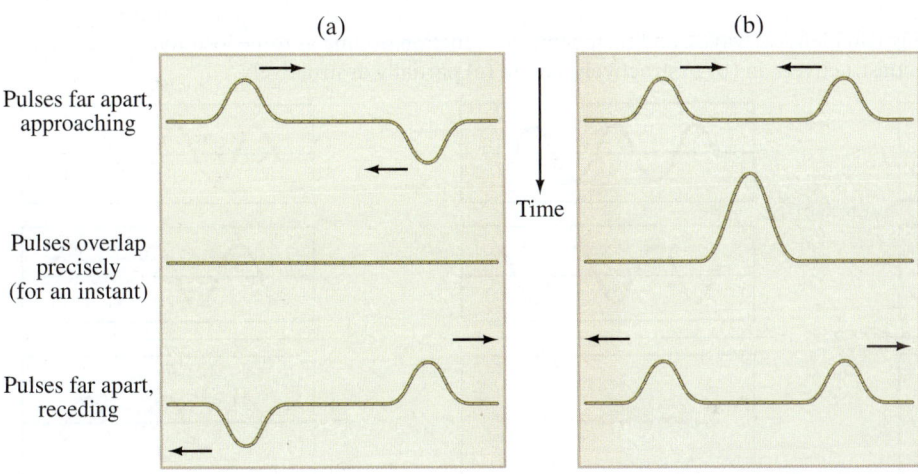

FIGURE 11–37 Two wave pulses pass each other. Where they overlap, interference occurs: (a) destructive, and (b) constructive. Read (a) and (b) downward (increasing time).

FIGURE 11–38 (a) Interference of water waves. (b) Constructive interference occurs where one wave's maximum (a crest) meets the other's maximum. Destructive interference ("flat water") occurs where one wave's maximum (a crest) meets the other's miminum (a trough).

When two rocks are thrown into a pond simultaneously, the two sets of circular waves that move outward interfere with one another as shown in Fig. 11–38a. In some areas of overlap, crests of one wave repeatedly meet crests of the other (and troughs meet troughs), Fig. 11–38b. Constructive interference is occurring at these points, and the water continuously oscillates up and down with greater amplitude than either wave separately. In other areas, destructive interference occurs where the water does not move up and down at all over time. This is where crests of one wave meet troughs of the other, and vice versa. Figure 11–39a shows the displacement of two identical waves graphically as a function of time, as well as their sum, for the case of constructive interference. For any two such waves, we use the term **phase** to describe the relative positions of their crests. When the crests and troughs are aligned as in Fig. 11–39a, for constructive interference, the two waves are **in phase**. At points where destructive interference occurs (Fig. 11–39b), crests of one wave repeatedly meet troughs of the other wave and the two waves are said to be completely **out of phase** or, more precisely, out of phase by one-half wavelength (or 180°).[†] That is, the crests of one wave occur a half wavelength behind the crests of the other wave. The relative phase of the two water waves in Fig. 11–38 in most areas is intermediate between these two extremes, resulting in *partially* destructive interference, as illustrated in Fig. 11–39c. If the amplitudes of two interfering waves are not equal, fully destructive interference (as in Fig. 11–39b) does not occur.

[†]One wavelength, or one full oscillation, corresponds to 360°—see Section 11–3, just after Eq. 11–8c, and also Fig. 11–7.

FIGURE 11–39 Graphs showing two identical waves, and their sum, as a function of time at three locations. In (a) the two waves interfere constructively, in (b) destructively, and in (c) partially destructively.

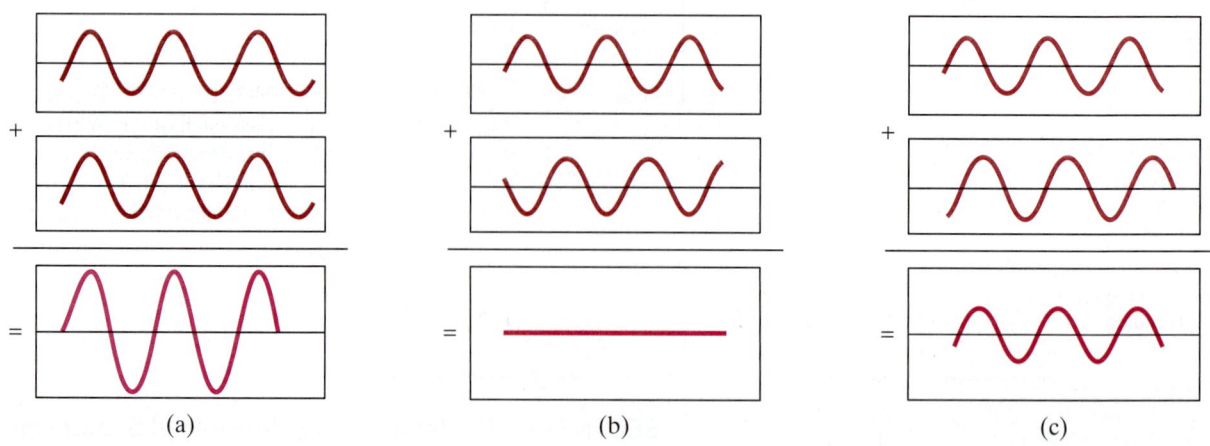

11–12 Standing Waves; Resonance

If you shake one end of a cord and the other end is kept fixed, a continuous wave will travel down to the fixed end and be reflected back, inverted, as we saw in Fig. 11–33a. As you continue to oscillate the cord, waves will travel in both directions, and the wave traveling along the cord, away from your hand, will interfere with the reflected wave coming back. Usually there will be quite a jumble. But if you oscillate the cord at just the right frequency, the two traveling waves will interfere in such a way that a large-amplitude **standing wave** will be produced, Fig. 11–40. It is called a "standing wave" because it does not appear to be traveling. The cord simply appears to have segments that oscillate up and down in a fixed pattern. The points of destructive interference, where the cord remains still at all times, are called **nodes**. Points of constructive interference, where the cord oscillates with maximum amplitude, are called **antinodes**. The nodes and antinodes remain in fixed positions for a particular frequency.

Standing waves can occur at more than one frequency. The lowest frequency of oscillation that produces a standing wave gives rise to the pattern shown in Fig. 11–40a. The standing waves shown in Figs. 11–40b and 11–40c are produced at precisely twice and three times the lowest frequency, respectively, assuming the tension in the cord is the same. The cord can also oscillate with four loops (four antinodes) at four times the lowest frequency, and so on.

The frequencies at which standing waves are produced are the **natural frequencies** or **resonant frequencies** of the cord, and the different standing wave patterns shown in Fig. 11–40 are different "resonant modes of vibration." A standing wave on a cord is the result of the interference of two waves traveling in opposite directions. A standing wave can also be considered a vibrating object at resonance. Standing waves represent the same phenomenon as the resonance of an oscillating spring or pendulum, which we discussed in Section 11–6. However, a spring or pendulum has only one resonant frequency, whereas the cord has an infinite number of resonant frequencies, each of which is a whole-number multiple of the lowest resonant frequency.

Consider a string stretched between two supports that is plucked like a guitar or violin string, Fig. 11–41a. Waves of a great variety of frequencies will travel in both directions along the string, will be reflected at the ends, and will travel back in the opposite direction. Most of these waves interfere with each other and quickly die out. However, those waves that correspond to the resonant frequencies of the string will persist. The ends of the string, since they are fixed, will be nodes. There may be other nodes as well. Some of the possible resonant modes of vibration (standing waves) are shown in Fig. 11–41b. Generally, the motion will be a combination of these different resonant modes, but only those frequencies that correspond to a resonant frequency will be present.

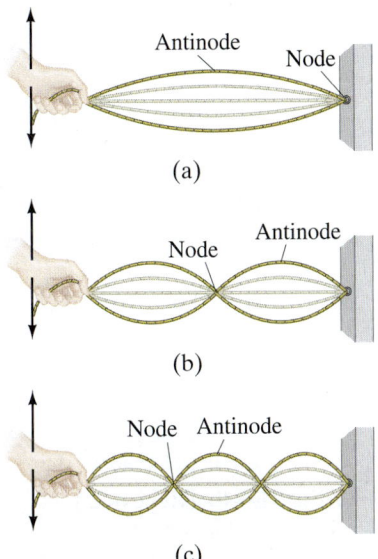

FIGURE 11–40 Standing waves corresponding to three resonant frequencies.

FIGURE 11–41 (a) A string is plucked. (b) Only standing waves corresponding to resonant frequencies persist for long.

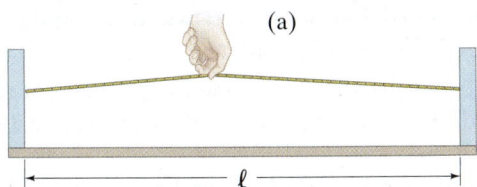

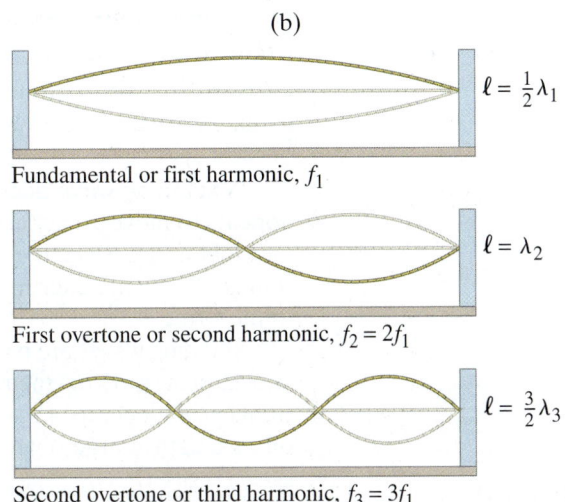

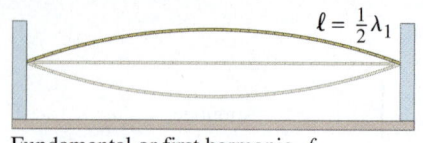

Fundamental or first harmonic, f_1

First overtone or second harmonic, $f_2 = 2f_1$

Second overtone or third harmonic, $f_3 = 3f_1$

FIGURE 11–41b (Repeated.)
(b) Only standing waves corresponding to resonant frequencies persist for long.

To determine the resonant frequencies, we first note that the wavelengths of the standing waves bear a simple relationship to the length ℓ of the string. The lowest frequency, called the **fundamental frequency**, corresponds to one antinode (or loop). And as can be seen in Fig. 11–41b, the whole length corresponds to one-half wavelength. Thus $\ell = \frac{1}{2}\lambda_1$, where λ_1 stands for the wavelength of the fundamental frequency. The other natural frequencies are called **overtones**; for a vibrating string they are whole-number (integral) multiples of the fundamental, and then are also called **harmonics**, with the fundamental being referred to as the **first harmonic**.† The next mode of vibration after the fundamental has two loops and is called the **second harmonic** (or first overtone), Fig. 11–41b. The length of the string ℓ at the second harmonic corresponds to one complete wavelength: $\ell = \lambda_2$. For the third and fourth harmonics, $\ell = \frac{3}{2}\lambda_3$, and $\ell = \frac{4}{2}\lambda_4 = 2\lambda_4$, respectively, and so on. In general, we can write

$$\ell = \frac{n\lambda_n}{2}, \quad \text{where } n = 1, 2, 3, \cdots.$$

The integer n labels the number of the harmonic: $n = 1$ for the fundamental, $n = 2$ for the second harmonic, and so on. We solve for λ_n and find

$$\lambda_n = \frac{2\ell}{n}, \quad n = 1, 2, 3, \cdots. \quad \begin{bmatrix} \text{string fixed} \\ \text{at both ends} \end{bmatrix} \quad (11\text{–}19\text{a})$$

To find the frequency f of each vibration we use Eq. 11–12, $f = v/\lambda$, and see that

$$f_n = \frac{v}{\lambda_n} = n\frac{v}{2\ell} = nf_1, \quad n = 1, 2, 3, \cdots, \quad (11\text{–}19\text{b})$$

where $f_1 = v/\lambda_1 = v/2\ell$ is the fundamental frequency. We see that each resonant frequency is an integer multiple of the fundamental frequency on a vibrating string.

Because a standing wave is equivalent to two traveling waves moving in opposite directions, the concept of wave velocity still makes sense and is given by Eq. 11–13 in terms of the tension F_T in the string and its mass per unit length ($\mu = m/\ell$). That is, $v = \sqrt{F_T/\mu}$ for waves traveling in either direction.

EXAMPLE 11–14 **Piano string.** A piano string 1.10 m long has mass 9.00 g. (a) How much tension must the string be under if it is to vibrate at a fundamental frequency of 131 Hz? (b) What are the frequencies of the first four harmonics?

APPROACH To determine the tension, we need to find the wave speed using Eq. 11–12 ($v = \lambda f$), and then use Eq. 11–13, solving it for F_T.

SOLUTION (a) The wavelength of the fundamental is $\lambda = 2\ell = 2.20$ m (Eq. 11–19a with $n = 1$). The speed of the wave on the string is $v = \lambda f = (2.20 \text{ m})(131 \text{ s}^{-1}) = 288$ m/s. Then we have (Eq. 11–13)

$$F_T = \mu v^2 = \frac{m}{\ell} v^2 = \left(\frac{9.00 \times 10^{-3} \text{ kg}}{1.10 \text{ m}}\right)(288 \text{ m/s})^2 = 679 \text{ N}.$$

(b) The first harmonic (the fundamental) has a frequency $f_1 = 131$ Hz. The frequencies of the second, third, and fourth harmonics are two, three, and four times the fundamental frequency: 262, 393, and 524 Hz, respectively.

NOTE The speed of the wave on the string is *not* the same as the speed of the sound wave that the piano string produces in the air (as we shall see in Chapter 12).

A standing wave does appear to be standing in place (and a traveling wave appears to move). The term "standing" wave is also meaningful from the point of view of energy. Since the string is at rest at the nodes, no energy flows past these points. Hence the energy is not transmitted down the string but "stands" in place in the string.

Standing waves are produced not only on strings, but also on any object that is struck, such as a drum membrane or an object made of metal or wood. The resonant frequencies depend on the dimensions of the object, just as for a string they depend on its length. Large objects have lower resonant frequencies than small objects.

†The term "harmonic" comes from music, because such integral multiples of frequencies "harmonize".

All musical instruments, from stringed to wind instruments (in which a column of air oscillates as a standing wave) to drums and other percussion instruments, depend on standing waves to produce their particular musical sounds, as we shall see in Chapter 12.

*11–13 Refraction[†]

When any wave strikes a boundary, some of the energy is reflected and some is transmitted or absorbed. When a two- or three-dimensional wave traveling in one medium crosses a boundary into a medium where its speed is different, the transmitted wave may move in a different direction than the incident wave, as shown in Fig. 11–42. This phenomenon is known as **refraction**. One example is a water wave; the velocity decreases in shallow water and the waves refract, as shown in Fig. 11–43. [When the wave velocity changes gradually, as in Fig. 11–43, without a sharp boundary, the waves change direction (refract) gradually.]

In Fig. 11–42, the velocity of the wave in medium 2 is less than in medium 1. In this case, the wave front bends so that it travels more nearly parallel to the boundary. That is, the *angle of refraction*, θ_r, is less than the *angle of incidence*, θ_i. To see why this is so, and to help us get a quantitative relation between θ_r and θ_i, let us think of each wave front as a row of soldiers. The soldiers are marching from firm ground (medium 1) into mud (medium 2) and hence are slowed down after the boundary. The soldiers that reach the mud first are slowed down first, and the row bends as shown in Fig. 11–44a. Let us consider the wave front (or row of soldiers) labeled A in Fig. 11–44b. In the same time t that A_1 moves a distance $\ell_1 = v_1 t$, we see that A_2 moves a distance $\ell_2 = v_2 t$. The two right triangles in Fig. 11–44b, shaded yellow and green, have the side labeled a in common. Thus

$$\sin\theta_1 = \frac{\ell_1}{a} = \frac{v_1 t}{a}$$

since a is the hypotenuse, and

$$\sin\theta_2 = \frac{\ell_2}{a} = \frac{v_2 t}{a}.$$

Dividing these two equations, we obtain the **law of refraction**:

$$\frac{\sin\theta_2}{\sin\theta_1} = \frac{v_2}{v_1}. \qquad (11\text{–}20)$$

Since θ_1 is the angle of incidence (θ_i), and θ_2 is the angle of refraction (θ_r), Eq. 11–20 gives the quantitative relation between the two. If the wave were going in the opposite direction, the geometry would not change; only θ_1 and θ_2 would change roles: θ_2 would be the angle of incidence and θ_1 the angle of refraction. Thus, if the wave travels into a medium where it can move faster, it will bend the opposite way, $\theta_r > \theta_i$. We see from Eq. 11–20 that if the velocity increases, the angle increases, and vice versa.

[†]This Section and the next are covered in more detail in Chapters 23 and 24 on optics.

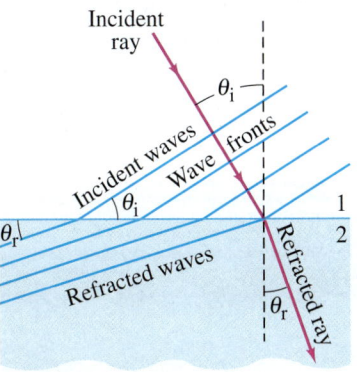

FIGURE 11–42 Refraction of waves passing a boundary.

FIGURE 11–43 Water waves refract gradually as they approach the shore, as their velocity decreases. There is no distinct boundary, as in Fig. 11–42, because the wave velocity changes gradually.

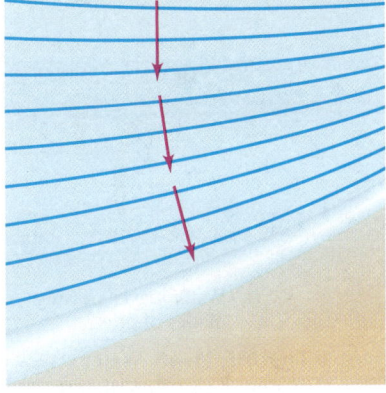

FIGURE 11–44 (a) Marching soldier analogy to derive (b) law of refraction for waves.

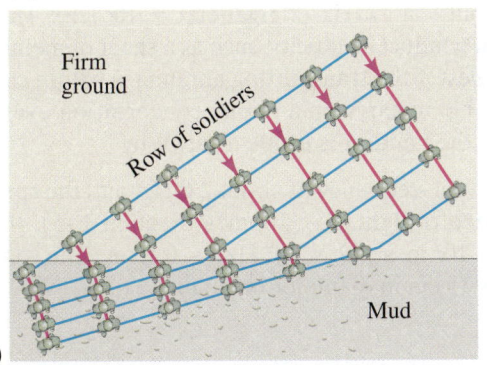

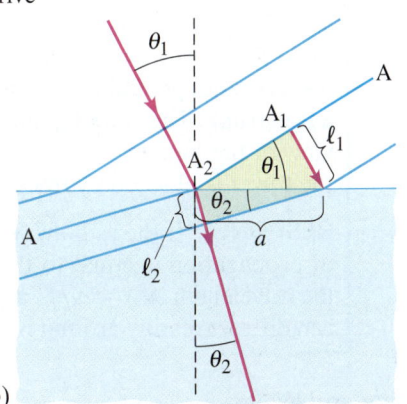

PHYSICS APPLIED
Earthquake wave refraction

Earthquake waves refract within the Earth as they travel through rock layers of different densities (which have different velocities) just as water waves do. Light waves refract as well, and when we discuss light, we shall find Eq. 11–20 very useful.

*11–14 Diffraction

Waves spread as they travel. When waves encounter an obstacle, they bend around it somewhat and pass into the region behind it, as shown in Fig. 11–45 for water waves. This phenomenon is called **diffraction**.

The amount of diffraction depends on the wavelength of the wave and on the size of the obstacle, as shown in Fig. 11–46. If the wavelength is much larger than the object, as with the grass blades of Fig. 11–46a, the wave bends around them almost as if they are not there. For larger objects, parts (b) and (c), there is more of a "shadow" region behind the obstacle where we might not expect the waves to penetrate—but they do, at least a little. Then notice in part (d), where the obstacle is the same as in part (c) but the wavelength is longer, that there is more diffraction into the shadow region. As a rule of thumb, *only if the wavelength is smaller than the size of the object will there be a significant shadow region.* This rule applies to *reflection* from an obstacle as well. Very little of a wave is reflected unless the wavelength is smaller than the size of the obstacle.

A rough guide to the amount of diffraction is

$$\theta(\text{radians}) \approx \frac{\lambda}{\ell},$$

where θ is roughly the angular spread of waves after they have passed through an opening of width ℓ or around an obstacle of width ℓ.

That waves can bend around obstacles, and thus can carry energy to areas behind obstacles, is very different from energy carried by material particles. A clear example is the following: if you are standing around a corner on one side of a building, you cannot be hit by a baseball thrown from the other side, but you can hear a shout or other sound because the sound waves diffract around the edges of the building.

(a)

(b)

FIGURE 11–45 Wave diffraction. In (a) the waves pass through a slit and into the "shadow region" behind. In (b) the waves are coming from the upper left. As they pass an obstacle, they bend around it into the shadow region behind it.

(a) Water waves passing blades of grass

(b) Stick in water

(c) Short-wavelength waves passing log

(d) Long-wavelength waves passing log

FIGURE 11–46 Water waves, coming from upper left, pass objects of various sizes. Note that the longer the wavelength compared to the size of the object, the more diffraction there is into the "shadow region."

CONCEPTUAL EXAMPLE 11–15 **Cell phones.** Cellular phones operate by radio waves with frequencies of about 1 or 2 GHz (1 gigahertz = 10^9 Hz). These waves cannot penetrate objects that conduct electricity, such as a sheet of metal or a tree trunk. The sound quality is best if the transmitting antenna is within clear view of the handset. Yet it is possible to carry on a phone conversation even if the tower is blocked by trees, or if the handset is inside a car. Why?

RESPONSE If the radio waves have a frequency of about 2 GHz, and the speed of propagation is equal to the speed of light, 3×10^8 m/s (Section 1–5), then the wavelength is $\lambda = v/f = (3 \times 10^8 \text{ m/s})/(2 \times 10^9 \text{ Hz}) = 0.15$ m. The waves can diffract readily around objects 15 cm in diameter or smaller.

*11–15 Mathematical Representation of a Traveling Wave

A simple wave with a single frequency, as in Fig. 11–47, is sinusoidal. To express such a wave mathematically, we assume it has a particular wavelength λ and frequency f. At $t = 0$, the wave shape shown is

$$y = A \sin \frac{2\pi}{\lambda} x, \qquad (11\text{–}21)$$

where y is the **displacement** of the wave (either longitudinal or transverse) at position x, λ is the wavelength, and A is the **amplitude** of the wave. [Equation 11–21 works because it repeats itself every wavelength: when $x = \lambda$, $y = \sin 2\pi = \sin 0$.]

Suppose the wave is moving to the right with speed v. After a time t, each part of the wave (indeed, the whole wave "shape") has moved to the right a distance vt. Figure 11–48 shows the wave at $t = 0$ as a solid curve, and at a later time t as a dashed curve. Consider any point on the wave at $t = 0$: say, a crest at some position x. After a time t, that crest will have traveled a distance vt, so its new position is a distance vt greater than its old position. To describe this crest (or other point on the wave shape), the argument of the sine function must have the same numerical value, so we replace x in Eq. 11–21 by $(x - vt)$:

$$y = A \sin\left[\frac{2\pi}{\lambda}(x - vt)\right]. \qquad (11\text{–}22)$$

Said another way, if you are on a crest, as t increases, x must increase at the same rate so that $(x - vt)$ remains constant.

For a wave traveling along the x axis to the left, toward decreasing values of x, v becomes $-v$, so

$$y = A \sin\left[\frac{2\pi}{\lambda}(x + vt)\right].$$

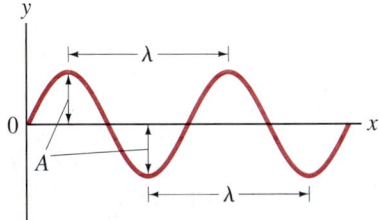

FIGURE 11–47 The characteristics of a single-frequency wave at $t = 0$ (just as in Fig. 11–24).

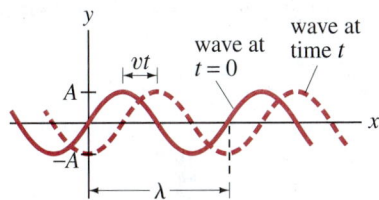

FIGURE 11–48 A traveling wave. In time t, the wave moves a distance vt.

Summary

An oscillating (or vibrating) object undergoes **simple harmonic motion** (SHM) if the restoring force is proportional to (the negative of) the displacement,

$$F = -kx. \qquad (11\text{–}1)$$

The maximum displacement from equilibrium is called the **amplitude**.

The **period**, T, is the time required for one complete cycle (back and forth), and the **frequency**, f, is the number of cycles per second; they are related by

$$f = \frac{1}{T}. \qquad (11\text{–}2)$$

The period of oscillation for a mass m on the end of a spring is given by

$$T = 2\pi\sqrt{\frac{m}{k}}. \qquad (11\text{–}6a)$$

SHM is **sinusoidal**, which means that the displacement as a function of time follows a sine curve.

During SHM, the total energy

$$E = \tfrac{1}{2}mv^2 + \tfrac{1}{2}kx^2 \qquad (11\text{–}3)$$

is continually changing from potential to kinetic and back again.

A **simple pendulum** of length ℓ approximates SHM if its amplitude is small and friction can be ignored. For small amplitudes, its period is given by

$$T = 2\pi\sqrt{\frac{\ell}{g}}, \qquad (11\text{–}11a)$$

where g is the acceleration of gravity.

When friction is present (for all real springs and pendulums), the motion is said to be **damped**. The maximum displacement decreases in time, and the mechanical energy is eventually all transformed to thermal energy.

If a varying force of frequency f is applied to a system capable of oscillating, the amplitude of oscillation can be very large if the frequency of the applied force is near the **natural** (or **resonant**) **frequency** of the oscillator. This is called **resonance**.

Vibrating objects act as sources of **waves** that travel outward from the source. Waves on water and on a cord are examples. The wave may be a **pulse** (a single crest), or it may be continuous (many crests and troughs).

The **wavelength** of a continuous sinusoidal wave is the distance between two successive crests.

The **frequency** is the number of full wavelengths (or crests) that pass a given point per unit time.

The **amplitude** of a wave is the maximum height of a crest, or depth of a trough, relative to the normal (or equilibrium) level.

The **wave speed** (how fast a crest moves) is equal to the product of wavelength and frequency,

$$v = \lambda f. \tag{11-12}$$

In a **transverse wave**, the oscillations are perpendicular to the direction in which the wave travels. An example is a wave on a cord.

In a **longitudinal wave**, the oscillations are along (parallel to) the line of travel; sound is an example.

Waves carry energy from place to place without matter being carried. The **intensity** of a wave is the energy per unit time carried across unit area (in watts/m^2). For three-dimensional waves traveling outward from a point source, the intensity decreases inversely as the square of the distance from the source (ignoring damping):

$$I \propto \frac{1}{r^2}. \tag{11-16b}$$

Wave intensity is proportional to the amplitude squared and to the frequency squared.

Waves reflect off objects in their path. When the **wave front** (of a two- or three-dimensional wave) strikes an object, the **angle of reflection** is equal to the **angle of incidence**. This is the **law of reflection**. When a wave strikes a boundary between two materials in which it can travel, part of the wave is reflected and part is transmitted.

When two waves pass through the same region of space at the same time, they **interfere**. The resultant displacement at any point and time is the sum of their separate displacements (= the **superposition principle**). This can result in **constructive interference**, **destructive interference**, or something in between, depending on the amplitudes and relative phases of the waves.

Waves traveling on a string of fixed length interfere with waves that have reflected off the end and are traveling back in the opposite direction. At certain frequencies, **standing waves** can be produced in which the waves seem to be standing still rather than traveling. The string (or other medium) is vibrating as a whole. This is a resonance phenomenon, and the frequencies at which standing waves occur are called **resonant frequencies**. Points of destructive interference (no oscillation) are called **nodes**. Points of constructive interference (maximum amplitude of vibration) are called **antinodes**.

[*Waves change direction, or **refract**, when traveling from one medium into a second medium where their speed is different. Waves spread, or **diffract**, as they travel and encounter obstacles. A rough guide to the amount of diffraction is $\theta \approx \lambda/\ell$, where λ is the wavelength and ℓ the width of an obstacle or opening. There is a significant "shadow region" only if the wavelength λ is smaller than the size of the obstacle.]

[*A traveling wave can be represented mathematically as $y = A \sin\{(2\pi/\lambda)(x \pm vt)\}$.]

Questions

1. Is the acceleration of a simple harmonic oscillator ever zero? If so, where?
2. Real springs have mass. Will the true period and frequency be larger or smaller than given by the equations for a mass oscillating on the end of an idealized massless spring? Explain.
3. How could you double the maximum speed of a simple harmonic oscillator (SHO)?
4. If a pendulum clock is accurate at sea level, will it gain or lose time when taken to high altitude? Why?
5. A tire swing hanging from a branch reaches nearly to the ground (Fig. 11–49). How could you estimate the height of the branch using only a stopwatch?

FIGURE 11–49 Question 5.

6. For a simple harmonic oscillator, when (if ever) are the displacement and velocity vectors in the same direction? When are the displacement and acceleration vectors in the same direction?
7. Two equal masses are attached to separate identical springs next to one another. One mass is pulled so its spring stretches 40 cm and the other is pulled so its spring stretches only 20 cm. The masses are released simultaneously. Which mass reaches the equilibrium point first?
8. What is the approximate period of your walking step?
9. What happens to the period of a playground swing if you rise up from sitting to a standing position?
10. Why can you make water slosh back and forth in a pan only if you shake the pan at a certain frequency?
11. Is the frequency of a simple periodic wave equal to the frequency of its source? Why or why not?
12. Explain the difference between the speed of a transverse wave traveling along a cord and the speed of a tiny piece of the cord.
13. What kind of waves do you think will travel along a horizontal metal rod if you strike its end (a) vertically from above and (b) horizontally parallel to its length?
14. Since the density of air decreases with an increase in temperature, but the bulk modulus B is nearly independent of temperature, how would you expect the speed of sound waves in air to vary with temperature?
15. If a rope has a free end, a pulse sent down the rope behaves differently on reflection than if the rope has that end fixed in position. What is this difference, and why does it occur?
16. How did geophysicists determine that part of the Earth's interior is liquid?

17. The speed of sound in most solids is somewhat greater than in air, yet the density of solids is much greater (10^3 to 10^4 times). Explain.

18. Give two reasons why circular water waves decrease in amplitude as they travel away from the source.

19. Two linear waves have the same amplitude and speed, and otherwise are identical, except one has half the wavelength of the other. Which transmits more energy? By what factor?

20. When a sinusoidal wave crosses the boundary between two sections of cord as in Fig. 11–34, the frequency does not change (although the wavelength and velocity do change). Explain why.

21. Is energy always conserved when two waves interfere? Explain.

22. If a string is vibrating as a standing wave in three loops, are there any places you could touch it with a knife blade without disturbing the motion?

23. Why do the strings used for the lowest-frequency notes on a piano normally have wire wrapped around them?

24. When a standing wave exists on a string, the vibrations of incident and reflected waves cancel at the nodes. Does this mean that energy was destroyed? Explain.

25. Can the amplitude of the standing waves in Fig. 11–40 be greater than the amplitude of the vibrations that cause them (up and down motion of the hand)?

26. "In a round bowl of water, waves move from the center to the rim, or from the rim to the center, depending on whether you strike at the center or at the rim." So wrote Dante Alighieri 700 years ago in his great poem *Paradiso* (Canto 14), the last part of his famous *Divine Comedy*. Try this experiment and discuss your results.

*27. AM radio signals can usually be heard behind a hill, but FM often cannot. That is, AM signals bend more than FM. Explain. (Radio signals, as we shall see, are carried by electromagnetic waves whose wavelength for AM is typically 200 to 600 m and for FM about 3 m.)

MisConceptual Questions

1. A mass on a spring in SHM (Fig. 11–1) has amplitude A and period T. At what point in the motion is the velocity zero and the acceleration zero simultaneously?
   (a) $x = A$.
   (b) $x > 0$ but $x < A$.
   (c) $x = 0$.
   (d) $x < 0$.
   (e) None of the above.

2. An object oscillates back and forth on the end of a spring. Which of the following statements are true at some time during the course of the motion?
   (a) The object can have zero velocity and, simultaneously, nonzero acceleration.
   (b) The object can have zero velocity and, simultaneously, zero acceleration.
   (c) The object can have zero acceleration and, simultaneously, nonzero velocity.
   (d) The object can have nonzero velocity and nonzero acceleration simultaneously.

3. An object of mass M oscillates on the end of a spring. To double the period, replace the object with one of mass:
   (a) $2M$.
   (b) $M/2$.
   (c) $4M$.
   (d) $M/4$.
   (e) None of the above.

4. An object of mass m rests on a frictionless surface and is attached to a horizontal ideal spring with spring constant k. The system oscillates with amplitude A. The oscillation frequency of this system can be increased by
   (a) decreasing k.
   (b) decreasing m.
   (c) increasing A.
   (d) More than one of the above.
   (e) None of the above will work.

5. When you use the approximation $\sin\theta \approx \theta$ for a pendulum, you must specify the angle θ in
   (a) radians only.
   (b) degrees only.
   (c) revolutions or radians.
   (d) degrees or radians.

6. Suppose you pull a simple pendulum to one side by an angle of 5°, let go, and measure the period of oscillation that ensues. Then you stop the oscillation, pull the pendulum to an angle of 10°, and let go. The resulting oscillation will have a period about _____ the period of the first oscillation.
   (a) four times
   (b) twice
   (c) half
   (d) one-fourth
   (e) the same as

7. At a playground, two young children are on identical swings. One child appears to be about twice as heavy as the other. If you pull them back together the same distance and release them to start them swinging, what will you notice about the oscillations of the two children?
   (a) The heavier child swings with a period twice that of the lighter one.
   (b) The lighter child swings with a period twice that of the heavier one.
   (c) Both children swing with the same period.

8. A grandfather clock is "losing" time because its pendulum moves too slowly. Assume that the pendulum is a massive bob at the end of a string. The motion of this pendulum can be sped up by (list all that work):
   (a) shortening the string.
   (b) lengthening the string.
   (c) increasing the mass of the bob.
   (d) decreasing the mass of the bob.

9. Consider a wave traveling down a cord and the transverse motion of a small piece of the cord. Which of the following is true?
   (a) The speed of the wave must be the same as the speed of a small piece of the cord.
   (b) The frequency of the wave must be the same as the frequency of a small piece of the cord.
   (c) The amplitude of the wave must be the same as the amplitude of a small piece of the cord.
   (d) All of the above are true.
   (e) Both (b) and (c) are true.

10. Two waves are traveling toward each other along a rope. When they meet, the waves
    (a) pass through each other.
    (b) bounce off of each other.
    (c) disappear.

11. Which of the following increases the speed of waves in a stretched elastic cord?
    (a) Increasing the wave amplitude.
    (b) Increasing the wave frequency.
    (c) Increasing the wavelength.
    (d) Stretching the elastic cord further.

12. Consider a wave on a string moving to the right, as shown in Fig. 11–50. What is the direction of the velocity of a particle of string at point B?
    (a) →
    (b) ↘
    (c) ↓
    (d) ↑
    (e) $\vec{v} = 0$, so no direction.

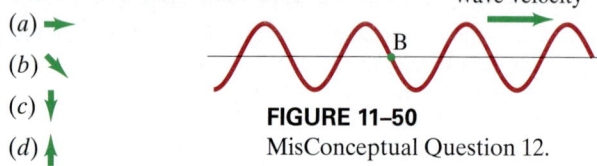

FIGURE 11–50
MisConceptual Question 12.

13. What happens when two waves, such as waves on a lake, come from different directions and run into each other?
    (a) They cancel each other out and disappear.
    (b) If they are the same size, they cancel each other out and disappear. If one wave is larger than the other, the smaller one disappears and the larger one shrinks but continues.
    (c) They get larger where they run into each other; then they continue in a direction between the direction of the two original waves and larger than either original wave.
    (d) They may have various patterns where they overlap, but each wave continues with its original pattern away from the region of overlap.
    (e) Waves cannot run into each other; they always come from the same direction and so are parallel.

14. A student attaches one end of a Slinky to the top of a table. She holds the other end in her hand, stretches it to a length ℓ, and then moves it back and forth to send a wave down the Slinky. If she next moves her hand faster while keeping the length of the Slinky the same, how does the wavelength down the Slinky change?
    (a) It increases.
    (b) It stays the same.
    (c) It decreases.

15. A wave transports
    (a) energy but not matter.
    (b) matter but not energy.
    (c) both energy and matter.

For assigned homework and other learning materials, go to the MasteringPhysics website.

Problems

11–1 to 11–3 Simple Harmonic Motion

1. (I) If a particle undergoes SHM with amplitude 0.21 m, what is the total distance it travels in one period?

2. (I) The springs of a 1700-kg car compress 5.0 mm when its 66-kg driver gets into the driver's seat. If the car goes over a bump, what will be the frequency of oscillations? Ignore damping.

3. (II) An elastic cord is 61 cm long when a weight of 75 N hangs from it but is 85 cm long when a weight of 210 N hangs from it. What is the "spring" constant k of this elastic cord?

4. (II) Estimate the stiffness of the spring in a child's pogo stick if the child has a mass of 32 kg and bounces once every 2.0 seconds.

5. (II) A fisherman's scale stretches 3.6 cm when a 2.4-kg fish hangs from it. (a) What is the spring stiffness constant and (b) what will be the amplitude and frequency of oscillation if the fish is pulled down 2.1 cm more and released so that it oscillates up and down?

6. (II) A small fly of mass 0.22 g is caught in a spider's web. The web oscillates predominantly with a frequency of 4.0 Hz. (a) What is the value of the effective spring stiffness constant k for the web? (b) At what frequency would you expect the web to oscillate if an insect of mass 0.44 g were trapped?

7. (II) A mass m at the end of a spring oscillates with a frequency of 0.83 Hz. When an additional 780-g mass is added to m, the frequency is 0.60 Hz. What is the value of m?

8. (II) A vertical spring with spring stiffness constant 305 N/m oscillates with an amplitude of 28.0 cm when 0.235 kg hangs from it. The mass passes through the equilibrium point ($y = 0$) with positive velocity at $t = 0$. (a) What equation describes this motion as a function of time? (b) At what times will the spring be longest and shortest?

9. (II) Figure 11–51 shows two examples of SHM, labeled A and B. For each, what is (a) the amplitude, (b) the frequency, and (c) the period?

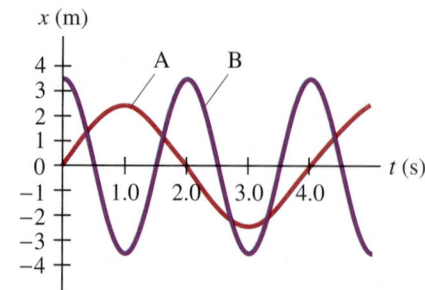

FIGURE 11–51
Problem 9.

10. (II) A balsa wood block of mass 52 g floats on a lake, bobbing up and down at a frequency of 3.0 Hz. (a) What is the value of the effective spring constant of the water? (b) A partially filled water bottle of mass 0.28 kg and almost the same size and shape of the balsa block is tossed into the water. At what frequency would you expect the bottle to bob up and down? Assume SHM.

11. (II) At what displacement of a SHO is the energy half kinetic and half potential?

12. (II) An object of unknown mass m is hung from a vertical spring of unknown spring constant k, and the object is observed to be at rest when the spring has stretched by 14 cm. The object is then given a slight push upward and executes SHM. Determine the period T of this oscillation.

13. (II) A 1.65-kg mass stretches a vertical spring 0.215 m. If the spring is stretched an additional 0.130 m and released, how long does it take to reach the (new) equilibrium position again?

14. (II) A 1.15-kg mass oscillates according to the equation $x = 0.650 \cos(8.40t)$ where x is in meters and t in seconds. Determine (a) the amplitude, (b) the frequency, (c) the total energy, and (d) the kinetic energy and potential energy when $x = 0.360$ m.

15. (II) A 0.25-kg mass at the end of a spring oscillates 2.2 times per second with an amplitude of 0.15 m. Determine (a) the speed when it passes the equilibrium point, (b) the speed when it is 0.10 m from equilibrium, (c) the total energy of the system, and (d) the equation describing the motion of the mass, assuming that at $t = 0$, x was a maximum.

16. (II) It takes a force of 91.0 N to compress the spring of a toy popgun 0.175 m to "load" a 0.160-kg ball. With what speed will the ball leave the gun if fired horizontally?

17. (II) If one oscillation has 3.0 times the energy of a second one of equal frequency and mass, what is the ratio of their amplitudes?

18. (II) A mass of 240 g oscillates on a horizontal frictionless surface at a frequency of 2.5 Hz and with amplitude of 4.5 cm. (a) What is the effective spring constant for this motion? (b) How much energy is involved in this motion?

19. (II) A mass resting on a horizontal, frictionless surface is attached to one end of a spring; the other end is fixed to a wall. It takes 3.6 J of work to compress the spring by 0.13 m. If the spring is compressed, and the mass is released from rest, it experiences a maximum acceleration of 12 m/s². Find the value of (a) the spring constant and (b) the mass.

20. (II) An object with mass 2.7 kg is executing simple harmonic motion, attached to a spring with spring constant $k = 310$ N/m. When the object is 0.020 m from its equilibrium position, it is moving with a speed of 0.55 m/s. (a) Calculate the amplitude of the motion. (b) Calculate the maximum speed attained by the object.

21. (II) At $t = 0$, an 885-g mass at rest on the end of a horizontal spring ($k = 184$ N/m) is struck by a hammer which gives it an initial speed of 2.26 m/s. Determine (a) the period and frequency of the motion, (b) the amplitude, (c) the maximum acceleration, (d) the total energy, and (e) the kinetic energy when $x = 0.40A$ where A is the amplitude.

22. (III) Agent Arlene devised the following method of measuring the muzzle velocity of a rifle (Fig. 11–52). She fires a bullet into a 4.148-kg wooden block resting on a smooth surface, and attached to a spring of spring constant $k = 162.7$ N/m. The bullet, whose mass is 7.870 g, remains embedded in the wooden block. She measures the maximum distance that the block compresses the spring to be 9.460 cm. What is the speed v of the bullet?

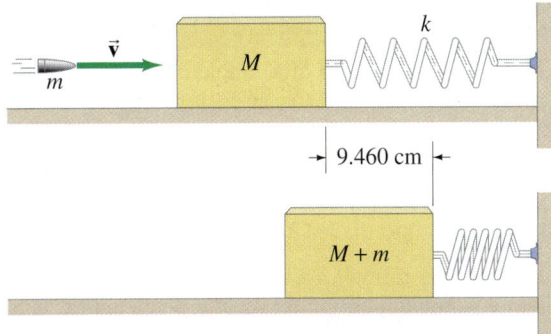

FIGURE 11–52 Problem 22.

23. (III) A bungee jumper with mass 65.0 kg jumps from a high bridge. After arriving at his lowest point, he oscillates up and down, reaching a low point seven more times in 43.0 s. He finally comes to rest 25.0 m below the level of the bridge. Estimate the spring stiffness constant and the unstretched length of the bungee cord assuming SHM.

24. (III) A block of mass m is supported by two identical parallel vertical springs, each with spring stiffness constant k (Fig. 11–53). What will be the frequency of vertical oscillation?

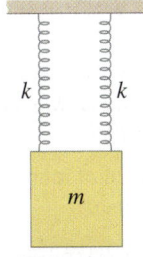

FIGURE 11–53
Problem 24.

25. (III) A 1.60-kg object oscillates at the end of a vertically hanging light spring once every 0.45 s. (a) Write down the equation giving its position y (+ upward) as a function of time t. Assume the object started by being compressed 16 cm from the equilibrium position (where $y = 0$), and released. (b) How long will it take to get to the equilibrium position for the first time? (c) What will be its maximum speed? (d) What will be the object's maximum acceleration, and where will it first be attained?

26. (III) Consider two objects, A and B, both undergoing SHM, but with different frequencies, as described by the equations $x_A = (2.0\text{ m}) \sin(4.0t)$ and $x_B = (5.0\text{ m}) \sin(3.0t)$, where t is in seconds. After $t = 0$, find the next three times t at which both objects simultaneously pass through the origin.

11–4 Simple Pendulum

27. (I) A pendulum has a period of 1.85 s on Earth. What is its period on Mars, where the acceleration of gravity is about 0.37 that on Earth?

28. (I) How long must a simple pendulum be if it is to make exactly one swing per second? (That is, one complete oscillation takes exactly 2.0 s.)

29. (I) A pendulum makes 28 oscillations in exactly 50 s. What is its (a) period and (b) frequency?

30. (II) What is the period of a simple pendulum 47 cm long (a) on the Earth, and (b) when it is in a freely falling elevator?

31. (II) Your grandfather clock's pendulum has a length of 0.9930 m. If the clock runs slow and loses 21 s per day, how should you adjust the length of the pendulum?

32. (II) Derive a formula for the maximum speed v_{max} of a simple pendulum bob in terms of g, the length ℓ, and the maximum angle of swing θ_{max}.

33. (III) A simple pendulum oscillates with an amplitude of 10.0°. What fraction of the time does it spend between $+5.0°$ and $-5.0°$? Assume SHM.

34. (III) A clock pendulum oscillates at a frequency of 2.5 Hz. At $t = 0$, it is released from rest starting at an angle of 12° to the vertical. Ignoring friction, what will be the position (angle in radians) of the pendulum at (a) $t = 0.25$ s, (b) $t = 1.60$ s, and (c) $t = 500$ s?

11–7 and 11–8 Waves

35. (I) A fisherman notices that wave crests pass the bow of his anchored boat every 3.0 s. He measures the distance between two crests to be 7.0 m. How fast are the waves traveling?

36. (I) A sound wave in air has a frequency of 282 Hz and travels with a speed of 343 m/s. How far apart are the wave crests (compressions)?

37. (I) Calculate the speed of longitudinal waves in (a) water, (b) granite, and (c) steel.

38. (I) AM radio signals have frequencies between 550 kHz and 1600 kHz (kilohertz) and travel with a speed of 3.0×10^8 m/s. What are the wavelengths of these signals? On FM the frequencies range from 88 MHz to 108 MHz (megahertz) and travel at the same speed. What are their wavelengths?

39. (II) P and S waves from an earthquake travel at different speeds, and this difference helps locate the earthquake "epicenter" (where the disturbance took place). (a) Assuming typical speeds of 8.5 km/s and 5.5 km/s for P and S waves, respectively, how far away did an earthquake occur if a particular seismic station detects the arrival of these two types of waves 1.5 min apart? (b) Is one seismic station sufficient to determine the position of the epicenter? Explain.

40. (II) A cord of mass 0.65 kg is stretched between two supports 8.0 m apart. If the tension in the cord is 120 N, how long will it take a pulse to travel from one support to the other?

41. (II) A 0.40-kg cord is stretched between two supports, 8.7 m apart. When one support is struck by a hammer, a transverse wave travels down the cord and reaches the other support in 0.85 s. What is the tension in the cord?

42. (II) A sailor strikes the side of his ship just below the surface of the sea. He hears the echo of the wave reflected from the ocean floor directly below 2.4 s later. How deep is the ocean at this point?

43. (II) Two children are sending signals along a cord of total mass 0.50 kg tied between tin cans with a tension of 35 N. It takes the vibrations in the string 0.55 s to go from one child to the other. How far apart are the children?

11–9 Energy Transported by Waves

44. (II) What is the ratio of (a) the intensities, and (b) the amplitudes, of an earthquake P wave passing through the Earth and detected at two points 15 km and 45 km from the source?

45. (II) The intensity of an earthquake wave passing through the Earth is measured to be 3.0×10^6 J/m²·s at a distance of 54 km from the source. (a) What was its intensity when it passed a point only 1.0 km from the source? (b) At what rate did energy pass through an area of 2.0 m² at 1.0 km?

46. (II) A bug on the surface of a pond is observed to move up and down a total vertical distance of 7.0 cm, from the lowest to the highest point, as a wave passes. If the ripples decrease to 4.5 cm, by what factor does the bug's maximum KE change?

47. (II) Two earthquake waves of the same frequency travel through the same portion of the Earth, but one is carrying 5.0 times the energy. What is the ratio of the amplitudes of the two waves?

11–11 Interference

48. (I) The two pulses shown in Fig. 11–54 are moving toward each other. (a) Sketch the shape of the string at the moment they directly overlap. (b) Sketch the shape of the string a few moments later. (c) In Fig. 11–37a, at the moment the pulses pass each other, the string is straight. What has happened to the energy at this moment?

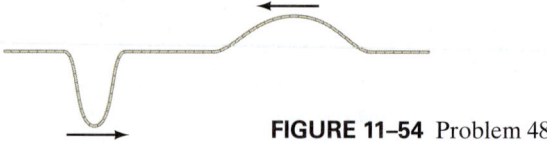

FIGURE 11–54 Problem 48.

11–12 Standing Waves; Resonance

49. (I) If a violin string vibrates at 440 Hz as its fundamental frequency, what are the frequencies of the first four harmonics?

50. (I) A violin string vibrates at 294 Hz when unfingered. At what frequency will it vibrate if it is fingered one-third of the way down from the end? (That is, only two-thirds of the string vibrates as a standing wave.)

51. (I) A particular string resonates in four loops at a frequency of 240 Hz. Give at least three other frequencies at which it will resonate. What is each called?

52. (II) The speed of waves on a string is 97 m/s. If the frequency of standing waves is 475 Hz, how far apart are two adjacent nodes?

53. (II) If two successive overtones of a vibrating string are 280 Hz and 350 Hz, what is the frequency of the fundamental?

54. (II) A guitar string is 92 cm long and has a mass of 3.4 g. The distance from the bridge to the support post is $\ell = 62$ cm, and the string is under a tension of 520 N. What are the frequencies of the fundamental and first two overtones?

55. (II) One end of a horizontal string is attached to a small-amplitude mechanical 60.0-Hz oscillator. The string's mass per unit length is 3.5×10^{-4} kg/m. The string passes over a pulley, a distance $\ell = 1.50$ m away, and weights are hung from this end, Fig. 11–55. What mass m must be hung from this end of the string to produce (a) one loop, (b) two loops, and (c) five loops of a standing wave? Assume the string at the oscillator is a node, which is nearly true.

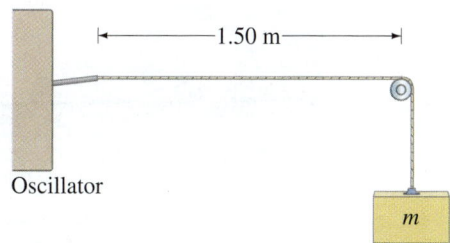

FIGURE 11–55 Problems 55 and 56.

56. (II) In Problem 55 (Fig. 11–55), the length ℓ of the string may be adjusted by moving the pulley. If the hanging mass m is fixed at 0.080 kg, how many different standing wave patterns may be achieved by varying ℓ between 10 cm and 1.5 m?

57. (II) When you slosh the water back and forth in a tub at just the right frequency, the water alternately rises and falls at each end, remaining relatively calm at the center. Suppose the frequency to produce such a standing wave in a 75-cm-wide tub is 0.85 Hz. What is the speed of the water wave?

*11–13 Refraction

*58. (I) An earthquake P wave traveling at 8.0 km/s strikes a boundary within the Earth between two kinds of material. If it approaches the boundary at an incident angle of 44° and the angle of refraction is 33°, what is the speed in the second medium?

*59. (II) A sound wave is traveling in warm air when it hits a layer of cold, dense air. If the sound wave hits the cold air interface at an angle of 25°, what is the angle of refraction? Assume that the cold air temperature is $-15°$C and the warm air temperature is $+15°$C. The speed of sound as a function of temperature can be approximated by $v = (331 + 0.60\,T)$ m/s, where T is in °C.

*60. (III) A longitudinal earthquake wave strikes a boundary between two types of rock at a 38° angle. As the wave crosses the boundary, the specific gravity changes from 3.6 to 2.5. Assuming that the elastic modulus is the same for both types of rock, determine the angle of refraction.

*11–14 Diffraction

*61. (II) What frequency of sound would have a wavelength the same size as a 0.75-m-wide window? (The speed of sound is 344 m/s at 20°C.) What frequencies would diffract through the window?

General Problems

62. A 62-kg person jumps from a window to a fire net 20.0 m directly below, which stretches the net 1.4 m. Assume that the net behaves like a simple spring. (a) Calculate how much it would stretch if the same person were lying in it. (b) How much would it stretch if the person jumped from 38 m?

63. An energy-absorbing car bumper has a spring constant of 410 kN/m. Find the maximum compression of the bumper if the car, with mass 1300 kg, collides with a wall at a speed of 2.0 m/s (approximately 5 mi/h).

64. The length of a simple pendulum is 0.72 m, the pendulum bob has a mass of 295 g, and it is released at an angle of 12° to the vertical. Assume SHM. (a) With what frequency does it oscillate? (b) What is the pendulum bob's speed when it passes through the lowest point of the swing? (c) What is the total energy stored in this oscillation assuming no losses?

65. A block of mass M is suspended from a ceiling by a spring with spring stiffness constant k. A penny of mass m is placed on top of the block. What is the maximum amplitude of oscillations that will allow the penny to just stay on top of the block? (Assume $m \ll M$.)

66. A block with mass $M = 6.0$ kg rests on a frictionless table and is attached by a horizontal spring ($k = 130$ N/m) to a wall. A second block, of mass $m = 1.25$ kg, rests on top of M. The coefficient of static friction between the two blocks is 0.30. What is the maximum possible amplitude of oscillation such that m will not slip off M?

67. A simple pendulum oscillates with frequency f. What is its frequency if the entire pendulum accelerates at $0.35\,g$ (a) upward, and (b) downward?

68. A 0.650-kg mass oscillates according to the equation $x = 0.25\sin(4.70\,t)$ where x is in meters and t is in seconds. Determine (a) the amplitude, (b) the frequency, (c) the period, (d) the total energy, and (e) the kinetic energy and potential energy when x is 15 cm.

69. An oxygen atom at a particular site within a DNA molecule can be made to execute simple harmonic motion when illuminated by infrared light. The oxygen atom is bound with a spring-like chemical bond to a phosphorus atom, which is rigidly attached to the DNA backbone. The oscillation of the oxygen atom occurs with frequency $f = 3.7 \times 10^{13}$ Hz. If the oxygen atom at this site is chemically replaced with a sulfur atom, the spring constant of the bond is unchanged (sulfur is just below oxygen in the Periodic Table). Predict the frequency after the sulfur substitution.

70. A rectangular block of wood floats in a calm lake. Show that, if friction is ignored, when the block is pushed gently down into the water and then released, it will then oscillate with SHM. Also, determine an equation for the force constant.

71. A 320-kg wooden raft floats on a lake. When a 68-kg man stands on the raft, it sinks 3.5 cm deeper into the water. When he steps off, the raft oscillates for a while. (*a*) What is the frequency of oscillation? (*b*) What is the total energy of oscillation (ignoring damping)?

72. A diving board oscillates with simple harmonic motion of frequency 2.8 cycles per second. What is the maximum amplitude with which the end of the board can oscillate in order that a pebble placed there (Fig. 11–56) does not lose contact with the board during the oscillation?

FIGURE 11–56 Problem 72.

73. A 950-kg car strikes a huge spring at a speed of 25 m/s (Fig. 11–57), compressing the spring 4.0 m. (*a*) What is the spring stiffness constant of the spring? (*b*) How long is the car in contact with the spring before it bounces off in the opposite direction?

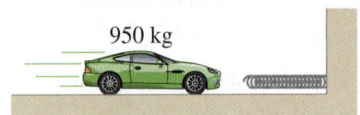

FIGURE 11–57 Problem 73.

74. A mass attached to the end of a spring is stretched a distance x_0 from equilibrium and released. At what distance from equilibrium will it have (*a*) velocity equal to half its maximum velocity, and (*b*) acceleration equal to half its maximum acceleration?

75. Carbon dioxide is a linear molecule. The carbon–oxygen bonds in this molecule act very much like springs. Figure 11–58 shows one possible way the oxygen atoms in this molecule can oscillate: the oxygen atoms oscillate symmetrically in and out, while the central carbon atom remains at rest. Hence each oxygen atom acts like a simple harmonic oscillator with a mass equal to the mass of an oxygen atom. It is observed that this oscillation occurs at a frequency $f = 2.83 \times 10^{13}$ Hz. What is the spring constant of the C—O bond?

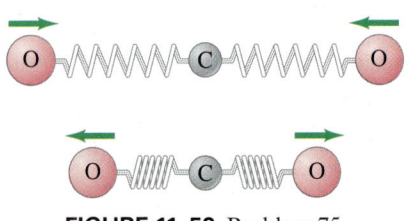

FIGURE 11–58 Problem 75, the CO_2 molecule.

76. A mass m is gently placed on the end of a freely hanging spring. The mass then falls 27.0 cm before it stops and begins to rise. What is the frequency of the oscillation?

77. Tall buildings are designed to sway in the wind. In a 100-km/h wind, suppose the top of a 110-story building oscillates horizontally with an amplitude of 15 cm at its natural frequency, which corresponds to a period of 7.0 s. Assuming SHM, find the maximum horizontal velocity and acceleration experienced by an employee as she sits working at her desk located on the top floor. Compare the maximum acceleration (as a percentage) with the acceleration due to gravity.

78. When you walk with a cup of coffee (diameter 8 cm) at just the right pace of about one step per second, the coffee sloshes higher and higher in your cup until eventually it starts to spill over the top, Fig 11–59. Estimate the speed of the waves in the coffee.

FIGURE 11–59 Problem 78.

79. A bug on the surface of a pond is observed to move up and down a total vertical distance of 0.12 m, lowest to highest point, as a wave passes. (*a*) What is the amplitude of the wave? (*b*) If the amplitude increases to 0.16 m, by what factor does the bug's maximum kinetic energy change?

80. An earthquake-produced surface wave can be approximated by a sinusoidal transverse wave. Assuming a frequency of 0.60 Hz (typical of earthquakes, which actually include a mixture of frequencies), what amplitude is needed so that objects begin to leave contact with the ground? [*Hint:* Set the acceleration $a > g$.]

81. Two strings on a musical instrument are tuned to play at 392 Hz (G) and 494 Hz (B). (*a*) What are the frequencies of the first two overtones for each string? (*b*) If the two strings have the same length and are under the same tension, what must be the ratio of their masses (m_G/m_B)? (*c*) If the strings, instead, have the same mass per unit length and are under the same tension, what is the ratio of their lengths (ℓ_G/ℓ_B)? (*d*) If their masses and lengths are the same, what must be the ratio of the tensions in the two strings?

82. A string can have a "free" end if that end is attached to a ring that can slide without friction on a vertical pole (Fig. 11–60). Determine the wavelengths of the resonant vibrations of such a string with one end fixed and the other free.

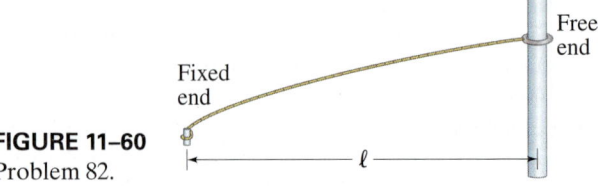

FIGURE 11–60 Problem 82.

83. The ripples in a certain groove 10.2 cm from the center of a $33\frac{1}{3}$-rpm phonograph record have a wavelength of 1.55 mm. What will be the frequency of the sound emitted?

84. A wave with a frequency of 180 Hz and a wavelength of 10.0 cm is traveling along a cord. The maximum speed of particles on the cord is the same as the wave speed. What is the amplitude of the wave?

85. Estimate the average power of a moving water wave that strikes the chest of an adult standing in the water at the seashore. Assume that the amplitude of the wave is 0.50 m, the wavelength is 2.5 m, and the period is 4.0 s.

86. A tsunami is a sort of pulse or "wave packet" consisting of several crests and troughs that become dramatically large as they enter shallow water at the shore. Suppose a tsunami of wavelength 235 km and velocity 550 km/h travels across the Pacific Ocean. As it approaches Hawaii, people observe an unusual decrease of sea level in the harbors. Approximately how much time do they have to run to safety? (In the absence of knowledge and warning, people have died during tsunamis, some of them attracted to the shore to see stranded fishes and boats.)

*87. For any type of wave that reaches a boundary beyond which its speed is increased, there is a maximum incident angle if there is to be a transmitted refracted wave. This maximum incident angle θ_{iM} corresponds to an angle of refraction equal to 90°. If $\theta_i > \theta_{iM}$, all the wave is reflected at the boundary and none is refracted, because refraction would correspond to $\sin\theta_r > 1$ (where θ_r is the angle of refraction), which is impossible. This phenomenon is referred to as *total internal reflection*. (a) Find a formula for θ_{iM} using the law of refraction, Eq. 11–20. (b) How far from the bank should a trout fisherman stand (Fig. 11–61) so trout won't be frightened by his voice (1.8 m above the ground)? The speed of sound is about 343 m/s in air and 1440 m/s in water.

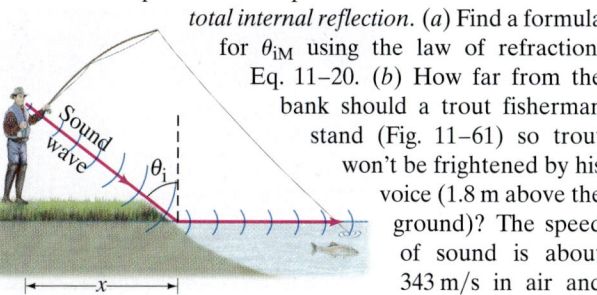

FIGURE 11–61 Problem 87b.

Search and Learn

1. Describe a procedure to measure the spring constant k of a car's springs. Assume that the owner's manual gives the car's mass M and that the shock absorbers are worn out so that the springs are underdamped. (See Sections 11–3 and 11–5.)

2. A particular unbalanced wheel of a car shakes when the car moves at 90.0 km/h. The wheel plus tire has mass 17.0 kg and diameter 0.58 m. By how much will the springs of this car compress when it is loaded with 280 kg? (Assume the 280 kg is split evenly among all four springs, which are identical.) [*Hint*: Reread Sections 11–1, 11–3, 11–6, and 8–3.]

3. Sometimes a car develops a pronounced rattle or vibration at a particular speed, especially if the road is hot enough that the tar between concrete slabs bumps up at regularly spaced intervals. Reread Sections 11–5 and 11–6, and decide whether each of the following is a factor and, if so, how: underdamping, overdamping, critical damping, and forced resonance.

4. Destructive interference occurs where two overlapping waves are $\frac{1}{2}$ wavelength or 180° out of phase. Explain why 180° is equivalent to $\frac{1}{2}$ wavelength.

5. Estimate the effective spring constant of a trampoline. [*Hint*: Go and jump, or watch, and give your data.]

6. A highway overpass was observed to resonate as one full loop $(\frac{1}{2}\lambda)$ when a small earthquake shook the ground vertically at 3.0 Hz. The highway department put a support at the center of the overpass, anchoring it to the ground as shown in Fig. 11–62. What resonant frequency would you now expect for the overpass? It is noted that earthquakes rarely do significant shaking above 5 or 6 Hz. Did the modifications do any good? Explain. (See Section 11–3.)

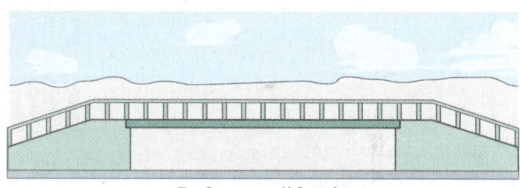

Before modification

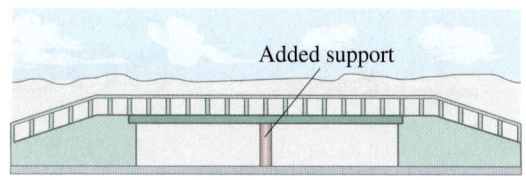

After modification

FIGURE 11–62 Search and Learn 6.

ANSWERS TO EXERCISES

A: (*b*).
B: (*c*).
C: (*a*) Increases; (*b*) increases; (*c*) increases.
D: (*c*).
E: (*c*).
F: (*a*).
G: (*c*).
H: (*d*).

"If music be the food of physics, play on." [Paraphrase of Shakespeare's *Twelfth Night*, line 1.]

Stringed instruments depend on transverse standing waves on strings to produce their harmonious sounds. The sound of wind instruments originates in longitudinal standing waves of an air column. Percussion instruments create more complicated standing waves.

Besides examining sources of sound, we also study the decibel scale of sound level, sound wave interference and beats, the Doppler effect, shock waves and sonic booms, and ultrasound imaging.

CHAPTER 12
Sound

CONTENTS

12–1 Characteristics of Sound
12–2 Intensity of Sound: Decibels
*12–3 The Ear and Its Response; Loudness
12–4 Sources of Sound: Vibrating Strings and Air Columns
*12–5 Quality of Sound, and Noise; Superposition
12–6 Interference of Sound Waves; Beats
12–7 Doppler Effect
*12–8 Shock Waves and the Sonic Boom
*12–9 Applications: Sonar, Ultrasound, and Medical Imaging

CHAPTER-OPENING QUESTION—Guess now!

A pianist plays the note "middle C." The sound is made by the vibration of the piano string and is propagated outward as a vibration of the air (which can reach your ear). How does the vibration on the string compare to the vibration in the air?

(a) The vibration on the string and the vibration in the air have the same wavelength.
(b) They have the same frequency.
(c) They have the same speed.
(d) Neither wavelength, frequency, nor speed are the same in the air as on the string.

Sound is associated with our sense of hearing and, therefore, with the physiology of our ears and the psychology of our brain, which interprets the sensations that reach our ears. The term *sound* also refers to the physical sensation that stimulates our ears: namely, longitudinal pressure waves.

We can distinguish three aspects of any sound. First, there must be a *source* for a sound; as with any mechanical wave, the source of a sound wave is a vibrating object. Second, the energy is transferred from the source in the form of longitudinal sound *waves* in air or other material. And third, the sound is *detected*, usually by an ear or by a microphone. We start by looking at sound waves themselves.

12–1 Characteristics of Sound

We saw in Chapter 11, Fig. 11–26, how a vibrating drumhead produces a sound wave in air. Indeed, we usually think of sound waves traveling in the air, because normally it is the vibrations of the air that force our eardrums to vibrate. But sound waves can also travel in other materials.

Two stones struck together under water can be heard by a swimmer beneath the surface, for the vibrations are carried to the ear by the water. When you put your ear flat against the ground, you can hear an approaching train or truck. In this case the ground does not actually touch your eardrum, but the longitudinal wave transmitted by the ground is called a sound wave just the same, since its vibrations cause the outer ear and the air within it to vibrate. Sound cannot travel in the absence of matter. For example, a bell ringing inside an evacuated jar cannot be heard, and sound cannot travel through the empty reaches of outer space.

The **speed of sound** is different in different materials. In air at 0°C and 1 atm, sound travels at a speed of 331 m/s. The speed of sound in various materials is given in Table 12–1. The values depend somewhat on temperature, especially for gases. For example, in air near room temperature, the speed increases approximately 0.60 m/s for each Celsius degree increase in temperature:

$$v \approx (331 + 0.60T) \text{ m/s}, \quad\quad\quad \text{[speed of sound in air]}$$

where T is the temperature in °C. Unless stated otherwise, we will assume in this Chapter that $T = 20$°C, so $v = [331 + (0.60)(20)]$ m/s = 343 m/s.

TABLE 12–1 Speed of Sound in Various Materials (20°C and 1 atm)

Material	Speed (m/s)
Air	343
Air (0°C)	331
Helium	1005
Hydrogen	1300
Water	1440
Sea water	1560
Iron and steel	≈ 5000
Glass	≈ 4500
Aluminum	≈ 5100
Hardwood	≈ 4000
Concrete	≈ 3000

> **CONCEPTUAL EXAMPLE 12–1** **Distance from a lightning strike.** A rule of thumb that tells how close lightning has struck is "one mile for every five seconds before the thunder is heard." Explain why this works, noting that the speed of light is so high (3×10^8 m/s, almost a million times faster than sound) that the time for light to travel to us is negligible compared to the time for the sound.
>
> **RESPONSE** The speed of sound in air is about 340 m/s, so to travel 1 km = 1000 m takes about 3 seconds. One mile is about 1.6 kilometers, so the time for the thunder to travel a mile is about $(1.6)(3) \approx 5$ seconds.

PHYSICS APPLIED
How far away is the lightning?

Two aspects of any sound are immediately evident to a human listener: "loudness" and "pitch." Each refers to a sensation in the consciousness of the listener. But to each of these subjective sensations there corresponds a physically measurable quantity. **Loudness** is related to the intensity (energy per unit time crossing unit area) in the sound wave, and we shall discuss it in the next Section.

The **pitch** of a sound refers to whether it is high, like the sound of a piccolo or violin, or low, like the sound of a bass drum or string bass. The physical quantity that determines pitch is the frequency, as was first noted by Galileo. The lower the frequency, the lower the pitch; the higher the frequency, the higher the pitch.[†] The best human ears can respond to frequencies from about 20 Hz to almost 20,000 Hz. (Recall that 1 Hz is 1 cycle per second.) This frequency range is called the **audible range**. These limits vary somewhat from one individual to another. One general trend is that as people age, they are less able to hear high frequencies, so the high-frequency limit may be 10,000 Hz or less.

Sound waves whose frequencies are outside the audible range may reach the ear, but we are not generally aware of them. Frequencies above 20,000 Hz are called **ultrasonic** (do not confuse with *supersonic*, which is used for an object moving with a speed faster than the speed of sound). Many animals can hear ultrasonic frequencies; dogs, for example, can hear sounds as high as 50,000 Hz, and bats can detect frequencies as high as 100,000 Hz. Ultrasonic waves have many useful applications in medicine and other fields, which we will discuss later in this Chapter.

CAUTION
Do not confuse ultrasonic (high frequency) with supersonic (high speed)

[†]Although pitch is determined mainly by frequency, it also depends to a slight extent on loudness. For example, a very loud sound may seem slightly lower in pitch than a quiet sound of the same frequency.

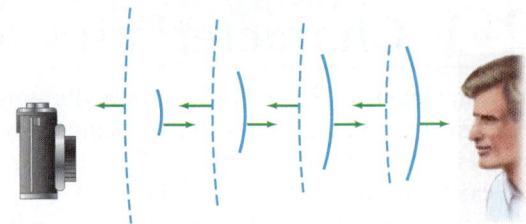

FIGURE 12–1 Example 12–2. Autofocusing camera emits an ultrasonic pulse. Solid lines represent the wave front of the outgoing wave pulse moving to the right; dashed lines represent the wave front of the pulse reflected off the person's face, returning to the camera. The time between emission and reception by the camera of these waves allows the camera mechanism to adjust the lens to focus at the proper distance.

PHYSICS APPLIED
Autofocusing camera

EXAMPLE 12–2 **Autofocusing with sound waves.** Autofocusing cameras emit a pulse of very high frequency (ultrasonic) sound that travels to the object being photographed, and include a sensor that detects the returning reflected sound, as shown in Fig. 12–1. To get an idea of the time sensitivity of the detector, calculate the travel time of the pulse for an object (a) 1.0 m away, and (b) 20 m away.

APPROACH If we assume the temperature is about 20°C, then the speed of sound is 343 m/s. Using this speed v and the total distance d back and forth in each case, we can obtain the time ($v = d/t$).

SOLUTION (a) The pulse travels 1.0 m to the object and 1.0 m back, for a total of 2.0 m. We solve for t in $v = d/t$:

$$t = \frac{d}{v} = \frac{2.0\,\text{m}}{343\,\text{m/s}} = 0.0058\,\text{s} = 5.8\,\text{ms}.$$

(b) The total distance now is $2 \times 20\,\text{m} = 40\,\text{m}$, so

$$t = \frac{40\,\text{m}}{343\,\text{m/s}} = 0.12\,\text{s} = 120\,\text{ms}.$$

NOTE Newer autofocus cameras use infrared light ($v = 3 \times 10^8\,\text{m/s}$) instead of ultrasound, and/or a digital sensor array that detects light intensity differences between adjacent receptors as the lens is automatically moved back and forth, choosing the lens position that provides maximum intensity differences (sharpest focus).

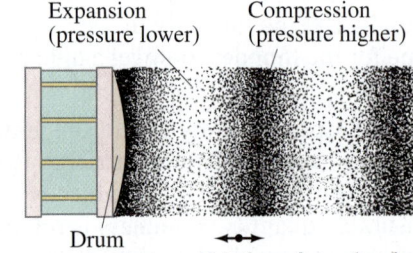

FIGURE 12–2 The membrane of a drum, as it vibrates, alternately compresses the air and, as it recedes (moves to the left), leaves a rarefaction or expansion of air. See also Fig. 11–26.

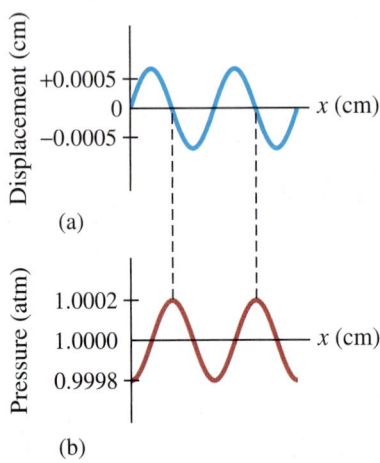

FIGURE 12–3 Representation of a sound wave in space at a given instant in terms of (a) displacement, and (b) pressure.

We often describe a sound wave in terms of the vibration of the molecules of the medium in which it travels—that is, in terms of the motion or displacement of the molecules. Sound waves can also be analyzed from the point of view of pressure. Indeed, longitudinal waves are often called **pressure waves**. The pressure variation is usually easier to measure than the displacement. As Fig. 12–2 shows, in a wave "compression" (where molecules are closest together), the pressure is higher than normal, whereas in an expansion (or *rarefaction*) the pressure is less than normal. Figure 12–3 shows a graphical representation of a sound wave in air in terms of (a) displacement and (b) pressure. Note that the displacement wave is a quarter wavelength out of phase with the pressure wave: where the pressure is a maximum or minimum, the displacement from equilibrium is zero; and where the pressure variation is zero, the displacement is a maximum or minimum.

Sound waves whose frequencies are below the audible range (that is, less than 20 Hz) are called **infrasonic**. Sources of infrasonic waves include earthquakes, thunder, volcanoes, and waves produced by vibrating heavy machinery. This last source can be particularly troublesome to workers, because infrasonic waves—even though inaudible—can cause damage to the human body. These low-frequency waves act in a resonant fashion, causing motion and irritation of the body's organs.

12–2 Intensity of Sound: Decibels

Loudness is a sensation in the consciousness of a human being and is related to a physically measurable quantity, the **intensity** of the wave. Intensity is defined as the energy transported by a wave per unit time across a unit area perpendicular to the energy flow. As we saw in Chapter 11, intensity is proportional to the square of the wave amplitude. Intensity has units of power per unit area, or watts/meter^2 (W/m^2).

An average human ear can detect sounds with an intensity as low as 10^{-12} W/m^2 and as high as 1 W/m^2 (and even higher, although above this it is painful). This is an incredibly wide range of intensity, spanning a factor of 10^{12} from lowest to highest. Presumably because of this wide range, what we perceive as loudness is not directly proportional to the intensity. To produce a sound that sounds about twice as loud requires a sound wave that has about 10 times the intensity. This is roughly valid at any sound level for frequencies near the middle of the audible range. For example, a sound wave of intensity 10^{-2} W/m^2 sounds to an average human being like it is about twice as loud as one with intensity of 10^{-3} W/m^2, and four times as loud as 10^{-4} W/m^2.

PHYSICS APPLIED
Wide range of human hearing

Sound Level

Because of this relationship between the subjective sensation of loudness and the physically measurable quantity "intensity," sound intensity levels are usually specified on a logarithmic scale. The unit on this scale is a **bel**, after the inventor Alexander Graham Bell, or much more commonly, the **decibel** (dB), which is $\frac{1}{10}$ bel (10 dB = 1 bel).† The **sound level**, β, of any sound is defined in terms of its intensity, I, as

$$\beta \text{ (in dB)} = 10 \log \frac{I}{I_0}, \tag{12-1}$$

where I_0 is the intensity of a chosen reference level, and the logarithm is to the base 10. I_0 is usually taken as the minimum intensity audible to a good ear—the "threshold of hearing," which is $I_0 = 1.0 \times 10^{-12}$ W/m^2. Thus, for example, the sound level of a sound whose intensity $I = 1.0 \times 10^{-10}$ W/m^2 will be

$$\beta = 10 \log \left(\frac{1.0 \times 10^{-10} \text{ W/m}^2}{1.0 \times 10^{-12} \text{ W/m}^2} \right) = 10 \log 100 = 20 \text{ dB},$$

since log 100 is equal to 2.0. (Appendix A has a brief review of logarithms.) Notice that the sound level at the threshold of hearing is 0 dB. That is, $\beta = 10 \log 10^{-12}/10^{-12} = 10 \log 1 = 0$ since log 1 = 0. Notice too that an increase in intensity by a factor of 10 corresponds to a sound level increase of 10 dB. An increase in intensity by a factor of 100 corresponds to a sound level increase of 20 dB. Thus a 50-dB sound is 100 times more intense than a 30-dB sound, and so on.

Intensities and sound levels for some common sounds are listed in Table 12–2.

CAUTION
0 dB does not mean zero intensity

EXAMPLE 12–3 Sound intensity on the street. At a busy street corner, the sound level is 75 dB. What is the intensity of sound there?

APPROACH We have to solve Eq. 12–1 for intensity I, remembering that $I_0 = 1.0 \times 10^{-12}$ W/m^2.

SOLUTION From Eq. 12–1

$$\log \frac{I}{I_0} = \frac{\beta}{10}.$$

Recalling that $x = \log y$ is the same as $y = 10^x$ (Appendix A–8), then

$$\frac{I}{I_0} = 10^{\beta/10}.$$

With $\beta = 75$, then

$$I = I_0 10^{\beta/10} = (1.0 \times 10^{-12} \text{ W/m}^2)(10^{7.5}) = 3.2 \times 10^{-5} \text{ W/m}^2.$$

TABLE 12–2 Intensity of Various Sounds

Source of the Sound	Sound Level (dB)	Intensity (W/m^2)
Jet plane at 30 m	140	100
Threshold of pain	120	1
Loud rock concert	120	1
Siren at 30 m	100	1×10^{-2}
Busy street traffic	80	1×10^{-4}
Noisy restaurant	70	1×10^{-5}
Talk, at 50 cm	65	3×10^{-6}
Quiet radio	40	1×10^{-8}
Whisper	30	1×10^{-9}
Rustle of leaves	10	1×10^{-11}
Threshold of hearing	0	1×10^{-12}

†The dB is dimensionless and so does not have to be included in calculations.

PHYSICS APPLIED
Loudspeaker response (± 3 dB)

EXAMPLE 12–4 **Loudspeaker response.** A high-quality loudspeaker is advertised to reproduce, at full volume, frequencies from 30 Hz to 18,000 Hz with uniform sound level ± 3 dB. That is, over this frequency range, the sound level output does not vary by more than 3 dB for a given input level. By what factor does the intensity change for a 3 dB change in output level?

APPROACH Let us call the average intensity I_1 and the average sound level β_1. Then the maximum intensity, I_2, corresponds to a level $\beta_2 = \beta_1 + 3$ dB. We then use the relation between intensity and sound level, Eq. 12–1.

SOLUTION Equation 12–1 gives

$$\beta_2 - \beta_1 = 10 \log \frac{I_2}{I_0} - 10 \log \frac{I_1}{I_0}$$

$$3 \text{ dB} = 10 \left(\log \frac{I_2}{I_0} - \log \frac{I_1}{I_0} \right) = 10 \log \frac{I_2}{I_1}$$

because $(\log a - \log b) = \log a/b$, as discussed in Appendix A–8. This last equation gives

$$\log \frac{I_2}{I_1} = 0.30,$$

or

$$\frac{I_2}{I_1} = 10^{0.30} = 2.0.$$

So ± 3 dB corresponds to a doubling or halving of the intensity.

NOTE From this last equation, we see also that $\log 2.0 = 0.30$. See also the short Table of Logarithms in Appendix A–8.

EXERCISE A If an increase of 3 dB means "twice as intense," what does an increase of 6 dB mean?

It is worth noting that a sound-level difference of 3 dB (which corresponds to a doubled intensity, as we just saw) corresponds to only a very small change in the subjective sensation of apparent loudness. Indeed, the average human can distinguish a difference in sound level of only about 1 or 2 dB.

Normally, the loudness or intensity of a sound decreases as you get farther from the source of the sound. Indoors, this effect is altered because of reflections from the walls. However, if a source is in the open so that sound can radiate out freely in all directions, the intensity decreases as the inverse square of the distance,

$$I \propto \frac{1}{r^2},$$

as we saw in Section 11–9. Over large distances, the intensity decreases faster than $1/r^2$ because some of the energy is transferred into irregular motion of air molecules. This loss happens more for higher frequencies, so any sound of mixed frequencies will be less "bright" at a distance. Far from an outdoor band, you hear mainly the boom of the drums.

CONCEPTUAL EXAMPLE 12–5 **Trumpet players.** A trumpeter plays at a sound level of 75 dB. Three equally loud trumpet players join in. What is the new sound level?

RESPONSE The intensity of four trumpets is four times the intensity of one trumpet $(= I_1)$ or $4I_1$. The sound level of the four trumpets would be

$$\beta = 10 \log \frac{4I_1}{I_0} = 10 \log 4 + 10 \log \frac{I_1}{I_0}$$

$$= 6.0 \text{ dB} + 75 \text{ dB} = 81 \text{ dB},$$

since $\log 4 = 0.60$.

EXERCISE B From Table 12–2, we see that ordinary conversation corresponds to a sound level of about 65 dB. If two people are talking at once, the sound level is (a) 65 dB, (b) 68 dB, (c) 75 dB, (d) 130 dB, (e) 62 dB.

EXAMPLE 12–6 **Airplane roar.** The sound level measured 30 m from a jet plane is 140 dB. Estimate the sound level at 300 m. (Ignore reflections from the ground.)

APPROACH Given the sound level, we can determine the intensity at 30 m using Eq. 12–1. Because intensity decreases as the square of the distance, ignoring reflections, we can find I at 300 m and again apply Eq. 12–1 to obtain the sound level.

SOLUTION The intensity I at 30 m is

$$140 \text{ dB} = 10 \log\left(\frac{I}{10^{-12} \text{ W/m}^2}\right)$$

or, dividing through by 10,

$$14 = \log\left(\frac{I}{10^{-12} \text{ W/m}^2}\right).$$

Recall that $y = \log x$ means $10^y = x$ (Appendix A–8). Then

$$10^{14} = \frac{I}{10^{-12} \text{ W/m}^2},$$

so $I = (10^{14})(10^{-12} \text{ W/m}^2) = 10^2 \text{ W/m}^2$. At 300 m, 10 times as far, the intensity, which decreases as $1/r^2$, will be $\left(\frac{1}{10}\right)^2 = 1/100$ as much, or 1 W/m^2. Hence, the sound level is

$$\beta = 10 \log\left(\frac{1 \text{ W/m}^2}{10^{-12} \text{ W/m}^2}\right) = 120 \text{ dB}.$$

Even at 300 m, the sound is at the threshold of pain. This is why workers at airports wear ear covers to protect their ears from damage (Fig. 12–4).

NOTE Here is a simpler approach that avoids Eq. 12–1. Because the intensity decreases as the square of the distance, at 10 times the distance the intensity decreases by $\left(\frac{1}{10}\right)^2 = \frac{1}{100}$. We can use the result that 10 dB corresponds to an intensity change by a factor of 10 (see just before Example 12–3). Then an intensity change by a factor of 100 corresponds to a sound-level change of $(2)(10 \text{ dB}) = 20 \text{ dB}$. This confirms our result above: $140 \text{ dB} - 20 \text{ dB} = 120 \text{ dB}$.

> **PHYSICS APPLIED**
> *Jet plane noise*

FIGURE 12–4 Example 12–6. Airport worker with sound-intensity-reducing ear covers (headphones).

Intensity Related to Amplitude

The intensity I of a wave is proportional to the square of the wave amplitude, A, as we saw in Section 11–9. We can therefore relate the amplitude quantitatively to the intensity I or sound level β, as the following Example shows.

EXAMPLE 12–7 **How tiny the displacement is.** Calculate the displacement of air molecules for a sound of frequency 1000 Hz at the threshold of hearing.

APPROACH In Section 11–9 we found a relation between intensity I and displacement amplitude A of a wave, Eq. 11–18. The amplitude of oscillation of air molecules is what we want to solve for, given the intensity. Assume the temperature is 20°C so the speed of sound is 343 m/s.

SOLUTION At the threshold of hearing, $I = 1.0 \times 10^{-12} \text{ W/m}^2$ (Table 12–2). We solve for the amplitude A in Eq. 11–18:

$$A = \frac{1}{\pi f}\sqrt{\frac{I}{2\rho v}}$$

$$= \frac{1}{(3.14)(1.0 \times 10^3 \text{ s}^{-1})}\sqrt{\frac{1.0 \times 10^{-12} \text{ W/m}^2}{(2)(1.29 \text{ kg/m}^3)(343 \text{ m/s})}} = 1.1 \times 10^{-11} \text{ m},$$

where we have taken the density of air to be 1.29 kg/m^3 (Table 10–1).

NOTE We see how incredibly sensitive the human ear is: it can detect displacements of air molecules which are less than the diameter of atoms (about 10^{-10} m).

> **PHYSICS APPLIED**
> *Incredible sensitivity of the ear*

*12–3 The Ear and Its Response; Loudness

The human ear is a remarkably sensitive detector of sound. Mechanical detectors of sound (microphones) can barely match the ear in detecting low-intensity sounds.

The function of the ear is to transform the vibrational energy of waves into electrical signals which are carried to the brain by way of nerves. A microphone performs a similar task. Sound waves striking the diaphragm of a microphone set it into vibration, and these vibrations are transformed into an electrical signal (Chapter 21) with the same frequencies, which can then be amplified and sent to a loudspeaker or recorder.

PHYSICS APPLIED
Human ear

Figure 12–5 is a diagram of the human ear. The ear consists of three main divisions: the outer ear, middle ear, and inner ear. In the outer ear, sound waves from the outside travel down the ear canal to the eardrum (the tympanum), which vibrates in response to the impinging waves. The middle ear consists of three small bones known as the hammer, anvil, and stirrup, which transfer the vibrations of the eardrum to the inner ear at the oval window. This delicate system of levers, coupled with the relatively large area of the eardrum compared to the area of the oval window, results in the pressure being amplified by a factor of about 20. The inner ear consists of the semicircular canals, which are important for controlling balance, and the liquid-filled cochlea where the vibrational energy of sound waves is transformed into electrical energy and sent to the brain.

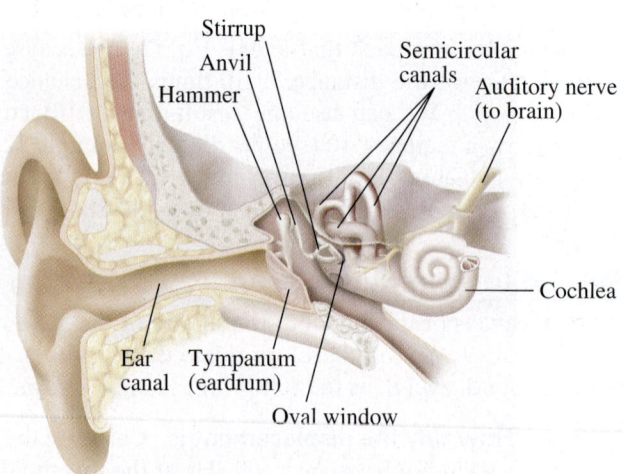

FIGURE 12–5 Diagram of the human ear.

*The Ear's Response

The ear is not equally sensitive to all frequencies. To hear the same loudness for sounds of different frequencies requires different intensities. Studies on large numbers of people have produced the averaged curves shown in Fig. 12–6 (top of next page). On this graph, each curve represents sounds that seemed to be equally loud. The number labeling each curve represents the **loudness level** (the units are called *phons*), which is numerically equal to the sound level in dB at 1000 Hz. For example, the curve labeled 40 represents sounds that are heard by an average person to have the same loudness as a 1000-Hz sound with a sound level of 40 dB. From this 40-phon curve, we see that a 100-Hz tone must be at a level of about 62 dB to be perceived as loud as a 1000-Hz tone of only 40 dB.

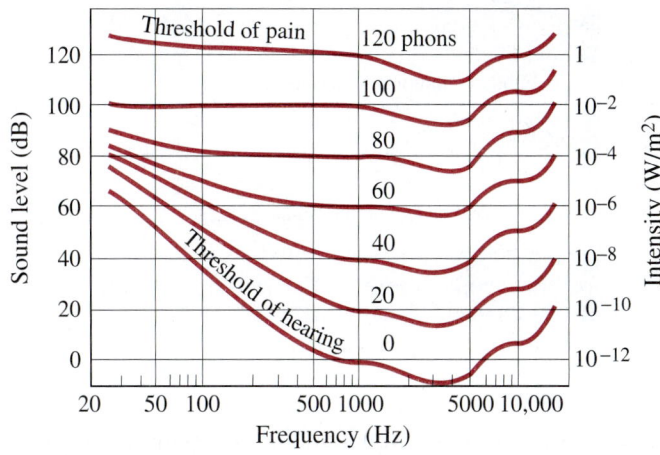

FIGURE 12–6 Sensitivity of the human ear as a function of frequency (see text). Note that the frequency scale is "logarithmic" in order to cover a wide range of frequencies.

The lowest curve in Fig. 12–6 (labeled 0) represents the sound level, as a function of frequency, for the *threshold of hearing*, the softest sound that is just audible by a very good ear. Note that the ear is most sensitive to sounds of frequency between 2000 and 4000 Hz, which are common in speech and music. Note too that whereas a 1000-Hz sound is audible at a level of 0 dB, a 100-Hz sound must be nearly 40 dB to be heard. The top curve in Fig. 12–6, labeled 120 phons, represents the *threshold of pain*. Sounds above this level can actually be felt and cause pain.

Figure 12–6 shows that at lower sound levels, our ears are less sensitive to the high and low frequencies relative to middle frequencies. The "loudness" control on some stereo systems is intended to compensate for this low-volume insensitivity. As the volume is turned down, the loudness control boosts the high and low frequencies relative to the middle frequencies so that the sound will have a more "normal-sounding" frequency balance. Many listeners, however, find the sound more pleasing or natural without the loudness control.

12–4 Sources of Sound: Vibrating Strings and Air Columns

The source of any sound is a vibrating object. Almost any object can vibrate and hence be a source of sound. We now discuss some simple sources of sound, particularly musical instruments. In musical instruments, the source is set into vibration by striking, plucking, bowing, or blowing. Standing waves are produced and the source vibrates at its natural resonant frequencies. The vibrating source is in contact with the air (or other medium) and pushes on it to produce sound waves that travel outward. The frequencies of the waves are the same as those of the source, but the speed and wavelengths can be different. A drum has a stretched membrane that vibrates. Xylophones and marimbas have metal or wood bars that can be set into vibration. Bells, cymbals, and gongs also make use of a vibrating metal. Many instruments make use of vibrating strings, such as the violin, guitar, and piano, or make use of vibrating columns of air, such as the flute, trumpet, and pipe organ. We have already seen that the pitch of a pure sound is determined by the frequency. Typical frequencies for musical notes on the "equally tempered chromatic scale" are given in Table 12–3 for the *octave* beginning with middle C. Note that one **octave** corresponds to a doubling of frequency. For example, middle C has frequency of 262 Hz whereas C′ (C above middle C) has twice that frequency, 524 Hz. [Middle C is the C or "do" note at the middle of a piano keyboard.]

TABLE 12–3 Equally Tempered Chromatic Scale[†]

Note	Name	Frequency (Hz)
C	do	262
C♯ or D♭		277
D	re	294
D♯ or E♭		311
E	mi	330
F	fa	349
F♯ or G♭		370
G	sol	392
G♯ or A♭		415
A	la	440
A♯ or B♭		466
B	ti	494
C′	do	524

[†] Only one octave is included.

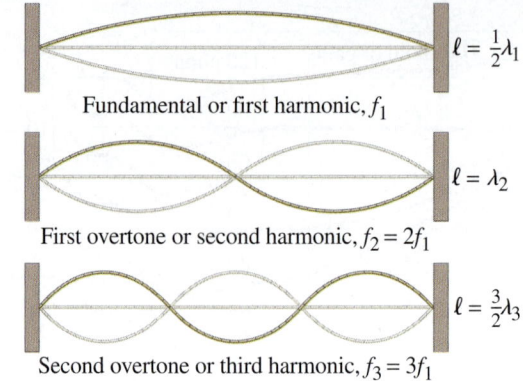

FIGURE 12–7 Standing waves on a string—only the lowest three frequencies are shown.

Stringed Instruments

PHYSICS APPLIED
Stringed instruments

We saw in Chapter 11, Fig. 11–41, how standing waves are established on a string, and we show this again here in Fig. 12–7. Such standing waves are the basis for all stringed instruments. The pitch is normally determined by the lowest resonant frequency, the **fundamental**, which corresponds to nodes occurring only at the ends. The string vibrating up and down as a whole corresponds to a half wavelength as shown at the top of Fig. 12–7; so the wavelength of the fundamental on the string is equal to twice the length ℓ of the string. Therefore, the fundamental frequency is $f_1 = v/\lambda = v/2\ell$, where v is the velocity of the wave on the string (*not* in the air). The possible frequencies for standing waves on a stretched string are whole-number multiples of the fundamental frequency:

$$f_n = nf_1 = n\frac{v}{2\ell}, \quad n = 1, 2, 3, \cdots$$

(just as in Eq. 11–19b), where $n = 1$ refers to the fundamental and $n = 2, 3, \cdots$ are the overtones. All of the standing waves, $n = 1, 2, 3, \cdots$, are called harmonics[†], as we saw in Section 11–12.

When a finger is placed on the string of a guitar or violin, the effective length of the vibrating string is shortened. So its fundamental frequency, and pitch, is higher since the wavelength of the fundamental is shorter (Fig. 12–8). The strings on a guitar or violin are all the same length. They sound at a different pitch because the strings have different mass per unit length, $\mu = m/\ell$, which affects the velocity on the string, Eq. 11–13,

$$v = \sqrt{F_T/\mu}. \quad \text{[stretched string]}$$

Thus the velocity on a heavier string is lower and the frequency will be lower for the same wavelength. The tension F_T also has an effect: indeed, adjusting the tension is the means for tuning the pitch of each string. In pianos and harps the strings are of different lengths. For the lower notes the strings are not only longer, but heavier as well, and the reason is illustrated in the following Example.

FIGURE 12–8 The wavelength of (a) an unfingered string is longer than that of (b) a fingered string. Hence, the frequency of the fingered string is higher. Only one string is shown on this guitar, and only the simplest standing wave, the fundamental, is shown.

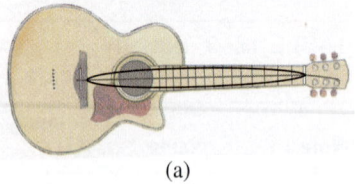
(a)

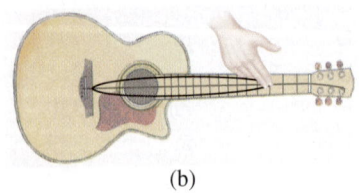
(b)

EXAMPLE 12–8 Piano strings. The highest key on a piano corresponds to a frequency about 150 times that of the lowest key. If the string for the highest note is 5.0 cm long, how long would the string for the lowest note have to be if it had the same mass per unit length and was under the same tension?

APPROACH Since $v = \sqrt{F_T/\mu}$, the velocity would be the same on each string. So the frequency is inversely proportional to the length ℓ of the string ($f = v/\lambda = v/2\ell$).

SOLUTION We can write, for the fundamental frequencies of each string, the ratio

$$\ell_L/\ell_H = f_H/f_L,$$

where the subscripts L and H refer to the lowest and highest notes, respectively. Thus $\ell_L = \ell_H(f_H/f_L) = (5.0 \text{ cm})(150) = 750 \text{ cm}$, or 7.5 m. This would be ridiculously long (≈ 25 ft) for a piano.

NOTE The longer strings of lower frequency are made heavier (higher mass per unit length), so even on grand pianos the strings are less than 3 m long.

[†]When the resonant frequencies above the fundamental (that is, the overtones) are integral multiples of the fundamental, as here, they are called harmonics. But if the overtones are not integral multiples of the fundamental, as is the case for a vibrating drumhead, for example, they are not harmonics.

EXERCISE C How many octaves does the piano of Example 12–8 cover?

EXAMPLE 12–9 **Frequencies and wavelengths in the violin.** A 0.32-m-long violin string is tuned to play A above middle C at 440 Hz. (*a*) What is the wavelength of the fundamental string vibration, and (*b*) what are the frequency and wavelength of the sound wave produced? (*c*) Why is there a difference?

APPROACH The wavelength of the fundamental string vibration equals twice the length of the string (Fig. 12–7). As the string vibrates, it pushes on the air, which is thus forced to oscillate at the same frequency as the string.

SOLUTION (*a*) From Fig. 12–7 the wavelength of the fundamental is

$$\lambda = 2\ell = 2(0.32 \text{ m}) = 0.64 \text{ m} = 64 \text{ cm}.$$

This is the wavelength of the standing wave on the string.

(*b*) The sound wave that travels outward in the air (to reach our ears) has the same frequency, 440 Hz. Its wavelength is

$$\lambda = \frac{v}{f} = \frac{343 \text{ m/s}}{440 \text{ Hz}} = 0.78 \text{ m} = 78 \text{ cm},$$

where v is the speed of sound in air (assumed at 20°C), Section 12–1.

(*c*) The wavelength of the sound wave is different from that of the standing wave on the string because the speed of sound in air (343 m/s at 20°C) is different from the speed of the wave on the string (= $f\lambda$ = 440 Hz × 0.64 m = 280 m/s) which depends on the tension in the string and its mass per unit length.

NOTE The frequencies on the string and in the air are the same: the string and air are in contact, and the string "forces" the air to vibrate at the same frequency. But the wavelengths are different because the wave speed on the string is different than that in air.

EXERCISE D Return to the Chapter-Opening Question, page 328, and answer it again now. Try to explain why you may have answered differently the first time.

Stringed instruments would not be very loud if they relied on their vibrating strings to produce the sound waves since the strings are too thin to compress and expand much air. Stringed instruments therefore make use of a kind of mechanical amplifier known as a *sounding board* (piano) or *sounding box* (guitar, violin), which acts to amplify the sound by putting a greater surface area in contact with the air (Fig. 12–9). When the strings are set into vibration, the sounding board or box is set into vibration as well. Since it has much greater area in contact with the air, it can produce a more intense sound wave. On an electric guitar, the sounding box is not so important since the vibrations of the strings are amplified electronically.

Wind Instruments

Instruments such as woodwinds, the brasses, and the pipe organ produce sound from the vibrations of standing waves in a column of air within a tube (Fig. 12–10). Standing waves can occur in the air of any cavity, including the human throat, but the frequencies present are complicated for any but very simple shapes such as the uniform, narrow tube of a flute or an organ pipe. In some instruments, a vibrating reed or the vibrating lip of the player helps to set up vibrations of the air column. In others, a stream of air is directed against one edge of the opening or mouthpiece, leading to turbulence which sets up the vibrations. Because of the disturbance, whatever its source, the air within the tube vibrates with a variety of frequencies, but only frequencies that correspond to standing waves will persist.

For a string fixed at both ends, Fig. 12–7, the standing waves have nodes (no movement) at the two ends, and one or more antinodes (large amplitude of vibration) in between. A node separates successive antinodes. The lowest-frequency standing wave, the *fundamental*, corresponds to a single antinode. The higher-frequency standing waves are called **overtones** or **harmonics**, as we saw in Section 11–12. Specifically, the first harmonic is the fundamental, the second harmonic (= first overtone) has twice the frequency of the fundamental, and so on.

(a)

(b)

FIGURE 12–9 (a) Piano, showing sounding board to which strings are attached; (b) sounding box (guitar).

FIGURE 12–10 Wind instruments: flute (left) and clarinet.

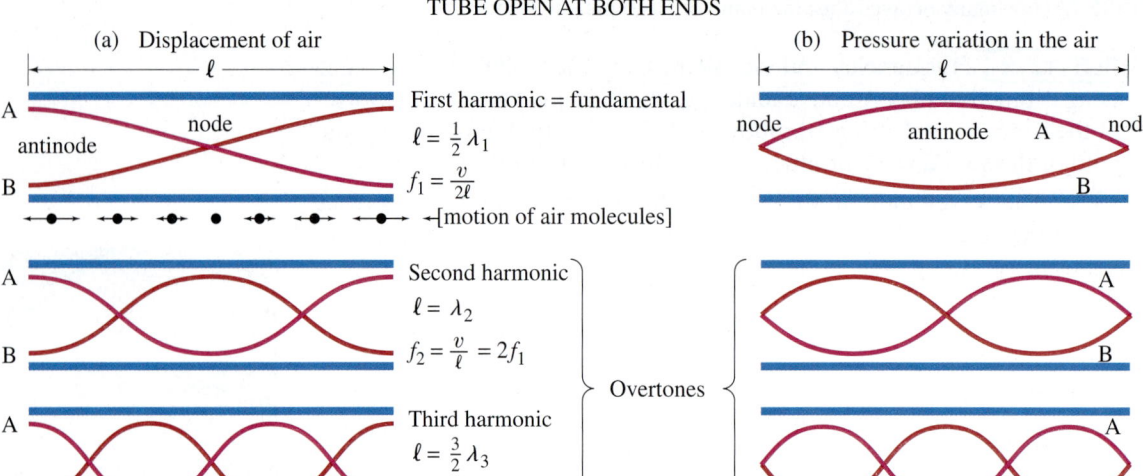

FIGURE 12–11 Graphs of the three simplest modes of vibration (standing waves) for a uniform tube open at both ends ("open tube"). These simplest modes of vibration are graphed in (a), on the left, in terms of the motion of the air (displacement), and in (b), on the right, in terms of air pressure. Each graph shows the wave format at two times, A and B, a half period apart. The actual motion of molecules for one case, the fundamental, is shown just below the tube at top left.

PHYSICS APPLIED
Wind instruments

The situation is similar for a column of air in a tube of uniform diameter, but we must remember that it is now air itself that is vibrating. We can describe the waves either in terms of the flow of the air—that is, in terms of the *displacement* of air—or in terms of the *pressure* in the air (see Figs. 12–2 and 12–3). In terms of displacement, the air at the closed end of a tube is a displacement node since the air is not free to move there, whereas near the open end of a tube there will be an antinode because the air can move freely in and out. The air within the tube vibrates in the form of longitudinal standing waves. A few of the possible modes of vibration for a tube open at both ends (called an **open tube**) are shown graphically in Fig. 12–11. Possible vibration modes for a tube that is open at one end but closed at the other (called a **closed tube**) are shown in Fig. 12–12. [A tube closed at *both* ends, having no connection to the outside air, would be useless as an instrument.] The graphs in part (a) of each Figure (left-hand sides) represent the displacement amplitude of the vibrating air in the tube. Note that these are graphs, and that the air molecules themselves oscillate *horizontally*, parallel to the tube length, as shown by the small arrows in the top diagram of Fig. 12–11a (on the left). The exact position of the antinode near the open end of a tube depends on the diameter of the tube, but if the diameter is small compared to the length, which is the usual case, the antinode occurs very close to the end as shown.† We assume this is the case in what follows. (The position of the antinode may also depend slightly on the wavelength and other factors.)

Let us look in detail at the open tube, in Fig. 12–11a, which might be an organ pipe or a flute. An open tube has displacement antinodes at both ends since the air is free to move at open ends. There must be at least one node within an open tube if there is to be a standing wave at all. A single node corresponds to the *fundamental frequency* of the tube. Since the distance between two successive nodes, or between two successive antinodes, is $\frac{1}{2}\lambda$, there is one-half of a wavelength within the length of the tube for the simplest case of the fundamental (top diagram in Fig. 12–11a): $\ell = \frac{1}{2}\lambda$, or $\lambda = 2\ell$. So the fundamental frequency is $f_1 = v/\lambda = v/2\ell$, where v is the velocity of sound in air (the air in the tube). The standing wave with two nodes is the *first overtone* or *second harmonic* and has half the wavelength ($\ell = \lambda$) and twice the frequency of the fundamental. Indeed, in a uniform tube open at both ends, the frequency of each overtone is an integral multiple of the fundamental frequency, as shown in Fig. 12–11a. This is just what is found for a string.

†The diameter D of a tube does affect the node at the open end of a tube. The end correction can be roughly approximated as adding $D/3$ to ℓ to give us an effective length for the tube in calculations.

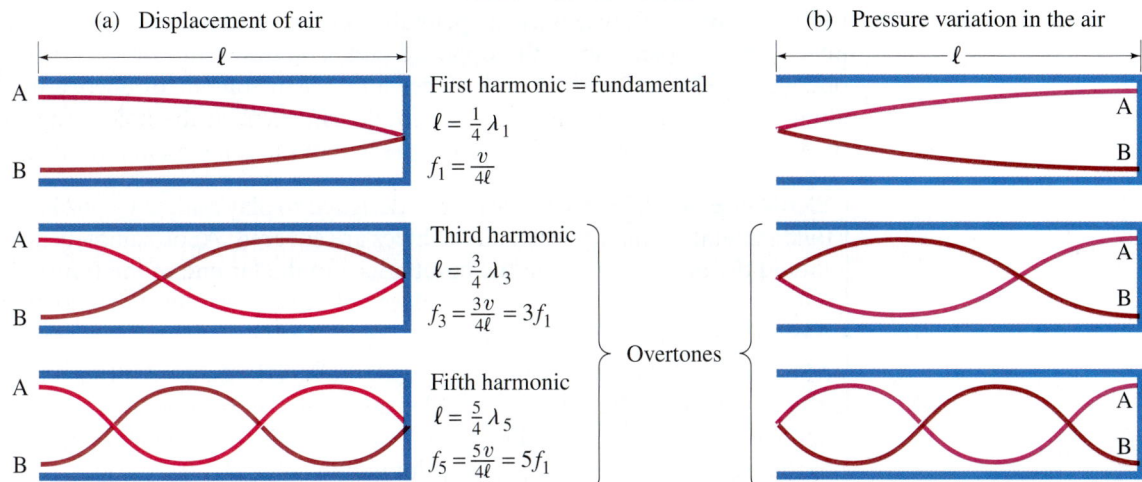

FIGURE 12–12 Modes of vibration (standing waves) for a tube closed at one end ("closed tube"). See caption for Fig. 12–11.

For a closed tube, shown in Fig. 12–12a, which could be an organ pipe, there is always a displacement node at the closed end (because the air is not free to move) and an antinode at the open end (where the air can move freely). Since the distance between a node and the nearest antinode is $\frac{1}{4}\lambda$, we see that the fundamental in a closed tube corresponds to only one-fourth of a wavelength within the length of the tube: $\ell = \lambda/4$, and $\lambda = 4\ell$. The fundamental frequency is thus $f_1 = v/4\ell$, or half that for an open pipe of the same length. There is another difference, for as we can see from Fig. 12–12a, only the odd harmonics are present in a closed tube: the overtones have frequencies equal to 3, 5, 7, ⋯ times the fundamental frequency. There is no way for waves with 2, 4, 6, ⋯ times the fundamental frequency to have a node at one end and an antinode at the other, and thus they cannot exist as standing waves in a closed tube.

Another way to analyze the vibrations in a uniform tube is to consider a description in terms of the *pressure* in the air, shown in part (b) of Figs. 12–11 and 12–12 (right-hand sides). Where the air in a wave is compressed, the pressure is higher, whereas in a wave expansion (or rarefaction), the pressure is less than normal. The open end of a tube is open to the atmosphere. Hence the pressure variation at an open end must be a *node*: the pressure does not alternate, but remains at the outside atmospheric pressure. If a tube has a closed end, the pressure at that closed end can readily alternate to be above or below atmospheric pressure. Hence there is a pressure *antinode* at a closed end of a tube. There can be pressure nodes and antinodes within the tube. Some of the possible vibrational modes in terms of pressure are shown in Fig. 12–11b for an open tube, and in Fig. 12–12b for a closed tube.

EXAMPLE 12–10 Organ pipes. What will be the fundamental frequency and first three overtones for a 26-cm-long organ pipe at 20°C if it is (a) open, (b) closed?

APPROACH All our calculations can be based on Figs. 12–11a and 12–12a.

SOLUTION (a) For the open pipe, Fig. 12–11a, the fundamental frequency is

$$f_1 = \frac{v}{2\ell} = \frac{343 \text{ m/s}}{2(0.26 \text{ m})} = 660 \text{ Hz}.$$

The speed v is the speed of sound in air (the air vibrating in the pipe). The overtones include all harmonics: 1320 Hz, 1980 Hz, 2640 Hz, and so on.

(b) For a closed pipe, Fig. 12–12a, the fundamental frequency is

$$f_1 = \frac{v}{4\ell} = \frac{343 \text{ m/s}}{4(0.26 \text{ m})} = 330 \text{ Hz}.$$

Only odd harmonics are present: the first three overtones are 990 Hz, 1650 Hz, and 2310 Hz.

NOTE The closed pipe plays 330 Hz, which, from Table 12–3, is E above middle C, whereas the open pipe of the same length plays 660 Hz, an octave higher.

Pipe organs use both open and closed pipes, with lengths from a few centimeters to 5 m or more. A flute acts as an open tube, for it is open not only where you blow into it, but is open also at the opposite end. The different notes on a flute are obtained by shortening the length of the vibrating air column, by uncovering holes along the tube (so a displacement antinode can occur at the hole). The shorter the length of the vibrating air column, the higher the fundamental frequency.

EXAMPLE 12–11 Flute. A flute is designed to play middle C (262 Hz) as the fundamental frequency when all the holes are covered. Approximately how long should the distance be from the mouthpiece to the far end of the flute? (This is only approximate since the antinode does not occur precisely at the mouthpiece.) Assume the temperature is 20°C.

APPROACH When all holes are covered, the length of the vibrating air column is the full length. The speed of sound in air at 20°C is 343 m/s. Because a flute is open at both ends, we use Fig. 12–11: the fundamental frequency f_1 is related to the length ℓ of the vibrating air column by $f = v/2\ell$.

SOLUTION Solving for ℓ, we find

$$\ell = \frac{v}{2f} = \frac{343 \text{ m/s}}{2(262 \text{ s}^{-1})} = 0.655 \text{ m}.$$

EXERCISE E To see why players of wind instruments "warm up" their instruments (so they will be in tune), determine the fundamental frequency of the flute of Example 12–11 when all holes are covered and the temperature is 10°C instead of 20°C.

EXAMPLE 12–12 ESTIMATE Wind noise frequencies. Wind can be noisy—it can "howl" in trees; it can "moan" in chimneys. What is causing the noise, and about what range of frequencies would you expect to hear?

APPROACH Gusts of air in the wind cause vibrations or oscillations of the tree limb (or air column in the chimney), which produce sound waves of the same frequency. The end of a tree limb fixed to the tree trunk is a node, whereas the other end is free to move and therefore is an antinode; the tree limb is thus about $\frac{1}{4}\lambda$ (Fig. 12–13).

SOLUTION We estimate $v \approx 4000 \text{ m/s}$ for the speed of sound in wood (Table 12–1). Suppose that a tree limb has length $\ell \approx 2 \text{ m}$; then $\lambda = 4\ell = 8 \text{ m}$ and $f = v/\lambda = (4000 \text{ m/s})/(8 \text{ m}) \approx 500 \text{ Hz}$.

NOTE Wind can excite air oscillations in a chimney, much like in an organ pipe or flute. A chimney is a fairly long tube, perhaps 3 m in length, acting like a tube open at either one end or even both ends. If open at both ends ($\lambda = 2\ell$), with $v \approx 340 \text{ m/s}$, we find $f_1 \approx v/2\ell \approx 57 \text{ Hz}$, which is a fairly low note—no wonder chimneys "moan"!

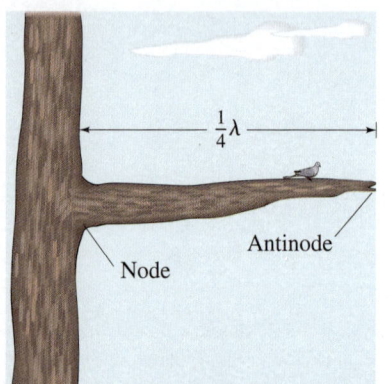

FIGURE 12–13 Example 12–12.

FIGURE 12–14 The amplitudes of the fundamental and first two overtones are added at each point to get the "sum," or composite waveform.

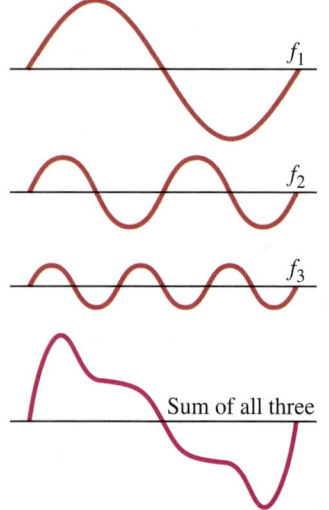

*12–5 Quality of Sound, and Noise; Superposition

Whenever we hear a sound, particularly a musical sound, we are aware of its loudness, its pitch, and also of a third aspect called "quality." For example, when a piano and then a flute play a note of the same loudness and pitch (say, middle C), there is a clear difference in the overall sound. We would never mistake a piano for a flute. This is what is meant by the **quality** of a sound. For musical instruments, the terms *timbre* and *tone color* are also used.

Just as loudness and pitch can be related to physically measurable quantities, so too can quality. The quality of a sound depends on the presence of overtones—their number and their relative amplitudes. Generally, when a note is played on a musical instrument, the fundamental as well as overtones are present simultaneously. Figure 12–14 illustrates how the *principle of superposition* (Section 11–11) applies to three wave forms, in this case the fundamental and first two overtones (with particular amplitudes): they add together at each point to give a composite **waveform**.

By "waveform" we mean the shape of the wave in space at a given moment. Normally, more than two overtones are present. [Any complex wave can be analyzed into a superposition of sinusoidal waves of appropriate amplitudes, wavelengths, and frequencies. Such an analysis is called a *Fourier analysis*.]

The relative amplitudes of the overtones for a given note are different for different musical instruments, which is what gives each instrument its characteristic quality or timbre. A bar graph showing the relative amplitudes of the harmonics for a given note produced by an instrument is called a *sound spectrum*. Several typical examples for different musical instruments are shown in Fig. 12–15. The fundamental usually has the greatest amplitude, and its frequency is what is heard as the pitch.

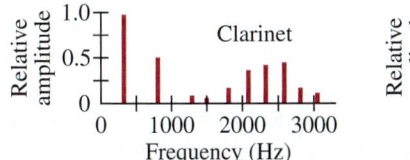
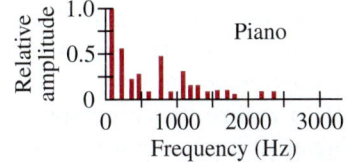
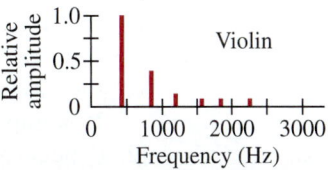

FIGURE 12–15 Sound spectra for different instruments. The spectra change when the instruments play different notes. The clarinet is a bit complicated: it acts like a closed tube at lower frequencies, having only odd harmonics, but at higher frequencies all harmonics occur as for an open tube.

The manner in which an instrument is played strongly influences the sound quality. Plucking a violin string, for example, makes a very different sound than pulling a bow across it. The sound spectrum at the very start (or end) of a note (as when a hammer strikes a piano string) can be very different from the subsequent sustained tone. This too affects the subjective tone quality of an instrument.

An ordinary sound, like that made by striking two stones together, is a noise that has a certain quality, but a clear pitch is not discernible. Such a noise is a mixture of many frequencies which bear little relation to one another. A sound spectrum made of that noise would not show discrete lines like those of Fig. 12–15. Instead it would show a continuous, or nearly continuous, spectrum of frequencies. Such a sound we call "noise" in comparison with the more harmonious sounds which contain frequencies that are simple multiples of the fundamental.

FIGURE 12–16 Sound waves from two loudspeakers interfere.

12–6 Interference of Sound Waves; Beats

Interference in Space

We saw in Section 11–11 that when two waves simultaneously pass through the same region of space, they interfere with one another. Interference also occurs with sound waves.

Consider two large loudspeakers, A and B, a distance d apart on the stage of an auditorium as shown in Fig. 12–16. Let us assume the two speakers are emitting sound waves of the same single frequency and that they are in phase: that is, when one speaker is forming a compression, so is the other. (We ignore reflections from walls, floor, etc.) The curved lines in the diagram represent the crests of sound waves from each speaker at one instant in time. We must remember that for a sound wave, a crest is a compression in the air whereas a trough—which falls between two crests—is a rarefaction. A human ear or detector at a point such as C, which is the same distance from each speaker, will experience a loud sound because the interference will be constructive—two crests reach it at one moment, two troughs reach it a moment later. On the other hand, at a point such as D in the diagram, little if any sound will be heard because destructive interference occurs—compressions of one wave meet rarefactions of the other and vice versa (see Fig. 11–38 and the related discussion on water waves in Section 11–11).

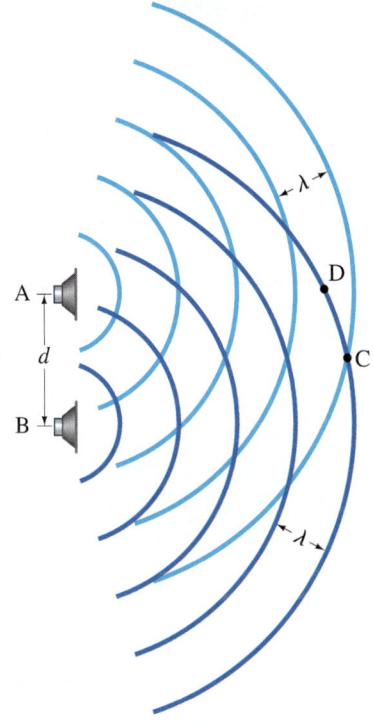

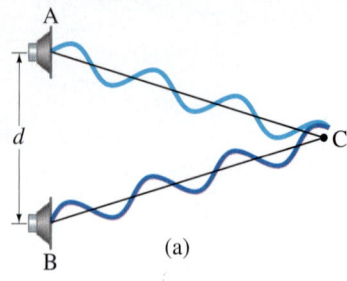
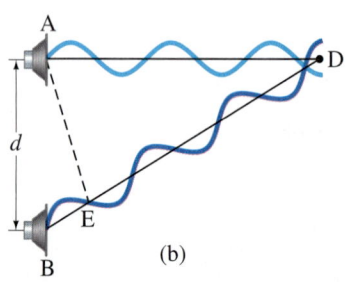

FIGURE 12–17 Sound waves of a single frequency from loudspeakers A and B (see Fig. 12–16) constructively interfere at C and destructively interfere at D. [Shown here are graphical representations, not the actual longitudinal sound waves.]

An analysis of this situation is perhaps clearer if we graphically represent the waveforms as in Fig. 12–17. In Fig. 12–17a it can be seen that at point C, constructive interference occurs because both waves simultaneously have crests or simultaneously have troughs when they arrive at C. In Fig. 12–17b we see that, to reach point D, the wave from speaker B must travel a greater distance than the wave from A. Thus the wave from B lags behind that from A. In this diagram, point E is chosen so that the distance ED is equal to AD. Thus we see that if the distance BE is equal to precisely one-half the wavelength of the sound, the two waves will be exactly out of phase when they reach D, and destructive interference occurs. This then is the criterion for determining at what points destructive interference occurs: destructive interference occurs at any point whose distance from one speaker is one-half wavelength greater than its distance from the other speaker. Notice that if this extra distance (BE in Fig. 12–17b) is equal to a whole wavelength (or 2, 3, ⋯ wavelengths), then the two waves will be in phase and *constructive interference* occurs. If the distance BE equals $\frac{1}{2}$, $1\frac{1}{2}$, $2\frac{1}{2}$, ⋯ wavelengths, *destructive interference* occurs.

It is important to realize that a person sitting at point D in Fig. 12–16 or 12–17 hears nothing at all (or nearly so), yet sound is coming from both speakers. Indeed, if one of the speakers is turned off, the sound from the other speaker will be clearly heard.

If a loudspeaker emits a whole range of frequencies, only specific wavelengths will destructively interfere completely at a given point.

EXAMPLE 12–13 Loudspeakers' interference. Two loudspeakers are 1.00 m apart. A person stands 4.00 m from one speaker. How far should this person be from the second speaker to detect destructive interference when the speakers emit an 1150-Hz sound? Assume the temperature is 20°C.

APPROACH To sense destructive interference, the person should be one-half wavelength closer to or farther from one speaker than from the other—that is, at a distance = 4.00 m ± λ/2. We can determine λ because we know f and v.

SOLUTION The speed of sound at 20°C is 343 m/s, so the wavelength of this sound is (Eq. 11–12)

$$\lambda = \frac{v}{f}$$

$$= \frac{343 \text{ m/s}}{1150 \text{ Hz}} = 0.30 \text{ m}.$$

For destructive interference to occur, the person must be one-half wavelength farther from one loudspeaker than from the other, or 0.15 m. Thus the person must be 3.85 m or 4.15 m from the second speaker.

NOTE If the speakers are less than 0.15 m apart, there will be no location that is 0.15 m farther from one speaker than the other, and there will be no place where destructive interference could occur.

Beats—Interference in Time

We have been discussing interference of sound waves that takes place in space. An interesting and important example of interference that occurs in time is the phenomenon known as **beats**: if two sources of sound—say, two tuning forks—are close in frequency but not exactly the same, sound waves from the two sources interfere with each other. The sound level at a given position alternately rises and falls in time, because the two waves are sometimes in phase and sometimes out of phase due to their different wavelengths. The regularly spaced intensity changes are called beats.

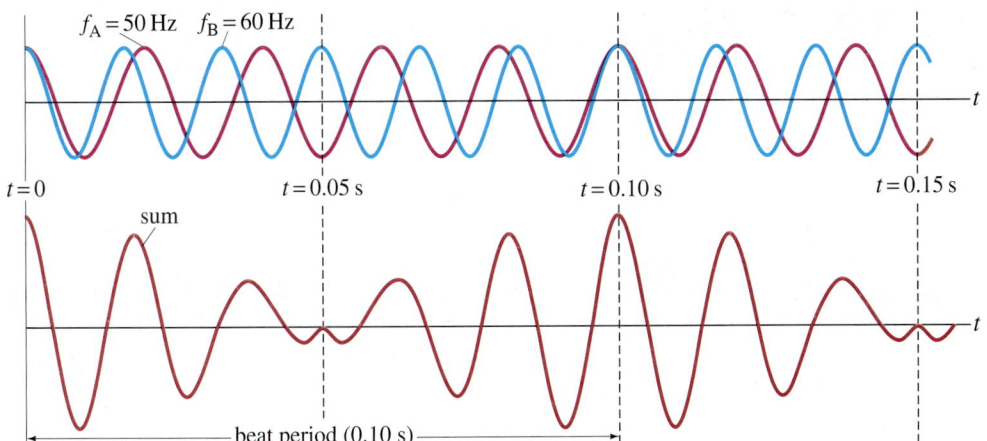

FIGURE 12–18 Beats occur as a result of the superposition of two sound waves of slightly different frequency.

To see how beats arise, consider two equal-amplitude sound waves of frequency $f_A = 50$ Hz and $f_B = 60$ Hz, respectively. In 1.00 s, the first source makes 50 vibrations whereas the second makes 60. We now examine the waves at one point in space equidistant from the two sources. The waveforms for each wave as a function of time, at a fixed position, are shown on the top graph of Fig. 12–18; the red line represents the 50-Hz wave, and the blue line represents the 60-Hz wave. The lower graph in Fig. 12–18 shows the sum of the two waves as a function of time. At time $t = 0$ the two waves are shown to be in phase and interfere constructively. Because the two waves vibrate at different rates, at time $t = 0.05$ s they are completely out of phase and interfere destructively. At $t = 0.10$ s, they are again in phase and the resultant amplitude again is large. Thus the resultant amplitude is large every 0.10 s and drops drastically in between. This rising and falling of the intensity is what is heard as beats.† In this case the beats are 0.10 s apart. That is, the **beat frequency** is ten per second, or 10 Hz. This result, that the beat frequency equals the difference in frequency of the two waves, is valid in general.

The phenomenon of beats can occur with any kind of wave and is a very sensitive method for comparing frequencies. For example, to tune a piano, a piano tuner listens for beats produced between his standard tuning fork and the frequency of a particular string on the piano, and knows it is in tune when the beats disappear. The members of an orchestra tune up by listening for beats between their instruments and the frequency of a standard tone (usually A above middle C at 440 Hz) produced by a piano or an oboe. Humans hear the individual beats if they are only a few per second. For higher beat frequencies, they run together and up to about 20 Hz or so you hear an intensity modulation (a wavering between loud and soft); above 20 Hz you hear a separate low tone (audible if the tones are strong enough).

PHYSICS APPLIED
Tuning a piano

EXAMPLE 12–14 **Beats.** A tuning fork produces a steady 400-Hz tone. When this tuning fork is struck and held near a vibrating guitar string, twenty beats are counted in five seconds. What are the possible frequencies produced by the guitar string?

APPROACH For beats to occur, the string must vibrate at a frequency different from 400 Hz by whatever the beat frequency is.

SOLUTION The beat frequency is

$$f_{beat} = 20 \text{ vibrations}/5 \text{ s} = 4 \text{ Hz}.$$

This is the difference of the frequencies of the two waves. Because one wave is known to be 400 Hz, the other must be either 404 Hz or 396 Hz.

---

†Beats will be heard even if the amplitudes are not equal, as long as the difference in amplitude is not great.

12–7 Doppler Effect

You may have noticed that you hear the pitch of the siren on a speeding fire truck drop abruptly as it passes you. Or you may have noticed the change in pitch of a blaring horn on a fast-moving car as it passes by you. The pitch of the engine noise of a race car changes as the car passes an observer. When a source of sound is moving toward an observer, the pitch the observer hears is higher than when the source is at rest; and when the source is traveling away from the observer, the pitch is lower. This phenomenon is known as the **Doppler effect**[†] and occurs for all types of waves. Let us now see why it occurs, and calculate the difference between the perceived and source sound frequencies when there is relative motion between source and observer.

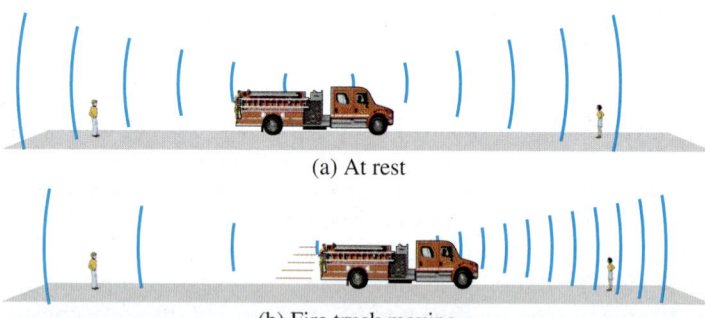

(a) At rest

(b) Fire truck moving

FIGURE 12–19 (a) Both observers on the sidewalk hear the same frequency from the fire truck at rest. (b) Doppler effect: observer toward whom the fire truck moves hears a higher-frequency sound, and observer behind the fire truck hears a lower-frequency sound.

Consider the siren of a fire truck at rest, which is emitting sound of a particular frequency in all directions as shown in Fig. 12–19a. The sound waves are moving at the speed of sound in air, v_{snd}, which is independent of the velocity of the source or observer. If our source, the fire truck, is moving, the siren emits sound at the same frequency as it does at rest. But the sound wavefronts it emits forward, in front of it, are closer together than when the fire truck is at rest, as shown in Fig. 12–19b. This is because the fire truck, as it moves, is partly "catching up" to the previously emitted wavefronts, and emits each crest closer to the previous one. Thus an observer on the sidewalk in front of the truck will detect more wave crests passing per second, so the frequency heard is higher. The wavefronts emitted behind the truck, on the other hand, are farther apart than when the truck is at rest because the truck is speeding away from them. Each new wavefront emitted is farther from the preceding one than when the truck is at rest. Hence, fewer wave crests per second pass by an observer behind the moving truck (Fig. 12–19b) and the perceived pitch is lower.

We can calculate the frequency shift by making use of Fig. 12–20. We assume the air (or other medium) is at rest in our reference frame. We consider first the stationary observer off to the right in Fig. 12–19. In Fig. 12–20a, the source of the sound is shown as a red dot, and is at rest. Two successive wave crests are shown, the second of which has just been emitted and is still near the source. The distance between these crests is λ, the wavelength. If the frequency of the source is f, then the time between emissions of wave crests is

$$T = \frac{1}{f} = \frac{\lambda}{v_{snd}}.$$

In Fig. 12–20b, the source is moving with a velocity v_{source} toward the observer.

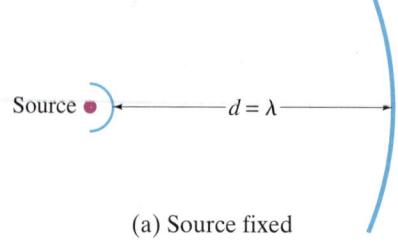

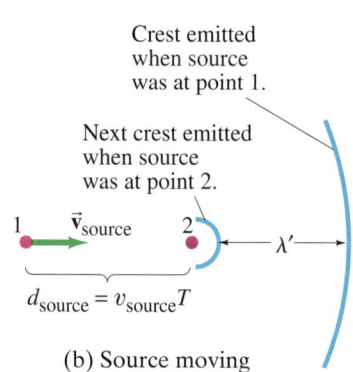

FIGURE 12–20 Determination of the frequency shift in the Doppler effect (see text). The red dot is the source.

(a) Source fixed

(b) Source moving

[†]After J. C. Doppler (1803–1853).

In a time T (just defined), the first wave crest has moved a distance $d = v_{snd}T = \lambda$, where v_{snd} is the velocity of the sound wave in air (which is the same whether the source is moving or not). In this same time, the source has moved a distance $d_{source} = v_{source}T$. Then the distance between successive wave crests, which is the wavelength λ' the observer on the right will perceive, is

$$\begin{aligned}\lambda' &= d - d_{source}\\ &= \lambda - v_{source}T\\ &= \lambda - v_{source}\frac{\lambda}{v_{snd}}\\ &= \lambda\left(1 - \frac{v_{source}}{v_{snd}}\right).\end{aligned}$$

We subtract λ from both sides of this equation and find that the shift in wavelength, $\Delta\lambda$, is

$$\Delta\lambda = \lambda' - \lambda = -\lambda\frac{v_{source}}{v_{snd}}.$$

So the shift in wavelength is directly proportional to the source speed v_{source}. The frequency f' that will be perceived by our stationary observer on the ground is given by (Eq. 11–12)

$$f' = \frac{v_{snd}}{\lambda'} = \frac{v_{snd}}{\lambda\left(1 - \frac{v_{source}}{v_{snd}}\right)}.$$

Because $v_{snd}/\lambda = f$, then

$$f' = \frac{f}{\left(1 - \frac{v_{source}}{v_{snd}}\right)}. \quad \begin{bmatrix}\text{source moving toward}\\ \text{stationary observer}\end{bmatrix} \quad (12\text{–}2\text{a})$$

Because the denominator is less than 1, the observed frequency f' is higher than the source frequency f. That is, $f' > f$. For example, if a source emits a sound of frequency 400 Hz when at rest, then when the source moves toward a fixed observer with a speed of 30 m/s, the observer hears a frequency (at 20°C) of

$$f' = \frac{400\text{ Hz}}{1 - \frac{30\text{ m/s}}{343\text{ m/s}}} = 438\text{ Hz}.$$

Now consider a source moving *away* from a stationary observer at a speed v_{source} (observer on the left in Fig. 12–19). Using the same arguments as above, the wavelength λ' perceived by our observer will have the minus sign on d_{source} (first equation on this page) changed to plus:

$$\begin{aligned}\lambda' &= d + d_{source}\\ &= \lambda\left(1 + \frac{v_{source}}{v_{snd}}\right).\end{aligned}$$

The difference between the observed and emitted wavelengths will be $\Delta\lambda = \lambda' - \lambda = +\lambda(v_{source}/v_{snd})$. The observed frequency of the wave is $f' = v_{snd}/\lambda'$, which equals

$$f' = \frac{f}{\left(1 + \frac{v_{source}}{v_{snd}}\right)}. \quad \begin{bmatrix}\text{source moving away from}\\ \text{stationary observer}\end{bmatrix} \quad (12\text{–}2\text{b})$$

If a source emitting at 400 Hz is moving away from a fixed observer at 30 m/s, the observer hears a frequency $f' = (400\text{ Hz})/[1 + (30\text{ m/s})/(343\text{ m/s})] = 368\text{ Hz}$.

FIGURE 12–21 Observer moving with speed v_obs toward a stationary source detects wave crests passing at speed $v' = v_\text{snd} + v_\text{obs}$ where v_snd is the speed of the sound waves in air.

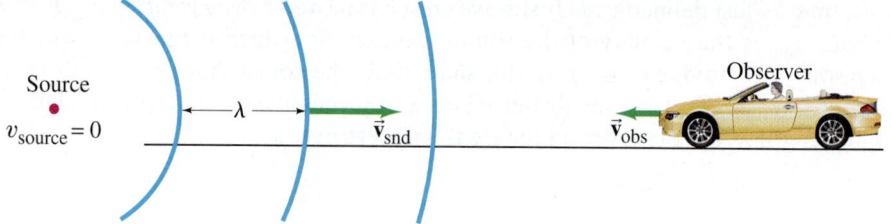

The Doppler effect also occurs when the source is at rest and the observer is in motion. If the observer is traveling *toward* the source, the pitch heard is higher than that of the emitted source frequency. If the observer is traveling *away* from the source, the pitch heard is lower. Quantitatively the change in frequency is different than for the case of a moving source. With a fixed source and a moving observer, the distance between wave crests, the wavelength λ, is not changed. But the velocity of the crests with respect to the observer *is* changed. If the observer is moving toward the source, Fig. 12–21, the speed v' of the waves relative to the observer is a simple addition of velocities: $v' = v_\text{snd} + v_\text{obs}$, where v_snd is the velocity of sound in air (we assume the air is still) and v_obs is the velocity of the observer. Hence, the frequency heard is

$$f' = \frac{v'}{\lambda} = \frac{v_\text{snd} + v_\text{obs}}{\lambda}.$$

Because $\lambda = v_\text{snd}/f$, then

$$f' = \frac{(v_\text{snd} + v_\text{obs})f}{v_\text{snd}},$$

or

$$f' = \left(1 + \frac{v_\text{obs}}{v_\text{snd}}\right)f. \quad \begin{bmatrix}\text{observer moving toward}\\\text{stationary source}\end{bmatrix} \quad (12\text{–}3\text{a})$$

If the observer is moving away from the source, the relative velocity is $v' = v_\text{snd} - v_\text{obs}$, so

$$f' = \left(1 - \frac{v_\text{obs}}{v_\text{snd}}\right)f. \quad \begin{bmatrix}\text{observer moving away}\\\text{from stationary source}\end{bmatrix} \quad (12\text{–}3\text{b})$$

EXAMPLE 12–15 A moving siren. The siren of a police car at rest emits at a predominant frequency of 1600 Hz. What frequency will you hear if you are at rest and the police car moves at 25.0 m/s (*a*) toward you, and (*b*) away from you?

APPROACH The observer is fixed, and the source moves, so we use Eqs. 12-2. The frequency you (the observer) hear is the emitted frequency f divided by the factor $(1 \pm v_\text{source}/v_\text{snd})$ where v_source is the speed of the police car. Use the minus sign when the car moves toward you (giving a higher frequency); use the plus sign when the car moves away from you (lower frequency).

SOLUTION (*a*) The car is moving toward you, so (Eq. 12–2a)

$$f' = \frac{f}{\left(1 - \dfrac{v_\text{source}}{v_\text{snd}}\right)} = \frac{1600 \text{ Hz}}{\left(1 - \dfrac{25.0 \text{ m/s}}{343 \text{ m/s}}\right)} = 1726 \text{ Hz} \approx 1730 \text{ Hz}.$$

(*b*) The car is moving away from you, so (Eq. 12–2b)

$$f' = \frac{f}{\left(1 + \dfrac{v_\text{source}}{v_\text{snd}}\right)} = \frac{1600 \text{ Hz}}{\left(1 + \dfrac{25.0 \text{ m/s}}{343 \text{ m/s}}\right)} = 1491 \text{ Hz} \approx 1490 \text{ Hz}.$$

EXERCISE F Suppose the police car of Example 12–15 is at rest and emits at 1600 Hz. What frequency would you hear if you were moving at 25.0 m/s (*a*) toward it, and (*b*) away from it?

When a sound wave is reflected from a moving obstacle, the frequency of the reflected wave will, because of the Doppler effect, be different from that of the incident wave. This is illustrated in the following Example.

EXAMPLE 12–16 Two Doppler shifts. A 5000-Hz sound wave is emitted by a stationary source. This sound wave reflects from an object moving 3.50 m/s toward the source (Fig. 12–22). What is the frequency of the wave reflected by the moving object as detected by a detector at rest near the source?

APPROACH There are actually two Doppler shifts in this situation. First, the moving object acts like an observer moving toward the source with speed $v_{obs} = 3.50$ m/s (Fig. 12–22a) and so "detects" a sound wave of frequency (Eq. 12–3a) $f' = f[1 + (v_{obs}/v_{snd})]$. Second, reflection of the wave from the moving object is equivalent to the object reemitting the wave at the same frequency, and thus acting effectively as a moving source with speed $v_{source} = 3.50$ m/s (Fig. 12–22b). The final frequency detected, f'', is given by $f'' = f'/[1 - (v_{source}/v_{snd})]$, Eq. 12–2a.

SOLUTION The frequency f' that is "detected" by the moving object is (Eq. 12–3a):

$$f' = \left(1 + \frac{v_{obs}}{v_{snd}}\right)f = \left(1 + \frac{3.50 \text{ m/s}}{343 \text{ m/s}}\right)(5000 \text{ Hz}) = 5051 \text{ Hz}.$$

The moving object now "emits" (reflects) a sound of frequency (Eq. 12–2a)

$$f'' = \frac{f'}{\left(1 - \dfrac{v_{source}}{v_{snd}}\right)} = \frac{5051 \text{ Hz}}{\left(1 - \dfrac{3.50 \text{ m/s}}{343 \text{ m/s}}\right)} = 5103 \text{ Hz}.$$

Thus the frequency shifts by 103 Hz.

NOTE Bats use this technique to be aware of their surroundings. This is also the principle behind Doppler radar as speed-measuring devices for vehicles, baseball pitches, tennis serves, storms such as tornadoes, and other objects.

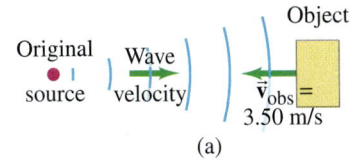

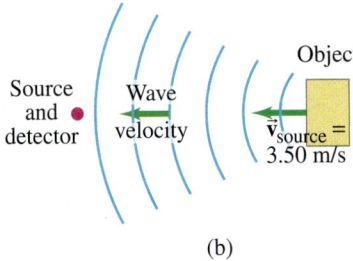

FIGURE 12–22 Example 12–16.

The emitted wave and the reflected wave in Example 12–16, when mixed together (say, electronically), interfere with one another and beats are produced. The beat frequency is equal to the difference in the two frequencies, 103 Hz. This Doppler technique is used in a variety of medical applications, usually with ultrasonic waves in the megahertz frequency range. For example, ultrasonic waves reflected from red blood cells can be used to determine the velocity of blood flow. Similarly, the technique can be used to detect the movement of the chest of a young fetus and to monitor its heartbeat.

For convenience, we can write Eqs. 12–2 and 12–3 as a single equation that covers all cases of both source and observer in motion:

$$f' = f\left(\frac{v_{snd} \pm v_{obs}}{v_{snd} \mp v_{source}}\right). \qquad \begin{bmatrix}\text{source and} \\ \text{observer moving}\end{bmatrix} \quad (12\text{–}4)$$

PHYSICS APPLIED
Doppler blood-flow meter and other medical uses

PROBLEM SOLVING
Getting the signs right

To get the signs right, recall from your own experience that the frequency is higher when observer and source approach each other, and lower when they move apart. Thus the upper signs in numerator and denominator apply if source and/or observer move toward each other; the lower signs apply if they are moving apart.

EXERCISE G How fast would a source have to approach an observer at rest for the observed frequency to be one octave above the produced frequency (frequency doubled)? (a) $\frac{1}{2}v_{snd}$, (b) v_{snd}, (c) $2v_{snd}$, (d) $4v_{snd}$.

PHYSICS APPLIED
Doppler effect for EM waves and weather forecasting

PHYSICS APPLIED
Redshift in cosmology

Doppler Effect for Light

The Doppler effect occurs for waves other than sound. Light and other types of electromagnetic waves (such as for radar) exhibit the Doppler effect: although the formulas for the frequency shift are not identical to Eqs. 12–2 and 12–3, the effect is similar (see Chapter 33). One important application is for weather forecasting using radar. The time delay between the emission of a radar pulse and its reception after being reflected off raindrops gives the position of precipitation. Measuring the Doppler shift in frequency (as in Example 12–16) tells how fast the storm is moving and in which direction. "Radar guns" used by police work similarly, measuring a car's speed by the Doppler shift of electromagnetic waves.

Another important application is to astronomy, where the velocities of galaxies can be estimated from the Doppler shift. Light from distant galaxies is shifted toward lower frequencies, indicating that the galaxies are moving away from us. This is called the **redshift** since red has the lowest frequency of visible light. The greater the frequency shift, the greater the velocity of recession. It is found that the farther the galaxies are from us, the faster they move away. This observation is the basis for the idea that the universe is expanding, and is one basis for the idea that the universe had a beginning affectionately called the "Big Bang" (Chapter 33).

*12–8 Shock Waves and the Sonic Boom

An object such as an airplane traveling faster than the speed of sound is said to have a **supersonic speed**. Such a speed is often given as a **Mach**[†] **number**, which is defined as the ratio of the speed of the object to the speed of sound in the surrounding medium. For example, a plane traveling 600 m/s high in the atmosphere, where the speed of sound is only 300 m/s, has a speed of Mach 2.

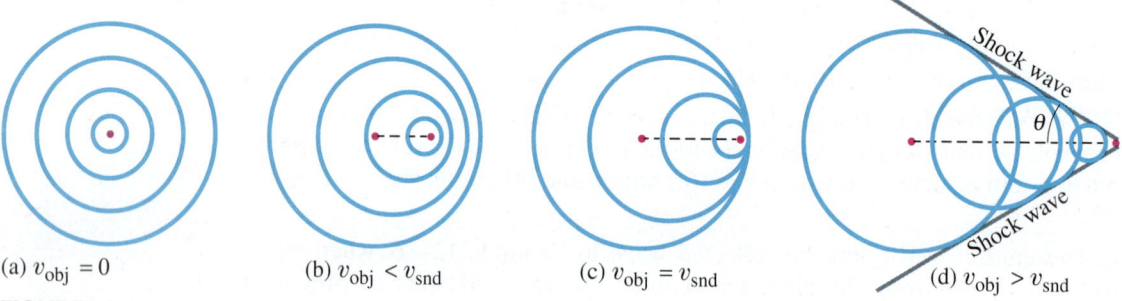

FIGURE 12–23 Sound waves emitted by an object (a) at rest or (b, c, and d) moving. (b) If the object's velocity is less than the velocity of sound, the Doppler effect occurs. (c) At $v_{obj} = v_{snd}$, the waves pile up in front, forming a "sound barrier." (d) If the object's velocity is greater than the velocity of sound, a shock wave is produced.

FIGURE 12–24 Bow waves produced by (a) a boat, (b) a duck.

(a)

(b)

When a source of sound moves at subsonic speeds (less than the speed of sound), the pitch of the sound is altered as we have seen (the Doppler effect); see also Figs. 12–23a and b. But if a source of sound moves faster than the speed of sound, a more dramatic effect known as a **shock wave** occurs. In this case, the source is actually "outrunning" the waves it produces. As shown in Fig. 12–23c, when the source is traveling *at* the speed of sound, the wave fronts it emits in the forward direction "pile up" directly in front of it. When the object moves faster, at a supersonic speed, the wave fronts pile up on one another along the sides, as shown in Fig. 12–23d. The different wave crests overlap one another and form a single very large crest which is the shock wave. Behind this very large crest there is usually a very large trough. A shock wave is essentially the result of constructive interference of a large number of wave fronts. A shock wave in air is analogous to the bow wave of a boat traveling faster than the speed of the water waves it produces, Fig. 12–24.

[†]After the Austrian physicist Ernst Mach (1838–1916).

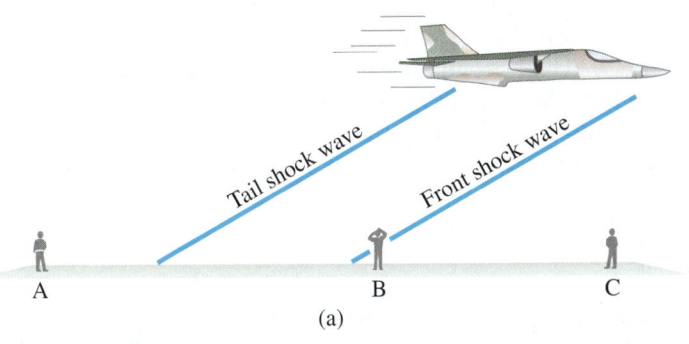
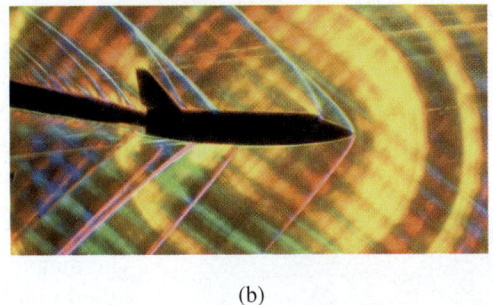

FIGURE 12–25 (a) The (double) sonic boom has already been heard by person A on the left. The front shock wave is just being heard by person B in the center. And both will shortly be heard by person C on the right. (b) Special photo of supersonic aircraft showing shock waves produced in the air. (Several closely spaced shock waves are produced by different parts of the aircraft.)

When an airplane travels at supersonic speeds, greater than the speed of sound in the air, the noise it makes and its disturbance of the air form into a shock wave containing a tremendous amount of sound energy. When the shock wave passes a listener, it is heard as a loud **sonic boom**. A sonic boom lasts only a fraction of a second, but the energy it contains is sometimes sufficient to break windows and cause other damage. Actually, a sonic boom is made up of two or more booms since major shock waves can form at the front and the rear of the aircraft (Fig. 12–25), as well as at the wings and other parts. Bow waves of a boat are also multiple, as can be seen in Fig. 12–24a.

When an aircraft accelerates toward the speed of sound, it encounters a barrier of sound waves in front of it (see Fig. 12–23c). To exceed the speed of sound, the aircraft needs extra thrust to pass through this **sound barrier**. This is called "breaking the sound barrier." Once a supersonic speed is attained, this barrier no longer impedes the motion. It is sometimes erroneously thought that a sonic boom is produced only at the moment an aircraft is breaking through the sound barrier. Actually, a shock wave follows the aircraft at all times it is traveling at supersonic speeds. A series of observers on the ground will each hear a loud "boom" as the shock wave passes, Fig. 12–25. The shock wave consists of a cone whose apex is at the aircraft. The angle of this cone, θ (see Fig. 12–23d), is given by

$$\sin\theta = \frac{v_{\text{snd}}}{v_{\text{obj}}}, \tag{12–5}$$

where v_{obj} is the velocity of the object (the aircraft) and v_{snd} is the velocity of sound in the medium (air for an airplane).

> **PHYSICS APPLIED**
> *Sonic boom*

*12–9 Applications: Sonar, Ultrasound, and Medical Imaging

*Sonar

The reflection of sound is used in many applications to determine distance. The **sonar**[†] or **pulse-echo** technique is used to locate underwater objects. A transmitter sends out a sound pulse through the water, and a detector receives its reflection, or echo, a short time later. This time interval is carefully measured, and from it the distance to the reflecting object can be determined since the speed of sound in water is known. The depth of the sea and the location of reefs, sunken ships, submarines, or schools of fish can be determined in this way. The interior structure of the Earth is studied in a similar way by detecting reflections of waves traveling through the Earth whose source was a deliberate explosion (called "soundings"). An analysis of waves reflected from various structures and boundaries within the Earth reveals characteristic patterns that are also useful in the exploration for oil and minerals. (*Radar* used at airports to track aircraft involves a similar pulse-echo technique except that it uses electromagnetic (EM) waves, which, like visible light, travel with a speed of 3×10^8 m/s.)

> **PHYSICS APPLIED**
> *Sonar: depth finding, Earth soundings*

[†]Sonar stands for "*so*und *na*vigation *r*anging."

Sonar generally makes use of **ultrasonic** frequencies: that is, sound waves whose frequencies are above 20 kHz, beyond the range of human detection. For sonar, the frequencies are typically in the range 20 kHz to 100 kHz. One reason for using ultrasound waves, other than the fact that they are inaudible, is that for shorter wavelengths there is less diffraction (Section 11–14) so the beam spreads less and smaller objects can be detected.

*Ultrasound Medical Imaging

The diagnostic use of ultrasound in medicine, in the form of images (sometimes called *sonograms*), is an important and interesting application of physical principles. A **pulse-echo technique** is used, much like sonar, except that the frequencies used are in the range of 1 to 10 MHz $(1\text{ MHz} = 10^6\text{ Hz})$. A high-frequency sound pulse is directed into the body, and its reflections from boundaries or interfaces between organs and other structures and lesions in the body are then detected. Tumors and other abnormal growths, or pockets of fluid, can be distinguished; the action of heart valves and the development of a fetus (Fig. 12–26) can be examined; and information about various organs of the body, such as the brain, heart, liver, and kidneys, can be obtained. Although ultrasound does not replace X-rays, for certain kinds of diagnosis it is more helpful. Some kinds of tissue or fluid are not detected in X-ray photographs, but ultrasound waves are reflected from their boundaries. "Real-time" ultrasound images are like a movie of a section of the interior of the body.

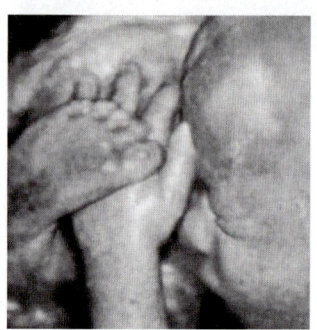

FIGURE 12–26 Ultrasound image of a human fetus within the uterus.

PHYSICS APPLIED
Ultrasound medical imaging

The pulse-echo technique for medical imaging works as follows. A brief pulse of ultrasound is emitted by a transducer that transforms an electrical pulse into a sound-wave pulse. Part of the pulse is reflected as echoes at each interface in the body, and most of the pulse (usually) continues on, Fig. 12–27a.

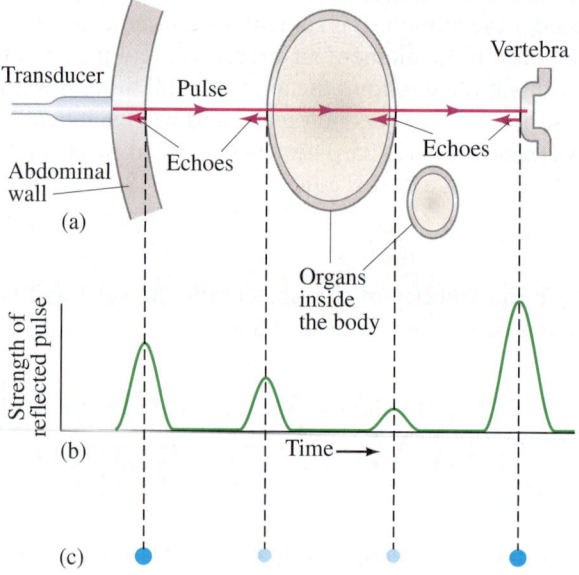

FIGURE 12–27 (a) Ultrasound pulse passes through the abdomen, reflecting from surfaces in its path. (b) Reflected pulses plotted as a function of time when received by transducer. The vertical dashed lines point out which reflected pulse goes with which surface. (c) "Dot display" for the same echoes: brightness of each dot is related to signal strength.

The detection of reflected pulses by the same transducer can then be displayed on the screen of a display monitor. The time elapsed from when the pulse is emitted to when each reflection (echo) is received is proportional to the distance to the reflecting surface. For example, if the distance from the transducer to the vertebra is 25 cm, the pulse travels a round-trip distance of $2 \times 25\text{ cm} = 0.50\text{ m}$. The speed of sound in human tissue is about 1540 m/s (close to that of sea water), so the time taken is

$$t = \frac{d}{v} = \frac{(0.50\text{ m})}{(1540\text{ m/s})} = 320\ \mu\text{s}.$$

The *strength* of a reflected pulse depends mainly on the difference in density of the two materials on either side of the interface and can be displayed as a pulse or as a dot (Figs. 12–27b and c). Each echo dot (Fig. 12–27c) can be represented as a point whose position is given by the time delay and whose brightness depends

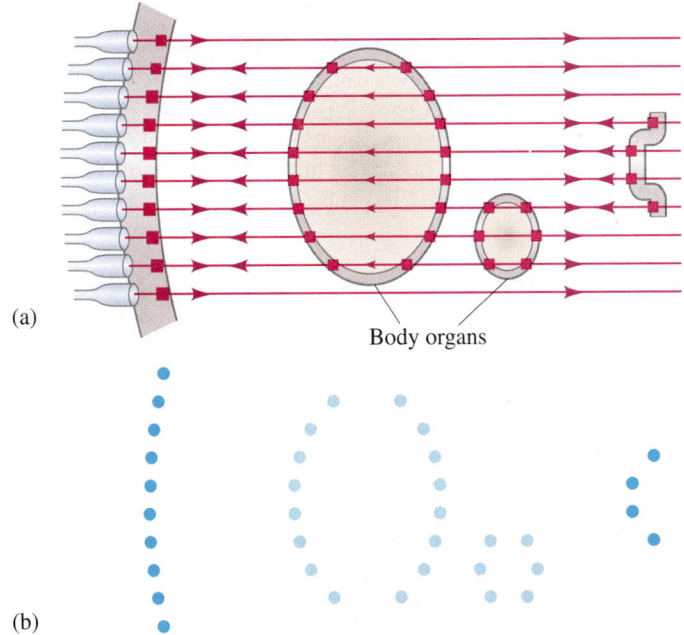

FIGURE 12–28 (a) Ten traces are made across the abdomen by moving the transducer, or by using an array of transducers. (b) The echoes from interfaces or boundaries (of organs) are plotted as dots to produce the image. More closely spaced traces would give a more detailed image.

on the strength of the echo. A two-dimensional image can then be formed out of these dots from a series of scans. The transducer is moved, or an array of transducers is used, each of which sends out a pulse at each position and receives echoes as shown in Fig. 12–28a. Each trace can be plotted, spaced appropriately one below the other, to form an image on a monitor screen as shown in Fig. 12–28b. Only 10 lines are shown in Fig. 12–28, so the image is crude. More lines give a more precise image.

Summary

Sound travels as a longitudinal wave in air and other materials. In air, the speed of sound increases with temperature. At 20°C, it is about 343 m/s.

The **pitch** of a sound is determined by the frequency; the higher the frequency, the higher the pitch.

The **audible range** of frequencies for humans is roughly 20 Hz to 20,000 Hz (1 Hz = 1 cycle per second).

The **loudness** or **intensity** of a sound is related to the amplitude squared of the wave. Because the human ear can detect sound intensities from 10^{-12} W/m² to over 1 W/m², sound levels are specified on a logarithmic scale. The **sound level** β, specified in decibels, is defined in terms of intensity I as

$$\beta \text{ (in dB)} = 10 \log\left(\frac{I}{I_0}\right), \quad (12\text{–}1)$$

where the reference intensity I_0 is usually taken to be 10^{-12} W/m².

Musical instruments are simple sources of sound in which *standing waves* are produced.

The strings of a stringed instrument may vibrate as a whole with nodes only at the ends; the frequency at which this standing wave occurs is called the **fundamental**. The fundamental frequency corresponds to a wavelength equal to twice the length of the string, $\lambda_1 = 2\ell$. The string can also vibrate at higher frequencies, called **overtones** or **harmonics**, in which there are one or more additional nodes. The frequency of each harmonic is a whole-number multiple of the fundamental.

In wind instruments, standing waves are set up in the column of air within the tube.

The vibrating air in an **open tube** (open at both ends) has displacement antinodes at both ends. The fundamental frequency corresponds to a wavelength equal to twice the tube length: $\lambda_1 = 2\ell$. The harmonics have frequencies that are 1, 2, 3, 4, ··· times the fundamental frequency, just as for strings.

For a **closed tube** (closed at one end), the fundamental corresponds to a wavelength four times the length of the tube: $\lambda_1 = 4\ell$. Only the odd harmonics are present, equal to 1, 3, 5, 7, ··· times the fundamental frequency.

Sound waves from different sources can interfere with each other. If two sounds are at slightly different frequencies, **beats** can be heard at a frequency equal to the difference in frequency of the two sources.

The **Doppler effect** refers to the change in pitch of a sound due to the motion either of the source or of the observer. If source and observer are approaching each other, the perceived pitch is higher. If they are moving apart, the perceived pitch is lower.

[*Shock waves and a sonic boom occur when an object moves at a **supersonic** speed—faster than the speed of sound. **Ultrasonic**-frequency (higher than 20 kHz) sound waves are used in many applications, including sonar and medical imaging.]

Questions

1. What is the evidence that sound travels as a wave?
2. What is the evidence that sound is a form of energy?
3. Children sometimes play with a homemade "telephone" by attaching a string to the bottoms of two paper cups. When the string is stretched and a child speaks into one cup, the sound can be heard at the other cup (Fig. 12–29). Explain clearly how the sound wave travels from one cup to the other.

FIGURE 12–29 Question 3.

4. When a sound wave passes from air into water, do you expect the frequency or wavelength to change?
5. What evidence can you give that the speed of sound in air does not depend significantly on frequency?
6. The voice of a person who has inhaled helium sounds very high-pitched. Why?
7. How will the air temperature in a room affect the pitch of organ pipes?
8. Explain how a tube might be used as a filter to reduce the amplitude of sounds in various frequency ranges. (An example is a car muffler.)
9. Why are the frets on a guitar (Fig. 12–30) spaced closer together as you move up the fingerboard toward the bridge?

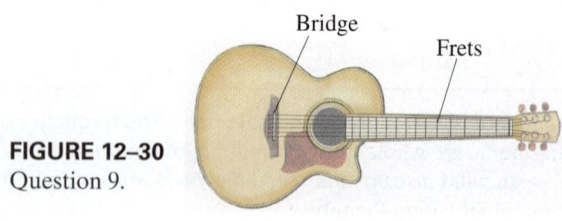

FIGURE 12–30 Question 9.

10. A noisy truck approaches you from behind a building. Initially you hear it but cannot see it. When it emerges and you do see it, its sound is suddenly "brighter"—you hear more of the high-frequency noise. Explain. [*Hint*: See Section 11–14 on diffraction.]
11. Standing waves can be said to be due to "interference in space," whereas beats can be said to be due to "interference in time." Explain.
12. In Fig. 12–16, if the frequency of the speakers is lowered, would the points D and C (where destructive and constructive interference occur) move farther apart or closer together? Explain.
13. Traditional methods of protecting the hearing of people who work in areas with very high noise levels have consisted mainly of efforts to block or reduce noise levels. With a relatively new technology, headphones are worn that do not block the ambient noise. Instead, a device is used which detects the noise, inverts it electronically, then feeds it to the headphones *in addition to* the ambient noise. How could adding *more* noise reduce the sound levels reaching the ears?

14. Consider the two waves shown in Fig. 12–31. Each wave can be thought of as a superposition of two sound waves with slightly different frequencies, as in Fig. 12–18. In which of the waves, (a) or (b), are the two component frequencies farther apart? Explain.

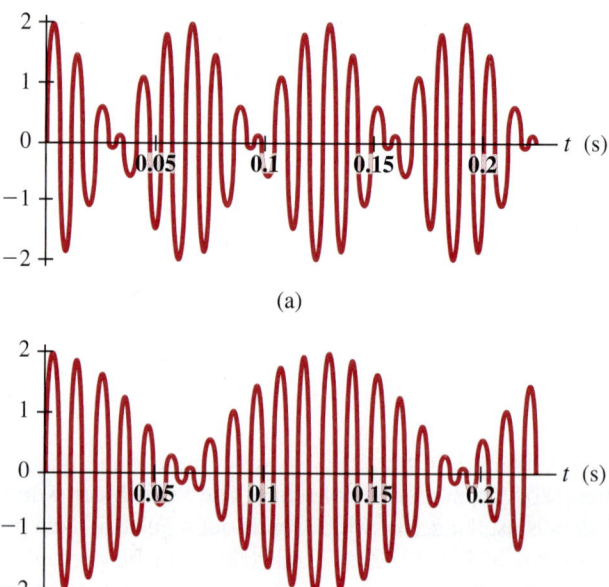

FIGURE 12–31 Question 14.

15. Is there a Doppler shift if the source and observer move in the same direction, with the same velocity? Explain.
16. If a wind is blowing, will this alter the frequency of the sound heard by a person at rest with respect to the source? Is the wavelength or velocity changed?
17. Figure 12–32 shows various positions of a child on a swing moving toward a person on the ground who is blowing a whistle. At which position, A through E, will the child hear the highest frequency for the sound of the whistle? Explain your reasoning.

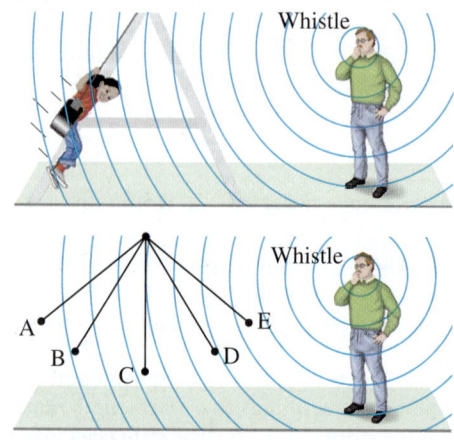

FIGURE 12–32 Question 17.

MisConceptual Questions

1. Do you expect an echo to return to you more quickly on a hot day or a cold day?
   (a) Hot day.
   (b) Cold day.
   (c) Same on both days.

2. Sound waves are
   (a) transverse waves characterized by the displacement of air molecules.
   (b) longitudinal waves characterized by the displacement of air molecules.
   (c) longitudinal waves characterized by pressure differences.
   (d) Both (b) and (c).
   (e) (a), (b), and (c).

3. The sound level near a noisy air conditioner is 70 dB. If two such units operate side by side, the sound level near them would be
   (a) 70 dB.
   (b) 73 dB.
   (c) 105 dB.
   (d) 140 dB.

4. To make a given sound seem twice as loud, how should a musician change the intensity of the sound?
   (a) Double the intensity.
   (b) Halve the intensity.
   (c) Quadruple the intensity.
   (d) Quarter the intensity.
   (e) Increase the intensity by a factor of 10.

5. A musical note that is two octaves higher than a second note
   (a) has twice the frequency of the second note.
   (b) has four times the frequency of the second note.
   (c) has twice the amplitude of the second note.
   (d) is 3 dB louder than the second note.
   (e) None of the above.

6. In which of the following is the wavelength of the lowest vibration mode the same as the length of the string or tube?
   (a) A string.
   (b) An open tube.
   (c) A tube closed at one end.
   (d) All of the above.
   (e) None of the above.

7. When a sound wave passes from air into water, what properties of the wave will change?
   (a) Frequency.
   (b) Wavelength.
   (c) Wave speed.
   (d) Both frequency and wavelength.
   (e) Both wave speed and wavelength.

8. A guitar string vibrates at a frequency of 330 Hz with wavelength 1.40 m. The frequency and wavelength of this sound in air (20°C) as it reaches our ears is
   (a) same frequency, same wavelength.
   (b) higher frequency, same wavelength.
   (c) lower frequency, same wavelength.
   (d) same frequency, longer wavelength.
   (e) same frequency, shorter wavelength.

9. A guitar player shortens the length of a guitar's vibrating string by pressing the string straight down onto a fret. The guitar then emits a higher-pitched note, because
   (a) the string's tension has been dramatically increased.
   (b) the string can vibrate with a much larger amplitude.
   (c) the string vibrates at a higher frequency.

10. An organ pipe with a fundamental frequency f is open at both ends. If one end is closed off, the fundamental frequency will
    (a) drop by half.
    (b) not change.
    (c) double.

11. Two loudspeakers are about 10 m apart in the front of a large classroom. If either speaker plays a pure tone at a single frequency of 400 Hz, the loudness seems pretty even as you wander around the room, and gradually decreases in volume as you move farther from the speaker. If both speakers then play the same tone together, what do you hear as you wander around the room?
    (a) The pitch of the sound increases to 800 Hz, and the sound is louder but not twice as loud. It is louder closer to the speakers and gradually decreases as you move away from the speakers—except near the back wall, where a slight echo makes the sound louder.
    (b) The sound is louder but maintains the same relative spatial pattern of gradually decreasing volume as you move away from the speakers.
    (c) As you move around the room, some areas seem to be dead spots with very little sound, whereas other spots seem to be louder than with only one speaker.
    (d) The sound is twice as loud—so loud that you cannot hear any difference as you move around the room.
    (e) At points equidistant from both speakers, the sound is twice as loud. In the rest of the room, the sound is the same as if a single speaker were playing.

12. You are driving at 75 km/h. Your sister follows in the car behind at 75 km/h. When you honk your horn, your sister hears a frequency
    (a) higher than the frequency you hear.
    (b) lower than the frequency you hear.
    (c) the same as the frequency you hear.
    (d) You cannot tell without knowing the horn's frequency.

13. A guitar string is vibrating at its fundamental frequency f. Which of the following is *not* true?
    (a) Each small section of the guitar string oscillates up and down at a frequency f.
    (b) The wavelength of the standing wave on the guitar string is $\lambda = v/f$, where v is the velocity of the wave on the string.
    (c) A sound wave created by this vibrating string propagates through the air with frequency f.
    (d) A sound wave created by this vibrating string propagates through the air with wavelength $\lambda = v/f$, where v is the velocity of sound in air.
    (e) The wavelength of the standing wave on the guitar string is $\lambda = \ell$, where ℓ is the length of the string.

For assigned homework and other learning materials, go to the MasteringPhysics website.

Problems

[Unless stated otherwise, assume $T = 20°C$ and $v_{sound} = 343$ m/s in air.]

12–1 Characteristics of Sound

1. (I) A hiker determines the length of a lake by listening for the echo of her shout reflected by a cliff at the far end of the lake. She hears the echo 2.5 s after shouting. Estimate the length of the lake.

2. (I) A sailor strikes the side of his ship just below the waterline. He hears the echo of the sound reflected from the ocean floor directly below 2.0 s later. How deep is the ocean at this point? Assume the speed of sound in sea water is 1560 m/s (Table 12–1) and does not vary significantly with depth.

3. (I) (a) Calculate the wavelengths in air at 20°C for sounds in the maximum range of human hearing, 20 Hz to 20,000 Hz. (b) What is the wavelength of an 18-MHz ultrasonic wave?

4. (II) On a warm summer day (31°C), it takes 4.80 s for an echo to return from a cliff across a lake. On a winter day, it takes 5.20 s. What is the temperature on the winter day?

5. (II) An ocean fishing boat is drifting just above a school of tuna on a foggy day. Without warning, an engine backfire occurs on another boat 1.55 km away (Fig. 12–33). How much time elapses before the backfire is heard (a) by the fish, and (b) by the fishermen?

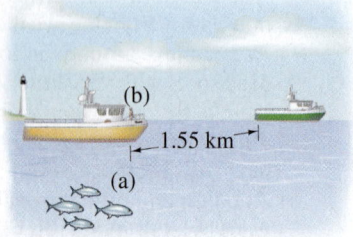

FIGURE 12–33 Problem 5.

6. (II) A person, with his ear to the ground, sees a huge stone strike the concrete pavement. A moment later two sounds are heard from the impact: one travels in the air and the other in the concrete, and they are 0.80 s apart. How far away did the impact occur? See Table 12–1.

7. (III) A stone is dropped from the top of a cliff. The splash it makes when striking the water below is heard 2.7 s later. How high is the cliff?

12–2 Intensity of Sound; Decibels

8. (I) What is the intensity of a sound at the pain level of 120 dB? Compare it to that of a whisper at 20 dB.

9. (I) What is the sound level of a sound whose intensity is 1.5×10^{-6} W/m²?

10. (II) You are trying to decide between two new stereo amplifiers. One is rated at 75 W per channel and the other is rated at 120 W per channel. In terms of dB, how much louder will the more powerful amplifier be when both are producing sound at their maximum levels?

11. (II) If two firecrackers produce a combined sound level of 85 dB when fired simultaneously at a certain place, what will be the sound level if only one is exploded? [Hint: Add intensities, not dBs.]

12. (II) A person standing a certain distance from an airplane with four equally noisy jet engines is experiencing a sound level of 140 dB. What sound level would this person experience if the captain shut down all but one engine? [Hint: Add intensities, not dBs.]

13. (II) One CD player is said to have a signal-to-noise ratio of 82 dB, whereas for a second CD player it is 98 dB. What is the ratio of intensities of the signal and the background noise for each device?

14. (II) A 55-dB sound wave strikes an eardrum whose area is 5.0×10^{-5} m². (a) How much energy is received by the eardrum per second? (b) At this rate, how long would it take your eardrum to receive a total energy of 1.0 J?

15. (II) At a rock concert, a dB meter registered 130 dB when placed 2.5 m in front of a loudspeaker on stage. (a) What was the power output of the speaker, assuming uniform spherical spreading of the sound and neglecting absorption in the air? (b) How far away would the sound level be 85 dB?

16. (II) A fireworks shell explodes 100 m above the ground, creating colorful sparks. How much greater is the sound level of the explosion for a person at a point directly below the explosion than for a person a horizontal distance of 200 m away (Fig. 12–34)?

FIGURE 12–34 Problem 16.

17. (II) If the amplitude of a sound wave is made 3.5 times greater, (a) by what factor will the intensity increase? (b) By how many dB will the sound level increase?

18. (II) Two sound waves have equal displacement amplitudes, but one has 2.2 times the frequency of the other. What is the ratio of their intensities?

19. (II) What would be the sound level (in dB) of a sound wave in air that corresponds to a displacement amplitude of vibrating air molecules of 0.13 mm at 440 Hz?

20. (II) (a) Estimate the power output of sound from a person speaking in normal conversation. Use Table 12–2. Assume the sound spreads roughly uniformly over a sphere centered on the mouth. (b) How many people would it take to produce a total sound output of 60 W of ordinary conversation? [Hint: Add intensities, not dBs.]

21. (III) Expensive amplifier A is rated at 220 W, while the more modest amplifier B is rated at 45 W. (a) Estimate the sound level in decibels you would expect at a point 3.5 m from a loudspeaker connected in turn to each amp. (b) Will the expensive amp sound twice as loud as the cheaper one?

*12–3 Loudness

*22. (I) A 5000-Hz tone must have what sound level to seem as loud as a 100-Hz tone that has a 50-dB sound level? (See Fig. 12–6.)

*23. (I) What are the lowest and highest frequencies that an ear can detect when the sound level is 40 dB? (See Fig. 12–6.)

*24. (II) Your ears can accommodate a huge range of sound levels. What is the ratio of highest to lowest intensity at (a) 100 Hz, (b) 5000 Hz? (See Fig. 12–6.)

12–4 Sources of Sound: Strings and Air Columns

25. (I) Estimate the number of octaves in the human audible range, 20 Hz to 20 kHz.
26. (I) What would you estimate for the length of a bass clarinet, assuming that it is modeled as a closed tube and that the lowest note that it can play is a D♭ whose frequency is 69 Hz?
27. (I) The A string on a violin has a fundamental frequency of 440 Hz. The length of the vibrating portion is 32 cm, and it has mass 0.35 g. Under what tension must the string be placed?
28. (I) An organ pipe is 116 cm long. Determine the fundamental and first three audible overtones if the pipe is (a) closed at one end, and (b) open at both ends.
29. (I) (a) What resonant frequency would you expect from blowing across the top of an empty soda bottle that is 24 cm deep, if you assumed it was a closed tube? (b) How would that change if it was one-third full of soda?
30. (I) If you were to build a pipe organ with open-tube pipes spanning the range of human hearing (20 Hz to 20 kHz), what would be the range of the lengths of pipes required?
31. (II) A tight guitar string has a frequency of 540 Hz as its third harmonic. What will be its fundamental frequency if it is fingered at a length of only 70% of its original length?
32. (II) Estimate the frequency of the "sound of the ocean" when you put your ear very near a 15-cm-diameter seashell (Fig. 12–35).

FIGURE 12–35 Problem 32.

33. (II) An unfingered guitar string is 0.68 m long and is tuned to play E above middle C (330 Hz). (a) How far from the end of this string must a fret (and your finger) be placed to play A above middle C (440 Hz)? (b) What is the wavelength on the string of this 440-Hz wave? (c) What are the frequency and wavelength of the sound wave produced in air at 22°C by this fingered string?
34. (II) (a) Determine the length of an open organ pipe that emits middle C (262 Hz) when the temperature is 18°C. (b) What are the wavelength and frequency of the fundamental standing wave in the tube? (c) What are λ and f in the traveling sound wave produced in the outside air?
35. (II) An organ is in tune at 22.0°C. By what percent will the frequency be off at 11°C?
36. (II) How far from the mouthpiece of the flute in Example 12–11 should the hole be that must be uncovered to play F above middle C at 349 Hz?
37. (II) (a) At $T = 22$°C, how long must an open organ pipe be to have a fundamental frequency of 294 Hz? (b) If this pipe is filled with helium, what is its fundamental frequency?
38. (II) A particular organ pipe can resonate at 264 Hz, 440 Hz, and 616 Hz, but not at any other frequencies in between. (a) Show why this is an open or a closed pipe. (b) What is the fundamental frequency of this pipe?
39. (II) A uniform narrow tube 1.70 m long is open at both ends. It resonates at two successive harmonics of frequencies 275 Hz and 330 Hz. What is (a) the fundamental frequency, and (b) the speed of sound in the gas in the tube?
40. (II) A pipe in air at 23.0°C is to be designed to produce two successive harmonics at 280 Hz and 320 Hz. How long must the pipe be, and is it open or closed?

41. (II) How many overtones are present within the audible range for a 2.18-m-long organ pipe at 20°C (a) if it is open, and (b) if it is closed?
42. (II) Determine the fundamental and first overtone frequencies when you are in a 9.0-m-long hallway with all doors closed. Model the hallway as a tube closed at both ends.
43. (III) When a player's finger presses a guitar string down onto a fret, the length of the vibrating portion of the string is shortened, thereby increasing the string's fundamental frequency (see Fig. 12–36). The string's tension and mass per unit length remain unchanged. If the unfingered length of the string is $\ell = 75.0$ cm, determine the positions x of the first six frets, if each fret raises the pitch of the fundamental by one musical note compared to the neighboring fret. On the equally tempered chromatic scale, the ratio of frequencies of neighboring notes is $2^{1/12}$.

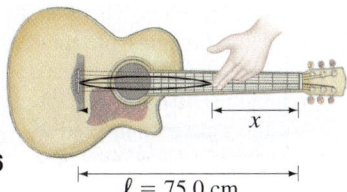

FIGURE 12–36 Problem 43.

$\ell = 75.0$ cm

44. (III) The human ear canal is approximately 2.5 cm long. It is open to the outside and is closed at the other end by the eardrum. Estimate the frequencies (in the audible range) of the standing waves in the ear canal. What is the relationship of your answer to the information in the graph of Fig. 12–6?

*12–5 Quality of Sound, Superposition

*45. (II) Approximately what are the intensities of the first two overtones of a violin compared to the fundamental? How many decibels softer than the fundamental are the first and second overtones? (See Fig. 12–15.)

12–6 Interference; Beats

46. (I) A piano tuner hears one beat every 2.0 s when trying to adjust two strings, one of which is sounding 350 Hz. How far off in frequency is the other string?
47. (I) A certain dog whistle operates at 23.5 kHz, while another (brand X) operates at an unknown frequency. If humans can hear neither whistle when played separately, but a shrill whine of frequency 5000 Hz occurs when they are played simultaneously, estimate the operating frequency of brand X.
48. (II) What is the beat frequency if middle C (262 Hz) and C♯ (277 Hz) are played together? What if each is played two octaves lower (each frequency reduced by a factor of 4)?
49. (II) A guitar string produces 3 beats/s when sounded with a 350-Hz tuning fork and 8 beats/s when sounded with a 355-Hz tuning fork. What is the vibrational frequency of the string? Explain your reasoning.
50. (II) Two violin strings are tuned to the same frequency, 294 Hz. The tension in one string is then decreased by 2.5%. What will be the beat frequency heard when the two strings are played together? [*Hint*: Recall Eq. 11–13.]
51. (II) The two sources of sound in Fig. 12–16 face each other and emit sounds of equal amplitude and equal frequency (305 Hz) but 180° out of phase. For what minimum separation of the two speakers will there be some point at which (a) complete constructive interference occurs and (b) complete destructive interference occurs. (Assume $T = 20$°C.)

52. (II) Two piano strings are supposed to be vibrating at 220 Hz, but a piano tuner hears three beats every 2.5 s when they are played together. (a) If one is vibrating at 220.0 Hz, what must be the frequency of the other (is there only one answer)? (b) By how much (in percent) must the tension be increased or decreased to bring them in tune?

53. (III) Two loudspeakers are 1.60 m apart. A person stands 3.00 m from one speaker and 3.50 m from the other. (a) What is the lowest frequency at which destructive interference will occur at this point if the speakers are in phase? (b) Calculate two other frequencies that also result in destructive interference at this point (give the next two highest). Let $T = 20°C$.

54. (III) Two loudspeakers are placed 3.00 m apart, as shown in Fig. 12–37. They emit 474-Hz sounds, in phase. A microphone is placed 3.20 m distant from a point midway between the two speakers, where an intensity maximum is recorded. (a) How far must the microphone be moved to the right to find the first intensity minimum? (b) Suppose the speakers are reconnected so that the 474-Hz sounds they emit are exactly out of phase. At what positions are the intensity maximum and minimum now?

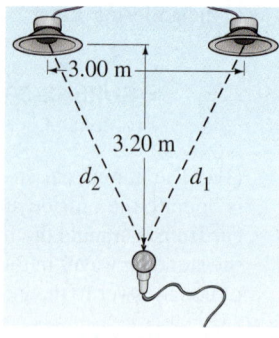

FIGURE 12–37
Problem 54.

55. (III) A source emits sound of wavelengths 2.54 m and 2.72 m in air. (a) How many beats per second will be heard? (Assume $T = 20°C$.) (b) How far apart in space are the regions of maximum intensity?

12–7 Doppler Effect

56. (I) The predominant frequency of a certain fire truck's siren is 1650 Hz when at rest. What frequency do you detect if you move with a speed of 30.0 m/s (a) toward the fire truck, and (b) away from it?

57. (II) A bat at rest sends out ultrasonic sound waves at 50.0 kHz and receives them returned from an object moving directly away from it at 27.5 m/s. What is the received sound frequency?

58. (II) Two automobiles are equipped with the same single-frequency horn. When one is at rest and the other is moving toward the first at 18 m/s, the driver at rest hears a beat frequency of 4.5 Hz. What is the frequency the horns emit? Assume $T = 20°C$.

59. (II) As a bat flies toward a wall at a speed of 6.0 m/s, the bat emits an ultrasonic sound wave with frequency 30.0 kHz. What frequency does the bat hear in the reflected wave?

60. (II) In one of the original Doppler experiments, a tuba was played at a frequency of 75 Hz on a moving flat train car, and a second identical tuba played the same tone while at rest in the railway station. What beat frequency was heard in the station if the train car approached the station at a speed of 14.0 m/s?

61. (II) A wave on the ocean surface with wavelength 44 m travels east at a speed of 18 m/s relative to the ocean floor. If, on this stretch of ocean, a powerboat is moving at 14 m/s (relative to the ocean floor), how often does the boat encounter a wave crest, if the boat is traveling (a) west, and (b) east?

62. (III) A police car sounding a siren with a frequency of 1580 Hz is traveling at 120.0 km/h. (a) What frequencies does an observer standing next to the road hear as the car approaches and as it recedes? (b) What frequencies are heard in a car traveling at 90.0 km/h in the opposite direction before and after passing the police car? (c) The police car passes a car traveling in the same direction at 80.0 km/h. What two frequencies are heard in this car?

63. (III) The Doppler effect using ultrasonic waves of frequency 2.25×10^6 Hz is used to monitor the heartbeat of a fetus. A (maximum) beat frequency of 240 Hz is observed. Assuming that the speed of sound in tissue is 1540 m/s, calculate the maximum velocity of the surface of the beating heart.

*12–8 Shock Waves; Sonic Boom

*64. (I) (a) How fast is an object moving on land if its speed at 24°C is Mach 0.33? (b) A high-flying jet cruising at 3000 km/h displays a Mach number of 3.1 on a screen. What is the speed of sound at that altitude?

*65. (I) The wake of a speedboat is 12° in a lake where the speed of the water wave is 2.2 km/h. What is the speed of the boat?

*66. (II) An airplane travels at Mach 2.1 where the speed of sound is 310 m/s. (a) What is the angle the shock wave makes with the direction of the airplane's motion? (b) If the plane is flying at a height of 6500 m, how long after it is directly overhead will a person on the ground hear the shock wave?

*67. (II) A space probe enters the thin atmosphere of a planet where the speed of sound is only about 42 m/s. (a) What is the probe's Mach number if its initial speed is 15,000 km/h? (b) What is the angle of the shock wave relative to the direction of motion?

*68. (II) A meteorite traveling 9200 m/s strikes the ocean. Determine the shock wave angle it produces (a) in the air just before entering the ocean, and (b) in the water just after entering. Assume $T = 20°C$.

*69. (III) You look directly overhead and see a plane exactly 1.45 km above the ground flying faster than the speed of sound. By the time you hear the sonic boom, the plane has traveled a horizontal distance of 2.0 km. See Fig. 12–38. Determine (a) the angle of the shock cone, θ, and (b) the speed of the plane and its Mach number. Assume the speed of sound is 330 m/s.

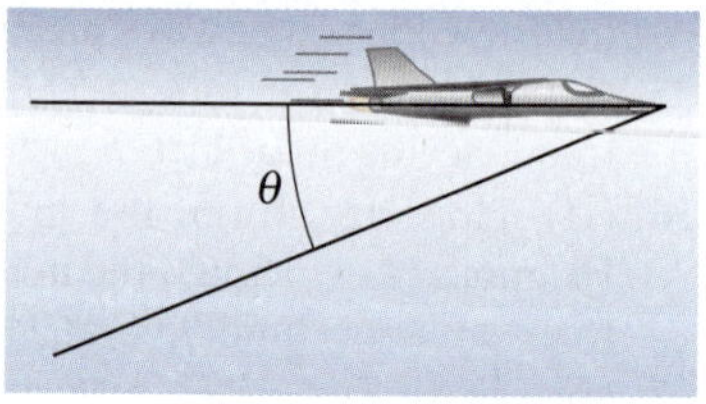

FIGURE 12–38
Problem 69.

*70. (III) A supersonic jet traveling at Mach 2.0 at an altitude of 9500 m passes directly over an observer on the ground. Where will the plane be relative to the observer when the latter hears the sonic boom? (See Fig. 12–39.)

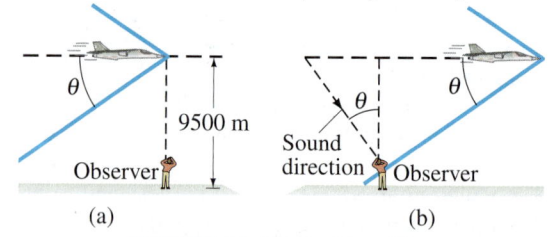

FIGURE 12–39 Problem 70.

General Problems

71. A fish finder uses a sonar device that sends 20,000-Hz sound pulses downward from the bottom of the boat, and then detects echoes. If the maximum depth for which it is designed to work is 85 m, what is the minimum time between pulses (in fresh water)?

72. A single mosquito 5.0 m from a person makes a sound close to the threshold of human hearing (0 dB). What will be the sound level of 200 such mosquitoes?

73. What is the resultant sound level when an 81-dB sound and an 87-dB sound are heard simultaneously?

74. The sound level 8.25 m from a loudspeaker, placed in the open, is 115 dB. What is the acoustic power output (W) of the speaker, assuming it radiates equally in all directions?

75. A stereo amplifier is rated at 225 W output at 1000 Hz. The power output drops by 12 dB at 15 kHz. What is the power output in watts at 15 kHz?

76. Workers around jet aircraft typically wear protective devices over their ears. Assume that the sound level of a jet airplane engine, at a distance of 30 m, is 130 dB, and that the average human ear has an effective radius of 2.0 cm. What would be the power intercepted by an unprotected ear at a distance of 30 m from a jet airplane engine?

77. In audio and communications systems, the **gain**, β, in decibels is defined for an amplifier as
$$\beta = 10 \log\left(\frac{P_{\text{out}}}{P_{\text{in}}}\right),$$
where P_{in} is the power input to the system and P_{out} is the power output. (a) A particular amplifier puts out 135 W of power for an input of 1.0 mW. What is its gain in dB? (b) If a signal-to-noise ratio of 93 dB is specified, what is the noise power if the output signal is 10 W?

78. Manufacturers typically offer a particular guitar string in a choice of diameters so that players can tune their instruments with a preferred string tension. For example, a nylon high-E string is available in a low- and high-tension model with diameter 0.699 mm and 0.724 mm, respectively. Assuming the density ρ of nylon is the same for each model, compare (as a ratio) the tension in a tuned high- and low-tension string.

79. A tuning fork is set into vibration above a vertical open tube filled with water (Fig. 12–40). The water level is allowed to drop slowly. As it does so, the air in the tube above the water level is heard to resonate with the tuning fork when the distance from the tube opening to the water level is 0.125 m and again at 0.395 m. What is the frequency of the tuning fork?

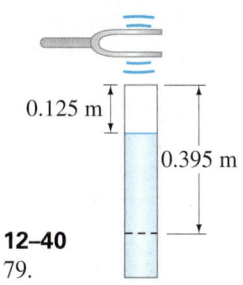

FIGURE 12–40 Problem 79.

80. Two identical tubes, each closed at one end, have a fundamental frequency of 349 Hz at 25.0°C. The air temperature is increased to 31.0°C in one tube. If the two pipes are now sounded together, what beat frequency results?

81. Each string on a violin is tuned to a frequency $1\frac{1}{2}$ times that of its neighbor. The four equal-length strings are to be placed under the same tension; what must be the mass per unit length of each string relative to that of the lowest string?

82. A particular whistle produces sound by setting up the fundamental standing wave in an air column 7.10 cm long. The tube is closed at one end. The whistle blower is riding in a car moving away from you at 25 m/s. What frequency do you hear?

83. The diameter D of a tube does affect the node at the open end of a tube. The end correction can be roughly approximated as adding $D/3$ to ℓ to give us an effective length for the tube in calculations. For a closed tube of length 0.55 m and diameter 3.0 cm, what are the frequencies of the first four harmonics, taking the end correction into consideration?

84. The frequency of a steam train whistle as it approaches you is 565 Hz. After it passes you, its frequency is measured as 486 Hz. How fast was the train moving (assume constant velocity)?

85. Two trains emit 508-Hz whistles. One train is stationary. The conductor on the stationary train hears a 3.5-Hz beat frequency when the other train approaches. What is the speed of the moving train?

86. Two loudspeakers are at opposite ends of a railroad car as it moves past a stationary observer at 12.0 m/s, as shown in Fig. 12–41. If the speakers have identical sound frequencies of 348 Hz, what is the beat frequency heard by the observer when (a) he listens from position A, in front of the car, (b) he is between the speakers, at B, and (c) he hears the speakers after they have passed him, at C?

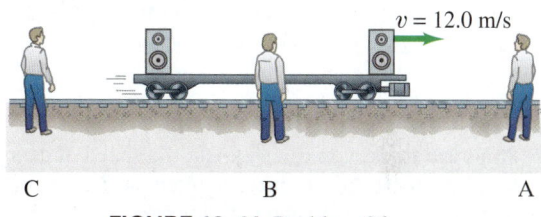

FIGURE 12–41 Problem 86.

87. Two open organ pipes, sounding together, produce a beat frequency of 6.0 Hz. The shorter one is 2.40 m long. How long is the other?

88. A bat flies toward a moth at speed 7.8 m/s while the moth is flying toward the bat at speed 5.0 m/s. The bat emits a sound wave of 51.35 kHz. What is the frequency of the wave detected by the bat after that wave reflects off the moth?

89. A bat emits a series of high-frequency sound pulses as it approaches a moth. The pulses are approximately 70.0 ms apart, and each is about 3.0 ms long. How far away can the moth be detected by the bat so that the echo from one pulse returns before the next pulse is emitted?

90. Two loudspeakers face each other at opposite ends of a long corridor. They are connected to the same source which produces a pure tone of 282 Hz. A person walks from one speaker toward the other at a speed of 1.6 m/s. What "beat" frequency does the person hear?

91. A sound-insulating door reduces the sound level by 30 dB. What fraction of the sound intensity passes through this door?

92. The "alpenhorn" (Fig. 12–42) was once used to send signals from one Alpine village to another. Since lower frequency sounds are less susceptible to intensity loss, long horns were used to create deep sounds. When played as a musical instrument, the alpenhorn must be blown in such a way that only one of the overtones is resonating. The most popular alpenhorn is about 3.4 m long, and it is called the F$^\#$ (or G$^\flat$) horn. What is the fundamental frequency of this horn, and which overtone is close to F$^\#$? (See Table 12–3.) Model as a tube open at both ends.

FIGURE 12–42 Problem 92.

93. Room acoustics for stereo listening can be compromised by the presence of standing waves, which can cause acoustic "dead spots" at the locations of the pressure nodes. Consider a living room 4.7 m long, 3.6 m wide, and 2.8 m high. Calculate the fundamental frequencies for the standing waves in this room.

94. A dramatic demonstration, called "singing rods," involves a long, slender aluminum rod held in the hand near the rod's midpoint. The rod is stroked with the other hand. With a little practice, the rod can be made to "sing," or emit a clear, loud, ringing sound. For an 80-cm-long rod, (a) what is the fundamental frequency of the sound? (b) What is its wavelength in the rod, and (c) what is the traveling wavelength of the sound in air at 20°C?

*95. The intensity at the threshold of hearing for the human ear at a frequency of about 1000 Hz is $I_0 = 1.0 \times 10^{-12}$ W/m^2, for which β, the sound level, is 0 dB. The threshold of pain at the same frequency is about 120 dB, or $I = 1.0$ W/m^2, corresponding to an increase of intensity by a factor of 10^{12}. By what factor does the displacement amplitude, A, vary?

96. A **Doppler flow meter** uses ultrasound waves to measure blood-flow speeds. Suppose the device emits sound at 3.5 MHz, and the speed of sound in human tissue is about 1540 m/s. What is the expected beat frequency if blood is flowing in large leg arteries at 3.0 cm/s directly away from the sound source?

Search and Learn

1. At a painfully loud concert, a 120-dB sound wave travels away from a loudspeaker at 343 m/s. How much sound wave energy is contained in each 1.0-cm^3 volume of air in the region near this loudspeaker? (See Sections 12–2 and 11–9.)

2. At a race track, you can estimate the speed of cars just by listening to the difference in pitch of the engine noise between approaching and receding cars. Suppose the sound of a certain car drops by a full octave (frequency halved) as it goes by on the straightaway. How fast is it going?

3. A person hears a pure tone in the 500 to 1000-Hz range coming from two sources. The sound is loudest at points equidistant from the two sources. To determine exactly what the frequency is, the person moves about and finds that the sound level is minimal at a point 0.25 m farther from one source than the other. What is the frequency of the sound?

4. A factory whistle emits sound of frequency 770 Hz. The wind velocity is 15.0 m/s from the north (heading south). What frequency will observers hear who are located, at rest, (a) due north, (b) due south, (c) due east, and (d) due west, of the whistle? What frequency is heard by a cyclist heading (e) north or (f) west, toward the whistle at 12.0 m/s? Assume $T = 20°C$.

5. A bugle is a tube of fixed length that behaves as if it is open at both ends. A bugler, by adjusting his lips correctly and blowing with proper air pressure, can cause a harmonic (usually other than the fundamental) of the air column within the tube to sound loudly. Standard military tunes like *Taps* and *Reveille* require only four musical notes: G4 (392 Hz), C5 (523 Hz), E5 (659 Hz), and G5 (784 Hz). (a) For a certain length ℓ, a bugle will have a sequence of four consecutive harmonics whose frequencies very nearly equal those associated with the notes G4, C5, E5, and G5. Determine this ℓ. (b) Which harmonic is each of the (approximate) notes G4, C5, E5, and G5 for the bugle?

ANSWERS TO EXERCISES

A: 4 times as intense.
B: (b).
C: 7 octaves, plus. [Note: This is like counting in binary, $2^7 = 128$; for more see Section 17–10.]
D: (b).
E: 257 Hz.
F: (a) 1717 Hz, (b) 1483 Hz.
G: (a).

Monument Valley, Arizona

Heating the air inside a "hot-air" balloon raises the air's temperature, causing it to expand, and forces some of the air out the opening at the bottom. The reduced amount of air inside means its density is lower than the outside air, so there is a net buoyant force upward on the balloon (Chapter 10). In this Chapter we study temperature and its effects on matter: thermal expansion and the gas laws. We examine the microscopic theory of matter as atoms or molecules that are continuously in motion, which we call kinetic theory. The temperature of a gas is directly related to the average translational kinetic energy of its molecules. We will consider ideal gases, but will also look at real gases and how they change phase, including evaporation, vapor pressure, and humidity.

Temperature and Kinetic Theory

CHAPTER 13

CONTENTS

13–1 Atomic Theory of Matter
13–2 Temperature and Thermometers
13–3 Thermal Equilibrium and the Zeroth Law of Thermodynamics
13–4 Thermal Expansion
13–5 The Gas Laws and Absolute Temperature
13–6 The Ideal Gas Law
13–7 Problem Solving with the Ideal Gas Law
13–8 Ideal Gas Law in Terms of Molecules: Avogadro's Number
13–9 Kinetic Theory and the Molecular Interpretation of Temperature
13–10 Distribution of Molecular Speeds
13–11 Real Gases and Changes of Phase
13–12 Vapor Pressure and Humidity
*13–13 Diffusion

CHAPTER-OPENING QUESTION—Guess now!
A hot-air balloon, open at one end (see photos above), rises when the air inside is heated by a flame. For the following properties, is the air inside the balloon higher, lower, or the same as for the air outside the balloon?
(i) Temperature. **(ii)** Pressure. **(iii)** Density.

This Chapter is the first of three (Chapters 13, 14, and 15) devoted to temperature, heat, and thermodynamics. Much of this Chapter discusses the theory that matter is made up of atoms and that these atoms are in continuous random motion. This theory is called the *kinetic theory*. ("Kinetic," you may recall from Chapter 6, is Greek for "moving.")

We also discuss the concept of temperature and how it is measured, as well as the measured properties of gases which serve as a foundation for kinetic theory.

13–1 Atomic Theory of Matter

The idea that matter is made up of atoms dates back to the ancient Greeks. According to the Greek philosopher Democritus, if a pure substance—say, a piece of iron—were cut into smaller and smaller bits, eventually a smallest piece of that substance would be obtained which could not be divided further. This smallest piece was called an **atom**, which in Greek means "indivisible." Today an atom is still the smallest piece of a substance, but we do not consider it indivisible. Rather it is viewed as consisting of a central nucleus (containing protons and neutrons) surrounded by electrons, Chapter 27.

Today the atomic theory is universally accepted. The experimental evidence in its favor, however, came mainly in the eighteenth, nineteenth, and twentieth centuries, and much of it was obtained from the analysis of chemical reactions.

We will often speak of the relative masses of individual atoms and molecules—what we call the **atomic mass** or **molecular mass**, respectively. (The terms *atomic weight* and *molecular weight* are sometimes used.) These masses are based on arbitrarily assigning the most abundant form of carbon atom, ^{12}C, the atomic mass of exactly 12.0000 **unified atomic mass units** (u). In terms of kilograms,

$$1\,u = 1.6605 \times 10^{-27}\,kg.$$

The average atomic mass of hydrogen is 1.0079 u, and the values for other atoms are as listed in the Periodic Table inside the back cover of this book, and also in Appendix B.† The molecular mass of a compound is the sum of atomic masses of the atoms making up the molecules of that compound.

[An **element** is a substance, such as neon, gold, iron, or copper, that cannot be broken down into simpler substances by chemical means. **Compounds** are substances made up of elements, and can be broken down into them; examples are carbon dioxide and water. The smallest piece of an element is an atom; the smallest piece of a compound is a **molecule**. Molecules are made up of atoms; a molecule of water, for example, is made up of two atoms of hydrogen and one of oxygen; its chemical formula is H_2O.]

An important piece of evidence for the atomic theory is called **Brownian motion**, named after the biologist Robert Brown, who is credited with its discovery in 1827. While he was observing tiny pollen grains suspended in water under his microscope, Brown noticed that the tiny grains moved about in erratic paths (Fig. 13–1), even though the water appeared to be perfectly still. The atomic theory easily explains Brownian motion if we assume that the atoms of any substance are continually in motion. Then Brown's tiny pollen grains are jostled about by the vigorous barrage of rapidly moving molecules of water.

In 1905, Albert Einstein examined Brownian motion from a theoretical point of view and was able to calculate from the experimental data the approximate size and mass of atoms and molecules. His calculations showed that the diameter of a typical atom is about 10^{-10} m.

At the start of Chapter 10, we distinguished the three common phases (or states) of matter—solid, liquid, gas—based on **macroscopic**, or "large-scale," properties. Now let us see how these three phases of matter differ, from the atomic or **microscopic** point of view. First of all, atoms and molecules must exert attractive forces on each other, because only this explains why a brick or a block of aluminum holds together in one piece. The attractive forces between molecules are of an electrical nature (more on this in later Chapters). When molecules come too close together, the force between them must become repulsive (electric repulsion between their outer electrons). We need this assumption to explain that matter takes up space. Thus molecules maintain a minimum distance from each other. In a solid material, the attractive forces are strong enough that the atoms or molecules move only slightly (oscillate) about relatively fixed positions, often in an array known as a crystal lattice, as shown in Fig. 13–2a. In a liquid, the atoms or molecules are moving more rapidly, or the forces between them are weaker, so that they are sufficiently free to pass around one another, as in Fig. 13–2b. In a gas, the forces are so weak, or

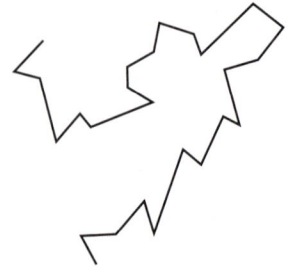

FIGURE 13–1 Path of a tiny particle (pollen grain, for example) suspended in water. The straight lines connect observed positions of the particle at equal time intervals.

FIGURE 13–2 Atomic arrangements in (a) a crystalline solid, (b) a liquid, and (c) a gas.

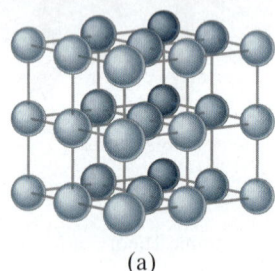

(a)

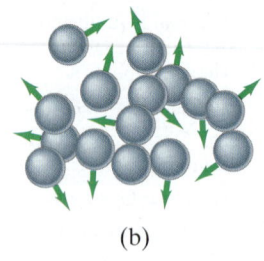

(b)

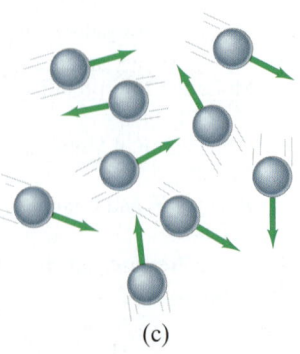

(c)

---

†The relative masses of different atoms came from analysis of chemical reactions, and the **law of definite proportions**. It states that when two or more elements combine to form a compound, they always do so in the same proportions by mass. For example, table salt is always formed from 23 parts sodium and 35 parts chlorine; and water from one part hydrogen and eight parts oxygen. A continuous theory of matter could not account for the law of definite proportions but atomic theory does: the proportions of each element that form a compound correspond to the relative masses of the combining atoms. One atom of sodium (Na) combines with one atom of chlorine (Cl) to form one molecule of salt (NaCl), and one atom of sodium has a mass 23/35 times as large as one of chlorine. Hydrogen, the lightest atom, was arbitrarily assigned the relative mass of 1. On this scale, carbon was about 12, oxygen 16, sodium 23, and so on. It was sometimes more complicated. For example, from the various compounds oxygen formed, its relative mass was judged to be 16; but this was inconsistent with the mass ratio in water of oxygen to hydrogen, only 8 to 1. This was explained by assuming two H atoms combine with one O atom to form a water molecule.

the speeds so high, that the molecules do not even stay close together. They move rapidly every which way, Fig. 13–2c, filling any container and occasionally colliding with one another. On average, the speeds are sufficiently high in a gas that when two molecules collide, the force of attraction is not strong enough to keep them close together and they fly off in new directions.

EXAMPLE 13–1 ESTIMATE **Distance between atoms.** The density of copper is 8.9×10^3 kg/m^3, and each copper atom has a mass of 63 u. Estimate the average distance between the centers of neighboring copper atoms.

APPROACH We consider a cube of copper 1 m on a side. From the given density ρ we can calculate the mass m of a cube of volume $V = 1$ m^3 ($m = \rho V$). We divide this mass m by the mass of one atom (63 u) to obtain the number of atoms in 1 m^3. We assume the atoms are in a uniform array, and we let N be the number of atoms in a 1-m length; then $(N)(N)(N) = N^3$ equals the total number of atoms in 1 m^3.

SOLUTION The mass of 1 copper atom is $63\,\text{u} = 63 \times 1.66 \times 10^{-27}$ kg $= 1.05 \times 10^{-25}$ kg. This means that in a cube of copper 1 m on a side (volume $= 1$ m^3), there are

$$\frac{8.9 \times 10^3 \text{ kg}}{1.05 \times 10^{-25} \text{ kg/atom}} = 8.5 \times 10^{28} \text{ atoms.}$$

The volume of a cube of side ℓ is $V = \ell^3$, so on one edge of the 1-m-long cube there are $(8.5 \times 10^{28})^{\frac{1}{3}}$ atoms $= 4.4 \times 10^9$ atoms. Hence the distance between neighboring atoms is

$$\frac{1 \text{ m}}{4.4 \times 10^9 \text{ atoms}} = 2.3 \times 10^{-10} \text{ m.}$$

NOTE Watch out for units. Even though "atoms" is not a unit, it is helpful to include it to make sure you calculate correctly.

NOTE The distance between atoms is essentially what we mean when we speak of the size or diameter of an atom. So we have calculated the size of a copper atom.

13–2 Temperature and Thermometers

In everyday life, **temperature** is a measure of how hot or cold something is. A hot oven is said to have a high temperature, whereas the ice of a frozen lake is said to have a low temperature.

Many properties of matter change with temperature. For example, most materials expand when their temperature is increased.[†] An iron beam is longer when hot than when cold. Concrete roads and sidewalks expand and contract slightly according to temperature, which is why compressible spacers or expansion joints (Fig. 13–3) are placed at regular intervals. The electrical resistance of matter changes with temperature (Chapter 18). So too does the color radiated by objects, at least at high temperatures: you may have noticed that the heating element of an electric stove glows with a red color when hot. At higher temperatures, solids such as iron glow orange or even white. The white light from an incandescent lightbulb comes from an extremely hot tungsten wire. The surface temperatures of the Sun and other stars can be measured by the predominant color (more precisely, wavelengths) of light they emit.

Instruments designed to measure temperature are called **thermometers**. There are many kinds of thermometers, but their operation always depends on some property of matter that changes with temperature. Many common thermometers rely on the expansion of a material with an increase in temperature. The first idea for a thermometer, by Galileo, made use of the expansion of a gas. Common thermometers today consist of a hollow glass tube filled with mercury or with alcohol colored with a red dye, as were the earliest usable thermometers (Fig. 13–4).

FIGURE 13–3 Expansion joint on a bridge. Note center white line of highway.

FIGURE 13–4 Thermometers built by the Accademia del Cimento (1657–1667) in Florence, Italy, are among the earliest known. These sensitive and exquisite instruments contained alcohol, sometimes colored, like many thermometers today.

[†]Most materials expand when their temperature is raised, but not all. Water, for example, in the range 0°C to 4°C contracts with an increase in temperature (see Section 13–4).

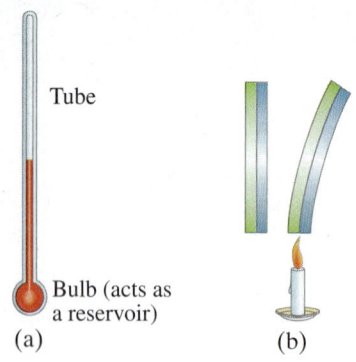

FIGURE 13–5 (a) Mercury- or alcohol-in-glass thermometer; (b) bimetallic strip.

FIGURE 13–6 Photograph of a thermometer using a coiled bimetallic strip.

FIGURE 13–7 Celsius and Fahrenheit scales compared.

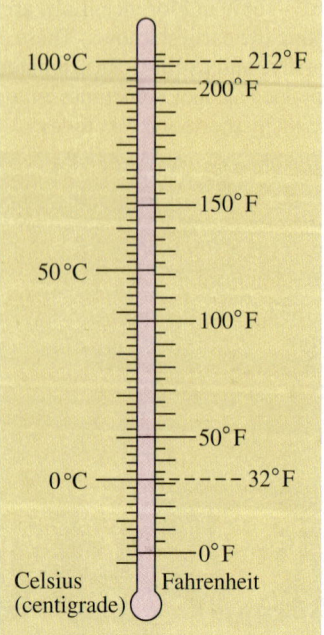

Inside a common liquid-in-glass thermometer, the liquid expands more than the glass when the temperature is increased, so the liquid level rises in the tube (Fig. 13–5a). Although metals also expand with temperature, the change in length of a metal rod, say, is generally too small to measure accurately for ordinary changes in temperature. However, a useful thermometer can be made by bonding together two different metals with different rates of expansion (Fig. 13–5b). When the temperature is increased, the different amounts of expansion cause the **bimetallic strip** to bend. Often the bimetallic strip is in the form of a coil, one end of which is fixed while the other is attached to a pointer, Fig. 13–6. Such thermometers are used as ordinary air thermometers, oven thermometers, automatic off switches in electric coffeepots, and in room thermostats for determining when the heater or air conditioner should go on or off. Very precise thermometers make use of electrical properties (Chapter 18), such as resistance thermometers, thermocouples, and thermistors, often with a digital readout.

Temperature Scales

In order to measure temperature quantitatively, some sort of numerical scale must be defined. The most common scale today is the **Celsius** or **centigrade** scale. In the United States, the **Fahrenheit** scale is common. The most important scale in scientific work is the absolute, or Kelvin, scale, and it will be discussed later in this Chapter.

One way to define a temperature scale is to assign arbitrary values to two readily reproducible temperatures. For both the Celsius and Fahrenheit scales these two fixed points are chosen to be the freezing point and the boiling point[†] of water, both taken at standard atmospheric pressure. On the Celsius scale, the freezing point of water is chosen to be 0°C ("zero degrees Celsius") and the boiling point 100°C. On the Fahrenheit scale, the freezing point is defined as 32°F and the boiling point 212°F. A practical thermometer is calibrated by placing it in carefully prepared environments at each of the two temperatures and marking the position of the liquid or pointer. For a Celsius scale, the distance between the two marks is divided into one hundred equal intervals representing each degree between 0°C and 100°C (hence the name "centigrade scale" meaning "hundred steps"). For the Fahrenheit scale, the two points are labeled 32°F and 212°F and the distance between them is divided into 180 equal intervals. For temperatures below the freezing point of water and above the boiling point of water, the scales may be extended using the same equally spaced intervals. However, thermometers can be used only over a limited temperature range because of their own limitations—for example, an alcohol-in-glass thermometer is rendered useless above temperatures where the alcohol vaporizes. For very low or very high temperatures, specialized thermometers are required, some of which we will mention later.

Every temperature on the Celsius scale corresponds to a particular temperature on the Fahrenheit scale, Fig. 13–7. To convert from one to the other, remember that 0°C corresponds to 32°F and that a range of 100° on the Celsius scale corresponds to a range of 180° on the Fahrenheit scale. Thus, one Fahrenheit degree (1 F°) corresponds to $100/180 = \frac{5}{9}$ of a Celsius degree (1 C°). That is, $1\,\text{F}° = \frac{5}{9}\text{C}°$. (Notice that when we refer to a specific temperature, we say "degrees Celsius," as in 20°C; but when we refer to a *change* in temperature or a temperature *interval*, we say "Celsius degrees," as in "2 C°.") The conversion between the two temperature scales can be written

$$T(°\text{C}) = \tfrac{5}{9}\bigl[T(°\text{F}) - 32\bigr]$$

or

$$T(°\text{F}) = \tfrac{9}{5}T(°\text{C}) + 32.$$

Rather than memorizing these relations, it may be simpler to remember that $0°\text{C} = 32°\text{F}$ and that a change of $5\,\text{C}° = $ a change of $9\,\text{F}°$.

[†]The freezing point of a substance is defined as that temperature at which the solid and liquid phases coexist in equilibrium—that is, without any net liquid changing into the solid or vice versa. Experimentally, this is found to occur at only one definite temperature, for a given pressure. Similarly, the boiling point is defined as that temperature at which the liquid and gas coexist in equilibrium. Since these points vary with pressure, the pressure must be specified (usually it is 1 atm).

EXAMPLE 13–2 **Taking your temperature.** Normal body temperature is 98.6°F. What is this on the Celsius scale?

APPROACH We recall that $0°C = 32°F$ and that a change of $5 C° = 9 F°$.

SOLUTION First we relate the given temperature to the freezing point of water (0°C). That is, 98.6°F is $98.6 - 32.0 = 66.6 F°$ above the freezing point of water. Since each F° is equal to $\frac{5}{9} C°$, this corresponds to $66.6 \times \frac{5}{9} = 37.0$ Celsius degrees above the freezing point. The freezing point of water is 0°C, so normal body temperature is 37.0°C.

CAUTION
Convert temperature by remembering $0°C = 32°F$ and a change of $5 C° = 9 F°$

*Standard Temperature Scale

Different materials do not expand in quite the same way over a wide temperature range. Consequently, if we calibrate different kinds of thermometers exactly as described above, they will not usually agree precisely.

Because of such discrepancies, some standard kind of thermometer must be chosen so that all temperatures can be precisely defined. The chosen standard for this purpose is the **constant-volume gas thermometer**. As shown in the simplified diagram of Fig. 13–8, this thermometer consists of a bulb filled with a low-pressure gas connected by a thin tube to a mercury manometer (Section 10–6). The volume of the gas is kept constant by raising or lowering the right-hand tube of the manometer so that the mercury in the left-hand tube coincides with the reference mark. An increase in temperature causes a proportional increase in pressure in the bulb. Thus the tube must be lifted higher to keep the gas volume constant. The height of the mercury in the right-hand column is then a measure of the temperature. This thermometer gives the same results for all gases in the limit of reducing the gas pressure in the bulb toward zero. The resulting scale serves as a basis for the **standard temperature scale**.

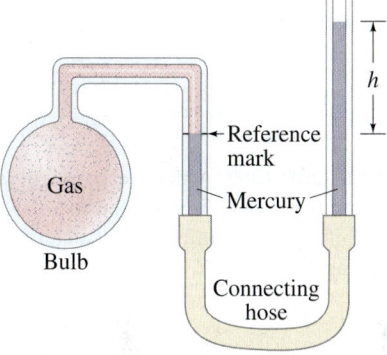

FIGURE 13–8 Constant-volume gas thermometer.

13–3 Thermal Equilibrium and the Zeroth Law of Thermodynamics

If two objects at different temperatures are placed in thermal contact (meaning thermal energy can transfer from one to the other), the two objects will eventually reach the same temperature. They are then said to be in **thermal equilibrium**. For example, you leave a fever thermometer in your mouth until it comes into thermal equilibrium with that environment; then you read it. Two objects are defined to be in thermal equilibrium if, when placed in thermal contact, no net energy flows from one to the other, and their temperatures don't change.

*The Zeroth Law of Thermodynamics

Experiments indicate that

> **if two systems are in thermal equilibrium with a third system, then they are in thermal equilibrium with each other.**

This postulate is called the **zeroth law of thermodynamics**. It has this unusual name because it was not until after the first and second laws of thermodynamics (Chapter 15) were worked out that scientists realized that this apparently obvious postulate needed to be stated first.

Temperature is a property of a system that determines whether the system will be in thermal equilibrium with other systems. When two systems are in thermal equilibrium, their temperatures are (by definition) equal, and no net thermal energy is exchanged between them. This is consistent with our everyday notion of temperature: when a hot object and a cold one are put into contact, they eventually come to the same temperature. Thus the importance of the zeroth law is that it allows a useful definition of temperature.

13–4 Thermal Expansion

Most substances expand when heated and contract when cooled. However, the amount of expansion or contraction varies, depending on the material.

Linear Expansion

Experiments indicate that the change in length $\Delta \ell$ of almost all solids is, to a good approximation, directly proportional to the change in temperature ΔT, as long as ΔT is not too large. The change in length is also proportional to the original length of the object, ℓ_0. That is, for the same temperature increase, a 4-m-long iron rod will increase in length twice as much as a 2-m-long iron rod. We can write this proportionality as an equation:

$$\Delta \ell = \alpha \ell_0 \Delta T, \tag{13-1a}$$

where α, the proportionality constant, is called the **coefficient of linear expansion** for the particular material and has units of $(C°)^{-1}$. We write $\ell = \ell_0 + \Delta \ell$, Fig. 13–9, and rewrite this equation as

$$\ell = \ell_0 (1 + \alpha \Delta T), \tag{13-1b}$$

where ℓ_0 is the length initially, at temperature T_0, and ℓ is the length after heating or cooling to a temperature T. If the temperature change $\Delta T = T - T_0$ is negative, then $\Delta \ell = \ell - \ell_0$ is also negative; the length shortens as the temperature decreases.

The values of α for various materials at 20°C are listed in Table 13–1. Actually, α does vary slightly with temperature (which is why thermometers made of different materials do not agree precisely). However, if the temperature range is not too great, the variation can usually be ignored.

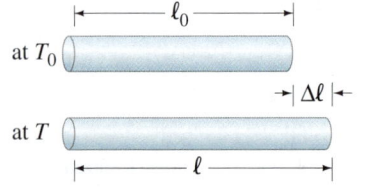

FIGURE 13–9 A thin rod of length ℓ_0 at temperature T_0 is heated to a new uniform temperature T and acquires length ℓ, where $\ell = \ell_0 + \Delta \ell$.

TABLE 13–1 Coefficients of Expansion, near 20°C		
Material	Coefficient of Linear Expansion, α $(C°)^{-1}$	Coefficient of Volume Expansion, β $(C°)^{-1}$
Solids		
Aluminum	25×10^{-6}	75×10^{-6}
Brass	19×10^{-6}	56×10^{-6}
Copper	17×10^{-6}	50×10^{-6}
Gold	14×10^{-6}	42×10^{-6}
Iron or steel	12×10^{-6}	35×10^{-6}
Lead	29×10^{-6}	87×10^{-6}
Glass (Pyrex®)	3×10^{-6}	9×10^{-6}
Glass (ordinary)	9×10^{-6}	27×10^{-6}
Quartz	0.4×10^{-6}	1×10^{-6}
Concrete and brick	$\approx 12 \times 10^{-6}$	$\approx 36 \times 10^{-6}$
Marble	$1.4–3.5 \times 10^{-6}$	$4–10 \times 10^{-6}$
Liquids		
Gasoline		950×10^{-6}
Mercury		180×10^{-6}
Ethyl alcohol		1100×10^{-6}
Glycerin		500×10^{-6}
Water		210×10^{-6}
Gases		
Air (and most other gases at atmospheric pressure)		3400×10^{-6}

EXAMPLE 13-3 **Bridge expansion.** The steel bed of a suspension bridge is 200 m long at 20°C. If the extremes of temperature to which it might be exposed are −30°C to +40°C, how much will it contract and expand?

APPROACH We assume the bridge bed will expand and contract linearly with temperature, as given by Eq. 13–1a.

SOLUTION From Table 13–1, we find that $\alpha = 12 \times 10^{-6} (C°)^{-1}$ for steel. The increase in length when it is at 40°C will be

$$\Delta \ell = \alpha \ell_0 \Delta T = (12 \times 10^{-6}/C°)(200 \text{ m})(40°C - 20°C) = 4.8 \times 10^{-2} \text{ m},$$

or 4.8 cm. When the temperature decreases to −30°C, $\Delta T = -50\, C°$. Then

$$\Delta \ell = (12 \times 10^{-6}/C°)(200 \text{ m})(-50\, C°) = -12.0 \times 10^{-2} \text{ m},$$

or a decrease in length of 12 cm. The total range the expansion joints must accommodate is 12 cm + 4.8 cm ≈ 17 cm (Fig. 13–3).

PHYSICS APPLIED
Expansion in structures

CONCEPTUAL EXAMPLE 13-4 **Do holes expand or contract?** If you heat a thin, circular ring (Fig. 13–10a) in the oven, does the ring's hole get larger or smaller?

RESPONSE If you guessed that the metal expands into the hole, making the hole smaller, it is not so. Imagine the ring is solid, like a coin (Fig. 13–10b). Draw a circle on it with a pen as shown. When the metal expands, the material inside the circle will expand along with the rest of the metal; so the dashed circle expands. Cutting the metal where the circle is shows that the hole in Fig. 13–10a increases in diameter.

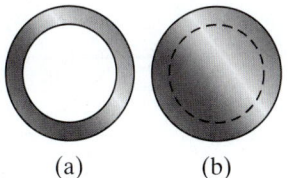
FIGURE 13-10 Example 13–4.

FIGURE 13-11 Example 13–5.

EXAMPLE 13-5 **Ring on a rod.** An iron ring is to fit snugly on a cylindrical iron rod (Fig. 13–11). At 20°C, the diameter of the rod is 6.445 cm and the inside diameter of the ring is 6.420 cm. To slip over the rod, the ring must be slightly larger than the rod diameter by about 0.008 cm. To what temperature must the ring be brought if its hole is to be large enough so it will slip over the rod?

APPROACH The hole in the ring must be increased from a diameter of 6.420 cm to 6.445 cm + 0.008 cm = 6.453 cm. The ring must be heated since the hole diameter will increase linearly with temperature (Example 13–4).

SOLUTION We solve for ΔT in Eq. 13–1a and find

$$\Delta T = \frac{\Delta \ell}{\alpha \ell_0} = \frac{6.453 \text{ cm} - 6.420 \text{ cm}}{(12 \times 10^{-6}/C°)(6.420 \text{ cm})} = 430\, C°.$$

So the ring must be raised at least to $T = (20°C + 430\, C°) = 450°C$.

NOTE In doing Problems, do not forget the last step, adding in the initial temperature (20°C here).

FIGURE 13-12 Example 13–6.

CONCEPTUAL EXAMPLE 13-6 **Opening a tight jar lid.** When the lid of a glass jar is tight, holding the lid under hot water for a short time will often make it easier to open (Fig. 13–12). Why?

RESPONSE The lid may be struck by the hot water more directly than the glass and so expand sooner. But even if not, metals generally expand more than glass for the same temperature change (α is greater—see Table 13–1).

PHYSICS APPLIED
Opening a tight lid

Volume Expansion

The change in *volume* of a material which undergoes a temperature change is given by a relation similar to Eq. 13–1a, namely,

$$\Delta V = \beta V_0 \Delta T, \qquad (13\text{–}2)$$

where V_0 is the original volume, ΔV is the change in volume when the temperature changes by ΔT, and β is the **coefficient of volume expansion**. The units of β are $(C°)^{-1}$.

Values of β for various materials are given in Table 13–1. Notice that for solids, β is normally equal to approximately 3α. Note also that linear expansion has no meaning for liquids and gases because they do not have fixed shapes.

Equations 13–1 and 13–2 are accurate only if $\Delta \ell$ (or ΔV) is small compared to ℓ_0 (or V_0). This is of particular concern for liquids and even more so for gases because of the large values of β. Furthermore, β itself varies substantially with temperature for gases. Therefore, a more convenient way of dealing with gases is needed, and will be discussed starting in Section 13–5.

EXAMPLE 13–7 **Gas tank in the Sun.** The 70-liter (L) steel gas tank of a car is filled to the top with gasoline at 20°C. The car sits in the Sun and the tank reaches a temperature of 40°C (104°F). How much gasoline do you expect to overflow from the tank?

PHYSICS APPLIED
Gas tank overflow

APPROACH Both the gasoline and the tank expand as the temperature increases, and we assume they do so linearly as described by Eq. 13–2. The volume of overflowing gasoline equals the volume increase of the gasoline minus the increase in volume of the tank.

SOLUTION The gasoline expands by

$$\Delta V = \beta V_0 \Delta T = (950 \times 10^{-6}/C°)(70\,L)(40°C - 20°C)$$
$$= 1.3\,L.$$

The tank also expands. We can think of it as a steel shell that undergoes volume expansion ($\beta = 35 \times 10^{-6}/C° \approx 3\alpha$). If the tank were solid, the surface layer (the shell) would expand just the same (as in Example 13–4). Thus the tank increases in volume by

$$\Delta V = (35 \times 10^{-6}/C°)(70\,L)(40°C - 20°C) = 0.049\,L,$$

so the tank expansion has little effect. More than a liter of gas could spill out.

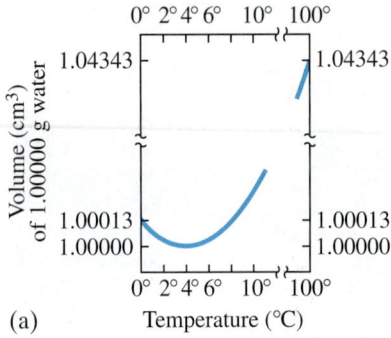
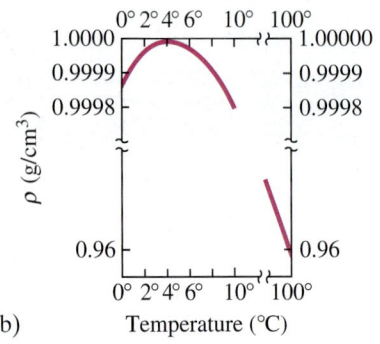

FIGURE 13–13 Behavior of water as a function of temperature near 4°C. (a) Volume of 1.00000 gram of water as a function of temperature. (b) Density vs. temperature. [Note the break in each axis.]

Anomalous Behavior of Water Below 4°C

Most substances expand more or less uniformly with an increase in temperature, as long as no phase change occurs. Water, however, does not follow the usual pattern. If water at 0°C is heated, it actually *decreases* in volume until it reaches 4°C. Above 4°C water behaves normally and expands in volume as the temperature is increased, Fig. 13–13. Water thus has its greatest density at 4°C. This anomalous behavior of water is of great importance for the survival of aquatic life during cold winters. When water in a lake (or river) is above 4°C and begins to cool by contact with cold air, the water at the surface sinks because it is denser. It is replaced by warmer water from below. This mixing continues until the temperature of the entire lake reaches 4°C. As the surface water cools further, it remains on the surface because it is less dense than the 4°C water below. Water thus freezes first at the surface, and the ice remains on the surface since ice (specific gravity = 0.917) is less dense than water. The water at the bottom remains liquid unless it is so cold that the whole body of water freezes. If water were like most substances, becoming more dense as it cools, the water at the bottom of a lake would be frozen first.

Lakes would freeze solid more easily because circulation would bring the warmer water to the surface to be efficiently cooled. The complete freezing of a lake would cause severe damage to its plant and animal life. Because of the unusual behavior of water below 4°C, it is rare for any large and deep body of water to freeze completely, and this is helped by the layer of ice on the surface which acts as an insulator to reduce the flow of heat out of the water into the cold air above. Without this peculiar but wonderful property of water, life on this planet as we know it might not have been possible.

PHYSICS APPLIED
Life under ice

Not only does water expand as it cools from 4°C to 0°C, it expands even more as it freezes to ice. This is why ice cubes float in water and pipes break when water inside them freezes.

*Thermal Stresses

In many situations, such as in buildings and roads, the ends of a beam or slab of material are rigidly fixed, which greatly limits expansion or contraction. If the temperature should change, large compressive or tensile stresses, called **thermal stresses**, will occur. The magnitude of such stresses can be calculated using the concept of elastic modulus developed in Chapter 9. To calculate the internal stress in a beam, we can think of this process as occurring in two steps: (1) the beam tries to expand (or contract) by an amount $\Delta \ell$ given by Eq. 13–1; (2) the solid in contact with the beam exerts a force to compress (or expand) it, keeping it at its original length. The force F required is given by Eq. 9–4:

$$\Delta \ell = \frac{1}{E} \frac{F}{A} \ell_0,$$

where E is Young's modulus for the material. To calculate the internal stress, F/A, we then set $\Delta \ell$ in Eq. 13–1a equal to $\Delta \ell$ in the equation above and find

$$\alpha \ell_0 \Delta T = \frac{1}{E} \frac{F}{A} \ell_0.$$

Hence, the stress is

$$\frac{F}{A} = \alpha E \, \Delta T.$$

For example, if 10-m-long concrete slabs are placed touching each other in a new park you are designing, a 30°C increase in temperature would produce a stress $F/A = \alpha E \, \Delta T = (12 \times 10^{-6}/\text{C}°)(20 \times 10^9 \, \text{N/m}^2)(30 \, \text{C}°) = 7.2 \times 10^6 \, \text{N/m}^2$. That stress would exceed the shear strength of concrete (Table 9–2), no doubt causing fracture and cracks. This is why soft spacers (or expansion joints) are placed between slabs on sidewalks and highways.

13–5 The Gas Laws and Absolute Temperature

Equation 13–2 is not useful for describing the expansion of a gas, partly because the expansion can be so great, and partly because gases generally expand to fill whatever container they are in. Indeed, Eq. 13–2 is meaningful only if the pressure is kept constant. The volume of a gas depends very much on the pressure as well as on the temperature. It is therefore valuable to determine a relation between the volume, the pressure, the temperature, and the quantity of a gas. Such a relation is called an **equation of state**. (By the word *state*, we mean the physical condition of the system.)

If the state of a system is changed, we will always wait until the pressure and temperature have reached the same values throughout. We thus consider only **equilibrium states** of a system—when the variables that describe it (such as temperature and pressure) are the same throughout the system and are not changing in time. We also note that the results of this Section are accurate only for gases that are not too dense (the pressure is not too high, on the order of an atmosphere or less) and not close to the liquefaction (boiling) point.

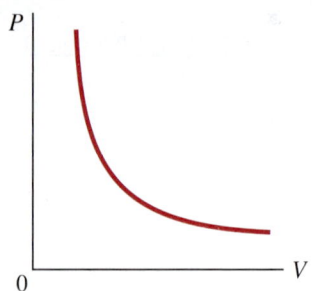

FIGURE 13–14 Pressure vs. volume of a fixed amount of gas at a constant temperature, showing the inverse relationship as given by Boyle's law: as the pressure decreases, the volume increases.

FIGURE 13–15 Volume of a fixed amount of gas as a function of (a) Celsius temperature, and (b) Kelvin temperature, when the pressure is kept constant.

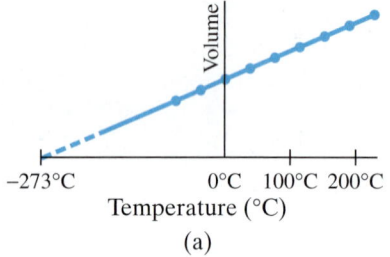

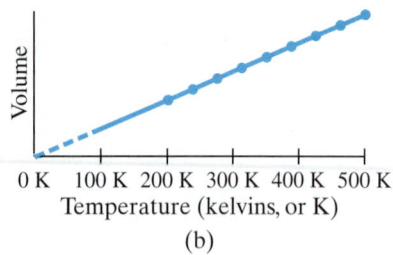

For a given quantity of gas it is found experimentally that, to a good approximation, *the volume of a gas is inversely proportional to the absolute pressure applied to it when the temperature is kept constant.* That is,

$$V \propto \frac{1}{P}, \qquad \text{[constant } T\text{]}$$

where P is the absolute pressure (*not* "gauge pressure"—see Section 10–4). For example, if the pressure on a gas is doubled, the volume is reduced to half its original volume. This relation is known as **Boyle's law**, after the Englishman Robert Boyle (1627–1691), who first stated it on the basis of his own experiments. A graph of P vs. V for a fixed temperature is shown in Fig. 13–14. Boyle's law can also be written

$$PV = \text{constant} \qquad \text{[constant } T\text{]}$$

for a fixed quantity of a gas kept at constant temperature. If either the pressure or volume of a fixed amount of gas is allowed to vary, the other variable also changes so that the product PV remains constant.

Temperature also affects the volume of a gas, but a quantitative relationship between V and T was not found until more than a century after Boyle's work. The Frenchman Jacques Charles (1746–1823) found that when the pressure is not too high and is kept constant, the volume of a gas increases with temperature at a nearly linear rate, as shown in Fig. 13–15a. However, all gases liquefy at low temperatures (for example, oxygen liquefies at $-183°C$), so the graph cannot be extended below the liquefaction point. Nonetheless, the graph is essentially a straight line and if projected to lower temperatures, as shown by the dashed line, it crosses the axis at about $-273°C$.

Such a graph can be drawn for any gas, and a straight line results which always projects back to $-273°C$ at zero volume. This seems to imply that if a gas could be cooled to $-273°C$, it would have zero volume, and at lower temperatures a negative volume, which makes no sense. It could be argued that $-273°C$ is the lowest temperature possible; indeed, many other more recent experiments indicate that this is so. This temperature is called the **absolute zero** of temperature. Its value has been determined to be $-273.15°C$.

Absolute zero forms the basis of a temperature scale known as the **absolute scale** or **Kelvin scale**, and it is used extensively in scientific work. On this scale the temperature is specified as degrees Kelvin or, preferably, simply as **kelvins** (K) without the degree sign. The intervals are the same as for the Celsius scale, but the zero on this scale (0 K) is chosen as absolute zero. Thus the freezing point of water ($0°C$) is 273.15 K, and the boiling point of water is 373.15 K. Indeed, any temperature on the Celsius scale can be changed to kelvins by adding 273.15 to it:

$$T(K) = T(°C) + 273.15.$$

Now let us look at Fig. 13–15b, where the graph of the volume of a gas versus absolute temperature is a straight line that passes through the origin. Thus, to a good approximation, *the volume of a fixed quantity of gas is directly proportional to the absolute temperature when the pressure is kept constant.* This is known as **Charles's law**, and is written

$$V \propto T. \qquad \text{[constant } P\text{]}$$

A third gas law, known as **Gay-Lussac's law**, after Joseph Gay-Lussac (1778–1850), states that *at constant volume, the absolute pressure of a fixed quantity of a gas is directly proportional to the absolute temperature*:

$$P \propto T. \qquad \text{[constant } V\text{]}$$

The laws of Boyle, Charles, and Gay-Lussac are not really laws in the sense that we use this term today (precise, deep, wide-ranging validity). They are really only approximations that are accurate for real gases only as long as the pressure and density of the gas are not too high, and the gas is not too close to liquefaction (condensation). The term *law* applied to these three relationships has become traditional, however, so we have stuck with that usage.

CONCEPTUAL EXAMPLE 13–8 Why you should not put a closed glass jar into a campfire. What could happen if you tossed an empty glass jar, with the lid on tight, into a fire, and why?

RESPONSE The inside of the jar is not empty. It is filled with air. As the fire heats the air inside, its temperature rises. The volume of the glass jar changes only slightly due to the heating. According to Gay-Lussac's law the pressure P of the air inside the jar can increase enough to cause the jar to explode, throwing glass pieces outward.

PHYSICS APPLIED
Throwing a jar into a campfire

13–6 The Ideal Gas Law

The gas laws of Boyle, Charles, and Gay-Lussac were obtained by means of an important scientific technique: namely, *considering one quantity and how it is affected by changing only one other variable, keeping all other variables constant.* These laws can now be combined into a single more general relation among all three variables—absolute pressure, volume, and absolute temperature of a fixed amount of gas:

$$PV \propto T.$$

This relation indicates how any of the quantities P, V, or T will vary when the other two quantities change. This relation reduces to Boyle's, Charles's, or Gay-Lussac's law when either T, P, or V, respectively, is held constant.

Finally, we must incorporate the effect of the amount of gas present. For example, when more air is forced into a balloon, the balloon gets bigger (Fig. 13–16). Indeed, careful experiments show that at constant temperature and pressure, the volume V of an enclosed gas increases in direct proportion to the mass m of gas present. Hence we write

$$PV \propto mT.$$

This proportion can be made into an equation by inserting a constant of proportionality. Experiment shows that this constant has a different value for different gases. However, the constant of proportionality turns out to be the same for all gases if, instead of the mass m, we use the number of *moles*.

The "mole" is an official SI unit for the amount of substance. One **mole** (abbreviated mol) is the amount of substance that contains 6.02×10^{23} objects (usually atoms, molecules, or ions, etc.). This number is called *Avogadro's number*, as discussed in Section 13–8. Its value comes from measurements. The mole's precise definition is the number of atoms in exactly 12 grams of carbon-12 (page 360).

Equivalently, 1 mol is that amount of substance whose mass in grams is numerically equal to the molecular mass of the substance (Section 13–1). For example, the mass of 1 mole of CO_2 is $[12 + (2 \times 16)] = 44$ g because carbon has atomic mass of 12 and oxygen 16 (see Periodic Table inside the rear cover).

In general, the number of moles, n, in a given sample of a pure substance is equal to the mass of the sample in grams divided by the molecular mass specified as grams per mole:

$$n \text{ (mole)} = \frac{\text{mass (grams)}}{\text{molecular mass (g/mol)}}.$$

For example, the number of moles in 132 g of CO_2 (molecular mass 44 u) is

$$n = \frac{132 \text{ g}}{44 \text{ g/mol}} = 3.0 \text{ mol.}$$

FIGURE 13–16 Blowing up a balloon means putting more air (more air molecules) into the balloon, which increases its volume. The pressure is nearly constant, at atmospheric pressure, except for the small effect of the balloon's elasticity.

IDEAL GAS LAW

We can now write the proportion above ($PV \propto mT$) as an equation:

$$PV = nRT, \quad (13\text{-}3)$$

where n represents the number of moles and R is the constant of proportionality. R is called the **universal gas constant** because its value is found experimentally to be the same for all gases. The value of R, in several sets of units (only the first is the proper SI unit), is

$$\begin{aligned} R &= 8.314 \, \text{J}/(\text{mol} \cdot \text{K}) &&\text{[SI units]} \\ &= 0.0821 \, (\text{L} \cdot \text{atm})/(\text{mol} \cdot \text{K}) \\ &= 1.99 \, \text{calories}/(\text{mol} \cdot \text{K}).^\dagger \end{aligned}$$

Equation 13–3 is called the **ideal gas law**, or the **equation of state for an ideal gas**. We use the term "ideal" because real gases do not follow Eq. 13–3 precisely, particularly at high pressure (and density) or when the gas is near the liquefaction point (= boiling point). However, at pressures less than an atmosphere or so, and when T is not close to the liquefaction point of the gas, Eq. 13–3 is quite accurate and useful for real gases.

CAUTION
Always give T in kelvins and P as absolute (not gauge) pressure

Always remember, when using the ideal gas law, that temperatures must be given in kelvins (K) and that the pressure P must always be *absolute* pressure, not gauge pressure (Section 10–4).

EXERCISE A Return to the Chapter-Opening Question, page 359, and answer it again now. Try to explain why you may have answered differently the first time.

EXERCISE B An ideal gas is contained in a steel sphere at 27.0°C and 1.00 atm absolute pressure. If no gas is allowed to escape and the temperature is raised to 127°C, what will be the new pressure? (*a*) 0.21 atm; (*b*) 0.75 atm; (*c*) 1.00 atm; (*d*) 1.33 atm; (*e*) 4.7 atm.

13–7 Problem Solving with the Ideal Gas Law

The ideal gas law is an extremely useful tool, and we now consider some Examples. We will often refer to "standard conditions" or **standard temperature and pressure** (**STP**), which means:

$$T = 273 \, \text{K} \, (0°\text{C}) \quad \text{and} \quad P = 1.00 \, \text{atm} = 1.013 \times 10^5 \, \text{N/m}^2 = 101.3 \, \text{kPa}.$$

EXAMPLE 13–9 Volume of one mole at STP. Determine the volume of 1.00 mol of any gas, assuming it behaves like an ideal gas, at STP.

APPROACH We use the ideal gas law, solving for V with $n = 1.00$ mol.

SOLUTION We solve for V in Eq. 13–3:

$$V = \frac{nRT}{P} = \frac{(1.00 \, \text{mol})(8.314 \, \text{J/mol} \cdot \text{K})(273 \, \text{K})}{(1.013 \times 10^5 \, \text{N/m}^2)} = 22.4 \times 10^{-3} \, \text{m}^3.$$

Since 1 liter (L) is $1000 \, \text{cm}^3 = 1.00 \times 10^{-3} \, \text{m}^3$, 1.00 mol of any (ideal) gas has volume $V = 22.4 \, \text{L}$ at STP.

PROBLEM SOLVING
1 mol of gas at STP has V = 22.4 L

The value of 22.4 L for the volume of 1 mol of an ideal gas at STP is worth remembering, for it sometimes makes calculation simpler.

EXERCISE C What is the volume of 1.00 mol of ideal gas at 546 K (= 2 × 273 K) and 2.0 atm absolute pressure? (*a*) 11.2 L; (*b*) 22.4 L; (*c*) 44.8 L; (*d*) 67.2 L; (*e*) 89.6 L.

---

†Sometimes it is useful to use R as given in terms of calories; calories will be defined in Section 14–1.

EXAMPLE 13-10 Helium balloon. A helium party balloon, assumed to be a perfect sphere, has a radius of 18.0 cm. At room temperature (20°C), its internal pressure is 1.05 atm. Find the number of moles of helium in the balloon and the mass of helium needed to inflate the balloon to these values.

APPROACH We can use the ideal gas law to find n, since we are given P and T, and can find V from the given radius.

SOLUTION We get the volume V from the formula for a sphere:
$$V = \tfrac{4}{3}\pi r^3$$
$$= \tfrac{4}{3}\pi (0.180 \text{ m})^3 = 0.0244 \text{ m}^3.$$

The pressure is given as 1.05 atm = 1.064×10^5 N/m². The temperature must be expressed in kelvins, so we change 20°C to $(20 + 273)$ K = 293 K. Finally, we use the value $R = 8.314$ J/(mol·K) because we are using SI units. Thus

$$n = \frac{PV}{RT} = \frac{(1.064 \times 10^5 \text{ N/m}^2)(0.0244 \text{ m}^3)}{(8.314 \text{ J/mol·K})(293 \text{ K})} = 1.066 \text{ mol}.$$

The mass of helium (atomic mass = 4.00 g/mol as given in the Periodic Table or Appendix B) can be obtained from

$$\text{mass} = n \times \text{molecular mass} = (1.066 \text{ mol})(4.00 \text{ g/mol}) = 4.26 \text{ g}$$

or 4.26×10^{-3} kg.

EXAMPLE 13-11 ESTIMATE Mass of air in a room. Estimate the mass of air in a room whose dimensions are 5 m × 3 m × 2.5 m high, at STP.

APPROACH First we determine the number of moles n using the given volume. Then we can multiply by the mass of one mole to get the total mass.

SOLUTION Example 13-9 told us that 1 mol of a gas at 0°C has a volume of 22.4 L = 22.4×10^{-3} m³. The room's volume is 5 m × 3 m × 2.5 m, so

$$n = \frac{(5 \text{ m})(3 \text{ m})(2.5 \text{ m})}{22.4 \times 10^{-3} \text{ m}^3/\text{mol}} \approx 1700 \text{ mol}.$$

Air is a mixture of about 20% oxygen (O_2) and 80% nitrogen (N_2). The molecular masses are 2×16 u = 32 u and 2×14 u = 28 u, respectively, for an average of about 29 u. Thus, 1 mol of air has a mass of about 29 g = 0.029 kg, so our room has a mass of air

$$m \approx (1700 \text{ mol})(0.029 \text{ kg/mol}) \approx 50 \text{ kg}.$$

NOTE That is roughly 100 lb of air!

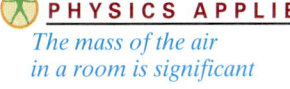
The mass of the air in a room is significant

EXERCISE D At 20°C, would there be (a) more, (b) less, or (c) the same mass of air in a room than at 0°C?

Frequently, volume is specified in liters and pressure in atmospheres. Rather than convert these to SI units, we can instead use the value of R given in Section 13-6 as 0.0821 L·atm/mol·K.

In many situations it is not necessary to use the value of R at all. For example, many problems involve a change in the pressure, temperature, and volume of a fixed amount of gas. In this case, $PV/T = nR$ = constant, since n and R remain constant. If we now let P_1, V_1, and T_1 represent the appropriate variables initially, and P_2, V_2, T_2 represent the variables after the change is made, then we can write

$$\frac{P_1 V_1}{T_1} = \frac{P_2 V_2}{T_2}. \qquad [\text{fixed } n]$$

Using the ideal gas law as a ratio

If we know any five of the quantities in this equation, we can solve for the sixth. Or, if one of the three variables is constant ($V_1 = V_2$, or $P_1 = P_2$, or $T_1 = T_2$) then we can use this equation to solve for one unknown when given the other three quantities.

PHYSICS APPLIED
Pressure in a hot tire

EXAMPLE 13-12 **Check tires cold.** An automobile tire is filled (Fig. 13–17) to a gauge pressure of 210 kPa (= 30 psi) at 10°C. After a drive of 100 km, the temperature within the tire rises to 40°C. What is the pressure within the tire now?

APPROACH We do not know the number of moles of gas, or the volume of the tire, but we assume they are constant. We use the ratio form of the ideal gas law.

SOLUTION Since $V_1 = V_2$, then

$$\frac{P_1}{T_1} = \frac{P_2}{T_2}.$$

This is, incidentally, a statement of Gay-Lussac's law. Since the pressure given is the gauge pressure (Section 10–4), we must add atmospheric pressure (= 101 kPa) to get the absolute pressure $P_1 = (210 \text{ kPa} + 101 \text{ kPa}) = 311$ kPa. We convert temperatures to kelvins by adding 273 and solve for P_2:

$$P_2 = P_1\left(\frac{T_2}{T_1}\right) = (3.11 \times 10^5 \text{ Pa})\left(\frac{313 \text{ K}}{283 \text{ K}}\right) = 344 \text{ kPa}.$$

Subtracting atmospheric pressure, we find the resulting gauge pressure to be 243 kPa, which is a 16% increase (= 35 psi).

NOTE This Example shows why car manuals emphasize checking tire pressure when the tires are cold.

FIGURE 13-17 Example 13–12.

13-8 Ideal Gas Law in Terms of Molecules: Avogadro's Number

The fact that the gas constant, R, has the same value for all gases is a remarkable reflection of simplicity in nature. It was first recognized, although in a slightly different form, by the Italian scientist Amedeo Avogadro (1776–1856). Avogadro stated that *equal volumes of gas at the same pressure and temperature contain equal numbers of molecules.* This is sometimes called **Avogadro's hypothesis**. That this is consistent with R being the same for all gases can be seen as follows. From Eq. 13–3, $PV = nRT$, we see that for the same number of moles, n, and the same pressure and temperature, the volume will be the same for all gases as long as R is the same. Second, the number of molecules in 1 mole is the same for all gases (see page 369). Thus Avogadro's hypothesis is equivalent to R being the same for all gases.

The number of molecules in one mole of any pure substance is known as **Avogadro's number**, N_A. Although Avogadro conceived the notion, he was not able to actually determine the value of N_A. Indeed, precise measurements were not done until the twentieth century.

A number of methods have been devised to measure N_A, and the accepted value today is (see inside front cover for more precise value)

Avogadro's number

$$N_A = 6.02 \times 10^{23}. \qquad \text{[molecules/mole]}$$

Since the total number of molecules, N, in a gas is equal to N_A times the number of moles $(N = nN_A)$, then the ideal gas law, Eq. 13–3, can be written in terms of the number of molecules present:

$$PV = nRT = \frac{N}{N_A}RT,$$

or

IDEAL GAS LAW
(in terms of molecules)

$$PV = NkT, \qquad (13\text{–}4)$$

where $k = R/N_A$ is called the **Boltzmann constant** and has the value

$$k = \frac{R}{N_A} = \frac{8.314 \text{ J/mol} \cdot \text{K}}{6.02 \times 10^{23}/\text{mol}} = 1.38 \times 10^{-23} \text{ J/K}.$$

EXAMPLE 13–13 Hydrogen atom mass. Use Avogadro's number to determine the mass of a hydrogen atom.

APPROACH The mass of one atom equals the mass of 1 mol divided by the number of atoms in 1 mol, N_A.

SOLUTION One mole of hydrogen atoms (atomic mass = 1.008 u, Section 13–1 or Appendix B) has a mass of 1.008×10^{-3} kg and contains 6.02×10^{23} atoms. Thus one atom has a mass

$$m = \frac{1.008 \times 10^{-3} \text{ kg}}{6.02 \times 10^{23}} = 1.67 \times 10^{-27} \text{ kg}.$$

NOTE Historically, the reverse was done: a precise value of N_A was obtained from a precise measurement of the mass of the hydrogen atom.

EXAMPLE 13–14 ESTIMATE How many molecules in one breath? Estimate how many molecules you breathe in with a 1.0-L breath of air.

Molecules in a breath

APPROACH We determine what fraction of a mole 1.0 L is using the result of Example 13–9 that 1 mole has a volume of 22.4 L at STP, and then multiply that by N_A to get the number of molecules in this number of moles.

SOLUTION One mole corresponds to 22.4 L at STP, so 1.0 L of air is $(1.0 \text{ L})/(22.4 \text{ L/mol}) = 0.045$ mol. Then 1.0 L of air contains

$$(0.045 \text{ mol})(6.02 \times 10^{23} \text{ molecules/mole}) \approx 3 \times 10^{22} \text{ molecules}.$$

13–9 Kinetic Theory and the Molecular Interpretation of Temperature

The analysis of matter in terms of atoms in continuous random motion is called **kinetic theory**. We now investigate the properties of a gas from the point of view of kinetic theory, which is based on the laws of classical mechanics. But to apply Newton's laws to each one of the vast number of molecules in a gas ($> 10^{25}/\text{m}^3$ at STP) is far beyond the capability of any present computer. Instead we take a statistical approach and determine averages of certain quantities, and connect these averages to macroscopic variables. We will demand that our microscopic description correspond to the macroscopic properties of gases; otherwise our theory would be of little value. Most importantly, we will arrive at an important relation between the average kinetic energy of molecules in a gas and the absolute temperature.

We make the following assumptions about the molecules in a gas. These assumptions reflect a simple view of a gas, but nonetheless the results they predict correspond well to the essential features of real gases that are at low pressure and are far from the liquefaction point. Under these conditions real gases follow the ideal gas law quite closely, and indeed the gas we now describe is referred to as an **ideal gas**. The assumptions representing the basic postulates of kinetic theory for an ideal gas are:

1. There are a large number of molecules, N, each of mass m, moving in random directions with a variety of speeds. This assumption agrees with our observation that a gas fills its container and, in the case of air on Earth, is kept from escaping only by the force of gravity.
2. The molecules are, on average, far apart from one another. That is, their average separation is much greater than the diameter of each molecule.
3. The molecules are assumed to obey the laws of classical mechanics, and are assumed to interact with one another only when they collide. Although molecules exert weak attractive forces on each other between collisions, the potential energy associated with these forces is small compared to the kinetic energy.
4. Collisions with another molecule or the wall of the vessel are assumed to be perfectly elastic, like the collisions of perfectly elastic billiard balls (Chapter 7). We assume the collisions are of very short duration compared to the time between collisions. Then we can ignore the potential energy associated with collisions in comparison to the kinetic energy between collisions.

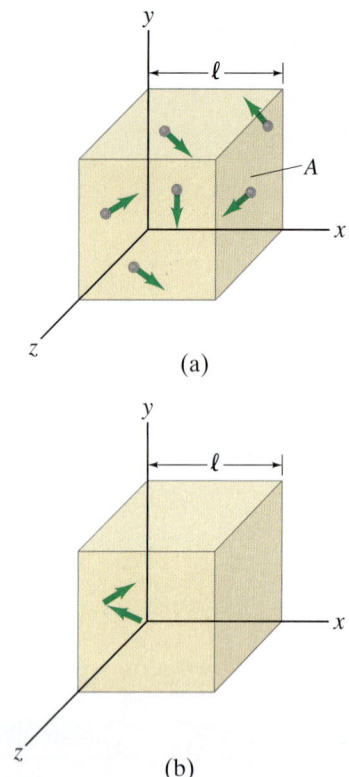

FIGURE 13–18 (a) Molecules of a gas moving about in a rectangular container. (b) Arrows indicate the momentum of one molecule as it rebounds from the end wall.

We can see how this kinetic view of a gas can explain Boyle's law (Section 13–5). The pressure exerted on a wall of a container of gas is due to the constant bombardment of molecules. If the volume is reduced by (say) half, the molecules are closer together and twice as many will be striking a given area of the wall per second. Hence we expect the pressure to be twice as great, in agreement with Boyle's law.

Now let us calculate quantitatively the pressure a gas exerts on its container in terms of microscopic quantities. We imagine that the molecules are inside a rectangular container (at rest) whose ends have area A and whose length is ℓ, as shown in Fig. 13–18a. The pressure exerted by the gas on the walls of its container is, according to our model, due to the collisions of the molecules with the walls. Let us focus our attention on the wall, of area A, at the left end of the container and examine what happens when one molecule strikes this wall, as shown in Fig. 13–18b. This molecule exerts a force on the wall, and according to Newton's third law the wall exerts an equal and opposite force back on the molecule. The magnitude of this force on the molecule, according to Newton's second law, is equal to the molecule's rate of change of momentum, $F = \Delta(mv)/\Delta t$ (Eq. 7–2). Assuming the collision is elastic, only the x component of the molecule's momentum changes, and it changes from $-mv_x$ (it is moving in the negative x direction) to $+mv_x$. Thus the change in the molecule's momentum, $\Delta(mv)$, which is the final momentum minus the initial momentum, is

$$\Delta(mv) = mv_x - (-mv_x) = 2mv_x$$

for one collision. This molecule will make many collisions with the wall, each separated by a time Δt, which is the time it takes the molecule to travel across the container and back again, a distance (x component) equal to 2ℓ. Thus $2\ell = v_x \Delta t$, or

$$\Delta t = \frac{2\ell}{v_x}.$$

The time Δt between collisions with a wall is very small, so the number of collisions per second is very large. Thus the average force—averaged over many collisions—will be equal to the momentum change during one collision divided by the time between collisions (Newton's second law, Eq. 7–2):

$$F = \frac{\Delta(mv)}{\Delta t} = \frac{2mv_x}{2\ell/v_x} = \frac{mv_x^2}{\ell}. \quad \text{[due to one molecule]}$$

During its passage back and forth across the container, the molecule may collide with the tops and sides of the container, but this does not alter its x component of momentum and thus does not alter our result. [It may also collide with other molecules, which may change its v_x. However, any loss (or gain) of momentum is acquired by other molecules, and because we will eventually sum over all the molecules, this effect will be included. So our result above is not altered.]

The actual force due to one molecule is intermittent, but because a huge number of molecules are striking the wall per second, the force is, on average, nearly constant. To calculate the force due to *all* the molecules in the container, we have to add the contributions of each. If all N molecules have the same mass m, the net force on the wall is

$$F = \frac{m}{\ell}(v_{x1}^2 + v_{x2}^2 + \cdots + v_{xN}^2),$$

where v_{x1} means v_x for molecule number 1 (we arbitrarily assign each molecule a number) and the sum extends over the total number of molecules N in the container. The average value of the square of the x component of velocity is

$$\overline{v_x^2} = \frac{v_{x1}^2 + v_{x2}^2 + \cdots + v_{xN}^2}{N}, \quad (13\text{–}5)$$

where the overbar (¯) means "average." Thus we can write the force as

$$F = \frac{m}{\ell} N \overline{v_x^2}. \quad (i)$$

We know that the square of any vector is equal to the sum of the squares of its components (theorem of Pythagoras). Thus $v^2 = v_x^2 + v_y^2 + v_z^2$ for any velocity v.

Taking averages, we obtain

$$\overline{v^2} = \overline{v_x^2} + \overline{v_y^2} + \overline{v_z^2}. \quad \text{(ii)}$$

Since the velocities of the molecules in our gas are assumed to be random, there is no preference to one direction or another. Hence

$$\overline{v_x^2} = \overline{v_y^2} = \overline{v_z^2}. \quad \text{(iii)}$$

Combining Eqs. (iii) and (ii), we get

$$\overline{v^2} = 3\overline{v_x^2}. \quad \text{(iv)}$$

We substitute Eq. (iv) into Eq. (i) for net force F (bottom of previous page):

$$F = \frac{m}{\ell} N \frac{\overline{v^2}}{3}.$$

The pressure on the wall is then

$$P = \frac{F}{A} = \frac{1}{3} \frac{Nm\overline{v^2}}{A\ell}$$

or

$$P = \frac{1}{3} \frac{Nm\overline{v^2}}{V}, \quad \begin{bmatrix} \text{pressure in an} \\ \text{ideal gas} \end{bmatrix} \quad \text{(13-6)}$$

where $V = \ell A$ is the volume of the container. This is the result we wanted, the pressure exerted by a gas on its container expressed in terms of molecular properties.

Equation 13–6 can be rewritten in a clearer form by multiplying both sides by V and rearranging the right-hand side:

$$PV = \tfrac{2}{3} N (\tfrac{1}{2} m \overline{v^2}). \quad \text{(13-7)}$$

The quantity $\tfrac{1}{2}m\overline{v^2}$ is the average translational kinetic energy ($\overline{KE}$) of the molecules in the gas. If we compare Eq. 13–7 with Eq. 13–4, the ideal gas law $PV = NkT$, we see that the two agree if

$$\tfrac{2}{3}(\tfrac{1}{2}m\overline{v^2}) = kT,$$

or

$$\boxed{\overline{KE} = \tfrac{1}{2} m \overline{v^2} = \tfrac{3}{2} kT.} \quad \text{[ideal gas]} \quad \text{(13-8)}$$

TEMPERATURE RELATED TO AVERAGE KINETIC ENERGY OF MOLECULES

This equation tells us that

the average translational kinetic energy of molecules in random motion in an ideal gas is directly proportional to the absolute temperature of the gas.

The higher the temperature, according to kinetic theory, the faster the molecules are moving on average. This relation is one of the triumphs of kinetic theory.

EXAMPLE 13–15 Molecular kinetic energy. What is the average translational kinetic energy of molecules in an ideal gas at 37°C?

APPROACH We use the absolute temperature in Eq. 13–8.

SOLUTION We change 37°C to 310 K and insert into Eq. 13–8:

$$\overline{KE} = \tfrac{3}{2} kT = \tfrac{3}{2}(1.38 \times 10^{-23} \text{ J/K})(310 \text{ K}) = 6.42 \times 10^{-21} \text{ J}.$$

NOTE A mole of molecules would have a total translational kinetic energy equal to $(6.42 \times 10^{-21} \text{ J})(6.02 \times 10^{23}) = 3860 \text{ J}$, which equals the kinetic energy of a 1-kg stone traveling almost 90 m/s.

EXERCISE E If molecules of hydrogen gas and oxygen gas were placed in the same balloon at room temperature, how would the average kinetic energies of the molecules compare? (*a*) They would be the same. (*b*) The hydrogen molecules would have greater kinetic energy. (*c*) The oxygen molecules would have greater kinetic energy. (*d*) Need more information.

Equation 13–8 holds not only for gases, but also applies reasonably accurately to liquids and solids. Thus the result of Example 13–15 would apply to molecules within living cells at body temperature (37°C).

We can use Eq. 13–8 to calculate how fast molecules are moving on average. Notice that the average in Eqs. 13–5 through 13–8 is over the *square* of the speed. The square root of $\overline{v^2}$ is called the **root-mean-square** speed, v_{rms} (since we are taking the square *root* of the *mean* of the *square* of the speed):

$$v_{rms} = \sqrt{\overline{v^2}} = \sqrt{\frac{3kT}{m}}. \qquad (13\text{–}9)$$

EXAMPLE 13–16 **Speeds of air molecules.** What is the rms speed of air molecules (O_2 and N_2) at room temperature (20°C)?

APPROACH To obtain v_{rms}, we need the masses of O_2 and N_2 molecules and then apply Eq. 13–9 to oxygen and nitrogen separately, since they have different masses.

SOLUTION The masses of one molecule of O_2 (molecular mass = 32 u) and N_2 (molecular mass = 28 u) are (where 1 u = 1.66×10^{-27} kg)

$$m(O_2) = (32)(1.66 \times 10^{-27} \text{ kg}) = 5.3 \times 10^{-26} \text{ kg},$$
$$m(N_2) = (28)(1.66 \times 10^{-27} \text{ kg}) = 4.6 \times 10^{-26} \text{ kg}.$$

Thus, for oxygen

$$v_{rms} = \sqrt{\frac{3kT}{m}} = \sqrt{\frac{(3)(1.38 \times 10^{-23} \text{ J/K})(293 \text{ K})}{(5.3 \times 10^{-26} \text{ kg})}} = 480 \text{ m/s},$$

and for nitrogen the result is $v_{rms} = 510$ m/s.

NOTE These speeds are more than 1700 km/h or 1000 mi/h, and are greater than the speed of sound, ≈ 340 m/s at 20°C (Chapter 12).

EXERCISE F By what factor must the absolute temperature change to double v_{rms}? (a) $\sqrt{2}$; (b) 2; (c) $2\sqrt{2}$; (d) 4; (e) 16.

*Kinetic Energy Near Absolute Zero

Equation 13–8, $\overline{KE} = \tfrac{3}{2}kT$, implies that as the temperature approaches absolute zero, the kinetic energy of molecules approaches zero. Modern quantum theory, however, tells us this is not quite so. Instead, as absolute zero is approached, the kinetic energy approaches a very small nonzero minimum value. Even though all real gases become liquid or solid near 0 K, molecular motion does not cease, even at absolute zero.

13–10 Distribution of Molecular Speeds

The molecules in a gas are assumed to be in random motion, which means that many molecules have speeds less than the rms speed and others have greater speeds. In 1859, James Clerk Maxwell (1831–1879) derived, on the basis of kinetic theory, that the speeds of molecules in a gas are distributed according to the graph shown in Fig. 13–19. This is known as the **Maxwell distribution of speeds**.[†] The speeds vary from zero to many times the rms speed, but as the graph shows, most molecules have speeds that are not far from the average. Less than 1% of the molecules exceed four times v_{rms}.

Experiments to determine the distribution of molecular speeds in real gases, starting in the 1920s, confirmed with considerable accuracy the Maxwell distribution and the direct proportion between average kinetic energy and absolute temperature, Eq. 13–8.

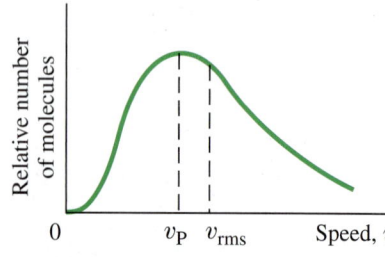

FIGURE 13–19 Distribution of speeds of molecules in an ideal gas. Note that v_{rms} is not at the peak of the curve (that speed is called the "most probable speed," v_P). This is because the curve is skewed to the right: it is not symmetrical.

[†]Mathematically, the distribution is given by $\Delta N = Cv^2 \exp(-\tfrac{1}{2}mv^2/kT)\Delta v$, where ΔN is the number of molecules with speed between v and $v + \Delta v$, C is a constant, and exp means the expression in parentheses is an exponent on the "natural number" $e = 2.718\ldots$.

Figure 13–20 shows the Maxwell distribution for two different temperatures. Just as v_{rms} increases with temperature, so the whole distribution curve shifts to the right at higher temperatures. Kinetic theory can be applied approximately to liquids and solutions. Figure 13–20 illustrates how kinetic theory can explain why many chemical reactions, including those in biological cells, take place more rapidly as the temperature increases. Most chemical reactions occur in a liquid solution, and the molecules have a speed distribution close to the Maxwell distribution. Two molecules may chemically react only if their kinetic energy is above some particular minimum value (called the *activation energy*), E_A, so that when they collide, they penetrate into each other somewhat. Figure 13–20 shows that at a higher temperature, many more molecules have a speed and kinetic energy KE above the needed threshold E_A.

PHYSICS APPLIED
How chemical reactions depend on T

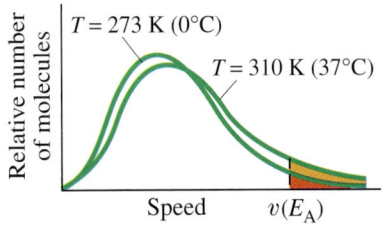

FIGURE 13–20 Distribution of molecular speeds for two different temperatures. Color shading shows proportions of molecules above a certain speed (corresponding to an activation energy $E_A = \frac{1}{2}mv^2$).

13–11 Real Gases and Changes of Phase

The ideal gas law, $PV = NkT$, is an accurate description of the behavior of a real gas as long as the pressure is not too high and the temperature is far from the liquefaction point. But what happens when these two criteria are not satisfied? First we discuss real gas behavior, and then we examine how kinetic theory can help us understand this behavior.

Let us look at a graph of pressure plotted against volume for a given amount of gas. On such a **PV diagram**, Fig. 13–21, each point represents the pressure and volume of an equilibrium state of the given substance. The various curves (labeled A, B, C, and D) show how the pressure varies as a function of volume for four different values of constant temperature T_A, T_B, T_C, and T_D. The red dashed curve A′ represents the behavior of a gas as predicted by the ideal gas law; that is, $PV =$ constant. The solid curve A represents the behavior of a real gas at the same temperature. Notice that at high pressure, the volume of a real gas is less than that predicted by the ideal gas law. The curves B and C in Fig. 13–21 represent the gas at successively lower temperatures, and we see that the behavior deviates even more from the curves predicted by the ideal gas law (for example, B′), and the deviation is greater the closer the gas is to liquefying.

To explain this behavior, note that at higher pressure we expect the molecules to be closer together. And at lower temperatures, the potential energy associated with attractive forces between the molecules (which we ignored before) is no longer negligible. These attractive forces tend to pull the molecules closer together so the volume is less than expected from the ideal gas law. At still lower temperatures, these forces cause liquefaction, and the molecules become very close together.

Curve D represents the situation when liquefaction occurs. At low pressure on curve D (on the right in Fig. 13–21), the substance is a gas and occupies a large volume. As the pressure is increased, the volume decreases until point b is reached. From point b to point a, the volume decreases with no change in pressure; the substance is gradually changing from the gas to the liquid phase. At point a, all of the substance has changed to liquid. Further increase in pressure reduces the volume only slightly—liquids are nearly incompressible—so on the left the curve is very steep as shown. The shaded area under the gold dashed line represents the region where the gas and liquid phases exist together in equilibrium.

Curve C in Fig. 13–21 represents the behavior of the substance at its **critical temperature**; the point c (the one point where curve C is horizontal) is called the **critical point**. At temperatures less than the critical temperature, a gas will change to the liquid phase if sufficient pressure is applied. Above the critical temperature (and this is the definition of the term), no amount of pressure can cause a gas to change phase and become a liquid. (Thus curves A and B represent the substance at temperatures where it can only be a gas.) The critical temperatures for various gases are given in Table 13–2. Scientists tried for many years to liquefy oxygen without success, which led to the idea that there must be a critical point. Oxygen can be liquefied only if first cooled below its critical temperature of $-118°C$.

FIGURE 13–21 *PV* diagram for a real substance. Curves A, B, C, and D represent the same substance at different temperatures ($T_A > T_B > T_C > T_D$).

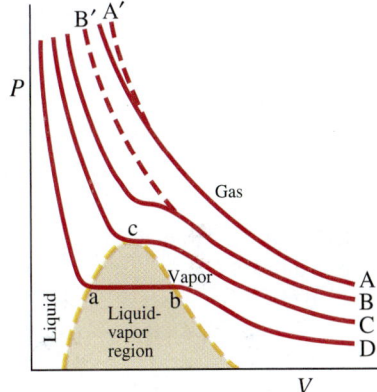

TABLE 13–2 Critical Temperatures and Pressures

Substance	Critical Temperature °C	K	Critical Pressure (atm)
Water	374	647	218
CO_2	31	304	72.8
Oxygen	−118	155	50
Nitrogen	−147	126	33.5
Hydrogen	−239.9	33.3	12.8
Helium	−267.9	5.3	2.3

Often a distinction is made between the terms "gas" and "vapor": a substance below its critical temperature in the gaseous state is called a **vapor**; above the critical temperature, it is called a **gas**.

The behavior of a substance can be diagrammed not only on a PV diagram but also on a PT diagram. A **PT diagram**, often called a **phase diagram**, is particularly convenient for comparing the different phases of a substance. Figure 13–22 is the phase diagram for water. The curve labeled ℓ-v represents those points where the liquid and vapor phases are in equilibrium—it is thus a graph of the boiling point versus pressure. Note that the curve correctly shows that at a pressure of 1 atm the boiling point is 100°C and that the boiling point is lowered for a decreased pressure. The curve s-ℓ represents points where solid and liquid exist in equilibrium and thus is a graph of the freezing point versus pressure.

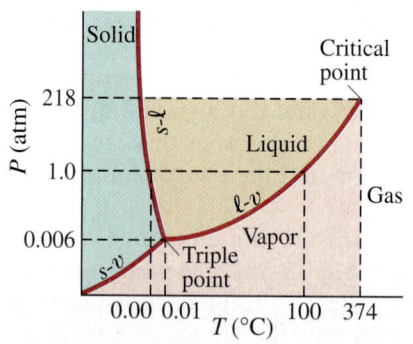

FIGURE 13–22 Phase diagram for water (note that the scales are not linear).

At 1 atm, the freezing point of water is 0°C, as shown. Notice also in Fig. 13–22 that at a pressure of 1 atm, the substance is in the liquid phase if the temperature is between 0°C and 100°C, but is in the solid or vapor phase if the temperature is below 0°C or above 100°C. The curve labeled s-v is the *sublimation point* versus pressure curve. **Sublimation** refers to the process whereby at low pressures a solid changes directly into the vapor phase without passing through the liquid phase. For water, sublimation occurs if the pressure of the water vapor is less than 0.0060 atm. Carbon dioxide, which in the solid phase is called dry ice, sublimates even at atmospheric pressure (Fig. 13–23).

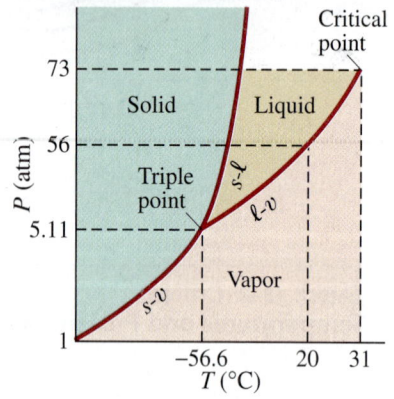

FIGURE 13–23 Phase diagram for carbon dioxide.

The intersection of the three curves (in Fig. 13–22) is the **triple point**. For water this occurs at $T = 273.16$ K and $P = 6.03 \times 10^{-3}$ atm. It is only at the triple point that the three phases can exist together in equilibrium. Because the triple point corresponds to a unique value of temperature and pressure, it is precisely reproducible and is often used as a point of reference. For example, the standard of temperature is usually specified as exactly 273.16 K at the triple point of water, rather than 273.15 K at the freezing point of water at 1 atm.

Notice that the solid-liquid (s-ℓ) curve for water (Fig. 13–22) slopes upward to the left. This is true only of substances that *expand* upon freezing: at a higher pressure, a lower temperature is needed to cause the liquid to freeze. More commonly, substances contract upon freezing and the s-ℓ curve slopes upward to the right, as shown for carbon dioxide (CO_2) in Fig. 13–23.

The phase transitions we have been discussing are the common ones. Some substances, however, can exist in several forms in the solid phase. A transition from one phase to another occurs at a particular temperature and pressure, just like ordinary phase changes. For example, ice has been observed in at least eight forms at very high pressure. Ordinary helium has two distinct liquid phases, called helium I and II. They exist only at temperatures within a few degrees of absolute zero. Helium II exhibits very unusual properties referred to as **superfluidity**. It has essentially zero viscosity and exhibits strange properties such as climbing up the sides of an open container. Also interesting are **liquid crystals** (used for computer and TV monitors, Section 24–11) which can be considered to be in a phase between liquid and solid.

13–12 Vapor Pressure and Humidity

Evaporation

If a glass of water is left out overnight, the water level will have dropped by morning. We say the water has evaporated, meaning that some of the water has changed to the vapor or gas phase.

This process of **evaporation** can be explained on the basis of kinetic theory. The molecules in a liquid move past one another with a variety of speeds that follow, approximately, the Maxwell distribution. There are strong attractive forces between these molecules, which is what keeps them close together in the liquid phase. A molecule near the surface of the liquid may, because of its speed, leave the liquid momentarily. But just as a rock thrown into the air returns to the Earth, so the attractive forces of the other molecules can pull the vagabond molecule back to the liquid surface—that is, if its velocity is not too large. A molecule with a high enough velocity, however, will escape the liquid entirely, like a rocket escaping the Earth, and become part of the gas phase. Only those molecules that have kinetic energy above a particular value can escape to the gas phase. We have already seen that kinetic theory predicts that the relative number of molecules with kinetic energy above a particular value (such as E_A in Fig. 13–20) increases with temperature. This is in accord with the well-known observation that the evaporation rate is greater at higher temperatures.

Because it is the fastest molecules that escape from the surface, the average speed of those remaining is less. When the average speed is less, the absolute temperature is less. Thus kinetic theory predicts that *evaporation is a cooling process*. You may have noticed this effect when you stepped out of a warm shower and felt cold as the water on your body began to evaporate; and after working up a sweat on a hot day, even a slight breeze makes you feel cool through evaporation. Try licking your finger and then blow on it.

Vapor Pressure

Air normally contains water vapor (water in the gas phase), and it comes mainly from evaporation. To look at this process in a little more detail, consider a closed container that is partially filled with water (or another liquid) and from which the air has been removed (Fig. 13–24). The fastest moving molecules quickly evaporate into the empty space above the liquid's surface. As they move about, some of these molecules strike the liquid surface and again become part of the liquid phase: this is called **condensation**. The number of molecules in the vapor increases until the number of molecules returning to the liquid equals the number leaving in the same time interval. Equilibrium then exists, and the space above the liquid surface is said to be *saturated*. The pressure of the vapor when it is saturated is called the **saturated vapor pressure** (or simply the vapor pressure).

The saturated vapor pressure of any substance depends on the temperature. At higher temperatures, more molecules have sufficient kinetic energy to break from the liquid surface into the vapor phase. Hence equilibrium will be reached at a higher vapor pressure. The saturated vapor pressure of water at various temperatures is given in Table 13–3. Notice that even solids—for example, ice—have a measurable saturated vapor pressure.

In everyday situations, evaporation from a liquid takes place into the air above it rather than into a vacuum. This does not materially alter the discussion above relating to Fig. 13–24. Equilibrium will still be reached when there are sufficient molecules in the gas phase that the number reentering the liquid equals the number leaving. The concentration of particular molecules (such as water) in the gas phase is not affected by the presence of air, although collisions with air molecules may lengthen the time needed to reach equilibrium. Thus equilibrium occurs at the same value of the saturated vapor pressure as if air were not there.

If the container is large or is not closed, all the liquid may evaporate before saturation is reached. And if the container is not sealed—as, for example, a room in your house—it is not likely that the air will become saturated with water vapor (unless it is raining outside).

FIGURE 13–24 Vapor appears above a liquid in a closed container.

PHYSICS APPLIED
Evaporation cools

TABLE 13–3 Saturated Vapor Pressure of Water

Temperature (°C)	Saturated Vapor Pressure	
	torr (= mm-Hg)	Pa (= N/m²)
−50	0.030	4.0
−10	1.95	2.60×10^2
0	4.58	6.11×10^2
5	6.54	8.72×10^2
10	9.21	1.23×10^3
15	12.8	1.71×10^3
20	17.5	2.33×10^3
25	23.8	3.17×10^3
30	31.8	4.24×10^3
40	55.3	7.37×10^3
50	92.5	1.23×10^4
60	149	1.99×10^4
70†	234	3.12×10^4
80	355	4.73×10^4
90	526	7.01×10^4
100‡	760	1.01×10^5
120	1489	1.99×10^5
150	3570	4.76×10^5

† Boiling point on summit of Mt. Everest.
‡ Boiling point at sea level.

FIGURE 13–25 Boiling: bubbles of water vapor float upward from the bottom (where the temperature is highest).

Boiling

The saturated vapor pressure of a liquid increases with temperature. When the temperature is raised to the point where the saturated vapor pressure at that temperature equals the external pressure, **boiling** occurs (Fig. 13–25). As the boiling point is approached, tiny bubbles tend to form in the liquid, which indicate a change from the liquid to the gas phase. However, if the vapor pressure inside the bubbles is less than the external pressure, the bubbles immediately are crushed. As the temperature is increased, the saturated vapor pressure inside a bubble eventually becomes equal to or exceeds the external air pressure. The bubble will then not collapse but can rise to the surface. Boiling has then begun. *A liquid boils when its saturated vapor pressure equals the external pressure.* This occurs for water at a pressure of 1 atm (760 torr) at 100°C, as can be seen from Table 13–3.

The boiling point of a liquid depends on the external pressure. At high elevations, the boiling point of water is somewhat less than at sea level because the air pressure is less up there. For example, on the summit of Mt. Everest (8850 m) the air pressure is about one-third of what it is at sea level, and from Table 13–3 we can see that water will boil at about 70°C. Cooking food by boiling takes longer at high elevations, because the boiling water is cooking at a lower temperature. Pressure cookers reduce cooking time because they build up a pressure as high as 2 atm, allowing a higher boiling (and cooking) temperature to be attained (Problem 64 and Fig. 13–32).

Partial Pressure and Humidity

When we refer to the weather as being dry or humid, we are referring to the water vapor content of the air. In a gas such as air, which is a mixture of several types of gases, the total pressure is the sum of the *partial pressures* of each gas present.† By **partial pressure**, we mean the pressure each gas would exert if it alone were present. The partial pressure of water in the air can be as low as zero and can vary up to a maximum equal to the saturated vapor pressure of water at the given temperature. Thus, at 20°C, the partial pressure of water cannot exceed 17.5 torr (see Table 13–3) or about 0.02 atm. The **relative humidity** is defined as the ratio of the partial pressure of water vapor to the saturated vapor pressure at a given temperature. It is usually expressed as a percentage:

$$\text{Relative humidity} = \frac{\text{partial pressure of H}_2\text{O}}{\text{saturated vapor pressure of H}_2\text{O}} \times 100\%.$$

Thus, when the humidity is close to 100%, the air holds nearly all the water vapor it can.

EXAMPLE 13–17 Relative humidity. On a particular hot day, the temperature is 30°C and the partial pressure of water vapor in the air is 21.0 torr. What is the relative humidity?

APPROACH From Table 13–3, we see that the saturated vapor pressure of water at 30°C is 31.8 torr.

SOLUTION The relative humidity is thus

$$\frac{21.0 \text{ torr}}{31.8 \text{ torr}} \times 100\% = 66\%.$$

PHYSICS APPLIED
Humidity and comfort

Humans are sensitive to humidity. A relative humidity of 40–50% is generally optimum for both health and comfort. High humidity, particularly on a hot day, reduces the evaporation of moisture from the skin, which is one of the body's vital mechanisms for regulating body temperature. Very low humidity, on the other hand, can dry the skin and mucous membranes.

---

†For example, 78% (by volume) of air molecules are nitrogen and 21% oxygen, with much smaller amounts of water vapor, argon, carbon dioxide, and other gases. At an air pressure of 1 atm, oxygen exerts a partial pressure of 0.21 atm and nitrogen 0.78 atm.

Air is saturated with water vapor when the partial pressure of water in the air is equal to the saturated vapor pressure at that temperature. If the partial pressure of water exceeds the saturated vapor pressure, the air is said to be **supersaturated**. This situation can occur when a temperature decrease occurs. For example, suppose the temperature is 30°C and the partial pressure of water is 21 torr, which represents a humidity of 66% as we saw in Example 13–17. Suppose now that the temperature falls to, say, 20°C, as might happen at nightfall. From Table 13–3 we see that the saturated vapor pressure of water at 20°C is 17.5 torr. Hence the relative humidity would be greater than 100%, and the supersaturated air cannot hold this much water vapor. The excess water may condense and appear as dew, clouds, or as fog or rain (Fig. 13–26).

When air containing a given amount of water is cooled, a temperature is reached where the partial pressure of water equals the saturated vapor pressure. This is called the **dew point**. Measurement of the dew point is the most accurate means of determining the relative humidity. One method uses a polished metal surface which is gradually cooled down while in contact with air. The temperature at which moisture begins to appear on the surface is the dew point, and the partial pressure of water can then be obtained from saturated vapor pressure Tables. If, for example, on a given day the temperature is 20°C and the dew point is 5°C, then the partial pressure of water (Table 13–3) in the 20°C air is 6.54 torr, whereas its saturated vapor pressure is 17.5 torr; hence the relative humidity is 6.54/17.5 = 37%.

PHYSICS APPLIED
Weather

(a)

(b)

(c)

FIGURE 13–26 (a) Fog or mist settling in a valley where the temperature has dropped below the dew point. (b) Dew drops on a leaf. (c) Clouds form on a sunny day at the beach due to air, nearly saturated with water vapor, rising to an altitude where the cooler temperature is at the dew point.

CONCEPTUAL EXAMPLE 13–18 **Dryness in winter.** Why does the air inside heated buildings seem very dry on a cold winter day?

RESPONSE Suppose the relative humidity outside on a −10°C day is 50%. Table 13–3 tells us the partial pressure of water in the air is about 1.0 torr. If this air is brought indoors and heated to +20°C, the relative humidity is (1.0 torr)/(17.5 torr) = 5.7%. Even if the outside air were saturated at a partial pressure of 1.95 torr, the inside relative humidity would still be at a low 11%.

*13–13 Diffusion

If you carefully place a few drops of food coloring in a glass of water as in Fig. 13–27, you will find that the color spreads throughout the water. The process may take some time (assuming you do not shake the glass), but eventually the color will become uniform. This mixing, known as **diffusion**, is readily explained by kinetic theory as due to the random movement of the molecules. Diffusion occurs in gases too. Common examples include perfume or smoke (or the odor of something cooking on a stove) diffusing in air, although convection (moving air currents) often plays a greater role in spreading odors than does diffusion. Diffusion depends on concentration, by which we mean the number of molecules or moles per unit volume. In general, *the diffusing substance moves from a region where its concentration is high to a region where its concentration is low.*

(a)

(b)

(c)

FIGURE 13–27 A few drops of food coloring (a) dropped into water, (b) spreads slowly throughout the water, eventually (c) becoming uniform.

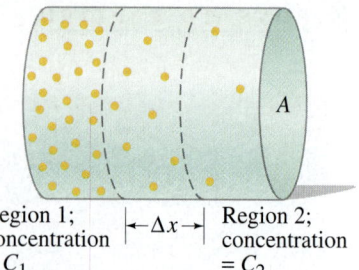

FIGURE 13–28 Diffusion occurs from a region of high concentration to one of lower concentration (only one type of molecule is shown).

TABLE 13–4 Diffusion Constants, D (20°C, 1 atm)

Diffusing Molecules	Medium	D (m²/s)
H_2	Air	6.3×10^{-5}
O_2	Air	1.8×10^{-5}
O_2	Water	100×10^{-11}
Glycine (an amino acid)	Water	95×10^{-11}
Blood hemoglobin	Water	6.9×10^{-11}
DNA (mass 6×10^6 u)	Water	0.13×10^{-11}

PHYSICS APPLIED
Diffusion time

Diffusion can be readily understood on the basis of kinetic theory and the random motion of molecules. Consider a tube of cross-sectional area A containing molecules in a higher concentration on the left than on the right, Fig. 13–28. We assume the molecules are in random motion. Yet there will be a net flow of molecules to the right. To see why, let us consider the small section of tube of length Δx as shown. Molecules from both regions 1 and 2 cross into this central section as a result of their random motion. The more molecules there are in a region, the more will strike a given area or cross a boundary. Since there is a greater concentration of molecules in region 1 than in region 2, more molecules cross into the central section from region 1 than from region 2. There is, then, a net flow of molecules from left to right, from high concentration (C_1) toward low concentration (C_2). The net flow becomes zero only when the concentrations become equal.

We might expect that the greater the difference in concentration, the greater the flow rate. Indeed, the rate of diffusion, J (number of molecules or moles or kg per second), is directly proportional to the difference in concentration per unit distance, $(C_1 - C_2)/\Delta x$ (which is called the **concentration gradient**), and to the cross-sectional area A (see Fig. 13–28):

$$J = DA \frac{C_1 - C_2}{\Delta x}. \qquad (13\text{–}10)$$

D is a constant of proportionality called the **diffusion constant**. Equation 13–10 is known as the **diffusion equation**, or **Fick's law**. If the concentrations are given in mol/m³, then J is the number of moles passing a given point per second. If the concentrations are given in kg/m³, then J is the mass movement per second (kg/s). The length Δx is given in meters, and area A in m². The values of D for a variety of substances diffusing in a particular medium are given in Table 13–4.

EXAMPLE 13–19 ESTIMATE Diffusion of ammonia in air. To get an idea of the time required for diffusion, estimate how long it might take for ammonia (NH_3) to be detected 10 cm from a bottle after it is opened, assuming only diffusion is occurring.

APPROACH This will be an order-of-magnitude calculation. The rate of diffusion J can be set equal to the number of molecules N diffusing across area A in a time t: $J = N/t$. Then the time $t = N/J$, where J is given by Eq. 13–10. We will have to make some assumptions and rough approximations about concentrations to use Eq. 13–10.

SOLUTION Using Eq. 13–10, we find

$$t = \frac{N}{J} = \frac{N}{DA} \frac{\Delta x}{\Delta C}.$$

The average concentration (midway between bottle and nose) can be approximated by $\overline{C} \approx N/V$, where V is the volume over which the molecules move and is roughly on the order of $V \approx A \Delta x$, where Δx is 10 cm = 0.10 m. We substitute $N = \overline{C} V = \overline{C} A \Delta x$ into the above equation:

$$t \approx \frac{(\overline{C} A \Delta x) \Delta x}{DA \Delta C} = \frac{\overline{C}}{\Delta C} \frac{(\Delta x)^2}{D}.$$

The concentration of ammonia is high near the bottle (C) and low near the detecting nose (≈ 0), so $\overline{C} \approx C/2 \approx \Delta C/2$, or $(\overline{C}/\Delta C) \approx \frac{1}{2}$. Since NH_3 molecules have a size somewhere between H_2 and O_2, from Table 13–4 we can estimate $D \approx 4 \times 10^{-5}$ m²/s. Then

$$t \approx \frac{1}{2} \frac{(0.10 \text{ m})^2}{(4 \times 10^{-5} \text{ m}^2/\text{s})} \approx 100 \text{ s},$$

or about a minute or two.

NOTE This result seems rather long from experience, suggesting that air currents (convection) are more important than diffusion for transmitting odors.

Diffusion is extremely important for living organisms. For example, molecules produced in certain chemical reactions within cells diffuse to other areas where they take part in other reactions.

Gas diffusion is important too. Plants require carbon dioxide for photosynthesis. The CO_2 diffuses into leaves from the outside air through tiny openings (stomata). As CO_2 is utilized by the cells, its concentration drops below that in the air outside, and more diffuses inward. Water vapor and oxygen produced by the cells diffuse outward into the air.

Animals also exchange oxygen and CO_2 with the environment. Oxygen is required for energy-producing reactions and must diffuse into cells. CO_2 is produced as an end product of many metabolic reactions and must diffuse out of cells. But diffusion is slow over longer distances, so only the smallest organisms in the animal world could survive without having developed complex respiratory and circulatory systems. In humans, oxygen is taken into the lungs, where it diffuses short distances across lung tissue and into the blood. Then the blood circulates it to cells throughout the body. The blood also carries CO_2 produced by the cells back to the lungs, where it diffuses outward.

PHYSICS APPLIED
Diffusion in living organisms

Summary

The atomic theory of matter postulates that all matter is made up of tiny entities called **atoms**, which are typically 10^{-10} m in diameter.

Atomic and **molecular masses** are specified on a scale where the most common form of carbon (^{12}C) is arbitrarily given the value 12.0000 u (atomic mass units), exactly.

The distinction between solids, liquids, and gases can be attributed to the strength of the attractive forces between the atoms or molecules and to their average speed.

Temperature is a measure of how hot or cold something is. **Thermometers** are used to measure temperature on the **Celsius** (°C), **Fahrenheit** (°F), and **Kelvin** (K) scales. Two standard points on each scale are the freezing point of water (0°C, 32°F, 273.15 K) and the boiling point of water (100°C, 212°F, 373.15 K). A one-kelvin change in temperature equals a change of one Celsius degree or $\frac{9}{5}$ Fahrenheit degrees. Kelvins are related to °C by

$$T(K) = T(°C) + 273.15.$$

When two objects at different temperatures are placed in contact, they eventually reach the same temperature and are then said to be in **thermal equilibrium**.

The change in length, $\Delta\ell$, of a solid, when its temperature changes by an amount ΔT, is directly proportional to the temperature change and to its original length ℓ_0. That is,

$$\Delta\ell = \alpha\ell_0 \Delta T, \quad (13\text{–}1a)$$

where α is the *coefficient of linear expansion*.

The change in volume of most solids, liquids, and gases is proportional to the temperature change and to the original volume V_0:

$$\Delta V = \beta V_0 \Delta T. \quad (13\text{–}2)$$

The *coefficient of volume expansion*, β, is approximately equal to 3α for uniform solids.

Water is unusual because, unlike most materials whose volume increases with temperature, its volume in the range from 0°C to 4°C actually decreases as the temperature increases.

The **ideal gas law**, or **equation of state for an ideal gas**, relates the pressure P, volume V, and temperature T (in kelvins) of n moles of gas by the equation

$$PV = nRT, \quad (13\text{–}3)$$

where $R = 8.314$ J/mol·K for all gases. Real gases obey the ideal gas law quite accurately if they are not at too high a pressure or near their liquefaction point.

One **mole** is that amount of a substance whose mass in grams is numerically equal to the atomic or molecular mass of that substance.

Avogadro's number, $N_A = 6.02 \times 10^{23}$, is the number of atoms or molecules in 1 mol of any pure substance.

The ideal gas law can be written in terms of the number of molecules N in the gas as

$$PV = NkT, \quad (13\text{–}4)$$

where $k = R/N_A = 1.38 \times 10^{-23}$ J/K is Boltzmann's constant.

According to the **kinetic theory** of gases, which is based on the idea that a gas is made up of molecules that are moving rapidly and randomly, the average translational kinetic energy of molecules is proportional to the Kelvin temperature T:

$$\overline{KE} = \tfrac{1}{2}m\overline{v^2} = \tfrac{3}{2}kT, \quad (13\text{–}8)$$

where k is Boltzmann's constant. At any moment, there exists a wide distribution of molecular speeds within a gas.

The behavior of real gases at high pressure, and/or when near their liquefaction point, deviates from the ideal gas law due to the attractive forces between molecules. Below the **critical temperature**, a gas can change to a liquid if sufficient pressure is applied; but if the temperature is higher than the critical temperature, no amount of pressure will cause a liquid surface to form. The **triple point** of a substance is that unique temperature and pressure at which all three phases—solid, liquid, and gas—can coexist in equilibrium.

Evaporation of a liquid is the result of the fastest moving molecules escaping from the surface. **Saturated vapor pressure** refers to the pressure of the vapor above a liquid when the two phases are in equilibrium. The vapor pressure of a substance (such as water) depends strongly on temperature, and at the boiling point is equal to atmospheric pressure. **Relative humidity** of air is the ratio of the actual partial pressure of water vapor in the air to the saturated vapor pressure at that temperature; it is usually expressed as a percentage.

[*Diffusion is the process whereby molecules of a substance move (on average) from one area to another because of a difference in that substance's concentration.]

Questions

1. Which has more atoms: 1 kg of lead or 1 kg of copper? (See the Periodic Table or Appendix B.) Explain why.
2. Name several properties of materials that could be used to make a thermometer.
3. Which is larger, 1 C° or 1 F°? Explain why.
4. In the relation $\Delta \ell = \alpha \ell_0 \Delta T$, should ℓ_0 be the initial length, the final length, or does it matter?
5. A flat bimetallic strip consists of a strip of aluminum riveted to a strip of iron. When heated, the strip will bend. Which metal will be on the outside of the curve? Why? [*Hint*: See Table 13–1.]
6. Long steam pipes that are fixed at the ends often have a section in the shape of a ∪. Why?
7. Figure 13–29 shows a diagram of a simple bimetallic **thermostat** used to control a furnace (or other heating or cooling system). The electric switch (attached to the bimetallic strip) is a glass vessel containing liquid mercury that conducts electricity when it touches both contact wires. Explain how this device controls the furnace and how it can be set at different temperatures.

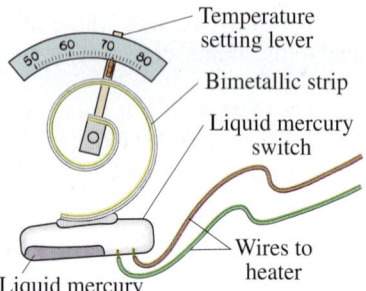

FIGURE 13–29 A thermostat (Question 7).

8. A glass container may break if one part of it is heated or cooled more rapidly than adjacent parts. Explain.
9. Explain why it is advisable to add water to an overheated automobile engine only slowly, and only with the engine running.
10. The units for the coefficient of linear expansion α are $(C°)^{-1}$, and there is no mention of a length unit such as meters. Would the expansion coefficient change if we used feet or millimeters instead of meters? Explain.
11. When a cold alcohol-in-glass thermometer is first placed in a hot tub of water, the alcohol initially descends a bit and then rises. Explain.
12. The principal virtue of Pyrex glass is that its coefficient of linear expansion is much smaller than that for ordinary glass (Table 13–1). Explain why this gives rise to the higher heat resistance of Pyrex.
13. Will a clock using a pendulum supported on a long thin brass rod that is accurate at 20°C run fast or slow on a hot day (30°C)? Explain.
14. Freezing a can of soda will cause its bottom and top to bulge so badly the can will not stand up. What has happened?
15. Will the buoyant force on an aluminum sphere submerged in water increase, decrease, or remain the same, if the temperature is increased from 20°C to 40°C? Explain.
16. Can you determine the temperature of a vacuum? Explain.
17. Escape velocity from the Earth refers to the minimum speed an object must have to leave the Earth and never return. (See also Fig 5-22, page 122). (*a*) The escape velocity from the Moon is about one-fifth what it is for the Earth, due to the Moon's smaller mass. Explain why the Moon has practically no atmosphere. (*b*) If hydrogen was once in the Earth's atmosphere, why would it have probably escaped?
18. What exactly does it mean when we say that oxygen boils at −183°C?
19. A length of thin wire is placed over a block of ice (or an ice cube) at 0°C. The wire hangs down both sides of the ice, and weights are hung from the ends of the wire. It is found that the wire cuts its way through the ice cube, but leaves a solid block of ice behind it. This process is called *regelation*. Explain how this happens by inferring how the freezing point of water depends on pressure.
20. (*a*) Why does food cook faster in a pressure cooker? (*b*) Why does pasta or rice need to boil longer at high altitudes? (*c*) Is it harder to boil water at high altitudes?
21. Is it possible to boil water at room temperature (20°C) without heating it? Explain.
22. Why does exhaled air appear as a little white cloud in the winter (Fig. 13–30)?

FIGURE 13–30 Question 22.

23. Explain why it is dangerous to open the radiator cap of an overheated automobile engine.

MisConceptual Questions

1. Rod A has twice the diameter of rod B, but both are made of iron and have the same initial length. Both rods are now subjected to the same change in temperature (but remain solid). How would the change in the rods' lengths compare?
   (*a*) Rod A > rod B.
   (*b*) Rod B > rod A.
   (*c*) Rod A = rod B.
   (*d*) Need to know whether the rods were cooled or heated.
2. The linear expansion of a material depends on which of the following?
   (*a*) The length of the material.
   (*b*) The change in temperature of the material.
   (*c*) The type of material.
   (*d*) All of the above.
   (*e*) Both (*b*) and (*c*).

3. A steel plate has a hole in it with a diameter of exactly 1.0 cm when the plate is at a temperature of 20°C. A steel ring has an inner diameter of exactly 1.0 cm at 20°C. Both the plate and the ring are heated to 100°C. Which statement is true?
   (a) The hole in the plate gets smaller, and the opening in the ring gets larger.
   (b) The opening in the ring gets larger, but we need the relative size of the plate and the hole to know what happens to the hole.
   (c) The hole in the plate and the opening in the ring get larger.
   (d) The hole in the plate and the opening in the ring get smaller.
   (e) The hole in the plate gets larger, and the opening in the ring gets smaller.

4. One mole of an ideal gas in a sealed rigid container is initially at a temperature of 100°C. The temperature is then increased to 200°C. The pressure in the gas
   (a) remains constant.
   (b) increases by about 25%.
   (c) doubles.
   (d) triples.

5. When an ideal gas is warmed from 20°C to 40°C, the gas's temperature T that appears in the ideal gas law increases by a factor
   (a) of 2.
   (b) of 1.07.
   (c) that depends on the temperature scale you use.

6. Two identical bottles at the same temperature contain the same gas. If bottle B has twice the volume and contains half the number of moles of gas as bottle A, how does the pressure in B compare with the pressure in A?
   (a) $P_B = \frac{1}{2} P_A$.
   (b) $P_B = 2 P_A$.
   (c) $P_B = \frac{1}{4} P_A$.
   (d) $P_B = 4 P_A$.
   (e) $P_B = P_A$.

7. The temperature of an ideal gas increases. Which of the following is true?
   (a) The pressure must decrease.
   (b) The pressure must increase.
   (c) The pressure must increase while the volume decreases.
   (d) The volume must increase while the pressure decreases.
   (e) The pressure, the volume, or both, may increase.

8. An ideal gas is in a sealed rigid container. The average kinetic energy of the gas molecules depends most on
   (a) the size of the container.
   (b) the number of molecules in the container.
   (c) the temperature of the gas.
   (d) the mass of the molecules.

9. Two ideal gases, A and B, are at the same temperature. If the molecular mass of the molecules in gas A is twice that of the molecules in gas B, the molecules' root-mean-square speed is
   (a) the same in both gases. (d) twice as great in B.
   (b) twice as great in A.   (e) 1.4 times greater in B.
   (c) 1.4 times greater in A.

10. In a mixture of the gases oxygen and helium, which statement is valid?
    (a) The helium atoms will be moving faster than the oxygen molecules, on average.
    (b) Both will be moving at the same speed.
    (c) The oxygen molecules will, on average, be moving more rapidly than the helium atoms.
    (d) The kinetic energy of helium atoms will exceed that of oxygen molecules.
    (e) None of the above.

11. Which of the following is *not* true about an ideal gas?
    (a) The average kinetic energy of the gas molecules increases as the temperature increases.
    (b) The volume of an ideal gas increases with temperature if the pressure is held constant.
    (c) The pressure of an ideal gas increases with temperature if the volume is held constant.
    (d) All gas molecules have the same speed at a particular temperature.
    (e) The molecules are assumed to be far apart compared to their size.

12. When using the ideal gas law, which of the following rules must be obeyed?
    (a) Always use temperature in kelvins and absolute pressure.
    (b) Always use volume in m³ and temperature in kelvins.
    (c) Always use gauge pressure and temperature in degrees Celsius.
    (d) Always use gauge pressure and temperature in kelvins.
    (e) Always use volume in m³ and gauge pressure.

13. The rms speed of the molecules of an ideal gas
    (a) is the same as the most probable speed of the molecules.
    (b) is always equal to $\sqrt{2}$ times the maximum molecular speed.
    (c) will increase as the temperature of a gas increases.
    (d) All of the above.

For assigned homework and other learning materials, go to the MasteringPhysics website.

Problems

13–1 Atomic Theory

1. (I) How does the number of atoms in a 27.5-gram gold ring compare to the number in a silver ring of the same mass?
2. (I) How many atoms are there in a 3.4-g copper coin?

13–2 Temperature and Thermometers

3. (I) (a) "Room temperature" is often taken to be 68°F. What is this on the Celsius scale? (b) The temperature of the filament in a lightbulb is about 1900°C. What is this on the Fahrenheit scale?

4. (I) Among the highest and lowest natural air temperatures claimed are 136°F in the Libyan desert and −129°F in Antarctica. What are these temperatures on the Celsius scale?

5. (I) A thermometer tells you that you have a fever of 38.9°C. What is this in Fahrenheit?

6. (I) (a) 18° below zero on the Celsius scale is what Fahrenheit temperature? (b) 18° below zero on the Fahrenheit scale is what Celsius temperature?

7. (II) Determine the temperature at which the Celsius and Fahrenheit scales give the same numerical reading ($T_C = T_F$).

8. (II) In an alcohol-in-glass thermometer, the alcohol column has length 12.61 cm at 0.0°C and length 22.79 cm at 100.0°C. What is the temperature if the column has length (a) 18.70 cm, and (b) 14.60 cm?

13–4 Thermal Expansion

9. (I) The Eiffel Tower (Fig. 13–31) is built of wrought iron approximately 300 m tall. Estimate how much its height changes between January (average temperature of 2°C) and July (average temperature of 25°C). Ignore the angles of the iron beams and treat the tower as a vertical beam.

FIGURE 13–31 Problem 9. The Eiffel Tower in Paris.

10. (I) A concrete highway is built of slabs 12 m long (15°C). How wide should the expansion cracks between the slabs be (at 15°C) to prevent buckling if the range of temperature is −30°C to +50°C?

11. (I) Super Invar™, an alloy of iron and nickel, is a strong material with a very low coefficient of thermal expansion ($0.20 \times 10^{-6}/C°$). A 1.8-m-long tabletop made of this alloy is used for sensitive laser measurements where extremely high tolerances are required. How much will this alloy table expand along its length if the temperature increases 6.0 C°? Compare to tabletops made of steel.

12. (II) To what temperature would you have to heat a brass rod for it to be 1.5% longer than it is at 25°C?

13. (II) To make a secure fit, rivets that are larger than the rivet hole are often used and the rivet is cooled (usually in dry ice) before it is placed in the hole. A steel rivet 1.872 cm in diameter is to be placed in a hole 1.870 cm in diameter in a metal at 22°C. To what temperature must the rivet be cooled if it is to fit in the hole?

14. (II) An ordinary glass is filled to the brim with 450.0 mL of water at 100.0°C. If the temperature of glass and water is decreased to 20.0°C, how much water could be added to the glass?

15. (II) An aluminum sphere is 8.75 cm in diameter. What will be its % change in volume if it is heated from 30°C to 160°C?

16. (II) It is observed that 55.50 mL of water at 20°C completely fills a container to the brim. When the container and the water are heated to 60°C, 0.35 g of water is lost. (a) What is the coefficient of volume expansion of the container? (b) What is the most likely material of the container? Density of water at 60°C is 0.98324 g/mL.

17. (II) A brass plug is to be placed in a ring made of iron. At 15°C, the diameter of the plug is 8.755 cm and that of the inside of the ring is 8.741 cm. They must both be brought to what common temperature in order to fit?

18. (II) A certain car has 14.0 L of liquid coolant circulating at a temperature of 93°C through the engine's cooling system. Assume that, in this normal condition, the coolant completely fills the 3.5-L volume of the aluminum radiator and the 10.5-L internal cavities within the aluminum engine. When a car overheats, the radiator, engine, and coolant expand and a small reservoir connected to the radiator catches any resultant coolant overflow. Estimate how much coolant overflows to the reservoir if the system goes from 93°C to 105°C. Model the radiator and engine as hollow shells of aluminum. The coefficient of volume expansion for coolant is $\beta = 410 \times 10^{-6}/C°$.

*19. (II) An aluminum bar has the desired length when at 12°C. How much stress is required to keep it at this length if the temperature increases to 35°C? [See Table 9–1.]

20. (III) The pendulum in a grandfather clock is made of brass and keeps perfect time at 17°C. How much time is gained or lost in a year if the clock is kept at 29°C? (Assume the frequency dependence on length for a simple pendulum applies; see Chapter 11.)

13–5 Gas Laws; Absolute Temperature

21. (I) Absolute zero is what temperature on the Fahrenheit scale?

22. (II) Typical temperatures in the interior of the Earth and Sun are about 4000°C and 15×10^6 °C, respectively. (a) What are these temperatures in kelvins? (b) What percent error is made in each case if a person forgets to change °C to K?

13–6 and 13–7 Ideal Gas Law

23. (I) If 3.50 m³ of a gas initially at STP is placed under a pressure of 3.20 atm, the temperature of the gas rises to 38.0°C. What is the volume?

24. (I) In an internal combustion engine, air at atmospheric pressure and a temperature of about 20°C is compressed in the cylinder by a piston to $\frac{1}{9}$ of its original volume (compression ratio = 9.0). Estimate the temperature of the compressed air, assuming the pressure reaches 40 atm.

25. (II) If 16.00 mol of helium gas is at 10.0°C and a gauge pressure of 0.350 atm, calculate (a) the volume of the helium gas under these conditions, and (b) the temperature if the gas is compressed to precisely half the volume at a gauge pressure of 1.00 atm.

26. (II) A storage tank contains 21.6 kg of nitrogen (N_2) at an absolute pressure of 3.45 atm. What will the pressure be if the nitrogen is replaced by an equal mass of CO_2 at the same temperature?

27. (II) A storage tank at STP contains 28.5 kg of nitrogen (N_2). (a) What is the volume of the tank? (b) What is the pressure if an additional 32.2 kg of nitrogen is added without changing the temperature?

28. (II) A scuba tank is filled with air to a gauge pressure of 204 atm when the air temperature is 29°C. A diver then jumps into the ocean and, after a short time on the ocean surface, checks the tank's gauge pressure and finds that it is only 191 atm. Assuming the diver has inhaled a negligible amount of air from the tank, what is the temperature of the ocean water?

29. (II) What is the pressure inside a 38.0-L container holding 105.0 kg of argon gas at 21.6°C?

30. (II) A sealed metal container contains a gas at 20.0°C and 1.00 atm. To what temperature must the gas be heated for the pressure to double to 2.00 atm? (Ignore expansion of the container.)

31. (II) A tire is filled with air at 15°C to a gauge pressure of 230 kPa. If the tire reaches a temperature of 38°C, what fraction of the original air must be removed if the original pressure of 230 kPa is to be maintained?

32. (II) If 61.5 L of oxygen at 18.0°C and an absolute pressure of 2.45 atm are compressed to 38.8 L and at the same time the temperature is raised to 56.0°C, what will the new pressure be?

33. (II) You buy an "airtight" bag of potato chips packaged at sea level, and take the chips on an airplane flight. When you take the potato chips out of your "carry-on" bag, you notice it has noticeably "puffed up." Airplane cabins are typically pressurized at 0.75 atm, and assuming the temperature inside an airplane is about the same as inside a potato chip processing plant, by what percentage has the bag "puffed up" in comparison to when it was packaged?

34. (II) A helium-filled balloon escapes a child's hand at sea level and 20.0°C. When it reaches an altitude of 3600 m, where the temperature is 5.0°C and the pressure only 0.68 atm, how will its volume compare to that at sea level?

35. (II) Compare the value for the density of water vapor at exactly 100°C and 1 atm (Table 10–1) with the value predicted from the ideal gas law. Why would you expect a difference?

36. (III) A sealed test tube traps 25.0 cm^3 of air at a pressure of 1.00 atm and temperature of 18°C. The test tube's stopper has a diameter of 1.50 cm and will "pop off" the test tube if a net upward force of 10.0 N is applied to it. To what temperature would you have to heat the trapped air in order to "pop off" the stopper? Assume the air surrounding the test tube is always at a pressure of 1.00 atm.

37. (III) An air bubble at the bottom of a lake 41.0 m deep has a volume of 1.00 cm^3. If the temperature at the bottom is 5.5°C and at the top 18.5°C, what is the radius of the bubble just before it reaches the surface?

13–8 Ideal Gas Law in Terms of Molecules; Avogadro's Number

38. (I) Calculate the number of molecules/m^3 in an ideal gas at STP.

39. (I) How many moles of water are there in 1.000 L at STP? How many molecules?

40. (II) Estimate the number of (a) moles and (b) molecules of water in all the Earth's oceans. Assume water covers 75% of the Earth to an average depth of 3 km.

41. (II) The lowest pressure attainable using the best available vacuum techniques is about 10^{-12} N/m^2. At such a pressure, how many molecules are there per cm^3 at 0°C?

42. (II) Is a gas mostly empty space? Check by assuming that the spatial extent of the gas molecules in air is about $\ell_0 = 0.3$ nm so one gas molecule occupies an approximate volume equal to ℓ_0^3. Assume STP.

13–9 Molecular Interpretation of Temperature

43. (I) (a) What is the average translational kinetic energy of a nitrogen molecule at STP? (b) What is the total translational kinetic energy of 1.0 mol of N$_2$ molecules at 25°C?

44. (I) Calculate the rms speed of helium atoms near the surface of the Sun at a temperature of about 6000 K.

45. (I) By what factor will the rms speed of gas molecules increase if the temperature is increased from 20°C to 160°C?

46. (I) A gas is at 20°C. To what temperature must it be raised to triple the rms speed of its molecules?

47. (I) What speed would a 1.0-g paper clip have if it had the same kinetic energy as a molecule at 22°C?

48. (II) The rms speed of molecules in a gas at 20.0°C is to be increased by 4.0%. To what temperature must it be raised?

49. (II) If the pressure in a gas is tripled while its volume is held constant, by what factor does v_{rms} change?

50. (II) Show that the rms speed of molecules in a gas is given by $v_{\text{rms}} = \sqrt{3P/\rho}$, where P is the pressure in the gas and ρ is the gas density.

51. (II) Show that for a mixture of two gases at the same temperature, the ratio of their rms speeds is equal to the inverse ratio of the square roots of their molecular masses, $v_1/v_2 = \sqrt{M_2/M_1}$.

52. (II) What is the rms speed of nitrogen molecules contained in an 8.5-m^3 volume at 2.9 atm if the total amount of nitrogen is 2100 mol?

53. (II) Two isotopes of uranium, ^{235}U and ^{238}U (the superscripts refer to their atomic masses), can be separated by a gas diffusion process by combining them with fluorine to make the gaseous compound UF$_6$. Calculate the ratio of the rms speeds of these molecules for the two isotopes, at constant T. Use Appendix B for masses.

54. (III) Calculate (a) the rms speed of an oxygen molecule at 0°C and (b) determine how many times per second it would move back and forth across a 5.0-m-long room on average, assuming it made no collisions with other molecules.

13–11 Real Gases; Phase Changes

55. (I) CO$_2$ exists in what phase when the pressure is 35 atm and the temperature is 35°C (Fig. 13–23)?

56. (I) (a) At atmospheric pressure, in what phases can CO$_2$ exist? (b) For what range of pressures and temperatures can CO$_2$ be a liquid? Refer to Fig. 13–23.

57. (I) Water is in which phase when the pressure is 0.01 atm and the temperature is (a) 90°C, (b) −20°C?

58. (II) You have a sample of water and are able to control temperature and pressure arbitrarily. (a) Using Fig. 13–22, describe the phase changes you would see if you started at a temperature of 85°C, a pressure of 180 atm, and decreased the pressure down to 0.004 atm while keeping the temperature fixed. (b) Repeat part (a) with the temperature at 0.0°C. Assume that you held the system at the starting conditions long enough for the system to stabilize before making further changes.

13–12 Vapor Pressure and Humidity

59. (I) What is the partial pressure of water vapor at 30°C if the humidity is 75%?

60. (I) What is the air pressure at a place where water boils at 80°C?

61. (II) What is the dew point if the humidity is 65% on a day when the temperature is 25°C?

62. (II) If the air pressure at a particular place in the mountains is 0.80 atm, estimate the temperature at which water boils.

63. (II) What is the mass of water in a closed room 5.0 m × 6.0 m × 2.4 m when the temperature is 25°C and the relative humidity is 55%?

64. (II) A **pressure cooker** is a sealed pot designed to cook food with the steam produced by boiling water somewhat above 100°C. The pressure cooker in Fig. 13–32 uses a weight of mass m to allow steam to escape at a certain pressure through a small hole (diameter d) in the cooker's lid. If $d = 3.0$ mm, what should m be in order to cook food at 120°C? Assume that atmospheric pressure outside the cooker is 1.01×10^5 Pa.

FIGURE 13–32 Problem 64.

65. (II) If the humidity in a sealed room of volume 420 m³ at 20°C is 65%, what mass of water can still evaporate from an open pan?

66. (III) Air that is at its dew point of 5°C is drawn into a building where it is heated to 22°C. What will be the relative humidity at this temperature? Assume constant pressure of 1.0 atm. Take into account the expansion of the air.

67. (III) When using a mercury barometer (Section 10–6), the vapor pressure of mercury is usually assumed to be zero. At room temperature mercury's vapor pressure is about 0.0015 mm-Hg. At sea level, the height h of mercury in a barometer is about 760 mm. (a) If the vapor pressure of mercury is neglected, is the true atmospheric pressure greater or less than the value read from the barometer? (b) What is the percent error? (c) What is the percent error if you use a water barometer and ignore water's saturated vapor pressure at STP?

*13–13 Diffusion

*68. (II) Estimate the time needed for a glycine molecule (see Table 13–4) to diffuse a distance of 25 μm in water at 20°C if its concentration varies over that distance from 1.00 mol/m³ to 0.50 mol/m³? Compare this "speed" to its rms (thermal) speed. The molecular mass of glycine is about 75 u.

*69. (II) Oxygen diffuses from the surface of insects to the interior through tiny tubes called tracheae. An average trachea is about 2 mm long and has cross-sectional area of 2×10^{-9} m². Assuming the concentration of oxygen inside is half what it is outside in the atmosphere, (a) show that the concentration of oxygen in the air (assume 21% is oxygen) at 20°C is about 8.7 mol/m³, then (b) calculate the diffusion rate J, and (c) estimate the average time for a molecule to diffuse in. Assume the diffusion constant is 1×10^{-5} m²/s.

General Problems

70. A Pyrex measuring cup was calibrated at normal room temperature. How much error will be made in a recipe calling for 375 mL of cool water, if the water and the cup are hot, at 95°C, instead of at room temperature? Neglect the glass expansion.

71. A precise steel tape measure has been calibrated at 14°C. At 37°C, (a) will it read high or low, and (b) what will be the percentage error?

72. A cubic box of volume 6.15×10^{-2} m³ is filled with air at atmospheric pressure at 15°C. The box is closed and heated to 165°C. What is the net force on each side of the box?

73. The gauge pressure in a helium gas cylinder is initially 32 atm. After many balloons have been blown up, the gauge pressure has decreased to 5 atm. What fraction of the original gas remains in the cylinder?

74. If a scuba diver fills his lungs to full capacity of 5.5 L when 9.0 m below the surface, to what volume would his lungs expand if he quickly rose to the surface? Is this advisable?

75. A house has a volume of 1200 m³. (a) What is the total mass of air inside the house at 15°C? (b) If the temperature drops to −15°C, what mass of air enters or leaves the house?

76. Estimate the number of air molecules in a room of length 6.0 m, width 3.0 m, and height 2.5 m. Assume the temperature is 22°C. How many moles does that correspond to?

77. An iron cube floats in a bowl of liquid mercury at 0°C. (a) If the temperature is raised to 25°C, will the cube float higher or lower in the mercury? (b) By what percent will the fraction of volume submerged change? [Hint: See Chapter 10.]

78. A helium balloon, assumed to be a perfect sphere, has a radius of 24.0 cm. At room temperature (20°C), its internal pressure is 1.08 atm. Determine the number of moles of helium in the balloon, and the mass of helium needed to inflate the balloon to these values.

79. A standard cylinder of oxygen used in a hospital has gauge pressure = 2000 psi (13,800 kPa) and volume = 14 L (0.014 m³) at $T = 295$ K. How long will the cylinder last if the flow rate, measured at atmospheric pressure, is constant at 2.1 L/min?

80. A brass lid screws tightly onto a glass jar at 15°C. To help open the jar, it can be placed into a bath of hot water. After this treatment, the temperatures of the lid and the jar are both 55°C. The inside diameter of the lid is 8.0 cm. Find the size of the gap (difference in radius) that develops by this procedure.

81. The density of gasoline at 0°C is 0.68×10^3 kg/m³. (a) What is the density on a hot day, when the temperature is 33°C? (b) What is the percent change in density?

82. The first real length standard, adopted more than 200 years ago, was a platinum bar with two very fine marks separated by what was defined to be exactly one meter. If this standard bar was to be accurate to within ± 1.0 μm, how carefully would the trustees have needed to control the temperature? The coefficient of linear expansion is 9×10^{-6}/C°.

83. If a steel band were to fit snugly around the Earth's equator at 25°C, but then was heated to 55°C, how high above the Earth would the band be (assume equal everywhere)?

84. In outer space the density of matter is about one atom per cm^3, mainly hydrogen atoms, and the temperature is about 2.7 K. Calculate the rms speed of these hydrogen atoms, and the pressure (in atmospheres).

85. (a) Estimate the rms speed of an amino acid, whose molecular mass is 89 u, in a living cell at 37°C. (b) What would be the rms speed of a protein of molecular mass 85,000 u at 37°C?

86. The escape speed from the Earth is 1.12×10^4 m/s, so that a gas molecule traveling away from Earth near the outer boundary of the Earth's atmosphere would, at this speed, be able to escape from the Earth's gravitational field and be lost to the atmosphere. At what temperature is the rms speed of (a) oxygen molecules, and (b) helium atoms equal to 1.12×10^4 m/s? (c) Can you explain why our atmosphere contains oxygen but not helium?

87. Consider a container of oxygen gas at a temperature of 23°C that is 1.00 m tall. Compare the gravitational potential energy of a molecule at the top of the container (assuming the potential energy is zero at the bottom) with the average kinetic energy of the molecules. Is it reasonable to neglect the potential energy?

88. A space vehicle returning from the Moon enters the Earth's atmosphere at a speed of about 42,000 km/h. Molecules (assume nitrogen) striking the nose of the vehicle with this speed correspond to what temperature? (Because of this high temperature, the nose of a space vehicle must be made of special materials; indeed, part of it does vaporize, and this is seen as a bright blaze upon reentry.)

89. A sauna has 8.5 m^3 of air volume, and the temperature is 85°C. The air is perfectly dry. How much water (in kg) should be evaporated if we want to increase the relative humidity from 0% to 10%? (See Table 13–3.)

90. A 0.50-kg trash-can lid is suspended against gravity by tennis balls thrown vertically upward at it. How many tennis balls per second must rebound from the lid elastically, assuming they have a mass of 0.060 kg and are thrown at 15 m/s?

91. In humid climates, people constantly **dehumidify** their cellars to prevent rot and mildew. If the cellar in a house (kept at 20°C) has 105 m^2 of floor space and a ceiling height of 2.4 m, what is the mass of water that must be removed from it in order to drop the humidity from 95% to a more reasonable 40%?

Search and Learn

1. This Chapter gives two ways to calculate the thermal expansion of a gas at a constant pressure of 1.0 atm. Use both methods to calculate the volume change of 1000 L of an ideal gas as it goes from −100°C to 0°C and from 0°C to 100°C. Why are the answers different?

2. A scuba tank when fully charged has a pressure of 180 atm at 18°C. The volume of the tank is 11.3 L. (a) What would the volume of the air be at 1.00 atm and at the same temperature? (b) Before entering the water, a person consumes 2.0 L of air in each breath, and breathes 12 times a minute. At this rate, how long would the tank last? (c) At a depth of 23.0 m in sea water at a temperature of 10°C, how long would the same tank last assuming the breathing rate does not change?

3. A 28.4-kg solid aluminum cylindrical wheel of radius 0.41 m is rotating about its axle in frictionless bearings with angular velocity $\omega = 32.8$ rad/s. If its temperature is then raised from 15.0°C to 95.0°C, what is the fractional change in ω?

4. A hot-air balloon achieves its buoyant lift by heating the air inside the balloon, which makes it less dense than the air outside. Suppose the volume of a balloon is 1800 m^3 and the required lift is 3300 N (rough estimate of the weight of the equipment and passenger). Calculate the temperature of the air inside the balloon which will produce the required lift. Assume the outside air is an ideal gas at 0°C. What factors limit the maximum altitude attainable by this method for a given load? [*Hint*: See Chapter 10.]

5. Estimate how many molecules of air are in each 2.0-L breath you inhale that were also in the last breath Galileo took. Assume the atmosphere is about 10 km high and of constant density. What other assumptions did you make?

6. (a) The second postulate of kinetic theory is that the molecules are, on average, far apart from one another. That is, their average separation is much greater than the diameter of each molecule. Is this assumption reasonable? To check, calculate the average distance between molecules of a gas at STP, and compare it to the diameter of a typical gas molecule, about 0.3 nm. (b) If the molecules were the diameter of ping-pong balls, say 4 cm, how far away would the next ping-pong ball be on average? (c) Repeat part *a*, but now assume the gas has been compressed so that the pressure is now 3 atm but still at 273 K. (d) Estimate what % of the total volume of gas is taken up by molecules themselves in parts *a* and *c*. [Note that the volume of the molecules themselves can become a significant part of the total volume at lower temperatures and higher pressures. Hence the actual volume the molecules have to bounce around in is less than the total volume. This contributes to the effect shown in Fig. 13–21 at high pressures where real gases (solid red lines) deviate from ideal gas behavior (dashed lines A' and B').]

ANSWERS TO EXERCISES

A: (i) Higher, (ii) same, (iii) lower.
B: (d).
C: (b).
D: (b) Less.
E: (a).
F: (d).

When it is cold, warm clothes act as insulators to reduce heat loss from the body to the environment by conduction and convection. Heat radiation from a campfire can warm you and your clothes. The fire can also transfer energy directly by heat convection and conduction to what you are cooking. Heat, like work, represents a transfer of energy. Heat is defined as a transfer of energy due to a difference of temperature. Internal energy U is the sum total of all the energy of all the molecules of the system.

14

Heat

CONTENTS

14–1 Heat as Energy Transfer
14–2 Internal Energy
14–3 Specific Heat
14–4 Calorimetry—Solving Problems
14–5 Latent Heat
14–6 Heat Transfer: Conduction
14–7 Heat Transfer: Convection
14–8 Heat Transfer: Radiation

CHAPTER-OPENING QUESTION—Guess now!

A 5-kg cube of warm iron (60°C) is put in thermal contact with a 10-kg cube of cold iron (15°C). Which statement is valid?

(a) Heat flows spontaneously from the warm cube to the cold cube until both cubes have the same heat content.
(b) Heat flows spontaneously from the warm cube to the cold cube until both cubes have the same temperature.
(c) Heat can flow spontaneously from the warm cube to the cold cube, but can also flow spontaneously from the cold cube to the warm cube.
(d) Heat flows from the larger cube to the smaller one because the larger one has more internal energy.

When a pot of cold water is placed on a hot burner of a stove, the temperature of the water increases. We say that heat "flows" from the hot burner to the cold water. When two objects at different temperatures are put in contact, heat spontaneously flows from the hotter one to the colder one. The spontaneous flow of heat is in the direction tending to equalize the temperature. If the two objects are kept in contact long enough for their temperatures to become equal, the objects are said to be in thermal equilibrium, and there is no further heat flow between them. For example, when a fever thermometer is first placed in your mouth, heat flows from your mouth to the thermometer. When the thermometer reaches the same temperature as the inside of your mouth, the thermometer and your mouth are then in equilibrium, and no more heat flows.

Heat and temperature are often confused. They are very different concepts, and in this Chapter we will make a clear distinction between them. We begin by defining and using the concept of heat. We also discuss how heat is used in calorimetry, how it is involved in changes of state of matter, and the processes of heat transfer—conduction, convection, and radiation.

14–1 Heat as Energy Transfer

We use the term "heat" in everyday life as if we knew what we meant. But the term is often used inconsistently, so it is important for us to define heat clearly, and to clarify the phenomena and concepts related to heat.

We commonly speak of the flow of heat—heat flows from a stove burner to a pot of soup, from the Sun to the Earth, from a person's mouth into a fever thermometer. Heat flows spontaneously from an object at higher temperature to one at lower temperature. Indeed, an eighteenth-century model of heat pictured heat flow as movement of a fluid substance called *caloric*. However, the caloric fluid could never be detected. In the nineteenth century, it was found that the various phenomena associated with heat could be described consistently using a new model that views heat as being akin to work, as we will discuss in a moment. First we note that a common unit for heat, still in use today, is named after caloric. It is called the **calorie** (cal) and is defined as *the amount of heat necessary to raise the temperature of 1 gram of water by 1 Celsius degree*. [To be precise, the particular temperature range from 14.5°C to 15.5°C is specified because the heat required is very slightly different at different temperatures. The difference is less than 1% over the range 0 to 100°C, and we will ignore it for most purposes.] More often used than the calorie is the **kilocalorie** (kcal), which is 1000 calories. Thus *1 kcal is the heat needed to raise the temperature of 1 kg of water by 1 C°*. Often a kilocalorie is called a **Calorie** (with a capital C), and this Calorie (or the kJ) is used to specify the energy value of food. In the British system of units, heat is measured in British thermal units (Btu). One **Btu** is defined as the heat needed to raise the temperature of 1 lb of water by 1 F°. It can be shown (Problem 5) that 1 Btu = 0.252 kcal = 1056 J. Also, one **therm** is 10^5 Btu.

CAUTION
Heat is not a fluid

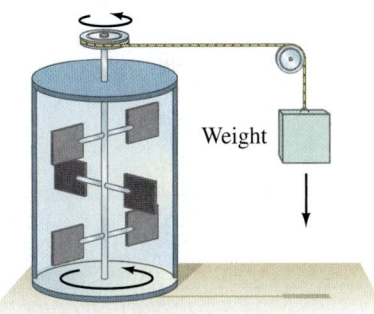

FIGURE 14–1 Joule's experiment on the mechanical equivalent of heat. [The work is transformed into internal energy (Section 14–2).]

The idea that heat is related to energy transfer was pursued by a number of scientists in the 1800s, particularly by an English brewer, James Prescott Joule (1818–1889). Joule and others performed a number of experiments that were crucial for establishing our present-day view that heat, like work, represents a transfer of energy. One of Joule's experiments is shown (simplified) in Fig. 14–1. The falling weight causes the paddle wheel to turn. The friction between the water and the paddle wheel causes the temperature of the water to rise slightly (barely measurable, in fact, by Joule). The same temperature rise could also be obtained by heating the water on a hot stove. In this and many other experiments (some involving electrical energy), Joule determined that a given amount of work done was always equivalent to a particular amount of heat input. Quantitatively, 4.186 joules (J) of work was found to be equivalent to 1 calorie (cal) of heat. This is known as the **mechanical equivalent of heat**:

$$4.186 \text{ J} = 1 \text{ cal};$$

$$4.186 \text{ kJ} = 1 \text{ kcal}.$$

As a result of these and other experiments, scientists came to interpret heat not as a substance, and not exactly as a form of energy. Rather, heat refers to a *transfer of energy*: when heat flows from a hot object to a cooler one, it is energy that is being transferred from the hot to the cold object. Thus, **heat** is *energy transferred from one object to another because of a difference in temperature*. In SI units, the unit for heat, as for any form of energy, is the joule. Nonetheless, calories and kcal are still sometimes used. Today the calorie is *defined* in terms of the joule (via the mechanical equivalent of heat, above), rather than in terms of the properties of water, as given previously. The latter is still handy to remember: 1 cal raises 1 g of water by 1 C°, or 1 kcal raises 1 kg of water by 1 C°.

CAUTION
Heat is energy transferred because of a ΔT

PHYSICS APPLIED
Working off Calories

EXAMPLE 14–1 ESTIMATE **Working off the extra Calories.** Suppose you throw caution to the wind and eat 500 Calories of ice cream and cake. To compensate, you want to do an equivalent amount of work climbing stairs or a mountain. How much total height must you climb?

APPROACH The work W you need to do in climbing stairs equals the change in gravitational potential energy: $W = \Delta PE = mgh$, where h is the vertical height climbed. For this estimate, let us approximate your mass as $m \approx 60$ kg.

SOLUTION 500 food Calories is 500 kcal, which in joules is

$$(500 \text{ kcal})(4.186 \times 10^3 \text{ J/kcal}) = 2.1 \times 10^6 \text{ J}.$$

The work done to climb a vertical height h is $W = mgh$. We solve for h:

$$h = \frac{W}{mg} = \frac{2.1 \times 10^6 \text{ J}}{(60 \text{ kg})(9.80 \text{ m/s}^2)} = 3600 \text{ m}.$$

This is a huge elevation change (over 11,000 ft).

NOTE The human body does not transform food energy with 100% efficiency—it is more like 20% efficient. As we shall discuss in the next Chapter, some energy is always "wasted," so you would actually have to climb only about $(0.2)(3600 \text{ m}) \approx 700$ m, which is more reasonable (about 2300 ft of elevation gain).

14–2 Internal Energy

The sum total of all the energy of all the molecules in an object is called its **internal energy**. (Sometimes **thermal energy** is used to mean the same thing.) We introduce the concept of internal energy now since it will help clarify ideas about heat.

Distinguishing Temperature, Heat, and Internal Energy

CAUTION
Distinguish heat from internal energy and from temperature

Using the kinetic theory, we can make a clear distinction between temperature, heat, and internal energy. Temperature (in kelvins) is a measure of the *average* kinetic energy of individual molecules (Eq. 13–8). Internal energy refers to the *total* energy of all the molecules within the object. (Thus two equal-mass hot ingots of iron may have the same temperature, but two of them have twice as much internal energy as one does.) Heat, finally, refers to a *transfer* of energy from one object to another because of a difference in temperature.

CAUTION
Direction of heat flow depends on temperatures (not on amount of internal energy)

Notice that the direction of heat flow between two objects depends on their temperatures, not on how much internal energy each has. Thus, if 50 g of water at 30°C is placed in contact (or mixed) with 200 g of water at 25°C, heat flows *from* the water at 30°C *to* the water at 25°C even though the internal energy of the 25°C water is much greater because there is so much more of it.

Internal Energy of an Ideal Gas

Let us calculate the internal energy of n moles of an ideal monatomic (one atom per molecule) gas. The internal energy, U, is the sum of the translational kinetic energies of all the atoms. This sum is equal to the average kinetic energy per molecule times the total number of molecules, N:

$$U = N(\tfrac{1}{2}m\overline{v^2}).$$

Using Eq. 13–8, $\overline{KE} = \tfrac{1}{2}m\overline{v^2} = \tfrac{3}{2}kT$, we can write this as

$$U = \tfrac{3}{2}NkT$$

or (recall Section 13–8)

$$U = \tfrac{3}{2}nRT, \qquad \begin{bmatrix}\text{internal energy of}\\ \text{ideal monatomic gas}\end{bmatrix} \quad (14\text{–}1)$$

where n is the number of moles. Thus, the internal energy of an ideal gas depends only on temperature and the number of moles (or molecules) of gas.

If the gas molecules contain more than one atom, then the rotational and vibrational energy of the molecules (Fig. 14–2) must also be taken into account. The internal energy will be greater at a given temperature than for a monatomic gas, but it will still be a function only of temperature for an ideal gas.

The internal energy of real gases also depends mainly on temperature, but where real gases deviate from ideal gas behavior, their internal energy depends also somewhat on pressure and volume (due to atomic potential energy).

The internal energy of liquids and solids is quite complicated, for it includes electrical potential energy associated with the forces (or "chemical" bonds) between atoms and molecules.

14–3 Specific Heat

If heat flows into an object, the object's temperature rises (assuming no phase change). But how much does the temperature rise? That depends. As early as the eighteenth century, experimenters had recognized that the amount of heat Q required to change the temperature of a given material is proportional to the mass m of the material present and to the temperature change ΔT. This remarkable simplicity in nature can be expressed in the equation

$$Q = mc\,\Delta T, \qquad (14\text{–}2)$$

where c is a quantity characteristic of the material called its **specific heat**. Because $c = Q/(m\,\Delta T)$, specific heat is specified in units[†] of J/kg·C° (the proper SI unit) or kcal/kg·C°. For water at 15°C and a constant pressure of 1 atm, $c = 4.186 \times 10^3$ J/kg·C° or 1.00 kcal/kg·C°, since, by definition of the cal and the joule, it takes 1 kcal of heat to raise the temperature of 1 kg of water by 1 C°. Table 14–1 gives the values of specific heat for other substances at 20°C. The values of c depend to some extent on temperature (as well as slightly on pressure), but for temperature changes that are not too great, c can often be considered constant.

EXAMPLE 14–2 How heat transferred depends on specific heat. (a) How much heat input is needed to raise the temperature of an empty 20-kg vat made of iron from 10°C to 90°C? (b) What if the vat is filled with 20 kg of water?

APPROACH We apply Eq. 14–2 to the different materials involved.

SOLUTION (a) Our system is the iron vat alone. From Table 14–1, the specific heat of iron is 450 J/kg·C°. The change in temperature is (90°C − 10°C) = 80 C°. Thus,

$$Q_{\text{vat}} = mc\,\Delta T = (20\text{ kg})(450\text{ J/kg·C°})(80\text{ C°}) = 7.2 \times 10^5\text{ J} = 720\text{ kJ}.$$

(b) Our system is the vat plus the water. The water alone would require

$$Q_{\text{w}} = mc\,\Delta T = (20\text{ kg})(4186\text{ J/kg·C°})(80\text{ C°}) = 6.7 \times 10^6\text{ J} = 6700\text{ kJ},$$

or almost 10 times what an equal mass of iron requires. The total, for the vat plus the water, is 720 kJ + 6700 kJ = 7400 kJ.

NOTE In (b), the iron vat and the water underwent the same temperature change, $\Delta T = 80$ C°, but their specific heats are different.

If the iron vat in part (a) of Example 14–2 had been *cooled* from 90°C to 10°C, 720 kJ of heat would have flowed *out* of the iron. In other words, Eq. 14–2 is valid for heat flow either in or out, with a corresponding increase or decrease in temperature.

We saw in part (b) of Example 14–2 that water requires almost 10 times as much heat as an equal mass of iron to make the same temperature change. Water has one of the highest specific heats of all substances, which makes it an ideal substance for hot-water space-heating systems and other uses that require a minimal drop in temperature for a given amount of heat transfer. It is the water content, too, that causes the apples rather than the crust in hot apple pie to burn our tongues, through heat transfer.

[†]Note that J/kg·C° means $\dfrac{\text{J}}{\text{kg·C°}}$ and *not* (J/kg)·C° = J·C°/kg (otherwise we would have written it that way).

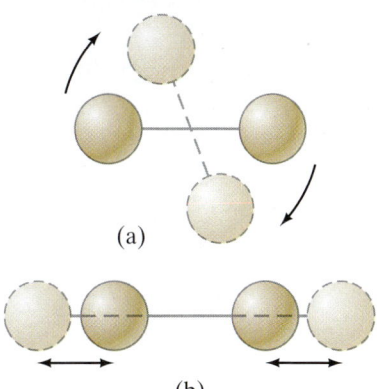

FIGURE 14–2 Besides translational kinetic energy, molecules can have (a) rotational kinetic energy, and (b) vibrational energy (both kinetic and potential).

TABLE 14–1 Specific Heats
(at 1 atm constant pressure and 20°C unless otherwise stated)

	Specific Heat, c	
Substance	J/kg·C°	kcal/kg·C° (= cal/g·C°)
Aluminum	900	0.22
Alcohol (ethyl)	2400	0.58
Copper	390	0.093
Glass	840	0.20
Iron or steel	450	0.11
Lead	130	0.031
Marble	860	0.21
Mercury	140	0.033
Silver	230	0.056
Wood	1700	0.4
Water		
Ice (−5°C)	2100	0.50
Liquid (15°C)	4186	1.00
Steam (110°C)	2010	0.48
Human body (average)	3470	0.83
Protein	1700	0.4

PHYSICS APPLIED
Practical effects of water's high specific heat

CONCEPTUAL EXAMPLE 14–3 **A very hot frying pan.** You accidentally let an empty iron frying pan get very hot on the stove (200°C or even more). What happens when you dunk it into a few inches of cool water in the bottom of the sink? Will the final temperature be midway between the initial temperatures of the water and pan? Will the water start boiling? Assume the mass of water is roughly the same as the mass of the frying pan.

RESPONSE Experience may tell you that the water warms up—perhaps by as much as 10 or 20 degrees. The water doesn't come close to boiling. The water's temperature increase is a lot less than the frying pan's temperature decrease. Why? Because the mass of water is roughly equal to that of the pan, and iron has a specific heat nearly 10 times smaller than that of water (Table 14–1). As heat leaves the frying pan and enters the water, the iron pan's temperature change will be about 10 times greater than that of the water. If, instead, you let a few drops of water fall onto the hot pan, that very small mass of water will sizzle and boil away (the pan's mass may be hundreds of times larger than that of the water drops).

EXERCISE A Return to the Chapter-Opening Question, page 390, and answer it again now. Try to explain why you may have answered differently the first time.

*Specific Heats for Gases

Specific heats for gases are more complicated than for solids and liquids, which change in volume only slightly with a change in temperature. Gases change strongly in volume with a change in temperature at constant pressure, as we saw in Chapter 13 with the gas laws; or, if kept at constant volume, the pressure in a gas changes strongly with temperature. The specific heat of a gas depends very much on how the process of changing its temperature is carried out. Most commonly, we deal with the specific heats of gases kept (a) at constant pressure (c_P) or (b) at constant volume (c_V). Some values are given in Table 14–2, where we see that c_P is always greater than c_V. For liquids and solids, this distinction is usually negligible. More details are given in Appendix D on molecular specific heats and the equipartition of energy.

TABLE 14–2 Specific Heats of Gases (kcal/kg·C°)

Gas	c_P (constant pressure)	c_V (constant volume)
Steam (100°C)	0.482	0.350
Oxygen	0.218	0.155
Helium	1.15	0.75
Carbon dioxide	0.199	0.153
Nitrogen	0.248	0.177

14–4 Calorimetry—Solving Problems

In discussing heat and thermodynamics, we often consider a particular **system**, which is any object or set of objects we choose to consider. Everything else in the universe is its "environment" (or the "surroundings"). There are several categories of systems. A **closed system** is one for which no mass enters or leaves (but energy may be exchanged with the environment). In an **open system**, mass may enter or leave (as may energy). Many (idealized) systems we study in physics are closed systems. But many systems, including plants and animals, are open systems since they exchange materials (food, oxygen, waste products) with the environment. A closed system is said to be **isolated** if no energy in any form (as well as no mass) passes across its boundaries; otherwise it is not isolated.

A perfectly isolated system is an ideal, but we often try to set up a system that can be closely approximated as an isolated system (preferably one we can deal with fairly easily). When different parts of an isolated system are at different temperatures, heat will flow (energy is transferred) from the part at higher temperature to the part at lower temperature—that is, within the system—until **thermal equilibrium** is reached, meaning the entire system is at the same temperature. For an isolated system, no energy is transferred into or out of it. So we can apply *conservation of energy* to such an isolated system. A simple intuitive way to set up a conservation of energy equation is to write that the heat lost by one part of the system is equal to the heat gained by the other part:

$$\text{heat lost} = \text{heat gained} \quad \text{[isolated system]}$$

or

$$\text{energy } out \text{ of one part} = \text{energy } into \text{ another part.} \quad \text{[isolated system]}$$

EXAMPLE 14-4 **The cup cools the tea.** If 200 cm³ of tea at 95°C is poured into a 150-g glass cup initially at 25°C (Fig. 14-3), what will be the common final temperature T of the tea and cup when equilibrium is reached, assuming no heat flows to the surroundings?

APPROACH We apply conservation of energy to our system of tea plus cup, which we are assuming is isolated: all of the heat that leaves the tea flows into the cup. We can use the specific heat equation, Eq. 14-2, to determine how the heat flow is related to the temperature changes.

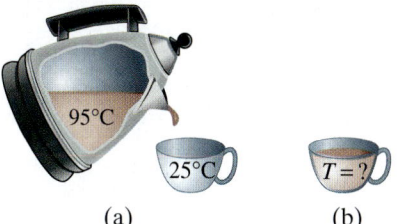

FIGURE 14-3 Example 14-4.

SOLUTION Because tea is mainly water, we can take its specific heat as 4186 J/kg·C° (Table 14-1), and its mass m is its density times its volume ($V = 200\,\text{cm}^3 = 200 \times 10^{-6}\,\text{m}^3$): $m = \rho V = (1.0 \times 10^3\,\text{kg/m}^3)(200 \times 10^{-6}\,\text{m}^3) = 0.20\,\text{kg}$. We use Eq. 14-2, apply conservation of energy, and let T be the as yet unknown final temperature:

$$\text{heat lost by tea} = \text{heat gained by cup}$$
$$m_{\text{tea}} c_{\text{tea}} (95°C - T) = m_{\text{cup}} c_{\text{cup}} (T - 25°C).$$

Putting in numbers and using Table 14-1 ($c_{\text{cup}} = 840\,\text{J/kg·C°}$ for glass), we solve for T, and find

$$(0.20\,\text{kg})(4186\,\text{J/kg·C°})(95°C - T) = (0.15\,\text{kg})(840\,\text{J/kg·C°})(T - 25°C)$$
$$79{,}500\,\text{J} - (837\,\text{J/C°})T = (126\,\text{J/C°})T - 3150\,\text{J}$$
$$T = 86°C.$$

The tea drops in temperature by 9 C° by coming into equilibrium with the cup.

NOTE The cup increases in temperature by 86°C − 25°C = 61 C°. Its much greater change in temperature (compared with that of the tea water) is due to its much smaller specific heat compared to that of water.

NOTE In this calculation, the ΔT (of Eq. 14-2, $Q = mc\,\Delta T$) is a positive quantity on both sides of our conservation of energy equation. On the left is "heat lost" and ΔT is the initial minus the final temperature ($95°C - T$), whereas on the right is "heat gained" and ΔT is the final minus the initial temperature.

⚠ **CAUTION**
When using heat lost = heat gained, ΔT is positive on both sides

Another, perhaps more general, way to set up the energy conservation equation for heat transfer within an isolated system is to write that the sum of all internal heat transfers within the system adds up to zero:

$$\Sigma Q = 0. \qquad [\text{isolated system}] \quad \textbf{(14-3)}$$

Each Q represents the heat entering or leaving one part of the system. Each term is written as $Q = mc(T_f - T_i)$, and $\Delta T = T_f - T_i$ is always the final minus the initial temperature. ΔT can be either positive or negative, depending on whether heat flows into or out of that part. Let us redo Example 14-4 using $\Sigma Q = 0$.

✏ **PROBLEM SOLVING**
Alternate approach: $\Sigma Q = 0$

EXAMPLE 14-4′ **Alternate Solution, $\Sigma Q = 0$.**

APPROACH We use Eq. 14-3, $\Sigma Q = 0$.

SOLUTION For each term, $\Delta T = T_f - T_i$. If $T_f < T_i$, then $\Delta T < 0$. Equation 14-3, $\Sigma Q = 0$, becomes

$$m_{\text{cup}} c_{\text{cup}} (T - 25°C) + m_{\text{tea}} c_{\text{tea}} (T - 95°C) = 0.$$

The second term is negative because T will be less than 95°C. Solving the algebra gives the same result, $T = 86°C$.

You are free to use either approach. They are entirely equivalent, algebraically. For example, if you move the first term in the displayed equation of the alternate Example 14-4 over to the other side of the equals sign, you obtain the "heat lost = heat gained" equation in the first version of Example 14-4.

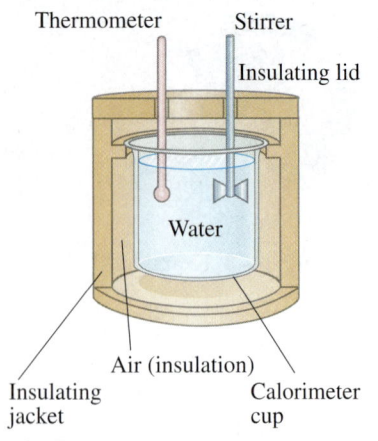

FIGURE 14–4 Simple water calorimeter.

The exchange of energy, as exemplified in Example 14–4, is the basis for a technique known as **calorimetry**, which is the quantitative measurement of heat exchange. To make such measurements, a **calorimeter** is used; a simple water calorimeter is shown in Fig. 14–4. It is very important that the calorimeter be well insulated so that almost no heat is exchanged with the surroundings. One important use of the calorimeter is in the determination of specific heats of substances. In the technique known as the "method of mixtures," a sample of a substance is heated to a high temperature, which is accurately measured, and then quickly placed in the cool water of the calorimeter. The heat lost by the sample will be gained by the water and the calorimeter cup. By measuring the final temperature of the mixture, the specific heat can be calculated, as illustrated in the following Example.

EXAMPLE 14–5 Unknown specific heat determined by calorimetry. An engineer wishes to determine the specific heat of a new metal alloy. A 0.150-kg sample of the alloy is heated to 540°C. It is then quickly placed in 0.400 kg of water at 10.0°C, which is contained in a 0.200-kg aluminum calorimeter cup. The final temperature of the system is 30.5°C. Calculate the specific heat of the alloy.

APPROACH We apply conservation of energy to our system, which we take to be the alloy sample, the water, and the calorimeter cup. We assume this system is isolated, and apply Eq. 14–3, $\Sigma Q = 0$.

SOLUTION Each term is of the form $Q = mc(T_f - T_i)$. Thus $\Sigma Q = 0$ gives

$$m_a c_a \Delta T_a + m_w c_w \Delta T_w + m_{cal} c_{cal} \Delta T_{cal} = 0$$

where the subscripts a, w, and cal refer to the alloy, water, and calorimeter, respectively, and each ΔT is the final temperature (30.5°C) minus the initial temperature for each object. When we put in values and use Table 14–1, this equation becomes

$$(0.150 \text{ kg})(c_a)(30.5°C - 540°C) + (0.400 \text{ kg})(4186 \text{ J/kg} \cdot C°)(30.5°C - 10.0°C)$$
$$+ (0.200 \text{ kg})(900 \text{ J/kg} \cdot C°)(30.5°C - 10.0°C) = 0,$$

or

$$-(76.4 \text{ kg} \cdot C°)c_a + 34{,}300 \text{ J} + 3690 \text{ J} = 0.$$

Solving for c_a we obtain:

$$c_a = 497 \text{ J/kg} \cdot C° \approx 500 \text{ J/kg} \cdot C°.$$

NOTE We rounded off because we have ignored any heat transferred to the thermometer and the stirrer (which is used to quicken the heat transfer process and thus reduce heat loss to the outside). To take them into account we would have to add (small) additional terms to the equation.

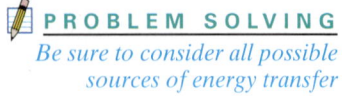

Be sure to consider all possible sources of energy transfer

In all Examples and Problems of this sort, be sure to include *all* objects that gain or lose heat (within reason). For simplicity, we have ignored very small masses, such as the thermometer and the stirrer, which will affect the energy balance only very slightly.

Bomb Calorimeter

Measuring Calorie content

A **bomb calorimeter** is used to measure the thermal energy released when a substance burns (including foods) to determine their Calorie content. A carefully weighed sample of the substance, with an excess amount of oxygen, is placed in a sealed container (the "bomb"). The bomb is placed in the water of the calorimeter and a fine wire passing into the bomb is then heated to ignite the mixture. The Calorie content of foods determined in this way can be unreliable because our bodies may not metabolize all the available energy (which would be excreted). Careful measurements and calculations need to take this into account.

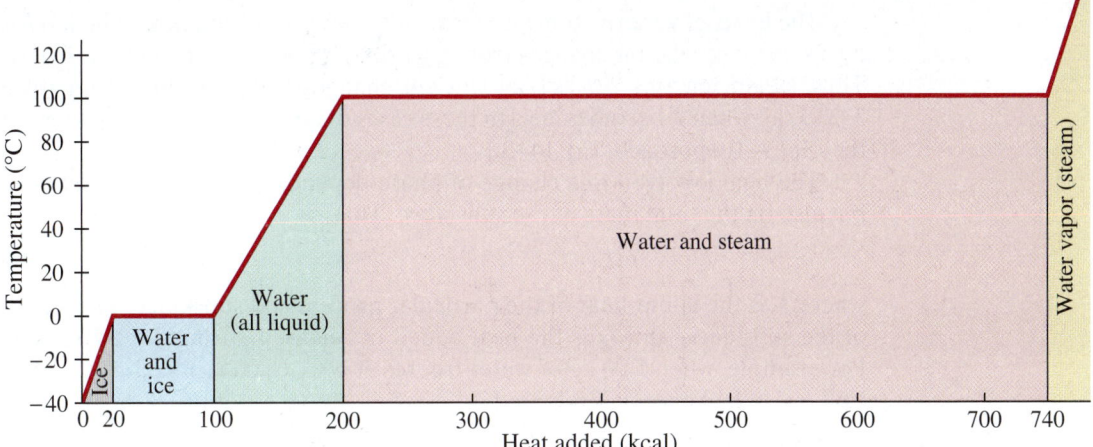

FIGURE 14–5 Temperature as a function of the heat added to bring 1.0 kg of ice at −40°C to steam above 100°C.

14–5 Latent Heat

When a material changes phase from solid to liquid, or from liquid to gas (see also Section 13–11), a certain amount of energy is involved in this **change of phase**. For example, let us trace what happens when a 1.0-kg block of ice at −40°C is heated at a slow steady rate until all the ice has changed to water, then the (liquid) water is heated to 100°C and changed to steam, and heated further above 100°C, all at 1 atm pressure. As shown in the graph of Fig. 14–5, as the ice is heated starting at −40°C, its temperature rises at a rate of about 2 C°/kcal of heat added (since for ice, $c \approx 0.50 \text{ kcal/kg} \cdot \text{C}°$). However, when 0°C is reached, the temperature stops increasing even though heat is still being added. The ice gradually changes to water in the liquid state, with no change in temperature. After about 40 kcal has been added at 0°C, half the ice remains and half has changed to water. After about 80 kcal, or 330 kJ, has been added, all the ice has changed to water, still at 0°C. Continued addition of heat causes the water's temperature to again increase, now at a rate of 1 C°/kcal. When 100°C is reached, the temperature again remains constant as the heat added changes the liquid water to vapor (steam). About 540 kcal (2260 kJ) is required to change the 1.0 kg of water completely to steam, after which the graph rises again, indicating that the temperature of the steam rises as heat is added.

The heat required to change 1.0 kg of a substance from the solid to the liquid state is called the **heat of fusion**; it is denoted by L_F. The heat of fusion of water is 79.7 kcal/kg or, in proper SI units, 333 kJ/kg ($= 3.33 \times 10^5$ J/kg). The heat required to change a substance from the liquid to the vapor phase is called the **heat of vaporization**, L_V. For water it is 539 kcal/kg or 2260 kJ/kg. Other substances follow graphs similar to Fig. 14–5, although the melting-point and boiling-point temperatures are different, as are the specific heats and heats of fusion and vaporization. Values for the heats of fusion and vaporization, which are also called the **latent heats**, are given in Table 14–3 for a number of substances.

TABLE 14–3 Latent Heats (at 1 atm)

Substance	Melting Point (°C)	Heat of Fusion kJ/kg	Heat of Fusion kcal/kg[†]	Boiling Point (°C)	Heat of Vaporization kJ/kg	Heat of Vaporization kcal/kg[†]
Oxygen	−218.8	14	3.3	−183	210	51
Nitrogen	−210.0	26	6.1	−195.8	200	48
Ethyl alcohol	−114	104	25	78	850	204
Ammonia	−77.8	33	8.0	−33.4	137	33
Water	0	333	79.7	100	2260	539
Lead	327	25	5.9	1750	870	208
Silver	961	88	21	2193	2300	558
Iron	1538	289	69.1	3023	6340	1520
Tungsten	3410	184	44	5900	4800	1150

[†]Numerical values in kcal/kg are the same in cal/g.

The heats of vaporization and fusion also refer to the amount of heat *released* by a substance when it changes from a gas to a liquid, or from a liquid to a solid. Thus, steam releases 2260 kJ/kg when it changes to water, and water releases 333 kJ/kg when it becomes ice. [In these cases of heat release, $Q < 0$ when using the $\Sigma Q = 0$ approach, Eq. 14–3.]

The heat involved in a change of phase depends not only on the latent heat but also on the total mass of the substance. That is,

$$Q = mL, \qquad (14\text{–}4)$$

where L is the latent heat of the particular process and substance, m is the mass of the substance, and Q is the heat added or released during the phase change. For example, when 5.00 kg of water freezes at 0°C, $(5.00 \text{ kg})(3.33 \times 10^5 \text{ J/kg}) = 1.67 \times 10^6 \text{ J}$ of energy is released.

EXERCISE B A pot of water is boiling on a gas stove, and then you turn up the heat. What happens? (*a*) The temperature of the water starts increasing. (*b*) There is a tiny decrease in the rate of water loss by evaporation. (*c*) The rate of water loss by evaporation increases. (*d*) There is an appreciable increase in both the rate of boiling and the temperature of the water. (*e*) None of these.

Calorimetry sometimes involves a change of state, as the following Examples show. Indeed, latent heats are often measured using calorimetry.

EXAMPLE 14–6 Making ice. How much energy does a freezer have to remove from 1.5 kg of water at 20°C to make ice at −12°C?

APPROACH We need to calculate the total energy removed by adding the heat outflow (1) to reduce the water temperature from 20°C to 0°C, (2) to change the liquid water to solid ice at 0°C, and (3) to lower the ice temperature from 0°C to −12°C.

SOLUTION The heat Q that needs to be removed from the 1.5 kg of water is

$$\begin{aligned} Q &= mc_w(20°\text{C} - 0°\text{C}) + mL_F + mc_{ice}[0° - (-12°\text{C})] \\ &= (1.5 \text{ kg})(4186 \text{ J/kg} \cdot \text{C}°)(20 \text{ C}°) + (1.5 \text{ kg})(3.33 \times 10^5 \text{ J/kg}) \\ &\quad + (1.5 \text{ kg})(2100 \text{ J/kg} \cdot \text{C}°)(12 \text{ C}°) \\ &= 6.6 \times 10^5 \text{ J} = 660 \text{ kJ}. \end{aligned}$$

EXERCISE C Which process in Example 14–6 required the greatest heat loss?

EXAMPLE 14–7 ESTIMATE Will all the ice melt? At a reception, a 0.50-kg chunk of ice at −10°C is placed in 3.0 kg of "iced" tea at 20°C. At what temperature and in what phase will the final mixture be? The tea can be considered as water. Ignore any heat flow to the surroundings, including the container.

First determine (or estimate) the final state

APPROACH Before we can write down an equation applying conservation of energy, we must first check to see if the final state will be all ice, a mixture of ice and water at 0°C, or all water. To bring the 3.0 kg of water at 20°C down to 0°C would require an energy release of

$$m_w c_w (20°\text{C} - 0°\text{C}) = (3.0 \text{ kg})(4186 \text{ J/kg} \cdot \text{C}°)(20 \text{ C}°) = 250{,}000 \text{ J}.$$

On the other hand, to raise the ice from −10°C to 0°C would require

$$\begin{aligned} m_{ice} c_{ice}[0°\text{C} - (-10°\text{C})] &= (0.50 \text{ kg})(2100 \text{ J/kg} \cdot \text{C}°)(10 \text{ C}°) \\ &= 10{,}500 \text{ J}, \end{aligned}$$

and to change the ice to water at 0°C would require

$$m_{ice} L_F = (0.50 \text{ kg})(333 \text{ kJ/kg}) = 167{,}000 \text{ J}.$$

The sum of the last two quantities is 10.5 kJ + 167 kJ = 177 kJ. This is not enough energy to bring the 3.0 kg of water at 20°C down to 0°C, so we see that the mixture must end up all water, somewhere between 0°C and 20°C.

SOLUTION To determine the final temperature T, we apply conservation of energy. We present both of the techniques discussed in Section 14–4.

Method 1: "$\Sigma Q = 0$" gives

$$\begin{pmatrix}\text{heat to raise}\\ \text{0.50 kg of ice}\\ \text{from }-10°C\\ \text{to }0°C\end{pmatrix} + \begin{pmatrix}\text{heat to change}\\ \text{0.50 kg}\\ \text{of ice}\\ \text{to water}\end{pmatrix} + \begin{pmatrix}\text{heat to raise}\\ \text{0.50 kg of water}\\ \text{from }0°C\\ \text{to }T\end{pmatrix} + \begin{pmatrix}\text{heat lost by}\\ \text{3.0 kg of}\\ \text{water cooling}\\ \text{from }20°C\text{ to }T\end{pmatrix} = 0.$$

PROBLEM SOLVING
Then determine the final temperature

Using some of the results from the "Approach" above, we obtain

$10{,}500\text{ J} + 167{,}000\text{ J} + (0.50\text{ kg})(4186\text{ J/kg}\cdot\text{C}°)(T - 0°C)$
$\qquad + (3.0\text{ kg})(4186\text{ J/kg}\cdot\text{C}°)(T - 20°C) = 0.$

Solving for T we obtain

$$T = 5.0°C.$$

Method 2: "heat gained = heat lost" produces a word equation like the one above (the last plus sign becomes an equals sign and we lose the "= 0"). That is,

$10{,}500\text{ J} + 167{,}000\text{ J} + (0.50\text{ kg})(4186\text{ J/kg}\cdot\text{C}°)(T - 0°C)$
$\qquad = (3.0\text{ kg})(4186\text{ J/kg}\cdot\text{C}°)(20°C - T).$

The term on the right is heat lost by 3.0 kg of water cooling from 20°C to T; here $\Delta T = T_i - T_f = 20°C - T$ for this approach. Algebraically, this equation is identical to the one above in the first method.[†]

EXERCISE D How much more ice at $-10°C$ would be needed in Example 14–7 to bring the tea down to 0°C, while just melting all the ice?

Calorimetry

1. Be sure you have sufficient information to apply energy conservation. Ask yourself: **is the system isolated** (or nearly so, enough to get a good estimate)? Do we know or can we calculate all significant sources of energy transfer?

2. Apply **conservation of energy**. Either write
$$\text{heat gained} = \text{heat lost},$$
or use
$$\Sigma Q = 0.$$

3. If **no phase changes** occur, each term in the energy conservation equation will have the form
$$Q = mc\,\Delta T.$$

4. If **phase changes** do or might occur, there may be terms in the energy conservation equation of the form $Q = mL$, where L is the latent heat. But *before* applying energy conservation, determine (or estimate) in which phase the final state will be, as we did in Example 14–7 by calculating the different contributing values for heat Q.

5. Note that when the system reaches thermal **equilibrium**, the final **temperature** of each substance will have the *same* value. There is only one T_f.

6. **Solve** your energy equation for the unknown.

Evaporation

The latent heat to change a liquid to a gas is needed not only at the boiling point. Water can change from the liquid to the gas phase even at room temperature. This process is called **evaporation** (see previous Chapter, Section 13–12). The value of the heat of vaporization of water increases slightly with a decrease in temperature: at 20°C, for example, it is 2450 kJ/kg (585 kcal/kg) compared to 2260 kJ/kg (= 539 kcal/kg) at 100°C. When water evaporates, the remaining liquid cools, because the energy required (the latent heat of vaporization) comes from the water itself; so its internal energy, and therefore its temperature, must drop.[‡]

[†]For any algebraic equation $A = B$, if you subtract B from both sides you get $A - B = B - B = 0$, or $A - B = 0$. Thus, if you move a term from one side of an equals sign to the other, the sign (+ or −) changes.

[‡]Also from the point of view of kinetic theory, evaporation is a cooling process because it is the fastest-moving molecules that escape from the surface (Section 13–12). Hence the average speed of the remaining molecules is less, so by Eq. 13–8 the temperature is less.

PHYSICS APPLIED
Body temperature

Evaporation of water from skin is one of the most important ways the body uses to control its temperature. When the temperature of the blood rises slightly above normal, the hypothalamus region of the brain detects this temperature increase and sends a signal to the sweat glands to increase their production. The energy (latent heat) needed to vaporize this water comes from the body, so the body cools.

Kinetic Theory of Latent Heats

We can make use of kinetic theory to see why energy is needed to melt or vaporize a substance. At the melting point, the latent heat of fusion does not act to increase the average kinetic energy (and the temperature) of the molecules in the solid, but instead is used to overcome the potential energy associated with the forces between the molecules. That is, work must be done against these attractive forces to break the molecules loose from their relatively fixed positions in the solid so they can freely roll over one another in the liquid phase. Similarly, energy is required for molecules held close together in the liquid phase to escape into the gaseous phase where they are far apart. This process is a more energetic reorganization of the molecules than is melting (the average distance between the molecules is greatly increased), and hence the heat of vaporization is generally much greater than the heat of fusion for a given substance.

14–6 Heat Transfer: Conduction

Heat is transferred from one place or object to another in three different ways: by *conduction,* by *convection,* and by *radiation.* We now discuss each of these in turn; but in practical situations, any two or all three may be operating at the same time. This Section deals with conduction.

When a metal poker is put in a hot fire, or a silver spoon is placed in a hot bowl of soup, the exposed end of the poker or spoon soon becomes hot as well, even though it is not directly in contact with the source of heat. We say that heat has been *conducted* from the hot end to the cold end.

Heat **conduction** in many materials can be visualized as being carried out via molecular collisions. As one end of an object is heated, the molecules there move faster and faster (= higher temperature). As these faster molecules collide with slower-moving neighbors, they transfer some of their kinetic energy to them, which in turn transfer some energy by collision with molecules still farther along the object. Thus the kinetic energy of thermal motion is transferred by molecular collision along the object. In metals, collisions of free electrons are mainly responsible for conduction. Conduction between objects in physical contact occurs similarly.

Heat conduction from one point to another takes place only if there is a difference in temperature between the two points. Indeed, it is found experimentally that the rate of heat flow through a substance is proportional to the difference in temperature between its ends. The rate of heat flow also depends on the size and shape of the object. To investigate this quantitatively, let us consider the heat flow through a uniform cylinder, as illustrated in Fig. 14–6. It is found experimentally that the heat flow Q over a time interval t is given by the relation

$$\frac{Q}{t} = kA\frac{T_1 - T_2}{\ell} \qquad (14\text{–}5)$$

where A is the cross-sectional area of the object, ℓ is the distance between the two ends, which are at temperatures T_1 and T_2, and k is a proportionality constant called the **thermal conductivity** which is characteristic of the material. From Eq. 14–5, we see that the rate of heat flow (units of J/s) is directly proportional to the cross-sectional area and to the temperature gradient[†] $(T_1 - T_2)/\ell$.

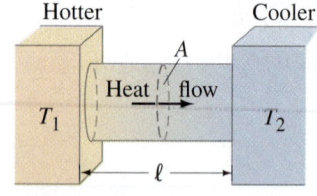

FIGURE 14–6 Heat conduction between regions at temperatures T_1 and T_2. If T_1 is greater than T_2, the heat flows to the right; the rate is given by Eq. 14–5.

[†]Equation 14–5 is quite similar to the relations describing diffusion (Section 13–13) and the flow of fluids through a pipe (Section 10–12). In those cases, the flow of matter was found to be proportional to the concentration gradient $(C_1 - C_2)/\ell$, or to the pressure gradient $(P_1 - P_2)/\ell$. This close similarity is one reason we speak of the "flow" of heat. Yet we must keep in mind that no substance is flowing in the case of heat—it is energy that is being transferred.

The thermal conductivities, k, for a variety of substances are given in Table 14–4. Substances for which k is large conduct heat rapidly and are said to be good **thermal conductors**. Most metals fall in this category, although there is a wide range even among them, as you may observe by holding the ends of a silver spoon and a stainless-steel spoon immersed in the same hot cup of soup. Substances for which k is small, such as wool, fiberglass, polyurethane, and goose down, are poor conductors of heat and are therefore good **thermal insulators**. The relative magnitudes of k can explain simple phenomena such as why a tile floor is much colder on the feet than a rug-covered floor at the same temperature. Tile is a better conductor of heat than the rug. Heat that flows from your foot to the rug is not conducted away rapidly, so the rug's surface quickly warms up to the temperature of your foot and feels good. But the tile conducts the heat away rapidly and thus can take more heat from your foot quickly, so your foot's surface temperature drops.

TABLE 14–4 Thermal Conductivities

Substance	Thermal Conductivity, k	
	$\dfrac{J}{(s \cdot m \cdot C°)}$	$\dfrac{kcal}{(s \cdot m \cdot C°)}$
Silver	420	10×10^{-2}
Copper	380	9.2×10^{-2}
Aluminum	200	5.0×10^{-2}
Steel	40	1.1×10^{-2}
Ice	2	5×10^{-4}
Glass	0.84	2.0×10^{-4}
Brick	0.84	2.0×10^{-4}
Concrete	0.84	2.0×10^{-4}
Water	0.56	1.4×10^{-4}
Human tissue	0.2	0.5×10^{-4}
Wood	0.1	0.3×10^{-4}
Fiberglass	0.048	0.12×10^{-4}
Cork	0.042	0.10×10^{-4}
Wool	0.040	0.10×10^{-4}
Goose down	0.025	0.060×10^{-4}
Polyurethane	0.024	0.057×10^{-4}
Air	0.023	0.055×10^{-4}

EXAMPLE 14–8 **Heat loss through windows.** A major source of heat loss from a house in cold weather is through the windows. Calculate the rate of heat flow through a glass window $2.0 \text{ m} \times 1.5 \text{ m}$ in area and 3.2 mm thick, if the temperatures at the inner and outer surfaces are 15.0°C and 14.0°C, respectively (Fig. 14–7).

APPROACH Heat flows by conduction through the 3.2-mm thickness of glass from the higher inside temperature to the lower outside temperature. We use the heat conduction equation, Eq. 14–5.

SOLUTION Here $A = (2.0 \text{ m})(1.5 \text{ m}) = 3.0 \text{ m}^2$ and $\ell = 3.2 \times 10^{-3}$ m. Using Table 14–4 to get k, we have

$$\frac{Q}{t} = kA\frac{T_1 - T_2}{\ell}$$

$$= \frac{(0.84 \text{ J/s} \cdot \text{m} \cdot \text{C°})(3.0 \text{ m}^2)(15.0°\text{C} - 14.0°\text{C})}{(3.2 \times 10^{-3} \text{ m})}$$

$$= 790 \text{ J/s}.$$

NOTE This rate of heat flow is equivalent to $(790 \text{ J/s})/(4.19 \times 10^3 \text{ J/kcal}) = 0.19 \text{ kcal/s}$, or $(0.19 \text{ kcal/s}) \times (3600 \text{ s/h}) = 680 \text{ kcal/h}$.

PHYSICS APPLIED
Heat loss through windows

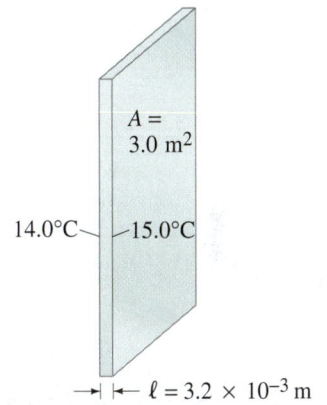

FIGURE 14–7 Example 14–8.

You might notice in Example 14–8 that 15°C is not very warm for the living room of a house. The room itself may indeed be much warmer, and the outside might be colder than 14°C. But the temperatures of 15°C and 14°C were specified as those at the window surfaces, and there is usually a considerable drop in temperature of the air in the vicinity of the window both on the inside and the outside. That is, the layer of air on either side of the window acts as an insulator, and normally the major part of the temperature drop between the inside and outside of the house takes place across the air layer. If there is a heavy wind, the air outside a window will constantly be replaced with cold air; the temperature gradient across the glass will be greater and there will be a much greater rate of heat loss. Increasing the width of the air layer, such as using two panes of glass separated by an air gap, will reduce the heat loss more than simply increasing the glass thickness, since the thermal conductivity of air is much less than that for glass. Such "double-pane windows" are often called *thermal windows*.

The insulating properties of clothing come from the insulating properties of air. Without clothes, our bodies in still air would heat the air in contact with the skin and would soon become reasonably comfortable because air is a very good insulator. But since air moves—there are breezes and drafts, and people move about—the warm air would be replaced by cold air, thus increasing the temperature difference and the heat loss from the body. Clothes keep us warm by trapping air so it cannot move readily. It is not the cloth that insulates us, but the air that the cloth traps. Goose down is a very good insulator because even a small amount of it fluffs up and traps a great amount of air.

| **EXERCISE E** Explain why drapes in front of a window reduce heat loss from a house.

R-values of thermal insulation

R-values for Building Materials

The insulating properties of building materials are often specified by *R*-values (or "thermal resistance"), defined for a given thickness ℓ of material as:

$$R = \frac{\ell}{k}.$$

The *R*-value of a given piece of material combines the thickness ℓ and the thermal conductivity k in one number. Larger *R* means better insulation from heat or cold. In the United States, *R*-values are given in British units as $\text{ft}^2 \cdot \text{h} \cdot \text{F}°/\text{Btu}$ (for example, *R*-19 means $R = 19\ \text{ft}^2 \cdot \text{h} \cdot \text{F}°/\text{Btu}$). Table 14–5 gives *R*-values for some common building materials. *R*-values increase directly with material thickness: for example, 2 inches of fiberglass is *R*-6, whereas 4 inches is *R*-12.

TABLE 14–5 *R*-values

Material	Thickness	*R*-value ($\text{ft}^2 \cdot \text{h} \cdot \text{F}°/\text{Btu}$)
Glass	$\frac{1}{8}$ inch	1
Brick	$3\frac{1}{2}$ inches	0.6–1
Plywood	$\frac{1}{2}$ inch	0.6
Fiberglass insulation	4 inches	12

14–7 Heat Transfer: Convection

Although liquids and gases are generally not very good conductors of heat, they can transfer heat rapidly by convection. **Convection** is the process whereby heat flows by the bulk movement of molecules from one place to another. Whereas conduction involves molecules (and/or electrons) moving only over small distances and colliding, convection involves the movement of large numbers of molecules over large distances.

A forced-air furnace, in which air is heated and then blown by a fan into a room, is an example of *forced convection*. *Natural convection* occurs as well, and one familiar example is that hot air rises. For instance, the air above a radiator (or other type of heater) expands as it is heated (Chapter 13), and hence its density decreases. Because its density is less than that of the surrounding cooler air, it rises via buoyancy, just as a log submerged in water floats upward because its density is less than that of water. Warm or cold ocean currents, such as the balmy Gulf Stream, represent natural convection on a global scale. Wind is another example of convection, and weather in general is strongly influenced by convective air currents.

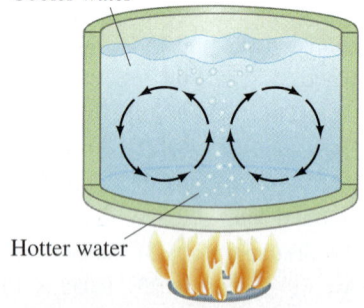
FIGURE 14–8 Convection currents in a pot of water being heated on a stove.

Ocean currents and wind

When a pot of water is heated (Fig. 14–8), convection currents are set up as the heated water at the bottom of the pot rises because of its reduced density. That heated water is replaced by cooler water from above. This principle is used in many heating systems, such as the hot-water radiator system shown in Fig. 14–9. Water is heated in the furnace, and as its temperature increases, it expands and rises as shown. This causes the water to circulate in the heating system. Hot water then enters the radiators, heat is transferred by conduction to the air, and the cooled water returns to the furnace. Thus, the water circulates because of convection; pumps are sometimes used to improve circulation. The air throughout the room also becomes heated as a result of convection. The air heated by the radiators rises and is replaced by cooler air, resulting in convective air currents, as shown by the green arrows in Fig. 14–9.

Convective home heating

FIGURE 14–9 Convection plays a role in heating a house. The circular arrows show convective air currents in the rooms.

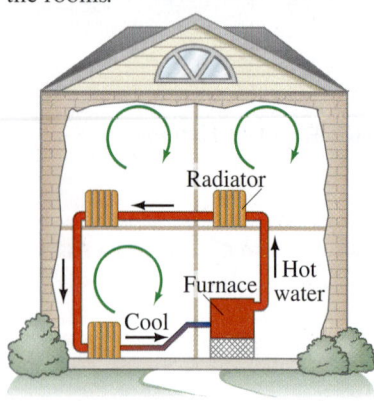

Other types of furnaces also depend on convection. Hot-air furnaces with registers (openings) near the floor often do not have fans but depend on natural convection, which can be appreciable. In other systems, a fan is used. In either case, it is important that cold air can return to the furnace so that convective currents circulate throughout the room if the room is to be uniformly heated.

The human body produces a great deal of thermal energy. Of the food energy transformed within the body, at best 20% is used to do work, so over 80% appears as thermal energy. During light activity, for example, if this thermal energy were not dissipated, the body temperature would rise about 3 C° per hour. Clearly, the heat generated by the body must be transferred to the outside. Is the heat transferred by conduction? The temperature of the skin in a comfortable environment is 33 to 35°C, whereas the interior of the body is at 37°C. Calculation shows (Problem 52) that, because of this small temperature difference, plus the low thermal conductivity of tissue, direct conduction is responsible for very little of the heat that must leave the body. Instead, the heat is carried to the skin by the blood. In addition to all its other important responsibilities, blood acts as a convective fluid to transfer heat to just beneath the surface of the skin. It is then conducted (over a very short distance) to the surface. Once at the surface, the heat is transferred to the environment by convection, evaporation, and radiation (Section 14–8).

Body heat: convection by blood

14–8 Heat Transfer: Radiation

Convection and conduction require the presence of matter as a medium to carry the heat from the hotter to the colder region. But a third type of heat transfer occurs without any medium at all. All life on Earth depends on the transfer of energy from the Sun, and this energy is transferred to the Earth over empty (or nearly empty) space. This form of energy transfer is heat—since the Sun's surface temperature (6000 K) is much higher than Earth's (≈ 300 K)—and is referred to as **radiation** (Fig. 14–10). The warmth we receive from a fire is mainly radiant energy.

As we shall see in later Chapters, radiation consists essentially of electromagnetic waves. Suffice it to say for now that radiation from the Sun consists of visible light plus many other wavelengths that the eye is not sensitive to, including infrared (IR) radiation, which is mainly responsible for heating the Earth.

The rate at which an object radiates energy has been found to be proportional to the fourth power of the Kelvin temperature, T. That is, an object at 2000 K, as compared to one at 1000 K, radiates energy at a rate $2^4 = 16$ times as much. The rate of radiation is also proportional to the area A of the emitting object, so the rate at which energy leaves the object, Q/t, is

$$\frac{Q}{t} = \epsilon \sigma A T^4. \tag{14–6}$$

This is called the **Stefan-Boltzmann equation**, and σ is a universal constant called the **Stefan-Boltzmann constant** which has the value

$$\sigma = 5.67 \times 10^{-8} \text{ W/m}^2 \cdot \text{K}^4.$$

The factor ϵ, called the **emissivity**, is a number[†] between 0 and 1 that is characteristic of the surface of the radiating material. Very black surfaces, such as charcoal, have emissivity close to 1, whereas shiny metal surfaces have ϵ close to zero and thus emit correspondingly less radiation. The value of ϵ depends somewhat on the temperature of the material.

Not only do shiny surfaces emit less radiation, but they absorb little of the radiation that falls upon them (most is reflected). Black and very dark objects are good emitters ($\epsilon \approx 1$); they also absorb nearly all the radiation that falls on them—which is why light-colored clothing is usually preferable to dark clothing on a hot day. Thus, **a good absorber is also a good emitter**.

Any object not only emits energy by radiation but also absorbs energy radiated by other objects. If an object of emissivity ϵ and area A is at a temperature T_1, it radiates energy at a rate $\epsilon \sigma A T_1^4$. If the object is surrounded by an environment at temperature T_2, the rate at which the surroundings radiate energy is proportional to T_2^4; then the rate that energy is *absorbed* by the object is proportional to T_2^4. The *net* rate of radiant heat flow from the object is given by the equation

$$\frac{Q}{t} = \epsilon \sigma A (T_1^4 - T_2^4), \tag{14–7}$$

where A is the surface area of the object, T_1 its temperature and ϵ its emissivity (at temperature T_1), and T_2 is the temperature of the surroundings. This equation is consistent with the experimental fact that equilibrium between the object and its surroundings is reached when they come to the same temperature. That is, Q/t must equal zero when $T_1 = T_2$, so ϵ must be the same for emission and absorption. This confirms the idea that a good emitter is a good absorber. Because both the object and its surroundings radiate energy, there is a net transfer of energy from one to the other unless everything is at the same temperature. From Eq. 14–7 it is clear that if $T_1 > T_2$, the net flow of heat is from the object to the surroundings, so the object cools. But if $T_1 < T_2$, the net heat flow is from the surroundings into the object, and the object's temperature rises. If different parts of the surroundings are at different temperatures, Eq. 14–7 becomes more complicated.

[†]ϵ is the Greek letter epsilon.

FIGURE 14–10 The Sun's surface radiates at 6000 K—much higher than the Earth's surface.

PHYSICS APPLIED
Dark vs. light clothing

PHYSICS APPLIED
The body's radiative heat loss

PROBLEM SOLVING
Must use the Kelvin temperature

PROBLEM SOLVING
$T_1^4 - T_2^4 \neq (T_1 - T_2)^4$

PHYSICS APPLIED
Room comfort

PHYSICS APPLIED
Prefer warm walls, cool air

EXAMPLE 14–9 ESTIMATE **Cooling by radiation.** An athlete is sitting unclothed in a locker room whose dark walls are at a temperature of 15°C. Estimate the body's rate of heat loss by radiation, assuming a skin temperature of 34°C and $\epsilon = 0.70$. Take the surface area of the body not in contact with the chair to be 1.5 m².

APPROACH We use Eq. 14–7, which requires Kelvin temperatures.

SOLUTION We have
$$\frac{Q}{t} = \epsilon \sigma A (T_1^4 - T_2^4)$$
$$= (0.70)(5.67 \times 10^{-8} \text{ W/m}^2 \cdot \text{K}^4)(1.5 \text{ m}^2)[(307 \text{ K})^4 - (288 \text{ K})^4] = 120 \text{ W}.$$

NOTE This person's "output" is a bit more than what a 100-W bulb uses.

NOTE Avoid a common error: $(T_1^4 - T_2^4) \neq (T_1 - T_2)^4$.

A resting person naturally produces heat internally at a rate of about 100 W, as we will see in Section 15–3, less than the heat loss by radiation as calculated in Example 14–9. Hence, our person's temperature would drop, causing considerable discomfort. The body responds to excessive heat loss by increasing its metabolic rate, and shivering is one method by which the body increases its metabolism. Naturally, clothes help a lot. Example 14–9 illustrates that a person may be uncomfortable even if the temperature of the air is, say, 25°C, which is quite a warm room. If the walls or floor are cold, radiation to them occurs no matter how warm the air is. Indeed, it is estimated that radiation accounts for about 50% of the heat loss from a sedentary person in a normal room. Rooms are most comfortable when the walls and floor are warm and the air is not so warm. Floors and walls can be heated by means of hot-water conduits or electric heating elements. Such first-rate heating systems are becoming more common today, and it is interesting to note that 2000 years ago the Romans, even in houses in the remote province of Great Britain, made use of hot-water and steam conduits in the floor to heat their houses.

EXAMPLE 14–10 ESTIMATE **Two teapots.** A ceramic teapot ($\epsilon = 0.70$) and a shiny one ($\epsilon = 0.10$) each hold 0.75 L of tea at 95°C. (a) Estimate the rate of heat loss from each, and (b) estimate the temperature drop after 30 min for each. Consider only radiation, and assume the surroundings are at 20°C.

APPROACH We are given all the information necessary to calculate the heat loss due to radiation, except for the area. The teapot holds 0.75 L, and we can approximate it as a cube 10 cm on a side (volume = 1.0 L), with five sides exposed. To estimate the temperature drop in (b), we use the concept of specific heat and ignore the contribution of the pots compared to that of the water.

SOLUTION (a) The teapot, approximated by a cube 10 cm on a side with five sides exposed, has a surface area of about $5 \times (0.1 \text{ m})^2 = 5 \times 10^{-2} \text{ m}^2$. The rate of heat loss would be about
$$\frac{Q}{t} = \epsilon \sigma A (T_1^4 - T_2^4)$$
$$= \epsilon (5.67 \times 10^{-8} \text{ W/m}^2 \cdot \text{K}^4)(5 \times 10^{-2} \text{ m}^2)[(368 \text{ K})^4 - (293 \text{ K})^4] \approx \epsilon (30) \text{ W},$$
or about 20 W for the ceramic pot ($\epsilon = 0.70$) and 3 W for the shiny one ($\epsilon = 0.10$).
(b) To estimate the temperature drop, we use the specific heat of water and ignore the contribution of the pots. The mass of 0.75 L of water is 0.75 kg. (Recall that $1.0 \text{ L} = 1000 \text{ cm}^3 = 1 \times 10^{-3} \text{ m}^3$ and $\rho = 1000 \text{ kg/m}^3$.) Using Eq. 14–2 and Table 14–1, we get
$$\frac{Q}{t} = mc \frac{\Delta T}{t}.$$
Then
$$\frac{\Delta T}{t} = \frac{Q/t}{mc} \approx \frac{\epsilon (30) \text{ J/s}}{(0.75 \text{ kg})(4186 \text{ J/kg} \cdot \text{C}°)} \approx \epsilon (0.01) \text{ C}°/\text{s}.$$
After 30 min (1800 s), $\Delta T = \epsilon (0.01 \text{ C}°/\text{s})t = \epsilon (0.01 \text{ C}°/\text{s})(1800 \text{ s}) \approx 18\epsilon \text{ C}°$, or about 12 C° for the ceramic pot ($\epsilon = 0.70$) and about 2 C° for the shiny one ($\epsilon = 0.10$). The shiny one clearly has an advantage, at least in terms of radiation.

NOTE Convection and conduction could play a greater role than radiation.

Heating of an object by radiation from the Sun cannot be calculated using Eq. 14–7 since this equation assumes a uniform temperature, T_2, of the environment surrounding the object, whereas the Sun is essentially a point source. Hence the Sun must be treated as a separate source of energy. About 1350 J of energy from the Sun strikes Earth's atmosphere per second per square meter of area at right angles to the Sun's rays. This number, 1350 W/m², is called the **solar constant**. The atmosphere may absorb as much as 70% of this energy before it reaches the ground, depending on the cloud cover. On a clear day, about 1000 W/m² reaches the Earth's surface. An object of emissivity ϵ with area A facing the Sun absorbs energy from the Sun at a rate, in watts, of about

$$\frac{Q}{t} = (1000 \text{ W/m}^2)\epsilon A \cos\theta, \quad (14\text{–}8)$$

where θ is the angle between the Sun's rays and a line perpendicular to the area A (Fig. 14–11). That is, $A \cos\theta$ is the "effective" area, at right angles to the Sun's rays.

The explanation for the **seasons** and the polar ice caps (see Fig. 14–12) depends on this $\cos\theta$ factor in Eq. 14–8. The seasons are *not* a result of how close the Earth is to the Sun—in fact, in the Northern Hemisphere, summer occurs when the Earth is farthest from the Sun. It is the angle (i.e., $\cos\theta$) that really matters. Furthermore, the reason the Sun heats the Earth more at midday than at sunrise or sunset is also related to this $\cos\theta$ factor.

EXAMPLE 14–11 ESTIMATE **Getting a tan—energy absorption.** What is the rate of energy absorption from the Sun by a person lying flat on the beach on a clear day if the Sun makes a 30° angle with the vertical? Assume that $\epsilon = 0.70$ and that 1000 W/m² reaches the Earth's surface.

APPROACH We use Eq. 14–8 and estimate a typical human to be roughly 2 m tall by 0.4 m wide, so $A \approx (2 \text{ m})(0.4 \text{ m}) = 0.8 \text{ m}^2$. See Fig. 14–11.

SOLUTION Since $\cos 30° = 0.866$, we have

$$\frac{Q}{t} = (1000 \text{ W/m}^2)\epsilon A \cos\theta$$
$$= (1000 \text{ W/m}^2)(0.70)(0.8 \text{ m}^2)(0.866) = 500 \text{ W}.$$

NOTE If a person wears light-colored clothing, ϵ is much smaller, so the energy absorbed is less.

An interesting application of thermal radiation to diagnostic medicine is **thermography**. A special instrument, the thermograph, scans the body, measuring the intensity of infrared[†] radiation from many points and forming a picture that resembles an X-ray (Fig. 14–13). Areas where metabolic activity is high, such as in tumors, can often be detected on a thermogram as a result of their higher temperature and consequent increased radiation.

[†]Infrared radiation is light whose wavelengths are longer than visible light (see Fig. 22–8).

PHYSICS APPLIED
Radiation from the Sun

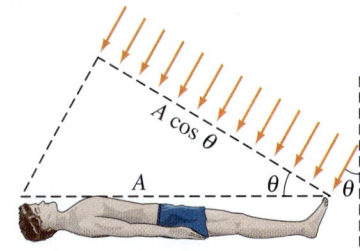

FIGURE 14–11 Radiant energy striking a body at an angle θ.

FIGURE 14–12 (a) Earth's seasons arise from the tilt of Earth's axis relative to its orbit around the Sun. (b) June sunlight makes an angle of about $23\frac{1}{2}°$ with the equator. Thus θ in the southern U.S. (label A) is near 0° (direct summer sunlight), but in the Southern Hemisphere (B), θ is 50° or 60°, and less heat can be absorbed—hence it is winter. Near the poles (C), there is never strong direct sunlight: $\cos\theta$ varies from about $\frac{1}{2}$ in summer to 0 in winter; so with little heating, ice can form.

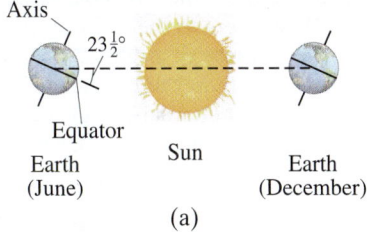

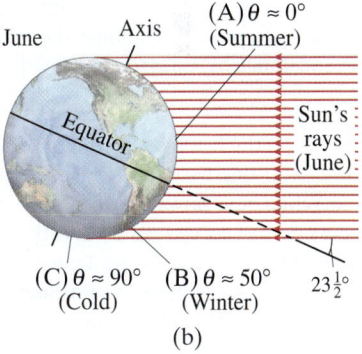

PHYSICS APPLIED
Thermography

FIGURE 14–13 Thermograms of a healthy person's arms and hands (a) before and (b) after smoking a cigarette, showing a temperature decrease due to impaired blood circulation associated with smoking. The thermograms have been color-coded according to temperature; the scale on the right goes from blue (cold) to white (hot).

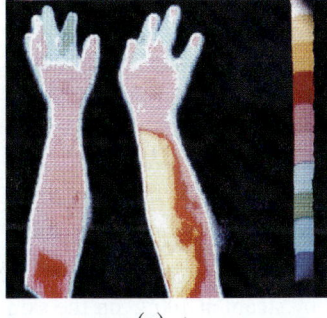

(a)

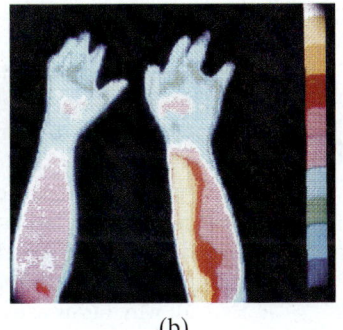

(b)

PHYSICS APPLIED
Astronomy—size of a star

EXAMPLE 14–12 ESTIMATE **Star radius.** The giant star Betelgeuse emits radiant energy at a rate 10^4 times greater than our Sun, whereas its surface temperature is only half (2900 K) that of our Sun. Estimate the radius of Betelgeuse, assuming $\epsilon = 1$. The Sun's radius is $r_S = 7 \times 10^8$ m.

APPROACH We assume both stars are spherical, with surface area $4\pi r^2$.

SOLUTION We solve Eq. 14–6 for A:
$$4\pi r^2 = A = \frac{(Q/t)}{\epsilon \sigma T^4}.$$

Then
$$\frac{r_B^2}{r_S^2} = \frac{(Q/t)_B}{(Q/t)_S} \cdot \frac{T_S^4}{T_B^4} = (10^4)(2^4) = 16 \times 10^4.$$

Hence $r_B = \sqrt{16 \times 10^4}\, r_S = (400)(7 \times 10^8\,\text{m}) \approx 3 \times 10^{11}$ m. If Betelgeuse were our Sun, it would envelop us (Earth is 1.5×10^{11} m from the Sun).

EXERCISE F Fanning yourself on a hot day cools you by (*a*) increasing the radiation rate of the skin; (*b*) increasing conductivity; (*c*) decreasing the mean free path of air; (*d*) increasing the evaporation of perspiration; (*e*) none of these.

Summary

Internal energy, U, refers to the total energy of all the molecules in an object. For an ideal monatomic gas,
$$U = \tfrac{3}{2}NkT = \tfrac{3}{2}nRT \qquad (14\text{–}1)$$
where N is the number of molecules or n is the number of moles.

Heat refers to the transfer of energy from one object to another because of a difference of temperature. Heat is thus measured in energy units, such as joules.

Heat and internal energy are also sometimes specified in calories or kilocalories (kcal), where
$$1\,\text{kcal} = 4.186\,\text{kJ}$$
is the amount of heat needed to raise the temperature of 1 kg of water by 1 C°.

The **specific heat**, c, of a substance is defined as the energy (or heat) required to change the temperature of unit mass of the substance by 1 degree; as an equation,
$$Q = mc\,\Delta T, \qquad (14\text{–}2)$$
where Q is the heat absorbed or given off, ΔT is the temperature increase or decrease, and m is the mass of the substance.

When heat flows between parts of an isolated system, conservation of energy tells us that the heat gained by one part of the system is equal to the heat lost by the other part of the system. This is the basis of **calorimetry**, which is the quantitative measurement of heat exchange.

Exchange of energy occurs, without a change in temperature, whenever a substance changes phase. The **heat of fusion** is the heat required to melt 1 kg of a solid into the liquid phase; it is also equal to the heat given off when the substance changes from liquid to solid. The **heat of vaporization** is the energy required to change 1 kg of a substance from the liquid to the vapor phase; it is also the energy given off when the substance changes from vapor to liquid.

Heat is transferred from one place (or object) to another in three different ways: conduction, convection, and radiation.

In **conduction**, energy is transferred through a substance by means of collisions between hotter (faster) molecules or electrons with their slower moving neighbors.

Convection is the transfer of energy by the mass movement of molecules even over considerable distances.

Radiation, which does not require the presence of matter, is energy transfer by electromagnetic waves, such as from the Sun. All objects radiate energy in an amount that is proportional to the fourth power of their Kelvin temperature (T^4) and to their surface area. The energy radiated (or absorbed) also depends on the nature of the surface (dark surfaces absorb and radiate more than do bright shiny ones), which is characterized by the emissivity, ϵ.

Radiation from the Sun arrives at the surface of the Earth on a clear day at a rate of about 1000 W/m².

Questions

1. What happens to the work done on a jar of orange juice when it is vigorously shaken?

2. When a hot object warms a cooler object, does temperature flow between them? Are the temperature changes of the two objects equal? Explain.

3. (*a*) If two objects of different temperatures are placed in contact, will heat naturally flow from the object with higher internal energy to the object with lower internal energy? (*b*) Is it possible for heat to flow even if the internal energies of the two objects are the same? Explain.

4. In warm regions where tropical plants grow but the temperature may drop below freezing a few times in the winter, the destruction of sensitive plants due to freezing can be reduced by watering them in the evening. Explain.

5. The specific heat of water is quite large. Explain why this fact makes water particularly good for heating systems (that is, hot-water radiators).

6. Why does water in a metal canteen stay cooler if the cloth jacket surrounding the canteen is kept moist?

7. Explain why burns caused by steam at 100°C on the skin are often more severe than burns caused by water at 100°C.

8. Explain why water cools (its temperature drops) when it evaporates, using the concepts of latent heat and internal energy.
9. Will pasta cook faster if the water boils more vigorously? Explain.
10. Very high in the Earth's atmosphere, the temperature can be 700°C. Yet an animal there would freeze to death rather than roast. Explain.
11. Explorers on failed Arctic expeditions have survived by covering themselves with snow. Why would they do that?
12. Why is wet sand at a beach cooler to walk on than dry sand?
13. If you hear that an object has "high heat content," does that mean that its temperature is high? Explain.
14. When hot-air furnaces are used to heat a house, why is it important that there be a vent for air to return to the furnace? What happens if this vent is blocked by a bookcase?
15. Ceiling fans are sometimes reversible, so that they drive the air down in one season and pull it up in another season. Explain which way you should set the fan (a) for summer, (b) for winter.
16. Goose down sleeping bags and parkas are often specified as so many inches or centimeters of *loft*, the actual thickness of the garment when it is fluffed up. Explain.
17. Microprocessor chips have a "heat sink" glued on top that looks like a series of fins. Why are they shaped like that?
18. Sea breezes are often encountered on sunny days at the shore of a large body of water. Explain, noting that the temperature of the land rises more rapidly than that of the nearby water.
19. The floor of a house on a foundation under which the air can flow is often cooler than a floor that rests directly on the ground (such as a concrete slab foundation). Explain.
20. A 22°C day is warm, while a swimming pool at 22°C feels cool. Why?
21. Explain why air temperature readings are always taken with the thermometer in the shade.
22. A premature baby in an incubator can be dangerously cooled even when the air temperature in the incubator is warm. Explain.

23. Does an ordinary electric fan cool the air? Why or why not? If not, why use it?
24. Heat loss occurs through windows by the following processes: (1) through the glass panes; (2) through the frame, particularly if it is metal; (3) ventilation around edges; and (4) radiation. (a) For the first three, what is (are) the mechanism(s): conduction, convection, or radiation? (b) Heavy curtains reduce which of these heat losses? Explain in detail.
25. A piece of wood lying in the Sun absorbs more heat than a piece of shiny metal. Yet the metal feels hotter than the wood when you pick it up. Explain.
26. The Earth cools off at night much more quickly when the weather is clear than when cloudy. Why?
27. An "emergency blanket" is a thin shiny (metal-coated) plastic foil. Explain how it can help to keep an immobile person warm.
28. Explain why cities situated by the ocean tend to have less extreme temperatures than inland cities at the same latitude.
29. A paper cup placed among hot coals will burn if empty (note burn spots at top of cup in Fig. 14–14), but won't burn if filled with water. Explain. Forget the marshmallows.

FIGURE 14–14 Question 29.

30. On a cold windy day, a window will feel colder than on an equally cold day with no wind. This is true even if no air leaks in near the window. Why?

MisConceptual Questions

1. When you put an ice cube in a glass of warm tea, which of the following happens?
   (a) Cold flows from the ice cube into the tea.
   (b) Cold flows from the ice cube into the tea and heat flows from the tea into the ice cube.
   (c) Heat flows from the tea into the ice cube.
   (d) Neither heat nor cold flows. Only temperature flows between the ice and the tea.

2. Both beakers A and B in Fig. 14–15 contain a mixture of ice and water at equilibrium. Which beaker is the coldest, or are they equal in temperature?
   (a) Beaker A.
   (b) Beaker B.
   (c) Equal.

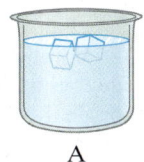 

FIGURE 14–15 MisConceptual Question 2. A   B

3. For objects at thermal equilibrium, which of the following is true?
   (a) Each is at the same temperature.
   (b) Each has the same internal energy.
   (c) Each has the same heat.
   (d) All of the above.
   (e) None of the above.

4. Which of the following happens when a material undergoes a phase change?
   (a) The temperature changes.
   (b) The chemical composition changes.
   (c) Heat flows into or out of the material.
   (d) The molecules break apart into atoms.

5. As heat is added to water, is it possible for the temperature measured by a thermometer in the water to remain constant?
   (a) Yes, the water could be changing phase.
   (b) No, adding heat will always change the temperature.
   (c) Maybe; it depends on the rate at which the heat is added.
   (d) Maybe; it depends on the initial water temperature.

6. A typical thermos bottle has a thin vacuum space between the shiny inner flask (which holds a liquid) and the shiny protective outer flask, often stainless steel. The vacuum space is excellent at preventing
   (a) conduction.
   (b) convection.
   (c) radiation.
   (d) conduction and convection.
   (e) conduction, convection, and radiation.

7. Heat is
   (a) a fluid called caloric.
   (b) a measure of the average kinetic energy of atoms.
   (c) the amount of energy transferred between objects as a result of a difference in temperature.
   (d) an invisible, odorless, weightless substance.
   (e) the total kinetic energy of an ideal gas.

8. Radiation is emitted
   (a) only by glowing objects such as the Sun.
   (b) only by objects whose temperature is greater than the temperature of the surroundings.
   (c) only by objects with more caloric than their surroundings.
   (d) by any object not at 0 K.
   (e) only by objects that have a large specific heat.

9. Ten grams of water is added to ten grams of ice in an insulated container. Will all of the ice melt?
   (a) Yes.   (b) No.   (c) More information is needed.

10. Two objects are made of the same material, but they have different masses and temperatures. If the objects are brought into thermal contact, which one will have the greater temperature change?
    (a) The one with the higher initial temperature.
    (b) The one with the lower initial temperature.
    (c) The one with the greater mass.
    (d) The one with the lesser mass.
    (e) The one with the higher specific heat.
    (f) Not enough information.

11. It has been a hot summer, so when you arrive at a lake, you decide to go for a swim even though it is nighttime. The water is cold! The next day, you go swimming again during the hottest part of the day, and even though the air is warmer the water is still almost as cold. Why?
    (a) Water is fairly dense compared with many other liquids.
    (b) Water remains in a liquid state for a wide range of temperatures.
    (c) Water has a high bulk modulus.
    (d) Water has a high specific heat.

12. Two equal-mass liquids, initially at the same temperature, are heated for the same time over the same stove. You measure the temperatures and find that one liquid has a higher temperature than the other. Which liquid has the higher specific heat?
    (a) The cooler one.
    (b) The hotter one.
    (c) Both are the same.

For assigned homework and other learning materials, go to the MasteringPhysics website.

Problems

14–1 Heat as Energy Transfer

1. (I) To what temperature will 8200 J of heat raise 3.0 kg of water that is initially at 10.0°C?

2. (I) How much heat (in joules) is required to raise the temperature of 34.0 kg of water from 15°C to 95°C?

3. (II) When a diver jumps into the ocean, water leaks into the gap region between the diver's skin and her wetsuit, forming a water layer about 0.5 mm thick. Assuming the total surface area of the wetsuit covering the diver is about 1.0 m², and that ocean water enters the suit at 10°C and is warmed by the diver to skin temperature of 35°C, estimate how much energy (in units of candy bars = 300 kcal) is required by this heating process.

4. (II) An average active person consumes about 2500 Cal a day. (a) What is this in joules? (b) What is this in kilowatt-hours? (c) If your power company charges about 10 ¢ per kilowatt-hour, how much would your energy cost per day if you bought it from the power company? Could you feed yourself on this much money per day?

5. (II) A British thermal unit (Btu) is a unit of heat in the British system of units. One Btu is defined as the heat needed to raise 1 lb of water by 1 F°. Show that
   $$1\ \text{Btu} = 0.252\ \text{kcal} = 1056\ \text{J}.$$

6. (II) How many joules and kilocalories are generated when the brakes are used to bring a 1300-kg car to rest from a speed of 95 km/h?

7. (II) A water heater can generate 32,000 kJ/h. How much water can it heat from 12°C to 42°C per hour?

8. (II) A small immersion heater is rated at 375 W. Estimate how long it will take to heat a cup of soup (assume this is 250 mL of water) from 15°C to 75°C.

14–3 and 14–4 Specific Heat; Calorimetry

9. (I) An automobile cooling system holds 18 L of water. How much heat does it absorb if its temperature rises from 15°C to 95°C?

10. (I) What is the specific heat of a metal substance if 135 kJ of heat is needed to raise 4.1 kg of the metal from 18.0°C to 37.2°C?

11. (II) (a) How much energy is required to bring a 1.0-L pot of water at 20°C to 100°C? (b) For how long could this amount of energy run a 60-W lightbulb?

12. (II) Samples of copper, aluminum, and water experience the same temperature rise when they absorb the same amount of heat. What is the ratio of their masses? [*Hint*: See Table 14–1.]

13. (II) How long does it take a 750-W coffeepot to bring to a boil 0.75 L of water initially at 11°C? Assume that the part of the pot which is heated with the water is made of 280 g of aluminum, and that no water boils away.

14. (II) What will be the equilibrium temperature when a 265-g block of copper at 245°C is placed in a 145-g aluminum calorimeter cup containing 825 g of water at 12.0°C?

15. (II) A 31.5-g glass thermometer reads 23.6°C before it is placed in 135 mL of water. When the water and thermometer come to equilibrium, the thermometer reads 41.8°C. What was the original temperature of the water? Ignore the mass of fluid inside the glass thermometer.

16. (II) A 0.40-kg iron horseshoe, just forged and very hot (Fig. 14–16), is dropped into 1.25 L of water in a 0.30-kg iron pot initially at 20.0°C. If the final equilibrium temperature is 25.0°C, estimate the initial temperature of the hot horseshoe.

FIGURE 14–16 Problem 16.

17. (II) When a 290-g piece of iron at 180°C is placed in a 95-g aluminum calorimeter cup containing 250 g of glycerin at 10°C, the final temperature is observed to be 38°C. Estimate the specific heat of glycerin.

18. (II) The *heat capacity*, C, of an object is defined as the amount of heat needed to raise its temperature by 1 C°. Thus, to raise the temperature by ΔT requires heat Q given by

$$Q = C \Delta T.$$

(a) Write the heat capacity C in terms of the specific heat, c, of the material. (b) What is the heat capacity of 1.0 kg of water? (c) Of 45 kg of water?

19. (II) The 1.20-kg head of a hammer has a speed of 7.5 m/s just before it strikes a nail (Fig. 14–17) and is brought to rest. Estimate the temperature rise of a 14-g iron nail generated by eight such hammer blows done in quick succession. Assume the nail absorbs all the energy.

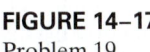

FIGURE 14–17 Problem 19.

20. (II) A 215-g sample of a substance is heated to 330°C and then plunged into a 105-g aluminum calorimeter cup containing 185 g of water and a 17-g glass thermometer at 10.5°C. The final temperature is 35.0°C. What is the specific heat of the substance? (Assume no water boils away.)

21. (II) A 0.095-kg aluminium sphere is dropped from the roof of a 55-m-high building. If 65% of the thermal energy produced when it hits the ground is absorbed by the sphere, what is its temperature increase?

22. (II) Estimate the Calorie content of 65 g of candy from the following measurements. A 15-g sample of the candy is placed in a small aluminum container of mass 0.325 kg filled with oxygen. This container is placed in 1.75 kg of water in an aluminum calorimeter cup of mass 0.624 kg at an initial temperature of 15.0°C. The oxygen–candy mixture in the small container (a "bomb calorimeter") is ignited, and the final temperature of the whole system is 53.5°C.

23. (II) Determine the energy content of 100 g of Karen's fudge cookies from the following measurements. A 10-g sample of a cookie is allowed to dry before putting it in a *bomb calorimeter* (page 396). The aluminum bomb has a mass of 0.615 kg and is placed in 2.00 kg of water contained in an aluminum calorimeter cup of mass 0.524 kg. The initial temperature of the system is 15.0°C, and its temperature after ignition is 36.0°C.

14–5 Latent Heat

24. (I) If 3.40×10^5 J of energy is supplied to a container of liquid oxygen at −183°C, how much oxygen can evaporate?

25. (II) How much heat is needed to melt 23.50 kg of silver that is initially at 25°C?

26. (II) During exercise, a person may give off 185 kcal of heat in 25 min by evaporation of water (at 20°C) from the skin. How much water has been lost? [*Hint*: See page 399.]

27. (II) What mass of steam at 100°C must be added to 1.00 kg of ice at 0°C to yield liquid water at 30°C?

28. (II) A 28-g ice cube at its melting point is dropped into an insulated container of liquid nitrogen. How much nitrogen evaporates if it is at its boiling point of 77 K and has a latent heat of vaporization of 200 kJ/kg? Assume for simplicity that the specific heat of ice is a constant and is equal to its value near its melting point.

29. (II) High-altitude mountain climbers do not eat snow, but always melt it first with a stove. To see why, calculate the energy absorbed from your body if you: (a) eat 1.0 kg of −15°C snow which your body warms to body temperature of 37°C; (b) melt 1.0 kg of −15°C snow using a stove and drink the resulting 1.0 kg of water at 2°C, which your body has to warm to 37°C.

30. (II) An iron boiler of mass 180 kg contains 730 kg of water at 18°C. A heater supplies energy at the rate of 58,000 kJ/h. How long does it take for the water (a) to reach the boiling point, and (b) to all have changed to steam?

31. (II) Determine the latent heat of fusion of mercury using the following calorimeter data: 1.00 kg of solid Hg at its melting point of −39.0°C is placed in a 0.620-kg aluminum calorimeter with 0.400 kg of water at 12.80°C; the resulting equilibrium temperature is 5.06°C.

32. (II) At a crime scene, the forensic investigator notes that the 6.2-g lead bullet that was stopped in a doorframe apparently melted completely on impact. Assuming the bullet was shot at room temperature (20°C), what does the investigator calculate as the minimum muzzle velocity of the gun?

33. (II) A 64-kg ice-skater moving at 7.5 m/s glides to a stop. Assuming the ice is at 0°C and that 50% of the heat generated by friction is absorbed by the ice, how much ice melts?

34. (II) A cube of ice is taken from the freezer at −8.5°C and placed in an 85-g aluminum calorimeter filled with 310 g of water at room temperature of 20.0°C. The final situation is all water at 17.0°C. What was the mass of the ice cube?

35. (II) A 55-g bullet traveling at 250 m/s penetrates a block of ice at 0°C and comes to rest within the ice. Assuming that the temperature of the bullet doesn't change appreciably, how much ice is melted as a result of the collision?

14–6 to 14–8 Conduction, Convection, Radiation

36. (I) Calculate the rate of heat flow by conduction through the windows of Example 14–8, assuming that there are strong gusty winds and the external temperature is −5°C.

37. (I) One end of a 56-cm-long copper rod with a diameter of 2.0 cm is kept at 460°C, and the other is immersed in water at 22°C. Calculate the heat conduction rate along the rod.

38. (II) (a) How much power is radiated by a tungsten sphere (emissivity $\epsilon = 0.35$) of radius 19 cm at a temperature of 25°C? (b) If the sphere is enclosed in a room whose walls are kept at −5°C, what is the *net* flow rate of energy out of the sphere?

Problems **409**

39. (II) How long does it take the Sun to melt a block of ice at 0°C with a flat horizontal area 1.0 m² and thickness 1.0 cm? Assume that the Sun's rays make an angle of 35° with the vertical and that the emissivity of ice is 0.050.

40. (II) *Heat conduction to skin.* Suppose 150 W of heat flows by conduction from the blood capillaries beneath the skin to the body's surface area of 1.5 m². If the temperature difference is 0.50 C°, estimate the average distance of capillaries below the skin surface.

41. (II) Two rooms, each a cube 4.0 m per side, share a 14-cm-thick brick wall. Because of a number of 100-W lightbulbs in one room, the air is at 30°C, while in the other room it is at 10°C. How many of the 100-W bulbs are needed to maintain the temperature difference across the wall?

42. (II) A 100-W lightbulb generates 95 W of heat, which is dissipated through a glass bulb that has a radius of 3.0 cm and is 0.50 mm thick. What is the difference in temperature between the inner and outer surfaces of the glass?

43. (III) Approximately how long should it take 8.2 kg of ice at 0°C to melt when it is placed in a carefully sealed Styrofoam ice chest of dimensions 25 cm × 35 cm × 55 cm whose walls are 1.5 cm thick? Assume that the conductivity of Styrofoam is double that of air and that the outside temperature is 34°C.

44. (III) A copper rod and an aluminum rod of the same length and cross-sectional area are attached end to end (Fig. 14–18). The copper end is placed in a furnace maintained at a constant temperature of 205°C. The aluminum end is placed in an ice bath held at a constant temperature of 0.0°C. Calculate the temperature at the point where the two rods are joined.

FIGURE 14–18 Problem 44.

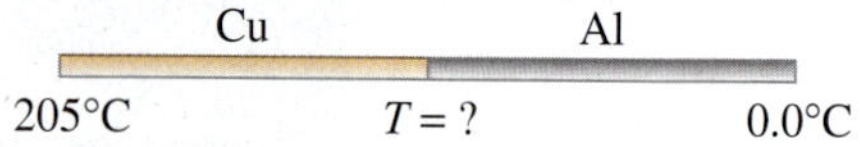

Cu Al
205°C $T = ?$ 0.0°C

*45. (III) Suppose the insulating qualities of the wall of a house come mainly from a 4.0-in. layer of brick and an R-19 layer of insulation, as shown in Fig. 14–19. What is the total rate of heat loss through such a wall, if its total area is 195 ft² and the temperature difference across it is 35 F°?

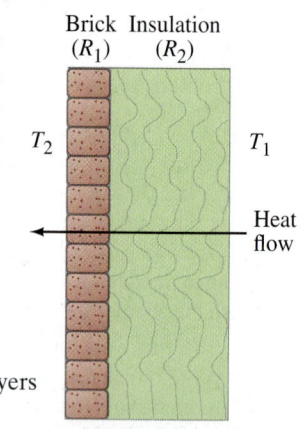

FIGURE 14–19 Problem 45. Two layers insulating a wall.

General Problems

46. A soft-drink can contains about 0.35 kg of liquid at 5°C. Drinking this liquid can actually consume some of the fat in the body, since energy is needed to warm the liquid to body temperature (37°C). How many food Calories should the drink have so that it is in perfect balance with the heat needed to warm the liquid (essentially water)?

47. (a) Estimate the total power radiated into space by the Sun, assuming it to be a perfect emitter at $T = 5500$ K. The Sun's radius is 7.0×10^8 m. (b) From this, determine the power per unit area arriving at the Earth, 1.5×10^{11} m away (Fig. 14–20).

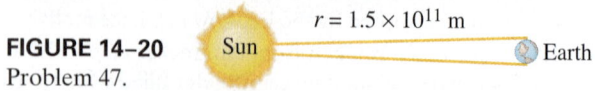

FIGURE 14–20 Problem 47.

48. To get an idea of how much thermal energy is contained in the world's oceans, estimate the heat liberated when a cube of ocean water, 1 km on each side, is cooled by 1 K. (Approximate the ocean water as pure water for this estimate.)

49. What will be the final result when equal masses of ice at 0°C and steam at 100°C are mixed together?

50. A mountain climber wears a goose-down jacket 3.5 cm thick with total surface area 0.95 m². The temperature at the surface of the clothing is −18°C and at the skin is 34°C. Determine the rate of heat flow by conduction through the jacket assuming (a) it is dry and the thermal conductivity k is that of goose down, and (b) the jacket is wet, so k is that of water and the jacket has matted to 0.50 cm thickness.

51. During light activity, a 70-kg person may generate 200 kcal/h. Assuming that 20% of this goes into useful work and the other 80% is converted to heat, estimate the temperature rise of the body after 45 min if none of this heat is transferred to the environment.

52. Estimate the rate at which heat can be conducted from the interior of the body to the surface. As a model, assume that the thickness of tissue is 4.0 cm, that the skin is at 34°C and the interior at 37°C, and that the surface area is 1.5 m². Compare this to the measured value of about 230 W that must be dissipated by a person working lightly. This clearly shows the necessity of convective cooling by the blood.

53. A bicyclist consumes 9.0 L of water over the span of 3.5 hours during a race. Making the approximation that 80% of the cyclist's energy goes into evaporating this water (at 20°C) as sweat, how much energy in kcal did the rider use during the ride? [*Hint:* See page 399.]

54. If coal gives off 30 MJ/kg when burned, how much coal is needed to heat a house requiring 2.0×10^5 MJ for the whole winter? Assume that 30% of the heat is lost up the chimney.

55. A 15-g lead bullet is tested by firing it into a fixed block of wood with a mass of 35 kg. The block and imbedded bullet together absorb all the heat generated. After thermal equilibrium has been reached, the system has a temperature rise measured as 0.020 C°. Estimate the bullet's entering speed.

56. A 310-kg marble boulder rolls off the top of a cliff and falls a vertical height of 120 m before striking the ground. Estimate the temperature rise of the rock if 50% of the heat generated remains in the rock.

57. A 2.3-kg lead ball is placed in a 2.5-L insulated pail of water initially at 20.0°C. If the final temperature of the water–lead combination is 32.0°C, what was the initial temperature of the lead ball?

58. A microwave oven is used to heat 250 g of water. On its maximum setting, the oven can raise the temperature of the liquid water from 20°C to 100°C in 1 min 45 s (= 105 s). (a) At what rate does the oven put energy into the liquid water? (b) If the power input from the oven to the water remains constant, determine how many grams of water will boil away if the oven is operated for 2 min (rather than just 1 min 45 s).

59. In a typical squash game (Fig. 14–21), two people hit a soft rubber ball at a wall. Assume that the ball hits the wall at a velocity of 22 m/s and bounces back at a velocity of 12 m/s, and that the kinetic energy lost in the process heats the ball. What will be the temperature increase of the ball after one bounce? (The specific heat of rubber is about 1200 J/kg·C°.)

FIGURE 14–21 Problem 59.

60. The temperature within the Earth's crust increases about 1.0 C° for each 30 m of depth. The thermal conductivity of the crust is 0.80 J/s·C°·m. (a) Determine the heat transferred from the interior to the surface for the entire Earth in 1.0 h. (b) Compare this heat to the 1000 W/m² that reaches the Earth's surface in 1.0 h from the Sun.

61. An iron meteorite melts when it enters the Earth's atmosphere. If its initial temperature was −105°C outside of Earth's atmosphere, calculate the minimum velocity the meteorite must have had before it entered Earth's atmosphere.

62. The temperature of the glass surface of a 75-W lightbulb is 75°C when the room temperature is 18°C. Estimate the temperature of a 150-W lightbulb with a glass bulb the same size. Consider only radiation, and assume that 90% of the energy is emitted as heat.

63. In a cold environment, a person can lose heat by conduction and radiation at a rate of about 200 W. Estimate how long it would take for the body temperature to drop from 36.6°C to 35.6°C if metabolism were nearly to stop. Assume a mass of 65 kg. (See Table 14–1.)

64. A 12-g lead bullet traveling at 220 m/s passes through a thin wall and emerges at a speed of 160 m/s. If the bullet absorbs 50% of the heat generated, (a) what will be the temperature rise of the bullet? (b) If the bullet's initial temperature was 20°C, will any of the bullet melt, and if so, how much?

65. A leaf of area 40 cm² and mass 4.5×10^{-4} kg directly faces the Sun on a clear day. The leaf has an emissivity of 0.85 and a specific heat of 0.80 kcal/kg·K. (a) Estimate the energy absorbed per second by the leaf from the Sun, and then (b) estimate the rate of rise of the leaf's temperature. (c) Will the temperature rise continue for hours? Why or why not? (d) Calculate the temperature the leaf would reach if it lost all its heat by radiation to the surroundings at 24°C. (e) In what other ways can the heat be dissipated by the leaf?

66. Using the result of part (a) in Problem 65, take into account radiation from the leaf to calculate how much water must be transpired (evaporated) by the leaf per hour to maintain a temperature of 35°C.

67. After a hot shower and dishwashing, there seems to be no hot water left in the 65-gal (245-L) water heater. This suggests that the tank has emptied and refilled with water at roughly 10°C. (a) How much energy does it take to reheat the water to 45°C? (b) How long would it take if the heater output is 9500 W?

68. A house thermostat is normally set to 22°C, but at night it is turned down to 16°C for 9.0 h. Estimate how much more heat would be needed (state as a percentage of daily usage) if the thermostat were not turned down at night. Assume that the outside temperature averages 0°C for the 9.0 h at night and 8°C for the remainder of the day, and that the heat loss from the house is proportional to the temperature difference inside and out. To obtain an estimate from the data, you must make other simplifying assumptions; state what these are.

Search and Learn

1. Create graphs similar to Fig. 14–5, but for lead and ethyl alcohol. Compare and contrast them with each other and with the graph for water. Are there any temperature ranges for which all three substances are liquids? All vapors? All solids? For convenience, use the specific heats given in Table 14–1 for all states of lead and ethyl alcohol.

2. (a) Using the solar constant, estimate the rate at which the whole Earth receives energy from the Sun. (b) Assume the Earth radiates an equal amount back into space (that is, the Earth is in equilibrium). Then, assuming the Earth is a perfect emitter ($\epsilon = 1.0$), estimate its average surface temperature. [Hint: Discuss why you use area $A = \pi r_E^2$ or $A = 4\pi r_E^2$ in each part.]

3. Calculate what will happen when 1000 J of heat is added to 100 grams of (a) ice at −20°C, (b) ice at 0°C, (c) water at 10°C, (d) water at 100°C, and (e) steam at 110°C.

4. A house has well-insulated walls 19.5 cm thick (assume conductivity of air) and area 410 m², a roof of wood 5.5 cm thick and area 250 m², and uncovered windows 0.65 cm thick and total area 33 m². (a) Assuming that heat is lost only by conduction, calculate the rate at which heat must be supplied to this house to maintain its inside temperature at 23°C if the outside temperature is −15°C. (b) If the house is initially at 15°C, estimate how much heat must be supplied to raise the temperature to 23°C within 30 min. Assume that only the air needs to be heated and that its volume is 750 m³. (c) If natural gas costs $0.080/kg and its heat of combustion is 5.4×10^7 J/kg, what is the monthly cost to maintain the house as in part (a) for 24 h each day, assuming 90% of the heat produced is used to heat the house? Take the specific heat of air to be 0.24 kcal/kg·C°.

ANSWERS TO EXERCISES

A: (b).
B: (c).
C: The phase change, liquid to ice (second process).
D: 0.21 kg.
E: The drapes trap a layer of air between the inside of the window and the room, which acts as an excellent insulator.
F: (d).

Thermodynamics is the study of heat and work. Heat is a transfer of energy due to a difference of temperature. Work is a transfer of energy by mechanical means, not due to a temperature difference. The first law of thermodynamics links the two in a general statement of energy conservation: the heat Q added to a system minus the net work W done by the system equals the change in internal energy ΔU of the system: $\Delta U = Q - W$.

There are many uses for a heat engine such as a modern coal-burning power plant, or a steam locomotive. The photograph shows a steam locomotive which produces steam that does work on a piston that moves linkage to turn locomotive wheels. The efficiency of any engine is limited by nature as described in the second law of thermodynamics. This great law is best stated in terms of a quantity called entropy, which is *not* conserved, but instead is constrained always to increase in any real process. Entropy is a measure of disorder. The second law of thermodynamics tells us that as time moves forward, the disorder in the universe increases. We also discuss practical matters such as heat engines, heat pumps, refrigerators, and air conditioners.

CHAPTER 15

The Laws of Thermodynamics

CONTENTS

15–1 The First Law of Thermodynamics

15–2 Thermodynamic Processes and the First Law

*15–3 Human Metabolism and the First Law

15–4 The Second Law of Thermodynamics—Introduction

15–5 Heat Engines

15–6 Refrigerators, Air Conditioners, and Heat Pumps

15–7 Entropy and the Second Law of Thermodynamics

15–8 Order to Disorder

15–9 Unavailability of Energy; Heat Death

*15–10 Statistical Interpretation of Entropy and the Second Law

*15–11 Thermal Pollution, Global Warming, and Energy Resources

CHAPTER-OPENING QUESTION—Guess now!

Fossil-fuel electric generating plants produce "thermal pollution." Part of the heat produced by the burning fuel is not converted to electric energy. The reason for this waste is

(a) the efficiency is higher if some heat is allowed to escape.
(b) engineering technology has not yet reached the point where 100% waste heat recovery is possible.
(c) some waste heat *must* be produced: this is a fundamental property of nature when converting heat to useful work.
(d) the plants rely on fossil fuels, not nuclear fuel.
(e) None of the above.

Thermodynamics is the name we give to the study of processes in which energy is transferred as heat and as work.

In Chapter 6 we saw that work is done when energy is transferred from one object to another by mechanical means. In Chapter 14 we saw that heat is a transfer of energy from one object to a second one at a lower temperature. Thus, heat is much like work. To distinguish them, *heat* is defined as *a transfer of energy due to a difference in temperature*, whereas work is a transfer of energy that is not due to a temperature difference.

In discussing thermodynamics, we often refer to particular systems. A **system** is any object or set of objects that we wish to consider (see Section 14–4). Everything else in the universe is referred to as its "environment" or the "surroundings."

In this Chapter, we examine the two great laws of thermodynamics. The first law of thermodynamics relates work and heat transfers to the change in internal energy of a system, and is a general statement of the conservation of energy. The second law of thermodynamics expresses limits on the ability to do useful work, and is often stated in terms of *entropy*, which is a measure of disorder. Besides these two great laws, we also discuss some important related practical devices: heat engines, refrigerators, heat pumps, and air conditioners.

15–1 The First Law of Thermodynamics

In Section 14–2, we defined the internal energy of a system as the sum total of all the energy of the molecules within the system. Then the internal energy of a system should increase if work is done on the system, or if heat is added to it. Similarly the internal energy should decrease if heat flows out of the system or if work is done by the system on something in the surroundings.

Thus it is reasonable to extend conservation of energy and propose an important law: the change in internal energy of a closed system, ΔU, will be equal to the energy added to the system by heating minus the work done by the system on the surroundings. In equation form we write

$$\Delta U = Q - W \qquad (15\text{–}1)$$

FIRST LAW OF THERMODYNAMICS

where Q is the net heat *added* to the system and W is the net work done *by* the system. We must be careful and consistent in following the sign conventions for Q and W. Because W in Eq. 15–1 is the work done *by* the system, then if work is done *on* the system, W will be negative and U will increase. Similarly, Q is positive for heat added to the system, so if heat leaves the system, Q is negative. [Caution: Elsewhere you may sometimes encounter the opposite convention for W where W is defined as the work done *on* the system; in that case Eq. 15–1 is written as $\Delta U = Q + W$.]

⚠ **CAUTION**
Heat added is +
Heat lost is −
Work on system is −
Work by system is +

Equation 15–1 is known as the **first law of thermodynamics**. It is one of the great laws of physics, and its validity rests on experiments (such as Joule's) to which no exceptions have been seen. Since Q and W represent energy transferred into or out of the system, the internal energy changes accordingly. Thus, the first law of thermodynamics is a general statement of the *law of conservation of energy*.

Note that the conservation of energy law was not able to be formulated until the 1800s, because it depended on the interpretation of heat as a transfer of energy.

A given system does not "have" a certain amount of heat or work. Rather, work and heat are involved in *thermodynamic processes* that can change the system from one state to another; they are not characteristic of the state itself. Quantities which describe the state of a system, such as internal energy U, pressure P, volume V, temperature T, and mass m or number of moles n, are called **state variables**. Q and W are *not* state variables.

⚠ **CAUTION**
P, V, T, U, m, n are state variables. W and Q are not: a system does not have an amount of heat or work

EXAMPLE 15–1 Using the first law. 2500 J of heat is added to a system, and 1800 J of work is done on the system. What is the change in internal energy of the system?

APPROACH We apply the first law of thermodynamics, Eq. 15–1, to our system.

SOLUTION The heat added to the system is $Q = 2500$ J. The work W done *by* the system is -1800 J. Why the minus sign? Because 1800 J done *on* the system (as given) equals -1800 J done *by* the system, and it is the latter we need for the sign conventions we used for Eq. 15–1. Hence

$$\Delta U = 2500 \text{ J} - (-1800 \text{ J}) = 2500 \text{ J} + 1800 \text{ J} = 4300 \text{ J}.$$

NOTE We did this calculation in detail to emphasize the importance of keeping careful track of signs. Both the heat and the work are inputs to the system, so we expect ΔU to be increased by both.

EXERCISE A What would be the internal energy change in Example 15–1 if 2500 J of heat is added to the system and 1800 J of work is done *by* the system (i.e., as output)?

*The First Law of Thermodynamics Extended

To write the first law of thermodynamics in a more complete form, consider a system that is moving so it has kinetic energy KE, and suppose there is also potential energy PE. Then the first law of thermodynamics would have to include these terms and would be written as

$$\Delta \text{KE} + \Delta \text{PE} + \Delta U = Q - W. \quad (15\text{-}2)$$

> **EXAMPLE 15–2 Kinetic energy transformed to thermal energy.** A 3.0-g bullet traveling at a speed of 400 m/s enters a tree and exits the other side with a speed of 200 m/s. Where did the bullet's lost KE go, and how much energy was transferred?
>
> **APPROACH** Take the bullet and tree as our system. No potential energy is involved. No work is done on (or by) the system by outside forces, nor is any heat added because no energy was transferred to or from the system due to a temperature difference. Thus the kinetic energy gets transformed into internal energy of the bullet and tree. This answers the first question.
>
> **SOLUTION** In the first law of thermodynamics as written in Eq. 15–2, we are given $Q = W = \Delta \text{PE} = 0$, so we have
>
> $$\Delta \text{KE} + \Delta U = 0$$
>
> or, using subscripts i and f for initial and final velocities,
>
> $$\Delta U = -\Delta \text{KE} = -(\text{KE}_f - \text{KE}_i) = \tfrac{1}{2} m(v_i^2 - v_f^2)$$
> $$= \tfrac{1}{2}(3.0 \times 10^{-3}\,\text{kg})[(400\,\text{m/s})^2 - (200\,\text{m/s})^2] = 180\,\text{J}.$$
>
> **NOTE** The internal energy of the bullet and tree both increase, as both experience a rise in temperature. If we had chosen the bullet alone as our system, work would be done on it and heat transfer would occur.

15–2 Thermodynamic Processes and the First Law

Let us analyze some thermodynamic processes in light of the first law of thermodynamics.

Isothermal Processes ($\Delta T = 0$)

To begin, we choose a very simple system: a fixed mass of an ideal gas enclosed in a container fitted with a movable piston as shown in Fig. 15–1.

First we consider an idealized process, such as adding heat or doing work, that is carried out at constant temperature. Such a process is called an **isothermal** process (from the Greek meaning "same temperature"). If the system is an ideal gas, then $PV = nRT$ (Eq. 13–3), so for a fixed amount of gas kept at constant temperature, $PV = $ constant. Thus a graph of pressure P vs. volume V, a **PV diagram**, would follow a curve like AB in Fig. 15–2 for an isothermal process. Each point on the curve, such as point A, represents the state of the system at a given moment—that is, its pressure P and volume V. At a lower temperature, another isothermal process would be represented by a curve like A′ B′ in Fig. 15–2 (the product $PV = nRT = $ constant is less when T is less). The curves shown in Fig. 15–2 are referred to as *isotherms*.

We assume the gas is in contact with a **heat reservoir** (a body whose mass is so large that, ideally, its temperature does not change significantly when heat is exchanged with our system). We also assume that a process of compression (volume decrease) or expansion (volume increase) is done very slowly, so that the process can be considered a *series of equilibrium states* all at the same constant temperature.†

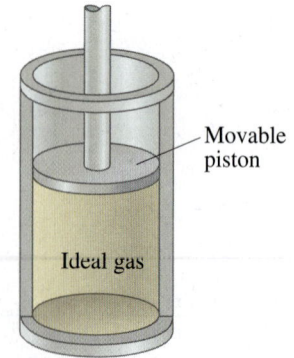

FIGURE 15–1 An ideal gas in a cylinder fitted with a movable piston.

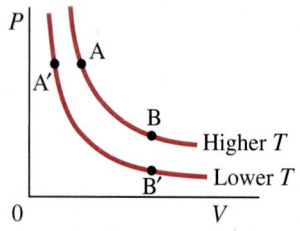

FIGURE 15–2 *PV* diagram for an ideal gas undergoing isothermal processes at two different temperatures.

---
†If a gas expands or is compressed quickly, there is turbulence and different parts of the gas in its container would be at different pressures and temperatures.

If the gas is initially in a state represented by point A in Fig. 15–2, and an amount of heat Q is added to the system, the pressure and volume will change and the state of the system will be represented by another point, B, on the diagram. If the temperature is to remain constant, the gas will expand and do an amount of work W on the environment (it exerts a force on the piston in Fig. 15–1 and moves it through a distance). The temperature and mass are kept constant so, from Eq. 14–1, the internal energy does not change: $\Delta U = \frac{3}{2} nR\, \Delta T = 0$. Hence, by the first law of thermodynamics, Eq. 15–1, $\Delta U = Q - W = 0$, so $W = Q$: the work done by the gas in an isothermal process equals the heat added to the gas.

Adiabatic Processes ($Q = 0$)

An **adiabatic** process is one in which no heat is allowed to flow into or out of the system: $Q = 0$. This situation can occur if the system is extremely well insulated, or the process happens so quickly that heat—which flows slowly—has no time to flow in or out. The very rapid expansion of gases in an internal combustion engine is one example of a process that is very nearly adiabatic. An adiabatic expansion of an ideal gas done very slowly can be represented by a curve like that labeled AC in Fig. 15–3. Since $Q = 0$, we have from Eq. 15–1 that $\Delta U = -W$. When a gas expands, it does work and W is positive, so the internal energy decreases; hence the temperature decreases as well (because $\Delta U = \frac{3}{2}nR\, \Delta T$). This is seen in Fig. 15–3 where the product PV ($= nRT$) is less at point C than at point B. (Compare to curve AB for an isothermal process, in which $\Delta U = 0$ and $\Delta T = 0$.) In the reverse operation, an adiabatic compression (going from C to A, for example), work is done *on* the gas, and hence the internal energy increases and the temperature rises. In a diesel engine, the fuel–air mixture is rapidly compressed adiabatically by a factor of 15 or more; the temperature rise is so great that the mixture ignites spontaneously, without spark plugs.

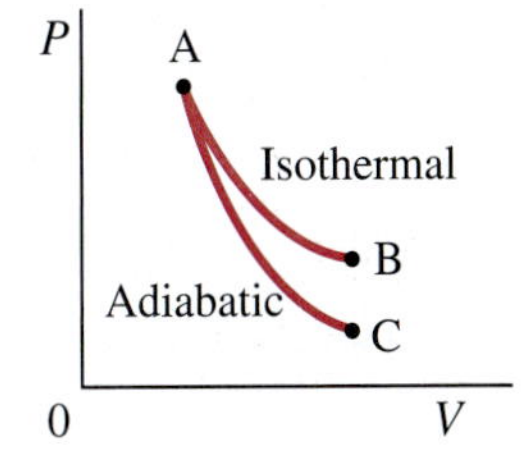

FIGURE 15–3 *PV* diagram for adiabatic (AC) and isothermal (AB) processes on an ideal gas.

Isobaric and Isovolumetric Processes

Isothermal and adiabatic processes are just two possible processes that can occur. Two other simple thermodynamic processes are illustrated on the *PV* diagrams of Fig. 15–4: (a) an **isobaric** process is one in which the pressure is kept constant, so the process is represented by a straight horizontal line on the *PV* diagram (Fig. 15–4a); (b) an **isovolumetric** (or *isochoric*) process is one in which the volume does not change (Fig. 15–4b). In these, and in all other processes, the first law of thermodynamics holds.

FIGURE 15–4 (a) Isobaric ("same pressure") process. (b) Isovolumetric ("same volume") process.

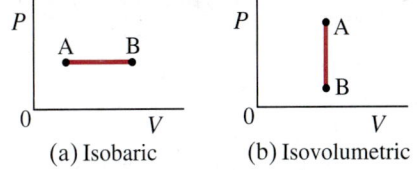

Work Done in Volume Changes

It is often valuable to calculate the work done in a process. If the pressure is kept constant during a process (isobaric), the work done is easily calculated. For example, if the gas in Fig. 15–5 expands very slowly against the piston, the work done by the gas to raise the piston is the force F times the distance d. But the force is just the pressure P of the gas times the area A of the piston, $F = PA$. Thus,

$$W = Fd = PAd.$$

Note that $Ad = \Delta V$, the change in volume of the gas, so

$$W = P\,\Delta V. \quad \text{[constant pressure]} \quad (15\text{–}3)$$

FIGURE 15–5 Work is done on the piston when the gas expands, moving the piston a distance d.

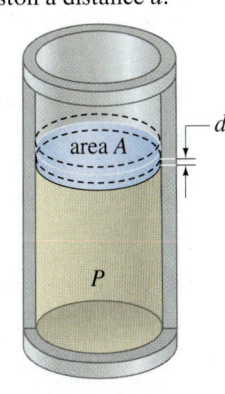

Equation 15–3 also holds if the gas is *compressed* at constant pressure, in which case ΔV is negative (since V decreases); W is then negative, which indicates that work is done *on* the gas. Equation 15–3 is also valid for liquids and solids, as long as the pressure is constant during the process.

In an isovolumetric process (Fig. 15–4b) the volume does not change, so no work is done, $W = 0$.

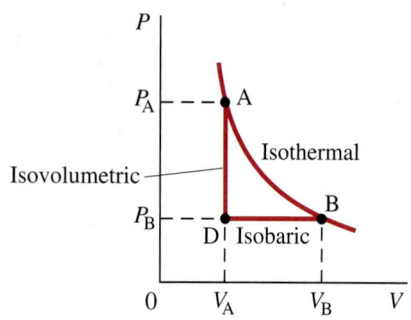

FIGURE 15–6 *PV* diagram for different processes (see the text), where the system changes from A to B.

Figure 15–6 shows the isotherm AB we saw in Fig. 15–2 as well as another possible process represented by the path ADB. In going from A to D, the gas does no work since the volume does not change. But in going from D to B, the gas does work equal to $P_B(V_B - V_A)$, and this is the total work done in the process ADB.

If the pressure varies during a process, such as for the isothermal process AB in Fig. 15–6 (and Fig. 15–2), Eq. 15–3 cannot be used directly to determine the work. A rough estimate can be obtained, however, by using an "average" value for P in Eq. 15–3. More accurately, the work done is equal to the area under the PV curve. This is obvious when the pressure is constant: as Fig. 15–7a shows, the shaded area is just $P_B(V_B - V_A)$, and this is the work done. Similarly, the work done during an isothermal process is equal to the shaded area shown in Fig. 15–7b. The calculation of work done in this case can be carried out using calculus, or by estimating the area on graph paper.

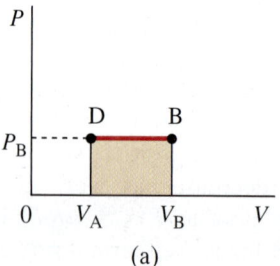
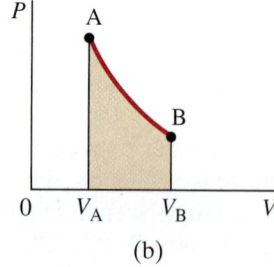

(a)   (b)

FIGURE 15–7 Work done by a gas is equal to the area under the *PV* curve.

CONCEPTUAL EXAMPLE 15–3 **Work in isothermal and adiabatic processes.** In Fig. 15–3 we saw the *PV* diagrams for a gas expanding in two ways, isothermally and adiabatically. The initial volume V_A was the same in each case, and the final volumes were the same $(V_B = V_C)$. In which process was more work done by the gas?

RESPONSE Our system is the gas. More work was done by the gas in the isothermal process, which we can see in two simple ways by looking at Fig. 15–3. First, the "average" pressure was higher during the isothermal process AB, so $W = P_{av}\Delta V$ was greater (ΔV is the same for both processes). Second, we can look at the area under each curve as we showed in Fig. 15–7b: the area under curve AB, which represents the work done, is greater (because curve AB is higher) than the area under AC in Fig. 15–3.

EXERCISE B Is the work done by the gas in process ADB of Fig. 15–6 greater than, less than, or equal to the work done in the isothermal process AB?

Table 15–1 gives a brief summary of the processes we have discussed. Many other types of processes can occur, but these "simple" ones are useful and can be dealt with by fairly simple means.

TABLE 15–1 Simple Thermodynamic Processes and the First Law

Process	What is constant:	The first law, $\Delta U = Q - W$, predicts:
Isothermal	T = constant	$\Delta T = 0$ makes $\Delta U = 0$, so $Q = W$
Isobaric	P = constant	$Q = \Delta U + W = \Delta U + P\Delta V$
Isovolumetric	V = constant	$\Delta V = 0$ makes $W = 0$, so $Q = \Delta U$
Adiabatic	$Q = 0$	$\Delta U = -W$

EXAMPLE 15–4 **First law in isobaric and isovolumetric processes.** An ideal gas is slowly compressed at a constant pressure of 2.0 atm from 10.0 L to 2.0 L. This process is represented in Fig. 15–8 as the path B to D. (In this process, some heat flows out of the gas and the temperature drops.) Heat is then added to the gas, holding the volume constant, and the pressure and temperature are allowed to rise (line DA) until the temperature reaches its original value ($T_A = T_B$). In the process BDA, calculate (a) the total work done by the gas, and (b) the total heat flow into the gas.

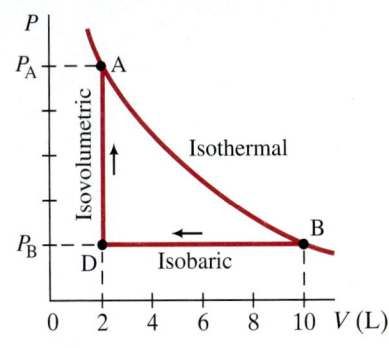

FIGURE 15–8 Example 15–4.

APPROACH (a) Work is done only in the compression process BD. In process DA, the volume is constant so $\Delta V = 0$ and no work is done (Eq. 15–3). (b) We use the first law of thermodynamics, Eq. 15–1.

SOLUTION (a) During the compression BD, the pressure is 2.0 atm $= 2(1.01 \times 10^5 \text{ N/m}^2)$ and the work done is $(1 \text{ L} = 10^3 \text{ cm}^3 = 10^{-3} \text{ m}^3)$

$$\begin{aligned} W &= P\,\Delta V \\ &= (2.02 \times 10^5 \text{ N/m}^2)[(2.0 \times 10^{-3} \text{ m}^3) - (10.0 \times 10^{-3} \text{ m}^3)] \\ &= -1.6 \times 10^3 \text{ J}. \end{aligned}$$

The total work done *by* the gas is -1.6×10^3 J, where the minus sign means that $+1.6 \times 10^3$ J of work is done *on* the gas.

(b) Because the temperature at the beginning and at the end of process BDA is the same, there is no change in internal energy: $\Delta U = 0$. From the first law of thermodynamics we have

$$0 = \Delta U = Q - W,$$

so

$$Q = W = -1.6 \times 10^3 \text{ J}.$$

Because Q is negative, 1600 J of heat flows out of the gas for the whole process, BDA.

EXERCISE C In Example 15–4, if the heat lost from the gas in the process BD is 8.4×10^3 J, what is the change in internal energy of the gas during process BD?

EXAMPLE 15–5 **Work done in an engine.** In an engine, 0.25 mol of an ideal monatomic gas in the cylinder expands rapidly and adiabatically against the piston. In the process, the temperature of the gas drops from 1150 K to 400 K. How much work does the gas do?

APPROACH We take the gas as our system (the piston is part of the surroundings). The pressure is not constant, so we can't use Eq. 15–3 ($W = P\,\Delta V$). Instead, we can use the first law of thermodynamics to find W because we can determine ΔU (from ΔT) and $Q = 0$ (the process is adiabatic).

SOLUTION We determine ΔU from Eq. 14–1 for the internal energy of an ideal monatomic gas, using subscripts f and i for final and initial states:

$$\begin{aligned} \Delta U &= U_f - U_i = \tfrac{3}{2} nR(T_f - T_i) \\ &= \tfrac{3}{2}(0.25 \text{ mol})(8.314 \text{ J/mol} \cdot \text{K})(400 \text{ K} - 1150 \text{ K}) \\ &= -2300 \text{ J}. \end{aligned}$$

Then, from the first law of thermodynamics, Eq. 15–1, the work done by the gas is

$$W = Q - \Delta U = 0 - (-2300 \text{ J}) = 2300 \text{ J}.$$

EXAMPLE 15–6 ΔU **for boiling water to steam.** Determine the change in internal energy of 1.00 liter of water (mass 1.00 kg) at 100°C when it is fully boiled from liquid to gas, which results in 1671 liters of steam at 100°C. Assume the process is done at atmospheric pressure.

APPROACH Our system is the water. The heat required here does not result in a temperature change; rather, a change in phase occurs. We can determine the heat Q required using the latent heat of water, as in Section 14–5. Work too will be done: $W = P\,\Delta V$. The first law of thermodynamics will then give us ΔU.

SOLUTION The latent heat of vaporization of water (Table 14–3) is $L_V = 22.6 \times 10^5$ J/kg. So the heat input required for this process is (Eq. 14–4)

$$Q = mL = (1.00\,\text{kg})(22.6 \times 10^5\,\text{J/kg}) = 22.6 \times 10^5\,\text{J}.$$

The work done by the water is (Eq. 15–3 since P is constant)

$$W = P\,\Delta V = (1.01 \times 10^5\,\text{N/m}^2)[(1671 \times 10^{-3}\,\text{m}^3) - (1 \times 10^{-3}\,\text{m}^3)]$$
$$= 1.69 \times 10^5\,\text{J},$$

where we used 1 atm = 1.01×10^5 N/m² and 1 L = 10^3 cm³ = 10^{-3} m³. Then

$$\Delta U = Q - W = (22.6 \times 10^5\,\text{J}) - (1.7 \times 10^5\,\text{J}) = 20.9 \times 10^5\,\text{J}.$$

NOTE Most of the heat added goes to increasing the internal energy of the water (increasing molecular energy to overcome the attraction that held the molecules close together in the liquid state). Only a small part (< 10%) goes into doing work.

NOTE Equation 14–1, $U = \tfrac{3}{2}nRT$, would tell us that $\Delta U = 0$ here because $\Delta T = 0$. Yet we determined that $\Delta U = 21 \times 10^5$ J. What is wrong? Equation 14–1 applies only to an ideal monatomic gas, not to liquid water.

*15–3 Human Metabolism and the First Law

PHYSICS APPLIED
Energy in the human body

Human beings and other animals do work. Work is done when a person walks or runs, or lifts a heavy object. Work requires energy. Energy is also needed for growth—to make new cells, and to replace old cells that have died. A great many energy-transforming processes occur within an organism, and they are referred to as *metabolism*.

We can apply the first law of thermodynamics,

$$\Delta U = Q - W,$$

to an organism: say, the human body. Work W is done by the body in its various activities; if this is not to result in a decrease in the body's internal energy (and temperature), energy must somehow be added to compensate. The body's internal energy is not maintained by a flow of heat Q into the body, however. Normally, the body is at a higher temperature than its surroundings, so heat usually flows *out* of the body. Even on a very hot day when heat is absorbed, the body has no way of utilizing this heat to support its vital processes. What then is the source of energy that allows us to do work? It is the internal energy (chemical potential energy) stored in foods (Fig. 15–9). In a closed system, the internal energy changes only as a result of heat flow or work done. In an open system, such as a human, internal energy itself can flow into or out of the system. When we eat food, we are bringing internal energy into our bodies directly, which thus increases the total internal energy U in our bodies. This energy eventually goes into work and heat flow from the body according to the first law.

The metabolic rate is the rate at which internal energy is transformed within the body. It is usually specified in kcal/h or in watts. Typical metabolic rates for a variety of human activities are given in Table 15–2 (top of next page) for an "average" 65-kg adult.

FIGURE 15–9 Bike rider getting an input of energy.

EXAMPLE 15–7 Energy transformation in the body. How much energy is transformed in 24 h by a 65-kg person who spends 8.0 h sleeping, 1.0 h at moderate physical labor, 4.0 h in light activity, and 11.0 h working at a desk or relaxing?

APPROACH The energy transformed during each activity equals the metabolic rate (Table 15–2) multiplied by the time.

SOLUTION Table 15–2 gives the metabolic rate in watts (J/s). Since there are 3600 s in an hour, the total energy transformed is

$$\begin{bmatrix}(8.0\,\text{h})(70\,\text{J/s}) + (1.0\,\text{h})(460\,\text{J/s})\\ + (4.0\,\text{h})(230\,\text{J/s}) + (11.0\,\text{h})(115\,\text{J/s})\end{bmatrix}(3600\,\text{s/h}) = 1.15 \times 10^7\,\text{J}.$$

NOTE Since $4.186 \times 10^3\,\text{J} = 1\,\text{kcal}$, this is equivalent to 2800 kcal; a food intake of 2800 Cal would compensate for this energy output. A 65-kg person who wanted to lose weight would have to eat less than 2800 Cal a day, or increase his or her level of activity. Exercise beats *any* diet technique.

TABLE 15–2 Metabolic Rates (65-kg human)

Activity	Metabolic Rate (approximate)	
	kcal/h	watts
Sleeping	60	70
Sitting upright	100	115
Light activity (eating, dressing, household chores)	200	230
Moderate work (tennis, walking)	400	460
Running (15 km/h)	1000	1150
Bicycling (race)	1100	1270

15–4 The Second Law of Thermodynamics—Introduction

The first law of thermodynamics states that energy is conserved. There are, however, many processes we can imagine that conserve energy but are not observed to occur in nature. For example, when a hot object is placed in contact with a cold object, heat flows from the hotter one to the colder one, never spontaneously from colder to hotter. If heat were to leave the colder object and pass to the hotter one, energy could still be conserved. Yet it does not happen spontaneously.[†] As a second example, consider what happens when you drop a rock and it hits the ground. The initial potential energy of the rock changes to kinetic energy as the rock falls. When the rock hits the ground, this energy in turn is transformed into internal energy of the rock and the ground in the vicinity of the impact; the molecules move faster and the temperature rises slightly. But have you seen the reverse happen—a rock at rest on the ground suddenly rise up in the air because the thermal energy of molecules is transformed into kinetic energy of the rock as a whole? Energy could be conserved in this process, yet we never see it happen.

There are many other examples of processes that occur in nature but whose reverse does not. Here are two more. (1) If you put a layer of salt in a jar and cover it with a layer of similar-sized grains of pepper, when you shake it you get a thorough mixture. But no matter how long you shake it, the mixture does not separate into two layers again. (2) Coffee cups and glasses break spontaneously if you drop them. But they do not go back together spontaneously (Fig. 15–10).

The first law of thermodynamics (conservation of energy) would not be violated if any of these processes occurred in reverse. To explain this lack of reversibility, scientists in the latter half of the nineteenth century formulated a new principle known as the second law of thermodynamics.

[†]By spontaneously, we mean by itself without input of work of some sort. (A refrigerator does move heat from a cold environment to a warmer one, but only because its motor does work—Section 15–6.)

(a) Initial state.　　(b) Later: cup reassembles and rises up.　　(c) Later still: cup lands on table.

FIGURE 15–10 Have you ever observed this process, a broken cup spontaneously reassembling and rising up onto a table? This process could conserve energy. But it never happens.

SECOND LAW OF THERMODYNAMICS
(Clausius statement)

The **second law of thermodynamics** is a statement about which processes occur in nature and which do not. It can be stated in a variety of ways, all of which are equivalent. One statement, due to R. J. E. Clausius (1822–1888), is that

heat can flow spontaneously from a hot object to a cold object; heat will not flow spontaneously from a cold object to a hot object.

Since this statement applies to one particular process, it is not obvious how it applies to other processes. A more general statement is needed that will include other possible processes in a more obvious way.

The development of a general statement of the second law of thermodynamics was based partly on the study of heat engines. A **heat engine** is any device that changes thermal energy into mechanical work, such as a steam engine or an automobile engine. We now examine heat engines, both from a practical point of view and to show their importance in developing the second law of thermodynamics.

15–5 Heat Engines

It is easy to produce thermal energy by doing work—for example, by simply rubbing your hands together briskly, or indeed by any frictional process. But to get work from thermal energy is more difficult, and a practical device to do so was invented only about 1700 with the development of the steam engine.

The basic idea behind any heat engine is that mechanical energy can be obtained from thermal energy only when heat is allowed to flow from a high temperature to a low temperature. In the process, some of the heat can then be transformed to mechanical work, as diagrammed schematically in Fig. 15–11. Useful heat engines run in a repeating *cycle*: that is, the system returns repeatedly to its starting point, and thus can run continuously. In each cycle the change in internal energy of the system is $\Delta U = 0$ because it returns to the starting state. Thus a heat input Q_H at a high temperature T_H is partly transformed into work W and partly exhausted as heat Q_L at a lower temperature T_L (Fig. 15–11). By conservation of energy, $Q_H = W + Q_L$. The high and low temperatures, T_H and T_L, are called the **operating temperatures** of the engine. Note carefully that we are now using a new (and intuitive) sign convention for heat engines: we take Q_H, Q_L, and W as always positive. The direction of each energy transfer is shown by the arrow on the applicable diagram, such as Fig. 15–11.

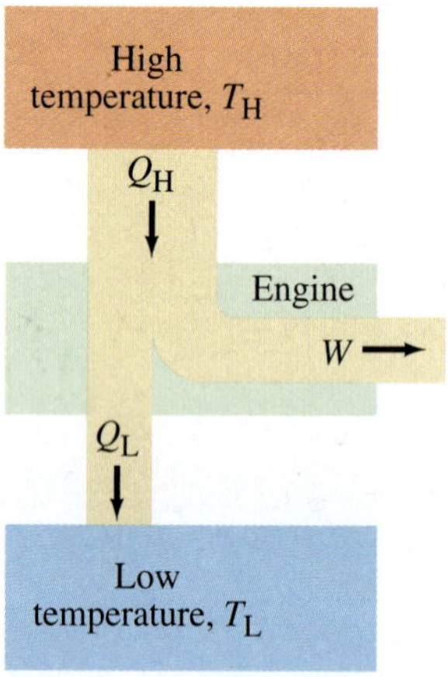

FIGURE 15–11 Schematic diagram of energy transfers for a heat engine.

CAUTION
Sign convention for heat engines:
$Q_H > 0$, $Q_L > 0$, $W > 0$

PHYSICS APPLIED
Engines

Steam Engine and Internal Combustion Engine

The operation of a steam engine is shown in Fig. 15–12. Steam engines are of two main types, each using steam heated by combustion of coal, oil, or gas, or by nuclear energy.

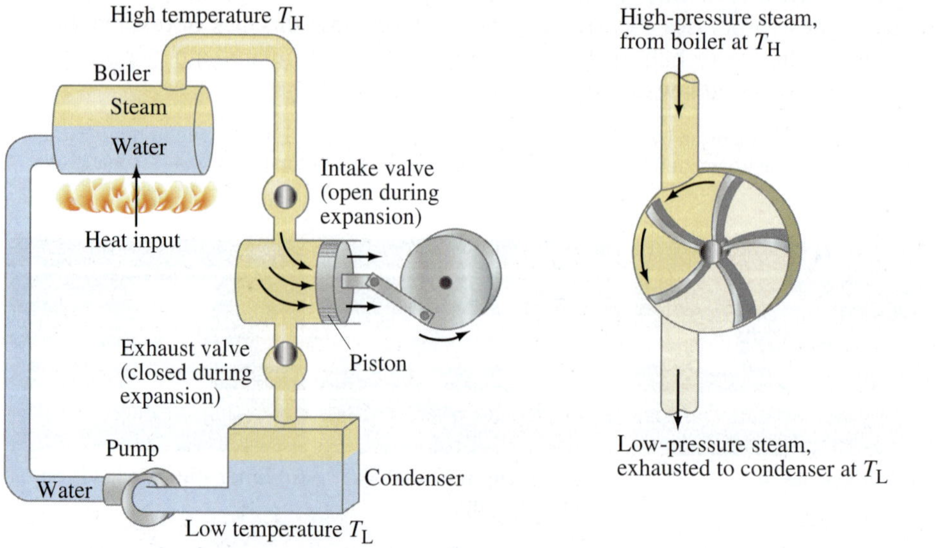

FIGURE 15–12 Steam engines.

(a) Reciprocating type

(b) Turbine (boiler and condenser not shown)

In a reciprocating engine, Fig. 15–12a, the heated steam passes through the intake valve and expands against a piston, forcing it to move. As the piston returns to its original position, it forces the gases out the exhaust valve which opens. A steam turbine, Fig. 15–12b, is very similar except that the reciprocating piston is replaced by a rotating turbine that resembles a paddlewheel with many sets of blades. Most of our electricity today is generated using steam turbines.[†] The material that is heated and cooled, steam in this case, is called the **working substance**. In an old-time steam engine (see page 412), the high temperature is obtained by burning coal, oil, or other fuel to heat the steam.

In an internal combustion engine (used in most automobiles), the high temperature is achieved by burning the gasoline–air mixture in the cylinder itself (ignited by the spark plug), as described in Fig. 15–13.

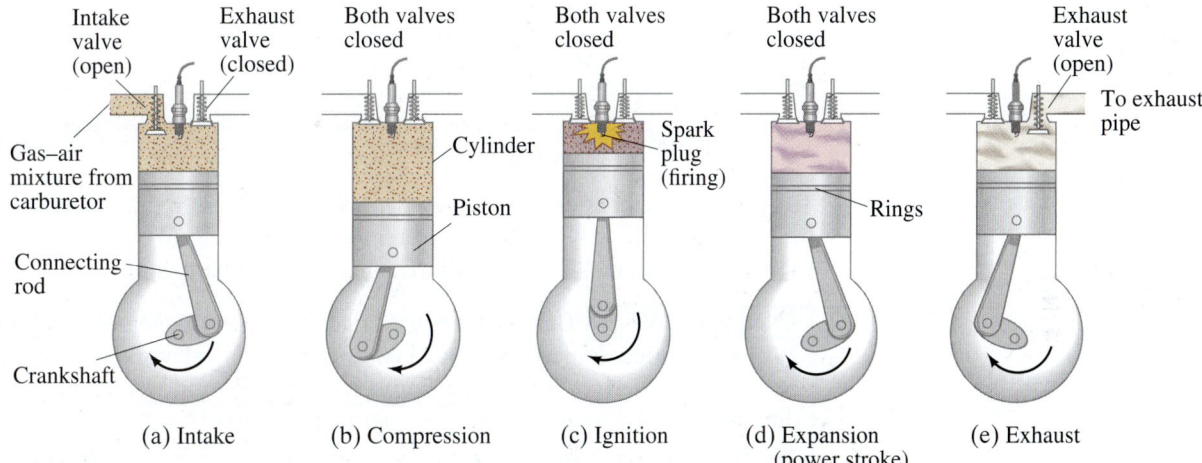

FIGURE 15–13 Four-stroke-cycle internal combustion engine: (a) the gasoline–air mixture flows into the cylinder as the piston moves down; (b) the piston moves upward and compresses the gas; (c) the brief instant when firing of the spark plug ignites the highly compressed gasoline–air mixture, raising it to a high temperature; (d) the gases, now at high temperature and pressure, expand against the piston in this, the power stroke; (e) the burned gases are pushed out to the exhaust pipe. When the piston reaches the top, the exhaust valve closes and the intake valve opens, and the whole cycle repeats. (a), (b), (d), and (e) are the four strokes of the cycle.

*Why a ΔT Is Needed to Drive a Heat Engine

To see why a *temperature difference* is required to run an engine, consider a steam engine. In the reciprocating engine, for example, suppose there were no condenser or pump (Fig. 15–12a), and that the steam was at the same temperature throughout the system. Then the pressure of the gas being exhausted would be the same as on intake. The work done by the gas *on* the piston when it expanded would equal the amount of work done *by* the piston to force the steam out the exhaust; hence, no net work would be done. In a real engine, the exhausted gas is cooled to a lower temperature and condensed so that the exhaust pressure is less than the intake pressure. Thus, the work the piston must do on the gas to expel it on the exhaust stroke is less than the work done by the gas on the piston during the intake. So a net amount of work can be obtained—but only if there is this difference of temperature. Similarly, in the gas turbine if the gas isn't cooled, the pressure on each side of the blades would be the same. By cooling the gas on the exhaust side, the pressure on the back side of the blade is less and hence the turbine turns.

[†]Even nuclear power plants utilize steam turbines; the nuclear fuel—uranium—serves as fuel to heat the steam.

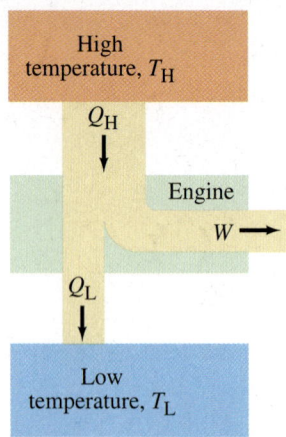

FIGURE 15–11 (Repeated.) Schematic diagram of energy transfers for a heat engine.

Efficiency

The **efficiency**, e, of any heat engine can be defined as the ratio of the work it does, W, to the heat input at the high temperature, Q_H (Fig. 15–11):

$$e = \frac{W}{Q_H}. \quad (15\text{–}4a)$$

This is a sensible definition since W is the output (what you get from the engine), whereas Q_H is what you put in and pay for in burned fuel. Since energy is conserved, the heat input Q_H must equal the work done plus the heat that flows out at the low temperature (Q_L):

$$Q_H = W + Q_L.$$

Thus $W = Q_H - Q_L$, and the efficiency of an engine is

$$e = \frac{W}{Q_H} = \frac{Q_H - Q_L}{Q_H}$$

or

$$e = 1 - \frac{Q_L}{Q_H}. \quad (15\text{–}4b)$$

To give the efficiency as a percent, we multiply Eq. 15-4 by 100. Note that e could be 1.0 (or 100%) only if Q_L were zero—that is, only if no heat were exhausted to the environment (which we will see shortly never happens).

EXAMPLE 15–8 Car efficiency. An automobile engine has an efficiency of 20% and produces an average of 23,000 J of mechanical work per second during operation. (a) How much heat input is required, and (b) how much heat is discharged as waste heat from this engine, per second?

APPROACH We want to find the heat input Q_H as well as the heat output Q_L, given $W = 23,000$ J each second and an efficiency $e = 0.20$. We can use the definition of efficiency, Eq. 15-4 in its various forms, to find first Q_H and then Q_L.

SOLUTION (a) From Eq. 15-4a, $e = W/Q_H$, we solve for Q_H:

$$Q_H = \frac{W}{e} = \frac{23{,}000\text{ J}}{0.20}$$
$$= 1.15 \times 10^5 \text{ J} = 115 \text{ kJ}.$$

The engine requires 115 kJ/s = 115 kW of heat input.

(b) Now we use Eq. 15-4b $(e = 1 - Q_L/Q_H)$ and solve for Q_L:

$$\frac{Q_L}{Q_H} = 1 - e$$

so

$$Q_L = (1-e)Q_H = (0.80)115 \text{ kJ}$$
$$= 92 \text{ kJ}.$$

The engine discharges heat to the environment at a rate of 92 kJ/s = 92 kW.

NOTE Of the 115 kJ that enters the engine per second, only 23 kJ (20%) does useful work whereas 92 kJ (80%) is wasted as heat output.

NOTE The problem was stated in terms of energy per unit time. We could just as well have stated it in terms of power, since 1 J/s = 1 watt.

Carnot Engine

To see how to increase efficiency, the French scientist Sadi Carnot (1796–1832) examined the characteristics of an ideal engine, now called a **Carnot engine**. No Carnot engine actually exists, but as a theoretical idea it played an important role in the development and understanding of the second law of thermodynamics.

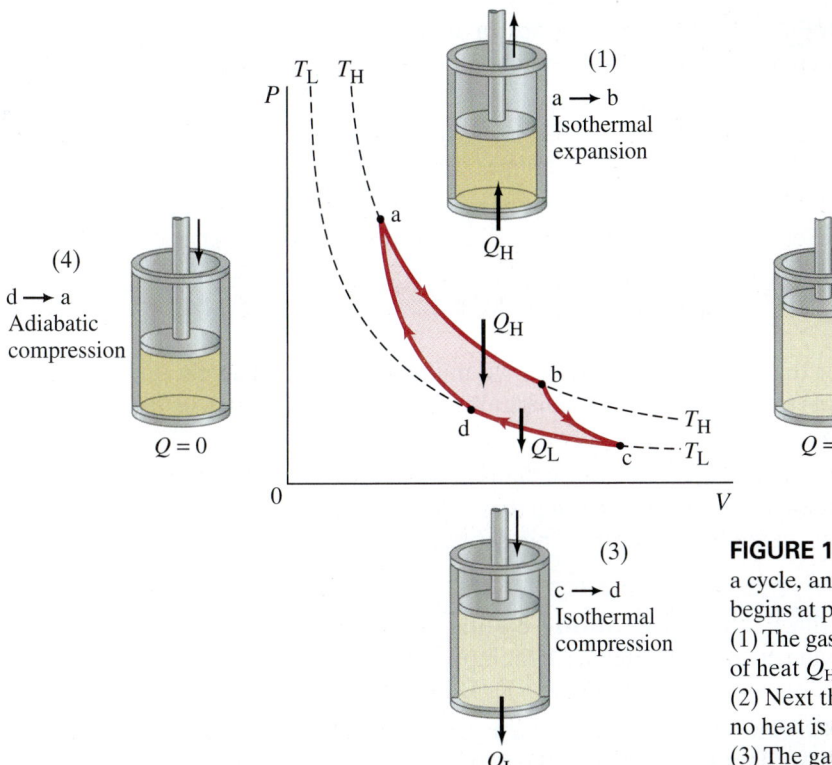

FIGURE 15–14 The Carnot cycle. Heat engines work in a cycle, and the cycle for the theoretical Carnot engine begins at point "a" on this PV diagram for an ideal gas. (1) The gas is first expanded isothermally, with the addition of heat Q_H, along the path "ab" at temperature T_H. (2) Next the gas expands adiabatically from "b" to "c"—no heat is exchanged, but the temperature drops to T_L. (3) The gas is then compressed at constant temperature T_L, path cd, and heat Q_L flows out. (4) Finally, the gas is compressed adiabatically, path da, back to its original state.

The idealized Carnot engine consisted of four processes done in a cycle, two of which are adiabatic ($Q = 0$) and two are isothermal ($\Delta T = 0$). This idealized cycle is shown in Fig. 15–14. Each of the processes was considered to be done **reversibly**. That is, each of the processes (say, during expansion of the gases against a piston) was done so slowly that the process could be considered a series of equilibrium states, and the whole process could be done in reverse with no change in the magnitude of work done or heat exchanged. A real process, on the other hand, would occur more quickly; there would be turbulence in the gas, friction would be present, and so on. Because of these factors, a real process cannot be done precisely in reverse—the turbulence would be different and the heat lost to friction would not reverse itself. Thus, real processes are **irreversible**.

The isothermal processes of a Carnot engine, where heats Q_H and Q_L are transferred, are assumed to be done at constant temperatures T_H and T_L. That is, the system is assumed to be in contact with idealized *heat reservoirs* (page 414) which are so large their temperatures don't change significantly when Q_H and Q_L are transferred.

Carnot showed that for an ideal reversible engine, the heats Q_H and Q_L are proportional to the operating temperatures T_H and T_L (in kelvins): $Q_H/Q_L = T_H/T_L$. So the efficiency can be written as

$$e_{\text{ideal}} = \frac{T_H - T_L}{T_H} = 1 - \frac{T_L}{T_H}. \qquad \begin{bmatrix} \text{Carnot (ideal)} \\ \text{efficiency} \end{bmatrix} \quad (15\text{-}5)$$

Equation 15–5 expresses the fundamental upper limit to the efficiency of any heat engine. A higher efficiency would violate the second law of thermodynamics.[†] Real engines always have an efficiency lower than this because of losses due to friction and the like. Real engines that are well designed reach 60 to 80% of the Carnot efficiency.

[†]If an engine had a higher efficiency than Eq. 15–5, it could be used in conjunction with a Carnot engine that is made to work in reverse as a refrigerator. If W was the same for both, the net result would be a flow of heat at a low temperature to a high temperature without work being done. That would violate the Clausius statement of the second law.

EXAMPLE 15–9 Steam engine efficiency. A steam engine operates between 500°C and 270°C. What is the maximum possible efficiency of this engine?

APPROACH The maximum possible efficiency is the idealized Carnot efficiency, Eq. 15–5. We must use kelvin temperatures.

SOLUTION We first change the temperature to kelvins by adding 273 to the given Celsius temperatures: $T_H = 773$ K and $T_L = 543$ K. Then

$$e_{ideal} = 1 - \frac{543}{773} = 0.30.$$

To get the efficiency in percent, we multiply by 100. Thus, the maximum (or Carnot) efficiency is 30%. Realistically, an engine might attain 0.70 of this value, or 21%.

NOTE In this Example the exhaust temperature is still rather high, 270°C. Steam engines are often arranged in series so that the exhaust of one engine is used as intake by a second or third engine.

EXAMPLE 15–10 A phony claim? An engine manufacturer makes the following claims: An engine's heat input per second is 9.0 kJ at 435 K. The heat output per second is 4.0 kJ at 285 K. Do you believe these claims?

APPROACH The engine's efficiency can be calculated from the definition, Eq. 15–4. It must be less than the maximum possible, Eq. 15–5.

SOLUTION The claimed efficiency of the engine is (Eq. 15–4)

$$e = \frac{Q_H - Q_L}{Q_H}$$

$$= \frac{9.0 \text{ kJ} - 4.0 \text{ kJ}}{9.0 \text{ kJ}} = 0.56,$$

or 56%. The maximum possible efficiency is given by the Carnot efficiency, Eq. 15–5:

$$e_{ideal} = \frac{T_H - T_L}{T_H}$$

$$= \frac{435 \text{ K} - 285 \text{ K}}{435 \text{ K}} = 0.34,$$

or 34%. The manufacturer's claims violate the second law of thermodynamics and cannot be believed.

EXERCISE D A motor is running with an intake temperature $T_H = 400$ K and an exhaust temperature $T_L = 300$ K. Which of the following are *not* possible efficiencies for the engine? (*a*) 0.10; (*b*) 0.16; (*c*) 0.24; (*d*) 0.30; (*e*) 0.33.

We can see from Eq. 15–5 that at normal temperatures, a 100% efficient engine is not possible. Only if the exhaust temperature, T_L, were at absolute zero would 100% efficiency be reachable. But getting to absolute zero is a practical (as well as theoretical) impossibility. [Careful experimentation suggests that absolute zero is unattainable. This result is known as the **third law of thermodynamics**.] Because no engine can be 100% efficient, it can be stated that

no device is possible whose sole effect is to transform a given amount of heat completely into work.

This is known as the **Kelvin-Planck statement of the second law of thermodynamics**. Figure 15–15 diagrams the ideal perfect heat engine, which can not exist.

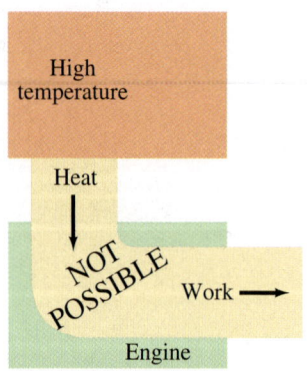

FIGURE 15–15 Diagram of an impossible perfect heat engine in which all heat input is used to do work.

SECOND LAW OF THERMODYNAMICS
(Kelvin-Planck statement)

If the second law were not true, so that a perfect engine could be built, rather remarkable things could happen. For example, if the engine of a ship did not need a low-temperature reservoir to exhaust heat into, the ship could sail across the ocean using the vast resources of the internal energy of the ocean water. Indeed, we would have no fuel problems at all!

> **EXERCISE E** Return to the Chapter-Opening Question, page 412, and answer it again now. Try to explain why you may have answered differently the first time.

15–6 Refrigerators, Air Conditioners, and Heat Pumps

The operating principle of refrigerators, air conditioners, and heat pumps is just the reverse of a heat engine. Each operates to transfer heat *out* of a cool environment into a warm environment. As diagrammed in Fig. 15–16, by doing work W, heat is taken from a low-temperature region, T_L (such as inside a refrigerator), and a greater amount of heat is exhausted at a high temperature, T_H (the room). Heat Q_L is removed from cooling coils *inside* the refrigerator and heat Q_H is given off by coils *outside* the rear of the refrigerator, Fig. 15–17. You can often feel this heated air coming out beneath the refrigerator. The work W is usually done by an electric motor which compresses a fluid, as illustrated in Fig. 15–17. (We assume Q_L, Q_H, and W are all positive, as in Section 15–5.)

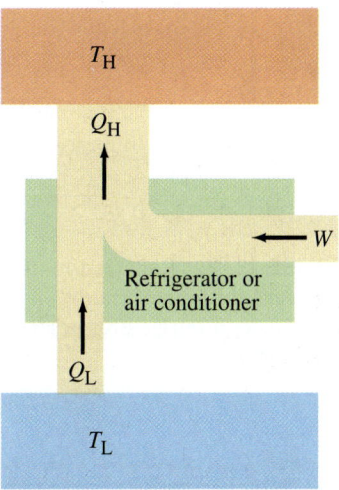

FIGURE 15–16 Schematic diagram of energy transfers for a refrigerator or air conditioner (AC).

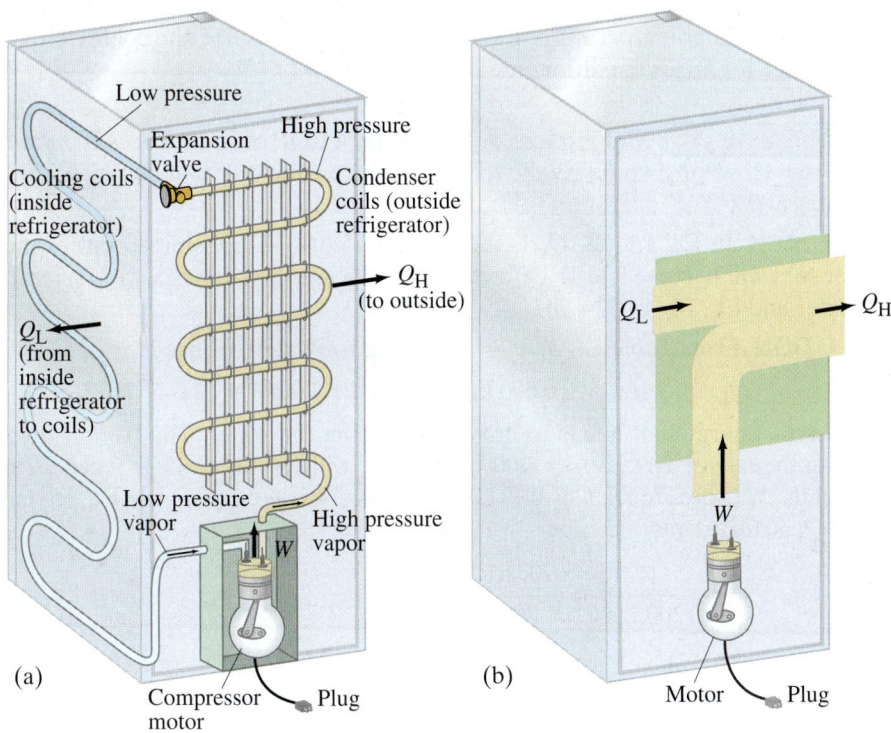

FIGURE 15–17 (a) Typical refrigerator system. The electric compressor motor forces a gas at high pressure through a heat exchanger (condenser) on the rear *outside* wall of the refrigerator, where Q_H is given off and the gas cools to become liquid. The liquid passes from a high-pressure region, via a valve, to low-pressure tubes on the *inside* walls of the refrigerator; the liquid evaporates at this lower pressure and thus absorbs heat (Q_L) from the inside of the refrigerator. The fluid returns to the compressor, where the cycle begins again. (b) Schematic diagram, like Fig. 15–16.

A perfect **refrigerator**—one in which no work is required to take heat from the low-temperature region to the high-temperature region—is not possible. This is the **Clausius statement of the second law of thermodynamics**, already mentioned in Section 15–4; it can be stated formally as

> **no device is possible whose sole effect is to transfer heat from one system at a temperature T_L into a second system at a higher temperature T_H.**

To make heat flow from a low-temperature object (or system) to one at a higher temperature, work must be done. Thus, *there can be no perfect refrigerator.*

PHYSICS APPLIED
Refrigerator

SECOND LAW OF THERMODYNAMICS
(Clausius statement)

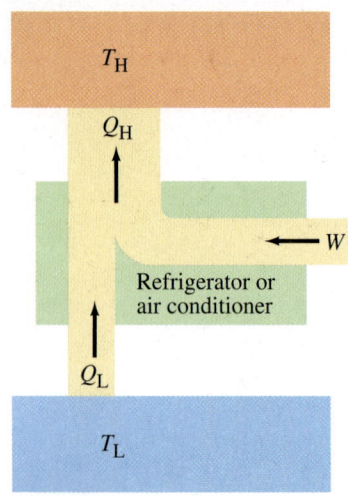

FIGURE 15-16 (Repeated.) Schematic diagram of energy transfers for a refrigerator or air conditioner (AC).

PHYSICS APPLIED
Air conditioner

The **coefficient of performance** (COP) of a refrigerator is defined as the heat Q_L removed from the low-temperature area (inside a refrigerator) divided by the work W done to remove the heat (Fig. 15–16):

$$\text{COP} = \frac{Q_L}{W}. \qquad \begin{bmatrix}\text{refrigerator and}\\ \text{air conditioner}\end{bmatrix} \quad (15\text{–}6a)$$

We use Q_L because it is the heat removed from inside that matters from a practical point of view. This makes sense because the more heat Q_L that can be removed from inside the refrigerator for a given amount of work, the better (more efficient) the refrigerator is. Energy is conserved, so from the first law of thermodynamics we can write $Q_L + W = Q_H$, or $W = Q_H - Q_L$ (see Fig. 15–16). Then Eq. 15-6a becomes

$$\text{COP} = \frac{Q_L}{W} = \frac{Q_L}{Q_H - Q_L}. \qquad \begin{bmatrix}\text{refrigerator and}\\ \text{air conditioner}\end{bmatrix} \quad (15\text{–}6b)$$

For an ideal refrigerator (not a perfect one, which is impossible), the best we could do would be

$$\text{COP}_{\text{ideal}} = \frac{T_L}{T_H - T_L}, \qquad \begin{bmatrix}\text{refrigerator and}\\ \text{air conditioner}\end{bmatrix} \quad (15\text{–}6c)$$

analogous to an ideal (Carnot) engine (Eq. 15–5).

An **air conditioner** works very much like a refrigerator, although the actual construction details are different: an air conditioner takes heat Q_L from inside a room or building at a low temperature, and deposits heat Q_H outside to the environment at a higher temperature. Equations 15–6 also describe the coefficient of performance for an air conditioner.

EXAMPLE 15–11 Making ice. A freezer has a COP of 2.8 and uses 200 watts of power. How long would it take to freeze an ice-cube tray that contains 600 g of water at 0°C?

APPROACH In Eq. 15–6b, Q_L is the heat that must be transferred out of the water so it will become ice. To determine Q_L, we use the latent heat of fusion L of water and Eq. 14–4, $Q = mL$.

SOLUTION From Table 14–3, $L = 333$ kJ/kg for water. Hence

$$Q_L = mL = (0.600\text{ kg})(3.33 \times 10^5\text{ J/kg}) = 2.0 \times 10^5\text{ J}$$

is the total energy that needs to be removed from the water. The freezer does work at the rate of 200 watts = 200 J/s = W/t, which is the work W it can do in t seconds. We solve for t: $t = W/(200\text{ J/s})$. For W, we can also use Eq. 15-6a: $W = Q_L/\text{COP}$. Thus

$$t = \frac{W}{200\text{ J/s}} = \frac{Q_L/\text{COP}}{200\text{ J/s}}$$

$$= \frac{2.0 \times 10^5\text{ J}}{(2.8)(200\text{ J/s})} = 360\text{ s},$$

or about 6 min.

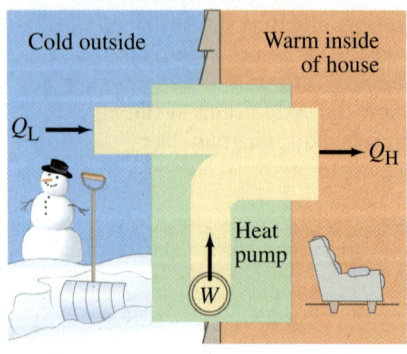

FIGURE 15–18 A heat pump uses an electric motor to "pump" heat from the cold outside to the warm inside of a house.

PHYSICS APPLIED
Heat pump

Heat naturally flows from high temperature to low temperature. Refrigerators and air conditioners do work to accomplish the opposite: to make heat flow from cold to hot. We might say they "pump" heat from cold areas to hotter areas, against the natural tendency of heat to flow from hot to cold, just as water can be pumped uphill, against the natural tendency to flow downhill. The term **heat pump** is usually reserved for a device that can heat a house in winter by using an electric motor that does work W to take heat Q_L from the outside at low temperature and delivers heat Q_H to the warmer inside of the house; see Fig. 15–18.

As in a refrigerator, there is an indoor and an outdoor heat exchanger (coils of the refrigerator) and an electric compressor motor. The operating principle is like that for a refrigerator or air conditioner; but the objective of a heat pump is to heat (deliver Q_H), rather than to cool (remove Q_L). Thus, the coefficient of performance of a heat pump is defined differently than for an air conditioner because it is the heat Q_H delivered to the inside of the house that is important now:

$$\text{COP} = \frac{Q_H}{W}. \qquad \text{[heat pump]} \quad (15\text{–}7)$$

The COP is necessarily greater than 1. Typical heat pumps today have COP ≈ 2.5 to 3. Most heat pumps can be "turned around" and used as air conditioners in the summer.

EXAMPLE 15–12 Heat pump. A heat pump has a coefficient of performance of 3.0 and is rated to do work at 1500 watts. (*a*) How much heat can it add to a room per second? (*b*) If the heat pump were turned around to act as an air conditioner in the summer, what would you expect its coefficient of performance to be, assuming all else stays the same?

APPROACH We use the definitions of coefficient of performance, which are different for the two devices in (*a*) and (*b*).

SOLUTION (*a*) We use Eq. 15–7 for the heat pump, and, since our device does 1500 J of work per second, it can pour heat into the room at a rate of

$$Q_H = \text{COP} \times W = 3.0 \times 1500 \text{ J} = 4500 \text{ J}$$

per second, or at a rate of 4500 W. [≈ 4 Btu/s.]

(*b*) If our device is turned around in summer, it can take heat Q_L from inside the house, doing 1500 J of work per second to then dump Q_H = 4500 J per second to the hot outside. Energy is conserved, so $Q_L + W = Q_H$ (see Fig. 15–18, but reverse the inside and outside of the house). Then

$$Q_L = Q_H - W = 4500 \text{ J} - 1500 \text{ J} = 3000 \text{ J}.$$

The coefficient of performance as an air conditioner would thus be (Eq. 15–6a)

$$\text{COP} = \frac{Q_L}{W} = \frac{3000 \text{ J}}{1500 \text{ J}} = 2.0.$$

NOTE The coefficients of performance are defined differently for heat pumps and air conditioners.

CAUTION
Heat pumps and air conditioners have different COP definitions

EXERCISE F The heat pump of Example 15–12 uses 1500 W of electric power to deliver 4500 W of heat. Does this sound like we're getting something for nothing? (*a*) Explain why we aren't. (*b*) Compare to the refrigerator of Example 15–11: the motor uses 200 W to extract how much heat?

A good heat pump can sometimes be a money saver and an energy saver, depending on the cost of the unit and installation, etc. Compare, for example, our heat pump in Example 15–12 to, say, a 1500-W electric heater. We plug the heater into the wall, it draws 1500 W of electricity, and it delivers 1500 W of heat to the room. Our heat pump when plugged into the wall also draws 1500 W of electricity (which is what we pay for), but it delivers 4500 W of heat!

*SEER Rating

Cooling devices such as refrigerators and air conditioners are often given a rating known as SEER (Seasonal Energy Efficiency Ratio), which is defined as

$$\text{SEER} = \frac{\text{(heat removed in Btu)}}{\text{(electrical input in watt-hours)}},$$

PHYSICS APPLIED
SEER rating

as measured by averaging over varying (seasonal) conditions. The definition of the SEER is basically the same as the COP except for the (unfortunate) mixed units. Given that 1 Btu = 1056 J (see Section 14–1 and Problem 5 in Chapter 14), then a SEER = 1 is a COP equal to (1 Btu/1 W·h) = (1056 J)/(1 J/s × 3600 s) = 0.29. A COP = 1 is a SEER = 1/0.29 = 3.4.

SECTION 15–6   **427**

15–7 Entropy and the Second Law of Thermodynamics

Thus far we have stated the second law of thermodynamics for specific situations. What we really need is a general statement of the second law of thermodynamics that will cover all situations, including processes discussed earlier in this Chapter that are not observed in nature even though they would not violate the first law of thermodynamics. It was not until the latter half of the nineteenth century that the second law of thermodynamics was finally stated in a general way—namely, in terms of a quantity called **entropy**, introduced by Clausius in the 1860s. Entropy, unlike heat, is a function of the state of a system. That is, a system in a given state has a temperature, a volume, a pressure, a mass, and also has a particular value of entropy. In the next Section, we will see that entropy can be interpreted as a measure of the order or disorder of a system.

When we deal with entropy—as with potential energy—it is the *change* in entropy during a process that is important, not the absolute amount. According to Clausius, the change in entropy S of a system, when an amount of heat Q is *added* to it by a reversible[†] process at constant temperature, is given by

$$\Delta S = \frac{Q}{T}, \qquad (15\text{–}8)$$

where T is the kelvin temperature. (If heat is lost, Q is negative in this equation, as per our original sign conventions on page 413.)

EXAMPLE 15–13 Entropy change in melting. An ice cube of mass 56 g is taken from a storage compartment at 0°C and placed in a paper cup. After a few minutes, exactly half of the mass of the ice cube has melted, becoming water at 0°C. Find the change in entropy of the ice/water.

APPROACH We consider the 56 g of water, initially in the form of ice, as our system. To determine the entropy change, we first must find the heat needed to melt the ice, which we do using the latent heat of fusion of water, $L = 333 \text{ kJ/kg}$ (Section 14–5). The heat Q required comes from the surroundings.

SOLUTION The heat required to melt 28 g of ice (half of the 56-g ice cube) is

$$Q = mL = (0.028 \text{ kg})(333 \text{ kJ/kg}) = 9.3 \text{ kJ}.$$

The temperature remains constant in our process, so we can find the change in entropy from Eq. 15–8:

$$\Delta S = \frac{Q}{T} = \frac{9.3 \text{ kJ}}{273 \text{ K}} = 34 \text{ J/K}.$$

NOTE The change in entropy of the surroundings (cup, air) has not been computed.

The temperature in Example 15–13 was constant, so the calculation was short. If the temperature varies during a process, a summation of the heat flow over the changing temperature can often be calculated using calculus or a computer. However, if the temperature change is not too great, a reasonable approximation can be made using the average value of the temperature, as indicated in the next Example.

EXAMPLE 15–14 ESTIMATE Entropy change when water samples are mixed. A sample of 50.0 kg of water at 20.00°C is mixed with 50.0 kg of water at 24.00°C. Estimate the change in entropy.

APPROACH The final temperature of the mixture will be 22.00°C, since we started with equal amounts of water. We use the specific heat of water and the methods of calorimetry (Sections 14–3 and 14–4) to determine the heat transferred. Then we use the average temperature of each sample of water to estimate the entropy change ($\Delta Q/T$).

[†]Real processes are irreversible. Because entropy is a state variable, the change in entropy ΔS for an irreversible process can be determined by calculating ΔS for a reversible process between the same two states.

SOLUTION A quantity of heat,

$$Q = mc\,\Delta T = (50.0\text{ kg})(4186\text{ J/kg}\cdot\text{C}°)(2.00\text{ C}°) = 4.186 \times 10^5\text{ J},$$

flows out of the hot water as it cools down from 24°C to 22°C, and this heat flows into the cold water as it warms from 20°C to 22°C. The total change in entropy, ΔS, will be the sum of the changes in entropy of the hot water, ΔS_H, and that of the cold water, ΔS_C:

$$\Delta S = \Delta S_H + \Delta S_C.$$

We estimate entropy changes by writing $\Delta S = Q/T_{av}$, where T_{av} is an "average" temperature for each process, which ought to give a reasonable estimate since the temperature change is small. For the hot water we use an average temperature of 23°C (296 K), and for the cold water an average temperature of 21°C (294 K). Thus

$$\Delta S_H \approx -\frac{4.186 \times 10^5\text{ J}}{296\text{ K}} = -1414\text{ J/K}$$

which is negative because this heat flows out (sign conventions, page 413), whereas heat is added to the cold water:

$$\Delta S_C \approx \frac{4.186 \times 10^5\text{ J}}{294\text{ K}} = 1424\text{ J/K}.$$

The entropy of the hot water (S_H) decreases because heat flows out of the hot water. But the entropy of the cold water (S_C) increases by a greater amount. The total change in entropy is

$$\Delta S = \Delta S_H + \Delta S_C \approx -1414\text{ J/K} + 1424\text{ J/K} \approx 10\text{ J/K}.$$

In Example 15–14, we saw that although the entropy of one part of the system decreased, the entropy of the other part increased by a greater amount; the net change in entropy of the whole system was positive. This result, which we have calculated for a specific case in Example 15–14, has been found to hold in all other cases tested. That is, the total entropy of an isolated system is found to increase in all natural processes. The second law of thermodynamics can be stated in terms of entropy as follows: *The entropy of an isolated system never decreases. It can only stay the same or increase.* Entropy can remain the same only for an idealized (reversible) process. For any real process, the change in entropy ΔS is greater than zero:

$$\Delta S > 0. \qquad\qquad \text{[real process]} \quad (15\text{–}9)$$

If the system is not isolated, then the change in entropy of the system, ΔS_{sys}, plus the change in entropy of the environment, ΔS_{env}, must be greater than or equal to zero:

$$\Delta S = \Delta S_{sys} + \Delta S_{env} \geq 0. \qquad\qquad (15\text{–}10)$$

Only idealized processes can have $\Delta S = 0$. Real processes always have $\Delta S > 0$. This, then, is the *general statement of the second law of thermodynamics*:

> **the total entropy of any system plus that of its environment increases as a result of any natural process.**

SECOND LAW OF THERMODYNAMICS *(general statement)*

Although the entropy of one part of the universe may decrease in any natural process (see Example 15–14), the entropy of some other part of the universe always increases by a greater amount, so the total entropy always increases.

Now that we finally have a quantitative general statement of the second law of thermodynamics, we can see that it is an unusual law. It differs considerably from other laws of physics, which are typically equalities (such as $F = ma$) or conservation laws (such as for energy and momentum). The second law of thermodynamics introduces a new quantity, the entropy S, but does not tell us it is conserved. Quite the opposite. Entropy is *not* conserved in natural processes. Entropy always increases in time for real processes.

15–8 Order to Disorder

The concept of entropy, as we have discussed it so far, may seem rather abstract. But we can relate it to the more ordinary concepts of *order* and *disorder*. In fact, the entropy of a system can be considered a *measure of the disorder of the system*. Then the second law of thermodynamics can be stated simply as:

natural processes tend to move toward a state of greater disorder.

> SECOND LAW OF THERMODYNAMICS
> *(general statement)*

Exactly what we mean by **disorder** may not always be clear, so we now consider a few examples. Some of these will show us how this very general statement of the second law applies beyond what we usually consider as thermodynamics.

Let us look at the simple processes mentioned in Section 15–4. First, a jar containing separate layers of salt and pepper is more orderly than a jar in which the salt and pepper are all mixed up. Shaking a jar containing separate layers results in a mixture, and no amount of shaking brings the orderly layers back again. The natural process is from a state of relative order (layers) to one of relative disorder (a mixture), not the reverse. That is, disorder increases. Second, a solid coffee cup is a more "orderly" and useful object than the pieces of a broken cup. Cups break when they fall, but they do not spontaneously mend themselves (as faked in Fig. 15–10). Again, the normal course of events is an increase of disorder.

Let us consider some processes for which we have actually calculated the entropy change, and see that an increase in entropy results in an increase in disorder (or vice versa). When ice melts to water at 0°C, the entropy of the water increases (Example 15–13). Intuitively, we can think of solid water, ice, as being more ordered than the less orderly fluid state which can flow all over the place. This change from order to disorder can be seen more clearly from the molecular point of view: the orderly arrangement of water molecules in an ice crystal has changed to the disorderly and somewhat random motion of the molecules in the fluid state.

When a hot substance is put in contact with a cold substance, heat flows from the high temperature to the low until the two substances reach the same intermediate temperature. Entropy increases, as we saw in Example 15–14. At the beginning of the process we can distinguish two classes of molecules: those with a high average kinetic energy (the hot object), and those with a low average kinetic energy (the cooler object). After the process in which heat flows, all the molecules are in one class with the same average kinetic energy; we no longer have the more orderly arrangement of molecules in two classes. Order has gone to disorder. Furthermore, the separate hot and cold objects could serve as the hot- and cold-temperature regions of a heat engine, and thus could be used to obtain useful work. But once the two objects are put in contact and reach the same temperature, no work can be obtained. Disorder has increased, because a system that has the ability to perform work must surely be considered to have a higher order than a system no longer able to do work.

When a stone falls to the ground, its macroscopic kinetic energy is transformed to thermal energy. (We noted earlier that the reverse never happens: a stone never absorbs thermal energy and rises into the air of its own accord.) This is another example of order changing to disorder. Thermal energy is associated with the disorderly random motion of molecules, but the molecules in the falling stone all have the same velocity downward in addition to their own random velocities. Thus, the more orderly kinetic energy of the stone as a whole (which could do useful work) is changed to disordered thermal energy when the stone strikes the ground. Disorder increases in this process, as it does in all processes that occur in nature.

Biological Development

> **PHYSICS APPLIED**
> *Biological development*

An interesting example of the increase in entropy relates to the biological development and growth of organisms. Clearly, a human being is a highly ordered organism. The development of an individual from a single cell to a grown person is a process of increasing order. Evolution too might be seen as an increase in order.

Do these processes violate the second law of thermodynamics? No, they do not. In the processes of growth and evolution, and even during the mature life of an individual, waste products are eliminated. These small molecules that remain as a result of metabolism are simple molecules without much order. Thus they represent relatively higher disorder or entropy. Indeed, the total entropy of the molecules cast aside by organisms during the processes of development and growth is greater than the decrease in entropy associated with the order of the growing individual or evolving species.

"Time's Arrow"

Another aspect of the second law of thermodynamics is that it tells us in which *direction* processes go. If you were to see a film being run backward, you would undoubtedly be able to tell that it *was* run backward. For you would see odd occurrences, such as a broken coffee cup rising from the floor and reassembling on a table, or a torn balloon suddenly becoming whole again and filled with air. We know these things don't happen in real life; they are processes in which order increases—or entropy decreases. They violate the second law of thermodynamics. When watching a movie (or imagining that time could go backward), we are tipped off to a reversal of time by observing whether entropy (and disorder) is increasing or decreasing. Hence, entropy has been called **time's arrow**, because it can tell us in which direction time is going.

15–9 Unavailability of Energy; Heat Death

In the process of heat conduction from a hot object to a cold one, we have seen that entropy increases and that order goes to disorder. The separate hot and cold objects could serve as the high- and low-temperature regions for a heat engine and thus could be used to obtain useful work. But after the two objects are put in contact with each other and reach the same uniform temperature, no work can be obtained from them. With regard to being able to do useful work, order has gone to disorder in this process.

The same can be said about a falling rock that comes to rest upon striking the ground. Before hitting the ground, all the kinetic energy of the rock could have been used to do useful work. But once the rock's mechanical kinetic energy becomes thermal energy, doing useful work is no longer possible.

Both these examples illustrate another important aspect of the second law of thermodynamics:

in any natural process, some energy becomes unavailable to do useful work.

In any process, no energy is ever lost (it is always conserved). Rather, energy becomes less useful—it can do less useful work. As time goes on, **energy is degraded**, in a sense. It goes from more orderly forms (such as mechanical) eventually to the least orderly form: internal, or thermal, energy. Entropy is a factor here because the amount of energy that becomes unavailable to do work is proportional to the change in entropy during any process.

A natural outcome of the degradation of energy is the prediction that as time goes on, the universe should approach a state of maximum disorder. Matter would become a uniform mixture, and heat would have flowed from high-temperature regions to low-temperature regions until the whole universe is at one temperature. No work could then be done. All the energy of the universe would have degraded to thermal energy. This prediction, called the **heat death** of the universe, has been much discussed, but it would lie *very* far in the future. It is a complicated subject, and some scientists question whether thermodynamics, as we now understand it, actually applies to the universe as a whole, which is at a much larger scale.[†]

[†]When a star, like our Sun, loses energy by radiation, it becomes *hotter* (not cooler). By losing energy, gravity is able to compress the gas of which the Sun is made—becoming smaller and denser means the Sun gets hotter. For astronomical objects, when heat flows from a hot object to a cooler object, the hot object gets hotter and the cool one cooler. That is, the temperature difference *increases*.

*15–10 Statistical Interpretation of Entropy and the Second Law

The ideas of entropy and disorder are made clearer with the use of a statistical or probabilistic analysis of the molecular state of a system. This statistical approach, which was first applied toward the end of the nineteenth century by Ludwig Boltzmann (1844–1906), makes a clear distinction between the "macrostate" and the "microstate" of a system. The **microstate** of a system would be specified by giving the position and velocity of every particle (or molecule). The **macrostate** of a system is specified by giving the far fewer macroscopic properties of the system—the temperature, pressure, number of moles, and so on. In reality, we can know only the macrostate of a system. There are generally far too many molecules in a system to be able to know the velocity and position of every one at a given moment. Nonetheless, we can hypothesize a great many different microstates that can correspond to the *same* macrostate.

Let us take a very simple example. Suppose you repeatedly shake four coins in your hand and drop them on a table. Specifying the *number* of heads and the number of tails that appear on a given throw is the *macrostate* of this system. Specifying *each coin* as being a head or a tail is the *microstate* of the system. In the following Table we see how many microstates correspond to each macrostate:

Macrostate	Possible Microstates (H = heads, T = tails)	Number of Microstates
4 heads	H H H H	1
3 heads, 1 tail	H H H T, H H T H, H T H H, T H H H	4
2 heads, 2 tails	H H T T, H T H T, T H H T, H T T H, T H T H, T T H H	6
1 head, 3 tails	T T T H, T T H T, T H T T, H T T T	4
4 tails	T T T T	1

Careful use of probability

CAUTION

A basic assumption behind the statistical approach is that *each microstate is equally probable*. Thus the number of microstates that give the same macrostate corresponds to the relative probability of that macrostate occurring. The macrostate of two heads and two tails is the most probable one in our case of tossing four coins; out of the total of 16 possible microstates, six correspond to two heads and two tails, so the probability of throwing two heads and two tails is 6 out of 16, or 38%. The probability of throwing one head and three tails is 4 out of 16, or 25%. The probability of four heads is only 1 in 16, or 6%. If you threw the coins 16 times, you might not find that two heads and two tails appear exactly 6 times, or four tails exactly once. These are only probabilities or averages. But if you made 1600 throws, very nearly 38% of them would be two heads and two tails. The greater the number of tries, the closer the percentages are to the calculated probabilities.

EXERCISE G In the Table above, what is the probability that there will be at least two heads? (a) $\frac{1}{2}$; (b) $\frac{1}{16}$; (c) $\frac{1}{8}$; (d) $\frac{3}{8}$; (e) $\frac{11}{16}$.

If we toss more coins—say, 100 all at the same time—the relative probability of throwing all heads (or all tails) is greatly reduced. There is only one microstate corresponding to all heads. For 99 heads and 1 tail, there are 100 microstates because each of the coins could be the one tail. The relative probabilities for other macrostates are given in Table 15–3 (top of next page). About 1.3×10^{30} microstates are possible.[†] Thus the relative probability of finding all heads is about 1 in 10^{30}, an incredibly unlikely event! The probability of obtaining 50 heads and 50 tails (see Table 15–3) is $(1.0 \times 10^{29})/(1.3 \times 10^{30}) = 0.08$ or 8%. The probability of obtaining anything between 45 and 55 heads is about 70%.

[†]Each coin has two possibilities, heads or tails. Then the possible number of microstates is $2 \times 2 \times 2 \times \cdots = 2^{100} = 1.3 \times 10^{30}$ (using a calculator or logarithms).

TABLE 15–3 Probabilities of Various Macrostates for 100 Coin Tosses

| Macrostate | | Number of | |
Heads	Tails	Microstates	Probability
100	0	1	7.9×10^{-31}
99	1	1.0×10^2	7.9×10^{-29}
90	10	1.7×10^{13}	1.4×10^{-17}
80	20	5.4×10^{20}	4.2×10^{-10}
60	40	1.4×10^{28}	0.011
55	45	6.1×10^{28}	0.047
50	50	1.0×10^{29}	0.077
45	55	6.1×10^{28}	0.047
40	60	1.4×10^{28}	0.011
20	80	5.4×10^{20}	4.2×10^{-10}
10	90	1.7×10^{13}	1.4×10^{-17}
1	99	1.0×10^2	7.9×10^{-29}
0	100	1	7.9×10^{-31}

Thus we see that as the number of coins increases, the probability of obtaining the most orderly arrangement (all heads or all tails) becomes extremely unlikely. The least orderly arrangement (half heads, half tails) is the most probable, and the probability of being within, say, 5% of the most probable arrangement greatly increases as the number of coins increases. These same ideas can be applied to the molecules of a system. For example, the most probable state of a gas (say, the air in a room) is one in which the molecules take up the whole space and move about randomly; this corresponds to the Maxwellian distribution, Fig. 15–19a (and see Section 13–10). On the other hand, the very orderly arrangement of all the molecules located in one corner of the room and all moving with the same velocity (Fig. 15–19b) is extremely unlikely.

From these examples, we can see that probability is directly related to disorder and hence to entropy. That is, the most probable state is the one with greatest entropy, or greatest disorder and randomness. The processes we actually observe are those where the entropy increase is greatest.

In terms of probability, the second law of thermodynamics—which tells us that entropy increases in any process—reduces to the statement that **those processes occur which are most probable**.

There is an additional element, however. The second law in terms of probability does not *forbid* a decrease in entropy. Rather, it says the probability is extremely low. It is not impossible that salt and pepper could separate spontaneously into layers, or that a broken cup could mend itself (Fig. 15–10). It is even possible that a lake could freeze over on a hot summer day (that is, for heat to flow out of the cold lake into the warmer surroundings). But the probability for such events occurring is almost always miniscule. In our coin examples, we saw that increasing the number of coins from 4 to 100 drastically reduced the probability of large deviations from the average, or most probable, arrangement. In ordinary systems we are not dealing with only 100 molecules, but with extremely large numbers of molecules: in 1 mole alone there are 6×10^{23} molecules. Hence the probability of deviation far from the average is incredibly tiny. For example, it has been calculated that the probability that a stone resting on the ground could transform 1 cal of thermal energy into mechanical energy and rise up into the air is much less likely than the probability that a group of monkeys typing randomly would by chance produce the complete works of Shakespeare. Put another way, the probability is less than once in the entire age of the universe; this could be what "never" means.

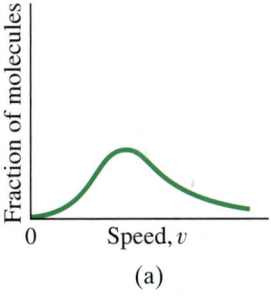

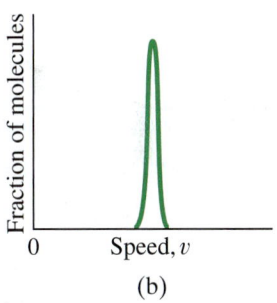

FIGURE 15–19 (a) Most probable distribution of molecular speeds in a gas (Maxwellian, or random). (b) Orderly, but highly unlikely, distribution of speeds in which all molecules have nearly the same speed.

 (a)
 (b)
 (c)

FIGURE 15–20 (a) A fossil-fuel steam plant (this one uses forest waste products = biomass). (b) An array of mirrors focuses sunlight on a boiler to produce steam at a solar energy installation. (c) Large cooling towers at an electric generating plant.

*15–11 Thermal Pollution, Global Warming, and Energy Resources

FIGURE 15–21 Mechanical energy is transformed to electric energy with a turbine and generator.

PHYSICS APPLIED
Heat engines and thermal pollution

Much of the energy we use in everyday life—from motor vehicles to most of the electricity produced by power plants—uses a heat engine. More than $\frac{2}{3}$ of the electric energy produced in the U.S. is generated at fossil-fuel steam plants (coal, oil, or gas—see Fig. 15–20a), and they make use of a heat engine (steam engines). In electric power plants, the steam drives the turbines and generators (Fig. 15–21) whose output is electric energy. The various means to turn the turbine are shown in Table 15–4, along with some of the advantages and disadvantages of each. Even nuclear power plants use a steam engine, run on nuclear fuel. Electricity produced by falling water at dams, by windmills, or by solar cells (Fig. 15–20b) does not involve a heat engine.

The heat output Q_L from every heat engine, from power plants to cars, is referred to as **thermal pollution** because this heat (Q_L) must be absorbed by the environment—such as by water from rivers or lakes, or by the air using large cooling towers (Fig. 15–20c). When water is the coolant, this heat raises the temperature of the water before returning it to its source, altering the natural ecology of aquatic life (largely because warmer water holds less oxygen). In the case of air cooling towers, the output heat Q_L raises the temperature of the atmosphere, which affects weather.

Air pollution—by which we mean the chemicals released in the burning of fossil fuels in cars, power plants, and industrial furnaces—gives rise to smog and other problems. Another issue is the buildup of CO_2 in the Earth's atmosphere due to the burning of fossil fuels; the carbon in them combines with O_2 of the air, forming CO_2. This CO_2 absorbs some of the infrared radiation that the Earth naturally emits (Section 14–8), and thus contributes to **global warming**. The **carbon footprint** of any activity (home appliances, steel production, manufacture of goods, transportation) refers to the negative impact on the environment due to the burning of fossil fuels which release CO_2 (and other noxious products) into the atmosphere, contributing to climate change. A carbon footprint can be expressed in kg or tons of carbon (in the CO_2). Limiting the burning of fossil fuels can help reduce these problems.

Thermal pollution, however, is unavoidable. Engineers can try to design and build engines that are more efficient, but they cannot surpass the Carnot efficiency and must live with T_L being at best the ambient temperature of water or air. The second law of thermodynamics tells us the limit imposed by nature. Even alternative energy sources like wind and solar cells can contribute to global warming (windmills slow down cooling winds, solar cells are "dark" and absorb more of the incident energy). What we can do, in light of the second law of thermodynamics, is use less energy and conserve our fuel resources. There is no other solution.

TABLE 15–4 Electric Energy Resources

Form of Electric Energy Production	% of Production (approx.) U.S.	% of Production (approx.) World	Advantages	Disadvantages
Fossil-fuel steam plants: burn coal, oil, or natural gas to boil water, producing high-pressure steam that turns a turbine of a generator (Figs. 15–12b, 15–21); uses heat engine.	72	66	We know how to build them; for now relatively inexpensive.	Air pollution; thermal pollution; limited efficiency; land devastation from extraction of raw materials (mining); global warming; accidents such as oil spills at sea; limited fuel supply (estimates range from a couple of decades to a few centuries).
Nuclear energy:				
Fission: nuclei of uranium or plutonium atoms split ("fission") with release of energy (Chapter 31) that heats steam; uses heat engine.	20	17	Normally almost no air pollution; less contribution to global warming; relatively inexpensive.	Thermal pollution; accidents can release damaging radioactivity; difficult disposal of radioactive by-products; possible diversion of nuclear material by terrorists; limited fuel supply.
Fusion: energy released when isotopes of hydrogen (or other small nuclei) combine or "fuse" (Chapter 31).	0	0	Relatively "clean"; vast fuel supply (hydrogen in water molecules in oceans); less contribution to global warming.	Not yet workable.
Hydroelectric: falling water turns turbines at the base of a dam.	7	16	No heat engine needed; no air, water, or thermal pollution; relatively inexpensive; high efficiency; dams can control flooding.	Reservoirs behind dams inundate scenic or inhabited land; dams block upstream migration of salmon and other fish for reproduction; few locations remain for new dams; drought.
Geothermal: natural steam from inside the Earth comes to the surface (hot springs, geysers, steam vents); or cold water passed down into contact with hot, dry rock is heated to steam.	<1	<1	No heat engine needed; little air pollution; good efficiency; relatively inexpensive and "clean."	Few appropriate sites; small production; mineral content of spent hot water can pollute.
Wind power: 3-kW to 5-MW windmills (vanes up to 50 m long) turn a generator.	≈1	<1	No heat engine; no air, water, or thermal pollution; relatively inexpensive.	Large array of big windmills might affect weather and be eyesores; slows down cooling winds; hazardous to migratory birds; winds not always strong.
Solar energy:	<1	<1		
Active solar heating: rooftop solar panels absorb the Sun's rays, which heat water in tubes for space heating and hot water supply.			No heat engine needed; no air or thermal pollution; unlimited fuel supply.	Space limitations; may require back-up; relatively expensive; less effective when cloudy.
Passive solar heating: architectural devices—windows along southern exposure, sunshade over windows to keep Sun's rays out in summer.			No heat engine needed; no air or thermal pollution; relatively inexpensive.	Almost none, but other methods needed too.
Solar cells (photovoltaic cells): convert sunlight directly into electricity without use of heat engine.			No heat engine; thermal, air, and water pollution very low; good efficiency (>30% and improving).	Expensive; chemical pollution at manufacture; much land needed as Sun's energy not concentrated; absorption (dark color) without reemission = global warming.

PROBLEM SOLVING

Thermodynamics

1. Define the **system** you are dealing with; distinguish the system under study from its surroundings.
2. When applying the first law of thermodynamics, be careful of **signs** associated with **work** and **heat**. In the first law, work done *by* the system is positive; work done *on* the system is negative. Heat *added* to the system is positive, but heat *removed* from it is negative. With heat engines, we usually consider the heat intake, the heat exhausted, and the work done as positive.
3. Watch the **units** used for work and heat; work is most often expressed in joules, and heat can be in calories, kilocalories, or joules. Be consistent: choose only one unit for use throughout a given problem.
4. **Temperatures** must generally be expressed in kelvins; temperature *differences* may be expressed in C° or K.
5. **Efficiency** (or coefficient of performance) is a ratio of two energy transfers: useful output divided by required input. Efficiency (but *not* coefficient of performance) is always less than 1 in value, and hence is often stated as a percentage.
6. The **entropy** of a system increases when heat is added to the system, and decreases when heat is removed. If heat is transferred from system A to system B, the change in entropy of A is negative and the change in entropy of B is positive.

Summary

The **first law of thermodynamics** states that the change in internal energy ΔU of a system is equal to the heat *added* to the system, Q, minus the work done *by* the system, W:

$$\Delta U = Q - W. \quad (15\text{-}1)$$

This important law is a statement of the conservation of energy, and is found to hold for all processes.

An **isothermal** process is a process carried out at constant temperature.

In an **adiabatic** process, no heat is exchanged ($Q = 0$).

The work W done by a gas at constant pressure P is given by

$$W = P \Delta V, \quad (15\text{-}3)$$

where ΔV is the change in volume of the gas.

A **heat engine** is a device for changing thermal energy, by means of heat flow, into useful work.

The **efficiency** e of a heat engine is defined as the ratio of the work W done by the engine to the high temperature heat input Q_H. Because of conservation of energy, the work output equals $Q_H - Q_L$, where Q_L is the heat exhausted at low temperature to the environment; hence

$$e = \frac{W}{Q_H} = \frac{Q_H - Q_L}{Q_H} = 1 - \frac{Q_L}{Q_H}. \quad (15\text{-}4)$$

Q_H, Q_L, and W, as defined for heat engines, are positive.

The *upper limit* on the efficiency (the *Carnot efficiency*) can be written in terms of the higher and lower operating temperatures (in kelvins) of the engine, T_H and T_L, as

$$e_{\text{ideal}} = 1 - \frac{T_L}{T_H}. \quad (15\text{-}5)$$

Real (irreversible) engines always have an efficiency less than this.

The operation of **refrigerators** and **air conditioners** is the reverse of a heat engine: work is done to extract heat Q_L from a cool region and exhaust it to a region at a higher temperature. The coefficient of performance (COP) for either is

$$\text{COP} = \frac{Q_L}{W}, \quad \begin{bmatrix} \text{refrigerator or} \\ \text{air conditioner} \end{bmatrix} \quad (15\text{-}6a)$$

where W is the work needed to remove heat Q_L from the area with the low temperature.

A **heat pump** uses work W to bring heat Q_L from the cold outside and deliver heat Q_H to warm the interior. The coefficient of performance of a heat pump is

$$\text{COP} = \frac{Q_H}{W}, \quad [\text{heat pump}] \quad (15\text{-}7)$$

because it is the heat Q_H delivered inside the building that counts.

The **second law of thermodynamics** can be stated in several equivalent ways:

(a) heat flows spontaneously from a hot object to a cold one, but not the reverse;

(b) there can be no 100% efficient heat engine—that is, one that can change a given amount of heat completely into work;

(c) natural processes tend to move toward a state of greater disorder or greater **entropy**.

Statement (c) is the most general statement of the second law of thermodynamics, and can be restated as: the total entropy, S, of any system plus that of its environment increases as a result of any natural process:

$$\Delta S > 0. \quad (15\text{-}9)$$

The change in entropy in a process that transfers heat Q at a constant temperature T is

$$\Delta S = \frac{Q}{T}. \quad (15\text{-}8)$$

Entropy is a quantitative measure of the **disorder** of a system. The second law of thermodynamics also indicates that as time goes on, energy is **degraded** to less useful forms—that is, it is less available to do useful work.

The second law of thermodynamics tells us in which direction processes tend to proceed; hence entropy is called "time's arrow."

[*Entropy can be examined from a statistical point of view, considering **macrostates** (for example, P, V, T) and **microstates** (state of each molecule). The most probable processes are the ones we observe. They are the ones that increase entropy the most. Processes that violate the second law "could" occur, but only with *extremely* low probability.]

[*All heat engines give rise to **thermal pollution** because they exhaust heat to the environment.]

Questions

1. In an isothermal process, 3700 J of work is done by an ideal gas. Is this enough information to tell how much heat has been added to the system? If so, how much? If not, why not?
2. Can mechanical energy ever be transformed completely into heat or internal energy? Can the reverse happen? In each case, if your answer is no, explain why not; if yes, give one or two examples.
3. Can the temperature of a system remain constant even though heat flows into or out of it? If so, give examples.
4. Explain why the temperature of a gas increases when it is compressed adiabatically.
5. An ideal monatomic gas expands slowly to twice its volume (1) isothermally; (2) adiabatically; (3) isobarically. Plot each on a PV diagram. In which process is ΔU the greatest, and in which is ΔU the least? In which is W the greatest and the least? In which is Q the greatest and the least?
6. (a) What happens if you remove the lid of a bottle containing chlorine gas? (b) Does the reverse process ever happen? Why or why not? (c) Can you think of two other examples of irreversibility?
7. Would a definition of heat engine efficiency as $e = W/Q_L$ be useful? Explain.
8. What are the high-temperature and the low-temperature areas for (a) an internal combustion engine, and (b) a steam engine? Are they, strictly speaking, heat reservoirs?
9. The oceans contain a tremendous amount of thermal (internal) energy. Why, in general, is it not possible to put this energy to useful work?
10. Can you warm a kitchen in winter by leaving the oven door open? Can you cool the kitchen on a hot summer day by leaving the refrigerator door open? Explain.
11. The COPs are defined differently for heat pumps and air conditioners. Explain why.
12. You are asked to test a machine that the inventor calls an "in-room air conditioner": a big box, standing in the middle of the room, with a cable that plugs into a power outlet. When the machine is switched on, you feel a stream of cold air coming out of it. How do you know that this machine cannot cool the room?
13. Think up several processes (other than those already mentioned) that would obey the first law of thermodynamics, but, if they actually occurred, would violate the second law.
14. Suppose a lot of papers are strewn all over the floor; then you stack them neatly. Does this violate the second law of thermodynamics? Explain.
15. The first law of thermodynamics is sometimes whimsically stated as, "You can't get something for nothing," and the second law as, "You can't even break even." Explain how these statements could be equivalent to the formal statements.
16. A gas is allowed to expand (a) adiabatically and (b) isothermally. In each process, does the entropy increase, decrease, or stay the same? Explain.
17. Which do you think has the greater entropy, 1 kg of solid iron or 1 kg of liquid iron? Why?
18. Give three examples, other than those mentioned in this Chapter, of naturally occurring processes in which order goes to disorder. Discuss the observability of the reverse process.
19. Entropy is often called "time's arrow" because it tells us in which direction natural processes occur. If a movie were run backward, name some processes that you might see that would tell you that time was "running backward."
20. Living organisms, as they grow, convert relatively simple food molecules into a complex structure. Is this a violation of the second law of thermodynamics? Explain your answer.

MisConceptual Questions

1. In an isobaric compression of an ideal gas,
   (a) no heat flows into the gas.
   (b) the internal energy of the gas remains constant.
   (c) no work is done on the gas.
   (d) work is done on the gas.
   (e) work is done by the gas.
2. Which is possible: converting (i) 100 J of work entirely into 100 J of heat, (ii) 100 J of heat entirely into 100 J of work?
   (a) Only (i) is possible.
   (b) Only (ii) is possible.
   (c) Both (i) and (ii) are possible.
   (d) Neither (i) nor (ii) is possible.
3. An ideal gas undergoes an isobaric compression and then an isovolumetric process that brings it back to its initial temperature. Had the gas undergone *one isothermal* process instead,
   (a) the work done on the gas would be the same.
   (b) the work done on the gas would be less.
   (c) the work done on the gas would be greater.
   (d) Need to know the temperature of the isothermal process.
4. An ideal gas undergoes an isothermal expansion from state A to state B. In this process (use sign conventions, page 413),
   (a) $Q = 0$, $\Delta U = 0$, $W > 0$.
   (b) $Q > 0$, $\Delta U = 0$, $W < 0$.
   (c) $Q = 0$, $\Delta U > 0$, $W > 0$.
   (d) $Q > 0$, $\Delta U = 0$, $W > 0$.
   (e) $Q = 0$, $\Delta U < 0$, $W < 0$.
5. An ideal gas undergoes an isothermal process. Which of the following statements are true? (i) No heat is added to or removed from the gas. (ii) The internal energy of the gas does not change. (iii) The average kinetic energy of the molecules does not change.
   (a) (i) only.
   (b) (i) and (ii) only.
   (c) (i) and (iii) only.
   (d) (ii) and (iii) only.
   (e) (i), (ii), and (iii).
   (f) None of the above.

6. An ideal gas undergoes an adiabatic expansion, a process in which no heat flows into or out of the gas. As a result,
   (a) the temperature of the gas remains constant and the pressure decreases.
   (b) both the temperature and pressure of the gas decrease.
   (c) the temperature of the gas decreases and the pressure increases.
   (d) both the temperature and volume of the gas increase.
   (e) both the temperature and pressure of the gas increase.

7. A heat engine operates between a high temperature of about 600°C and a low temperature of about 300°C. What is the maximum theoretical efficiency for this engine?
   (a) = 100%.   (b) ≈ 66%.   (c) ≈ 50%.   (d) ≈ 34%.
   (e) Cannot be determined from the given information.

8. On a very hot day, could you cool your kitchen by leaving the refrigerator door open?
   (a) Yes, but it would be very expensive.
   (b) Yes, but only if the humidity is below 50%.
   (c) No, the refrigerator would exhaust the same amount of heat into the room as it takes out of the room.
   (d) No, the heat exhausted by the refrigerator into the room is more than the heat the refrigerator takes out of the room.

9. Which of the following possibilities could increase the efficiency of a heat engine or an internal combustion engine?
   (a) Increase the temperature of the hot part of the system and reduce the temperature of the exhaust.
   (b) Increase the temperatures of both the hot part and the exhaust part of the system by the same amount.
   (c) Decrease the temperatures of both the hot part and the exhaust part of the system by the same amount.
   (d) Decrease the temperature of the hot part and increase the temperature of the exhaust part by the same amount.
   (e) None of the above; only redesigning the engine or using better gas could improve the engine's efficiency.

10. About what percentage of the heat produced by burning gasoline is turned into useful work by a typical automobile?
    (a) 20%.   (b) 50%.   (c) 80%.   (d) 90%.   (e) Nearly 100%.

11. Which statement is true regarding the entropy change of an ice cube that melts?
    (a) Since melting occurs at the melting point temperature, there is no temperature change so there is no entropy change.
    (b) Entropy increases.
    (c) Entropy decreases.

For assigned homework and other learning materials, go to the MasteringPhysics website.

Problems

15–1 and 15–2 First Law of Thermodynamics
[Recall sign conventions, page 413.]

1. (I) An ideal gas expands isothermally, performing 4.30×10^3 J of work in the process. Calculate (a) the change in internal energy of the gas, and (b) the heat absorbed during this expansion.

2. (I) One liter of air is cooled at constant pressure until its volume is halved, and then it is allowed to expand isothermally back to its original volume. Draw the process on a PV diagram.

3. (II) Sketch a PV diagram of the following process: 2.5 L of ideal gas at atmospheric pressure is cooled at constant pressure to a volume of 1.0 L, and then expanded isothermally back to 2.5 L, whereupon the pressure is increased at constant volume until the original pressure is reached.

4. (II) A gas is enclosed in a cylinder fitted with a light frictionless piston and maintained at atmospheric pressure. When 254 kcal of heat is added to the gas, the volume is observed to increase slowly from 12.0 m³ to 16.2 m³. Calculate (a) the work done by the gas and (b) the change in internal energy of the gas.

5. (II) A 1.0-L volume of air initially at 3.5 atm of (gauge) pressure is allowed to expand isothermally until the pressure is 1.0 atm. It is then compressed at constant pressure to its initial volume, and lastly is brought back to its original pressure by heating at constant volume. Draw the process on a PV diagram, including numbers and labels for the axes.

6. (II) The pressure in an ideal gas is cut in half slowly, while being kept in a container with rigid walls. In the process, 465 kJ of heat left the gas. (a) How much work was done during this process? (b) What was the change in internal energy of the gas during this process?

7. (II) In an engine, an almost ideal gas is compressed adiabatically to half its volume. In doing so, 2630 J of work is done on the gas. (a) How much heat flows into or out of the gas? (b) What is the change in internal energy of the gas? (c) Does its temperature rise or fall?

8. (II) An ideal gas expands at a constant total pressure of 3.0 atm from 410 mL to 690 mL. Heat then flows out of the gas at constant volume, and the pressure and temperature are allowed to drop until the temperature reaches its original value. Calculate (a) the total work done by the gas in the process, and (b) the total heat flow into the gas.

9. (II) 8.5 moles of an ideal monatomic gas expand adiabatically, performing 8300 J of work in the process. What is the change in temperature of the gas during this expansion?

10. (II) Consider the following two-step process. Heat is allowed to flow out of an ideal gas at constant volume so that its pressure drops from 2.2 atm to 1.4 atm. Then the gas expands at constant pressure, from a volume of 5.9 L to 9.3 L, where the temperature reaches its original value. See Fig. 15–22. Calculate (a) the total work done by the gas in the process, (b) the change in internal energy of the gas in the process, and (c) the total heat flow into or out of the gas.

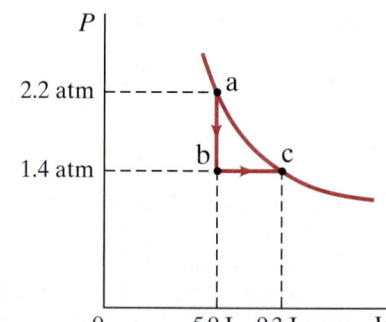

FIGURE 15–22
Problem 10.

11. (II) Use the conservation of energy to explain why the temperature of a well-insulated gas increases when it is compressed—say, by pushing down on a piston—whereas the temperature decreases when the gas expands. Show your reasoning.

12. (III) The PV diagram in Fig. 15–23 shows two possible states of a system containing 1.75 moles of a monatomic ideal gas. ($P_1 = P_2 = 425$ N/m², $V_1 = 2.00$ m³, $V_2 = 8.00$ m³.) (a) Draw the process which depicts an isobaric expansion from state 1 to state 2, and label this process A. (b) Find the work done by the gas and the change in internal energy of the gas in process A. (c) Draw the two-step process which depicts an isothermal expansion from state 1 to the volume V_2, followed by an isovolumetric increase in temperature to state 2, and label this process B. (d) Find the change in internal energy of the gas for the two-step process B.

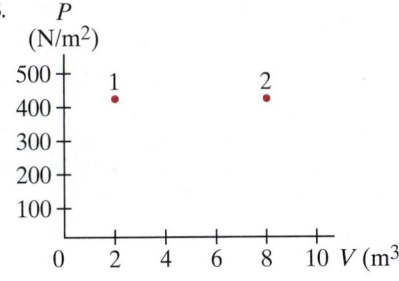

FIGURE 15–23 Problem 12.

13. (III) When a gas is taken from a to c along the curved path in Fig. 15–24, the work done by the gas is $W = -35$ J and the heat added to the gas is $Q = -175$ J. Along path abc, the work done by the gas is $W = -56$ J. (That is, 56 J of work is done *on* the gas.) (a) What is Q for path abc? (b) If $P_c = \frac{1}{2}P_b$, what is W for path cda? (c) What is Q for path cda? (d) What is $U_a - U_c$? (e) If $U_d - U_c = 42$ J, what is Q for path da?

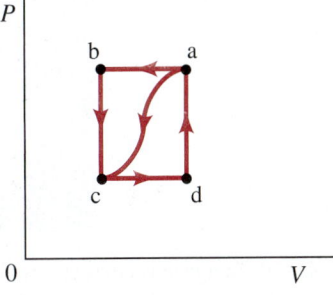

FIGURE 15–24 Problem 13.

*15–3 Human Metabolism

*14. (I) How much energy would the person of Example 15–7 transform if instead of working 11.0 h she took a noontime break and ran at 15 km/h for 1.0 h?

*15. (I) Calculate the average metabolic rate of a 65-kg person who sleeps 8.0 h, sits at a desk 6.0 h, engages in light activity 6.0 h, watches TV 2.0 h, plays tennis 1.5 h, and runs 0.50 h daily.

*16. (II) A 65-kg person decides to lose weight by sleeping one hour less per day, using the time for light activity. How much weight (or mass) can this person expect to lose in 1 year, assuming no change in food intake? Assume that 1 kg of fat stores about 40,000 kJ of energy.

*17. (II) (a) How much energy is transformed by a typical 65-kg person who runs at 15 km/h for 30 min/day in one week (Table 15–2)? (b) How many food calories would the person have to eat to make up for this energy loss?

15–5 Heat Engines

18. (I) A heat engine exhausts 8200 J of heat while performing 2600 J of useful work. What is the efficiency of this engine?

19. (I) What is the maximum efficiency of a heat engine whose operating temperatures are 560°C and 345°C?

20. (I) The exhaust temperature of a heat engine is 230°C. What is the high temperature if the Carnot efficiency is 34%?

21. (I) A heat engine does 9200 J of work per cycle while absorbing 25.0 kcal of heat from a high-temperature reservoir. What is the efficiency of this engine?

22. (I) A heat engine's high temperature T_H could be ambient temperature, because liquid nitrogen at 77 K could be T_L and is cheap. What would be the efficiency of a Carnot engine that made use of heat transferred from air at room temperature (293 K) to the liquid nitrogen "fuel" (Fig. 15–25)?

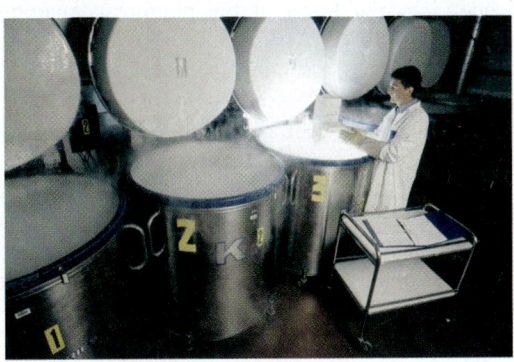

FIGURE 15–25 Problem 22.

23. (II) Which will improve the efficiency of a Carnot engine more: a 10 C° increase in the high-temperature reservoir, or a 10 C° decrease in the low-temperature reservoir? Give detailed results. Can you state a generalization?

24. (II) A certain power plant puts out 580 MW of electric power. Estimate the heat discharged per second, assuming that the plant has an efficiency of 32%.

25. (II) A nuclear power plant operates at 65% of its maximum theoretical (Carnot) efficiency between temperatures of 660°C and 330°C. If the plant produces electric energy at the rate of 1.4 GW, how much exhaust heat is discharged per hour?

26. (II) A heat engine exhausts its heat at 340°C and has a Carnot efficiency of 36%. What exhaust temperature would enable it to achieve a Carnot efficiency of 42%?

27. (II) A Carnot engine's operating temperatures are 210°C and 45°C. The engine's power output is 910 W. Calculate the rate of heat output.

28. (II) A four-cylinder gasoline engine has an efficiency of 0.22 and delivers 180 J of work per cycle per cylinder. If the engine runs at 25 cycles per second (1500 rpm), determine (a) the work done per second, and (b) the total heat input per second from the gasoline. (c) If the energy content of gasoline is 130 MJ per gallon, how long does one gallon last?

29. (II) A Carnot engine performs work at the rate of 520 kW with an input of 950 kcal of heat per second. If the temperature of the heat source is 520°C, at what temperature is the waste heat exhausted?

30. (II) A heat engine uses a heat source at 580°C and has an ideal (Carnot) efficiency of 22%. To increase the ideal efficiency to 42%, what must be the temperature of the heat source?

31. (III) A typical compact car experiences a total drag force of about 350 N at 55 mi/h. If this car gets 32 miles per gallon of gasoline at this speed, and a liter of gasoline (1 gal = 3.8 L) releases about 3.2×10^7 J when burned, what is the car's efficiency?

Problems 439

15–6 Refrigerators, Air Conditioners, Heat Pumps

32. (I) If an ideal refrigerator keeps its contents at 2.5°C when the house temperature is 22°C, what is its COP?

33. (I) The low temperature of a freezer cooling coil is −8°C and the discharge temperature is 33°C. What is the maximum theoretical coefficient of performance?

34. (II) What is the temperature inside an ideal refrigerator-freezer that operates with a COP = 7.0 in a 22°C room?

35. (II) A heat pump is used to keep a house warm at 22°C. How much work is required of the pump to deliver 3100 J of heat into the house if the outdoor temperature is (a) 0°C, (b) −15°C? Assume a COP of 3.0. (c) Redo for both temperatures, assuming an ideal (Carnot) coefficient of performance COP = $T_H/(T_H - T_L)$.

36. (II) (a) What is the coefficient of performance of an ideal heat pump that extracts heat from 6°C air outside and deposits heat inside a house at 24°C? (b) If this heat pump operates on 1200 W of electrical power, what is the maximum heat it can deliver into the house each hour? See Problem 35.

37. (II) What volume of water at 0°C can a freezer make into ice cubes in 1.0 h, if the coefficient of performance of the cooling unit is 6.0 and the power input is 1.2 kilowatt?

38. (II) How much less per year would it cost a family to operate a heat pump that has a coefficient of performance of 2.9 than an electric heater that costs $2000 to heat their home for a year? If the conversion to the heat pump costs $15,000, how long would it take the family to break even on heating costs? How much would the family save in 20 years?

15–7 Entropy

39. (I) What is the change in entropy of 320 g of steam at 100°C when it is condensed to water at 100°C?

40. (I) 1.0 kg of water is heated from 0°C to 100°C. Estimate the change in entropy of the water.

41. (I) What is the change in entropy of 1.00 m³ of water at 0°C when it is frozen to ice at 0°C?

42. (II) A 5.8-kg box having an initial speed of 4.0 m/s slides along a rough table and comes to rest. Estimate the total change in entropy of the universe. Assume all objects are at room temperature (293 K).

43. (II) If 1.00 m³ of water at 0°C is frozen and cooled to −8.0°C by being in contact with a great deal of ice at −8.0°C, estimate the total change in entropy of the process.

44. (II) An aluminum rod conducts 8.40 cal/s from a heat source maintained at 225°C to a large body of water at 22°C. Calculate the rate at which entropy increases in this process.

45. (II) A 2.8-kg piece of aluminum at 28.5°C is placed in 1.0 kg of water in a Styrofoam container at room temperature (20.0°C). Estimate the net change in entropy of the system.

46. (II) A falling rock has kinetic energy KE just before striking the ground and coming to rest. What is the total change in entropy of rock plus environment as a result of this collision?

47. (II) 1.0 kg of water at 35°C is mixed with 1.0 kg of water at 45°C in a well-insulated container. Estimate the net change in entropy of the system.

48. (III) A real heat engine working between heat reservoirs at 970 K and 650 K produces 550 J of work per cycle for a heat input of 2500 J. (a) Compare the efficiency of this real engine to that of an ideal (Carnot) engine. (b) Calculate the total entropy change of the universe per cycle of the real engine, and (c) also if the engine is ideal (Carnot).

*15–10 Statistical Interpretation

*49. (II) Calculate the probabilities, when you throw two dice, of obtaining (a) a 4, and (b) a 10.

*50. (II) Suppose that you repeatedly shake six coins in your hand and drop them on the floor. Construct a table showing the number of microstates that correspond to each macrostate. What is the probability of obtaining (a) three heads and three tails, and (b) six heads?

*51. (III) A bowl contains many red, orange, and green jelly beans, in equal numbers. You are to make a line of 3 jelly beans by randomly taking 3 beans from the bowl. (a) Construct a table showing the number of microstates that correspond to each macrostate. Then determine the probability of (b) all 3 beans red, and (c) 2 greens, 1 orange.

*52. (III) Rank the following five-card hands in order of increasing probability: (a) four aces and a king; (b) six of hearts, eight of diamonds, queen of clubs, three of hearts, jack of spades; (c) two jacks, two queens, and an ace; and (d) any hand having no two equal-value cards (no pairs, etc.). Discuss your ranking in terms of microstates and macrostates.

*15–11 Energy Resources

*53. (I) Solar cells (Fig. 15–26) can produce about 40 W of electricity per square meter of surface area if directly facing the Sun. How large an area is required to supply the needs of a house that requires 24 kWh/day? Would this fit on the roof of an average house? (Assume the Sun shines about 9 h/day.)

FIGURE 15–26 Problem 53.

*54. (II) Energy may be stored by pumping water to a high reservoir when demand is low and then releasing it to drive turbines during peak demand. Suppose water is pumped to a lake 115 m above the turbines at a rate of 1.00 × 10⁵ kg/s for 10.0 h at night. (a) How much energy (kWh) is needed to do this each night? (b) If all this energy is released during a 14-h day, at 75% efficiency, what is the average power output?

*55. (II) Water is stored in an artificial lake created by a dam (Fig. 15–27). The water depth is 48 m at the dam, and a steady flow rate of 32 m³/s is maintained through hydroelectric turbines installed near the base of the dam. How much electrical power can be produced?

FIGURE 15–27 Problem 55: Flaming Gorge Dam on the Green River in Utah.

General Problems

56. An inventor claims to have built an engine that produces 2.00 MW of usable work while taking in 3.00 MW of thermal energy at 425 K, and rejecting 1.00 MW of thermal energy at 215 K. Is there anything fishy about his claim? Explain.

57. When 5.80×10^5 J of heat is added to a gas enclosed in a cylinder fitted with a light frictionless piston maintained at atmospheric pressure, the volume is observed to increase from 1.9 m^3 to 4.1 m^3. Calculate (a) the work done by the gas, and (b) the change in internal energy of the gas. (c) Graph this process on a PV diagram.

58. A restaurant refrigerator has a coefficient of performance of 4.6. If the temperature in the kitchen outside the refrigerator is 32°C, what is the lowest temperature that could be obtained inside the refrigerator if it were ideal?

59. A particular car does work at the rate of about 7.0 kJ/s when traveling at a steady 21.8 m/s along a level road. This is the work done against friction. The car can travel 17 km on 1.0 L of gasoline at this speed (about 40 mi/gal). What is the minimum value for T_H if T_L is 25°C? The energy available from 1.0 L of gas is 3.2×10^7 J.

60. A "Carnot" refrigerator (the reverse of a Carnot engine) absorbs heat from the freezer compartment at a temperature of −17°C and exhausts it into the room at 25°C. (a) How much work would the refrigerator do to change 0.65 kg of water at 25°C into ice at −17°C? (b) If the compressor output is 105 W and runs 25% of the time, how long will this take?

61. It has been suggested that a heat engine could be developed that made use of the temperature difference between water at the surface of the ocean and water several hundred meters deep. In the tropics, the temperatures may be 27°C and 4°C, respectively. (a) What is the maximum efficiency such an engine could have? (b) Why might such an engine be feasible in spite of the low efficiency? (c) Can you imagine any adverse environmental effects that might occur?

62. A cooling unit for a new freezer has an inner surface area of 8.0 m^2, and is bounded by walls 12 cm thick with a thermal conductivity of 0.050 W/m·K. The inside must be kept at −15°C in a room that is at 22°C. The motor for the cooling unit must run no more than 15% of the time. What is the minimum power requirement of the cooling motor?

63. Refrigeration units can be rated in "tons." A 1-ton air conditioning system can remove sufficient energy to freeze 1 ton (2000 pounds = 909 kg) of 0°C water into 0°C ice in one 24-h day. If, on a 35°C day, the interior of a house is maintained at 22°C by the continuous operation of a 5-ton air conditioning system, how much does this cooling cost the homeowner per hour? Assume the work done by the refrigeration unit is powered by electricity that costs $0.10 per kWh and that the unit's coefficient of performance is 18% that of an ideal refrigerator. 1 kWh = 3.60×10^6 J.

64. Two 1100-kg cars are traveling 85 km/h in opposite directions when they collide and are brought to rest. Estimate the change in entropy of the universe as a result of this collision. Assume $T = 20°C$.

65. A 110-g insulated aluminum cup at 35°C is filled with 150 g of water at 45°C. After a few minutes, equilibrium is reached. (a) Determine the final temperature, and (b) estimate the total change in entropy.

66. The burning of gasoline in a car releases about 3.0×10^4 kcal/gal. If a car averages 41 km/gal when driving 110 km/h, which requires 25 hp, what is the efficiency of the engine under those conditions?

67. A Carnot engine operates with $T_L = 20°C$ and has an efficiency of 25%. By how many kelvins should the high operating temperature T_H be increased to achieve an efficiency of 35%?

68. Calculate the work done by an ideal gas in going from state A to state C in Fig. 15–28 for each of the following processes: (a) ADC, (b) ABC, and (c) AC directly.

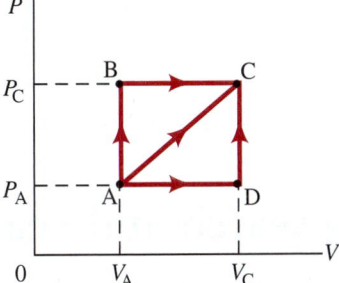

FIGURE 15–28 Problem 68.

69. A 38% efficient power plant puts out 850 MW of electrical power. Cooling towers take away the exhaust heat. (a) If the air temperature is allowed to rise 7.0 C°, estimate what volume of air (km^3) is heated per day. Will the local climate be heated significantly? (b) If the heated air were to form a layer 180 m thick, estimate how large an area it would cover for 24 h of operation. Assume the air has density 1.3 kg/m^3 and has specific heat of about 1.0 kJ/kg·C° at constant pressure.

70. Suppose a power plant delivers energy at 880 MW using steam turbines. The steam goes into the turbines superheated at 625 K and deposits its unused heat in river water at 285 K. Assume that the turbine operates as an ideal Carnot engine. (a) If the river flow rate is 37 m^3/s, estimate the average temperature increase of the river water immediately downstream from the power plant. (b) What is the entropy increase per kilogram of the downstream river water in J/kg·K?

71. A car engine whose output power is 135 hp operates at about 15% efficiency. Assume the engine's water temperature of 85°C is its cold-temperature (exhaust) reservoir and 495°C is its thermal "intake" temperature (the temperature of the exploding gas–air mixture). (a) What is the ratio of its efficiency relative to its maximum possible (Carnot) efficiency? (b) Estimate how much power (in watts) goes into moving the car, and how much heat, in joules and in kcal, is exhausted to the air in 1.0 h.

72. An ideal monatomic gas is contained in a tall cylindrical jar of cross-sectional area 0.080 m^2 fitted with an airtight frictionless 0.15-kg movable piston. When the gas is heated (at constant pressure) from 25°C to 55°C, the piston rises 1.0 cm. How much heat was required for this process? Assume atmospheric pressure outside. [*Hint*: See Section 14–2.]

73. Metabolizing 1.0 kg of fat results in about 3.7×10^7 J of internal energy in the body. (a) In one day, how much fat does the body burn to maintain the body temperature of a person staying in bed and metabolizing at an average rate of 95 W? (b) How long would it take to burn 1.0 kg of fat this way assuming there is no food intake?

74. (a) At a steam power plant, steam engines work in pairs, the heat output of the first one being the approximate heat input of the second. The operating temperatures of the first are 750°C and 440°C, and of the second 415°C and 270°C. If the heat of combustion of coal is 2.8×10^7 J/kg, at what rate must coal be burned if the plant is to put out 950 MW of power? Assume the efficiency of the engines is 65% of the ideal (Carnot) efficiency. (b) Water is used to cool the power plant. If the water temperature is allowed to increase by no more than 4.5 C°, estimate how much water must pass through the plant per hour.

75. Suppose a heat pump has a stationary bicycle attachment that allows *you* to provide the work instead of using an electrical wall outlet. If your heat pump has a coefficient of performance of 2.0 and you can cycle at a racing pace (Table 15–2) at 20% efficiency for a half hour, how much heat can you provide?

76. An ideal air conditioner keeps the temperature inside a room at 21°C when the outside temperature is 32°C. If 4.8 kW of power enters a room through the windows in the form of direct radiation from the Sun, how much electrical power would be saved if the windows were shaded so only 500 W enters?

77. An ideal heat pump is used to maintain the inside temperature of a house at $T_{in} = 22°C$ when the outside temperature is T_{out}. Assume that when it is operating, the heat pump does work at a rate of 1500 W. Also assume that the house loses heat via conduction through its walls and other surfaces at a rate given by $(650 \text{ W/C°})(T_{in} - T_{out})$. (a) For what outside temperature would the heat pump have to operate all the time in order to maintain the house at an inside temperature of 22°C? (b) If the outside temperature is 8°C, what percentage of the time does the heat pump have to operate in order to maintain the house at an inside temperature of 22°C?

Search and Learn

1. What happens to the internal energy of water vapor in the air that condenses on the outside of a cold glass of water? Is work done or heat exchanged? Explain in detail.

2. Draw a *PV* diagram for an ideal gas which undergoes a three-step cyclic thermodynamic process in which the first step has $\Delta U = 0$ and $W > 0$, the second step has $W = 0$, and the third step has $Q = 0$ and $W < 0$.

3. What exactly is a Carnot engine and why is it important? How practical is it?

4. (a) Make up an advertisement for a refrigerator or air conditioner that violates the first law of thermodynamics (see Section 15–6). (b) Make up an ad for a car engine that violates the second law of thermodynamics (see Section 15–5).

*5. One day a person sleeps for 7.0 h, goes running for an hour, gets dressed and eats breakfast for an hour, sits at work for the next 9.0 h, does household chores for a couple of hours, eats dinner for an hour, surfs the Internet and watches TV for a couple of hours, and finally takes an hour for a bath and getting ready for bed. If all of the energy associated with these activities is considered as heat that the body outputs to the environment, estimate the change in entropy the person has provided. (See Sections 15–3 and 15–7.)

6. A particular 1.5-m² photovoltaic panel operating in direct sunlight produces electricity at 20% efficiency. The resulting electricity is used to operate an electric stove that can be used to heat water. A second system uses a 1.5-m² curved mirror to concentrate the Sun's energy directly onto a container of water. Estimate how long it takes each system to heat 1.0 kg of water from 25°C to 95°C. (See also Chapter 14.)

7. A dehumidifier removes water vapor from air and has been referred to as a "refrigerator with an open door." The humid air is pulled in by a fan and passes over a cold coil, whose temperature is less than the dew point, and some of the air's water condenses. After this water is extracted, the air is warmed back to its original temperature and sent into the room. In a well-designed dehumidifier, the heat that is removed by the cooling coil mostly comes from the condensation of water vapor to liquid, and this heat is used to re-warm the air. Estimate how much water is removed in 1.0 h by an ideal dehumidifier, if the temperature of the room is 25°C, the water condenses at 8°C, and the dehumidifier does work at the rate of 600 W of electrical power. (See Sections 15–6, 13–12, and 14–5.)

*8. **Trees offsetting CO_2.** Trees can help offset the buildup of CO_2 due to burning coal and other fossil fuels. CO_2 can be absorbed by tree foliage. Trees use the carbon to grow, and release O_2 into the atmosphere. Suppose a refrigerator uses 600 kWh of electricity per year (about 2×10^9 J) from a 33% efficient coal-fired power plant. Burning 1 kg of coal releases about 2×10^7 J of energy. Assume coal is all carbon, which when burned in air becomes CO_2. (a) How much coal is burned per year to run this refrigerator? (b) Assuming a forest can capture 1700 kg of carbon per hectare (= 10,000 m²) per year, estimate how many square meters of forest are needed to capture the carbon (in the form now of CO_2) emitted in (a).

ANSWERS TO EXERCISES

A: 700 J.
B: Less.
C: -6.8×10^3 J.
D: (d), (e).
E: (c).
F: (a) Heat Q_L comes from outside to conserve energy; (b) 560 W.
G: (e).

This comb has acquired a static electric charge, either from passing through hair, or being rubbed by a cloth or paper towel. The electrical charge on the comb induces a polarization (separation of charge) in scraps of paper, and thus attracts them.

Our introduction to electricity in this Chapter covers conductors and insulators, and Coulomb's law which relates the force between two point charges as a function of their distance apart. We also introduce the powerful concept of electric field.

Electric Charge and Electric Field

CHAPTER 16

CONTENTS

16–1 Static Electricity; Electric Charge and Its Conservation
16–2 Electric Charge in the Atom
16–3 Insulators and Conductors
16–4 Induced Charge; the Electroscope
16–5 Coulomb's Law
16–6 Solving Problems Involving Coulomb's Law and Vectors
16–7 The Electric Field
16–8 Electric Field Lines
16–9 Electric Fields and Conductors
*16–10 Electric Forces in Molecular Biology: DNA Structure and Replication
*16–11 Photocopy Machines and Computer Printers Use Electrostatics
*16–12 Gauss's Law

CHAPTER-OPENING QUESTION—Guess now!
Two identical tiny spheres have the same electric charge. If their separation is doubled, the force each exerts on the other will be
(a) half.
(b) double.
(c) four times larger.
(d) one-quarter as large.
(e) unchanged.

The word "electricity" may evoke an image of complex modern technology: lights, motors, electronics, and computers. But the electric force plays an even deeper role in our lives. According to atomic theory, electric forces between atoms and molecules hold them together to form liquids and solids, and electric forces are also involved in the metabolic processes that occur within our bodies. Many of the forces we have dealt with so far, such as elastic forces, the normal force, and friction and other contact forces (pushes and pulls), are now considered to result from electric forces acting at the atomic level. Gravity, on the other hand, is a separate force.[†]

[†]As we discussed in Section 5–9, physicists in the twentieth century came to recognize four different fundamental forces in nature: (1) gravitational force, (2) electromagnetic force (we will see later that electric and magnetic forces are intimately related), (3) strong nuclear force, and (4) weak nuclear force. The last two forces operate at the level of the nucleus of an atom. Recent theory has combined the electromagnetic and weak nuclear forces so they are now considered to have a common origin known as the electroweak force. We discuss the other forces in later Chapters.

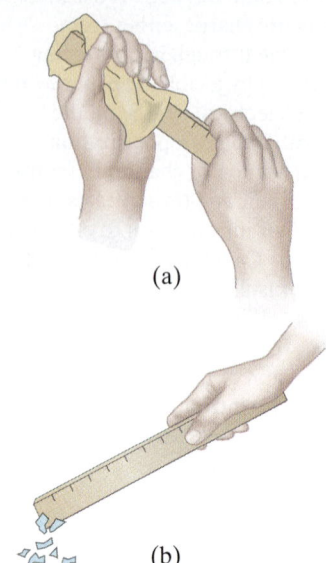

FIGURE 16–1 (a) Rub a plastic ruler with a cloth or paper towel, and (b) bring it close to some tiny pieces of paper.

FIGURE 16–2 Like charges repel one another; unlike charges attract. (Note color coding: we color positive charged objects pink or red, and negative charges blue-green. We use these colors especially for point charges, but not always for real objects.)

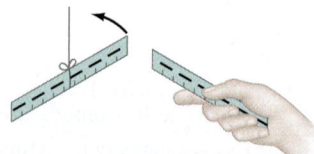

(a) Two charged plastic rulers repel

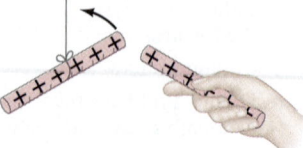

(b) Two charged glass rods repel

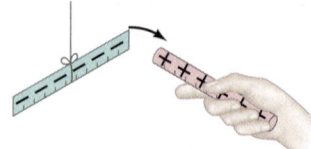

(c) Charged glass rod attracts charged plastic ruler

LAW OF CONSERVATION OF ELECTRIC CHARGE

The earliest studies on electricity date back to the ancients, but only since the late 1700s has electricity been studied in detail. We will discuss the development of ideas about electricity, including practical devices, as well as its relation to magnetism, in the next seven Chapters.

16–1 Static Electricity; Electric Charge and Its Conservation

The word *electricity* comes from the Greek word *elektron*, which means "amber." Amber is petrified tree resin, and the ancients knew that if you rub a piece of amber with a cloth, the amber attracts small pieces of leaves or dust. A piece of hard rubber, a glass rod, or a plastic ruler rubbed with a cloth will also display this "amber effect," or **static electricity** as we call it today. You can readily pick up small pieces of paper with a plastic comb or ruler that you have just vigorously rubbed with even a paper towel. See the photo on the previous page and Fig. 16–1. You have probably experienced static electricity when combing your hair or when taking a synthetic blouse or shirt from a clothes dryer. And you may have felt a shock when you touched a metal doorknob after sliding across a car seat or walking across a synthetic carpet. In each case, an object becomes "charged" as a result of rubbing, and is said to possess a net **electric charge**.

Is all electric charge the same, or is there more than one type? In fact, there are *two* types of electric charge, as the following simple experiments show. A plastic ruler suspended by a thread is vigorously rubbed with a cloth to charge it. When a second plastic ruler, which has been charged in the same way, is brought close to the first, it is found that one ruler *repels* the other. This is shown in Fig. 16–2a. Similarly, if a rubbed glass rod is brought close to a second charged glass rod, again a repulsive force is seen to act, Fig. 16–2b. However, if the charged glass rod is brought close to the charged plastic ruler, it is found that they *attract* each other, Fig. 16–2c. The charge on the glass must therefore be different from that on the plastic. Indeed, it is found experimentally that all charged objects fall into one of two categories. Either they are attracted to the plastic and repelled by the glass; or they are repelled by the plastic and attracted to the glass. Thus there seem to be two, and only two, types of electric charge. Each type of charge repels the same type but attracts the opposite type. That is: **unlike charges attract; like charges repel**.

The two types of electric charge were referred to as **positive** and **negative** by the American statesman, philosopher, and scientist Benjamin Franklin (1706–1790). The choice of which name went with which type of charge was arbitrary. Franklin's choice set the charge on the rubbed glass rod to be positive charge, so the charge on a rubbed plastic ruler (or amber) is called negative charge. We still follow this convention today.

Franklin argued that whenever a certain amount of charge is produced on one object, an equal amount of the opposite type of charge is produced on another object. The positive and negative are to be treated *algebraically*, so during any process, the net change in the amount of charge produced is zero. For example, when a plastic ruler is rubbed with a paper towel, the plastic acquires a negative charge and the towel acquires an equal amount of positive charge. The charges are separated, but the sum of the two is zero.

This is an example of a law that is now well established: the **law of conservation of electric charge**, which states that

the net amount of electric charge produced in any process is zero;

or, said another way,

no net electric charge can be created or destroyed.

If one object (or a region of space) acquires a positive charge, then an equal amount of negative charge will be found in neighboring areas or objects. No violations have ever been found, and the law of conservation of electric charge is as firmly established as those for energy and momentum.

16–2 Electric Charge in the Atom

Only within the past century has it become clear that an understanding of electricity originates inside the atom itself. In later Chapters we will discuss atomic structure and the ideas that led to our present view of the atom in more detail. But it will help our understanding of electricity if we discuss it briefly now.

A simplified model of an atom shows it as having a tiny but massive, positively charged nucleus surrounded by one or more negatively charged electrons (Fig. 16–3). The nucleus contains protons, which are positively charged, and neutrons, which have no net electric charge. All protons and all electrons have exactly the same magnitude of electric charge; but their signs are opposite. Hence neutral atoms, having no net charge, contain equal numbers of protons and electrons. Sometimes an atom may lose one or more of its electrons, or may gain extra electrons, in which case it will have a net positive or negative charge and is called an **ion**.

In solid materials the nuclei tend to remain close to fixed positions, whereas some of the electrons may move quite freely. When an object is *neutral*, it contains equal amounts of positive and negative charge. The charging of a solid object by rubbing can be explained by the transfer of electrons from one object to the other. When a plastic ruler becomes negatively charged by rubbing with a paper towel, electrons are transferred from the towel to the plastic, leaving the towel with a positive charge equal in magnitude to the negative charge acquired by the plastic. In liquids and gases, nuclei or ions can move as well as electrons.

Normally when objects are charged by rubbing, they hold their charge only for a limited time and eventually return to the neutral state. Where does the charge go? Usually the excess charge "leaks off" onto water molecules in the air. This is because water molecules are **polar**—that is, even though they are neutral, their charge is not distributed uniformly, Fig. 16–4. Thus the extra electrons on, say, a charged plastic ruler can "leak off" into the air because they are attracted to the positive end of water molecules. A positively charged object, on the other hand, can be neutralized by transfer of loosely held electrons from water molecules in the air. On dry days, static electricity is much more noticeable since the air contains fewer water molecules to allow leakage of charge. On humid or rainy days, it is difficult to make any object hold a net charge for long.

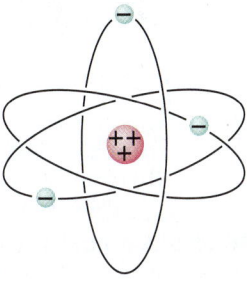

FIGURE 16–3 Simple model of the atom.

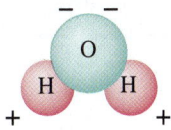

FIGURE 16–4 Diagram of a water molecule. Because it has opposite charges on different ends, it is called a "polar" molecule.

16–3 Insulators and Conductors

Suppose we have two metal spheres, one highly charged and the other electrically neutral (Fig. 16–5a). If we now place a metal object, such as a nail, so that it touches both spheres (Fig. 16–5b), the previously uncharged sphere quickly becomes charged. If, instead, we had connected the two spheres by a wooden rod or a piece of rubber (Fig. 16–5c), the uncharged ball would not become noticeably charged. Materials like the iron nail are said to be **conductors** of electricity, whereas wood and rubber are **nonconductors** or **insulators**.

Metals are generally good conductors, whereas most other materials are insulators (although even insulators conduct electricity very slightly). Nearly all natural materials fall into one or the other of these two distinct categories. However, a few materials (notably silicon and germanium) fall into an intermediate category known as **semiconductors**.

From the atomic point of view, the electrons in an insulating material are bound very tightly to the nuclei. In a good metal conductor, on the other hand, some of the electrons are bound very loosely and can move about freely within the metal (although they cannot *leave* the metal easily) and are often referred to as **free electrons** or **conduction electrons**. When a positively charged object is brought close to or touches a conductor, the free electrons in the conductor are attracted by this positively charged object and move quickly toward it. If a negatively charged object is brought close to the conductor, the free electrons in the conductor move swiftly away from it. In a semiconductor, there are many fewer free electrons, and in an insulator, almost none.

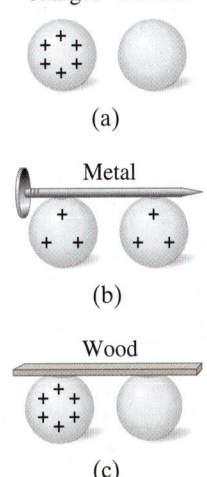

FIGURE 16–5 (a) A charged metal sphere and a neutral metal sphere. (b) The two spheres connected by a conductor (a metal nail), which conducts charge from one sphere to the other. (c) The original two spheres connected by an insulator (wood); almost no charge is conducted.

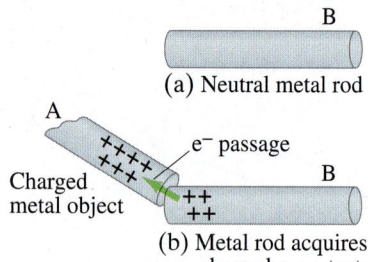

FIGURE 16–6 A neutral metal rod in (a) will acquire a positive charge if placed in contact (b) with a positively charged metal object. (Electrons move as shown by the green arrow.) This is called charging by conduction.

16–4 Induced Charge; the Electroscope

Suppose a positively charged metal object A is brought close to an uncharged metal object B. If the two touch, the free electrons in the neutral one are attracted to the positively charged object and some of those electrons will pass over to it, Fig. 16–6. Since object B, originally neutral, is now missing some of its negative electrons, it will have a net positive charge. This process is called **charging by conduction**, or "by contact," and the two objects end up with the same sign of charge.

Now suppose a positively charged object is brought close to a neutral metal rod, but does not touch it. Although the free electrons of the metal rod do not leave the rod, they still move within the metal toward the external positive charge, leaving a positive charge at the opposite end of the rod (Fig. 16–7b). A charge is said to have been *induced* at the two ends of the metal rod. No net charge has been created in the rod: charges have merely been *separated*. The net charge on the metal rod is still zero. However, if the metal is separated into two pieces, we would have two charged objects: one charged positively and one charged negatively. This is **charging by induction**.

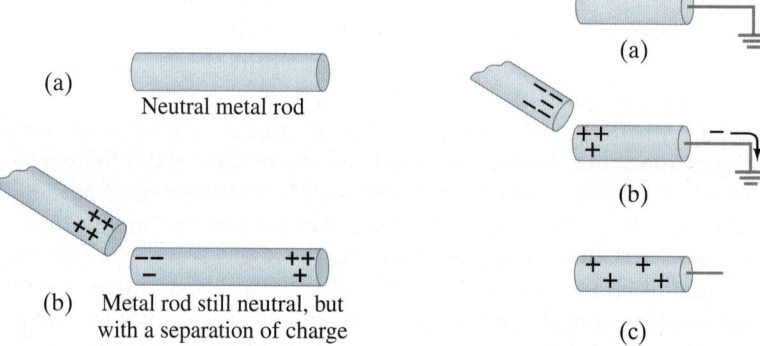

FIGURE 16–7 Charging by induction: if the rod in (b) is cut into two parts, each part will have a net charge.

FIGURE 16–8 Inducing a charge on an object connected to ground.

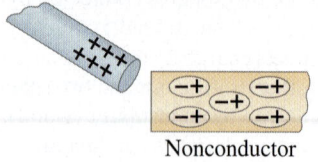

FIGURE 16–9 A charged object brought near a nonconductor causes a charge separation within the nonconductor's molecules.

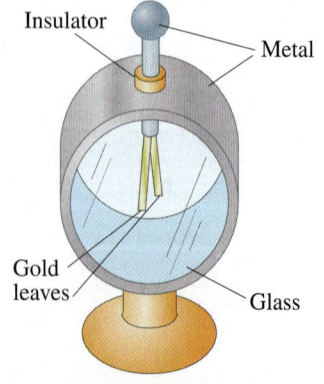

FIGURE 16–10 Electroscope.

Another way to induce a net charge on a metal object is to first connect it with a conducting wire to the ground (or a conducting pipe leading into the ground) as shown in Fig. 16–8a (the symbol ⏚ means connected to "ground"). The object is then said to be **grounded** or "earthed." The Earth, because it is so large and can conduct, easily accepts or gives up electrons; hence it acts like a reservoir for charge. If a charged object—say negative this time—is brought up close to the metal object, free electrons in the metal are repelled and many of them move down the wire into the Earth, Fig. 16–8b. This leaves the metal positively charged. If the wire is now cut, the metal object will have a positive induced charge on it (Fig. 16–8c). If the wire is cut *after* the negative object is moved away, the electrons would all have moved from the ground back into the metal object and it would be neutral again.

Charge separation can also be done in nonconductors. If you bring a positively charged object close to a neutral nonconductor as shown in Fig. 16–9, almost no electrons can move about freely within the nonconductor. But they can move slightly within their own atoms and molecules. Each oval in Fig. 16–9 represents a molecule (not to scale); the negatively charged electrons, attracted to the external positive charge, tend to move in its direction within their molecules. Because the negative charges in the nonconductor are nearer to the external positive charge, the nonconductor as a whole is attracted to the external positive charge (see the Chapter-Opening Photo, page 443).

An **electroscope** is a device that can be used for detecting charge. As shown in Fig. 16–10, inside a case are two movable metal leaves, often made of gold foil, connected to a metal knob on the outside. (Sometimes only one leaf is movable.)

If a positively charged object is brought close to the knob, a separation of charge is induced: electrons are attracted up into the knob, and the leaves become positively charged, Fig. 16–11a. The two leaves repel each other as shown, because they are both positively charged. If, instead, the knob is charged by conduction (touching), the whole apparatus acquires a net charge as shown in Fig. 16–11b. In either case, the greater the amount of charge, the greater the separation of the leaves.

Note that you cannot tell the sign of the charge in this way, since negative charge will cause the leaves to separate just as much as an equal amount of positive charge; in either case, the two leaves repel each other. An electroscope can, however, be used to determine the sign of the charge if it is first charged by conduction: say, negatively, as in Fig. 16–12a. Now if a negative object is brought close, as in Fig. 16–12b, more electrons are induced to move down into the leaves and they separate further. If a positive charge is brought close instead, the electrons are induced to flow upward, so the leaves are less negative and their separation is reduced, Fig. 16–12c.

The electroscope was used in the early studies of electricity. The same principle, aided by some electronics, is used in much more sensitive modern **electrometers**.

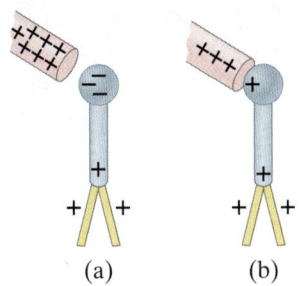

FIGURE 16–11 Electroscope charged (a) by induction, (b) by conduction.

FIGURE 16–12 A previously charged electroscope can be used to determine the sign of a charged object.

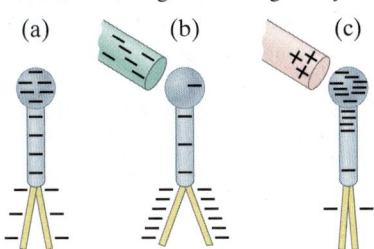

16–5 Coulomb's Law

We have seen that an electric charge exerts a force of attraction or repulsion on other electric charges. What factors affect the magnitude of this force? To find an answer, the French physicist Charles Coulomb (1736–1806) investigated electric forces in the 1780s using a torsion balance (Fig. 16–13) much like that used by Cavendish for his studies of the gravitational force (Chapter 5).

Precise instruments for the measurement of electric charge were not available in Coulomb's time. Nonetheless, Coulomb was able to prepare small spheres with different magnitudes of charge in which the *ratio* of the charges was known.[†] Although he had some difficulty with induced charges, Coulomb was able to argue that the electric force one tiny charged object exerts on a second tiny charged object is directly proportional to the charge on each of them. That is, if the charge on either one of the objects is doubled, the force is doubled; and if the charge on both of the objects is doubled, the force increases to four times the original value. This was the case when the distance between the two charges remained the same. If the distance between them was allowed to increase, he found that the force decreased with the *square of the distance* between them. That is, if the distance was doubled, the force fell to one-fourth of its original value. Thus, Coulomb concluded, the magnitude of the force F that one small charged object exerts on a second one is proportional to the product of the magnitude of the charge on one, Q_1, times the magnitude of the charge on the other, Q_2, and inversely proportional to the square of the distance r between them (Fig. 16–14). As an equation, we can write **Coulomb's law** as

$$F = k\frac{Q_1 Q_2}{r^2}, \qquad \text{[magnitudes]} \quad (16\text{–}1)$$

where k is a proportionality constant.[‡]

FIGURE 16–13 Coulomb's apparatus: when an external charged sphere is placed close to the charged one on the suspended bar, the bar rotates slightly. The suspending fiber resists the twisting motion, and the angle of twist is proportional to the force applied. With this apparatus, Coulomb investigated how the electric force varies as a function of the magnitude of the charges and of the distance between them.

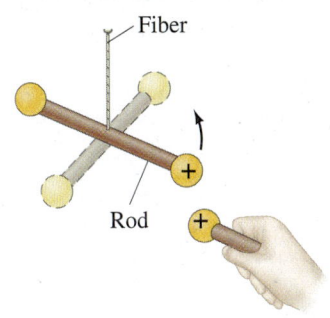

COULOMB'S LAW

FIGURE 16–14 Coulomb's law, Eq. 16–1, gives the force between two point charges, Q_1 and Q_2, a distance r apart.

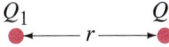

[†]Coulomb reasoned that if a charged conducting sphere is placed in contact with an identical uncharged sphere, the charge on the first would be shared equally by the two of them because of symmetry. He thus had a way to produce charges equal to $\frac{1}{2}$, $\frac{1}{4}$, and so on, of the original charge.

[‡]The validity of Coulomb's law today rests on precision measurements that are much more sophisticated than Coulomb's original experiment. The exponent 2 on r in Coulomb's law has been shown to be accurate to 1 part in 10^{16} [that is, $2 \pm (1 \times 10^{-16})$].

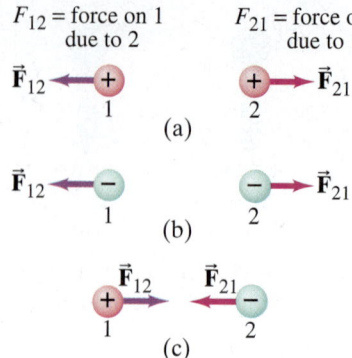

FIGURE 16–15 The direction of the static electric force one point charge exerts on another is always along the line joining the two charges, and depends on whether the charges have the same sign as in (a) and (b), or opposite signs (c).

As we just saw, Coulomb's law, Eq. 16–1,

$$F = k\frac{Q_1 Q_2}{r^2},\qquad \text{[magnitudes]} \quad (16\text{–}1)$$

gives the *magnitude* of the electric force that either charge exerts on the other. The *direction* of the electric force *is always along the line joining the two charges.* If the two charges have the same sign, the force on either charge is directed away from the other (they repel each other). If the two charges have opposite signs, the force on one is directed toward the other (they attract). See Fig. 16–15. Notice that the force one charge exerts on the second is equal but opposite to that exerted by the second on the first, in accord with Newton's third law.

The SI unit of charge is the **coulomb** (C). The precise definition of the coulomb today is in terms of electric current and magnetic field, and will be discussed later (Section 20–6). In SI units, the constant k in Coulomb's law has the value

$$k = 8.988 \times 10^9 \,\text{N} \cdot \text{m}^2/\text{C}^2$$

or, when we only need two significant figures,

$$k \approx 9.0 \times 10^9 \,\text{N} \cdot \text{m}^2/\text{C}^2.$$

Thus, 1 C is that amount of charge which, if placed on each of two point objects that are 1.0 m apart, will result in each object exerting a force of $(9.0 \times 10^9 \,\text{N} \cdot \text{m}^2/\text{C}^2)(1.0\,\text{C})(1.0\,\text{C})/(1.0\,\text{m})^2 = 9.0 \times 10^9 \,\text{N}$ on the other. This would be an enormous force, equal to the weight of almost a million tons. We rarely encounter charges as large as a coulomb.[†]

Charges produced by rubbing ordinary objects (such as a comb or plastic ruler) are typically around a microcoulomb ($1\,\mu\text{C} = 10^{-6}\,\text{C}$) or less. Objects that carry a positive charge have a deficit of electrons, whereas negatively charged objects have an excess of electrons. The charge on one electron has been determined to have a magnitude of about $1.6022 \times 10^{-19}\,\text{C}$, and is negative. This is the smallest charge observed in nature,[‡] and because it is fundamental, it is given the symbol e and is often referred to as the **elementary charge**:

$$e = 1.6022 \times 10^{-19}\,\text{C} \approx 1.6 \times 10^{-19}\,\text{C}.$$

Note that e is defined as a positive number, so the charge on the electron is $-e$. (The charge on a proton, on the other hand, is $+e$.) Since an object cannot gain or lose a fraction of an electron, the net charge on any object must be an integral multiple of this charge. Electric charge is thus said to be **quantized** (existing only in discrete amounts: $1e, 2e, 3e$, etc.). Because e is so small, however, we normally do not notice this discreteness in macroscopic charges ($1\,\mu\text{C}$ requires about 10^{13} electrons), which thus seem continuous.

Coulomb's law looks a lot like the *law of universal gravitation*, $F = Gm_1 m_2/r^2$, which expresses the magnitude of the gravitational force a mass m_1 exerts on a mass m_2 (Eq. 5–4). Both are **inverse square laws** ($F \propto 1/r^2$). Both also have a proportionality to a property of each object—mass for gravity, electric charge for electricity. And both act over a distance (that is, there is no need for contact). A major difference between the two laws is that gravity is always an attractive force, whereas the electric force can be either attractive or repulsive. Electric charge comes in two types, positive and negative; gravitational mass is only positive.

The constant k in Eq. 16–1 is often written in terms of another constant, ϵ_0, called the **permittivity of free space**. It is related to k by $k = 1/4\pi\epsilon_0$. Coulomb's law can be written

COULOMB'S LAW
(in terms of ϵ_0)

$$F = \frac{1}{4\pi\epsilon_0}\frac{Q_1 Q_2}{r^2}, \qquad (16\text{–}2)$$

where

$$\epsilon_0 = \frac{1}{4\pi k} = 8.85 \times 10^{-12}\,\text{C}^2/\text{N} \cdot \text{m}^2.$$

[†]In the once common cgs system of units, k is set equal to 1, and the unit of electric charge is called the *electrostatic unit* (esu) or the statcoulomb. One esu is defined as that charge, on each of two point objects 1 cm apart, that gives rise to a force of 1 dyne.

[‡]According to the Standard Model of elementary particle physics, subnuclear particles called quarks have a smaller charge than the electron, equal to $\tfrac{1}{3}e$ or $\tfrac{2}{3}e$. Quarks have not been detected directly as isolated objects, and theory indicates that free quarks may not be detectable.

Equation 16–2 looks more complicated than Eq. 16–1, but other fundamental equations we haven't seen yet are simpler in terms of ϵ_0 rather than k. It doesn't matter which form we use since Eqs. 16–1 and 16–2 are equivalent. (The latest precise values of e and ϵ_0 are given inside the front cover.)[†]

Equations 16–1 and 16–2 apply to objects whose size is much smaller than the distance between them. Ideally, it is precise for **point charges** (spatial size negligible compared to other distances). For finite-sized objects, it is not always clear what value to use for r, particularly since the charge may not be distributed uniformly on the objects. If the two objects are spheres and the charge is known to be distributed uniformly on each, then r is the distance between their centers.

Coulomb's law describes the force between two charges when they are at rest. Additional forces come into play when charges are in motion, and will be discussed in later Chapters. In this Chapter we discuss only charges at rest, the study of which is called **electrostatics**, and Coulomb's law gives the **electrostatic force**.

When calculating with Coulomb's law, we usually use magnitudes, ignoring signs of the charges, and determine the direction of a force separately based on whether the force is attractive or repulsive.

> **PROBLEM SOLVING**
> *Use magnitudes in Coulomb's law; find force direction from signs of charges*

EXERCISE A Return to the Chapter-Opening Question, page 443, and answer it again now. Try to explain why you may have answered differently the first time.

EXAMPLE 16–1 **Electric force on electron by proton.** Determine the magnitude and direction of the electric force on the electron of a hydrogen atom exerted by the single proton $(Q_2 = +e)$ that is the atom's nucleus. Assume the average distance between the revolving electron and the proton is $r = 0.53 \times 10^{-10}$ m, Fig. 16–16.

APPROACH To find the force magnitude we use Coulomb's law, $F = kQ_1Q_2/r^2$ (Eq. 16–1), with $r = 0.53 \times 10^{-10}$ m. The electron and proton have the same magnitude of charge, e, so $Q_1 = Q_2 = 1.6 \times 10^{-19}$ C.

SOLUTION The magnitude of the force is

$$F = k\frac{Q_1Q_2}{r^2} = \frac{(9.0 \times 10^9 \text{ N}\cdot\text{m}^2/\text{C}^2)(1.6 \times 10^{-19} \text{ C})(1.6 \times 10^{-19} \text{ C})}{(0.53 \times 10^{-10} \text{ m})^2}$$

$$= 8.2 \times 10^{-8} \text{ N}.$$

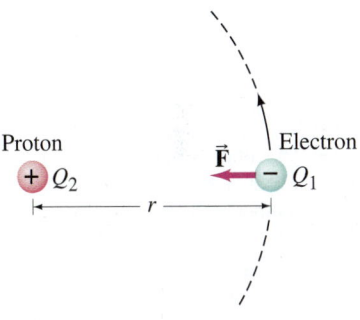

FIGURE 16–16 Example 16–1.

The direction of the force on the electron is toward the proton, because the charges have opposite signs so the force is attractive.

CONCEPTUAL EXAMPLE 16–2 **Which charge exerts the greater force?** Two positive point charges, $Q_1 = 50\,\mu\text{C}$ and $Q_2 = 1\,\mu\text{C}$, are separated by a distance ℓ, Fig. 16–17. Which is larger in magnitude, the force that Q_1 exerts on Q_2, or the force that Q_2 exerts on Q_1?

FIGURE 16–17 Example 16–2.

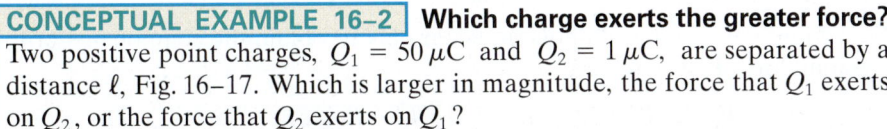

RESPONSE From Coulomb's law, the force on Q_1 exerted by Q_2 is

$$F_{12} = k\frac{Q_1Q_2}{\ell^2}.$$

The force on Q_2 exerted by Q_1 is

$$F_{21} = k\frac{Q_2Q_1}{\ell^2}$$

which is the same magnitude. The equation is symmetric with respect to the two charges, so $F_{21} = F_{12}$.

NOTE Newton's third law also tells us these two forces must have equal magnitude.

EXERCISE B In Example 16–2, how is the direction of F_{12} related to the direction of F_{21}?

[†]Our convention for units, such as $C^2/N\cdot m^2$ for ϵ_0, means m^2 is in the denominator. That is, $C^2/N\cdot m^2$ means $C^2/(N\cdot m^2)$ and does *not* mean $(C^2/N)\cdot m^2 = C^2\cdot m^2/N$.

Keep in mind that Coulomb's law, Eq. 16–1 or 16–2, gives the force on a charge due to only *one* other charge. If several (or many) charges are present, the *net force on any one of them will be the vector sum of the forces due to each of the others*. This **principle of superposition** is based on experiment, and tells us that electric force vectors add like any other vector. For example, if you have a system of four charges, the net force on charge 1, say, is the sum of the forces exerted on charge 1 by charges 2, 3, and 4. The magnitudes of these three forces are determined from Coulomb's law, and then are added vectorially.

16–6 Solving Problems Involving Coulomb's Law and Vectors

The electric force between charged particles at rest (sometimes referred to as the **electrostatic force** or as the **Coulomb force**) is, like all forces, a vector: it has both magnitude and direction. When several forces act on an object (call them $\vec{F}_1, \vec{F}_2$, etc.), the net force $\vec{F}_{net}$ on the object is the vector sum of all the forces acting on it:

$$\vec{F}_{net} = \vec{F}_1 + \vec{F}_2 + \cdots.$$

This is the principle of superposition for forces. We studied how to add vectors in Chapter 3; then in Chapter 4 we used the rules for adding vectors to obtain the net force on an object by adding the different vector forces acting on it. It might be useful now to review Sections 3–2, 3–3, and 3–4. Here is a brief review of vectors.

Vector Addition Review

Suppose two vector forces, $\vec{F}_1$ and $\vec{F}_2$, act on an object (Fig. 16–18a). They can be added using the tail-to-tip method (Fig. 16–18b) or by the parallelogram method (Fig. 16–18c), as discussed in Section 3–2. These two methods are useful for *understanding* a given problem (for getting a picture in your mind of what is going on). But for *calculating* the direction and magnitude of the resultant sum, it is more precise to use the method of adding components. Figure 16–18d shows the forces $\vec{F}_1$ and $\vec{F}_2$ resolved into components along chosen x and y axes (for more details, see Section 3–4). From the definitions of the trigonometric functions (Figs. 3–11 and 3–12), we have

$$F_{1x} = F_1 \cos\theta_1 \qquad F_{2x} = F_2 \cos\theta_2$$
$$F_{1y} = F_1 \sin\theta_1 \qquad F_{2y} = -F_2 \sin\theta_2.$$

We add up the x and y components separately to obtain the components of the resultant force $\vec{F}$, which are

$$F_x = F_{1x} + F_{2x} = F_1 \cos\theta_1 + F_2 \cos\theta_2,$$
$$F_y = F_{1y} + F_{2y} = F_1 \sin\theta_1 - F_2 \sin\theta_2.$$

The magnitude of the resultant (or *net*) force $\vec{F}$ is

$$F = \sqrt{F_x^2 + F_y^2}.$$

The direction of $\vec{F}$ is specified by the angle θ that $\vec{F}$ makes with the x axis, which is given by

$$\tan\theta = \frac{F_y}{F_x}.$$

Adding Electric Forces; Principle of Superposition

When dealing with several charges, it is helpful to use double subscripts on each of the forces involved. The first subscript refers to the particle *on* which the force acts; the second refers to the particle that exerts the force. For example, if we have three charges, $\vec{F}_{31}$ means the force exerted *on* particle 3 *by* particle 1.

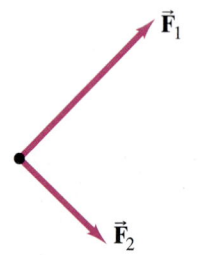

(a) Two forces acting on an object.

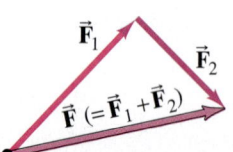

(b) The total, or net, force is $\vec{F} = \vec{F}_1 + \vec{F}_2$ by the tail-to-tip method of adding vectors.

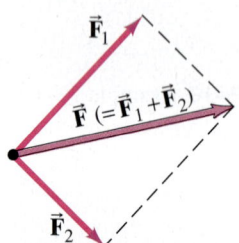

(c) $\vec{F} = \vec{F}_1 + \vec{F}_2$ by the parallelogram method.

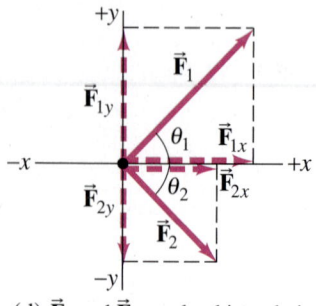

(d) $\vec{F}_1$ and $\vec{F}_2$ resolved into their x and y components.

FIGURE 16–18 Review of vector addition.

As in all problem solving, it is very important to draw a diagram, in particular a free-body diagram (Chapter 4) for each object, showing all the forces acting *on* that object. In applying Coulomb's law, we can deal with charge magnitudes only (leaving out minus signs) to get the magnitude of each force. Then determine separately the direction of the force physically (along the line joining the two particles: like charges repel, unlike charges attract), and show the force on the diagram. (You could determine direction first if you like.) Finally, add all the forces on one object together as vectors to obtain the net force on that object.

EXAMPLE 16–3 **Three charges in a line.** Three charged particles are arranged in a line, as shown in Fig. 16–19a. Calculate the net electrostatic force on particle 3 (the $-4.0\,\mu\text{C}$ on the right) due to the other two charges.

APPROACH The net force on particle 3 is the vector sum of the force $\vec{F}_{31}$ exerted on particle 3 by particle 1 and the force $\vec{F}_{32}$ exerted on 3 by particle 2:

$$\vec{F} = \vec{F}_{31} + \vec{F}_{32}.$$

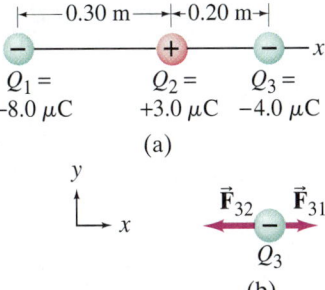

FIGURE 16–19 Example 16–3.

SOLUTION The magnitudes of these two forces are obtained using Coulomb's law, Eq. 16–1:

$$F_{31} = k\frac{Q_3 Q_1}{r_{31}^2}$$

$$= \frac{(9.0 \times 10^9\,\text{N}\cdot\text{m}^2/\text{C}^2)(4.0 \times 10^{-6}\,\text{C})(8.0 \times 10^{-6}\,\text{C})}{(0.50\,\text{m})^2} = 1.2\,\text{N},$$

where $r_{31} = 0.50\,\text{m}$ is the distance from Q_3 to Q_1. Similarly,

$$F_{32} = k\frac{Q_3 Q_2}{r_{32}^2}$$

$$= \frac{(9.0 \times 10^9\,\text{N}\cdot\text{m}^2/\text{C}^2)(4.0 \times 10^{-6}\,\text{C})(3.0 \times 10^{-6}\,\text{C})}{(0.20\,\text{m})^2} = 2.7\,\text{N}.$$

Since we were calculating the magnitudes of the forces, we omitted the signs of the charges. But we must be aware of them to get the direction of each force. Let the line joining the particles be the x axis, and we take it positive to the right. Then, because $\vec{F}_{31}$ is repulsive and $\vec{F}_{32}$ is attractive, the directions of the forces are as shown in Fig. 16–19b: F_{31} points in the positive x direction (away from Q_1) and F_{32} points in the negative x direction (toward Q_2). The net force on particle 3 is then

$$F = -F_{32} + F_{31}$$
$$= -2.7\,\text{N} + 1.2\,\text{N} = -1.5\,\text{N}.$$

The magnitude of the net force is 1.5 N, and it points to the left.

NOTE Charge Q_1 acts on charge Q_3 just as if Q_2 were not there (this is the principle of superposition). That is, the charge in the middle, Q_2, in no way blocks the effect of charge Q_1 acting on Q_3. Naturally, Q_2 exerts its own force on Q_3.

CAUTION
Each charge exerts its own force. No charge blocks the effect of the others

EXERCISE C Determine the magnitude and direction of the net force on charge Q_2 in Fig. 16–19a.

FIGURE 16–20 Determining the forces for Example 16–4. (a) The directions of the individual forces are as shown because $\vec{F}_{32}$ is repulsive (the force on Q_3 is in the direction away from Q_2 because Q_3 and Q_2 are both positive) whereas $\vec{F}_{31}$ is attractive (Q_3 and Q_1 have opposite signs), so $\vec{F}_{31}$ points toward Q_1. (b) Adding $\vec{F}_{32}$ to $\vec{F}_{31}$ to obtain the net force $\vec{F}$.

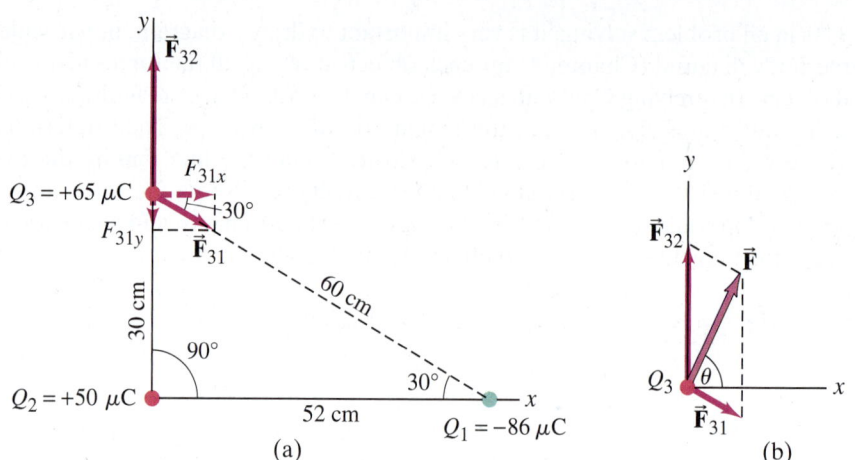

EXAMPLE 16–4 Electric force using vector components. Calculate the net electrostatic force on charge Q_3 shown in Fig. 16–20a due to the charges Q_1 and Q_2.

APPROACH We use Coulomb's law to find the magnitudes of the individual forces. The direction of each force will be along the line connecting Q_3 to Q_1 or Q_2. The forces $\vec{F}_{31}$ and $\vec{F}_{32}$ have the directions shown in Fig. 16–20a, since Q_1 exerts an attractive force on Q_3, and Q_2 exerts a repulsive force. The forces $\vec{F}_{31}$ and $\vec{F}_{32}$ are *not* along the same line, so to find the resultant force on Q_3 we resolve $\vec{F}_{31}$ and $\vec{F}_{32}$ into x and y components and perform the vector addition.

SOLUTION The magnitudes of $\vec{F}_{31}$ and $\vec{F}_{32}$ are (ignoring signs of the charges since we know the directions)

$$F_{31} = k\frac{Q_3 Q_1}{r_{31}^2} = \frac{(9.0 \times 10^9\,\text{N}\cdot\text{m}^2/\text{C}^2)(6.5 \times 10^{-5}\,\text{C})(8.6 \times 10^{-5}\,\text{C})}{(0.60\,\text{m})^2} = 140\,\text{N},$$

$$F_{32} = k\frac{Q_3 Q_2}{r_{32}^2} = \frac{(9.0 \times 10^9\,\text{N}\cdot\text{m}^2/\text{C}^2)(6.5 \times 10^{-5}\,\text{C})(5.0 \times 10^{-5}\,\text{C})}{(0.30\,\text{m})^2} = 325\,\text{N}.$$

(We keep 3 significant figures until the end, and then keep 2 because only 2 are given.) We resolve $\vec{F}_{31}$ into its components along the x and y axes, as shown in Fig. 16–20a:

$$F_{31x} = F_{31}\cos 30° = (140\,\text{N})\cos 30° = 120\,\text{N},$$
$$F_{31y} = -F_{31}\sin 30° = -(140\,\text{N})\sin 30° = -70\,\text{N}.$$

The force $\vec{F}_{32}$ has only a y component. So the net force $\vec{F}$ on Q_3 has components

$$F_x = F_{31x} = 120\,\text{N},$$
$$F_y = F_{32} + F_{31y} = 325\,\text{N} - 70\,\text{N} = 255\,\text{N}.$$

The magnitude of the net force is

$$F = \sqrt{F_x^2 + F_y^2} = \sqrt{(120\,\text{N})^2 + (255\,\text{N})^2} = 280\,\text{N};$$

and it acts at an angle θ (see Fig. 16–20b) given by

$$\tan\theta = \frac{F_y}{F_x} = \frac{255\,\text{N}}{120\,\text{N}} = 2.13,$$

so $\theta = \tan^{-1}(2.13) = 65°$.

NOTE Because $\vec{F}_{31}$ and $\vec{F}_{32}$ are not along the same line, the magnitude of $\vec{F}_3$ is not equal to the sum (or difference as in Example 16–3) of the separate magnitudes. That is, F_3 is not equal to $F_{31} + F_{32}$; nor does it equal $F_{32} - F_{31}$. Instead we had to do vector addition.

CONCEPTUAL EXAMPLE 16–5 **Make the force on Q_3 zero.** In Fig. 16–20, where could you place a fourth charge, $Q_4 = -50\,\mu\text{C}$, so that the net force on Q_3 would be zero?

RESPONSE By the principle of superposition, we need a force in exactly the opposite direction to the resultant $\vec{F}$ due to Q_2 and Q_1 that we calculated in Example 16–4, Fig. 16–20b. Our force must have magnitude 280 N, and must point down and to the left of Q_3 in Fig. 16–20b, in the direction opposite to $\vec{F}$. So Q_4 must be along this line. See Fig. 16–21.

| **EXERCISE D** In Example 16–5, what distance r must Q_4 be from Q_3?

| **EXERCISE E** (a) Consider two point charges, $+Q$ and $-Q$, which are fixed a distance d apart. Can you find a location where a third positive charge Q could be placed so that the net electric force on this third charge is zero? (b) What if the first two charges were both $+Q$?

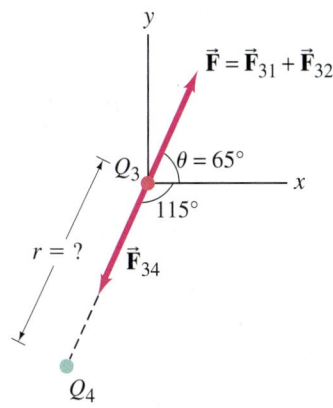

FIGURE 16–21 Example 16–5 and Exercise D: Q_4 exerts force ($\vec{F}_{34}$) that makes the net force on Q_3 zero.

16–7 The Electric Field

Many common forces might be referred to as "contact forces," such as your hands pushing or pulling a cart, or a tennis racket hitting a tennis ball.

In contrast, both the gravitational force and the electrical force act over a distance: there is a force between two objects even when the objects are not touching. The idea of a force *acting at a distance* was a difficult one for early thinkers. Newton himself felt uneasy with this idea when he published his law of universal gravitation. A helpful way to look at the situation uses the idea of the **field**, developed by the British scientist Michael Faraday (1791–1867). In the electrical case, according to Faraday, an *electric field* extends outward from every charge and permeates all of space (Fig. 16–22). If a second charge (call it Q_2) is placed near the first charge, it feels a force exerted by the electric field that is there (say, at point P in Fig. 16–22). The electric field at point P is considered to interact directly with charge Q_2 to produce the force on Q_2.

We can in principle investigate the electric field surrounding a charge or group of charges by measuring the force on a small positive **test charge** which is at rest. By a test charge we mean a charge so small that the force it exerts does not significantly affect the charges that create the field. If a tiny positive test charge q is placed at various locations in the vicinity of a single positive charge Q as shown in Fig. 16–23 (points A, B, C), the force exerted on q is as shown. The force at B is less than at A because B's distance from Q is greater (Coulomb's law); and the force at C is smaller still. In each case, the force on q is directed radially away from Q. The electric field is defined in terms of the force on such a positive test charge. In particular, the **electric field**, $\vec{E}$, at any point in space is defined as the force $\vec{F}$ exerted on a tiny positive test charge placed at that point divided by the magnitude of the test charge q:

$$\vec{E} = \frac{\vec{F}}{q}. \qquad (16\text{–}3)$$

More precisely, $\vec{E}$ is defined as the limit of $\vec{F}/q$ as q is taken smaller and smaller, approaching zero. That is, q is so tiny that it exerts essentially no force on the other charges which created the field. From this definition (Eq. 16–3), we see that the electric field at any point in space is a vector whose direction is the direction of the force on a tiny positive test charge at that point, and whose magnitude is the *force per unit charge*. Thus $\vec{E}$ has SI units of newtons per coulomb (N/C).

The reason for defining $\vec{E}$ as $\vec{F}/q$ (with $q \to 0$) is so that $\vec{E}$ does not depend on the magnitude of the test charge q. This means that $\vec{E}$ describes only the effect of the charges creating the electric field at that point.

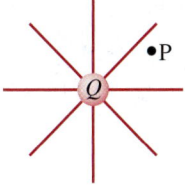

FIGURE 16–22 An electric field surrounds every charge. The red lines indicate the electric field extending out from charge Q, and P is an arbitrary point.

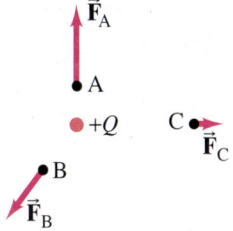

FIGURE 16–23 Force exerted by charge $+Q$ on a small test charge, q, placed at points A, B, and C.

The electric field at any point in space can be measured, based on the definition, Eq. 16–3. For simple situations with one or several point charges, we can calculate $\vec{E}$. For example, the electric field at a distance r from a single point charge Q would have magnitude

$$E = \frac{F}{q} = \frac{kqQ/r^2}{q}$$

$$E = k\frac{Q}{r^2}; \qquad \text{[single point charge]} \quad (16\text{–}4a)$$

or, in terms of ϵ_0 as in Eq. 16–2 ($k = 1/4\pi\epsilon_0$):

$$E = \frac{1}{4\pi\epsilon_0}\frac{Q}{r^2}. \qquad \text{[single point charge]} \quad (16\text{–}4b)$$

Notice that E is independent of the test charge q—that is, E depends only on the charge Q which produces the field, and not on the value of the test charge q. Equations 16–4 are referred to as the electric field form of Coulomb's law.

If we are given the electric field $\vec{E}$ at a given point in space, then we can calculate the force $\vec{F}$ on any charge q placed at that point by writing (see Eq. 16–3):

$$\vec{F} = q\vec{E}. \qquad (16\text{–}5)$$

This is valid even if q is not small as long as q does not cause the charges creating $\vec{E}$ to move. If q is positive, $\vec{F}$ and $\vec{E}$ point in the same direction. If q is negative, $\vec{F}$ and $\vec{E}$ point in opposite directions. See Fig. 16–24.

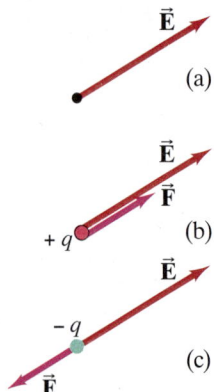

FIGURE 16–24 (a) Electric field at a given point in space. (b) Force on a positive charge at that point. (c) Force on a negative charge at that point.

EXAMPLE 16–6 Photocopy machine. A photocopy machine works by arranging positive charges (in the pattern to be copied) on the surface of a drum, then gently sprinkling negatively charged dry toner (ink) particles onto the drum. The toner particles temporarily stick to the pattern on the drum (Fig. 16–25) and are later transferred to paper and "melted" to produce the copy. Suppose each toner particle has a mass of 9.0×10^{-16} kg and carries an average of 20 extra electrons to provide an electric charge. Assuming that the electric force on a toner particle must exceed twice its weight in order to ensure sufficient attraction, compute the required electric field strength near the surface of the drum.

APPROACH The electric force on a toner particle of charge $q = 20e$ is $F = qE$, where E is the needed electric field. This force needs to be at least as great as twice the weight (mg) of the particle.

SOLUTION The minimum value of electric field satisfies the relation

$$qE = 2mg$$

where $q = 20e$. Hence

$$E = \frac{2mg}{q} = \frac{2(9.0 \times 10^{-16}\,\text{kg})(9.8\,\text{m/s}^2)}{20(1.6 \times 10^{-19}\,\text{C})} = 5.5 \times 10^3\,\text{N/C}.$$

PHYSICS APPLIED
Photocopier

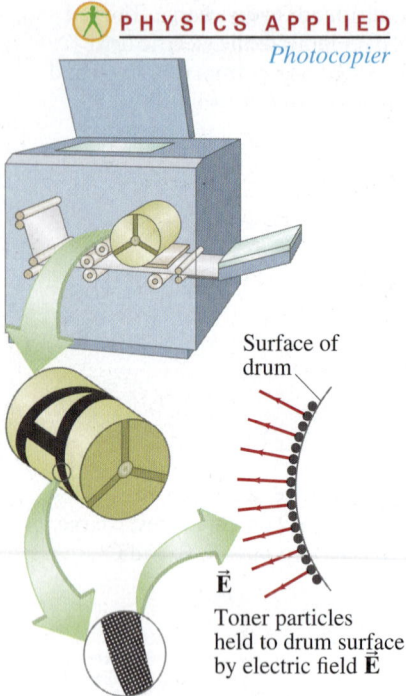

FIGURE 16–25 Example 16–6.

FIGURE 16–26 Example 16–7. Electric field at point P (a) due to a negative charge Q, and (b) due to a positive charge Q, each 30 cm from P.

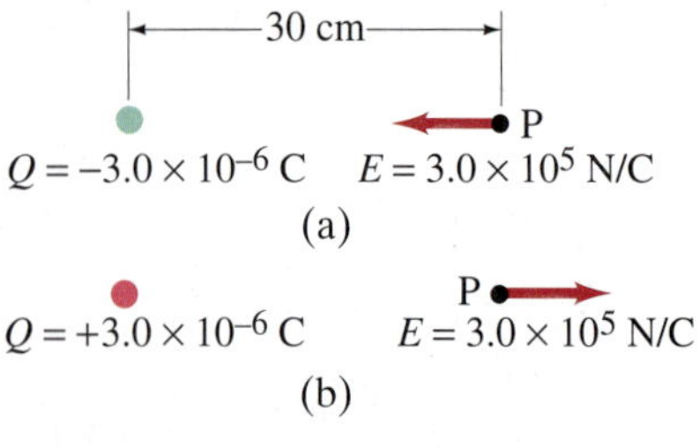

EXAMPLE 16–7 Electric field of a single point charge. Calculate the magnitude and direction of the electric field at a point P which is 30 cm to the right of a point charge $Q = -3.0 \times 10^{-6}$ C.

APPROACH The magnitude of the electric field due to a single point charge is given by Eq. 16–4. The direction is found using the sign of the charge Q.

SOLUTION The magnitude of the electric field is:

$$E = k\frac{Q}{r^2} = \frac{(9.0 \times 10^9\,\text{N}\cdot\text{m}^2/\text{C}^2)(3.0 \times 10^{-6}\,\text{C})}{(0.30\,\text{m})^2} = 3.0 \times 10^5\,\text{N/C}.$$

The direction of the electric field is *toward* the charge Q, to the left as shown in Fig. 16–26a, since we defined the direction as that of the force on a positive test charge which here would be attractive. If Q had been positive, the electric field would have pointed away, as in Fig. 16–26b.

NOTE There is no electric charge at point P. But there is an electric field there. The only real charge is Q.

This Example illustrates a general result: The electric field $\vec{E}$ due to a positive charge points away from the charge, whereas $\vec{E}$ due to a negative charge points toward that charge.

EXERCISE F Find the magnitude and direction of the electric field due to a $-2.5\,\mu C$ charge 50 cm below it.

If the electric field at a given point in space is due to more than one charge, the individual fields (call them $\vec{E}_1$, $\vec{E}_2$, etc.) due to each charge are added vectorially to get the total field at that point:

$$\vec{E} = \vec{E}_1 + \vec{E}_2 + \cdots.$$

The validity of this **superposition principle** for electric fields is fully confirmed by experiment.

EXAMPLE 16–8 $\vec{E}$ **at a point between two charges.** Two point charges are separated by a distance of 10.0 cm. One has a charge of $-25\,\mu C$ and the other $+50\,\mu C$. (a) Determine the direction and magnitude of the electric field at a point P between the two charges that is 2.0 cm from the negative charge (Fig. 16–27a). (b) If an electron (mass = 9.11×10^{-31} kg) is placed at rest at P and then released, what will be its initial acceleration (direction and magnitude)?

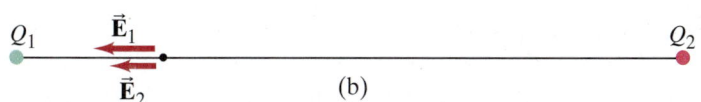

(a)

(b)

FIGURE 16–27 Example 16–8. In (b), we don't know the relative lengths of $\vec{E}_1$ and $\vec{E}_2$ until we do the calculation.

APPROACH The electric field at P will be the vector sum of the fields created separately by Q_1 and Q_2. The field due to the negative charge Q_1 points toward Q_1, and the field due to the positive charge Q_2 points away from Q_2. Thus both fields point to the left as shown in Fig. 16–27b, and we can add the magnitudes of the two fields together algebraically, ignoring the signs of the charges. In (b) we use Newton's second law ($\Sigma \vec{F} = m\vec{a}$) to find the acceleration, where $\Sigma \vec{F} = q\Sigma\vec{E}$.

SOLUTION (a) Each field is due to a point charge as given by Eq. 16–4, $E = kQ/r^2$. The total field points to the left and has magnitude

$$E = k\frac{Q_1}{r_1^2} + k\frac{Q_2}{r_2^2} = k\left(\frac{Q_1}{r_1^2} + \frac{Q_2}{r_2^2}\right)$$

$$= (9.0 \times 10^9\,\text{N}\cdot\text{m}^2/\text{C}^2)\left(\frac{25 \times 10^{-6}\,\text{C}}{(2.0 \times 10^{-2}\,\text{m})^2} + \frac{50 \times 10^{-6}\,\text{C}}{(8.0 \times 10^{-2}\,\text{m})^2}\right)$$

$$= 6.3 \times 10^8\,\text{N/C}.$$

(b) The electric field points to the left, so the electron will feel a force to the *right* since it is negatively charged. Therefore the acceleration $a = F/m$ (Newton's second law) will be to the right. The force on a charge q in an electric field E is $F = qE$ (Eq. 16–5). Hence the magnitude of the electron's initial acceleration is

$$a = \frac{F}{m} = \frac{qE}{m} = \frac{(1.60 \times 10^{-19}\,\text{C})(6.3 \times 10^8\,\text{N/C})}{9.11 \times 10^{-31}\,\text{kg}} = 1.1 \times 10^{20}\,\text{m/s}^2.$$

NOTE By considering the directions of *each* field ($\vec{E}_1$ and $\vec{E}_2$) before doing any calculations, we made sure our calculation could be done simply and correctly.

EXERCISE G Four charges of equal magnitude, but possibly different sign, are placed on the corners of a square. What arrangement of charges will produce an electric field with the greatest magnitude at the center of the square? (a) All four positive charges; (b) all four negative charges; (c) three positive and one negative; (d) two positive and two negative; (e) three negative and one positive.

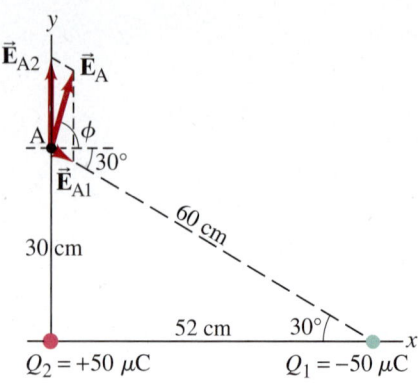

FIGURE 16–28 Calculation of the electric field at point A, Example 16–9.

PROBLEM SOLVING
Ignore signs of charges and determine direction physically, showing directions on diagram

EXAMPLE 16–9 $\vec{E}$ **above two point charges.** Calculate the total electric field at point A in Fig. 16–28 due to both charges, Q_1 and Q_2.

APPROACH The calculation is much like that of Example 16–4, except now we are dealing with electric fields instead of force. The electric field at point A is the vector sum of the fields $\vec{E}_{A1}$ due to Q_1, and $\vec{E}_{A2}$ due to Q_2. We find the magnitude of the field produced by each point charge, then we add their components to find the total field at point A.

SOLUTION The magnitude of the electric field produced at point A by each of the charges Q_1 and Q_2 is given by $E = kQ/r^2$, so

$$E_{A1} = \frac{(9.0 \times 10^9 \text{ N} \cdot \text{m}^2/\text{C}^2)(50 \times 10^{-6} \text{ C})}{(0.60 \text{ m})^2} = 1.25 \times 10^6 \text{ N/C},$$

$$E_{A2} = \frac{(9.0 \times 10^9 \text{ N} \cdot \text{m}^2/\text{C}^2)(50 \times 10^{-6} \text{ C})}{(0.30 \text{ m})^2} = 5.0 \times 10^6 \text{ N/C}.$$

The direction of E_{A1} points from A toward Q_1 (negative charge), whereas E_{A2} points from A away from Q_2, as shown; so the total electric field at A, $\vec{E}_A$, has components

$$E_{Ax} = E_{A1} \cos 30° = 1.1 \times 10^6 \text{ N/C},$$
$$E_{Ay} = E_{A2} - E_{A1} \sin 30° = 4.4 \times 10^6 \text{ N/C}.$$

Thus the magnitude of $\vec{E}_A$ is

$$E_A = \sqrt{(1.1)^2 + (4.4)^2} \times 10^6 \text{ N/C} = 4.5 \times 10^6 \text{ N/C},$$

and its direction is ϕ (Fig. 16–28) given by $\tan \phi = E_{Ay}/E_{Ax} = 4.4/1.1 = 4.0$, so $\phi = 76°$.

It is worthwhile summarizing here what we have learned about solving electrostatics problems.

PROBLEM SOLVING

Electrostatics: Electric Forces and Electric Fields

Whether you use electric field or electrostatic forces, the procedure for solving electrostatics problems is similar:

1. **Draw** a careful **diagram**—namely, a free-body diagram for each object, showing all the forces acting on that object, or showing the electric field at a point due to all significant charges present. Determine the **direction** of each force or electric field physically: like charges repel each other, unlike charges attract; fields point away from a + charge, and toward a − charge. Show and label each vector force or field on your diagram.

2. **Apply Coulomb's law** to calculate the magnitude of the force that each contributing charge exerts on a charged object, or the magnitude of the electric field each charge produces at a given point. Deal only with magnitudes of charges (leaving out minus signs), and obtain the magnitude of each force or electric field.

3. **Add vectorially** all the forces on an object, or the contributing fields at a point, to get the resultant. Use **symmetry** (say, in the geometry) whenever possible.

EXAMPLE 16–10 $\vec{E}$ **equidistant above two point charges.** Figure 16–29 (top of next page) is the same as Fig. 16–28 but includes point B, which is equidistant (40 cm) from Q_1 and Q_2. Calculate the total electric field at point B in Fig. 16–29 due to both charges, Q_1 and Q_2.

APPROACH We explicitly follow the steps of the Problem Solving Strategy above.

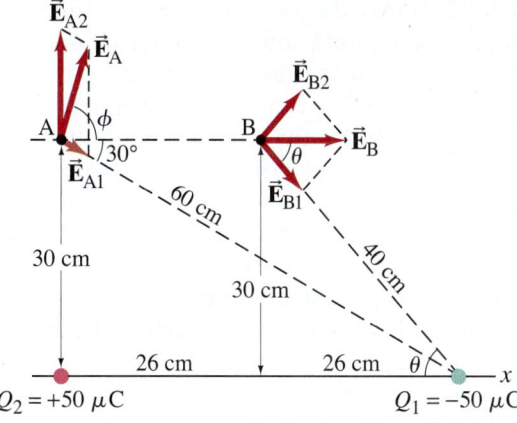

FIGURE 16–29 Same as Fig. 16–28 but with point B added. Calculation of the electric field at points A and B for Examples 16–9 and 16–10.

SOLUTION

1. **Draw** a careful **diagram**. The **directions** of the electric fields $\vec{E}_{B1}$ and $\vec{E}_{B2}$, as well as the net field $\vec{E}_B$, are shown in Fig. 16–29. $\vec{E}_{B2}$ points away from the positive charge Q_2; $\vec{E}_{B1}$ points toward the negative charge Q_1.

2. **Apply Coulomb's law** to find the magnitudes of the contributing electric fields. Because B is equidistant from the two equal charges (40 cm by the Pythagorean theorem), the magnitudes of E_{B1} and E_{B2} are the same; that is,

$$E_{B1} = E_{B2} = \frac{kQ}{r^2} = \frac{(9.0 \times 10^9 \,\text{N} \cdot \text{m}^2/\text{C}^2)(50 \times 10^{-6} \,\text{C})}{(0.40 \,\text{m})^2} = 2.8 \times 10^6 \,\text{N/C}.$$

3. **Add vectorially**, and use **symmetry** when possible. The y components of $\vec{E}_{B1}$ and $\vec{E}_{B2}$ are equal and opposite. Because of this symmetry, the total field E_B is horizontal and equals $E_{B1} \cos\theta + E_{B2} \cos\theta = 2E_{B1} \cos\theta$. From Fig. 16–29, $\cos\theta = 26 \,\text{cm}/40 \,\text{cm} = 0.65$. Then

$$E_B = 2E_{B1} \cos\theta = 2(2.8 \times 10^6 \,\text{N/C})(0.65) = 3.6 \times 10^6 \,\text{N/C},$$

and the direction of $\vec{E}_B$ is along the $+x$ direction.

PROBLEM SOLVING
Use symmetry to save work, when possible

16–8 Electric Field Lines

Since the electric field is a vector, it is sometimes referred to as a *vector field*. We could indicate the electric field with arrows at various points in a given situation, such as at A, B, and C in Fig. 16–30. The directions of $\vec{E}_A$, $\vec{E}_B$, and $\vec{E}_C$ are the same as for the forces shown earlier in Fig. 16–23, but the magnitudes (arrow lengths) are different since we divide $\vec{F}$ by q to get $\vec{E}$. However, the relative lengths of $\vec{E}_A$, $\vec{E}_B$, and $\vec{E}_C$ are the same as for the forces since we divide by the same q each time. To indicate the electric field in such a way at *many* points, however, would result in many arrows, which would quickly become cluttered and confusing. To avoid this, we use another technique, that of field lines.

To visualize the electric field, we draw a series of lines to indicate the direction of the electric field at various points in space. These **electric field lines** (or **lines of force**) are drawn to indicate the direction of the force due to the given field on a positive test charge. The lines of force due to a single isolated positive charge are shown in Fig. 16–31a, and for a single isolated negative charge in Fig. 16–31b. In part (a) the lines point radially outward from the charge, and in part (b) they point radially inward toward the charge because that is the direction the force would be on a positive test charge in each case (as in Fig. 16–26). Only a few representative lines are shown. We could draw lines in between those shown since the electric field exists there as well. We can draw the lines so that the *number of lines starting on a positive charge, or ending on a negative charge, is proportional to the magnitude of the charge*. Notice that nearer the charge, where the electric field is greater ($F \propto 1/r^2$), the lines are closer together. This is a general property of electric field lines: *the closer together the lines are, the stronger the electric field in that region*. In fact, field lines can be drawn so that the number of lines crossing unit area perpendicular to $\vec{E}$ is proportional to the magnitude of the electric field.

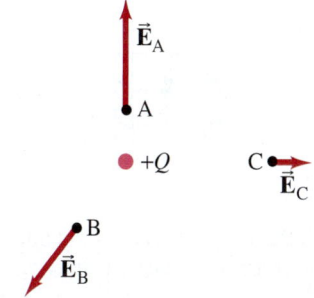

FIGURE 16–30 Electric field vector, shown at three points, due to a single point charge Q. (Compare to Fig. 16–23.)

FIGURE 16–31 Electric field lines (a) near a single positive point charge, (b) near a single negative point charge.

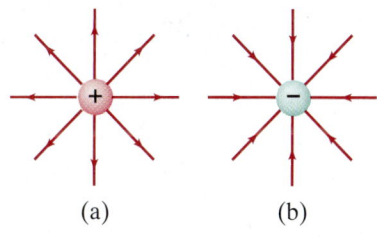

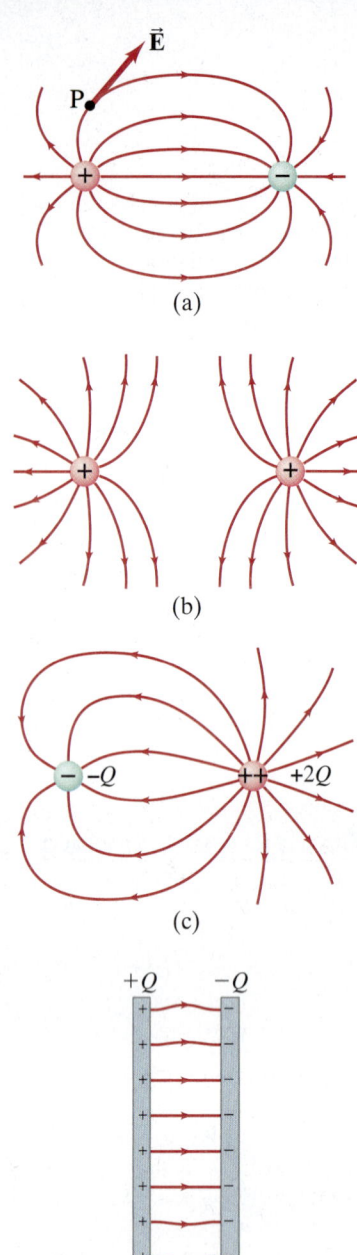

FIGURE 16–32 Electric field lines for four arrangements of charges.

FIGURE 16–33 The Earth's gravitational field, which at any point is directed toward the Earth's center (the force on any mass points toward the Earth's center).

Figure 16–32a shows the electric field lines due to two equal charges of opposite sign, a combination known as an **electric dipole**. The electric field lines are curved in this case and are directed from the positive charge to the negative charge. The direction of the electric field at any point is tangent to the field line at that point as shown by the vector arrow $\vec{E}$ at point P. To satisfy yourself that this is the correct pattern for the electric field lines, you can make a few calculations such as those done in Examples 16–9 and 16–10 for just this case (see Fig. 16–29). Figure 16–32b shows the electric field lines for two equal positive charges, and Fig. 16–32c for unequal charges, $-Q$ and $+2Q$. Note that twice as many lines leave $+2Q$ as enter $-Q$ (number of lines is proportional to magnitude of Q). Finally, in Fig. 16–32d, we see in cross section the field lines between two flat parallel plates carrying equal but opposite charges. Notice that the electric field lines between the two plates start out perpendicular to the surface of the metal plates (we will see why this is true in the next Section) and go directly from one plate to the other, as we expect because a positive test charge placed between the plates would feel a strong repulsion from the positive plate and a strong attraction to the negative plate. The field lines between two close plates are parallel and equally spaced in the central region, but fringe outward near the edges. Thus, in the central region, the electric field has the same magnitude at all points, and we can write

$$E = \text{constant.} \quad \begin{bmatrix} \text{between two closely spaced, oppositely} \\ \text{charged, flat parallel plates} \end{bmatrix} \quad (16\text{–}6)$$

The fringing of the field near the edges can often be ignored, particularly if the separation of the plates is small compared to their height and width.[†] We summarize the properties of field lines as follows:

1. Electric field lines indicate the direction of the electric field; the field points in the direction tangent to the field line at any point.
2. The lines are drawn so that the magnitude of the electric field, E, is proportional to the number of lines crossing unit area perpendicular to the lines. The closer together the lines, the stronger the field.
3. Electric field lines start on positive charges and end on negative charges; and the number starting or ending is proportional to the magnitude of the charge.

Also note that field lines never cross. Why not? Because it would not make sense for the electric field to have two directions at the same point.

Gravitational Field

The field concept can also be applied to the gravitational force (Chapter 5). Thus we can say that a **gravitational field** exists for every object that has mass. One object attracts another by means of the gravitational field. The Earth, for example, can be said to possess a gravitational field (Fig. 16–33) which is responsible for the gravitational force on objects. The *gravitational field* is defined as the *force per unit mass*. The magnitude of the Earth's gravitational field at any point above the Earth's surface is thus GM_E/r^2, where M_E is the mass of the Earth, r is the distance of the point from the Earth's center, and G is the gravitational constant (Chapter 5). At the Earth's surface, r is the radius of the Earth and the gravitational field is equal to g, the acceleration due to gravity. Beyond the Earth, the gravitational field can be calculated at any point as a sum of terms due to Earth, Sun, Moon, and other bodies that contribute significantly.

[†]The magnitude of the constant electric field between two parallel plates is given by $E = Q/\epsilon_0 A$, where Q is the magnitude of the charge on each plate and A is the area of one plate. We show this in the optional Section 16–12 on Gauss's law.

16–9 Electric Fields and Conductors

We now discuss some properties of conductors. First, *the electric field inside a conductor is zero in the static situation*—that is, when the charges are at rest. If there were an electric field within a conductor, there would be a force on the free electrons. The electrons would move until they reached positions where the electric field, and therefore the electric force on them, did become zero.

This reasoning has some interesting consequences. For one, *any net charge on a conductor distributes itself on the surface*. For a negatively charged conductor, you can imagine that the negative charges repel one another and race to the surface to get as far from one another as possible. Another consequence is the following. Suppose that a positive charge Q is surrounded by an isolated uncharged metal conductor whose shape is a spherical shell, Fig. 16–34. Because there can be no field within the metal, the lines leaving the central positive charge must end on negative charges on the inner surface of the metal. That is, the encircled charge $+Q$ induces an equal amount of negative charge, $-Q$, on the inner surface of the spherical shell. Since the shell is neutral, a positive charge of the same magnitude, $+Q$, must exist on the outer surface of the shell. Thus, although no field exists in the metal itself, an electric field exists outside of it, as shown in Fig. 16–34, as if the metal were not even there.

A related property of static electric fields and conductors is that *the electric field is always perpendicular to the surface outside of a conductor*. If there were a component of $\vec{E}$ parallel to the surface (Fig. 16–35), it would exert a force on free electrons at the surface, causing the electrons to move along the surface until they reached positions where no net force was exerted on them parallel to the surface—that is, until the electric field was perpendicular to the surface.

These properties apply only to conductors. Inside a nonconductor, which does not have free electrons, a static electric field can exist as we will see in the next Chapter. Also, the electric field outside a nonconductor does not necessarily make an angle of 90° to the surface.

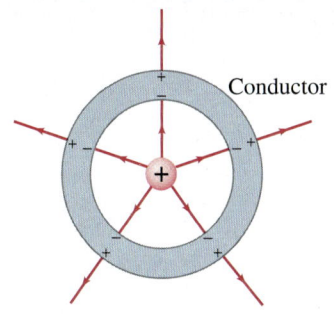

FIGURE 16–34 A charge inside a neutral spherical metal shell induces charge on its surfaces. The electric field exists even beyond the shell, but not within the conductor itself.

FIGURE 16–35 If the electric field $\vec{E}$ at the surface of a conductor had a component parallel to the surface, $\vec{E}_\parallel$, the latter would accelerate electrons into motion. In the static case, $\vec{E}_\parallel$ must be zero, and the electric field must be perpendicular to the conductor's surface: $\vec{E} = \vec{E}_\perp$.

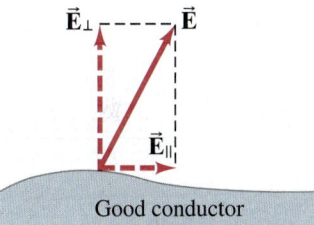

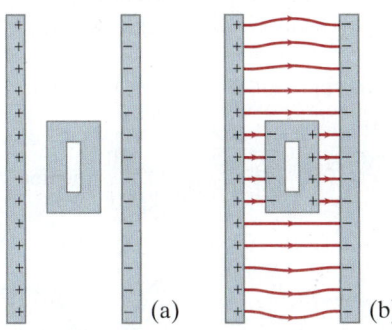

FIGURE 16–36 Example 16–11.

FIGURE 16–37 High-voltage "Van de Graaff" generators create strong electric fields in the vicinity of the "Faraday cage" below. The strong field accelerates stray electrons in the atmosphere to the KE needed to knock electrons out of air atoms, causing an avalanche of charge which flows to (or from) the metal cage. The metal cage protects the person inside it.

CONCEPTUAL EXAMPLE 16–11 Shielding, and safety in a storm. A neutral hollow metal box is placed between two parallel charged plates as shown in Fig. 16–36a. What is the field like inside the box?

RESPONSE If our metal box had been solid, and not hollow, free electrons in the box would have redistributed themselves along the surface until all their individual fields would have canceled each other inside the box. The net field inside the box would have been zero. For a hollow box, the external field is not changed since the electrons in the metal can move just as freely as before to the surface. Hence the field inside the hollow metal box is also zero, and the field lines are shown in Fig. 16–36b. A conducting box is an effective device for shielding delicate instruments and electronic circuits from unwanted external electric fields. We also can see that a relatively safe place to be during a lightning storm is inside a parked car, surrounded by metal. See also Fig. 16–37, where a person inside a porous "cage" is protected from a strong electric discharge. (It is not safe in a lightning storm to be near a tree which can conduct, or out in the open where you are taller than the surroundings.)

PHYSICS APPLIED
Electrical shielding

*16–10 Electric Forces in Molecular Biology: DNA Structure and Replication

PHYSICS APPLIED
Inside a cell: kinetic theory plus electrostatic force

The study of the structure and functioning of a living cell at the molecular level is known as molecular biology. It is an important area for application of physics. The interior of every biological cell is mainly water. We can imagine a cell as a thick soup of molecules continually in motion (kinetic theory, Chapter 13), colliding with one another with various amounts of kinetic energy. These molecules interact with one another because of the *electrostatic force* between molecules.

Indeed, cellular processes are now considered to be the result of *random ("thermal") molecular motion plus the ordering effect of the electrostatic force.* As an example, we look at DNA structure and replication. The picture we present is a model of what happens based on physical theories and experiment.

The genetic information that is passed on from generation to generation in all living cells is contained in the chromosomes, which are made up of genes. Each gene contains the information needed to produce a particular type of protein molecule, and that information is built into the principal molecule of a chromosome, DNA (deoxyribonucleic acid), Fig. 16–38. DNA molecules are made up of many small molecules known as nucleotide bases which are each *polar* (Section 16–2) due to unequal sharing of electrons. There are four types of nucleotide bases in DNA: adenine (A), cytosine (C), guanine (G), and thymine (T).

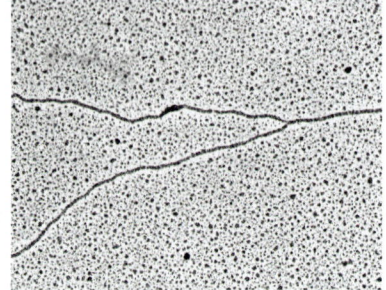

FIGURE 16–38 Image of DNA replicating, made by a transmission electron microscope.

The DNA of a chromosome generally consists of two long DNA strands wrapped about one another in the shape of a "double helix." The genetic information is contained in the specific order of the four bases (A, C, G, T) along each strand. As shown in Fig. 16–39, the two strands are attracted by electrostatic forces—that is, by the attraction of positive charges to negative charges that exist on parts of the molecules. We see in Fig. 16–39a that an A (adenine) on one strand is always opposite a T on the other strand; similarly, a G is always opposite a C. This important ordering effect occurs because the shapes of A, T, C, and G are such that a T fits closely only into an A, and a G into a C. Only in the case of this close proximity of the charged portions is the electrostatic force great enough to hold them together even for a short time (Fig. 16–39b), forming what are referred to as "weak bonds."

PHYSICS APPLIED
DNA structure

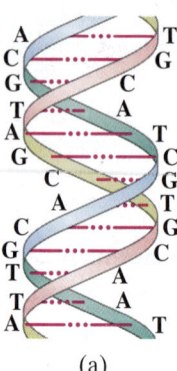

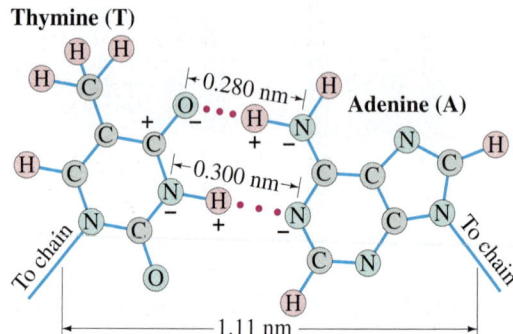

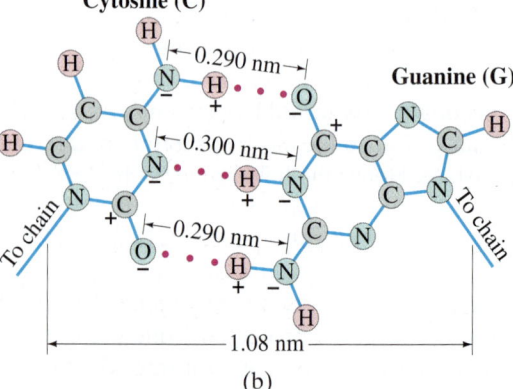

(a)

(b)

FIGURE 16–39 (a) Schematic diagram of a section of DNA double helix. (b) "Close-up" view of the helix, showing how A and T attract each other and how G and C attract each other through electrostatic forces. The + and − signs indicated on certain atoms represent net charges, usually a fraction of e, due to uneven sharing of electrons. The red dots indicate the electrostatic attraction (often called a "weak bond" or "hydrogen bond"—Section 29–3). Note that there are two weak bonds between A and T, and three between C and G.

The electrostatic force between A and T, and between C and G, exists because these molecules have charged parts. These charges are due to some electrons in each of these molecules spending more time orbiting one atom than another. For example, the electron normally on the H atom of adenine (upper part of Fig. 16–39b) spends some of its time orbiting the adjacent N atom (more on this in Chapter 29), so the N has a net negative charge and the H is left with a net positive charge. This H^+ atom of adenine is then attracted to the O^- atom of thymine. These net + and − charges usually have magnitudes of a fraction of e (charge on the electron) such as $0.2e$ or $0.4e$. (This is what we mean by "polar" molecules.)

[When H^+ is involved, the weak bond it can make with a nearby negative charge, such as O^-, is relatively strong (partly because H^+ is so small) and is referred to as a **hydrogen bond** (Section 29–3).]

When the DNA replicates (duplicates) itself just before cell division, the arrangement of A opposite T and G opposite C is crucial for ensuring that the genetic information is passed on accurately to the next generation, Fig. 16–40. The two strands of DNA separate (with the help of enzymes, which also operate via the electrostatic force), leaving the charged parts of the bases exposed.

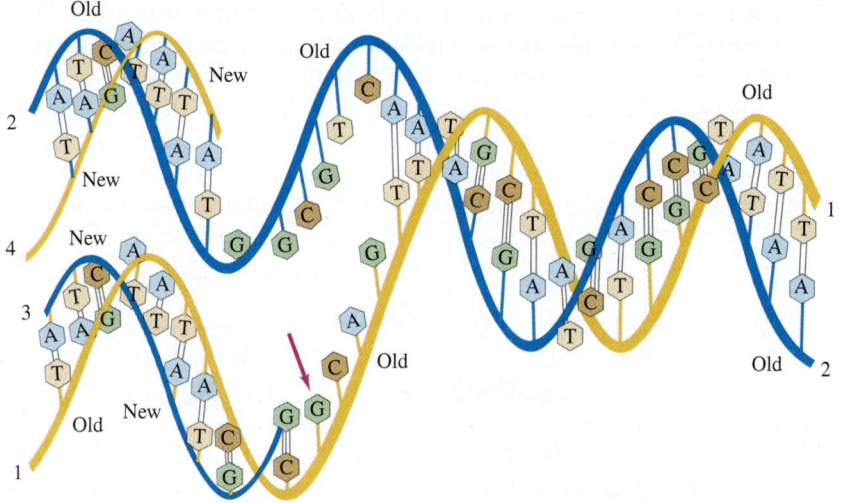

FIGURE 16–40 Replication of DNA.

Once replication starts, let us see how the correct order of bases occurs by looking at the G molecule indicated by the red arrow in Fig. 16–40. Many unattached nucleotide bases of all four kinds are bouncing around in the cellular fluid, and the only type that will experience attraction to our G, if it comes close to it, will be a C. The charges on the other three bases can not get close enough to those on the G to provide a significant attractive force—remember that the electrostatic (Coulomb) force decreases rapidly with distance $(\propto 1/r^2)$. Because the G does not attract an A, T, or G appreciably, an A, T, or G will be knocked away by collisions with other molecules before enzymes can attach it to the growing chain (number 3 in Fig. 16–40). But the electrostatic force will often hold a C opposite our G long enough so that an enzyme can attach the C to the growing end of the new chain. Thus electrostatic forces are responsible for selecting the bases in the proper order during replication. Note in Fig. 16–40 that the new number 4 strand has the same order of bases as the old number 1 strand; and the new number 3 strand is the same as the old number 2. So the two new double helixes, 1–3 and 2–4, are identical to the original 1–2 helix. Hence the genetic information is passed on accurately to the next generation.

This process of DNA replication is often presented as if it occurred in clockwork fashion—as if each molecule knew its role and went to its assigned place. But this is not the case. The forces of attraction are rather weak and become significant only when charged parts of the two molecules have "complementary shapes," meaning they can get close enough so that the electrostatic force $(\propto 1/r^2)$ is strong enough to form weak bonds. If the molecular shapes are not just right, there is almost no electrostatic attraction, which is why there are so few mistakes. Thus, out of the random motion of the molecules, the electrostatic force acts to bring order out of chaos.

The random (thermal) velocities of molecules in a cell affect *cloning*. When a bacterial cell divides, the two new bacteria have nearly identical DNA. Even if the DNA were perfectly identical, the two new bacteria would not end up behaving in exactly the same way. Long protein, DNA, and RNA molecules get bumped into different shapes, and even the expression[†] of genes can thus be different. Loosely held parts of large molecules such as a methyl group (CH_3) can also be knocked off by a strong collision with another molecule. Hence, cloned organisms are not identical, even if their DNA were identical. Indeed, there can not really be genetic determinism.

*16–11 Photocopy Machines and Computer Printers Use Electrostatics

Photocopy machines and laser printers use electrostatic attraction to print an image. They each use a different technique to project an image onto a special cylindrical drum (or rotating conveyor belt). The drum is typically made of aluminum, a good conductor; its surface is coated with a thin layer of selenium, which has the interesting property (called "photoconductivity") of being an electrical nonconductor in the dark, but a conductor when exposed to light.

In a **photocopier**, lenses and mirrors focus an image of the original sheet of paper onto the drum, much like a camera lens focuses an image on an electronic detector or film. Step 1, done in the dark, is the placing of a uniform positive charge on the drum's selenium layer by a charged roller or rod: see Fig. 16–41.

Photocopy machines

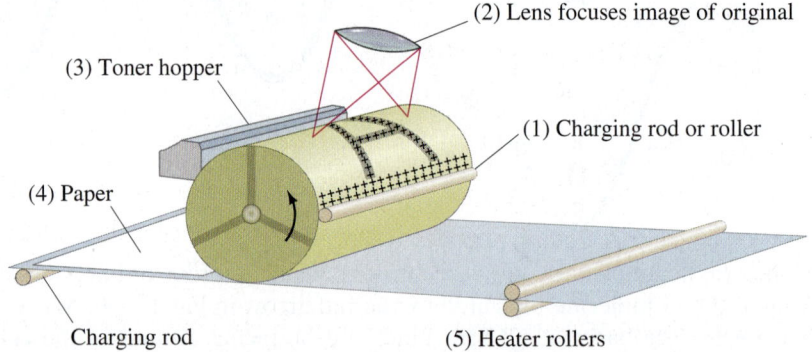

FIGURE 16–41 Inside a photocopy machine: (1) the selenium drum is given a + charge; (2) the lens focuses image on drum—only dark spots stay charged; (3) toner particles (negatively charged) are attracted to positive areas on drum; (4) the image is transferred to paper; (5) heat binds the image to the paper.

In step 2, the image to be copied is projected onto the drum. For simplicity, let us assume the image is a dark letter A on a white background (as on the page of a book) as shown in Fig. 16–41. The letter A on the drum is dark, but all around it is light. At all these light places, the selenium becomes conducting and electrons flow in from the aluminum beneath, neutralizing those positive areas. In the dark areas of the letter A, the selenium is nonconducting and so retains the positive charge already put on it, Fig. 16–41. In step 3, a fine dark powder known as *toner* is given a negative charge, and is brushed on the drum as it rotates. The negatively charged toner particles are attracted to the positive areas on the drum (the A in our case) and stick only there. In step 4, the rotating drum presses against a piece of paper which has been positively charged more strongly than the selenium, so the toner particles are transferred to the paper, forming the final image. Finally, step 5, the paper is heated to fix the toner particles firmly on the paper.

In a color copier (or printer), this process is repeated for each color—black, cyan (blue), magenta (red), and yellow. Combining these four colors in different proportions produces any desired color.

[†]The separate genes of a DNA double helix can be covered by protein molecules, keeping those genes from being "expressed"—that is, translated into the proteins they code for (see Section 29–3).

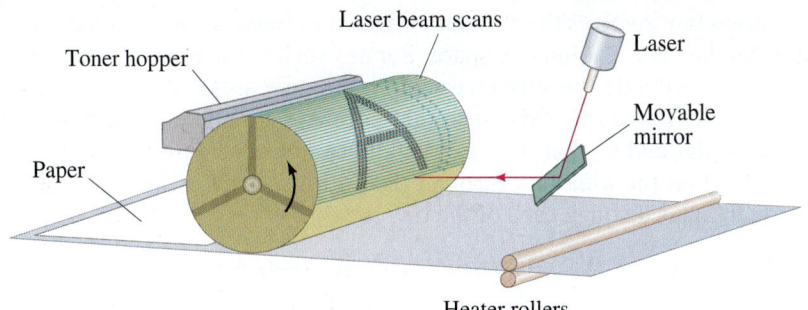

FIGURE 16–42 Inside a laser printer: a movable mirror sweeps the laser beam in horizontal lines across the drum.

A **laser printer** uses a computer output to program the intensity of a laser beam onto the selenium-coated drum of Fig. 16–42. The thin beam of light from the laser is scanned (by a movable mirror) from side to side across the drum in a series of horizontal lines, each line just below the previous line. As the beam sweeps across the drum, the intensity of the beam is varied by the computer output, being strong for a point that is meant to be white or bright, and weak or zero for points that are meant to come out dark. After each sweep, the drum rotates very slightly for additional sweeps, Fig. 16–42, until a complete image is formed on it. The light parts of the selenium become conducting and lose their (previously given) positive electric charge, and the toner sticks only to the dark, electrically charged areas. The drum then transfers the image to paper, as in a photocopier.

An **inkjet printer** does not use a drum. Instead nozzles spray tiny droplets of ink directly at the paper. The nozzles are swept across the paper, each sweep just above the previous one as the paper moves down. On each sweep, the ink makes dots on the paper, except for those points where no ink is desired, as directed by the computer. The image consists of a huge number of very tiny dots. The quality or resolution of a printer is usually specified in dots per inch (dpi) in each (linear) direction.

PHYSICS APPLIED
Laser printer

PHYSICS APPLIED
Inkjet printer

*16–12 Gauss's Law

An important relation in electricity is Gauss's law, developed by the great mathematician Karl Friedrich Gauss (1777–1855). It relates electric charge and electric field, and is a more general and elegant form of Coulomb's law.

Gauss's law involves the concept of **electric flux**, which refers to the electric field passing through a given area. For a uniform electric field $\vec{E}$ passing through an area A, as shown in Fig. 16–43a, the electric flux Φ_E is defined as

$$\Phi_E = EA \cos\theta,$$

where θ is the angle between the electric field direction and a line drawn perpendicular to the area. The flux can be written equivalently as

$$\Phi_E = E_\perp A = EA_\perp = EA \cos\theta, \qquad (16\text{–}7)$$

where $E_\perp = E \cos\theta$ is the component of $\vec{E}$ perpendicular to the area (Fig. 16–43b) and, similarly, $A_\perp = A \cos\theta$ is the projection of the area A perpendicular to the field $\vec{E}$ (Fig. 16–43c).

Electric flux can be interpreted in terms of field lines. We mentioned in Section 16–8 that field lines can always be drawn so that the number (N) passing through unit area perpendicular to the field $(A_\perp)$ is proportional to the magnitude of the field (E): that is, $E \propto N/A_\perp$. Hence,

$$N \propto EA_\perp = \Phi_E, \qquad (16\text{–}8)$$

so the flux through an area is proportional to the number of lines passing through that area.

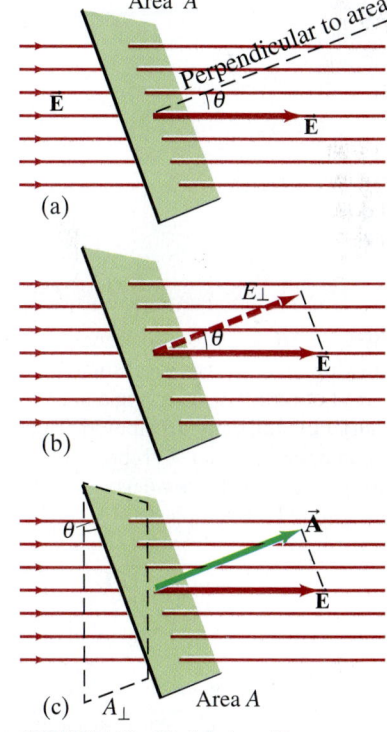

FIGURE 16–43 (a) A uniform electric field $\vec{E}$ passing through a flat square area A. (b) $E_\perp = E \cos\theta$ is the component of $\vec{E}$ perpendicular to the plane of area A. (c) $A_\perp = A \cos\theta$ is the projection (dashed) of the area A perpendicular to the field $\vec{E}$.

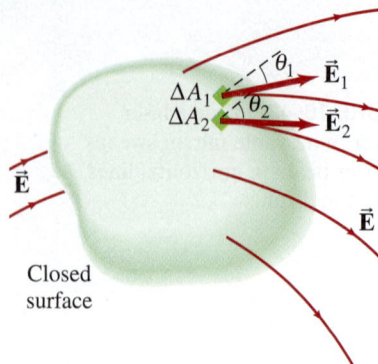

FIGURE 16–44 Electric field lines passing through a closed surface. The surface is divided up into many tiny areas, ΔA_1, ΔA_2, $\cdots$, and so on, of which only two are shown.

Gauss's law involves the *total* flux through a closed surface—a surface of any shape that encloses a volume of space. For any such surface, such as that shown in Fig. 16–44, we divide the surface up into many tiny areas, $\Delta A_1, \Delta A_2, \Delta A_3, \cdots$, and so on. We make the division so that each ΔA is small enough that it can be considered flat and so that the electric field can be considered constant through each ΔA. Then the *total* flux through the entire surface is the sum over all the individual fluxes through each of the tiny areas:

$$\Phi_E = E_1 \Delta A_1 \cos\theta_1 + E_2 \Delta A_2 \cos\theta_2 + \cdots$$
$$= \sum E \Delta A \cos\theta = \sum E_\perp \Delta A,$$

where the symbol Σ means "sum of." We saw in Section 16–8 that the number of field lines starting on a positive charge or ending on a negative charge is proportional to the magnitude of the charge. Hence, the *net* number of lines N pointing out of any closed surface (number of lines pointing out minus the number pointing in) must be proportional to the net charge enclosed by the surface, Q_{encl}. But from Eq. 16–8, we have that the net number of lines N is proportional to the total flux Φ_E. Therefore,

$$\Phi_E = \sum_{\substack{\text{closed}\\\text{surface}}} E_\perp \Delta A \propto Q_{\text{encl}}.$$

The constant of proportionality, to be consistent with Coulomb's law, is $1/\epsilon_0$, so we have

GAUSS'S LAW

$$\sum_{\substack{\text{closed}\\\text{surface}}} E_\perp \Delta A = \frac{Q_{\text{encl}}}{\epsilon_0}, \tag{16–9}$$

where the sum (Σ) is over any closed surface, and Q_{encl} is the net charge enclosed within that surface. This is **Gauss's law**.

Coulomb's law and Gauss's law can be used to determine the electric field due to a given (static) charge distribution. Gauss's law is useful when the charge distribution is simple and symmetrical. However, we must choose the closed "gaussian" surface very carefully so we can determine $\vec{\mathbf{E}}$. We normally choose a surface that has just the **symmetry** needed so that E will be constant on all or on parts of its surface.

EXAMPLE 16–12 Charged spherical conducting shell. A thin spherical shell of radius r_0 possesses a total net charge Q that is uniformly distributed on it, Fig. 16–45. Determine the electric field at points (*a*) outside the shell, and (*b*) inside the shell.

APPROACH Because the charge is distributed symmetrically, the electric field must be *symmetric*. Thus the field outside the shell must be directed radially outward (inward if $Q < 0$) and must depend only on r.

SOLUTION (*a*) We choose our imaginary gaussian surface as a sphere of radius r ($r > r_0$) concentric with the shell, shown in Fig. 16–45 as the dashed circle A_1. Then, by symmetry, the electric field will have the same magnitude at all points on this gaussian surface. Because $\vec{\mathbf{E}}$ is perpendicular to this surface, Gauss's law gives (with $Q_{\text{encl}} = Q$ in Eq. 16–9)

$$\sum E_\perp \Delta A = E \sum \Delta A = E(4\pi r^2) = \frac{Q}{\epsilon_0},$$

where $4\pi r^2$ is the surface area of our sphere (gaussian surface) of radius r. Thus

$$E = \frac{1}{4\pi\epsilon_0}\frac{Q}{r^2}. \qquad [r > r_0]$$

We see that the field outside a uniformly charged spherical shell is the same as if all the charge were concentrated at the center as a point charge.

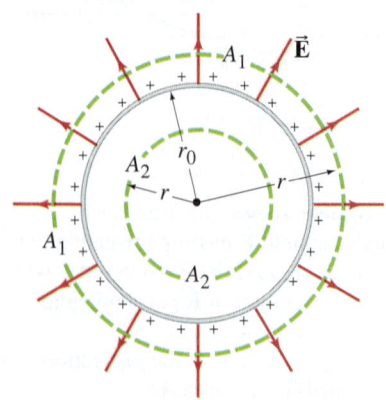

FIGURE 16–45 Cross-sectional drawing of a thin spherical shell (gray) of radius r_0, carrying a net charge Q uniformly distributed. The green circles A_1 and A_2 represent two gaussian surfaces we use to determine $\vec{\mathbf{E}}$. Example 16–12.

(b) Inside the shell, the electric field must also be symmetric. So E must again have the same value at all points on a spherical gaussian surface (A_2 in Fig. 16–45) concentric with the shell. Thus, E can be factored out of the sum and, with $Q_{encl} = 0$ because the charge inside surface A_2 is zero, we have

$$\sum E_\perp \Delta A = E \sum \Delta A$$
$$= E(4\pi r^2) = \frac{Q_{encl}}{\epsilon_0} = 0.$$

Hence

$$E = 0 \qquad [r < r_0]$$

inside a uniform spherical shell of charge (as claimed in Section 16–9).

The results of Example 16–12 also apply to a uniform *solid* spherical conductor that is charged, since all the charge would lie in a thin layer at the surface (Section 16–9). In particular

$$E = \frac{1}{4\pi\epsilon_0} \frac{Q}{r^2}$$

outside a spherical conductor. Thus, the electric field outside a spherically symmetric distribution of charge is the same as for a point charge of the same magnitude at the center of the sphere. This result applies also outside a uniformly charged nonconductor, because we can use the same gaussian surface A_1 (Fig. 16–45) and the same *symmetry* argument. We can also consider this a demonstration of our statement in Chapter 5 about the **gravitational force**, which is also a perfect $1/r^2$ force: The gravitational force exerted by a uniform sphere is the same as if all the mass were at the center, as stated on page 120.

EXAMPLE 16–13 E **near any conducting surface.** Show that the magnitude of the electric field just outside the surface of a good conductor of any shape is given by

$$E = \frac{\sigma}{\epsilon_0},$$

where σ is defined as the surface charge density, Q/A, on the conductor's surface at that point.

APPROACH We choose as our gaussian surface a small cylindrical box, very small in height so that one of its circular ends is just above the conductor (Fig. 16–46). The other end is just below the conductor's surface, and the very short sides are perpendicular to it.

SOLUTION The electric field is zero inside a conductor and is perpendicular to the surface just outside it (Section 16–9), so electric flux passes only through the outside end of our cylindrical box; no flux passes through the very short sides or through the inside end of our gaussian box. We choose the area A (of the flat cylinder end above the conductor surface) small enough so that E is essentially uniform over it. Then Gauss's law gives

$$\sum E_\perp \Delta A = EA = \frac{Q_{encl}}{\epsilon_0} = \frac{\sigma A}{\epsilon_0},$$

and therefore

$$E = \frac{\sigma}{\epsilon_0}. \qquad \text{[at surface of conductor]}$$

This useful result applies for any shape conductor, including a large, uniformly charged flat sheet: the electric field will be constant and equal to σ/ϵ_0.

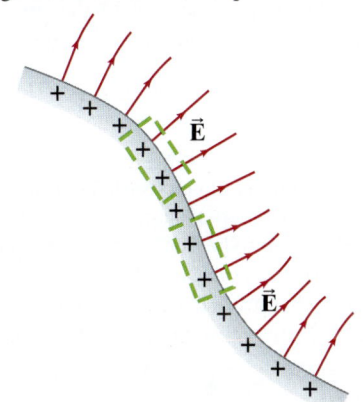

FIGURE 16–46 Electric field near the surface of a conductor. Two small cylindrical boxes are shown dashed. Either one can serve as our gaussian surface. Example 16–13.

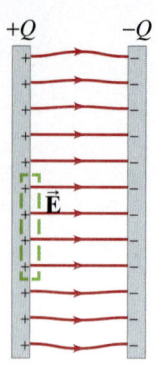

FIGURE 16–47 The electric field between two closely spaced parallel plates is uniform and equal to $E = \sigma/\epsilon_0$.

This last Example also gives us the field between the two parallel plates we discussed in Fig. 16–32d. If the plates are large compared to their separation, then the field lines are perpendicular to the plates and, except near the edges, they are parallel to each other. Therefore the electric field (see Fig. 16–47, which shows a similar very thin gaussian surface as Fig. 16–46) is also

$$E = \frac{\sigma}{\epsilon_0} = \frac{Q/A}{\epsilon_0} = \frac{Q}{\epsilon_0 A}, \quad \begin{bmatrix} \text{between two closely spaced,} \\ \text{oppositely charged, parallel plates} \end{bmatrix} \quad (16\text{–}10)$$

where $Q = \sigma A$ is the charge on one of the plates.

Summary

There are two kinds of **electric charge**, positive and negative. These designations are to be taken algebraically—that is, any charge is plus or minus so many coulombs (C), in SI units.

Electric charge is **conserved**: if a certain amount of one type of charge is produced in a process, an equal amount of the opposite type is also produced; thus the *net* charge produced is zero.

According to atomic theory, electricity originates in the atom, which consists of a positively charged nucleus surrounded by negatively charged electrons. Each electron has a charge $-e = -1.60 \times 10^{-19}$ C.

Electric **conductors** are those materials in which many electrons are relatively free to move, whereas electric **insulators** or **nonconductors** are those in which very few electrons are free to move.

An object is negatively charged when it has an excess of electrons, and positively charged when it has less than its normal number of electrons. The net charge on any object is a whole number times $+e$ or $-e$. That is, charge is **quantized**.

An object can become charged by rubbing (in which electrons are transferred from one material to another), **by conduction** (which is transfer of charge from one charged object to another by touching), or **by induction** (the separation of charge within an object because of the close approach of another charged object but without touching).

Electric charges exert a force on each other. If two charges are of opposite types, one positive and one negative, they each exert an attractive force on the other. If the two charges are the same type, each repels the other.

The magnitude of the force one point charge exerts on another is proportional to the product of their charges, and inversely proportional to the square of the distance between them:

$$F = k\frac{Q_1 Q_2}{r^2} = \frac{1}{4\pi\epsilon_0}\frac{Q_1 Q_2}{r^2}; \quad (16\text{–}1, 16\text{–}2)$$

this is **Coulomb's law**.

We think of an **electric field** as existing in space around any charge or group of charges. The force on another charged object is then said to be due to the electric field present at its location.

The *electric field*, $\vec{E}$, at any point in space due to one or more charges, is defined as the force per unit charge that would act on a tiny positive test charge q placed at that point:

$$\vec{E} = \frac{\vec{F}}{q}. \quad (16\text{–}3)$$

The magnitude of the electric field a distance r from a point charge Q is

$$E = k\frac{Q}{r^2}. \quad (16\text{–}4a)$$

The total electric field at a point in space is equal to the vector sum of the individual fields due to each contributing charge. This is the **principle of superposition**.

Electric fields are represented by **electric field lines** that start on positive charges and end on negative charges. Their direction indicates the direction the force would be on a tiny positive test charge placed at each point. The lines can be drawn so that the number per unit area is proportional to the magnitude of E.

The static electric field inside a conductor is zero, and the electric field lines just outside a charged conductor are perpendicular to its surface.

[*In the replication of DNA, the electrostatic force plays a crucial role in selecting the proper molecules so that the genetic information is passed on accurately from generation to generation.]

[*Photocopiers and computer printers use electric charge placed on toner particles and a drum to form an image.]

[*The **electric flux** passing through a small area A for a uniform electric field $\vec{E}$ is

$$\Phi_E = E_\perp A, \quad (16\text{–}7)$$

where $E_\perp$ is the component of $\vec{E}$ perpendicular to the surface. The flux through a surface is proportional to the number of field lines passing through it.]

[***Gauss's law** states that the total flux summed over any closed surface (considered as made up of many small areas ΔA) is equal to the net charge Q_{encl} enclosed by the surface divided by ϵ_0:

$$\sum_{\substack{\text{closed}\\\text{surface}}} E_\perp \Delta A = \frac{Q_{\text{encl}}}{\epsilon_0}. \quad (16\text{–}9)$$

Gauss's law can be used to determine the electric field due to given charge distributions, but its usefulness is mainly limited to cases where the charge distribution displays much symmetry. The real importance of Gauss's law is that it is a general and elegant statement of the relation between electric charge and electric field.]

Questions

1. If you charge a pocket comb by rubbing it with a silk scarf, how can you determine if the comb is positively or negatively charged?

2. Why does a shirt or blouse taken from a clothes dryer sometimes cling to your body?

3. Explain why fog or rain droplets tend to form around ions or electrons in the air.

4. Why does a plastic ruler that has been rubbed with a cloth have the ability to pick up small pieces of paper? Why is this difficult to do on a humid day?

5. A positively charged rod is brought close to a neutral piece of paper, which it attracts. Draw a diagram showing the separation of charge in the paper, and explain why attraction occurs.

6. Contrast the *net charge* on a conductor to the "free charges" in the conductor.

7. Figures 16–7 and 16–8 show how a charged rod placed near an uncharged metal object can attract (or repel) electrons. There are a great many electrons in the metal, yet only some of them move as shown. Why not all of them?

8. When an electroscope is charged, its two leaves repel each other and remain at an angle. What balances the electric force of repulsion so that the leaves don't separate further?

9. The balloon in Fig. 16–48 was rubbed on a student's hair. Explain why the water drip curves instead of falling vertically.

FIGURE 16–48 Question 9.

10. The form of Coulomb's law is very similar to that for Newton's law of universal gravitation. What are the differences between these two laws? Compare also gravitational mass and electric charge.

11. When a charged ruler attracts small pieces of paper, sometimes a piece jumps quickly away after touching the ruler. Explain.

12. We are not normally aware of the gravitational or electric force between two ordinary objects. What is the reason in each case? Give an example where we are aware of each one and why.

13. Explain why the test charges we use when measuring electric fields must be small.

14. When determining an electric field, must we use a *positive* test charge, or would a negative one do as well? Explain.

15. Draw the electric field lines surrounding two negative electric charges a distance ℓ apart.

16. Assume that the two opposite charges in Fig. 16–32a are 12.0 cm apart. Consider the magnitude of the electric field 2.5 cm from the positive charge. On which side of this charge—top, bottom, left, or right—is the electric field the strongest? The weakest? Explain.

17. Consider the electric field at the three points indicated by the letters A, B, and C in Fig. 16–49. First draw an arrow at each point indicating the direction of the net force that a positive test charge would experience if placed at that point, then list the letters in order of *decreasing* field strength (strongest first). Explain.

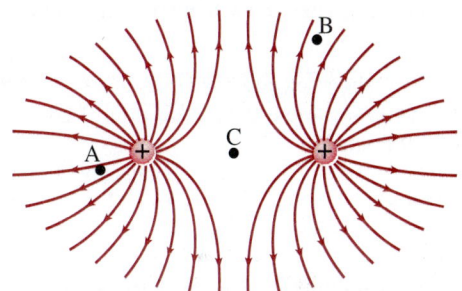

FIGURE 16–49 Question 17.

18. Why can electric field lines never cross?

19. Show, using the three rules for field lines given in Section 16–8, that the electric field lines starting or ending on a single point charge must be symmetrically spaced around the charge.

20. Given two point charges, Q and $2Q$, a distance ℓ apart, is there a point along the straight line that passes through them where $E = 0$ when their signs are (*a*) opposite, (*b*) the same? If yes, state roughly where this point will be.

21. Consider a small positive test charge located on an electric field line at some point, such as point P in Fig. 16–32a. Is the direction of the velocity and/or acceleration of the test charge along this line? Discuss.

*22. A point charge is surrounded by a spherical gaussian surface of radius r. If the sphere is replaced by a cube of side r, will Φ_E be larger, smaller, or the same? Explain.

MisConceptual Questions

1. $Q_1 = -0.10 \,\mu\text{C}$ is located at the origin. $Q_2 = +0.10 \,\mu\text{C}$ is located on the positive x axis at $x = 1.0$ m. Which of the following is true of the force on Q_1 due to Q_2?
   (a) It is attractive and directed in the $+x$ direction.
   (b) It is attractive and directed in the $-x$ direction.
   (c) It is repulsive and directed in the $+x$ direction.
   (d) It is repulsive and directed in the $-x$ direction.

2. Swap the positions of Q_1 and Q_2 of MisConceptual Question 1. Which of the following is true of the force on Q_1 due to Q_2?
   (a) It does not change.
   (b) It changes from attractive to repulsive.
   (c) It changes from repulsive to attractive.
   (d) It changes from the $+x$ direction to the $-x$ direction.
   (e) It changes from the $-x$ direction to the $+x$ direction.

3. Fred the lightning bug has a mass m and a charge $+q$. Jane, his lightning-bug wife, has a mass of $\frac{3}{4}m$ and a charge $-2q$. Because they have charges of opposite sign, they are attracted to each other. Which is attracted more to the other, and by how much?
   (a) Fred, twice as much.
   (b) Jane, twice as much.
   (c) Fred, four times as much.
   (d) Jane, four times as much.
   (e) They are attracted to each other by the same amount.

4. Figure 16–50 shows electric field lines due to a point charge. What can you say about the field at point 1 compared with the field at point 2?
   (a) The field at point 2 is larger, because point 2 is on a field line.
   (b) The field at point 1 is larger, because point 1 is not on a field line.
   (c) The field at point 1 is zero, because point 1 is not on a field line.
   (d) The field at point 1 is larger, because the field lines are closer together in that region.

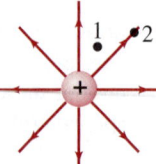

FIGURE 16–50
MisConceptual Question 4.

5. A negative point charge is in an electric field created by a positive point charge. Which of the following is true?
   (a) The field points toward the positive charge, and the force on the negative charge is in the same direction as the field.
   (b) The field points toward the positive charge, and the force on the negative charge is in the opposite direction to the field.
   (c) The field points away from the positive charge, and the force on the negative charge is in the same direction as the field.
   (d) The field points away from the positive charge, and the force on the negative charge is in the opposite direction to the field.

6. As an object acquires a positive charge, its mass usually
   (a) decreases.
   (b) increases.
   (c) stays the same.
   (d) becomes negative.

7. Refer to Fig. 16–32d. If the two charged plates were moved until they are half the distance shown without changing the charge on the plates, the electric field near the center of the plates would
   (a) remain almost exactly the same.
   (b) increase by a factor of 2.
   (c) increase, but not by a factor of 2.
   (d) decrease by a factor of 2.
   (e) decrease, but not by a factor of 2.

8. We wish to determine the electric field at a point near a positively charged metal sphere (a good conductor). We do so by bringing a small positive test charge, q_0, to this point and measure the force F_0 on it. F_0/q_0 will be _____ the electric field $\vec{E}$ as it was at that point before the test charge was present.
   (a) greater than
   (b) less than
   (c) equal to

9. We are usually not aware of the electric force acting between two everyday objects because
   (a) the electric force is one of the weakest forces in nature.
   (b) the electric force is due to microscopic-sized particles such as electrons and protons.
   (c) the electric force is invisible.
   (d) most everyday objects have as many plus charges as minus charges.

10. To be safe during a lightning storm, it is best to be
    (a) in the middle of a grassy meadow.
    (b) inside a metal car.
    (c) next to a tall tree in a forest.
    (d) inside a wooden building.
    (e) on a metal observation tower.

11. Which are the worst places in MisConceptual Question 10?

12. Which vector best represents the direction of the electric field at the fourth corner of the square due to the three charges shown in Fig. 16–51?

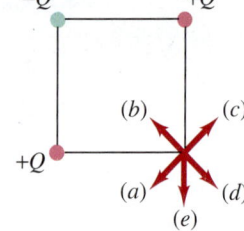

FIGURE 16–51
MisConceptual Question 12.

13. A small metal ball hangs from the ceiling by an insulating thread. The ball is attracted to a positively charged rod held near the ball. The charge of the ball must be
    (a) positive.
    (b) negative.
    (c) neutral.
    (d) positive or neutral.
    (e) negative or neutral.

For assigned homework and other learning materials, go to the MasteringPhysics website.

Problems

16–5 and 16–6 Coulomb's Law

$[1 \text{ mC} = 10^{-3} \text{ C}, \ 1 \mu\text{C} = 10^{-6} \text{ C}, \ 1 \text{ nC} = 10^{-9} \text{ C.}]$

1. (I) What is the magnitude of the electric force of attraction between an iron nucleus ($q = +26e$) and its innermost electron if the distance between them is 1.5×10^{-12} m?

2. (I) How many electrons make up a charge of $-48.0 \mu\text{C}$?

3. (I) What is the magnitude of the force a $+25 \mu\text{C}$ charge exerts on a $+2.5$ mC charge 16 cm away?

4. (I) What is the repulsive electrical force between two protons 4.0×10^{-15} m apart from each other in an atomic nucleus?

5. (II) When an object such as a plastic comb is charged by rubbing it with a cloth, the net charge is typically a few microcoulombs. If that charge is $3.0 \mu\text{C}$, by what percentage does the mass of a 9.0-g comb change during charging?

6. (II) Two charged dust particles exert a force of 4.2×10^{-2} N on each other. What will be the force if they are moved so they are only one-eighth as far apart?

7. (II) Two small charged spheres are 6.52 cm apart. They are moved, and the force each exerts on the other is found to have tripled. How far apart are they now?

8. (II) A person scuffing her feet on a wool rug on a dry day accumulates a net charge of $-28 \mu\text{C}$. How many excess electrons does she get, and by how much does her mass increase?

9. (II) What is the total charge of all the electrons in a 12-kg bar of gold? What is the net charge of the bar? (Gold has 79 electrons per atom and an atomic mass of 197 u.)

10. (II) Compare the electric force holding the electron in orbit $(r = 0.53 \times 10^{-10} \text{ m})$ around the proton nucleus of the hydrogen atom, with the gravitational force between the same electron and proton. What is the ratio of these two forces?

11. (II) Particles of charge $+65$, $+48$, and $-95 \mu\text{C}$ are placed in a line (Fig. 16–52). The center one is 0.35 m from each of the others. Calculate the net force on each charge due to the other two.

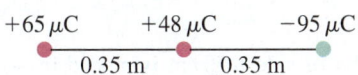

FIGURE 16–52 Problem 11.

12. (II) Three positive particles of equal charge, $+17.0 \mu\text{C}$, are located at the corners of an equilateral triangle of side 15.0 cm (Fig. 16–53). Calculate the magnitude and direction of the net force on each particle due to the other two.

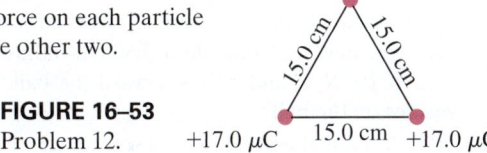

FIGURE 16–53 Problem 12.

13. (II) A charge Q is transferred from an initially uncharged plastic ball to an identical ball 24 cm away. The force of attraction is then 17 mN. How many electrons were transferred from one ball to the other?

14. (II) A charge of 6.15 mC is placed at each corner of a square 0.100 m on a side. Determine the magnitude and direction of the force on each charge.

15. (II) At each corner of a square of side ℓ there are point charges of magnitude Q, $2Q$, $3Q$, and $4Q$ (Fig. 16–54). Determine the magnitude and direction of the force on the charge $2Q$.

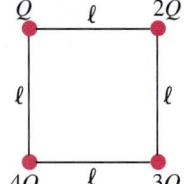

FIGURE 16–54 Problem 15.

16. (II) A large electroscope is made with "leaves" that are 78-cm-long wires with tiny 21-g spheres at the ends. When charged, nearly all the charge resides on the spheres. If the wires each make a 26° angle with the vertical (Fig. 16–55), what total charge Q must have been applied to the electroscope? Ignore the mass of the wires.

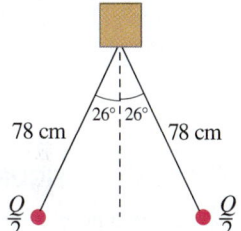

FIGURE 16–55 Problem 16.

17. (III) Two small nonconducting spheres have a total charge of $90.0 \mu\text{C}$. (a) When placed 28.0 cm apart, the force each exerts on the other is 12.0 N and is repulsive. What is the charge on each? (b) What if the force were attractive?

18. (III) Two charges, $-Q$ and $-3Q$, are a distance ℓ apart. These two charges are free to move but do not because there is a third (fixed) charge nearby. What must be the magnitude of the third charge and its placement in order for the first two to be in equilibrium?

16–7 and 16–8 Electric Field, Field Lines

19. (I) Determine the magnitude and direction of the electric force on an electron in a uniform electric field of strength 2460 N/C that points due east.

20. (I) A proton is released in a uniform electric field, and it experiences an electric force of 1.86×10^{-14} N toward the south. Find the magnitude and direction of the electric field.

21. (I) Determine the magnitude and direction of the electric field 21.7 cm directly above an isolated 33.0×10^{-6} C charge.

22. (I) A downward electric force of 6.4 N is exerted on a $-7.3 \mu\text{C}$ charge. Find the magnitude and direction of the electric field at the position of this charge.

23. (II) Determine the magnitude of the acceleration experienced by an electron in an electric field of 756 N/C. How does the direction of the acceleration depend on the direction of the field at that point?

24. (II) Determine the magnitude and direction of the electric field at a point midway between a $-8.0 \mu\text{C}$ and a $+5.8 \mu\text{C}$ charge 6.0 cm apart. Assume no other charges are nearby.

25. (II) Draw, approximately, the electric field lines about two point charges, $+Q$ and $-3Q$, which are a distance ℓ apart.

26. (II) What is the electric field strength at a point in space where a proton experiences an acceleration of 2.4 million "g's"?

27. (II) An electron is released from rest in a uniform electric field and accelerates to the north at a rate of 105 m/s^2. Find the magnitude and direction of the electric field.

28. (II) The electric field midway between two equal but opposite point charges is 386 N/C, and the distance between the charges is 16.0 cm. What is the magnitude of the charge on each?

29. (II) Calculate the electric field at one corner of a square 1.22 m on a side if the other three corners are occupied by 3.25×10^{-6} C charges.

30. (II) Calculate the electric field at the center of a square 42.5 cm on a side if one corner is occupied by a $-38.6 \text{ }\mu\text{C}$ charge and the other three are occupied by $-27.0 \text{ }\mu\text{C}$ charges.

31. (II) Determine the direction and magnitude of the electric field at the point P in Fig. 16–56. The charges are separated by a distance $2a$, and point P is a distance x from the midpoint between the two charges. Express your answer in terms of Q, x, a, and k.

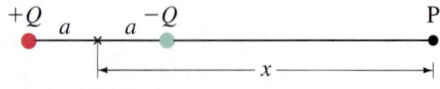

FIGURE 16–56 Problem 31.

32. (II) Two point charges, $Q_1 = -32 \text{ }\mu\text{C}$ and $Q_2 = +45 \text{ }\mu\text{C}$, are separated by a distance of 12 cm. The electric field at the point P (see Fig. 16–57) is zero. How far from Q_1 is P?

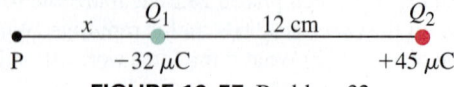

FIGURE 16–57 Problem 32.

33. (II) Determine the electric field $\vec{E}$ at the origin 0 in Fig. 16–58 due to the two charges at A and B.

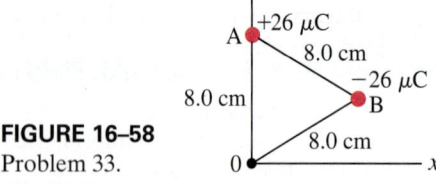

FIGURE 16–58 Problem 33.

34. (II) You are given two unknown point charges, Q_1 and Q_2. At a point on the line joining them, one-third of the way from Q_1 to Q_2, the electric field is zero (Fig. 16–59). What is the ratio Q_1/Q_2?

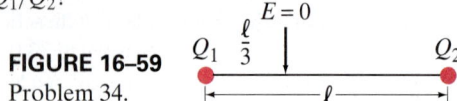

FIGURE 16–59 Problem 34.

35. (III) Use Coulomb's law to determine the magnitude and direction of the electric field at points A and B in Fig. 16–60 due to the two positive charges ($Q = 4.7 \text{ }\mu\text{C}$) shown. Are your results consistent with Fig. 16–32b?

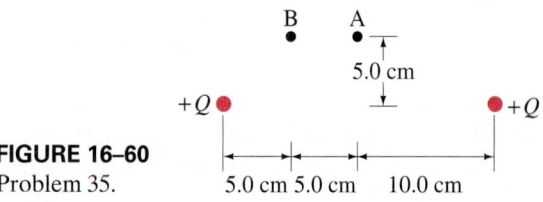

FIGURE 16–60 Problem 35.

36. (III) An electron (mass $m = 9.11 \times 10^{-31}$ kg) is accelerated in the uniform field $\vec{E}$ ($E = 1.45 \times 10^4$ N/C) between two thin parallel charged plates. The separation of the plates is 1.60 cm. The electron is accelerated from rest near the negative plate and passes through a tiny hole in the positive plate, Fig. 16–61. (a) With what speed does it leave the hole? (b) Show that the gravitational force can be ignored.

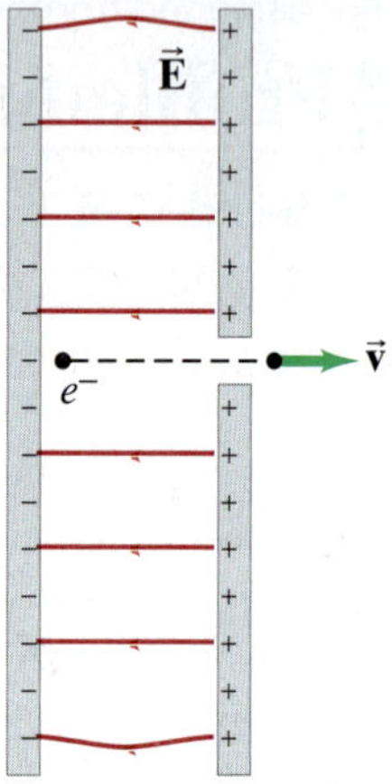

FIGURE 16–61 Problem 36.

*16–10 DNA

*37. (III) The two strands of the helix-shaped DNA molecule are held together by electrostatic forces as shown in Fig. 16–39. Assume that the net average charge (due to electron sharing) indicated on H and N atoms has magnitude $0.2e$ and on the indicated C and O atoms is $0.4e$. Assume also that atoms on each molecule are separated by 1.0×10^{-10} m. Estimate the net force between (a) a thymine and an adenine; and (b) a cytosine and a guanine. For each bond (red dots) consider only the three atoms in a line (two atoms on one molecule, one atom on the other). (c) Estimate the total force for a DNA molecule containing 10^5 pairs of such molecules. Assume half are A–T pairs and half are C–G pairs.

*16–12 Gauss's Law

*38. (I) The total electric flux from a cubical box of side 28.0 cm is $1.85 \times 10^3 \text{ N} \cdot \text{m}^2/\text{C}$. What charge is enclosed by the box?

*39. (II) In Fig. 16–62, two objects, O_1 and O_2, have charges $+1.0 \text{ }\mu\text{C}$ and $-2.0 \text{ }\mu\text{C}$, respectively, and a third object, O_3, is electrically neutral. (a) What is the electric flux through the surface A_1 that encloses all three objects? (b) What is the electric flux through the surface A_2 that encloses the third object only?

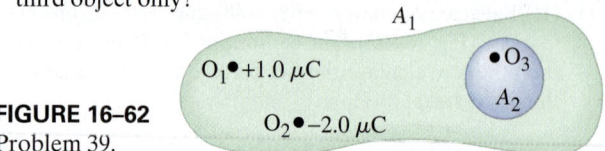

FIGURE 16–62 Problem 39.

*40. (II) A cube of side 8.50 cm is placed in a uniform field $E = 7.50 \times 10^3$ N/C with edges parallel to the field lines. (a) What is the net flux through the cube? (b) What is the flux through each of its six faces?

*41. (II) The electric field between two parallel square metal plates is 130 N/C. The plates are 0.85 m on a side and are separated by 3.0 cm. What is the charge on each plate (assume equal and opposite)? Neglect edge effects.

*42. (II) The field just outside a 3.50-cm-radius metal ball is 3.75×10^2 N/C and points toward the ball. What charge resides on the ball?

*43. (III) A point charge Q rests at the center of an uncharged thin spherical conducting shell. (See Fig. 16–34.) What is the electric field E as a function of r (a) for r less than the inner radius of the shell, (b) inside the shell, and (c) beyond the shell? (d) How does the shell affect the field due to Q alone? How does the charge Q affect the shell?

General Problems

44. How close must two electrons be if the magnitude of the electric force between them is equal to the weight of either at the Earth's surface?

45. Given that the human body is mostly made of water, estimate the total amount of positive charge in a 75-kg person.

46. A 3.0-g copper penny has a net positive charge of 32 μC. What fraction of its electrons has it lost?

47. Measurements indicate that there is an electric field surrounding the Earth. Its magnitude is about 150 N/C at the Earth's surface and points inward toward the Earth's center. What is the magnitude of the electric charge on the Earth? Is it positive or negative? [*Hint*: The electric field outside a uniformly charged sphere is the same as if all the charge were concentrated at its center.]

48. (a) The electric field near the Earth's surface has magnitude of about 150 N/C. What is the acceleration experienced by an electron near the surface of the Earth? (b) What about a proton? (c) Calculate the ratio of each acceleration to $g = 9.8$ m/s^2.

49. A water droplet of radius 0.018 mm remains stationary in the air. If the downward-directed electric field of the Earth is 150 N/C, how many excess electron charges must the water droplet have?

50. Estimate the net force between the CO group and the HN group shown in Fig. 16–63. The C and O have charges $\pm 0.40e$, and the H and N have charges $\pm 0.20e$, where $e = 1.6 \times 10^{-19}$ C. [*Hint*: Do not include the "internal" forces between C and O, or between H and N.]

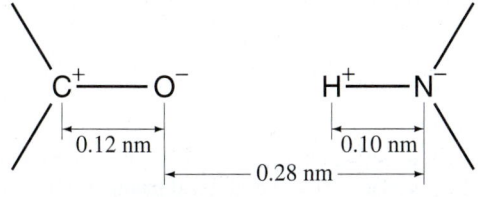

FIGURE 16–63 Problem 50.

51. In a simple model of the hydrogen atom, the electron revolves in a circular orbit around the proton with a speed of 2.2×10^6 m/s. Determine the radius of the electron's orbit. [*Hint*: See Chapter 5 on circular motion.]

52. Two small charged spheres hang from cords of equal length ℓ as shown in Fig. 16–64 and make small angles θ_1 and θ_2 with the vertical. (a) If $Q_1 = Q$, $Q_2 = 2Q$, and $m_1 = m_2 = m$, determine the ratio θ_1/θ_2. (b) Estimate the distance between the spheres.

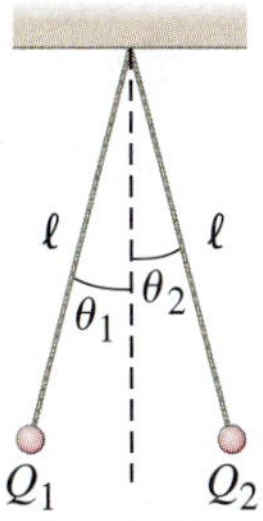

FIGURE 16–64 Problem 52.

53. A positive point charge $Q_1 = 2.5 \times 10^{-5}$ C is fixed at the origin of coordinates, and a negative point charge $Q_2 = -5.0 \times 10^{-6}$ C is fixed to the x axis at $x = +2.4$ m. Find the location of the place(s) along the x axis where the electric field due to these two charges is zero.

54. Dry air will break down and generate a spark if the electric field exceeds about 3×10^6 N/C. How much charge could be packed onto a green pea (diameter 0.75 cm) before the pea spontaneously discharges? [*Hint*: Eqs. 16–4 work outside a sphere if r is measured from its center.]

55. Two point charges, $Q_1 = -6.7 \mu$C and $Q_2 = 1.8 \mu$C, are located between two oppositely charged parallel plates, as shown in Fig. 16–65. The two charges are separated by a distance of $x = 0.47$ m. Assume that the electric field produced by the charged plates is uniform and equal to $E = 53{,}000$ N/C. Calculate the net electrostatic force on Q_1 and give its direction.

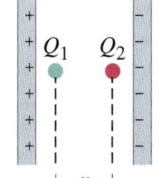

FIGURE 16–65 Problem 55.

56. Packing material made of pieces of foamed polystyrene can easily become charged and stick to each other. Given that the density of this material is about 35 kg/m^3, estimate how much charge might be on a 2.0-cm-diameter foamed polystyrene sphere, assuming the electric force between two spheres stuck together is equal to the weight of one sphere.

57. A point charge ($m = 1.0$ gram) at the end of an insulating cord of length 55 cm is observed to be in equilibrium in a uniform horizontal electric field of 9500 N/C, when the pendulum's position is as shown in Fig. 16–66, with the charge 12 cm above the lowest (vertical) position. If the field points to the right in Fig. 16–66, determine the magnitude and sign of the point charge.

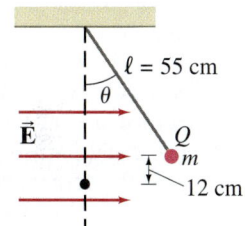

FIGURE 16–66 Problem 57.

58. Two small, identical conducting spheres A and B are a distance R apart; each carries the same charge Q. (a) What is the force sphere B exerts on sphere A? (b) An identical sphere with zero charge, sphere C, makes contact with sphere B and is then moved very far away. What is the net force now acting on sphere A? (c) Sphere C is brought back and now makes contact with sphere A and is then moved far away. What is the force on sphere A in this third case?

59. For an experiment, a colleague of yours says he smeared toner particles uniformly over the surface of a sphere 1.0 m in diameter and then measured an electric field of 5000 N/C near its surface. (a) How many toner particles (Example 16–6) would have to be on the surface to produce these results? (b) What is the total mass of the toner particles?

60. A proton ($m = 1.67 \times 10^{-27}$ kg) is suspended at rest in a uniform electric field $\vec{\mathbf{E}}$. Take into account gravity at the Earth's surface, and determine $\vec{\mathbf{E}}$.

61. A point charge of mass 0.185 kg, and net charge $+0.340\ \mu C$, hangs at rest at the end of an insulating cord above a large sheet of charge. The horizontal sheet of fixed uniform charge creates a uniform vertical electric field in the vicinity of the point charge. The tension in the cord is measured to be 5.18 N. Calculate the magnitude and direction of the electric field due to the sheet of charge (Fig. 16–67).

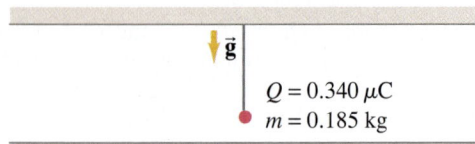

FIGURE 16–67 Problem 61.

62. An electron with speed $v_0 = 5.32 \times 10^6$ m/s is traveling parallel to an electric field of magnitude $E = 9.45 \times 10^3$ N/C. (a) How far will the electron travel before it stops? (b) How much time will elapse before it returns to its starting point?

63. Given the two charges shown in Fig. 16–68, at what position(s) x is the electric field zero?

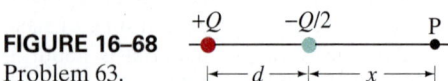

FIGURE 16–68 Problem 63.

64. What is the total charge of all the electrons in a 25-kg bar of aluminum? (Aluminum has 13 electrons per atom and an atomic mass of 27 u.)

65. Two point charges, $+Q$ and $-Q$ of mass m, are placed on the ends of a massless rod of length ℓ, which is fixed to a table by a pin through its center. If the apparatus is then subjected to a uniform electric field E parallel to the table and perpendicular to the rod, find the net torque on the system of rod plus charges.

66. Determine the direction and magnitude of the electric field at point P, Fig. 16–69. The two charges are separated by a distance of $2a$. Point P is on the perpendicular bisector of the line joining the charges, a distance x from the midpoint between them. Express your answers in terms of Q, x, a, and k.

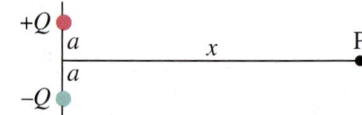

FIGURE 16–69 Problem 66.

67. Two moles of carbon contain 7.22×10^{24} electrons. Two electrically neutral carbon spheres, each containing 1 mole of carbon, are separated by 15.0 cm (center to center). What fraction of electrons would have to be transferred from one sphere to the other for the electric force and the gravitational force between the spheres to be equal?

Search and Learn

1. Referring to Section 16–4 and Figs. 16–11 and 16–12, what happens to the separation of the leaves of an electroscope when the charging object is removed from an electroscope (a) charged by induction and (b) charged by conduction? (c) Is it possible to tell whether the electroscope in Fig. 16–12a has been charged by induction or by conduction? If so, which way was it charged? (d) Draw electric field lines (Section 16–8) for the electroscopes in Figs. 16–11a and 16–11b, omitting the fields around the charging rod. How do the fields differ?

2. Four equal positive point charges, each of charge $6.4\ \mu C$, are at the corners of a square of side 9.2 cm. What charge should be placed at the center of the square so that all charges are at equilibrium? Is this a stable or an unstable equilibrium (Section 9–4) in the plane?

3. Suppose electrons enter a uniform electric field midway between two plates at an angle θ_0 to the horizontal, as shown in Fig. 16–70. The path is symmetrical, so they leave at the same angle θ_0 and just barely miss the top plate. What is θ_0? Ignore fringing of the field.

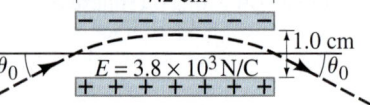

FIGURE 16–70 Search and Learn 3.

4. What experimental observations mentioned in the text rule out the possibility that the numerator in Coulomb's law contains the sum $(Q_1 + Q_2)$ rather than the product $Q_1 Q_2$?

5. Near the surface of the Earth, there is a downward electric field of 150 N/C and a downward gravitational field of 9.8 N/kg. A charged 1.0-kg mass is observed to fall with acceleration 8.0 m/s². What are the magnitude and sign of its charge?

6. Identical negative charges $(Q = -e)$ are located at two of the three vertices of an equilateral triangle. The length of a side of the triangle is ℓ. What is the magnitude of the net electric field at the third vertex? If a third identical negative charge was located at the third vertex, then what would be the net electrostatic force on it due to the other two charges? Use symmetry and explain how you used it.

7. Suppose that electrical attraction, rather than gravity, were responsible for holding the Moon in orbit around the Earth. If equal and opposite charges Q were placed on the Earth and the Moon, what should be the value of Q to maintain the present orbit? Use data given on the inside front cover of this book. Treat the Earth and Moon as point particles.

ANSWERS TO EXERCISES

A: (d).
B: Opposite.
C: 0.3 N, to the right.
D: 0.32 m.
E: (a) No; (b) yes, midway between them.
F: 9.0×10^4 N/C, vertically upward.
G: (d), if the two + charges are not at opposite corners (use symmetry).

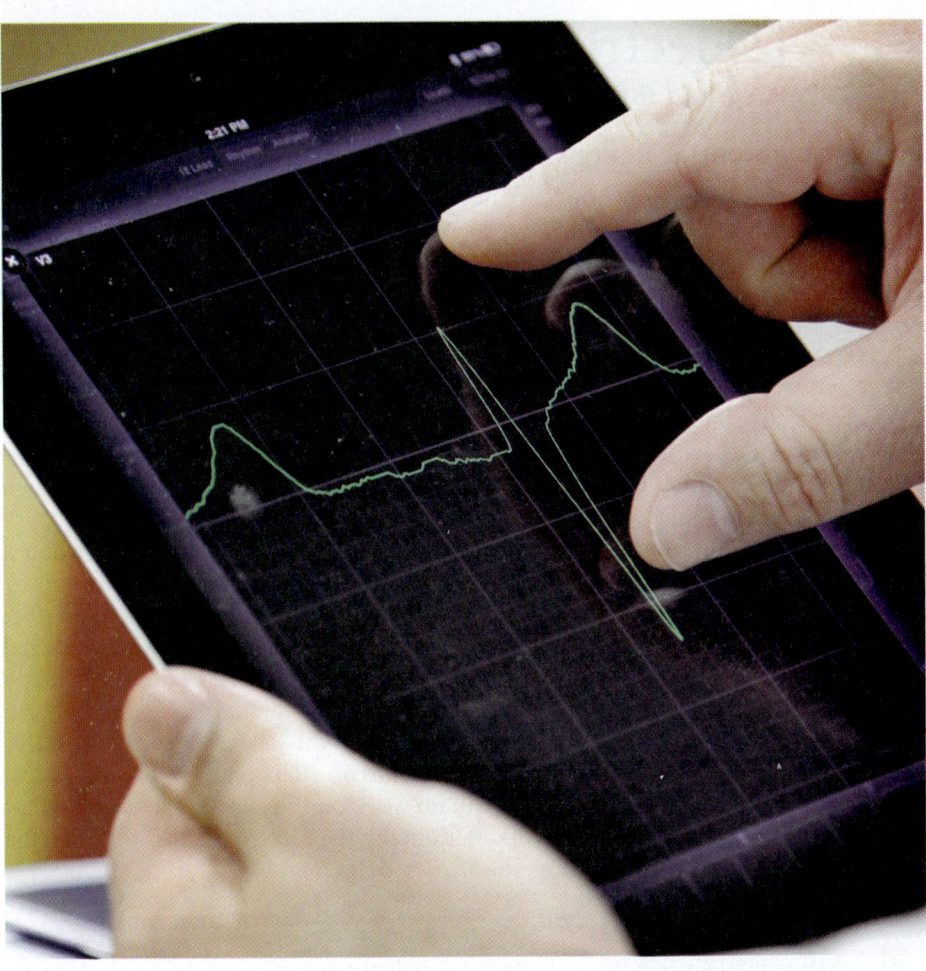

We are used to voltage in our lives—a 12-volt car battery, 110 V or 220 V at home, 1.5-volt flashlight batteries, and so on. Here we see displayed the voltage produced across a human heart, known as an electrocardiogram. Voltage is the same as electric potential difference between two points. Electric potential is defined as the potential energy per unit charge.

We discuss voltage and its relation to electric field, as well as electric energy storage, capacitors, and applications including the ECG shown here, binary numbers and digital electronics, TV and computer monitors, and digital TV.

CHAPTER 17

Electric Potential

CHAPTER-OPENING QUESTION—Guess now!
When two positively charged small spheres are pushed toward each other, what happens to their potential energy?
- (a) It remains unchanged.
- (b) It decreases.
- (c) It increases.
- (d) There is no potential energy in this situation.

We saw in Chapter 6 that the concept of energy was extremely valuable in dealing with the subject of mechanics. For one thing, energy is a conserved quantity and is thus an important tool for understanding nature. Furthermore, we saw that many Problems could be solved using the energy concept even though a detailed knowledge of the forces involved was not possible, or when a calculation involving Newton's laws would have been too difficult.

The energy point of view can be used in electricity, and it is especially useful. It not only extends the law of conservation of energy, but it gives us another way to view electrical phenomena. The energy concept is also a tool in solving Problems more easily in many cases than by using forces and electric fields.

CONTENTS

- 17–1 Electric Potential Energy and Potential Difference
- 17–2 Relation between Electric Potential and Electric Field
- 17–3 Equipotential Lines and Surfaces
- 17–4 The Electron Volt, a Unit of Energy
- 17–5 Electric Potential Due to Point Charges
- *17–6 Potential Due to Electric Dipole; Dipole Moment
- 17–7 Capacitance
- 17–8 Dielectrics
- 17–9 Storage of Electric Energy
- 17–10 Digital; Binary Numbers; Signal Voltage
- *17–11 TV and Computer Monitors: CRTs, Flat Screens
- *17–12 Electrocardiogram (ECG or EKG)

17–1 Electric Potential Energy and Potential Difference

Electric Potential Energy

To apply conservation of energy, we need to define electric potential energy as we did for other types of potential energy. As we saw in Chapter 6, potential energy can be defined only for a conservative force. The work done by a conservative force in moving an object between any two positions is independent of the path taken. The electrostatic force between any two charges (Eq. 16–1, $F = kQ_1Q_2/r^2$) is conservative because the dependence on position is just like the gravitational force (Eq. 5–4), which is conservative. Hence we can define potential energy PE for the electrostatic force.

We saw in Chapter 6 that the change in potential energy between any two points, a and b, equals the negative of the work done by the conservative force on an object as it moves from point a to point b: $\Delta \text{PE} = -W$.

Thus we define the change in electric potential energy, $\text{PE}_b - \text{PE}_a$, when a point charge q moves from some point a to another point b, as the negative of the work done by the electric force on the charge as it moves from point a to point b. For example, consider the electric field between two equally but oppositely charged parallel plates; we assume their separation is small compared to their width and height, so the field $\vec{E}$ will be uniform over most of the region, Fig. 17–1. Now consider a tiny positive point charge q placed at the point "a" very near the positive plate as shown. This charge q is so small that it has no effect on $\vec{E}$. If this charge q at point a is released, the electric force will do work on the charge and accelerate it toward the negative plate. The work W done by the electric field E to move the charge a distance d is (using Eq. 16–5, $F = qE$)

$$W = Fd = qEd. \quad [\text{uniform } \vec{E}]$$

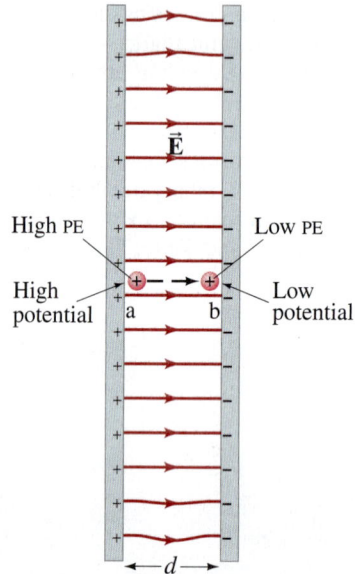

FIGURE 17–1 Work is done by the electric field $\vec{E}$ in moving the positive charge from position a to position b.

The change in electric potential energy equals the negative of the work done by the electric force:

$$\text{PE}_b - \text{PE}_a = -qEd \quad [\text{uniform } \vec{E}] \quad \textbf{(17–1)}$$

for this case of uniform electric field $\vec{E}$. In the case illustrated, the potential energy decreases (ΔPE is negative); and as the charged particle accelerates from point a to point b in Fig. 17–1, the particle's kinetic energy KE increases—by an equal amount. In accord with the conservation of energy, electric potential energy is transformed into kinetic energy, and the total energy is conserved. Note that the positive charge q has its greatest potential energy at point a, near the positive plate.[†] The reverse is true for a negative charge: its potential energy is greatest near the negative plate.

Electric Potential and Potential Difference

In Chapter 16, we found it useful to define the electric field as the force per unit charge. Similarly, it is useful to define the **electric potential** (or simply the **potential** when "electric" is understood) as the *electric potential energy per unit charge*. Electric potential is given the symbol V. If a positive test charge q in an electric field has electric potential energy PE_a at some point a (relative to some zero potential energy), the electric potential V_a at this point is

$$V_a = \frac{\text{PE}_a}{q}. \quad \textbf{(17–2a)}$$

As we discussed in Chapter 6, only differences in potential energy are physically meaningful. Hence only the **difference in potential**, or the **potential difference**, between two points a and b (such as those shown in Fig. 17–1) is measurable.

[†]At point a, the positive charge q has its greatest ability to do work (on some other object or system).

When the electric force does positive work on a charge, the kinetic energy increases and the potential energy decreases. The difference in potential energy, $PE_b - PE_a$, is equal to the negative of the work, W_{ba}, done by the electric field to move the charge from a to b; so the potential difference V_{ba} is

$$V_{ba} = V_b - V_a = \frac{PE_b - PE_a}{q} = -\frac{W_{ba}}{q}. \quad (17\text{--}2b)$$

Note that electric potential, like electric field, does not depend on our test charge q. V depends on the other charges that create the field, not on the test charge q; q acquires potential energy by being in the potential V due to the other charges.

We can see from our definition that the positive plate in Fig. 17–1 is at a higher potential than the negative plate. Thus a positively charged object moves naturally from a high potential to a low potential. A negative charge does the reverse.

The unit of electric potential, and of potential difference, is joules/coulomb and is given a special name, the **volt**, in honor of Alessandro Volta (1745–1827) who is best known for inventing the electric battery. The volt is abbreviated V, so 1 V = 1 J/C. Potential difference, since it is measured in volts, is often referred to as **voltage**. (Be careful not to confuse V for volts, with italic V for voltage.)

If we wish to speak of the potential V_a at some point a, we must be aware that V_a depends on where the potential is chosen to be zero. The zero for electric potential in a given situation can be chosen arbitrarily, just as for potential energy, because only differences in potential energy can be measured. Often the ground, or a conductor connected directly to the ground (the Earth), is taken as zero potential, and other potentials are given with respect to ground. (Thus, a point where the voltage is 50 V is one where the difference of potential between it and ground is 50 V.) In other cases, as we shall see, we may choose the potential to be zero at an infinite distance.

CONCEPTUAL EXAMPLE 17–1 **A negative charge.** Suppose a negative charge, such as an electron, is placed near the negative plate in Fig. 17–1, at point b, shown here in Fig. 17–2. If the electron is free to move, will its electric potential energy increase or decrease? How will the electric potential change?

RESPONSE An electron released at point b will be attracted to the positive plate. As the electron accelerates toward the positive plate, its kinetic energy increases, so its potential energy *decreases*: $PE_a < PE_b$ and $\Delta PE = PE_a - PE_b < 0$. But note that the electron moves from point b at low potential to point a at higher potential: $\Delta V = V_a - V_b > 0$. (Potentials V_a and V_b are due to the charges on the plates, not due to the electron.) The signs of ΔPE and ΔV are opposite because of the negative charge of the electron.

NOTE A positive charge placed next to the negative plate at b would stay there, with no acceleration. A positive charge tends to move from high potential to low.

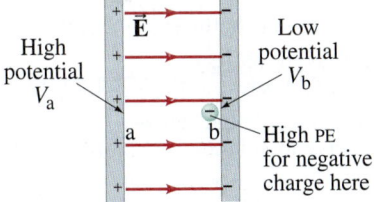

FIGURE 17–2 Central part of Fig. 17–1, showing a negative point charge near the negative plate. Example 17–1.

⚠️ **CAUTION**
A negative charge has high PE when potential V is low

Because the electric potential difference is defined as the potential energy difference per unit charge, then the change in potential energy of a charge q when it moves from point a to point b is

$$\Delta PE = PE_b - PE_a = q(V_b - V_a) = qV_{ba}. \quad (17\text{--}3)$$

That is, if an object with charge q moves through a potential difference V_{ba}, its potential energy changes by an amount qV_{ba}. For example, if the potential difference between the two plates in Fig. 17–1 is 6 V, then a +1 C charge moved from point b to point a will gain (1 C)(6 V) = 6 J of electric potential energy. (And it will lose 6 J of electric potential energy if it moves from a to b.) Similarly, a +2 C charge will gain ΔPE = (2 C)(6 V) = 12 J, and so on. Thus, electric potential difference is a measure of how much energy an electric charge can acquire in a given situation. And, since energy is the ability to do work, the electric potential difference is also a measure of how much *work* a given charge can do. The exact amount of energy or work depends both on the potential difference and on the charge.

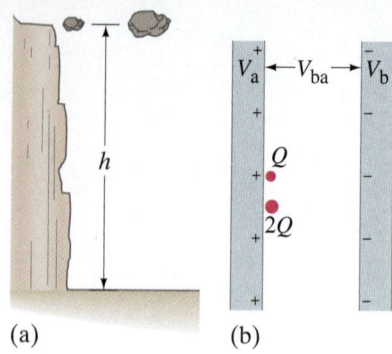

FIGURE 17–3 (a) Two rocks are at the same height. The larger rock has more potential energy. (b) Two positive charges have the same electric potential. The $2Q$ charge has more potential energy.

TABLE 17–1 Some Typical Potential Differences (Voltages)

Source	Voltage (approx.)
Thundercloud to ground	10^8 V
High-voltage power line	10^5–10^6 V
Automobile ignition	10^4 V
Household outlet	10^2 V
Automobile battery	12 V
Flashlight battery (AA, AAA, C, D)	1.5 V
Resting potential across nerve membrane	10^{-1} V
Potential changes on skin (ECG and EEG)	10^{-4} V

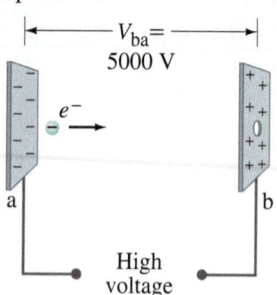

FIGURE 17–4 Electron accelerated, Example 17–2.

To better understand electric potential, let's make a comparison to the gravitational case when a rock falls from the top of a cliff. The greater the height, h, of a cliff, the more potential energy ($= mgh$) the rock has at the top of the cliff relative to the bottom, and the more kinetic energy it will have when it reaches the bottom. The actual amount of kinetic energy it will acquire, and the amount of work it can do, depends both on the height of the cliff and the mass m of the rock. A large rock and a small rock can be at the same height h (Fig. 17–3a) and thus have the same "gravitational potential," but the larger rock has the greater potential energy (it has more mass). The electrical case is similar (Fig. 17–3b): the potential energy change, or the work that can be done, depends both on the potential difference (corresponding to the height of the cliff) and on the charge (corresponding to mass), Eq. 17–3. But note a significant difference: electric charge comes in two types, $+$ and $-$, whereas gravitational mass is always $+$.

Sources of electrical energy such as batteries and electric generators are meant to maintain a potential difference. The actual amount of energy transformed by such a device depends on how much charge flows, as well as the potential difference (Eq. 17–3). For example, consider an automobile headlight connected to a 12.0-V battery. The amount of energy transformed (into light and thermal energy) is proportional to how much charge flows, which in turn depends on how long the light is on. If over a given period of time 5.0 C of charge flows through the light, the total energy transformed is $(5.0\,\text{C})(12.0\,\text{V}) = 60\,\text{J}$. If the headlight is left on twice as long, 10.0 C of charge will flow and the energy transformed is $(10.0\,\text{C})(12.0\,\text{V}) = 120\,\text{J}$. Table 17–1 presents some typical voltages.

EXAMPLE 17–2 Electron in TV tube. Suppose an electron is accelerated from rest through a potential difference $V_b - V_a = V_{ba} = +5000$ V (Fig. 17–4). (a) What is the change in electric potential energy of the electron? What is (b) the kinetic energy, and (c) the speed of the electron ($m = 9.1 \times 10^{-31}$ kg) as a result of this acceleration?

APPROACH The electron, accelerated toward the positive plate, will change in potential energy by an amount $\Delta \text{PE} = qV_{ba}$ (Eq. 17–3). The loss in potential energy will equal its gain in kinetic energy (energy conservation).

SOLUTION (a) The charge on an electron is $q = -e = -1.6 \times 10^{-19}$ C. Therefore its change in potential energy is

$$\Delta \text{PE} = qV_{ba} = (-1.6 \times 10^{-19}\,\text{C})(+5000\,\text{V}) = -8.0 \times 10^{-16}\,\text{J}.$$

The minus sign indicates that the potential energy decreases. The potential difference, V_{ba}, has a positive sign because the final potential V_b is higher than the initial potential V_a. Negative electrons are attracted toward a positive electrode (or plate) and repelled away from a negative electrode.

(b) The potential energy lost by the electron becomes kinetic energy KE. From conservation of energy (Eq. 6–11a), $\Delta \text{KE} + \Delta \text{PE} = 0$, so

$$\Delta \text{KE} = -\Delta \text{PE}$$
$$\tfrac{1}{2}mv^2 - 0 = -q(V_b - V_a) = -qV_{ba},$$

where the initial kinetic energy is zero since we are given that the electron started from rest. So the final $\text{KE} = -qV_{ba} = 8.0 \times 10^{-16}\,\text{J}$.

(c) In the equation just above we solve for v:

$$v = \sqrt{-\frac{2qV_{ba}}{m}} = \sqrt{-\frac{2(-1.6 \times 10^{-19}\,\text{C})(5000\,\text{V})}{9.1 \times 10^{-31}\,\text{kg}}} = 4.2 \times 10^7\,\text{m/s}.$$

NOTE The electric potential energy does not depend on the mass, only on the charge and voltage. The speed *does* depend on m.

EXERCISE A Instead of the electron in Example 17–2, suppose a proton ($m = 1.67 \times 10^{-27}$ kg) was accelerated from rest by a potential difference $V_{ba} = -5000$ V. What would be the proton's (a) change in PE, and (b) final speed?

17–2 Relation between Electric Potential and Electric Field

The effects of any charge distribution can be described either in terms of electric field or in terms of electric potential. Electric potential is often easier to use because it is a scalar, whereas electric field is a vector. There is an intimate connection between the potential and the field. Let us consider the case of a uniform electric field, such as that between the parallel plates of Fig. 17–1 whose difference of potential is V_{ba}. The work done by the electric field to move a positive charge q from point a to point b is equal to the negative of the change in potential energy (Eq. 17–2b), so

$$W = -q(V_b - V_a) = -qV_{ba}.$$

We can also write the work done as the force times distance, where the force on q is $F = qE$, so

$$W = Fd = qEd,$$

where d is the distance (parallel to the field lines) between points a and b. We now set these two expressions for W equal and find $qV_{ba} = -qEd$, or

$$V_{ba} = -Ed. \qquad \text{[uniform } \vec{E}\text{]} \quad \textbf{(17–4a)}$$

If we solve for E, we find

$$E = -\frac{V_{ba}}{d}. \qquad \text{[uniform } \vec{E}\text{]} \quad \textbf{(17–4b)}$$

From Eq. 17–4b we see that the units for electric field can be written as volts per meter (V/m), as well as newtons per coulomb (N/C, from $E = F/q$). These are equivalent because $1\,\text{N/C} = 1\,\text{N·m/C·m} = 1\,\text{J/C·m} = 1\,\text{V/m}$. The minus sign in Eq. 17–4b tells us that $\vec{E}$ points in the direction of decreasing potential V.

EXAMPLE 17–3 **Electric field obtained from voltage.** Two parallel plates are charged to produce a potential difference of 50 V. If the separation between the plates is 0.050 m, calculate the magnitude of the electric field in the space between the plates (Fig. 17–5).

APPROACH We apply Eq. 17–4b to obtain the magnitude of E, assumed uniform.

SOLUTION The magnitude of the electric field is

$$E = V_{ba}/d = (50\,\text{V}/0.050\,\text{m}) = 1000\,\text{V/m}.$$

NOTE Equations 17–4 apply only for a uniform electric field. The general relationship between $\vec{E}$ and V is more complicated.

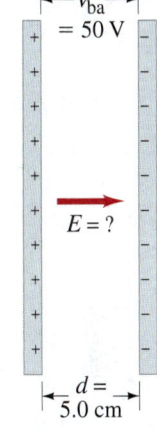

FIGURE 17–5 Example 17–3.

*General Relation between $\vec{E}$ and V

In a region where $\vec{E}$ is not uniform, the connection between $\vec{E}$ and V takes on a different form than Eqs. 17–4. In general, it is possible to show that the electric field in a given direction at any point in space is equal to the *rate at which the electric potential decreases over distance in that direction*. For example, the x component of the electric field is given by $E_x = -\Delta V/\Delta x$, where ΔV is the change in potential over a very short distance Δx.

Breakdown Voltage

When very high voltages are present, air can become ionized due to the high electric fields. Any odd free electron can be accelerated to sufficient kinetic energy to knock electrons out of O_2 and N_2 molecules of the air. This **breakdown** of air occurs when the electric field exceeds about 3×10^6 V/m. When electrons recombine with their molecules, light is emitted. Such breakdown of air is the source of lightning, the spark of a car's spark plug, and even short sparks between your fingers and a doorknob after you walk across a synthetic rug or slide across a car seat (which can result in a significant transfer of charge to you).

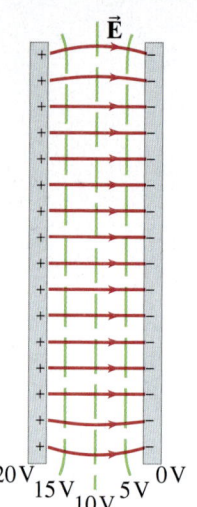

17–3 Equipotential Lines and Surfaces

The electric potential can be represented by drawing **equipotential lines** or, in three dimensions, **equipotential surfaces**. An equipotential surface is one on which all points are at the same potential. That is, the potential difference between any two points on the surface is zero, so no work is required to move a charge from one point on the surface to the other. An *equipotential surface must be perpendicular to the electric field* at any point. If this were not so—that is, if there were a component of $\vec{E}$ parallel to the surface—it would require work to move the charge along the surface against this component of $\vec{E}$; and this would contradict the idea that it is an *equi*potential surface.

The fact that the electric field lines and equipotential surfaces are mutually perpendicular helps us locate the equipotentials when the electric field lines are known. In a normal two-dimensional drawing, we show equipotential *lines*, which are the intersections of equipotential surfaces with the plane of the drawing. In Fig. 17–6, a few of the equipotential lines are drawn (dashed green lines) for the electric field (red lines) between two parallel plates at a potential difference of 20 V. The negative plate is arbitrarily chosen to be zero volts and the potential of each equipotential line is indicated. Note that $\vec{E}$ points toward lower values of V. The equipotential lines for the case of two equal but oppositely charged particles are shown in Fig. 17–7 as green dashed lines. (This combination of equal + and − charges is called an "electric dipole," as we saw in Section 16–8; see Fig. 16–32a.)

Unlike electric field lines, which start and end on electric charges, equipotential lines and surfaces are always continuous and never end, and so continue beyond the borders of Figs. 17–6 and 17–7. A useful analogy for equipotential lines is a topographic map: the contour lines are gravitational equipotential lines (Fig. 17–8).

We saw in Section 16–9 that there can be no electric field within a conductor in the static case, for otherwise the free electrons would feel a force and would move. Indeed the *entire volume of a conductor must be entirely at the same potential in the static case*. The surface of a conductor is thus an equipotential surface. (If it weren't, the free electrons at the surface would move, because whenever there is a potential difference between two points, free charges will move.) This is fully consistent with our result in Section 16–9 that the electric field at the surface of a conductor must be perpendicular to the surface.

FIGURE 17–6 Equipotential lines (the green dashed lines) between two charged parallel plates are always perpendicular to the electric field (solid red lines).

FIGURE 17–7 Equipotential lines (green, dashed) are always perpendicular to the electric field lines (solid red), shown here for two equal but oppositely charged particles (an "electric dipole").

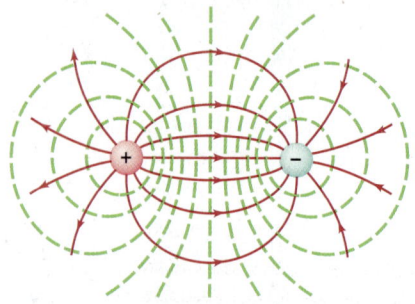

FIGURE 17–8 A topographic map (here, a portion of the Sierra Nevada in California) shows continuous contour lines, each of which is at a fixed height above sea level. Here they are at 80-ft (25-m) intervals. If you walk along one contour line, you neither climb nor descend. If you cross lines, and especially if you climb perpendicular to the lines, you will be changing your gravitational potential (rapidly, if the lines are close together).

17–4 The Electron Volt, a Unit of Energy

The joule is a very large unit for dealing with energies of electrons, atoms, or molecules. For this purpose, the unit **electron volt** (eV) is used. One electron volt is defined as the energy acquired by a particle carrying a charge whose magnitude equals that on the electron ($q = e$) as a result of moving through a potential difference of 1 V. The charge on an electron has magnitude 1.6022×10^{-19} C, and the change in potential energy equals qV. So 1 eV is equal to $(1.6022 \times 10^{-19} \text{ C})(1.00 \text{ V}) = 1.6022 \times 10^{-19}$ J:

$$1 \text{ eV} = 1.6022 \times 10^{-19} \text{ J} \approx 1.60 \times 10^{-19} \text{ J}.$$

Electron volt

An electron that accelerates through a potential difference of 1000 V will lose 1000 eV of potential energy and thus gain 1000 eV or 1 keV (kiloelectron volt) of kinetic energy.

On the other hand, if a particle with a charge equal to twice the magnitude of the charge on the electron ($= 2e = 3.2 \times 10^{-19}$ C) moves through a potential difference of 1000 V, its kinetic energy will increase by 2000 eV = 2 keV.

Although the electron volt is handy for *stating* the energies of molecules and elementary particles, it is *not* a proper SI unit. For calculations, electron volts should be converted to joules using the conversion factor just given. In Example 17–2, for example, the electron acquired a kinetic energy of 8.0×10^{-16} J. We can quote this energy as 5000 eV ($= 8.0 \times 10^{-16}$ J$/1.6 \times 10^{-19}$ J/eV), but when determining the speed of a particle in SI units, we must use the KE in joules (J).

EXERCISE B What is the kinetic energy of a He^{2+} ion released from rest and accelerated through a potential difference of 2.5 kV? (*a*) 2500 eV, (*b*) 500 eV, (*c*) 5000 eV, (*d*) 10,000 eV, (*e*) 250 eV.

17–5 Electric Potential Due to Point Charges

The electric potential at a distance r from a single point charge Q can be derived from the expression for its electric field (Eq. 16–4, $E = kQ/r^2$) using calculus. The potential in this case is usually taken to be zero at infinity ($= \infty$, which means extremely, indefinitely, far away); this is also where the electric field ($E = kQ/r^2$) is zero. The result is

$$V = k\frac{Q}{r}$$
$$= \frac{1}{4\pi\epsilon_0}\frac{Q}{r}, \qquad \begin{bmatrix}\text{single point charge}\\ V = 0 \text{ at } r = \infty\end{bmatrix} \quad (17\text{–}5)$$

where $k = 8.99 \times 10^9$ N·m^2/C$^2 \approx 9.0 \times 10^9$ N·m^2/C^2. We can think of V here as representing the absolute potential at a distance r from the charge Q, where $V = 0$ at $r = \infty$; or we can think of V as the potential difference between r and infinity. (The symbol ∞ means infinitely far away.) Notice that the potential V decreases with the first power of the distance, whereas the electric field (Eq. 16–4) decreases as the *square* of the distance. The potential near a positive charge is large and positive, and it decreases toward zero at very large distances, Fig. 17–9a. The potential near a negative charge is negative and increases toward zero at large distances, Fig. 17–9b. Equation 17–5 is sometimes called the **Coulomb potential** (it has its origin in Coulomb's law).

⚠ CAUTION

$V \propto \frac{1}{r}, \quad E \propto \frac{1}{r^2}$ *for a point charge*

FIGURE 17–9 Potential V as a function of distance r from a single point charge Q when the charge is (a) positive, (b) negative.

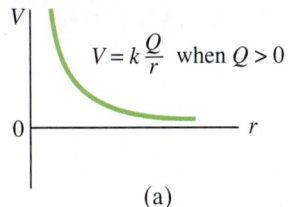
(a)

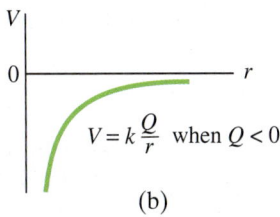
(b)

EXAMPLE 17–4 Potential due to a positive or a negative charge. Determine the potential at a point 0.50 m (*a*) from a $+20\ \mu$C point charge, (*b*) from a $-20\ \mu$C point charge.

APPROACH The potential due to a point charge is given by Eq. 17–5, $V = kQ/r$.

SOLUTION (*a*) At a distance of 0.50 m from a positive 20 μC charge, the potential is

$$V = k\frac{Q}{r}$$
$$= (9.0 \times 10^9 \text{ N·m}^2/\text{C}^2)\left(\frac{20 \times 10^{-6} \text{ C}}{0.50 \text{ m}}\right) = 3.6 \times 10^5 \text{ V}.$$

(*b*) For the negative charge,

$$V = (9.0 \times 10^9 \text{ N·m}^2/\text{C}^2)\left(\frac{-20 \times 10^{-6} \text{ C}}{0.50 \text{ m}}\right) = -3.6 \times 10^5 \text{ V}.$$

NOTE Potential can be positive or negative, and we always include a charge's sign when we find electric potential.

✎ PROBLEM SOLVING
Keep track of charge signs for electric potential

EXAMPLE 17–5 Work required to bring two positive charges close together. What minimum work must be done by an external force to bring a charge $q = 3.00\ \mu C$ from a great distance away (take $r = \infty$) to a point 0.500 m from a charge $Q = 20.0\ \mu C$?

APPROACH To find the work we cannot simply multiply the force times distance because the force is proportional to $1/r^2$ and so is not constant. Instead we can set the change in potential energy equal to the (positive of the) work required of an *external* force (Chapter 6, Eq. 6–7a), and Eq. 17–3: $W_{ext} = \Delta PE = q(V_b - V_a)$. We get the potentials V_b and V_a using Eq. 17–5.

SOLUTION The external work required is equal to the change in potential energy:

$$W_{ext} = q(V_b - V_a) = q\left(\frac{kQ}{r_b} - \frac{kQ}{r_a}\right),$$

where $r_b = 0.500$ m and $r_a = \infty$. The right-hand term within the parentheses is zero ($1/\infty = 0$) so

$$W_{ext} = (3.00 \times 10^{-6}\ C)\frac{(8.99 \times 10^9\ N \cdot m^2/C^2)(2.00 \times 10^{-5}\ C)}{(0.500\ m)} = 1.08\ J.$$

NOTE We could not use Eqs. 17–4 here because they apply *only* to uniform fields. But we did use Eq. 17–3 because it is always valid.

> **CAUTION**
> *We cannot use $W = Fd$ if F is not constant*

EXERCISE C What work is required to bring a charge $q = 3.00\ \mu C$ originally a distance of 1.50 m from a charge $Q = 20.0\ \mu C$ until it is 0.50 m away?

To determine the electric field at points near a collection of two or more point charges requires adding up the electric fields due to each charge. Since the electric field is a vector, this can be time consuming or complicated. To find the electric potential at a point due to a collection of point charges is far easier, because the electric potential is a scalar, and hence you only need to add numbers (with appropriate signs) without concern for direction.

EXAMPLE 17–6 Potential above two charges. Calculate the electric potential (*a*) at point A in Fig. 17–10 due to the two charges shown, and (*b*) at point B. [This is the same situation as Examples 16–9 and 16–10, Fig. 16–29, where we calculated the electric field at these points.]

APPROACH The total potential at point A (or at point B) is the algebraic sum of the potentials at that point due to each of the two charges Q_1 and Q_2. The potential due to each single charge is given by Eq. 17–5. We do not have to worry about directions because electric potential is a scalar quantity. But we do have to keep track of the signs of charges.

> **CAUTION**
> *Potential is a scalar and has no components*

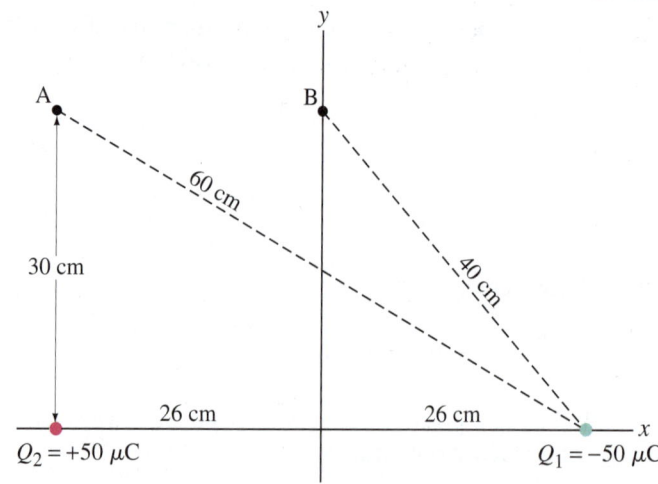

FIGURE 17–10 Example 17–6. (See also Examples 16–9 and 16–10, Fig. 16–29.)

SOLUTION (a) We add the potentials at point A due to each charge Q_1 and Q_2, and we use Eq. 17–5 for each:

$$V_A = V_{A2} + V_{A1}$$
$$= k\frac{Q_2}{r_{2A}} + k\frac{Q_1}{r_{1A}}$$

where $r_{1A} = 60\text{ cm}$ and $r_{2A} = 30\text{ cm}$. Then

$$V_A = \frac{(9.0 \times 10^9 \text{ N} \cdot \text{m}^2/\text{C}^2)(5.0 \times 10^{-5}\text{ C})}{0.30 \text{ m}} + \frac{(9.0 \times 10^9 \text{ N} \cdot \text{m}^2/\text{C}^2)(-5.0 \times 10^{-5}\text{ C})}{0.60 \text{ m}}$$

$$= 1.50 \times 10^6 \text{ V} - 0.75 \times 10^6 \text{ V}$$
$$= 7.5 \times 10^5 \text{ V}.$$

(b) At point B, $r_{1B} = r_{2B} = 0.40$ m, so

$$V_B = V_{B2} + V_{B1}$$
$$= \frac{(9.0 \times 10^9 \text{ N} \cdot \text{m}^2/\text{C}^2)(5.0 \times 10^{-5}\text{ C})}{0.40 \text{ m}} + \frac{(9.0 \times 10^9 \text{ N} \cdot \text{m}^2/\text{C}^2)(-5.0 \times 10^{-5}\text{ C})}{0.40 \text{ m}}$$
$$= 0 \text{ V}.$$

NOTE The two terms in the sum in (b) cancel for any point equidistant from Q_1 and Q_2 $(r_{1B} = r_{2B})$. Thus the potential will be zero everywhere on the plane equidistant between the two opposite charges. This plane is an equipotential surface with $V = 0$.

Simple summations like these can be performed for any number of point charges.

CONCEPTUAL EXAMPLE 17–7 Potential energies. Consider the three pairs of charges shown in Fig. 17–11. Call them Q_1 and Q_2. (a) Which set has a positive potential energy? (b) Which set has the most negative potential energy? (c) Which set requires the most work to separate the charges to infinity? Assume the charges all have the same magnitude.

RESPONSE The potential energy equals the work required to bring the two charges near each other, starting at a great distance (∞). Assume the left (+) charge Q_1 is already there. To bring a second charge Q_2 close to the first from a great distance away (∞) requires external work

$$W_{\text{ext}} = Q_2 V = k\frac{Q_1 Q_2}{r}$$

where r is the final distance between them. Thus the potential energy of the two charges is

$$\text{PE} = k\frac{Q_1 Q_2}{r}.$$

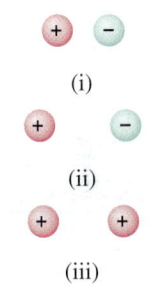

FIGURE 17–11 Example 17–7.

(a) Set (iii) has a positive potential energy because the charges have the same sign. (b) Both (i) and (ii) have opposite signs of charge and negative PE. Because r is smaller in (i), the PE is most negative for (i). (c) Set (i) will require the most work for separation to infinity. The more negative the potential energy, the more work required to separate the charges and bring the PE up to zero ($r = \infty$), as in Fig. 17–9b.

EXERCISE D Return to the Chapter-Opening Question, page 473, and answer it again now. Try to explain why you may have answered differently the first time.

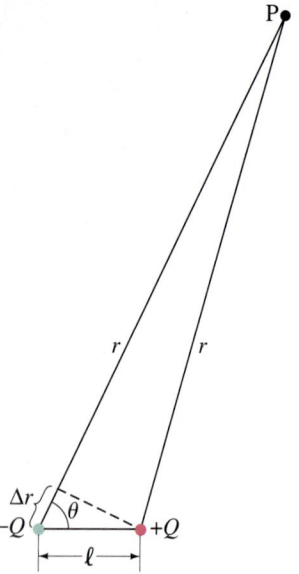

FIGURE 17–12 Electric dipole. Calculation of potential V at point P.

*17–6 Potential Due to Electric Dipole; Dipole Moment

Two equal point charges Q, of opposite sign, separated by a distance ℓ, are called an **electric dipole**. The electric field lines and equipotential surfaces for a dipole were shown in Fig. 17–7. Because electric dipoles occur often in physics, as well as in other disciplines such as molecular biology, it is useful to examine them more closely.

The electric potential at an arbitrary point P due to a dipole, Fig. 17–12, is the sum of the potentials due to each of the two charges:

$$V = \frac{kQ}{r} + \frac{k(-Q)}{r + \Delta r} = kQ\left(\frac{1}{r} - \frac{1}{r + \Delta r}\right) = kQ\frac{\Delta r}{r(r + \Delta r)},$$

where r is the distance from P to the positive charge and $r + \Delta r$ is the distance to the negative charge. This equation becomes simpler if we consider points P whose distance from the dipole is much larger than the separation of the two charges—that is, for $r \gg \ell$. From Fig. 17–12 we see that $\Delta r = \ell \cos\theta$; since $r \gg \Delta r = \ell \cos\theta$, we can neglect Δr in the denominator as compared to r. Then we obtain

$$V \approx \frac{kQ\ell \cos\theta}{r^2}. \qquad \text{[dipole; } r \gg \ell\text{]} \quad (17\text{–}6\text{a})$$

We see that the potential decreases as the *square* of the distance from the dipole, whereas for a single point charge the potential decreases with the first power of the distance (Eq. 17–5). It is not surprising that the potential should fall off faster for a dipole: when you are far from a dipole, the two equal but opposite charges appear so close together as to tend to neutralize each other.

The product $Q\ell$ in Eq. 17–6a is referred to as the **dipole moment**, p, of the dipole. Equation 17–6a in terms of the dipole moment is

$$V \approx \frac{kp\cos\theta}{r^2}. \qquad \text{[dipole; } r \gg \ell\text{]} \quad (17\text{–}6\text{b})$$

A dipole moment has units of coulomb-meters ($C \cdot m$), although for molecules a smaller unit called a *debye* is sometimes used: 1 debye = $3.33 \times 10^{-30}\,C \cdot m$.

In many molecules, even though they are electrically neutral, the electrons spend more time in the vicinity of one atom than another, which results in a separation of charge. Such molecules have a dipole moment and are called **polar molecules**. We already saw that water (Fig. 16–4) is a polar molecule, and we have encountered others in our discussion of molecular biology (Section 16–10). Table 17–2 gives the dipole moments for several molecules. The + and − signs indicate on which atoms these charges lie. The last two entries are a part of many organic molecules and play an important role in molecular biology.

Dipoles in molecular biology

TABLE 17–2 Dipole Moments of Selected Molecules

Molecule	Dipole Moment ($C \cdot m$)
$H_2^{(+)}O^{(-)}$	6.1×10^{-30}
$H^{(+)}Cl^{(-)}$	3.4×10^{-30}
$N^{(-)}H_3^{(+)}$	5.0×10^{-30}
$>N^{(-)}\!-\!H^{(+)}$	$\approx 3.0 \times 10^{-30}$ ‡
$>C^{(+)}\!=\!O^{(-)}$	$\approx 8.0 \times 10^{-30}$ ‡

‡ These last two groups often appear on larger molecules; hence the value for the dipole moment will vary somewhat, depending on the rest of the molecule.

17–7 Capacitance

A **capacitor** is a device that can store electric charge, and normally consists of two conducting objects (usually plates or sheets) placed near each other but not touching. Capacitors are widely used in electronic circuits and sometimes are called **condensers**. Capacitors store charge for later use, such as in a camera flash, and as energy backup in devices like computers if the power fails. Capacitors also block surges of charge and energy to protect circuits. Very tiny capacitors serve as memory for the "ones" and "zeros" of the binary code in the random access memory (RAM) of computers and other electronic devices (as in Fig. 17–35). Capacitors serve many other applications as well, some of which we will discuss.

Uses of capacitors

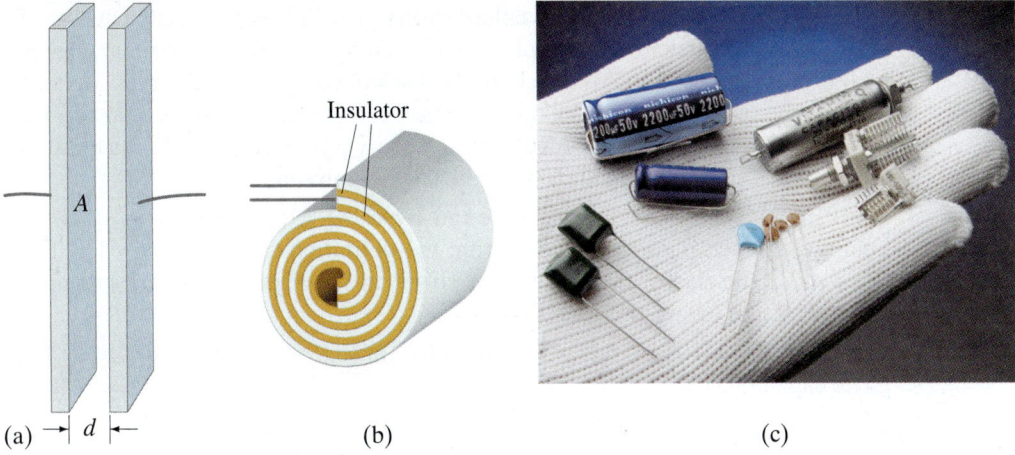

FIGURE 17–13 Capacitors: diagrams of (a) parallel plate, (b) cylindrical (rolled up parallel plate). (c) Photo of some real capacitors.

A simple capacitor consists of a pair of parallel plates of area A separated by a small distance d (Fig. 17–13a). Often the two plates are rolled into the form of a cylinder with paper or other insulator separating the plates, Fig. 17–13b; Fig. 17–13c is a photo of some actual capacitors used for various applications. In circuit diagrams, the symbol

$$\dashv\vdash \quad \text{or} \quad \dashv\mid\vdash \qquad \text{[capacitor symbol]}$$

represents a capacitor. A battery, which is a source of voltage, is indicated by the symbol

$$\dashv\mid\vdash \qquad \text{[battery symbol]}$$

with unequal arms.

If a voltage is applied across a capacitor by connecting the capacitor to a battery with conducting wires as in Fig. 17–14, the two plates quickly become charged: one plate acquires a negative charge, the other an equal amount of positive charge. Each battery terminal and the capacitor plate connected to it are at the same potential. Hence the full battery voltage appears across the capacitor. For a given capacitor, it is found that the amount of charge Q acquired by each plate is proportional to the magnitude of the potential difference V between the plates:

$$Q = CV. \qquad (17\text{–}7)$$

The constant of proportionality, C, in Eq. 17–7 is called the **capacitance** of the capacitor. The unit of capacitance is coulombs per volt, and this unit is called a **farad** (F). Common capacitors have capacitance in the range of 1 pF (picofarad = 10^{-12} F) to 10^3 μF (microfarad = 10^{-6} F). The relation, Eq. 17–7, was first suggested by Volta in the late eighteenth century.

In Eq. 17–7 and from now on, we will use simply V (in italics) to represent a potential difference, such as that produced by a battery, rather than V_{ba}, ΔV, or $V_b - V_a$, as previously.

Also, be sure not to confuse *italic* letters V and C which stand for voltage and capacitance, with non-italic V and C which stand for the units volts and coulombs.

The capacitance C does not in general depend on Q or V. Its value depends only on the size, shape, and relative position of the two conductors, and also on the material that separates them. For a parallel-plate capacitor whose plates have area A and are separated by a distance d of air (Fig. 17–13a), the capacitance is given by

$$C = \epsilon_0 \frac{A}{d}. \qquad \text{[parallel-plate capacitor]} \qquad (17\text{–}8)$$

We see that C depends only on geometric factors, A and d, and not on Q or V. We derive this useful relation in the optional subsection at the end of this Section. The constant ϵ_0 is the *permittivity of free space*, which, as we saw in Chapter 16, has the value 8.85×10^{-12} C^2/N·m^2.

EXERCISE E Graphs for charge versus voltage are shown in Fig. 17–15 for three capacitors, A, B, and C. Which has the greatest capacitance?

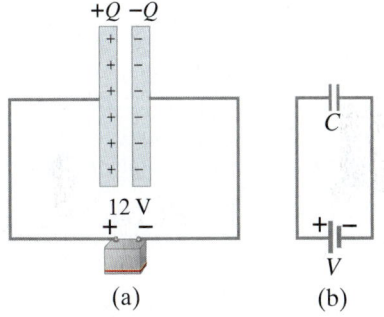

FIGURE 17–14 (a) Parallel-plate capacitor connected to a battery. (b) Same circuit shown using symbols.

⚠ **CAUTION**
V = potential difference from here on

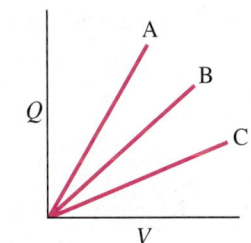

FIGURE 17–15 Exercise E.

SECTION 17–7 Capacitance **483**

EXAMPLE 17-8 **Capacitor calculations.** (a) Calculate the capacitance of a parallel-plate capacitor whose plates are 20 cm × 3.0 cm and are separated by a 1.0-mm air gap. (b) What is the charge on each plate if a 12-V battery is connected across the two plates? (c) What is the electric field between the plates? (d) Estimate the area of the plates needed to achieve a capacitance of 1 F, assuming the air gap d is 100 times smaller, or 10 microns (1 **micron** = 1 μm = 10^{-6} m).

APPROACH The capacitance is found by using Eq. 17-8, $C = \epsilon_0 A/d$. The charge on each plate is obtained from the definition of capacitance, Eq. 17-7, $Q = CV$. The electric field is uniform, so we can use Eq. 17-4b for the magnitude $E = V/d$. In (d) we use Eq. 17-8 again.

SOLUTION (a) The area $A = (20 \times 10^{-2}\,\text{m})(3.0 \times 10^{-2}\,\text{m}) = 6.0 \times 10^{-3}\,\text{m}^2$. The capacitance C is then

$$C = \epsilon_0 \frac{A}{d} = (8.85 \times 10^{-12}\,\text{C}^2/\text{N} \cdot \text{m}^2)\frac{6.0 \times 10^{-3}\,\text{m}^2}{1.0 \times 10^{-3}\,\text{m}} = 53\,\text{pF}.$$

(b) The charge on each plate is

$$Q = CV = (53 \times 10^{-12}\,\text{F})(12\,\text{V}) = 6.4 \times 10^{-10}\,\text{C}.$$

(c) From Eq. 17-4b for a uniform electric field, the magnitude of E is

$$E = \frac{V}{d} = \frac{12\,\text{V}}{1.0 \times 10^{-3}\,\text{m}} = 1.2 \times 10^4\,\text{V/m}.$$

(d) We solve for A in Eq. 17-8 and substitute $C = 1.0\,\text{F}$ and $d = 1.0 \times 10^{-5}\,\text{m}$ to find that we need plates with an area

$$A = \frac{Cd}{\epsilon_0} \approx \frac{(1\,\text{F})(1.0 \times 10^{-5}\,\text{m})}{(9 \times 10^{-12}\,\text{C}^2/\text{N} \cdot \text{m}^2)} \approx 10^6\,\text{m}^2.$$

NOTE This is the area of a square 10^3 m or 1 km on a side. That is inconveniently large. Large-capacitance capacitors will not be simple parallel plates.

PHYSICS APPLIED
Capacitor as power backup; condenser microphone; computer keyboard

Not long ago, a capacitance greater than a few mF was unusual. Today capacitors are available that are 1 or 2 F, yet they are just a few cm on a side. Such capacitors are used as power backups, for example, in computer memory and electronics where the time and date can be maintained through tiny charge flow. [Capacitors are superior to rechargable batteries for this purpose because they can be recharged more than 10^5 times with no degradation.] Such high-capacitance capacitors can be made of *activated carbon* which has very high porosity, so that the surface area is very large; one-tenth of a gram of activated carbon can have a surface area of 100 m². Furthermore, the equal and opposite charges exist in an electric "double layer" about 10^{-9} m thick. Thus, the capacitance of 0.1 g of activated carbon, whose internal area can be 10^2 m², is equivalent to a parallel-plate capacitor with $C \approx \epsilon_0 A/d = (8.85 \times 10^{-12}\,\text{C}^2/\text{N} \cdot \text{m}^2)(10^2\,\text{m}^2)/(10^{-9}\,\text{m}) \approx 1\,\text{F}$.

The proportionality, $C \propto A/d$ in Eq. 17-8, is valid also for a parallel-plate capacitor that is rolled up into a spiral cylinder, as in Fig. 17-13b. However, the constant factor, ϵ_0, must be replaced if an insulator such as paper separates the plates, as is usual, as discussed in the next Section.

One type of microphone is a **condenser**, or capacitor, **microphone**, diagrammed in Fig. 17-16. The changing air pressure in a sound wave causes one plate of the capacitor C to move back and forth. The voltage across the capacitor changes at the same frequency as the sound wave.

Some computer keyboards operate by capacitance. As shown in Fig. 17-17, each key is connected to the upper plate of a capacitor. The upper plate moves down when the key is pressed, reducing the spacing between the capacitor plates, and increasing the capacitance (Eq. 17-8: smaller d, larger C). The *change* in capacitance results in an electric signal that is detected by an electronic circuit.

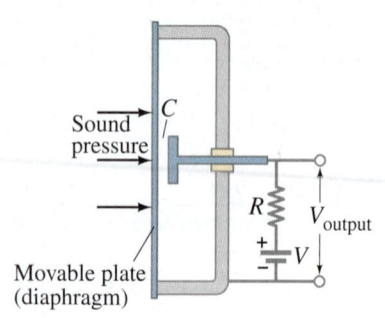

FIGURE 17-16 Diagram of a condenser microphone.

FIGURE 17-17 Key on a computer keyboard. Pressing the key reduces the plate spacing, increasing the capacitance.

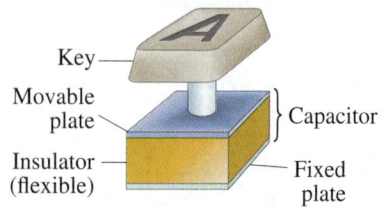

*Derivation of Capacitance for Parallel-Plate Capacitor

Equation 17–8 can be derived using the result from Section 16–12 on Gauss's law, namely that the electric field between two parallel plates is given by Eq. 16–10:

$$E = \frac{Q/A}{\epsilon_0}.$$

We combine this with Eq. 17–4a, using magnitudes, $V = Ed$, to obtain

$$V = \left(\frac{Q}{A\epsilon_0}\right)d.$$

Then, from Eq. 17–7, the definition of capacitance,

$$C = \frac{Q}{V} = \frac{Q}{(Q/A\epsilon_0)d} = \epsilon_0 \frac{A}{d}$$

which is Eq. 17–8.

17–8 Dielectrics

In most capacitors there is an insulating sheet of material, such as paper or plastic, called a **dielectric** between the plates (Fig. 17–18). This serves several purposes. First, dielectrics break down (allowing electric charge to flow) less readily than air, so higher voltages can be applied without charge passing across the gap. Furthermore, a dielectric allows the plates to be placed closer together without touching, thus allowing an increased capacitance because d is smaller in Eq. 17–8. Thirdly, it is found experimentally that if the dielectric fills the space between the two conductors, it increases the capacitance by a factor K, known as the **dielectric constant**. Thus, for a parallel-plate capacitor,

$$C = K\epsilon_0 \frac{A}{d}. \qquad (17\text{–}9)$$

This can be written

$$C = \epsilon \frac{A}{d},$$

where $\epsilon = K\epsilon_0$ is called the **permittivity** of the material.

The values of the dielectric constant for various materials are given in Table 17–3. Also shown in Table 17–3 is the **dielectric strength**, the maximum electric field before breakdown (charge flow) occurs.

FIGURE 17–18 A cylindrical capacitor, unrolled from its case to show the dielectric between the plates. See also Fig. 17–13b.

> **CONCEPTUAL EXAMPLE 17–9** **Inserting a dielectric at constant V.** An air-filled capacitor consisting of two parallel plates separated by a distance d is connected to a battery of constant voltage V and acquires a charge Q. While it is still connected to the battery, a slab of dielectric material with $K = 3$ is inserted between the plates of the capacitor. Will Q increase, decrease, or stay the same?
>
> **RESPONSE** Since the capacitor remains connected to the battery, the voltage stays constant and equal to the battery voltage V. The capacitance C increases when the dielectric material is inserted because K in Eq. 17–9 has increased. From the relation $Q = CV$, if V stays constant, but C increases, Q must increase as well. As the dielectric is inserted, more charge will be pulled from the battery and deposited onto the plates of the capacitor as its capacitance increases.

> **EXERCISE F** If the dielectric in Example 17–9 fills the space between the plates, by what factor does (a) the capacitance change, (b) the charge on each plate change?

> **CONCEPTUAL EXAMPLE 17–10** **Inserting a dielectric into an isolated capacitor.** Suppose the air-filled capacitor of Example 17–9 is charged (to Q) and then disconnected from the battery. Next a dielectric is inserted between the plates. Will Q, C, or V change?
>
> **RESPONSE** The charge Q remains the same—the capacitor is isolated, so there is nowhere for the charge to go. The capacitance increases as a result of inserting the dielectric (Eq. 17–9). The voltage across the capacitor also changes—it *decreases* because, by Eq. 17–7, $Q = CV$, so $V = Q/C$; if Q stays constant and C increases (it is in the denominator), then V decreases.

TABLE 17–3
Dielectric Constants (at 20°C)

Material	Dielectric constant K	Dielectric strength (V/m)
Vacuum	1.0000	
Air (1 atm)	1.0006	3×10^6
Paraffin	2.2	10×10^6
Polystyrene	2.6	24×10^6
Vinyl (plastic)	2–4	50×10^6
Paper	3.7	15×10^6
Quartz	4.3	8×10^6
Oil	4	12×10^6
Glass, Pyrex	5	14×10^6
Rubber, neoprene	6.7	12×10^6
Porcelain	6–8	5×10^6
Mica	7	150×10^6
Water (liquid)	80	
Strontium titanate	300	8×10^6

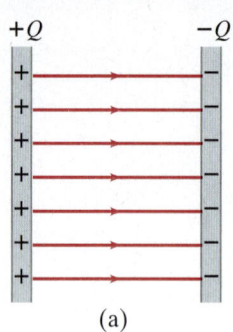
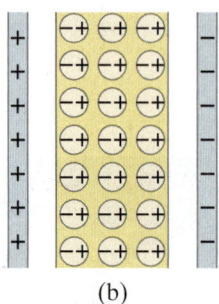
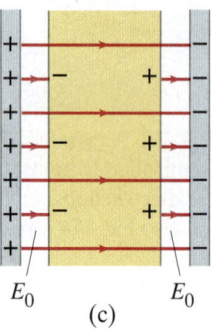

FIGURE 17-19 Molecular view of the effects of a dielectric.

*Molecular Description of Dielectrics

Let us examine, from the molecular point of view, why the capacitance of a capacitor should be larger when a dielectric is between the plates. A capacitor C_0 whose plates are separated by an air gap has a charge $+Q$ on one plate and $-Q$ on the other (Fig. 17–19a). Assume it is isolated (not connected to a battery) so charge cannot flow to or from the plates. The potential difference between the plates, V_0, is given by Eq. 17–7:

$$Q = C_0 V_0,$$

where the subscripts refer to air between the plates. Now we insert a dielectric between the plates (Fig. 17–19b). Because of the electric field between the capacitor plates, the dielectric molecules will tend to become oriented as shown in Fig. 17–19b. If the dielectric molecules are *polar*, the positive end is attracted to the negative plate and vice versa. Even if the dielectric molecules are not polar, electrons within them will tend to move slightly toward the positive capacitor plate, so the effect is the same. The net effect of the aligned dipoles is a net negative charge on the outer edge of the dielectric facing the positive plate, and a net positive charge on the opposite side, as shown in Fig. 17–19c.

Some of the electric field lines, then, do not pass through the dielectric but instead end on charges induced on the surface as shown in Fig. 17–19c. Hence the electric field within the dielectric is less than in air. That is, the electric field in the space between the capacitor plates, assumed filled by the dielectric, has been reduced by some factor K. The voltage across the capacitor is reduced by the same factor K because $V = Ed$ (Eq. 17–4) and hence, by Eq. 17–7, $Q = CV$, the capacitance C must increase by that same factor K to keep Q constant.

17–9 Storage of Electric Energy

A charged capacitor stores electric energy by separating + and − charges. The energy stored in a capacitor will be equal to the work done to charge it. The net effect of charging a capacitor is to remove charge from one plate and add it to the other plate. This is what a battery does when it is connected to a capacitor. A capacitor does not become charged instantly. It takes some time, often very little (Section 19–6). Initially, when the capacitor is uncharged, no work is required to move the first bit of charge over. As more charge is transferred, work is needed to move charge against the increasing voltage V. The work needed to add a small amount of charge Δq, when a potential difference V is across the plates, is $\Delta W = V \Delta q$. The total work needed to move total charge Q is equivalent to moving all the charge Q across a voltage equal to the *average* voltage during the process. (This is just like calculating the work done to compress a spring, Section 6–4, page 148.) The average voltage is $(V_f - 0)/2 = V_f/2$, where V_f is the final voltage; so the work to move the total charge Q from one plate to the other is

$$W = Q \frac{V_f}{2}.$$

Thus we can say that the electric potential energy, PE, stored in a capacitor is

$$\text{PE} = \text{energy} = \tfrac{1}{2} QV,$$

where V is the potential difference between the plates (we dropped the subscript), and Q is the charge on each plate. Since $Q = CV$, we can also write

$$\text{PE} = \tfrac{1}{2} QV = \tfrac{1}{2} CV^2 = \tfrac{1}{2} \frac{Q^2}{C}. \qquad (17\text{--}10)$$

PHYSICS APPLIED
Camera flash

EXAMPLE 17–11 Energy stored in a capacitor. A camera flash unit (Fig. 17–20) stores energy in a 660-μF capacitor at 330 V. (*a*) How much electric energy can be stored? (*b*) What is the power output if nearly all this energy is released in 1.0 ms?

APPROACH We use Eq. 17–10 in the form $\text{PE} = \tfrac{1}{2} CV^2$ because we are given C and V.

SOLUTION (*a*) The energy stored is

$$\text{PE} = \tfrac{1}{2}CV^2 = \tfrac{1}{2}(660 \times 10^{-6}\,\text{F})(330\,\text{V})^2 = 36\,\text{J}.$$

(*b*) If this energy is released in $\tfrac{1}{1000}$ of a second ($= 1.0\,\text{ms} = 1.0 \times 10^{-3}\,\text{s}$), the power output is $P = \text{PE}/t = (36\,\text{J})/(1.0 \times 10^{-3}\,\text{s}) = 36{,}000\,\text{W}.$

| **EXERCISE G** A capacitor stores 0.50 J of energy at 9.0 V. What is its capacitance?

CONCEPTUAL EXAMPLE 17–12 **Capacitor plate separation increased.**
A parallel-plate capacitor carries charge Q and is then disconnected from a battery. The two plates are initially separated by a distance d. Suppose the plates are pulled apart until the separation is $2d$. How has the energy stored in this capacitor changed?

RESPONSE If we increase the plate separation d, we decrease the capacitance according to Eq. 17–8, $C = \epsilon_0 A/d$, by a factor of 2. The charge Q hasn't changed. So according to Eq. 17–10, where we choose the form $\text{PE} = \tfrac{1}{2}Q^2/C$ because we know Q is the same and C has been halved, the reduced C means the PE stored increases by a factor of 2.

NOTE We can see why the energy stored increases from a physical point of view: the two plates are charged equal and opposite, so they attract each other. If we pull them apart, we must do work, so we raise the potential energy.

FIGURE 17–20 A camera flash unit. The 660-μF capacitor is the black cylinder.

It is useful to think of the energy stored in a capacitor as being stored in the electric field between the plates. As an example let us calculate the energy stored in a parallel-plate capacitor in terms of the electric field.

We have seen that the electric field $\vec{E}$ between two close parallel plates is nearly uniform and its magnitude is related to the potential difference by $V = Ed$ (Eq. 17–4), where d is the separation. Also, Eq. 17–8 tells us $C = \epsilon_0 A/d$ for a parallel-plate capacitor. Thus

$$\text{PE} = \tfrac{1}{2}CV^2 = \tfrac{1}{2}\left(\frac{\epsilon_0 A}{d}\right)(E^2 d^2)$$
$$= \tfrac{1}{2}\epsilon_0 E^2 A d.$$

The quantity Ad is the volume between the plates in which the electric field E exists. If we divide both sides of this equation by the volume, we obtain an expression for the energy per unit volume or **energy density**:

$$\text{energy density} = \frac{\text{PE}}{\text{volume}} = \tfrac{1}{2}\epsilon_0 E^2. \qquad (17\text{–}11)$$

The *electric energy stored per unit volume in any region of space is proportional to the square of the electric field* in that region. We derived Eq. 17–11 for the special case of a parallel-plate capacitor. But it can be shown to be true for any region of space where there is an electric field. Indeed, we will use this result when we discuss electromagnetic radiation (Chapter 22).

Health Effects

The energy stored in a large capacitance can give you a burn or a shock. One reason you are warned not to touch a circuit, or open an electronic device, is because capacitors may still be carrying charge even if the external power is turned off.

On the other hand, the basis of a heart *defibrillator* is a capacitor charged to a high voltage. A heart attack can be characterized by fast irregular beating of the heart, known as *ventricular* (or *cardiac*) *fibrillation*. The heart then does not pump blood to the rest of the body properly, and if the interruption lasts for long, death results. A sudden, brief jolt of charge through the heart from a defibrillator can cause complete heart stoppage, sometimes followed by a resumption of normal beating. The defibrillator capacitor is charged to a high voltage, typically a few thousand volts, and is allowed to discharge very rapidly through the heart via a pair of wide contacts known as "pads" or "paddles" that spread out the current over the chest (Fig. 17–21).

PHYSICS APPLIED
Shocks, burns, defibrillators

FIGURE 17–21 Heart defibrillator.

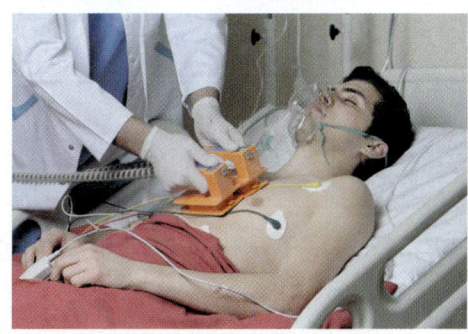

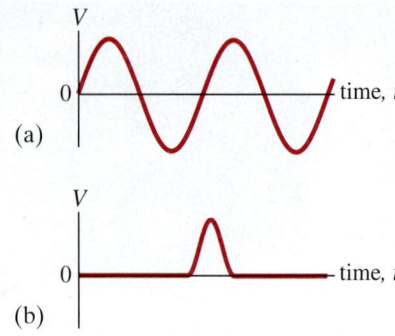

FIGURE 17–22 Two kinds of signal voltage: (a) sinusoidal, (b) a pulse, both analog. Many other shapes are possible.

TABLE 17–4 Binary to Decimal

Binary† number	Decimal number
00000000	0
00000001	1
00000010	2
00000011	3
00000100	4
00000111	7
00001000	8
00100101	37
11111111	255

†Note that we start counting from right to left: the 1's digit is on the far right, then the 2's, the 4's, the 8's, the 16's, the 32's, the 64's, and the 128's.

17–10 Digital; Binary Numbers; Signal Voltage

Batteries and a wall plug are meant to provide a constant **supply voltage** as power to operate a flashlight, an electric heater, and other electric and electronic devices.

A **signal voltage**, on the other hand, is a voltage intended to affect something else. A signal voltage varies in time and can also be very brief. For example, a sound such as a pure tone, which may be sinusoidal as we discussed in Chapters 11 and 12 (see Figs. 11–24 and 12–14), will produce an output voltage from a high quality microphone that is also sinusoidal (Fig. 17–22a). That signal voltage is amplified and reaches a loudspeaker, making it produce the sound we hear. A signal voltage is sometimes a simple pulse (Fig. 17–22b; see also Figs. 11–23 and 11–33), and often acts to change some aspect of an electronic device.

Signal voltages are sent to cell phones ("I've got signal"), to computers from the Internet, or to TV sets with the information on the picture and sound. Not long ago, signal voltages were **analog**—the voltage varied continuously, as in Fig. 17–22.

Today, television and computer signals are **digital** and use a binary number system to represent a numerical value. In a normal number, such as 609, there are *ten* choices for each digit—from 0 to 9—and normal numbers are called **decimal** (Latin for ten). In a **binary** number, each digit or **bit** has only *two* possibilities, 0 or 1 (sometimes referred to as "off" or "on"). In binary, 0001 means "one," 0010 means 2, 0011 means 3, and 1101 means $8 + 4 + 0 + 1 = 13$ in decimal. See Table 17–4, and note that counting starts from the right, just as in regular decimal (the "ones" digit is last, on the far right, then to the left is the "tens" and then "hundreds": for 609, the "ones" are 9, the "hundreds" are 6). Any value can be represented by a voltage pattern something like that shown in Fig. 17–23.

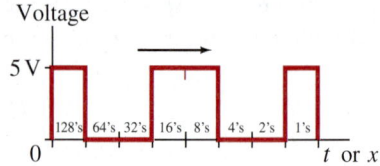

FIGURE 17–23 A traveling digital signal: voltage vs. position x or time t. If standing alone, this sequence would represent 10011001 or 153 ($= 128 + 0 + 0 + 16 + 8 + 0 + 0 + 1$).

A "1" is a positive voltage such as $+5\,\text{V}$, whereas a "0" is $0\,\text{V}$. The brightness signal, for example, that goes to each of the millions of tiny picture elements or "subpixels" of a TV or computer screen (Fig. 17–31, Section 17–11), is contained in a **byte**. One byte is 8 bits, which means

each byte of 8 bits allows $2^8 = 256$ possibilities

(that is, 0 to 255) or 256 shades for each of 3 colors: red, green, blue. The full color of each pixel (the three subpixel colors) has $(256)^3 = 17 \times 10^6$ possibilities. Digital television signals, which we discuss in the next Section, are transmitted at about $19\,\text{Mb/s} = 19$ Megabits per second. So 19×10^6 bits pass a given point per second, or one bit every 53 nanoseconds. We could write this in terms of bytes as $2.4\,\text{MB/s}$, where for bytes we use capital B (and lower-case b for bits).

When an analog signal, such as the pure sine wave of Fig. 17–22a, is converted to digital (**analog-to-digital converter**, ADC), the digital signal may look like the blue squared-off curve of Fig. 17–24. The digital signal has a limited number of discrete values. The difference between the original continuous analog signal and its digital approximation is called the **quantization error** or **quantization loss**. To minimize that loss, there are two important factors: (i) the **resolution** or **bit depth**, which is the number of bits or values for the voltage of each sample (= measurement); (ii) the **sampling rate**, which is the number of times per second the original analog voltage is measured ("sampled").

Consider a digital approximation for a 100-Hz sine wave: Figure 17–24 shows (i) a 0 to 6-V, 2-bit depth, measuring only 4 possible voltages (00, 01, 10, 11, or 0, 1, 2, 3 in decimal), and (ii) a sampling rate of (9 samples in one cycle or wavelength) $\times$ (100 cycles/s = 100 Hz) which is 900 samples/s or 900 Hz. This is very poor quality. For high quality reproduction, a greater bit depth and higher sampling rate are needed, which requires more memory, and more data to be transmitted.

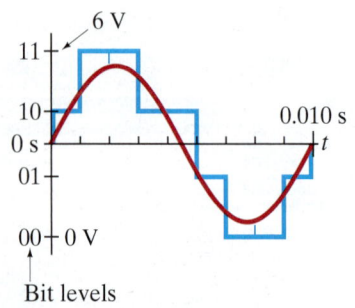

FIGURE 17–24 The red analog sine wave, which is at a 100-Hz frequency (1 wavelength is done in 0.010 s), has been converted to a 2-bit (4 level) digital signal (blue).

For audio CDs, the sampling rate is 44.1 kHz (44,100 samplings every second) and 16-bit resolution, meaning each sampled voltage can have $2^8 \times 2^8 = 2^{16} \approx 65,000$ different voltage levels between, say, 0 and 5 volts. See Fig. 17–25 for details. Audio recording today typically uses 96 kHz and 24-bit ($2^{24} \approx 17 \times 10^6$ voltage levels) to give a better approximation of the original analog signal (on super-CDs or solid-state memory), but must be transferred down to 44.1 kHz and 16-bit to produce ordinary CDs. (DVDs can use 192 kHz sampling rate for sound.) But iPods and MP3 players have lower sampling rates and much less detail, which many listeners can notice.

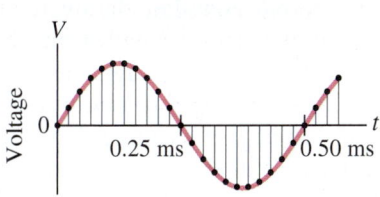

FIGURE 17–25 The sine wave shown could represent the analog electric signal from a microphone due to a pure 2000-Hz tone. (See Chapters 11 and 12.) The analog-to-digital electronics **samples** the signal—that is, measures and records the signal's voltage at intervals, many times per second. Each dot on the curve represents the voltage measured (sampled) at that point. The sampling rate in this diagram is 44,100 each second, or 44.1 kHz, like a CD. That is, a sample is taken every $(1\text{ s})/44{,}100 = 0.000023\text{ s} = 0.023$ ms. In 0.50 ms, as shown here, 22 samples (black dots) are taken. This is an alternate way to represent sampling compared to Fig. 17-24, and shows that we cannot see any changes that might happen between the samplings (dots).

Figure 17–25 gives some details about a pure 2000-Hz sound sampled at 44.1 kHz. Normal musical sounds are a complex summation of many such sine waves of different frequencies and amplitudes. A simple summation was shown in Fig. 12–14. Another example is shown in Fig. 17–26, where we can see that the fine details may be missed by a digital conversion. Look at Fig. 17–25: if that were 20,000 Hz (highest frequency of human hearing), it would be sampled only about two times per wavelength. Both those samples might be zero volts—obviously missing the entire waveform. Over many wavelengths, it might eventually reproduce the waveform somewhat well. But many sounds only last milliseconds, like the initial attack of a piano note or plucked guitar string. Many audiophiles hear the difference between an original vinyl record and its subsequent release as a CD at 44.1 kHz.

Digital audio signals must be converted back to analog (**digital-to-analog converter**, DAC) before being sent to a loudspeaker or headset. Even in a TV, the digital signals are converted to analog voltages before addressing the pixels (next Section), although the picture itself might be said to be digital since it is made up of separate pixels.

Digital photographs are made up of millions of "pixels" to produce a sharp image that is not "pixelated" or blurry. Also important (and complicated) are the number of bits provided for colors, plus the ability of the sensors (Chapter 25) to sustain a wide range of brightnesses under dim and bright light conditions.

Digital data has some real advantages: for one, it can be **compressed**, in the sense that repeated information can be reduced so that less memory space is required—fewer bits and bytes. For example, adjacent "pixels" on a photograph that includes a blue sky may be essentially identical. If 200 almost identical pixels can be coded as identical, that takes up less memory (or "size") than to specify all the 200 pixels individually. Compression schemes, like **jpeg** for photos, lose some information and may be noticeable. In audio, MP3 players use one-tenth the space that a CD does, but many listeners don't notice. Compression is one reason that more data or "information" can be transmitted digitally for a given **bandwidth**. [Bandwidth is the fixed range of frequencies allotted to each radio or TV station or Internet connection, and limits the number of bits transmitted per second.]

In audio, many listeners claim that digital does not match analog in full sound quality. And what about movies? Will digital ever match Technicolor?

*Noise

Digital information transmission has another advantage: any distortion or unwanted (external) electrical signal that intrudes from outside, broadly called **noise**, can badly corrupt an analog signal: Fig. 17–27a shows a time-varying analog signal, and Fig. 17–27b shows nasty outside noise interfering with it. But a digital signal is still readable unless the noise is very large, on the order of half the bit signal itself (Figs. 17–27c and d).

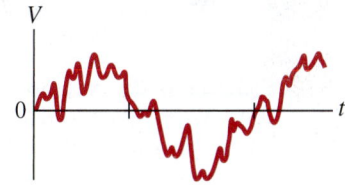

FIGURE 17–26 This type of complex signal is much more normal than the pure sine wave of Fig. 17–25. Sampling may not catch all the details, especially because the waveform is changing very fast in time.

FIGURE 17–27 (a) Original analog signal and (b) the same signal dirtied up by outside signals (= noise). (c) A digital signal is still readable (d) without error if the noise is not too great.

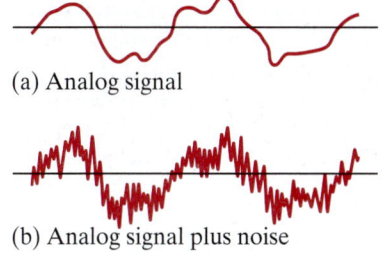

(a) Analog signal

(b) Analog signal plus noise

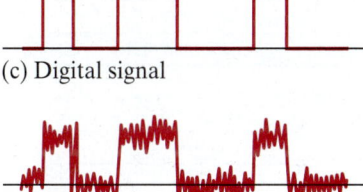

(c) Digital signal

(d) Digital signal plus noise

*17–11 TV and Computer Monitors: CRTs, Flat Screens

The first television receivers used a **cathode ray tube** (**CRT**), and as recently as 2008 they accounted for half of all new TV sales. Two years later it was tough to find a new CRT set to buy. Even though new TV sets are flat screen plasma or **liquid crystal displays** (**LCD**), an understanding of how a CRT works is useful.

*CRT

The operation of a CRT depends on **thermionic emission**, discovered by Thomas Edison (1847–1931). Consider a voltage applied to two small electrodes inside an evacuated glass "tube" as shown in Fig. 17–28: the **cathode** is negative, and the **anode** is positive. If the cathode is heated (usually by an electric current) so that it becomes hot and glowing, it is found that negative charges leave the cathode and flow to the positive anode. These negative charges are now called electrons, but originally they were called **cathode rays** because they seemed to come from the cathode (more detail in Section 27–1 on the discovery of the electron).

Figure 17–29 is a simplified sketch of a CRT which is contained in an evacuated glass tube. A beam of electrons, emitted by the heated cathode, is accelerated by the high-voltage anode and passes through a small hole in that anode. The inside of the tube face on the right (the screen) is coated with a fluorescent material that glows at the spot where the electrons hit. Voltage applied across the horizontal and vertical deflection plates, Fig. 17–29, can be varied to deflect the electron beam to different spots on the screen.

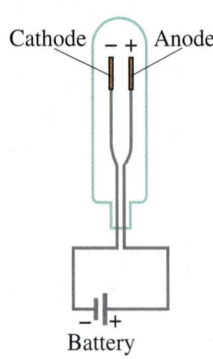

FIGURE 17–28 If the cathode inside the evacuated glass tube is heated to glowing (by an electric current, not shown), negatively charged "cathode rays" (= electrons) are "boiled off" and flow across to the anode (+), to which they are attracted.

PHYSICS APPLIED
CRT

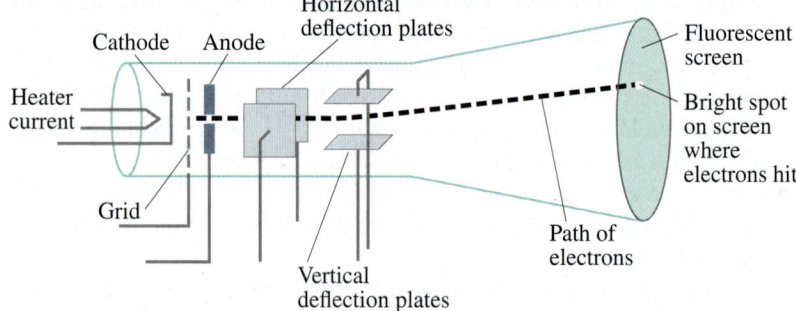

FIGURE 17–29 A cathode-ray tube. Magnetic deflection coils are commonly used in place of the electric deflection plates shown here. The relative positions of the elements have been exaggerated for clarity.

PHYSICS APPLIED
TV and computer monitors

FIGURE 17–30 Electron beam sweeps across a CRT television screen in a succession of horizontal lines, referred to as a **raster**. Each horizontal sweep is made by varying the voltage on the horizontal deflection plates (Fig. 17–29). Then the electron beam is moved down a short distance by a change in voltage on the vertical deflection plates, and the process is repeated.

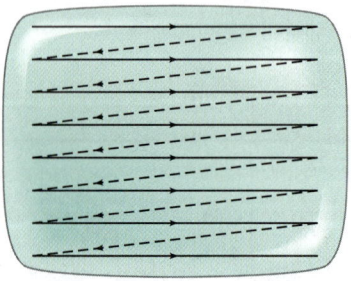

In TV and computer monitors, the CRT electron beam sweeps over the screen in the manner shown in Fig. 17–30 by carefully synchronized voltages applied to the deflection plates (more commonly by magnetic deflection coils—Chapter 20). During each horizontal sweep of the electron beam, the **grid** (Fig. 17–29) receives a signal voltage that limits the flow of electrons at each instant during the sweep; the more negative the grid voltage is, the more electrons are repelled and fewer pass through, producing a less bright spot on the screen. Thus the varying grid voltage is responsible for the brightness of each spot on the screen. At the end of each horizontal sweep of the electron beam, the horizontal deflection voltage changes dramatically to bring the beam back to the opposite side of the screen, and the vertical voltage changes slightly so the beam begins a new horizontal sweep slightly below the previous one. The difference in brightness of the spots on the screen forms the "picture." **Color screens** have red, green, and blue phosphors which glow when struck by the electron beam. The various brightnesses of adjacent red, green, and blue phosphors (so close together we don't distinguish them) produce almost any color. Analog TV for the U.S. provided 480 visible horizontal sweeps[†] to form a complete picture every $\frac{1}{30}$ s. With 30 new frames or pictures every second (25 in countries with 50-Hz line voltage), a "moving picture" is displayed on the TV screen. (Note: commercial movies on film are 24 frames per second.)

[†]525 lines in total, but only 480 form the picture; the other 45 lines contain other information such as synchronization. The sweep is **interlaced**: that is, every $\frac{1}{60}$ s every other line is traced, and in the next $\frac{1}{60}$ s, the lines in between are traced.

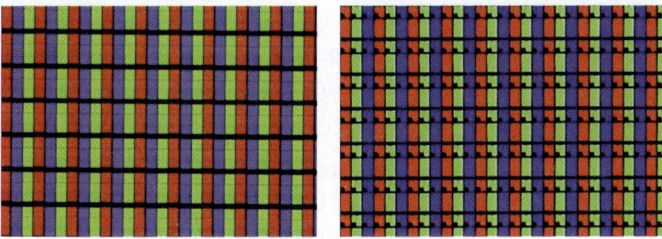

FIGURE 17–31 Close up of a tiny section of two typical LCD screens. You can even make out wires and transistors in the one on the right.

*Flat Screens and Addressing Pixels

Today's flat screens contain millions of tiny *picture elements*, or **pixels**. Each pixel consists of 3 **subpixels**, a red, a green, and a blue. A close up of a common arrangement of pixels is shown in Fig. 17–31 for an LCD screen. (How liquid crystals work in an LCD screen is described in Section 24–11.) Subpixels are so small that at normal viewing distances we don't distinguish them and the separate red (R), green (G), and blue (B) subpixels blend to produce almost any color, depending on the relative brightnesses of the three subpixels. Liquid crystals act as filters (R, G, and B) that filter the light from a white **backlight**, usually fluorescent lamps or *light-emitting diodes* (LED, Section 29–9).[†] The picture you see on the screen depends on the level of brightness of each subpixel, as suggested in Fig. 17–32 for a simple black and white picture.

High definition (HD) television screens have 1080 horizontal rows of pixels, each row consisting of 1920 pixels across the screen. That is, there are 1920 vertical columns, for a total of nearly 2 million pixels. Today, television in the U.S. is transmitted digitally at a rate of 60 Hz—that is, 60 frames or pictures per second (50 Hz in many countries) which makes the "moving picture." To form one frame, each subpixel must have the correct brightness. We now describe one way of doing this.

The brightness of each LCD subpixel (Section 24–11) depends on the voltage between its front and its back: if this voltage ΔV is zero, that subpixel is at maximum brightness; if ΔV is at its maximum (which might be +5 volts), that subpixel is dark.

Giving the correct voltage (to provide the correct brightness) is called **addressing** the subpixel. Typically the front of the subpixel is maintained at a positive voltage, such as +5 V. On the back of the display, the voltage at each subpixel is provided at the intersection of the 1080 horizontal wires (rows) and 1920 × 3 (colors) ≈ 6000 vertical wires (columns). See Fig. 17–33, which shows the array, or **matrix**, of wires. Each intersection of one vertical and one horizontal wire lies behind one subpixel. Because many frames are shown per second, the signal voltages applied are brief, like a pulse (see Fig. 17-22b or 11-23).

PHYSICS APPLIED
How flat screens work

FIGURE 17–32 Example of an image made up of many small squares or *pixels* (picture elements). This one has rather low resolution.

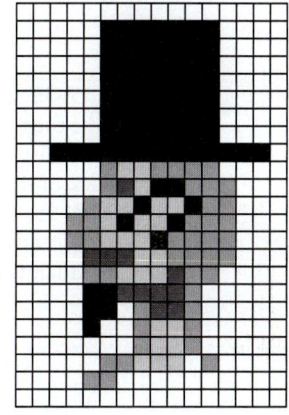

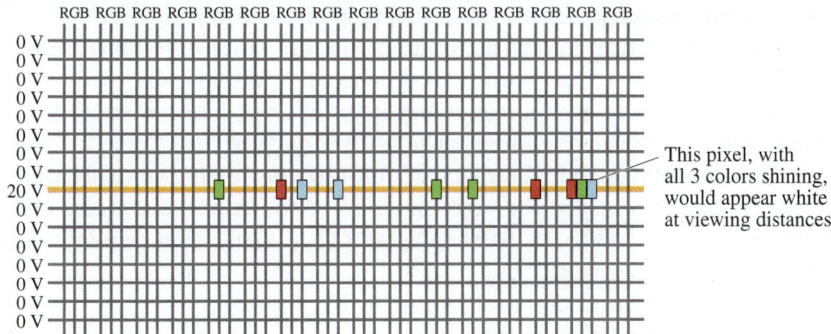

FIGURE 17–33 Array of wires (a matrix) behind all the pixels on an LCD screen. Each intersection of two wires is at a subpixel (red, green, or blue). One horizontal wire is activated at a time (the orange one at the moment shown) meaning it is at a positive voltage (+20 V) which allows that one row of pixels to be addressed at that moment; all other horizontal wires are at 0 V. At this moment, the data stream arrives to all the vertical wires, presenting the needed voltage (between 0 and 5 V) to produce the correct brightness for each of the nearly 6000 subpixels along the activated row.

The video signal that arrives at the display **activates** only one horizontal wire at a time (the orange one in Fig. 17–33): that one horizontal wire has a voltage (let's say +20 V) whereas all the others are at 0 V. That 20 V is not applied directly to the pixels, but *allows* the vertical wires to apply briefly the proper "signal voltage" to each subpixel along that row (via a transistor, see below). These signal voltages, known as the **data stream**, are applied to all the vertical wires just as that one row is activated: they provide the correct brightness for each subpixel in that activated row. A few subpixels are highlighted in Fig. 17–33. Immediately afterward, the other rows are activated, one by one, until the entire frame has been completed (in $\frac{1}{60}$ s).

[†]LEDs are discussed in Section 29–9. Home TVs advertised as LED generally mean an LCD screen with an LED backlight. LED pixels small enough for home screens are difficult to make, but actual LED screens are found in very large displays such as at stadiums.

*SECTION 17–11 491

Then a new frame is started. The addressing of subpixels for each row of each frame serves the same purpose as the sweep of the electron beam in a CRT, Fig. 17–30.

*Active Matrix (advanced)

High-definition displays use an **active matrix**, meaning that a tiny **thin-film transistor** (**TFT**) is attached to a corner of the back of each subpixel. (Transistors are discussed in Section 29–10.) One electrode of each TFT, called the "source," is connected to the vertical wire which addresses that subpixel, Fig. 17–34, and the "drain" electrode is connected to the back of the subpixel. The horizontal wire that serves the subpixel is connected to the transistor's **gate** electrode. The gate's voltage, by attracting charge or not, functions as a switch to connect or disconnect the source voltage to the drain and to the back of the subpixel (its front is fixed at +5 V). The potential difference ΔV across a subpixel determines if that subpixel will be bright in color ($\Delta V = 0$), black (ΔV = maximum), or something in between. See Fig. 17–35. All the subpixel TFTs along the one activated horizontal wire (the orange one in Fig. 17–33) will have +20 V at the gate: the TFTs are turned "on," like a switch. That allows electric charge to flow, connecting the vertical wire signal voltage at each TFT source to its drain and to the back of the subpixel. Thus all subpixels along one row receive the brightness needed for that line of the frame.

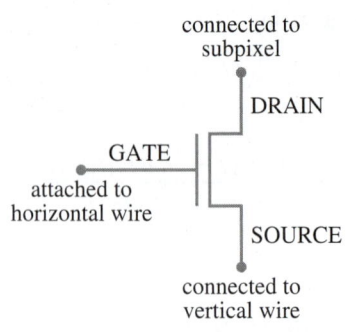

FIGURE 17–34 Thin-film transistor. One is attached to each screen subpixel.

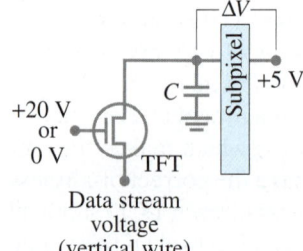

FIGURE 17–35 Circuit diagram for one subpixel. The front of the subpixel is at +5 V. If the TFT gate is at 20 V (horizontal wire activated), the data stream voltage is applied to the back of the subpixel, and determines the brightness of that subpixel. If the gate is at 0 V (horizontal wire not activated), the TFT is "off": no charge passes through, and the capacitance helps maintain ΔV until the subpixel is updated $\frac{1}{60}$ s later.

Within a subpixel's electronics is a capacitance that helps maintain the ΔV until that subpixel is **updated** with a new signal for the next frame, $\frac{1}{60}$ s later (≈ 17 ms). The row below the orange one shown in Fig. 17–33 is activated about 15 μs later $\left[= \left(\frac{1}{60}\text{s}\right)\left(\frac{1}{1080 \text{ lines}}\right)\right]$. The 6000 vertical wires (**data lines**) get their signal voltages (data stream) updated just before each row is activated in order to establish the brightness of each subpixel in that next row. All 1080 rows are activated, one-by-one, within $\frac{1}{60}$ s (≈ 17 ms) to complete that frame. Then a new frame is started.

New TV sets today can often refresh the screen at a higher rate. A **refresh rate** of 120 Hz (or 240 Hz) means that frames are interpolated between the normal ones, by averaging, which produces less blurring in fast action scenes.

Digital TV is transmitted at about 19 Mb/s (19 megabits/s) as mentioned in Section 17–10. (This rate is way too slow to do a full refresh every $\frac{1}{60}$ s—try the calculation and see—so a lot of compression is done and the areas where most movement occurs get refreshed.) The TV set or "box" that receives the digital video signal has to decode the signal in order to send analog voltages to the pixels of the screen, and at just the right time. TV stations in the U.S. are allowed to broadcast HD at 1080 × 1920 pixels or at 720 × 1280, or in standard definition (SD) of 480 × 704 pixels.

[When you read 1080p or 1080i for a TV, the "p" stands for "progressive," meaning an entire frame is made in $\frac{1}{60}$ s as described above. The "i" stands for "interlaced," meaning all the odd rows (half the picture) are done in $\frac{1}{60}$ s and then all the even rows are done in the next $\frac{1}{60}$ s, so a full picture is done at 30 per second or 30 Hz, thus reducing the data (or bit) rate. Analog TV (US) was 480i.]

*Oscilloscopes

An **oscilloscope** is a device for amplifying, measuring, and visually displaying an electrical signal as a function of time on an LCD or CRT monitor, or computer screen. The visible "trace" on the screen, which could be an electrocardiogram (Fig. 17–36), or a signal from an experiment on nerve conduction, is a plot of the signal voltage (vertically) versus time (horizontally). [In a CRT, the electron beam is swept horizontally at a uniform rate in time by the horizontal deflection plates, Figs. 17–29 and 17–30. The signal to be displayed is applied (after amplification) to the vertical deflection plates.]

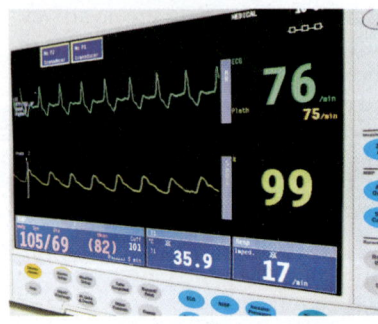

FIGURE 17–36 An electrocardiogram (ECG) trace displayed on a CRT.

PHYSICS APPLIED
Oscilloscope

*17–12 Electrocardiogram (ECG or EKG)

Each time the heart beats, changes in electrical potential occur on its surface that can be detected using *electrodes* (metal contacts), which are attached to the skin. The changes in potential are small, on the order of millivolts (mV), and must be amplified. They are displayed with a chart recorder on paper, or on a monitor (CRT or LCD), Fig. 17–36. An **electrocardiogram** (ECG or EKG) is the record of the potential changes for a given person's heart. An example is shown in Fig. 17–37. We now look at the source of these potential changes and their relation to heart activity.

PHYSICS APPLIED
Electrocardiogram

FIGURE 17–37 Typical ECG. Two heart beats are shown.

Both muscle and nerve cells have an electric dipole layer across the cell wall. That is, in the normal situation there is a net positive charge on the exterior surface and a net negative charge on the interior surface, Fig. 17–38a. The amount of charge depends on the size of the cell, but is approximately 10^{-3} C/m² of surface. For a cell whose surface area is 10^{-5} m², the total charge on either surface is thus $\approx 10^{-8}$ C. Just before the contraction of heart muscles, changes occur in the cell wall, so that positive ions on the exterior of the cell are able to pass through the wall and neutralize charge on the inside, or even make the inside surface slightly positive compared to the exterior. This "depolarization" starts at one end of the cell and progresses toward the opposite end, as indicated by the arrow in Fig. 17–38b, until the whole muscle is depolarized; the muscle then repolarizes to its original state (Fig. 17–38a), all in less than a second. Figure 17–38c shows rough graphs of the potential V as a function of time at the two points P and P′ (on either side of this cell) as the depolarization moves across the cell. The path of depolarization within the heart as a whole is more complicated, and produces the complex potential difference as a function of time, Fig. 17–37.

It is standard procedure to divide a typical electrocardiogram into regions corresponding to the various deflections (or "waves"), as shown in Fig. 17–37. Each of the deflections corresponds to the activity of a particular part of the heart beat (Fig. 10–42). The P wave corresponds to contraction of the atria. The QRS group corresponds to contraction of the ventricles as the depolarization follows a very complicated path. The T wave corresponds to recovery (repolarization) of the heart in preparation for the next cycle.

The ECG is a powerful tool in identifying heart defects. For example, the right side of the heart enlarges if the right ventricle must push against an abnormally large load (as when blood vessels become hardened or clogged). This problem is readily observed on an ECG, because the S wave becomes very large (negatively). *Infarcts*, which are dead regions of the heart muscle that result from heart attacks, are also detected on an ECG because they reflect the depolarization wave.

FIGURE 17–38 Heart muscle cell showing (a) charge dipole layer in resting state; (b) depolarization of cell progressing as muscle begins to contract; and (c) potential V at points P and P′ as a function of time.

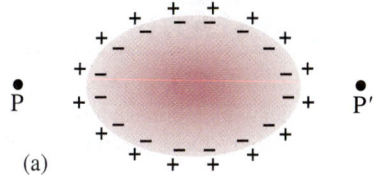

(a)

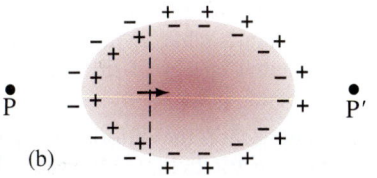

(b)

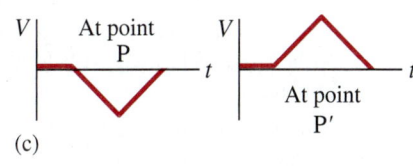

(c)

Summary

The **electric potential** V at any point in space is defined as the electric potential energy per unit charge:

$$V_a = \frac{PE_a}{q}. \quad (17\text{–}2a)$$

The **electric potential difference** between any two points is defined as the work done to move a 1 C electric charge between the two points. Potential difference is measured in volts (1 V = 1 J/C) and is often referred to as **voltage**.

The change in potential energy when a charge q moves through a potential difference V_{ba} is

$$\Delta PE = qV_{ba}. \quad (17\text{–}3)$$

The potential difference V_{ba} between two points a and b where a uniform electric field E exists is given by

$$V_{ba} = -Ed, \quad (17\text{–}4a)$$

where d is the distance between the two points.

An **equipotential line** or **surface** is all at the same potential, and is perpendicular to the electric field at all points.

The electric potential at a position P due to a single point charge Q, relative to zero potential at infinity, is given by

$$V = \frac{kQ}{r}, \quad (17\text{–}5)$$

where r is the distance from Q to position P and $k = 1/4\pi\epsilon_0$.

[*The potential due to an **electric dipole** drops off as $1/r^2$. The **dipole moment** is $p = Q\ell$, where ℓ is the distance between the two equal but opposite charges of magnitude Q.]

A **capacitor** is a device used to store charge (and electric energy), and consists of two nontouching conductors. The two conductors hold equal and opposite charges, of magnitude Q. The ratio of this charge Q to the potential difference V between the conductors is called the **capacitance**, C:

$$C = \frac{Q}{V}, \quad \text{or} \quad Q = CV. \quad (17\text{-}7)$$

The capacitance of a parallel-plate capacitor is proportional to the area A of each plate and inversely proportional to their separation d:

$$C = \epsilon_0 \frac{A}{d}. \quad (17\text{-}8)$$

The space between the two conductors of a capacitor contains a nonconducting material such as air, paper, or plastic. These materials are referred to as **dielectrics**, and the capacitance is proportional to a property of dielectrics called the **dielectric constant**, K (equal to 1 for air).

A charged capacitor stores an amount of electric energy given by

$$\text{PE} = \tfrac{1}{2}QV = \tfrac{1}{2}CV^2 = \tfrac{1}{2}\frac{Q^2}{C}. \quad (17\text{-}10)$$

This energy can be thought of as stored in the electric field between the plates.

The energy stored in any electric field E has a density

$$\frac{\text{electric PE}}{\text{volume}} = \tfrac{1}{2}\epsilon_0 E^2. \quad (17\text{-}11)$$

Digital electronics converts an analog **signal voltage** into an approximate digital voltage based on a **binary code**: each **bit** has two possibilities, 1 or 0 (also "on" or "off"). The binary number 1101 equals 13. A **byte** is 8 bits and provides $2^8 = 256$ voltage levels. **Sampling rate** is the number of voltage measurements done on the analog signal per second. The **bit depth** is the number of digital voltage levels available at each sampling. CDs are 44.1 kHz, 16-bit.

[*TV and computer monitors traditionally used a **cathode ray tube** (CRT) which accelerates electrons by high voltage, and sweeps them across the screen in a regular way using magnetic coils or electric deflection plates. **LCD flat screens** contain millions of **pixels**, each with a red, green, and blue **subpixel** whose brightness is addressed every $\frac{1}{60}$ s via a **matrix** of horizontal and vertical wires using a **digital** (**binary**) code.]

[*An **electrocardiogram** (ECG or EKG) records the potential changes of each heart beat as the cells depolarize and repolarize.]

Questions

1. If two points are at the same potential, does this mean that no net work is done in moving a test charge from one point to the other? Does this imply that no force must be exerted? Explain.

2. If a negative charge is initially at rest in an electric field, will it move toward a region of higher potential or lower potential? What about a positive charge? How does the potential energy of the charge change in each instance? Explain.

3. State clearly the difference (a) between electric potential and electric field, (b) between electric potential and electric potential energy.

4. An electron is accelerated from rest by a potential difference of 0.20 V. How much greater would its final speed be if it is accelerated with four times as much voltage? Explain.

5. Is there a point along the line joining two equal positive charges where the electric field is zero? Where the electric potential is zero? Explain.

6. Can a particle ever move from a region of low electric potential to one of high potential and yet have its electric potential energy decrease? Explain.

7. If $V = 0$ at a point in space, must $\vec{E} = 0$? If $\vec{E} = 0$ at some point, must $V = 0$ at that point? Explain. Give examples for each.

8. Can two equipotential lines cross? Explain.

9. Draw in a few equipotential lines in Fig. 16–32b and c.

10. When a battery is connected to a capacitor, why do the two plates acquire charges of the same magnitude? Will this be true if the two plates are different sizes or shapes?

11. A conducting sphere carries a charge Q and a second identical conducting sphere is neutral. The two are initially isolated, but then they are placed in contact. (a) What can you say about the potential of each when they are in contact? (b) Will charge flow from one to the other? If so, how much?

12. The parallel plates of an isolated capacitor carry opposite charges, Q. If the separation of the plates is increased, is a force required to do so? Is the potential difference changed? What happens to the work done in the pulling process?

13. If the electric field $\vec{E}$ is uniform in a region, what can you infer about the electric potential V? If V is uniform in a region of space, what can you infer about $\vec{E}$?

14. Is the electric potential energy of two isolated unlike charges positive or negative? What about two like charges? What is the significance of the sign of the potential energy in each case?

15. If the voltage across a fixed capacitor is doubled, the amount of energy it stores (a) doubles; (b) is halved; (c) is quadrupled; (d) is unaffected; (e) none of these. Explain.

16. How does the energy stored in a capacitor change when a dielectric is inserted if (a) the capacitor is isolated so Q does not change; (b) the capacitor remains connected to a battery so V does not change? Explain.

17. A dielectric is pulled out from between the plates of a capacitor which remains connected to a battery. What changes occur to (a) the capacitance, (b) the charge on the plates, (c) the potential difference, (d) the energy stored in the capacitor, and (e) the electric field? Explain your answers.

18. We have seen that the capacitance C depends on the size and position of the two conductors, as well as on the dielectric constant K. What then did we mean when we said that C is a constant in Eq. 17-7?

MisConceptual Questions

1. A $+0.2\,\mu C$ charge is in an electric field. What happens if that charge is replaced by a $+0.4\,\mu C$ charge?
   (a) The electric potential doubles, but the electric potential energy stays the same.
   (b) The electric potential stays the same, but the electric potential energy doubles.
   (c) Both the electric potential and electric potential energy double.
   (d) Both the electric potential and electric potential energy stay the same.

2. Two identical positive charges are placed near each other. At the point halfway between the two charges,
   (a) the electric field is zero and the potential is positive.
   (b) the electric field is zero and the potential is zero.
   (c) the electric field is not zero and the potential is positive.
   (d) the electric field is not zero and the potential is zero.
   (e) None of these statements is true.

3. Four identical point charges are arranged at the corners of a square [Hint: Draw a figure]. The electric field E and potential V at the center of the square are
   (a) $E = 0, V = 0$.
   (b) $E = 0, V \neq 0$.
   (c) $E \neq 0, V \neq 0$.
   (d) $E \neq 0, V = 0$.
   (e) $E = V$ regardless of the value.

4. Which of the following statements is valid?
   (a) If the potential at a particular point is zero, the field at that point must be zero.
   (b) If the field at a particular point is zero, the potential at that point must be zero.
   (c) If the field throughout a particular region is constant, the potential throughout that region must be zero.
   (d) If the potential throughout a particular region is constant, the field throughout that region must be zero.

5. If it takes an amount of work W to move two $+q$ point charges from infinity to a distance d apart from each other, then how much work should it take to move three $+q$ point charges from infinity to a distance d apart from each other?
   (a) $2W$.
   (b) $3W$.
   (c) $4W$.
   (d) $6W$.

6. A proton ($Q = +e$) and an electron ($Q = -e$) are in a constant electric field created by oppositely charged plates. You release the proton from near the positive plate and the electron from near the negative plate. Which feels the larger electric force?
   (a) The proton.
   (b) The electron.
   (c) Neither—there is no force.
   (d) The magnitude of the force is the same for both and in the same direction.
   (e) The magnitude of the force is the same for both but in opposite directions.

7. When the proton and electron in MisConceptual Question 6 strike the opposite plate, which one has more kinetic energy?
   (a) The proton.
   (b) The electron.
   (c) Both acquire the same kinetic energy.
   (d) Neither—there is no change in kinetic energy.
   (e) They both acquire the same kinetic energy but with opposite signs.

8. Which of the following do not affect capacitance?
   (a) Area of the plates.
   (b) Separation of the plates.
   (c) Material between the plates.
   (d) Charge on the plates.
   (e) Energy stored in the capacitor.

9. A battery establishes a voltage V on a parallel-plate capacitor. After the battery is disconnected, the distance between the plates is doubled without loss of charge. Accordingly, the capacitance _____ and the voltage between the plates _____.
   (a) increases; decreases.
   (b) decreases; increases.
   (c) increases; increases.
   (d) decreases; decreases.
   (e) stays the same; stays the same.

10. Which of the following is a vector?
    (a) Electric potential.
    (b) Electric potential energy.
    (c) Electric field.
    (d) Equipotential lines.
    (e) Capacitance.

11. A $+0.2\,\mu C$ charge is in an electric field. What happens if that charge is replaced by a $-0.2\,\mu C$ charge?
    (a) The electric potential changes sign, but the electric potential energy stays the same.
    (b) The electric potential stays the same, but the electric potential energy changes sign.
    (c) Both the electric potential and electric potential energy change sign.
    (d) Both the electric potential and electric potential energy stay the same.

Problems

17–1 to 17–4 Electric Potential

1. (I) How much work does the electric field do in moving a $-7.7\,\mu C$ charge from ground to a point whose potential is $+65$ V higher?

2. (I) How much work does the electric field do in moving a proton from a point at a potential of $+125$ V to a point at -45 V? Express your answer both in joules and electron volts.

3. (I) What potential difference is needed to stop an electron that has an initial velocity $v = 6.0 \times 10^5$ m/s?

4. (I) How much kinetic energy will an electron gain (in joules and eV) if it accelerates through a potential difference of 18,500 V?

5. (I) An electron acquires 6.45×10^{-16} J of kinetic energy when it is accelerated by an electric field from plate A to plate B. What is the potential difference between the plates, and which plate is at the higher potential?

6. (I) How strong is the electric field between two parallel plates 6.8 mm apart if the potential difference between them is 220 V?

7. (I) An electric field of 525 V/m is desired between two parallel plates 11.0 mm apart. How large a voltage should be applied?

8. (I) The electric field between two parallel plates connected to a 45-V battery is 1900 V/m. How far apart are the plates?

9. (I) What potential difference is needed to give a helium nucleus ($Q = 2e$) 85.0 keV of kinetic energy?

10. (II) Two parallel plates, connected to a 45-V power supply, are separated by an air gap. How small can the gap be if the air is not to become conducting by exceeding its breakdown value of $E = 3 \times 10^6$ V/m?

11. (II) The work done by an external force to move a $-6.50\,\mu C$ charge from point A to point B is 15.0×10^{-4} J. If the charge was started from rest and had 4.82×10^{-4} J of kinetic energy when it reached point B, what must be the potential difference between A and B?

12. (II) What is the speed of an electron with kinetic energy (a) 850 eV, and (b) 0.50 keV?

13. (II) What is the speed of a proton whose KE is 4.2 keV?

14. (II) An alpha particle (which is a helium nucleus, $Q = +2e$, $m = 6.64 \times 10^{-27}$ kg) is emitted in a radioactive decay with KE = 5.53 MeV. What is its speed?

15. (II) An electric field greater than about 3×10^6 V/m causes air to break down (electrons are removed from the atoms and then recombine, emitting light). See Section 17–2 and Table 17–3. If you shuffle along a carpet and then reach for a doorknob, a spark flies across a gap you estimate to be 1 mm between your finger and the doorknob. Estimate the voltage between your finger and the doorknob. Why is no harm done?

16. (II) An electron starting from rest acquires 4.8 keV of KE in moving from point A to point B. (a) How much KE would a proton acquire, starting from rest at B and moving to point A? (b) Determine the ratio of their speeds at the end of their respective trajectories.

17. (II) Draw a conductor in the oblong shape of a football. This conductor carries a net negative charge, $-Q$. Draw in a dozen or so electric field lines and equipotential lines.

17–5 Potential Due to Point Charges

[Let $V = 0$ at $x = \infty$.]

18. (I) What is the electric potential 15.0 cm from a 3.00 μC point charge?

19. (I) A point charge Q creates an electric potential of $+165$ V at a distance of 15 cm. What is Q?

20. (II) A $+35\,\mu C$ point charge is placed 46 cm from an identical $+35\,\mu C$ charge. How much work would be required to move a $+0.50\,\mu C$ test charge from a point midway between them to a point 12 cm closer to either of the charges?

21. (II) (a) What is the electric potential 2.5×10^{-15} m away from a proton (charge $+e$)? (b) What is the electric potential energy of a system that consists of two protons 2.5×10^{-15} m apart—as might occur inside a typical nucleus?

22. (II) Three point charges are arranged at the corners of a square of side ℓ as shown in Fig. 17–39. What is the potential at the fourth corner (point A)?

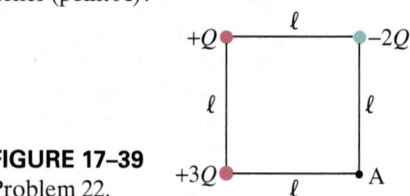

FIGURE 17–39 Problem 22.

23. (II) An electron starts from rest 24.5 cm from a fixed point charge with $Q = -6.50$ nC. How fast will the electron be moving when it is very far away?

24. (II) Two identical $+9.5\,\mu C$ point charges are initially 5.3 cm from each other. If they are released at the same instant from rest, how fast will each be moving when they are very far away from each other? Assume they have identical masses of 1.0 mg.

25. (II) Two point charges, $3.0\,\mu C$ and $-2.0\,\mu C$, are placed 4.0 cm apart on the x axis. At what points along the x axis is (a) the electric field zero and (b) the potential zero?

26. (II) How much work must be done to bring three electrons from a great distance apart to 1.0×10^{-10} m from one another (at the corners of an equilateral triangle)?

27. (II) Point a is 62 cm north of a $-3.8\,\mu C$ point charge, and point b is 88 cm west of the charge (Fig. 17–40). Determine (a) $V_b - V_a$ and (b) $\vec{E}_b - \vec{E}_a$ (magnitude and direction).

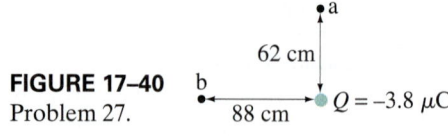

FIGURE 17–40 Problem 27.

28. (II) Many chemical reactions release energy. Suppose that at the beginning of a reaction, an electron and proton are separated by 0.110 nm, and their final separation is 0.100 nm. How much electric potential energy was lost in this reaction (in units of eV)?

29. (III) How much voltage must be used to accelerate a proton (radius 1.2×10^{-15} m) so that it has sufficient energy to just "touch" a silicon nucleus? A silicon nucleus has a charge of $+14e$, and its radius is about 3.6×10^{-15} m. Assume the potential is that for point charges.

30. (III) Two equal but opposite charges are separated by a distance d, as shown in Fig. 17–41. Determine a formula for $V_{BA} = V_B - V_A$ for points B and A on the line between the charges situated as shown.

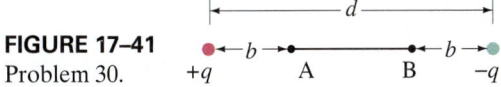

FIGURE 17–41 Problem 30.

31. (III) In the Bohr model of the hydrogen atom, an electron orbits a proton (the nucleus) in a circular orbit of radius 0.53×10^{-10} m. (a) What is the electric potential at the electron's orbit due to the proton? (b) What is the kinetic energy of the electron? (c) What is the total energy of the electron in its orbit? (d) What is the *ionization energy*—that is, the energy required to remove the electron from the atom and take it to $r = \infty$, at rest? Express the results of parts (b), (c), and (d) in joules and eV.

*17–6 Electric Dipoles

*32. (I) An electron and a proton are 0.53×10^{-10} m apart. What is their dipole moment if they are at rest?

*33. (II) Calculate the electric potential due to a dipole whose dipole moment is 4.2×10^{-30} C·m at a point 2.4×10^{-9} m away if this point is (a) along the axis of the dipole nearer the positive charge; (b) 45° above the axis but nearer the positive charge; (c) 45° above the axis but nearer the negative charge.

*34. (III) The dipole moment, considered as a vector, points from the negative to the positive charge. The water molecule, Fig. 17–42, has a dipole moment $\vec{p}$ which can be considered as the vector sum of the two dipole moments, $\vec{p}_1$ and $\vec{p}_2$, as shown. The distance between each H and the O is about 0.96×10^{-10} m. The lines joining the center of the O atom with each H atom make an angle of 104°, as shown, and the net dipole moment has been measured to be $p = 6.1 \times 10^{-30}$ C·m. Determine the charge q on each H atom.

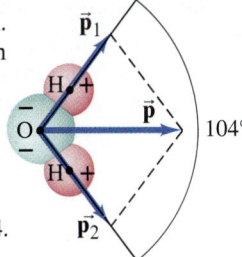

FIGURE 17–42 Problem 34. A water molecule, H₂O.

17–7 Capacitance

35. (I) The two plates of a capacitor hold $+2500\ \mu C$ and $-2500\ \mu C$ of charge, respectively, when the potential difference is 960 V. What is the capacitance?

36. (I) An 8500-pF capacitor holds plus and minus charges of 16.5×10^{-8} C. What is the voltage across the capacitor?

37. (I) How much charge flows from each terminal of a 12.0-V battery when it is connected to a 5.00-μF capacitor?

38. (I) A 0.20-F capacitor is desired. What area must the plates have if they are to be separated by a 3.2-mm air gap?

39. (II) The charge on a capacitor increases by 15 μC when the voltage across it increases from 97 V to 121 V. What is the capacitance of the capacitor?

40. (II) An electric field of 8.50×10^5 V/m is desired between two parallel plates, each of area 45.0 cm² and separated by 2.45 mm of air. What charge must be on each plate?

41. (II) If a capacitor has opposite 4.2 μC charges on the plates, and an electric field of 2.0 kV/mm is desired between the plates, what must each plate's area be?

42. (II) It takes 18 J of energy to move a 0.30-mC charge from one plate of a 15-μF capacitor to the other. How much charge is on each plate?

43. (II) To get an idea how big a farad is, suppose you want to make a 1-F air-filled parallel-plate capacitor for a circuit you are building. To make it a reasonable size, suppose you limit the plate area to 1.0 cm². What would the gap have to be between the plates? Is this practically achievable?

44. (II) How strong is the electric field between the plates of a 0.80-μF air-gap capacitor if they are 2.0 mm apart and each has a charge of 62 μC?

45. (III) A 2.50-μF capacitor is charged to 746 V and a 6.80-μF capacitor is charged to 562 V. These capacitors are then disconnected from their batteries. Next the positive plates are connected to each other and the negative plates are connected to each other. What will be the potential difference across each and the charge on each? [*Hint*: Charge is conserved.]

46. (III) A 7.7-μF capacitor is charged by a 165-V battery (Fig. 17–43a) and then is disconnected from the battery. When this capacitor (C_1) is then connected (Fig. 17–43b) to a second (initially uncharged) capacitor, C_2, the final voltage on each capacitor is 15 V. What is the value of C_2? [*Hint*: Charge is conserved.]

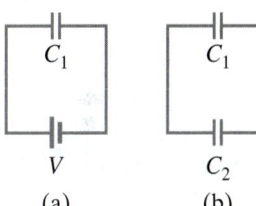

FIGURE 17–43 Problems 46 and 58. (a) (b)

17–8 Dielectrics

47. (I) What is the capacitance of two square parallel plates 6.6 cm on a side that are separated by 1.8 mm of paraffin?

48. (I) What is the capacitance of a pair of circular plates with a radius of 5.0 cm separated by 2.8 mm of mica?

49. (II) An uncharged capacitor is connected to a 21.0-V battery until it is fully charged, after which it is disconnected from the battery. A slab of paraffin is then inserted between the plates. What will now be the voltage between the plates?

50. (II) A 3500-pF air-gap capacitor is connected to a 32-V battery. If a piece of mica is placed between the plates, how much charge will flow from the battery?

51. (II) The electric field between the plates of a paper-separated ($K = 3.75$) capacitor is 8.24×10^4 V/m. The plates are 1.95 mm apart, and the charge on each is 0.675 μC. Determine the capacitance of this capacitor and the area of each plate.

17–9 Electric Energy Storage

52. (I) 650 V is applied to a 2800-pF capacitor. How much energy is stored?

53. (I) A cardiac defibrillator is used to shock a heart that is beating erratically. A capacitor in this device is charged to 5.0 kV and stores 1200 J of energy. What is its capacitance?

54. (II) How much energy is stored by the electric field between two square plates, 8.0 cm on a side, separated by a 1.5-mm air gap? The charges on the plates are equal and opposite and of magnitude 370 μC.

55. (II) A homemade capacitor is assembled by placing two 9-in. pie pans 4 cm apart and connecting them to the opposite terminals of a 9-V battery. Estimate (a) the capacitance, (b) the charge on each plate, (c) the electric field halfway between the plates, and (d) the work done by the battery to charge them. (e) Which of the above values change if a dielectric is inserted?

56. (II) A parallel-plate capacitor has fixed charges $+Q$ and $-Q$. The separation of the plates is then halved. (a) By what factor does the energy stored in the electric field change? (b) How much work must be done to reduce the plate separation from d to $\frac{1}{2}d$? The area of each plate is A.

57. (II) There is an electric field near the Earth's surface whose magnitude is about 150 V/m. How much energy is stored per cubic meter in this field?

58. (III) A 3.70-μF capacitor is charged by a 12.0-V battery. It is disconnected from the battery and then connected to an uncharged 5.00-μF capacitor (Fig. 17–43). Determine the total stored energy (a) before the two capacitors are connected, and (b) after they are connected. (c) What is the change in energy?

17–10 Digital

59. (I) Write the decimal number 116 in binary.

60. (I) Write the binary number 01010101 as a decimal number.

61. (I) Write the binary number 1010101010101010 as a decimal number.

62. (II) Consider a rather coarse 4-bit analog-to-digital conversion where the maximum voltage is 5.0 V. (a) What voltage does 1011 represent? (b) What is the 4-bit representation for 2.0 V?

63. (II) (a) 16-bit sampling provides how many different possible voltages? (b) 24-bit sampling provides how many different possible voltages? (c) For color TV, 3 subpixels, each 8 bits, provides a total of how many different colors?

64. (II) A few extraterrestrials arrived. They had two hands, but claimed that $3 + 2 = 11$. How many fingers did they have on their two hands? Note that our decimal system (and ten characters: 0, 1, 2, $\cdots$, 9) surely has its origin because we have ten fingers. [*Hint*: 11 is in their system. In our decimal system, the result would be written as 5.]

*17–11 TV and Computer Monitors

*65. (II) Figure 17–44 is a photograph of a computer screen shot by a camera set at an exposure time of $\frac{1}{4}$ s. During the exposure the cursor arrow was moved around by the mouse, and we see it 15 times. (a) Explain why we see the cursor 15 times. (b) What is the refresh rate of the screen?

FIGURE 17–44 Problem 65.

*66. (III) In a given CRT, electrons are accelerated horizontally by 9.0 kV. They then pass through a uniform electric field E for a distance of 2.8 cm, which deflects them upward so they travel 22 cm to the top of the screen, 11 cm above the center. Estimate the value of E.

*67. (III) Electrons are accelerated by 6.0 kV in a CRT. The screen is 30 cm wide and is 34 cm from the 2.6-cm-long deflection plates. Over what range must the horizontally deflecting electric field vary to sweep the beam fully across the screen?

General Problems

68. A lightning flash transfers 4.0 C of charge and 5.2 MJ of energy to the Earth. (a) Across what potential difference did it travel? (b) How much water could this boil and vaporize, starting from room temperature? (See also Chapter 14.)

69. In an older television tube, electrons are accelerated by thousands of volts through a vacuum. If a television set were laid on its back, would electrons be able to move upward against the force of gravity? What potential difference, acting over a distance of 2.4 cm, would be needed to balance the downward force of gravity so that an electron would remain stationary? Assume that the electric field is uniform.

70. How does the energy stored in a capacitor change, as the capacitor remains connected to a battery, if the separation of the plates is doubled?

71. How does the energy stored in an isolated capacitor change if (a) the potential difference is doubled, or (b) the separation of the plates is doubled?

72. A huge 4.0-F capacitor has enough stored energy to heat 2.8 kg of water from 21°C to 95°C. What is the potential difference across the plates?

73. A proton ($q = +e$) and an alpha particle ($q = +2e$) are accelerated by the same voltage V. Which gains the greater kinetic energy, and by what factor?

74. Dry air will break down if the electric field exceeds 3.0×10^6 V/m. What amount of charge can be placed on a parallel-plate capacitor if the area of each plate is 65 cm^2?

75. Three charges are at the corners of an equilateral triangle (side ℓ) as shown in Fig. 17–45. Determine the potential at the midpoint of each of the sides. Let $V = 0$ at $r = \infty$.

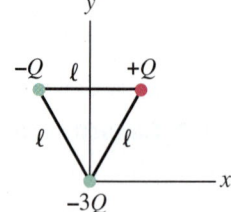

FIGURE 17–45 Problem 75.

76. It takes 15.2 J of energy to move a 13.0-mC charge from one plate of a 17.0-μF capacitor to the other. How much charge is on each plate? Assume constant voltage.

77. A 3.4 μC and a −2.6 μC charge are placed 2.5 cm apart. At what points along the line joining them is (a) the electric field zero, and (b) the electric potential zero?

78. Near the surface of the Earth there is an electric field of about 150 V/m which points downward. Two identical balls with mass $m = 0.670$ kg are dropped from a height of 2.00 m, but one of the balls is positively charged with $q_1 = 650$ μC, and the second is negatively charged with $q_2 = -650$ μC. Use conservation of energy to determine the difference in the speed of the two balls when they hit the ground. (Neglect air resistance.)

79. The power supply for a pulsed nitrogen laser has a 0.050-μF capacitor with a maximum voltage rating of 35 kV. (a) Estimate how much energy could be stored in this capacitor. (b) If 12% of this stored electrical energy is converted to light energy in a pulse that is 6.2 microseconds long, what is the power of the laser pulse?

80. In a **photocell**, ultraviolet (UV) light provides enough energy to some electrons in barium metal to eject them from the surface at high speed. To measure the maximum energy of the electrons, another plate above the barium surface is kept at a negative enough potential that the emitted electrons are slowed down and stopped, and return to the barium surface. See Fig. 17–46. If the plate voltage is −3.02 V (compared to the barium) when the fastest electrons are stopped, what was the speed of these electrons when they were emitted?

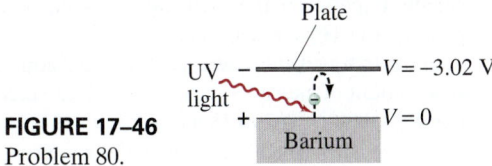

FIGURE 17–46
Problem 80.

81. A +38 μC point charge is placed 36 cm from an identical +38 μC charge. A −1.5 μC charge is moved from point A to point B as shown in Fig. 17–47. What is the change in potential energy?

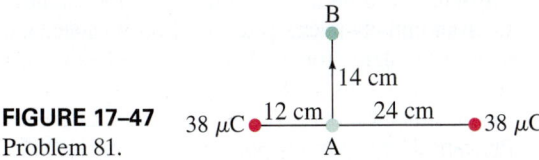

FIGURE 17–47
Problem 81.

82. Paper has a dielectric constant $K = 3.7$ and a dielectric strength of 15×10^6 V/m. Suppose that a typical sheet of paper has a thickness of 0.11 mm. You make a "homemade" capacitor by placing a sheet of 21×14 cm paper between two aluminum foil sheets (Fig. 17–48) of the same size. (a) What is the capacitance C of your device? (b) About how much charge could you store on your capacitor before it would break down?

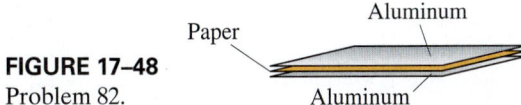

FIGURE 17–48
Problem 82.

83. A capacitor is made from two 1.1-cm-diameter coins separated by a 0.10-mm-thick piece of paper ($K = 3.7$). A 12-V battery is connected to the capacitor. How much charge is on each coin?

84. A +3.5 μC charge is 23 cm to the right of a −7.2 μC charge. At the midpoint between the two charges, (a) determine the potential and (b) the electric field.

85. A parallel-plate capacitor with plate area 3.0 cm² and air-gap separation 0.50 mm is connected to a 12-V battery, and fully charged. The battery is then disconnected. (a) What is the charge on the capacitor? (b) The plates are now pulled to a separation of 0.75 mm. What is the charge on the capacitor now? (c) What is the potential difference between the plates now? (d) How much work was required to pull the plates to their new separation?

86. A 2.1-μF capacitor is fully charged by a 6.0-V battery. The battery is then disconnected. The capacitor is not ideal and the charge slowly leaks out from the plates. The next day, the capacitor has lost half its stored energy. Calculate the amount of charge lost.

87. Two point charges are fixed 4.0 cm apart from each other. Their charges are $Q_1 = Q_2 = 6.5$ μC, and their masses are $m_1 = 1.5$ mg and $m_2 = 2.5$ mg. (a) If Q_1 is released from rest, what will be its speed after a very long time? (b) If both charges are released from rest at the same time, what will be the speed of Q_1 after a very long time?

88. Two charges are placed as shown in Fig. 17–49 with $q_1 = 1.2$ μC and $q_2 = -3.3$ μC. Find the potential difference between points A and B.

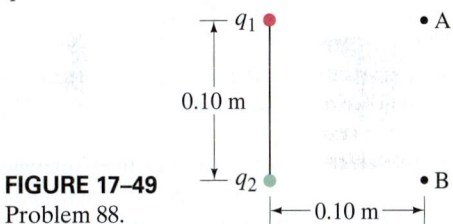

FIGURE 17–49
Problem 88.

89. If the electrons in a single raindrop, 3.5 mm in diameter, could be removed from the Earth (without removing the atomic nuclei), by how much would the potential of the Earth increase?

90. Thunderclouds may develop a voltage difference of about 5×10^7 V. Given that an electric field of 3×10^6 V/m is required to produce an electrical spark within a volume of air, estimate the length of a thundercloud lightning bolt. [Can you see why, when lightning strikes from a cloud to the ground, the bolt has to propagate as a sequence of steps?]

91. A manufacturer claims that a carpet will not generate more than 6.0 kV of static electricity. What magnitude of charge would have to be transferred between a carpet and a shoe for there to be a 6.0-kV potential difference between the shoe and the carpet? Approximate the area of the shoe and assume the shoe and carpet are large sheets of charge separated by a small distance $d = 1.0$ mm.

92. Compact "ultracapacitors" with capacitance values up to several thousand farads are now commercially available. One application for ultracapacitors is in providing power for electrical circuits when other sources (such as a battery) are turned off. To get an idea of how much charge can be stored in such a component, assume a 1200-F ultracapacitor is initially charged to 12.0 V by a battery and is then disconnected from the battery. If charge is then drawn off the plates of this capacitor at a rate of 1.0 mC/s, say, to power the backup memory of some electrical device, how long (in days) will it take for the potential difference across this capacitor to drop to 6.0 V?

93. An electron is accelerated horizontally from rest by a potential difference of 2200 V. It then passes between two horizontal plates 6.5 cm long and 1.3 cm apart that have a potential difference of 250 V (Fig. 17–50). At what angle θ will the electron be traveling after it passes between the plates?

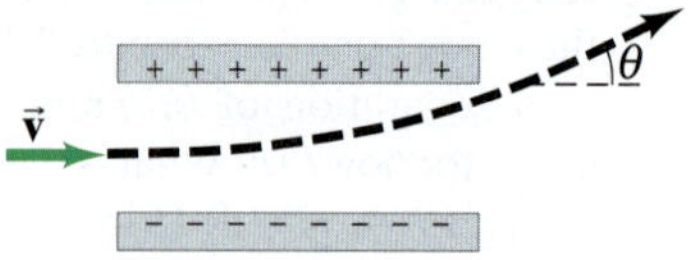

FIGURE 17–50 Problem 93.

94. In the **dynamic random access memory (DRAM)** of a computer, each memory cell contains a capacitor for charge storage. Each of these cells represents a single binary-bit value of "1" when its 35-fF capacitor $(1\ \text{fF} = 10^{-15}\ \text{F})$ is charged at 1.5 V, or "0" when uncharged at 0 V. (a) When fully charged, how many excess electrons are on a cell capacitor's negative plate? (b) After charge has been placed on a cell capacitor's plate, it slowly "leaks" off at a rate of about 0.30 fC/s. How long does it take for the potential difference across this capacitor to decrease by 2.0% from its fully charged value? (Because of this leakage effect, the charge on a DRAM capacitor is "refreshed" many times per second.) Note: A DRAM cell is shown in Fig. 21–29.

95. In the DRAM computer chip of Problem 94, suppose the two parallel plates of one cell's 35-fF capacitor are separated by a 2.0-nm-thick insulating material with dielectric constant $K = 25$. (a) Determine the area A (in μm^2) of the cell capacitor's plates. (b) If the plate area A accounts for half of the area of each cell, estimate how many megabytes of memory can be placed on a 3.0-cm^2 silicon wafer. (1 byte = 8 bits.)

96. A parallel-plate capacitor with plate area $A = 2.0\ m^2$ and plate separation $d = 3.0\ mm$ is connected to a 35-V battery (Fig. 17–51a). (a) Determine the charge on the capacitor, the electric field, the capacitance, and the energy stored in the capacitor. (b) With the capacitor still connected to the battery, a slab of plastic with dielectric strength $K = 3.2$ is placed between the plates of the capacitor, so that the gap is completely filled with the dielectric (Fig. 17–51b). What are the new values of charge, electric field, capacitance, and the energy stored in the capacitor?

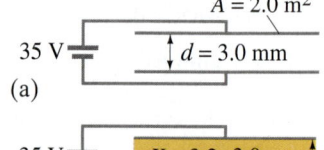

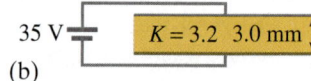

FIGURE 17–51 Problem 96.

Search and Learn

1. Make a list of rules for and properties of equipotential surfaces or lines. You should be able to find eight distinct rules in the text.
2. Figure 17–8 shows contour lines (elevations). Just for fun, assume they are equipotential lines on a flat 2-dimensional surface with the values shown being in volts. Estimate the magnitude and direction of the "electric field" (a) between Iceberg Lake and Cecile Lake and (b) at the Minaret Mine. Assume that up is $+y$, right is $+x$, and that Cecile Lake is about 1.0 km wide in the middle.
3. In lightning storms, the potential difference between the Earth and the bottom of thunderclouds may be 35,000,000 V. The bottoms of the thunderclouds are typically 1500 m above the Earth, and can have an area of 110 km^2. Modeling the Earth–cloud system as a huge capacitor, calculate (a) the capacitance of the Earth–cloud system, (b) the charge stored in the "capacitor," and (c) the energy stored in the "capacitor."
4. The potential energy stored in a capacitor (Section 17–9) can be written as either $CV^2/2$ or $Q^2/2C$. In the first case the energy is proportional to C; in the second case the energy is proportional to $1/C$. (a) Explain how both of these equations can be correct. (b) When might you use the first equation and when might you use the second equation? (c) If a paper dielectric is inserted into a parallel-plate capacitor that is attached to a battery (V does not change), by what factor will the energy stored in the capacitor change? (d) If a quartz dielectric is inserted into a charged parallel-plate capacitor that is isolated from any battery, by what factor will the energy stored in the capacitor change?

5. Suppose it takes 75 kW of power for your car to travel at a constant speed on the highway. (a) What is this in horsepower? (b) How much energy in joules would it take for your car to travel at highway speed for 5.0 hours? (c) Suppose this amount of energy is to be stored in the electric field of a parallel-plate capacitor (Section 17–9). If the voltage on the capacitor is to be 850 V, what is the required capacitance? (d) If this capacitor were to be made from activated carbon (Section 17–7), the voltage would be limited to no more than 10 V. In this case, how many grams of activated carbon would be required? (e) Is this practical?

6. Capacitors can be used as "electric charge counters." Consider an initially uncharged capacitor of capacitance C with its bottom plate grounded and its top plate connected to a source of electrons. (a) If N electrons flow onto the capacitor's top plate, show that the resulting potential difference V across the capacitor is directly proportional to N. (b) Assume the voltage-measuring device can accurately resolve voltage changes of about 1 mV. What value of C would be necessary to resolve the arrival of an individual electron? (c) Using modern semiconductor technology, a micron-size capacitor can be constructed with parallel conducting plates separated by an insulator of dielectric constant $K = 3$ and thickness $d = 100$ nm. What side length ℓ should the square plates have (in μm)?

ANSWERS TO EXERCISES

A: (a) -8.0×10^{-16} J; (b) 9.8×10^5 m/s.
B: (c).
C: 0.72 J.
D: (c).
E: A.
F: (a) 3 times greater; (b) 3 times greater.
G: 12 mF.

The glow of the thin wire filament of incandescent lightbulbs is caused by the electric current passing through it. Electric energy is transformed to thermal energy (via collisions between moving electrons and atoms of the wire), which causes the wire's temperature to become so high that it glows. In halogen lamps (tungsten–halogen), shown on the right, the tungsten filament is surrounded by a halogen gas such as bromine or iodine in a clear tube. Halogens, via chemical reactions, restore many of the tungsten atoms that were evaporated from the hot filament, allowing longer life, higher temperature (typically 2900 K versus 2700 K), better efficiency, and whiter light.

Electric current and electric power in electric circuits are of basic importance in everyday life. We examine both dc and ac in this Chapter, and include the microscopic analysis of electric current.

Electric Currents

CHAPTER 18

CHAPTER-OPENING QUESTION—Guess now!

The conductors shown are all made of copper and are at the same temperature. Which conductor would have the greatest resistance to the flow of charge entering from the left? Which would offer the least resistance?

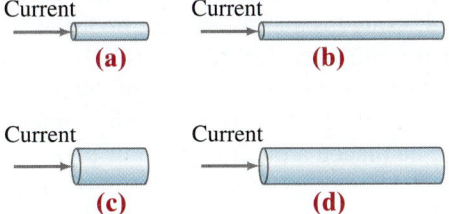

CONTENTS

- 18–1 The Electric Battery
- 18–2 Electric Current
- 18–3 Ohm's Law: Resistance and Resistors
- 18–4 Resistivity
- 18–5 Electric Power
- 18–6 Power in Household Circuits
- 18–7 Alternating Current
- *18–8 Microscopic View of Electric Current
- *18–9 Superconductivity
- *18–10 Electrical Conduction in the Human Nervous System

In the previous two Chapters we have been studying static electricity: electric charges at rest. In this Chapter we begin our study of charges in motion, and we call a flow of charge an electric current.

In everyday life we are familiar with electric currents in wires and other conductors. Most practical electrical devices depend on electric current: current through a lightbulb, current in the heating element of a stove, hair dryer, or electric heater, as well as currents in electronic devices. Electric currents can exist in conductors such as wires, but also in semiconductor devices, human cells and their membranes (Section 18–10), and in empty space.

In electrostatic situations, we saw in Section 16–9 that the electric field must be zero inside a conductor (if it weren't, the charges would move). But when charges are *moving* along a conductor, an electric field is needed to set charges into motion, and to keep them in motion against even low resistance in any normal conductor. We can control the flow of charge using electric fields and electric potential (voltage), concepts we have just been discussing. In order to have a current in a wire, a potential difference is needed, which can be provided by a battery.

We first look at electric current from a macroscopic point of view. Later in the Chapter we look at currents from a microscopic (theoretical) point of view as a flow of electrons in a wire.

18–1 The Electric Battery

Until the year 1800, the technical development of electricity consisted mainly of producing a static charge by friction. It all changed in 1800 when Alessandro Volta (1745–1827; Fig. 18–1) invented the electric battery, and with it produced the first steady flow of electric charge—that is, a steady electric current.

The events that led to the discovery of the battery are interesting. Not only was this an important discovery, but it also gave rise to a famous scientific debate.

In the 1780s, Luigi Galvani (1737–1798), professor at the University of Bologna, carried out a series of experiments on the contraction of a frog's leg muscle by using static electricity. Galvani found that the muscle also contracted when dissimilar metals were inserted into the frog. Galvani believed that the source of the electric charge was in the frog muscle or nerve itself, and that the metal merely transmitted the charge to the proper points. When he published his work in 1791, he termed this charge "animal electricity." Many wondered, including Galvani himself, if he had discovered the long-sought "life-force."

Volta, at the University of Pavia 200 km away, was skeptical of Galvani's results, and came to believe that the source of the electricity was not in the animal itself, but rather in the *contact between the dissimilar metals*. Volta realized that a moist conductor, such as a frog muscle or moisture at the contact point of two dissimilar metals, was necessary in the circuit if it was to be effective. He also saw that the contracting frog muscle was a sensitive instrument for detecting electric "tension" or "electromotive force" (his words for what we now call voltage), in fact more sensitive than the best available electroscopes that he and others had developed.[†]

FIGURE 18–1 Alessandro Volta. In this portrait, Volta demonstrates his battery to Napoleon in 1801.

Volta's research found that certain combinations of metals produced a greater effect than others, and, using his measurements, he listed them in order of effectiveness. (This "electrochemical series" is still used by chemists today.) He also found that carbon could be used in place of one of the metals.

Volta then conceived his greatest contribution to science. Between a disc of zinc and one of silver, he placed a piece of cloth or paper soaked in salt solution or dilute acid and piled a "battery" of such couplings, one on top of another, as shown in Fig. 18–2. This "pile" or "battery" produced a much increased potential difference. Indeed, when strips of metal connected to the two ends of the pile were brought close, a spark was produced. Volta had designed and built the first electric battery. He published his discovery in 1800.

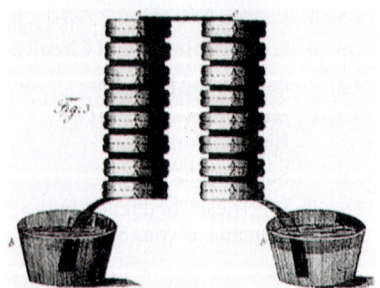

FIGURE 18–2 A voltaic battery, from Volta's original publication.

[†]Volta's most sensitive electroscope (Section 16–4) measured about 40 V per degree (angle of leaf separation). Nonetheless, he was able to estimate the potential differences produced by combinations of dissimilar metals in contact. For a silver–zinc contact he got about 0.7 V, remarkably close to today's value of 0.78 V.

Electric Cells and Batteries

A battery produces electricity by transforming chemical energy into electrical energy. Today a great variety of electric cells and batteries are available, from flashlight batteries to the storage battery of a car. The simplest batteries contain two plates or rods made of dissimilar metals (one can be carbon) called **electrodes**. The electrodes are immersed in a solution or paste, such as a dilute acid, called the **electrolyte**. Such a device is properly called an **electric cell**, and several cells connected together is a **battery**, although today even a single cell is called a battery. The chemical reactions involved in most electric cells are quite complicated. Here we describe how one very simple cell works, emphasizing the physical aspects.

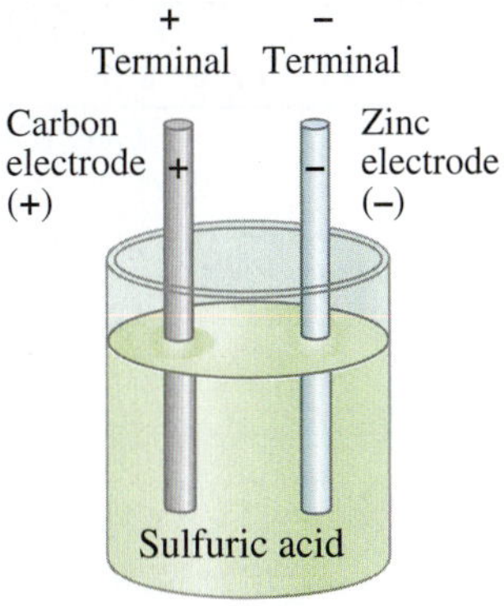

The cell shown in Fig. 18–3 uses dilute sulfuric acid as the electrolyte. One of the electrodes is made of carbon, the other of zinc. The part of each electrode outside the solution is called the **terminal**, and connections to wires and circuits are made here. The acid tends to dissolve the zinc electrode. Each zinc atom leaves two electrons behind on the electrode and enters the solution as a positive ion. The zinc electrode thus acquires a negative charge. The electrolyte becomes positively charged, and can pull electrons off the carbon electrode. Thus the carbon electrode becomes positively charged. Because there is an opposite charge on the two electrodes, there is a potential difference between the two terminals.

In a cell whose terminals are not connected, only a small amount of the zinc is dissolved, for as the zinc electrode becomes increasingly negative, any new positive zinc ions produced are attracted back to the electrode. Thus, a particular potential difference (or voltage) is maintained between the two terminals. If charge is allowed to flow between the terminals, say, through a wire (or a lightbulb), then more zinc can be dissolved. After a time, one or the other electrode is used up and the cell becomes "dead."

The voltage that exists between the terminals of a battery depends on what the electrodes are made of and their relative ability to be dissolved or give up electrons.

When two or more cells are connected so that the positive terminal of one is connected to the negative terminal of the next, they are said to be connected in *series* and their voltages add up. Thus, the voltage between the ends of two 1.5-V AA flashlight batteries connected in series is 3.0 V, whereas the six 2-V cells of an automobile storage battery give 12 V. Figure 18–4a shows a diagram of a common "dry cell" or "flashlight battery" used not only in flashlights but in many portable electronic devices, and Fig. 18–4b shows two smaller ones connected in series to a flashlight bulb. An incandescent lightbulb consists of a thin, coiled wire (filament) inside an evacuated glass bulb, as shown in Fig. 18–5 and in the Chapter-Opening Photos, page 501. When charge passes through the filament, it gets very hot (≈ 2800 K) and glows. Other bulb types, such as fluorescent, work differently.

FIGURE 18–3 Simple electric cell.

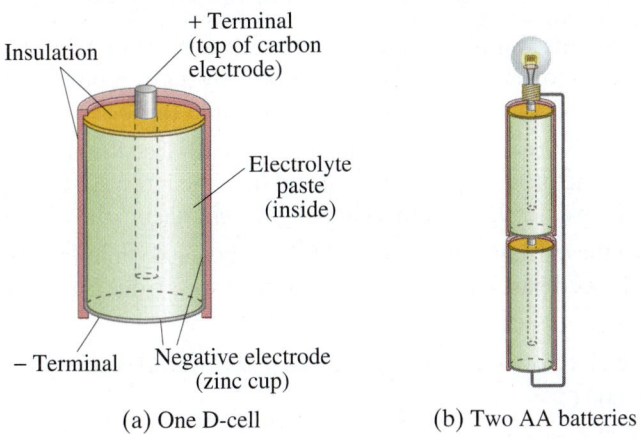

FIGURE 18–4 (a) Diagram of an ordinary dry cell (like a D-cell or AA). The cylindrical zinc cup is covered on the sides; its flat bottom is the negative terminal. (b) Two dry cells (AA type) connected in series. Note that the positive terminal of one cell pushes against the negative terminal of the other.

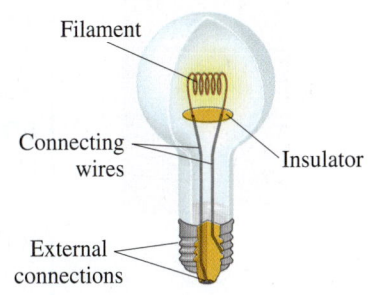

FIGURE 18–5 An ordinary incandescent lightbulb: the fine wire of the filament becomes so hot that it glows. Incandescent halogen bulbs enclose the filament in a small quartz tube filled with a halogen gas (bromine or iodine) which allows longer filament life and higher filament temperature for greater efficiency and whiteness.

SECTION 18–1 The Electric Battery 503

PHYSICS APPLIED
Electric cars

Electric Cars

Considerable research is being done to improve batteries for electric cars and for hybrids (which use both a gasoline internal combustion engine and an electric motor). One type of battery is lithium-ion, in which the anode contains lithium and the cathode is carbon. Electric cars need no gear changes and can develop full torque starting from rest, and so can accelerate quickly and smoothly. The distance an electric car can go between charges of the battery (its "range") is an important parameter because each recharging of an electric car battery may take hours, not minutes like a gas fill-up. Because charging an electric car can draw a large current over a period of several hours, electric power companies may need to upgrade their power grids so they won't fail when many electric cars are being charged at the same time in a small urban area.

18–2 Electric Current

The purpose of a battery is to produce a potential difference, which can then make charges move. When a continuous conducting path is connected between the terminals of a battery, we have an electric **circuit**, Fig. 18–6a. On any diagram of a circuit, as in Fig. 18–6b, we use the symbol

$$\underset{+\ -}{\dashv\vdash} \quad \text{or} \quad \dashv\vdash \qquad \text{[battery symbol]}$$

to represent a battery. The device connected to the battery could be a lightbulb, a heater, a radio, or some other device. When such a circuit is formed, charge can move (or flow) through the wires of the circuit, from one terminal of the battery to the other, as long as the conducting path is continuous. Any flow of charge such as this is called an **electric current**.

More precisely, the electric current in a wire is defined as the net amount of charge that passes through the wire's full cross section at any point per unit time. Thus, the current I is defined as

$$I = \frac{\Delta Q}{\Delta t}, \tag{18-1}$$

where ΔQ is the amount of charge that passes through the conductor at any location during the time interval Δt.

Electric current is measured in coulombs per second; this is given a special name, the **ampere** (abbreviated amp or A), after the French physicist André Ampère (1775–1836). Thus, 1 A = 1 C/s. Smaller units of current are often used, such as the milliampere $(1\,\text{mA} = 10^{-3}\,\text{A})$ and microampere $(1\,\mu\text{A} = 10^{-6}\,\text{A})$.

A current can flow in a circuit only if there is a *continuous* conducting path. We then have a **complete circuit**. If there is a break in the circuit, say, a cut wire, we call it an **open circuit** and no current flows. In any single circuit, with only a single path for current to follow such as in Fig. 18–6b, a steady current at any instant is the same at one point (say, point A) as at any other point (such as B). This follows from the conservation of electric charge: charge doesn't disappear. A battery does not create (or destroy) any net charge, nor does a lightbulb absorb or destroy charge.

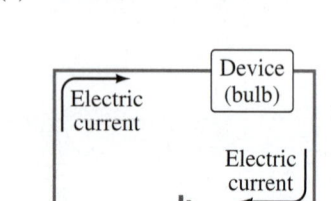

FIGURE 18–6 (a) A simple electric circuit. (b) Schematic drawing of the same circuit, consisting of a battery, connecting wires (thick gray lines), and a lightbulb or other device.

CAUTION
A battery does not create charge; a lightbulb does not destroy charge

> **EXAMPLE 18–1 Current is flow of charge.** A steady current of 2.5 A exists in a wire for 4.0 min. (a) How much total charge passes by a given point in the circuit during those 4.0 min? (b) How many electrons would this be?
>
> **APPROACH** (a) Current is flow of charge per unit time, Eq. 18–1, so the amount of charge passing a point is the product of the current and the time interval. (b) To get the number of electrons, we divide the total charge by the charge on one electron.
>
> **SOLUTION** (a) Since the current was 2.5 A, or 2.5 C/s, then in 4.0 min (= 240 s) the total charge that flowed past a given point in the wire was, from Eq. 18–1,
>
> $$\Delta Q = I\,\Delta t = (2.5\,\text{C/s})(240\,\text{s}) = 600\,\text{C}.$$
>
> (b) The charge on one electron is 1.60×10^{-19} C, so 600 C would consist of
>
> $$\frac{600\,\text{C}}{1.6 \times 10^{-19}\,\text{C/electron}} = 3.8 \times 10^{21}\,\text{electrons}.$$

EXERCISE A If 1 million electrons per second pass a point in a wire, what is the current?

CONCEPTUAL EXAMPLE 18–2 **How to connect a battery.** What is wrong with each of the schemes shown in Fig. 18–7 for lighting a flashlight bulb with a flashlight battery and a single wire?

RESPONSE (*a*) There is no closed path for charge to flow around. Charges might briefly start to flow from the battery toward the lightbulb, but there they run into a "dead end," and the flow would immediately come to a stop.
(*b*) Now there is a closed path passing to and from the lightbulb; but the wire touches only one battery terminal, so there is no potential difference in the circuit to make the charge move. Neither here, nor in (*a*), does the bulb light up.
(*c*) Nothing is wrong here. This is a complete circuit: charge can flow out from one terminal of the battery, through the wire and the bulb, and into the other terminal. This scheme will light the bulb.

In many real circuits, wires are connected to a common conductor that provides continuity. This common conductor is called **ground**, usually represented as ⏚ or ⏚, and really is connected to the ground for a building or house. In a car, one terminal of the battery is called "ground," but is not connected to the earth itself—it is connected to the frame of the car, as is one connection to each lightbulb and other devices. Thus the car frame is a conductor in each circuit, ensuring a continuous path for charge flow, and is called "ground" for the car's circuits. (Note that the car frame is well insulated from the earth by the rubber tires.)

We saw in Chapter 16 that conductors contain many free electrons. Thus, if a continuous conducting wire is connected to the terminals of a battery, negatively charged electrons flow in the wire. When the wire is first connected, the potential difference between the terminals of the battery sets up an electric field inside the wire and parallel to it. Free electrons at one end of the wire are attracted into the positive terminal, and at the same time other electrons enter the other end of the wire at the negative terminal of the battery. There is a continuous flow of electrons throughout the wire that begins as soon as the wire is connected to *both* terminals.

When the conventions of positive and negative charge were invented two centuries ago, however, it was assumed that positive charge flowed in a wire. For nearly all purposes, positive charge flowing in one direction is exactly equivalent to negative charge flowing in the opposite direction, as shown in Fig. 18–8. Today, we still use the historical convention of positive charge flow when discussing the direction of a current. So when we speak of the current direction in a circuit, we mean the direction positive charge would flow. This is sometimes referred to as **conventional current**. When we want to speak of the direction of electron flow, we will specifically state it is the electron current. In liquids and gases, both positive and negative charges (ions) can move.

In practical life, such as rating the total charge of a car battery, you may see the unit **ampere-hour** (A·h): from Eq. 18–1, $\Delta Q = I \Delta t$.

EXERCISE B How many coulombs is 1.00 A·h?

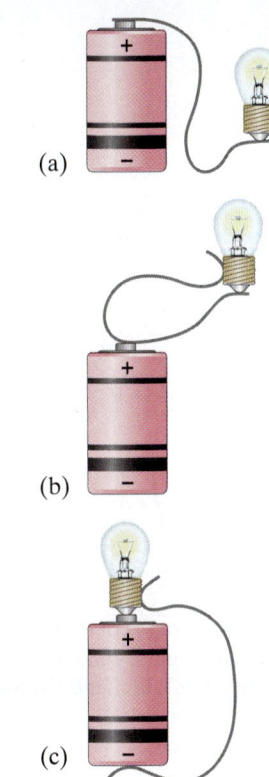

FIGURE 18–7 Example 18–2.

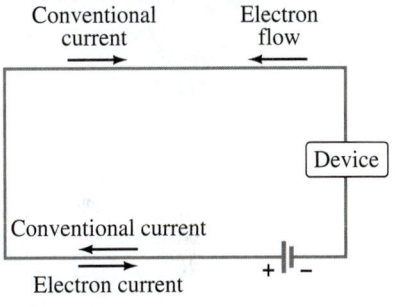

FIGURE 18–8 Conventional current from + to − is equivalent to a negative electron flow from − to +.

CAUTION
Distinguish conventional current from electron flow

18–3 Ohm's Law: Resistance and Resistors

To produce an electric current in a circuit, a difference in potential is required. One way of producing a potential difference along a wire is to connect its ends to the opposite terminals of a battery. It was Georg Simon Ohm (1787–1854) who established experimentally that the current in a metal wire is proportional to the potential difference V applied to its two ends:

$$I \propto V.$$

If, for example, we connect a wire to the two terminals of a 6-V battery, the current in the wire will be twice what it would be if the wire were connected to a 3-V battery. It is also found that reversing the sign of the voltage does not affect the magnitude of the current.

Exactly how large the current is in a wire depends not only on the voltage between its ends, but also on the resistance the wire offers to the flow of electrons. Electron flow is impeded because of collisions with the atoms of the wire.† We define electrical **resistance** R as the proportionality factor between the voltage V (between the ends of the wire) and the current I (passing through the wire):

$$V = IR. \qquad (18\text{-}2)$$

Ohm found experimentally that in metal conductors R is a constant independent of V, a result known as **Ohm's law**. Equation 18-2, $V = IR$, is itself sometimes called Ohm's law, but only when referring to materials or devices for which R is a constant independent of V. But R is not a constant for many substances other than metals, nor for devices such as diodes, vacuum tubes, transistors, and so on. Even for metals, R is not constant if the temperature changes much: for a lightbulb filament the measured resistance is low for small currents, but is much higher at the filament's normal large operating current that puts it at the high temperature needed to make it glow ($\approx 3000\,\text{K}$). Thus Ohm's "law" is not a fundamental law of nature, but rather a description of a certain class of materials: metal conductors, whose temperature does not change much. Such materials are said to be "ohmic." Materials or devices that do not follow Ohm's law are said to be *nonohmic*. See Fig. 18-9.

The unit for resistance is called the **ohm** and is abbreviated Ω (Greek capital letter omega). Because $R = V/I$, we see that $1.0\,\Omega$ is equivalent to $1.0\,\text{V/A}$.

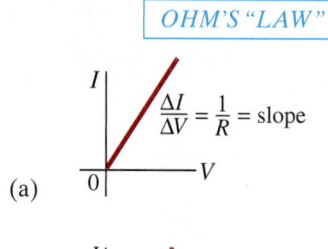
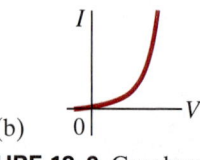

FIGURE 18–9 Graphs of current vs. voltage (a) for a metal conductor which obeys Ohm's law, and (b) for a nonohmic device, in this case a semiconductor diode.

FIGURE 18–10 Flashlight (Example 18-3). Note how the circuit is completed along the side strip.

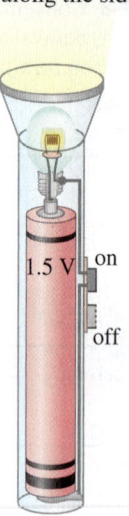

EXAMPLE 18–3 Flashlight bulb resistance. A small flashlight bulb (Fig. 18-10) draws 300 mA from its 1.5-V battery. (*a*) What is the resistance of the bulb? (*b*) If the battery becomes weak and the voltage drops to 1.2 V, how would the current change? Assume the bulb is approximately ohmic.

APPROACH We apply Ohm's law to the bulb, where the voltage applied across it is the battery voltage.

SOLUTION (*a*) We change 300 mA to 0.30 A and use Eq. 18-2:

$$R = \frac{V}{I} = \frac{1.5\,\text{V}}{0.30\,\text{A}} = 5.0\,\Omega.$$

(*b*) If the resistance stays the same, the current would be

$$I = \frac{V}{R} = \frac{1.2\,\text{V}}{5.0\,\Omega} = 0.24\,\text{A} = 240\,\text{mA},$$

or a decrease of 60 mA.

NOTE With the smaller current in part (*b*), the bulb filament's temperature would be lower and the bulb less bright. Also, resistance does depend on temperature (Section 18-4), so our calculation is only a rough approximation.

EXERCISE C What is the resistance of a lightbulb if 0.50 A flows through it when 120 V is connected across it?

All electric devices, from heaters to lightbulbs to stereo amplifiers, offer resistance to the flow of current. The filaments of lightbulbs (Fig. 18-5) and electric heaters are special types of wires whose resistance results in their becoming very hot. Generally, the connecting wires have very low resistance in comparison to the resistance of the wire filaments or coils, so the connecting wires usually have a minimal effect on the magnitude of the current.

---

†A useful analogy compares the flow of electric charge in a wire to the flow of water in a river, or in a pipe, acted on by gravity. If the river (or pipe) is nearly level, the flow rate is small. But if one end is somewhat higher than the other, the water flow rate—or current—is greater. The greater the difference in height, the swifter the current. We saw in Chapter 17 that electric potential is analogous to the height of a cliff for gravity. Just as an increase in height can cause a greater flow of water, so a greater electric potential difference, or voltage, causes a greater electric current. Resistance in a wire is analogous to rocks in a river that retard water flow.

In many circuits, particularly in electronic devices, **resistors** are used to control the amount of current. Resistors have resistances ranging from less than an ohm to millions of ohms (see Figs. 18–11 and 18–12). The main types are "wire-wound" resistors which consist of a coil of fine wire, "composition" resistors which are usually made of carbon, resistors made of thin carbon or metal films, and (on tiny integrated circuit "chips") undoped semiconductors.

When we draw a diagram of a circuit, we use the symbol

—⋀⋀⋀— [resistor symbol]

to indicate a resistance. Wires whose resistance is negligible, however, are shown simply as straight lines. Figure 18–12 and its Table show one way to specify the resistance of a resistor.

FIGURE 18–11 Photo of resistors (striped), plus other devices on a circuit board.

Resistor Color Code

Color	Number	Multiplier	Tolerance
Black	0	1	
Brown	1	10^1	1%
Red	2	10^2	2%
Orange	3	10^3	
Yellow	4	10^4	
Green	5	10^5	
Blue	6	10^6	
Violet	7	10^7	
Gray	8	10^8	
White	9	10^9	
Gold		10^{-1}	5%
Silver		10^{-2}	10%
No color			20%

FIGURE 18–12 The resistance value of a given resistor is written on the exterior, or may be given as a color code as shown below and in the Table: the first two colors represent the first two digits in the value of the resistance, the third color represents the power of ten that it must be multiplied by, and the fourth is the manufactured tolerance. For example, a resistor whose four colors are red, green, yellow, and silver has a resistance of $25 \times 10^4 \, \Omega = 250{,}000 \, \Omega = 250 \, \text{k}\Omega$, plus or minus 10%. [An alternative code is a number such as 104, which means $R = 1.0 \times 10^4 \, \Omega$.]

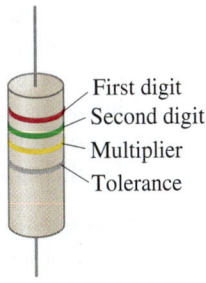

First digit
Second digit
Multiplier
Tolerance

CONCEPTUAL EXAMPLE 18–4 **Current and potential.** Current I enters a resistor R as shown in Fig. 18–13. (a) Is the potential higher at point A or at point B? (b) Is the current greater at point A or at point B?

RESPONSE (a) Positive charge always flows from + to −, from high potential to low potential. So if current I is conventional (positive) current, point A is at a higher potential than point B.

(b) Conservation of charge requires that whatever charge flows into the resistor at point A, an equal amount of charge emerges at point B. Charge or current does not get "used up" by a resistor. So the current is the same at A and B.

FIGURE 18–13 Example 18–4.

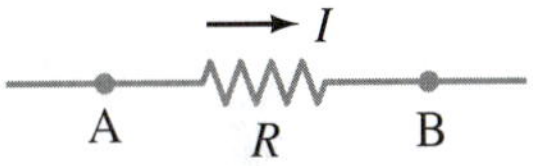

An electric potential decrease, as from point A to point B in Example 18–4, is often called a **potential drop** or a **voltage drop**.

Some Helpful Clarifications

Here we briefly summarize some possible misunderstandings and clarifications. Batteries do not put out a constant current. Instead, batteries are intended to maintain a constant potential difference, or very nearly so. (Details in the next Chapter.) Thus a battery should be considered a source of voltage. The voltage is applied *across* a wire or device.

Electric current passes *through* a wire or device (connected to a battery), and its magnitude depends on that device's resistance. The resistance is a *property* of the wire or device. The voltage, on the other hand, is external to the wire or device, and is applied across the two ends of the wire or device. The current through the device might be called the "response": the current increases if the voltage increases or the resistance decreases, as $I = V/R$.

⚠ **CAUTION**
Voltage is applied across *a device; current passes* through *a device*

Current is *not* a vector, even though current does have a direction. In a thin wire, the direction of the current is always parallel to the wire at each point, no matter how the wire curves, just like water in a pipe. The direction of conventional (positive) current is from high potential (+) toward lower potential (−).

Current and charge do not increase or decrease or get "used up" when going through a wire or other device. The amount of charge that goes in at one end comes out at the other end.

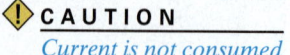
CAUTION
Current is not consumed

18–4 Resistivity

It is found experimentally that the resistance R of a uniform wire is directly proportional to its length ℓ and inversely proportional to its cross-sectional area A. That is,

$$R = \rho \frac{\ell}{A}, \tag{18-3}$$

where ρ (Greek letter "rho"), the constant of proportionality, is called the **resistivity** and depends on the material used. Typical values of ρ, whose units are $\Omega \cdot m$ (see Eq. 18–3), are given for various materials in the middle column of Table 18–1 which is divided into the categories *conductors*, *insulators*, and *semiconductors* (Section 16–3). The values depend somewhat on purity, heat treatment, temperature, and other factors. Notice that silver has the lowest resistivity and is thus the best conductor (although it is expensive). Copper is close, and much less expensive, which is why most wires are made of copper. Aluminum, although it has a higher resistivity, is much less dense than copper; it is thus preferable to copper in some situations, such as for transmission lines, because its resistance for the same weight is less than that for copper.[†]

EXERCISE D Return to the Chapter-Opening Question, page 501, and answer it again now. Try to explain why you may have answered differently the first time.

[†]The reciprocal of the resistivity, called the **electrical conductivity**, is $\sigma = 1/\rho$ and has units of $(\Omega \cdot m)^{-1}$.

TABLE 18–1 Resistivity and Temperature Coefficients (at 20°C)

Material	Resistivity, ρ ($\Omega \cdot$ m)	Temperature Coefficient, α (C°)$^{-1}$
Conductors		
Silver	1.59×10^{-8}	0.0061
Copper	1.68×10^{-8}	0.0068
Gold	2.44×10^{-8}	0.0034
Aluminum	2.65×10^{-8}	0.00429
Tungsten	5.6×10^{-8}	0.0045
Iron	9.71×10^{-8}	0.00651
Platinum	10.6×10^{-8}	0.003927
Mercury	98×10^{-8}	0.0009
Nichrome (Ni, Fe, Cr alloy)	100×10^{-8}	0.0004
Semiconductors[‡]		
Carbon (graphite)	$(3-60) \times 10^{-5}$	−0.0005
Germanium	$(1-500) \times 10^{-3}$	−0.05
Silicon	0.1–60	−0.07
Insulators		
Glass	$10^9 - 10^{12}$	
Hard rubber	$10^{13} - 10^{15}$	

[‡] Values depend strongly on the presence of even slight amounts of impurities.

EXERCISE E A copper wire has a resistance of 10 Ω. What would its resistance be if it had the same diameter but was only half as long? (a) 20 Ω, (b) 10 Ω, (c) 5 Ω, (d) 1 Ω, (e) none of these.

EXAMPLE 18–5 Speaker wires. Suppose you want to connect your stereo to remote speakers (Fig. 18–14). (a) If each wire must be 20 m long, what diameter copper wire should you use to keep the resistance less than 0.10 Ω per wire? (b) If the current to each speaker is 4.0 A, what is the potential difference, or voltage drop, across each wire?

APPROACH We solve Eq. 18–3 to get the area A, from which we can calculate the wire's radius using $A = \pi r^2$. The diameter is $2r$. In (b) we can use Ohm's law, $V = IR$.

SOLUTION (a) We solve Eq. 18–3 for the area A and find ρ for copper in Table 18–1:

$$A = \rho \frac{\ell}{R} = \frac{(1.68 \times 10^{-8}\,\Omega\cdot\text{m})(20\,\text{m})}{(0.10\,\Omega)} = 3.4 \times 10^{-6}\,\text{m}^2.$$

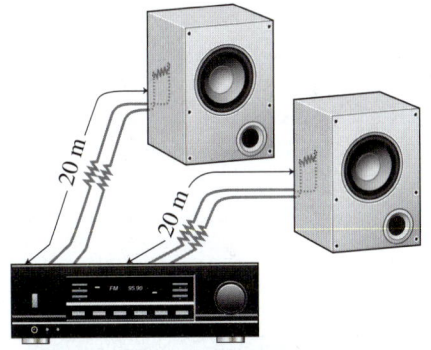

FIGURE 18–14 Example 18–5.

The cross-sectional area A of a circular wire is $A = \pi r^2$. The radius must then be at least

$$r = \sqrt{\frac{A}{\pi}} = 1.04 \times 10^{-3}\,\text{m} = 1.04\,\text{mm}.$$

The diameter is twice the radius and so must be at least $2r = 2.1$ mm.
(b) From $V = IR$ we find that the voltage drop across each wire is

$$V = IR = (4.0\,\text{A})(0.10\,\Omega) = 0.40\,\text{V}.$$

NOTE The voltage drop across the wires reduces the voltage that reaches the speakers from the stereo amplifier, thus reducing the sound level a bit.

CONCEPTUAL EXAMPLE 18–6 Stretching changes resistance. Suppose a wire of resistance R could be stretched uniformly until it was twice its original length. What would happen to its resistance? Assume the amount of material, and therefore its volume, doesn't change.

RESPONSE If the length ℓ doubles, then the cross-sectional area A is halved, because the volume $(V = A\ell)$ of the wire remains the same. From Eq. 18–3 we see that the resistance would increase by a factor of four $(2/\frac{1}{2} = 4)$.

EXERCISE F Copper wires in houses typically have a diameter of about 1.5 mm. How long a wire would have a 1.0-Ω resistance?

Temperature Dependence of Resistivity

The resistivity of a material depends somewhat on temperature. The resistance of metals generally increases with temperature. This is not surprising, because at higher temperatures, the atoms are moving more rapidly and are arranged in a less orderly fashion. So they might be expected to interfere more with the flow of electrons. If the temperature change is not too great, the resistivity of metals usually increases nearly linearly with temperature. That is,

$$\rho_T = \rho_0[1 + \alpha(T - T_0)] \tag{18–4}$$

where ρ_0 is the resistivity at some reference temperature T_0 (such as 0°C or 20°C), ρ_T is the resistivity at a temperature T, and α is the **temperature coefficient of resistivity**. Values for α are given in Table 18–1. Note that the temperature coefficient for semiconductors can be negative. Why? It seems that at higher temperatures, some of the electrons that are normally not free in a semiconductor become free and can contribute to the current. Thus, the resistance of a semiconductor can decrease with an increase in temperature.

PHYSICS APPLIED
Resistance thermometer

EXAMPLE 18-7 **Resistance thermometer.** The variation in electrical resistance with temperature can be used to make precise temperature measurements. Platinum is commonly used since it is relatively free from corrosive effects and has a high melting point. Suppose at 20.0°C the resistance of a platinum resistance thermometer is 164.2 Ω. When placed in a particular solution, the resistance is 187.4 Ω. What is the temperature of this solution?

APPROACH Since the resistance R is directly proportional to the resistivity ρ, we can combine Eq. 18–3 with Eq. 18–4 to find R as a function of temperature T, and then solve that equation for T.

SOLUTION Equation 18–3 tells us $R = \rho \ell / A$, so we multiply Eq. 18–4 by (ℓ/A) to obtain

$$R = R_0[1 + \alpha(T - T_0)].$$

Here $R_0 = \rho_0 \ell / A$ is the resistance of the wire at $T_0 = 20.0°C$. We solve this equation for T and find (see Table 18–1 for α)

$$T = T_0 + \frac{R - R_0}{\alpha R_0} = 20.0°C + \frac{187.4\,\Omega - 164.2\,\Omega}{(3.927 \times 10^{-3}(C°)^{-1})(164.2\,\Omega)} = 56.0°C.$$

NOTE Resistance thermometers have the advantage that they can be used at very high or low temperatures where gas or liquid thermometers would be useless.

NOTE More convenient for some applications is a **thermistor** (Fig. 18–15), which consists of a metal oxide or semiconductor whose resistance also varies in a repeatable way with temperature. Thermistors can be made quite small and respond very quickly to temperature changes.

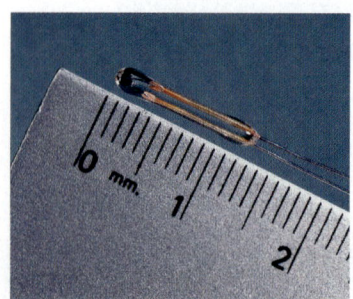

FIGURE 18-15 A thermistor only 13 mm long, shown next to a millimeter ruler.

EXERCISE G The resistance of the tungsten filament of a common incandescent lightbulb is how many times greater at its operating temperature of 2800 K than its resistance at room temperature? (*a*) Less than 1% greater; (*b*) roughly 10% greater; (*c*) about 2 times greater; (*d*) roughly 10 times greater; (*e*) more than 100 times greater.

The value of α in Eq. 18–4 can itself depend on temperature, so it is important to check the temperature range of validity of any value (say, in a handbook of physical data). If the temperature range is wide, Eq. 18–4 is not adequate and terms proportional to the square and cube of the temperature are needed, but these terms are generally very small except when $T - T_0$ is large.

18-5 Electric Power

Electric energy is useful to us because it can be easily transformed into other forms of energy. Motors transform electric energy into mechanical energy, and are examined in Chapter 20.

In other devices such as electric heaters, stoves, toasters, and hair dryers, electric energy is transformed into thermal energy in a wire resistance known as a "heating element." And in an ordinary lightbulb, the tiny wire filament (Fig. 18–5 and Chapter-Opening Photo) becomes so hot it glows; only a few percent of the energy is transformed into visible light, and the rest, over 90%, into thermal energy. Lightbulb filaments and heating elements (Fig. 18–16) in household appliances have resistances typically of a few ohms to a few hundred ohms.

Electric energy is transformed into thermal energy or light in such devices, and there are many collisions between the moving electrons and the atoms of the wire. In each collision, part of the electron's kinetic energy is transferred to the atom with which it collides. As a result, the kinetic energy of the wire's atoms increases and hence the temperature (Section 13–9) of the wire element increases. The increased thermal energy can be transferred as heat by conduction and convection to the air in a heater or to food in a pan, by radiation to bread in a toaster, or radiated as light.

FIGURE 18-16 Hot electric stove burner glows because of energy transformed by electric current.

To find the power transformed by an electric device, recall that the energy transformed when a charge Q moves through a potential difference V is QV (Eq. 17–3). Then the power P, which is the rate energy is transformed, is

$$P = \frac{\text{energy transformed}}{\text{time}} = \frac{QV}{t}.$$

The charge that flows per second, Q/t, is the electric current I. Thus we have

$$P = IV. \qquad (18\text{–}5)$$

This general relation gives us the power transformed by any device, where I is the current passing through it and V is the potential difference across it. It also gives the power delivered by a source such as a battery. The SI unit of electric power is the same as for any kind of power, the **watt** (1 W = 1 J/s).

The rate of energy transformation in a resistance R can be written in two other ways, starting with the general relation $P = IV$ and substituting in Ohm's law, $V = IR$:

$$P = IV = I(IR) = I^2 R \qquad (18\text{–}6\text{a})$$

$$P = IV = \left(\frac{V}{R}\right)V = \frac{V^2}{R}. \qquad (18\text{–}6\text{b})$$

Equations 18–6a and b apply only to resistors, whereas Eq. 18–5, $P = IV$, is more general and applies to any device.

EXAMPLE 18–8 **Headlights.** Calculate the resistance of a 40-W automobile headlight designed for 12 V (Fig. 18–17).

APPROACH We solve for R in Eq. 18–6b, which has the given variables.

SOLUTION From Eq. 18–6b,

$$R = \frac{V^2}{P} = \frac{(12\ \text{V})^2}{(40\ \text{W})} = 3.6\ \Omega.$$

NOTE This is the resistance when the bulb is burning brightly at 40 W. When the bulb is cold, the resistance is much lower, as we saw in Eq. 18–4 (see also Exercise G). Since the current is high when the resistance is low, lightbulbs burn out most often when first turned on.

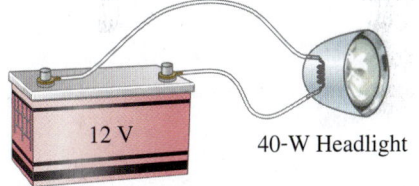

FIGURE 18–17 Example 18–8.

PHYSICS APPLIED
Why lightbulbs burn out when first turned on

CAUTION
You pay for energy, which is power × time, not for power

It is energy, not power, that you pay for on your electric bill. Since power is the *rate* energy is transformed, the total energy used by any device is simply its power consumption multiplied by the time it is on. If the power is in watts and the time is in seconds, the energy will be in joules since 1 W = 1 J/s. Electric companies usually specify the energy with a much larger unit, the **kilowatt-hour** (kWh). One kWh = $(1000\ \text{W})(3600\ \text{s})$ = 3.60×10^6 J.

EXAMPLE 18–9 **Electric heater.** An electric heater draws a steady 15.0 A on a 120-V line. How much power does it require and how much does it cost per month (30 days) if it operates 3.0 h per day and the electric company charges 9.2 cents per kWh?

APPROACH We use Eq. 18–5, $P = IV$, to find the power. We multiply the power (in kW) by the time (h) used in a month and by the cost per energy unit, $0.092 per kWh, to get the cost per month.

SOLUTION The power is

$$P = IV = (15.0\ \text{A})(120\ \text{V})$$
$$= 1800\ \text{W} = 1.80\ \text{kW}.$$

The time (in hours) the heater is used per month is $(3.0\ \text{h/d})(30\ \text{d}) = 90\ \text{h}$, which at 9.2¢/kWh would cost $(1.80\ \text{kW})(90\ \text{h})(\$0.092/\text{kWh}) = \$15$, just for this heater.

NOTE Household current is actually alternating (ac), but our solution is still valid assuming the given values for V and I are the proper averages (rms) as we discuss in Section 18–7.

PHYSICS APPLIED
Lightning

FIGURE 18–18 Example 18–10. A lightning bolt.

EXAMPLE 18–10 ESTIMATE Lightning bolt. Lightning is a spectacular example of electric current in a natural phenomenon (Fig. 18–18). There is much variability to lightning bolts, but a typical event might transfer 10^9 J of energy across a potential difference of perhaps 5×10^7 V during a time interval of about 0.2 s. Use this information to estimate (a) the total amount of charge transferred between cloud and ground, (b) the current in the lightning bolt, and (c) the average power delivered over the 0.2 s.

APPROACH We estimate the charge Q, recalling that potential energy change equals the potential difference ΔV times the charge Q, Eq. 17–3. We equate ΔPE with the energy transferred, $\Delta \text{PE} \approx 10^9$ J. Next, the current I is Q/t (Eq. 18–1) and the power P is energy/time.

SOLUTION (a) From Eq. 17–3, the energy transformed is $\Delta \text{PE} = Q \Delta V$. We solve for Q:

$$Q = \frac{\Delta \text{PE}}{\Delta V} \approx \frac{10^9 \text{ J}}{5 \times 10^7 \text{ V}} = 20 \text{ coulombs}.$$

(b) The current during the 0.2 s is about

$$I = \frac{Q}{t} \approx \frac{20 \text{ C}}{0.2 \text{ s}} = 100 \text{ A}.$$

(c) The average power delivered is

$$P = \frac{\text{energy}}{\text{time}} = \frac{10^9 \text{ J}}{0.2 \text{ s}} = 5 \times 10^9 \text{ W} = 5 \text{ GW}.$$

We can also use Eq. 18–5:

$$P = IV = (100 \text{ A})(5 \times 10^7 \text{ V}) = 5 \text{ GW}.$$

NOTE Since most lightning bolts consist of several stages, it is possible that individual parts could carry currents much higher than the 100 A calculated above.

EXERCISE H Since 1 kWh = 3.6×10^6 J, how much mass must be lifted against gravity through one meter to do the equivalent amount of work?

18–6 Power in Household Circuits

PHYSICS APPLIED
Safety—wires getting hot

PHYSICS APPLIED
Fuses, circuit breakers, and shorts

The electric wires that carry electricity to lights and other electric appliances in houses and buildings have some resistance, although usually it is quite small. Nonetheless, if the current is large enough, the wires will heat up and produce thermal energy at a rate equal to $I^2 R$, where R is the wire's resistance. One possible hazard is that the current-carrying wires in the wall of a building may become so hot as to start a fire. Thicker wires have less resistance (see Eq. 18–3) and thus can carry more current without becoming too hot. When a wire carries more current than is safe, it is said to be "overloaded." To prevent overloading, **fuses** or **circuit breakers** are installed in circuits. They are basically switches (Fig. 18–19, top of next page) that open the circuit when the current exceeds a safe value. A 20-A fuse or circuit breaker, for example, opens when the current passing through it exceeds 20 A. If a circuit repeatedly burns out a fuse or opens a circuit breaker, and no connected device requires more than 20 A, there are two possibilities: there may be too many devices drawing current in that circuit; or there is a fault somewhere, such as a "short." A short, or "short circuit," means that two wires have touched that should not have (perhaps because the insulation has worn through) so the path of the current is shortened through a path of very low resistance. With reduced resistance, the current becomes very large and can make a wire hot enough to start a fire. Short circuits should be remedied immediately.

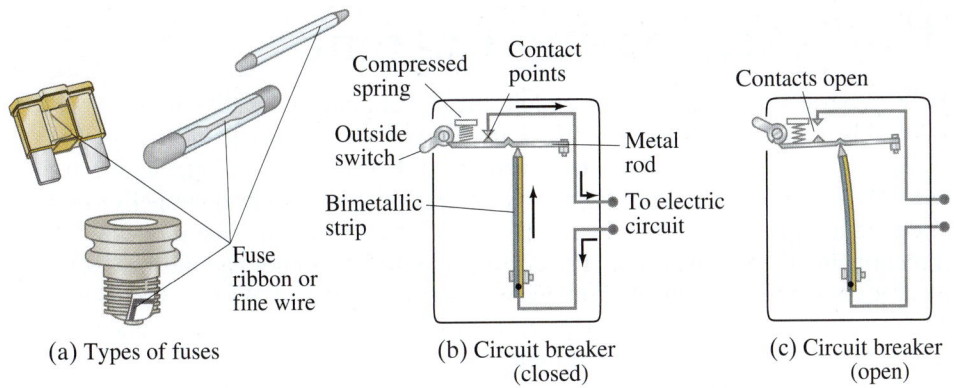

(a) Types of fuses   (b) Circuit breaker (closed)   (c) Circuit breaker (open)

FIGURE 18–19 (a) Fuses. When current exceeds a certain value, the metallic ribbon or wire inside melts and the circuit opens. Then the fuse must be replaced. (b) One type of circuit breaker. Current passes through a bimetallic strip. When the current exceeds a safe level, the heating of the bimetallic strip causes the strip to bend so far to the left that the notch in the spring-loaded metal rod drops down over the end of the bimetallic strip (c) and the circuit opens at the contact points (one is attached to the rod) and the outside switch is also flipped. When the bimetallic strip cools, it can be reset using the outside switch. Better magnetic-type circuit breakers are discussed in Chapters 20 and 21.

Household circuits are designed with the various devices connected so that each receives the standard voltage (Fig. 18–20) from the electric company (usually 120 V in the United States). Circuits with the devices arranged as in Fig. 18–20 are called *parallel circuits*, as we will discuss in the next Chapter. When a fuse blows or circuit breaker opens, it is important to check the total current being drawn on that circuit, which is the sum of the currents in each device.

FIGURE 18–20 Connection of household appliances.

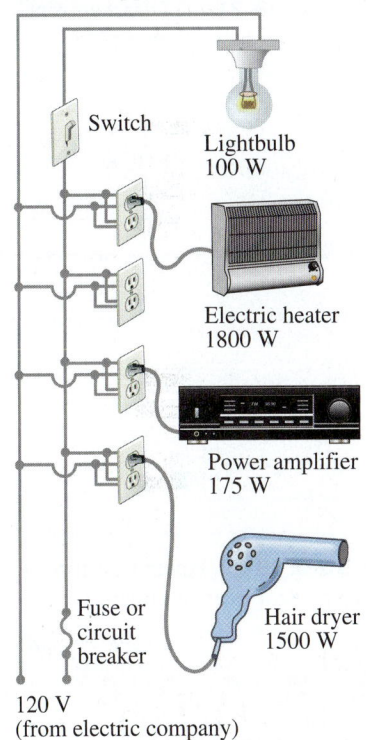

EXAMPLE 18–11 Will a fuse blow? Determine the total current drawn by all the devices in the circuit of Fig. 18–20.

APPROACH Each device has the same 120-V voltage across it. The current each draws from the source is found from $I = P/V$, Eq. 18–5.

SOLUTION The circuit in Fig. 18–20 draws the following currents: the lightbulb draws $I = P/V = 100 \text{ W}/120 \text{ V} = 0.8 \text{ A}$; the heater draws $1800 \text{ W}/120 \text{ V} = 15.0 \text{ A}$; the power amplifier draws a maximum of $175 \text{ W}/120 \text{ V} = 1.5 \text{ A}$; and the hair dryer draws $1500 \text{ W}/120 \text{ V} = 12.5 \text{ A}$. The total current drawn, if all devices are used at the same time, is

$$0.8 \text{ A} + 15.0 \text{ A} + 1.5 \text{ A} + 12.5 \text{ A} = 29.8 \text{ A}.$$

NOTE The heater draws as much current as 18 100-W lightbulbs. For safety, the heater should probably be on a circuit by itself.

If the circuit in Fig. 18–20 is designed for a 20-A fuse, the fuse should blow, and we hope it will, to prevent overloaded wires from getting hot enough to start a fire. Something will have to be turned off to get this circuit below 20 A. (Houses and apartments usually have several circuits, each with its own fuse or circuit breaker; try moving one of the devices to another circuit.) If the circuit is designed with heavier wire and a 30-A fuse, the fuse shouldn't blow—if it does, a short may be the problem. (The most likely place for a short is in the cord of one of the devices.) Proper fuse size is selected according to the wire used to supply the current. A properly rated fuse should *never* be replaced by a higher-rated one, even in a car. A fuse blowing or a circuit breaker opening is acting like a switch, making an "open circuit." By an open circuit, we mean that there is no longer a complete conducting path, so no current can flow; it is as if $R = \infty$.

PHYSICS APPLIED
Proper fuses and shorts

CONCEPTUAL EXAMPLE 18–12 A dangerous extension cord. Your 1800-W portable electric heater is too far from your desk to warm your feet. Its cord is too short, so you plug it into an extension cord rated at 11 A. Why is this dangerous?

RESPONSE 1800 W at 120 V draws a 15-A current. The wires in the extension cord rated at 11 A could become hot enough to melt the insulation and cause a fire.

PHYSICS APPLIED
Extension cords and possible danger

EXERCISE I How many 60-W 120-V lightbulbs can operate on a 20-A line? (*a*) 2; (*b*) 3; (*c*) 6; (*d*) 20; (*e*) 40.

SECTION 18–6 Power in Household Circuits   **513**

18–7 Alternating Current

When a battery is connected to a circuit, the current moves steadily in one direction. This is called a **direct current**, or **dc**. Electric generators at electric power plants, however, produce **alternating current**, or **ac**. (Sometimes capital letters are used, DC and AC.) An alternating current reverses direction many times per second and is commonly sinusoidal, Fig. 18–21. The electrons in a wire first move in one direction and then in the other. The current supplied to homes and businesses by electric companies is ac throughout virtually the entire world. We will discuss and analyze ac circuits in detail in Chapter 21. But because ac circuits are so common in real life, we will discuss some of their basic aspects here.

The voltage produced by an ac electric generator is sinusoidal, as we shall see later. The current it produces is thus sinusoidal (Fig. 18–21b). We can write the voltage as a function of time as

$$V = V_0 \sin 2\pi f t = V_0 \sin \omega t. \quad (18\text{–}7\text{a})$$

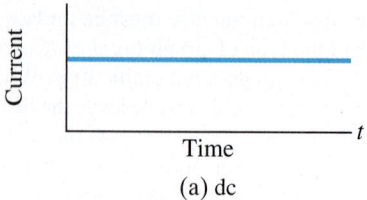
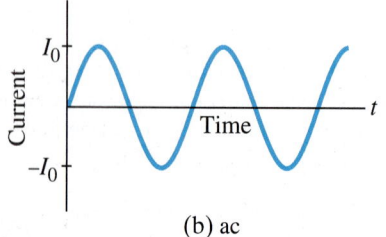

FIGURE 18–21 (a) Direct current, and (b) alternating current, as functions of time.

The potential V oscillates between $+V_0$ and $-V_0$, and V_0 is referred to as the **peak voltage**. The frequency f is the number of complete oscillations made per second, and $\omega = 2\pi f$. In most areas of the United States and Canada, f is 60 Hz (the unit "hertz," as we saw in Chapters 8 and 11, means cycles per second). In many countries, 50 Hz is used.

Equation 18–2, $V = IR$, works also for ac: if a voltage V exists across a resistance R, then the current I through the resistance is

$$I = \frac{V}{R} = \frac{V_0}{R} \sin \omega t = I_0 \sin \omega t. \quad (18\text{–}7\text{b})$$

The quantity $I_0 = V_0/R$ is the **peak current**. The current is considered positive when the electrons flow in one direction and negative when they flow in the opposite direction. It is clear from Fig. 18–21b that an alternating current is as often positive as it is negative. Thus, the average current is zero. This does not mean, however, that no power is needed or that no heat is produced in a resistor. Electrons do move back and forth, and do produce heat. Indeed, the power transformed in a resistance R at any instant is (Eq. 18–7b)

$$P = I^2 R = I_0^2 R \sin^2 \omega t.$$

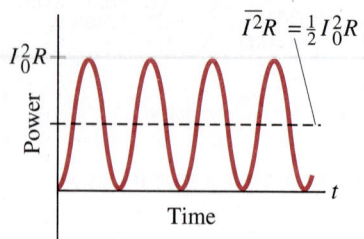

FIGURE 18–22 Power transformed in a resistor in an ac circuit.

Because the current is squared, we see that the power is always positive, as graphed in Fig. 18–22. The quantity $\sin^2 \omega t$ varies between 0 and 1; and it is not too difficult to show[†] that its average value is $\frac{1}{2}$, as indicated in Fig. 18–22. Thus, the *average power* transformed, $\overline{P}$, is

$$\overline{P} = \tfrac{1}{2} I_0^2 R.$$

Since power can also be written $P = V^2/R = (V_0^2/R) \sin^2 \omega t$, we also have that the average power is

$$\overline{P} = \tfrac{1}{2} \frac{V_0^2}{R}.$$

The average or mean value of the *square* of the current or voltage is thus what is important for calculating average power: $\overline{I^2} = \tfrac{1}{2} I_0^2$ and $\overline{V^2} = \tfrac{1}{2} V_0^2$. The square root of each of these is the **rms** (root-mean-square) value of the current or voltage:

$$I_{\text{rms}} = \sqrt{\overline{I^2}} = \frac{I_0}{\sqrt{2}} = 0.707 I_0, \quad (18\text{–}8\text{a})$$

$$V_{\text{rms}} = \sqrt{\overline{V^2}} = \frac{V_0}{\sqrt{2}} = 0.707 V_0. \quad (18\text{–}8\text{b})$$

[†]A graph of $\cos^2 \omega t$ versus t is identical to that for $\sin^2 \omega t$ in Fig. 18–22, except that the points are shifted (by $\tfrac{1}{4}$ cycle) on the time axis. Thus the average value of $\sin^2$ and $\cos^2$, averaged over one or more full cycles, will be the same. From the trigonometric identity $\sin^2 \theta + \cos^2 \theta = 1$, we can write

$$\overline{(\sin^2 \omega t)} + \overline{(\cos^2 \omega t)} = 2\overline{(\sin^2 \omega t)} = 1.$$

Hence the average value of $\sin^2 \omega t$ is $\tfrac{1}{2}$.

The rms values of V and I are sometimes called the *effective values*. They are useful because they can be substituted directly into the power formulas, Eqs. 18–5 and 18–6, to get the average power:

$$\overline{P} = I_{rms} V_{rms} \tag{18-9a}$$

$$\overline{P} = \tfrac{1}{2} I_0^2 R = I_{rms}^2 R \tag{18-9b}$$

$$\overline{P} = \tfrac{1}{2} \frac{V_0^2}{R} = \frac{V_{rms}^2}{R}. \tag{18-9c}$$

Thus, a direct current whose values of I and V equal the rms values of I and V for an alternating current will produce the same power. Hence it is usually the rms value of current and voltage that is specified or measured. For example, in the United States and Canada, standard line voltage is 120-V ac. The 120 V is V_{rms}; the peak voltage V_0 is (Eq. 18–8b)

$$V_0 = \sqrt{2}\, V_{rms} = 170\,\text{V}.$$

In much of the world (Europe, Australia, Asia) the rms voltage is 240 V, so the peak voltage is 340 V. The line voltage can vary, depending on the total load; the frequency of 60 Hz or 50 Hz, however, remains extremely steady.

EXAMPLE 18–13 **Hair dryer.** (*a*) Calculate the resistance and the peak current in a 1500-W hair dryer (Fig. 18–23) connected to a 120-V ac line. (*b*) What happens if it is connected to a 240-V ac line in Britain?

APPROACH We are given $\overline{P}$ and V_{rms}, so $I_{rms} = \overline{P}/V_{rms}$ (Eq. 18–9a or 18–5), and $I_0 = \sqrt{2}\, I_{rms}$. Then we find R from $V = IR$.

SOLUTION (*a*) We solve Eq. 18–9a for the rms current:

$$I_{rms} = \frac{\overline{P}}{V_{rms}} = \frac{1500\,\text{W}}{120\,\text{V}} = 12.5\,\text{A}.$$

Then

$$I_0 = \sqrt{2}\, I_{rms} = 17.7\,\text{A}.$$

The resistance is

$$R = \frac{V_{rms}}{I_{rms}} = \frac{120\,\text{V}}{12.5\,\text{A}} = 9.6\,\Omega.$$

The resistance could equally well be calculated using peak values:

$$R = \frac{V_0}{I_0} = \frac{170\,\text{V}}{17.7\,\text{A}} = 9.6\,\Omega.$$

(*b*) When connected to a 240-V line, more current would flow and the resistance would change with the increased temperature (Section 18–4). But let us make an estimate of the power transformed based on the same 9.6-Ω resistance. The average power would be

$$\overline{P} = \frac{V_{rms}^2}{R}$$

$$= \frac{(240\,\text{V})^2}{(9.6\,\Omega)} = 6000\,\text{W}.$$

This is four times the dryer's power rating and would undoubtedly melt the heating element or the wire coils of the motor.

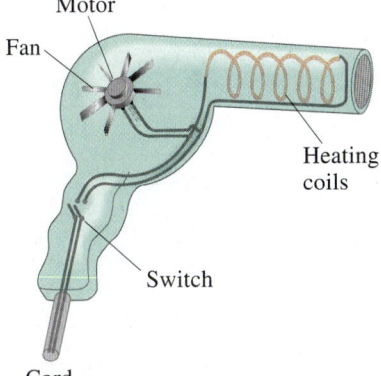

FIGURE 18–23 A hair dryer. Most of the current goes through the heating coils, a pure resistance; a small part goes to the motor to turn the fan. Example 18–13.

This Section has given a brief introduction to the simpler aspects of alternating currents. We will discuss ac circuits in more detail in Chapter 21. In Chapter 19 we will deal with the details of dc circuits only.

FIGURE 18–24 Electric field $\vec{E}$ in a wire gives electrons in random motion a drift velocity $\vec{v}_d$. Note $\vec{v}_d$ is in the opposite direction of $\vec{E}$ because electrons have a negative charge ($\vec{F} = q\vec{E}$).

FIGURE 18–25 Electrons in the volume $A\ell$ will all pass through the cross section indicated in a time Δt, where $\ell = v_d \Delta t$.

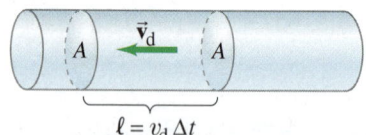

*18–8 Microscopic View of Electric Current

It can be useful to analyze a simple model of electric current at the microscopic level of atoms and electrons. In a conducting wire, for example, we can imagine the free electrons as moving about randomly at high speeds, bouncing off the atoms of the wire (somewhat like the molecules of a gas—Sections 13–8 to 13–10). When an electric field exists in the wire, Fig. 18–24, the electrons feel a force and initially begin to accelerate. But they soon reach a more or less steady average velocity known as their **drift velocity**, v_d (collisions with atoms in the wire keep them from accelerating further). The drift velocity is normally very much smaller than the electrons' average random speed.

We can relate v_d to the macroscopic current I in the wire. In a time Δt, the electrons will travel a distance $\ell = v_d \Delta t$ on average. Suppose the wire has cross-sectional area A. Then in time Δt, electrons in a volume $V = A\ell = Av_d \Delta t$ will pass through the cross section A of wire, as shown in Fig. 18–25. If there are n free electrons (each with magnitude of charge e) per unit volume, then the total number of electrons is $N = nV$ (V is volume, not voltage) and the total charge ΔQ that passes through the area A in a time Δt is

$$\Delta Q = \text{(number of charges, } N) \times \text{(charge per particle)}$$
$$= (nV)(e) = (nAv_d \Delta t)(e).$$

The magnitude of the current I in the wire is thus

$$I = \frac{\Delta Q}{\Delta t} = neAv_d. \quad (18\text{–}10)$$

EXAMPLE 18–14 **Electron speed in wire.** A copper wire 3.2 mm in diameter carries a 5.0-A current. Determine the drift velocity of the free electrons. Assume that one electron per Cu atom is free to move (the others remain bound to the atom).

APPROACH We apply Eq. 18–10 to find the drift velocity v_d if we can determine the number n of free electrons per unit volume. Since we assume there is one free electron per atom, the density of free electrons, n, is the same as the number of Cu atoms per unit volume. The atomic mass of Cu is 63.5 u (see Periodic Table inside the back cover), so 63.5 g of Cu contains one mole or 6.02×10^{23} free electrons. To find the volume V of this amount of copper, and then $n = N/V$, we use the mass density of copper (Table 10–1), $\rho_D = 8.9 \times 10^3$ kg/m³, where $\rho_D = m/V$. (We use ρ_D to distinguish it here from ρ for resistivity.)

SOLUTION The number of free electrons per unit volume, $n = N/V$ (where $V = \text{volume} = m/\rho_D$), is

$$n = \frac{N}{V} = \frac{N}{m/\rho_D} = \frac{N(1 \text{ mole})}{m(1 \text{ mole})} \rho_D$$

$$n = \left(\frac{6.02 \times 10^{23} \text{ electrons}}{63.5 \times 10^{-3} \text{ kg}}\right)(8.9 \times 10^3 \text{ kg/m}^3) = 8.4 \times 10^{28} \text{ m}^{-3}.$$

The cross-sectional area of the wire is $A = \pi r^2 = \pi(1.6 \times 10^{-3} \text{ m})^2 = 8.0 \times 10^{-6}$ m². Then, by Eq. 18–10, the drift velocity has magnitude

$$v_d = \frac{I}{neA} = \frac{5.0 \text{ A}}{(8.4 \times 10^{28} \text{ m}^{-3})(1.6 \times 10^{-19} \text{ C})(8.0 \times 10^{-6} \text{ m}^2)}$$
$$= 4.6 \times 10^{-5} \text{ m/s} \approx 0.05 \text{ mm/s}.$$

NOTE The actual speed of electrons bouncing around inside the metal is estimated to be about 1.6×10^6 m/s at 20°C, very much greater than the drift velocity.

The drift velocity of electrons in a wire is slow, only about 0.05 mm/s in Example 18–14, which means it takes an electron about 20×10^3 s, or $5\frac{1}{2}$ h, to travel only 1 m. This is not how fast "electricity travels": when you flip a light switch, the light—even if many meters away—goes on nearly instantaneously. Why? Because electric fields travel essentially at the speed of light (3×10^8 m/s). We can think of electrons in a wire as being like a pipe full of water: when a little water enters one end of the pipe, some water immediately comes out the other end.

*18–9 Superconductivity

At very low temperatures, well below 0°C, the resistivity (Section 18–4) of certain metals and certain compounds or alloys becomes zero as measured by the highest-precision techniques. Materials in such a state are said to be **superconducting**. This phenomenon was first observed by H. K. Onnes (1853–1926) in 1911 when he cooled mercury below 4.2 K (−269°C) and found that the resistance of mercury suddenly dropped to zero. In general, superconductors become superconducting only below a certain *transition temperature* or *critical temperature*, T_C, which is usually within a few degrees of absolute zero. Current in a ring-shaped superconducting material has been observed to flow for years in the absence of a potential difference, with no measurable decrease. Measurements show that the resistivity ρ of superconductors is less than $4 \times 10^{-25}\ \Omega\cdot\text{m}$, which is over 10^{16} times smaller than that for copper, and is considered to be zero in practice. See Fig. 18–26.

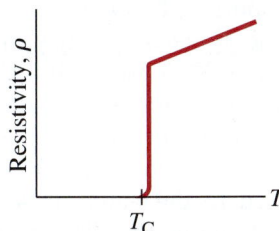

FIGURE 18–26 A superconducting material has zero resistivity when its temperature is below T_C, its "critical temperature." At temperatures above T_C, the resistivity jumps to a "normal" nonzero value and increases with temperature as most materials do (Eq. 18–4).

Before 1986 the highest temperature at which a material was found to superconduct was 23 K, which required liquid helium to keep the material cold. In 1987, a compound of yttrium, barium, copper, and oxygen (YBCO) was developed that can be superconducting at 90 K. Since this is above the boiling temperature of liquid nitrogen, 77 K, liquid nitrogen is sufficiently cold to keep the material superconducting. This was an important breakthrough because liquid nitrogen is much more easily and cheaply obtained than is the liquid helium needed for earlier superconductors. Superconductivity at temperatures as high as 160 K has been reported, though in fragile compounds.

To develop high-T_C superconductors for use as wires (such as for wires in "superconducting electromagnets"—Section 20–7), many applications today utilize a bismuth-strontium-calcium-copper oxide (BSCCO). A major challenge is how to make a useable, bendable wire out of the BSCCO, which is very brittle. (One solution is to embed tiny filaments of the high-T_C superconductor in a metal alloy, which is not resistanceless but has resistance much less than a conventional copper cable.)

*18–10 Electrical Conduction in the Human Nervous System

An interesting example of the flow of electric charge is in the human nervous system, which provides us with the means for being aware of the world, for communication within the body, and for controlling the body's muscles. Although the detailed functioning of the hugely complex nervous system still is not well understood, we do have a reasonable understanding of how messages are transmitted within the nervous system: they are electrical signals passing along the basic element of the nervous system, the **neuron**.

Neurons are living cells of unusual shape (Fig. 18–27). Attached to the main cell body are several small appendages known as *dendrites* and a long tail called the *axon*. Signals are received by the dendrites and are propagated along the axon. When a signal reaches the nerve endings, it is transmitted to the next neuron or to a muscle at a connection called a *synapse*.

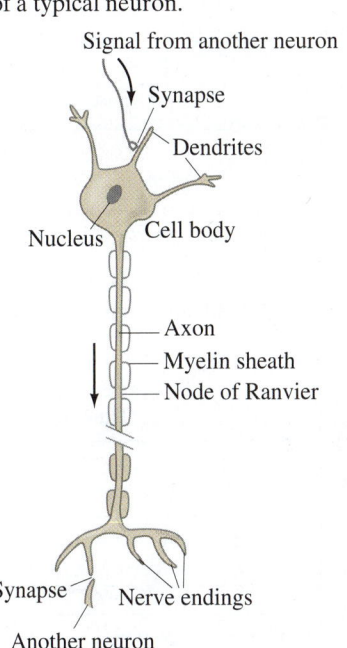

FIGURE 18–27 A simplified sketch of a typical neuron.

TABLE 18–2 Concentrations of Ions Inside and Outside a Typical Axon		
	Concentration inside axon (mol/m³)	Concentration outside axon (mol/m³)
K^+	140	5
Na^+	15	140
Cl^-	9	125

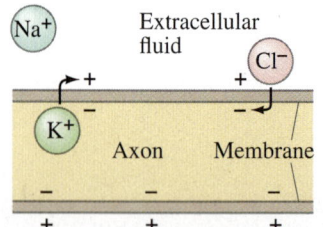

FIGURE 18–28 How a dipole layer of charge forms on a cell membrane.

FIGURE 18–29 Measuring the potential difference between the inside and outside of a nerve cell.

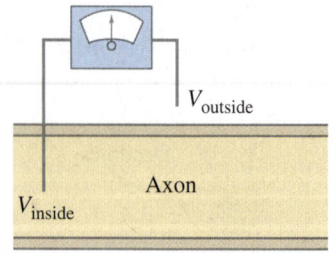

FIGURE 18–30 Action potential.

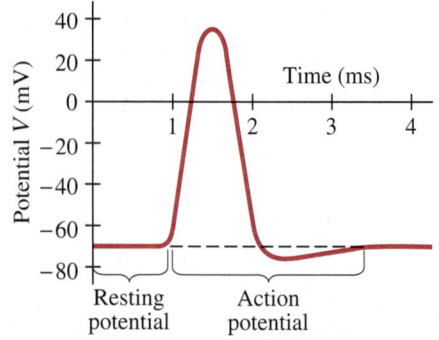

A neuron, before transmitting an electrical signal, is in the so-called "resting state." Like nearly all living cells, neurons have a net positive charge on the outer surface of the cell membrane and a negative charge on the inner surface. This difference in charge, or **dipole layer**, means that a potential difference exists across the cell membrane. When a neuron is not transmitting a signal, this **resting potential**, normally stated as

$$V_{\text{inside}} - V_{\text{outside}},$$

is typically -60 mV to -90 mV, depending on the type of organism. The most common ions in a cell are K^+, Na^+, and Cl^-. There are large differences in the concentrations of these ions inside and outside an axon, as indicated by the typical values given in Table 18–2. Other ions are also present, so the fluids both inside and outside the axon are electrically neutral. Because of the differences in concentration, there is a tendency for ions to diffuse across the membrane (see Section 13–13 on diffusion). However, in the resting state the cell membrane prevents any net flow of Na^+ through a mechanism of active transport[‡] of Na^+ ions out of the cell by a particular protein to which Na^+ attach; energy needed comes from ATP. But it does allow the flow of Cl^- ions, and less so of K^+ ions, and it is these two ions that produce the dipole charge layer on the membrane. Because there is a greater concentration of K^+ inside the cell than outside, more K^+ ions tend to diffuse outward across the membrane than diffuse inward. A K^+ ion that passes through the membrane becomes attached to the outer surface of the membrane, and leaves behind an equal negative charge that lies on the inner surface of the membrane (Fig. 18–28). The fluids themselves remain neutral. What keeps the ions on the membrane is their attraction for each other across the membrane. Independently, Cl^- ions tend to diffuse *into* the cell since their concentration outside is higher. Both K^+ and Cl^- diffusion tends to charge the interior surface of the membrane negative and the outside positive. As charge accumulates on the membrane surface, it becomes increasingly difficult for more ions to diffuse: K^+ ions trying to move outward, for example, are repelled by the positive charge already there. Equilibrium is reached when the tendency to diffuse because of the concentration difference is just balanced by the electrical potential difference across the membrane. The greater the concentration difference, the greater the potential difference across the membrane (-60 mV to -90 mV).

The most important aspect of a neuron is not that it has a resting potential (most cells do), but rather that it can respond to a stimulus and conduct an electrical signal along its length. The stimulus could be thermal (when you touch a hot stove) or chemical (as in taste buds); it could be pressure (as on the skin or at the eardrum), or light (as in the eye); or it could be the electric stimulus of a signal coming from the brain or another neuron. In the laboratory, the stimulus is usually electrical and is applied by a tiny probe at some point on the neuron. If the stimulus exceeds some threshold, a voltage pulse will travel down the axon. This voltage pulse can be detected at a point on the axon using a voltmeter or an oscilloscope connected as in Fig. 18–29. This voltage pulse has the shape shown in Fig. 18–30, and is called an **action potential**. As can be seen, the potential increases from a resting potential of about -70 mV and becomes a positive 30 mV or 40 mV. The action potential lasts for about 1 ms and travels down an axon with a speed of 30 m/s to 150 m/s. When an action potential is stimulated, the nerve is said to have "fired."

What causes the action potential? At the point where the stimulus occurs, the membrane suddenly alters its permeability, becoming much more permeable to Na^+ than to K^+ and Cl^- ions. Thus, Na^+ ions rush into the cell and the inner surface of the wall becomes positively charged, and the potential difference quickly swings positive ($\approx +30$ mV in Fig. 18–30). Just as suddenly, the membrane returns to its original characteristics; it becomes impermeable to Na^+ and in fact pumps out Na^+ ions. The diffusion of Cl^- and K^+ ions again predominates and the original resting potential is restored (-70 mV in Fig. 18–30).

[‡]This transport mechanism is sometimes referred to as the "sodium pump."

What causes the action potential to travel along the axon? The action potential occurs at the point of stimulation, as shown in Fig. 18–31a. The membrane momentarily is positive on the inside and negative on the outside at this point. Nearby charges are attracted toward this region, as shown in Fig. 18–31b. The potential in these adjacent regions then drops, causing an action potential there. Thus, as the membrane returns to normal at the original point, nearby it experiences an action potential, so the action potential moves down the axon (Figs. 18–31c and d).

You may wonder if the number of ions that pass through the membrane would significantly alter the concentrations. The answer is no; and we can show why (and again show the power and usefulness of physics) by treating the axon as a capacitor as we do in Search and Learn Problem 8 (the concentration changes by less than 1 part in 10^4).

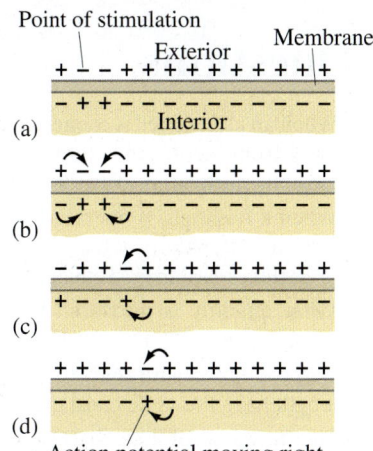

FIGURE 18–31 Propagation of action potential along axon membrane.

Summary

An electric **battery** serves as a source of nearly constant potential difference by transforming chemical energy into electric energy. A simple battery consists of two electrodes made of different metals immersed in a solution or paste known as an electrolyte.

Electric current, I, refers to the rate of flow of electric charge and is measured in **amperes** (A): 1 A equals a flow of 1 C/s past a given point.

The direction of **conventional current** is that of positive charge flow. In a wire, it is actually negatively charged electrons that move, so they flow in a direction opposite to the conventional current. A positive charge flow in one direction is almost always equivalent to a negative charge flow in the opposite direction. Positive conventional current always flows from a high potential to a low potential.

The **resistance** R of a device is defined by the relation

$$V = IR, \qquad (18\text{–}2)$$

where I is the current in the device when a potential difference V is applied across it. For materials such as metals, R is a constant independent of V (thus $I \propto V$), a result known as **Ohm's law**. Thus, the current I coming from a battery of voltage V depends on the resistance R of the circuit connected to it.

Voltage is applied *across* a device or between the ends of a wire. Current passes *through* a wire or device. Resistance is a property *of* the wire or device.

The unit of resistance is the **ohm** (Ω), where $1\,\Omega = 1\,\text{V/A}$. See Table 18–3.

TABLE 18–3 Summary of Units

Current	$1\,\text{A} = 1\,\text{C/s}$
Potential difference	$1\,\text{V} = 1\,\text{J/C}$
Power	$1\,\text{W} = 1\,\text{J/s}$
Resistance	$1\,\Omega = 1\,\text{V/A}$

The resistance R of a wire is inversely proportional to its cross-sectional area A, and directly proportional to its length ℓ and to a property of the material called its resistivity:

$$R = \frac{\rho \ell}{A}. \qquad (18\text{–}3)$$

The **resistivity**, ρ, increases with temperature for metals, but for semiconductors it may decrease.

The rate at which energy is transformed in a resistance R from electric to other forms of energy (such as heat and light) is equal to the product of current and voltage. That is, the **power** transformed, measured in watts, is given by

$$P = IV, \qquad (18\text{–}5)$$

which for resistors can be written as

$$P = I^2 R = \frac{V^2}{R}. \qquad (18\text{–}6)$$

The SI unit of power is the **watt** ($1\,\text{W} = 1\,\text{J/s}$).

The total electric energy transformed in any device equals the product of the power and the time during which the device is operated. In SI units, energy is given in joules ($1\,\text{J} = 1\,\text{W}\cdot\text{s}$), but electric companies use a larger unit, the **kilowatt-hour** ($1\,\text{kWh} = 3.6 \times 10^6\,\text{J}$).

Electric current can be **direct current** (**dc**), in which the current is steady in one direction; or it can be **alternating current** (**ac**), in which the current reverses direction at a particular frequency f, typically 60 Hz. Alternating currents are typically sinusoidal in time,

$$I = I_0 \sin \omega t, \qquad (18\text{–}7\text{b})$$

where $\omega = 2\pi f$, and are produced by an alternating voltage.

The **rms** values of sinusoidally alternating currents and voltages are given by

$$I_\text{rms} = \frac{I_0}{\sqrt{2}} \quad \text{and} \quad V_\text{rms} = \frac{V_0}{\sqrt{2}}, \qquad (18\text{–}8)$$

respectively, where I_0 and V_0 are the **peak** values. The power relationship, $P = IV = I^2 R = V^2/R$, is valid for the average power in alternating currents when the rms values of V and I are used.

[*The current in a wire, at the microscopic level, is considered to be a slow **drift velocity** of electrons, $\vec{v}_\text{d}$. The current I is given by

$$I = neAv_\text{d}, \qquad (18\text{–}10)$$

where n is the number of free electrons per unit volume, e is the magnitude of the charge on an electron, and A is the cross-sectional area of the wire.]

[*At very low temperatures certain materials become **superconducting**, which means their electrical resistance becomes zero.]

[*The human nervous system operates via electrical conduction: when a nerve "fires," an electrical signal travels as a voltage pulse known as an **action potential**.]

Questions

1. When an electric cell is connected to a circuit, electrons flow away from the negative terminal in the circuit. But within the cell, electrons flow *to* the negative terminal. Explain.
2. When a flashlight is operated, what is being used up: battery current, battery voltage, battery energy, battery power, or battery resistance? Explain.
3. What quantity is measured by a battery rating given in ampere-hours (A·h)? Explain.
4. Can a copper wire and an aluminum wire of the same length have the same resistance? Explain.
5. One terminal of a car battery is said to be connected to "ground." Since it is not really connected to the ground, what is meant by this expression?
6. The equation $P = V^2/R$ indicates that the power dissipated in a resistor decreases if the resistance is increased, whereas the equation $P = I^2R$ implies the opposite. Is there a contradiction here? Explain.
7. What happens when a lightbulb burns out?
8. If the resistance of a small immersion heater (to heat water for tea or soup, Fig. 18–32) was increased, would it speed up or slow down the heating process? Explain.

FIGURE 18–32 Question 8.

9. If a rectangular solid made of carbon has sides of lengths a, $2a$, and $3a$, to which faces would you connect the wires from a battery so as to obtain (*a*) the least resistance, (*b*) the greatest resistance?
10. Explain why lightbulbs almost always burn out just as they are turned on and not after they have been on for some time.
11. Which draws more current, a 100-W lightbulb or a 75-W bulb? Which has the higher resistance?
12. Electric power is transferred over large distances at very high voltages. Explain how the high voltage reduces power losses in the transmission lines.
13. A 15-A fuse blows out repeatedly. Why is it dangerous to replace this fuse with a 25-A fuse?
14. When electric lights are operated on low-frequency ac (say, 5 Hz), they flicker noticeably. Why?
15. Driven by ac power, the same electrons pass back and forth through your reading lamp over and over again. Explain why the light stays lit instead of going out after the first pass of electrons.
16. The heating element in a toaster is made of Nichrome wire. Immediately after the toaster is turned on, is the current magnitude (I_{rms}) in the wire increasing, decreasing, or staying constant? Explain.
17. Is current used up in a resistor? Explain.
18. Why is it more dangerous to turn on an electric appliance when you are standing outside in bare feet than when you are inside wearing shoes with thick soles?
*19. Compare the drift velocities and electric currents in two wires that are geometrically identical and the density of atoms is similar, but the number of free electrons per atom in the material of one wire is twice that in the other.
*20. A voltage V is connected across a wire of length ℓ and radius r. How is the electron drift speed affected if (*a*) ℓ is doubled, (*b*) r is doubled, (*c*) V is doubled, assuming in each case that other quantities stay the same?

MisConceptual Questions

1. When connected to a battery, a lightbulb glows brightly. If the battery is reversed and reconnected to the bulb, the bulb will glow
   (*a*) brighter.   (*c*) with the same brightness.
   (*b*) dimmer.   (*d*) not at all.
2. When a battery is connected to a lightbulb properly, current flows through the lightbulb and makes it glow. How much current flows through the battery compared with the lightbulb?
   (*a*) More.
   (*b*) Less.
   (*c*) The same amount.
   (*d*) No current flows through the battery.
3. Which of the following statements about Ohm's law is true?
   (*a*) Ohm's law relates the current through a wire to the voltage across the wire.
   (*b*) Ohm's law holds for all materials.
   (*c*) Any material that obeys Ohm's law does so independently of temperature.
   (*d*) Ohm's law is a fundamental law of physics.
   (*e*) Ohm's law is valid for superconductors.
4. Electrons carry energy from a battery to a lightbulb. What happens to the electrons when they reach the lightbulb?
   (*a*) The electrons are used up.
   (*b*) The electrons stay in the lightbulb.
   (*c*) The electrons are emitted as light.
   (*d*) Fewer electrons leave the bulb than enter it.
   (*e*) None of the above.
5. Where in the circuit of Fig. 18–33 is the current the largest, (*a*), (*b*), (*c*), or (*d*)? Or (*e*) it is the same at all points?

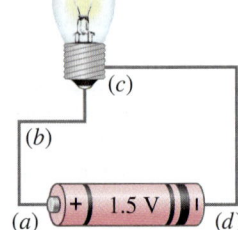

FIGURE 18–33 MisConceptual Question 5.

6. When you double the *voltage* across a certain material or device, you observe that the *current* increases by a factor of 3. What can you conclude?
   (*a*) Ohm's law is obeyed, because the current increases when *V* increases.
   (*b*) Ohm's law is not obeyed in this case.
   (*c*) This situation has nothing to do with Ohm's law.

7. When current flows through a resistor,
   (a) some of the charge is used up by the resistor.
   (b) some of the current is used up by the resistor.
   (c) Both (a) and (b) are true.
   (d) Neither (a) nor (b) is true.

8. The unit kilowatt-hour is a measure of
   (a) the rate at which energy is transformed.
   (b) power.
   (c) an amount of energy.
   (d) the amount of power used per second.

9. Why might a circuit breaker open if you plug too many electrical devices into a single circuit?
   (a) The voltage becomes too high.
   (b) The current becomes too high.
   (c) The resistance becomes too high.
   (d) A circuit breaker will not "trip" no matter how many electrical devices you plug into the circuit.

10. Nothing happens when birds land on a power line, yet we are warned not to touch a power line with a ladder. What is the difference?
    (a) Birds have extremely high internal resistance compared to humans.
    (b) There is little to no voltage drop between a bird's two feet, but there is a significant voltage drop between the top of a ladder touching a power line and the bottom of the ladder on the ground.
    (c) Dangerous current comes from the ground only.
    (d) Most birds don't understand the situation.

11. When a light switch is turned on, the light comes on immediately because
    (a) the electrons coming from the power source move through the initially empty wires very fast.
    (b) the electrons already in the wire are instantly "pushed" by a voltage difference.
    (c) the lightbulb may be old with low resistance. It would take longer if the bulb were new and had high resistance.
    (d) the electricity bill is paid. The electric company can make it take longer when the bill is unpaid.

For assigned homework and other learning materials, go to the MasteringPhysics website.

Problems

18–2 and 18–3 Electric Current, Resistance, Ohm's Law

(*Note*: The charge on one electron is 1.60×10^{-19} C.)

1. (I) A current of 1.60 A flows in a wire. How many electrons are flowing past any point in the wire per second?

2. (I) A service station charges a battery using a current of 6.7 A for 5.0 h. How much charge passes through the battery?

3. (I) What is the current in amperes if 1200 Na$^+$ ions flow across a cell membrane in 3.1 μs? The charge on the sodium is the same as on an electron, but positive.

4. (I) What is the resistance of a toaster if 120 V produces a current of 4.6 A?

5. (I) What voltage will produce 0.25 A of current through a 4800-Ω resistor?

6. (I) How many coulombs are there in a 75 ampere-hour car battery?

7. (II) (a) What is the current in the element of an electric clothes dryer with a resistance of 8.6 Ω when it is connected to 240 V? (b) How much charge passes through the element in 50 min? (Assume direct current.)

8. (II) A bird stands on a dc electric transmission line carrying 4100 A (Fig. 18–34). The line has 2.5×10^{-5} Ω resistance per meter, and the bird's feet are 4.0 cm apart. What is the potential difference between the bird's feet?

FIGURE 18–34 Problem 8.

9. (II) A hair dryer draws 13.5 A when plugged into a 120-V line. (a) What is its resistance? (b) How much charge passes through it in 15 min? (Assume direct current.)

10. (II) A 4.5-V battery is connected to a bulb whose resistance is 1.3 Ω. How many electrons leave the battery per minute?

11. (II) An electric device draws 5.60 A at 240 V. (a) If the voltage drops by 15%, what will be the current, assuming nothing else changes? (b) If the resistance of the device were reduced by 15%, what current would be drawn at 240 V?

18–4 Resistivity

12. (I) What is the diameter of a 1.00-m length of tungsten wire whose resistance is 0.32 Ω?

13. (I) What is the resistance of a 5.4-m length of copper wire 1.5 mm in diameter?

14. (II) Calculate the ratio of the resistance of 10.0 m of aluminum wire 2.2 mm in diameter, to 24.0 m of copper wire 1.8 mm in diameter.

15. (II) Can a 2.2-mm-diameter copper wire have the same resistance as a tungsten wire of the same length? Give numerical details.

16. (II) A certain copper wire has a resistance of 15.0 Ω. At what point along its length must the wire be cut so that the resistance of one piece is 4.0 times the resistance of the other? What is the resistance of each piece?

17. (II) Compute the voltage drop along a 21-m length of household no. 14 copper wire (used in 15-A circuits). The wire has diameter 1.628 mm and carries a 12-A current.

18. (II) Two aluminum wires have the same resistance. If one has twice the length of the other, what is the ratio of the diameter of the longer wire to the diameter of the shorter wire?

19. (II) A rectangular solid made of carbon has sides of lengths 1.0 cm, 2.0 cm, and 4.0 cm, lying along the x, y, and z axes, respectively (Fig. 18–35). Determine the resistance for current that passes through the solid in (a) the x direction, (b) the y direction, and (c) the z direction. Assume the resistivity is $\rho = 3.0 \times 10^{-5}\,\Omega\cdot m$.

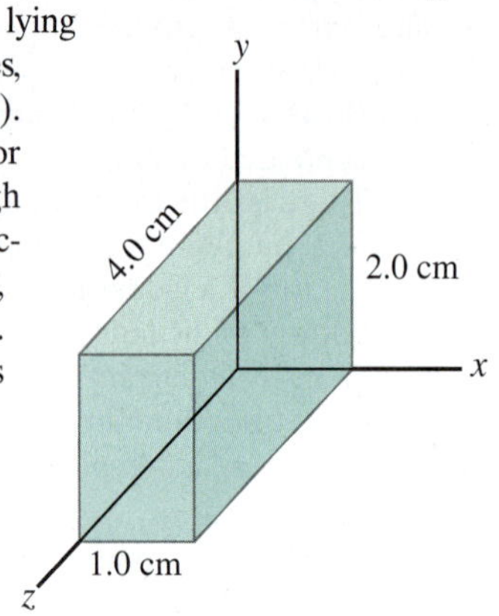

FIGURE 18–35
Problem 19.

20. (II) A length of wire is cut in half and the two lengths are wrapped together side by side to make a thicker wire. How does the resistance of this new combination compare to the resistance of the original wire?

21. (II) How much would you have to raise the temperature of a copper wire (originally at 20°C) to increase its resistance by 12%?

22. (II) Determine at what temperature aluminum will have the same resistivity as tungsten does at 20°C.

23. (II) A 100-W lightbulb has a resistance of about 12 Ω when cold (20°C) and 140 Ω when on (hot). Estimate the temperature of the filament when hot assuming an average temperature coefficient of resistivity $\alpha = 0.0045\,(C°)^{-1}$.

24. (III) A length of aluminum wire is connected to a precision 10.00-V power supply, and a current of 0.4212 A is precisely measured at 23.5°C. The wire is placed in a new environment of unknown temperature where the measured current is 0.3818 A. What is the unknown temperature?

25. (III) For some applications, it is important that the value of a resistance not change with temperature. For example, suppose you made a 3.20-kΩ resistor from a carbon resistor and a Nichrome wire-wound resistor connected together so the total resistance is the sum of their separate resistances. What value should each of these resistors have (at 0°C) so that the combination is temperature independent?

26. (III) A 10.0-m length of wire consists of 5.0 m of copper followed by 5.0 m of aluminum, both of diameter 1.4 mm. A voltage difference of 95 mV is placed across the composite wire. (a) What is the total resistance (sum) of the two wires? (b) What is the current through the wire? (c) What are the voltages across the aluminum part and across the copper part?

18–5 and 18–6 Electric Power

27. (I) What is the maximum power consumption of a 3.0-V portable CD player that draws a maximum of 240 mA of current?

28. (I) The heating element of an electric oven is designed to produce 3.3 kW of heat when connected to a 240-V source. What must be the resistance of the element?

29. (I) What is the maximum voltage that can be applied across a 3.9-kΩ resistor rated at $\frac{1}{4}$ watt?

30. (I) (a) Determine the resistance of, and current through, a 75-W lightbulb connected to its proper source voltage of 110 V. (b) Repeat for a 250-W bulb.

31. (I) An electric car has a battery that can hold 16 kWh of energy (approximately 6×10^7 J). If the battery is designed to operate at 340 V, how many coulombs of charge would need to leave the battery at 340 V and return at 0 V to equal the stored energy of the battery?

32. (I) An electric car uses a 45-kW (160-hp) motor. If the battery pack is designed for 340 V, what current would the motor need to draw from the battery? Neglect any energy losses in getting energy from the battery to the motor.

33. (II) A 120-V hair dryer has two settings: 950 W and 1450 W. (a) At which setting do you guess the resistance to be higher? After making a guess, determine the resistance at (b) the lower setting, and (c) the higher setting.

34. (II) A 12-V battery causes a current of 0.60 A through a resistor. (a) What is its resistance, and (b) how many joules of energy does the battery lose in a minute?

35. (II) A 120-V fish-tank heater is rated at 130 W. Calculate (a) the current through the heater when it is operating, and (b) its resistance.

36. (II) You buy a 75-W lightbulb in Europe, where electricity is delivered at 240 V. If you use the bulb in the United States at 120 V (assume its resistance does not change), how bright will it be relative to 75-W 120-V bulbs? [Hint: Assume roughly that brightness is proportional to power consumed.]

37. (II) How many kWh of energy does a 550-W toaster use in the morning if it is in operation for a total of 5.0 min? At a cost of 9.0 cents/kWh, estimate how much this would add to your monthly electric energy bill if you made toast four mornings per week.

38. (II) At $0.095/kWh, what does it cost to leave a 25-W porch light on day and night for a year?

39. (II) What is the total amount of energy stored in a 12-V, 65 A·h car battery when it is fully charged?

40. (II) An ordinary flashlight uses two D-cell 1.5-V batteries connected in series to provide 3.0 V across the bulb, as in Fig. 18–4b (Fig. 18–36). The bulb draws 380 mA when turned on. (a) Calculate the resistance of the bulb and the power dissipated. (b) By what factor would the power increase if four D-cells in series (total 6.0 V) were used with the same bulb? (Neglect heating effects of the filament.) Why shouldn't you try this?

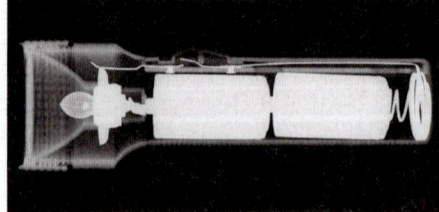

FIGURE 18–36
Problem 40
(X-ray of a flashlight).

41. (II) How many 75-W lightbulbs, connected to 120 V as in Fig. 18–20, can be used without blowing a 15-A fuse?

42. (II) An extension cord made of two wires of diameter 0.129 cm (no. 16 copper wire) and of length 2.7 m (9 ft) is connected to an electric heater which draws 18.0 A on a 120-V line. How much power is dissipated in the cord?

43. (II) You want to design a portable electric blanket that runs on a 1.5-V battery. If you use a 0.50-mm-diameter copper wire as the heating element, how long should the wire be if you want to generate 18 W of heating power? What happens if you accidentally connect the blanket to a 9.0-V battery?

44. (II) A power station delivers 750 kW of power at 12,000 V to a factory through wires with total resistance 3.0 Ω. How much less power is wasted if the electricity is delivered at 50,000 V rather than 12,000 V?

45. (III) A small immersion heater can be used in a car to heat a cup of water for coffee or tea. If the heater can heat 120 mL of water from 25°C to 95°C in 8.0 min, (a) approximately how much current does it draw from the car's 12-V battery, and (b) what is its resistance? Assume the manufacturer's claim of 85% efficiency.

46. (III) The current in an electromagnet connected to a 240-V line is 21.5 A. At what rate must cooling water pass over the coils for the water temperature to rise no more than 6.50 C°?

18–7 Alternating Current

47. (I) Calculate the peak current in a 2.7-kΩ resistor connected to a 220-V rms ac source.

48. (I) An ac voltage, whose peak value is 180 V, is across a 310-Ω resistor. What are the rms and peak currents in the resistor?

49. (II) Estimate the resistance of the 120-V_{rms} circuits in your house as seen by the power company, when (a) everything electrical is unplugged, and (b) two 75-W lightbulbs are on.

50. (II) The peak value of an alternating current in a 1500-W device is 6.4 A. What is the rms voltage across it?

51. (II) An 1800-W arc welder is connected to a 660-V_{rms} ac line. Calculate (a) the peak voltage and (b) the peak current.

52. (II) Each channel of a stereo receiver is capable of an average power output of 100 W into an 8-Ω loudspeaker (see Fig. 18–14). What are the rms voltage and the rms current fed to the speaker (a) at the maximum power of 100 W, and (b) at 1.0 W when the volume is turned down?

53. (II) Determine (a) the maximum instantaneous power dissipated by a 2.2-hp pump connected to a 240-V_{rms} ac power source, and (b) the maximum current passing through the pump.

54. (II) A heater coil connected to a 240-V_{rms} ac line has a resistance of 38 Ω. (a) What is the average power used? (b) What are the maximum and minimum values of the instantaneous power?

*18–8 Microscopic View of Electric Current

*55. (II) A 0.65-mm-diameter copper wire carries a tiny dc current of 2.7 μA. Estimate the electron drift velocity.

*56. (II) A 4.80-m length of 2.0-mm-diameter wire carries a 750-mA dc current when 22.0 mV is applied to its ends. If the drift velocity is 1.7×10^{-5} m/s, determine (a) the resistance R of the wire, (b) the resistivity ρ, and (c) the number n of free electrons per unit volume.

*57. (III) At a point high in the Earth's atmosphere, He^{2+} ions in a concentration of $2.4 \times 10^{12}/m^3$ are moving due north at a speed of 2.0×10^6 m/s. Also, a $7.0 \times 10^{11}/m^3$ concentration of O_2^- ions is moving due south at a speed of 6.2×10^6 m/s. Determine the magnitude and direction of the net current passing through unit area (A/m^2).

*18–10 Nerve Conduction

*58. (I) What is the magnitude of the electric field across an axon membrane 1.0×10^{-8} m thick if the resting potential is −70 mV?

*59. (II) A neuron is stimulated with an electric pulse. The action potential is detected at a point 3.70 cm down the axon 0.0052 s later. When the action potential is detected 7.20 cm from the point of stimulation, the time required is 0.0063 s. What is the speed of the electric pulse along the axon? (Why are two measurements needed instead of only one?)

*60. (III) During an action potential, Na^+ ions move into the cell at a rate of about 3×10^{-7} $mol/m^2 \cdot s$. How much power must be produced by the "active Na^+ pumping" system to produce this flow against a +30-mV potential difference? Assume that the axon is 10 cm long and 20 μm in diameter.

General Problems

61. A person accidentally leaves a car with the lights on. If each of the two headlights uses 40 W and each of the two taillights 6 W, for a total of 92 W, how long will a fresh 12-V battery last if it is rated at 75 A·h? Assume the full 12 V appears across each bulb.

62. A sequence of potential differences V is applied across a wire (diameter = 0.32 mm, length = 11 cm) and the resulting currents I are measured as follows:

V (V)	0.100	0.200	0.300	0.400	0.500
I (mA)	72	142	218	290	357

(a) If this wire obeys Ohm's law, graphing I vs. V will result in a straight-line plot. Explain why this is so and determine the theoretical predictions for the straight line's slope and y-intercept. (b) Plot I vs. V. Based on this plot, can you conclude that the wire obeys Ohm's law (i.e., did you obtain a straight line with the expected y-intercept, within the values of the significant figures)? If so, determine the wire's resistance R. (c) Calculate the wire's resistivity and use Table 18–1 to identify the solid material from which it is composed.

63. What is the average current drawn by a 1.0-hp 120-V motor? (1 hp = 746 W.)

64. The **conductance** G of an object is defined as the reciprocal of the resistance R; that is, G = 1/R. The unit of conductance is a *mho* (= ohm^{-1}), which is also called the *siemens* (S). What is the conductance (in siemens) of an object that draws 440 mA of current at 3.0 V?

65. The heating element of a 110-V, 1500-W heater is 3.8 m long. If it is made of iron, what must its diameter be?

66. (a) A particular household uses a 2.2-kW heater 2.0 h/day ("on" time), four 100-W lightbulbs 6.0 h/day, a 3.0-kW electric stove element for a total of 1.0 h/day, and miscellaneous power amounting to 2.0 kWh/day. If electricity costs $0.115 per kWh, what will be their monthly bill (30 d)? (b) How much coal (which produces 7500 kcal/kg) must be burned by a 35%-efficient power plant to provide the yearly needs of this household?

67. A small city requires about 15 MW of power. Suppose that instead of using high-voltage lines to supply the power, the power is delivered at 120 V. Assuming a two-wire line of 0.50-cm-diameter copper wire, estimate the cost of the energy lost to heat per hour per meter. Assume the cost of electricity is about 12 cents per kWh.

68. A 1600-W hair dryer is designed for 117 V. (a) What will be the percentage change in power output if the voltage drops to 105 V? Assume no change in resistance. (b) How would the actual change in resistivity with temperature affect your answer?

69. The wiring in a house must be thick enough so it does not become so hot as to start a fire. What diameter must a copper wire be if it is to carry a maximum current of 35 A and produce no more than 1.5 W of heat per meter of length?

70. Determine the resistance of the tungsten filament in a 75-W 120-V incandescent lightbulb (a) at its operating temperature of about 2800 K, (b) at room temperature.

71. Suppose a current is given by the equation $I = 1.40 \sin 210t$, where I is in amperes and t in seconds. (a) What is the frequency? (b) What is the rms value of the current? (c) If this is the current through a 24.0-Ω resistor, write the equation that describes the voltage as a function of time.

72. A microwave oven running at 65% efficiency delivers 950 W to the interior. Find (a) the power drawn from the source, and (b) the current drawn. Assume a source voltage of 120 V.

73. A 1.00-Ω wire is stretched uniformly to 1.50 times its original length. What is its resistance now?

74. 220 V is applied to two different conductors made of the same material. One conductor is twice as long and twice the diameter of the second. What is the ratio of the power transformed in the first relative to the second?

75. An electric power plant can produce electricity at a fixed power P, but the plant operator is free to choose the voltage V at which it is produced. This electricity is carried as an electric current I through a transmission line (resistance R) from the plant to the user, where it provides the user with electric power P'. (a) Show that the reduction in power $\Delta P = P - P'$ due to transmission losses is given by $\Delta P = P^2 R / V^2$. (b) In order to reduce power losses during transmission, should the operator choose V to be as large or as small as possible?

76. A 2800-W oven is connected to a 240-V source. (a) What is the resistance of the oven? (b) How long will it take to bring 120 mL of 15°C water to 100°C assuming 65% efficiency? (c) How much will this cost at 11 cents/kWh?

77. A proposed electric vehicle makes use of storage batteries as its source of energy. It is powered by 24 batteries, each 12 V, 95 A·h. Assume that the car is driven on level roads at an average speed of 45 km/h, and the average friction force is 440 N. Assume 100% efficiency and neglect energy used for acceleration. No energy is consumed when the vehicle is stopped, since the engine doesn't need to idle. (a) Determine the horsepower required. (b) After approximately how many kilometers must the batteries be recharged?

78. A 15.2-Ω resistor is made from a coil of copper wire whose total mass is 15.5 g. What is the diameter of the wire, and how long is it?

79. A fish-tank heater is rated at 95 W when connected to 120 V. The heating element is a coil of Nichrome wire. When uncoiled, the wire has a total length of 3.5 m. What is the diameter of the wire?

80. A 100-W, 120-V lightbulb has a resistance of 12 Ω when cold (20°C) and 140 Ω when on (hot). Calculate its power consumption (a) at the instant it is turned on, and (b) after a few moments when it is hot.

81. In an automobile, the system voltage varies from about 12 V when the car is off to about 13.8 V when the car is on and the charging system is in operation, a difference of 15%. By what percentage does the power delivered to the headlights vary as the voltage changes from 12 V to 13.8 V? Assume the headlight resistance remains constant.

82. A tungsten filament used in a flashlight bulb operates at 0.20 A and 3.0 V. If its resistance at 20°C is 1.5 Ω, what is the temperature of the filament when the flashlight is on?

83. Lightbulb A is rated at 120 V and 40 W for household applications. Lightbulb B is rated at 12 V and 40 W for automotive applications. (a) What is the current through each bulb? (b) What is the resistance of each bulb? (c) In one hour, how much charge passes through each bulb? (d) In one hour, how much energy does each bulb use? (e) Which bulb requires larger diameter wires to connect its power source and the bulb?

84. An air conditioner draws 18 A at 220-V ac. The connecting cord is copper wire with a diameter of 1.628 mm. (a) How much power does the air conditioner draw? (b) If the length of the cord (containing two wires) is 3.5 m, how much power is dissipated in the wiring? (c) If no. 12 wire, with a diameter of 2.053 mm, was used instead, how much power would be dissipated in the wiring? (d) Assuming that the air conditioner is run 12 h per day, how much money per month (30 days) would be saved by using no. 12 wire? Assume that the cost of electricity is 12 cents per kWh.

85. An electric wheelchair is designed to run on a single 12-V battery rated to provide 100 ampere-hours (100 A·h). (a) How much energy is stored in this battery? (b) If the wheelchair experiences an average total retarding force (mainly friction) of 210 N, how far can the wheelchair travel on one charge?

86. If a wire of resistance R is stretched uniformly so that its length doubles, by what factor does the power dissipated in the wire change, assuming it remains hooked up to the same voltage source? Assume the wire's volume and density remain constant.

87. Copper wire of diameter 0.259 cm is used to connect a set of appliances at 120 V, which draw 1450 W of power total. (a) What power is wasted in 25.0 m of this wire? (b) What is your answer if wire of diameter 0.412 cm is used?

88. Battery-powered electricity is very expensive compared with that available from a wall outlet. Estimate the cost per kWh of (a) an alkaline D-cell (cost $1.70) and (b) an alkaline AA-cell (cost $1.25). These batteries can provide a continuous current of 25 mA for 820 h and 120 h, respectively, at 1.5 V. (c) Compare to the cost of a normal 120-V ac house source at $0.10/kWh.

89. A copper pipe has an inside diameter of 3.00 cm and an outside diameter of 5.00 cm (Fig. 18–37). What is the resistance of a 10.0-m length of this pipe?

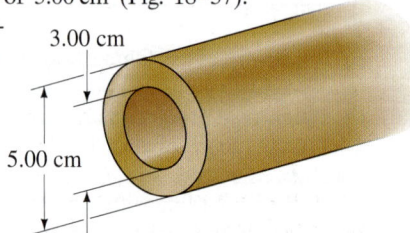

FIGURE 18–37 Problem 89.

*90. The Tevatron accelerator at Fermilab (Illinois) is designed to carry an 11-mA beam of protons ($q = 1.6 \times 10^{-19}$ C) traveling at very nearly the speed of light (3.0×10^8 m/s) around a ring 6300 m in circumference. How many protons are in the beam?

*91. The level of liquid helium (temperature ≈ 4 K) in its storage tank can be monitored using a vertically aligned niobium–titanium (NbTi) wire, whose length ℓ spans the height of the tank. In this level-sensing setup, an electronic circuit maintains a constant electrical current I at all times in the NbTi wire and a voltmeter monitors the voltage difference V across this wire. Since the superconducting critical temperature for NbTi is 10 K, the portion of the wire immersed in the liquid helium is in the superconducting state, while the portion above the liquid (in helium vapor with temperature above 10 K) is in the normal state. Define $f = x/\ell$ to be the fraction of the tank filled with liquid helium (Fig. 18–38) and V_0 to be the value of V when the tank is empty ($f = 0$). Determine the relation between f and V (in terms of V_0).

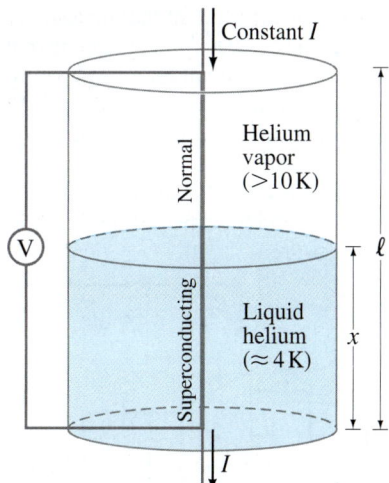

FIGURE 18–38 Problem 91.

Search and Learn

1. Why is Ohm's law less of a law than Newton's laws?
2. A traditional incandescent lamp filament may have been lit to a temperature of 2700 K. A contemporary halogen incandescent lamp filament may be at around 2900 K. (a) Estimate the percent improvement of the halogen bulb over the traditional one. [*Hint*: See Section 14–8.] (b) To produce the same amount of light as a traditional 100-W bulb, estimate what wattage a halogen bulb should use.
3. You find a small cylindrical resistor that measures 9.00 mm in length and 2.15 mm in diameter, and it has a color code of red, yellow, brown, and gold. What is the resistor made of primarily?
4. Small changes in the length of an object can be measured using a **strain gauge** sensor, which is a wire that when undeformed has length ℓ_0, cross-sectional area A_0, and resistance R_0. This sensor is rigidly affixed to the object's surface, aligning its length in the direction in which length changes are to be measured. As the object deforms, the length of the wire sensor changes by $\Delta \ell$, and the resulting change ΔR in the sensor's resistance is measured. Assuming that as the solid wire is deformed to a length ℓ, its density and volume remain constant (only approximately valid), show that the strain ($= \Delta\ell/\ell_0$) of the wire sensor, and thus of the object to which it is attached, is approximately $\Delta R/2R_0$. [See Sections 18–4 and 9–5.]
5. An electric heater is used to heat a room of volume 65 m³. Air is brought into the room at 5°C and is completely replaced twice per hour. Heat loss through the walls amounts to approximately 850 kcal/h. If the air is to be maintained at 22°C, what minimum wattage must the heater have? (The specific heat of air is about 0.17 kcal/kg·C°. Reread parts of Chapter 14 and Section 18–5.)
6. Household wiring has sometimes used aluminium instead of copper. (a) Using Table 18–1, find the ratio of the resistance of a copper wire to that of an aluminum wire of the same length and diameter. (b) Typical copper wire used for home wiring in the U.S. has a diameter of 1.63 mm. What is the resistance of 125 m of this wire? (c) What would be the resistance of the same wire if it were made of aluminum? (d) How much power would be dissipated in each wire if it carried 18 A of current? (e) What should be the diameter of the aluminum wire for it to have the same resistance as the copper wire? (f) In Section 18–4, a statement is made about the resistance of copper and aluminum wires of the same weight. Using Table 10–1 for the densities of copper and aluminum, find the resistance of an aluminum wire of the same mass and length as the copper wire in part (b). Is the statement true?
*7. How far can an average electron move along the wires of a 650-W toaster during an alternating current cycle? The power cord has copper wires of diameter 1.7 mm and is plugged into a standard 60-Hz 120-V ac outlet. [*Hint*: The maximum current in the cycle is related to the maximum drift velocity. The maximum velocity in an oscillation is related to the maximum displacement; see Chapter 11.]
*8. **Capacitance of an axon.** (a) Do an order-of-magnitude estimate for the capacitance of an axon 10 cm long of radius 10 μm. The thickness of the membrane is about 10^{-8} m, and the dielectric constant is about 3. (b) By what factor does the concentration (number of ions per volume) of Na^+ ions in the cell change as a result of one action potential?

ANSWERS TO EXERCISES

A: 1.6×10^{-13} A.
B: 3600 C.
C: 240 Ω.
D: (b), (c).
E: (c).
F: 110 m.
G: (d).
H: 370,000 kg, or about 5000 people.
I: (e) 40.

This cell phone has an attachment that measures a person's blood sugar level, and plots it over a period of days. All electronic devices contain circuits that are dc, at least in part. The circuit diagram below shows a possible amplifier circuit for an audio output (cell phone ear piece). We have already met two of the circuit elements shown: resistors and capacitors, and we discuss them in circuits in this Chapter. (The large triangle is an amplifier chip containing transistors.) We also discuss how voltmeters and ammeters work, and how measurements affect the quantity being measured.

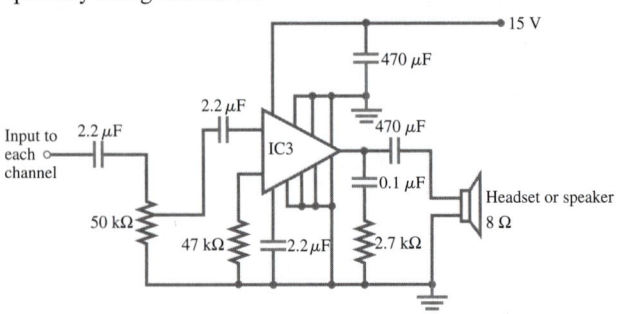

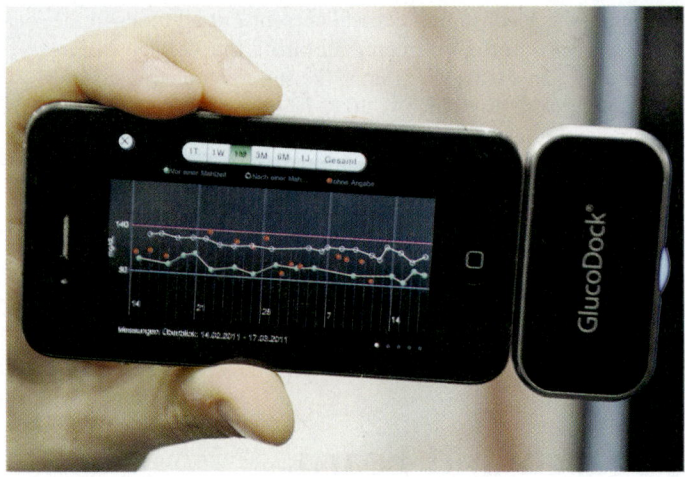

19 DC Circuits

CONTENTS

- 19–1 EMF and Terminal Voltage
- 19–2 Resistors in Series and in Parallel
- 19–3 Kirchhoff's Rules
- 19–4 EMFs in Series and in Parallel; Charging a Battery
- 19–5 Circuits Containing Capacitors in Series and in Parallel
- 19–6 RC Circuits—Resistor and Capacitor in Series
- 19–7 Electric Hazards
- 19–8 Ammeters and Voltmeters—Measurement Affects the Quantity Being Measured

TABLE 19–1 Symbols for Circuit Elements

Symbol	Device
─┤├─	Battery
─┤├─ or ─┤(─	Capacitor
─\/\/\─	Resistor
───	Wire with negligible resistance
─/─	Switch
⏚ or ↓	Ground

CHAPTER-OPENING QUESTION—Guess now!

The automobile headlight bulbs shown in the circuits here are identical. The battery connection which produces more light is

(a) circuit 1.
(b) circuit 2.
(c) both the same.
(d) not enough information.

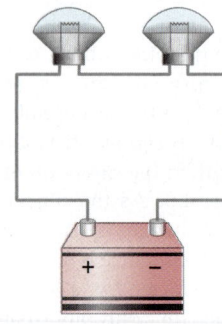

Circuit 1

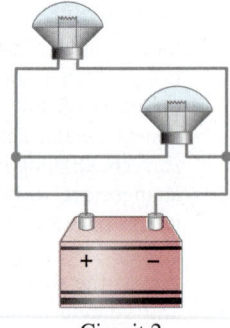

Circuit 2

Electric circuits are basic parts of all electronic devices from cell phones and TV sets to computers and automobiles. Scientific measurements—whether in physics, biology, or medicine—make use of electric circuits. In Chapter 18, we discussed the basic principles of electric current. Now we apply these principles to analyze dc circuits involving combinations of batteries, resistors, and capacitors. We also study the operation of some useful instruments.[†]

When we draw a diagram for a circuit, we represent batteries, capacitors, and resistors by the symbols shown in Table 19–1. Wires whose resistance is negligible compared with other resistance in the circuit are drawn as straight lines. A ground symbol (⏚ or ↓) may mean a real connection to the ground, perhaps via a metal pipe, or it may mean a common connection, such as the frame of a car.

For the most part in this Chapter, except in Section 19–6 on RC circuits, we will be interested in circuits operating in their steady state. We won't be looking at a circuit at the moment a change is made in it, such as when a battery or resistor is connected or disconnected, but only when the currents have reached their steady values.

[†]AC circuits that contain only a voltage source and resistors can be analyzed like the dc circuits in this Chapter. However, ac circuits that contain capacitors and other circuit elements are more complicated, and we discuss them in Chapter 21.

19–1 EMF and Terminal Voltage

To have current in an electric circuit, we need a device such as a battery or an electric generator that transforms one type of energy (chemical, mechanical, or light, for example) into electric energy. Such a device is called a **source** of **electromotive force**[†] or of **emf**. The *potential difference* between the terminals of such a source, when no current flows to an external circuit, is called the **emf** of the source. The symbol $\mathcal{E}$ is usually used for emf (don't confuse $\mathcal{E}$ with E for electric field), and its unit is volts.

A battery is not a source of constant current—the current out of a battery varies according to the resistance in the circuit. A battery *is*, however, a nearly constant voltage source, but not perfectly constant as we now discuss. For example, if you start a car with the headlights on, you may notice the headlights dim. This happens because the starter draws a large current, and the battery voltage drops below its rated emf as a result. The voltage drop occurs because the chemical reactions in a battery cannot supply charge fast enough to maintain the full emf. For one thing, charge must move (within the electrolyte) between the electrodes of the battery, and there is always some hindrance to completely free flow. Thus, a battery itself has some resistance, which is called its **internal resistance**; it is usually designated r.

CAUTION
Why battery voltage isn't perfectly constant

A real battery is modeled as if it were a perfect emf $\mathcal{E}$ in series with a resistor r, as shown in Fig. 19–1. Since this resistance r is inside the battery, we can never separate it from the battery. The two points a and b in Fig. 19–1 represent the two terminals of the battery. What we measure is the **terminal voltage** $V_{ab} = V_a - V_b$. When no current is drawn from the battery, the terminal voltage equals the emf, which is determined by the chemical reactions in the battery: $V_{ab} = \mathcal{E}$. However, when a current I flows from the battery there is an internal drop in voltage equal to Ir. Thus the terminal voltage (the actual voltage applied to a circuit) is

$$V_{ab} = \mathcal{E} - Ir. \qquad \text{[current } I \text{ flows from battery]} \quad (19\text{–}1)$$

For example, if a 12-V battery has an internal resistance of $0.1\,\Omega$, then when 10 A flows from the battery, the terminal voltage is $12\,\text{V} - (10\,\text{A})(0.1\,\Omega) = 11\,\text{V}$. The internal resistance of a battery is usually small. For example, an ordinary flashlight battery when fresh may have an internal resistance of perhaps $0.05\,\Omega$. (However, as it ages and the electrolyte dries out, the internal resistance increases to many ohms.)

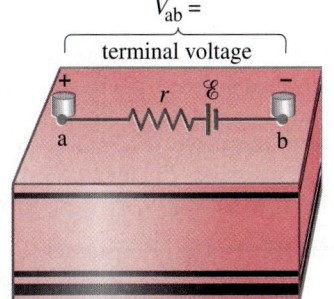
FIGURE 19–1 Diagram for an electric cell or battery.

EXAMPLE 19–1 Battery with internal resistance. A $65.0\text{-}\Omega$ resistor is connected to the terminals of a battery whose emf is 12.0 V and whose internal resistance is $0.5\,\Omega$, Fig. 19–2. Calculate (*a*) the current in the circuit, (*b*) the terminal voltage of the battery, V_{ab}, and (*c*) the power dissipated in the resistor R and in the battery's internal resistance r.

APPROACH We first consider the battery as a whole, which is shown in Fig. 19–2 as an emf $\mathcal{E}$ and internal resistance r between points a and b. Then we apply $V = IR$ to the circuit itself.

SOLUTION (*a*) From Eq. 19–1, we have $V_{ab} = \mathcal{E} - Ir$. We apply Ohm's law (Eq. 18–2) to this battery and the resistance R of the circuit: $V_{ab} = IR$. Hence

$$\mathcal{E} - Ir = IR$$

or $\mathcal{E} = I(R + r)$. So

$$I = \frac{\mathcal{E}}{R + r} = \frac{12.0\,\text{V}}{65.0\,\Omega + 0.5\,\Omega} = \frac{12.0\,\text{V}}{65.5\,\Omega} = 0.183\,\text{A}.$$

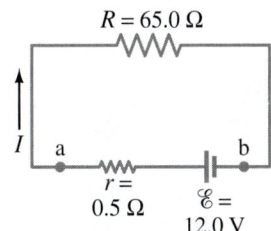
FIGURE 19–2 Example 19–1.

(*b*) The terminal voltage is

$$V_{ab} = \mathcal{E} - Ir = 12.0\,\text{V} - (0.183\,\text{A})(0.5\,\Omega) = 11.9\,\text{V}.$$

(*c*) The power dissipated in R (Eq. 18–6) is

$$P_R = I^2 R = (0.183\,\text{A})^2 (65.0\,\Omega) = 2.18\,\text{W},$$

and in the battery's resistance r it is

$$P_r = I^2 r = (0.183\,\text{A})^2 (0.5\,\Omega) = 0.02\,\text{W}.$$

[†]The term "electromotive force" is a misnomer—it does not refer to a "force" that is measured in newtons. To avoid confusion, we use the abbreviation, emf.

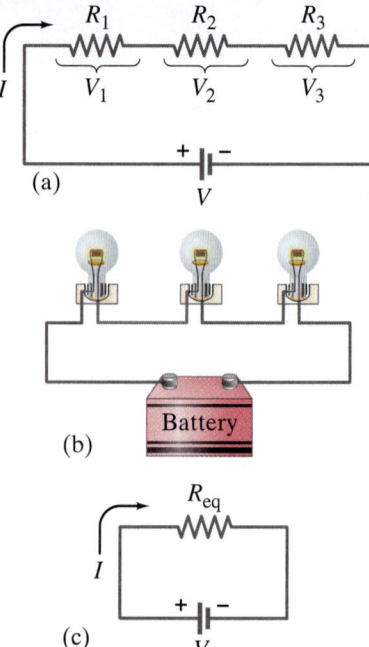

FIGURE 19–3 (a) Resistances connected in series. (b) Resistances could be lightbulbs, or any other type of resistance. (c) Equivalent single resistance R_{eq} that draws the same current: $R_{eq} = R_1 + R_2 + R_3$.

FIGURE 19–4 (a) Resistances connected in parallel. (b) Resistances could be lightbulbs. (c) The equivalent circuit with R_{eq} obtained from Eq. 19–4:
$$\frac{1}{R_{eq}} = \frac{1}{R_1} + \frac{1}{R_2} + \frac{1}{R_3}.$$

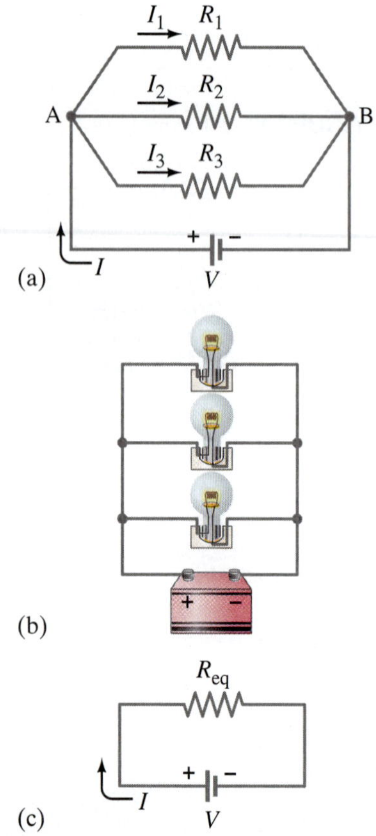

Unless stated otherwise, we assume the battery's internal resistance is negligible, and the battery voltage given is its terminal voltage, which we will usually write as V rather than V_{ab}. Do not confuse V (italic) for voltage, with V (not italic) for the volt unit.

19–2 Resistors in Series and in Parallel

When two or more resistors are connected end to end along a single path as shown in Fig. 19–3a, they are said to be connected in **series**. The resistors could be simple resistors as were pictured in Fig. 18–11, or they could be lightbulbs (Fig. 19–3b), or heating elements, or other resistive devices. Any charge that passes through R_1 in Fig. 19–3a will also pass through R_2 and then R_3. Hence the same current I passes through each resistor. (If it did not, this would imply that either charge was not conserved, or that charge was accumulating at some point in the circuit, which does not happen in the steady state.)

We let V represent the potential difference (voltage) across all three resistors in Fig. 19–3a. We assume all other resistance in the circuit can be ignored, so V equals the terminal voltage supplied by the battery. We let V_1, V_2, and V_3 be the potential differences across each of the resistors, R_1, R_2, and R_3, respectively. From Ohm's law, $V = IR$, we can write $V_1 = IR_1$, $V_2 = IR_2$, and $V_3 = IR_3$. Because the resistors are connected end to end, energy conservation tells us that the total voltage V is equal to the sum of the voltages across each resistor:

$$V = V_1 + V_2 + V_3 = IR_1 + IR_2 + IR_3. \quad \text{[series]} \quad (19\text{–}2)$$

Now let us determine the equivalent single resistance R_{eq} that would draw the same current I as our combination of three resistors in series; see Fig. 19–3c. Such a single resistance R_{eq} would be related to V by

$$V = IR_{eq}.$$

We equate this expression with Eq. 19–2, $V = I(R_1 + R_2 + R_3)$, and find

$$R_{eq} = R_1 + R_2 + R_3. \quad \text{[series]} \quad (19\text{–}3)$$

When we put several resistances in series, the total or equivalent resistance is the sum of the separate resistances. (Sometimes we call it "net resistance.") This sum applies to any number of resistances in series. Note that when you add more resistance to the circuit, the current through the circuit will decrease. For example, if a 12-V battery is connected to a 4-Ω resistor, the current will be 3 A. But if the 12-V battery is connected to three 4-Ω resistors in series, the total resistance is 12 Ω and the current through the entire circuit will be only 1 A.

Another way to connect resistors is in **parallel**, so that the current from the source splits into separate branches or paths (Fig. 19–4a). Wiring in houses and buildings is arranged so all electric devices are in parallel, as we saw in Chapter 18, Fig. 18–20. With parallel wiring, if you disconnect one device (say, R_1 in Fig. 19–4a), the current to the other devices is not interrupted. Compare to a series circuit, where if one device (say, R_1 in Fig. 19–3a) is disconnected, the current *is* stopped to all others.

In a parallel circuit, Fig. 19–4a, the total current I that leaves the battery splits into three separate paths. We let I_1, I_2, and I_3 be the currents through each of the resistors, R_1, R_2, and R_3, respectively. Because *electric charge is conserved*, the current I flowing into junction A (where the different wires or conductors meet, Fig. 19–4a) must equal the current flowing out of the junction. Thus

$$I = I_1 + I_2 + I_3. \quad \text{[parallel]}$$

When resistors are connected in parallel, each has the same voltage across it. (Indeed, any two points in a circuit connected by a wire of negligible resistance are at the same potential.) Hence the full voltage of the battery is applied to each resistor in Fig. 19–4a. Applying Ohm's law to each resistor, we have

$$I_1 = \frac{V}{R_1}, \quad I_2 = \frac{V}{R_2}, \quad \text{and} \quad I_3 = \frac{V}{R_3}.$$

Let us now determine what single resistor R_{eq} (Fig. 19–4c) will draw the same current I as these three resistances in parallel. This equivalent resistance R_{eq} must satisfy Ohm's law too:

$$I = \frac{V}{R_{eq}}.$$

We now combine the equations above:

$$I = I_1 + I_2 + I_3,$$

$$\frac{V}{R_{eq}} = \frac{V}{R_1} + \frac{V}{R_2} + \frac{V}{R_3}.$$

When we divide out the V from each term, we have

$$\frac{1}{R_{eq}} = \frac{1}{R_1} + \frac{1}{R_2} + \frac{1}{R_3}. \quad \text{[parallel]} \quad (19\text{--}4)$$

For example, suppose you connect two 4-Ω loudspeakers in parallel to a single set of output terminals of an amplifier. The equivalent resistance of the two 4-Ω "resistors" in parallel is

$$\frac{1}{R_{eq}} = \frac{1}{4\,\Omega} + \frac{1}{4\,\Omega} = \frac{2}{4\,\Omega} = \frac{1}{2\,\Omega},$$

and so $R_{eq} = 2\,\Omega$. Thus the net (or equivalent) resistance is *less* than each single resistance. This may at first seem surprising. But remember that when you connect resistors in parallel, you are giving the current additional paths to follow. Hence the net resistance will be less.†

Equations 19–3 and 19–4 make good sense. Recalling Eq. 18–3 for resistivity, $R = \rho \ell / A$, we see that placing resistors in series effectively increases the length and therefore the resistance; putting resistors in parallel effectively increases the area through which current flows, thus reducing the overall resistance.

Note that whenever a group of resistors is replaced by the equivalent resistance, current and voltage and power in the rest of the circuit are unaffected.

| **EXERCISE A** You have a 10-Ω and a 15-Ω resistor. What is the smallest and largest equivalent resistance that you can make with these two resistors?

CONCEPTUAL EXAMPLE 19–2 **Series or parallel?** (a) The lightbulbs in Fig. 19–5 are identical. Which configuration produces more light? (b) Which way do you think the headlights of a car are wired? Ignore change of filament resistance R with current.

RESPONSE (a) The equivalent resistance of the parallel circuit is found from Eq. 19–4, $1/R_{eq} = 1/R + 1/R = 2/R$. Thus $R_{eq} = R/2$. The parallel combination then has lower resistance ($= R/2$) than the series combination ($R_{eq} = R + R = 2R$). There will be more total current in the parallel configuration (2), since $I = V/R_{eq}$ and V is the same for both circuits. The total power transformed, which is related to the light produced, is $P = IV$, so the greater current in (2) means more light is produced.

(b) Headlights are wired in parallel (2), because if one bulb goes out, the other bulb can stay lit. If they were in series (1), when one bulb burned out (the filament broke), the circuit would be open and no current would flow, so neither bulb would light.

FIGURE 19–5 Example 19–2.

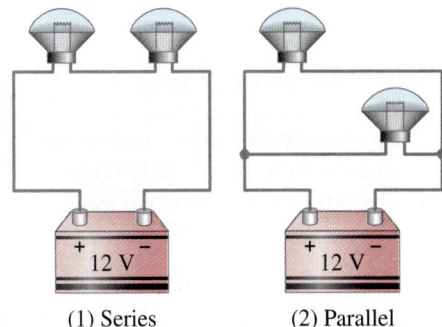

(1) Series (2) Parallel

| **EXERCISE B** Return to the Chapter-Opening Question, page 526, and answer it again now. Try to explain why you may have answered differently the first time.

†An analogy may help. Consider two identical pipes taking in water near the top of a dam and releasing it at the bottom as shown in the figure to the right. If both pipes are open, rather than only one, twice as much water will flow through. That is, the net resistance to the flow of water will be reduced by half with two equal pipes open, just as for electrical resistors in parallel.

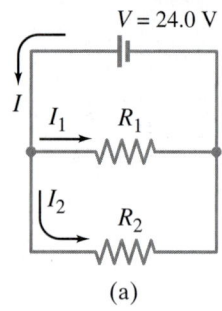

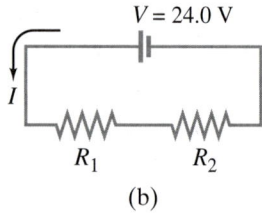

FIGURE 19–6 Example 19–3.

EXAMPLE 19–3 Series and parallel resistors. Two 100-Ω resistors are connected (a) in parallel, and (b) in series, to a 24.0-V battery (Fig. 19–6). What is the current through each resistor and what is the equivalent resistance of each circuit?

APPROACH We use Ohm's law and the ideas just discussed for series and parallel connections to get the current in each case. We can also use Eqs. 19–3 and 19–4.

SOLUTION (a) Any given charge (or electron) can flow through only one or the other of the two resistors in Fig. 19–6a. Just as a river may break into two streams when going around an island, here too the total current I from the battery (Fig. 19–6a) splits to flow through each resistor, so I equals the sum of the separate currents through the two resistors:

$$I = I_1 + I_2.$$

The potential difference across each resistor is the battery voltage $V = 24.0$ V. Applying Ohm's law to each resistor gives

$$I = I_1 + I_2 = \frac{V}{R_1} + \frac{V}{R_2} = \frac{24.0 \text{ V}}{100 \text{ }\Omega} + \frac{24.0 \text{ V}}{100 \text{ }\Omega} = 0.24 \text{ A} + 0.24 \text{ A} = 0.48 \text{ A}.$$

The equivalent resistance is

$$R_{eq} = \frac{V}{I} = \frac{24.0 \text{ V}}{0.48 \text{ A}} = 50 \text{ }\Omega.$$

We could also have obtained this result from Eq. 19–4:

$$\frac{1}{R_{eq}} = \frac{1}{100 \text{ }\Omega} + \frac{1}{100 \text{ }\Omega} = \frac{2}{100 \text{ }\Omega} = \frac{1}{50 \text{ }\Omega},$$

so $R_{eq} = 50 \text{ }\Omega$.

(b) All the current that flows out of the battery passes first through R_1 and then through R_2 because they lie along a single path, Fig. 19–6b. So the current I is the same in both resistors; the potential difference V across the battery equals the total change in potential across the two resistors:

$$V = V_1 + V_2.$$

Ohm's law gives $V = IR_1 + IR_2 = I(R_1 + R_2)$. Hence

$$I = \frac{V}{R_1 + R_2} = \frac{24.0 \text{ V}}{100 \text{ }\Omega + 100 \text{ }\Omega} = 0.120 \text{ A}.$$

The equivalent resistance, using Eq. 19–3, is $R_{eq} = R_1 + R_2 = 200 \text{ }\Omega$. We can also get R_{eq} by thinking from the point of view of the battery: the total resistance R_{eq} must equal the battery voltage divided by the current it delivers:

$$R_{eq} = \frac{V}{I} = \frac{24.0 \text{ V}}{0.120 \text{ A}} = 200 \text{ }\Omega.$$

NOTE The voltage across R_1 is $V_1 = IR_1 = (0.120 \text{ A})(100 \text{ }\Omega) = 12.0$ V, and that across R_2 is $V_2 = IR_2 = 12.0$ V, each being half of the battery voltage. A simple circuit like Fig. 19–6b is thus often called a simple **voltage divider**.

EXAMPLE 19–4 Circuit with series and parallel resistors. How much current is drawn from the battery shown in Fig. 19–7a?

APPROACH The current I that flows out of the battery all passes through the 400-Ω resistor, but then it splits into I_1 and I_2 passing through the 500-Ω and 700-Ω resistors. The latter two resistors are in parallel with each other. We look for something that we already know how to treat. So let's start by finding the equivalent resistance, R_P, of the parallel resistors, 500 Ω and 700 Ω. Then we can consider this R_P to be in series with the 400-Ω resistor.

SOLUTION The equivalent resistance, R_P, of the 500-Ω and 700-Ω resistors in parallel is

$$\frac{1}{R_P} = \frac{1}{500 \text{ }\Omega} + \frac{1}{700 \text{ }\Omega} = 0.0020 \text{ }\Omega^{-1} + 0.0014 \text{ }\Omega^{-1} = 0.0034 \text{ }\Omega^{-1}.$$

This is $1/R_P$, so we take the reciprocal to find R_P.

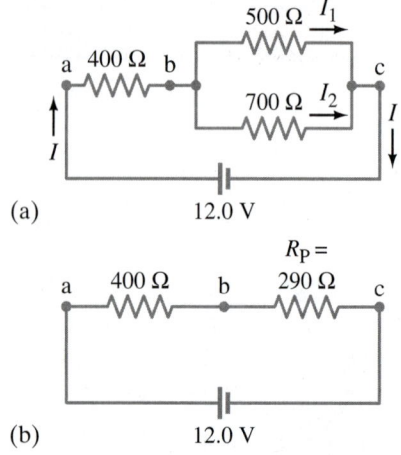

FIGURE 19–7 (a) Circuit for Examples 19–4 and 19–5. (b) Equivalent circuit, showing the equivalent resistance of 290 Ω for the two parallel resistors in (a).

It is a common mistake to forget to take this reciprocal. The units of reciprocal ohms, Ω^{-1}, are a reminder. Thus

$$R_P = \frac{1}{0.0034\ \Omega^{-1}} = 290\ \Omega.$$

This 290 Ω is the equivalent resistance of the two parallel resistors, and is in series with the 400-Ω resistor (see equivalent circuit, Fig. 19–7b). To find the total equivalent resistance R_{eq}, we add the 400-Ω and 290-Ω resistances, since they are in series:

$$R_{eq} = 400\ \Omega + 290\ \Omega = 690\ \Omega.$$

The total current flowing from the battery is then

$$I = \frac{V}{R_{eq}} = \frac{12.0\ V}{690\ \Omega} = 0.0174\ A \approx 17\ mA.$$

NOTE This I is also the current flowing through the 400-Ω resistor, but not through the 500-Ω and 700-Ω resistors (both currents are less—see the next Example).

CAUTION
Remember to take the reciprocal

EXAMPLE 19–5 Current in one branch. What is the current I_1 through the 500-Ω resistor in Fig. 19–7a?

APPROACH We need the voltage across the 500-Ω resistor, which is the voltage between points b and c in Fig. 19–7a, and we call it V_{bc}. Once V_{bc} is known, we can apply Ohm's law, $V = IR$, to get the current. First we find the voltage across the 400-Ω resistor, V_{ab}, since we know that 17.4 mA passes through it (Example 19–4).

SOLUTION V_{ab} can be found using $V = IR$:

$$V_{ab} = (0.0174\ A)(400\ \Omega) = 7.0\ V.$$

The total voltage across the network of resistors is $V_{ac} = 12.0\ V$, so V_{bc} must be $12.0\ V - 7.0\ V = 5.0\ V$. Ohm's law gives the current I_1 through the 500-Ω resistor:

$$I_1 = \frac{5.0\ V}{500\ \Omega} = 1.0 \times 10^{-2}\ A = 10\ mA.$$

This is the answer we wanted. We can also calculate the current I_2 through the 700-Ω resistor since the voltage across it is also 5.0 V:

$$I_2 = \frac{5.0\ V}{700\ \Omega} = 7\ mA.$$

NOTE When I_1 combines with I_2 to form the total current I (at point c in Fig. 19–7a), their sum is $10\ mA + 7\ mA = 17\ mA$. This equals the total current I as calculated in Example 19–4, as it should.

CONCEPTUAL EXAMPLE 19–6 Bulb brightness in a circuit. The circuit in Fig. 19–8 has three identical lightbulbs, each of resistance R. (a) When switch S is closed, how will the brightness of bulbs A and B compare with that of bulb C? (b) What happens when switch S is opened? Use a minimum of mathematics.

RESPONSE (a) With switch S closed, the current that passes through bulb C must split into two equal parts when it reaches the junction leading to bulbs A and B because the resistance of bulb A equals that of B. Thus, A and B each receive half of C's current; A and B will be equally bright, but less bright than C ($P = I^2R$).
(b) When the switch S is open, no current can flow through bulb A, so it will be dark. Now, the same current passes through bulbs B and C, so B and C will be equally bright. The equivalent resistance of this circuit ($= R + R$) is greater than that of the circuit with the switch closed, so the current leaving the battery is reduced. Thus, bulb C will be dimmer when we open the switch, but bulb B will be brighter because it gets more current when the switch is open (you may want to use some mathematics here).

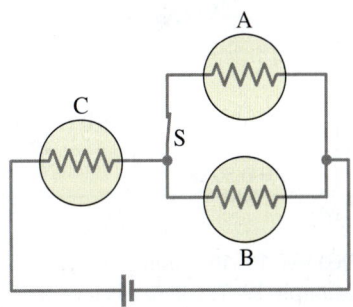

FIGURE 19–8 Example 19–6, three identical lightbulbs. Each yellow circle with —⋀⋀— inside represents a lightbulb and its resistance.

FIGURE 19–9 Exercise C.

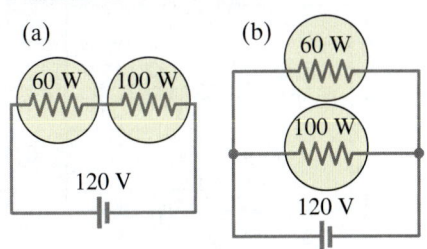

EXERCISE C A 100-W, 120-V lightbulb and a 60-W, 120-V lightbulb are connected in two different ways as shown in Fig. 19–9. In each case, which bulb glows more brightly? Ignore change of filament resistance with current (and temperature).

SECTION 19–2 Resistors in Series and in Parallel   **531**

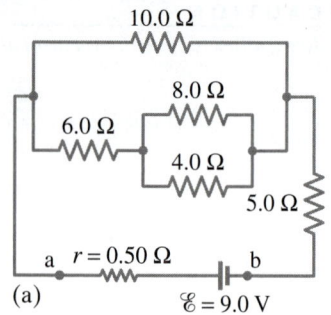

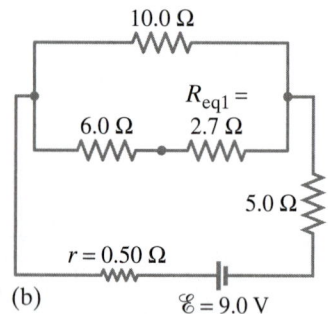

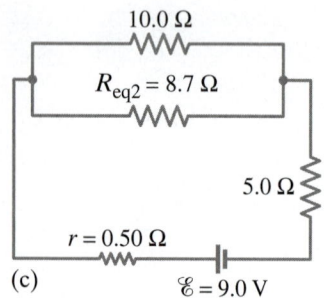

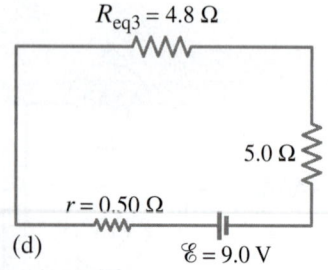

FIGURE 19–10 Circuit for Example 19–7, where r is the internal resistance of the battery.

FIGURE 19–11 Currents can be calculated using Kirchhoff's rules.

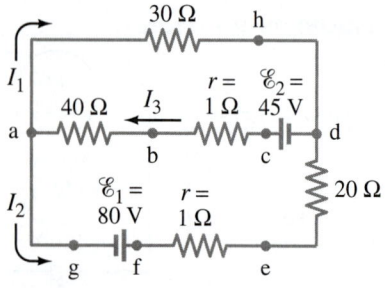

EXAMPLE 19–7 Analyzing a circuit. A 9.0-V battery whose internal resistance r is $0.50\,\Omega$ is connected in the circuit shown in Fig. 19–10a. (a) How much current is drawn from the battery? (b) What is the terminal voltage of the battery? (c) What is the current in the 6.0-Ω resistor?

APPROACH To find the current out of the battery, we first need to determine the equivalent resistance R_{eq} of the entire circuit, including r, which we do by identifying and isolating simple series or parallel combinations of resistors. Once we find I from Ohm's law, $I = \mathscr{E}/R_{eq}$, we get the terminal voltage using $V_{ab} = \mathscr{E} - Ir$. For (c) we apply Ohm's law to the 6.0-Ω resistor.

SOLUTION (a) We want to determine the equivalent resistance of the circuit. But where do we start? We note that the 4.0-Ω and 8.0-Ω resistors are in parallel, and so have an equivalent resistance R_{eq1} given by

$$\frac{1}{R_{eq1}} = \frac{1}{8.0\,\Omega} + \frac{1}{4.0\,\Omega} = \frac{3}{8.0\,\Omega};$$

so $R_{eq1} = 2.7\,\Omega$. This $2.7\,\Omega$ is in series with the 6.0-Ω resistor, as shown in the equivalent circuit of Fig. 19–10b. The net resistance of the lower arm of the circuit is then

$$R_{eq2} = 6.0\,\Omega + 2.7\,\Omega = 8.7\,\Omega,$$

as shown in Fig. 19–10c. The equivalent resistance R_{eq3} of the 8.7-Ω and 10.0-Ω resistances in parallel is given by

$$\frac{1}{R_{eq3}} = \frac{1}{10.0\,\Omega} + \frac{1}{8.7\,\Omega} = 0.21\,\Omega^{-1},$$

so $R_{eq3} = (1/0.21\,\Omega^{-1}) = 4.8\,\Omega$. This $4.8\,\Omega$ is in series with the 5.0-Ω resistor and the 0.50-Ω internal resistance of the battery (Fig. 19–10d), so the total equivalent resistance R_{eq} of the circuit is $R_{eq} = 4.8\,\Omega + 5.0\,\Omega + 0.50\,\Omega = 10.3\,\Omega$. Hence the current drawn is

$$I = \frac{\mathscr{E}}{R_{eq}} = \frac{9.0\,\text{V}}{10.3\,\Omega} = 0.87\,\text{A}.$$

(b) The terminal voltage of the battery is

$$V_{ab} = \mathscr{E} - Ir = 9.0\,\text{V} - (0.87\,\text{A})(0.50\,\Omega) = 8.6\,\text{V}.$$

(c) Now we can work back and get the current in the 6.0-Ω resistor. It must be the same as the current through the $8.7\,\Omega$ shown in Fig. 19–10c (why?). The voltage across that $8.7\,\Omega$ will be the emf of the battery minus the voltage drops across r and the 5.0-Ω resistor: $V_{8.7} = 9.0\,\text{V} - (0.87\,\text{A})(0.50\,\Omega + 5.0\,\Omega)$. Applying Ohm's law, we get the current (call it I')

$$I' = \frac{9.0\,\text{V} - (0.87\,\text{A})(0.50\,\Omega + 5.0\,\Omega)}{8.7\,\Omega} = 0.48\,\text{A}.$$

This is the current through the 6.0-Ω resistor.

19–3 Kirchhoff's Rules

In the last few Examples we have been able to find the currents in circuits by combining resistances in series and parallel, and using Ohm's law. This technique can be used for many circuits. However, some circuits are too complicated for that analysis. For example, we cannot find the currents in each part of the circuit shown in Fig. 19–11 simply by combining resistances as we did before.

To deal with complicated circuits, we use Kirchhoff's rules, devised by G. R. Kirchhoff (1824–1887) in the mid-nineteenth century. There are two rules, and they are simply convenient applications of the laws of conservation of charge and energy.

Kirchhoff's first rule or **junction rule** is based on the conservation of electric charge (we already used it to derive the equation for parallel resistors). It states that

> **at any junction point, the sum of all currents entering the junction must equal the sum of all currents leaving the junction.**

Junction rule (conservation of charge)

That is, whatever charge goes in must come out. For example, at the junction point a in Fig. 19–11, I_3 is entering whereas I_1 and I_2 are leaving. Thus Kirchhoff's junction rule states that $I_3 = I_1 + I_2$. We already saw an instance of this in the NOTE at the end of Example 19–5.

Kirchhoff's second rule or **loop rule** is based on the conservation of energy. It states that

> **the sum of the changes in potential around any closed loop of a circuit must be zero.**

Loop rule (conservation of energy)

To see why this rule should hold, consider a rough analogy with the potential energy of a roller coaster on its track. When the roller coaster starts from the station, it has a particular potential energy. As it is pulled up the first hill, its gravitational potential energy increases and reaches a maximum at the top. As it descends the other side, its potential energy decreases and reaches a local minimum at the bottom of the hill. As the roller coaster continues on its up and down path, its potential energy goes through more changes. But when it arrives back at the starting point, it has exactly as much potential energy as it had when it started at this point. Another way of saying this is that there was as much uphill as there was downhill.

Similar reasoning can be applied to an electric circuit. We will analyze the circuit of Fig. 19–11 shortly, but first we consider the simpler circuit in Fig. 19–12. We have chosen it to be the same as the equivalent circuit of Fig. 19–7b already discussed. The current in this circuit is $I = (12.0\text{ V})/(690\ \Omega) = 0.0174\text{ A}$, as we calculated in Example 19–4. (We keep an extra digit in I to reduce rounding errors.) The positive side of the battery, point e in Fig. 19–12a, is at a high potential compared to point d at the negative side of the battery. That is, point e is like the top of a hill for a roller coaster. We follow the current around the circuit starting at any point. We choose to start at point d and follow a small positive test charge completely around this circuit. As we go, we note all changes in potential. When the test charge returns to point d, the potential will be the same as when we started (total change in potential around the circuit is zero). We plot the changes in potential around the circuit in Fig. 19–12b; point d is arbitrarily taken as zero.

FIGURE 19–12 Changes in potential around the circuit in (a) are plotted in (b).

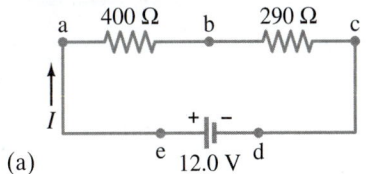

As our positive test charge goes from point d, which is the negative or low potential side of the battery, to point e, which is the positive terminal (high potential side) of the battery, the potential increases by 12.0 V. (This is like the roller coaster being pulled up the first hill.) That is,

$$V_{ed} = +12.0\text{ V}.$$

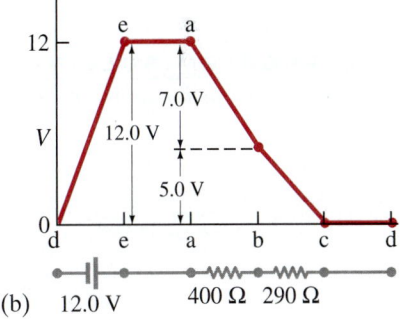

When our test charge moves from point e to point a, there is no change in potential because there is no source of emf and negligible resistance in the connecting wires.

Next, as the charge passes through the 400-Ω resistor to get to point b, there is a decrease in potential of $V = IR = (0.0174\text{ A})(400\ \Omega) = 7.0\text{ V}$. The positive test charge is flowing "downhill" since it is heading toward the negative terminal of the battery, as indicated in the graph of Fig. 19–12b. Because this is a *decrease* in potential, we use a *negative* sign:

$$V_{ba} = V_b - V_a = -7.0\text{ V}.$$

PROBLEM SOLVING
Be consistent with signs when applying the loop rule

As the charge proceeds from b to c there is another potential decrease (a "voltage drop") of $(0.0174\text{ A}) \times (290\ \Omega) = 5.0\text{ V}$, and this too is a decrease in potential:

$$V_{cb} = -5.0\text{ V}.$$

There is no change in potential as our test charge moves from c to d as we assume negligible resistance in the wires.

The sum of all the changes in potential around the circuit of Fig. 19–12 is

$$+12.0\text{ V} - 7.0\text{ V} - 5.0\text{ V} = 0.$$

This is exactly what Kirchhoff's loop rule said it would be.

PROBLEM SOLVING

Kirchhoff's Rules

1. **Label the current** in each separate branch of the given circuit with a different subscript, such as I_1, I_2, I_3 (see Fig. 19–11 or 19–13). Each current refers to a segment between two junctions. Choose the direction of each current, using an arrow. The direction can be chosen arbitrarily: if the current is actually in the opposite direction, it will come out with a minus sign in the solution.

2. **Identify the unknowns**. You will need as many independent equations as there are unknowns. You may write down more equations than this, but you will find that some of the equations will be redundant (that is, not be independent in the sense of providing new information). You may use $V = IR$ for each resistor, which sometimes will reduce the number of unknowns.

3. **Apply Kirchhoff's junction rule** at one or more junctions.

4. **Apply Kirchhoff's loop rule** for one or more loops: follow each loop in one direction only. Pay careful attention to subscripts, and to signs:
   (a) For a resistor, apply Ohm's law; the potential difference is negative (a decrease) if your chosen loop direction is the same as the chosen current direction through that resistor. The potential difference is positive (an increase) if your chosen loop direction is *opposite* to the chosen current direction.
   (b) For a battery, the potential difference is positive if your chosen loop direction is from the negative terminal toward the positive terminal; the potential difference is negative if the loop direction is from the positive terminal toward the negative terminal.

5. **Solve the equations** algebraically for the unknowns. Be careful with signs. At the end, check your answers by plugging them into the original equations, or even by using any additional loop or junction rule equations not used previously.

EXAMPLE 19–8 Using Kirchhoff's rules. Calculate the currents I_1, I_2, and I_3 in the three branches of the circuit in Fig. 19–13 (which is the same as Fig. 19–11).

APPROACH and SOLUTION

1. **Label the currents** and their directions. Figure 19–13 uses the labels I_1, I_2, and I_3 for the current in the three separate branches. Since (positive) current tends to move away from the positive terminal of a battery, we choose I_2 and I_3 to have the directions shown in Fig. 19–13. The direction of I_1 is not obvious in advance, so we arbitrarily chose the direction indicated. If the current actually flows in the opposite direction, our answer will have a negative sign.

2. **Identify the unknowns.** We have three unknowns (I_1, I_2, and I_3) and therefore we need three equations, which we get by applying Kirchhoff's junction and loop rules.

3. **Junction rule**: We apply Kirchhoff's junction rule to the currents at point a, where I_3 enters and I_2 and I_1 leave:

$$I_3 = I_1 + I_2. \quad \text{(i)}$$

This same equation holds at point d, so we get no new information by writing an equation for point d.

📝 **PROBLEM SOLVING**
Choose current directions arbitrarily

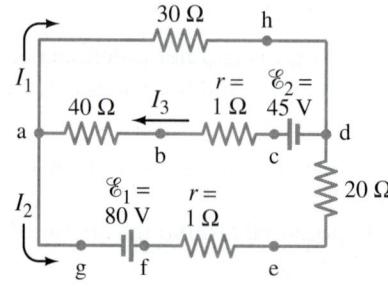

FIGURE 19–13 Currents can be calculated using Kirchhoff's rules. See Example 19–8.

4. **Loop rule**: We apply Kirchhoff's loop rule to two different closed loops. First we apply it to the upper loop ahdcba. We start (and end) at point a. From a to h we have a potential decrease $V_{ha} = -(I_1)(30\,\Omega)$. From h to d there is no change, but from d to c the potential increases by 45 V: that is, $V_{cd} = +45$ V. From c to a the potential decreases through the two resistances by an amount $V_{ac} = -(I_3)(40\,\Omega + 1\,\Omega) = -(41\,\Omega)I_3$. Thus we have $V_{ha} + V_{cd} + V_{ac} = 0$, or

$$-30I_1 + 45 - 41I_3 = 0, \qquad \text{(ii)}$$

where we have omitted the units (volts and amps) so we can more easily see the algebra. For our second loop, we take the outer loop ahdefga. (We could have chosen the lower loop abcdefga instead.) Again we start at point a, and going to point h we have $V_{ha} = -(I_1)(30\,\Omega)$. Next, $V_{dh} = 0$. But when we take our positive test charge from d to e, it actually is going uphill, against the current—or at least against the *assumed* direction of the current, which is what counts in this calculation. Thus $V_{ed} = +I_2(20\,\Omega)$ has a *positive* sign. Similarly, $V_{fe} = +I_2(1\,\Omega)$. From f to g there is a decrease in potential of 80 V because we go from the high potential terminal of the battery to the low. Thus $V_{gf} = -80$ V. Finally, $V_{ag} = 0$, and the sum of the potential changes around this loop is

$$-30I_1 + (20 + 1)I_2 - 80 = 0. \qquad \text{(iii)}$$

> ✏️ **PROBLEM SOLVING**
> *Be consistent with signs when applying the loop rule*

Our major work is done. The rest is algebra.

5. **Solve the equations**. We have three equations—labeled (i), (ii), and (iii)—and three unknowns. From Eq. (iii) we have

$$I_2 = \frac{80 + 30I_1}{21} = 3.8 + 1.4I_1. \qquad \text{(iv)}$$

From Eq. (ii) we have

$$I_3 = \frac{45 - 30I_1}{41} = 1.1 - 0.73I_1. \qquad \text{(v)}$$

We substitute Eqs. (iv) and (v) into Eq. (i):

$$I_1 = I_3 - I_2 = 1.1 - 0.73I_1 - 3.8 - 1.4I_1.$$

We solve for I_1, collecting terms:

$$3.1I_1 = -2.7$$
$$I_1 = -0.87 \text{ A}.$$

The negative sign indicates that the direction of I_1 is actually opposite to that initially assumed and shown in Fig. 19–13. The answer automatically comes out in amperes because our voltages and resistances were in volts and ohms. From Eq. (iv) we have

> ✏️ **PROBLEM SOLVING**
> *I_1 is in the opposite direction from that assumed in Fig. 19–13*

$$I_2 = 3.8 + 1.4I_1 = 3.8 + 1.4(-0.87) = 2.6 \text{ A},$$

and from Eq. (v)

$$I_3 = 1.1 - 0.73I_1 = 1.1 - 0.73(-0.87) = 1.7 \text{ A}.$$

This completes the solution.

NOTE The unknowns in different situations are not necessarily currents. It might be that the currents are given and we have to solve for unknown resistance or voltage. The variables are then different, but the technique is the same.

EXERCISE D Write the Kirchhoff equation for the lower loop abcdefga of Example 19–8 and show, assuming the currents calculated in this Example, that the potentials add to zero for this lower loop.

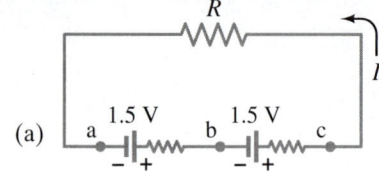
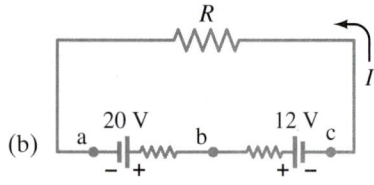
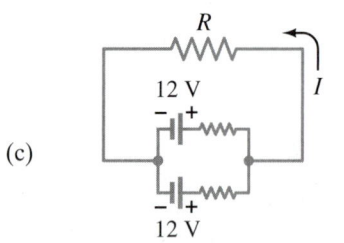

FIGURE 19–14 Batteries in series, (a) and (b), and in parallel (c).

FIGURE 19–15 Example 19–9, a jump start.

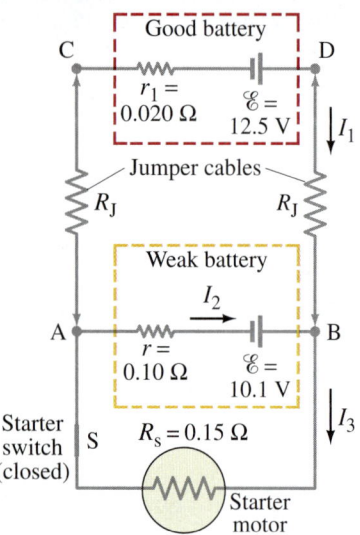

19–4 EMFs in Series and in Parallel; Charging a Battery

When two or more sources of emf, such as batteries, are arranged in series as in Fig. 19–14a, the total voltage is the algebraic sum of their respective voltages. On the other hand, when a 20-V and a 12-V battery are connected oppositely, as shown in Fig. 19–14b, the net voltage V_{ca} is 8 V (ignoring voltage drop across internal resistances). That is, a positive test charge moved from a to b gains in potential by 20 V, but when it passes from b to c it drops by 12 V. So the net change is 20 V − 12 V = 8 V. You might think that connecting batteries in reverse like this would be wasteful. For most purposes that would be true. But such a reverse arrangement is precisely how a battery charger works. In Fig. 19–14b, the 20-V source is charging up the 12-V battery. Because of its greater voltage, the 20-V source is forcing charge back into the 12-V battery: electrons are being forced into its negative terminal and removed from its positive terminal.

An automobile alternator keeps the car battery charged in the same way. A voltmeter placed across the terminals of a (12-V) car battery with the engine running fairly fast can tell you whether or not the alternator is charging the battery. If it is, the voltmeter reads 13 or 14 V. If the battery is not being charged, the voltage will be 12 V, or less if the battery is discharging. Car batteries can be recharged, but other batteries may not be rechargeable because the chemical reactions in many cannot be reversed. In such cases, the arrangement of Fig. 19–14b would simply waste energy.

Sources of emf can also be arranged in parallel, Fig. 19–14c, which—if the emfs are the same—can provide more energy when large currents are needed. Each of the cells in parallel has to produce only a fraction of the total current, so the energy loss due to internal resistance is less than for a single cell; and the batteries will go dead less quickly.

EXAMPLE 19–9 Jump starting a car. A good car battery is being used to jump start a car with a weak battery. The good battery has an emf of 12.5 V and internal resistance 0.020 Ω. Suppose the weak battery has an emf of 10.1 V and internal resistance 0.10 Ω. Each copper jumper cable is 3.0 m long and 0.50 cm in diameter, and can be attached as shown in Fig. 19–15. Assume the starter motor can be represented as a resistor $R_s = 0.15\ \Omega$. Determine the current through the starter motor (a) if only the weak battery is connected to it, and (b) if the good battery is also connected, as shown in Fig. 19–15.

APPROACH We apply Kirchhoff's rules, but in (b) we will first need to determine the resistance of the jumper cables using their dimensions and the resistivity ($\rho = 1.68 \times 10^{-8}\ \Omega \cdot$m for copper) as discussed in Section 18–4.

SOLUTION (a) The circuit with only the weak battery and no jumper cables is simple: an emf of 10.1 V connected to two resistances in series, 0.10 Ω + 0.15 Ω = 0.25 Ω. Hence the current is $I = V/R = (10.1\ \text{V})/(0.25\ \Omega) = 40$ A.

(b) We need to find the resistance of the jumper cables that connect the good battery to the weak one. From Eq. 18–3, each has resistance

$$R_J = \frac{\rho \ell}{A} = \frac{(1.68 \times 10^{-8}\ \Omega \cdot \text{m})(3.0\ \text{m})}{(\pi)(0.25 \times 10^{-2}\ \text{m})^2} = 0.0026\ \Omega.$$

Kirchhoff's loop rule for the full outside loop gives

$$12.5\ \text{V} - I_1(2R_J + r_1) - I_3 R_S = 0$$
$$12.5\ \text{V} - I_1(0.025\ \Omega) - I_3(0.15\ \Omega) = 0 \qquad \textbf{(i)}$$

since $(2R_J + r) = (0.0052\ \Omega + 0.020\ \Omega) = 0.025\ \Omega$.

536 CHAPTER 19 DC Circuits

The loop rule for the lower loop, including the weak battery and the starter, gives

$$10.1 \text{ V} - I_3(0.15 \, \Omega) - I_2(0.10 \, \Omega) = 0. \quad \text{(ii)}$$

The junction rule at point B gives

$$I_1 + I_2 = I_3. \quad \text{(iii)}$$

We have three equations in three unknowns. From Eq. (iii),

$$I_1 = I_3 - I_2$$

and we substitute this into Eq. (i):

$$12.5 \text{ V} - (I_3 - I_2)(0.025 \, \Omega) - I_3(0.15 \, \Omega) = 0,$$
$$12.5 \text{ V} - I_3(0.175 \, \Omega) + I_2(0.025 \, \Omega) = 0.$$

Combining this last equation with Eq. (ii) gives

$$12.5 \text{ V} - I_3(0.175 \, \Omega) + \left(\frac{10.1 \text{ V} - I_3(0.15 \, \Omega)}{0.10 \, \Omega}\right)(0.025 \, \Omega) = 0$$

or

$$I_3 = \frac{12.5 \text{ V} + 2.5 \text{ V}}{(0.175 \, \Omega + 0.0375 \, \Omega)} = 71 \text{ A},$$

quite a bit better than in part (*a*).
The other currents are $I_2 = -5$ A and $I_1 = 76$ A. Note that $I_2 = -5$ A is in the opposite direction from what we assumed in Fig. 19–15. The terminal voltage of the weak 10.1-V battery when being charged is

$$V_{BA} = 10.1 \text{ V} - (-5 \text{ A})(0.10 \, \Omega) = 10.6 \text{ V}.$$

NOTE The circuit in Fig. 19–15, without the starter motor, is how a battery can be charged. The stronger battery pushes charge back into the weaker battery.

EXERCISE E If the jumper cables of Example 19–9 were mistakenly connected in reverse, the positive terminal of each battery would be connected to the negative terminal of the other battery (Fig. 19–16). What would be the current *I* even before the starter motor is engaged (the switch S in Fig. 19–16 is open)? Why could this cause the batteries to explode?

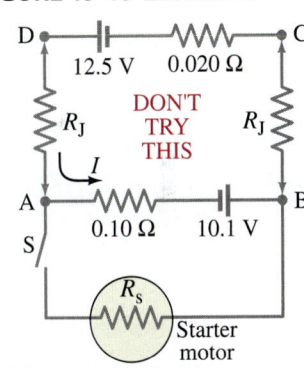

FIGURE 19–16 Exercise E.

Safety when Jump Starting

Before jump starting a car's weak battery, be sure both batteries are 12 V and check the polarity of both batteries. The following (cautious) procedure applies if the negative (−) terminal is ground (attached by a cable to the metal car frame and motor), and the "hot" terminal is positive (+) on both batteries, as is the case for most modern cars. The + terminal is usually marked by a red color, often a red cover. The safest procedure is to first connect the hot (+) terminal of the weak battery to the hot terminal of the good battery (using the cable with red clamps). Spread apart the handles of each clamp to squeeze the contact tightly. Then connect the black cable, first to the ground terminal of the good battery, and the other end to a clean exposed metal part (i.e., at ground) on the car with the weak battery. (This last connection should preferably be not too close to the battery, which in rare cases might leak H_2 gas that could ignite at the spark that may accompany the final connection.) This is safer than connecting directly to the ground terminal. When you are ready to start the disabled car, it helps to have the good car running (to keep its battery fully charged). As soon as the disabled car starts, immediately detach the cables in the exact reverse order (ground cable first).

In the photo of Fig. 19–15, the above procedure is not being followed. Note the safety error: with ground terminals connected, if the red clamp (+12 V) touches a metal part (= ground), even if dropped by the person, a short circuit with damaging high electric current can occur (hundreds of amps).

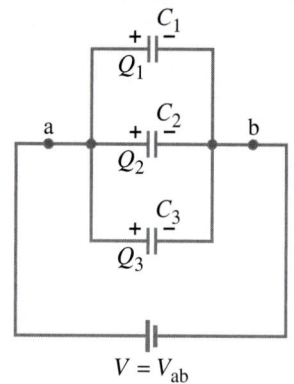

FIGURE 19–17 Capacitors in parallel: $C_{eq} = C_1 + C_2 + C_3$.

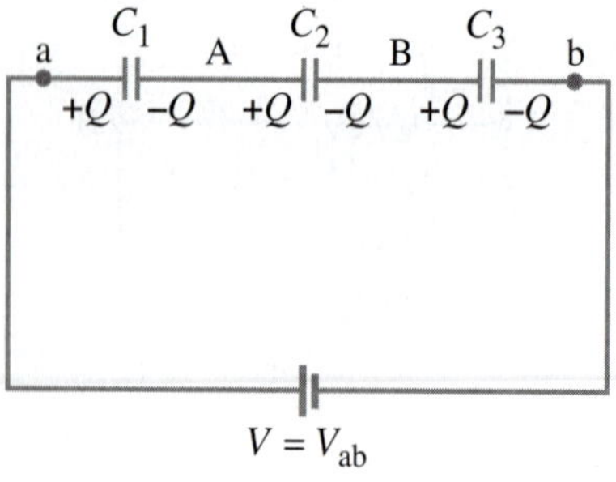

19–5 Circuits Containing Capacitors in Series and in Parallel

Just as resistors can be placed in series or in parallel in a circuit, so can capacitors (Chapter 17). We first consider a **parallel** connection as shown in Fig. 19–17. If a battery supplies a potential difference V to points a and b, this same potential difference $V = V_{ab}$ exists across each of the capacitors. That is, since the left-hand plates of all the capacitors are connected by conductors, they all reach the same potential V_a when connected to the battery; and the right-hand plates each reach potential V_b. Each capacitor plate acquires a charge given by $Q_1 = C_1 V$, $Q_2 = C_2 V$, and $Q_3 = C_3 V$. The total charge Q that must leave the battery is then

$$Q = Q_1 + Q_2 + Q_3 = C_1 V + C_2 V + C_3 V.$$

Let us try to find a single equivalent capacitor that will hold the same charge Q at the same voltage $V = V_{ab}$. It will have a capacitance C_{eq} given by

$$Q = C_{eq} V.$$

Combining the two previous equations, we have

$$C_{eq} V = C_1 V + C_2 V + C_3 V = (C_1 + C_2 + C_3) V$$

or

$$C_{eq} = C_1 + C_2 + C_3. \qquad \text{[parallel]} \quad (19\text{–}5)$$

The net effect of connecting capacitors in parallel is thus to *increase* the capacitance. Connecting capacitors in parallel is essentially increasing the area of the plates where charge can accumulate (see, for example, Eq. 17–8).

Capacitors can also be connected in **series**: that is, end to end as shown in Fig. 19–18. A charge $+Q$ flows from the battery to one plate of C_1, and $-Q$ flows to one plate of C_3. The regions A and B between the capacitors were originally neutral, so the net charge there must still be zero. The $+Q$ on the left plate of C_1 attracts a charge of $-Q$ on the opposite plate. Because region A must have a zero net charge, there is $+Q$ on the left plate of C_2. The same considerations apply to the other capacitors, so we see that the charge on each capacitor plate has the same magnitude Q. A single capacitor that could replace these three in series without affecting the circuit (that is, Q and V the same) would have a capacitance C_{eq} where

$$Q = C_{eq} V.$$

FIGURE 19–18 Capacitors in series: $\dfrac{1}{C_{eq}} = \dfrac{1}{C_1} + \dfrac{1}{C_2} + \dfrac{1}{C_3}$.

The total voltage V across the three capacitors in series must equal the sum of the voltages across each capacitor:

$$V = V_1 + V_2 + V_3.$$

We also have for each capacitor $Q = C_1 V_1$, $Q = C_2 V_2$, and $Q = C_3 V_3$, so we substitute for V_1, V_2, V_3, and V into the last equation and get

$$\frac{Q}{C_{eq}} = \frac{Q}{C_1} + \frac{Q}{C_2} + \frac{Q}{C_3} = Q\left(\frac{1}{C_1} + \frac{1}{C_2} + \frac{1}{C_3}\right)$$

or

$$\frac{1}{C_{eq}} = \frac{1}{C_1} + \frac{1}{C_2} + \frac{1}{C_3}. \qquad \text{[series]} \quad (19\text{–}6)$$

> **⚠ CAUTION**
> *Formula for capacitors in series resembles formula for resistors in parallel*

Notice that the equivalent capacitance C_{eq} is *smaller* than the smallest contributing capacitance. Notice also that the forms of the equations for capacitors in series or in parallel are the reverse of their counterparts for resistance. That is, the formula for capacitors in series resembles the formula for resistors in parallel.

EXAMPLE 19–10 Equivalent capacitance. Determine the capacitance of a single capacitor that will have the same effect as the combination shown in Fig. 19–19a. Take $C_1 = C_2 = C_3 = C$.

APPROACH First we find the equivalent capacitance of C_2 and C_3 in parallel, and then consider that capacitance in series with C_1.

SOLUTION Capacitors C_2 and C_3 are connected in parallel, so they are equivalent to a single capacitor having capacitance

$$C_{23} = C_2 + C_3 = C + C = 2C.$$

This C_{23} is in series with C_1, Fig. 19–19b, so the equivalent capacitance of the entire circuit, C_{eq}, is given by

$$\frac{1}{C_{eq}} = \frac{1}{C_1} + \frac{1}{C_{23}} = \frac{1}{C} + \frac{1}{2C} = \frac{3}{2C}.$$

Hence the equivalent capacitance of the entire combination is $C_{eq} = \tfrac{2}{3}C$, and it is smaller than any of the contributing capacitances, $C_1 = C_2 = C_3 = C$.

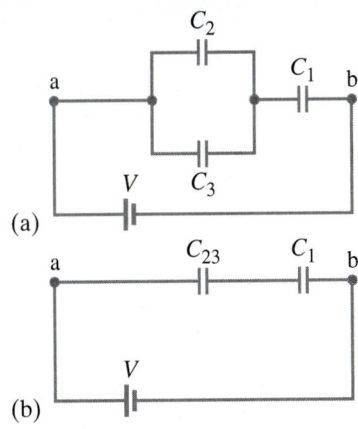

FIGURE 19–19
Examples 19–10 and 19–11.

PROBLEM SOLVING
Remember to take the reciprocal

EXERCISE F Consider two identical capacitors $C_1 = C_2 = 10\,\mu F$. What are the smallest and largest capacitances that can be obtained by connecting these in series or parallel combinations? (a) $0.2\,\mu F$, $5\,\mu F$; (b) $0.2\,\mu F$, $10\,\mu F$; (c) $0.2\,\mu F$, $20\,\mu F$; (d) $5\,\mu F$, $10\,\mu F$; (e) $5\,\mu F$, $20\,\mu F$; (f) $10\,\mu F$, $20\,\mu F$.

EXAMPLE 19–11 Charge and voltage on capacitors. Determine the charge on each capacitor in Fig. 19–19a of Example 19–10 and the voltage across each, assuming $C = 3.0\,\mu F$ and the battery voltage is $V = 4.0\,V$.

APPROACH We have to work "backward" through Example 19–10. That is, we find the charge Q that leaves the battery, using the equivalent capacitance. Then we find the charge on each separate capacitor and the voltage across each. Each step uses Eq. 17–7, $Q = CV$.

SOLUTION The 4.0-V battery behaves as if it is connected to a capacitance $C_{eq} = \tfrac{2}{3}C = \tfrac{2}{3}(3.0\,\mu F) = 2.0\,\mu F$. Therefore the charge Q that leaves the battery, by Eq. 17–7, is

$$Q = CV = (2.0\,\mu F)(4.0\,V) = 8.0\,\mu C.$$

From Fig. 19–19a, this charge arrives at the negative plate of C_1, so $Q_1 = 8.0\,\mu C$. The charge Q that leaves the positive plate of the battery is split evenly between C_2 and C_3 (symmetry: $C_2 = C_3$) and is $Q_2 = Q_3 = \tfrac{1}{2}Q = 4.0\,\mu C$. Next, the voltages across C_2 and C_3 have to be the same. The voltage across each capacitor is obtained using $V = Q/C$. So

$$V_1 = Q_1/C_1 = (8.0\,\mu C)/(3.0\,\mu F) = 2.7\,V$$
$$V_2 = Q_2/C_2 = (4.0\,\mu C)/(3.0\,\mu F) = 1.3\,V$$
$$V_3 = Q_3/C_3 = (4.0\,\mu C)/(3.0\,\mu F) = 1.3\,V.$$

19–6 RC Circuits—Resistor and Capacitor in Series

Capacitor Charging

Capacitors and resistors are often found together in a circuit. Such **RC circuits** are common in everyday life. They are used to control the speed of a car's windshield wipers and the timing of traffic lights; they are used in camera flashes, in heart pacemakers, and in many other electronic devices. In *RC* circuits, we are not so interested in the final "steady state" voltage and charge on the capacitor, but rather in how these variables change in time.

A simple *RC* circuit is shown in Fig. 19–20a. When the switch S is closed, current immediately begins to flow through the circuit. Electrons will flow out from the negative terminal of the battery, through the resistor *R*, and accumulate on the upper plate of the capacitor. And electrons will flow into the positive terminal of the battery, leaving a positive charge on the other plate of the capacitor. As charge accumulates on the capacitor, the potential difference across it increases ($V_C = Q/C$), and the current is reduced until eventually the voltage across the capacitor equals the emf of the battery, $\mathcal{E}$. There is then no further current flow, and no potential difference across the resistor. The potential difference V_C across the capacitor, which is proportional to the charge on it ($V_C = Q/C$, Eq. 17–7), thus increases in time as shown in Fig. 19–20b. The shape of this curve is a type of exponential, and is given by the formula[†]

$$V_C = \mathcal{E}(1 - e^{-t/RC}), \qquad (19\text{–}7\text{a})$$

where we use the subscript C to remind us that V_C is the voltage across the capacitor and is given here as a function of time *t*. [The constant *e*, known as the base for natural logarithms, has the value $e = 2.718\cdots$. Do not confuse this *e* with *e* for the charge on the electron.]

We can write a similar formula for the charge $Q (= CV_C)$ on the capacitor:

$$Q = Q_0(1 - e^{-t/RC}), \qquad (19\text{–}7\text{b})$$

where Q_0 represents the maximum charge.

The product of the resistance *R* times the capacitance *C*, which appears in the exponent, is called the **time constant** τ of the circuit:

$$\tau = RC. \qquad (19\text{–}7\text{c})$$

The time constant is a measure of how quickly the capacitor becomes charged. [The units of *RC* are $\Omega \cdot F = (V/A)(C/V) = C/(C/s) = s$.] Specifically, it can be shown that the product *RC* gives the time required for the capacitor's voltage (and charge) to reach 63% of the maximum. This can be checked[‡] using any calculator with an e^x key: $e^{-1} = 0.37$, so for $t = RC$, then $(1 - e^{-t/RC}) = (1 - e^{-1}) = (1 - 0.37) = 0.63$. In a circuit, for example, where $R = 200\,\text{k}\Omega$ and $C = 3.0\,\mu\text{F}$, the time constant is $(2.0 \times 10^5\,\Omega)(3.0 \times 10^{-6}\,\text{F}) = 0.60\,\text{s}$. If the resistance is much smaller, the time constant is much smaller and the capacitor becomes charged much more quickly. This makes sense, because a lower resistance will retard the flow of charge less. All circuits contain some resistance (if only in the connecting wires), so a capacitor can never be charged instantaneously when connected to a battery.

Finally, what is the voltage V_R across the resistor in Fig. 19–20a? The imposed battery voltage is $\mathcal{E}$, so

$$V_R = \mathcal{E} - V_C = \mathcal{E}(1 - 1 + e^{-t/RC}) = \mathcal{E}e^{-t/RC}.$$

This is called an **exponential decay**. The current *I* flowing in the circuit is that flowing through the resistor and is also an exponential decay:

$$I = \frac{V_R}{R} = \frac{\mathcal{E}}{R}e^{-t/RC}. \qquad (19\text{–}7\text{d})$$

When the switch of the circuit in Fig. 19–20a is closed, the current is largest at first because there is no charge on the capacitor to impede it. As charge builds on the capacitor, the current decreases in time. That is exactly what Eq. 19–7d and Fig. 19–20c tell us.

> **CAUTION**
> *Don't confuse e for exponential with e for electron charge*

FIGURE 19–20 After the switch S closes in the *RC* circuit shown in (a), the voltage V_C across the capacitor increases with time as shown in (b), and the current through the resistor decreases with time as shown in (c).

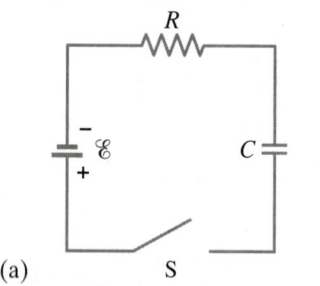
(a)

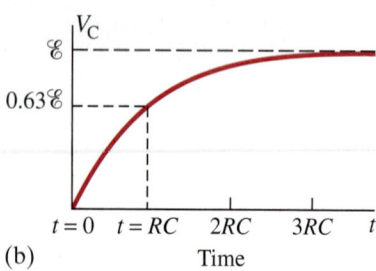
(b)

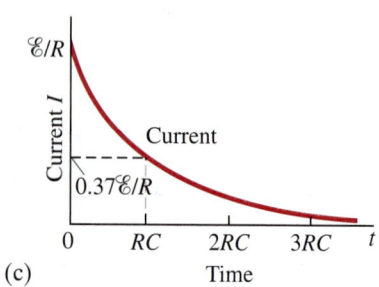
(c)

[†]The derivation uses calculus.

[‡]More simply, since $e = 2.718\cdots$, then $e^{-1} = 1/e = 1/2.718 = 0.37$. Note that *e* is the inverse operation to the natural logarithm ln: $\ln(e) = 1$, and $\ln(e^x) = x$.

EXAMPLE 19–12 **RC circuit, with emf.** The capacitance in the circuit of Fig. 19–20a is $C = 0.30\ \mu\text{F}$, the total resistance is $R = 20\ \text{k}\Omega$, and the battery emf is 12 V. Determine (a) the time constant, (b) the maximum charge the capacitor could acquire, (c) the time it takes for the charge to reach 99% of this value, and (d) the maximum current.

APPROACH We use Fig. 19–20 and Eqs. 19–7a, b, c, and d.

SOLUTION (a) The time constant is $RC = (2.0 \times 10^4\ \Omega)(3.0 \times 10^{-7}\ \text{F}) = 6.0 \times 10^{-3}\ \text{s} = 6.0\ \text{ms}$.

(b) The maximum charge would occur when no further current flows, so
$Q_0 = C\mathcal{E} = (3.0 \times 10^{-7}\ \text{F})(12\ \text{V}) = 3.6\ \mu\text{C}$.

(c) In Eq. 19–7b, we set $Q = 0.99 C\mathcal{E}$:

$$0.99 C\mathcal{E} = C\mathcal{E}(1 - e^{-t/RC}),$$

or

$$e^{-t/RC} = 1 - 0.99 = 0.01.$$

We take the natural logarithm of both sides (Appendix A–8), recalling that $\ln e^x = x$:

$$\frac{t}{RC} = -\ln(0.01) = 4.6$$

so

$$t = 4.6 RC = (4.6)(6.0 \times 10^{-3}\ \text{s}) = 28 \times 10^{-3}\ \text{s}$$

or 28 ms (less than $\tfrac{1}{30}$ s).

(d) The current is a maximum at $t = 0$ (the moment when the switch is closed) and there is no charge yet on the capacitor ($Q = 0$):

$$I_\text{max} = \frac{\mathcal{E}}{R} = \frac{12\ \text{V}}{2.0 \times 10^4\ \Omega} = 600\ \mu\text{A}.$$

Capacitor Discharging

The circuit just discussed involved the *charging* of a capacitor by a battery through a resistance. Now let us look at another situation: a capacitor is already charged to a voltage V_0 and charge Q_0, and it is then allowed to *discharge* through a resistance R as shown in Fig. 19–21a. In this case there is no battery. When the switch S is closed, charge begins to flow through resistor R from one side of the capacitor toward the other side, until the capacitor is fully discharged. The voltage across the capacitor decreases, as shown in Fig. 19–21b. This "exponential decay" curve is given by

$$V_C = V_0 e^{-t/RC},$$

where V_0 is the initial voltage across the capacitor. The voltage falls 63% of the way to zero (to $0.37 V_0$) in a time $\tau = RC$. Because the charge Q on the capacitor is $Q = CV$ (and $Q_0 = CV_0$), we can write

$$Q = Q_0 e^{-t/RC}$$

for a discharging capacitor, where Q_0 is the initial charge.

The voltage across the resistor will have the same magnitude as that across the capacitor at any instant, but the opposite sign, because there is zero applied emf: $V_C + V_R = 0$ so $V_R = -V_C = -V_0 e^{-t/RC}$. A graph of V_R vs. time would just be Fig. 19–21b upside down. The current $I = V_R/R = -(V_0/R)e^{-t/RC} = -I_0 e^{-t/RC}$. The current has its greatest magnitude at $t = 0$ and decreases exponentially in time. (The current has a minus sign because in Fig. 19–21a it flows in the opposite direction as compared to the current in Fig. 19–20a.)

FIGURE 19–21 For the *RC* circuit shown in (a), the voltage V_C across the capacitor decreases with time t, as shown in (b), after the switch S is closed at $t = 0$. The charge on the capacitor follows the same curve since $Q \propto V_C$.

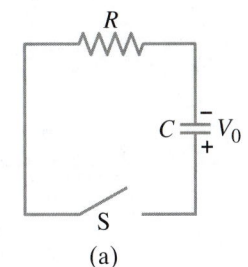

(a)

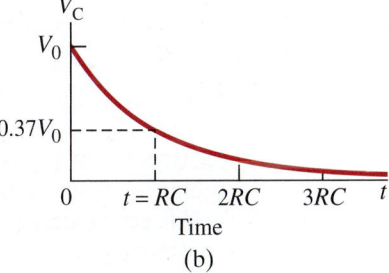

(b)

EXAMPLE 19–13 **A discharging RC circuit.** If a charged capacitor, $C = 35\ \mu F$, is connected to a resistance $R = 120\ \Omega$ as in Fig. 19–21a, how much time will elapse until the voltage falls to 10% of its original (maximum) value?

APPROACH The voltage across the capacitor decreases according to $V_C = V_0 e^{-t/RC}$. We set $V_C = 0.10 V_0$ (10% of V_0), but first we need to calculate $\tau = RC$.

SOLUTION The time constant for this circuit is given by

$$\tau = RC = (120\ \Omega)(35 \times 10^{-6}\ F) = 4.2 \times 10^{-3}\ s.$$

After a time t the voltage across the capacitor will be

$$V_C = V_0 e^{-t/RC}.$$

We want to know the time t for which $V_C = 0.10 V_0$. We substitute into the above equation

$$0.10 V_0 = V_0 e^{-t/RC}$$

so

$$e^{-t/RC} = 0.10.$$

The inverse operation to the exponential e is the natural log, ln. Thus

$$\ln(e^{-t/RC}) = -\frac{t}{RC} = \ln 0.10 = -2.3.$$

Solving for t, we find the elapsed time is

$$t = 2.3(RC) = (2.3)(4.2 \times 10^{-3}\ s) = 9.7 \times 10^{-3}\ s = 9.7\ ms.$$

NOTE We can find the time for any specified voltage across a capacitor by using $t = RC \ln(V_0/V_C)$.

CONCEPTUAL EXAMPLE 19–14 **Bulb in RC circuit.** In the circuit of Fig. 19–22, the capacitor is originally uncharged. Describe the behavior of the lightbulb from the instant switch S is closed until a long time later.

RESPONSE When the switch is first closed, the current in the circuit is high and the lightbulb burns brightly. As the capacitor charges, the voltage across the capacitor increases, causing the current to be reduced, and the lightbulb dims. As the potential difference across the capacitor approaches the same voltage as the battery, the current decreases toward zero and the lightbulb goes out.

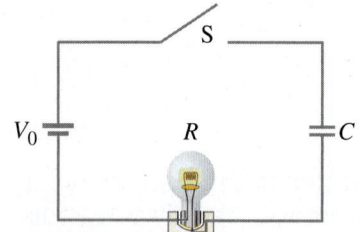

FIGURE 19–22 Example 19–14.

Medical and Other Applications of RC Circuits

The charging and discharging in an RC circuit can be used to produce voltage pulses at a regular frequency. The charge on the capacitor increases to a particular voltage, and then discharges. One way of initiating the discharge of the capacitor is by the use of a gas-filled tube which has an electrical breakdown when the voltage across it reaches a certain value V_0. After the discharge is finished, the tube no longer conducts current and the recharging process repeats itself, starting at a lower voltage V_0'. Figure 19–23 shows a possible circuit, and the **sawtooth voltage** it produces.

PHYSICS APPLIED
Sawtooth voltage

PHYSICS APPLIED
Blinking flashers

A simple blinking light can be an application of a sawtooth oscillator circuit. Here the emf is supplied by a battery; the neon bulb flashes on at a rate of perhaps 1 cycle per second. The main component of a "flasher unit" is a moderately large capacitor.

FIGURE 19–23 (a) An RC circuit, coupled with a gas-filled tube as a switch, can produce (b) a repeating "sawtooth" voltage.

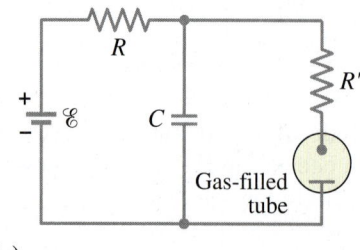
(a)

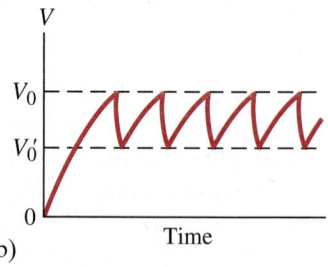
(b)

The intermittent windshield wipers of a car can also use an *RC* circuit. The *RC* time constant, which can be changed using a multi-positioned switch for different values of *R* with fixed *C*, determines the rate at which the wipers come on.

PHYSICS APPLIED
Windshield wipers on "intermittent"

> **EXERCISE G** A typical turn signal flashes perhaps twice per second, so its time constant is on the order of 0.5 s. Estimate the resistance in the circuit, assuming a moderate capacitor of $C = 1\,\mu\text{F}$.

An important medical use of an *RC* circuit is the electronic heart pacemaker, which can make a stopped heart start beating again by applying an electric stimulus through electrodes attached to the chest. The stimulus can be repeated at the normal heartbeat rate if necessary. The heart itself contains *pacemaker* cells, which send out tiny electric pulses at a rate of 60 to 80 per minute. These signals induce the start of each heartbeat. In some forms of heart disease, the natural pacemaker fails to function properly, and the heart loses its beat. Such patients use *electronic pacemakers* which produce a regular voltage pulse that starts and controls the frequency of the heartbeat. The electrodes are implanted in or near the heart (Fig. 19–24), and the circuit contains a capacitor and a resistor. The charge on the capacitor increases to a certain point and then discharges a pulse to the heart. Then it starts charging again. The pulsing rate depends on the time constant *RC*.

PHYSICS APPLIED
Heart pacemaker

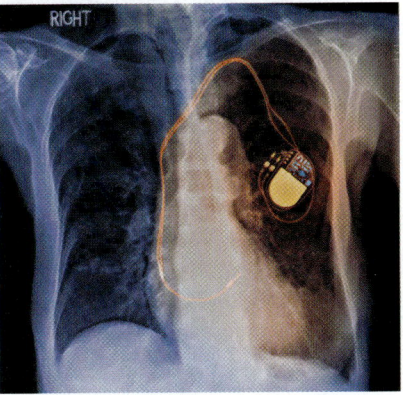
FIGURE 19–24 Electronic battery-powered pacemaker can be seen on the rib cage in this X-ray (color added).

19–7 Electric Hazards

Excess electric current can overheat wires in buildings and cause fires, as discussed in Section 18–6. Electric current can also damage the human body or even be fatal. Electric current through the human body can cause damage in two ways: (1) heating tissue and causing burns; (2) stimulating nerves and muscles, and we feel a "shock." The severity of a shock depends on the magnitude of the current, how long it acts, and through what part of the body it passes. A current passing through vital organs such as the heart or brain is especially damaging.

PHYSICS APPLIED
Dangers of electricity

A current of about 1 mA or more can be felt and may cause pain. Currents above 10 mA cause severe contraction of the muscles, and a person may not be able to let go of the source of the current (say, a faulty appliance or wire). Death from paralysis of the respiratory system can occur. Artificial respiration can sometimes revive a victim. If a current above about 80 to 100 mA passes across the torso, so that a portion passes through the heart for more than a second or two, the heart muscles will begin to contract irregularly and blood will not be properly pumped. This condition is called **ventricular fibrillation**. If it lasts for long, death results. Strangely enough, if the current is much larger, on the order of 1 A, death by heart failure may be less likely,[†] but such currents can cause serious burns if concentrated through a small area of the body.

It is current that harms, but it is voltage that drives the current. The seriousness of an electric shock depends on the current and thus on the applied voltage and the effective resistance of the body. Living tissue has low resistance because the fluid of cells contains ions that can conduct quite well. However, the outer layer of skin, when dry, offers high resistance and is thus protective. The effective resistance between two points on opposite sides of the body when the skin is dry is on the order of 10^4 to $10^6\,\Omega$. But when the skin is wet, the resistance may be $10^3\,\Omega$ or less. A person who is barefoot or wearing thin-soled shoes will be in good contact with the ground, and touching a 120-V line with a wet hand can result in a current

$$I = \frac{120\,\text{V}}{1000\,\Omega} = 120\,\text{mA}.$$

As we saw, this could be lethal.

[†]Larger currents apparently bring the entire heart to a standstill. Upon release of the current, the heart returns to its normal rhythm. This may not happen when fibrillation occurs because, once started, it can be hard to stop. Fibrillation may also occur as a result of a heart attack or during heart surgery. A device known as a *defibrillator* (described in Section 17–9) can apply a brief high current to the heart, causing complete heart stoppage which is often followed by resumption of normal beating.

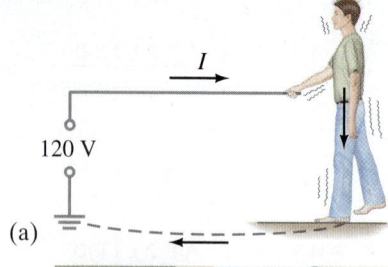

FIGURE 19–25 You can receive a shock when the circuit is completed.

⚠️ **CAUTION**
Keep one hand in your pocket when other touches electricity

🚶 **PHYSICS APPLIED**
Grounding and shocks

You can get a shock by becoming part of a complete circuit. Figure 19–25 shows two ways the circuit might be completed when you accidentally touch a "hot" electric wire—"hot" meaning a high potential relative to ground such as 120 V (normal U.S. household voltage) or 240 V (many other countries). The other wire of building wiring is connected to ground—either by a wire connected to a buried conductor, or via a metal water pipe into the ground. The current in Fig. 19–25a passes from the high-voltage wire through you to ground through your bare feet, and back along the ground (a fair conductor) to the ground terminal of the source. If you stand on a good insulator—thick rubber-soled shoes or a dry wood floor—there will be much more resistance in the circuit and much less current through you. If you stand with bare feet on the ground, or in a bathtub, there is lethal danger because the resistance is much less and the current greater. In a bathtub (or swimming pool), not only are you wet, which reduces your resistance, but the water is in contact with the drain pipe (typically metal) that leads to the ground. It is strongly recommended that you not touch anything electrical when wet or in bare feet. The use of non-metal pipes would be protective.

In Fig. 19–25b, a person touches a faulty "hot" wire with one hand, and the other hand touches a sink faucet (connected to ground via the pipe or even by water in a non-metal pipe). The current is particularly dangerous because it passes across the chest, through the heart and lungs. A useful rule: if one hand is touching something electrical, keep your other hand in your back pocket (don't use it!), and wear thick rubber-soled shoes. Also remove metal jewelry, especially rings (your finger is usually moist under a ring).

You can come into contact with a hot wire by touching a bare wire whose insulation has worn off, or from a bare wire inside an appliance when you're tinkering with it. (Always unplug an electrical device before investigating its insides!)† Also, a wire inside a device may break or lose its insulation and come in contact with the case. If the case is metal, it will conduct electricity. A person could then suffer a severe shock merely by touching the case, as shown in Fig. 19–26b. To prevent

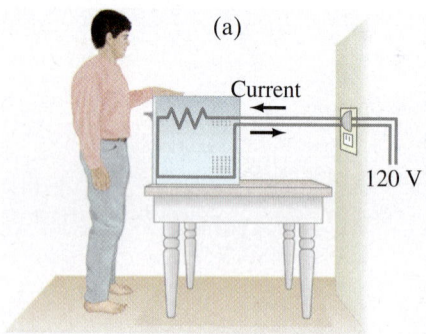
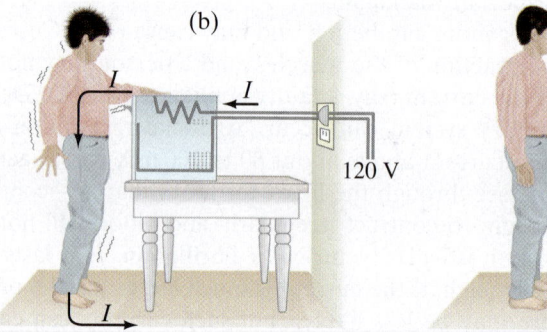
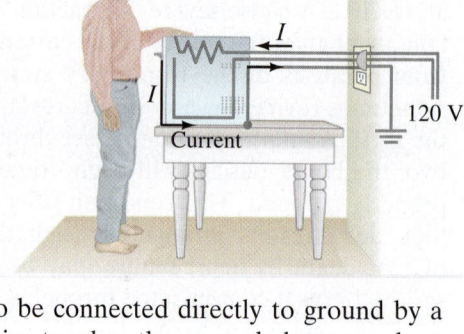

FIGURE 19–26 (a) An electric oven operating normally with a 2-prong plug. (b) A short to a metal case which is ungrounded, causing a shock. (c) A short to the case which is grounded by a 3-prong plug; almost no current goes through the person.

an accident, metal cases are supposed to be connected directly to ground by a separate ground wire. Then if a "hot" wire touches the grounded case, a short circuit to ground immediately occurs internally, as shown in Fig. 19–26c, and most of the current passes through the low-resistance ground wire rather than through the person. Furthermore, the high current should open a fuse or circuit breaker. Grounding a metal case is done by a separate ground wire connected to the third (round) prong of a 3-prong plug. Never cut off the third prong of a plug—it could save your life. A three-prong plug, and an adapter, are shown in Figs. 19–27a and b.

†Even then you can get a bad shock from a capacitor that hasn't been discharged until you touch it.

FIGURE 19–27 (a) A 3-prong plug, and (b) an adapter (white) for old-fashioned 2-prong outlets—be sure to screw down the ground tab (green color in photo).

Safe Wiring

Why is a third wire needed? The 120 V is carried by the other two wires—one **hot** (120 V ac), the other **neutral**, which is itself grounded. The third "dedicated" ground wire with the round prong may seem redundant. But it is protection for two reasons: (1) It protects against internal wiring that may have been done incorrectly or is faulty as discussed above, Fig. 19–26. (2) The *neutral* wire carries the full normal current ("return" current from the hot 120 V) and it does have resistance—so there can be a voltage drop along the neutral wire, normally small; but if connections are poor or corroded, or the plug is loose, the resistance could be large enough that you might feel that voltage if you touched the neutral wire some distance from its grounding point.

Some electrical devices come with only two wires, and the plug's two prongs are of different widths; the plug can be inserted only one way into the outlet so that the intended neutral (wider prong) in the device is connected to neutral in the wiring (Fig. 19–28). For example, the screw threads of a lightbulb are meant to be connected to neutral (and the base contact to hot), to avoid shocks when changing a bulb in a possibly protruding socket. Devices with 2-prong plugs do *not* have their cases grounded; they are supposed to have double electric insulation (or have a nonmetal case). Take extra care anyway.

The insulation on a wire may be color coded. Hand-held meters (Section 19–8) may have red (hot) and black (ground) lead wires. But in a U.S. house, the hot wire is often black (though it may be red), whereas white is neutral and green (or bare) is the dedicated ground, Fig. 19–29. But beware: these color codes cannot always be trusted.

[In the U.S., three wires normally enter a house: two *hot* wires at 120 V each (which add together to 240 V for appliances or devices that run on 240 V) plus the grounded *neutral* (carrying return current for the two hot wires). See Fig. 19–29. The "dedicated" *ground* wire (non-current carrying) is a fourth wire that does not come from the electric company but enters the house from a nearby heavy stake in the ground or a buried metal pipe. The two hot wires can feed separate 120-V circuits in the house, so each 120-V circuit inside the house has only three wires, including the dedicated ground.]

Normal circuit breakers (Sections 18–6 and 20–7) protect equipment and buildings from overload and fires. They protect humans only in some circumstances, such as the very high currents that result from a short, if they respond quickly enough. *Ground fault circuit interrupters* (GFCI or GFI), described in Section 21–9, are designed to protect people from the much lower currents (10 mA to 100 mA) that are lethal but would not throw a 15-A circuit breaker or blow a 20-A fuse.

Another danger is **leakage current**, by which we mean a current along an unintended path. Leakage currents are often "capacitively coupled." For example, a wire in a lamp forms a capacitor with the metal case; charges moving in one conductor attract or repel charge in the other, so there is a current. Typical electrical codes limit leakage currents to 1 mA for any device, which is usually harmless. It could be dangerous, however, to a hospital patient with implanted electrodes, due to the absence of the protective skin layer and because the current can pass directly through the heart. Although 100 mA may be needed to cause heart fibrillation when entering through the hands and spreading out through the body (very little of it actually passing through the heart), but as little as 0.02 mA can cause fibrillation when passing directly to the heart. Thus, a "wired" patient is in considerable danger from leakage current even from as simple an act as touching a lamp.

Finally, don't touch a downed power line (lethal!) or even get near it. A hot power line is at thousands of volts. A huge current can flow along the ground from the point where the high-voltage wire touches the ground. This current is great enough that the voltage between your two feet could be large and dangerous. Tip: stand on one foot, or run so only one foot touches the ground at a time.

CAUTION
Necessity of third (ground) wire

FIGURE 19–28 A polarized 2-prong plug.

CAUTION
Black wire may be either ground or hot. Beware!

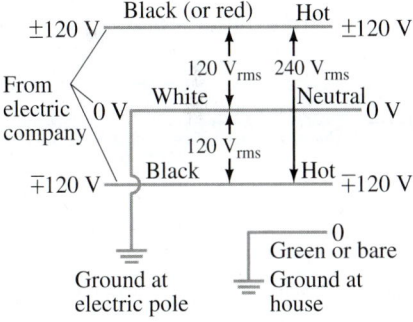

FIGURE 19–29 Four wires entering a typical house. The color codes for wires are not always as shown here—be careful!

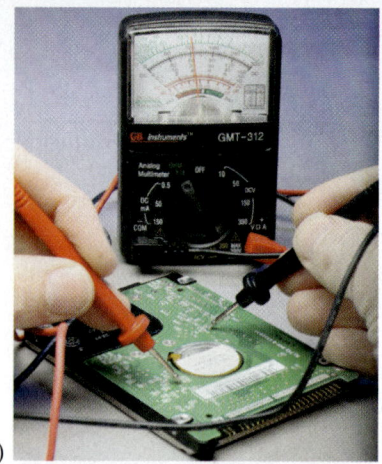

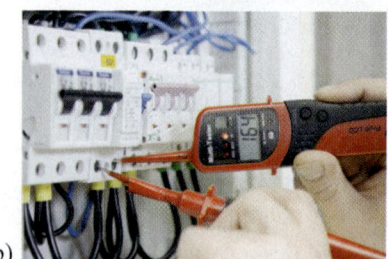

FIGURE 19–30 (a) An analog multimeter. (b) An electronic digital meter measuring voltage at a circuit breaker.

Ammeters use shunt resistor in parallel

FIGURE 19–31 An ammeter is a galvanometer in parallel with a (shunt) resistor with low resistance, R_{sh}.

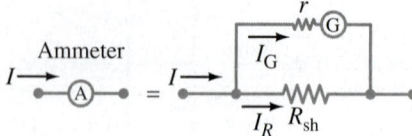

19–8 Ammeters and Voltmeters—Measurement Affects the Quantity Being Measured

Measurement is a fundamental part of physics, and is not as simple as you might think. Measuring instruments can not be taken for granted; their results are not perfect and often need to be interpreted. As an illustration of measurement "theory" we examine here how electrical quantities are measured using meters. We also examine how meters affect the quantity they attempt to measure.

An **ammeter** measures current, and a **voltmeter** measures potential difference or voltage. Each can be either: (1) an *analog* meter, which displays numerical values by the position of a pointer that can move across a scale (Fig. 19–30a); or (2) a *digital* meter, which displays the numerical value in numbers (Fig. 19–30b). We now examine how analog meters work.

An analog ammeter or voltmeter, in which the reading is by a pointer on a scale (Fig. 19–30a), uses a *galvanometer*. A galvanometer works on the principle of the force between a magnetic field and a current-carrying coil of wire; it is straightforward to understand and will be discussed in Chapter 20. For now, we only need to know that the deflection of the needle of a galvanometer is proportional to the current flowing through it. The *full-scale current sensitivity* of a galvanometer, I_m, is the electric current needed to make the needle deflect full scale, typically about 50 μA.

A galvanometer whose sensitivity I_m is 50 μA can measure currents from about 1 μA (currents smaller than this would be hard to read on the scale) up to 50 μA. To measure larger currents, a resistor is placed in parallel with the galvanometer. An analog **ammeter**, represented by the symbol •–Ⓐ–•, consists of a galvanometer (•–Ⓖ–•) in parallel with a resistor called the **shunt resistor**, as shown in Fig. 19–31. ("Shunt" is a synonym for "in parallel.") The shunt resistance is R_{sh}, and the resistance of the galvanometer coil is r. The value of R_{sh} is chosen according to the full-scale deflection desired; R_{sh} is normally very small—giving an ammeter a very small net resistance—so most of the current passes through R_{sh} and very little ($\lesssim 50\,\mu\text{A}$) passes through the galvanometer to deflect the needle.

EXAMPLE 19–15 Ammeter design. Design an ammeter to read 1.0 A at full scale using a galvanometer with a full-scale sensitivity of 50 μA and a resistance $r = 30\,\Omega$. Check if the scale is linear.

APPROACH Only 50 μA $(= I_G = 0.000050\,\text{A})$ of the 1.0-A current passes through the galvanometer to give full-scale deflection. The rest of the current $(I_R = 0.999950\,\text{A})$ passes through the small shunt resistor, R_{sh}, Fig. 19–31. The potential difference across the galvanometer equals that across the shunt resistor (they are in parallel). We apply Ohm's law to find R_{sh}.

SOLUTION Because $I = I_G + I_R$, when $I = 1.0$ A flows into the meter, we want I_R through the shunt resistor to be $I_R = 0.999950$ A. The potential difference across the shunt is the same as across the galvanometer, so

$$I_R R_{sh} = I_G r.$$

Then

$$R_{sh} = \frac{I_G r}{I_R} = \frac{(5.0 \times 10^{-5}\,\text{A})(30\,\Omega)}{(0.999950\,\text{A})} = 1.5 \times 10^{-3}\,\Omega,$$

or 0.0015 Ω. The shunt resistor must thus have a *very* low resistance and most of the current passes through it.

Because $I_G = I_R(R_{sh}/r)$ and (R_{sh}/r) is constant, we see that the scale is linear (needle deflection is proportional to I_G). If the current $I \approx I_R$ into the meter is half of full scale, 0.50 A, the current to the galvanometer will be $I_G = I_R(R_{sh}/r) = (0.50\,\text{A})(1.5 \times 10^{-3}\,\Omega)/(30\,\Omega) = 25\,\mu\text{A}$, which would make the needle deflect halfway, as it should.

An analog **voltmeter** (•–Ⓥ–•) consists of a galvanometer and a resistor R_{ser} connected in series, Fig. 19–32. R_{ser} is usually large, giving a voltmeter a high internal resistance.

Voltmeters use series resistor

EXAMPLE 19–16 Voltmeter design. Using a galvanometer with internal resistance $r = 30\,\Omega$ and full-scale current sensitivity of $50\,\mu A$, design a voltmeter that reads from 0 to 15 V. Is the scale linear?

APPROACH When a potential difference of 15 V exists across the terminals of our voltmeter, we want $50\,\mu A$ to be passing through it so as to give a full-scale deflection.

SOLUTION From Ohm's law, $V = IR$, we have (Fig. 19–32)
$$15\,V = (50\,\mu A)(r + R_{ser}),$$
so
$$R_{ser} = (15\,V)/(5.0 \times 10^{-5}\,A) - r = 300\,k\Omega - 30\,\Omega \approx 300\,k\Omega.$$

Notice that $r = 30\,\Omega$ is so small compared to the value of R_{ser} that it doesn't influence the calculation significantly. The scale will again be linear: if the voltage to be measured is 6.0 V, the current passing through the voltmeter will be $(6.0\,V)/(3.0 \times 10^5\,\Omega) = 2.0 \times 10^{-5}\,A$, or $20\,\mu A$. This will produce two-fifths of full-scale deflection, as required $(6.0\,V/15.0\,V = 2/5)$.

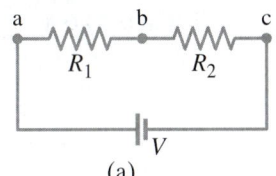

FIGURE 19–32 A voltmeter is a galvanometer in series with a resistor with high resistance, R_{ser}.

FIGURE 19–33 Measuring current and voltage.

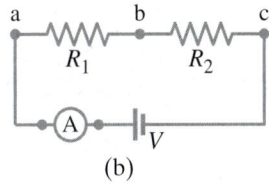
(a)

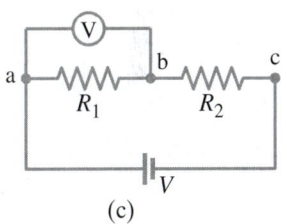
(b)

(c)

How to Connect Meters

Suppose you wish to determine the current I in the circuit shown in Fig. 19–33a, and the voltage V across the resistor R_1. How exactly are ammeters and voltmeters connected to the circuit being measured?

Because an ammeter is used to measure the current flowing in the circuit, it must be inserted directly into the circuit, in series with the other elements, as shown in Fig. 19–33b. The smaller its internal resistance, the less it affects the circuit.

A voltmeter is connected "externally," in parallel with the circuit element across which the voltage is to be measured. It measures the potential difference between two points. Its two wire leads (connecting wires) are connected to the two points, as shown in Fig. 19–33c, where the voltage across R_1 is being measured. The larger its internal resistance ($R_{ser} + r$, Fig. 19–32), the less it affects the circuit being measured.

Effects of Meter Resistance

It is important to know the sensitivity of a meter, for in many cases the resistance of the meter can seriously affect your results. Consider the following Example.

PHYSICS APPLIED
Correcting for meter resistance

EXAMPLE 19–17 Voltage reading vs. true voltage. An electronic circuit has two 15-kΩ resistors, R_1 and R_2, connected in series, Fig. 19–34a. The battery voltage is 8.0 V and it has negligible internal resistance. A voltmeter has resistance of 50 kΩ on the 5.0-V scale. What voltage does the meter read when connected across R_1, Fig. 19–34b, and what error is caused by the meter's finite resistance?

APPROACH The meter acts as a resistor in parallel with R_1. We use parallel and series resistor analyses and Ohm's law to find currents and voltages.

SOLUTION The voltmeter resistance of 50,000 Ω is in parallel with $R_1 = 15\,k\Omega$, Fig. 19–34b. The net resistance R_{eq} of these two is
$$\frac{1}{R_{eq}} = \frac{1}{50\,k\Omega} + \frac{1}{15\,k\Omega} = \frac{13}{150\,k\Omega};$$
so $R_{eq} = 11.5\,k\Omega$. This $R_{eq} = 11.5\,k\Omega$ is in series with $R_2 = 15\,k\Omega$, so the total resistance is now $26.5\,k\Omega$ (not the original $30\,k\Omega$). Hence the current from the battery is
$$I = \frac{8.0\,V}{26.5\,k\Omega} = 3.0 \times 10^{-4}\,A = 0.30\,mA.$$
Then the voltage drop across R_1, which is the same as that across the voltmeter, is $(3.0 \times 10^{-4}\,A)(11.5 \times 10^3\,\Omega) = 3.5\,V$. [The voltage drop across R_2 is $(3.0 \times 10^{-4}\,A)(15 \times 10^3\,\Omega) = 4.5\,V$, for a total of 8.0 V.] If we assume the meter is accurate, it reads 3.5 V. In the original circuit, without the meter, $R_1 = R_2$ so the voltage across R_1 is half that of the battery, or 4.0 V. Thus the voltmeter, because of its internal resistance, gives a low reading. It is off by 0.5 V, or more than 10%.

NOTE Often the **sensitivity** of a voltmeter is specified on its face as, for example, 10,000 Ω/V. Then on a 5.0-V scale, the voltmeter would have a resistance given by $(5.0\,V)(10,000\,\Omega/V) = 50,000\,\Omega$. The meter's resistance depends on the scale used.

FIGURE 19–34 Example 19–17.

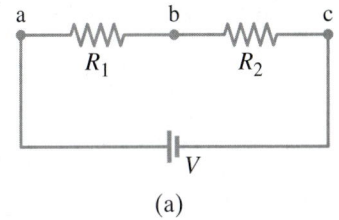
(a)

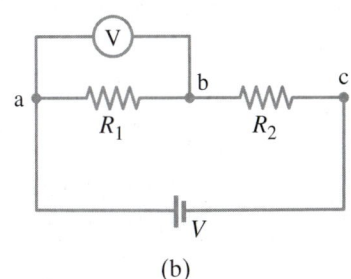
(b)

Example 19–17 illustrates how seriously a meter can affect a circuit and give a misleading reading. If the resistance of a voltmeter is much higher than the resistance of the circuit, however, it will have little effect and its readings can be more accurate, at least to the manufactured precision of the meter, which for analog meters is typically 3% to 4% of full-scale deflection. Even an ammeter can interfere with a circuit, but the effect is minimal if its resistance is much less than that of the circuit as a whole. For both voltmeters and ammeters, the more sensitive the galvanometer, the less effect it will have on the circuit. [A 50,000-Ω/V meter is far better than a 1000-Ω/V meter.]

> **⚠ CAUTION**
> *Measurements affect the quantity being measured*

Whenever we make a measurement on a circuit, to some degree we affect that circuit (Example 19–17). This is true for other types of measurement as well: when we make a measurement on a system, we affect that system in some way. On a temperature measurement, for example, the thermometer can exchange heat with the system, thus altering the temperature it is measuring. It is important to be able to make needed corrections, as we saw in Example 19–17.

Other Meters

The meters described above are for direct current. A dc meter can be modified to measure ac (alternating current, Section 18–7) with the addition of diodes (Chapter 29), which allow current to flow in one direction only. An ac meter can be calibrated to read rms or peak values.

Voltmeters and ammeters can have several series or shunt resistors to offer a choice of range. **Multimeters** can measure voltage, current, and resistance. Sometimes a multimeter is called a VOM (Volt-Ohm-Meter or Volt-Ohm-Milliammeter).

An **ohmmeter** measures resistance, and must contain a battery of known voltage connected in series to a resistor (R_{ser}) and to an ammeter which contains a shunt R_{sh} (Fig. 19–35). The resistor whose resistance is to be measured completes the circuit, and must not be connected in a circuit containing a voltage source. The needle deflection of the meter is inversely proportional to the resistance. The scale calibration depends on the value of its series resistor, which is changeable in a multimeter. Because an ohmmeter sends a current through the device whose resistance is to be measured, it should not be used on very delicate devices that could be damaged by the current.

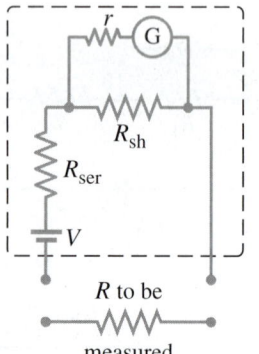

FIGURE 19–35 An ohmmeter.

Digital Meters

Digital meters (see Fig. 19–30b) are used in the same way as analog meters: they are inserted directly into the circuit, in series, to measure current (Fig. 19–33b), and connected "outside," in parallel with the circuit, to measure voltage (Fig. 19–33c).

The internal construction of digital meters is very different from analog meters. First, digital meters do not use a galvanometer, but rather semiconductor devices (Chapter 29). The electronic circuitry and digital readout are more sensitive than a galvanometer, and have less effect on the circuit to be measured. When we measure dc voltages, a digital meter's resistance is very high, commonly on the order of 10 MΩ to 100 MΩ $(10^7$–$10^8\ \Omega)$, and doesn't change significantly when different voltage scales are selected. A 100-MΩ digital meter draws very little current when connected across even a 1-MΩ resistance.

The precision of digital meters is exceptional, often one part in 10^4 $(= 0.01\%)$ or better. This precision is not the same as accuracy, however. A precise meter of internal resistance $10^8\ \Omega$ will not give accurate results if used to measure a voltage across a 10^8-Ω resistor—in which case it is necessary to do a calculation like that in Example 19–17.

Summary

A device that transforms another type of energy into electrical energy is called a **source** of **emf**. A battery behaves like a source of emf in series with an **internal resistance**. The emf is the potential difference determined by the chemical reactions in the battery and equals the terminal voltage when no current is drawn. When a current is drawn, the voltage at the battery's terminals is less than its emf by an amount equal to the potential decrease Ir across the internal resistance.

When resistances are connected in **series** (end to end in a single linear path), the equivalent resistance is the sum of the individual resistances:

$$R_{eq} = R_1 + R_2 + \cdots. \qquad (19\text{--}3)$$

In a series combination, R_{eq} is greater than any component resistance.

When resistors are connected in **parallel**, it is the reciprocals that add up:

$$\frac{1}{R_{eq}} = \frac{1}{R_1} + \frac{1}{R_2} + \cdots. \qquad (19\text{--}4)$$

In a parallel connection, the net resistance is less than any of the individual resistances.

Kirchhoff's rules are useful in determining the currents and voltages in circuits. Kirchhoff's **junction rule** is based on conservation of electric charge and states that the sum of all currents entering any junction equals the sum of all currents leaving that junction. The second, or **loop rule**, is based on conservation of energy and states that the algebraic sum of the changes in potential around any closed path of the circuit must be zero.

When capacitors are connected in **parallel**, the equivalent capacitance is the sum of the individual capacitances:

$$C_{eq} = C_1 + C_2 + \cdots. \qquad (19\text{--}5)$$

When capacitors are connected in **series**, it is the reciprocals that add up:

$$\frac{1}{C_{eq}} = \frac{1}{C_1} + \frac{1}{C_2} + \cdots. \qquad (19\text{--}6)$$

When an **RC circuit** containing a resistance R in series with a capacitance C is connected to a dc source of emf, the voltage across the capacitor rises gradually in time characterized by an exponential of the form $(1 - e^{-t/RC})$, where the **time constant**

$$\tau = RC \qquad (19\text{--}7)$$

is the time it takes for the voltage to reach 63% of its maximum value.

A capacitor discharging through a resistor is characterized by the same time constant: in a time $\tau = RC$, the voltage across the capacitor drops to 37% of its initial value. The charge on the capacitor, and the voltage across it, decrease as $e^{-t/RC}$.

Electric shocks are caused by current passing through the body. To avoid shocks, the body must not become part of a complete circuit by allowing different parts of the body to touch objects at different potentials. Commonly, shocks are caused by one part of the body touching ground ($V = 0$) and another part touching a nonzero electric potential.

An **ammeter** measures current. An analog ammeter consists of a galvanometer and a parallel **shunt resistor** that carries most of the current. An analog **voltmeter** consists of a galvanometer and a series resistor. An ammeter is inserted *into* the circuit whose current is to be measured. A voltmeter is external, being connected in parallel to the element whose voltage is to be measured. Digital meters have greater internal resistance and affect the circuit to be measured less than do analog meters.

Questions

1. Explain why birds can sit on power lines safely, even though the wires have no insulation around them, whereas leaning a metal ladder up against a power line is extremely dangerous.
2. Discuss the advantages and disadvantages of Christmas tree lights connected in parallel versus those connected in series.
3. If all you have is a 120-V line, would it be possible to light several 6-V lamps without burning them out? How?
4. Two lightbulbs of resistance R_1 and R_2 $(R_2 > R_1)$ and a battery are all connected in series. Which bulb is brighter? What if they are connected in parallel? Explain.
5. Household outlets are often double outlets. Are these connected in series or parallel? How do you know?
6. With two identical lightbulbs and two identical batteries, explain how and why you would arrange the bulbs and batteries in a circuit to get the maximum possible total power to the lightbulbs. (Ignore internal resistance of batteries.)
7. If two identical resistors are connected in series to a battery, does the battery have to supply more power or less power than when only one of the resistors is connected? Explain.
8. You have a single 60-W bulb lit in your room. How does the overall resistance of your room's electric circuit change when you turn on an additional 100-W bulb? Explain.
9. Suppose three identical capacitors are connected to a battery. Will they store more energy if connected in series or in parallel?
10. When applying Kirchhoff's loop rule (such as in Fig. 19–36), does the sign (or direction) of a battery's emf depend on the direction of current through the battery? What about the terminal voltage?

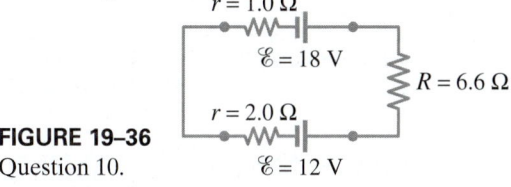

FIGURE 19–36 Question 10.

11. Different lamps might have batteries connected in either of the two arrangements shown in Fig. 19–37. What would be the advantages of each scheme?

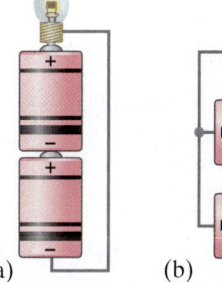 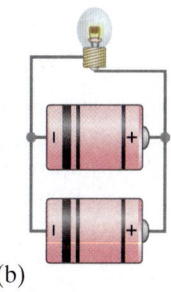

FIGURE 19–37 Question 11. (a) (b)

12. For what use are batteries connected in series? For what use are they connected in parallel? Does it matter if the batteries are nearly identical or not in either case?

13. Can the terminal voltage of a battery ever exceed its emf? Explain.
14. Explain in detail how you could measure the internal resistance of a battery.
15. In an RC circuit, current flows from the battery until the capacitor is completely charged. Is the total energy supplied by the battery equal to the total energy stored by the capacitor? If not, where does the extra energy go?
16. Given the circuit shown in Fig. 19–38, use the words "increases," "decreases," or "stays the same" to complete the following statements:
    (a) If R_7 increases, the potential difference between A and E _____. Assume no resistance in Ⓐ and ℰ.
    (b) If R_7 increases, the potential difference between A and E _____. Assume Ⓐ and ℰ have resistance.
    (c) If R_7 increases, the voltage drop across R_4 _____.
    (d) If R_2 decreases, the current through R_1 _____.
    (e) If R_2 decreases, the current through R_6 _____.
    (f) If R_2 decreases, the current through R_3 _____.
    (g) If R_5 increases, the voltage drop across R_2 _____.
    (h) If R_5 increases, the voltage drop across R_4 _____.
    (i) If R_2, R_5, and R_7 increase, $\mathcal{E}\;(r=0)$ _____.

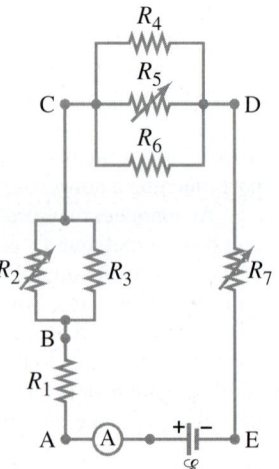

FIGURE 19–38 Question 16. R_2, R_5, and R_7 are *variable* resistors (you can change their resistance), given the symbol ⏦.

17. Design a circuit in which two different switches of the type shown in Fig. 19–39 can be used to operate the same lightbulb from opposite sides of a room.

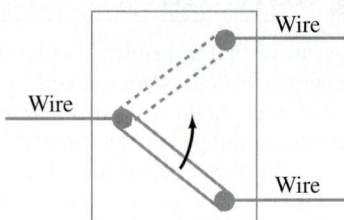

FIGURE 19–39 Question 17.

18. Why is it more dangerous to turn on an electric appliance when you are standing outside in bare feet than when you are inside wearing shoes with thick soles?
19. What is the main difference between an analog voltmeter and an analog ammeter?
20. What would happen if you mistakenly used an ammeter where you needed to use a voltmeter?
21. Explain why an ideal ammeter would have zero resistance and an ideal voltmeter infinite resistance.
22. A voltmeter connected across a resistor always reads *less* than the actual voltage (i.e., when the meter is not present). Explain.
23. A small battery-operated flashlight requires a single 1.5-V battery. The bulb is barely glowing. But when you take the battery out and check it with a digital voltmeter, it registers 1.5 V. How would you explain this?

MisConceptual Questions

1. In which circuits shown in Fig. 19–40 are resistors connected in series?

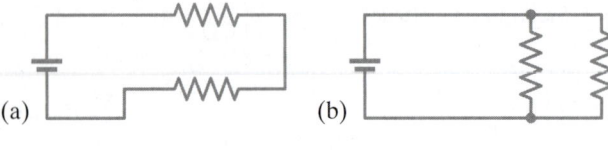

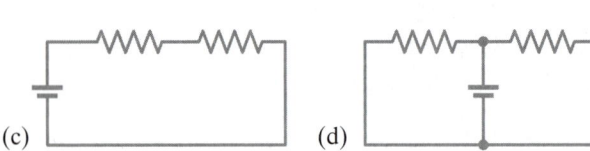

FIGURE 19–40 MisConceptual Question 1.

2. Which resistors in Fig. 19–41 are connected in parallel?
    (a) All three.
    (b) R_1 and R_2.
    (c) R_2 and R_3.
    (d) R_1 and R_3.
    (e) None of the above.

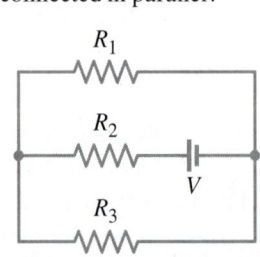

FIGURE 19–41 MisConceptual Question 2.

3. A 10,000-Ω resistor is placed in series with a 100-Ω resistor. The current in the 10,000-Ω resistor is 10 A. If the resistors are swapped, how much current flows through the 100-Ω resistor?
    (a) >10 A.   (b) <10 A.   (c) 10 A.
    (d) Need more information about the circuit.

4. Two identical 10-V batteries and two identical 10-Ω resistors are placed in series as shown in Fig. 19–42. If a 10-Ω lightbulb is connected with one end connected between the batteries and other end between the resistors, how much current will flow through the lightbulb?
    (a) 0 A.
    (b) 1 A.
    (c) 2 A.
    (d) 4 A.

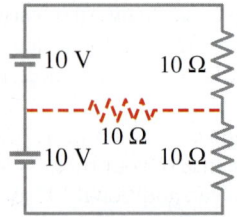

FIGURE 19–42 MisConceptual Question 4.

5. Which resistor shown in Fig. 19–43 has the greatest current going through it? Assume that all the resistors are equal.
   (a) R_1.
   (b) R_1 and R_2.
   (c) R_3 and R_4.
   (d) R_5.
   (e) All of them the same.

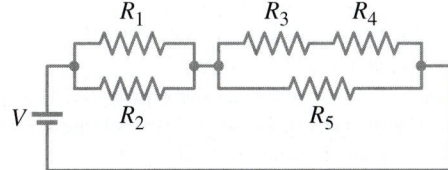

FIGURE 19–43 MisConceptual Question 5.

6. Figure 19–44 shows three identical bulbs in a circuit. What happens to the brightness of bulb A if you replace bulb B with a short circuit?
   (a) Bulb A gets brighter.
   (b) Bulb A gets dimmer.
   (c) Bulb A's brightness does not change.
   (d) Bulb A goes out.

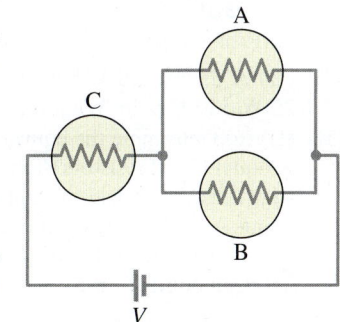

FIGURE 19–44 MisConceptual Question 6.

7. When the switch shown in Fig. 19–45 is closed, what will happen to the voltage across resistor R_4? It will
   (a) increase.   (b) decrease.   (c) stay the same.

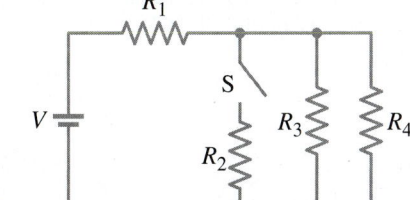

FIGURE 19–45 MisConceptual Questions 7 and 8.

8. When the switch shown in Fig. 19–45 is closed, what will happen to the voltage across resistor R_1? It will
   (a) increase.   (b) decrease.   (c) stay the same.

9. As a capacitor is being charged in an RC circuit, the current flowing through the resistor is
   (a) increasing.   (c) constant.
   (b) decreasing.   (d) zero.

10. For the circuit shown in Fig. 19–46, what happens when the switch S is closed?
    (a) Nothing. Current cannot flow through the capacitor.
    (b) The capacitor immediately charges up to the battery emf.
    (c) The capacitor eventually charges up to the full battery emf at a rate determined by R and C.
    (d) The capacitor charges up to a fraction of the battery emf determined by R and C.
    (e) The capacitor charges up to a fraction of the battery emf determined by R only.

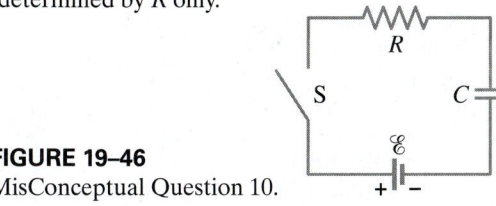

FIGURE 19–46 MisConceptual Question 10.

11. The capacitor in the circuit shown in Fig. 19–47 is charged to an initial value Q. When the switch is closed, it discharges through the resistor. It takes 2.0 seconds for the charge to drop to $\frac{1}{2}Q$. How long does it take to drop to $\frac{1}{4}Q$?
    (a) 3.0 seconds.
    (b) 4.0 seconds.
    (c) Between 2.0 and 3.0 seconds.
    (d) Between 3.0 and 4.0 seconds.
    (e) More than 4.0 seconds.

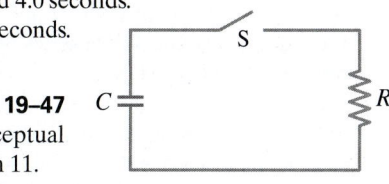

FIGURE 19–47 MisConceptual Question 11.

12. A resistor and a capacitor are used in series to control the timing in the circuit of a heart pacemaker. To design a pacemaker that can double the heart rate when the patient is exercising, which statement below is true? The capacitor
    (a) needs to discharge faster, so the resistance should be decreased.
    (b) needs to discharge faster, so the resistance should be increased.
    (c) needs to discharge slower, so the resistance should be decreased.
    (d) needs to discharge slower, so the resistance should be increased.
    (e) does not affect the timing, regardless of the resistance.

13. Why is an appliance cord with a three-prong plug safer than one with two prongs?
    (a) The 120 V from the outlet is split among three wires, so it isn't as high a voltage as when it is only split between two wires.
    (b) Three prongs fasten more securely to the wall outlet.
    (c) The third prong grounds the case, so the case cannot reach a high voltage.
    (d) The third prong acts as a ground wire, so the electrons have an easier time leaving the appliance. As a result, fewer electrons build up in the appliance.
    (e) The third prong controls the capacitance of the appliance, so it can't build up a high voltage.

14. When capacitors are connected in series, the effective capacitance is _____ the smallest capacitance; when capacitors are connected in parallel, the effective capacitance is _____ the largest capacitance.
    (a) greater than; equal to.
    (b) greater than; less than.
    (c) less than; greater than.
    (d) equal to; less than.
    (e) equal to; equal to.

15. If ammeters and voltmeters are not to significantly alter the quantities they are measuring,
    (a) the resistance of an ammeter and a voltmeter should be much higher than that of the circuit element being measured.
    (b) the resistance of an ammeter should be much lower, and the resistance of a voltmeter should be much higher, than those of the circuit being measured.
    (c) the resistance of an ammeter should be much higher, and the resistance of a voltmeter should be much lower, than those of the circuit being measured.
    (d) the resistance of an ammeter and a voltmeter should be much lower than that of the circuit being measured.
    (e) None of the above.

Problems

19–1 Emf and Terminal Voltage

1. (I) Calculate the terminal voltage for a battery with an internal resistance of 0.900 Ω and an emf of 6.00 V when the battery is connected in series with (a) a 71.0-Ω resistor, and (b) a 710-Ω resistor.

2. (I) Four 1.50-V cells are connected in series to a 12.0-Ω lightbulb. If the resulting current is 0.45 A, what is the internal resistance of each cell, assuming they are identical and neglecting the resistance of the wires?

3. (II) What is the internal resistance of a 12.0-V car battery whose terminal voltage drops to 8.8 V when the starter motor draws 95 A? What is the resistance of the starter?

19–2 Resistors in Series and Parallel

[In these Problems neglect the internal resistance of a battery unless the Problem refers to it.]

4. (I) A 650-Ω and an 1800-Ω resistor are connected in series with a 12-V battery. What is the voltage across the 1800-Ω resistor?

5. (I) Three 45-Ω lightbulbs and three 65-Ω lightbulbs are connected in series. (a) What is the total resistance of the circuit? (b) What is the total resistance if all six are wired in parallel?

6. (II) Suppose that you have a 580-Ω, a 790-Ω, and a 1.20-kΩ resistor. What is (a) the maximum, and (b) the minimum resistance you can obtain by combining these?

7. (II) How many 10-Ω resistors must be connected in series to give an equivalent resistance to five 100-Ω resistors connected in parallel?

8. (II) Design a "voltage divider" (see Example 19–3) that would provide one-fifth (0.20) of the battery voltage across R_2, Fig. 19–6. What is the ratio R_1/R_2?

9. (II) Suppose that you have a 9.0-V battery and wish to apply a voltage of only 3.5 V. Given an unlimited supply of 1.0-Ω resistors, how could you connect them to make a "voltage divider" that produces a 3.5-V output for a 9.0-V input?

10. (II) Three 1.70-kΩ resistors can be connected together in four different ways, making combinations of series and/or parallel circuits. What are these four ways, and what is the net resistance in each case?

11. (II) A battery with an emf of 12.0 V shows a terminal voltage of 11.8 V when operating in a circuit with two lightbulbs, each rated at 4.0 W (at 12.0 V), which are connected in parallel. What is the battery's internal resistance?

12. (II) Eight identical bulbs are connected in series across a 120-V line. (a) What is the voltage across each bulb? (b) If the current is 0.45 A, what is the resistance of each bulb, and what is the power dissipated in each?

13. (II) Eight bulbs are connected in parallel to a 120-V source by two long leads of total resistance 1.4 Ω. If 210 mA flows through each bulb, what is the resistance of each, and what fraction of the total power is wasted in the leads?

14. (II) A close inspection of an electric circuit reveals that a 480-Ω resistor was inadvertently soldered in the place where a 350-Ω resistor is needed. How can this be fixed without removing anything from the existing circuit?

15. (II) Eight 7.0-W Christmas tree lights are connected in series to each other and to a 120-V source. What is the resistance of each bulb?

16. (II) Determine (a) the equivalent resistance of the circuit shown in Fig. 19–48, (b) the voltage across each resistor, and (c) the current through each resistor.

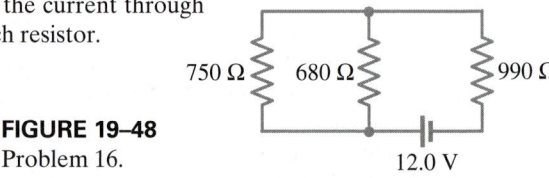

FIGURE 19–48 Problem 16.

17. (II) A 75-W, 120-V bulb is connected in parallel with a 25-W, 120-V bulb. What is the net resistance?

18. (II) (a) Determine the equivalent resistance of the "ladder" of equal 175-Ω resistors shown in Fig. 19–49. In other words, what resistance would an ohmmeter read if connected between points A and B? (b) What is the current through each of the three resistors on the left if a 50.0-V battery is connected between points A and B?

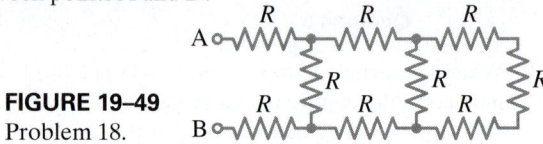

FIGURE 19–49 Problem 18.

19. (II) What is the net resistance of the circuit connected to the battery in Fig. 19–50?

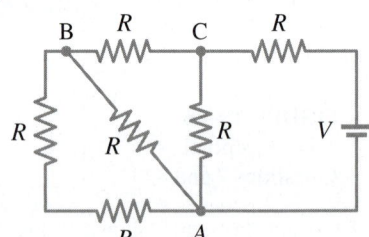

FIGURE 19–50 Problems 19 and 20.

20. (II) Calculate the current through each resistor in Fig. 19–50 if each resistance R = 3.25 kΩ and V = 12.0 V. What is the potential difference between points A and B?

21. (III) Two resistors when connected in series to a 120-V line use one-fourth the power that is used when they are connected in parallel. If one resistor is 4.8 kΩ, what is the resistance of the other?

22. (III) Three equal resistors (R) are connected to a battery as shown in Fig. 19–51. Qualitatively, what happens to (a) the voltage drop across each of these resistors, (b) the current flow through each, and (c) the terminal voltage of the battery, when the switch S is opened, after having been closed for a long time? (d) If the emf of the battery is 9.0 V, what is its terminal voltage when the switch is closed if the internal resistance r is 0.50 Ω and R = 5.50 Ω? (e) What is the terminal voltage when the switch is open?

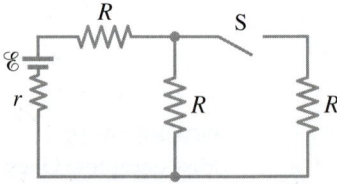

FIGURE 19–51 Problem 22.

23. (III) A 2.5-kΩ and a 3.7-kΩ resistor are connected in parallel; this combination is connected in series with a 1.4-kΩ resistor. If each resistor is rated at 0.5 W (maximum without overheating), what is the maximum voltage that can be applied across the whole network?

24. (III) Consider the network of resistors shown in Fig. 19–52. Answer qualitatively: (a) What happens to the voltage across each resistor when the switch S is closed? (b) What happens to the current through each when the switch is closed? (c) What happens to the power output of the battery when the switch is closed? (d) Let $R_1 = R_2 = R_3 = R_4 = 155\,\Omega$ and $V = 22.0\,V$. Determine the current through each resistor before and after closing the switch. Are your qualitative predictions confirmed?

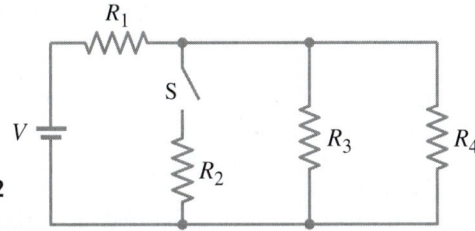

FIGURE 19–52 Problem 24.

19–3 Kirchhoff's Rules

25. (I) Calculate the current in the circuit of Fig. 19–53, and show that the sum of all the voltage changes around the circuit is zero.

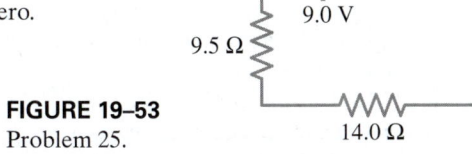

FIGURE 19–53 Problem 25.

26. (II) Determine the terminal voltage of each battery in Fig. 19–54.

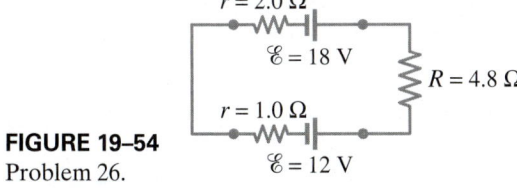

FIGURE 19–54 Problem 26.

27. (II) For the circuit shown in Fig. 19–55, find the potential difference between points a and b. Each resistor has $R = 160\,\Omega$ and each battery is 1.5 V.

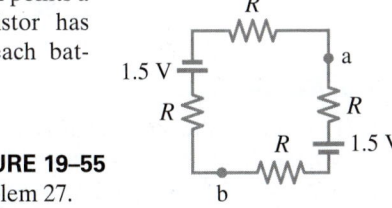

FIGURE 19–55 Problem 27.

28. (II) Determine the magnitudes and directions of the currents in each resistor shown in Fig. 19–56. The batteries have emfs of $\mathcal{E}_1 = 9.0\,V$ and $\mathcal{E}_2 = 12.0\,V$ and the resistors have values of $R_1 = 25\,\Omega$, $R_2 = 68\,\Omega$, and $R_3 = 35\,\Omega$. (a) Ignore internal resistance of the batteries. (b) Assume each battery has internal resistance $r = 1.0\,\Omega$.

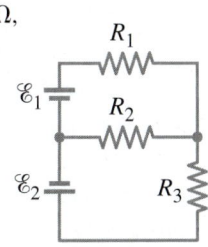

FIGURE 19–56 Problem 28.

29. (II) (a) What is the potential difference between points a and d in Fig. 19–57 (similar to Fig. 19–13, Example 19–8), and (b) what is the terminal voltage of each battery?

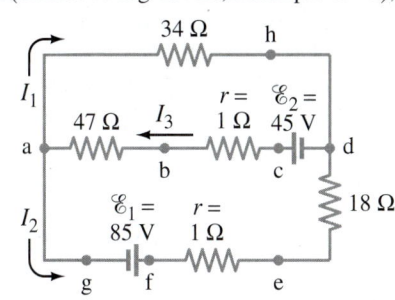

FIGURE 19–57 Problem 29.

30. (II) Calculate the magnitude and direction of the currents in each resistor of Fig. 19–58.

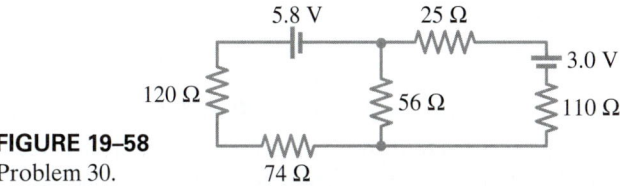

FIGURE 19–58 Problem 30.

31. (II) Determine the magnitudes and directions of the currents through R_1 and R_2 in Fig. 19–59.

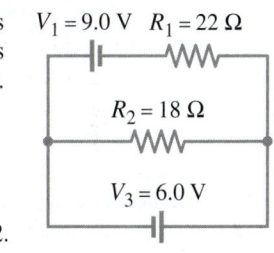

FIGURE 19–59 Problems 31 and 32.

32. (II) Repeat Problem 31, now assuming that each battery has an internal resistance $r = 1.4\,\Omega$.

33. (III) (a) A network of five equal resistors R is connected to a battery $\mathcal{E}$ as shown in Fig. 19–60. Determine the current I that flows out of the battery. (b) Use the value determined for I to find the single resistor R_{eq} that is equivalent to the five-resistor network.

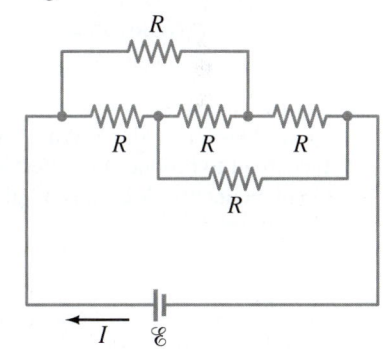

FIGURE 19–60 Problem 33.

34. (III) (a) Determine the currents I_1, I_2, and I_3 in Fig. 19–61. Assume the internal resistance of each battery is $r = 1.0\,\Omega$. (b) What is the terminal voltage of the 6.0-V battery?

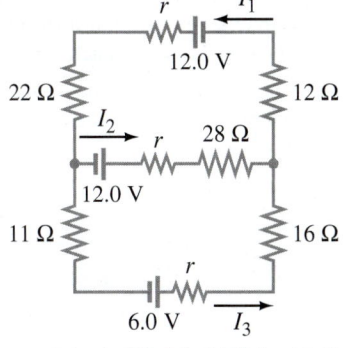

FIGURE 19–61 Problems 34 and 35.

35. (III) What would the current I_1 be in Fig. 19–61 if the 12-Ω resistor is shorted out (resistance = 0)? Let $r = 1.0\,\Omega$.

19–4 Emfs Combined, Battery Charging

36. (II) Suppose two batteries, with unequal emfs of 2.00 V and 3.00 V, are connected as shown in Fig. 19–62. If each internal resistance is $r = 0.350\ \Omega$, and $R = 4.00\ \Omega$, what is the voltage across the resistor R?

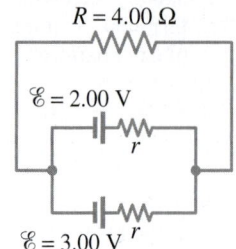

FIGURE 19–62 Problem 36.

37. (II) A battery for a proposed electric car is to have three hundred 3-V lithium ion cells connected such that the total voltage across all of the cells is 300 V. Describe a possible connection configuration (using series and parallel connections) that would meet these battery specifications.

19–5 Capacitors in Series and Parallel

38. (I) (a) Six 4.8-μF capacitors are connected in parallel. What is the equivalent capacitance? (b) What is their equivalent capacitance if connected in series?

39. (I) A 3.00-μF and a 4.00-μF capacitor are connected in series, and this combination is connected in parallel with a 2.00-μF capacitor (see Fig. 19–63). What is the net capacitance?

40. (II) If 21.0 V is applied across the whole network of Fig. 19–63, calculate (a) the voltage across each capacitor and (b) the charge on each capacitor.

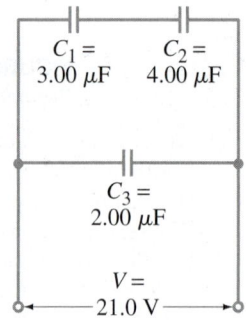

FIGURE 19–63 Problems 39 and 40.

41. (II) The capacitance of a portion of a circuit is to be reduced from 2900 pF to 1200 pF. What capacitance can be added to the circuit to produce this effect without removing existing circuit elements? Must any existing connections be broken to accomplish this?

42. (II) An electric circuit was accidentally constructed using a 7.0-μF capacitor instead of the required 16-μF value. Without removing the 7.0-μF capacitor, what can a technician add to correct this circuit?

43. (II) Consider three capacitors, of capacitance 3200 pF, 5800 pF, and 0.0100 μF. What maximum and minimum capacitance can you form from these? How do you make the connection in each case?

44. (II) Determine the equivalent capacitance between points a and b for the combination of capacitors shown in Fig. 19–64.

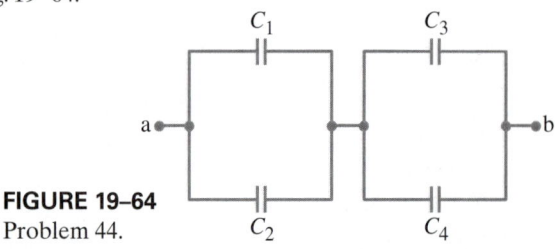

FIGURE 19–64 Problem 44.

45. (II) What is the ratio of the voltage V_1 across capacitor C_1 in Fig. 19–65 to the voltage V_2 across capacitor C_2?

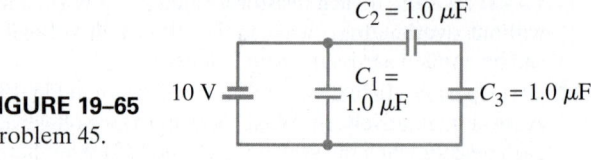

FIGURE 19–65 Problem 45.

46. (II) A 0.50-μF and a 1.4-μF capacitor are connected in series to a 9.0-V battery. Calculate (a) the potential difference across each capacitor and (b) the charge on each. (c) Repeat parts (a) and (b) assuming the two capacitors are in parallel.

47. (II) A circuit contains a single 250-pF capacitor hooked across a battery. It is desired to store four times as much energy in a combination of two capacitors by adding a single capacitor to this one. How would you hook it up, and what would its value be?

48. (II) Suppose three parallel-plate capacitors, whose plates have areas A_1, A_2, and A_3 and separations d_1, d_2, and d_3, are connected in parallel. Show, using only Eq. 17–8, that Eq. 19–5 is valid.

49. (II) Two capacitors connected in parallel produce an equivalent capacitance of 35.0 μF but when connected in series the equivalent capacitance is only 4.8 μF. What is the individual capacitance of each capacitor?

50. (III) Given three capacitors, $C_1 = 2.0\ \mu F$, $C_2 = 1.5\ \mu F$, and $C_3 = 3.0\ \mu F$, what arrangement of parallel and series connections with a 12-V battery will give the minimum voltage drop across the 2.0-μF capacitor? What is the minimum voltage drop?

51. (III) In Fig. 19–66, suppose $C_1 = C_2 = C_3 = C_4 = C$. (a) Determine the equivalent capacitance between points a and b. (b) Determine the charge on each capacitor and the potential difference across each in terms of V.

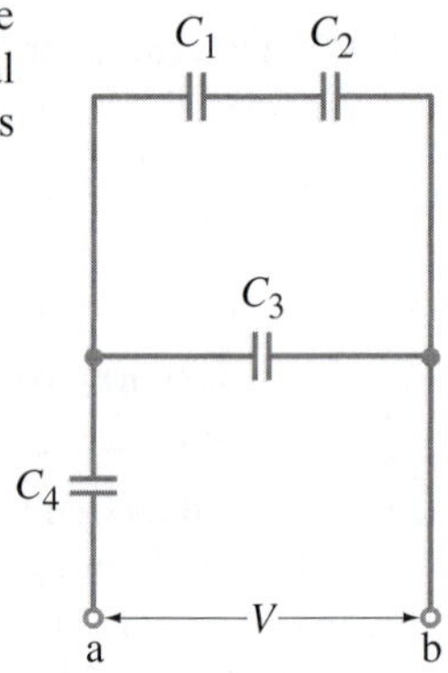

FIGURE 19–66 Problem 51.

19–6 RC Circuits

52. (I) Estimate the value of resistances needed to make a variable timer for intermittent windshield wipers: one wipe every 15 s, 8 s, 4 s, 2 s, 1 s. Assume the capacitor used is on the order of 1 μF. See Fig. 19–67.

FIGURE 19–67 Problem 52.

53. (II) Electrocardiographs are often connected as shown in Fig. 19–68. The lead wires to the legs are said to be capacitively coupled. A time constant of 3.0 s is typical and allows rapid changes in potential to be recorded accurately. If $C = 3.0\,\mu\text{F}$, what value must R have? [*Hint*: Consider each leg as a separate circuit.]

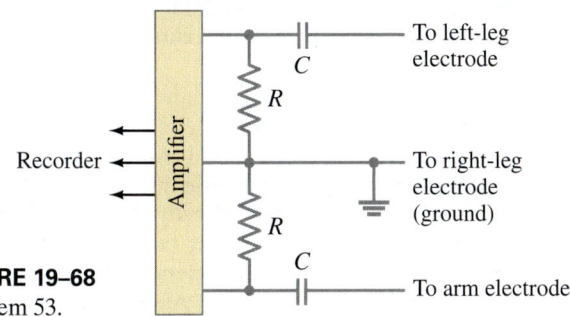

FIGURE 19–68 Problem 53.

54. (II) In Fig. 19–69 (same as Fig. 19–20a), the total resistance is 15.0 kΩ, and the battery's emf is 24.0 V. If the time constant is measured to be 18.0 μs, calculate (*a*) the total capacitance of the circuit and (*b*) the time it takes for the voltage across the resistor to reach 16.0 V after the switch is closed.

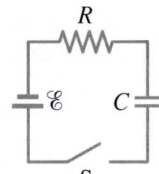

FIGURE 19–69 Problem 54.

55. (II) Two 3.8-μF capacitors, two 2.2-kΩ resistors, and a 16.0-V source are connected in series. Starting from the uncharged state, how long does it take for the current to drop from its initial value to 1.50 mA?

56. (II) The RC circuit of Fig. 19–70 (same as Fig. 19–21a) has $R = 8.7$ kΩ and $C = 3.0\,\mu$F. The capacitor is at voltage V_0 at $t = 0$, when the switch is closed. How long does it take the capacitor to discharge to 0.25% of its initial voltage?

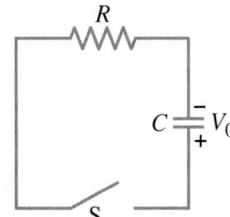

FIGURE 19–70 Problem 56.

57. (III) Consider the circuit shown in Fig. 19–71, where all resistors have the same resistance R. At $t = 0$, with the capacitor C uncharged, the switch is closed. (*a*) At $t = 0$, the three currents can be determined by analyzing a simpler, but equivalent, circuit. Draw this simpler circuit and use it to find the values of I_1, I_2, and I_3 at $t = 0$. (*b*) At $t = \infty$, the currents can be determined by analyzing a simpler, equivalent circuit. Draw this simpler circuit and implement it in finding the values of I_1, I_2, and I_3 at $t = \infty$. (*c*) At $t = \infty$, what is the potential difference across the capacitor?

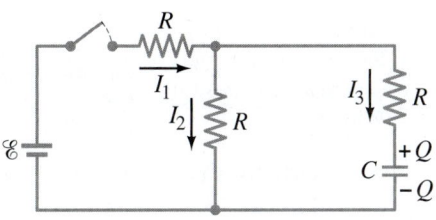

FIGURE 19–71 Problem 57.

58. (III) Two resistors and two uncharged capacitors are arranged as shown in Fig. 19–72. Then a potential difference of 24 V is applied across the combination as shown. (*a*) What is the potential at point a with switch S open? (Let $V = 0$ at the negative terminal of the source.) (*b*) What is the potential at point b with the switch open? (*c*) When the switch is closed, what is the final potential of point b? (*d*) How much charge flows through the switch S after it is closed?

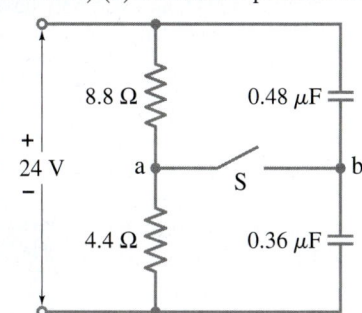

FIGURE 19–72 Problem 58.

19–8 Ammeters and Voltmeters

59. (I) (*a*) An ammeter has a sensitivity of 35,000 Ω/V. What current in the galvanometer produces full-scale deflection? (*b*) What is the resistance of a voltmeter on the 250-V scale if the meter sensitivity is 35,000 Ω/V?

60. (II) An ammeter whose internal resistance is 53 Ω reads 5.25 mA when connected in a circuit containing a battery and two resistors in series whose values are 720 Ω and 480 Ω. What is the actual current when the ammeter is absent?

61. (II) A milliammeter reads 35 mA full scale. It consists of a 0.20-Ω resistor in parallel with a 33-Ω galvanometer. How can you change this ammeter to a voltmeter giving a full-scale reading of 25 V without taking the ammeter apart? What will be the sensitivity (Ω/V) of your voltmeter?

62. (II) A galvanometer has an internal resistance of 32 Ω and deflects full scale for a 55-μA current. Describe how to use this galvanometer to make (*a*) an ammeter to read currents up to 25 A, and (*b*) a voltmeter to give a full-scale deflection of 250 V.

63. (III) A battery with $\mathscr{E} = 12.0$ V and internal resistance $r = 1.0\,\Omega$ is connected to two 7.5-kΩ resistors in series. An ammeter of internal resistance 0.50 Ω measures the current, and at the same time a voltmeter with internal resistance 15 kΩ measures the voltage across one of the 7.5-kΩ resistors in the circuit. What do the ammeter and voltmeter read? What is the % "error" from the current and voltage *without* meters?

64. (III) What internal resistance should the voltmeter of Example 19–17 have to be in error by less than 5%?

65. (III) Two 9.4-kΩ resistors are placed in series and connected to a battery. A voltmeter of sensitivity 1000 Ω/V is on the 3.0-V scale and reads 1.9 V when placed across either resistor. What is the emf of the battery? (Ignore its internal resistance.)

66. (III) When the resistor R in Fig. 19–73 is 35 Ω, the high-resistance voltmeter reads 9.7 V. When R is replaced by a 14.0-Ω resistor, the voltmeter reading drops to 8.1 V. What are the emf and internal resistance of the battery?

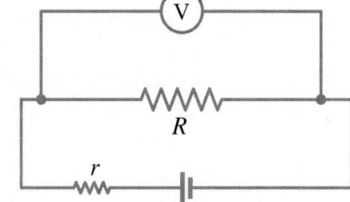

FIGURE 19–73 Problem 66.

General Problems

67. Suppose that you wish to apply a 0.25-V potential difference between two points on the human body. The resistance is about 1800 Ω, and you only have a 1.5-V battery. How can you connect up one or more resistors to produce the desired voltage?

68. A **three-way lightbulb** can produce 50 W, 100 W, or 150 W, at 120 V. Such a bulb contains two filaments that can be connected to the 120 V individually or in parallel (Fig. 19–74). (a) Describe how the connections to the two filaments are made to give each of the three wattages. (b) What must be the resistance of each filament?

FIGURE 19–74 Problem 68.

69. What are the values of effective capacitance which can be obtained by connecting four identical capacitors, each having a capacitance C?

70. Electricity can be a hazard in hospitals, particularly to patients who are connected to electrodes, such as an ECG. Suppose that the motor of a motorized bed shorts out to the bed frame, and the bed frame's connection to a ground has broken (or was not there in the first place). If a nurse touches the bed and the patient at the same time, the nurse becomes a conductor and a complete circuit can be made through the patient to ground through the ECG apparatus. This is shown schematically in Fig. 19–75. Calculate the current through the patient.

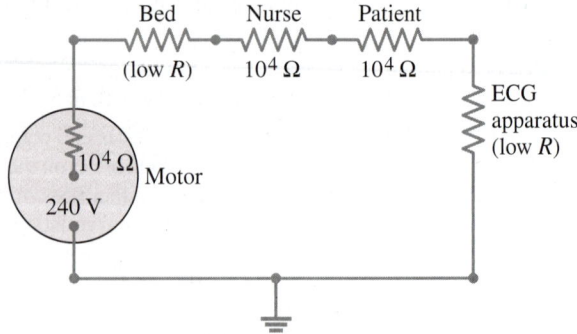

FIGURE 19–75 Problem 70.

71. A heart pacemaker is designed to operate at 72 beats/min using a 6.5-μF capacitor in a simple RC circuit. What value of resistance should be used if the pacemaker is to fire (capacitor discharge) when the voltage reaches 75% of maximum and then drops to 0 V (72 times a minute)?

72. Suppose that a person's body resistance is 950 Ω (moist skin). (a) What current passes through the body when the person accidentally is connected to 120 V? (b) If there is an alternative path to ground whose resistance is 25 Ω, what then is the current through the body? (c) If the voltage source can produce at most 1.5 A, how much current passes through the person in case (b)?

73. One way a multiple-speed ventilation fan for a car can be designed is to put resistors in series with the fan motor. The resistors reduce the current through the motor and make it run more slowly. Suppose the current in the motor is 5.0 A when it is connected directly across a 12-V battery. (a) What series resistor should be used to reduce the current to 2.0 A for low-speed operation? (b) What power rating should the resistor have? Assume that the motor's resistance is roughly the same at all speeds.

74. A **Wheatstone bridge** is a type of "bridge circuit" used to make measurements of resistance. The unknown resistance to be measured, R_x, is placed in the circuit with accurately known resistances R_1, R_2, and R_3 (Fig. 19–76). One of these, R_3, is a variable resistor which is adjusted so that when the switch is closed momentarily, the ammeter Ⓐ shows zero current flow. The bridge is then said to be balanced. (a) Determine R_x in terms of R_1, R_2, and R_3. (b) If a Wheatstone bridge is "balanced" when $R_1 = 590\ \Omega$, $R_2 = 972\ \Omega$, and $R_3 = 78.6\ \Omega$, what is the value of the unknown resistance?

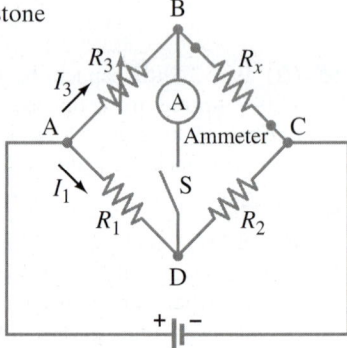

FIGURE 19–76 Problem 74. Wheatstone bridge.

75. The internal resistance of a 1.35-V mercury cell is 0.030 Ω, whereas that of a 1.5-V dry cell is 0.35 Ω. Explain why three mercury cells can more effectively power a 2.5-W hearing aid that requires 4.0 V than can three dry cells.

76. How many $\frac{1}{2}$-W resistors, each of the same resistance, must be used to produce an equivalent 3.2-kΩ, 3.5-W resistor? What is the resistance of each, and how must they be connected? Do not exceed $P = \frac{1}{2}$ W in each resistor.

77. A **solar cell**, 3.0 cm square, has an output of 350 mA at 0.80 V when exposed to full sunlight. A solar panel that delivers close to 1.3 A of current at an emf of 120 V to an external load is needed. How many cells will you need to create the panel? How big a panel will you need, and how should you connect the cells to one another?

78. The current through the 4.0-kΩ resistor in Fig. 19–77 is 3.10 mA. What is the terminal voltage V_{ba} of the "unknown" battery? (There are two answers. Why?)

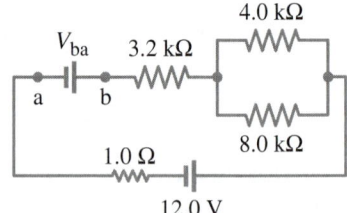

FIGURE 19–77 Problem 78.

79. A power supply has a fixed output voltage of 12.0 V, but you need $V_T = 3.5$ V output for an experiment. (a) Using the voltage divider shown in Fig. 19–78, what should R_2 be if R_1 is 14.5 Ω? (b) What will the terminal voltage V_T be if you connect a load to the 3.5-V output, assuming the load has a resistance of 7.0 Ω?

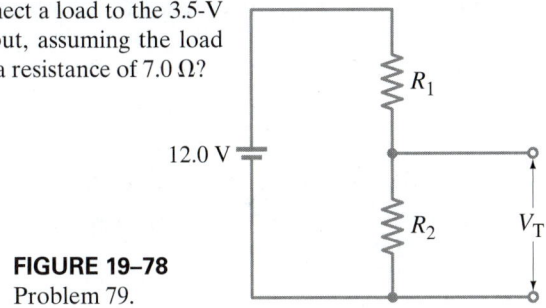

FIGURE 19–78
Problem 79.

80. A battery produces 40.8 V when 8.40 A is drawn from it, and 47.3 V when 2.80 A is drawn. What are the emf and internal resistance of the battery?

81. In the circuit shown in Fig. 19–79, the 33-Ω resistor dissipates 0.80 W. What is the battery voltage?

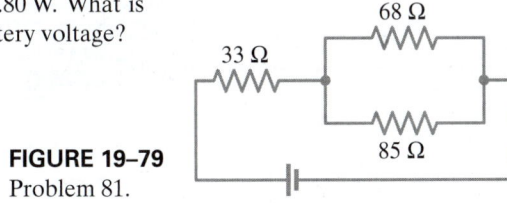

FIGURE 19–79
Problem 81.

82. For the circuit shown in Fig. 19–80, determine (a) the current through the 16-V battery and (b) the potential difference between points a and b, $V_a - V_b$.

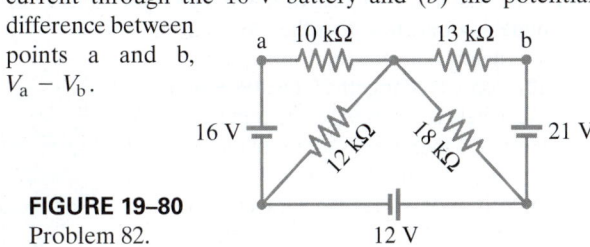

FIGURE 19–80
Problem 82.

83. The current through the 20-Ω resistor in Fig. 19–81 does not change whether the two switches S_1 and S_2 are both open or both closed. Use this clue to determine the value of the unknown resistance R.

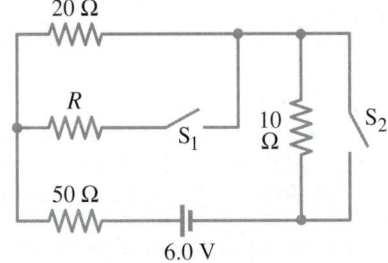

FIGURE 19–81
Problem 83.

84. (a) What is the equivalent resistance of the circuit shown in Fig. 19–82? [*Hint*: Redraw the circuit to see series and parallel better.] (b) What is the current in the 14-Ω resistor? (c) What is the current in the 12-Ω resistor? (d) What is the power dissipation in the 4.5-Ω resistor?

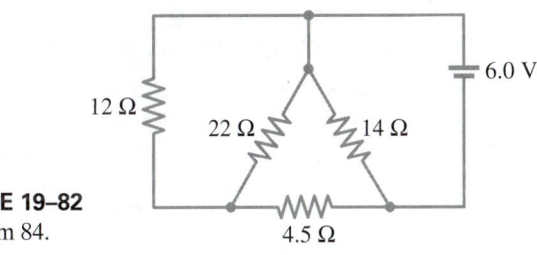

FIGURE 19–82
Problem 84.

85. (a) A voltmeter and an ammeter can be connected as shown in Fig. 19–83a to measure a resistance R. If V is the voltmeter reading, and I is the ammeter reading, the value of R will not quite be V/I (as in Ohm's law) because some current goes through the voltmeter. Show that the actual value of R is

$$\frac{1}{R} = \frac{I}{V} - \frac{1}{R_V},$$

where R_V is the voltmeter resistance. Note that $R \approx V/I$ if $R_V \gg R$. (b) A voltmeter and an ammeter can also be connected as shown in Fig. 19–83b to measure a resistance R. Show in this case that

$$R = \frac{V}{I} - R_A,$$

where V and I are the voltmeter and ammeter readings and R_A is the resistance of the ammeter. Note that $R \approx V/I$ if $R_A \ll R$.

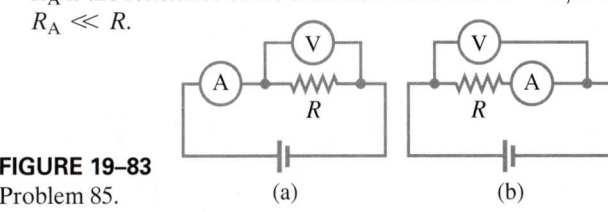

FIGURE 19–83
Problem 85.

86. The circuit shown in Fig. 19–84 uses a neon-filled tube as in Fig. 19–23a. This neon lamp has a threshold voltage V_0 for conduction, because no current flows until the neon gas in the tube is ionized by a sufficiently strong electric field. Once the threshold voltage is exceeded, the lamp has negligible resistance. The capacitor stores electrical energy, which can be released to flash the lamp. Assume that $C = 0.150\ \mu\text{F}$, $R = 2.35 \times 10^6\ \Omega$, $V_0 = 90.0$ V, and $\mathcal{E} = 105$ V. (a) Assuming the circuit is hooked up to the emf at time $t = 0$, at what time will the light first flash? (b) If the value of R is increased, will the time you found in part (a) increase or decrease? (c) The flashing of the lamp is very brief. Why? (d) Explain what happens after the lamp flashes for the first time.

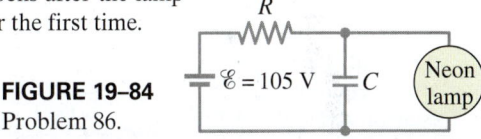

FIGURE 19–84
Problem 86.

87. A flashlight bulb rated at 2.0 W and 3.0 V is operated by a 9.0-V battery. To light the bulb at its rated voltage and power, a resistor R is connected in series as shown in Fig. 19–85. What value should the resistor have?

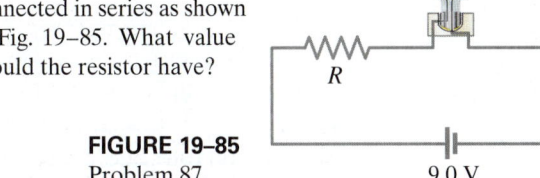

FIGURE 19–85
Problem 87.

88. In Fig. 19–86, let $V = 10.0$ V and $C_1 = C_2 = C_3 = 25.4\ \mu\text{F}$. How much energy is stored in the capacitor network (a) as shown, (b) if the capacitors were all in series, and (c) if the capacitors were all in parallel?

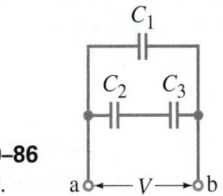

FIGURE 19–86
Problem 88.

89. A 12.0-V battery, two resistors, and two capacitors are connected as shown in Fig. 19–87. After the circuit has been connected for a long time, what is the charge on each capacitor?

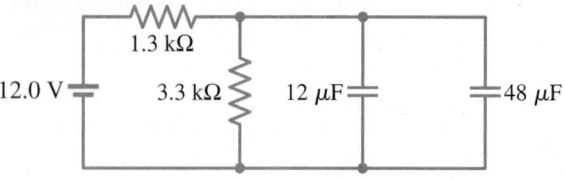

FIGURE 19–87 Problem 89.

90. Determine the current in each resistor of the circuit shown in Fig. 19–88.

FIGURE 19–88 Problem 90.

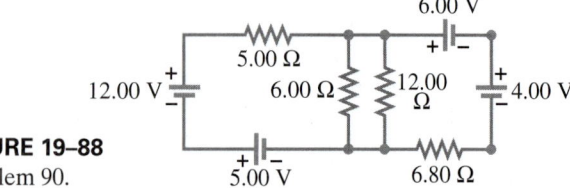

91. How much energy must a 24-V battery expend to charge a 0.45-μF and a 0.20-μF capacitor fully when they are placed (a) in parallel, (b) in series? (c) How much charge flowed from the battery in each case?

92. Two capacitors, $C_1 = 2.2\,\mu$F and $C_2 = 1.2\,\mu$F, are connected in parallel to a 24-V source as shown in Fig. 19–89a. After they are charged they are disconnected from the source and from each other, and then reconnected directly to each other with plates of opposite sign connected together (see Fig. 19–89b). Find the charge on each capacitor and the potential across each after equilibrium is established (Fig. 19–89c).

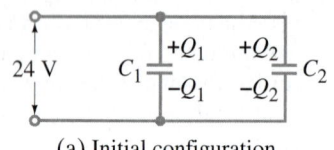

(a) Initial configuration.

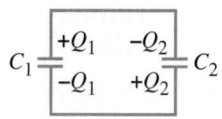

(b) At the instant of reconnection only.

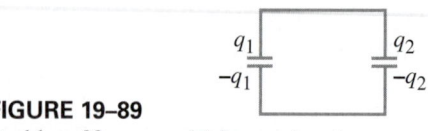

FIGURE 19–89 Problem 92.

(c) Later, after charges move.

93. The switch S in Fig. 19–90 is connected downward so that capacitor C_2 becomes fully charged by the battery of voltage V_0. If the switch is then connected upward, determine the charge on each capacitor after the switching.

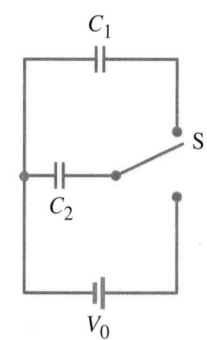

FIGURE 19–90 Problem 93.

94. The performance of the starter circuit in a car can be significantly degraded by a small amount of corrosion on a battery terminal. Figure 19–91a depicts a properly functioning circuit with a battery (12.5-V emf, 0.02-Ω internal resistance) attached via corrosion-free cables to a starter motor of resistance $R_S = 0.15\,\Omega$. Sometime later, corrosion between a battery terminal and a starter cable introduces an extra series resistance of only $R_C = 0.10\,\Omega$ into the circuit as suggested in Fig. 19–91b. Let P_0 be the power delivered to the starter in the circuit free of corrosion, and let P be the power delivered to the circuit with corrosion. Determine the ratio P/P_0.

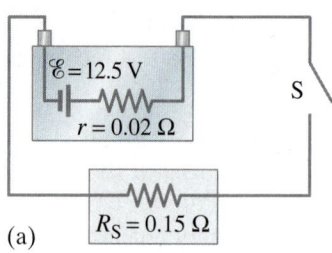

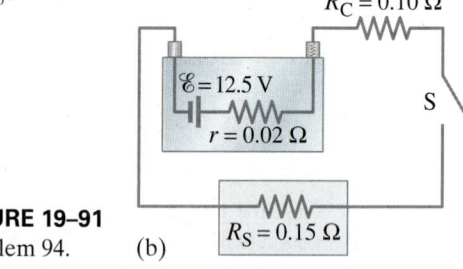

FIGURE 19–91 Problem 94.

95. The variable capacitance of an old radio tuner consists of four plates connected together placed alternately between four other plates, also connected together (Fig. 19–92). Each plate is separated from its neighbor by 1.6 mm of air. One set of plates can move so that the area of overlap of each plate varies from 2.0 cm^2 to 9.0 cm^2. (a) Are these seven capacitors connected in series or in parallel? (b) Determine the range of capacitance values.

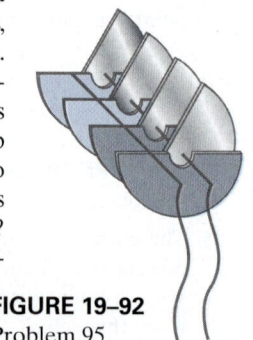

FIGURE 19–92 Problem 95.

96. A 175-pF capacitor is connected in series with an unknown capacitor, and as a series combination they are connected to a 25.0-V battery. If the 175-pF capacitor stores 125 pC of charge on its plates, what is the unknown capacitance?

97. In the circuit shown in Fig. 19–93, $C_1 = 1.0\,\mu$F, $C_2 = 2.0\,\mu$F, $C_3 = 2.4\,\mu$F, and a voltage $V_{ab} = 24$ V is applied across points a and b. After C_1 is fully charged, the switch is thrown to the right. What is the final charge and potential difference on each capacitor?

FIGURE 19–93 Problem 97.

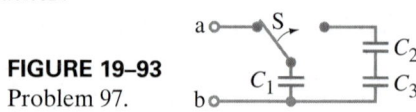

Search and Learn

1. Compare the formulas for resistors and for capacitors when connected in series and in parallel by filling in the Table below. Discuss and explain the differences. Consider the role of voltage V.

	R_{eq}	C_{eq}
Series		
Parallel		

2. Fill in the Table below for a combination of two unequal resistors of resistance R_1 and R_2. Assume the electric potential on the low-voltage end of the combination is V_A volts and the potential at the high-voltage end of the combination is V_B volts. First draw diagrams.

Property	Resistors in Series	Resistors in Parallel
Equivalent resistance		
Current through equivalent resistance		
Voltage across equivalent resistance		
Voltage across the pair of resistors		
Voltage across each resistor	$V_1 =$ $V_2 =$	$V_1 =$ $V_2 =$
Voltage at a point between the resistors		Not applicable
Current through each resistor	$I_1 =$ $I_2 =$	$I_1 =$ $I_2 =$

3. Cardiac defibrillators are discussed in Section 17–9. (a) Choose a value for the resistance so that the 1.0-μF capacitor can be charged to 3000 V in 2.0 seconds. Assume that this 3000 V is 95% of the full source voltage. (b) The effective resistance of the human body is given in Section 19–7. If the defibrillator discharges with a time constant of 10 ms, what is the effective capacitance of the human body?

4. A **potentiometer** is a device to precisely measure potential differences or emf, using a **null** technique. In the simple potentiometer circuit shown in Fig. 19–94, R' represents the total resistance of the resistor from A to B (which could be a long uniform "slide" wire), whereas R represents the resistance of only the part from A to the movable contact at C. When the unknown emf to be measured, $\mathcal{E}_x$, is placed into the circuit as shown, the movable contact C is moved until the galvanometer G gives a null reading (i.e., zero) when the switch S is closed. The resistance between A and C for this situation we call R_x. Next, a standard emf, $\mathcal{E}_s$, which is known precisely, is inserted into the circuit in place of $\mathcal{E}_x$ and again the contact C is moved until zero current flows through the galvanometer when the switch S is closed. The resistance between A and C now is called R_s. Show that the unknown emf is given by

$$\mathcal{E}_x = \left(\frac{R_x}{R_s}\right)\mathcal{E}_s$$

where R_x, R_s, and $\mathcal{E}_s$ are all precisely known. The working battery is assumed to be fresh and to give a constant voltage.

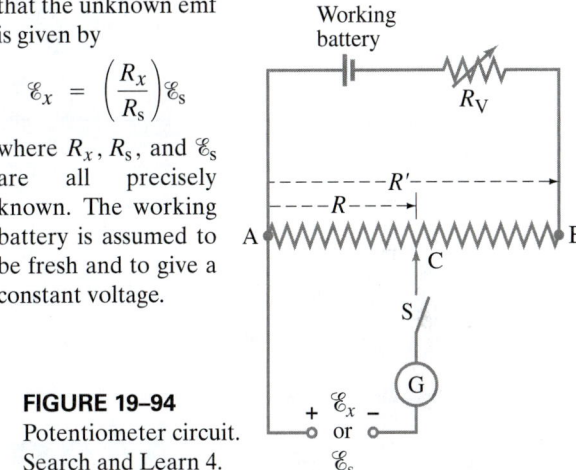

FIGURE 19–94 Potentiometer circuit. Search and Learn 4.

5. The circuit shown in Fig. 19–95 is a primitive 4-bit **digital-to-analog converter (DAC)**. In this circuit, to represent each digit (2^n) of a binary number, a "1" has the nth switch closed whereas zero ("0") has the switch open. For example, 0010 is represented by closing switch $n = 1$, while all other switches are open. Show that the voltage V across the 1.0-Ω resistor for the binary numbers 0001, 0010, 0100, and 1001 (which in decimal represent 1, 2, 4, 9) follows the pattern that you expect for a 4-bit DAC. (Section 17–10 may help.)

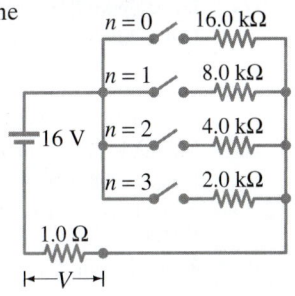

FIGURE 19–95 Search and Learn 5.

ANSWERS TO EXERCISES

A: 6 Ω and 25 Ω.
B: (b).
C: (a) 60-W bulb; (b) 100-W bulb. [Can you explain why? In (a), recall $P = I^2R$.]
D: $41I_3 - 45 + 21I_2 - 80 = 0$.
E: 180 A; this high current through the batteries could cause them to become very hot (and dangerous—possibly exploding): the power dissipated in the weak battery would be $P = I^2r = (180\text{ A})^2(0.10\text{ }\Omega) = 3200\text{ W}$!
F: (e).
G: $\approx$ 500 kΩ.

Magnets produce magnetic fields, but so do electric currents. Compass needles are magnets, and they align along the direction of any magnetic field present. Here, the compasses show the presence (and direction) of a magnetic field near a current-carrying wire. We shall see in this Chapter how magnetic field is defined, and how magnetic fields exert forces on electric currents and on charged particles. We also discuss useful applications of the interaction between magnetic fields and electric currents and moving electric charges, such as motors and loudspeakers.

Magnetism

CONTENTS

- 20–1 Magnets and Magnetic Fields
- 20–2 Electric Currents Produce Magnetic Fields
- 20–3 Force on an Electric Current in a Magnetic Field; Definition of $\vec{B}$
- 20–4 Force on an Electric Charge Moving in a Magnetic Field
- 20–5 Magnetic Field Due to a Long Straight Wire
- 20–6 Force between Two Parallel Wires
- 20–7 Solenoids and Electromagnets
- 20–8 Ampère's Law
- 20–9 Torque on a Current Loop; Magnetic Moment
- 20–10 Applications: Motors, Loudspeakers, Galvanometers
- *20–11 Mass Spectrometer
- *20–12 Ferromagnetism: Domains and Hysteresis

CHAPTER-OPENING QUESTION—Guess now!

Which of the following can experience a force when placed in the magnetic field of a magnet?

(a) An electric charge at rest.
(b) An electric charge moving.
(c) An electric current in a wire.
(d) Another magnet.

The history of magnetism began thousands of years ago, when in a region of Asia Minor known as Magnesia, rocks were found that could attract each other. These rocks were called "magnets" after their place of discovery.

Not until the nineteenth century, however, was it seen that magnetism and electricity are closely related. A crucial discovery was that electric currents produce magnetic effects (we will say "magnetic fields") like magnets do. All kinds of practical devices depend on magnetism, from compasses to motors, loudspeakers, computer memory, and electric generators.

20–1 Magnets and Magnetic Fields

You probably have observed a magnet attract paper clips, nails, and other objects made of iron, as in Fig. 20–1. Any magnet, whether it is in the shape of a bar or a horseshoe, has two ends or faces, called **poles**, which is where the magnetic effect is strongest. If a bar magnet is suspended from a fine thread, it is found that one pole of the magnet will always point toward the north. It is not known for sure when this fact was discovered, but it is known that the Chinese were making use of it as an aid to navigation by the eleventh century and perhaps earlier.

This is the principle of a compass. A compass needle is simply a bar magnet which is supported at its center of gravity so that it can rotate freely. The pole of a freely suspended magnet that points toward geographic north is called the **north pole** of the magnet. The other pole points toward the south and is called the **south pole**.

It is a familiar observation that when two magnets are brought near one another, each exerts a force on the other. The force can be either attractive or repulsive and can be felt even when the magnets don't touch. If the north pole of one bar magnet is brought near the north pole of a second magnet, the force is repulsive. Similarly, if the south poles are brought close, the force is repulsive. But when the north pole of one magnet is brought near the south pole of another magnet, the force is attractive. These results are shown in Fig. 20–2, and are reminiscent of the forces between electric charges: like poles repel, and unlike poles attract. But *do not confuse magnetic poles with electric charge*. They are very different. One important difference is that a positive or negative electric charge can easily be isolated. But an isolated single magnetic pole has never been observed. If a bar magnet is cut in half, you do not obtain isolated north and south poles. Instead, two new magnets are produced, Fig. 20–3, each with north (N) and south (S) poles. If the cutting operation is repeated, more magnets are produced, each with a north and a south pole. Physicists have searched for isolated single magnetic poles (monopoles), but no **magnetic monopole** has ever been observed.

Besides iron, a few other materials, such as cobalt, nickel, gadolinium, and some of their oxides and alloys, show strong magnetic effects. They are said to be **ferromagnetic** (from the Latin word *ferrum* for iron). Other materials show some slight magnetic effect, but it is very weak and can be detected only with delicate instruments. We will look in more detail at ferromagnetism in Section 20–12.

In Chapter 16, we used the concept of an electric field surrounding an electric charge. In a similar way, we can picture a **magnetic field** surrounding a magnet. The force one magnet exerts on another can then be described as the interaction between one magnet and the magnetic field of the other. Just as we drew electric field lines, we can also draw **magnetic field lines**. They can be drawn, as for electric field lines, so that

1. the direction of the magnetic field is tangent to a field line at any point, and
2. the number of lines per unit area is proportional to the strength of the magnetic field.

The *direction* of the magnetic field at a given location can be defined as the direction that the north pole of a compass needle would point if placed at that location. (We will give a more precise definition of magnetic field shortly.) Figure 20–4a shows how thin iron filings (acting like tiny magnets) reveal the magnetic field lines by lining up like the compass needles. The magnetic field determined in this way for the field surrounding a bar magnet is shown in Fig. 20–4b. Notice that because of our definition, the lines always point out from the north pole and in toward the south pole of a magnet (the north pole of a magnetic compass needle is attracted to the south pole of the magnet).

Magnetic field lines continue inside a magnet, as indicated in Fig. 20–4b. Indeed, given the lack of single magnetic poles, magnetic field lines always form closed loops, unlike electric field lines that begin on positive charges and end on negative charges.

FIGURE 20–1 A horseshoe magnet attracts pins made of iron.

FIGURE 20–2 Like poles of two magnets repel; unlike poles attract.

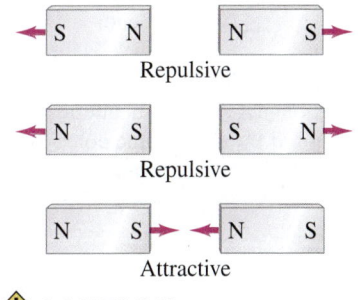

> **CAUTION**
> *Magnets do not attract all metals*

FIGURE 20–3 If you split a magnet, you won't get isolated north and south poles; instead, two new magnets are produced, each with a north and a south pole.

> **CAUTION**
> *Magnetic field lines form closed loops, unlike electric field lines*

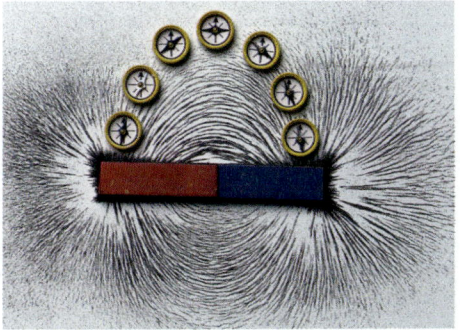

(a)

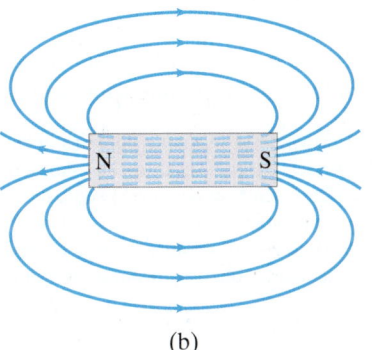

(b)

FIGURE 20–4 (a) Visualizing magnetic field lines around a bar magnet, using iron filings and compass needles. The red end of the bar magnet is its north pole. The N pole of a nearby compass needle points away from the north pole of the magnet. (b) Diagram of magnetic field lines for a bar magnet.

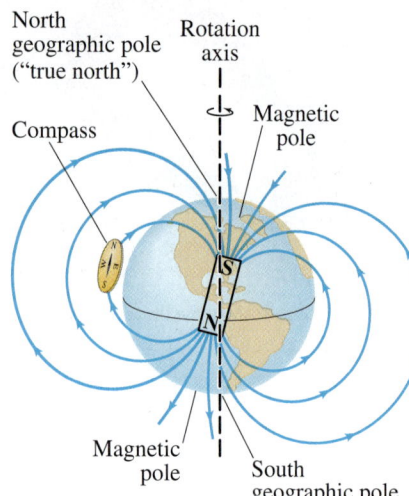

FIGURE 20–5 The Earth acts like a huge magnet. But its magnetic poles are not at the geographic poles (on the Earth's rotation axis).

Earth's Magnetic Field

The Earth's magnetic field is shown in Fig. 20–5, and is thought to be produced by electric currents in the Earth's molten iron outer core. The pattern of field lines is almost as though there were an imaginary bar magnet inside the Earth. Since the north pole (N) of a compass needle points north, the Earth's **magnetic pole** which is in the geographic north is magnetically a south pole, as indicated in Fig. 20–5 by the S on the schematic bar magnet inside the Earth. Remember that the north pole of one magnet is attracted to the south pole of another magnet. Nonetheless, Earth's pole in the north is still often called the "north magnetic pole," or "geomagnetic north," simply because it is in the north. Similarly, the Earth's southern magnetic pole, which is near the geographic south pole, is magnetically a north pole (N). The Earth's magnetic poles do not coincide with the **geographic poles**, which are on the Earth's axis of rotation. The north magnetic pole, for example, is in the Canadian Arctic, now on the order of 1000 km[†] from the geographic north pole, or **true north**. This difference must be taken into account for accurate use of a compass (Fig. 20–6). The angular difference between the direction of a compass needle (which points along the magnetic field lines) at any location and true (geographical) north is called the **magnetic declination**. In the U.S. it varies from 0° to about 20°, depending on location.

Notice in Fig. 20–5 that the Earth's magnetic field at most locations is not tangent to the Earth's surface. The angle that the Earth's magnetic field makes with the horizontal at any point is referred to as the **angle of dip**, or the "inclination." It is 67° at New York, for example, and 55° at Miami.

EXERCISE A Does the Earth's magnetic field have a greater magnitude near the poles or near the equator? [How can you tell using the field lines in Fig. 20–5?]

PHYSICS APPLIED
Use of a compass

FIGURE 20–6 Using a map and compass in the wilderness. First you align the compass case so the needle points away from true north (N) exactly the number of degrees of declination stated on the map (for this topographic map, it is 17° as shown just to the left of the compass). Then align the map with true north, as shown, *not* with the compass needle. [This is an old map (1953) of a part of California; on new maps (2012) the declination is only 13°, telling us the position of magnetic north has moved—see footnote below.]

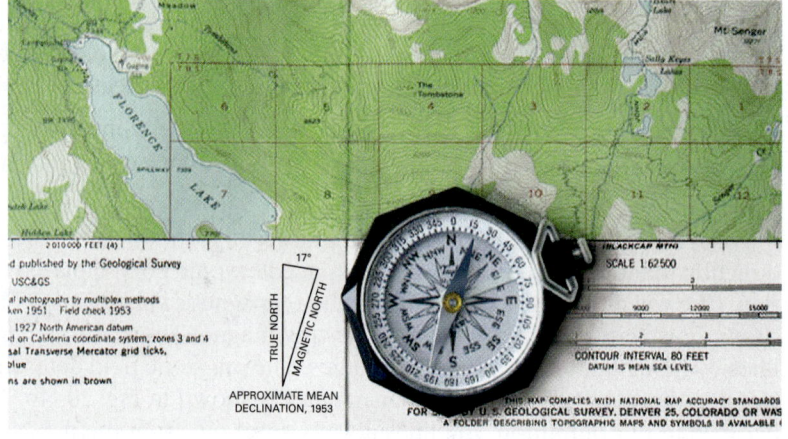

Uniform Magnetic Field

The simplest magnetic field is one that is uniform—it doesn't change in magnitude or direction from one point to another. A perfectly uniform field over a large area is not easy to produce. But the field between two flat parallel pole pieces of a magnet is nearly uniform if the area of the pole faces is large compared to their separation, as shown in Fig. 20–7. At the edges, the field "fringes" out somewhat: the magnetic field lines are no longer quite parallel and uniform. The parallel evenly spaced field lines in the central region of the gap indicate that the field is uniform at points not too near the edges, much like the electric field between two parallel plates (Fig. 17–1).

FIGURE 20–7 Magnetic field between two wide poles of a magnet is nearly uniform, except near the edges.

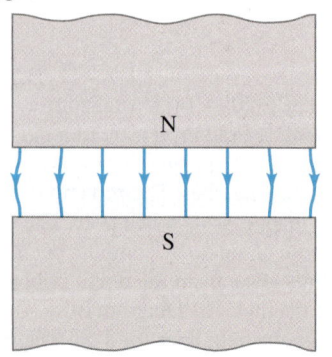

[†]Earth's north magnetic pole has been moving over time, on the order of 10 km per year in recent decades. Magnetism in rocks solidified at various times in the past (age determined by radioactive dating—see Section 30–11) suggests that Earth's magnetic poles have not only moved significantly over geologic time, but have also reversed direction 400 times over the last 330 million years. Also note that a compass gives a false reading if you are standing on rock containing magnetized iron ore (as you move around, the compass needle is inconsistent).

  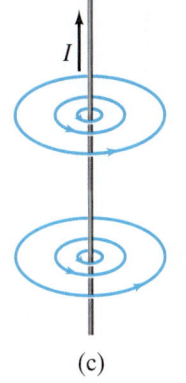 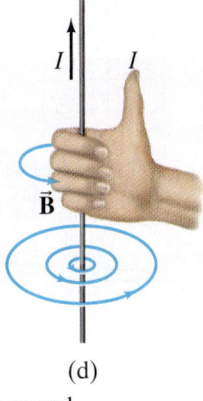

(a) (b) (c) (d)

FIGURE 20–8 (a) Deflection of compass needles near a current-carrying wire, showing the presence and direction of the magnetic field. (b) Iron filings also align along the direction of the magnetic field lines near a straight current-carrying wire. (c) Diagram of the magnetic field lines around an electric current in a straight wire. (d) Right-hand rule for remembering the direction of the magnetic field: when the thumb points in the direction of the conventional current, the fingers wrapped around the wire point in the direction of the magnetic field. ($\vec{B}$ is the symbol for magnetic field.)

20–2 Electric Currents Produce Magnetic Fields

During the eighteenth century, many scientists sought to find a connection between electricity and magnetism. A stationary electric charge and a magnet were shown to have no influence on each other. But in 1820, Hans Christian Oersted (1777–1851) found that when a compass is placed near a wire, the compass needle deflects if (and only if) the wire carries an electric current. As we have seen, a compass needle is deflected by a magnetic field. So Oersted's experiment showed that **an electric current produces a magnetic field**. He had found a connection between electricity and magnetism.

A compass needle placed near a straight section of current-carrying wire experiences a force, causing the needle to align tangent to a circle around the wire, Fig. 20–8a. Thus, the magnetic field lines produced by a current in a straight wire are in the form of circles with the wire at their center, Figs. 20–8b and c. The direction of these lines is indicated by the north pole of the compasses in Fig. 20–8a. There is a simple way to remember the direction of the magnetic field lines in this case. It is called a **right-hand rule**: grasp the wire with your right hand so that your thumb points in the direction of the conventional (positive) current; then your fingers will encircle the wire in the direction of the magnetic field, Fig. 20–8d.

Right-Hand-Rule-1:
Magnetic field direction produced by electric current

The magnetic field lines due to a circular loop of current-carrying wire can be determined in a similar way by placing a compass at various locations near the loop. The result is shown in Fig. 20–9. Again the right-hand rule can be used, as shown in Fig. 20–10. Unlike the uniform field shown in Fig. 20–7, the magnetic fields shown in Figs. 20–8 and 20–9 are *not* uniform—the fields are different in magnitude and direction at different locations.

FIGURE 20–9 Magnetic field lines due to a circular loop of wire.

EXERCISE B A straight wire carries a current directly toward you. In what direction are the magnetic field lines surrounding the wire?

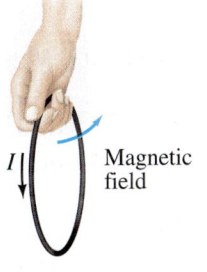

FIGURE 20–10 Right-hand rule for determining the direction of the magnetic field relative to the current in a loop of wire.

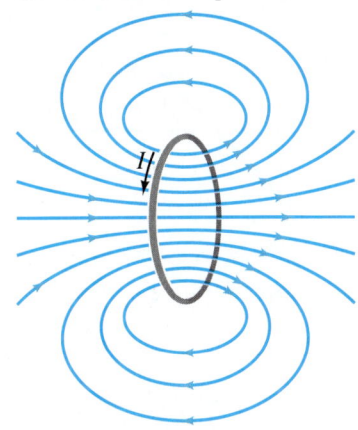

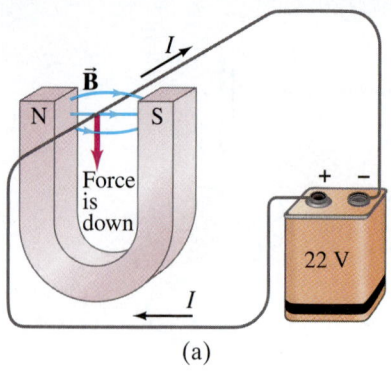
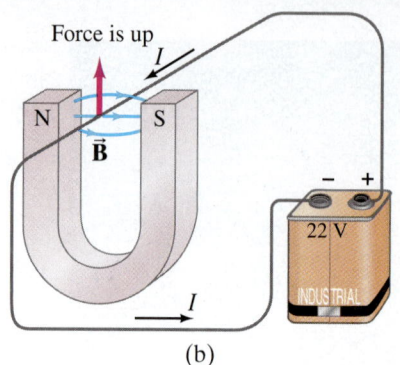
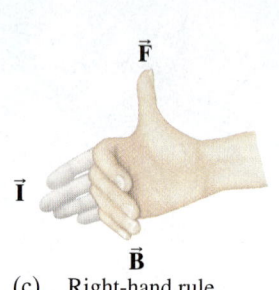

(a)   (b)   (c)   Right-hand rule

FIGURE 20–11 (a) Force on a current-carrying wire placed in a magnetic field $\vec{B}$; (b) same, but current reversed; (c) right-hand rule for setup in (b), with current $\vec{I}$ shown as if a vector with direction.

20–3 Force on an Electric Current in a Magnetic Field; Definition of $\vec{B}$

In Section 20–2 we saw that an electric current exerts a force on a magnet, such as a compass needle. By Newton's third law, we might expect the reverse to be true as well: we should expect that *a magnet exerts a force on a current-carrying wire*. Experiments indeed confirm this effect, and it too was first observed by Oersted.

Suppose a straight wire is placed in the magnetic field between the poles of a horseshoe magnet as shown in Fig. 20–11, where the vector symbol $\vec{B}$ represents the magnitude and direction of the magnetic field. When a current flows in the wire, experiment shows that a force is exerted on the wire. But this force is *not* toward one or the other pole of the magnet. Instead, the force is directed at right angles to the magnetic field direction, downward in Fig. 20–11a. If the current is reversed in direction, the force is in the opposite direction, upward as shown in Fig. 20–11b. Experiments show that *the direction of the force is always perpendicular to the direction of the current and also perpendicular to the direction of the magnetic field*, $\vec{B}$.

Right-Hand-Rule-2: *Force on current exerted by $\vec{B}$*

The direction of the force is given by another **right-hand rule**, as illustrated in Fig. 20–11c. Orient your right hand until your outstretched fingers can point in the direction of the conventional current I, and when you bend your fingers they point in the direction of the magnetic field lines, $\vec{B}$. Then your outstretched thumb will point in the direction of the force $\vec{F}$ on the wire.

This right-hand rule describes the direction of the force. What about the magnitude of the force on the wire? It is found experimentally that the magnitude of the force is directly proportional to the current I in the wire, to the magnetic field B (assumed uniform), and to the length ℓ of wire exposed to the magnetic field. The force also depends on the angle θ between the current direction and the magnetic field (Fig. 20–12), being proportional to $\sin\theta$. Thus, the force on a wire carrying a current I with length ℓ in a uniform magnetic field B is given by

$$F \propto I\ell B \sin\theta.$$

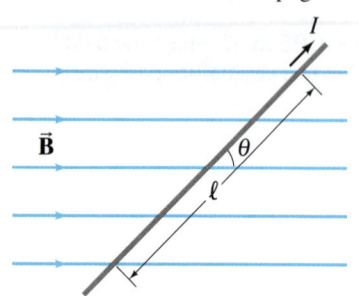

FIGURE 20–12 Current-carrying wire in a magnetic field. Force on the wire is directed into the page.

When the current is perpendicular to the field lines ($\theta = 90°$ and $\sin 90° = 1$), the force is strongest. When the wire is parallel to the magnetic field lines ($\theta = 0°$), there is no force at all.

Up to now we have not defined the magnetic field strength precisely. In fact, the magnetic field B can be conveniently defined in terms of the above proportion so that the proportionality constant is precisely 1. Thus we have

$$F = I\ell B \sin\theta. \qquad (20\text{–}1)$$

If the current's direction is perpendicular to the field $\vec{B}$ ($\theta = 90°$), then the force is

$$F_{\max} = I\ell B. \qquad [\text{current} \perp \vec{B}] \quad (20\text{–}2)$$

(If B is not uniform, then B in Eqs. 20–1 and 20–2 can be the average field over the length ℓ of the wire.)

The magnitude of $\vec{B}$ can be defined using Eq. 20–2 as $B = F_{\max}/I\ell$, where $F_{\max}$ is the magnitude of the force on a straight length ℓ of wire carrying a current I when the wire is perpendicular to $\vec{B}$.

EXERCISE C A wire carrying current I is perpendicular to a magnetic field of strength B. Assuming a fixed length of wire, which of the following changes will result in decreasing the force on the wire by a factor of 2? (a) Decrease the angle from 90° to 45°; (b) decrease the angle from 90° to 30°; (c) decrease the current in the wire to $I/2$; (d) decrease the magnetic field to $B/2$; (e) none of these will do it.

The SI unit for magnetic field B is the **tesla** (T). From Eq. 20–1 or 20–2, we see that $1\,\text{T} = 1\,\text{N/A}\cdot\text{m}$. An older name for the tesla is the "weber per meter squared" ($1\,\text{Wb/m}^2 = 1\,\text{T}$). Another unit sometimes used to specify magnetic field is a cgs unit, the **gauss** (G): $1\,\text{G} = 10^{-4}\,\text{T}$. A field given in gauss should always be changed to teslas before using with other SI units. To get a "feel" for these units, we note that the magnetic field of the Earth at its surface is about $\frac{1}{2}\,\text{G}$ or $0.5 \times 10^{-4}\,\text{T}$. On the other hand, strong electromagnets can produce fields on the order of 2 T and superconducting magnets can produce over 10 T.

EXAMPLE 20–1 Magnetic force on a current-carrying wire. A wire carrying a steady (dc) 30-A current has a length $\ell = 12$ cm between the pole faces of a magnet. The wire is at an angle $\theta = 60°$ to the field (Fig. 20–13). The magnetic field is approximately uniform at 0.90 T. We ignore the field beyond the pole pieces. Determine the magnitude and direction of the force on the wire.

APPROACH We use Eq. 20–1, $F = I\ell B \sin\theta$.

SOLUTION The force F on the 12-cm length of wire within the uniform field B is
$$F = I\ell B \sin\theta = (30\,\text{A})(0.12\,\text{m})(0.90\,\text{T})(\sin 60°) = 2.8\,\text{N}.$$
We use right-hand-rule-2 to find the direction of $\vec{F}$. Hold your right hand flat, pointing your fingers in the direction of the current. Then bend your fingers (maybe needing to rotate your hand) so they point along $\vec{B}$, Fig. 20–13. Your thumb then points into the page, which is thus the direction of the force F.

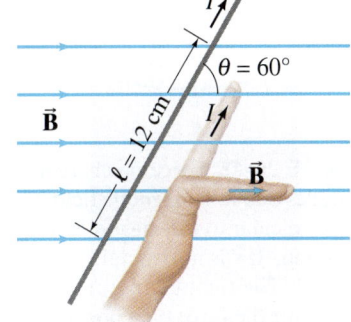

FIGURE 20–13 Example 20–1. For right-hand-rule-2, the thumb points into the page. See Fig. 20–11c.

EXERCISE D A straight power line carries 30 A and is perpendicular to the Earth's magnetic field of $0.50 \times 10^{-4}\,\text{T}$. What magnitude force is exerted on 100 m of this power line?

On a diagram, when we want to represent an electric current or a magnetic field that is pointing out of the page (toward us) or into the page, we use $\odot$ or $\times$, respectively. The $\odot$ is meant to resemble the tip of an arrow pointing directly toward the reader, whereas the $\times$ or $\otimes$ resembles the tail of an arrow pointing away. See Fig. 20–14.

EXAMPLE 20–2 Measuring a magnetic field. A rectangular loop of wire hangs vertically as shown in Fig. 20–14. A magnetic field $\vec{B}$ is directed horizontally, perpendicular to the plane of the loop, and points out of the page as represented by the symbol $\odot$. The magnetic field $\vec{B}$ is very nearly uniform along the horizontal portion of wire ab (length $\ell = 10.0$ cm) which is near the center of the gap of a large magnet producing the field. The top portion of the wire loop is out of the field. The loop hangs from a balance (reads 0 when $B = 0$) which measures a downward magnetic force of $F = 3.48 \times 10^{-2}\,\text{N}$ when the wire carries a current $I = 0.245$ A. What is the magnitude of the magnetic field B?

APPROACH Three straight sections of the wire loop are in the magnetic field: a horizontal section and two vertical sections. We apply Eq. 20–1 to each section and use the right-hand rule.

SOLUTION Using right-hand-rule-2 (page 564), we see that the magnetic force on the left vertical section of wire points to the left, and the force on the vertical section on the right points to the right. These two forces are equal and in opposite directions and so add up to zero. Hence, the net magnetic force on the loop is that on the horizontal section ab, whose length is $\ell = 0.100$ m. The angle θ between $\vec{B}$ and the wire is $\theta = 90°$, so $\sin\theta = 1$. Thus Eq. 20–1 gives
$$B = \frac{F}{I\ell} = \frac{3.48 \times 10^{-2}\,\text{N}}{(0.245\,\text{A})(0.100\,\text{m})} = 1.42\,\text{T}.$$

NOTE This technique can be a precise means of determining magnetic field strength.

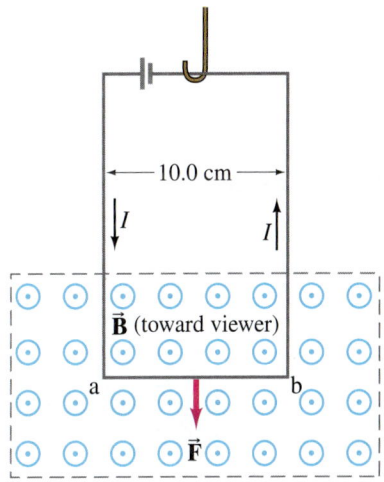

FIGURE 20–14 Measuring a magnetic field $\vec{B}$. Example 20–2.

20–4 Force on an Electric Charge Moving in a Magnetic Field

We have seen that a current-carrying wire experiences a force when placed in a magnetic field. Since a current in a wire consists of moving electric charges, we might expect that freely moving charged particles (not in a wire) would also experience a force when passing through a magnetic field. Free electric charges are not as easy to produce in the lab as a current in a wire, but it can be done, and experiments do show that moving electric charges experience a force in a magnetic field.

From what we already know, we can predict the force on a single electric charge moving in a magnetic field $\vec{B}$. If N such particles of charge q pass by a given point in time t, they constitute a current $I = Nq/t$. We let t be the time for a charge q to travel a distance ℓ in a magnetic field $\vec{B}$; then $\ell = vt$ where v is the magnitude of the velocity $\vec{v}$ of the particle. Thus, the force on these N particles is, by Eq. 20–1, $F = I\ell B \sin\theta = (Nq/t)(vt)B\sin\theta = NqvB\sin\theta$. The force on *one* of the N particles is then

$$F = qvB\sin\theta. \qquad [\theta \text{ between } \vec{v} \text{ and } \vec{B}] \quad (20\text{–}3)$$

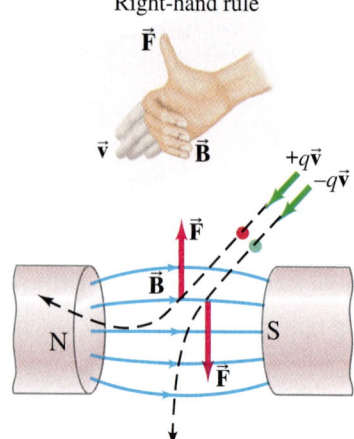

FIGURE 20–15 Force on charged particles due to a magnetic field is perpendicular to the magnetic field direction. If $\vec{v}$ is horizontal, then $\vec{F}$ is vertical. The right-hand rule is shown for the force on a positive charge, $+q$.

This equation gives the magnitude of the force exerted by a magnetic field on a particle of charge q moving with velocity v at a point where the magnetic field has magnitude B. The angle between $\vec{v}$ and $\vec{B}$ is θ. The force is greatest when the particle moves perpendicular to $\vec{B}$ ($\theta = 90°$):

$$F_{\text{max}} = qvB. \qquad [\vec{v} \perp \vec{B}] \quad (20\text{–}4)$$

Right-hand-rule-3:
Force on moving charge exerted by $\vec{B}$

The force is *zero* if the particle moves *parallel* to the field lines ($\theta = 0°$). The *direction* of the force is perpendicular to the magnetic field $\vec{B}$ and to the velocity $\vec{v}$ of the particle. For a positive charge, the force direction is given by another **right-hand rule**: you orient your right hand so that your outstretched fingers point along the direction of the particle's velocity ($\vec{v}$), and when you bend your fingers they must point along the direction of $\vec{B}$. Then your thumb will point in the direction of the force. This is true only for *positively* charged particles, and will be "up" for the positive particle shown in Fig. 20–15. For negatively charged particles, the force is in exactly the opposite direction, "down" in Fig. 20–15.

CONCEPTUAL EXAMPLE 20–3 **Negative charge near a magnet.** A negative charge $-Q$ is placed at rest near a magnet. Will the charge begin to move? Will it feel a force? What if the charge were positive, $+Q$?

RESPONSE No to all questions. A charge at rest has velocity equal to zero. Magnetic fields exert a force only on moving electric charges (Eq. 20–3).

EXERCISE E Return to the Chapter-Opening Question, page 560, and answer it again now. Try to explain why you may have answered differently the first time.

EXAMPLE 20–4 **Magnetic force on a proton.** A magnetic field exerts a force of 8.0×10^{-14} N toward the west on a proton moving vertically upward at a speed of 5.0×10^6 m/s (Fig. 20–16a). When moving horizontally in a northerly direction, the force on the proton is zero (Fig. 20–16b). Determine the magnitude and direction of the magnetic field in this region. (The charge on a proton is $q = +e = 1.6 \times 10^{-19}$ C.)

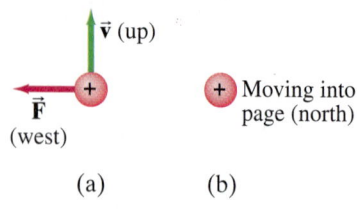

FIGURE 20–16 Example 20–4.

APPROACH Since the force on the proton is zero when moving north, the field must be in a north–south direction ($\theta = 0°$ in Eq. 20–3). To produce a force to the west when the proton moves upward, right-hand-rule-3 tells us that $\vec{B}$ must point toward the north. (Your thumb points west and the outstretched fingers of your right hand point upward only when your bent fingers point north.) The magnitude of $\vec{B}$ is found using Eq. 20–3.

SOLUTION Equation 20–3 with $\theta = 90°$ gives

$$B = \frac{F}{qv} = \frac{8.0 \times 10^{-14}\,\text{N}}{(1.6 \times 10^{-19}\,\text{C})(5.0 \times 10^6\,\text{m/s})} = 0.10\,\text{T}.$$

EXERCISE F Determine the force on the proton of Example 20–4 if it heads horizontally south.

EXAMPLE 20–5 ESTIMATE **Magnetic force on ions during a nerve pulse.**
Estimate the magnitude of the magnetic force due to the Earth's magnetic field on ions crossing a cell membrane during an action potential (Section 18–10). Assume the speed of the ions is 10^{-2} m/s.

APPROACH Using $F = qvB$, set the magnetic field of the Earth to be roughly $B \approx 10^{-4}$ T, and the charge $q \approx e \approx 10^{-19}$ C.

SOLUTION $F \approx (10^{-19}\,\text{C})(10^{-2}\,\text{m/s})(10^{-4}\,\text{T}) = 10^{-25}$ N.

NOTE This is an extremely small force. Yet it is thought that migrating animals do somehow detect the Earth's magnetic field, and this is an area of active research.

The path of a charged particle moving in a plane perpendicular to a uniform magnetic field is a circle as we shall now show. In Fig. 20–17 the magnetic field is directed *into* the paper, as represented by ×'s. An electron at point P is moving to the right, and the force on it at this point is toward the bottom of the page as shown (use the right-hand rule and reverse the direction for negative charge). The electron is thus deflected toward the page bottom. A moment later, say, when it reaches point Q, the force is still perpendicular to the velocity and is in the direction shown. Because the force is always perpendicular to $\vec{v}$, the magnitude of $\vec{v}$ does not change—the electron moves at constant speed. We saw in Chapter 5 that if the force on a particle is always perpendicular to its velocity $\vec{v}$, the particle moves in a circle and has a centripetal acceleration of magnitude $a = v^2/r$ (Eq. 5–1). Thus a charged particle moves in a circular path with a constant magnitude of centripetal acceleration in a uniform magnetic field (see Fig. 20–18). The electron moves clockwise in Fig. 20–17. A positive particle in this field would feel a force in the opposite direction and would thus move counterclockwise.

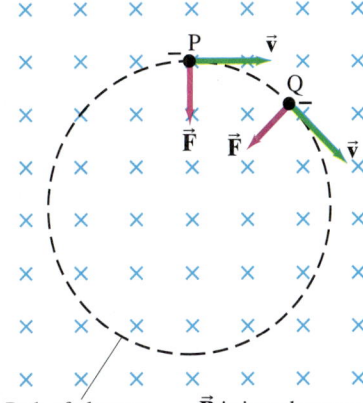

Path of electron $\vec{B}$ is into the page

FIGURE 20–17 Force exerted by a uniform magnetic field on a moving charged particle (in this case, an electron) produces a circular path.

FIGURE 20–18 The white ring inside the glass tube is the glow of a beam of electrons that ionize the gas molecules. The red coils of current-carrying wire produce a nearly uniform magnetic field, illustrating the circular path of charged particles in a uniform magnetic field.

EXAMPLE 20–6 **Electron's path in a uniform magnetic field.** An electron travels at 1.5×10^7 m/s in a plane perpendicular to a uniform 0.010-T magnetic field. Describe its path quantitatively. Ignore gravity (= very small in comparison).

APPROACH The electron moves at speed v in a curved path and so must have a centripetal acceleration $a = v^2/r$ (Eq. 5–1). We find the radius of curvature using Newton's second law. The force is given by Eq. 20–3 with $\sin \theta = 1$: $F = qvB$.

SOLUTION We insert F and a into Newton's second law:

$$\Sigma F = ma$$
$$qvB = \frac{mv^2}{r}.$$

We solve for r and find

$$r = \frac{mv}{qB}.$$

Since $\vec{F}$ is perpendicular to $\vec{v}$, the magnitude of $\vec{v}$ doesn't change. From this equation we see that if $\vec{B}$ = constant, then r = constant, and the curve must be a circle as we claimed above. To get r we put in the numbers:

$$r = \frac{(9.1 \times 10^{-31}\,\text{kg})(1.5 \times 10^7\,\text{m/s})}{(1.6 \times 10^{-19}\,\text{C})(0.010\,\text{T})} = 0.85 \times 10^{-2}\,\text{m} = 8.5\,\text{mm}.$$

NOTE See Fig. 20–18. If the magnetic field B is larger, is the radius larger or smaller?

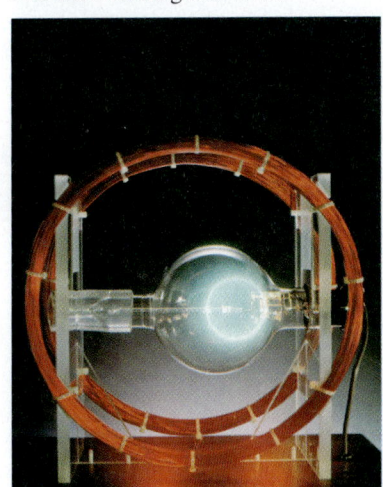

The time T required for a particle of charge q moving with constant speed v to make one circular revolution in a uniform magnetic field $\vec{B}$ ($\perp \vec{v}$) is $T = 2\pi r/v$, where $2\pi r$ is the circumference of its circular path. From Example 20–6, $r = mv/qB$, so

$$T = \frac{2\pi m}{qB}.$$

Since T is the period of rotation, the frequency of rotation is

$$f = \frac{1}{T} = \frac{qB}{2\pi m}. \quad (20\text{–}5)$$

This is often called the **cyclotron frequency** of a particle in a field because this is the frequency at which particles revolve in a cyclotron (see Problem 88).

CONCEPTUAL EXAMPLE 20–7 **Stopping charged particles.** An electric charge q moving in an electric field $\vec{E}$ can be decelerated to a stop if the force $\vec{F} = q\vec{E}$ (Eq. 16–5) acts in the direction opposite to the charge's velocity. Can a magnetic field be used to stop a charged particle?

RESPONSE No, because the force is always *perpendicular* to the velocity of the particle and thus can only change the direction but not the magnitude of its velocity. Also the magnetic force cannot do work on the particle (force and displacement are perpendicular, Eq. 6–1) and so cannot change the kinetic energy of the particle, Eq. 6–4.

PROBLEM SOLVING

Magnetic Fields

Magnetic fields are somewhat analogous to the electric fields of Chapter 16, but there are several important differences to recall:

1. The force experienced by a charged particle moving in a magnetic field is *perpendicular* to the direction of the magnetic field (and to the direction of the velocity of the particle), whereas the force exerted by an electric field is *parallel* to the direction of the field (and independent of the velocity of the particle).

2. The *right-hand rule*, in its different forms, is intended to help you determine the directions of magnetic field, and the forces they exert, and/or the directions of electric current or charged particle velocity. The right-hand rules (Table 20–1) are designed to deal with the "perpendicular" nature of these quantities.

3. The equations in this Chapter are generally not printed as vector equations, but involve magnitudes only. Right-hand rules are to be used to find directions of vector quantities.

Physical Situation	Example	How to Orient Right Hand	Result
1. Magnetic field produced by current (RHR-1)	Fig. 20–8d	Wrap fingers around wire with thumb pointing in direction of current I	Fingers curl in direction of $\vec{B}$
2. Force on electric current I due to magnetic field (RHR-2)	Fig. 20–11c	Fingers first point straight along current I, then bend along magnetic field $\vec{B}$	Thumb points in direction of the force $\vec{F}$
3. Force on electric charge $+q$ due to magnetic field (RHR-3)	Fig. 20–15	Fingers point along particle's velocity $\vec{v}$, then along $\vec{B}$	Thumb points in direction of the force $\vec{F}$

TABLE 20–1 Summary of Right-hand Rules (= RHR)

CONCEPTUAL EXAMPLE 20–8 **A helical path.** What is the path of a charged particle in a uniform magnetic field if its velocity is *not* perpendicular to the magnetic field?

RESPONSE The velocity vector can be broken down into components parallel and perpendicular to the field. The velocity component parallel to the field lines experiences no force ($\theta = 0$), so this component remains constant. The velocity component perpendicular to the field results in circular motion about the field lines. Putting these two motions together produces a helical (spiral) motion around the field lines as shown in Fig. 20–19.

EXERCISE G What is the sign of the charge in Fig. 20–19? How would you modify the drawing if the charge had the opposite sign?

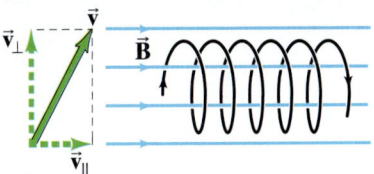

FIGURE 20–19 Example 20–8.

FIGURE 20–20 (a) Diagram showing a charged particle that approaches the Earth and is "captured" by the magnetic field of the Earth. Such particles follow the field lines toward the poles as shown. (b) Photo of aurora borealis.

*Aurora Borealis

Charged ions approach the Earth from the Sun (the "solar wind") and enter the atmosphere mainly near the poles, sometimes causing a phenomenon called the **aurora borealis** or "northern lights" in northern latitudes. To see why, consider Example 20–8 and Fig. 20–20 (see also Fig. 20–19). In Fig. 20–20 we imagine a stream of charged particles approaching the Earth. The velocity component *perpendicular* to the field for each particle becomes a circular orbit around the field lines, whereas the velocity component *parallel* to the field carries the particle along the field lines toward the poles. As a particle approaches the Earth's North Pole, the magnetic field is stronger and the radius of the helical path becomes smaller (see Example 20–6, $r \propto 1/B$).

A high concentration of high-speed charged particles ionizes the air, and as the electrons recombine with atoms, light is emitted (Chapter 27) which is the aurora. Auroras are especially spectacular during periods of high sunspot activity when more charged particles are emitted and more come toward Earth.

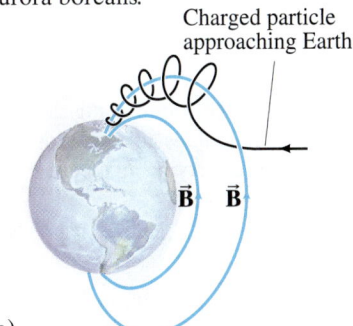

*The Hall Effect

When a current-carrying conductor is held fixed in a magnetic field, the field exerts a sideways force on the charges moving in the conductor. For example, if electrons move to the right in the rectangular conductor shown in Fig. 20–21a, the inward magnetic field will exert a downward force on the electrons of magnitude $F = ev_dB$, where v_d is the drift velocity of the electrons (Section 18–8). Thus the electrons will tend to move nearer to side D than side C, causing a potential difference between sides C and D of the conductor. This potential difference builds up until the electric field $\vec{E}_H$ that it produces exerts a force ($= e\vec{E}_H$) on the moving charges that is equal and opposite to the magnetic force ($= ev_dB$). This is the **Hall effect**, named after E. H. Hall who discovered it in 1879. The difference of potential produced is called the **Hall emf**. Its magnitude is $V_{Hall} = E_Hd = (F/e)d = v_dBd$, where d is the width of the conductor.

A current of negative charges moving to the right is equivalent to positive charges moving to the left, at least for most purposes. But the Hall effect can distinguish these two. As can be seen in Fig. 20–21b, positive particles moving to the left are deflected downward, so that the bottom surface is positive relative to the top surface. This is the reverse of part (a). Indeed, the direction of the emf in the Hall effect first revealed that it is negative particles that move in metal conductors, and that positive "holes" move in *p*-type semiconductors.

Because the Hall emf is proportional to B, the Hall effect can be used to measure magnetic fields. A device to do so is called a *Hall probe*. When B is known, the Hall emf can be used to determine the drift velocity of charge carriers.

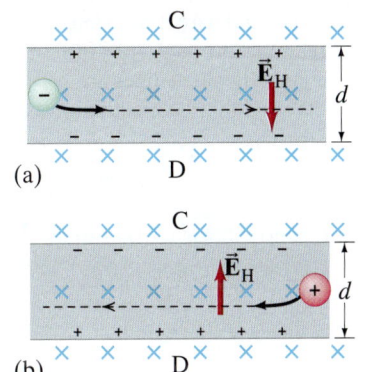

FIGURE 20–21 The Hall effect. (a) Negative charges moving to the right as the current. (b) Positive charges moving to the left as the current.

20–5 Magnetic Field Due to a Long Straight Wire

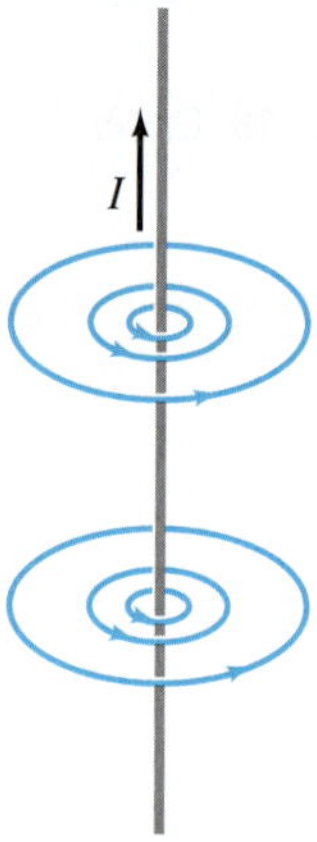

We saw in Section 20–2, Fig. 20–8, that the magnetic field lines due to the electric current in a long straight wire form circles with the wire at the center (Fig. 20–22). You might expect that the field strength at a given point would be greater if the current flowing in the wire were greater; and that the field would be less at points farther from the wire. This is indeed the case. Careful experiments show that the magnetic field B due to the current in a long straight wire is directly proportional to the current I in the wire and inversely proportional to the distance r from the wire:

$$B \propto \frac{I}{r}.$$

This relation is valid as long as r, the perpendicular distance to the wire, is much less than the distance to the ends of the wire (i.e., the wire is long).

The proportionality constant is written as $\mu_0/2\pi$, so

$$B = \frac{\mu_0}{2\pi}\frac{I}{r}. \qquad \text{[near a long straight wire]} \quad (20\text{–}6)$$

The value of the constant μ_0, which is called the **permeability of free space**, is[†]
$\mu_0 = 4\pi \times 10^{-7}\,\text{T}\cdot\text{m/A}$.

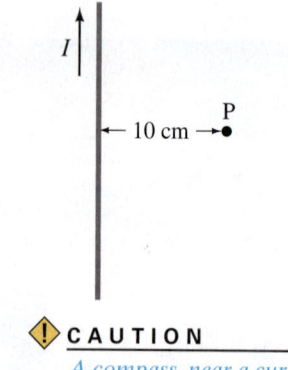

FIGURE 20–22 Same as Fig. 20–8c, magnetic field lines around a long straight wire carrying an electric current I.

FIGURE 20–23 Example 20–9.

EXAMPLE 20–9 Calculation of $\vec{B}$ near a wire. An electric wire in the wall of a building carries a dc current of 25 A vertically upward. What is the magnetic field due to this current at a point P, 10 cm due north of the wire (Fig. 20–23)?

APPROACH We assume the wire is much longer than the 10-cm distance to the point P so we can apply Eq. 20–6.

SOLUTION According to Eq. 20–6:

$$B = \frac{\mu_0 I}{2\pi r} = \frac{(4\pi \times 10^{-7}\,\text{T}\cdot\text{m/A})(25\,\text{A})}{(2\pi)(0.10\,\text{m})} = 5.0 \times 10^{-5}\,\text{T},$$

or 0.50 G. By right-hand-rule-1 (page 568), the field points to the west (into the page in Fig. 20–23) at point P.

NOTE The magnetic field at point P produced by the wire has about the same magnitude as Earth's, so a compass at P would not point north but to the northwest.

NOTE Most electrical wiring in buildings consists of cables with two wires in each cable. Since the two wires carry current in opposite directions, their magnetic fields cancel to a large extent, but may still affect sensitive electronic devices.

⚠ **CAUTION**
A compass, near a current, may not point north

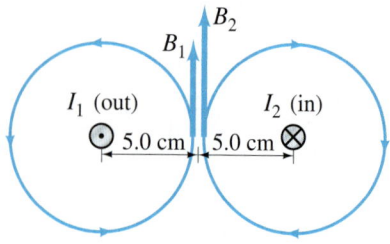

FIGURE 20–24 Example 20–10. Wire 1 carrying current I_1 out towards us, and wire 2 carrying current I_2 into the page, produce magnetic fields whose lines are circles around their respective wires.

EXAMPLE 20–10 Magnetic field midway between two currents. Two parallel straight wires 10.0 cm apart carry currents in opposite directions (Fig. 20–24). Current $I_1 = 5.0\,\text{A}$ is out of the page, and $I_2 = 7.0\,\text{A}$ is into the page. Determine the magnitude and direction of the magnetic field halfway between the two wires.

APPROACH The magnitude of the field produced by each wire is calculated from Eq. 20–6. The direction of *each* wire's field is determined with the right-hand rule. The total field is the vector sum of the two fields at the midway point.

SOLUTION The magnetic field lines due to current I_1 form circles around the wire of I_1, and right-hand-rule-1 (Fig. 20–8d) tells us they point counterclockwise around the wire. The field lines due to I_2 form circles around the wire of I_2 and point clockwise, Fig. 20–24. At the midpoint, both fields point upward in Fig. 20–24 as shown, and so add together. The midpoint is 0.050 m from each wire.

[†]The constant is chosen in this complicated way so that Ampère's law (Section 20–8), which is considered more fundamental, will have a simple and elegant form.

From Eq. 20–6 the magnitudes of B_1 and B_2 are

$$B_1 = \frac{\mu_0 I_1}{2\pi r} = \frac{(4\pi \times 10^{-7}\,\text{T·m/A})(5.0\,\text{A})}{2\pi(0.050\,\text{m})} = 2.0 \times 10^{-5}\,\text{T};$$

$$B_2 = \frac{\mu_0 I_2}{2\pi r} = \frac{(4\pi \times 10^{-7}\,\text{T·m/A})(7.0\,\text{A})}{2\pi(0.050\,\text{m})} = 2.8 \times 10^{-5}\,\text{T}.$$

The total field is *up* with a magnitude of

$$B = B_1 + B_2 = 4.8 \times 10^{-5}\,\text{T}.$$

EXERCISE H Suppose both I_1 and I_2 point into the page in Fig. 20–24. What then is the field B midway between the wires?

CONCEPTUAL EXAMPLE 20–11 **Magnetic field due to four wires.** Figure 20–25 shows four long parallel wires which carry equal currents into or out of the page as shown. In which configuration, (*a*) or (*b*), is the magnetic field greater at the center of the square?

RESPONSE It is greater in (*a*). The arrows illustrate the directions of the field produced by each wire; check it out, using the right-hand rule to confirm these results. The net field at the center is the superposition of the four fields (which are of equal magnitude), which will point to the left in (*a*) and is zero in (*b*).

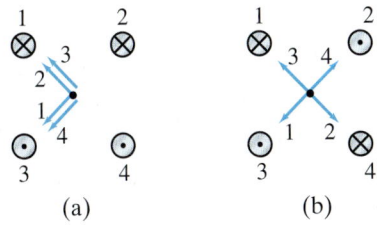

FIGURE 20–25 Example 20–11.

20–6 Force between Two Parallel Wires

We have seen that a wire carrying a current produces a magnetic field (magnitude given by Eq. 20–6 for a long straight wire). Also, a current-carrying wire feels a force when placed in a magnetic field (Section 20–3, Eq. 20–1). Thus, we expect that two current-carrying wires will exert a force on each other.

Consider two long parallel wires separated by a distance d, as in Fig. 20–26a. They carry currents I_1 and I_2, respectively. Each current produces a magnetic field that is "felt" by the other, so each must exert a force on the other. For example, the magnetic field B_1 produced by I_1 in Fig 20–26 is given by Eq. 20–6, which at the location of wire 2 points into the page and has magnitude

$$B_1 = \frac{\mu_0}{2\pi}\frac{I_1}{d}.$$

See Fig. 20–26b, where the field due *only* to I_1 is shown. According to Eq. 20–2, the force F_2 exerted by B_1 on a length ℓ_2 of wire 2, carrying current I_2, has magnitude

$$F_2 = I_2 B_1 \ell_2.$$

Note that the force on I_2 is due only to the field produced by I_1. Of course, I_2 also produces a field, but it does not exert a force on itself. We substitute B_1 into the formula for F_2 and find that the force on a length ℓ_2 of wire 2 is

$$F_2 = \frac{\mu_0}{2\pi}\frac{I_1 I_2}{d}\ell_2. \qquad \text{[parallel wires]} \quad (20\text{–}7)$$

If we use right-hand-rule-1 of Fig. 20–8d, we see that the lines of B_1 are as shown in Fig. 20–26b. Then using right-hand-rule-2 of Fig. 20–11c, we see that the force exerted on I_2 will be to the left in Fig. 20–26b. That is, I_1 exerts an attractive force on I_2 (Fig. 20–27a). This is true as long as the currents are in the same direction. If I_2 is in the opposite direction from I_1, right-hand-rule-2 indicates that the force is in the opposite direction. That is, I_1 exerts a repulsive force on I_2 (Fig. 20–27b).

Reasoning similar to that above shows that the magnetic field produced by I_2 exerts an equal but opposite force on I_1. We expect this to be true also from Newton's third law. Thus, as shown in Fig. 20–27, parallel currents in the same direction attract each other, whereas parallel currents in opposite directions repel.

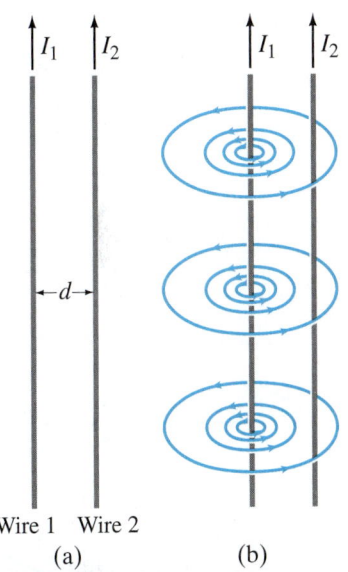

FIGURE 20–26 (a) Two parallel conductors carrying currents I_1 and I_2. (b) Magnetic field $\vec{B}_1$ produced by I_1. (Field produced by I_2 is not shown.) $\vec{B}_1$ points into page at position of I_2.

FIGURE 20–27 (a) Parallel currents in the same direction exert an attractive force on each other. (b) Antiparallel currents (in opposite directions) exert a repulsive force on each other.

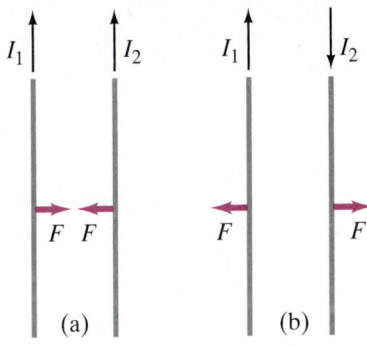

EXAMPLE 20–12 **Force between two current-carrying wires.** The two wires of a 2.0-m-long appliance cord are 3.0 mm apart and carry a current of 8.0 A. Calculate the force one wire exerts on the other.

APPROACH Each wire is in the magnetic field of the other when the current is on, so we can apply Eq. 20–7.

SOLUTION Equation 20–7 gives

$$F = \frac{\mu_0}{2\pi} \frac{I_1 I_2}{d} \ell_2 = \frac{(4\pi \times 10^{-7}\,\text{T·m/A})(8.0\,\text{A})^2(2.0\,\text{m})}{(2\pi)(3.0 \times 10^{-3}\,\text{m})} = 8.5 \times 10^{-3}\,\text{N}.$$

The currents are in opposite directions (one toward the appliance, the other away from it), so the force would be repulsive and tend to spread the wires apart.

Definition of the Ampere and the Coulomb

You may have wondered how the constant μ_0 in Eq. 20–6 could be exactly $4\pi \times 10^{-7}\,\text{T·m/A}$. Here is how it happened. With an older definition of the ampere, μ_0 was measured experimentally to be very close to this value. Today, μ_0 is *defined* to be exactly $4\pi \times 10^{-7}\,\text{T·m/A}$. This could not be done if the ampere were defined independently. The ampere, the unit of current, is now defined in terms of the magnetic field it produces using the defined value of μ_0.

In particular, we use the force between two parallel current-carrying wires, Eq. 20–7, to define the ampere precisely. If $I_1 = I_2 = 1$ A exactly, and the two wires are exactly 1 m apart, then

$$\frac{F}{\ell} = \frac{\mu_0}{2\pi} \frac{I_1 I_2}{d} = \frac{(4\pi \times 10^{-7}\,\text{T·m/A})}{(2\pi)} \frac{(1\,\text{A})(1\,\text{A})}{(1\,\text{m})} = 2 \times 10^{-7}\,\text{N/m}.$$

Thus, one **ampere** is defined as *that current flowing in each of two long parallel wires, 1 m apart, which results in a force of exactly 2×10^{-7} N per meter of length of each wire.*

This is the precise definition of the ampere, and because it is readily reproducible, is called an **operational definition**. The **coulomb** is defined in terms of the ampere as being *exactly* one ampere-second: 1 C = 1 A·s.

20–7 Solenoids and Electromagnets

PHYSICS APPLIED
Solenoids and electromagnets

A long coil of wire consisting of many loops (or turns) of wire is called a **solenoid**. The current in each loop produces a magnetic field, as we saw in Fig. 20–9. The magnetic field within a solenoid can be fairly large because it is the sum of the fields due to the current in each loop (Fig. 20–28). A solenoid acts like a magnet; one end can be considered the north pole and the other the south pole, depending on the direction of the current in the loops (use the right-hand rule). Since the magnetic field lines leave the north pole of a magnet, the north pole of the solenoid in Fig. 20–28 is on the right. As we will see in the next Section, the magnetic field inside a tightly wrapped solenoid with N turns of wire in a length ℓ, each carrying current I, is

$$B = \frac{\mu_0 N I}{\ell}. \qquad (20\text{–}8)$$

If a piece of iron is placed inside a solenoid, the magnetic field is increased greatly because the iron becomes a magnet. The resulting magnetic field is the sum of the field due to the current and the field due to the iron, and can be hundreds or thousands of times the field due to the current alone (see Section 20–12). Such an iron-core solenoid is an **electromagnet**.

Electromagnets have many practical applications, from use in motors and generators to producing large magnetic fields for research. Sometimes an iron core is not present—the magnetic field then comes only from the current in the wire coils. A large field B in this case requires a large current I, which produces a large amount of waste heat $(P = I^2 R)$. But if the current-carrying wires are made of superconducting material kept below the transition temperature (Section 18–9), very high fields can be produced, and no electric power is needed to maintain the large current in the superconducting coils. Energy is required, however, to refrigerate the coils at the low temperatures where they superconduct.

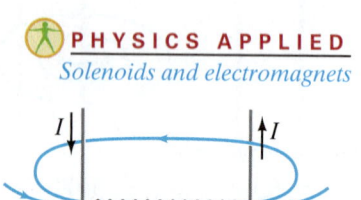
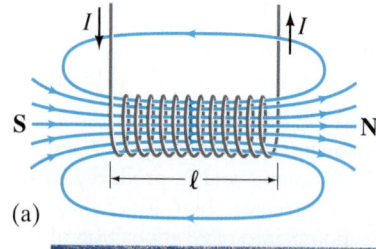

FIGURE 20–28 (a) Magnetic field of a solenoid. The north pole of this solenoid, thought of as a magnet, is on the right, and the south pole is on the left. (b) Photo of iron filings aligning along $\vec{B}$ field lines of a solenoid with loosely spaced loops. The field is smoother if the loops are closely spaced.

Another useful device consists of a solenoid into which a rod of iron is partially inserted. This combination is also referred to as a **solenoid**. One simple use is as a doorbell (Fig. 20–29). When the circuit is closed by pushing the button, the coil effectively becomes a magnet and exerts a force on the iron rod. The rod is pulled into the coil and strikes the bell. A large solenoid is used for the starter of a car: when you engage the starter, you are closing a circuit that not only turns the starter motor, but first activates a solenoid that moves the starter into direct contact with the gears on the engine's flywheel. Solenoids are used a lot as switches in cars and many other devices. They have the advantage of moving mechanical parts quickly and accurately.

PHYSICS APPLIED
Doorbell, car starter

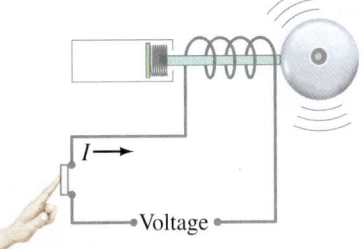

Magnetic Circuit Breakers

Modern circuit breakers that protect houses and buildings from overload and fire contain not only a "thermal" part (bimetallic strip as described in Section 18–6, Fig. 18–19) but also a magnetic sensor. If the current is above a certain level, the magnetic field the current produces pulls an iron plate that breaks the same contact points as in Figs. 18–19b and c. Magnetic circuit breakers react quickly (< 10 ms), and for buildings are designed to react to the high currents of short circuits (but not shut off for the start-up surges of motors).

In more sophisticated circuit breakers, including ground fault circuit interrupters (GFCIs—discussed in Section 21–9), a solenoid is used. The iron rod of Fig. 20–29, instead of striking a bell, strikes one side of a pair of electric contact points, opening them and opening the circuit. They react very quickly (≈ 1 ms) and to very small currents (≈ 5 mA) and thus protect humans (not just property) and save lives.

FIGURE 20–29 Solenoid used as a doorbell.

PHYSICS APPLIED
Magnetic circuit breakers

20–8 Ampère's Law

The relation between the current in a long straight wire and the magnetic field it produces is given by Eq. 20–6, Section 20–5. This equation is valid *only* for a long straight wire. Is there a general relation between a current in a wire of any shape and the magnetic field around it? Yes: the French scientist André Marie Ampère (1775–1836) proposed such a relation shortly after Oersted's discovery. Consider any (arbitrary) closed path around a current, as shown in Fig. 20–30, and imagine this path as being made up of short segments each of length $\Delta \ell$. We take the product of the length of each segment times the component of magnetic field $\vec{B}$ parallel to that segment. If we now sum all these terms, the result (according to Ampère) will be equal to μ_0 times the net current I_{encl} that passes through the surface *enclosed* by the path. This is known as **Ampère's law** and can be written

$$\Sigma B_\parallel \Delta \ell = \mu_0 I_{encl}. \qquad (20\text{–}9)$$

AMPÈRE'S LAW

The symbol Σ means "the sum of" and $B_\parallel$ means the component of $\vec{B}$ parallel to that particular $\Delta \ell$. The lengths $\Delta \ell$ are chosen small enough so that $B_\parallel$ is essentially constant along each length. The sum must be made over a closed path, and I_{encl} is the total net current enclosed by this closed path.

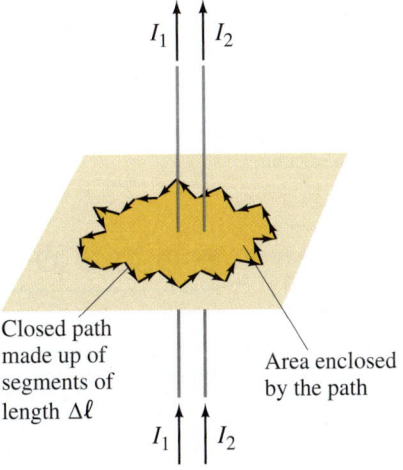

FIGURE 20–30 Arbitrary path enclosing electric currents, for Ampère's law. The path is broken down into segments of equal length $\Delta \ell$. The total current enclosed by the path shown is $I_{encl} = I_1 + I_2$.

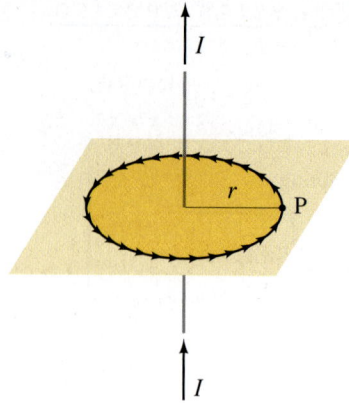

FIGURE 20–31 Circular path of radius r.

Field Due to a Straight Wire

We can check Ampère's law by applying it to the simple case of a long straight wire carrying a current I. Let us find the magnitude of B at point P, a distance r from the wire in Fig. 20–31. The magnetic field lines are circles with the wire at their center (as in Fig. 20–8). As the path to be used in Eq. 20–9, we choose a convenient one: a circle of radius r, because at any point on this path, $\vec{B}$ will be tangent to this circle. For any short segment of the circle (Fig. 20–31), $\vec{B}$ will be parallel to that segment, so $B_\parallel = B$. Suppose we break the circular path down into 100 segments.[†] Then Ampère's law states that

$$(B\,\Delta\ell)_1 + (B\,\Delta\ell)_2 + (B\,\Delta\ell)_3 + \cdots + (B\,\Delta\ell)_{100} = \mu_0 I.$$

The dots represent all the terms we did not write down. All the segments are the same distance from the wire, so by *symmetry* we expect B to be the same at each segment. We can then factor out B from the sum:

$$B(\Delta\ell_1 + \Delta\ell_2 + \Delta\ell_3 + \cdots + \Delta\ell_{100}) = \mu_0 I.$$

The sum of the segment lengths $\Delta\ell$ equals the circumference of the circle, $2\pi r$. Thus we have

$$B(2\pi r) = \mu_0 I,$$

or

$$B = \frac{\mu_0 I}{2\pi r}.$$

This is just Eq. 20–6 for the magnetic field near a long straight wire, so Ampère's law agrees with experiment in this case.

A great many experiments indicate that Ampère's law is valid in general. Practically, it can be used to calculate the magnetic field mainly for simple or symmetric situations. Its importance is that it relates the magnetic field to the current in a direct and mathematically elegant way. Ampère's law is considered one of the basic laws of electricity and magnetism. It is valid for any situation where the currents and fields are not changing in time.

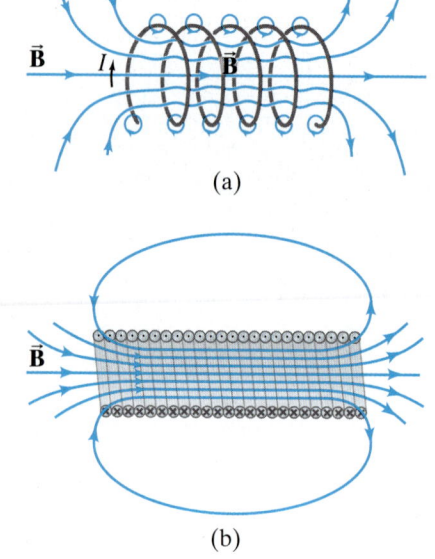

FIGURE 20–32 (a) Magnetic field due to several loops of a solenoid. (b) For many closely spaced loops, the field is very nearly uniform.

Field Inside a Solenoid

We now use Ampère's law to calculate the magnetic field inside a *solenoid* (Section 20–7), a long coil of wire with many loops or turns, Fig. 20–32. Each loop produces a magnetic field as was shown in Fig. 20–9, and the total field inside the solenoid will be the sum of the fields due to each current loop as shown in Fig. 20–32a for a few loops. If the solenoid has many loops and they are close together, the field inside will be nearly uniform and parallel to the solenoid axis except at the ends, as shown in Fig. 20–32b. Outside the solenoid, the field lines spread out in space, so the magnetic field is much weaker than inside. For applying Ampère's law, we choose the path abcd shown in Fig. 20–33 far from either end. We consider this path as made up of four straight segments, the sides of the rectangle: ab, bc, cd, da. Then Ampère's law, Eq. 20–9, becomes

$$(B_\parallel \Delta\ell)_{ab} + (B_\parallel \Delta\ell)_{bc} + (B_\parallel \Delta\ell)_{cd} + (B_\parallel \Delta\ell)_{da} = \mu_0 I_{encl}.$$

The first term in the sum on the left will be (nearly) zero because the field outside the solenoid is negligible compared to the field inside. Furthermore, $\vec{B}$ is perpendicular to the segments bc and da, so these terms are zero, too.

[†] Actually, Ampère's law is precisely accurate when there is an infinite number of infinitesimally short segments, but that leads into calculus.

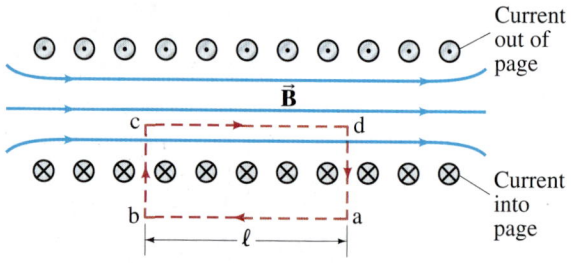

FIGURE 20–33 Cross-sectional view into a solenoid. The magnetic field inside is straight except at the ends. Red dashed lines indicate the path chosen for use in Ampère's law. ⊙ and ⊗ are electric current direction (in the wire loops) out of the page and into the page.

Thus the left side of our Ampère equation we just wrote becomes $(B_\parallel \Delta\ell)_{cd} = B\ell$, where B is the field inside the solenoid, and ℓ is the length cd. We set $B\ell$ equal to μ_0 times the current enclosed by our chosen rectangular loop: if a current I flows in the wire of the solenoid, the total current enclosed by our path abcd is NI, where N is the number of loops (or turns) our path encircles (five in Fig. 20–33). Thus Ampère's law gives us

$$B\ell = \mu_0 NI,$$

so

$$B = \frac{\mu_0 IN}{\ell}, \qquad \text{[solenoid]} \quad \textbf{(20–8 } again\textbf{)}$$

as we quoted in the previous Section. This is the magnetic field magnitude inside a solenoid. B depends only on the number of loops per unit length, N/ℓ, and the current I. The field does not depend on the position within the solenoid, so B is uniform inside the solenoid. This is strictly true only for an infinite solenoid, but it is a good approximation for real ones at points not close to the ends.

The direction of the magnetic field inside the solenoid is found by applying right-hand-rule-1, Fig. 20–8d (see also Figs. 20–9 and 20–10), and is as shown in Fig. 20–33.

20–9 Torque on a Current Loop; Magnetic Moment

When an electric current flows in a closed loop of wire placed in an external magnetic field, as shown in Fig. 20–34, the magnetic force on the current can produce a torque. This is the principle behind a number of important practical devices, including motors and analog voltmeters and ammeters, which we discuss in the next Section.

Current flows through the rectangular loop in Fig. 20–34a, whose face we assume is parallel to $\vec{\mathbf{B}}$. $\vec{\mathbf{B}}$ exerts no force and no torque on the horizontal segments of wire because they are parallel to the field and $\sin\theta = 0$ in Eq. 20–1. But the magnetic field does exert a force on each of the vertical sections of wire as shown, $\vec{\mathbf{F}}_1$ and $\vec{\mathbf{F}}_2$ (see also top view, Fig. 20–34b). By right-hand-rule-2 (Fig. 20–11c or Table 20–1) the direction of the force on the upward current on the left is in the opposite direction from the equal magnitude force $\vec{\mathbf{F}}_2$ on the downward current on the right. These forces give rise to a net torque that acts to rotate the coil about its vertical axis.

Let us calculate the magnitude of this torque. From Eq. 20–2 (current $\perp \vec{\mathbf{B}}$), the force $F = IaB$, where a is the length of the vertical arm of the coil (Fig. 20–34a). The lever arm for each force is $b/2$, where b is the width of the coil and the "axis" is at the midpoint. The torques around this axis produced by $\vec{\mathbf{F}}_1$ and $\vec{\mathbf{F}}_2$ act in the same direction (Fig. 20–34b), so the total torque τ is the sum of the two torques:

$$\tau = IaB\frac{b}{2} + IaB\frac{b}{2} = IabB = IAB,$$

where $A = ab$ is the area of the coil. If the coil consists of N loops of wire, the current is then NI, so the torque becomes

$$\tau = NIAB.$$

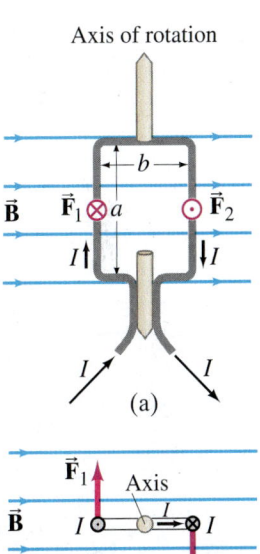

If the coil makes an angle with the magnetic field, as shown in Fig. 20–34c, the forces are unchanged, but each lever arm is reduced from $\frac{1}{2}b$ to $\frac{1}{2}b\sin\theta$. Note that the angle θ is taken to be the angle between $\vec{\mathbf{B}}$ and the perpendicular to the face of the coil, Fig. 20–34c. So the torque becomes

$$\tau = NIAB\sin\theta. \qquad \textbf{(20–10)}$$

This formula, derived here for a rectangular coil, is valid for any shape of flat coil.

The quantity NIA is called the **magnetic dipole moment** of the coil:

$$M = NIA \qquad \textbf{(20–11)}$$

and is considered a vector perpendicular to the coil.

FIGURE 20–34 Calculating the torque on a current loop in a magnetic field $\vec{\mathbf{B}}$. (a) Loop face parallel to $\vec{\mathbf{B}}$ field lines; (b) top view; (c) loop makes an angle to $\vec{\mathbf{B}}$, reducing the torque since the lever arm is reduced.

EXAMPLE 20–13 **Torque on a coil.** A circular loop of wire has a diameter of 20.0 cm and contains 10 loops. The current in each loop is 3.00 A, and the coil is placed in a 2.00-T external magnetic field. Determine the maximum and minimum torque exerted on the coil by the field.

APPROACH Equation 20–10 is valid for any shape of coil, including circular loops. Maximum and minimum torque are determined by the angle θ the coil makes with the magnetic field.

SOLUTION The area of one loop of the coil is
$$A = \pi r^2 = \pi (0.100 \text{ m})^2 = 3.14 \times 10^{-2} \text{ m}^2.$$

The maximum torque occurs when the coil's face is parallel to the magnetic field, so $\theta = 90°$ in Fig. 20–34c, and $\sin \theta = 1$ in Eq. 20–10:
$$\tau = NIAB \sin \theta = (10)(3.00 \text{ A})(3.14 \times 10^{-2} \text{ m}^2)(2.00 \text{ T})(1) = 1.88 \text{ N} \cdot \text{m}.$$

The minimum torque occurs if $\sin \theta = 0$, for which $\theta = 0°$, and then $\tau = 0$ from Eq. 20–10.

NOTE If the coil is free to turn, it will rotate toward the orientation with $\theta = 0°$.

20–10 Applications: Motors, Loudspeakers, Galvanometers

There are many practical applications of the forces related to magnetism. Among the most common are motors and loudspeakers. First we look at the galvanometer, which is the easiest to explain, and which you find on the instrument panels of automobiles and other devices whose readout is via a pointer or needle.

Galvanometer

The basic component of analog meters (those with pointer and dial), including analog ammeters, voltmeters, and ohmmeters, including gauges on car dashboards, is a galvanometer. We have already seen how these meters are designed (Section 19–8), and now we can examine how the crucial element, a galvanometer, works. As shown in Fig. 20–35, a **galvanometer** consists of a coil of wire (with attached pointer) suspended in the magnetic field of a permanent magnet. When current flows through the loop of wire, the magnetic field B exerts a torque τ on the loop, as given by Eq. 20–10,
$$\tau = NIAB \sin \theta.$$

This torque is opposed by a spring which exerts a torque τ_s approximately proportional to the angle ϕ through which it is turned (Hooke's law). That is,
$$\tau_s = k\phi,$$
where k is the stiffness constant of the spring. The coil and attached pointer rotate to the angle where the torques balance. When the needle is in equilibrium at rest, the torques have equal magnitude: $k\phi = NIAB \sin \theta$, so
$$\phi = \frac{NIAB \sin \theta}{k}.$$

The deflection of the pointer, ϕ, is directly proportional to the current I flowing in the coil, but also depends on the angle θ the coil makes with $\vec{B}$. For a useful meter we need ϕ to depend only on the current I, independent of θ. To solve this problem, magnets with curved pole pieces are used and the galvanometer coil is wrapped around a cylindrical iron core as shown in Fig. 20–36. The iron tends to concentrate the magnetic field lines so that $\vec{B}$ always points parallel to the face of the coil at the wire outside the core. The force is then always perpendicular to the face of the coil, and the torque will not vary with angle. Thus ϕ will be proportional to I, as required for a useful meter.

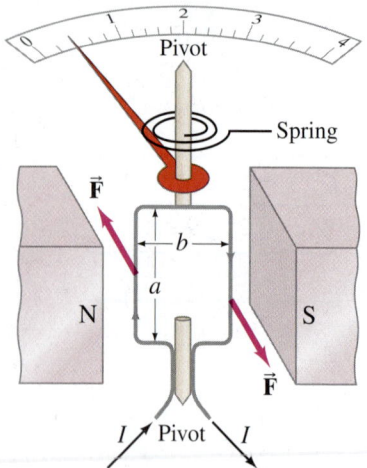

FIGURE 20–35 Galvanometer.

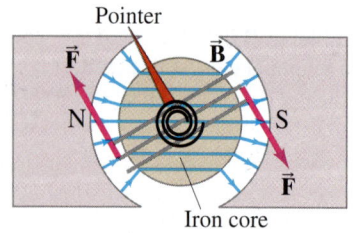

FIGURE 20–36 Galvanometer coil (3 loops shown) wrapped on an iron core.

Electric Motors

An **electric motor** changes electric energy into (rotational) mechanical energy. A motor works on the same principle as a galvanometer (a torque is exerted on a current-carrying loop in a magnetic field) except that the coil must turn continuously in one direction. The coil is mounted on an iron cylinder called the **rotor** or **armature**, Fig. 20–37. Actually, there are several coils, although only one is indicated in Fig. 20–37. The armature is mounted on a shaft or axle. When the armature is in the position shown in Fig. 20–37, the magnetic field exerts forces on the current in the loop as shown (perpendicular to $\vec{B}$ and to the current direction). However, when the coil, which is rotating clockwise in Fig. 20–37, passes beyond the vertical position, the forces would then act to return the coil back toward the vertical if the current remained the same. But if the current could be reversed at that critical moment, the forces would reverse, and the coil would continue rotating in the same direction. Thus, alternation of the current is necessary if a motor is to turn continuously in one direction. This can be achieved in a **dc motor** with the use of **commutators** and **brushes**: as shown in Fig. 20–38, input current passes through stationary brushes that rub against the conducting commutators mounted on the motor shaft. At every half revolution, each commutator changes its connection over to the other brush. Thus the current in the coil reverses every half revolution as required for continuous rotation.

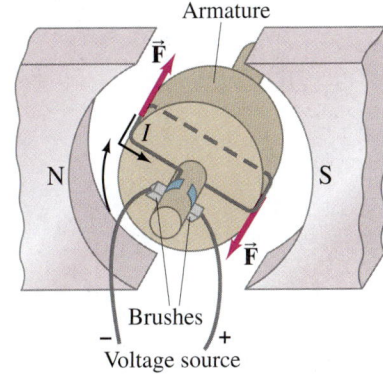

FIGURE 20–37 Diagram of a simple dc motor. (Magnetic field lines are as shown in Fig. 20–36.)

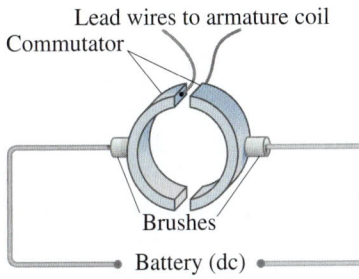

FIGURE 20–38 Commutator-brush arrangement in a dc motor ensures alternation of the current in the armature to keep rotation continuous in one direction. The commutators are attached to the motor shaft and turn with it, whereas the brushes remain stationary.

Most motors contain several coils, called *windings*, each connected to a different portion of the armature, Fig. 20–39. Current flows through each coil only during a small part of a revolution, at the time when its orientation results in the maximum torque. In this way, a motor produces a much steadier torque than can be obtained from a single coil.

An **ac motor**, with ac current as input, can work without commutators since the current itself alternates. Many motors use wire coils to produce the magnetic field (electromagnets) instead of a permanent magnet. Indeed the design of most motors is more complex than described here, but the general principles remain the same.

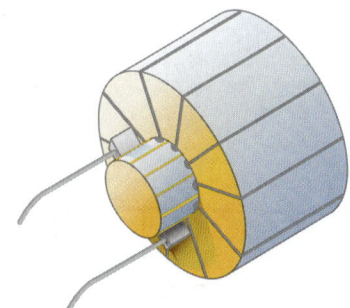

FIGURE 20–39 Motor with many windings.

FIGURE 20–40 Loudspeaker.

Loudspeakers and Headsets

Loudspeakers and audio headsets also work on the principle that a magnet exerts a force on a current-carrying wire. The electrical output of a stereo or TV set is connected to the wire leads of the speaker or earbuds. The speaker leads are connected internally to a coil of wire, which is itself attached to the speaker cone, Fig. 20–40. The speaker cone is usually made of stiffened cardboard and is mounted so that it can move back and forth freely (except at its attachment on the outer edges). A permanent magnet is mounted directly in line with the coil of wire. When the alternating current of an audio signal flows through the wire coil, which is free to move within the magnet, the coil experiences a force due to the magnetic field of the magnet. (The force is to the right at the instant shown in Fig. 20–40, RHR-2, page 568.) As the current alternates at the frequency of the audio signal, the coil and attached speaker cone move back and forth at the same frequency, causing alternate compressions and rarefactions of the adjacent air, and sound waves are produced. A speaker thus changes electrical energy into sound energy, and the frequencies and intensities of the emitted sound waves can be an accurate reproduction of the electrical input.

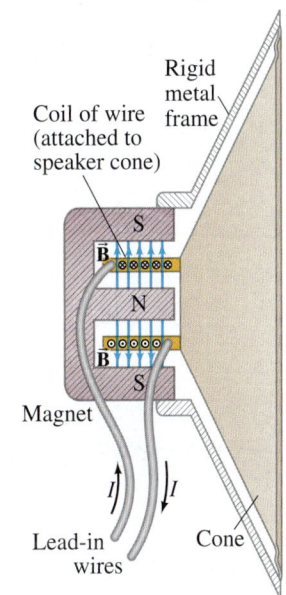

*20–11 Mass Spectrometer

PHYSICS APPLIED
The mass spectrometer

A **mass spectrometer** is a device to measure masses of atoms. It is used today not only in physics but also in chemistry, geology, and medicine, often to identify atoms (and their concentration) in given samples. Ions are produced by heating the sample, or by using an electric current. As shown in Fig. 20–41, the ions (mass m, charge q) pass through slit S_1 and enter a region (before S_2) where there are crossed ($\perp$) electric and magnetic fields. Ions follow a straight-line path in this region if the electric force qE (upward on a positive ion) is just balanced by the magnetic force qvB (downward on a positive ion): that is, if $qE = qvB$, or

$$v = \frac{E}{B}.$$

Only those ions whose speed is $v = E/B$ will pass through undeflected and emerge through slit S_2. (This arrangement is called a **velocity selector**.) In the semicircular region, after S_2, there is only a magnetic field, B', so the ions follow a circular path. The radius of the circular path is found from their mark on film, or by detectors, if B' is fixed. If instead r is fixed by the position of a detector, then B' is varied until detection occurs. Newton's second law, $\Sigma F = ma$, applied to an ion moving in a circle under the influence only of the magnetic field B' gives $qvB' = mv^2/r$. Since $v = E/B$, we have

$$m = \frac{qB'r}{v} = \frac{qBB'r}{E}. \quad (20\text{–}12)$$

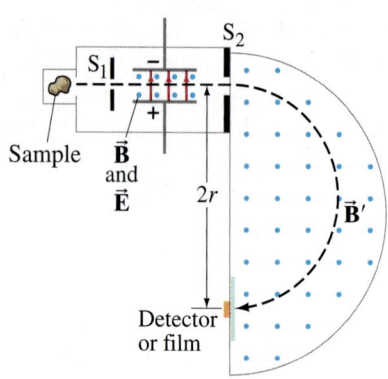

FIGURE 20–41 Bainbridge-type mass spectrometer. The magnetic fields B and B' point out of the paper (indicated by the dots).

All the quantities on the right side are known or can be measured, and thus m can be determined.

Historically, the masses of many atoms were measured this way. When a pure substance was used, it was sometimes found that two or more closely spaced marks would appear on the film. For example, neon produced two marks whose radii corresponded to atoms of 20 and 22 atomic mass units (u). Impurities were ruled out and it was concluded that there must be two types of neon with different masses. These different forms were called **isotopes**. It was soon found that most elements are mixtures of isotopes, and the difference in mass is due to different numbers of neutrons (discussed in Chapter 30).

EXAMPLE 20–14 **Mass spectrometry.** Carbon atoms of atomic mass 12.0 u are found to be mixed with an unknown element. In a mass spectrometer with fixed B', the carbon traverses a path of radius 22.4 cm and the unknown's path has a 26.2-cm radius. What is the unknown element? Assume the ions of both elements have the same charge.

APPROACH The carbon and unknown atoms pass through the same electric and magnetic fields. Hence their masses are proportional to the radius of their respective paths (see Eq. 20–12).

SOLUTION We write a ratio for the masses, using Eq. 20–12:

$$\frac{m_x}{m_C} = \frac{qBB'r_x/E}{qBB'r_C/E} = \frac{r_x}{r_C}$$

$$= \frac{26.2 \text{ cm}}{22.4 \text{ cm}} = 1.17.$$

Thus $m_x = 1.17 \times 12.0\text{ u} = 14.0\text{ u}$. The other element is probably nitrogen (see the Periodic Table, inside the back cover).

NOTE The unknown could also be an isotope such as carbon-14 ($^{14}_{6}\text{C}$). See Appendix B. Further physical or chemical analysis would be needed.

*20–12 Ferromagnetism: Domains and Hysteresis

We saw in Section 20–1 that iron (and a few other materials) can be made into strong magnets. These materials are said to be **ferromagnetic**.

*Sources of Ferromagnetism

Microscopic examination reveals that a piece of iron is made up of tiny regions known as **domains**, less than 1 mm in length or width. Each domain behaves like a tiny magnet with a north and a south pole. In an unmagnetized piece of iron, the domains are arranged randomly, Fig. 20–42a. The magnetic effects of the domains cancel each other out, so this piece of iron is not a magnet. In a magnet, the domains are preferentially aligned in one direction as shown in Fig. 20–42b (downward in this case). A magnet can be made from an unmagnetized piece of iron by placing it in a strong magnetic field. (You can make a needle magnetic, for example, by stroking it with one pole of a strong magnet.) The magnetization direction of domains may actually rotate slightly to be more nearly parallel to the external field, and the borders of domains may move so domains with magnetic orientation parallel to the external field grow larger (compare Figs. 20–42a and b).

We can now explain how a magnet can pick up unmagnetized pieces of iron like paper clips. The magnet's field causes a slight realignment of the domains in the unmagnetized object so that it becomes a temporary magnet with its north pole facing the south pole of the permanent magnet; thus, attraction results. Similarly, elongated iron filings in a magnetic field acquire aligned domains and align themselves to reveal the shape of the magnetic field, Fig. 20–43.

An iron magnet can remain magnetized for a long time, and is referred to as a "permanent magnet." But if you drop a magnet on the floor or strike it with a hammer, you can jar the domains into randomness and the magnet loses some or all of its magnetism. Heating a permanent magnet can also cause loss of magnetism, for raising the temperature increases the random thermal motion of atoms, which tends to randomize the domains. Above a certain temperature known as the **Curie temperature** (1043 K for iron), a magnet cannot be made at all.

The striking similarity between the fields produced by a bar magnet and by a loop of electric current (Figs. 20–4b, 20–9) offers a clue that perhaps magnetic fields produced by electric currents may have something to do with ferromagnetism. According to modern atomic theory, atoms can be roughly visualized as having electrons that orbit around a central nucleus. The electrons are charged, and so constitute an electric current and therefore produce a magnetic field. But the fields due to orbiting electrons end up adding to zero. Electrons themselves produce an additional magnetic field, almost as if they and their electric charge were spinning about their own axes. And it is this magnetic field due to electron **spin**[†] that is believed to produce ferromagnetism in most ferromagnetic materials.

It is believed today that *all* magnetic fields are caused by electric currents. This means that magnetic field lines always form closed loops, unlike electric field lines which begin on positive charges and end on negative charges.

*Magnetic Permeability

If a piece of ferromagnetic material like iron is placed inside a solenoid to form an electromagnet (Section 20–7), the magnetic field increases greatly over that produced by the current in the solenoid coils alone, often by hundreds or thousands of times. This happens because the domains in the iron become aligned by the external field produced by the current in the solenoid coil.

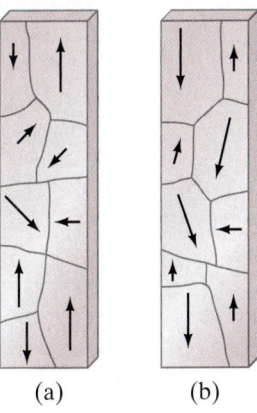

(a)   (b)

FIGURE 20–42 (a) An unmagnetized piece of iron is made up of domains that are randomly arranged. Each domain is like a tiny magnet; the arrows represent the magnetization direction, with the arrowhead being the N pole. (b) In a magnet, the domains are preferentially aligned in one direction (down in this case), and may be altered in size by the magnetization process.

FIGURE 20–43 Iron filings line up along magnetic field lines due to a permanent magnet.

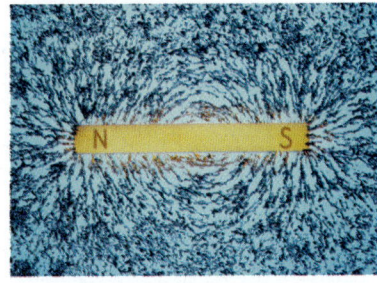

⚠ **CAUTION**
$\vec{B}$ *lines form closed loops,*
$\vec{E}$ *lines start on* ⊕ *and end on* ⊖

[†]The name "spin" comes from an early suggestion that this intrinsic magnetic field arises from the electron "spinning" on its axis (as well as "orbiting" the nucleus) to produce the extra field. However, this view of a spinning electron is oversimplified and not valid (see Chapter 28).

The total magnetic field $\vec{B}$ is then the sum of two terms,
$$\vec{B} = \vec{B}_0 + \vec{B}_M.$$
$\vec{B}_0$ is the field due to the current in the solenoid coil and $\vec{B}_M$ is the additional field due to the iron. Often $B_M \gg B_0$. The total field can also be written by replacing the constant μ_0 in Eq. 20–8 ($B = \mu_0 NI/\ell$ for a solenoid) by another constant called the **magnetic permeability** μ, which is characteristic of the magnetic material inside the coil. Then $B = \mu NI/\ell$. For ferromagnetic materials, μ is much greater than μ_0. For all other materials, its value is very close to μ_0.[†] The value of μ, however, is not constant for ferromagnetic materials; it depends on the strength of the "external" field B_0, as the following experiment shows.

*Hysteresis

Measurements on magnetic materials often use a **torus** or **toroid**, which is like a long solenoid bent into the shape of a donut (Fig. 20–44), so practically all the lines of $\vec{B}$ remain within the toroid. Consider a toroid with an iron core that is initially unmagnetized and there is no current in the wire loops. Then the current I is slowly increased, and B_0 (which is due only to I) increases linearly with I. The total field B also increases, but follows the curved line shown in Fig. 20–45 which is a graph of total B vs. B_0. Initially, point a, the domains are randomly oriented. As B_0 increases, the domains become more and more aligned until at point b, nearly all are aligned. The iron is said to be approaching **saturation**.

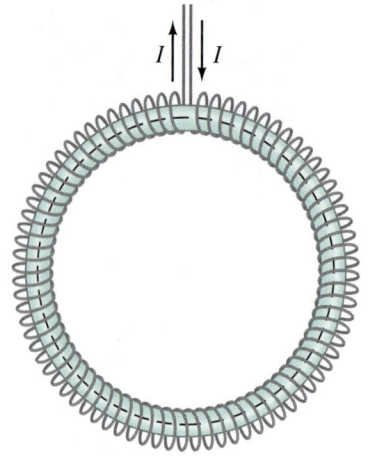

FIGURE 20–44 Iron-core toroid.

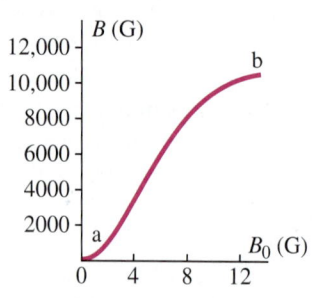

FIGURE 20–45 Total magnetic field B in an iron-core toroid as a function of the external field B_0 (B_0 is caused by the current I in the coil). We use gauss ($1\,G = 10^{-4}\,T$) so that labels are clear.

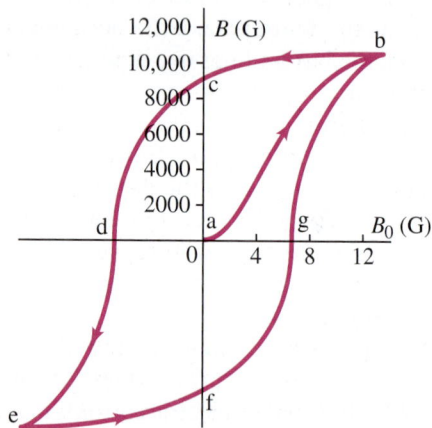

FIGURE 20–46 Hysteresis curve.

Next, suppose current in the coil is reduced, so the field B_0 decreases. If the current (and B_0) is reduced to zero, point c in Fig. 20–46, the domains do *not* become completely random. Instead, some permanent magnetism remains in the iron core. If the current is increased in the opposite direction, enough domains can be turned around so the total B becomes zero at point d. As the reverse current is increased further, the iron approaches saturation in the opposite direction, point e. Finally, if the current is again reduced to zero (point f) and then increased in the original direction, the total field follows the path efgb, again approaching saturation at point b.

Notice that the field did not pass through the origin (point a) in this cycle. The fact that the curve does not retrace itself on the same path is called **hysteresis**. The curve bcdefgb is called a **hysteresis loop**. In such a cycle, much energy is transformed to thermal energy (friction) due to realigning of the domains. Note that at points c and f, the iron core is magnetized even though there is no current in the coils. These points correspond to a permanent magnet.

[†]All materials are slightly magnetic. Nonferromagnetic materials fall into two principal classes: (1) **paramagnetic** materials consist of atoms that have a net magnetic dipole moment which can align slightly with an external field, just as the galvanometer coil in Fig. 20–35 experiences a torque that tends to align it; (2) **diamagnetic** materials have atoms with no net dipole moment, but in the presence of an external field electrons revolving in one direction increase in speed slightly whereas electrons revolving in the opposite direction are reduced in speed; the result is a slight net magnetic effect that opposes the external field.

Summary

A magnet has two **poles**, north and south. The north pole is that end which points toward geographic north when the magnet is freely suspended. Like poles of two magnets repel each other, whereas unlike poles attract.

We can picture that a **magnetic field** surrounds every magnet. The SI unit for magnetic field is the **tesla** (T).

Electric currents produce magnetic fields. For example, the lines of magnetic field due to a current in a straight wire form circles around the wire, and the field exerts a force on magnets (or currents) near it.

A magnetic field exerts a force on an electric current. For a straight wire of length ℓ carrying a current I, the force has magnitude

$$F = I\ell B \sin\theta, \quad (20\text{-}1)$$

where θ is the angle between the magnetic field $\vec{B}$ and the current direction. The direction of the force is perpendicular to the current-carrying wire and to the magnetic field, and is given by a right-hand rule. Equation 20–1 serves as the definition of magnetic field $\vec{B}$.

Similarly, a magnetic field exerts a force on a charge q moving with velocity v of magnitude

$$F = qvB \sin\theta, \quad (20\text{-}3)$$

where θ is the angle between $\vec{v}$ and $\vec{B}$. The direction of $\vec{F}$ is perpendicular to $\vec{v}$ and to $\vec{B}$ (again a right-hand rule). The path of a charged particle moving perpendicular to a uniform magnetic field is a circle.

The magnitude of the magnetic field produced by a current I in a long straight wire, at a distance r from the wire, is

$$B = \frac{\mu_0}{2\pi}\frac{I}{r}. \quad (20\text{-}6)$$

Two currents exert a force on each other via the magnetic field each produces. Parallel currents in the same direction attract each other; currents in opposite directions repel.

The magnetic field inside a long tightly wound solenoid is

$$B = \mu_0 NI/\ell, \quad (20\text{-}8)$$

where N is the number of loops in a length ℓ of coil, and I is the current in each loop.

Ampère's law states that around any chosen closed loop path, the sum of each path segment $\Delta\ell$ times the component of $\vec{B}$ parallel to the segment equals μ_0 times the current I enclosed by the closed path:

$$\Sigma B_\parallel \Delta\ell = \mu_0 I_{encl}. \quad (20\text{-}9)$$

The torque τ on N loops of current I in a magnetic field $\vec{B}$ is

$$\tau = NIAB \sin\theta. \quad (20\text{-}10)$$

The force or torque exerted on a current-carrying wire by a magnetic field is the basis for operation of many devices, such as **motors**, **loudspeakers**, and **galvanometers** used in analog electric meters.

[*A **mass spectrometer** uses electric and magnetic fields to determine the mass of ions.]

[*Iron and a few other materials that are **ferromagnetic** can be made into strong permanent magnets. Ferromagnetic materials are made up of tiny **domains**—each a tiny magnet—which are preferentially aligned in a permanent magnet. When iron or another ferromagnetic material is placed in a magnetic field B_0 due to a current, the iron becomes magnetized. When the current is turned off, the material remains magnetized; when the current is increased in the opposite direction, a graph of the total field B versus B_0 is a **hysteresis loop**, and the fact that the curve does not retrace itself is called **hysteresis**.]

Questions

1. A compass needle is not always balanced parallel to the Earth's surface, but one end may dip downward. Explain.
2. Explain why the Earth's "north pole" is really a magnetic south pole. Indicate how north and south magnetic poles were defined and how we can tell experimentally that the north pole is really a south magnetic pole.
3. In what direction are the magnetic field lines surrounding a straight wire carrying a current that is moving directly away from you? Explain.
4. A horseshoe magnet is held vertically with the north pole on the left and south pole on the right. A wire passing between the poles, equidistant from them, carries a current directly away from you. In what direction is the force on the wire? Explain.
5. Will a magnet attract any metallic object, such as those made of aluminum or copper? (Try it and see.) Why is this so?
6. Two iron bars attract each other no matter which ends are placed close together. Are both magnets? Explain.
7. The magnetic field due to current in wires in your home can affect a compass. Discuss the effect in terms of currents, including if they are ac or dc.
8. If a negatively charged particle enters a region of uniform magnetic field which is perpendicular to the particle's velocity, will the kinetic energy of the particle increase, decrease, or stay the same? Explain your answer. (Neglect gravity and assume there is no electric field.)
9. In Fig. 20–47, charged particles move in the vicinity of a current-carrying wire. For each charged particle, the arrow indicates the initial direction of motion of the particle, and the + or − indicates the sign of the charge. For each of the particles, indicate the direction of the magnetic force due to the magnetic field produced by the wire. Explain.

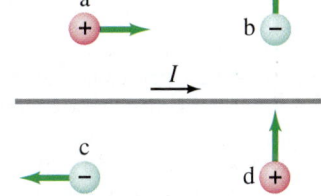

FIGURE 20–47 Question 9.

10. Three particles, a, b, and c, enter a magnetic field and follow paths as shown in Fig. 20–48. What can you say about the charge on each particle? Explain.

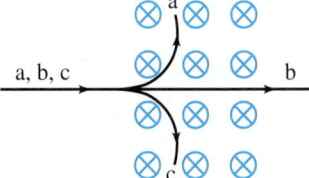

FIGURE 20–48 Question 10.

11. Can an iron rod attract a magnet? Can a magnet attract an iron rod? What must you consider to answer these questions?

12. A positively charged particle in a nonuniform magnetic field follows the trajectory shown in Fig. 20–49. Indicate the direction of the magnetic field at points near the path, assuming the path is always in the plane of the page, and indicate the relative magnitudes of the field in each region. Explain your answers.

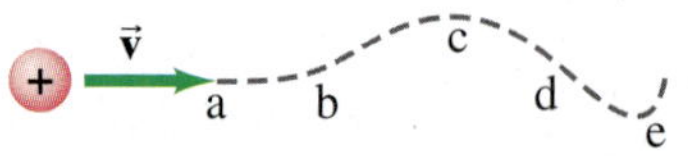

FIGURE 20–49
Question 12.

13. Explain why a strong magnet held near a CRT television screen (Section 17–11) causes the picture to become distorted. Also, explain why the picture sometimes goes completely black where the field is the strongest. [But don't risk damage to your TV by trying this.]

14. Suppose you have three iron rods, two of which are magnetized but the third is not. How would you determine which two are the magnets without using any additional objects?

15. Can you set a resting electron into motion with a magnetic field? With an electric field? Explain.

16. A charged particle is moving in a circle under the influence of a uniform magnetic field. If an electric field that points in the same direction as the magnetic field is turned on, describe the path the charged particle will take.

17. A charged particle moves in a straight line through a particular region of space. Could there be a nonzero magnetic field in this region? If so, give two possible situations.

18. If a moving charged particle is deflected sideways in some region of space, can we conclude, for certain, that $\vec{B} \neq 0$ in that region? Explain.

19. Two insulated long wires carrying equal currents I cross at right angles to each other. Describe the magnetic force one exerts on the other.

20. A horizontal current-carrying wire, free to move in Earth's gravitational field, is suspended directly above a parallel, current-carrying wire. (a) In what direction is the current in the lower wire? (b) Can the lower wire be held in stable equilibrium due to the magnetic force of the upper wire? Explain.

21. What would be the effect on B inside a long solenoid if (a) the diameter of all the loops was doubled, (b) the spacing between loops was doubled, or (c) the solenoid's length was doubled along with a doubling in the total number of loops?

22. A type of magnetic switch similar to a solenoid is a **relay** (Fig. 20–50). A relay is an electromagnet (the iron rod inside the coil does not move) which, when activated, attracts a strip of iron on a pivot. Design a relay to close an electrical switch. A relay is used when you need to switch on a circuit carrying a very large current but do not want that large current flowing through the main switch. For example, a car's starter switch is connected to a relay so that the large current needed for the starter doesn't pass to the dashboard switch.

FIGURE 20–50
Question 22.

*23. Two ions have the same mass, but one is singly ionized and the other is doubly ionized. How will their positions on the film of a mass spectrometer (Fig. 20–41) differ? Explain.

*24. Why will either pole of a magnet attract an unmagnetized piece of iron?

*25. An unmagnetized nail will not attract an unmagnetized paper clip. However, if one end of the nail is in contact with a magnet, the other end *will* attract a paper clip. Explain.

MisConceptual Questions

1. Indicate which of the following will produce a magnetic field:
   (a) A magnet.
   (b) The Earth.
   (c) An electric charge at rest.
   (d) A moving electric charge.
   (e) An electric current.
   (f) The voltage of a battery not connected to anything.
   (g) An ordinary piece of iron.
   (h) A piece of any metal.

2. A current in a wire points into the page as shown at the right. In which direction is the magnetic field at point A (choose below)?

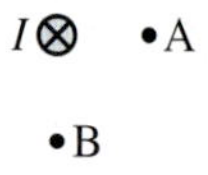

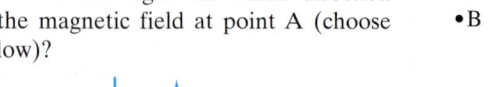

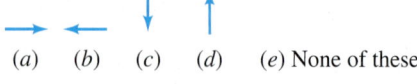

   (a)   (b)   (c)   (d)   (e) None of these.

3. In which direction (see above) is the magnetic field at point B?

4. When a charged particle moves parallel to the direction of a magnetic field, the particle travels in a
   (a) straight line.   (c) helical path.
   (b) circular path.   (d) hysteresis loop.

5. As a proton moves through space, it creates
   (a) an electric field only.
   (b) a magnetic field only.
   (c) both an electric field and magnetic field.
   (d) nothing; the electric field and magnetic fields cancel each other out.

6. Which statements about the force on a charged particle placed in a magnetic field are true?
   (a) A magnetic force is exerted only if the particle is moving.
   (b) The force is a maximum if the particle is moving in the direction of the field.
   (c) The force causes the particle to gain kinetic energy.
   (d) The direction of the force is along the magnetic field.
   (e) A magnetic field always exerts a force on a charged particle.

7. Which of the following statements is false? The magnetic field of a current-carrying wire
   (a) is directed circularly around the wire.
   (b) decreases inversely with the distance from the wire.
   (c) exists only if the current in the wire is changing.
   (d) depends on the magnitude of the current.

8. A wire carries a current directly away from you. Which way do the magnetic field lines produced by this wire point?
   (a) They point parallel to the wire in the direction of the current.
   (b) They point parallel to the wire opposite the direction of the current.
   (c) They point toward the wire.
   (d) They point away from the wire.
   (e) They make circles around the wire.

9. A proton enters a uniform magnetic field that is perpendicular to the proton's velocity (Fig. 20–51). What happens to the kinetic energy of the proton?
   (a) It increases.
   (b) It decreases.
   (c) It stays the same.
   (d) It depends on the velocity direction.
   (e) It depends on the B field direction.

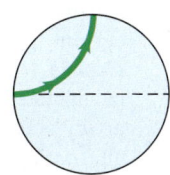

FIGURE 20–51
MisConceptual Question 9.

10. For a charged particle, a constant magnetic field can be used to change
    (a) only the direction of the particle's velocity.
    (b) only the magnitude of the particle's velocity.
    (c) both the magnitude and direction of the particle's velocity.
    (d) None of the above.

11. Which of the following statements about the force on a charged particle due to a magnetic field are not valid?
    (a) It depends on the particle's charge.
    (b) It depends on the particle's velocity.
    (c) It depends on the strength of the external magnetic field.
    (d) It acts at right angles to the direction of the particle's motion.
    (e) None of the above; all of these statements are valid.

12. Two parallel wires are vertical. The one on the left carries a 10-A current upward. The other carries 5-A current downward. Compare the magnitude of the force that each wire exerts on the other.
    (a) The wire on the left carries twice as much current, so it exerts twice the force on the right wire as the right one exerts on the left one.
    (b) The wire on the left exerts a smaller force. It creates a magnetic field twice that due to the wire on the right; and therefore has less energy to cause a force on the wire on the right.
    (c) The two wires exert the same force on each other.
    (d) Not enough information; we need the length of the wire.

For assigned homework and other learning materials, go to the MasteringPhysics website.

Problems

20–3 Force on Electric Current in Magnetic Field

1. (I) (a) What is the force per meter of length on a straight wire carrying a 6.40-A current when perpendicular to a 0.90-T uniform magnetic field? (b) What if the angle between the wire and field is 35.0°?

2. (I) How much current is flowing in a wire 4.80 m long if the maximum force on it is 0.625 N when placed in a uniform 0.0800-T field?

3. (I) A 240-m length of wire stretches between two towers and carries a 120-A current. Determine the magnitude of the force on the wire due to the Earth's magnetic field of 5.0×10^{-5} T which makes an angle of 68° with the wire.

4. (I) A 2.6-m length of horizontal wire carries a 4.5-A current toward the south. The dip angle of the Earth's magnetic field makes an angle of 41° to the wire. Estimate the magnitude of the magnetic force on the wire due to the Earth's magnetic field of 5.5×10^{-5} T.

5. (I) The magnetic force per meter on a wire is measured to be only 45% of its maximum possible value. What is the angle between the wire and the magnetic field?

6. (II) The force on a wire carrying 6.45 A is a maximum of 1.28 N when placed between the pole faces of a magnet. If the pole faces are 55.5 cm in diameter, what is the approximate strength of the magnetic field?

7. (II) The force on a wire is a maximum of 8.50×10^{-2} N when placed between the pole faces of a magnet. The current flows horizontally to the right and the magnetic field is vertical. The wire is observed to "jump" toward the observer when the current is turned on. (a) What type of magnetic pole is the top pole face? (b) If the pole faces have a diameter of 10.0 cm, estimate the current in the wire if the field is 0.220 T. (c) If the wire is tipped so that it makes an angle of 10.0° with the horizontal, what force will it now feel? [*Hint*: What length of wire will now be in the field?]

8. (II) Suppose a straight 1.00-mm-diameter copper wire could just "float" horizontally in air because of the force due to the Earth's magnetic field $\vec{B}$, which is horizontal, perpendicular to the wire, and of magnitude 5.0×10^{-5} T. What current would the wire carry? Does the answer seem feasible? Explain briefly.

20–4 Force on Charge Moving in Magnetic Field

9. (I) Determine the magnitude and direction of the force on an electron traveling 7.75×10^5 m/s horizontally to the east in a vertically upward magnetic field of strength 0.45 T.

10. (I) An electron is projected vertically upward with a speed of 1.70×10^6 m/s into a uniform magnetic field of 0.640 T that is directed horizontally away from the observer. Describe the electron's path in this field.

11. (I) Alpha particles (charge $q = +2e$, mass $m = 6.6 \times 10^{-27}$ kg) move at 1.6×10^6 m/s. What magnetic field strength would be required to bend them into a circular path of radius $r = 0.14$ m?

12. (I) Find the direction of the force on a negative charge for each diagram shown in Fig. 20–52, where $\vec{v}$ (green) is the velocity of the charge and $\vec{B}$ (blue) is the direction of the magnetic field. ($\otimes$ means the vector points inward. $\odot$ means it points outward, toward you.)

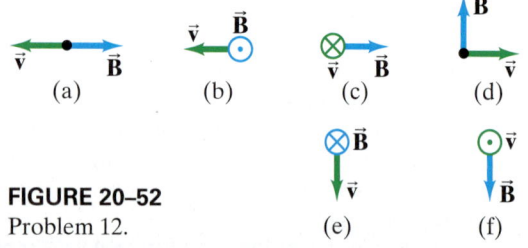

FIGURE 20–52
Problem 12.

13. (I) Determine the direction of $\vec{B}$ for each case in Fig. 20–53, where $\vec{F}$ represents the maximum magnetic force on a positively charged particle moving with velocity $\vec{v}$.

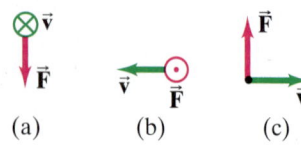

FIGURE 20–53
Problem 13.

14. (II) Determine the velocity of a beam of electrons that goes undeflected when moving perpendicular to an electric and to a magnetic field. $\vec{E}$ and $\vec{B}$ are also perpendicular to each other and have magnitudes 7.7×10^3 V/m and 7.5×10^{-3} T, respectively. What is the radius of the electron orbit if the electric field is turned off?

15. (II) A helium ion ($Q = +2e$) whose mass is 6.6×10^{-27} kg is accelerated by a voltage of 3700 V. (a) What is its speed? (b) What will be its radius of curvature if it moves in a plane perpendicular to a uniform 0.340-T field? (c) What is its period of revolution?

16. (II) For a particle of mass m and charge q moving in a circular path in a magnetic field B, (a) show that its kinetic energy is proportional to r^2, the square of the radius of curvature of its path. (b) Show that its angular momentum is $L = qBr^2$, around the center of the circle.

17. (II) A 1.5-MeV (kinetic energy) proton enters a 0.30-T field, in a plane perpendicular to the field. What is the radius of its path? See Section 17–4.

18. (II) An electron experiences the greatest force as it travels 2.8×10^6 m/s in a magnetic field when it is moving northward. The force is vertically upward and of magnitude 6.2×10^{-13} N. What is the magnitude and direction of the magnetic field?

19. (II) A proton and an electron have the same kinetic energy upon entering a region of constant magnetic field. What is the ratio of the radii of their circular paths?

20. (III) A proton (mass m_p), a deuteron ($m = 2m_p$, $Q = e$), and an alpha particle ($m = 4m_p$, $Q = 2e$) are accelerated by the same potential difference V and then enter a uniform magnetic field $\vec{B}$, where they move in circular paths perpendicular to $\vec{B}$. Determine the radius of the paths for the deuteron and alpha particle in terms of that for the proton.

21. (III) A 3.40-g bullet moves with a speed of 155 m/s perpendicular to the Earth's magnetic field of 5.00×10^{-5} T. If the bullet possesses a net charge of 18.5×10^{-9} C, by what distance will it be deflected from its path due to the Earth's magnetic field after it has traveled 1.50 km?

*22. (III) A **Hall probe**, consisting of a thin rectangular slab of current-carrying material, is calibrated by placing it in a known magnetic field of magnitude 0.10 T. When the field is oriented normal to the slab's rectangular face, a Hall emf of 12 mV is measured across the slab's width. The probe is then placed in a magnetic field of unknown magnitude B, and a Hall emf of 63 mV is measured. Determine B assuming that the angle θ between the unknown field and the plane of the slab's rectangular face is (a) $\theta = 90°$, and (b) $\theta = 60°$.

*23. (III) The Hall effect can be used to measure **blood flow rate** because the blood contains ions that constitute an electric current. (a) Does the sign of the ions influence the emf? Explain. (b) Determine the flow velocity in an artery 3.3 mm in diameter if the measured emf across the width of the artery is 0.13 mV and B is 0.070 T. (In actual practice, an alternating magnetic field is used.)

*24. (III) A long copper strip 1.8 cm wide and 1.0 mm thick is placed in a 1.2-T magnetic field as in Fig. 20–21a. When a steady current of 15 A passes through it, the Hall emf is measured to be 1.02 μV. Determine (a) the drift velocity of the electrons and (b) the density of free (conducting) electrons (number per unit volume) in the copper. [Hint: See also Section 18–8.]

20–5 and 20–6 Magnetic Field of Straight Wire, Force between Two Wires

25. (I) Jumper cables used to start a stalled vehicle often carry a 65-A current. How strong is the magnetic field 4.5 cm from one cable? Compare to the Earth's magnetic field (5.0×10^{-5} T).

26. (I) If an electric wire is allowed to produce a magnetic field no larger than that of the Earth (0.50×10^{-4} T) at a distance of 12 cm from the wire, what is the maximum current the wire can carry?

27. (I) Determine the magnitude and direction of the force between two parallel wires 25 m long and 4.0 cm apart, each carrying 25 A in the same direction.

28. (I) A vertical straight wire carrying an upward 28-A current exerts an attractive force per unit length of 7.8×10^{-4} N/m on a second parallel wire 9.0 cm away. What current (magnitude and direction) flows in the second wire?

29. (II) In Fig. 20–54, a long straight wire carries current I out of the page toward you. Indicate, with appropriate arrows, the direction and (relative) magnitude of $\vec{B}$ at each of the points C, D, and E in the plane of the page.

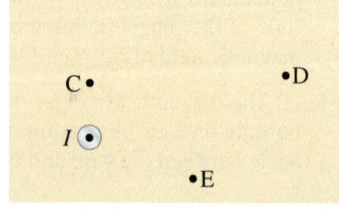

FIGURE 20–54
Problem 29.

30. (II) An experiment on the Earth's magnetic field is being carried out 1.00 m from an electric cable. What is the maximum allowable current in the cable if the experiment is to be accurate to ±3.0%?

31. (II) A rectangular loop of wire is placed next to a straight wire, as shown in Fig. 20–55. There is a current of 3.5 A in both wires. Determine the magnitude and direction of the net force on the loop.

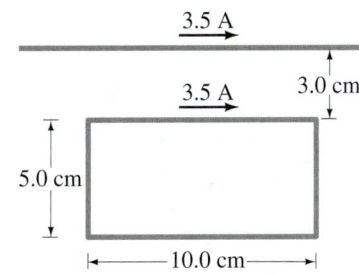

FIGURE 20–55 Problem 31.

32. (II) A horizontal compass is placed 18 cm due south from a straight vertical wire carrying a 48-A current downward. In what direction does the compass needle point at this location? Assume the horizontal component of the Earth's field at this point is 0.45×10^{-4} T and the magnetic declination is 0°.

33. (II) A long horizontal wire carries 24.0 A of current due north. What is the net magnetic field 20.0 cm due west of the wire if the Earth's field there points downward, 44° below the horizontal, and has magnitude 5.0×10^{-5} T?

34. (II) A straight stream of protons passes a given point in space at a rate of 2.5×10^9 protons/s. What magnetic field do they produce 1.5 m from the beam?

35. (II) Determine the magnetic field midway between two long straight wires 2.0 cm apart in terms of the current I in one when the other carries 25 A. Assume these currents are (a) in the same direction, and (b) in opposite directions.

36. (II) Two straight parallel wires are separated by 7.0 cm. There is a 2.0-A current flowing in the first wire. If the magnetic field strength is found to be zero between the two wires at a distance of 2.2 cm from the first wire, what is the magnitude and direction of the current in the second wire?

37. (II) Two long straight wires each carry a current I out of the page toward the viewer, Fig. 20–56. Indicate, with appropriate arrows, the direction of $\vec{B}$ at each of the points 1 to 6 in the plane of the page. State if the field is zero at any of the points.

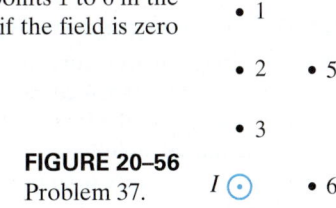

FIGURE 20–56 Problem 37.

38. (II) A power line carries a current of 95 A west along the tops of 8.5-m-high poles. (a) What is the magnitude and direction of the magnetic field produced by this wire at the ground directly below? How does this compare with the Earth's magnetic field of about $\frac{1}{2}$ G? (b) Where would the wire's magnetic field cancel the Earth's field?

39. (II) A compass needle points 17° E of N outdoors. However, when it is placed 12.0 cm to the east of a vertical wire inside a building, it points 32° E of N. What is the magnitude and direction of the current in the wire? The Earth's field there is 0.50×10^{-4} T and is horizontal.

40. (II) A long pair of insulated wires serves to conduct 24.5 A of dc current to and from an instrument. If the wires are of negligible diameter but are 2.8 mm apart, what is the magnetic field 10.0 cm from their midpoint, in their plane (Fig. 20–57)? Compare to the magnetic field of the Earth.

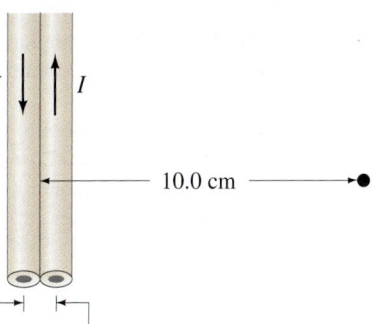

FIGURE 20–57 Problems 40 and 41.

41. (II) A third wire is placed in the plane of the two wires shown in Fig. 20–57 parallel and just to the right. If it carries 25.0 A upward, what force per meter of length does it exert on each of the other two wires? Assume it is 2.8 mm from the nearest wire, center to center.

42. (III) Two long thin parallel wires 13.0 cm apart carry 28-A currents in the same direction. Determine the magnetic field vector at a point 10.0 cm from one wire and 6.0 cm from the other (Fig. 20–58).

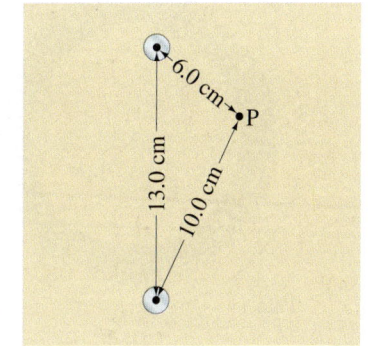

FIGURE 20–58 Problem 42.

43. (III) Two long wires are oriented so that they are perpendicular to each other. At their closest, they are 20.0 cm apart (Fig. 20–59). What is the magnitude of the magnetic field at a point midway between them if the top one carries a current of 20.0 A and the bottom one carries 12.0 A?

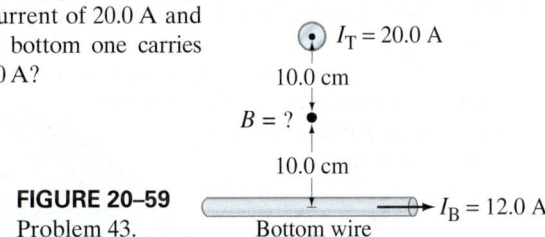

FIGURE 20–59 Problem 43.

20–7 Solenoids and Electromagnets

44. (I) A thin 12-cm-long solenoid has a total of 460 turns of wire and carries a current of 2.0 A. Calculate the field inside the solenoid near the center.

45. (I) A 30.0-cm-long solenoid 1.25 cm in diameter is to produce a field of 4.65 mT at its center. How much current should the solenoid carry if it has 935 turns of the wire?

46. (I) A 42-cm-long solenoid, 1.8 cm in diameter, is to produce a 0.030-T magnetic field at its center. If the maximum current is 4.5 A, how many turns must the solenoid have?

47. (II) A 550-turn horizontal solenoid is 15 cm long. The current in its coils is 38 A. A straight wire cuts through the center of the solenoid, along a 3.0-cm diameter. This wire carries a 22-A current downward (and is connected by other wires that don't concern us). What is the force on this wire assuming the solenoid's magnetic field points due east?

48. (III) You have 1.0 kg of copper and want to make a practical solenoid that produces the greatest possible magnetic field for a given voltage. Should you make your copper wire long and thin, short and fat, or something else? Consider other variables, such as solenoid diameter, length, and so on. Explain your reasoning.

20–8 Ampère's Law

49. (III) A *toroid* is a solenoid in the shape of a donut (Fig. 20–60). Use Ampère's law along the circular paths, shown dashed in Fig. 20–60a, to determine that the magnetic field (a) inside the toroid is $B = \mu_0 NI/2\pi R$, where N is the total number of turns, and (b) outside the toroid is $B = 0$. (c) Is the field inside a toroid uniform like a solenoid's? If not, how does it vary?

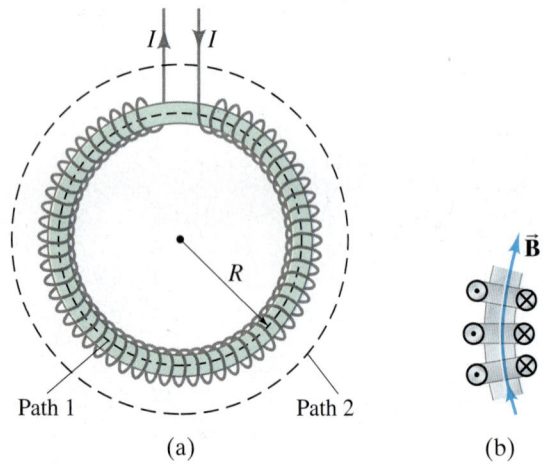

FIGURE 20–60 Problem 49. (a) A toroid or torus. (b) A section of the toroid showing direction of the current for three loops: ⊙ means current toward you, and ⊗ means current away from you.

50. (III) (a) Use Ampère's law to show that the magnetic field between the conductors of a **coaxial cable** (Fig. 20–61) is $B = \mu_0 I/2\pi r$ if r (distance from center) is greater than the radius of the inner wire and less than the radius of the outer cylindrical braid (= ground). (b) Show that $B = 0$ outside the coaxial cable.

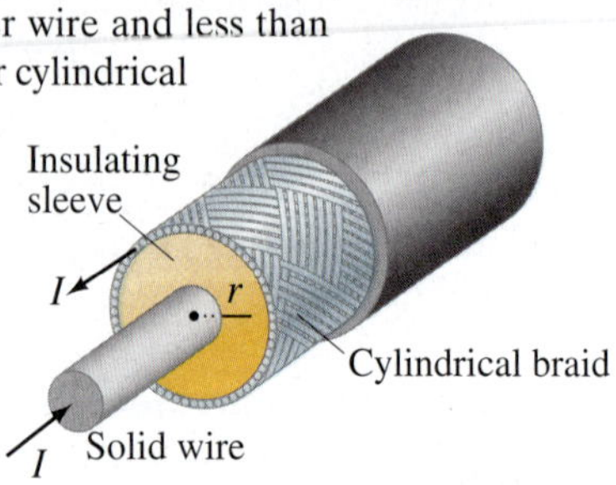

FIGURE 20–61 Coaxial cable. Problem 50.

20–9 and 20–10 Torque on Current Loop, Motors, Galvanometers

51. (I) A single square loop of wire 22.0 cm on a side is placed with its face parallel to the magnetic field as in Fig. 20–34b. When 5.70 A flows in the coil, the torque on it is 0.325 m·N. What is the magnetic field strength?

52. (I) If the current to a motor drops by 12%, by what factor does the output torque change?

53. (I) A galvanometer needle deflects full scale for a 53.0-μA current. What current will give full-scale deflection if the magnetic field weakens to 0.760 of its original value?

54. (II) A circular coil 12.0 cm in diameter and containing nine loops lies flat on the ground. The Earth's magnetic field at this location has magnitude 5.50×10^{-5} T and points into the Earth at an angle of 56.0° below a line pointing due north. If a 7.20-A clockwise current passes through the coil, (a) determine the torque on the coil, and (b) which edge of the coil rises up: north, east, south, or west?

*20–11 Mass Spectrometer

*55. (I) Protons move in a circle of radius 6.10 cm in a 0.566-T magnetic field. What value of electric field could make their paths straight? In what direction must the electric field point?

*56. (I) In a mass spectrometer, germanium atoms have radii of curvature equal to 21.0, 21.6, 21.9, 22.2, and 22.8 cm. The largest radius corresponds to an atomic mass of 76 u. What are the atomic masses of the other isotopes?

*57. (II) Suppose the electric field between the electric plates in the mass spectrometer of Fig. 20–41 is 2.88×10^4 V/m and the magnetic fields are $B = B' = 0.68$ T. The source contains carbon isotopes of mass numbers 12, 13, and 14 from a long-dead piece of a tree. (To estimate masses of the atoms, multiply by 1.67×10^{-27} kg.) How far apart are the lines formed by the singly charged ions of each type on the photographic film? What if the ions were doubly charged?

*58. (II) One form of mass spectrometer accelerates ions by a voltage V before they enter a magnetic field B. The ions are assumed to start from rest. Show that the mass of an ion is $m = qB^2R^2/2V$, where R is the radius of the ions' path in the magnetic field and q is their charge.

*59. (II) An unknown particle moves in a straight line through crossed electric and magnetic fields with $E = 1.5$ kV/m and $B = 0.034$ T. If the electric field is turned off, the particle moves in a circular path of radius $r = 2.7$ cm. What might the particle be?

*60. (III) A mass spectrometer is monitoring air pollutants. It is difficult, however, to separate molecules of nearly equal mass such as CO (28.0106 u) and N_2 (28.0134 u). How large a radius of curvature must a spectrometer have (Fig. 20–41) if these two molecules are to be separated on the film by 0.50 mm?

*20–12 Ferromagnetism, Hysteresis

*61. (I) A long thin iron-core solenoid has 380 loops of wire per meter, and a 350-mA current flows through the wire. If the permeability of the iron is $3000\mu_0$, what is the total field B inside the solenoid?

*62. (II) An iron-core solenoid is 38 cm long and 1.8 cm in diameter, and has 780 turns of wire. The magnetic field inside the solenoid is 2.2 T when 48 A flows in the wire. What is the permeability μ at this high field strength?

*63. (II) The following are some values of B and B_0 for a piece of iron as it is being magnetized (note different units):

$B_0(10^{-4}$ T)	0.0	0.13	0.25	0.50	0.63	0.78	1.0	1.3
B(T)	0.0	0.0042	0.010	0.028	0.043	0.095	0.45	0.67

$B_0(10^{-4}$ T)	1.9	2.5	6.3	13.0	130	1300	10,000
B(T)	1.01	1.18	1.44	1.58	1.72	2.26	3.15

Determine the magnetic permeability μ for each value and plot a graph of μ versus B_0.

General Problems

64. Two long straight parallel wires are 15 cm apart. Wire A carries 2.0-A current. Wire B's current is 4.0 A in the same direction. (*a*) Determine the magnetic field magnitude due to wire A at the position of wire B. (*b*) Determine the magnetic field due to wire B at the position of wire A. (*c*) Are these two magnetic fields equal and opposite? Why or why not? (*d*) Determine the force on wire A due to wire B, and the force on wire B due to wire A. Are these two forces equal and opposite? Why or why not?

65. Protons with momentum 4.8×10^{-21} kg·m/s are magnetically steered clockwise in a circular path 2.2 m in diameter. Determine the magnitude and direction of the field in the magnets surrounding the beam pipe.

66. A small but rigid ∪-shaped wire carrying a 5.0-A current (Fig. 20–62) is placed inside a solenoid. The solenoid is 15.0 cm long and has 700 loops of wire, and the current in each loop is 7.0 A. What is the net force on the ∪-shaped wire?

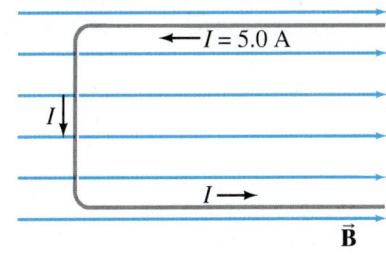

FIGURE 20–62
Problem 66.

67. The power cable for an electric trolley (Fig. 20–63) carries a horizontal current of 330 A toward the east. The Earth's magnetic field has a strength 5.0×10^{-5} T and makes an angle of dip of 22° at this location. Calculate the magnitude and direction of the magnetic force on an 18-m length of this cable.

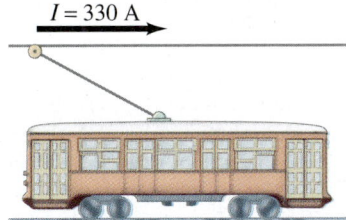

FIGURE 20–63
Problem 67.

68. A particle of charge q moves in a circular path of radius r perpendicular to a uniform magnetic field B. Determine its linear momentum in terms of the quantities given.

69. An airplane has acquired a net charge of 1280 μC. If the Earth's magnetic field of 5.0×10^{-5} T is perpendicular to the airplane's velocity of magnitude 120 m/s, determine the force on the airplane.

70. A 32-cm-long solenoid, 1.8 cm in diameter, is to produce a 0.050-T magnetic field at its center. If the maximum current is 6.4 A, how many turns must the solenoid have?

71. Near the equator, the Earth's magnetic field points almost horizontally to the north and has magnitude $B = 0.50 \times 10^{-4}$ T. What should be the magnitude and direction for the velocity of an electron if its weight is to be exactly balanced by the magnetic force?

72. A doubly charged helium atom, whose mass is 6.6×10^{-27} kg, is accelerated by a voltage of 3200 V. (*a*) What will be its radius of curvature in a uniform 0.240-T field? (*b*) What is its period of revolution?

73. Four very long straight parallel wires, located at the corners of a square of side ℓ, carry equal currents I_0 perpendicular to the page as shown in Fig. 20–64. Determine the magnitude and direction of $\vec{B}$ at the center C of the square.

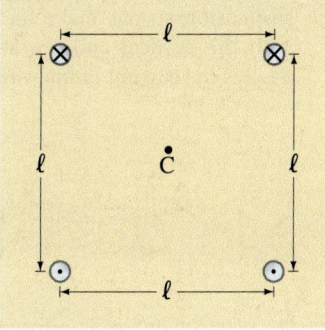

FIGURE 20–64
Problem 73.

74. (*a*) What value of magnetic field would make a beam of electrons, traveling to the west at a speed of 4.8×10^6 m/s, go undeflected through a region where there is a uniform electric field of 12,000 V/m pointing south? (*b*) What is the direction of the magnetic field if it is perpendicular to the electric field? (*c*) What is the frequency of the circular orbit of the electrons if the electric field is turned off?

75. Magnetic fields are very useful in particle accelerators for "beam steering"; that is, the magnetic fields can be used to change the direction of the beam of charged particles without altering their speed (Fig. 20–65). Show how this could work with a beam of protons. What happens to protons that are not moving with the speed for which the magnetic field was designed? If the field extends over a region 5.0 cm wide and has a magnitude of 0.41 T, by approximately what angle θ will a beam of protons traveling at 2.5×10^6 m/s be bent?

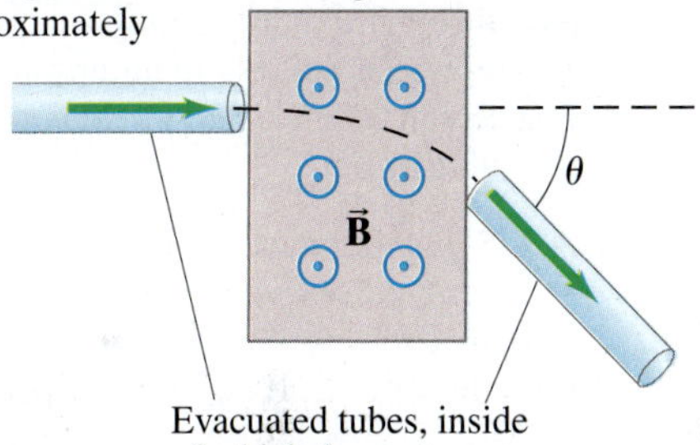

FIGURE 20–65
Problem 75.

Evacuated tubes, inside of which the protons move with velocity indicated by the green arrows

76. The magnetic field B at the center of a circular coil of wire carrying a current I (as in Fig. 20–9) is

$$B = \frac{\mu_0 N I}{2r},$$

where N is the number of loops in the coil and r is its radius. Imagine a simple model in which the Earth's magnetic field of about 1 G ($= 1 \times 10^{-4}$ T) near the poles is produced by a single current loop around the equator. Roughly estimate the current this loop would carry.

77. A proton follows a spiral path through a gas in a uniform magnetic field of 0.010 T, perpendicular to the plane of the spiral, as shown in Fig. 20–66. In two successive loops, at points P and Q, the radii are 10.0 mm and 8.5 mm, respectively. Calculate the change in the kinetic energy of the proton as it travels from P to Q.

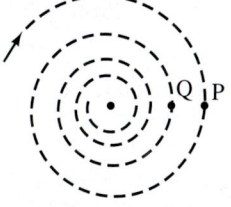

FIGURE 20–66
Problem 77.

78. Two long straight aluminum wires, each of diameter 0.42 mm, carry the same current but in opposite directions. They are suspended by 0.50-m-long strings as shown in Fig. 20–67. If the suspension strings make an angle of 3.0° with the vertical and are hanging freely, what is the current in the wires?

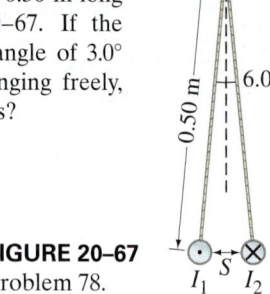

FIGURE 20–67
Problem 78.

79. An electron enters a uniform magnetic field $B = 0.23$ T at a 45° angle to $\vec{B}$. Determine the radius r and pitch p (distance between loops) of the electron's helical path assuming its speed is 3.0×10^6 m/s. See Fig. 20–68.

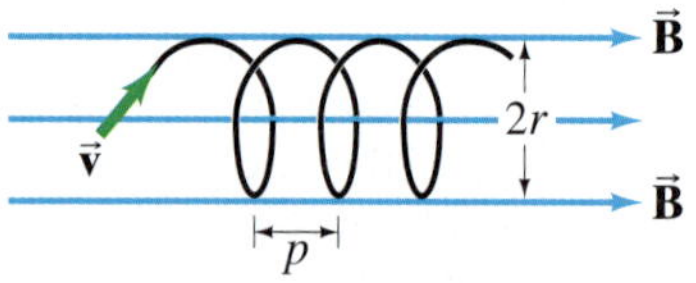

FIGURE 20–68
Problem 79.

80. A motor run by a 9.0-V battery has a 20-turn square coil with sides of length 5.0 cm and total resistance 28 Ω. When spinning, the magnetic field felt by the wire in the coil is 0.020 T. What is the maximum torque on the motor?

81. Electrons are accelerated horizontally by 2.2 kV. They then pass through a uniform magnetic field B for a distance of 3.8 cm, which deflects them upward so they reach the top of a screen 22 cm away, 11 cm above the center. Estimate the value of B.

82. A 175-g model airplane charged to 18.0 mC and traveling at 3.4 m/s passes within 8.6 cm of a wire, nearly parallel to its path, carrying a 25-A current. What acceleration (in g's) does this interaction give the airplane?

83. A uniform conducting rod of length ℓ and mass m sits atop a fulcrum, which is placed a distance $\ell/4$ from the rod's left-hand end and is immersed in a uniform magnetic field of magnitude B directed into the page (Fig. 20–69). An object whose mass M is 6.0 times greater than the rod's mass is hung from the rod's left-hand end. What current (direction and magnitude) should flow through the rod in order for it to be "balanced" (i.e., be at rest horizontally) on the fulcrum? (Flexible connecting wires which exert negligible force on the rod are not shown.)

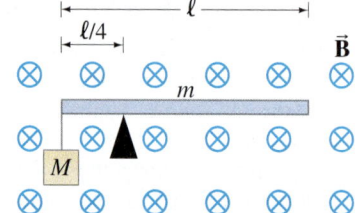

FIGURE 20–69
Problem 83.

84. Suppose the Earth's magnetic field at the equator has magnitude 0.50×10^{-4} T and a northerly direction at all points. Estimate the speed a singly ionized uranium ion ($m = 238$ u, $q = +e$) would need to circle the Earth 6.0 km above the equator. Can you ignore gravity? [Ignore relativity.]

85. A particle with charge q and momentum p, initially moving along the x axis, enters a region where a uniform magnetic field B_0 extends over a width $x = \ell$ as shown in Fig. 20–70. The particle is deflected a distance d in the $+y$ direction as it traverses the field. Determine (a) whether q is positive or negative, and (b) the magnitude of its momentum p in terms of q, B_0, ℓ, and d.

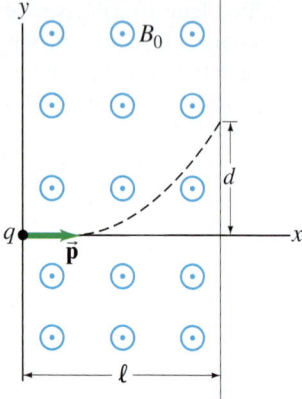

FIGURE 20–70
Problem 85.

86. A bolt of lightning strikes a metal flag pole, one end of which is anchored in the ground. Estimate the force the Earth's magnetic field can exert on the flag pole while the lightning-induced current flows. See Example 18–10.

87. A sort of "projectile launcher" is shown in Fig. 20–71. A large current moves in a closed loop composed of fixed rails, a power supply, and a very light, almost frictionless bar (pale green) touching the rails. A magnetic field is perpendicular to the plane of the circuit. If the bar has a length $\ell = 28$ cm, a mass of 1.5 g, and is placed in a field of 1.7 T, what constant current flow is needed to accelerate the bar from rest to 28 m/s in a distance of 1.0 m? In what direction must the magnetic field point?

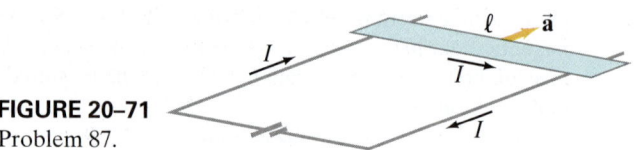

FIGURE 20–71
Problem 87.

88. The **cyclotron** (Fig. 20–72) is a device used to accelerate elementary particles such as protons to high speeds. Particles starting at point A with some initial velocity travel in semicircular orbits in the magnetic field B. The particles are accelerated to higher speeds each time they pass through the gap between the metal "dees," where there is an electric field E. (There is no electric field inside the hollow metal dees where the electrons move in circular paths.) The electric field changes direction each half-cycle, owing to an ac voltage $V = V_0 \sin 2\pi ft$, so that the particles are increased in speed at each passage through the gap. (a) Show that the frequency f of the voltage must be $f = Bq/2\pi m$, where q is the charge on the particles and m their mass. (b) Show that the kinetic energy of the particles increases by $2qV_0$ each revolution, assuming that the gap is small. (c) If the radius of the cyclotron is 2.0 m and the magnetic field strength is 0.50 T, what will be the maximum kinetic energy of accelerated protons in MeV?

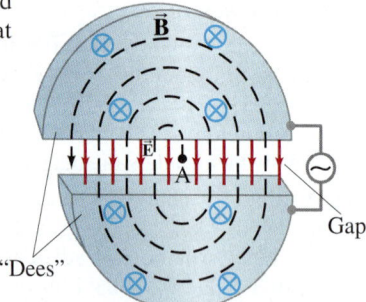

FIGURE 20–72
A cyclotron.
Problem 88.

89. Three long parallel wires are 3.8 cm from one another. (Looking along them, they are at three corners of an equilateral triangle.) The current in each wire is 8.00 A, but its direction in wire M is opposite to that in wires N and P (Fig. 20–73). (*a*) Determine the magnetic force per unit length on each wire due to the other two. (*b*) In Fig. 20–73, determine the magnitude and direction of the magnetic field at the midpoint of the line between wire M and wire N.

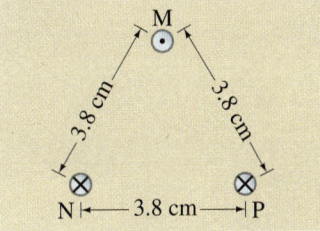

FIGURE 20–73 Problems 89 and 90.

90. In Fig. 20–73 the top wire is 1.00-mm-diameter copper wire and is suspended in air due to the two magnetic forces from the bottom two wires. The current flow through the two bottom wires is 75 A in each. Calculate the required current flow in the suspended wire (M).

91. You want to get an idea of the magnitude of magnetic fields produced by overhead power lines. You estimate that a transmission wire is about 13 m above the ground. The local power company tells you that the lines operate at 240 kV and provide a maximum power of 46 MW. Estimate the magnetic field you might experience walking under one such power line, and compare to the Earth's field.

92. Two long parallel wires 8.20 cm apart carry 19.2-A currents in the same direction. Determine the magnetic field vector at a point P, 12.0 cm from one wire and 13.0 cm from the other (Fig. 20–74). [*Hint*: Use the law of cosines; see Appendix A or inside rear cover.]

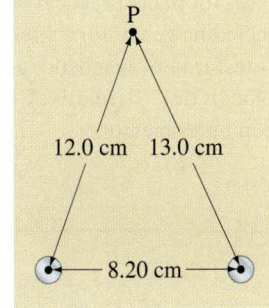

FIGURE 20–74 Problem 92.

Search and Learn

1. How many magnetic force equations are there in Chapter 20? List each one and explain when it applies. For each magnetic force equation, show how the units work out to give force in newtons.

2. An electron is moving north at a constant speed of 3.0×10^4 m/s. (*a*) In what direction should an electric field point if the electron is to be accelerated to the east? (*b*) In what direction should a magnetic field point if the electron is to be accelerated to the west? (*c*) If the electric field of part *a* has a strength of 330 V/m, what magnetic field (magnitude and direction) will produce zero net force on the electron? (*d*) If the electron in part *c* is moving faster than 3.0×10^4 m/s, in which direction will it be accelerated? What if it is moving slower than 3.0×10^4 m/s? (*e*) Now consider electrons that move perpendicular to both a magnetic field and to an electric field, which are perpendicular to each other. If only electrons with speeds of 5.5×10^4 m/s go straight through undeflected, what is the ratio of the magnitudes of electric field to magnetic field? Without knowing the value of the electric field, can you know the value of the magnetic field?

3. (*a*) A particle of charge *q* moves in a circular path of radius *r* in a uniform magnetic field $\vec{B}$. If the magnitude of the magnetic field is double, and the kinetic energy of the particle is the same, how does the angular momentum of the particle differ? (*b*) Show that the magnetic dipole moment *M* (Section 20–9) of an electron orbiting the proton nucleus of a hydrogen atom is related to the orbital angular momentum *L* of the electron by

$$M = \frac{e}{2m}L.$$

4. (*a*) Two long parallel wires, each 2.0 mm in diameter and 9.00 cm apart, carry equal 1.0-A currents in the same direction, Fig. 20–75. Determine $\vec{B}$ along the *x* axis between the wires as a function of *x*. (*b*) Graph *B* vs. *x* from $x = 1.0$ mm to $x = 89.0$ mm.

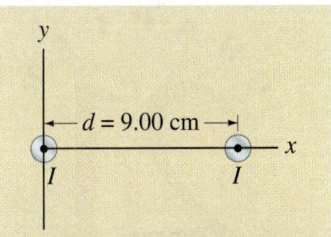

FIGURE 20–75 Search and Learn 4.

5. The force on a moving particle in a magnetic field is the idea behind **electromagnetic pumping**. It can be used to pump metallic fluids (such as sodium) and to pump blood in artificial heart machines. A basic design is shown in Fig. 20–76. For blood, an electric field is applied perpendicular to a blood vessel and to the magnetic field. Explain in detail how ions in the blood are caused to move. Do positive and negative ions feel a force in the same direction?

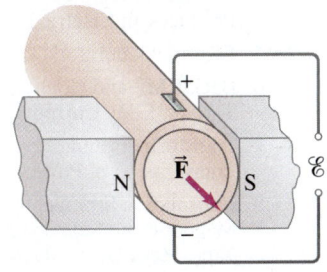

FIGURE 20–76 Electromagnetic pumping in a blood vessel. Search and Learn 5.

ANSWERS TO EXERCISES

A: Near the poles, where the field lines are closer together.
B: Circles, pointing counterclockwise.
C: (*b*), (*c*), (*d*).
D: 0.15 N.
E: (*b*), (*c*), (*d*).
F: Zero.
G: Negative; the helical path would rotate in the opposite direction (still going to the right).
H: 0.8×10^{-5} T, up.

One of the great laws of physics is Faraday's law of induction, which says that a changing magnetic flux produces an induced emf. This photo shows a bar magnet moving into (or out of) a coil of wire, and the galvanometer registers an induced current. This phenomenon of electromagnetic induction is the basis for many practical devices, including generators, alternators, transformers, magnetic recording on tape or disk (hard drive), and computer memory.

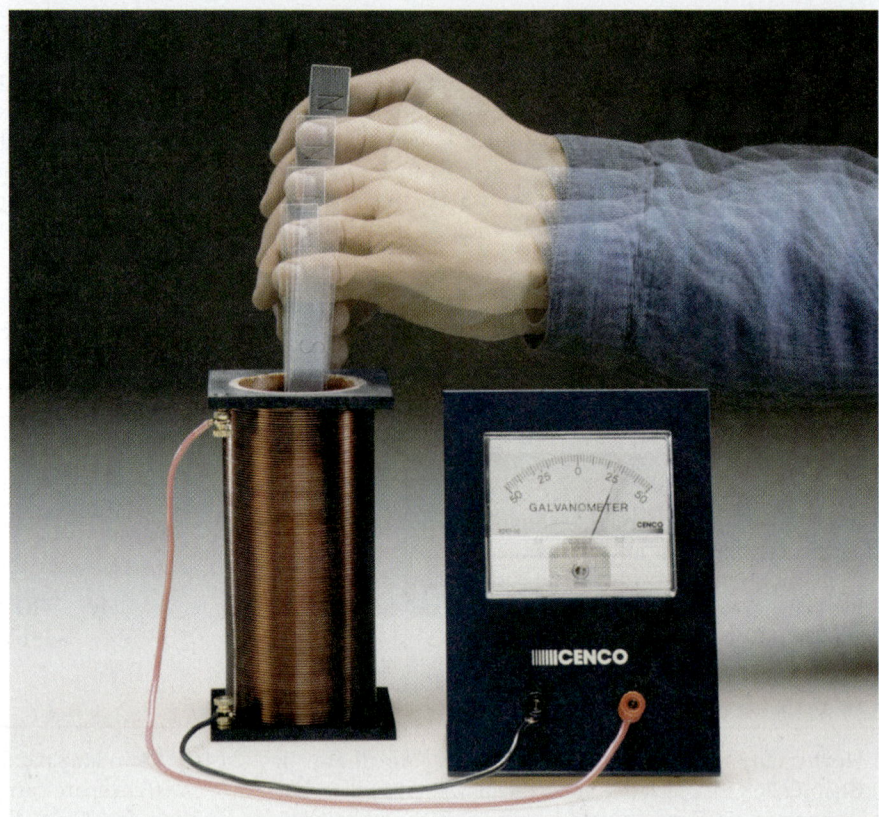

CHAPTER 21

Electromagnetic Induction and Faraday's Law

CONTENTS

21–1 Induced EMF
21–2 Faraday's Law of Induction; Lenz's Law
21–3 EMF Induced in a Moving Conductor
21–4 Changing Magnetic Flux Produces an Electric Field
21–5 Electric Generators
21–6 Back EMF and Counter Torque; Eddy Currents
21–7 Transformers and Transmission of Power
*21–8 Information Storage: Magnetic and Semiconductor; Tape, Hard Drive, RAM
*21–9 Applications of Induction: Microphone, Seismograph, GFCI
*21–10 Inductance
*21–11 Energy Stored in a Magnetic Field
*21–12 LR Circuit
*21–13 AC Circuits and Reactance
*21–14 LRC Series AC Circuit
*21–15 Resonance in AC Circuits

CHAPTER-OPENING QUESTION—Guess now!

In the photograph above, the bar magnet is inserted down into the coil of wire, and is left there for 1 minute; then it is pulled up and out from the coil. What would an observer watching the galvanometer see?

(a) No change (pointer stays on zero): without a battery there is no current to detect.
(b) A small current flows while the magnet is inside the coil of wire.
(c) A current spike as the magnet enters the coil, and then nothing.
(d) A current spike as the magnet enters the coil, and then a steady small current.
(e) A current spike as the magnet enters the coil, then nothing (pointer at zero), then a current spike in the opposite direction as the magnet exits the coil.

In Chapter 20, we discussed two ways in which electricity and magnetism are related: (1) an electric current produces a magnetic field; and (2) a magnetic field exerts a force on an electric current or on a moving electric charge. These discoveries were made in 1820–1821. Scientists then began to wonder: if electric currents produce a magnetic field, is it possible that a magnetic field can produce an electric current? Ten years later the American Joseph Henry (1797–1878) and the Englishman Michael Faraday (1791–1867) independently found that it was possible. Henry actually made the discovery first. But Faraday published his results earlier and investigated the subject in more detail. We now discuss this phenomenon and some of its world-changing applications including the electric generator.

21–1 Induced EMF

In his attempt to produce an electric current from a magnetic field, Faraday used an apparatus like that shown in Fig. 21–1. A coil of wire, X, was connected to a battery. The current that flowed through X produced a magnetic field that was intensified by the ring-shaped iron core around which the wire was wrapped. Faraday hoped that a strong steady current in X would produce a great enough magnetic field to produce a current in a second coil Y wrapped on the same iron ring.

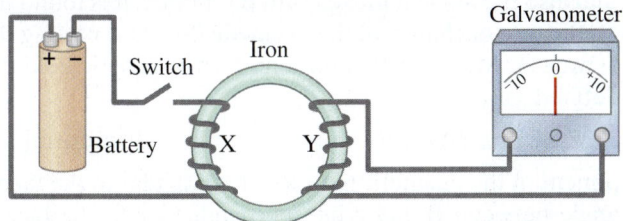

FIGURE 21–1 Faraday's experiment to induce an emf.

This second circuit, Y, contained a galvanometer to detect any current but contained no battery. He met no success with constant currents. But the long-sought effect was finally observed when Faraday noticed the galvanometer in circuit Y deflect strongly at the moment he closed the switch in circuit X. And the galvanometer deflected strongly in the opposite direction when he opened the switch in X. A constant current in X produced a constant magnetic field which produced *no* current in Y. Only when the current in X was starting or stopping was a current produced in Y.

Faraday concluded that although a constant magnetic field produces no current in a conductor, a *changing* magnetic field can produce an electric current. Such a current is called an **induced current**. When the magnetic field through coil Y changes, a current occurs in Y as if there were a source of emf in circuit Y. We therefore say that

a changing magnetic field induces an emf.

Faraday did further experiments on **electromagnetic induction**, as this phenomenon is called. For example, Fig. 21–2 shows that if a magnet is moved quickly into a coil of wire, a current is induced in the wire. If the magnet is quickly removed, a current is induced in the opposite direction ($\vec{B}$ through the coil decreases). Furthermore, if the magnet is held steady and the coil of wire is moved toward or away from the magnet, again an emf is induced and a current flows. Motion or change is required to induce an emf. It doesn't matter whether the magnet or the coil moves. It is their *relative motion* that counts.

> **CAUTION**
> *Changing $\vec{B}$, not $\vec{B}$ itself, induces current*

> **CAUTION**
> *Relative motion—magnet or coil moving induces current*

FIGURE 21–2 (a) A current is induced when a magnet is moved toward a coil, momentarily increasing the magnetic field through the coil. (b) The induced current is opposite when the magnet is moved away from the coil ($\vec{B}$ decreases). Note that the galvanometer zero is at the center of the scale and the needle deflects left or right, depending on the direction of the current. In (c), no current is induced if the magnet does not move relative to the coil. It is the relative motion that counts here: the magnet can be held steady and the coil moved, which also induces an emf.

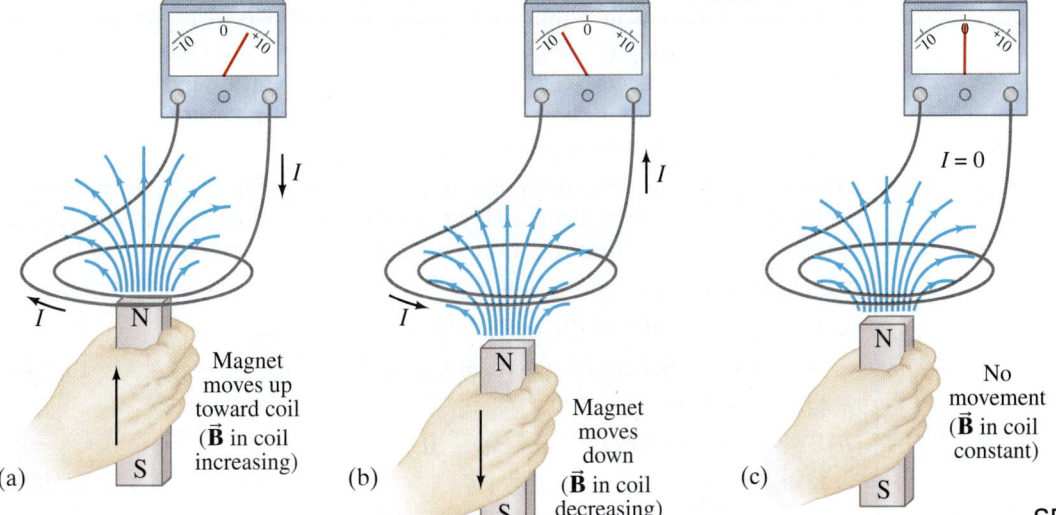

EXERCISE A Return to the Chapter-Opening Question, page 590, and answer it again now. Try to explain why you may have answered differently the first time.

21–2 Faraday's Law of Induction; Lenz's Law

Faraday investigated quantitatively what factors influence the magnitude of the emf induced. He found first of all that the more rapidly the magnetic field changes, the greater the induced emf. He also found that the induced emf depends on the area of the circuit loop (and also the angle it makes with $\vec{B}$). In fact, it is found that the emf is proportional to the rate of change of the **magnetic flux**, Φ_B, passing through the circuit or loop of area A. Magnetic flux for a uniform magnetic field through a loop of area A is defined as

$$\Phi_B = B_\perp A = BA\cos\theta. \qquad [B \text{ uniform}] \quad (21\text{–}1)$$

Here $B_\perp$ is the component of the magnetic field $\vec{B}$ perpendicular to the face of the loop, and θ is the angle between $\vec{B}$ and a line perpendicular to the face of the loop. These quantities are shown in Fig. 21–3 for a square loop of side ℓ whose area is $A = \ell^2$. When the face of the loop is parallel to $\vec{B}$, $\theta = 90°$ and $\Phi_B = 0$. When $\vec{B}$ is perpendicular to the face of the loop, $\theta = 0°$, and

$$\Phi_B = BA. \qquad [\text{uniform } \vec{B} \perp \text{loop face}]$$

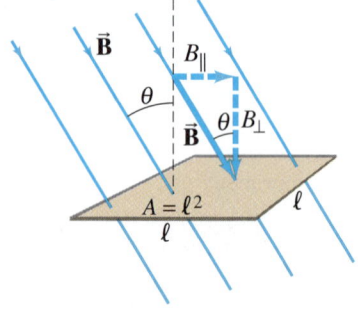

FIGURE 21–3 Determining the flux through a flat loop of wire. This loop is square, of side ℓ and area $A = \ell^2$.

As we saw in Chapter 20, the lines of $\vec{B}$ (like lines of $\vec{E}$) can be drawn such that the number of lines per unit area is proportional to the field strength. Then the flux Φ_B can be thought of as being proportional to the *total number of lines passing through the area enclosed by the loop*. This is illustrated in Fig. 21–4, where three wire loops of a coil are viewed from the side (on edge). For $\theta = 90°$, no magnetic field lines pass through the loops and $\Phi_B = 0$, whereas Φ_B is a maximum when $\theta = 0°$. The unit of magnetic flux is the tesla-meter^2; this is called a **weber**: $1 \text{ Wb} = 1 \text{ T}\cdot\text{m}^2$.

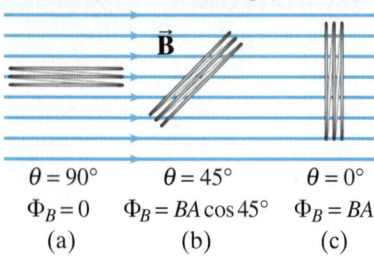

FIGURE 21–4 Magnetic flux Φ_B is proportional to the number of lines of $\vec{B}$ that pass through the loops of a coil (here with 3 loops).

With our definition of flux, Eq. 21–1, we can write down the results of Faraday's investigations: The emf $\mathscr{E}$ induced in a circuit is equal to the rate of change of magnetic flux through the circuit:

$$\mathscr{E} = -\frac{\Delta\Phi_B}{\Delta t}. \qquad [1 \text{ loop}] \quad (21\text{–}2\text{a})$$

FARADAY'S LAW OF INDUCTION

This fundamental result is known as **Faraday's law of induction**, and it is one of the basic laws of electromagnetism.

If the circuit contains N loops that are closely wrapped so the same flux passes through each, the emfs induced in each loop add together, so the total emf is

$$\mathscr{E} = -N\frac{\Delta\Phi_B}{\Delta t}. \qquad [N \text{ loops}] \quad (21\text{–}2\text{b})$$

FARADAY'S LAW OF INDUCTION

EXAMPLE 21–1 A loop of wire in a magnetic field. A square loop of wire of side $\ell = 5.0$ cm is in a uniform magnetic field $B = 0.16$ T. What is the magnetic flux in the loop (a) when $\vec{B}$ is perpendicular to the face of the loop and (b) when $\vec{B}$ is at an angle of 30° to the area of the loop? (c) What is the magnitude of the average current in the loop if it has a resistance of 0.012 Ω and it is rotated from position (b) to position (a) in 0.14 s?

APPROACH We use the definition $\Phi_B = BA\cos\theta$, Eq. 21–1, to calculate the magnetic flux. Then we use Faraday's law of induction to find the induced emf in the coil, and from that the induced current ($I = \mathscr{E}/R$).

SOLUTION The area of the coil is $A = \ell^2 = (5.0 \times 10^{-2} \text{ m})^2 = 2.5 \times 10^{-3} \text{ m}^2$.
(a) $\vec{B}$ is perpendicular to the coil's face, so $\theta = 0°$ and

$$\Phi_B = BA\cos 0° = (0.16 \text{ T})(2.5 \times 10^{-3} \text{ m}^2)(1) = 4.0 \times 10^{-4} \text{ T}\cdot\text{m}^2$$

or 4.0×10^{-4} Wb.
(b) The angle θ is 30° and $\cos 30° = 0.866$, so

$$\Phi_B = BA\cos\theta = (0.16 \text{ T})(2.5 \times 10^{-3} \text{ m}^2)\cos 30° = 3.5 \times 10^{-4} \text{ T}\cdot\text{m}^2$$

or 3.5×10^{-4} Wb, a bit less than in part (a).

(c) The magnitude of the induced emf (Eq. 21–2a) during the 0.14-s time interval is

$$\mathcal{E} = \frac{\Delta \Phi_B}{\Delta t} = \frac{(4.0 \times 10^{-4}\,\text{T}\cdot\text{m}^2) - (3.5 \times 10^{-4}\,\text{T}\cdot\text{m}^2)}{0.14\,\text{s}} = 3.6 \times 10^{-4}\,\text{V}.$$

Before and after the loop rotates, when it is at rest, the emf is zero. The current in the wire loop (Ohm's law) while it is rotating is

$$I = \frac{\mathcal{E}}{R} = \frac{3.6 \times 10^{-4}\,\text{V}}{0.012\,\Omega} = 0.030\,\text{A} = 30\,\text{mA}.$$

The minus signs in Eqs. 21–2a and b are there to remind us in which direction the induced emf acts. Experiments show that

a current produced by an induced emf moves in a direction so that the magnetic field created by that current opposes the original change in flux.

This is known as **Lenz's law**. Be aware that we are now discussing two distinct magnetic fields: (1) the changing magnetic field or flux that induces the current, and (2) the magnetic field produced by the induced current (all currents produce a magnetic field). The second (induced) field opposes the *change* in the first.

Lenz's law can be said another way, valid even if no current can flow (as when a circuit is not complete):

An induced emf is always in a direction that opposes the original change in flux that caused it.

Let us apply Lenz's law to the relative motion between a magnet and a coil, Fig. 21–2. The changing flux through the coil induces an emf in the coil, producing a current. This induced current produces its own magnetic field. In Fig. 21–2a the distance between the coil and the magnet decreases. The magnet's magnetic field (and number of field lines) through the coil increases, and therefore the flux increases. The magnetic field of the magnet points upward. To oppose the upward increase, the magnetic field produced by the induced current needs to point *downward* inside the coil. Thus, Lenz's law tells us the current moves as shown in Fig. 21–2a (use the right-hand rule). In Fig. 21–2b, the flux *decreases* (because the magnet is moved away and B decreases), so the induced current in the coil produces an *upward* magnetic field through the coil that is "trying" to maintain the status quo. Thus the current in Fig. 21–2b is in the opposite direction from Fig. 21–2a.

It is important to note that an emf is induced whenever there is a change in *flux* through the coil, and we now consider some more possibilities.

> **CAUTION**
> *Distinguish two different magnetic fields*

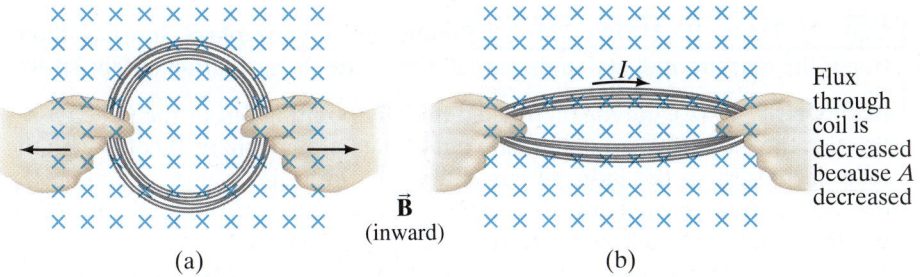

Magnetic flux $\Phi_B = BA\cos\theta$, so an emf can be induced in three ways: (1) by a changing magnetic field B; (2) by changing the area A of the loop in the field; or (3) by changing the loop's orientation θ with respect to the field. Figures 21–1 and 21–2 showed case 1. Cases 2 and 3 are illustrated in Figs. 21–5 and 21–6.

FIGURE 21–5 A current can be induced by changing the coil's area, even though B doesn't change. Here the area A is reduced by pulling on the sides of the coil: the *flux* through the coil is reduced as we go from (a) to (b). The brief induced current acts in the direction shown so as to try to maintain the original flux ($\Phi = BA$) by producing its own magnetic field into the page. That is, as area A decreases, the current acts to increase B in the original (inward) direction.

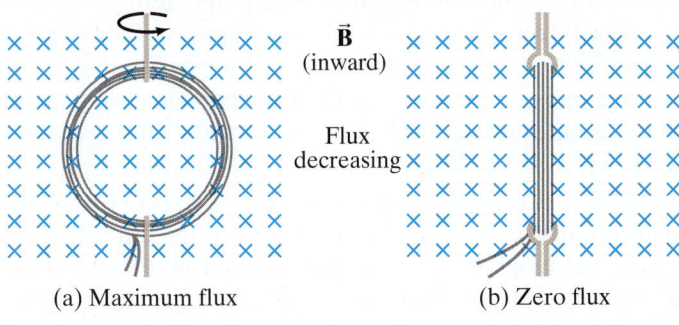

(a) Maximum flux (b) Zero flux

FIGURE 21–6 A current can be induced by rotating a coil in a magnetic field. The flux through the coil changes from (a) to (b) because θ (in Eq. 21–1, $\Phi = BA\cos\theta$) went from 0° ($\cos\theta = 1$) to 90° ($\cos\theta = 0$).

SECTION 21–2 593

FIGURE 21–7 Example 21–2: An induction stove.

CONCEPTUAL EXAMPLE 21–2 **Induction stove.** In an induction stove (Fig. 21–7), an ac current exists in a coil that is the "burner" (a burner that never gets hot). Why will it heat a metal pan, usually iron, but not a glass container?

RESPONSE The ac current sets up a changing magnetic field that passes through the pan bottom. This changing magnetic field induces a current in the pan, and since the pan offers resistance, electric energy is transformed to thermal energy which heats the pan and its contents. If the pan is iron, magnetic hysteresis due to the changing current produces additional heating. A glass container offers such high resistance that little current is induced and little energy is transferred $(P = V^2/R)$.

PROBLEM SOLVING — Lenz's Law

Lenz's law is used to determine the direction of the (conventional) electric current induced in a loop due to a change in magnetic flux inside the loop. To produce an induced current you need

(a) a closed conducting loop, and

(b) an external magnetic flux through the loop that is changing in time.

1. Determine whether the magnetic flux ($\Phi_B = BA\cos\theta$) inside the loop is decreasing, increasing, or unchanged.

2. The magnetic field due to the induced current: (a) points in the same direction as the external field if the flux is decreasing; (b) points in the opposite direction from the external field if the flux is increasing; or (c) is zero if the flux is not changing.

3. Once you know the direction of the induced magnetic field, use right-hand-rule-1 (page 563, Chapter 20) to find the direction of the induced current.

4. Always keep in mind that there are two magnetic fields: (1) an external field whose flux must be changing if it is to induce an electric current, and (2) a magnetic field produced by the induced current.

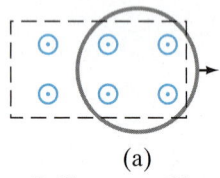
(a) Pulling a round loop to the right out of a magnetic field which points out of the page

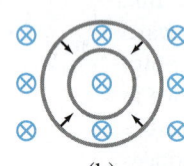
(b) Shrinking a loop in a magnetic field pointing into the page

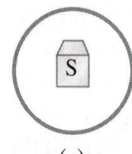
(c) S magnetic pole moving from below, up toward the loop

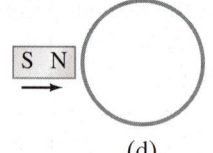
(d) N magnetic pole moving toward loop in the plane of the loop

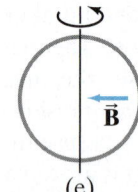
(e) Rotating the loop by pulling the left side toward us and pushing the right side in; the magnetic field points from right to left

FIGURE 21–8 Example 21–3.

⚠ **CAUTION**
Magnetic field created by induced current opposes change in external flux, not necessarily opposing the external field

CONCEPTUAL EXAMPLE 21–3 **Practice with Lenz's law.** In which direction is the current, induced in the circular loop for each situation in Fig. 21–8?

RESPONSE In (a), the magnetic field initially pointing out of the page passes through the loop. If you pull the loop out of the field, magnetic flux through the loop decreases; so the induced current will be in a direction to maintain the decreasing flux through the loop: the current will be counterclockwise to produce a magnetic field outward (toward the reader).

(b) The external field is into the page. The coil area gets smaller, so the flux will decrease; hence the induced current will be clockwise, producing its own field into the page to make up for the flux decrease.

(c) Magnetic field lines point into the S pole of a magnet, so as the magnet moves toward us and the loop, the magnet's field points into the page and is getting stronger. The current in the loop will be induced in the counterclockwise direction in order to produce a field $\vec{B}$ out of the page.

(d) The field is in the plane of the loop, so no magnetic field lines pass through the loop and the flux through the loop is zero throughout the process; hence there is no change in flux and no induced emf or current in the loop.

(e) Initially there is no flux through the loop. When you start to rotate the loop, the external field through the loop begins increasing to the left. To counteract this change in flux, the loop will have current induced in a counterclockwise direction so as to produce its own field to the right.

EXAMPLE 21–4 **Pulling a coil from a magnetic field.** A 100-loop square coil of wire, with side $\ell = 5.00$ cm and total resistance $R = 100\,\Omega$, is positioned perpendicular to a uniform magnetic field $B = 0.600$ T, as shown in Fig. 21–9. It is quickly pulled from the field at constant speed (moving perpendicular to $\vec{B}$) to a region where B drops abruptly to zero. At $t = 0$, the right edge of the coil is at the edge of the field. It takes 0.100 s for the whole coil to reach the field-free region. Determine (a) the rate of change in flux through one loop of the coil, and (b) the total emf and current induced in the 100-loop coil. (c) How much energy is dissipated in the coil? (d) What was the average force required (F_{ext})?

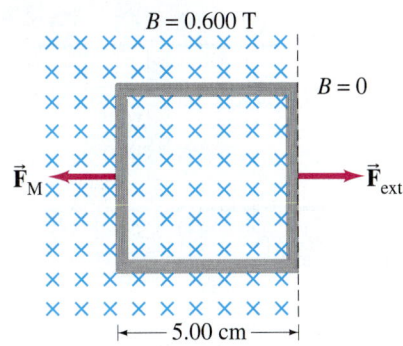

FIGURE 21–9 Example 21–4. The square coil in a magnetic field $B = 0.600$ T is pulled abruptly to the right to a region where $B = 0$. (The forces shown are discussed in the alternate solution at the end of Example 21–4.)

APPROACH We start by finding how the magnetic flux, $\Phi_B = BA\cos 0° = BA$, changes during the time interval $\Delta t = 0.100$ s. Faraday's law then gives the induced emf and Ohm's law gives the current.

SOLUTION (a) The coil's area is $A = \ell^2 = (5.00 \times 10^{-2}\,\text{m})^2 = 2.50 \times 10^{-3}\,\text{m}^2$. The flux through one loop is initially $\Phi_B = BA = (0.600\,\text{T})(2.50 \times 10^{-3}\,\text{m}^2) = 1.50 \times 10^{-3}$ Wb. After 0.100 s, the flux is zero. The rate of change in flux is constant (because the coil is square), and for one loop is equal to

$$\frac{\Delta \Phi_B}{\Delta t} = \frac{0 - (1.50 \times 10^{-3}\,\text{Wb})}{0.100\,\text{s}} = -1.50 \times 10^{-2}\,\text{Wb/s}.$$

(b) The emf induced (Eq. 21–2) in the 100-loop coil during this 0.100-s interval is

$$\mathscr{E} = -N\frac{\Delta \Phi_B}{\Delta t} = -(100)(-1.50 \times 10^{-2}\,\text{Wb/s}) = 1.50\,\text{V}.$$

The current is found by applying Ohm's law to the 100-Ω coil:

$$I = \frac{\mathscr{E}}{R} = \frac{1.50\,\text{V}}{100\,\Omega} = 1.50 \times 10^{-2}\,\text{A} = 15.0\,\text{mA}.$$

By Lenz's law, the current must be clockwise to produce more $\vec{B}$ into the page and thus oppose the decreasing flux into the page.

(c) The total energy dissipated in the coil is the product of the power ($= I^2R$) and the time:

$$E = Pt = I^2Rt = (1.50 \times 10^{-2}\,\text{A})^2(100\,\Omega)(0.100\,\text{s}) = 2.25 \times 10^{-3}\,\text{J}.$$

(d) We can use the result of part (c) and apply the work-energy principle: the energy dissipated E is equal to the work W needed to pull the coil out of the field (Chapter 6). Because $W = \bar{F}_{\text{ext}}\,d$ where $d = 5.00$ cm, then

$$\bar{F}_{\text{ext}} = \frac{W}{d} = \frac{2.25 \times 10^{-3}\,\text{J}}{5.00 \times 10^{-2}\,\text{m}} = 0.0450\,\text{N}.$$

Alternate Solution (d) We can also calculate the force directly using Eq. 20–2 for constant $\vec{B}$, $F = I\ell B$. The force the magnetic field exerts on the top and bottom sections of the square coil of Fig. 21–9 are in opposite directions and cancel each other. The magnetic force $\vec{F}_M$ exerted on the left vertical section of the square coil acts to the left as shown because the current is up (clockwise). The right side of the loop is in the region where $\vec{B} = 0$. Hence the external force to the right, $\vec{F}_{\text{ext}}$, needed to just overcome the magnetic force to the left (on $N = 100$ loops), is

$$F_{\text{ext}} = NI\ell B = (100)(0.0150\,\text{A})(0.0500\,\text{m})(0.600\,\text{T}) = 0.0450\,\text{N},$$

which is the same answer, confirming our use of energy conservation above.

FIGURE 21–10 Exercise B.

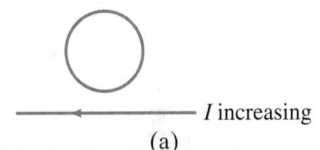
(a)

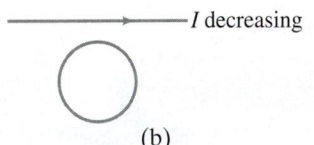
(b)

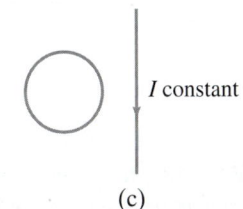
(c)

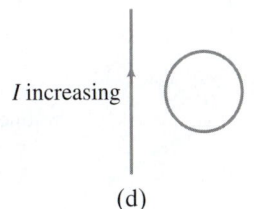
(d)

| **EXERCISE B** What is the direction of the induced current in the circular loop due to the current shown in each part of Fig. 21–10?

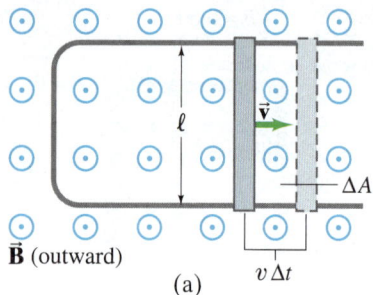

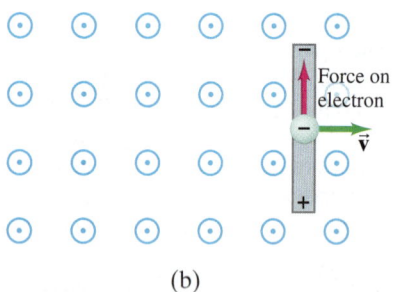

FIGURE 21–11 (a) A conducting rod is moved to the right on a U-shaped conductor in a uniform magnetic field $\vec{B}$ that points out of the page. The induced current is clockwise. (b) Upward force on an electron in the metal rod (moving to the right) due to $\vec{B}$ pointing out of the page; hence electrons can collect at the top of the rod, leaving + charge at the bottom.

FIGURE 21–12 Example 21–5.

PHYSICS APPLIED
Blood-flow measurement

FIGURE 21–13 Measurement of blood velocity from the induced emf. Example 21–6.

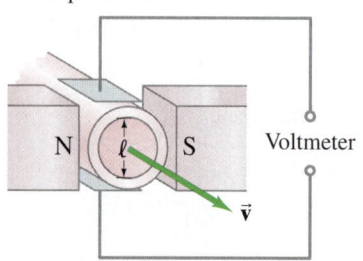

596 CHAPTER 21

21–3 EMF Induced in a Moving Conductor

Another way to induce an emf is shown in Fig. 21–11a, and this situation helps illuminate the nature of the induced emf. Assume that a uniform magnetic field $\vec{B}$ is perpendicular to the area bounded by the U-shaped conductor and the movable rod resting on it. If the rod is made to move at a speed v to the right, it travels a distance $\Delta x = v\,\Delta t$ in a time Δt. Therefore, the area of the loop increases by an amount $\Delta A = \ell\,\Delta x = \ell v\,\Delta t$ in a time Δt. By Faraday's law there is an induced emf $\mathcal{E}$ whose magnitude is given by

$$\mathcal{E} = \frac{\Delta \Phi_B}{\Delta t} = \frac{B\,\Delta A}{\Delta t} = \frac{B\ell v\,\Delta t}{\Delta t} = B\ell v. \quad (21\text{–}3)$$

The induced current is clockwise (to counter the increasing flux).

Equation 21–3 is valid as long as B, ℓ, and v are mutually perpendicular. (If they are not, we use only the components of each that are mutually perpendicular.) An emf induced on a conductor moving in a magnetic field is sometimes called **motional emf**.

We can also obtain Eq. 21–3 without using Faraday's law. We saw in Chapter 20 that a charged particle moving with speed v perpendicular to a magnetic field B experiences a force $F = qvB$ (Eq. 20–4). When the rod of Fig. 21–11a moves to the right with speed v, the electrons in the rod also move with this speed. Therefore, since $\vec{v} \perp \vec{B}$, each electron feels a force $F = qvB$, which acts up the page as the red arrow in Fig. 21–11b shows. If the rod is not in contact with the U-shaped conductor, electrons would collect at the upper end of the rod, leaving the lower end positive (see signs in Fig. 21–11b). There must thus be an induced emf. If the rod is in contact with the U-shaped conductor (Fig. 21–11a), the electrons will flow into the U. There will then be a clockwise (conventional) current in the loop. To calculate the emf, we determine the work W needed to move a charge q from one end of the rod to the other against this potential difference: $W = $ force × distance $= (qvB)(\ell)$. The emf equals the work done per unit charge, so $\mathcal{E} = W/q = qvB\ell/q = B\ell v$, the same result as from Faraday's law above, Eq. 21–3.

> **EXERCISE C** In what direction will the electrons flow in Fig. 21–11 if the rod moves to the left, decreasing the area of the current loop?

> **EXAMPLE 21–5 ESTIMATE Does a moving airplane develop a large emf?**
> An airplane travels 1000 km/h in a region where the Earth's magnetic field is about 5×10^{-5} T and is nearly vertical (Fig. 21–12). What is the potential difference induced between the wing tips that are 70 m apart?
> **APPROACH** We consider the wings to be a 70-m-long conductor moving through the Earth's magnetic field. We use Eq. 21–3 to get the emf.
> **SOLUTION** Since $v = 1000$ km/h $= 280$ m/s, and $\vec{v} \perp \vec{B}$, we have
> $$\mathcal{E} = B\ell v = (5 \times 10^{-5}\,\text{T})(70\,\text{m})(280\,\text{m/s}) \approx 1\,\text{V}.$$
> **NOTE** Not much to worry about.

> **EXAMPLE 21–6 Electromagnetic blood-flow measurement.** The rate of blood flow in our body's vessels can be measured using the apparatus shown in Fig. 21–13, since blood contains charged ions. Suppose that the blood vessel is 2.0 mm in diameter, the magnetic field is 0.080 T, and the measured emf is 0.10 mV. What is the flow velocity v of the blood?
> **APPROACH** The magnetic field $\vec{B}$ points horizontally from left to right (N pole toward S pole). The induced emf acts over the width $\ell = 2.0$ mm of the blood vessel, perpendicular to $\vec{B}$ and $\vec{v}$ (Fig. 21–13), just as in Fig. 21–11. We can then use Eq. 21–3 to get v.
> **SOLUTION** We solve for v in Eq. 21–3:
> $$v = \frac{\mathcal{E}}{B\ell} = \frac{(1.0 \times 10^{-4}\,\text{V})}{(0.080\,\text{T})(2.0 \times 10^{-3}\,\text{m})} = 0.63\,\text{m/s}.$$
> **NOTE** In actual practice, an alternating current is used to produce an alternating magnetic field. The induced emf is then alternating.

21–4 Changing Magnetic Flux Produces an Electric Field

We have seen that a changing magnetic flux induces an emf. In a closed loop of wire there will also be an induced current, which implies there is an electric field in the wire causing the electrons to start moving. Indeed, this and other results suggest the important conclusion that

a changing magnetic flux produces an electric field.

This result applies not only to wires and other conductors, but is a general result that applies to any region in space. Indeed, an electric field will be produced (= induced) at any point in space where there is a changing magnetic field.

We can get a simple formula for E in terms of B for the case of electrons in a moving conductor, as in Fig. 21–11. The electrons feel a force (upwards in Fig. 21–11b); and if we put ourselves in the reference frame of the conductor, this force accelerating the electrons implies that there is an electric field in the conductor. Electric field is defined as the force per unit charge, $E = F/q$, where here $F = qvB$ (Eq. 20–4). Thus the effective field E in the rod must be

$$E = \frac{F}{q} = \frac{qvB}{q} = vB, \qquad (21\text{–}4)$$

which is a useful result.

21–5 Electric Generators

We discussed alternating currents (ac) briefly in Section 18–7. Now we examine how ac is generated: by an **electric generator** or **dynamo**. A generator transforms mechanical energy into electric energy, just the opposite of what a motor does (Section 20–10). A simplified diagram of an **ac generator** is shown in Fig. 21–14. A generator consists of many loops of wire (only one is shown) wound on an **armature** that can rotate in a magnetic field. The axle is turned by some mechanical means (falling water, steam turbine, car motor belt), and an emf is induced in the rotating coil. An electric current is thus the *output* of a generator. Suppose in Fig. 21–14 that the armature is rotating clockwise; then right-hand-rule-3 (p. 568) applied to charged particles in the wire (or Lenz's law) tells us that the (conventional) current in the wire labeled b on the armature is outward towards us; therefore the current is outward through brush b. (Each brush is fixed and presses against a continuous slip ring that rotates with the armature.) After one-half revolution, wire b will be where wire a is now in Fig. 21–14, and the current then at brush b will be inward. Thus the current produced is alternating.

The frequency f is 60 Hz for general use in the United States and Canada, whereas 50 Hz is used in many countries. Most of the power generated in the United States is done at steam plants, where the burning of fossil fuels (coal, oil, natural gas) boils water to produce high-pressure steam that turns a turbine connected to the generator axle (Fig. 15–21). Turbines can also be turned by water pressure at a dam (hydroelectric). At nuclear power plants, the nuclear energy released is used to produce steam to turn turbines. Indeed, a heat engine (Chapter 15) connected to a generator is the principal means of generating electric power. The frequency of 60 Hz or 50 Hz is maintained very precisely by power companies.

A **dc generator** is much like an ac generator, except the slip rings are replaced by split-ring commutators, Fig. 21–15a, just as in a dc motor (Figs. 20–37 and 20–38). The output of such a generator is as shown and can be smoothed out by placing a capacitor in parallel with the output.[†] More common is the use of many armature windings, as in Fig. 21–15b, which produces a smoother output.

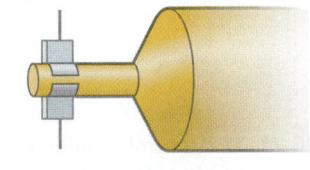

FIGURE 21–14 An ac generator.

FIGURE 21–15 (a) A dc generator with one set of commutators, and (b) a dc generator with many sets of commutators and windings.

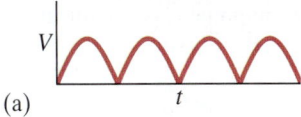
(a)

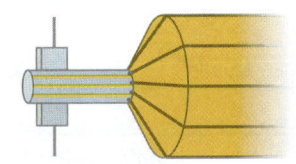

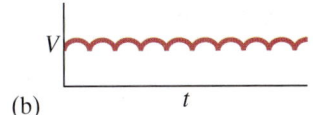
(b)

[†]A capacitor tends to store charge and, if the time constant RC is long enough, helps to smooth out the voltage as shown in the figure to the right.

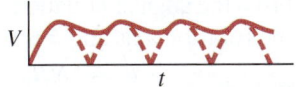

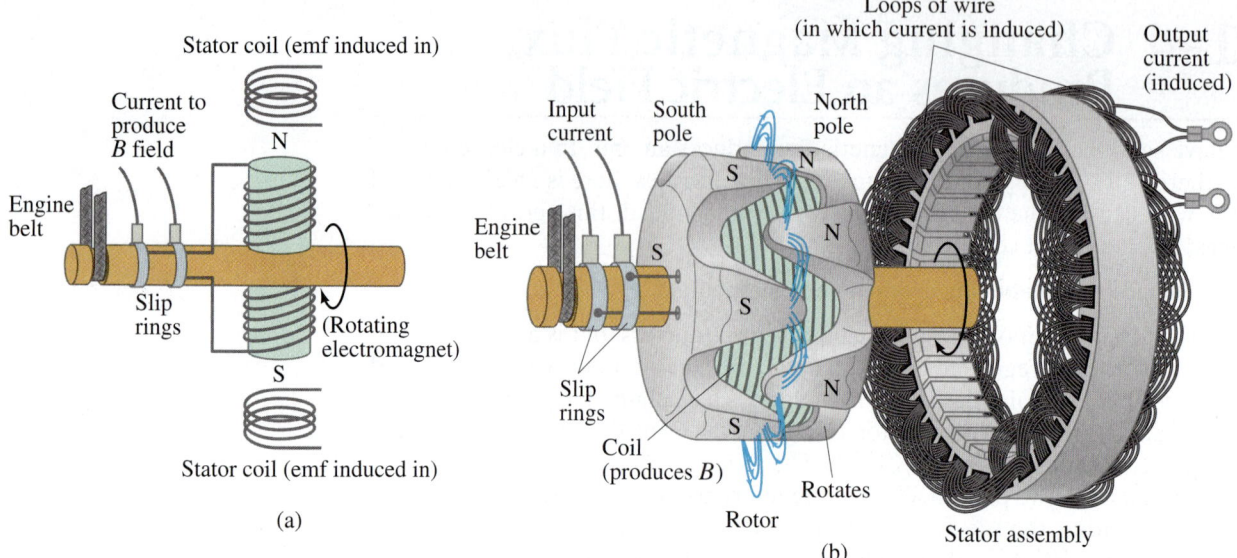

FIGURE 21–16 (a) Simplified schematic diagram of an alternator. The input current to the rotor from the battery is connected through continuous slip rings. Sometimes the rotor electromagnet is replaced by a permanent magnet (no input current). (b) Actual shape of an alternator. The rotor is made to turn by a belt from the engine. The current in the wire coil of the rotor produces a magnetic field inside it on its axis that points horizontally from left to right (not shown), thus making north and south poles of the plates attached at either end. These end plates are made with triangular fingers that are bent over the coil—hence there are alternating N and S poles quite close to one another, with magnetic field lines between them as shown by the blue lines. As the rotor turns, these field lines pass through the fixed stator coils (shown on the right for clarity, but in operation the rotor rotates within the stator), inducing a current in them, which is the output.

*Alternators

PHYSICS APPLIED
Alternators

Automobiles used to use dc generators. Today they mainly use **alternators**, which avoid the problems of wear and electrical arcing (sparks) across the split-ring commutators of dc generators. Alternators differ from generators in that an electromagnet, called the **rotor**, is fed by current from the battery and is made to rotate by a belt from the engine. The magnetic field of the turning rotor passes through a surrounding set of stationary coils called the **stator** (Fig. 21–16), inducing an alternating current in the stator coils, which is the output. This ac output is changed to dc for charging the battery by the use of semiconductor diodes, which allow current flow in one direction only.

Deriving the Generator Equation

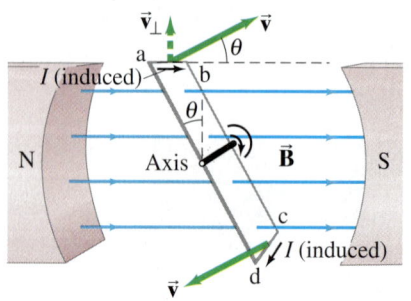

FIGURE 21–17 The emf is induced in the segments ab and cd, whose velocity components perpendicular to the field $\vec{B}$ are $v \sin \theta$.

Figure 21–17 shows the wire loop on a generator armature. The loop is being made to rotate clockwise in a uniform magnetic field $\vec{B}$. The velocity of the two lengths ab and cd at this instant are shown. Although the sections of wire bc and da are moving, the force on electrons in these sections is toward the side of the wire, not along the wire's length. The emf generated is thus due only to the force on charges in the sections ab and cd. From right-hand-rule-3, we see that the direction of the induced current in ab is from a toward b. And in the lower section, it is from c to d; so the flow is continuous in the loop. The magnitude of the emf generated in ab is given by Eq. 21–3, except that we must take the component of the velocity perpendicular to B:

$$\mathscr{E} = B\ell v_\perp,$$

where ℓ is the length of ab. From Fig. 21–17 we see that $v_\perp = v \sin \theta$, where θ is the angle the loop's face makes with the vertical. The emf induced in cd has the same magnitude and is in the same direction. Therefore their emfs add, and the total emf is

$$\mathscr{E} = 2NB\ell v \sin \theta,$$

where we have multiplied by N, the number of loops in the coil.

If the coil is rotating with constant angular velocity ω, then the angle $\theta = \omega t$. From the angular equations (Eq. 8–4), $v = \omega r = \omega(h/2)$, where r is the distance from the rotation axis and h is the length of bc or ad. Thus $\mathscr{E} = 2NB\omega\ell(h/2) \sin \omega t$, or

$$\mathscr{E} = NB\omega A \sin \omega t, \quad (21\text{–}5)$$

where $A = \ell h$ is the area of the loop. This equation holds for any shape coil, not just for a rectangle as derived. Thus, the output emf of the generator is sinusoidally alternating (see Fig. 21–18 and Section 18–7). Since ω is expressed in radians per second, we can write $\omega = 2\pi f$, where f is the frequency (in Hz = s^{-1}). The rms output (see Section 18–7, Eq. 18–8b) is

$$V_{rms} = \frac{NB\omega A}{\sqrt{2}}.$$

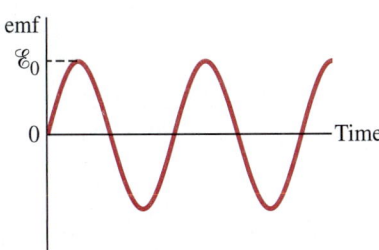

FIGURE 21–18 An ac generator produces an alternating current. The output emf $\mathcal{E} = \mathcal{E}_0 \sin \omega t$, where $\mathcal{E}_0 = NA\omega B$ (Eq. 21–5).

EXAMPLE 21–7 **An ac generator.** The armature of a 60-Hz ac generator rotates in a 0.15-T magnetic field. If the area of the coil is 2.0×10^{-2} m^2, how many loops must the coil contain if the peak output is to be $\mathcal{E}_0 = 170$ V?

APPROACH From Eq. 21–5 we see that the maximum emf is $\mathcal{E}_0 = NBA\omega$.

SOLUTION We solve Eq. 21–5 for N with $\omega = 2\pi f = (6.28)(60 \text{ s}^{-1}) = 377 \text{ s}^{-1}$:

$$N = \frac{\mathcal{E}_0}{BA\omega} = \frac{170 \text{ V}}{(0.15 \text{ T})(2.0 \times 10^{-2} \text{ m}^2)(377 \text{ s}^{-1})} = 150 \text{ turns}.$$

21–6 Back EMF and Counter Torque; Eddy Currents

Back EMF, in a Motor

A motor turns and produces mechanical energy when a current is made to flow in it. From our description in Section 20–10 of a simple dc motor, you might expect that the armature would accelerate indefinitely due to the torque on it. However, as the armature of the motor turns, the magnetic flux through the coil changes and an emf is generated. This induced emf acts to oppose the motion (Lenz's law) and is called the **back emf** or **counter emf**. The greater the speed of the motor, the greater the back emf. A motor normally turns and does work on something, but if there were no load to push (or rotate), the motor's speed would increase until the back emf equaled the input voltage. When there is a mechanical load, the speed of the motor may be limited also by the load. The back emf will then be less than the external applied voltage. The greater the mechanical load, the slower the motor rotates and the lower is the back emf ($\mathcal{E} \propto \omega$, Eq. 21–5).

EXAMPLE 21–8 **Back emf in a motor.** The armature windings of a dc motor have a resistance of 5.0 Ω. The motor is connected to a 120-V line, and when the motor reaches full speed against its normal load, the back emf is 108 V. Calculate (a) the current into the motor when it is just starting up, and (b) the current when the motor reaches full speed.

APPROACH As the motor is just starting up, it is turning very slowly, so there is negligible back emf. The only voltage is the 120-V line. The current is given by Ohm's law with $R = 5.0$ Ω. At full speed, we must include as emfs both the 120-V applied emf and the opposing back emf.

SOLUTION (a) At start up, the current is controlled by the 120 V applied to the coil's 5.0-Ω resistance. By Ohm's law,

$$I = \frac{V}{R} = \frac{120 \text{ V}}{5.0 \text{ Ω}} = 24 \text{ A}.$$

(b) When the motor is at full speed, the back emf must be included in the equivalent circuit shown in Fig. 21–19. In this case, Ohm's law (or Kirchhoff's rule) gives

$$120 \text{ V} - 108 \text{ V} = I(5.0 \text{ Ω}).$$

Therefore

$$I = \frac{12 \text{ V}}{5.0 \text{ Ω}} = 2.4 \text{ A}.$$

NOTE This result shows that the current can be very high when a motor first starts up. This is why the lights in your house may dim when the motor of the refrigerator (or other large motor) starts up. The large initial refrigerator current causes the voltage to the lights to drop because the house wiring has resistance and there is some voltage drop across it when large currents are drawn.

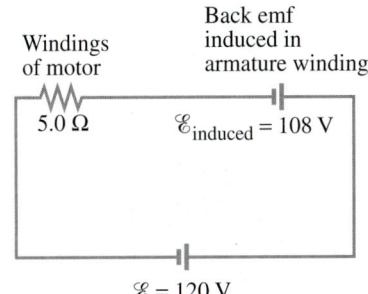

FIGURE 21–19 Circuit of a motor showing induced back emf. Example 21–8.

PHYSICS APPLIED
Burning out a motor

CONCEPTUAL EXAMPLE 21–9 **Motor overload.** When using an appliance such as a blender, electric drill, or sewing machine, if the appliance is overloaded or jammed so that the motor slows appreciably or stops while the power is still connected, the motor can burn out and be ruined. Explain why this happens.

RESPONSE The motors are designed to run at a certain speed for a given applied voltage, and the designer must take the expected back emf into account. If the rotation speed is reduced, the back emf will not be as high as expected ($\mathscr{E} \propto \omega$, Eq. 21–5). The current will increase and may become large enough that the windings of the motor heat up and may melt, ruining the motor.

Counter Torque, in a Generator

In a generator, the situation is the reverse of that for a motor. As we saw, the mechanical turning of the armature induces an emf in the loops, which is the output. If the generator rotates but is not connected to an external circuit, the emf exists at the terminals but there is no current. In this case, it takes little effort to turn the armature. But if the generator *is* connected to a device that draws current, then a current flows in the coils of the armature. Because this current-carrying coil is in an external magnetic field, there will be a torque exerted on it (as in a motor), and this torque opposes the motion (use right-hand-rule-2, page 568, for the force on a wire in Fig. 21–14 or 21–17). This is called a **counter torque**. The greater the electrical load—that is, the more current that is drawn—the greater will be the counter torque. Hence the external applied torque will have to be greater to keep the generator turning. This makes sense from the conservation of energy principle. More mechanical energy input is needed to produce more electric energy output.

EXERCISE D A bicycle headlight is powered by a generator that is turned by the bicycle wheel. (*a*) If you speed up, how does the power to the light change? (*b*) Does the generator resist being turned as the bicycle's speed increases, and if so how?

Eddy Currents

Induced currents are not always confined to well-defined paths such as in wires. Consider, for example, the rotating metal wheel in Fig. 21–20a. An external magnetic field is applied to a limited area of the wheel as shown and points into the page. The section of wheel in the magnetic field has an emf induced in it because the conductor is moving, carrying electrons with it. The flow of induced (conventional) current in the wheel is upward in the region of the magnetic field (Fig. 21–20b), and the current follows a downward return path outside that region. Why? According to Lenz's law, the induced currents oppose the change that causes them. Consider the part of the rotating wheel labeled c in Fig. 21–20b, where the magnetic field is zero but is just about to enter a region where $\vec{B}$ points into the page. To oppose this inward increase in magnetic field, the induced current is counterclockwise to produce a field pointing out of the page (right-hand-rule-1). Similarly, region d is about to move to e, where $\vec{B}$ is zero; hence the current is clockwise to produce an inward field opposed to this decreasing flux inward. These currents are referred to as **eddy currents**. They can be present in any conductor that is moving across a magnetic field or through which the magnetic flux is changing.

In Fig. 21–20b, the magnetic field exerts a force $\vec{F}$ on the induced currents it has created, and that force opposes the rotational motion. Eddy currents can be used in this way as a smooth braking device on, say, a rapid-transit car. In order to stop the car, an electromagnet can be turned on that applies its field either to the wheels or to the moving steel rail below. Eddy currents can also be used to dampen (reduce) the oscillation of a vibrating system, which is referred to as **magnetic damping**.

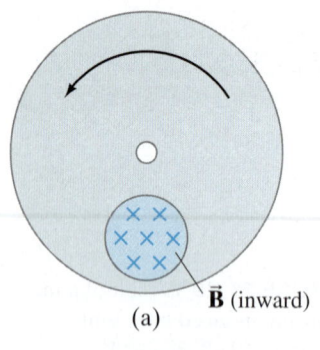

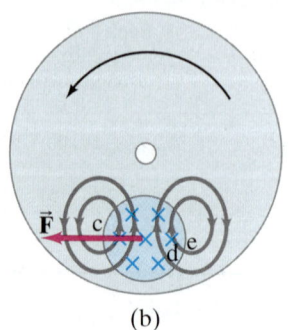

FIGURE 21–20 Production of eddy currents in a rotating wheel. The gray lines in (b) indicate induced current.

Eddy currents, however, can be a problem. For example, eddy currents induced in the armature of a motor or generator produce heat ($P = I\mathscr{E}$) and waste energy. To reduce the eddy currents, the armatures are *laminated*; that is, they are made of very thin sheets of iron that are well insulated from one another (used also in transformers, Fig. 21–23). The total path length of the eddy currents is confined to each slab, which increases the total resistance; hence the current is less and there is less wasted energy.

Walk-through metal detectors (Fig. 21–21) use electromagnetic induction and eddy currents to detect metal objects. Several coils are situated in the walls of the walk-through at different heights. In one technique, the coils are given brief pulses of current, hundreds or thousands of times per second. When a person passes through the walk-through, any metal object being carried will have eddy currents induced in it, and the small magnetic field produced by that eddy current can be detected, setting off an alert or alarm.

FIGURE 21–21 Metal detector.

21–7 Transformers and Transmission of Power

A transformer is a device for increasing or decreasing an ac voltage. Transformers are found everywhere: on utility poles (Fig. 21–22) to reduce the high voltage from the electric company to a usable voltage in houses (120 V or 240 V), in chargers for cell phones, laptops, and other electrical devices, in your car to give the needed high voltage to the spark plugs, and in many other applications. A **transformer** consists of two coils of wire known as the **primary** and **secondary** coils. The two coils can be interwoven (with insulated wire); or they can be linked by an iron core which is laminated to minimize eddy-current losses (Section 21–6), as shown in Fig. 21–23. Transformers are designed so that (nearly) all the magnetic flux produced by the current in the primary coil also passes through the secondary coil, and we assume this is true in what follows. We also assume that energy losses (in resistance and hysteresis) can be ignored—a good approximation for real transformers, which are often better than 99% efficient.

FIGURE 21–22 Repairing a step-down transformer on a utility pole.

When an ac voltage is applied to the primary coil, the changing magnetic field it produces will induce an ac voltage of the same frequency in the secondary coil. However, the voltage will be different according to the number of "turns" or loops in each coil. From Faraday's law, the voltage or emf induced in the secondary coil is

$$V_S = N_S \frac{\Delta \Phi_B}{\Delta t},$$

where N_S is the number of turns in the secondary coil, and $\Delta \Phi_B / \Delta t$ is the rate at which the magnetic flux changes.

The input primary voltage, V_P, is related to the rate at which the flux changes through it,

$$V_P = N_P \frac{\Delta \Phi_B}{\Delta t},$$

where N_P is the number of turns in the primary coil. This follows because the changing flux produces a back emf, $N_P \Delta \Phi_B / \Delta t$, in the primary that balances the applied voltage V_P if the resistance of the primary can be ignored (Kirchhoff's rules). We divide these two equations, assuming little or no flux is lost, to find

$$\frac{V_S}{V_P} = \frac{N_S}{N_P}. \qquad (21\text{–}6)$$

FIGURE 21–23 Step-up transformer ($N_P = 4$, $N_S = 12$).

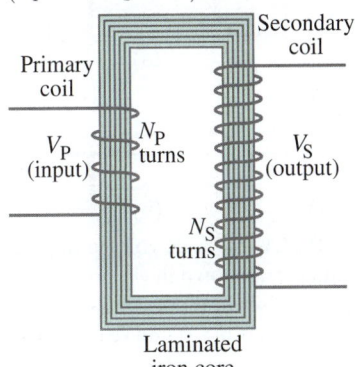

This *transformer equation* tells how the secondary (output) voltage is related to the primary (input) voltage; V_S and V_P in Eq. 21–6 can be the rms values (Section 18–7) for both, or peak values for both. Steady dc voltages don't work in a transformer because there would be no changing magnetic flux.

If the secondary coil contains more loops than the primary coil ($N_S > N_P$), we have a **step-up transformer**. The secondary voltage is greater than the primary voltage. For example, if the secondary coil has twice as many turns as the primary coil, then the secondary voltage will be twice that of the primary voltage. If N_S is less than N_P, we have a **step-down transformer**.

Although ac voltage can be increased (or decreased) with a transformer, we don't get something for nothing. Energy conservation tells us that the power output can be no greater than the power input. A well-designed transformer can be greater than 99% efficient, so little energy is lost to heat. The power output thus essentially equals the power input. Since power $P = IV$ (Eq. 18–5), we have

$$I_P V_P = I_S V_S,$$

or (remembering Eq. 21–6),

$$\frac{I_S}{I_P} = \frac{N_P}{N_S}. \tag{21–7}$$

EXAMPLE 21–10 Cell phone charger. The charger for a cell phone contains a transformer that reduces 120-V (or 240-V) ac to 5.0-V ac to charge the 3.7-V battery (Section 19–4). (It also contains diodes to change the 5.0-V ac to 5.0-V dc.) Suppose the secondary coil contains 30 turns and the charger supplies 700 mA. Calculate (a) the number of turns in the primary coil, (b) the current in the primary, and (c) the power transformed.

APPROACH We assume the transformer is ideal, with no flux loss, so we can use Eq. 21–6 and then Eq. 21–7.

SOLUTION (a) This is a step-down transformer, and from Eq. 21–6 we have

$$N_P = N_S \frac{V_P}{V_S} = \frac{(30)(120\text{ V})}{(5.0\text{ V})} = 720 \text{ turns.}$$

(b) From Eq. 21–7

$$I_P = I_S \frac{N_S}{N_P} = (0.70\text{ A})\left(\frac{30}{720}\right) = 29 \text{ mA.}$$

(c) The power transformed is

$$P = I_S V_S = (0.70\text{ A})(5.0\text{ V}) = 3.5 \text{ W.}$$

NOTE The power in the primary coil, $P = (0.029\text{ A})(120\text{ V}) = 3.5$ W, is the same as the power in the secondary coil. There is 100% efficiency in power transfer for our ideal transformer.

EXERCISE E How many turns would you want in the secondary coil of a transformer having $N_P = 400$ turns if it were to reduce the voltage from 120-V ac to 3.0-V ac?

A transformer operates only on ac. A dc current in the primary coil does not produce a changing flux and therefore induces no emf in the secondary. However, if a dc voltage is applied to the primary through a switch, at the instant the switch is opened or closed there will be an induced voltage in the secondary. For example, if the dc is turned on and off as shown in Fig. 21–24a, the voltage induced in the secondary is as shown in Fig. 21–24b. Notice that the secondary voltage drops to zero when the dc voltage is steady. This is basically how, in the **ignition system** of an automobile, the high voltage is created to produce the spark across the gap of a spark plug that ignites the gas-air mixture. The transformer is referred to as an "ignition coil," and transforms the 12 V dc of the battery (when switched off in the primary) into a spike of as much as 30 kV in the secondary.

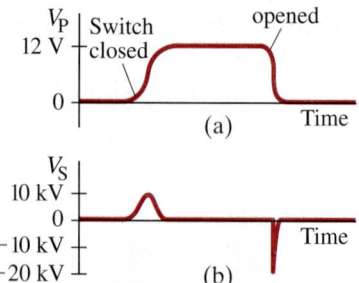

FIGURE 21–24 A dc voltage turned on and off as shown in (a) produces voltage pulses in the secondary (b). Voltage scales in (a) and (b) are not the same.

PHYSICS APPLIED
Car ignition system

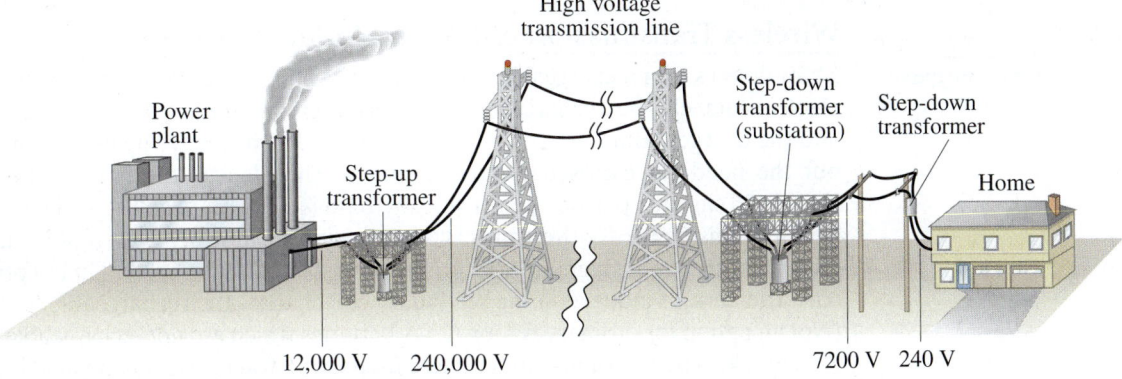

FIGURE 21–25 The transmission of electric power from power plants to homes makes use of transformers at various stages.

Transformers play an important role in the transmission of electricity. Power plants are often situated some distance from metropolitan areas, so electricity must then be transmitted over long distances (Fig. 21–25). There is always some power loss in the transmission lines, and this loss can be minimized if the power is transmitted at high voltage, using transformers, as the following Example shows.

PHYSICS APPLIED
Transformers help power transmission

EXAMPLE 21–11 Transmission lines. An average of 120 kW of electric power is sent to a small town from a power plant 10 km away. The transmission lines have a total resistance of 0.40 Ω. Calculate the power loss if the power is transmitted at (*a*) 240 V and (*b*) 24,000 V.

APPROACH We cannot use $P = V^2/R$ because if R is the resistance of the transmission lines, we don't know the voltage drop along them. The given voltages are applied across the lines plus the load (the town). But we can determine the current I in the lines ($= P/V$), and then find the power loss from $P_L = I^2R$, for both cases (*a*) and (*b*).

SOLUTION (*a*) If 120 kW is sent at 240 V, the total current will be

$$I = \frac{P}{V} = \frac{1.2 \times 10^5 \text{ W}}{2.4 \times 10^2 \text{ V}} = 500 \text{ A}.$$

The power loss in the lines, P_L, is then

$$P_L = I^2R = (500 \text{ A})^2 (0.40 \text{ Ω}) = 100 \text{ kW}.$$

Thus, over 80% of all the power would be wasted as heat in the power lines!
(*b*) If 120 kW is sent at 24,000 V, the total current will be

$$I = \frac{P}{V} = \frac{1.2 \times 10^5 \text{ W}}{2.4 \times 10^4 \text{ V}} = 5.0 \text{ A}.$$

The power loss in the lines is then

$$P_L = I^2R = (5.0 \text{ A})^2 (0.40 \text{ Ω}) = 10 \text{ W},$$

which is less than $\frac{1}{100}$ of 1%: a far better efficiency.

NOTE We see that the higher voltage results in less current, and thus less power is wasted as heat in the transmission lines. It is for this reason that power is usually transmitted at very high voltages, as high as 700 kV.

The great advantage of ac, and a major reason it is in nearly universal use[†], is that the voltage can easily be stepped up or down by a transformer. The output voltage of an electric generating plant is stepped up prior to transmission. Upon arrival in a city, it is stepped down in stages at electric substations prior to distribution. The voltage in lines along city streets is typically 2400 V or 7200 V and is stepped down to 240 V or 120 V for home use by transformers (Figs. 21–22 and 21–25).

[†]DC transmission along wires does exist, and has some advantages (if the current is constant, there is no induced current in nearby conductors as there is with ac). But boosting to high voltage and down again at the receiving end requires more complicated electronics.

SECTION 21–7 Transformers and Transmission of Power

PHYSICS APPLIED
Charging phones, cars, etc., by induction

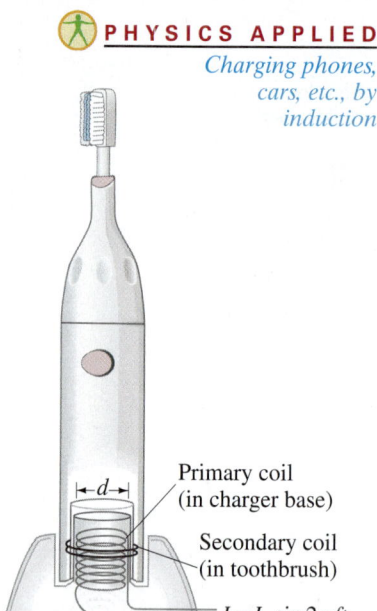

FIGURE 21–26 This electric toothbrush contains rechargeable batteries which are being recharged as it sits on its base. Charging occurs from a primary coil in the base to a secondary coil in the toothbrush. The toothbrush can be lifted from its base when you want to brush your teeth.

PHYSICS APPLIED
Hard drive

Wireless Transmission of Power—Inductive Charging

Many devices with rechargeable batteries, like cell phones, cordless phones, and even electric cars, can be recharged using a direct metal contact between the device and the charger. But devices can also be charged "wirelessly" by induction, without the need for exposed electric contacts. The electric toothbrush shown in Fig. 21–26 sits on a plastic base. Inside the base is a "primary coil" connected to an ac outlet. Inside the toothbrush is a "secondary coil" in which a current is induced due to the changing magnetic field produced by the changing current in the primary coil. The current induced in the secondary coil charges the rechargeable batteries. (Not an option for ordinary AA or AAA batteries which are *not* rechargeable.) The effect is like a transformer—except here there is no iron to contain the field lines, so there is less efficiency. But you can separate the two parts (toothbrush and charger) and brush your teeth. Many heart pacemakers are given power inductively: power in an external coil is transmitted to a secondary coil in the pacemaker (Fig. 19–25) inside the person's body near the heart. Inductive charging is also a possible means for recharging an electric car's batteries.

Wireless transmission of power must be done over short distances to maintain a reasonable efficiency. Wireless transmission of signals (information) can be done over great distances (Section 22–7) because even fairly low power signals can be detected, and it is the information in the signal voltages that counts, not power.

*21–8 Information Storage: Magnetic and Semiconductor; Tape, Hard Drive, RAM

*Magnetic Storage: Read/Write on Tape and Disks

Recording and playback on tape or disk is done by magnetic **heads**. Magnetic tapes contain a thin layer of ferromagnetic oxide on a thin plastic tape. Computer **hard drives** (HD) store digital information (applications and data): they have a thin layer of ferromagnetic material on the surface of each rotating disk or platter, Fig. 21–27a. During recording of an audio or video signal on tape, or "writing" on a hard drive, the voltage is sent to the recording head which acts as a tiny electromagnet (Fig. 21–27b) that magnetizes the tiny section of tape or disk passing the narrow gap in the head at each instant. During playback, or "reading" of an HD, the changing magnetism of the moving tape or disk at the gap causes corresponding changes in the magnetic field within the soft-iron head, which in turn induces an emf in the coil (Faraday's law). This induced emf is the output signal that can be processed by the computer, or for audio can be amplified and sent to a loudspeaker (for video to a monitor or TV).

FIGURE 21–27 (a) Photo of a hard drive showing several platters and read/write heads that can quickly move from the edge of the disk to the center. (b) Read/Write (playback/recording) head for disk or tape. In writing or recording, the electric input signal to the head, which acts as an electromagnet, magnetizes the passing tape or disk. In reading or playback, the changing magnetic field of the passing tape or disk induces a changing magnetic field in the head, which in turn induces in the coil an emf that is the output signal.

(a)

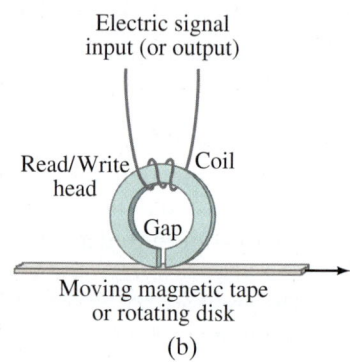
(b)

Audio and video signals may be **analog**, varying continuously in amplitude over time: the variation in degree of magnetization at sequential points reflects the variation in amplitude and frequency of the audio or video signal. In modern equipment, analog signals (say from a microphone) are electronically converted to **digital**—which means a series of **bits**, each of which is a "1" or a "0", that forms a **binary code** as discussed in Section 17–10. (Recall also, 8 bits in a row = 1 **byte**.) Computers process only digital information.

604 CHAPTER 21

CD-ROMs, CDs (audio compact discs), and DVDs (digital video discs) are read by an **optical drive** (not magnetic): a laser emits a narrow beam of light that reflects off the "grooves" of the rotating disc containing "pits" as described in Section 28–11.

> **PHYSICS APPLIED**
> *CDs and DVDs*

*Semiconductor Memory: DRAM, Flash

Basic to your computer is its **random access memory** (**RAM**). This is where the information you are working with at any given time is temporarily stored and manipulated by you. Each data storage location can be accessed and read (or written) directly and quickly, so you don't have to wait. In contrast, hard drives, tape, flash and external devices are more permanent **storage**, and they are much slower to access because the data must be searched for, sequentially, such as along the circular tracks of hard drives (Fig. 21–27a). Programs, applications, and data that you want to use are imported by the computer into the RAM from their more permanent (and more slowly accessed) storage area.[†]

RAM is based on semiconductor technology, storing the binary bits ("0" or "1") as electric charge or voltage. Some computers may use semiconductors also for long-term storage ("flash memory") in place of a hard drive.

A common type of RAM is **dynamic random access memory** or **DRAM**, which uses arrays of transistors known as MOSFETs (metal-oxide semiconductor field-effect transistors). Transistors will be discussed in Section 29–10, but we already encountered them in Section 17–11 about TV screen addressing, Fig. 17–34, which we show again here, Fig. 21–28. A MOSFET transistor in RAM serves basically as an on–off switch: the voltage on the **gate** terminal acts to control the conductivity between the **source** and the **drain** terminals, thus allowing current to flow (or not) between them.

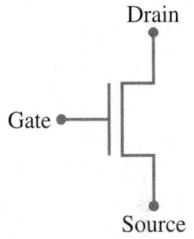

FIGURE 21–28 Symbol for a MOSFET transistor. The gate acts to attract or repel charge, and thus open or close the connection along the semiconductor that connects source and drain.

Each memory "cell," which in DRAM consists of one transistor and a capacitor, stores one bit (= a "0" or a "1"). Each cell is extremely small physically, less than 100 nm across.[‡] Typical DRAM chips (integrated circuits) contain billions of these memory cells. To see how they work, we look at a tiny part, the simple four-cell array shown in Fig. 21–29. One side of each cell capacitor is grounded; the other side is connected to the transistor source. The drain of each transistor is connected to a very thin conducting wire or "line," a **bit-line**, that runs across the array of cells. Each gate is connected to a **word-line**. A particular bit is a "1" or a "0" depending on whether the capacitor of the cell is charged to a voltage V (maybe 5 V) or is at zero (uncharged, or at a very small voltage).

To **write** data, say on the upper left cell in Fig. 21–29, word-line-1 is given a high enough pulse of voltage to "turn on" the transistor. That is, the high gate voltage attracts charge and allows bit-line-1 and the capacitor to be connected. Thus charge can flow from bit-line-1 to the capacitor, charging it either to V or to 0, depending on the bit-line-1 voltage at that moment, thus writing a "1" or a "0".

The lower left cell in Fig. 21–29 can be written at the same time by setting bit-line-2 voltage to V or zero.

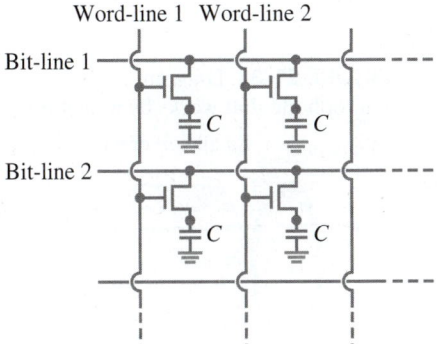

FIGURE 21–29 A tiny 2 × 2 cell, part of a simple DRAM array. The word-lines and bit-lines do not touch each other where they cross.

Now let us see a simple way to **read** a cell. In order to read the data stored ("1" or "0") on the upper left cell, a voltage of about $\frac{1}{2}V$ is given to bit-line-1. Then word-line-1 is given enough voltage to turn on the transistor and connect bit-line-1 to the capacitor. The capacitor, if uncharged (= "0"), will now drag charge from bit-line-1 and the bit-line voltage will drop *below* $\frac{1}{2}V$. If the capacitor is already charged to V (= "1"), the connection to bit-line-1 will raise bit-line-1's voltage to *above* $\frac{1}{2}V$. A sensor at the end of bit-line-1 will detect either change in voltage (increase means it reads a "1", decrease a "0"). All cells connected to one word-line are read at the same moment. The capacitor voltage has been altered by the small charge flow during the reading process. So that cell or bit which has just been read needs to be written again, or "refreshed."

[†]Computer specifications may use "memory" for the random access (fast) memory, and "storage" for the long-term (and slower access) information on hard drives, flash drives, and related devices.

[‡]At 100 nm, 10^5 bits can fit along a 1-cm line, (1 cm/100 nm) = 10^{-2} m/10^{-7} m. So a square, 1 cm on a side, can hold $10^5 \times 10^5 = 10^{10}$ = 10 Gbits ≈ 1 GB (gigabyte). Today, cells are even smaller than 100 nm: a (30 nm)^2 cell can hold ≈ 10 GB in a 1 cm^2 area.

The transistors are imperfect switches and allow the charge on the tiny capacitors in each cell to be "leaky" and lose charge fairly quickly, so every cell has to be read and rewritten (refreshed) many times per second. The D in DRAM stands for this "dynamic" refreshing action. If the power is turned off, the capacitors lose their charge and the data are lost. DRAM is thus referred to as being **volatile** memory, whereas a hard drive keeps its (magnetic) memory even when the electric power is off and is called **nonvolatile** memory (doesn't "evaporate").

Flash memory is also made of semiconductor material on tiny "chips." The transistor structures are more complicated, and are able to keep the data even without power so they are nonvolatile. Each MOSFET contains a second gate (the **floating gate**) insulated on both faces, and can hold charge for many years. Charged or not corresponds to a "1" or a "0" bit. Figure 21–30 is a diagram of such an **NVM** (nonvolatile memory) cell. The floating gate is insulated from the standard gate and the semiconductor connecting source and drain. A high positive voltage on the gate (+20 V) forces electrons in the semiconductor (at 0 V) to pass through the thin insulator into the floating gate by a process of quantum mechanical **tunneling** (discussed in Section 30–12). This charge is stored on the floating gate as a "1" bit. The erase process is done by applying the opposite (−20 V) voltage to force electrons to tunnel out of the floating gate, returning it to the uncharged state (= a "0" bit). The erase process is slow (milliseconds vs. ns for DRAM), so erasure is done in large blocks of memory. Flash memory[†] is slower to read or write, and is too slow to use as RAM. Instead, flash memory can be used in place of a hard drive as general storage in computers and tablets, and may be called a "solid state device" (**SSD**). Flash is also used for flash drives, memory cards (such as SD cards), thumb drives, cell phone and portable player memory, and external computer memory.

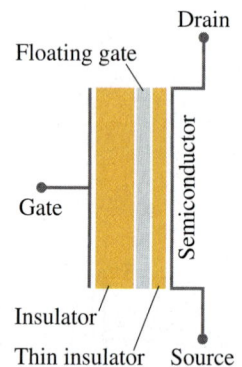

FIGURE 21–30 A floating gate nonvolatile memory cell (NVM).

Magnetoresistive RAM (**MRAM**) is a recent development, involving (again) magnetic properties. One cell (storing one bit) consists of two tiny ferromagnetic plates (separated by an insulator), one of which is permanently magnetized. The other plate can be magnetized in one direction or the other, for a "1" or a "0", by current in nearby wires. Cell size is a bit large, but MRAM is fast and nonvolatile (no power and no refresh needed) and therefore has the potential to be used as any type of memory.

*21–9 Applications of Induction: Microphone, Seismograph, GFCI

*Microphone

The condenser microphone was discussed in Section 17–7. Many other types operate on the principle of induction. In one form, a microphone is just the inverse of a loudspeaker (Section 20–10). A small coil connected to a membrane is suspended close to a small permanent magnet, as shown in Fig. 21–31. The coil moves in the magnetic field when sound waves strike the membrane, and this motion induces an emf in the moving coil. The frequency of the induced emf will be just that of the impinging sound waves, and this emf is the "signal" that can be amplified and sent to loudspeakers or recorder.

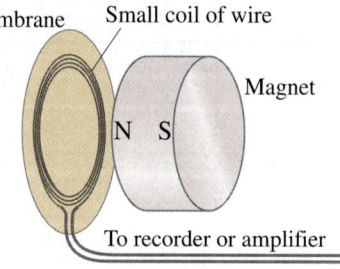

FIGURE 21–31 Diagram of a microphone that works by induction.

PHYSICS APPLIED
Credit card

*Credit Card Reader

When you pass a credit card through a reader at a store, the magnetic stripe on the back of the card passes over a read head just as for a computer hard drive. The magnetic stripe contains personal information about your account and connects by telephone line for approval from your credit card company. Newer cards use semiconductor chips that are more difficult to fraudulently copy.

[†]Why the name "Flash"? It may come from the erase process: large blocks erased "in a flash," and/or because the earliest floating gate memories were erased by a flash of UV light which ejected the stored electrons.

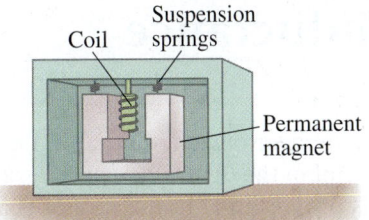

FIGURE 21–32 One type of seismograph, in which the coil is fixed to the case and moves with the Earth's surface. The magnet, suspended by springs, has inertia and does not move instantaneously with the coil (and case), so there is relative motion between magnet and coil.

*Seismograph

In geophysics, a **seismograph** measures the intensity of earthquake waves using a magnet and a coil of wire. Either the magnet or the coil is fixed to the case, and the other is inertial (suspended by a spring; Fig. 21–32). The relative motion of magnet and coil when the surface of the Earth shakes induces an emf output.

PHYSICS APPLIED
Seismograph

*Ground Fault Circuit Interrupter (GFCI)

Fuses and circuit breakers (Sections 18–6 and 20–7) protect buildings from electricity-induced fire, and apparatus from damage, due to undesired high currents. But they do not turn off the current until it is very much greater than that which can cause permanent damage to humans or death (≈ 100 mA). If fast enough, they may protect humans in some cases, such as very high currents due to short circuits. A **ground fault circuit interrupter** (GFCI) is meant above all to protect humans.

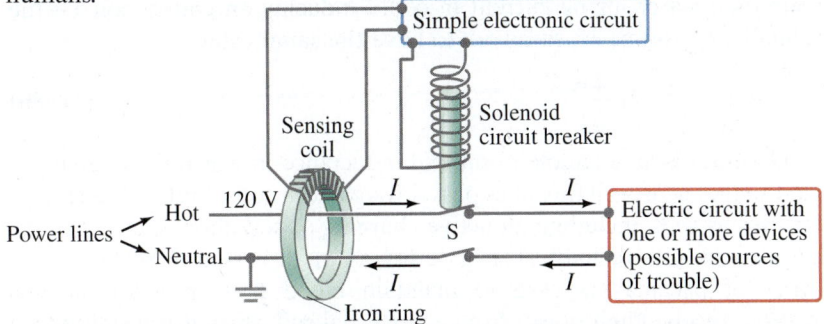

FIGURE 21–33 A ground fault circuit interrupter (GFCI).

PHYSICS APPLIED
GFCI

FIGURE 21–34 (a) A GFCI wall outlet. GFCIs can be recognized by their "test" and "reset" buttons. (b) Add-on GFCI that plugs into outlet.

Electromagnetic induction is the physical basis of a GFCI. As shown in Fig. 21–33, the two conductors of a power line connected to an electric circuit or device (red) pass through a small iron ring. Around the ring are many loops of thin wire that serve as a sensing coil. Under normal conditions (no ground fault), the current moving in the hot power wire is exactly balanced by the returning current in the neutral wire. If something goes wrong and the hot wire touches the ungrounded metal case of the device or appliance, some of the entering current can pass through a person who touches the case and then to ground (a **ground fault**). Then the return current in the neutral wire will be less than the entering current in the hot wire, so there is a *net current* passing through the GFCI's iron ring. Because the current is ac, it is changing and that current difference produces a changing magnetic field in the iron, thus inducing an emf in the sensing coil wrapped around the iron. For example, if a device draws 8.0 A, and there is a ground fault through a person of 100 mA (= 0.1 A), then 7.9 A will appear in the neutral wire. The emf induced in the sensing coil by this 100-mA difference is amplified by a simple transistor circuit and sent to its own solenoid circuit breaker that opens the circuit at the switch S, thus protecting your life.

If the case of the faulty device is grounded, the difference in current is even higher when there is a fault, and the GFCI trips very quickly.

GFCIs can sense current differences as low as 5 mA and react in 1 ms, saving lives. They can be small to fit as a wall outlet (Fig. 21–34a), or as a plug-in unit into which you plug a hair dryer or toaster (Fig. 21–34b). It is especially important to have GFCIs installed in kitchens, in bathrooms, outdoors, and near swimming pools, where people are most in danger of touching ground. GFCIs always have a "test" button (to be sure the GFCI itself works) and a "reset" button (after it goes off).

(a)

(b)

*21–10 Inductance

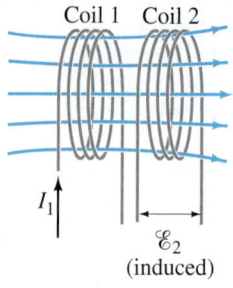

FIGURE 21–35 A changing current in one coil will induce a current in the second coil.

*Mutual Inductance

If two coils of wire are near one another, as in Fig. 21–35, a changing current in one will induce an emf in the other. We apply Faraday's law to coil 2: the emf $\mathcal{E}_2$ induced in coil 2 is proportional to the rate of change of magnetic flux passing through it. A changing flux in coil 2 is produced by a changing current I_1 in coil 1. So $\mathcal{E}_2$ is proportional to the rate of change of the current in coil 1:

$$\mathcal{E}_2 = -M \frac{\Delta I_1}{\Delta t}, \quad (21\text{–}8a)$$

where we assume the time interval Δt is very small, and the constant of proportionality, M, is called the **mutual inductance**. (The minus sign is because of Lenz's law, the induced emf opposes the changing flux.) Mutual inductance has units of $V \cdot s/A = \Omega \cdot s$, which is called the **henry** (H), after Joseph Henry: $1\,H = 1\,\Omega \cdot s$.

The mutual inductance M is a "constant" in that it does not depend on I_1; M depends on "geometric" factors such as the size, shape, number of turns, and relative positions of the two coils, and also on whether iron (or other ferromagnetic material) is present. For example, the farther apart the two coils are in Fig. 21–35, the fewer lines of flux can pass through coil 2, so M will be less. If we consider the inverse situation—a changing current in coil 2 inducing an emf in coil 1—the proportionality constant, M, turns out to have the same value,

$$\mathcal{E}_1 = -M \frac{\Delta I_2}{\Delta t}. \quad (21\text{–}8b)$$

A transformer is an example of mutual inductance in which the coupling is maximized so that nearly all flux lines pass through both coils. Mutual inductance has other uses as well, including inductive charging of cell phones, electric cars, and other devices with rechargeable batteries, as we discussed in Section 21–7. Some types of pacemakers used to maintain blood flow in heart patients (Section 19–6) receive their power from an external coil which is transmitted via mutual inductance to a second coil in the pacemaker near the heart. This type has the advantage over battery-powered pacemakers in that surgery is not needed to replace a battery when it wears out.

PHYSICS APPLIED
Pacemaker

*Self-Inductance

The concept of inductance applies also to an isolated single coil. When a changing current passes through a coil or solenoid, a changing magnetic flux is produced inside the coil, and this in turn induces an emf. This induced emf opposes the change in flux (Lenz's law); it is much like the back emf generated in a motor. (For example, if the current through the coil is increasing, the increasing magnetic flux induces an emf that opposes the original current and tends to retard its increase.) The induced emf $\mathcal{E}$ is proportional to the rate of change in current (and is in the direction opposed to the change, hence the minus sign):

$$\mathcal{E} = -L \frac{\Delta I}{\Delta t}. \quad (21\text{–}9)$$

The constant of proportionality L is called the **self-inductance**, or simply the **inductance** of the coil. It, too, is measured in henrys. The magnitude of L depends on the size and shape of the coil and on the presence of an iron core.

An ac circuit (Section 18–7) always contains some inductance, but often it is quite small unless the circuit contains a coil of many loops or turns. A coil that has significant self-inductance L is called an **inductor**. It is shown on circuit diagrams by the symbol

 . [inductor symbol]

CONCEPTUAL EXAMPLE 21-12 **Direction of emf in inductor.** Current passes through the coil in Fig. 21–36 from left to right as shown. (a) If the current is increasing with time, in which direction is the induced emf? (b) If the current is decreasing in time, what then is the direction of the induced emf?

RESPONSE (a) From Lenz's law we know that the induced emf must oppose the change in magnetic flux. If the current is increasing, so is the magnetic flux. The induced emf acts to oppose the increasing flux, which means it acts like a source of emf that opposes the outside source of emf driving the current. So the induced emf in the coil acts to oppose I in Fig. 21–36a. In other words, the inductor might be thought of as a battery with a positive terminal at point A (tending to block the current entering at A), and negative at point B.

(b) If the current is decreasing, then by Lenz's law the induced emf acts to bolster the flux—like a source of emf reinforcing the external emf. The induced emf acts to increase I in Fig. 21–36b, so in this situation you can think of the induced emf as a battery with its negative terminal at point A to attract more current (conventional, +) to move to the right.

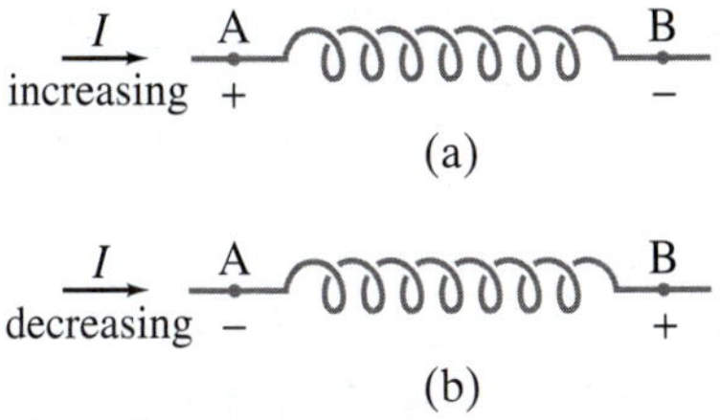

FIGURE 21–36 Example 21–12.

EXAMPLE 21-13 **Solenoid inductance.** (a) Determine a formula for the self-inductance L of a long tightly wrapped solenoid coil of length ℓ and cross-sectional area A, that contains N turns (or loops) of wire. (b) Calculate the value of L if $N = 100$, $\ell = 5.0$ cm, $A = 0.30$ cm^2, and the solenoid is air filled.

APPROACH The induced emf in a coil can be determined either from Faraday's law ($\mathcal{E} = -N \Delta\Phi_B/\Delta t$) or the self-inductance ($\mathcal{E} = -L \Delta I/\Delta t$). If we equate these two expressions, we can solve for the inductance L since we know how to calculate the flux Φ_B for a solenoid using Eq. 20–8 ($B = \mu_0 IN/\ell$).

SOLUTION (a) We equate Faraday's law (Eq. 21–2b) and Eq. 21–9 for the inductance:

$$\mathcal{E} = -N \frac{\Delta\Phi_B}{\Delta t} = -L \frac{\Delta I}{\Delta t},$$

and solve for L:

$$L = N \frac{\Delta\Phi_B}{\Delta I}.$$

We know $\Phi_B = BA$ (Eq. 21–1), and Eq. 20–8 gives us the magnetic field B for a solenoid, $B = \mu_0 NI/\ell$, so the magnetic flux inside the solenoid is

$$\Phi_B = \frac{\mu_0 NIA}{\ell}.$$

Any change in current, ΔI, causes a change in flux

$$\Delta\Phi_B = \frac{\mu_0 N \Delta I A}{\ell}.$$

We put this into our equation above for L:

$$L = N \frac{\Delta\Phi_B}{\Delta I} = \frac{\mu_0 N^2 A}{\ell}.$$

(b) Using $\mu_0 = 4\pi \times 10^{-7}$ T·m/A, and putting in values given,

$$L = \frac{(4\pi \times 10^{-7} \text{ T·m/A})(100)^2(3.0 \times 10^{-5} \text{ m}^2)}{(5.0 \times 10^{-2} \text{ m})} = 7.5 \,\mu\text{H}.$$

*21–11 Energy Stored in a Magnetic Field

In Section 17–9 we saw that the energy stored in a capacitor is equal to $\frac{1}{2}CV^2$. By using a similar argument, it can be shown that the energy U stored in an inductance L, carrying a current I, is

$$U = \text{energy} = \tfrac{1}{2}LI^2.$$

Just as the energy stored in a capacitor can be considered to reside in the electric field between its plates, so the energy in an inductor can be considered to be stored in its magnetic field.

To write the energy in terms of the magnetic field, we quote the result of Example 21–13 that the inductance of a solenoid is $L = \mu_0 N^2 A/\ell$. The magnetic field B in a solenoid is related to the current I (see Eq. 20–8) by $B = \mu_0 NI/\ell$. Thus, $I = B\ell/\mu_0 N$, and

$$U = \text{energy} = \tfrac{1}{2}LI^2 = \tfrac{1}{2}\left(\frac{\mu_0 N^2 A}{\ell}\right)\left(\frac{B\ell}{\mu_0 N}\right)^2 = \tfrac{1}{2}\frac{B^2}{\mu_0}A\ell.$$

We can think of this energy as residing in the volume enclosed by the windings, which is $A\ell$. Then the energy per unit volume, or **energy density**, is

$$u = \text{energy density} = \tfrac{1}{2}\frac{B^2}{\mu_0}. \qquad (21\text{–}10)$$

This formula, which was derived for the special case of a solenoid, can be shown to be valid for any region of space where a magnetic field exists. If a ferromagnetic material is present, μ_0 is replaced by μ. This equation is analogous to that for an electric field, $\tfrac{1}{2}\epsilon_0 E^2$, Section 17–9.

*21–12 LR Circuit

Any inductor will have some resistance. We represent an inductor by drawing the inductance L and its resistance R separately, as in Fig. 21–37. The resistance R could also include any other resistance in the circuit. Now we ask, what happens when a battery of voltage V_0 is connected in series to such an LR circuit? At the instant the switch connecting the battery is closed, the current starts to flow. It is opposed by the induced emf in the inductor because of the changing current. However, as soon as current starts to flow, there is a voltage drop across the resistance ($V = IR$). Hence, the voltage drop across the inductance is reduced and the current increases less rapidly. The current thus rises gradually, as shown in Fig. 21–38a, and approaches the steady value $I_0 = V_0/R$ when there is no more emf across the inductor (I is no longer changing) so all the voltage drop is across the resistance. The shape of the curve for I as a function of time is

$$I = \left(\frac{V_0}{R}\right)(1 - e^{-t/\tau}), \qquad [LR \text{ circuit with emf}]$$

where e is the number $e = 2.718\cdots$ (see Section 19–6) and

$$\tau = \frac{L}{R}$$

is the **time constant** of the circuit. When $t = \tau$, then $(1 - e^{-1}) = 0.63$, so τ is the time required for the current to reach $0.63 I_{\max}$.

Next, if the battery is suddenly switched out of the circuit (dashed line in Fig. 21–37), it takes time for the current to drop to zero, as shown in Fig. 21–38b. This is an exponential decay curve given by

$$I = I_0 e^{-t/\tau}. \qquad [LR \text{ circuit without emf}]$$

The time constant τ is the time for the current to decrease to $0.37 I_{\max}$ (37% of the original value), and again equals L/R.

These graphs show that there is always some "lag time" or "reaction time" when an electromagnet, for example, is turned on or off. We also see that an LR circuit has properties similar to an RC circuit (Section 19–6). Unlike the capacitor case, however, the time constant here is *inversely* proportional to R.

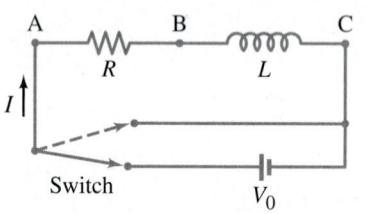

FIGURE 21–37 LR circuit.

FIGURE 21–38 (a) Growth of current in an LR circuit when connected to a battery. (b) Decay of current when the LR circuit is shorted out (battery is out of the circuit).

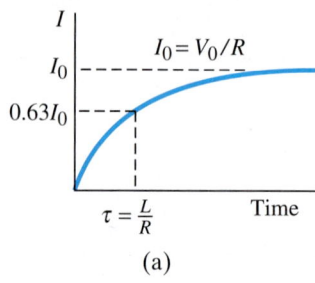

(a)

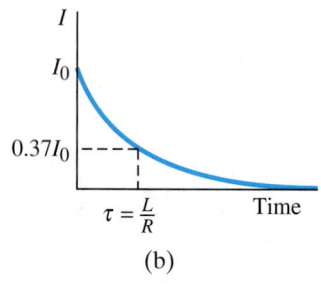

(b)

*21–13 AC Circuits and Reactance

We have previously discussed circuits that contain combinations of resistor, capacitor, and inductor, but only when they are connected to a dc source of emf or to zero voltage. Now we discuss these circuit elements when they are connected to a source of alternating voltage that produces an alternating current (ac).

First we examine, one at a time, how a resistor, a capacitor, and an inductor behave when connected to a source of alternating voltage, represented by the symbol

●—(∼)—● [alternating voltage]

which produces a sinusoidal voltage of frequency f. We assume in each case that the emf gives rise to a current

$$I = I_0 \cos 2\pi f t,$$

where t is time and I_0 is the peak current. Remember (Section 18–7) that $V_{rms} = V_0/\sqrt{2}$ and $I_{rms} = I_0/\sqrt{2}$ (Eq. 18–8).

*Resistor

When an ac source is connected to a resistor as in Fig. 21–39a, the current increases and decreases with the alternating voltage according to Ohm's law,

$$V = IR = I_0 R \cos 2\pi f t = V_0 \cos 2\pi f t$$

where $V_0 = I_0 R$ is the peak voltage. Figure 21–39b shows the voltage (red curve) and the current (blue curve) as a function of time. Because the current is zero when the voltage is zero and the current reaches a peak when the voltage does, we say that the current and voltage are **in phase**. Energy is transformed into heat (Section 18–7), at an average rate $\overline{P} = \overline{IV} = I_{rms}^2 R = V_{rms}^2/R$.

*Inductor

In Fig. 21–40a an inductor of inductance L (symbol ⎯⏣⎯) is connected to the ac source. We ignore any resistance it might have (it is usually small). The voltage applied to the inductor will be equal to the "back" emf generated in the inductor by the changing current as given by Eq. 21–9. This is because the sum of the electric potential changes around any closed circuit must add up to zero, by Kirchhoff's rule. Thus

$$V - L\frac{\Delta I}{\Delta t} = 0$$

or

$$\frac{\Delta I}{\Delta t} = \frac{V}{L}$$

where V is the sinusoidally varying voltage of the source and $L\,\Delta I/\Delta t$ is the voltage induced in the inductor. According to the last equation, I is increasing most rapidly when V has its maximum value, $V = V_0$. And I will be decreasing most rapidly when $V = -V_0$. These two instants correspond to points d and b on the graph of voltage versus time in Fig. 21–40b. By going point by point in this manner, the curve of V versus t as compared to that for I versus t can be constructed, and they are shown by the blue and red lines, respectively, in Fig. 21–40b. Notice that the current reaches its peaks (and troughs) $\frac{1}{4}$ cycle after the voltage does. We say that the

current lags the voltage by 90° for an inductor.

Because the current and voltage in an inductor are *out of phase* by 90°, the product IV (= power) is as often positive as it is negative (Fig. 21–40b). So no energy is transformed in an inductor on average; and no energy is dissipated as thermal energy.

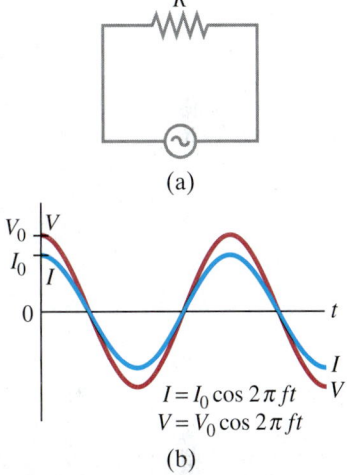

FIGURE 21–39 (a) Resistor connected to an ac source. (b) Current (blue curve) is in phase with the voltage (red) across a resistor.

FIGURE 21–40 (a) Inductor connected to an ac source. (b) Current (blue curve) lags voltage (red curve) by a quarter cycle, or 90°.

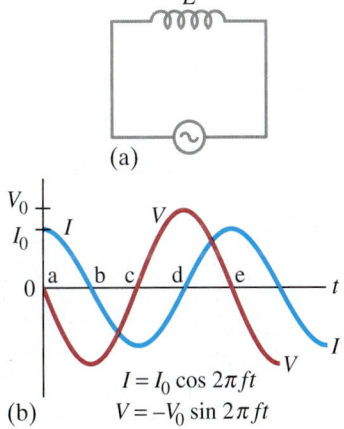

Just as a resistor impedes the flow of charge, so too an inductor impedes the flow of charge in an alternating current due to the back emf produced. For a resistor R, the current and voltage are related by $V = IR$. We can write a similar relation for an inductor:

$$V = IX_L, \quad \begin{bmatrix}\text{rms or peak values,} \\ \text{not at any instant}\end{bmatrix} \quad \text{(21–11a)}$$

where X_L is called the **inductive reactance**. X_L has units of ohms. The quantities V and I in Eq. 21–11a can refer either to rms for both, or to peak values for both (see Section 18–7). Although this equation can relate the peak values, the peak current and voltage are not reached at the same time; so Eq. 21–11a is *not valid at a particular instant*, as is the case for a resistor ($V = IR$). Careful calculation (using calculus), as well as experiment, shows that

$$X_L = \omega L = 2\pi f L, \quad \text{(21–11b)}$$

where $\omega = 2\pi f$ and f is the frequency of the ac.

For example, the inductive reactance of a 0.300-H inductor at 120 V and 60.0 Hz is $X_L = 2\pi f L = (6.28)(60.0\text{ s}^{-1})(0.300\text{ H}) = 113\text{ }\Omega$.

*Capacitor

When a capacitor is connected to a battery, the capacitor plates quickly acquire equal and opposite charges; but no steady current flows in the circuit. A capacitor prevents the flow of a dc current. But if a capacitor is connected to an alternating source of voltage, as in Fig. 21–41a, an alternating current will flow continuously. This can happen because when the ac voltage is first turned on, charge begins to flow and one plate acquires a negative charge and the other a positive charge. But when the voltage reverses itself, the charges flow in the opposite direction. Thus, for an alternating applied voltage, an ac current is present in the circuit continuously.

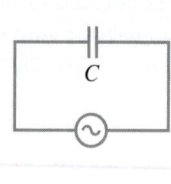
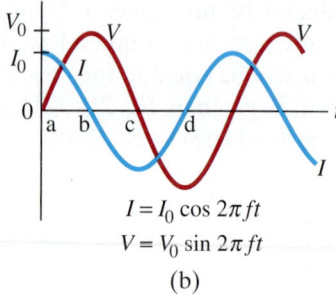

FIGURE 21–41 (a) Capacitor connected to an ac source. (b) Current leads voltage by a quarter cycle, or 90°.

The applied voltage must equal the voltage across the capacitor: $V = Q/C$, where C is the capacitance and Q the charge on the plates. Thus the charge Q on the plates is in phase with the voltage. But what about the current I? At point a in Fig. 21–41b, when the voltage is zero and starts increasing, the charge on the plates is zero. Thus charge flows readily toward the plates and the current I is large. As the voltage approaches its maximum of V_0 (point b in Fig. 21–41b), the charge that has accumulated on the plates tends to prevent more charge from flowing, so the current I drops to zero at point b. Thus the current follows the blue curve in Fig. 21–41b. Like an inductor, the voltage and current are out of phase by 90°. But for a capacitor, the current reaches its peaks $\frac{1}{4}$ cycle before the voltage does, so we say that the

current leads the voltage by 90° for a capacitor.

Because the current and voltage are out of phase, the average power dissipated is zero, just as for an inductor. Thus *only a resistance will dissipate energy* as thermal energy in an ac circuit.

A relationship between the applied voltage and the current in a capacitor can be written just as for an inductance:

$$V = IX_C, \quad \begin{bmatrix}\text{rms or peak}\\ \text{values}\end{bmatrix} \quad (21\text{–}12\text{a})$$

where X_C is called the **capacitive reactance** and has units of ohms. V and I can both be rms or both maximum (V_0 and I_0); X_C depends on both the capacitance C and the frequency f:

$$X_C = \frac{1}{\omega C} = \frac{1}{2\pi f C}, \quad (21\text{–}12\text{b})$$

where $\omega = 2\pi f$. For dc conditions, $f = 0$ and X_C becomes infinite, as it should because a capacitor does not pass dc current.

EXAMPLE 21–14 Capacitor reactance. What is the rms current in the circuit of Fig. 21–41a if $C = 1.00\,\mu\text{F}$ and $V_{\text{rms}} = 120\,\text{V}$? Calculate for (a) $f = 60.0\,\text{Hz}$, and then for (b) $f = 6.00 \times 10^5\,\text{Hz}$.

APPROACH We find the reactance using Eq. 21–12b, and solve for current in the equivalent form of Ohm's law, Eq. 21–12a.

SOLUTION (a) $X_C = 1/2\pi f C = 1/(2\pi)(60.0\,\text{s}^{-1})(1.00 \times 10^{-6}\,\text{F}) = 2.65\,\text{k}\Omega$. The rms current is (Eq. 21–12a):

$$I_{\text{rms}} = \frac{V_{\text{rms}}}{X_C} = \frac{120\,\text{V}}{2.65 \times 10^3\,\Omega} = 45.2\,\text{mA}.$$

(b) For $f = 6.00 \times 10^5\,\text{Hz}$, X_C will be $0.265\,\Omega$ and $I_{\text{rms}} = 452\,\text{A}$, vastly larger!

NOTE The dependence on f is dramatic. For high frequencies, the capacitive reactance is very small.

Two common applications of capacitors are illustrated in Figs. 21–42a and b. In Fig. 21–42a, circuit A is said to be capacitively coupled to circuit B. The purpose of the capacitor is to prevent a dc voltage from passing from A to B but allowing an ac signal to pass relatively unimpeded (if C is sufficiently large). In Fig. 21–42b, the

PHYSICS APPLIED
Capacitors as filters

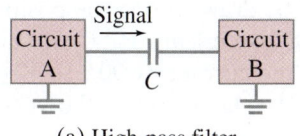
(a) High-pass filter

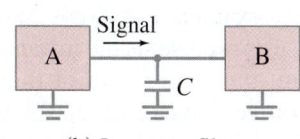
(b) Low-pass filter

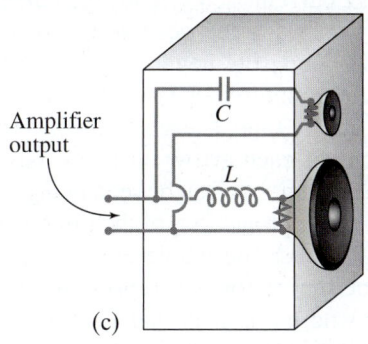

FIGURE 21–42 (a) and (b): Two common uses for a capacitor as a filter. (c) Simple loudspeaker cross-over.

capacitor also passes ac and not dc. In this case, a dc voltage can be maintained between circuits A and B, but an ac signal leaving A passes to ground instead of into B. Thus the capacitor in Fig. 21–42b acts like a **filter** when a constant dc voltage is required; any sharp variation in voltage passes to ground instead of into circuit B.

EXERCISE F The capacitor C in Fig. 21–42a is often called a "high-pass" filter, and the one in Fig. 21–42b a "low-pass" filter. Explain why.

Loudspeakers having separate "woofer" (low-frequency speaker) and "tweeter" (high-frequency speaker) may use a simple "cross-over" that consists of a capacitor in the tweeter circuit to impede low-frequency signals, and an inductor in the woofer circuit to impede high-frequency signals ($X_L = 2\pi f L$). Hence mainly low-frequency sounds reach and are emitted by the woofer. See Fig. 21–42c.

PHYSICS APPLIED
Loudspeaker cross-over

*21–14 LRC Series AC Circuit

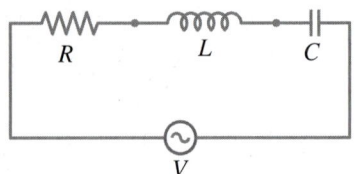

FIGURE 21–43 An *LRC* circuit.

Let us examine a circuit containing all three elements in series: a resistor R, an inductor L, and a capacitor C, Fig. 21–43. If a given circuit contains only two of these elements, we can still use the results of this Section by setting $R = 0$, $X_L = 0$, or $X_C = 0$, as needed. We let V_R, V_L, and V_C represent the voltage across each element at a *given instant* in time; and V_{R0}, V_{L0}, and V_{C0} represent the *maximum* (peak) values of these voltages. The voltage across each of the elements will follow the phase relations we discussed in the previous Section. At any instant the voltage V supplied by the source will be, by Kirchhoff's loop rule,

$$V = V_R + V_L + V_C. \qquad (21\text{–}13\text{a})$$

CAUTION
Peak voltages do not add to yield source voltage

Because the various voltages are not in phase, they do not reach their peak values at the same time, so the peak voltage of the source V_0 will *not* equal $V_{R0} + V_{L0} + V_{C0}$.

*Phasor Diagrams

The current in an *LRC* circuit at any instant is the same at all points in the circuit (charge does not pile up in the wires). Thus the currents in each element are in phase with each other, even though the voltages are not. We choose our origin in time $(t = 0)$ so that the current I at any time t is

$$I = I_0 \cos 2\pi f t.$$

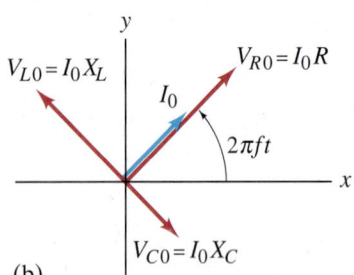

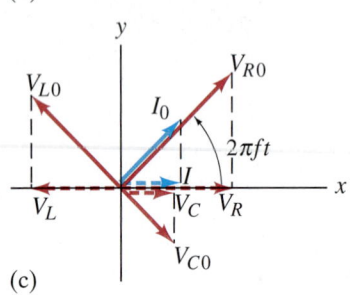

AC circuits are complicated to analyze. The easiest approach is to use a sort of vector device known as a **phasor diagram**. Arrows (treated like vectors) are drawn in an xy coordinate system to represent each voltage. The length of each arrow represents the magnitude of the peak voltage across each element:

$$V_{R0} = I_0 R, \quad V_{L0} = I_0 X_L, \quad \text{and} \quad V_{C0} = I_0 X_C. \qquad (21\text{–}13\text{b})$$

V_{R0} is in phase with the current and is initially $(t = 0)$ drawn along the positive x axis, as is the current I_0. V_{L0} leads the current by 90°, so it leads V_{R0} by 90° and is initially drawn along the positive y axis. V_{C0} lags the current by 90°, so V_{C0} is drawn initially along the negative y axis. See Fig. 21–44a. If we let the vector diagram rotate counterclockwise at frequency f, we get the diagram shown in Fig. 21–44b; after a time, t, each arrow has rotated through an angle $2\pi f t$. Then the projections of each arrow on the x axis represent the voltages across each element at the instant t, as can be seen in Fig. 21–44c. For example $I = I_0 \cos 2\pi f t$.

The sum of the projections of the three voltage vectors represents the instantaneous voltage across the whole circuit, V. Therefore, the vector sum of these vectors will be the vector that represents the peak source voltage, V_0, as shown in Fig. 21–45 where it is seen that V_0 makes an angle ϕ with I_0 and V_{R0}. As time passes, V_0 rotates with the other vectors, so the instantaneous voltage V (projection of V_0 on the x axis) is (see Fig. 21–45)

$$V = V_0 \cos(2\pi f t + \phi).$$

FIGURE 21–44 Phasor diagram for a series *LRC* circuit at (a) $t = 0$, (b) time t later. (c) Projections on the x axis which give I, V_R, V_C, V_L at time t.

FIGURE 21–45 Phasor diagram for a series *LRC* circuit showing the sum vector, V_0.

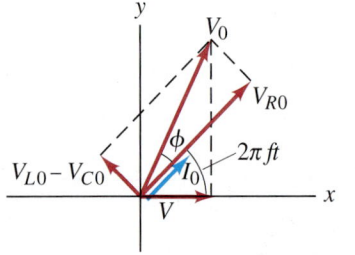

The voltage V across the whole circuit must equal the source voltage (Fig. 21–43). Thus the voltage from the source is out of phase with the current by an angle ϕ.

From this analysis we can now determine the total **impedance** Z of the circuit, which is defined in analogy to resistance and reactance as

$$V_{\text{rms}} = I_{\text{rms}} Z, \quad \text{or} \quad V_0 = I_0 Z. \qquad (21\text{–}14)$$

From Fig. 21–45 we see, using the Pythagorean theorem (V_0 is the hypotenuse of a

right triangle), that (use Eq. 21–13b)

$$V_0 = \sqrt{V_{R0}^2 + (V_{L0} - V_{C0})^2}$$
$$= I_0\sqrt{R^2 + (X_L - X_C)^2}.$$

Thus, from Eq. 21–14, the total impedance Z is

$$Z = \sqrt{R^2 + (X_L - X_C)^2}. \qquad (21\text{–}15)$$

Also from Fig. 21–45, we can find the phase angle ϕ between voltage and current:

$$\tan\phi = \frac{V_{L0} - V_{C0}}{V_{R0}} = \frac{I_0(X_L - X_C)}{I_0 R} = \frac{X_L - X_C}{R} \qquad (21\text{–}16a)$$

and

$$\cos\phi = \frac{V_{R0}}{V_0} = \frac{I_0 R}{I_0 Z} = \frac{R}{Z}. \qquad (21\text{–}16b)$$

Figure 21–45 was drawn for the case $X_L > X_C$, and the current lags the source voltage by ϕ. When the reverse is true, $X_L < X_C$, then ϕ in Eqs. 21–16 is less than zero, and the current leads the source voltage.

We saw earlier that power is dissipated only by a resistance; none is dissipated by inductance or capacitance. Therefore, the average power is given by $\overline{P} = I_{\text{rms}}^2 R$. But from Eq. 21–16b, $R = Z\cos\phi$. Therefore

$$\overline{P} = I_{\text{rms}}^2 Z\cos\phi = I_{\text{rms}} V_{\text{rms}} \cos\phi. \qquad (21\text{–}17)$$

The factor $\cos\phi$ is referred to as the **power factor** of the circuit.

EXAMPLE 21–15 *LRC circuit.* Suppose $R = 25.0\,\Omega$, $L = 30.0\,\text{mH}$, and $C = 12.0\,\mu\text{F}$ in Fig. 21–43, and they are connected in series to a 90.0-V ac (rms) 500-Hz source. Calculate (*a*) the current in the circuit, and (*b*) the voltmeter readings (rms) across each element.

APPROACH To obtain the current, we determine the impedance (Eq. 21–15 plus Eqs. 21–11b and 21–12b), and then use $I_{\text{rms}} = V_{\text{rms}}/Z$ (Eq. 21–14). Voltage drops across each element are found using Ohm's law or equivalent for each element: $V_R = IR$, $V_L = IX_L$, and $V_C = IX_C$.

SOLUTION (*a*) First, we find the reactance of the inductor and capacitor at $f = 500\,\text{Hz} = 500\,\text{s}^{-1}$:

$$X_L = 2\pi f L = 94.2\,\Omega, \qquad X_C = \frac{1}{2\pi f C} = 26.5\,\Omega.$$

Then the total impedance is

$$Z = \sqrt{R^2 + (X_L - X_C)^2} = \sqrt{(25.0\,\Omega)^2 + (94.2\,\Omega - 26.5\,\Omega)^2} = 72.2\,\Omega.$$

From the impedance version of Ohm's law, Eq. 21–14,

$$I_{\text{rms}} = \frac{V_{\text{rms}}}{Z} = \frac{90.0\,\text{V}}{72.2\,\Omega} = 1.25\,\text{A}.$$

(*b*) The rms voltage across each element is

$$(V_R)_{\text{rms}} = I_{\text{rms}} R = (1.25\,\text{A})(25.0\,\Omega) = 31.2\,\text{V}$$
$$(V_L)_{\text{rms}} = I_{\text{rms}} X_L = (1.25\,\text{A})(94.2\,\Omega) = 118\,\text{V}$$
$$(V_C)_{\text{rms}} = I_{\text{rms}} X_C = (1.25\,\text{A})(26.5\,\Omega) = 33.1\,\text{V}.$$

NOTE These voltages do *not* add up to the source voltage, 90.0 V (rms). Indeed, the rms voltage across the inductance *exceeds* the source voltage. This can happen because the different voltages are out of phase with each other: so, at any instant the capacitor's voltage might be negative which compensates for a large positive inductor voltage. The rms voltages, however, are always positive by definition. Although the rms voltages need not add up to the source voltage, the instantaneous voltages at any time must add up to the source voltage at that instant.

CAUTION
Individual peak or rms voltages do NOT add up to source voltage (due to phase differences)

*21–15 Resonance in AC Circuits

The rms current in an LRC series circuit is given by (see Eqs. 21–14, 21–15, 21–11b, and 21–12b):

$$I_{rms} = \frac{V_{rms}}{Z} = \frac{V_{rms}}{\sqrt{R^2 + \left(2\pi f L - \frac{1}{2\pi f C}\right)^2}}. \quad (21\text{–}18)$$

Because the reactance of inductors and capacitors depends on the frequency f of the source, the current in an LRC circuit depends on frequency. From Eq. 21–18 we see that the current will be maximum at a frequency that satisfies

$$2\pi f L - \frac{1}{2\pi f C} = 0.$$

We solve this for f, and call the solution f_0:

$$f_0 = \frac{1}{2\pi}\sqrt{\frac{1}{LC}}. \quad \text{[resonance]} \quad (21\text{–}19)$$

When $f = f_0$, the circuit is in **resonance**, and f_0 is the **resonant frequency** of the circuit. At this frequency, $X_C = X_L$, so the impedance is purely resistive. A graph of I_{rms} versus f is shown in Fig. 21–46 for particular values of R, L, and C. For smaller R compared to X_L and X_C, the resonance peak will be higher and sharper.

When R is very small, we speak of an **LC circuit**. The energy in an LC circuit oscillates, at frequency f_0, between the inductor and the capacitor, with some being dissipated in R (some resistance is unavoidable). This is called an **LC oscillation** or an **electromagnetic oscillation**. Not only does the charge oscillate back and forth, but so does the energy, which oscillates between being stored in the electric field of the capacitor and in the magnetic field of the inductor.

Electric resonance is used in many circuits. Radio and TV sets, for example, use resonant circuits for tuning in a station. Many frequencies reach the circuit from the antenna, but a significant current flows only for frequencies at or near the resonant frequency chosen (the station you want). Either L or C is variable so that different stations can be tuned in (more on this in Chapter 22).

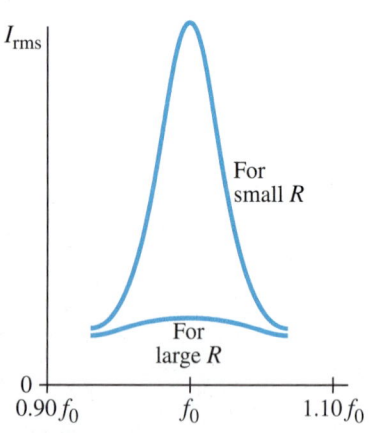

FIGURE 21–46 Current in an LRC circuit as a function of frequency, showing resonance peak at $f = f_0 = 1/(2\pi\sqrt{LC})$.

Summary

The **magnetic flux** passing through a loop is equal to the product of the area of the loop times the perpendicular component of the magnetic field:

$$\Phi_B = B_\perp A = BA\cos\theta. \quad (21\text{–}1)$$

If the magnetic flux through a coil of wire changes in time, an emf is induced in the coil. The magnitude of the induced emf equals the time rate of change of the magnetic flux through the loop times the number N of loops in the coil:

$$\mathcal{E} = -N\frac{\Delta\Phi_B}{\Delta t}. \quad (21\text{–}2b)$$

This is **Faraday's law of induction**.

The induced emf can produce a current whose magnetic field opposes the original change in flux (**Lenz's law**).

Faraday's law also tells us that a changing magnetic field produces an electric field; and that a straight wire of length ℓ moving with speed v perpendicular to a magnetic field of strength B has an emf induced between its ends equal to

$$\mathcal{E} = B\ell v. \quad (21\text{–}3)$$

An electric **generator** changes mechanical energy into electric energy. Its operation is based on Faraday's law: a coil of wire is made to rotate uniformly by mechanical means in a magnetic field, and the changing flux through the coil induces a sinusoidal current, which is the output of the generator.

A motor, which operates in the reverse of a generator, acts like a generator in that a **back emf** is induced in its rotating coil. Because this back emf opposes the input voltage, it can act to limit the current in a motor coil. Similarly, a generator acts somewhat like a motor in that a **counter torque** acts on its rotating coil.

A **transformer**, which is a device to change the magnitude of an ac voltage, consists of a primary coil and a secondary coil. The changing flux due to an ac voltage in the primary coil induces an ac voltage in the secondary coil. In a 100% efficient transformer, the ratio of output to input voltages (V_S/V_P) equals the ratio of the number of turns N_S in the secondary to the number N_P in the primary:

$$\frac{V_S}{V_P} = \frac{N_S}{N_P}. \quad (21\text{–}6)$$

The ratio of secondary to primary current is in the inverse ratio of turns:

$$\frac{I_S}{I_P} = \frac{N_P}{N_S}. \quad (21\text{–}7)$$

[*Read/write heads for computer hard drives and tape, as well as microphones, ground fault circuit interrupters, and seismographs, are all applications of electromagnetic induction.]

[*A changing current in a coil of wire will produce a changing magnetic field that induces an emf in a second coil placed nearby. The **mutual inductance**, M, is defined by

$$\mathcal{E}_2 = -M \frac{\Delta I_1}{\Delta t}. \quad (21\text{–}8)]$$

[*Within a single coil, the changing B due to a changing current induces an opposing emf, $\mathcal{E}$, so a coil has a **self-inductance** L defined by

$$\mathcal{E} = -L \frac{\Delta I}{\Delta t}. \quad (21\text{–}9)]$$

[*The energy stored in an inductance L carrying current I is given by $U = \frac{1}{2} L I^2$. This energy can be thought of as being stored in the magnetic field of the inductor. The energy density u in any magnetic field B is given by

$$u = \frac{1}{2} \frac{B^2}{\mu_0}. \quad (21\text{–}10)]$$

[*When an inductance L and resistor R are connected in series to a source of emf, V_0, the current rises as

$$I = \frac{V_0}{R}(1 - e^{-t/\tau}),$$

where $\tau = L/R$ is the **time constant**. If the battery is suddenly switched out of the LR circuit, the current drops exponentially, $I = I_{\max} e^{-t/\tau}$.]

[*Inductive and capacitive **reactance**, X, defined as for resistors, is the proportionality constant between voltage and current (either the rms or peak values). Across an inductor,

$$V = I X_L, \quad (21\text{–}11\text{a})$$

and across a capacitor,

$$V = I X_C. \quad (21\text{–}12\text{a})$$

The reactance of an inductor increases with frequency f,

$$X_L = 2\pi f L, \quad (21\text{–}11\text{b})$$

whereas the reactance of a capacitor decreases with frequency f,

$$X_C = \frac{1}{2\pi f C}. \quad (21\text{–}12\text{b})$$

The current through a resistor is always in phase with the voltage across it, but in an inductor the current lags the voltage by 90°, and in a capacitor the current leads the voltage by 90°.]

[*In an LRC series circuit, the total **impedance** Z is defined by the equivalent of $V = IR$ for resistance, namely,

$$V_0 = I_0 Z \quad \text{or} \quad V_{\text{rms}} = I_{\text{rms}} Z; \quad (21\text{–}14)$$

Z is given by

$$Z = \sqrt{R^2 + (X_L - X_C)^2}. \quad (21\text{–}15)]$$

[*An LRC series circuit **resonates** at a frequency given by

$$f_0 = \frac{1}{2\pi} \sqrt{\frac{1}{LC}}. \quad (21\text{–}19)$$

The rms current in the circuit is largest when the applied voltage has a frequency equal to f_0.]

Questions

1. What would be the advantage, in Faraday's experiments (Fig. 21–1), of using coils with many turns?
2. What is the difference between magnetic flux and magnetic field?
3. Suppose you are holding a circular ring of wire in front of you and (a) suddenly thrust a magnet, south pole first, away from you toward the center of the circle. Is a current induced in the wire? (b) Is a current induced when the magnet is held steady within the ring? (c) Is a current induced when you withdraw the magnet? For each yes answer, specify the direction. Explain your answers.
4. (a) A wire loop is pulled away from a current-carrying wire (Fig. 21–47). What is the direction of the induced current in the loop: clockwise or counterclockwise?
   (b) What if the wire loop stays fixed as the current I decreases? Explain your answers. 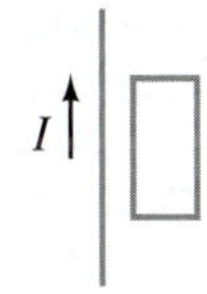
   FIGURE 21–47 Question 4.
5. (a) If the north pole of a thin flat magnet moves on a table toward a loop also on the table (Fig. 21–48), in what direction is the induced current in the loop? Assume the magnet is the same thickness as the wire. (b) What if the magnet is four times thicker than the wire loop? Explain your answers. 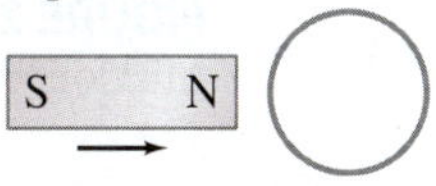
   FIGURE 21–48 Question 5.
6. Suppose you are looking along a line through the centers of two circular (but separate) wire loops, one behind the other. A battery is suddenly connected to the front loop, establishing a clockwise current. (a) Will a current be induced in the second loop? (b) If so, when does this current start? (c) When does it stop? (d) In what direction is this current? (e) Is there a force between the two loops? (f) If so, in what direction?
7. The battery mentioned in Question 6 is disconnected. Will a current be induced in the second loop? If so, when does it start and stop? In what direction is this current?
8. In Fig. 21–49, determine the direction of the induced current in resistor R_A (a) when coil B is moved toward coil A, (b) when coil B is moved away from A, (c) when the resistance R_B is increased but the coils remain fixed. Explain your answers.

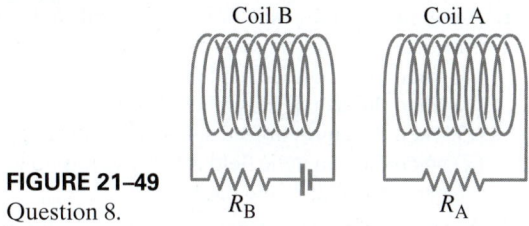

FIGURE 21–49 Question 8.

9. In situations where a small signal must travel over a distance, a *shielded cable* is used in which the signal wire is surrounded by an insulator and then enclosed by a cylindrical conductor (shield) carrying the return current. Why is a "shield" necessary?
10. What is the advantage of placing the two insulated electric wires carrying ac close together or even twisted about each other?
11. Explain why, exactly, the lights may dim briefly when a refrigerator motor starts up. When an electric heater is turned on, the lights may stay dimmed as long as the heater is on. Explain the difference.
12. Use Figs. 21–14 and 21–17 plus the right-hand rules to show why the counter torque in a generator *opposes* the motion.
13. Will an eddy current brake (Fig. 21–20) work on a copper or aluminum wheel, or must the wheel be ferromagnetic? Explain.

14. A bar magnet falling inside a vertical metal tube reaches a terminal velocity even if the tube is evacuated so that there is no air resistance. Explain.
15. It has been proposed that eddy currents be used to help sort solid waste for recycling. The waste is first ground into tiny pieces and iron removed with a magnet. The waste then is allowed to slide down an incline over permanent magnets. How will this aid in the separation of nonferrous metals (Al, Cu, Pb, brass) from nonmetallic materials?
16. The pivoted metal bar with slots in Fig. 21–50 falls much more quickly through a magnetic field than does a solid bar. Explain.

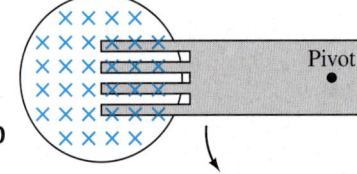

FIGURE 21–50 Question 16.

17. If an aluminum sheet is held between the poles of a large bar magnet, it requires some force to pull it out of the magnetic field even though the sheet is not ferromagnetic and does not touch the pole faces. Explain.
18. A bar magnet is held above the floor and dropped (Fig. 21–51). In case (a), the magnet falls through a wire loop. In case (b), there is nothing between the magnet and the floor. How will the speeds of the magnets compare? Explain.

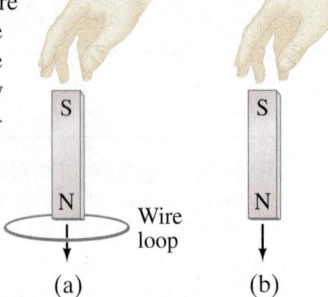

FIGURE 21–51 Question 18 and MisConceptual Question 5.

19. A metal bar, pivoted at one end, oscillates freely in the absence of a magnetic field; but in a magnetic field, its oscillations are quickly damped out. Explain. (This *magnetic damping* is used in a number of practical devices.)
20. An enclosed transformer has four wire leads coming from it. How could you determine the ratio of turns on the two coils without taking the transformer apart? How would you know which wires paired with which?
21. The use of higher-voltage lines in homes—say, 600 V or 1200 V—would reduce energy waste. Why are they not used?
22. A transformer designed for a 120-V ac input will often "burn out" if connected to a 120-V dc source. Explain. [*Hint*: The resistance of the primary coil is usually very low.]
*23. How would you arrange two flat circular coils so that their mutual inductance was (a) greatest, (b) least (without separating them by a great distance)? Explain.
*24. Does the emf of the battery in Fig. 21–37 affect the time needed for the *LR* circuit to reach (a) a given fraction of its maximum possible current, (b) a given value of current? Explain.
*25. In an *LRC* circuit, can the rms voltage across (a) an inductor, (b) a capacitor, be greater than the rms voltage of the ac source? Explain.
*26. Describe briefly how the frequency of the source emf affects the impedance of (a) a pure resistance, (b) a pure capacitance, (c) a pure inductance, (d) an *LRC* circuit near resonance (R small), (e) an *LRC* circuit far from resonance (R small).
*27. Describe how to make the impedance in an *LRC* circuit a minimum.
*28. An *LRC* resonant circuit is often called an *oscillator* circuit. What is it that oscillates?
*29. Is the ac current in the inductor always the same as the current in the resistor of an *LRC* circuit? Explain.

MisConceptual Questions

1. A coil rests in the plane of the page while a magnetic field is directed into the page. A clockwise current is induced
   (a) when the magnetic field gets stronger.
   (b) when the size of the coil decreases.
   (c) when the coil is moved sideways across the page.
   (d) when the magnetic field is tilted so it is no longer perpendicular to the page.
2. A wire loop moves at constant velocity without rotation through a constant magnetic field. The induced current in the loop will be
   (a) clockwise.   (b) counterclockwise.   (c) zero.
   (d) We need to know the orientation of the loop relative to the magnetic field.
3. A square loop moves to the right from an area where $\vec{B} = 0$, completely through a region containing a uniform magnetic field directed into the page (Fig. 21–52), and then out to $B = 0$ after point L. A current is induced in the loop
   (a) only as it passes line J.   (d) as it passes line J or line L.
   (b) only as it passes line K.   (e) as it passes all three lines.
   (c) only as it passes line L.

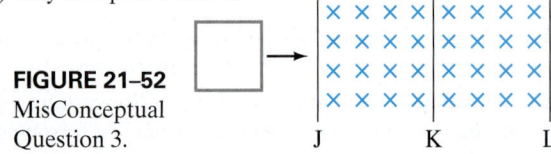

FIGURE 21–52 MisConceptual Question 3.

4. Two loops of wire are moving in the vicinity of a very long straight wire carrying a steady current (Fig. 21–53). Find the direction of the induced current in each loop.

   For C: For D:
   (a) clockwise.   (a) clockwise.
   (b) counterclockwise. (b) counterclockwise.
   (c) zero.   (c) zero.
   (d) alternating (ac). (d) alternating (ac).

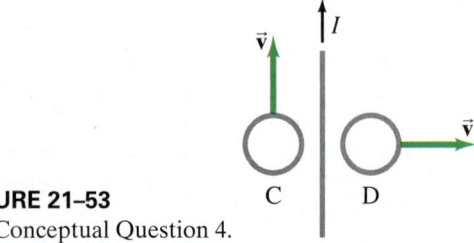

FIGURE 21–53 MisConceptual Question 4.

5. If there is induced current in Question 18 (see Fig. 21–51), wouldn't that cost energy? Where would that energy come from in case (a)?
   (a) Induced current doesn't need energy.
   (b) Energy conservation is violated.
   (c) There is less kinetic energy.
   (d) There is more gravitational potential energy.

6. A nonconducting plastic hoop is held in a magnetic field that points out of the page (Fig. 21–54). As the strength of the field increases,
   (a) an induced emf will be produced that causes a clockwise current.
   (b) an induced emf will be produced that causes a counterclockwise current.
   (c) an induced emf will be produced but no current.
   (d) no induced emf will be produced.

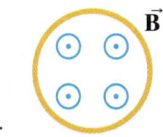

FIGURE 21–54 MisConceptual Question 6.

7. A long straight wire carries a current I as shown in Fig. 21–55. A small loop of wire rests in the plane of the page. Which of the following will *not* induce a current in the loop?
   (a) Increasing the current in the straight wire.
   (b) Moving the loop in a direction parallel to the wire.
   (c) Rotating the loop so that it becomes perpendicular to the plane of the page.
   (d) Moving the loop farther from the wire without rotating it.
   (e) Moving the loop farther from the wire while rotating it.

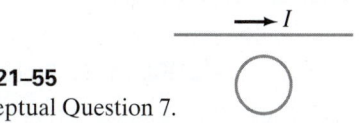

FIGURE 21–55 MisConceptual Question 7.

8. Two separate but nearby coils are mounted along the same axis. A power supply controls the flow of current in the first coil, and thus the magnetic field it produces. The second coil is connected only to an ammeter. The ammeter will indicate that a current is flowing in the second coil
   (a) whenever a current flows in the first coil.
   (b) only when a steady current flows in the first coil.
   (c) only when the current in the first coil changes.
   (d) only if the second coil is connected to the power supply by rewiring it to be in series with the first coil.

9. When a generator is used to produce electric current, the resulting electric energy originates from which source?
   (a) The generator's magnetic field.
   (b) Whatever rotates the generator's axle.
   (c) The resistance of the generator's coil.
   (d) Back emf.
   (e) Empty space.

10. Which of the following will *not* increase a generator's voltage output?
    (a) Rotating the generator faster.
    (b) Increasing the area of the coil.
    (c) Rotating the magnetic field so that it is more closely parallel to the generator's rotation axis.
    (d) Increasing the magnetic field through the coil.
    (e) Increasing the number of turns in the coil.

11. Which of the following can a transformer accomplish?
    (a) Changing voltage but not current.
    (b) Changing current but not voltage.
    (c) Changing power.
    (d) Changing both current and voltage.

12. A laptop computer's charger unit converts 120 V from a wall power outlet to the lower voltage required by the laptop. Inside the charger's plastic case is a diode or rectifier (discussed in Chapter 29) that changes ac to dc plus a
    (a) battery.
    (b) motor.
    (c) generator.
    (d) transformer.
    (e) transmission line.

13. Which of the following statements about transformers is false?
    (a) Transformers work using ac current or dc current.
    (b) If the current in the secondary is higher, the voltage is lower.
    (c) If the voltage in the secondary is higher, the current is lower.
    (d) If no flux is lost, the product of the voltage and the current is the same in the primary and secondary coils.

14. A 10-V, 1.0-A dc current is run through a step-up transformer that has 10 turns on the input side and 20 turns on the output side. What is the output?
    (a) 10 V, 0.5 A.
    (b) 20 V, 0.5 A.
    (c) 20 V, 1 A.
    (d) 10 V, 1 A.
    (e) 0 V, 0 A.

15. The alternating electric current at a wall outlet is most commonly produced by
    (a) a connection to rechargeable batteries.
    (b) a rotating coil that is immersed in a magnetic field.
    (c) accelerating electrons between oppositely charged capacitor plates.
    (d) using an electric motor.
    (e) alternately heating and cooling a wire.

*16. When you swipe a credit card, the machine sometimes fails to read the card. What can you do differently?
    (a) Swipe the card more slowly so that the reader has more time to read the magnetic stripe.
    (b) Swipe the card more quickly so that the induced emf is higher.
    (c) Swipe the card more quickly so that the induced currents are reduced.
    (d) Swipe the card more slowly so that the magnetic fields don't change so fast.

*17. Which of the following is true about all series ac circuits?
    (a) The voltage across any circuit element is a maximum when the current is a maximum in that circuit element.
    (b) The current at any point in the circuit is always the same as the current at any other point in the circuit.
    (c) The current in the circuit is a maximum when the source ac voltage is a maximum.
    (d) Resistors, capacitors, and inductors can all change the phase of the current.

For assigned homework and other learning materials, go to the MasteringPhysics website.

Problems

21–1 to 21–4 Faraday's Law of Induction

1. (I) The magnetic flux through a coil of wire containing two loops changes at a constant rate from -58 Wb to $+38$ Wb in 0.34 s. What is the emf induced in the coil?

2. (I) The north pole of the magnet in Fig. 21–57 is being inserted into the coil. In which direction is the induced current flowing through resistor R? Explain.

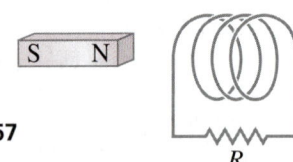

FIGURE 21–57 Problem 2.

3. (I) The rectangular loop in Fig. 21–58 is being pushed to the right, where the magnetic field points inward. In what direction is the induced current? Explain your reasoning.

FIGURE 21–58 Problem 3.

4. (I) If the solenoid in Fig. 21–59 is being pulled away from the loop shown, in what direction is the induced current in the loop? Explain.

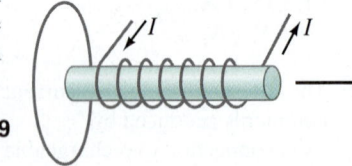

FIGURE 21–59 Problem 4.

5. (II) An 18.5-cm-diameter loop of wire is initially oriented perpendicular to a 1.5-T magnetic field. The loop is rotated so that its plane is parallel to the field direction in 0.20 s. What is the average induced emf in the loop?

6. (II) A fixed 10.8-cm-diameter wire coil is perpendicular to a magnetic field 0.48 T pointing up. In 0.16 s, the field is changed to 0.25 T pointing down. What is the average induced emf in the coil?

7. (II) A 16-cm-diameter circular loop of wire is placed in a 0.50-T magnetic field. (a) When the plane of the loop is perpendicular to the field lines, what is the magnetic flux through the loop? (b) The plane of the loop is rotated until it makes a 42° angle with the field lines. What is the angle θ in Eq. 21–1 for this situation? (c) What is the magnetic flux through the loop at this angle?

8. (II) (a) If the resistance of the resistor in Fig. 21–60 is slowly increased, what is the direction of the current induced in the small circular loop inside the larger loop? (b) What would it be if the small loop were placed outside the larger one, to the left? Explain your answers.

FIGURE 21–60 Problem 8.

9. (II) The moving rod in Fig. 21–11 is 12.0 cm long and is pulled at a speed of 18.0 cm/s. If the magnetic field is 0.800 T, calculate (a) the emf developed, and (b) the electric field felt by electrons in the rod.

10. (II) A circular loop in the plane of the paper lies in a 0.65-T magnetic field pointing into the paper. The loop's diameter changes from 20.0 cm to 6.0 cm in 0.50 s. What is (a) the direction of the induced current, (b) the magnitude of the average induced emf, and (c) the average induced current if the coil resistance is 2.5 Ω?

11. (II) What is the direction of the induced current in the circular loop due to the current shown in each part of Fig. 21–61? Explain why.

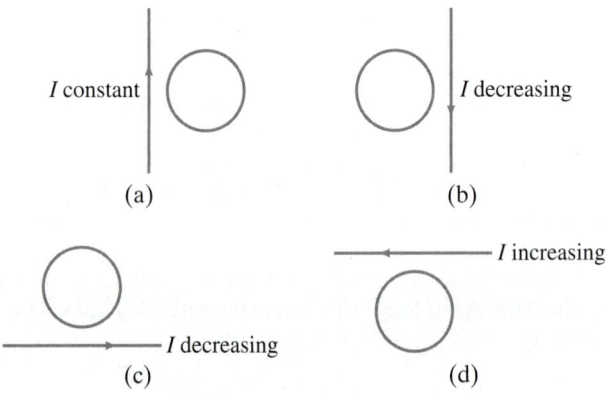

FIGURE 21–61 Problem 11.

12. (II) A 600-turn solenoid, 25 cm long, has a diameter of 2.5 cm. A 14-turn coil is wound tightly around the center of the solenoid. If the current in the solenoid increases uniformly from 0 to 5.0 A in 0.60 s, what will be the induced emf in the short coil during this time?

13. (II) When a car drives through the Earth's magnetic field, an emf is induced in its vertical 55-cm-long radio antenna. If the Earth's field $(5.0 \times 10^{-5}\,\text{T})$ points north with a dip angle of 38°, what is the maximum emf induced in the antenna and which direction(s) will the car be moving to produce this maximum value? The car's speed is 30.0 m/s on a horizontal road.

14. (II) Part of a single rectangular loop of wire with dimensions shown in Fig. 21–62 is situated inside a region of uniform magnetic field of 0.550 T. The total resistance of the loop is 0.230 Ω. Calculate the force required to pull the loop from the field (to the right) at a constant velocity of 3.10 m/s. Neglect gravity.

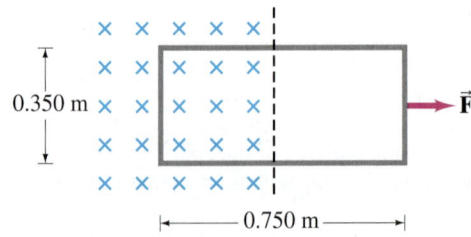

FIGURE 21–62 Problem 14.

15. (II) In order to make the rod of Fig. 21–11a move to the right at speed v, you need to apply an external force on the rod to the right. (a) Explain and determine the magnitude of the required force. (b) What external power is needed to move the rod? (Do not confuse this external force on the rod with the upward force on the electrons shown in Fig. 21–11b.)

620 CHAPTER 21

16. (II) In Fig. 21–11, the moving rod has a resistance of 0.25 Ω and moves on rails 20.0 cm apart. The stationary U-shaped conductor has negligible resistance. When a force of 0.350 N is applied to the rod, it moves to the right at a constant speed of 1.50 m/s. What is the magnetic field?

17. (III) In Fig. 21–11, the rod moves with a speed of 1.6 m/s on rails 30.0 cm apart. The rod has a resistance of 2.5 Ω. The magnetic field is 0.35 T, and the resistance of the U-shaped conductor is 21.0 Ω at a given instant. Calculate (a) the induced emf, (b) the current in the U-shaped conductor, and (c) the external force needed to keep the rod's velocity constant at that instant.

18. (III) A 22.0-cm-diameter coil consists of 30 turns of circular copper wire 2.6 mm in diameter. A uniform magnetic field, perpendicular to the plane of the coil, changes at a rate of 8.65×10^{-3} T/s. Determine (a) the current in the loop, and (b) the rate at which thermal energy is produced.

19. (III) The magnetic field perpendicular to a single 13.2-cm-diameter circular loop of copper wire decreases uniformly from 0.670 T to zero. If the wire is 2.25 mm in diameter, how much charge moves past a point in the coil during this operation?

21–5 Generators

20. (II) The generator of a car idling at 1100 rpm produces 12.7 V. What will the output be at a rotation speed of 2500 rpm, assuming nothing else changes?

21. (II) A 550-loop circular armature coil with a diameter of 8.0 cm rotates at 120 rev/s in a uniform magnetic field of strength 0.55 T. (a) What is the rms voltage output of the generator? (b) What would you do to the rotation frequency in order to double the rms voltage output?

22. (II) A generator rotates at 85 Hz in a magnetic field of 0.030 T. It has 950 turns and produces an rms voltage of 150 V and an rms current of 70.0 A. (a) What is the peak current produced? (b) What is the area of each turn of the coil?

23. (III) A simple generator has a square armature 6.0 cm on a side. The armature has 85 turns of 0.59-mm-diameter copper wire and rotates in a 0.65-T magnetic field. The generator is used to power a lightbulb rated at 12.0 V and 25.0 W. At what rate should the generator rotate to provide 12.0 V to the bulb? Consider the resistance of the wire on the armature.

21–6 Back EMF and Torque

24. (I) A motor has an armature resistance of 3.65 Ω. If it draws 8.20 A when running at full speed and connected to a 120-V line, how large is the back emf?

25. (I) The back emf in a motor is 72 V when operating at 1800 rpm. What would be the back emf at 2300 rpm if the magnetic field is unchanged?

26. (II) What will be the current in the motor of Example 21–8 if the load causes it to run at half speed?

21–7 Transformers

[Assume 100% efficiency, unless stated otherwise.]

27. (I) A transformer is designed to change 117 V into 13,500 V, and there are 148 turns in the primary coil. How many turns are in the secondary coil?

28. (I) A transformer has 360 turns in the primary coil and 120 in the secondary coil. What kind of transformer is this, and by what factor does it change the voltage? By what factor does it change the current?

29. (I) A step-up transformer increases 25 V to 120 V. What is the current in the secondary coil as compared to the primary coil?

30. (I) Neon signs require 12 kV for their operation. To operate from a 240-V line, what must be the ratio of secondary to primary turns of the transformer? What would the voltage output be if the transformer were connected in reverse?

31. (II) A model-train transformer plugs into 120-V ac and draws 0.35 A while supplying 6.8 A to the train. (a) What voltage is present across the tracks? (b) Is the transformer step-up or step-down?

32. (II) The output voltage of a 95-W transformer is 12 V, and the input current is 25 A. (a) Is this a step-up or a step-down transformer? (b) By what factor is the voltage multiplied?

33. (II) A transformer has 330 primary turns and 1240 secondary turns. The input voltage is 120 V and the output current is 15.0 A. What are the output voltage and input current?

34. (II) If 35 MW of power at 45 kV (rms) arrives at a town from a generator via 4.6-Ω transmission lines, calculate (a) the emf at the generator end of the lines, and (b) the fraction of the power generated that is wasted in the lines.

35. (II) For the transmission of electric power from power plant to home, as depicted in Fig. 21–25, where the electric power sent by the plant is 100 kW, about how far away could the house be from the power plant before power loss is 50%? Assume the wires have a resistance per unit length of 5×10^{-5} Ω/m.

36. (II) For the electric power transmission system shown in Fig. 21–25, what is the ratio N_S/N_P for (a) the step-up transformer, (b) the step-down transformer next to the home?

37. (III) Suppose 2.0 MW is to arrive at a large shopping mall over two 0.100-Ω lines. Estimate how much power is saved if the voltage is stepped up from 120 V to 1200 V and then down again, rather than simply transmitting at 120 V. Assume the transformers are each 99% efficient.

38. (III) Design a dc transmission line that can transmit 925 MW of electricity 185 km with only a 2.5% loss. The wires are to be made of aluminum and the voltage is 660 kV.

*21–10 Inductance

*39. (I) If the current in a 160-mH coil changes steadily from 25.0 A to 10.0 A in 350 ms, what is the magnitude of the induced emf?

*40. (I) What is the inductance of a coil if the coil produces an emf of 2.50 V when the current in it changes from −28.0 mA to +31.0 mA in 14.0 ms?

*41. (I) Determine the inductance L of a 0.60-m-long air-filled solenoid 2.9 cm in diameter containing 8500 loops.

*42. (I) How many turns of wire would be required to make a 130-mH inductor out of a 30.0-cm-long air-filled solenoid with a diameter of 5.8 cm?

*43. (II) An air-filled cylindrical inductor has 2600 turns, and it is 2.5 cm in diameter and 28.2 cm long. (a) What is its inductance? (b) How many turns would you need to generate the same inductance if the core were iron-filled instead? Assume the magnetic permeability of iron is about 1200 times that of free space.

*44. (II) A coil has 2.25-Ω resistance and 112-mH inductance. If the current is 3.00 A and is increasing at a rate of 3.80 A/s, what is the potential difference across the coil at this moment?

*45. (III) A physics professor wants to demonstrate the large size of the henry unit. On the outside of a 12-cm-diameter plastic hollow tube, she wants to wind an air-filled solenoid with self-inductance of 1.0 H using copper wire with a 0.81-mm diameter. The solenoid is to be tightly wound with each turn touching its neighbor (the wire has a thin insulating layer on its surface so the neighboring turns are not in electrical contact). How long will the plastic tube need to be and how many kilometers of copper wire will be required? What will be the resistance of this solenoid?

*46. (III) A long thin solenoid of length ℓ and cross-sectional area A contains N_1 closely packed turns of wire. Wrapped tightly around it is an insulated coil of N_2 turns, Fig. 21–63. Assume all the flux from coil 1 (the solenoid) passes through coil 2, and calculate the mutual inductance.

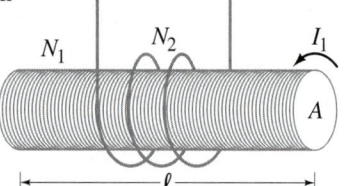

FIGURE 21–63 Problem 46.

*21–11 Magnetic Energy Storage

*47. (I) The magnetic field inside an air-filled solenoid 36 cm long and 2.0 cm in diameter is 0.72 T. Approximately how much energy is stored in this field?

*48. (II) At $t = 0$, the current through a 45.0-mH inductor is 50.0 mA and is increasing at the rate of 115 mA/s. What is the initial energy stored in the inductor, and how long does it take for the energy to increase by a factor of 5.0 from the initial value?

*49. (II) Assuming the Earth's magnetic field averages about 0.50×10^{-4} T near Earth's surface, estimate the total energy stored in this field in the first 10 km above Earth's surface.

*21–12 LR Circuit

*50. (II) It takes 2.56 ms for the current in an LR circuit to increase from zero to 0.75 its maximum value. Determine (a) the time constant of the circuit, (b) the resistance of the circuit if $L = 31.0$ mH.

*51. (II) How many time constants does it take for the potential difference across the resistor in an LR circuit like that in Fig. 21–37 to drop to 2.5% of its original value, after the switch is moved to the upper position, removing V_0 from the circuit?

*52. (III) Determine $\Delta I/\Delta t$ at $t = 0$ (when the battery is connected) for the LR circuit of Fig. 21–37 and show that if I continued to increase at this rate, it would reach its maximum value in one time constant.

*53. (III) After how many time constants does the current in Fig. 21–37 reach within (a) 10%, (b) 1.0%, and (c) 0.1% of its maximum value?

*21–13 AC Circuits and Reactance

*54. (I) What is the reactance of a 6.20-μF capacitor at a frequency of (a) 60.0 Hz, (b) 1.00 MHz?

*55. (I) At what frequency will a 32.0-mH inductor have a reactance of 660 Ω?

*56. (I) At what frequency will a 2.40-μF capacitor have a reactance of 6.10 kΩ?

*57. (II) Calculate the reactance of, and rms current in, a 260-mH radio coil connected to a 240-V (rms) 10.0-kHz ac line. Ignore resistance.

*58. (II) An inductance coil operates at 240 V and 60.0 Hz. It draws 12.2 A. What is the coil's inductance?

*59. (II) (a) What is the reactance of a well-insulated 0.030-μF capacitor connected to a 2.0-kV (rms) 720-Hz line? (b) What will be the peak value of the current?

*21–14 LRC Circuits

*60. (II) For a 120-V rms 60-Hz voltage, an rms current of 70 mA passing through the human body for 1.0 s could be lethal. What must be the impedance of the body for this to occur?

*61. (II) A 36-kΩ resistor is in series with a 55-mH inductor and an ac source. Calculate the impedance of the circuit if the source frequency is (a) 50 Hz, and (b) 3.0×10^4 Hz.

*62. (II) A 3.5-kΩ resistor and a 3.0-μF capacitor are connected in series to an ac source. Calculate the impedance of the circuit if the source frequency is (a) 60 Hz, and (b) 60,000 Hz.

*63. (II) Determine the resistance of a coil if its impedance is 235 Ω and its reactance is 115 Ω.

*64. (II) Determine the total impedance, phase angle, and rms current in an LRC circuit connected to a 10.0-kHz, 725-V (rms) source if $L = 28.0$ mH, $R = 8.70$ kΩ, and $C = 6250$ pF.

*65. (II) An ac voltage source is connected in series with a 1.0-μF capacitor and a 650-Ω resistor. Using a digital ac voltmeter, the amplitude of the voltage source is measured to be 4.0 V rms, while the voltages across the resistor and across the capacitor are found to be 3.0 V rms and 2.7 V rms, respectively. Determine the frequency of the ac voltage source. Why is the voltage measured across the voltage source not equal to the sum of the voltages measured across the resistor and across the capacitor?

*66. (III) (a) What is the rms current in an LR circuit when a 60.0-Hz 120-V rms ac voltage is applied, where $R = 2.80$ kΩ and $L = 350$ mH? (b) What is the phase angle between voltage and current? (c) How much power is dissipated? (d) What are the rms voltage readings across R and L?

*67. (III) (a) What is the rms current in an RC circuit if $R = 6.60$ kΩ, $C = 1.80$ μF, and the rms applied voltage is 120 V at 60.0 Hz? (b) What is the phase angle between voltage and current? (c) What are the voltmeter readings across R and C?

*68. (III) Suppose circuit B in Fig. 21–42a consists of a resistance $R = 520$ Ω. The filter capacitor has capacitance $C = 1.2$ μF. Will this capacitor act to eliminate 60-Hz ac but pass a high-frequency signal of frequency 6.0 kHz? To check this, determine the voltage drop across R for a 130-mV signal of frequency (a) 60 Hz; (b) 6.0 kHz.

*21–15 Resonance in AC Circuits

*69. (I) A 3500-pF capacitor is connected in series to a 55.0-μH coil of resistance 4.00 Ω. What is the resonant frequency of this circuit?

*70. (II) The variable capacitor in the tuner of an AM radio has a capacitance of 2800 pF when the radio is tuned to a station at 580 kHz. (a) What must be the capacitance for a station at 1600 kHz? (b) What is the inductance (assumed constant)?

*71. (II) An LRC circuit has $L = 14.8$ mH and $R = 4.10$ Ω. (a) What value must C have to produce resonance at 3600 Hz? (b) What will be the maximum current at resonance if the peak external voltage is 150 V?

*72. (III) A resonant circuit using a 260-nF capacitor is to resonate at 18.0 kHz. The air-core inductor is to be a solenoid with closely packed coils made from 12.0 m of insulated wire 1.1 mm in diameter. How many loops will the inductor contain?

*73. (III) A 2200-pF capacitor is charged to 120 V and then quickly connected to an inductor. The frequency of oscillation is observed to be 19 kHz. Determine (a) the inductance, (b) the peak value of the current, and (c) the maximum energy stored in the magnetic field of the inductor.

General Problems

74. Suppose you are looking at two wire loops in the plane of the page as shown in Fig. 21–64. When switch S is closed in the left-hand coil, (a) what is the direction of the induced current in the other loop? (b) What is the situation after a "long" time? (c) What is the direction of the induced current in the right-hand loop if that loop is quickly pulled horizontally to the right? (d) Suppose the right-hand loop also has a switch like the left-hand loop. The switch in the left-hand loop has been closed a long time when the switch in the right-hand loop is closed. What happens in this case? Explain each answer.

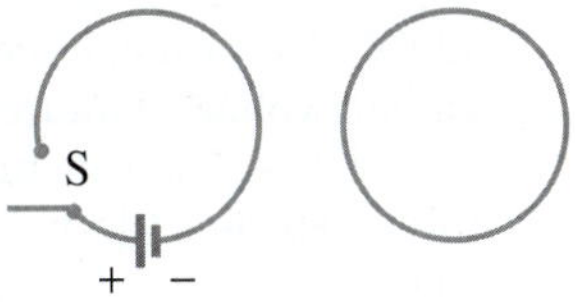

FIGURE 21–64
Problem 74.

75. A square loop 24.0 cm on a side has a resistance of 6.10 Ω. It is initially in a 0.665-T magnetic field, with its plane perpendicular to $\vec{B}$, but is removed from the field in 40.0 ms. Calculate the electric energy dissipated in this process.

76. A high-intensity desk lamp is rated at 45 W but requires only 12 V. It contains a transformer that converts 120-V household voltage. (a) Is the transformer step-up or step-down? (b) What is the current in the secondary coil when the lamp is on? (c) What is the current in the primary coil? (d) What is the resistance of the bulb when on?

77. A flashlight can be made that is powered by the induced current from a magnet moving through a coil of wire. The coil and magnet are inside a plastic tube that can be shaken causing the magnet to move back and forth through the coil. Assume the magnet has a maximum field strength of 0.05 T. Make reasonable assumptions and specify the size of the coil and the number of turns necessary to light a standard 1-watt, 3-V flashlight bulb.

78. Conceptual Example 21–9 states that an overloaded motor may burn out due to high currents. Suppose you have a blender with an internal resistance of 3.0 Ω. (a) At 120 V, what is the initial current through the blender? (b) The blender is rated at 2.0 A for continuous use. What is the back emf of the blender? (c) At what rate is heat dissipated in the blender during normal use? (d) If the blender jams and stops turning, at what rate is heat dissipated in the motor coils?

79. Power is generated at 24 kV at a generating plant located 56 km from a town that requires 55 MW of power at 12 kV. Two transmission lines from the plant to the town each have a resistance of 0.10 Ω/km. What should the output voltage of the transformer at the generating plant be for an overall transmission efficiency of 98.5%, assuming a perfect transformer?

80. The primary windings of a transformer which has an 88% efficiency are connected to 110-V ac. The secondary windings are connected across a 2.4-Ω, 75-W lightbulb. (a) Calculate the current through the primary windings of the transformer. (b) Calculate the ratio of the number of primary windings of the transformer to the number of secondary windings of the transformer.

81. A pair of power transmission lines each have a 0.95-Ω resistance and carry 740 A over 9.0 km. If the rms input voltage is 42 kV, calculate (a) the voltage at the other end, (b) the power input, (c) power loss in the lines, and (d) the power output.

82. Two resistanceless rails rest 32 cm apart on a 6.0° ramp. They are joined at the bottom by a 0.60-Ω resistor. At the top a copper bar of mass 0.040 kg (ignore its resistance) is laid across the rails. Assuming a vertical 0.45-T magnetic field, what is the terminal (steady) velocity of the bar as it slides frictionlessly down the rails?

83. Show that the power loss in transmission lines, P_L, is given by $P_L = (P_T)^2 R_L / V^2$, where P_T is the power transmitted to the user, V is the delivered voltage, and R_L is the resistance of the power lines.

84. A coil with 190 turns, a radius of 5.0 cm, and a resistance of 12 Ω surrounds a solenoid with 230 turns/cm and a radius of 4.5 cm (Fig. 21–65). The current in the solenoid changes at a constant rate from 0 to 2.0 A in 0.10 s. Calculate the magnitude and direction of the induced current in the outer coil.

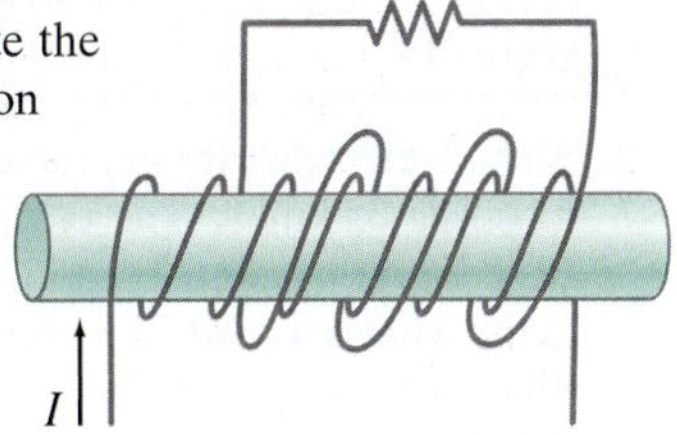

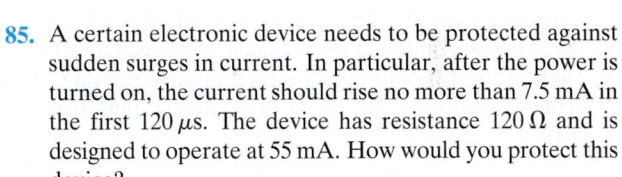

FIGURE 21–65
Problem 84.

85. A certain electronic device needs to be protected against sudden surges in current. In particular, after the power is turned on, the current should rise no more than 7.5 mA in the first 120 μs. The device has resistance 120 Ω and is designed to operate at 55 mA. How would you protect this device?

86. A 35-turn 12.5-cm-diameter coil is placed between the pole pieces of an electromagnet. When the electromagnet is turned on, the flux through the coil changes, inducing an emf. At what rate (in T/s) must the magnetic field change if the emf is to be 120 V?

87. Calculate the peak output voltage of a simple generator whose square armature windings are 6.60 cm on a side; the armature contains 125 loops and rotates in a field of 0.200 T at a rate of 120 rev/s.

*88. Typical large values for electric and magnetic fields attained in laboratories are about 1.0×10^4 V/m and 2.0 T. (a) Determine the energy density for each field and compare. (b) What magnitude electric field would be needed to produce the same energy density as the 2.0-T magnetic field?

*89. Determine the inductance L of the primary of a transformer whose input is 220 V at 60.0 Hz if the current drawn is 6.3 A. Assume no current in the secondary.

*90. A 130-mH coil whose resistance is 15.8 Ω is connected to a capacitor C and a 1360-Hz source voltage. If the current and voltage are to be in phase, what value must C have?

*91. The wire of a tightly wound solenoid is unwound and used to make another tightly wound solenoid of twice the diameter. By what factor does the inductance change?

*92. The **Q factor** of a resonant ac circuit (Section 21–15) can be defined as the ratio of the voltage across the capacitor (or inductor) to the voltage across the resistor, at resonance. The larger the Q factor, the sharper the resonance curve will be and the sharper the tuning. (a) Show that the Q factor is given by the equation $Q = (1/R)\sqrt{L/C}$. (b) At a resonant frequency $f_0 = 1.0$ MHz, what must be the values of L and R to produce a Q factor of 650? Assume that $C = 0.010\,\mu\text{F}$.

Search and Learn

1. (a) Sections 19–7 and 21–9 discuss conditions when and where it is especially important to have ground fault circuit interrupters (GFCIs) installed. What is it about those places that makes "touching ground" especially risky? (b) Describe how a GFCI works and compare to fuses and circuit breakers (see also Section 18–6).

2. While demonstrating Faraday's law to her class, a physics professor inadvertently moves the gold ring on her finger from a location where a 0.68-T magnetic field points along her finger to a zero-field location in 45 ms. The 1.5-cm-diameter ring has a resistance and mass of 55 $\mu\Omega$ and 15 g, respectively. (a) Estimate the thermal energy produced in the ring due to the flow of induced current. (b) Find the temperature rise of the ring, assuming all of the thermal energy produced goes into increasing the ring's temperature. The specific heat of gold is 129 J/kg·C°.

3. A small electric car overcomes a 250-N friction force when traveling 35 km/h. The electric motor is powered by ten 12-V batteries connected in series and is coupled directly to the wheels whose diameters are 58 cm. The 290 armature coils are rectangular, 12 cm by 15 cm, and rotate in a 0.65-T magnetic field. (a) How much current does the motor draw to produce the required torque? (b) What is the back emf? (c) How much power is dissipated in the coils? (d) What percent of the input power is used to drive the car? [*Hint*: Check Sections 6–10, 18–5, 20–9, 20–10, and 21–6.]

4. Explain the advantage of using ac rather than dc current when electric power needs to be transported long distances. (See Section 21–7.)

5. A power line carrying a sinusoidally varying current with frequency $f = 60$ Hz and peak value $I_0 = 155$ A runs at a height of 7.0 m across a farmer's land (Fig. 21–66). The farmer constructs a vertical 2.0-m-high 2000-turn rectangular wire coil below the power line. The farmer hopes to use the induced voltage in this coil to power 120-V electrical equipment, which requires a sinusoidally varying voltage with frequency $f = 60$ Hz and peak value $V_0 = 170$ V. Estimate the length ℓ of the coil needed. Would this be stealing? [*Hint*: Consider ΔB over one-quarter of a cycle $(\frac{1}{240}\,\text{s})$. See Sections 20–5 and 18–7.]

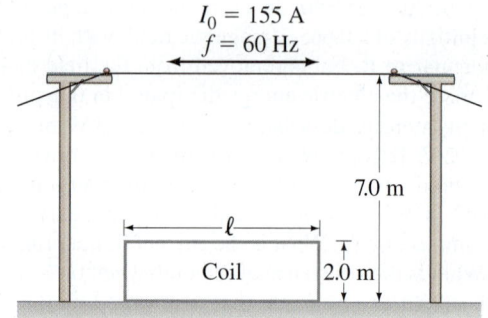

FIGURE 21–66 Search and Learn 5.

6. A **ballistic galvanometer** is a device that measures the total charge Q that passes through it in a short time. It is connected to a **search coil** that measures B (also called a **flip coil**) which is a small coil with N turns, each of cross-sectional area A. The flip coil is placed in the magnetic field to be measured with its face perpendicular to the field. It is then quickly rotated 180° about a diameter. Show that the total charge Q that flows in the induced current during this short "flip" time is proportional to the magnetic field B. In particular, show that

$$B = \frac{QR}{2NA}$$

where R is the total resistance of the circuit including the coil and ballistic galvanometer which measures charge Q.

ANSWERS TO EXERCISES

A: (e).
B: (a) Counterclockwise; (b) clockwise; (c) zero; (d) counterclockwise.
C: Clockwise (conventional current counterclockwise).
D: (a) Increase (brighter); (b) yes; resists more (counter torque).
E: 10 turns.
F: From Eq. 21–11b, the higher the frequency the lower the reactance, so in (a) more high frequency current flows to circuit B. In (b) higher frequencies pass to ground whereas lower frequencies pass more easily to circuit B.

Wireless technology is all around us: radio and television, cell phones, wi-fi, Bluetooth, and all wireless communication. These devices work by electromagnetic waves traveling through space. Wireless devices are applications of Marconi's development of long-distance transmission of information a century ago.

In this photo we see the first humans to land on the Moon. In the background is a television camera that sent live moving images through empty space to Earth where it was shown live.

We will see in this Chapter that Maxwell predicted the existence of EM waves from his famous equations. Maxwell's equations themselves are a magnificent summary of electromagnetism. We will also see that EM waves carry energy and momentum, and that light itself is an electromagnetic wave.

Electromagnetic Waves

CHAPTER 22

CHAPTER-OPENING QUESTION—Guess now!

Which of the following best describes the difference between radio waves and X-rays?
- **(a)** X-rays are radiation whereas radio waves are electromagnetic waves.
- **(b)** Both can be thought of as electromagnetic waves. They differ only in wavelength and frequency.
- **(c)** X-rays are pure energy. Radio waves are made of fields, not energy.
- **(d)** Radio waves come from electric currents in an antenna. X-rays are not related to electric charge.
- **(e)** X-rays are made up of particles called photons whereas radio waves are oscillations in space.

CONTENTS

- 22–1 Changing Electric Fields Produce Magnetic Fields; Maxwell's Equations
- 22–2 Production of Electromagnetic Waves
- 22–3 Light as an Electromagnetic Wave and the Electromagnetic Spectrum
- 22–4 Measuring the Speed of Light
- 22–5 Energy in EM Waves
- 22–6 Momentum Transfer and Radiation Pressure
- 22–7 Radio and Television; Wireless Communication

The culmination of electromagnetic theory in the nineteenth century was the prediction, and the experimental verification, that waves of electromagnetic fields could travel through space. This achievement opened a whole new world of communication: first the wireless telegraph, then radio and television, and more recently cell phones, remote-control devices, wi-fi, and Bluetooth. Most important was the spectacular prediction that light is an electromagnetic wave.

The theoretical prediction of electromagnetic waves was the work of the Scottish physicist James Clerk Maxwell (1831–1879; Fig. 22–1), who unified, in one magnificent theory, all the phenomena of electricity and magnetism.

FIGURE 22–1 James Clerk Maxwell.

22-1 Changing Electric Fields Produce Magnetic Fields; Maxwell's Equations

The development of electromagnetic theory in the early part of the nineteenth century by Oersted, Ampère, and others was not actually done in terms of electric and magnetic fields. The idea of the field was introduced somewhat later by Faraday, and was not generally used until Maxwell showed that all electric and magnetic phenomena could be described using only four equations involving electric and magnetic fields. These equations, known as **Maxwell's equations**, are the basic equations for all electromagnetism. They are fundamental in the same sense that Newton's three laws of motion and the law of universal gravitation are for mechanics. In a sense, they are even more fundamental, because they are consistent with the theory of relativity (Chapter 26), whereas Newton's laws are not. Because all of electromagnetism is contained in this set of four equations, Maxwell's equations are considered one of the great triumphs of the human intellect.

Although we will not present Maxwell's equations in mathematical form since they involve calculus, we will summarize them here in words. They are:

(1) a generalized form of Coulomb's law that relates electric field to its source, electric charge (= Gauss's law, Section 16–12);

(2) a similar law for the magnetic field, except that magnetic field lines are always continuous—they do not begin or end (as electric field lines do, on charges);

(3) an electric field is produced by a changing magnetic field (Faraday's law);

(4) a magnetic field is produced by an electric current (Ampère's law), or by a changing electric field.

Law (3) is Faraday's law (see Chapter 21, especially Section 21–4). The first part of law (4), that a magnetic field is produced by an electric current, was discovered by Oersted, and the mathematical relation is given by Ampère's law (Section 20–8). But the second part of law (4) is an entirely new aspect predicted by Maxwell: Maxwell argued that if a changing magnetic field produces an electric field, as given by Faraday's law, then the reverse might be true as well: **a changing electric field will produce a magnetic field**. This was an *hypothesis* by Maxwell. It is based on the idea of *symmetry* in nature. Indeed, the size of the effect in most cases is so small that Maxwell recognized it would be difficult to detect it experimentally.

*Maxwell's Fourth Equation (Ampère's Law Extended)

To back up the idea that a changing electric field might produce a magnetic field, we use an indirect argument that goes something like this. According to Ampère's law (Section 20–8), $\Sigma B_{\parallel} \Delta \ell = \mu_0 I$. That is, divide any closed path you choose into short segments $\Delta \ell$, multiply each segment by the parallel component of the magnetic field B at that segment, and then sum all these products over the complete closed path. That sum will then equal μ_0 times the total current I that passes through a surface bounded by the path. When we applied Ampère's law to the field around a straight wire (Section 20–8), we imagined the current as passing through the circular area enclosed by our circular loop. That area is the flat surface 1 shown in Fig. 22–2. However, we could just as well use the sack-shaped surface 2 in Fig. 22–2 as the surface for Ampère's law because the same current I passes through it.

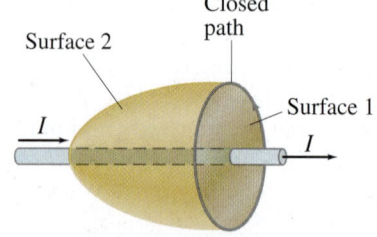

FIGURE 22–2 Ampère's law applied to two different surfaces bounded by the same closed path.

Now consider the closed path for the situation of Fig. 22–3 where a capacitor is being discharged. Ampère's law works for surface 1 (current I passes through surface 1), but it does not work for surface 2 because no current passes through surface 2. There is a magnetic field around the wire, so the left side of Ampère's law

$$\Sigma B_\parallel \Delta \ell = \mu_0 I$$

is not zero around the circular closed path; yet no current flows through surface 2, so the right side *is* zero for surface 2. We seem to have a contradiction of Ampère's law. There is a magnetic field present in Fig. 22–3, however, only if charge is flowing to or away from the capacitor plates. The changing charge on the plates means that the electric field between the plates is changing in time. Maxwell resolved the problem of no current through surface 2 in Fig. 22–3 by proposing that the changing electric field between the plates is *equivalent to* an electric current. He called it a **displacement current**, I_D. An ordinary current I is then called a "conduction current," and Ampère's law, as generalized by Maxwell, becomes

$$\Sigma B_\parallel \Delta \ell = \mu_0 (I + I_D).$$

Ampère's law will now apply also for surface 2 in Fig. 22–3, where I_D refers to the changing electric field.

Combining Eq. 17–7 for the charge on a capacitor, $Q = CV$, with Eq. 17–8, $C = \epsilon_0 A/d$, and with the magnitudes in Eq. 17–4a, $V = Ed$, we can write $Q = CV = (\epsilon_0 A/d)(Ed) = \epsilon_0 AE$. Then the current I_D becomes

$$I_D = \frac{\Delta Q}{\Delta t} = \epsilon_0 \frac{\Delta \Phi_E}{\Delta t},$$

where $\Phi_E = EA$ is the **electric flux**, defined in analogy to magnetic flux (Section 21–2). Then, Ampère's law becomes

$$\Sigma B_\parallel \Delta \ell = \mu_0 I + \mu_0 \epsilon_0 \frac{\Delta \Phi_E}{\Delta t}. \tag{22-1}$$

Ampère's law (general form)

This equation embodies Maxwell's idea that a magnetic field can be caused not only by a normal electric current, but also by a changing electric field or changing electric flux.

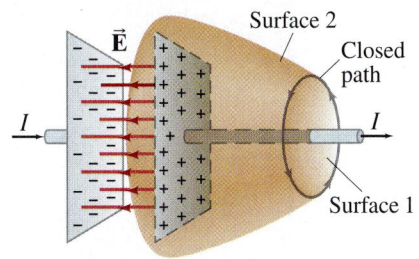

FIGURE 22–3 A capacitor discharging. A conduction current passes through surface 1, but no conduction current passes through the sacklike surface 2. An extra term is needed in Ampère's law.

22–2 Production of Electromagnetic Waves

According to Maxwell, a magnetic field will be produced in empty space if there is a changing electric field. From this, Maxwell derived another startling conclusion. If a changing magnetic field produces an electric field, that electric field is itself changing. This changing electric field will, in turn, produce a magnetic field, which will be changing, and so it too will produce a changing electric field; and so on. When Maxwell worked with his equations, he found that the net result of these interacting changing fields was a *wave* of electric and magnetic fields that can propagate (travel) through space! We now examine, in a simplified way, how such **electromagnetic waves** can be produced.

Consider two conducting rods that will serve as an "antenna" (Fig. 22–4a). Suppose these two rods are connected by a switch to the opposite terminals of a battery. When the switch is closed, the upper rod quickly becomes positively charged and the lower one negatively charged. Electric field lines are formed as indicated by the lines in Fig. 22–4b. While the charges are flowing, a current exists whose direction is indicated by the black arrows. A magnetic field is therefore produced near the antenna. The magnetic field lines encircle the rod-like antenna and therefore, in Fig. 22–4, $\vec{B}$ points into the page ($\otimes$) on the right and out of the page ($\odot$) on the left. Now we ask, how far out do these electric and magnetic fields extend? In the static case, the fields would extend outward indefinitely far. However, when the switch in Fig. 22–4 is closed, the fields quickly appear nearby, but it takes time for them to reach distant points. Both electric and magnetic fields store energy, and this energy cannot be transferred to distant points at infinite speed.

FIGURE 22–4 Fields produced by charge flowing into conductors. It takes time for the $\vec{E}$ and $\vec{B}$ fields to travel outward to distant points. The fields are shown to the right of the antenna, but they move out in all directions, symmetrically about the (vertical) antenna.

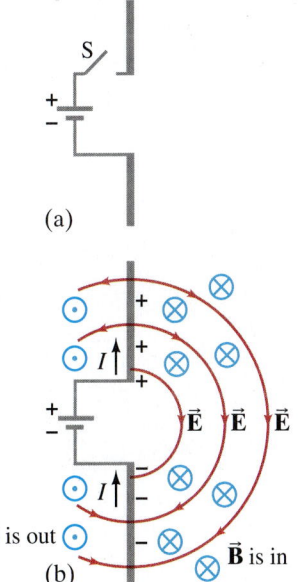

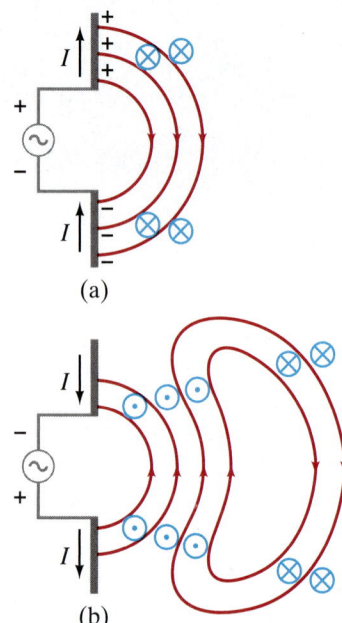

FIGURE 22–5 Sequence showing electric and magnetic fields that spread outward from oscillating charges on two conductors (the antenna) connected to an ac source (see the text).

Now we look at the situation of Fig. 22–5, where our antenna is connected to an ac generator. In Fig. 22–5a, the connection has just been completed. Charge starts building up, and fields form just as in Fig. 22–4b. The + and − signs in Fig. 22–5a indicate the net charge on each rod at a given instant. The black arrows indicate the direction of the current. The electric field is represented by red lines in the plane of the page; and the magnetic field, according to the right-hand rule, is into (⊗) or out of (⊙) the page in blue. In Fig. 22–5b, the voltage of the ac generator has reversed in direction; the current is reversed and the new magnetic field is in the opposite direction. Because the new fields have changed direction, the old lines fold back to connect up to some of the new lines and form closed loops as shown.† The old fields, however, don't suddenly disappear; they are on their way to distant points. Indeed, because a changing magnetic field produces an electric field, and a changing electric field produces a magnetic field, this combination of changing electric and magnetic fields moving outward is self-supporting, no longer depending on the antenna charges.

The fields not far from the antenna, referred to as the *near field*, become quite complicated, but we are not so interested in them. We are mainly interested in the fields far from the antenna (they are generally what we detect), which we refer to as the **radiation field**. The electric field lines form loops, as shown in Fig. 22–6a, and continue moving outward. The magnetic field lines also form closed loops, but are not shown because they are perpendicular to the page. Although the lines are shown only on the right of the source, fields also travel in other directions. The field strengths are greatest in directions perpendicular to the oscillating charges; and they drop to zero along the direction of oscillation—above and below the antenna in Fig. 22–6a.

FIGURE 22–6 (a) The radiation fields (far from the antenna) produced by a sinusoidal signal on the antenna. The red closed loops represent electric field lines. The magnetic field lines, perpendicular to the page and represented by blue ⊗ and ⊙, also form closed loops. (b) Very far from the antenna, the wave fronts (field lines) are essentially flat over a fairly large area, and are referred to as *plane waves*.

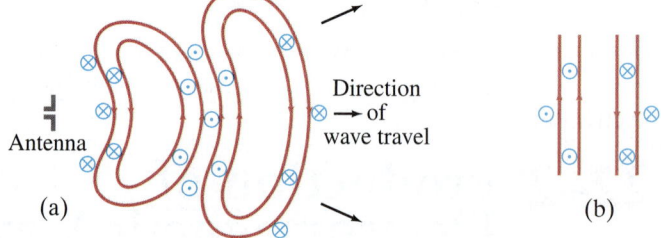

The magnitudes of both $\vec{E}$ and $\vec{B}$ in the radiation field are found to decrease with distance as $1/r$. (Compare this to the static electric field given by Coulomb's law where $\vec{E}$ decreases as $1/r^2$.) The energy carried by the electromagnetic wave is proportional (as for any wave, Chapter 11) to the square of the amplitude, E^2 or B^2, as will be discussed further in Section 22–7, so the intensity of the wave decreases as $1/r^2$. Thus the energy carried by EM waves follows the **inverse square law** just as for sound waves (Eqs. 11–16).

Several things about the radiation field can be noted from Fig. 22–6. First, *the electric and magnetic fields at any point are perpendicular to each other, and to the direction of wave travel*. Second, we can see that the fields alternate in direction ($\vec{B}$ is into the page at some points and out of the page at others; $\vec{E}$ points up at some points and down at others). Thus, the field strengths vary from a maximum in one direction, to zero, to a maximum in the other direction. The electric and magnetic fields are "in phase": that is, they each are zero at the same points and reach their maxima at the same points in space. Finally, very far from the antenna (Fig. 22–6b) the field lines are quite flat over a reasonably large area, and the waves are referred to as **plane waves**.

---

†We are considering waves traveling through empty space. There are no charges for lines of $\vec{E}$ to start or stop on, so they form closed loops. Magnetic field lines always form closed loops.

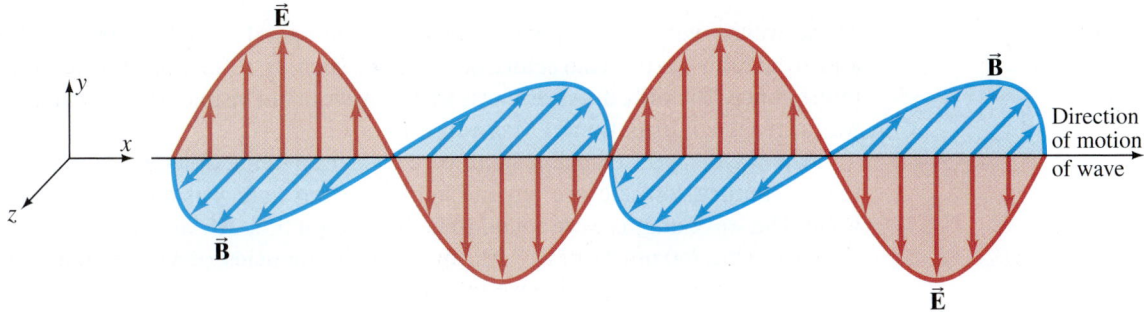

FIGURE 22–7 Electric and magnetic field strengths in an electromagnetic wave. $\vec{E}$ and $\vec{B}$ are at right angles to each other. The entire pattern moves in a direction perpendicular to both $\vec{E}$ and $\vec{B}$.

If the source voltage varies sinusoidally, then the electric and magnetic field strengths in the radiation field will also vary sinusoidally. The sinusoidal character of the waves is shown in Fig. 22–7, which displays the field directions and magnitudes plotted as a function of position along the direction of wave travel. Notice that $\vec{B}$ and $\vec{E}$ are perpendicular to each other and to the direction of wave travel.

We call these waves electromagnetic (EM) waves. They are *transverse* waves because the amplitude is perpendicular to the direction of wave travel. However, EM waves are always waves of *fields*, not of matter (like waves on water or a rope). Because they are fields, EM waves can propagate in empty space.

As we have seen, EM waves are produced by electric charges that are oscillating, and hence are undergoing acceleration. In fact, we can say in general that

accelerating electric charges give rise to electromagnetic waves.

Maxwell derived a formula for the speed of EM waves:

$$v = c = \frac{E}{B}, \tag{22-2}$$

where c is the special symbol for the speed of electromagnetic waves in empty space, and E and B are the magnitudes of electric and magnetic fields at the same point in space. More specifically, it was also shown that

$$c = \frac{1}{\sqrt{\epsilon_0 \mu_0}}. \quad \text{[speed of EM waves]} \tag{22-3}$$

When Maxwell put in the values for ϵ_0 and μ_0, he found

$$c = \frac{1}{\sqrt{\epsilon_0 \mu_0}} = \frac{1}{\sqrt{(8.85 \times 10^{-12}\,\text{C}^2/\text{N} \cdot \text{m}^2)(4\pi \times 10^{-7}\,\text{N} \cdot \text{s}^2/\text{C}^2)}} = 3.00 \times 10^8\,\text{m/s},$$

which is exactly equal to the measured speed of light in vacuum (Section 22–4).

> **EXERCISE A** At a particular instant in time, a wave has its electric field pointing north and its magnetic field pointing up. In which direction is the wave traveling? (*a*) South, (*b*) west, (*c*) east, (*d*) down, (*e*) not enough information. [See Fig. 22–7.]

22–3 Light as an Electromagnetic Wave and the Electromagnetic Spectrum

Maxwell's prediction that EM waves should exist was startling. Equally remarkable was the speed at which EM waves were predicted to travel—3.00×10^8 m/s, the same as the measured speed of light.

Light had been shown some 60 years before Maxwell's work to behave like a wave (we'll discuss this in Chapter 24). But nobody knew what kind of wave it was. What is it that is oscillating in a light wave? Maxwell, on the basis of the calculated speed of EM waves, argued that light must be an electromagnetic wave. This idea soon came to be generally accepted by scientists, but not fully until after EM waves were experimentally detected. EM waves were first generated and detected experimentally by Heinrich Hertz (1857–1894) in 1887, eight years after Maxwell's death. Hertz used a spark-gap apparatus in which charge was made to rush back and forth for a short time, generating waves whose frequency was about 10^9 Hz. He detected them some distance away using a loop of wire in which an emf was induced when a changing magnetic field passed through.

These waves were later shown to travel at the speed of light, 3.00×10^8 m/s, and to exhibit all the characteristics of light such as reflection, refraction, and interference. The only difference was that they were not visible. Hertz's experiment was a strong confirmation of Maxwell's theory.

The wavelengths of visible light were measured in the first decade of the nineteenth century, long before anyone imagined that light was an electromagnetic wave. The wavelengths were found to lie between 4.0×10^{-7} m and 7.5×10^{-7} m, or 400 nm to 750 nm $(1 \text{ nm} = 10^{-9} \text{ m})$. The frequencies of visible light can be found using Eq. 11–12, which we rewrite here:

$$c = \lambda f, \tag{22-4}$$

where f and λ are the frequency and wavelength, respectively, of the wave. Here, c is the speed of light, 3.00×10^8 m/s; it gets the special symbol c because of its universality for all EM waves in free space. Equation 22–4 tells us that the frequencies of visible light are between 4.0×10^{14} Hz and 7.5×10^{14} Hz. (Recall that 1 Hz = 1 cycle per second = 1 s^{-1}.)

But visible light is only one kind of EM wave. As we have seen, Hertz produced EM waves of much lower frequency, about 10^9 Hz. These are now called **radio waves**, because frequencies in this range are used to transmit radio and TV signals. Electromagnetic waves, or EM radiation as we sometimes call it, have been produced or detected over a wide range of frequencies. They are usually categorized as shown in Fig. 22–8, which is known as the **electromagnetic spectrum**.

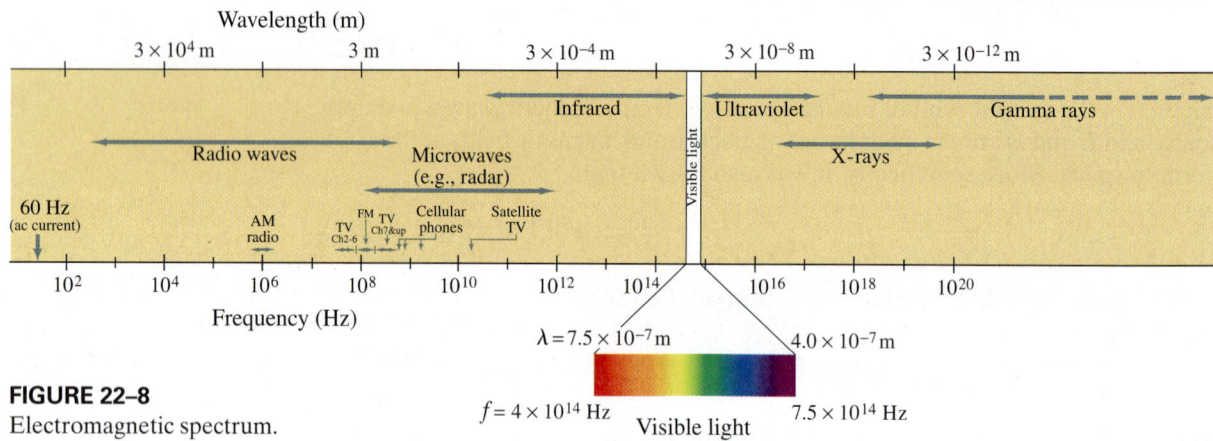

FIGURE 22–8 Electromagnetic spectrum.

Radio waves and microwaves can be produced in the laboratory using electronic equipment (Fig. 22–5). Higher-frequency waves are very difficult to produce electronically. These and other types of EM waves are produced in natural processes, as emission from atoms, molecules, and nuclei (more on this later). EM waves can be produced by the acceleration of electrons or other charged particles, such as electrons in the antenna of Fig. 22–5. X-rays have very short wavelengths (and very high frequencies), and they are produced (Chapters 25 and 28) when fast-moving electrons are rapidly decelerated upon striking a metal target. Even the visible light emitted by an ordinary incandescent bulb is due to electrons undergoing acceleration within the hot filament.

EXERCISE B Return to the Chapter-Opening Question, page 625, and answer it again now. Try to explain why you may have answered differently the first time.

We will meet various types of EM waves later. However, it is worth mentioning here that infrared (IR) radiation (EM waves whose frequency is just less than that of visible light) is mainly responsible for the heating effect of the Sun. The Sun emits not only visible light but substantial amounts of IR and UV (ultraviolet) as well. The molecules of our skin tend to "resonate" at infrared frequencies, so it is these that are preferentially absorbed and thus warm us. We humans experience EM waves differently depending on their wavelengths: Our eyes detect wavelengths between about 4×10^{-7} m and 7.5×10^{-7} m (visible light), whereas our skin detects longer wavelengths (IR). Many EM wavelengths we don't detect directly at all.

Light and other electromagnetic waves travel at a speed of 3×10^8 m/s. Compare this to sound, which travels (see Chapter 12) at a speed of about 300 m/s in air, a million times slower; or to typical freeway speeds of a car, 30 m/s (100 km/h, or 60 mi/h), 10 million times slower than light. EM waves differ from sound waves in another big way: sound waves travel in a medium such as air, and involve motion of air molecules; EM waves do not involve any material—only fields, and they can travel in empty space.

> **CAUTION**
> *Sound and EM waves are different*

EXAMPLE 22–1 Wavelengths of EM waves. Calculate the wavelength (a) of a 60-Hz EM wave, (b) of a 93.3-MHz FM radio wave, and (c) of a beam of visible red light from a laser at frequency 4.74×10^{14} Hz.

APPROACH All of these waves are electromagnetic waves, so their speed is $c = 3.00 \times 10^8$ m/s. We solve for λ in Eq. 22–4: $\lambda = c/f$.

SOLUTION (a)
$$\lambda = \frac{c}{f} = \frac{3.00 \times 10^8 \text{ m/s}}{60 \text{ s}^{-1}} = 5.0 \times 10^6 \text{ m},$$

or 5000 km. 60 Hz is the frequency of ac current in the United States, and, as we see here, one wavelength stretches all the way across the continental USA.

(b)
$$\lambda = \frac{3.00 \times 10^8 \text{ m/s}}{93.3 \times 10^6 \text{ s}^{-1}} = 3.22 \text{ m}.$$

The length of an FM radio antenna is often about half this ($\frac{1}{2}\lambda$), or $1\frac{1}{2}$ m.

(c)
$$\lambda = \frac{3.00 \times 10^8 \text{ m/s}}{4.74 \times 10^{14} \text{ s}^{-1}} = 6.33 \times 10^{-7} \text{ m } (= 633 \text{ nm}).$$

EXERCISE C What are the frequencies of (a) an 80-m-wavelength radio wave, and (b) an X-ray of wavelength 5.5×10^{-11} m?

EXAMPLE 22–2 ESTIMATE Cell phone antenna. The antenna of a cell phone is often $\frac{1}{4}$ wavelength long. A particular cell phone has an 8.5-cm-long straight rod for its antenna. Estimate the operating frequency of this phone.

APPROACH The basic equation relating wave speed, wavelength, and frequency is $c = \lambda f$; the wavelength λ equals four times the antenna's length.

SOLUTION The antenna is $\frac{1}{4}\lambda$ long, so $\lambda = 4(8.5 \text{ cm}) = 34 \text{ cm} = 0.34$ m. Then $f = c/\lambda = (3.0 \times 10^8 \text{ m/s})/(0.34 \text{ m}) = 8.8 \times 10^8$ Hz = 880 MHz.

NOTE Radio antennas are not always straight conductors. The conductor may be a round loop to save space. See Fig. 22–18b.

EXERCISE D How long should a $\frac{1}{4}$-λ antenna be for an aircraft radio operating at 165 MHz?

Electromagnetic waves can travel along transmission lines as well as in empty space. When a source of emf is connected to a transmission line—be it two parallel wires or a coaxial cable (Fig. 22–9)—the electric field within the wire is not set up immediately at all points along the wires. This is based on the same argument we used in Section 22–2 with reference to Fig. 22–5. Indeed, it can be shown that if the wires are separated by empty space or air, the electrical signal travels along the wires at the speed $c = 3.0 \times 10^8$ m/s. For example, when you flip a light switch, the light actually goes on a tiny fraction of a second later. If the wires are in a medium whose electric permittivity is ϵ and magnetic permeability is μ (Sections 17–8 and 20–12, respectively), the speed is not given by Eq. 22–3, but by

$$v = \frac{1}{\sqrt{\epsilon \mu}}$$

instead.

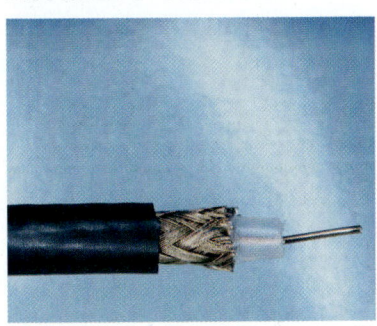

FIGURE 22–9 Coaxial cable.

EXAMPLE 22–3 ESTIMATE **Phone call time lag.** You make a telephone call from New York to a friend in London. Estimate how long it will take the electrical signal generated by your voice to reach London, assuming the signal is (*a*) carried on a telephone cable under the Atlantic Ocean, and (*b*) sent via satellite 36,000 km above the ocean. Would there be a noticeable delay in either case?

APPROACH The signal is carried on a telephone wire or in the air via satellite. In either case it is an electromagnetic wave. Electronics as well as the wire or cable slow things down, but as a rough estimate we take the speed to be $c = 3.0 \times 10^8$ m/s.

SOLUTION The distance from New York to London is about 5000 km.
(*a*) The time delay via the cable is $t = d/c \approx (5 \times 10^6 \text{ m})/(3.0 \times 10^8 \text{ m/s}) = 0.017$ s.
(*b*) Via satellite the time would be longer because communications satellites, which are usually geosynchronous (Example 5–12), move at a height of 36,000 km. The signal would have to go up to the satellite and back down, or about 72,000 km. The actual distance the signal would travel would be a little more than this as the signal would go up and down on a diagonal (5000 km New York to London, small compared to the distance up to the satellite). Thus $t = d/c \approx (7.2 \times 10^7 \text{ m})/(3 \times 10^8 \text{ m/s}) \approx 0.24$ s, one way. Both directions $\approx \frac{1}{2}$ s.

NOTE When the signal travels via the underwater cable, there is only a hint of a delay and conversations are fairly normal. When the signal is sent via satellite, the delay *is* noticeable. The length of time between the end of when you speak and your friend receives it and replies, and then you hear the reply, would be about a half second beyond the normal time in a conversation, as we just calculated. This is enough to be noticeable, and you have to adjust for it so you don't start talking again while your friend's reply is on the way back to you.

EXERCISE E If you are on the phone via satellite to someone only 100 km away, would you notice the same effect discussed in the NOTE above?

EXERCISE F If your voice traveled as a sound wave, how long would it take to go from New York to London?

22–4 Measuring the Speed of Light

Galileo attempted to measure the speed of light by trying to measure the time required for light to travel a known distance between two hilltops. He stationed an assistant on one hilltop and himself on another, and ordered the assistant to lift the cover from a lamp the instant he saw a flash from Galileo's lamp. Galileo measured the time between the flash of his lamp and when he received the light from his assistant's lamp. The time was so short that Galileo concluded it merely represented human reaction time, and that the speed of light must be extremely high.

The first successful determination that the speed of light is finite was made by the Danish astronomer Ole Roemer (1644–1710). Roemer had noted that the carefully measured orbital period of Io, a moon of Jupiter with an average period of 42.5 h, varied slightly, depending on the relative position of Earth and Jupiter. He attributed this variation in the apparent period to the change in distance between the Earth and Jupiter during one of Io's periods, and the time it took light to travel the extra distance. Roemer concluded that the speed of light—though great—is finite.

Since then a number of techniques have been used to measure the speed of light. Among the most important were those carried out by the American Albert A. Michelson (1852–1931). Michelson used the rotating mirror apparatus diagrammed in Fig. 22–10 for a series of high-precision experiments carried out from 1880 to the 1920s. Light from a source would hit one face of a rotating eight-sided mirror. The reflected light traveled to a stationary mirror a large distance away and back again as shown. If the rotating mirror was turning at just the right rate, the returning beam of light would reflect from one face of the mirror into a small telescope through which the observer looked. If the speed of rotation was only slightly different, the beam would be deflected to one side and would not be seen by the observer. From the required speed of the rotating mirror and the known distance to the stationary mirror, the speed of light could be calculated. In the 1920s, Michelson set up the rotating mirror on the top of Mt. Wilson in southern California and the stationary mirror on Mt. Baldy (Mt. San Antonio) 35 km away. He later measured the speed of light in vacuum using a long evacuated tube.

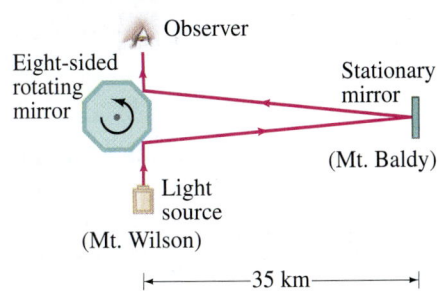

FIGURE 22–10 Michelson's speed-of-light apparatus (not to scale).

Today the speed of light, c, in vacuum is taken as

$$c = 2.99792458 \times 10^8 \, \text{m/s},$$

and is *defined* to be this value. This means that the standard for length, the meter, is no longer defined separately. Instead, as we noted in Section 1–5, the meter is now formally defined as the distance light travels in vacuum in 1/299,792,458 of a second.

We usually round off c to

$$c = 3.00 \times 10^8 \, \text{m/s}$$

when extremely precise results are not required. In air, the speed is only slightly less.

22–5 Energy in EM Waves

Electromagnetic waves carry energy from one region of space to another. This energy is associated with the moving electric and magnetic fields. In Section 17–9, we saw that the energy density u_E (J/m³) stored in an electric field E is $u_E = \frac{1}{2}\epsilon_0 E^2$ (Eq. 17–11). The energy density stored in a magnetic field B, as we discussed in Section 21–11, is given by $u_B = \frac{1}{2}B^2/\mu_0$ (Eq. 21–10). Thus, the total energy stored per unit volume in a region of space where there is an electromagnetic wave is

$$u = u_E + u_B = \frac{1}{2}\epsilon_0 E^2 + \frac{1}{2}\frac{B^2}{\mu_0}. \tag{22-5}$$

In this equation, E and B represent the electric and magnetic field strengths of the wave at any instant in a small region of space. We can write Eq. 22–5 in terms of the E field only using Eqs. 22–2 ($B = E/c$) and 22–3 ($c = 1/\sqrt{\epsilon_0 \mu_0}$) to obtain

$$u = \frac{1}{2}\epsilon_0 E^2 + \frac{1}{2}\frac{\epsilon_0 \mu_0 E^2}{\mu_0} = \epsilon_0 E^2. \tag{22-6a}$$

Note here that the energy density associated with the B field equals that due to the E field, and each contributes half to the total energy. We can also write the energy density in terms of the B field only:

$$u = \epsilon_0 E^2 = \epsilon_0 c^2 B^2 = \frac{B^2}{\mu_0}, \tag{22-6b}$$

or in one term containing both E and B,

$$u = \epsilon_0 E^2 = \epsilon_0 EcB = \frac{\epsilon_0 EB}{\sqrt{\epsilon_0 \mu_0}}$$

or

$$u = \sqrt{\frac{\epsilon_0}{\mu_0}} EB. \tag{22-6c}$$

Equations 22–6 give the energy density of EM waves in any region of space at any instant.

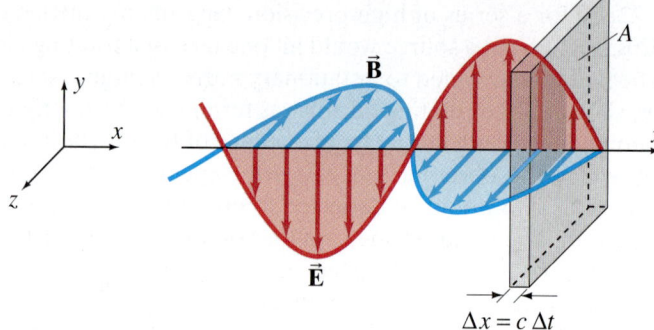

FIGURE 22–11 Electromagnetic wave carrying energy through area A.

The energy a wave transports per unit time per unit area is the **intensity** I, as defined in Sections 11–9 and 12–2.† The units of I are W/m². The energy ΔU is the energy density (Eq. 22–6a) times the volume V. Hence the energy passing through an area A in a time Δt (see Fig. 22–11) is

$$\Delta U = u\,\Delta V = (u)(A\,\Delta x) = (\epsilon_0 E^2)(Ac\,\Delta t)$$

because $\Delta x = c\,\Delta t$. Therefore, the magnitude of the intensity (energy per unit area per time Δt, or power per unit area) is

$$I = \frac{\Delta U}{A\,\Delta t} = \frac{(\epsilon_0 E^2)(Ac\,\Delta t)}{A\,\Delta t} = \epsilon_0 c E^2.$$

From Eqs. 22–2 and 22–3, this can also be written

$$I = \epsilon_0 c E^2 = \frac{c}{\mu_0} B^2 = \frac{EB}{\mu_0}. \quad (22\text{–}7)$$

We can also find the *average intensity* over an extended period of time, if E and B are sinusoidal. Then $\overline{E^2} = E_0^2/2$, just as for electric currents and voltages, Section 18–7, Eqs. 18–8. Thus

$$\overline{I} = \frac{1}{2}\epsilon_0 c E_0^2 = \frac{1}{2}\frac{c}{\mu_0} B_0^2 = \frac{E_0 B_0}{2\mu_0}. \quad (22\text{–}8)$$

Here E_0 and B_0 are the maximum values of E and B. We can also write

$$\overline{I} = \frac{E_{\text{rms}} B_{\text{rms}}}{\mu_0},$$

where E_{rms} and B_{rms} are the rms values ($E_{\text{rms}} = \sqrt{\overline{E^2}} = E_0/\sqrt{2}$, and $B_{\text{rms}} = \sqrt{\overline{B^2}} = B_0/\sqrt{2}$).

EXAMPLE 22–4 *E and B from the Sun.* Radiation from the Sun reaches the Earth (above the atmosphere) with an intensity of about 1350 W/m² = 1350 J/s·m². Assume that this is a single EM wave, and calculate the maximum values of E and B.

APPROACH We solve Eq. 22–8 ($\overline{I} = \frac{1}{2}\epsilon_0 c E_0^2$) for E_0 in terms of $\overline{I}$ and use $\overline{I} = 1350$ J/s·m².

SOLUTION $E_0 = \sqrt{\dfrac{2\overline{I}}{\epsilon_0 c}} = \sqrt{\dfrac{2(1350\text{ J/s·m}^2)}{(8.85 \times 10^{-12}\text{ C}^2/\text{N·m}^2)(3.00 \times 10^8\text{ m/s})}}$

$= 1.01 \times 10^3$ V/m.

From Eq. 22–2, $B = E/c$, so

$$B_0 = \frac{E_0}{c} = \frac{1.01 \times 10^3 \text{ V/m}}{3.00 \times 10^8 \text{ m/s}} = 3.37 \times 10^{-6}\text{ T}.$$

NOTE Although B has a small numerical value compared to E (because of the way the different units for E and B are defined), B contributes the same energy to the wave as E does, as we saw earlier.

⚠ **CAUTION**
E and B have very different values (due to how units are defined), but E and B contribute equal energy

---

†The intensity I for EM waves is often called the **Poynting vector** and given the symbol $\vec{S}$. Its direction is that in which the energy is being transported, which is the direction the wave is traveling, and its magnitude is the intensity ($S = I$).

22-6 Momentum Transfer and Radiation Pressure

If electromagnetic waves carry energy, then we would expect them to also carry linear momentum. When an electromagnetic wave encounters the surface of an object, a force will be exerted on the surface as a result of the momentum transfer ($F = \Delta p/\Delta t$) just as when a moving object strikes a surface. The force per unit area exerted by the waves is called **radiation pressure**, and its existence was predicted by Maxwell. He showed that if a beam of EM radiation (light, for example) is completely absorbed by an object, then the momentum transferred is

$$\Delta p = \frac{\Delta U}{c}, \qquad \begin{bmatrix}\text{radiation}\\ \text{fully}\\ \text{absorbed}\end{bmatrix} \quad (22\text{-}9\text{a})$$

where ΔU is the energy absorbed by the object in a time Δt and c is the speed of light. If, instead, the radiation is fully reflected (suppose the object is a mirror), then the momentum transferred is twice as great, just as when a ball bounces elastically off a surface:

$$\Delta p = \frac{2\,\Delta U}{c}. \qquad \begin{bmatrix}\text{radiation}\\ \text{fully}\\ \text{reflected}\end{bmatrix} \quad (22\text{-}9\text{b})$$

If a surface absorbs some of the energy, and reflects some of it, then $\Delta p = a\,\Delta U/c$, where a has a value between 1 and 2.

Using Newton's second law we can calculate the force and the pressure exerted by EM radiation on an object. The force F is given by

$$F = \frac{\Delta p}{\Delta t}.$$

The radiation pressure P (assuming full absorption) is given by (see Eq. 22-9a)

$$P = \frac{F}{A} = \frac{1}{A}\frac{\Delta p}{\Delta t} = \frac{1}{Ac}\frac{\Delta U}{\Delta t}.$$

We discussed in Section 22-5 that the average intensity $\bar{I}$ is defined as energy per unit time per unit area:

$$\bar{I} = \frac{\Delta U}{A\,\Delta t}.$$

Hence the radiation pressure is

$$P = \frac{\bar{I}}{c}. \qquad \begin{bmatrix}\text{fully}\\ \text{absorbed}\end{bmatrix} \quad (22\text{-}10\text{a})$$

If the light is fully reflected, the radiation pressure is twice as great (Eq. 22-9b):

$$P = \frac{2\bar{I}}{c}. \qquad \begin{bmatrix}\text{fully}\\ \text{reflected}\end{bmatrix} \quad (22\text{-}10\text{b})$$

EXAMPLE 22-5 ESTIMATE Solar pressure. Radiation from the Sun that reaches the Earth's surface (after passing through the atmosphere) transports energy at a rate of about 1000 W/m². Estimate the pressure and force exerted by the Sun on your outstretched hand.

APPROACH The radiation is partially reflected and partially absorbed, so let us estimate simply $P = \bar{I}/c$.

SOLUTION $P \approx \dfrac{\bar{I}}{c} = \dfrac{1000\text{ W/m}^2}{3\times 10^8\text{ m/s}} \approx 3\times 10^{-6}\text{ N/m}^2.$

An estimate of the area of your outstretched hand might be about 10 cm by 20 cm, so $A \approx 0.02\text{ m}^2$. Then the force is

$$F = PA \approx (3\times 10^{-6}\text{ N/m}^2)(0.02\text{ m}^2) \approx 6\times 10^{-8}\text{ N}.$$

NOTE These numbers are tiny. The force of gravity on your hand, for comparison, is maybe a half pound, or with $m = 0.2$ kg, $mg \approx (0.2\text{ kg})(9.8\text{ m/s}^2) \approx 2\text{ N}$. The radiation pressure on your hand is imperceptible compared to gravity.

EXAMPLE 22–6 ESTIMATE **A solar sail.** Proposals have been made to use the radiation pressure from the Sun to help propel spacecraft around the solar system. (a) About how much force would be applied on a 1 km × 1 km highly reflective sail when about the same distance from the Sun as the Earth is? (b) By how much would this increase the speed of a 5000-kg spacecraft in one year? (c) If the spacecraft started from rest, about how far would it travel in a year?

APPROACH (a) Pressure P is force per unit area, so $F = PA$. We use the estimate of Example 22–5, doubling it for a reflecting surface $P = 2\bar{I}/c$. (b) We find the acceleration from Newton's second law, and assume it is constant, and then find the speed from $v = v_0 + at$. (c) The distance traveled is given by $x = \frac{1}{2}at^2$.

SOLUTION (a) Doubling the result of Example 22–5, we get a solar pressure that is about $2\bar{I}/c \approx 10^{-5} \text{ N/m}^2$, rounding off. Then the force is $F \approx PA = (10^{-5} \text{ N/m}^2)(10^3 \text{ m})(10^3 \text{ m}) \approx 10 \text{ N}$.

(b) The acceleration is $a \approx F/m \approx (10 \text{ N})/(5000 \text{ kg}) \approx 2 \times 10^{-3} \text{ m/s}^2$. One year has $(365 \text{ days})(24 \text{ h/day})(3600 \text{ s/h}) \approx 3 \times 10^7 \text{ s}$. The speed increase is $v - v_0 = at \approx (2 \times 10^{-3} \text{ m/s}^2)(3 \times 10^7 \text{ s}) \approx 6 \times 10^4 \text{ m/s}$ ($\approx 200{,}000$ km/h!).

(c) Starting from rest, this acceleration would result in a distance traveled of about $d = \frac{1}{2}at^2 \approx \frac{1}{2}(2 \times 10^{-3} \text{ m/s}^2)(3 \times 10^7 \text{ s})^2 \approx 10^{12} \text{ m}$ in a year, about seven times the Sun–Earth distance. This result would apply if the spacecraft was far from the Earth so the Earth's gravitational force is small compared to 10 N.

NOTE A large sail providing a small force over a long time could result in a lot of motion. [Gravity due to the Sun and planets has been ignored, but in reality would have to be considered.]

Although you cannot directly feel the effects of radiation pressure, the phenomenon is quite dramatic when applied to atoms irradiated by a finely focused laser beam. An atom has a mass on the order of 10^{-27} kg, and a laser beam can deliver energy at a rate of 1000 W/m². This is the same intensity used in Example 22–5, but here a radiation pressure of 10^{-6} N/m² would be very significant on a molecule whose mass might be 10^{-23} to 10^{-26} kg. It is possible to move atoms and molecules around by steering them with a laser beam, in a device called **optical tweezers**. Optical tweezers have some remarkable applications. They are of great interest to biologists, especially since optical tweezers can manipulate live microorganisms, and components within a cell, without damaging them. Optical tweezers have been used to measure the elastic properties of DNA by pulling each end of the molecule with such a laser "tweezers."

Optical tweezers (move cell parts, DNA elasticity)

22–7 Radio and Television; Wireless Communication

Wireless transmission

Electromagnetic waves offer the possibility of transmitting information over long distances. Among the first to realize this and put it into practice was Guglielmo Marconi (1874–1937) who, in the 1890s, invented and developed wireless communication. With it, messages could be sent at the speed of light without the use of wires. The first signals were merely long and short pulses that could be translated into words by a code, such as the "dots" and "dashes" of the Morse code: they were digital wireless, believe it or not. In 1895 Marconi sent wireless signals a kilometer or two in Italy. By 1901 he had sent test signals 3000 km across the ocean from Newfoundland, Canada, to Cornwall, England (Fig. 22–12). In 1903 he sent the first practical commercial messages from Cape Cod, Massachusetts, to England: the London *Times* printed news items sent from its New York correspondent. 1903 was also the year of the first powered airplane flight by the Wright brothers. The hallmarks of the modern age—wireless communication and flight—date from the same year. Our modern world of wireless communication, including radio, television, cordless phones, cell phones, Bluetooth, wi-fi, and satellite communication, are based on Marconi's pioneering work.

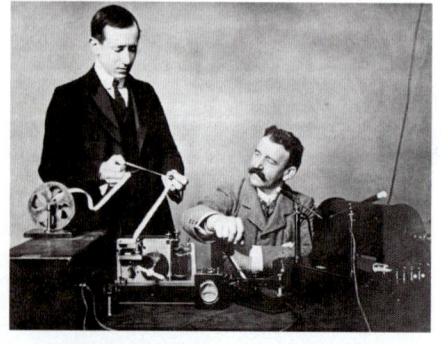
FIGURE 22–12 Guglielmo Marconi (1874–1937), on the left, receiving signals in Cornwall, 1901.

The next decade saw the development of vacuum tubes. Out of this early work radio and television were born. We now discuss briefly (1) how radio and TV signals are transmitted, and (2) how they are received at home.

The process by which a radio station transmits information (words and music) is outlined in Fig. 22–13. The audio (sound) information is changed into an electrical signal of the same frequencies by, say, a microphone, a laser, or a magnetic read/write head. This electrical signal is called an audiofrequency (AF) signal, because the frequencies are in the audio range (20 to 20,000 Hz). The signal is amplified electronically and is then mixed with a radio-frequency (RF) signal called its **carrier frequency**, which represents that station. AM radio stations have carrier frequencies from about 530 kHz to 1700 kHz. For example, "710 on your dial" means a station whose carrier frequency is 710 kHz. FM radio stations have much higher carrier frequencies, between 88 MHz and 108 MHz. The carrier frequencies for broadcast TV stations in the United States lie between 54 MHz and 72 MHz, between 76 MHz and 88 MHz, between 174 MHz and 216 MHz, and between 470 MHz and 698 MHz. Today's digital broadcasting (see Sections 17–10 and 17–11) uses the same frequencies as the pre-2009 analog transmission.

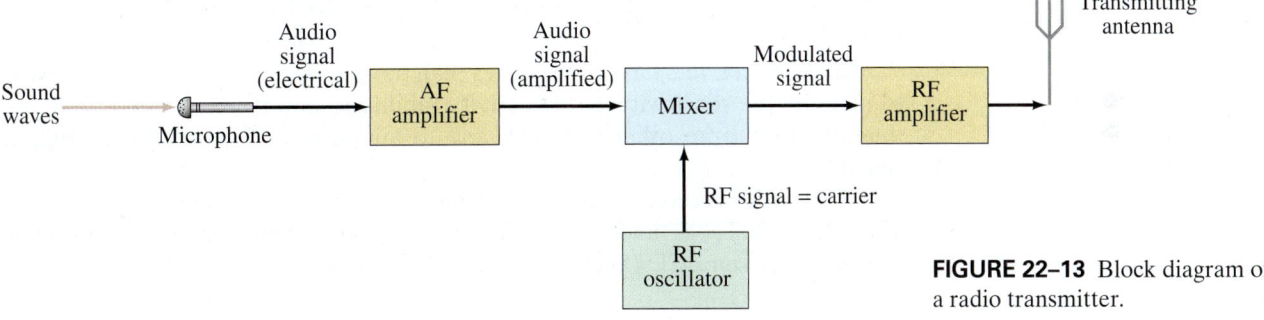

FIGURE 22–13 Block diagram of a radio transmitter.

The mixing of the audio and carrier frequencies is done in two ways. In **amplitude modulation** (AM), the amplitude of the high-frequency carrier wave is made to vary in proportion to the amplitude of the audio signal, as shown in Fig. 22–14. It is called "amplitude modulation" because the *amplitude* of the carrier is altered ("modulate" means to change or alter). In **frequency modulation** (FM), the *frequency* of the carrier wave is made to change in proportion to the audio signal's amplitude, as shown in Fig. 22–15. The mixed signal is amplified further and sent to the transmitting antenna (Fig. 22–13), where the complex mixture of frequencies is sent out in the form of EM waves. In digital communication, the signal is put into digital form (Section 17–10) which modulates the carrier.

A television transmitter works in a similar way, using FM for audio and AM for video; both audio and video signals are mixed with carrier frequencies.

PHYSICS APPLIED
AM and FM

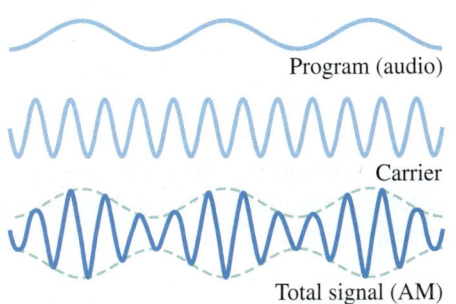

FIGURE 22–14 In amplitude modulation (AM), the amplitude of the carrier signal is made to vary in proportion to the audio signal's amplitude.

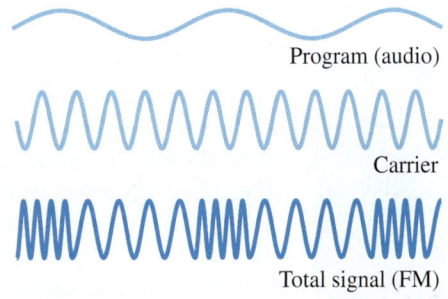

FIGURE 22–15 In frequency modulation (FM), the frequency of the carrier signal is made to change in proportion to the audio signal's amplitude. This method is used by FM radio and television.

FIGURE 22–16 Block diagram of a simple radio receiver.

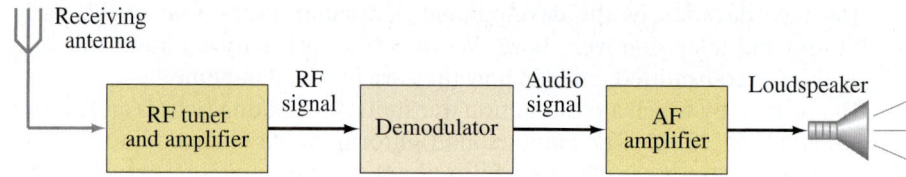

PHYSICS APPLIED
Radio and TV receivers

FIGURE 22–17 Simple tuning stage of a radio.

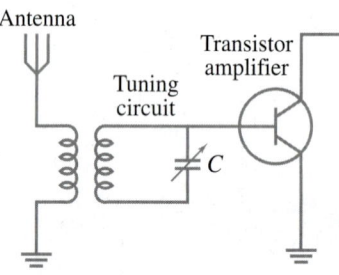

Now let us look at the other end of the process, the reception of radio and TV programs at home. A simple radio receiver is diagrammed in Fig. 22–16. The EM waves sent out by all stations are received by the antenna. The signals the antenna detects and sends to the receiver are very small and contain frequencies from many different stations. The receiver uses a resonant LC circuit (Section 21–15) to select out a particular RF frequency (actually a narrow range of frequencies) corresponding to a particular station. A simple way of tuning a station is shown in Fig. 22–17. A particular station is "tuned in" by adjusting C and/or L so that the resonant frequency of the circuit ($f_0 = 1/(2\pi\sqrt{LC})$, Eq. 21–19) equals that of the station's carrier frequency. The signal, containing both audio and carrier frequencies, next goes to the *demodulator*, or *detector* (Fig. 22–16), where "demodulation" takes place—that is, the audio signal is separated from the RF carrier frequency. The audio signal is amplified and sent to a loudspeaker or headphones.

Modern receivers have more stages than those shown. Various means are used to increase the sensitivity and selectivity (ability to detect weak signals and distinguish them from other stations), and to minimize distortion of the original signal.[†]

A television receiver does similar things to both the audio and the video signals. The audio signal goes finally to the loudspeaker, and the video signal to the monitor screen, such as an LCD (Sections 17–11 and 24–11).

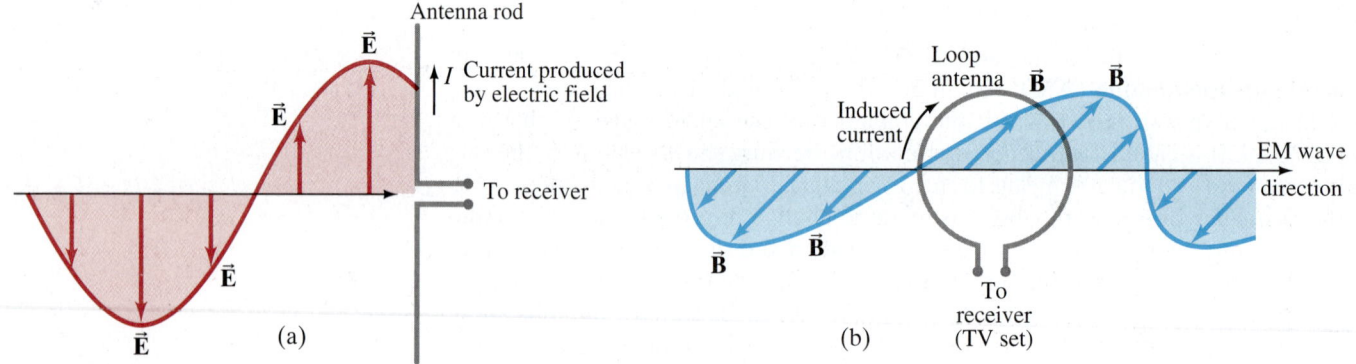

FIGURE 22–18 Antennas. (a) Electric field of EM wave produces a current in an antenna consisting of straight wire or rods. (b) The moving and changing magnetic field induces an emf and current in a loop antenna.

FIGURE 22–19 A satellite dish.

One kind of antenna consists of one or more conducting rods; the electric field in the EM waves exerts a force on the electrons in the conductor, causing them to move back and forth at the frequencies of the waves (Fig. 22–18a). A second type of antenna consists of a tubular coil of wire which detects the magnetic field of the wave: the changing B field induces an emf in the coil (Fig. 22–18b). A satellite dish (Fig. 22–19) consists of a parabolic reflector that focuses the EM waves onto a "horn," similar to a concave mirror telescope (Fig. 25–22).

[†]For *FM stereo broadcasting*, two signals are carried by the carrier wave. One signal contains frequencies up to about 15 kHz, which includes most audio frequencies. The other signal includes the same range of frequencies, but 19 kHz is added to it. A stereo receiver subtracts this 19,000-Hz signal and distributes the two signals to the left and right channels. The first signal consists of the sum of left and right channels (L + R), so monophonic radios (one speaker) detect all the sound. The second signal is the difference between left and right (L − R). Hence a stereo receiver must add and subtract the two signals to get pure left and right signals for each channel.

EXAMPLE 22–7 **Tuning a station.** Calculate the transmitting wavelength of an FM radio station that transmits at 100.1 MHz.

APPROACH Radio is transmitted as an EM wave, so the speed is $c = 3.0 \times 10^8$ m/s. The wavelength is found from Eq. 22–4, $\lambda = c/f$.

SOLUTION The carrier frequency is $f = 100.1$ MHz $\approx 1.0 \times 10^8$ s^{-1}, so

$$\lambda = \frac{c}{f} = \frac{(3.0 \times 10^8 \text{ m/s})}{(1.0 \times 10^8 \text{ s}^{-1})} = 3.0 \text{ m}.$$

NOTE The wavelengths of other FM signals (88 MHz to 108 MHz) are close to the 3.0-m wavelength of this station. FM antennas are typically 1.5 m long, or about a half wavelength. This length is chosen so that the antenna reacts in a resonant fashion and thus is more sensitive to FM frequencies. AM radio antennas would have to be very long and impractical to be either $\frac{1}{2}\lambda$ or $\frac{1}{4}\lambda$.

Other EM Wave Communications

The various regions of the radio-wave spectrum are assigned by governmental agencies for various purposes. Besides those mentioned above, there are "bands" assigned for use by ships, airplanes, police, military, amateurs, satellites and space, and radar. Cell phones, for example, are complete radio transmitters and receivers. In the U.S., CDMA cell phones function on two different bands: 800 MHz and 1900 MHz (= 1.9 GHz). Europe, Asia, and much of the rest of the world use a different system: the international standard called GSM (Global System for Mobile Communication), on 900-MHz and 1800-MHz bands. The U.S. now also has the GSM option (at 850 MHz and 1.9 GHz), as does much of the rest of the Americas. A 700-MHz band is being made available for cell phones (it used to carry TV broadcast channels, no longer used). Radio-controlled toys (cars, sailboats, robotic animals, etc.) can use various frequencies from 27 MHz to 75 MHz. Automobile remote entry (keyless) may operate around 300 MHz or 400 MHz.

Cable TV channels are carried as electromagnetic waves along a coaxial cable (Fig. 22–9) rather than being broadcast and received through the "air." The channels are in the same part of the EM spectrum, hundreds of MHz, but some are at frequencies not available for TV broadcast. Digital satellite TV and radio are carried in the microwave portion of the spectrum (12 to 14 GHz and 2.3 GHz, respectively).

PHYSICS APPLIED
Cell phones, remote controls, cable TV, satellite TV and radio

Wireless from the Moon

In 1969, astronauts first landed on the Moon. It was shown live on television (Fig. 22–20). The transmitting TV camera can be seen in the Chapter-Opening photo, page 625. At that time, someone pointed out that Columbus and other early navigators could have imagined that humans might one day reach the Moon. But they would never have believed possible that moving images could be sent from the Moon to the Earth through empty space.

FIGURE 22–20 The first person on the Moon, Neil Armstrong, July 20, 1969, pointed out "One small step for a man, one giant leap for mankind."

Summary

James Clerk Maxwell synthesized an elegant theory in which all electric and magnetic phenomena could be described using four equations, now called **Maxwell's equations**. They are based on earlier ideas, but Maxwell added one more—that a changing electric field produces a magnetic field.

Maxwell's theory predicted that transverse **electromagnetic (EM) waves** would be produced by accelerating electric charges, and these waves would propagate (move) through space at the speed of light:

$$c = \frac{1}{\sqrt{\epsilon_0 \mu_0}} = 3.00 \times 10^8 \text{ m/s}. \quad (22\text{--}3)$$

The oscillating electric and magnetic fields in an EM wave are perpendicular to each other and to the direction of propagation. These EM waves are waves of fields, not matter, and can propagate in empty space.

The wavelength λ and frequency f of EM waves are related to their speed c by

$$c = \lambda f \quad (22\text{--}4)$$

just as for other waves.

After EM waves were experimentally detected, it became generally accepted that light is an EM wave. The **electromagnetic spectrum** includes EM waves of a wide variety of wavelengths, from microwaves and radio waves to visible light to X-rays and gamma rays, all of which travel through space at a speed $c = 3.0 \times 10^8$ m/s.

The average *intensity* (W/m²) of an EM wave is

$$\bar{I} = \frac{1}{2}\epsilon_0 c E_0^2 = \frac{1}{2}\frac{c}{\mu_0}B_0^2 = \frac{1}{2}\frac{E_0 B_0}{\mu_0}, \quad (22\text{-}8)$$

where E_0 and B_0 are the peak values of the electric and magnetic fields, respectively, in the wave.

EM waves carry momentum and exert a **radiation pressure** proportional to the intensity I of the wave.

Radio, TV, cell phone, and other wireless signals are transmitted through space in the radio-wave or microwave part of the EM spectrum.

Questions

1. The electric field in an EM wave traveling north oscillates in an east–west plane. Describe the direction of the magnetic field vector in this wave. Explain.
2. Is sound an EM wave? If not, what kind of wave is it?
3. Can EM waves travel through a perfect vacuum? Can sound waves?
4. When you flip a light switch on, does the light go on immediately? Explain.
5. Are the wavelengths of radio and television signals longer or shorter than those detectable by the human eye?
6. When you connect two loudspeakers to the output of a stereo amplifier, should you be sure the lead-in wires are equal in length to avoid a time lag between speakers? Explain.
7. In the electromagnetic spectrum, what type of EM wave would have a wavelength of 10^3 km? 1 km? 1 m? 1 cm? 1 mm? 1 μm?
8. Can radio waves have the same frequencies as sound waves (20 Hz–20,000 Hz)?
9. If a radio transmitter has a vertical antenna, should a receiver's antenna (rod type) be vertical or horizontal to obtain best reception?
10. The carrier frequencies of FM broadcasts are much higher than for AM broadcasts. On the basis of what you learned about diffraction in Chapter 11, explain why AM signals can be detected more readily than FM signals behind low hills or buildings.
11. Discuss how cordless telephones make use of EM waves. What about cell phones?
12. A lost person may signal by switching a flashlight on and off using Morse code. This is actually a modulated EM wave. Is it AM or FM? What is the frequency of the carrier, approximately?

MisConceptual Questions

1. In a vacuum, what is the difference between a radio wave and an X-ray?
   (*a*) Wavelength.   (*b*) Frequency.   (*c*) Speed.
2. The radius of an atom is on the order of 10^{-10} m. In comparison, the wavelength of visible light is
   (*a*) much smaller. (*b*) about the same size. (*c*) much larger.
3. Which of the following travel at the same speed as light? (Choose all that apply.)
   (*a*) Radio waves.   (*d*) Ultrasonic waves.   (*g*) Gamma rays.
   (*b*) Microwaves.   (*e*) Infrared radiation.   (*h*) X-rays.
   (*c*) Radar.   (*f*) Cell phone signals.
4. Which of the following types of electromagnetic radiation travels the fastest?
   (*a*) Radio waves.
   (*b*) Visible light waves.
   (*c*) X-rays.
   (*d*) Gamma rays.
   (*e*) All the above travel at the same speed.
5. In empty space, which quantity is always larger for X-ray radiation than for a radio wave?
   (*a*) Amplitude.   (*c*) Frequency.
   (*b*) Wavelength.   (*d*) Speed.
6. If electrons in a wire vibrate up and down 1000 times per second, they will create an electromagnetic wave having
   (*a*) a wavelength of 1000 m.   (*c*) a speed of 1000 m/s.
   (*b*) a frequency of 1000 Hz.   (*d*) an amplitude of 1000 m.
7. If the Earth–Sun distance were doubled, the intensity of radiation from the Sun that reaches the Earth's surface would
   (*a*) quadruple. (*b*) double. (*c*) drop to $\frac{1}{2}$. (*d*) drop to $\frac{1}{4}$.
8. An electromagnetic wave is traveling straight down toward the center of the Earth. At a certain moment in time the electric field points west. In which direction does the magnetic field point at this moment?
   (*a*) North.   (*d*) West.   (*g*) Either (*a*) or (*b*).
   (*b*) South.   (*e*) Up.   (*h*) Either (*c*) or (*d*).
   (*c*) East.   (*f*) Down.   (*i*) Either (*e*) or (*f*).
9. If the intensity of an electromagnetic wave doubles,
   (*a*) the electric field must also double.
   (*b*) the magnetic field must also double.
   (*c*) both the magnetic field and the electric field must increase by a factor of $\sqrt{2}$.
   (*d*) Any of the above.
10. If all else is the same, for which surface would the radiation pressure from light be the greatest?
    (*a*) A black surface.
    (*b*) A gray surface.
    (*c*) A yellow surface.
    (*d*) A white surface.
    (*e*) All experience the same radiation pressure, because they are exposed to the same light.
11. Starting in 2009, TV stations in the U.S. switched to digital signals. [See Sections 22–7, 17–10, and 17–11.] To watch today's digital broadcast TV, could you use a pre-2009 TV antenna meant for analog? Explain.
    (*a*) No; analog antennas do not receive digital signals.
    (*b*) No; digital signals are broadcast at different frequencies, so you need a different antenna.
    (*c*) Yes; digital signals are broadcast with the same carrier frequencies, so your old antenna will be fine.
    (*d*) No; you cannot receive digital signals through an antenna and need to switch to cable or satellite.

Problems

22–1 B Produced by Changing E

*1. (II) Determine the rate at which the electric field changes between the round plates of a capacitor, 8.0 cm in diameter, if the plates are spaced 1.1 mm apart and the voltage across them is changing at a rate of 120 V/s.

*2. (II) Calculate the displacement current I_D between the square plates, 5.8 cm on a side, of a capacitor if the electric field is changing at a rate of 1.6×10^6 V/m·s.

*3. (II) At a given instant, a 3.8-A current flows in the wires connected to a parallel-plate capacitor. What is the rate at which the electric field is changing between the plates if the square plates are 1.60 cm on a side?

*4. (III) A 1500-nF capacitor with circular parallel plates 2.0 cm in diameter is accumulating charge at the rate of 32.0 mC/s at some instant in time. What will be the induced magnetic field strength 10.0 cm radially outward from the center of the plates? What will be the value of the field strength after the capacitor is fully charged?

22–2 EM Waves

5. (I) If the electric field in an EM wave has a peak magnitude of 0.72×10^{-4} V/m, what is the peak magnitude of the magnetic field strength?

6. (I) If the magnetic field in a traveling EM wave has a peak magnitude of 10.5 nT, what is the peak magnitude of the electric field?

7. (I) In an EM wave traveling west, the B field oscillates up and down vertically and has a frequency of 90.0 kHz and an rms strength of 7.75×10^{-9} T. Determine the frequency and rms strength of the electric field. What is the direction of its oscillations?

8. (I) How long does it take light to reach us from the Sun, 1.50×10^8 km away?

9. (II) How long should it take the voices of astronauts on the Moon to reach the Earth? Explain in detail.

22–3 Electromagnetic Spectrum

10. (I) An EM wave has a wavelength of 720 nm. What is its frequency, and how would we classify it?

11. (I) An EM wave has frequency 7.14×10^{14} Hz. What is its wavelength, and how would we classify it?

12. (I) A widely used "short-wave" radio broadcast band is referred to as the 49-m band. What is the frequency of a 49-m radio signal?

13. (I) What is the frequency of a microwave whose wavelength is 1.50 cm?

14. (II) Electromagnetic waves and sound waves can have the same frequency. (a) What is the wavelength of a 1.00-kHz electromagnetic wave? (b) What is the wavelength of a 1.00-kHz sound wave? (The speed of sound in air is 341 m/s.) (c) Can you hear a 1.00-kHz electromagnetic wave?

15. (II) (a) What is the wavelength of a 22.75×10^9 Hz radar signal? (b) What is the frequency of an X-ray with wavelength 0.12 nm?

16. (II) How long would it take a message sent as radio waves from Earth to reach Mars when Mars is (a) nearest Earth, (b) farthest from Earth? Assume that Mars and Earth are in the same plane and that their orbits around the Sun are circles (Mars is $\approx 230 \times 10^6$ km from the Sun).

17. (II) Our nearest star (other than the Sun) is 4.2 light-years away. That is, it takes 4.2 years for the light it emits to reach Earth. How far away is it in meters?

18. (II) A light-year is a measure of distance (not time). How many meters does light travel in a year?

19. (II) Pulsed lasers used for science and medicine produce very brief bursts of electromagnetic energy. If the laser light wavelength is 1062 nm (Neodymium–YAG laser), and the pulse lasts for 34 picoseconds, how many wavelengths are found within the laser pulse? How brief would the pulse need to be to fit only one wavelength?

22–4 Measuring the Speed of Light

20. (II) What is the minimum angular speed at which Michelson's eight-sided mirror would have had to rotate to reflect light into an observer's eye by succeeding mirror faces (1/8 of a revolution, Fig. 22–10)?

21. (II) A student wants to scale down Michelson's light-speed experiment to a size that will fit in one room. An eight-sided mirror is available, and the stationary mirror can be mounted 12 m from the rotating mirror. If the arrangement is otherwise as shown in Fig. 22–10, at what minimum rate must the mirror rotate?

22–5 Energy in EM Wave

22. (I) The $\vec{E}$ field in an EM wave has a peak of 22.5 mV/m. What is the average rate at which this wave carries energy across unit area per unit time?

23. (II) The magnetic field in a traveling EM wave has an rms strength of 22.5 nT. How long does it take to deliver 365 J of energy to 1.00 cm² of a wall that it hits perpendicularly?

24. (II) How much energy is transported across a 1.00-cm² area per hour by an EM wave whose E field has an rms strength of 30.8 mV/m?

25. (II) A spherically spreading EM wave comes from an 1800-W source. At a distance of 5.0 m, what is the intensity, and what is the rms value of the electric field?

26. (II) If the amplitude of the B field of an EM wave is 2.2×10^{-7} T, (a) what is the amplitude of the E field? (b) What is the average power transported across unit area by the EM wave?

27. (II) What is the average energy contained in a 1.00-m³ volume near the Earth's surface due to radiant energy from the Sun? See Example 22–4.

28. (II) A 15.8-mW laser puts out a narrow beam 2.40 mm in diameter. What are the rms values of E and B in the beam?

29. (II) Estimate the average power output of the Sun, given that about 1350 W/m² reaches the upper atmosphere of the Earth.

30. (II) A high-energy pulsed laser emits a 1.0-ns-long pulse of average power 1.5×10^{11} W. The beam is nearly a cylinder 2.2×10^{-3} m in radius. Determine (a) the energy delivered in each pulse, and (b) the rms value of the electric field.

22–6 Radiation Pressure

31. (II) Estimate the radiation pressure due to a bulb that emits 25 W of EM radiation at a distance of 9.5 cm from the center of the bulb. Estimate the force exerted on your fingertip if you place it at this point.

32. (II) What size should the solar panel on a satellite orbiting Jupiter be if it is to collect the same amount of radiation from the Sun as a 1.0-m^2 solar panel on a satellite orbiting Earth? [*Hint*: Assume the inverse square law (Eq. 11–16b).]

33. (III) Suppose you have a car with a 100-hp engine. How large a solar panel would you need to replace the engine with solar power? Assume that the solar panels can utilize 20% of the maximum solar energy that reaches the Earth's surface (1000 W/m^2).

22–7 Radio, TV

34. (I) What is the range of wavelengths for (*a*) FM radio (88 MHz to 108 MHz) and (*b*) AM radio (535 kHz to 1700 kHz)?

35. (I) Estimate the wavelength for a 1.9-GHz cell phone transmitter.

36. (I) Compare 980 on the AM dial to 98.1 on FM. Which has the longer wavelength, and by what factor is it larger?

37. (I) What are the wavelengths for two TV channels that broadcast at 54.0 MHz (Channel 2) and 692 MHz (Channel 51)?

38. (I) The variable capacitor in the tuner of an AM radio has a capacitance of 2500 pF when the radio is tuned to a station at 550 kHz. What must the capacitance be for a station near the other end of the dial, 1610 kHz?

39. (I) The oscillator of a 98.3-MHz FM station has an inductance of 1.8 μH. What value must the capacitance be?

40. (II) A certain FM radio tuning circuit has a fixed capacitor $C = 810$ pF. Tuning is done by a variable inductance. What range of values must the inductance have to tune stations from 88 MHz to 108 MHz?

41. (II) An amateur radio operator wishes to build a receiver that can tune a range from 14.0 MHz to 15.0 MHz. A variable capacitor has a minimum capacitance of 86 pF. (*a*) What is the required value of the inductance? (*b*) What is the maximum capacitance used on the variable capacitor?

42. (II) A satellite beams microwave radiation with a power of 13 kW toward the Earth's surface, 550 km away. When the beam strikes Earth, its circular diameter is about 1500 m. Find the rms electric field strength of the beam.

43. (III) A 1.60-m-long FM antenna is oriented parallel to the electric field of an EM wave. How large must the electric field be to produce a 1.00-mV (rms) voltage between the ends of the antenna? What is the rate of energy transport per m^2?

General Problems

44. Who will hear the voice of a singer first: a person in the balcony 50.0 m away from the stage (see Fig. 22–21), or a person 1200 km away at home whose ear is next to the radio listening to a live broadcast? Roughly how much sooner? Assume the microphone is a few centimeters from the singer and the temperature is 20°C.

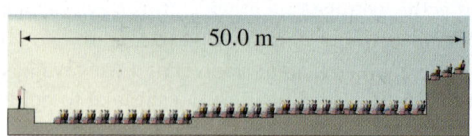

FIGURE 22–21 Problem 44.

45. A global positioning system (GPS) functions by determining the travel times for EM waves from various satellites to a land-based GPS receiver. If the receiver is to detect a change in travel distance on the order of 3 m, what is the associated change in travel time (in ns) that must be measured?

46. Light is emitted from an ordinary lightbulb filament in wave-train bursts about 10^{-8} s in duration. What is the length in space of such wave trains?

47. The voice from an astronaut on the Moon (Fig. 22–22) was beamed to a listening crowd on Earth. If you were standing 28 m from the loudspeaker on Earth, what was the total time lag between when you heard the sound and when the sound entered a microphone on the Moon? Explain whether the microphone was inside the space helmet, or outside, and why.

FIGURE 22–22 Problem 47.

48. Radio-controlled clocks throughout the United States receive a radio signal from a transmitter in Fort Collins, Colorado, that accurately (within a microsecond) marks the beginning of each minute. A slight delay, however, is introduced because this signal must travel from the transmitter to the clocks. Assuming Fort Collins is no more than 3000 km from any point in the U.S., what is the longest travel-time delay?

49. If the Sun were to disappear or radically change its output, how long would it take for us on Earth to learn about it?

50. Cosmic microwave background radiation fills space with an average energy density of about 4 × 10^{-14} J/m^3. (*a*) Find the rms value of the electric field associated with this radiation. (*b*) How far from a 7.5-kW radio transmitter emitting uniformly in all directions would you find a comparable value?

51. What are E_0 and B_0 at a point 2.50 m from a light source whose output is 18 W? Assume the bulb emits radiation of a single frequency uniformly in all directions.

52. Estimate the rms electric field in the sunlight that hits Mars, knowing that the Earth receives about 1350 W/m^2 and that Mars is 1.52 times farther from the Sun (on average) than is the Earth.

53. The average intensity of a particular TV station's signal is 1.0 × 10^{-13} W/m^2 when it arrives at a 33-cm-diameter satellite TV antenna. (*a*) Calculate the total energy received by the antenna during 4.0 hours of viewing this station's programs. (*b*) Estimate the amplitudes of the E and B fields of the EM wave.

54. What length antenna would be appropriate for a portable device that could receive satellite TV?

55. A radio station is allowed to broadcast at an average power not to exceed 25 kW. If an electric field amplitude of 0.020 V/m is considered to be acceptable for receiving the radio transmission, estimate how many kilometers away you might be able to detect this station.

56. The radiation pressure (Section 22–6) created by electromagnetic waves might someday be used to power spacecraft through the use of a "solar sail," Example 22–6. (a) Assuming total reflection, what would be the pressure on a solar sail located at the same distance from the Sun as the Earth (where $I = 1350$ W/m^2)? (b) Suppose the sail material has a mass of 1 g/m^2. What would be the acceleration of the sail due to solar radiation pressure? (c) A realistic solar sail would have a payload. How big a sail would you need to accelerate a 100-kg payload at 1×10^{-3} m/s^2?

57. Suppose a 35-kW radio station emits EM waves uniformly in all directions. (a) How much energy per second crosses a 1.0-m^2 area 1.0 km from the transmitting antenna? (b) What is the rms magnitude of the $\vec{E}$ field at this point, assuming the station is operating at full power? What is the rms voltage induced in a 1.0-m-long vertical car antenna (c) 1.0 km away, (d) 50 km away?

58. A point source emits light energy uniformly in all directions at an average rate P_0 with a single frequency f. Show that the peak electric field in the wave is given by

$$E_0 = \sqrt{\frac{\mu_0 c P_0}{2\pi r^2}}.$$

[*Hint*: The surface area of a sphere is $4\pi r^2$.]

59. What is the maximum power level of a radio station so as to avoid electrical breakdown of air at a distance of 0.65 m from the transmitting antenna? Assume the antenna is a point source. Air breaks down in an electric field of about 3×10^6 V/m.

60. Estimate how long an AM antenna would have to be if it were (a) $\frac{1}{2}\lambda$ or (b) $\frac{1}{4}\lambda$. AM radio is roughly 1 MHz (530 kHz to 1.7 MHz).

61. 12 km from a radio station's transmitting antenna, the amplitude of the electric field is 0.12 V/m. What is the average power output of the radio station?

Search and Learn

1. How practical is solar power for various devices? Assume that on a sunny day, sunlight has an intensity of 1000 W/m^2 at the surface of Earth and that a solar-cell panel can convert 20% of that sunlight into electric power. Calculate the area A of solar panel needed to power (a) a calculator that consumes 50 mW, (b) a hair dryer that consumes 1500 W, (c) a car that would require 40 hp. (d) In each case, would the area A be small enough to be mounted on the device itself, or in the case of (b) on the roof of a house?

2. A powerful laser portrayed in a movie provides a 3-mm diameter beam of green light with a power of 3 W. A good agent inside the Space Shuttle aims the laser beam at an enemy astronaut hovering outside. The mass of the enemy astronaut is 120 kg and the Space Shuttle 103,000 kg. (a) Determine the "radiation-pressure" force exerted on the enemy by the laser beam assuming her suit is perfectly reflecting. (b) If the enemy is 30 m from the Shuttle's center of mass, estimate the gravitational force the Shuttle exerts on the enemy. (c) Which of the two forces is larger, and by what factor?

3. The Arecibo radio telescope in Puerto Rico can detect a radio wave with an intensity as low as 1×10^{-23} W/m^2. Consider a "best-case" scenario for communication with extraterrestrials: suppose an advanced civilization a distance x away from Earth is able to transform the entire power output of a Sun-like star completely into a radio-wave signal which is transmitted uniformly in all directions. (a) In order for Arecibo to detect this radio signal, what is the maximum value for x in light-years (1 ly $\approx 10^{16}$ m)? (b) How does this maximum value compare with the 100,000-ly size of our Milky Way galaxy? The intensity of sunlight at Earth's orbital distance from the Sun is 1350 W/m^2. [*Hint*: Assume the inverse square law (Eq. 11–16b).]

4. Laser light can be focused (at best) to a spot with a radius r equal to its wavelength λ. Suppose a 1.0-W beam of green laser light ($\lambda = 5 \times 10^{-7}$ m) forms such a spot and illuminates a cylindrical object of radius r and length r (Fig. 22–23). Estimate (a) the radiation pressure and force on the object, and (b) its acceleration, if its density equals that of water and it absorbs all the radiation. [This order-of-magnitude calculation convinced researchers of the feasibility of "optical tweezers," page 636.]

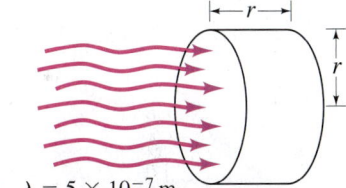

FIGURE 22–23
Search and Learn 4. $\lambda = 5 \times 10^{-7}$ m

ANSWERS TO EXERCISES

A: (c).
B: (b).
C: (a) 3.8×10^6 Hz; (b) 5.5×10^{18} Hz.
D: 45 cm.
E: Yes; the signal still travels 72,000 km.
F: Over 4 hours.

Reflection from still water, as from a glass mirror, can be analyzed using the ray model of light.

Is this picture right side up, or upside down? How can you tell? What are the clues? Notice the people and position of the Sun. Ray diagrams, which we will learn to draw in this Chapter, can provide the answer. See Example 23–3.

In this first Chapter on light and optics, we use the ray model of light to understand the formation of images by mirrors, both plane and curved (spherical). We also study refraction—how light rays bend when they go from one medium to another—and how, via refraction, images are formed by lenses, which are the crucial part of so many optical instruments.

Light: Geometric Optics

CONTENTS

23–1 The Ray Model of Light
23–2 Reflection; Image Formation by a Plane Mirror
23–3 Formation of Images by Spherical Mirrors
23–4 Index of Refraction
23–5 Refraction: Snell's Law
23–6 Total Internal Reflection; Fiber Optics
23–7 Thin Lenses; Ray Tracing
23–8 The Thin Lens Equation
*23–9 Combinations of Lenses
*23–10 Lensmaker's Equation

CHAPTER-OPENING QUESTIONS—Guess now!

1. A 2.0-m-tall person is standing 2.0 m from a flat vertical mirror staring at her image. What minimum height must the mirror's reflecting glass have if the person is to see her entire body, from the top of her head to her feet?
  (a) 0.50 m. **(b)** 1.0 m. **(c)** 1.5 m. **(d)** 2.0 m. **(e)** 2.5 m.

2. The focal length of a lens is
  (a) the diameter of the lens.
  (b) the thickness of the lens.
  (c) the distance from the lens at which incoming parallel rays bend to intersect at a point.
  (d) the distance from the lens at which all real images are formed.

The sense of sight is extremely important to us, for it provides us with a large part of our information about the world. How do we see? What is the something called *light* that enters our eyes and causes the sensation of sight? How does light behave so that we can see everything that we do? We saw in Chapter 22 that light can be considered a form of electromagnetic radiation. We now examine the subject of light in detail in the next three Chapters.

We see an object in one of two ways: (1) the object may be a *source* of light, such as a lightbulb, a flame, or a star, in which case we see the light emitted directly from the source; or, more commonly, (2) we see an object by light *reflected* from it.

In the latter case, the light may have originated from the Sun, artificial lights, or a campfire. An understanding of how objects *emit* light was not achieved until the 1920s, and will be discussed in Chapter 27. How light is *reflected* from objects was understood much earlier, and will be discussed in Section 23–2.

23–1 The Ray Model of Light

A great deal of evidence suggests that *light travels in straight lines* under a wide variety of circumstances.[†] For example, a source of light like the Sun (which at its great distance from us is nearly a "point source") casts distinct shadows, and the beam from a laser pointer appears to be a straight line. In fact, we infer the positions of objects in our environment by assuming that light moves from the object to our eyes in straight-line paths. Our orientation to the physical world is based on this assumption.

This reasonable assumption is the basis of the **ray model** of light. This model assumes that light travels in straight-line paths called light **rays**. Actually, a ray is an idealization; it is meant to represent an extremely narrow beam of light. When we see an object, according to the ray model, light reaches our eyes from each point on the object. Although light rays leave each point in many different directions, normally only a small bundle of these rays can enter the pupil of an observer's eye, as shown in Fig. 23–1. If the person's head moves to one side, a different bundle of rays will enter the eye from each point.

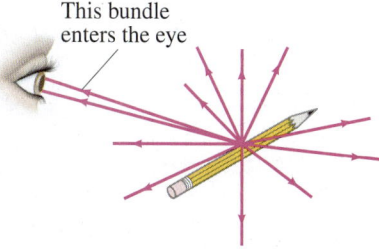

FIGURE 23–1 Light rays come from each single point on an object. A small bundle of rays leaving one point is shown entering a person's eye.

We saw in Chapter 22 that light can be considered as an electromagnetic wave. Although the ray model of light does not deal with this aspect of light (we discuss the wave nature of light in Chapter 24), the ray model has been very successful in describing many aspects of light such as reflection, refraction, and the formation of images by mirrors and lenses. Because these explanations involve straight-line rays at various angles, this subject is referred to as **geometric optics**.

23–2 Reflection; Image Formation by a Plane Mirror

When light strikes the surface of an object, some of the light is reflected. The rest can be absorbed by the object (and transformed to thermal energy) or, if the object is transparent like glass or water, part can be transmitted through. For a very smooth shiny object such as a silvered mirror, over 95% of the light may be reflected.

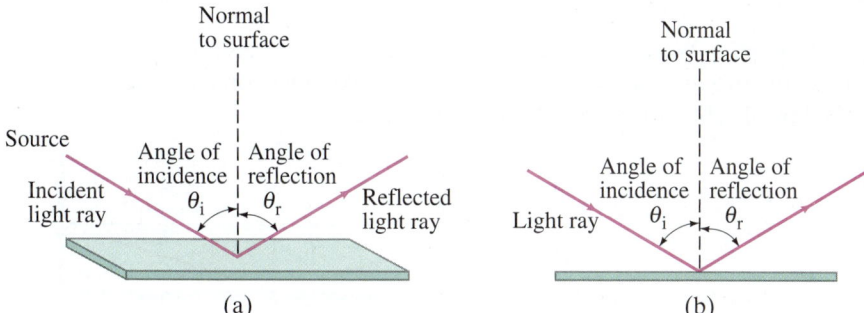

FIGURE 23–2 Law of reflection: (a) shows a 3-D view of an incident ray being reflected at the top of a flat surface; (b) shows a side or "end-on" view, which we will usually use because of its clarity.

When a narrow beam of light strikes a flat surface (Fig. 23–2), we define the **angle of incidence**, θ_i, to be the angle an incident ray makes with the normal (perpendicular) to the surface, and the **angle of reflection**, θ_r, to be the angle the reflected ray makes with the normal. It is found that the *incident and reflected rays lie in the same plane with the normal to the surface*, and that

the angle of reflection equals the angle of incidence, $\theta_r = \theta_i$.

This is the **law of reflection**, and it is depicted in Fig. 23–2. It was known to the ancient Greeks, and you can confirm it yourself by shining a narrow flashlight beam or a laser pointer at a mirror in a darkened room.

LAW OF REFLECTION

[†]In a uniform transparent medium such as air or glass: But not always, such as for nonuniform air that allows optical illusions and mirages which we discuss in Section 24–2 (Fig. 24–4).

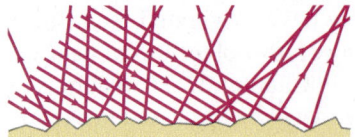

FIGURE 23–3 Diffuse reflection from a rough surface.

When light is incident upon a rough surface, even microscopically rough such as this page, it is reflected in many directions, as shown in Fig. 23–3. This is called **diffuse reflection**. The law of reflection still holds, however, at each small section of the surface. Because of diffuse reflection in all directions, an ordinary object can be seen at many different angles by the light reflected from it. When you move your head to the side, different reflected rays reach your eye from each point on the object (such as this page), Fig. 23–4a. Let us compare diffuse reflection to reflection from a mirror, which is known as **specular reflection**. ("Speculum" is Latin for mirror.) When a narrow beam of light shines on a mirror, the light will not reach your eye unless your eye is positioned at just the right place where the law of reflection is satisfied, as shown in Fig. 23–4b. This is what gives rise to the special image-forming properties of mirrors.

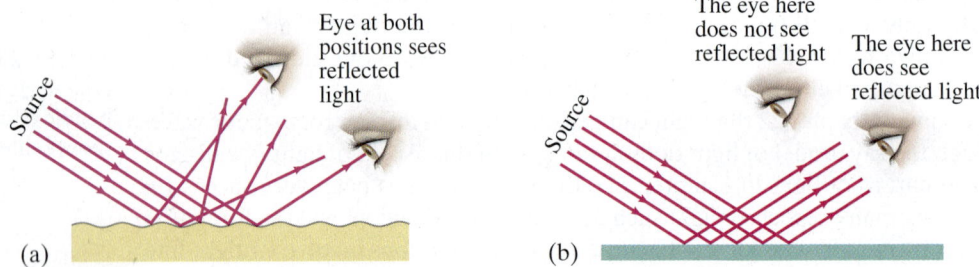

FIGURE 23–4 A narrow beam of light shines on (a) white paper, and (b) a mirror. In part (a), you can see with your eye the white light (and printed words) reflected at various positions because of diffuse reflection. But in part (b), you see the reflected light only when your eye is placed correctly $(\theta_r = \theta_i)$; mirror reflection is also known as specular reflection. (Galileo, using similar arguments, showed that the Moon must have a rough surface rather than a highly polished surface like a mirror, as some people thought.)

EXAMPLE 23–1 Reflection from flat mirrors. Two flat mirrors are perpendicular to each other. An incoming beam of light makes an angle of 15° with the first mirror as shown in Fig. 23–5a. What angle will the outgoing beam make with the second mirror?

APPROACH We sketch the path of the beam as it reflects off the two mirrors, and draw the two normals to the mirrors for the two reflections. We use geometry and the law of reflection to find the various angles.

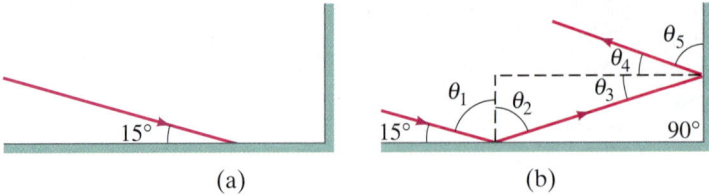

FIGURE 23–5 Example 23–1.

SOLUTION In Fig. 23–5b, $\theta_1 + 15° = 90°$, so $\theta_1 = 75°$; by the law of reflection $\theta_2 = \theta_1 = 75°$ too. Using the fact that the sum of the three angles of a triangle is always 180°, and noting that the two normals to the two mirrors are perpendicular to each other, we have $\theta_2 + \theta_3 + 90° = 180°$. Thus $\theta_3 = 180° - 90° - 75° = 15°$. By the law of reflection, $\theta_4 = \theta_3 = 15°$, so $\theta_5 = 75°$ is the angle the reflected ray makes with the second mirror surface.

NOTE The outgoing ray is parallel to the incoming ray. Reflectors on bicycles, cars, and other applications use this principle.

When you look straight into a mirror, you see what appears to be yourself as well as various objects around and behind you, Fig. 23–6. Your face and the other objects look as if they are in front of you, beyond the mirror. But what you see in the mirror is an **image** of the objects, including yourself, that are in front of the mirror. Also, you don't see yourself as others see you, because left and right appear reversed in the image.

A **plane mirror** is one with a smooth flat reflecting surface. Figure 23–7 shows how an image is formed by a plane mirror according to the ray model. We are viewing the mirror, on edge, in the diagram of Fig. 23–7, and the rays are shown reflecting from the front surface. (Good mirrors are generally made by putting a highly reflective metallic coating on one surface of a very flat piece of glass.) Rays from two different points on an object (the bottle on the left in Fig. 23–7) are shown: two rays are shown leaving from a point on the top of the bottle, and two more from a point on the bottom. Rays leave each point on the object going in many directions (as in Fig. 23–1), but only those that enclose the bundle of rays that enter the eye from each of the two points are shown. Each set of diverging rays that reflect from the mirror and enter the eye *appear to come from a single point* behind the mirror, called the **image point**, as shown by the dashed lines. That is, our eyes and brain interpret any rays that enter an eye as having traveled straight-line paths. The point from which each bundle of rays seems to come is one point on the image. For each point on the object, there is a corresponding image point. (This analysis of how a plane mirror forms an image was published by Kepler in 1604.)

FIGURE 23–6 When you look in a mirror, you see an image of yourself and objects around you. You don't see yourself as others see you, because left and right appear reversed in the image.

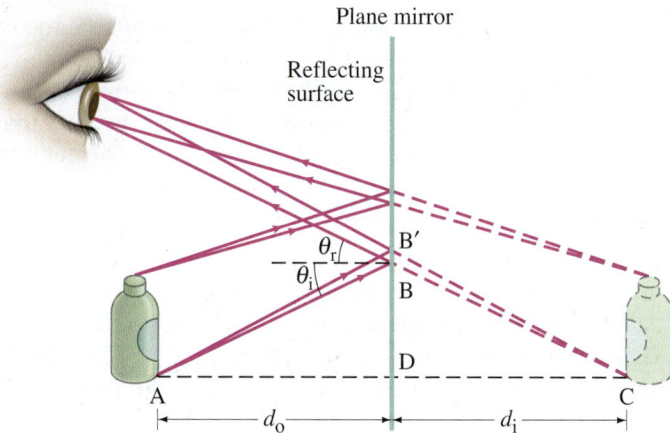

FIGURE 23–7 Formation of a virtual image by a plane mirror. Only the bundle of rays from the top and bottom of the object which reach the eye is shown.

Let us concentrate on the two rays that leave point A on the object in Fig. 23–7, and strike the mirror at points B and B'. We use geometry now, for the rays at B. The angles ADB and CDB are right angles; and because of the law of reflection, $\theta_i = \theta_r$ at point B. Therefore, by geometry, angles ABD and CBD are also equal. The two triangles ABD and CBD are thus congruent, and the length AD = CD. That is, the image appears as far behind the mirror as the object is in front. The **image distance**, d_i (perpendicular distance from mirror to image, Fig. 23–7), equals the **object distance**, d_o (perpendicular distance from object to mirror). From the geometry, we also can see that the height of the image is the same as that of the object.

The light rays do not actually pass through the image location itself in Fig. 23–7. (Note where the red lines are dashed to show they are our projections, not rays.) The image would not appear on paper or film placed at the location of the image. Therefore, it is called a **virtual image**. This is to distinguish it from a **real image** in which the light does pass through the image and which therefore could appear on a white surface, or on film or on an electronic sensor placed at the image position. Our eyes can see both real and virtual images, as long as the diverging rays enter our pupils. We will see that curved mirrors and lenses can form real images, as well as virtual. A movie projector lens, for example, produces a real image that is visible on the screen.

SECTION 23–2 Reflection; Image Formation by a Plane Mirror

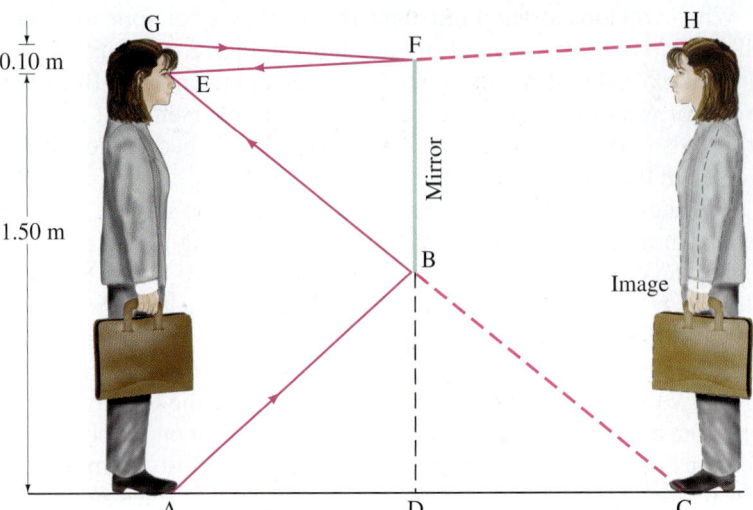

FIGURE 23-8 Seeing oneself in a mirror. Example 23-2.

PHYSICS APPLIED
How tall a mirror do you need to see a reflection of your entire self?

EXAMPLE 23-2 How tall must a full-length mirror be? A woman 1.60 m tall stands in front of a vertical plane mirror. What is the minimum height of the mirror, and how high must its lower edge be above the floor, if she is to be able to see her whole body? Assume that her eyes are 10 cm below the top of her head.

APPROACH For her to see her whole body, light rays from the top of her head (point G) and from the bottom of her foot (A) must reflect from the mirror and enter her eye, Fig. 23-8. We don't show two rays diverging from each point as we did in Fig. 23-7, where we wanted to find where the image is. Now that we know the image is the same distance behind a plane mirror as the object is in front, we only need to show one ray leaving point G (top of head) and one ray leaving point A (her toe), and then use geometry.

SOLUTION First consider the ray that leaves her foot at A, reflects at B, and enters the eye at E. The mirror needs to extend no lower than B. The angle of reflection equals the angle of incidence, so the height BD is half of the height AE. Because AE = 1.60 m − 0.10 m = 1.50 m, then BD = 0.75 m. Similarly, if the woman is to see the top of her head, the top edge of the mirror only needs to reach point F, which is 5 cm below the top of her head (half of GE = 10 cm). Thus, DF = 1.55 m, and the mirror needs to have a vertical height of only (1.55 m − 0.75 m) = 0.80 m. And the mirror's bottom edge must be 0.75 m above the floor.

NOTE We see that a mirror, if positioned at the correct height (as in Fig. 23-8), need be only half as tall as a person for that person to be able to see all of himself or herself.

EXERCISE A Does the result of Example 23-2 depend on your distance from the mirror? (Try it and see, it's fun.)

EXERCISE B Return to Chapter-Opening Question 1, page 644, and answer it again now. Try to explain why you may have answered differently the first time.

CONCEPTUAL EXAMPLE 23-3 Is the photo upside down? Close examination of the photograph on the first page of this Chapter reveals that in the top portion, the image of the Sun is seen clearly, whereas in the lower portion, the image of the Sun is partially blocked by the tree branches. Show why the reflection is not the same as the real scene by drawing a sketch of this situation, showing the Sun, the camera, the branch, and two rays going from the Sun to the camera (one direct and one reflected). Is the photograph right side up?

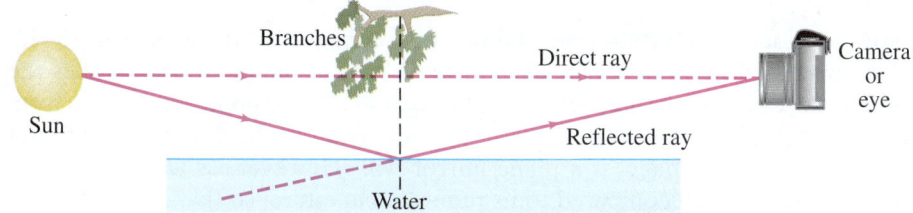

FIGURE 23–9 Example 23–3.

RESPONSE We need to draw two diagrams, one assuming the photo on p. 644 is right side up, and another assuming it is upside down. Figure 23–9 is drawn assuming the photo is upside down. In this case, the Sun blocked by the tree would be the direct view, and the full view of the Sun the reflection: the ray which reflects off the water and into the camera travels at an angle below the branch, whereas the ray that travels directly to the camera passes through the branches. This works. Try to draw a diagram assuming the photo is right side up (thus assuming that the image of the Sun in the reflection is higher above the horizon than it is as viewed directly). It won't work. The photo on p. 644 is upside down.

Also, what about the people in the photo? Try to draw a diagram showing why they don't appear in the reflection. [*Hint*: Assume they are not sitting at the edge of the pool, but back from the edge.] Then try to draw a diagram of the reverse (i.e., assume the photo is right side up so the people are visible only in the reflection). Reflected images are not perfect replicas when different planes (distances) are involved.

23–3 Formation of Images by Spherical Mirrors

Reflecting surfaces can also be *curved*, usually *spherical*, which means they form a section of a sphere. A **spherical mirror** is called **convex** if the reflection takes place on the outer surface of the spherical shape so that the center of the mirror surface bulges out toward the viewer, Fig. 23–10a. A mirror is called **concave** if the reflecting surface is on the inner surface of the sphere so that the mirror surface curves away from the viewer (like a "cave"), Fig. 23–10b. Concave mirrors are used as shaving or cosmetic mirrors (**magnifying mirrors**), Fig. 23–11a, because they magnify. Convex mirrors are sometimes used on cars and trucks (rearview mirrors) and in shops (to watch for theft), because they take in a wide field of view, Fig. 23–11b.

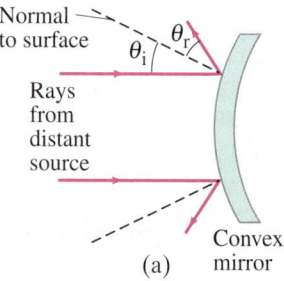
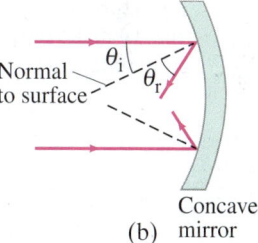

FIGURE 23–10 Mirrors with convex and concave spherical surfaces. Note that $\theta_r = \theta_i$ for each ray. (The dashed lines are perpendicular to the mirror surface at each point shown.)

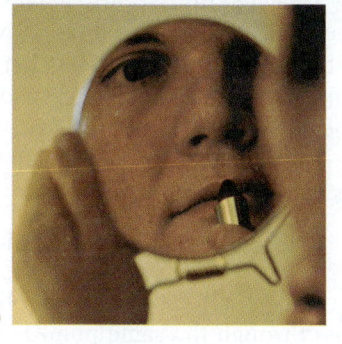

(a) (b)

FIGURE 23–11 (a) A concave cosmetic mirror gives a magnified image. (b) A convex mirror in a store reduces image size and so includes a wide field of view. Note the extreme distortion—this mirror has a large curved surface and does not fit the "paraxial ray" approximation discussed on the next page.

Focal Point and Focal Length

To see how spherical mirrors form images, we first consider an object that is very far from a concave mirror. For a distant object, as shown in Fig. 23–12, the rays from each point on the object that strike the mirror will be nearly parallel. *For an object infinitely far away* (the Sun and stars approach this), *the rays would be precisely parallel.*

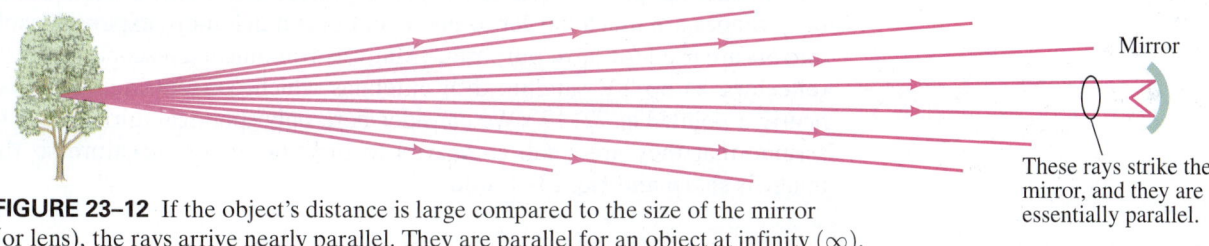

FIGURE 23–12 If the object's distance is large compared to the size of the mirror (or lens), the rays arrive nearly parallel. They are parallel for an object at infinity (∞).

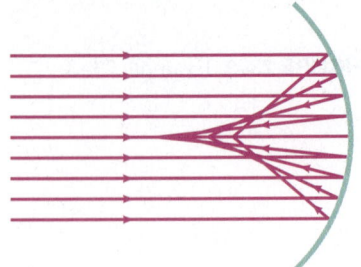

FIGURE 23–13 Parallel rays striking a concave spherical mirror do not intersect (or focus) at precisely a single point. (This "defect" is referred to as "spherical aberration.")

Now consider such parallel rays falling on a concave mirror as in Fig. 23–13. The law of reflection holds for each of these rays at the point each strikes the mirror. As can be seen, they are not all brought to a single point. In order to form a sharp image, the rays must come to a point. Thus a spherical mirror will not make as sharp an image as a plane mirror will. However, as we show below, if the mirror is small compared to its radius of curvature, so that a reflected ray makes only a *small angle* with the incident ray (2θ in Fig. 23–14), then the rays will cross each other at very nearly a single point, or **focus**. In the case shown in Fig. 23–14, the incoming rays are parallel to the **principal axis**, which is defined as the straight line perpendicular to the curved surface at its center (line CA in Fig. 23–14). The point F, where incident parallel rays come to a focus after reflection, is called the **focal point** of the mirror. The distance between F and the center of the mirror, length FA, is called the **focal length**, f, of the mirror. The focal point is also the *image point for an object infinitely far away* along the principal axis. The image of the Sun, for example, would be at F.

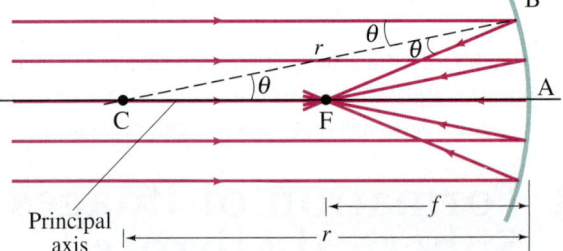

FIGURE 23–14 Rays parallel to the principal axis of a concave spherical mirror come to a focus at F, the focal point, as long as the mirror is small in width as compared to its radius of curvature, r, so that the rays are "paraxial"—that is, make only small angles with the horizontal axis.

Now we will show, for a mirror whose reflecting surface is small compared to its radius of curvature, that the rays very nearly meet at a common point, F, and we will also determine the focal length f. In this approximation, we consider only rays that make a small angle with the principal axis; such rays are called **paraxial rays**, and their angles are exaggerated in Fig. 23–14 to make the labels clear. First we consider a ray that strikes the mirror at B in Fig. 23–14. The point C is the center of curvature of the mirror (the center of the sphere of which the mirror is a part). So the dashed line CB is equal to r, the radius of curvature, and CB is normal to the mirror's surface at B. The incoming ray that hits the mirror at B makes an angle θ with this normal, and hence the reflected ray, BF, also makes an angle θ with the normal (law of reflection). The angle BCF is also θ, as shown. The triangle CBF is isosceles because two of its angles are equal. Thus length CF = FB. We assume the mirror surface is small compared to the mirror's radius of curvature, so the angles are small, and the length FB is nearly equal to length FA. In this approximation, FA = FC. But FA = f, the focal length, and CA = $2 \times$ FA = r. Thus the focal length is half the radius of curvature:

$$f = \frac{r}{2}. \qquad \text{[spherical mirror]} \quad (23\text{–}1)$$

We assumed only that the angle θ was small, so this result applies for all other incident paraxial rays. Thus all paraxial rays pass through the same point F, the focal point.

Since it is only approximately true that the rays come to a perfect focus at F, the more curved the mirror, the worse the approximation (Fig. 23–13) and the more blurred the image. This "defect" of spherical mirrors is called **spherical aberration**; we will discuss it more with regard to lenses in Chapter 25. A **parabolic reflector**, on the other hand, will reflect the rays to a perfect focus. However, because parabolic shapes are much harder to make and thus much more expensive, spherical mirrors are used for most purposes. (Many astronomical telescopes use parabolic reflectors, as do TV satellite dish antennas which concentrate radio waves to nearly a point, Fig. 22–19.) We consider here only spherical mirrors and we will assume that they are small compared to their radius of curvature so that the image is sharp and Eq. 23–1 holds.

Image Formation—Ray Diagrams

We saw that for an object at infinity, the image is located at the focal point of a concave spherical mirror, where $f = r/2$. But where does the image lie for an object not at infinity? First consider the object shown as an arrow in Fig. 23–15a, which is placed between F and C at point O (O for object). Let us determine where the image will be for a given point O′ at the top of the object, by finding the point where rays drawn from the tip of the arrow converge after reflecting from the mirror. To do this we can draw several rays and make sure these reflect from the mirror such that the angle of reflection equals the angle of incidence.

(a) Ray 1 goes out from O′ parallel to the axis and reflects through F.

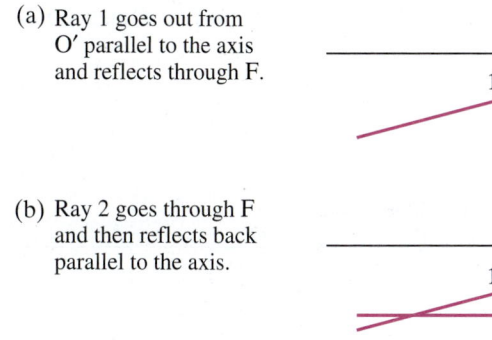

(b) Ray 2 goes through F and then reflects back parallel to the axis.

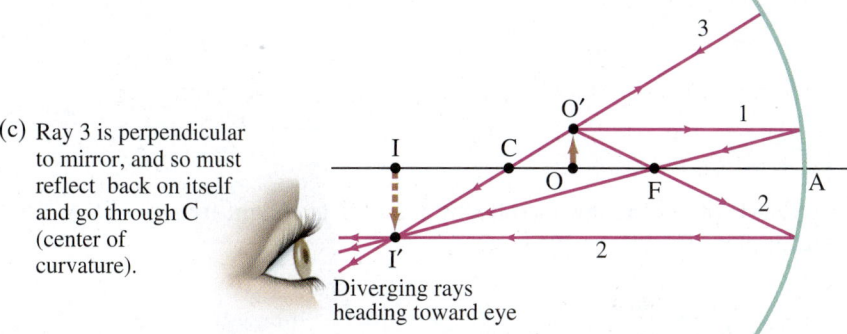

FIGURE 23–15 Rays leave point O′ on the object (an arrow). Shown are the three most useful rays for determining where the image I′ is formed. [Note that our mirror is not small compared to f, so our diagram will not give the precise position of the image.]

(c) Ray 3 is perpendicular to mirror, and so must reflect back on itself and go through C (center of curvature).

Many rays could be drawn leaving any point on an object, but determining the image position is faster if we deal with three particular rays. These are the rays labeled 1, 2, and 3 in Fig. 23–15 and we draw them leaving object point O′ as follows:

Ray 1 leaving O′ is drawn parallel to the axis; therefore after reflection it must pass along a line through F, Fig. 23–15a (just as parallel rays did in Fig. 23–14).

Ray 2 leaves O′ and is made to pass through F (Fig. 23–15b); therefore it must reflect so it is parallel to the axis. (In reverse, a parallel ray passes through F.)

Ray 3 is drawn along a radius of the spherical surface (Fig. 23–15c) and is perpendicular to the mirror, so it is reflected back on itself and passes through C, the center of curvature.

RAY DIAGRAM
Finding the image position for a curved mirror

All three rays leave a single point O′ on the object. After reflection from a (small) mirror, the point at which these rays cross is the image point I′. All other rays from the same object point will also pass through this image point. To find the image point for any object point, only these three types of rays need to be drawn. Only two of these rays are needed, but the third serves as a check.

PROBLEM SOLVING
Image point is where reflected rays intersect

We have shown the image point in Fig. 23–15 only for a single point on the object. Other points on the object are imaged nearby. For instance, the bottom of the arrow, on the principal axis at point O, is imaged on the axis at point I. So a complete image of the object is formed (dashed arrow in Fig. 23–15c). Because the light actually passes through the image, this is a **real image** that will appear on a white surface or film placed there. This can be compared to the virtual image formed by a plane mirror (the light does not pass through that image, Fig. 23–7).

The image in Fig. 23–15 can be seen by the eye only when the eye is placed to the left of the image, so that some of the rays *diverging* from each point on the image (as point I′) can enter the eye as shown in Fig. 23–15c (just as in Figs. 23–1 and 23–7).

Mirror Equation and Magnification

Image points can be determined, roughly, by drawing the three rays as just described, Fig. 23–15. But it is difficult to draw small angles for the "paraxial" rays as we assumed. For more accurate results, we now derive an equation that gives the image distance if the object distance and radius of curvature of the mirror are known. To do this, we refer to Fig. 23–16. The **object distance**, d_o, is the distance of the object (point O) from the center of the mirror. The **image distance**, d_i, is the distance of the image (point I) from the center of the mirror. The height of the object OO' is called h_o and the height of the image, I'I, is h_i. Two rays leaving O' are shown: O'FBI' (same as ray 2 in Fig. 23–15) and O'AI', which is a fourth type of ray that reflects at the center of the mirror and can also be used to find an image point.

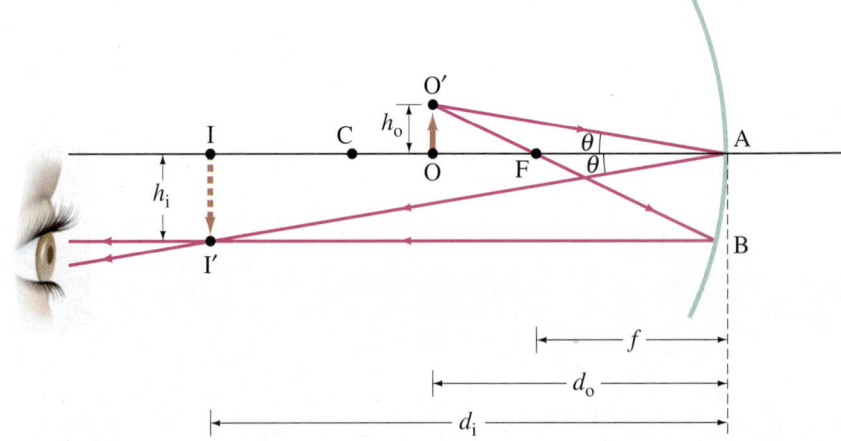

FIGURE 23–16 Diagram for deriving the mirror equation. For the derivation, we assume the mirror size is small compared to its radius of curvature.

The ray O'AI' obeys the law of reflection, so the two right triangles O'AO and I'AI are similar. Therefore, we have

$$\frac{h_o}{h_i} = \frac{d_o}{d_i}.$$

For the other ray shown, O'FBI', the triangles O'FO and AFB are also similar because the angles at F are equal and we use the approximation $AB = h_i$ (mirror small compared to its radius). Furthermore $FA = f$, the focal length of the mirror, so

$$\frac{h_o}{h_i} = \frac{OF}{FA} = \frac{d_o - f}{f}.$$

The left sides of the two preceding expressions are the same, so we can equate the right sides:

$$\frac{d_o}{d_i} = \frac{d_o - f}{f}.$$

We now divide both sides by d_o and rearrange to obtain

Mirror equation
$$\frac{1}{d_o} + \frac{1}{d_i} = \frac{1}{f}. \tag{23-2}$$

This is the equation we were seeking. It is called the **mirror equation** and relates the object and image distances to the focal length f (where $f = r/2$).

The mirror equation also holds for a plane mirror: the focal length is $f = r/2 = \infty$ (Eq. 23–1), and Eq. 23–2 gives $d_i = -d_o$.

The **magnification**, m, of a mirror is defined as the height of the image divided by the height of the object. From our first set of similar triangles in Fig. 23–16, or the first equation just below Fig. 23–16, we can write:

$$m = \frac{h_i}{h_o} = -\frac{d_i}{d_o}. \qquad (23\text{–}3)$$

The minus sign in Eq. 23–3 is inserted as a convention. Indeed, we must be careful about the signs of all quantities in Eqs. 23–2 and 23–3. Sign conventions are chosen so as to give the correct locations and orientations of images, as predicted by ray diagrams. The **sign conventions** we use are:

Sign conventions for mirrors

1. the image height h_i is positive if the image is upright, and negative if inverted, relative to the object (assuming h_o is taken as positive);
2. d_i or d_o is positive if image or object is in front of the mirror (as in Fig. 23–16); if either image or object is behind the mirror, the corresponding distance is negative. [An example of $d_i < 0$ can be seen in Fig. 23–17, Example 23–6.]†

Thus the magnification (Eq. 23–3) is positive for an upright image and negative for an inverted image (upside down). We summarize sign conventions more fully in the Problem Solving Strategy following our discussion of convex mirrors later in this Section.

Concave Mirror Examples

EXAMPLE 23–4 **Image in a concave mirror.** A 1.50-cm-high object is placed 20.0 cm from a concave mirror with radius of curvature 30.0 cm. Determine (*a*) the position of the image, and (*b*) its size.

APPROACH We determine the focal length from the radius of curvature (Eq. 23–1), $f = r/2 = 15.0$ cm. The ray diagram is basically the same as Fig. 23–16, since the object is between F and C. The position and size of the image are found from Eqs. 23–2 and 23–3.

SOLUTION Referring to Fig. 23–16, we have CA = r = 30.0 cm, FA = f = 15.0 cm, and OA = d_o = 20.0 cm.
(*a*) We start with the mirror equation, Eq. 23–2, rearranging it (subtracting $(1/d_o)$ from both sides):

$$\frac{1}{d_i} = \frac{1}{f} - \frac{1}{d_o} = \frac{1}{15.0 \text{ cm}} - \frac{1}{20.0 \text{ cm}} = 0.0167 \text{ cm}^{-1}.$$

So $d_i = 1/(0.0167 \text{ cm}^{-1}) = 60.0$ cm. Because d_i is positive, the image is 60.0 cm in front of the mirror, on the same side as the object.

Remember to take the reciprocal

(*b*) From Eq. 23–3, the magnification is

$$m = -\frac{d_i}{d_o} = -\frac{60.0 \text{ cm}}{20.0 \text{ cm}} = -3.00.$$

The image is 3.0 times larger than the object, and its height is

$$h_i = mh_o = (-3.00)(1.5 \text{ cm}) = -4.5 \text{ cm}.$$

The minus sign reminds us that the image is inverted, as shown in Fig. 23–16.

NOTE When an object is further from a concave mirror than the focal point, we can see from Fig. 23–15 or 23–16 that the image is always inverted and real.

CONCEPTUAL EXAMPLE 23–5 **Reversible rays.** If the object in Example 23–4 is placed instead where the image is (see Fig. 23–16), where will the new image be?

RESPONSE The mirror equation is *symmetric* in d_o and d_i. Thus the new image will be where the old object was. Indeed, in Fig. 23–16 we need only reverse the direction of the rays to get our new situation.

---

†d_o is always positive for a real object; $d_o < 0$ can happen only if the object is an image formed by another mirror or lens—see Example 23–16.

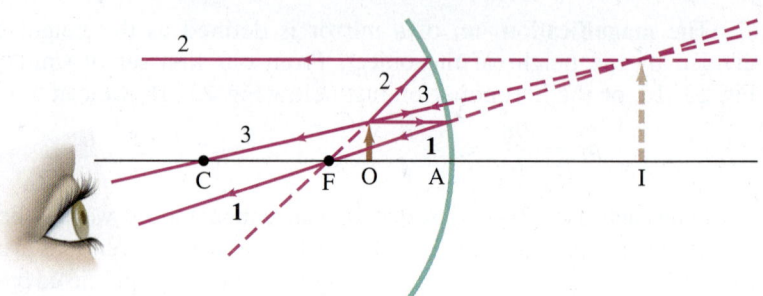

FIGURE 23–17 Object placed within the focal point F. The image is *behind* the mirror and is *virtual*, Example 23–6. [Note that the vertical scale (height of object = 1.0 cm) is different from the horizontal (OA = 10.0 cm) for ease of drawing, and reduces the precision of the drawing.]

EXAMPLE 23–6 Object closer to concave mirror than focal point.
A 1.00-cm-high object is placed 10.0 cm from a concave mirror whose radius of curvature is 30.0 cm. (*a*) Draw a ray diagram to locate (approximately) the position of the image. (*b*) Determine the position of the image and the magnification analytically.

APPROACH We draw the ray diagram using the rays as in Fig. 23–15, page 651. An analytic solution uses Eqs. 23–1, 23–2, and 23–3.

SOLUTION (*a*) Since $f = r/2 = 15.0$ cm, the object is between the mirror and the focal point. We draw the three rays as described earlier (Fig. 23–15); they are shown leaving the tip of the object in Fig. 23–17. Ray 1 leaves the tip of our object heading toward the mirror parallel to the axis, and reflects through F. Ray 2 cannot head toward F because it would not strike the mirror; so ray 2 must point as if it started at F (dashed line in Fig. 23–17) and heads to the mirror, and then is reflected parallel to the principal axis. Ray 3 is perpendicular to the mirror and reflects back on itself. The rays reflected from the mirror diverge and so never meet at a point. They appear to be coming from a point behind the mirror (dashed lines). This point locates the image of the tip of the arrow. The image is thus behind the mirror and is *virtual*.

(*b*) We use Eq. 23–2 to find d_i when $d_o = 10.0$ cm:

$$\frac{1}{d_i} = \frac{1}{f} - \frac{1}{d_o} = \frac{1}{15.0\ \text{cm}} - \frac{1}{10.0\ \text{cm}} = \frac{2 - 3}{30.0\ \text{cm}} = -\frac{1}{30.0\ \text{cm}}.$$

Therefore, $d_i = -30.0$ cm. The minus sign means the image is behind the mirror, which our diagram also showed us. The magnification is $m = -d_i/d_o = -(-30.0\ \text{cm})/(10.0\ \text{cm}) = +3.00$. So the image is 3.00 times larger than the object. The plus sign indicates that the image is upright (same as object), which is consistent with the ray diagram, Fig. 23–17.

NOTE The image distance cannot be obtained accurately by measuring on Fig. 23–17, because our diagram violates the paraxial ray assumption (we draw rays at steeper angles to make them clearly visible).

NOTE When the object is located inside the focal point of a concave mirror ($d_o < f$), the image is always upright and virtual. If the object O in Fig. 23–17 is you, you see yourself clearly, because the reflected rays at point O (you) are diverging. Your image is upright and enlarged. This is how a shaving or cosmetic mirror is used—you must place your head closer to the mirror than the focal point if you are to see yourself right-side up (see the photograph, Fig. 23–11a). [If the object is *beyond* the focal point, as in Fig. 23–15, the image is real and inverted: upside down—and hard to use!]

PHYSICS APPLIED
Magnifying mirror: Seeing yourself upright and magnified in a concave mirror

Seeing the Image; Seeing Yourself

For a person's eye to see a sharp image, the eye must be at a place where it intercepts diverging rays from points on the image, as is the case for the eye's position in Figs. 23–15, 23–16, and 23–17. When we look at normal objects, we always detect rays diverging toward the eye as shown in Fig. 23–1. (Or, for very distant objects like stars, the rays become essentially parallel, as in Fig. 23–12.)

If you placed your eye between points O and I in Fig. 23–16, for example, *converging* rays from the object OO′ would enter your eye and the lens of your eye could not bring them to a focus; you would see a blurry image or no perceptible image at all. [We will discuss the eye more in Chapter 25.]

If *you* are the object OO′ in Fig. 23–16, situated between F and C, and are trying to see yourself in the mirror, you would see a blur; but the person whose eye is shown in Fig. 23–16 could see you clearly. If you are to the left of C in Fig. 23–16, where $d_o > 2f$, you can see yourself clearly, but upside down. Why? Because then the rays arriving from the image will be *diverging* at your position (Fig. 23–18), and your eye can then focus them. You can also see yourself clearly, and right side up, if you are closer to the mirror than its focal point ($d_o < f$), as we saw in Example 23–6, Fig. 23–17.

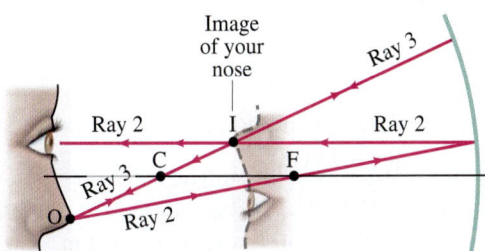

FIGURE 23–18 You can see a clear inverted image of your face in a concave mirror when you are beyond C ($d_o > 2f$), because the rays that arrive at your eye are *diverging*. Standard rays 2 and 3 are shown leaving point O on your nose. Ray 2 (and other nearby rays) enters your eye. Notice that rays are diverging as they move to the left of image point I.

Convex Mirrors

The analysis used for concave mirrors can be applied to **convex** mirrors. Even the mirror equation (Eq. 23–2) holds for a convex mirror, although the quantities involved must be carefully defined. Figure 23–19a shows parallel rays falling on a convex mirror. Again spherical aberration is significant (Fig. 23–13), unless we assume the mirror is small compared to its radius of curvature. The reflected rays diverge, but seem to come from point F behind the mirror, Fig. 23–19a. This is the **focal point**, and its distance from the center of the mirror (point A) is the **focal length**, f. The equation $f = r/2$ is valid also for a convex mirror. We see that an object at infinity produces a virtual image in a convex mirror. Indeed, no matter where the object is placed on the reflecting side of a convex mirror, the image will be virtual and upright, as indicated in Fig. 23–19b. To find the image we draw rays 1 and 3 according to the rules used before on the concave mirror, as shown in Fig. 23–19b. Note that although rays 1 and 3 don't actually pass through points F and C, the line along which each is drawn does (shown dashed).

The mirror equation, Eq. 23–2, holds for convex mirrors but the focal length f and radius of curvature must be considered negative. The proof is left as a Problem. It is also left as a Problem to show that Eq. 23–3 for the magnification is also valid.

FIGURE 23–19 Convex mirror: (a) the focal point is at F, behind the mirror; (b) the image I of the object at O is virtual, upright, and smaller than the object. [Not to scale for Example 23–7.]

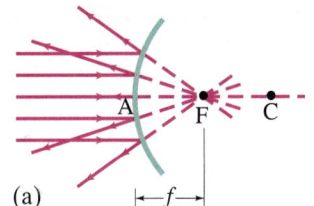
(a)

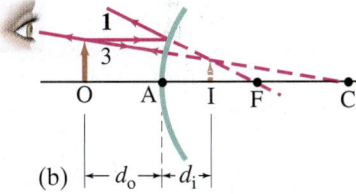
(b)

PROBLEM SOLVING

Spherical Mirrors

1. Always **draw a ray diagram** even though you are going to make an analytic calculation—the diagram serves as a check, even if not precise. From one point on the object, draw at least two, preferably three, of the easy-to-draw rays using the rules described in Fig. 23–15. The image point is where the reflected rays intersect (real image) or appear to intersect (virtual).

2. Apply the **mirror equation**, Eq. 23–2, and the **magnification equation**, Eq. 23–3. It is crucially important to follow the sign conventions—see the next point.

3. **Sign Conventions**
   (a) When the object, image, or focal point is on the reflecting side of the mirror (on the left in our drawings), the corresponding distance is positive. If any of these points is behind the mirror (on the right) the corresponding distance is negative.[†]
   (b) The image height h_i is positive if the image is upright, and negative if inverted, relative to the object (h_o is always taken as positive).

4. **Check** that the analytic solution is consistent with the ray diagram.

[†]Object distances are positive for material objects, but can be negative in systems with more than one mirror or lens—see Section 23–9.

PHYSICS APPLIED
Convex rearview mirror

EXAMPLE 23–7 **Convex rearview mirror.** An external rearview car mirror is convex with a radius of curvature of 16.0 m (Fig. 23–20). Determine the location of the image and its magnification for an object 10.0 m from the mirror.

APPROACH We follow the steps of the Problem Solving Strategy explicitly.

SOLUTION

1. **Draw a ray diagram.** The ray diagram will be like Fig. 23–19b, but the large object distance ($d_o = 10.0$ m) makes a precise drawing difficult. We have a convex mirror, so r is negative by convention.

2. **Mirror and magnification equations.** The center of curvature of a convex mirror is behind the mirror, as is its focal point, so we set $r = -16.0$ m so that the focal length is $f = r/2 = -8.0$ m. The object is in front of the mirror, $d_o = 10.0$ m. Solving the mirror equation, Eq. 23–2, for $1/d_i$ gives

$$\frac{1}{d_i} = \frac{1}{f} - \frac{1}{d_o} = \frac{1}{-8.0\text{ m}} - \frac{1}{10.0\text{ m}} = \frac{-10.0 - 8.0}{80.0\text{ m}} = -\frac{18}{80.0\text{ m}}.$$

Thus $d_i = -80.0\text{ m}/18 = -4.4$ m. Equation 23–3 gives the magnification

$$m = -\frac{d_i}{d_o} = -\frac{(-4.4\text{ m})}{(10.0\text{ m})} = +0.44.$$

3. **Sign conventions.** The image distance is negative, -4.4 m, so the image is *behind* the mirror. The magnification is $m = +0.44$, so the image is *upright* (same orientation as object, which is useful) and about half what it would be in a plane mirror.

4. **Check.** Our results are consistent with Fig. 23–19b.

FIGURE 23–20 Example 23–7.

Convex rearview mirrors on vehicles sometimes come with a warning that objects are closer than they appear in the mirror. The fact that d_i may be smaller than d_o (as in Example 23–7) seems to contradict this observation. The real reason the object seems farther away is that its image in the convex mirror is *smaller* than it would be in a plane mirror, and we judge distance of ordinary objects such as other cars mostly by their size.

23–4 Index of Refraction

We saw in Chapter 22 that the speed of light in vacuum (like other EM waves) is

$$c = 2.99792458 \times 10^8 \text{ m/s},$$

which is usually rounded off to

$$3.00 \times 10^8 \text{ m/s}$$

when extremely precise results are not required.

In air, the speed is only slightly less. In other transparent materials, such as glass and water, the speed is always less than that in vacuum. For example, in water light travels at about $\tfrac{3}{4}c$. The ratio of the speed of light in vacuum to the speed v in a given material is called the **index of refraction**, n, of that material:

$$n = \frac{c}{v}. \quad (23\text{–}4)$$

The index of refraction is never less than 1, and values for various materials are given in Table 23–1. For example, since $n = 1.33$ for water, the speed of light in water is

$$v = \frac{c}{n} = \frac{(3.00 \times 10^8 \text{ m/s})}{1.33} = 2.26 \times 10^8 \text{ m/s}.$$

As we shall see later, n varies somewhat with the wavelength of the light—except in vacuum—so a particular wavelength is specified in Table 23–1, that of yellow light with wavelength $\lambda = 589$ nm.

That light travels more slowly in matter than in vacuum can be explained at the atomic level as being due to the absorption and reemission of light by atoms and molecules of the material.

TABLE 23–1 Indices of Refraction[†]

Material	$n = \dfrac{c}{v}$
Vacuum	1.0000
Air (at STP)	1.0003
Water	1.33
Ethyl alcohol	1.36
Glass	
Fused quartz	1.46
Crown glass	1.52
Light flint	1.58
Plastic	
Acrylic, Lucite, CR-39	1.50
Polycarbonate	1.59
"High-index"	1.6–1.7
Sodium chloride	1.53
Diamond	2.42

[†]$\lambda = 589$ nm.

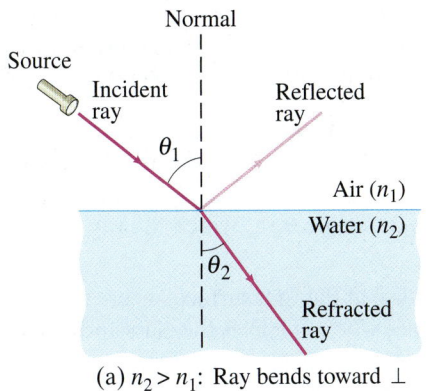
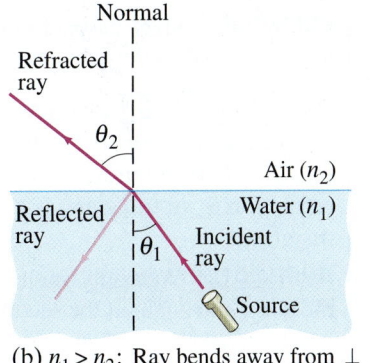

(a) $n_2 > n_1$: Ray bends toward ⊥

(b) $n_1 > n_2$: Ray bends away from ⊥

FIGURE 23–21 Refraction.
(a) Light refracted when passing from air (n_1) into water (n_2): $n_2 > n_1$.
(b) Light refracted when passing from water (n_1) into air (n_2): $n_1 > n_2$.

23–5 Refraction: Snell's Law

When light passes from one transparent medium into another with a different index of refraction, some or all of the incident light is reflected at the boundary. The rest passes into the new medium. If a ray of light is incident at an angle to the surface (other than perpendicular), the ray changes direction as it enters the new medium. This change in direction, or bending, of the light ray is called **refraction**.

Figure 23–21a shows a ray passing from air into water. Angle θ_1 is the angle the incident ray makes with the normal (perpendicular) to the surface and is called the **angle of incidence**. Angle θ_2 is the **angle of refraction**, the angle the refracted ray makes with the normal to the surface. Notice that the ray bends toward the normal when entering the water. This is always the case when the ray enters a medium where the speed of light is *less* (and the index of refraction is greater, Eq. 23–4). If light travels from one medium into a second where its speed is *greater*, the ray bends away from the normal; this is shown in Fig. 23–21b for a ray traveling from water to air.

⚠ **CAUTION**
Angles of incidence and refraction are measured from the perpendicular, not from the surface

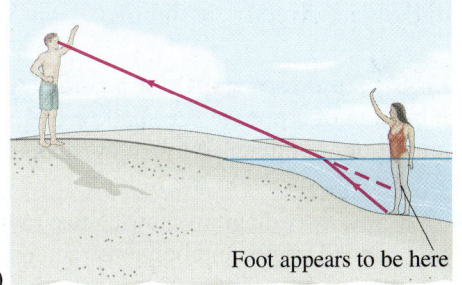

Foot appears to be here

(a) (b)

FIGURE 23–22 (a) Photograph, and (b) ray diagram showing why a person's legs look shorter standing in water: a ray from the bather's foot to the observer's eye bends at the water's surface, and our brain interprets the light as traveling in a straight line, from higher up (dashed line).

Refraction is responsible for a number of common optical illusions. For example, a person standing in waist-deep water appears to have shortened legs (Fig. 23–22). The rays leaving the person's foot are bent at the surface. The observer's brain assumes the rays to have traveled a straight-line path (dashed red line), and so the feet appear to be higher than they really are. Similarly, when you put a straw in water, it appears to be bent (Fig. 23–23). This also means that water is deeper than it appears.

FIGURE 23–23 A straw in water looks bent even when it isn't.

Snell's Law

The angle of refraction depends on the speed of light in the two media and on the incident angle. An analytic relation between θ_1 and θ_2 in Fig. 23–21 was arrived at experimentally about 1621 by Willebrord Snell (1591–1626). Known as **Snell's law**, it is written:

$$n_1 \sin \theta_1 = n_2 \sin \theta_2. \qquad (23\text{–}5)$$

θ_1 is the angle of incidence and θ_2 is the angle of refraction; n_1 and n_2 are the respective indices of refraction in the materials. See Fig. 23–21. The incident and refracted rays lie in the same plane, which also includes the perpendicular to the surface. Snell's law is the **law of refraction**. (Snell's law was derived in Section 11–13 for water waves where Eq. 11–20 is just a combination of Eqs. 23–5 and 23–4, and we derive it again in Chapter 24 using the wave theory of light.)

Snell's law shows that if $n_2 > n_1$, then $\theta_2 < \theta_1$. Thus, if light enters a medium where n is greater (and its speed is less), the ray is bent toward the normal. And if $n_2 < n_1$, then $\theta_2 > \theta_1$, so the ray bends away from the normal. See Fig. 23–21.

SNELL'S LAW
(LAW OF REFRACTION)

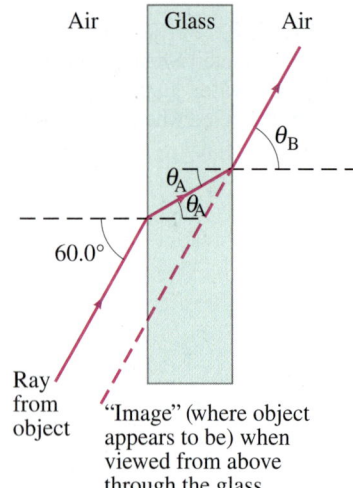

FIGURE 23–24 Light passing through a piece of glass (Example 23–8).

EXERCISE C Light passes from a medium with $n = 1.3$ (water) into a medium with $n = 1.5$ (glass). Is the light bent toward or away from the perpendicular to the interface?

EXAMPLE 23–8 Refraction through flat glass. Light traveling in air strikes a flat piece of uniformly thick glass at an incident angle of 60.0°, as shown in Fig. 23–24. If the index of refraction of the glass is 1.50, (*a*) what is the angle of refraction θ_A in the glass; (*b*) what is the angle θ_B at which the ray emerges from the glass?

APPROACH We apply Snell's law twice: at the first surface, where the light enters the glass, and again at the second surface where it leaves the glass and enters the air.

SOLUTION (*a*) The incident ray is in air, so $n_1 = 1.00$ and $n_2 = 1.50$. Applying Snell's law where the light enters the glass ($\theta_1 = 60.0°$, $\theta_2 = \theta_A$) gives

$$(1.00) \sin 60.0° = (1.50) \sin \theta_A$$

or

$$\sin \theta_A = \frac{1.00}{1.50} \sin 60.0° = 0.5774,$$

and $\theta_A = 35.3°$.

(*b*) Since the faces of the glass are parallel, the incident angle at the second surface is also θ_A (geometry), so $\sin \theta_A = 0.5774$. At this second interface, $n_1 = 1.50$ and $n_2 = 1.00$. Thus the ray re-enters the air at an angle θ_B given by

$$\sin \theta_B = \frac{1.50}{1.00} \sin \theta_A = 0.866,$$

and $\theta_B = 60.0°$. The direction of a light ray is thus unchanged by passing through a flat piece of glass of uniform thickness.

NOTE This result is valid for any angle of incidence. The ray is displaced slightly to one side, however. You can observe this by looking through a piece of glass (near its edge) at some object and then moving your head to the side slightly so that you see the object directly. It "jumps."

⚠ CAUTION (real life)
Water is deeper than it looks

FIGURE 23–25 Example 23–9.

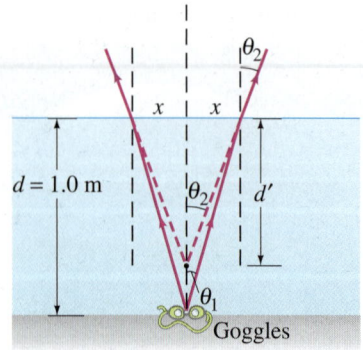

EXAMPLE 23–9 Apparent depth of a pool. A swimmer has dropped her goggles to the bottom of a pool at the shallow end, marked as 1.0 m deep. But the goggles don't look that deep. Why? How deep do the goggles appear to be when you look straight down into the water?

APPROACH We draw a ray diagram showing two rays going upward from a point on the goggles at a small angle, and being refracted at the water's (flat) surface, Fig. 23–25. The two rays traveling upward from the goggles are refracted *away* from the normal as they exit the water, and so appear to be diverging from a point above the goggles (dashed lines), which is why the water seems less deep than it actually is. We are looking straight down, so all angles are small (but exaggerated in Fig. 23–25 for clarity).

SOLUTION To calculate the apparent depth d' (Fig. 23–25), given a real depth $d = 1.0$ m, we use Snell's law with $n_1 = 1.33$ for water and $n_2 = 1.0$ for air:

$$\sin \theta_2 = n_1 \sin \theta_1.$$

We are considering only small angles, so $\sin \theta \approx \tan \theta \approx \theta$, with θ in radians. So Snell's law becomes

$$\theta_2 \approx n_1 \theta_1.$$

From Fig. 23–25, we see that $\theta_2 \approx \tan \theta_2 = x/d'$ and $\theta_1 \approx \tan \theta_1 = x/d$. Putting these into Snell's law, $\theta_2 \approx n_1 \theta_1$, we get

$$\frac{x}{d'} \approx n_1 \frac{x}{d}$$

or

$$d' \approx \frac{d}{n_1} = \frac{1.0 \text{ m}}{1.33} = 0.75 \text{ m}.$$

The pool seems only three-fourths as deep as it actually is.

NOTE Water in general is deeper than it looks—a useful safety guideline.

23-6 Total Internal Reflection; Fiber Optics

When light passes from one material into a second material where the index of refraction is less (say, from water into air), the refracted light ray bends away from the normal, as for rays I and J in Fig. 23–26. At a particular incident angle, the angle of refraction will be 90°, and the refracted ray would skim the surface (ray K).

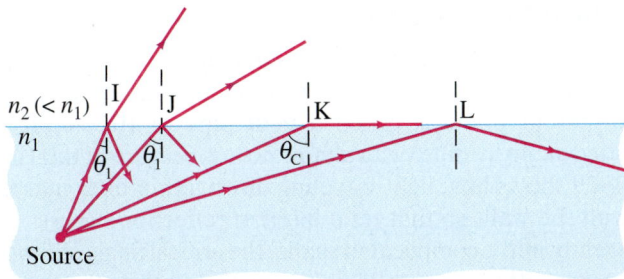

FIGURE 23–26 Since $n_2 < n_1$, light rays are totally internally reflected if the incident angle $\theta_1 > \theta_C$, as for ray L. If $\theta_1 < \theta_C$, as for rays I and J, only a part of the light is reflected, and the rest is refracted.

The incident angle at which this occurs is called the **critical angle**, θ_C. From Snell's law, θ_C is given by

$$\sin \theta_C = \frac{n_2}{n_1} \sin 90° = \frac{n_2}{n_1}. \quad (23\text{-}6)$$

For any incident angle less than θ_C, there will be a refracted ray, although part of the light will also be reflected at the boundary. However, for incident angles θ_1 greater than θ_C, Snell's law would tell us that $\sin \theta_2 \; (= n_1 \sin \theta_1 / n_2)$ would be greater than 1.00 when $n_2 < n_1$. Yet the sine of an angle can never be greater than 1.00. In this case there is no refracted ray at all, and *all of the light is reflected*, as for ray L in Fig. 23–26. This effect is called **total internal reflection**. Total internal reflection occurs only when light strikes a boundary where the medium beyond has a *lower* index of refraction.

> **CAUTION**
> *Total internal reflection (occurs only if refractive index is smaller beyond boundary)*

CONCEPTUAL EXAMPLE 23–10 View up from under water. Describe what a person would see who looked up at the world from beneath the perfectly smooth surface of a lake or swimming pool.

RESPONSE For an air–water interface, the critical angle is given by

$$\sin \theta_C = \frac{1.00}{1.33} = 0.750.$$

Therefore, $\theta_C = 49°$. Thus the person would see the outside world compressed into a circle whose edge makes a 49° angle with the vertical. Beyond this angle, the person would see reflections from the sides and bottom of the lake or pool (Fig. 23–27).

EXERCISE D Light traveling in air strikes a glass surface with $n = 1.48$. For what range of angles will total internal reflection occur?

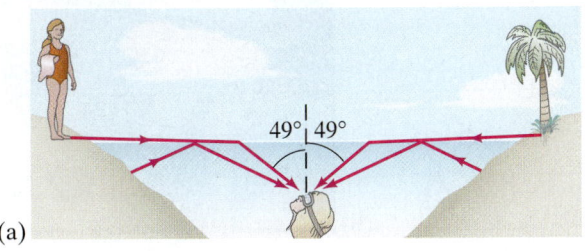
(a)

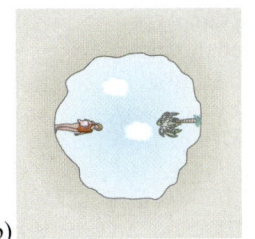
(b)

FIGURE 23–27 (a) Light rays entering submerged person's eye, and (b) view looking upward from beneath the water (the surface of the water must be very smooth). Example 23–10.

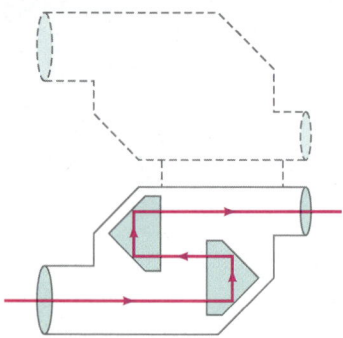

FIGURE 23–28 Total internal reflection of light by prisms in binoculars.

PHYSICS APPLIED
Fiber optics in communications and medicine—bronchoscopes, colonoscopes, endoscopes

FIGURE 23–29 Light reflected totally at the interior surface of a glass or transparent plastic fiber.

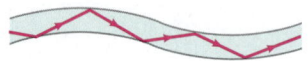

Many optical instruments, such as binoculars, use total internal reflection within a prism to reflect light. The advantage is that very nearly 100% of the light is reflected, whereas even the best mirrors reflect somewhat less than 100%. Thus the image is brighter, especially after several reflections. For glass with $n = 1.50$, $\theta_C = 41.8°$. Therefore, 45° prisms will reflect all the light internally, if oriented as shown in the binoculars of Fig. 23–28.

| **EXERCISE E** What would happen if we immersed the 45° glass prisms in Fig. 23–28 in water?

Fiber Optics; Medical Instruments

Total internal reflection is the principle behind **fiber optics**. Glass and plastic fibers as thin as a few micrometers in diameter are commonly used. A bundle of such slender transparent fibers is called a **light pipe** or **fiber-optic cable**. Light[†] can be transmitted along the fiber with almost no loss because of total internal reflection. Figure 23–29 shows how light traveling down a thin fiber makes only glancing collisions with the walls so that total internal reflection occurs. Even if the light pipe is bent gently into a complicated shape, the critical angle still won't be exceeded, so light is transmitted practically undiminished to the other end. Very small losses do occur, mainly by reflection at the ends and absorption within the fiber.

Important applications of fiber-optic cables are in communications and medicine. They are used in place of wire to carry telephone calls, video signals, and computer data. The signal is a modulated light beam (a light beam whose intensity can be varied) and data is transmitted at a much higher rate and with less loss and less interference than an electrical signal in a copper wire. Fibers have been developed that can support over one hundred separate wavelengths, each modulated to carry more than 10 gigabits (10^{10} bits) of information per second. That amounts to a terabit (10^{12} bits) per second for one hundred wavelengths.

The use of fiber optics to transmit a clear picture is particularly useful in medicine, Fig. 23–30. For example, a patient's lungs can be examined by inserting a fiber-optic cable known as a bronchoscope through the mouth and down the bronchial tube. Light is sent down an outer set of fibers to illuminate the lungs. The reflected light returns up a central core set of fibers. Light directly in front of each fiber travels up that fiber. At the opposite end, a viewer sees a series of bright and dark spots, much like a TV screen—that is, a picture of what lies at the opposite end. Lenses are used at each end of the cable. The image may be viewed directly or on a monitor screen or film. The fibers must be optically insulated from one another, usually by a thin coating of material with index of refraction less than that of the fiber. The more fibers there are, and the smaller they are, the more detailed the picture. Such instruments, including bronchoscopes, colonoscopes (for viewing the colon), and endoscopes (stomach or other organs), are extremely useful for examining hard-to-reach places.

[†]Fiber-optic devices use not only visible light but also infrared light, ultraviolet light, and microwaves.

FIGURE 23–30 (a) How a fiber-optic image is made. (b) Example of a fiber-optic device inserted through the mouth to view the vocal cords, with the image on screen.

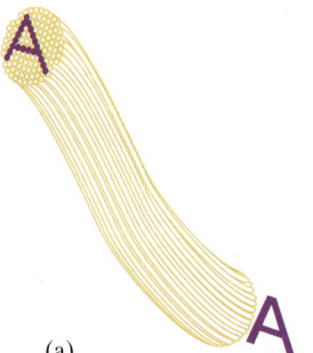
(a)

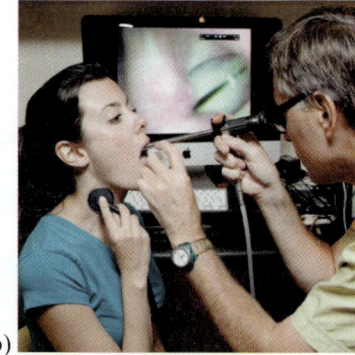
(b)

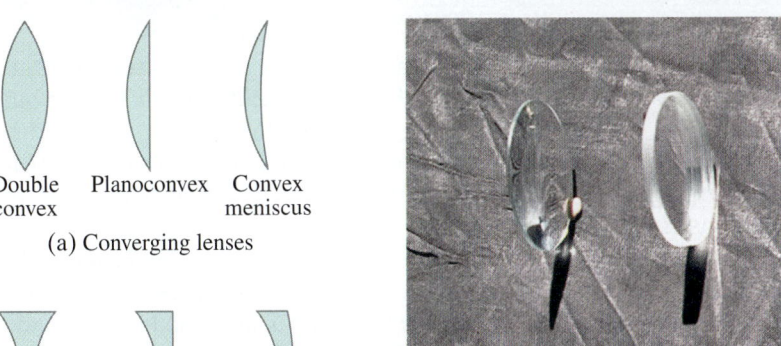
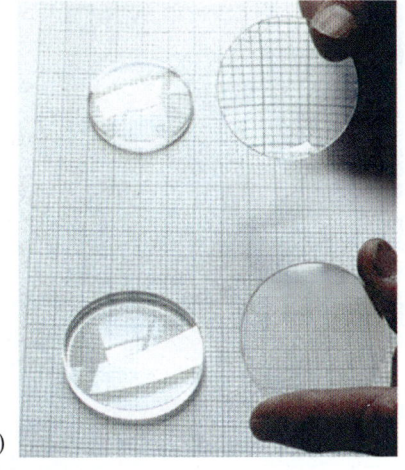

(a) Converging lenses

(b) Diverging lenses

FIGURE 23–31 (a) Converging lenses and (b) diverging lenses, shown in cross section. Converging lenses are thicker at the center whereas diverging lenses are thicker at the edges. (c) Photo of a converging lens (on the left) and a diverging lens (right). (d) Converging lenses (above), and diverging lenses (below), lying flat, and raised off the paper to form images.

23–7 Thin Lenses; Ray Tracing

The most important simple optical device is the thin lens. The development of optical devices using lenses dates to the sixteenth and seventeenth centuries, although the earliest record of eyeglasses dates from the late thirteenth century. Today we find lenses in eyeglasses, cameras, magnifying glasses, telescopes, binoculars, microscopes, and medical instruments. A thin lens is usually circular, and its two faces are portions of a sphere. (Cylindrical faces are also possible, but we will concentrate on spherical.) The two faces can be concave, convex, or plane. Several types are shown in Figs. 23–31a and b in cross section. The importance of lenses is that they form images of objects—see Fig. 23–32.

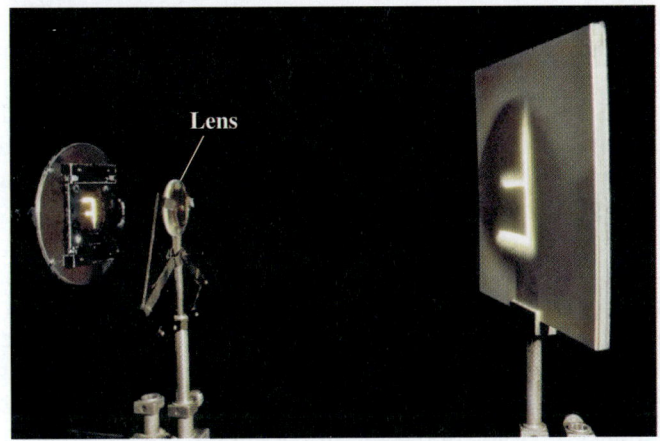

FIGURE 23–32 Converging lens (in holder) forms an image (large "F" on screen at right) of a bright object (illuminated "F" at the left).

Consider parallel rays striking the double convex lens shown in cross section in Fig. 23–33. We assume the lens is made of transparent material such as glass or transparent plastic with index of refraction greater than that of the air outside. The **axis** of a lens is a straight line passing through the center of the lens and perpendicular to its two surfaces (Fig. 23–33). From Snell's law, we can see that each ray in Fig. 23–33 is bent toward the axis when the ray enters the lens and again when it leaves the lens at the back surface. (Note the dashed lines indicating the normals to each surface for the top ray.) If rays parallel to the axis fall on a thin lens, they will be focused to a point called the **focal point**, F. This will not be precisely true for a lens with spherical surfaces. But it will be very nearly true—that is, parallel rays will be focused to a tiny region that is nearly a point—if the diameter of the lens is small compared to the radii of curvature of the two lens surfaces. This criterion is satisfied by a **thin lens**, one that is very thin compared to its diameter, and we consider only thin lenses here.

FIGURE 23–33 Parallel rays are brought to a focus by a converging thin lens.

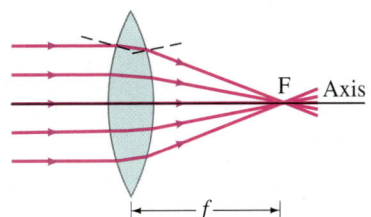

FIGURE 23–34 Image of the Sun burning wood.

FIGURE 23–35 Parallel rays at an angle are focused on the focal plane.

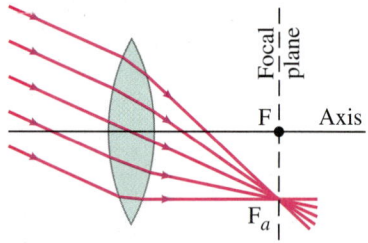

FIGURE 23–36 Diverging lens.

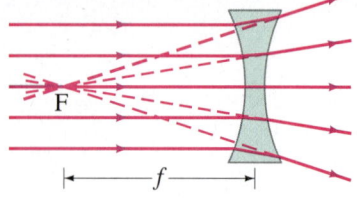

The rays from a point on a distant object are essentially parallel—see Fig. 23–12. Therefore we can say that *the focal point is the image point for an object at infinity on the lens axis*, Fig. 23–33. Thus, the focal point of a lens can be found by locating the point where the Sun's rays (or those from some other distant object) are brought to a sharp image, Fig. 23–34. The distance of the focal point from the center of the lens is called the **focal length**, f, Fig. 23–33. A lens can be turned around so that light can pass through it from the opposite side. The *focal length is the same on both sides*, as we shall see later, even if the curvatures of the two lens surfaces are different. If parallel rays fall on a lens at an angle, as in Fig. 23–35, they focus at a point F_a. The plane containing all focus points, such as F and F_a in Fig. 23–35, is called the **focal plane** of the lens.

Any lens (in air) that is thicker in the center than at the edges will make parallel rays converge to a point, and is called a **converging lens** (see Fig. 23–31a). Lenses that are thinner in the center than at the edges (Fig. 23–31b) are called **diverging lenses** because they make parallel light diverge, as shown in Fig. 23–36. The focal point, F, of a diverging lens is defined as that point from which refracted rays, originating from parallel incident rays, seem to emerge as shown in Fig. 23–36. And the distance from F to the center of the lens is called the **focal length**, f, just as for a converging lens.

EXERCISE F Return to Chapter-Opening Question 2, page 644, and answer it again now. Try to explain why you may have answered differently the first time.

Optometrists and ophthalmologists, instead of using the focal length, use the reciprocal of the focal length to specify the strength of eyeglass (or contact) lenses. This is called the **power**, P, of a lens:

$$P = \frac{1}{f}. \tag{23-7}$$

The unit for lens power is the **diopter** (D), which is an inverse meter: $1\,\text{D} = 1\,\text{m}^{-1}$. For example, a 20-cm-focal-length lens has a power $P = 1/(0.20\,\text{m}) = 5.0\,\text{D}$. We will mainly use the focal length, but we will refer again to the power of a lens when we discuss eyeglass lenses in Chapter 25.

The most important parameter of a lens is its focal length f, which is the same on both sides of the lens. For a converging lens, f can be measured by finding the image point for the Sun or other distant objects. Once f is known, the image position can be determined for any object. To find the image point by drawing rays would be difficult if we had to determine the refractive angles at the front surface of the lens and again at the back surface where the ray exits. We can save ourselves a lot of effort by making use of certain facts we already know, such as that a ray parallel to the axis of the lens passes (after refraction) through the focal point. To determine an image point, we can consider only the three rays indicated in Fig. 23–37, which uses an arrow (on the left) as the object, and a converging lens forming an image (dashed arrow) to the right. These rays, emanating from a single point on the object, are drawn as if the lens were infinitely thin, and we show only a single sharp bend at the center line of the lens instead of the refractions at each surface. These three rays are drawn as follows:

RAY DIAGRAM
Finding the image position formed by a thin lens

Ray 1 is drawn parallel to the axis, Fig. 23–37a; therefore it is refracted by the lens so that it passes along a line through the focal point F behind the lens.

Ray 2 is drawn to pass through the other focal point F' (front side of lens in Fig. 23–37) and emerge from the lens parallel to the axis, Fig. 23–37b. (In reverse it would be a parallel ray going left and passing through F'.)

Ray 3 is directed toward the very center of the lens, where the two surfaces are essentially parallel to each other, Fig. 23–37c. This ray therefore emerges from the lens at the same angle as it entered. The ray would be displaced slightly to one side, as we saw in Example 23–8; but since we assume the lens is thin, we draw ray 3 straight through as shown.

The point where these three rays cross is the image point for that object point. Actually, any two of these rays will suffice to locate the image point, but drawing the third ray can serve as a check.

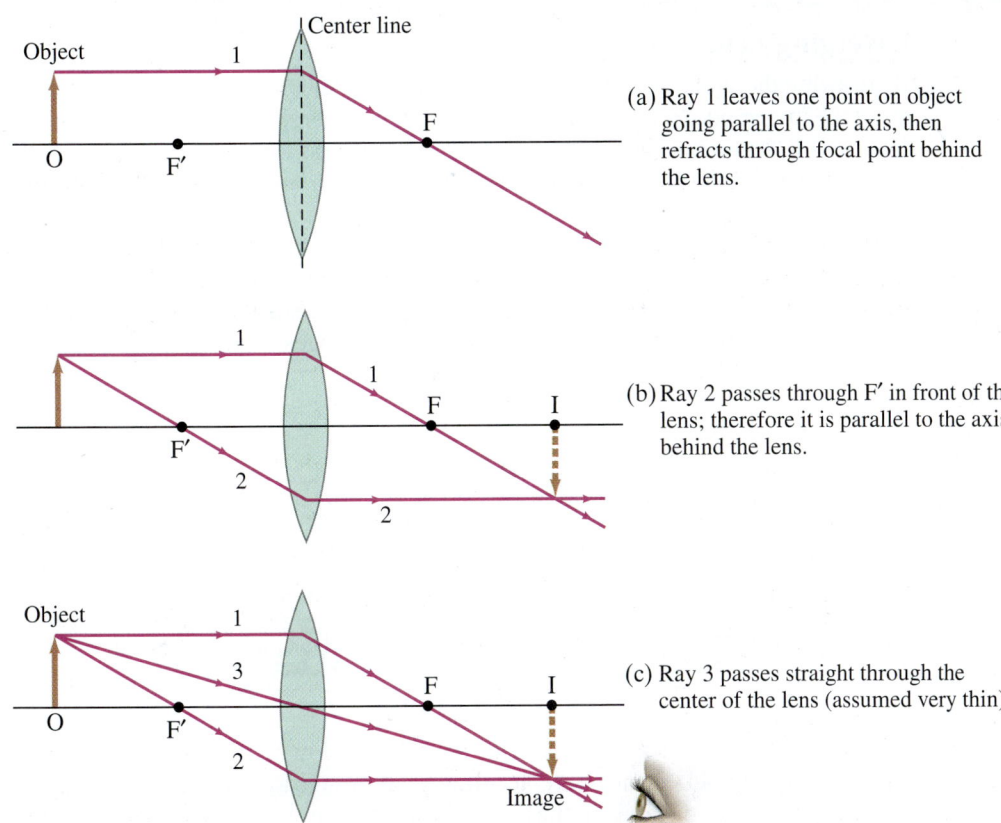

(a) Ray 1 leaves one point on object going parallel to the axis, then refracts through focal point behind the lens.

(b) Ray 2 passes through F′ in front of the lens; therefore it is parallel to the axis behind the lens.

(c) Ray 3 passes straight through the center of the lens (assumed very thin).

FIGURE 23–37 Finding the image by ray tracing for a converging lens. Rays are shown leaving one point on the object (an arrow). Shown are the three most useful rays, leaving the tip of the object, for determining where the image of that point is formed. (Note that the focal points F and F′ on either side of the lens are the same distance f from the center of the lens.)

Using these three rays for one object point, we can find the image point for that point of the object (the top of the arrow in Fig. 23–37). The image points for all other points on the object can be found similarly to determine the complete image of the object. Because the rays actually pass through the image for the case shown in Fig. 23–37, it is a **real image** (see pages 647 and/or 651). The image could be detected by film or electronic sensor, and actually be seen on a white surface or screen placed at the position of the image (Fig. 23–38).

CONCEPTUAL EXAMPLE 23–11 Half-blocked lens. What happens to the image of an object if the top half of a lens is covered by a piece of cardboard?

RESPONSE Let us look at the rays in Fig. 23–37. If the top half (or any half of the lens) is blocked, you might think that half the image is blocked. But in Fig. 23–37c, we see how the rays used to create the "top" of the image pass through both the top and the bottom of the lens. Only three of many rays are shown—many more rays pass through the lens, and they can form the image. You don't lose the image. But covering part of the lens cuts down on the total light received and reduces the brightness of the image.

NOTE If the lens is partially blocked by your thumb, you may notice an out of focus image of part of that thumb.

Seeing the Image

The image can also be seen directly by the eye when the eye is placed behind the image, as shown in Fig. 23–37c, so that some of the rays diverging from each point on the image can enter the eye. We can see a sharp image only for rays *diverging* from each point on the image, because we see normal objects when diverging rays from each point enter the eye as shown in Fig. 23–1. A normal eye cannot focus converging rays; if your eye was positioned between points F and I in Fig. 23–37c, it would not see a clear image. (More about our eyes in Section 25–2.) Figure 23–38 shows an image seen (a) on a white surface and (b) directly by the eye (and a camera) placed behind the image. The eye can see both real and virtual images (see next page) as long as the eye is positioned so rays diverging from the image enter it.

FIGURE 23–38 (a) A converging lens can form a real image (here of a distant building, upside down) on a white wall. (b) That same real image is also directly visible to the eye. [Figure 23–31d shows images (graph paper) seen by the eye made by both diverging and converging lenses.]

(a)

(b)

SECTION 23–7 Thin Lenses; Ray Tracing 663

Diverging Lens

By drawing the same three rays emerging from a single object point, we can determine the image position formed by a diverging lens, as shown in Fig. 23–39. Note that ray 1 is drawn parallel to the axis, but does not pass through the focal point F' behind the lens. Instead it seems to come (dashed line) from the focal point F in front of the lens. Ray 2 is directed toward F' and is refracted parallel to the lens axis by the lens. Ray 3 passes directly through the center of the lens. The three refracted rays seem to emerge from a point on the left of the lens. This is the image point, I. Because the rays do not pass through the image, it is a **virtual image**. Note that the eye does not distinguish between real and virtual images—both are visible.

FIGURE 23–39 Finding the image by ray tracing for a diverging lens.

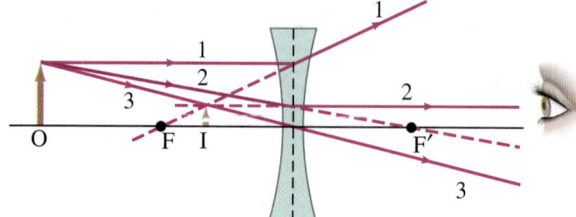

23–8 The Thin Lens Equation

We now derive an equation that relates the image distance to the object distance and the focal length of a thin lens. This equation will make the determination of image position quicker and more accurate than doing ray tracing. Let d_o be the object distance, the distance of the object from the center of the lens, and d_i be the image distance, the distance of the image from the center of the lens, Fig. 23–40.

FIGURE 23–40 Deriving the lens equation for a converging lens.

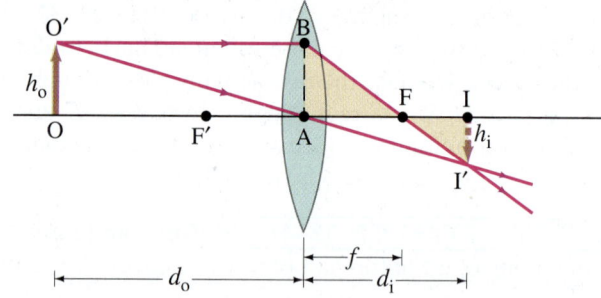

Let h_o and h_i refer to the heights of the object and image. Consider the two rays shown in Fig. 23–40 for a converging lens, assumed to be very thin. The right triangles FI'I and FBA (highlighted in yellow) are similar because angle AFB equals angle IFI'; so

$$\frac{h_i}{h_o} = \frac{d_i - f}{f},$$

since length $AB = h_o$. Triangles OAO' and IAI' are similar as well. Therefore,

$$\frac{h_i}{h_o} = \frac{d_i}{d_o}.$$

We equate the right sides of these two equations (the left sides are the same), and divide by d_i to obtain

$$\frac{1}{f} - \frac{1}{d_i} = \frac{1}{d_o}$$

or

THIN LENS EQUATION

$$\frac{1}{d_o} + \frac{1}{d_i} = \frac{1}{f}. \tag{23–8}$$

This is called the **thin lens equation**. It relates the image distance d_i to the object distance d_o and the focal length f. It is the most useful equation in geometric optics. (Interestingly, it is exactly the same as the mirror equation, Eq. 23–2.)

If the object is at infinity, then $1/d_o = 0$, so $d_i = f$. Thus the focal length is the image distance for an object at infinity, as mentioned earlier.

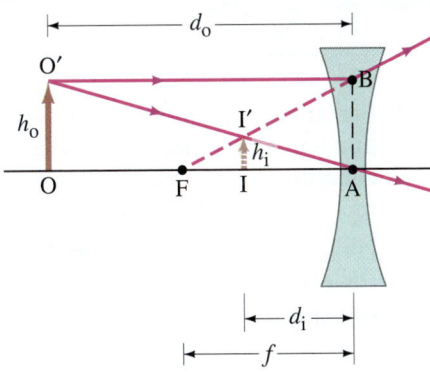

FIGURE 23–41 Deriving the lens equation for a diverging lens.

We can derive the lens equation for a diverging lens using Fig. 23–41. Triangles IAI' and OAO' are similar; and triangles IFI' and AFB are similar. Thus (noting that length $AB = h_o$)

$$\frac{h_i}{h_o} = \frac{d_i}{d_o}$$

and

$$\frac{h_i}{h_o} = \frac{f - d_i}{f}.$$

When we equate the right sides of these two equations and simplify, we obtain

$$\frac{1}{d_o} - \frac{1}{d_i} = -\frac{1}{f}.$$

This equation becomes the same as Eq. 23–8 if we make f and d_i negative. That is, we take f to be *negative for a diverging lens*, and d_i negative when the image is on the same side of the lens as the light comes from. Thus Eq. 23–8 will be valid for both converging and diverging lenses, and for *all* situations, if we use the following **sign conventions**:

> **CAUTION**
> *Focal length is negative for diverging lens*

> **PROBLEM SOLVING**
> *SIGN CONVENTIONS for lenses*

1. The focal length is positive for converging lenses and negative for diverging lenses.
2. The object distance is positive if the object is on the side of the lens from which the light is coming (this is always the case for real objects; but when lenses are used in combination, it might not be so: see Example 23–16); otherwise, it is negative.
3. The image distance is positive if the image is on the opposite side of the lens from where the light is coming; if it is on the same side, d_i is negative. Equivalently, the image distance is positive for a real image (Fig. 23–40) and negative for a virtual image (Fig. 23–41).
4. The height of the image, h_i, is positive if the image is upright, and negative if the image is inverted relative to the object. (h_o is always taken as upright and positive.)

The **magnification**, m, of a lens is defined as the ratio of the image height to object height, $m = h_i/h_o$. From Figs. 23–40 and 23–41 and the conventions just stated (for which we will need a minus sign), we have

$$m = \frac{h_i}{h_o} = -\frac{d_i}{d_o}. \quad (23\text{–}9)$$

For an upright image the magnification is positive, and for an inverted image the magnification is negative.

From sign convention 1, it follows that the power (Eq. 23–7) of a converging lens, in diopters, is positive, whereas the power of a diverging lens is negative. A converging lens is sometimes referred to as a **positive lens**, and a diverging lens as a **negative lens**.

Diverging lenses (see Fig. 23–41) always produce an upright virtual image for any real object, no matter where that object is. Converging lenses can produce real (inverted) images as in Fig. 23–40, or virtual (upright) images, depending on object position, as we shall see.

SECTION 23–8 The Thin Lens Equation **665**

PROBLEM SOLVING

Thin Lenses

1. Draw a **ray diagram**, as precise as possible, but even a rough one can serve as confirmation of analytic results. Choose one point on the object and draw at least two, or preferably three, of the easy-to-draw rays described in Figs. 23–37 and 23–39. The image point is where the rays intersect.

2. For analytic solutions, solve for unknowns in the **thin lens equation** (Eq. 23–8) and the **magnification equation** (Eq. 23–9). The thin lens equation involves reciprocals—don't forget to take the reciprocal.

3. Follow the **sign conventions** listed just above.

4. Check that your analytic answers are **consistent** with your ray diagram.

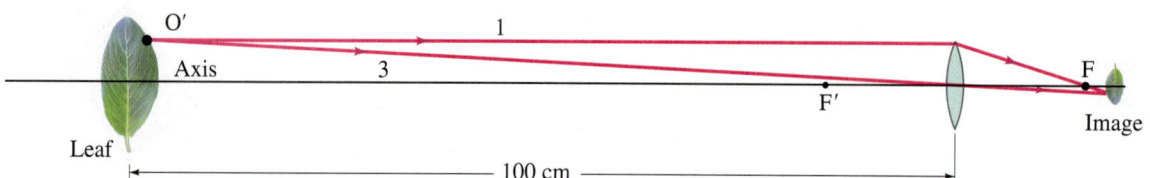

FIGURE 23–42 Example 23–12. (Not to scale.)

EXAMPLE 23–12 Image formed by converging lens. What is (a) the position, and (b) the size, of the image of a 7.6-cm-high leaf placed 1.00 m from a +50.0-mm-focal-length camera lens?

APPROACH We follow the steps of the Problem Solving Strategy explicitly.

SOLUTION

1. **Ray diagram.** Figure 23–42 is an approximate ray diagram, showing only rays 1 and 3 for a single point on the leaf. We see that the image ought to be a little behind the focal point F, to the right of the lens.

2. **Thin lens and magnification equations.** (a) We find the image position analytically using the thin lens equation, Eq. 23–8. The camera lens is converging, with $f = +5.00$ cm, and $d_o = 100$ cm, and so the thin lens equation gives

$$\frac{1}{d_i} = \frac{1}{f} - \frac{1}{d_o} = \frac{1}{5.00 \text{ cm}} - \frac{1}{100 \text{ cm}} = \frac{20.0 - 1.0}{100 \text{ cm}} = \frac{19.0}{100 \text{ cm}}.$$

Then, taking the reciprocal,

$$d_i = \frac{100 \text{ cm}}{19.0} = 5.26 \text{ cm},$$

or 52.6 mm behind the lens.
(b) The magnification is

$$m = -\frac{d_i}{d_o} = -\frac{5.26 \text{ cm}}{100 \text{ cm}} = -0.0526,$$

so

$$h_i = mh_o = (-0.0526)(7.6 \text{ cm}) = -0.40 \text{ cm}.$$

The image is 4.0 mm high.

3. **Sign conventions.** The image distance d_i came out positive, so the image is behind the lens. The image height is $h_i = -0.40$ cm; the minus sign means the image is inverted.

4. **Consistency.** The analytic results of steps 2 and 3 are consistent with the ray diagram, Fig. 23–42: the image is behind the lens and inverted.

NOTE Part (a) tells us that the image is 2.6 mm farther from the lens than the image for an object at infinity, which would equal the focal length, 50.0 mm. Indeed, when focusing a camera lens, the closer the object is to the camera, the farther the lens must be from the sensor or film.

EXERCISE G If the leaf (object) of Example 23–12 is moved farther from the lens, does the image move closer to or farther from the lens? (Don't calculate!)

EXAMPLE 23–13 **Object close to converging lens.** An object is placed 10 cm from a 15-cm-focal-length converging lens. Determine the image position and size (a) analytically, and (b) using a ray diagram.

APPROACH The object is within the focal point—closer to the lens than the focal point F as $d_o < f$. We first use Eqs. 23–8 and 23–9 to obtain an analytic solution, and then confirm with a ray diagram using the special rays 1, 2, and 3 for a single object point.

SOLUTION (a) Given $f = 15\text{ cm}$ and $d_o = 10\text{ cm}$, then

$$\frac{1}{d_i} = \frac{1}{15\text{ cm}} - \frac{1}{10\text{ cm}} = \frac{2-3}{30\text{ cm}} = -\frac{1}{30\text{ cm}},$$

and $d_i = -30\text{ cm}$. (Remember to take the reciprocal!) Because d_i is negative, the image must be virtual and on the same side of the lens as the object (sign convention 3, page 665). The magnification

$$m = -\frac{d_i}{d_o} = -\frac{-30\text{ cm}}{10\text{ cm}} = 3.0.$$

CAUTION
Don't forget to take the reciprocal

The image is three times as large as the object and is upright. This lens is being used as a magnifying glass, which we discuss in more detail in Section 25–3.

(b) The ray diagram is shown in Fig. 23–43 and confirms the result in part (a). We choose point O' on the top of the object and draw ray 1. Ray 2, however, may take some thought: if we draw it heading toward F', it is going the wrong way—so we have to draw it as if coming from F' (and so dashed), striking the lens, and then going out parallel to the lens axis. We project it backward, with a dashed line, as we must do also for ray 1, in order to find where they cross. Ray 3 is drawn through the lens center, and it crosses the other two rays at the image point, I'.

NOTE From Fig. 23–43 we can see that, when an object is placed between a converging lens and its focal point, the image is virtual.

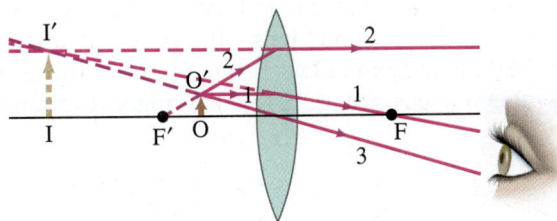

FIGURE 23–43 An object placed within the focal point of a converging lens produces a virtual image. Example 23–13.

EXAMPLE 23–14 **Diverging lens.** Where must a small insect be placed if a 25-cm-focal-length diverging lens is to form a virtual image 20 cm from the lens, on the same side as the object?

APPROACH The ray diagram is basically that of Fig. 23–41 because our lens here is diverging and our image is given as in front of the lens within the focal distance. (It would be a valuable exercise to draw the ray diagram to scale, precisely, now.) The insect's distance, d_o, can be calculated using the thin lens equation.

SOLUTION The lens is diverging, so f is negative: $f = -25\text{ cm}$. The image distance must be negative too because the image is in front of the lens (sign conventions), so $d_i = -20\text{ cm}$. The lens equation, Eq. 23–8, gives

$$\frac{1}{d_o} = \frac{1}{f} - \frac{1}{d_i} = -\frac{1}{25\text{ cm}} + \frac{1}{20\text{ cm}} = \frac{-4+5}{100\text{ cm}} = \frac{1}{100\text{ cm}}.$$

So the object must be 100 cm in front of the lens.

*23–9 Combinations of Lenses

Many optical instruments use lenses in combination. When light passes through more than one lens, we find the image formed by the first lens as if it were alone. Then this image becomes the *object* for the second lens. Next we find the image formed by this second lens using the first image as object. This second image is the final image if there are only two lenses. The total magnification will be the product of the separate magnifications of each lens. Even if the second lens intercepts the light from the first lens before it forms an image, this technique still works.

EXAMPLE 23–15 A two-lens system. Two converging lenses, A and B, with focal lengths $f_A = 20.0$ cm and $f_B = 25.0$ cm, are placed 80.0 cm apart, as shown in Fig. 23–44a. An object is placed 60.0 cm in front of the first lens as shown in Fig. 23–44b. Determine (*a*) the position, and (*b*) the magnification, of the final image formed by the combination of the two lenses.

APPROACH Starting at the tip of our object O, we draw rays 1, 2, and 3 for the first lens, A, and also a ray 4 which, after passing through lens A, acts for the second lens, B, as ray 3′ (through the center). We use primes now for the standard rays relative to lens B. Ray 2 for lens A exits parallel, and so is ray 1′ for lens B. To determine the position of the image I_A formed by lens A, we use Eq. 23–8 with $f_A = 20.0$ cm and $d_{oA} = 60.0$ cm. The distance of I_A (lens A's image) from lens B is the object distance d_{oB} for lens B. The final image is found using the thin lens equation, this time with all distances relative to lens B. For (*b*) the magnifications are found from Eq. 23–9 for each lens in turn.

SOLUTION (*a*) The object is a distance $d_{oA} = +60.0$ cm from the first lens, A, and this lens forms an image whose position can be calculated using the thin lens equation:

$$\frac{1}{d_{iA}} = \frac{1}{f_A} - \frac{1}{d_{oA}} = \frac{1}{20.0 \text{ cm}} - \frac{1}{60.0 \text{ cm}} = \frac{3-1}{60.0 \text{ cm}} = \frac{1}{30.0 \text{ cm}}.$$

> **CAUTION**
> *Object distance for second lens is **not** equal to the image distance for first lens*

So the first image I_A is at $d_{iA} = 30.0$ cm behind the first lens. This image becomes the object for the second lens, B. It is a distance $d_{oB} = 80.0$ cm $-$ 30.0 cm $=$ 50.0 cm in front of lens B (Fig. 23–44b). The image formed by lens B, again using the thin lens equation, is at a distance d_{iB} from the lens B:

$$\frac{1}{d_{iB}} = \frac{1}{f_B} - \frac{1}{d_{oB}} = \frac{1}{25.0 \text{ cm}} - \frac{1}{50.0 \text{ cm}} = \frac{2-1}{50.0 \text{ cm}} = \frac{1}{50.0 \text{ cm}}.$$

Hence $d_{iB} = 50.0$ cm behind lens B. This is the final image—see Fig. 23–44b.

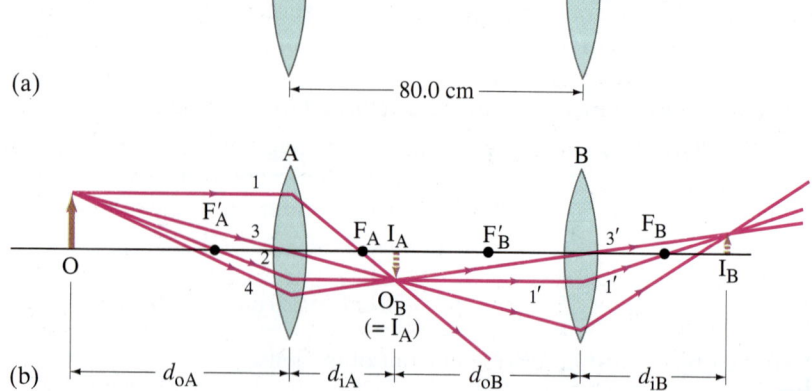

FIGURE 23–44 Two lenses, A and B, used in combination, Example 23–15. The small numbers refer to the easily drawn rays.

(b) Lens A has a magnification (Eq. 23–9)
$$m_A = -\frac{d_{iA}}{d_{oA}} = -\frac{30.0\text{ cm}}{60.0\text{ cm}} = -0.500.$$
Thus, the first image is inverted and is half as high as the object (again Eq. 23–9):
$$h_{iA} = m_A h_{oA} = -0.500 h_{oA}.$$
Lens B takes this first image as object and changes its height by a factor
$$m_B = -\frac{d_{iB}}{d_{oB}} = -\frac{50.0\text{ cm}}{50.0\text{ cm}} = -1.000.$$
The second lens reinverts the image (the minus sign) but doesn't change its size. The final image height is (remember h_{oB} is the same as h_{iA})
$$h_{iB} = m_B h_{oB} = m_B h_{iA} = m_B m_A h_{oA} = (m_{\text{total}}) h_{oA}.$$
The total magnification is the product of m_A and m_B, which here equals $m_{\text{total}} = m_A m_B = (-1.000)(-0.500) = +0.500$, or half the original height, and the final image is upright.

PROBLEM SOLVING
Total magnification is
$m_{\text{total}} = m_A m_B$

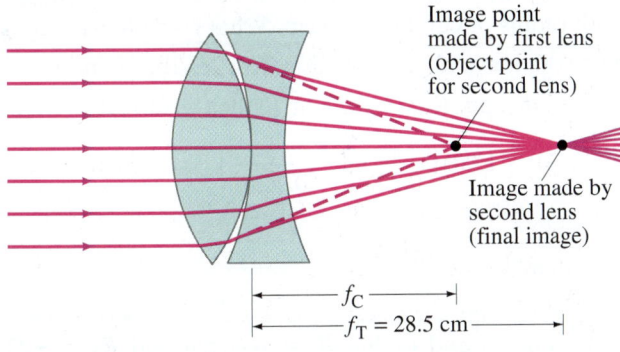

FIGURE 23–45 Determining the focal length of a diverging lens. Example 23–16.

EXAMPLE 23–16 Measuring f for a diverging lens. To measure the focal length of a diverging lens, a converging lens is placed in contact with it, as shown in Fig. 23–45. The Sun's rays are focused by this combination at a point 28.5 cm behind the lenses as shown. If the converging lens has a focal length f_C of 16.0 cm, what is the focal length f_D of the diverging lens? Assume both lenses are thin and the space between them is negligible.

APPROACH The image distance for the first lens equals its focal length (16.0 cm) since the object distance is infinity (∞). The position of this image, even though it is never actually formed, acts as the object for the second (diverging) lens. We apply the thin lens equation to the diverging lens to find its focal length, given that the final image is at $d_i = 28.5$ cm.

SOLUTION Rays from the Sun are focused 28.5 cm behind the combination, so the focal length of the total combination is $f_T = 28.5$ cm. If the diverging lens was absent, the converging lens would form the image at its focal point—that is, at a distance $f_C = 16.0$ cm behind it (dashed lines in Fig. 23–45). When the diverging lens is placed next to the converging lens, we treat the image formed by the first lens as the *object* for the second lens. Since this object lies to the right of the diverging lens, this is a situation where d_o is negative (see the sign conventions, page 665). Thus, for the diverging lens, the object is virtual and $d_o = -16.0$ cm. The diverging lens forms the image of this virtual object at a distance $d_i = 28.5$ cm away (given). Thus,
$$\frac{1}{f_D} = \frac{1}{d_o} + \frac{1}{d_i} = \frac{1}{-16.0\text{ cm}} + \frac{1}{28.5\text{ cm}} = -0.0274\text{ cm}^{-1}.$$
We take the reciprocal to find $f_D = -1/(0.0274\text{ cm}^{-1}) = -36.5$ cm.

CAUTION
$d_o < 0$

NOTE If this technique is to work, the converging lens must be "stronger" than the diverging lens—that is, it must have a focal length whose magnitude is less than that of the diverging lens.

*23–10 Lensmaker's Equation

A useful equation, called the **lensmaker's equation**, relates the focal length of a lens to the radii of curvature R_1 and R_2 of its two surfaces and its index of refraction n:

Lensmaker's equation

$$\frac{1}{f} = (n - 1)\left(\frac{1}{R_1} + \frac{1}{R_2}\right). \quad (23\text{–}10)$$

If both surfaces are convex, R_1 and R_2 are considered positive.† For a concave surface, the radius must be considered *negative*.

Notice that Eq. 23-10 is *symmetrical* in R_1 and R_2. Thus, if a lens is turned around so that light impinges on the other surface, the focal length is the same even if the two lens surfaces are different. This confirms what we said earlier: a lens' focal length is the same on both sides of the lens.

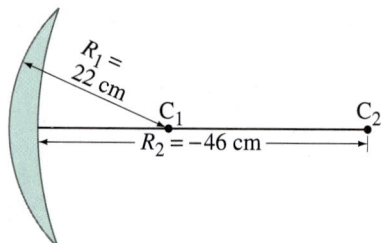

FIGURE 23–46 Example 23–17. The left surface is convex (center bulges outward); the right surface is concave.

> **EXAMPLE 23–17 Calculating f for a converging lens.** A convex meniscus lens (Figs. 23-31a and 23-46) is made from glass with $n = 1.50$. The radius of curvature of the convex surface (left in Fig. 23-46) is 22 cm. The surface on the right is concave with radius of curvature 46 cm. What is the focal length?
>
> **APPROACH** We use the lensmaker's equation, Eq. 23-10, to find f.
>
> **SOLUTION** $R_1 = 0.22 \text{ m}$ and $R_2 = -0.46 \text{ m}$ (concave surface). Then
>
> $$\frac{1}{f} = (1.50 - 1.00)\left(\frac{1}{0.22 \text{ m}} - \frac{1}{0.46 \text{ m}}\right) = 1.19 \text{ m}^{-1}.$$
>
> So
>
> $$f = \frac{1}{1.19 \text{ m}^{-1}} = 0.84 \text{ m},$$
>
> and the lens is converging since $f > 0$.
>
> **NOTE** If we turn the lens around so that $R_1 = -46$ cm and $R_2 = +22$ cm, we get the same result.
>
> **NOTE** Because Eq. 23-10 gives $1/f$, it gives directly the power of a lens in diopters, Eq. 23-7. The power of this lens is about 1.2 D.

†Some books use a different convention: R_1 and R_2 may be considered positive if their centers of curvature are to the right of the lens; then a minus sign replaces the + sign in their version of Eq. 23-10.

Summary

Light appears to travel along straight-line paths, called **rays**, through uniform transparent materials including air and glass. When light reflects from a flat surface, the *angle of reflection equals the angle of incidence*. This **law of reflection** explains why mirrors can form **images**.

In a **plane mirror**, the image is virtual, upright, the same size as the object, and as far behind the mirror as the object is in front.

A **spherical mirror** can be concave or convex. A **concave** spherical mirror focuses parallel rays of light (light from a very distant object) to a point called the **focal point**. The distance of this point from the mirror is the **focal length** f of the mirror and

$$f = \frac{r}{2} \quad (23\text{–}1)$$

where r is the radius of curvature of the mirror.

Parallel rays falling on a **convex mirror** reflect from the mirror as if they diverged from a common point behind the mirror. The distance of this point from the mirror is the focal length and is considered negative for a convex mirror.

For a given object, the approximate position and size of the image formed by a mirror can be found by ray tracing. Algebraically, the relation between image and object distances, d_i and d_o, and the focal length f, is given by the **mirror equation**:

$$\frac{1}{d_o} + \frac{1}{d_i} = \frac{1}{f}. \quad (23\text{–}2)$$

The ratio of image height h_i to object height h_o, which equals the magnification m of a mirror, is

$$m = \frac{h_i}{h_o} = -\frac{d_i}{d_o}. \quad (23\text{–}3)$$

If the rays that converge to form an image actually pass through the image, so the image would appear on a screen or film placed there, the image is said to be a **real image**. If the light rays do not actually pass through the image, the image is a **virtual image**.

The speed of light v depends on the **index of refraction**, n, of the material:

$$n = \frac{c}{v}, \quad (23\text{–}4)$$

where c is the speed of light in vacuum.

When light passes from one transparent medium into another, the rays bend or refract. The **law of refraction** (**Snell's law**) states that

$$n_1 \sin \theta_1 = n_2 \sin \theta_2, \quad (23\text{–}5)$$

where n_1 and θ_1 are the index of refraction and angle with the normal (perpendicular) to the surface for the incident ray, and n_2 and θ_2 are for the refracted ray.

When light rays reach the boundary of a material where the index of refraction decreases, the rays will be **totally internally reflected** if the incident angle, θ_1, is such that Snell's law would

670 CHAPTER 23 Light: Geometric Optics

predict $\sin \theta_2 > 1$. This occurs if θ_1 exceeds the critical angle θ_C given by

$$\sin \theta_C = \frac{n_2}{n_1}. \qquad (23\text{-}6)$$

A lens uses refraction to produce a real or virtual image. Parallel rays of light are focused to a point, the **focal point**, by a **converging** lens. The distance of the focal point from the lens is the **focal length** f of the lens. It is the same on both sides of the lens.

After parallel rays pass through a **diverging** lens, they appear to diverge from a point in front of the lens, which is its focal point; and the corresponding focal length is considered negative.

The **power** P of a lens, which is $P = 1/f$ (Eq. 23-7), is given in diopters, which are units of inverse meters (m^{-1}).

For a given object, the position and size of the image formed by a lens can be found approximately by ray tracing. Algebraically, the relation between image and object distances, d_i and d_o, and the focal length f, is given by the **thin lens equation**:

$$\frac{1}{d_o} + \frac{1}{d_i} = \frac{1}{f}. \qquad (23\text{-}8)$$

The ratio of image height to object height, which equals the **magnification** m for a lens, is

$$m = \frac{h_i}{h_o} = -\frac{d_i}{d_o}. \qquad (23\text{-}9)$$

When using the various equations of geometric optics, you must remember the **sign conventions** for all quantities involved: carefully review them (pages 655 and 665) when doing Problems.

[*When two (or more) thin lenses are used in combination to produce an image, the thin lens equation can be used for each lens in sequence. The image produced by the first lens acts as the object for the second lens.]

[*The **lensmaker's equation** relates the radii of curvature of the lens surfaces and the lens' index of refraction to the focal length of the lens.]

Questions

1. Archimedes is said to have burned the whole Roman fleet in the harbor of Syracuse, Italy, by focusing the rays of the Sun with a huge spherical mirror. Is this[†] reasonable?
2. What is the focal length of a plane mirror? What is the magnification of a plane mirror?
3. Although a plane mirror appears to reverse left and right, it doesn't reverse up and down. Discuss why this happens, noting that front to back is also reversed. Also discuss what happens if, while standing, you look up vertically at a horizontal mirror on the ceiling.
4. An object is placed along the principal axis of a spherical mirror. The magnification of the object is -2.0. Is the image real or virtual, inverted or upright? Is the mirror concave or convex? On which side of the mirror is the image located?
5. If a concave mirror produces a real image, is the image necessarily inverted? Explain.
6. How might you determine the speed of light in a solid, rectangular, transparent object?
7. When you look at the Moon's reflection from a ripply sea, it appears elongated (Fig. 23-47). Explain.

FIGURE 23-47 Question 7.

8. What is the angle of refraction when a light ray is incident perpendicular to the boundary between two transparent materials?

9. When you look down into a swimming pool or a lake, are you likely to overestimate or underestimate its depth? Explain. How does the apparent depth vary with the viewing angle? (Use ray diagrams.)
10. Draw a ray diagram to show why a stick or straw looks bent when part of it is under water (Fig. 23-23).
11. When a wide beam of parallel light enters water at an angle, the beam broadens. Explain.
12. You look into an aquarium and view a fish inside. One ray of light from the fish is shown emerging from the tank in Fig. 23-48. The apparent position of the fish is also shown (dashed ray). In the drawing, indicate the approximate position of the actual fish. Briefly justify your answer.

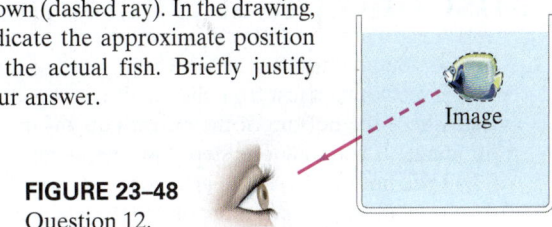

FIGURE 23-48 Question 12.

13. How can you "see" a round drop of water on a table even though the water is transparent and colorless?
14. A ray of light is refracted through three different materials (Fig. 23-49). Which material has (a) the largest index of refraction, (b) the smallest?

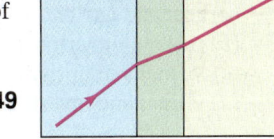

FIGURE 23-49 Question 14.

15. A child looks into a pool to see how deep it is. She then drops a small toy into the pool to help decide how deep the pool is. After this careful investigation, she decides it is safe to jump in—only to discover the water is over her head. What went wrong with her interpretation of her experiment?
16. Can a light ray traveling in air be totally reflected when it strikes a smooth water surface if the incident angle is chosen correctly? Explain.

[†]Students at MIT did a feasibility study. See www.mit.edu/2.009/www/experiments/deathray/10_ArchimedesResult.html.

17. What type of mirror is shown in Fig. 23–50? Explain.

FIGURE 23–50 Question 17 and Problem 15.

18. Light rays from stars (including our Sun) always bend toward the vertical direction as they pass through the Earth's atmosphere. (a) Why does this make sense? (b) What can you conclude about the apparent positions of stars as viewed from Earth? Draw a circle for Earth, a dot for you, and 3 or 4 stars at different angles.

19. Where must the film be placed if a camera lens is to make a sharp image of an object far away? Explain.

20. A photographer moves closer to his subject and then refocuses. Does the camera lens move farther away from or closer to the camera film or sensor? Explain.

21. Can a diverging lens form a real image under any circumstances? Explain.

22. Light rays are said to be "reversible." Is this consistent with the thin lens equation? Explain.

23. Can real images be projected on a screen? Can virtual images? Can either be photographed? Discuss carefully.

24. A thin converging lens is moved closer to a nearby object. Does the real image formed change (a) in position, (b) in size? If yes, describe how.

25. If a glass converging lens is placed in water, its focal length in water will be (a) longer, (b) shorter, or (c) the same as in air. Explain.

26. Compare the mirror equation with the thin lens equation. Discuss similarities and differences, especially the sign conventions for the quantities involved.

27. A lens is made of a material with an index of refraction $n = 1.25$. In air, it is a converging lens. Will it still be a converging lens if placed in water? Explain, using a ray diagram.

28. (a) Does the focal length of a lens depend on the fluid in which it is immersed? (b) What about the focal length of a spherical mirror? Explain.

29. An underwater lens consists of a carefully shaped thin-walled plastic container filled with air. What shape should it have in order to be (a) converging, (b) diverging? Use ray diagrams to support your answer.

30. The thicker a double convex lens is in the center as compared to its edges, the shorter its focal length for a given lens diameter. Explain.

*31. A non-symmetrical lens (say, planoconvex) forms an image of a nearby object. Use the lensmaker's equation to explain if the image point changes when the lens is turned around.

*32. Example 23–16 shows how to use a converging lens to measure the focal length of a diverging lens. (a) Why can't you measure the focal length of a diverging lens directly? (b) It is said that for this to work, the converging lens must be stronger than the diverging lens. What is meant by "stronger," and why is this statement true?

MisConceptual Questions

1. Suppose you are standing about 3 m in front of a mirror. You can see yourself just from the top of your head to your waist, where the bottom of the mirror cuts off the rest of your image. If you walk one step closer to the mirror
   (a) you will not be able to see any more of your image.
   (b) you will be able to see more of your image, below your waist.
   (c) you will see less of your image, with the cutoff rising to be above your waist.

2. When the reflection of an object is seen in a flat mirror, the image is
   (a) real and upright.
   (b) real and inverted.
   (c) virtual and upright.
   (d) virtual and inverted.

3. You want to create a spotlight that will shine a bright beam of light with all of the light rays parallel to each other. You have a large concave spherical mirror and a small lightbulb. Where should you place the lightbulb?
   (a) At the focal point of the mirror.
   (b) At the radius of curvature of the mirror.
   (c) At any point, because all rays bouncing off the mirror will be parallel.
   (d) None of the above; you can't make parallel rays with a concave mirror.

4. When you look at a fish in a still stream from the bank, the fish appears shallower than it really is due to refraction. From directly above, it appears
   (a) deeper than it really is.
   (b) at its actual depth.
   (c) shallower than its real depth.
   (d) It depends on your height above the water.

5. Parallel light rays cross interfaces from medium 1 into medium 2 and then into medium 3 as shown in Fig. 23–51. What can we say about the relative sizes of the indices of refraction of these media?
   (a) $n_1 > n_2 > n_3$.
   (b) $n_3 > n_2 > n_1$.
   (c) $n_2 > n_3 > n_1$.
   (d) $n_1 > n_3 > n_2$.
   (e) $n_2 > n_1 > n_3$.
   (f) None of the above.

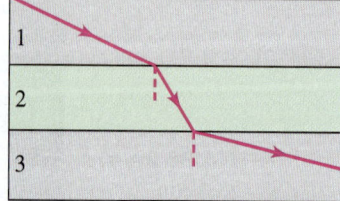

FIGURE 23–51 MisConceptual Question 5.

6. To shoot a swimming fish with an intense light beam from a *laser gun*, you should aim
   (a) directly at the image.
   (b) slightly above the image.
   (c) slightly below the image.

7. When moonlight strikes the surface of a calm lake, what happens to this light?
   (a) All of it reflects from the water surface back to the air.
   (b) Some of it reflects back to the air; some enters the water.
   (c) All of it enters the water.
   (d) All of it disappears via absorption by water molecules.

8. If you shine a light through an optical fiber, why does it come out the end but not out the sides?
   (a) It does come out the sides, but this effect is not obvious because the sides are so much longer than the ends.
   (b) The sides are mirrored, so the light reflects.
   (c) Total internal reflection makes the light reflect from the sides.
   (d) The light flows along the length of the fiber, never touching the sides.

9. A converging lens, such as a typical magnifying glass,
   (a) always produces a magnified image (taller than object).
   (b) always produces an image smaller than the object.
   (c) always produces an upright image.
   (d) always produces an inverted image (upside down).
   (e) None of these statements are true.

10. Virtual images can be formed by
    (a) only mirrors.
    (b) only lenses.
    (c) only plane mirrors.
    (d) only curved mirrors or lenses.
    (e) plane and curved mirrors, and lenses.

11. A lens can be characterized by its *power*, which
    (a) is the same as the magnification.
    (b) tells how much light the lens can focus.
    (c) depends on where the object is located.
    (d) is the reciprocal of the focal length.

12. You cover half of a lens that is forming an image on a screen. Compare what happens when you cover the top half of the lens versus the bottom half.
    (a) When you cover the top half of the lens, the top half of the image disappears; when you cover the bottom half of the lens, the bottom half of the image disappears.
    (b) When you cover the top half of the lens, the bottom half of the image disappears; when you cover the bottom half of the lens, the top half of the image disappears.
    (c) The image becomes half as bright in both cases.
    (d) Nothing happens in either case.
    (e) The image disappears in both cases.

13. Which of the following can form an image?
    (a) A plane mirror.
    (b) A curved mirror.
    (c) A lens curved on both sides.
    (d) A lens curved on only one side.
    (e) All of the above.

14. As an object moves from just outside the focal point of a converging lens to just inside it, the image goes from _____ and _____ to _____ and _____.
    (a) large; inverted; large; upright.
    (b) large; upright; large; inverted.
    (c) small; inverted; small; upright.
    (d) small; upright; small; inverted.

For assigned homework and other learning materials, go to the MasteringPhysics website.

Problems

23–2 Reflection; Plane Mirrors

1. (I) When you look at yourself in a 60-cm-tall plane mirror, you see the same amount of your body whether you are close to the mirror or far away. (Try it and see.) Use ray diagrams to show why this should be true.

2. (I) Suppose that you want to take a photograph of yourself as you look at your image in a mirror 3.1 m away. For what distance should the camera lens be focused?

3. (II) Two plane mirrors meet at a 135° angle, Fig. 23–52. If light rays strike one mirror at 34° as shown, at what angle ϕ do they leave the second mirror?

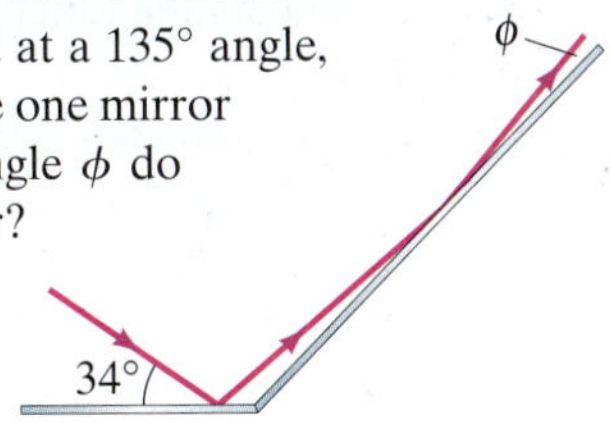

FIGURE 23–52 Problem 3.

4. (II) A person whose eyes are 1.72 m above the floor stands 2.20 m in front of a vertical plane mirror whose bottom edge is 38 cm above the floor, Fig. 23–53. What is the horizontal distance x to the base of the wall supporting the mirror of the nearest point on the floor that can be seen reflected in the mirror?

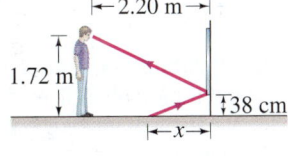

FIGURE 23–53 Problem 4.

5. (II) Stand up two plane mirrors so they form a 90.0° angle as in Fig. 23–54. When you look into this double mirror, you see yourself as others see you, instead of reversed as in a single mirror. Make a ray diagram to show how this occurs.

FIGURE 23–54 Problem 5.

6. (II) Two plane mirrors, nearly parallel, are facing each other 2.3 m apart as in Fig. 23–55. You stand 1.6 m away from one of these mirrors and look into it. You will see multiple images of yourself. (a) How far away from you are the first three images of yourself in the mirror in front of you? (b) Are these first three images facing toward you or away from you?

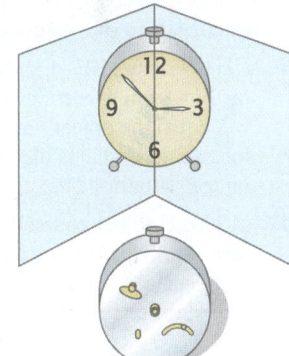

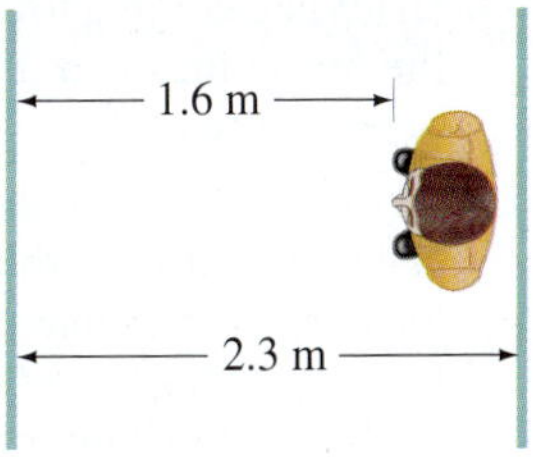

FIGURE 23–55 Problem 6.

7. (III) Suppose you are 94 cm from a plane mirror. What area of the mirror is used to reflect the rays entering one eye from a point on the tip of your nose if your pupil diameter is 4.5 mm?

23–3 Spherical Mirrors

8. (I) A solar cooker, really a concave mirror pointed at the Sun, focuses the Sun's rays 18.8 cm in front of the mirror. What is the radius of the spherical surface from which the mirror was made?

9. (I) How far from a concave mirror (radius 21.0 cm) must an object be placed if its image is to be at infinity?

10. (II) A small candle is 38 cm from a concave mirror having a radius of curvature of 24 cm. (a) What is the focal length of the mirror? (b) Where will the image of the candle be located? (c) Will the image be upright or inverted?

11. (II) An object 3.0 mm high is placed 16 cm from a convex mirror of radius of curvature 16 cm. (a) Show by ray tracing that the image is virtual, and estimate the image distance. (b) Show that the (negative) image distance can be computed from Eq. 23–2 using a focal length of -8.0 cm. (c) Compute the image size, using Eq. 23–3.

12. (II) A dentist wants a small mirror that, when 2.00 cm from a tooth, will produce a 4.0× upright image. What kind of mirror must be used and what must its radius of curvature be?

13. (II) You are standing 3.4 m from a convex security mirror in a store. You estimate the height of your image to be half of your actual height. Estimate the radius of curvature of the mirror.

14. (II) The image of a distant tree is virtual and very small when viewed in a curved mirror. The image appears to be 19.0 cm behind the mirror. What kind of mirror is it, and what is its radius of curvature?

15. (II) A mirror at an amusement park shows an upright image of any person who stands 1.9 m in front of it. If the image is three times the person's height, what is the radius of curvature of the mirror? (See Fig. 23–50.)

16. (II) In Example 23–4, show that if the object is moved 10.0 cm farther from the concave mirror, the object's image size will equal the object's actual size. Stated as a multiple of the focal length, what is the object distance for this "actual-sized image" situation?

17. (II) You look at yourself in a shiny 8.8-cm-diameter Christmas tree ball. If your face is 25.0 cm away from the ball's front surface, where is your image? Is it real or virtual? Is it upright or inverted?

18. (II) Some rearview mirrors produce images of cars to your rear that are smaller than they would be if the mirror were flat. Are the mirrors concave or convex? What is a mirror's radius of curvature if cars 16.0 m away appear 0.33 their normal size?

19. (II) When walking toward a concave mirror you notice that the image flips at a distance of 0.50 m. What is the radius of curvature of the mirror?

20. (II) (a) Where should an object be placed in front of a concave mirror so that it produces an image at the same location as the object? (b) Is the image real or virtual? (c) Is the image inverted or upright? (d) What is the magnification of the image?

21. (II) A shaving or makeup mirror is designed to magnify your face by a factor of 1.40 when your face is placed 20.0 cm in front of it. (a) What type of mirror is it? (b) Describe the type of image that it makes of your face. (c) Calculate the required radius of curvature for the mirror.

22. (II) Use two techniques, (a) a ray diagram, and (b) the mirror equation, to show that the magnitude of the magnification of a concave mirror is less than 1 if the object is beyond the center of curvature C ($d_o > r$), and is greater than 1 if the object is within C ($d_o < r$).

23. (III) Show, using a ray diagram, that the magnification m of a convex mirror is $m = -d_i/d_o$, just as for a concave mirror. [Hint: Consider a ray from the top of the object that reflects at the center of the mirror.]

24. (III) An object is placed a distance r in front of a wall, where r exactly equals the radius of curvature of a certain concave mirror. At what distance from the wall should this mirror be placed so that a real image of the object is formed on the wall? What is the magnification of the image?

23–4 Index of Refraction

25. (I) The speed of light in ice is 2.29×10^8 m/s. What is the index of refraction of ice?

26. (I) What is the speed of light in (a) ethyl alcohol, (b) lucite, (c) crown glass?

27. (II) The speed of light in a certain substance is 82% of its value in water. What is the index of refraction of that substance?

23–5 Refraction; Snell's Law

28. (I) A flashlight beam strikes the surface of a pane of glass ($n = 1.56$) at a 67° angle to the normal. What is the angle of refraction?

29. (I) A diver shines a flashlight upward from beneath the water at a 35.2° angle to the vertical. At what angle does the light leave the water?

30. (I) A light beam coming from an underwater spotlight exits the water at an angle of 56.0°. At what angle of incidence did it hit the air–water interface from below the surface?

31. (I) Rays of the Sun are seen to make a 36.0° angle to the vertical beneath the water. At what angle above the horizon is the Sun?

32. (II) An aquarium filled with water has flat glass sides whose index of refraction is 1.54. A beam of light from outside the aquarium strikes the glass at a 43.5° angle to the perpendicular (Fig. 23–56). What is the angle of this light ray when it enters (a) the glass, and then (b) the water? (c) What would be the refracted angle if the ray entered the water directly?

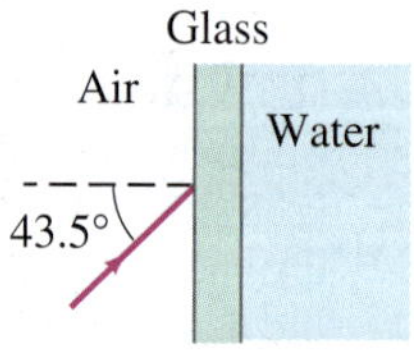

FIGURE 23–56
Problem 32.

33. (II) A beam of light in air strikes a slab of glass ($n = 1.51$) and is partially reflected and partially refracted. Determine the angle of incidence if the angle of reflection is twice the angle of refraction.

34. (II) In searching the bottom of a pool at night, a watchman shines a narrow beam of light from his flashlight, 1.3 m above the water level, onto the surface of the water at a point 2.5 m from his foot at the edge of the pool (Fig. 23–57). Where does the spot of light hit the bottom of the 2.1-m-deep pool? Measure from the bottom of the wall beneath his foot.

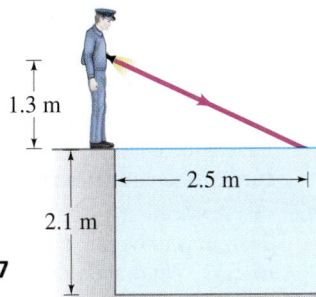

FIGURE 23–57 Problem 34.

23–6 Total Internal Reflection

35. (I) What is the critical angle for the interface between water and crown glass? To be internally reflected, the light must start in which material?
36. (I) The critical angle for a certain liquid–air surface is 47.2°. What is the index of refraction of the liquid?
37. (II) A beam of light is emitted in a pool of water from a depth of 82.0 cm. Where must it strike the air–water interface, relative to the spot directly above it, in order that the light does *not* exit the water?
38. (II) A beam of light is emitted 8.0 cm beneath the surface of a liquid and strikes the air surface 7.6 cm from the point directly above the source. If total internal reflection occurs, what can you say about the index of refraction of the liquid?
39. (III) (a) What is the minimum index of refraction for a glass or plastic prism to be used in binoculars (Fig. 23–28) so that total internal reflection occurs at 45°? (b) Will binoculars work if their prisms (assume $n = 1.58$) are immersed in water? (c) What minimum n is needed if the prisms are immersed in water?
40. (III) A beam of light enters the end of an optic fiber as shown in Fig. 23–58. (a) Show that we can guarantee total internal reflection at the side surface of the material (at point A), if the index of refraction is greater than about 1.42. In other words, regardless of the angle α, the light beam reflects back into the material at point A, assuming air outside. (b) What if the fiber were immersed in water?

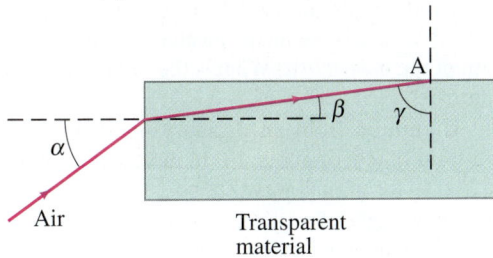

FIGURE 23–58 Problem 40.

23–7 and 23–8 Thin Lenses

41. (I) A sharp image is located 391 mm behind a 215-mm-focal-length converging lens. Find the object distance (a) using a ray diagram, (b) by calculation.
42. (I) Sunlight is observed to focus at a point 16.5 cm behind a lens. (a) What kind of lens is it? (b) What is its power in diopters?
43. (I) (a) What is the power of a 32.5-cm-focal-length lens? (b) What is the focal length of a −6.75-D lens? Are these lenses converging or diverging?

44. (II) A certain lens focuses light from an object 1.55 m away as an image 48.3 cm on the other side of the lens. What type of lens is it and what is its focal length? Is the image real or virtual?
45. (II) A 105-mm-focal-length lens is used to focus an image on the sensor of a camera. The maximum distance allowed between the lens and the sensor plane is 132 mm. (a) How far in front of the sensor should the lens (assumed thin) be positioned if the object to be photographed is 10.0 m away? (b) 3.0 m away? (c) 1.0 m away? (d) What is the closest object this lens could photograph sharply?
46. (II) Use ray diagrams to show that a real image formed by a thin lens is always inverted, whereas a virtual image is always upright if the object is real.
47. (II) A stamp collector uses a converging lens with focal length 28 cm to view a stamp 16 cm in front of the lens. (a) Where is the image located? (b) What is the magnification?
48. (II) It is desired to magnify reading material by a factor of 3.0× when a book is placed 9.0 cm behind a lens. (a) Draw a ray diagram and describe the type of image this would be. (b) What type of lens is needed? (c) What is the power of the lens in diopters?
49. (II) A −7.00-D lens is held 12.5 cm from an ant 1.00 mm high. Describe the position, type, and height of the image.
50. (II) An object is located 1.50 m from a 6.5-D lens. By how much does the image move if the object is moved (a) 0.90 m closer to the lens, and (b) 0.90 m farther from the lens?
51. (II) (a) How far from a 50.0-mm-focal-length lens must an object be placed if its image is to be magnified 2.50× and be real? (b) What if the image is to be virtual and magnified 2.50×?
52. (II) Repeat Problem 51 for a −50.0-mm-focal-length lens. [*Hint*: Consider objects real or virtual (formed by some other piece of optics).]
53. (II) How far from a converging lens with a focal length of 32 cm should an object be placed to produce a real image which is the same size as the object?
54. (II) (a) A 2.40-cm-high insect is 1.30 m from a 135-mm-focal-length lens. Where is the image, how high is it, and what type is it? (b) What if $f = -135$ mm?
55. (III) A bright object and a viewing screen are separated by a distance of 86.0 cm. At what location(s) between the object and the screen should a lens of focal length 16.0 cm be placed in order to produce a sharp image on the screen? [*Hint*: First draw a diagram.]
56. (III) How far apart are an object and an image formed by an 85-cm-focal-length converging lens if the image is 3.25× larger than the object and is real?
57. (III) In a film projector, the film acts as the object whose image is projected on a screen (Fig. 23–59). If a 105-mm-focal-length lens is to project an image on a screen 25.5 m away, how far from the lens should the film be? If the film is 24 mm wide, how wide will the picture be on the screen?

FIGURE 23–59 Film projector, Problem 57.

*23–9 Lens Combinations

*58. (II) A diverging lens with $f = -36.5$ cm is placed 14.0 cm behind a converging lens with $f = 20.0$ cm. Where will an object at infinity be focused?

*59. (II) Two 25.0-cm-focal-length converging lenses are placed 16.5 cm apart. An object is placed 35.0 cm in front of one lens. Where will the final image formed by the second lens be located? What is the total magnification?

*60. (II) A 38.0-cm-focal-length converging lens is 28.0 cm behind a diverging lens. Parallel light strikes the diverging lens. After passing through the converging lens, the light is again parallel. What is the focal length of the diverging lens? [*Hint*: First draw a ray diagram.]

*61. (II) Two lenses, one converging with focal length 20.0 cm and one diverging with focal length -10.0 cm, are placed 25.0 cm apart. An object is placed 60.0 cm in front of the converging lens. Determine (*a*) the position and (*b*) the magnification of the final image formed. (*c*) Sketch a ray diagram for this system.

*62. (II) A lighted candle is placed 36 cm in front of a converging lens of focal length $f_1 = 13$ cm, which in turn is 56 cm in front of another converging lens of focal length $f_2 = 16$ cm (see Fig. 23–60). (*a*) Draw a ray diagram and estimate the location and the relative size of the final image. (*b*) Calculate the position and relative size of the final image.

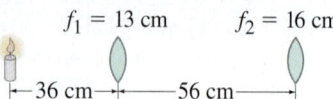

FIGURE 23–60 Problem 62.

*23–10 Lensmaker's Equation

*63. (I) A double concave lens has surface radii of 33.4 cm and 28.8 cm. What is the focal length if $n = 1.52$?

*64. (I) Both surfaces of a double convex lens have radii of 34.1 cm. If the focal length is 28.9 cm, what is the index of refraction of the lens material?

*65. (I) A planoconvex lens (Fig. 23–31a) with $n = 1.55$ is to have a focal length of 16.3 cm. What is the radius of curvature of the convex surface?

*66. (II) A symmetric double convex lens with a focal length of 22.0 cm is to be made from glass with an index of refraction of 1.52. What should be the radius of curvature for each surface?

*67. (II) A prescription for an eyeglass lens calls for +3.50 diopters. The lensmaker grinds the lens from a "blank" with $n = 1.56$ and convex front surface of radius of curvature of 30.0 cm. What should be the radius of curvature of the other surface?

*68. (III) An object is placed 96.5 cm from a glass lens ($n = 1.52$) with one concave surface of radius 22.0 cm and one convex surface of radius 18.5 cm. Where is the final image? What is the magnification?

General Problems

69. Sunlight is reflected off the Moon. How long does it take that light to reach us from the Moon?

70. You hold a small flat mirror 0.50 m in front of you and can see your reflection twice in that mirror because there is a full-length mirror 1.0 m behind you (Fig. 23–61). Determine the distance of each image from you.

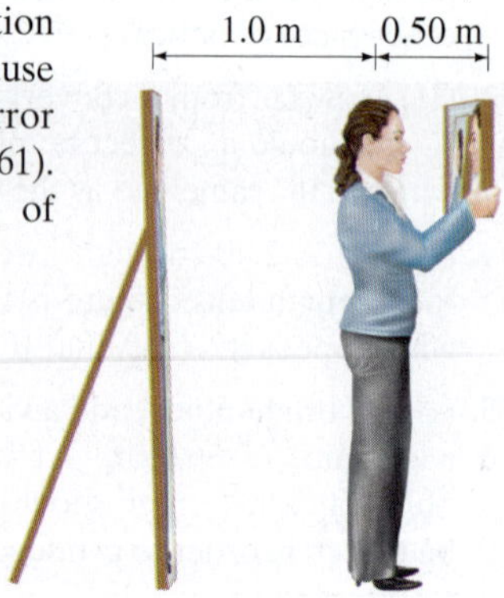

FIGURE 23–61 Problem 70.

71. We wish to determine the depth of a swimming pool filled with water by measuring the width ($x = 6.50$ m) and then noting that the far bottom edge of the pool is just visible at an angle of 13.0° above the horizontal as shown in Fig. 23–62. Calculate the depth of the pool.

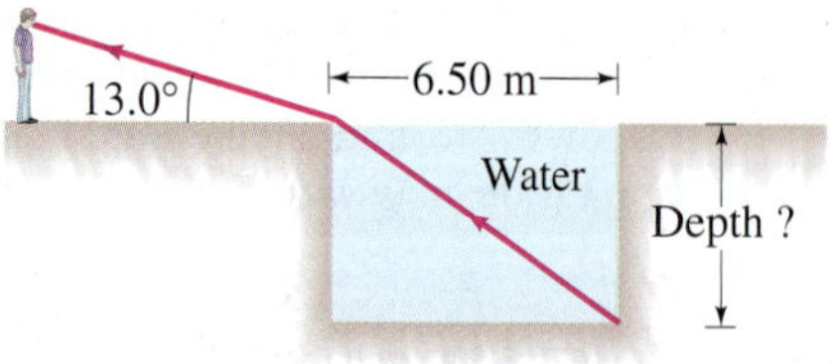

FIGURE 23–62 Problem 71.

72. The critical angle of a certain piece of plastic in air is $\theta_C = 37.8°$. What is the critical angle of the same plastic if it is immersed in water?

73. A pulse of light takes 2.63 ns (see Table 1–4) to travel 0.500 m in a certain material. Determine the material's index of refraction, and identify this material.

74. When an object is placed 60.0 cm from a certain converging lens, it forms a real image. When the object is moved to 40.0 cm from the lens, the image moves 10.0 cm farther from the lens. Find the focal length of this lens.

75. A 4.5-cm-tall object is placed 32 cm in front of a spherical mirror. It is desired to produce a virtual image that is upright and 3.5 cm tall. (*a*) What type of mirror should be used? (*b*) Where is the image located? (*c*) What is the focal length of the mirror? (*d*) What is the radius of curvature of the mirror?

76. Light is emitted from an ordinary lightbulb filament in wave-train bursts of about 10^{-8} s in duration. What is the length in space of such wave trains?

77. If the apex angle of a prism is $\phi = 75°$ (see Fig. 23–63), what is the minimum incident angle for a ray if it is to emerge from the opposite side (i.e., not be totally internally reflected), given $n = 1.58$?

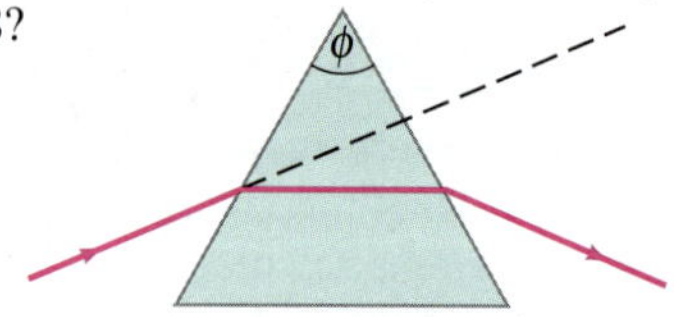

FIGURE 23–63 Problem 77.

78. (*a*) A plane mirror can be considered a limiting case of a spherical mirror. Specify what this limit is. (*b*) Determine an equation that relates the image and object distances in this limit of a plane mirror. (*c*) Determine the magnification of a plane mirror in this same limit. (*d*) Are your results in parts (*b*) and (*c*) consistent with the discussion of Section 23–2 on plane mirrors?

79. An object is placed 18 cm from a certain mirror. The image is half the height of the object, inverted, and real. How far is the image from the mirror, and what is the radius of curvature of the mirror?

80. Light is incident on an equilateral glass prism at a 45.0° angle to one face, Fig. 23–64. Calculate the angle at which light emerges from the opposite face. Assume that $n = 1.54$.

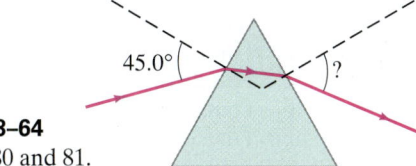

FIGURE 23–64 Problems 80 and 81.

81. Suppose a ray strikes the left face of the prism in Fig. 23–64 at 45.0° as shown, but is totally internally reflected at the opposite side. If the apex angle (at the top) is $\theta = 65.0°$, what can you say about the index of refraction of the prism?

82. (a) An object 37.5 cm in front of a certain lens is imaged 8.20 cm in front of that lens (on the same side as the object). What type of lens is this, and what is its focal length? Is the image real or virtual? (b) If the image were located, instead, 44.5 cm in front of the lens, what type of lens would it be and what focal length would it have?

83. How large is the image of the Sun on a camera sensor with (a) a 35-mm-focal-length lens, (b) a 50-mm-focal-length lens, and (c) a 105-mm-focal-length lens? The Sun has diameter 1.4×10^6 km, and it is 1.5×10^8 km away.

84. Figure 23–65 is a photograph of an eyeball with the image of a boy in a doorway. (a) Is the eye here acting as a lens or as a mirror? (b) Is the eye being viewed right side up or is the camera taking this photo upside down? (c) Explain, based on all possible images made by a convex mirror or lens.

FIGURE 23–65 Problem 84.

85. Which of the two lenses shown in Fig. 23–66 is converging, and which is diverging? Explain using ray diagrams and show how each image is formed.

FIGURE 23–66 Problem 85.

86. Figure 23–67 shows a liquid-detecting prism device that might be used inside a washing machine. If no liquid covers the prism's hypotenuse, total internal reflection of the beam from the light source produces a large signal in the light sensor. If liquid covers the hypotenuse, some light escapes from the prism into the liquid and the light sensor's signal decreases. Thus a large signal from the light sensor indicates the absence of liquid in the reservoir. Determine the allowable range for the prism's index of refraction n.

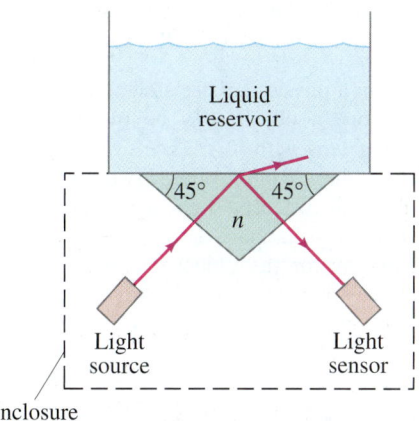

FIGURE 23–67 Problem 86.

*87. (a) Show that if two thin lenses of focal lengths f_1 and f_2 are placed in contact with each other, the focal length of the combination is given by $f_T = f_1 f_2 / (f_1 + f_2)$. (b) Show that the power P of the combination of two lenses is the sum of their separate powers, $P = P_1 + P_2$.

*88. Two converging lenses are placed 30.0 cm apart. The focal length of the lens on the right is 20.0 cm, and the focal length of the lens on the left is 15.0 cm. An object is placed to the left of the 15.0-cm-focal-length lens. A final image from both lenses is inverted and located halfway between the two lenses. How far to the left of the 15.0-cm-focal-length lens is the original object?

*89. An object is placed 30.0 cm from a +5.0-D lens. A spherical mirror with focal length 25 cm is placed 75 cm behind the lens. Where is the final image? (Note that the mirror reflects light back through the lens.) Be sure to draw a diagram.

*90. A small object is 25.0 cm from a diverging lens as shown in Fig. 23–68. A converging lens with a focal length of 12.0 cm is 30.0 cm to the right of the diverging lens. The two-lens system forms a real inverted image 17.0 cm to the right of the converging lens. What is the focal length of the diverging lens?

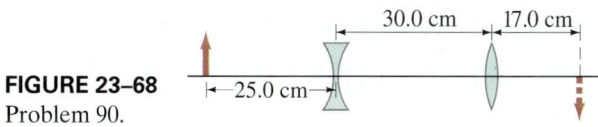

FIGURE 23–68 Problem 90.

Search and Learn

1. (a) Describe the difference between a real image and a virtual image? (b) Can your eyes tell the difference? (c) How can you tell the difference on a ray diagram? (d) How could you tell the difference between a virtual image and a real image experimentally? (e) If you were to take a photograph of a virtual image, would you see the image in the photograph? (f) If you were to put a piece of photographic film at the location of a virtual image, would the image be captured on the film? (g) Explain any differences in your answers to parts (e) and (f).

2. Students in a physics lab are assigned to find the location where a bright object may be placed in order that a converging lens with $f = 12$ cm will produce an image three times the size of the object. Two students complete the assignment at different times using identical equipment, but when they compare notes later, they discover that their answers for the object distance are not the same. Explain why they do not necessarily need to repeat the lab, and justify your response with a calculation.

3. Both a converging lens and a concave mirror can produce virtual images that are larger than the object. Concave mirrors can be used as makeup mirrors, but converging lenses cannot be. (a) Draw ray diagrams to explain why not. (b) If a concave mirror has the same focal length as a converging lens, and an object is placed first at a distance of $\frac{1}{2}f$ from the lens and then at a distance of $\frac{1}{2}f$ from the mirror, how will the magnification of the object compare in the two cases?

4. (a) Did the person we see in Fig. 23–69 shoot the picture we are looking at? We see her in three different mirrors. Describe (b) what type of mirror each is, and (c) her position relative to the focal point and center of curvature.

5. Justify the second part of sign convention 3, page 665, starting "Equivalently." Use ray diagrams for all possible situations. Cite Figures already in the text and draw any others needed.

6. The only means to create a real image with a single lens would be to place
   (a) the object inside the focal length of a converging lens;
   (b) the object inside the focal length of a diverging lens;
   (c) the object outside the focal length of a converging lens;
   (d) the object outside the focal length of a diverging lens;
   (e) any of the above, given the correct distance from the focal point.

7. Make a table showing the sign conventions for mirrors and lenses. Include the sign convention for the mirrors and lenses themselves and for the image and object heights and distances for each.

8. Figure 23–70 shows a converging lens held above three equal-sized letters A. In (a) the lens is 5 cm from the paper, and in (b) the lens is 15 cm from the paper. Estimate the focal length of the lens. What is the image position for each case?

(a)

(b)

FIGURE 23–70 Search and Learn 8.

FIGURE 23–69 Search and Learn 4.

ANSWERS TO EXERCISES

A: No.
B: (b).
C: Toward.
D: None.
E: No total internal reflection, $\theta_C > 45°$.
F: (c).
G: Closer to it.

The beautiful colors from the surface of this soap bubble can be nicely explained by the wave theory of light. A soap bubble is a very thin spherical film filled with air. Light reflected from the outer and inner surfaces of this thin film of soapy water interferes constructively to produce the bright colors. Which color we see at any point depends on the thickness of the soapy water film at that point and also on the viewing angle. Near the top of the bubble, we see a small black area surrounded by a silver or white area. The bubble's thickness is smallest at that black spot, perhaps only about 30 nm thick, and is fully transparent (we see the black background).

We cover fundamental aspects of the wave nature of light, including two-slit interference and interference in thin films.

The Wave Nature of Light

CHAPTER 24

CHAPTER-OPENING QUESTION—Guess now!

When a thin layer of oil lies on top of water or wet pavement, you can often see swirls of color. We also see swirls of color on the soap bubble shown above. What causes these colors?

(a) Additives in the oil or soap reflect various colors.
(b) Chemicals in the oil or soap absorb various colors.
(c) Dispersion due to differences in index of refraction in the oil or soap.
(d) The interactions of the light with a thin boundary layer where the oil (or soap) and the water have mixed irregularly.
(e) Light waves reflected from the top and bottom surfaces of the thin oil or soap film can add up constructively for particular wavelengths.

L ight carries energy. Evidence for this can come from focusing the Sun's rays with a magnifying glass on a piece of paper and burning a hole in it. But how does light travel, and in what form is this energy carried? In our discussion of waves in Chapter 11, we noted that energy can be carried from place to place in basically two ways: by particles or by waves. In the first case, material objects or particles can carry energy, such as an avalanche of rocks or rushing water. In the second case, water waves and sound waves, for example, can carry energy over long distances even though the oscillating particles of the medium do not travel these distances. In view of this, what can we say about the nature of light: does light travel as a stream of particles away from its source, or does light travel in the form of waves that spread outward from the source?

CONTENTS

24–1 Waves vs. Particles; Huygens' Principle and Diffraction
*24–2 Huygens' Principle and the Law of Refraction
24–3 Interference—Young's Double-Slit Experiment
24–4 The Visible Spectrum and Dispersion
24–5 Diffraction by a Single Slit or Disk
24–6 Diffraction Grating
24–7 The Spectrometer and Spectroscopy
24–8 Interference in Thin Films
*24–9 Michelson Interferometer
24–10 Polarization
*24–11 Liquid Crystal Displays (LCD)
*24–12 Scattering of Light by the Atmosphere

679

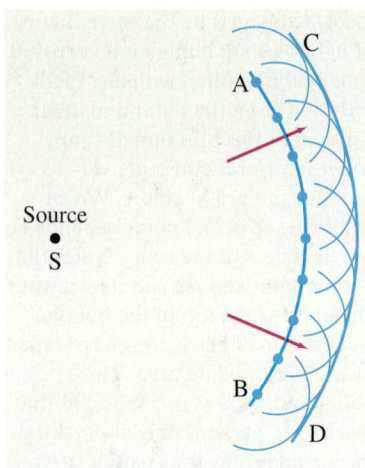

FIGURE 24–1 Huygens' principle, used to determine wave front CD when wave front AB is given.

FIGURE 24–2 Huygens' principle is consistent with diffraction (a) around the edge of an obstacle, (b) through a large hole, (c) through a small hole whose size is on the order of the wavelength of the wave.

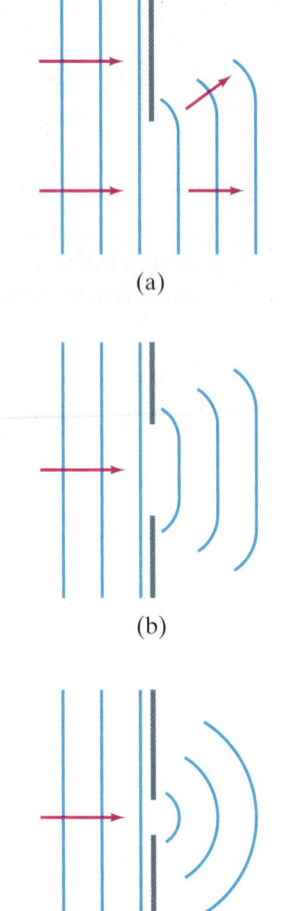

Historically, this question has turned out to be a difficult one. For one thing, light does not reveal itself in any obvious way as being made up of tiny particles; nor do we see tiny light waves passing by as we do water waves. The evidence seemed to favor first one side and then the other until about 1830, when most physicists had accepted the wave theory. By the end of the nineteenth century, light was considered to be an *electromagnetic wave* (Chapter 22). In the early twentieth century, light was shown to have a particle nature as well, as we shall discuss in Chapter 27. We now speak of the wave–particle duality of light. The wave theory of light remains valid and has proved very successful. In this Chapter we investigate the evidence for the wave theory and how it has been used to explain a wide range of phenomena.

24–1 Waves vs. Particles; Huygens' Principle and Diffraction

The Dutch scientist Christian Huygens (1629–1695), a contemporary of Newton, proposed a wave theory of light that had much merit. Still useful today is a technique Huygens developed for predicting the future position of a wave front when an earlier position is known. By a **wave front**, we mean all the points along a two- or three-dimensional wave that form a wave crest—what we simply call a "wave" as seen on the ocean. Wave fronts are perpendicular to rays as discussed in Chapter 11 (Fig. 11–35). **Huygens' principle** can be stated as follows: *Every point on a wave front can be considered as a source of tiny wavelets that spread out in the forward direction at the speed of the wave itself. The new wave front is the envelope of all the wavelets—that is, the tangent to all of them.*

As an example of the use of Huygens' principle, consider the wave front AB in Fig. 24–1, which is traveling away from a source S. We assume the medium is *isotropic*—that is, the speed v of the waves is the same in all directions. To find the wave front a short time t after it is at AB, tiny circles are drawn at points along AB with radius $r = vt$. The centers of these tiny circles are shown as blue dots on the original wave front AB, and the circles represent Huygens' (imaginary) wavelets. The tangent to all these wavelets, the curved line CD, is the new position of the wave front after a time t.

Huygens' principle is particularly useful for analyzing what happens when waves run into an obstacle and the wave fronts are partially interrupted. Huygens' principle predicts that waves bend in behind an obstacle, as shown in Fig. 24–2. This is just what water waves do, as we saw in Chapter 11 (Figs. 11–45 and 11–46). The bending of waves behind obstacles into the "shadow region" is known as **diffraction**. Since diffraction occurs for waves, but not for particles, it can serve as one means for distinguishing the nature of light.

Note, as shown in Fig. 24–2, that diffraction is most prominent when the size of the opening is on the order of the wavelength of the wave. If the opening is much larger than the wavelength, diffraction may go unnoticed.

Does light exhibit diffraction? In the mid-seventeenth century, the Jesuit priest Francesco Grimaldi (1618–1663) had observed that when sunlight entered a darkened room through a tiny hole in a screen, the spot on the opposite wall was larger than would be expected from geometric rays. He also observed that the border of the image was not clear but was surrounded by colored fringes. Grimaldi attributed this to the diffraction of light.

The wave model of light nicely accounts for diffraction. But the ray model (Chapter 23) cannot account for diffraction, and it is important to be aware of such limitations to the ray model. Geometric optics using rays is successful in a wide range of situations only because normal openings and obstacles are much larger than the wavelength of the light, and so relatively little diffraction or bending occurs.

*24–2 Huygens' Principle and the Law of Refraction

The laws of reflection and refraction were well known in Newton's time. The law of reflection could not distinguish between the two theories we just discussed: waves versus particles. When waves reflect from an obstacle, the angle of incidence equals the angle of reflection (Fig. 11–36). The same is true of particles—think of a tennis ball without spin striking a flat surface.

The law of refraction is another matter. Consider a ray of light entering a medium where it is bent toward the normal, as when traveling from air into water. As shown in Fig. 24–3, this bending can be constructed using Huygens' principle if we assume the speed of light is less in the second medium ($v_2 < v_1$). In time t, point B on wave front AB (perpendicular to the incoming ray) travels a distance $v_1 t$ to reach point D. Point A on the wave front, traveling in the second medium, goes a distance $v_2 t$ to reach point C, and $v_2 t < v_1 t$. Huygens' principle is applied to points A and B to obtain the curved wavelets shown at C and D. The wave front is tangent to these two wavelets, so the new wave front is the line CD. Hence the rays, which are perpendicular to the wave fronts, bend toward the normal if $v_2 < v_1$, as drawn. (This is basically the same discussion as we used around Fig. 11–42.)

Newton favored a particle theory of light which predicted the opposite result, that the speed of light would be greater in the second medium ($v_2 > v_1$). Thus the wave theory predicts that the speed of light in water, for example, is less than in air; and Newton's particle theory predicts the reverse. An experiment to actually measure the speed of light in water was performed in 1850 by the French physicist Jean Foucault, and it confirmed the wave-theory prediction. By then, however, the wave theory was already fully accepted, as we shall see in the next Section.

Snell's law of refraction follows directly from Huygens' principle, given that the speed of light v in any medium is related to the speed in a vacuum, c, and the index of refraction, n, by Eq. 23–4: that is, $v = c/n$. From the Huygens' construction of Fig. 24–3, angle ADC is equal to θ_2 and angle BAD is equal to θ_1. Then for the two triangles that have the common side AD, we have

$$\sin \theta_1 = \frac{v_1 t}{AD}, \quad \sin \theta_2 = \frac{v_2 t}{AD}.$$

We divide these two equations and obtain

$$\frac{\sin \theta_1}{\sin \theta_2} = \frac{v_1}{v_2}.$$

Then, by Eq. 23–4, $v_1 = c/n_1$ and $v_2 = c/n_2$, so we have

$$n_1 \sin \theta_1 = n_2 \sin \theta_2,$$

which is Snell's law of refraction, Eq. 23–5. (The law of reflection can be derived from Huygens' principle in a similar way.)

When a light wave travels from one medium to another, its frequency does not change, but its wavelength does. This can be seen from Fig. 24–3, where each of the blue lines representing a wave front corresponds to a crest (peak) of the wave. Then

$$\frac{\lambda_2}{\lambda_1} = \frac{v_2 t}{v_1 t} = \frac{v_2}{v_1} = \frac{n_1}{n_2},$$

where, in the last step, we used Eq. 23–4, $v = c/n$. If medium 1 is a vacuum (or air), so $n_1 = 1$, $v_1 = c$, and we call λ_1 simply λ, then the wavelength in another medium of index of refraction n ($= n_2$) will be

$$\lambda_n = \frac{\lambda}{n}. \qquad (24\text{–}1)$$

This result is consistent with the frequency f being unchanged no matter what medium the wave is traveling in, since $c = f\lambda$.

> **EXERCISE A** A light beam in air with wavelength = 500 nm, frequency = 6.0×10^{14} Hz, and speed = 3.0×10^8 m/s goes into glass which has an index of refraction = 1.5. What are the wavelength, frequency, and speed of the light in the glass?

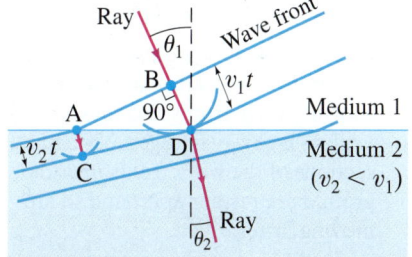

FIGURE 24–3 Refraction explained, using Huygens' principle. Wave fronts are perpendicular to the rays.

⚠️ **CAUTION**
Frequency is fixed, wavelength can change

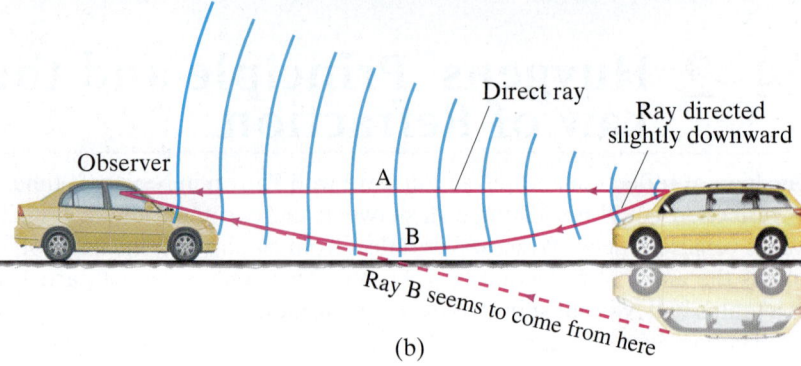

(a)   (b)

FIGURE 24–4 (a) A highway mirage. (b) Drawing (greatly exaggerated) showing wave fronts and rays to explain highway mirages. Note how sections of the wave fronts near the ground are farther apart and so are moving faster.

PHYSICS APPLIED
Highway mirages

Wave fronts can be used to explain how mirages are produced by refraction of light. For example, on a hot day motorists sometimes see a mirage of water on the highway ahead of them, with distant vehicles seemingly reflected in it (Fig. 24–4a). On a hot day, there can be a layer of very hot air next to the roadway (made hot by the Sun beating down on the road). Hot air is less dense than cooler air, so the index of refraction is slightly lower in the hot air. In Fig. 24–4b, we see a diagram of light coming from one point on a distant car (on the right) heading left toward the observer. Wave fronts and two rays (perpendicular to the wave fronts) are shown. Ray A heads directly at the observer and follows a straight-line path, and represents the normal view of the distant car. Ray B is a ray initially directed slightly downward but, instead of hitting the road, it bends slightly as it moves through layers of air of different index of refraction. The wave fronts, shown in blue in Fig. 24–4b, move slightly faster in the layers of (less dense) air nearer the ground. Thus ray B is bent as shown, and seems to the observer to be coming from below (dashed line) as if reflected off the road. Hence the mirage.

24–3 Interference—Young's Double-Slit Experiment

In 1801, the Englishman Thomas Young (1773–1829) obtained convincing evidence for the wave nature of light and was even able to measure wavelengths for visible light. Figure 24–5a shows a schematic diagram of Young's famous double-slit experiment. To have light from a single source, Young used the sunlight passing through a very narrow slit in a window covering. This beam of parallel rays falls on a screen containing two closely spaced slits, S_1 and S_2. (The slits and their separation are very narrow, not much larger than the wavelength of the light.) If light consists of tiny particles, we would expect to see two bright lines on a screen placed behind the slits as in (b). But instead, a series of bright lines are seen as in (c). Young was able to explain this result as a **wave-interference** phenomenon.

FIGURE 24–5 (a) Young's double-slit experiment. (b) If light consists of particles, we would expect to see two bright lines on the screen behind the slits. (c) In fact, many lines are observed. The slits and their separation need to be very thin.

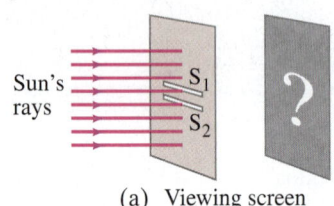

(a) Viewing screen

(b) Viewing screen (particle theory prediction)

(c) Viewing screen (actual)

To understand why, consider **plane waves**† of light of a single wavelength—called **monochromatic**, meaning "one color"—falling on the two slits as shown in Fig. 24–6. Because of diffraction, the waves leaving the two small slits spread out as shown. This is equivalent to the interference pattern produced when two rocks are thrown into a lake (Fig. 11–38), or when sound from two loudspeakers interferes (Fig. 12–16). Recall Section 11–11 on wave interference.

†See pages 312 and 628.

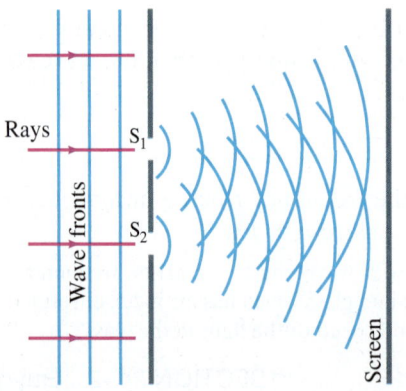

FIGURE 24–6 Plane waves (parallel flat wave fronts) fall on two slits. If light is a wave, light passing through one of two slits should interfere with light passing through the other slit.

682 CHAPTER 24

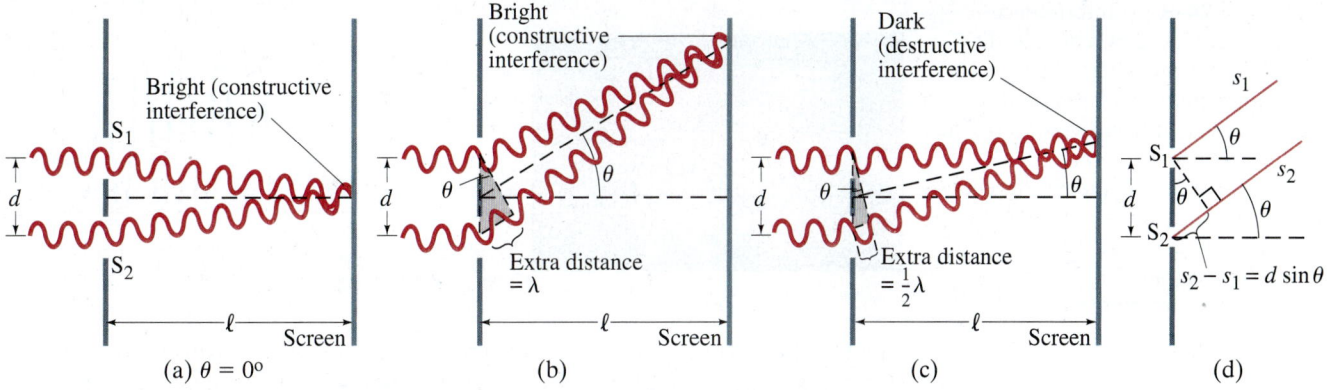

FIGURE 24–7 How the wave theory explains the pattern of lines seen in the double-slit experiment. (a) At the center of the screen, waves from each slit travel the same distance and are in phase. [Assume $\ell \gg d$.] (b) At this angle θ, the lower wave travels an extra distance of one whole wavelength, and the waves are in phase; note from the shaded triangle that the path difference equals $d \sin \theta$. (c) For this angle θ, the lower wave travels an extra distance equal to one-half wavelength, so the two waves arrive at the screen fully out of phase. (d) A more detailed diagram showing the geometry for parts (b) and (c).

To see how an interference pattern is produced on the screen, we make use of Fig. 24–7. Waves of wavelength λ are shown entering the slits S_1 and S_2, which are a distance d apart. The waves spread out in all directions after passing through the slits (Fig. 24–6), but they are shown in Figs. 24–7a, b, and c only for three different angles θ. In Fig. 24–7a, the waves reaching the center of the screen are shown ($\theta = 0°$). Waves from the two slits travel the same distance, so they are **in phase**: a crest of one wave arrives at the same time as a crest of the other wave. Hence the amplitudes of the two waves add to form a larger amplitude as shown in Fig. 24–8a. This is **constructive interference**, and there is a bright line at the center of the screen. Constructive interference also occurs when the paths of the two rays differ by one wavelength (or any whole number of wavelengths), as shown in Fig. 24–7b; also here there will be a bright line on the screen. But if one ray travels an extra distance of one-half wavelength (or $\frac{3}{2}\lambda$, $\frac{5}{2}\lambda$, and so on), the two waves are exactly **out of phase** (Section 11–11) when they reach the screen: the crests of one wave arrive at the same time as the troughs of the other wave, and so they add to produce zero amplitude (Fig. 24–8b). This is **destructive interference**, and the screen is dark, Fig. 24–7c. Thus, there will be a series of bright and dark lines (or **fringes**) on the viewing screen.

To determine exactly where the bright lines fall, first note that Fig. 24–7 is somewhat exaggerated; in real situations, the distance d between the slits is very small compared to the distance ℓ to the screen. The rays from each slit for each case will therefore be essentially parallel, and θ is the angle they make with the horizontal as shown in Fig. 24–7d. From the shaded right triangles shown in Figs. 24–7b and c, we can see that the extra distance traveled by the lower ray is $d \sin \theta$ (seen more clearly in Fig. 24–7d). Constructive interference will occur, and a bright fringe will appear on the screen, when the *path difference*, $d \sin \theta$, equals a whole number of wavelengths:

$$d \sin \theta = m\lambda, \qquad m = 0, 1, 2, \cdots. \quad \begin{bmatrix}\text{constructive}\\ \text{interference}\\ \text{(bright)}\end{bmatrix} \quad (24\text{–}2a)$$

The value of m is called the **order** of the interference fringe. The first order ($m = 1$), for example, is the first fringe on each side of the central fringe (which is at $\theta = 0$, $m = 0$). Destructive interference occurs when the path difference $d \sin \theta$ is $\frac{1}{2}\lambda$, $\frac{3}{2}\lambda$, and so on:

$$d \sin \theta = (m + \tfrac{1}{2})\lambda, \qquad m = 0, 1, 2, \cdots. \quad \begin{bmatrix}\text{destructive}\\ \text{interference}\\ \text{(dark)}\end{bmatrix} \quad (24\text{–}2b)$$

The bright fringes are peaks or maxima of light intensity, the dark fringes are minima.

FIGURE 24–8 Two traveling waves are shown undergoing (a) constructive interference, (b) destructive interference. (See also Section 11–11.)

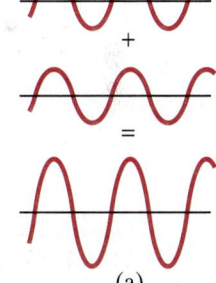

(a)

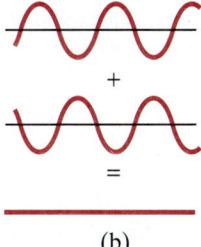

(b)

FIGURE 24–9 (a) Interference fringes produced by a double-slit experiment and detected by photographic film placed on the viewing screen. The arrow marks the central fringe. (b) Graph of the intensity of light in the interference pattern. Also shown are values of m for Eq. 24–2a (constructive interference) and Eq. 24–2b (destructive interference).

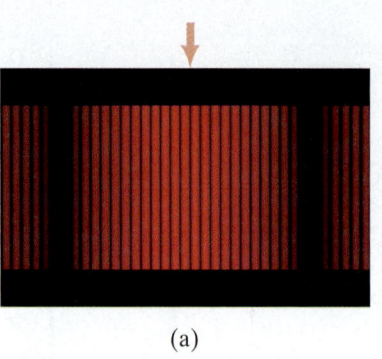
(a)

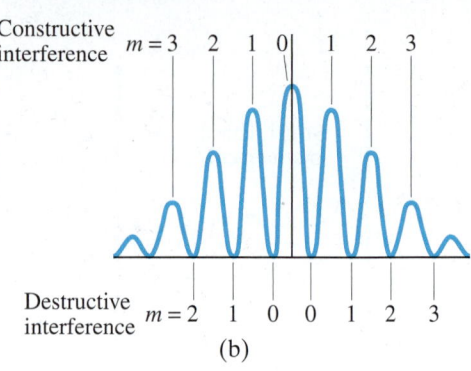
(b)

The intensity of the bright fringes is greatest for the central fringe ($m = 0$) and decreases for higher orders, as shown in Fig. 24–9. How much the intensity decreases with increasing order depends on the width of the two slits.

EXAMPLE 24–1 Line spacing for double-slit interference. A screen containing two slits 0.100 mm apart is 1.20 m from the viewing screen. Light of wavelength $\lambda = 500$ nm falls on the slits from a distant source. Approximately how far apart will adjacent bright interference fringes be on the screen?

APPROACH The angular position of bright (constructive interference) fringes is found using Eq. 24–2a. The distance between the first two fringes (say) can be found using right triangles as shown in Fig. 24–10.

CAUTION
Use the approximation $\theta \approx \tan\theta$ or $\theta \approx \sin\theta$ only if θ is small and in radians

SOLUTION Given $d = 0.100$ mm $= 1.00 \times 10^{-4}$ m, $\lambda = 500 \times 10^{-9}$ m, and $\ell = 1.20$ m, the first-order fringe ($m = 1$) occurs at an angle θ given by

$$\sin\theta_1 = \frac{m\lambda}{d} = \frac{(1)(500 \times 10^{-9}\,\text{m})}{1.00 \times 10^{-4}\,\text{m}} = 5.00 \times 10^{-3}.$$

This is a very small angle, so we can take $\sin\theta \approx \theta$, with θ in radians. The first-order fringe will occur a distance x_1 above the center of the screen (see Fig. 24–10), given by $x_1/\ell = \tan\theta_1 \approx \theta_1$, so

$$x_1 \approx \ell\theta_1 = (1.20\,\text{m})(5.00 \times 10^{-3}) = 6.00\,\text{mm}.$$

The second-order fringe ($m = 2$) will occur at

$$x_2 \approx \ell\theta_2 = \ell\frac{2\lambda}{d} = 12.0\,\text{mm}$$

above the center, and so on. Thus the lower-order fringes are 6.00 mm apart.

NOTE The spacing between fringes is essentially uniform until the approximation $\sin\theta \approx \theta$ is no longer valid.

FIGURE 24–10 Examples 24–1 and 24–2. For small angles θ (give θ in radians), the interference fringes occur at distance $x = \theta\ell$ above the center fringe ($m = 0$); θ_1 and x_1 are for the first-order fringe ($m = 1$); θ_2 and x_2 are for $m = 2$.

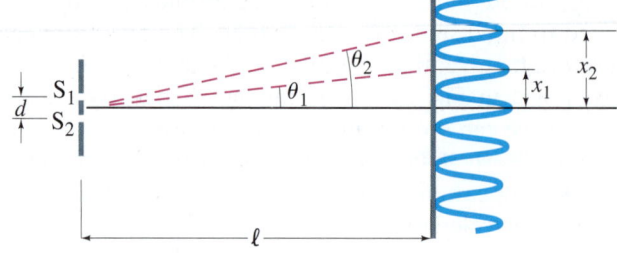

CONCEPTUAL EXAMPLE 24–2 Changing the wavelength. (a) What happens to the interference pattern shown in Fig. 24–10, Example 24–1, if the incident light (500 nm) is replaced by light of wavelength 700 nm? (b) What happens instead if the wavelength stays at 500 nm but the slits are moved farther apart?

RESPONSE (a) When λ increases in Eq. 24–2a but d stays the same, the angle θ for bright fringes increases and the interference pattern spreads out. (b) Increasing the slit spacing d reduces θ for each order, so the lines are closer together.

From Eqs. 24–2 we can see that, except for the zeroth-order fringe at the center, the position of the fringes depends on wavelength. When white light falls on the two slits, as Young found in his experiments, the central fringe is white, but the first (and higher) order fringes contain a spectrum of colors like a rainbow.

Using Eq. 24–2a, we can see that θ is smallest for violet light and largest for red (Fig. 24–11). By measuring the position of these fringes, Young was the first to determine the wavelengths of visible light. In doing so, he showed that what distinguishes different colors physically is their wavelength (or frequency), an idea put forward earlier by Grimaldi in 1665.

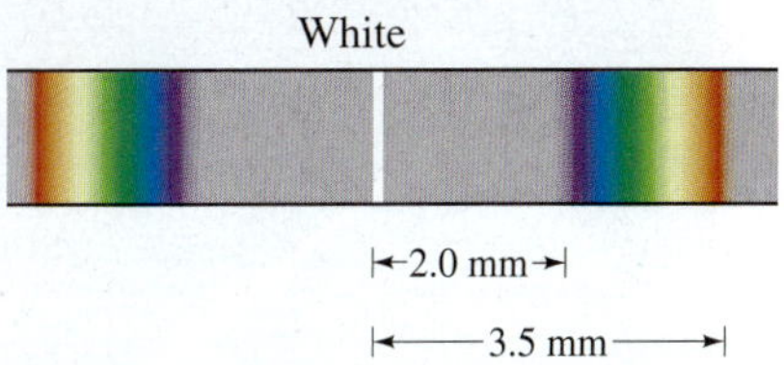

EXAMPLE 24–3 **Wavelengths from double-slit interference.** White light passes through two slits 0.50 mm apart, and an interference pattern is observed on a screen 2.5 m away. The first-order fringe resembles a rainbow with violet and red light at opposite ends. The violet light is about 2.0 mm and the red 3.5 mm from the center of the central white fringe (Fig. 24–11). Estimate the wavelengths for the violet and red light.

APPROACH We find the angles for violet and red light from the distances given and the diagram of Fig. 24–10. Then we use Eq. 24–2a to obtain the wavelengths. Because 3.5 mm is much less than 2.5 m, we can use the small-angle approximation.

SOLUTION We use Eq. 24–2a ($d \sin \theta = m\lambda$) with $m = 1$, $d = 5.0 \times 10^{-4}$ m, and $\sin \theta \approx \tan \theta \approx \theta$. Also $\theta \approx x/\ell$ (Fig. 24–10), so for violet light, $x = 2.0$ mm, and

$$\lambda = \frac{d \sin \theta}{m} \approx \frac{d\theta}{m} \approx \frac{d}{m}\frac{x}{\ell} = \left(\frac{5.0 \times 10^{-4} \text{ m}}{1}\right)\left(\frac{2.0 \times 10^{-3} \text{ m}}{2.5 \text{ m}}\right) = 4.0 \times 10^{-7} \text{ m},$$

or 400 nm. For red light, $x = 3.5$ mm, so

$$\lambda \approx \frac{d}{m}\frac{x}{\ell} = \left(\frac{5.0 \times 10^{-4} \text{ m}}{1}\right)\left(\frac{3.5 \times 10^{-3} \text{ m}}{2.5 \text{ m}}\right) = 7.0 \times 10^{-7} \text{ m} = 700 \text{ nm}.$$

EXERCISE B For the setup in Example 24–3, how far from the central white fringe is the first-order fringe for green light $\lambda = 500$ nm?

Coherence

The two slits in Figs. 24–6 and 24–7 act as if they were two sources of radiation. They are called **coherent sources** because the waves leaving them have the same wavelength and frequency, and bear the same phase relationship to each other at all times. This happens because the waves come from a single source to the left of the two slits. An interference pattern is observed only when the sources are coherent. If two tiny lightbulbs replaced the two slits, an interference pattern would not be seen. The light emitted by one lightbulb would have a random phase with respect to the second bulb, and the screen would be more or less uniformly illuminated. Two such sources, whose output waves have phases that bear no fixed relationship to each other over time, are called **incoherent sources**.

24–4 The Visible Spectrum and Dispersion

Two of the most important properties of light are readily describable in terms of the wave theory of light: intensity (or brightness) and color. The **intensity** of light is the energy it carries per unit area per unit time, and is related to the square of the amplitude of the wave, as for any wave (see Section 11–9, or Eqs. 22–7 and 22–8). The **color** of light is related to the frequency f or wavelength λ of the light. (Recall $\lambda f = c = 3.0 \times 10^8$ m/s, Eq. 22–4.) Visible light—that to which our eyes are sensitive—consists of frequencies from 4×10^{14} Hz to 7.5×10^{14} Hz, corresponding to wavelengths in air of about 400 nm to 750 nm.[†] This is the **visible spectrum**, and within it lie the different colors from violet to red, as shown in Fig. 24–12.

[†]Sometimes the angstrom (Å) unit is used when referring to light: $1 \text{ Å} = 1 \times 10^{-10}$ m. Visible light has wavelengths in air of 4000 Å to 7500 Å.

FIGURE 24–11 First-order fringes for a double slit are a full spectrum, like a rainbow. Also Example 24–3.

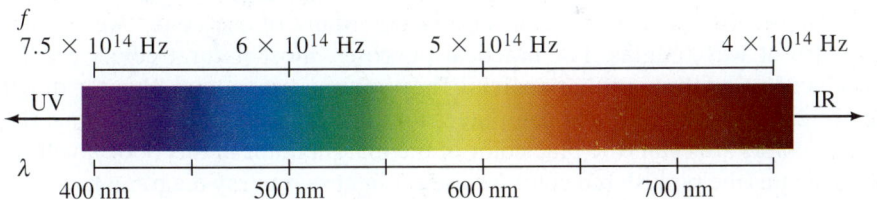

FIGURE 24–12 The spectrum of visible light, showing the range of frequencies and wavelengths in air for the various colors. Many colors, such as brown, do not appear in the spectrum; they are made from a mixture of wavelengths.

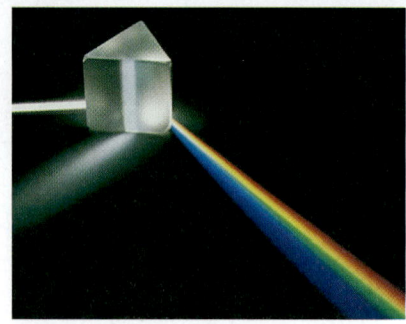

FIGURE 24–13 White light passing through a prism is spread out into its constituent colors.

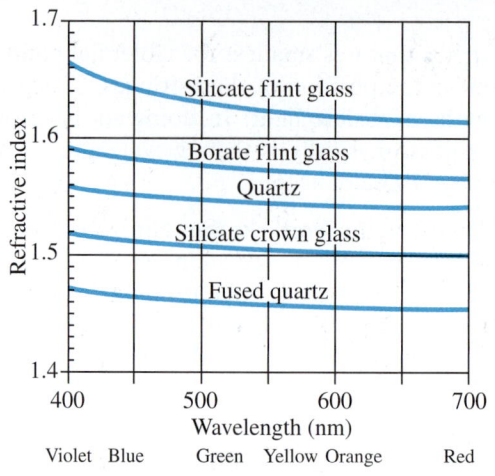

FIGURE 24–14 Index of refraction as a function of wavelength for various transparent solids.

FIGURE 24–15 White light dispersed by a prism into the visible spectrum.

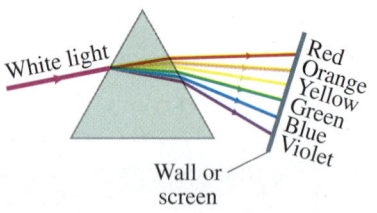

Light with wavelength (in air) shorter than 400 nm (= violet) is called **ultraviolet** (UV), and light with wavelength longer than 750 nm (= red) is called **infrared** (IR).† Although human eyes are not sensitive to UV or IR, some types of photographic film and other detectors do respond to them.

A prism can separate white light into a rainbow of colors, as shown in Fig. 24–13. This happens when the index of refraction of a material depends on the wavelength, as shown for several materials in Fig. 24–14. White light is a mixture of all visible wavelengths, and when incident on a prism, as in Fig. 24–15, the different wavelengths are bent to varying degrees. Because the index of refraction is greater for the shorter wavelengths, violet light is bent the most and red the least, as shown in Fig. 24–15. This spreading of white light into the full spectrum is called **dispersion**.

Rainbows are a spectacular example of dispersion—by drops of water. You can see rainbows when you look at falling water droplets with the Sun behind you. Figure 24–16 shows how red and violet rays are bent by spherical water droplets and are reflected off the back surface of the droplet. Red is bent the least and so reaches the observer's eyes from droplets higher in the sky, as shown in Fig. 24–16a. Thus the top of the rainbow is red.

FIGURE 24–16 (a) Ray diagram showing how a rainbow (b) is formed.

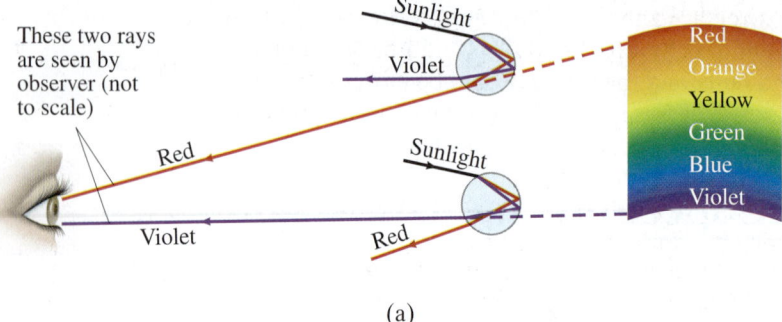

(a) (b)

FIGURE 24–17 Diamond.

Diamonds achieve their brilliance (Fig. 24–17) from a combination of dispersion and total internal reflection. Because diamonds have a very high index of refraction of about 2.4, the critical angle for total internal reflection is only 25°. The light dispersed into a spectrum inside the diamond therefore strikes many of the internal surfaces of the diamond before it strikes one at less than 25° and emerges. After many such reflections, the light has traveled far enough that the colors have become sufficiently separated to be seen individually and brilliantly by the eye after leaving the diamond.

The visible spectrum, Fig. 24–12, does not show all the colors seen in nature. For example, there is no brown in Fig. 24–12. Many of the colors we see are a mixture of wavelengths. For practical purposes, most natural colors can be reproduced using three primary colors. They are red, green, and blue for direct source viewing such as TV and computer monitors. For inks used in printing, the primary colors are cyan (the blue color of the margin notes in this book), yellow, and magenta (the pinkish red color we use for light rays in ray diagrams).

†The complete electromagnetic spectrum is illustrated in Fig. 22–8.

CONCEPTUAL EXAMPLE 24–4 **Observed color of light under water.** We said that color depends on wavelength. For example, light of wavelength $\lambda_0 = 650$ nm in air, we see red. If we observe the same object when under water, it still looks red. But the wavelength in water λ_w is (Eq. 24–1) $\lambda_w = \lambda_0/n_w = 650$ nm$/1.33 = 489$ nm. Light with wavelength 489 nm in air would appear blue in air. Can you explain why the light appears red rather than blue when observed under water?

RESPONSE Today we have little doubt that it is our brains that express colors, based on the wavelengths of light that strike the receptor cells within the retina (at the rear of the eyeball, as diagrammed in the next Chapter, Fig. 25–9). For objects under water, the water does nothing to change the frequency, but does change the wavelength to λ_0/n_w. When that light enters the eye, the frequency is still unchanged, but the speed is changed to c/n_{eye} where n_{eye} is the index of refraction of the fluid that fills the interior of the eye and is in contact with the retina. The wavelength of light that reaches the retina is $\lambda_{eye} = \lambda_0/n_{eye}$, and is the same whether the light enters from the air or from water.

24–5 Diffraction by a Single Slit or Disk

Young's double-slit experiment put the wave theory of light on a firm footing. But full acceptance came only with studies on diffraction (Section 24–1) more than a decade later, in the 1810s and 1820s.

We have already discussed diffraction briefly with regard to water waves (Section 11–14) as well as for light (Section 24–1). We have seen that diffraction refers to the spreading or bending of waves around edges. Let's look in more detail.

In 1819 Augustin Fresnel (1788–1827) presented to the French Academy a wave theory of light that predicted and explained interference and diffraction effects. Almost immediately Siméon Poisson (1781–1840) pointed out a counter-intuitive inference: according to Fresnel's wave theory, if light from a point source were to fall on a solid disk, part of the incident light would be diffracted around the edges and would constructively interfere at the center of the shadow (Fig. 24–18). That prediction seemed very unlikely. But when the experiment was actually carried out by Francois Arago, the bright spot was seen at the very center of the shadow (Fig. 24–19a). This was strong evidence for the wave theory.

Figure 24–19a is a photograph of the shadow cast by a coin using a coherent point source of light, a laser in this case. The bright spot is clearly present at the center. Note also the bright and dark fringes beyond the shadow. These resemble the interference fringes of a double slit. Indeed, they are due to interference of waves diffracted around the outer edge of the disk, and the group of fringes is referred to as a **diffraction pattern**. A diffraction pattern exists around any sharp-edged object illuminated by a point source, as shown in Figs. 24–19b and c. We are not always aware of diffraction because most sources of light in everyday life are not points, so light from different parts of the source washes out the pattern.

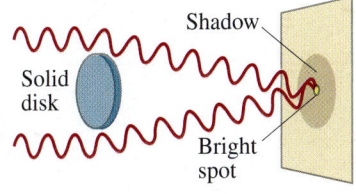

FIGURE 24–18 If light is a wave, a bright spot will appear at the center of the shadow of a solid disk illuminated by a point source of monochromatic light.

FIGURE 24–19 Diffraction pattern of (a) a circular disk (a coin), (b) scissors, (c) a single slit, each illuminated by a (nearly) point source of coherent monochromatic light.

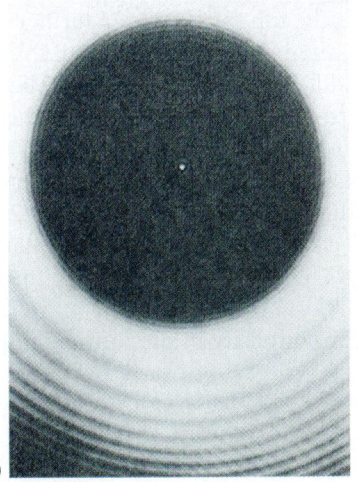
(a)

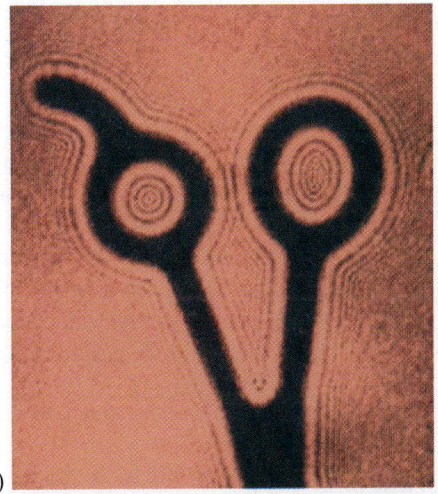
(b)

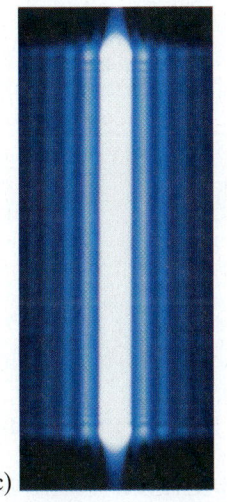
(c)

To see how a diffraction pattern arises, we analyze the important case of monochromatic light passing through a narrow slit (as for Fig. 24–19c). We assume that parallel rays (plane waves) of light pass straight through a slit of width D to a viewing screen very far away.† As we know from studying water waves and from Huygens' principle, waves passing through a slit spread out in all directions. We will now examine how the waves passing through different parts of the slit interfere with each other.

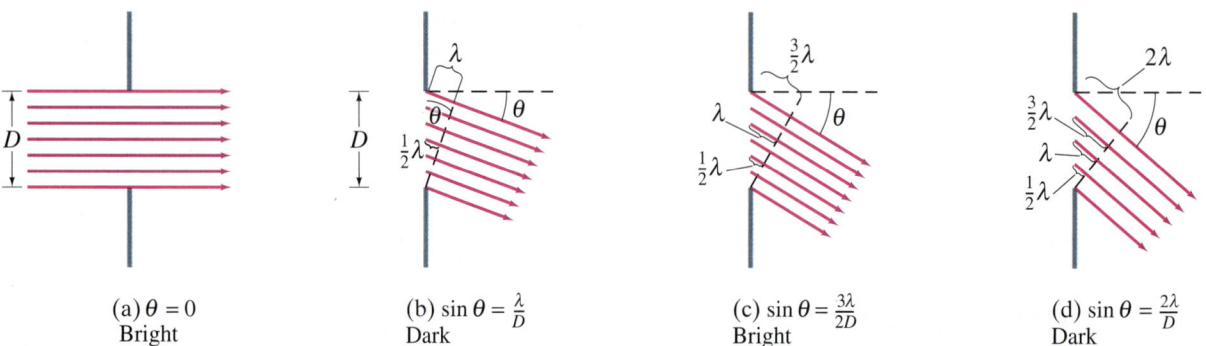

(a) $\theta = 0$
Bright

(b) $\sin \theta = \frac{\lambda}{D}$
Dark

(c) $\sin \theta = \frac{3\lambda}{2D}$
Bright

(d) $\sin \theta = \frac{2\lambda}{D}$
Dark

FIGURE 24–20 Analysis of diffraction pattern formed by light passing through a narrow slit of width D.

Parallel rays of monochromatic light pass through the narrow slit as shown in Fig. 24–20a. The slit width D is on the order of the wavelength λ of the light, but the slit's length (into and out of page) may be large compared to λ. The light falls on a screen which is assumed to be very far away, so the rays heading toward any point are very nearly parallel before they meet at the screen. First we consider rays that pass straight through as in Fig. 24–20a. They are all in phase, so there will be a central bright spot on the screen (see Fig. 24–19c). In Fig. 24–20b, we consider rays moving at an angle θ such that the ray from the top of the slit travels exactly one wavelength farther than the ray from the bottom edge to reach the screen. The ray passing through the very center of the slit will travel one-half wavelength farther than the ray at the bottom of the slit. These two rays will be exactly out of phase with one another and so will destructively interfere when they overlap at the screen. Similarly, a ray slightly above the bottom one will cancel a ray that is the same distance above the central one. Indeed, each ray passing through the lower half of the slit will cancel with a corresponding ray passing through the upper half. Thus, all the rays destructively interfere in pairs, and so the light intensity will be zero on the viewing screen at this angle. The angle θ at which this takes place can be seen from Fig. 24–20b to occur when $\lambda = D \sin \theta$, so

$$\sin \theta = \frac{\lambda}{D}. \qquad \text{[first minimum]} \quad (24\text{–}3a)$$

The light intensity is a maximum at $\theta = 0°$ and decreases to a minimum (intensity = zero) at the angle θ given by Eq. 24–3a.

Now consider a larger angle θ such that the top ray travels $\frac{3}{2}\lambda$ farther than the bottom ray, as in Fig. 24–20c. In this case, the rays from the bottom third of the slit will cancel in pairs with those in the middle third because they will be $\lambda/2$ out of phase. However, light from the top third of the slit will still reach the screen, so there will be a bright spot (or fringe) centered near $\sin \theta \approx 3\lambda/2D$, but it will not be nearly as bright as the central spot at $\theta = 0°$. For an even larger angle θ such that the top ray travels 2λ farther than the bottom ray, Fig. 24–20d, rays from the bottom quarter of the slit will cancel with those in the quarter just above it because the path lengths differ by $\lambda/2$. And the rays through the quarter of the slit just above center will cancel with those through the top quarter. At this angle there will again be a minimum of zero intensity in the diffraction pattern.

†If the viewing screen is not far away, lenses can be used to make the rays parallel.

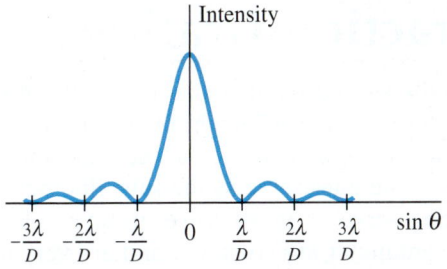

FIGURE 24–21 Intensity in the diffraction pattern of a single slit as a function of sin θ. Note that the central maximum is not only much higher than the maxima to each side, but it is also twice as wide ($2\lambda/D$ wide) as any of the others (each only λ/D wide).

A plot of the intensity as a function of angle is shown in Fig. 24–21. This corresponds well with the photo of Fig. 24–19c. Notice that minima (zero intensity) occur on both sides of center at

$$D \sin \theta = m\lambda, \quad m = \pm 1, \pm 2, \pm 3, \cdots, \quad \text{[minima]} \quad \textbf{(24–3b)}$$

but *not* at $m = 0$ where there is the strongest maximum. Between the minima, smaller intensity maxima occur at approximately (not exactly) $m \approx \frac{3}{2}, \frac{5}{2}, \cdots$.

Note that the *minima* for a diffraction pattern, Eq. 24–3b, satisfy a criterion that looks very similar to that for the *maxima* (bright spots or fringes) for double-slit interference, Eq. 24–2a. Also note that D is a single slit width, whereas d in Eqs. 24–2 is the distance between two slits.

> ⚠ **CAUTION**
> *Don't confuse Eqs. 24–2 for interference with Eqs. 24–3 for diffraction; note the differences*

EXAMPLE 24–5 Single-slit diffraction maximum. Light of wavelength 750 nm passes through a slit 1.0×10^{-3} mm wide. How wide is the central maximum (*a*) in degrees, and (*b*) in centimeters, on a screen 20 cm away?

APPROACH The width of the central maximum goes from the first minimum on one side to the first minimum on the other side. We use Eq. 24–3a to find the angular position of the first single-slit diffraction minimum.

SOLUTION (*a*) The first minimum occurs at

$$\sin \theta = \frac{\lambda}{D} = \frac{7.5 \times 10^{-7}\,\text{m}}{1.0 \times 10^{-6}\,\text{m}} = 0.75.$$

So $\theta = 49°$. This is the angle between the center and the first minimum, Fig. 24–22. The angle subtended by the whole central maximum, between the minima above and below the center, is twice this, or 98°.
(*b*) The width of the central maximum is $2x$, where $\tan \theta = x/20\,\text{cm}$. So $2x = 2(20\,\text{cm})(\tan 49°) = 46\,\text{cm}.$

NOTE A large width of the screen will be illuminated, but it will not normally be very bright since the amount of light that passes through such a small slit will be small and it is spread over a large area. Note also that we *cannot* use the small-angle approximation here ($\theta \approx \sin \theta \approx \tan \theta$) because θ is large.

FIGURE 24–22 Example 24–5.

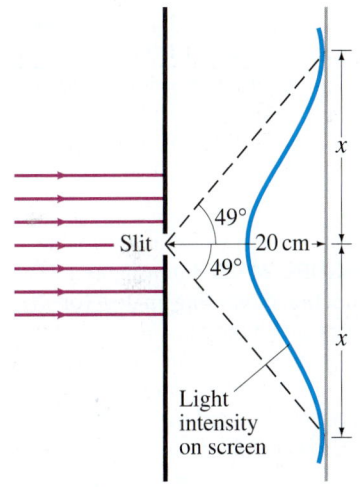

EXERCISE C In Example 24–5, red light ($\lambda = 750$ nm) was used. If instead yellow light ($\lambda = 550$ nm) had been used, would the central maximum be wider or narrower?

CONCEPTUAL EXAMPLE 24–6 Diffraction spreads. Light shines through a small rectangular slit that is narrower in the vertical direction than the horizontal, Fig. 24–23. (*a*) Would you expect the diffraction pattern to be more spread out in the vertical direction or in the horizontal direction? (*b*) Should a rectangular loudspeaker horn at a stadium be tall and narrow, or wide and flat?

RESPONSE (*a*) From Eq. 24–3a we can see that if we make the slit width D smaller, the pattern spreads out more (θ will be larger in Eq. 24–3a). This is consistent with our study of waves in Chapter 11. The diffraction through the rectangular hole will be wider vertically, since the opening is smaller in that direction.
(*b*) For a stadium loudspeaker, the sound pattern desired is one spread out horizontally, so the horn should be tall and narrow (rotate Fig. 24–23 by 90°).

FIGURE 24–23 Example 24–6.

24–6 Diffraction Grating

A large number of equally spaced parallel slits is called a **diffraction grating**, although the term "interference grating" might be as appropriate. Gratings can be made by precision machining of very fine parallel lines on a glass plate. The untouched spaces between the lines serve as the slits. Photographic transparencies of an original grating serve as inexpensive gratings. Gratings containing 10,000 lines or slits per centimeter are common, and are very useful for precise measurements of wavelengths. A diffraction grating containing slits is called a **transmission grating**. Another type of diffraction grating is the **reflection grating**, made by ruling fine lines on a metallic or glass surface from which light is reflected and analyzed. The analysis is basically the same as for a transmission grating, which we now discuss.

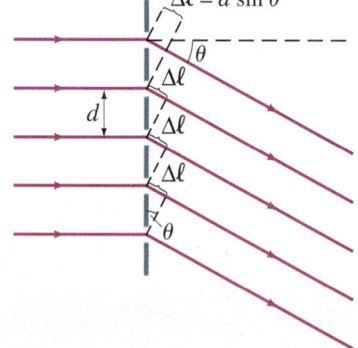

FIGURE 24–24 Diffraction grating.

The analysis of a diffraction grating is much like that of Young's double-slit experiment. We assume parallel rays of light are incident on the grating as shown in Fig. 24–24. We also assume that the slits are narrow enough so that diffraction by each of them spreads light over a very wide angle on a distant screen beyond the grating, and interference can occur with light from all the other slits. Light rays that pass through each slit without deviation ($\theta = 0°$) interfere constructively to produce a bright maximum at the center of the screen. Constructive interference also occurs at an angle θ such that rays from adjacent slits travel an extra distance of $\Delta\ell = m\lambda$, where m is an integer. If d is the distance *between* slits, then we see from Fig. 24–24 that $\Delta\ell = d \sin\theta$, and

$$\sin\theta = \frac{m\lambda}{d}, \qquad m = 0, 1, 2, \cdots \qquad \begin{bmatrix}\text{diffraction grating}\\ \text{principal maxima}\end{bmatrix} \quad (24\text{–}4)$$

CAUTION
Diffraction grating is analyzed using interference formulas, not diffraction formulas

is the criterion to have a brightness maximum. This is the same equation as for the double-slit situation, and again m is called the **order** of the pattern.

There is an important difference between a double-slit and a multiple-slit pattern. The bright maxima are much *sharper* and *narrower* for a grating. Why? Suppose the angle θ in Fig. 24–24 is increased just slightly beyond θ required for a maximum. For only two slits, the two waves will be only slightly out of phase, so nearly full constructive interference occurs. This means the maxima are wide (see Fig. 24–9). For a grating, the waves from two adjacent slits will also not be significantly out of phase. But waves from one slit and those from a second one a few hundred slits away may be exactly out of phase; all or nearly all the light can cancel in pairs in this way. For example, suppose the angle θ is very slightly different from its first-order maximum, so that the extra path length for a pair of adjacent slits is not exactly λ but rather 1.0010λ. The wave through one slit and another one 500 slits below will have a path difference of $1\lambda + (500)(0.0010\lambda) = 1.5000\lambda$, or $1\tfrac{1}{2}$ wavelengths, so the two will be out of phase and cancel. A pair of slits, one below each of these, will also cancel. That is, the light from slit 1 cancels with light from slit 501; light from slit 2 cancels with light from slit 502, and so on. Thus even for a tiny angle[†] corresponding to an extra path length of $\tfrac{1}{1000}\lambda$, there is much destructive interference, and so the maxima of a diffraction grating are very narrow. The more slits there are in a grating, the sharper will be the peaks (see Fig. 24–25). Because a grating produces much sharper maxima than two slits alone, and also much brighter maxima because there are many more slits, a grating is a far more precise device for measuring wavelengths.

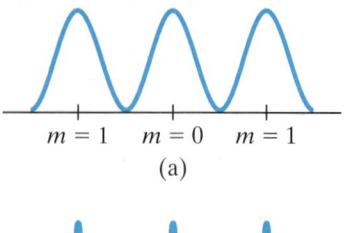
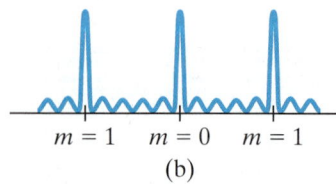

FIGURE 24–25 Intensity as a function of viewing angle θ (or position on the screen) for (a) two slits, (b) six slits. For a diffraction grating, the number of slits is very large ($\approx 10^4$) and the peaks are narrower still.

Suppose the light striking a diffraction grating is not monochromatic, but consists of two or more distinct wavelengths. Then for all orders other than $m = 0$, each wavelength will produce a maximum at a different angle (Eq. 24–4), forming a line on the screen as shown in Fig. 24–26a.

[†]Depending on the total number of slits, there may or may not be complete cancellation for such an angle, so there will be very tiny peaks between the main maxima (see Fig. 24–25b), but they are usually much too small to be seen.

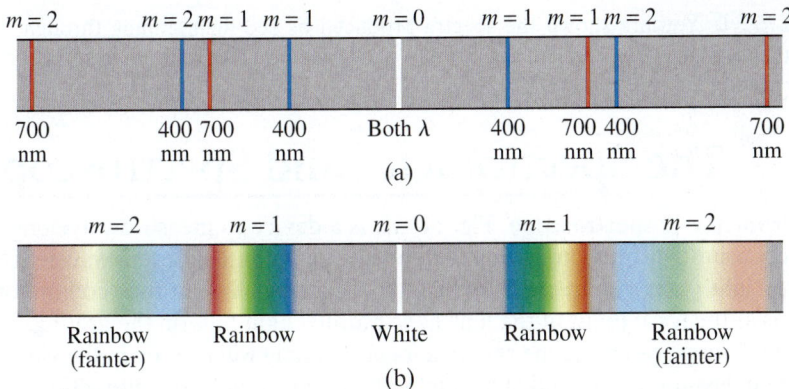

FIGURE 24–26 Spectra produced by a grating: (a) two wavelengths, 400 nm and 700 nm; (b) white light. The second order will normally be dimmer than the first order. (Higher orders are not shown.) If the grating spacing is small enough, the second and higher orders will be missing.

If white light strikes a grating, the central ($m = 0$) maximum will be a sharp white line. But for all other orders, there will be a distinct spectrum of colors spread out over a certain angular width, Fig. 24–26b. Because a diffraction grating spreads out light into its component wavelengths, the resulting pattern is called a **spectrum**.

EXAMPLE 24–7 Diffraction grating: line positions. Determine the angular positions of the first- and second-order lines (maxima) for light of wavelength 400 nm and 700 nm incident on a grating containing 10,000 slits per centimeter.

APPROACH First we find the distance d between grating slits: if the grating has N slits in 1 m, then the distance between slits is $d = 1/N$ meters. Then we use Eq. 24–4, $\sin\theta = m\lambda/d$, to get the angles for the two wavelengths for $m = 1$ and 2.

SOLUTION The grating contains 1.00×10^4 slits/cm = 1.00×10^6 slits/m, which means the distance between slits is $d = (1/1.00 \times 10^6)$ m = 1.00×10^{-6} m = $1.00\ \mu$m. In first order ($m = 1$), the angles are

$$\sin\theta_{400} = \frac{m\lambda}{d} = \frac{(1)(4.00 \times 10^{-7}\,\text{m})}{1.00 \times 10^{-6}\,\text{m}} = 0.400$$

$$\sin\theta_{700} = \frac{(1)(7.00 \times 10^{-7}\,\text{m})}{1.00 \times 10^{-6}\,\text{m}} = 0.700$$

so $\theta_{400} = 23.6°$ and $\theta_{700} = 44.4°$. In second order,

$$\sin\theta_{400} = \frac{2\lambda}{d} = \frac{(2)(4.00 \times 10^{-7}\,\text{m})}{1.00 \times 10^{-6}\,\text{m}} = 0.800$$

$$\sin\theta_{700} = \frac{(2)(7.00 \times 10^{-7}\,\text{m})}{1.00 \times 10^{-6}\,\text{m}} = 1.40$$

so $\theta_{400} = 53.1°$. But the second order does not exist for $\lambda = 700$ nm because $\sin\theta$ cannot exceed 1. No higher orders will appear.

EXAMPLE 24–8 Spectra overlap. White light containing wavelengths from 400 nm to 750 nm strikes a grating containing 4000 slits/cm. Show that the blue at $\lambda = 450$ nm of the third-order spectrum overlaps the red at 700 nm of the second order.

APPROACH We use $\sin\theta = m\lambda/d$ to calculate the angular positions of the $m = 3$ blue maximum and the $m = 2$ red one.

SOLUTION The grating spacing is $d = (1/4000)$ cm = 2.50×10^{-6} m. The blue of the third order occurs at an angle θ given by

$$\sin\theta = \frac{m\lambda}{d} = \frac{(3)(4.50 \times 10^{-7}\,\text{m})}{(2.50 \times 10^{-6}\,\text{m})} = 0.540.$$

Red in second order occurs at

$$\sin\theta = \frac{(2)(7.00 \times 10^{-7}\,\text{m})}{(2.50 \times 10^{-6}\,\text{m})} = 0.560,$$

which is a greater angle; so the second order overlaps into the beginning of the third-order spectrum.

SECTION 24–6 Diffraction Grating

EXERCISE D You are shown the spectra produced by red light shining through two different gratings. The maxima in spectrum A are farther apart than those in spectrum B. Which grating has more slits/cm?

24–7 The Spectrometer and Spectroscopy

A **spectrometer** or **spectroscope**, Fig. 24–27, is a device to measure wavelengths accurately using a diffraction grating (or a prism) to separate different wavelengths of light. Light from a source passes through a narrow slit S in the "collimator." The slit is at the focal point of the lens L, so parallel light falls on the grating. The movable telescope can bring the rays to a focus. Nothing will be seen in the viewing telescope unless it is positioned at an angle θ that corresponds to a diffraction peak (first order is usually used) of a wavelength emitted by the source. The angle θ can be measured to very high accuracy, so the wavelength can be determined to high accuracy using Eq. 24-4:

$$\lambda = \frac{d}{m}\sin\theta,$$

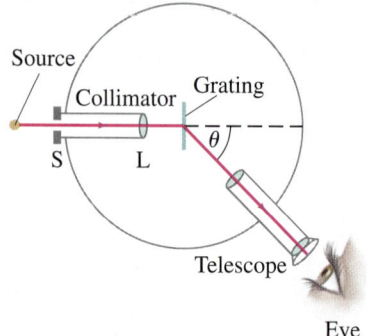

FIGURE 24–27 Spectrometer or spectroscope.

where m is an integer representing the order, and d is the distance between grating slits. The bright line you see in a spectrometer corresponding to a discrete particular wavelength is actually an image of the slit S. A narrower slit results in dimmer light, but we can measure the angular position more precisely. If the light contains a continuous range of wavelengths, then a continuous spectrum is seen in the spectroscope.

The spectrometer in Fig. 24–27 uses a transmission grating. Others may use a reflection grating, or sometimes a prism. A prism works because of dispersion (Section 24–4), bending light of different wavelengths into different angles. A prism is not a linear device and must be calibrated because λ is not $\propto \sin\theta$; see Fig. 24–14.

An important use of a spectrometer is for the identification of atoms or molecules. When a gas is heated or an electric current is passed through it, the gas emits a characteristic **line spectrum**. That is, only certain discrete wavelengths of light are emitted, and these are different for different elements and compounds.[†] Figure 24–28 shows the line spectra for a number of elements in the gas state. Line spectra occur only for gases at high temperatures and low pressure and density. The light from heated solids, such as a lightbulb filament, and even from a dense gaseous object such as the Sun, produces a **continuous spectrum** including a wide range of wavelengths.

Figure 24–28 also shows the Sun's "continuous spectrum," which contains a number of *dark* lines (only the most prominent are shown), called **absorption lines**. Atoms and molecules can absorb light at the same wavelengths at which they emit light.

FIGURE 24–28 Line spectra for the gases indicated, and the spectrum from the Sun showing absorption lines.

[†]Why atoms and molecules emit line spectra was a great mystery for many years and played a central role in the development of modern quantum theory, as we shall see in Chapter 27.

The Sun's absorption lines are due to absorption by atoms and molecules in the cooler outer atmosphere of the Sun, as well as by atoms and molecules in the Earth's atmosphere. A careful analysis of all the Sun's thousands of absorption lines reveals that at least two-thirds of all elements are present in the Sun's atmosphere. The presence of elements in the atmosphere of nearby planets, in interstellar space, and in stars, is also determined by spectroscopy.

Spectroscopy is useful for determining the presence of certain types of molecules in laboratory specimens where chemical analysis would be difficult. For example, biological DNA and different types of protein absorb light in particular regions of the spectrum (such as in the UV). The material to be examined, which is often in solution, is placed in a monochromatic light beam whose wavelength is selected by the placement angle of a diffraction grating or prism. The amount of absorption, as compared to a standard solution without the specimen, can reveal not only the presence of a particular type of molecule, but also its concentration.

Light emission and absorption also occur outside the visible part of the spectrum, such as in the UV and IR regions. Glass absorbs light in these regions, so reflection gratings and mirrors (in place of lenses) are used. Special types of film or sensors are used for detection.

PHYSICS APPLIED
Chemical and biochemical analysis by spectroscopy

EXAMPLE 24–9 Hydrogen spectrum. Light emitted by hot hydrogen gas is observed with a spectroscope using a diffraction grating having 1.00×10^4 slits/cm. The spectral lines nearest to the center (0°) are a violet line at 24.2°, a blue line at 25.7°, a blue-green line at 29.1°, and a red line at 41.0° from the center. What are the wavelengths of these spectral lines of hydrogen?

APPROACH We get the wavelengths from the angles by using $\lambda = (d/m) \sin \theta$ where d is the spacing between slits, and m is the order of the spectrum (Eq. 24–4).

SOLUTION Since these are the closest lines to $\theta = 0°$, this is the first-order spectrum ($m = 1$). The slit spacing is $d = 1/(1.00 \times 10^4 \text{ cm}^{-1}) = 1.00 \times 10^{-6}$ m. The wavelength of the violet line is

$$\lambda = \left(\frac{d}{m}\right) \sin \theta = \left(\frac{1.00 \times 10^{-6} \text{ m}}{1}\right) \sin 24.2° = 4.10 \times 10^{-7} \text{ m} = 410 \text{ nm}.$$

The other wavelengths are:

blue: $\lambda = (1.00 \times 10^{-6} \text{ m}) \sin 25.7° = 434$ nm,
blue-green: $\lambda = (1.00 \times 10^{-6} \text{ m}) \sin 29.1° = 486$ nm,
red: $\lambda = (1.00 \times 10^{-6} \text{ m}) \sin 41.0° = 656$ nm.

NOTE In an unknown mixture of gases, these four spectral lines need to be seen to identify that the mixture contains hydrogen.

24–8 Interference in Thin Films

Interference of light gives rise to many everyday phenomena such as the bright colors reflected from soap bubbles and from thin oil or gasoline films on water, Fig. 24–29. In these and other cases, the colors are a result of constructive interference between light reflected from the two surfaces of the thin film. The effect is observed only if the thickness of the film is on the order of the wavelength of the light. If the film thickness is greater than a few wavelengths, the effect gets washed out.

FIGURE 24–29 Thin-film interference patterns seen in (a) a soap bubble, (b) a thin film of soapy water, and (c) a thin layer of oil on wet pavement.

(a)

(b)
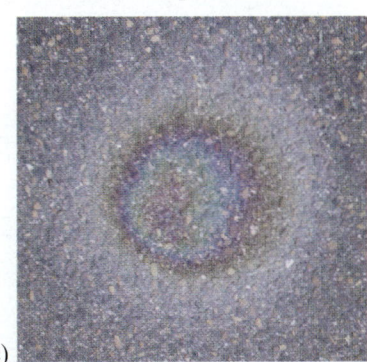
(c)

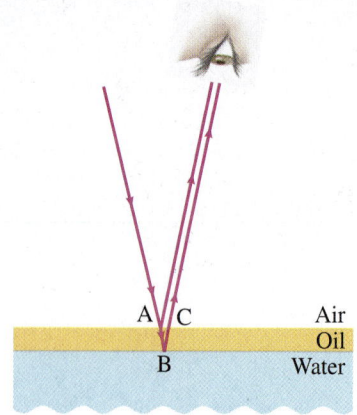

FIGURE 24–30 Light reflected from the upper and lower surfaces of a thin film of oil lying on water.

To see how this **thin-film interference** happens, consider a smooth surface of water on top of which is a thin uniform layer of another substance, say an oil whose index of refraction is less than that of water (we'll see why we assume this shortly); see Fig. 24–30. Assume for now that the incident light is of a single wavelength. Part of the incident light is reflected at A on the top surface, and part of the light transmitted is reflected at B on the lower surface. The part reflected at the lower surface must travel the extra distance ABC. If this *path difference* ABC equals one or a whole number of wavelengths in the film (λ_n), the two waves will reach the eye in phase and interfere constructively. Hence the region AC on the surface film will appear bright. But if ABC equals $\frac{1}{2}\lambda_n, \frac{3}{2}\lambda_n$, and so on, the two waves will be exactly out of phase and destructive interference occurs: the area AC on the film will show no reflection—it will be dark (transparent to the dark material below). The wavelength λ_n is *the wavelength in the film*: $\lambda_n = \lambda/n$, where n is the index of refraction in the film and λ is the wavelength in vacuum. See Eq. 24–1.

When white light falls on such a film, the path difference ABC will equal λ_n (or $m\lambda_n$, with $m =$ an integer) for only one wavelength at a given viewing angle. The color corresponding to λ (λ in air) will be seen as very bright. For light viewed at a slightly different angle, the path difference ABC will be longer or shorter and a different color will undergo constructive interference. Thus, for an extended (nonpoint) source emitting white light, a series of bright colors will be seen next to one another. Variations in thickness of the film will also alter the path difference ABC and therefore affect the color of light that is most strongly reflected.

> **EXERCISE E** Return to the Chapter-Opening Question, page 679, and answer it again now. Try to explain why you may have answered differently the first time.

When a curved glass surface is placed in contact with a flat glass surface, Fig. 24–31, a series of concentric rings is seen when illuminated from above by either white light (as shown) or by monochromatic light. These are called **Newton's rings**[†] and they are due to interference between waves reflected by the top and bottom surfaces of the very thin *air gap* between the two pieces of glass. Because this gap (which is equivalent to a thin film) increases in width from the central contact point out to the edges, the extra path length for the lower ray (equal to BCD) varies. Where it equals 0, $\frac{1}{2}\lambda$, λ, $\frac{3}{2}\lambda$, 2λ, and so on, it corresponds to constructive and destructive interference; and this gives rise to the series of bright colored circles seen in Fig. 24–31b. The color you see at a given radius corresponds to constructive interference; at that radius, other colors partially or fully destructively interfere. (If monochromatic light is used, the rings are alternately bright and dark.)

The point of contact of the two glass surfaces (A in Fig. 24–31a) is not bright in Fig. 24–31b. Since the path difference is zero here, our previous analysis would suggest that the waves reflected from each surface are in phase—so this central area ought to be bright. But it is dark, which tells us the two waves must be completely

[†]Although Newton gave an elaborate description of them, they had been first observed and described by his contemporary, Robert Hooke.

FIGURE 24–31 Newton's rings. (a) Light rays reflected from upper and lower surfaces of the thin air gap can interfere. (b) Photograph of interference patterns using white light.

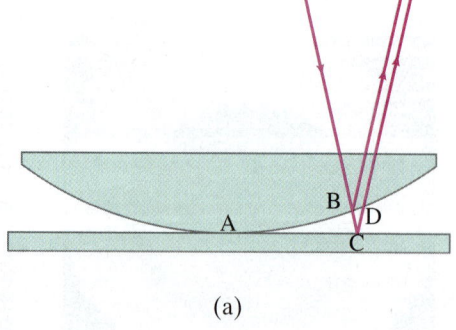

(a)

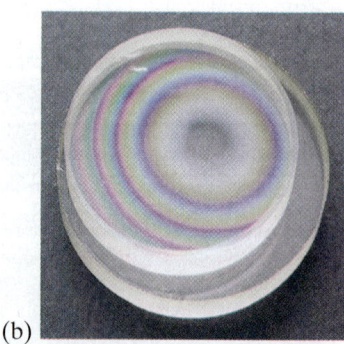

(b)

out of phase. This can happen only if one of the waves, upon reflection, flips over—a crest becomes a trough—see Fig. 24–32. We say that the reflected wave has undergone a **phase shift** of 180°, or of half a wave cycle ($\frac{1}{2}\lambda$). Indeed, this and other experiments reveal that, at normal incidence,

a beam of light, reflected by a material with index of refraction greater than that of the material in which it is traveling, changes phase by 180° or $\frac{1}{2}$ cycle;

see Fig. 24–32. This phase shift acts just like a path difference of $\frac{1}{2}\lambda$. If the index of refraction of the reflecting material is less than that of the material in which the light is traveling, no phase shift occurs.[†]

Thus the wave reflected at the curved surface above the air gap in Fig. 24–31a undergoes no change in phase. But the wave reflected at the lower surface, where the beam in air strikes the glass, undergoes a $\frac{1}{2}$-cycle phase shift, equivalent to a $\frac{1}{2}\lambda$ path difference. Thus the two waves reflected near the point of contact A of the two glass surfaces (where the air gap approaches zero thickness) will be a half cycle (or 180°) out of phase, and a dark spot occurs. Bright colored rings will occur when the path difference is $\frac{1}{2}\lambda$, $\frac{3}{2}\lambda$, and so on, because the phase shift at one surface effectively adds a path difference of $\frac{1}{2}\lambda$ ($= \frac{1}{2}$ cycle). (If monochromatic light is used, the bright Newton's rings will be separated by dark bands which occur when the path difference BCD in Fig. 24–31a is equal to an integral number of wavelengths.)

Returning for a moment to Fig. 24–30, the light reflecting at both interfaces, air–oil and oil–water, *each* underwent a phase shift of 180° equivalent to a path difference of $\frac{1}{2}\lambda$, since we assumed $n_{water} > n_{oil} > n_{air}$. Because the two phase shifts were equal, they didn't affect our analysis.

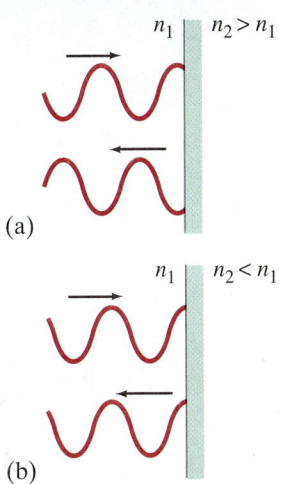

(a)

(b)

FIGURE 24–32 (a) Reflected ray changes phase by 180° or $\frac{1}{2}$ cycle if $n_2 > n_1$, but (b) does not if $n_2 < n_1$.

FIGURE 24–33 (a) Light rays reflected from the upper and lower surfaces of a thin wedge of air (between two glass plates) interfere to produce bright and dark bands. (b) Pattern observed when glass plates are optically flat; (c) pattern when plates are not so flat. See Example 24–10.

EXAMPLE 24–10 Thin film of air, wedge-shaped. A very fine wire 7.35×10^{-3} mm in diameter is placed between two flat glass plates as in Fig. 24–33a. Light whose wavelength in air is 600 nm falls (and is viewed) perpendicular to the plates and a series of bright and dark bands is seen, Fig. 24–33b. How many light and dark bands will there be in this case? Will the area next to the wire be bright or dark?

APPROACH We need to consider two effects: (1) path differences for rays reflecting from the two close surfaces (thin wedge of air between the two glass plates), and (2) the $\frac{1}{2}$-cycle phase shift at the lower surface (point E in Fig. 24–33a), where rays in air can enter glass (or be reflected). Because there is a phase shift only at the lower surface, there will be a dark band (no reflection) when the path difference is 0, λ, 2λ, 3λ, and so on. Since the light rays are perpendicular to the plates, the extra path length (DEF) equals $2t$, where t is the thickness of the air gap at any point.

SOLUTION Dark bands will occur where

$$2t = m\lambda, \qquad m = 0, 1, 2, \cdots.$$

Bright bands occur when $2t = (m + \frac{1}{2})\lambda$, where m is an integer. At the position of the wire, $t = 7.35 \times 10^{-6}$ m. At this point there will be $2t/\lambda = (2)(7.35 \times 10^{-6} \text{ m})/(6.00 \times 10^{-7} \text{ m}) = 24.5$ wavelengths. This is a "half integer," so the area next to the wire will be bright. There will be a total of 25 dark lines along the plates, corresponding to path lengths DEF of $0\lambda, 1\lambda, 2\lambda, 3\lambda, \cdots, 24\lambda$, including the one at the point of contact A ($m = 0$). Between them, there will be 24 bright lines plus the one at the end, or 25.

NOTE The bright and dark bands will be straight only if the glass plates are extremely flat. If they are not, the pattern is uneven, as in Fig. 24–33c. Thus we see a very precise way of testing a glass surface for flatness. Spherical lens surfaces can be tested for precision by placing the lens on a flat glass surface and observing Newton's rings (Fig. 24–31b) for perfect circularity.

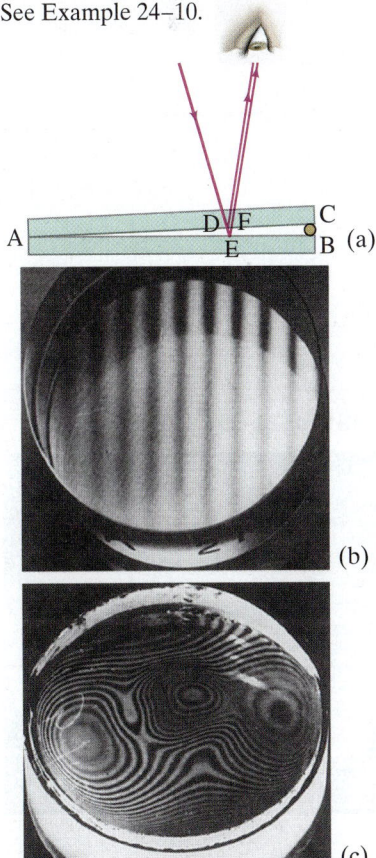

(a)

(b)

(c)

PHYSICS APPLIED
Testing glass for flatness

[†]This result corresponds to the reflection of a wave traveling along a cord when it reaches the end. As we saw in Fig. 11–33, if the end is tied down, the wave changes phase and the pulse flips over, but if the end is free, no phase shift occurs.

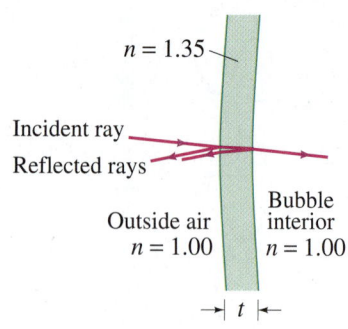

FIGURE 24–34 Soap bubble, Example 24–11. The incident and reflected rays are assumed to be perpendicular to the bubble's surface. They are shown at a slight angle so we can distinguish them.

CAUTION

A formula is not enough: you must also check for phase changes at surfaces

When white light (rather than monochromatic light) is incident on the thin wedge of air in Fig. 24–31a or 24–33a, a colorful series of fringes is seen because constructive interference occurs for different wavelengths in the reflected light at different thicknesses along the wedge.

A soap bubble (Fig. 24–29a and Chapter-Opening Photo) is a thin spherical shell (or film) with air inside. The variations in thickness of a soap bubble film give rise to bright colors reflected from the soap bubble. (There is air on both sides of the bubble film.) Similar variations in film thickness produce the bright colors seen reflecting from a thin layer of oil or gasoline on a puddle or lake (Fig. 24–29c). Which wavelengths appear brightest also depends on the viewing angle.

EXAMPLE 24–11 **Thickness of soap bubble skin.** A soap bubble appears green ($\lambda = 540$ nm) at the point on its front surface nearest the viewer. What is the smallest thickness the soap bubble film could have? Assume $n = 1.35$.

APPROACH Assume the light is reflected perpendicularly from the point on a spherical surface nearest the viewer, Fig. 24–34. The light rays also reflect from the inner surface of the soap bubble film as shown. The path difference of these two reflected rays is $2t$, where t is the thickness of the soap film. Light reflected from the first (outer) surface undergoes a 180° phase change (index of refraction of soap is greater than that of air), whereas reflection at the second (inner) surface does not. To determine the thickness t for an interference maximum, we must use the wavelength of light in the soap ($n = 1.35$).

SOLUTION The 180° phase change at only one surface is equivalent to a $\frac{1}{2}\lambda$ path difference. Therefore, green light is bright when the minimum path difference equals $\frac{1}{2}\lambda_n$. Thus, $2t = \lambda_n/2$, so

$$t = \frac{\lambda_n}{4} = \frac{\lambda}{4n} = \frac{(540 \text{ nm})}{(4)(1.35)} = 100 \text{ nm}.$$

This is the smallest thickness.

NOTE At this small thickness, blue (450 nm) and red (600 nm) also would reflect fairly constructively, so the bubble would appear almost white. The green color is more likely to be seen at the *next* thickness that gives constructive interference, $2t = 3\lambda/2n$, because other colors would be more fully cancelled by destructive interference. Then t would be $t = 3\lambda/4n = 300$ nm. Note that green is seen in air, so $\lambda = 540$ nm (not λ/n).

*Colors in a Thin Soap Film

FIGURE 24–29b (Repeated.)

The thin film of soapy water (in a plastic loop) shown in Fig. 24–29b (repeated here) has stood vertically for a long time. Gravity has pulled the soapy water downward, so the film increases in thickness going toward the bottom. The top section is so thin (perhaps 30 nm thick $\ll \lambda$) that light reflected from the front and back surfaces have almost zero path difference. Thus the 180° phase change at the front surface assures that the two reflected waves are 180° out of phase for all wavelengths of visible light. The white light incident on this thin film does not reflect at the top part of the film, so the top is transparent and we see the background which is black.

Below the black area at the top, there is a thin blue line, and then a white band. The film has thickened to perhaps 75 to 100 nm, so the shortest wavelength (blue) light begins to partially interfere constructively. But just below, where the thickness is slightly greater (100 nm), the path difference is reasonably close to $\lambda/2$ for much of the spectrum and we see white or silver.[†]

Immediately below the white band in this Figure we see a brown band, where $t \approx 200$ nm, and many wavelengths (not all) are close to λ—and those colors destructively interfere, leaving only a few colors to partially interfere constructively, giving us murky brown.

[†]Why? Recall that red starts at 600 nm in air; so most colors in the spectrum lie between 450 nm and 600 nm in air; but in water the wavelengths are $n = 1.33$ times smaller, 340 nm to 450 nm, so a 100-nm thickness is a 200-nm path difference, not far from $\lambda/2$ for most colors.

Farther down in Fig. 24–29b, with increasing thickness t, a path difference $2t = 510$ nm corresponds nicely to $\frac{3}{2}\lambda$ for blue, but not for other colors, so we see blue ($\frac{3}{2}\lambda$ path difference plus $\frac{1}{2}\lambda$ phase change = constructive interference). Other colors experience constructive interference (at $\frac{3}{2}\lambda$ and then at $\frac{5}{2}\lambda$) at still greater thicknesses, so going down we see a series of separated colors something like a rainbow.

In the soap bubble of our Chapter-Opening Photo (page 679), similar things happen: at the top (where the film is thinnest) we see black and then silver-white, just as in the soap film shown in Fig. 24–29b.

Also examine the oil film on wet pavement shown in Fig. 24–29c (repeated here). The oil film is thickest at the center and thins out toward the edges. Notice the whitish outer ring where most colors constructively interfere, which would suggest a thickness on the order of 100 nm as discussed above for the white band in the soap film. Beyond the outer white band of the oil film, Fig. 24–29c, there is still some oil, but the film is so thin that reflected light from upper and lower surfaces destructively interfere and you can see right through this very thin oil film.

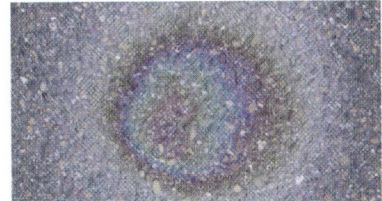

FIGURE 24–29c (Repeated.)

Lens Coatings

An important application of thin-film interference is in the coating of glass to make it "nonreflecting," particularly for lenses. A glass surface reflects about 4% of the light incident upon it. Good-quality cameras, microscopes, and other optical devices may contain six to ten thin lenses. Reflection from all these surfaces can reduce the light level considerably, and multiple reflections produce a background haze that reduces the quality of the image. By reducing reflection, transmission and sharpness are increased.

PHYSICS APPLIED
Lens coatings

A very thin coating on the lens surfaces can reduce reflections considerably. The thickness of the coating is chosen so that light (at least for one wavelength) reflecting from the front and rear surfaces of the film destructively interferes. Destructive interference can occur nearly completely for one particular wavelength depending on the thickness of the coating. Nearby wavelengths will at least partially destructively interfere, but a single coating cannot eliminate reflections for all wavelengths. Nonetheless, a single coating can reduce total reflection from 4% to 1% of the incident light. Often the coating is designed to eliminate the center of the reflected spectrum (around 550 nm). The extremes of the spectrum—red and violet—will not be reduced as much. Since a mixture of red and violet produces purple, the light seen reflected from such coated lenses is purple (Fig. 24–35). Lenses containing two or three separate coatings can more effectively reduce a wider range of reflecting wavelengths.

FIGURE 24–35 A coated lens. Note color of light reflected from the front lens surface.

PROBLEM SOLVING

Interference

1. **Interference effects** depend on the simultaneous arrival of two or more waves at the same point in space.
2. **Constructive interference** occurs when waves with the same wavelength arrive in phase with each other: a crest of one wave arrives at the same time as a crest of the other wave(s). The amplitudes of the waves then add to form a larger amplitude. Constructive interference also occurs when the path difference is exactly one full wavelength or any integer multiple of a full wavelength: $1\lambda, 2\lambda, 3\lambda, \cdots$.
3. **Destructive interference** occurs when a crest of one wave arrives at the same time as a trough of the other wave. The amplitudes add, but they are of opposite sign, so the total amplitude is reduced to zero if the two amplitudes are equal. Destructive interference occurs whenever the phase difference is half a wave cycle, or the path difference is a half-integral number of wavelengths. Thus, the total amplitude will be zero if two identical waves arrive one-half wavelength out of phase, or $(m + \frac{1}{2})\lambda$ out of phase, where m is an integer.
4. For thin-film interference, an extra half-wavelength **phase shift** occurs when light **reflects** from an optically more dense medium (going from a material of lesser toward greater index of refraction).

SECTION 24–8 Interference in Thin Films

EXAMPLE 24–12 **Nonreflective coating.** What is the thickness of an optical coating of MgF$_2$ whose index of refraction is $n = 1.38$ and which is designed to eliminate reflected light at wavelengths (in air) around 550 nm when incident normally on glass for which $n = 1.50$?

APPROACH We explicitly follow the procedure outlined in the Problem Solving Strategy on page 697.

SOLUTION

1. **Interference effects.** Consider two rays reflected from the front and rear surfaces of the coating on the lens as shown in Fig. 24–36. The rays are drawn not quite perpendicular to the lens so we can see each of them. These two reflected rays will interfere with each other.
2. **Constructive interference.** We want to eliminate reflected light, so we do not consider constructive interference.
3. **Destructive interference.** To eliminate reflection, we want reflected rays 1 and 2 to be $\frac{1}{2}$ cycle out of phase with each other so that they destructively interfere. The phase difference is due to the path difference $2t$ traveled by ray 2, as well as any phase change in either ray due to reflection.
4. **Reflection phase shift.** Rays 1 and 2 *both* undergo a change of phase by $\frac{1}{2}$ cycle when they reflect from the coating's front and rear surfaces, respectively (at both surfaces the index of refraction increases). Thus there is no net change in phase due to the reflections. The net phase difference will be due to the extra path $2t$ taken by ray 2 in the coating, where $n = 1.38$. We want $2t$ to equal $\frac{1}{2}\lambda_n$ so that destructive interference occurs, where $\lambda_n = \lambda/n$ is the wavelength in the coating. With $2t = \lambda_n/2 = \lambda/2n$, then

$$t = \frac{\lambda_n}{4} = \frac{\lambda}{4n} = \frac{(550 \text{ nm})}{(4)(1.38)} = 99.6 \text{ nm}.$$

NOTE We could have set $2t = \left(m + \frac{1}{2}\right)\lambda_n$, where m is an integer. The smallest thickness ($m = 0$) is usually chosen because destructive interference will occur over the widest angle.

NOTE Complete destructive interference occurs only for the given wavelength of visible light. Longer and shorter wavelengths will have only partial cancellation.

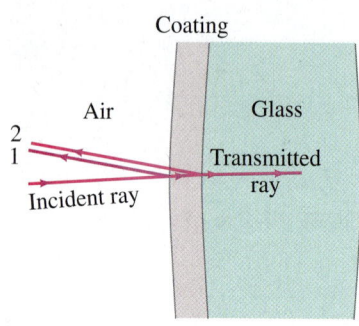

FIGURE 24–36 Lens coating, Example 24–12. Incident ray of light is partially reflected at the front surface of a lens coating (ray 1) and again partially reflected at the rear surface of the coating (ray 2), with most of the energy passing through as the transmitted ray into the glass.

*24–9 Michelson Interferometer

A useful instrument involving wave interference is the **Michelson interferometer** (Fig. 24–37),[†] invented by the American Albert A. Michelson (Section 22–4). Monochromatic light from a single point on an extended source is shown striking a half-silvered mirror M$_S$. This **beam splitter** mirror M$_S$ has a thin layer of silver that reflects only half the light that hits it, so that half of the beam passes through to a fixed mirror M$_2$, where it is reflected back. The other half is reflected by M$_S$ to a mirror M$_1$ that is movable (by a fine-thread screw), where it is also reflected back. Upon its return, part of beam 1 passes through M$_S$ and reaches a sensor or the eye; and part of beam 2, on its return, is reflected by M$_S$ into the eye. If the two path lengths are identical, the two coherent beams entering the eye constructively interfere and brightness will be seen. If the movable mirror is moved a distance $\lambda/4$, one beam will travel an extra distance equal to $\lambda/2$ (because it travels back and forth over the distance $\lambda/4$). In this case, the two beams will destructively interfere and darkness will be seen. As M$_1$ is moved farther, brightness will recur (when the path difference is λ), then darkness, and so on.

Very precise length measurements can be made with an interferometer. The motion of mirror M$_1$ by only $\frac{1}{4}\lambda$ produces a clear difference between brightness and darkness. For $\lambda = 400$ nm, this means a precision of 100 nm, or 10^{-4} mm! If mirror M$_1$ is tilted very slightly, the bright or dark spots are seen instead as a series of bright and dark lines or "fringes" that move as M$_1$ moves. By counting the number of fringes (or fractions thereof) that pass a reference line, extremely precise length measurements can be made.

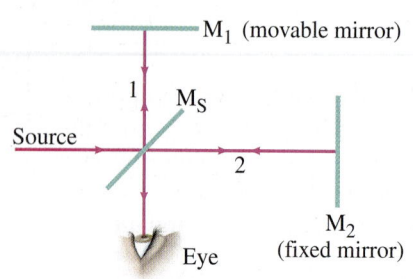

FIGURE 24–37 Michelson interferometer.

[†]There are other types of interferometer, but Michelson's is the best known.

24–10 Polarization

An important and useful property of light is that it can be *polarized*. To see what this means, let us examine waves traveling on a rope. A rope can oscillate in a vertical plane, Fig. 24–38a, or in a horizontal plane, Fig. 24–38b. In either case, the wave is said to be **linearly polarized** or **plane-polarized**—*the oscillations are in a plane*.

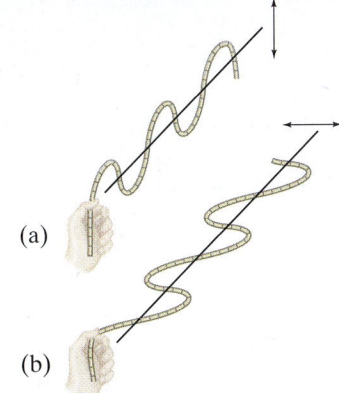

FIGURE 24–38 Transverse waves on a rope polarized (a) in a vertical plane and (b) in a horizontal plane.

If we now place an obstacle containing a vertical slit in the path of the wave, Fig. 24–39, a vertically polarized wave passes through the vertical slit, but a horizontally polarized wave will not. If a horizontal slit were used, the vertically polarized wave would be stopped. If both types of slit were used, both types of wave would be stopped by one slit or the other. Note that polarization can exist *only* for *transverse waves*, and not for longitudinal waves such as sound. The latter oscillate only along the direction of motion, and neither orientation of slit would stop them.

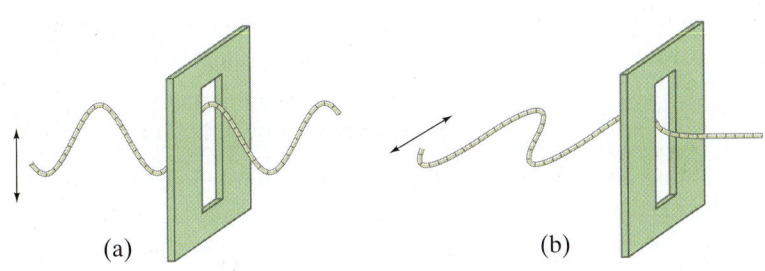

FIGURE 24–39 (a) A vertically polarized wave passes through a vertical slit, but (b) a horizontally polarized wave will not.

Maxwell's theory of light as electromagnetic (EM) waves predicted that light can be polarized since an EM wave is a transverse wave. The direction of polarization in a plane-polarized EM wave is taken as the direction of the electric field vector $\vec{E}$.

Light is not necessarily polarized. It can also be **unpolarized**, which means that the source has oscillations in many planes at once, as shown in Fig. 24–40. Ordinary lightbulbs emit unpolarized light, as does the Sun.

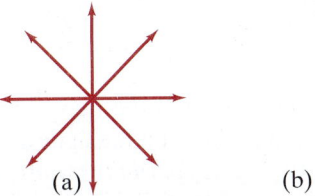

FIGURE 24–40 (a) Oscillation of the electric field vectors in unpolarized light. The light is traveling into or out of the page. (b) Electric field in linear polarized light.

Polaroids (Polarization by Absorption)

Plane-polarized light can be obtained from unpolarized light using certain crystals such as tourmaline. Or, more commonly, we use a **Polaroid sheet**. (Polaroid materials were invented in 1929 by Edwin Land.) A Polaroid sheet consists of long complex molecules arranged parallel to one another. Such a Polaroid acts like a series of parallel slits to allow one orientation of polarization to pass through nearly undiminished. This direction is called the *transmission axis* of the Polaroid. Polarization perpendicular to this direction is absorbed almost completely by the Polaroid.

Absorption by a Polaroid can be explained at the molecular level. An electric field $\vec{E}$ that oscillates parallel to the long molecules can set electrons into motion along the molecules, thus doing work on them and transferring energy. Hence, if $\vec{E}$ is parallel to the molecules, it gets absorbed. An electric field $\vec{E}$ perpendicular to the long molecules does not have this possibility of doing work and transferring its energy, and so passes through freely. When we speak of the *transmission axis* of a Polaroid, we mean the direction for which $\vec{E}$ is passed, so a Polaroid axis is *perpendicular* to the long molecules. [If we want to think of there being slits between the parallel molecules in the sense of Fig. 24–39, then Fig. 24–39 would apply for the $\vec{B}$ field in the EM wave, not the $\vec{E}$ field.]

FIGURE 24–41 Vertical Polaroid transmits only the vertical component of a wave (electric field) incident upon it.

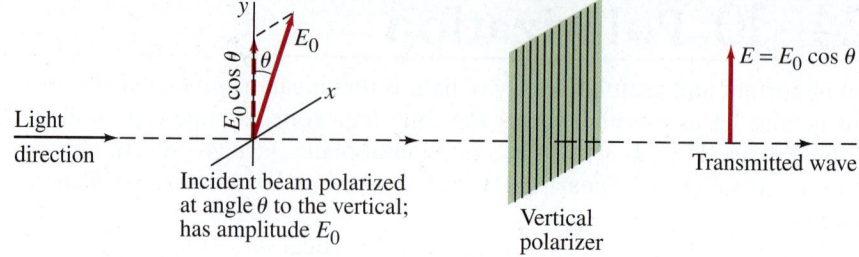

If a beam of plane-polarized light strikes a Polaroid whose transmission axis is at an angle θ to the incident polarization direction, the beam will emerge plane-polarized parallel to the Polaroid transmission axis, and the amplitude of E will be reduced to $E \cos \theta$, Fig. 24–41. Thus, a Polaroid passes only that component of polarization (the electric field vector, $\vec{E}$) that is parallel to its transmission axis. Because the intensity of a light beam is proportional to the square of the amplitude (Sections 11–9 and 22–5), the intensity of a plane-polarized beam transmitted by a polarizer is proportional to $(E_0 \cos \theta)^2$, a relation called Malus' law,

$$I = I_0 \cos^2 \theta, \qquad \begin{bmatrix} \text{intensity of plane-polarized} \\ \text{wave passed by polarizer} \end{bmatrix} \quad (24\text{–}5)$$

where I_0 is the incoming intensity and θ is the angle between the polarizer transmission axis and the plane of polarization of the incoming wave.

A Polaroid can be used as a **polarizer** to *produce* plane-polarized light from unpolarized light, since only the component of light parallel to the axis is transmitted. A Polaroid can also be used as an **analyzer** to determine (1) if light is polarized and (2) the plane of polarization. A Polaroid acting as an analyzer will pass the same amount of light independent of the orientation of its axis if the light is unpolarized; try rotating one lens of a pair of Polaroid sunglasses while looking through it at a lightbulb. If the light is polarized, however, when you rotate the Polaroid the transmitted light will be a maximum when the plane of polarization is parallel to the Polaroid's transmission axis, and a minimum when perpendicular to it. If you do this while looking at the sky, preferably at right angles to the Sun's direction, you will see that skylight is polarized. (Direct sunlight is unpolarized, but don't look directly at the Sun, even through a polarizer, for damage to the eye may occur.) If the light transmitted by an analyzer Polaroid falls to zero at one orientation, then the light is 100% plane-polarized. If it merely reaches a minimum, the light is *partially polarized*.

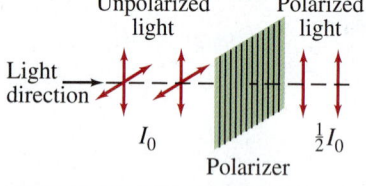

FIGURE 24–42 Unpolarized light has equal intensity vertical and horizontal components. After passing through a polarizer, one of these components is eliminated. The intensity of the light is reduced to half.

Unpolarized light consists of light with random directions of polarization. Each of these polarization directions can be resolved into components along two mutually perpendicular directions. On average, an unpolarized beam can be thought of as two plane-polarized beams of equal magnitude perpendicular to one another. When unpolarized light passes through a polarizer, one component is eliminated. So the intensity of the light passing through is reduced by half because half the light is eliminated: $I = \frac{1}{2} I_0$ (Fig. 24–42).

When two Polaroids are *crossed*—that is, their polarizing axes are perpendicular to one another—unpolarized light can be entirely stopped. As shown in Fig. 24–43, unpolarized light is made plane-polarized by the first Polaroid (the polarizer). The second Polaroid, the analyzer, then eliminates this component since its transmission axis is perpendicular to the first.

FIGURE 24–43 Crossed Polaroids completely eliminate light.

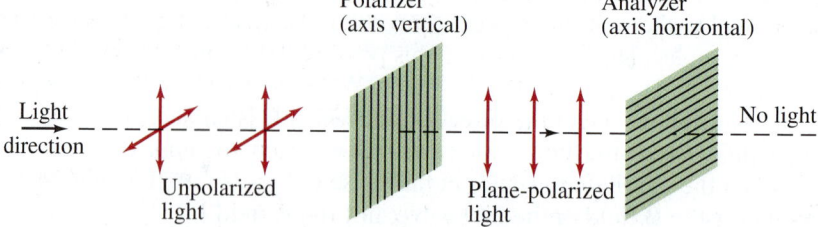

You can try this with Polaroid sunglasses (Fig. 24–44). Note that Polaroid sunglasses eliminate 50% of unpolarized light because of their polarizing property; they absorb even more because they are colored. Plane-polarized light in any direction is also stopped by crossed Polaroids.

EXAMPLE 24–13 **Two Polaroids at 60°.** Unpolarized light passes through two Polaroids; the axis of the first is vertical and that of the second is at 60° to the vertical. Describe the orientation and intensity of the transmitted light.

APPROACH Half of the unpolarized light is absorbed by the first Polaroid, and the remaining light emerges plane-polarized vertically. When that light passes through the second Polaroid, the intensity is further reduced according to Eq. 24–5, and the plane of polarization is then along the axis of the second Polaroid.

SOLUTION The first Polaroid eliminates half the light, so the intensity is reduced by half: $I_1 = \frac{1}{2}I_0$. The light reaching the second polarizer is vertically polarized and so is reduced in intensity (Eq. 24–5) to

$$I_2 = I_1(\cos 60°)^2 = \tfrac{1}{4}I_1.$$

Thus, $I_2 = \frac{1}{8}I_0$. The transmitted light has an intensity one-eighth that of the original and is plane-polarized at a 60° angle to the vertical.

FIGURE 24–44 Crossed Polaroids. When the two polarized sunglass lenses overlap, with axes perpendicular, almost no light passes through.

CONCEPTUAL EXAMPLE 24–14 **Three Polaroids.** We saw in Fig. 24–43 that when unpolarized light falls on two crossed Polaroids (axes at 90°), no light passes through. What happens if a third Polaroid, with axis at 45° to each of the other two, is placed between them (Fig. 24–45a)?

RESPONSE We start just as in Example 24–13 and recall again that light emerging from each Polaroid is polarized parallel to that Polaroid's axis. Thus the angle in Eq. 24–5 is that between the transmission axes of each pair of Polaroids taken in turn. The first Polaroid changes the unpolarized light to plane-polarized and reduces the intensity from I_0 to $I_1 = \frac{1}{2}I_0$. The second polarizer further reduces the intensity by $(\cos 45°)^2$, Eq. 24–5:

$$I_2 = I_1(\cos 45°)^2 = \tfrac{1}{2}I_1 = \tfrac{1}{4}I_0.$$

The light leaving the second polarizer is plane-polarized at 45° (Fig. 24–45b) relative to the third polarizer, so the third one reduces the intensity to

$$I_3 = I_2(\cos 45°)^2 = \tfrac{1}{2}I_2,$$

or $I_3 = \frac{1}{8}I_0$. Thus $\frac{1}{8}$ of the original intensity gets transmitted.

NOTE If we don't insert the 45° Polaroid, zero intensity results (Fig. 24–43).

EXERCISE F How much light would pass through if the 45° polarizer in Example 24–14 was placed not between the other two polarizers but (a) before the vertical (first) polarizer, or (b) after the horizontal polarizer?

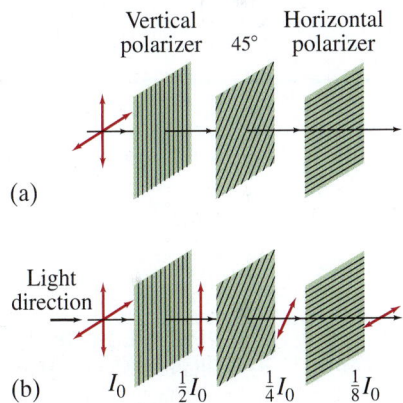

FIGURE 24–45 Example 24–14.

Polarization by Reflection

Another means of producing polarized light from unpolarized light is by reflection. When light strikes a nonmetallic surface at any angle other than perpendicular, the reflected beam is polarized preferentially in the plane parallel to the surface, Fig. 24–46. In other words, the component with polarization in the plane perpendicular to the surface is preferentially transmitted or absorbed. You can check this by rotating Polaroid sunglasses while looking through them at a flat surface of a lake or road. Since most outdoor surfaces are horizontal, Polaroid sunglasses are made with their axes vertical to eliminate the more strongly reflected horizontal component, and thus reduce glare.

FIGURE 24–46 Light reflected from a nonmetallic surface, such as the smooth surface of water in a lake, is partially polarized parallel to the surface.

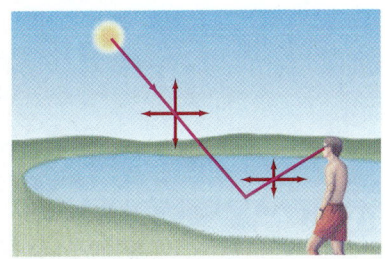

FIGURE 24–47 Photographs of a lake, (a) allowing all light into the camera lens, and (b) using a polarizer. The polarizer is adjusted to absorb most of the (polarized) light reflected from the water's surface, allowing the dimmer light from the bottom of the lake, and any fish lying there, to be seen more readily.

(a) (b)

People who go fishing wear Polaroids to eliminate reflected glare from the surface of a lake or stream and thus see beneath the water more clearly (Fig. 24–47).

The amount of polarization in the reflected beam depends on the angle, varying from no polarization at normal incidence to 100% polarization at an angle known as the **polarizing angle** θ_p.[†] This angle is related to the index of refraction of the two materials on either side of the boundary by the equation

$$\tan \theta_p = \frac{n_2}{n_1}, \qquad (24\text{-}6a)$$

where n_1 is the index of refraction of the material in which the incident beam is traveling, and n_2 is that of the medium beyond the reflecting boundary. If the beam is traveling in air, $n_1 = 1$, and Eq. 24–6a becomes

$$\tan \theta_p = n. \qquad (24\text{-}6b)$$

The polarizing angle θ_p is also called **Brewster's angle**, and Eqs. 24–6 *Brewster's law*, after the Scottish physicist David Brewster (1781–1868), who worked it out experimentally in 1812. Equations 24–6 can be derived from the electromagnetic wave theory of light. It is interesting that at Brewster's angle, the reflected ray and the transmitted (refracted) ray make a 90° angle to each other; that is, $\theta_p + \theta_r = 90°$, where θ_r is the refraction angle (Fig. 24–48). This can be seen

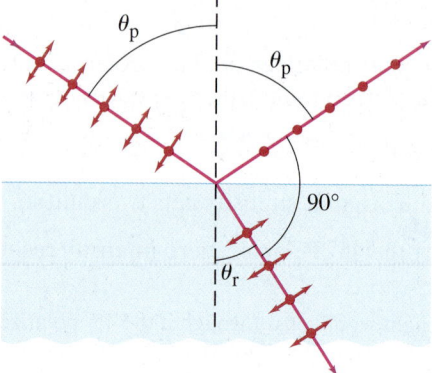

FIGURE 24–48 At θ_p the reflected light is plane-polarized parallel to the surface, and $\theta_p + \theta_r = 90°$, where θ_r is the refraction angle. (The large dots represent vibrations perpendicular to the page.)

by substituting Eq. 24–6a, $n_2 = n_1 \tan \theta_p = n_1 \sin \theta_p / \cos \theta_p$, into Snell's law, $n_1 \sin \theta_p = n_2 \sin \theta_r$, which gives $\cos \theta_p = \sin \theta_r$ which can only hold if $\theta_p = 90° - \theta_r$ (see Trigonometric identities inside back cover or Appendix page A-8).

EXAMPLE 24–15 **Polarizing angle.** (a) At what incident angle is sunlight reflected from a lake perfectly plane-polarized? (b) What is the refraction angle?

APPROACH The polarizing angle at the surface is Brewster's angle, Eq. 24–6b. We find the angle of refraction from Snell's law.

SOLUTION (a) We use Eq. 24–6b with $n = 1.33$, so $\tan \theta_p = 1.33$ giving $\theta_p = 53.1°$.
(b) From Snell's law, $\sin \theta_r = \sin \theta_p / n = \sin 53.1° / 1.33 = 0.601$ giving $\theta_r = 36.9°$.

NOTE $\theta_p + \theta_r = 53.1° + 36.9° = 90.0°$, as expected.

[†]Only a fraction of the incident light is reflected at the surface of a transparent medium. Although this reflected light is 100% polarized (if $\theta = \theta_p$), the remainder of the light, which is transmitted into the new medium, is only partially polarized.

*24–11 Liquid Crystal Displays (LCD)

A wonderful use of polarization is in a **liquid crystal display** (LCD). LCDs are used as the display in cell phones, other hand-held electronic devices, and flat-panel computer and television screens.

A liquid crystal display is made up of many tiny rectangles called **pixels**, or "picture elements." The picture you see depends on which pixels are dark or light and of what color, as suggested in Fig. 24–49 for a simple black and white picture.

Liquid crystals are organic materials that at room temperature exist in a phase that is neither fully solid nor fully liquid. They are sort of gooey, and their molecules display a randomness of position characteristic of liquids, as discussed in Section 13–1 and Fig. 13–2b. They also show some of the orderliness of a solid crystal (Fig. 13–2a), but only in one dimension.

The liquid crystals we find useful are made up of relatively rigid rod-like molecules that interact weakly with each other and tend to align parallel to each other, as shown in Fig. 24–50.

In a simple LCD, each pixel (picture element) contains liquid crystal material sandwiched between two glass plates whose inner surfaces have been brushed to form nanometer-wide parallel scratches. The rod-like liquid crystal molecules in contact with the scratches tend to line up along the scratches. The two plates typically have their scratches at 90° to each other, and the weak electric forces between the rod-like molecules tend to keep them nearly aligned with their nearest neighbors, resulting in the twisted pattern shown in Fig. 24–51a.

The outer surfaces of the glass plates each have a thin film polarizer, they too oriented at 90° to each other. Unpolarized light incident from the left becomes plane-polarized, and the liquid crystal molecules keep this polarization aligned with their rod-like shape. That is, the plane of polarization of the light rotates with the molecules as the light passes through the liquid crystal. The light emerges with its plane of polarization rotated by 90°, and readily passes through the second polarizer, Fig. 24–51a. A tiny LCD pixel in this situation will appear bright.

FIGURE 24–49 Example of an image made up of many small squares or *pixels* (picture elements).

FIGURE 24–50 Liquid crystal molecules tend to align in one dimension (parallel to each other) but have random positions (left-right, up-down).

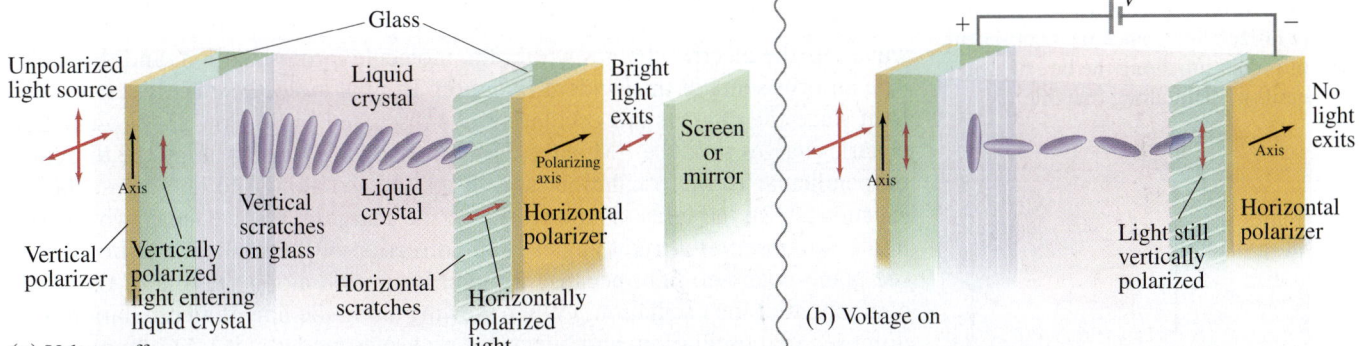

(a) Voltage off

(b) Voltage on

Now suppose a voltage is applied to transparent electrodes on each glass plate of the pixel. The rod-like molecules are polar (or can acquire an internal separation of charge due to the applied electric field). The applied voltage tends to align the molecules end-to-end, and they no longer follow the careful twisted pattern shown in Fig. 24–51a. Instead the applied electric field tends to align the molecules end-to-end, left to right (perpendicular to the glass plates), Fig. 24–51b, and then they don't affect the light polarization significantly. The entering plane-polarized light no longer has its plane of polarization rotated as it passes through the liquid crystal, and no light can exit through the second (horizontal) polarizer (Fig. 24–51b). With the voltage on, the pixel appears dark.[†]

FIGURE 24–51 (a) "Twisted" form of liquid crystal. Light polarization plane is rotated 90°, and so is transmitted by the horizontal polarizer. Only one line of molecules is shown. (b) Molecules disoriented by electric field. The plane of polarization is not changed, so light does not pass through the horizontal polarizer. (The transparent electrodes are not shown.)

[†]Some displays use an opposite system: the polarizers are parallel to each other (the scratches remain at 90° to maintain the twist). Then voltage *off* results in *black* (no light), and voltage *on* results in bright light.

FIGURE 24–52 Watch-face LCD display with altimeter. The black segments or pixels have a voltage applied to them. Note that the 8 uses all seven segments (pixels); other numbers use fewer.

FIGURE 24–53 Arrangement of subpixels on a TV or computer display (enlarged).

Simple display screens, such as for watches and calculators, use ambient light as the source (you can't see the display in the dark), with a mirror behind the LCD to reflect the light back. There are only a few pixels, corresponding to the elongated segments needed to form the numbers from 0 to 9 (and letters in some displays), as seen in Fig. 24–52. Any pixels to which a voltage is applied appear dark and form part of a number. With no voltage, pixels pass light through the polarizers to the mirror and back out, which forms a bright background. Displays with white numbers on a dark background have the voltages reversed.

Television, cell phone, and computer LCDs are more sophisticated. A color pixel consists of three cells, or subpixels, each covered with a red, green, or blue filter (Fig. 24–53). Varying brightnesses of these three primary colors can yield almost any natural color. A good-quality screen consists of millions of pixels. Behind this array of pixels is a light source, often thin fluorescent tubes the diameter of a straw, or light-emitting diodes (LEDs). The light passes through the liquid crystal subpixels, or not, depending on the voltage applied to each, as we discussed in detail in Section 17–11. See especially Figs. 17–31 and 17–33.

[To obtain a range of gray scale or range of color brightness, each subpixel cannot simply go on or off as in Fig. 24–51. Several techniques can be used depending on the construction of the LCD. If the voltage applied in Fig. 24–51b is small enough, the disorientation of the molecules may be small, allowing some rotation of the polarization vector and thus some light can pass through, the actual amount depending on the voltage. Alternatively, each subpixel can be pulsed—the length of time it is *on* affects the perceived brightness. The effect of stronger or weaker brightness can instead be provided by the number of nearby subpixels of the same color that are turned on or off; this third system lets the eye "average" over many pixels, but reduces the sharpness or resolution of the picture.]

*24–12 Scattering of Light by the Atmosphere

Sunsets are red, the sky is blue, and skylight is polarized (at least partially). These phenomena can be explained on the basis of the *scattering* of light by the molecules of the atmosphere. In Fig. 24–54 we see unpolarized light from the Sun impinging on a molecule of the Earth's atmosphere. The electric field of the EM wave sets the electric charges within the molecule into oscillation, and the molecule absorbs some of the incident radiation. But the molecule quickly reemits this light since the charges are oscillating. As discussed in Section 22–2, oscillating electric charges produce EM waves. The intensity is strongest along the direction perpendicular to the oscillation, and drops to zero along the line of oscillation (Section 22–2). In Fig. 24–54 the motion of the charges is resolved into two components. An observer at right angles to the direction of the sunlight, as shown, will see plane-polarized light because no light is emitted along the line of the other component of the oscillation. (When viewing along the line of an oscillation, you don't see that oscillation, and hence see no waves made by it.) At other viewing angles, both components will be present; one will be stronger, however, so the light appears partially polarized. Thus, the process of scattering explains the polarization of skylight.

Scattering of light by the Earth's atmosphere depends on wavelength λ. For particles much smaller than the wavelength of light (such as molecules of air), the particles will be less of an obstruction to long wavelengths than to short ones. The scattering decreases, in fact, as $1/\lambda^4$. Blue and violet light are thus scattered much more than red and orange, which is why the sky looks blue. At sunset, the Sun's rays pass through a maximum length of atmosphere. Much of the blue has been taken out by scattering. The light that reaches us at this low angle where the Sun is near the horizon, and reflects off clouds and haze, is thus lacking in blue. That is why sunsets appear reddish.

The dependence of scattering on $1/\lambda^4$ is valid only if the scattering objects are much smaller than the wavelength of the light. This is valid for oxygen and nitrogen molecules whose diameters are about 0.2 nm. Clouds, however, contain water droplets or crystals that are much larger than λ. They scatter all frequencies of light nearly uniformly. Hence clouds appear white (or gray, if shadowed).

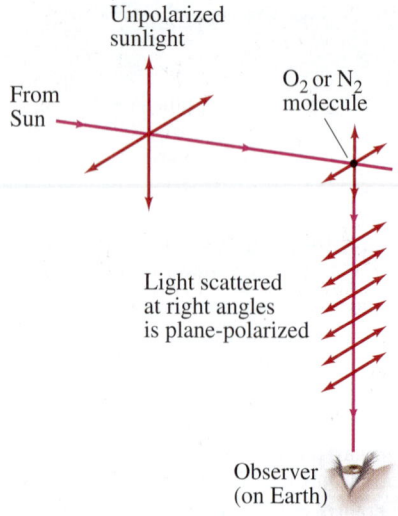

FIGURE 24–54 Unpolarized sunlight scattered by molecules of the air. An observer at right angles sees plane-polarized light, since the component of oscillation along the line of sight emits no light along that line.

PHYSICS APPLIED
Why the sky is blue
Why sunsets are red
Why clouds are white

Summary

The wave theory of light is strongly supported by observations that light exhibits **interference** and **diffraction**. Wave theory also explains the refraction of light and the fact that light travels more slowly in transparent solids and liquids than it does in air.

[*An aid to predicting wave behavior is **Huygens' principle**, which states that every point on a wave front can be considered as a source of tiny wavelets that spread out in the forward direction at the speed of the wave itself. The new wave front is the envelope (the common tangent) of all the wavelets.]

The wavelength of light in a medium with index of refraction n is

$$\lambda_n = \frac{\lambda}{n}, \qquad (24\text{-}1)$$

where λ is the wavelength in vacuum; the frequency is not changed.

Young's double-slit experiment demonstrated the interference of light. The observed bright spots of the interference pattern are explained as constructive interference between the beams coming through the two slits, where the beams differ in path length by an integral number of wavelengths. The dark areas in between are due to destructive interference when the path lengths differ by $\frac{1}{2}\lambda$, $\frac{3}{2}\lambda$, and so on. The angles θ at which **constructive interference** occurs are given by

$$\sin\theta = m\frac{\lambda}{d}, \qquad (24\text{-}2a)$$

where λ is the wavelength of the light, d is the separation of the slits, and m is an integer (0, 1, 2, ···). **Destructive interference** occurs at angles θ given by

$$\sin\theta = \left(m + \tfrac{1}{2}\right)\frac{\lambda}{d}, \qquad (24\text{-}2b)$$

where m is an integer (0, 1, 2, ···).

Two sources of light are perfectly **coherent** if the waves leaving them are of the same single frequency and maintain the same phase relationship at all times. If the light waves from the two sources have a random phase with respect to each other over time (as for two lightbulbs), the two sources are **incoherent**.

The frequency or wavelength of light determines its color. The **visible spectrum** in air extends from about 400 nm (violet) to about 750 nm (red).

Glass prisms spread white light into its constituent colors because the index of refraction varies with wavelength, a phenomenon known as **dispersion**.

The formula $\sin\theta = m\lambda/d$ for constructive interference also holds for a **diffraction grating**, which consists of many parallel slits or lines, separated from each other by a distance d.

The peaks of constructive interference are much brighter and sharper for a diffraction grating than for a two-slit apparatus.

A diffraction grating (or a prism) is used in a **spectrometer** to separate different colors and observe **line spectra**. For a given order m, θ depends on λ. Precise determination of wavelength can be done with a spectrometer by careful measurement of θ.

Diffraction refers to the fact that light, like other waves, bends around objects it passes, and spreads out after passing through narrow slits. This bending gives rise to a **diffraction pattern** due to interference between rays of light that travel different distances.

Light passing through a very narrow slit of width D (on the order of the wavelength λ) will produce a pattern with a bright central maximum of half-width θ given by

$$\sin\theta = \frac{\lambda}{D}, \qquad (24\text{-}3a)$$

flanked by fainter lines to either side.

Light reflected from the front and rear surfaces of a thin film of transparent material can interfere constructively or destructively, depending on the path difference. A phase change of 180° or $\frac{1}{2}\lambda$ occurs when the light reflects at a surface where the index of refraction increases. Such **thin-film interference** has many practical applications, such as lens coatings and using Newton's rings to check uniformity of glass surfaces.

In **unpolarized light**, the electric field vectors oscillate in all transverse directions. If the electric vector oscillates only in one plane, the light is said to be **plane-polarized**. Light can also be partially polarized.

When an unpolarized light beam passes through a **Polaroid** sheet, the emerging beam is plane-polarized. When a light beam is polarized and passes through a Polaroid, the intensity varies as the Polaroid is rotated. Thus a Polaroid can act as a **polarizer** or as an **analyzer**.

The intensity I_0 of a plane-polarized light beam incident on a Polaroid is reduced to

$$I = I_0 \cos^2\theta \qquad (24\text{-}5)$$

where θ is the angle between the axis of the Polaroid and the initial plane of polarization.

Light can also be partially or fully **polarized by reflection**. If light traveling in air is reflected from a medium of index of refraction n, the reflected beam will be *completely* plane-polarized if the incident angle θ_p is given by

$$\tan\theta_p = n. \qquad (24\text{-}6b)$$

The fact that light can be polarized shows that it must be a transverse wave.

Questions

1. Does Huygens' principle apply to sound waves? To water waves? Explain how Huygens' principle makes sense for water waves, where each point vibrates up and down.
2. Why is light sometimes described as rays and sometimes as waves?
3. We can hear sounds around corners but we cannot see around corners; yet both sound and light are waves. Explain the difference.
4. Two rays of light from the same source destructively interfere if their path lengths differ by how much?
5. Monochromatic red light is incident on a double slit, and the interference pattern is viewed on a screen some distance away. Explain how the fringe pattern would change if the red light source is replaced by a blue light source.
6. If Young's double-slit experiment were submerged in water, how would the fringe pattern be changed?

7. Why doesn't the light from the two headlights of a distant car produce an interference pattern?

8. Why are interference fringes noticeable only for a *thin* film like a soap bubble and not for a thick piece of glass?

9. Why are the fringes of Newton's rings (Fig. 24–31) closer together as you look farther from the center?

10. Some coated lenses appear greenish yellow when seen by reflected light. What reflected wavelengths do you suppose the coating is designed to eliminate completely?

11. A drop of oil on a pond appears bright at its edges, where its thickness is much less than the wavelengths of visible light. What can you say about the index of refraction of the oil compared to that of water?

12. Radio waves and visible light are both electromagnetic waves. Why can a radio receive a signal behind a hill when we cannot see the transmitting antenna?

13. Hold one hand close to your eye and focus on a distant light source through a narrow slit between two fingers. (Adjust your fingers to obtain the best pattern.) Describe the pattern that you see.

14. For diffraction by a single slit, what is the effect of increasing (*a*) the slit width, (*b*) the wavelength?

15. Describe the single-slit diffraction pattern produced when white light falls on a slit having a width of (*a*) 60 nm, (*b*) 60,000 nm.

16. What happens to the diffraction pattern of a single slit if the whole apparatus is immersed in (*a*) water, (*b*) a vacuum, instead of in air.

17. What is the difference in the interference patterns formed by two slits 10^{-4} cm apart as compared to a diffraction grating containing 10^4 slits/cm?

18. For a diffraction grating, what is the advantage of (*a*) many slits, (*b*) closely spaced slits?

19. White light strikes (*a*) a diffraction grating and (*b*) a prism. A rainbow appears on a wall just below the direction of the horizontal incident beam in each case. What is the color of the top of the rainbow in each case? Explain.

20. What does polarization tell us about the nature of light?

21. Explain the advantage of polarized sunglasses over plain tinted sunglasses.

22. How can you tell if a pair of sunglasses is polarizing or not?

*23. What would be the color of the sky if the Earth had no atmosphere?

*24. If the Earth's atmosphere were 50 times denser than it is, would sunlight still be white, or would it be some other color?

MisConceptual Questions

1. Light passing through a double-slit arrangement is viewed on a distant screen. The interference pattern observed on the screen would have the widest spaced fringes for the case of
   (*a*) red light and a small slit spacing.
   (*b*) blue light and a small slit spacing.
   (*c*) red light and a large slit spacing.
   (*d*) blue light and a large slit spacing.

2. Light from a green laser of wavelength 530 nm passes through two slits that are 400 nm apart. The resulting pattern formed on a screen in front of the slits is shown in Fig. 24–55. If point A is the same distance from both slits, how much closer is point B to one slit than to the other?
   (*a*) 530 nm.
   (*b*) 265 nm.
   (*c*) 400 nm.
   (*d*) 0 nm.
   (*e*) It depends on the distance to the screen.

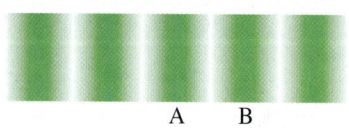

FIGURE 24–55 MisConceptual Question 2.

3. The colors in a rainbow are caused by
   (*a*) the interaction of the light reflected from different raindrops.
   (*b*) different amounts of absorption for light of different colors by the water in the raindrops.
   (*c*) different amounts of refraction for light of different colors by the water in the raindrops.
   (*d*) the downward motion of the raindrops.

4. A double-slit experiment yields an interference pattern due to the path length difference from light traveling through one slit versus the other. Why does a single slit show a diffraction pattern?
   (*a*) There is a path length difference from waves originating at different parts of the slit.
   (*b*) The wavelength of the light is shorter than the slit.
   (*c*) The light passing through the slit interferes with light that does not pass through.
   (*d*) The single slit must have something in the middle of it, causing it to act like a double slit.

5. If you hold two fingers very close together and look at a bright light, you see lines between the fingers. What is happening?
   (*a*) You are holding your fingers too close to your eye to be able to focus on it.
   (*b*) You are seeing a diffraction pattern.
   (*c*) This is a quantum-mechanical tunneling effect.
   (*d*) The brightness of the light is overwhelming your eye.

6. Light passes through a slit that is about 5×10^{-3} m high and 5×10^{-7} m wide. The central bright light visible on a distant screen will be
   (*a*) about 5×10^{-3} m high and about 5×10^{-7} m wide.
   (*b*) about 5×10^{-3} m high and wider than 5×10^{-7} m.
   (*c*) about 5×10^{-3} m high and narrower than 5×10^{-7} m.
   (*d*) taller than 5×10^{-3} m high and wider than 5×10^{-7} m.
   (*e*) taller than 5×10^{-3} m high and about 5×10^{-7} m wide.

7. Blue light of wavelength λ passes through a single slit of width d and forms a diffraction pattern on a screen. If we replace the blue light by red light of wavelength 2λ, we can retain the original diffraction pattern if we change the slit width
   (a) to $d/4$.
   (b) to $d/2$.
   (c) not at all.
   (d) to $2d$.
   (e) to $4d$.

8. Imagine holding a circular disk in a beam of monochromatic light (Fig. 24–56). If diffraction occurs at the edge of the disk, the center of the shadow is
   (a) darker than the rest of the shadow.
   (b) a bright spot.
   (c) bright or dark, depending on the wavelength.
   (d) bright or dark, depending on the distance to the screen.

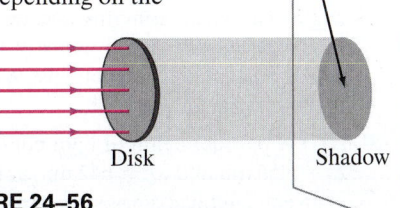

FIGURE 24–56 MisConceptual Question 8.

9. If someone is around a corner from you, what is the main reason you can hear him speaking but can't see him?
   (a) Sound travels farther in air than light does.
   (b) Sound can travel through walls, but light cannot.
   (c) Sound waves have long enough wavelengths to bend around a corner; light wavelengths are too short to bend much.
   (d) Sound waves reflect off walls, but light cannot.

10. When a CD is held at an angle, the reflected light contains many colors. What causes these colors?
    (a) An anti-theft encoding intended to prevent copying of the CD.
    (b) The different colors correspond to different data bits.
    (c) Light reflected from the closely spaced grooves adds constructively for different wavelengths at different angles.
    (d) It is part of the decorative label on the CD.

11. If a thin film has a thickness that is
    (a) $\frac{1}{4}$ of a wavelength, constructive interference will always occur.
    (b) $\frac{1}{4}$ of a wavelength, destructive interference will always occur.
    (c) $\frac{1}{2}$ of a wavelength, constructive interference will always occur.
    (d) $\frac{1}{2}$ of a wavelength, destructive interference will always occur.
    (e) None of the above is always true.

12. If unpolarized light is incident from the left on three polarizers as shown in Fig. 24–57, in which case will some light get through?
    (a) Case 1 only.
    (b) Case 2 only.
    (c) Case 3 only.
    (d) Cases 1 and 3.
    (e) All three cases.

FIGURE 24–57 MisConceptual Question 12.

For assigned homework and other learning materials, go to the MasteringPhysics website.

Problems

24–3 Double-Slit Interference

1. (I) Monochromatic light falling on two slits 0.018 mm apart produces the fifth-order bright fringe at an 8.6° angle. What is the wavelength of the light used?

2. (I) The third-order bright fringe of 610-nm light is observed at an angle of 31° when the light falls on two narrow slits. How far apart are the slits?

3. (II) Monochromatic light falls on two very narrow slits 0.048 mm apart. Successive fringes on a screen 6.50 m away are 8.5 cm apart near the center of the pattern. Determine the wavelength and frequency of the light.

4. (II) If 720-nm and 660-nm light passes through two slits 0.62 mm apart, how far apart are the second-order fringes for these two wavelengths on a screen 1.0 m away?

5. (II) Water waves having parallel crests 4.5 cm apart pass through two openings 7.5 cm apart in a board. At a point 3.0 m beyond the board, at what angle relative to the "straight-through" direction would there be little or no wave action?

6. (II) A red laser from the physics lab is marked as producing 632.8-nm light. When light from this laser falls on two closely spaced slits, an interference pattern formed on a wall several meters away has bright red fringes spaced 5.00 mm apart near the center of the pattern. When the laser is replaced by a small laser pointer, the fringes are 5.14 mm apart. What is the wavelength of light produced by the laser pointer?

7. (II) Light of wavelength 680 nm falls on two slits and produces an interference pattern in which the third-order bright red fringe is 38 mm from the central fringe on a screen 2.8 m away. What is the separation of the two slits?

8. (II) Light of wavelength λ passes through a pair of slits separated by 0.17 mm, forming a double-slit interference pattern on a screen located a distance 37 cm away. Suppose that the image in Fig. 24–9a is an actual-size reproduction of this interference pattern. Use a ruler to measure a pertinent distance on this image; then utilize this measured value to determine λ (nm).

9. (II) A parallel beam of light from a He–Ne laser, with a wavelength 633 nm, falls on two very narrow slits 0.068 mm apart. How far apart are the fringes in the center of the pattern on a screen 3.3 m away?

10. (II) A physics professor wants to perform a lecture demonstration of Young's double-slit experiment for her class using the 633-nm light from a He–Ne laser. Because the lecture hall is very large, the interference pattern will be projected on a wall that is 5.0 m from the slits. For easy viewing by all students in the class, the professor wants the distance between the $m = 0$ and $m = 1$ maxima to be 35 cm. What slit separation is required in order to produce the desired interference pattern?

11. (II) Suppose a thin piece of glass is placed in front of the lower slit in Fig. 24–7 so that the two waves enter the slits 180° out of phase (Fig. 24–58). Draw in detail the interference pattern seen on the screen.

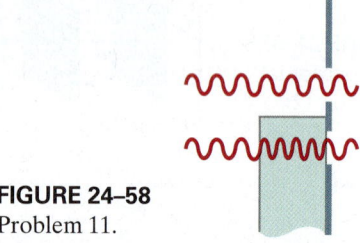

FIGURE 24–58
Problem 11.

12. (II) In a double-slit experiment it is found that blue light of wavelength 480 nm gives a second-order maximum at a certain location on the screen. What wavelength of visible light would have a minimum at the same location?

13. (II) Two narrow slits separated by 1.0 mm are illuminated by 544-nm light. Find the distance between adjacent bright fringes on a screen 4.0 m from the slits.

14. (II) Assume that light of a single color, rather than white light, passes through the two-slit setup described in Example 24–3. If the distance from the central fringe to a first-order fringe is measured to be 2.9 mm on the screen, determine the light's wavelength (in nm) and color (see Fig. 24–11).

15. (II) In a double-slit experiment, the third-order maximum for light of wavelength 480 nm is located 16 mm from the central bright spot on a screen 1.6 m from the slits. Light of wavelength 650 nm is then projected through the same slits. How far from the central bright spot will the second-order maximum of this light be located?

16. (II) Light of wavelength 470 nm in air shines on two slits 6.00×10^{-2} mm apart. The slits are immersed in water, as is a viewing screen 40.0 cm away. How far apart are the fringes on the screen?

17. (III) A very thin sheet of plastic ($n = 1.60$) covers one slit of a double-slit apparatus illuminated by 680-nm light. The center point on the screen, instead of being a maximum, is dark. What is the (minimum) thickness of the plastic?

24–4 Visible Spectrum; Dispersion

18. (I) By what percent is the speed of blue light (450 nm) less than the speed of red light (680 nm), in silicate flint glass (see Fig. 24–14)?

19. (II) A light beam strikes a piece of glass at a 65.00° incident angle. The beam contains two wavelengths, 450.0 nm and 700.0 nm, for which the index of refraction of the glass is 1.4831 and 1.4754, respectively. What is the angle between the two refracted beams?

20. (III) A parallel beam of light containing two wavelengths, $\lambda_1 = 455$ nm and $\lambda_2 = 642$ nm, enters the silicate flint glass of an equilateral prism as shown in Fig. 24–59. At what angles, θ_1 and θ_2, does each beam leave the prism (give angle with normal to the face)? See Fig. 24–14.

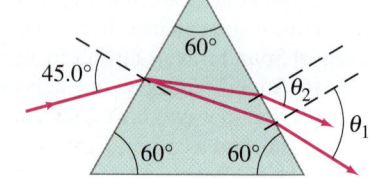

FIGURE 24–59
Problem 20.

24–5 Single-Slit Diffraction

21. (I) If 680-nm light falls on a slit 0.0425 mm wide, what is the angular width of the central diffraction peak?

22. (I) Monochromatic light falls on a slit that is 2.60×10^{-3} mm wide. If the angle between the first dark fringes on either side of the central maximum is 28.0° (dark fringe to dark fringe), what is the wavelength of the light used?

23. (II) When blue light of wavelength 440 nm falls on a single slit, the first dark bands on either side of center are separated by 51.0°. Determine the width of the slit.

24. (II) A single slit 1.0 mm wide is illuminated by 450-nm light. What is the width of the central maximum (in cm) in the diffraction pattern on a screen 6.0 m away?

25. (II) How wide is the central diffraction peak on a screen 2.30 m behind a 0.0348-mm-wide slit illuminated by 558-nm light?

26. (II) Consider microwaves which are incident perpendicular to a metal plate which has a 1.6-cm slit in it. Discuss the angles at which there are diffraction minima for wavelengths of (a) 0.50 cm, (b) 1.0 cm, and (c) 3.0 cm.

27. (II) (a) For a given wavelength λ, what is the minimum slit width for which there will be no diffraction minima? (b) What is the minimum slit width so that no visible light exhibits a diffraction minimum?

28. (II) Light of wavelength 620 nm falls on a slit that is 3.80×10^{-3} mm wide. Estimate how far the first bright diffraction fringe is from the strong central maximum if the screen is 10.0 m away.

29. (II) Monochromatic light of wavelength 633 nm falls on a slit. If the angle between the first two bright fringes on either side of the central maximum is 32°, estimate the slit width.

30. (II) Coherent light from a laser diode is emitted through a rectangular area $3.0\,\mu\text{m} \times 1.5\,\mu\text{m}$ (horizontal-by-vertical). If the laser light has a wavelength of 780 nm, determine the angle between the first diffraction minima (a) above and below the central maximum, (b) to the left and right of the central maximum.

31. (III) If parallel light falls on a single slit of width D at a 28.0° angle to the normal, describe the diffraction pattern.

24–6 and 24–7 Diffraction Gratings

32. (I) At what angle will 510-nm light produce a second-order maximum when falling on a grating whose slits are 1.35×10^{-3} cm apart?

33. (I) A grating that has 3800 slits per cm produces a third-order fringe at a 22.0° angle. What wavelength of light is being used?

34. (I) A grating has 7400 slits/cm. How many spectral orders can be seen (400 to 700 nm) when it is illuminated by white light?

35. (II) Red laser light from a He–Ne laser ($\lambda = 632.8$ nm) creates a second-order fringe at 53.2° after passing through the grating. What is the wavelength λ of light that creates a first-order fringe at 20.6°?

36. (II) How many slits per centimeter does a grating have if the third order occurs at a 15.0° angle for 620-nm light?

37. (II) A source produces first-order lines when incident normally on a 9800-slit/cm diffraction grating at angles 28.8°, 36.7°, 38.6°, and 41.2°. What are the wavelengths?

38. (II) White light containing wavelengths from 410 nm to 750 nm falls on a grating with 7800 slits/cm. How wide is the first-order spectrum on a screen 3.40 m away?

39. (II) A diffraction grating has 6.5×10^5 slits/m. Find the angular spread in the second-order spectrum between red light of wavelength 7.0×10^{-7} m and blue light of wavelength 4.5×10^{-7} m.

40. (II) Two first-order spectrum lines are measured by a 9650-slit/cm spectroscope at angles, on each side of center, of $+26°38'$, $+41°02'$ and $-26°18'$, $-40°27'$. Calculate the wavelengths based on these data.

41. (II) What is the highest spectral order that can be seen if a grating with 6500 slits per cm is illuminated with 633-nm laser light? Assume normal incidence.

42. (II) The first-order line of 589-nm light falling on a diffraction grating is observed at a 14.5° angle. How far apart are the slits? At what angle will the third order be observed?

43. (II) Two (and only two) full spectral orders can be seen on either side of the central maximum when white light is sent through a diffraction grating. What is the maximum number of slits per cm for the grating?

24–8 Thin-Film Interference

44. (I) If a soap bubble is 120 nm thick, what wavelength is most strongly reflected at the center of the outer surface when illuminated normally by white light? Assume that $n = 1.32$.

45. (I) How far apart are the dark bands in Example 24–10 if the glass plates are each 21.5 cm long?

46. (II) (a) What is the smallest thickness of a soap film ($n = 1.33$) that would appear black if illuminated with 480-nm light? Assume there is air on both sides of the soap film. (b) What are two other possible thicknesses for the film to appear black? (c) If the thickness t was much less than λ, why would the film also appear black?

47. (II) A lens appears greenish yellow ($\lambda = 570$ nm is strongest) when white light reflects from it. What minimum thickness of coating ($n = 1.25$) do you think is used on such a glass lens ($n = 1.52$), and why?

48. (II) A thin film of oil ($n_o = 1.50$) with varying thickness floats on water ($n_w = 1.33$). When it is illuminated from above by white light, the reflected colors are as shown in Fig. 24–60. In air, the wavelength of yellow light is 580 nm. (a) Why are there no reflected colors at point A? (b) What is the oil's thickness t at point B?

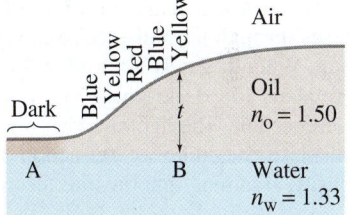

FIGURE 24–60 Problem 48.

49. (II) How many uncoated thin lenses in an optical instrument would reduce the amount of light passing through the instrument to 50% or less? (Assume the same transmission percent at each of the two surfaces—see page 697.)

50. (II) A total of 35 bright and 35 dark Newton's rings (not counting the dark spot at the center) are observed when 560-nm light falls normally on a planoconvex lens resting on a flat glass surface (Fig. 24–31). How much thicker is the lens at the center than the edges?

51. (II) If the wedge between the glass plates of Example 24–10 is filled with some transparent substance other than air—say, water—the pattern shifts because the wavelength of the light changes. In a material where the index of refraction is n, the wavelength is $\lambda_n = \lambda/n$, where λ is the wavelength in vacuum (Eq. 24–1). How many dark bands would there be if the wedge of Example 24–10 were filled with water?

52. (II) A fine metal foil separates one end of two pieces of optically flat glass, as in Fig. 24–33. When light of wavelength 670 nm is incident normally, 24 dark bands are observed (with one at each end). How thick is the foil?

53. (II) How thick (minimum) should the air layer be between two flat glass surfaces if the glass is to appear bright when 450-nm light is incident normally? What if the glass is to appear dark?

54. (III) A thin oil slick $(n_o = 1.50)$ floats on water $(n_w = 1.33)$. When a beam of white light strikes this film at normal incidence from air, the only enhanced reflected colors are red (650 nm) and violet (390 nm). From this information, deduce the (minimum) thickness t of the oil slick.

55. (III) A uniform thin film of alcohol ($n = 1.36$) lies on a flat glass plate ($n = 1.56$). When monochromatic light, whose wavelength can be changed, is incident normally, the reflected light is a minimum for $\lambda = 525$ nm and a maximum for $\lambda = 655$ nm. What is the minimum thickness of the film?

*24–9 Michelson Interferometer

*56. (II) How far must the mirror M_1 in a Michelson interferometer be moved if 680 fringes of 589-nm light are to pass by a reference line?

*57. (II) What is the wavelength of the light entering an interferometer if 362 bright fringes are counted when the movable mirror moves 0.125 mm?

*58. (II) A micrometer is connected to the movable mirror of an interferometer. When the micrometer is tightened down on a thin metal foil, the net number of bright fringes that move, compared to closing the empty micrometer, is 296. What is the thickness of the foil? The wavelength of light used is 589 nm.

*59. (III) One of the beams of an interferometer (Fig. 24–61) passes through a small evacuated glass container 1.155 cm deep. When a gas is allowed to slowly fill the container, a total of 158 dark fringes are counted to move past a reference line. The light used has a wavelength of 632.8 nm. Calculate the index of refraction of the gas at its final density, assuming that the interferometer is in vacuum.

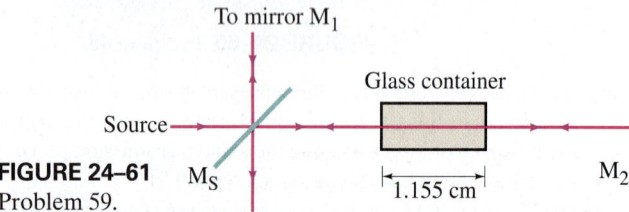

FIGURE 24–61 Problem 59.

24–10 Polarization

60. (I) Two polarizers are oriented at 72° to one another. Unpolarized light falls on them. What fraction of the light intensity is transmitted?

61. (I) What is Brewster's angle for an air–glass ($n = 1.56$) surface?

62. (II) At what angle should the axes of two Polaroids be placed so as to reduce the intensity of the incident unpolarized light to (a) $\frac{1}{3}$, (b) $\frac{1}{10}$?

63. (II) Two polarizers are oriented at 42.0° to one another. Light polarized at a 21.0° angle to each polarizer passes through both. What is the transmitted intensity (%)?

64. (II) Three perfectly polarizing sheets are spaced 2 cm apart and in parallel planes. The transmission axis of the second sheet is 30° relative to the first one. The transmission axis of the third sheet is 90° relative to the *first* one. Unpolarized light impinges on the first polarizing sheet. What percent of this light is transmitted out through the third polarizer?

65. (II) A piece of material, suspected of being a stolen diamond ($n = 2.42$), is submerged in oil of refractive index 1.43 and illuminated by unpolarized light. It is found that the reflected light is completely polarized at an angle of 62°. Is it diamond? Explain.

66. (II) Two Polaroids are aligned so that the initially unpolarized light passing through them is a maximum. At what angle should one of them be placed so the transmitted intensity is subsequently reduced by half?

67. (II) What is Brewster's angle for a diamond submerged in water?

68. (II) The critical angle for total internal reflection at a boundary between two materials is 58°. What is Brewster's angle at this boundary? Give two answers, one for each material.

69. (II) What would Brewster's angle be for reflections off the surface of water for light coming from beneath the surface? Compare to the angle for total internal reflection, and to Brewster's angle from above the surface.

70. (II) Unpolarized light of intensity I_0 passes through six successive Polaroid sheets each of whose axis makes a 35° angle with the previous one. What is the intensity of the transmitted beam?

71. (III) Two polarizers are oriented at 48° to each other and plane-polarized light is incident on them. If only 35% of the light gets through both of them, what was the initial polarization direction of the incident light?

72. (III) Four polarizers are placed in succession with their axes vertical, at 30.0° to the vertical, at 60.0° to the vertical, and at 90.0° to the vertical. (a) Calculate what fraction of the incident unpolarized light is transmitted by the four polarizers. (b) Can the transmitted light be *decreased* by removing one of the polarizers? If so, which one? (c) Can the transmitted light intensity be extinguished by removing polarizers? If so, which one(s)?

General Problems

73. Light of wavelength 5.0×10^{-7} m passes through two parallel slits and falls on a screen 5.0 m away. Adjacent bright bands of the interference pattern are 2.0 cm apart. (a) Find the distance between the slits. (b) The same two slits are next illuminated by light of a different wavelength, and the fifth-order minimum for this light occurs at the same point on the screen as the fourth-order minimum for the previous light. What is the wavelength of the second source of light?

74. Television and radio waves reflecting from mountains or airplanes can interfere with the direct signal from the station. (a) What kind of interference will occur when 75-MHz television signals arrive at a receiver directly from a distant station, and are reflected from a nearby airplane 122 m directly above the receiver? Assume $\frac{1}{2}\lambda$ change in phase of the signal upon reflection. (b) What kind of interference will occur if the plane is 22 m closer to the receiver?

75. Red light from three separate sources passes through a diffraction grating with 3.60×10^5 slits/m. The wavelengths of the three lines are 6.56×10^{-7} m (hydrogen), 6.50×10^{-7} m (neon), and 6.97×10^{-7} m (argon). Calculate the angles for the first-order diffraction line of each source.

76. What is the index of refraction of a clear material if a minimum thickness of 125 nm, when laid on glass, is needed to reduce reflection to nearly zero when light of 675 nm is incident normally upon it? Do you have a choice for an answer?

77. Light of wavelength 650 nm passes through two narrow slits 0.66 mm apart. The screen is 2.40 m away. A second source of unknown wavelength produces its second-order fringe 1.23 mm closer to the central maximum than the 650-nm light. What is the wavelength of the unknown light?

78. Monochromatic light of variable wavelength is incident normally on a thin sheet of plastic film in air. The reflected light is a maximum only for $\lambda = 491.4$ nm and $\lambda = 688.0$ nm in the visible spectrum. What is the thickness of the film ($n = 1.58$)? [Hint: Assume successive values of m.]

79. Show that the second- and third-order spectra of white light produced by a diffraction grating always overlap. What wavelengths overlap?

80. A radio station operating at 90.3 MHz broadcasts from two identical antennas at the same elevation but separated by a 9.0-m horizontal distance d, Fig. 24–62. A maximum signal is found along the midline, perpendicular to d at its midpoint and extending horizontally in both directions. If the midline is taken as 0°, at what other angle(s) θ is a maximum signal detected? A minimum signal? Assume all measurements are made much farther than 9.0 m from the antenna towers.

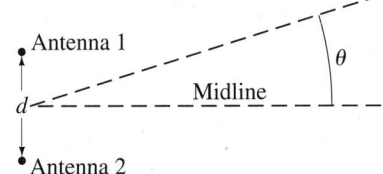

FIGURE 24–62 Problem 80.

81. Calculate the minimum thickness needed for an antireflective coating ($n = 1.38$) applied to a glass lens in order to eliminate (a) blue (450 nm), or (b) red (720 nm) reflections for light at normal incidence.

82. Stealth aircraft are designed to not reflect radar, whose wavelength is typically 2 cm, by using an antireflecting coating. Ignoring any change in wavelength in the coating, estimate its thickness.

83. A laser beam passes through a slit of width 1.0 cm and is pointed at the Moon, which is approximately 380,000 km from the Earth. Assume the laser emits waves of wavelength 633 nm (the red light of a He–Ne laser). Estimate the width of the beam when it reaches the Moon due to diffraction.

84. A thin film of soap ($n = 1.34$) coats a piece of flat glass ($n = 1.52$). How thick is the film if it reflects 643-nm red light most strongly when illuminated normally by white light?

85. When violet light of wavelength 415 nm falls on a single slit, it creates a central diffraction peak that is 8.20 cm wide on a screen that is 3.15 m away. How wide is the slit?

86. A series of polarizers are each rotated 10° from the previous polarizer. Unpolarized light is incident on this series of polarizers. How many polarizers does the light have to go through before it is $\frac{1}{5}$ of its original intensity?

87. The wings of a certain beetle have a series of parallel lines across them. When normally incident 480-nm light is reflected from the wing, the wing appears bright when viewed at an angle of 56°. How far apart are the lines?

88. A teacher stands well back from an outside doorway 0.88 m wide, and blows a whistle of frequency 950 Hz. Ignoring reflections, estimate at what angle(s) it is *not* possible to hear the whistle clearly on the playground outside the doorway. Assume 340 m/s for the speed of sound.

89. Light is incident on a diffraction grating with 7200 slits/cm and the pattern is viewed on a screen located 2.5 m from the grating. The incident light beam consists of two wavelengths, $\lambda_1 = 4.4 \times 10^{-7}$ m and $\lambda_2 = 6.8 \times 10^{-7}$ m. Calculate the linear distance between the first-order bright fringes of these two wavelengths on the screen.

90. How many slits per centimeter must a grating have if there is to be no second-order spectrum for any visible wavelength?

91. When yellow sodium light, $\lambda = 589$ nm, falls on a diffraction grating, its first-order peak on a screen 72.0 cm away falls 3.32 cm from the central peak. Another source produces a line 3.71 cm from the central peak. What is its wavelength? How many slits/cm are on the grating?

92. Two of the lines of the atomic hydrogen spectrum have wavelengths of 656 nm and 410 nm. If these fall at normal incidence on a grating with 7700 slits/cm, what will be the angular separation of the two wavelengths in the first-order spectrum?

93. A tungsten–halogen bulb emits a continuous spectrum of ultraviolet, visible, and infrared light in the wavelength range 360 nm to 2000 nm. Assume that the light from a tungsten–halogen bulb is incident on a diffraction grating with slit spacing d and that the first-order brightness maximum for the wavelength of 1200 nm occurs at angle θ. What other wavelengths within the spectrum of incident light will produce a brightness maximum at this same angle θ? [Optical filters are used to deal with this bothersome effect when a continuous spectrum of light is measured by a spectrometer.]

94. At what angle above the horizon is the Sun when light reflecting off a smooth lake is polarized most strongly?

95. Unpolarized light falls on two polarizer sheets whose axes are at right angles. (a) What fraction of the incident light intensity is transmitted? (b) What fraction is transmitted if a third polarizer is placed between the first two so that its axis makes a 56° angle with the axis of the first polarizer? (c) What if the third polarizer is in front of the other two?

96. At what angle should the axes of two Polaroids be placed so as to reduce the intensity of the incident unpolarized light by an additional factor (after the first Polaroid cuts it in half) of (a) 4, (b) 10, (c) 100?

Search and Learn

1. Compare Figs. 24–5, 24–6, and 24–7, which are different representations of the double-slit experiment. For each figure state the direction the light is traveling. Where are the wave crests in terms of this direction? How are they represented in each figure? Give one advantage of each figure in helping you understand the double-slit experiment and interference.

2. Discuss the similarities, and differences, of double-slit interference and single-slit diffraction.

3. Describe why the various colors of visible light appear as they do in Fig. 24–16, where red is at the top and violet at the bottom, and in Fig. 24–26, where violet is closest to the central maximum and red is farthest from the central maximum.

4. When can we use geometric optics as in Chapter 23, and when do we need to use the more complicated wave model of light discussed in Chapter 24? In particular, what are the physical characteristics that matter in making this decision?

5. A parallel beam of light containing two wavelengths, 420 nm and 650 nm, enters a borate flint glass equilateral prism (Fig. 24–63). (a) What is the angle between the two beams leaving the prism? (b) Repeat part (a) for a diffraction grating with 5800 slits/cm. (c) Discuss two advantages of a diffraction grating, including one that you see from your results.

6. Suppose you viewed the light *transmitted* through a thin coating layered on a flat piece of glass. Draw a diagram, similar to Fig. 24–30 or 24–36, and describe the conditions required for maxima and minima. Consider all possible values of index of refraction. Discuss the relative intensity of the minima compared to the maxima and to zero.

7. What percent of visible light is reflected from plain glass? Assume your answer refers to transmission through each surface, front and back. How does the presence of multiple lenses in a good camera degrade the image? What is suggested in Section 24–8 to reduce this reflection? Explain in words, and sketch how this solution works. For a glass lens in air, about how much improvement does this solution provide?

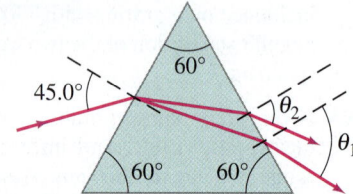

FIGURE 24–63 Search and Learn 5.

ANSWERS TO EXERCISES

A: 333 nm; 6.0×10^{14} Hz; 2.0×10^8 m/s.
B: 2.5 mm.
C: Narrower.
D: A.
E: (e).
F: Zero for both (a) and (b), because the two successive polarizers at 90° cancel all light. The 45° Polaroid must be inserted *between* the other two if transmission is to occur.

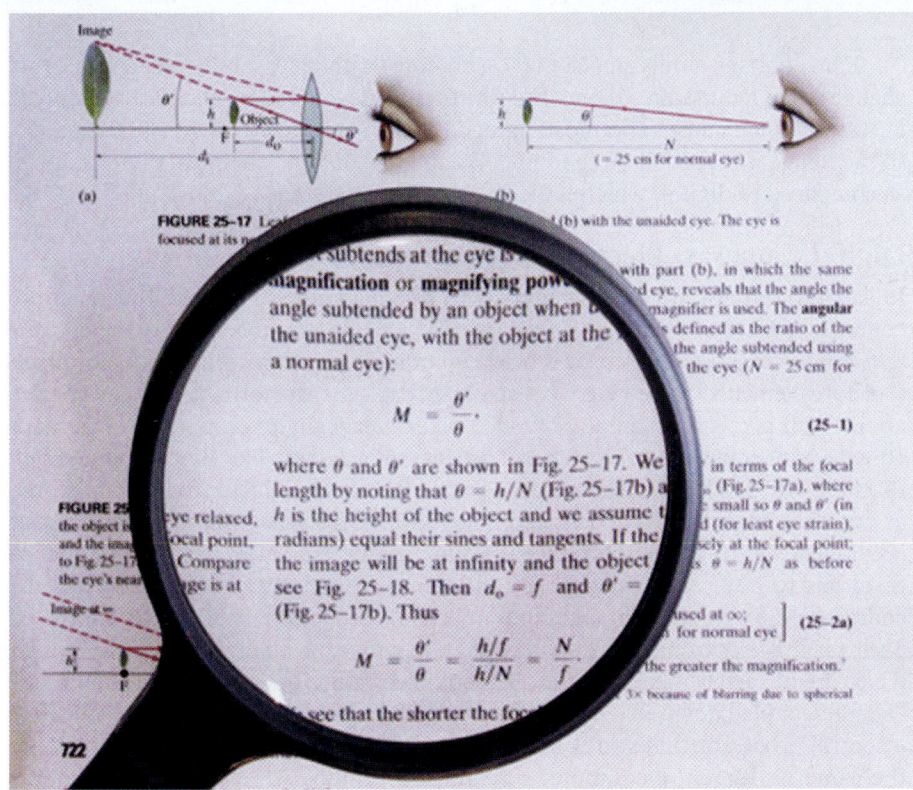

Of the many optical devices we discuss in this Chapter, the magnifying glass is the simplest. Here it is magnifying part of page 722 of this Chapter, which describes how the magnifying glass works according to the ray model. In this Chapter we examine film and digital cameras, the human eye, eyeglasses, telescopes, and microscopes as well as image resolution, X-rays, and CT scans.

Optical Instruments

CHAPTER 25

CHAPTER-OPENING QUESTION—Guess now!

Because of diffraction, a light microscope has a useful magnification of about
(a) 50×; (b) 100×; (c) 500×; (d) 2000×; (e) 5000×;
and the smallest objects it can resolve have a size of about
(a) 10 nm; (b) 100 nm; (c) 500 nm; (d) 2500 nm; (e) 5500 nm.

In our discussion of the behavior of light in the two previous Chapters, we also described a few instruments such as the spectrometer and the Michelson interferometer. In this Chapter, we will discuss some more common instruments, most of which use lenses, including the camera, telescope, microscope, and the human eye. To describe their operation, we will use ray diagrams as we did in Chapter 23. However, we will see that understanding some aspects of their operation will require the wave nature of light.

25–1 Cameras: Film and Digital

The basic elements of a **camera** are a lens, a light-tight box, a shutter to let light pass through the lens only briefly, and in a digital camera an electronic sensor or in a traditional camera a piece of film (Fig. 25–1). When the shutter is opened for a brief "exposure," light from external objects in the field of view is focused by the lens as an image on the sensor or film.

You can see the image yourself if you remove the back of a conventional camera, keeping the shutter open, and view through a piece of tissue paper (on which an image can form) placed where the film should be.

CONTENTS

25–1 Cameras: Film and Digital
25–2 The Human Eye; Corrective Lenses
25–3 Magnifying Glass
25–4 Telescopes
25–5 Compound Microscope
25–6 Aberrations of Lenses and Mirrors
25–7 Limits of Resolution; Circular Apertures
25–8 Resolution of Telescopes and Microscopes; the λ Limit
25–9 Resolution of the Human Eye and Useful Magnification
*25–10 Specialty Microscopes and Contrast
25–11 X-Rays and X-Ray Diffraction
*25–12 X-Ray Imaging and Computed Tomography (CT Scan)

FIGURE 25–1 A simple camera.

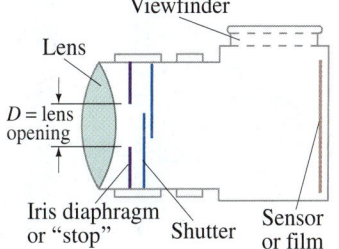

PHYSICS APPLIED
Cameras, film and digital

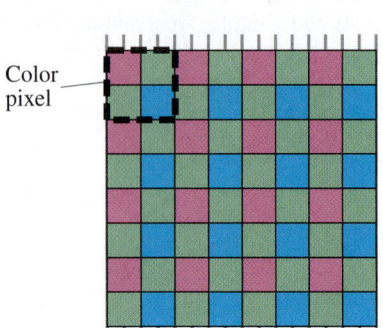

FIGURE 25–2 Portion of a typical Bayer array sensor. A square group of four pixels $^{RG}_{GB}$ is sometimes called a "color pixel."

FIGURE 25–3 A layered or "Foveon" tri-pixel that includes all three colors, arranged vertically so light can pass through all three subpixels.

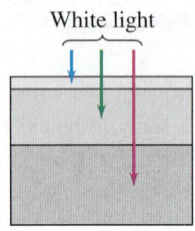

FIGURE 25–4 Suppose we take a picture that includes a thin black line (our object) on a white background. The *image* of this black line has a colored "halo" (red above, blue below) due to the mosaic arrangement of color filter pixels, as shown by the colors transmitted to the image. Computer averaging minimizes such color problems (the green at top and bottom of image may average with nearby pixels to give white or nearly so) but the image is consequently "softened" or blurred.

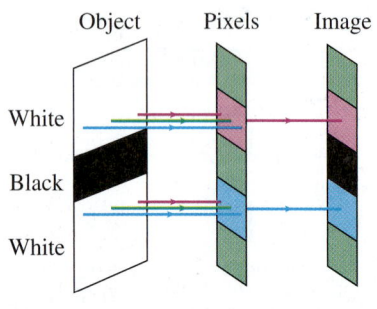

Film contains a thin **emulsion** (= a coating) with light-sensitive chemicals that change when light strikes them. The film is then developed by chemicals dissolved in water, which causes the most changed areas (brightest light) to turn opaque, so the image is recorded on the film.† We might call film "chemical photography" as compared to digital, which is electronic.

Digital Cameras, Electronic Sensors (CCD, CMOS)

In a **digital camera**, the film is replaced by a semiconductor sensor. Two types are common: **CCD** (*charge-coupled device*) and **CMOS** (*complementary metal oxide semiconductor*). A CCD sensor is made up of millions of tiny semiconductor **pixels** ("picture elements")—see Fig. 24–49. A 12-MP (12-megapixel) sensor might contain about 4000 pixels horizontally by 3000 pixels vertically over an area of perhaps 16×12 mm, and preferably larger such as 36×24 mm like 35-mm film. Light reaching any pixel liberates electrons within the semiconductor‡ which are stored as charge in that pixel's capacitance. The more intense the light, the more charge accumulates during the brief exposure time. After exposure, the charge on each pixel has to be "read" (measured) and stored. A reader circuit first reads the charge on the pixel capacitance right next to it. Immediately after, the charge on each pixel is electronically transferred to its adjacent pixel, towards the reader which reads each pixel charge in sequence, one-by-one. Hence the name "charge-coupled device." All this information (the brightness of each pixel) goes to a central processor that stores it and allows re-formation of the image later onto the camera's screen, a computer screen, or a printer. After all the pixel charge information is transferred to memory (Section 21–8), a new picture can be taken.

A CMOS sensor also uses a silicon semiconductor, and incorporates transistor electronics within each pixel, allowing parallel readout, somewhat like the similar MOSFET array that was shown in Fig. 21–29.

Sensor sizes are typically in the ratio 4:3 or 3:2. A larger sensor is better because it can hold more pixels, and/or each pixel can be larger and hold more charge (free electrons) to provide a wider range of brightness, better color accuracy, and better sensitivity in low-light conditions.

In the most common array of pixels, referred to as a **Bayer mosaic**, color is achieved by red, green, and blue filters over alternating pixels as shown in Fig. 25–2, similar to what a color LCD or CRT screen does (Sections 17–11 and 24–11). The sensor type shown in Fig. 25–2 contains twice as many green pixels as red or blue (because green seems to have a stronger influence on the human eye's sensation of sharpness). The computer-analyzed color at many pixels is often an average with nearest-neighbor colors to reduce memory size (= compression, see page 489).

Each different color of pixel in a Bayer array is counted as a *separate* pixel. In contrast, in an LCD screen (Sections 17–11 and 24–11), a group of three subpixels is counted as one pixel, a more conservative count.

An alternative technology, called "Foveon," uses a semiconductor layer system. Different wavelengths of light penetrate silicon to different depths, as shown in Fig. 25–3: blue wavelengths are absorbed in the top layer, allowing green and red light to pass through. Longer wavelengths (green) are absorbed in the second layer, and the bottom layer detects the longest wavelengths (red). All three colors are detected by each "tri-pixel" site, resulting in better color resolution and fewer artifacts.

Digital Artifacts

Digital cameras can produce image artifacts (artificial effects in the image not present in the original) resulting from the electronic sensing of the image. One example using the Bayer mosaic of pixels (Fig. 25–2) is described in Fig. 25–4.

---

†This is called a *negative*, because the black areas correspond to bright objects and vice versa. The same process occurs during printing to produce a black-and-white "positive" picture from the negative. Color film has three emulsion layers (or dyes) corresponding to the three primary colors.

‡Specifically, photons of light knock electrons in the valence band up to the conduction band. This material on semiconductors is covered in Chapter 29.

Camera Adjustments

There are three main adjustments on good-quality cameras: shutter speed, *f*-stop, and focusing. Although most cameras today make these adjustments automatically, it is valuable to understand these adjustments to use a camera effectively. For special or top-quality work, a manual camera is indispensable (Fig. 25–5).

Exposure time or shutter speed This refers to how quickly the digital sensor can make an accurate reading of the images, or how long the shutter of a camera is open and the film or sensor is exposed. It could vary from a second or more ("time exposures") to $\frac{1}{1000}$ s or faster. To avoid blurring from camera movement, exposure times shorter than $\frac{1}{100}$ s are normally needed. If the object is moving, even shorter exposure times are needed to "stop" the action ($\frac{1}{1000}$ s or less). If the exposure time is not fast enough, the image will be blurred by camera shake. Blurring in low light conditions is more of a problem with cell-phone cameras whose inexpensive sensors need to have the shutter open longer to collect enough light. Digital still cameras or cell phones that take short videos must have a fast enough "sampling" time and fast "clearing" (of the charge) so as to take pictures at least 15 frames per second; preferable is 24 fps (like film) or 30, 60, or 120 fps like TV refresh rates (25, 50, or 100 in areas like Europe where 50 Hz is the normal line voltage frequency).

FIGURE 25–5 On this camera, the *f*-stops and the focusing ring are on the camera lens. Shutter speeds are selected on the small wheel on top of the camera body.

***f*-stop** The amount of light reaching the sensor or film depends on the area of the **lens opening** as well as shutter speed, and must be carefully controlled to avoid **underexposure** (too little light so the picture is dark and only the brightest objects show up) or **overexposure** (too much light, so that bright objects have a lack of contrast and "washed-out" appearance). A high quality camera controls the exposure with a "stop" or iris diaphragm, whose opening is of variable diameter, placed behind the lens (Fig. 25–1). The lens opening is controlled (automatically or manually) to compensate for bright or dark lighting conditions, the sensitivity of the sensor or film,[†] and for different shutter speeds. The size of the opening is specified by the ***f*-stop** or ***f*-number**, defined as

$$f\text{-stop} = \frac{f}{D},$$

where f is the focal length of the lens and D is the diameter of the lens opening (see Fig. 25–1). For example, when a 50-mm-focal-length lens has an opening $D = 25$ mm, then $f/D = 50 \text{ mm}/25 \text{ mm} = 2$, so we say it is set at $f/2$. When this lens is set at $f/5.6$, the opening is only 9 mm ($50/9 = 5.6$). For faster shutter speeds, or low light conditions, a wider lens opening must be used to get a proper exposure, which corresponds to a smaller *f*-stop number. The smaller the *f*-stop number, the larger the opening and the more light passes through the lens to the sensor or film. The smallest *f*-number of a lens (largest opening) is referred to as the *speed* of the lens. The best lenses may have a speed of $f/2.0$, or even faster. The advantage of a fast lens is that it allows pictures to be taken under poor lighting conditions. Good quality lenses consist of several elements to reduce the defects present in simple thin lenses (Section 25–6). Standard *f*-stops are

1.0, 1.4, 2.0, 2.8, 4.0, 5.6, 8, 11, 16, 22, and 32

(Fig. 25–5). Each of these stops corresponds to a diameter reduction by a factor of $\sqrt{2} \approx 1.4$. Because the amount of light reaching the film is proportional to the *area* of the opening, and therefore proportional to the diameter squared, each standard *f*-stop corresponds to a factor of 2 change in light intensity reaching the film.

[†]Different films have different sensitivities to light, referred to as the "film speed" and specified as an "ISO (or ASA) number." A "faster" film is more sensitive and needs less light to produce a good image, but is grainier which you see when the image is enlarged. Digital cameras may have a "gain" or "ISO" adjustment for sensitivity. A typical everyday ISO might be 200 or so. Adjusting a CCD to be "faster" (high ISO like 3200) for low light conditions results in "noise," resulting in graininess just as in film cameras.

SECTION 25–1 Cameras: Film and Digital **715**

Focusing Focusing is the operation of placing the lens at the correct position relative to the sensor or film for the sharpest image. The image distance is smallest for objects at infinity (the symbol ∞ is used for infinity) and is equal to the focal length, as we saw in Section 23–7. For closer objects, the image distance is greater than the focal length, as can be seen from the lens equation, $1/f = 1/d_o + 1/d_i$ (Eq. 23–8). To focus on nearby objects, the lens must therefore be moved away from the sensor or film, and this is usually done on a manual camera by turning a ring on the lens.

If the lens is focused on a nearby object, a sharp image of it will be formed, but the image of distant objects may be blurry (Fig. 25–6). The rays from a point on the distant object will be out of focus—instead of a point, they will form a circle on the sensor or film as shown (exaggerated) in Fig. 25–7. The distant object will thus produce an image consisting of overlapping circles and will be blurred. These circles are called **circles of confusion**. To have near and distant objects sharp in the same photo, you (or the camera) can try setting the lens focus at an intermediate position. For a given distance setting, there is a range of distances over which the circles of confusion will be small enough that the images will be reasonably sharp. This is called the **depth of field**. For a sensor or film width of 36 mm (including 35-mm film cameras), the depth of field is usually based on a maximum circle of confusion diameter of 0.030 mm, even 0.02 mm or 0.01 mm for critical work or very large photographs. The depth of field varies with the lens opening. If the lens opening is smaller, only rays through the central part of the lens are accepted, and these form smaller circles of confusion for a given object distance. Hence, at smaller lens openings, a greater range of object distances will fit within the circle of confusion criterion, so the depth of field is greater. Smaller lens openings, however, result in reduced resolution due to diffraction (discussed later in this Chapter). Best resolution is typically found around $f/5.6$ or $f/8$.

FIGURE 25–6 Photos taken with a camera lens (a) focused on a nearby object with distant object blurry, and (b) focused on a more distant object with nearby object blurry.

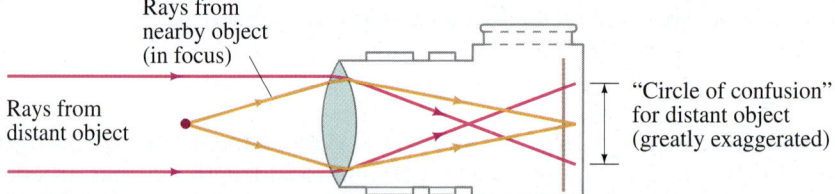

FIGURE 25–7 When the lens is positioned to focus on a nearby object, points on a distant object produce circles and are therefore blurred. (The effect is shown greatly exaggerated.)

EXAMPLE 25–1 Camera focus. How far must a 50.0-mm-focal-length camera lens be moved from its infinity setting to sharply focus an object 3.00 m away?

APPROACH For an object at infinity, the image is at the focal point, by definition (Section 23–7). For an object distance of 3.00 m, we use the thin lens equation, Eq. 23–8, to find the image distance (distance of lens to film or sensor).

SOLUTION When focused at infinity, the lens is 50.0 mm from the film. When focused at $d_o = 3.00$ m, the image distance is given by the lens equation,

$$\frac{1}{d_i} = \frac{1}{f} - \frac{1}{d_o} = \frac{1}{50.0 \text{ mm}} - \frac{1}{3000 \text{ mm}} = \frac{3000 - 50}{(3000)(50.0) \text{ mm}} = \frac{2950}{150{,}000 \text{ mm}}.$$

We solve for d_i and find $d_i = 50.8$ mm, so the lens needs to move 0.8 mm away from the film or digital sensor.

EXERCISE A If the lens of Example 25–1 is 50.4 mm from the film or sensor, what is the object distance for sharp focus?

CONCEPTUAL EXAMPLE 25–2 **Shutter speed.** To improve the depth of field, you "stop down" your camera lens by two f-stops from $f/4$ to $f/8$. What should you do to the shutter speed to maintain the same exposure?

RESPONSE The amount of light admitted by the lens is proportional to the area of the lens opening. Reducing the lens opening by two f-stops reduces the diameter by a factor of 2, and the area by a factor of 4. To maintain the same exposure, the shutter must be open four times as long. If the shutter speed had been $\frac{1}{500}$ s, you would have to increase the exposure time to $\frac{1}{125}$ s.

*Picture Sharpness

The sharpness of a picture depends not only on accurate focusing and short exposure times, but also on the graininess of the film, or number of pixels on a digital sensor. Fine-grained films and tiny pixels are "slower," meaning they require longer exposures for a given light level. All pixels are rarely used because digital cameras have averaging (or "compression") programs, such as JPEG, which reduce memory size by averaging over pixels where little contrast is detected. But some detail is inevitably lost. For example, a small blue lake may seem uniform, and coding 600 pixels as identical takes less memory than specifying all 600. Any slight variation of the water surface is lost. Full "RAW" data uses more memory. Film records everything (down to its grain size), as does RAW if your camera offers it. The processor also averages over pixels in low light conditions, resulting in a less sharp photo.

The quality of the lens strongly affects the image quality, and we discuss lens resolution and diffraction effects in Sections 25–6 and 25–7. The sharpness, or **resolution**, of a lens is often given as so many lines per millimeter, measured by photographing a standard set of parallel black lines on a white background (sometimes said as "line pairs/mm") on fine-grain film or high quality sensor, or as so many dots per inch (dpi). The minimum spacing of distinguishable lines or dots gives the resolution. A lens that can give 50 lines/mm is reasonable, 100 lines/mm is very good (= 100 dots/mm ≈ 2500 dpi). Electronic sensors also have a resolution and it is sometimes given as line pairs across the full sensor width.

A "full" Bayer pixel (upper left in Fig. 25–2) is 4 regular pixels: for example, to make a white dot as part of a white line (between two black lines when determining lens resolution), all 4 Bayer pixels (RGGB) would have to be bright. For a Foveon, all three colors of one pixel need to be bright to produce a white dot.[†]

EXAMPLE 25–3 **Pixels and resolution.** A digital camera offers a maximum resolution of 4000 × 3000 pixels on a 32 mm × 24 mm sensor. How sharp should the lens be to make use of this sensor resolution in RAW?

APPROACH We find the number of pixels per millimeter and require the lens to be at least that good.

SOLUTION We can either take the image height (3000 pixels in 24 mm) or the width (4000 pixels in 32 mm):

$$\frac{3000 \text{ pixels}}{24 \text{ mm}} = 125 \text{ pixels/mm}.$$

We would like the lens to match this resolution of 125 lines or dots per mm, which would be a quite good lens. If the lens is not this good, fewer pixels and less memory could be used.

NOTE Increasing lens resolution is a tougher problem today than is squeezing more pixels on a CCD or CMOS sensor. The sensor for high quality cameras must also be physically larger for better image accuracy and greater light sensitivity in low light conditions.

[†]Consider a 4000 × 3000 pixel array. For a Foveon, each "full pixel" (Fig. 25–3) has all 3 colors, each of which can be counted as a pixel, so it may be considered as 4000 × 3000 × 3 = 36 MP. For a Bayer sensor, Fig. 25–2, 4000 × 3000 is 12 MP (6 MP of green, 3 MP each of red and blue). There are more green pixels because they are most important in our eyes' ability to note resolution. So the distance between green pixels is a rough guide to the sharpness of a Bayer. To match a 4000 × 3000 Foveon (36 MP, or 12 MP of tri-pixel sites), a Bayer would need to have about 24 MP (because it would then have 12 MP of green). This "equivalence" is only a rough approximation.

PHYSICS APPLIED
When is a photo sharp?

EXAMPLE 25–4 Blown-up photograph. A photograph looks sharp at normal viewing distances if the dots or lines are resolved to perhaps 10 dots/mm. Would an 8 × 10-inch enlargement of a photo taken by the camera in Example 25–3 seem sharp?

APPROACH We assume the image is 4000 × 3000 pixels on a 32 × 24-mm sensor as in Example 25–3, or 125 pixels/mm. We make an enlarged photo 8 × 10 in. = 20 cm × 25 cm.

SOLUTION The short side of the sensor is 24 mm = 2.4 cm long, and that side of the photograph is 8 inches or 20 cm. Thus the size is increased by a factor of 20 cm/2.4 cm ≈ 8× (or 25 cm/3.2 cm ≈ 8×). To fill the 8 × 10-in. paper, we assume the enlargement is 8×. The pixels are thus enlarged 8×. So the pixel count of 125/mm on the sensor becomes 125/8 = 15 per mm on the print. Hence an 8 × 10-inch print would be a sharp photograph. We could go 50% larger—11 × 14 or maybe even 12 × 18 inches.

In order to make very large photographic prints, large-format cameras are used such as 6 cm × 6 cm ($2\frac{1}{4}$ inch square)—either film or sensor—and even 4 × 5 inch and 8 × 10 inch (using sheet film or glass plates).

EXERCISE B The criterion of 0.030 mm as the diameter of a circle of confusion as acceptable sharpness is how many dots per mm on the sensor?

Telephotos and Wide-angles

Camera lenses are categorized into normal, telephoto, and wide angle, according to focal length and film size. A **normal lens** covers the sensor or film with a field of view that corresponds approximately to that of normal vision. A "normal" lens for 35-mm film has a focal length of 50 mm. The best digital cameras aim for a sensor of the same size† (24 mm × 36 mm). (If the sensor is smaller, digital cameras sometimes specify focal lengths to correspond with classic 35-mm cameras.) **Telephoto lenses** act like telescopes to magnify images. They have longer focal lengths than a normal lens: as we saw in Section 23–8 (Eq. 23–9), the height of the image for a given object distance is proportional to the image distance, and the image distance will be greater for a lens with longer focal length. For distant objects, the image height is very nearly proportional to the focal length. Thus a 200-mm telephoto lens for use with a 35-mm camera gives a 4× magnification over the normal 50-mm lens. A **wide-angle lens** has a shorter focal length than normal: a wider field of view is included, and objects appear smaller. A **zoom lens** is one whose focal length can be changed (by changing the distance between the thin lenses that make up the compound lens) so that you seem to zoom up to, or away from, the subject as you change the focal length.

Digital cameras may have an **optical zoom** meaning the lens can change focal length and maintain resolution. But an "electronic" or **digital zoom** just enlarges the dots (pixels) with loss of sharpness.

Different types of viewing systems are used in cameras. In some cameras, you view through a small window just above the lens as in Fig. 25–1. In a **single-lens reflex** camera (SLR), you actually view through the lens with the use of prisms and mirrors (Fig. 25–8). A mirror hangs at a 45° angle behind the lens and flips up out of the way just before the shutter opens. SLRs have the advantage that you can see almost exactly what you will get. Digital cameras use an LCD display, and it too can show what you will get on the photo if it is carefully designed.

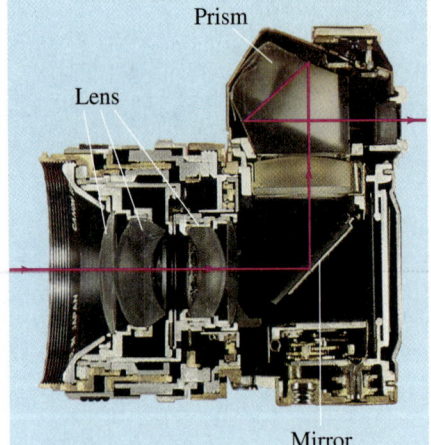

FIGURE 25–8 Single-lens reflex (SLR) camera, showing how the image is viewed through the lens with the help of a movable mirror and prism.

---

†A "35-mm camera" uses film that is physically 35 mm wide; that 35 mm is not to be confused with a focal length. 35-mm film has sprocket holes, so only 24 mm of its height is used for the photo; the width is usually 36 mm for stills. Thus one frame is 36 mm × 24 mm. Movie frames on 35-mm film are 24 mm × 18 mm.

25–2 The Human Eye; Corrective Lenses

PHYSICS APPLIED
The eye

The human eye resembles a camera in its basic structure (Fig. 25–9), but is far more sophisticated. The interior of the eye is filled with a transparent gel-like substance called the *vitreous humor* with index of refraction $n = 1.337$. Light enters this enclosed volume through the cornea and lens. Between the cornea and lens is a watery fluid, the aqueous humor (*aqua* is "water" in Latin) with $n = 1.336$. A diaphragm, called the **iris** (the colored part of your eye), adjusts automatically to control the amount of light entering the eye, similar to a camera. The hole in the iris through which light passes (the **pupil**) is black because no light is reflected from it (it's a hole), and very little light is reflected back out from the interior of the eye. The **retina**, which plays the role of the film or sensor in a camera, is on the curved rear surface of the eye. The retina consists of a complex array of nerves and receptors known as *rods* and *cones* which act to change light energy into electrical signals that travel along the nerves. The reconstruction of the image from all these tiny receptors is done mainly in the brain, although some analysis may also be done in the complex interconnected nerve network at the retina itself. At the center of the retina is a small area called the **fovea**, about 0.25 mm in diameter, where the cones are very closely packed and the sharpest image and best color discrimination are found.

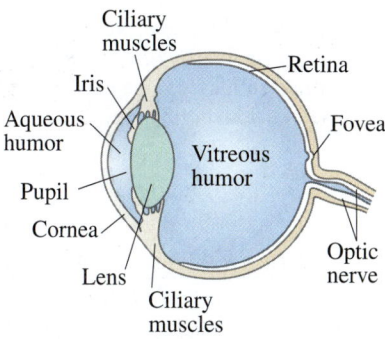
FIGURE 25–9 Diagram of a human eye.

Unlike a camera, the eye contains no shutter. The equivalent operation is carried out by the nervous system, which analyzes the signals to form images at the rate of about 30 per second. This can be compared to motion picture or television cameras, which operate by taking a series of still pictures at a rate of 24 (movies) and 60 or 30 (U.S. television) per second. Their rapid projection on the screen gives the appearance of motion.

The lens of the eye ($n = 1.386$ to 1.406) does little of the bending of the light rays. Most of the refraction is done at the front surface of the **cornea** ($n = 1.376$) at its interface with air ($n = 1.0$). The lens acts as a fine adjustment for focusing at different distances. This is accomplished by the ciliary muscles (Fig. 25–9), which change the curvature of the lens so that its focal length is changed. To focus on a distant object, the ciliary muscles of the eye are relaxed and the lens is thin, as shown in Fig. 25–10a, and parallel rays focus at the focal point (on the retina). To focus on a nearby object, the muscles contract, causing the center of the lens to thicken, Fig. 25–10b, thus shortening the focal length so that images of nearby objects can be focused on the retina, behind the new focal point. This focusing adjustment is called **accommodation**.

FIGURE 25–10 Accommodation by a normal eye: (a) lens relaxed, focused at infinity; (b) lens thickened, focused on a nearby object.

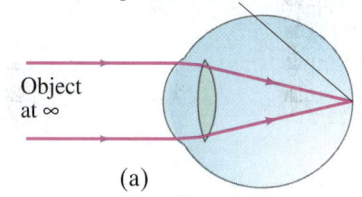

The closest distance at which the eye can focus clearly is called the **near point** of the eye. For young adults it is typically 25 cm, although younger children can often focus on objects as close as 10 cm. As people grow older, the ability to accommodate is reduced and the near point increases. A given person's **far point** is the farthest distance at which an object can be seen clearly. For some purposes it is useful to speak of a **normal eye** (a sort of average over the population), defined as an eye having a near point of 25 cm and a far point of infinity. To check your own near point, place this book close to your eye and slowly move it away until the type is sharp.

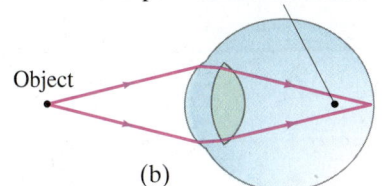

The "normal" eye is sort of an ideal. Many people have eyes that do not accommodate within the "normal" range of 25 cm to infinity, or have some other defect. Two common defects are nearsightedness and farsightedness. Both can be corrected to a large extent with lenses—either eyeglasses or contact lenses.

In **nearsightedness**, or **myopia**, the human eye can focus only on nearby objects. The far point is not infinity but some shorter distance, so distant objects are not seen clearly. Nearsightedness is usually caused by an eyeball that is too long, although sometimes it is the curvature of the cornea that is too great. In either case, images of distant objects are focused in front of the retina.

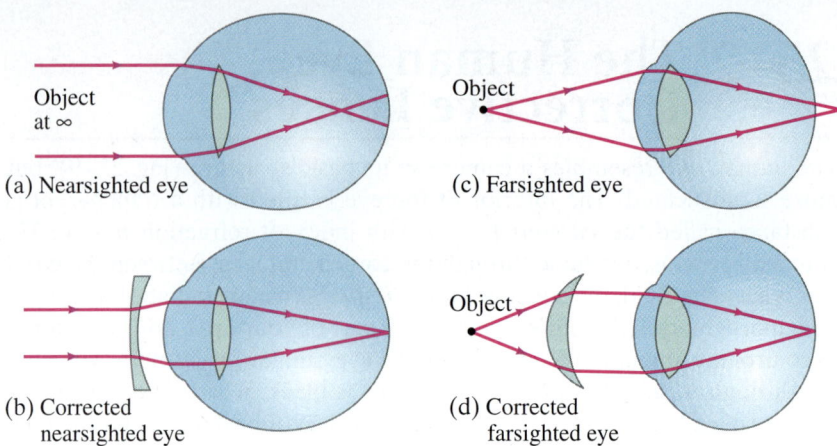

FIGURE 25–11 Correcting eye defects with lenses. (a) A nearsighted eye, which cannot focus clearly on distant objects (focal point is in front of retina), can be corrected (b) by use of a diverging lens. (c) A farsighted eye, which cannot focus clearly on nearby objects (focus point behind retina), can be corrected (d) by use of a converging lens.

PHYSICS APPLIED
Corrective lenses

A diverging lens, because it causes parallel rays to diverge, allows the rays to be focused at the retina (Figs. 25–11a and b) and thus can correct nearsightedness.

In **farsightedness**, or *hyperopia*, the eye cannot focus on nearby objects. Although distant objects are usually seen clearly, the near point is somewhat greater than the "normal" 25 cm, which makes reading difficult. This defect is caused by an eyeball that is too short or (less often) by a cornea that is not sufficiently curved. It is corrected by a converging lens, Figs. 25–11c and d. Similar to hyperopia is *presbyopia*, which is the lessening ability of the eye to accommodate as a person ages, and the near point moves out. Converging lenses also compensate for this.

Astigmatism is usually caused by an out-of-round cornea or lens so that point objects are focused as short lines, which blurs the image. It is as if the cornea were spherical with a cylindrical section superimposed. As shown in Fig. 25–12, a cylindrical lens focuses a point into a line parallel to its axis. An astigmatic eye may focus rays in one plane, such as the vertical plane, at a shorter distance than it does for rays in a horizontal plane. Astigmatism is corrected with the use of a compensating cylindrical lens. Lenses for eyes that are nearsighted or farsighted as well as astigmatic are ground with superimposed spherical and cylindrical surfaces, so that the radius of curvature of the correcting lens is different in different planes.

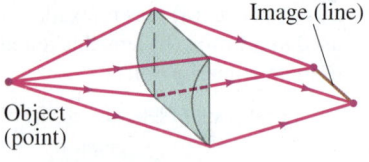

FIGURE 25–12 A cylindrical lens forms a line image of a point object because it is converging in one plane only.

EXAMPLE 25–5 Farsighted eye. A farsighted eye has a near point of 100 cm. Reading glasses must have what lens power so that a newspaper can be read at a distance of 25 cm? Assume the lens is very close to the eye.

APPROACH When the object is placed 25 cm from the lens ($= d_o$), we want the image to be 100 cm away on the *same* side of the lens (so the eye can focus it), and so the image is virtual, as shown in Fig. 25–13, and $d_i = -100$ cm will be negative. We use the thin lens equation (Eq. 23–8) to determine the needed focal length. Optometrists' prescriptions specify the power ($P = 1/f$, Eq. 23–7) given in diopters ($1\ \text{D} = 1\ \text{m}^{-1}$).

SOLUTION Given that $d_o = 25$ cm and $d_i = -100$ cm, the thin lens equation gives

$$\frac{1}{f} = \frac{1}{d_o} + \frac{1}{d_i} = \frac{1}{25\ \text{cm}} + \frac{1}{-100\ \text{cm}} = \frac{4-1}{100\ \text{cm}} = \frac{1}{33\ \text{cm}}.$$

So $f = 33$ cm $= 0.33$ m. The power P of the lens is $P = 1/f = +3.0$ D. The plus sign indicates that it is a converging lens.

NOTE We chose the image position to be where the eye can actually focus. The lens needs to put the image there, given the desired placement of the object (newspaper).

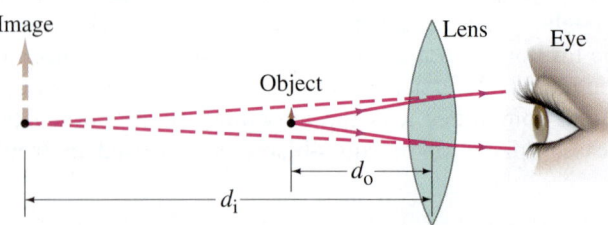

FIGURE 25–13 Lens of reading glasses (Example 25–5).

EXAMPLE 25–6 Nearsighted eye. A nearsighted eye has near and far points of 12 cm and 17 cm, respectively. (a) What lens power is needed for this person to see distant objects clearly, and (b) what then will be the near point? Assume that the lens is 2.0 cm from the eye (typical for eyeglasses).

APPROACH For a distant object ($d_o = \infty$), the lens must put the image at the far point of the eye as shown in Fig. 25–14a, 17 cm in front of the eye. We can use the thin lens equation to find the focal length of the lens, and from this its lens power. The new near point (as shown in Fig. 25–14b) can be calculated for the lens by again using the thin lens equation.

SOLUTION (a) For an object at infinity ($d_o = \infty$), the image must be in front of the lens 17 cm from the eye or (17 cm − 2 cm) = 15 cm from the lens; hence $d_i = -15$ cm. We use the thin lens equation to solve for the focal length of the needed lens:

$$\frac{1}{f} = \frac{1}{d_o} + \frac{1}{d_i} = \frac{1}{\infty} + \frac{1}{-15 \text{ cm}} = -\frac{1}{15 \text{ cm}}.$$

So $f = -15$ cm $= -0.15$ m or $P = 1/f = -6.7$ D. The minus sign indicates that it must be a diverging lens for the myopic eye.

(b) The near point when glasses are worn is where an object is placed (d_o) so that the lens forms an image at the "near point of the naked eye," namely 12 cm from the eye. That image point is (12 cm − 2 cm) = 10 cm in front of the lens, so $d_i = -0.10$ m and the thin lens equation gives

$$\frac{1}{d_o} = \frac{1}{f} - \frac{1}{d_i} = -\frac{1}{0.15 \text{ m}} + \frac{1}{0.10 \text{ m}} = \frac{-2 + 3}{0.30 \text{ m}} = \frac{1}{0.30 \text{ m}}.$$

So $d_o = 30$ cm, which means the near point when the person is wearing glasses is 30 cm in front of the lens, or 32 cm from the eye.

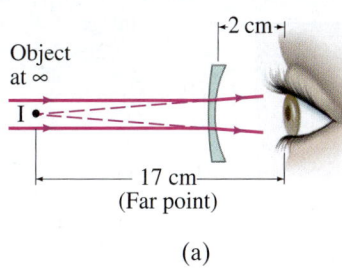

(a)

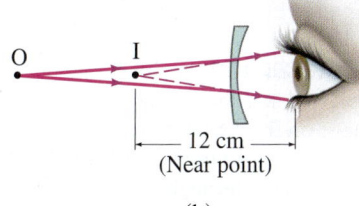

(b)

FIGURE 25–14 Example 25–6.

Contact Lenses

Suppose contact lenses are used to correct the eye in Example 25–6. Since contacts are placed directly on the cornea, we would not subtract out the 2.0 cm for the image distances. That is, for distant objects $d_i = f = -17$ cm, so $P = 1/f = -5.9$ D. The new near point would be 41 cm. Thus we see that a contact lens and an eyeglass lens will require slightly different powers, or focal lengths, for the same eye because of their different placements relative to the eye. We also see that glasses in this case give a better near point than contacts.

PHYSICS APPLIED
Contact lenses—different f and P

EXERCISE C What power of contact lens is needed for an eye to see distant objects if its far point is 25 cm?

Underwater Vision

When your eyes are under water, distant underwater objects look blurry because at the water–cornea interface, the difference in indices of refraction is very small: $n = 1.33$ for water, 1.376 for the cornea. Hence light rays are bent very little and are focused far behind the retina, Fig. 25–15a. If you wear goggles or a face mask, you restore an air–cornea interface ($n = 1.0$ and 1.376, respectively) and the rays can be focused, Fig. 25–15b.

PHYSICS APPLIED
Underwater vision

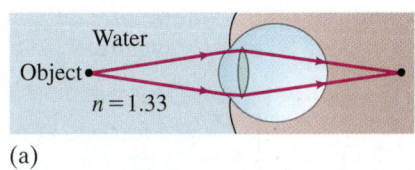

(a)

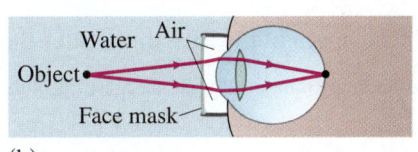

(b)

FIGURE 25–15 (a) Under water, we see a blurry image because light rays are bent much less than in air. (b) If we wear goggles, we again have an air–cornea interface and can see clearly.

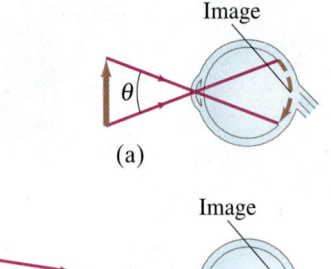

(a)

(b)

FIGURE 25–16 When the same object is viewed at a shorter distance, the image on the retina is greater, so the object appears larger and more detail can be seen. The angle θ that the object subtends in (a) is greater than in (b). *Note*: This is not a normal ray diagram because we are showing only one ray from each point.

25–3 Magnifying Glass

Much of the remainder of this Chapter will deal with optical devices that are used to produce magnified images of objects. We first discuss the **simple magnifier**, or **magnifying glass**, which is simply a converging lens (see Chapter-Opening Photo, page 713).

How large an object appears, and how much detail we can see on it, depends on the size of the image it makes on the retina. This, in turn, depends on the angle subtended by the object at the eye. For example, a penny held 30 cm from the eye looks twice as tall as one held 60 cm away because the angle it subtends is twice as great (Fig. 25–16). When we want to examine detail on an object, we bring it up close to our eyes so that it subtends a greater angle. However, our eyes can accommodate only up to a point (the near point), and we will assume a standard distance of $N = 25$ cm as the near point in what follows.

A magnifying glass allows us to place the object closer to our eye so that it subtends a greater angle. As shown in Fig. 25–17a, the object is placed at the focal point or just within it. Then the converging lens produces a virtual image, which must be at least 25 cm from the eye if the eye is to focus on it. If the eye is relaxed, the image will be at infinity, and in this case the object is exactly at the focal point. (You make this slight adjustment yourself when you "focus" on the object by moving the magnifying glass.)

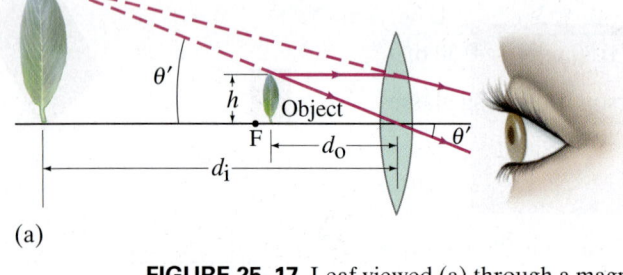

(a)

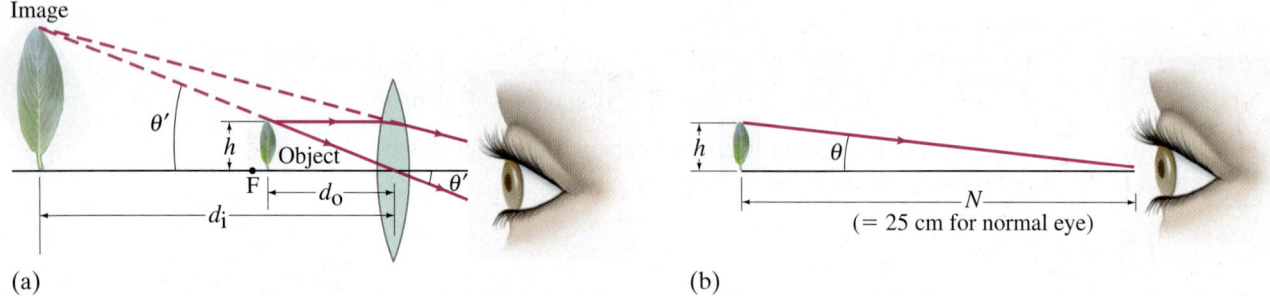

(b)

FIGURE 25–17 Leaf viewed (a) through a magnifying glass, and (b) with the unaided eye. The eye is focused at its near point in both cases.

A comparison of part (a) of Fig. 25–17 with part (b), in which the same object is viewed at the near point with the unaided eye, reveals that the angle the object subtends at the eye is much larger when the magnifier is used. The **angular magnification** or **magnifying power**, M, of the lens is defined as the ratio of the angle subtended by an object when using the lens, to the angle subtended using the unaided eye, with the object at the near point N of the eye ($N = 25$ cm for a normal eye):

$$M = \frac{\theta'}{\theta}, \quad (25\text{–}1)$$

where θ and θ' are shown in Fig. 25–17. We can write M in terms of the focal length by noting that $\theta = h/N$ (Fig. 25–17b) and $\theta' = h/d_o$ (Fig. 25–17a), where h is the height of the object and we assume the angles are small so θ and θ' (in radians) equal their sines and tangents. If the eye is relaxed (for least eye strain), the image will be at infinity and the object will be precisely at the focal point; see Fig. 25–18. Then $d_o = f$ and $\theta' = h/f$, whereas $\theta = h/N$ as before (Fig. 25–17b). Thus

$$M = \frac{\theta'}{\theta} = \frac{h/f}{h/N} = \frac{N}{f}. \quad \begin{bmatrix}\text{eye focused at }\infty; \\ N = 25\text{ cm for normal eye}\end{bmatrix} \quad (25\text{–}2a)$$

We see that the shorter the focal length of the lens, the greater the magnification.[†]

FIGURE 25–18 With the eye relaxed, the object is placed at the focal point, and the image is at infinity. Compare to Fig. 25–17a where the image is at the eye's near point.

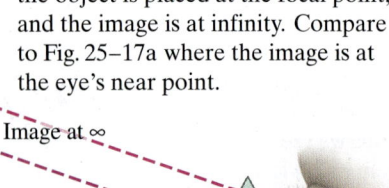

[†]Simple single-lens magnifiers are limited to about 2 or 3× because of blurring due to spherical aberration (Section 25–6).

The magnification of a given lens can be increased a bit by moving the lens and adjusting your eye so it focuses on the image at the eye's near point. In this case, $d_i = -N$ (see Fig. 25–17a) if your eye is very near the magnifier. Then the object distance d_o is given by

$$\frac{1}{d_o} = \frac{1}{f} - \frac{1}{d_i} = \frac{1}{f} + \frac{1}{N}.$$

We see from this equation that $d_o = fN/(f + N) < f$, as shown in Fig. 25–17a, since $N/(f + N)$ must be less than 1. With $\theta' = h/d_o$ the magnification is

$$M = \frac{\theta'}{\theta} = \frac{h/d_o}{h/N}$$

$$= N\left(\frac{1}{d_o}\right) = N\left(\frac{1}{f} + \frac{1}{N}\right)$$

or

$$M = \frac{N}{f} + 1. \qquad \begin{bmatrix} \text{eye focused at near point, } N; \\ N = 25 \text{ cm for normal eye} \end{bmatrix} \quad \text{(25–2b)}$$

We see that the magnification is slightly greater when the eye is focused at its near point, as compared to when it is relaxed.

EXAMPLE 25–7 ESTIMATE **A jeweler's "loupe."** An 8-cm-focal-length converging lens is used as a "jeweler's loupe," which is a magnifying glass. Estimate (a) the magnification when the eye is relaxed, and (b) the magnification if the eye is focused at its near point $N = 25$ cm.

APPROACH The magnification when the eye is relaxed is given by Eq. 25–2a. When the eye is focused at its near point, we use Eq. 25–2b and we assume the lens is near the eye.

SOLUTION (a) With the relaxed eye focused at infinity,

$$M = \frac{N}{f} = \frac{25 \text{ cm}}{8 \text{ cm}} \approx 3\times.$$

(b) The magnification when the eye is focused at its near point ($N = 25$ cm), and the lens is near the eye, is

$$M = 1 + \frac{N}{f} = 1 + \frac{25}{8} \approx 4\times.$$

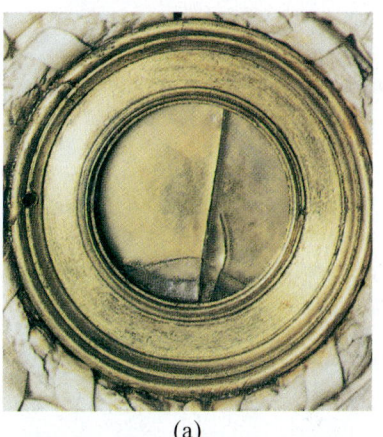

FIGURE 25–19 (a) Objective lens (mounted now in an ivory frame) from the telescope with which Galileo made his world-shaking discoveries, including the moons of Jupiter. (b) Telescopes made by Galileo (1609).

(a)

25–4 Telescopes

A telescope is used to magnify objects that are very far away. In most cases, the object can be considered to be at infinity.

Galileo, although he did not invent it,[†] developed the telescope into a usable and important instrument. He was the first to examine the heavens with the telescope (Fig. 25–19), and he made world-shaking discoveries, including the moons of Jupiter, the phases of Venus, sunspots, the structure of the Moon's surface, and that the Milky Way is made up of a huge number of individual stars.

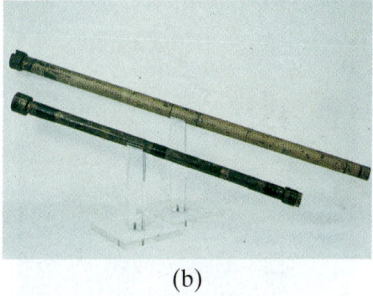

(b)

[†]Galileo built his first telescope in 1609 after having heard of such an instrument existing in Holland. The first telescopes magnified only three to four times, but Galileo soon made a 30-power instrument. The first Dutch telescopes date from about 1604 and probably were copies of an Italian telescope built around 1590. Kepler (see Chapter 5) gave a ray description (in 1611) of the Keplerian telescope, which is named for him because he first described it, although he did not build it.

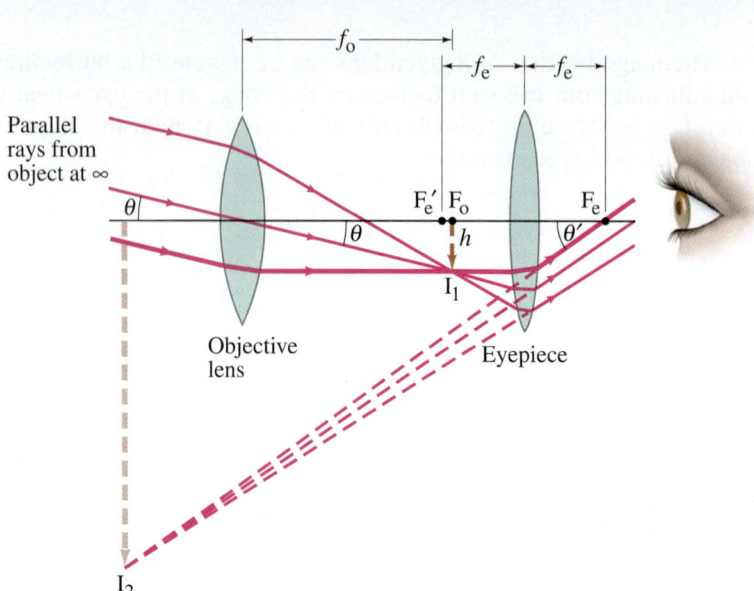

FIGURE 25–20 Astronomical telescope (refracting). Parallel light from one point on a distant object $(d_o = \infty)$ is brought to a focus by the objective lens in its focal plane. This image (I_1) is magnified by the eyepiece to form the final image I_2. Only two of the rays shown entering the objective are standard rays (2 and 3) as described in Fig. 23–37.

Several types of **astronomical telescope** exist. The common **refracting** type, sometimes called **Keplerian**, contains two converging lenses located at opposite ends of a long tube, as illustrated in Fig. 25–20. The lens closest to the object is called the **objective lens** (focal length f_o) and forms a real image I_1 of the distant object in the plane of its focal point F_o (or near it if the object is not at infinity). The second lens, called the **eyepiece** (focal length f_e), acts as a magnifier. That is, the eyepiece magnifies the image I_1 formed by the objective lens to produce a second, greatly magnified image, I_2, which is virtual and inverted. If the viewing eye is relaxed, the eyepiece is adjusted so the image I_2 is at infinity. Then the real image I_1 is at the focal point F'_e of the eyepiece, and the distance between the lenses is $f_o + f_e$ for an object at infinity.

PROBLEM SOLVING
Distance between lenses = $f_o + f_e$ for relaxed eye

To find the total angular magnification of this telescope, we note that the angle an object subtends as viewed by the unaided eye is just the angle θ subtended at the telescope objective. From Fig. 25–20 we can see that $\theta \approx h/f_o$, where h is the height of the image I_1 and we assume θ is small so that $\tan\theta \approx \theta$. Note, too, that the thickest of the three rays drawn in Fig. 25–20 is parallel to the axis before it strikes the eyepiece and therefore is refracted through the eyepiece focal point F_e on the far side. Thus, $\theta' \approx h/f_e$ and the **total magnifying power** (that is, angular magnification, which is what is always quoted) of this telescope is

$$M = \frac{\theta'}{\theta} = \frac{(h/f_e)}{(h/f_o)} = -\frac{f_o}{f_e}, \qquad \begin{bmatrix}\text{telescope}\\\text{magnification}\end{bmatrix} \quad (25\text{–}3)$$

where we used Eq. 25–1 and we inserted a minus sign to indicate that the image is inverted. To achieve a large magnification, the objective lens should have a long focal length and the eyepiece a short focal length.

FIGURE 25–21 This large refracting telescope was built in 1897 and is housed at Yerkes Observatory in Wisconsin. The objective lens is 102 cm (40 inches) in diameter, and the telescope tube is about 19 m long. Example 25–8.

EXAMPLE 25–8 Telescope magnification. The largest optical refracting telescope in the world is located at the Yerkes Observatory in Wisconsin, Fig. 25–21. It is referred to as a "40-inch" telescope, meaning that the diameter of the objective is 40 in., or 102 cm. The objective lens has a focal length of 19 m, and the eyepiece has a focal length of 10 cm. (*a*) Calculate the total magnifying power of this telescope. (*b*) Estimate the length of the telescope.

APPROACH Equation 25–3 gives the magnification. The length of the telescope is the distance between the two lenses.

SOLUTION (*a*) From Eq. 25–3 we find

$$M = -\frac{f_o}{f_e} = -\frac{19\text{ m}}{0.10\text{ m}} = -190\times.$$

(*b*) For a relaxed eye, the image I_1 is at the focal point of both the eyepiece and the objective lenses. The distance between the two lenses is thus $f_o + f_e \approx 19$ m, which is essentially the length of the telescope.

EXERCISE D A 40× telescope has a 1.2-cm focal length eyepiece. What is the focal length of the objective lens?

For an astronomical telescope to produce bright images of faint stars, the objective lens must be large to allow in as much light as possible. Indeed, the diameter of the objective lens (and hence its "light-gathering power") is an important parameter for an astronomical telescope, which is why the largest ones are specified by giving the objective diameter (such as the 10-meter Keck telescope in Hawaii). The construction and grinding of large lenses is very difficult. Therefore, the largest telescopes are **reflecting telescopes** which use a curved mirror as the objective, Fig. 25–22. A mirror has only one surface to be ground and can be supported along its entire surface[†] (a large lens, supported at its edges, would sag under its own weight). Often, the eyepiece lens or mirror (see Fig. 25–22) is removed so that the real image formed by the objective mirror can be recorded directly on film or on an electronic sensor (CCD or CMOS, Section 25–1).

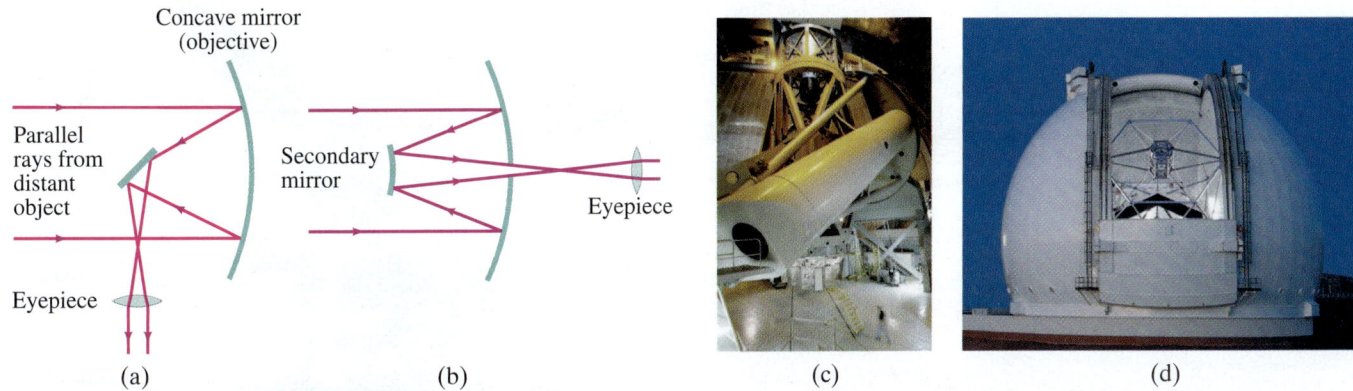

FIGURE 25–22 A concave mirror can be used as the objective of an astronomical telescope. Arrangement (a) is called the Newtonian focus, and (b) the Cassegrainian focus. Other arrangements are also possible. (c) The 200-inch (mirror diameter) Hale telescope on Palomar Mountain in California. (d) The 10-meter Keck telescope on Mauna Kea, Hawaii. The Keck combines thirty-six 1.8-meter six-sided mirrors into the equivalent of a very large single reflector, 10 m in diameter.

A **terrestrial telescope**, for viewing objects on Earth, must provide an upright image—seeing normal objects upside down would be difficult (much less important for viewing stars). Two designs are shown in Fig. 25–23. The **Galilean** type, which Galileo used for his great astronomical discoveries, has a diverging lens as eyepiece which intercepts the converging rays from the objective lens before they reach a focus, and acts to form a virtual upright image, Fig. 25–23a. This design is still used in opera glasses. The tube is reasonably short, but the field of view is small. The second type, shown in Fig. 25–23b, is often called a **spyglass** and makes use of a third convex lens that acts to make the image upright as shown. A spyglass must be quite long. The most practical design today is the **prism binocular** which was shown in Fig. 23–28. The objective and eyepiece are converging lenses. The prisms reflect the rays by total internal reflection and shorten the physical size of the device, and they also act to produce an upright image. One prism reinverts the image in the vertical plane, the other in the horizontal plane.

[†]Another advantage of mirrors is that they exhibit no chromatic aberration because the light doesn't pass through them; and they can be ground into a parabolic shape to correct for spherical aberration (Section 25–6). The reflecting telescope was first proposed by Newton.

FIGURE 25–23 Terrestrial telescopes that produce an upright image: (a) Galilean; (b) spyglass, or erector type.

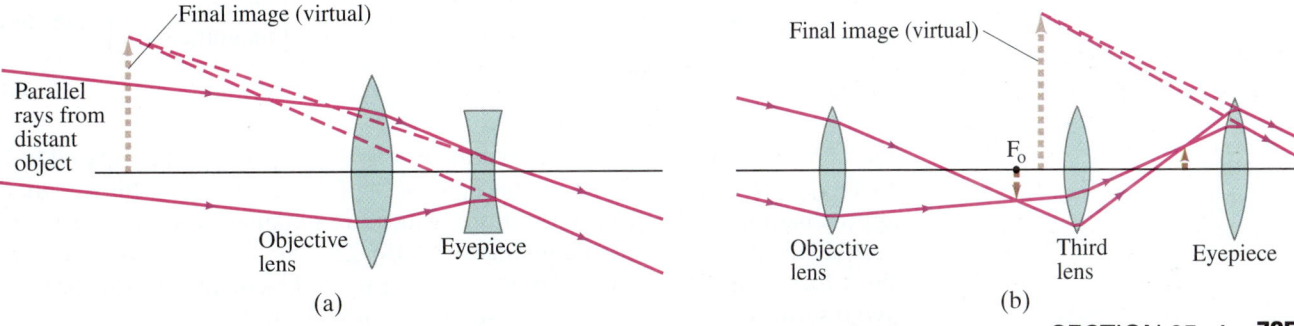

25–5 Compound Microscope

PHYSICS APPLIED
Microscopes

The compound **microscope**, like the telescope, has both objective and eyepiece (or ocular) lenses, Fig. 25–24. The design is different from that for a telescope because a microscope is used to view objects that are very close, so the object distance is very small. The object is placed just beyond the objective's focal point as shown in Fig. 25–24a. The image I_1 formed by the objective lens is real, quite far from the objective lens, and much enlarged. The eyepiece is positioned so that this image is near the eyepiece focal point F_e. The image I_1 is magnified by the eyepiece into a very large virtual image, I_2, which is seen by the eye and is inverted. Modern microscopes use a third "tube" lens behind the objective, but we will analyze the simpler arrangement shown in Fig. 25–24a.

FIGURE 25–24 Compound microscope: (a) ray diagram, (b) photograph (illumination comes from below, outlined in red, then up through the slide holding the sample or object).

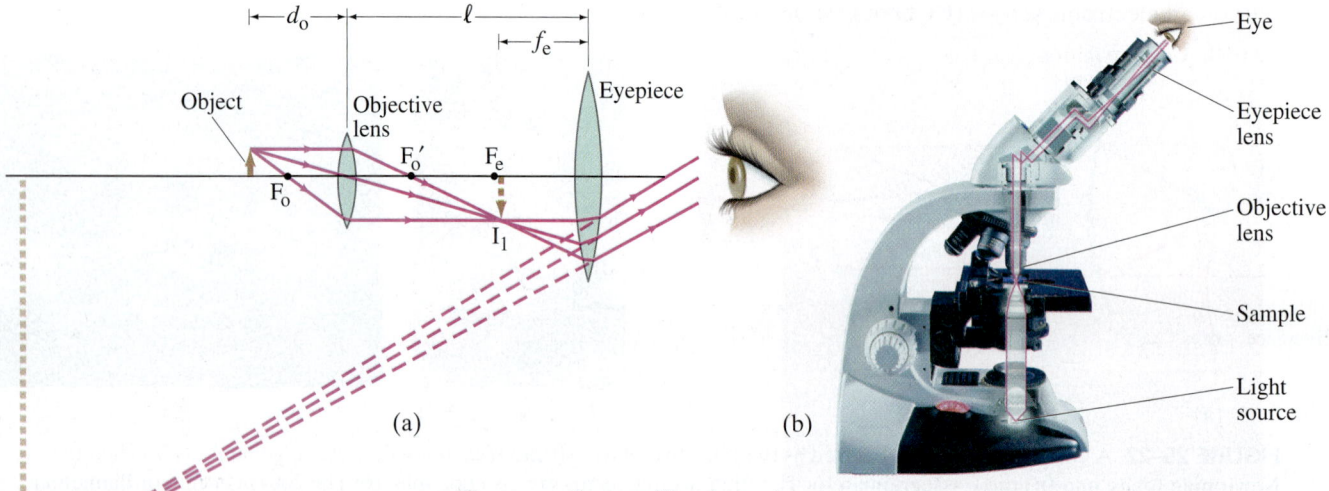

The overall magnification of a microscope is the product of the magnifications produced by the two lenses. The image I_1 formed by the objective lens is a factor m_o greater than the object itself. From Fig. 25–24a and Eq. 23–9 for the magnification of a simple lens, we have

$$m_o = \frac{h_i}{h_o} = \frac{d_i}{d_o} = \frac{\ell - f_e}{d_o}, \quad (25\text{–}4)$$

where d_o and d_i are the object and image distances for the objective lens, ℓ is the distance between the lenses (equal to the length of the barrel), and we ignored the minus sign in Eq. 23–9 which only tells us that the image is inverted. We set $d_i = \ell - f_e$, which is exact only if the eye is relaxed, so that the image I_1 is at the eyepiece focal point F_e. The eyepiece acts like a simple magnifier. If we assume that the eye is relaxed, the eyepiece angular magnification M_e is (from Eq. 25–2a)

$$M_e = \frac{N}{f_e}, \quad (25\text{–}5)$$

where the near point $N = 25$ cm for the normal eye. Since the eyepiece enlarges the image formed by the objective, the overall angular magnification M is the product of the magnification of the objective lens, m_o, times the angular magnification, M_e, of the eyepiece lens (Eqs. 25–4 and 25–5):

$$M = M_e m_o = \left(\frac{N}{f_e}\right)\left(\frac{\ell - f_e}{d_o}\right) \quad \begin{bmatrix}\text{microscope}\\\text{magnification}\end{bmatrix} \quad (25\text{–}6a)$$

$$\approx \frac{N\ell}{f_e f_o}. \quad [f_o \text{ and } f_e \ll \ell] \quad (25\text{–}6b)$$

The approximation, Eq. 25–6b, is accurate when f_e and f_o are small compared to ℓ, so $\ell - f_e \approx \ell$, and the object is near F_o so $d_o \approx f_o$ (Fig. 25–24a). This is a good approximation for large magnifications, which are obtained when f_o and f_e are very small (they are in the denominator of Eq. 25–6b). To make lenses of very short focal length, compound lenses involving several elements must be used to avoid serious aberrations, as discussed in the next Section.

EXAMPLE 25–9 Microscope. A compound microscope consists of a 10× eyepiece and a 50× objective 17.0 cm apart. Determine (a) the overall magnification, (b) the focal length of each lens, and (c) the position of the object when the final image is in focus with the eye relaxed. Assume a normal eye, so $N = 25$ cm.

APPROACH The overall magnification is the product of the eyepiece magnification and the objective magnification. The focal length of the eyepiece is found from Eq. 25–2a or 25–5 for the magnification of a simple magnifier. For the objective lens, it is easier to next find d_o (part c) using Eq. 25–4 before we find f_o.

SOLUTION (a) The overall magnification is $(10\times)(50\times) = 500\times$.
(b) The eyepiece focal length is (Eq. 25–5) $f_e = N/M_e = 25 \text{ cm}/10 = 2.5$ cm.
Next we solve Eq. 25–4 for d_o, and find

$$d_o = \frac{\ell - f_e}{m_o} = \frac{(17.0 \text{ cm} - 2.5 \text{ cm})}{50} = 0.29 \text{ cm}.$$

Then, from the thin lens equation for the objective with $d_i = \ell - f_e = 14.5$ cm (see Fig. 25–24a),

$$\frac{1}{f_o} = \frac{1}{d_o} + \frac{1}{d_i} = \frac{1}{0.29 \text{ cm}} + \frac{1}{14.5 \text{ cm}} = 3.52 \text{ cm}^{-1};$$

so $f_o = 1/(3.52 \text{ cm}^{-1}) = 0.28$ cm.
(c) We just calculated $d_o = 0.29$ cm, which is very close to f_o.

25–6 Aberrations of Lenses and Mirrors

In Chapter 23 we developed a theory of image formation by a thin lens. We found, for example, that all rays from each point on an object are brought to a single point as the image point. This result, and others, were based on approximations for a thin lens, mainly that all rays make small angles with the axis and that we can use $\sin\theta \approx \theta$. Because of these approximations, we expect deviations from the simple theory, which are referred to as **lens aberrations**. There are several types of aberration; we will briefly discuss each of them separately, but all may be present at one time.

Consider an object at any point (even at infinity) on the axis of a lens with spherical surfaces. Rays from this point that pass through the outer regions of the lens are brought to a focus at a different point from those that pass through the center of the lens. This is called **spherical aberration**, and is shown exaggerated in Fig. 25–25. Consequently, the image seen on a screen or film will not be a point but a tiny circular patch of light. If the sensor or film is placed at the point C, as indicated, the circle will have its smallest diameter, which is referred to as the **circle of least confusion**. Spherical aberration is present whenever spherical surfaces are used. It can be reduced by using nonspherical (= aspherical) lens surfaces, but grinding such lenses is difficult and expensive. Spherical aberration can be reduced by the use of several lenses in combination, and by using primarily the central part of lenses.

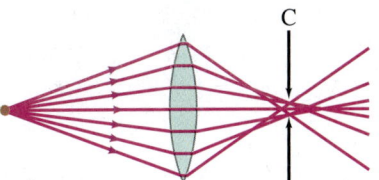

FIGURE 25–25 Spherical aberration (exaggerated). Circle of least confusion is at C.

For object points off the lens axis, additional aberrations occur. Rays passing through the different parts of the lens cause spreading of the image that is noncircular. There are two effects: **coma** (because the image of a point is comet-shaped rather than a tiny circle) and **off-axis astigmatism**.[†] Furthermore, the image points for objects off the axis but at the same distance from the lens do not fall on a flat plane but on a curved surface—that is, the focal plane is not flat. (We expect this because the points on a flat plane, such as the film in a camera, are not equidistant from the lens.) This aberration is known as **curvature of field** and is a problem in cameras and other devices where the sensor or film is a flat plane. In the eye, however, the retina is curved, which compensates for this effect.

[†]Although the effect is the same as for astigmatism in the eye (Section 25–2), the cause is different. Off-axis astigmatism is no problem in the eye because objects are clearly seen only at the fovea, on the lens axis.

Another aberration, **distortion**, is a result of variation of magnification at different distances from the lens axis. Thus a straight-line object some distance from the axis may form a curved image. A square grid of lines may be distorted to produce "barrel distortion," or "pincushion distortion," Fig. 25–26. The former is common in extreme wide-angle lenses.

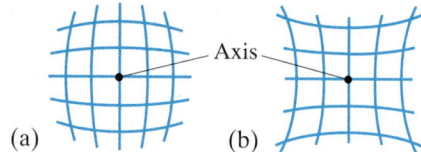

FIGURE 25–26 Distortion: lenses may image a square grid of perpendicular lines to produce (a) barrel distortion or (b) pincushion distortion. These distortions can be seen in the photograph of Fig. 23–31d for a simple lens.

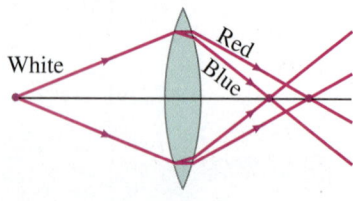

FIGURE 25–27 Chromatic aberration. Different colors are focused at different points.

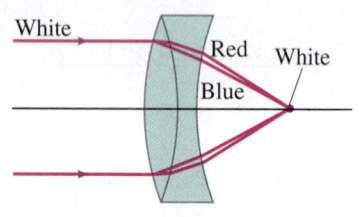

FIGURE 25–28 Achromatic doublet.

PHYSICS APPLIED
Aberration in the human eye

All the above aberrations occur for monochromatic light and hence are referred to as *monochromatic aberrations*. Normal light is not monochromatic, and there will also be **chromatic aberration**. This aberration arises because of dispersion—the variation of index of refraction of transparent materials with wavelength (Section 24–4). For example, blue light is bent more than red light by glass. So if white light is incident on a lens, the different colors are focused at different points, Fig. 25–27, and have slightly different magnifications resulting in colored fringes in the image. Chromatic aberration can be eliminated for any two colors (and reduced greatly for all others) by the use of two lenses made of different materials with different indices of refraction and dispersion. Normally one lens is converging and the other diverging, and they are often cemented together (Fig. 25–28). Such a lens combination is called an **achromatic doublet** (or "color-corrected" lens).

To reduce aberrations, high-quality lenses are **compound lenses** consisting of many simple lenses, referred to as **elements**. A typical high-quality camera lens may contain six to eight (or more) elements. For simplicity we will usually indicate lenses in diagrams as if they were simple lenses.

The human eye is also subject to aberrations, but they are minimal. Spherical aberration, for example, is minimized because (1) the cornea is less curved at the edges than at the center, and (2) the lens is less dense at the edges than at the center. Both effects cause rays at the outer edges to be bent less strongly, and thus help to reduce spherical aberration. Chromatic aberration is partially compensated for because the lens absorbs the shorter wavelengths appreciably and the retina is less sensitive to the blue and violet wavelengths. This is just the region of the spectrum where dispersion—and thus chromatic aberration—is greatest (Fig. 24–14).

Spherical mirrors (Section 23–3) also suffer aberrations including spherical aberration (see Fig. 23–13). Mirrors can be ground in a parabolic shape to correct for aberrations, but they are much harder to make and therefore very expensive. Spherical mirrors do not, however, exhibit chromatic aberration because the light does not pass through them (no refraction, no dispersion).

25–7 Limits of Resolution; Circular Apertures

The ability of a lens to produce distinct images of two point objects very close together is called the **resolution** of the lens. The closer the two images can be and still be seen as distinct (rather than overlapping blobs), the higher the resolution. The resolution of a camera lens, for example, is often specified as so many dots or lines per millimeter, as mentioned in Section 25–1.

Two principal factors limit the resolution of a lens. The first is lens aberrations. As we just saw, because of spherical and other aberrations, a point object is not a point on the image but a tiny blob. Careful design of compound lenses can reduce aberrations significantly, but they cannot be eliminated entirely. The second factor that limits resolution is *diffraction*, which cannot be corrected for because it is a natural result of the wave nature of light. We discuss it now.

In Section 24–5, we saw that because light travels as a wave, light from a point source passing through a slit is spread out into a diffraction pattern (Figs. 24–19 and 24–21). A lens, because it has edges, acts like a round slit. When a lens forms the image of a point object, the image is actually a tiny diffraction pattern. Thus *an image would be blurred even if aberrations were absent.*

In the analysis that follows, we assume that the lens is free of aberrations, so we can concentrate on diffraction effects and how much they limit the resolution of a lens. In Fig. 24–21 we saw that the diffraction pattern produced by light passing through a rectangular slit has a central maximum in which most of the light falls. This central peak falls to a minimum on either side of its center at an angle θ given by

$$\sin\theta = \lambda/D$$

(this is Eq. 24–3a), where D is the slit width and λ the wavelength of light used. θ is the angular half-width of the central maximum, and for small angles (in radians) can be written

$$\theta \approx \sin\theta = \frac{\lambda}{D}.$$

There are also low-intensity fringes beyond.

For a lens, or any circular hole, the image of a point object will consist of a *circular* central peak (called the *diffraction spot* or *Airy disk*) surrounded by faint circular fringes, as shown in Fig. 25–29a. The central maximum has an angular half-width given by

$$\theta = \frac{1.22\lambda}{D},$$

where D is the diameter of the circular opening. [This is a theoretical result for a perfect circle or lens. For real lenses or circles, the factor is on the order of 1 to 2.] This formula differs from that for a slit (Eq. 24–3) by the factor 1.22. This factor appears because the width of a circular hole is not uniform (like a rectangular slit) but varies from its diameter D to zero. A mathematical analysis shows that the "average" width is $D/1.22$. Hence we get the equation above rather than Eq. 24–3. The intensity of light in the diffraction pattern from a point source of light passing through a circular opening is shown in Fig. 25–30. The image for a non-point source is a superposition of such patterns. For most purposes we need consider only the central spot, since the concentric rings are so much dimmer.

If two point objects are very close, the diffraction patterns of their images will overlap as shown in Fig. 25–29b. As the objects are moved closer, a separation is reached where you can't tell if there are two overlapping images or a single image. The separation at which this happens may be judged differently by different observers. However, a generally accepted criterion is that proposed by Lord Rayleigh (1842–1919). This **Rayleigh criterion** states that *two images are just resolvable when the center of the diffraction disk of one image is directly over the first minimum in the diffraction pattern of the other.* This is shown in Fig. 25–31. Since the first minimum is at an angle $\theta = 1.22\lambda/D$ from the central maximum, Fig. 25–31 shows that two objects can be considered *just resolvable* if they are separated by at least an angle θ given by

$$\theta = \frac{1.22\lambda}{D}. \qquad \begin{bmatrix}\text{2 points just resolvable;}\\ \theta \text{ in radians}\end{bmatrix} \quad (25\text{–}7)$$

In this equation, D is the diameter of the lens, and applies also to a mirror diameter. This is the limit on resolution set by the wave nature of light due to diffraction. A smaller angle means better resolution: you can make out closer objects. We see from Eq. 25–7 that using a shorter wavelength λ can reduce θ and thus increase resolution.

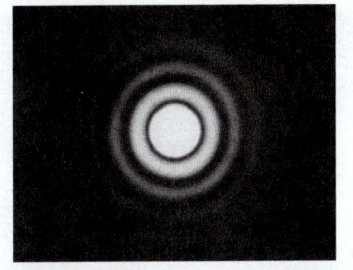
(a)

(b)

FIGURE 25–29 Photographs of images (greatly magnified) formed by a lens, showing the diffraction pattern of an image for: (a) a single point object; (b) two point objects whose images are barely resolved.

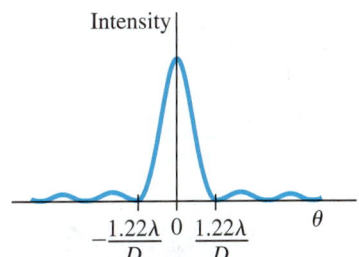

FIGURE 25–30 Intensity of light across the diffraction pattern of a circular hole.

FIGURE 25–31 The *Rayleigh criterion*. Two images are just resolvable when the center of the diffraction peak of one is directly over the first minimum in the diffraction pattern of the other. The two point objects O and O' subtend an angle θ at the lens; only one ray (it passes through the center of the lens) is drawn for each object, to indicate the center of the diffraction pattern of its image.

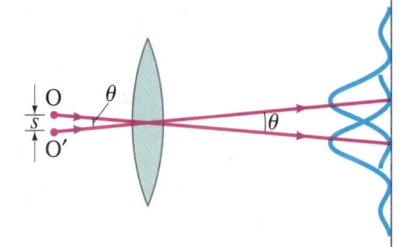

FIGURE 25–32 Hubble Space Telescope, with Earth in the background. The flat orange panels are solar cells that collect energy from the Sun to power the equipment.

How well the eye can see

EXAMPLE 25–10 Hubble Space Telescope. The Hubble Space Telescope (HST) is a reflecting telescope that was placed in orbit above the Earth's atmosphere, so its resolution would not be limited by turbulence in the atmosphere (Fig. 25–32). Its objective diameter is 2.4 m. For visible light, say $\lambda = 550$ nm, estimate the improvement in resolution the Hubble offers over Earth-bound telescopes, which are limited in resolution by movement of the Earth's atmosphere to about half an arc second. (Each degree is divided into 60 minutes each containing 60 seconds, so $1° = 3600$ arc seconds.)

APPROACH Angular resolution for the Hubble is given (in radians) by Eq. 25–7. The resolution for Earth telescopes is given, and we first convert it to radians so we can compare.

SOLUTION Earth-bound telescopes are limited to an angular resolution of

$$\theta = \frac{1}{2}\left(\frac{1}{3600}\right)°\left(\frac{2\pi \text{ rad}}{360°}\right) = 2.4 \times 10^{-6} \text{ rad}.$$

The Hubble, on the other hand, is limited by diffraction (Eq. 25–7) which for $\lambda = 550$ nm is

$$\theta = \frac{1.22\lambda}{D} = \frac{1.22(550 \times 10^{-9} \text{ m})}{2.4 \text{ m}} = 2.8 \times 10^{-7} \text{ rad},$$

thus giving almost ten times better resolution (2.4×10^{-6} rad$/2.8 \times 10^{-7}$ rad $\approx 9\times$).

EXAMPLE 25–11 ESTIMATE Eye resolution. You are in an airplane at an altitude of 10,000 m. If you look down at the ground, estimate the minimum separation s between objects that you could distinguish. Could you count cars in a parking lot? Consider only diffraction, and assume your pupil is about 3.0 mm in diameter and $\lambda = 550$ nm.

APPROACH We use the Rayleigh criterion, Eq. 25–7, to estimate θ. The separation s of objects is $s = \ell\theta$, where $\ell = 10^4$ m and θ is in radians.

SOLUTION In Eq. 25–7, we set $D = 3.0$ mm for the opening of the eye:

$$s = \ell\theta = \ell\frac{1.22\lambda}{D}$$

$$= (10^4 \text{ m})\frac{(1.22)(550 \times 10^{-9} \text{ m})}{3.0 \times 10^{-3} \text{ m}} = 2.2 \text{ m}.$$

Yes, you could just resolve a car (roughly 2 m wide by 3 or 4 m long) and so could count the number of cars in the lot.

EXERCISE E Someone claims a spy satellite camera can see 3-cm-high newspaper headlines from an altitude of 100 km. If diffraction were the only limitation ($\lambda = 550$ nm), use Eq. 25–7 to determine what diameter lens the camera would have.

25–8 Resolution of Telescopes and Microscopes; the λ Limit

You might think that a microscope or telescope could be designed to produce any desired magnification, depending on the choice of focal lengths and quality of the lenses. But this is not possible, because of diffraction. An increase in magnification above a certain point merely results in magnification of the diffraction patterns. This can be highly misleading since we might think we are seeing details of an object when we are really seeing details of the diffraction pattern.

To examine this problem, we apply the Rayleigh criterion: two objects (or two nearby points on one object) are just resolvable if they are separated by an angle θ (Fig. 25–31) given by Eq. 25–7:

$$\theta = \frac{1.22\lambda}{D}.$$

This formula is valid for either a microscope or a telescope, where D is the diameter of the objective lens or mirror. For a telescope, the resolution is specified by stating θ as given by this equation.[†]

EXAMPLE 25–12 **Telescope resolution (radio wave vs. visible light).** What is the theoretical minimum angular separation of two stars that can just be resolved by (a) the 200-inch telescope on Palomar Mountain (Fig. 25–22c); and (b) the Arecibo radiotelescope (Fig. 25–33), whose diameter is 300 m and whose radius of curvature is also 300 m. Assume $\lambda = 550$ nm for the visible-light telescope in part (a), and $\lambda = 4$ cm (the shortest wavelength at which the radiotelescope has operated) in part (b).

APPROACH We apply the Rayleigh criterion (Eq. 25–7) for each telescope.

SOLUTION (a) With $D = 200$ in. $= 5.1$ m, we have from Eq. 25–7 that

$$\theta = \frac{1.22\lambda}{D} = \frac{(1.22)(5.50 \times 10^{-7}\,\text{m})}{(5.1\,\text{m})} = 1.3 \times 10^{-7}\,\text{rad},$$

or 0.75×10^{-5} deg. (Note that this is equivalent to resolving two points less than 1 cm apart from a distance of 100 km!)

(b) For radio waves with $\lambda = 0.04$ m emitted by stars, the resolution is

$$\theta = \frac{(1.22)(0.04\,\text{m})}{(300\,\text{m})} = 1.6 \times 10^{-4}\,\text{rad}.$$

The resolution is less because the wavelength is so much larger, but the larger objective collects more radiation and thus detects fainter objects.

NOTE In both cases, we determined the limit set by diffraction. The resolution for a visible-light Earth-bound telescope is not this good because of aberrations and, more importantly, turbulence in the atmosphere. In fact, large-diameter objectives are not justified by increased resolution, but by their greater light-gathering ability—they allow more light in, so fainter objects can be seen. Radiotelescopes are not hindered by atmospheric turbulence, and the resolution found in (b) is a good estimate.

FIGURE 25–33 The 300-meter radiotelescope in Arecibo, Puerto Rico, uses radio waves (Fig. 22–8) instead of visible light.

PHYSICS APPLIED
Why large-diameter objectives

For a microscope, it is more convenient to specify the actual distance, s, between two points that are just barely resolvable: see Fig. 25–31. Since objects are normally placed near the focal point of the microscope objective, the angle subtended by two objects is $\theta = s/f$, so $s = f\theta$. If we combine this with Eq. 25–7, we obtain the **resolving power** (**RP**) of a microscope

$$\text{RP} = s = f\theta = \frac{1.22\lambda f}{D}, \qquad \text{[microscope]} \quad (25\text{–}8)$$

where f is the objective lens' focal length (not frequency) and D its diameter. The distance s is called the resolving power of the lens because it is the minimum separation of two object points that can just be resolved—assuming the highest quality lens since this limit is imposed by the wave nature of light. A smaller RP means better resolution, better detail.

[†]Earth-bound telescopes with large-diameter objectives are usually limited not by diffraction but by other effects such as turbulence in the atmosphere. The resolution of a high-quality microscope, on the other hand, normally *is* limited by diffraction; microscope objectives are complex compound lenses containing many elements of small diameter (since f is small), thus reducing aberrations.

Diffraction sets an ultimate limit on the detail that can be seen on any object. In Eq. 25–8 for the resolving power of a microscope, the focal length of the lens cannot practically be made less than (approximately) the radius of the lens (= $D/2$), and even that is very difficult (see the lensmaker's equation, Eq. 23–10). In this best case, Eq. 25–8 gives, with $f \approx D/2$,

$$\text{RP} \approx \frac{\lambda}{2}. \tag{25-9}$$

Thus we can say, to within a factor of 2 or so, that

> **it is not possible to resolve detail of objects smaller than the wavelength of the radiation being used.**

Wavelength limits resolution

This is an important and useful rule of thumb.

Compound lenses in microscopes are now designed so well that the actual limit on resolution is often set by diffraction—that is, by the wavelength of the light used. To obtain greater detail, one must use radiation of shorter wavelength. The use of UV radiation can increase the resolution by a factor of perhaps 2. Far more important, however, was the discovery in the early twentieth century that electrons have wave properties (Chapter 27) and that their wavelengths can be very small. The wave nature of electrons is utilized in the electron microscope (Section 27–9), which can magnify 100 to 1000 times more than a visible-light microscope because of the much shorter wavelengths. X-rays, too, have very short wavelengths and are often used to study objects in great detail (Section 25–11).

25–9 Resolution of the Human Eye and Useful Magnification

The resolution of the human eye is limited by several factors, all of roughly the same order of magnitude. The resolution is best at the fovea, where the cone spacing is smallest, about 3 μm (= 3000 nm). The diameter of the pupil varies from about 0.1 cm to about 0.8 cm. So for $\lambda = 550$ nm (where the eye's sensitivity is greatest), the diffraction limit is about $\theta \approx 1.22\lambda/D \approx 8 \times 10^{-5}$ rad to 6×10^{-4} rad. The eye is about 2 cm long, giving a resolving power (Eq. 25–8) of $s \approx (2 \times 10^{-2}\text{ m})(8 \times 10^{-5}\text{ rad}) \approx 2\ \mu\text{m}$ at best, to about 10 μm at worst (pupil small). Spherical and chromatic aberration also limit the resolution to about 10 μm. The net result is that the eye can just resolve objects whose angular separation is around

$$5 \times 10^{-4}\text{ rad.} \qquad \begin{bmatrix}\text{best eye}\\ \text{resolution}\end{bmatrix}$$

This corresponds to objects separated by 1 cm at a distance of about 20 m.

The typical near point of a human eye is about 25 cm. At this distance, the eye can just resolve objects that are $(25\text{ cm})(5 \times 10^{-4}\text{ rad}) \approx 10^{-4}\text{ m} = \frac{1}{10}$ mm apart.† Since the best light microscopes can resolve objects no smaller than about 200 nm at best (Eq. 25–9 for violet light, $\lambda = 400$ nm), the useful magnification [= (resolution by naked eye)/(resolution by microscope)] is limited to about

$$\frac{10^{-4}\text{ m}}{200 \times 10^{-9}\text{ m}} \approx 500\times. \qquad \begin{bmatrix}\text{maximum useful}\\ \text{microscope magnification}\end{bmatrix}$$

In practice, magnifications of about 1000× are often used to minimize eyestrain. Any greater magnification would simply make visible the diffraction pattern produced by the microscope objective lens.

EXERCISE F Return to the Chapter-Opening Question, page 713, and answer it again now. Try to explain why you may have answered differently the first time.

---

†A nearsighted eye that needs -8 or -10 D lenses can have a near point of 8 or 10 cm, and a higher resolution up close (without glasses) of a factor of $2\frac{1}{2}$ or 3, or $\approx \frac{1}{25}$ mm $\approx 40\ \mu$m.

*25–10 Specialty Microscopes and Contrast

All the resolving power a microscope can attain will be useless if the object to be seen cannot be distinguished from the background. The difference in brightness between the image of an object and the image of the surroundings is called **contrast**. Achieving high contrast is an important problem in microscopy and other forms of imaging. The problem arises in biology, for example, because cells consist largely of water and are almost uniformly transparent to light. We now briefly discuss two special types of microscope that can increase contrast: the interference and phase-contrast microscopes.

An **interference microscope** makes use of the wave properties of light in a direct way to increase contrast in a transparent object. Consider a transparent object—say, a bacterium in water (Fig. 25–34). Coherent light enters uniformly from the left and is in phase at all points such as a and b. If the object is as transparent as the water, the beam leaving at d will be as bright as that at c. There will be no contrast and the object will not be seen. However, if the object's refractive index is slightly different from that of the surrounding medium, the wavelength within the object will be altered as shown. Hence light waves at points c and d will differ in phase, if not in amplitude. The interference microscope changes this difference in phase into a difference of amplitude which our eyes can detect. Light that passes through the sample is superimposed onto a reference beam that does not pass through the object, so that they interfere. One way of doing this is shown in Fig. 25–35. Light from a source is split into two equal beams by a half-silvered mirror, MS_1. One beam passes through the object, and the second (comparison beam) passes through an identical system without the object. The two meet again and are superposed by the half-silvered mirror MS_2 before entering the eyepiece and eye. The path length (and amplitude) of the comparison beam is adjustable so that the background can be dark; that is, full destructive interference occurs. Light passing through the object (beam bd in Fig. 25–34) will also interfere with the comparison beam. But because of its different phase, the interference will not be completely destructive. Thus it will appear brighter than the background. Where the object varies in thickness, the phase difference between beams ac and bd in Fig. 25–34 will be different, thus affecting the amount of interference. Hence *variation in the thickness of the object will appear as variations in brightness in the image.*

A **phase-contrast microscope** also makes use of interference and differences in phase to produce a high-contrast image. Contrast is achieved by a circular glass *phase plate* that has a groove (or a raised portion) in the shape of a ring, positioned so undeviated source rays pass through it, but rays deviated by the object do not pass through this ring. Because the rays deviated by the object travel through a different thickness of glass than the undeviated source rays, the two can be out of phase and can interfere destructively at the object image plane. Thus the image of the object can contrast sharply with the background. Phase-contrast microscope images tend to have "halos" around them (as a result of diffraction from the phase-plate opening), so care must be taken in the interpretation of images.

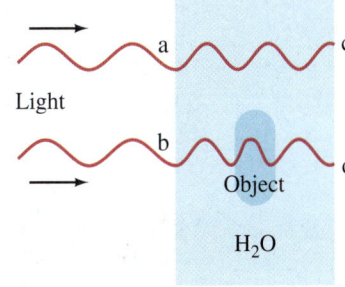

FIGURE 25–34 Object—say, a bacterium—in a water solution.

Interference microscope

FIGURE 25–35 Diagram of an interference microscope.

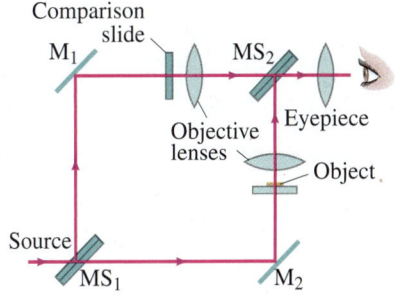

Phase-contrast microscope

FIGURE 25–36 X-ray tube. Electrons emitted by a heated filament in a vacuum tube are accelerated by a high voltage. When they strike the surface of the anode, the "target," X-rays are emitted.

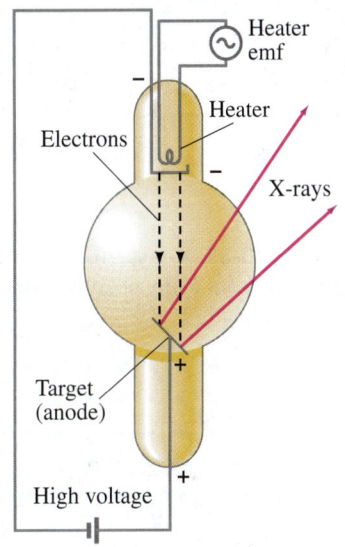

25–11 X-Rays and X-Ray Diffraction

In 1895, W. C. Roentgen (1845–1923) discovered that when electrons were accelerated by a high voltage in a vacuum tube and allowed to strike a glass or metal surface inside the tube, fluorescent minerals some distance away would glow, and photographic film would become exposed. Roentgen attributed these effects to a new type of radiation (different from cathode rays). They were given the name **X-rays** after the algebraic symbol x, meaning an unknown quantity. He soon found that X-rays penetrated through some materials better than through others, and within a few weeks he presented the first X-ray photograph (of his wife's hand). The production of X-rays today is usually done in a tube (Fig. 25–36) similar to Roentgen's, using voltages of typically 30 kV to 150 kV.

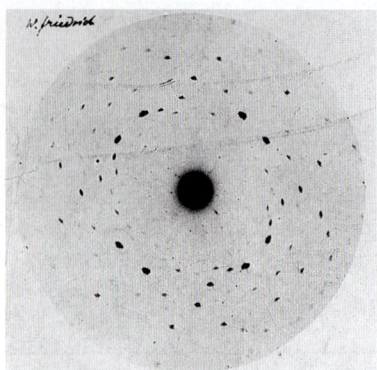

FIGURE 25–37 This X-ray diffraction pattern is one of the first observed by Max von Laue in 1912 when he aimed a beam of X-rays at a zinc sulfide crystal. The diffraction pattern was detected directly on a photographic plate.

Investigations into the nature of X-rays indicated they were not charged particles (such as electrons) since they could not be deflected by electric or magnetic fields. It was suggested that they might be a form of invisible light. However, they showed no diffraction or interference effects using ordinary gratings. Indeed, if their wavelengths were much smaller than the typical grating spacing of 10^{-6} m $(= 10^3$ nm), no effects would be expected. Around 1912, Max von Laue (1879–1960) suggested that if the atoms in a crystal were arranged in a regular array (see Fig. 13–2a), such a crystal might serve as a diffraction grating for very short wavelengths on the order of the spacing between atoms, estimated to be about 10^{-10} m $(= 10^{-1}$ nm). Experiments soon showed that X-rays scattered from a crystal did indeed show the peaks and valleys of a diffraction pattern (Fig. 25–37). Thus it was shown, in a single blow, that X-rays have a wave nature and that atoms are arranged in a regular way in crystals. Today, X-rays are recognized as electromagnetic radiation with wavelengths in the range of about 10^{-2} nm to 10 nm, the range readily produced in an X-ray tube.

*X-Ray Diffraction

We saw in Sections 25–7 and 25–8 that light of shorter wavelength provides greater resolution when we are examining an object microscopically. Since X-rays have much shorter wavelengths than visible light, they should in principle offer much greater resolution. However, there seems to be no effective material to use as lenses for the very short wavelengths of X-rays. Instead, the clever but complicated technique of **X-ray diffraction** (or **crystallography**) has proved very effective for examining the microscopic world of atoms and molecules. In a simple crystal such as NaCl, the atoms are arranged in an orderly cubical fashion, Fig. 25–38, with atoms spaced a distance d apart. Suppose that a beam of X-rays is incident on the crystal at an angle ϕ to the surface, and that the two rays shown are reflected from two subsequent planes of atoms as shown. The two rays will constructively interfere if the extra distance ray I travels is a whole number of wavelengths farther than the distance ray II travels. This extra distance is $2d \sin \phi$. Therefore, constructive interference will occur when

$$m\lambda = 2d \sin \phi, \qquad m = 1, 2, 3, \cdots, \qquad (25\text{--}10)$$

where m can be any integer. (Notice that ϕ is *not* the angle with respect to the normal to the surface.) This is called the **Bragg equation** after W. L. Bragg (1890–1971), who derived it and who, together with his father, W. H. Bragg (1862–1942), developed the theory and technique of X-ray diffraction by crystals in 1912–1913. If the X-ray wavelength is known and the angle ϕ is measured, the distance d between atoms can be obtained. This is the basis for X-ray crystallography.

FIGURE 25–38 X-ray diffraction by a crystal.

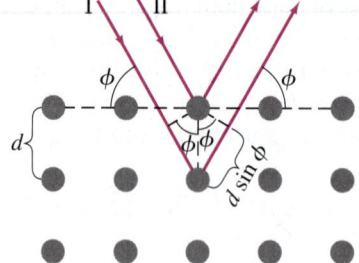

FIGURE 25–39 X-rays can be diffracted from many possible planes within a crystal.

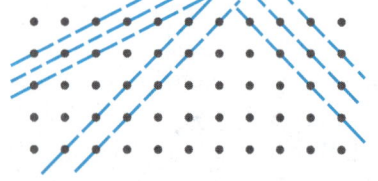

EXERCISE G When X-rays of wavelength 0.10×10^{-9} m are scattered from a sodium chloride crystal, a second-order diffraction peak is observed at 21°. What is the spacing between the planes of atoms for this scattering?

Actual X-ray diffraction patterns are quite complicated. First of all, a crystal is a three-dimensional object, and X-rays can be diffracted from different planes at different angles within the crystal, as shown in Fig. 25–39. Although the analysis is complex, a great deal can be learned from X-ray diffraction about any substance that can be put in crystalline form.

X-ray diffraction has been very useful in determining the structure of biologically important molecules, such as the double helix structure of DNA, worked out by James Watson and Francis Crick in 1953. See Fig. 25–40, and for models of the double helix, Figs. 16–39a and 16–40. Around 1960, the first detailed structure of a protein molecule, myoglobin, was elucidated with the aid of X-ray diffraction. Soon the structure of an important constituent of blood, hemoglobin, was worked out, and since then the structures of a great many molecules have been determined with the help of X-rays.

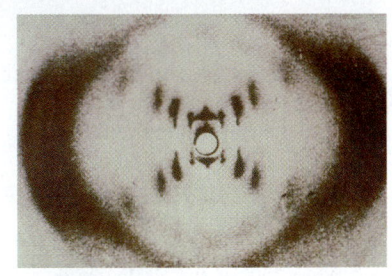

FIGURE 25–40 X-ray diffraction photo of DNA molecules taken by Rosalind Franklin in the early 1950s. The cross of spots suggested that DNA is a helix.

*25–12 X-Ray Imaging and Computed Tomography (CT Scan)

*Normal X-Ray Image

For a conventional medical or dental X-ray photograph, the X-rays emerging from the tube (Fig. 25–36) pass through the body and are detected on photographic film, a digital sensor, or a fluorescent screen, Fig. 25–41. The rays travel in very nearly straight lines through the body with minimal deviation since at X-ray wavelengths there is little diffraction or refraction. There is absorption (and scattering), however; and the difference in absorption by different structures in the body is what gives rise to the image produced by the transmitted rays. The less the absorption, the greater the transmission and the darker the film. The image is, in a sense, a "shadow" of what the rays have passed through. The X-ray image is *not* produced by focusing rays with lenses as for the instruments discussed earlier in this Chapter.

Normal X-ray image

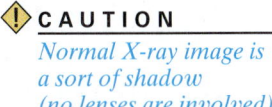
Normal X-ray image is a sort of shadow (no lenses are involved)

FIGURE 25–41 Conventional X-ray imaging, which is essentially shadowing.

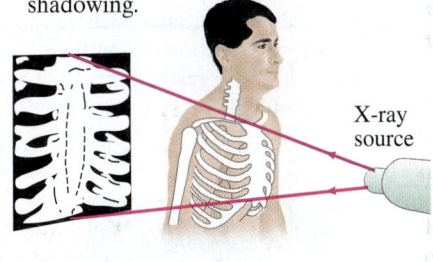

*Tomography Images (CT)

In conventional X-ray images, a body's thickness is projected onto film or a sensor; structures overlap and in many cases are difficult to distinguish. In the 1970s, a revolutionary X-ray technique was developed called **computed tomography** (CT), which produces an image of a *slice* through the body. (The word **tomography** comes from the Greek: *tomos* = slice, *graph* = picture.) Structures and lesions previously impossible to visualize can now be seen with remarkable clarity. The principle behind CT is shown in Fig. 25–42: a thin collimated (parallel) beam of X-rays passes through the body to a detector that measures the transmitted intensity. Measurements are made at a large number of points as the source and detector are moved past the body together. The apparatus is then rotated slightly about the body axis and again scanned; this is repeated at (perhaps) 1° intervals for 180°. The intensity of the transmitted beam for the many points of each scan, and for each angle, is sent to a computer that reconstructs the image of the slice. Note that the imaged slice is perpendicular to the long axis of the body. For this reason, CT is sometimes called **computerized axial tomography** (CAT), although the abbreviation CAT, as in CAT scan, can also be read as **computer-assisted tomography**.

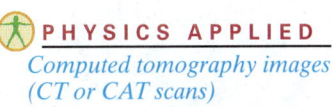
Computed tomography images (CT or CAT scans)

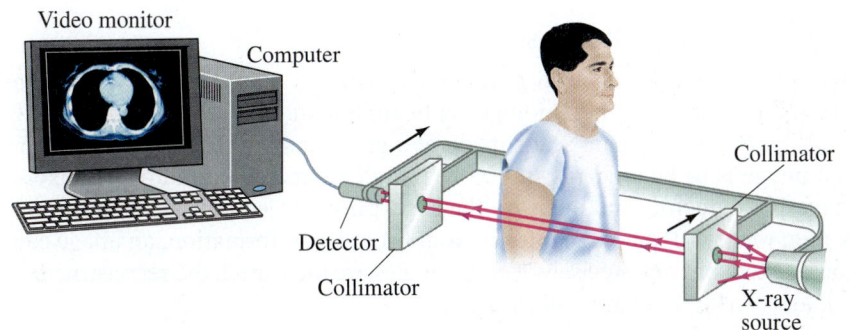

FIGURE 25–42 Tomographic imaging: the X-ray source and detector move together across the body, the transmitted intensity being measured at a large number of points. Then the "source-detector" assembly is rotated slightly (say, 1°) around a vertical axis, and another scan is made. This process is repeated for perhaps 180°. The computer reconstructs the image of the slice and it is presented on a TV or computer monitor.

FIGURE 25–43 (a) Fan-beam scanner. Rays transmitted through the entire body are measured simultaneously at each angle. The source and detector rotate to take measurements at different angles. In another type of fan-beam scanner, there are detectors around the entire 360° of the circle which remain fixed as the source moves. (b) In still another type, a beam of electrons from a source is directed by magnetic fields at tungsten targets surrounding the patient.

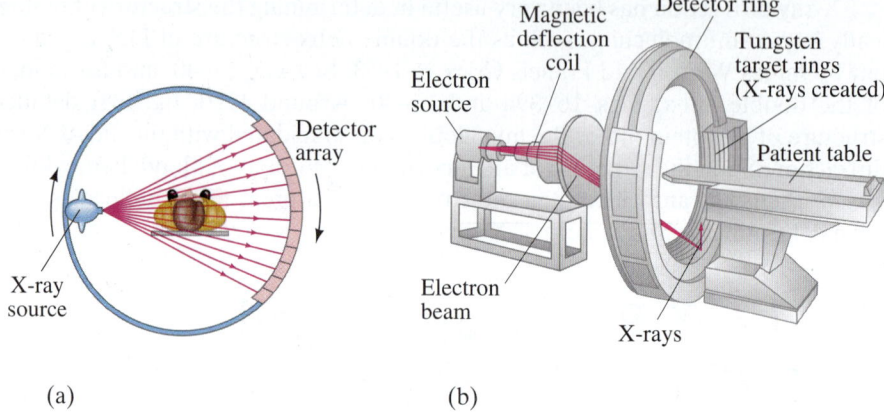

The use of a single detector as in Fig. 25–42 would require a few minutes for the many scans needed to form a complete image. Much faster scanners use a fan beam, Fig. 25–43a, in which beams passing through the entire cross section of the body are detected simultaneously by many detectors. The source and detectors are then rotated about the patient, and an image requires only a few seconds. Even faster, and therefore useful for heart scans, are fixed source machines wherein an electron beam is directed (by magnetic fields) to tungsten targets surrounding the patient, creating the X-rays. See Fig. 25–43b.

*Image Formation

But how is the image formed? We can think of the slice to be imaged as being divided into many tiny picture elements (or **pixels**), which could be squares (as in Fig. 24–49). For CT, the width of each pixel is chosen according to the width of the detectors and/or the width of the X-ray beams, and this determines the resolution of the image, which might be 1 mm. An X-ray detector measures the intensity of the transmitted beam. Subtracting this value from the intensity of the beam at the source yields the total absorption (called a "projection") along that beam line. Complicated mathematical techniques are used to analyze all the absorption projections for the huge number of beam scans measured (see the next Subsection), obtaining the absorption at each pixel and assigning each a "grayness value" according to how much radiation was absorbed. The image is made up of tiny spots (pixels) of varying shades of gray. Often the amount of absorption is color-coded. The colors in the resulting **false-color** image have nothing to do, however, with the actual color of the object. The actual images are monochromatic (various shades of gray, depending on the absorption). Only *visible* light has color; X-rays do not.

Figure 25–44 illustrates what actual CT images look like. It is generally agreed that CT scanning has revolutionized some areas of medicine by providing much less invasive, and/or more accurate, diagnosis.

Computed tomography can also be applied to ultrasound imaging (Section 12–9) and to emissions from radioisotopes and nuclear magnetic resonance (Sections 31–8 and 31–9).

FIGURE 25–44 Two CT images, with different resolutions, each showing a cross section of a brain. Photo (a) is of low resolution; photo (b), of higher resolution, shows a brain tumor, and uses false color to highlight it.

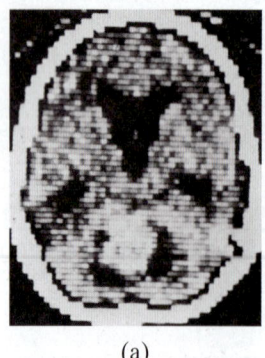

(a)

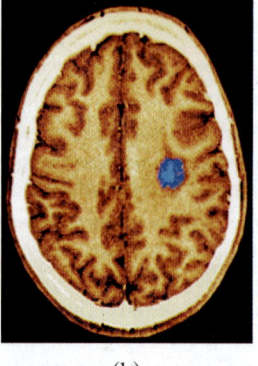

(b)

*Tomographic Image Reconstruction

How can the "grayness" of each pixel be determined even though all we can measure is the total absorption along each beam line in the slice? It can be done only by using the many beam scans made at a great many different angles. Suppose the image is to be an array of 100×100 elements for a total of 10^4 pixels. If we have 100 detectors and measure the absorption projections at 100 different angles, then we get 10^4 pieces of information. From this information, an image can be reconstructed, but not precisely. If more angles are measured, the reconstruction of the image can be done more accurately.

To suggest how mathematical reconstruction is done, we consider a very simple case using the **iterative** technique ("to iterate" is from the Latin "to repeat"). Suppose our sample slice is divided into the simple 2 × 2 pixels as shown in Fig. 25–45. The number inside each pixel represents the amount of absorption by the material in that area (say, in tenths of a percent): that is, 4 represents twice as much absorption as 2. But we cannot directly measure these values—they are the unknowns we want to solve for. All we can measure are the projections—the total absorption along each beam line—and these are shown in Fig. 25–45 outside the yellow squares as the sum of the absorptions for the pixels along each line at four different angles. These projections (given at the tip of each arrow) are what we can measure, and we now want to work back from them to see how close we can get to the true absorption value for each pixel. We start our analysis with each pixel being assigned a zero value, Fig. 25–46a. In the iterative technique, we use the projections to estimate the absorption value in each square, and repeat for each angle. The angle 1 projections are 7 and 13 (Fig. 25–45). We divide each of these equally between their two squares: each square in the left column of Fig. 25–46a gets $3\frac{1}{2}$ (half of 7), and each square in the right column gets $6\frac{1}{2}$ (half of 13).

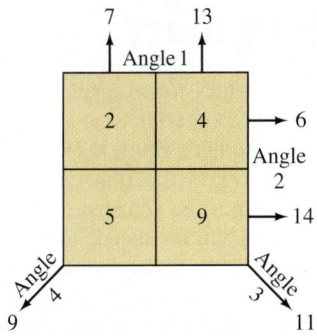

FIGURE 25–45 A simple 2 × 2 image showing true absorption values (inside the squares) and measured projections.

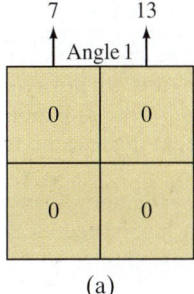

FIGURE 25–46 Reconstructing the image using projections in an iterative procedure.

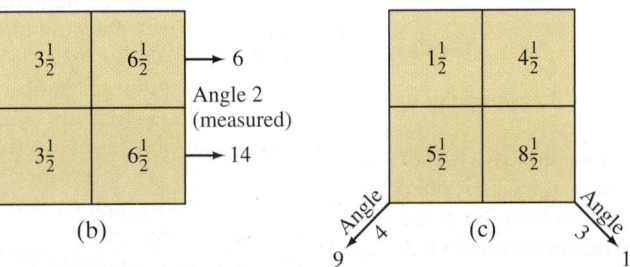

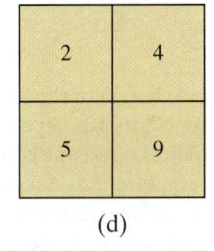

(a)   (b)   (c)   (d)

Next we use the projections at angle 2. We calculate the difference between the measured projections at angle 2 (6 and 14) and the projections based on the previous estimate (top row: $3\frac{1}{2} + 6\frac{1}{2} = 10$; same for bottom row). Then we distribute this difference equally to the squares in that row. For the top row, we have

$$3\frac{1}{2} + \frac{6-10}{2} = 1\frac{1}{2} \quad \text{and} \quad 6\frac{1}{2} + \frac{6-10}{2} = 4\frac{1}{2};$$

and for the bottom row,

$$3\frac{1}{2} + \frac{14-10}{2} = 5\frac{1}{2} \quad \text{and} \quad 6\frac{1}{2} + \frac{14-10}{2} = 8\frac{1}{2}.$$

These values are inserted as shown in Fig. 25–46c. Next, the projection at angle 3 (= 11), combined with the difference as above, gives

(upper left) $\quad 1\frac{1}{2} + \dfrac{11-10}{2} = 2 \quad$ and $\quad$ (lower right) $\quad 8\frac{1}{2} + \dfrac{11-10}{2} = 9;$

and then for angle 4 we have

(lower left) $\quad 5\frac{1}{2} + \dfrac{9-10}{2} = 5 \quad$ and $\quad$ (upper right) $\quad 4\frac{1}{2} + \dfrac{9-10}{2} = 4.$

The result, shown in Fig. 25–46d, corresponds exactly to the true values. (In real situations, the true values are not known, which is why these computer techniques are required.) To obtain these numbers exactly, we used six pieces of information (two each at angles 1 and 2, one each at angles 3 and 4). For the much larger number of pixels used for actual images, exact values are generally not attained. Many iterations may be needed, and the calculation is considered sufficiently precise when the difference between calculated and measured projections is sufficiently small. The above example illustrates the "convergence" of the process: the first iteration (b to c in Fig. 25–46) changed the values by 2, the last iteration (c to d) by only $\frac{1}{2}$.

Summary

A **camera** lens forms an image on film, or on an electronic sensor (CCD or CMOS) in a digital camera. Light is allowed in briefly through a shutter. The image is focused by moving the lens relative to the film or sensor, and the **f-stop** (or lens opening) must be adjusted for the brightness of the scene and the chosen shutter speed. The f-stop is defined as the ratio of the focal length to the diameter of the lens opening.

The human **eye** also adjusts for the available light—by opening and closing the iris. It focuses not by moving the lens, but by adjusting the shape of the lens to vary its focal length. The image is formed on the retina, which contains an array of receptors known as rods and cones.

Diverging eyeglass or contact lenses are used to correct the defect of a nearsighted eye, which cannot focus well on distant objects. Converging lenses are used to correct for defects in which the eye cannot focus on close objects.

A **simple magnifier** is a converging lens that forms a virtual image of an object placed at (or within) the focal point. The **angular magnification**, when viewed by a relaxed normal eye, is

$$M = \frac{N}{f}, \quad (25\text{–}2a)$$

where f is the focal length of the lens and N is the near point of the eye (25 cm for a "normal" eye).

An **astronomical telescope** consists of an **objective lens** or mirror, and an **eyepiece** that magnifies the real image formed by the objective. The **magnification** is equal to the ratio of the objective and eyepiece focal lengths, and the image is inverted:

$$M = -\frac{f_o}{f_e}. \quad (25\text{–}3)$$

A compound **microscope** also uses objective and eyepiece lenses, and the final image is inverted. The total magnification is the product of the magnifications of the two lenses and is approximately

$$M \approx \frac{N\ell}{f_e f_o}, \quad (25\text{–}6b)$$

where ℓ is the distance between the lenses, N is the near point of the eye, and f_o and f_e are the focal lengths of objective and eyepiece, respectively.

Microscopes, telescopes, and other optical instruments are limited in the formation of sharp images by **lens aberrations**. These include **spherical aberration**, in which rays passing through the edge of a lens are not focused at the same point as those that pass near the center; and **chromatic aberration**, in which different colors are focused at different points. Compound lenses, consisting of several elements, can largely correct for aberrations.

The wave nature of light also limits the sharpness, or **resolution**, of images. Because of diffraction, it is *not possible to discern details smaller than the wavelength* of the radiation being used. The useful magnification of a light microscope is limited by diffraction to about 500×.

[*__X-rays__ are a form of electromagnetic radiation of very short wavelength. They are produced when high-speed electrons, accelerated by high voltage in an evacuated tube, strike a glass or metal target.]

[*__Computed tomography__ (CT or CAT scan) uses many narrow X-ray beams through a section of the body to construct an image of that section.]

Questions

1. Why must a camera lens be moved farther from the sensor or film to focus on a closer object?
2. Why is the depth of field greater, and the image sharper, when a camera lens is "stopped down" to a larger f-number? Ignore diffraction.
3. Describe how diffraction affects the statement of Question 2. [*Hint*: See Eq. 24–3 or 25–7.]
4. Why are bifocals needed mainly by older persons and not generally by younger people?
5. Will a nearsighted person who wears corrective lenses in her glasses be able to see clearly underwater when wearing those glasses? Use a diagram to show why or why not.
6. You can tell whether people are nearsighted or farsighted by looking at the width of their face through their glasses. If a person's face appears narrower through the glasses (Fig. 25–47), is the person farsighted or nearsighted? Try to explain, but also check experimentally with friends who wear glasses.

FIGURE 25–47 Question 6.

7. In attempting to discern distant details, people will sometimes squint. Why does this help?
8. Is the image formed on the retina of the human eye upright or inverted? Discuss the implications of this for our perception of objects.
9. The human eye is much like a camera—yet, when a camera shutter is left open and the camera is moved, the image will be blurred. But when you move your head with your eyes open, you still see clearly. Explain.
10. Reading glasses use converging lenses. A simple magnifier is also a converging lens. Are reading glasses therefore magnifiers? Discuss the similarities and differences between converging lenses as used for these two different purposes.
11. Nearsighted people often look over (or under) their glasses when they want to see something small up close, like a cell phone screen. Why?
12. Spherical aberration in a thin lens is minimized if rays are bent equally by the two surfaces. If a planoconvex lens is used to form a real image of an object at infinity, which surface should face the object? Use ray diagrams to show why.
13. Explain why chromatic aberration occurs for thin lenses but not for mirrors.
14. Inexpensive microscopes for children's use usually produce images that are colored at the edges. Why?

15. Which aberrations present in a simple lens are not present (or are greatly reduced) in the human eye?
16. By what factor can you improve resolution, other things being equal, if you use blue light ($\lambda = 450$ nm) rather than red (700 nm)?
17. Atoms have diameters of about 10^{-8} cm. Can visible light be used to "see" an atom? Explain.
18. Which color of visible light would give the best resolution in a microscope? Explain.
19. For both converging and diverging lenses, discuss how the focal length for red light differs from that for violet light.
20. The 300-meter radiotelescope in Arecibo, Puerto Rico (Fig. 25–33), is the world's largest radiotelescope, but many other radiotelescopes are also very large. Why are radiotelescopes so big? Why not make optical telescopes that are equally large? (The largest optical telescopes have diameters of about 10 meters.)

MisConceptual Questions

1. The image of a nearby object formed by a camera lens is
   (a) at the lens' focal point.
   (b) always blurred.
   (c) at the same location as the image of an object at infinity.
   (d) farther from the lens than the lens' focal point.
2. What is a megapixel in a digital camera?
   (a) A large spot on the detector where the image is focused.
   (b) A special kind of lens that gives a sharper image.
   (c) A number related to how many photographs the camera can store.
   (d) A million light-sensitive spots on the detector.
   (e) A number related to how fast the camera can take pictures.
3. When a nearsighted person looks at a distant object through her glasses, the image produced by the glasses should be
   (a) about 25 cm from her eye.
   (b) at her eye's far point.
   (c) at her eye's near point.
   (d) at the far point for a normal eye.
4. If the distance from your eye's lens to the retina is shorter than for a normal eye, you will struggle to see objects that are
   (a) nearby.   (c) colorful.
   (b) far away.   (d) moving fast.
5. The image produced on the retina of the eye is _____ compared to the object being viewed.
   (a) inverted.   (c) sideways.
   (b) upright.   (d) enlarged.
6. How do eyeglasses help a nearsighted person see more clearly?
   (a) Diverging lenses bend light entering the eye, so the image focuses farther from the front of the eye.
   (b) Diverging lenses bend light entering the eye, so the image focuses closer to the front of the eye.
   (c) Converging lenses bend light entering the eye, so the image focuses farther from the front of the eye.
   (d) Converging lenses bend light entering the eye, so the image focuses closer to the front of the eye.
   (e) Lenses adjust the distance from the cornea to the back of the eye.
7. When you closely examine an object through a magnifying glass, the magnifying glass
   (a) makes the object bigger.
   (b) makes the object appear closer than it actually is.
   (c) makes the object appear farther than it actually is.
   (d) causes additional light rays to be emitted by the object.
8. It would be impossible to build a microscope that could use visible light to see the molecular structure of a crystal because.
   (a) lenses with enough magnification cannot be made.
   (b) lenses cannot be ground with fine enough precision.
   (c) lenses cannot be placed in the correct place with enough precision.
   (d) diffraction limits the resolving power to about the size of the wavelength of the light used.
   (e) More than one of the above is correct.
9. Why aren't white-light microscopes made with a magnification of 3000×?
   (a) Lenses can't be made large enough.
   (b) Lenses can't be made small enough.
   (c) Lenses can't be made with short enough focal lengths.
   (d) Lenses can't be made with long enough focal lengths.
   (e) Diffraction limits useful magnification to several times less than this.
10. The resolving power of a microscope is greatest when the object being observed is illuminated by
    (a) ultraviolet light.   (c) visible light.
    (b) infrared light.   (d) radio waves.
11. Which of the following statements is true?
    (a) A larger-diameter lens can better resolve two distant points.
    (b) Red light can better resolve two distant points than blue light can.
    (c) It is easier to resolve distant objects than nearer objects.
    (d) Objects that are closer together are easier to resolve than objects that are farther apart.
12. While you are photographing a dog, it begins to move away. What must you do to keep it in focus?
    (a) Increase the f-stop value.
    (b) Decrease the f-stop value.
    (c) Move the lens away from the sensor or film.
    (d) Move the lens closer to the sensor or film.
    (e) None of the above.
13. A converging lens, like the type used in a magnifying glass,
    (a) always produces a magnified image (image taller than the object).
    (b) can also produce an image smaller than the object.
    (c) always produces an upright image.
    (d) can also produce an inverted image (upside down).
    (e) None of these statements are true.

Problems

25–1 Camera

1. (I) A properly exposed photograph is taken at $f/16$ and $\frac{1}{100}$ s. What lens opening is required if the shutter speed is $\frac{1}{400}$ s?

2. (I) A television camera lens has a 17-cm focal length and a lens diameter of 6.0 cm. What is its f-number?

3. (I) A 65-mm-focal-length lens has f-stops ranging from $f/1.4$ to $f/22$. What is the corresponding range of lens diaphragm diameters?

4. (I) A light meter reports that a camera setting of $\frac{1}{500}$ s at $f/5.6$ will give a correct exposure. But the photographer wishes to use $f/11$ to increase the depth of field. What should the shutter speed be?

5. (II) For a camera equipped with a 55-mm-focal-length lens, what is the object distance if the image height equals the object height? How far is the object from the image on the film?

6. (II) A nature photographer wishes to shoot a 34-m-tall tree from a distance of 65 m. What focal-length lens should be used if the image is to fill the 24-mm height of the sensor?

7. (II) A 200-mm-focal-length lens can be adjusted so that it is 200.0 mm to 208.2 mm from the film. For what range of object distances can it be adjusted?

8. (II) How large is the image of the Sun on film used in a camera with (a) a 28-mm-focal-length lens, (b) a 50-mm-focal-length lens, and (c) a 135-mm-focal-length lens? (d) If the 50-mm lens is considered normal for this camera, what relative magnification does each of the other two lenses provide? The Sun has diameter 1.4×10^6 km, and it is 1.5×10^8 km away.

9. (II) If a 135-mm telephoto lens is designed to cover object distances from 1.30 m to ∞, over what distance must the lens move relative to the plane of the sensor or film?

10. (III) Show that for objects very far away (assume infinity), the magnification of any camera lens is proportional to its focal length.

25–2 Eye and Corrective Lenses

11. (I) A human eyeball is about 2.0 cm long and the pupil has a maximum diameter of about 8.0 mm. What is the "speed" of this lens?

12. (II) A person struggles to read by holding a book at arm's length, a distance of 52 cm away. What power of reading glasses should be prescribed for her, assuming they will be placed 2.0 cm from the eye and she wants to read at the "normal" near point of 25 cm?

13. (II) Reading glasses of what power are needed for a person whose near point is 125 cm, so that he can read a computer screen at 55 cm? Assume a lens–eye distance of 1.8 cm.

14. (II) An eye is corrected by a -5.50-D lens, 2.0 cm from the eye. (a) Is this eye near- or farsighted? (b) What is this eye's far point without glasses?

15. (II) A person's right eye can see objects clearly only if they are between 25 cm and 85 cm away. (a) What power of contact lens is required so that objects far away are sharp? (b) What will be the near point with the lens in place?

16. (II) About how much longer is the nearsighted eye in Example 25–6 than the 2.0 cm of a normal eye?

17. (II) A person has a far point of 14 cm. What power glasses would correct this vision if the glasses were placed 2.0 cm from the eye? What power contact lenses, placed on the eye, would the person need?

18. (II) One lens of a nearsighted person's eyeglasses has a focal length of -26.0 cm and the lens is 1.8 cm from the eye. If the person switches to contact lenses placed directly on the eye, what should be the focal length of the corresponding contact lens?

19. (II) What is the focal length of the eye–lens system when viewing an object (a) at infinity, and (b) 34 cm from the eye? Assume that the lens–retina distance is 2.0 cm.

20. (III) The closely packed cones in the fovea of the eye have a diameter of about 2 μm. For the eye to discern two images on the fovea as distinct, assume that the images must be separated by at least one cone that is not excited. If these images are of two point-like objects at the eye's 25-cm near point, how far apart are these barely resolvable objects? Assume the eye's diameter (cornea-to-fovea distance) is 2.0 cm.

21. (III) A nearsighted person has near and far points of 10.6 and 20.0 cm, respectively. If she puts on contact lenses with power $P = -4.00$ D, what are her new near and far points?

25–3 Magnifying Glass

22. (I) What is the focal length of a magnifying glass of 3.2$\times$ magnification for a relaxed normal eye?

23. (I) What is the magnification of a lens used with a relaxed eye if its focal length is 16 cm?

24. (I) A magnifier is rated at 3.5$\times$ for a normal eye focusing on an image at the near point. (a) What is its focal length? (b) What is its focal length if the 3.5$\times$ refers to a relaxed eye?

25. (II) Sherlock Holmes is using an 8.20-cm-focal-length lens as his magnifying glass. To obtain maximum magnification, where must the object be placed (assume a normal eye), and what will be the magnification?

26. (II) A small insect is placed 4.85 cm from a +5.00-cm-focal-length lens. Calculate (a) the position of the image, and (b) the angular magnification.

27. (II) A 3.80-mm-wide bolt is viewed with a 9.60-cm-focal-length lens. A normal eye views the image at its near point. Calculate (a) the angular magnification, (b) the width of the image, and (c) the object distance from the lens.

28. (II) A magnifying glass with a focal length of 9.2 cm is used to read print placed at a distance of 8.0 cm. Calculate (a) the position of the image; (b) the angular magnification.

29. (III) A writer uses a converging lens of focal length $f = 12$ cm as a magnifying glass to read fine print on his book contract. Initially, the writer holds the lens above the fine print so that its image is at infinity. To get a better look, he then moves the lens so that the image is at his 25-cm near point. How far, and in what direction (toward or away from the fine print) did the writer move the lens? Assume his eye is adjusted to remain always very near the magnifying glass.

30. (III) A magnifying glass is rated at 3.0$\times$ for a normal eye that is relaxed. What would be the magnification for a relaxed eye whose near point is (a) 75 cm, and (b) 15 cm? Explain the differences.

25–4 Telescopes

31. (I) What is the magnification of an astronomical telescope whose objective lens has a focal length of 82 cm, and whose eyepiece has a focal length of 2.8 cm? What is the overall length of the telescope when adjusted for a relaxed eye?

32. (I) The overall magnification of an astronomical telescope is desired to be 25×. If an objective of 88-cm focal length is used, what must be the focal length of the eyepiece? What is the overall length of the telescope when adjusted for use by the relaxed eye?

33. (II) A 7.0× binocular has 3.5-cm-focal-length eyepieces. What is the focal length of the objective lenses?

34. (II) An astronomical telescope has an objective with focal length 75 cm and a +25-D eyepiece. What is the total magnification?

35. (II) An astronomical telescope has its two lenses spaced 82.0 cm apart. If the objective lens has a focal length of 78.5 cm, what is the magnification of this telescope? Assume a relaxed eye.

36. (II) A Galilean telescope adjusted for a relaxed eye is 36.8 cm long. If the objective lens has a focal length of 39.0 cm, what is the magnification?

37. (II) What is the magnifying power of an astronomical telescope using a reflecting mirror whose radius of curvature is 6.1 m and an eyepiece whose focal length is 2.8 cm?

38. (II) The Moon's image appears to be magnified 150× by a reflecting astronomical telescope with an eyepiece having a focal length of 3.1 cm. What are the focal length and radius of curvature of the main (objective) mirror?

39. (II) A 120× astronomical telescope is adjusted for a relaxed eye when the two lenses are 1.10 m apart. What is the focal length of each lens?

40. (II) An astronomical telescope longer than about 50 cm is not easy to hold by hand. Estimate the maximum angular magnification achievable for a telescope designed to be handheld. Assume its eyepiece lens, if used as a magnifying glass, provides a magnification of 5× for a relaxed eye with near point $N = 25$ cm.

41. (III) A reflecting telescope (Fig. 25–22b) has a radius of curvature of 3.00 m for its objective mirror and a radius of curvature of -1.50 m for its eyepiece mirror. If the distance between the two mirrors is 0.90 m, how far in front of the eyepiece should you place the electronic sensor to record the image of a star?

42. (III) A 6.5× pair of binoculars has an objective focal length of 26 cm. If the binoculars are focused on an object 4.0 m away (from the objective), what is the magnification? (The 6.5× refers to objects at infinity; Eq. 25–3 holds only for objects at infinity and not for nearby ones.)

25–5 Microscopes

43. (I) A microscope uses an eyepiece with a focal length of 1.70 cm. Using a normal eye with a final image at infinity, the barrel length is 17.5 cm and the focal length of the objective lens is 0.65 cm. What is the magnification of the microscope?

44. (I) A 720× microscope uses a 0.40-cm-focal-length objective lens. If the barrel length is 17.5 cm, what is the focal length of the eyepiece? Assume a normal eye and that the final image is at infinity.

45. (I) A 17-cm-long microscope has an eyepiece with a focal length of 2.5 cm and an objective with a focal length of 0.33 cm. What is the approximate magnification?

46. (II) A microscope has a 14.0× eyepiece and a 60.0× objective lens 20.0 cm apart. Calculate (a) the total magnification, (b) the focal length of each lens, and (c) where the object must be for a normal relaxed eye to see it in focus.

47. (II) Repeat Problem 46 assuming that the final image is located 25 cm from the eyepiece (near point of a normal eye).

48. (II) A microscope has a 1.8-cm-focal-length eyepiece and a 0.80-cm objective. Assuming a relaxed normal eye, calculate (a) the position of the object if the distance between the lenses is 14.8 cm, and (b) the total magnification.

49. (II) The eyepiece of a compound microscope has a focal length of 2.80 cm and the objective lens has $f = 0.740$ cm. If an object is placed 0.790 cm from the objective lens, calculate (a) the distance between the lenses when the microscope is adjusted for a relaxed eye, and (b) the total magnification.

50. (III) An inexpensive instructional lab microscope allows the user to select its objective lens to have a focal length of 32 mm, 15 mm, or 3.9 mm. It also has two possible eyepieces with magnifications 5× and 15×. Each objective forms a real image 160 mm beyond its focal point. What are the largest and smallest overall magnifications obtainable with this instrument?

25–6 Lens Aberrations

51. (II) An achromatic lens is made of two very thin lenses, placed in contact, that have focal lengths $f_1 = -27.8$ cm and $f_2 = +25.3$ cm. (a) Is the combination converging or diverging? (b) What is the net focal length?

*52. (III) A planoconvex lens (Fig. 23–31a) has one flat surface and the other has $R = 14.5$ cm. This lens is used to view a red and yellow object which is 66.0 cm away from the lens. The index of refraction of the glass is 1.5106 for red light and 1.5226 for yellow light. What are the locations of the red and yellow images formed by the lens? [*Hint*: See Section 23–10.]

25–7 to 25–9 Resolution Limits

53. (I) What is the angular resolution limit (degrees) set by diffraction for the 100-inch (254-cm mirror diameter) Mt. Wilson telescope ($\lambda = 560$ nm)?

54. (I) What is the resolving power of a microscope ($\lambda = 550$ nm) with a 5-mm-diameter objective which has $f = 9$ mm?

55. (II) Two stars 18 light-years away are barely resolved by a 66-cm (mirror diameter) telescope. How far apart are the stars? Assume $\lambda = 550$ nm and that the resolution is limited by diffraction.

56. (II) The nearest neighboring star to the Sun is about 4 light-years away. If a planet happened to be orbiting this star at an orbital radius equal to that of the Earth–Sun distance, what minimum diameter would an Earth-based telescope's aperture have to be in order to obtain an image that resolved this star–planet system? Assume the light emitted by the star and planet has a wavelength of 550 nm.

57. (II) If you could shine a very powerful flashlight beam toward the Moon, estimate the diameter of the beam when it reaches the Moon. Assume that the beam leaves the flashlight through a 5.0-cm aperture, that its white light has an average wavelength of 550 nm, and that the beam spreads due to diffraction only.

58. (II) The normal lens on a 35-mm camera has a focal length of 50.0 mm. Its aperture diameter varies from a maximum of 25 mm ($f/2$) to a minimum of 3.0 mm ($f/16$). Determine the resolution limit set by diffraction for ($f/2$) and ($f/16$). Specify as the number of lines per millimeter resolved on the detector or film. Take $\lambda = 560$ nm.

59. (III) Suppose that you wish to construct a telescope that can resolve features 6.5 km across on the Moon, 384,000 km away. You have a 2.0-m-focal-length objective lens whose diameter is 11.0 cm. What focal-length eyepiece is needed if your eye can resolve objects 0.10 mm apart at a distance of 25 cm? What is the resolution limit set by the size of the objective lens (that is, by diffraction)? Use $\lambda = 560$ nm.

*25–11 X-Ray Diffraction

*60. (II) X-rays of wavelength 0.138 nm fall on a crystal whose atoms, lying in planes, are spaced 0.285 nm apart. At what angle ϕ (relative to the surface, Fig. 25–38) must the X-rays be directed if the first diffraction maximum is to be observed?

*61. (II) First-order Bragg diffraction is observed at 23.8° relative to the crystal surface, with spacing between atoms of 0.24 nm. (a) At what angle will second order be observed? (b) What is the wavelength of the X-rays?

*62. (II) If X-ray diffraction peaks corresponding to the first three orders ($m = 1$, 2, and 3) are measured, can both the X-ray wavelength λ and lattice spacing d be determined? Prove your answer.

*25–12 Imaging by Tomography

*63. (II) (a) Suppose for a conventional X-ray image that the X-ray beam consists of parallel rays. What would be the magnification of the image? (b) Suppose, instead, that the X-rays come from a point source (as in Fig. 25–41) that is 15 cm in front of a human body which is 25 cm thick, and the film is pressed against the person's back. Determine and discuss the range of magnifications that result.

General Problems

64. A **pinhole** camera uses a tiny pinhole instead of a lens. Show, using ray diagrams, how reasonably sharp images can be formed using such a pinhole camera. In particular, consider two point objects 2.0 cm apart that are 1.0 m from a 1.0-mm-diameter pinhole. Show that on a piece of film 7.0 cm behind the pinhole the two objects produce two separate circles that do not overlap.

65. Suppose that a correct exposure is $\frac{1}{250}$ s at $f/11$. Under the same conditions, what exposure time would be needed for a *pinhole* camera (Problem 64) if the pinhole diameter is 1.0 mm and the film is 7.0 cm from the hole?

66. An astronomical telescope has a magnification of 7.5×. If the two lenses are 28 cm apart, determine the focal length of each lens.

67. (a) How far away can a human eye distinguish two car headlights 2.0 m apart? Consider only diffraction effects and assume an eye pupil diameter of 6.0 mm and a wavelength of 560 nm. (b) What is the minimum angular separation an eye could resolve when viewing two stars, considering only diffraction effects? In reality, it is about 1' of arc. Why is it not equal to your answer in (b)?

68. Figure 25–48 was taken from the NIST Laboratory (National Institute of Standards and Technology) in Boulder, CO, 2.0 km from the hiker in the photo. The Sun's image was 15 mm across on the film. Estimate the focal length of the camera lens (actually a telescope). The Sun has diameter 1.4×10^6 km, and it is 1.5×10^8 km away.

69. A 1.0-cm-diameter lens with a focal length of 35 cm uses blue light to image two objects 15 m away that are very close together. What is the closest those objects can be to each other and still be imaged as separate objects?

70. A movie star catches a reporter shooting pictures of her at home. She claims the reporter was trespassing. To prove her point, she gives as evidence the film she seized. Her 1.65-m height is 8.25 mm high on the film, and the focal length of the camera lens was 220 mm. How far away from the subject was the reporter standing?

71. As early morning passed toward midday, and the sunlight got more intense, a photographer noted that, if she kept her shutter speed constant, she had to change the f-number from $f/5.6$ to $f/16$. By what factor had the sunlight intensity increased during that time?

72. A child has a near point of 15 cm. What is the maximum magnification the child can obtain using a 9.5-cm-focal-length magnifier? What magnification can a normal eye obtain with the same lens? Which person sees more detail?

73. A woman can see clearly with her right eye only when objects are between 45 cm and 135 cm away. Prescription bifocals should have what powers so that she can see distant objects clearly (upper part) and be able to read a book 25 cm away (lower part) with her right eye? Assume that the glasses will be 2.0 cm from the eye.

74. What is the magnifying power of a $+4.0$-D lens used as a magnifier? Assume a relaxed normal eye.

75. A physicist lost in the mountains tries to make a telescope using the lenses from his reading glasses. They have powers of $+2.0$ D and $+5.5$ D, respectively. (a) What maximum magnification telescope is possible? (b) Which lens should be used as the eyepiece?

76. A person with normal vision adjusts a microscope for a good image when her eye is relaxed. She then places a camera where her eye was. For what object distance should the camera be set? Explain.

77. A 50-year-old man uses $+2.5$-D lenses to read a newspaper 25 cm away. Ten years later, he must hold the paper 38 cm away to see clearly with the same lenses. What power lenses does he need now in order to hold the paper 25 cm away? (Distances are measured from the lens.)

FIGURE 25–48
Problem 68.

78. Two converging lenses, one with $f = 4.0$ cm and the other with $f = 48$ cm, are made into a telescope. (a) What are the length and magnification? Which lens should be the eyepiece? (b) Assume these lenses are now combined to make a microscope; if the magnification needs to be 25×, how long would the microscope be?

79. An X-ray tube operates at 95 kV with a current of 25 mA and nearly all the electron energy goes into heat. If the specific heat of the 0.065-kg anode plate is 0.11 kcal/kg·C°, what will be the temperature rise per minute if no cooling water is used? (See Fig. 25–36.)

80. Human vision normally covers an angle of roughly 40° horizontally. A "normal" camera lens then is defined as follows: When focused on a distant horizontal object which subtends an angle of 40°, the lens produces an image that extends across the full horizontal extent of the camera's light-recording medium (film or electronic sensor). Determine the focal length f of the "normal" lens for the following types of cameras: (a) a 35-mm camera that records images on film 36 mm wide; (b) a digital camera that records images on a charge-coupled device (CCD) 1.60 cm wide.

81. The objective lens and the eyepiece of a telescope are spaced 85 cm apart. If the eyepiece is +19 D, what is the total magnification of the telescope?

82. Sam purchases +3.50-D eyeglasses which correct his faulty vision to put his near point at 25 cm. (Assume he wears the lenses 2.0 cm from his eyes.) Calculate (a) the focal length of Sam's glasses, (b) Sam's near point without glasses. (c) Pam, who has normal eyes with near point at 25 cm, puts on Sam's glasses. Calculate Pam's near point with Sam's glasses on.

83. Spy planes fly at extremely high altitudes (25 km) to avoid interception. If their cameras are to discern features as small as 5 cm, what is the minimum aperture of the camera lens to afford this resolution? (Use $\lambda = 580$ nm.)

84. X-rays of wavelength 0.0973 nm are directed at an unknown crystal. The second diffraction maximum is recorded when the X-rays are directed at an angle of 21.2° relative to the crystal surface. What is the spacing between crystal planes?

85. The Hubble Space Telescope, with an objective diameter of 2.4 m, is viewing the Moon. Estimate the minimum distance between two objects on the Moon that the Hubble can distinguish. Consider diffraction of light with wavelength 550 nm. Assume the Hubble is near the Earth.

86. The Earth and Moon are separated by about 400×10^6 m. When Mars is 8×10^{10} m from Earth, could a person standing on Mars resolve the Earth and its Moon as two separate objects without a telescope? Assume a pupil diameter of 5 mm and $\lambda = 550$ nm.

87. You want to design a spy satellite to photograph license plate numbers. Assuming it is necessary to resolve points separated by 5 cm with 550-nm light, and that the satellite orbits at a height of 130 km, what minimum lens aperture (diameter) is required?

88. Given two 12-cm-focal-length lenses, you attempt to make a crude microscope using them. While holding these lenses a distance 55 cm apart, you position your microscope so that its objective lens is distance d_o from a small object. Assume your eye's near point $N = 25$ cm. (a) For your microscope to function properly, what should d_o be? (b) Assuming your eye is relaxed when using it, what magnification M does your microscope achieve? (c) Since the length of your microscope is not much greater than the focal lengths of its lenses, the approximation $M \approx N\ell/f_e f_o$ is not valid. If you apply this approximation to your microscope, what % error do you make in your microscope's true magnification?

*89. The power of one lens in a pair of eyeglasses is -3.5 D. The radius of curvature of the outside surface is 16.0 cm. What is the radius of curvature of the inside surface? The lens is made of plastic with $n = 1.62$.

Search and Learn

1. Digital cameras may offer an optical zoom or a digital zoom. An optical zoom uses a variable focal-length lens, so only the central part of the field of view fills the entire sensor; a digital zoom electronically includes only the central pixels of the sensor, so objects are larger in the final picture. Discuss which is better, and why.

2. Which of the following statements is true? (See Section 25–2.) Write a brief explanation why each is true or false. (a) Contact lenses and eyeglasses for the same person would have the same power. (b) Farsighted people can see far clearly but not near. (c) Nearsighted people cannot see near or far clearly. (d) Astigmatism in vision is corrected by using different spherical lenses for each eye.

3. Redo Examples 25–3 and 25–4 assuming the sensor has only 6 MP. Explain the different results and their impact on finished photographs.

4. Describe at least four advantages of using mirrors rather than lenses for an astronomical telescope.

5. An astronomical telescope, Fig. 25–20, produces an inverted image. One way to make a telescope that produces an upright image is to insert a third lens between the objective and the eyepiece, Fig. 25–23b. To have the same magnification, the non-inverting telescope will be longer. Suppose lenses of focal length 150 cm, 1.5 cm, and 10 cm are available. Where should these three lenses be placed to make a non-inverting telescope with magnification 100×?

6. Mizar, the second star from the end of the Big Dipper's handle, appears to have a companion star, Alcor. From Earth, Mizar and Alcor have an angular separation of 12 arc minutes (1 arc min = $\frac{1}{60}$ of 1°). Using Examples 25–10 and 25–11, estimate the angular resolution of the human eye (in arc min). From your estimate, explain if these two stars can be resolved by the naked eye.

ANSWERS TO EXERCISES

A: 6.3 m.
B: 33 dots/mm.
C: $P = -4.0$ D.
D: 48 cm.
E: 2 m.
F: (c) as stated on page 732; (c) by the λ rule.
G: 0.28 nm.

A science fantasy book called *Mr Tompkins in Wonderland* (1940), by physicist George Gamow, imagined a world in which the speed of light was only 10 m/s (20 mi/h). Mr Tompkins had studied relativity and when he began "speeding" on a bicycle, he "expected that he would be immediately shortened, and was very happy about it as his increasing figure had lately caused him some anxiety. To his great surprise, however, nothing happened to him or to his cycle. On the other hand, the picture around him completely changed. The streets grew shorter, the windows of the shops began to look like narrow slits, and the policeman on the corner became the thinnest man he had ever seen. 'By Jove!' exclaimed Mr Tompkins excitedly, 'I see the trick now. This is where the word *relativity* comes in.'"

Relativity does indeed predict that objects moving relative to us at high speed, close to the speed of light c, are shortened in length. We don't notice it as Mr Tompkins did, because $c = 3 \times 10^8$ m/s is incredibly fast. We will study length contraction, time dilation, simultaneity non-agreement, and how energy and mass are equivalent ($E = mc^2$).

CHAPTER 26
The Special Theory of Relativity

CONTENTS

26–1 Galilean–Newtonian Relativity
26–2 Postulates of the Special Theory of Relativity
26–3 Simultaneity
26–4 Time Dilation and the Twin Paradox
26–5 Length Contraction
26–6 Four-Dimensional Space–Time
26–7 Relativistic Momentum
26–8 The Ultimate Speed
26–9 $E = mc^2$; Mass and Energy
26–10 Relativistic Addition of Velocities
26–11 The Impact of Special Relativity

CHAPTER-OPENING QUESTION—Guess now!

A rocket is headed away from Earth at a speed of $0.80c$. The rocket fires a small payload at a speed of $0.70c$ (relative to the rocket) aimed away from Earth. How fast is the payload moving relative to Earth?

(a) $1.50c$;
(b) a little less than $1.50c$;
(c) a little over c;
(d) a little under c;
(e) $0.75c$.

Physics at the end of the nineteenth century looked back on a period of great progress. The theories developed over the preceding three centuries had been very successful in explaining a wide range of natural phenomena. Newtonian mechanics beautifully explained the motion of objects on Earth and in the heavens. Furthermore, it formed the basis for successful treatments of fluids, wave motion, and sound. Kinetic theory explained the behavior of gases and other materials. Maxwell's theory of electromagnetism embodied all of electric and magnetic phenomena, and it predicted the existence of electromagnetic waves that would behave just like light—so light came to be thought of as an electromagnetic wave. Indeed, it seemed that the natural world, as seen through the eyes of physicists, was very well explained. A few puzzles remained, but it was felt that these would soon be explained using already known principles.

It did not turn out so simply. Instead, these puzzles were to be solved only by the introduction, in the early part of the twentieth century, of two revolutionary new theories that changed our whole conception of nature: the *theory of relativity* and *quantum theory*.

Physics as it was known at the end of the nineteenth century (what we've covered up to now in this book) is referred to as **classical physics**. The new physics that grew out of the great revolution at the turn of the twentieth century is now called **modern physics**. In this Chapter, we present the special theory of relativity, which was first proposed by Albert Einstein (1879–1955; Fig. 26–1) in 1905. In Chapter 27, we introduce the equally momentous quantum theory.

FIGURE 26–1 Albert Einstein (1879–1955), one of the great minds of the twentieth century, was the creator of the special and general theories of relativity.

26–1 Galilean–Newtonian Relativity

Einstein's special theory of relativity deals with how we observe events, particularly how objects and events are observed from different frames of reference.[†] This subject had already been explored by Galileo and Newton.

The special theory of relativity deals with events that are observed and measured from so-called **inertial reference frames** (Section 4–2 and Appendix C), which are reference frames in which Newton's first law is valid: if an object experiences no net force, the object either remains at rest or continues in motion with constant speed in a straight line. It is usually easiest to analyze events when they are observed and measured by observers at rest in an inertial frame. The Earth, though not quite an inertial frame (it rotates), is close enough that for most purposes we can approximate it as an inertial frame. Rotating or otherwise accelerating frames of reference are noninertial frames,[‡] and won't concern us in this Chapter (they are dealt with in Einstein's general theory of relativity, as we will see in Chapter 33).

A reference frame that moves with constant velocity with respect to an inertial frame is itself also an inertial frame, since Newton's laws hold in it as well. When we say that we observe or make measurements from a certain reference frame, it means that we are at rest in that reference frame.

[†]A reference frame is a set of coordinate axes fixed to some object such as the Earth, a train, or the Moon. See Section 2–1.

[‡]On a rotating platform (say a merry-go-round), for example, a ball at rest starts moving outward even though no object exerts a force on it. This is therefore not an inertial frame. See Appendix C, Fig. C–1.

(a) Reference frame = car

(b) Reference frame = Earth

FIGURE 26–2 A coin is dropped by a person in a moving car. The upper views show the moment of the coin's release, the lower views are a short time later. (a) In the reference frame of the car, the coin falls straight down (and the tree moves to the left). (b) In a reference frame fixed on the Earth, the coin has an initial velocity (= to car's) and follows a curved (parabolic) path.

Both Galileo and Newton were aware of what we now call the **relativity principle** applied to mechanics: that *the basic laws of physics are the same in all inertial reference frames.* You may have recognized its validity in everyday life. For example, objects move in the same way in a smoothly moving (constant-velocity) train or airplane as they do on Earth. (This assumes no vibrations or rocking which would make the reference frame noninertial.) When you walk, drink a cup of soup, play pool, or drop a pencil on the floor while traveling in a train, airplane, or ship moving at constant velocity, the objects move just as they do when you are at rest on Earth. Suppose you are in a car traveling rapidly at constant velocity. If you drop a coin from above your head inside the car, how will it fall? It falls straight downward with respect to the car, and hits the floor directly below the point of release, Fig. 26–2a. This is just how objects fall on the Earth—straight down—and thus our experiment in the moving car is in accord with the relativity principle. (If you drop the coin out the car's window, this won't happen because the moving air drags the coin backward relative to the car.)

Note in this example, however, that to an observer on the Earth, the coin follows a curved path, Fig. 26–2b. The actual path followed by the coin is different as viewed from different frames of reference. This does not violate the relativity principle because this principle states that the *laws* of physics are the same in all inertial frames. The same law of gravity, and the same laws of motion, apply in both reference frames. The acceleration of the coin is the same in both reference frames. The difference in Figs. 26–2a and b is that in the Earth's frame of reference, the coin has an initial velocity (equal to that of the car). The laws of physics therefore predict it will follow a parabolic path like any projectile (Chapter 3). In the car's reference frame, there is no initial velocity, and the laws of physics predict that the coin will fall straight down. The laws are the same in both reference frames, although the specific paths are different.

> **CAUTION**
> *Laws are the same, but paths may be different in different reference frames*

Galilean–Newtonian relativity involves certain unprovable assumptions that make sense from everyday experience. It is assumed that the lengths of objects are the same in one reference frame as in another, and that time passes at the same rate in different reference frames. In classical mechanics, then, space and time intervals are considered to be **absolute**: their measurement does not change from one reference frame to another. The mass of an object, as well as all forces, are assumed to be unchanged by a change in inertial reference frame.

> **CAUTION**
> *Length and time intervals are absolute (pre-relativity)*

The position of an object, however, is different when specified in different reference frames, and so is velocity. For example, a person may walk inside a bus toward the front with a speed of 2 m/s. But if the bus moves 10 m/s with respect to the Earth, the person is then moving with a speed of 12 m/s with respect to the Earth. The acceleration of an object, however, is the same in any inertial reference frame according to classical mechanics. This is because the change in velocity, and the time interval, will be the same. For example, the person in the bus may accelerate from 0 to 2 m/s in 1.0 seconds, so $a = 2$ m/s^2 in the reference frame of the bus. With respect to the Earth, the acceleration is

$$(12 \text{ m/s} - 10 \text{ m/s})/(1.0 \text{ s}) = 2 \text{ m/s}^2,$$

> **CAUTION**
> *Position and velocity are different in different reference frames, but length is the same (classical)*

which is the same.

Since neither F, m, nor a changes from one inertial frame to another, Newton's second law, $F = ma$, does not change. Thus Newtons' second law satisfies the relativity principle. The other laws of mechanics also satisfy the relativity principle.

That the laws of mechanics are the same in all inertial reference frames implies that no one inertial frame is special in any sense. We express this important conclusion by saying that **all inertial reference frames are equivalent** for the description of mechanical phenomena. No one inertial reference frame is any better than another. A reference frame fixed to a car or an aircraft traveling at constant velocity is as good as one fixed on the Earth. When you travel smoothly at constant velocity in a car or airplane, it is just as valid to say you are at rest and the Earth is moving as it is to say the reverse.[†] There is no experiment you can do to tell which frame is "really" at rest and which is moving. Thus, there is no way to single out one particular reference frame as being at absolute rest.

A complication arose, however, in the last half of the nineteenth century. Maxwell's comprehensive and successful theory of electromagnetism (Chapter 22) predicted that light is an electromagnetic wave. Maxwell's equations gave the velocity of light c as 3.00×10^8 m/s; and this is just what is measured. The question then arose: in what reference frame does light have precisely the value predicted by Maxwell's theory? It was assumed that light would have a different speed in different frames of reference. For example, if observers could travel on a rocket ship at a speed of 1.0×10^8 m/s away from a source of light, we might expect them to measure the speed of the light reaching them to be $(3.0 \times 10^8 \text{ m/s}) - (1.0 \times 10^8 \text{ m/s}) = 2.0 \times 10^8$ m/s. But Maxwell's equations have no provision for relative velocity. They predicted the speed of light to be $c = 3.0 \times 10^8$ m/s, which seemed to imply that there must be some preferred reference frame where c would have this value.

We discussed in Chapters 11 and 12 that waves can travel on water and along ropes or strings, and sound waves travel in air and other materials. Nineteenth-century physicists viewed the material world in terms of the laws of mechanics, so it was natural for them to assume that light too must travel in some *medium*. They called this transparent medium the **ether** and assumed it permeated all space.[‡] It was therefore assumed that the velocity of light given by Maxwell's equations must be with respect to the ether.[§]

Scientists soon set out to determine the speed of the Earth relative to this absolute frame, whatever it might be. A number of clever experiments were designed. The most direct were performed by A. A. Michelson and E. W. Morley in the 1880s. They measured the difference in the speed of light in different directions using Michelson's interferometer (Section 24–9). They expected to find a difference depending on the orientation of their apparatus with respect to the ether. For just as a boat has different speeds relative to the land when it moves upstream, downstream, or across the stream, so too light would be expected to have different speeds depending on the velocity of the ether past the Earth.

Strange as it may seem, they detected no difference at all. This was a great puzzle. A number of explanations were put forth over a period of years, but they led to contradictions or were otherwise not generally accepted. This **null result** was one of the great puzzles at the end of the nineteenth century.

Then in 1905, Albert Einstein proposed a radical new theory that reconciled these many problems in a simple way. But at the same time, as we shall see, it completely changed our ideas of space and time.

[†]We use the reasonable approximation that Earth is an inertial reference frame.

[‡]The medium for light waves could not be air, since light travels from the Sun to Earth through nearly empty space. Therefore, another medium was postulated, the ether. The ether was not only transparent but, because of difficulty in detecting it, was assumed to have zero density.

[§]Also, it appeared that Maxwell's equations did *not* satisfy the relativity principle: They were simplest in the frame where $c = 3.00 \times 10^8$ m/s, in a reference frame at rest in the ether. In any other reference frame, extra terms were needed to account for relative velocity. Although other laws of physics obeyed the relativity principle, the laws of electricity and magnetism apparently did not. Einstein's second postulate (next Section) resolved this problem: Maxwell's equations do satisfy relativity.

26–2 Postulates of the Special Theory of Relativity

The problems that existed at the start of the twentieth century with regard to electromagnetic theory and Newtonian mechanics were beautifully resolved by Einstein's introduction of the special theory of relativity in 1905. Unaware of the Michelson–Morley null result, Einstein was motivated by certain questions regarding electromagnetic theory and light waves. For example, he asked himself: "What would I see if I rode a light beam?" The answer was that instead of a traveling electromagnetic wave, he would see alternating electric and magnetic fields at rest whose magnitude changed in space, but did not change in time. Such fields, he realized, had never been detected and indeed were not consistent with Maxwell's electromagnetic theory. He argued, therefore, that it was unreasonable to think that the speed of light relative to any observer could be reduced to zero, or in fact reduced at all. This idea became the second postulate of his theory of relativity.

In his famous 1905 paper, Einstein proposed doing away with the idea of the ether and the accompanying assumption of a preferred or absolute reference frame at rest. This proposal was embodied in two postulates. The first was an extension of the Galilean–Newtonian relativity principle to include not only the laws of mechanics but also those of the rest of physics, including electricity and magnetism:

> RELATIVITY PRINCIPLE

First postulate (the relativity principle): **The laws of physics have the same form in all inertial reference frames.**

The first postulate can also be stated as: *there is no experiment you can do in an inertial reference frame to determine if you are at rest or moving uniformly at constant velocity.*

The second postulate is consistent with the first:

> SPEED OF LIGHT PRINCIPLE

Second postulate (constancy of the speed of light): **Light propagates through empty space with a definite speed c independent of the speed of the source or observer.**

These two postulates form the foundation of Einstein's **special theory of relativity**. It is called "special" to distinguish it from his later "general theory of relativity," which deals with noninertial (accelerating) reference frames (Chapter 33). The special theory, which is what we discuss here, deals only with inertial frames.

The second postulate may seem hard to accept, for it seems to violate common sense. First of all, we have to think of light traveling through empty space. Giving up the ether is not too hard, however, since it had never been detected. But the second postulate also tells us that the speed of light in vacuum is always the same, 3.00×10^8 m/s, no matter what the speed of the observer or the source. Thus, a person traveling toward or away from a source of light will measure the same speed for that light as someone at rest with respect to the source. This conflicts with our everyday experience: we would expect to have to add in the velocity of the observer. On the other hand, perhaps we can't expect our everyday experience to be helpful when dealing with the high velocity of light. Furthermore, the null result of the Michelson–Morley experiment is fully consistent with the second postulate.[†]

Einstein's proposal has a certain beauty. By doing away with the idea of an absolute reference frame, it was possible to reconcile classical mechanics with Maxwell's electromagnetic theory. The speed of light predicted by Maxwell's equations *is* the speed of light in vacuum in *any* reference frame.

Einstein's theory required us to give up common sense notions of space and time, and in the following Sections we will examine some strange but interesting consequences of special relativity. Our arguments for the most part will be simple ones.

[†]The Michelson–Morley experiment can also be considered as evidence for the first postulate, since it was intended to measure the motion of the Earth relative to an absolute reference frame. Its failure to do so implies the absence of any such preferred frame.

We will use a technique that Einstein himself did: we will imagine very simple experimental situations in which little mathematics is needed. In this way, we can see many of the consequences of relativity theory without getting involved in detailed calculations. Einstein called these **thought experiments**.

26–3 Simultaneity

An important consequence of the theory of relativity is that we can no longer regard time as an absolute quantity. No one doubts that time flows onward and never turns back. But according to relativity, the time interval between two events, and even whether or not two events are simultaneous, depends on the observer's reference frame. By an **event**, which we use a lot here, we mean something that happens at a particular place and at a particular time.

Two events are said to occur simultaneously if they occur at exactly the same time. But how do we know if two events occur precisely at the same time? If they occur at the same point in space—such as two apples falling on your head at the same time—it is easy. But if the two events occur at widely separated places, it is more difficult to know whether the events are simultaneous since we have to take into account the time it takes for the light from them to reach us. Because light travels at finite speed, a person who sees two events must calculate back to find out when they actually occurred. For example, if two events are *observed* to occur at the same time, but one actually took place farther from the observer than the other, then the more distant one must have occurred earlier, and the two events were not simultaneous.

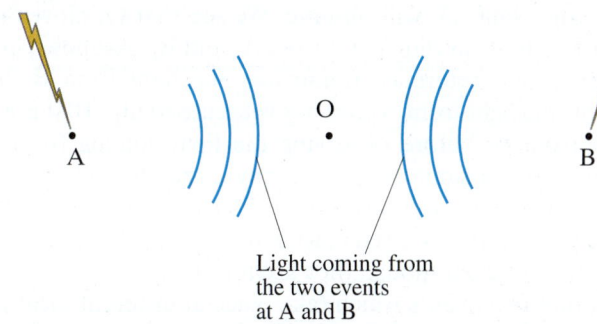

FIGURE 26–3 A moment after lightning strikes at points A and B, the pulses of light (shown as blue waves) are traveling toward the observer O, but O "sees" the lightning only when the light reaches O.

We now imagine a simple thought experiment. Assume an observer, called O, is located exactly halfway between points A and B where two events occur, Fig. 26–3. Suppose the two events are lightning that strikes the points A and B, as shown. For brief events like lightning, only short pulses of light (blue in Fig. 26–3) will travel outward from A and B and reach O. Observer O "sees" the events when the pulses of light reach point O. If the two pulses reach O at the same time, then the two events had to be simultaneous. This is because (i) the two light pulses travel at the same speed (postulate 2), and (ii) the distance OA equals OB, so the time for the light to travel from A to O and from B to O must be the same. Observer O can then definitely state that the two events occurred simultaneously. On the other hand, if O sees the light from one event before that from the other, then the former event occurred first.

The question we really want to examine is this: if two events are simultaneous to an observer in one reference frame, are they also simultaneous to another observer moving with respect to the first? Let us call the observers O_1 and O_2 and assume they are fixed in reference frames 1 and 2 that move with speed v relative to one another. These two reference frames can be thought of as two rockets or two trains (Fig. 26–4). O_2 says that O_1 is moving to the right with speed v, as in Fig. 26–4a; and O_1 says O_2 is moving to the left with speed v, as in Fig. 26–4b. Both viewpoints are legitimate according to the relativity principle. [There is no third point of view that will tell us which one is "really" moving.]

FIGURE 26–4 Observers O_1 and O_2, on two different trains (two different reference frames), are moving with relative speed v. (a) O_2 says that O_1 is moving to the right. (b) O_1 says that O_2 is moving to the left. Both viewpoints are legitimate: it all depends on your reference frame.

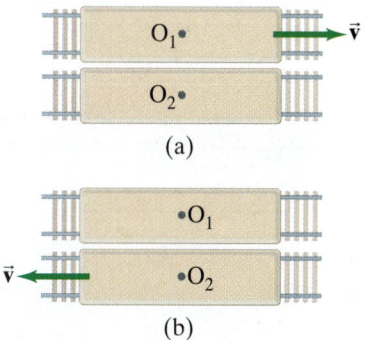

Now suppose that observers O_1 and O_2 observe and measure two lightning strikes. The lightning bolts mark both trains where they strike: at A_1 and B_1 on O_1's train, and at A_2 and B_2 on O_2's train, Fig. 26–5a. For simplicity, we assume that O_1 is exactly halfway between A_1 and B_1, and O_2 is halfway between A_2 and B_2. Let us first put ourselves in O_2's reference frame, so we observe O_1 moving to the right with speed v. Let us also assume that the two events occur *simultaneously* in O_2's frame, and just at the instant when O_1 and O_2 are opposite each other, Fig. 26–5a. A short time later, Fig. 26–5b, light from A_2 and from B_2 reach O_2 at the same time (we assumed this). Since O_2 knows (or measures) the distances O_2A_2 and O_2B_2 as equal, O_2 knows the two events are simultaneous in the O_2 reference frame.

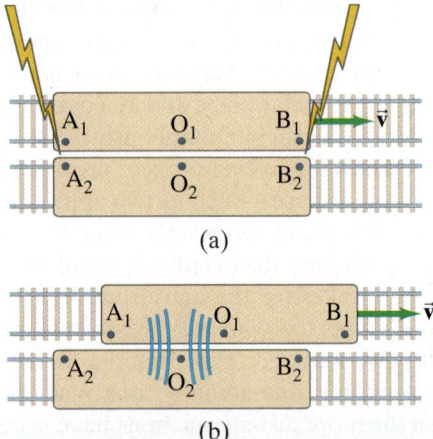

FIGURE 26–5 Thought experiment on simultaneity. In both (a) and (b) we are in the reference frame of observer O_2, who sees the reference frame of O_1 moving to the right. In (a), one lightning bolt strikes the two reference frames at A_1 and A_2, and a second lightning bolt strikes at B_1 and B_2. (b) A moment later, the light (shown in blue) from the two events reaches O_2 at the same time. So according to observer O_2, the two bolts of lightning struck simultaneously. But in O_1's reference frame, the light from B_1 has already reached O_1, whereas the light from A_1 has not yet reached O_1. So in O_1's reference frame, the event at B_1 must have preceded the event at A_1. Simultaneity in time is not absolute.

But what does observer O_1 observe and measure? From our (O_2) reference frame, we can predict what O_1 will observe. We see that O_1 moves to the right during the time the light is traveling to O_1 from A_1 and B_1. As shown in Fig. 26–5b, we can see from our O_2 reference frame that the light from B_1 has already passed O_1, whereas the light from A_1 has not yet reached O_1. That is, O_1 observes the light coming from B_1 before observing the light coming from A_1. Given (i) that light travels at the same speed c in any direction and in any reference frame, and (ii) that the distance O_1A_1 equals O_1B_1, then observer O_1 can only conclude that the event at B_1 occurred before the event at A_1. The two events are *not* simultaneous for O_1, even though they are for O_2.

We thus find that two events which take place at different locations and are simultaneous to one observer, are actually not simultaneous to a second observer who moves relative to the first.

It may be tempting to ask: "Which observer is right, O_1 or O_2?" The answer, according to relativity, is that they are *both* right. There is no "best" reference frame we can choose to determine which observer is right. Both frames are equally good. We can only conclude that *simultaneity is not an absolute concept*, but is relative. We are not aware of this lack of agreement on simultaneity in everyday life because the effect is noticeable only when the relative speed of the two reference frames is very large (near c), or the distances involved are very large.

26–4 Time Dilation and the Twin Paradox

The fact that two events simultaneous to one observer may not be simultaneous to a second observer suggests that time itself is not absolute. Could it be that time passes differently in one reference frame than in another? This is, indeed, just what Einstein's theory of relativity predicts, as the following thought experiment shows.

Figure 26–6 shows a spaceship traveling past Earth at high speed. The point of view of an observer on the spaceship is shown in part (a), and that of an observer on Earth in part (b). Both observers have accurate clocks. The person on the spaceship (Fig. 26–6a) flashes a light and measures the time it takes the light to travel directly across the spaceship and return after reflecting from a mirror (the rays are drawn at a slight angle for clarity). In the reference frame of the spaceship, the

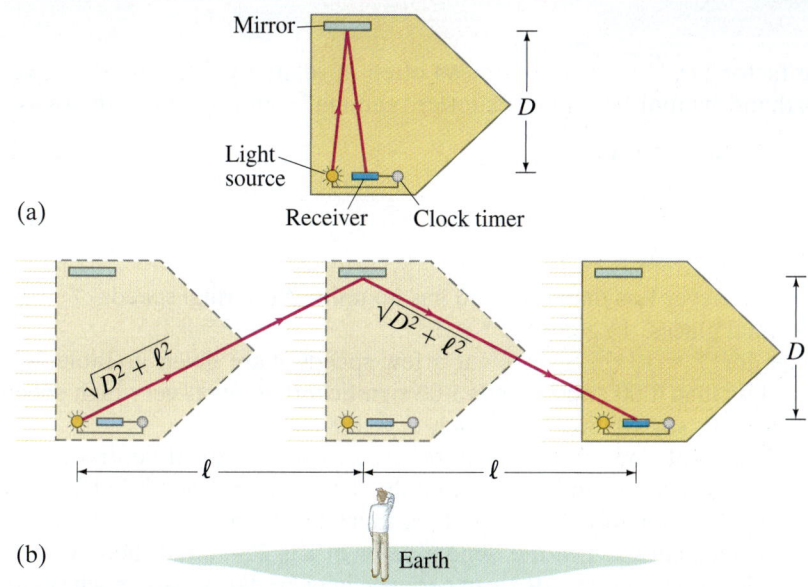

FIGURE 26–6 Time dilation can be shown by a thought experiment: the time it takes for light to travel across a spaceship and back is longer for the observer on Earth (b) than for the observer on the spaceship (a).

light travels a distance $2D$ at speed c, Fig. 26–6a; so the time required to go across and back, Δt_0, is

$$\Delta t_0 = \frac{2D}{c}.$$

The observer on Earth, Fig. 26–6b, observes the same process. But to this observer, the spaceship is moving. So the light travels the diagonal path shown going across the spaceship, reflecting off the mirror, and returning to the sender. Although the light travels at the same speed to this observer (the second postulate), it travels a greater distance. Hence the time required, as measured by the observer on Earth, will be *greater* than that measured by the observer on the spaceship.

Let us determine the time interval Δt measured by the observer on Earth between sending and receiving the light. In time Δt, the spaceship travels a distance $2\ell = v\,\Delta t$ where v is the speed of the spaceship (Fig. 26–6b). The light travels a total distance on its diagonal path (Pythagorean theorem) of $2\sqrt{D^2 + \ell^2} = c\,\Delta t$, where $\ell = v\,\Delta t/2$. Therefore

$$c\,\Delta t = 2\sqrt{D^2 + \ell^2} = 2\sqrt{D^2 + v^2(\Delta t)^2/4}.$$

We square both sides to find $c^2(\Delta t)^2 = 4D^2 + v^2(\Delta t)^2$, and solve for $(\Delta t)^2$:

$$(\Delta t)^2 = 4D^2/(c^2 - v^2)$$

so

$$\Delta t = \frac{2D}{c\sqrt{1 - v^2/c^2}}.$$

We combine this equation for Δt with the formula for Δt_0 above, $\Delta t_0 = 2D/c$:

$$\Delta t = \frac{\Delta t_0}{\sqrt{1 - v^2/c^2}}. \qquad (26\text{–}1a) \quad \boxed{\text{TIME DILATION}}$$

Since $\sqrt{1 - v^2/c^2}$ is always less than 1, we see that $\Delta t > \Delta t_0$. That is, the time interval between the two events (the sending of the light, and its reception on the spaceship) is *greater* for the observer on Earth than for the observer on the spaceship. This is a general result of the theory of relativity, and is known as **time dilation**. The time dilation effect can be stated as

> **clocks moving relative to an observer are measured to run more slowly, as compared to clocks at rest.**

However, we should not think that the clocks are somehow at fault.

> **Time is actually measured to pass more slowly in any moving reference frame as compared to your own.**

This remarkable result is an inevitable outcome of the two postulates of the special theory of relativity.

The factor $1/\sqrt{1 - v^2/c^2}$ occurs so often in relativity that we often give it the shorthand symbol γ (the Greek letter "gamma"), and write Eq. 26–1a as

$$\Delta t = \gamma \, \Delta t_0 \tag{26–1b}$$

where

$$\gamma = \frac{1}{\sqrt{1 - v^2/c^2}}. \tag{26–2}$$

Note that γ is never less than one, and has no units. At normal speeds, $\gamma = 1$ to many decimal places. In general, $\gamma \geq 1$.

Values for $\gamma = 1/\sqrt{1 - v^2/c^2}$ at a few speeds v are given in Table 26–1. γ is never less than 1.00 and exceeds 1.00 significantly only at very high speeds, much above let's say 10^6 m/s (for which $\gamma = 1.000006$).

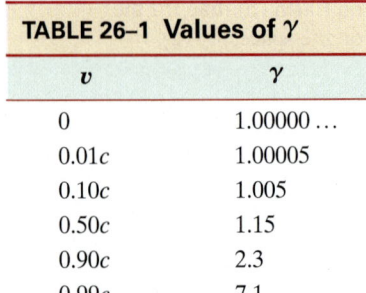

TABLE 26–1 Values of γ

v	γ
0	1.00000...
0.01c	1.00005
0.10c	1.005
0.50c	1.15
0.90c	2.3
0.99c	7.1

The concept of time dilation may be hard to accept, for it contradicts our experience. We can see from Eq. 26–1 that the time dilation effect is indeed negligible unless v is reasonably close to c. If v is much less than c, then the term v^2/c^2 is much smaller than the 1 in the denominator of Eq. 26–1, and then $\Delta t \approx \Delta t_0$ (see Example 26–2). The speeds we experience in everyday life are much smaller than c, so it is little wonder we don't ordinarily notice time dilation. But experiments that have tested the time dilation effect have confirmed Einstein's predictions. In 1971, for example, extremely precise atomic clocks were flown around the Earth in jet planes. The speed of the planes (10^3 km/h) was much less than c, so the clocks had to be accurate to nanoseconds (10^{-9} s) in order to detect any time dilation. They were this accurate, and they confirmed Eqs. 26–1 to within experimental error. Time dilation had been confirmed decades earlier, however, by observations on "elementary particles" which have very small masses (typically 10^{-30} to 10^{-27} kg) and so require little energy to be accelerated to speeds close to the speed of light, c. Many of these elementary particles are not stable and decay after a time into lighter particles. One example is the muon, whose mean lifetime is 2.2 μs when at rest. Careful experiments showed that when a muon is traveling at high speeds, its lifetime is measured to be longer than when it is at rest, just as predicted by the time dilation formula.

EXAMPLE 26–1 **Lifetime of a moving muon.** (a) What will be the mean lifetime of a muon as measured in the laboratory if it is traveling at $v = 0.60c = 1.80 \times 10^8$ m/s with respect to the laboratory? A muon's mean lifetime at rest is $2.20\,\mu\text{s} = 2.20 \times 10^{-6}$ s. (b) How far does a muon travel in the laboratory, on average, before decaying?

APPROACH If an observer were to move along with the muon (the muon would be at rest to this observer), the muon would have a mean life of 2.20×10^{-6} s. To an observer in the lab, the muon lives longer because of time dilation. We find the mean lifetime using Eq. 26–1 and the average distance using $d = v \, \Delta t$.

SOLUTION (a) From Eq. 26–1 with $v = 0.60c$, we have

$$\Delta t = \frac{\Delta t_0}{\sqrt{1 - v^2/c^2}}$$

$$= \frac{2.20 \times 10^{-6}\,\text{s}}{\sqrt{1 - 0.36c^2/c^2}} = \frac{2.20 \times 10^{-6}\,\text{s}}{\sqrt{0.64}} = 2.8 \times 10^{-6}\,\text{s}.$$

(b) Relativity predicts that a muon with speed 1.80×10^8 m/s would travel an average distance $d = v \, \Delta t = (1.80 \times 10^8 \text{ m/s})(2.8 \times 10^{-6}\,\text{s}) = 500$ m, and this is the distance that is measured experimentally in the laboratory.

NOTE At a speed of 1.8×10^8 m/s, classical physics would tell us that with a mean life of 2.2 μs, an average muon would travel $d = vt = (1.8 \times 10^8 \text{ m/s})(2.2 \times 10^{-6}\,\text{s}) = 400$ m. This is shorter than the distance measured.

EXERCISE A What is the muon's mean lifetime (Example 26–1) if it is traveling at $v = 0.90c$? (a) 0.42 μs; (b) 2.3 μs; (c) 5.0 μs; (d) 5.3 μs; (e) 12.0 μs.

We need to clarify how to use Eq. 26–1, $\Delta t = \gamma \Delta t_0$, and the meaning of Δt and Δt_0. The equation is true only when Δt_0 represents the time interval between the two events *in a reference frame where an observer at rest sees the two events occur at the same point in space* (as in Fig. 26–6a where the two events are the light flash being sent and being received). This time interval, Δt_0, is called the **proper time**. Then Δt in Eqs. 26–1 represents the time interval between the two events as measured in a reference frame *moving* with speed v with respect to the first. In Example 26–1 above, Δt_0 (and not Δt) was set equal to 2.2×10^{-6} s because it is only in the rest frame of the muon that the two events ("birth" and "decay") occur at the same point in space. The proper time Δt_0 is the shortest time between the events any observer can measure. In any other moving reference frame, the time Δt is greater.

CAUTION
Proper time Δt_0 is for 2 events at the same point in space

CAUTION
Proper time is shortest: $\Delta t > \Delta t_0$

EXAMPLE 26–2 **Time dilation at 100 km/h.** Let us check time dilation for everyday speeds. A car traveling 100 km/h covers a certain distance in 10.00 s according to the driver's watch. What does an observer at rest on Earth measure for the time interval?

APPROACH The car's speed relative to Earth, written in meters per second, is 100 km/h = $(1.00 \times 10^5 \text{ m})/(3600 \text{ s}) = 27.8$ m/s. The driver is at rest in the reference frame of the car, so we set $\Delta t_0 = 10.00$ s in the time dilation formula.

SOLUTION We use Eq. 26–1a:

$$\Delta t = \frac{\Delta t_0}{\sqrt{1 - \frac{v^2}{c^2}}} = \frac{10.00 \text{ s}}{\sqrt{1 - \left(\frac{27.8 \text{ m/s}}{3.00 \times 10^8 \text{ m/s}}\right)^2}}$$

$$= \frac{10.00 \text{ s}}{\sqrt{1 - (8.59 \times 10^{-15})}}.$$

If you put these numbers into a calculator, you will obtain $\Delta t = 10.00$ s, because the denominator differs from 1 by such a tiny amount. The time measured by an observer fixed on Earth would show no difference from that measured by the driver, even with the best instruments. A computer that could calculate to a large number of decimal places would reveal a slight difference between Δt and Δt_0.

NOTE We can estimate the difference using the binomial expansion (Appendix A–5),

$$(1 \pm x)^n \approx 1 \pm nx. \qquad [\text{for } x \ll 1]$$

In our time dilation formula, we have the factor $\gamma = (1 - v^2/c^2)^{-\frac{1}{2}}$. Thus[†]

$$\Delta t = \gamma \Delta t_0 = \Delta t_0 \left(1 - \frac{v^2}{c^2}\right)^{-\frac{1}{2}} \approx \Delta t_0 \left(1 + \frac{1}{2} \frac{v^2}{c^2}\right)$$

$$\approx 10.00 \text{ s} \left[1 + \frac{1}{2} \left(\frac{27.8 \text{ m/s}}{3.00 \times 10^8 \text{ m/s}}\right)^2\right]$$

$$\approx 10.00 \text{ s} + 4 \times 10^{-14} \text{ s}.$$

So the difference between Δt and Δt_0 is predicted to be 4×10^{-14} s, an extremely small amount.

PROBLEM SOLVING
Use of the binomial expansion

EXERCISE B A certain atomic clock keeps precise time on Earth. If the clock is taken on a spaceship traveling at a speed $v = 0.60c$, does this clock now run slow according to the people (*a*) on the spaceship, (*b*) on Earth?

[†]Recall that $1/x^n$ is written as x^{-n}, such as $1/x^2 = x^{-2}$, Appendix A–2.

EXAMPLE 26–3 Reading a magazine on a spaceship. A passenger on a fictional high-speed spaceship traveling between Earth and Jupiter at a steady speed of $0.75c$ reads a magazine which takes 10.0 min according to her watch. (a) How long does this take as measured by Earth-based clocks? (b) How much farther is the spaceship from Earth at the end of reading the article than it was at the beginning?

APPROACH (a) The time interval in one reference frame is related to the time interval in the other by Eq. 26–1a or b. (b) At constant speed, distance is speed × time. Because there are two time intervals (Δt and Δt_0) we will get two distances, one for each reference frame. [This surprising result is explored in the next Section (26–5).]

SOLUTION (a) The given 10.0-min time interval is the proper time Δt_0—starting and finishing the magazine happen at the same place on the spaceship. Earth clocks measure

$$\Delta t = \frac{\Delta t_0}{\sqrt{1 - (v^2/c^2)}} = \frac{10.00 \text{ min}}{\sqrt{1 - (0.75)^2}} = 15.1 \text{ min}.$$

(b) In the Earth frame, the rocket travels a distance $D = v\,\Delta t = (0.75c)(15.1 \text{ min}) = (0.75)(3.0 \times 10^8 \text{ m/s})(15.1 \text{ min} \times 60 \text{ s/min}) = 2.04 \times 10^{11}$ m. In the spaceship's frame, the Earth is moving away from the spaceship at $0.75c$, but the time is only 10.0 min, so the distance is measured to be $D_0 = v\,\Delta t_0 = (2.25 \times 10^8 \text{ m/s})(600 \text{ s}) = 1.35 \times 10^{11}$ m.

Space Travel?

Time dilation has aroused interesting speculation about space travel. According to classical (Newtonian) physics, to reach a star 100 light-years away would not be possible for ordinary mortals (1 light-year is the distance light can travel in 1 year $= 3.0 \times 10^8$ m/s $\times 3.16 \times 10^7$ s $= 9.5 \times 10^{15}$ m). Even if a spaceship could travel at close to the speed of light, it would take over 100 years to reach such a star. But time dilation tells us that the time involved could be less. In a spaceship traveling at $v = 0.999c$, the time for such a trip would be only about $\Delta t_0 = \Delta t \sqrt{1 - v^2/c^2} = (100 \text{ yr})\sqrt{1 - (0.999)^2} = 4.5$ yr. Thus time dilation allows such a trip, but the enormous practical problems of achieving such speeds may not be possible to overcome, certainly not in the near future.

When we talk in this Chapter and in the Problems about spaceships moving at speeds close to c, it is for understanding and for fun, but not realistic, although for tiny elementary particles such high speeds *are* realistic.

In this example, 100 years would pass on Earth, whereas only 4.5 years would pass for the astronaut on the trip. Is it just the clocks that would slow down for the astronaut? No.

> All processes, including aging and other life processes, run more slowly for the astronaut as measured by an Earth observer. But to the astronaut, time would pass in a normal way.

The astronaut would experience 4.5 years of normal sleeping, eating, reading, and so on. And people on Earth would experience 100 years of ordinary activity.

Twin Paradox

Not long after Einstein proposed the special theory of relativity, an apparent paradox was pointed out. According to this **twin paradox**, suppose one of a pair of 20-year-old twins takes off in a spaceship traveling at very high speed to a distant star and back again, while the other twin remains on Earth. According to the Earth twin, the astronaut twin will age less. Whereas 20 years might pass for the Earth twin, perhaps only 1 year (depending on the spacecraft's speed) would pass for the traveler. Thus, when the traveler returns, the earthbound twin could expect to be 40 years old whereas the traveling twin would be only 21.

This is the viewpoint of the twin on the Earth. But what about the traveling twin? If all inertial reference frames are equally good, won't the traveling twin make all the claims the Earth twin does, only in reverse? Can't the astronaut twin claim that since the Earth is moving away at high speed, time passes more slowly on Earth and the twin on Earth will age less? This is the opposite of what the Earth twin predicts. They cannot both be right, for after all the spacecraft returns to Earth and a direct comparison of ages and clocks can be made.

There is, however, no contradiction here. One of the viewpoints is indeed incorrect. The consequences of the special theory of relativity—in this case, time dilation—can be applied only by observers in an inertial reference frame. The Earth is such a frame (or nearly so), whereas the spacecraft is not. The spacecraft accelerates at the start and end of its trip and when it turns around at the far point of its journey. Part of the time, the astronaut twin may be in an inertial frame (and is justified in saying the Earth twin's clocks run slow). But during the accelerations, the twin on the spacecraft is not in an inertial frame. So she cannot use special relativity to predict their relative ages when she returns to Earth. The Earth twin stays in the same inertial frame, and we can thus trust her predictions based on special relativity. Thus, there is no paradox. The prediction of the Earth twin that the traveling twin ages less is the correct one.

*Global Positioning System (GPS)

PHYSICS APPLIED
Global positioning system (GPS)

Airplanes, cars, boats, and hikers use **global positioning system** (**GPS**) receivers to tell them quite accurately where they are at a given moment (Fig. 26–7). There are more than 30 global positioning system satellites that send out precise time signals using atomic clocks. Your receiver compares the times received from at least four satellites, all of whose times are carefully synchronized to within 1 part in 10^{13}. By comparing the time differences with the known satellite positions and the fixed speed of light, the receiver can determine how far it is from each satellite and thus where it is on the Earth. It can do this to an accuracy of a few meters, if it has been constructed to make corrections such as the one below due to relativity.

CONCEPTUAL EXAMPLE 26–4 **A relativity correction to GPS.** GPS satellites move at about 4 km/s = 4000 m/s. Show that a good GPS receiver needs to correct for time dilation if it is to produce results consistent with atomic clocks accurate to 1 part in 10^{13}.

RESPONSE Let us calculate the magnitude of the time dilation effect by inserting $v = 4000$ m/s into Eq. 26–1a:

$$\Delta t = \frac{1}{\sqrt{1 - \frac{v^2}{c^2}}} \Delta t_0 = \frac{1}{\sqrt{1 - \left(\frac{4 \times 10^3 \text{ m/s}}{3 \times 10^8 \text{ m/s}}\right)^2}} \Delta t_0$$

$$= \frac{1}{\sqrt{1 - 1.8 \times 10^{-10}}} \Delta t_0.$$

We use the binomial expansion: $(1 \pm x)^n \approx 1 \pm nx$ for $x \ll 1$ (see Appendix A–5) which here is $(1 - x)^{-\frac{1}{2}} \approx 1 + \frac{1}{2}x$. That is

$$\Delta t = \left(1 + \frac{1}{2}(1.8 \times 10^{-10})\right) \Delta t_0 = (1 + 9 \times 10^{-11}) \Delta t_0.$$

The time "error" divided by the time interval is

$$\frac{(\Delta t - \Delta t_0)}{\Delta t_0} = 1 + 9 \times 10^{-11} - 1 = 9 \times 10^{-11} \approx 1 \times 10^{-10}.$$

Time dilation, if not accounted for, would introduce an error of about 1 part in 10^{10}, which is 1000 times greater than the precision of the atomic clocks. Not correcting for time dilation means a receiver could give much poorer position accuracy.

NOTE GPS devices must make other corrections as well, including effects associated with general relativity.

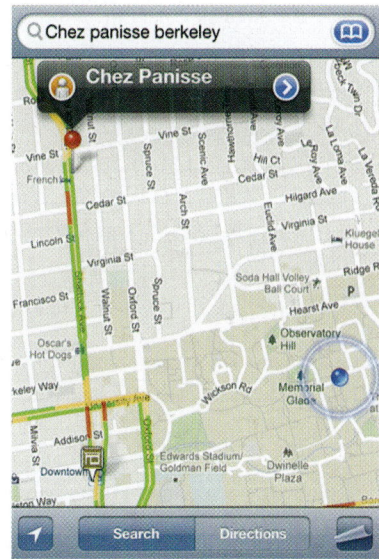

FIGURE 26–7 A visiting professor of physics uses the GPS on her smart phone to find a restaurant (red dot). Her location in the physics department is the blue dot. Traffic on some streets is also shown (green = good, orange = slow, red = heavy traffic) which comes in part by tracking cell phone movements.

26–5 Length Contraction

Time intervals are not the only things different in different reference frames. Space intervals—lengths and distances—are different as well, according to the special theory of relativity, and we illustrate this with a thought experiment.

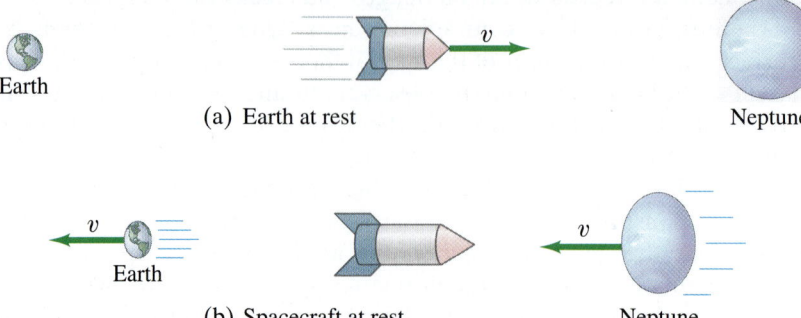

FIGURE 26–8 (a) A spaceship traveling at very high speed v from Earth to the planet Neptune, as seen from Earth's frame of reference. (b) According to an observer on the spaceship, Earth and Neptune are moving at the very high speed v: Earth leaves the spaceship, and a time Δt_0 later Neptune arrives at the spaceship.

Observers on Earth watch a spacecraft traveling at speed v from Earth to, say, Neptune, Fig. 26–8a. The distance between the planets, as measured by the Earth observers, is ℓ_0. The time required for the trip, measured from Earth, is

$$\Delta t = \frac{\ell_0}{v}. \qquad \text{[Earth observer]}$$

In Fig. 26–8b we see the point of view of observers on the spacecraft. In this frame of reference, the spaceship is at rest; Earth and Neptune move† with speed v. The time between departure of Earth and arrival of Neptune, as observed from the spacecraft, is the "proper time" Δt_0 (page 753), because these two events occur at the same point in space (i.e., at the spacecraft). Therefore the time interval is less for the spacecraft observers than for the Earth observers. That is, because of time dilation (Eq. 26–1a), the time for the trip as viewed by the spacecraft is

$$\Delta t_0 = \Delta t \sqrt{1 - v^2/c^2}$$
$$= \Delta t / \gamma. \qquad \text{[spacecraft observer]}$$

Because the spacecraft observers measure the same speed but less time between these two events, they also measure the distance as less. If we let ℓ be the distance between the planets as viewed by the spacecraft observers, then $\ell = v \Delta t_0$, which we can rewrite as $\ell = v \Delta t_0 = v \Delta t \sqrt{1 - v^2/c^2} = \ell_0 \sqrt{1 - v^2/c^2}$. Thus we have the important result that

LENGTH CONTRACTION

$$\ell = \ell_0 \sqrt{1 - v^2/c^2} \qquad (26\text{–}3\text{a})$$

or, using γ (Eq. 26–2),

$$\ell = \frac{\ell_0}{\gamma}. \qquad (26\text{–}3\text{b})$$

This is a general result of the special theory of relativity and applies to lengths of objects as well as to distance between objects. The result can be stated most simply in words as:

> the length of an object moving relative to an observer is measured to be shorter along its direction of motion than when it is at rest.

CAUTION
Proper length is measured in reference frame where the two positions are at rest

This is called **length contraction**. The length ℓ_0 in Eqs. 26–3 is called the **proper length**. It is the length of the object (or distance between two points whose positions are measured at the same time) as determined by *observers at rest* with respect to the object. Equations 26–3 give the length ℓ that will be measured by observers when the object travels past them at speed v.

†We assume v is much greater than the relative speed of Neptune and Earth (which we thus ignore).

It is important to note that length contraction occurs *only along the direction of motion*. For example, the moving spaceship in Fig. 26–8a is shortened in length, but its height is the same as when it is at rest.

Length contraction, like time dilation, is not noticeable in everyday life because the factor $\sqrt{1 - v^2/c^2}$ in Eq. 26–3a differs significantly from 1.00 only when v is very large.

EXAMPLE 26–5 Painting's contraction. A rectangular painting measures 1.00 m tall and 1.50 m wide, Fig. 26–9a. It is hung on the side wall of a spaceship which is moving past the Earth at a speed of $0.90c$. (*a*) What are the dimensions of the picture according to the captain of the spaceship? (*b*) What are the dimensions as seen by an observer on the Earth?

APPROACH We apply the length contraction formula, Eq. 26–3a, to the dimension parallel to the motion; v is the speed of the painting relative to the Earth observer.

SOLUTION (*a*) The painting is at rest ($v = 0$) on the spaceship so it (as well as everything else in the spaceship) looks perfectly normal to everyone on the spaceship. The captain sees a 1.00-m by 1.50-m painting.

(*b*) Only the dimension in the direction of motion is shortened, so the height is unchanged at 1.00 m, Fig. 26–9b. The length, however, is contracted to

$$\ell = \ell_0 \sqrt{1 - \frac{v^2}{c^2}}$$
$$= (1.50\,\text{m})\sqrt{1 - (0.90)^2} = 0.65\,\text{m}.$$

So the picture has dimensions 1.00 m × 0.65 m to an observer on Earth.

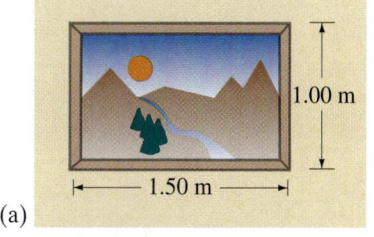

(a)

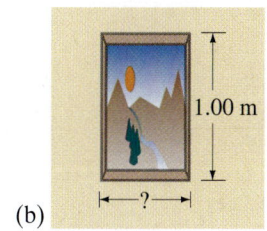

(b)

FIGURE 26–9 Example 26–5.

EXAMPLE 26–6 A fantasy supertrain. A very fast train with a "proper length" of $\ell_0 = 500$ m (measured by people at rest on the train) is passing through a tunnel that is 200 m long according to observers on the ground. Let us imagine the train's speed to be so great that the train fits completely within the tunnel as seen by observers on the ground. That is, the engine is just about to emerge from one end of the tunnel at the time the last car disappears into the other end. What is the train's speed?

APPROACH Since the train just fits inside the tunnel, its length measured by the person on the ground is $\ell = 200$ m. The length contraction formula, Eq. 26–3a or b, can thus be used to solve for v.

SOLUTION Substituting $\ell = 200$ m and $\ell_0 = 500$ m into Eq. 26–3a gives

$$200\,\text{m} = 500\,\text{m}\sqrt{1 - \frac{v^2}{c^2}};$$

dividing both sides by 500 m and squaring, we get

$$(0.40)^2 = 1 - \frac{v^2}{c^2}$$

or

$$\frac{v}{c} = \sqrt{1 - (0.40)^2}$$

and

$$v = 0.92c.$$

NOTE No real train could go this fast. But it is fun to think about.

NOTE An observer on the *train* would *not* see the two ends of the train inside the tunnel at the same time. Recall that observers moving relative to each other do not agree about simultaneity. (See Example 26–7, next.)

EXERCISE C What is the length of the tunnel as measured by observers on the train in Example 26–6?

CONCEPTUAL EXAMPLE 26–7 **Resolving the train and tunnel length.** Observers at rest on the Earth see a very fast 200-m-long train pass through a 200-m-long tunnel (as in Example 26–6) so that the train momentarily disappears from view inside the tunnel. Observers on the train measure the train's length to be 500 m and the tunnel's length to be only 80 m (Exercise C, using Eq. 26–3a). Clearly a 500-m-long train cannot fit inside an 80-m-long tunnel. How is this apparent inconsistency explained?

RESPONSE Events simultaneous in one reference frame may not be simultaneous in another. Let the engine emerging from one end of the tunnel be "event A," and the last car disappearing into the other end of the tunnel "event B." To observers in the Earth frame, events A and B are simultaneous. To observers on the train, however, the events are not simultaneous. In the train's frame, event A occurs before event B. As the engine emerges from the tunnel, observers on the train observe the last car as still $500\,\text{m} - 80\,\text{m} = 420\,\text{m}$ from the entrance to the tunnel.

26–6 Four-Dimensional Space–Time

Let us imagine a person is on a train moving at a very high speed, say $0.65c$, Fig. 26–10. This person begins a meal at 7:00 and finishes at 7:15, according to a clock on the train. The two events, beginning and ending the meal, take place at the same point on the train, so the "proper time" between these two events is 15 min. To observers on Earth, the plate is moving and the meal will take longer—20 min according to Eqs. 26–1. Let us assume that the meal was served on a 20-cm-diameter plate (its "proper length"). To observers on the Earth, the plate is moving and is only 15 cm wide (length contraction). Thus, to observers on the Earth, the meal looks smaller but lasts longer.

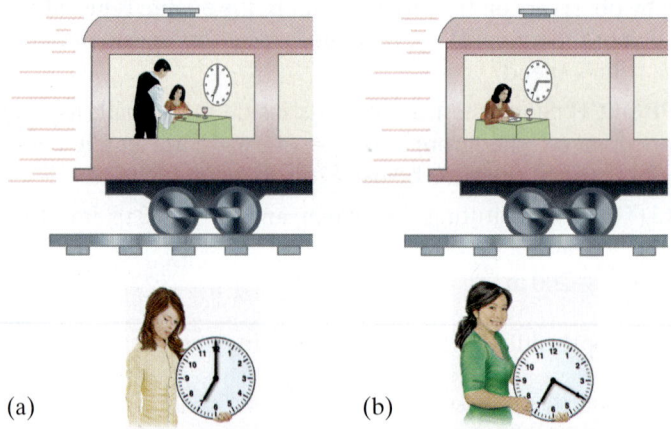

FIGURE 26–10 According to an accurate clock on a fast-moving train, a person (a) begins dinner at 7:00 and (b) finishes at 7:15. At the beginning of the meal, two observers on Earth set their watches to correspond with the clock on the train. These observers measure the eating time as 20 minutes.

In a sense the two effects, time dilation and length contraction, balance each other. When viewed from the Earth, what an object seems to lose in size it gains in length of time it lasts. Space, or length, is exchanged for time.

Considerations like this led to the idea of **four-dimensional space–time**: space takes up three dimensions and time is a fourth dimension. Space and time are intimately connected. Just as when we squeeze a balloon we make one dimension larger and another smaller, so when we examine objects and events from different reference frames, a certain amount of space is exchanged for time, or vice versa.

Although the idea of four dimensions may seem strange, it refers to the idea that any object or event is specified by four quantities—three to describe where in space, and one to describe when in time. The really unusual aspect of four-dimensional space–time is that space and time can intermix: a little of one can be exchanged for a little of the other when the reference frame is changed.

[In Galilean–Newtonian relativity, the time interval between two events, Δt, and the distance between two events or points, Δx, are invariant quantities no matter what inertial reference frame they are viewed from. Neither of these quantities is invariant according to Einstein's relativity. But there is an invariant quantity in four-dimensional space–time, called the **space–time interval**, which is $(\Delta s)^2 = (c\,\Delta t)^2 - (\Delta x)^2$.]

26–7 Relativistic Momentum

So far in this Chapter, we have seen that two basic mechanical quantities, length and time intervals, need modification because they are relative—their value depends on the reference frame from which they are measured. We might expect that other physical quantities might need some modification according to the theory of relativity, such as momentum and energy.

The analysis of collisions between two particles shows that if we want to preserve the law of conservation of momentum in relativity, we must redefine momentum as

$$p = \frac{mv}{\sqrt{1 - v^2/c^2}} = \gamma mv. \qquad (26\text{–}4)$$

Here γ is shorthand for $1/\sqrt{1 - v^2/c^2}$ as before (Eq. 26–2). For speeds much less than the speed of light, Eq. 26–4 gives the classical momentum, $p = mv$.

Relativistic momentum has been tested many times on tiny elementary particles (such as muons), and it has been found to behave in accord with Eq. 26–4.

EXAMPLE 26–8 Momentum of moving electron. Compare the momentum of an electron to its classical value when it has a speed of (a) 4.00×10^7 m/s in the CRT of an old TV set, and (b) $0.98c$ in an accelerator used for cancer therapy.

APPROACH We use Eq. 26–4 for the momentum of a moving electron.

SOLUTION (a) At $v = 4.00 \times 10^7$ m/s, the electron's momentum is

$$p = \frac{mv}{\sqrt{1 - \dfrac{v^2}{c^2}}} = \frac{mv}{\sqrt{1 - \dfrac{(4.00 \times 10^7 \text{ m/s})^2}{(3.00 \times 10^8 \text{ m/s})^2}}} = 1.01 mv.$$

The factor $\gamma = 1/\sqrt{1 - v^2/c^2} \approx 1.01$, so the momentum is only about 1% greater than the classical value. (If we put in the mass of an electron, $m = 9.11 \times 10^{-31}$ kg, the momentum is $p = 1.01 mv = 3.68 \times 10^{-23}$ kg·m/s, compared to 3.64×10^{-23} kg·m/s classically.)

(b) With $v = 0.98c$, the momentum is

$$p = \frac{mv}{\sqrt{1 - \dfrac{v^2}{c^2}}} = \frac{mv}{\sqrt{1 - \dfrac{(0.98c)^2}{c^2}}} = \frac{mv}{\sqrt{1 - (0.98)^2}} = 5.0 mv.$$

An electron traveling at 98% the speed of light has $\gamma = 5.0$ and a momentum 5.0 times its classical value.

*Rest Mass and Relativistic Mass

CAUTION
Most physicists prefer to consider the mass of a particle as fixed

The relativistic definition of momentum, Eq. 26–4, has sometimes been interpreted as an increase in the mass of an object. In this interpretation, a particle can have a **relativistic mass**, m_{rel}, which increases with speed according to

$$m_{\text{rel}} = \frac{m}{\sqrt{1 - v^2/c^2}}.$$

In this "mass-increase" formula, m is referred to as the **rest mass** of the object. With this interpretation, *the mass of an object appears to increase as its speed increases.* But there are problems with relativistic mass. If we plug it into formulas like $F = ma$ or $\text{KE} = \frac{1}{2}mv^2$, we obtain formulas that do not agree with experiment. (If we write Newton's second law in its more general form, $\vec{F} = \Delta\vec{p}/\Delta t$, that would get a correct result.) Also, be careful *not* to think a mass acquires more particles or more molecules as its speed becomes very large. It doesn't. Today, most physicists prefer not to use relativistic mass, so an object has only one mass (its rest mass), and it is only the momentum that increases with speed.

Whenever we talk about the mass of an object, we will always mean its rest mass (a fixed value). [But see Problem 46.]

26–8 The Ultimate Speed

A basic result of the special theory of relativity is that the speed of an object cannot equal or exceed the speed of light. That the speed of light is a natural speed limit in the universe can be seen from any of Eqs. 26–1, 26–3, or 26–4. It is perhaps easiest to see from Eq. 26–4. As an object is accelerated to greater and greater speeds, its momentum becomes larger and larger. Indeed, if v were to equal c, the denominator in this equation would be zero, and the momentum would be infinite. To accelerate an object up to $v = c$ would thus require infinite energy, and so is not possible.

26–9 $E = mc^2$; Mass and Energy

If momentum needs to be modified to fit with relativity as we just saw in Eq. 26–4, then we might expect that energy would also need to be rethought. Indeed, Einstein not only developed a new formula for kinetic energy, but also found a new relation between mass and energy, and the startling idea that mass is a form of energy.

We start with the work-energy principle (Chapter 6), hoping it is still valid in relativity and will give verifiable results. That is, we assume the net work done on a particle is equal to its change in kinetic energy (KE). Using this principle, Einstein showed that at high speeds the formula $\text{KE} = \frac{1}{2}mv^2$ is not correct. Instead, Einstein showed that the kinetic energy of a particle of mass m traveling at speed v is given by

$$\text{KE} = \frac{mc^2}{\sqrt{1 - v^2/c^2}} - mc^2. \quad (26\text{–}5\text{a})$$

In terms of $\gamma = 1/\sqrt{1 - v^2/c^2}$ we can rewrite Eq. 26–5a as

$$\text{KE} = \gamma mc^2 - mc^2 = (\gamma - 1)mc^2. \quad (26\text{–}5\text{b})$$

Equation 26–5a requires some interpretation. The first term increases with the speed v of the particle. The second term, mc^2, is constant; it is called the **rest energy** of the particle, and represents a form of energy that a particle has even when at rest. Note that if a particle is at rest ($v = 0$) the first term in Eq. 26–5a becomes mc^2, so $\text{KE} = 0$ as it should.

We can rearrange Eq. 26–5b to get

$$\gamma mc^2 = mc^2 + \text{KE}.$$

We call γmc^2 the *total energy* E of the particle (assuming no potential energy), because it equals the rest energy plus the kinetic energy:

$$E = \text{KE} + mc^2. \qquad (26\text{–}6\text{a})$$

The total energy[†] can also be written, using Eqs. 26–5, as

$$E = \gamma mc^2 = \frac{mc^2}{\sqrt{1 - v^2/c^2}}. \qquad (26\text{–}6\text{b})$$

For a particle at rest in a given reference frame, KE is zero in Eq. 26–6a, so the total energy is its rest energy:

$$E = mc^2. \qquad (26\text{–}7)$$

MASS RELATED TO ENERGY

Here we have Einstein's famous formula, $E = mc^2$. This formula mathematically relates the concepts of energy and mass. But if this idea is to have any physical meaning, then mass ought to be convertible to other forms of energy and vice versa. Einstein suggested that this might be possible, and indeed changes of mass to other forms of energy, and vice versa, have been experimentally confirmed countless times in nuclear and elementary particle physics. For example, an electron and a positron (= a positive electron, see Section 32–3) have often been observed to collide and disappear, producing pure electromagnetic radiation. The amount of electromagnetic energy produced is found to be exactly equal to that predicted by Einstein's formula, $E = mc^2$. The reverse process is also commonly observed in the laboratory: electromagnetic radiation under certain conditions can be converted into material particles such as electrons (see Section 27–6 on pair production). On a larger scale, the energy produced in nuclear power plants is a result of the loss in mass of the uranium fuel as it undergoes the process called fission (Chapter 31). Even the radiant energy we receive from the Sun is an example of $E = mc^2$; the Sun's mass is continually decreasing as it radiates electromagnetic energy outward.

The relation $E = mc^2$ is now believed to apply to all processes, although the changes are often too small to measure. That is, when the energy of a system changes by an amount ΔE, the mass of the system changes by an amount Δm given by

$$\Delta E = (\Delta m)(c^2). \qquad (26\text{–}8)$$

In a nuclear reaction where an energy E is required or released, the masses of the reactants and the products will be different by $\Delta m = \Delta E/c^2$.

EXAMPLE 26–9 Pion's kinetic energy. A π^0 meson ($m = 2.4 \times 10^{-28}$ kg) travels at a speed $v = 0.80c = 2.4 \times 10^8$ m/s. What is its kinetic energy? Compare to a classical calculation.

APPROACH We use Eq. 26–5 and compare to $\frac{1}{2}mv^2$.

SOLUTION We substitute values into Eq. 26–5a

$$\begin{aligned}\text{KE} &= mc^2\left(\frac{1}{\sqrt{1 - v^2/c^2}} - 1\right) \\ &= (2.4 \times 10^{-28}\text{ kg})(3.0 \times 10^8\text{ m/s})^2\left(\frac{1}{(1 - 0.64)^{\frac{1}{2}}} - 1\right) \\ &= 1.4 \times 10^{-11}\text{ J}.\end{aligned}$$

PROBLEM SOLVING
Relativistic kinetic energy

Notice that the units of mc^2 are kg·m²/s², which is the joule.

NOTE Classically KE $= \frac{1}{2}mv^2 = \frac{1}{2}(2.4 \times 10^{-28}\text{ kg})(2.4 \times 10^8\text{ m/s})^2 = 6.9 \times 10^{-12}$ J, about half as much, but this is not a correct result. Note that $\frac{1}{2}\gamma mv^2$ also does not work.

[†]This is for a "free particle," without forces and potential energy. Potential energy terms can be added.

EXAMPLE 26–10 Energy from nuclear decay. The energy required or released in nuclear reactions and decays comes from a change in mass between the initial and final particles. In one type of radioactive decay (Chapter 30), an atom of uranium ($m = 232.03716$ u) decays to an atom of thorium ($m = 228.02874$ u) plus an atom of helium ($m = 4.00260$ u) where the masses given are in atomic mass units ($1\,\text{u} = 1.6605 \times 10^{-27}$ kg). Calculate the energy released in this decay.

APPROACH The initial mass minus the total final mass gives the mass loss in atomic mass units (u); we convert that to kg, and multiply by c^2 to find the energy released, $\Delta E = \Delta m c^2$.

SOLUTION The initial mass is 232.03716 u, and after the decay the mass is 228.02874 u + 4.00260 u = 232.03134 u, so there is a loss of mass of 0.00582 u. This mass, which equals $(0.00582\,\text{u})(1.66 \times 10^{-27}\,\text{kg}) = 9.66 \times 10^{-30}$ kg, is changed into energy. By $\Delta E = \Delta m c^2$, we have

$$\Delta E = (9.66 \times 10^{-30}\,\text{kg})(3.0 \times 10^8\,\text{m/s})^2$$
$$= 8.70 \times 10^{-13}\,\text{J}.$$

Since $1\,\text{MeV} = 1.60 \times 10^{-13}$ J (Section 17–4), the energy released is 5.4 MeV.

In the tiny world of atoms and nuclei, it is common to quote energies in eV (electron volts) or multiples such as MeV (10^6 eV). Momentum (see Eq. 26–4) can be quoted in units of eV/c (or MeV/c). And mass can be quoted (from $E = mc^2$) in units of eV/c^2 (or MeV/c^2). Note the use of c to keep the units correct. The masses of the electron and the proton can be shown to be 0.511 MeV/c^2 and 938 MeV/c^2, respectively. For example, for the electron, $mc^2 = (9.11 \times 10^{-31}\,\text{kg})(2.998 \times 10^8\,\text{m/s})^2/(1.602 \times 10^{-13}\,\text{J/MeV}) = 0.511$ MeV. See also the Table inside the front cover.

EXAMPLE 26–11 A 1-TeV proton. The Tevatron accelerator at Fermilab in Illinois can accelerate protons to a kinetic energy of 1.0 TeV (10^{12} eV). What is the speed of such a proton?

APPROACH We solve the kinetic energy formula, Eq. 26–5a, for v.

SOLUTION The rest energy of a proton is $mc^2 = 938$ MeV or 9.38×10^8 eV. Compared to the kinetic energy of 10^{12} eV, the rest energy can be neglected, so we simplify Eq. 26–5a to

$$\text{KE} \approx \frac{mc^2}{\sqrt{1 - v^2/c^2}}.$$

We solve this for v in the following steps:

$$\sqrt{1 - \frac{v^2}{c^2}} = \frac{mc^2}{\text{KE}};$$

$$1 - \frac{v^2}{c^2} = \left(\frac{mc^2}{\text{KE}}\right)^2;$$

$$\frac{v^2}{c^2} = 1 - \left(\frac{mc^2}{\text{KE}}\right)^2 = 1 - \left(\frac{9.38 \times 10^8\,\text{eV}}{1.0 \times 10^{12}\,\text{eV}}\right)^2;$$

$$v = \sqrt{1 - (9.38 \times 10^{-4})^2}\,c$$
$$= 0.99999956\,c.$$

So the proton is traveling at a speed very nearly equal to c.

At low speeds, $v \ll c$, the relativistic formula for kinetic energy reduces to the classical one, as we now show by using the binomial expansion (Appendix A): $(1 \pm x)^n = 1 \pm nx + \cdots$, keeping only two terms because $x = v/c$ is very much less than 1. With $n = -\frac{1}{2}$ we expand the square root in Eq. 26–5a

$$\text{KE} = mc^2\left(\frac{1}{\sqrt{1 - v^2/c^2}} - 1\right)$$

so that

$$\text{KE} \approx mc^2\left(1 + \frac{1}{2}\frac{v^2}{c^2} + \cdots - 1\right) \approx \tfrac{1}{2}mv^2.$$

The dots in the first expression represent very small terms in the expansion which we neglect since we assumed that $v \ll c$. Thus at low speeds, the relativistic form for kinetic energy reduces to the classical form, $\text{KE} = \tfrac{1}{2}mv^2$. This makes relativity a viable theory in that it can predict accurate results at low speed as well as at high. Indeed, the other equations of special relativity also reduce to their classical equivalents at ordinary speeds: length contraction, time dilation, and modifications to momentum as well as kinetic energy, all disappear for $v \ll c$ since $\sqrt{1 - v^2/c^2} \approx 1$.

A useful relation between the total energy E of a particle and its momentum p can also be derived. The momentum of a particle of mass m and speed v is given by Eq. 26–4

$$p = \gamma mv = \frac{mv}{\sqrt{1 - v^2/c^2}}.$$

The total energy is

$$E = \text{KE} + mc^2$$

or

$$E = \gamma mc^2 = \frac{mc^2}{\sqrt{1 - v^2/c^2}}.$$

We square this equation (and we insert "$v^2 - v^2$" which is zero, but will help us):

$$E^2 = \frac{m^2c^2c^2}{1 - v^2/c^2} = \frac{m^2c^2(c^2 - v^2 + v^2)}{1 - v^2/c^2} = \frac{m^2c^2v^2}{1 - v^2/c^2} + \frac{m^2c^2(c^2 - v^2)}{1 - v^2/c^2}$$

$$= p^2c^2 + \frac{m^2c^4(1 - v^2/c^2)}{1 - v^2/c^2}$$

or

$$E^2 = p^2c^2 + m^2c^4. \tag{26–9}$$

Thus, the total energy can be written in terms of the momentum p, or in terms of the kinetic energy (Eq. 26–6a), where we have assumed there is no potential energy.

*Invariant Energy–Momentum

We can rewrite Eq. 26–9 as $E^2 - p^2c^2 = m^2c^4$. Since the mass m of a given particle is the same in any reference frame, we see that the quantity $E^2 - p^2c^2$ must also be the same in any reference frame. Thus, at any given moment the total energy E and momentum p of a particle will be different in different reference frames, but the quantity $E^2 - p^2c^2$ will have the same value in all inertial reference frames. We say that the quantity $E^2 - p^2c^2$ is **invariant**.

When Do We Use Relativistic Formulas?

From a practical point of view, we do not have much opportunity in our daily lives to use the mathematics of relativity. For example, the γ factor, $\gamma = 1/\sqrt{1 - v^2/c^2}$, has a value of 1.005 when $v = 0.10c$. Thus, for speeds even as high as $0.10c = 3.0 \times 10^7$ m/s, the factor $\sqrt{1 - v^2/c^2}$ in relativistic formulas gives a numerical correction of less than 1%. For speeds less than $0.10c$, or unless mass and energy are interchanged, we don't usually need the more complicated relativistic formulas, and can use the simpler classical formulas.

If you are given a particle's mass m and its kinetic energy KE, you can do a quick calculation to determine if you need to use relativistic formulas or if classical ones are good enough. You simply compute the ratio KE/mc^2 because (Eq. 26–5b)

$$\frac{\text{KE}}{mc^2} = \gamma - 1 = \frac{1}{\sqrt{1 - v^2/c^2}} - 1.$$

If this ratio comes out to be less than, say, 0.01, then $\gamma \leq 1.01$ and relativistic equations will correct the classical ones by about 1%. If your expected precision is no better than 1%, classical formulas are good enough. But if your precision is 1 part in 1000 (0.1%) then you would want to use relativistic formulas. If your expected precision is only 10%, you need relativity if $(\text{KE}/mc^2) \gtrsim 0.1$.

EXERCISE D For 1% accuracy, does an electron with KE = 100 eV need to be treated relativistically? [*Hint*: The mass of an electron is 0.511 MeV.]

26–10 Relativistic Addition of Velocities

Consider a rocket ship that travels away from the Earth with speed v, and assume that this rocket has fired off a second rocket that travels at speed u' with respect to the first (Fig. 26–11). We might expect that the speed u of rocket 2 with respect to Earth is $u = v + u'$, which in the case shown in Fig. 26–11 is $u = 0.60c + 0.60c = 1.20c$. But, as discussed in Section 26–8, no object can travel faster than the speed of light in any reference frame. Indeed, Einstein showed that since length and time are different in different reference frames, the classical addition-of-velocities formula is no longer valid. Instead, the correct formula is

$$u = \frac{v + u'}{1 + vu'/c^2} \quad \left[\begin{array}{l}\vec{u} \text{ and } \vec{v} \text{ along} \\ \text{the same direction}\end{array}\right] \quad (26\text{–}10)$$

for motion along a straight line. We derive this formula in Appendix E. If u' is in the opposite direction from v, then u' must have a minus sign in the above equation so $u = (v - u')/(1 - vu'/c^2)$.

EXAMPLE 26–12 **Relative velocity, relativistically.** Calculate the speed of rocket 2 in Fig. 26–11 with respect to Earth.

APPROACH We combine the speed of rocket 2 relative to rocket 1 with the speed of rocket 1 relative to Earth, using the relativistic Eq. 26–10 because the speeds are high and they are along the same line.

SOLUTION Rocket 2 moves with speed $u' = 0.60c$ with respect to rocket 1. Rocket 1 has speed $v = 0.60c$ with respect to Earth. The speed of rocket 2 with respect to Earth is (Eq. 26–10)

$$u = \frac{0.60c + 0.60c}{1 + \frac{(0.60c)(0.60c)}{c^2}} = \frac{1.20c}{1.36} = 0.88c.$$

NOTE The speed of rocket 2 relative to Earth is less than c, as it must be.

We can see that Eq. 26–10 reduces to the classical form for velocities small compared to the speed of light since $1 + vu'/c^2 \approx 1$ for v and $u' \ll c$. Thus, $u \approx v + u'$, as in classical physics (Chapter 3).

Let us test our formula at the other extreme, that of the speed of light. Suppose that rocket 1 in Fig. 26–11 sends out a beam of light so that $u' = c$. Equation 26–10 tells us that the speed of this light relative to Earth is

$$u = \frac{0.60c + c}{1 + \frac{(0.60c)(c)}{c^2}} = \frac{1.60c}{1.60} = c,$$

which is fully consistent with the second postulate of relativity.

EXERCISE E Use Eq. 26–10 to calculate the speed of rocket 2 in Fig. 26–11 relative to Earth if it was shot from rocket 1 at a speed $u' = 3000$ km/s $= 0.010c$. Assume rocket 1 had a speed $v = 6000$ km/s $= 0.020c$.

EXERCISE F Return to the Chapter-Opening Question, page 744, and answer it again now. Try to explain why you may have answered differently the first time.

Relative velocities do not add simply, as in classical mechanics ($v \ll c$)

Relativistic addition of velocities formula ($\vec{u}$ and $\vec{v}$ along same line)

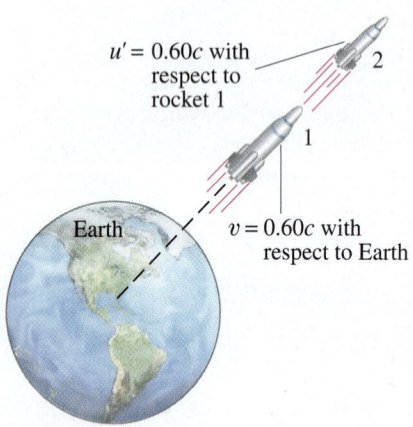

FIGURE 26–11 Rocket 1 leaves Earth at $v = 0.60c$. Rocket 2 is fired from rocket 1 with speed $u' = 0.60c$. What is the speed of rocket 2 with respect to the Earth? Example 26–12.

26–11 The Impact of Special Relativity

A great many experiments have been performed to test the predictions of the special theory of relativity. Within experimental error, no contradictions have been found. Scientists have therefore accepted relativity as an accurate description of nature.

At speeds much less than the speed of light, the relativistic formulas reduce to the old classical ones, as we have discussed. We would, of course, hope—or rather, insist—that this be true since Newtonian mechanics works so well for objects moving with speeds $v \ll c$. This insistence that a more general theory (such as relativity) give the same results as a more restricted theory (such as classical mechanics which works for $v \ll c$) is called the **correspondence principle**. The two theories must correspond where their realms of validity overlap. Relativity thus does not contradict classical mechanics. Rather, it is a more general theory, of which classical mechanics is now considered to be a limiting case.

The importance of relativity is not simply that it gives more accurate results, especially at very high speeds. Much more than that, it has changed the way we view the world. The concepts of space and time are now seen to be relative, and intertwined with one another, whereas before they were considered absolute and separate. Even our concepts of matter and energy have changed: either can be converted to the other. The impact of relativity extends far beyond physics. It has influenced the other sciences, and even the world of art and literature; it has, indeed, entered the general culture.

The special theory of relativity we have studied in this Chapter deals with inertial (nonaccelerating) reference frames. In Chapter 33 we will discuss briefly the more complicated "general theory of relativity" which can deal with non-inertial reference frames.

Summary

An **inertial reference frame** is one in which Newton's law of inertia holds. Inertial reference frames move at constant velocity relative to one another. Accelerating reference frames are **noninertial**.

The **special theory of relativity** is based on two principles: the **relativity principle**, which states that the laws of physics are the same in all inertial reference frames, and the principle of the **constancy of the speed of light**, which states that the speed of light in empty space has the same value in all inertial reference frames.

One consequence of relativity theory is that two events that are simultaneous in one reference frame may not be simultaneous in another. Other effects are **time dilation**: moving clocks are measured to run slow; and **length contraction**: the length of a moving object is measured to be shorter (in its direction of motion) than when it is at rest. Quantitatively,

$$\Delta t = \frac{\Delta t_0}{\sqrt{1 - v^2/c^2}} = \gamma \, \Delta t_0 \qquad (26\text{-}1)$$

$$\ell = \ell_0 \sqrt{1 - v^2/c^2} = \frac{\ell_0}{\gamma} \qquad (26\text{-}3)$$

where ℓ and Δt are the length and time interval of objects (or events) observed as they move by at the speed v; ℓ_0 and Δt_0 are the **proper length** and **proper time**—that is, the same quantities as measured in the rest frame of the objects or events. The quantity γ is shorthand for

$$\gamma = \frac{1}{\sqrt{1 - v^2/c^2}}. \qquad (26\text{-}2)$$

The theory of relativity has changed our notions of space and time, and of momentum, energy, and mass. Space and time are seen to be intimately connected, with time being the fourth dimension in addition to the three dimensions of space.

The **momentum** of an object is given by

$$p = \gamma m v = \frac{mv}{\sqrt{1 - v^2/c^2}}. \qquad (26\text{-}4)$$

Mass and energy are interconvertible. The equation

$$E = mc^2 \qquad (26\text{-}7)$$

tells how much energy E is needed to create a mass m, or vice versa. Said another way, $E = mc^2$ is the amount of energy an object has because of its mass m. The law of conservation of energy must include mass as a form of energy.

The kinetic energy KE of an object moving at speed v is given by

$$\text{KE} = \frac{mc^2}{\sqrt{1 - v^2/c^2}} - mc^2 = (\gamma - 1)mc^2 \qquad (26\text{-}5)$$

where m is the mass of the object. The total energy E, if there is no potential energy, is

$$\begin{aligned} E &= \text{KE} + mc^2 \\ &= \gamma mc^2. \end{aligned} \qquad (26\text{-}6)$$

The momentum p of an object is related to its total energy E (assuming no potential energy) by

$$E^2 = p^2c^2 + m^2c^4. \qquad (26\text{-}9)$$

Velocity addition also must be done in a special way. All these relativistic effects are significant only at high speeds, close to the speed of light, which itself is the ultimate speed in the universe.

Questions

1. You are in a windowless car in an exceptionally smooth train moving at constant velocity. Is there any physical experiment you can do in the train car to determine whether you are moving? Explain.
2. You might have had the experience of being at a red light when, out of the corner of your eye, you see the car beside you creep forward. Instinctively you stomp on the brake pedal, thinking that you are rolling backward. What does this say about absolute and relative motion?
3. A worker stands on top of a railroad car moving at constant velocity and throws a heavy ball straight up (from his point of view). Ignoring air resistance, explain whether the ball will land back in his hand or behind him.
4. Does the Earth really go around the Sun? Or is it also valid to say that the Sun goes around the Earth? Discuss in view of the relativity principle (that there is no best reference frame). Explain. See Section 5–8.
5. If you were on a spaceship traveling at $0.6c$ away from a star, at what speed would the starlight pass you?
6. The time dilation effect is sometimes expressed as "moving clocks run slowly." Actually, this effect has nothing to do with motion affecting the functioning of clocks. What then does it deal with?
7. Does time dilation mean that time actually passes more slowly in moving reference frames or that it only *seems* to pass more slowly?
8. A young-looking woman astronaut has just arrived home from a long trip. She rushes up to an old gray-haired man and in the ensuing conversation refers to him as her son. How might this be possible?
9. If you were traveling away from Earth at speed $0.6c$, would you notice a change in your heartbeat? Would your mass, height, or waistline change? What would observers on Earth using telescopes say about you?
10. Do time dilation and length contraction occur at ordinary speeds, say 90 km/h?
11. Suppose the speed of light were infinite. What would happen to the relativistic predictions of length contraction and time dilation?
12. Explain how the length contraction and time dilation formulas might be used to indicate that c is the limiting speed in the universe.
13. Discuss how our everyday lives would be different if the speed of light were only 25 m/s.
14. The drawing at the start of this Chapter shows the street as seen by Mr Tompkins, for whom the speed of light is $c = 20$ mi/h. What does Mr Tompkins look like to the people standing on the street (Fig. 26–12)? Explain.

FIGURE 26–12 Question 14. Mr Tompkins as seen by people on the sidewalk. See also Chapter-Opening Figure on page 744.

15. An electron is limited to travel at speeds less than c. Does this put an upper limit on the momentum of an electron? If so, what is this upper limit? If not, explain.
16. Can a particle of nonzero mass attain the speed of light? Explain.
17. Does the equation $E = mc^2$ conflict with the conservation of energy principle? Explain.
18. If mass is a form of energy, does this mean that a spring has more mass when compressed than when relaxed? Explain.
19. It is not correct to say that "matter can neither be created nor destroyed." What must we say instead?
20. Is our intuitive notion that velocities simply add, as in Section 3–8, completely wrong?

MisConceptual Questions

1. The fictional rocket ship *Adventure* is measured to be 50 m long by the ship's captain inside the rocket. When the rocket moves past a space dock at $0.5c$, space-dock personnel measure the rocket ship to be 43.3 m long. What is its proper length?
    (*a*) 50 m.   (*b*) 43.3 m.   (*c*) 93.3 m.   (*d*) 13.3 m.
2. As rocket ship *Adventure* (MisConceptual Question 1) passes by the space dock, the ship's captain flashes a flashlight at 1.00-s intervals as measured by space-dock personnel. How often does the flashlight flash relative to the captain?
    (*a*) Every 1.15 s.   (*b*) Every 1.00 s.   (*c*) Every 0.87 s.
    (*d*) We need to know the distance between the ship and the space dock.
3. For the flashing of the flashlight in MisConceptual Question 2, what time interval is the proper time interval?
    (*a*) 1.15 s.   (*b*) 1.00 s.   (*c*) 0.87 s.   (*d*) 0.13 s.
4. The rocket ship of MisConceptual Question 1 travels to a star many light-years away, then turns around and returns at the same speed. When it returns to the space dock, who would have aged less: the space-dock personnel or ship's captain?
    (*a*) The space-dock personnel.
    (*b*) The ship's captain.
    (*c*) Both the same amount, because both sets of people were moving relative to each other.
    (*d*) We need to know how far away the star is.
5. An Earth observer notes that clocks on a passing spacecraft run slowly. The person on the spacecraft
    (*a*) agrees her clocks move slower than those on Earth.
    (*b*) feels normal, and her heartbeat and eating habits are normal.
    (*c*) observes that Earth clocks are moving slowly.
    (*d*) The real time is in between the times measured by the two observers.
    (*e*) Both (*a*) and (*b*).
    (*f*) Both (*b*) and (*c*).

6. Spaceships A and B are traveling directly toward each other at a speed 0.5c relative to the Earth, and each has a headlight aimed toward the other ship. What value do technicians on ship B get by measuring the speed of the light emitted by ship A's headlight?
   (a) 0.5c. (b) 0.75c. (c) 1.0c. (d) 1.5c.

7. Relativistic formulas for time dilation, length contraction, and mass are valid
   (a) only for speeds less than 0.10c.
   (b) only for speeds greater than 0.10c.
   (c) only for speeds very close to c.
   (d) for all speeds.

8. Which of the following will two observers in inertial reference frames always agree on? (Choose all that apply.)
   (a) The time an event occurred.
   (b) The distance between two events.
   (c) The time interval between the occurence of two events.
   (d) The speed of light.
   (e) The validity of the laws of physics.
   (f) The simultaneity of two events.

9. Two observers in different inertial reference frames moving relative to each other at nearly the speed of light see the same two events but, using precise equipment, record different time intervals between the two events. Which of the following is true of their measurements?
   (a) One observer is incorrect, but it is impossible to tell which one.
   (b) One observer is incorrect, and it is possible to tell which one.
   (c) Both observers are incorrect.
   (d) Both observers are correct.

10. You are in a rocket ship going faster and faster. As your speed increases and your velocity gets closer to the speed of light, which of the following do you observe in your frame of reference?
    (a) Your mass increases.
    (b) Your length shortens in the direction of motion.
    (c) Your wristwatch slows down.
    (d) All of the above.
    (e) None of the above.

11. You are in a spaceship with no windows, radios, or other means to check outside. How could you determine whether your spaceship is at rest or moving at constant velocity?
    (a) By determining the apparent velocity of light in the spaceship.
    (b) By checking your precision watch. If it's running slow, then the ship is moving.
    (c) By measuring the lengths of objects in the spaceship. If they are shortened, then the ship is moving.
    (d) Give up, because you can't tell.

12. The period of a pendulum attached in a spaceship is 2 s while the spaceship is parked on Earth. What is the period to an observer on Earth when the spaceship moves at 0.6c with respect to the Earth?
    (a) Less than 2 s.
    (b) More than 2 s.
    (c) 2 s.

13. Two spaceships, each moving at a speed 0.75c relative to the Earth, are headed directly toward each other. What do occupants of one ship measure the speed of other ship to be?
    (a) 0.96c. (b) 1.0c. (c) 1.5c. (d) 1.75c. (e) 0.75c.

For assigned homework and other learning materials, go to the MasteringPhysics website.

Problems

26–4 and 26–5 Time Dilation, Length Contraction

1. (I) A spaceship passes you at a speed of 0.850c. You measure its length to be 44.2 m. How long would it be when at rest?

2. (I) A certain type of elementary particle travels at a speed of 2.70×10^8 m/s. At this speed, the average lifetime is measured to be 4.76×10^{-6} s. What is the particle's lifetime at rest?

3. (II) You travel to a star 135 light-years from Earth at a speed of 2.90×10^8 m/s. What do you measure this distance to be?

4. (II) What is the speed of a pion if its average lifetime is measured to be 4.40×10^{-8} s? At rest, its average lifetime is 2.60×10^{-8} s.

5. (II) In an Earth reference frame, a star is 49 light-years away. How fast would you have to travel so that to you the distance would be only 35 light-years?

6. (II) At what speed v will the length of a 1.00-m stick look 10.0% shorter (90.0 cm)?

7. (II) At what speed do the relativistic formulas for (a) length and (b) time intervals differ from classical values by 1.00%? (This is a reasonable way to estimate when to use relativistic calculations rather than classical.)

8. (II) You decide to travel to a star 62 light-years from Earth at a speed that tells you the distance is only 25 light-years. How many years would it take you to make the trip?

9. (II) A friend speeds by you in her spacecraft at a speed of 0.720c. It is measured in your frame to be 4.80 m long and 1.35 m high. (a) What will be its length and height at rest? (b) How many seconds elapsed on your friend's watch when 20.0 s passed on yours? (c) How fast did you appear to be traveling according to your friend? (d) How many seconds elapsed on your watch when she saw 20.0 s pass on hers?

10. (II) A star is 21.6 light-years from Earth. How long would it take a spacecraft traveling 0.950c to reach that star as measured by observers: (a) on Earth, (b) on the spacecraft? (c) What is the distance traveled according to observers on the spacecraft? (d) What will the spacecraft occupants compute their speed to be from the results of (b) and (c)?

11. (II) A fictional news report stated that starship *Enterprise* had just returned from a 5-year voyage while traveling at 0.70c. (a) If the report meant 5.0 years of *Earth time*, how much time elapsed on the ship? (b) If the report meant 5.0 years of *ship time*, how much time passed on Earth?

12. (II) A box at rest has the shape of a cube 2.6 m on a side. This box is loaded onto the flat floor of a spaceship and the spaceship then flies past us with a horizontal speed of 0.80c. What is the volume of the box as we observe it?

13. (III) Escape velocity from the Earth is 11.2 km/s. What would be the percent decrease in length of a 68.2-m-long spacecraft traveling at that speed as seen from Earth?

14. (III) An unstable particle produced in an accelerator experiment travels at constant velocity, covering 1.00 m in 3.40 ns in the lab frame before changing ("decaying") into other particles. In the rest frame of the particle, determine (a) how long it lived before decaying, (b) how far it moved before decaying.

15. (III) How fast must a pion be moving on average to travel 32 m before it decays? The average lifetime, at rest, is 2.6×10^{-8} s.

26–7 Relativistic Momentum

16. (I) What is the momentum of a proton traveling at $v = 0.68c$?

17. (II) (a) A particle travels at $v = 0.15c$. By what percentage will a calculation of its momentum be wrong if you use the classical formula? (b) Repeat for $v = 0.75c$.

18. (II) A particle of mass m travels at a speed $v = 0.22c$. At what speed will its momentum be doubled?

19. (II) An unstable particle is at rest and suddenly decays into two fragments. No external forces act on the particle or its fragments. One of the fragments has a speed of $0.60c$ and a mass of 6.68×10^{-27} kg, while the other has a mass of 1.67×10^{-27} kg. What is the speed of the less massive fragment?

20. (II) What is the percent change in momentum of a proton that accelerates from (a) $0.45c$ to $0.85c$, (b) $0.85c$ to $0.98c$?

26–9 $E = mc^2$; Mass and Energy

21. (I) Calculate the rest energy of an electron in joules and in MeV (1 MeV $= 1.60 \times 10^{-13}$ J).

22. (I) When a uranium nucleus at rest breaks apart in the process known as *fission* in a nuclear reactor, the resulting fragments have a total kinetic energy of about 200 MeV. How much mass was lost in the process?

23. (I) The total annual energy consumption in the United States is about 1×10^{20} J. How much mass would have to be converted to energy to fuel this need?

24. (I) Calculate the mass of a proton (1.67×10^{-27} kg) in MeV/c^2.

25. (I) A certain chemical reaction requires 4.82×10^4 J of energy input for it to go. What is the increase in mass of the products over the reactants?

26. (II) Calculate the kinetic energy and momentum of a proton traveling 2.90×10^8 m/s.

27. (II) What is the momentum of a 950-MeV proton (that is, its kinetic energy is 950 MeV)?

28. (II) What is the speed of an electron whose kinetic energy is 1.12 MeV?

29. (II) (a) How much work is required to accelerate a proton from rest up to a speed of $0.985c$? (b) What would be the momentum of this proton?

30. (II) At what speed will an object's kinetic energy be 33% of its rest energy?

31. (II) Determine the speed and the momentum of an electron ($m = 9.11 \times 10^{-31}$ kg) whose KE equals its rest energy.

32. (II) A proton is traveling in an accelerator with a speed of 1.0×10^8 m/s. By what factor does the proton's kinetic energy increase if its speed is doubled?

33. (II) How much energy can be obtained from conversion of 1.0 gram of mass? How much mass could this energy raise to a height of 1.0 km above the Earth's surface?

34. (II) To accelerate a particle of mass m from rest to speed $0.90c$ requires work W_1. To accelerate the particle from speed $0.90c$ to $0.99c$ requires work W_2. Determine the ratio W_2/W_1.

35. (II) Suppose there was a process by which two photons, each with momentum 0.65 MeV/c, could collide and make a single particle. What is the maximum mass that the particle could possess?

36. (II) What is the speed of a proton accelerated by a potential difference of 165 MV?

37. (II) What is the speed of an electron after being accelerated from rest by 31,000 V?

38. (II) The kinetic energy of a particle is 45 MeV. If the momentum is 121 MeV/c, what is the particle's mass?

39. (II) Calculate the speed of a proton ($m = 1.67 \times 10^{-27}$ kg) whose kinetic energy is exactly half (a) its total energy, (b) its rest energy.

40. (II) Calculate the kinetic energy and momentum of a proton ($m = 1.67 \times 10^{-27}$ kg) traveling 8.65×10^7 m/s. By what percentages would your calculations have been in error if you had used classical formulas?

41. (II) Suppose a spacecraft of mass 17,000 kg is accelerated to $0.15c$. (a) How much kinetic energy would it have? (b) If you used the classical formula for kinetic energy, by what percentage would you be in error?

42. (II) A negative muon traveling at 53% the speed of light collides head on with a positive muon traveling at 65% the speed of light. The two muons (each of mass 105.7 MeV/c^2) annihilate, and produce how much electromagnetic energy?

43. (II) Two identical particles of mass m approach each other at equal and opposite speeds, v. The collision is completely inelastic and results in a single particle at rest. What is the mass of the new particle? How much energy was lost in the collision? How much kinetic energy was lost in this collision?

44. (III) The americium nucleus, $^{241}_{95}$Am, decays to a neptunium nucleus, $^{237}_{93}$Np, by emitting an alpha particle of mass 4.00260 u and kinetic energy 5.5 MeV. Estimate the mass of the neptunium nucleus, ignoring its recoil, given that the americium mass is 241.05682 u.

45. (III) Show that the kinetic energy KE of a particle of mass m is related to its momentum p by the equation
$$p = \sqrt{\text{KE}^2 + 2\text{KE}\, mc^2}/c.$$

*46. (III) What magnetic field B is needed to keep 998-GeV protons revolving in a circle of radius 1.0 km? Use the relativistic mass. The proton's "rest mass" is 0.938 GeV/c^2. (1 GeV $= 10^9$ eV.) [*Hint*: In relativity, $m_{\text{rel}} v^2/r = qvB$ is still valid in a magnetic field, where $m_{\text{rel}} = \gamma m$.]

26–10 Relativistic Addition of Velocities

47. (I) A person on a rocket traveling at $0.40c$ (with respect to the Earth) observes a meteor come from behind and pass her at a speed she measures as $0.40c$. How fast is the meteor moving with respect to the Earth?

48. (II) Two spaceships leave Earth in opposite directions, each with a speed of $0.60c$ with respect to Earth. (a) What is the velocity of spaceship 1 relative to spaceship 2? (b) What is the velocity of spaceship 2 relative to spaceship 1?

49. (II) A spaceship leaves Earth traveling at 0.65c. A second spaceship leaves the first at a speed of 0.82c with respect to the first. Calculate the speed of the second ship with respect to Earth if it is fired (a) in the same direction the first spaceship is already moving, (b) directly backward toward Earth.

50. (II) An observer on Earth sees an alien vessel approach at a speed of 0.60c. The fictional starship *Enterprise* comes to the rescue (Fig. 26–13), overtaking the aliens while moving directly toward Earth at a speed of 0.90c relative to Earth. What is the relative speed of one vessel as seen by the other?

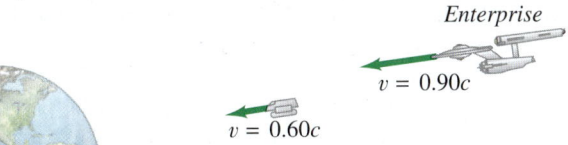

FIGURE 26–13 Problem 50.

51. (II) A spaceship in distress sends out two escape pods in opposite directions. One travels at a speed $v_1 = +0.70c$ in one direction, and the other travels at a speed $v_2 = -0.80c$ in the other direction, as observed from the spaceship. What speed does the first escape pod measure for the second escape pod?

52. (II) Rocket A passes Earth at a speed of 0.65c. At the same time, rocket B passes Earth moving 0.95c relative to Earth in the same direction as A. How fast is B moving relative to A when it passes A?

53. (II) Your spaceship, traveling at 0.90c, needs to launch a probe out the forward hatch so that its speed relative to the planet that you are approaching is 0.95c. With what speed must it leave your ship?

General Problems

54. What is the speed of a particle when its kinetic energy equals its rest energy? Does the mass of the particle affect the result?

55. The nearest star to Earth is Proxima Centauri, 4.3 light-years away. (a) At what constant velocity must a spacecraft travel from Earth if it is to reach the star in 4.9 years, as measured by travelers on the spacecraft? (b) How long does the trip take according to Earth observers?

56. According to the special theory of relativity, the factor γ that determines the length contraction and the time dilation is given by $\gamma = 1/\sqrt{1 - v^2/c^2}$. Determine the numerical values of γ for an object moving at speed $v = 0.01c, 0.05c, 0.10c, 0.20c, 0.30c, 0.40c, 0.50c, 0.60c, 0.70c, 0.80c, 0.90c, 0.95c,$ and $0.99c$. Make a graph of γ versus v.

57. A healthy astronaut's heart rate is 60 beats/min. Flight doctors on Earth can monitor an astronaut's vital signs remotely while in flight. How fast would an astronaut be flying away from Earth if the doctor measured her having a heart rate of 25 beats/min?

58. (a) What is the speed v of an electron whose kinetic energy is 14,000 times its rest energy? You can state the answer as the difference $c - v$. Such speeds are reached in the Stanford Linear Accelerator, SLAC. (b) If the electrons travel in the lab through a tube 3.0 km long (as at SLAC), how long is this tube in the electrons' reference frame? [*Hint*: Use the binomial expansion.]

59. What minimum amount of electromagnetic energy is needed to produce an electron and a positron together? A positron is a particle with the same mass as an electron, but has the opposite charge. (Note that electric charge is conserved in this process. See Section 27–6.)

60. How many grams of matter would have to be totally destroyed to run a 75-W lightbulb for 1.0 year?

61. A free neutron can decay into a proton, an electron, and a neutrino. Assume the neutrino's mass is zero; the other masses can be found in the Table inside the front cover. Determine the total kinetic energy shared among the three particles when a neutron decays at rest.

62. An electron ($m = 9.11 \times 10^{-31}$ kg) is accelerated from rest to speed v by a conservative force. In this process, its potential energy decreases by 6.20×10^{-14} J. Determine the electron's speed, v.

63. The Sun radiates energy at a rate of about 4×10^{26} W. (a) At what rate is the Sun's mass decreasing? (b) How long does it take for the Sun to lose a mass equal to that of Earth? (c) Estimate how long the Sun could last if it radiated constantly at this rate.

64. How much energy would be required to break a helium nucleus into its constituents, two protons and two neutrons? The masses of a proton (including an electron), a neutron, and neutral helium are, respectively, 1.00783 u, 1.00867 u, and 4.00260 u. (This energy difference is called the *total binding energy* of the $^4_2$He nucleus.)

65. Show analytically that a particle with momentum p and energy E has a speed given by
$$v = \frac{pc^2}{E} = \frac{pc}{\sqrt{m^2c^2 + p^2}}.$$

66. Two protons, each having a speed of 0.990c in the laboratory, are moving toward each other. Determine (a) the momentum of each proton in the laboratory, (b) the total momentum of the two protons in the laboratory, and (c) the momentum of one proton as seen by the other proton.

67. When two moles of hydrogen molecules (H_2) and one mole of oxygen molecules (O_2) react to form two moles of water (H_2O), the energy released is 484 kJ. How much does the mass decrease in this reaction? What % of the total original mass is this?

68. The fictional starship *Enterprise* obtains its power by combining matter and antimatter, achieving complete conversion of mass into energy. If the mass of the *Enterprise* is approximately 6×10^9 kg, how much mass must be converted into kinetic energy to accelerate it from rest to one-tenth the speed of light?

69. Make a graph of the kinetic energy versus momentum for (a) a particle of nonzero mass, and (b) a particle with zero mass.

70. A spaceship and its occupants have a total mass of 160,000 kg. The occupants would like to travel to a star that is 35 light-years away at a speed of 0.70c. To accelerate, the engine of the spaceship changes mass directly to energy. (a) Estimate how much mass will be converted to energy to accelerate the spaceship to this speed. (b) Assuming the acceleration is rapid, so the speed for the entire trip can be taken to be 0.70c, determine how long the trip will take according to the astronauts on board.

71. In a nuclear reaction two identical particles are created, traveling in opposite directions. If the speed of each particle is 0.82c, relative to the laboratory frame of reference, what is one particle's speed relative to the other particle?

72. A 36,000-kg spaceship is to travel to the vicinity of a star 6.6 light-years from Earth. Passengers on the ship want the (one-way) trip to take no more than 1.0 year. How much work must be done on the spaceship to bring it to the speed necessary for this trip?

73. Suppose a 14,500-kg spaceship left Earth at a speed of 0.90c. What is the spaceship's kinetic energy? Compare with the total U.S. annual energy consumption (about 10^{20} J).

74. A pi meson of mass m_π decays at rest into a muon (mass m_μ) and a neutrino of negligible or zero mass. Show that the kinetic energy of the muon is $\text{KE}_\mu = (m_\pi - m_\mu)^2 c^2 / (2m_\pi)$.

75. An astronaut on a spaceship traveling at 0.75c relative to Earth measures his ship to be 23 m long. On the ship, he eats his lunch in 28 min. (a) What length is the spaceship according to observers on Earth? (b) How long does the astronaut's lunch take to eat according to observers on Earth?

76. Astronomers measure the distance to a particular star to be 6.0 light-years (1 ly = distance light travels in 1 year). A spaceship travels from Earth to the vicinity of this star at steady speed, arriving in 3.50 years as measured by clocks on the spaceship. (a) How long does the trip take as measured by clocks in Earth's reference frame? (b) What distance does the spaceship travel as measured in its own reference frame?

77. An electron is accelerated so that its kinetic energy is greater than its rest energy mc^2 by a factor of (a) 5.00, (b) 999. What is the speed of the electron in each case?

78. You are traveling in a spaceship at a speed of 0.70c away from Earth. You send a laser beam toward the Earth traveling at velocity c relative to you. What do observers on the Earth measure for the speed of the laser beam?

79. A farm boy studying physics believes that he can fit a 13.0-m-long pole into a 10.0-m-long barn if he runs fast enough, carrying the pole. Can he do it? Explain in detail. How does this fit with the idea that when he is running the barn looks even shorter than 10.0 m?

80. An atomic clock is taken to the North Pole, while another stays at the Equator. How far will they be out of synchronization after 2.0 years has elapsed? [*Hint*: Use the binomial expansion, Appendix A.]

81. An airplane travels 1300 km/h around the Earth in a circle of radius essentially equal to that of the Earth, returning to the same place. Using special relativity, estimate the difference in time to make the trip as seen by Earth and by airplane observers. [*Hint*: Use the binomial expansion, Appendix A.]

Search and Learn

1. Determine about how fast Mr Tompkins is traveling in the Chapter-Opening Photograph. Do you agree with the picture in terms of the way Mr Tompkins would see the world? Explain. [*Hint*: Assume the bank clock and Stop sign facing us are round according to the people on the sidewalk.]

2. Examine the experiment of Fig. 26–5 from O_1's reference frame. In this case, O_1 will be at rest and will see the lightning bolt at B_1 and B_2, before the lightning bolt at A_1 and A_2. Will O_1 recognize that O_2, who is moving with speed v to the left, will see the two events as simultaneous? Explain in detail, drawing diagrams equivalent to Fig. 26–5. [*Hint*: Include length contraction.]

3. Using Example 26–2 as a guide, show that for objects that move slowly in comparison to c, the length contraction formula is roughly $\ell \approx \ell_0(1 - \frac{1}{2}v^2/c^2)$. Use this approximation to find the "length shortening" $\Delta \ell = \ell_0 - \ell$ of the train in Example 26–6 if the train travels at 100 km/h (rather than 0.92c).

4. In Example 26–5, the spaceship is moving at 0.90c in the horizontal direction relative to an observer on the Earth. If instead the spaceship moved at 0.90c directed at 30° above the horizontal, what would be the painting's dimensions as seen by the observer on Earth?

5. Protons from outer space crash into the Earth's atmosphere at a high rate. These protons create particles that eventually decay into other particles called *muons*. This cosmic debris travels through the atmosphere. Every second, dozens of muons pass through your body. If a muon is created 30 km above the Earth's surface, what minimum speed and kinetic energy must the muon have in order to hit Earth's surface? A muon's mean lifetime (at rest) is 2.20 μs and its mass is 105.7 MeV/c^2.

6. As a rough rule, anything traveling faster than about 0.1c is called *relativistic*—that is, special relativity is a significant effect. Determine the speed of an electron in a hydrogen atom (radius 0.53×10^{-10} m) and state whether or not it is relativistic. (Treat the electron as though it were in a circular orbit around the proton. See hint for Problem 46.)

ANSWERS TO EXERCISES

A: (c).
B: (a) No; (b) yes.
C: 80 m.
D: No: $\text{KE}/mc^2 \approx 2 \times 10^{-4}$.

E: 0.030c, same as classical, to an accuracy of better than 0.1%.
F: (d).

Electron microscopes (EM) produce images using electrons which have wave properties just as light does. Because the wavelength of electrons can be much smaller than that of visible light, much greater resolution and magnification can be obtained. A scanning electron microscope (SEM) can produce images with a three-dimensional quality.

All EM images are monochromatic (black and white). Artistic coloring has been added here, as is common. On the left is an SEM image of a blood clot forming (yellow-color web) due to a wound. White blood cells are colored green here for visibility. On the right, red blood cells in a small artery. A red blood cell travels about 15 km a day inside our bodies and lives roughly 4 months before damage or rupture. Humans contain 4 to 6 liters of blood, and 2 to 3×10^{13} red blood cells.

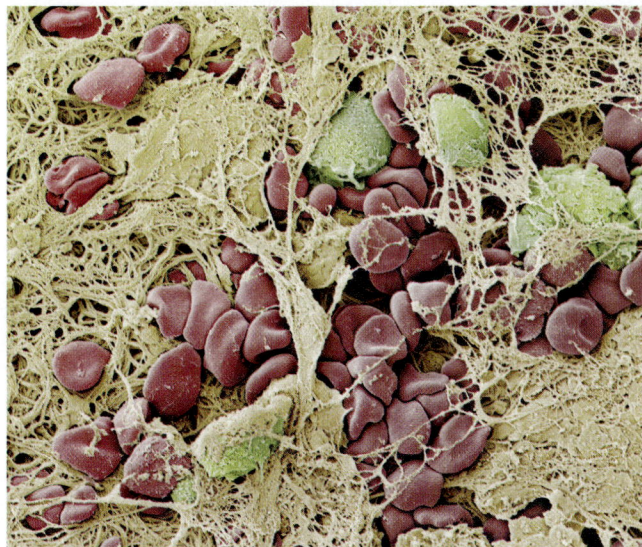 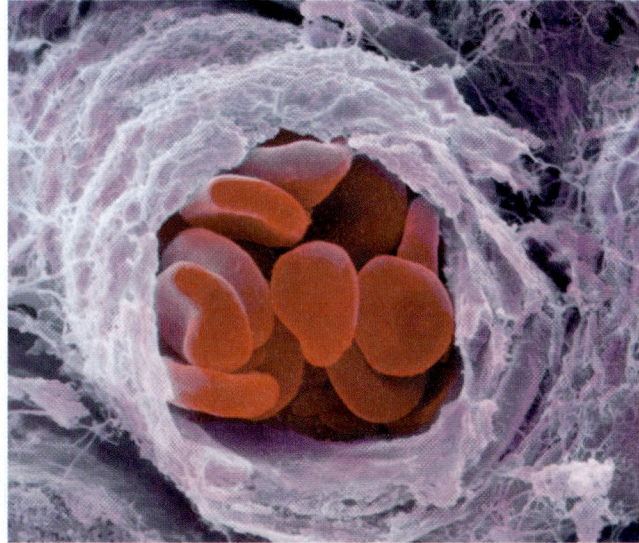

Early Quantum Theory and Models of the Atom

CHAPTER 27

CHAPTER-OPENING QUESTION—Guess now!

It has been found experimentally that
- **(a)** light behaves as a wave.
- **(b)** light behaves as a particle.
- **(c)** electrons behave as particles.
- **(d)** electrons behave as waves.
- **(e)** all of the above are true.
- **(f)** only (a) and (b) are true.
- **(g)** only (a) and (c) are true.
- **(h)** none of the above are true.

The second aspect of the revolution that shook the world of physics in the early part of the twentieth century was the quantum theory (the other was Einstein's theory of relativity). Unlike the special theory of relativity, the revolution of quantum theory required almost three decades to unfold, and many scientists contributed to its development. It began in 1900 with Planck's quantum hypothesis, and culminated in the mid-1920s with the theory of quantum mechanics of Schrödinger and Heisenberg which has been so effective in explaining the structure of matter. The discovery of the electron in the 1890s, with which we begin this Chapter, might be said to mark the beginning of modern physics, and is a sort of precursor to the quantum theory.

CONTENTS

- 27–1 Discovery and Properties of the Electron
- 27–2 Blackbody Radiation; Planck's Quantum Hypothesis
- 27–3 Photon Theory of Light and the Photoelectric Effect
- 27–4 Energy, Mass, and Momentum of a Photon
- *27–5 Compton Effect
- 27–6 Photon Interactions; Pair Production
- 27–7 Wave–Particle Duality; the Principle of Complementarity
- 27–8 Wave Nature of Matter
- 27–9 Electron Microscopes
- 27–10 Early Models of the Atom
- 27–11 Atomic Spectra: Key to the Structure of the Atom
- 27–12 The Bohr Model
- 27–13 de Broglie's Hypothesis Applied to Atoms

27–1 Discovery and Properties of the Electron

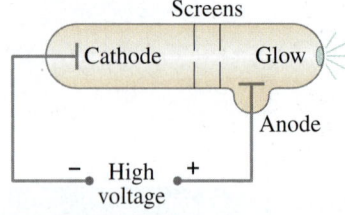

FIGURE 27–1 Discharge tube. In some models, one of the screens is the anode (positive plate).

Toward the end of the nineteenth century, studies were being done on the discharge of electricity through rarefied gases. One apparatus, diagrammed in Fig. 27–1, was a glass tube fitted with electrodes and evacuated so only a small amount of gas remained inside. When a very high voltage was applied to the electrodes, a dark space seemed to extend outward from the cathode (negative electrode) toward the opposite end of the tube; and that far end of the tube would glow. If one or more screens containing a small hole were inserted as shown, the glow was restricted to a tiny spot on the end of the tube. It seemed as though something being emitted by the cathode traveled across to the opposite end of the tube. These "somethings" were named **cathode rays**.

There was much discussion at the time about what these rays might be. Some scientists thought they might resemble light. But the observation that the bright spot at the end of the tube could be deflected to one side by an electric or magnetic field suggested that cathode rays were charged particles; and the direction of the deflection was consistent with a negative charge. Furthermore, if the tube contained certain types of rarefied gas, the path of the cathode rays was made visible by a slight glow.

Estimates of the charge e of the cathode-ray particles, as well as of their charge-to-mass ratio e/m, had been made by 1897. But in that year, J. J. Thomson (1856–1940) was able to measure e/m directly, using the apparatus shown in Fig. 27–2. Cathode rays are accelerated by a high voltage and then pass between a pair of parallel plates built into the tube. Another voltage applied to the parallel plates produces an electric field $\vec{E}$, and a pair of coils produces a magnetic field $\vec{B}$. If $E = B = 0$, the cathode rays follow path b in Fig. 27–2.

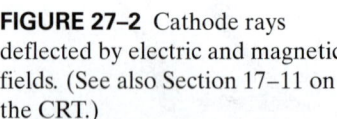

FIGURE 27–2 Cathode rays deflected by electric and magnetic fields. (See also Section 17–11 on the CRT.)

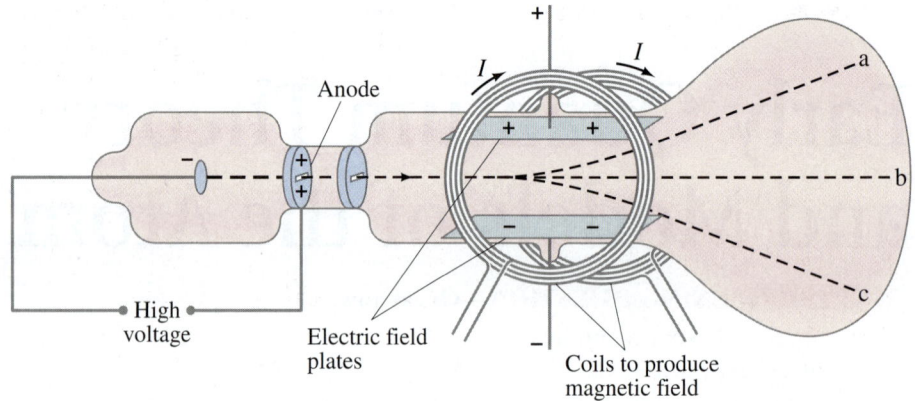

When only the electric field is present, say with the upper plate positive, the cathode rays are deflected upward as in path a in Fig. 27–2. If only a magnetic field exists, say inward, the rays are deflected downward along path c. These observations are just what is expected for a negatively charged particle. The force on the rays due to the magnetic field is $F = evB$, where e is the charge and v is the velocity of the cathode rays (Eq. 20–4). In the absence of an electric field, the rays are bent into a curved path, and applying Newton's second law $F = ma$ with $a =$ centripetal acceleration gives

$$evB = m\frac{v^2}{r},$$

and thus

$$\frac{e}{m} = \frac{v}{Br}.$$

The radius of curvature r can be measured and so can B. The velocity v can be found by applying an electric field in addition to the magnetic field. The electric field E is adjusted so that the cathode rays are undeflected and follow path b in Fig. 27–2. In this situation the upward force due to the electric field, $F = eE$, is balanced by the downward force due to the magnetic field, $F = evB$. We equate the two forces, $eE = evB$, and find

$$v = \frac{E}{B}.$$

Combining this with the above equation we have

$$\frac{e}{m} = \frac{E}{B^2 r}. \tag{27–1}$$

The quantities on the right side can all be measured, and although e and m could not be determined separately, the ratio e/m could be determined. The accepted value today is $e/m = 1.76 \times 10^{11}$ C/kg. Cathode rays soon came to be called **electrons**.

Discovery in Science

The "discovery" of the electron, like many others in science, is not quite so obvious as discovering gold or oil. Should the discovery of the electron be credited to the person who first saw a glow in the tube? Or to the person who first called them cathode rays? Perhaps neither one, for they had no conception of the electron as we know it today. In fact, the credit for the discovery is generally given to Thomson, but not because he was the first to see the glow in the tube. Rather it is because he believed that this phenomenon was due to tiny negatively charged particles and made careful measurements on them. Furthermore he argued that these particles were constituents of atoms, and not ions or atoms themselves as many thought, and he developed an electron theory of matter. His view is close to what we accept today, and this is why Thomson is credited with the "discovery." Note, however, that neither he nor anyone else ever actually saw an electron itself. We discuss this briefly, for it illustrates the fact that discovery in science is not always a clear-cut matter. In fact some philosophers of science think the word "discovery" is often not appropriate, such as in this case.

Electron Charge Measurement

Thomson believed that an electron was not an atom, but rather a constituent, or part, of an atom. Convincing evidence for this came soon with the determination of the charge and the mass of the cathode rays. Thomson's student J. S. Townsend made the first direct (but rough) measurements of e in 1897. But it was the more refined **oil-drop experiment** of Robert A. Millikan (1868–1953) that yielded a precise value for the charge on the electron and showed that charge comes in discrete amounts. In this experiment, tiny droplets of mineral oil carrying an electric charge were allowed to fall under gravity between two parallel plates, Fig. 27–3. The electric field E between the plates was adjusted until the drop was suspended in midair. The downward pull of gravity, mg, was then just balanced by the upward force due to the electric field. Thus $qE = mg$ so the charge $q = mg/E$. The mass of the droplet was determined by measuring its terminal velocity in the absence of the electric field. Often the droplet was charged negatively, but sometimes it was positive, suggesting that the droplet had acquired or lost electrons (by friction, leaving the atomizer). Millikan's painstaking observations and analysis presented convincing evidence that any charge was an integral multiple of a smallest charge, e, that was ascribed to the electron, and that the value of e was 1.6×10^{-19} C. This value of e, combined with the measurement of e/m, gives the mass of the electron to be $(1.6 \times 10^{-19}$ C$)/(1.76 \times 10^{11}$ C/kg$) = 9.1 \times 10^{-31}$ kg. This mass is less than a thousandth the mass of the smallest atom, and thus confirmed the idea that the electron is only a part of an atom. The accepted value today for the mass of the electron is

$$m_e = 9.11 \times 10^{-31} \text{ kg}.$$

The experimental result that any charge is an integral multiple of e means that electric charge is *quantized* (exists only in discrete amounts).

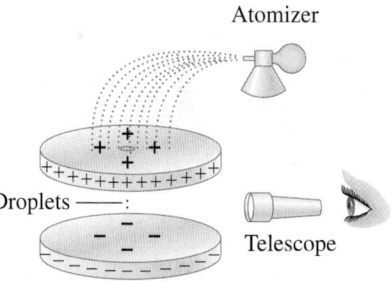

FIGURE 27–3 Millikan's oil-drop experiment.

27–2 Blackbody Radiation; Planck's Quantum Hypothesis

Blackbody Radiation

One of the observations that was unexplained at the end of the nineteenth century was the spectrum of light emitted by hot objects. We saw in Section 14–8 that all objects emit radiation whose total intensity is proportional to the fourth power of the Kelvin (absolute) temperature (T^4). At normal temperatures (≈ 300 K), we are not aware of this electromagnetic radiation because of its low intensity. At higher temperatures, there is sufficient infrared radiation that we can feel heat if we are close to the object. At still higher temperatures (on the order of 1000 K), objects actually glow, such as a red-hot electric stove burner or the heating element in a toaster. At temperatures above 2000 K, objects glow with a yellow or whitish color, such as white-hot iron and the filament of a lightbulb. The light emitted contains a continuous range of wavelengths or frequencies, and the spectrum is a plot of intensity vs. wavelength or frequency. As the temperature increases, the electromagnetic radiation emitted by objects not only increases in total intensity but has its peak intensity at higher and higher frequencies.

The spectrum of light emitted by a hot dense object is shown in Fig. 27–4 for an idealized **blackbody**. A blackbody is a body that, when cool, would absorb all the radiation falling on it (and so would appear black under reflection when illuminated by other sources). The radiation such an idealized blackbody would emit when hot and luminous, called **blackbody radiation** (though not necessarily black in color), approximates that from many real objects. The 6000-K curve in Fig. 27–4, corresponding to the temperature of the surface of the Sun, peaks in the visible part of the spectrum. For lower temperatures, the total intensity drops considerably and the peak occurs at longer wavelengths (or lower frequencies). This is why objects glow with a red color at around 1000 K. It is found experimentally that the wavelength at the peak of the spectrum, λ_P, is related to the Kelvin temperature T by

$$\lambda_P T = 2.90 \times 10^{-3} \, \text{m} \cdot \text{K}. \tag{27–2}$$

This is known as **Wien's law**.

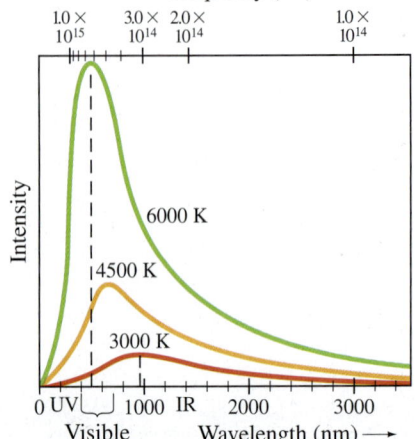

FIGURE 27–4 Measured spectra of wavelengths and frequencies emitted by a blackbody at three different temperatures.

EXAMPLE 27–1 **The Sun's surface temperature.** Estimate the temperature of the surface of our Sun, given that the Sun emits light whose peak intensity occurs in the visible spectrum at around 500 nm.

APPROACH We assume the Sun acts as a blackbody, and use $\lambda_P = 500$ nm in Wien's law (Eq. 27–2).

SOLUTION Wien's law gives

$$T = \frac{2.90 \times 10^{-3} \, \text{m} \cdot \text{K}}{\lambda_P} = \frac{2.90 \times 10^{-3} \, \text{m} \cdot \text{K}}{500 \times 10^{-9} \, \text{m}} \approx 6000 \, \text{K}.$$

EXAMPLE 27–2 **Star color.** Suppose a star has a surface temperature of 32,500 K. What color would this star appear?

APPROACH We assume the star emits radiation as a blackbody, and solve for λ_P in Wien's law, Eq. 27–2.

SOLUTION From Wien's law we have

$$\lambda_P = \frac{2.90 \times 10^{-3} \, \text{m} \cdot \text{K}}{T} = \frac{2.90 \times 10^{-3} \, \text{m} \cdot \text{K}}{3.25 \times 10^4 \, \text{K}} = 89.2 \, \text{nm}.$$

The peak is in the UV range of the spectrum, and will be way to the left in Fig. 27–4. In the visible region, the curve will be descending, so the shortest visible wavelengths will be strongest. Hence the star will appear bluish (or blue-white).

NOTE This example helps us to understand why stars have different colors (reddish for the coolest stars; orangish, yellow, white, bluish for "hotter" stars.)

EXERCISE A What is the color of an object at 4000 K?

Planck's Quantum Hypothesis

In the year 1900, Max Planck (1858–1947) proposed a theory that was able to reproduce the graphs of Fig. 27–4. His theory, still accepted today, made a new and radical assumption: that the energy of the oscillations of atoms within molecules cannot have just any value; instead each has energy which is a multiple of a minimum value related to the frequency of oscillation by

$$E = hf.$$

Here h is a new constant, now called **Planck's constant**, whose value was estimated by Planck by fitting his formula for the blackbody radiation curve to experiment. The value accepted today is

$$h = 6.626 \times 10^{-34} \, \text{J} \cdot \text{s}.$$

Planck's assumption suggests that the energy of any molecular vibration could be only a whole number multiple of hf:

$$E = nhf, \qquad n = 1, 2, 3, \cdots, \qquad (27\text{–}3)$$

where n is called a **quantum number** ("quantum" means "discrete amount" as opposed to "continuous"). This idea is often called **Planck's quantum hypothesis**, although little attention was brought to this point at the time. In fact, it appears that Planck considered it more as a mathematical device to get the "right answer" rather than as an important discovery. Planck himself continued to seek a classical explanation for the introduction of h. The recognition that this was an important and radical innovation did not come until later, after about 1905 when others, particularly Einstein, entered the field.

The quantum hypothesis, Eq. 27–3, states that the energy of an oscillator can be $E = hf$, or $2hf$, or $3hf$, and so on, but there cannot be vibrations with energies between these values. That is, energy would not be a continuous quantity as had been believed for centuries; rather it is **quantized**—it exists only in discrete amounts. The smallest amount of energy possible (hf) is called the **quantum of energy**. Recall from Chapter 11 that the energy of an oscillation is proportional to the amplitude squared. Another way of expressing the quantum hypothesis is that not just any amplitude of vibration is possible. The possible values for the amplitude are related to the frequency f.

A simple analogy may help. Compare a ramp, on which a box can be placed at any height, to a flight of stairs on which the box can have only certain discrete amounts of potential energy, as shown in Fig. 27–5.

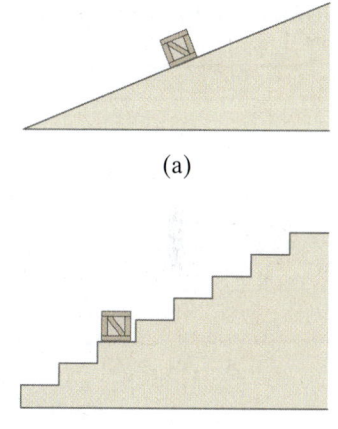

FIGURE 27–5 Ramp versus stair analogy. (a) On a ramp, a box can have continuous values of potential energy. (b) But on stairs, the box can have only discrete (quantized) values of energy.

27–3 Photon Theory of Light and the Photoelectric Effect

In 1905, the same year that he introduced the special theory of relativity, Einstein made a bold extension of the quantum idea by proposing a new theory of light. Planck's work had suggested that the vibrational energy of molecules in a radiating object is quantized with energy $E = nhf$, where n is an integer and f is the frequency of molecular vibration. Einstein argued that when light is emitted by a molecular oscillator, the molecule's vibrational energy of nhf must decrease by an amount hf (or by $2hf$, etc.) to another integer times hf, such as $(n - 1)hf$. Then to conserve energy, the light ought to be emitted in packets, or *quanta*, each with an energy

$$E = hf, \qquad (27\text{–}4)$$

Photon energy

where f is here the frequency of the emitted light. Again h is Planck's constant. Because all light ultimately comes from a radiating source, this idea suggests that *light is transmitted as tiny particles*, or **photons** as they are now called, as well as via the waves predicted by Maxwell's electromagnetic theory. The photon theory of light was also a radical departure from classical ideas. Einstein proposed a test of the quantum theory of light: quantitative measurements on the photoelectric effect.

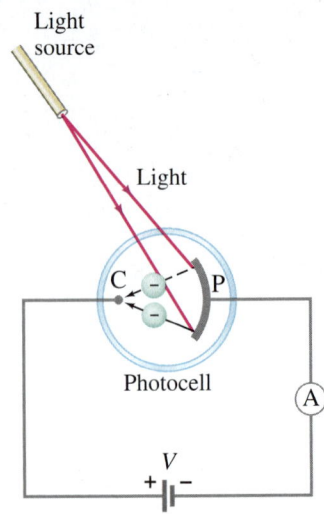

FIGURE 27–6 The photoelectric effect.

When light shines on a metal surface, electrons are found to be emitted from the surface. This effect is called the **photoelectric effect** and it occurs in many materials, but is most easily observed with metals. It can be observed using the apparatus shown in Fig. 27–6. A metal plate P and a smaller electrode C are placed inside an evacuated glass tube, called a **photocell**. The two electrodes are connected to an ammeter and a source of emf, as shown. When the photocell is in the dark, the ammeter reads zero. But when light of sufficiently high frequency illuminates the plate, the ammeter indicates a current flowing in the circuit. We explain completion of the circuit by imagining that electrons, ejected from the plate by the impinging light, flow across the tube from the plate to the "collector" C as indicated in Fig. 27–6.

That electrons should be emitted when light shines on a metal is consistent with the electromagnetic (EM) wave theory of light: the electric field of an EM wave could exert a force on electrons in the metal and eject some of them. Einstein pointed out, however, that the wave theory and the photon theory of light give very different predictions on the details of the photoelectric effect. For example, one thing that can be measured with the apparatus of Fig. 27–6 is the maximum kinetic energy (KE_{max}) of the emitted electrons. This can be done by using a variable voltage source and reversing the terminals so that electrode C is negative and P is positive. The electrons emitted from P will be repelled by the negative electrode, but if this reverse voltage is small enough, the fastest electrons will still reach C and there will be a current in the circuit. If the reversed voltage is increased, a point is reached where the current reaches zero—no electrons have sufficient kinetic energy to reach C. This is called the *stopping potential*, or *stopping voltage*, V_0, and from its measurement, KE_{max} can be determined using conservation of energy (loss of kinetic energy = gain in potential energy):

$$KE_{max} = eV_0.$$

Now let us examine the details of the photoelectric effect from the point of view of the wave theory versus Einstein's particle theory.

First the wave theory, assuming monochromatic light. The two important properties of a light wave are its intensity and its frequency (or wavelength). When these two quantities are varied, the wave theory makes the following predictions:

Wave theory predictions

1. If the light intensity is increased, the number of electrons ejected and their maximum kinetic energy should be increased because the higher intensity means a greater electric field amplitude, and the greater electric field should eject electrons with higher speed.
2. The frequency of the light should not affect the kinetic energy of the ejected electrons. Only the intensity should affect KE_{max}.

The photon theory makes completely different predictions. First we note that in a monochromatic beam, all photons have the same energy ($= hf$). Increasing the intensity of the light beam means increasing the number of photons in the beam, but does not affect the energy of each photon as long as the frequency is not changed. According to Einstein's theory, an electron is ejected from the metal by a collision with a single photon. In the process, all the photon energy is transferred to the electron and the photon ceases to exist. Since electrons are held in the metal by attractive forces, some minimum energy W_0 is required just to get an electron out through the surface. W_0 is called the **work function**, and is a few electron volts ($1\,eV = 1.6 \times 10^{-19}\,J$) for most metals. If the frequency f of the incoming light is so low that hf is less than W_0, then the photons will not have enough energy to eject any electrons at all. If $hf > W_0$, then electrons will be ejected and energy will be conserved in the process. That is, the input energy (of the photon), hf, will equal the outgoing kinetic energy KE of the electron plus the energy required to get it out of the metal, W:

$$hf = KE + W. \qquad (27\text{–}5a)$$

The least tightly held electrons will be emitted with the most kinetic energy (KE_{max}), in which case W in this equation becomes the work function W_0, and KE becomes KE_{max}:

$$hf = \text{KE}_{\text{max}} + W_0. \qquad \text{[least bound electrons]} \quad \textbf{(27-5b)}$$

Many electrons will require more energy than the bare minimum (W_0) to get out of the metal, and thus the kinetic energy of such electrons will be less than the maximum.

From these considerations, the photon theory makes the following predictions:

1. An increase in intensity of the light beam means more photons are incident, so more electrons will be ejected; but since the energy of each photon is not changed, the maximum kinetic energy of electrons is not changed by an increase in intensity.

2. If the frequency of the light is increased, the maximum kinetic energy of the electrons increases linearly, according to Eq. 27–5b. That is,

$$\text{KE}_{\text{max}} = hf - W_0.$$

This relationship is plotted in Fig. 27–7.

3. If the frequency f is less than the "cutoff" frequency f_0, where $hf_0 = W_0$, no electrons will be ejected, no matter how great the intensity of the light.

These predictions of the photon theory are very different from the predictions of the wave theory. In 1913–1914, careful experiments were carried out by R. A. Millikan. The results were fully in agreement with Einstein's photon theory.

One other aspect of the photoelectric effect also confirmed the photon theory. If extremely low light intensity is used, the wave theory predicts a time delay before electron emission so that an electron can absorb enough energy to exceed the work function. The photon theory predicts no such delay—it only takes one photon (if its frequency is high enough) to eject an electron—and experiments showed no delay. This too confirmed Einstein's photon theory.

Photon theory predictions

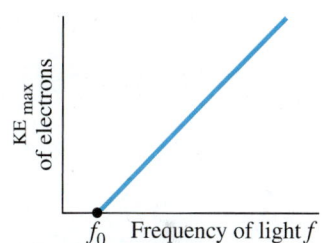

FIGURE 27–7 Photoelectric effect: the maximum kinetic energy of ejected electrons increases linearly with the frequency of incident light. No electrons are emitted if $f < f_0$.

EXAMPLE 27-3 Photon energy. Calculate the energy of a photon of blue light, $\lambda = 450$ nm in air (or vacuum).

APPROACH The photon has energy $E = hf$ (Eq. 27-4) where $f = c/\lambda$ (Eq. 22-4).

SOLUTION Since $f = c/\lambda$, we have

$$E = hf = \frac{hc}{\lambda} = \frac{(6.63 \times 10^{-34}\,\text{J}\cdot\text{s})(3.00 \times 10^8\,\text{m/s})}{(4.5 \times 10^{-7}\,\text{m})} = 4.4 \times 10^{-19}\,\text{J},$$

or $(4.4 \times 10^{-19}\,\text{J})/(1.60 \times 10^{-19}\,\text{J/eV}) = 2.8\,\text{eV}$. (See definition of eV in Section 17-4, $1\,\text{eV} = 1.60 \times 10^{-19}\,\text{J}$.)

EXAMPLE 27-4 ESTIMATE Photons from a lightbulb. Estimate how many visible light photons a 100-W lightbulb emits per second. Assume the bulb has a typical efficiency of about 3% (that is, 97% of the energy goes to heat).

APPROACH Let's assume an average wavelength in the middle of the visible spectrum, $\lambda \approx 500$ nm. The energy of each photon is $E = hf = hc/\lambda$. Only 3% of the 100-W power is emitted as visible light, or $3\,\text{W} = 3\,\text{J/s}$. The number of photons emitted per second equals the light output of $3\,\text{J/s}$ divided by the energy of each photon.

SOLUTION The energy emitted in one second ($= 3\,\text{J}$) is $E = Nhf$ where N is the number of photons emitted per second and $f = c/\lambda$. Hence

$$N = \frac{E}{hf} = \frac{E\lambda}{hc} = \frac{(3\,\text{J})(500 \times 10^{-9}\,\text{m})}{(6.63 \times 10^{-34}\,\text{J}\cdot\text{s})(3.00 \times 10^8\,\text{m/s})} \approx 8 \times 10^{18}$$

per second, or almost 10^{19} photons emitted per second, an enormous number.

EXERCISE B A beam contains infrared light of a single wavelength, 1000 nm, and monochromatic UV at 100 nm, both of the same intensity. Are there more 100-nm photons or more 1000-nm photons?

EXAMPLE 27–5 Photoelectron speed and energy. What is the kinetic energy and the speed of an electron ejected from a sodium surface whose work function is $W_0 = 2.28$ eV when illuminated by light of wavelength (a) 410 nm, (b) 550 nm?

APPROACH We first find the energy of the photons ($E = hf = hc/\lambda$). If the energy is greater than W_0, then electrons will be ejected with varying amounts of KE, with a maximum of $KE_{max} = hf - W_0$.

SOLUTION (a) For $\lambda = 410$ nm,

$$hf = \frac{hc}{\lambda} = 4.85 \times 10^{-19} \text{ J} \quad \text{or} \quad 3.03 \text{ eV}.$$

The maximum kinetic energy an electron can have is given by Eq. 27–5b, $KE_{max} = 3.03$ eV $- 2.28$ eV $= 0.75$ eV, or $(0.75 \text{ eV})(1.60 \times 10^{-19} \text{ J/eV}) = 1.2 \times 10^{-19}$ J. Since $KE = \frac{1}{2}mv^2$ where $m = 9.1 \times 10^{-31}$ kg,

$$v_{max} = \sqrt{\frac{2KE}{m}} = 5.1 \times 10^5 \text{ m/s}.$$

Most ejected electrons will have less KE and less speed than these maximum values.

(b) For $\lambda = 550$ nm, $hf = hc/\lambda = 3.61 \times 10^{-19}$ J $= 2.26$ eV. Since this photon energy is less than the work function, no electrons are ejected.

NOTE In (a) we used the nonrelativistic equation for kinetic energy. If v had turned out to be more than about $0.1c$, our calculation would have been inaccurate by more than a percent or so, and we would probably prefer to redo it using the relativistic form (Eq. 26–5).

EXERCISE C Determine the lowest frequency and the longest wavelength needed to emit electrons from sodium.

By converting units, we can show that the energy of a photon in electron volts, when given the wavelength λ in nm, is

$$E \text{ (eV)} = \frac{1.240 \times 10^3 \text{ eV} \cdot \text{nm}}{\lambda \text{ (nm)}}. \quad \text{[photon energy in eV]}$$

Applications of the Photoelectric Effect

The photoelectric effect, besides playing an important historical role in confirming the photon theory of light, also has many practical applications. Burglar alarms and automatic doors often make use of the photocell circuit of Fig. 27–6. When a person interrupts the beam of light, the sudden drop in current in the circuit activates a switch—often a solenoid—which operates a bell or opens the door. UV or IR light is sometimes used in burglar alarms because of its invisibility. Many smoke detectors use the photoelectric effect to detect tiny amounts of smoke that interrupt the flow of light and so alter the electric current. Photographic light meters use this circuit as well. Photocells are used in many other devices, such as absorption spectrophotometers, to measure light intensity. One type of film sound track is a variably shaded narrow section at the side of the film, Fig. 27–8. Light passing through the film is thus "modulated," and the output electrical signal of the photocell detector follows the frequencies on the sound track. For many applications today, the vacuum-tube photocell of Fig. 27–6 has been replaced by a semiconductor device known as a **photodiode** (Section 29–9). In these semiconductors, the absorption of a photon liberates a bound electron so it can move freely, which changes the conductivity of the material and the current through a photodiode is altered.

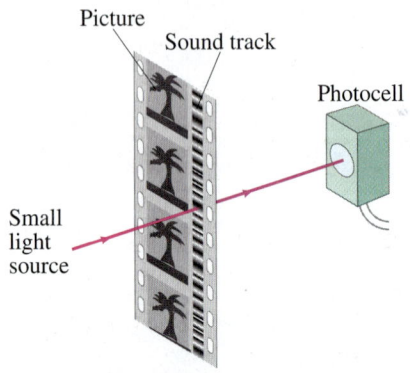

FIGURE 27–8 Optical sound track on movie film. In the projector, light from a small source (different from that for the picture) passes through the sound track on the moving film.

27–4 Energy, Mass, and Momentum of a Photon

We have just seen (Eq. 27–4) that the total energy of a single photon is given by $E = hf$. Because a photon always travels at the speed of light, it is truly a relativistic particle. Thus we must use relativistic formulas for dealing with its mass, energy, and momentum. The momentum of any particle of mass m is given by $p = mv/\sqrt{1 - v^2/c^2}$. Since $v = c$ for a photon, the denominator is zero. To avoid having an infinite momentum, we conclude that the photon's mass must be zero: $m = 0$. This makes sense too because a photon can never be at rest (it always moves at the speed of light). A photon's kinetic energy is its total energy:

$$\text{KE} = E = hf. \qquad \text{[photon]}$$

The momentum of a photon can be obtained from the relativistic formula (Eq. 26–9) $E^2 = p^2c^2 + m^2c^4$ where we set $m = 0$, so $E^2 = p^2c^2$ or

$$p = \frac{E}{c}. \qquad \text{[photon]}$$

Since $E = hf$ for a photon, its momentum is related to its wavelength by

$$p = \frac{E}{c} = \frac{hf}{c} = \frac{h}{\lambda}. \qquad (27\text{–}6)$$

CAUTION
Momentum of photon is not mv

EXAMPLE 27–6 ESTIMATE **Photon momentum and force.** Suppose the 10^{19} photons emitted per second from the 100-W lightbulb in Example 27–4 were all focused onto a piece of black paper and absorbed. (*a*) Calculate the momentum of one photon and (*b*) estimate the force all these photons could exert on the paper.

APPROACH Each photon's momentum is obtained from Eq. 27–6, $p = h/\lambda$. Next, each absorbed photon's momentum changes from $p = h/\lambda$ to zero. We use Newton's second law, $F = \Delta p/\Delta t$, to get the force. Let $\lambda = 500$ nm.

SOLUTION (*a*) Each photon has a momentum

$$p = \frac{h}{\lambda} = \frac{6.63 \times 10^{-34} \text{ J} \cdot \text{s}}{500 \times 10^{-9} \text{ m}} = 1.3 \times 10^{-27} \text{ kg} \cdot \text{m/s}.$$

(*b*) Using Newton's second law for $N = 10^{19}$ photons (Example 27–4) whose momentum changes from h/λ to 0, we obtain

$$F = \frac{\Delta p}{\Delta t} = \frac{Nh/\lambda - 0}{1 \text{ s}} = N\frac{h}{\lambda} \approx (10^{19} \text{ s}^{-1})(10^{-27} \text{ kg} \cdot \text{m/s}) \approx 10^{-8} \text{ N}.$$

NOTE This is a tiny force, but we can see that a very strong light source could exert a measurable force, and near the Sun or a star the force due to photons in electromagnetic radiation could be considerable. See Section 22–6.

EXAMPLE 27–7 **Photosynthesis.** In *photosynthesis*, pigments such as chlorophyll in plants capture the energy of sunlight to change CO_2 to useful carbohydrate. About nine photons are needed to transform one molecule of CO_2 to carbohydrate and O_2. Assuming light of wavelength $\lambda = 670$ nm (chlorophyll absorbs most strongly in the range 650 nm to 700 nm), how efficient is the photosynthetic process? The reverse chemical reaction releases an energy of 4.9 eV/molecule of CO_2, so 4.9 eV is needed to transform CO_2 to carbohydrate.

APPROACH The efficiency is the minimum energy required (4.9 eV) divided by the actual energy absorbed, nine times the energy (hf) of one photon.

SOLUTION The energy of nine photons, each of energy $hf = hc/\lambda$, is $(9)(6.63 \times 10^{-34} \text{ J} \cdot \text{s})(3.00 \times 10^8 \text{ m/s})/(6.7 \times 10^{-7} \text{ m}) = 2.7 \times 10^{-18}$ J or 17 eV. Thus the process is about $(4.9 \text{ eV}/17 \text{ eV}) = 29\%$ efficient.

PHYSICS APPLIED
Photosynthesis

*27–5 Compton Effect

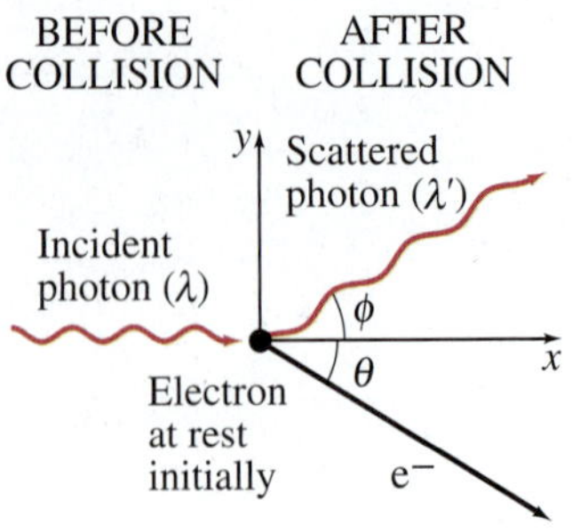

FIGURE 27–9 The Compton effect. A single photon of wavelength λ strikes an electron in some material, knocking it out of its atom. The scattered photon has less energy (some energy is given to the electron) and hence has a longer wavelength λ' (shown exaggerated). Experiments found scattered X-rays of just the wavelengths predicted by conservation of energy and momentum using the photon model.

Besides the photoelectric effect, a number of other experiments were carried out in the early twentieth century which also supported the photon theory. One of these was the **Compton effect** (1923) named after its discoverer, A. H. Compton (1892–1962). Compton aimed short-wavelength light (actually X-rays) at various materials, and detected light scattered at various angles. He found that the scattered light had a slightly longer wavelength than did the incident light, and therefore a slightly lower frequency indicating a loss of energy. He explained this result on the basis of the photon theory as incident photons colliding with electrons of the material, Fig. 27–9. Using Eq. 27–6 for momentum of a photon, Compton applied the laws of conservation of momentum and energy to the collision of Fig. 27–9 and derived the following equation for the wavelength of the scattered photons:

$$\lambda' = \lambda + \frac{h}{m_e c}(1 - \cos\phi), \qquad (27\text{--}7)$$

where m_e is the mass of the electron. (The quantity $h/m_e c$, which has the dimensions of length, is called the **Compton wavelength** of the electron.) We see that the predicted wavelength of scattered photons depends on the angle ϕ at which they are detected. Compton's measurements of 1923 were consistent with this formula. The wave theory of light predicts no such shift: an incoming electromagnetic wave of frequency f should set electrons into oscillation at frequency f; and such oscillating electrons would reemit EM waves of this same frequency f (Section 22–2), which would not change with angle (ϕ). Hence the Compton effect adds to the firm experimental foundation for the photon theory of light.

EXERCISE D When a photon scatters off an electron by the Compton effect, which of the following increases: its energy, frequency, wavelength?

EXAMPLE 27–8 X-ray scattering. X-rays of wavelength 0.140 nm are scattered from a very thin slice of carbon. What will be the wavelengths of X-rays scattered at (a) 0°, (b) 90°, (c) 180°?

APPROACH This is an example of the Compton effect, and we use Eq. 27–7 to find the wavelengths.

SOLUTION (a) For $\phi = 0°$, $\cos\phi = 1$ and $1 - \cos\phi = 0$. Then Eq. 27–7 gives $\lambda' = \lambda = 0.140$ nm. This makes sense since for $\phi = 0°$, there really isn't any collision as the photon goes straight through without interacting.
(b) For $\phi = 90°$, $\cos\phi = 0$, and $1 - \cos\phi = 1$. So

$$\lambda' = \lambda + \frac{h}{m_e c} = 0.140 \text{ nm} + \frac{6.63 \times 10^{-34} \text{ J·s}}{(9.11 \times 10^{-31} \text{ kg})(3.00 \times 10^8 \text{ m/s})}$$

$$= 0.140 \text{ nm} + 2.4 \times 10^{-12} \text{ m} = 0.142 \text{ nm};$$

that is, the wavelength is longer by one Compton wavelength ($= h/m_e c^2$ = 0.0024 nm for an electron).
(c) For $\phi = 180°$, which means the photon is scattered backward, returning in the direction from which it came (a direct "head-on" collision), $\cos\phi = -1$, and $1 - \cos\phi = 2$. So

$$\lambda' = \lambda + 2\frac{h}{m_e c} = 0.140 \text{ nm} + 2(0.0024 \text{ nm}) = 0.145 \text{ nm}.$$

NOTE The maximum shift in wavelength occurs for backward scattering, and it is twice the Compton wavelength.

PHYSICS APPLIED
Measuring bone density

The Compton effect has been used to diagnose bone disease such as osteoporosis. Gamma rays, which are photons of even shorter wavelength than X-rays, coming from a radioactive source are scattered off bone material. The total intensity of the scattered radiation is proportional to the density of electrons, which is in turn proportional to the bone density. A low bone density may indicate osteoporosis.

27–6 Photon Interactions; Pair Production

When a photon passes through matter, it interacts with the atoms and electrons. There are four important types of interactions that a photon can undergo:

1. The *photoelectric effect*: A photon may knock an electron out of an atom and in the process the photon disappears.
2. The photon may knock an atomic electron to a higher energy state in the atom if its energy is not sufficient to knock the electron out altogether. In this process the photon also disappears, and all its energy is given to the atom. Such an atom is then said to be in an *excited state*, and we shall discuss it more later.
3. The photon can be scattered from an electron (or a nucleus) and in the process lose some energy; this is the *Compton effect* (Fig. 27–9). But notice that the photon is not slowed down. It still travels with speed c, but its frequency will be lower because it has lost some energy.
4. *Pair production*: A photon can actually create matter, such as the production of an electron and a positron, Fig. 27–10. (A positron has the same mass as an electron, but the opposite charge, $+e$.)

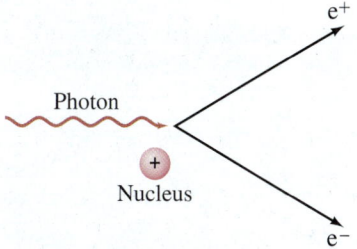

FIGURE 27–10 Pair production: a photon disappears and produces an electron and a positron.

In process 4, **pair production**, the photon disappears in the process of creating the electron–positron pair. This is an example of mass being created from pure energy, and it occurs in accord with Einstein's equation $E = mc^2$. Notice that a photon cannot create an electron alone since electric charge would not then be conserved. The inverse of pair production also occurs: if a positron comes close to an electron, the two quickly **annihilate** each other and their energy, including their mass, appears as electromagnetic energy of photons. Because positrons are not as plentiful in nature as electrons, they usually do not last long.

Electron–positron annihilation is the basis for the type of medical imaging known as PET, as discussed in Section 31–8.

EXAMPLE 27–9 Pair production. (a) What is the minimum energy of a photon that can produce an electron–positron pair? (b) What is this photon's wavelength?

APPROACH The minimum photon energy E equals the rest energy (mc^2) of the two particles created, via Einstein's famous equation $E = mc^2$ (Eq. 26–7). There is no energy left over, so the particles produced will have zero kinetic energy. The wavelength is $\lambda = c/f$ where $E = hf$ for the original photon.

SOLUTION (a) Because $E = mc^2$, and the mass created is equal to two electron masses, the photon must have energy

$$E = 2(9.11 \times 10^{-31} \text{ kg})(3.00 \times 10^8 \text{ m/s})^2 = 1.64 \times 10^{-13} \text{ J} = 1.02 \text{ MeV}$$

($1 \text{ MeV} = 10^6 \text{ eV} = 1.60 \times 10^{-13} \text{ J}$). A photon with less energy cannot undergo pair production.

(b) Since $E = hf = hc/\lambda$, the wavelength of a 1.02-MeV photon is

$$\lambda = \frac{hc}{E} = \frac{(6.63 \times 10^{-34} \text{ J} \cdot \text{s})(3.00 \times 10^8 \text{ m/s})}{(1.64 \times 10^{-13} \text{ J})} = 1.2 \times 10^{-12} \text{ m},$$

which is 0.0012 nm. Such photons are in the gamma-ray (or very short X-ray) region of the electromagnetic spectrum (Fig. 22–8).

NOTE Photons of higher energy (shorter wavelength) can also create an electron–positron pair, with the excess energy becoming kinetic energy of the particles.

Pair production cannot occur in empty space, for momentum could not be conserved. In Example 27–9, for instance, energy is conserved, but only enough energy was provided to create the electron–positron pair at rest and thus with zero momentum, which could not equal the initial momentum of the photon. Indeed, it can be shown that at any energy, an additional massive object, such as an atomic nucleus (Fig. 27–10), must take part in the interaction to carry off some of the momentum.

27–7 Wave–Particle Duality; the Principle of Complementarity

The photoelectric effect, the Compton effect, and other experiments have placed the particle theory of light on a firm experimental basis. But what about the classic experiments of Young and others (Chapter 24) on interference and diffraction which showed that the wave theory of light also rests on a firm experimental basis?

We seem to be in a dilemma. Some experiments indicate that light behaves like a wave; others indicate that it behaves like a stream of particles. These two theories seem to be incompatible, but both have been shown to have validity. Physicists finally came to the conclusion that this duality of light must be accepted as a fact of life. It is referred to as the **wave–particle duality**. Apparently, light is a more complex phenomenon than just a simple wave or a simple beam of particles.

To clarify the situation, the great Danish physicist Niels Bohr (1885–1962, Fig. 27–11) proposed his famous **principle of complementarity**. It states that to understand an experiment, sometimes we find an explanation using wave theory and sometimes using particle theory. Yet we must be aware of both the wave and particle aspects of light if we are to have a full understanding of light. Therefore these two aspects of light complement one another.

It is not easy to "visualize" this duality. We cannot readily picture a combination of wave and particle. Instead, we must recognize that the two aspects of light are different "faces" that light shows to experimenters.

Part of the difficulty stems from how we think. Visual pictures (or models) in our minds are based on what we see in the everyday world. We apply the concepts of waves and particles to light because in the macroscopic world we see that energy is transferred from place to place by these two methods. We cannot see directly whether light is a wave or particle, so we do indirect experiments. To explain the experiments, we apply the models of waves or of particles to the nature of light. But these are abstractions of the human mind. When we try to conceive of what light really "is," we insist on a visual picture. Yet there is no reason why light should conform to these models (or visual images) taken from the macroscopic world. The "true" nature of light—if that means anything—is not possible to visualize. The best we can do is recognize that our knowledge is limited to the indirect experiments, and that in terms of everyday language and images, light reveals both wave and particle properties.

It is worth noting that Einstein's equation $E = hf$ itself links the particle and wave properties of a light beam. In this equation, E refers to the energy of a particle; and on the other side of the equation, we have the frequency f of the corresponding wave.

FIGURE 27–11 Niels Bohr (right), walking with Enrico Fermi along the Appian Way outside Rome. This photo shows one important way physics is done.

CAUTION

*Not correct to say light is a wave and/or a particle. Light can **act** like a wave or like a particle*

27–8 Wave Nature of Matter

In 1923, Louis de Broglie (1892–1987) extended the idea of the wave–particle duality. He appreciated the *symmetry* in nature, and argued that if light sometimes behaves like a wave and sometimes like a particle, then perhaps those things in nature thought to be particles—such as electrons and other material objects—might also have wave properties. De Broglie proposed that the wavelength of a material particle would be related to its momentum in the same way as for a photon, Eq. 27–6, $p = h/\lambda$. That is, for a particle having linear momentum $p = mv$, the wavelength λ is given by

de Broglie wavelength

$$\lambda = \frac{h}{p}, \quad (27\text{–}8)$$

and is valid classically ($p = mv$ for $v \ll c$) and relativistically ($p = \gamma mv = mv/\sqrt{1 - v^2/c^2}$). This is sometimes called the **de Broglie wavelength** of a particle.

EXAMPLE 27–10 Wavelength of a ball. Calculate the de Broglie wavelength of a 0.20-kg ball moving with a speed of 15 m/s.

APPROACH We use Eq. 27–8.

SOLUTION $\lambda = \dfrac{h}{p} = \dfrac{h}{mv} = \dfrac{(6.6 \times 10^{-34}\,\text{J}\cdot\text{s})}{(0.20\,\text{kg})(15\,\text{m/s})} = 2.2 \times 10^{-34}\,\text{m}.$

Ordinary objects, such as the ball of Example 27–10, have unimaginably small wavelengths. Even if the speed is extremely small, say 10^{-4} m/s, the wavelength would be about 10^{-29} m. Indeed, the wavelength of any ordinary object is much too small to be measured and detected. The problem is that the properties of waves, such as interference and diffraction, are significant only when the size of objects or slits is not much larger than the wavelength. And there are no known objects or slits to diffract waves only 10^{-30} m long, so the wave properties of ordinary objects go undetected.

But tiny elementary particles, such as electrons, are another matter. Since the mass m appears in the denominator of Eq. 27–8, a very small mass should have a much larger wavelength.

EXAMPLE 27–11 Wavelength of an electron. Determine the wavelength of an electron that has been accelerated through a potential difference of 100 V.

APPROACH If the kinetic energy is much less than the rest energy, we can use the classical formula, $\text{KE} = \tfrac{1}{2}mv^2$ (see end of Section 26–9). For an electron, $mc^2 = 0.511$ MeV. We then apply conservation of energy: the kinetic energy acquired by the electron equals its loss in potential energy. After solving for v, we use Eq. 27–8 to find the de Broglie wavelength.

SOLUTION The gain in kinetic energy equals the loss in potential energy: $\Delta \text{PE} = eV - 0$. Thus $\text{KE} = eV$, so $\text{KE} = 100$ eV. The ratio $\text{KE}/mc^2 = 100\,\text{eV}/(0.511 \times 10^6\,\text{eV}) \approx 10^{-4}$, so relativity is not needed. Thus

$$\tfrac{1}{2}mv^2 = eV$$

and

$$v = \sqrt{\dfrac{2eV}{m}} = \sqrt{\dfrac{(2)(1.6 \times 10^{-19}\,\text{C})(100\,\text{V})}{(9.1 \times 10^{-31}\,\text{kg})}} = 5.9 \times 10^6\,\text{m/s}.$$

Then

$$\lambda = \dfrac{h}{mv} = \dfrac{(6.63 \times 10^{-34}\,\text{J}\cdot\text{s})}{(9.1 \times 10^{-31}\,\text{kg})(5.9 \times 10^6\,\text{m/s})} = 1.2 \times 10^{-10}\,\text{m},$$

or 0.12 nm.

EXERCISE E As a particle travels faster, does its de Broglie wavelength decrease, increase, or remain the same?

EXERCISE F Return to the Chapter-Opening Question, page 771, and answer it again now. Try to explain why you may have answered differently the first time.

Electron Diffraction

From Example 27–11, we see that electrons can have wavelengths on the order of 10^{-10} m, and even smaller. Although small, this wavelength can be detected: the spacing of atoms in a crystal is on the order of 10^{-10} m and the orderly array of atoms in a crystal could be used as a type of diffraction grating, as was done earlier for X-rays (see Section 25–11). C. J. Davisson and L. H. Germer performed the crucial experiment: they scattered electrons from the surface of a metal crystal and, in early 1927, observed that the electrons were scattered into a pattern of regular peaks. When they interpreted these peaks as a diffraction pattern, the wavelength of the diffracted electron wave was found to be just that predicted by de Broglie, Eq. 27–8. In the same year, G. P. Thomson (son of J. J. Thomson) used a different experimental arrangement and also detected diffraction of electrons. (See Fig. 27–12. Compare it to X-ray diffraction, Section 25–11.) Later experiments showed that protons, neutrons, and other particles also have wave properties.

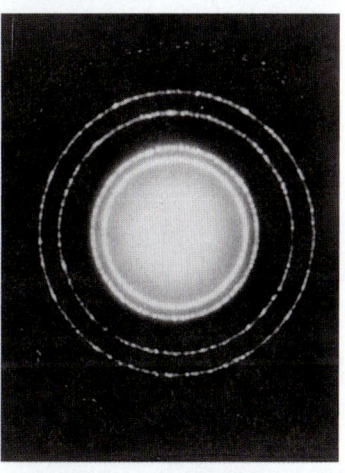

FIGURE 27–12 Diffraction pattern of electrons scattered from aluminum foil, as recorded on film.

Thus the wave–particle duality applies to material objects as well as to light. The principle of complementarity applies to matter as well. That is, we must be aware of both the particle and wave aspects in order to have an understanding of matter, including electrons. But again we must recognize that a visual picture of a "wave–particle" is not possible.

PHYSICS APPLIED
Electron diffraction

EXAMPLE 27–12 Electron diffraction. The wave nature of electrons is manifested in experiments where an electron beam interacts with the atoms on the surface of a solid, especially crystals. By studying the angular distribution of the diffracted electrons, one can indirectly measure the geometrical arrangement of atoms. Assume that the electrons strike perpendicular to the surface of a solid (see Fig. 27–13), and that their energy is low, $KE = 100\,\text{eV}$, so that they interact only with the surface layer of atoms. If the smallest angle at which a diffraction maximum occurs is at $24°$, what is the separation d between the atoms on the surface?

SOLUTION Treating the electrons as waves, we need to determine the condition where the difference in path traveled by the wave diffracted from adjacent atoms is an integer multiple of the de Broglie wavelength, so that constructive interference occurs. The path length difference is $d \sin \theta$ (Fig. 27–13); so for the smallest value of θ we must have

$$d \sin \theta = \lambda.$$

However, λ is related to the (non-relativistic) kinetic energy KE by

$$KE = \frac{p^2}{2m_e} = \frac{h^2}{2m_e \lambda^2}.$$

Thus

$$\lambda = \frac{h}{\sqrt{2m_e\, KE}}$$

$$= \frac{(6.63 \times 10^{-34}\,\text{J}\cdot\text{s})}{\sqrt{2(9.11 \times 10^{-31}\,\text{kg})(100\,\text{eV})(1.6 \times 10^{-19}\,\text{J/eV})}} = 0.123\,\text{nm}.$$

The surface inter-atomic spacing is

$$d = \frac{\lambda}{\sin \theta} = \frac{0.123\,\text{nm}}{\sin 24°} = 0.30\,\text{nm}.$$

NOTE Experiments of this type verify both the wave nature of electrons and the orderly array of atoms in crystalline solids.

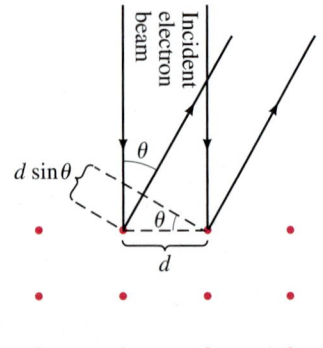
FIGURE 27–13 Example 27–12. The red dots represent atoms in an orderly array in a solid.

What Is an Electron?

We might ask ourselves: "What is an electron?" The early experiments of J. J. Thomson (Section 27–1) indicated a glow in a tube, and that glow moved when a magnetic field was applied. The results of these and other experiments were best interpreted as being caused by tiny negatively charged particles which we now call electrons. No one, however, has actually seen an electron directly. The drawings we sometimes make of electrons as tiny spheres with a negative charge on them are merely convenient pictures (now recognized to be inaccurate). Again we must rely on experimental results, some of which are best interpreted using the particle model and others using the wave model. These models are mere pictures that we use to extrapolate from the macroscopic world to the tiny microscopic world of the atom. And there is no reason to expect that these models somehow reflect the reality of an electron. We thus use a wave or a particle model (whichever works best in a situation) so that we can talk about what is happening. But we should not be led to believe that an electron *is* a wave or a particle. Instead we could say that an electron is the set of its properties that we can measure. Bertrand Russell said it well when he wrote that an electron is "a logical construction."

27–9 Electron Microscopes

PHYSICS APPLIED
Electron microscope

The idea that electrons have wave properties led to the development of the **electron microscope** (EM), which can produce images of much greater magnification than a light microscope. Figures 27–14 and 27–15 are diagrams of two types, developed around the middle of the twentieth century: the **transmission electron microscope** (TEM), which produces a two-dimensional image, and the **scanning electron microscope** (SEM), which produces images with a three-dimensional quality.

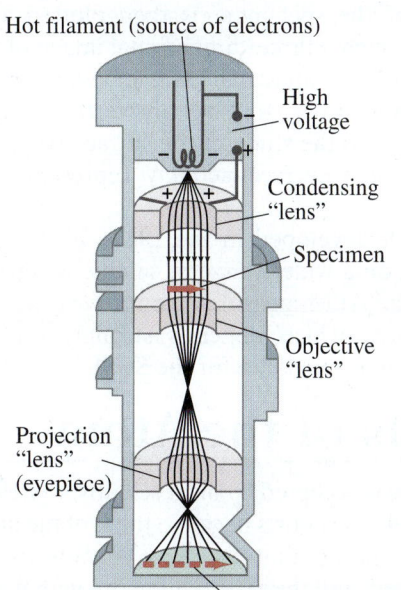

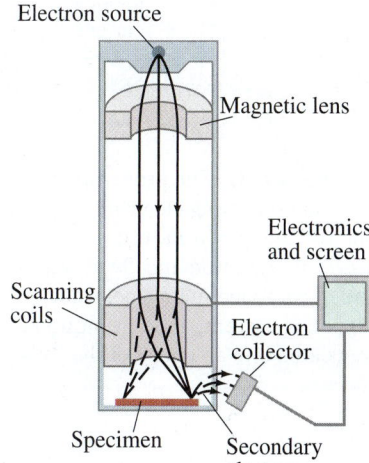

FIGURE 27–14 Transmission electron microscope. The magnetic field coils are designed to be "magnetic lenses," which bend the electron paths and bring them to a focus, as shown. The sensors of the image measure electron intensity only, no color.

FIGURE 27–15 Scanning electron microscope. Scanning coils move an electron beam back and forth across the specimen. Secondary electrons produced when the beam strikes the specimen are collected and their intensity affects the brightness of pixels in a monitor to produce a picture.

In both types, the objective and eyepiece lenses are actually magnetic fields that exert forces on the electrons to bring them to a focus. The fields are produced by carefully designed current-carrying coils of wire. Photographs using each type are shown in Fig. 27–16. EMs measure the intensity of electrons, producing monochromatic photos. Color is often added artificially to highlight.

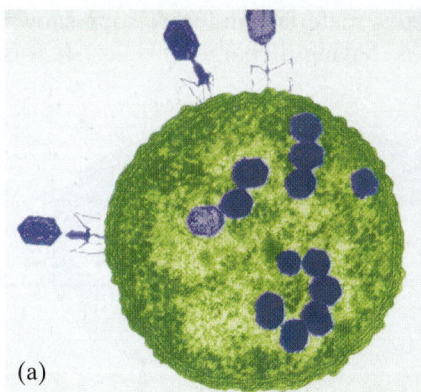
(a)

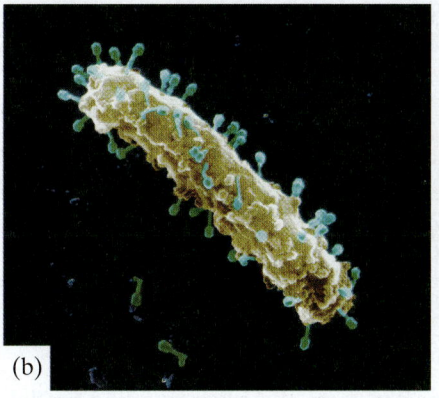
(b)

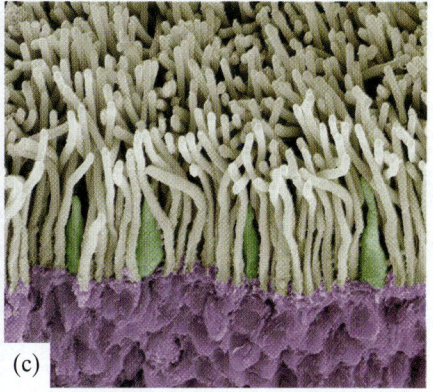
(c)

As discussed in Sections 25–7 and 25–8, the maximum resolution of details on an object is about the size of the wavelength of the radiation used to view it. Electrons accelerated by voltages on the order of 10^5 V have wavelengths of about 0.004 nm. The maximum resolution obtainable would be on this order, but in practice, aberrations in the magnetic lenses limit the resolution in transmission electron microscopes to about 0.1 to 0.5 nm. This is still 1000 times better than a visible-light microscope, and corresponds to a useful magnification of about a million. Such magnifications are difficult to achieve, and more common magnifications are 10^4 to 10^5. The maximum resolution of a scanning electron microscope is less, typically 5 to 10 nm although new high-resolution SEMs approach 1 nm.

FIGURE 27–16 Electron micrographs, in false color, of (a) viruses attacking a cell of the bacterium *Escherichia coli* (TEM, $\approx 50,000\times$). (b) Same subject by an SEM ($\approx 35,000\times$). (c) SEM image of an eye's retina (Section 25–2); the rods and cones have been colored beige and green, respectively. Part (c) is also on the cover of this book.

SECTION 27–9

PHYSICS APPLIED
STM and AFM

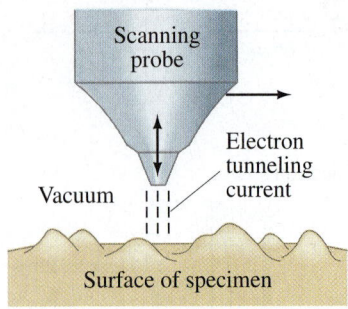

FIGURE 27–17 The probe tip of a scanning tunneling electron microscope, as it is moved horizontally, automatically moves up and down to maintain a constant tunneling current, and this motion is translated into an image of the surface.

FIGURE 27–18 Plum-pudding model of the atom.

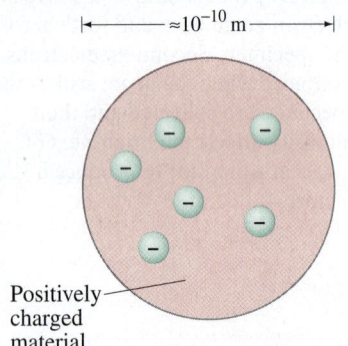

The **scanning tunneling electron microscope** (STM), developed in the 1980s, contains a tiny probe, whose tip may be only one (or a few) atoms wide, that is moved across the specimen to be examined in a series of linear passes. The tip, as it scans, remains very close to the surface of the specimen, about 1 nm above it, Fig. 27–17. A small voltage applied between the probe and the surface causes electrons to leave the surface and pass through the vacuum to the probe, by a process known as *tunneling* (discussed in Section 30–12). This "tunneling" current is very sensitive to the gap width, so a feedback mechanism can be used to raise and lower the probe to maintain a constant electron current. The probe's vertical motion, following the surface of the specimen, is then plotted as a function of position, scan after scan, producing a three-dimensional image of the surface. Surface features as fine as the size of an atom can be resolved: a resolution better than 50 pm (0.05 nm) laterally and 0.01 to 0.001 nm vertically. This kind of resolution has given a great impetus to the study of the surface structure of materials. The "topographic" image of a surface actually represents the distribution of electron charge.

The **atomic force microscope** (AFM), developed in the 1980s, is in many ways similar to an STM, but can be used on a wider range of sample materials. Instead of detecting an electric current, the AFM measures the force between a cantilevered tip and the sample, a force which depends strongly on the tip–sample separation at each point. The tip is moved as for the STM.

27–10 Early Models of the Atom

The idea that matter is made up of atoms was accepted by most scientists by 1900. With the discovery of the electron in the 1890s, scientists began to think of the atom itself as having a structure with electrons as part of that structure. We now discuss how our modern view of the atom developed, and the quantum theory with which it is intertwined.[†]

A typical model of the atom in the 1890s visualized the atom as a homogeneous sphere of positive charge inside of which there were tiny negatively charged electrons, a little like plums in a pudding, Fig. 27–18.

Around 1911, Ernest Rutherford (1871–1937) and his colleagues performed experiments whose results contradicted the plum-pudding model of the atom. In these experiments a beam of positively charged alpha (α) particles was directed at a thin sheet of metal foil such as gold, Fig. 27–19. (These newly discovered α particles were emitted by certain radioactive materials and were soon shown to be doubly ionized helium atoms—that is, having a charge of $+2e$.) It was

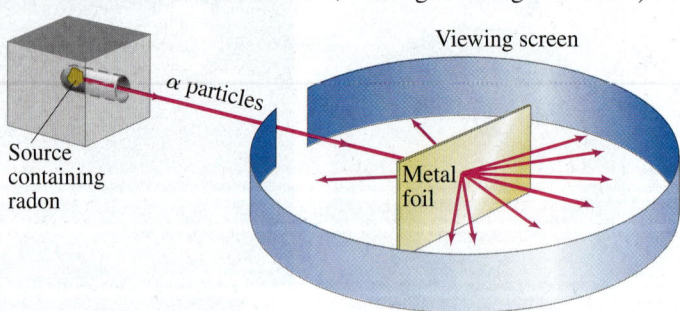

FIGURE 27–19 Experimental setup for Rutherford's experiment: α particles emitted by radon are deflected by the atoms of a thin metal foil and a few rebound backward.

expected from the plum-pudding model that the alpha particles would not be deflected significantly because electrons are so much lighter than alpha particles, and the alpha particles should not have encountered any massive concentration of positive charge to strongly repel them. The experimental results completely contradicted these predictions. It was found that most of the alpha particles passed through the foil unaffected, as if the foil were mostly empty space.

[†]Some readers may say: "Tell us the facts as we know them today, and don't bother us with the historical background and its outmoded theories." Such an approach would ignore the creative aspect of science and thus give a false impression of how science develops. Moreover, it is not really possible to understand today's view of the atom without insight into the concepts that led to it.

And of those deflected, a few were deflected at very large angles—some even backward, nearly in the direction from which they had come. This could happen, Rutherford reasoned, only if the positively charged alpha particles were being repelled by a massive positive charge concentrated in a very small region of space (see Fig. 27–20). He hypothesized that the atom must consist of a tiny but massive positively charged nucleus, containing over 99.9% of the mass of the atom, surrounded by much lighter electrons some distance away. The electrons would be moving in orbits about the nucleus—much as the planets move around the Sun—because if they were at rest, they would fall into the nucleus due to electrical attraction. See Fig. 27–21. Rutherford's experiments suggested that the nucleus must have a radius of about 10^{-15} to 10^{-14} m. From kinetic theory, and especially Einstein's analysis of Brownian motion (see Section 13–1), the radius of atoms was estimated to be about 10^{-10} m. Thus the electrons would seem to be at a distance from the nucleus of about 10,000 to 100,000 times the radius of the nucleus itself. (If the nucleus were the size of a baseball, the atom would have the diameter of a big city several kilometers across.) So an atom would be mostly empty space.

Rutherford's **planetary model** of the atom (also called the **nuclear model** of the atom) was a major step toward how we view the atom today. It was not, however, a complete model and presented some major problems, as we shall see.

27–11 Atomic Spectra: Key to the Structure of the Atom

Earlier in this Chapter we saw that heated solids (as well as liquids and dense gases) emit light with a continuous spectrum of wavelengths. This radiation is assumed to be due to oscillations of atoms and molecules, which are largely governed by the interaction of each atom or molecule with its neighbors.

Rarefied gases can also be excited to emit light. This is done by intense heating, or more commonly by applying a high voltage to a "discharge tube" containing the gas at low pressure, Fig. 27–22. The radiation from excited gases had been observed early in the nineteenth century, and it was found that the spectrum was not continuous. Rather, excited gases emit light of only certain wavelengths, and when this light is analyzed through the slit of a spectroscope or spectrometer, a **line spectrum** is seen rather than a continuous spectrum. The line spectra emitted by a number of elements in the visible region are shown below in Fig. 27–23, and in Chapter 24, Fig. 24–28. The **emission spectrum** is characteristic of the material and can serve as a type of "fingerprint" for identification of the gas.

We also saw (Chapter 24) that if a continuous spectrum passes through a rarefied gas, dark lines are observed in the emerging spectrum, at wavelengths corresponding to lines normally emitted by the gas. This is called an **absorption spectrum** (Fig. 27–23c), and it became clear that gases can absorb light at the same frequencies at which they emit. Using film sensitive to ultraviolet and to infrared light, it was found that gases emit and absorb discrete frequencies in these regions as well as in the visible.

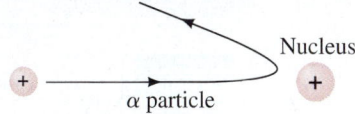

FIGURE 27–20 Backward rebound of α particles in Fig. 27–19 explained as the repulsion from a heavy positively charged nucleus.

FIGURE 27–21 Rutherford's model of the atom: electrons orbit a tiny positive nucleus (not to scale). The atom is visualized as mostly empty space.

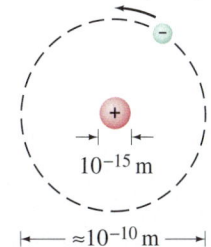

FIGURE 27–22 Gas-discharge tube: (a) diagram; (b) photo of an actual discharge tube for hydrogen.

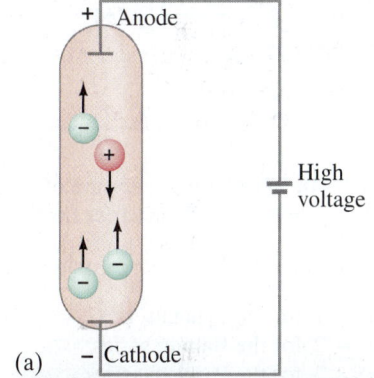

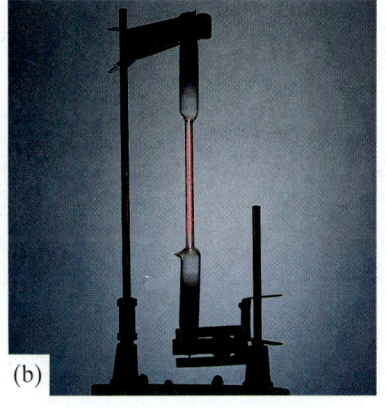

FIGURE 27–23 Emission spectra of the gases (a) atomic hydrogen, (b) helium, and (c) the *solar absorption* spectrum.

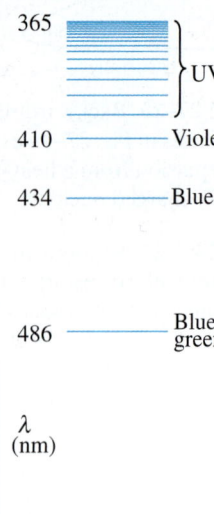

FIGURE 27–24 Balmer series of lines for hydrogen.

In low-density gases, the atoms are far apart on average and hence the light emitted or absorbed is assumed to be by *individual atoms* rather than through interactions between atoms, as in a solid, liquid, or dense gas. Thus the line spectra serve as a key to the structure of the atom: any theory of atomic structure must be able to explain why atoms emit light only of discrete wavelengths, and it should be able to predict what these wavelengths are.

Hydrogen is the simplest atom—it has only one electron orbiting its nucleus. It also has the simplest spectrum. The spectrum of most atoms shows little apparent regularity. But the spacing between lines in the hydrogen spectrum decreases in a regular way, Fig. 27–24. Indeed, in 1885, J. J. Balmer (1825–1898) showed that the four lines in the visible portion of the hydrogen spectrum (with measured wavelengths 656 nm, 486 nm, 434 nm, and 410 nm) have wavelengths that fit the formula

$$\frac{1}{\lambda} = R\left(\frac{1}{2^2} - \frac{1}{n^2}\right), \qquad n = 3, 4, \cdots. \qquad (27\text{–}9)$$

Here n takes on the values 3, 4, 5, 6 for the four visible lines, and R, called the **Rydberg constant**, has the value $R = 1.0974 \times 10^7 \text{ m}^{-1}$. Later it was found that this **Balmer series** of lines extended into the UV region, ending at $\lambda = 365$ nm, as shown in Fig. 27–24. Balmer's formula, Eq. 27–9, also worked for these lines with higher integer values of n. The lines near 365 nm become too close together to distinguish, but the limit of the series at 365 nm corresponds to $n = \infty$ (so $1/n^2 = 0$ in Eq. 27–9).

Later experiments on hydrogen showed that there were similar series of lines in the UV and IR regions, and each series had a pattern just like the Balmer series, but at different wavelengths, Fig. 27–25. Each of these series was found to

FIGURE 27–25 Line spectrum of atomic hydrogen. Each series fits the formula $\frac{1}{\lambda} = R\left(\frac{1}{n'^2} - \frac{1}{n^2}\right)$ where $n' = 1$ for the Lyman series, $n' = 2$ for the Balmer series, $n' = 3$ for the Paschen series, and so on; n can take on all integer values from $n = n' + 1$ up to infinity. The only lines in the visible region of the electromagnetic spectrum are part of the Balmer series.

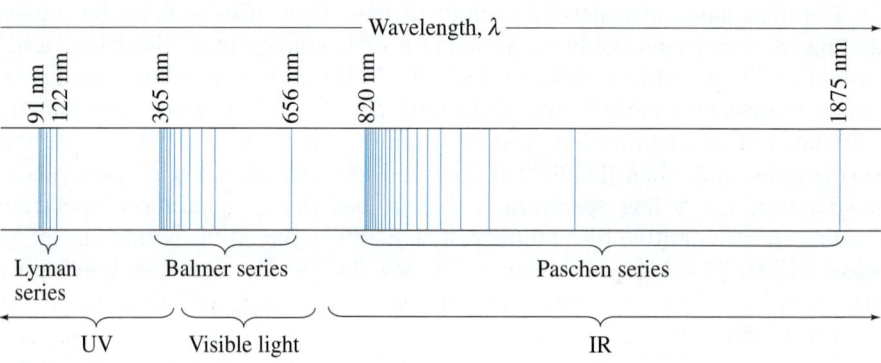

fit a formula with the same form as Eq. 27–9 but with the $1/2^2$ replaced by $1/1^2$, $1/3^2$, $1/4^2$, and so on. For example, the **Lyman series** contains lines with wavelengths from 91 nm to 122 nm (in the UV region) and fits the formula

$$\frac{1}{\lambda} = R\left(\frac{1}{1^2} - \frac{1}{n^2}\right), \qquad n = 2, 3, \cdots.$$

The wavelengths of the **Paschen series** (in the IR region) fit

$$\frac{1}{\lambda} = R\left(\frac{1}{3^2} - \frac{1}{n^2}\right), \qquad n = 4, 5, \cdots.$$

The Rutherford model was unable to explain why atoms emit line spectra. It had other difficulties as well. According to the Rutherford model, electrons orbit the nucleus, and since their paths are curved the electrons are accelerating. Hence they should give off light like any other accelerating electric charge (Chapter 22).

Since light carries off energy and energy is conserved, the electron's own energy must decrease to compensate. Hence electrons would be expected to spiral into the nucleus. As they spiraled inward, their frequency would increase in a short time and so too would the frequency of the light emitted. Thus the two main difficulties of the Rutherford model are these: (1) it predicts that light of a continuous range of frequencies will be emitted, whereas experiment shows line spectra; (2) it predicts that atoms are unstable—electrons would quickly spiral into the nucleus—but we know that atoms in general are stable, because there is stable matter all around us.

Clearly Rutherford's model was not sufficient. Some sort of modification was needed, and Niels Bohr provided it in a model that included the quantum hypothesis. Although the Bohr model has been superseded, it did provide a crucial stepping stone to our present understanding. And some aspects of the Bohr model are still useful today, so we examine it in detail in the next Section.

27–12 The Bohr Model

Bohr had studied in Rutherford's laboratory for several months in 1912 and was convinced that Rutherford's planetary model of the atom had validity. But in order to make it work, he felt that the newly developing quantum theory would somehow have to be incorporated in it. The work of Planck and Einstein had shown that in heated solids, the energy of oscillating electric charges must change discontinuously—from one discrete energy state to another, with the emission of a quantum of light. Perhaps, Bohr argued, the electrons in an atom also cannot lose energy continuously, but must do so in quantum "jumps." In working out his model during the next year, Bohr postulated that electrons move about the nucleus in circular orbits, but that only certain orbits are allowed. He further postulated that an electron in each orbit would have a definite energy and would move in the orbit *without radiating energy* (even though this violated classical ideas since accelerating electric charges are supposed to emit EM waves; see Chapter 22). He thus called the possible orbits **stationary states**. In this **Bohr model**, light is emitted only when an electron jumps from a higher (upper) stationary state to another of lower energy, Fig. 27–26. When such a transition occurs, a single photon of light is emitted whose energy, by energy conservation, is given by

$$hf = E_u - E_\ell, \qquad (27\text{–}10)$$

where E_u refers to the energy of the upper state and E_ℓ the energy of the lower state.

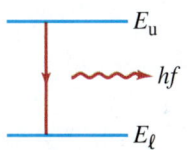

FIGURE 27–26 An atom emits a photon (energy $= hf$) when its energy changes from E_u to a lower energy E_ℓ.

In 1912–13, Bohr set out to determine what energies these orbits would have in the simplest atom, hydrogen; the spectrum of light emitted could then be predicted from Eq. 27–10. In the Balmer formula he had the key he was looking for. Bohr quickly found that his theory would agree with the Balmer formula if he assumed that the electron's angular momentum L is quantized and equal to an integer n times $h/2\pi$. As we saw in Chapter 8 angular momentum is given by $L = I\omega$, where I is the moment of inertia and ω is the angular velocity. For a single particle of mass m moving in a circle of radius r with speed v, $I = mr^2$ and $\omega = v/r$; hence, $L = I\omega = (mr^2)(v/r) = mvr$. Bohr's **quantum condition** is

$$L = mvr_n = n\frac{h}{2\pi}, \qquad n = 1, 2, 3, \cdots, \qquad (27\text{–}11)$$

where n is an integer and r_n is the radius of the nth possible orbit. The allowed orbits are numbered $1, 2, 3, \cdots$, according to the value of n, which is called the **principal quantum number** of the orbit.

Equation 27–11 did not have a firm theoretical foundation. Bohr had searched for some "quantum condition," and such tries as $E = hf$ (where E represents the energy of the electron in an orbit) did not give results in accord with experiment. Bohr's reason for using Eq. 27–11 was simply that it worked; and we now look at how. In particular, let us determine what the Bohr theory predicts for the measurable wavelengths of emitted light.

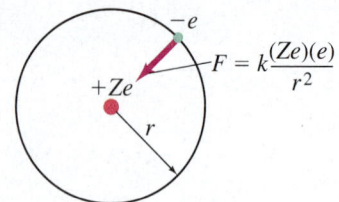

FIGURE 27–27 Electric force (Coulomb's law) keeps the negative electron in orbit around the positively charged nucleus.

An electron in a circular orbit of radius r (Fig. 27–27) would have a centripetal acceleration v^2/r produced by the electrical force of attraction between the negative electron and the positive nucleus. This force is given by Coulomb's law,

$$F = k\frac{(Ze)(e)}{r^2},$$

where $k = 1/4\pi\epsilon_0 = 8.99 \times 10^9 \,\text{N}\cdot\text{m}^2/\text{C}^2$. The charge on the electron is $q_1 = -e$, and that on the nucleus is $q_2 = +Ze$, where Ze is the charge on the nucleus: $+e$ is the charge on a proton, Z is the number of protons in the nucleus (called "atomic number," Section 28–7).[†] For the hydrogen atom, $Z = +1$.

In Newton's second law, $F = ma$, we substitute Coulomb's law for F and $a = v^2/r_n$ for a particular allowed orbit of radius r_n, and obtain

$$F = ma$$

$$k\frac{Ze^2}{r_n^2} = \frac{mv^2}{r_n}.$$

We solve this for r_n,

$$r_n = \frac{kZe^2}{mv^2},$$

and then substitute for v from Eq. 27–11 (which says $v = nh/2\pi m r_n$):

$$r_n = \frac{kZe^2 4\pi^2 m r_n^2}{n^2 h^2}.$$

We solve for r_n (it appears on both sides, so we cancel one of them) and find

$$r_n = \frac{n^2 h^2}{4\pi^2 m k Z e^2} = \frac{n^2}{Z} r_1 \qquad n = 1, 2, 3 \cdots, \qquad (27\text{–}12)$$

where n is an integer (Eq. 27–11), and

$$r_1 = \frac{h^2}{4\pi^2 m k e^2}.$$

Equation 27–12 gives the radii of all possible orbits. The smallest orbit is for $n = 1$, and for hydrogen ($Z = 1$) has the value

$$r_1 = \frac{(1)^2 (6.626 \times 10^{-34} \,\text{J}\cdot\text{s})^2}{4\pi^2 (9.11 \times 10^{-31}\,\text{kg})(8.99 \times 10^9\,\text{N}\cdot\text{m}^2/\text{C}^2)(1.602 \times 10^{-19}\,\text{C})^2}$$

$$r_1 = 0.529 \times 10^{-10}\,\text{m}. \qquad (27\text{–}13)$$

The radius of the smallest orbit in hydrogen, r_1, is sometimes called the **Bohr radius**. From Eq. 27–12, we see that the radii of the larger orbits[‡] increase as n^2, so

$$r_2 = 4r_1 = 2.12 \times 10^{-10}\,\text{m},$$
$$r_3 = 9r_1 = 4.76 \times 10^{-10}\,\text{m},$$
$$\vdots$$
$$r_n = n^2 r_1, \qquad n = 1, 2, 3, \cdots.$$

The first four orbits are shown in Fig. 27–28. Notice that, according to Bohr's model, an electron can exist only in the orbits given by Eq. 27–12. There are no allowable orbits in between.

For an atom with $Z \neq 1$, we can write the orbital radii, r_n, using Eq. 27–12:

$$r_n = \frac{n^2}{Z}(0.529 \times 10^{-10}\,\text{m}), \qquad n = 1, 2, 3, \cdots. \qquad (27\text{–}14)$$

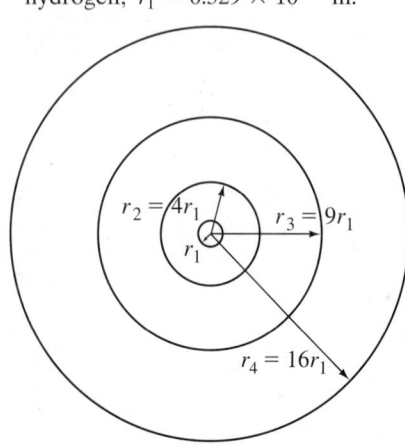

FIGURE 27–28 The four smallest orbits in the Bohr model of hydrogen; $r_1 = 0.529 \times 10^{-10}$ m.

[†]We include Z in our derivation so that we can treat other single-electron ("hydrogenlike") atoms such as the ions He$^+$ ($Z = 2$) and Li^{2+} ($Z = 3$). Helium in the neutral state has two electrons; if one electron is missing, the remaining He$^+$ ion consists of one electron revolving around a nucleus of charge $+2e$. Similarly, doubly ionized lithium, Li^{2+}, also has a single electron, and in this case $Z = 3$.

[‡]Be careful not to believe that these well-defined orbits actually exist. Today electrons are better thought of as forming "clouds," as discussed in Chapter 28.

In each of its possible orbits, the electron in a Bohr model atom would have a definite energy, as the following calculation shows. The total energy equals the sum of the kinetic and potential energies. The potential energy of the electron is given by $\text{PE} = qV = -eV$, where V is the potential due to a point charge $+Ze$ as given by Eq. 17–5: $V = kQ/r = kZe/r$. So

$$\text{PE} = -eV = -k\frac{Ze^2}{r}.$$

The total energy E_n for an electron in the nth orbit of radius r_n is the sum of the kinetic and potential energies:

$$E_n = \tfrac{1}{2}mv^2 - \frac{kZe^2}{r_n}.$$

When we substitute v from Eq. 27–11 and r_n from Eq. 27–12 into this equation, we obtain

$$E_n = -\frac{2\pi^2 Z^2 e^4 m k^2}{h^2}\frac{1}{n^2} \qquad n = 1, 2, 3, \cdots. \qquad \textbf{(27–15a)}$$

If we evaluate the constant term in Eq. 27–15a and convert it to electron volts, as is customary in atomic physics, we obtain

$$E_n = -(13.6\,\text{eV})\frac{Z^2}{n^2}, \qquad n = 1, 2, 3, \cdots. \qquad \textbf{(27–15b)}$$

The lowest energy level ($n = 1$) for hydrogen ($Z = 1$) is

$$E_1 = -13.6\,\text{eV}.$$

Since n^2 appears in the denominator of Eq. 27–15b, the energies of the larger orbits in hydrogen ($Z = 1$) are given by

$$E_n = \frac{-13.6\,\text{eV}}{n^2}.$$

For example,

$$E_2 = \frac{-13.6\,\text{eV}}{4} = -3.40\,\text{eV},$$

$$E_3 = \frac{-13.6\,\text{eV}}{9} = -1.51\,\text{eV}.$$

We see that not only are the orbit radii quantized, but from Eqs. 27–15, so is the energy. The quantum number n that labels the orbit radii also labels the energy levels. The lowest **energy level** or **energy state** has energy E_1, and is called the **ground state**. The higher states, E_2, E_3, and so on, are called **excited states**. The fixed energy levels are also called **stationary states**.

Notice that although the energy for the larger orbits has a smaller numerical value, all the energies are less than zero. Thus, $-3.4\,\text{eV}$ is a higher energy than $-13.6\,\text{eV}$. Hence the orbit closest to the nucleus (r_1) has the lowest energy (the most negative). The reason the energies have negative values has to do with the way we defined the zero for potential energy. For two point charges, $\text{PE} = kq_1q_2/r$ corresponds to zero potential energy when the two charges are infinitely far apart (Section 17–5). Thus, an electron that can just barely be free from the atom by reaching $r = \infty$ (or, at least, far from the nucleus) with zero kinetic energy will have $E = \text{KE} + \text{PE} = 0 + 0 = 0$, corresponding to $n = \infty$ in Eqs. 27–15. If an electron is free and has kinetic energy, then $E > 0$. To remove an electron that is part of an atom requires an energy input (otherwise atoms would not be stable). Since $E \geq 0$ for a free electron, then an electron bound to an atom needs to have $E < 0$. That is, energy must be added to bring its energy up, from a negative value to at least zero in order to free it.

The minimum energy required to remove an electron from an atom initially in the ground state is called the **binding energy** or **ionization energy**. The ionization energy for hydrogen has been measured to be 13.6 eV, and this corresponds precisely to removing an electron from the lowest state, $E_1 = -13.6\,\text{eV}$, up to $E = 0$ where it can be free.

Spectra Lines Explained

It is useful to show the various possible energy values as horizontal lines on an energy-level diagram. This is shown for hydrogen in Fig. 27–29. The electron in a hydrogen atom can be in any one of these levels according to Bohr theory. But it could never be in between, say at $-9.0\,\text{eV}$. At room temperature, nearly all H atoms will be in the ground state ($n = 1$). At higher temperatures, or during an electric discharge when there are many collisions between free electrons and atoms, many atoms can be in excited states ($n > 1$). Once in an excited state, an atom's electron can jump down to a lower state, and give off a photon in the process. This is, according to the Bohr model, the origin of the emission spectra of excited gases.

Note that above $E = 0$, an electron is free and can have any energy (E is not quantized). Thus there is a continuum of energy states above $E = 0$, as indicated in the energy-level diagram of Fig. 27–29.

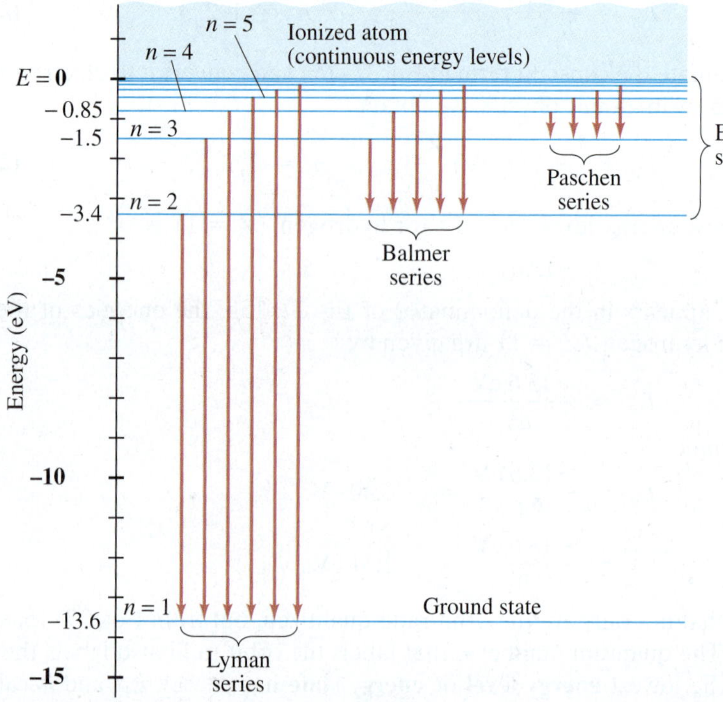

FIGURE 27–29 Energy-level diagram for the hydrogen atom, showing the transitions for the spectral lines of the Lyman, Balmer, and Paschen series (Fig. 27–25). Each vertical arrow represents an atomic transition that gives rise to the photons of one spectral line (a single wavelength or frequency).

The vertical arrows in Fig. 27–29 represent the transitions or jumps that correspond to the various observed spectral lines. For example, an electron jumping from the level $n = 3$ to $n = 2$ would give rise to the 656-nm line in the Balmer series, and the jump from $n = 4$ to $n = 2$ would give rise to the 486-nm line (see Fig. 27–24). We can predict wavelengths of the spectral lines emitted according to Bohr theory by combining Eq. 27–10 with Eq. 27–15. Since $hf = hc/\lambda$, we have from Eq. 27–10

$$\frac{1}{\lambda} = \frac{hf}{hc} = \frac{1}{hc}(E_n - E_{n'}),$$

where n refers to the upper state and n' to the lower state. Then using Eq. 27–15,

$$\frac{1}{\lambda} = \frac{2\pi^2 Z^2 e^4 m k^2}{h^3 c}\left(\frac{1}{n'^2} - \frac{1}{n^2}\right). \tag{27–16}$$

This theoretical formula has the same form as the experimental Balmer formula, Eq. 27–9, with $n' = 2$. Thus we see that the Balmer series of lines corresponds to transitions or "jumps" that bring the electron down to the second energy level. Similarly, $n' = 1$ corresponds to the Lyman series and $n' = 3$ to the Paschen series (see Fig. 27–29).

When the constant in Eq. 27–16 is evaluated with $Z = 1$, it is found to have the measured value of the Rydberg constant, $R = 1.0974 \times 10^7\,\text{m}^{-1}$ in Eq. 27–9, in accord with experiment (see Problem 54).

The great success of Bohr's model is that it gives an explanation for why atoms emit line spectra, and accurately predicts the wavelengths of emitted light for hydrogen. The Bohr model also explains absorption spectra: photons of just the right wavelength can knock an electron from one energy level to a higher one. To conserve energy, only photons that have just the right energy will be absorbed. This explains why a continuous spectrum of light entering a gas will emerge with dark (absorption) lines at frequencies that correspond to emission lines (Fig. 27–23c).

The Bohr theory also ensures the stability of atoms. It establishes stability by decree: the ground state is the lowest state for an electron and there is no lower energy level to which it can go and emit more energy. Finally, as we saw above, the Bohr theory accurately predicts the ionization energy of 13.6 eV for hydrogen. However, the Bohr model was not so successful for other atoms, and has been superseded as we shall discuss in the next Chapter. We discuss the Bohr model because it *was* an important start and because we still use the concept of stationary states, the ground state, and transitions between states. Also, the terminology used in the Bohr model is still used by chemists and spectroscopists.

EXAMPLE 27–13 **Wavelength of a Lyman line.** Use Fig. 27–29 to determine the wavelength of the first Lyman line, the transition from $n = 2$ to $n = 1$. In what region of the electromagnetic spectrum does this lie?

APPROACH We use Eq. 27–10, $hf = E_u - E_\ell$, with the energies obtained from Fig. 27–29 to find the energy and the wavelength of the transition. The region of the electromagnetic spectrum is found using the EM spectrum in Fig. 22–8.

SOLUTION In this case, $hf = E_2 - E_1 = \{-3.4 \text{ eV} - (-13.6 \text{ eV})\} = 10.2 \text{ eV} = (10.2 \text{ eV})(1.60 \times 10^{-19} \text{ J/eV}) = 1.63 \times 10^{-18}$ J. Since $\lambda = c/f$, we have

$$\lambda = \frac{c}{f} = \frac{hc}{E_2 - E_1} = \frac{(6.63 \times 10^{-34} \text{ J} \cdot \text{s})(3.00 \times 10^8 \text{ m/s})}{1.63 \times 10^{-18} \text{ J}} = 1.22 \times 10^{-7} \text{ m},$$

or 122 nm, which is in the UV region of the EM spectrum, Fig. 22–8. See also Fig. 27–25, where this value is confirmed experimentally.

NOTE An alternate approach: use Eq. 27–16 to find λ, and get the same result.

EXAMPLE 27–14 **Wavelength of a Balmer line.** Use the Bohr model to determine the wavelength of light emitted when a hydrogen atom makes a transition from the $n = 6$ to the $n = 2$ energy level.

APPROACH We can use Eq. 27–16 or its equivalent, Eq. 27–9, with $R = 1.097 \times 10^7 \text{ m}^{-1}$.

SOLUTION We find

$$\frac{1}{\lambda} = (1.097 \times 10^7 \text{ m}^{-1})\left(\frac{1}{4} - \frac{1}{36}\right) = 2.44 \times 10^6 \text{ m}^{-1}.$$

So $\lambda = 1/(2.44 \times 10^6 \text{ m}^{-1}) = 4.10 \times 10^{-7}$ m or 410 nm. This is the fourth line in the Balmer series, Fig. 27–24, and is violet in color.

EXAMPLE 27–15 **Absorption wavelength.** Use Fig. 27–29 to determine the maximum wavelength that hydrogen in its ground state can absorb. What would be the next smaller wavelength that would work?

APPROACH Maximum wavelength corresponds to minimum energy, and this would be the jump from the ground state up to the first excited state (Fig. 27–29). The next smaller wavelength occurs for the jump from the ground state to the second excited state.

SOLUTION The energy needed to jump from the ground state to the first excited state is $13.6 \text{ eV} - 3.4 \text{ eV} = 10.2 \text{ eV}$; the required wavelength, as we saw in Example 27–13, is 122 nm. The energy to jump from the ground state to the second excited state is $13.6 \text{ eV} - 1.5 \text{ eV} = 12.1 \text{ eV}$, which corresponds to a wavelength

$$\lambda = \frac{c}{f} = \frac{hc}{hf} = \frac{hc}{E_3 - E_1} = \frac{(6.63 \times 10^{-34} \text{ J} \cdot \text{s})(3.00 \times 10^8 \text{ m/s})}{(12.1 \text{ eV})(1.60 \times 10^{-19} \text{ J/eV})} = 103 \text{ nm}.$$

EXAMPLE 27–16 **He^+ ionization energy.** (a) Use the Bohr model to determine the ionization energy of the He^+ ion, which has a single electron. (b) Also calculate the maximum wavelength a photon can have to cause ionization. The helium atom is the second atom, after hydrogen, in the Periodic Table (next Chapter); its nucleus contains 2 protons and normally has 2 electrons circulating around it, so $Z = 2$.

APPROACH We want to determine the minimum energy required to lift the electron from its ground state and to barely reach the free state at $E = 0$. The ground state energy of He^+ is given by Eq. 27–15b with $n = 1$ and $Z = 2$.

SOLUTION (a) Since all the symbols in Eq. 27–15b are the same as for the calculation for hydrogen, except that Z is 2 instead of 1, we see that E_1 will be $Z^2 = 2^2 = 4$ times the E_1 for hydrogen:
$$E_1 = 4(-13.6 \text{ eV}) = -54.4 \text{ eV}.$$
Thus, to ionize the He^+ ion should require 54.4 eV, and this value agrees with experiment.

(b) The maximum wavelength photon that can cause ionization will have energy $hf = 54.4$ eV and wavelength
$$\lambda = \frac{c}{f} = \frac{hc}{hf} = \frac{(6.63 \times 10^{-34} \text{ J·s})(3.00 \times 10^8 \text{ m/s})}{(54.4 \text{ eV})(1.60 \times 10^{-19} \text{ J/eV})} = 22.8 \text{ nm}.$$
If $\lambda > 22.8$ nm, ionization can not occur.

NOTE If the atom absorbed a photon of greater energy (wavelength shorter than 22.8 nm), the atom could still be ionized and the freed electron would have kinetic energy of its own.

In this Example 27–16, we saw that E_1 for the He^+ ion is four times more negative than that for hydrogen. Indeed, the energy-level diagram for He^+ looks just like that for hydrogen, Fig. 27–29, except that the numerical values for each energy level are four times larger. Note, however, that we are talking here about the He^+ *ion*. Normal (neutral) helium has two electrons and its energy level diagram is entirely different.

CONCEPTUAL EXAMPLE 27–17 **Hydrogen at 20°C.** (a) Estimate the average kinetic energy of whole hydrogen atoms (not just the electrons) at room temperature. (b) Use the result to explain why, at room temperature, very few H atoms are in excited states and nearly all are in the ground state, and hence emit no light.

RESPONSE According to kinetic theory (Chapter 13), the average kinetic energy of atoms or molecules in a gas is given by Eq. 13–8:
$$\overline{KE} = \tfrac{3}{2} kT,$$
where $k = 1.38 \times 10^{-23}$ J/K is Boltzmann's constant, and T is the kelvin (absolute) temperature. Room temperature is about $T = 300$ K, so
$$\overline{KE} = \tfrac{3}{2}(1.38 \times 10^{-23} \text{ J/K})(300 \text{ K}) = 6.2 \times 10^{-21} \text{ J},$$
or, in electron volts:
$$\overline{KE} = \frac{6.2 \times 10^{-21} \text{ J}}{1.6 \times 10^{-19} \text{ J/eV}} = 0.04 \text{ eV}.$$
The average KE of an atom as a whole is thus very small compared to the energy between the ground state and the next higher energy state (13.6 eV − 3.4 eV = 10.2 eV). Any atoms in excited states quickly fall to the ground state and emit light. Once in the ground state, collisions with other atoms can transfer energy of only 0.04 eV on the average. A small fraction of atoms can have much more energy (see Section 13–10 on the distribution of molecular speeds), but even a kinetic energy that is 10 times the average is not nearly enough to excite atoms into states above the ground state. Thus, at room temperature, practically all atoms are in the ground state. Atoms can be excited to upper states by very high temperatures, or by applying a high voltage so a current of high energy electrons passes through the gas as in a discharge tube (Fig. 27–22).

Correspondence Principle

We should note that Bohr made some radical assumptions that were at variance with classical ideas. He assumed that electrons in fixed orbits do not radiate light even though they are accelerating (moving in a circle), and he assumed that angular momentum is quantized. Furthermore, he was not able to say how an electron moved when it made a transition from one energy level to another. On the other hand, there is no real reason to expect that in the tiny world of the atom electrons would behave as ordinary-sized objects do. Nonetheless, he felt that where quantum theory overlaps with the macroscopic world, it should predict classical results. This is the **correspondence principle**, already mentioned in regard to relativity (Section 26–11). This principle does work for Bohr's theory of the hydrogen atom. The orbit sizes and energies are quite different for $n = 1$ and $n = 2$, say. But orbits with $n = 100,000,000$ and $100,000,001$ would be very close in radius and energy (see Fig. 27–29). Indeed, transitions between such large orbits, which would approach macroscopic sizes, would be imperceptible. Such orbits would thus appear to be continuously spaced, which is what we expect in the everyday world.

Finally, it must be emphasized that the well-defined orbits of the Bohr model do not actually exist. The Bohr model is only a model, not reality. The idea of electron orbits was rejected a few years later, and today electrons are thought of (Chapter 28) as forming "probability clouds."

27–13 de Broglie's Hypothesis Applied to Atoms

Bohr's theory was largely of an *ad hoc* nature. Assumptions were made so that theory would agree with experiment. But Bohr could give no reason why the orbits were quantized, nor why there should be a stable ground state. Finally, ten years later, a reason was proposed by Louis de Broglie. We saw in Section 27–8 that in 1923, de Broglie proposed that material particles, such as electrons, have a wave nature; and that this hypothesis was confirmed by experiment several years later.

One of de Broglie's original arguments in favor of the wave nature of electrons was that it provided an explanation for Bohr's theory of the hydrogen atom. According to de Broglie, a particle of mass m moving with a nonrelativistic speed v would have a wavelength (Eq. 27–8) of

$$\lambda = \frac{h}{mv}.$$

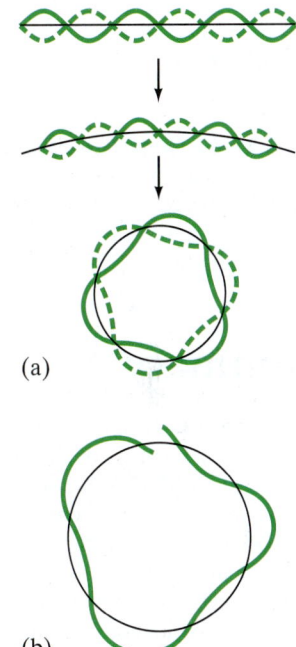

FIGURE 27–30 (a) An ordinary standing wave compared to a circular standing wave. (b) When a wave does not close (and hence interferes destructively with itself), it rapidly dies out.

FIGURE 27–31 Standing circular waves for two, three, and five wavelengths on the circumference; n, the number of wavelengths, is also the quantum number.

Each electron orbit in an atom, he proposed, is actually a standing wave. As we saw in Chapter 11, when a violin or guitar string is plucked, a vast number of wavelengths are excited. But only certain ones—those that have nodes at the ends—are sustained. These are the *resonant* modes of the string. Waves with other wavelengths interfere with themselves upon reflection and their amplitudes quickly drop to zero. With electrons moving in circles, according to Bohr's theory, de Broglie argued that the electron wave was a *circular* standing wave that closes on itself, Fig. 27–30a. If the wavelength of a wave does not close on itself, as in Fig. 27–30b, destructive interference takes place as the wave travels around the loop, and the wave quickly dies out. Thus, the only waves that persist are those for which the circumference of the circular orbit contains a whole number of wavelengths, Fig. 27–31. The circumference of a Bohr orbit of radius r_n is $2\pi r_n$, so to have constructive interference, we need

$$2\pi r_n = n\lambda, \quad n = 1, 2, 3, \cdots.$$

When we substitute $\lambda = h/mv$, we get $2\pi r_n = nh/mv$, or

$$mvr_n = \frac{nh}{2\pi}.$$

This is just the *quantum condition* proposed by Bohr on an *ad hoc* basis, Eq. 27–11. It is from this equation that the discrete orbits and energy levels were derived.

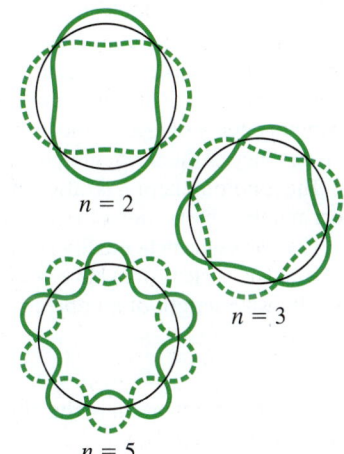

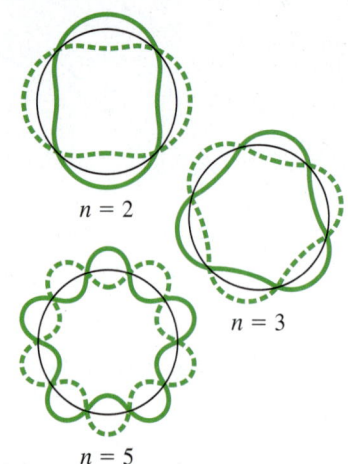

FIGURE 27–31 (Repeated.) Standing circular waves for two, three, and five wavelengths on the circumference; n, the number of wavelengths, is also the quantum number.

Now we have a first explanation for the quantized orbits and energy states in the Bohr model: they are due to the wave nature of the electron, and only resonant "standing" waves can persist.† This implies that the *wave–particle duality* is at the root of atomic structure.

In viewing the circular electron waves of Fig. 27–31, the electron is not to be thought of as following the oscillating wave pattern. In the Bohr model of hydrogen, the electron moves in a circle. The circular wave, on the other hand, represents the *amplitude* of the electron "matter wave," and in Fig. 27–31 the wave amplitude is shown superimposed on the circular path of the particle orbit for convenience.

Bohr's theory worked well for hydrogen and for one-electron ions. But it did not prove successful for multi-electron atoms. Bohr theory could not predict line spectra even for the next simplest atom, helium. It could not explain why some emission lines are brighter than others, nor why some lines are split into two or more closely spaced lines ("fine structure"). A new theory was needed and was indeed developed in the 1920s. This new and radical theory is called *quantum mechanics*. It finally solved the problem of atomic structure, but it gives us a very different view of the atom: the idea of electrons in well-defined orbits was replaced with the idea of electron "clouds." This new theory of quantum mechanics has given us a wholly different view of the basic mechanisms underlying physical processes.

†We note, however, that Eq. 27–11 is no longer considered valid, as discussed in the next Chapter.

Summary

The electron was discovered using an evacuated cathode ray tube. The measurement of the charge-to-mass ratio (e/m) of the electron was done using magnetic and electric fields. The charge e on the electron was first measured in the Millikan oil-drop experiment and then its mass was obtained from the measured value of the e/m ratio.

Quantum theory has its origins in **Planck's quantum hypothesis** that molecular oscillations are **quantized**: their energy E can only be integer (n) multiples of hf, where h is Planck's constant and f is the natural frequency of oscillation:

$$E = nhf. \quad (27\text{–}3)$$

This hypothesis explained the spectrum of radiation emitted by a **blackbody** at high temperature.

Einstein proposed that for some experiments, light could be pictured as being emitted and absorbed as **quanta** (particles), which we now call **photons**, each with energy

$$E = hf \quad (27\text{–}4)$$

and momentum

$$p = \frac{E}{c} = \frac{hf}{c} = \frac{h}{\lambda}. \quad (27\text{–}6)$$

He proposed the photoelectric effect as a test for the photon theory of light. In the **photoelectric effect**, the photon theory says that each incident photon can strike an electron in a material and eject it if the photon has sufficient energy. The maximum energy of ejected electrons is then linearly related to the frequency of the incident light.

The photon theory is also supported by the **Compton effect** and the observation of electron–positron **pair production**.

The **wave–particle duality** refers to the idea that light and matter (such as electrons) have both wave and particle properties. The wavelength of an object is given by

$$\lambda = \frac{h}{p}, \quad (27\text{–}8)$$

where p is the momentum of the object ($p = mv$ for a particle of mass m and speed v).

The **principle of complementarity** states that we must be aware of both the particle and wave properties of light and of matter for a complete understanding of them.

Electron microscopes (EM) make use of the wave properties of electrons to form an image: their "lenses" are magnetic. Various types of EM exist: some can magnify 100,000× (1000× better than a light microscope); others can give a 3-D image.

Early models of the atom include Rutherford's planetary (or nuclear) model of an atom which consists of a tiny but massive positively charged nucleus surrounded (at a relatively great distance) by electrons.

To explain the **line spectra** emitted by atoms, as well as the stability of atoms, the **Bohr model** postulated that: (1) electrons bound in an atom can only occupy orbits for which the angular momentum is quantized, which results in discrete values for the radius and energy; (2) an electron in such a **stationary state** emits no radiation; (3) if an electron jumps to a lower state, it emits a photon whose energy equals the difference in energy between the two states; (4) the angular momentum L of atomic electrons is quantized by the rule $L = nh/2\pi$, where n is an integer called the **quantum number**. The $n = 1$ state is the **ground state**, which in hydrogen has an energy $E_1 = -13.6 \text{ eV}$. Higher values of n correspond to **excited states**, and their energies are

$$E_n = -(13.6 \text{ eV})\frac{Z^2}{n^2}, \quad (27\text{–}15b)$$

where Ze is the charge on the nucleus. Atoms are excited to these higher states by collisions with other atoms or electrons, or by absorption of a photon of just the right frequency.

De Broglie's hypothesis that electrons (and other matter) have a wavelength $\lambda = h/mv$ gave an explanation for Bohr's quantized orbits by bringing in the wave–particle duality: the orbits correspond to circular standing waves in which the circumference of the orbit equals a whole number of wavelengths.

Questions

1. Does a lightbulb at a temperature of 2500 K produce as white a light as the Sun at 6000 K? Explain.

2. If energy is radiated by all objects, why can we not see them in the dark? (See also Section 14–8.)

3. What can be said about the relative temperatures of whitish-yellow, reddish, and bluish stars? Explain.

4. Darkrooms for developing black-and-white film were sometimes lit by a red bulb. Why red? Explain if such a bulb would work in a darkroom for developing color film.

5. If the threshold wavelength in the photoelectric effect increases when the emitting metal is changed to a different metal, what can you say about the work functions of the two metals?

6. Explain why the existence of a cutoff frequency in the photoelectric effect more strongly favors a particle theory rather than a wave theory of light.

7. UV light causes sunburn, whereas visible light does not. Suggest a reason.

8. The work functions for sodium and cesium are 2.28 eV and 2.14 eV, respectively. For incident photons of a given frequency, which metal will give a higher maximum kinetic energy for the electrons? Explain.

9. Explain how the photoelectric circuit of Fig. 27–6 could be used in (a) a burglar alarm, (b) a smoke detector, (c) a photographic light meter.

10. (a) Does a beam of infrared photons always have less energy than a beam of ultraviolet photons? Explain. (b) Does a single photon of infrared light always have less energy than a single photon of ultraviolet light? Why?

11. Light of 450-nm wavelength strikes a metal surface, and a stream of electrons emerges from the metal. If light of the same intensity but of wavelength 400 nm strikes the surface, are more electrons emitted? Does the energy of the emitted electrons change? Explain.

*12. If an X-ray photon is scattered by an electron, does the photon's wavelength change? If so, does it increase or decrease? Explain.

*13. In both the photoelectric effect and in the Compton effect, a photon collides with an electron causing the electron to fly off. What is the difference between the two processes?

14. Why do we say that light has wave properties? Why do we say that light has particle properties?

15. Why do we say that electrons have wave properties? Why do we say that electrons have particle properties?

16. What are the differences between a photon and an electron? Be specific: make a list.

17. If an electron and a proton travel at the same speed, which has the shorter wavelength? Explain.

18. An electron and a proton are accelerated through the same voltage. Which has the longer wavelength? Explain why.

19. In Rutherford's planetary model of the atom, what keeps the electrons from flying off into space?

20. When a wide spectrum of light passes through hydrogen gas at room temperature, absorption lines are observed that correspond only to the Lyman series. Why don't we observe the other series?

21. How can you tell if there is oxygen near the surface of the Sun?

22. (a) List at least three successes of the Bohr model of the atom, according to Section 27–12. (b) List at least two observations that the Bohr model could not explain, according to Section 27–13.

23. According to Section 27–11, what were the two main difficulties of the Rutherford model of the atom?

24. Is it possible for the de Broglie wavelength of a "particle" to be greater than the dimensions of the particle? To be smaller? Is there any direct connection? Explain.

25. How can the spectrum of hydrogen contain so many lines when hydrogen contains only one electron?

26. Explain how the closely spaced energy levels for hydrogen near the top of Fig. 27–29 correspond to the closely spaced spectral lines at the top of Fig. 27–24.

27. In a helium atom, which contains two electrons, do you think that on average the electrons are closer to the nucleus or farther away than in a hydrogen atom? Why?

28. The Lyman series is brighter than the Balmer series, because this series of transitions ends up in the most common state for hydrogen, the ground state. Why then was the Balmer series discovered first?

29. Use conservation of momentum to explain why photons emitted by hydrogen atoms have slightly less energy than that predicted by Eq. 27–10.

30. State if a continuous or a line spectrum is produced by each of the following: (a) a hot solid object; (b) an excited, rarefied gas; (c) a hot liquid; (d) light from a hot solid that passes through a cooler rarefied gas; (e) a hot dense gas. For each, if a line spectrum is produced, is it an emission or an absorption spectrum?

31. Suppose we obtain an emission spectrum for hydrogen at very high temperature (when some of the atoms are in excited states), and an absorption spectrum at room temperature, when all atoms are in the ground state. Will the two spectra contain identical lines?

MisConceptual Questions

1. Which of the following statements is true regarding how blackbody radiation changes as the temperature of the radiating object increases?
   (a) Both the maximum intensity and the peak wavelength increase.
   (b) The maximum intensity increases, and the peak wavelength decreases.
   (c) Both the maximum intensity and the peak wavelength decrease.
   (d) The maximum intensity decreases, and the peak wavelength increases.

2. As red light shines on a piece of metal, no electrons are released. When the red light is slowly changed to shorter-wavelength light (basically progressing through the rainbow), nothing happens until yellow light shines on the metal, at which point electrons are released from the metal. If this metal is replaced with a metal having a higher work function, which light would have the best chance of releasing electrons from the metal?
   (a) Blue.
   (b) Red.
   (c) Yellow would still work fine.
   (d) We need to know more about the metals involved.

3. A beam of red light and a beam of blue light have equal intensities. Which statement is true?
   (a) There are more photons in the blue beam.
   (b) There are more photons in the red beam.
   (c) Both beams contain the same number of photons.
   (d) The number of photons is not related to intensity.

4. Which of the following is necessarily true?
   (a) Red light has more energy than violet light.
   (b) Violet light has more energy than red light.
   (c) A single photon of red light has more energy than a single photon of violet light.
   (d) A single photon of violet light has more energy than a single photon of red light.
   (e) None of the above.
   (f) A combination of the above (specify).

5. If a photon of energy E ejects electrons from a metal with kinetic energy KE, then a photon with energy $E/2$
   (a) will eject electrons with kinetic energy KE/2.
   (b) will eject electrons with an energy greater than KE/2.
   (c) will eject electrons with an energy less than KE/2.
   (d) might not eject any electrons.

6. If the momentum of an electron were doubled, how would its wavelength change?
   (a) No change.
   (b) It would be halved.
   (c) It would double.
   (d) It would be quadrupled.
   (e) It would be reduced to one-fourth.

7. Which of the following can be thought of as either a wave or a particle?
   (a) Light.
   (b) An electron.
   (c) A proton.
   (d) All of the above.

8. When you throw a baseball, its de Broglie wavelength is
   (a) the same size as the ball.
   (b) about the same size as an atom.
   (c) about the same size as an atom's nucleus.
   (d) much smaller than the size of an atom's nucleus.

9. Electrons and photons of light are similar in that
   (a) both have momentum given by h/λ.
   (b) both exhibit wave–particle duality.
   (c) both are used in diffraction experiments to explore structure.
   (d) All of the above.
   (e) None of the above.

10. In Rutherford's famous set of experiments described in Section 27–10, the fact that some alpha particles were deflected at large angles indicated that (choose all that apply)
    (a) the nucleus was positive.
    (b) charge was quantized.
    (c) the nucleus was concentrated in a small region of space.
    (d) most of the atom is empty space.
    (e) None of the above.

11. Which of the following electron transitions between two energy states (n) in the hydrogen atom corresponds to the emission of a photon with the longest wavelength?
    (a) $2 \rightarrow 5$.
    (b) $5 \rightarrow 2$.
    (c) $5 \rightarrow 8$.
    (d) $8 \rightarrow 5$.

12. If we set the potential energy of an electron and a proton to be zero when they are an infinite distance apart, then the lowest energy a bound electron in a hydrogen atom can have is
    (a) 0.
    (b) -13.6 eV.
    (c) any possible value.
    (d) any value between -13.6 eV and 0.

13. Which of the following is the currently accepted model of the atom?
    (a) The plum-pudding model.
    (b) The Rutherford atom.
    (c) The Bohr atom.
    (d) None of the above.

14. Light has all of the following except:
    (a) mass.
    (b) momentum.
    (c) kinetic energy.
    (d) frequency.
    (e) wavelength.

For assigned homework and other learning materials, go to the MasteringPhysics website.

Problems

27–1 Discovery of the Electron

1. (I) What is the value of e/m for a particle that moves in a circle of radius 14 mm in a 0.86-T magnetic field if a perpendicular 640-V/m electric field will make the path straight?

2. (II) (a) What is the velocity of a beam of electrons that go undeflected when passing through crossed (perpendicular) electric and magnetic fields of magnitude 1.88×10^4 V/m and 2.60×10^{-3} T, respectively? (b) What is the radius of the electron orbit if the electric field is turned off?

3. (II) An oil drop whose mass is 2.8×10^{-15} kg is held at rest between two large plates separated by 1.0 cm (Fig. 27–3), when the potential difference between the plates is 340 V. How many excess electrons does this drop have?

27–2 Blackbodies; Planck's Quantum Hypothesis

4. (I) How hot is a metal being welded if it radiates most strongly at 520 nm?

5. (I) Estimate the peak wavelength for radiation emitted from (a) ice at 0°C, (b) a floodlamp at 3100 K, (c) helium at 4 K, assuming blackbody emission. In what region of the EM spectrum is each?

6. (I) (a) What is the temperature if the peak of a blackbody spectrum is at 18.0 nm? (b) What is the wavelength at the peak of a blackbody spectrum if the body is at a temperature of 2200 K?

7. (I) An HCl molecule vibrates with a natural frequency of 8.1×10^{13} Hz. What is the difference in energy (in joules and electron volts) between successive values of the oscillation energy?

8. (II) The steps of a flight of stairs are 20.0 cm high (vertically). If a 62.0-kg person stands with both feet on the same step, what is the gravitational potential energy of this person, relative to the ground, on (a) the first step, (b) the second step, (c) the third step, (d) the nth step? (e) What is the change in energy as the person descends from step 6 to step 2?

9. (II) Estimate the peak wavelength of light emitted from the pupil of the human eye (which approximates a blackbody) assuming normal body temperature.

27–3 and 27–4 Photons and the Photoelectric Effect

10. (I) What is the energy of photons (joules) emitted by a 91.7-MHz FM radio station?

11. (I) What is the energy range (in joules and eV) of photons in the visible spectrum, of wavelength 400 nm to 750 nm?

12. (I) A typical gamma ray emitted from a nucleus during radioactive decay may have an energy of 320 keV. What is its wavelength? Would we expect significant diffraction of this type of light when it passes through an everyday opening, such as a door?

13. (I) Calculate the momentum of a photon of yellow light of wavelength 5.80×10^{-7} m.

14. (I) What is the momentum of a $\lambda = 0.014$ nm X-ray photon?

15. (I) For the photoelectric effect, make a table that shows expected observations for a particle theory of light and for a wave theory of light. Circle the actual observed effects. (See Section 27–3.)

16. (II) About 0.1 eV is required to break a "hydrogen bond" in a protein molecule. Calculate the minimum frequency and maximum wavelength of a photon that can accomplish this.

17. (II) What minimum frequency of light is needed to eject electrons from a metal whose work function is 4.8×10^{-19} J?

18. (II) The human eye can respond to as little as 10^{-18} J of light energy. For a wavelength at the peak of visual sensitivity, 550 nm, how many photons lead to an observable flash?

19. (II) What is the longest wavelength of light that will emit electrons from a metal whose work function is 2.90 eV?

20. (II) The work functions for sodium, cesium, copper, and iron are 2.3, 2.1, 4.7, and 4.5 eV, respectively. Which of these metals will not emit electrons when visible light shines on it?

21. (II) In a photoelectric-effect experiment it is observed that no current flows unless the wavelength is less than 550 nm. (a) What is the work function of this material? (b) What stopping voltage is required if light of wavelength 400 nm is used?

22. (II) What is the maximum kinetic energy of electrons ejected from barium ($W_0 = 2.48$ eV) when illuminated by white light, $\lambda = 400$ to 750 nm?

23. (II) Barium has a work function of 2.48 eV. What is the maximum kinetic energy of electrons if the metal is illuminated by UV light of wavelength 365 nm? What is their speed?

24. (II) When UV light of wavelength 255 nm falls on a metal surface, the maximum kinetic energy of emitted electrons is 1.40 eV. What is the work function of the metal?

25. (II) The threshold wavelength for emission of electrons from a given surface is 340 nm. What will be the maximum kinetic energy of ejected electrons when the wavelength is changed to (a) 280 nm, (b) 360 nm?

26. (II) A certain type of film is sensitive only to light whose wavelength is less than 630 nm. What is the energy (eV and kcal/mol) needed for the chemical reaction to occur which causes the film to change?

27. (II) When 250-nm light falls on a metal, the current through a photoelectric circuit (Fig. 27–6) is brought to zero at a stopping voltage of 1.64 V. What is the work function of the metal?

28. (II) In a photoelectric experiment using a clean sodium surface, the maximum energy of the emitted electrons was measured for a number of different incident frequencies, with the following results.

Frequency ($\times 10^{14}$ Hz)	Energy (eV)
11.8	2.60
10.6	2.11
9.9	1.81
9.1	1.47
8.2	1.10
6.9	0.57

Plot the graph of these results and find: (a) Planck's constant; (b) the cutoff frequency of sodium; (c) the work function.

29. (II) Show that the energy E (in electron volts) of a photon whose wavelength is λ (nm) is given by

$$E = \frac{1.240 \times 10^3 \text{ eV} \cdot \text{nm}}{\lambda \text{ (nm)}}.$$

Use at least 4 significant figures for values of h, c, e (see inside front cover).

*27–5 Compton Effect

*30. (I) A high-frequency photon is scattered off of an electron and experiences a change of wavelength of 1.7×10^{-4} nm. At what angle must a detector be placed to detect the scattered photon (relative to the direction of the incoming photon)?

*31. (II) The quantity h/mc, which has the dimensions of length, is called the *Compton wavelength*. Determine the Compton wavelength for (a) an electron, (b) a proton. (c) Show that if a photon has wavelength equal to the Compton wavelength of a particle, the photon's energy is equal to the rest energy of the particle, mc^2.

*32. (II) X-rays of wavelength $\lambda = 0.140$ nm are scattered from carbon. What is the expected Compton wavelength shift for photons detected at angles (relative to the incident beam) of exactly (a) 45°, (b) 90°, (c) 180°?

27–6 Pair Production

33. (I) How much total kinetic energy will an electron–positron pair have if produced by a 3.64-MeV photon?

34. (II) What is the longest wavelength photon that could produce a proton–antiproton pair? (Each has a mass of 1.67×10^{-27} kg.)

35. (II) What is the minimum photon energy needed to produce a $\mu^+\mu^-$ pair? The mass of each μ (muon) is 207 times the mass of an electron. What is the wavelength of such a photon?

36. (II) An electron and a positron, each moving at 3.0×10^5 m/s, collide head on, disappear, and produce two photons, each with the same energy and momentum moving in opposite directions. Determine the energy and momentum of each photon.

37. (II) A gamma-ray photon produces an electron and a positron, each with a kinetic energy of 285 keV. Determine the energy and wavelength of the photon.

27–8 Wave Nature of Matter

38. (I) Calculate the wavelength of a 0.21-kg ball traveling at 0.10 m/s.

39. (I) What is the wavelength of a neutron ($m = 1.67 \times 10^{-27}$ kg) traveling at 8.5×10^4 m/s?

40. (II) Through how many volts of potential difference must an electron, initially at rest, be accelerated to achieve a wavelength of 0.27 nm?

41. (II) Calculate the ratio of the kinetic energy of an electron to that of a proton if their wavelengths are equal. Assume that the speeds are nonrelativistic.

42. (II) An electron has a de Broglie wavelength $\lambda = 4.5 \times 10^{-10}$ m. (a) What is its momentum? (b) What is its speed? (c) What voltage was needed to accelerate it from rest to this speed?

43. (II) What is the wavelength of an electron of energy (a) 10 eV, (b) 100 eV, (c) 1.0 keV?

44. (II) Show that if an electron and a proton have the same nonrelativistic kinetic energy, the proton has the shorter wavelength.

45. (II) Calculate the de Broglie wavelength of an electron if it is accelerated from rest by 35,000 V as in Fig. 27–2. Is it relativistic? How does its wavelength compare to the size of the "neck" of the tube, typically 5 cm? Do we have to worry about diffraction problems blurring the picture on the CRT screen?

46. (III) A Ferrari with a mass of 1400 kg approaches a freeway underpass that is 12 m across. At what speed must the car be moving, in order for it to have a wavelength such that it might somehow "diffract" after passing through this "single slit"? How do these conditions compare to normal freeway speeds of 30 m/s?

27–9 Electron Microscope

47. (II) What voltage is needed to produce electron wavelengths of 0.26 nm? (Assume that the electrons are nonrelativistic.)

48. (II) Electrons are accelerated by 2850 V in an electron microscope. Estimate the maximum possible resolution of the microscope.

27–11 and 27–12 Spectra and the Bohr Model

49. (I) For the three hydrogen transitions indicated below, with n being the initial state and n' being the final state, is the transition an absorption or an emission? Which is higher, the initial state energy or the final state energy of the atom? Finally, which of these transitions involves the largest energy photon? (a) $n = 1$, $n' = 3$; (b) $n = 6$, $n' = 2$; (c) $n = 4$, $n' = 5$.

50. (I) How much energy is needed to ionize a hydrogen atom in the $n = 3$ state?

51. (I) The second longest wavelength in the Paschen series in hydrogen (Fig. 27–29) corresponds to what transition?

52. (I) Calculate the ionization energy of doubly ionized lithium, Li^{2+}, which has $Z = 3$ (and is in the ground state).

53. (I) (a) Determine the wavelength of the second Balmer line ($n = 4$ to $n = 2$ transition) using Fig. 27–29. Determine likewise (b) the wavelength of the second Lyman line and (c) the wavelength of the third Balmer line.

54. (I) Evaluate the Rydberg constant R using the Bohr model (compare Eqs. 27–9 and 27–16) and show that its value is $R = 1.0974 \times 10^7$ m^{-1}. (Use values inside front cover to 5 or 6 significant figures.)

55. (II) What is the longest wavelength light capable of ionizing a hydrogen atom in the ground state?

56. (II) What wavelength photon would be required to ionize a hydrogen atom in the ground state and give the ejected electron a kinetic energy of 11.5 eV?

57. (II) In the Sun, an ionized helium (He^+) atom makes a transition from the $n = 6$ state to the $n = 2$ state, emitting a photon. Can that photon be absorbed by hydrogen atoms present in the Sun? If so, between what energy states will the hydrogen atom transition occur?

58. (II) Construct the energy-level diagram for the He^+ ion (like Fig. 27–29).

59. (II) Construct the energy-level diagram for doubly ionized lithium, Li^{2+}.

60. (II) Determine the electrostatic potential energy and the kinetic energy of an electron in the ground state of the hydrogen atom.

61. (II) A hydrogen atom has an angular momentum of 5.273×10^{-34} kg·m^2/s. According to the Bohr model, what is the energy (eV) associated with this state?

62. (II) An excited hydrogen atom could, in principle, have a radius of 1.00 cm. What would be the value of n for a Bohr orbit of this size? What would its energy be?

63. (II) Is the use of nonrelativistic formulas justified in the Bohr atom? To check, calculate the electron's velocity, v, in terms of c, for the ground state of hydrogen, and then calculate $\sqrt{1 - v^2/c^2}$.

64. (III) Show that the magnitude of the electrostatic potential energy of an electron in any Bohr orbit of a hydrogen atom is twice the magnitude of its kinetic energy in that orbit.

65. (III) Suppose an electron was bound to a proton, as in the hydrogen atom, but by the gravitational force rather than by the electric force. What would be the radius, and energy, of the first Bohr orbit?

General Problems

66. The Big Bang theory (Chapter 33) states that the beginning of the universe was accompanied by a huge burst of photons. Those photons are still present today and make up the so-called cosmic microwave background radiation. The universe radiates like a blackbody with a temperature today of about 2.7 K. Calculate the peak wavelength of this radiation.

67. At low temperatures, nearly all the atoms in hydrogen gas will be in the ground state. What minimum frequency photon is needed if the photoelectric effect is to be observed?

68. A beam of 72-eV electrons is scattered from a crystal, as in X-ray diffraction, and a first-order peak is observed at $\theta = 38°$. What is the spacing between planes in the diffracting crystal? (See Section 25–11.)

69. A microwave oven produces electromagnetic radiation at $\lambda = 12.2$ cm and produces a power of 720 W. Calculate the number of microwave photons produced by the microwave oven each second.

70. Sunlight reaching the Earth's atmosphere has an intensity of about 1300 W/m². Estimate how many photons per square meter per second this represents. Take the average wavelength to be 550 nm.

71. A beam of red laser light ($\lambda = 633$ nm) hits a black wall and is fully absorbed. If this light exerts a total force $F = 5.8$ nN on the wall, how many photons per second are hitting the wall?

72. A flashlight emits 2.5 W of light. As the light leaves the flashlight in one direction, a reaction force is exerted on the flashlight in the opposite direction. Estimate the size of this reaction force.

73. A **photomultiplier tube** (a very sensitive light sensor), is based on the photoelectric effect: incident photons strike a metal surface and the resulting ejected electrons are collected. By counting the number of collected electrons, the number of incident photons (i.e., the incident light intensity) can be determined. (a) If a photomultiplier tube is to respond properly for incident wavelengths throughout the visible range (410 nm to 750 nm), what is the maximum value for the work function W_0 (eV) of its metal surface? (b) If W_0 for its metal surface is above a certain threshold value, the photomultiplier will only function for incident ultraviolet wavelengths and be unresponsive to visible light. Determine this threshold value (eV).

74. If a 100-W lightbulb emits 3.0% of the input energy as visible light (average wavelength 550 nm) uniformly in all directions, estimate how many photons per second of visible light will strike the pupil (4.0 mm diameter) of the eye of an observer, (a) 1.0 m away, (b) 1.0 km away.

75. An electron and a positron collide head on, annihilate, and create two 0.85-MeV photons traveling in opposite directions. What were the initial kinetic energies of electron and positron?

76. By what potential difference must (a) a proton ($m = 1.67 \times 10^{-27}$ kg), and (b) an electron ($m = 9.11 \times 10^{-31}$ kg), be accelerated from rest to have a wavelength $\lambda = 4.0 \times 10^{-12}$ m?

77. In some of Rutherford's experiments (Fig. 27–19) the α particles (mass = 6.64×10^{-27} kg) had a kinetic energy of 4.8 MeV. How close could they get to the surface of a gold nucleus (radius $\approx 7.0 \times 10^{-15}$ m, charge = $+79e$)? Ignore the recoil motion of the nucleus.

78. By what fraction does the mass of an H atom decrease when it makes an $n = 3$ to $n = 1$ transition?

79. Calculate the ratio of the gravitational force to the electric force for the electron in the ground state of a hydrogen atom. Can the gravitational force be reasonably ignored?

80. Electrons accelerated from rest by a potential difference of 12.3 V pass through a gas of hydrogen atoms at room temperature. What wavelengths of light will be emitted?

81. In a particular photoelectric experiment, a stopping potential of 2.10 V is measured when ultraviolet light of wavelength 270 nm is incident on the metal. Using the same setup, what will the new stopping potential be if blue light of wavelength 440 nm is used, instead?

82. Neutrons can be used in diffraction experiments to probe the lattice structure of crystalline solids. Since the neutron's wavelength needs to be on the order of the spacing between atoms in the lattice, about 0.3 nm, what should the speed of the neutrons be?

83. In Chapter 22, the intensity of light striking a surface was related to the electric field of the associated electromagnetic wave. For photons, the intensity is the number of photons striking a 1-m² area per second. Suppose 1.0×10^{12} photons of 497-nm light are incident on a 1-m² surface every second. What is the intensity of the light? Using the wave model of light, what is the maximum electric field of the electromagnetic wave?

84. The intensity of the Sun's light in the vicinity of the Earth is about 1350 W/m². Imagine a spacecraft with a mirrored square sail of dimension 1.0 km. Estimate how much thrust (in newtons) this craft will experience due to collisions with the Sun's photons. [Hint: Assume the photons bounce off the sail with no change in the magnitude of their momentum.]

85. Light of wavelength 280 nm strikes a metal whose work function is 2.2 eV. What is the shortest de Broglie wavelength for the electrons that are produced as photoelectrons?

86. Photons of energy 6.0 eV are incident on a metal. It is found that current flows from the metal until a stopping potential of 3.8 V is applied. If the wavelength of the incident photons is doubled, what is the maximum kinetic energy of the ejected electrons? What would happen if the wavelength of the incident photons was tripled?

87. What would be the theoretical limit of resolution for an electron microscope whose electrons are accelerated through 110 kV? (Relativistic formulas should be used.)

88. Assume hydrogen atoms in a gas are initially in their ground state. If free electrons with kinetic energy 12.75 eV collide with these atoms, what photon wavelengths will be emitted by the gas?

89. Visible light incident on a diffraction grating with slit spacing of 0.010 mm has the first maximum at an angle of 3.6° from the central peak. If electrons could be diffracted by the same grating, what electron velocity would produce the same diffraction pattern as the visible light?

90. (a) Suppose an unknown element has an absorption spectrum with lines corresponding to 2.5, 4.7, and 5.1 eV above its ground state and an ionization energy of 11.5 eV. Draw an energy level diagram for this element. (b) If a 5.1-eV photon is absorbed by an atom of this substance, in which state was the atom before absorbing the photon? What will be the energies of the photons that can subsequently be emitted by this atom?

91. A photon of momentum 3.53×10^{-28} kg·m/s is emitted from a hydrogen atom. To what spectrum series does this photon belong, and from what energy level was it ejected?

92. Light of wavelength 464 nm falls on a metal which has a work function of 2.28 eV. (a) How much voltage should be applied to bring the current to zero? (b) What is the maximum speed of the emitted electrons? (c) What is the de Broglie wavelength of these electrons?

93. An electron accelerated from rest by a 96-V potential difference is injected into a 3.67×10^{-4} T magnetic field where it travels in an 18-cm-diameter circle. Calculate e/m from this information.

94. Estimate the number of photons emitted by the Sun in a year. (Take the average wavelength to be 550 nm and the intensity of sunlight reaching the Earth (outer atmosphere) as 1350 W/m².)

95. Apply Bohr's assumptions to the Earth–Moon system to calculate the allowed energies and radii of motion. Given the known distance between the Earth and Moon, is the quantization of the energy and radius apparent?

96. At what temperature would the average kinetic energy (Chapter 13) of a molecule of hydrogen gas (H_2) be sufficient to excite a hydrogen atom out of the ground state?

Search and Learn

1. Name the person or people who did each of the following: (a) made the first direct measurement of the charge-to-mass ratio of the electron (Section 27–1); (b) measured the charge on the electron and showed that it is quantized (Section 27–1); (c) proposed the radical assumption that the vibrational energy of molecules in a radiating object is quantized (Sections 27–2, 27–3); (d) found that light (X-rays) scattered off electrons in a material will decrease the energy of the photons (Section 27–5); (e) proposed that the wavelength of a material particle would be related to its momentum in the same way as for a photon (Section 27–8); (f) performed the first crucial experiment illustrating electron diffraction (Section 27–8); (g) deciphered the nuclear model of the atom by aiming α particles at gold foil (Section 27–10).

2. State the principle of complementarity, and give at least two experimental results that support this principle for electrons and for photons. (See Section 27–7 and also Sections 27–3 and 27–8.)

3. Imagine the following Young's double-slit experiment using matter rather than light: electrons are accelerated through a potential difference of 12 V, pass through two closely spaced slits separated by a distance d, and create an interference pattern. (a) Using Example 27–11 and Section 24–3 as guides, find the required value for d if the first-order interference fringe is to be produced at an angle of 10°. (b) Given the approximate size of atoms, would it be possible to construct the required two-slit set-up for this experiment?

4. Does each of the following support the wave nature or the particle nature of light? (a) The existence of the cutoff frequency in the photoelectric effect; (b) Young's double-slit experiment; (c) the shift in the photon frequency in Compton scattering; (d) the diffraction of light.

5. (a) From Sections 22–3, 24–4, and 27–3, estimate the minimum energy (eV) that initiates the chemical process on the retina responsible for vision. (b) Estimate the threshold photon energy above which the eye registers no sensation of sight.

6. (a) A rubidium atom ($m = 85$ u) is at rest with one electron in an excited energy level. When the electron jumps to the ground state, the atom emits a photon of wavelength $\lambda = 780$ nm. Determine the resulting (nonrelativistic) recoil speed v of the atom. (b) The recoil speed sets the lower limit on the temperature to which an ideal gas of rubidium atoms can be cooled in a laser-based **atom trap**. Using the kinetic theory of gases (Chapter 13), estimate this "lowest achievable" temperature.

7. Suppose a particle of mass m is confined to a one-dimensional box of width L. According to quantum theory, the particle's wave (with $\lambda = h/mv$) is a standing wave with nodes at the edges of the box. (a) Show the possible modes of vibration on a diagram. (b) Show that the kinetic energy of the particle has quantized energies given by $\text{KE} = n^2h^2/8mL^2$, where n is an integer. (c) Calculate the ground-state energy ($n = 1$) for an electron confined to a box of width 0.50×10^{-10} m. (d) What is the ground-state energy, and speed, of a baseball ($m = 140$ g) in a box 0.65 m wide? (e) An electron confined to a box has a ground-state energy of 22 eV. What is the width of the box? [*Hint*: See Sections 27–8, 27–13, and 11–12.]

ANSWERS TO EXERCISES

A: $\lambda_p = 725$ nm, so red.
B: More 1000-nm photons (each has lower energy).
C: 5.50×10^{14} Hz, 545 nm.
D: Only λ.
E: Decrease.
F: (e).

A neon tube is a thin glass tube, moldable into various shapes, filled with neon (or other) gas that glows with a particular color when a current at high voltage passes through it. Gas atoms, excited to upper energy levels, jump down to lower energy levels and emit light (photons) whose wavelengths (color) are characteristic of the type of gas.

In this Chapter we study what quantum mechanics tells us about atoms, their energy levels, and the effect of the exclusion principle for atoms with more than one electron. We also discuss interesting applications such as lasers and holography.

Quantum Mechanics of Atoms

CHAPTER 28

CONTENTS

28–1 Quantum Mechanics—A New Theory

28–2 The Wave Function and Its Interpretation; the Double-Slit Experiment

28–3 The Heisenberg Uncertainty Principle

28–4 Philosophic Implications; Probability versus Determinism

28–5 Quantum-Mechanical View of Atoms

28–6 Quantum Mechanics of the Hydrogen Atom; Quantum Numbers

28–7 Multielectron Atoms; the Exclusion Principle

28–8 The Periodic Table of Elements

*28–9 X-Ray Spectra and Atomic Number

*28–10 Fluorescence and Phosphorescence

28–11 Lasers

*28–12 Holography

CHAPTER-OPENING QUESTION—Guess now!

The uncertainty principle states that
 (a) no measurement can be perfect because it is technologically impossible to make perfect measuring instruments.
 (b) it is impossible to measure exactly where a particle is, unless it is at rest.
 (c) it is impossible to simultaneously know both the position and the momentum of a particle with complete certainty.
 (d) a particle cannot actually have a completely certain value of momentum.

Bohr's model of the atom gave us a first (though rough) picture of what an atom is like. It proposed explanations for why there is emission and absorption of light by atoms at only certain wavelengths. The wavelengths of the line spectra and the ionization energy for hydrogen (and one-electron ions) are in excellent agreement with experiment. But the Bohr model had important limitations. It was not able to predict line spectra for more complex atoms—atoms with more than one electron—not even for the neutral helium atom, which has only two electrons. Nor could it explain why emission lines, when viewed with great precision, consist of two or more very closely spaced lines (referred to as *fine structure*). The Bohr model also did not explain why some spectral lines were brighter than others. And it could not explain the bonding of atoms in molecules or in solids and liquids.

From a theoretical point of view, too, the Bohr model was not satisfactory: it was a strange mixture of classical and quantum ideas. Moreover, the wave–particle duality was not really resolved.

803

FIGURE 28–1 Erwin Schrödinger with Lise Meitner. (She was a codiscoverer of nuclear fission, Chapter 31.)

FIGURE 28–2 Werner Heisenberg (center) on Lake Como (Italy) with Enrico Fermi (left) and Wolfgang Pauli (right).

We mention these limitations of the Bohr model not to disparage it—for it was a landmark in the history of science. Rather, we mention them to show why, in the early 1920s, it became increasingly evident that a new, more comprehensive theory was needed. It was not long in coming. Less than two years after de Broglie gave us his matter–wave hypothesis, Erwin Schrödinger (1887–1961; Fig. 28–1) and Werner Heisenberg (1901–1976; Fig. 28–2) independently developed a new comprehensive theory.

28–1 Quantum Mechanics—A New Theory

The new theory, called **quantum mechanics**, has been extremely successful. It unifies the wave–particle duality into a single consistent theory and has successfully dealt with the spectra emitted by complex atoms, even the fine details. It explains the relative brightness of spectral lines and how atoms form molecules. It is also a much more general theory that covers all quantum phenomena from blackbody radiation to atoms and molecules. It has explained a wide range of natural phenomena and from its predictions many new practical devices have become possible. Indeed, it has been so successful that it is accepted today by nearly all physicists as the fundamental theory underlying physical processes.

Quantum mechanics deals mainly with the microscopic world of atoms and light. But this new theory, when it is applied to macroscopic phenomena, must be able to produce the old classical laws. This, the **correspondence principle** (already mentioned in Section 27–12), is satisfied fully by quantum mechanics.

This doesn't mean we should throw away classical theories such as Newton's laws. In the everyday world, classical laws are far easier to apply and they give sufficiently accurate descriptions. But when we deal with high speeds, close to the speed of light, we must use the theory of relativity; and when we deal with the tiny world of the atom, we use quantum mechanics.

Although we won't go into the detailed mathematics of quantum mechanics, we will discuss the main ideas and how they involve the wave and particle properties of matter to explain atomic structure and other applications.

28–2 The Wave Function and Its Interpretation; the Double-Slit Experiment

The important properties of any wave are its wavelength, frequency, and amplitude. For an electromagnetic wave, the frequency (or wavelength) determines whether the light is in the visible spectrum or not, and if so, what color it is. We also have seen that the frequency is a measure of the energy of the corresponding photon, $E = hf$ (Eq. 27–4). The amplitude or displacement of an electromagnetic wave at any point is the strength of the electric (or magnetic) field at that point, and is related to the intensity of the wave (the brightness of the light).

For material particles such as electrons, quantum mechanics relates the wavelength to momentum according to de Broglie's formula, $\lambda = h/p$, Eq. 27–8. But what corresponds to the *amplitude* or *displacement* of a matter wave? The amplitude of an electromagnetic wave is represented by the electric and magnetic fields, E and B. In quantum mechanics, this role is played by the **wave function**, which is given the symbol Ψ (the Greek capital letter psi, pronounced "sigh"). Thus Ψ represents the wave displacement, as a function of time and position, of a new kind of field which we might call a "matter" field or a matter wave.

To understand how to interpret the wave function Ψ, we make an analogy with light using the wave–particle duality.

We saw in Chapter 11 that the intensity I of any wave is proportional to the square of the amplitude. This holds true for light waves as well, as we saw in Chapter 22. That is,

$$I \propto E^2,$$

where E is the electric field strength. From the *particle* point of view, the intensity of a light beam (of given frequency) is proportional to the number of photons, N, that pass through a given area per unit time. The more photons there are, the greater the intensity. Thus

$$I \propto E^2 \propto N.$$

This proportion can be turned around so that we have

$$N \propto E^2.$$

That is, the number of photons (striking a page of this book, say) is proportional to the square of the electric field strength.

If the light beam is very weak, only a few photons will be involved. Indeed, it is possible to "build up" a photograph in a camera using very weak light so the effect of photons arriving can be seen. If we are dealing with only one photon, the relationship above $\left(N \propto E^2\right)$ can be interpreted in a slightly different way. At any point, the square of the electric field strength E^2 is a measure of the *probability* that a photon will be at that location. At points where E^2 is large, there is a high probability the photon will be there; where E^2 is small, the probability is low.

We can interpret matter waves in the same way, as was first suggested by Max Born (1882–1970) in 1927. The wave function Ψ may vary in magnitude from point to point in space and time. If Ψ describes a collection of many electrons, then Ψ^2 at any point will be proportional to the number of electrons expected to be found at that point. When dealing with small numbers of electrons we can't make very exact predictions, so Ψ^2 takes on the character of a probability. If Ψ, which depends on time and position, represents a single electron (say, in an atom), then Ψ^2 is interpreted like this: Ψ^2 *at a certain point in space and time represents the probability of finding the electron at the given position and time.* Thus Ψ^2 is often referred to as the **probability density** or **probability distribution**.

Double-Slit Interference Experiment for Electrons

To understand this better, we take as a thought experiment the familiar double-slit experiment, and consider it both for light and for electrons.

Consider two slits whose size and separation are on the order of the wavelength of whatever we direct at them, either light or electrons, Fig. 28–3. We know very well what would happen in this case for light, since this is just Young's double-slit experiment (Section 24–3): an interference pattern would be seen on the screen behind. If light were replaced by electrons with wavelength comparable to the slit size, they too would produce an interference pattern (recall Fig. 27–12). In the case of light, the pattern would be visible to the eye or could be recorded on film, semiconductor sensor, or screen. For electrons, a fluorescent screen could be used (it glows where an electron strikes).

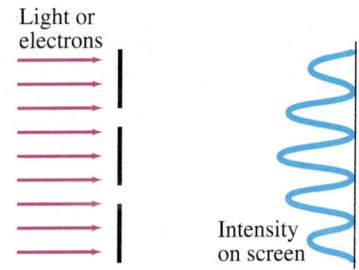

FIGURE 28–3 Parallel beam, of light or electrons, falls on two slits whose sizes are comparable to the wavelength. An interference pattern is observed.

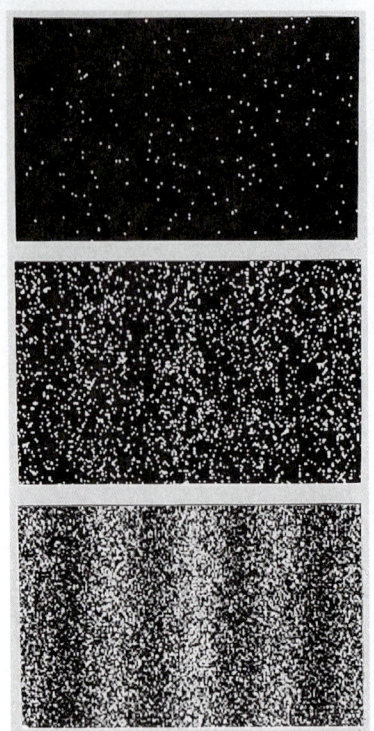

FIGURE 28–4 Young's double-slit experiment done with electrons—note that the pattern is not evident with only a few electrons (top photo), but with more and more electrons (second and third photos), the familiar double-slit interference pattern (Chapter 24) is seen.

If we reduced the flow of electrons (or photons) so they passed through the slits one at a time, we would see a flash each time one struck the screen. At first, the flashes would seem random. Indeed, there is no way to predict just where any one electron would hit the screen. If we let the experiment run for a long time, and kept track of where each electron hit the screen, we would soon see a pattern emerging—the interference pattern predicted by the wave theory; see Fig. 28–4. Thus, although we could not predict where a given electron would strike the screen, we could predict probabilities. (The same can be said for photons.) The probability, as we saw, is proportional to Ψ^2. Where Ψ^2 is zero, we would get a minimum in the interference pattern. And where Ψ^2 is a maximum, we would get a peak in the interference pattern.

The interference pattern would thus occur even when electrons (or photons) passed through the slits one at a time. So the interference pattern could not arise from the interaction of one electron with another. It is as if an electron passed through both slits at the same time, interfering with itself. This is possible because an electron is not precisely a particle. It is as much a wave as it is a particle, and a wave could travel through both slits at once. But what would happen if we covered one of the slits so we knew that the electron passed through the other slit, and a little later we covered the second slit so the electron had to have passed through the first slit? The result would be that no interference pattern would be seen. We would see, instead, two bright areas (or diffraction patterns) on the screen behind the slits.

If both slits are open, the screen shows an interference pattern as if each electron passed through both slits, like a wave. Yet each electron would make a tiny spot on the screen as if it were a particle.

The main point of this discussion is this: if we treat electrons (and other particles) as if they were waves, then Ψ represents the wave amplitude. If we treat them as particles, then we must treat them on a *probabilistic* basis. The square of the wave function, Ψ^2, gives the probability of finding a given electron at a given point. We cannot predict—or even follow—the path of a single electron precisely through space and time.

28–3 The Heisenberg Uncertainty Principle

Whenever a measurement is made, some uncertainty is always involved. For example, you cannot make an absolutely exact measurement of the length of a table. Even with a measuring stick that has markings 1 mm apart, there will be an inaccuracy of perhaps $\frac{1}{2}$ mm or so. More precise instruments will produce more precise measurements. But there is always some uncertainty involved in a measurement, no matter how good the measuring device. We expect that by using more precise instruments, the uncertainty in a measurement can be made indefinitely small.

But according to quantum mechanics, there is actually a limit to the precision of certain measurements. This limit is not a restriction on how well instruments can be made; rather, it is inherent in nature. It is the result of two factors: the wave–particle duality, and the unavoidable interaction between the thing observed and the observing instrument. Let us look at this in more detail.

To make a measurement on an object without disturbing it, at least a little, is not possible. Consider trying to locate a lost Ping-pong ball in a dark room: you could probe about with your hand or a stick, or you could shine a light and detect the photons reflecting off the ball. When you search with your hand or a stick, you find the ball's position when you touch it, but at the same time you unavoidably bump it, and give it some momentum. Thus you won't know its *future* position. If you search for the Ping-pong ball using light, in order to "see" the ball at least one photon (really, quite a few) must scatter from it, and the reflected photon must enter your eye or some other detector. When a photon strikes an ordinary-sized object, it only slightly alters the motion or position of the object.

But a photon striking a tiny object like an electron transfers enough momentum to greatly change the electron's motion and position in an unpredictable way. The mere act of measuring the position of an object at one time makes our knowledge of its future position imprecise.

Now let us see where the wave–particle duality comes in. Imagine a thought experiment in which we are trying to measure the position of an object, say an electron, with photons, Fig. 28–5. (The arguments would be similar if we were using, instead, an electron microscope.) As we saw in Chapter 25, objects can be seen to a precision at best of about the wavelength of the radiation used due to diffraction. If we want a precise position measurement, we must use a short wavelength. But a short wavelength corresponds to high frequency and large momentum ($p = h/\lambda$); and the more momentum the photons have, the more momentum they can give the object when they strike it. If we use photons of longer wavelength, and correspondingly smaller momentum, the object's motion when struck by the photons will not be affected as much. But the longer wavelength means lower resolution, so the object's position will be less accurately known. Thus the act of observing produces an uncertainty in both the *position* and the *momentum* of the electron. This is the essence of the *uncertainty principle* first enunciated by Heisenberg in 1927.

Quantitatively, we can make an approximate calculation of the magnitude of the uncertainties. If we use light of wavelength λ, the position can be measured at best to a precision of about λ. That is, the uncertainty in the position measurement, Δx, is approximately

$$\Delta x \approx \lambda.$$

Suppose that the object can be detected by a single photon. The photon has a momentum $p_x = h/\lambda$ (Eq. 27–6). When the photon strikes our object, it will give some or all of this momentum to the object, Fig. 28–5. Therefore, the final x momentum of our object will be uncertain in the amount

$$\Delta p_x \approx \frac{h}{\lambda}$$

since we can't tell how much momentum will be transferred. The product of these uncertainties is

$$(\Delta x)(\Delta p_x) \approx (\lambda)\left(\frac{h}{\lambda}\right) \approx h.$$

The uncertainties could be larger than this, depending on the apparatus and the number of photons needed for detection. A more careful mathematical calculation shows the product of the uncertainties as, at best, about

$$(\Delta x)(\Delta p_x) \gtrsim \frac{h}{2\pi}. \tag{28-1}$$

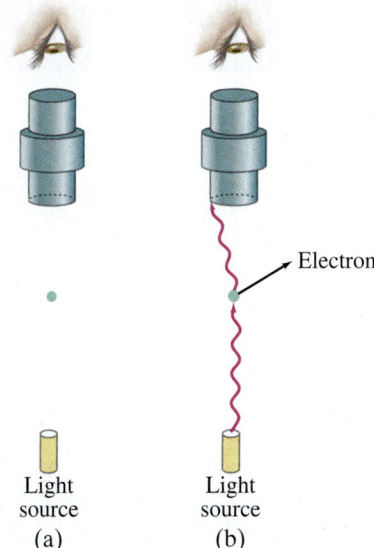

FIGURE 28–5 Thought experiment for observing an electron with a powerful light microscope. At least one photon must scatter from the electron (transferring some momentum to it) and enter the microscope.

UNCERTAINTY PRINCIPLE
(position and momentum)

This is a mathematical statement of the **Heisenberg uncertainty principle**, or, as it is sometimes called, the **indeterminancy principle**. It tells us that we cannot measure both the position *and* momentum of an object precisely at the same time. The more accurately we try to measure the position, so that Δx is small, the greater will be the uncertainty in momentum, Δp_x. If we try to measure the momentum very accurately, then the uncertainty in the position becomes large. The uncertainty principle does not forbid individual precise measurements, however. For example, in principle we could measure the position of an object exactly. But then its momentum would be completely unknown. Thus, although we might know the position of the object exactly at one instant, we could have no idea at all where it would be a moment later. The uncertainties expressed here are inherent in nature, and reflect the best precision theoretically attainable even with the best instruments.

 **CAUTION**
Uncertainties not due to instrument deficiency, but inherent in nature (wave–particle)

SECTION 28–3 The Heisenberg Uncertainty Principle **807**

EXERCISE A Return to the Chapter-Opening Question, page 803, and answer it again now. Try to explain why you may have answered differently the first time.

Another useful form of the uncertainty principle relates energy and time, and we examine this as follows. The object to be detected has an uncertainty in position $\Delta x \approx \lambda$. The photon that detects it travels with speed c, and it takes a time $\Delta t \approx \Delta x/c \approx \lambda/c$ to pass through the distance of uncertainty. Hence, the measured time when our object is at a given position is uncertain by about

$$\Delta t \approx \frac{\lambda}{c}.$$

Since the photon can transfer some or all of its energy ($= hf = hc/\lambda$) to our object, the uncertainty in energy of our object as a result is

$$\Delta E \approx \frac{hc}{\lambda}.$$

The product of these two uncertainties is

$$(\Delta E)(\Delta t) \approx \left(\frac{hc}{\lambda}\right)\left(\frac{\lambda}{c}\right) \approx h.$$

A more careful calculation gives

UNCERTAINTY PRINCIPLE
(energy and time)

$$(\Delta E)(\Delta t) \gtrsim \frac{h}{2\pi}. \quad (28\text{-}2)$$

This form of the uncertainty principle tells us that the energy of an object can be uncertain (or can be interpreted as briefly nonconserved) by an amount ΔE for a time $\Delta t \approx h/(2\pi \Delta E)$.

The quantity $(h/2\pi)$ appears so often in quantum mechanics that for convenience it is given the symbol $\hbar$ ("h-bar"). That is,

$$\hbar = \frac{h}{2\pi} = \frac{6.626 \times 10^{-34} \text{ J} \cdot \text{s}}{2\pi} = 1.055 \times 10^{-34} \text{ J} \cdot \text{s}.$$

By using this notation, Eqs. 28–1 and 28–2 for the uncertainty principle can be written

$$(\Delta x)(\Delta p_x) \gtrsim \hbar$$

and

$$(\Delta E)(\Delta t) \gtrsim \hbar.$$

We have been discussing the position and velocity of an electron as if it were a particle. But it isn't simply a particle. Indeed, we have the uncertainty principle because an electron—and matter in general—has wave as well as particle properties. What the uncertainty principle really tells us is that if we insist on thinking of the electron as a particle, then there are certain limitations on this simplified view—namely, that the position and velocity cannot both be known precisely at the same time; and even that the electron does not *have* a precise position and momentum at the same time (because it is not simply a particle). Similarly, the energy can be uncertain (or nonconserved) by an amount ΔE for a time $\Delta t \approx \hbar/\Delta E$.

Because Planck's constant, h, is so small, the uncertainties expressed in the uncertainty principle are usually negligible on the macroscopic level. But at the level of atomic sizes, the uncertainties are significant. Because we consider ordinary objects to be made up of atoms containing nuclei and electrons, the uncertainty principle is relevant to our understanding of all of nature. The uncertainty principle expresses, perhaps most clearly, the probabilistic nature of quantum mechanics. It thus is often used as a basis for philosophic discussion.

EXAMPLE 28-1 Position uncertainty of electron. An electron moves in a straight line with a constant speed $v = 1.10 \times 10^6$ m/s which has been measured to a precision of 0.10%. What is the maximum precision with which its position could be simultaneously measured?

APPROACH The momentum is $p = mv$, and the uncertainty in p is $\Delta p = 0.0010p$. The uncertainty principle (Eq. 28-1) gives us the smallest uncertainty in position Δx using the equals sign.

SOLUTION The momentum of the electron is

$$p = mv = (9.11 \times 10^{-31}\,\text{kg})(1.10 \times 10^6\,\text{m/s}) = 1.00 \times 10^{-24}\,\text{kg}\cdot\text{m/s}.$$

The uncertainty in the momentum is 0.10% of this, or $\Delta p = 1.0 \times 10^{-27}$ kg·m/s. From the uncertainty principle, the best simultaneous position measurement will have an uncertainty of

$$\Delta x \approx \frac{\hbar}{\Delta p} = \frac{1.055 \times 10^{-34}\,\text{J}\cdot\text{s}}{1.0 \times 10^{-27}\,\text{kg}\cdot\text{m/s}} = 1.1 \times 10^{-7}\,\text{m},$$

or 110 nm.

NOTE This is about 1000 times the diameter of an atom.

EXERCISE B An electron's position is measured with a precision of 0.50×10^{-10} m. Find the minimum uncertainty in its momentum and velocity.

EXAMPLE 28-2 Position uncertainty of a baseball. What is the uncertainty in position, imposed by the uncertainty principle, on a 150-g baseball thrown at (93 ± 2) mi/h $= (42 \pm 1)$ m/s?

APPROACH The uncertainty in the speed is $\Delta v = 1$ m/s. We multiply Δv by m to get Δp and then use the uncertainty principle, solving for Δx.

SOLUTION The uncertainty in the momentum is

$$\Delta p = m\,\Delta v = (0.150\,\text{kg})(1\,\text{m/s}) = 0.15\,\text{kg}\cdot\text{m/s}.$$

Hence the uncertainty in a position measurement could be as small as

$$\Delta x = \frac{\hbar}{\Delta p} = \frac{1.055 \times 10^{-34}\,\text{J}\cdot\text{s}}{0.15\,\text{kg}\cdot\text{m/s}} = 7 \times 10^{-34}\,\text{m}.$$

NOTE This distance is far smaller than any we could imagine observing or measuring. It is trillions of trillions of times smaller than an atom. Indeed, the uncertainty principle sets no relevant limit on measurement for macroscopic objects.

EXAMPLE 28-3 ESTIMATE J/ψ lifetime calculated. The J/ψ meson, discovered in 1974, was measured to have an average mass of 3100 MeV/c^2 (note the use of energy units since $E = mc^2$) and a mass "width" of 63 keV/c^2. By this we mean that the masses of different J/ψ mesons were actually measured to be slightly different from one another. This mass "width" is related to the very short lifetime of the J/ψ before it decays into other particles. From the uncertainty principle, if the particle exists for only a time Δt, its mass (or rest energy) will be uncertain by $\Delta E \approx \hbar/\Delta t$. Estimate the J/$\psi$ lifetime.

APPROACH We use the energy–time version of the uncertainty principle, Eq. 28-2.

SOLUTION The uncertainty of 63 keV/c^2 in the J/ψ's mass is an uncertainty in its rest energy, which in joules is

$$\Delta E = (63 \times 10^3\,\text{eV})(1.60 \times 10^{-19}\,\text{J/eV}) = 1.01 \times 10^{-14}\,\text{J}.$$

Then we expect its lifetime τ ($= \Delta t$ using Eq. 28-2) to be

$$\tau \approx \frac{\hbar}{\Delta E} = \frac{1.055 \times 10^{-34}\,\text{J}\cdot\text{s}}{1.01 \times 10^{-14}\,\text{J}} \approx 1 \times 10^{-20}\,\text{s}.$$

Lifetimes this short are difficult to measure directly, and the assignment of very short lifetimes depends on this use of the uncertainty principle.

28–4 Philosophic Implications; Probability versus Determinism

The classical Newtonian view of the world is a deterministic one (see Section 5–8). One of its basic ideas is that once the position and velocity of an object are known at a particular time, its future position can be predicted if the forces on it are known. For example, if a stone is thrown a number of times with the same initial velocity and angle, and the forces on it remain the same, the path of the projectile will always be the same. If the forces are known (gravity and air resistance, if any), the stone's path can be precisely predicted. This mechanistic view implies that the future unfolding of the universe, assumed to be made up of particulate objects, is completely determined.

This classical deterministic view of the physical world has been radically altered by quantum mechanics. As we saw in the analysis of the double-slit experiment (Section 28–2), electrons all treated in the same way will not all end up in the same place. According to quantum mechanics, certain probabilities exist that an electron will arrive at different points. This is very different from the classical view, in which the path of a particle is precisely predictable from the initial position and velocity and the forces exerted on it. According to quantum mechanics, the position and velocity of an object cannot even be known accurately at the same time. This is expressed in the uncertainty principle, and arises because basic entities, such as electrons, are not considered simply as particles: they have wave properties as well. Quantum mechanics allows us to calculate only the probability[†] that, say, an electron (when thought of as a particle) will be observed at various places. Quantum mechanics says there is some inherent unpredictability in nature. This is very different from the deterministic view of classical mechanics.

Because matter is considered to be made up of atoms, even ordinary-sized objects are expected to be governed by probability, rather than by strict determinism. For example, quantum mechanics predicts a finite (but negligibly small) probability that when you throw a stone, its path might suddenly curve upward instead of following the downward-curved parabola of normal projectile motion. Quantum mechanics predicts with extremely high probability that ordinary objects will behave just as the classical laws of physics predict. But these predictions are considered probabilities, not absolute certainties. The reason that macroscopic objects behave in accordance with classical laws with such high probability is due to the large number of molecules involved: when large numbers of objects are present in a statistical situation, deviations from the average (or most probable) approach zero. It is the average configuration of vast numbers of molecules that follows the so-called fixed laws of classical physics with such high probability, and gives rise to an apparent "determinism." Deviations from classical laws are observed when small numbers of molecules are dealt with. We can say, then, that although there are no precise deterministic laws in quantum mechanics, there are statistical laws based on probability.

It is important to note that there is a difference between the probability imposed by quantum mechanics and that used in the nineteenth century to understand thermodynamics and the behavior of gases in terms of molecules (Chapters 13 and 15). In thermodynamics, probability is used because there are far too many particles to keep track of. But the molecules are still assumed to move and interact in a deterministic way following Newton's laws. Probability in quantum mechanics is quite different; it is seen as *inherent* in nature, and not as a limitation on our abilities to calculate or to measure.

[†]Note that these probabilities can be calculated precisely, just like predictions of probabilities at rolling dice or dealing cards; but they are unlike predictions of probabilities at sporting events, which are only estimates.

The view presented here is the generally accepted one and is called the **Copenhagen interpretation** of quantum mechanics in honor of Niels Bohr's home, since it was largely developed there through discussions between Bohr and other prominent physicists.

Because electrons are not simply particles, they cannot be thought of as following particular paths in space and time. This suggests that a description of matter in space and time may not be completely correct. This deep and far-reaching conclusion has been a lively topic of discussion among philosophers. Perhaps the most important and influential philosopher of quantum mechanics was Bohr. He argued that a space–time description of actual atoms and electrons is not possible. Yet a description of experiments on atoms or electrons must be given in terms of space and time and other concepts familiar to ordinary experience, such as waves and particles. We must not let our *descriptions* of experiments lead us into believing that atoms or electrons themselves actually move in space and time as classical particles.

28–5 Quantum-Mechanical View of Atoms

At the beginning of this Chapter, we discussed the limitations of the Bohr model of atomic structure. Now we examine the quantum-mechanical theory of atoms, which is a far more complete theory than the old Bohr model. Although the Bohr model has been discarded as an accurate description of nature, nonetheless, quantum mechanics reaffirms certain aspects of the older theory, such as that electrons in an atom exist only in discrete states of definite energy, and that a photon of light is emitted (or absorbed) when an electron makes a transition from one state to another. But quantum mechanics is a much deeper theory, and has provided us with a very different view of the atom. According to quantum mechanics, electrons do not exist in well-defined circular orbits as in the Bohr model. Rather, the electron (because of its wave nature) can be thought of as spread out in space as a "**cloud**." The size and shape of the electron cloud can be calculated for a given state of an atom. For the ground state in the hydrogen atom, the electron cloud is spherically symmetric, as shown in Fig. 28–6. The electron cloud at its higher densities roughly indicates the "size" of an atom. But just as a cloud may not have a distinct border, atoms do not have a precise boundary or a well-defined size. Not all electron clouds have a spherical shape, as we shall see later in this Chapter.

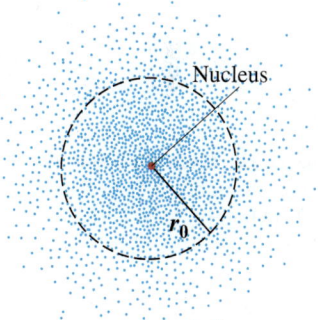

FIGURE 28–6 Electron cloud or "probability distribution" for the ground state of the hydrogen atom, as seen from afar. The dots represent a hypothetical detection of an electron at each point: dots closer together represent more probable presence of an electron (denser cloud). The dashed circle represents the Bohr radius r_0.

The electron cloud can be interpreted from either the particle or the wave viewpoint. Remember that by a particle we mean something that is localized in space—it has a definite position at any given instant. By contrast, a wave is spread out in space. The electron cloud, spread out in space as in Fig. 28–6, is a result of the wave nature of electrons. Electron clouds can also be interpreted as **probability distributions** (or **probability density**) for a particle. As we saw in Section 28–3, we cannot predict the path an electron will follow (thinking of it as a particle). After one measurement of its position we cannot predict exactly where it will be at a later time. We can only calculate the probability that it will be found at different points. If you were to make 500 different measurements of the position of an electron in a hydrogen atom, the majority of the results would show the electron at points where the probability is high (dark area in Fig. 28–6). Only occasionally would the electron be found where the probability is low. The electron cloud or probability distribution becomes small (or thin) at places, especially far away, but never becomes zero. So quantum mechanics suggests that an atom is *not* mostly empty space, and that there is no truly empty space in the universe.

28–6 Quantum Mechanics of the Hydrogen Atom; Quantum Numbers

We now look more closely at what quantum mechanics tells us about the hydrogen atom. Much of what we say here also applies to more complex atoms, which are discussed in the next Section.

Quantum mechanics is a much more sophisticated and successful theory than Bohr's. Yet in a few details they agree. Quantum mechanics predicts the same basic energy levels (Fig. 27–29) for the hydrogen atom as does the Bohr model. That is,

$$E_n = -\frac{13.6 \text{ eV}}{n^2}, \qquad n = 1, 2, 3, \cdots,$$

where n is an integer. In the simple Bohr model, there was only one quantum number, n. In quantum mechanics, four different quantum numbers are needed to specify each state in the atom:

(1) The quantum number, n, from the Bohr model is found also in quantum mechanics and is called the **principal quantum number**. It can have any integer value from 1 to ∞. The total energy of a state in the hydrogen atom depends on n, as we saw above.

(2) The **orbital quantum number**, ℓ, is related to the magnitude of the angular momentum of the electron; ℓ can take on integer values from 0 to $(n-1)$. For the ground state, $n = 1$, ℓ can only be zero.† For $n = 3$, ℓ can be 0, 1, or 2. The actual magnitude of the angular momentum L is related to the quantum number ℓ by

$$L = \sqrt{\ell(\ell+1)}\, \hbar \tag{28–3}$$

(where again $\hbar = h/2\pi$). The value of ℓ has almost no effect on the total energy in the hydrogen atom; only n does to any appreciable extent (but see *fine structure* below). In atoms with two or more electrons, the energy does depend on ℓ as well as n, as we shall see.

(3) The **magnetic quantum number**, m_ℓ, is related to the *direction* of the electron's angular momentum, and it can take on integer values ranging from $-\ell$ to $+\ell$. For example, if $\ell = 2$, then m_ℓ can be $-2, -1, 0, +1$, or $+2$. Since angular momentum is a vector, it is not surprising that both its magnitude and its direction would be quantized. For $\ell = 2$, the five different directions allowed can be represented by the diagram of Fig. 28–7. This limitation on the direction of $\vec{L}$ is often called **space quantization**. In quantum mechanics, the direction of the angular momentum is usually specified by giving its component along the z axis (this choice is arbitrary). Then L_z is related to m_ℓ by the equation

$$L_z = m_\ell \hbar.$$

The values of L_x and L_y are not definite, however. The name for m_ℓ derives not from theory (which relates it to L_z), but from experiment. It was found that when a gas-discharge tube was placed in a magnetic field, the spectral lines were split into several very closely spaced lines. This splitting, known as the **Zeeman effect**, implies that the energy levels must be split (Fig. 28–8), and thus that the energy of a state depends not only on n but also on m_ℓ when a magnetic field is applied—hence the name "magnetic quantum number."

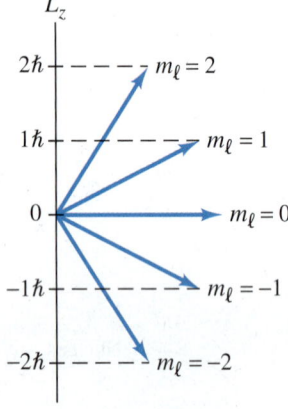

FIGURE 28–7 Quantization of angular momentum direction for $\ell = 2$. (Magnitude of $\vec{L}$ is $L = \sqrt{6}\,\hbar$.)

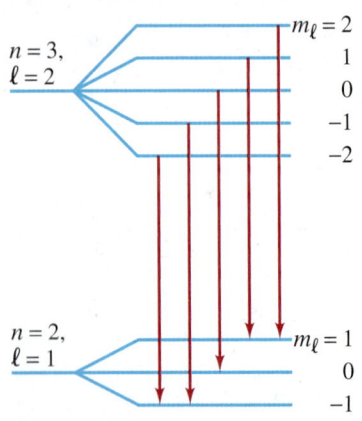

FIGURE 28–8 Energy levels (not to scale). When a magnetic field is applied, the $n = 3$, $\ell = 2$ energy level is split into five separate levels, corresponding to the five values of m_ℓ (2, 1, 0, −1, −2). An $n = 2$, $\ell = 1$ level is split into three levels ($m_\ell = 1, 0, -1$). Transitions can occur between levels (not all transitions are shown), with photons of several slightly different frequencies being given off (the Zeeman effect).

†This replaces Bohr theory, which assigned $\ell = 1$ to the ground state (Eq. 27–11).

(4) Finally, there is the **spin quantum number**, m_s, which for an electron can have only two values, $m_s = +\frac{1}{2}$ and $m_s = -\frac{1}{2}$. The existence of this quantum number did not come out of Schrödinger's original wave theory, as did n, ℓ, and m_ℓ. Instead, a subsequent modification by P. A. M. Dirac (1902–1984) explained its presence as a relativistic effect. The first hint that m_s was needed, however, came from experiment. A careful study of the spectral lines of hydrogen showed that each actually consisted of two (or more) very closely spaced lines even in the absence of an external magnetic field. It was at first hypothesized that this tiny splitting of energy levels, called **fine structure**, was due to angular momentum associated with a spinning of the electron. That is, the electron might spin on its axis as well as orbit the nucleus, just as the Earth spins on its axis as it orbits the Sun. The interaction between the tiny current of the spinning electron could then interact with the magnetic field due to the orbiting charge and cause the small observed splitting of energy levels. (The energy thus depends slightly on m_ℓ and m_s.)† Today we consider the picture of a spinning electron as not legitimate. We cannot even view an electron as a localized object, much less a spinning one. What is important is that the electron can have two different states due to some intrinsic property that behaves like an angular momentum, and we still call this property "spin." The two possible values of m_s ($+\frac{1}{2}$ and $-\frac{1}{2}$) are often said to be "spin up" and "spin down," referring to the two possible directions of the spin angular momentum.

The possible values of the four quantum numbers for an electron in the hydrogen atom are summarized in Table 28–1.

TABLE 28–1 Quantum Numbers for an Electron

Name	Symbol	Possible Values
Principal	n	$1, 2, 3, \cdots, \infty$.
Orbital	ℓ	For a given n: ℓ can be $0, 1, 2, \cdots, n-1$.
Magnetic	m_ℓ	For given n and ℓ: m_ℓ can be $\ell, \ell-1, \cdots, 0, \cdots, -\ell$.
Spin	m_s	For each set of n, ℓ, and m_ℓ: m_s can be $+\frac{1}{2}$ or $-\frac{1}{2}$.

CONCEPTUAL EXAMPLE 28–4 Possible states for $n = 3$. How many different states are possible for an electron with principal quantum number $n = 3$?

RESPONSE For $n = 3$, ℓ can have the values $\ell = 2, 1, 0$. For $\ell = 2$, m_ℓ can be $2, 1, 0, -1, -2$, which is five different possibilities. For each of these, m_s can be either up or down ($+\frac{1}{2}$ or $-\frac{1}{2}$); so for $\ell = 2$, there are $2 \times 5 = 10$ states. For $\ell = 1$, m_ℓ can be $1, 0, -1$, and since m_s can be $+\frac{1}{2}$ or $-\frac{1}{2}$ for each of these, we have 6 more possible states. Finally, for $\ell = 0$, m_ℓ can only be 0, and there are only 2 states corresponding to $m_s = +\frac{1}{2}$ and $-\frac{1}{2}$. The total number of states is $10 + 6 + 2 = 18$, as detailed in the following Table:

n	ℓ	m_ℓ	m_s	n	ℓ	m_ℓ	m_s
3	2	2	$\frac{1}{2}$	3	1	1	$\frac{1}{2}$
3	2	2	$-\frac{1}{2}$	3	1	1	$-\frac{1}{2}$
3	2	1	$\frac{1}{2}$	3	1	0	$\frac{1}{2}$
3	2	1	$-\frac{1}{2}$	3	1	0	$-\frac{1}{2}$
3	2	0	$\frac{1}{2}$	3	1	-1	$\frac{1}{2}$
3	2	0	$-\frac{1}{2}$	3	1	-1	$-\frac{1}{2}$
3	2	-1	$\frac{1}{2}$	3	0	0	$\frac{1}{2}$
3	2	-1	$-\frac{1}{2}$	3	0	0	$-\frac{1}{2}$
3	2	-2	$\frac{1}{2}$				
3	2	-2	$-\frac{1}{2}$				

EXERCISE C An electron has $n = 4$, $\ell = 2$. Which of the following values of m_ℓ are possible: 4, 3, 2, 1, 0, -1, -2, -3, -4?

†Fine structure is said to be due to a **spin–orbit interaction**.

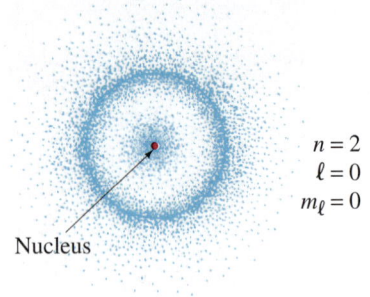

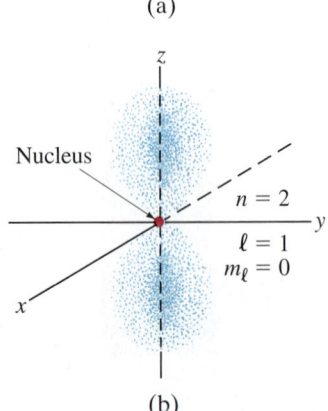

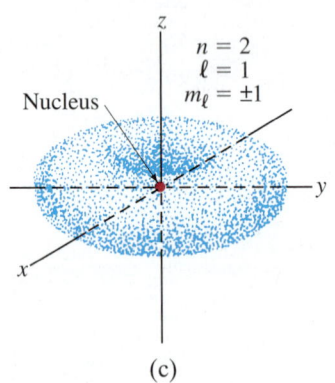

FIGURE 28–9 Electron cloud, or probability distribution, for $n = 2$ states in hydrogen. [The donut-shaped orbit in (c) is the sum of two dumbbell-shaped orbits, as in (b), along the x and y axes added together.]

EXAMPLE 28–5 E **and** L **for** $n = 3$. Determine (a) the energy and (b) the orbital angular momentum for an electron in each of the hydrogen atom states with $n = 3$, as in Example 28–4.

APPROACH The energy of a state depends only on n, except for the very small corrections mentioned above, which we will ignore. Energy is calculated as in the Bohr model, $E_n = -13.6 \text{ eV}/n^2$. For angular momentum we use Eq. 28–3.

SOLUTION (a) Since $n = 3$ for all these states, they all have the same energy,

$$E_3 = -\frac{13.6 \text{ eV}}{(3)^2} = -1.51 \text{ eV}.$$

(b) For $\ell = 0$, Eq. 28–3 gives

$$L = \sqrt{\ell(\ell + 1)}\,\hbar = 0.$$

For $\ell = 1$,

$$L = \sqrt{1(1+1)}\,\hbar = \sqrt{2}\,\hbar = 1.49 \times 10^{-34} \text{ J·s}.$$

For $\ell = 2$, $L = \sqrt{2(2+1)}\,\hbar = \sqrt{6}\,\hbar$.

NOTE Atomic angular momenta are generally given as a multiple of $\hbar$ ($\sqrt{2}\,\hbar$ or $\sqrt{6}\,\hbar$ in this case), rather than in SI units.

EXERCISE D What are the energy and angular momentum of the electron in a hydrogen atom with $n = 6$, $\ell = 4$?

Although ℓ and m_ℓ do not significantly affect the energy levels in hydrogen, they do affect the electron probability distribution in space. For $n = 1$, ℓ and m_ℓ can only be zero and the electron distribution is as shown in Fig. 28–6. For $n = 2$, ℓ can be 0 or 1. The distribution for $n = 2$, $\ell = 0$ is shown in Fig. 28–9a, and it is seen to differ from that for the ground state (Fig. 28–6), although it is still spherically symmetric. For $n = 2$, $\ell = 1$, the distributions are not spherically symmetric as shown in Figs. 28–9b (for $m_\ell = 0$) and 28–9c (for $m_\ell = +1$ or -1).

Although the spatial distributions of the electron can be calculated for the various states, it is difficult to measure them experimentally. Most of the experimental information about atoms has come from a careful examination of the emission spectra under various conditions as in Figs. 27–23 and 24–28.

[Chemists refer to atomic states, and especially the shape in space of their probability distributions, as **orbitals**. Each atomic orbital is characterized by its quantum numbers n, ℓ, and m_ℓ, and can hold one or two electrons ($m_s = +\frac{1}{2}$ or $m_s = -\frac{1}{2}$); s-orbitals ($\ell = 0$) are spherically symmetric, Figs. 28–6 and 28–9a; p-orbitals ($\ell = 1$) can be dumbbell shaped with lobes, Fig. 28–9b, or donut shaped if combining $m_\ell = +1$ and $m_\ell = -1$, Fig. 28–9c.]

Selection Rules: Allowed and Forbidden Transitions

Another prediction of quantum mechanics is that when a photon is emitted or absorbed, transitions can occur only between states with values of ℓ that differ by exactly one unit:

$$\Delta \ell = \pm 1.$$

According to this **selection rule**, an electron in an $\ell = 2$ state can jump only to a state with $\ell = 1$ or $\ell = 3$. It cannot jump to a state with $\ell = 2$ or $\ell = 0$. A transition such as $\ell = 2$ to $\ell = 0$ is called a **forbidden transition**. Actually, such a transition is not absolutely forbidden and can occur, but only with very low probability compared to **allowed transitions**—those that satisfy the selection rule $\Delta \ell = \pm 1$. Since the orbital angular momentum of an H atom must change by one unit when it emits a photon, conservation of angular momentum tells us that the photon must carry off angular momentum. Indeed, experimental evidence of many sorts shows that the photon can be assigned a spin angular momentum of $1\hbar$.

28–7 Multielectron Atoms; the Exclusion Principle

We have discussed the hydrogen atom in detail because it is the simplest to deal with. Now we briefly discuss more complex atoms, those that contain more than one electron. Their energy levels can be determined experimentally from an analysis of their emission spectra. The energy levels are *not* the same as in the H atom, because the electrons interact with each other as well as with the nucleus. Each electron in a complex atom still occupies a particular state characterized by the quantum numbers n, ℓ, m_ℓ, and m_s. For atoms with more than one electron, the energy levels depend on both n and ℓ.

The number of electrons in a neutral atom is called its **atomic number**, Z; Z is also the number of positive charges (protons) in the nucleus, and determines what kind of atom it is. That is, Z determines the fundamental properties that distinguish one type of atom from another.

To understand the possible arrangements of electrons in an atom, a new principle was needed. It was introduced by Wolfgang Pauli (1900–1958; Fig. 28–2) and is called the **Pauli exclusion principle**. It states:

No two electrons in an atom can occupy the same quantum state.

Thus, no two electrons in an atom can have exactly the same set of the quantum numbers n, ℓ, m_ℓ, and m_s. The Pauli exclusion principle forms the basis not only for understanding atoms, but also for understanding molecules and bonding, and other phenomena as well. (See also note at end of this Section.)

Let us now look at the structure of some of the simpler atoms when they are in the ground state. After hydrogen, the next simplest atom is *helium* with two electrons. Both electrons can have $n = 1$, because one can have spin up $(m_s = +\tfrac{1}{2})$ and the other spin down $(m_s = -\tfrac{1}{2})$, thus satisfying the exclusion principle. Since $n = 1$, then ℓ and m_ℓ must be zero (Table 28–1, page 813). Thus the two electrons have the quantum numbers indicated at the top of Table 28–2.

Lithium has three electrons, two of which can have $n = 1$. But the third cannot have $n = 1$ without violating the exclusion principle. Hence the third electron must have $n = 2$. It happens that the $n = 2$, $\ell = 0$ level has a lower energy than $n = 2$, $\ell = 1$. So the electrons in the ground state have the quantum numbers indicated in Table 28–2. The quantum numbers of the third electron could also be, say, $(n, \ell, m_\ell, m_s) = (3, 1, -1, \tfrac{1}{2})$. But the atom in this case would be in an excited state, because it would have greater energy. It would not be long before it jumped to the ground state with the emission of a photon. At room temperature, unless extra energy is supplied (as in a discharge tube), the vast majority of atoms are in the ground state.

We can continue in this way to describe the quantum numbers of each electron in the ground state of larger and larger atoms. The quantum numbers for sodium, with its eleven electrons, are shown in Table 28–2.

EXERCISE E Construct a Table of the ground-state quantum numbers for beryllium, $Z = 4$ (like those in Table 28–2).

Figure 28–10 shows a simple energy level diagram where occupied states are shown as up or down arrows $(m_s = +\tfrac{1}{2}$ or $-\tfrac{1}{2})$, and possible empty states are shown as a small circle.

TABLE 28–2 Ground-State Quantum Numbers

Helium, $Z = 2$

n	ℓ	m_ℓ	m_s
1	0	0	$\tfrac{1}{2}$
1	0	0	$-\tfrac{1}{2}$

Lithium, $Z = 3$

n	ℓ	m_ℓ	m_s
1	0	0	$\tfrac{1}{2}$
1	0	0	$-\tfrac{1}{2}$
2	0	0	$\tfrac{1}{2}$

Sodium, $Z = 11$

n	ℓ	m_ℓ	m_s
1	0	0	$\tfrac{1}{2}$
1	0	0	$-\tfrac{1}{2}$
2	0	0	$\tfrac{1}{2}$
2	0	0	$-\tfrac{1}{2}$
2	1	1	$\tfrac{1}{2}$
2	1	1	$-\tfrac{1}{2}$
2	1	0	$\tfrac{1}{2}$
2	1	0	$-\tfrac{1}{2}$
2	1	-1	$\tfrac{1}{2}$
2	1	-1	$-\tfrac{1}{2}$
3	0	0	$\tfrac{1}{2}$

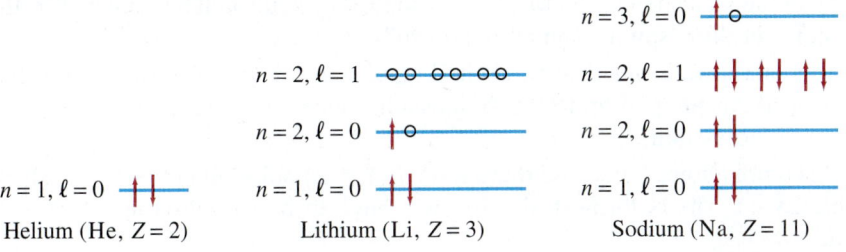

FIGURE 28–10 Energy level diagrams (not to scale) showing occupied states (arrows) and unoccupied states (o) for the ground states of He, Li, and Na. Note that we have shown the $n = 2$, $\ell = 1$ level of Li even though it is empty.

The ground-state configuration for all atoms is given in the **Periodic Table**, which is displayed inside the back cover of this book, and discussed in the next Section.

[The *exclusion principle* applies to identical particles whose spin quantum number is a half-integer ($\frac{1}{2}, \frac{3}{2}$, and so on), including electrons, protons, and neutrons; such particles are called **fermions**, after Enrico Fermi who derived a statistical theory describing them. A basic assumption is that all electrons are **identical**, indistinguishable one from another. Similarly, all protons are identical, all neutrons are identical, and so on. The exclusion principle does not apply to particles with integer spin (0, 1, 2, and so on), such as the photon and π meson, all of which are referred to as **bosons** (after Satyendranath Bose, who derived a statistical theory for them).]

28–8 The Periodic Table of Elements

More than a century ago, Dmitri Mendeleev (1834–1907) arranged the (then) known elements into what we now call the **Periodic Table** of the elements. The atoms were arranged according to increasing mass, but also so that elements with similar chemical properties would fall in the same column. Today's version is shown inside the back cover of this book. Each square contains the atomic number Z, the symbol for the element, and the atomic mass (in atomic mass units). Finally, the lower left corner shows the configuration of the ground state of the atom. This requires some explanation. Electrons with the same value of n are referred to as being in the same **shell**. Electrons with $n = 1$ are in one shell (the K shell), those with $n = 2$ are in a second shell (the L shell), those with $n = 3$ are in the third (M) shell, and so on. Electrons with the same values of n and ℓ are referred to as being in the same **subshell**. Letters are often used to specify the value of ℓ as shown in Table 28–3. That is, $\ell = 0$ is the s subshell; $\ell = 1$ is the p subshell; $\ell = 2$ is the d subshell; beginning with $\ell = 3$, the letters follow the alphabet, f, g, h, i, and so on. (The first letters s, p, d, and f were originally abbreviations of "sharp," "principal," "diffuse," and "fundamental," terms referring to the experimental spectra.)

The Pauli exclusion principle limits the number of electrons possible in each shell and subshell. For any value of ℓ, there are $2\ell + 1$ possible m_ℓ values (m_ℓ can be any integer from 1 to ℓ, from -1 to $-\ell$, or zero), and two possible m_s values. There can be, therefore, at most $2(2\ell + 1)$ electrons in any ℓ subshell. For example, for $\ell = 2$, five m_ℓ values are possible (2, 1, 0, -1, -2), and for each of these, m_s can be $+\frac{1}{2}$ or $-\frac{1}{2}$ for a total of 2(5) = 10 states. Table 28–3 lists the maximum number of electrons that can occupy each subshell.

Because the energy levels depend almost entirely on the values of n and ℓ, it is customary to specify the electron configuration simply by giving the n value and the appropriate letter for ℓ, with the number of electrons in each subshell given as a superscript. The ground-state configuration of sodium, for example, is written as $1s^2 2s^2 2p^6 3s^1$. This is simplified in the Periodic Table by specifying the configuration only of the outermost electrons and any other nonfilled subshells (see Table 28–4 here, and the Periodic Table inside the back cover).

TABLE 28–3 Value of ℓ

Value of ℓ	Letter Symbol	Maximum Number of Electrons in Subshell
0	s	2
1	p	6
2	d	10
3	f	14
4	g	18
5	h	22
⋮	⋮	⋮

TABLE 28–4 Electron Configuration of Some Elements

Z (Number of Electrons)	Element[†]	Ground State Configuration (outer electrons)
1	H	$1s^1$
2	He	$1s^2$
3	Li	$2s^1$
4	Be	$2s^2$
5	B	$2s^2 2p^1$
6	C	$2s^2 2p^2$
7	N	$2s^2 2p^3$
8	O	$2s^2 2p^4$
9	F	$2s^2 2p^5$
10	Ne	$2s^2 2p^6$
11	Na	$3s^1$
12	Mg	$3s^2$
13	Al	$3s^2 3p^1$
14	Si	$3s^2 3p^2$
15	P	$3s^2 3p^3$
16	S	$3s^2 3p^4$
17	Cl	$3s^2 3p^5$
18	Ar	$3s^2 3p^6$
19	K	$3d^0 4s^1$
20	Ca	$3d^0 4s^2$
21	Sc	$3d^1 4s^2$
22	Ti	$3d^2 4s^2$
23	V	$3d^3 4s^2$
24	Cr	$3d^5 4s^1$
25	Mn	$3d^5 4s^2$
26	Fe	$3d^6 4s^2$

[†]Names of elements can be found in Appendix B.

CONCEPTUAL EXAMPLE 28–6 **Electron configurations.** Which of the following electron configurations are possible, and which are not: (a) $1s^2 2s^2 2p^6 3s^3$; (b) $1s^2 2s^2 2p^6 3s^2 3p^5 4s^2$; (c) $1s^2 2s^2 2p^6 2d^1$?

RESPONSE (a) This is not allowed, because too many electrons (three) are shown in the s subshell of the M ($n = 3$) shell. The s subshell has $m_\ell = 0$, with two slots only, for "spin up" and "spin down" electrons.

(b) This is allowed, but it is an excited state. One of the electrons from the $3p$ subshell has jumped up to the $4s$ subshell. Since there are 19 electrons, the element is potassium.

(c) This is not allowed, because there is no d ($\ell = 2$) subshell in the $n = 2$ shell (Table 28–1). The outermost electron will have to be (at least) in the $n = 3$ shell.

EXERCISE F Write the complete ground-state configuration for gallium, with its 31 electrons.

The grouping of atoms in the Periodic Table is according to increasing atomic number, Z. It was designed to also show regularity according to chemical properties. Although this is treated in chemistry textbooks, we discuss it here briefly because it is a result of quantum mechanics. See the Periodic Table inside the back cover.

All the **noble gases** (in column VIII of the Periodic Table) have completely filled shells or subshells. That is, their outermost subshell is completely full, and the electron distribution is spherically symmetric. With such full spherical symmetry, other electrons are not attracted nor are electrons readily lost (ionization energy is high). This is why the noble gases are chemically inert (more on this when we discuss molecules and bonding in Chapter 29). Column VII contains the **halogens**, which lack one electron from a filled shell. Because of the shapes of the orbits (see Section 29–1), an additional electron can be accepted from another atom, and hence these elements are quite reactive. They have a valence of -1, meaning that when an extra electron is acquired, the resulting ion has a net charge of $-1e$. Column I of the Periodic Table contains the **alkali metals**, all of which have a single outer s electron. This electron spends most of its time outside the inner closed shells and subshells which shield it from most of the nuclear charge. Indeed, it is relatively far from the nucleus and is attracted to it by a net charge of only about $+1e$, because of the shielding effect of the other electrons. Hence this outer electron is easily removed and can spend much of its time around another atom, forming a molecule. This is why the alkali metals are very chemically reactive and have a valence of $+1$. The other columns of the Periodic Table can be treated similarly.

The presence of the **transition elements** in the center of the Periodic Table, as well as the lanthanides (rare earths) and actinides below, is a result of incomplete inner shells. For the lowest Z elements, the subshells are filled in a simple order: first $1s$, then $2s$, followed by $2p$, $3s$, and $3p$. You might expect that $3d$ ($n = 3$, $\ell = 2$) would be filled next, but it isn't. Instead, the $4s$ level actually has a slightly lower energy than the $3d$ (due to electrons interacting with each other), so it fills first (K and Ca). Only then does the $3d$ shell start to fill up, beginning with Sc, as can be seen in Table 28–4. (The $4s$ and $3d$ levels are close, so some elements have only one $4s$ electron, such as Cr.) Most of the chemical properties of these transition elements are governed by the relatively loosely held $4s$ electrons, and hence they usually have valences of $+1$ or $+2$. A similar effect is responsible for the *lanthanides* and *actinides*, which are shown at the bottom of the Periodic Table for convenience. All have very similar chemical properties, which are determined by their two outer $6s$ or $7s$ electrons, whereas the different numbers of electrons in the unfilled inner shells have little effect.

⚠️ **CAUTION**
Subshells are not always filled in "order"

*28–9 X-Ray Spectra and Atomic Number

The line spectra of atoms in the visible, UV, and IR regions of the EM spectrum are mainly due to transitions between states of the outer electrons. Much of the positive charge of the nucleus is shielded from these electrons by the negative charge on the inner electrons. But the innermost electrons in the $n = 1$ shell "see" the full charge of the nucleus. Since the energy of a level is proportional to Z^2 (see Eq. 27–15), for an atom with $Z = 50$, we would expect wavelengths about $50^2 = 2500$ times shorter than those found in the Lyman series of hydrogen (around 100 nm), or $(100 \text{ nm})/(2500) \approx 10^{-2}$ to 10^{-1} nm. Such short wavelengths lie in the X-ray region of the spectrum.

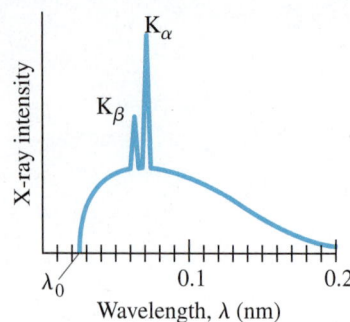

FIGURE 28–11 Spectrum of X-rays emitted from a molybdenum target in an X-ray tube operated at 50 kV.

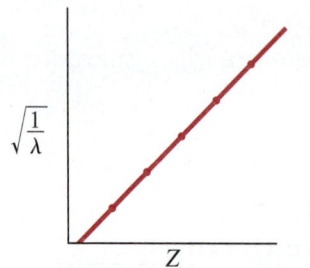

FIGURE 28–12 Plot of $\sqrt{1/\lambda}$ vs. Z for K_α X-ray lines.

X-rays are produced when electrons accelerated by a high voltage strike the metal target inside an X-ray tube (Section 25–11). If we look at the spectrum of wavelengths emitted by an X-ray tube, we see that the spectrum consists of two parts: a continuous spectrum with a cutoff at some λ_0 which depends only on the voltage across the tube, and a series of peaks superimposed. A typical example is shown in Fig. 28–11. The smooth curve and the cutoff wavelength λ_0 move to the left as the voltage across the tube increases. The sharp lines or peaks (labeled K_α and K_β in Fig. 28–11), however, remain at the same wavelength when the voltage is changed, although they are located at different wavelengths when different target materials are used. This observation suggests that the peaks are characteristic of the target material used. Indeed, we can explain the peaks by imagining that the electrons accelerated by the high voltage of the tube can reach sufficient energies that, when they collide with the atoms of the target, they can knock out one of the very tightly held inner electrons. Then we explain these **characteristic X-rays** (the peaks in Fig. 28–11) as photons emitted when an electron in an upper state drops down to fill the vacated lower state. The K lines result from transitions *into* the K shell ($n = 1$). The K_α line consists of photons emitted in a transition that originates from the $n = 2$ (L) shell and drops to the $n = 1$ (K) shell. On the other hand, the K_β line reflects a transition from the $n = 3$ (M) shell down to the K shell. An L line is due to a transition into the L shell, and so on.

Measurement of the characteristic X-ray spectra has allowed a determination of the inner energy levels of atoms. It has also allowed the determination of Z values for many atoms, because (as we have seen) the wavelength of the shortest characteristic X-rays emitted will be inversely proportional to Z^2. Actually, for an electron jumping from, say, the $n = 2$ to the $n = 1$ level (K_α line), the wavelength is inversely proportional to $(Z - 1)^2$ because the nucleus is shielded by the one electron that still remains in the 1s level. In 1914, H. G. J. Moseley (1887–1915) found that a plot of $\sqrt{1/\lambda}$ vs. Z produced a straight line, Fig. 28–12, where λ is the wavelength of the K_α line. The Z values of a number of elements were determined by fitting them to such a **Moseley plot**. The work of Moseley put the concept of atomic number on a firm experimental basis.

> **EXAMPLE 28–7 X-ray wavelength.** Estimate the wavelength for an $n = 2$ to $n = 1$ transition in molybdenum ($Z = 42$). What is the energy of such a photon?
>
> **APPROACH** We use the Bohr formula, Eq. 27–16 for $1/\lambda$, with Z^2 replaced by $(Z - 1)^2 = (41)^2$.
>
> **SOLUTION** Equation 27–16 gives
>
> $$\frac{1}{\lambda} = \left(\frac{2\pi^2 e^4 m k^2}{h^3 c}\right)(Z - 1)^2 \left(\frac{1}{n'^2} - \frac{1}{n^2}\right)$$
>
> where $n = 2$, $n' = 1$, and $k = 8.99 \times 10^9 \, \text{N} \cdot \text{m}^2/\text{c}^2$. We substitute in values:
>
> $$\frac{1}{\lambda} = (1.097 \times 10^7 \, \text{m}^{-1})(41)^2 \left(\frac{1}{1} - \frac{1}{4}\right) = 1.38 \times 10^{10} \, \text{m}^{-1}.$$
>
> So
>
> $$\lambda = \frac{1}{1.38 \times 10^{10} \, \text{m}^{-1}} = 0.072 \, \text{nm}.$$
>
> This is close to the measured value (Fig. 28–11) of 0.071 nm. Each of these photons would have energy (in eV) of:
>
> $$E = hf = \frac{hc}{\lambda} = \frac{(6.63 \times 10^{-34} \, \text{J} \cdot \text{s})(3.00 \times 10^8 \, \text{m/s})}{(7.2 \times 10^{-11} \, \text{m})(1.60 \times 10^{-19} \, \text{J/eV})} = 17 \, \text{keV}.$$
>
> The denominator includes the conversion factor from joules to eV.

EXAMPLE 28–8 Determining atomic number. High-energy electrons are used to bombard an unknown material. The strongest peak is found for X-rays emitted with an energy of 7.5 keV. Guess what the material is.

APPROACH The highest intensity X-rays are generally for the K_α line (see Fig. 28–11) which occurs when high-energy electrons knock out K shell electrons (the innermost orbit, $n = 1$) and their place is taken by electrons from the L shell ($n = 2$). We use the Bohr model, and assume the electrons of the unknown atoms (Z) "see" a nuclear charge of $Z - 1$ (screened by one electron).

SOLUTION The hydrogen transition $n = 2$ to $n = 1$ would yield $E_H = 13.6\,\text{eV} - 3.4\,\text{eV} = 10.2\,\text{eV}$ (see Fig. 27–29 or Example 27–13). Energy of our unknown E_Z is proportional to Z^2 (Eq. 27–15), or rather $(Z - 1)^2$ because the nucleus is shielded by the one electron in a $1s$ state (see above), so we can use ratios:

$$\frac{E_Z}{E_H} = \frac{(Z-1)^2}{1^2} = \frac{7500\,\text{eV}}{10.2\,\text{eV}} = 735,$$

so $Z - 1 = \sqrt{735} = 27$, and $Z = 28$, which makes it cobalt.

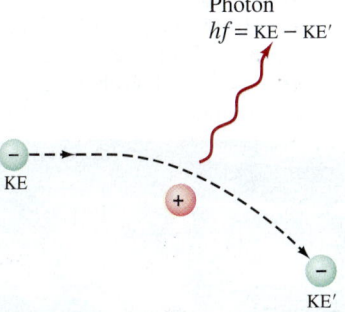

FIGURE 28–13 Bremsstrahlung photon produced by an electron decelerated by interaction with a target atom.

Now we briefly analyze the continuous part of an X-ray spectrum (Fig. 28–11) based on the photon theory of light. When electrons strike the target, they collide with atoms of the material and give up most of their energy as heat (about 99%, so X-ray tubes must be cooled). Electrons can also give up energy by emitting a photon of light: an electron decelerated by interaction with atoms of the target (Fig. 28–13) emits radiation because of its deceleration (Chapter 22), and in this case it is called **bremsstrahlung** (German for "braking radiation"). Because energy is conserved, the energy of the emitted photon, hf, equals the loss of kinetic energy of the electron, $\Delta\text{KE} = \text{KE} - \text{KE}'$, so

$$hf = \Delta\text{KE}.$$

An electron may lose all or a part of its energy in such a collision. The continuous X-ray spectrum (Fig. 28–11) is explained as being due to such bremsstrahlung collisions in which varying amounts of energy are lost by the electrons. The shortest-wavelength X-ray (the highest frequency) must be due to an electron that gives up *all* its kinetic energy to produce one photon in a single collision. Since the initial kinetic energy of an electron is equal to the energy given it by the accelerating voltage, V, then $\text{KE} = eV$. In a single collision in which the electron is brought to rest ($\text{KE}' = 0$), then $\Delta\text{KE} = eV$ and

$$hf_0 = eV.$$

We set $f_0 = c/\lambda_0$ where λ_0 is the cutoff wavelength (Fig. 28–11) and find

$$\lambda_0 = \frac{hc}{eV}. \quad (28\text{–}4)$$

This prediction for λ_0 corresponds precisely with that observed experimentally. This result is further evidence that X-rays are a form of electromagnetic radiation (light) and that the photon theory of light is valid.

EXAMPLE 28–9 Cutoff wavelength. What is the shortest-wavelength X-ray photon emitted in an X-ray tube subjected to 50 kV?

APPROACH The electrons striking the target will have a KE of 50 keV. The shortest-wavelength photons are due to collisions in which all of the electron's KE is given to the photon so $\text{KE} = eV = hf_0$.

SOLUTION From Eq. 28–4,

$$\lambda_0 = \frac{hc}{eV} = \frac{(6.63 \times 10^{-34}\,\text{J}\cdot\text{s})(3.0 \times 10^8\,\text{m/s})}{(1.6 \times 10^{-19}\,\text{C})(5.0 \times 10^4\,\text{V})} = 2.5 \times 10^{-11}\,\text{m},$$

or 0.025 nm.

NOTE This result agrees well with experiment, Fig. 28–11.

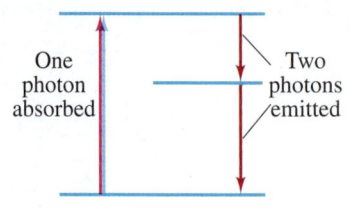

FIGURE 28–14 Fluorescence.

PHYSICS APPLIED
Fluorescence analysis and fluorescent lightbulbs

FIGURE 28–15 When UV light (a range of wavelengths) illuminates these various "fluorescent" rocks, they fluoresce in the visible region of the spectrum.

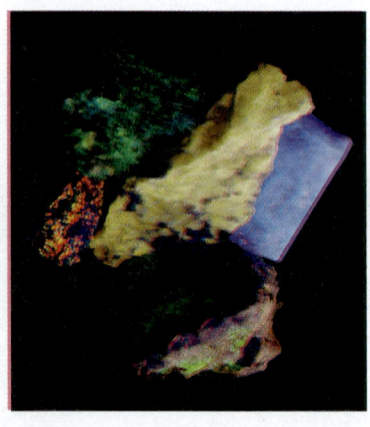

FIGURE 28–16 (a) Absorption of a photon. (b) Stimulated emission. E_u and E_ℓ refer to "upper" and "lower" energy states.

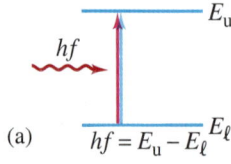

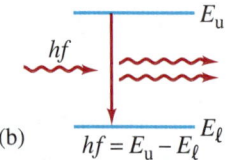

*28–10 Fluorescence and Phosphorescence

When an atom is excited from one energy state to a higher one by the absorption of a photon, it may return to the lower level in a series of two (or more) transitions if there is at least one energy level in between (Fig. 28–14). The photons emitted will consequently have lower energy and frequency than the absorbed photon. When the absorbed photon is in the UV and the emitted photons are in the visible region of the spectrum, this phenomenon is called **fluorescence** (Fig. 28–15).

The wavelength for which fluorescence will occur depends on the energy levels of the particular atoms. Because the frequencies are different for different substances, and because many substances fluoresce readily, fluorescence is a powerful tool for identification of compounds. It is also used for assaying—determining how much of a substance is present—and for following substances along a natural *metabolic pathway* in biological organisms. For detection of a given compound, the stimulating light must be monochromatic, and solvents or other materials present must not fluoresce in the same region of the spectrum. Sometimes the observation of fluorescent light being emitted is sufficient to detect a compound. In other cases, spectrometers are used to measure the wavelengths and intensities of the emitted light.

Fluorescent lightbulbs work in a two-step process. The applied voltage accelerates electrons that strike atoms of the gas in the tube and cause them to be excited. When the excited atoms jump down to their normal levels, they emit UV photons which strike a fluorescent coating on the inside of the tube. The light we see is a result of this material fluorescing in response to the UV light striking it.

Materials such as those used for luminous watch dials, and other glow-in-the-dark products, are said to be **phosphorescent**. When an atom is raised to a normal excited state, it drops back down within about 10^{-8} s. In phosphorescent substances, atoms can be excited by photon absorption to energy levels called **metastable**, which are states that last much longer because to jump down is a "forbidden" transition (Section 28–6). Metastable states can last even a few seconds or longer. In a collection of such atoms, many of the atoms will descend to the lower state fairly soon, but many will remain in the excited state for over an hour. Hence light will be emitted even after long periods. When you put a luminous watch dial close to a bright lamp, many atoms are excited to metastable states, and you can see the glow for a long time afterward.

28–11 Lasers

A **laser** is a device that can produce a very narrow intense beam of monochromatic coherent light. (By **coherent**, we mean that across any cross section of the beam, all parts have the same phase.[†]) The emitted beam is a nearly perfect plane wave. An ordinary light source, on the other hand, emits light in all directions (so the intensity decreases rapidly with distance), and the emitted light is incoherent (the different parts of the beam are not in phase with each other). The excited atoms that emit the light in an ordinary lightbulb act independently, so each photon emitted can be considered as a short wave train lasting about 10^{-8} s. Different wave trains bear no phase relation to one another. Just the opposite is true of lasers.

The action of a laser is based on quantum theory. We have seen that a photon can be absorbed by an atom if (and only if) the photon energy hf corresponds to the energy difference between an occupied energy level of the atom and an available excited state, Fig. 28–16a. If the atom is already in the excited state, it may jump down spontaneously (i.e., no stimulus) to the lower state with the emission of a photon. However, if a photon with this same energy strikes the excited atom, it can stimulate the atom to make the transition sooner to the lower state, Fig. 28–16b. This phenomenon is called **stimulated emission**: not only do we still have the original photon, but also a second one of the same frequency as a result

[†]See also Section 24–3.

of the atom's transition. These two photons are exactly *in phase*, and they are moving in the same direction. This is how coherent light is produced in a laser. The name "laser" is an acronym for **L**ight **A**mplification by **S**timulated **E**mission of **R**adiation.

Normally, most atoms are in the lower state, so the majority of incident photons will be absorbed. To obtain the coherent light from stimulated emission, two conditions must be satisfied. First, the atoms must be excited to the higher state so that an **inverted population** is produced in which more atoms are in the upper state than in the lower one (Fig. 28–17). Then *emission* of photons will dominate over absorption. And second, the higher state must be a **metastable state**—a state in which the electrons remain longer than usual[†] so that the transition to the lower state occurs by stimulated emission rather than spontaneously.

Figure 28–18 is a schematic diagram of a laser: the "lasing" material is placed in a long narrow tube at the ends of which are two mirrors, one of which is partially transparent (transmitting perhaps 1 or 2%). Some of the excited atoms drop down fairly soon after being excited. One of these is the blue atom shown on the far left in Fig. 28–18. If the emitted photon strikes another atom in the excited state, it stimulates this atom to emit a photon of the *same* frequency, moving in the *same* direction, and *in phase* with it. These two photons then move on to strike other atoms causing more stimulated emission. As the process continues, the number of photons multiplies. When the photons strike the end mirrors, most are reflected back, and as they move in the opposite direction, they continue to stimulate more atoms to emit photons. As the photons move back and forth between the mirrors, a small percentage passes through the partially transparent mirror at one end. These photons make up the narrow coherent external laser beam. (Inside the tube, some spontaneously emitted photons will be emitted at an angle to the axis, and these will merely go out the side of the tube and not affect the narrow width of the main beam.)

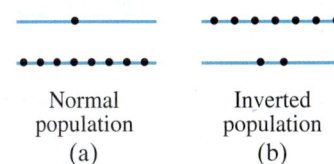

FIGURE 28–17 Two energy levels for a collection of atoms. Each dot represents the energy state of one atom. (a) A normal situation; (b) an inverted population.

⚠️ **CAUTION**
Laser: photons have same frequency and direction, and are in phase

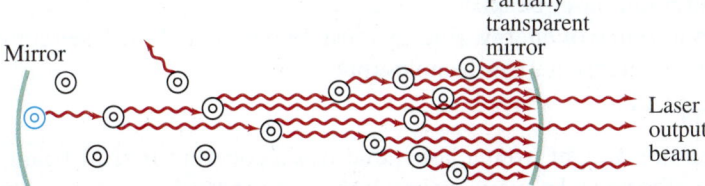

FIGURE 28–18 Laser diagram, showing excited atoms stimulated to emit light.

In a well-designed laser, the spreading of the beam is limited only by diffraction, so the angular spread is $\approx \lambda/D$ (see Eq. 24–3 or 25–7) where D is the diameter of the end mirror. The diffraction spreading can be incredibly small. The light energy, instead of spreading out in space as it does for an ordinary light source, can be a pencil-thin beam.

Creating an Inverted Population

The excitation of the atoms in a laser can be done in several ways to produce the necessary inverted population. In a **ruby laser**, the lasing material is a ruby rod consisting of Al_2O_3 with a small percentage of aluminum (Al) atoms replaced by chromium (Cr) atoms. The Cr atoms are the ones involved in lasing. In a process called **optical pumping**, the atoms are excited by strong flashes of light of wavelength 550 nm, which corresponds to a photon energy of 2.2 eV. As shown in Fig. 28–19, the atoms are excited from state E_0 to state E_2. The atoms quickly decay either back to E_0 or to the intermediate state E_1, which is metastable with a lifetime of about 3×10^{-3} s (compared to 10^{-8} s for ordinary levels). With strong pumping action, more atoms can be found in the E_1 state than are in the E_0 state. Thus we have the inverted population needed for lasing. As soon as a few atoms in the E_1 state jump down to E_0, they emit photons that produce stimulated emission of the other atoms, and the lasing action begins. A ruby laser thus emits a beam whose photons have energy 1.8 eV and a wavelength of 694.3 nm (or "ruby-red" light).

FIGURE 28–19 Energy levels of chromium in a ruby crystal. Photons of energy 2.2 eV "pump" atoms from E_0 to E_2, which then decay to metastable state E_1. Lasing action occurs by stimulated emission of photons in transition from E_1 to E_0.

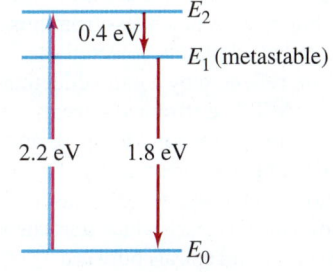

[†]An excited atom may land in such a state and can jump to a lower state only by a so-called forbidden transition (Section 28–6), which is why its lifetime is longer than normal.

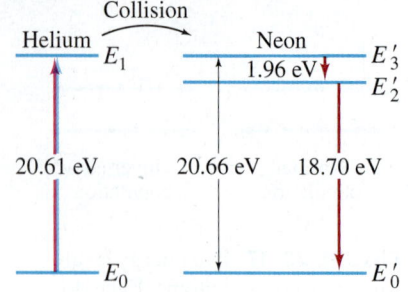

FIGURE 28–20 Energy levels for He and Ne. He is excited in the electric discharge to the E_1 state. This energy is transferred to the E'_3 level of the Ne by collision. E'_3 is metastable and decays to E'_2 by stimulated emission.

In a **helium–neon laser** (He–Ne), the lasing material is a gas, a mixture of about 85% He and 15% Ne. The atoms are excited by applying a high voltage to the tube so that an electric discharge takes place within the gas. In the process, some of the He atoms are raised to the metastable state E_1 shown in Fig. 28–20, which corresponds to a jump of 20.61 eV, almost exactly equal to an excited state in neon, 20.66 eV. The He atoms do not quickly return to the ground state by spontaneous emission, but instead often give their excess energy to a Ne atom when they collide—see Fig. 28–20. In such a collision, the He drops to the ground state and the Ne atom is excited to the state E'_3 (the prime refers to neon states). The slight difference in energy (0.05 eV) is supplied by the kinetic energy of the moving atoms. In this manner, the E'_3 state in Ne—which is metastable—becomes more populated than the E'_2 level. This inverted population between E'_3 and E'_2 is what is needed for lasing.

Very common now are **semiconductor diode lasers**, also called **pn junction lasers**, which utilize an inverted population of electrons between the conduction band and the lower-energy valence band (Section 29–9). When an electron jumps down, a photon can be emitted, which in turn can stimulate another electron to make the transition and emit another photon, in phase. The needed mirrors (as in Fig. 28–18) are made by the polished ends of the *pn* crystal. Semiconductor lasers are used in CD and DVD players (see below), and in many other applications.

Other types of laser include: *chemical lasers*, in which the energy input comes from the chemical reaction of highly reactive gases; *dye lasers*, whose frequency is tunable; CO_2 *gas lasers*, capable of high power output in the infrared; and *rare-earth solid-state lasers* such as the high-power Nd:YAG laser.

The excitation of the atoms in a laser can be done continuously or in pulses. In a **pulsed laser**, the atoms are excited by periodic inputs of energy. In a **continuous laser**, the energy input is continuous: as atoms are stimulated to jump down to the lower level, they are soon excited back up to the upper level so the output is a continuous laser beam.

No laser is a source of energy. Energy must be put in, and the laser converts a part of it into an intense narrow beam output.

Laser is not an energy source

PHYSICS APPLIED
DVD and CD players, bar codes

*Applications

The unique feature of light from a laser, that it is a coherent narrow beam, has found many applications. In everyday life, lasers are used as bar-code readers (at store checkout stands) and in compact disc (CD) and digital video disc (DVD) players. The laser beam reflects off the stripes and spaces of a bar code, or off the tiny pits of a CD or DVD as shown in Fig. 28–21a. The recorded information on a CD or DVD is a series of pits and spaces representing 0s and 1s (or "off" and "on") of a binary code (Section 17–10) that is decoded electronically before being sent to the audio or video system. The laser of a CD player starts reading at the inside of the disc which rotates at about 500 rpm at the start. As the disc rotates, the laser follows the spiral track (Fig. 28–21b), and as it moves outward the disc must slow down because each successive circumference ($C = 2\pi r$) is slightly longer as r increases; at the outer edge, the disc is rotating about 200 rpm. A 1-hour CD has a track roughly 5 km long; the track width is about 1600 nm ($= 1.6\,\mu\text{m}$) and the distance between pits is about 800 nm. DVDs contain much more information.

FIGURE 28–21 (a) Reading a CD (or DVD). The fine beam of a laser, focused even more finely with lenses, is directed at the undersurface of a rotating compact disc. The beam is reflected back from the areas between pits but reflects much less from pits. The reflected light is detected as shown, reflected by a half-reflecting mirror MS. The strong and weak reflections correspond to the 0s and 1s of the binary code representing the audio or video signal. (b) A laser follows the CD track which starts near the center and spirals outward.

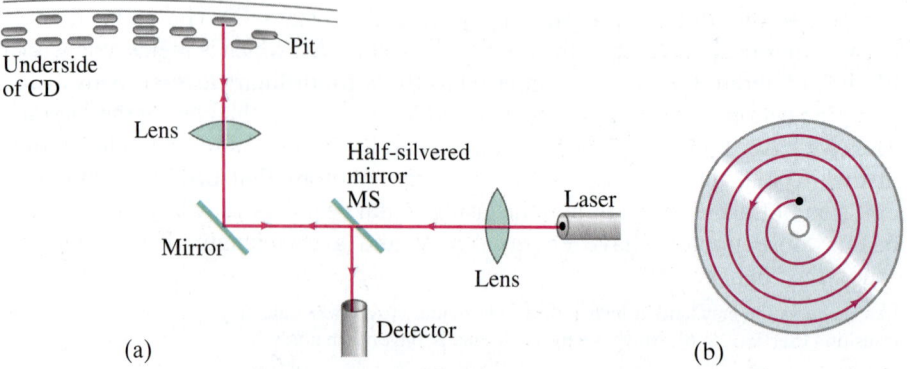

Standard DVDs use a thinner track (0.7 μm) and shorter pit length (400 nm). Blu-ray discs use a "blue" laser with a short wavelength (405 nm) and narrower beam, allowing a narrower track (0.3 μm) that can store much more data for high definition. DVDs can also have two layers, one below the other. When the laser focuses on the second layer, the light passes through the semitransparent surface layer. The second layer may start reading at the outer edge instead of inside. DVDs can also have a single or double layer on *both* surfaces of the disc.

Lasers are a useful surgical tool. The narrow intense beam can be used to destroy tissue in a localized area, or to break up gallstones and kidney stones. Because of the heat produced, a laser beam can be used to "weld" broken tissue, such as a detached retina, Fig. 28–22, or to mold the cornea of the eye (by vaporizing tiny bits of material) to correct myopia and other eye defects (LASIK surgery). The laser beam can be carried by an optical fiber (Section 23–6) to the surgical point, sometimes as an additional fiber-optic path on an endoscope (again Section 23–6). An example is the removal of plaque clogging human arteries. Lasers have been used to destroy tiny organelles within a living cell by researchers studying how the absence of that organelle affects the behavior of the cell. Laser beams are used to destroy cancerous and precancerous cells; and the heat seals off capillaries and lymph vessels, thus "cauterizing" the wound to prevent spread of the disease.

The intense heat produced in a small area by a laser beam is used for welding and machining metals and for drilling tiny holes in hard materials. Because a laser beam is coherent, monochromatic, narrow, and essentially parallel, lenses can be used to focus the light into even smaller areas. The precise straightness of a laser beam is also useful to surveyors for lining up equipment accurately, especially in inaccessible places.

PHYSICS APPLIED
Medical and other uses of lasers

FIGURE 28–22 Laser being used in eye surgery.

*28–12 Holography

One of the most interesting applications of laser light is the production of three-dimensional images called **holograms** (see Fig. 28–23). In an ordinary photograph, the film simply records the intensity of light reaching it at each point. When the photograph or transparency is viewed, light reflecting from it or passing through it gives us a two-dimensional picture. In holography, the images are formed by interference, without lenses. A laser hologram is typically made on a photographic emulsion (film). A broadened laser beam is split into two parts by a half-silvered mirror, Fig. 28–24. One part goes directly to the film; the rest passes to the object to be photographed, from which it is reflected to the film. Light from every point on the object reaches each point on the film, and the interference of the two beams allows the film to record both the intensity and relative phase of the light at each point. It is crucial that the incident light be coherent—that is, in phase at all points—which is why a laser is used. After the film is developed, it is placed again in a laser beam and a three-dimensional image of the object is created. You can walk around such an image and see it from different sides as if it were the original object. Yet, if you try to touch it with your hand, there will be nothing material there.

PHYSICS APPLIED
Holography

FIGURE 28–23 An athlete (race-car driver) puts his hand on, or through, a holographic image of himself.

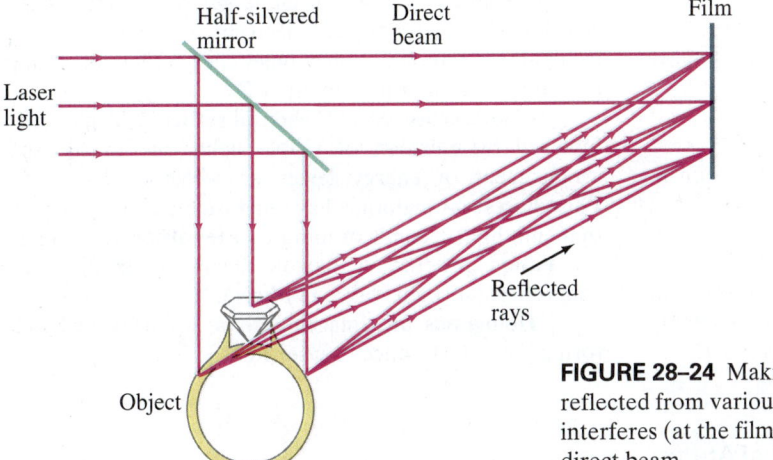

FIGURE 28–24 Making a hologram. Light reflected from various points on the object interferes (at the film) with light from the direct beam.

Volume or **white-light holograms** do not require a laser to see the image, but can be viewed with ordinary white light (preferably a nearly point source, such as the Sun or a clear bulb with a small bright filament). Such holograms must be made, however, with a laser. They are made not on thin film, but on a *thick* emulsion. The interference pattern in the film emulsion can be thought of as an array of bands or ribbons where constructive interference occurred. This array, and the reconstruction of the image, can be compared to Bragg scattering of X-rays from the atoms in a crystal (see Section 25–11). White light can reconstruct the image because the Bragg condition ($m\lambda = 2d\sin\theta$) selects out the appropriate single wavelength. If the hologram is originally produced by lasers emitting the three additive primary colors (red, green, and blue), the three-dimensional image can be seen in full color when viewed with white light.

Summary

In 1925, Schrödinger and Heisenberg separately worked out a new theory, **quantum mechanics**, which is now considered to be the fundamental theory at the atomic level. It is a statistical theory rather than a deterministic one.

An important aspect of quantum mechanics is the Heisenberg **uncertainty principle**. It results from the wave–particle duality and the unavoidable interaction between an observed object and the observer.

One form of the uncertainty principle states that the position x and momentum p_x of an object cannot both be measured precisely at the same time. The products of the uncertainties, $(\Delta x)(\Delta p_x)$, can be no less than $\hbar$ ($= h/2\pi$):

$$(\Delta p_x)(\Delta x) \gtrsim \hbar. \quad (28\text{–}1)$$

Another form of the uncertainty principle states that the energy can be uncertain (or nonconserved) by an amount ΔE for a time Δt, where

$$(\Delta E)(\Delta t) \gtrsim \hbar. \quad (28\text{–}2)$$

According to quantum mechanics, the state of an electron in an atom is specified by four **quantum numbers**: n, ℓ, m_ℓ, and m_s:

(1) n, the **principal quantum number**, can take on any integer value (1, 2, 3, $\cdots$) and corresponds to the quantum number of the old Bohr model;

(2) ℓ, the **orbital quantum number**, can take on values from 0 up to $n - 1$;

(3) m_ℓ, the **magnetic quantum number**, can take on integer values from $-\ell$ to $+\ell$;

(4) m_s, the **spin quantum number**, can be $+\tfrac{1}{2}$ or $-\tfrac{1}{2}$.

The energy levels in the hydrogen atom depend on n, whereas in other atoms they depend on n and ℓ.

The orbital angular momentum of an atom has magnitude $L = \sqrt{\ell(\ell+1)}\,\hbar$ and z component $L_z = m_\ell \hbar$.

When an external magnetic field is applied, the spectral lines are split (the **Zeeman effect**), indicating that the energy depends also on m_ℓ in this case.

Even in the absence of a magnetic field, precise measurements of spectral lines show a tiny splitting of the lines called **fine structure**, whose explanation is that the energy depends very slightly on m_ℓ and m_s.

Transitions between states that obey the **selection rule** $\Delta \ell = \pm 1$ are far more probable than other so-called **forbidden transitions**.

The arrangement of electrons in multi-electron atoms is governed by the Pauli **exclusion principle**, which states that no two electrons can occupy the same quantum state—that is, they cannot have the same set of quantum numbers n, ℓ, m_ℓ, and m_s.

As a result, electrons in multi-electron atoms are grouped into **shells** (according to the value of n) and **subshells** (according to ℓ).

Electron configurations are specified using the numerical values of n, and using letters for ℓ: s, p, d, f, etc., for $\ell = 0, 1, 2, 3$, and so on, plus a superscript for the number of electrons in that subshell. Thus, the ground state of hydrogen is $1s^1$, whereas that for oxygen is $1s^2 2s^2 2p^4$.

In the **Periodic Table**, the elements are arranged in horizontal rows according to increasing **atomic number** (= number of electrons in the neutral atom). The shell structure gives rise to a periodicity in the properties of the elements, so that each vertical column can contain elements with similar chemical properties.

X-rays, which are a form of electromagnetic radiation of very short wavelength, are produced when high-speed electrons strike a target. The spectrum of X-rays so produced consists of two parts, a *continuous* spectrum produced when the electrons are decelerated by atoms of the target, and *peaks* representing photons emitted by atoms of the target after being excited by collision with the high-speed electrons. Measurement of these peaks allows determination of inner energy levels of atoms and determination of atomic number Z.

[*****Fluorescence** occurs when absorbed UV photons are followed by emission of visible light, due to the special arrangement of energy levels of atoms of the material. **Phosphorescent** materials have **metastable** states (long-lived) that emit light seconds or minutes after absorption of light.]

Lasers produce a narrow beam of monochromatic coherent light (light waves *in phase*).

[*****Holograms** are images with a 3-dimensional quality, formed by interference of laser light.]

Questions

1. Compare a matter wave Ψ to (a) a wave on a string, (b) an EM wave. Discuss similarities and differences.
2. Explain why Bohr's theory of the atom is not compatible with quantum mechanics, particularly the uncertainty principle.
3. Explain why it is that the more massive an object is, the easier it becomes to predict its future position.
4. In view of the uncertainty principle, why does a baseball seem to have a well-defined position and speed, whereas an electron does not?
5. Would it ever be possible to balance a very sharp needle precisely on its point? Explain.
6. A cold thermometer is placed in a hot bowl of soup. Will the temperature reading of the thermometer be the same as the temperature of the hot soup before the measurement was made? Explain.
7. Does the uncertainty principle set a limit to how well you can make any single measurement of position? Explain.
8. If you knew the position of an object precisely, with no uncertainty, how well would you know its momentum?
9. When you check the pressure in a tire, doesn't some air inevitably escape? Is it possible to avoid this escape of air altogether? What is the relation to the uncertainty principle?
10. It has been said that the ground-state energy in the hydrogen atom can be precisely known but the excited states have some uncertainty in their values (an "energy width"). Is this consistent with the uncertainty principle in its energy form? Explain.
11. Which model of the hydrogen atom, the Bohr model or the quantum-mechanical model, predicts that the electron spends more time near the nucleus? Explain.
12. The size of atoms varies by only a factor of three or so, from largest to smallest, yet the number of electrons varies from one to over 100. Explain.
13. Excited hydrogen and excited helium atoms both radiate light as they jump down to the $n = 1$, $\ell = 0$, $m_\ell = 0$ state. Why do the two elements have very different emission spectra?
14. How would the Periodic Table look if there were no electron spin but otherwise quantum mechanics were valid? Consider the first 20 elements or so.
15. Which of the following electron configurations are not allowed: (a) $1s^2 2s^2 2p^4 3s^2 4p^2$; (b) $1s^2 2s^2 2p^8 3s^1$; (c) $1s^2 2s^2 2p^6 3s^2 3p^5 4s^2 4d^5 4f^1$? If not allowed, explain why.
16. In what column of the Periodic Table would you expect to find the atom with each of the following configurations: (a) $1s^2 2s^2 2p^5$; (b) $1s^2 2s^2 2p^6 3s^2$; (c) $1s^2 2s^2 2p^6 3s^2 3p^6$; (d) $1s^2 2s^2 2p^6 3s^2 3p^6 4s^1$?
17. Why do chlorine and iodine exhibit similar properties?
18. Explain why potassium and sodium exhibit similar properties.
19. Why are the chemical properties of the rare earths so similar? [Hint: Examine the Periodic Table.]
20. The ionization energy for neon ($Z = 10$) is 21.6 eV, and that for sodium ($Z = 11$) is 5.1 eV. Explain the large difference.
21. Why do we expect electron transitions deep within an atom to produce shorter wavelengths than transitions by outer electrons?
22. Does the Bohr model of the atom violate the uncertainty principle? Explain.
23. Briefly explain why noble gases are nonreactive and why alkali metals are highly reactive. (See Section 28–8.)
24. Compare spontaneous emission to stimulated emission.
25. How does laser light differ from ordinary light? How is it the same?
26. Explain how a 0.0005-W laser beam, photographed at a distance, can seem much stronger than a 1000-W street lamp at the same distance.
27. Does the intensity of light from a laser fall off as the inverse square of the distance? Explain.
*28. Why does the cutoff wavelength in Fig. 28–11 imply a photon nature for light?
*29. Why do we not expect perfect agreement between measured values of characteristic X-ray line wavelengths and those calculated using the Bohr model, as in Example 28–7?
*30. How would you figure out which lines in an X-ray spectrum correspond to K_α, K_β, L, etc., transitions?

MisConceptual Questions

1. An atom has the electron configuration $1s^2 2s^2 2p^6 3s^2 3p^6 4s^1$. How many electrons does this atom have?
   (a) 15. (b) 19. (c) 30. (d) 46.
2. For the electron configuration of MisConceptual Question 1, what orbital quantum numbers do the electrons have?
   (a) 0.
   (b) 0 and 1.
   (c) 0 and 1 and 2.
   (d) 0 and 1 and 2 and 3.
   (e) 0 and 1 and 2 and 3 and 4.
3. If a beam of electrons is fired through a slit,
   (a) the electrons can be deflected because of their wave properties.
   (b) only electrons that hit the edge of the slit are deflected.
   (c) electrons can interact with electromagnetic waves in the slit, forming a diffraction pattern.
   (d) the probability of an electron making it through the slit depends on the uncertainty principle.
4. What is meant by the ground state of an atom?
   (a) All of the quantum numbers have their lowest values ($n = 1$, $\ell = m_\ell = 0$).
   (b) The principal quantum number of the electrons in the outer shell is 1.
   (c) All of the electrons are in the lowest energy state, consistent with the exclusion principle.
   (d) The electrons are in the lowest state allowed by the uncertainty principle.
5. The Pauli exclusion principle applies to all electrons
   (a) in the same shell, but not electrons in different shells.
   (b) in the same container of atoms.
   (c) in the same column of the Periodic Table.
   (d) in incomplete shells.
   (e) in the same atom.

6. Which of the following is the best paraphrasing of the Heisenberg uncertainty principle?
   (a) Only if you know the exact position of a particle can you know the exact momentum of the particle.
   (b) The larger the momentum of a particle, the smaller the position of the particle.
   (c) The more precisely you know the position of a particle, the less well you can know the momentum of the particle.
   (d) The better you know the position of a particle, the better you can know the momentum of the particle.
   (e) How well you can determine the position and momentum of a particle depends on the particle's quantum numbers.

7. Which of the following is required by the Pauli exclusion principle?
   (a) No electron in an atom can have the same set of quantum numbers as any other electron in that atom.
   (b) Each electron in an atom must have the same n value.
   (c) Each electron in an atom must have different m_ℓ values.
   (d) Only two electrons can be in any particular shell of an atom.
   (e) No two electrons in a collection of atoms can have the exact same set of quantum numbers.

8. Under what condition(s) can the exact location and velocity of an electron be measured at the same time?
   (a) The electron is in the ground state of the atom.
   (b) The electron is in an excited state of the atom.
   (c) The electron is free (not bound to an atom).
   (d) Both (a) and (b).
   (e) Never.

9. According to the uncertainty principle,
   (a) there is always an uncertainty in a measurement of the position of a particle.
   (b) there is always an uncertainty in a measurement of the momentum of a particle.
   (c) there is always an uncertainty in a simultaneous measurement of both the position and momentum of a particle.
   (d) All of the above.

10. Which of the following is *not* always a property of lasers?
    (a) All of the photons in laser light have the same phase.
    (b) All laser photons have nearly identical frequencies.
    (c) Laser light moves as a beam, spreading out very slowly.
    (d) Laser light is always brighter than other sources of light.
    (e) Lasers depend on an inverted population of atoms where more atoms occupy a higher energy state than some lower energy state.

For assigned homework and other learning materials, go to the MasteringPhysics website.

Problems

28–2 Wave Function, Double-Slit

1. (II) The neutrons in a parallel beam, each having kinetic energy 0.025 eV, are directed through two slits 0.40 mm apart. How far apart will the interference peaks be on a screen 1.0 m away? [*Hint*: First find the wavelength of the neutron.]

2. (II) Pellets of mass 2.0 g are fired in parallel paths with speeds of 120 m/s through a hole 3.0 mm in diameter. How far from the hole must you be to detect a 1.0-cm-diameter spread in the beam of pellets?

28–3 Uncertainty Principle

3. (I) A proton is traveling with a speed of $(8.660 \pm 0.012) \times 10^5$ m/s. With what maximum precision can its position be ascertained? [*Hint*: $\Delta p = m \Delta v$.]

4. (I) If an electron's position can be measured to a precision of 2.4×10^{-8} m, how precisely can its speed be known?

5. (I) An electron remains in an excited state of an atom for typically 10^{-8} s. What is the minimum uncertainty in the energy of the state (in eV)?

6. (II) The Z^0 boson, discovered in 1985, is the mediator of the weak nuclear force, and it typically decays very quickly. Its average rest energy is 91.19 GeV, but its short lifetime shows up as an intrinsic width of 2.5 GeV. What is the lifetime of this particle? [*Hint*: See Example 28–3.]

7. (II) What is the uncertainty in the mass of a muon ($m = 105.7$ MeV/c^2), specified in eV/c^2, given its lifetime of 2.20 μs?

8. (II) A free neutron ($m = 1.67 \times 10^{-27}$ kg) has a mean life of 880 s. What is the uncertainty in its mass (in kg)?

9. (II) An electron and a 140-g baseball are each traveling 120 m/s measured to a precision of 0.065%. Calculate and compare the uncertainty in position of each.

10. (II) A radioactive element undergoes an alpha decay with a lifetime of 12 μs. If alpha particles are emitted with 5.5-MeV kinetic energy, find the percent uncertainty $\Delta E/E$ in the particle energy.

11. (II) If an electron's position can be measured to a precision of 15 nm, what is the uncertainty in its speed? Assuming the minimum speed must be at least equal to its uncertainty, what is the electron's minimum kinetic energy?

12. (II) Estimate the lowest possible energy of a neutron contained in a typical nucleus of radius 1.2×10^{-15} m. [*Hint*: Assume a particle can have an energy as large as its uncertainty.]

13. (III) How precisely can the position of a 5.00-keV electron be measured assuming its energy is known to 1.00%?

14. (III) Use the uncertainty principle to show that if an electron were present in the nucleus ($r \approx 10^{-15}$ m), its kinetic energy (use relativity) would be hundreds of MeV. (Since such electron energies are not observed, we conclude that electrons are not present in the nucleus.) [*Hint*: Assume a particle can have an energy as large as its uncertainty.]

28–6 to 28–8 Quantum Numbers, Exclusion Principle

15. (I) For $n = 6$, what values can ℓ have?
16. (I) For $n = 6$, $\ell = 3$, what are the possible values of m_ℓ and m_s?
17. (I) How many electrons can be in the $n = 5$, $\ell = 3$ subshell?
18. (I) How many different states are possible for an electron whose principal quantum number is $n = 4$? Write down the quantum numbers for each state.
19. (I) List the quantum numbers for each electron in the ground state of (a) carbon ($Z = 6$), (b) aluminum ($Z = 13$).
20. (I) List the quantum numbers for each electron in the ground state of oxygen ($Z = 8$).
21. (I) Calculate the magnitude of the angular momentum of an electron in the $n = 5$, $\ell = 3$ state of hydrogen.
22. (I) If a hydrogen atom has $\ell = 4$, what are the possible values for n, m_ℓ, and m_s?
23. (II) If a hydrogen atom has $m_\ell = -3$, what are the possible values of n, ℓ, and m_s?
24. (II) Show that there can be 18 electrons in a "g" subshell.
25. (II) What is the full electron configuration in the ground state for elements with Z equal to (a) 26, (b) 34, (c) 38? [Hint: See the Periodic Table inside the back cover.]
26. (II) What is the full electron configuration for (a) silver (Ag), (b) gold (Au), (c) uranium (U)? [Hint: See the Periodic Table inside the back cover.]
27. (II) A hydrogen atom is in the $5d$ state. Determine (a) the principal quantum number, (b) the energy of the state, (c) the orbital angular momentum and its quantum number ℓ, and (d) the possible values for the magnetic quantum number.
28. (II) Estimate the binding energy of the third electron in lithium using the Bohr model. [Hint: This electron has $n = 2$ and "sees" a net charge of approximately $+1e$.] The measured value is 5.36 eV.
29. (II) Show that the total angular momentum is zero for a filled subshell.
30. (II) For each of the following atomic transitions, state whether the transition is allowed or forbidden, and why: (a) $4p \to 3p$; (b) $3p \to 1s$; (c) $4d \to 2d$; (d) $5d \to 3s$; (e) $4s \to 2p$.
31. (II) An electron has $m_\ell = 2$ and is in its lowest possible energy state. What are the values of n and ℓ for this electron?
32. (II) An excited H atom is in a $6d$ state. (a) Name all the states (n, ℓ) to which the atom is "allowed" to make a transition with the emission of a photon. (b) How many different wavelengths are there (ignoring fine structure)?

*28–9 X-Rays

*33. (I) What are the shortest-wavelength X-rays emitted by electrons striking the face of a 28.5-kV TV picture tube? What are the longest wavelengths?
*34. (I) If the shortest-wavelength bremsstrahlung X-rays emitted from an X-ray tube have $\lambda = 0.035$ nm, what is the voltage across the tube?
*35. (I) Show that the cutoff wavelength λ_0 in an X-ray spectrum is given by
$$\lambda_0 = \frac{1240 \text{ nm}}{V},$$
where V is the X-ray tube voltage in volts.
*36. (I) For the spectrum of X-rays emitted from a molybdenum target discussed relative to Fig. 28–11, determine the maximum and minimum energy.
*37. (II) Use the result of Example 28–7 ($Z = 42$) to estimate the X-ray wavelength emitted when a cobalt atom ($Z = 27$) makes a transition from $n = 2$ to $n = 1$.
*38. (II) Estimate the wavelength for an $n = 3$ to $n = 2$ transition in iron ($Z = 26$).
*39. (II) Use the Bohr model to estimate the wavelength for an $n = 3$ to $n = 1$ transition in molybdenum ($Z = 42$). The measured value is 0.063 nm. Why do we not expect perfect agreement?
*40. (II) A mixture of iron and an unknown material is bombarded with electrons. The wavelengths of the K_α lines are 194 pm for iron and 229 pm for the unknown. What is the unknown material?

28–11 Lasers

41. (II) A laser used to weld detached retinas puts out 25-ms-long pulses of 640-nm light which average 0.68-W output during a pulse. How much energy can be deposited per pulse and how many photons does each pulse contain? [Hint: See Example 27–4.]
42. (II) A low-power laser used in a physics lab might have a power of 0.50 mW and a beam diameter of 3.0 mm. Calculate (a) the average light intensity of the laser beam, and (b) compare it to the intensity of a very powerful lightbulb emitting 100-W of light as viewed from 2.0 m.
43. (II) Calculate the wavelength of the He–Ne laser (see Fig. 28–20).
44. (II) Estimate the angular spread of a laser beam due to diffraction if the beam emerges through a 3.0-mm-diameter mirror. Assume that $\lambda = 694$ nm. What would be the diameter of this beam if it struck (a) a satellite 340 km above the Earth, or (b) the Moon? [Hint: See Sections 24–5 and 25–7.]

General Problems

45. The magnitude of the orbital angular momentum in an excited state of hydrogen is 6.84×10^{-34} J·s and the z component is 2.11×10^{-34} J·s. What are all the possible values of n, ℓ, and m_ℓ for this state?
46. An electron in the $n = 2$ state of hydrogen remains there on average about 10^{-8} s before jumping to the $n = 1$ state. (a) Estimate the uncertainty in the energy of the $n = 2$ state. (b) What fraction of the transition energy is this? (c) What is the wavelength, and width (in nm), of this line in the spectrum of hydrogen?
47. What are the largest and smallest possible values for the angular momentum L of an electron in the $n = 6$ shell?
48. A 12-g bullet leaves a rifle at a speed of 150 m/s. (a) What is the wavelength of this bullet? (b) If the position of the bullet is known to a precision of 0.60 cm (radius of the barrel), what is the minimum uncertainty in its momentum?
49. If an electron's position can be measured to a precision of 2.0×10^{-8} m, what is the uncertainty in its momentum? Assuming its momentum must be at least equal to its uncertainty, estimate the electron's wavelength.
50. The ionization (binding) energy of the outermost electron in boron is 8.26 eV. (a) Use the Bohr model to estimate the "effective charge," Z_{eff}, seen by this electron. (b) Estimate the average orbital radius.

51. Using the Bohr formula for the radius of an electron orbit, estimate the average distance from the nucleus for an electron in the innermost ($n = 1$) orbit of a uranium atom ($Z = 92$). Approximately how much energy would be required to remove this innermost electron?

52. Protons are accelerated from rest across 480 V. They are then directed at two slits 0.70 mm apart. How far apart will the interference peaks be on a screen 28 m away?

53. How many electrons can there be in an "h" subshell?

54. (a) Show that the number of different states possible for a given value of ℓ is equal to $2(2\ell + 1)$. (b) What is this number for $\ell = 0, 1, 2, 3, 4, 5,$ and 6?

55. Show that the number of different electron states possible for a given value of n is $2n^2$. (See Problem 54.)

56. A beam of electrons with kinetic energy 45 keV is shot through two narrow slits in a barrier. The slits are a distance 2.0×10^{-6} m apart. If a screen is placed 45.0 cm behind the barrier, calculate the spacing between the "bright" fringes of the interference pattern produced on the screen.

57. The angular momentum in the hydrogen atom is given both by the Bohr model and by quantum mechanics. Compare the results for $n = 2$.

58. The lifetime of a typical excited state in an atom is about 10 ns. Suppose an atom falls from one such excited state to a lower one, and emits a photon of wavelength about 500 nm. Find the fractional energy uncertainty $\Delta E/E$ and wavelength uncertainty $\Delta \lambda/\lambda$ of this photon.

59. A 1300-kg car is traveling with a speed of (22 ± 0.22) m/s. With what maximum precision can its position be determined?

60. An atomic spectrum contains a line with a wavelength centered at 488 nm. Careful measurements show the line is really spread out between 487 and 489 nm. Estimate the lifetime of the excited state that produced this line.

61. An electron and a proton, each initially at rest, are accelerated across the same voltage. Assuming that the uncertainty in their position is given by their de Broglie wavelength, find the ratio of the uncertainty in their momentum.

62. If the principal quantum number n were limited to the range from 1 to 6, how many elements would we find in nature?

63. If your de Broglie wavelength were 0.50 m, how fast would you be moving if your mass is 68.0 kg? Would you notice diffraction effects as you walk through a doorway? Approximately how long would it take you to walk through the doorway?

64. Suppose that the spectrum of an unknown element shows a series of lines with one out of every four matching a line from the Lyman series of hydrogen. Assuming that the unknown element is an ion with Z protons and one electron, determine Z and the element in question.

*65. Photons of wavelength 0.154 nm are emitted from the surface of a certain metal when it is bombarded with high-energy radiation. If this photon wavelength corresponds to the K_α line, what is the element?

Search and Learn

1. Use the uncertainty principle to estimate the position uncertainty for the electron in the ground state of the hydrogen atom. [*Hint*: Determine the momentum using the Bohr model of Section 27–12 and assume the momentum can be anywhere between this value and zero.] How does this result compare to the Bohr radius?

2. On what factors does the periodicity of the Periodic Table depend? Consider the exclusion principle, quantization of angular momentum, spin, and any others you can think of.

3. As discussed in Section 28–5: (a) List two aspects of the Bohr model that the quantum-mechanical theory of the atom retained. (b) Give one major difference between the Bohr model and the quantum-mechanical theory of the atom.

4. Estimate (a) the quantum number ℓ for the orbital angular momentum of the Earth about the Sun, and (b) the number of possible orientations for the plane of Earth's orbit.

5. Show that the diffraction spread of a laser beam, $\approx \lambda/D$ (Section 28–11), is precisely what you might expect from the uncertainty principle. See also Chapters 24 and 25. [*Hint*: Since the beam's width is constrained by the dimension of the aperture D, the component of the light's momentum perpendicular to the laser axis is uncertain.]

6. For noble gases, the halogens, and the alkali metals, explain the atomic structure that is common to each group and how that structure explains a common property of the group. (See Section 28–8.)

7. Imagine a line whose length equals the diameter of the smallest orbit in the Bohr model. If we are told only that an electron is located somewhere on this line, then the electron's position can be specified as $x = 0 \pm r_1$, where the origin is at the line's center and r_1 is the Bohr radius. Such an electron can never be observed at rest, but instead at a minimum will have a speed somewhere in the range from $v = 0 - \Delta v$ to $0 + \Delta v$. Determine Δv.

8. What is uncertain in the Heisenberg uncertainty principle? Explain. (See Section 28–3.)

ANSWERS TO EXERCISES

A: (c).
B: 2.1×10^{-24} kg·m/s, 2.3×10^6 m/s.
C: 2, 1, 0, −1, −2.
D: −0.38 eV, $\sqrt{20}\,\hbar$.
E: Add one line to Li in Table 28–2: 2, 0, 0, $-\frac{1}{2}$.
F: $1s^2 2s^2 2p^6 3s^2 3p^6 3d^{10} 4s^2 4p^1$.

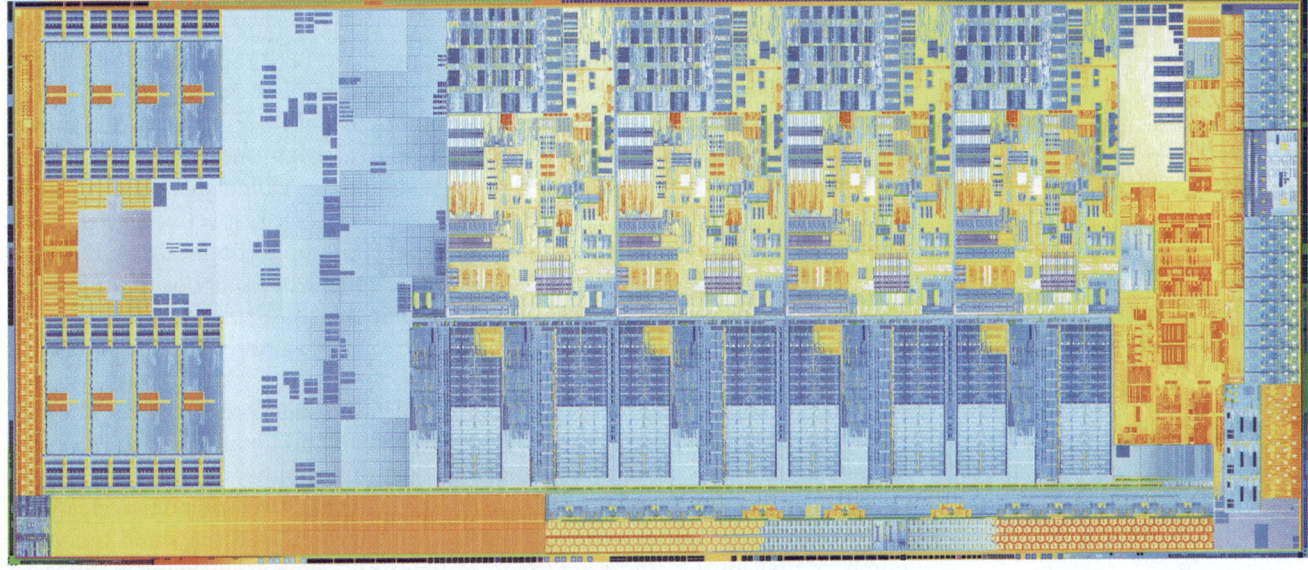

This computer processor chip contains over 1.4 billion transistors, plus diodes and other semiconductor electronic elements, all in a space of about 1 cm². It uses 22-nm technology, meaning the "wires" (conducting lines) are 22 nm wide.

Before discussing semiconductors and their applications, we study the quantum theory description of bonding between atoms to form molecules, and how it explains molecular behavior. We then examine how atoms and molecules form solids, with emphasis on metals as well as on semiconductors and their use in electronics.

Molecules and Solids

CHAPTER-OPENING QUESTION—Guess now!

What holds a solid together?
(a) Gravitational forces.
(b) Magnetic forces.
(c) Electric forces.
(d) Glue.
(e) Nuclear forces.

CONTENTS

*29–1 Bonding in Molecules
*29–2 Potential-Energy Diagrams for Molecules
*29–3 Weak (van der Waals) Bonds
*29–4 Molecular Spectra
*29–5 Bonding in Solids
*29–6 Free-Electron Theory of Metals; Fermi Energy
*29–7 Band Theory of Solids
*29–8 Semiconductors and Doping
*29–9 Semiconductor Diodes, LEDs, OLEDs
*29–10 Transistors: Bipolar and MOSFETs
*29–11 Integrated Circuits, 22-nm Technology

Since its development in the 1920s, quantum mechanics has had a profound influence on our lives, both intellectually and technologically. Even the way we view the world has changed, as we have seen in the last few Chapters. Now we discuss how quantum mechanics has given us an understanding of the structure of molecules and matter in bulk, as well as a number of important applications including semiconductor devices and applications to biology. Semiconductor devices, like transistors, now may be only a few atoms thick, which is the realm of quantum mechanics.

*29–1 Bonding in Molecules

One of the great successes of quantum mechanics was to give scientists, at last, an understanding of the nature of chemical bonds. Because it is based in physics, and because this understanding is so important in many fields, we discuss it here.

By a molecule, we mean a group of two or more atoms that are strongly held together so as to function as a single unit. When atoms make such an attachment, we say that a chemical **bond** has been formed. There are two main types of strong chemical bond: covalent and ionic. Many bonds are actually intermediate between these two types.

*Covalent Bonds

To understand how *covalent bonds* are formed, we take the simplest case, the bond that holds two hydrogen atoms together to form the hydrogen molecule, H_2. The mechanism is basically the same for other covalent bonds. As two H atoms approach each other, the electron clouds begin to overlap, and the electrons from each atom can "orbit" both nuclei. (This is sometimes called **sharing** electrons.) If both electrons are in the ground state ($n = 1$) of their respective atoms, there are two possibilities: their spins can be parallel (both up or both down), in which case the total spin is $S = \frac{1}{2} + \frac{1}{2} = 1$; or their spins can be opposite ($m_s = +\frac{1}{2}$ for one, and $m_s = -\frac{1}{2}$ for the other), so that the total spin $S = 0$. We shall now see that a bond is formed only for the $S = 0$ state, when the spins are opposite.

First we consider the $S = 1$ state, for which the spins are the same. The two electrons cannot both be in the lowest energy state and be attached to the same atom, for then they would have identical quantum numbers in violation of the exclusion principle. The exclusion principle tells us that, because no two electrons can occupy the same quantum state, if two electrons have the same quantum numbers, they must be different in some other way—namely, by being in different places in space (for example, attached to different atoms). Thus, for $S = 1$, when the two atoms approach each other, the electrons will stay away from each other as shown by the probability distribution of Fig. 29–1. The electrons spend very little time between the two nuclei, so the positively charged nuclei repel each other and no bond is formed.

For the $S = 0$ state, on the other hand, the spins are opposite and the two electrons are consequently in different quantum states (m_s is different, $+\frac{1}{2}$ for one, $-\frac{1}{2}$ for the other). Hence the two electrons can come close together, and the probability distribution looks like Fig. 29–2: the electrons can spend much of their time between the two nuclei. The two positively charged nuclei are attracted to the negatively charged electron cloud between them, and this is the attraction that holds the two hydrogen atoms together to form a hydrogen molecule. This is a **covalent bond**.

The probability distributions of Figs. 29–1 and 29–2 can perhaps be better understood on the basis of waves. What the exclusion principle requires is that when the spins are the same, there is destructive interference of the electron wave functions in the region between the two atoms. But when the spins are opposite, constructive interference occurs in the region between the two atoms, resulting in a large amount of negative charge there. Thus a covalent bond can be said to be the result of constructive interference of the electron wave functions in the space between the two atoms, and of the electrostatic attraction of the two positive nuclei for the negative charge concentration between them.

Why a bond is formed can also be understood from the energy point of view. When the two H atoms approach close to one another, if the spins of their electrons are opposite, the electrons can occupy the same space, as discussed above. This means that each electron can now move about in the space of two atoms instead of in the volume of only one. Because each electron now occupies more space, it is less well localized. From the uncertainty principle with Δx larger, we see that Δp and the minimum momentum can be less. With less momentum, each electron has less energy when the two atoms combine than when they are separate. That is, the molecule has less energy than the two separate atoms, and so is more stable. An energy input is required to break the H_2 molecule into two separate H atoms, so the H_2 molecule is a stable entity. This is what we mean by a *bond*. The energy required to break a bond is called the **bond energy**, the **binding energy**, or the **dissociation energy**. For the hydrogen molecule, H_2, the bond energy is 4.5 eV.

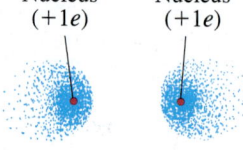

FIGURE 29–1 Electron probability distribution (electron cloud) for two H atoms when the spins are the same: $S = \frac{1}{2} + \frac{1}{2} = 1$.

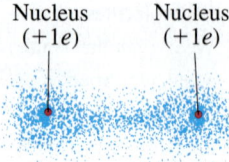

FIGURE 29–2 Electron probability distribution (cloud) around two H atoms when the spins are opposite ($S = 0$): in this case, a bond is formed because the positive nuclei are attracted to the concentration of the electron cloud's negative charge between them. This is a hydrogen molecule, H_2.

*Ionic Bonds

An *ionic bond* is, in a sense, a special case of the covalent bond. Instead of the electrons being shared equally, they are shared unequally. For example, in sodium chloride (NaCl), the outer electron of the sodium spends nearly all its time around the chlorine (Fig. 29–3). The chlorine atom acquires a net negative charge as a result of the extra electron, whereas the sodium atom is left with a net positive charge. The electrostatic attraction between these two charged atoms holds them together. The resulting bond is called an **ionic bond** because it is created by the attraction between the two ions (Na^+ and Cl^-). But to understand the ionic bond, we must understand why the extra electron from the sodium spends so much of its time around the chlorine. After all, the chlorine atom is neutral; why should it attract another electron?

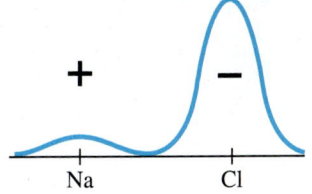

FIGURE 29–3 Probability distribution for the outermost electron of Na in NaCl.

The answer lies in the probability distributions of the electrons in the two neutral atoms. Sodium contains 11 electrons, 10 of which are in spherically symmetric closed shells (Fig. 29–4). The last electron spends most of its time beyond these closed shells. Because the closed shells have a total charge of $-10e$ and the nucleus has charge $+11e$, the outermost electron in sodium "feels" a net attraction due to $+1e$. It is not held very strongly. On the other hand, 12 of chlorine's 17 electrons form closed shells, or subshells (corresponding to $1s^2 2s^2 2p^6 3s^2$). These 12 electrons form a spherically symmetric shield around the nucleus. The other five electrons are in $3p$ states whose probability distributions are not spherically symmetric and have a form similar to those for the $2p$ states in hydrogen shown in Figs. 28–9b and c. Four of these $3p$ electrons can have "doughnut-shaped" distributions symmetric about the z axis, as shown in Fig. 29–5. The fifth can have a "barbell-shaped" distribution (as for $m_\ell = 0$ in Fig. 28–9b), which in Fig. 29–5 is shown only in dashed outline because it is half empty. That is, the exclusion principle allows one more electron to be in this state (it will have spin opposite to that of the electron already there). If an extra electron—say from a Na atom—happens to be in the vicinity, it can be in this state, perhaps at point x in Fig. 29–5. It could experience an attraction due to as much as $+5e$ because the $+17e$ of the nucleus is partly shielded at this point by the 12 inner electrons. Thus, the outer electron of a sodium atom will be more strongly attracted by the $+5e$ of the chlorine atom than by the $+1e$ of its own atom. This, combined with the strong attraction between the two ions when the extra electron stays with the Cl^-, produces the charge distribution of Fig. 29–3, and hence the ionic bond.

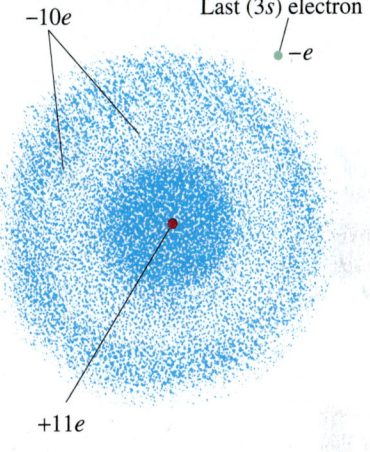

FIGURE 29–4 In a neutral sodium atom, the 10 inner electrons shield the nucleus, so the single outer electron is attracted by a net charge of $+1e$.

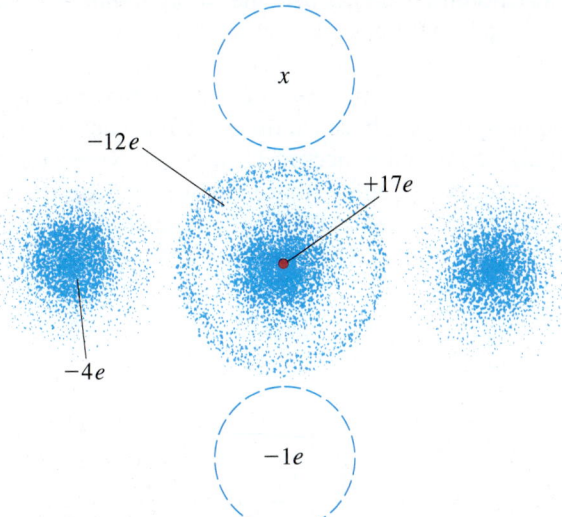

FIGURE 29–5 Neutral chlorine atom. The $+17e$ of the nucleus is shielded by the 12 electrons in the inner shells and subshells. Four of the five $3p$ electrons are shown in doughnut-shaped clouds (seen in cross section at left and right), and the fifth is in the dashed-line cloud concentrated about the z axis (vertical). An extra electron at x will be attracted by a net charge that can be as much as $+5e$.

*Partial Ionic Character of Covalent Bonds

A pure covalent bond in which the electrons are shared equally occurs mainly in symmetrical molecules such as H_2, O_2, and Cl_2. When the atoms involved are different from each other, usually the shared electrons are more likely to be in the vicinity of one atom than the other. The extreme case is an ionic bond. In intermediate cases the *covalent bond* is said to have a **partial ionic character**.

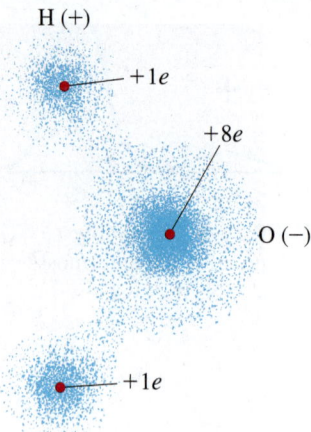

FIGURE 29–6 The water molecule H_2O is polar.

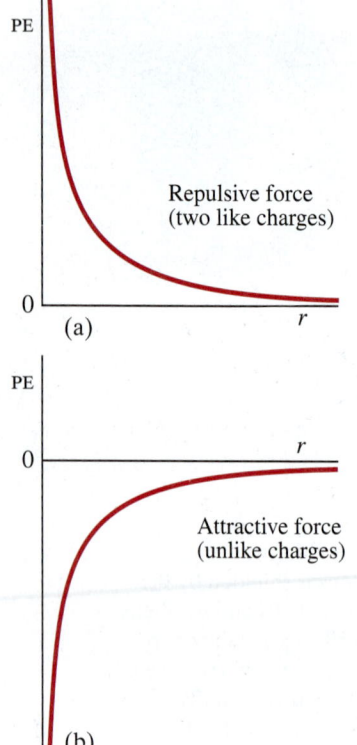

FIGURE 29–7 Potential energy PE as a function of separation r for two point charges of (a) like sign and (b) opposite sign.

The molecules themselves are **polar**—that is, one part (or parts) of the molecule has a net positive charge and other parts a net negative charge. An example is the water molecule, H_2O (Fig. 29–6). Covalent bonds have shared electrons, which in H_2O are more likely to be found around the oxygen atom than around the two hydrogens. The reason is similar to that discussed above in connection with ionic bonds. Oxygen has eight electrons ($1s^2 2s^2 2p^4$), of which four form a spherically symmetric core and the other four could have, for example, a doughnut-shaped distribution. The barbell-shaped distribution on the z axis (like that shown dashed in Fig. 29–5) could be empty, so electrons from hydrogen atoms can be attracted by a net charge of $+4e$. They are also attracted by the H nuclei, so they partly orbit the H atoms as well as the O atom. The net effect is that there is a net positive charge on each H atom (less than $+1e$), because the electrons spend only part of their time there. And, there is a net negative charge on the O atom.

*29–2 Potential-Energy Diagrams for Molecules

It is useful to analyze the interaction between two objects—say, between two atoms or molecules—with the use of a potential-energy diagram, which is a plot of the potential energy versus the separation distance.

For the simple case of two point charges, q_1 and q_2, the potential energy PE is given by (we combine Eqs. 17–2a and 17–5)

$$\text{PE} = k\frac{q_1 q_2}{r},$$

where r is the distance between the charges, and the constant k ($= 1/4\pi\epsilon_0$) is equal to $9.0 \times 10^9 \, \text{N} \cdot \text{m}^2/\text{C}^2$. If the two charges have the same sign, the potential energy is positive for all values of r, and a graph of PE versus r in this case is shown in Fig. 29–7a. The force is repulsive (the charges have the *same* sign) and the curve rises as r decreases; this makes sense because if one particle moves freely toward the other (r getting smaller), the repulsion slows it down so its KE gets smaller, meaning PE gets larger. If, on the other hand, the two charges are of the *opposite* sign, the potential energy is negative because the product $q_1 q_2$ is negative. The force is attractive in this case, and the graph of PE ($\propto -1/r$) versus r looks like Fig. 29–7b. The potential energy becomes more *negative* as r decreases.

Now let us look at the potential-energy diagram for the formation of a covalent bond, such as for the hydrogen molecule, H_2. The potential energy PE of one H atom in the presence of the other is plotted in Fig. 29–8. Starting at large r, the PE decreases as the atoms approach, because the electrons concentrate between the two nuclei (Fig. 29–2), so attraction occurs. However, at very short distances, the electrons would be "squeezed out"—there is no room for them between the two nuclei. Without the electrons between them, each nucleus would feel a repulsive force due to the other, so the curve rises as r decreases further.

FIGURE 29–8 Potential-energy diagram for the H_2 molecule; r is the separation of the two H atoms. The binding energy (the energy difference between PE = 0 and the lowest energy state near the bottom of the well) is 4.5 eV, and $r_0 = 0.074$ nm.

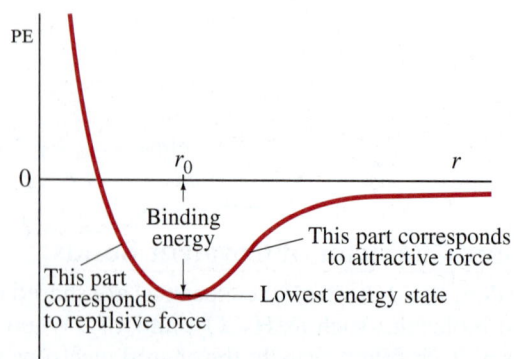

There is an optimum separation of the atoms, r_0 in Fig. 29–8, at which the energy is lowest. This is the point of greatest stability for the hydrogen molecule, and r_0 is the average separation of atoms in the H_2 molecule. The depth of this "well" is the *binding energy*,[†] as shown. This is how much energy must be put into the system to separate the two atoms to infinity, where the PE = 0. For the H_2 molecule, the binding energy is about 4.5 eV and $r_0 = 0.074$ nm.

For many bonds, the potential-energy curve has the shape shown in Fig. 29–9. There is still an optimum distance r_0 at which the molecule is stable. But when the atoms approach from a large distance, the force is initially repulsive rather than attractive. The atoms thus do not form a bond spontaneously. Some additional energy must be injected into the system to get it over the "hump" (or barrier) in the potential-energy diagram. This required energy is called the **activation energy**.

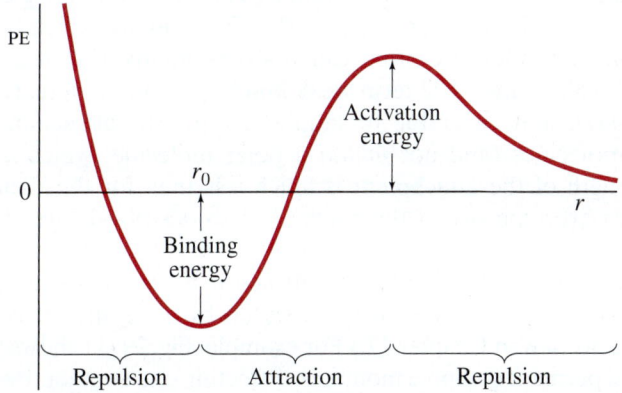

FIGURE 29–9 Potential-energy diagram for a bond requiring an activation energy.

The curve of Fig. 29–9 is much more common than that of Fig. 29–8. The activation energy often reflects a need to break other bonds, before the one under discussion can be made. For example, to make water from O_2 and H_2, the H_2 and O_2 molecules must first be broken into H and O atoms by an input of energy; this is what the activation energy represents. Then the H and O atoms can combine to form H_2O with the release of a great deal more energy than was put in initially. The initial activation energy can be provided by applying an electric spark to a mixture of H_2 and O_2, breaking a few of these molecules into H and O atoms. When these atoms combine to form H_2O, a lot of energy is released (the ground state is near the bottom of the well) which provides the activation energy needed for further reactions: additional H_2 and O_2 molecules are broken up and recombined to form H_2O.

The potential-energy diagrams for ionic bonds, such as NaCl, may be more like Fig. 29–8: the Na^+ and Cl^- ions attract each other at distances a bit larger than some r_0, but at shorter distances the overlapping of inner electron shells gives rise to repulsion. The two atoms thus are most stable at some intermediate separation, r_0. For partially ionic bonds, there is usually an activation energy, Fig. 29–9.

Sometimes the potential energy of a bond looks like that of Fig. 29–10. In this case, the energy of the bonded molecule, at a separation r_0, is greater than when there is no bond ($r = \infty$). That is, an energy *input* is required to make the bond (hence the binding energy is negative), and there is energy release when the bond is broken. Such a bond is stable only because there is the barrier of the activation energy. This type of bond is important in living cells, for it is in such bonds that energy can be stored efficiently in certain molecules, particularly ATP (adenosine triphosphate). The bond that connects the last phosphate group (designated Ⓟ in Fig. 29–10) to the rest of the molecule (ADP, meaning adenosine diphosphate, since it contains only two phosphates) has potential energy of the shape shown in Fig. 29–10. Energy is stored in this bond. When the bond is broken (ATP → ADP + Ⓟ), energy is released and this energy can be used to make other chemical reactions "go."

PHYSICS APPLIED
ATP and energy in the cell

FIGURE 29–10 Potential-energy diagram for the formation of ATP from ADP and phosphate (Ⓟ).

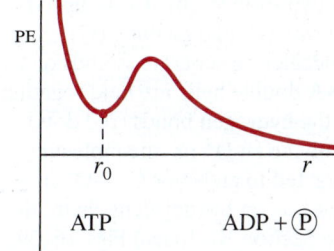

[†]The binding energy corresponds not quite to the bottom of the potential-energy curve, but to the lowest quantum energy state, slightly above the bottom, as shown in Fig. 29–8.

In living cells, many chemical reactions have activation energies that are often on the order of several eV. Such energy barriers are not easy to overcome in the cell. This is where enzymes come in. They act as *catalysts*, which means they act to lower the activation energy so that reactions can occur that otherwise would not. Enzymes act via the electrostatic force to distort the bonding electron clouds, so that the initial bonds are easily broken.

*29–3 Weak (van der Waals) Bonds

Once a bond between two atoms or ions is made, energy must normally be supplied to break the bond and separate the atoms. As mentioned in Section 29–1, this energy is called the *bond energy* or *binding energy*. The binding energy for covalent and ionic bonds is typically 2 to 5 eV. These bonds, which hold atoms together to *form* molecules, are often called **strong bonds** to distinguish them from so-called "weak bonds." The term **weak bond**, as we use it here, refers to an attachment *between* molecules due to simple electrostatic attraction—such as *between* polar molecules (and not *within* a polar molecule, which is a strong bond). The strength of the attachment is much less than for the strong bonds. Binding energies are typically in the range 0.04 to 0.3 eV—hence their name "weak bonds."

Weak bonds are generally the result of attraction between dipoles. (A pair of equal point charges q of opposite sign, separated by a distance ℓ, is called an **electric dipole**, as we saw in Chapter 17.) For example, Fig. 29–11 shows two molecules, which have permanent dipole moments, attracting one another. Besides such **dipole–dipole bonds**, there can also be **dipole–induced dipole bonds**, in which a polar molecule with a permanent dipole moment can induce a dipole moment in an otherwise electrically balanced (nonpolar) molecule, just as a single charge can induce a separation of charge in a nearby object (see Fig. 16–7). There can even be an attraction between two nonpolar molecules, because their electrons are moving about: at any instant there may be a transient separation of charge, creating a brief dipole moment and weak attraction. All these weak bonds are referred to as **van der Waals bonds**, and the forces involved **van der Waals forces**. The potential energy has the general shape shown in Fig. 29–8, with the attractive van der Waals potential energy varying as $1/r^6$. The force decreases greatly with increased distance.

When one of the atoms in a dipole–dipole bond is hydrogen, as in Fig. 29–11, it is called a **hydrogen bond**. A hydrogen bond is generally the strongest of the weak bonds, because the hydrogen atom is the smallest atom and can be approached more closely. Hydrogen bonds also have a partial "covalent" character: that is, electrons between the two dipoles may be shared to a small extent, making a stronger, more lasting bond.

Weak bonds are very important for understanding the activities of cells, such as the double helix shape of DNA (Fig. 29–12), and DNA replication

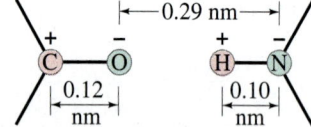

FIGURE 29–11 The C^+—O^- and H^+—N^- dipoles attract each other. (These dipoles may be part of, for example, the nucleotide bases cytosine and guanine in DNA molecules. See Fig. 29–12.) The + and − charges typically have magnitudes of a fraction of e.

PHYSICS APPLIED
DNA

FIGURE 29–12 (a) Model of part of a DNA double helix. The red dots represent hydrogen bonds between the two strands. (b) "Close-up" view: cytosine (C) and guanine (G) molecules on separate strands of a DNA double helix are held together by the hydrogen bonds (red dots) involving an H^+ on one molecule attracted to an N^- or C^+—O^- of a molecule on the adjacent chain. See also Section 16–10 and Figs. 16–39 and 16–40.

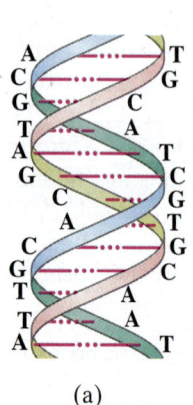

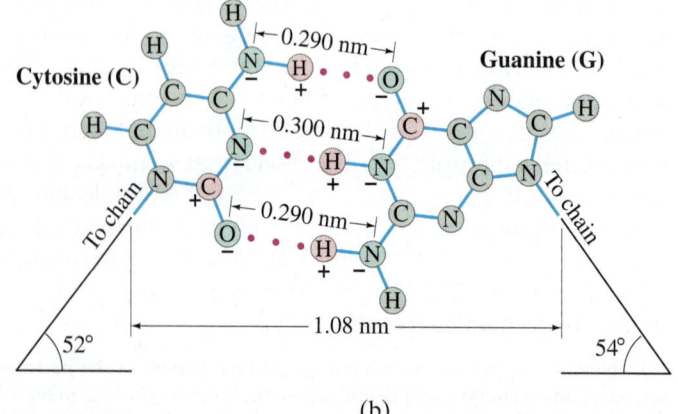

(a)   (b)

(see Section 16–10). The average kinetic energy of molecules in a living cell at normal temperatures ($T \approx 300$ K) is around $\frac{3}{2}kT \approx 0.04$ eV (kinetic theory, Chapter 13), about the magnitude of weak bonds. This means that a weak bond can readily be broken just by a molecular collision. Hence weak bonds are not very permanent—they are, instead, brief attachments. This helps them play particular roles in the cell. On the other hand, strong bonds—those that hold molecules together—are almost never broken simply by molecular collision because their binding energies are much higher (≈ 2 to 5 eV). Thus they are relatively permanent. They can be broken by chemical action (the making of even stronger bonds), and this usually happens in the cell with the aid of an enzyme, which is a protein molecule.

EXAMPLE 29–1 **Nucleotide energy.** Calculate the potential energy between a C^+–O^- dipole of the nucleotide base cytosine and the nearby H^+–N^- dipole of guanine, assuming that the two dipoles are lined up as shown in Fig. 29–11. Dipole moment ($= q\ell$) measurements (see Table 17–2 and Fig. 29–11) give

$$q_H = -q_N = \frac{3.0 \times 10^{-30} \, \text{C} \cdot \text{m}}{0.10 \times 10^{-9} \, \text{m}} = 3.0 \times 10^{-20} \, \text{C} = 0.19e,$$

and

$$q_C = -q_O = \frac{8.0 \times 10^{-30} \, \text{C} \cdot \text{m}}{0.12 \times 10^{-9} \, \text{m}} = 6.7 \times 10^{-20} \, \text{C} = 0.42e.$$

APPROACH We want to find the potential energy of the two charges in one dipole due to the two charges in the other, because this will be equal to the work needed to pull the two dipoles infinitely far apart. The potential energy of a charge q_1 in the presence of a charge q_2 is $\text{PE} = k(q_1 q_2/r_{12})$ where $k = 9.0 \times 10^9 \, \text{N} \cdot \text{m}^2/\text{C}^2$ and r_{12} is the distance between the two charges. (See Eqs. 17–2 and 17–5.)

SOLUTION The potential energy consists of four terms:

$$\text{PE} = \text{PE}_{CH} + \text{PE}_{CN} + \text{PE}_{OH} + \text{PE}_{ON}$$

where PE_{CH} means the potential energy of C in the presence of H, and similarly for the other terms. We do not have terms corresponding to C and O, or N and H, because the two dipoles are assumed to be stable entities. Then, using the distances shown in Fig. 29–11, we get:

$$\text{PE} = k \left[\frac{q_C q_H}{r_{CH}} + \frac{q_C q_N}{r_{CN}} + \frac{q_O q_H}{r_{OH}} + \frac{q_O q_N}{r_{ON}} \right]$$

$$= (9.0 \times 10^9 \, \text{N} \cdot \text{m}^2/\text{C}^2)(6.7 \times 10^{-20} \, \text{C})(3.0 \times 10^{-20} \, \text{C}) \left(\frac{1}{r_{CH}} - \frac{1}{r_{CN}} - \frac{1}{r_{OH}} + \frac{1}{r_{ON}} \right)$$

$$= (9.0 \times 10^9 \, \text{N} \cdot \text{m}^2/\text{C}^2)(6.7)(3.0) \frac{(10^{-20} \, \text{C})^2}{(10^{-9} \, \text{m})} \left(\frac{1}{0.31} - \frac{1}{0.41} - \frac{1}{0.19} + \frac{1}{0.29} \right)$$

$$= -1.86 \times 10^{-20} \, \text{J} = -0.12 \, \text{eV}.$$

The potential energy is negative, meaning 0.12 eV of work (or energy input) is required to separate the dipoles. That is, the binding energy of this "weak" or hydrogen bond is 0.12 eV. This is only an estimate, of course, since other charges in the vicinity would have an influence too.

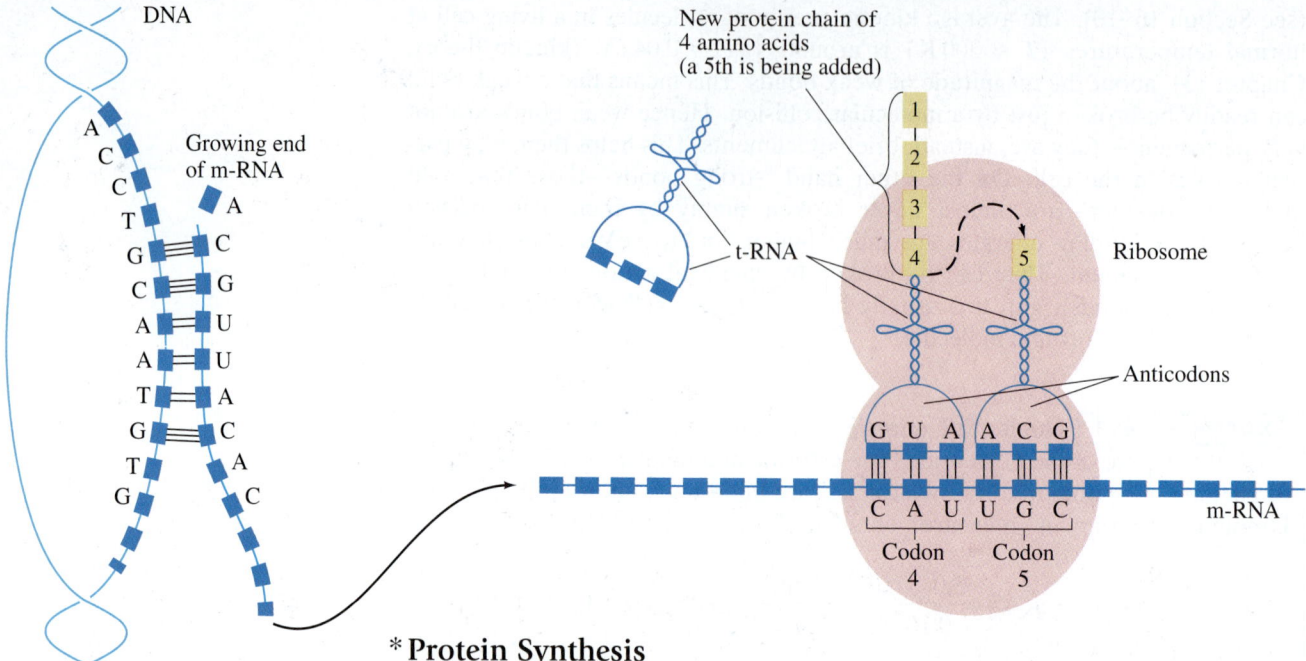

FIGURE 29–13 Protein synthesis. The yellow rectangles represent amino acids. See text for details.

PHYSICS APPLIED
Protein synthesis

*Protein Synthesis

Weak bonds, especially hydrogen bonds, are crucial to the process of protein synthesis. Proteins serve as structural parts of the cell and as enzymes to catalyze chemical reactions needed for the growth and survival of the organism. A protein molecule consists of one or more chains of small molecules known as *amino acids*. There are 20 different amino acids, and a single protein chain may contain hundreds of them in a specific order. The standard model for how amino acids are connected together in the correct order to form a protein molecule is shown schematically in Fig. 29–13.

We begin at the DNA double helix: each gene on a chromosome contains the information for producing one protein. The ordering of the four bases, A, C, G, and T, provides the "code," the **genetic code**, for the order of amino acids in the protein. First, the DNA double helix unwinds and a new molecule called *messenger*-RNA (m-RNA) is synthesized using one strand of the DNA as a "template." m-RNA is a chain molecule containing four different bases, like those of DNA (Section 16–10) except that thymine (T) is replaced by the similar uracil molecule (U). Near the top left in Fig. 29–13, a C has just been added to the growing m-RNA chain in much the same way that DNA replicates (Fig. 16–40); and an A, attracted and held close to the T on the DNA chain by the electrostatic force, will soon be attached to the C by an enzyme. The order of the bases, and thus the genetic information, is preserved in the m-RNA because the shapes of the molecules only allow the "proper" one to get close enough so the electrostatic force can act to form weak bonds.

Next, the m-RNA is buffeted about in the cell (recall kinetic theory, Chapter 13) until it gets close to a tiny organelle known as a *ribosome*, to which it can attach by electrostatic attraction (on the right in Fig. 29–13), because their shapes allow the charged parts to get close enough to form weak bonds. (Recall that force decreases greatly with separation distance.) Also held by the electrostatic force to the ribosome are one or two *transfer*-RNA (t-RNA) molecules. These t-RNA molecules "translate" the genetic code of nucleotide bases into amino acids in the following way. There is a different t-RNA molecule for each amino acid and each combination of three bases. On one end of a t-RNA molecule is an amino acid. On the other end of the t-RNA molecule is the appropriate "anticodon," a set of three nucleotide bases that "code" for that amino acid. If all three bases of an anticodon match the three bases of the "codon" on the m-RNA (in the sense of G to C and A to U), the anticodon is attracted electrostatically to the m-RNA codon and that t-RNA molecule is held there briefly. The ribosome has two particular attachment sites which hold two t-RNA molecules while enzymes bond the two amino acids together to lengthen the amino acid chain (yellow in Fig. 29–13). As each amino acid is connected by an enzyme (four are already connected in Fig. 29–13, top right, and a fifth is about to be connected), the old t-RNA molecule is removed—perhaps by a random collision with some molecule in the cellular fluid. A new one soon becomes attracted as the ribosome moves along the m-RNA.

This process of protein synthesis is often presented as if it occurred in clockwork fashion—as if each molecule knew its role and went to its assigned place. But this is not the case. The forces of attraction between the electric charges of the molecules are rather weak and become significant only when the molecules can come close together, and when several weak bonds can be made. Indeed, if the shapes are not just right, the electrostatic attraction is nearly zero, which is why there are few mistakes. The fact that weak bonds are weak is very important. If they were strong, collisions with other molecules would not allow a t-RNA molecule to be released from the ribosome, or the m-RNA to be released from the DNA. If they were not temporary encounters, metabolism would grind to a halt.

As each amino acid is added to the next, the protein molecule grows in length until it is complete. Even as it is being made, this chain is being buffeted about in the cell—we might think of a wiggling worm. But a protein molecule has electrically charged polar groups along its length. And as it takes on various shapes, the electric forces of attraction between different parts of the molecule will eventually lead to a particular shape of the protein which is quite stable. Each type of protein has its own special shape, depending on the location of charged atoms. In the last analysis, the final shape depends on the order of the amino acids.

*29–4 Molecular Spectra

When atoms combine to form molecules, the probability distributions of the outer electrons overlap and this interaction alters the energy levels. Nonetheless, molecules can undergo transitions between electron energy levels just as atoms do. For example, the H_2 molecule can absorb a photon of just the right frequency to excite one of its ground-state electrons to an excited state. The excited electron can then return to the ground state, emitting a photon. The energy of photons emitted by molecules can be of the same order of magnitude as for atoms, typically 1 to 10 eV, or less.

Additional energy levels become possible for molecules (but not for atoms) because the molecule as a whole can rotate, and the atoms of the molecule can vibrate relative to each other. The energy levels for both rotational and vibrational levels are quantized, and are generally spaced much more closely (10^{-3} to 10^{-1} eV) than the electronic levels. Each atomic energy level thus becomes a set of closely spaced levels corresponding to the vibrational and rotational motions, Fig. 29–14. Transitions from one level to another appear as many very closely spaced lines. In fact, the lines are not always distinguishable, and these spectra are called **band spectra**. Each type of molecule has its own characteristic spectrum, which can be used for identification and for determination of structure. We now look in more detail at rotational and vibrational states in molecules.

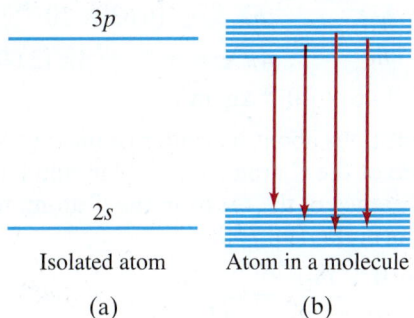

FIGURE 29–14 (a) The individual energy levels of an isolated atom become (b) bands of closely spaced levels in molecules, as well as in solids and liquids.

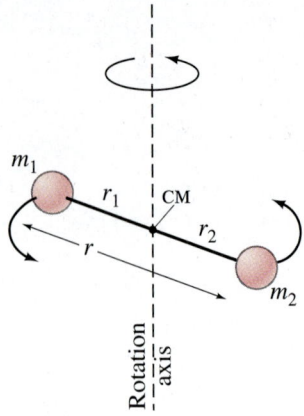

FIGURE 29–15 Diatomic molecule rotating about a vertical axis.

FIGURE 29–16 Rotational energy levels and allowed transitions (emission and absorption) for a diatomic molecule. Upward-pointing arrows represent absorption of a photon, and downward arrows represent emission of a photon.

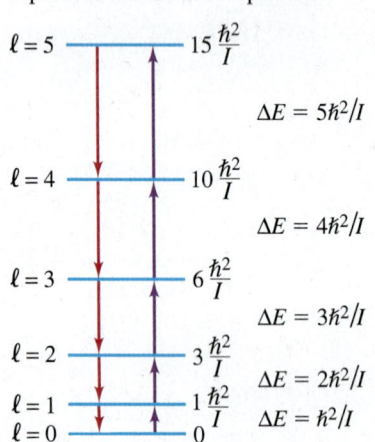

*Rotational Energy Levels in Molecules

We consider only diatomic molecules, although the analysis can be extended to polyatomic molecules. When a diatomic molecule rotates about its center of mass as shown in Fig. 29–15, its kinetic energy of rotation (see Section 8–7) is

$$E_{\text{rot}} = \frac{1}{2}I\omega^2 = \frac{(I\omega)^2}{2I},$$

where $I\omega$ is the angular momentum (Section 8–8). Quantum mechanics predicts quantization of angular momentum just as in atoms (see Eq. 28–3):

$$I\omega = \sqrt{\ell(\ell+1)}\,\hbar, \qquad \ell = 0, 1, 2, \cdots,$$

where ℓ is an integer called the **rotational angular momentum quantum number**. Thus the rotational energy is quantized:

$$E_{\text{rot}} = \frac{(I\omega)^2}{2I} = \ell(\ell+1)\frac{\hbar^2}{2I}. \qquad \ell = 0, 1, 2, \cdots. \quad (29\text{–}1)$$

Transitions between rotational energy levels are subject to the **selection rule** (as in Section 28–6):

$$\Delta\ell = \pm 1.$$

The energy of a photon emitted or absorbed for a transition between rotational states with angular momentum quantum number ℓ and $\ell - 1$ will be

$$\Delta E_{\text{rot}} = E_\ell - E_{\ell-1} = \frac{\hbar^2}{2I}\ell(\ell+1) - \frac{\hbar^2}{2I}(\ell-1)(\ell)$$

$$= \frac{\hbar^2}{I}\ell. \qquad \begin{bmatrix}\ell \text{ is for upper} \\ \text{energy state}\end{bmatrix} \quad (29\text{–}2)$$

We see that the transition energy increases directly with ℓ. Figure 29–16 shows some of the allowed rotational energy levels and transitions. Measured absorption lines fall in the microwave or far-infrared regions of the spectrum (energies $\approx 10^{-3}$ eV), and their frequencies are generally $2, 3, 4, \cdots$ times higher than the lowest one, as predicted by Eq. 29–2.

EXERCISE A Determine the three lowest rotational energy states (in eV) for a nitrogen molecule which has a moment of inertia $I = 1.39 \times 10^{-46}$ kg·m².

EXAMPLE 29–2 Rotational transition. A rotational transition $\ell = 1$ to $\ell = 0$ for the molecule CO has a measured absorption wavelength $\lambda_1 = 2.60$ mm (microwave region). Use this to calculate (a) the moment of inertia of the CO molecule, and (b) the CO bond length, r.

APPROACH The absorption wavelength is used to find the energy of the absorbed photon, and we can then calculate the moment of inertia, I, from Eq. 29–2. The moment of inertia is related to the CO separation (bond length r).

SOLUTION (a) The photon energy, $E = hf = hc/\lambda$, equals the rotational energy level difference, ΔE_{rot}. From Eq. 29–2, we can write

$$\frac{\hbar^2}{I}\ell = \Delta E_{\text{rot}} = hf = \frac{hc}{\lambda_1}.$$

With $\ell = 1$ (the upper state) in this case, we solve for I:

$$I = \frac{\hbar^2\ell}{hc}\lambda_1 = \frac{h\lambda_1}{4\pi^2 c} = \frac{(6.63 \times 10^{-34}\,\text{J·s})(2.60 \times 10^{-3}\,\text{m})}{4\pi^2(3.00 \times 10^8\,\text{m/s})}$$

$$= 1.46 \times 10^{-46}\,\text{kg·m}^2.$$

(b) The molecule rotates about its center of mass (CM) as shown in Fig. 29–15. Let m_1 be the mass of the C atom, $m_1 = 12$ u, and let m_2 be the mass of the O, $m_2 = 16$ u. The distance of the CM from the C atom, which is r_1 in Fig. 29–15, is given by the CM formula, Eq. 7–9:

$$r_1 = \frac{0 + m_2 r}{m_1 + m_2} = \frac{16}{12 + 16}r = 0.57r.$$

The O atom is a distance $r_2 = r - r_1 = 0.43r$ from the CM. The moment of inertia of the CO molecule about its CM is then (see Example 8–9)

$$\begin{aligned} I &= m_1 r_1^2 + m_2 r_2^2 \\ &= [(12\,\text{u})(0.57r)^2 + (16\,\text{u})(0.43r)^2][1.66 \times 10^{-27}\,\text{kg/u}] \\ &= (1.14 \times 10^{-26}\,\text{kg})r^2. \end{aligned}$$

We solve for r and use the result of part (a) for I:

$$r = \sqrt{\frac{1.46 \times 10^{-46}\,\text{kg}\cdot\text{m}^2}{1.14 \times 10^{-26}\,\text{kg}}} = 1.13 \times 10^{-10}\,\text{m} = 0.113\,\text{nm} \approx 0.11\,\text{nm}.$$

| **EXERCISE B** What are the wavelengths of the next three rotational transitions for CO?

*Vibrational Energy Levels in Molecules

The potential energy of the two atoms in a typical diatomic molecule has the shape shown in Fig. 29–8 or 29–9, and Fig. 29–17 again shows the PE for the H_2 molecule (solid curve). This PE curve, at least in the vicinity of the equilibrium separation r_0, closely resembles the potential energy of a harmonic oscillator, PE $= \tfrac{1}{2}kx^2$, which is shown superposed in dashed lines. Thus, for small displacements from r_0, each atom experiences a restoring force approximately proportional to the displacement, and the molecule vibrates as a simple harmonic oscillator (SHO)—see Chapter 11. According to quantum mechanics, the possible quantized energy levels are

$$E_{\text{vib}} = (\nu + \tfrac{1}{2})hf, \qquad \nu = 0, 1, 2, \cdots, \qquad (29\text{–}3)$$

where f is the classical frequency (see Chapter 11—f depends on the mass of the atoms and on the bond strength or "stiffness") and ν is an integer called the **vibrational quantum number**. The lowest energy state ($\nu = 0$) is not zero (as for rotation), but has $E = \tfrac{1}{2}hf$. This is called the **zero-point energy**. Higher states have energy $\tfrac{3}{2}hf$, $\tfrac{5}{2}hf$, and so on, as shown in Fig. 29–18. Transitions between vibrational energy levels are subject to the **selection rule**

$$\Delta\nu = \pm 1,$$

so allowed transitions occur only between adjacent states[†], and all give off (or absorb) photons of energy

$$\Delta E_{\text{vib}} = hf. \qquad (29\text{–}4)$$

This is very close to experimental values for small ν. But for higher energies, the PE curve (Fig. 29–17) begins to deviate from a perfect SHO curve, which affects the wavelengths and frequencies of the transitions. Typical transition energies are on the order of 10^{-1} eV, roughly 10 to 100 times larger than for rotational transitions, with wavelengths in the infrared region of the spectrum ($\approx 10^{-5}$ m).

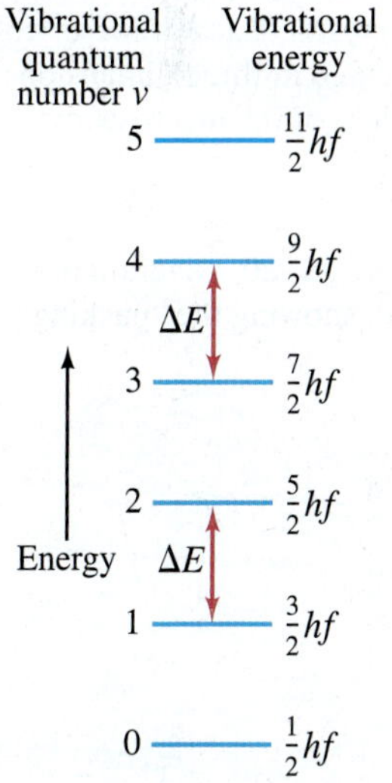

EXAMPLE 29–3 Vibrational energy levels in hydrogen. Hydrogen molecule vibrations emit infrared radiation of wavelength around 2300 nm. (a) What is the separation in energy between adjacent vibrational levels? (b) What is the lowest vibrational energy state?

APPROACH The energy separation between adjacent vibrational levels is (Eq. 29–4) $\Delta E_{\text{vib}} = hf = hc/\lambda$. The lowest energy (Eq. 29–3) has $\nu = 0$.

SOLUTION

(a) $\Delta E_{\text{vib}} = hf = \dfrac{hc}{\lambda} = \dfrac{(6.63 \times 10^{-34}\,\text{J}\cdot\text{s})(3.00 \times 10^8\,\text{m/s})}{(2300 \times 10^{-9}\,\text{m})(1.60 \times 10^{-19}\,\text{J/eV})} = 0.54\,\text{eV},$

where the denominator includes the conversion factor from joules to eV.

(b) The lowest vibrational energy has $\nu = 0$ in Eq. 29–3:

$$E_{\text{vib}} = (\nu + \tfrac{1}{2})hf = \tfrac{1}{2}hf = 0.27\,\text{eV}.$$

| **EXERCISE C** What is the energy of the first vibrational state above the ground state in the hydrogen molecule?

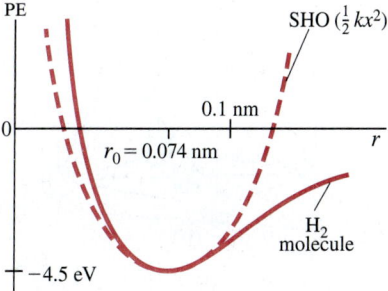

FIGURE 29–17 Potential energy for the H_2 molecule and for a simple harmonic oscillator (PE $= \tfrac{1}{2}kx^2$, with $|x| = |r - r_0|$).

FIGURE 29–18 Allowed vibrational energies for a diatomic molecule, where f is the fundamental frequency of vibration (see Chapter 11). The energy levels are equally spaced. Transitions are allowed only between adjacent levels ($\Delta\nu = \pm 1$).

[†]Forbidden transitions with $\Delta\nu = 2$ are emitted with much lower probability, but their observation can be important in some cases, such as in astronomy.

*29–5 Bonding in Solids

Quantum mechanics has been a great tool for understanding the structure of solids. This active field of research today is called **solid-state physics**, or **condensed-matter physics** so as to include liquids as well. The rest of this Chapter is devoted to this subject, and we begin with a brief look at the structure of solids and the bonds that hold them together.

Although some solid materials are *amorphous* in structure (such as glass), in that the atoms and molecules show no long-range order, we are interested here in the large class of *crystalline* substances whose atoms, ions, or molecules are generally accepted to form an orderly array known as a **lattice**. Figure 29–19 shows three of the possible arrangements of atoms in a crystal: simple cubic, face-centered cubic, and body-centered cubic. The NaCl crystal lattice is shown in Fig. 29–20.

The molecules of a solid are held together in a number of ways. The most common are by *covalent* bonding (such as between the carbon atoms of the diamond crystal) and by *ionic* bonding (as in a NaCl crystal). Often the bonds are partially covalent and partially ionic. Our discussion of these bonds earlier in this Chapter for molecules applies equally well to solids.

Let us look for a moment at the NaCl crystal of Fig. 29–20. Each Na^+ ion feels an attractive Coulomb potential due to each of the six "nearest neighbor" Cl^- ions surrounding it. Note that one Na^+ does not "belong" exclusively to one Cl^-, so we must not think of ionic solids as consisting of individual molecules. Each Na^+ also feels a repulsive Coulomb potential due to other Na^+ ions, although this is weaker since the Na^+ ions are farther away.

A different type of bond occurs in metals. Metal atoms have relatively loosely held outer electrons. **Metallic bond** theories propose that in a metallic solid, these outer electrons roam rather freely among all the metal atoms which, without their outer electrons, act like positive ions. According to the theory, the electrostatic attraction between the metal ions and this negative electron "gas" is responsible, at least in part, for holding the solid together. The binding energy of metal bonds is typically 1 to 3 eV, somewhat weaker than ionic or covalent bonds (5 to 10 eV in solids). The "free electrons" are responsible for the high electrical and thermal conductivity of metals. This theory also nicely accounts for the shininess of smooth metal surfaces: the free electrons can vibrate at any frequency, so when light of a range of frequencies falls on a metal, the electrons can vibrate in response and re-emit light of those same frequencies. Hence, the reflected light will consist largely of the same frequencies as the incident light. Compare this to nonmetallic materials that have a distinct color—the atomic electrons exist only in certain energy states, and when white light falls on them, the atoms absorb at certain frequencies, and reflect other frequencies which make up the color we see.

Here is a brief comparison of important strong bonds:

- ionic: an electron is "grabbed" from one atom by another;
- covalent: electrons are shared by atoms within a single molecule;
- metallic: electrons are shared by all atoms in the metal.

The atoms or molecules of some materials, such as the noble gases, can form only **weak bonds** with each other. As we saw in Section 29–3, weak bonds have very low binding energies and would not be expected to hold atoms together as a liquid or solid at room temperature. The noble gases condense only at very low temperatures, where the atomic (thermal) kinetic energy is small and the weak attraction can then hold the atoms together.

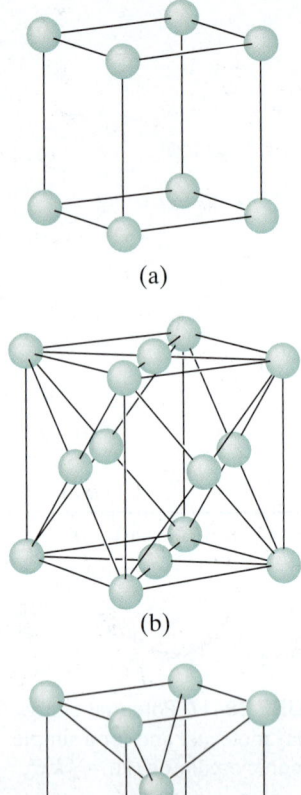

FIGURE 29–19 Arrangement of atoms in (a) a simple cubic crystal, (b) face-centered cubic crystal (note the atom at the center of each face), and (c) body-centered cubic crystal. Each of these "cells" is repeated in three dimensions to the edges of the macroscopic crystal.

FIGURE 29–20 Diagram of an NaCl crystal, showing the "packing" of atoms.

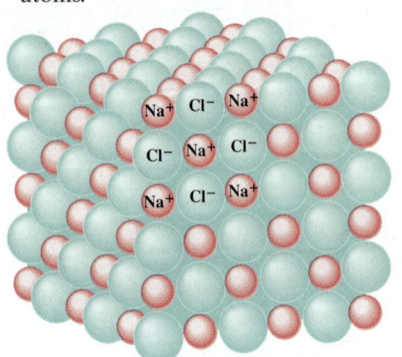

EXERCISE D Return to the Chapter-Opening Question, page 829, and answer it again now. Try to explain why you may have answered differently the first time.

*29–6 Free-Electron Theory of Metals; Fermi Energy

The free-electron theory of metals considers electrons in a metal as being in constant motion like an ideal gas, which we discussed in Chapter 13. For a classical ideal gas, at very low temperatures near absolute zero, $T = 0\,\text{K}$, all the particles would be in the lowest state, with zero kinetic energy $\left(= \frac{3}{2}kT = 0\right)$. But the situation is vastly different for an electron gas because, according to quantum mechanics, electrons obey the exclusion principle and can be only in certain possible energy levels or states. Electrons also obey a quantum statistics called **Fermi–Dirac statistics**[†] that takes into account the exclusion principle. All particles that have spin $\frac{1}{2}$ (or other half-integral spin: $\frac{3}{2}, \frac{5}{2}$, etc.), such as electrons, protons, and neutrons, obey Fermi–Dirac statistics and are referred to as **fermions** (see Section 28–7). The electron gas in a metal is often called a **Fermi gas**. According to the exclusion principle, no two electrons in the metal can have the same set of quantum numbers. Therefore, in each of the energy states available for the electrons in our "gas," there can be at most two electrons: one with spin up $\left(m_s = +\frac{1}{2}\right)$ and one with spin down $\left(m_s = -\frac{1}{2}\right)$. Thus, at $T = 0\,\text{K}$, the possible energy levels will be filled, two electrons each, up to a maximum level called the **Fermi level**. This is shown in Fig. 29–21, where the vertical axis is the "density of occupied states," whose meaning is similar to the Maxwell distribution for a classical gas (Section 13–10). The energy of the state at the Fermi level is called the **Fermi energy**, E_F. For copper, $E_F = 7.0\,\text{eV}$. This is very much greater than the energy of thermal motion at room temperature ($\overline{KE} = \frac{3}{2}kT \approx 0.04\,\text{eV}$, Eq. 13–8). Clearly, all motion does not stop at absolute zero.

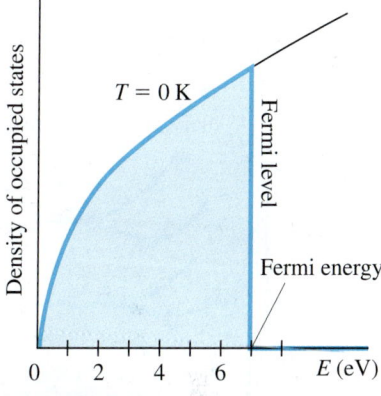

FIGURE 29–21 At $T = 0\,\text{K}$, all states up to energy E_F, called the Fermi energy, are filled. (Shown here for copper.)

At $T = 0$, all states with energy below E_F are occupied, and all states above E_F are empty. What happens for $T > 0$? We expect that at least some of the electrons will increase in energy due to thermal motion. Figure 29–22 shows the density of occupied states for $T = 1200\,\text{K}$, a temperature at which a metal is so hot it would glow. We see that the distribution differs very little from that at $T = 0$. We see also that the changes that do occur are concentrated about the Fermi level. A few electrons from slightly below the Fermi level move to energy states slightly above it. The average energy of the electrons increases only very slightly when the temperature is increased from $T = 0\,\text{K}$ to $T = 1200\,\text{K}$. This is very different from the behavior of an ideal gas, for which kinetic energy increases directly with T. Nonetheless, this behavior is readily understood as follows. Energy of thermal motion at $T = 1200\,\text{K}$ is about $\frac{3}{2}kT \approx 0.1\,\text{eV}$. The Fermi level, on the other hand, is on the order of several eV: for copper it is $E_F \approx 7.0\,\text{eV}$. An electron at $T = 1200\,\text{K}$ may have 7 eV of energy, but it can acquire at most only a few times 0.1 eV of energy by a (thermal) collision with the lattice. Only electrons very near the Fermi level would find vacant states close enough to make such a transition. Essentially none of the electrons could increase in energy by, say, 3 eV, so electrons farther down in the electron gas are unaffected. Only electrons near the top of the energy distribution can be thermally excited to higher states. And their new energy is on the average only slightly higher than their old energy. This model of free electrons in a metal as a "gas," though incomplete, provides good explanations for the thermal and electrical conductivity of metals.

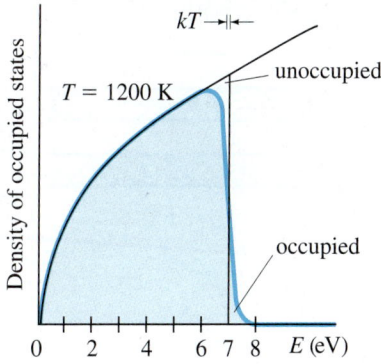

FIGURE 29–22 The density of occupied states for the electron gas in copper. The width kT shown above the graph represents thermal energy at $T = 1200\,\text{K}$.

[†]Developed independently by Enrico Fermi (Figs. 1–13, 27–11, 28–2, 30–7) in early 1926 and by P. A. M. Dirac a few months later. See Section 28–7.

*29–7 Band Theory of Solids

We saw in Section 29–1 that when two hydrogen atoms approach each other, the wave functions overlap, and the two 1s states (one for each atom) divide into two states of different energy. (As we saw, only one of these states, $S = 0$, has low enough energy to give a bound H_2 molecule.) Figure 29–23a shows this situation for 1s and 2s states for two atoms: as the two atoms get closer (toward the left in Fig. 29–23a), the 1s and 2s states split into two levels. If six atoms come together, as in Fig. 29–23b, each of the states splits into six levels. If a large number of atoms come together to form a solid, then each of the original atomic levels becomes a **band** as shown in Fig. 29–23c. The energy levels are so close together in each band that they seem essentially continuous. This is why the spectrum of heated solids (Section 27–2) appears continuous. (See also Fig. 29–14 and its discussion at start of Section 29–4.)

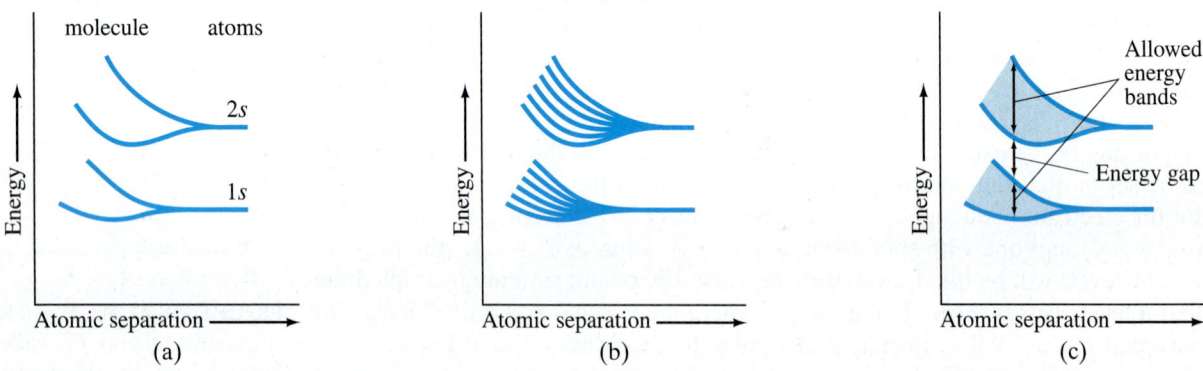

FIGURE 29–23 The splitting of 1s and 2s atomic energy levels as (a) two atoms approach each other (the atomic separation decreases toward the left on the graph); (b) the same for six atoms, and (c) for many atoms when they come together to form a solid.

FIGURE 29–24 Energy bands for sodium (Na).

The crucial aspect of a good **conductor** is that the highest energy band containing electrons is only partially filled. Consider sodium metal, for example, whose energy bands are shown in Fig. 29–24. The 1s, 2s, and 2p bands are full (just as in a sodium atom) and don't concern us. The 3s band, however, is only half full. To see why, recall that the exclusion principle stipulates that in an atom, only two electrons can be in the 3s state, one with spin up and one with spin down. These two states have slightly different energy. For a solid consisting of N atoms, the 3s band will contain $2N$ possible energy states. A sodium atom has a single 3s electron, so in a sample of sodium metal containing N atoms, there are N electrons in the 3s band, and N unoccupied states. When a potential difference is applied across the metal, electrons can respond by accelerating and increasing their energy, since there are plenty of unoccupied states of slightly higher energy available. Hence, a current flows readily and sodium is a good conductor. The characteristic of all good conductors is that the highest energy band is only partially filled, or two bands overlap so that unoccupied states are available. An example of the latter is magnesium, which has two 3s electrons, so its 3s band is filled. But the unfilled 3p band overlaps the 3s band in energy, so there are lots of available states for the electrons to move into. Thus magnesium, too, is a good conductor.

In a material that is a good **insulator**, on the other hand, the highest band containing electrons, called the **valence band**, is completely filled. The next highest energy band, called the **conduction band**, is separated from the valence band by a "forbidden" **energy gap** (or **band gap**), E_g, of typically 5 to 10 eV. So at room temperature (300 K), where thermal energies (that is, average kinetic energy—see Chapter 13) are on the order of $\frac{3}{2}kT \approx 0.04$ eV, almost no electrons can acquire the 5 eV needed to reach the conduction band. When a potential difference is applied across the material, no available states are accessible to the electrons, and no current flows. Hence, the material is a good insulator.

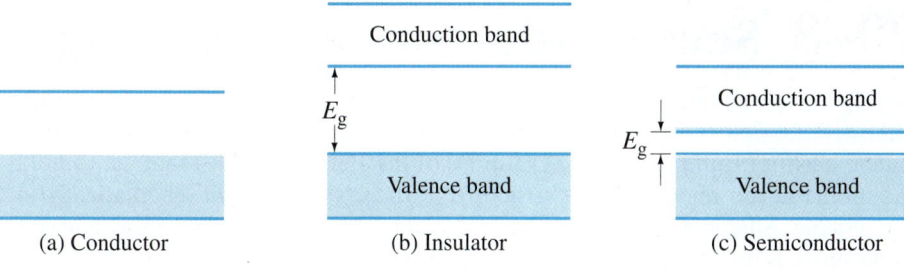

FIGURE 29–25 Energy bands for (a) a conductor, (b) an insulator, which has a large energy gap E_g, and (c) a semiconductor, which has a small energy gap E_g. Shading represents occupied states. Pale shading in (c) represents electrons that can pass from the top of the valence band to the bottom of the conduction band due to thermal agitation at room temperature (exaggerated).

Figure 29–25 compares the relevant energy bands (a) for conductors, (b) for insulators, and also (c) for the important class of materials known as **semiconductors**. The bands for a pure (or **intrinsic**) semiconductor, such as silicon or germanium, are like those for an insulator, except that the unfilled conduction band is separated from the filled valence band by a much smaller energy gap, E_g, which for silicon is $E_g = 1.12 \text{ eV}$. At room temperature, electrons are moving about with varying amounts of kinetic energy $(\overline{KE} = \frac{3}{2}kT)$, according to kinetic theory, Chapter 13. A few electrons can acquire enough thermal energy to reach the conduction band, and so a very small current may flow when a voltage is applied. At higher temperatures, more electrons have enough energy to jump the gap (top end of thermal distribution—see Fig. 13–20). Often this effect can more than offset the effects of more frequent collisions due to increased disorder at higher temperature, so the resistivity of semiconductors can *decrease* with increasing temperature (see Table 18–1). But this is not the whole story of semiconductor conduction. When a potential difference is applied to a semiconductor, the few electrons in the conduction band move toward the positive electrode. Electrons in the valence band try to do the same thing, and a few can because there are a small number of unoccupied states which were left empty by the electrons reaching the conduction band. Such unfilled electron states are called **holes**. Each electron in the valence band that fills a hole in this way as it moves toward the positive electrode leaves behind its own hole, so the holes migrate toward the negative electrode. As the electrons tend to accumulate at one side of the material, the holes tend to accumulate on the opposite side. We will look at this phenomenon in more detail in the next Section.

EXAMPLE 29–4 Calculating the energy gap. It is found that the conductivity of a certain semiconductor increases when light of wavelength 345 nm or shorter strikes it, suggesting that electrons are being promoted from the valence band to the conduction band. What is the energy gap, E_g, for this semiconductor?

APPROACH The longest wavelength (lowest energy) photon to cause an increase in conductivity has $\lambda = 345$ nm, and its energy $(= hf)$ equals the energy gap.

SOLUTION The gap energy equals the energy of a $\lambda = 345$-nm photon:

$$E_g = hf = \frac{hc}{\lambda} = \frac{(6.63 \times 10^{-34} \text{ J} \cdot \text{s})(3.00 \times 10^8 \text{ m/s})}{(345 \times 10^{-9} \text{ m})(1.60 \times 10^{-19} \text{ J/eV})} = 3.6 \text{ eV}.$$

CONCEPTUAL EXAMPLE 29–5 Which is transparent? The energy gap for silicon is 1.12 eV at room temperature, whereas that of zinc sulfide (ZnS) is 3.6 eV. Which one of these is opaque to visible light, and which is transparent?

RESPONSE Visible-light photons span energies from roughly 1.8 eV to 3.1 eV. ($E = hf = hc/\lambda$ where $\lambda = 400$ nm to 700 nm and $1 \text{ eV} = 1.6 \times 10^{-19}$ J.) Light is absorbed by the electrons in a material. Silicon's energy gap is small enough to absorb these photons, thus bumping electrons well up into the conduction band, so silicon is opaque. On the other hand, zinc sulfide's energy gap is so large that no visible-light photons would be absorbed; they would pass right through the material which would thus be transparent.

PHYSICS APPLIED
Transparency

*29–8 Semiconductors and Doping

Nearly all electronic devices today use semiconductors—mainly silicon (Si), although the first transistor (1948) was made with germanium (Ge). An atom of silicon has four outer electrons (group IV of the Periodic Table) that act to hold the atoms in the regular lattice structure of the crystal, shown schematically in Fig. 29–26a. Silicon acquires properties useful for electronics when a tiny amount of impurity is introduced into the crystal structure (perhaps 1 part in 10^6 or 10^7). This is called **doping** the semiconductor. Two kinds of doped semiconductor can be made, depending on the type of impurity used. The impurity can be an element whose atoms have five outer electrons (group V in the Periodic Table), such as arsenic. Then we have the situation shown in Fig. 29–26b, with a few arsenic atoms holding positions in the crystal lattice where normally silicon atoms are. Only four of arsenic's electrons fit into the bonding structure. The fifth does not fit in and can move relatively freely, somewhat like the electrons in a conductor. Because of this small number of extra electrons, a doped semiconductor becomes slightly conducting. The density of conduction electrons in an **intrinsic** (= undoped) semiconductor at room temperature is very low, usually less than 1 per 10^9 atoms. With an impurity concentration of 1 in 10^6 or 10^7 when doped, the conductivity will be much higher and it can be controlled with great precision. An arsenic-doped silicon crystal is an ***n*-type semiconductor** because *negative* charges (electrons) carry the electric current.

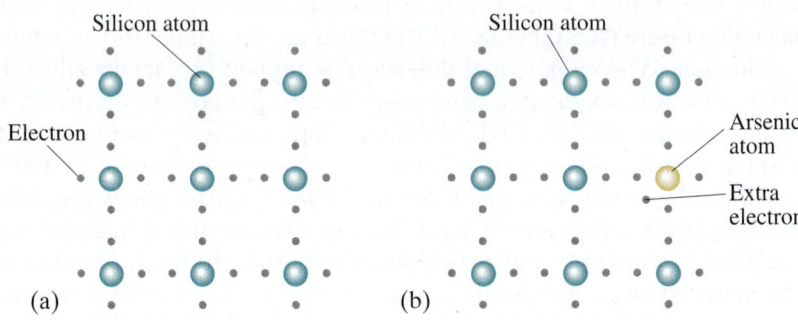

FIGURE 29–26 Two-dimensional representation of a silicon crystal. (a) Four (outer) electrons surround each silicon atom. (b) Silicon crystal doped with a small percentage of arsenic atoms: the extra electron doesn't fit into the crystal lattice and so is free to move about. This is an *n*-type semiconductor.

In a ***p*-type semiconductor**, a small percentage of semiconductor atoms are replaced by atoms with three outer electrons (group III in the Periodic Table), such as boron. As shown in Fig. 29–27a, there is a **hole** in the lattice structure near a boron atom because it has only three outer electrons. An electron from a nearby silicon atom can jump into this hole and fill it. But this leaves a hole where that electron had previously been, Fig. 29–27b. The vast majority of atoms are silicon, so holes are almost always next to a silicon atom. Since silicon atoms require four outer electrons to be neutral, this means there is a net positive charge at the hole. Whenever an electron moves to fill a hole, the positive hole is then at the previous position of that electron. Another electron can then fill this hole, and the hole thus moves to a new location; and so on. This type of semiconductor is called *p*-type because it is the positive holes that carry the electric current.[†] Note, however, that both *p*-type and *n*-type semiconductors have *no net charge* on them.

> **CAUTION**
> *p-type semiconductors act as though + charges move—but electrons actually do the moving*

[†]Each electron that fills a hole moves a very short distance (∼1 atom < 1 nm) whereas holes move much larger distances and so are the real carriers of the current. We can tell the current is carried by positive charges (holes) by using the Hall effect, Section 20–4.

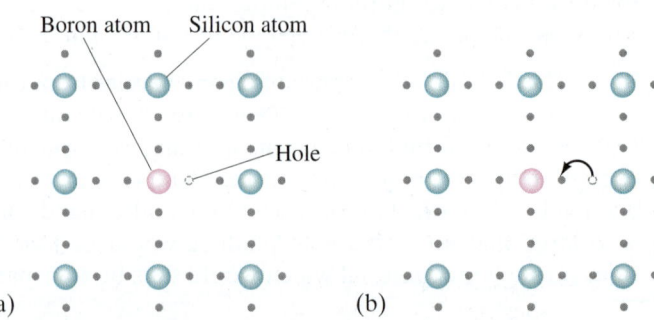

FIGURE 29–27 A *p*-type semiconductor, boron-doped silicon. (a) Boron has only three outer electrons, so there is an empty spot, or *hole* in the structure. (b) An electron from a nearby silicon atom can jump into the hole and fill it. As a result, the hole moves to a new location (to the right in this diagram), to where the electron used to be.

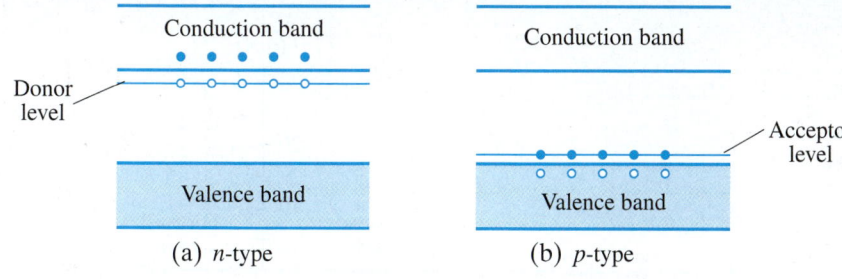

FIGURE 29–28 Impurity energy levels in doped semiconductors.

According to the band theory (Section 29–7), in a doped semiconductor the impurity provides additional energy states between the bands as shown in Fig. 29–28. In an *n*-type semiconductor, the impurity energy level lies just below the conduction band, Fig. 29–28a. Electrons in this energy level need only about 0.05 eV in Si to reach the conduction band which is on the order of the thermal energy, $\frac{3}{2}kT$ ($\approx$ 0.04 eV at 300 K). At room temperature, the tiny % of electrons in this donor level (roughly 1 in 10^6) can readily make the transition upward. This energy level can thus supply electrons to the conduction band, so it is called a **donor** level. In *p*-type semiconductors, the impurity energy level is just above the valence band (Fig. 29–28b). It is called an **acceptor** level because electrons from the valence band can jump into it with only average thermal energy. Positive holes are left behind in the valence band, and as other electrons move into these holes, the holes move as discussed earlier.

> **EXERCISE E** Which of the following impurity atoms in silicon would produce a *p*-type semiconductor? (*a*) Ge; (*b*) Ne; (*c*) Al; (*d*) As; (*e*) Ga; (*f*) none of the above.

*29–9 Semiconductor Diodes, LEDs, OLEDs

Semiconductor diodes and transistors are essential components of modern electronic devices. The miniaturization achieved today allows many billions of diodes, transistors, resistors, etc., to be fabricated (adding doping atoms) on a single *chip* less than a millimeter on a side.

At the interface between an *n*-type and a *p*-type semiconductor, a ***pn* junction diode** is formed. Separately, the two semiconductors are electrically neutral. But near the junction, a few electrons diffuse from the *n*-type into the *p*-type semiconductor, where they fill a few of the holes. The *n*-type is left with a positive charge, and the *p*-type acquires a net negative charge. Thus an "intrinsic" *potential difference* is established, with the *n* side positive relative to the *p* side, and this prevents further diffusion of electrons. The "junction" is actually a very thin layer between the charged *n* and *p* semiconductors where all holes are filled with electrons. This junction region is called the **depletion layer** (depleted of electrons and holes).†

If a battery is connected to a diode with the positive terminal to the *p* side and the negative terminal to the *n* side as in Fig. 29–29a, the externally applied voltage opposes the intrinsic potential difference and the diode is said to be **forward biased**. If the voltage is great enough, about 0.6 V for Si at room temperature, it overcomes that intrinsic potential difference and a large current can flow. The positive holes in the *p*-type semiconductor are repelled by the positive terminal of the battery, and the electrons in the *n*-type are repelled by the negative terminal of the battery. The holes and electrons meet at the junction, and the electrons cross over and fill the holes. A current is flowing. The positive terminal of the battery is continually pulling electrons off the *p* end, forming new holes, and electrons are being supplied by the negative terminal at the *n* end.

When the diode is **reverse biased**, as in Fig. 29–29b, the holes in the *p* end are attracted to the battery's negative terminal and the electrons in the *n* end are attracted to the positive terminal. Almost no current carriers meet near the junction and, ideally, no current flows.

FIGURE 29–29 Schematic diagram showing how a semiconductor diode operates. Current flows when the voltage is connected in forward bias, as in (a), but not when connected in reverse bias, as in (b).

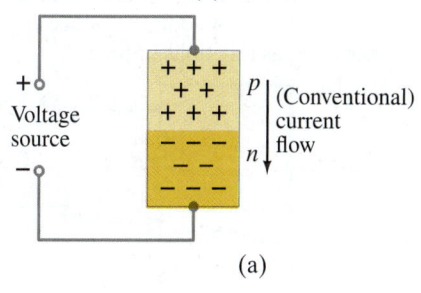

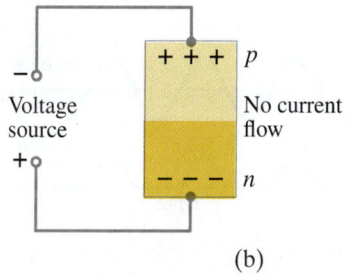

†One way to form the *pn* boundary at the **nanometer** thicknesses on chips is to implant (or diffuse) *n*-type donor atoms into the surface of a *p*-type semiconductor, converting a layer of the *p*-type semiconductor into *n*-type.

FIGURE 29–30 Current through a silicon *pn* diode as a function of applied voltage.

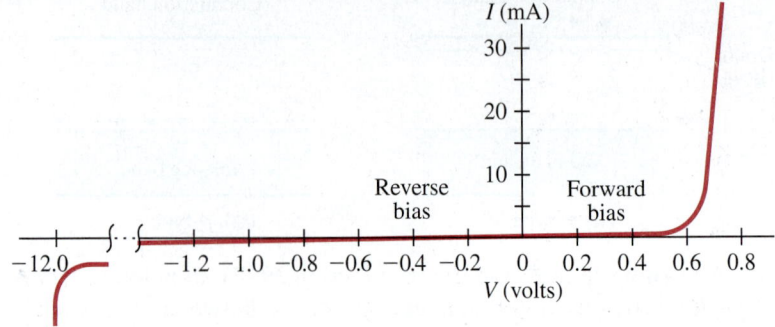

A graph of current versus voltage for a typical diode is shown in Fig. 29–30. A forward bias greater than 0.6 V allows a large current to flow. In reverse bias, a real diode allows a small amount of reverse current to flow; for most practical purposes, it is negligible.[†]

The symbol for a diode is

⟶▷|⟵ [diode]

where the arrow represents the direction conventional (+) current flows readily.

EXAMPLE 29–6 A diode. The diode whose current–voltage characteristics are shown in Fig. 29–30 is connected in series with a 4.0-V battery in forward bias and a resistor. If a current of 15 mA is to pass through the diode, what resistance must the resistor have?

APPROACH We use Fig. 29–30, where we see that the voltage drop across the diode is about 0.7 V when the current is 15 mA. Then we use simple circuit analysis and Ohm's law (Chapters 18 and 19).

SOLUTION The voltage drop across the resistor is 4.0 V − 0.7 V = 3.3 V, so $R = V/I = (3.3\,\text{V})/(1.5 \times 10^{-2}\,\text{A}) = 220\,\Omega$.

If the voltage across a diode connected in reverse bias is increased greatly, breakdown occurs. The electric field across the junction becomes so large that ionization of atoms results. The electrons thus pulled off their atoms contribute to a larger and larger current as breakdown continues. The voltage remains constant over a wide range of currents. This is shown on the far left in Fig. 29–30. This property of diodes can be used to accurately regulate a voltage supply. A diode designed for this purpose is called a **zener diode**. When placed across the output of an unregulated power supply, a zener diode can maintain the voltage at its own breakdown voltage as long as the supply voltage is always above this point. Zener diodes can be obtained corresponding to voltages of a few volts to hundreds of volts.

A diode is called a **nonlinear device** because the current is not proportional to the voltage. That is, a graph of current versus voltage (Fig. 29–30) is not a straight line, as it is for a resistor (which ideally *is* linear).

*Rectifiers

Since a *pn* junction diode allows current to flow only in one direction (as long as the voltage is not too high), it can serve as a **rectifier**—to change ac into dc. A simple rectifier circuit is shown in Fig. 29–31a. The ac source applies a voltage across the diode alternately positive and negative. Only during half of each cycle will a current pass through the diode; only then is there a current through the resistor R. Hence, a graph of the voltage V_{ab} across R as a function of time looks like the output voltage shown in Fig. 29–31b. This **half-wave rectification** is not exactly dc, but it is unidirectional. More useful is a **full-wave rectifier** circuit, which uses two diodes (or sometimes four) as shown in Fig. 29–32a (top of next page). At any given instant, either one diode or the other will conduct current to the right.

FIGURE 29–31 (a) A simple (half-wave) rectifier circuit using a semiconductor diode. (b) AC source input voltage, and output voltage across R, as functions of time.

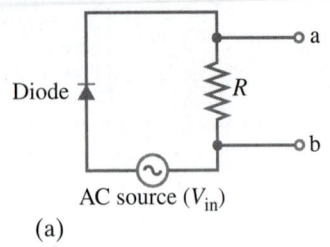

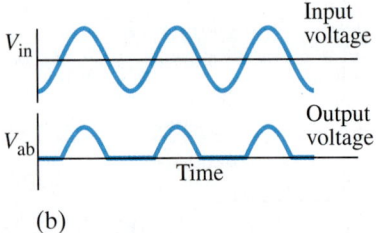

[†]At room temperature, the reverse current is a few pA in Si; but it increases rapidly with temperature, and may render a diode ineffective above 200°C.

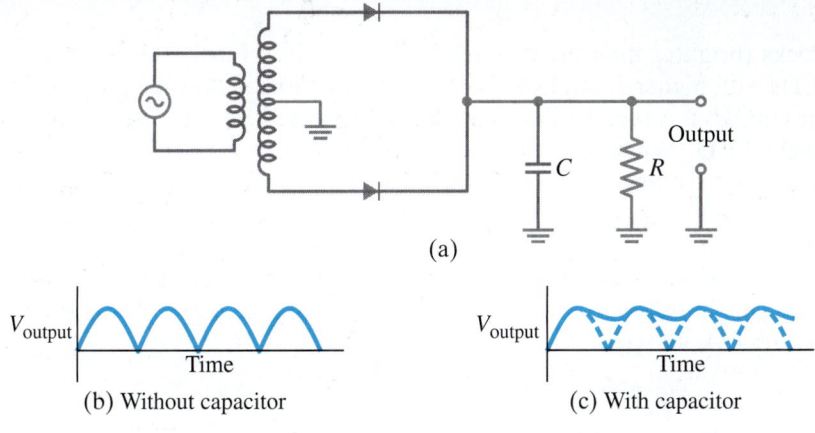

FIGURE 29–32 (a) Full-wave rectifier circuit (including a transformer so the magnitude of the voltage can be changed). (b) Output voltage in the absence of capacitor C. (c) Output voltage with the capacitor in the circuit.

Therefore, the output across the load resistor R will be as shown in Fig. 29–32b. Actually this is the voltage if the capacitor C were not in the circuit. The capacitor tends to store charge and, if the time constant RC is sufficiently long, helps to smooth out the current as shown in Fig. 29–32c. (The variation in output shown in Fig. 29–32c is called **ripple voltage**.)

Rectifier circuits are important because most line voltage in buildings is ac, and most electronic devices require a dc voltage for their operation. Hence, diodes are found in nearly all electronic devices including radios, TV sets, computers, and chargers for cell phones and other devices.

*Photovoltaic Cells

Solar cells, also called **photovoltaic cells**, are rather heavily doped pn junction diodes used to convert sunlight into electric energy. Photons are absorbed, creating electron–hole pairs if the photon energy is greater than the band gap energy, E_g (see Figs. 29–25c and 29–28). That is, the absorbed photon excites an electron from the valence band up to the conduction band, leaving behind a hole in the valence band. The created electrons and holes produce a current that, when connected to an external circuit, becomes a source of emf and power. A typical silicon pn junction may produce about 0.6 V. Many are connected in series to produce a higher voltage. Such series strings are connected in parallel within a **photovoltaic panel**. Research includes experimenting with combinations of semiconductors. A good photovoltaic panel can have an output of perhaps 50 W/m^2, averaged over day and night, sunny and cloudy. The world's total electricity demand is on the order of 10^{12} W, which could be met with solar cells covering an area of only about 200 km × 200 km of Earth's surface.[†]

Photodiodes (Section 27–3) and **semiconductor particle detectors** (Section 30–13) operate similarly.

*LEDs

A **light-emitting diode (LED)** is sort of the reverse of a photovoltaic cell. When a pn junction is forward biased, a current begins to flow. Electrons cross from the n-region into the p-region, recombining with holes, and a photon can be emitted with an energy about equal to the band gap energy, E_g. This does not work well with silicon diodes.[‡] But high light-emission is achieved with **compound semiconductors**, typically involving a group III and a group V element such as gallium and arsenic (= gallium arsenide = GaAs). Remarkably, GaAs has a crystal structure very similar to Si. See Fig. 29–33. For doping of GaAs, group VI atoms (like Se) can serve as donors, and group II atoms (valence +2, such as Zn) as acceptors. The energy gap for GaAs is $E_g = 1.42$ eV, corresponding to near-infrared photons with wavelength 870 nm (almost visible). Such infrared LEDs are suitable for use in remote-control devices for TVs, DVD players, stereos, car door locks, and so on.

The first visible-light LED, developed in the early 1960s, was made of a semiconductor compound of gallium, arsenic, and phosphorus (= GaAsP) which emitted red light. The red LED soon found use as the familiar indicator lights (on–off) on electronic devices, and as the bright red read-out on calculators and

PHYSICS APPLIED
LEDs and applications
Car safety (brakes)

FIGURE 29–33 (a) Two Si atoms forming the covalent bond showing the electrons in different colors for each of the two separate atoms. (In Fig. 29–26a we showed each atom separately to emphasize the four outer electrons in each.) (b) A gallium–arsenic pair, also covalently bonded.

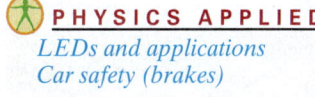

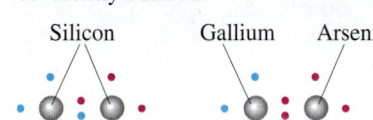

[†]Electricity makes up about 5% of total global energy use.
[‡]Electron-hole recombination in silicon results mostly in heat, as lattice vibrations called **phonons**.

digital clocks (brighter than the dimmer LCD readouts). Further development led to LEDs with higher E_g and shorter wavelengths: first yellow, then finally in 1995, blue (InGaN). A blue LED was important because it gave the possibility of a **white-light LED**. White light can be approximated by LEDs in two ways: (1) using a red, a yellow–green, and a blue LED; (2) using a blue LED with coatings of "powders" or "phosphors" that are fluorescent (Section 28–10). For the latter, the high-energy blue LED photons are themselves emitted, plus they can excite the various phosphors to excited states which decay in two or more steps, emitting light of lower energy and longer wavelengths. Figure 29–34 shows typical spectra of both types.

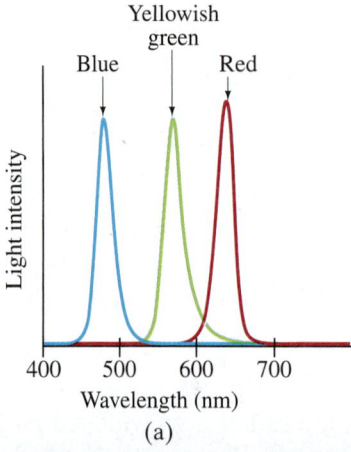
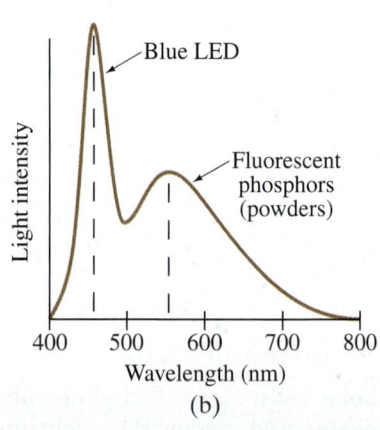

FIGURE 29–34 (a) A combination of three LEDs of three different colors gives a sort of white color, but there are large wavelength gaps, so some colors would not be reflected and would appear black; this type is rarely used now. (b) A blue LED with fluorescent phosphors or powders gives a better approximation of white light. (Thanks to M. Vannoni and G. Molesini for (b).)

FIGURE 29–35 LED flashlights. Note the tiny LEDs, each maybe $\frac{1}{2}$ cm in diameter.

LED "bulbs" are available to replace other types of lighting in applications such as flashlights (Fig. 29–35), street lighting, traffic signals, car brake lights, billboards, backlighting for LCD screens, and large display screens at stadiums. LED lights, sometimes called **solid-state** lighting, are longer-lived (50,000 hours vs. 1000–2000 for ordinary bulbs), more efficient (up to 5 times), and rugged. A small town in Italy, Torraca, was the first to have all its street lighting be LED (2007). LEDs can be as small as 1 or 2 mm wide, and are individual units with wires connected directly to them. They can be used for large TV screens in stadiums, but a home TV would require much smaller LED size, meaning fabrication of many on a crystalline semiconductor, and the pixels would be addressed as discussed in Section 17–11 for LCD screens.

*Pulse Oximeter

A **pulse oximeter** uses two LEDs to measure the % oxygen (O_2) saturation in your blood. One LED is red, 660 nm, and the other IR (900–940 nm). The LED beams pass through a finger (Fig. 29–36) or earlobe and are detected by a photodiode. Oxygenated red blood cells absorb less red and more infrared light than deoxygenated cells. A ratio of absorbed light (red/IR) of 0.5 corresponds to nearly 100% O_2 saturation; a ratio of 1.0 is about 85% and 2.0 corresponds to about 50% (bad). The LED measures during complete pulses, including blood surges, and the device can also count your heartbeat rate.

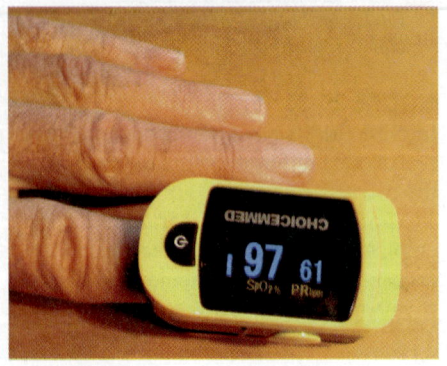

FIGURE 29–36 A pulse oximeter.

*pn Diode Lasers

Diode lasers, using a *pn*-junction in forward bias like an LED, are the most compact of lasers and are very common: they read CDs and DVDs and are used as pointers and in laser printers. They emit photons like an LED but, like all lasers (Section 28–11), need to have an *inverted population* of states for the lasing frequency. This is achieved by applying a high forward-bias voltage. The large current brings many electrons into the conduction band at the junction layer, and holes into the valence band, and before the electrons have time to combine with holes, they form an inverted population. When one electron drops down into a hole and emits a photon, that photon stimulates other electrons to drop down as well, *in phase*, creating coherent laser light. Opposite ends of the crystal are made parallel and very smooth so they act as the mirrors needed for lasing, as shown in our laser diagram, Fig. 28–18.

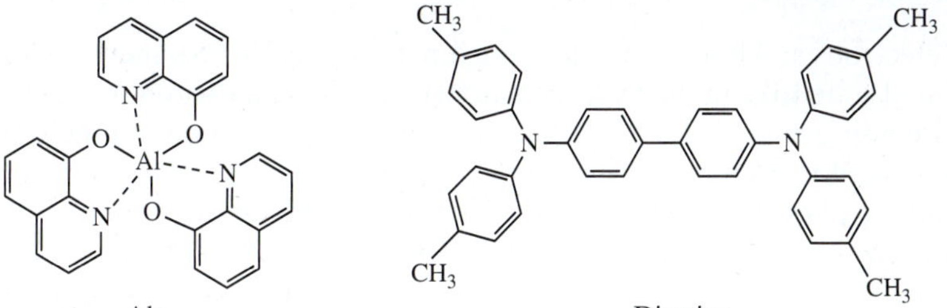

FIGURE 29–37 These two organic molecules were used in the first OLEDs (1987). The hexagons have carbon at each corner, and an attached hydrogen, unless otherwise noted.

*OLED (Organic LED)

Many organic compounds have semiconductor properties. Useful ones can have mobile electrons and holes. A practical organic **electroluminescent** (**EL**) device, an **organic light-emitting diode** (**OLED**) was first described in the late 1980s.

Organic compounds contain carbon (C), hydrogen (H), often nitrogen and oxygen, and sometimes other atoms. We usually think of them as coming from life—plants and animals. They are also found in petroleum, and some can be synthesized in the lab. Organic compounds can be complex, and often contain the familiar hexagonal "benzene ring" with C atoms at all (or most) of the six corners. The two organic compounds shown in Fig. 29–37 were used as n-type and p-type layers in the earliest useful OLED. **Polymers**, long organic molecules with repeating structural units, can also be used for an OLED.

The simplest OLED consists of two organic layers, the **emissive layer** and the **conductive layer**, each 20 to 50 nm thick, sandwiched between two electrodes, Fig. 29–38. The anode is typically transparent, to let the light out. It can be made of a very thin layer of indium–tin oxide (**ITO**), which is transparent and conductive, coated on a glass slab. The cathode is often metallic, but could also be made of transparent material.

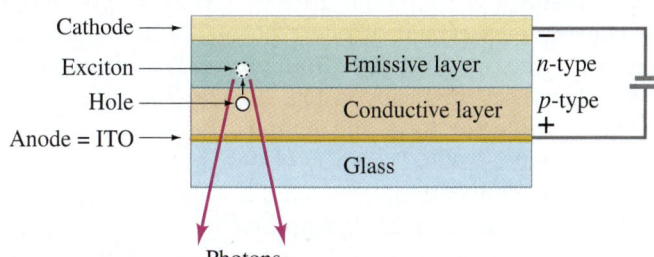

FIGURE 29–38 An OLED with two organic layers. Hole–electron recombination into an exciton (dashed circle) occurs in the emissive layer, followed by photon emission. Photons emitted in the wrong direction (upward in the diagram) reduce efficiency.

OLEDs can be smaller and thinner than ordinary inorganic LEDs. They can be more easily constructed as a unit for a screen display (i.e., more cheaply, but still quite expensive) than for inorganic LEDs. Their use as screens on cell phones, cameras, and TVs produces brighter light and greater contrast, and they need less power (important for battery life of portable devices) than LCD screens. Why? They need no backlight (like LCDs) because they emit the light themselves. OLEDs can be fabricated as a matrix, usually active matrix (**AMOLED**), using the same type of addressing described in Section 17–11 for LCDs. OLED displays are much thinner than LCDs and retain brightness at larger viewing angles. They can even be fabricated on curved or flexible substrates—try the windshield of your car (Fig. 29–39). The array may be RGBG (similar to a Bayer mosaic, Fig. 25–2) or RGBW where W = white is meant to give greater brightness. The subpixels can also be stacked, one above the other (similar to the Foveon, Fig. 25–3).

FIGURE 29–39 Head up displays on curved windshields can use curved OLEDs to show, for example, your speed without having to look down at the speedometer.

*OLED Functioning (advanced)

According to band theory, when a voltage is applied ($\approx$ 2 to 5 V), electrons are "injected" (engineering term) into energy states of the **lowest unoccupied molecular orbitals** (**LUMO**) of the emissive layer. At the same time, electrons are withdrawn from the **highest occupied molecular orbitals** (**HOMO**) of the conductive layer at the cathode—which is equivalent to **holes** being "injected" into the conductive layer. The LUMO and HOMO energy levels are analogous to the conduction and valence bands of inorganic silicon diodes (Fig. 29–28). Holes travel in the HOMO, electrons in the LUMO. ("Orbital" is a chemistry word for the states occupied by the electrons in a molecule.)

When electrons and holes meet near the junction (Fig. 29–38), they can form a sort of bound state (like in the hydrogen atom) known as an **exciton**. An exciton has a small binding energy (0.1 to 1 eV), and a very short lifetime on the order of nanoseconds. When an exciton "decays" (the negative electron and positive hole combine), a photon is emitted. These photons are the useful output.

The energy hf of the photon, and its frequency corresponding to the color, depends on the energy structure of the exciton. The energy gap, LUMO–HOMO, sets an upper limit on hf, but the vibrational energy levels of the molecules reduce that by varying amounts, as does the binding energy of the exciton. The spectrum has a peak, like those in Fig. 29–34a, but is wider, 100–200 nm at half maximum. The organic molecules are chosen so that the photons have frequencies in the color range desired, say for a display subpixel: bluish (B), greenish (G), or red (R).

The conductive layer is also called the **hole transport layer** (**HTL**), which name expresses its purpose. The emissive layer, on the other hand (Fig. 29–38), serves two purposes: (1) it serves to transport electrons toward the junction, and (2) it is in this layer (near the junction) that holes meet electrons to form excitons and then combine and emit light. These two functions can be divided in a more sophisticated OLED that has three layers: Adjacent to the cathode is the **electron transport layer** (**ETL**), plus there is an **emissive layer** (**EML**) sandwiched between the ETL and the HTL. The emissive layer can be complex, containing a **host** material plus a **guest** compound in small concentration—a kind of doping—to fine-tune energy levels and efficiency.

*29–10 Transistors: Bipolar and MOSFETs

The **bipolar junction transistor** was invented in 1948 by J. Bardeen, W. Shockley, and W. Brattain. It consists of a crystal of one type of doped semiconductor sandwiched between two of the opposite type. Both *npn* and *pnp* transistors can be made, and they are shown schematically in Fig. 29–40a. The three semiconductors are given the names **collector**, **base**, and **emitter**. The symbols for *npn* and *pnp* transistors are shown in Fig. 29–40b. The arrow is always placed on the emitter and indicates the direction of (conventional) current flow in normal operation.

The operation of an *npn* transistor as an **amplifier** is shown in Fig. 29–41. A dc voltage V_{CE} is maintained between the collector and emitter by battery $\mathscr{E}_C$. The voltage applied to the base is called the *base bias voltage*, V_{BE}. If V_{BE} is positive, conduction electrons in the emitter are attracted into the base. The base region is very thin, much less than 1 μm, so most of these electrons flow right across into the collector which is maintained at a positive voltage. A large current, I_C, flows between collector and emitter and a much smaller current, I_B, through the base. In the steady state, I_B and I_C can be considered dc. But a small variation (= ac) in the base voltage due to an input signal attracts (or repels)

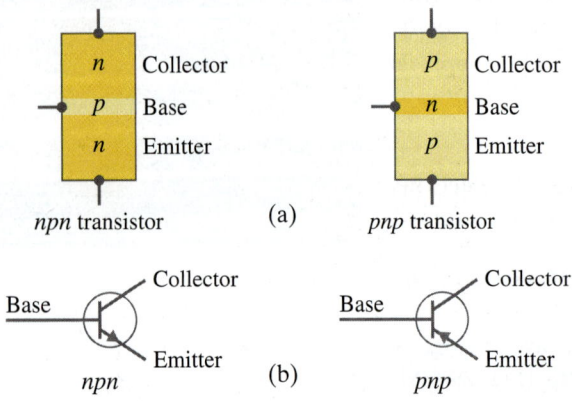

FIGURE 29–40 (a) Schematic diagram of *npn* and *pnp* transistors. (b) Symbols for *npn* and *pnp* transistors.

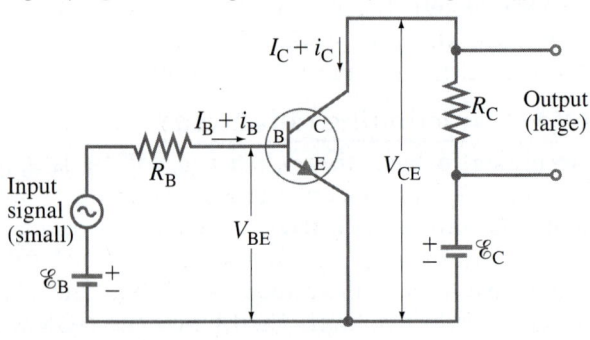

FIGURE 29–41 An *npn* transistor used as an amplifier. I_B is the current produced by $\mathscr{E}_B$ (in the absence of a signal), i_B is the ac signal current (= change in I_B).

charge that passes through into the collector and thus can cause a large *change* in the collector current and a large change in the voltage drop across the output resistor R_C. Hence a transistor can *amplify* a small signal into a larger one.

Typically a small ac signal (call it i_B) is to be amplified, and when added to the base bias current (and voltage) causes the current and voltage at the collector to vary at the same rate but magnified. Thus, what is important for amplification is the *change* in collector current for a given input *change* in base current. We label these ac signal currents (= changes in I_C and I_B) as i_C and i_B. The **current gain** is defined as the ratio

$$\beta_I = \frac{\text{output (collector) ac current}}{\text{input (base) ac current}} = \frac{i_C}{i_B}.$$

β_I may be on the order of 10 to 100. Similarly, the **voltage gain** is

$$\beta_V = \frac{\text{output (collector) ac voltage}}{\text{input (base) ac voltage}}.$$

Transistors are the basic elements in modern electronic amplifiers of all sorts.

A *pnp* transistor operates like an *npn*, except that holes move instead of electrons. The collector voltage is negative, and so is the base voltage in normal operation.

Another kind of transistor, very important, is the **MOSFET** (metal-oxide semiconductor field-effect transistor) common in **digital circuits** as a type of switch. Its construction is shown in Fig. 29–42a, and its symbol in Fig. 29–42b. What is called the emitter in a bipolar transistor is called the **source** in a MOSFET, and the collector is called the **drain**. The base is called the **gate**. The gate acts to let a current flow, or not, from the source to the drain, depending on the electric field it (the gate) provides across an insulator that separates it from the *p*-type semiconductor below, Fig. 29–42a. Hence the name "field-effect transistor" (FET).[†] MOSFETs are often used like switches, on or off, which in digital circuits can allow the storage of a binary bit, a "1" or a "0". We discussed uses of MOSFETs relative to digital TV (Section 17–11) and computer memory storage (Section 21–8).

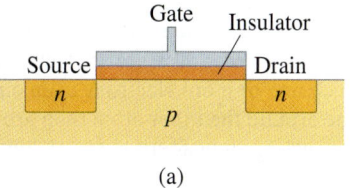

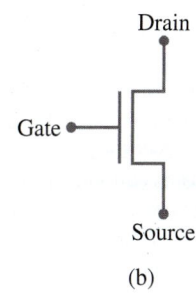

FIGURE 29–42 (a) Construction of a MOSFET of *n*- and *p*-type semiconductors and a gate of metal or heavily doped silicon (= a good conductor). (b) Symbol for a MOSFET which suggests its function.

*29–11 Integrated Circuits, 22-nm Technology

Although individual transistors are very small compared to the once-used vacuum tubes, they are huge compared to **integrated circuits** or **chips** (photo at start of this Chapter), invented in 1959 independently by Jack Kilby and Robert Noyce. Tiny amounts of impurities can be inserted or injected at particular locations within a single silicon crystal or wafer. These can be arranged to form diodes, transistors, resistors (undoped semiconductors), and very thin connecting "wires" (= conductors) which are heavily doped thin lines. Capacitors and inductors can also be formed, but also can be connected separately. Integrated circuits are the heart of computers, televisions, calculators, cameras, and the electronic instruments that control aircraft, space vehicles, and automobiles.

A tiny chip, a few millimeters on a side, may contain billions of transistors and other circuit elements. The number of elements/mm^2 has been doubling every 2 or 3 years. We often hear of the **technology generation**, which is a number that refers to the minimum width of a conducting line ("wire"). The gate of a MOSFET may be even smaller. Since 2003 we have passed from 90-nm technology to 65-nm, to 45-nm, to 32-nm, to 22-nm, every 2 to 3 years, and now 16-nm technology which—being only a few atoms wide—may involve new structures and quantum-mechanical effects. Smaller means more diodes and transistors per mm^2 and therefore greater speed (faster response time) because the distance signals have to travel is less. Smaller also means lower power consumption. Size, speed, and power have all been improved 10 to 100 million times in the last 40 years.

[†]The "MOS" comes from a version with a **M**etal gate, silicon di**O**xide insulator, and a **S**emiconductor (*p*-type shown in Fig. 29–42a). The gate can also be heavily doped silicon (= good conductor).

Summary

Quantum mechanics explains the bonding together of atoms to form **molecules**. In a **covalent bond**, the atoms share electrons. The electron clouds of two or more atoms overlap because of constructive interference between the electron waves. The positive nuclei are attracted to this concentration of negative charge between them, forming the bond.

An **ionic bond** is an extreme case of a covalent bond in which one or more electrons from one atom spend much more time around the other atom than around their own. The atoms then act as oppositely charged ions that attract each other, forming the bond.

These **strong bonds** hold molecules together, and also hold atoms and molecules together in solids. Also important are **weak bonds** (or **van der Waals bonds**), which are generally dipole attractions between molecules.

When atoms combine to form molecules, the energy levels of the outer electrons are altered because they now interact with each other. Additional energy levels also become possible because the atoms can vibrate with respect to each other, and the molecule as a whole can rotate. The energy levels for both vibrational and rotational motion are quantized, and are very close together (typically, 10^{-1} eV to 10^{-3} eV apart). Each atomic energy level thus becomes a set of closely spaced levels corresponding to the vibrational and rotational motions. Transitions from one level to another appear as many very closely spaced lines. The resulting spectra are called **band spectra**.

The quantized rotational energy levels are given by

$$E_{\text{rot}} = \ell(\ell + 1)\frac{\hbar^2}{2I}, \quad \ell = 0, 1, 2, \cdots, \quad \text{(29–1)}$$

where I is the moment of inertia of the molecule.

The energy levels for vibrational motion are given by

$$E_{\text{vib}} = \left(v + \tfrac{1}{2}\right)hf, \quad v = 0, 1, 2, \cdots, \quad \text{(29–3)}$$

where f is the classical natural frequency of vibration for the molecule. Transitions between energy levels are subject to the selection rules $\Delta\ell = \pm 1$ and $\Delta v = \pm 1$.

Some **solids** are bound together by covalent and ionic bonds, just as molecules are. In metals, the electrostatic force between free electrons and positive ions helps form the **metallic bond**.

In the free-electron theory of metals, electrons occupy the possible energy states according to the exclusion principle. At $T = 0\,\text{K}$, all possible states are filled up to a maximum energy level called the **Fermi energy**, E_F, the magnitude of which is typically a few eV. All states above E_F are vacant at $T = 0\,\text{K}$.

In a crystalline solid, the possible energy states for electrons are arranged in **bands**. Within each band the levels are very close together, but between the bands there may be forbidden **energy gaps**. Good conductors are characterized by the highest occupied band (the **conduction band**) being only partially full, so lots of states are available to electrons to move about and accelerate when a voltage is applied. In a good insulator, the highest occupied energy band (the **valence band**) is completely full, and there is a large energy gap (5 to 10 eV) to the next highest band, the *conduction band*. At room temperature, molecular kinetic energy (thermal energy) available due to collisions is only about 0.04 eV, so almost no electrons can jump from the valence to the conduction band in an insulator. In a **semiconductor**, the gap between valence and conduction bands is much smaller, on the order of 1 eV, so a few electrons can make the transition from the essentially full valence band to the nearly empty conduction band, allowing a small amount of conductivity.

In a **doped** semiconductor, a small percentage of impurity atoms with five or three valence electrons replace a few of the normal silicon atoms with their four valence electrons. A five-electron impurity produces an **n-type** semiconductor with negative electrons as carriers of current. A three-electron impurity produces a **p-type** semiconductor in which positive **holes** carry the current. The energy level of impurity atoms lies slightly below the conduction band in an *n*-type semiconductor, and acts as a **donor** from which electrons readily pass into the conduction band. The energy level of impurity atoms in a *p*-type semiconductor lies slightly above the valence band and acts as an **acceptor** level, since electrons from the valence band easily reach it, leaving holes behind to act as charge carriers.

A semiconductor **diode** consists of a **pn junction** and allows current to flow in one direction only; *pn* junction diodes are used as **rectifiers** to change ac to dc, as photovoltaic cells to produce electricity from sunlight, and as lasers. **Light-emitting diodes** (**LED**) use compound semiconductors which can emit light when a forward-bias voltage is applied; uses include readouts, infrared remote controls, visible lighting (flashlights, street lights), and very large TV screens. LEDs using organic molecules or polymers (**OLED**) are used as screens on cell phones and other displays. Common **transistors** consist of three semiconductor sections, either as *pnp* or *npn*. Transistors can amplify electrical signals and in computers serve as switches or **gates** for the 1s and 0s of digital bits. An integrated circuit consists of a tiny semiconductor crystal or **chip** on which many transistors, diodes, resistors, and other circuit elements are constructed by placement of impurities.

Questions

1. What type of bond would you expect for (*a*) the N_2 molecule, (*b*) the HCl molecule, (*c*) Fe atoms in a solid?
2. Describe how the molecule $CaCl_2$ could be formed.
3. Does the H_2 molecule have a permanent dipole moment? Does O_2? Does H_2O? Explain.
4. Although the molecule H_3 is not stable, the ion H_3^+ is. Explain, using the Pauli exclusion principle.
5. Would you expect the molecule H_2^+ to be stable? If so, where would the single electron spend most of its time?
6. Explain why the carbon atom ($Z = 6$) usually forms four bonds with hydrogen-like atoms.
7. The energy of a molecule can be divided into four categories. What are they?
8. If conduction electrons are free to roam about in a metal, why don't they leave the metal entirely?
9. Explain why the resistivity of metals increases with increasing temperature whereas the resistivity of semiconductors may decrease with increasing temperature.
10. Compare the resistance of a *pn* junction diode connected in forward bias to its resistance when connected in reverse bias.
11. Explain how a transistor can be used as a switch.

12. Figure 29–43 shows a "bridge-type" full-wave rectifier. Explain how the current is rectified and how current flows during each half cycle.

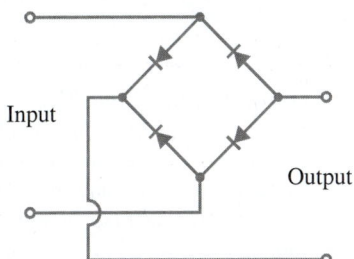

FIGURE 29–43 Question 12.

13. What is the main difference between *n*-type and *p*-type semiconductors?

14. Explain on the basis of energy bands why the sodium chloride crystal is a good insulator. [*Hint*: Consider the shells of Na$^+$ and Cl$^-$ ions.]

15. In a transistor, the base–emitter junction and the base–collector junction are essentially diodes. Are these junctions reverse-biased or forward-biased in the application shown in Fig. 29–41?

16. A transistor can amplify an electronic signal, meaning it can increase the power of an input signal. Where does it get the energy to increase the power?

17. A silicon semiconductor is doped with phosphorus. Will these atoms be donors or acceptors? What type of semiconductor will this be?

18. Do diodes and transistors obey Ohm's law? Explain.

19. Can a diode be used to amplify a signal? Explain.

MisConceptual Questions

1. What holds molecules together?
   (*a*) Gravitational forces.
   (*b*) Magnetic forces.
   (*c*) Electric forces.
   (*d*) Glue.
   (*e*) Nuclear forces.

2. Which of the following is true for covalently bound diatomic molecules such as H$_2$?
   (*a*) All electrons in the two atoms have identical quantum numbers.
   (*b*) The molecule has fewer electrons than the two separate atoms do.
   (*c*) The molecule has less energy than two separate atoms.
   (*d*) The energy of the molecule is greatest when the atoms are separated by one bond length.

3. A hydrogen atom ($Z = 1$) is bonded to a lithium atom ($Z = 3$) in lithium hydride, LiH. Which of the following are possible spin states of the two shared electrons?
   (*a*) $+\frac{1}{2}, +\frac{1}{2}$.
   (*b*) $-\frac{1}{2}, -\frac{1}{2}$.
   (*c*) $+\frac{1}{2}, -\frac{1}{2}$.
   (*d*) Both (*a*) and (*b*).
   (*e*) Any of the above.

4. Ionic bonding is related to
   (*a*) magnetic dipole interactions.
   (*b*) the transfer of one or more electrons from one atom to another.
   (*c*) the sharing of electrons between atoms.
   (*d*) the transfer of electrons to the solid.
   (*e*) oscillation dipoles.

5. Consider Fig. 29–10. As the last phosphate group approaches and then bonds to the ADP molecule, which of the following is true? Choose all that apply.
   (*a*) The phosphate group is first repelled and then attracted to the ADP molecule.
   (*b*) The phosphate group is always attracted to the ADP molecule.
   (*c*) The phosphate group is always repelled by the ADP molecule.
   (*d*) The system first loses and then stores potential energy.
   (*e*) Both binding energy and activation energy are negative.
   (*f*) Both binding energy and activation energy are positive.

6. Which type of bond holds the molecules of the DNA double helix together?
   (*a*) Covalent bond.
   (*b*) Ionic bond.
   (*c*) Einstein bond.
   (*d*) Van der Waals bond.

7. In a *p*-type semiconductor, a hole is
   (*a*) a region in the molecular structure where an atom is missing.
   (*b*) an extra electron from one of the donor atoms.
   (*c*) an extra positively charged particle in the molecular structure.
   (*d*) a region missing an electron relative to the rest of the molecular structure.

8. The electrical resistance of a semiconductor may decrease with increasing temperature because, at elevated temperature, more electrons
   (*a*) collide with the crystal lattice.
   (*b*) move faster.
   (*c*) are able to jump across the energy gap.
   (*d*) form weak van der Waals bonds.

9. Which of the following would *not* be used as an impurity in doping silicon?
   (*a*) Germanium.
   (*b*) Gallium.
   (*c*) Boron.
   (*d*) Phosphorus.
   (*e*) Arsenic.

10. Why are metals good conductors?
    (*a*) Gaining a tiny bit of energy allows their electrons to move.
    (*b*) They have more electrons than protons, so some of the electrons are extra and free to move.
    (*c*) They have more protons than electrons, so some of the protons are extra and free to move.
    (*d*) Gaining a tiny bit of energy allows their protons to move.
    (*e*) Electrons are tightly bound to their atoms.

Problems

*29–1 to 29–3 Molecular Bonds

1. (I) Estimate the binding energy of a KCl molecule by calculating the electrostatic potential energy when the K^+ and Cl^- ions are at their stable separation of 0.28 nm. Assume each has a charge of magnitude $1.0e$.

2. (II) The measured binding energy of KCl is 4.43 eV. From the result of Problem 1, estimate the contribution to the binding energy of the repelling electron clouds at the equilibrium distance $r_0 = 0.28$ nm.

3. (II) The equilibrium distance r_0 between two atoms in a molecule is called the **bond length**. Using the bond lengths of homogeneous molecules (like H_2, O_2, and N_2), one can estimate the bond length of heterogeneous molecules (like CO, CN, and NO). This is done by summing half of each bond length of the homogenous molecules to estimate that of the heterogeneous molecule. Given the following bond lengths: H_2 (= 74 pm), N_2 (= 145 pm), O_2 (= 121 pm), C_2 (= 154 pm), estimate the bond lengths for: HN, CN, and NO.

4. (II) Binding energies are often measured experimentally in kcal per mole, and then the binding energy in eV per molecule is calculated from that result. What is the conversion factor in going from kcal per mole to eV per molecule? What is the binding energy of KCl (= 4.43 eV) in kcal per mole?

5. (III) Estimate the binding energy of the H_2 molecule, assuming the two H nuclei are 0.074 nm apart and the two electrons spend 33% of their time midway between them.

6. (III) (a) Apply reasoning similar to that in the text for the $S = 0$ and $S = 1$ states in the formation of the H_2 molecule to show why the molecule He_2 is *not* formed. (b) Explain why the He_2^+ molecular ion *could* form. (Experiment shows it has a binding energy of 3.1 eV at $r_0 = 0.11$ nm.)

*29–4 Molecular Spectra

7. (I) Show that the quantity $\hbar^2/I$ has units of energy.

8. (II) (a) Calculate the "characteristic rotational energy," $\hbar^2/2I$, for the O_2 molecule whose bond length is 0.121 nm. (b) What are the energy and wavelength of photons emitted in an $\ell = 3$ to $\ell = 2$ transition?

9. (II) The "characteristic rotational energy," $\hbar^2/2I$, for N_2 is 2.48×10^{-4} eV. Calculate the N_2 bond length.

10. (II) The equilibrium separation of H atoms in the H_2 molecule is 0.074 nm (Fig. 29–8). Calculate the energies and wavelengths of photons for the rotational transitions (a) $\ell = 1$ to $\ell = 0$, (b) $\ell = 2$ to $\ell = 1$, and (c) $\ell = 3$ to $\ell = 2$.

11. (II) Determine the wavelength of the photon emitted when the CO molecule makes the rotational transition $\ell = 5$ to $\ell = 4$. [*Hint*: See Example 29–2.]

12. (II) Calculate the bond length for the NaCl molecule given that three successive wavelengths for rotational transitions are 23.1 mm, 11.6 mm, and 7.71 mm.

13. (II) (a) Use the curve of Fig. 29–17 to estimate the stiffness constant k for the H_2 molecule. (Recall that $PE = \frac{1}{2}kx^2$.) (b) Then estimate the fundamental wavelength for vibrational transitions using the classical formula (Chapter 11), but use only $\frac{1}{2}$ the mass of an H atom (because both H atoms move).

*29–5 Bonding in Solids

14. (II) Common salt, NaCl, has a density of 2.165 g/cm³. The molecular weight of NaCl is 58.44. Estimate the distance between nearest neighbor Na and Cl ions. [*Hint*: Each ion can be considered to be at the corner of a cube.]

15. (II) Repeat Problem 14 for KCl whose density is 1.99 g/cm³.

16. (II) The spacing between "nearest neighbor" Na and Cl ions in a NaCl crystal is 0.24 nm. What is the spacing between two nearest neighbor Na ions?

*29–7 Band Theory of Solids

17. (I) A semiconductor is struck by light of slowly increasing frequency and begins to conduct when the wavelength of the light is 620 nm. Estimate the energy gap E_g.

18. (I) Calculate the longest-wavelength photon that can cause an electron in silicon ($E_g = 1.12$ eV) to jump from the valence band to the conduction band.

19. (II) The energy gap between valence and conduction bands in germanium is 0.72 eV. What range of wavelengths can a photon have to excite an electron from the top of the valence band into the conduction band?

20. (II) The band gap of silicon is 1.12 eV. (a) For what range of wavelengths will silicon be transparent? (See Example 29–5.) In what region of the electromagnetic spectrum does this transparent range begin? (b) If window glass is transparent for all visible wavelengths, what is the minimum possible band gap value for glass (assume $\lambda = 400$ nm to 700 nm)? [*Hint*: If the photon has less energy than the band gap, the photon will pass through the solid without being absorbed.]

21. (II) The energy gap E_g in germanium is 0.72 eV. When used as a photon detector, roughly how many electrons can be made to jump from the valence to the conduction band by the passage of an 830-keV photon that loses all its energy in this fashion?

22. (III) We saw that there are $2N$ possible electron states in the 3s band of Na, where N is the total number of atoms. How many possible electron states are there in the (a) 2s band, (b) 2p band, and (c) 3p band? (d) State a general formula for the total number of possible states in any given electron band.

*29–8 Semiconductors and Doping

23. (III) Suppose that a silicon semiconductor is doped with phosphorus so that one silicon atom in 1.5×10^6 is replaced by a phosphorus atom. Assuming that the "extra" electron in every phosphorus atom is donated to the conduction band, by what factor is the density of conduction electrons increased? The density of silicon is 2330 kg/m³, and the density of conduction electrons in pure silicon is about 10^{16} m⁻³ at room temperature.

*29–9 Diodes

24. (I) At what wavelength will an LED radiate if made from a material with an energy gap $E_g = 1.3\,\text{eV}$?

25. (I) If an LED emits light of wavelength $\lambda = 730\,\text{nm}$, what is the energy gap (in eV) between valence and conduction bands?

26. (I) A semiconductor diode laser emits 1.3-μm light. Assuming that the light comes from electrons and holes recombining, what is the band gap in this laser material?

27. (II) A silicon diode, whose current–voltage characteristics are given in Fig. 29–30, is connected in series with a battery and a 960-Ω resistor. What battery voltage is needed to produce a 14-mA current?

28. (II) An ac voltage of 120-V rms is to be rectified. Estimate very roughly the average current in the output resistor R (= 31 kΩ) for (a) a half-wave rectifier (Fig. 29–31), and (b) a full-wave rectifier (Fig. 29–32) without capacitor.

29. (III) Suppose that the diode of Fig. 29–30 is connected in series to a 180-Ω resistor and a 2.0-V battery. What current flows in the circuit? [*Hint*: Draw a line on Fig. 29–30 representing the current in the resistor as a function of the voltage across the diode; the intersection of this line with the characteristic curve will give the answer.]

30. (III) Sketch the resistance as a function of current, for $V > 0$, for the diode shown in Fig. 29–30.

31. (III) A 120-V rms 60-Hz voltage is to be rectified with a full-wave rectifier as in Fig. 29–32, where $R = 33\,\text{k}\Omega$, and $C = 28\,\mu\text{F}$. (a) Make a rough estimate of the average current. (b) What happens if $C = 0.10\,\mu\text{F}$? [*Hint*: See Section 19–6.]

*29–10 Transistors

32. (I) From Fig. 29–41, write an equation for the relationship between the base current (I_B), the collector current (I_C), and the emitter current (I_E, not labeled in Fig. 29–41). Assume $i_B = i_C = 0$.

33. (I) Draw a circuit diagram showing how a *pnp* transistor can operate as an amplifier, similar to Fig. 29–41 showing polarities, etc.

34. (II) If the current gain of the transistor amplifier in Fig. 29–41 is $\beta = i_C/i_B = 95$, what value must R_C have if a 1.0-μA ac base current is to produce an ac output voltage of 0.42 V?

35. (II) Suppose that the current gain of the transistor in Fig. 29–41 is $\beta = i_C/i_B = 85$. If $R_C = 3.8\,\text{k}\Omega$, calculate the ac output voltage for an ac input current of 2.0 μA.

36. (II) An amplifier has a voltage gain of 75 and a 25-kΩ load (output) resistance. What is the peak output current through the load resistor if the input voltage is an ac signal with a peak of 0.080 V?

37. (II) A transistor, whose current gain $\beta = i_C/i_B = 65$, is connected as in Fig. 29–41 with $R_B = 3.8\,\text{k}\Omega$ and $R_C = 7.8\,\text{k}\Omega$. Calculate (a) the voltage gain, and (b) the power amplification.

General Problems

38. Use the uncertainty principle to estimate the binding energy of the H_2 molecule by calculating the difference in kinetic energy of the electrons between (i) when they are in separate atoms and (ii) when they are in the molecule. Take Δx for the electrons in the separated atoms to be the radius of the first Bohr orbit, 0.053 nm, and for the molecule take Δx to be the separation of the nuclei, 0.074 nm. [*Hint*: Let $\Delta p \approx \Delta p_x$.]

39. The average translational kinetic energy of an atom or molecule is about $KE = \frac{3}{2}kT$ (see Section 13–9), where $k = 1.38 \times 10^{-23}\,\text{J/K}$ is Boltzmann's constant. At what temperature T will KE be on the order of the bond energy (and hence the bond easily broken by thermal motion) for (a) a covalent bond (say H_2) of binding energy 4.0 eV, and (b) a "weak" hydrogen bond of binding energy 0.12 eV?

40. A diatomic molecule is found to have an activation energy of 1.3 eV. When the molecule is disassociated, 1.6 eV of energy is released. Draw a potential energy curve for this molecule.

41. In the ionic salt KF, the separation distance between ions is about 0.27 nm. (a) Estimate the electrostatic potential energy between the ions assuming them to be point charges (magnitude 1e). (b) When F "grabs" an electron, it releases 3.41 eV of energy, whereas 4.34 eV is required to ionize K. Find the binding energy of KF relative to free K and F atoms, neglecting the energy of repulsion.

42. The rotational absorption spectrum of a molecule displays peaks about 8.9×10^{11} Hz apart. Determine the moment of inertia of this molecule.

43. For O_2 with a bond length of 0.121 nm, what is the moment of inertia about an axis through the center of mass perpendicular to the line joining the 2 atoms?

44. Must we consider quantum effects for everyday rotating objects? Estimate the differences between rotational energy levels for a spinning baton compared to the energy of the baton. Assume the baton consists of a uniform 32-cm-long bar with a mass of 230 g and two small end masses, each of mass 380 g, and it rotates at 1.8 rev/s about the bar's center.

45. For a certain semiconductor, the longest wavelength radiation that can be absorbed is 2.06 mm. What is the energy gap in this semiconductor?

46. When EM radiation is incident on diamond, it is found that light with wavelengths shorter than 226 nm will cause the diamond to conduct. What is the energy gap between the valence band and the conduction band for diamond?

47. The energy gap between valence and conduction bands in zinc sulfide is 3.6 eV. What range of wavelengths can a photon have to excite an electron from the top of the valence band into the conduction band?

48. Most of the Sun's radiation has wavelengths shorter than 1100 nm. For a solar cell to absorb all this, what energy gap ought the material have?

49. A TV remote control emits IR light. If the detector on the TV set is *not* to react to visible light, could it make use of silicon as a "window" with its energy gap $E_g = 1.12$ eV? What is the shortest-wavelength light that can strike silicon without causing electrons to jump from the valence band to the conduction band?

50. Green and blue LEDs became available many years after red LEDs were first developed. Approximately what energy gaps would you expect to find in green (525 nm) and in blue (465 nm) LEDs?

51. Consider a monatomic solid with a weakly bound cubic lattice, with each atom connected to six neighbors, each bond having a binding energy of 3.4×10^{-3} eV. When this solid melts, its latent heat of fusion goes directly into breaking the bonds between the atoms. Estimate the latent heat of fusion for this solid, in J/mol. [*Hint*: Show that in a simple cubic lattice (Fig. 29–44), there are *three* times as many bonds as there are atoms, when the number of atoms is large.]

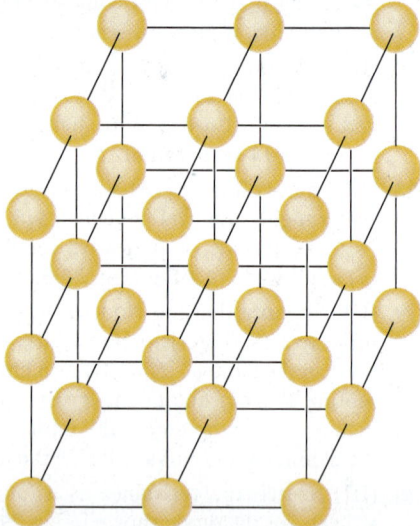

FIGURE 29–44 Problem 51.

Search and Learn

1. Explain why metals are shiny. (See Section 29–5.)
2. Compare the potential energy diagram for an H_2 molecule with the potential energy diagram for ATP formation from ADP and Ⓟ. Explain the significance of the difference in shapes of the two diagrams. (See Section 29–2.)
3. (*a*) Why are weak bonds important in cells? (*b*) Explain why heating proteins too much may cause them to denature—that is, lose the specific shape they need to function. (See Section 29–3.) (*c*) What is the strongest weak bond, and why? (*d*) If this bond, and the other weak bonds, were stronger (that is, too strong), what would be the consequence for protein synthesis?
4. Assume conduction electrons in a semiconductor behave as an ideal gas. (This is not true for conduction electrons in a metal.) (*a*) Taking mass $m = 9 \times 10^{-31}$ kg and temperature $T = 300$ K, determine the de Broglie wavelength of a semiconductor's conduction electrons. (*b*) Given that the spacing between atoms in a semiconductor's atomic lattice is on the order of 0.3 nm, would you expect room-temperature conduction electrons to travel in straight lines or diffract when traveling through this lattice? Explain.

5. A strip of silicon 1.6 cm wide and 1.0 mm thick is immersed in a magnetic field of strength 1.5 T perpendicular to the strip (Fig. 29–45). When a current of 0.28 mA is run through the strip, there is a resulting Hall effect voltage of 18 mV across the strip (Section 20–4). How many electrons per silicon atom are in the conduction band? The density of silicon is 2330 kg/m³.

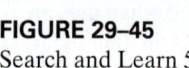

FIGURE 29–45 Search and Learn 5.

6. For an arsenic donor atom in a doped silicon semiconductor, assume that the "extra" electron moves in a Bohr orbit about the arsenic ion. For this electron in the ground state, take into account the dielectric constant $K = 12$ of the Si lattice (which represents the weakening of the Coulomb force due to all the other atoms or ions in the lattice), and estimate (*a*) the binding energy, and (*b*) the orbit radius for this extra electron. [*Hint*: Substitute $\epsilon = K\epsilon_0$ in Coulomb's law; see Section 17–8 and also 27–12.]

ANSWERS TO EXERCISES

A: 0; 5.00×10^{-4} eV; 1.50×10^{-3} eV.
B: 1.30 mm, 0.87 mm, 0.65 mm.
C: 0.81 eV.
D: (*c*).
E: (*c*), (*e*).

In this Chapter we begin our discussion of nuclear physics. We study the properties of nuclei, the various forms of radioactivity, and how radioactive decay can be used in a variety of fields to determine the age of old objects, from bones and trees to rocks and other mineral substances, and obtain information on the history of the Earth.

Shown is one version of a **Chart of the Nuclides**. Each horizontal row has a square for each known isotope (nuclide) of one element with a particular Z value (= number of electrons in the neutral atom = number of protons in the nucleus). At the far left is a white box with the average atomic weight (or a range if uncertain) of the naturally occurring isotopes of that element. Each vertical column contains nuclides with the same neutron number N. For $N = 1$ (to right of pencil), starting at the bottom, there is a lone neutron, then above it $^2_1$H, then $^3_2$He and $^4_3$Li. Each square is color coded: black means a stable nuclide. Radioactive nuclides are blue green for β^- decay, pink for β^+ decay or electron capture (ε) such as $^7_4$Be, yellow for α decay, and so on. Thus $^1_1$H and $^2_1$H are stable but $^3_1$H (tritium) undergoes β^- decay with half-life = 12.3 years ("a" is for Latin "anno" = year). The squares contain the atomic mass of that isotope, or half-life and energy released if radioactive. Other details may be alternate decay modes and certain cross sections (σ).

Nuclear Physics and Radioactivity

CHAPTER 30

CHAPTER-OPENING QUESTION—Guess now!

If half of an 80-μg sample of $^{60}_{27}$Co decays in 5.3 years, how much $^{60}_{27}$Co is left in 10.6 years?

(a) 10 μg.
(b) 20 μg.
(c) 30 μg.
(d) 40 μg.
(e) 0 μg.

CONTENTS

30–1 Structure and Properties of the Nucleus
30–2 Binding Energy and Nuclear Forces
30–3 Radioactivity
30–4 Alpha Decay
30–5 Beta Decay
30–6 Gamma Decay
30–7 Conservation of Nucleon Number and Other Conservation Laws
30–8 Half-Life and Rate of Decay
30–9 Calculations Involving Decay Rates and Half-Life
30–10 Decay Series
30–11 Radioactive Dating
*30–12 Stability and Tunneling
30–13 Detection of Particles

I n the early part of the twentieth century, Rutherford's experiments (Section 27–10) led to the idea that at the center of an atom there is a tiny but massive nucleus with a positive charge. At the same time that the quantum theory was being developed and scientists were attempting to understand the structure of the atom and its electrons, investigations into the nucleus itself had also begun. In this Chapter and the next, we take a brief look at *nuclear physics*.

30–1 Structure and Properties of the Nucleus

An important question for physicists was whether the nucleus had a structure, and what that structure might be. By the early 1930s, a model of the nucleus had been developed that is still useful. According to this model, a nucleus is made up of two types of particles: protons and neutrons. [These "particles" also have wave properties, but for ease of visualization and language, we usually refer to them simply as "particles."] A **proton** is the nucleus of the simplest atom, hydrogen. The proton has a positive charge ($= +e = +1.60 \times 10^{-19}$ C, the same magnitude as for the electron) and its mass is measured to be

$$m_p = 1.67262 \times 10^{-27} \text{ kg}.$$

The **neutron**, whose existence was ascertained in 1932 by the English physicist James Chadwick (1891–1974), is electrically neutral ($q = 0$), as its name implies. Its mass is very slightly larger than that of the proton:

$$m_n = 1.67493 \times 10^{-27} \text{ kg}.$$

These two constituents of a nucleus, neutrons and protons, are referred to collectively as **nucleons**.

Although a normal hydrogen nucleus consists of a single proton alone, the nuclei of all other elements consist of both neutrons and protons. The different nuclei are often referred to as **nuclides**. The number of protons in a nucleus (or nuclide) is called the **atomic number** and is designated by the symbol Z. The total number of nucleons, neutrons plus protons, is designated by the symbol A and is called the **atomic mass number**, or sometimes simply **mass number**. This name is used since the mass of a nucleus is very closely A times the mass of one nucleon. A nuclide with 7 protons and 8 neutrons thus has $Z = 7$ and $A = 15$. The **neutron number** N is $N = A - Z$.

To specify a given nuclide, we need give only A and Z. A special symbol is commonly used which takes the form

$$^A_Z X,$$

where X is the chemical symbol for the element (see Appendix B, and the Periodic Table inside the back cover), A is the atomic mass number, and Z is the atomic number. For example, $^{15}_{7}N$ means a nitrogen nucleus containing 7 protons and 8 neutrons for a total of 15 nucleons. In a neutral atom, the number of electrons orbiting the nucleus is equal to the atomic number Z (since the charge on an electron has the same magnitude but opposite sign to that of a proton). The main properties of an atom, and how it interacts with other atoms, are largely determined by the number of electrons. Hence Z determines what kind of atom it is: carbon, oxygen, gold, or whatever. It is redundant to specify both the symbol of a nucleus and its atomic number Z as described above. If the nucleus is nitrogen, for example, we know immediately that $Z = 7$. The subscript Z is thus sometimes dropped and $^{15}_{7}N$ is then written simply ^{15}N; in words we say "nitrogen fifteen."

For a particular type of atom (say, carbon), nuclei are found to contain different numbers of neutrons, although they all have the same number of protons. For example, carbon nuclei always have 6 protons, but they may have 5, 6, 7, 8, 9, or 10 neutrons. Nuclei that contain the same number of protons but different numbers of neutrons are called **isotopes**. Thus, $^{11}_{6}C$, $^{12}_{6}C$, $^{13}_{6}C$, $^{14}_{6}C$, $^{15}_{6}C$, and $^{16}_{6}C$ are all isotopes of carbon. The isotopes of a given element are not all equally common. For example, 98.9% of naturally occurring carbon (on Earth) is the isotope $^{12}_{6}C$, and about 1.1% is $^{13}_{6}C$. These percentages are referred to as the **natural abundances**.[†] Even hydrogen has isotopes: 99.99% of natural hydrogen is $^{1}_{1}H$, a simple proton, as the nucleus; there are also $^{2}_{1}H$, called **deuterium**, and $^{3}_{1}H$, **tritium**, which besides the proton contain 1 or 2 neutrons. (The bare nucleus in each case is called the **deuteron** and **triton**.)

[†]The mass value for each element as given in the Periodic Table (inside back cover) is an average weighted according to the natural abundances of its isotopes.

Many isotopes that do not occur naturally can be produced in the laboratory by means of nuclear reactions (more on this later). Indeed, all elements beyond uranium ($Z > 92$) do not occur naturally on Earth and are only produced artificially (in the laboratory), as are many nuclides with $Z \leq 92$.

The approximate size of nuclei was determined originally by Rutherford from the scattering of charged particles by thin metal foils. We cannot speak about a definite size for nuclei because of the wave–particle duality (Section 27–7): their spatial extent must remain somewhat fuzzy. Nonetheless a rough "size" can be measured by scattering high-speed electrons off nuclei. It is found that nuclei have a roughly spherical shape with a radius that increases with A according to the approximate formula

$$r \approx (1.2 \times 10^{-15}\,\text{m})(A^{\frac{1}{3}}). \tag{30-1}$$

Since the volume of a sphere is $V = \frac{4}{3}\pi r^3$, we see that the volume of a nucleus is approximately proportional to the number of nucleons, $V \propto A$ (because $(A^{\frac{1}{3}})^3 = A$). This is what we would expect if nucleons were like impenetrable billiard balls: if you double the number of balls, you double the total volume. Hence, all nuclei have nearly the same density, and it is enormous (see Example 30–2).

The metric abbreviation for 10^{-15} m is the fermi (after Enrico Fermi, Fig. 30–7) or the femtometer, fm (see Table 1–4 or inside the front cover). Thus $1.2 \times 10^{-15}\,\text{m} = 1.2\,\text{fm}$ or 1.2 fermis.

EXAMPLE 30–1 ESTIMATE Nuclear sizes. Estimate the diameter of the smallest and largest naturally occurring nuclei: (a) $^{1}_{1}\text{H}$, (b) $^{238}_{92}\text{U}$.

APPROACH The radius r of a nucleus is related to its number of nucleons A by Eq. 30–1. The diameter $d = 2r$.

SOLUTION (a) For hydrogen, $A = 1$, Eq. 30–1 gives

$$d = \text{diameter} = 2r \approx 2(1.2 \times 10^{-15}\,\text{m})(A^{\frac{1}{3}}) = 2.4 \times 10^{-15}\,\text{m}$$

since $A^{\frac{1}{3}} = 1^{\frac{1}{3}} = 1$.

(b) For uranium $d \approx (2.4 \times 10^{-15}\,\text{m})(238)^{\frac{1}{3}} = 15 \times 10^{-15}\,\text{m}$.

The range of nuclear diameters is only from 2.4 fm to 15 fm.

NOTE Because nuclear radii vary as $A^{\frac{1}{3}}$, the largest nuclei (such as uranium with $A = 238$) have a radius only about $\sqrt[3]{238} \approx 6$ times that of the smallest, hydrogen ($A = 1$).

EXAMPLE 30–2 ESTIMATE Nuclear and atomic densities. Compare the density of nuclear matter to the density of normal solids.

APPROACH The density of normal liquids and solids is on the order of 10^3 to $10^4\,\text{kg/m}^3$ (see Table 10–1), and because the atoms are close packed, atoms have about this density too. We therefore compare the density (mass per volume) of a nucleus to that of its atom as a whole.

SOLUTION The mass of a proton is greater than the mass of an electron by a factor

$$\frac{1.67 \times 10^{-27}\,\text{kg}}{9.1 \times 10^{-31}\,\text{kg}} \approx 2000.$$

Thus, over 99.9% of the mass of an atom is in the nucleus, and for our estimate we can say the mass of the atom equals the mass of the nucleus, $m_{\text{nucl}}/m_{\text{atom}} = 1$. Atoms have a radius of about 10^{-10} m (Chapter 27) and nuclei on the order of 10^{-15} m (Eq. 30–1). Thus the ratio of nuclear density to atomic density is about

$$\frac{\rho_{\text{nucl}}}{\rho_{\text{atom}}} = \frac{(m_{\text{nucl}}/V_{\text{nucl}})}{(m_{\text{atom}}/V_{\text{atom}})} = \left(\frac{m_{\text{nucl}}}{m_{\text{atom}}}\right)\frac{\frac{4}{3}\pi r_{\text{atom}}^3}{\frac{4}{3}\pi r_{\text{nucl}}^3} \approx (1)\frac{(10^{-10})^3}{(10^{-15})^3} = 10^{15}.$$

The nucleus is 10^{15} times more dense than ordinary matter.

The masses of nuclei can be determined from the radius of curvature of fast-moving nuclei (as ions) in a known magnetic field using a mass spectrometer, as discussed in Section 20–11. Indeed the existence of different isotopes of the same element (different number of neutrons) was discovered using this device.

CAUTION
Masses are for neutral atom (nucleus plus electrons)

Nuclear masses can be specified in **unified atomic mass units** (u). On this scale, a neutral $^{12}_{6}C$ atom is given the exact value 12.000000 u. A neutron then has a measured mass of 1.008665 u, a proton 1.007276 u, and a neutral hydrogen atom $^{1}_{1}H$ (proton plus electron) 1.007825 u. The masses of many nuclides are given in Appendix B. It should be noted that the masses in this Table, as is customary, are for the *neutral atom* (including electrons), and not for a bare nucleus.

Masses may be specified using the electron-volt energy unit, $1\,eV = 1.6022 \times 10^{-19}\,J$ (Section 17–4). This can be done because mass and energy are related, and the precise relationship is given by Einstein's equation $E = mc^2$ (Chapter 26). Since the mass of a proton is $1.67262 \times 10^{-27}\,kg$, or 1.007276 u, then 1 u is equal to

$$1.0000\,u = (1.0000\,u)\left(\frac{1.67262 \times 10^{-27}\,kg}{1.007276\,u}\right) = 1.66054 \times 10^{-27}\,kg;$$

this is equivalent to an energy (see Table inside front cover) in MeV ($= 10^6\,eV$) of

$$E = mc^2 = \frac{(1.66054 \times 10^{-27}\,kg)(2.9979 \times 10^8\,m/s)^2}{(1.6022 \times 10^{-19}\,J/eV)} = 931.5\,MeV.$$

Thus,

$$1\,u = 1.6605 \times 10^{-27}\,kg = 931.5\,MeV/c^2.$$

The rest masses of some of the basic particles are given in Table 30–1. As a rule of thumb, to remember, the masses of neutron and proton are about $1\,GeV/c^2$ ($= 1000\,MeV/c^2$) which is about 2000 times the mass of an electron ($\approx \frac{1}{2}\,MeV/c^2$).

TABLE 30–1
Rest Masses in Kilograms, Unified Atomic Mass Units, and MeV/c^2

Object	Mass		
	kg	u	MeV/c^2
Electron	9.1094×10^{-31}	0.00054858	0.51100
Proton	1.67262×10^{-27}	1.007276	938.27
$^{1}_{1}H$ atom	1.67353×10^{-27}	1.007825	938.78
Neutron	1.67493×10^{-27}	1.008665	939.57

Just as an electron has intrinsic spin and angular momentum quantum numbers, so too do nuclei and their constituents, the proton and neutron. Both the proton and the neutron are spin $\frac{1}{2}$ particles, just like the electron. A nucleus, made up of protons and neutrons, has a **nuclear spin** quantum number, I, that can be either integer or half integer, depending on whether it is made up of an even or an odd number of nucleons.

30–2 Binding Energy and Nuclear Forces

Binding Energies

The total mass of a stable nucleus is always less than the sum of the masses of its separate protons and neutrons, as the following Example shows.

EXAMPLE 30–3 $^{4}_{2}He$ **mass compared to its constituents.** Compare the mass of a $^{4}_{2}He$ atom to the total mass of its constituent particles.

PROBLEM SOLVING
Keep track of electron masses

APPROACH The $^{4}_{2}He$ nucleus contains 2 protons and 2 neutrons. Tables normally give the masses of neutral atoms—that is, nucleus plus its Z electrons. We must therefore be sure to balance out the electrons when we compare masses. Thus we use the mass of $^{1}_{1}H$ rather than that of a proton alone. We look up the mass of the $^{4}_{2}He$ atom in Appendix B (it includes the mass of 2 electrons), as well as the mass for the 2 neutrons and 2 hydrogen atoms (= 2 protons + 2 electrons).

SOLUTION The mass of a neutral $^4_2$He atom, from Appendix B, is 4.002603 u. The mass of two neutrons and two H atoms (2 protons including the 2 electrons) is

$$2m_n = 2(1.008665 \text{ u}) = 2.017330 \text{ u}$$
$$2m(^1_1\text{H}) = 2(1.007825 \text{ u}) = \underline{2.015650 \text{ u}}$$
$$\text{sum} = 4.032980 \text{ u.}$$

Thus the mass of $^4_2$He is measured to be less than the masses of its constituents by an amount $4.032980 \text{ u} - 4.002603 \text{ u} = 0.030377 \text{ u}$.

Where has this lost mass of 0.030377 u disappeared to? It must be $E = mc^2$.

If the four nucleons suddenly came together to form a $^4_2$He nucleus, the mass "loss" would appear as energy of another kind (such as radiation, or kinetic energy). The mass (or energy) difference in the case of $^4_2$He, given in energy units, is $(0.030377 \text{ u})(931.5 \text{ MeV/u}) = 28.30 \text{ MeV}$. This difference is referred to as the **total binding energy** of the nucleus. The total binding energy represents the amount of energy that must be put *into* a nucleus in order to break it apart into its constituents. If the mass of, say, a $^4_2$He nucleus were exactly equal to the mass of two neutrons plus two protons, the nucleus could fall apart without any input of energy. To be stable, the mass of a nucleus *must* be less than that of its constituent nucleons, so that energy input *is* needed to break it apart.

Binding energy is not something a nucleus has—it is energy it "lacks" relative to the total mass of its separate constituents.

[As a comparison, we saw in Chapter 27 that the binding energy of the one electron in the hydrogen atom is 13.6 eV; so the mass of a $^1_1$H atom is less than that of a single proton plus a single electron by $13.6 \text{ eV}/c^2$. The binding energies of nuclei are on the order of MeV, so the eV binding energies of electrons can be ignored. Nuclear binding energies, compared to nuclear masses, are on the order of $(28 \text{ MeV}/4000 \text{ MeV}) \approx 1 \times 10^{-2}$, where we used helium's binding energy of 28.3 MeV (see above) and mass $\approx 4 \times 940 \text{ MeV} \approx 4000 \text{ MeV}$.]

EXERCISE A Determine how much less the mass of the $^7_3$Li nucleus is compared to that of its constituents. See Appendix B.

The **binding energy per nucleon** is defined as the total binding energy of a nucleus divided by A, the total number of nucleons. We calculated above that the binding energy of $^4_2$He is 28.3 MeV, so its binding energy per nucleon is $28.3 \text{ MeV}/4 = 7.1 \text{ MeV}$. Figure 30–1 shows the measured binding energy per nucleon as a function of A for stable nuclei. The curve rises as A increases and reaches a plateau at about 8.7 MeV per nucleon above $A \approx 40$. Beyond $A \approx 80$, the curve decreases slowly, indicating that larger nuclei are held together less tightly than those in the middle of the Periodic Table. We will see later that these characteristics allow the release of nuclear energy in the processes of fission and fusion.

> **CAUTION**
> *Mass of nucleus must be less than mass of constituents*

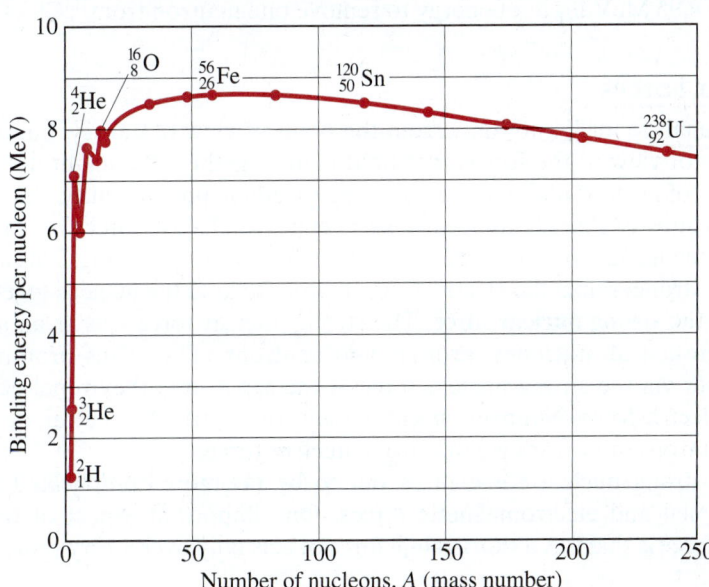

FIGURE 30–1 Binding energy per nucleon for the more stable nuclides as a function of mass number A.

EXAMPLE 30–4 **Binding energy for iron.** Calculate the total binding energy and the binding energy per nucleon for $^{56}_{26}$Fe, the most common stable isotope of iron.

APPROACH We subtract the mass of a $^{56}_{26}$Fe atom from the total mass of 26 hydrogen atoms and 30 neutrons, all found in Appendix B. Then we convert mass units to energy units; finally we divide by $A = 56$, the total number of nucleons.

SOLUTION $^{56}_{26}$Fe has 26 protons and 30 neutrons whose separate masses are

$$26m(^1_1\text{H}) = (26)(1.007825 \text{ u}) = 26.20345 \text{ u} \quad \text{(includes 26 electrons)}$$
$$30m_n = (30)(1.008665 \text{ u}) = \underline{30.25995 \text{ u}}$$
$$\text{sum} = 56.46340 \text{ u}.$$
$$\text{Subtract mass of } ^{56}_{26}\text{Fe:} = \underline{-55.93494 \text{ u}} \quad \text{(Appendix B)}$$
$$\Delta m = 0.52846 \text{ u}.$$

The total binding energy is thus

$$(0.52846 \text{ u})(931.5 \text{ MeV/u}) = 492.26 \text{ MeV}$$

and the binding energy per nucleon is

$$\frac{492.26 \text{ MeV}}{56 \text{ nucleons}} = 8.79 \text{ MeV}.$$

NOTE The binding energy per nucleon graph (Fig. 30–1) peaks about here, for iron. So the iron nucleus, and its neighbors, are the most stable of nuclei.

EXERCISE B Determine the binding energy per nucleon for $^{16}_{8}$O.

EXAMPLE 30–5 **Binding energy of last neutron.** What is the binding energy of the last neutron in $^{13}_{6}$C?

APPROACH If $^{13}_{6}$C lost one neutron, it would be $^{12}_{6}$C. We subtract the mass of $^{13}_{6}$C from the masses of $^{12}_{6}$C and a free neutron.

SOLUTION Obtaining the masses from Appendix B, we have

$$\text{Mass } ^{12}_{6}\text{C} = 12.000000 \text{ u}$$
$$\text{Mass } ^1_0\text{n} = \underline{1.008665 \text{ u}}$$
$$\text{Total} = 13.008665 \text{ u}.$$
$$\text{Subtract mass of } ^{13}_{6}\text{C:} \quad \underline{-13.003355 \text{ u}}$$
$$\Delta m = 0.005310 \text{ u}$$

which in energy is $(931.5 \text{ MeV/u})(0.005310 \text{ u}) = 4.95 \text{ MeV}$. That is, it would require 4.95 MeV input of energy to remove one neutron from $^{13}_{6}$C.

Nuclear Forces

We can analyze nuclei not only from the point of view of energy, but also from the point of view of the forces that hold them together. We might not expect a collection of protons and neutrons to come together spontaneously, since protons are all positively charged and thus exert repulsive electric forces on each other. Since stable nuclei *do* stay together, another force must be acting. This new force has to be stronger than the electric force in order to hold the nucleus together, and is called the **strong nuclear force**. The strong nuclear force acts as an attractive force between all nucleons, protons and neutrons alike. Thus protons attract each other via the strong nuclear force at the same time they repel each other via the electric force. Neutrons, because they are electrically neutral, only attract other neutrons or protons via the strong nuclear force.

The strong nuclear force turns out to be far more complicated than the gravitational and electromagnetic forces. One important aspect of the strong nuclear force is that it is a **short-range** force: it acts only over a very short distance.

It is very strong between two nucleons if they are less than about 10^{-15} m apart, but it is essentially zero if they are separated by a distance greater than this. Compare this to electric and gravitational forces, which decrease as $1/r^2$ but continue acting over any distances and are therefore called **long-range** forces.

The strong nuclear force has some strange features. For example, if a nuclide contains too many or too few neutrons relative to the number of protons, the binding of the nucleons is reduced; nuclides that are too unbalanced in this regard are unstable. As shown in Fig. 30–2, stable nuclei tend to have the same number of protons as neutrons ($N = Z$) up to about $A = 30$. Beyond this, stable nuclei contain more neutrons than protons. This makes sense since, as Z increases, the electrical repulsion increases, so a greater number of neutrons—which exert only the attractive strong nuclear force—are required to maintain stability. For very large Z, no number of neutrons can overcome the greatly increased electric repulsion. Indeed, there are no completely stable nuclides above $Z = 82$.

What we mean by a *stable nucleus* is one that stays together indefinitely. What then is an *unstable nucleus*? It is one that comes apart; and this results in radioactive decay. Before we discuss the important subject of radioactivity (next Section), we note that there is a second type of nuclear force that is much weaker than the strong nuclear force. It is called the **weak nuclear force**, and we are aware of its existence only because it shows itself in certain types of radioactive decay. These two nuclear forces, the strong and the weak, together with the gravitational and electromagnetic forces, comprise the four fundamental types of force in nature.

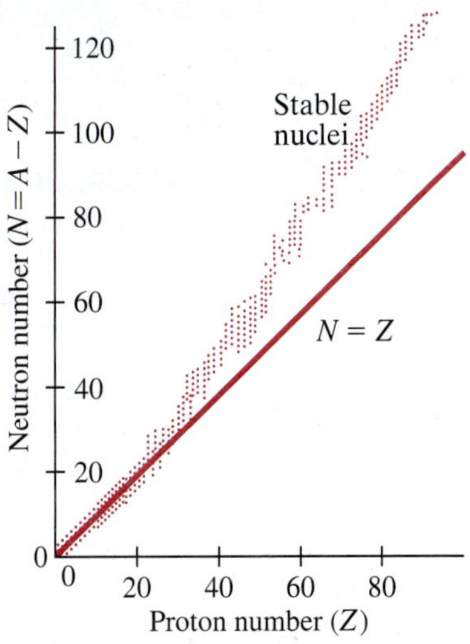

FIGURE 30–2 Number of neutrons versus number of protons for stable nuclides, which are represented by dots. The straight line represents $N = Z$.

30–3 Radioactivity

Nuclear physics had its beginnings in 1896. In that year, Henri Becquerel (1852–1908) made an important discovery: in his studies of phosphorescence, he found that a certain mineral (which happened to contain uranium) would darken a photographic plate even when the plate was wrapped to exclude light. It was clear that the mineral emitted some new kind of radiation that, unlike X-rays (Section 25–11), occurred without any external stimulus. This new phenomenon eventually came to be called **radioactivity**.

Soon after Becquerel's discovery, Marie Curie (1867–1934) and her husband, Pierre Curie (1859–1906), isolated two previously unknown elements that were very highly radioactive (Fig. 30–3). These were named polonium and radium. Other radioactive elements were soon discovered as well. The radioactivity was found in every case to be unaffected by the strongest physical and chemical treatments, including strong heating or cooling or the action of strong chemicals. It was suspected that the source of radioactivity must be deep within the atom, coming from the nucleus. It became apparent that radioactivity is the result of the **disintegration** or **decay** of an unstable nucleus. Certain isotopes are not stable, and they decay with the emission of some type of radiation or "rays."

Many unstable isotopes occur in nature, and such radioactivity is called "natural radioactivity." Other unstable isotopes can be produced in the laboratory by nuclear reactions (Section 31–1); these are said to be produced "artificially" and to have "artificial radioactivity." Radioactive isotopes are sometimes referred to as **radioisotopes** or **radionuclides**.

Rutherford and others began studying the nature of the rays emitted in radioactivity about 1898. They classified the rays into three distinct types according to their penetrating power. One type of radiation could barely penetrate a piece of paper. The second type could pass through as much as 3 mm of aluminum. The third was extremely penetrating: it could pass through several centimeters of lead and still be detected on the other side. They named these three types of radiation alpha (α), beta (β), and gamma (γ), respectively, after the first three letters of the Greek alphabet.

FIGURE 30–3 Marie and Pierre Curie in their laboratory (about 1906) where radium was discovered.

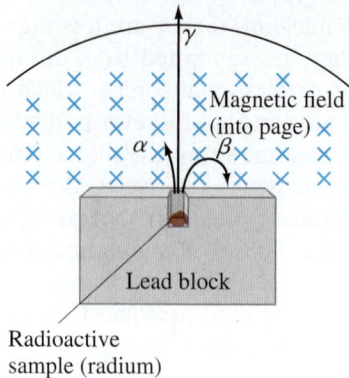

FIGURE 30–4 Alpha and beta rays are bent in opposite directions by a magnetic field, whereas gamma rays are not bent at all.

Each type of ray was found to have a different charge and hence is bent differently in a magnetic field, Fig. 30–4; α rays are positively charged, β rays are negatively charged, and γ rays are neutral. It was soon found that all three types of radiation consisted of familiar kinds of particles. Gamma rays are very high-energy *photons* whose energy is even higher than that of X-rays. Beta rays were found to be identical to *electrons* that orbit the nucleus, but they are created within the nucleus itself. Alpha rays (or α particles) are simply the nuclei of *helium* atoms, ^4_2He; that is, an α ray consists of two protons and two neutrons bound together.

We now discuss each of these three types of radioactivity, or decay, in more detail.

30–4 Alpha Decay

Experiments show that when nuclei decay, the number of nucleons (= mass number A) is conserved, as well as electric charge (= Ze). When a nucleus emits an α particle (^4_2He), the remaining nucleus will be different from the original: it has lost two protons and two neutrons. Radium 226 ($^{226}_{88}\text{Ra}$), for example, is an α emitter. It decays to a nucleus with $Z = 88 - 2 = 86$ and $A = 226 - 4 = 222$. The nucleus with $Z = 86$ is radon (Rn)—see Appendix B or the Periodic Table. Thus radium decays to radon with the emission of an α particle. This is written

$$^{226}_{88}\text{Ra} \rightarrow {}^{222}_{86}\text{Rn} + {}^4_2\text{He}.$$

FIGURE 30–5 Radioactive decay of radium to radon with emission of an alpha particle.

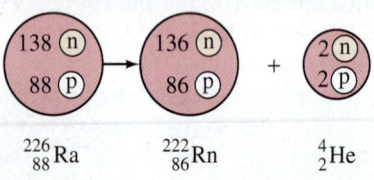

See Fig. 30–5.

When α decay occurs, a different element is formed. The **daughter** nucleus ($^{222}_{86}\text{Rn}$ in this case) is different from the **parent** nucleus ($^{226}_{88}\text{Ra}$ in this case). This changing of one element into another is called **transmutation** of the elements.

Alpha decay can be written in general as

$$^A_Z\text{N} \rightarrow {}^{A-4}_{Z-2}\text{N}' + {}^4_2\text{He} \qquad [\alpha \text{ decay}]$$

where N is the parent, N' the daughter, and Z and A are the atomic number and atomic mass number, respectively, of the parent.

> **EXERCISE C** $^{154}_{66}\text{Dy}$ decays by α emission to what element? (*a*) Pb, (*b*) Gd, (*c*) Sm, (*d*) Er, (*e*) Yb.

Alpha decay occurs because the strong nuclear force is unable to hold very large nuclei together. The nuclear force is a short-range force: it acts only between neighboring nucleons. But the electric force acts all the way across a large nucleus. For very large nuclei, the large Z means the repulsive electric force becomes so large (Coulomb's law) that the strong nuclear force is unable to hold the nucleus together.

We can express the instability of the parent nucleus in terms of energy (or mass): the mass of the parent nucleus is greater than the mass of the daughter nucleus plus the mass of the α particle. The mass difference appears as kinetic energy, which is carried away by the α particle and the recoiling daughter nucleus. The total energy released is called the **disintegration energy**, Q, or the **Q-value** of the decay. From conservation of energy,

$$M_P c^2 = M_D c^2 + m_\alpha c^2 + Q,$$

where Q equals the kinetic energy of the daughter and α particle, and M_P, M_D, and m_α are the masses of the parent, daughter, and α particle, respectively. Thus

$$Q = M_P c^2 - (M_D + m_\alpha)c^2. \qquad \text{(30–2)}$$

If the parent had *less* mass than the daughter plus the α particle (so $Q < 0$), the decay would violate conservation of energy. Such decays have never been observed, another confirmation of this great conservation law.

EXAMPLE 30–6 Uranium decay energy release. Calculate the disintegration energy when $^{232}_{92}U$ (mass = 232.037156 u) decays to $^{228}_{90}Th$ (228.028741 u) with the emission of an α particle. (As always, masses given are for neutral atoms.)

APPROACH We use conservation of energy as expressed in Eq. 30–2. $^{232}_{92}U$ is the parent, $^{228}_{90}Th$ is the daughter.

SOLUTION Since the mass of the $^4_2$He is 4.002603 u (Appendix B), the total mass in the final state $(m_{Th} + m_{He})$ is

$$228.028741 \text{ u} + 4.002603 \text{ u} = 232.031344 \text{ u}.$$

The mass lost when the $^{232}_{92}U$ decays $(m_U - m_{Th} - m_{He})$ is

$$232.037156 \text{ u} - 232.031344 \text{ u} = 0.005812 \text{ u}.$$

Because 1 u = 931.5 MeV, the energy Q released is

$$Q = (0.005812 \text{ u})(931.5 \text{ MeV/u}) = 5.4 \text{ MeV}$$

and this energy appears as kinetic energy of the α particle and the daughter nucleus.

Additional Example

EXAMPLE 30–7 Kinetic energy of the α in $^{232}_{92}U$ decay. For the $^{232}_{92}U$ decay of Example 30–6, how much of the 5.4-MeV disintegration energy will be carried off by the α particle?

APPROACH In any reaction, momentum must be conserved as well as energy.

SOLUTION Before disintegration, the nucleus can be assumed to be at rest, so the total momentum was zero. After disintegration, the total vector momentum must still be zero so the magnitude of the α particle's momentum must equal the magnitude of the daughter's momentum (Fig. 30–6):

$$m_\alpha v_\alpha = m_D v_D.$$

Thus $v_\alpha = m_D v_D / m_\alpha$ and the α's kinetic energy is

$$KE_\alpha = \tfrac{1}{2} m_\alpha v_\alpha^2 = \tfrac{1}{2} m_\alpha \left(\frac{m_D v_D}{m_\alpha}\right)^2 = \tfrac{1}{2} m_D v_D^2 \left(\frac{m_D}{m_\alpha}\right) = \left(\frac{m_D}{m_\alpha}\right) KE_D$$

$$= \left(\frac{228.028741 \text{ u}}{4.002603 \text{ u}}\right) KE_D = 57 \, KE_D.$$

The total disintegration energy is $Q = KE_\alpha + KE_D = 57 \, KE_D + KE_D = 58 \, KE_D$. Hence

$$KE_\alpha = 57 \, KE_D = \frac{57}{58} Q = 5.3 \text{ MeV}.$$

The lighter α particle carries off (57/58) or 98% of the total kinetic energy. The total energy released is 5.4 MeV, so the daughter nucleus, which recoils in the opposite direction, carries off only 0.1 MeV.

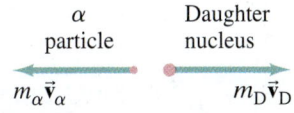

FIGURE 30–6 Momentum conservation in Example 30–7.

Why α Particles?

Why, you may wonder, do nuclei emit this combination of four nucleons called an α particle? Why not just four separate nucleons, or even one? The answer is that the α particle is very strongly bound, so that its mass is significantly less than that of four separate nucleons. That helps the final state in α decay to have less total mass, thus allowing certain nuclides to decay which could not decay to, say, 2 protons plus 2 neutrons. For example, $^{232}_{92}$U could not decay to 2p + 2n because the masses of the daughter $^{228}_{90}$Th plus four separate nucleons is 228.028741 u + 2(1.007825 u) + 2(1.008665 u) = 232.061721 u, which is greater than the mass of the $^{232}_{92}$U parent (232.037156 u). Such a decay would violate the conservation of energy. Indeed, we have never seen $^{232}_{92}$U → $^{228}_{90}$Th + 2p + 2n. Similarly, it is almost always true that the emission of a single nucleon is energetically not possible; see Example 30–5.

Smoke Detectors—An Application

PHYSICS APPLIED
Smoke detector

One widespread application of nuclear physics is present in nearly every home in the form of an ordinary **smoke detector**. One type of smoke detector contains about 0.2 mg of the radioactive americium isotope, $^{241}_{95}$Am, in the form of AmO_2. The radiation continually ionizes the nitrogen and oxygen molecules in the air space between two oppositely charged plates. The resulting conductivity allows a small steady electric current. If smoke enters, the radiation is absorbed by the smoke particles rather than by the air molecules, thus reducing the current. The current drop is detected by the device's electronics and sets off the alarm. The radiation dose that escapes from an intact americium smoke detector is much less than the natural radioactive background, and so can be considered relatively harmless. There is no question that smoke detectors save lives and reduce property damage.

30–5 Beta Decay

β⁻ Decay

Transmutation of elements also occurs when a nucleus decays by β decay—that is, with the emission of an electron or β⁻ particle. The nucleus $^{14}_{6}$C, for example, emits an electron when it decays:

$$^{14}_{6}\text{C} \rightarrow {}^{14}_{7}\text{N} + e^- + \text{neutrino},$$

where e⁻ is the symbol for the electron. The particle known as the neutrino has charge $q = 0$ and a very small mass, long thought to be zero. It was not initially detected and was only later hypothesized to exist, as we shall discuss later in this Section. No nucleons are lost when an electron is emitted, and the total number of nucleons, A, is the same in the daughter nucleus as in the parent. But because an electron has been emitted from the nucleus itself, the charge on the daughter nucleus is +1e greater than that on the parent. The parent nucleus in the decay written above had $Z = +6$, so from charge conservation the nucleus remaining behind must have a charge of +7e. So the daughter nucleus has $Z = 7$, which is nitrogen.

CAUTION
β-decay e⁻ comes from nucleus (it is not an orbital electron)

It must be carefully noted that the electron emitted in β decay is *not* an orbital electron. Instead, the electron is created *within the nucleus itself*. What happens is that one of the neutrons changes to a proton and in the process (to conserve charge) emits an electron. Indeed, free neutrons actually do decay in this fashion:

$$n \rightarrow p + e^- + \text{neutrino}.$$

To remind us of their origin in the nucleus, the electrons emitted in β decay are often referred to as "β particles." They are, nonetheless, indistinguishable from orbital electrons.

EXAMPLE 30–8 Energy release in $^{14}_{6}C$ decay. How much energy is released when $^{14}_{6}C$ decays to $^{14}_{7}N$ by β emission?

APPROACH We find the mass difference before and after decay, Δm. The energy released is $E = (\Delta m)c^2$. The masses given in Appendix B are those of the neutral atom, and we have to keep track of the electrons involved. Assume the parent nucleus has six orbiting electrons so it is neutral; its mass is 14.003242 u. The daughter in this decay, $^{14}_{7}N$, is not neutral because it has the same six orbital electrons circling it but the nucleus has a charge of $+7e$. However, the mass of this daughter with its six electrons, plus the mass of the emitted electron (which makes a total of seven electrons), is just the mass of a neutral nitrogen atom.

SOLUTION The total mass in the final state is

$$(\text{mass of } {}^{14}_{7}N \text{ nucleus } + 6 \text{ electrons}) + (\text{mass of 1 electron}),$$

and this is equal to

$$\text{mass of neutral } {}^{14}_{7}N \text{ (includes 7 electrons)},$$

which from Appendix B is a mass of 14.003074 u. So the mass difference is $14.003242\,u - 14.003074\,u = 0.000168\,u$, which is equivalent to an energy change $\Delta m\,c^2 = (0.000168\,u)(931.5\,\text{MeV/u}) = 0.156\,\text{MeV}$ or 156 keV.

NOTE The neutrino doesn't contribute to either the mass or charge balance because it has $q = 0$ and $m \approx 0$.

> **CAUTION**
> *Be careful with atomic and electron masses in β decay*

According to Example 30–8, we would expect the emitted electron to have a kinetic energy of 156 keV. (The daughter nucleus, because its mass is very much larger than that of the electron, recoils with very low velocity and hence gets very little of the kinetic energy—see Example 30–7.) Indeed, very careful measurements indicate that a few emitted β particles do have kinetic energy close to this calculated value. But the vast majority of emitted electrons have somewhat less energy. In fact, the energy of the emitted electron can be anywhere from zero up to the maximum value as calculated above. This range of electron kinetic energy was found for any β decay. It was as if the law of conservation of energy was being violated, and Bohr actually considered this possibility. Careful experiments indicated that linear momentum and angular momentum also did not seem to be conserved. Physicists were troubled at the prospect of giving up these laws, which had worked so well in all previous situations.

In 1930, Wolfgang Pauli proposed an alternate solution: perhaps a new particle that was very difficult to detect was emitted during β decay in addition to the electron. This hypothesized particle could be carrying off the energy, momentum, and angular momentum required to maintain the conservation laws. This new particle was named the **neutrino**—meaning "little neutral one"—by the great Italian physicist Enrico Fermi (1901–1954; Fig. 30–7), who in 1934 worked out a detailed theory of β decay. (It was Fermi who, in this theory, postulated the existence of the fourth force in nature which we call the *weak nuclear force*.) The neutrino has zero charge, spin of $\frac{1}{2}\hbar$, and was long thought to have zero mass, although today we are quite sure that it has a very tiny mass $(< 0.14\,\text{eV}/c^2)$. If its mass were zero, it would be much like a photon in that it is neutral and would travel at the speed of light. But the neutrino is very difficult to detect. In 1956, complex experiments produced further evidence for the existence of the neutrino; but by then, most physicists had already accepted its existence.

The symbol for the neutrino is the Greek letter nu (ν). The correct way of writing the decay of $^{14}_{6}C$ is then

$$^{14}_{6}C \rightarrow {}^{14}_{7}N + e^- + \bar{\nu}.$$

The bar ($^-$) over the neutrino symbol is to indicate that it is an "antineutrino." (Why this is called an antineutrino rather than simply a neutrino is discussed in Chapter 32.)

FIGURE 30–7 Enrico Fermi, as portrayed on a US postage stamp. Fermi contributed significantly to both theoretical and experimental physics, a feat almost unique in modern times: statistical theory of identical particles that obey the exclusion principle (= fermions); theory of the weak interaction and β decay; neutron physics; induced radioactivity and new elements; first nuclear reactor; first resonance of particle physics; led and inspired a vast amount of other nuclear research.

In β decay, it is the weak nuclear force that plays the crucial role. The neutrino is unique in that it interacts with matter only via the weak force, which is why it is so hard to detect.

β⁺ Decay

Many isotopes decay by electron emission. They are always isotopes that have too many neutrons compared to the number of protons. That is, they are isotopes that lie above the stable isotopes plotted in Fig. 30–2. But what about unstable isotopes that have too few neutrons compared to their number of protons—those that fall below the stable isotopes of Fig. 30–2? These, it turns out, decay by emitting a **positron** instead of an electron. A positron (sometimes called an e^+ or $β^+$ particle) has the same mass as the electron, but it has a positive charge of $+1e$. Because it is so like an electron, except for its charge, the positron is called the **antiparticle**† to the electron. An example of a $β^+$ decay is that of $^{19}_{10}\text{Ne}$:

$$^{19}_{10}\text{Ne} \rightarrow ^{19}_{9}\text{F} + e^+ + \nu,$$

where e^+ stands for a positron. Note that the ν emitted here is a neutrino, whereas that emitted in $β^-$ decay is called an antineutrino. Thus an antielectron (= positron) is emitted with a neutrino, whereas an antineutrino is emitted with an electron; this gives a certain balance as discussed in Chapter 32.

We can write $β^-$ and $β^+$ decay, in general, as follows:

$$^{A}_{Z}\text{N} \rightarrow ^{A}_{Z+1}\text{N}' + e^- + \bar{\nu} \qquad [β^- \text{ decay}]$$
$$^{A}_{Z}\text{N} \rightarrow ^{A}_{Z-1}\text{N}' + e^+ + \nu, \qquad [β^+ \text{ decay}]$$

where N is the parent nucleus and N' is the daughter.

Electron Capture

Besides $β^-$ and $β^+$ emission, there is a third related process. This is **electron capture** (abbreviated EC in Appendix B) and occurs when a nucleus absorbs one of its orbiting electrons. An example is $^{7}_{4}\text{Be}$, which as a result becomes $^{7}_{3}\text{Li}$. The process is written

$$^{7}_{4}\text{Be} + e^- \rightarrow ^{7}_{3}\text{Li} + \nu,$$

or, in general,

$$^{A}_{Z}\text{N} + e^- \rightarrow ^{A}_{Z-1}\text{N}' + \nu. \qquad [\text{electron capture}]$$

Usually it is an electron in the innermost (K) shell that is captured, in which case the process is called **K-capture**. The electron disappears in the process, and a proton in the nucleus becomes a neutron; a neutrino is emitted as a result. This process is inferred experimentally by detection of emitted X-rays (due to other electrons jumping down to fill the state of the captured e^-).

30–6 Gamma Decay

Gamma rays are photons having very high energy. They have their origin in the decay of a nucleus, much like emission of photons by excited atoms. Like an atom, a nucleus itself can be in an excited state. When it jumps down to a lower energy state, or to the ground state, it emits a photon which we call a γ ray. The possible states of a nucleus are much farther apart in energy than those of an atom: on the order of keV or MeV, as compared to a few eV for electrons in an atom. Hence, the emitted photons have energies that can range from a few keV to several MeV. For a given decay, the γ ray always has the same energy. Since a γ ray carries no charge, there is no change in the element as a result of a γ decay.

How does a nucleus get into an excited state? It may occur because of a violent collision with another particle. More commonly, the nucleus remaining after a previous radioactive decay may be in an excited state. A typical example is shown

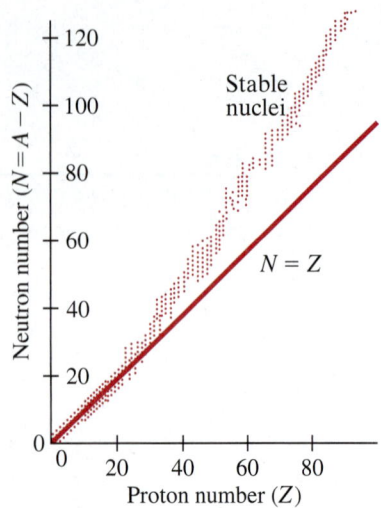

FIGURE 30–2 (Repeated.) Number of neutrons versus number of protons for stable nuclides, which are represented by dots. The straight line represents $N = Z$.

---

†Discussed in Chapter 32. Briefly, an antiparticle has the same mass as its corresponding particle, but opposite charge. A particle and its antiparticle can quickly annihilate each other, releasing energy in the form of two γ rays: $e^+ + e^- \rightarrow 2\gamma$.

in the energy-level diagram of Fig. 30–8. $^{12}_{5}$B can decay by β decay directly to the ground state of $^{12}_{6}$C; or it can go by β decay to an excited state of $^{12}_{6}$C, written $^{12}_{6}$C*, which itself decays by emission of a 4.4-MeV γ ray to the ground state of $^{12}_{6}$C.

We can write γ decay as

$$^{A}_{Z}N^* \rightarrow {}^{A}_{Z}N + \gamma, \qquad [\gamma \text{ decay}]$$

where the asterisk means "excited state" of that nucleus.

What, you may wonder, is the difference between a γ ray and an X-ray? They both are electromagnetic radiation (photons) and, though γ rays usually have higher energy than X-rays, their range of energies overlap to some extent. The difference is not intrinsic. We use the term X-ray if the photon is produced by an electron–atom interaction, and γ ray if the photon is produced in a nuclear process.

*Isomers; Internal Conversion

In some cases, a nucleus may remain in an excited state for some time before it emits a γ ray. The nucleus is then said to be in a **metastable state** and is called an **isomer**.

An excited nucleus can sometimes return to the ground state by another process known as **internal conversion** with no γ ray emitted. In this process, the excited nucleus interacts with one of the orbital electrons and ejects this electron from the atom with the same kinetic energy (minus the binding energy of the electron) that an emitted γ ray would have had.

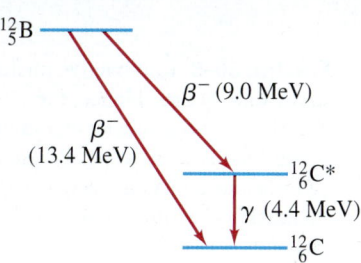

FIGURE 30–8 Energy-level diagram showing how $^{12}_{5}$B can decay to the ground state of $^{12}_{6}$C by β decay (total energy released = 13.4 MeV), or can instead β decay to an excited state of $^{12}_{6}$C (indicated by *), which subsequently decays to its ground state by emitting a 4.4-MeV γ ray.

30–7 Conservation of Nucleon Number and Other Conservation Laws

In all three types of radioactive decay, the classical conservation laws hold. Energy, linear momentum, angular momentum, and electric charge are all conserved. These quantities are the same before the decay as after. But a new conservation law is also revealed, the **law of conservation of nucleon number**. According to this law, the total number of nucleons (A) remains constant in any process, although one type can change into the other type (protons into neutrons or vice versa). This law holds in all three types of decay. [In Chapter 32 we will generalize this and call it conservation of baryon number.]

Table 30–2 gives a summary of α, β, and γ decay.

TABLE 30–2 The Three Types of Radioactive Decay

α decay:
$${}^{A}_{Z}N \rightarrow {}^{A-4}_{Z-2}N' + {}^{4}_{2}He$$
β decay:
$${}^{A}_{Z}N \rightarrow {}^{A}_{Z+1}N' + e^{-} + \bar{\nu}$$
$${}^{A}_{Z}N \rightarrow {}^{A}_{Z-1}N' + e^{+} + \nu$$
$${}^{A}_{Z}N + e^{-} \rightarrow {}^{A}_{Z-1}N' + \nu \quad [EC]^{\dagger}$$
γ decay:
$${}^{A}_{Z}N^* \rightarrow {}^{A}_{Z}N + \gamma$$

†Electron capture.
*Indicates the excited state of a nucleus.

30–8 Half-Life and Rate of Decay

A macroscopic sample of any radioactive isotope consists of a vast number of radioactive nuclei. These nuclei do not all decay at one time. Rather, they decay one by one over a period of time. This is a random process: we can not predict exactly when a given nucleus will decay. But we can determine, on a probabilistic basis, approximately how many nuclei in a sample will decay over a given time period, by assuming that each nucleus has the same probability of decaying in each second that it exists.

The number of decays ΔN that occur in a very short time interval Δt is then proportional to Δt and to the total number N of radioactive (parent) nuclei present:

$$\Delta N = -\lambda N \Delta t \qquad (30\text{–}3a)$$

where the minus sign means N is decreasing. We rewrite this to get the **rate of decay** (number of decays per second):

$$\frac{\Delta N}{\Delta t} = -\lambda N. \qquad (30\text{–}3b)$$

In these equations, λ is a measurable constant called the **decay constant**, which is different for different isotopes. The greater λ is, the greater the rate of decay, $\Delta N/\Delta t$, and the more "radioactive" that isotope is said to be.

FIGURE 30–9 Radioactive nuclei decay one by one. Hence, the number of parent nuclei in a sample is continually decreasing. When a $^{14}_{6}C$ nucleus emits an electron (b), the nucleus becomes a $^{14}_{7}N$ nucleus. Another decays in (c).

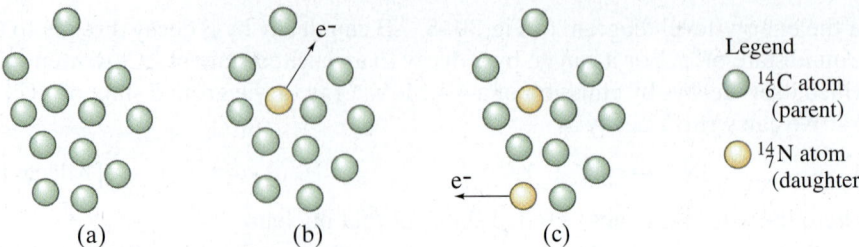

The number of decays that occur in the short time interval Δt is designated ΔN because each decay that occurs corresponds to a decrease by one in the number N of parent nuclei present. That is, radioactive decay is a "one-shot" process, Fig. 30–9. Once a particular parent nucleus decays into its daughter, it cannot do it again.

Exponential Decay

Equation 30–3a or b can be solved for N (using calculus) and the result is

$$N = N_0 e^{-\lambda t}, \qquad (30\text{--}4)$$

where N_0 is the number of parent nuclei present at any chosen time $t = 0$, and N is the number remaining after a time t. The symbol e is the natural exponential (encountered earlier in Sections 19–6 and 21–12) whose value is $e = 2.718\cdots$. Thus the number of parent nuclei in a sample decreases exponentially in time. This is shown in Fig. 30–10a for the decay of $^{14}_{6}C$. Equation 30–4 is called the **radioactive decay law**.

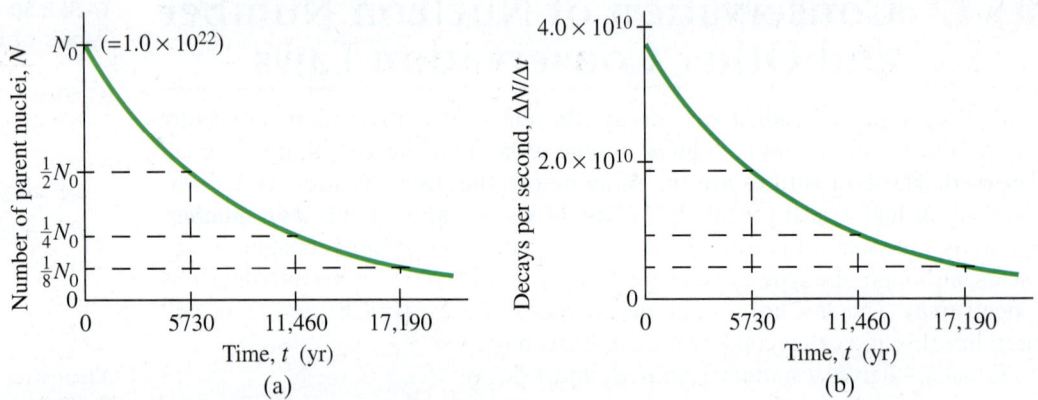

FIGURE 30–10 (a) The number N of parent nuclei in a given sample of $^{14}_{6}C$ decreases exponentially. We assume a sample that has $N_0 = 1.00 \times 10^{22}$ nuclei. (b) The number of decays per second also decreases exponentially. The half-life of $^{14}_{6}C$ is 5730 yr, which means that the number of parent nuclei, N, and the rate of decay, $\Delta N/\Delta t$, decrease by half every 5730 yr.

The number of decays per second, or decay rate R, is the magnitude of $\Delta N/\Delta t$, and is also called the **activity** of the sample. The magnitude (always positive) of a quantity is often indicated using vertical lines. The magnitude of $\Delta N/\Delta t$ is written $|\Delta N/\Delta t|$ and it is proportional to N (see Eq. 30–3b). So it too decreases exponentially in time at the same rate (Fig. 30–10b). The activity of a pure sample at time t is

$$R = \left|\frac{\Delta N}{\Delta t}\right| = R_0 e^{-\lambda t}, \qquad (30\text{--}5)$$

where $R_0 = |\Delta N/\Delta t|_0$ is the activity at $t = 0$.

Equation 30–5 is also referred to as the **radioactive decay law** (as is Eq. 30–4).

Half-Life

The rate of decay of any isotope is often specified by giving its "half-life" rather than the decay constant λ. The **half-life** of an isotope is defined as the time it takes for half the original amount of parent isotope in a given sample to decay.

For example, the half-life of $^{14}_{6}C$ is about 5730 years. If at some time a piece of petrified wood contains, say, 1.00×10^{22} nuclei of $^{14}_{6}C$, then 5730 years later it will contain half as many, 0.50×10^{22} nuclei. After another 5730 years it will contain 0.25×10^{22} nuclei, and so on. This is shown in Fig. 30–10a. Since the rate of decay $\Delta N/\Delta t$ is proportional to N, it, too, decreases by a factor of 2 every half-life (Fig. 30–10b).

The half-lives of known radioactive isotopes vary from very short ($\approx 10^{-22}$ s) to more than 10^{23} yr ($> 10^{30}$ s). The half-lives of many isotopes are given in Appendix B. It should be clear that the half-life (which we designate $T_{\frac{1}{2}}$) bears an inverse relationship to the decay constant. The longer the half-life of an isotope, the more slowly it decays, and hence λ is smaller. Conversely, very active isotopes (large λ) have very short half-lives. The precise relationship between half-life and decay constant is

$$T_{\frac{1}{2}} = \frac{\ln 2}{\lambda} = \frac{0.693}{\lambda}. \qquad (30\text{–}6)$$

We derive this in the next (optional) subsection.

EXERCISE D The half-life of $^{22}_{11}Na$ is 2.6 years. How much $^{22}_{11}Na$ will be left of a pure 1.0-μg sample after 7.8 yr? (a) None. (b) $\frac{1}{8}$ μg. (c) $\frac{1}{4}$ μg. (d) $\frac{1}{2}$ μg. (e) 0.693 μg.

EXERCISE E Return to the Chapter-Opening Question, page 857, and answer it again now. Try to explain why you may have answered differently the first time.

*Deriving the Half-Life Formula

We can derive Eq. 30–6 starting from Eq. 30–4 by setting $N = N_0/2$ at $t = T_{\frac{1}{2}}$:

$$\frac{N_0}{2} = N_0 e^{-\lambda T_{\frac{1}{2}}}$$

so

$$\frac{1}{2} = e^{-\lambda T_{\frac{1}{2}}}$$

and

$$e^{\lambda T_{\frac{1}{2}}} = 2.$$

We take natural logs of both sides ("ln" and "e" are inverse operations, meaning $\ln(e^x) = x$) and find

$$\ln\left(e^{\lambda T_{\frac{1}{2}}}\right) = \ln 2,$$

so

$$\lambda T_{\frac{1}{2}} = \ln 2 = 0.693$$

and

$$T_{\frac{1}{2}} = \frac{\ln 2}{\lambda} = \frac{0.693}{\lambda},$$

which is Eq. 30–6.

*Mean Life

Sometimes the **mean life** τ of an isotope is quoted, which is defined as $\tau = 1/\lambda$. Then Eq. 30–4 can be written $N = N_0 e^{-t/\tau}$, just as for RC and LR circuits (Chapters 19 and 21 where τ was called the time constant). The mean life of an isotope is then given by (see also Eq. 30–6)

$$\tau = \frac{1}{\lambda} = \frac{T_{\frac{1}{2}}}{0.693}. \qquad \text{[mean life]} \quad (30\text{–}7)$$

The mean life and half-life differ by a factor of 0.693, so confusing them can cause serious error (and has). The radioactive decay law, Eq. 30–5, can then be written as $R = R_0 e^{-t/\tau}$.

CAUTION
Do not confuse half-life and mean life

30–9 Calculations Involving Decay Rates and Half-Life

Let us now consider Examples of what we can determine about a sample of radioactive material if we know the half-life.

EXAMPLE 30–9 **Sample activity.** The isotope $^{14}_{6}C$ has a half-life of 5730 yr. If a sample contains 1.00×10^{22} carbon-14 nuclei, what is the activity of the sample?

APPROACH We first use the half-life to find the decay constant (Eq. 30–6), and use that to find the activity, Eq. 30–3b. The number of seconds in a year is $(60)(60)(24)(365\frac{1}{4}) = 3.156 \times 10^7$ s.

SOLUTION The decay constant λ from Eq. 30–6 is

$$\lambda = \frac{0.693}{T_{\frac{1}{2}}} = \frac{0.693}{(5730 \text{ yr})(3.156 \times 10^7 \text{ s/yr})} = 3.83 \times 10^{-12} \text{ s}^{-1}.$$

From Eqs. 30–3b and 30–5, the activity or rate of decay is

$$R = \left|\frac{\Delta N}{\Delta t}\right| = \lambda N = (3.83 \times 10^{-12} \text{ s}^{-1})(1.00 \times 10^{22}) = 3.83 \times 10^{10} \text{ decays/s}.$$

Notice that the graph of Fig. 30–10b starts at this value, corresponding to the original value of $N = 1.0 \times 10^{22}$ nuclei in Fig. 30–10a.

NOTE The unit "decays/s" is often written simply as s^{-1} since "decays" is not a unit but refers only to the number. This simple unit of activity is called the becquerel: 1 Bq = 1 decay/s, as discussed in Chapter 31.

CONCEPTUAL EXAMPLE 30–10 **Safety: Activity versus half-life.** One might think that a short half-life material is safer than a long half-life material because it will not last as long. Is that true?

RESPONSE No. A shorter half-life means the activity is higher and thus more "radioactive" and can cause more biological damage. In contrast, a longer half-life for the same sample size N means a lower activity but we have to worry about it for longer and find safe storage until it reaches a safe (low) level of activity.

EXAMPLE 30–11 **A sample of radioactive $^{13}_{7}N$.** A laboratory has 1.49 µg of pure $^{13}_{7}N$, which has a half-life of 10.0 min (600 s). (a) How many nuclei are present initially? (b) What is the rate of decay (activity) initially? (c) What is the activity after 1.00 h? (d) After approximately how long will the activity drop to less than one per second $(= 1\text{ s}^{-1})$?

APPROACH We use the definition of the mole and Avogadro's number (Sections 13–6 and 13–8) to find (a) the number of nuclei. For (b) we get λ from the given half-life and use Eq. 30–3b for the rate of decay. For (c) and (d) we use Eq. 30–5.

SOLUTION (a) The atomic mass is 13.0, so 13.0 g will contain 6.02×10^{23} nuclei (Avogadro's number). We have only 1.49×10^{-6} g, so the number of nuclei N_0 that we have initially is given by the ratio

$$\frac{N_0}{6.02 \times 10^{23}} = \frac{1.49 \times 10^{-6} \text{ g}}{13.0 \text{ g}}.$$

Solving for N_0, we find $N_0 = 6.90 \times 10^{16}$ nuclei.
(b) From Eq. 30–6,

$$\lambda = 0.693/T_{\frac{1}{2}} = (0.693)/(600 \text{ s}) = 1.155 \times 10^{-3} \text{ s}^{-1}.$$

Then, at $t = 0$ (see Eqs. 30–3b and 30–5)

$$R_0 = \left|\frac{\Delta N}{\Delta t}\right|_0 = \lambda N_0 = (1.155 \times 10^{-3} \text{ s}^{-1})(6.90 \times 10^{16}) = 7.97 \times 10^{13} \text{ decays/s}.$$

(c) After $1.00\,\text{h} = 3600\,\text{s}$, the magnitude of the activity will be (Eq. 30–5)

$$R = R_0 e^{-\lambda t} = (7.97 \times 10^{13}\,\text{s}^{-1}) e^{-(1.155 \times 10^{-3}\,\text{s}^{-1})(3600\,\text{s})} = 1.25 \times 10^{12}\,\text{s}^{-1}.$$

(d) We want to determine the time t when $R = 1.00\,\text{s}^{-1}$. From Eq. 30–5, we have

$$e^{-\lambda t} = \frac{R}{R_0} = \frac{1.00\,\text{s}^{-1}}{7.97 \times 10^{13}\,\text{s}^{-1}} = 1.25 \times 10^{-14}.$$

We take the natural log (ln) of both sides ($\ln e^{-\lambda t} = -\lambda t$) and divide by λ to find

$$t = -\frac{\ln(1.25 \times 10^{-14})}{\lambda} = 2.77 \times 10^4\,\text{s} = 7.70\,\text{h}.$$

Easy Alternate Solution to (c) $1.00\,\text{h} = 60.0$ minutes is 6 half-lives, so the activity will decrease to $(\tfrac{1}{2})(\tfrac{1}{2})(\tfrac{1}{2})(\tfrac{1}{2})(\tfrac{1}{2})(\tfrac{1}{2}) = (\tfrac{1}{2})^6 = \tfrac{1}{64}$ of its original value, or $(7.97 \times 10^{13})/(64) = 1.25 \times 10^{12}$ per second.

30–10 Decay Series

It is often the case that one radioactive isotope decays to another isotope that is also radioactive. Sometimes this daughter decays to yet a third isotope which also is radioactive. Such successive decays are said to form a **decay series**. An important example is illustrated in Fig. 30–11. As can be seen, $^{238}_{92}\text{U}$ decays by α emission to $^{234}_{90}\text{Th}$, which in turn decays by β decay to $^{234}_{91}\text{Pa}$. The series continues as shown, with several possible branches near the bottom, ending at the stable lead isotope, $^{206}_{82}\text{Pb}$. The two last decays can be

$$^{206}_{81}\text{Tl} \rightarrow {}^{206}_{82}\text{Pb} + e^- + \bar{\nu}, \qquad (T_{\frac{1}{2}} = 4.2\,\text{min})$$

or

$$^{210}_{84}\text{Po} \rightarrow {}^{206}_{82}\text{Pb} + \alpha. \qquad (T_{\frac{1}{2}} = 138\,\text{days})$$

Other radioactive series also exist.

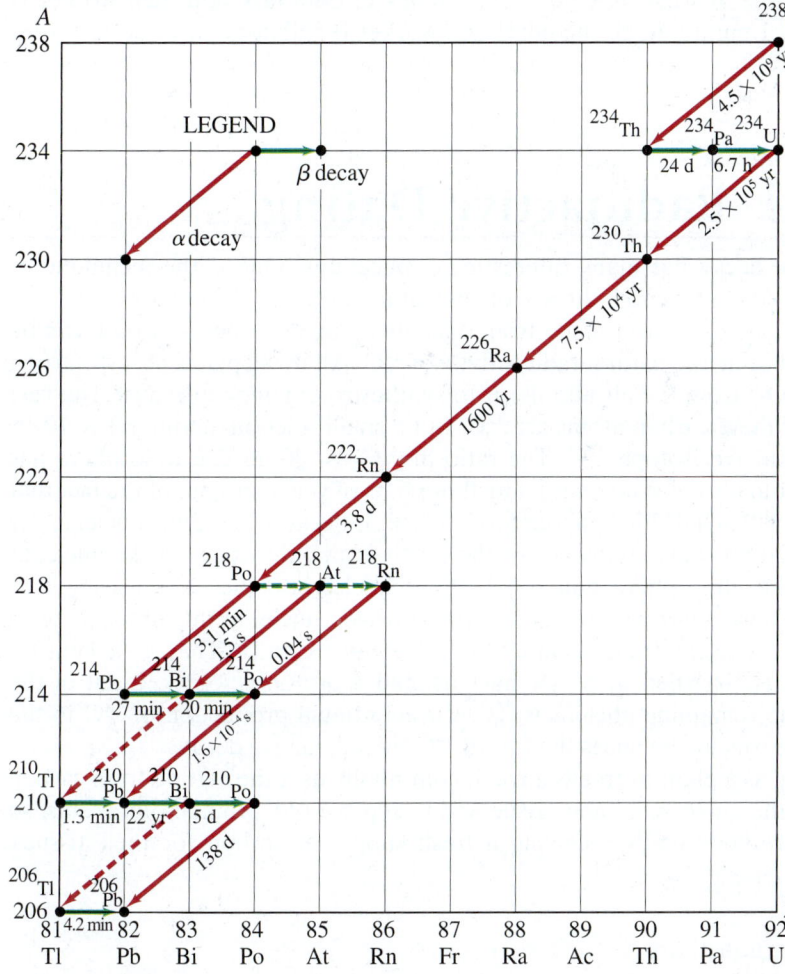

FIGURE 30–11 Decay series beginning with $^{238}_{92}\text{U}$. Nuclei in the series are specified by a dot representing A and Z values. Half-lives are given in seconds (s), minutes (min), hours (h), days (d), or years (yr). Note that a horizontal arrow represents β decay (A does not change), whereas a diagonal line represents α decay (A changes by 4, Z changes by 2). For the four nuclides shown that can decay by both α and β decay, the more prominent decay (in these four cases, $>99.9\%$) is shown as a solid arrow and the less common decay ($<0.1\%$) as a dashed arrow.

Because of such decay series, certain radioactive elements are found in nature that otherwise would not be. When the solar system (including Earth) was formed about 5 billion years ago, it is believed that nearly all nuclides were present, having been formed (by fusion and neutron capture, Sections 31–3 and 33–2) in a nearby supernova explosion (Section 33–2). Many isotopes with short half-lives decayed quickly and no longer are detected in nature today. But long-lived isotopes, such as $^{238}_{92}U$ with a half-life of 4.5×10^9 yr, still do exist in nature today. Indeed, about half of the original $^{238}_{92}U$ still remains. We might expect, however, that radium ($^{226}_{88}Ra$), with a half-life of 1600 yr, would have disappeared from the Earth long ago. Indeed, the original $^{226}_{88}Ra$ nuclei must by now have all decayed. However, because $^{238}_{92}U$ decays (in several steps, Fig. 30–11) to $^{226}_{88}Ra$, the supply of $^{226}_{88}Ra$ is continually replenished, which is why it is still found on Earth today. The same can be said for many other radioactive nuclides.

CONCEPTUAL EXAMPLE 30–12 **Decay chain.** In the decay chain of Fig. 30–11, if we look at the decay of $^{234}_{92}U$, we see four successive nuclides with half-lives of 250,000 yr, 75,000 yr, 1600 yr, and a little under 4 days. Each decay in the chain has an alpha particle of a characteristic energy, and so we can monitor the radioactive decay rate of each nuclide. Given a sample that was pure $^{234}_{92}U$ a million years ago, which alpha decay would you expect to have the highest activity rate in the sample?

RESPONSE The first instinct is to say that the process with the shortest half-life would show the highest activity. Surprisingly, perhaps, the activities of the four nuclides in this sample are all the same. The reason is that in each case the decay of the parent acts as a bottleneck to the decay of the daughter. Compared to the 1600-yr half-life of $^{226}_{88}Ra$, for example, its daughter $^{222}_{86}Rn$ decays almost immediately, but it cannot decay until it is made. (This is like an automobile assembly line: if worker A takes 20 minutes to do a task and then worker B takes only 1 minute to do the next task, worker B still does only one car every 20 minutes.)

30–11 Radioactive Dating

Radioactive decay has many interesting applications. One is the technique of *radioactive dating* by which the age of ancient materials can be determined.

PHYSICS APPLIED
Carbon-14 dating

The age of any object made from once-living matter, such as wood, can be determined using the natural radioactivity of $^{14}_{6}C$. All living plants absorb carbon dioxide (CO_2) from the air and use it to synthesize organic molecules. The vast majority of these carbon atoms are $^{12}_{6}C$, but a small fraction, about 1.3×10^{-12}, is the radioactive isotope $^{14}_{6}C$. The ratio of $^{14}_{6}C$ to $^{12}_{6}C$ in the atmosphere has remained roughly constant over many thousands of years, in spite of the fact that $^{14}_{6}C$ decays with a half-life of about 5730 yr. This is because energetic nuclei in the cosmic radiation, which impinges on the Earth from outer space, strike nuclei of atoms in the atmosphere and break those nuclei into pieces, releasing free neutrons. Those neutrons can collide with nitrogen nuclei in the atmosphere to produce the nuclear transformation $n + ^{14}_{7}N \rightarrow ^{14}_{6}C + p$. That is, a neutron strikes and is absorbed by a $^{14}_{7}N$ nucleus, and a proton is knocked out in the process. The remaining nucleus is $^{14}_{6}C$. This continual production of $^{14}_{6}C$ in the atmosphere roughly balances the loss of $^{14}_{6}C$ by radioactive decay.

As long as a plant or tree is alive, it continually uses the carbon from carbon dioxide in the air to build new tissue and to replace old. Animals eat plants, so they too are continually receiving a fresh supply of carbon for their tissues.

Organisms cannot distinguish† $^{14}_{6}C$ from $^{12}_{6}C$, and because the ratio of $^{14}_{6}C$ to $^{12}_{6}C$ in the atmosphere remains nearly constant, the ratio of the two isotopes within the living organism remains nearly constant as well. When an organism dies, carbon dioxide is no longer taken in and utilized. Because the $^{14}_{6}C$ decays radioactively, the ratio of $^{14}_{6}C$ to $^{12}_{6}C$ in a dead organism decreases over time. The half-life of $^{14}_{6}C$ is about 5730 yr, so the $^{14}_{6}C/^{12}_{6}C$ ratio decreases by half every 5730 yr. If, for example, the $^{14}_{6}C/^{12}_{6}C$ ratio of an ancient wooden tool is half of what it is in living trees, then the object must have been made from a tree that was felled about 5730 years ago.

Actually, corrections must be made for the fact that the $^{14}_{6}C/^{12}_{6}C$ ratio in the atmosphere has not remained precisely constant over time. The determination of what this ratio has been over the centuries has required techniques such as comparing the expected ratio to the actual ratio for objects whose age is known, such as very old trees whose annual rings can be counted reasonably accurately.

EXAMPLE 30–13 **An ancient animal.** The mass of carbon in an animal bone fragment found in an archeological site is 200 g. If the bone registers an activity of 16 decays/s, what is its age?

PHYSICS APPLIED
Archeological dating

APPROACH First we determine how many $^{14}_{6}C$ atoms there were in our 200-g sample when the animal was alive, given the known fraction of $^{14}_{6}C$ to $^{12}_{6}C$, 1.3×10^{-12}. Then we use Eq. 30–3b to find the activity back then, and Eq. 30–5 to find out how long ago that was by solving for the time t.

SOLUTION The 200 g of carbon is nearly all $^{12}_{6}C$; 12.0 g of $^{12}_{6}C$ contains 6.02×10^{23} atoms, so 200 g contains

$$\left(\frac{6.02 \times 10^{23} \text{ atoms/mol}}{12.0 \text{ g/mol}}\right)(200 \text{ g}) = 1.00 \times 10^{25} \text{ atoms.}$$

When the animal was alive, the ratio of $^{14}_{6}C$ to $^{12}_{6}C$ in the bone was 1.3×10^{-12}. The number of $^{14}_{6}C$ nuclei at that time was

$$N_0 = (1.00 \times 10^{25} \text{ atoms})(1.3 \times 10^{-12}) = 1.3 \times 10^{13} \text{ atoms.}$$

From Eq. 30–3b with $\lambda = 3.83 \times 10^{-12} \text{ s}^{-1}$ (Example 30–9) the magnitude of the activity when the animal was still alive ($t = 0$) was

$$R_0 = \left|\frac{\Delta N}{\Delta t}\right|_0 = \lambda N_0 = (3.83 \times 10^{-12} \text{ s}^{-1})(1.3 \times 10^{13}) = 50 \text{ s}^{-1}.$$

From Eq. 30–5

$$R = R_0 e^{-\lambda t}$$

where R, its activity now, is given as 16 s^{-1}. Then

$$16 \text{ s}^{-1} = (50 \text{ s}^{-1}) e^{-\lambda t}$$

or

$$e^{\lambda t} = \frac{50}{16}.$$

We take the natural logs of both sides (and divide by λ) to get

$$t = \frac{1}{\lambda} \ln\left(\frac{50}{16}\right) = \frac{1}{3.83 \times 10^{-12} \text{ s}^{-1}} \ln\left(\frac{50}{16}\right)$$

$$= 2.98 \times 10^{11} \text{ s} = 9400 \text{ yr,}$$

which is the time elapsed since the death of the animal.

---

†Organisms operate almost exclusively via chemical reactions—which involve only the outer orbital electrons of the atom; extra neutrons in the nucleus have essentially no effect.

Geological Time Scale Dating

Carbon dating is useful only for determining the age of objects less than about 60,000 years old. The amount of $^{14}_{6}C$ remaining in objects older than that is usually too small to measure accurately, although new techniques are allowing detection of even smaller amounts of $^{14}_{6}C$, pushing the time frame further back. On the other hand, radioactive isotopes with longer half-lives can be used in certain circumstances to obtain the age of older objects. For example, the decay of $^{238}_{92}U$, because of its long half-life of 4.5×10^9 years, is useful in determining the ages of rocks on a geologic time scale. When molten material on Earth long ago solidified into rock as the temperature dropped, different compounds solidified according to the melting points, and thus different compounds separated to some extent. Uranium present in a material became fixed in position and the daughter nuclei that result from the decay of uranium were also fixed in that position. Thus, by measuring the amount of $^{238}_{92}U$ remaining in the material relative to the amount of daughter nuclei, the time when the rock solidified can be determined.

PHYSICS APPLIED
Geological dating

PHYSICS APPLIED
Oldest Earth rocks and earliest life

Radioactive dating methods using $^{238}_{92}U$ and other isotopes have shown the age of the oldest Earth rocks to be about 4×10^9 yr. The age of rocks in which the oldest fossilized organisms are embedded indicates that life appeared more than $3\frac{1}{2}$ billion years ago. The earliest fossilized remains of mammals are found in rocks 200 million years old, and humanlike creatures seem to have appeared more than 2 million years ago. Radioactive dating has been indispensable for the reconstruction of Earth's history.

*30–12 Stability and Tunneling

Radioactive decay occurs only if the mass of the parent nucleus is greater than the sum of the masses of the daughter nucleus and all particles emitted. For example, $^{238}_{92}U$ can decay to $^{234}_{90}Th$ because the mass of $^{238}_{92}U$ is greater than the mass of the $^{234}_{90}Th$ plus the mass of the α particle. Because systems tend to go in the direction that reduces their internal or potential energy (a ball rolls downhill, a positive charge moves toward a negative charge), you may wonder why an unstable nucleus doesn't fall apart immediately. In other words, why do $^{238}_{92}U$ nuclei ($T_{\frac{1}{2}} = 4.5 \times 10^9$ yr) and other isotopes have such long half-lives? Why don't parent nuclei all decay at once?

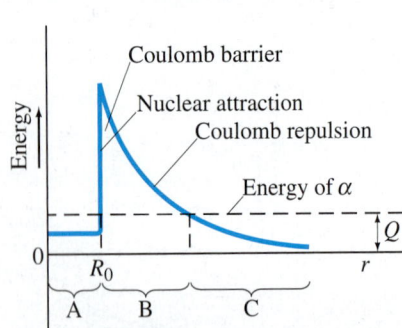

FIGURE 30–12 Potential energy for alpha particle and nucleus, showing the "Coulomb barrier" through which the α particle must tunnel to escape. The Q-value of the reaction is also indicated.

The answer has to do with quantum theory and the nature of the forces involved. One way to view the situation is with the aid of a potential-energy diagram, as in Fig. 30–12. Let us consider the particular case of the decay $^{238}_{92}U \rightarrow ^{234}_{90}Th + ^{4}_{2}He$. The blue line represents the potential energy, including rest mass, where we imagine the α particle as a separate entity within the $^{238}_{92}U$ nucleus. The region labeled A in Fig. 30–12 represents the PE of the α particle when it is held within the uranium nucleus by the strong nuclear force (R_0 is the nuclear radius). Region C represents the PE when the α particle is free of the nucleus. The downward-curving PE (proportional to $1/r$) represents the electrical repulsion (Coulomb's law) between the positively charged α and the $^{234}_{90}Th$ nucleus. To get to region C, the α particle has to get through the "**Coulomb barrier**" shown. Since the PE just beyond $r = R_0$ (region B) is greater than the energy of the alpha particle (dashed line), the α particle could not escape the nucleus if it were governed by classical physics. It could escape only if there were an input of energy equal to the height of the barrier. Nuclei decay spontaneously, however, without any input of energy. How, then, does the α particle get from region A to region C? It actually passes through the barrier in a process known as quantum-mechanical **tunneling**. Classically, this could not happen, because an α particle in region B (within the barrier) would be violating the conservation-of-energy principle.[†]

[†]The total energy E (dashed line in Fig. 30–12) would be less than the PE; because KE $= \frac{1}{2}mv^2 > 0$, then classically, $E =$ KE $+$ PE could not be less than the PE.

The uncertainty principle, however, tells us that energy conservation can be violated by an amount ΔE for a length of time Δt given by

$$(\Delta E)(\Delta t) \approx \frac{h}{2\pi}.$$

We saw in Section 28–3 that this is a result of the wave–particle duality. Thus quantum mechanics allows conservation of energy to be violated for brief periods that may be long enough for an α particle to "tunnel" through the barrier. ΔE would represent the energy difference between the average barrier height and the particle's energy, and Δt the time to pass through the barrier. The higher and wider the barrier, the less time the α particle has to escape and the less likely it is to do so. It is therefore the height and width of this barrier that control the rate of decay and half-life of an isotope.

30–13 Detection of Particles

Individual particles such as electrons, protons, α particles, neutrons, and γ rays are not detected directly by our senses. Consequently, a variety of instruments have been developed to detect them.

Counters

One of the most common detectors is the **Geiger counter**. As shown in Fig. 30–13, it consists of a cylindrical metal tube filled with a certain type of gas. A long wire runs down the center and is kept at a high positive voltage ($\approx 10^3$ V) with respect to the outer cylinder. The voltage is just slightly less than that required to ionize the gas atoms. When a charged particle enters through the thin "window" at one end of the tube, it ionizes a few atoms of the gas. The freed electrons are attracted toward the positive wire, and as they are accelerated they strike and ionize additional atoms. An "avalanche" of electrons is quickly produced, and when it reaches the wire anode, it produces a voltage pulse. The pulse, after being amplified, can be sent to an electronic counter, which counts how many particles have been detected. Or the pulses can be sent to a loudspeaker and each detection of a particle is heard as a "click." Only a fraction of the radiation emitted by a sample is detected by any detector.

A **scintillation counter** makes use of a solid, liquid, or gas known as a **scintillator** or **phosphor**. The atoms of a scintillator are easily excited when struck by an incoming particle and emit visible light when they return to their ground states. Typical scintillators are crystals of NaI and certain plastics. One face of a solid scintillator is cemented to a photomultiplier tube, and the whole is wrapped with opaque material to keep it light-tight (in the dark) or is placed within a light-tight container. The **photomultiplier (PM) tube** converts the energy of the scintillator-emitted photon(s) into an electric signal. A PM tube is a vacuum tube containing several electrodes (typically 8 to 14), called *dynodes*, which are maintained at successively higher voltages as shown in Fig. 30–14. At its top surface is a photoelectric surface, called the *photocathode*, whose work function (Section 27–3) is low enough that an electron is easily released when struck by a photon from the scintillator. Such an electron is accelerated toward the positive voltage of the first dynode. When it strikes the first dynode, the electron has acquired sufficient kinetic energy so that it can eject two to five more electrons. These, in turn, are accelerated toward the higher voltage second dynode, and a multiplication process begins. The number of electrons striking the last dynode may be 10^6 or more. Thus the passage of a particle through the scintillator results in an electric signal at the output of the PM tube that can be sent to an electronic counter just as for a Geiger tube. Solid scintillators are much more dense than the gas of a Geiger counter, and so are much more efficient detectors—especially for γ rays, which interact less with matter than do α or β particles. Scintillators that can measure the total energy deposited are much used today and are called **calorimeters**.

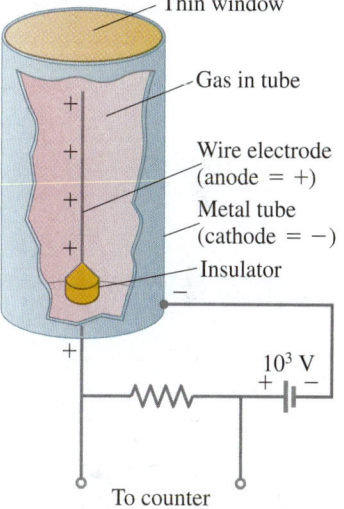

FIGURE 30–13 Diagram of a Geiger counter.

FIGURE 30–14 Scintillation counter with a photomultiplier tube.

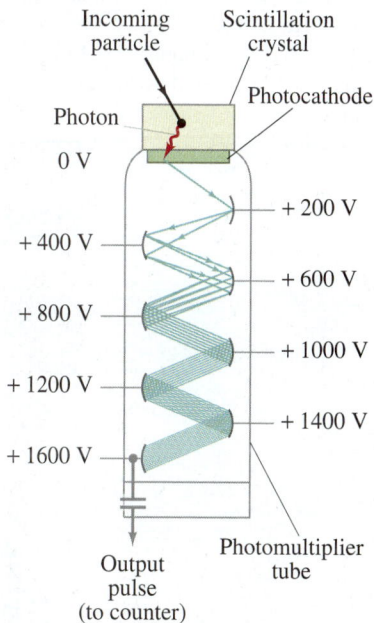

In tracer work (Section 31–7), **liquid scintillators** are often used. Radioactive samples taken at different times or from different parts of an organism are placed directly in small bottles containing the liquid scintillator. This is particularly convenient for detection of β rays from $^3_1$H and $^{14}_6$C, which have very low energies and have difficulty passing through the outer covering of a crystal scintillator or Geiger tube. A PM tube is still used to produce the electric signal from the liquid scintillator.

A **semiconductor detector** consists of a reverse-biased *pn* junction diode (Section 29–9). A charged particle passing through the junction can excite electrons into the conduction band, leaving holes in the valence band. The freed charges produce a short electric current pulse that can be counted as for Geiger and scintillation counters.

Hospital workers and others who work around radiation may carry *film badges* which detect the accumulation of radiation exposure. The film inside is periodically replaced and developed, the darkness of the developed film being related to total exposure (see Section 31–5).

Visualization

The devices discussed so far are used for counting the number of particles (or decays of a radioactive isotope). There are also devices that allow the track of charged particles to be *seen*. Very important are semiconductor detectors. **Silicon wafer semiconductors** have their surface etched into separate tiny pixels, each providing particle position information. They are much used in elementary particle physics (Chapter 32) to track the positions of particles produced and to determine their point of origin and/or their momentum (with the help of a magnetic field). The pixel arrangement can be CCD or CMOS (Section 25–1), the latter able to incorporate electronics inside, allowing fast readout.

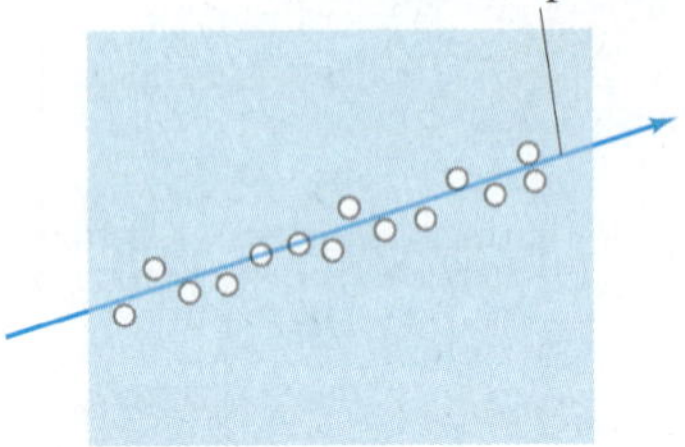

One of the oldest tracking devices is the **photographic emulsion**, which can be small and portable, used now particularly for cosmic-ray studies from balloons. A charged particle passing through an emulsion ionizes the atoms along its path. These points undergo a chemical change, and when the emulsion is developed (like film) the particle's path is revealed.

In a **cloud chamber**, used in the early days of nuclear physics, a gas is cooled to a temperature slightly below its usual condensation point ("supercooled"). Tiny droplets form around ions produced when a charged particle passes through (Fig. 30–15). Light scattering from these droplets reveals the track of the particle.

The **bubble chamber**, invented in 1952 by Donald A. Glaser (1926–2013), makes use of a superheated liquid kept close to its normal boiling point. Bubbles characteristic of boiling form around ions produced by the passage of a charged particle, revealing paths of particles that recently passed through. Because a bubble chamber uses a liquid, often liquid hydrogen, many more interactions can occur than in a cloud chamber. A magnetic field applied across the chamber makes charged particle paths curve (Chapter 20) and allows the momentum of charged particles to be determined from the radius of curvature of their paths.

A **multiwire**[†] **chamber** consists of a set of closely spaced fine wires immersed in a gas (Fig. 30–16). Many wires are grounded, and the others between are kept at very high voltage. A charged particle passing through produces ions in the gas. Freed electrons drift toward the nearest high voltage wire, creating an "avalanche" of many more ions, and producing an electric pulse or signal at that wire. The positions of the particles are determined electronically by the position of the wire and by the time it takes the pulses to reach "readout" electronics at the ends of the wires. The paths of the particles are reconstructed electronically by computers which can "draw" a picture of the tracks, as shown in Fig. 32–15, Chapter 32. An external magnetic field curves the paths, allowing the momentum of the particles to be measured.

Many detectors are also **calorimeters** which measure the energy of the particles.

FIGURE 30–15 In a cloud chamber or bubble chamber, droplets or bubbles are formed around ions produced by the passage of a charged particle.

FIGURE 30–16 Multiwire chamber inside the Collider Detector at Fermilab (CDF). Figure 32–15 in Chapter 32 was made with this detector.

[†]Also called *wire drift chamber* or *wire proportional chamber*.

Summary

Nuclear physics is the study of atomic nuclei. Nuclei contain **protons** and **neutrons**, which are collectively known as **nucleons**. The total number of nucleons, A, is the nucleus's **atomic mass number**. The number of protons, Z, is the **atomic number**. The number of neutrons equals $A - Z$. **Isotopes** are nuclei with the same Z, but with different numbers of neutrons. For an element X, an isotope of given Z and A is represented by

$$^A_Z X.$$

The nuclear radius is approximately proportional to $A^{\frac{1}{3}}$, indicating that all nuclei have about the same density. Nuclear masses are specified in **unified atomic mass units** (u), where the mass of $^{12}_{6}C$ (including its 6 electrons) is defined as exactly 12.000000 u. In terms of the energy equivalent (because $E = mc^2$),

$$1\,u = 931.5\,MeV/c^2 = 1.66 \times 10^{-27}\,kg.$$

The mass of a stable nucleus is less than the sum of the masses of its constituent nucleons. The difference in mass (times c^2) is the **total binding energy**. It represents the energy needed to break the nucleus into its constituent nucleons. The **binding energy per nucleon** averages about 8 MeV per nucleon, and is lowest for low mass and high mass nuclei.

Unstable nuclei undergo **radioactive decay**; they change into other nuclei with the emission of an α, β, or γ particle. An α particle is a $^4_2 He$ nucleus; a β particle is an electron or positron; and a γ ray is a high-energy photon. In β decay, a **neutrino** is also emitted. The transformation of **parent** nuclei into **daughter** nuclei is called **transmutation** of the elements. Radioactive decay occurs spontaneously only when the mass of the products is less than the mass of the parent nucleus. The loss in mass appears as kinetic energy of the products.

Nuclei are held together by the **strong nuclear force**. The **weak nuclear force** makes itself apparent in β decay. These two forces, plus the gravitational and electromagnetic forces, are the four known types of force.

Electric charge, linear and angular momentum, mass–energy, and **nucleon number** are **conserved** in all decays.

Radioactive decay is a statistical process. For a given type of radioactive nucleus, the number of nuclei that decay (ΔN) in a time Δt is proportional to the number N of parent nuclei present:

$$\Delta N = -\lambda N\,\Delta t; \qquad (30\text{–}3a)$$

the minus sign means N *decreases* in time.

The proportionality constant λ is called the **decay constant** and is characteristic of the given nucleus. The number N of nuclei remaining after a time t decreases exponentially,

$$N = N_0 e^{-\lambda t}, \qquad (30\text{–}4)$$

as does the **activity**, $R = $ magnitude of $\Delta N/\Delta t$:

$$R = \left|\frac{\Delta N}{\Delta t}\right|_0 e^{-\lambda t}. \qquad (30\text{–}5)$$

The **half-life**, $T_{\frac{1}{2}}$, is the time required for half the nuclei of a radioactive sample to decay. It is related to the decay constant by

$$T_{\frac{1}{2}} = \frac{0.693}{\lambda}. \qquad (30\text{–}6)$$

Radioactive dating is the use of radioactive decay to determine the age of certain objects, such as carbon dating.

[*Alpha decay occurs via a purely quantum-mechanical process called **tunneling** through a barrier.]

Particle detectors include **Geiger counters**, **scintillators** with attached **photomultiplier tubes**, and **semiconductor detectors**. Detectors that can image particle tracks include **semiconductors**, photographic **emulsions**, **bubble chambers**, and **multiwire chambers**.

Questions

1. What do different isotopes of a given element have in common? How are they different?

2. What are the elements represented by the X in the following: (a) $^{232}_{92}X$; (b) $^{18}_{7}X$; (c) $^{1}_{1}X$; (d) $^{86}_{38}X$; (e) $^{252}_{100}X$?

3. How many protons and how many neutrons do each of the isotopes in Question 2 have?

4. Identify the element that has 87 nucleons and 50 neutrons.

5. Why are the atomic masses of many elements (see the Periodic Table) not close to whole numbers?

6. Why are atoms much more likely to emit an alpha particle than to emit separate neutrons and protons?

7. What are the similarities and the differences between the strong nuclear force and the electric force?

8. What is the experimental evidence in favor of radioactivity being a nuclear process?

9. The isotope $^{64}_{29}Cu$ is unusual in that it can decay by γ, β^-, and β^+ emission. What is the resulting nuclide for each case?

10. A $^{238}_{92}U$ nucleus decays via α decay to a nucleus containing how many neutrons?

11. Describe, in as many ways as you can, the difference between α, β, and γ rays.

12. Fill in the missing particle or nucleus:
    (a) $^{45}_{20}Ca \rightarrow ? + e^- + \bar{\nu}$
    (b) $^{58}_{29}Cu^* \rightarrow ? + \gamma$
    (c) $^{46}_{24}Cr \rightarrow ^{46}_{23}V + ?$
    (d) $^{234}_{94}Pu \rightarrow ? + \alpha$
    (e) $^{239}_{93}Np \rightarrow ^{239}_{94}Pu + ?$

13. Immediately after a $^{238}_{92}U$ nucleus decays to $^{234}_{90}Th + ^{4}_{2}He$, the daughter thorium nucleus may still have 92 electrons circling it. Since thorium normally holds only 90 electrons, what do you suppose happens to the two extra ones?

14. When a nucleus undergoes either β^- or β^+ decay, what happens to the energy levels of the atomic electrons? What is likely to happen to these electrons following the decay?

15. The alpha particles from a given alpha-emitting nuclide are generally monoenergetic; that is, they all have the same kinetic energy. But the beta particles from a beta-emitting nuclide have a spectrum of energies. Explain the difference between these two cases.

16. Do isotopes that undergo electron capture generally lie above or below the stable nuclides in Fig. 30–2?
17. Can hydrogen or deuterium emit an α particle? Explain.
18. Why are many artificially produced radioactive isotopes rare in nature?
19. An isotope has a half-life of one month. After two months, will a given sample of this isotope have completely decayed? If not, how much remains?
20. Why are none of the elements with $Z > 92$ stable?
21. A proton strikes a $^{6}_{3}\text{Li}$ nucleus. As a result, an α particle and another particle are released. What is the other particle?
22. Can $^{14}_{6}\text{C}$ dating be used to measure the age of stone walls and tablets of ancient civilizations? Explain.
23. Explain the absence of β^+ emitters in the radioactive decay series of Fig. 30–11.
24. As $^{222}_{86}\text{Rn}$ decays into $^{206}_{82}\text{Pb}$, how many alpha and beta particles are emitted? Does it matter which path in the decay series is chosen? Why or why not?
25. A ^{238}U nucleus (initially at rest) decays into a ^{234}Th nucleus and an alpha particle. Which has the greater (i) momentum, (ii) velocity, (iii) kinetic energy? Explain.
    (a) The ^{234}Th nucleus.
    (b) The alpha particle.
    (c) Both the same.

MisConceptual Questions

1. Elements of the Periodic Table are distinguished by
   (a) the number of protons in the nucleus.
   (b) the number of neutrons in the nucleus.
   (c) the number of electrons in the atom.
   (d) Both (a) and (b).
   (e) (a), (b), and (c).

2. A nucleus has
   (a) more energy than its component neutrons and protons have.
   (b) less energy than its component neutrons and protons have.
   (c) the same energy as its component neutrons and protons have.
   (d) more energy than its component neutrons and protons have when the nucleus is at rest but less energy than when it is moving.

3. Which of the following will generally create a more stable nucleus?
   (a) Having more nucleons.
   (b) Having more protons than neutrons.
   (c) Having a larger binding energy per nucleon.
   (d) Having the same number of electrons as protons.
   (e) Having a larger total binding energy.

4. There are 82 protons in a lead nucleus. Why doesn't the lead nucleus burst apart?
   (a) Coulomb repulsive force doesn't act inside the nucleus.
   (b) Gravity overpowers the Coulomb repulsive force inside the nucleus.
   (c) The negatively charged neutrons balance the positively charged protons.
   (d) Protons lose their positive charge inside the nucleus.
   (e) The strong nuclear force holds the nucleus together.

5. The half-life of a radioactive nucleus is
   (a) half the time it takes for the entire substance to decay.
   (b) the time it takes for half of the substance to decay.
   (c) the same as the decay constant.
   (d) Both (a) and (b) (they are the same).
   (e) All of the above.

6. As a radioactive sample decays,
   (a) the half-life increases.
   (b) the half-life decreases.
   (c) the activity remains the same.
   (d) the number of radioactive nuclei increases.
   (e) None of the above.

7. If the half-life of a radioactive sample is 10 years, then it should take _____ years for the sample to decay completely.
   (a) 10. (b) 20. (c) 40.
   (d) Cannot be determined.

8. A sample's half-life is 1 day. What fraction of the original sample will have decayed after 3 days?
   (a) $\frac{1}{8}$. (b) $\frac{1}{4}$. (c) $\frac{1}{2}$. (d) $\frac{3}{4}$. (e) $\frac{7}{8}$. (f) All of it.

9. After three half-lives, what fraction of the original radioactive material is left?
   (a) None. (b) $\frac{1}{16}$. (c) $\frac{1}{8}$. (d) $\frac{1}{4}$. (e) $\frac{3}{4}$. (f) $\frac{7}{8}$.

10. Technetium $^{98}_{43}\text{Tc}$ has a half-life of 4.2×10^6 yr. Strontium $^{90}_{38}\text{Sr}$ has a half-life of 28.79 yr. Which statements are true?
    (a) The decay constant of Sr is greater than the decay constant of Tc.
    (b) The activity of 100 g of Sr is less than the activity of 100 g of Tc.
    (c) The long half-life of Tc means that it decays by alpha decay.
    (d) A Tc atom has a higher probability of decaying in 1 yr than a Sr atom.
    (e) 28.79 g of Sr has the same activity as 4.2×10^6 g of Tc.

11. A material having which decay constant would have the shortest half-life?
    (a) 100/second.
    (b) 5/year.
    (c) 8/century.
    (d) 10^9/day.

12. Uranium-238 decays to lead-206 through a series of
    (a) alpha decays.
    (b) beta decays.
    (c) gamma decays.
    (d) some combination of alpha, beta, and gamma decays.

13. Carbon dating is useful only for determining the age of objects less than about _____ years old.
    (a) 4.5 million.
    (b) 1.2 million.
    (c) 600,000.
    (d) 60,000.
    (e) 6000.
14. Radon has a half-life of about 1600 years. The Earth is several billion years old, so why do we still find radon on this planet?
    (a) Ice-age temperatures preserved some of it.
    (b) Heavier unstable isotopes decay into it.
    (c) It is created in lightning strikes.
    (d) It is replenished by cosmic rays.
    (e) Its half-life has increased over time.
    (f) Its half-life has decreased over time.
15. How does an atom's nucleus stay together and remain stable?
    (a) The attractive gravitational force between the protons and neutrons overcomes the repulsive electrostatic force between the protons.
    (b) Having just the right number of neutrons overcomes the electrostatic force between the protons.
    (c) A strong covalent bond develops between the neutrons and protons, because they are so close to each other.
    (d) None of the above.
16. What has greater mass?
    (a) A neutron and a proton that are far from each other (unbound).
    (b) A neutron and a proton that are bound together in a hydrogen (deuterium) nucleus.
    (c) Both the same.

For assigned homework and other learning materials, go to the MasteringPhysics website.

Problems

[See Appendix B for masses]

30–1 Nuclear Properties

1. (I) A pi meson has a mass of 139 MeV/c^2. What is this in atomic mass units?
2. (I) What is the approximate radius of an α particle ($^4_2$He)?
3. (I) By what % is the radius of $^{238}_{92}$U greater than the radius of $^{232}_{92}$U?
4. (II) (a) What is the approximate radius of a $^{112}_{48}$Cd nucleus? (b) Approximately what is the value of A for a nucleus whose radius is 3.7×10^{-15} m?
5. (II) What is the mass of a bare α particle (without electrons) in MeV/c^2?
6. (II) Suppose two alpha particles were held together so they were just "touching" (use Eq. 30–1). Estimate the electrostatic repulsive force each would exert on the other. What would be the acceleration of an alpha particle subjected to this force?
7. (II) (a) What would be the radius of the Earth if it had its actual mass but had the density of nuclei? (b) By what factor would the radius of a $^{238}_{92}$U nucleus increase if it had the Earth's density?
8. (II) What stable nucleus has approximately half the radius of a uranium nucleus? [Hint: Find A and use Appendix B to get Z.]
9. (II) If an alpha particle were released from rest near the surface of a $^{257}_{100}$Fm nucleus, what would its kinetic energy be when far away?
10. (II) (a) What is the fraction of the hydrogen atom's mass ($^1_1$H) that is in the nucleus? (b) What is the fraction of the hydrogen atom's volume that is occupied by the nucleus?
11. (II) Approximately how many nucleons are there in a 1.0-kg object? Does it matter what the object is made of? Why or why not?
12. (III) How much kinetic energy, in MeV, must an α particle have to just "touch" the surface of a $^{232}_{92}$U nucleus?

30–2 Binding Energy

13. (I) Estimate the total binding energy for $^{63}_{29}$Cu, using Fig. 30–1.
14. (I) Use Fig. 30–1 to estimate the total binding energy of (a) $^{238}_{92}$U, and (b) $^{84}_{36}$Kr.
15. (II) Calculate the binding energy per nucleon for a $^{15}_{7}$N nucleus, using Appendix B.
16. (II) Use Appendix B to calculate the binding energy of $^2_1$H (deuterium).
17. (II) Determine the binding energy of the last neutron in a $^{23}_{11}$Na nucleus.
18. (II) Calculate the total binding energy, and the binding energy per nucleon, for (a) $^7_3$Li, (b) $^{195}_{78}$Pt. Use Appendix B.
19. (II) Compare the average binding energy of a nucleon in $^{23}_{11}$Na to that in $^{24}_{11}$Na, using Appendix B.
20. (III) How much energy is required to remove (a) a proton, (b) a neutron, from $^{15}_{7}$N? Explain the difference in your answers.
21. (III) (a) Show that the nucleus $^8_4$Be (mass = 8.005305 u) is unstable and will decay into two α particles. (b) Is $^{12}_{6}$C stable against decay into three α particles? Show why or why not.

30–3 to 30–7 Radioactive Decay

22. (I) The $^7_3$Li nucleus has an excited state 0.48 MeV above the ground state. What wavelength gamma photon is emitted when the nucleus decays from the excited state to the ground state?
23. (II) Show that the decay $^{11}_{6}$C → $^{10}_{5}$B + p is not possible because energy would not be conserved.
24. (II) Calculate the energy released when tritium, $^3_1$H, decays by β^- emission.
25. (II) What is the maximum kinetic energy of an electron emitted in the β decay of a free neutron?
26. (II) Give the result of a calculation that shows whether or not the following decays are possible:
    (a) $^{233}_{92}$U → $^{232}_{92}$U + n;
    (b) $^{14}_{7}$N → $^{13}_{7}$N + n;
    (c) $^{40}_{19}$K → $^{39}_{19}$K + n.
27. (II) $^{24}_{11}$Na is radioactive. (a) Is it a β^- or β^+ emitter? (b) Write down the decay reaction, and estimate the maximum kinetic energy of the emitted β.

28. (II) A $^{238}_{92}$U nucleus emits an α particle with kinetic energy = 4.20 MeV. (a) What is the daughter nucleus, and (b) what is the approximate atomic mass (in u) of the daughter atom? Ignore recoil of the daughter nucleus.

29. (II) Calculate the maximum kinetic energy of the β particle emitted during the decay of $^{60}_{27}$Co.

30. (II) How much energy is released in electron capture by beryllium: $^{7}_{4}$Be + e^- → $^{7}_{3}$Li + ν?

31. (II) The isotope $^{218}_{84}$Po can decay by either α or β^- emission. What is the energy release in each case? The mass of $^{218}_{84}$Po is 218.008973 u.

32. (II) The nuclide $^{32}_{15}$P decays by emitting an electron whose maximum kinetic energy can be 1.71 MeV. (a) What is the daughter nucleus? (b) Calculate the daughter's atomic mass (in u).

33. (II) A photon with a wavelength of 1.15×10^{-13} m is ejected from an atom. Calculate its energy and explain why it is a γ ray from the nucleus or a photon from the atom.

34. (II) How much recoil energy does a $^{40}_{19}$K nucleus get when it emits a 1.46-MeV gamma ray?

35. (II) Determine the maximum kinetic energy of β^+ particles released when $^{11}_{6}$C decays to $^{11}_{5}$B. What is the maximum energy the neutrino can have? What is the minimum energy of each?

36. (III) Show that when a nucleus decays by β^+ decay, the total energy released is equal to
$$(M_P - M_D - 2m_e)c^2,$$
where M_P and M_D are the masses of the parent and daughter atoms (neutral), and m_e is the mass of an electron or positron.

37. (III) When $^{238}_{92}$U decays, the α particle emitted has 4.20 MeV of kinetic energy. Calculate the recoil kinetic energy of the daughter nucleus and the Q-value of the decay.

30–8 to 30–11 Half-Life, Decay Rates, Decay Series, Dating

38. (I) (a) What is the decay constant of $^{238}_{92}$U whose half-life is 4.5×10^9 yr? (b) The decay constant of a given nucleus is 3.2×10^{-5} s^{-1}. What is its half-life?

39. (I) A radioactive material produces 1120 decays per minute at one time, and 3.6 h later produces 140 decays per minute. What is its half-life?

40. (I) What fraction of a sample of $^{68}_{32}$Ge, whose half-life is about 9 months, will remain after 2.5 yr?

41. (I) What is the activity of a sample of $^{14}_{6}$C that contains 6.5×10^{20} nuclei?

42. (I) What fraction of a radioactive sample is left after exactly 5 half-lives?

43. (II) The iodine isotope $^{131}_{53}$I is used in hospitals for diagnosis of thyroid function. If 782 μg are ingested by a patient, determine the activity (a) immediately, (b) 1.50 h later when the thyroid is being tested, and (c) 3.0 months later. Use Appendix B.

44. (II) How many nuclei of $^{238}_{92}$U remain in a rock if the activity registers 420 decays per second?

45. (II) In a series of decays, the nuclide $^{235}_{92}$U becomes $^{207}_{82}$Pb. How many α and β^- particles are emitted in this series?

46. (II) $^{124}_{55}$Cs has a half-life of 30.8 s. (a) If we have 8.7 μg initially, how many Cs nuclei are present? (b) How many are present 2.6 min later? (c) What is the activity at this time? (d) After how much time will the activity drop to less than about 1 per second?

47. (II) Calculate the mass of a sample of pure $^{40}_{19}$K with an initial decay rate of 2.4×10^5 s^{-1}. The half-life of $^{40}_{19}$K is 1.248×10^9 yr.

48. (II) Calculate the activity of a pure 6.7-μg sample of $^{32}_{15}$P ($T_{\frac{1}{2}} = 1.23 \times 10^6$ s).

49. (II) A sample of $^{233}_{92}$U ($T_{\frac{1}{2}} = 1.59 \times 10^5$ yr) contains 4.50×10^{18} nuclei. (a) What is the decay constant? (b) Approximately how many disintegrations will occur per minute?

50. (II) The activity of a sample drops by a factor of 6.0 in 9.4 minutes. What is its half-life?

51. (II) A 345-g sample of pure carbon contains 1.3 parts in 10^{12} (atoms) of $^{14}_{6}$C. How many disintegrations occur per second?

52. (II) A sample of $^{238}_{92}$U is decaying at a rate of 4.20×10^2 decays/s. What is the mass of the sample?

53. (II) **Rubidium–strontium dating.** The rubidium isotope $^{87}_{37}$Rb, a β emitter with a half-life of 4.75×10^{10} yr, is used to determine the age of rocks and fossils. Rocks containing fossils of ancient animals contain a ratio of $^{87}_{38}$Sr to $^{87}_{37}$Rb of 0.0260. Assuming that there was no $^{87}_{38}$Sr present when the rocks were formed, estimate the age of these fossils.

54. (II) Two of the naturally occurring radioactive decay sequences start with $^{232}_{90}$Th and with $^{235}_{92}$U. The first five decays of these two sequences are:

and
$$\alpha, \beta, \beta, \alpha, \alpha$$
$$\alpha, \beta, \alpha, \beta, \alpha.$$

Determine the resulting intermediate daughter nuclei in each case.

55. (II) An ancient wooden club is found that contains 73 g of carbon and has an activity of 7.0 decays per second. Determine its age assuming that in living trees the ratio of ^{14}C/^{12}C atoms is about 1.3×10^{-12}.

56. (II) Use Fig. 30–11 and calculate the relative decay rates for α decay of $^{218}_{84}$Po and $^{214}_{84}$Po.

57. (III) The activity of a radioactive source decreases by 5.5% in 31.0 hours. What is the half-life of this source?

58. (III) $^{7}_{4}$Be decays with a half-life of about 53 d. It is produced in the upper atmosphere, and filters down onto the Earth's surface. If a plant leaf is detected to have 350 decays/s of $^{7}_{4}$Be, (a) how long do we have to wait for the decay rate to drop to 25 per second? (b) Estimate the initial mass of $^{7}_{4}$Be on the leaf.

59. (III) At $t = 0$, a pure sample of radioactive nuclei contains N_0 nuclei whose decay constant is λ. Determine a formula for the number of daughter nuclei, N_D, as a function of time; assume the daughter is stable and that $N_D = 0$ at $t = 0$.

General Problems

60. Which radioactive isotope of lead is being produced if the measured activity of a sample drops to 1.050% of its original activity in 4.00 h?

61. An old wooden tool is found to contain only 4.5% of the $^{14}_{6}C$ that an equal mass of fresh wood would. How old is the tool?

62. A neutron star consists of neutrons at approximately nuclear density. Estimate, for a 10-km-diameter neutron star, (a) its mass number, (b) its mass (kg), and (c) the acceleration of gravity at its surface.

63. **Tritium dating.** The $^{3}_{1}H$ isotope of hydrogen, which is called *tritium* (because it contains three nucleons), has a half-life of 12.3 yr. It can be used to measure the age of objects up to about 100 yr. It is produced in the upper atmosphere by cosmic rays and brought to Earth by rain. As an application, determine approximately the age of a bottle of wine whose $^{3}_{1}H$ radiation is about $\frac{1}{10}$ that present in new wine.

64. Some elementary particle theories (Section 32–11) suggest that the proton may be unstable, with a half-life $\geq 10^{33}$ yr. (a) How long would you expect to wait for one proton in your body to decay (approximate your body as all water)? (b) Of the roughly 7 billion people on Earth, about how many would have a proton in their body decay in a 70 yr lifetime?

65. The original experiments which established that an atom has a heavy, positive nucleus were done by shooting alpha particles through gold foil. The alpha particles had a kinetic energy of 7.7 MeV. What is the closest they could get to the center of a gold nucleus? How does this compare with the size of the nucleus?

66. How long must you wait (in half-lives) for a radioactive sample to drop to 2.00% of its original activity?

67. If the potassium isotope $^{40}_{19}K$ gives 42 decays/s in a liter of milk, estimate how much $^{40}_{19}K$ and regular $^{39}_{19}K$ are in a liter of milk. Use Appendix B.

68. Strontium-90 is produced as a nuclear fission product of uranium in both reactors and atomic bombs. Look at its location in the Periodic Table to see what other elements it might be similar to chemically, and tell why you think it might be dangerous to ingest. It has too many neutrons to be stable, and it decays with a half-life of about 29 yr. How long will we have to wait for the amount of $^{90}_{38}Sr$ on the Earth's surface to reach 1% of its current level, assuming no new material is scattered about? Write down the decay reaction, including the daughter nucleus. The daughter is radioactive: write down its decay.

69. The activity of a sample of $^{35}_{16}S$ ($T_{\frac{1}{2}} = 87.37$ days) is 4.28×10^{4} decays per second. What is the mass of the sample?

70. The nuclide $^{191}_{76}Os$ decays with β^{-} energy of 0.14 MeV accompanied by γ rays of energy 0.042 MeV and 0.129 MeV. (a) What is the daughter nucleus? (b) Draw an energy-level diagram showing the ground states of the parent and daughter and excited states of the daughter. (c) To which of the daughter states does β^{-} decay of $^{191}_{76}Os$ occur?

71. Determine the activities of (a) 1.0 g of $^{131}_{53}I$ ($T_{\frac{1}{2}} = 8.02$ days) and (b) 1.0 g of $^{238}_{92}U$ ($T_{\frac{1}{2}} = 4.47 \times 10^{9}$ yr).

72. Use Fig. 30–1 to estimate the total binding energy for copper and then estimate the energy, in joules, needed to break a 3.0-g copper penny into its constituent nucleons.

73. Instead of giving atomic masses for nuclides as in Appendix B, some Tables give the **mass excess**, Δ, defined as $\Delta = M - A$, where A is the atomic mass number and M is the mass in u. Determine the mass excess, in u and in MeV/c^2, for: (a) $^{4}_{2}He$; (b) $^{12}_{6}C$; (c) $^{86}_{38}Sr$; (d) $^{235}_{92}U$. (e) From a glance at Appendix B, can you make a generalization about the sign of Δ as a function of Z or A?

74. When water is placed near an intense neutron source, the neutrons can be slowed down to almost zero speed by collisions with the water molecules, and are eventually captured by a hydrogen nucleus to form the stable isotope called **deuterium**, $^{2}_{1}H$, giving off a gamma ray. What is the energy of the gamma ray?

75. The practical limit for carbon-14 dating is about 60,000 years. If a bone contains 1.0 kg of carbon, and the animal died 60,000 years ago, what is the activity today?

76. Using Section 30–2 and Appendix B, determine the energy required to remove one neutron from $^{4}_{2}He$. How many times greater is this energy than the binding energy of the last neutron in $^{13}_{6}C$?

77. (a) If all of the atoms of the Earth were to collapse and simply become nuclei, what would be the Earth's new radius? (b) If all of the atoms of the Sun were to collapse and simply become nuclei, what would be the Sun's new radius?

78. (a) A 72-gram sample of natural carbon contains the usual fraction of $^{14}_{6}C$. Estimate roughly how long it will take before there is only one $^{14}_{6}C$ nucleus left. (b) How does the answer in (a) change if the sample is 340 grams? What does this tell you about the limits of carbon dating?

79. If the mass of the proton were just a little closer to the mass of the neutron, the following reaction would be possible even at low collision energies:

$$e^{-} + p \rightarrow n + \nu.$$

(a) Why would this situation be catastrophic? (See last paragraph of Chapter 33.) (b) By what percentage would the proton's mass have to be increased to make this reaction possible?

80. What is the ratio of the kinetic energies for an alpha particle and a beta particle if both make tracks with the same radius of curvature in a magnetic field, oriented perpendicular to the paths of the particles?

81. A 1.00-g sample of natural samarium emits α particles at a rate of 120 s^{-1} due to the presence of $^{147}_{62}Sm$. The natural abundance of $^{147}_{62}Sm$ is 15%. Calculate the half-life for this decay process.

82. Almost all of naturally occurring uranium is $^{238}_{92}U$ with a half-life of 4.468×10^{9} yr. Most of the rest of natural uranium is $^{235}_{92}U$ with a half-life of 7.04×10^{8} yr. Today a sample contains 0.720% $^{235}_{92}U$. (a) What was this percentage 1.0 billion years ago? (b) What percentage of uranium will be $^{235}_{92}U$ 100 million years from now?

83. A banana contains about 420 mg of potassium, of which a small fraction is the radioactive isotope $^{40}_{19}$K (Appendix B). Estimate the activity of an average banana due to $^{40}_{19}$K.

84. When $^{23}_{10}$Ne (mass = 22.9947 u) decays to $^{23}_{11}$Na (mass = 22.9898 u), what is the maximum kinetic energy of the emitted electron? What is its minimum energy? What is the energy of the neutrino in each case? Ignore recoil of the daughter nucleus.

85. (a) In α decay of, say, a $^{226}_{88}$Ra nucleus, show that the nucleus carries away a fraction $1/(1 + \frac{1}{4}A_D)$ of the total energy available, where A_D is the mass number of the daughter nucleus. [Hint: Use conservation of momentum as well as conservation of energy.] (b) Approximately what percentage of the energy available is thus carried off by the α particle when $^{226}_{88}$Ra decays?

86. Decay series, such as that shown in Fig. 30–11, can be classified into four families, depending on whether the mass numbers have the form $4n$, $4n + 1$, $4n + 2$, or $4n + 3$, where n is an integer. Justify this statement and show that for a nuclide in any family, all its daughters will be in the same family.

Search and Learn

1. Describe in detail why we think there is a strong nuclear force.

2. (a) Under what circumstances could a fermium nucleus decay into an einsteinium nucleus? (b) What about the reverse, an Es nucleus decaying into Fm?

3. Using the uncertainty principle and the radius of a nucleus, estimate the minimum possible kinetic energy of a nucleon in, say, iron. Ignore relativistic corrections. [Hint: A particle can have a momentum at least as large as its momentum uncertainty.]

4. In Fig. 30–17, a nucleus decays and emits a particle that enters a region with a uniform magnetic field of 0.012 T directed into the page. The path of the detected particle is shown. (a) What type of radioactive decay is this? (b) If the radius of the circular arc is 4.7 mm, what is the velocity of the particle?

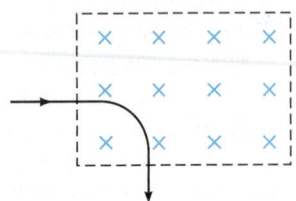

FIGURE 30–17 Search and Learn 4.

5. In both internal conversion and β decay, an electron is emitted. How could you determine which decay process occurred?

6. Suppose we discovered that several thousand years ago cosmic rays had bombarded the Earth's atmosphere a lot more than we had thought. Compared to previous calculations of the carbon-dated age of organic matter, we would now calculate it to be older, younger, or the same age as previously calculated? Explain.

7. In 1991, the frozen remains of a Neolithic-age man, nicknamed Otzi, were found in the Italian Alps by hikers. The body was well preserved, as were his bow, arrows, knife, axe, other tools, and clothing. The date of his death can be determined using carbon-14 dating. (a) What is the decay constant for $^{14}_{6}$C? (b) How many $^{14}_{6}$C atoms per gram of $^{12}_{6}$C are there in a living organism? (c) What is the activity per gram in naturally occurring carbon for a living organism? (d) For Otzi, the activity per gram of carbon was measured to be 0.121. How long ago did he live?

8. Some radioactive isotopes have half-lives that are greater than the age of the universe (like gadolinium or samarium). The only way to determine these half-lives is to monitor the decay rate of a sample that contains these isotopes. For example, suppose we find an asteroid that currently contains about 15,000 kg of $^{152}_{64}$Gd (gadolinium) and we detect an activity of 1 decay/s. Estimate the half-life of gadolinium (in years).

ANSWERS TO EXERCISES

A: 0.042130 u.
B: 7.98 MeV/nucleon.
C: (b) Gd.
D: (b) $\frac{1}{8}\mu$g.
E: (b) 20 μg.

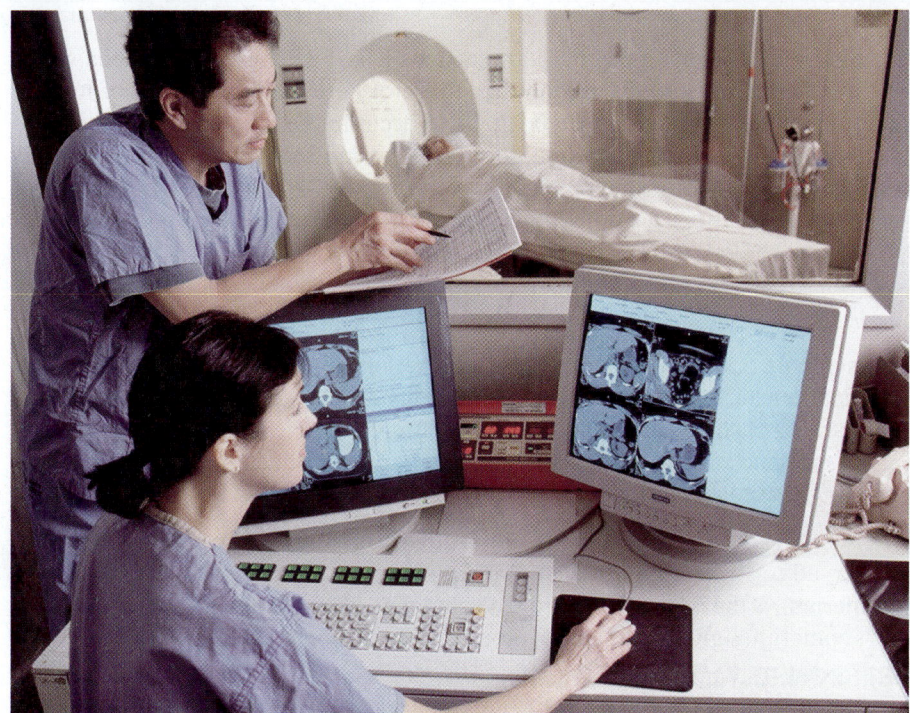

Technicians are looking at an MRI image of sections through a patient's body. MRI is one of several powerful types of medical imaging based on physics used by doctors to diagnose illnesses.

This Chapter opens with basic and important physics topics of nuclear reactions, nuclear fission, and nuclear fusion, and how we obtain nuclear energy. Then we examine the health aspects of radiation—dosimetry, therapy, and imaging: MRI, PET, and SPECT.

Nuclear Energy; Effects and Uses of Radiation

CHAPTER 31

CHAPTER-OPENING QUESTION—Guess now!

The Sun is powered by
(a) nuclear alpha decay.
(b) nuclear beta decay.
(c) nuclear gamma decay.
(d) nuclear fission.
(e) nuclear fusion.

CONTENTS

31–1 Nuclear Reactions and the Transmutation of Elements
31–2 Nuclear Fission; Nuclear Reactors
31–3 Nuclear Fusion
31–4 Passage of Radiation Through Matter; Biological Damage
31–5 Measurement of Radiation— Dosimetry
*31–6 Radiation Therapy
*31–7 Tracers in Research and Medicine
*31–8 Emission Tomography: PET and SPECT
31–9 Nuclear Magnetic Resonance (NMR) and Magnetic Resonance Imaging (MRI)

We continue our study of nuclear physics in this Chapter. We begin with a discussion of nuclear reactions, and then we examine the important huge energy-releasing processes of fission and fusion. We also deal with the effects of nuclear radiation passing through matter, particularly biological matter, and how radiation is used medically for therapy, diagnosis, and imaging techniques.

31–1 Nuclear Reactions and the Transmutation of Elements

When a nucleus undergoes α or β decay, the daughter nucleus is a different element from the parent. The transformation of one element into another, called **transmutation**, also occurs via nuclear reactions. A **nuclear reaction** is said to occur when a nucleus is struck by another nucleus, or by a simpler particle such as a γ ray, neutron, or proton, and an interaction takes place. Ernest Rutherford was the first to report seeing a nuclear reaction. In 1919 he observed that some of the α particles passing through nitrogen gas were absorbed and protons emitted. He concluded that nitrogen nuclei had been transformed into oxygen nuclei via the reaction

$$^{4}_{2}\text{He} + ^{14}_{7}\text{N} \rightarrow ^{17}_{8}\text{O} + ^{1}_{1}\text{H},$$

where $^{4}_{2}\text{He}$ is an α particle, and $^{1}_{1}\text{H}$ is a proton.

885

Since then, a great many nuclear reactions have been observed. Indeed, many of the radioactive isotopes used in the laboratory are made by means of nuclear reactions. Nuclear reactions can be made to occur in the laboratory, but they also occur regularly in nature. In Chapter 30 we saw an example: $^{14}_{6}C$ is continually being made in the atmosphere via the reaction $n + ^{14}_{7}N \rightarrow ^{14}_{6}C + p$.

Nuclear reactions are sometimes written in a shortened form: for example,
$$n + ^{14}_{7}N \rightarrow ^{14}_{6}C + p$$
can be written
$$^{14}_{7}N\,(n,\,p)\,^{14}_{6}C.$$

The symbols outside the parentheses on the left and right represent the initial and final nuclei, respectively. The symbols inside the parentheses represent the bombarding particle (first) and the emitted small particle (second).

In any nuclear reaction, both electric charge and nucleon number are conserved. These conservation laws are often useful, as the following Example shows.

CONCEPTUAL EXAMPLE 31–1 Deuterium reaction. A neutron is observed to strike an $^{16}_{8}O$ nucleus, and a deuteron is given off. (A **deuteron**, or **deuterium**, is the isotope of hydrogen containing one proton and one neutron, $^{2}_{1}H$; it is sometimes given the symbol d or D.) What is the nucleus that results?

RESPONSE We have the reaction $n + ^{16}_{8}O \rightarrow ? + ^{2}_{1}H$. The total number of nucleons initially is $1 + 16 = 17$, and the total charge is $0 + 8 = 8$. The same totals apply after the reaction. Hence the product nucleus must have $Z = 7$ and $A = 15$. From the Periodic Table, we find that it is nitrogen that has $Z = 7$, so the nucleus produced is $^{15}_{7}N$.

EXERCISE A Determine the resulting nucleus in the reaction $n + ^{137}_{56}Ba \rightarrow ? + \gamma$.

Energy and momentum are also conserved in nuclear reactions, and can be used to determine whether or not a given reaction can occur. For example, if the total mass of the final products is less than the total mass of the initial particles, this decrease in mass (recall $\Delta E = \Delta m\,c^2$) is converted to kinetic energy (KE) of the outgoing particles. But if the total mass of the products is greater than the total mass of the initial reactants, the reaction requires energy. The reaction will then not occur unless the bombarding particle has sufficient kinetic energy. Consider a nuclear reaction of the general form
$$a + X \rightarrow Y + b, \qquad (31\text{–}1)$$
where particle a is a moving projectile particle (or small nucleus) that strikes nucleus X, producing nucleus Y and particle b (typically, p, n, α, γ). We define the **reaction energy**, or **Q-value**, in terms of the masses involved, as
$$Q = (M_a + M_X - M_b - M_Y)c^2. \qquad (31\text{–}2a)$$
For a γ ray, $M = 0$. If energy is released by the reaction, $Q > 0$. If energy is required, $Q < 0$.

Because energy is conserved, Q has to be equal to the change in kinetic energy (final minus initial):
$$Q = KE_b + KE_Y - KE_a - KE_X. \qquad (31\text{–}2b)$$
If X is a target nucleus at rest (or nearly so) struck by incoming particle a, then $KE_X = 0$. For $Q > 0$, the reaction is said to be *exothermic* or *exoergic*; energy is released in the reaction, so the total kinetic energy is greater after the reaction than before. If Q is negative, the reaction is said to be *endothermic* or *endoergic*: an energy input is required to make the reaction happen. The energy input comes from the kinetic energy of the initial colliding particles (a and X).

EXAMPLE 31–2 A slow-neutron reaction. The nuclear reaction
$$n + ^{10}_{5}B \rightarrow ^{7}_{3}Li + ^{4}_{2}He$$
is observed to occur even when very slow-moving neutrons (mass $M_n = 1.0087\,u$) strike boron atoms at rest. For a particular reaction in which $KE_n \approx 0$, the outgoing helium ($M_{He} = 4.0026\,u$) is observed to have a speed of 9.30×10^6 m/s. Determine (a) the kinetic energy of the lithium ($M_{Li} = 7.0160\,u$), and (b) the Q-value of the reaction.

APPROACH Since the neutron and boron are both essentially at rest, the total momentum before the reaction is zero; momentum is conserved and so must be zero afterward as well. Thus,

$$M_{Li} v_{Li} = M_{He} v_{He}.$$

We solve this for v_{Li} and substitute it into the equation for kinetic energy. In (b) we use Eq. 31–2b.

SOLUTION (a) We can use classical kinetic energy with little error, rather than relativistic formulas, because $v_{He} = 9.30 \times 10^6$ m/s is not close to the speed of light c. And v_{Li} will be even less because $M_{Li} > M_{He}$. Thus we can write the KE of the lithium, using the momentum equation just above, as

$$KE_{Li} = \frac{1}{2} M_{Li} v_{Li}^2 = \frac{1}{2} M_{Li} \left(\frac{M_{He} v_{He}}{M_{Li}} \right)^2 = \frac{M_{He}^2 v_{He}^2}{2 M_{Li}}.$$

We put in numbers, changing the mass in u to kg and recall that 1.60×10^{-13} J = 1 MeV:

$$KE_{Li} = \frac{(4.0026 \text{ u})^2 (1.66 \times 10^{-27} \text{ kg/u})^2 (9.30 \times 10^6 \text{ m/s})^2}{2(7.0160 \text{ u})(1.66 \times 10^{-27} \text{ kg/u})}$$

$$= 1.64 \times 10^{-13} \text{ J} = 1.02 \text{ MeV}.$$

(b) We are given the data $KE_a = KE_X = 0$ in Eq. 31–2b, so $Q = KE_{Li} + KE_{He}$, where

$$KE_{He} = \tfrac{1}{2} M_{He} v_{He}^2 = \tfrac{1}{2} (4.0026 \text{ u})(1.66 \times 10^{-27} \text{ kg/u})(9.30 \times 10^6 \text{ m/s})^2$$

$$= 2.87 \times 10^{-13} \text{ J} = 1.80 \text{ MeV}.$$

Hence, $Q = 1.02$ MeV + 1.80 MeV = 2.82 MeV.

EXAMPLE 31–3 Will the reaction "go"? Can the reaction

$$p + {}^{13}_{6}C \rightarrow {}^{13}_{7}N + n$$

occur when ${}^{13}_{6}C$ is bombarded by 2.0-MeV protons?

APPROACH The reaction will "go" if the reaction is exothermic ($Q > 0$) and even if $Q < 0$ if the input momentum and kinetic energy are sufficient. First we calculate Q from the difference between final and initial masses using Eq. 31–2a, and look up the masses in Appendix B.

SOLUTION The total masses before and after the reaction are:

Before	After
$M({}^{13}_{6}C) = 13.003355$	$M({}^{13}_{7}N) = 13.005739$
$M({}^{1}_{1}H) = 1.007825$	$M(n) = 1.008665$
14.011180	14.014404

(We must use the mass of the ${}^{1}_{1}H$ atom rather than that of the bare proton because the masses of ${}^{13}_{6}C$ and ${}^{13}_{7}N$ include the electrons, and we must include an equal number of electron masses on each side of the equation.) The products have an excess mass of

$$(14.014404 - 14.011180) \text{u} = 0.003224 \text{ u} \times 931.5 \text{ MeV/u} = 3.00 \text{ MeV}.$$

Thus $Q = -3.00$ MeV, and the reaction is endothermic. This reaction requires energy, and the 2.0-MeV protons do not have enough to make it go.

NOTE The incoming proton in this Example would need more than 3.00 MeV of kinetic energy to make this reaction go; 3.00 MeV would be enough to conserve energy, but a proton of this energy would produce the ${}^{13}_{7}N$ and n with no kinetic energy and hence no momentum. Since an incident 3.0-MeV proton has momentum, conservation of momentum would be violated. A calculation using conservation of energy *and* of momentum, as we did in Examples 30–7 and 31–2, shows that the minimum proton energy, called the **threshold energy**, is 3.23 MeV in this case.

(a)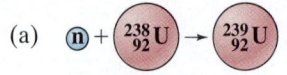

Neutron captured by $^{238}_{92}$U.

(b)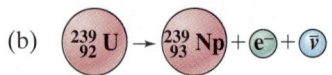

$^{239}_{92}$U decays by β decay to neptunium-239.

(c)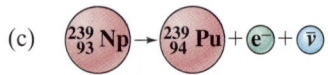

$^{239}_{93}$Np itself decays by β decay to produce plutonium-239.

FIGURE 31–1 Neptunium and plutonium are produced in this series of reactions, after bombardment of $^{238}_{92}$U by neutrons.

FIGURE 31–2 Projectile particles strike a target of area A and thickness ℓ made up of n nuclei per unit volume.

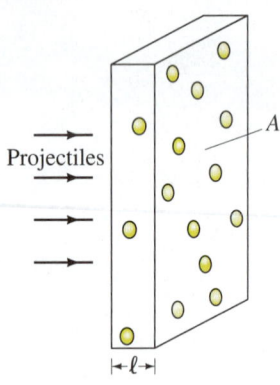

Neutron Physics

The artificial transmutation of elements took a great leap forward in the 1930s when Enrico Fermi realized that neutrons would be the most effective projectiles for causing nuclear reactions and in particular for producing new elements. Because neutrons have no net electric charge, they are not repelled by positively charged nuclei as are protons or alpha particles. Hence the probability of a neutron reaching the nucleus and causing a reaction is much greater than for charged projectiles,[†] particularly at low energies. Between 1934 and 1936, Fermi and his co-workers in Rome produced many previously unknown isotopes by bombarding different elements with neutrons. Fermi realized that if the heaviest known element, uranium, is bombarded with neutrons, it might be possible to produce new elements with atomic numbers greater than that of uranium. After several years of hard work, it was suspected that two new elements had been produced, neptunium ($Z = 93$) and plutonium ($Z = 94$). The full confirmation that such "transuranic" elements could be produced came several years later at the University of California, Berkeley. The reactions are shown in Fig. 31–1.

It was soon shown that what Fermi had actually observed when he bombarded uranium was an even stranger process—one that was destined to play an extraordinary role in the world at large. We discuss it in Section 31–2.

*Cross Section

Some reactions have a higher probability of occurring than others. The reaction probability is specified by a quantity called the collision **cross section**. Although the size of a nucleus, like that of an atom, is not a clearly defined quantity since the edges are not distinct like those of a tennis ball or baseball, we can nonetheless define a *cross section* for nuclei undergoing collisions by using an analogy. Suppose that projectile particles strike a stationary target of total area A and thickness ℓ, as shown in Fig. 31–2. Assume also that the target is made up of identical objects (such as marbles or nuclei), each of which has a cross-sectional area σ, and we assume the incoming projectiles are small by comparison. We assume that the target objects are fairly far apart and the thickness ℓ is so small that we don't have to worry about overlapping. This is often a reasonable assumption because nuclei have diameters on the order of 10^{-14} m but are at least 10^{-10} m (atomic size) apart even in solids. If there are n nuclei per unit volume, the total cross-sectional area of all these tiny targets is

$$A' = nA\ell\sigma$$

since $nA\ell = (n)(\text{volume})$ is the total number of targets and σ is the cross-sectional area of each. If $A' \ll A$, most of the incident projectile particles will pass through the target without colliding. If R_0 is the rate at which the projectile particles strike the target (number/second), the rate at which collisions occur, R, is

$$R = R_0 \frac{A'}{A} = R_0 \frac{nA\ell\sigma}{A}$$

so

$$R = R_0 n\ell\sigma.$$

Thus, by measuring the collision rate, R, we can determine σ:

$$\sigma = \frac{R}{R_0 n\ell}.$$

The cross section σ is an "effective" target area. It is a *measure of the probability of a collision or of a particular reaction occurring* per target nucleus, independent of the dimensions of the entire target. The concept of cross section is useful

---

[†]That is, positively charged particles. Electrons rarely cause nuclear reactions because they do not interact via the strong nuclear force.

because σ depends only on the properties of the interacting particles, whereas R depends on the thickness and area of the physical (macroscopic) target, on the number of particles in the incident beam, and so on.

31–2 Nuclear Fission; Nuclear Reactors

In 1938, the German scientists Otto Hahn and Fritz Strassmann made an amazing discovery. Following up on Fermi's work, they found that uranium bombarded by neutrons sometimes produced smaller nuclei that were roughly half the size of the original uranium nucleus. Lise Meitner and Otto Frisch quickly realized what had happened: the uranium nucleus, after absorbing a neutron, actually had split into two roughly equal pieces. This was startling, for until then the known nuclear reactions involved knocking out only a tiny fragment (for example, n, p, or α) from a nucleus.

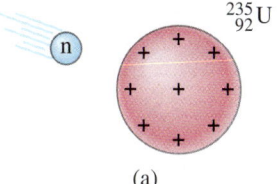
(a)

Nuclear Fission and Chain Reactions

This new phenomenon was named **nuclear fission** because of its resemblance to biological fission (cell division). It occurs much more readily for $^{235}_{92}\text{U}$ than for the more common $^{238}_{92}\text{U}$. The process can be visualized by imagining the uranium nucleus to be like a liquid drop. According to this **liquid-drop model**, the neutron absorbed by the $^{235}_{92}\text{U}$ nucleus (Fig. 31–3a) gives the nucleus extra internal energy (like heating a drop of water). This intermediate state, or **compound nucleus**, is $^{236}_{92}\text{U}$ (because of the absorbed neutron), Fig. 31–3b. The extra energy of this nucleus—it is in an excited state—appears as increased motion of the individual nucleons inside, which causes the nucleus to take on abnormal elongated shapes. When the nucleus elongates (in this model) into the shape shown in Fig. 31–3c, the attraction of the two ends via the short-range nuclear force is greatly weakened by the increased separation distance. Then the electric repulsive force becomes dominant, and the nucleus splits in two (Fig. 31–3d). The two resulting nuclei, X_1 and X_2, are called **fission fragments**, and in the process a number of neutrons (typically two or three) are also given off. The reaction can be written

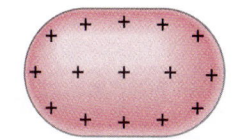
(b) $^{236}_{92}\text{U}$ (compound nucleus)

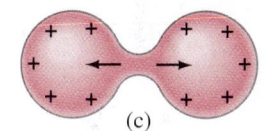
(c)

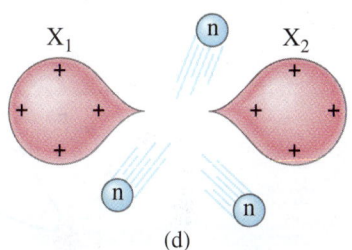
(d)

$$\text{n} + {}^{235}_{92}\text{U} \rightarrow {}^{236}_{92}\text{U} \rightarrow X_1 + X_2 + \text{neutrons}. \quad (31\text{–}3)$$

The compound nucleus, $^{236}_{92}\text{U}$, exists for less than 10^{-12} s, so the process occurs very quickly. The two fission fragments, X_1 and X_2, rarely split the original uranium mass precisely half and half, but more often as about 40%–60%. A typical fission reaction is

$$\text{n} + {}^{235}_{92}\text{U} \rightarrow {}^{141}_{56}\text{Ba} + {}^{92}_{36}\text{Kr} + 3\text{n}, \quad (31\text{–}4)$$

although many others also occur.

FIGURE 31–3 Fission of a $^{235}_{92}\text{U}$ nucleus after capture of a neutron, according to the liquid-drop model.

FIGURE 31–4 Mass distribution of fission fragments from $^{235}_{92}\text{U}$ + n. The small arrow indicates equal mass fragments ($\frac{1}{2} \times (236 - 2) = 117$, assuming 2 neutrons are liberated). Note that the vertical scale is logarithmic.

CONCEPTUAL EXAMPLE 31–4 **Counting nucleons.** Identify the element X in the fission reaction $\text{n} + {}^{235}_{92}\text{U} \rightarrow {}^{A}_{Z}X + {}^{93}_{38}\text{Sr} + 2\text{n}$.

RESPONSE The number of nucleons is conserved (Section 30–7). The uranium nucleus with 235 nucleons plus the incoming neutron make $235 + 1 = 236$ nucleons. So there must be 236 nucleons after the reaction. The Sr has 93 nucleons, and the two neutrons make 95 nucleons, so X has $A = 236 - 95 = 141$. Electric charge is also conserved: before the reaction, the total charge is $92e$. After the reaction the total charge is $(Z + 38)e$ and must equal $92e$. Thus $Z = 92 - 38 = 54$. The element with $Z = 54$ is xenon (see Appendix B or the Periodic Table inside the back cover), so the isotope is $^{141}_{54}\text{Xe}$.

EXERCISE B In the fission reaction $\text{n} + {}^{235}_{92}\text{U} \rightarrow {}^{137}_{53}\text{I} + {}^{96}_{39}\text{Y} + \text{neutrons}$, how many neutrons are produced?

Figure 31–4 shows the measured distribution of $^{235}_{92}\text{U}$ fission fragments according to mass. Only rarely (about 1 in 10^4) does a fission result in equal mass fragments (arrow in Fig. 31–4).

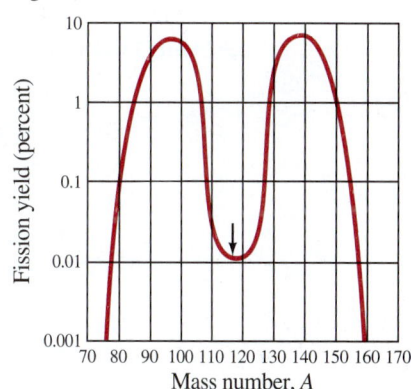

A tremendous amount of energy is released in a fission reaction because the mass of $^{235}_{92}\text{U}$ is considerably greater than the total mass of the fission fragments plus released neutrons. This can be seen from the binding-energy-per-nucleon curve of Fig. 30–1; the binding energy per nucleon for uranium is about 7.6 MeV/nucleon, but for fission fragments that have intermediate mass (in the center portion of the graph, $A \approx 100$), the average binding energy per nucleon is about 8.5 MeV/nucleon. Since the fission fragments are more tightly bound, the sum of their masses is less than the mass of the uranium. The difference in mass, or energy, between the original uranium nucleus and the fission fragments is about $8.5 - 7.6 = 0.9$ MeV per nucleon. Because there are 236 nucleons involved in each fission, the total energy released per fission is

$$(0.9 \,\text{MeV/nucleon})(236 \,\text{nucleons}) \approx 200 \,\text{MeV}. \qquad (31\text{–}5)$$

This is an enormous amount of energy for one single nuclear event. At a practical level, the energy from one fission is tiny. But if many such fissions could occur in a short time, an enormous amount of energy at the macroscopic level would be available. A number of physicists, including Fermi, recognized that the neutrons released in each fission (Eqs. 31–3 and 31–4) could be used to create a **chain reaction**. That is, one neutron initially causes one fission of a uranium nucleus; the two or three neutrons released can go on to cause additional fissions, so the process multiplies as shown schematically in Fig. 31–5.

If a **self-sustaining chain reaction** was actually possible in practice, the enormous energy available in fission could be released on a larger scale. Fermi and his co-workers (at the University of Chicago) showed it was possible by constructing the first **nuclear reactor** in 1942 (Fig. 31–6).

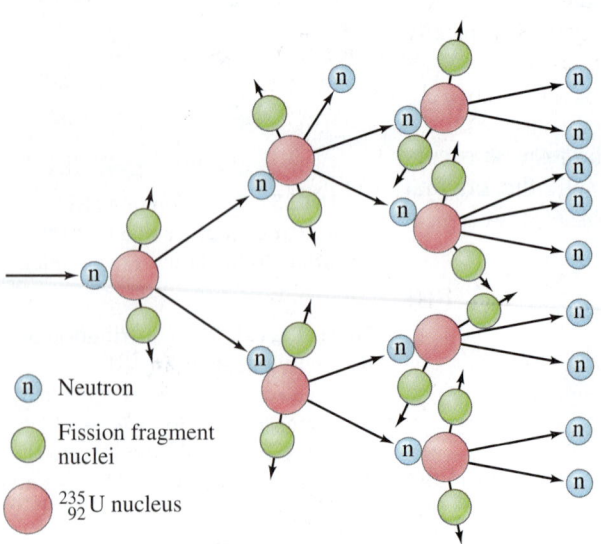

FIGURE 31–5 Chain reaction.

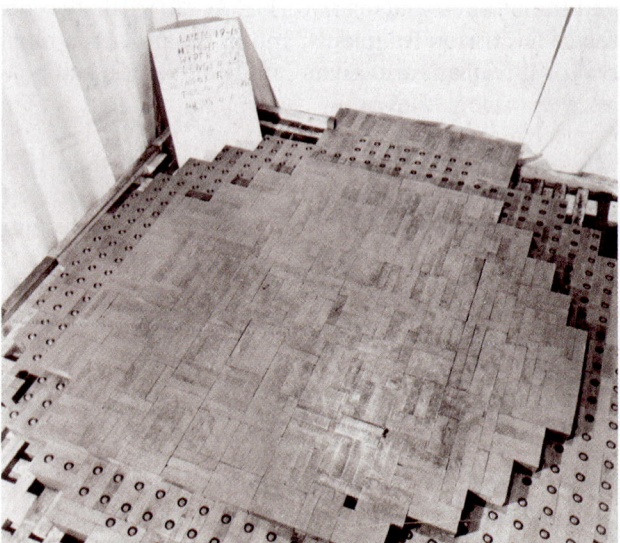

FIGURE 31–6 This is the only photograph of the first nuclear reactor, built by Fermi under the grandstand of Stagg Field at the University of Chicago. It is shown here under construction as a layer of graphite (used as moderator) was being placed over a layer of natural uranium. On December 2, 1942, Fermi slowly withdrew the cadmium control rods and the reactor went critical. This first self-sustaining chain reaction was announced to Washington, via telephone, by Arthur Compton who witnessed the event and reported: "The Italian navigator has just landed in the new world."

Nuclear Reactors

Several problems have to be overcome to make any nuclear reactor function. First, the probability that a $^{235}_{92}$U nucleus will absorb a neutron is large only for slow neutrons, but the neutrons emitted during a fission (which are needed to sustain a chain reaction) are moving very fast. A substance known as a **moderator** must be used to slow down the neutrons. The most effective moderator will consist of atoms whose mass is as close as possible to that of the neutrons. (To see why this is true, recall from Chapter 7 that a billiard ball striking an equal mass ball at rest can itself be stopped in one collision; but a billiard ball striking a heavy object bounces off with nearly unchanged speed.) The best moderator would thus contain $^1_1$H atoms. Unfortunately, $^1_1$H tends to absorb neutrons. But the isotope of hydrogen called *deuterium*, $^2_1$H, does not absorb many neutrons and is thus an almost ideal moderator. Either $^1_1$H or $^2_1$H can be used in the form of water. In the latter case, it is **heavy water**, in which the hydrogen atoms have been replaced by deuterium. Another common moderator is *graphite*, which consists of $^{12}_6$C atoms.

A second problem is that the neutrons produced in one fission may be absorbed and produce other nuclear reactions with other nuclei in the reactor, rather than produce further fissions. In a "light-water" reactor, the $^1_1$H nuclei absorb neutrons, as does $^{238}_{92}$U to form $^{239}_{92}$U in the reaction n + $^{238}_{92}$U → $^{239}_{92}$U + γ. Naturally occurring uranium[†] contains 99.3% $^{238}_{92}$U and only 0.7% fissionable $^{235}_{92}$U. To increase the probability of fission of $^{235}_{92}$U nuclei, natural uranium can be **enriched** to increase the percentage of $^{235}_{92}$U by using processes such as diffusion or centrifugation. Enrichment is not usually necessary for reactors using heavy water as moderator because heavy water doesn't absorb neutrons.[‡]

The third problem is that some neutrons will escape through the surface of the reactor core before they can cause further fissions (Fig. 31–7). Thus the mass of fuel must be sufficiently large for a self-sustaining chain reaction to take place. The minimum mass of uranium needed is called the **critical mass**. The value of the critical mass depends on the moderator, the fuel ($^{239}_{94}$Pu may be used instead of $^{235}_{92}$U), and how much the fuel is enriched, if at all. Typical values are on the order of a few kilograms (that is, neither grams nor thousands of kilograms). Critical mass depends also on the average number of neutrons released per fission: 2.5 for $^{235}_{92}$U, 2.9 for $^{239}_{94}$Pu so the critical mass for $^{239}_{94}$Pu is smaller.

To have a self-sustaining chain reaction, on average at least one neutron produced in each fission must go on to produce another fission. The average number of neutrons per fission that do go on to produce further fissions is called the **neutron multiplication factor**, f. For a self-sustaining chain reaction, we must have $f \geq 1$. If $f < 1$, the reactor is "subcritical." If $f > 1$, it is "supercritical" (and could become dangerously explosive). Reactors are equipped with movable **control rods** (good neutron absorbers like cadmium or boron), whose function is to absorb neutrons and maintain the reactor at just barely "critical," $f = 1$.

The release of neutrons and subsequent fissions occur so quickly that manipulation of the control rods to maintain $f = 1$ would not be possible if it weren't for the small percentage ($\approx 1\%$) of so-called **delayed neutrons**. They come from the decay of neutron-rich fission fragments (or their daughters) having lifetimes on the order of seconds—sufficient to allow enough reaction time to operate the control rods and maintain $f = 1$.

Nuclear reactors have been built for use in research and to produce electric power. Fission produces many neutrons and a "research reactor" is basically an intense source of neutrons. These neutrons can be used as projectiles in nuclear reactions to produce nuclides not found in nature, including isotopes used as tracers and for therapy. A "power reactor" is used to produce electric power.

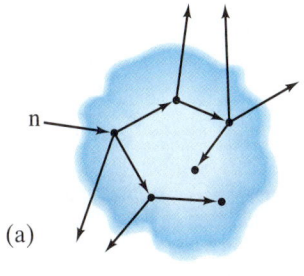

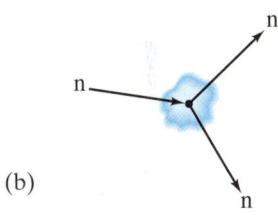

FIGURE 31–7 If the amount of uranium exceeds the critical mass, as in (a), a sustained chain reaction is possible. If the mass is less than critical, as in (b), too many neutrons escape before additional fissions occur, and the chain reaction is not sustained.

[†]$^{238}_{92}$U will fission, but only with fast neutrons ($^{238}_{92}$U is more stable than $^{235}_{92}$U). The probability of absorbing a fast neutron and producing a fission is too low to produce a self-sustaining chain reaction.

[‡]Other types of power or research reactors need 3% to 20% of $^{235}_{92}$U. Atom bombs need greater enrichment: $\geq 80\%$.

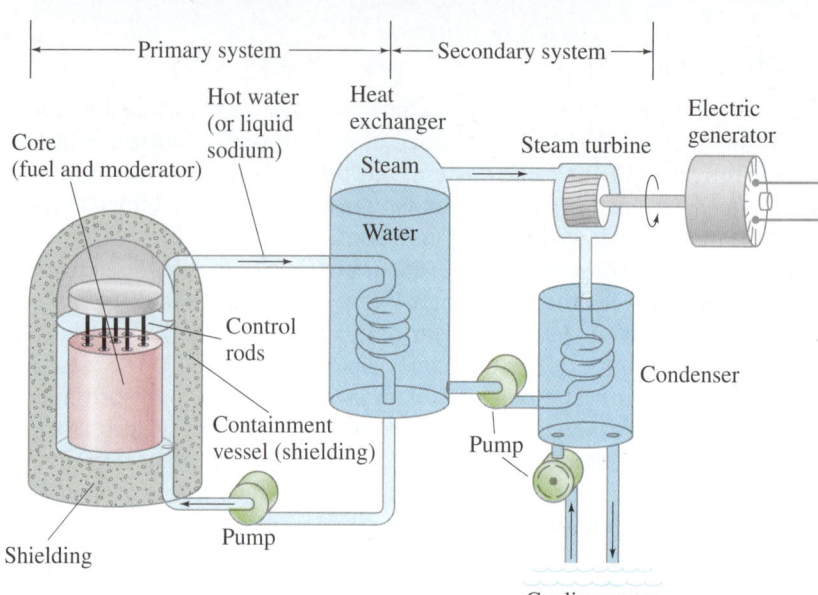

FIGURE 31–8 A nuclear reactor. The heat generated by the fission process in the fuel rods is carried off by hot water or liquid sodium and is used to boil water to steam in the heat exchanger. The steam drives a turbine to generate electricity and is then cooled in the condenser (to reduce pressure on the back side of the turbine blades).

The energy released in the fission process appears as heat, which is used to boil water and produce steam to drive a turbine connected to an electric generator (Fig. 31–8). The **core** of a nuclear reactor consists of the fuel and a moderator (water in most U.S. commercial reactors). The fuel is usually uranium enriched so that it contains 2 to 4 percent $^{235}_{92}\text{U}$. Water at high pressure or other liquid (such as liquid sodium) is allowed to flow through the core. The thermal energy it absorbs is used to produce steam in the heat exchanger, so the fissionable fuel acts as the heat input for a heat engine (Chapter 15).

There are problems associated with nuclear power plants. Besides the usual thermal pollution associated with any heat engine (Section 15–11), there is the serious problem of disposal of the radioactive fission fragments produced in the reactor, plus radioactive nuclides produced by neutrons interacting with the structural parts of the reactor. Fission fragments, like their uranium or plutonium parents, have about 50% more neutrons than protons. Nuclei with atomic number in the typical range for fission fragments ($Z \approx 30$ to 60) are stable only if they have more nearly equal numbers of protons and neutrons (see Fig. 30–2). Hence the highly neutron-rich fission fragments are very unstable and decay radioactively. The accidental release of highly radioactive fission fragments into the atmosphere poses a serious threat to human health (Section 31–4), as does possible leakage of the radioactive wastes when they are disposed of. The accidents at Three Mile Island, Pennsylvania (1979), at Chernobyl, Russia (1986), and at Fukushima, Japan (2011), have illustrated some of the dangers and have shown that nuclear plants must be located, constructed, maintained, and operated with great care and precision (Fig. 31–9).

FIGURE 31–9 Smoke rising from Fukushima, Japan, after the nuclear power plant meltdown in 2011.

Finally, the lifetime of nuclear power plants is limited to 30-some years, due to buildup of radioactivity and the fact that the structural materials themselves are weakened by the intense conditions inside. The cost of "decommissioning" a power plant is very great.

So-called **breeder reactors** were proposed as a solution to the problem of limited supplies of fissionable uranium, $^{235}_{92}\text{U}$. A breeder reactor is one in which some of the neutrons produced in the fission of $^{235}_{92}\text{U}$ are absorbed by $^{238}_{92}\text{U}$, and $^{239}_{94}\text{Pu}$ is produced via the set of reactions shown in Fig. 31–1. $^{239}_{94}\text{Pu}$ is fissionable with slow neutrons, so after separation it can be used as a fuel in a nuclear reactor. Thus a breeder reactor "breeds" new fuel[†] ($^{239}_{94}\text{Pu}$) from otherwise useless $^{238}_{92}\text{U}$. Natural uranium is 99.3 percent $^{238}_{92}\text{U}$, which in a breeder becomes useful fissionable $^{239}_{94}\text{Pu}$, thus increasing the supply of fissionable fuel by more than a factor of 100. But breeder reactors have the same problems as other reactors, plus other serious problems. Not only is plutonium a serious health hazard in itself (radioactive with a half-life of 24,000 years), but plutonium produced in a reactor can readily be used in a bomb, increasing the danger of nuclear proliferation and theft of fuel to produce a bomb.

[†] A breeder reactor does *not* produce more fuel than it uses.

Nuclear power presents risks. Other large-scale energy-conversion methods, such as conventional oil and coal-burning steam plants, also present health and environmental hazards; some of them were discussed in Section 15–11, and include air pollution, oil spills, and the release of CO_2 gas which can trap heat as in a greenhouse to raise the Earth's temperature. The solution to the world's needs for energy is not only technological, but also economic and political. A major factor surely is to "conserve"—to minimize our energy use. "Reduce, reuse, recycle."

EXAMPLE 31–5 **Uranium fuel amount.** Estimate the minimum amount of $^{235}_{92}U$ that needs to undergo fission in order to run a 1000-MW power reactor per year of continuous operation. Assume an efficiency (Chapter 15) of about 33%.

APPROACH At 33% efficiency, we need 3×1000 MW $= 3000 \times 10^6$ J/s input. Each fission releases about 200 MeV (Eq. 31–5), so we divide the energy for a year by 200 MeV to get the number of fissions needed per year. Then we multiply by the mass of one uranium atom.

SOLUTION For 1000 MW output, the total power generation needs to be 3000 MW, of which 2000 MW is dumped as "waste" heat. Thus the total energy release in 1 yr $(3 \times 10^7 \text{ s})$ from fission needs to be about

$$(3 \times 10^9 \text{ J/s})(3 \times 10^7 \text{ s}) \approx 10^{17} \text{ J}.$$

If each fission releases 200 MeV of energy, the number of fissions required for a year is

$$\frac{(10^{17} \text{ J})}{(2 \times 10^8 \text{ eV/fission})(1.6 \times 10^{-19} \text{ J/eV})} \approx 3 \times 10^{27} \text{ fissions}.$$

The mass of a single uranium atom is about $(235 \text{ u})(1.66 \times 10^{-27} \text{ kg/u}) \approx 4 \times 10^{-25}$ kg, so the total uranium mass needed is

$$(4 \times 10^{-25} \text{ kg/fission})(3 \times 10^{27} \text{ fissions}) \approx 1000 \text{ kg},$$

or about a ton of $^{235}_{92}U$.

NOTE Because $^{235}_{92}U$ makes up only 0.7% of natural uranium, the yearly requirement for uranium is on the order of a hundred tons. This is orders of magnitude less than coal, both in mass and volume. Coal releases 2.8×10^7 J/kg, whereas $^{235}_{92}U$ can release 10^{17} J per ton, as we just calculated, or 10^{17} J/10^3 kg $= 10^{14}$ J/kg. For natural uranium, the figure is 100 times less, 10^{12} J/kg.

EXERCISE C A nuclear-powered submarine needs 6000-kW input power. How many $^{235}_{92}U$ fissions is this per second?

Atom Bomb

The first use of fission, however, was not to produce electric power. Instead, it was first used as a fission bomb (called the "atomic bomb"). In early 1940, with Europe already at war, Germany's leader, Adolf Hitler, banned the sale of uranium from the Czech mines he had recently taken over. Research into the fission process suddenly was enshrouded in secrecy. Physicists in the United States were alarmed. A group of them approached Einstein—a man whose name was a household word—to send a letter to President Franklin Roosevelt about the possibilities of using nuclear fission for a bomb far more powerful than any previously known, and inform him that Germany might already have begun development of such a bomb. Roosevelt responded by authorizing the program known as the Manhattan Project, to see if a bomb could be built. Work began in earnest after Fermi's demonstration in 1942 that a sustained chain reaction was possible. A new secret laboratory was developed on an isolated mesa in New Mexico known as Los Alamos. Under the direction of J. Robert Oppenheimer (1904–1967; Fig. 31–10), it became the home of famous scientists from all over Europe and the United States.

FIGURE 31–10 J. Robert Oppenheimer, on the left, with General Leslie Groves, who was the administrative head of Los Alamos during World War II. The photograph was taken at the Trinity site in the New Mexico desert, where the first atomic bomb was exploded.

FIGURE 31–11 Photo taken a month after the bomb was dropped on Nagasaki. The shacks were constructed afterwards from debris in the ruins. The bombs dropped on Hiroshima and Nagasaki were each equivalent to about 20,000 tons of the common explosive TNT ($\sim 10^{14}$ J).

To build a bomb that was subcritical during transport but that could be made supercritical (to produce a chain reaction) at just the right moment, two pieces of uranium were used, each less than the critical mass but together greater than the critical mass. The two masses, kept separate until the moment of detonation, were then forced together quickly by a kind of gun, and a chain reaction of explosive proportions occurred. An alternate bomb detonated conventional explosives (TNT) surrounding a plutonium sphere to compress it by implosion to double its density, making it more than critical and causing a nuclear explosion. The first fission bomb was tested in the New Mexico desert in July 1945. It was successful. In early August, a fission bomb using uranium was dropped on Hiroshima and a second, using plutonium, was dropped on Nagasaki (Fig. 31–11), both in Japan. World War II ended shortly thereafter.

Besides its destructive power, a fission bomb produces many highly radioactive fission fragments, as does a nuclear reactor. When a fission bomb explodes, these radioactive isotopes are released into the atmosphere as **radioactive fallout**.

Testing of nuclear bombs in the atmosphere after World War II was a cause of concern, because the movement of air masses spread the fallout all over the globe. Radioactive fallout eventually settles to the Earth, particularly in rainfall, and is absorbed by plants and grasses and enters the food chain. This is a far more serious problem than the same radioactivity on the exterior of our bodies, because α and β particles are largely absorbed by clothing and the outer (dead) layer of skin. But inside our bodies as food, the isotopes are in contact with living cells. One particularly dangerous radioactive isotope is $^{90}_{38}\text{Sr}$, which is chemically much like calcium and becomes concentrated in bone, where it causes bone cancer and destroys bone marrow. The 1963 treaty signed by over 100 nations that bans nuclear weapons testing in the atmosphere was motivated because of the hazards of fallout.

31–3 Nuclear Fusion

The mass of every stable nucleus is less than the sum of the masses of its constituent protons and neutrons. For example, the mass of the helium isotope $^{4}_{2}\text{He}$ is less than the mass of two protons plus two neutrons, Example 30–3. If two protons and two neutrons were to come together to form a helium nucleus, there would be a loss of mass. This mass loss is manifested in the release of energy.

Nuclear Fusion; Stars

The process of building up nuclei by bringing together individual protons and neutrons, or building larger nuclei by combining small nuclei, is called **nuclear fusion**. In Fig. 31–12 (same as Fig. 30–1), we can see why small nuclei can combine to form larger ones with the release of energy: it is because the binding energy per nucleon is less for light nuclei than it is for heavier nuclei (up to about $A \approx 60$).

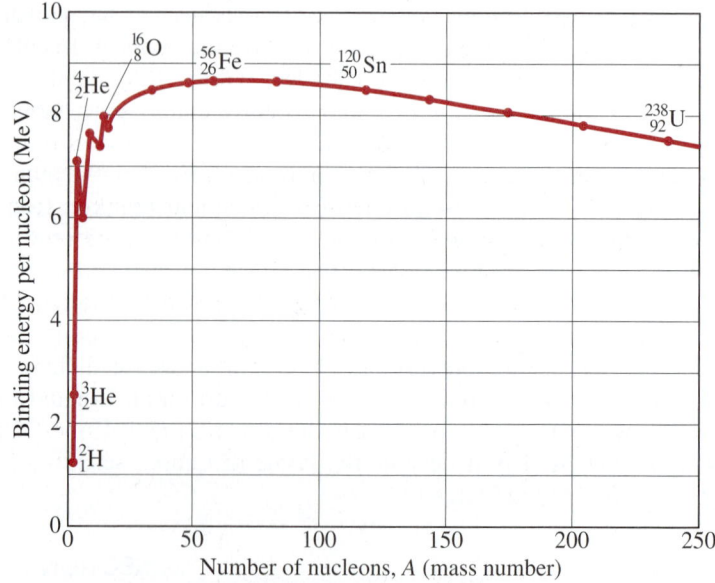

FIGURE 31–12 Average binding energy per nucleon as a function of mass number A for stable nuclei. Same as Fig. 30–1.

For two positively charged nuclei to get close enough to fuse, they must have very high kinetic energy to overcome the electric repulsion. It is believed that many of the elements in the universe were originally formed through the process of fusion in stars (see Chapter 33) where the temperature is extremely high, corresponding to high KE (Eq. 13–8). Today fusion is still producing the prodigious amounts of light energy (EM waves) stars emit, including our Sun.

EXAMPLE 31–6 **Fusion energy release.** One of the simplest fusion reactions involves the production of deuterium, ^2_1H, from a neutron and a proton: $^1_1\text{H} + \text{n} \rightarrow {^2_1\text{H}} + \gamma$. How much energy is released in this reaction?

APPROACH The energy released equals the difference in mass (times c^2) between the initial and final masses.

SOLUTION From Appendix B, the initial mass is
$$1.007825\ \text{u} + 1.008665\ \text{u} = 2.016490\ \text{u},$$
and after the reaction the mass is that of the ^2_1H, namely 2.014102 u (the γ is massless). The mass difference is
$$2.016490\ \text{u} - 2.014102\ \text{u} = 0.002388\ \text{u},$$
so the energy released is
$$(\Delta m)c^2 = (0.002388\ \text{u})(931.5\ \text{MeV/u}) = 2.22\ \text{MeV},$$
and it is carried off by the ^2_1H nucleus and the γ ray.

The energy output of our Sun is believed to be due principally to the following sequence of fusion reactions:

$$^1_1\text{H} + {^1_1\text{H}} \rightarrow {^2_1\text{H}} + e^+ + \nu \qquad (0.42\ \text{MeV}) \qquad \textbf{(31–6a)}$$
$$^1_1\text{H} + {^2_1\text{H}} \rightarrow {^3_2\text{He}} + \gamma \qquad (5.49\ \text{MeV}) \qquad \textbf{(31–6b)}$$
$$^3_2\text{He} + {^3_2\text{He}} \rightarrow {^4_2\text{He}} + {^1_1\text{H}} + {^1_1\text{H}}. \qquad (12.86\ \text{MeV}) \qquad \textbf{(31–6c)}$$

Proton–proton chain

where the energy released (Q-value) for each reaction is given in parentheses. (They include keeping track of atomic electrons as for β-decay, Section 30–5.) These reactions are between nuclei (without electrons at these very high temperatures); the first reaction can be written as
$$\text{p} + \text{p} \rightarrow \text{d} + e^+ + \nu$$
where p = proton and d = deuteron. The net effect of this sequence, which is called the **proton–proton chain**, is for four protons to combine to form one ^4_2He nucleus plus two positrons, two neutrinos, and two gamma rays:
$$4\,^1_1\text{H} \rightarrow {^4_2\text{He}} + 2e^+ + 2\nu + 2\gamma. \qquad \textbf{(31–7)}$$

Note that it takes two of each of the first two reactions (Eqs. 31–6a and b) to produce the two ^3_2He for the third reaction. Also, each of the two e^+ (Eq. 31–6a) quickly annihilates with an electron to produce 2 γ rays (Section 27–6) with total energy $2m_e c^2 = 1.02$ MeV. So the total energy release for the net reaction, Eq. 31–7, is
$$2(0.42\ \text{MeV}) + 2(1.02\ \text{MeV}) + 2(5.49\ \text{MeV}) + 12.86\ \text{MeV} = 26.7\ \text{MeV}.$$

The first reaction, the formation of deuterium from two protons (Eq. 31–6a), has a very low probability, and so limits the rate at which the Sun produces energy. (Thank goodness; this is why the Sun and other stars have long lifetimes and are still shining brightly.)

EXERCISE D Return to the Chapter-Opening Question, page 885, and answer it again now. Try to explain why you may have answered it differently the first time.

EXAMPLE 31–7 ESTIMATE **Estimating fusion energy.** Estimate the energy released if the following reaction occurred:
$$^2_1\text{H} + {^2_1\text{H}} \rightarrow {^4_2\text{He}}.$$

APPROACH We use Fig. 31–12 for a quick estimate.

SOLUTION We see in Fig. 31–12 that each ^2_1H has a binding energy of about $1\tfrac{1}{4}$ MeV/nucleon, which for 2 nuclei of mass 2 is $4 \times (1\tfrac{1}{4}) \approx 5$ MeV. The ^4_2He has a binding energy per nucleon (Fig. 31–12) of about 7 MeV for a total of 4×7 MeV $\approx$ 28 MeV. Hence the energy release is about 28 MeV $-$ 5 MeV $\approx$ 23 MeV.

In stars hotter than the Sun, it is more likely that the energy output comes principally from the **carbon** (or **CNO**) **cycle**, which comprises the following sequence of reactions:

Carbon cycle

$$^{12}_{6}C + {}^{1}_{1}H \rightarrow {}^{13}_{7}N + \gamma$$
$$^{13}_{7}N \rightarrow {}^{13}_{6}C + e^+ + \nu$$
$$^{13}_{6}C + {}^{1}_{1}H \rightarrow {}^{14}_{7}N + \gamma$$
$$^{14}_{7}N + {}^{1}_{1}H \rightarrow {}^{15}_{8}O + \gamma$$
$$^{15}_{8}O \rightarrow {}^{15}_{7}N + e^+ + \nu$$
$$^{15}_{7}N + {}^{1}_{1}H \rightarrow {}^{12}_{6}C + {}^{4}_{2}He.$$

No net carbon is consumed in this cycle and the net effect is the same as the proton–proton chain, Eq. 31–7 (plus one extra γ). The theory of the proton–proton chain and of the carbon cycle as the source of energy for the Sun and stars was first worked out by Hans Bethe (1906–2005) in 1939.

CONCEPTUAL EXAMPLE 31–8 **Stellar fusion.** What is the heaviest element likely to be produced in fusion processes in stars?

RESPONSE Fusion is possible if the final products have more binding energy (less mass) than the reactants, because then there is a net release of energy. Since the binding energy curve in Fig. 31–12 (or Fig. 30–1) peaks near $A \approx 56$ to 58 which corresponds to iron or nickel, it would not be energetically favorable to produce elements heavier than that. Nevertheless, in the center of massive stars or in supernova explosions, there is enough initial kinetic energy available to drive endothermic reactions that produce heavier elements as well.

EXERCISE E If the Sun is generating a constant amount of energy via fusion, the mass of the Sun must be (*a*) increasing, (*b*) decreasing, (*c*) constant, (*d*) irregular.

Possible Fusion Reactors

PHYSICS APPLIED
Fusion energy reactors

The possibility of utilizing the energy released in fusion to make a power reactor is very attractive. The fusion reactions most likely to succeed in a reactor involve the isotopes of hydrogen, ${}^{2}_{1}H$ (deuterium) and ${}^{3}_{1}H$ (tritium), and are as follows, with the energy released given in parentheses:

$$^{2}_{1}H + {}^{2}_{1}H \rightarrow {}^{3}_{1}H + {}^{1}_{1}H \quad (4.03 \text{ MeV}) \quad \textbf{(31–8a)}$$
$$^{2}_{1}H + {}^{2}_{1}H \rightarrow {}^{3}_{2}He + n \quad (3.27 \text{ MeV}) \quad \textbf{(31–8b)}$$
$$^{2}_{1}H + {}^{3}_{1}H \rightarrow {}^{4}_{2}He + n. \quad (17.59 \text{ MeV}) \quad \textbf{(31–8c)}$$

Comparing these energy yields with that for the fission of ${}^{235}_{92}U$, we can see that the energy released in fusion reactions can be greater for a given mass of fuel than in fission. Furthermore, as fuel, a fusion reactor could use deuterium, which is very plentiful in the water of the oceans (the natural abundance of ${}^{2}_{1}H$ is 0.0115% on average, or about 1 g of deuterium per 80 L of water). The simple proton–proton reaction of Eq. 31–6a, which could use a much more plentiful source of fuel, ${}^{1}_{1}H$, has such a small probability of occurring that it cannot be considered a possibility on Earth.

Although a useful fusion reactor has not yet been achieved, considerable progress has been made in overcoming the inherent difficulties. The problems are associated with the fact that all nuclei have a positive charge and repel each other. However, if they can be brought close enough together so that the short-range attractive strong nuclear force can come into play, it can pull the nuclei together and fusion can occur. For the nuclei to get close enough together, they must have large kinetic energy to overcome the electric repulsion. High kinetic energies are readily attainable with particle accelerators (Chapter 32), but the number of particles involved is too small. To produce realistic amounts of energy, we must deal with matter in bulk, for which high kinetic energy means higher temperatures.

Indeed, very high temperatures are required for sustained fusion to occur, and fusion devices are often referred to as **thermonuclear devices**. The interiors of the Sun and other stars are very hot, many millions of degrees, so the nuclei are moving fast enough for fusion to take place, and the energy released keeps the temperature high so that further fusion reactions can occur. The Sun and the stars represent huge self-sustaining thermonuclear reactors that stay together because of their great gravitational mass. But on Earth, containment of the fast-moving nuclei at the high temperatures and densities required has proven difficult.

It was realized after World War II that the temperature produced within a fission (or "atomic") bomb was close to 10^8 K. This suggested that a fission bomb could be used to ignite a fusion bomb (popularly known as a thermonuclear or hydrogen bomb) to release the vast energy of fusion. The uncontrollable release of fusion energy in an H-bomb (in 1952) was relatively easy to obtain. But to realize usable energy from fusion at a slow and controlled rate has turned out to be a serious challenge.

EXAMPLE 31–9 ESTIMATE Temperature needed for d–t fusion. Estimate the temperature required for deuterium–tritium fusion (d–t) to occur.

APPROACH We assume the nuclei approach head-on, each with kinetic energy KE, and that the nuclear force comes into play when the distance between their centers equals the sum of their nuclear radii. The electrostatic potential energy (Chapter 17) of the two particles at this distance equals the minimum total kinetic energy of the two particles when far apart. The average kinetic energy is related to Kelvin temperature by Eq. 13–8.

SOLUTION The radii of the two nuclei ($A_d = 2$ and $A_t = 3$) are given by Eq. 30–1: $r_d \approx 1.5$ fm, $r_t \approx 1.7$ fm, so $r_d + r_t = 3.2 \times 10^{-15}$ m. We equate the kinetic energy of the two initial particles to the potential energy when at this distance:

$$2\text{KE} \approx \frac{1}{4\pi\epsilon_0} \frac{e^2}{(r_d + r_t)}$$

$$\approx \left(9.0 \times 10^9 \frac{\text{N} \cdot \text{m}^2}{\text{C}^2}\right) \frac{(1.6 \times 10^{-19}\,\text{C})^2}{(3.2 \times 10^{-15}\,\text{m})(1.6 \times 10^{-19}\,\text{J/eV})} \approx 0.45\,\text{MeV}.$$

Thus, KE ≈ 0.22 MeV, and if we ask that the average kinetic energy be this high, then from Eq. 13–8, $\frac{3}{2}kT = \overline{\text{KE}}$, we have a temperature of

$$T = \frac{2\overline{\text{KE}}}{3k} = \frac{2(0.22\,\text{MeV})(1.6 \times 10^{-13}\,\text{J/MeV})}{3(1.38 \times 10^{-23}\,\text{J/K})} \approx 2 \times 10^9\,\text{K}.$$

NOTE More careful calculations show that the temperature required for fusion is actually about an order of magnitude less than this rough estimate, partly because it is not necessary that the *average* kinetic energy be 0.22 MeV—a small percentage of nuclei with this much energy (in the high-energy tail of the Maxwell distribution, Fig. 13–20) would be sufficient. Reasonable estimates for a usable fusion reactor are in the range $T \gtrsim 1$ to 4×10^8 K.

A high temperature is required for a fusion reactor. But there must also be a high density of nuclei to ensure a sufficiently high collision rate. A real difficulty with controlled fusion is to contain nuclei long enough and at a high enough density for sufficient reactions to occur so that a usable amount of energy is obtained. At the temperatures needed for fusion, the atoms are ionized, and the resulting collection of nuclei and electrons is referred to as a **plasma**. Ordinary materials vaporize at a few thousand degrees at most, and hence cannot be used to contain a high-temperature plasma. Two major containment techniques are *magnetic confinement* and *inertial confinement*.

In **magnetic confinement**, magnetic fields are used to try to contain the hot plasma. A simple approach is the "magnetic bottle" shown in Fig. 31–13. The paths of the charged particles in the plasma are bent by the magnetic field; where magnetic field lines are close together, the force on the particles reflects them

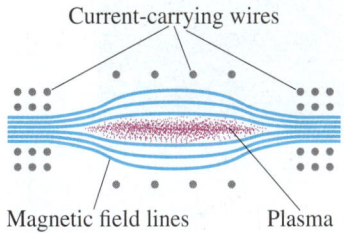

FIGURE 31–13 "Magnetic bottle" used to confine a plasma.

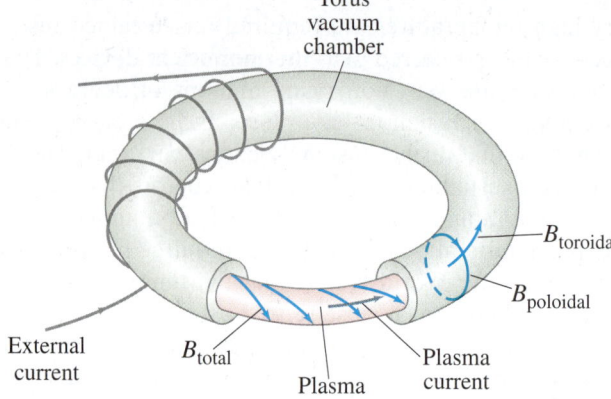

FIGURE 31–14 Tokamak configuration, showing the total $\vec{B}$ field due to external current plus current in the plasma itself.

FIGURE 31–15 (a) Tokamak: split image view of the Joint European Torus (JET) located near Oxford, England. Interior, on the left, and an actual plasma in there ($T \approx 1 \times 10^8$ K) on the right. (b) A 2-mm-diameter round d–t (deuterium–tritium) inertial target, being filled through a thin glass tube from above, at the National Ignition Facility (NIF), Lawrence Livermore National Laboratory, California.

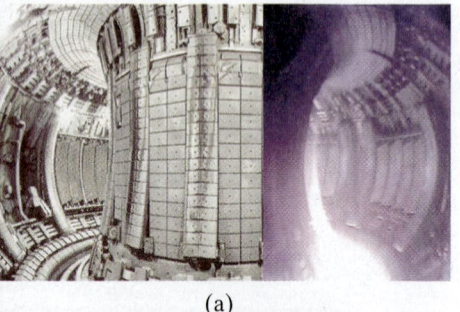

(a)

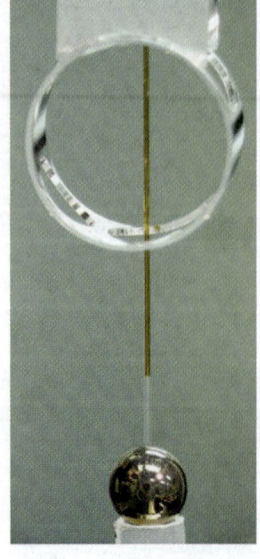

(b)

back toward the center. Unfortunately, magnetic bottles develop "leaks" and the charged particles slip out before sufficient fusion takes place. The most promising design today is the **tokamak**, first developed in Russia. A tokamak (Fig. 31–14) is toroid-shaped (a torus, which is like a donut) and involves complicated magnetic fields: current-carrying conductors produce a magnetic field directed along the axis of the toroid ("toroidal" field); an additional field is produced by currents within the plasma itself ("poloidal" field). The combination produces a helical field as shown in Fig. 31–14, confining the plasma, at least briefly, so it doesn't touch the vacuum chamber's metal walls (Fig. 31–15a).

In 1957, J. D. Lawson showed that the product of ion density n (= ions/m^3) and confinement time τ must exceed a minimum value of approximately

$$n\tau \gtrsim 3 \times 10^{20}\, \text{s/m}^3.$$

This **Lawson criterion** must be reached to produce **ignition**, meaning fusion that continues after all external heating is turned off. Practically, it is expected to be achieved with $n \approx 1$ to $3 \times 10^{20}\, \text{m}^{-3}$ and $\tau \approx 1$ to 3 s. To reach **break-even**, the point at which the energy output due to fusion is equal to the energy input to heat the plasma, requires an $n\tau$ about an order of magnitude less. The break-even point was very closely approached in the 1990s at the Tokamak Fusion Test Reactor (TFTR) at Princeton, and the very high temperature needed for ignition (4×10^8 K) was exceeded—although not both of these at the same time.

Magnetic confinement fusion research continues throughout the world. This research will help us in developing the huge multinational test device (European Union, India, Japan, South Korea, Russia, China, and the U.S.), called ITER (International Thermonuclear Experimental Reactor), situated in France. It is expected that ITER will produce temperatures above 10^8 K for extended periods (minutes or hours) and to begin running by the 2020s, with an expected power output of about 500 MW, 10 times the input energy. ITER is planned to be the final research step before building a commercial reactor.

The second method for containing the fuel for fusion is **inertial confinement fusion** (ICF): a small pellet or capsule of deuterium and tritium (Fig. 31–15b) is struck simultaneously from hundreds of directions by very intense laser beams. The intense influx of energy heats and ionizes the pellet into a plasma, compressing it and heating it to temperatures at which fusion can occur ($> 10^8$ K). The confinement time is on the order of 10^{-11} to 10^{-9} s, during which time the ions do not move appreciably because of their own inertia, and fusion can take place.

31–4 Passage of Radiation Through Matter; Biological Damage

When we speak of *radiation*, we include α, β, γ, and X-rays, as well as protons, neutrons, and other particles such as pions (see Chapter 32). Because charged particles can ionize the atoms or molecules of any material they pass through, they are referred to as **ionizing radiation**. And because radiation produces ionization, it can cause considerable damage to materials, particularly to biological tissue.

Charged particles, such as α and β rays and protons, cause ionization because of electric forces. That is, when they pass through a material, they can attract or repel electrons strongly enough to remove them from the atoms of the material. Since the α and β rays emitted by radioactive substances have energies on the order of 1 MeV (10^4 to 10^7 eV), whereas ionization of atoms and molecules requires on the order of 10 eV (Chapter 27), we see that a single α or β particle can cause thousands of ionizations.

Neutral particles also give rise to ionization when they pass through materials. For example, X-ray and γ-ray photons can ionize atoms by knocking out electrons by means of the photoelectric and Compton effects (Chapter 27). Furthermore, if a γ ray has sufficient energy (greater than 1.02 MeV), it can undergo pair production: an electron and a positron are produced (Section 27–6). The charged particles produced in all of these processes can themselves go on to produce further ionization. Neutrons, on the other hand, interact with matter mainly by collisions with nuclei, with which they interact strongly. Often the nucleus is broken apart by such a collision, altering the molecule of which it was a part. The fragments produced can in turn cause ionization.

Radiation passing through matter can do considerable damage. Metals and other structural materials become brittle and their strength can be weakened if the radiation is very intense, as in nuclear reactor power plants and for space vehicles that must pass through areas of intense cosmic radiation.

Biological Damage

PHYSICS APPLIED
Biological radiation damage

The radiation damage produced in biological organisms is due primarily to ionization produced in cells. Several related processes can occur. Ions or radicals are produced that are highly reactive and take part in chemical reactions that interfere with the normal operation of the cell. All forms of radiation can ionize atoms by knocking out electrons. If these are bonding electrons, the molecule may break apart, or its structure may be altered so that it does not perform its normal function or may perform a harmful function. In the case of proteins, the loss of one molecule is not serious if there are other copies of the protein in the cell and additional copies can be made from the gene that codes for it. However, large doses of radiation may damage so many molecules that new copies cannot be made quickly enough, and the cell dies.

Damage to the DNA is more serious, since a cell may have only one copy. Each alteration in the DNA can affect a gene and alter the molecule that gene codes for (Section 29–3), so that needed proteins or other molecules may not be made at all. Again the cell may die. The death of a single cell is not normally a problem, since the body can replace it with a new one. (There are exceptions, such as neurons, which are mostly not replaceable, so their loss is serious.) But if many cells die, the organism may not be able to recover. On the other hand, a cell may survive but be defective. It may go on dividing and produce many more defective cells, to the detriment of the whole organism. Thus radiation can cause cancer—the rapid uncontrolled production of cells.

The possible damage done by the medical use of X-rays and other radiation must be balanced against the medical benefits and prolongation of life as a result of their diagnostic use.

31–5 Measurement of Radiation—Dosimetry

Although the passage of ionizing radiation through the human body can cause considerable damage, radiation can also be used to treat certain diseases, particularly cancer, often by using very narrow beams directed at a cancerous tumor in order to destroy it (Section 31–6). It is therefore important to be able to quantify the amount, or **dose**, of radiation. This is the subject of **dosimetry**.

PHYSICS APPLIED
Dosimetry

The strength of a source can be specified at a given time by stating the **source activity**: how many nuclear decays (or disintegrations) occur per second. The traditional unit is the **curie** (Ci), defined as

$$1 \text{ Ci} = 3.70 \times 10^{10} \text{ decays per second.}$$

(This number comes from the original definition as the activity of exactly one gram of radium.) Although the curie is still in common use, the SI unit for source activity is the **becquerel** (Bq), defined as

$$1 \text{ Bq} = 1 \text{ decay/s.}$$

> ⚠ **CAUTION**
> *In the lab, activity will be less than written on the bottle—note the date*

Commercial suppliers of **radionuclides** (radioactive nuclides) used as tracers specify the activity at a given time. Because the activity decreases over time, more so for short-lived isotopes, it is important to take this decrease into account.

The magnitude of the source activity, $\Delta N/\Delta t$, is related to the number of radioactive nuclei present, N, and to the half-life, $T_{\frac{1}{2}}$, by (see Section 30–8):

$$\frac{\Delta N}{\Delta t} = \lambda N = \frac{0.693}{T_{\frac{1}{2}}} N.$$

EXAMPLE 31–10 Radioactivity taken up by cells. In a certain experiment, 0.016 μCi of $^{32}_{15}\text{P}$ is injected into a medium containing a culture of bacteria. After 1.0 h the cells are washed and a 70% efficient detector (counts 70% of emitted β rays) records 720 counts per minute from the cells. What percentage of the original $^{32}_{15}\text{P}$ was taken up by the cells?

APPROACH The half-life of $^{32}_{15}\text{P}$ is about 14 days (Appendix B), so we can ignore any loss of activity over 1 hour. From the given activity, we find how many β rays are emitted. We can compare 70% of this to the $(720/\text{min})/(60 \text{ s/min}) = 12$ per second detected.

SOLUTION The total number of decays per second originally was $(0.016 \times 10^{-6})(3.7 \times 10^{10}) = 590$. The counter could be expected to count 70% of this, or 410 per second. Since it counted $720/60 = 12$ per second, then $12/410 = 0.029$ or 2.9% was incorporated into the cells.

Another type of measurement is the exposure or **absorbed dose**—that is, the *effect* the radiation has on the absorbing material. The earliest unit of dosage was the **roentgen** (R), defined in terms of the amount of ionization produced by the radiation (1 R = 1.6×10^{12} ion pairs per gram of dry air at standard conditions). Today, 1 R is defined as the amount of X-ray or γ radiation that deposits 0.878×10^{-2} J of energy per kilogram of air. The roentgen was largely superseded by another unit of absorbed dose applicable to any type of radiation, the **rad**: *1 rad is that amount of radiation which deposits energy per unit mass of 1.00×10^{-2} J/kg in any absorbing material.* (This is quite close to the roentgen for X- and γ rays.) The proper SI unit for absorbed dose is the **gray** (Gy):

$$1 \text{ Gy} = 1 \text{ J/kg} = 100 \text{ rad.} \tag{31–9}$$

The absorbed dose depends not only on the energy per particle and on the strength of a given source or of a radiation beam (number of particles per second), but also on the type of material that is absorbing the radiation. Bone, for example, absorbs more of X-ray or γ radiation normally used than does flesh, so the same beam passing through a human body deposits a greater dose (in rads or grays) in bone than in flesh.

The gray and the rad are physical units of dose—the energy deposited per unit mass of material. They are, however, not the most meaningful units for measuring the biological damage produced by radiation because equal doses of different types of radiation cause differing amounts of damage. For example, 1 rad of α radiation does 10 to 20 times the amount of damage as 1 rad of β or γ rays. This difference arises largely because α rays (and other heavy particles such as protons and neutrons) move much more slowly than β and γ rays of equal energy due to their greater mass. Hence, ionizing collisions occur closer together, so more irreparable damage can be done. The **relative biological effectiveness** (RBE) of a given type of radiation is defined as the number of rads of X-ray or γ radiation that produces the same biological damage as 1 rad of the given radiation. For example, 1 rad of slow neutrons does the same damage as 5 rads of X-rays. Table 31–1 gives the RBE for several types of radiation. The numbers are approximate because they depend somewhat on the energy of the particles and on the type of damage that is used as the criterion.

The **effective dose** can be given as the product of the dose in rads and the RBE, and this unit is known as the **rem** (which stands for *rad equivalent man*):

$$\text{effective dose (in rem)} = \text{dose (in rad)} \times \text{RBE}. \quad (31\text{–}10\text{a})$$

This unit is being replaced by the SI unit for "effective dose," the **sievert** (Sv):

$$\text{effective dose (Sv)} = \text{dose (Gy)} \times \text{RBE} \quad (31\text{–}10\text{b})$$

so

$$1\,\text{Sv} = 100\,\text{rem} \quad \text{or} \quad 1\,\text{rem} = 10\,\text{mSv}.$$

By these definitions, 1 rem (or 1 Sv) of any type of radiation does approximately the same amount of biological damage. For example, 50 rem of fast neutrons does the same damage as 50 rem of γ rays. But note that 50 rem of fast neutrons is only 5 rads, whereas 50 rem of γ rays is 50 rads.

TABLE 31–1 Relative Biological Effectiveness (RBE)

Type	RBE
X- and γ rays	1
β (electrons)	1
Protons	2
Slow neutrons	5
Fast neutrons	≈ 10
α particles and heavy ions	≈ 20

Human Exposure to Radiation

We are constantly exposed to low-level radiation from natural sources: cosmic rays, natural radioactivity in rocks and soil, and naturally occurring radioactive isotopes in our food, such as $^{40}_{19}\text{K}$. **Radon**, $^{222}_{86}\text{Rn}$, is of considerable concern today. It is the product of radium decay and is an intermediate in the decay series from uranium (see Fig. 30–11). Most intermediates remain in the rocks where formed, but radon is a gas that can escape from rock (and from building material like concrete) to enter the air we breathe, and damage the interior of the lung.

PHYSICS APPLIED
Radon

The **natural radioactive background** averages about 0.30 rem (300 mrem) per year per person in the U.S., although there are large variations. From medical X-rays and scans, the average person receives about 50 to 60 mrem per year, giving an average total dose of about 360 mrem (3.6 mSv) per person. U.S. government regulators suggest an upper limit of allowed radiation for an individual in the general populace at about 100 mrem (1 mSv) per year in addition to natural background. It is believed that even low doses of radiation increase the chances of cancer or genetic defects; there is *no safe level* or threshold of radiation exposure.

PHYSICS APPLIED
Human radiation exposure

The upper limit for people who work around radiation—in hospitals, in power plants, in research—has been set higher, a maximum of 20 mSv (2 rem) whole-body dose, averaged over some years (a maximum of 50 mSv (5 rem/yr) in any one year). To monitor exposure, those people who work around radiation generally carry some type of dosimeter, one common type being a **radiation film badge** which is a piece of film wrapped in light-tight material. The passage of ionizing radiation through the film changes it so that the film is darkened upon development, and thus indicates the received dose. Newer types include the *thermoluminescent dosimeter* (TLD). Dosimeters and badges do not protect the worker, but high levels detected suggest reassignment or modified work practices to reduce radiation exposure to acceptable levels.

PHYSICS APPLIED
Radiation worker exposure
Film badge

Large doses of radiation can cause unpleasant symptoms such as nausea, fatigue, and loss of body hair, because of cellular damage. Such effects are sometimes referred to as **radiation sickness**. Very large doses can be fatal, although the time span of the dose is important. A brief dose of 10 Sv (1000 rem) is nearly always fatal. A 3-Sv (300-rem) dose in a short period of time is fatal in about 50% of patients within a month. However, the body possesses remarkable repair processes, so that a 3-Sv dose spread over several weeks is usually not fatal. It will, nonetheless, cause considerable damage to the body.

PHYSICS APPLIED
Radiation sickness

The effects of low doses over a long time are difficult to determine and are not well known as yet.

CONCEPTUAL EXAMPLE 31–11 **Limiting the dose.** A worker in an environment with a radioactive source is warned that she is accumulating a dose too quickly and will have to lower her exposure by a factor of ten to continue working for the rest of the year. If the worker is able to work farther away from the source, how much farther away is necessary?

RESPONSE If the energy is radiated uniformly in all directions, then the intensity (dose/area) should decrease as the distance squared, just as it does for sound and light. If she can work four times farther away, the exposure lowers by a factor of sixteen, enough to make her safe.

EXAMPLE 31–12 **Whole-body dose.** What whole-body dose is received by a 70-kg laboratory worker exposed to a 40-mCi $^{60}_{27}$Co source, assuming the person's body has cross-sectional area 1.5 m² and is normally about 4.0 m from the source for 4.0 h per day? $^{60}_{27}$Co emits γ rays of energy 1.33 MeV and 1.17 MeV in quick succession. Approximately 50% of the γ rays interact in the body and deposit all their energy. (The rest pass through.)

APPROACH Of the given energy emitted, only a fraction passes through the worker, equal to *her* area divided by the total area (or $4\pi r^2$) over a full sphere of radius $r = 4.0$ m (Fig. 31–16).

SOLUTION The total γ-ray energy per decay is $(1.33 + 1.17)$ MeV $= 2.50$ MeV, so the total energy emitted by the source per second is

$$(0.040 \text{ Ci})(3.7 \times 10^{10} \text{ decays/Ci·s})(2.50 \text{ MeV}) = 3.7 \times 10^9 \text{ MeV/s}.$$

The proportion of this energy intercepted by the body is its 1.5-m² area divided by the area of a sphere of radius 4.0 m (Fig. 31–16):

$$\frac{1.5 \text{ m}^2}{4\pi r^2} = \frac{1.5 \text{ m}^2}{4\pi(4.0 \text{ m})^2} = 7.5 \times 10^{-3}.$$

So the rate energy is deposited in the body (remembering that only 50% of the γ rays interact in the body) is

$$E = \left(\tfrac{1}{2}\right)(7.5 \times 10^{-3})(3.7 \times 10^9 \text{ MeV/s})(1.6 \times 10^{-13} \text{ J/MeV}) = 2.2 \times 10^{-6} \text{ J/s}.$$

Since 1 Gy = 1 J/kg, the whole-body dose rate for this 70-kg person is $(2.2 \times 10^{-6} \text{ J/s})/(70 \text{ kg}) = 3.1 \times 10^{-8}$ Gy/s. In 4.0 h, this amounts to a dose of

$$(4.0 \text{ h})(3600 \text{ s/h})(3.1 \times 10^{-8} \text{ Gy/s}) = 4.5 \times 10^{-4} \text{ Gy}.$$

RBE ≈ 1 for gammas, so the effective dose is 450 μSv (Eqs. 31–10b and 31–9) or:

$$(100 \text{ rad/Gy})(4.5 \times 10^{-4} \text{ Gy})(1 \text{ rem/rad}) = 45 \text{ mrem} = 0.45 \text{ mSv}.$$

NOTE This 45-mrem effective dose is almost 50% of the normal allowed dose for a whole year (100 mrem/yr), or 1% of the maximum one-year allowance for radiation workers. This worker should not receive such a large dose every day and should seek ways to reduce it (shield the source, vary the work, work farther from the source, work less time this close to source, etc.).

FIGURE 31–16 Radiation spreads out in all directions. A person 4.0 m away intercepts only a fraction: her cross-sectional area divided by the area of a sphere of radius 4.0 m. Example 31–12.

We have assumed that the intensity of radiation decreases as the square of the distance. It actually falls off faster than $1/r^2$ because of absorption in the air, so our answers are a slight overestimate of dose received.

PHYSICS APPLIED
Radon exposure

EXAMPLE 31–13 **Radon exposure.** In the U.S., yearly deaths from radon exposure (the second leading cause of lung cancer) are estimated to be on the order of 20,000 and maybe much more. The Environmental Protection Agency recommends taking action to reduce the radon concentration in living areas if it exceeds 4 pCi/L of air. In some areas 50% of houses exceed this level from naturally occurring radon in the soil. Estimate (*a*) the number of decays/s in 1.0 m³ of air and (*b*) the mass of radon that emits 4.0 pCi of $^{222}_{86}$Rn radiation.

APPROACH We can use the definition of the curie to determine how many decays per second correspond to 4 pCi, then Eq. 30–3b to determine how many nuclei of radon it takes to have this activity $\Delta N/\Delta t$.

SOLUTION (a) We saw at the start of this Section that $1\,\text{Ci} = 3.70 \times 10^{10}$ decays/s. Thus

$$\frac{\Delta N}{\Delta t} = 4.0\,\text{pCi} = (4.0 \times 10^{-12}\,\text{Ci})(3.70 \times 10^{10}\,\text{decays/s/Ci})$$

$$= 0.148\,\text{s}^{-1}$$

per liter of air. In $1\,\text{m}^3$ of air $(1\,\text{m}^3 = 10^6\,\text{cm}^3 = 10^3\,\text{L})$ there would be $(0.148\,\text{s}^{-1})(1000) = 150$ decays/s.

(b) From Eqs. 30–3b and 30–6

$$\frac{\Delta N}{\Delta t} = \lambda N = \frac{0.693}{T_{\frac{1}{2}}} N.$$

Appendix B tells us $T_{\frac{1}{2}} = 3.8235$ days for radon, so

$$N = \left(\frac{\Delta N}{\Delta t}\right) \frac{T_{\frac{1}{2}}}{0.693}$$

$$= (0.148\,\text{s}^{-1}) \frac{(3.8235\,\text{days})(8.64 \times 10^4\,\text{s/day})}{0.693}$$

$$= 7.06 \times 10^4 \text{ atoms of radon-222}.$$

The molar mass (222 u) and Avogadro's number are used to find the mass:

$$m = \frac{(7.06 \times 10^4\,\text{atoms})(222\,\text{g/mol})}{6.02 \times 10^{23}\,\text{atoms/mol}} = 2.6 \times 10^{-17}\,\text{g}$$

or 26 attograms in 1 L of air at the limit of 4 pCi/L. This 2.6×10^{-17} g/L is 2.6×10^{-14} grams of radon per m³ of air.

NOTE Each radon atom emits 4 α particles and 4 β particles, each one capable of causing many harmful ionizations, before the sequence of decays reaches a stable element.

*31–6 Radiation Therapy

The medical application of radioactivity and radiation to human beings involves two basic aspects: (1) **radiation therapy**—the treatment of disease (mainly cancer)—which we discuss in this Section; and (2) the *diagnosis* of disease, which we discuss in the following Sections of this Chapter.

Radiation can cause cancer. It can also be used to treat it. Rapidly growing cancer cells are especially susceptible to destruction by radiation. Nonetheless, large doses are needed to kill the cancer cells, and some of the surrounding normal cells are inevitably killed as well. It is for this reason that cancer patients receiving radiation therapy often suffer side effects characteristic of radiation sickness. To minimize the destruction of normal cells, a narrow beam of γ or X-rays is often used when a cancerous tumor is well localized. The beam is directed at the tumor, and the source (or body) is rotated so that the beam passes through various parts of the body to keep the dose at any one place as low as possible—except at the tumor and its immediate surroundings, where the beam passes at all times (Fig. 31–17). The radiation may be from a radioactive source such as $^{60}_{27}\text{Co}$, or it may be from an X-ray machine that produces photons in the range 200 keV to 5 MeV. Protons, neutrons, electrons, and pions, which are produced in particle accelerators (Section 32–1), are also being used in cancer therapy.

PHYSICS APPLIED
Radiation therapy

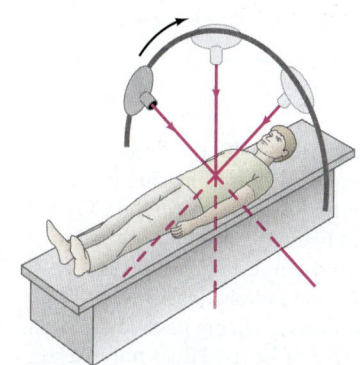

FIGURE 31–17 Radiation source rotates so that the beam always passes through the diseased tissue, but minimizes the dose in the rest of the body.

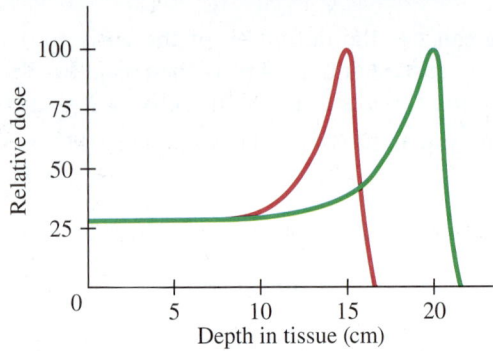

FIGURE 31–18 Energy deposited in tissue as a function of depth for 170-MeV protons (red curve) and 190-MeV protons (green). The peak of each curve is often called the Bragg peak.

PHYSICS APPLIED
Proton therapy

Protons used to kill tumors have a special property that makes them particularly useful. As shown in Fig. 31–18, when protons enter tissue, most of their energy is deposited at the end of their path. The protons' initial kinetic energy can be chosen so that most of the energy is deposited at the depth of the tumor itself, to destroy it. The incoming protons deposit only a small amount of energy in the tissue in front of the tumor, and none at all behind the tumor, thus having less negative effect on healthy tissue than X- or γ rays. Because tumors have physical size, even several centimeters in diameter, a range of proton energies is often used. Heavier ions, such as α particles or carbon ions, are similarly useful. This **proton therapy** technique is more than a half century old, but the necessity of having a large accelerator has meant that few hospitals have used the technique until now. Many such "proton centers" are now being built.

Another form of treatment is to insert a tiny radioactive source directly inside a tumor, which will eventually kill the majority of the cells. A similar technique is used to treat cancer of the thyroid with the radioactive isotope $^{131}_{53}\text{I}$. The thyroid gland concentrates iodine present in the bloodstream, particularly in any area where abnormal growth is taking place. Its intense radioactivity can destroy the defective cells.

Another application of radiation is for sterilizing bandages, surgical equipment, and even packaged foods such as ground beef, chicken, and produce, because bacteria and viruses can be killed or deactivated by large doses of radiation.

(a)

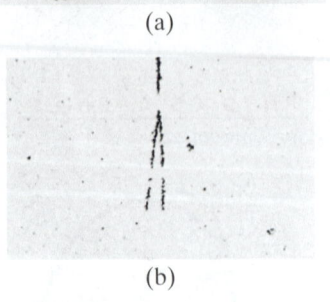

(b)

FIGURE 31–19 (a) Autoradiograph of a leaf exposed for 30 s to $^{14}\text{CO}_2$. Only the tissue where the CO_2 has been taken up, to be used in photosynthesis (Example 27–7), has become radioactive. The non-metabolizing tissue of the veins is free of $^{14}_{6}\text{C}$ and does not blacken the X-ray sheet. (b) Autoradiograph of chromosomal DNA. The dashed arrays of film grains show the Y-shaped growing point of replicating DNA.

*31–7 Tracers in Research and Medicine

Radioactive isotopes are used in biological and medical research as **tracers**. A given compound is artificially synthesized incorporating a radioactive isotope such as $^{14}_{6}\text{C}$ or $^{3}_{1}\text{H}$. Such "tagged" molecules can then be traced as they move through an organism or as they undergo chemical reactions. The presence of these tagged molecules (or parts of them, if they undergo chemical change) can be detected by a Geiger or scintillation counter, which detects emitted radiation (see Section 30–13). How food molecules are digested, and to what parts of the body they are diverted, can be traced in this way.

Radioactive tracers have been used to determine how amino acids and other essential compounds are synthesized by organisms. The permeability of cell walls to various molecules and ions can be determined using radioactive tracers: the tagged molecule or ion is injected into the extracellular fluid, and the radioactivity present inside and outside the cells is measured as a function of time.

In a technique known as **autoradiography**, the position of the radioactive isotopes is detected on film. For example, the distribution of carbohydrates produced in the leaves of plants from absorbed CO_2 can be observed by keeping the plant in an atmosphere where the carbon atom in the CO_2 is $^{14}_{6}\text{C}$. After a time, a leaf is placed firmly on a photographic plate and the emitted radiation darkens the film most strongly where the isotope is most strongly concentrated (Fig. 31–19a). Autoradiography using labeled nucleotides (components of DNA) has revealed much about the details of DNA replication (Fig. 31–19b). Today gamma cameras are used in a similar way—see next page.

For medical diagnosis, the radionuclide commonly used today is $^{99m}_{43}$Tc, a long-lived excited state of technetium-99 (the "m" in the symbol stands for "metastable" state). It is formed when $^{99}_{42}$Mo decays. The great usefulness of $^{99m}_{43}$Tc derives from its convenient half-life of 6 h (short, but not too short) and the fact that it can combine with a large variety of compounds. The compound to be labeled with the radionuclide is so chosen because it concentrates in the organ or region of the anatomy to be studied. Detectors outside the body then record, or image, the distribution of the radioactively labeled compound. The detection could be done by a single detector (Fig. 31–20a) which is moved across the body, measuring the intensity of radioactivity at a large number of points. The image represents the relative intensity of radioactivity at each point. The relative radioactivity is a diagnostic tool. For example, high or low radioactivity may represent overactivity or underactivity of an organ or part of an organ, or in another case may represent a lesion or tumor. More complex **gamma cameras** make use of many detectors which simultaneously record the radioactivity at many points. The measured intensities can be displayed on a TV or computer monitor. The image is sometimes called a scintigram (after scintillator), Fig. 31–20b. Gamma cameras are relatively inexpensive, but their resolution is limited—by non-perfect collimation[†]. Yet they allow "dynamic" studies: images that change in time, like a movie.

PHYSICS APPLIED
Medical diagnosis

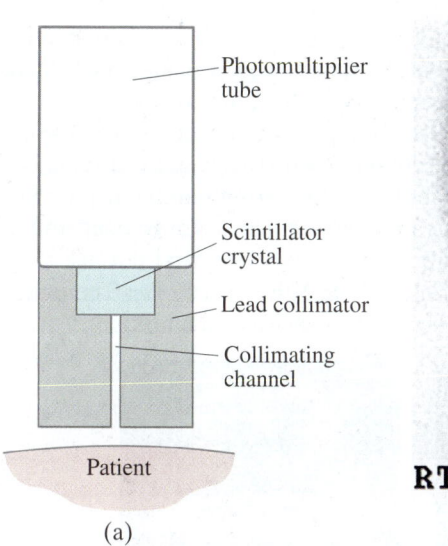
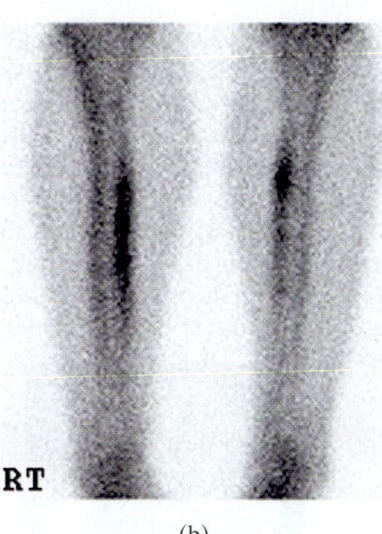

FIGURE 31–20 (a) Collimated gamma-ray detector for scanning (moving) over a patient. The collimator selects γ rays that come in a (nearly) straight line from the patient. Without the collimator, γ rays from all parts of the body could strike the scintillator, producing a poor image. Detectors today usually have many collimator tubes and are called *gamma cameras*. (b) Gamma camera image (scintigram), of both legs of a patient with shin splints, detecting γs from $^{99m}_{43}$Tc.

*31–8 Emission Tomography: PET and SPECT

The images formed using the standard techniques of nuclear medicine, as briefly discussed in the previous Section, are produced from radioactive tracer sources within the *volume* of the body. It is also possible to image the radioactive emissions from a single plane or slice through the body using the computed tomography techniques discussed in Section 25–12. A gamma camera measures the radioactive intensity from the tracer at many points and angles around the patient. The data are processed in much the same way as for X-ray CT scans (Section 25–12). This technique is referred to as **single photon emission computed tomography** (SPECT), or simply SPET (single photon emission tomography).

Another important technique is **positron emission tomography** (PET), which makes use of positron emitters such as $^{11}_{6}$C, $^{13}_{7}$N, $^{15}_{8}$O, and $^{18}_{9}$F whose half-lives are short. These isotopes are incorporated into molecules that, when inhaled or injected, accumulate in the organ or region of the body to be studied.

PHYSICS APPLIED
Medical imaging

PHYSICS APPLIED
Emission tomography (SPECT, PET)

[†]To "collimate" means to "make parallel," usually by blocking non-parallel rays with a narrow tube inside lead, as in Fig. 31–20a.

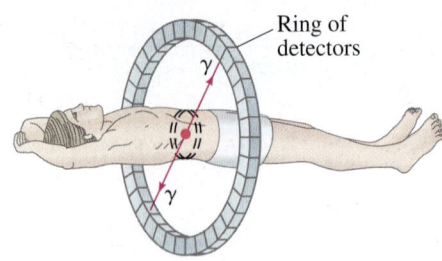

FIGURE 31–21 Positron emission tomography (PET) system showing a ring of detectors to detect the two annihilation γ rays $(e^+ + e^- \to 2\gamma)$ emitted at $180°$ to each other.

When such a nuclide undergoes β^+ decays, the emitted positron travels at most a few millimeters before it collides with a normal electron. In this collision, the positron and electron are annihilated, producing two γ rays $(e^+ + e^- \to 2\gamma)$, each having an energy of 511 keV $(= m_e c^2)$. The two γ rays fly off in opposite directions $(180° \pm 0.25°)$ since they must have almost exactly equal and opposite momenta to conserve momentum (the momenta of the initial e^+ and e^- are essentially zero compared to the momenta afterward of the γ rays). Because the photons travel along the same line in opposite directions, their detection in coincidence by rings of detectors surrounding the patient (Fig. 31–21) readily establishes the line along which the emission took place. If the difference in time of arrival of the two photons could be determined accurately, the actual position of the emitting nuclide along that line could be calculated. Present-day electronics can measure times to at best ± 300 ps, so at the γ ray's speed $(c = 3 \times 10^8 \text{ m/s})$, the actual position could be determined to an accuracy on the order of about $d = vt \approx (3 \times 10^8 \text{ m/s})(300 \times 10^{-12} \text{ s}) \approx 10$ cm, which is not very useful. Although there may be future potential for *time-of-flight* measurements to determine position, today computed tomography techniques are used instead, similar to those for X-ray CT, which can reconstruct PET images with a resolution on the order of 2–5 mm. One big advantage of PET is that no collimators are needed (as for detection of a single photon—see Fig. 31–20a). Thus, fewer photons are "wasted" and lower doses can be administered to the patient with PET.

Both PET and SPECT systems can give images that relate to biochemistry, metabolism, and function. This is to be compared to X-ray CT scans, whose images reflect shape and structure—that is, the anatomy of the imaged region.

Figure 31–22 shows PET scans of the same person's brain (a) when using a cell phone near the ear and (b) with the cell phone off. The bright red spots in (a) indicate a higher rate of glucose metabolism, suggesting excitability of brain tissue (the glucose was tagged with a radioactive tracer). Emfs from the cell phone antenna thus seem to affect metabolism and may be harmful to us!

The colors shown here are faked (only visible light has colors). The original images are various shades of gray, representing intensity (or counts).

FIGURE 31–22 False-color PET scans of a horizontal section through a brain showing glucose metabolism rates (red is high) by a person (a) using a cell phone near the ear, and (b) with the cell phone off.

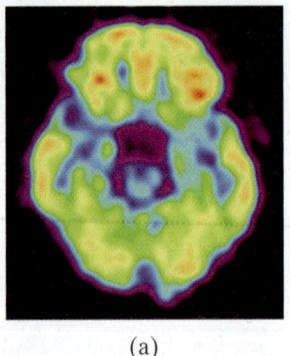
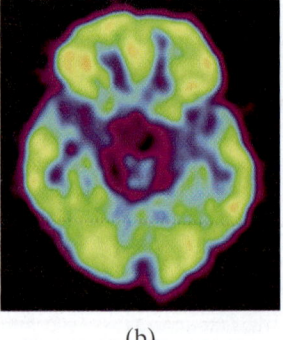

(a)   (b)

31–9 Nuclear Magnetic Resonance (NMR) and Magnetic Resonance Imaging (MRI)

Nuclear magnetic resonance (NMR) is a phenomenon which soon after its discovery in 1946 became a powerful research tool in a variety of fields from physics to chemistry and biochemistry. It is also an important medical imaging technique. We first briefly discuss the phenomenon, and then look at its applications.

*Nuclear Magnetic Resonance (NMR)

We saw in Chapter 28 (Section 28–6) that when atoms are placed in a magnetic field, atomic energy levels split into several closely spaced levels (see Fig. 28–8). Nuclei, too, exhibit these magnetic properties. We examine only the simplest, the hydrogen (H) nucleus, since it is the one most used, even for medical imaging.

The $^1_1$H nucleus consists of a single proton. Its spin angular momentum (and its magnetic moment), like that of the electron, can take on only two values when placed in a magnetic field: we call these "spin up" (parallel to the field) and "spin down" (antiparallel to the field), as suggested in Fig. 31–23. When a magnetic field is present, the energy of the nucleus splits into two levels as shown in Fig. 31–24, with the spin up (parallel to field) having the lower energy. (This is like the Zeeman effect for atomic levels, Fig. 28–8.) The difference in energy ΔE between these two levels is proportional to the total magnetic field B_T at the nucleus:

$$\Delta E = kB_T,$$

where k is a proportionality constant that is different for different nuclides.

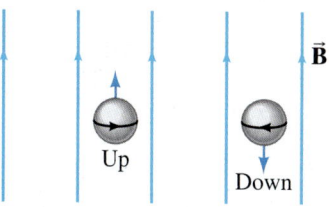

FIGURE 31–23 Schematic picture of a proton in a magnetic field $\vec{B}$ (pointing upward) with the two possible states of proton spin, up and down.

In a standard **nuclear magnetic resonance** (NMR) setup, the sample to be examined is placed in a static magnetic field. A radiofrequency (RF) pulse of electromagnetic radiation (that is, photons) is applied to the sample. If the frequency, f, of this pulse corresponds precisely to the energy difference between the two energy levels (Fig. 31–24), so that

$$hf = \Delta E = kB_T, \quad (31\text{–}11)$$

then the photons of the RF beam will be absorbed, exciting many of the nuclei from the lower state to the upper state. This is a resonance phenomenon because there is significant absorption only if f is very near $f = kB_T/h$. Hence the name "nuclear magnetic resonance." For free $^1_1$H nuclei, the frequency is 42.58 MHz for a magnetic field $B_T = 1.0$ T. If the H atoms are bound in a molecule the total magnetic field B_T at the H nuclei will be the sum of the external applied field (B_{ext}) plus the local magnetic field (B_{local}) due to electrons and nuclei of neighboring atoms. Since f is proportional to B_T, the value of f for a given external field will be slightly different for bound H atoms than for free atoms:

$$hf = k(B_{\text{ext}} + B_{\text{local}}).$$

FIGURE 31–24 Energy E_0 in the absence of a magnetic field splits into two levels in the presence of a magnetic field.

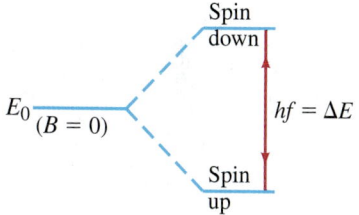

This small change in frequency can be measured, and is called the "chemical shift." A great deal has been learned about the structure of molecules and bonds using this NMR technique.

*Magnetic Resonance Imaging (MRI)

For producing medically useful NMR images—now commonly called MRI, or **magnetic resonance imaging**—the element most used is hydrogen since it is the commonest element in the human body and gives the strongest NMR signals. The experimental apparatus is shown in Fig. 31–25. The large coils set up the static magnetic field, and the RF coils produce the RF pulse of electromagnetic waves (photons) that cause the nuclei to jump from the lower state to the upper one (Fig. 31–24). These same coils (or another coil) can detect the absorption of energy or the emitted radiation (also of frequency $f = \Delta E/h$, Eq. 31–11) when the nuclei jump back down to the lower state.

PHYSICS APPLIED
NMR imaging (MRI)

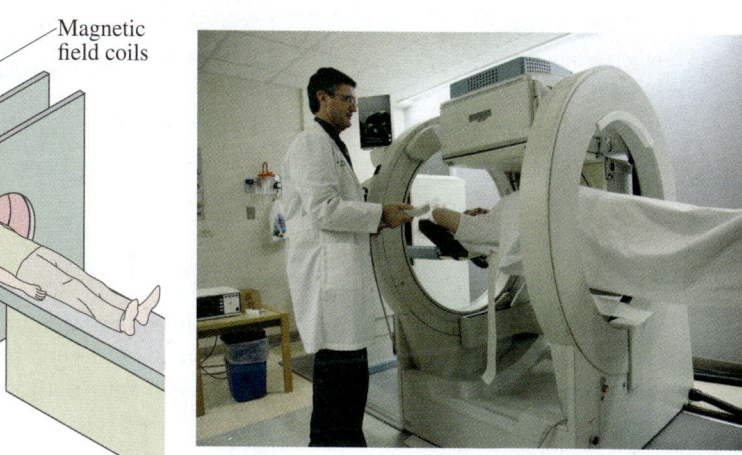

FIGURE 31–25 NMR imaging setup: (a) diagram; (b) photograph.

SECTION 31–9 Nuclear Magnetic Resonance (NMR) and Magnetic Resonance Imaging (MRI)

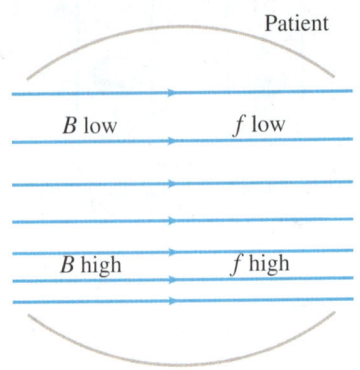

FIGURE 31–26 A static field that is stronger at the bottom than at the top. The frequency of absorbed or emitted radiation is proportional to B in NMR.

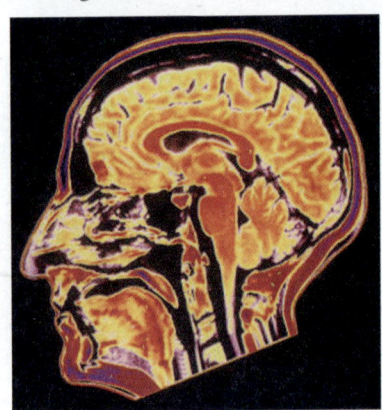

FIGURE 31–27 False-color NMR image (MRI) through the head showing structures in the brain.

The formation of a two-dimensional or three-dimensional image can be done using techniques similar to those for computed tomography (Section 25–12). The simplest thing to measure for creating an image is the intensity of absorbed and/or reemitted radiation from many different points of the body, and this would be a measure of the density of H atoms at each point. But how do we determine from what part of the body a given photon comes? One technique is to give the static magnetic field a gradient; that is, instead of applying a uniform magnetic field, B_T, the field is made to vary with position across the width of the sample (or patient), Fig. 31–26. Because the frequency absorbed by the H nuclei is proportional to B_T (Eq. 31–11), only one plane within the body will have the proper value of B_T to absorb photons of a particular frequency f. By varying f, absorption by different planes can be measured. Alternately, if the field gradient is applied *after* the RF pulse, the frequency of the emitted photons will be a measure of where they were emitted. If a magnetic field gradient in one direction is applied during excitation (absorption of photons) and photons of a single frequency are transmitted, only H nuclei in one thin slice will be excited. By applying a gradient during reemission in a direction perpendicular to the first, the frequency f of the reemitted radiation will represent depth in that slice. Other ways of varying the magnetic field throughout the volume of the body can be used in order to correlate NMR frequency with position.

A reconstructed image based on the density of H atoms (that is, the intensity of absorbed or emitted radiation) is not very interesting. More useful are images based on the rate at which the nuclei decay back to the ground state, and such images can produce resolution of 1 mm or better. This NMR technique (sometimes called **spin-echo**) produces images of great diagnostic value, both in the delineation of structure (anatomy) and in the study of metabolic processes. An NMR image is shown in Fig. 31–27.

NMR imaging is considered to be noninvasive. We can calculate the energy of the photons involved: as mentioned above, in a 1.0-T magnetic field, $f = 42.58$ MHz for $^1_1$H. This corresponds to an energy of $hf = (6.6 \times 10^{-34} \text{ J} \cdot \text{s})(43 \times 10^6 \text{ Hz}) \approx 3 \times 10^{-26}$ J or about 10^{-7} eV. Since molecular bonds are on the order of 1 eV, the RF photons can cause little cellular disruption. This should be compared to X- or γ rays, whose energies are 10^4 to 10^6 eV and thus can cause significant damage. The static magnetic fields, though often large (as high as 1.0 to 1.5 T), are believed to be harmless (except for people who wear heart pacemakers).

TABLE 31–2 Medical Imaging Techniques

Technique	Where Discussed in This Book	Optimal Resolution
Conventional X-ray	Section 25–12	$\frac{1}{2}$ mm
CT scan, X-ray	Section 25–12	$\frac{1}{2}$ mm
Nuclear medicine (tracers)	Section 31–7	1 cm
SPECT (single photon emission)	Section 31–8	1 cm
PET (positron emission)	Section 31–8	2–5 mm
MRI (NMR)	Section 31–9	$\frac{1}{2}$–1 mm
Ultrasound	Section 12–9	0.3–2 mm

Table 31–2 lists the major techniques we have discussed for imaging the interior of the human body, along with the optimum resolution attainable today. Resolution is only one factor that must be considered, because the different imaging techniques provide different types of information that are useful for different types of diagnosis.

Summary

A **nuclear reaction** occurs when two nuclei collide and two or more other nuclei (or particles) are produced. In this process, as in radioactivity, **transmutation** (change) of elements occurs.

The **reaction energy** or **Q-value** of a reaction $a + X \rightarrow Y + b$ is

$$Q = (M_a + M_X - M_b - M_Y)c^2 \quad \textbf{(31–2a)}$$
$$= KE_b + KE_Y - KE_a - KE_X. \quad \textbf{(31–2b)}$$

In **fission**, a heavy nucleus such as uranium splits into two intermediate-sized nuclei after being struck by a neutron. $^{235}_{92}U$ is fissionable by slow neutrons, whereas some fissionable nuclei require fast neutrons. Much energy is released in fission (≈ 200 MeV per fission) because the binding energy per nucleon is lower for heavy nuclei than it is for intermediate-sized nuclei, so the mass of a heavy nucleus is greater than the total mass of its fission products. The fission process releases neutrons, so that a **chain reaction** is possible. The **critical mass** is the minimum mass of fuel needed so that enough emitted neutrons go on to produce more fissions and sustain a chain reaction. In a **nuclear reactor** or nuclear weapon, a **moderator** is used to slow down the released neutrons.

The **fusion** process, in which small nuclei combine to form larger ones, also releases energy. The energy from our Sun originates in the fusion reactions known as the **proton–proton chain** in which four protons fuse to form a ^4_2He nucleus producing 25 MeV of energy. A useful fusion reactor for power generation has not yet proved possible because of the difficulty in containing the fuel (e.g., deuterium) long enough at the extremely high temperature required ($\approx 10^8$ K). Nonetheless, progress has been made in confining the collection of charged ions known as a **plasma**. The two main methods are **magnetic confinement**, using a magnetic field in a device such as the donut-shaped **tokamak**, and **inertial confinement** in which intense laser beams compress a fuel pellet of deuterium and tritium.

Radiation can cause damage to materials, including biological tissue. Quantifying amounts of radiation is the subject of **dosimetry**. The **curie** (Ci) and the **becquerel** (Bq) are units that measure the **source activity** or rate of decay of a sample: 1 Ci = 3.70×10^{10} decays per second, whereas 1 Bq = 1 decay/s. The **absorbed dose**, often specified in **rads**, measures the amount of energy deposited per unit mass of absorbing material: 1 rad is the amount of radiation that deposits energy at the rate of 10^{-2} J/kg of material. The SI unit of absorbed dose is the **gray**: 1 Gy = 1 J/kg = 100 rad. The **effective dose** is often specified by the **rem** = rad × RBE, where RBE is the "relative biological effectiveness" of a given type of radiation; 1 rem of any type of radiation does approximately the same amount of biological damage. The average dose received per person per year in the United States is about 360 mrem. The SI unit for effective dose is the **sievert**: 1 Sv = 100 rem.

[*Nuclear radiation is used in medicine for cancer therapy, and for imaging of biological structure and processes. Tomographic imaging of the human body, which can provide 3-dimensional detail, includes several types: PET, SPET (= SPECT), MRI, and CT scans (discussed in Chapter 25). MRI makes use of **nuclear magnetic resonance** (NMR).]

Questions

1. Fill in the missing particles or nuclei:
   (a) $n + ^{232}_{90}Th \rightarrow ? + \gamma$;
   (b) $n + ^{137}_{56}Ba \rightarrow ^{137}_{55}Cs + ?$;
   (c) $d + ^2_1H \rightarrow ^4_2He + ?$;
   (d) $\alpha + ^{197}_{79}Au \rightarrow ? + d$
   where d stands for deuterium.

2. When $^{22}_{11}Na$ is bombarded by deuterons (^2_1H), an α particle is emitted. What is the resulting nuclide? Write down the reaction equation.

3. Why are neutrons such good projectiles for producing nuclear reactions?

4. What is the Q-value for radioactive decay reactions?
   (a) $Q < 0$.   (b) $Q > 0$.   (c) $Q = 0$.
   (d) The sign of Q depends on the nucleus.

5. The energy from nuclear fission appears in the form of thermal energy—but the thermal energy of what?

6. (a) If $^{235}_{92}U$ released only 1.5 neutrons per fission on average (instead of 2.5), would a chain reaction be possible? (b) If so, how would the chain reaction be different than if 3 neutrons were released per fission?

7. Why can't uranium be enriched by chemical means?

8. How can a neutron, with practically no kinetic energy, excite a nucleus to the extent shown in Fig. 31–3?

9. Why would a porous block of uranium be more likely to explode if kept under water rather than in air?

10. A reactor that uses highly enriched uranium can use ordinary water (instead of heavy water) as a moderator and still have a self-sustaining chain reaction. Explain.

11. Why must the fission process release neutrons if it is to be useful?

12. Why are neutrons released in a fission reaction?

13. What is the reason for the "secondary system" in a nuclear reactor, Fig. 31–8? That is, why is the water heated by the fuel in a nuclear reactor not used directly to drive the turbines?

14. What is the basic difference between fission and fusion?

15. Discuss the relative merits and disadvantages, including pollution and safety, of power generation by fossil fuels, nuclear fission, and nuclear fusion.

16. Why do gamma particles penetrate matter more easily than beta particles do?

17. Light energy emitted by the Sun and stars comes from the fusion process. What conditions in the interior of stars make this possible?

18. How do stars, and our Sun, maintain confinement of the plasma for fusion?

19. People who work around metals that emit alpha particles are trained that there is little danger from proximity or touching the material, but they must take extreme precautions against ingesting it. Why? (Eating and drinking while working are forbidden.)

20. What is the difference between absorbed dose and effective dose? What are the SI units for each?

21. Radiation is sometimes used to sterilize medical supplies and even food. Explain how it works.

*22. How might radioactive tracers be used to find a leak in a pipe?

MisConceptual Questions

1. In a nuclear reaction, which of the following is *not* conserved?
   (a) Energy.
   (b) Momentum.
   (c) Electric charge.
   (d) Nucleon number.
   (e) None of the above.

2. Fission fragments are typically
   (a) β^+ emitters.
   (b) β^- emitters.
   (c) Both.
   (d) Neither.

3. Which of the following properties would decrease the critical mass needed to sustain a nuclear chain reaction?
   (a) Low boiling point.
   (b) High melting point.
   (c) More neutrons released per fission.
   (d) Low nuclear density.
   (e) Filled valence shell.
   (f) All of the above.

4. Rather than having a maximum at about $A \approx 60$, as shown in Fig. 31–12, suppose the average binding energy per nucleon continually increased with increasing mass number. Then,
   (a) fission would still be possible, but not fusion.
   (b) fusion would still be possible, but not fission.
   (c) both fission and fusion would still be possible.
   (d) neither fission nor fusion would be possible.

5. Why is a moderator needed in a normal uranium fission reactor?
   (a) To increase the rate of neutron capture by uranium-235.
   (b) To increase the rate of neutron capture by uranium-238.
   (c) To increase the rate of production of plutonium-239.
   (d) To increase the critical mass of the fission fuel.
   (e) To provide more neutrons for the reaction.
   (f) All of the above.

6. What is the difference between nuclear fission and nuclear fusion?
   (a) Nuclear fission is used for bombs; nuclear fusion is used in power plants.
   (b) There is no difference. Fission and fusion are different names for the same physical phenomenon.
   (c) Nuclear fission refers to using deuterium to create a nuclear reaction.
   (d) Nuclear fusion occurs spontaneously, as happens to the C^{14} used in carbon dating.
   (e) In nuclear fission, a nucleus splits; in nuclear fusion, nucleons or nuclei and nucleons join to form a new nucleus.

7. A primary difficulty in energy production by fusion is
   (a) the scarcity of necessary fuel.
   (b) the disposal of radioactive by-products produced.
   (c) the high temperatures necessary to overcome the electrical repulsion of protons.
   (d) the fact that it is possible in volcanic regions only.

8. If two hydrogen nuclei, $^2_1$H, each of mass m_H, fuse together and form a helium nucleus of mass m_{He},
   (a) $m_{He} < 2m_H$.
   (b) $m_{He} = 2m_H$.
   (c) $m_{He} > 2m_H$.
   (d) All of the above are possible.

9. Which radiation induces the most biological damage for a given amount of energy deposited in tissue?
   (a) Alpha particles.
   (b) Gamma radiation.
   (c) Beta radiation.
   (d) All do the same damage for the same deposited energy.

10. Which would produce the most energy in a single reaction?
    (a) The fission reaction associated with uranium-235.
    (b) The fusion reaction of the Sun (two hydrogen nuclei fused to one helium nucleus).
    (c) Both (a) and (b) are about the same.
    (d) Need more information.

11. The fuel necessary for fusion-produced energy could be derived from
    (a) water.
    (b) superconductors.
    (c) uranium.
    (d) helium.
    (e) sunlight.

12. Which of the following is true?
    (a) Any amount of radiation is harmful to living tissue.
    (b) Radiation is a natural part of the environment.
    (c) All forms of radiation will penetrate deep into living tissue.
    (d) None of the above is true.

13. Which of the following would reduce the cell damage due to radiation for a lab technician who works with radioactive isotopes in a hospital or lab?
    (a) Increase the worker's distance from the radiation source.
    (b) Decrease the time the worker is exposed to the radiation.
    (c) Use shielding to reduce the amount of radiation that strikes the worker.
    (d) Have the worker wear a radiation badge when working with the radioactive isotopes.
    (e) All of the above.

14. If the same dose of each type of radiation was provided over the same amount of time, which type would be most harmful?
    (a) X-rays.
    (b) γ rays.
    (c) β rays.
    (d) α particles.

15. $^{235}_{92}$U releases an average of 2.5 neutrons per fission compared to 2.9 for $^{239}_{94}$Pu. Which has the smaller critical mass?
    (a) $^{235}_{92}$U.
    (b) $^{239}_{94}$Pu.
    (c) Both the same.

For assigned homework and other learning materials, go to the MasteringPhysics website.

Problems

(NOTE: Masses are found in Appendix B.)

31–1 Nuclear Reactions, Transmutation

1. (I) Natural aluminum is all $^{27}_{13}$Al. If it absorbs a neutron, what does it become? Does it decay by β^+ or β^-? What will be the product nucleus?

2. (I) Determine whether the reaction $^2_1$H + $^2_1$H → $^3_2$He + n requires a threshold energy, and why.

3. (I) Is the reaction n + $^{238}_{92}$U → $^{239}_{92}$U + γ possible with slow neutrons? Explain.

4. (II) (a) Complete the following nuclear reaction, p + ? → $^{32}_{16}$S + γ. (b) What is the Q-value?

5. (II) The reaction p + $^{18}_8$O → $^{18}_9$F + n requires an input of energy equal to 2.438 MeV. What is the mass of $^{18}_9$F?

6. (II) (a) Can the reaction n + $^{24}_{12}$Mg → $^{23}_{11}$Na + d occur if the bombarding particles have 18.00 MeV of kinetic energy? (d stands for deuterium, $^2_1$H.) (b) If so, how much energy is released? If not, what kinetic energy is needed?

7. (II) (a) Can the reaction p + $^7_3$Li → $^4_2$He + α occur if the incident proton has kinetic energy = 3100 keV? (b) If so, what is the total kinetic energy of the products? If not, what kinetic energy is needed?

8. (II) In the reaction α + $^{14}_7$N → $^{17}_8$O + p, the incident α particles have 9.85 MeV of kinetic energy. The mass of $^{17}_8$O is 16.999132 u. (a) Can this reaction occur? (b) If so, what is the total kinetic energy of the products? If not, what kinetic energy is needed?

9. (II) Calculate the Q-value for the "capture" reaction α + $^{16}_8$O → $^{20}_{10}$Ne + γ.

10. (II) Calculate the total kinetic energy of the products of the reaction d + $^{13}_6$C → $^{14}_7$N + n if the incoming deuteron has kinetic energy KE = 41.4 MeV.

11. (II) Radioactive $^{14}_6$C is produced in the atmosphere when a neutron is absorbed by $^{14}_7$N. Write the reaction and find its Q-value.

12. (II) An example of a **stripping** nuclear reaction is d + $^6_3$Li → X + p. (a) What is X, the resulting nucleus? (b) Why is it called a "stripping" reaction? (c) What is the Q-value of this reaction? Is the reaction endothermic or exothermic?

13. (II) An example of a **pick-up** nuclear reaction is $^3_2$He + $^{12}_6$C → X + α. (a) Why is it called a "pick-up" reaction? (b) What is the resulting nucleus? (c) What is the Q-value of this reaction? Is the reaction endothermic or exothermic?

14. (II) Does the reaction p + $^7_3$Li → $^4_2$He + α require energy, or does it release energy? How much energy?

15. (II) Calculate the energy released (or energy input required) for the reaction α + $^9_4$Be → $^{12}_6$C + n.

31–2 Nuclear Fission

16. (I) What is the energy released in the fission reaction of Eq. 31–4? (The masses of $^{141}_{56}$Ba and $^{92}_{36}$Kr are 140.914411 u and 91.926156 u, respectively.)

17. (I) Calculate the energy released in the fission reaction n + $^{235}_{92}$U → $^{88}_{38}$Sr + $^{136}_{54}$Xe + 12n. Use Appendix B, and assume the initial kinetic energy of the neutron is very small.

18. (I) How many fissions take place per second in a 240-MW reactor? Assume 200 MeV is released per fission.

19. (I) The energy produced by a fission reactor is about 200 MeV per fission. What fraction of the mass of a $^{235}_{92}$U nucleus is this?

20. (II) Suppose that the average electric power consumption, day and night, in a typical house is 960 W. What initial mass of $^{235}_{92}$U would have to undergo fission to supply the electrical needs of such a house for a year? (Assume 200 MeV is released per fission, as well as 100% efficiency.)

21. (II) Consider the fission reaction

$$^{235}_{92}\text{U} + \text{n} \rightarrow {}^{133}_{51}\text{Sb} + {}^{98}_{41}\text{Nb} + ?\text{n}.$$

(a) How many neutrons are produced in this reaction? (b) Calculate the energy release. The atomic masses for Sb and Nb isotopes are 132.915250 u and 97.910328 u, respectively.

22. (II) How much mass of $^{235}_{92}$U is required to produce the same amount of energy as burning 1.0 kg of coal (about 3×10^7 J)?

23. (II) What initial mass of $^{235}_{92}$U is required to operate a 950-MW reactor for 1 yr? Assume 34% efficiency.

24. (II) If a 1.0-MeV neutron emitted in a fission reaction loses one-half of its kinetic energy in each collision with moderator nuclei, how many collisions must it make to reach thermal energy ($\frac{3}{2}kT = 0.040$ eV)?

25. (II) Assuming a fission of $^{236}_{92}$U into two roughly equal fragments, estimate the electric potential energy just as the fragments separate from each other. Assume that the fragments are spherical (see Eq. 30–1) and compare your calculation to the nuclear fission energy released, about 200 MeV.

26. (III) Suppose that the neutron multiplication factor is 1.0004. If the average time between successive fissions in a chain of reactions is 1.0 ms, by what factor will the reaction rate increase in 1.0 s?

31–3 Nuclear Fusion

27. (I) What is the average kinetic energy of protons at the center of a star where the temperature is 2×10^7 K? [Hint: See Eq. 13–8.]

28. (II) Show that the energy released in the fusion reaction $^2_1$H + $^3_1$H → $^4_2$He + n is 17.59 MeV.

29. (II) Show that the energy released when two deuterium nuclei fuse to form $^3_2$He with the release of a neutron is 3.27 MeV (Eq. 31–8b).

Problems 911

30. (II) Verify the Q-value stated for each of the reactions of Eqs. 31–6. [*Hint*: Use Appendix B; be careful with electrons (included in mass values except for p, d, t).]

31. (II) (*a*) Calculate the energy release per gram of fuel for the reactions of Eqs. 31–8a, b, and c. (*b*) Calculate the energy release per gram of uranium $^{235}_{92}$U in fission, and give its ratio to each reaction in (*a*).

32. (II) How much energy is released when $^{238}_{92}$U absorbs a slow neutron (kinetic energy ≈ 0) and becomes $^{239}_{92}$U?

33. (II) If a typical house requires 960 W of electric power on average, what minimum amount of deuterium fuel would have to be used in a year to supply these electrical needs? Assume the reaction of Eq. 31–8b.

34. (II) If $^6_3$Li is struck by a slow neutron, it can form $^4_2$He and another nucleus. (*a*) What is the second nucleus? (This is a method of generating this isotope.) (*b*) How much energy is released in the process?

35. (II) Suppose a fusion reactor ran on "d–d" reactions, Eqs. 31–8a and b in equal amounts. Estimate how much natural water, for fuel, would be needed per hour to run a 1150-MW reactor, assuming 33% efficiency.

36. (III) Show that the energies carried off by the $^4_2$He nucleus and the neutron for the reaction of Eq. 31–8c are about 3.5 MeV and 14 MeV, respectively. Are these fixed values, independent of the plasma temperature?

37. (III) How much energy (J) is contained in 1.00 kg of water if its natural deuterium is used in the fusion reaction of Eq. 31–8a? Compare to the energy obtained from the burning of 1.0 kg of gasoline, about 5×10^7 J.

38. (III) (*a*) Give the ratio of the energy needed for the first reaction of the *carbon cycle* to the energy needed for a deuterium–tritium reaction (Example 31–9). (*b*) If a deuterium–tritium reaction actually requires a temperature $T \approx 3 \times 10^8$ K, estimate the temperature needed for the first carbon-cycle reaction.

31–5 Dosimetry

39. (I) 350 rads of α-particle radiation is equivalent to how many rads of X-rays in terms of biological damage?

40. (I) A dose of 4.0 Sv of γ rays in a short period would be lethal to about half the people subjected to it. How many grays is this?

41. (I) How many rads of slow neutrons will do as much biological damage as 72 rads of fast neutrons?

42. (II) How much energy is deposited in the body of a 65-kg adult exposed to a 2.5-Gy dose?

43. (II) A cancer patient is undergoing radiation therapy in which protons with an energy of 1.2 MeV are incident on a 0.20-kg tumor. (*a*) If the patient receives an effective dose of 1.0 rem, what is the absorbed dose? (*b*) How many protons are absorbed by the tumor? Assume RBE ≈ 1.

44. (II) A 0.035-μCi sample of $^{32}_{15}$P is injected into an animal for tracer studies. If a Geiger counter intercepts 35% of the emitted β particles, what will be the counting rate, assumed 85% efficient?

45. (II) About 35 eV is required to produce one ion pair in air. Show that this is consistent with the two definitions of the roentgen given in the text.

46. (II) A 1.6-mCi source of $^{32}_{15}$P (in NaHPO$_4$), a β emitter, is implanted in a tumor where it is to administer 32 Gy. The half-life of $^{32}_{15}$P is 14.3 days, and 1.0 mCi delivers about 10 mGy/min. Approximately how long should the source remain implanted?

47. (II) What is the mass of a 2.50-μCi $^{14}_6$C source?

48. (II) $^{57}_{27}$Co emits 122-keV γ rays. If a 65-kg person swallowed 1.55 μCi of $^{57}_{27}$Co, what would be the dose rate (Gy/day) averaged over the whole body? Assume that 50% of the γ-ray energy is deposited in the body. [*Hint*: Determine the rate of energy deposited in the body and use the definition of the gray.]

49. (II) Ionizing radiation can be used on meat products to reduce the levels of microbial pathogens. Refrigerated meat is limited to 4.5 kGy. If 1.6-MeV electrons irradiate 5 kg of beef, how many electrons would it take to reach the allowable limit?

50. (III) Huge amounts of radioactive $^{131}_{53}$I were released in the accident at Chernobyl in 1986. Chemically, iodine goes to the human thyroid. (It can be used for diagnosis and treatment of thyroid problems.) In a normal thyroid, $^{131}_{53}$I absorption can cause damage to the thyroid. (*a*) Write down the reaction for the decay of $^{131}_{53}$I. (*b*) Its half-life is 8.0 d; how long would it take for ingested $^{131}_{53}$I to become 5.0% of the initial value? (*c*) Absorbing 1 mCi of $^{131}_{53}$I can be harmful; what mass of iodine is this?

51. (III) Assume a liter of milk typically has an activity of 2000 pCi due to $^{40}_{19}$K. If a person drinks two glasses (0.5 L) per day, estimate the total effective dose (in Sv and in rem) received in a year. As a crude model, assume the milk stays in the stomach 12 hr and is then released. Assume also that roughly 10% of the 1.5 MeV released per decay is absorbed by the body. Compare your result to the normal allowed dose of 100 mrem per year. Make your estimate for (*a*) a 60-kg adult, and (*b*) a 6-kg baby.

52. (III) Radon gas, $^{222}_{86}$Rn, is considered a serious health hazard (see discussion in text). It decays by α-emission. (*a*) What is the daughter nucleus? (*b*) Is the daughter nucleus stable or radioactive? If the latter, how does it decay, and what is its half-life? (See Fig. 30–11.) (*c*) Is the daughter nucleus also a noble gas, or is it chemically reactive? (*d*) Suppose 1.4 ng of $^{222}_{86}$Rn seeps into a basement. What will be its activity? If the basement is then sealed, what will be the activity 1 month later?

31–9 NMR

53. (II) Calculate the wavelength of photons needed to produce NMR transitions in free protons in a 1.000-T field. In what region of the spectrum is this wavelength?

General Problems

54. Consider a system of nuclear power plants that produce 2100 MW. (a) What total mass of $^{235}_{92}$U fuel would be required to operate these plants for 1 yr, assuming that 200 MeV is released per fission? (b) Typically 6% of the $^{235}_{92}$U nuclei that fission produce strontium-90, $^{90}_{38}$Sr, a β^- emitter with a half-life of 29 yr. What is the total radioactivity of the $^{90}_{38}$Sr, in curies, produced in 1 yr? (Neglect the fact that some of it decays during the 1-yr period.)

55. J. Chadwick discovered the neutron by bombarding $^9_4$Be with the popular projectile of the day, alpha particles. (a) If one of the reaction products was the then unknown neutron, what was the other product? (b) What is the Q-value of this reaction?

56. Fusion temperatures are often given in keV. Determine the conversion factor from kelvins to keV using, as is common in this field, $\overline{KE} = kT$ without the factor $\frac{3}{2}$.

57. One means of enriching uranium is by diffusion of the gas UF$_6$. Calculate the ratio of the speeds of molecules of this gas containing $^{235}_{92}$U and $^{238}_{92}$U, on which this process depends.

58. (a) What mass of $^{235}_{92}$U was actually fissioned in the first atomic bomb, whose energy was the equivalent of about 20 kilotons of TNT (1 kiloton of TNT releases 5×10^{12} J)? (b) What was the actual mass transformed to energy?

59. The average yearly background radiation in a certain town consists of 32 mrad of X-rays and γ rays plus 3.4 mrad of particles having a RBE of 10. How many rem will a person receive per year on average?

60. A shielded γ-ray source yields a dose rate of 0.048 rad/h at a distance of 1.0 m for an average-sized person. If workers are allowed a maximum dose of 5.0 rem in 1 year, how close to the source may they operate, assuming a 35-h work week? Assume that the intensity of radiation falls off as the square of the distance. (It actually falls off more rapidly than $1/r^2$ because of absorption in the air, so your answer will give a better-than-permissible value.)

61. Radon gas, $^{222}_{86}$Rn, is formed by α decay. (a) Write the decay equation. (b) Ignoring the kinetic energy of the daughter nucleus (it's so massive), estimate the kinetic energy of the α particle produced. (c) Estimate the momentum of the alpha and of the daughter nucleus. (d) Estimate the kinetic energy of the daughter, and show that your approximation in (b) was valid.

62. In the net reaction, Eq. 31–7, for the proton–proton chain in the Sun, the neutrinos escape from the Sun with energy of about 0.5 MeV. The remaining energy, 26.2 MeV, is available to heat the Sun. Use this value to calculate the "heat of combustion" per kilogram of hydrogen fuel and compare it to the heat of combustion of coal, about 3×10^7 J/kg.

63. Energy reaches Earth from the Sun at a rate of about 1300 W/m^2. Calculate (a) the total power output of the Sun, and (b) the number of protons consumed per second in the reaction of Eq. 31–7, assuming that this is the source of all the Sun's energy. (c) Assuming that the Sun's mass of 2.0×10^{30} kg was originally all protons and that all could be involved in nuclear reactions in the Sun's core, how long would you expect the Sun to "glow" at its present rate? See Problem 62. [Hint: Use $1/r^2$ law.]

64. Estimate how many solar neutrinos pass through a 180-m^2 ceiling of a room, at latitude 44°, for an hour around midnight on midsummer night. [Hint: See Problems 62 and 63.]

65. Estimate how much total energy would be released via fission if 2.0 kg of uranium were enriched to 5% of the isotope $^{235}_{92}$U.

66. Some stars, in a later stage of evolution, may begin to fuse two $^{12}_6$C nuclei into one $^{24}_{12}$Mg nucleus. (a) How much energy would be released in such a reaction? (b) What kinetic energy must two carbon nuclei each have when far apart, if they can then approach each other to within 6.0 fm, center-to-center? (c) Approximately what temperature would this require?

67. An average adult body contains about 0.10 μCi of $^{40}_{19}$K, which comes from food. (a) How many decays occur per second? (b) The potassium decay produces beta particles with energies of around 1.4 MeV. Estimate the dose per year in sieverts for a 65-kg adult. Is this a significant fraction of the 3.6-mSv/yr background rate?

68. When the nuclear reactor accident occurred at Chernobyl in 1986, 2.0×10^7 Ci were released into the atmosphere. Assuming that this radiation was distributed uniformly over the surface of the Earth, what was the activity per square meter? (The actual activity was not uniform; even within Europe wet areas received more radioactivity from rainfall.)

69. A star with a large helium abundance can burn helium in the reaction $^4_2$He + $^4_2$He + $^4_2$He $\rightarrow$ $^{12}_6$C. What is the Q-value for this reaction?

70. A 1.2-μCi $^{137}_{55}$Cs source is used for 1.4 hours by a 62-kg worker. Radioactive $^{137}_{55}$Cs decays by β^- decay with a half-life of 30 yr. The average energy of the emitted betas is about 190 keV per decay. The β decay is quickly followed by a γ with an energy of 660 keV. Assuming the person absorbs *all* emitted energy, what effective dose (in rem) is received?

71. Suppose a future fusion reactor would be able to put out 1000 MW of electrical power continuously. Assume the reactor will produce energy solely through the reaction given in Eq. 31–8a and will convert this energy to electrical energy with an efficiency of 33%. Estimate the minimum amount of deuterium needed to run this facility per year.

72. If a 65-kg power plant worker has been exposed to the maximum slow-neutron radiation for a given year, how much total energy (in J) has that worker absorbed? What if he were exposed to fast protons?

73. Consider the fission reaction

$$n + {}^{235}_{92}U \rightarrow {}^{92}_{38}Sr + X + 3n.$$

(a) What is X? (b) If this were part of a chain reaction in a fission power reactor running at "barely critical," what would happen on average to the three produced neutrons? (c) (optional) What is the Q-value of this reaction? [Hint: Mass values can be found at www.nist.gov/pml/data/comp.cfm.]

74. A large amount of $^{90}_{38}$Sr was released during the Chernobyl nuclear reactor accident in 1986. The $^{90}_{38}$Sr enters the body through the food chain. How long will it take for 85% of the $^{90}_{38}$Sr released during the accident to decay? See Appendix B.

75. Three radioactive sources have the same activity, 35 mCi. Source A emits 1.0-MeV γ rays, source B emits 2.0-MeV γ rays, and source C emits 2.0-MeV alphas. What is the relative danger of these sources?

76. A 55-kg patient is to be given a medical test involving the ingestion of $^{99m}_{43}$Tc (Section 31–7) which decays by emitting a 140-keV gamma. The half-life for this decay is 6 hours. Assuming that about half the gamma photons exit the body without interacting with anything, what must be the initial activity of the Tc sample if the whole-body dose cannot exceed 50 mrem? Make the rough approximation that biological elimination of Tc can be ignored.

Search and Learn

1. Referring to Section 31–2, (a) state three problems that must be overcome to make a functioning fission nuclear reactor; (b) state three environmental problems or dangers that do or could result from the operation of a nuclear fission reactor; (c) describe an additional problem or danger associated with a breeder reactor.

2. Referring to Section 31–3, (a) why can small nuclei combine to form larger ones, releasing energy in the process? (b) Why does the first reaction in the proton–proton chain limit the rate at which the Sun produces energy? (c) What are the heaviest elements for which energy is released if the elements are created by fusion of lighter elements? (d) What keeps the Sun and stars together, allowing them to sustain fusion? (e) What two methods are currently being investigated to contain high-temperature plasmas on the Earth to create fusion in the laboratory?

3. Deuterium makes up 0.0115% of natural hydrogen on average. Make a rough estimate of the total deuterium in the Earth's oceans and estimate the total energy released if all of it were used in fusion reactors.

4. The energy output of massive stars is believed to be due to the *carbon cycle* (see text). (a) Show that no carbon is consumed in this cycle and that the net effect is the same as for the proton–proton chain. (b) What is the total energy release? (c) Determine the energy output for each reaction and decay. (d) Why might the carbon cycle require a higher temperature ($\approx 2 \times 10^7$ K) than the proton–proton chain ($\approx 1.5 \times 10^7$ K)?

5. Consider the effort by humans to harness nuclear fusion as a viable energy source. (a) What are some advantages of using nuclear fusion rather than nuclear fission? (b) What is the major technological problem with using controlled nuclear fusion as a source of energy? (c) Discuss two different approaches to solving this problem. (d) What fuel is necessary in a nuclear fusion reaction? (e) Write a nuclear reaction using two nuclei of the fuel in part (d) to create a third nucleus. (f) Calculate the Q-value of the reaction in part (e).

6. (a) Explain how each of the following can cause damage to materials: beta particles, alpha particles, energetic neutrons, and gamma rays. (b) How might metals be damaged? (c) How can the damage affect living cells?

ANSWERS TO EXERCISES

A: $^{138}_{56}$Ba.
B: 3 neutrons.
C: 2×10^{17}.
D: (e).
E: (b).

This photo is a computer reconstruction of particles produced due to a 7 TeV proton–proton collision at the Large Hadron Collider (LHC). It is a candidate for having produced the long-sought Higgs boson (plus other particles). The Higgs in this case could have decayed (very quickly $\sim 10^{-22}$ s) into two Z bosons (which are carriers of the weak force):

$$H^0 \rightarrow Z^0 + Z^0.$$

We don't see the tracks of the Z^0 particles because (1) they are neutral and (2) they decay too quickly ($\sim 10^{-24}$ s), in this case:

$$Z^0 \rightarrow e^- + e^+.$$

The tracks of the 2 electrons and 2 positrons are shown as green lines. The Higgs is thought to play a fundamental role in the Standard Model of particle physics, importantly providing mass to fundamental particles.

The CMS detector of this photo uses a combination of the detector types discussed in Section 30–13. A magnetic field causes particles to move in curved paths so the momentum of each can be measured (Section 20–4). Tracks of particles with very large momentum, such as our electrons here, are barely curved.

In this Chapter we will study elementary particle physics from its beginnings until today, including antiparticles, neutrinos, quarks, the Standard Model, and theories that go beyond. We start with the great machines that accelerate particles so they can collide at high energies.

CHAPTER 32

Elementary Particles

CHAPTER-OPENING QUESTIONS—Guess now!

1. Physicists reserve the term "fundamental particle" for particles with a special property. What do you think that special property is?
   (a) Particles that are massless.
   (b) Particles that possess the minimum allowable electric charge.
   (c) Particles that have no internal structure.
   (d) Particles that produce no force on other objects.

2. The fundamental particles as we see them today, besides the long-sought-for Higgs boson, are
   (a) atoms and electrons.
   (b) protons, neutrons, and electrons.
   (c) protons, neutrons, electrons, and photons.
   (d) quarks, leptons, and gauge bosons (carriers of force).
   (e) hadrons, leptons, and gauge bosons.

I n the final two Chapters of this book we discuss two of the most exciting areas of contemporary physics: elementary particles in this Chapter, and cosmology and astrophysics in Chapter 33. These are subjects at the forefront of knowledge—elementary particles treats the smallest objects in the universe; cosmology treats the largest (and oldest) aspects of the universe. The reader who wants an understanding of the great beauties of present-day science (and its limits) will want to read these Chapters. So will those who want to be good citizens, even if there is not time to cover them in a physics course.

CONTENTS

- 32–1 High-Energy Particles and Accelerators
- 32–2 Beginnings of Elementary Particle Physics—Particle Exchange
- 32–3 Particles and Antiparticles
- 32–4 Particle Interactions and Conservation Laws
- 32–5 Neutrinos
- 32–6 Particle Classification
- 32–7 Particle Stability and Resonances
- 32–8 Strangeness? Charm? Towards a New Model
- 32–9 Quarks
- 32–10 The Standard Model: QCD and Electroweak Theory
- 32–11 Grand Unified Theories
- 32–12 Strings and Supersymmetry

In this penultimate Chapter we discuss *elementary particle* physics, which represents the human endeavor to understand the basic building blocks of all matter, and the fundamental forces that govern their interactions.

Almost a century ago, by the 1930s, it was accepted that all atoms can be considered to be made up of neutrons, protons, and electrons. The basic constituents of the universe were no longer considered to be atoms (as they had been for 2000 years) but rather the proton, neutron, and electron. Besides these three "elementary particles," several others were also known: the positron (a positive electron), the neutrino, and the γ particle (or photon), for a total of six elementary particles.

By the 1950s and 1960s many new types of particles similar to the neutron and proton were discovered, as well as many "midsized" particles called *mesons* whose masses were mostly less than nucleon masses but more than the electron mass. (Other mesons, found later, have masses greater than nucleons.) Physicists felt that these particles could not all be fundamental, and must be made up of even smaller constituents (later confirmed by experiment), which were given the name *quarks*.

By the term **fundamental particle**, we mean a particle that is so simple, so basic, that it has no internal structure[†] (is not made up of smaller subunits)—see Chapter-Opening Question 1.

Today, the fundamental constituents of matter are considered to be **quarks** (they make up protons and neutrons as well as mesons) and **leptons** (a class that includes electrons, positrons, and neutrinos). There are also the "carriers of force" known as **gauge bosons**, including the photon, gluons, and W and Z bosons. In addition there is the elusive **Higgs** boson, predicted in the 1960s but whose first suggestions of experimental detection came only in 2011–2013. The theory that describes our present view is called the **Standard Model**. How we came to our present understanding of elementary particles is the subject of this Chapter.

One of the exciting developments of the last few years is an emerging synthesis between the study of elementary particles and astrophysics (Chapter 33). In fact, recent observations in astrophysics have led to the conclusion that the greater part of the mass–energy content of the universe is not ordinary matter but two mysterious and invisible forms known as "dark matter" and "dark energy" which cannot be explained by the Standard Model in its present form.

Indeed, we are now aware that the Standard Model is not sufficient. There are problems and important questions still unanswered, and we will mention some of them in this Chapter and how we hope to answer them.

32–1 High-Energy Particles and Accelerators

In the late 1940s, after World War II, it was found that if the incoming particle in a nuclear reaction (Section 31–1) has sufficient energy, new types of particles can be produced. The earliest experiments used **cosmic rays**—particles that impinge on the Earth from space. In the laboratory, various types of particle accelerators have been constructed to accelerate protons or electrons to high energies so they can collide with other particles—often protons (the hydrogen nucleus). Heavy ions, up to lead (Pb), have also been accelerated. These **high-energy accelerators** have been used to probe more deeply into matter, to produce and study new particles, and to give us information about the basic forces and constituents of nature. The particles produced in high-energy collisions can be detected by a variety of special detectors, discussed in Section 30–13, including scintillation counters, bubble chambers, multiwire chambers, and semiconductors. The rate of production of

---

[†]Recall from Section 13–1 that the word "atom" comes from the Greek meaning "indivisible." Atoms have a substructure (protons, neutrons) so are not fundamental. Yet an atom is still the smallest "piece" of an element that has the characteristics of that material.

any group of particles is quantified using the concept of *cross section*, Section 31–1. Because the projectile particles are at high energy, this field is sometimes called **high-energy physics**.

Wavelength and Resolution

Particles accelerated to high energy can probe the interior of nuclei and nucleons or other particles they strike. An important factor is that faster-moving projectiles can reveal more detail. The wavelength of projectile particles is given by de Broglie's wavelength formula (Eq. 27–8),

$$\lambda = \frac{h}{p}, \tag{32-1}$$

showing that the greater the momentum p of the bombarding particle, the shorter its wavelength. As discussed in Chapter 25 on optical instruments, resolution of details in images is limited by the wavelength: the shorter the wavelength, the finer the detail that can be obtained. This is one reason why particle accelerators of higher and higher energy have been built in recent years: to probe ever deeper into the structure of matter, to smaller and smaller size.

EXAMPLE 32–1 **High resolution with electrons.** What is the wavelength, and hence the expected resolution, for 1.3-GeV electrons?

APPROACH Because 1.3 GeV is much larger than the electron mass, we must be dealing with relativistic speeds. The momentum of the electrons is found from Eq. 26–9, and the wavelength is $\lambda = h/p$.

SOLUTION Each electron has $KE = 1.3 \text{ GeV} = 1300 \text{ MeV}$, which is about 2500 times the rest energy of the electron ($mc^2 = 0.51 \text{ MeV}$). Thus we can ignore the term $(mc^2)^2$ in Eq. 26–9, $E^2 = p^2c^2 + m^2c^4$, and we solve for p:

$$p = \sqrt{\frac{E^2 - m^2c^4}{c^2}} \approx \sqrt{\frac{E^2}{c^2}} = \frac{E}{c}.$$

Therefore the de Broglie wavelength is

$$\lambda = \frac{h}{p} = \frac{hc}{E},$$

where $E = 1.3 \text{ GeV}$. Hence

$$\lambda = \frac{(6.63 \times 10^{-34} \text{ J} \cdot \text{s})(3.0 \times 10^8 \text{ m/s})}{(1.3 \times 10^9 \text{ eV})(1.6 \times 10^{-19} \text{ J/eV})} = 0.96 \times 10^{-15} \text{ m},$$

or 0.96 fm. This resolution of about 1 fm is on the order of the size of nuclei (see Eq. 30–1).

NOTE The maximum possible resolution of this beam of electrons is far greater than for a light beam in a light microscope ($\lambda \approx 500 \text{ nm}$).

EXERCISE A What is the wavelength of a proton with $KE = 1.00 \text{ TeV}$?

A major reason today for building high-energy accelerators is that new particles of greater mass can be produced at higher collision energies, transforming the kinetic energy of the colliding particles into massive particles by $E = mc^2$, as we will discuss shortly. Now we look at particle accelerators.

Cyclotron

The cyclotron was developed in 1930 by E. O. Lawrence (1901–1958; Fig. 32–1) at the University of California, Berkeley. It uses a magnetic field to maintain charged ions—usually protons—in nearly circular paths. Although particle physicists no longer use simple cyclotrons, they are used in medicine for treating cancer, and their operating principles are useful for understanding modern accelerators.

FIGURE 32–1 Ernest O. Lawrence, left, with Donald Cooksey and the "dees" of an early cyclotron.

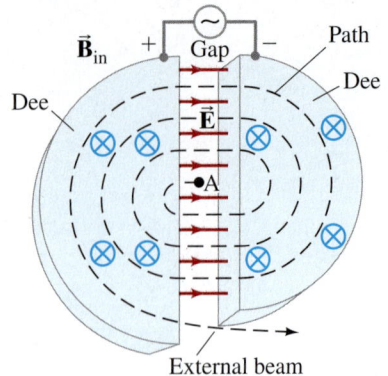

FIGURE 32–2 Diagram of a cyclotron. The magnetic field, applied by a large electromagnet, points into the page. The protons start at A, the ion source. The red electric field lines shown are for the alternating electric field in the gap at a certain moment.

The protons move in a vacuum inside two D-shaped cavities, as shown in Fig. 32–2. Each time they pass into the gap between the "dees," a voltage accelerates them (the electric force), increasing their speed and increasing the radius of curvature of their path in the magnetic field. After many revolutions, the protons acquire high kinetic energy and reach the outer edge of the cyclotron where they strike a target. The protons speed up only when they are in the gap *between* the dees, and the voltage must be alternating. When protons are moving to the right across the gap in Fig. 32–2, the right dee must be electrically negative and the left one positive. A half-cycle later, the protons are moving to the left, so the left dee must be negative in order to accelerate them.

The frequency, f, of the applied voltage must be equal to the frequency of the circulating protons. When ions of charge q are circulating *within* the hollow dees, the net force F on each ion is due to the magnetic field B, so $F = qvB$, where v is the speed of the ion at a given moment (Eq. 20–4). The magnetic force is perpendicular to both $\vec{v}$ and $\vec{B}$, and does not speed up the ions but causes them to move in circles; the acceleration within the dees is centripetal and equals v^2/r, where r is the radius of the ion's path at a given moment. We use Newton's second law, $F = ma$, and find that

$$F = ma$$
$$qvB = \frac{mv^2}{r}$$

when the protons are within the dees (not the gap), so their (constant) speed at radius r is

$$v = \frac{qBr}{m}.$$

The time required for a complete revolution is the period T and is equal to

$$T = \frac{\text{distance}}{\text{speed}} = \frac{2\pi r}{qBr/m} = \frac{2\pi m}{qB}.$$

Hence the frequency of revolution f is

$$f = \frac{1}{T} = \frac{qB}{2\pi m}. \tag{32–2}$$

This is known as the **cyclotron frequency**.

EXAMPLE 32–2 Cyclotron. A small cyclotron of maximum radius $R = 0.25$ m accelerates protons in a 1.7 T magnetic field. Calculate (a) the frequency needed for the applied alternating voltage, and (b) the kinetic energy of protons when they leave the cyclotron.

APPROACH The frequency of the protons revolving within the dees (Eq. 32–2) must equal the frequency of the voltage applied across the gap if the protons are going to increase in speed.

SOLUTION (a) From Eq. 32–2,

$$f = \frac{qB}{2\pi m} = \frac{(1.6 \times 10^{-19}\text{ C})(1.7\text{ T})}{(6.28)(1.67 \times 10^{-27}\text{ kg})} = 2.6 \times 10^7\text{ Hz} = 26\text{ MHz},$$

which is in the radio-wave region of the EM spectrum (Fig. 22–8).
(b) The protons leave the cyclotron at $r = R = 0.25$ m. From $qvB = mv^2/r$ (see above), we have $v = qBr/m$, so their kinetic energy is

$$\text{KE} = \frac{1}{2}mv^2 = \frac{1}{2}m\frac{q^2B^2R^2}{m^2} = \frac{q^2B^2R^2}{2m}$$

$$= \frac{(1.6 \times 10^{-19}\text{ C})^2(1.7\text{ T})^2(0.25\text{ m})^2}{(2)(1.67 \times 10^{-27}\text{ kg})} = 1.4 \times 10^{-12}\text{ J} = 8.7\text{ MeV}.$$

NOTE The kinetic energy is much less than the rest energy of the proton (938 MeV), so relativity is not needed.

NOTE The magnitude of the voltage applied to the dees does not appear in the formula for KE, and so does not affect the final energy. But the higher this voltage, the fewer the revolutions required to bring the protons to full energy.

An important aspect of the cyclotron is that the frequency of the applied voltage, as given by Eq. 32–2, does not depend on the radius r of the particle's path. Thus the frequency does not have to be changed as the protons or ions start from the source and are accelerated to paths of larger and larger radii. But this is only true at nonrelativistic energies. At higher speeds, the momentum (Eq. 26–4) is $p = \gamma mv = mv/\sqrt{1 - v^2/c^2}$, so m in Eq. 32–2 has to be replaced by γm and the cyclotron frequency f (Eq. 32–2) depends on speed v because γ does. To keep the particles in sync, machines called **synchrocyclotrons** reduce the frequency in time to correspond to the increase of γm (in Eq. 32–2) as a packet of charged particles increases in speed more slowly at larger orbits.

Synchrotron

Another way to accelerate relativistic particles is to increase the magnetic field B in time so as to keep f (Eq. 32–2) constant as the particles speed up. Such devices are called **synchrotrons**; the particles move in a circle of fixed radius, which can be very large. The larger the radius, the greater the KE of the particles can be for a given magnetic field strength (see argument on previous page). The biggest synchrotron of all is at the European Center for Nuclear Research (CERN) in Geneva, Switzerland, the Large Hadron Collider (LHC). It is 4.3 km in radius, and 27 km in circumference, and accelerates protons to 4 TeV (soon to be 7 TeV).

The *Tevatron* accelerator at Fermilab (Fermi National Accelerator Laboratory, near Chicago, Illinois, has a radius of 1.0 km.[†] The Tevatron accelerated protons to about 1000 GeV = 1 TeV (hence its name, 1 TeV = 10^{12} eV). It was shut down in 2011.

FIGURE 32–3 The interior of the tunnel of the main accelerator at Fermilab, showing (red) the ring of superconducting magnets used to keep particles moving in a circular path at the 1-TeV Tevatron.

These large synchrotrons use a narrow ring of magnets (see Fig. 32–3) with each magnet placed at the same radius from the center of the circle. The magnets are interrupted by gaps where high voltage accelerates the particles to higher speeds. Another way to describe the acceleration is to say the particles "surf" on a traveling electromagnetic wave within radiofrequency (RF) cavities. (The particles are first given considerable energy in smaller accelerators, "injectors," before being injected into the large ring of the large synchrotron.)

One problem of any accelerator is that accelerating electric charges radiate electromagnetic energy (see Chapter 22). Since ions or electrons are accelerated in an accelerator, we can expect considerable energy to be lost by radiation. The effect increases with energy and is especially important in circular machines where centripetal acceleration is present, such as synchrotrons, and hence is called **synchrotron radiation**. Synchrotron radiation can be useful, however. Intense beams of photons (γ rays) are sometimes needed, and they are often obtained from an electron synchrotron. Strong sources of such photons are referred to as **light sources**.

[†]Robert Wilson, who helped design the Tevatron, and founded the field of proton therapy (Section 31–6), expressed his views on accelerators and national security in this exchange with Senator John Pastore during testimony before a Congressional Committee in 1969:

> Pastore: "Is there anything connected with the hopes of this accelerator [the Tevatron] that in any way involves the security of the country?"
>
> Robert Wilson: "No sir, I don't believe so."
>
> Pastore: "Nothing at all?"
>
> Wilson: "Nothing at all. ... "
>
> Pastore: "It has no value in that respect?"
>
> Wilson: "It has only to do with the respect with which we regard one another, the dignity of men, our love of culture. ... It has to do with are we good painters, good sculptors, great poets? I mean all the things we really venerate in our country and are patriotic about ... it has nothing to do directly with defending our country except to make it worth defending."

Linear Accelerators

In a **linear accelerator** (linac), electrons or ions are accelerated along a straight-line path, Fig. 32–4, passing through a series of tubular conductors. Voltage applied to the tubes is alternating so that when electrons (say) reach a gap, the tube in front of them is positive and the one they just left is negative. At low speeds, the particles cover less distance in the same amount of time, so the tubes are shorter at first. Electrons, with their small mass, get close to the speed of light quickly, $v \approx c$, and the tubes are nearly equal in length. Linear accelerators are particularly important for accelerating electrons to avoid loss of energy due to synchrotron radiation. The largest electron linear accelerator has been at Stanford University (Stanford Linear Accelerator Center, or SLAC), about 3 km (2 mi) long, accelerating electrons to 50 GeV. Linacs accelerating protons are used as injectors into circular machines to provide initial kinetic energy. Many hospitals have 10-MeV electron linacs that strike a metal foil to produce γ ray photons to irradiate tumors.

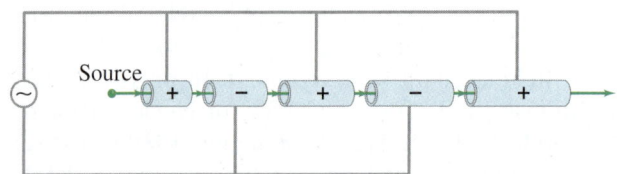

FIGURE 32–4 Diagram of a simple linear accelerator.

Colliding Beams

High-energy physics experiments were once done by aiming a beam of particles from an accelerator at a stationary target. But to obtain the maximum possible collision energy from a given accelerator, two beams of particles are now accelerated to very high energy and are steered so that they collide head-on. One way to accomplish such **colliding beams** with a single accelerator is through the use of **storage rings**, in which oppositely circulating beams can be repeatedly brought into collision with one another at particular points. For example, in the experiments that provided strong evidence for the top quark (Section 32–9 and Fig. 32–15), the Fermilab Tevatron accelerated protons and antiprotons each to 900 GeV, so that the combined energy of head-on collisions was 1.8 TeV.

The largest collider is the Large Hadron Collider (LHC) at CERN, with a circumference of 26.7 km (Fig. 32–5). The two colliding beams are designed to each carry 7-TeV protons for a total interaction energy of 14 TeV. For the experiments in 2011 and 2012 the total interaction energy was 7 TeV and 8 TeV. The protons for each of the beams, moving in opposite directions, are accelerated in several stages. The penultimate is SPS (Super Proton Synchrotron), seen in Fig. 32–5a which accelerates protons from 28 GeV to the 450 GeV at which they are injected into the LHC itself.

FIGURE 32–5 (a) The large circle represents the position of the tunnel, about 100 m below the ground at CERN (near Geneva) on the French–Swiss border, which houses the LHC. The smaller circle shows the position of the Super Proton Synchrotron used for accelerating protons prior to injection into the LHC. (b) Circulating proton beams, in opposite directions, inside the vacuum tube within the LHC tunnel.

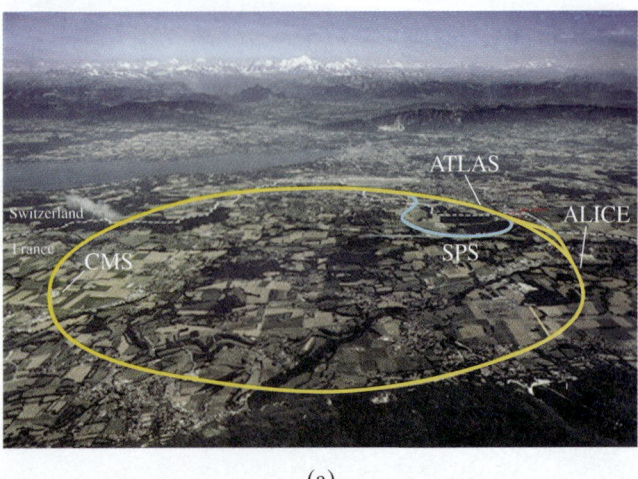

(a)

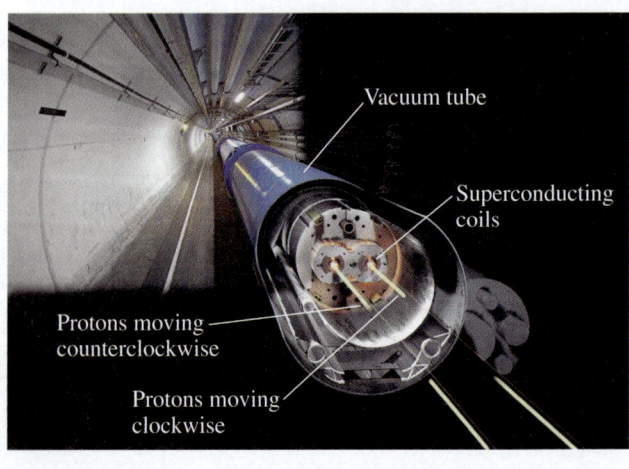

(b)

Figure 32–6 shows part of one of the detectors (ATLAS) as it was being constructed at the LHC. The detectors within ATLAS include silicon semiconductor detectors with huge numbers of pixels used to track particle paths and find their point of interaction, and to measure their radius of curvature in a magnetic field and thus determine their momentum (Section 20–4). Their energy is determined in "calorimeters" utilizing plastic, liquid, or dense metal compound crystal scintillators (Section 30–13).

In the planning stage is the International Linear Collider (ILC) which would have colliding beams of e^- and e^+ at around 0.3 to 1 TeV. It would utilize semiconductor detectors using CMOS (Section 25–1) with embedded transistors to allow fast readout.

FIGURE 32–6 ATLAS, one of the large complex detectors at the LHC, is shown here as it was being built. In 2012 it was used to provide evidence for the Higgs boson. Note the people near the bottom. From the outside, the CMS detector at the LHC looks similar.

EXAMPLE 32–3 **Protons at relativistic speeds.** Determine the energy required to accelerate a proton in a high-energy accelerator (*a*) from rest to $v = 0.900c$, and (*b*) from $v = 0.900c$ to $v = 0.999c$. (*c*) What is the kinetic energy achieved by the proton in each case?

APPROACH We use the work-energy principle, which is still valid relativistically as mentioned in Section 26–9: $W = \Delta KE$.

SOLUTION The kinetic energy of a proton of mass m is given by Eq. 26–5,

$$KE = (\gamma - 1)mc^2,$$

where the relativistic factor γ is

$$\gamma = \frac{1}{\sqrt{1 - v^2/c^2}}.$$

The work-energy theorem becomes

$$W = \Delta KE = (\gamma_2 - 1)mc^2 - (\gamma_1 - 1)mc^2 = (\gamma_2 - \gamma_1)mc^2$$

where γ_1 and γ_2 are for the initial and final speeds, $v_1 = 0$, $v_2 = 0.900c$.
(*a*) For $v = v_1 = 0$, $\gamma_1 = 1$; and for $v_2 = 0.900c$

$$\gamma_2 = \frac{1}{\sqrt{1 - (0.900)^2}} = 2.29.$$

For a proton, $mc^2 = 938$ MeV, so the work (or energy) needed to accelerate it from rest to $v_2 = 0.900c$ is

$$W = \Delta KE = (\gamma_2 - \gamma_1)mc^2$$
$$= (2.29 - 1.00)(938 \text{ MeV}) = 1.21 \text{ GeV}.$$

(*b*) To go from $v_2 = 0.900c$ to $v_3 = 0.999c$, we need

$$\gamma_3 = \frac{1}{\sqrt{1 - (0.999)^2}} = 22.4.$$

So the work needed to accelerate a proton from $0.900c$ to $0.999c$ is

$$W = \Delta KE = (\gamma_3 - \gamma_2)mc^2$$
$$= (22.4 - 2.29)(938 \text{ MeV}) = 18.9 \text{ GeV},$$

which is 15 times as much.
(*c*) The kinetic energy reached by the proton in (*a*) is just equal to the work done on it, $KE = 1.21$ GeV. The final kinetic energy of the proton in (*b*), moving at $v_3 = 0.999c$, is

$$KE = (\gamma_3 - 1)mc^2 = (21.4)(938 \text{ MeV}) = 20.1 \text{ GeV}.$$

NOTE This result makes sense because, starting from rest, we did work

$$W = 1.21 \text{ GeV} + 18.9 \text{ GeV} = 20.1 \text{ GeV}$$

on it.

32–2 Beginnings of Elementary Particle Physics—Particle Exchange

The accepted model for elementary particles today views *quarks* and *leptons* as the fundamental constituents of ordinary matter. To understand our present-day view of elementary particles, it is necessary to understand the ideas leading up to its formulation.

Elementary particle physics might be said to have begun in 1935 when the Japanese physicist Hideki Yukawa (1907–1981) predicted the existence of a new particle that would in some way mediate the strong nuclear force. To understand Yukawa's idea, we first consider the electromagnetic force. When we first discussed electricity, we saw that the electric force acts over a distance, without contact. To better perceive how a force can act over a distance, we used the idea of a **field**. The force that one charged particle exerts on a second can be said to be due to the electric field set up by the first. Similarly, the magnetic field can be said to carry the magnetic force. Later (Chapter 22), we saw that electromagnetic (EM) fields can travel through space as waves. Finally, in Chapter 27, we saw that electromagnetic radiation (light) can be considered as either a wave or as a collection of particles called *photons*. Because of this wave–particle duality, it is possible to imagine that the electromagnetic force between charged particles is due to

(1) the EM field set up by one charged particle and felt by the other, or

(2) an exchange of photons (γ particles) between them.

(a) Repulsive force (children throwing pillows)

(b) Attractive force (children grabbing pillows from each other's hands)

FIGURE 32–7 Forces equivalent to particle exchange. (a) Repulsive force (children on roller skates throwing pillows at each other). (b) Attractive force (children grabbing pillows from each other's hands).

It is (2) that we want to concentrate on here, and a crude analogy for how an exchange of particles could give rise to a force is suggested in Fig. 32–7. In part (a), two children start throwing heavy pillows at each other; each throw and each catch results in the child being pushed backward by the impulse. This is the equivalent of a repulsive force. On the other hand, if the two children exchange pillows by grabbing them out of the other person's hand, they will be pulled toward each other, as when an attractive force acts.

For the electromagnetic force, it is photons exchanged between two charged particles that give rise to the force between them. A simple diagram describing this photon exchange is shown in Fig. 32–8. Such a diagram, called a **Feynman diagram** after its inventor, the American physicist Richard Feynman (1918–1988), is based on the theory of **quantum electrodynamics** (QED).

FIGURE 32–8 Feynman diagram showing a photon acting as the carrier of the electromagnetic force between two electrons. This is sort of an *x* vs. *t* graph, with *t* increasing upward. Starting at the bottom, two electrons approach each other. As they get close, momentum and energy get transferred from one to the other, carried by a photon (or more than one), and the two electrons bounce apart.

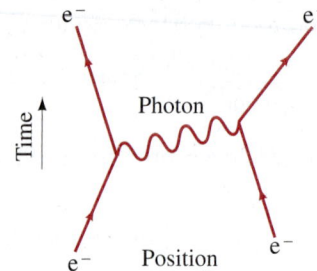

Figure 32–8 represents the simplest case in QED, in which a single photon is exchanged. One of the charged particles emits the photon and recoils somewhat as a result; and the second particle absorbs the photon. In such a collision or *interaction*, energy and momentum are transferred from one charged particle to the other, carried by the photon. The photon is absorbed by the second particle after it is emitted by the first particle and is not observable. Hence the photon is referred to as a **virtual** photon, in contrast to one that is free and can be detected by instruments. The photon is said to **mediate**, or **carry**, the electromagnetic force.

By analogy with photon exchange that mediates the electromagnetic force, Yukawa argued that there ought to be a particle that mediates the strong nuclear force—the force that holds nucleons together in the nucleus. Yukawa called this predicted particle a **meson** (meaning "medium mass"). Figure 32–9 is a Feynman diagram showing the original model of meson exchange: a meson carrying the strong force between a neutron and a proton.

A rough estimate of the mass of the meson can be made as follows. Suppose the proton on the left in Fig. 32–9 is at rest. For it to emit a meson would require energy (to make the meson's mass) which, coming from nowhere, would violate conservation of energy. But the uncertainty principle allows nonconservation of energy by an amount ΔE if it occurs only for a time Δt given by $(\Delta E)(\Delta t) \approx h/2\pi$. We set ΔE equal to the energy needed to create the mass m of the meson: $\Delta E = mc^2$. Conservation of energy is violated only as long as the meson exists, which is the time Δt required for the meson to pass from one nucleon to the other, where it is absorbed and disappears. If we assume the meson travels at relativistic speed, close to the speed of light c, then Δt need be at most about $\Delta t = d/c$, where d is the maximum distance that can separate the interacting nucleons. Thus we can write

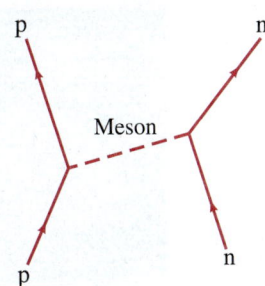

FIGURE 32–9 Early model showing meson exchange when a proton and neutron interact via the strong nuclear force. (Today, as we shall see shortly, we view the strong force as carried by gluons between quarks.)

$$\Delta E \, \Delta t \approx \frac{h}{2\pi}$$

$$mc^2\left(\frac{d}{c}\right) \approx \frac{h}{2\pi}$$

or

$$mc^2 \approx \frac{hc}{2\pi d}. \quad (32\text{–}3)$$

The range of the strong nuclear force (the maximum distance away it can be felt) is small—not much more than the size of a nucleon or small nucleus (see Eq. 30–1)—so let us take $d \approx 1.5 \times 10^{-15}$ m. Then from Eq. 32–3,

$$mc^2 \approx \frac{hc}{2\pi d} = \frac{(6.6 \times 10^{-34}\,\text{J}\cdot\text{s})(3.0 \times 10^8\,\text{m/s})}{(6.28)(1.5 \times 10^{-15}\,\text{m})} \approx 2.1 \times 10^{-11}\,\text{J} = 130\,\text{MeV}.$$

The mass of the predicted meson, roughly $130\,\text{MeV}/c^2$, is about 250 times the electron mass of $0.51\,\text{MeV}/c^2$.

EXERCISE B What effect does an increase in the mass of the virtual exchange particle have on the range of the force it mediates? (*a*) Decreases it; (*b*) increases it; (*c*) has no appreciable effect; (*d*) decreases the range for charged particles and increases the range for neutral particles.

Note that since the electromagnetic force has infinite range, Eq. 32–3 with $d = \infty$ tells us that the exchanged particle for the electromagnetic force, the photon, will have zero mass, which it does.

The particle predicted by Yukawa was discovered in cosmic rays by C. F. Powell and G. Occhialini in 1947, and is called the "π" or pi meson, or simply the **pion**. It comes in three charge states: $+e$, $-e$, or 0, where $e = 1.6 \times 10^{-19}$ C. The π^+ and π^- have mass of $139.6\,\text{MeV}/c^2$ and the π^0 a mass of $135.0\,\text{MeV}/c^2$, all close to Yukawa's prediction. All three interact strongly with matter. Reactions observed in the laboratory, using a particle accelerator, include

$$p + p \rightarrow p + p + \pi^0,$$
$$p + p \rightarrow p + n + \pi^+. \quad (32\text{–}4)$$

The incident proton from the accelerator must have sufficient energy to produce the additional mass of the free pion.

Yukawa's theory of pion exchange as carrier of the strong force has been superseded by *quantum chromodynamics* in which protons, neutrons, and other strongly interacting particles are made up of basic entities called *quarks*, and the basic carriers of the strong force are *gluons*, as we shall discuss shortly. But the basic idea of the earlier theory, that forces can be understood as the exchange of particles, remains valid.

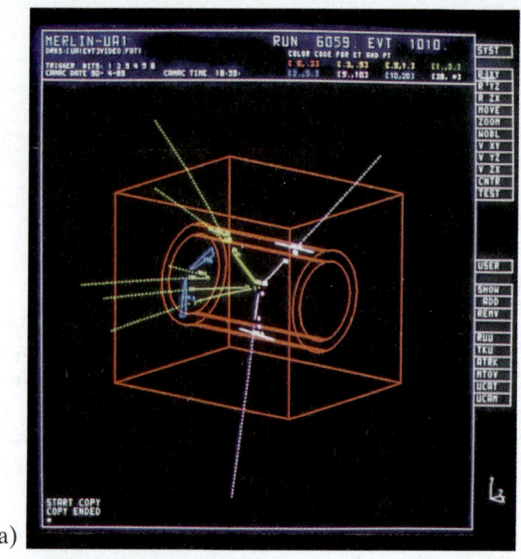

FIGURE 32–10 (a) Computer reconstruction of a Z-particle decay into an electron and a positron ($Z^0 \rightarrow e^+ + e^-$) whose tracks are shown in white, which took place in the UA1 detector at CERN. (b) Photo of the UA1 detector at CERN as it was being built. 1980s.

There are four known types of force—or interactions—in nature. The electromagnetic force is carried by the photon, the strong force by gluons. What about the other two: the weak force and gravity? These too are believed to be mediated by particles. The particles that transmit the weak force are referred to as the W^+, W^-, and Z^0, and were detected in 1983 (Fig. 32–10). The quantum (or carrier) of the gravitational force has been named the **graviton**, but its existence has not been detected and it may not be detectable.

A comparison of the four forces is given in Table 32–1, where they are listed according to their (approximate) relative strengths. Although gravity may be the most obvious force in daily life (because of the huge mass of the Earth), on a nuclear scale gravity is by far the weakest of the four forces and its effect at the particle level can nearly always be ignored.

TABLE 32–1 The Four Forces in Nature

Type	Relative Strength (approx., for 2 protons in nucleus)	Field Particle
Strong	1	Gluons (= g)
Electromagnetic	10^{-2}	Photon (= γ)
Weak	10^{-6}	$W^\pm$ and Z^0
Gravitational	10^{-38}	Graviton (?)

32–3 Particles and Antiparticles

The positron, as we discussed in Sections 27–6 (pair production) and 30–5 (β^+ decay), is basically a positive electron. That is, many of its properties are the same as for the electron, such as mass, but it has the opposite electric charge ($+e$). Other quantum numbers that we will discuss shortly are also reversed. The positron is said to be the **antiparticle** to the electron.

The positron was first detected as a curved path in a cloud chamber in a magnetic field by Carl Anderson in 1932. It was predicted that other particles also would have antiparticles. It was decades before another type was found.

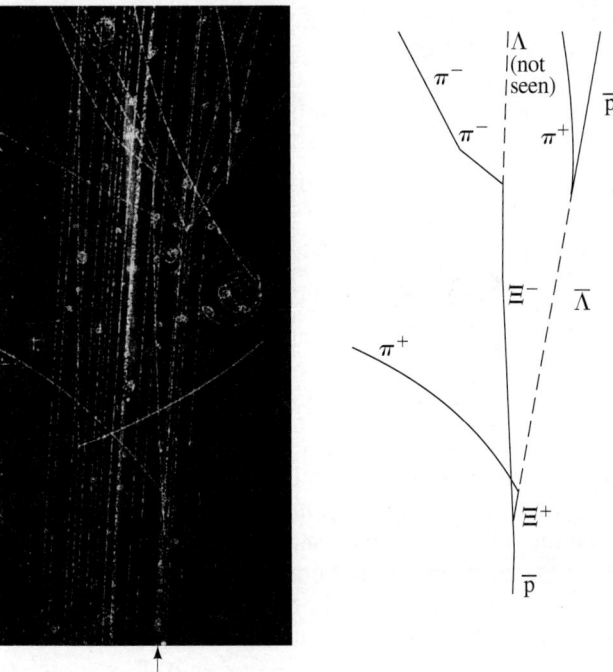

FIGURE 32–11 Liquid-hydrogen bubble-chamber photograph of an antiproton ($\bar{p}$) colliding with a proton at rest, producing a Xi—anti-Xi pair ($\bar{p} + p \rightarrow \Xi^- + \bar{\Xi}^+$) that subsequently decay into other particles. The drawing indicates the assignment of particles to each track, which is based on how or if that particle decays, and on mass values estimated from measurement of momentum (curvature of track in magnetic field) and energy (thickness of track, for example). Neutral particle paths are shown by dashed lines since neutral particles rarely ionize atoms, around which bubbles form, and hence leave no tracks. 1950s.

Finally, in 1955 the antiparticle to the proton, the **antiproton** ($\bar{p}$), which carries a negative charge (Fig. 32–11), was discovered at the University of California, Berkeley, by Emilio Segrè (1905–1989, Fig. 32–12) and Owen Chamberlain (1920–2006). A bar, such as over the p, is used to indicate the antiparticle ($\bar{p}$). Soon after, the antineutron ($\bar{n}$) was found. All particles have antiparticles. But a few, like the photon, the π^0, and the Higgs, do not have distinct antiparticles—we say that they are their own antiparticles.

Antiparticles are produced in nuclear reactions when there is sufficient energy available to produce the required mass, and they do not live very long in the presence of matter. For example, a positron is stable when by itself but rarely survives for long; as soon as it encounters an electron, the two annihilate each other. The energy of their vanished mass, plus any kinetic energy they possessed, is usually converted into the energy of two γ rays. Annihilation also occurs for all other particle–antiparticle pairs.

Antimatter is a term referring to material that would be made up of "antiatoms" in which antiprotons and antineutrons would form the nucleus around which positrons (antielectrons) would move. The term is also used for antiparticles in general. If there were pockets of antimatter in the universe, a huge explosion would occur if it should encounter normal matter. It is believed that antimatter was prevalent in the very early universe (Section 33–7).

*Negative Sea of Electrons; Vacuum State

The original idea for antiparticles came from a relativistic wave equation developed in 1928 by the Englishman P. A. M. Dirac (1902–1984). Recall that, as we saw in Chapter 26, the total energy E of a particle with mass m and momentum p and zero potential energy is given by Eq. 26–9, $E^2 = p^2c^2 + m^2c^4$. Thus

$$E = \pm\sqrt{p^2c^2 + m^2c^4}.$$

Dirac applied his new equation and found that it included solutions with both + and − signs. He could not ignore the solution with the negative sign, which we might have thought unphysical. If those negative energy states are real, then we would expect normal free electrons to drop down into those states, emitting

FIGURE 32–12 Emilio Segrè: he worked with Fermi in the 1930s, later discovered the first "man-made" element, technetium, and other elements, and then the antiproton. The inscription below the photo is from a book by Segrè given to this book's author.

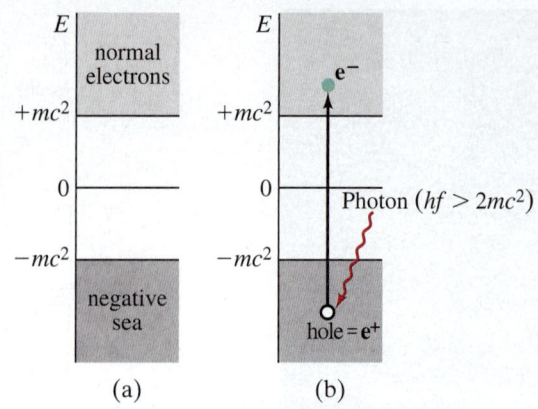

FIGURE 32–13 (a) Possible energy states for an electron. Note the vast sea of fully occupied electron states at $E < -mc^2$. (b) An electron in the negative sea is hit by a photon $(E > 2mc^2)$ and knocks it up to a normal positive energy state. The positive "hole" left behind acts like a positive electron—it is a positron.

photons—never experimentally seen. To deal with this difficulty, Dirac postulated that all those negative energy states are *normally occupied*. That is, what we thought was the **vacuum** is instead a vast **sea of electrons** in negative energy states (Fig. 32–13a). These electrons are not normally observable. But if a photon strikes one of these negative energy electrons, that electron can be knocked up to a normal $(E > mc^2)$ energy state as shown in Fig. 32–13b. Note in Fig. 32–13 that there are no energy states between $E = -mc^2$ and $E = +mc^2$ (because p^2 cannot be negative in the equation $E = \pm\sqrt{p^2c^2 + m^2c^4}$). The photon that knocks an e^- from the negative sea up to a normal state (Fig. 32–13b) must have an energy greater than $2mc^2$. What is left behind is a hole (as in semiconductors, Sections 29–7 and 29–8) with positive charge. We call that "hole" a **positron**, and it can move around as a free particle with positive energy. Thus Fig. 32–13b represents (Section 27–6) **pair production**: $\gamma \rightarrow e^-e^+$.

The vast sea of electrons with negative energy in Fig. 32–13 is the vacuum (or **vacuum state**). According to quantum mechanics, the vacuum is not empty, but contains electrons and other particles as well. The uncertainty principle allows a particle to jump briefly up to a normal energy, thus creating a **particle–antiparticle** pair. It is possible that they could be the source of the recently discovered *dark energy* that fills the universe (Chapter 33). We still have a lot to learn.

32–4 Particle Interactions and Conservation Laws

One of the important uses of high-energy accelerators and colliders is to study the interactions of elementary particles with each other. As a means of ordering this subnuclear world, the conservation laws are indispensable. The laws of conservation of energy, of momentum, of angular momentum, and of electric charge are found to hold precisely in all particle interactions.

A study of particle interactions has revealed a number of new conservation laws which (just like the old ones) are ordering principles: they help to explain why some reactions occur and others do not. For example, the following reaction has never been observed:

$$p + n \not\rightarrow p + p + \bar{p}$$

even though charge, energy, and so on, are conserved ($\not\rightarrow$ means the reaction does not occur). To understand why such a reaction does not occur, physicists hypothesized a new conservation law, the conservation of **baryon number**. (Baryon number is a generalization of nucleon number, which we saw earlier is conserved in nuclear reactions and decays.) All nucleons are defined to have baryon number $B = +1$, and all antinucleons (antiprotons, antineutrons) have $B = -1$. All other types of particles, such as photons, mesons, and electrons and other leptons, have $B = 0$. The reaction shown at the start of this paragraph does not conserve baryon number since the left side has $B = (+1) + (+1) = +2$, and the right has $B = (+1) + (+1) + (-1) = +1$. On the other hand, the following reaction does conserve B and *does* occur if the incoming proton has sufficient energy:

$$p + p \rightarrow p + p + \bar{p} + p,$$
$$B = +1 + 1 = +1 + 1 - 1 + 1.$$

As indicated, $B = +2$ on both sides of this equation. From these and other reactions, the **conservation of baryon number** has been established as a basic principle of physics.

Also useful are conservation laws for the three **lepton numbers**, associated with weak interactions including decays. In ordinary β decay, an electron or positron is emitted along with a neutrino or antineutrino. In another type of decay, a particle known as a "μ" or **muon**, can be emitted instead of an electron. The muon (discovered in 1937) seems to be much like an electron, except its mass is 207 times larger $(106 \text{ MeV}/c^2)$. The neutrino (ν_e) that accompanies an emitted electron is found to be different from the neutrino (ν_μ) that accompanies an emitted muon. Each of these neutrinos has an antiparticle: $\bar{\nu}_e$ and $\bar{\nu}_\mu$. In ordinary β decay we have, for example,

$$n \rightarrow p + e^- + \bar{\nu}_e$$

but not $n \not\rightarrow p + e^- + \bar{\nu}_\mu$. To explain why these do not occur, the concept of **electron lepton number**, L_e, was invented. If the electron (e^-) and the electron neutrino (ν_e) are assigned $L_e = +1$, and e^+ and $\bar{\nu}_e$ are assigned $L_e = -1$, whereas all other particles have $L_e = 0$, then all observed decays conserve L_e. For example, in $n \rightarrow p + e^- + \bar{\nu}_e$, initially $L_e = 0$, and afterward $L_e = 0 + (+1) + (-1) = 0$. Decays that do not conserve L_e, even though they would obey the other conservation laws, are not observed to occur.

In a decay involving muons, such as

$$\pi^+ \rightarrow \mu^+ + \nu_\mu,$$

a second quantum number, **muon lepton number** (L_μ), is conserved. The μ^- and ν_μ are assigned $L_\mu = +1$, and their antiparticles μ^+ and $\bar{\nu}_\mu$ have $L_\mu = -1$, whereas all other particles have $L_\mu = 0$. L_μ too is conserved in interactions and decays. Similar assignments can be made for the **tau lepton number**, L_τ, associated with the τ lepton (discovered in 1976 with mass more than 3000 times the electron mass) and its neutrino, ν_τ.

Antiparticles have not only opposite electric charge from their particles, but also opposite B, L_e, L_μ, and L_τ. For example, a neutron has $B = +1$, an antineutron has $B = -1$ (and all the L's are zero).

CAUTION

The different types of neutrinos are not identical

CONCEPTUAL EXAMPLE 32–4 **Lepton number in muon decay.** Which of the following decay schemes is possible for muon decay: (a) $\mu^- \rightarrow e^- + \bar{\nu}_e$; (b) $\mu^- \rightarrow e^- + \bar{\nu}_e + \nu_\mu$; (c) $\mu^- \rightarrow e^- + \nu_e$? All of these particles have $L_\tau = 0$.

RESPONSE A μ^- has $L_\mu = +1$ and $L_e = 0$. This is the initial state for all decays given, and the final state must also have $L_\mu = +1$, $L_e = 0$. In (a), the final state has $L_\mu = 0 + 0 = 0$, and $L_e = +1 - 1 = 0$; L_μ would not be conserved and indeed this decay is not observed to occur. The final state of (b) has $L_\mu = 0 + 0 + 1 = +1$ and $L_e = +1 - 1 + 0 = 0$, so both L_μ and L_e are conserved. This is in fact the most common decay mode of the μ^-. Lastly, (c) does not occur because L_e $(= +2$ in the final state) is not conserved, nor is L_μ.

EXAMPLE 32–5 **Energy and momentum are conserved.** In addition to the "number" conservation laws which help explain the decay schemes of particles, we can also apply the laws of conservation of energy and momentum. The decay of a Σ^+ particle at rest with mass $1189 \text{ MeV}/c^2$ (Table 32–2 in Section 32–6) can yield a proton ($m_p = 938 \text{ MeV}/c^2$) and a neutral pion, π^0 ($m_{\pi^0} = 135 \text{ MeV}/c^2$):
$$\Sigma^+ \to p + \pi^0.$$
Determine the kinetic energies of the proton and π^0.

APPROACH We find the energy release from the change in mass ($E = mc^2$) as we did for nuclear processes (Eq. 30–2 or 31–2a). Then we apply conservation of energy and momentum, using relativistic formulas as the energies are large.

SOLUTION The energy released, $Q = \text{KE}_p + \text{KE}_{\pi^0}$, is the change in mass $\times c^2$:
$$Q = [m_{\Sigma^+} - (m_p + m_{\pi^0})]c^2 = [1189 - (938 + 135)] \text{ MeV} = 116 \text{ MeV}.$$
Next we apply conservation of momentum: the initial particle Σ^+ is at rest, so the π^0 and p have opposite momentum but are equal in magnitude: $p_{\pi^0} = p_p$. We square this equation, $p_{\pi^0}^2 = p_p^2$, which becomes, using Eq. 26–9 ($p^2c^2 = E^2 - m^2c^4$),
$$E_{\pi^0}^2 - m_{\pi^0}^2 c^4 = E_p^2 - m_p^2 c^4.$$
Solving for $E_{\pi^0}^2$:
$$E_{\pi^0}^2 = E_p^2 - m_p^2 c^4 + m_{\pi^0}^2 c^4.$$
We substitute Eq. 26–6a, $E = \text{KE} + mc^2$, for both the π^0 and the p:
$$(\text{KE}_{\pi^0} + m_{\pi^0} c^2)^2 = (\text{KE}_p + m_p c^2)^2 - m_p^2 c^4 + m_{\pi^0}^2 c^4$$
$$\text{KE}_{\pi^0}^2 + 2\text{KE}_{\pi^0} m_{\pi^0} c^2 + \cancel{m_{\pi^0}^2 c^4} = \text{KE}_p^2 + 2\text{KE}_p m_p c^2 + \cancel{m_p^2 c^4} - \cancel{m_p^2 c^4} + \cancel{m_{\pi^0}^2 c^4}.$$
Next (after cancelling as shown) we substitute $\text{KE}_p = Q - \text{KE}_{\pi^0}$:
$$\cancel{\text{KE}_{\pi^0}^2} + 2\text{KE}_{\pi^0} m_{\pi^0} c^2 = Q^2 - 2Q\text{KE}_{\pi^0} + \cancel{\text{KE}_{\pi^0}^2} + 2Qm_p c^2 - 2\text{KE}_{\pi^0} m_p c^2.$$
After cancelling as shown, we solve for KE_{π^0}:
$$\text{KE}_{\pi^0} = \frac{Q^2 + 2Qm_p c^2}{2m_{\pi^0} c^2 + 2Q + 2m_p c^2} = \frac{(116 \text{ MeV})^2 + 2(116 \text{ MeV})(938 \text{ MeV})}{2(135 \text{ MeV}) + 2(116 \text{ MeV}) + 2(938 \text{ MeV})}$$
which gives $\text{KE}_{\pi^0} = 97 \text{ MeV}$. Then $\text{KE}_p = 116 \text{ MeV} - 97 \text{ MeV} = 19 \text{ MeV}$.

32–5 Neutrinos

We first met neutrinos with regard to β decay in Section 30–5. The study of neutrinos is a "hot" subject today. Experiments are being carried out in deep underground laboratories, sometimes in deep mine shafts. The thick layer of earth above is meant to filter out all other "background" particles, leaving mainly the very weakly interacting neutrinos to arrive at the detectors.

Some very important results have come to the fore in recent years. First there was the **solar neutrino problem**. The energy output of the Sun is believed to be due to the nuclear fusion reactions discussed in Chapter 31, Eqs. 31–6 and 31–7. The neutrinos emitted in these reactions are all ν_e (accompanied by e^+). But the rate at which ν_e arrive at Earth was measured starting in the late 1960s to be much less than expected based on the power output of the Sun. It was then proposed that, on the long trip between Sun and Earth, ν_e might turn into ν_μ or ν_τ. Subsequent experiments, definitive only in 2001, confirmed this hypothesis. Thus the three neutrinos, ν_e, ν_μ, ν_τ, can change into one another in certain circumstances, a phenomenon called **neutrino flavor oscillation**[†]. (Each of the three neutrino types is called, whimsically, a different "flavor.") This result suggests that the lepton numbers L_e, L_μ, and L_τ are not perfectly conserved. But the sum, $L_e + L_\mu + L_\tau$, is believed to be always conserved.

[†]Neutrino oscillations had first been proposed in 1957 by Bruno Pontecorvo. He also proposed that the electron and muon neutrinos are different species; and he also suggested a way to confirm the existence of neutrinos by detecting $\bar{\nu}_e$ emitted in huge numbers by a nuclear reactor, an experiment carried out by Frederick Reines and Clyde Cowan in the 1950s. The experimentalists who confirmed these two predictions were awarded the Nobel Prize, but not the theorist who proposed them.

The second exceptional result has long been speculated on: are neutrinos massless as originally thought, or do they have a nonzero mass? Rough upper limits on the masses have been made. Today astrophysical experiments show that the sum of all three neutrino masses combined is less than about $0.14\,\text{eV}/c^2$. But can all the masses be zero? Not if there are the flavor oscillations discussed above. It seems that at most, one type could have zero mass, and it is likely that at least one neutrino type has a mass of at least $0.04\,\text{eV}/c^2$.

As a result of neutrino oscillations, the three types of neutrino may not be exactly what we thought they were (e, μ, τ). If not, the three basic neutrinos, called 1, 2, and 3, are combinations of ν_e, ν_μ, and ν_τ.

Another outstanding question is whether or not neutrinos are in the category called **Majorana particles**,[†] meaning they would be their own antiparticles, like γ, π^0, and Higgs. If so, a lot of other questions (and answers) would appear.

*Neutrino Mass Estimate from a Supernova

The explosion of a supernova in the outer parts of our Galaxy in 1987 (Section 33–2) released lots of neutrinos and offered an opportunity to estimate electron neutrino mass. If neutrinos do have mass, then their speed would be less than c, and neutrinos of different energy would take different times to travel the 170,000 light-years from the supernova to Earth. To get an idea of how such a measurement could be done, suppose two neutrinos from "SN1987A" were emitted at the same time and were actually detected on Earth (via the reaction $\bar\nu_e + p \to n + e^+$) 10 seconds apart, with measured kinetic energies of about 20 MeV and 10 MeV. From other laboratory measurements we expect the neutrino mass to be less than 100 eV; and since our neutrinos have kinetic energy of 20 MeV and 10 MeV, we can make the approximation $m_\nu c^2 \ll E$, so that E (the total energy) is essentially equal to the kinetic energy. We use Eq. 26–6b, which tells us

$$E = \frac{m_\nu c^2}{\sqrt{1 - v^2/c^2}}.$$

We solve this for v, the velocity of a neutrino with energy E:

$$v = c\left(1 - \frac{m_\nu^2 c^4}{E^2}\right)^{\frac{1}{2}} = c\left(1 - \frac{m_\nu^2 c^4}{2E^2} + \cdots\right),$$

where we have used the binomial expansion $(1-x)^{\frac{1}{2}} = 1 - \frac{1}{2}x + \cdots$, and we ignore higher-order terms since $m_\nu^2 c^4 \ll E^2$. The time t for a neutrino to travel a distance d ($= 170{,}000$ ly) is

$$t = \frac{d}{v} = \frac{d}{c\left(1 - \frac{m_\nu^2 c^4}{2E^2}\right)} \approx \frac{d}{c}\left(1 + \frac{m_\nu^2 c^4}{2E^2}\right),$$

where again we used the binomial expansion $[(1-x)^{-1} = 1 + x + \cdots]$. The difference in arrival times for our two neutrinos of energies $E_1 = 20\,\text{MeV}$ and $E_2 = 10\,\text{MeV}$ is

$$t_2 - t_1 = \frac{d}{c}\frac{m_\nu^2 c^4}{2}\left(\frac{1}{E_2^2} - \frac{1}{E_1^2}\right).$$

We solve this for $m_\nu c^2$ and set $t_2 - t_1 = 10\,\text{s}$:

$$m_\nu c^2 = \left[\frac{2c(t_2 - t_1)}{d}\frac{E_1^2 E_2^2}{E_1^2 - E_2^2}\right]^{\frac{1}{2}} = 22 \times 10^{-6}\,\text{MeV} = 22\,\text{eV}.$$

This calculation, with its optimistic assumptions, estimates the mass of the neutrino to be $22\,\text{eV}/c^2$. But there would be experimental uncertainties, and even worse there is the unwarranted assumption that the two neutrinos were emitted at the same time.

[†]The brilliant young physicist Ettore Majorana (1906–1938) disappeared from a ship under mysterious circumstances in 1938 at the age of 31.

Theoretical models of supernova explosions suggest that the neutrinos are emitted in a burst that lasts from a second or two up to perhaps 10 s. If we assume the neutrinos are not emitted simultaneously but rather at any time over a 10-s interval, then that 10-s difference in arrival times could be due to a 10-s difference in their emission time. In this case the data would be consistent with zero neutrino mass, and it put an approximate *upper limit* of $22\ \text{eV}/c^2$.

The actual detection of these neutrinos was brilliant—it was a rare event that allowed us to detect something other than EM radiation from beyond the solar system, and was an exceptional confirmation of theory. In the experiments, the most sensitive detector consisted of several thousand tons of water in an underground chamber. It detected 11 events in 12 seconds, probably via the reaction $\bar{\nu}_e + \text{p} \rightarrow \text{n} + \text{e}^+$. There was not a clear correlation between energy and time of arrival. Nonetheless, a careful analysis of that experiment set a rough upper limit on the electron antineutrino mass of about $4\ \text{eV}/c^2$. The more recent results mentioned above are much more definitive—they provide evidence that mass is much smaller, and that it is *not zero*—but precise neutrino masses still elude us.

32–6 Particle Classification

In the decades after the discovery of the π meson in the late 1940s, hundreds of other subnuclear particles were discovered. One way to categorize the particles is according to their interactions, since not all particles interact via all four of the forces known in nature (though all interact via gravity). Table 32–2 (next page) lists some of the more common particles classified in this way along with many of their properties. At the top of Table 32–2 are the so-called "fundamental" particles which we believe have no internal structure. Below them are some of the "composite" particles which are made up of quarks, according to the Standard Model.

The **fundamental particles** include the **gauge bosons** (so-named after the theory that describes them, *gauge theory*), which include the gluons, the photon, and the W and Z particles; these are the particles that mediate (or "carry") the strong, electromagnetic, and weak interactions, respectively.

Also fundamental are the **leptons**, which are particles that do not interact via the strong force but do interact via the weak nuclear force. Leptons that carry electric charge also interact via the electromagnetic force. The leptons include the electron, the muon, and the tau, and three types of neutrino: the electron neutrino (ν_e), the muon neutrino (ν_μ), and the tau neutrino (ν_τ). Each lepton has an antiparticle. Finally, the recently detected Higgs boson is also considered to be fundamental, with no internal structure.

The second category of particle in Table 32–2 is the **hadrons**, which are **composite** particles (made up of quarks as we will discuss shortly). Hadrons are particles that interact via the strong nuclear force and are said to be **strongly interacting particles**. They also interact via the other forces, but the strong force predominates at short distances. The hadrons include the proton, neutron, pion, and many other particles. They are divided into two subgroups: **baryons**, which are particles that have baryon number $+1$ (or -1 in the case of their antiparticles) and, as we shall see, are each made up of three quarks; and **mesons**, which have baryon number $= 0$, and are made up of a quark and an antiquark.

Only a few of the hundreds of hadrons (a veritable "zoo") are included in Table 32–2. Notice that the baryons Λ, Σ, Ξ, and Ω all decay to lighter-mass baryons, and eventually to a proton or neutron. All these processes conserve baryon number. Since there is no particle lighter than the proton with $B = +1$, if baryon number is strictly conserved, the proton itself cannot decay and is stable. (But see Section 32–11.) Note that Table 32–2 gives the **mean life** (τ) of each particle (as is done in particle physics), not the half-life ($T_{\frac{1}{2}}$). Recall that they differ by a factor 0.693: $\tau = T_{\frac{1}{2}}/\ln 2 = T_{\frac{1}{2}}/0.693$, Eq. 30–7. The term **lifetime** in particle physics means the mean life τ ($=$ mean lifetime).

The baryon and lepton numbers (B, L_e, L_μ, L_τ), as well as strangeness S (Section 32–8), as given in Table 32–2 are for particles; their antiparticles have opposite sign for these numbers.

TABLE 32–2 Particles (selected)†

Category	Forces involved	Particle name	Symbol	Anti-particle	Spin	Mass (MeV/c^2)	B	L_e	L_μ	L_τ	S	Mean life (s)	Principal Decay Modes
Fundamental													
Gauge bosons (force carriers)	str	Gluons	g	Self	1	0	0	0	0	0	0	Stable	
	em	Photon	γ	Self	1	0	0	0	0	0	0	Stable	
	w, em	W	W^+	W^-	1	80.385×10^3	0	0	0	0	0	3×10^{-25}	$e\nu_e, \mu\nu_\mu, \tau\nu_\tau$, hadrons
	w	Z	Z^0	Self	1	91.19×10^3	0	0	0	0	0	3×10^{-25}	$e^+e^-, \mu^+\mu^-, \tau^+\tau^-$, hadrons
Higgs boson	w, str	Higgs	H^0	Self	0	125×10^3	0	0	0	0	0	1.6×10^{-22}	$b\bar{b}, Z^0Z^0, W^+W^-, g\bar{g}, \tau\bar{\tau}, \gamma\gamma$
Leptons	w, em‡	Electron	e^-	e^+	$\frac{1}{2}$	0.511	0	+1	0	0	0	Stable	
		Neutrino (e)	ν_e	$\bar{\nu}_e$	$\frac{1}{2}$	0 (<0.14 eV/c^2)‡	0	+1	0	0	0	Stable	
		Muon	μ^-	μ^+	$\frac{1}{2}$	105.7	0	0	+1	0	0	2.20×10^{-6}	$e^-\bar{\nu}_e\nu_\mu$
		Neutrino (μ)	ν_μ	$\bar{\nu}_\mu$	$\frac{1}{2}$	0 (<0.14 eV/c^2)‡	0	0	+1	0	0	Stable	
		Tau	τ^-	τ^+	$\frac{1}{2}$	1777	0	0	0	+1	0	2.91×10^{-13}	$\mu^-\bar{\nu}_\mu\nu_\tau, e^-\bar{\nu}_e\nu_\tau$, hadrons $+\nu_\tau$
		Neutrino (τ)	ν_τ	$\bar{\nu}_\tau$	$\frac{1}{2}$	0 (<0.14 eV/c^2)‡	0	0	0	+1	0	Stable	
Quarks	w, em, str	(see Table 32–3)											
Hadrons (composite), selected													
Mesons (quark–antiquark)	str, em, w	Pion	π^+	π^-	0	139.6	0	0	0	0	0	2.60×10^{-8}	$\mu^+\nu_\mu$
			π^0	Self	0	135.0	0	0	0	0	0	0.85×10^{-16}	2γ
		Kaon	K^+	K^-	0	493.7	0	0	0	0	+1	1.24×10^{-8}	$\mu^+\nu_\mu, \pi^+\pi^0$
			K^0_S	$\bar{K}^0_S$	0	497.6	0	0	0	0	+1	0.895×10^{-10}	$\pi^+\pi^-, 2\pi^0$
			K^0_L	$\bar{K}^0_L$	0	497.6	0	0	0	0	+1	5.12×10^{-8}	$\pi^\pm e^\mp \overset{(-)}{\nu}_e, \pi^\pm \mu^\mp \overset{(-)}{\nu}_\mu, 3\pi$
		Eta	η^0	Self	0	547.9	0	0	0	0	0	5.1×10^{-19}	$2\gamma, 3\pi^0, \pi^+\pi^-\pi^0$
		Rho	ρ^0	Self	1	775	0	0	0	0	0	4.4×10^{-24}	$\pi^+\pi^-, 2\pi^0$
			ρ^+	ρ^-	1	775	0	0	0	0	0	4.4×10^{-24}	$\pi^+\pi^0$
		and others											
Baryons (3 quarks)	str, em, w	Proton	p	$\bar{p}$	$\frac{1}{2}$	938.3	+1	0	0	0	0	Stable	
		Neutron	n	$\bar{n}$	$\frac{1}{2}$	939.6	+1	0	0	0	0	882	$pe^-\bar{\nu}_e$
		Lambda	Λ^0	$\bar{\Lambda}^0$	$\frac{1}{2}$	1115.7	+1	0	0	0	−1	2.63×10^{-10}	$p\pi^-, n\pi^0$
		Sigma	Σ^+	$\bar{\Sigma}^-$	$\frac{1}{2}$	1189.4	+1	0	0	0	−1	0.80×10^{-10}	$p\pi^0, n\pi^+$
			Σ^0	$\bar{\Sigma}^0$	$\frac{1}{2}$	1192.6	+1	0	0	0	−1	7.4×10^{-20}	$\Lambda^0\gamma$
			Σ^-	$\bar{\Sigma}^+$	$\frac{1}{2}$	1197.4	+1	0	0	0	−1	1.48×10^{-10}	$n\pi^-$
		Xi	Ξ^0	$\bar{\Xi}^0$	$\frac{1}{2}$	1314.9	+1	0	0	0	−2	2.90×10^{-10}	$\Lambda^0\pi^0$
			Ξ^-	$\bar{\Xi}^+$	$\frac{1}{2}$	1321.7	+1	0	0	0	−2	1.64×10^{-10}	$\Lambda^0\pi^-$
		Omega	Ω^-	Ω^+	$\frac{3}{2}$	1672.5	+1	0	0	0	−3	0.82×10^{-10}	$\Xi^0\pi^-, \Lambda^0 K^-, \Xi^-\pi^0$
		and others											

†See also Table 32–4 for particles with charm and bottom. S in this Table stands for "strangeness" (see Section 32–8). More detail online at: pdg.lbl.gov.
‡Neutrinos partake only in the weak interaction. Experimental upper limits on neutrino masses are given in parentheses, as obtained mainly from the WMAP survey (Chapter 33). Detection of neutrino oscillations suggests that at least one type of neutrino has a nonzero mass greater than 0.04 eV/c^2.

EXAMPLE 32–6 Baryon decay. Show that the decay modes of the Σ^+ baryon given in Table 32–2 do not violate the conservation laws we have studied up to now: energy, charge, baryon number, lepton numbers.

APPROACH Table 32–2 shows two possible decay modes, (a) $\Sigma^+ \to p + \pi^0$, (b) $\Sigma^+ \to n + \pi^+$. All the particles have lepton numbers equal to zero.

SOLUTION (a) Energy: for $\Sigma^+ \to p + \pi^0$ the change in mass-energy is

$$\Delta(Mc^2) = m_\Sigma c^2 - m_p c^2 - m_{\pi^0} c^2$$
$$= 1189.4 \text{ MeV} - 938.3 \text{ MeV} - 135.0 \text{ MeV} = +116.1 \text{ MeV},$$

so energy can be conserved with the resulting particles having kinetic energy.

Charge: $+e = +e + 0$, so charge is conserved.

Baryon number: $+1 = +1 + 0$, so baryon number is conserved.

(b) Energy: for $\Sigma^+ \to n + \pi^+$, the mass-energy change is

$$\Delta(Mc^2) = m_\Sigma c^2 - m_n c^2 - m_{\pi^+} c^2$$
$$= 1189.4 \text{ MeV} - 939.6 \text{ MeV} - 139.6 \text{ MeV} = +110.2 \text{ MeV}.$$

This reaction releases 110.2 MeV of energy as kinetic energy of the products.

Charge: $+e = 0 + e$, so charge is conserved.

Baryon number: $+1 = +1 + 0$, so baryon number is conserved.

32–7 Particle Stability and Resonances

Many particles listed in Table 32–2 are unstable. The lifetime of an unstable particle depends on which force is most active in causing the decay. When a stronger force influences a decay, that decay occurs more quickly. Decays caused by the weak force typically have lifetimes of 10^{-13} s or longer (W and Z decay directly and more quickly). Decays via the electromagnetic force have much shorter lifetimes, typically about 10^{-16} to 10^{-19} s, and normally involve a γ (photon). Most of the unstable particles included in Table 32–2 decay either via the weak or the electromagnetic interaction.

Many particles have been found that decay via the strong interaction, with very short lifetimes, typically about 10^{-23} s. Their lifetimes are so short they do not travel far enough to be detected before decaying. The existence of such short-lived particles is inferred from their decay products. Consider the first such particle discovered (by Fermi), using a beam of π^+ particles with varying amounts of energy directed through a hydrogen target (protons). The number of interactions (π^+ scattered) plotted versus the pion's kinetic energy is shown in Fig. 32–14. The large number of interactions around 200 MeV led Fermi to conclude that the π^+ and proton combined momentarily to form a short-lived particle before coming apart again, or at least that they resonated together for a short time. Indeed, the large peak in Fig. 32–14 resembles a resonance curve (see Figs. 11–18 and 21–46), and this new "particle"—now called the Δ—is referred to as a **resonance**. Hundreds of other resonances have been found, and are regarded as excited states of lighter mass particles such as a nucleon.

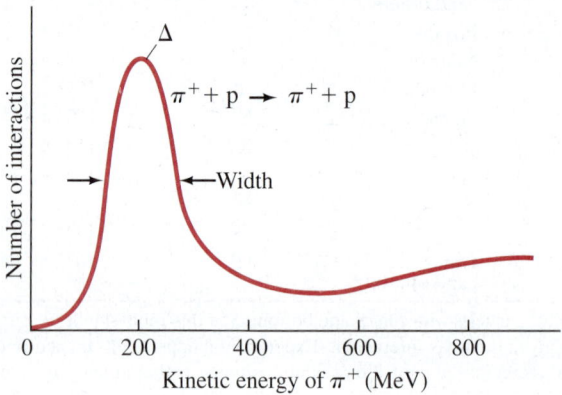

FIGURE 32–14 Number of π^+ particles scattered elastically by a proton target as a function of the incident π^+ kinetic energy. The resonance shape represents the formation of a short-lived particle, the Δ, which has a charge in this case of $+2e$ (Δ^{++}).

The **width** of a resonance—in Fig. 32–14 the full width of the Δ peak at half the maximum is on the order of 100 MeV—is an interesting application of the uncertainty principle. If a particle lives only 10^{-23} s, then its mass (i.e., its rest energy) will be uncertain by an amount

$$\Delta E \approx \frac{h}{2\pi \, \Delta t} \approx \frac{(6.6 \times 10^{-34} \, \text{J} \cdot \text{s})}{(6)(10^{-23} \, \text{s})} \approx 10^{-11} \, \text{J} \approx 100 \, \text{MeV},$$

which is what is observed. Actually, the lifetimes of $\approx 10^{-23}$ s for such resonances are inferred by the reverse process: from the measured width being ≈ 100 MeV.

32–8 Strangeness? Charm? Towards a New Model

In the early 1950s, the newly found particles K, Λ, and Σ were found to behave rather strangely in two ways. First, they were always produced in pairs. For example, the reaction

$$\pi^- + p \rightarrow K^0 + \Lambda^0$$

occurred with high probability, but the similar reaction $\pi^- + p \not\rightarrow K^0 + n$ was never observed to occur even though it did not violate any known conservation law. The second feature of these **strange particles**, as they came to be called, was that they were produced via the strong interaction (that is, at a high interaction rate), but did not decay at a fast rate characteristic of the strong interaction (even though they decayed into strongly interacting particles).

To explain these observations, a new quantum number, **strangeness**, and a new conservation law, **conservation of strangeness**, were introduced. By assigning the strangeness numbers (S) indicated in Table 32–2, the production of strange particles in pairs was explained. Antiparticles were assigned opposite strangeness from their particles. For example, in the reaction $\pi^- + p \rightarrow K^0 + \Lambda^0$, the initial state has strangeness $S = 0 + 0 = 0$, and the final state has $S = +1 - 1 = 0$, so strangeness is conserved. But for $\pi^- + p \not\rightarrow K^0 + n$, the initial state has $S = 0$ and the final state has $S = +1 + 0 = +1$, so strangeness would not be conserved; and this reaction is not observed.

To explain the decay of strange particles, it is assumed that strangeness is conserved in the strong interaction but is *not conserved in the weak interaction*. Thus, strange particles were forbidden by strangeness conservation to decay to nonstrange particles of lower mass via the strong interaction, but could decay by means of the weak interaction at the observed longer lifetimes of 10^{-10} to 10^{-8} s.

The conservation of strangeness was the first example of a **partially conserved** quantity. In this case, the quantity strangeness is conserved by strong interactions but not by weak.

CAUTION
Partially conserved quantities

CONCEPTUAL EXAMPLE 32–7 **Guess the missing particle.** Using the conservation laws for particle interactions, determine the possibilities for the missing particle in the reaction

$$\pi^- + p \rightarrow K^0 + ?$$

in addition to $K^0 + \Lambda^0$ mentioned above.

RESPONSE We write equations for the conserved numbers in this reaction, with B, L_e, S, and Q as unknowns whose determination will reveal what the possible particle might be:

Baryon number: $0 + 1 = 0 + B$
Lepton number: $0 + 0 = 0 + L_e$
Charge: $-1 + 1 = 0 + Q$
Strangeness: $0 + 0 = 1 + S$.

The unknown product particle would have to have these characteristics:

$$B = +1 \quad L_e = 0 \quad Q = 0 \quad S = -1.$$

In addition to Λ^0, a neutral sigma particle, Σ^0, is also consistent with these numbers.

In the next Section we will discuss another partially conserved quantity which was given the name **charm**. The discovery in 1974 of a particle with charm helped solidify a new theory involving quarks, which we now discuss.

32–9 Quarks

One difference between leptons and hadrons is that the hadrons interact via the strong interaction, whereas the leptons do not. There is an even more fundamental difference. The six leptons (e^-, μ^-, τ^-, ν_e, ν_μ, ν_τ) are considered to be truly fundamental particles because they do not show any internal structure, and have no measurable size. (Attempts to determine the size of leptons have put an upper limit of about 10^{-18} m.) On the other hand, there are hundreds of hadrons, and experiments indicate they do have an internal structure. When an electron collides with another electron, it scatters off as per Coulomb's law. But electrons scattering off a proton reveal a more complex pattern, implying internal parts within the proton (= quarks).

FIGURE 32–15 This computer-generated reconstruction of a proton–antiproton collision at Fermilab occurred at an energy of nearly 2 TeV. It is one of the events that provided evidence for the top quark (1995). The multiwire chamber (Section 30–13) is in a magnetic field, and the radius of curvature of the charged particle tracks is a measure of each particle's momentum (Chapter 20). The white dots represent signals seen on the electric wires of the multiwire chamber. The colored lines are particle paths. The top quark (t) has too brief a lifetime $(\approx 10^{-23}\text{ s})$ to be detected itself, so we look for its possible decay products. Analysis indicates the following interaction and subsequent decays:

$$p + \bar{p} \longrightarrow t + \bar{t}$$
$$\quad\quad\quad\quad\quad\quad \hookrightarrow W^- + \bar{b}$$
$$\quad\quad\quad\quad\quad\quad\quad\quad \hookrightarrow \text{jet}$$
$$\quad\quad\quad\quad\quad\quad\quad\quad\quad \to \mu^- + \bar{\nu}_\mu$$
$$\quad\quad\quad\quad \hookrightarrow W^+ + b$$
$$\quad\quad\quad\quad\quad\quad \hookrightarrow \text{jet}$$
$$\quad\quad\quad\quad\quad \hookrightarrow u + \bar{d}$$
$$\quad\quad\quad\quad\quad\quad\quad \hookrightarrow \text{jet}$$
$$\quad\quad\quad\quad\quad\quad\quad \to \text{jet}$$

The tracks in the photo include jets (groups of particles moving in roughly the same direction), and a muon (μ^-) whose track is the pink one enclosed by a yellow rectangle to make it stand out.

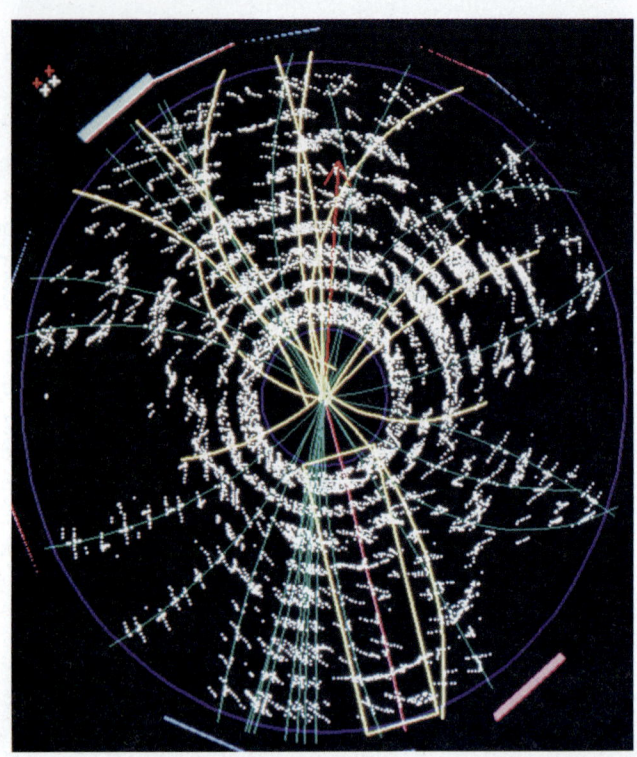

In 1963, M. Gell-Mann and G. Zweig proposed that none of the hadrons, not even the proton and neutron, are truly fundamental, but instead are made up of combinations of three more fundamental pointlike entities called (somewhat whimsically) **quarks**.[†] Today, the quark theory is well-accepted, and quarks are considered truly fundamental particles, like leptons. The three quarks originally proposed were named **up**, **down**, and **strange**, with abbreviations u, d, s. The theory today has six quarks, just as there are six leptons—based on a presumed *symmetry* in nature. The other three quarks are called **charm**, **bottom**, and **top** (c, b, t). The names apply also to new properties of each quark (quantum numbers c, b, t) that distinguish these new quarks from the 3 original quarks (see Table 32–3). These properties (like strangeness) are conserved in strong, but not weak, interactions. Figure 32–15 shows one of the events that provided evidence for the top quark.

All quarks have spin $\frac{1}{2}$ and an electric charge of either $+\frac{2}{3}e$ or $-\frac{1}{3}e$ (that is, a fraction of the previously thought smallest charge e). Antiquarks have opposite sign of electric charge Q, baryon number B, strangeness S, charm c, bottom b, and top t. Other properties of quarks are shown in Table 32–3.

[†]Gell-Mann chose the word from a phrase in James Joyce's *Finnegans Wake*.

TABLE 32–3 Properties of Quarks (Antiquarks have opposite sign Q, B, S, c, b, t)

			Quarks					
Name	Symbol	Mass (MeV/c^2)	Charge Q	Baryon Number B	Strangeness S	Charm c	Bottom b	Top t
Up	u	2.3	$+\frac{2}{3}e$	$\frac{1}{3}$	0	0	0	0
Down	d	4.8	$-\frac{1}{3}e$	$\frac{1}{3}$	0	0	0	0
Strange	s	95	$-\frac{1}{3}e$	$\frac{1}{3}$	-1	0	0	0
Charm	c	1275	$+\frac{2}{3}e$	$\frac{1}{3}$	0	$+1$	0	0
Bottom	b	4180	$-\frac{1}{3}e$	$\frac{1}{3}$	0	0	-1	0
Top[†]	t	173,500	$+\frac{2}{3}e$	$\frac{1}{3}$	0	0	0	$+1$

[†]The top quark, with its extremely short lifetime of 5×10^{-25} s, does not live long enough to form hadrons.

TABLE 32–4 Partial List of Heavy Hadrons, with Charm and Bottom ($L_e = L_\mu = L_\tau = 0$)

Category	Particle	Anti-particle	Spin	Mass (MeV/c^2)	Baryon Number B	Strangeness S	Charm c	Bottom b	Mean life (s)	Principal Decay Modes
Mesons	D^+	D^-	0	1869.6	0	0	+1	0	10.4×10^{-13}	K + others, e + others
	D^0	$\overline{D}^0$	0	1864.9	0	0	+1	0	4.1×10^{-13}	K + others, μ or e + others
	D_S^+	D_S^-	0	1968.5	0	+1	+1	0	5.0×10^{-13}	K + others
	J/ψ (3097)	Self	1	3096.9	0	0	0	0	0.71×10^{-20}	Hadrons, e^+e^-, $\mu^+\mu^-$
	Υ (9460)	Self	1	9460.3	0	0	0	0	1.2×10^{-20}	Hadrons, $\mu^+\mu^-$, e^+e^-, $\tau^+\tau^-$
	B^-	B^+	0	5279.3	0	0	0	-1	1.6×10^{-12}	D^0 + others
	B^0	$\overline{B}^0$	0	5279.6	0	0	0	-1	1.5×10^{-12}	D^0 + others
Baryons	Λ_c^+	Λ_c^-	$\tfrac{1}{2}$	2286	+1	0	+1	0	2.0×10^{-13}	Hadrons (e.g., Λ + others)
	Σ_c^{++}	Σ_c^{--}	$\tfrac{1}{2}$	2454	+1	0	+1	0	2.9×10^{-22}	$\Lambda_c^+ \pi^+$
	Σ_c^+	Σ_c^-	$\tfrac{1}{2}$	2453	+1	0	+1	0	$>1.4 \times 10^{-22}$	$\Lambda_c^+ \pi^0$
	Σ_c^0	$\overline{\Sigma}_c^0$	$\tfrac{1}{2}$	2454	+1	0	+1	0	3.0×10^{-22}	$\Lambda_c^+ \pi^-$
	Λ_b^0	$\overline{\Lambda}_b^0$	$\tfrac{1}{2}$	5619	+1	0	0	-1	1.4×10^{-12}	$J/\psi \Lambda^0$, $pD^0\pi^-$, $\Lambda_c^+\pi^+\pi^-\pi^-$

All hadrons are considered to be made up of combinations of quarks (plus the gluons that hold them together), and their properties are described by looking at their quark content. Mesons consist of a quark–antiquark pair. For example, a π^+ meson is a $u\overline{d}$ combination: note that for the $u\overline{d}$ pair (Table 32–3), $Q = \tfrac{2}{3}e + \tfrac{1}{3}e = +1e$, $B = \tfrac{1}{3} - \tfrac{1}{3} = 0$, $S = 0 + 0 = 0$, as they must for a π^+; and a $K^+ = u\overline{s}$, with $Q = +1$, $B = 0$, $S = +1$. A π^0 can be made of $u\overline{u}$ or $d\overline{d}$.

Baryons, on the other hand, consist of three quarks. For example, a neutron is n = ddu, whereas an antiproton is $\overline{p} = \overline{u}\,\overline{u}\,\overline{d}$. See Fig. 32–16. Strange particles all contain an s or $\overline{s}$ quark, whereas charm particles contain a c or $\overline{c}$ quark. A few of these hadrons are listed in Table 32–4.

Current models suggest that quarks may be so tightly bound together that they may not ever exist singly in the free state. But quarks can be detected indirectly when they turn into narrow **jets** of other particles, as in Fig. 32–15. Also, observations of very high energy electrons scattered off protons suggest that protons are indeed made up of constituents.

Today, the truly **fundamental particles** are considered to be the six quarks, the six leptons, the gauge bosons that carry the fundamental forces, and the Higgs (page 939). See Table 32–5, where the quarks and leptons are arranged in three "families" or "generations." Ordinary matter—atoms made of protons, neutrons, and electrons—is contained in the "first generation." The others are thought to have existed in the very early universe, but are seen by us today only at powerful colliders or in cosmic rays. All of the hundreds of hadrons can be accounted for by combinations of the six quarks and six antiquarks.

EXERCISE C Return to the Chapter-Opening Questions, page 915, and answer them again now. Try to explain why you may have answered differently the first time.

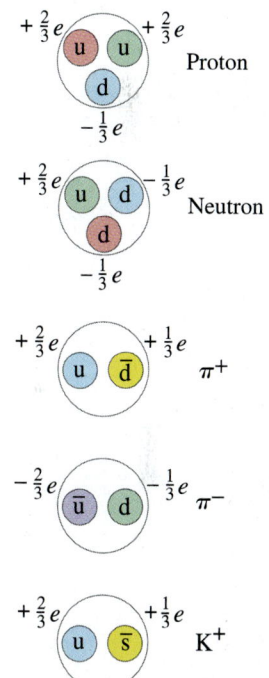

FIGURE 32–16 Quark compositions for several particles.

TABLE 32–5 The Fundamental Particles[†] of the Standard Model

Gauge bosons	γ	g	Z^0	$W^\pm$
Higgs boson	H^0			

	Quarks		and	Leptons	
First generation	u	d		e	ν_e
Second generation	s	c		μ	ν_μ
Third generation	b	t		τ	ν_τ

[†]The graviton (G^0) has been hypothesized but not detected and may not be detectable. It is not part of the Standard Model.

CONCEPTUAL EXAMPLE 32–8 **Quark combinations.** Find the baryon number, charge, and strangeness for the following quark combinations, and identify the hadron particle that is made up of these quark combinations: (a) udd, (b) u$\bar{u}$, (c) uss, (d) sdd, and (e) b$\bar{u}$.

RESPONSE We use Table 32–3 to get the properties of the quarks, then Table 32–2 or 32–4 to find the particle that has these properties.

(a) udd has
$$Q = +\tfrac{2}{3}e - \tfrac{1}{3}e - \tfrac{1}{3}e = 0,$$
$$B = \tfrac{1}{3} + \tfrac{1}{3} + \tfrac{1}{3} = 1,$$
$$S = 0 + 0 + 0 = 0,$$

as well as $c = 0$, bottom $= 0$, top $= 0$. The only baryon ($B = +1$) that has $Q = 0$, $S = 0$, etc., is the neutron (Table 32–2).

(b) u$\bar{u}$ has $Q = \tfrac{2}{3}e - \tfrac{2}{3}e = 0$, $B = \tfrac{1}{3} - \tfrac{1}{3} = 0$, and all other quantum numbers $= 0$. Sounds like a π^0 (d$\bar{d}$ also gives a π^0).

(c) uss has $Q = 0$, $B = +1$, $S = -2$, others $= 0$. This is a Ξ^0.

(d) sdd has $Q = -1$, $B = +1$, $S = -1$, so must be a Σ^-.

(e) b$\bar{u}$ has $Q = -1$, $B = 0$, $S = 0$, $c = 0$, bottom $= -1$, top $= 0$. This must be a B$^-$ meson (Table 32–4).

EXERCISE D What is the quark composition of a K$^-$ meson?

32–10 The Standard Model: QCD and Electroweak Theory

Not long after the quark theory was proposed, it was suggested that quarks have another property (or quality) called **color**, or "color charge" (analogous to electric charge). The distinction between the six types of quark (u, d, s, c, b, t) was referred to as **flavor**. According to theory, each flavor of quark can have one of three colors, usually designated red, green, and blue. (These are the three primary colors which, when added together in appropriate amounts, as on a TV screen, produce white.) Note that the names "color" and "flavor" have nothing to do with our senses, but are purely whimsical—as are other names, such as charm, in this new field. (We did, however, "color" the quarks in Fig. 32–16.) The antiquarks are colored antired, antigreen, and antiblue. Baryons are made up of three quarks, one of each color. Mesons consist of a quark–antiquark pair of a particular color and its anticolor. Both baryons and mesons are thus colorless or white.

Originally, the idea of quark color was proposed to preserve the Pauli exclusion principle (Section 28–7). Not all particles obey the exclusion principle. Those that do, such as electrons, protons, and neutrons, are called **fermions**. Those that don't are called **bosons**. These two categories are distinguished also in their spin: bosons have integer spin (0, 1, 2, etc.) whereas fermions have half-integer spin, usually $\tfrac{1}{2}$ as for electrons and nucleons, but other fermions have spin $\tfrac{3}{2}, \tfrac{5}{2}$, etc. Matter is made up mainly of fermions, but the carriers of the forces (γ, W, Z, and gluons) are all bosons. Quarks are fermions (they have spin $\tfrac{1}{2}$) and therefore should obey the exclusion principle. Yet for three particular baryons (uuu, ddd, and sss), all three quarks would have the same quantum numbers, and at least two quarks have their spin in the same direction (since there are only two choices, spin up $[m_s = +\tfrac{1}{2}]$ or spin down $[m_s = -\tfrac{1}{2}]$). This would seem to violate the exclusion principle; but if quarks have that additional quantum number *color*, which is different for each quark, it would serve to distinguish them and allow the exclusion principle to hold. Although quark color, and the resulting threefold increase in the number of quarks, was originally an *ad hoc* idea, it also served to bring the theory into better agreement with experiment, such as predicting the correct lifetime of the π^0 meson, and the measured rate of hadron production in observed e^+e^- collisions at accelerators. The idea of color soon became a central feature of the theory as determining the force binding quarks together in a hadron.

Each quark is assumed to carry a *color charge*, analogous to electric charge, and the strong force between quarks is referred to as the **color force**. This theory of the strong force is called **quantum chromodynamics** (*chroma* = color in Greek), or **QCD**, to indicate that the force acts between color charges (and not between, say, electric charges). The strong force between two hadrons is considered to be a force between the quarks that make them up, as suggested in Fig. 32–17.

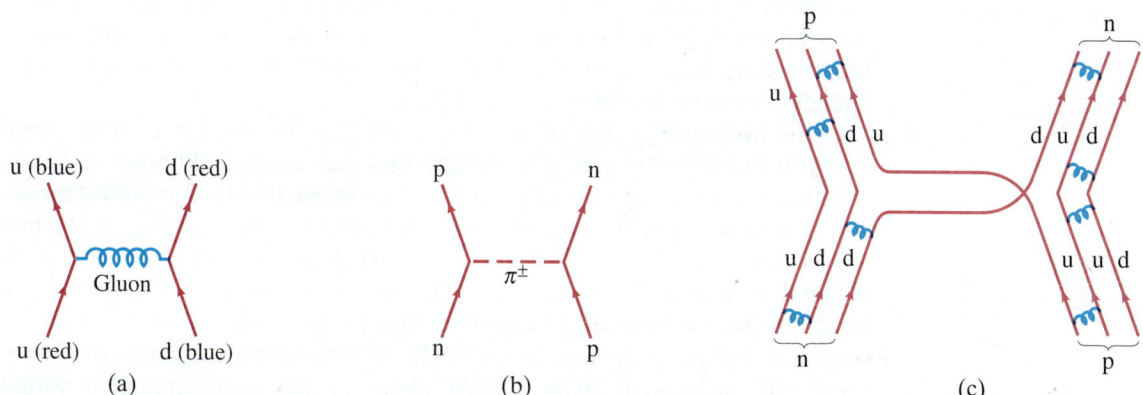

FIGURE 32–17 (a) The force between two quarks holding them together as part of a proton, for example, is carried by a gluon, which in this case involves a change in color. (b) Strong interaction $n + p \rightarrow n + p$ with the exchange of a charged π meson (+ or −, depending on whether it is considered moving to the left or to the right). (c) Quark representation of the same interaction $n + p \rightarrow n + p$. The blue coiled lines between quarks represent gluon exchanges holding the hadrons together. (The exchanged meson may be regarded as $\bar{u}d$ emitted by the n and absorbed by the p, or as $u\bar{d}$ emitted by p and absorbed by n, because a u (or d) quark going to the left in the diagram is equivalent to a $\bar{u}$ (or $\bar{d}$) going to the right.)

The particles that transmit the color force (analogous to photons for the EM force) are called **gluons** (a play on "glue"). They are included in Tables 32–2 and 32–5. There are eight gluons, according to the theory, all massless and all have color charge.[†]

You might ask what would happen if we try to see a single quark with color by reaching deep inside a hadron and extracting a single quark. Quarks are so tightly bound to other quarks that extracting one would require a tremendous amount of energy, so much that it would be sufficient to create more quarks ($E = mc^2$). Indeed, such experiments are done at modern particle colliders and all we get is more hadrons (quark–antiquark pairs, or triplets, which we observe as mesons or baryons), never an isolated quark. This property of quarks, that they are always bound in groups that are colorless, is called **confinement**.

The color force has the interesting property that, as two quarks approach each other very closely (equivalently, have high energy), the force between them becomes small. This aspect is referred to as **asymptotic freedom**.

The weak force, as we have seen, is thought to be mediated by the W^+, W^-, and Z^0 particles. It acts between the "weak charges" that each particle has. Each fundamental particle can thus have electric charge, weak charge, color charge, and gravitational mass, although one or more of these could be zero. For example, all leptons have color charge of zero, so they do not interact via the strong force.

[†]Compare to the EM interaction, where the photon has no electric charge. Because gluons have color charge, they could attract each other and form composite particles (photons cannot). Such "glueballs" are being searched for.

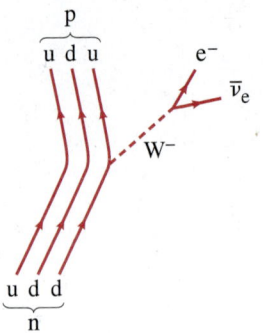

FIGURE 32–18 Quark representation of the Feynman diagram for β decay of a neutron into a proton. Example 32–9.

CONCEPTUAL EXAMPLE 32–9 **Beta decay.** Draw a Feynman diagram, showing what happens in beta decay using quarks.

RESPONSE Beta decay is a result of the weak interaction, and the mediator is either a $W^\pm$ or Z^0 particle. What happens, in part, is that a neutron (udd quarks) decays into a proton (uud). Apparently a d quark (charge $-\frac{1}{3}e$) has turned into a u quark (charge $+\frac{2}{3}e$). Charge conservation means that a negatively charged particle, namely a W^-, was emitted by the d quark. Since an electron and an antineutrino appear in the final state, they must have come from the decay of the virtual W^-, as shown in Fig. 32–18.

To summarize, the Standard Model says that the truly fundamental particles (Table 32–5) are the leptons, the quarks, the gauge bosons (photon, W and Z, and the gluons), and the Higgs boson. The photon, leptons, W^+, W^-, and Z^0 have all been observed in experiments. But only combinations of quarks (baryons and mesons) have been observed in the free state, and it seems likely that free quarks and gluons cannot be observed in isolation.

One important aspect of theoretical work is the attempt to find a **unified** basis for the different forces in nature. This was a long-held hope of Einstein, which he was never able to fulfill. A so-called **gauge theory** that unifies the weak and electromagnetic interactions was put forward in the 1960s by S. Weinberg, S. Glashow, and A. Salam. In this **electroweak theory**, the weak and electromagnetic forces are seen as two different manifestations of a single, more fundamental, *electroweak* interaction. The electroweak theory has had many successes, including the prediction of the $W^\pm$ particles as carriers of the weak force, with masses of $80.38 \pm 0.02 \text{ GeV}/c^2$ in excellent agreement with the measured values of $80.385 \pm 0.015 \text{ GeV}/c^2$ (and similar accuracy for the Z^0).

The combination of electroweak theory plus QCD for the strong interaction is referred to today as the **Standard Model (SM)** of fundamental particles.

EXAMPLE 32–10 **ESTIMATE** **Range of weak force.** The weak nuclear force is of very short range, meaning it acts over only a very short distance. Estimate its range using the masses (Table 32–2) of the $W^\pm$ and Z: $m \approx 80$ or $90 \text{ GeV}/c^2 \approx 10^2 \text{ GeV}/c^2$.

APPROACH We assume the $W^\pm$ or Z^0 exchange particles can exist for a time Δt given by the uncertainty principle (Section 28–3), $\Delta t \approx \hbar/\Delta E$, where $\Delta E \approx mc^2$ is the energy needed to create the virtual particle ($W^\pm$, Z) that carries the weak force.

SOLUTION Let Δx be the distance the virtual W or Z can move before it must be reabsorbed within the time $\Delta t \approx \hbar/\Delta E$. To find an upper limit on Δx, and hence the maximum range of the weak force, we let the W or Z travel close to the speed of light, so $\Delta x \lesssim c\,\Delta t$. Recalling that $1 \text{ GeV} = 1.6 \times 10^{-10}$ J, then

$$\Delta x \lesssim c\,\Delta t \approx \frac{c\hbar}{\Delta E} \approx \frac{(3 \times 10^8 \text{ m/s})(10^{-34} \text{ J} \cdot \text{s})}{(10^2 \text{ GeV})(1.6 \times 10^{-10} \text{ J/GeV})} \approx 10^{-18} \text{ m}.$$

This is indeed a very small range.

NOTE Compare this to the range of the electromagnetic force whose range is infinite ($1/r^2$ never becomes zero for any finite r), which makes sense because the mass of its virtual exchange particle, the photon, is zero (in the denominator of the above equation).

[We did a similar calculation for the strong force in Section 32–2, estimating the mass of the π meson as exchange particle. In our deeper view of the strong force, namely the color force between quarks within a nucleon, the gluons have zero mass, which implies infinite range (see formula in Example 32–10). We might have expected a range of about 10^{-15} m (nuclear size). But according to the Standard Model, the color force is weak at very close distances and increases greatly with distance (causing quark confinement). Thus its range could be infinite.]

Theoreticians have wondered why the W and Z have large masses rather than being massless like the photon. Peter Higgs and others in 1964 used electroweak theory to suggest an explanation by means of a **Higgs field** and its particle, the **Higgs boson**, which interact with the W and Z to "slow them down." In being forced to go slower than the speed of light, they would have to have mass ($m = 0$ only if $v = c$). Indeed, the Higgs field is thought to permeate the vacuum ("empty space") and to perhaps confer mass on particles that now have mass by slowing them down. In 2012 strong evidence was announced at CERN's Large Hadron Collider (Section 32–1) for a particle of mass 125 GeV/c^2 that is thought to be the long-sought Higgs boson of the Standard Model. But intense research continues, not only to better understand this particle, but to search for additional Higgs-like particles suggested by theories that go beyond the Standard Model such as *supersymmetry* (Section 32–12).

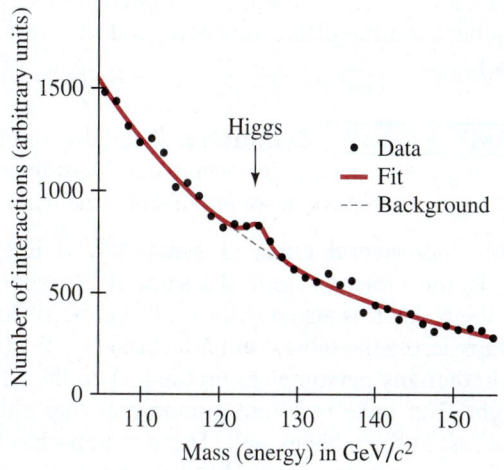

FIGURE 32–19 Evidence for the Higgs boson.

FIGURE 32–20 Fabiola Gianotti, leader of the ATLAS team (3000 physicists), at the LHC with theorist Peter Higgs, July 4, 2012, when the long hoped-for boson was announced.

Figure 32–19 shows the "resonance" bump (Section 32–7) that represents the Higgs boson as detected by the CMS team at the LHC. A second experiment, ATLAS, came up with the same mass. ATLAS is considered the largest scientific experiment ever (see Fig. 32–20).

There is no way to know if the Chapter-Opening Photo of a possible Higgs event, page 915, is actually a Higgs or is a background event. As can be seen in Fig. 32–19, there are many more background events around 125 GeV than there are in the resonance bump representing the Higgs boson.

32–11 Grand Unified Theories

The Standard Model, for all its success, cannot explain some important issues—such as why the charge on the electron has *exactly* the same magnitude as the charge on the proton. This is crucial, because if the charge magnitudes were even a little different, atoms would not be neutral and the resulting large electric forces would surely have made life impossible. Indeed, the Standard Model is now considered to be a low-energy approximation to a more complete theory.

With the success of unified electroweak theory, theorists are trying to incorporate it and QCD for the strong (color) force into a so-called **grand unified theory** (**GUT**).

One type of such a grand unified theory of the electromagnetic, weak, and strong forces has been proposed in which there is only one class of particle—leptons and quarks belong to the same family and are able to change freely from one type to the other—and the three forces are different aspects of a single underlying force. The unity is predicted to occur, however, only on a scale of less than about 10^{-31} m, corresponding to a typical particle energy of about 10^{16} GeV.

If two elementary particles (leptons or quarks) approach each other to within this **unification scale**, the apparently fundamental distinction between them would not exist at this level, and a quark could readily change to a lepton, or vice versa. Baryon and lepton numbers would not be conserved. The weak, electromagnetic, and strong (color) force would blend to a force of a single strength.

What happens between the unification distance of 10^{-31} m and more normal (larger) distances is referred to as **symmetry breaking**. As an analogy, consider an atom in a crystal. Deep within the atom, there is much symmetry—in the innermost regions the electron cloud is spherically symmetric (Chapter 28). Farther out, this symmetry breaks down—the electron clouds are distributed preferentially along the lines (bonds) joining the atoms in the crystal. In a similar way, at 10^{-31} m the force between elementary particles is theorized to be a single force—it is symmetrical and does not single out one type of "charge" over another. But at larger distances, that symmetry is broken and we see three distinct forces. (In the "Standard Model" of electroweak interactions, Section 32–10, the symmetry breaking between the electromagnetic and the weak interactions occurs at about 10^{-18} m.)

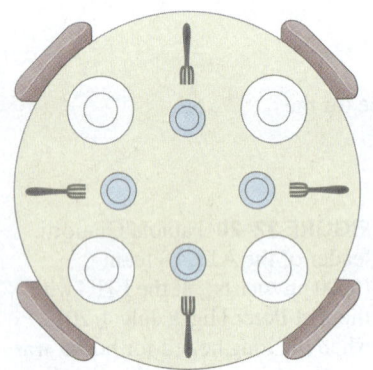

FIGURE 32–21 Symmetry around a table. Example 32–11.

CONCEPTUAL EXAMPLE 32–11 Symmetry. The table in Fig. 32–21 has four identical place settings. Four people sit down to eat. Describe the symmetry of this table and what happens to it when someone starts the meal.

RESPONSE The table has several kinds of symmetry. It is symmetric to rotations of 90°: that is, the table will look the same if everyone moved one chair to the left or to the right. It is also north–south symmetric and east–west symmetric, so that swaps across the table don't affect the way the table looks. It also doesn't matter whether any person picks up the fork to the left of the plate or the fork to the right. But once that first person picks up either fork, the choice is set for all the rest at the table as well. The symmetry has been *broken*. The underlying symmetry is still there—the blue glasses could still be chosen either way—but some choice must get made and at that moment the symmetry of the diners is broken.

Another example of symmetry breaking is a pencil standing on its point before falling. Standing, it looks the same from any horizontal direction. From above, it is a tiny circle. But when it falls to the table, it points in one particular direction—the symmetry is broken.

Proton Decay

Since unification is thought to occur at such tiny distances and huge energies, the theory is difficult to test experimentally. But it is not completely impossible. One testable prediction is the idea that the proton might decay (via, for example, $p \rightarrow \pi^0 + e^+$) and violate conservation of baryon number. This could happen if two quarks within a proton approached to within 10^{-31} m of each other. But it is very unlikely at normal temperature and energy, so the decay of a proton can only be an unlikely process. In the simplest form of GUT, the theoretical estimate of the proton mean life for the decay mode $p \rightarrow \pi^0 + e^+$ is about 10^{31} yr, and this is now within the realm of testability.[†] Proton decays have still not been seen, and experiments put the lower limit on the proton mean life for the above mode to be about 10^{33} yr, somewhat greater than this prediction. This may seem a disappointment, but on the other hand, it presents a challenge. Indeed more complex GUTs may resolve this conflict.

[†]This is much larger than the age of the universe ($\approx 14 \times 10^9$ yr). But we don't have to wait 10^{31} yr to see. Instead we can wait for one decay among 10^{31} protons over a year (see Eqs. 30–3a and 30–7, $\Delta N = \lambda N \Delta t = N \Delta t / \tau$, and Example 32–12).

EXAMPLE 32-12 ESTIMATE **Proton decay.** An experiment uses 3300 tons of water waiting to see a proton decay of the type $p \rightarrow \pi^0 + e^+$. If the experiment is run for 4 years without detecting a decay, estimate the lower limit on the proton mean life.

APPROACH As with radioactive decay, the number of decays is proportional to the number of parent species (N), the time interval (Δt), and the decay constant (λ) which is related to the mean life τ by (see Eqs. 30–3 and 30–7):

$$\Delta N = -\lambda N \Delta t = -\frac{N \Delta t}{\tau}.$$

SOLUTION Dealing only with magnitudes, we solve for τ:

$$\tau = \frac{N \Delta t}{\Delta N}.$$

Thus for $\Delta N < 1$ (we don't see even one decay) over the four-year trial,

$$\tau > N(4 \, \text{yr}),$$

where N is the number of protons in 3300 tons of water. To determine N, we note that each molecule of H_2O contains $2 + 8 = 10$ protons. So one mole of water ($18 \, g$, 6×10^{23} molecules) contains $10 \times 6 \times 10^{23} = 6 \times 10^{24}$ protons in 18 g of water ($= 18 \, g/1000 \, g = 1/56$ of a kg), or about 3×10^{26} protons per kilogram. One ton is 10^3 kg, so the 3300 tons contains $(3.3 \times 10^6 \, \text{kg})(3 \times 10^{26} \, \text{protons/kg}) \approx 1 \times 10^{33}$ protons. Then our very rough estimate for a lower limit on the proton mean life is $\tau > (10^{33})(4 \, \text{yr}) \approx 4 \times 10^{33}$ yr.

*GUT and Cosmology

An interesting prediction of unified theories relates to cosmology (Chapter 33). It was thought by many theorists that during the first 10^{-35} s after the theorized Big Bang that created the universe, the temperature was so extremely high that particles had energies corresponding to the unification scale. Baryon number would not have been conserved then, perhaps allowing an imbalance that might account for the observed predominance of matter ($B > 0$) over antimatter ($B < 0$) in the universe. The fact that we are surrounded by matter, with no significant antimatter in sight, is considered a problem in search of an explanation (not given by the Standard Model). We call this the **matter–antimatter problem**. To understand it may require still undiscovered phenomena—perhaps related to quarks or neutrinos, or the Higgs boson or supersymmetry (next Section).

Many theorists no longer think the Big Bang was sufficiently hot to create unification. Nonetheless we see that there is a deep connection between investigations at either end of the size scale: theories about the tiniest objects (elementary particles) have a strong bearing on the understanding of the universe on a large scale. We look at this more in the next Chapter.

Figure 32–22 is a rough diagram indicating how the four fundamental forces in nature might have "condensed out" (a symmetry was broken) as time went on after the Big Bang (Chapter 33), and as the mean temperature of the universe and the typical particle energy decreased.

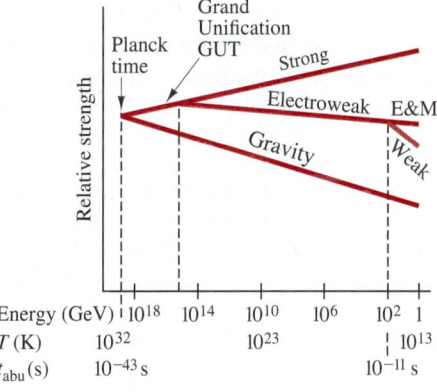

FIGURE 32-22 Time and energy plot of the four fundamental forces, perhaps unified at the "Planck time" (10^{-43} s after the birth of the universe), and how each condensed out, assuming a very hot Big Bang. The symbol t_{abu} means time after the birth of the universe. Note that the typical particle energy (and average temperature of the universe) decreases to the right, as time increases.

32–12 Strings and Supersymmetry

We have seen that the Standard Model is unable to address important experimental issues, and that theoreticians are attacking the problem as experimenters search for new data, new particles, new concepts.

Even more ambitious than grand unified theories are attempts to also incorporate gravity, and thus unify all four forces in nature into a single theory. (Such theories are sometimes referred to misleadingly as **theories of everything**.) A major attempt to unify all four forces is called **string theory**, introduced by Gabriele Veneziano in 1968: Each fundamental particle (Table 32–5) is imagined not as a point but as a one-dimensional **string**, perhaps 10^{-35} m long, which vibrates in a particular standing wave pattern. (You might say each particle is a different note on a tiny stretched string.) More sophisticated theories propose the fundamental entities as being multidimensional **branes** (after 2-D membranes).

A related idea that also goes way beyond the Standard Model is **supersymmetry**, which applied to strings is known as **superstring theory**. Supersymmetry, developed by Bruno Zumino (1923–) and Julius Wess (1934–2007), predicts that interactions exist that would change fermions into bosons and vice versa, and that each known fermion would have a supersymmetric boson partner. Thus, for each quark (a fermion), there would be a **squark** (a boson) or "supersymmetric" quark. For every lepton there would be a **slepton**. Likewise, for every known boson (photons and gluons, for example), there would be a supersymmetric fermion (**photinos** and **gluinos**). Supersymmetry predicts also that a *graviton*, which transmits the gravity force, has a partner, the **gravitino**. Supersymmetry (often abbreviated SUSY) offers solutions to a number of important theoretical problems. Supersymmetric particles are a candidate for the "dark matter" of the universe (discussed in Chapter 33). But why hasn't this "missing part" of the universe ever been detected? The best guess is that supersymmetric particles might be heavier than their conventional counterparts, perhaps too heavy to have been produced in today's accelerators. A search for supersymmetric particles is already being done at CERN's new Large Hadron Collider.

Versions of these theories predict other interesting properties, such as that space has 11 dimensions, but 7 of them are "coiled up" so we normally only notice the 4-D of space–time. We would like to know if and how many extra dimensions there are, and how and why they are hidden. We hope to have some answers from the new LHC (Section 32–1).

Some theorists think SUSY and other theories are approximations to a more fundamental, still undiscovered, **M-theory**. Edward Witten coined the term when proposing an 11 dimensional approximation, but never said what "M" stands for.

The world of elementary particles is opening new vistas. What happens in the future is bound to be exciting.

Summary

Particle accelerators are used to accelerate charged particles, such as electrons and protons, so they can have very high energy collisions with other particles. High-energy particles have short wavelength and so can be used to probe the structure of matter in great detail (very small distances). High kinetic energy also allows the creation of new particles through collisions (via $E = mc^2$).

Cyclotrons and **synchrotrons** use a magnetic field to keep the particles in a circular path and accelerate them at intervals by high voltage. **Linear accelerators** accelerate particles along a line. **Colliding beams** allow higher interaction energy.

An **antiparticle** has the same mass as a particle but opposite charge. Certain other properties may also be opposite: for example, the antiproton has **baryon number** (nucleon number) opposite ($B = -1$) to that for the proton ($B = +1$).

In all nuclear and particle reactions, the following conservation laws hold: momentum, angular momentum, mass–energy, electric charge, baryon number, and **lepton numbers**.

Certain particles have a property called **strangeness**, which is conserved by the strong force but not by the weak force. The properties **charm**, **bottom**, and **top** also are conserved by the strong force but not by the weak force.

Just as the electromagnetic force can be said to be due to an exchange of photons, the strong nuclear force is carried by massless **gluons**. The W and Z particles carry the weak force. These fundamental force carriers (photon, W and Z, gluons) are called **gauge bosons**.

Other particles can be classified as either *leptons* or *hadrons*. **Leptons** participate only in gravity, the weak, and the electromagnetic interactions. **Hadrons**, which today are considered **composite** particles, are made up of **quarks**, and participate in all four interactions, including the strong interaction. The hadrons can be classified as **mesons**, with baryon number zero, and **baryons**, with nonzero baryon number.

Most particles, except for the photon, electron, neutrinos, and proton, decay with measurable mean lives varying from 10^{-25} s to 10^3 s. The mean life depends on which force is predominant. Weak decays usually have mean lives greater than about 10^{-13} s. Electromagnetic decays typically have mean lives on the order of 10^{-16} to 10^{-19} s. The shortest lived particles, called **resonances**, decay via the strong interaction and live typically for only about 10^{-23} s.

Today's **Standard Model** of elementary particles considers **quarks** as the basic building blocks of the hadrons. The six quark "flavors" are called **up**, **down**, **strange**, **charm**, **bottom**, and **top**. It is expected that there are the same number of quarks as leptons (six of each), and that quarks and leptons are truly fundamental particles along with the gauge bosons (γ, W, Z, gluons) and the Higgs boson.

Quarks are said to have **color**, and, according to **quantum chromodynamics** (QCD), the strong color force acts between their color charges and is transmitted by **gluons**. **Electroweak theory** views the weak and electromagnetic forces as two aspects of a single underlying interaction. QCD plus the electroweak theory are referred to as the *Standard Model* of the fundamental particles.

Grand unified theories of forces suggest that at very short distance (10^{-31} m) and very high energy, the weak, electromagnetic, and strong forces would appear as a single force, and the fundamental difference between quarks and leptons would disappear.

According to **string theory**, the fundamental particles may be tiny strings, 10^{-35} m long, distinguished by their standing wave pattern. **Supersymmetry** predicts that each fermion (or boson) has a corresponding boson (or fermion) partner.

Questions

1. Give a reaction between two nucleons, similar to Eq. 32–4, that could produce a π^-.
2. If a proton is moving at very high speed, so that its kinetic energy is much greater than its rest energy (mc^2), can it then decay via $p \rightarrow n + \pi^+$?
3. What would an "antiatom," made up of the antiparticles to the constituents of normal atoms, consist of? What might happen if *antimatter*, made of such antiatoms, came in contact with our normal world of matter?
4. What particle in a decay signals the electromagnetic interaction?
5. (*a*) Does the presence of a neutrino among the decay products of a particle necessarily mean that the decay occurs via the weak interaction? (*b*) Do all decays via the weak interaction produce a neutrino? Explain.
6. Why is it that a neutron decays via the weak interaction even though the neutron and one of its decay products (proton) are strongly interacting?
7. Which of the four interactions (strong, electromagnetic, weak, gravitational) does an electron take part in? A neutrino? A proton?
8. Verify that charge and baryon number are conserved in each of the decays shown in Table 32–2.
9. Which of the particle decays listed in Table 32–2 occur via the electromagnetic interaction?
10. Which of the particle decays listed in Table 32–2 occur by the weak interaction?
11. The Δ baryon has spin $\frac{3}{2}$, baryon number 1, and charge $Q = +2, +1, 0,$ or -1. Why is there no charge state $Q = -2$?
12. Which of the particle decays in Table 32–4 occur via the electromagnetic interaction?
13. Which of the particle decays in Table 32–4 occur by the weak interaction?
14. Quarks have spin $\frac{1}{2}$. How do you account for the fact that baryons have spin $\frac{1}{2}$ or $\frac{3}{2}$, and mesons have spin 0 or 1?
15. Suppose there were a kind of "neutrinolet" that was massless, had no color charge or electrical charge, and did not feel the weak force. Could you say that this particle even exists?
16. Is it possible for a particle to be both (*a*) a lepton and a baryon? (*b*) a baryon and a hadron? (*c*) a meson and a quark? (*d*) a hadron and a lepton? Explain.
17. Using the ideas of quantum chromodynamics, would it be possible to find particles made up of two quarks and no antiquarks? What about two quarks and two antiquarks?
18. Why can neutrons decay when they are free, but not when they are inside a stable nucleus?
19. Is the reaction $e^- + p \rightarrow n + \bar{\nu}_e$ possible? Explain.
20. Occasionally, the Λ will decay by the following reaction: $\Lambda^0 \rightarrow p^+ + e^- + \bar{\nu}_e$. Which of the four forces in nature is responsible for this decay? How do you know?

MisConceptual Questions

1. There are six kinds (= flavors) of quarks: up, down, strange, charm, bottom, and top. Which flavors make up most of the known matter in the universe?
   (*a*) Up and down quarks.
   (*b*) Strange and charm quarks.
   (*c*) Bottom and top quarks.
   (*d*) All of the above.
2. Which of the following particles can not be composed of quarks?
   (*a*) Proton.
   (*b*) Electron.
   (*c*) π meson.
   (*d*) Neutron.
   (*e*) Higgs boson.

3. If gravity is the weakest force, why is it the one we notice most?
    (a) Our bodies are not sensitive to the other forces.
    (b) The other forces act only within atoms and therefore have no effect on us.
    (c) Gravity may be "very weak" but always attractive, and the Earth has enormous mass. The strong and weak nuclear forces have very short range. The electromagnetic force has a long range, but most matter is electrically neutral.
    (d) At long distances, the gravitational force is actually stronger than the other forces.
    (e) The other forces act only on elementary particles, not on objects our size.
4. Is it possible for a tau lepton (whose mass is almost twice that of a proton) to decay into only hadrons?
    (a) Yes, because it is so massive it could decay into a proton and pions.
    (b) Yes, it could decay into pions and nothing else.
    (c) No, such a decay would violate lepton number; all of its decay products must be leptons.
    (d) No, its decay products must include a tau neutrino but could include hadrons such as pions.
    (e) No, the tau lepton is too massive to decay.
5. Many particle accelerators are circular because:
    (a) particles accelerate faster around circles.
    (b) in order to move in a circle, acceleration is required.
    (c) a circular accelerator has a shorter length than a square one.
    (d) the particles can be accelerated through the same potential difference many times, making the accelerator more compact.
    (e) a particle moving in a circle needs more energy than a particle moving in a straight line.
6. Which of the following are today considered fundamental particles (that is, not composed of smaller components)? Choose as many as apply.
    (a) Atoms. (b) Electrons. (c) Protons. (d) Neutrons. (e) Quarks. (f) Photon. (g) Higgs boson.
7. The electron's antiparticle is called the positron. Which of the following properties, if any, are the same for electrons and positrons?
    (a) Mass.
    (b) Charge.
    (c) Lepton number.
    (d) None of the above.
8. The strong nuclear force between a neutron and a proton is due to
    (a) the exchange of π mesons between the neutron and the proton.
    (b) the conservation of baryon number.
    (c) the beta decay of the neutron into the proton.
    (d) the exchange of gluons between the quarks within the neutron and the proton.
    (e) Both (a) and (d) at different scales.
9. Electrons are still considered fundamental particles (in the group called leptons). But protons and neutrons are no longer considered fundamental; they have substructure and are made up of
    (a) pions. (b) leptons. (c) quarks. (d) bosons. (e) photons.
10. Which of the following will interact via the weak nuclear force *only*?
    (a) Quarks. (b) Gluons. (c) Neutrons. (d) Neutrinos. (e) Electrons. (f) Muons. (g) Higgs boson.

For assigned homework and other learning materials, go to the MasteringPhysics website.

Problems

32–1 Particles and Accelerators

1. (I) What is the total energy of a proton whose kinetic energy is 4.65 GeV?
2. (I) Calculate the wavelength of 28-GeV electrons.
3. (I) If α particles are accelerated by the cyclotron of Example 32–2, what must be the frequency of the voltage applied to the dees?
4. (I) What is the time for one complete revolution for a very high-energy proton in the 1.0-km-radius Fermilab accelerator?
5. (II) What strength of magnetic field is used in a cyclotron in which protons make 3.1×10^7 revolutions per second?
6. (II) (a) If the cyclotron of Example 32–2 accelerated α particles, what maximum energy could they attain? What would their speed be? (b) Repeat for deuterons ($^2_1$H). (c) In each case, what frequency of voltage is required?
7. (II) Which is better for resolving details of the nucleus: 25-MeV alpha particles or 25-MeV protons? Compare each of their wavelengths with the size of a nucleon in a nucleus.
8. (II) What is the wavelength (= minimum resolvable size) of 7.0-TeV protons at the LHC?
9. (II) The 1.0-km radius Fermilab Tevatron took about 20 seconds to bring the energies of the stored protons from 150 GeV to 1.0 TeV. The acceleration was done once per turn. Estimate the energy given to the protons on each turn. (You can assume that the speed of the protons is essentially c the whole time.)
10. (II) A cyclotron with a radius of 1.0 m is to accelerate deuterons ($^2_1$H) to an energy of 12 MeV. (a) What is the required magnetic field? (b) What frequency is needed for the voltage between the dees? (c) If the potential difference between the dees averages 22 kV, how many revolutions will the particles make before exiting? (d) How much time does it take for one deuteron to go from start to exit? (e) Estimate how far it travels during this time.
11. (III) Show that the energy of a particle (charge e) in a synchrotron, in the relativistic limit ($v \approx c$), is given by E (in eV) = Brc, where B is the magnetic field and r is the radius of the orbit (SI units).

32–2 to 32–6 Particle Interactions, Particle Exchange

12. (I) About how much energy is released when a Λ^0 decays to n + π^0? (See Table 32–2.)
13. (I) How much energy is released in the decay
    $$\pi^+ \rightarrow \mu^+ + \nu_\mu?$$
    See Table 32–2.

14. (I) Estimate the range of the strong force if the mediating particle were the kaon instead of a pion.

15. (I) How much energy is required to produce a neutron–antineutron pair?

16. (II) Determine the total energy released when Σ^0 decays to Λ^0 and then to a proton.

17. (II) Two protons are heading toward each other with equal speeds. What minimum kinetic energy must each have if a π^0 meson is to be created in the process? (See Table 32–2.)

18. (II) What minimum kinetic energy must a proton and an antiproton each have if they are traveling at the same speed toward each other, collide, and produce a K^+K^- pair in addition to themselves? (See Table 32–2.)

19. (II) What are the wavelengths of the two photons produced when a proton and antiproton at rest annihilate?

20. (II) The Λ^0 cannot decay by the following reactions. What conservation laws are violated in each of the reactions?
    (a) $\Lambda^0 \nrightarrow n + \pi^-$
    (b) $\Lambda^0 \nrightarrow p + K^-$
    (c) $\Lambda^0 \nrightarrow \pi^+ + \pi^-$

21. (II) What would be the wavelengths of the two photons produced when an electron and a positron, each with 420 keV of kinetic energy, annihilate in a head-on collision?

22. (II) Which of the following reactions and decays are possible? For those forbidden, explain what laws are violated.
    (a) $\pi^- + p \to n + \eta^0$
    (b) $\pi^+ + p \to n + \pi^0$
    (c) $\pi^+ + p \to p + e^+$
    (d) $p \to e^+ + \nu_e$
    (e) $\mu^+ \to e^+ + \bar{\nu}_\mu$
    (f) $p \to n + e^+ + \nu_e$

23. (II) Antiprotons can be produced when a proton with sufficient energy hits a stationary proton. Even if there is enough energy, which of the following reactions will not happen?
    $p + p \to p + \bar{p}$
    $p + p \to p + p + \bar{p}$
    $p + p \to p + p + p + \bar{p}$
    $p + p \to p + e^+ + e^+ + \bar{p}$

24. (III) In the rare decay $\pi^+ \to e^+ + \nu_e$, what is the kinetic energy of the positron? Assume the π^+ decays from rest and $m_\nu = 0$.

25. (III) For the decay $\Lambda^0 \to p + \pi^-$, calculate (a) the Q-value (energy released), and (b) the kinetic energy of the p and π^-, assuming the Λ^0 decays from rest. (Use relativistic formulas.)

26. (III) Calculate the maximum kinetic energy of the electron when a muon decays from rest via $\mu^- \to e^- + \bar{\nu}_e + \nu_\mu$. [*Hint*: In what direction do the two neutrinos move relative to the electron in order to give the electron the maximum kinetic energy? Both energy and momentum are conserved; use relativistic formulas.]

32–7 to 32–11 Resonances, Standard Model, Quarks, QCD, GUT

27. (I) The mean life of the Σ^0 particle is 7×10^{-20} s. What is the uncertainty in its rest energy? Express your answer in MeV.

28. (I) The measured width of the ψ (3686) meson is about 300 keV. Estimate its mean life.

29. (I) The measured width of the J/ψ meson is 88 keV. Estimate its mean life.

30. (I) The B^- meson is a $b\bar{u}$ quark combination. (a) Show that this is consistent for all quantum numbers. (b) What are the quark combinations for B^+, B^0, $\bar{B}^0$?

31. (I) What is the energy width (or uncertainty) of (a) η^0, and (b) ρ^+? See Table 32–2.

32. (II) Which of the following decays are possible? For those that are forbidden, explain which laws are violated.
    (a) $\Xi^0 \to \Sigma^+ + \pi^-$
    (b) $\Omega^- \to \Sigma^0 + \pi^- + \nu$
    (c) $\Sigma^0 \to \Lambda^0 + \gamma + \gamma$

33. (II) In ordinary radioactive decay, a W particle may be created even though the decaying particle has less mass than the W particle. If you assume $\Delta E \approx$ mass of the virtual W, what is the expected lifetime of the W?

34. (II) What quark combinations produce (a) a Ξ^0 baryon and (b) a Ξ^- baryon?

35. (II) What are the quark combinations that can form (a) a neutron, (b) an antineutron, (c) a Λ^0, (d) a $\bar{\Sigma}^0$?

36. (II) What particles do the following quark combinations produce: (a) uud, (b) $\bar{u}\bar{u}\bar{s}$, (c) $\bar{u}s$, (d) $d\bar{u}$, (e) $\bar{c}s$?

37. (II) What is the quark combination needed to produce a D^0 meson ($Q = B = S = 0$, $c = +1$)?

38. (II) The D_S^+ meson has $S = c = +1$, $B = 0$. What quark combination would produce it?

39. (II) Draw a possible Feynman diagram using quarks (as in Fig. 32–17c) for the reaction $\pi^- + p \to \pi^0 + n$.

40. (II) Draw a Feynman diagram for the reaction $n + \nu_\mu \to p + \mu^-$.

General Problems

41. What is the total energy of a proton whose kinetic energy is 15 GeV? What is its wavelength?

42. The mean lifetimes listed in Table 32–2 are in terms of *proper time*, measured in a reference frame where the particle is at rest. If a tau lepton is created with a kinetic energy of 950 MeV, how long would its track be as measured in the lab, on average, ignoring any collisions?

43. (a) How much energy is released when an electron and a positron annihilate each other? (b) How much energy is released when a proton and an antiproton annihilate each other? (All particles have KE ≈ 0.)

44. If 2×10^{14} protons moving at $v \approx c$, with KE $= 4.0$ TeV, are stored in the 4.3-km-radius ring of the LHC, (a) how much current (amperes) is carried by this beam? (b) How fast would a 1500-kg car have to move to carry the same kinetic energy as this beam?

45. Protons are injected into the 4.3-km-radius Large Hadron Collider with an energy of 450 GeV. If they are accelerated by 8.0 MV each revolution, how far do they travel and approximately how much time does it take for them to reach 4.0 TeV?

46. Which of the following reactions are possible, and by what interaction could they occur? For those forbidden, explain why.
    (a) $\pi^- + p \to K^0 + p + \pi^0$
    (b) $K^- + p \to \Lambda^0 + \pi^0$
    (c) $K^+ + n \to \Sigma^+ + \pi^0 + \gamma$
    (d) $K^+ \to \pi^0 + \pi^0 + \pi^+$
    (e) $\pi^+ \to e^+ + \nu_e$

47. Which of the following reactions are possible, and by what interaction could they occur? For those forbidden, explain why.
    (a) $\pi^- + p \to K^+ + \Sigma^-$
    (b) $\pi^+ + p \to K^+ + \Sigma^+$
    (c) $\pi^- + p \to \Lambda^0 + K^0 + \pi^0$
    (d) $\pi^+ + p \to \Sigma^0 + \pi^0$
    (e) $\pi^- + p \to p + e^- + \bar{\nu}_e$

48. One decay mode for a π^+ is $\pi^+ \to \mu^+ + \nu_\mu$. What would be the equivalent decay for a π^-? Check conservation laws.

49. Symmetry breaking occurs in the electroweak theory at about 10^{-18} m. Show that this corresponds to an energy that is on the order of the mass of the $W^\pm$.

50. Calculate the Q-value for each of the reactions, Eq. 32-4, for producing a pion.

51. How many fundamental fermions are there in a water molecule?

52. The mass of a π^0 can be measured by observing the reaction $\pi^- + p \to \pi^0 + n$ with initial kinetic energies near zero. The neutron is observed to be emitted with a kinetic energy of 0.60 MeV. Use conservation of energy and momentum to determine the π^0 mass.

53. (a) Show that the so-called unification distance of 10^{-31} m in grand unified theory is equivalent to an energy of about 10^{16} GeV. Use the uncertainty principle, and also de Broglie's wavelength formula, and explain how they apply. (b) Calculate the temperature corresponding to 10^{16} GeV.

54. Calculate the Q-value for the reaction $\pi^- + p \to \Lambda^0 + K^0$, when negative pions strike stationary protons. Estimate the minimum pion kinetic energy needed to produce this reaction. [Hint: Assume Λ^0 and K^0 move off with the same velocity.]

55. A proton and an antiproton annihilate each other at rest and produce two pions, π^- and π^+. What is the kinetic energy of each pion?

56. For the reaction $p + p \to 3p + \bar{p}$, where one of the initial protons is at rest, use relativistic formulas to show that the threshold energy is $6m_p c^2$, equal to three times the magnitude of the Q-value of the reaction, where m_p is the proton mass. [Hint: Assume all final particles have the same velocity.]

57. At about what kinetic energy (in eV) can the rest energy of a proton be ignored when calculating its wavelength, if the wavelength is to be within 1.0% of its true value? What are the corresponding wavelength and speed of the proton?

58. Use the quark model to describe the reaction
$$\bar{p} + n \to \pi^- + \pi^0.$$

59. Identify the missing particle in the following reactions.
    (a) $p + p \to p + n + \pi^+ + ?$ (b) $p + ? \to n + \mu^+$

60. What fraction of the speed of light c is the speed of a 7.0-TeV proton?

61. Using the information in Section 32–1, show that the Large Hadron Collider's two colliding proton beams can resolve details that are less than 1/10,000 the size of a nucleus.

62. Searches are underway for a process called **neutrinoless double beta decay**, in which a nucleus decays by emitting two electrons. (a) If the parent nucleus is $^{96}_{40}$Zr, what would the daughter nucleus be? (b) What conservation laws would be violated during this decay? (c) How could $^{96}_{40}$Zr decay to the same daughter nucleus without violating any conservation laws?

63. Estimate the lifetime of the Higgs boson from the width of the "bump" in Fig. 32–19, using the uncertainty principle. [Note: This is not a realistic estimate because the underlying processes are very complicated.]

Search and Learn

1. (a) What are the two major classes of particles that make up the matter of the universe? (b) Name six types, or flavors, of each class of particles. (c) What are the four known fundamental forces in the universe? (d) Name the particles that carry the forces in part c. Which force is much weaker than the other three?

2. (a) What property characterizes all hadrons? (b) What property characterizes all baryons? (c) What property characterizes all mesons?

3. Show that all conservation laws hold for all the decays described in Fig. 32–15 for the decays of the top quark.

4. The Higgs boson, Section 32–10, has very probably been detected at the CERN LHC. (a) If a Higgs boson at rest decays into two tau leptons, what is the kinetic energy of each tau? Follow the analysis of Example 32–5. See Table 32–2. (b) What are the signs of the electric charges of the two tau leptons? (c) Could a Higgs boson decay into two Z bosons (Table 32–2)?

5. (a) Show, by conserving momentum and energy, that it is impossible for an isolated electron to radiate only a single photon. (b) With this result in mind, how can you defend the photon exchange diagram in Fig. 32–8?

6. What magnetic field is required for the 4.25-km-radius Large Hadron Collider (LHC) to accelerate protons to 7.0 TeV? [Hint: Use relativity, Chapter 26.]

ANSWERS TO EXERCISES

A: 1.24×10^{-18} m = 1.24 am.
B: (a).
C: (c); (d).
D: $s\bar{u}$.

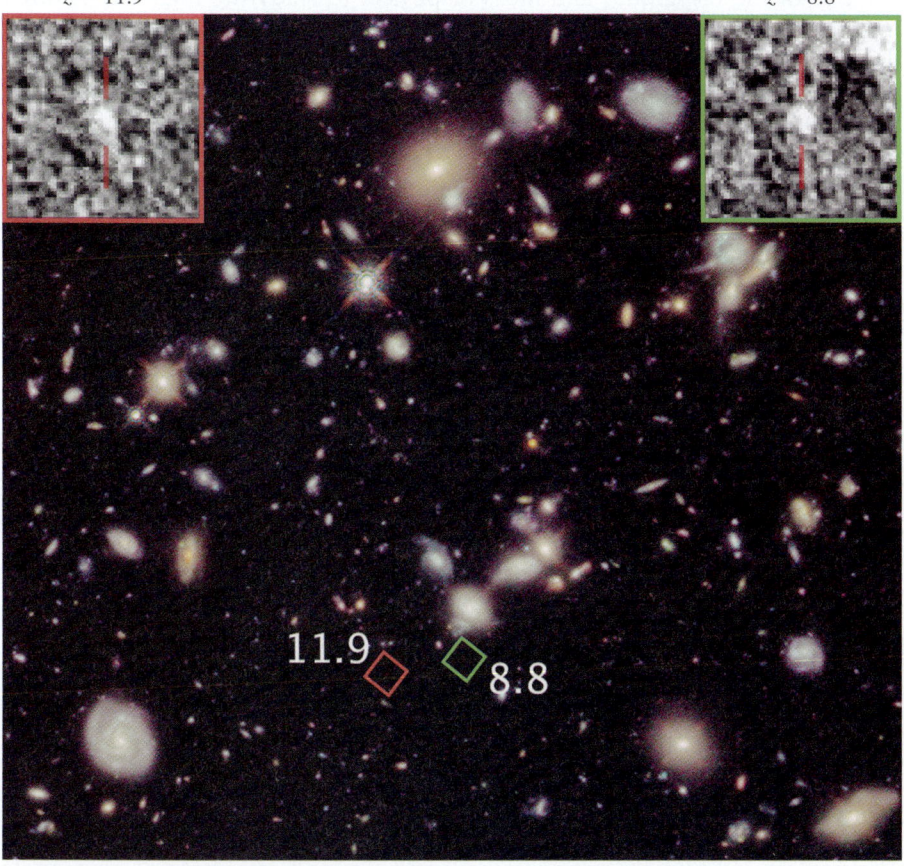

This Hubble eXtreme Deep Field (XDF) photograph is of a very small part of the sky. It includes what may be the most distant galaxies observable by us (small red and green squares, and shown enlarged in the corners), with $z \approx 8.8$ and 11.9, that already existed when the universe was about 0.4 billion years old. We see these galaxies as they appeared then, 13.4 billion years ago, which is when they emitted this light. The most distant galaxies were young and small and grew to become large galaxies by colliding and merging with other small galaxies.

We examine the latest theories on how stars and galaxies form and evolve, including the role of nucleosynthesis, as well as Einstein's general theory of relativity which deals with gravity and curvature of space. We take a thorough look at the evidence for the expansion of the universe, and the Standard Model of the universe evolving from an initial Big Bang. We point out some unsolved problems, including the nature of dark matter and dark energy that make up most of our universe.

Astrophysics and Cosmology

CHAPTER 33

CHAPTER-OPENING QUESTIONS—Guess now!

1. Until recently, astronomers expected the expansion rate of the universe would be decreasing. Why?
   (a) Friction.
   (b) The second law of thermodynamics.
   (c) Gravity.
   (d) The electromagnetic force.

2. The universe began expanding right at the beginning. How long will it continue to expand?
   (a) Until it runs out of room.
   (b) Until friction slows it down and brings it to a stop.
   (c) Until all galaxies are moving at the speed of light relative to the center.
   (d) Possibly forever.

CONTENTS

- 33–1 Stars and Galaxies
- 33–2 Stellar Evolution: Birth and Death of Stars, Nucleosynthesis
- 33–3 Distance Measurements
- 33–4 General Relativity: Gravity and the Curvature of Space
- 33–5 The Expanding Universe: Redshift and Hubble's Law
- 33–6 The Big Bang and the Cosmic Microwave Background
- 33–7 The Standard Cosmological Model: Early History of the Universe
- 33–8 Inflation: Explaining Flatness, Uniformity, and Structure
- 33–9 Dark Matter and Dark Energy
- 33–10 Large-Scale Structure of the Universe
- 33–11 Finally …

In the previous Chapter, we studied the tiniest objects in the universe—the elementary particles. Now we leap to the grandest objects in the universe—stars, galaxies, and clusters of galaxies—plus the history and structure of the universe itself. These two extreme realms, elementary particles and the cosmos, are among the most intriguing and exciting subjects in science. And, surprisingly, these two extreme realms are related in a fundamental way, as was already hinted in Chapter 32.

947

Use of the techniques and ideas of physics to study the night sky is often referred to as **astrophysics**. Central to our present theoretical understanding of the universe (or cosmos) is Einstein's *general theory of relativity* which represents our most complete understanding of gravitation. Many other aspects of physics are involved, from electromagnetism and thermodynamics to atomic and nuclear physics as well as elementary particles. General Relativity serves also as the foundation for modern **cosmology**, which is the study of the universe as a whole. Cosmology deals especially with the search for a theoretical framework to understand the observed universe, its origin, and its future. The questions posed by cosmology are profound and difficult; the possible answers stretch the imagination. They are questions like "Has the universe always existed, or did it have a beginning in time?" Either alternative is difficult to imagine: time going back indefinitely into the past, or an actual moment when the universe began (but, then, what was there before?). And what about the size of the universe? Is it infinite in size? It is hard to imagine infinity. Or is it finite in size? This is also hard to imagine, for if the universe is finite, it does not make sense to ask what is beyond it, because the universe is all there is.

In the last 10 to 20 years, so much progress has occurred in astrophysics and cosmology that many scientists are calling recent work a "Golden Age" for cosmology. Our survey will be qualitative, but we will nonetheless touch on the major ideas. We begin with a look at what can be seen beyond the Earth.

FIGURE 33–1 Sections of the Milky Way. In (a), the thin line is the trail of an artificial Earth satellite in this long time exposure. The dark diagonal area is due to dust absorption of visible light, blocking the view. In (b) the view is toward the center of the Galaxy (taken in summer from Arizona).

(a)

(b)

33–1 Stars and Galaxies

According to the ancients, the stars, except for the few that seemed to move relative to the others (the planets), were fixed on a sphere beyond the last planet. The universe was neatly self-contained, and we on Earth were at or near its center. But in the centuries following Galileo's first telescopic observations of the night sky in 1609, our view of the universe has changed dramatically. We no longer place ourselves at the center, and we view the universe as vastly larger. The distances involved are so great that we specify them in terms of the time it takes light to travel the given distance: for example,

$$1 \text{ light-second} = (3.0 \times 10^8 \text{ m/s})(1.0 \text{ s}) = 3.0 \times 10^8 \text{ m} = 300{,}000 \text{ km};$$
$$1 \text{ light-minute} = (3.0 \times 10^8 \text{ m/s})(60 \text{ s}) = 18 \times 10^6 \text{ km}.$$

The most common unit is the **light-year** (ly):

$$1 \text{ ly} = (2.998 \times 10^8 \text{ m/s})(3.156 \times 10^7 \text{ s/yr})$$
$$= 9.46 \times 10^{15} \text{ m} \approx 10^{13} \text{ km} \approx 10^{16} \text{ m}.$$

For specifying distances to the Sun and Moon, we usually use meters or kilometers, but we could specify them in terms of light seconds or minutes. The Earth–Moon distance is 384,000 km, which is 1.28 light-seconds. The Earth–Sun distance is 1.50×10^{11} m, or 150,000,000 km; this is equal to 8.3 light-minutes (it takes 8.3 min for light emitted by the Sun to reach us). Far out in our solar system, Pluto is about 6×10^9 km from the Sun, or 6×10^{-4} ly.[†] The nearest star to us, other than the Sun, is Proxima Centauri, about 4.2 ly away.

On a clear moonless night, thousands of stars of varying degrees of brightness can be seen, as well as the long cloudy stripe known as the Milky Way (Fig. 33–1). Galileo first observed, with his telescope, that the Milky Way is comprised of countless individual stars. A century and a half later (about 1750), Thomas Wright suggested that the Milky Way was a flat disk of stars extending to great distances in a plane, which we call the **Galaxy** (Greek for "milky way").

[†]We can also say this is about 5 light-hours.

Our Galaxy has a diameter of almost 100,000 light-years and a thickness of roughly 2000 ly. It has a central bulge and spiral arms (Fig. 33–2). Our Sun, which is a star like many others, is located about halfway from the galactic center to the edge, some 26,000 ly from the center. Our Galaxy contains roughly 400 billion (4×10^{11}) stars. The Sun orbits the galactic center approximately once every 250 million years, so its speed is roughly 200 km/s relative to the center of the Galaxy. The total mass of all the stars in our Galaxy is estimated to be about 4×10^{41} kg of ordinary matter. There is also strong evidence that our Galaxy is permeated and surrounded by a massive invisible "halo" of "dark matter" (Section 33–9).

FIGURE 33–2 Our Galaxy, as it would appear from the outside: (a) "edge view," in the plane of the disk; (b) "top view," looking down on the disk. (If only we could see it like this— from the outside!) (c) Infrared photograph of the inner reaches of the Milky Way, showing the central bulge and disk of our Galaxy. This very wide angle photo taken from the COBE satellite (Section 33–6) extends over 360° of sky. The white dots are nearby stars.

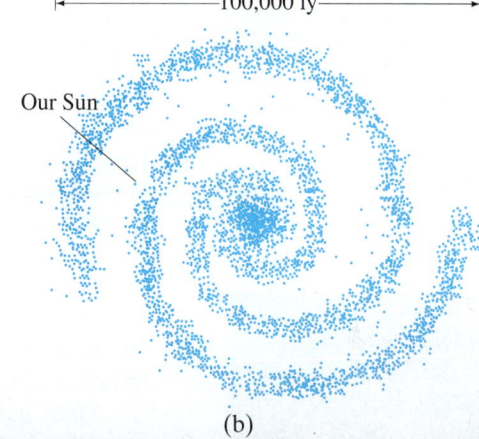

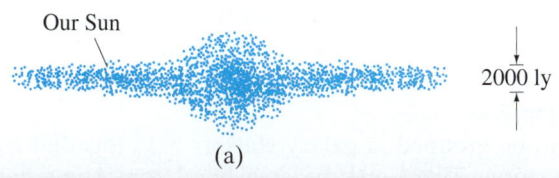

EXAMPLE 33–1 ESTIMATE **Our Galaxy's mass.** Estimate the total mass of our Galaxy using the orbital data above for the Sun about the center of the Galaxy. Assume the mass of the Galaxy is concentrated in the central bulge.

APPROACH We assume that the Sun (including our solar system) has total mass m and moves in a circular orbit about the center of the Galaxy (total mass M), and that the mass M can be considered as being located at the center of the Galaxy. We then apply Newton's second law, $F = ma$, with a being the centripetal acceleration, $a = v^2/r$, and for F we use the universal law of gravitation (Chapter 5).

SOLUTION Our Sun and solar system orbit the center of the Galaxy, according to the best measurements as mentioned above, with a speed of about $v = 200$ km/s at a distance from the Galaxy center of about $r = 26,000$ ly. We use Newton's second law:

$$F = ma$$
$$G\frac{Mm}{r^2} = m\frac{v^2}{r}$$

where M is the mass of the Galaxy and m is the mass of our Sun and solar system. Solving this, we find

$$M = \frac{rv^2}{G} \approx \frac{(26{,}000 \text{ ly})(10^{16} \text{ m/ly})(2 \times 10^5 \text{ m/s})^2}{6.67 \times 10^{-11} \text{ N} \cdot \text{m}^2/\text{kg}^2} \approx 2 \times 10^{41} \text{ kg}.$$

NOTE In terms of *numbers* of stars, if they are like our Sun ($m = 2.0 \times 10^{30}$ kg), there would be about $(2 \times 10^{41} \text{ kg})/(2 \times 10^{30} \text{ kg}) \approx 10^{11}$ or very roughly on the order of 100 billion stars.

FIGURE 33–3 This globular star cluster is located in the constellation Hercules.

FIGURE 33–4 This gaseous nebula, found in the constellation Carina, is about 9000 light-years from us.

In addition to stars both within and outside the Milky Way, we can see by telescope many faint cloudy patches in the sky which were all referred to once as "nebulae" (Latin for "clouds"). A few of these, such as those in the constellations Andromeda and Orion, can actually be discerned with the naked eye on a clear night. Some are **star clusters** (Fig. 33–3), groups of stars that are so numerous they appear to be a cloud. Others are glowing clouds of gas or dust (Fig. 33–4), and it is for these that we now mainly reserve the word **nebula**.

Most fascinating are those that belong to a third category: they often have fairly regular elliptical shapes. Immanuel Kant (about 1755) guessed they are faint because they are a great distance beyond our Galaxy. At first it was not universally accepted that these objects were **extragalactic**—that is, outside our Galaxy. But the very large telescopes constructed in the twentieth century revealed that individual stars could be resolved within these extragalactic objects and that many contain spiral arms. Edwin Hubble (1889–1953) did much of this observational work in the 1920s using the 2.5-m (100-inch) telescope[†] on Mt. Wilson near Los Angeles, California, then the world's largest. Hubble demonstrated that these objects were indeed extragalactic because of their great distances. The distance to our nearest large galaxy,[‡] Andromeda, is over 2 million light-years, a distance 20 times greater than the diameter of our Galaxy. It seemed logical that these nebulae must be **galaxies** similar to ours. (Note that it is usual to capitalize the word "galaxy" only when it refers to our own.) Today it is thought there are roughly 10^{11} galaxies in the observable universe—that is, roughly as many galaxies as there are stars in a galaxy. See Fig. 33–5.

Many galaxies tend to be grouped in **galaxy clusters** held together by their mutual gravitational attraction. There may be anywhere from a few dozen to many thousands of galaxies in each cluster. Furthermore, clusters themselves seem to be organized into even larger aggregates: clusters of clusters of galaxies, or **superclusters**. The farthest detectable galaxies are more than 10^{10} ly distant. See Table 33–1 (top of next page).

[†]2.5 m (= 100 inches) refers to the diameter of the curved objective mirror. The bigger the mirror, the more light it collects (greater brightness) and the less diffraction there is (better resolution), so more and fainter stars can be seen. See Chapter 25. Until recently, photographic films or plates were used to take long time exposures. Now large solid-state CCD or CMOS sensors (Section 25–1) are available containing hundreds of millions of pixels (compared to 10 million pixels in a good-quality digital camera).

[‡]The *Magellanic clouds* are much closer than Andromeda, but are small and are usually considered small satellite galaxies of our own Galaxy.

FIGURE 33–5 Photographs of galaxies. (a) Spiral galaxy in the constellation Hydra. (b) Two galaxies: the larger and more dramatic one is known as the Whirlpool galaxy. (c) An infrared image (given "false" colors) of the same galaxies as in (b), here showing the arms of the spiral as having more substance than in the visible light photo (b); the different colors correspond to different light intensities. Visible light is scattered and absorbed by interstellar dust much more than infrared is, so infrared gives us a clearer image.

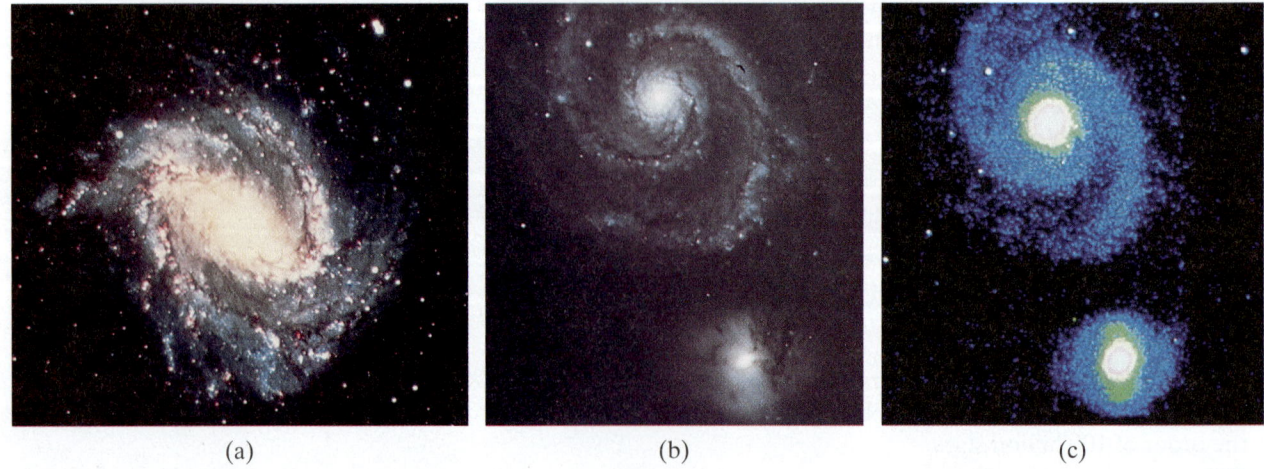

(a)   (b)   (c)

CONCEPTUAL EXAMPLE 33–2 **Looking back in time.** Astronomers often think of their telescopes as time machines, looking back toward the origin of the universe. How far back do they look?

RESPONSE The distance in light-years measures how long in years the light has been traveling to reach us, so Table 33–1 tells us also how far back in time we are looking. For example, if we saw Proxima Centauri explode into a supernova today, then the event would have really occurred about 4.2 years ago. The most distant galaxies emitted the light we see now roughly 13×10^9 years ago. What we see was how they were then, 13×10^9 yr ago.

EXERCISE A Suppose we could place a huge mirror 1 light-year away from us. What would we see in this mirror if it is facing us on Earth? When did what we see in the mirror take place? (This might be called a "time machine.")

Table 33–1 Astronomical Distances

Object	Approx. Distance from Earth (ly)
Moon	4×10^{-8}
Sun	1.6×10^{-5}
Size of solar system (distance to Pluto)	6×10^{-4}
Nearest star (Proxima Centauri)	4.2
Center of our Galaxy	2.6×10^4
Nearest large galaxy	2.4×10^6
Farthest galaxies	13.4×10^9

Besides the usual stars, clusters of stars, galaxies, and clusters and superclusters of galaxies, the universe contains many other interesting objects. Among these are stars known as *red giants*, *white dwarfs*, *neutron stars*, exploding stars called *novae* and *supernovae*, and *black holes* whose gravity is so strong that even light cannot escape them. In addition, there is electromagnetic radiation that reaches the Earth but does not come from the bright pointlike objects we call stars: particularly important is the microwave background radiation that arrives nearly uniformly from all directions in the universe.

Finally, there are **active galactic nuclei** (**AGN**), which are very luminous pointlike sources of light in the centers of distant galaxies. The most dramatic examples of AGN are **quasars** ("quasistellar objects" or QSOs), which are so luminous that the surrounding starlight of the galaxy is drowned out. Their luminosity is thought to come from matter falling into a giant black hole at a galaxy's center.

33–2 Stellar Evolution: Birth and Death of Stars, Nucleosynthesis

The stars appear unchanging. Night after night the night sky reveals no significant variations. Indeed, on a human time scale, the vast majority of stars change very little (except for novae, supernovae, and certain variable stars). Although stars *seem* fixed in relation to each other, many move sufficiently for the motion to be detected. Speeds of stars relative to neighboring stars can be hundreds of km/s, but at their great distance from us, this motion is detectable only by careful measurement. There is also a great range of brightness among stars, due to differences in the rate stars emit energy and to their different distances from us.

Luminosity and Brightness of Stars

Any star or galaxy has an **intrinsic luminosity**, L (or simply **luminosity**), which is its total power radiated in watts. Also important is the **apparent brightness**, b, defined as the power crossing unit area at the Earth perpendicular to the path of the light. Given that energy is conserved, and ignoring any absorption in space, the total emitted power L when it reaches a distance d from the star will be spread over a sphere of surface area $4\pi d^2$. If d is the distance from the star to the Earth, then L must be equal to $4\pi d^2$ times b (power per unit area at Earth). That is,

$$b = \frac{L}{4\pi d^2}. \quad (33–1)$$

EXAMPLE 33–3 **Apparent brightness.** Suppose a star has luminosity equal to that of our Sun. If it is 10 ly away from Earth, how much dimmer will it appear?

APPROACH We use the inverse square law in Eq. 33–1 to determine the relative brightness $(b \propto 1/d^2)$ since the luminosity L is the same for both stars.

SOLUTION Using the inverse square law, the star appears dimmer by a factor

$$\frac{b_{\text{star}}}{b_{\text{Sun}}} = \frac{d_{\text{Sun}}^2}{d_{\text{star}}^2} = \frac{(1.5 \times 10^8 \text{ km})^2}{(10 \text{ ly})^2 (10^{13} \text{ km/ly})^2} \approx 2 \times 10^{-12}.$$

Careful study of nearby stars has shown that the luminosity for most stars depends on the mass: *the more massive the star, the greater its luminosity*[†]. Another important parameter of a star is its surface temperature, which can be determined from the spectrum of electromagnetic frequencies it emits. As we saw in Chapter 27, as the temperature of a body increases, the spectrum shifts from predominantly lower frequencies (and longer wavelengths, such as red) to higher frequencies (and shorter wavelengths such as blue). Quantitatively, the relation is given by Wien's law (Eq. 27–2): the wavelength λ_P at the peak of the spectrum of light emitted by a blackbody (we often approximate stars as blackbodies) is inversely proportional to its Kelvin temperature T; that is, $\lambda_P T = 2.90 \times 10^{-3}$ m·K. The surface temperatures of stars typically range from about 3000 K (reddish) to about 50,000 K (UV).

EXAMPLE 33–4 Determining star temperature and star size. Suppose that the distances from Earth to two nearby stars can be reasonably estimated, and that their measured apparent brightnesses suggest the two stars have about the same luminosity, L. The spectrum of one of the stars peaks at about 700 nm (so it is reddish). The spectrum of the other peaks at about 350 nm (bluish). Use Wien's law (Eq. 27–2) and the Stefan-Boltzmann equation (Section 14–8) to determine (*a*) the surface temperature of each star, and (*b*) how much larger one star is than the other.

APPROACH We determine the surface temperature T for each star using Wien's law and each star's peak wavelength. Then, using the Stefan-Boltzmann equation (power output or luminosity $\propto AT^4$ where A = surface area of emitter), we can find the surface area ratio and relative sizes of the two stars.

SOLUTION (*a*) Wien's law (Eq. 27–2) states that $\lambda_P T = 2.90 \times 10^{-3}$ m·K. So the temperature of the reddish star is

$$T_r = \frac{2.90 \times 10^{-3}\,\text{m·K}}{\lambda_P} = \frac{2.90 \times 10^{-3}\,\text{m·K}}{700 \times 10^{-9}\,\text{m}} = 4140\,\text{K}.$$

The temperature of the bluish star will be double this because its peak wavelength is half (350 nm vs. 700 nm):

$$T_b = 8280\,\text{K}.$$

(*b*) The Stefan-Boltzmann equation, Eq. 14–6, states that the power radiated *per unit area* of surface from a blackbody is proportional to the fourth power of the Kelvin temperature, T^4. The temperature of the bluish star is double that of the reddish star, so the bluish one must radiate $(T_b/T_r)^4 = 2^4 = 16$ times as much energy per unit area. But we are given that they have the same luminosity (the same total power output); so the surface area of the blue star must be $\frac{1}{16}$ that of the red one. The surface area of a sphere is $4\pi r^2$, so the radius of the reddish star is $\sqrt{16} = 4$ times larger than the radius of the bluish star (or $4^3 = 64$ times the volume).

H–R Diagram

An important astronomical discovery, made around 1900, was that for most stars, the color is related to the intrinsic luminosity and therefore to the mass. A useful way to present this relationship is by the so-called Hertzsprung–Russell (H–R) diagram. On the H–R diagram, the horizontal axis shows the surface temperature T and the vertical axis is the luminosity L; each star is represented by a point

---

[†]Applies to "main-sequence" stars (see next page). The mass of a star can be determined by observing its gravitational effects on other visible objects. Many stars are part of a cluster, the simplest being a binary star in which two stars orbit around each other, allowing their masses to be determined using rotational mechanics.

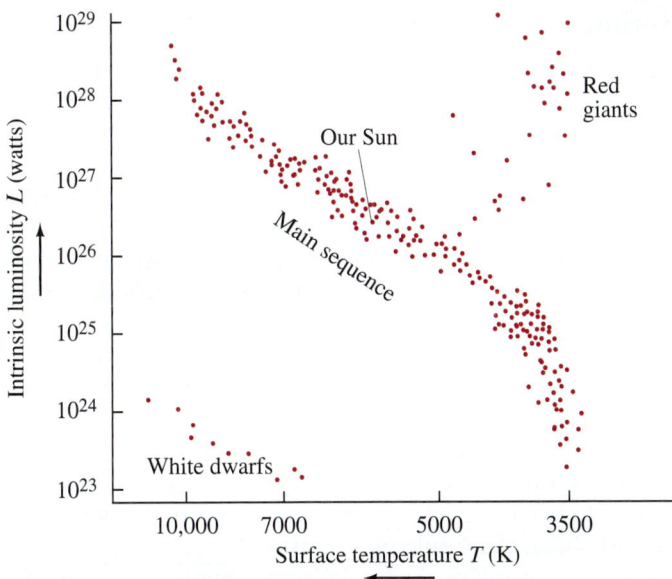

FIGURE 33–6 Hertzsprung–Russell (H–R) diagram is a logarithmic graph of luminosity vs. surface temperature T of stars (note that T increases to the left).

on the diagram, Fig. 33–6. Most stars fall along the diagonal band termed the **main sequence**. Starting at the lower right we find the coolest stars: by Wien's law, $\lambda_P T = $ constant, their light output peaks at long wavelengths, so they are reddish in color. They are also the least luminous and therefore of low mass. Farther up toward the left we find hotter and more luminous stars that are whitish, like our Sun. Still farther up we find even more luminous and more massive stars, bluish in color. Stars that fall on this diagonal band are called *main-sequence stars*. There are also stars that fall outside the main sequence. Above and to the right we find extremely large stars, with high luminosities but with low (reddish) color temperature: these are called **red giants**. At the lower left, there are a few stars of low luminosity but with high temperature: these are the **white dwarfs**.

EXAMPLE 33–5 ESTIMATE **Distance to a star using the H–R diagram and color.** Suppose that detailed study of a certain star suggests that it most likely fits on the main sequence of an H–R diagram. Its measured apparent brightness is $b = 1.0 \times 10^{-12}$ W/m², and the peak wavelength of its spectrum is $\lambda_P \approx 600$ nm. Estimate its distance from us.

APPROACH We find the temperature using Wien's law, Eq. 27–2. The luminosity is estimated for a main-sequence star on the H–R diagram of Fig. 33–6, and then the distance is found using the relation between brightness and luminosity, Eq. 33–1.

SOLUTION The star's temperature, from Wien's law (Eq. 27–2), is

$$T \approx \frac{2.90 \times 10^{-3} \text{ m} \cdot \text{K}}{600 \times 10^{-9} \text{ m}} \approx 4800 \text{ K}.$$

A star on the main sequence of an H–R diagram at this temperature has luminosity of about $L \approx 1 \times 10^{26}$ W, read off of Fig. 33–6. Then, from Eq. 33–1,

$$d = \sqrt{\frac{L}{4\pi b}} \approx \sqrt{\frac{1 \times 10^{26} \text{ W}}{4(3.14)(1.0 \times 10^{-12} \text{ W/m}^2)}} \approx 3 \times 10^{18} \text{ m}.$$

Its distance from us in light-years is

$$d = \frac{3 \times 10^{18} \text{ m}}{10^{16} \text{ m/ly}} \approx 300 \text{ ly}.$$

EXERCISE B Estimate the distance to a 6000-K main-sequence star with an apparent brightness of 2.0×10^{-12} W/m².

Stellar Evolution; Nucleosynthesis

Why are there different types of stars, such as red giants and white dwarfs, as well as main-sequence stars? Were they all born this way, in the beginning? Or might each different type represent a different age in the life cycle of a star? Astronomers and astrophysicists today believe the latter is the case. Note, however, that we cannot actually follow any but the tiniest part of the life cycle of any given star because they live for ages vastly greater than ours, on the order of millions or billions of years. Nonetheless, let us follow the process of **stellar evolution** from the birth to the death of a star, as astrophysicists have theoretically reconstructed it today.

Stars are born, it is believed, when gaseous clouds (mostly hydrogen) contract due to the pull of gravity. A huge gas cloud might fragment into numerous contracting masses, each mass centered in an area where the density is only slightly greater than that at nearby points. Once such "globules" form, gravity causes each to contract in toward its center of mass. As the particles of such a *protostar* accelerate inward, their kinetic energy increases. Eventually, when the kinetic energy is sufficiently high, the Coulomb repulsion between the positive charges is not strong enough to keep all the hydrogen nuclei apart, and nuclear fusion can take place.

In a star like our Sun, the fusion of hydrogen (sometimes referred to as "burning")[†] occurs via the *proton–proton chain* (Section 31–3, Eqs. 31–6), in which four protons fuse to form a ^4_2He nucleus with the release of γ rays, positrons, and neutrinos: $4\,^1_1\text{H} \rightarrow\, ^4_2\text{He} + 2\,e^+ + 2\nu_e + 2\gamma$. These reactions require a temperature of about 10^7 K, corresponding to an average kinetic energy ($\approx kT$) of about 1 keV (Eq. 13–8). In more massive stars, the carbon cycle produces the same net effect: four ^1_1H produce a ^4_2He—see Section 31–3. The fusion reactions take place primarily in the core of a star, where T may be on the order of 10^7 to 10^8 K. (The surface temperature is much lower—on the order of a few thousand kelvins.) The tremendous release of energy in these fusion reactions produces an outward pressure sufficient to halt the inward gravitational contraction. Our protostar, now really a young *star*, stabilizes on the *main sequence*. Exactly where the star falls along the main sequence depends on its mass. The more massive the star, the farther up (and to the left) it falls on the H–R diagram of Fig. 33–6. Our Sun required perhaps 30 million years to reach the main sequence, and is expected to remain there about 10 billion years (10^{10} yr). Although most stars are billions of years old, evidence is strong that stars are actually being born at this moment. More massive stars have shorter lives, because they are hotter and the Coulomb repulsion is more easily overcome, so they use up their fuel faster. Our Sun may remain on the main sequence for 10^{10} years, but a star ten times more massive may reside there for only 10^7 years.

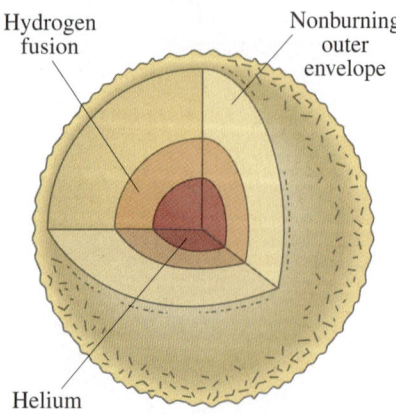

FIGURE 33–7 A shell of "burning" hydrogen (fusing to become helium) surrounds the core where the newly formed helium gravitates.

As hydrogen fuses to form helium, the helium that is formed is denser and tends to accumulate in the central core where it was formed. As the core of helium grows, hydrogen continues to fuse in a shell around it: see Fig. 33–7. When much of the hydrogen within the core has been consumed, the production of energy decreases at the center and is no longer sufficient to prevent the huge gravitational forces from once again causing the core to contract and heat up. The hydrogen in the shell around the core then fuses even more fiercely because of this rise in temperature, allowing the outer envelope of the star to expand and to cool. The surface temperature, thus reduced, produces a spectrum of light that peaks at longer wavelength (reddish).

This process marks a new step in the evolution of a star. The star has become redder, it has grown in size, and it has become more luminous, which means it has left the main sequence. It will have moved to the right and upward on the

---

[†]The word "burn," meaning fusion, is put in quotation marks because these high-temperature fusion reactions occur via a *nuclear* process, and must not be confused with ordinary burning (of, say, paper, wood, or coal) in air, which is a *chemical* reaction, occurring at the *atomic* level (and at a much lower temperature).

H–R diagram, as shown in Fig. 33–8. As it moves upward, it enters the **red giant** stage. Thus, theory explains the origin of red giants as a natural step in a star's evolution. Our Sun, for example, has been on the main sequence for about $4\frac{1}{2}$ billion years. It will probably remain there another 5 or 6 billion years. When our Sun leaves the main sequence, it is expected to grow in diameter (as it becomes a red giant) by a factor of 100 or more, possibly swallowing up inner planets such as Mercury and possibly Venus and even Earth.

If the star is like our Sun, or larger, further fusion can occur. As the star's outer envelope expands, its core continues to shrink and heat up. When the temperature reaches about 10^8 K, even helium nuclei, in spite of their greater charge and hence greater electrical repulsion, can come close enough to each other to undergo fusion. The reactions are

$$\begin{aligned} {}^{4}_{2}\text{He} + {}^{4}_{2}\text{He} &\rightarrow {}^{8}_{4}\text{Be} \\ {}^{4}_{2}\text{He} + {}^{8}_{4}\text{Be} &\rightarrow {}^{12}_{6}\text{C} \end{aligned} \qquad (33\text{–}2)$$

with the emission of two γ rays. These two reactions must occur in quick succession (because ${}^{8}_{4}\text{Be}$ is very unstable), and the net effect is

$$3\,{}^{4}_{2}\text{He} \rightarrow {}^{12}_{6}\text{C} + 2\gamma. \qquad (Q = 7.3\,\text{MeV})$$

This fusion of helium causes a change in the star which moves rapidly to the "horizontal branch" on the H–R diagram (Fig. 33–8). Further fusion reactions are possible, with ${}^{4}_{2}\text{He}$ fusing with ${}^{12}_{6}\text{C}$ to form ${}^{16}_{8}\text{O}$. In more massive stars, higher Z elements like ${}^{20}_{10}\text{Ne}$ or ${}^{24}_{12}\text{Mg}$ can be made. This process of creating heavier nuclei from lighter ones (or by absorption of neutrons which tends to occur at higher Z) is called **nucleosynthesis**.

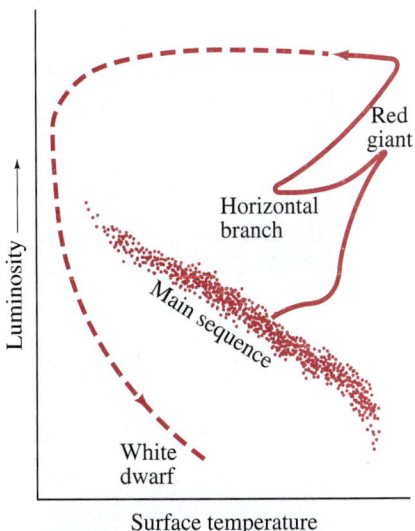

FIGURE 33–8 Evolutionary "track" of a star like our Sun represented on an H–R diagram.

Low Mass Stars—White Dwarfs

The final fate of a star depends on its mass. Stars can lose mass as parts of their outer envelope move off into space. Stars born with a mass less than about 8 solar masses (8× the mass of our Sun) eventually end up with a residual mass less than about 1.4 solar masses. A residual mass of 1.4 solar masses is known as the **Chandrasekhar limit**. For stars smaller than this, no further fusion energy can be obtained because of the large Coulomb repulsion between nuclei. The core of such a "low mass" star (original mass ≲ 8 solar masses) contracts under gravity. The outer envelope expands again and the star becomes an even brighter and larger red giant, Fig. 33–8. Eventually the outer layers escape into space, and the newly revealed surface is hotter than before. So the star moves to the left in the H–R diagram (horizontal dashed line in Fig. 33–8). Then, as the core shrinks the star cools, and typically follows the downward dashed route shown on the left in Fig. 33–8, becoming a **white dwarf**. A white dwarf with a residual mass equal to that of the Sun would be about the size of the Earth. A white dwarf contracts to the point at which the electrons start to overlap, but no further because, by the Pauli exclusion principle, no two electrons can be in the same quantum state. At this point the star is supported against further collapse by this **electron degeneracy** pressure. A white dwarf continues to lose internal energy by radiation, decreasing in temperature and becoming dimmer until it glows no more. It has then become a cold dark chunk of extremely dense material.

High Mass Stars—Supernovae, Neutron Stars, Black Holes

Stars whose original mass is greater than about 8 solar masses are thought to follow a very different scenario. A star with this great a mass can contract under gravity and heat up even further. At temperatures $T \approx 3$ or 4×10^9 K, nuclei as heavy as ${}^{56}_{26}\text{Fe}$ and ${}^{56}_{28}\text{Ni}$ can be made. But here the formation of heavy nuclei from lighter ones, by fusion, ends. As we saw in Fig. 30–1, the average binding energy per nucleon begins to decrease for A greater than about 60. Further fusions would *require* energy, rather than release it.

At these extremely high temperatures, well above 10^9 K, high-energy collisions can cause the breaking apart of iron and nickel nuclei into He nuclei, and eventually into protons and neutrons:

$$^{56}_{26}\text{Fe} \rightarrow 13\,^4_2\text{He} + 4\text{n}$$
$$^4_2\text{He} \rightarrow 2\text{p} + 2\text{n}.$$

These are energy-requiring (endothermic) reactions, which rob energy from the core, allowing gravitational contraction to begin. This then can force electrons and protons together to form neutrons in **inverse β decay**:

$$\text{e}^- + \text{p} \rightarrow \text{n} + \nu.$$

As a result of these reactions, the pressure in the core drops precipitously. As the core collapses under the huge gravitational forces, the tremendous mass becomes essentially an enormous nucleus made up almost exclusively of neutrons. The size of the star is no longer limited by the exclusion principle applied to electrons, but rather by **neutron degeneracy** pressure, and the star contracts rapidly to form an enormously dense **neutron star**. The core of a neutron star contracts to the point at which all neutrons are as close together as they are in an atomic nucleus. That is, the density of a neutron star is on the order of 10^{14} times greater than normal solids and liquids on Earth. A cupful of such dense matter would weigh billions of tons. A neutron star that has a mass 1.5 times that of our Sun would have a diameter of only about 20 km. (Compare this to a white dwarf with 1 solar mass whose diameter would be $\approx 10^4$ km, as mentioned on the previous page.)

The contraction of the core of a massive star would mean a great reduction in gravitational potential energy. Somehow this energy would have to be released. Indeed, it was suggested in the 1930s that the final core collapse to a neutron star could be accompanied by a catastrophic explosion known as a **supernova** (plural = supernovae). The tremendous energy release (Fig. 33–9) could form virtually all elements of the Periodic Table (see below) and blow away the entire outer envelope of the star, spreading its contents into interstellar space. The presence of heavy elements on Earth and in our solar system suggests that our solar system formed from the debris of many such supernova explosions.

The elements heavier than Ni are thought to form mainly by **neutron capture** in these exploding supernovae (rather than by fusion, as for elements up to Ni). Large numbers of free neutrons, resulting from nuclear reactions, are present inside those highly evolved stars and they can readily combine with, say, a $^{56}_{26}\text{Fe}$ nucleus to form (if three are captured) $^{59}_{26}\text{Fe}$, which decays to $^{59}_{27}\text{Co}$. The $^{59}_{27}\text{Co}$ can capture neutrons, also becoming neutron rich and decaying by β^- to the next higher Z element, and so on to the highest Z elements.

The final state of a neutron star depends on its mass. If the final mass is less than about three solar masses, the subsequent evolution of the neutron star is thought to resemble that of a white dwarf. If the mass is greater than this (original mass $\gtrsim 40$ solar masses), the neutron star collapses under gravity, overcoming even neutron degeneracy. Gravity would then be so strong that emitted light could not escape—it would be pulled back in by the force of gravity. Since no radiation could escape from such a "star," we could not see it— it would be black. An object may pass by it and be deflected by its gravitational field, but if the object came too close it would be swallowed up, never to escape. This is a **black hole**.

FIGURE 33–9 The star indicated by the arrow in (a) exploded as a supernova, as shown in (b). It was detected in 1987, and named SN1987A. The large bright spot in (b) indicates a huge release of energy but does not represent the physical size.

Novae and Supernovae

Novae (singular is *nova*, meaning "new" in Latin) are faint stars that have suddenly increased in brightness by as much as a factor of 10^6 and last for a month or two before fading. Novae are thought to be faint white dwarfs that have pulled mass from a nearby companion (they make up a *binary* system), as illustrated in Fig. 33–10. The captured mass of hydrogen suddenly fuses into helium at a high rate for a few weeks. Many novae (maybe all) are *recurrent*—they repeat their bright glow years later.

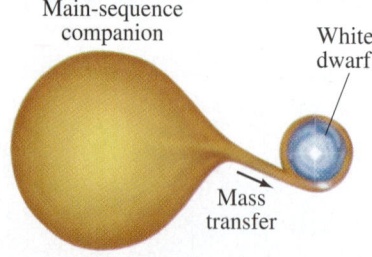

FIGURE 33–10 Hypothetical model for novae and Type Ia supernovae, showing how a white dwarf could pull mass from its normal companion.

Supernovae are also brief explosive events, but release millions of times more energy than novae, up to 10^{10} times more luminous than our Sun. The peak of brightness may exceed that of the entire galaxy in which they are located, but lasts only a few days or weeks. They slowly fade over a few months. Many supernovae form by core collapse to a neutron star as described above. See Fig. 33–9.

Type Ia supernovae are different. They all seem to have very nearly the same luminosity. They are believed to be binary stars, one of which is a white dwarf that pulls mass from its companion, much like for a nova, Fig. 33–10. The mass is higher, and as mass is captured and the total mass approaches the Chandrasekhar limit of 1.4 solar masses, it explodes as a "white-dwarf" supernova by undergoing a "thermonuclear runaway"—an uncontrolled chain of nuclear reactions that entirely destroys the white dwarf. Type Ia supernovae are useful to us as "standard candles" in the night sky to help us determine distance—see next Section.

33–3 Distance Measurements

Parallax

We have talked about the vast distances of objects in the universe. But how do we measure these distances? One basic technique employs simple geometry to measure the **parallax** of a star. By parallax we mean the apparent motion of a star, against the background of much more distant stars, due to the Earth's motion around the Sun. As shown in Fig. 33–11, we can measure the angle 2ϕ that the star appears to shift, relative to very distant stars, when viewed 6 months apart. If we know the distance d from Earth to Sun, we can reconstruct the right triangles shown in Fig. 33–11 and can then determine the distance D to the star. This is essentially the way the heights of mountains are determined, by "triangulation": see Example 1–8.

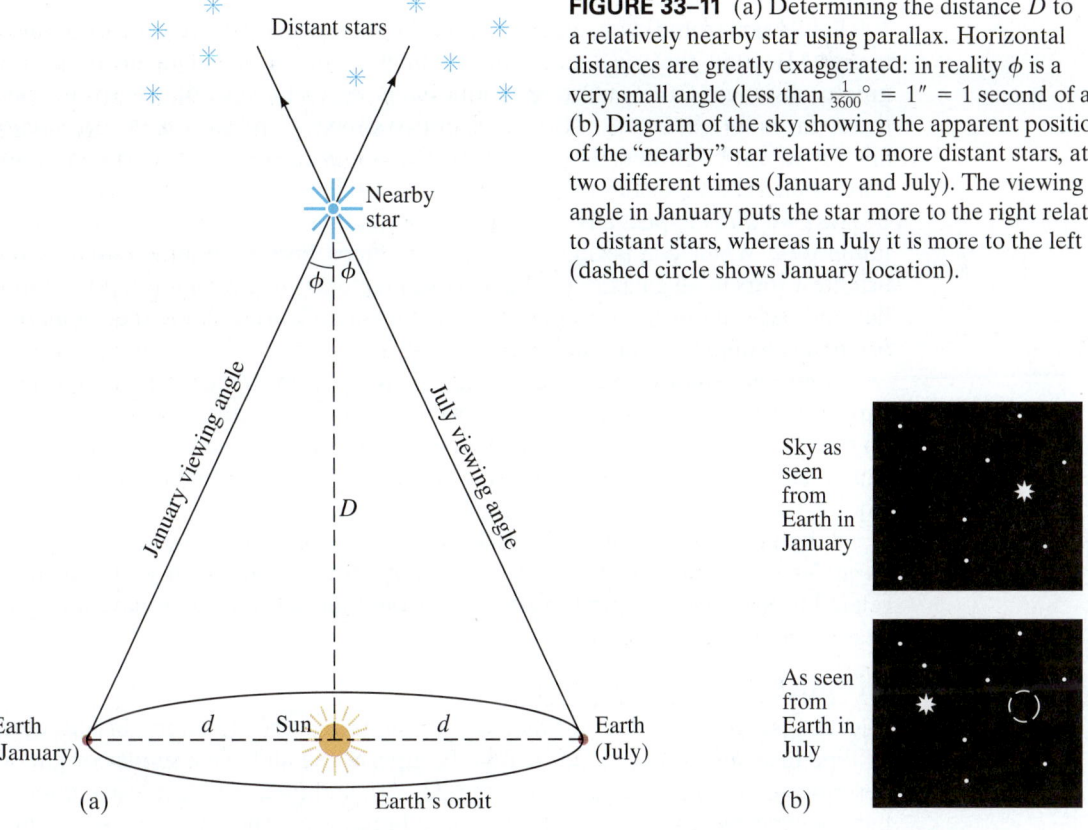

FIGURE 33–11 (a) Determining the distance D to a relatively nearby star using parallax. Horizontal distances are greatly exaggerated: in reality ϕ is a very small angle (less than $\frac{1}{3600}° = 1'' = 1$ second of arc). (b) Diagram of the sky showing the apparent position of the "nearby" star relative to more distant stars, at two different times (January and July). The viewing angle in January puts the star more to the right relative to distant stars, whereas in July it is more to the left (dashed circle shows January location).

EXAMPLE 33–6 ESTIMATE **Distance to a star using parallax.** Estimate the distance D to a star if the angle 2ϕ in Fig. 33–11a is measured to be $2\phi = 0.00012°$.

APPROACH From trigonometry, $\tan\phi = d/D$ in Fig. 33–11a. The Sun–Earth distance is $d = 1.5 \times 10^8$ km (inside front cover).

SOLUTION The angle $\phi = 0.00006°$, or about $(0.00006°)(2\pi\,\text{rad}/360°) = 1.0 \times 10^{-6}$ radians. We can use $\tan\phi \approx \phi$ because ϕ is very small. We solve for D in $\tan\phi = d/D$. The distance D to the star is

$$D = \frac{d}{\tan\phi} \approx \frac{d}{\phi} = \frac{1.5 \times 10^8\,\text{km}}{1.0 \times 10^{-6}\,\text{rad}} = 1.5 \times 10^{14}\,\text{km},$$

or about 15 ly.

*Parsec

Distances to stars are often specified in terms of parallax angle (ϕ in Fig. 33–11a) given in seconds of arc: 1 second (1″) is $\frac{1}{60}$ of one minute (1′) of arc, which is $\frac{1}{60}$ of a degree, so $1″ = \frac{1}{3600}$ of a degree. The distance is then specified in **parsecs** (pc) (meaning *par*allax angle in *sec*onds of arc): $D = 1/\phi$ with ϕ in seconds of arc. In Example 33–6, $\phi = (6 \times 10^{-5})°(3600) = 0.22″$ of arc, so we would say the star is at a distance of $1/0.22″ = 4.5$ pc. One parsec is given by (recall $D = d/\phi$, and we set the Sun–Earth distance (Fig. 33–11a) as $d = 1.496 \times 10^{11}$ m):

$$1\,\text{pc} = \frac{d}{1″} = \frac{1.496 \times 10^{11}\,\text{m}}{(1″)\left(\frac{1'}{60''}\right)\left(\frac{1°}{60'}\right)\left(\frac{2\pi\,\text{rad}}{360°}\right)} = 3.086 \times 10^{16}\,\text{m}$$

$$1\,\text{pc} = (3.086 \times 10^{16}\,\text{m})\left(\frac{1\,\text{ly}}{9.46 \times 10^{15}\,\text{m}}\right) = 3.26\,\text{ly}.$$

Distant Stars and Galaxies

Parallax can be used to determine the distance to stars as far away as about 100 light-years from Earth, and from an orbiting spacecraft perhaps 5 to 10 times farther. Beyond that distance, parallax angles are too small to measure. For greater distances, more subtle techniques must be employed. We might compare the apparent brightnesses of two stars, or two galaxies, and use the *inverse square law* (apparent brightness drops off as the square of the distance) to roughly estimate their relative distances. We can't expect this technique to be very precise because we don't expect any two stars, or two galaxies, to have the same intrinsic luminosity. When comparing galaxies, a perhaps better estimate assumes the brightest stars in all galaxies (or the brightest galaxies in galaxy clusters) are similar and have about the same intrinsic luminosity. Consequently, their *apparent brightness* would be a measure of how far away they were.

Another technique makes use of the H–R diagram. Measurement of a star's surface temperature (from its spectrum) places it at a certain point (within 20%) on the H–R diagram, assuming it is a main-sequence star, and then its luminosity can be estimated from the vertical axis (Fig. 33–6). Its apparent brightness and Eq. 33–1 give its approximate distance; see Example 33–5.

A better estimate comes from comparing *variable stars*, especially *Cepheid variables* whose luminosity varies over time with a period that is found to be related to their average luminosity. Thus, from their period and apparent brightness we get their distance.

Distance via SNIa, Redshift

The largest distances are estimated by comparing the apparent brightnesses of Type Ia supernovae ("SNIa"). Type Ia supernovae all have a similar origin (as described on the previous page and Fig. 33–10), and their brief explosive burst of light is expected to be of nearly the same luminosity. They are thus sometimes referred to as "standard candles."

Another important technique for estimating the distance of very distant galaxies is from the "redshift" in the line spectra of elements and compounds. The amount of redshift is related to the expansion of the universe, as we shall discuss in Section 33–5. It is useful for objects farther than 10^7 to 10^8 ly away.

As we look farther and farther away, measurement techniques are less and less reliable, so there is more uncertainty in the measurements of large distances.

33–4 General Relativity: Gravity and the Curvature of Space

We have seen that the force of gravity plays an important role in the processes that occur in stars. Gravity too is important for the evolution of the universe as a whole. The reasons gravity plays a dominant role in the universe, and not one of the other of the four forces in nature, are (1) it is long-range and (2) it is always attractive. The strong and weak nuclear forces act over very short distances only, on the order of the size of a nucleus; hence they do not act over astronomical distances (they do act between nuclei and nucleons in stars to produce nuclear reactions). The electromagnetic force, like gravity, acts over great distances. But it can be either attractive or repulsive. And since the universe does not seem to contain large areas of net electric charge, a large net force does not occur. But gravity acts only as an *attractive* force between *all* masses, and there are large accumulations of mass in the universe. The force of gravity as Newton described it in his law of universal gravitation was modified by Einstein. In his general theory of relativity, Einstein developed a theory of gravity that now forms the basis of cosmological dynamics.

In the *special theory of relativity* (Chapter 26), Einstein concluded that there is no way for an observer to determine whether a given frame of reference is at rest or is moving at constant velocity in a straight line. Thus the laws of physics must be the same in different inertial reference frames. But what about the more general case of motion where reference frames can be *accelerating*?

Einstein tackled the problem of accelerating reference frames in his **general theory of relativity** and in it also developed a theory of gravity. The mathematics of General Relativity is complex, so our discussion will be mainly qualitative.

We begin with Einstein's **principle of equivalence**, which states that

no experiment can be performed that could distinguish between a uniform gravitational field and an equivalent uniform acceleration.

If observers sensed that they were accelerating (as in a vehicle speeding around a sharp curve), they could not prove by any experiment that in fact they weren't simply experiencing the pull of a gravitational field. Conversely, we might think we are being pulled by gravity when in fact we are undergoing an acceleration having nothing to do with gravity.

As a thought experiment, consider a person in a freely falling elevator near the Earth's surface. If our observer held out a book and let go of it, what would happen? Gravity would pull it downward toward the Earth, but at the same rate ($g = 9.8 \text{ m/s}^2$) at which the person and elevator were falling. So the book would hover right next to the person's hand (Fig. 33–12). The effect is exactly the same as if this reference frame was at rest and *no* forces were acting. On the other hand, if the elevator was out in space where the gravitational field is essentially zero, the released book would float, just as it does in Fig. 33–12. Next, if the elevator (out in space) is accelerated upward (using rockets) at an acceleration of 9.8 m/s^2, the book as seen by our observer would fall to the floor with an acceleration of 9.8 m/s^2, just as if it were falling due to gravity at the surface of the Earth. According to the principle of equivalence, the observer could not determine whether the book fell because the elevator was accelerating upward, or because a gravitational field was acting downward and the elevator was at rest. The two descriptions are equivalent.

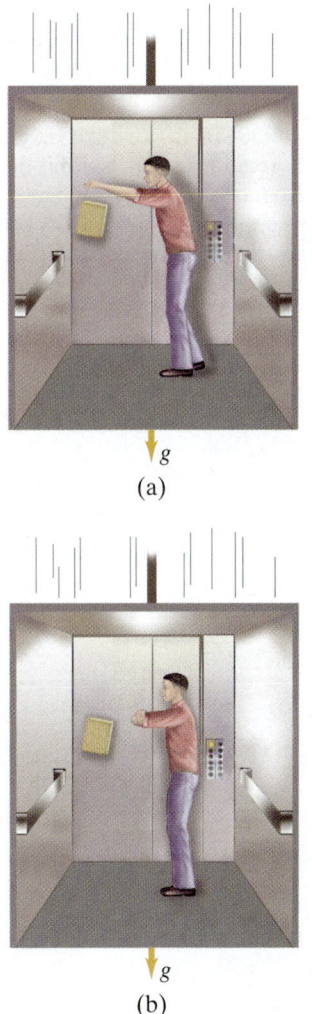

FIGURE 33–12 In an elevator falling freely under gravity, (a) a person releases a book; (b) the released book hovers next to the owner's hand; (b) is a few moments after (a).

The principle of equivalence is related to the concept that there are two types of mass. Newton's second law, $F = ma$, uses **inertial mass**. We might say that inertial mass represents "resistance" to any type of force. The second type of mass is **gravitational mass**. When one object attracts another by the gravitational force (Newton's law of universal gravitation, $F = Gm_1m_2/r^2$, Chapter 5), the strength of the force is proportional to the product of the *gravitational masses* of the two objects. This is much like Coulomb's law for the electric force between two objects which is proportional to the product of their electric charges. The electric charge on an object is not related to its inertial mass; so why should we expect that an object's gravitational mass (call it gravitational charge if you like) be related to its inertial mass? All along we have assumed they were the same. Why? Because no experiment—not even of high precision—has been able to discern any measurable difference between inertial mass and gravitational mass. (For example, in the absence of air resistance, all objects fall at the same acceleration, g, on Earth.) This is another way to state the equivalence principle: *gravitational mass is equivalent to inertial mass.*

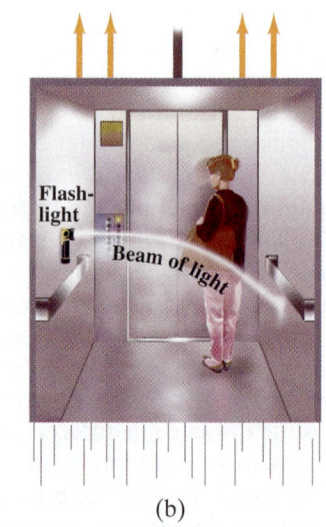

FIGURE 33–13 (a) Light beam goes straight across an elevator which is not accelerating. (b) The light beam bends (exaggerated) according to an observer in an accelerating elevator whose speed increases in the upward direction.

The principle of equivalence can be used to show that light ought to be deflected by the gravitational force due to a massive object. Consider another thought experiment, in which an elevator is in free space where virtually no gravity acts. If a light beam is emitted by a flashlight attached to the side of the elevator, the beam travels straight across the elevator and makes a spot on the opposite side if the elevator is at rest or moving at constant velocity (Fig. 33–13a). If instead the elevator is accelerating upward, as in Fig. 33–13b, the light beam still travels straight across in a reference frame at rest. In the upwardly accelerating elevator, however, the beam is observed to curve downward. Why? Because during the time the light travels from one side of the elevator to the other, the elevator is moving upward at a vertical speed that is increasing relative to the light. Next we note that according to the equivalence principle, an upwardly accelerating reference frame is equivalent to a downward gravitational field. Hence, we can picture the curved light path in Fig. 33–13b as being due to the effect of a gravitational field. Thus, from the principle of equivalence, we expect gravity to exert a force on a beam of light and to bend it out of a straight-line path!

That light is affected by gravity is an important prediction of Einstein's general theory of relativity. And it can be tested. The amount a light beam would be deflected from a straight-line path must be small even when passing a massive object. (For example, light near the Earth's surface after traveling 1 km is predicted to drop only about 10^{-10} m, which is equal to the diameter of a small atom and not detectable.) The most massive object near us is the Sun, and it was calculated that light from a distant star would be deflected by 1.75″ of arc (tiny but detectable) as it passed by the edge of the Sun (Fig. 33–14). However, such a measurement could be made only during a total eclipse of the Sun, so that the Sun's tremendous brightness would not obscure the starlight passing near its edge.

FIGURE 33–14 (a) Two stars in the sky observed from Earth. (b) If the light from one of these stars passes very near the Sun, whose gravity bends the rays, the star will appear higher than it actually is (follow the ray backwards). [Not to scale.]

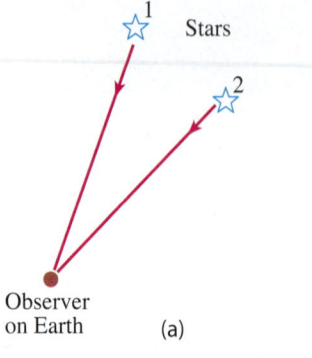
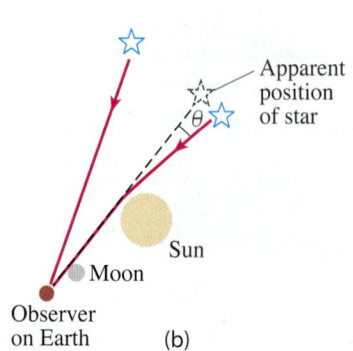

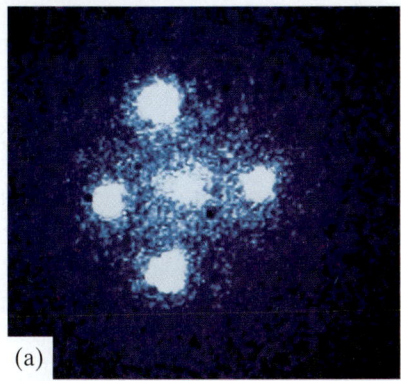
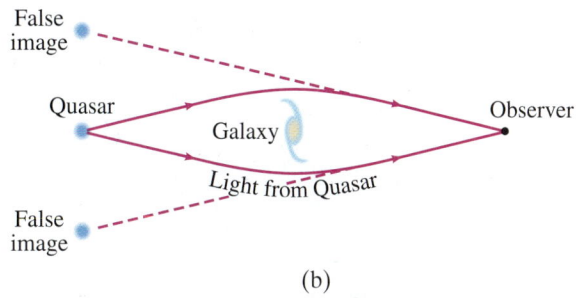

FIGURE 33–15 (a) Hubble Space Telescope photograph of the so-called "Einstein cross," thought to represent "gravitational lensing": the central spot is a relatively nearby galaxy, whereas the four other spots are thought to be images of a single quasar *behind* the galaxy. (b) Diagram showing how the galaxy could bend the light coming from the quasar behind it to produce the four images. See also Fig. 33–14. [If the shape of the nearby galaxy and distant quasar were perfect spheres and perfectly aligned, we would expect the "image" of the distant quasar to be a circular ring or halo instead of the four separate images seen here. Such a ring is called an "Einstein ring."]

An opportune eclipse occurred in 1919, and scientists journeyed to the South Atlantic to observe it. Their photos of stars just behind the Sun revealed shifts in accordance with Einstein's prediction. Another example of gravitational deflection of light is **gravitational lensing**, as described in Fig. 33–15. The very distant galaxies shown in the XDF photo at the start of this Chapter, page 947, are thought to be visible only because of gravitational lensing (and magnification of their emitted light) by nearer galaxies—as if the nearby galaxies acted as a magnifying glass.

The mathematician Fermat showed in the 1600s that optical phenomena, including reflection, refraction, and effects of lenses, can be derived from a simple principle: that light traveling between two points follows the shortest path in space. Thus if gravity curves the path of light, then gravity must be able to curve space itself. That is, *space itself can be curved*, and it is gravitational mass that causes the curvature. Indeed, the curvature of space—or rather, of four-dimensional space-time—is a basic aspect of Einstein's General Relativity.

What is meant by **curved space**? To understand, recall that our normal method of viewing the world is via Euclidean plane geometry. In Euclidean geometry, there are many axioms and theorems we take for granted, such as that the sum of the angles of any triangle is 180°. Non-Euclidean geometries, which involve curved space, have also been imagined by mathematicians. It is hard enough to imagine three-dimensional curved space, much less curved four-dimensional space-time. So let us try to understand the idea of curved space by using two-dimensional surfaces.

Consider, for example, the two-dimensional surface of a sphere. It is clearly curved, Fig. 33–16, at least to us who view it from the outside—from our three-dimensional world. But how would hypothetical two-dimensional creatures determine whether their two-dimensional space was flat (a plane) or curved? One way would be to measure the sum of the angles of a triangle. If the surface is a plane, the sum of the angles is 180°, as we learn in plane geometry. But if the space is curved, and a sufficiently large triangle is constructed, the sum of the angles will *not* be 180°. To construct a triangle on a curved surface, say the sphere of Fig. 33–16, we must use the equivalent of a straight line: that is, the shortest distance between two points, which is called a **geodesic**. On a sphere, a geodesic is an arc of a great circle (an arc in a plane passing through the center of the sphere) such as the Earth's equator and the Earth's longitude lines. Consider, for example, the large triangle of Fig. 33–16: its sides are two longitude lines passing from the north pole to the equator, and the third side is a section of the equator as shown. The two longitude lines make 90° angles with the equator (look at a world globe to see this more clearly). They make an angle with each other at the north pole, which could be, say, 90° as shown; the sum of these angles is 90° + 90° + 90° = 270°. This is clearly *not* a Euclidean space. Note, however, that if the triangle is small in comparison to the radius of the sphere, the angles will add up to nearly 180°, and the triangle (and space) will seem flat.

FIGURE 33–16 On a two-dimensional curved surface, the sum of the angles of a triangle may not be 180°.

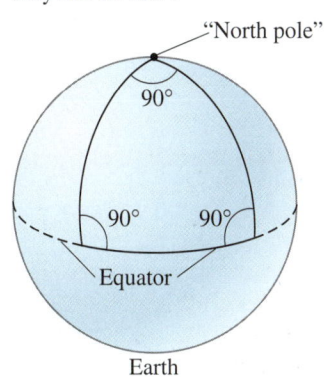

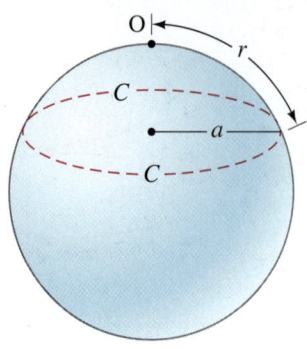
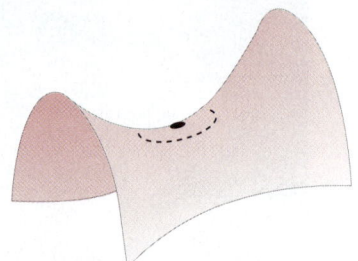

FIGURE 33–17 On a spherical surface (a two-dimensional world) a circle of circumference C is drawn (red) about point O as the center. The radius of the circle (not the sphere) is the distance r along the surface. (Note that in our three-dimensional view, we can tell that $C = 2\pi a$. Since $r > a$, then $C < 2\pi r$.)

FIGURE 33–18 Example of a two-dimensional surface with negative curvature.

Another way to test the curvature of space is to measure the radius r and circumference C of a large circle. On a plane surface, $C = 2\pi r$. But on a two-dimensional spherical surface, C is *less* than $2\pi r$, as can be seen in Fig. 33–17. The proportionality between C and r is *less* than 2π. Such a surface is said to have *positive curvature*. On the saddlelike surface of Fig. 33–18, the circumference of a circle is greater than $2\pi r$, and the sum of the angles of a triangle is less than 180°. Such a surface is said to have a *negative curvature*.

Curvature of the Universe

What about our universe? On a large scale (not just near a large mass), what is the overall curvature of the universe? Does it have positive curvature, negative curvature, or is it flat (zero curvature)? We perceive our world as Euclidean (flat), but we can not exclude the possibility that space could have a curvature so slight that we don't normally notice it. This is a crucial question in cosmology, and it can be answered only by precise experimentation.

If the universe had a positive curvature, the universe would be *closed*, or *finite* in volume. This would *not* mean that the stars and galaxies extended out to a certain boundary, beyond which there is empty space. There is no boundary or edge in such a universe. The universe is all there is. If a particle were to move in a straight line in a particular direction, it would eventually return to the starting point—perhaps eons of time later.

On the other hand, if the curvature of space was zero or negative, the universe would be *open*. It could just go on forever. An open universe could be *infinite*; but according to recent research, even that may not necessarily be so.

Today the evidence is very strong that the universe on a large scale is very close to being flat. Indeed, it is so close to being flat that we can't tell if it might have very slightly positive or very slightly negative curvature.

Black Holes

According to Einstein's theory of general relativity (sometimes abbreviated GR), space-time is curved near massive objects. We might think of space as being like a thin rubber sheet: if a heavy weight is placed on the sheet, it sags as shown in Fig. 33–19a (top of next page). The weight corresponds to a huge mass that causes space (space itself!) to curve. Thus, in the context of general relativity† we do not speak of the "force" of gravity acting on objects. Instead we say that objects and light rays move as they do because space-time is curved. An object starting at rest or moving slowly near the great mass of Fig. 33–19a would follow a geodesic (the equivalent of a straight line in plane geometry) toward that great mass.

The extreme curvature of space-time shown in Fig. 33–19b could be produced by a **black hole**. A black hole, as we mentioned in Section 33–2, has such strong gravity that even light cannot escape from it. To become a black hole, an object of mass M must undergo **gravitational collapse**, contracting by gravitational self-attraction to within a radius called the **Schwarzschild radius**,

$$R = \frac{2GM}{c^2},$$

where G is the gravitational constant and c the speed of light. If an object collapses to within this radius, it is predicted by general relativity to collapse to a point at $r = 0$, forming an infinitely dense singularity. This prediction is uncertain, however, because in this realm we need to combine quantum mechanics with gravity, a unification of theories not yet achieved (Section 32–12).

| **EXERCISE C** What is the Schwarzschild radius for an object with 10 solar masses?

The Schwarzschild radius also represents the event horizon of a black hole. By **event horizon** we mean the surface beyond which no emitted signals can ever reach us, and thus inform us of events that happen beyond that surface. As a star collapses toward a black hole, the light it emits is pulled harder and harder by gravity, but we can still see it. Once the matter passes within the event horizon, the emitted light cannot escape but is pulled back in by gravity (= curvature of space-time).

All we can know about a black hole is its mass, its angular momentum (rotating black holes), and its electric charge. No other information, no details of its structure or the kind of matter it was formed of, can be known because no information can escape.

How might we observe black holes? We cannot see them because no light can escape from them. They would be black objects against a black sky. But they do exert a gravitational force on nearby objects, and also on light rays (or photons) that pass nearby (just like in Fig. 33–15). The black hole believed to be at the center of our Galaxy ($M \approx 4 \times 10^6 \, M_{Sun}$) was discovered by examining the motion of matter in its vicinity. Another technique is to examine stars which appear to move as if they were one member of a *binary system* (two stars rotating about their common center of mass), but without a visible companion. If the unseen star is a black hole, it might be expected to pull off gaseous material from its visible companion (as in Fig. 33–10). As this matter approached the black hole, it would be highly accelerated and should emit X-rays of a characteristic type before plunging inside the event horizon. Such X-rays, plus a sufficiently high mass estimate from the rotational motion, can provide evidence for a black hole. One of the many candidates for a black hole is in the binary-star system Cygnus X-1. It is widely believed that the center of most galaxies is occupied by a black hole with a mass 10^6 to 10^9 times the mass of a typical star like our Sun.

| **EXERCISE D** A black hole has radius R. Its mass is proportional to (*a*) R, (*b*) R^2, (*c*) R^3. Justify your answer.

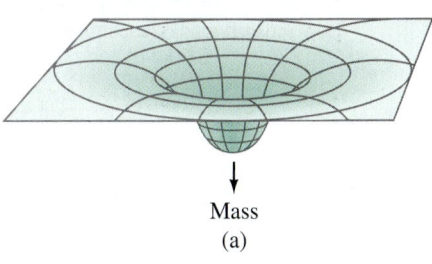

Mass
(a)

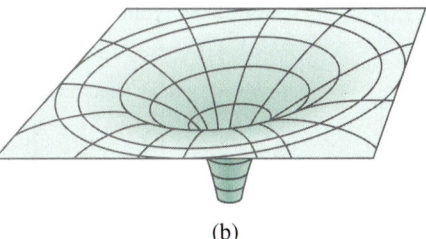

(b)

FIGURE 33–19 (a) Rubber-sheet analogy for space-time curved by matter. (b) Same analogy for a black hole, which can "swallow up" objects that pass near.

---

†Alexander Pope (1688–1744) wrote an epitaph for Newton:
  "Nature, and Nature's laws lay hid in night:
  God said, *Let Newton be!* and all was light."
Sir John Squire (1884–1958), perhaps uncomfortable with Einstein's profound thoughts, added:
  "It did not last: the Devil howling '*Ho!*
  Let Einstein be!*'* restored the status quo."

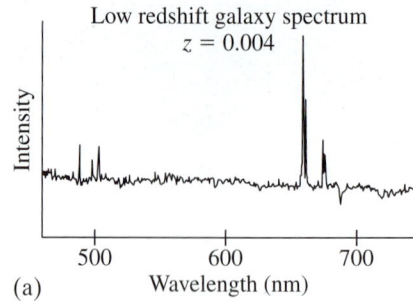

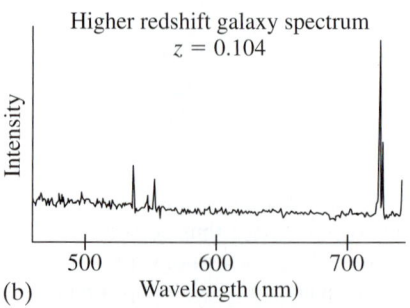

FIGURE 33–20 Atoms and molecules emit and absorb light of particular frequencies depending on the spacing of their energy levels, as we saw in Chapters 27 to 29. (a) The spectrum of light received from a relatively slow-moving galaxy. (b) Spectrum of a galaxy moving away from us at a much higher speed. Note how the peaks (or lines) in the spectrum have moved to longer wavelengths. The redshift is $z = (\lambda_{obs} - \lambda_{rest})/\lambda_{rest}$.

HUBBLE'S LAW

33–5 The Expanding Universe: Redshift and Hubble's Law

We discussed in Section 33–2 how individual stars evolve from their birth to their death as white dwarfs, neutron stars, or black holes. But what about the universe as a whole: is it static, or does it change? One of the most important scientific discoveries of the twentieth century was that distant galaxies are racing away from us, and that the farther they are from us at a given time, the faster they are moving away. How astronomers arrived at this astonishing idea, and what it means for the past history of the universe as well as its future, will occupy us for the remainder of the book.

Observational evidence that the universe is expanding was first put forth by Edwin Hubble in 1929. This idea was based on distance measurements of galaxies (Section 33–3), and determination of their velocities by the Doppler shift of spectral lines in the light received from them (Fig. 33–20). In Chapter 12 we saw how the frequency of sound is higher and the wavelength shorter if the source and observer move toward each other. If the source moves away from the observer, the frequency is lower and the wavelength longer. The **Doppler effect** occurs also for light, but the formula for light is slightly different than for sound and is given by[†]

$$\lambda_{obs} = \lambda_{rest}\sqrt{\frac{1 + v/c}{1 - v/c}}, \quad \begin{bmatrix} \text{source and observer moving} \\ \text{away from each other} \end{bmatrix} \quad (33\text{–}3)$$

where λ_{rest} is the emitted wavelength as seen in a reference frame at rest with respect to the source, and λ_{obs} is the wavelength observed in a frame moving with velocity v away from the source along the line of sight. (For relative motion *toward* each other, $v < 0$ in this formula.) When a distant source emits light of a particular wavelength, and the source is moving away from us, the wavelength appears longer to us: the color of the light (if it is visible) is shifted toward the red end of the visible spectrum, an effect known as a **redshift**. (If the source moves toward us, the color shifts toward the blue or shorter wavelength.)

In the spectra of stars in other galaxies, lines are observed that correspond to lines in the known spectra of particular atoms (see Section 27–11 and Figs. 24–28 and 27–23). What Hubble found was that the lines seen in the spectra from distant galaxies were generally *redshifted*, and that the amount of shift seemed to be approximately proportional to the distance of the galaxy from us. That is, the velocity v of a galaxy moving away from us is proportional to its distance d from us:

$$v = H_0 d. \quad (33\text{–}4)$$

This is **Hubble's law**, one of the most fundamental astronomical ideas. It was first suggested, in 1927, by Georges Lemaître, a Belgian physics professor and priest, who also first proposed what later came to be called the Big Bang. The constant H_0 is called the **Hubble parameter**.

The value of H_0 until recently was uncertain by over 20%, and thought to be between 15 and 25 km/s/Mly. But recent measurements now put its value more precisely at

$$H_0 = 21 \text{ km/s/Mly}$$

(that is, 21 km/s per million light-years of distance). The current uncertainty is about 2%, or ±0.5 km/s/Mly. [H_0 can be written in terms of parsecs (Section 33–3) as $H_0 = 67$ km/s/Mpc (that is, 67 km/s per megaparsec of distance) with an uncertainty of about ±1.2 km/s/Mpc.]

[†]For light there is no medium and we can make no distinction between motion of the source and motion of the observer (special relativity), as we did for sound which travels in a medium.

Redshift Origins

Galaxies very near us seem to be moving randomly relative to us: some move towards us (blueshifted), others away from us (redshifted); their speeds are on the order of $0.001c$. But for more distant galaxies, the velocity of recession is much greater than the velocity of local random motion, and so is dominant and Hubble's law (Eq. 33–4) holds very well. More distant galaxies have higher recession velocity and a larger redshift, and we call their redshift a **cosmological redshift**. We interpret this redshift today as due to the *expansion of space* itself. We can think of the originally emitted wavelength λ_{rest} as being stretched out (becoming longer) along with the expanding space around it, as suggested in Fig. 33–21. Although Hubble thought of the redshift as a Doppler shift, now we prefer to understand it in this sense of expanding space. (But note that atoms in galaxies do not expand as space expands; they keep their regular size.)

There is a third way to produce a redshift, which we mention for completeness: a **gravitational redshift**. Light leaving a massive star is gaining in gravitational potential energy (just like a stone thrown upward from Earth). So the kinetic energy of each photon, hf, must be getting smaller (to conserve energy). A smaller frequency f means a larger (longer) wavelength λ ($= c/f$), which is a redshift.

The amount of a redshift is specified by the **redshift parameter**, z, defined as

$$z = \frac{\lambda_{obs} - \lambda_{rest}}{\lambda_{rest}} = \frac{\Delta\lambda}{\lambda_{rest}}, \qquad (33\text{–}5\text{a})$$

where λ_{rest} is a wavelength as seen by an observer at rest relative to the source, and λ_{obs} is the wavelength measured by a moving observer. Equation 33–5a can be written as

$$z = \frac{\lambda_{obs}}{\lambda_{rest}} - 1 \qquad (33\text{–}5\text{b})$$

and

$$z + 1 = \frac{\lambda_{obs}}{\lambda_{rest}}. \qquad (33\text{–}5\text{c})$$

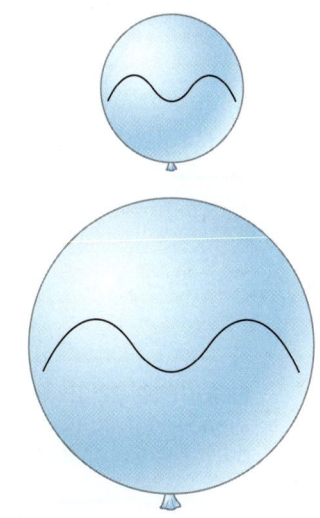

FIGURE 33–21 Simplified model of a 2-dimensional universe, imagined as a balloon. As you blow up the balloon (= expanding universe), the wavelength of a wave on its surface gets longer (redshifted).

For low speeds not close to the speed of light ($v \lesssim 0.1\,c$), the Doppler formula (Eq. 33–3) can be used to show (Problem 32) that z is proportional to the speed of the source toward or away from us:

$$z = \frac{\lambda_{obs} - \lambda_{rest}}{\lambda_{rest}} = \frac{\Delta\lambda}{\lambda_{rest}} \approx \frac{v}{c}. \qquad [v \ll c] \quad (33\text{–}6)$$

But redshifts are not always small, in which case the approximation of Eq. 33–6 is not valid. For high z galaxies, not even Eq. 33–3 applies because the redshift is due to the expansion of space (cosmological redshift), not the Doppler effect. Our Chapter-Opening Photograph, page 947, shows two very distant high z galaxies, $z = 8.8$ and 11.9, which are also shown enlarged.

*Scale Factor (advanced)

The expansion of space can be described as a scaling of the typical distance between two points or objects in the universe. If two distant galaxies are a distance d_0 apart at some initial time, then a time t later they will be separated by a greater distance $d(t)$. The **scale factor** is the same as for light, expressed in Eq. 33–5a:

$$\frac{d(t) - d_0}{d_0} = \frac{\Delta\lambda}{\lambda} = z$$

or

$$\frac{d(t)}{d_0} = 1 + z.$$

Thus, for example, if a galaxy has $z = 3$, then the scale factor is now $(1 + 3) = 4$ times larger than when the light was emitted from that galaxy. That is, the average distance between galaxies has become 4 times larger. Thus the factor by which the wavelength has increased since it was emitted tells us by what factor the universe (or the typical distance between objects) has increased.

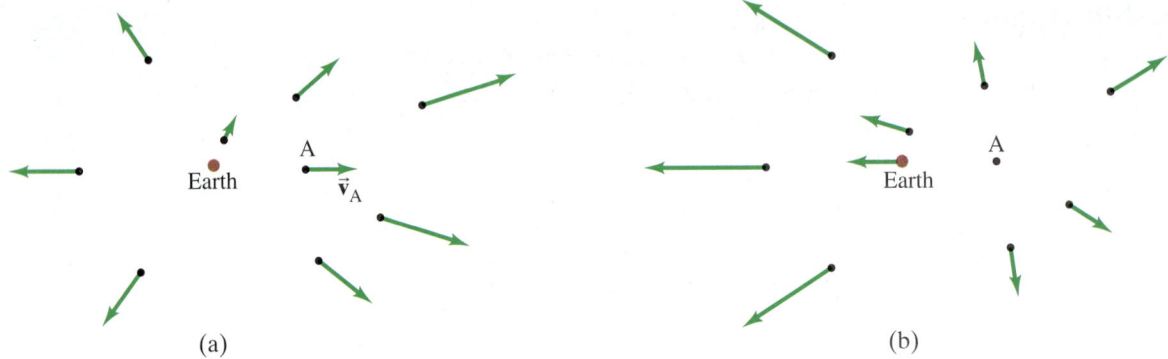

FIGURE 33–22 Expansion of the universe looks the same from any point in the universe. If you are on Earth as shown in part (a), or you are instead at galaxy A (which is at rest in the reference frame shown in (b)), all other galaxies appear to be racing away from you.

Expansion, and the Cosmological Principle

What does it mean that distant galaxies are all moving away from us, and with ever greater speed the farther they are from us? It seems to suggest some kind of explosive expansion that started at some very distant time in the past. And at first sight we seem to be in the middle of it all. But we aren't. The expansion appears the same from any other point in the universe. To understand why, see Fig. 33–22. In Fig. 33–22a we have the view from Earth (or from our Galaxy). The velocities of surrounding galaxies are indicated by arrows, pointing away from us, and the arrows are longer (faster speeds) for galaxies more distant from us. Now, what if we were on the galaxy labeled A in Fig. 33–22a? From Earth, galaxy A appears to be moving to the right at a velocity, call it $\vec{v}_A$, represented by the arrow pointing to the right. If we were *on* galaxy A, Earth would appear to be moving to the left at velocity $-\vec{v}_A$. To determine the velocities of other galaxies relative to A, we vectorially add the velocity vector, $-\vec{v}_A$, to all the velocity arrows shown in Fig. 33–22a. This yields Fig. 33–22b, where we see that the universe is expanding away from galaxy A as well; and the velocities of galaxies receding from A are proportional to their current distance from A. *The universe looks pretty much the same from different points*.

Thus the expansion of the universe can be stated as follows: all galaxies are racing away from *each other* at an average rate of about 21 km/s per million light-years of distance between them. The ramifications of this idea are profound, and we discuss them in a moment.

A basic assumption in cosmology has been that on a large scale, the universe would look the same to observers at different places at the same time. In other words, the universe is both *isotropic* (looks the same in all directions) and *homogeneous* (would look the same if we were located elsewhere, say in another galaxy). This assumption is called the **cosmological principle**. On a local scale, say in our solar system or within our Galaxy, it clearly does not apply (the sky looks different in different directions). But it has long been thought to be valid if we look on a large enough scale, so that the average population density of galaxies and clusters of galaxies ought to be the same in different areas of the sky. This seems to be valid on distances greater than about 700 Mly. The expansion of the universe (Fig. 33–22) is consistent with the cosmological principle; and the near uniformity of the cosmic microwave background radiation (discussed in Section 33–6) supports it. Another way to state the cosmological principle is that *our place in the universe is not special*.

The expansion of the universe, as described by Hubble's law, strongly suggests that galaxies must have been closer together in the past than they are now. This is, in fact, the basis of the *Big Bang* theory of the origin of the universe, which pictures the universe as a relentless expansion starting from a very hot and compressed beginning. We discuss the Big Bang in detail shortly, but first let us see what can be said about the age of the universe.

One way to estimate the age of the universe uses the Hubble parameter. With $H_0 \approx 21$ km/s per 10^6 light-years, the time required for the galaxies to arrive at their present separations would be approximately (starting with $v = d/t$ and using Hubble's law, Eq. 33–4),

$$t = \frac{d}{v} = \frac{d}{H_0 d} = \frac{1}{H_0} \approx \frac{(10^6 \text{ ly})(0.95 \times 10^{13} \text{ km/ly})}{(21 \text{ km/s})(3.16 \times 10^7 \text{ s/yr})} \approx 14 \times 10^9 \text{ yr,}$$

or 14 billion years. The age of the universe calculated in this way is called the *characteristic expansion time* or "Hubble age." It is a very rough estimate and assumes the rate of expansion of the universe was constant (which today we are quite sure is not true). Today's best measurements give the age of the universe as about 13.8×10^9 yr, in remarkable agreement with the rough Hubble age estimate.

*Steady-State Model

Before discussing the Big Bang in detail, we mention one alternative to the Big Bang—the **steady-state model**—which assumed that the universe is infinitely old and on average looks the same now as it always has. (This assumed uniformity in time as well as space was called the *perfect cosmological principle*.) According to the steady-state model, no large-scale changes have taken place in the universe as a whole, particularly no Big Bang. To maintain this view in the face of the recession of galaxies away from each other, matter would need to be created continuously to maintain the assumption of uniformity. The rate of mass creation required is very small—about one nucleon per cubic meter every 10^9 years.

The steady-state model provided the Big Bang model with healthy competition in the mid-twentieth century. But the discovery of the cosmic microwave background radiation (next Section), as well as other observations of the universe, has made the Big Bang model universally accepted.

33–6 The Big Bang and the Cosmic Microwave Background

The expansion of the universe suggests that typical objects in the universe were once much closer together than they are now. This is the basis for the idea that the universe began about 14 billion years ago as an expansion from a state of very high density and temperature known affectionately as the **Big Bang**.

The birth of the universe was not an explosion, because an explosion blows pieces out into the surrounding space. Instead, the Big Bang was the start of an expansion of space itself. The observable universe was relatively very small at the start and has been expanding, getting ever larger, ever since. The initial tiny universe of extremely dense matter is not to be thought of as a concentrated mass in the midst of a much larger space around it. The initial tiny but dense universe was the *entire universe*. There wouldn't have been anything else. When we say that the universe was once smaller than it is now, we mean that the average separation between objects (such as electrons or galaxies) was less. The universe may have been infinite in extent even then, and it may still be now (only bigger). The **observable universe** (that which we have the possibility of observing because light has had time to reach us) is, however, finite.

A major piece of evidence supporting the Big Bang is the **cosmic microwave background** radiation (or CMB) whose discovery came about as follows.

In 1964, Arno Penzias and Robert Wilson pointed their horn antenna for detecting radio waves (Fig. 33–23) into the sky. With it they detected widespread emission, and became convinced that it was coming from outside our Galaxy. They made precise measurements at a wavelength $\lambda = 7.35$ cm, in the microwave region of the electromagnetic spectrum (Fig. 22–8). The intensity of this radiation was found initially not to vary by day or night or time of year, nor to depend on direction. It came from all directions in the universe with equal intensity, to a precision of better than 1%. It could only be concluded that this radiation came from the universe as a whole.

FIGURE 33–23 Photo of Arno Penzias (right, who signed it "Arno") and Robert Wilson. Behind them their "horn antenna."

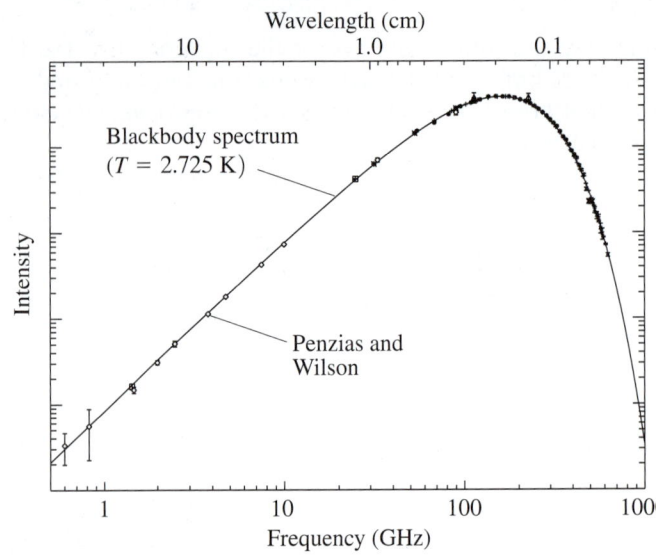

FIGURE 33–24 Spectrum of cosmic microwave background radiation, showing blackbody curve and experimental measurements including at the frequency detected by Penzias and Wilson. (Thanks to G. F. Smoot and D. Scott. The vertical bars represent the most recent experimental uncertainty in a measurement.)

The intensity of this CMB measured at $\lambda = 7.35$ cm corresponds to blackbody radiation (see Section 27–2) at a temperature of about 3 K. When radiation at other wavelengths was measured by the COBE satellite (COsmic Background Explorer), the intensities were found to fall on a nearly perfect blackbody curve as shown in Fig. 33–24, corresponding to a temperature of 2.725 K (± 0.002 K).

The remarkable uniformity of the CMB was in accordance with the cosmological principle. But theorists felt that there needed to be some small inhomogeneities, or "anisotropies," in the CMB that would have provided "seeds" at which galaxy formation could have started. Small areas of slightly higher density, which could have contracted under gravity to form clusters of galaxies, were indeed found. These tiny inhomogeneities in density and temperature were detected first by the COBE satellite experiment in 1992, led by George Smoot and John Mather (Fig. 33–25).

FIGURE 33–25 COBE scientists John Mather (chief scientist and responsible for measuring the blackbody form of the spectrum) and George Smoot (chief investigator for anisotropy experiment) shown here during celebrations for their Dec. 2006 Nobel Prize, given for their discovery of the spectrum and anisotropy of the CMB using the COBE instrument.

This discovery of the **anisotropy** of the CMB ranks with the discovery of the CMB itself in the history of cosmology. The blackbody fit and the anisotropy were the culmination of decades of research by pioneers such as Richard Muller, Paul Richards, and David Wilkinson. Subsequent experiments gave us greater detail in 2003, 2006, and 2012 with the WMAP (Wilkinson Microwave Anisotropy Probe) results, Fig. 33–26, and even more recently with the European Planck satellite results in 2013.

The CMB provides strong evidence in support of the Big Bang, and gives us information about conditions in the very early universe. In fact, in the late 1940s, George Gamow and his collaborators calculated that a Big Bang origin of the universe should have generated just such a microwave background radiation.

To understand why, let us look at what a Big Bang might have been like. (Today we usually use the term "Big Bang" to refer to the *process*, starting from a moment after the birth of the universe through the subsequent expansion.) The temperature must have been extremely high at the start, so high that there could not have been any atoms in the very early stages of the universe (high energy collisions would have broken atoms apart into nuclei and free electrons). Instead, the universe would have consisted solely of radiation (photons) and a plasma of charged electrons and other elementary particles. The universe would have been

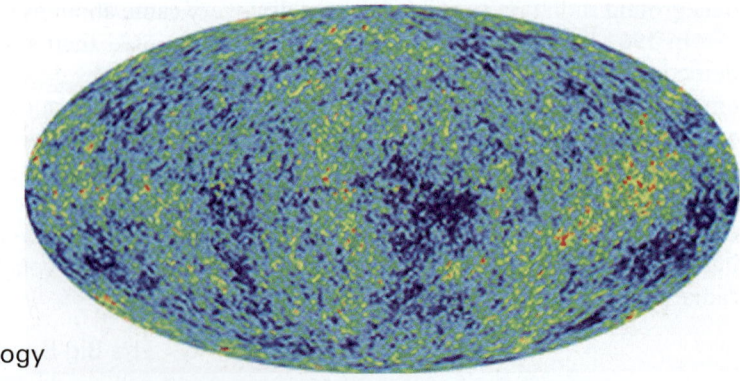

FIGURE 33–26 Measurements of the cosmic microwave background radiation over the entire sky, color-coded to represent differences in temperature from the average 2.725 K: the color scale ranges from +200 μK (red) to −200 μK (dark blue), representing slightly hotter and colder spots (associated with variations in density). Results are from the WMAP satellite in 2012: the angular resolution is 0.2°.

opaque—the photons in a sense "trapped," traveling very short distances before being scattered again, primarily by electrons. Indeed, the details of the microwave background radiation provide strong evidence that matter and radiation were once in equilibrium at a very high temperature. As the universe expanded, the energy spread out over an increasingly larger volume and the temperature dropped. Not long before the temperature had fallen to ~3000 K, some 380,000 years later, could nuclei and electrons combine together as stable atoms. With the disappearance of free electrons, as they combined with nuclei to form atoms, the radiation would have been freed—**decoupled** from matter, we say. The universe became *transparent* because photons were now free to travel nearly unimpeded straight through the universe.

It is this radiation, from 380,000 years after the birth of the universe, that we now see as the CMB. As the universe expanded, so too the wavelengths of the radiation lengthened, thus redshifting to longer wavelengths that correspond to lower temperature (recall Wien's law, $\lambda_P T$ = constant, Section 27–2), until they would have reached the 2.7-K background radiation we observe today.

Looking Back toward the Big Bang—Lookback Time

Figure 33–27 shows our Earth point of view, looking out in all directions back toward the Big Bang and the brief (380,000-year-long) period when radiation was trapped in the early plasma (yellow band). The time it takes light to reach us from an event is called its **lookback time**. The "close-up" insert in Fig. 33–27 shows a photon scattering repeatedly inside that early plasma and then exiting the plasma in a straight line. No matter what direction we look, our view of the very early universe is blocked by this wall of plasma. It is like trying to look into a very thick fog or into the surface of the Sun—we can see only as far as its surface, called the **surface of last scattering**, but not into it. Wavelengths from there are redshifted by $z \approx 1100$. Time $\Delta t'$ in Fig. 33–27 is the lookback time (not real time that goes forward).

Recall that when we view an object far away, we are seeing it as it was then, when the light was emitted, not as it would appear today.

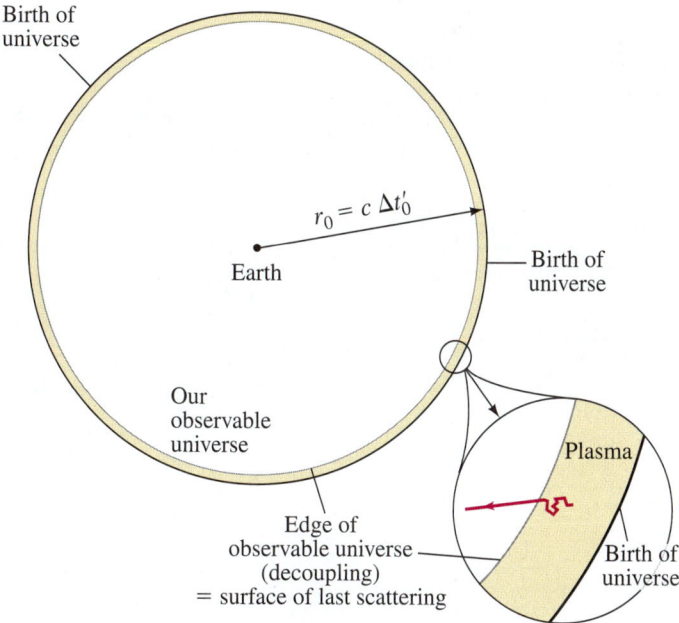

FIGURE 33–27 When we look out from the Earth, we look back in time. Any other observer in the universe would see more or less the same thing. The farther an object is from us, the longer ago the light we see had to have left it. We cannot see quite as far as the Big Bang because our view is blocked—we can see only as far as the "surface of last scattering," which radiated the CMB. The insert on the lower right shows the earliest 380,000 years of the universe when it was opaque: a photon is shown scattering many times and then (at decoupling, 380,000 yr after the birth of the universe) becoming free to travel in a straight line. If this photon wasn't heading our way when "liberated," many others were. Galaxies are not shown, but would be concentrated close to Earth in this diagram because they were created relatively recently. *Note:* This diagram is not a normal map. Maps show a section of the world as might be seen all *at a given time*. This diagram shows space (like a map), but each point is *not* at the same time. The light coming from a point a distance r from Earth took a time $\Delta t' = r/c$ to reach Earth, and thus shows an event that took place long ago, a time $\Delta t' = r/c$ in the past, which we call its "lookback time." The universe began $\Delta t'_0 = 13.8$ Gyr ago.

The Observable Universe

Figure 33–27 can easily be misinterpreted: it is not a picture of the universe at a given instant, but is intended to suggest how we look out in all directions from our observation point (the Earth, or near it). Be careful not to think that the birth of the universe took place in a circle or a sphere surrounding us as if Fig. 33–27 were a photo taken at a given moment. What Fig. 33–27 does show is what we can see, the *observable universe*. Better yet, it shows the *most* we could see.

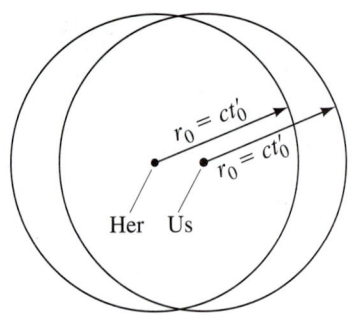

FIGURE 33–28 Two observers, on widely separated galaxies, have different horizons, different observable universes.

We would undoubtedly be arrogant to think that we could see the entire universe. Indeed, theories assume that we cannot see everything, that the **entire universe** is greater than the **observable universe**, which is a sphere of radius $r_0 = ct_0$ centered on the observer, with t_0 being the age of the universe. We can never see further back than the time it takes light to reach us.

Consider, for example, an observer in another galaxy, very far from us, located to the left of our observation point in Fig. 33–27. That observer would not yet have seen light coming from the far right of the large circle in Fig. 33–27 that we see—it will take some time for that light to reach her. But she will have already, some time ago, seen the light coming from the left that we are seeing now. In fact, her observable universe, superimposed on ours, is suggested by Fig. 33–28.

The edge of our observable universe is called the **horizon**. We could, in principle, see as far as the horizon, but not beyond it. An observer in another galaxy, far from us, will have a different horizon.

33–7 The Standard Cosmological Model: Early History of the Universe

In the last decade or two, a convincing theory of the origin and evolution of the universe has been developed, now called the **Standard Cosmological Model**. Part of this theory is based on recent theoretical and experimental advances in elementary particle physics, and part from observations of the universe including COBE, WMAP, and Planck satellites. Indeed, cosmology and elementary particle physics have cross-fertilized to a surprising extent.

Let us go back to the earliest of times—as close as possible to the Big Bang—and follow a Standard Model theoretical scenario of events as the universe expanded and cooled after the Big Bang. Initially we talk of extremely small time intervals as well as extremely high temperatures, far higher than any temperature in the universe today. Figure 33–29 is a compressed graphical representation of the events, and it may be helpful to consult it as we go along.

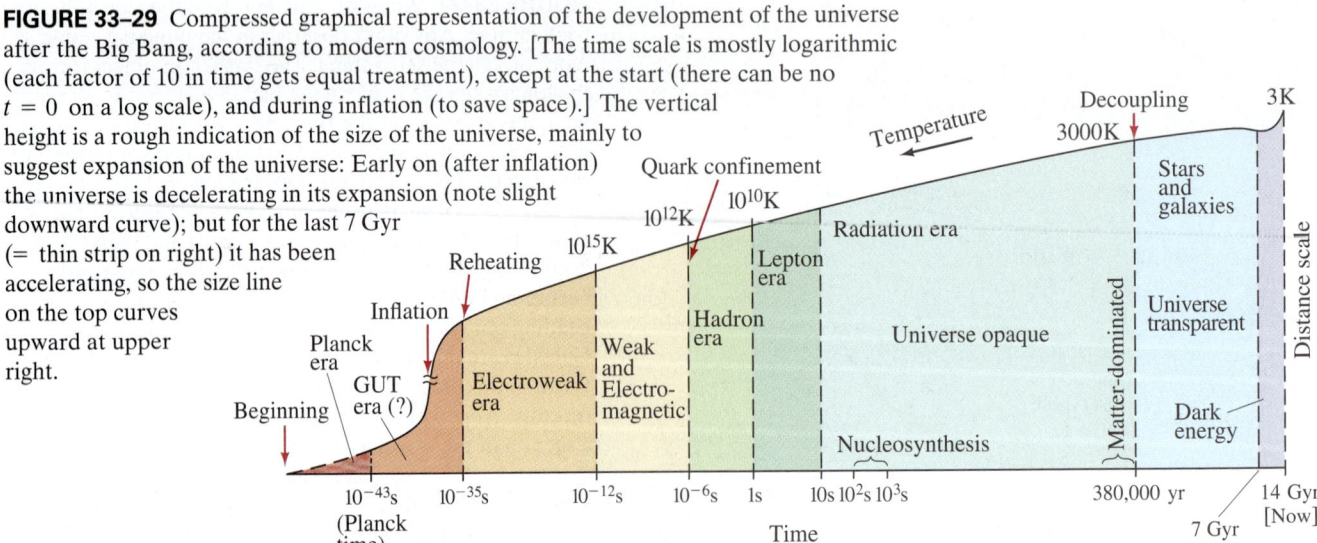

FIGURE 33–29 Compressed graphical representation of the development of the universe after the Big Bang, according to modern cosmology. [The time scale is mostly logarithmic (each factor of 10 in time gets equal treatment), except at the start (there can be no $t = 0$ on a log scale), and during inflation (to save space).] The vertical height is a rough indication of the size of the universe, mainly to suggest expansion of the universe: Early on (after inflation) the universe is decelerating in its expansion (note slight downward curve); but for the last 7 Gyr (= thin strip on right) it has been accelerating, so the size line on the top curves upward at upper right.

The History

We begin at a time only a minuscule fraction of a second after the "beginning" of the universe, 10^{-43} s. This time (sometimes referred to as the **Planck time**) is an unimaginably short time, and predictions can be only speculative. Earlier, we can say nothing because we do not have a theory of quantum gravity which would be needed for the incredibly high densities and temperatures during this "Planck era."

The first theories of the Big Bang assumed the universe was extremely hot in the beginning, maybe 10^{32} K, and then gradually cooled down while expanding. In those first moments after 10^{-43} s, the four forces of nature were thought to be united—there was only one force (Chapter 32, Fig. 32–22). Then a kind of "phase transition" would have occurred during which the gravitational force would have "condensed out" as a separate force. This and subsequent phase transitions, as shown in Fig. 32–22, are analogous to phase transitions water undergoes as it cools from a gas condensing into a liquid, and with further cooling freezes into ice.† The *symmetry* of the four forces would have been broken leaving the strong, weak, and electromagnetic forces still unified, and the universe would have entered the **grand unified era** (GUT—see Section 32–11).

This scenario of a *hot* Big Bang is now doubted by some important theorists, such as Andreí Linde, whose theories suggest the universe was much cooler at the Planck time. But what happened next to the universe, though very strange, is accepted by most cosmologists: a brilliant idea, suggested by Linde and Alan Guth in the early 1980s, proposed that the universe underwent an incredible exponential expansion, increasing in size by a factor of 10^{30} or maybe much more, in a tiny fraction of a second, perhaps 10^{-35} s or 10^{-32} s. The usefulness of this **inflationary scenario** is that it solved major problems with earlier Big Bang models, such as explaining why the universe is flat, as well as the thermal equilibrium to provide the nearly uniform CMB, as discussed below.

When inflation ended, whatever energy caused it then ended up being transformed into elementary particles with very high kinetic energy, corresponding to very high temperature (Eq. 13–8, $\overline{KE} = \frac{3}{2}kT$). That process is referred to as **reheating**, and the universe was now a "soup" of leptons, quarks, and other particles. We can think of this "soup" as a plasma of particles and antiparticles, as well as photons—all in roughly equal numbers—colliding with one another frequently and exchanging energy.

The temperature of the universe at the end of inflation was much lower than that expected by the hot Big Bang theory. But it would have been high enough so that the weak and electromagnetic forces were unified into a single force, and this stage of the universe is sometimes called the **electroweak era**. Approximately 10^{-12} s after the Big Bang, the temperature dropped to about 10^{15} K corresponding to randomly moving particles with an average kinetic energy KE of about 100 GeV (see Eq. 13–8):

$$\text{KE} \approx kT \approx \frac{(1.4 \times 10^{-23}\,\text{J/K})(10^{15}\,\text{K})}{1.6 \times 10^{-19}\,\text{J/eV}} \approx 10^{11}\,\text{eV} = 100\,\text{GeV}.$$

(As an estimate, we usually ignore the factor $\frac{3}{2}$ in Eq. 13–8.) At that time, symmetry between weak and electromagnetic forces would have broken down, and the weak force separated from the electromagnetic.

As the universe cooled down to about 10^{12} K (KE $\approx$ 100 MeV), approximately 10^{-6} s after the Big Bang, quarks stop moving freely and begin to "condense" into more normal particles: nucleons and the other hadrons and their antiparticles. With this **confinement of quarks**, the universe entered the **hadron era**. But it did not last long. Very soon the vast majority of hadrons disappeared. To see why, let us focus on the most familiar hadrons: nucleons and their antiparticles. When the average kinetic energy of particles was somewhat higher than 1 GeV, protons, neutrons, and their antiparticles were continually being created out of the energies of collisions involving photons and other particles, such as

$$\text{photons} \rightarrow \text{p} + \overline{\text{p}}$$
$$\rightarrow \text{n} + \overline{\text{n}}.$$

But just as quickly, particles and antiparticles would annihilate: for example

$$\text{p} + \overline{\text{p}} \rightarrow \text{photons or leptons}.$$

So the processes of creation and annihilation of nucleons were in equilibrium. The numbers of nucleons and antinucleons were high—roughly as many as there were electrons, positrons, or photons. But as the universe expanded and cooled, and the average kinetic energy of particles dropped below about 1 GeV, which is the minimum energy needed in a typical collision to create nucleons and antinucleons (about 940 MeV each), the process of nucleon creation could not continue.

---

†It may be interesting to point out that this story of origins here bears some resemblance to ancient accounts (nonscientific) that mention the "void," "formless wasteland" (or "darkness over the deep"), "abyss," "divide the waters" (= a phase transition?), not to mention the sudden appearance of light.

Annihilation could continue, however, with antinucleons annihilating nucleons, until almost no nucleons were left. But not quite zero. Somehow we need to explain our present world of matter (nucleons and electrons) with very little antimatter in sight.

To explain our world of matter, we might suppose that earlier in the universe, after the inflationary period, a slight excess of quarks over antiquarks was formed.† This would have resulted in a slight excess of nucleons over antinucleons. And it is these "leftover" nucleons that we are made of today. The excess of nucleons over antinucleons was probably about one part in 10^9. During the hadron era, there should have been about as many nucleons as photons. After it ended, the "leftover" nucleons thus numbered only about one nucleon per 10^9 photons, and this ratio has persisted to this day. Protons, neutrons, and all other heavier particles were thus tremendously reduced in number by about 10^{-6} s after the Big Bang. The lightest hadrons, the pions, soon disappeared, about 10^{-4} s after the Big Bang; because they are the lightest mass hadrons (140 MeV), pions were the last hadrons able to be created as the temperature (and average kinetic energy) dropped. Lighter particles, including electrons and neutrinos, were the dominant form of matter, and the universe entered the **lepton era**.

By the time the first full second had passed (clearly the most eventful second in history!), the universe had cooled to about 10 billion degrees, 10^{10} K. The average kinetic energy was about 1 MeV. This was still sufficient energy to create electrons and positrons and balance their annihilation reactions, since their masses correspond to about 0.5 MeV. So there were about as many e^+ and e^- as there were photons. But within a few more seconds, the temperature had dropped sufficiently so that e^+ and e^- could no longer be formed. Annihilation ($e^+ + e^- \rightarrow$ photons) continued. And, like nucleons before them, electrons and positrons all but disappeared from the universe—except for a slight excess of electrons over positrons (later to join with nuclei to form atoms). Thus, about $t = 10$ s after the Big Bang, the universe entered the **radiation era** (Fig. 33–29). Its major constituents were photons and neutrinos. But the neutrinos, partaking only in the weak force, rarely interacted. So the universe, until then experiencing significant amounts of energy in matter and in radiation, now became **radiation-dominated**: much more energy was contained in radiation than in matter, a situation that would last more than 50,000 years.

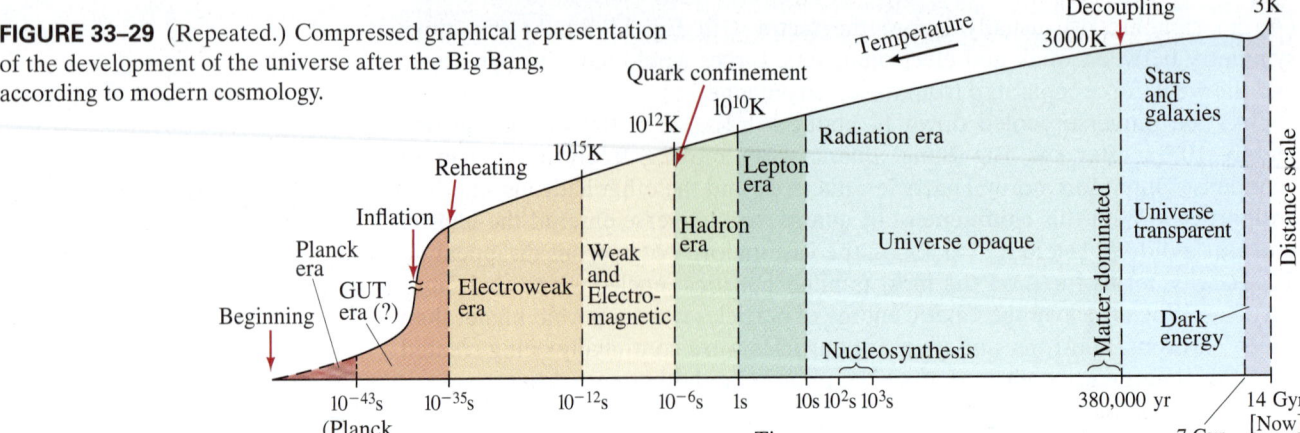

FIGURE 33–29 (Repeated.) Compressed graphical representation of the development of the universe after the Big Bang, according to modern cosmology.

Meanwhile, during the next few minutes, crucial events were taking place. Beginning about 2 or 3 minutes after the Big Bang, nuclear fusion began to occur. The temperature had dropped to about 10^9 K, corresponding to an average kinetic energy $\overline{KE} \approx 100$ keV, where nucleons could strike each other and be able to fuse (Section 31–3), but now cool enough so newly formed nuclei would not be immediately broken apart by subsequent collisions. Deuterium, helium, and very tiny amounts of lithium nuclei were made. But the universe was cooling too quickly, and larger nuclei were not made. After only a few minutes, probably not even a quarter of an hour after the Big Bang, the temperature dropped far enough that nucleosynthesis stopped, not to start again for millions of years (in stars).

†Why this could have happened is a question for which we are seeking an answer today.

Thus, after the first quarter hour or so of the universe, matter consisted mainly of bare nuclei of hydrogen (about 75%) and helium (about 25%)† as well as electrons. But radiation (photons) continued to dominate.

Our story is almost complete. The next important event is thought to have occurred 380,000 years later. The universe had expanded to about $\frac{1}{1000}$ of its present scale, and the temperature had cooled to about 3000 K. The average kinetic energy of nuclei, electrons, and photons was less than an electron volt. Since ionization energies of atoms are on the order of eV, then as the temperature dropped below this point, electrons could orbit the bare nuclei and remain there (without being ejected by collisions), thus forming atoms. This period is often called the **recombination** epoch (a misnomer since electrons had never before been combined with nuclei to form atoms). With the disappearance of free electrons and the birth of atoms, the photons—which had been continually scattering from the free electrons—now became free to spread throughout the universe. As mentioned in the previous Section, we say that the photons became **decoupled** from matter. Thus *decoupling* occurred at *recombination*. The energy contained in radiation had been decreasing (lengthening in wavelength as the universe expanded); and at about $t = 56,000$ yr (even before decoupling) the energy contained in matter became dominant over radiation. The universe was said to have become **matter-dominated** (marked on Fig. 33–29). As the universe continued to expand, the electromagnetic radiation cooled further, to 2.7 K today, forming the cosmic microwave background radiation we detect from everywhere in the universe.

After the birth of atoms, then stars and galaxies could begin to form: by self-gravitation around mass concentrations (inhomogeneities). Stars began to form about 200 million years after the Big Bang, galaxies after almost 10^9 years. The universe continued to evolve until today, some 14 billion years after it started.

* * *

This scenario, like other scientific models, cannot be said to be "proven." Yet this model is remarkably effective in explaining the evolution of the universe we live in, and makes predictions which can be tested against the next generation of observations.

A major event, and something only discovered recently, is that when the universe was about half as old as it is now (about 7 Gyr ago), its expansion began to accelerate. This was a big surprise because it was assumed the expansion of the universe would slow down due to gravitational attraction of all objects toward each other. This acceleration in the expansion of the universe is said to be due to "dark energy," as we discuss in Section 33–9. On the right in Fig. 33–29 is a narrow vertical strip that represents the most recent 7 billion years of the universe, during which *dark energy* seems to have dominated.

33–8 Inflation: Explaining Flatness, Uniformity, and Structure

The idea that the universe underwent a period of exponential inflation early in its life, expanding by a factor of 10^{30} or more (previous Section), was first put forth by Alan Guth and Andreí Linde. Many sophisticated models based on this general idea have since been proposed. The energy required for this wild expansion may have been due to fields somewhat like the Higgs field (Section 32–10). So far, the evidence for inflation is indirect; yet it is a feature of most viable cosmological models because it alone is able to provide natural explanations for several remarkable features of our universe.

---

†This Standard Model prediction of a 25% primordial production of helium agrees with what we observe today—the universe *does* contain about 25% He—and it is strong evidence in support of the Standard Big Bang Model. Furthermore, the theory says that 25% He abundance is fully consistent with there being three neutrino types, which is the number we observe. And it sets an upper limit of four to the maximum number of possible neutrino types. This is a striking example of the powerful connection between particle physics and cosmology.

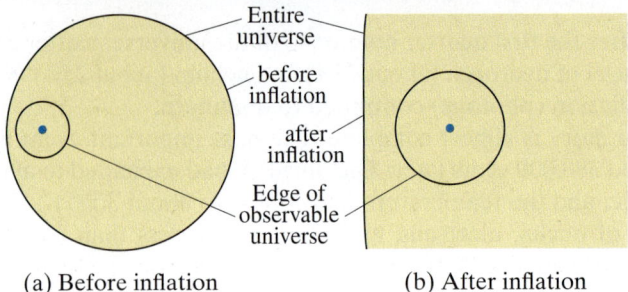

FIGURE 33–30 (a) Simple 2-D model of the entire universe; the observable universe is suggested by the small circle centered on us (blue dot). (b) Edge of entire universe is essentially flat after the 10^{30}-fold expansion during inflation.

Flatness

First of all, our best measurements suggest that the universe is flat, that it has zero curvature. As scientists, we would like some reason for this remarkable result. To see how inflation explains flatness, consider a simple 2-dimensional model of the universe as we did earlier in Figs. 33–16 and 33–21. A circle in this 2-dimensional universe (= surface of a sphere, Fig. 33–30a) represents the *observable* universe as seen by an observer at the blue dot. A possible hypothesis is that inflation occurred over a time interval that very roughly doubled the age of the universe from, let us say, $t = 1 \times 10^{-35}$ s to $t = 2 \times 10^{-35}$ s. The size of the *observable* universe ($r = ct$) would have increased by a factor of two during inflation, while the radius of curvature of the *entire* universe increased by an enormous factor of 10^{30} or more. Thus the edge of our 2-D sphere representing the entire universe would have seemed flat to a high degree of precision, as shown in Fig. 33–30b. Even if the time of inflation was a factor of 10 or 100 (instead of 2), the expansion factor of 10^{30} or more would have blotted out any possibility of observing anything but a flat universe.

CMB Uniformity

Inflation also explains why the CMB is so uniform. Without inflation, the tiny universe at 10^{-35} s would not have been small enough for all parts of it to have been in contact and so reach the same temperature (information cannot travel faster than c). To see this, suppose that the currently observable universe came from a region of space about 1 cm in diameter at $t = 10^{-36}$ s, as per original Big Bang theory. In that 10^{-36} s, light could have traveled $d = ct = (3 \times 10^8 \text{ m/s})(10^{-36} \text{ s}) = 10^{-27}$ m, way too small for the opposite sides of a 1-cm-wide "universe" to have been in communication. But if that region had been 10^{30} times smaller ($= 10^{-32}$ m), as proposed by the inflation model, there could have been contact and thermal equilibrium to produce the observed nearly uniform CMB. Inflation, by making the very early universe extremely small, assures that all parts of that region which is today's observable universe could have been in thermal equilibrium. And after inflation the universe could be large enough to give us today's observable universe.

Galaxy Seeds, Fluctuations, Magnetic Monopoles

Inflation also gives us a clue as to how the present structure of the universe (galaxies and clusters of galaxies) came about. We saw earlier that, according to the uncertainty principle, energy might be not conserved by an amount ΔE for a time $\Delta t \approx \hbar / \Delta E$. Forces, whether electromagnetic or other types, can undergo such tiny **quantum fluctuations** according to quantum theory, but they are so tiny they are not detectable unless magnified in some way. That is what inflation might have done: it could have magnified those fluctuations perhaps 10^{30} times in size, which would give us the density irregularities seen in the cosmic microwave background (WMAP, Fig. 33–26). That would be very nice, because the density variations we see in the CMB are what we believe were the seeds that later coalesced under gravity into galaxies and galaxy clusters, and our models fit the data extremely well.

Sometimes it is said that the quantum fluctuations occurred in the **vacuum state** or vacuum energy. This could be possible because the vacuum is no longer considered to be empty, as we discussed in Section 32–3 relative to positrons as holes in a negative energy sea of electrons. Indeed, the vacuum is thought to be filled with fields and particles occupying all the possible negative energy states.

Also, the virtual exchange particles that carry the forces, as discussed in Chapter 32, could leave their brief virtual states and actually become real as a result of the 10^{30} magnification of space (according to inflation) and the very short time over which it occurred ($\Delta t = \hbar/\Delta E$).

Inflation helps us too with the puzzle of why **magnetic monopoles** (Section 20–1) have never been observed, yet isolated magnetic poles may well have been copiously produced at the start. After inflation, they would have been so far apart that we have never stumbled on one.

Inflation may solve outstanding problems, but we may need new physics to understand how inflation occurred. Many predictions of inflationary theory have been confirmed by recent cosmological observations.

33–9 Dark Matter and Dark Energy

According to the Standard Big Bang Model, the universe is evolving and changing. Individual stars are being created, evolving, and then dying to become white dwarfs, neutron stars, or black holes. At the same time, the universe as a whole is expanding. One important question is whether the universe will continue to expand forever. Until the late 1990s, the universe was thought to be dominated by matter which interacts by gravity, and the fate of the universe was connected to the curvature of space-time (Section 33–4). If the universe had *negative* curvature, the expansion of the universe would never stop, although the rate of expansion would decrease due to the gravitational attraction of its parts. Such a universe would be *open* and infinite. If the universe is *flat* (no curvature), it would still be open and infinite but its expansion would slowly approach a zero rate. If the universe had *positive* curvature, it would be *closed* and finite; the effect of gravity would be strong enough that the expansion would eventually stop and the universe would begin to contract, collapsing back onto itself in a **big crunch**.

Critical Density

According to the above scenario (which does not include inflation or the recently discovered acceleration of the universe), the fate of the universe would depend on the average mass–energy density in the universe. For an average mass density greater than a critical value known as the **critical density**, estimated to be about

$$\rho_c \approx 10^{-26} \text{ kg/m}^3$$

(i.e., a few nucleons/m³ on average throughout the universe), space-time would have a positive curvature and gravity would prevent expansion from continuing forever. Eventually (if $\rho > \rho_c$) gravity would pull the universe back into a big crunch. If instead the actual density was equal to the critical density, $\rho = \rho_c$, the universe would be flat and open, just barely expanding forever. If the actual density was less than the critical density, $\rho < \rho_c$, the universe would have negative curvature and would easily expand forever. See Fig. 33–31. Today we believe the universe is very close to flat. But recent evidence suggests the universe is expanding at an *accelerating* rate, as discussed below.

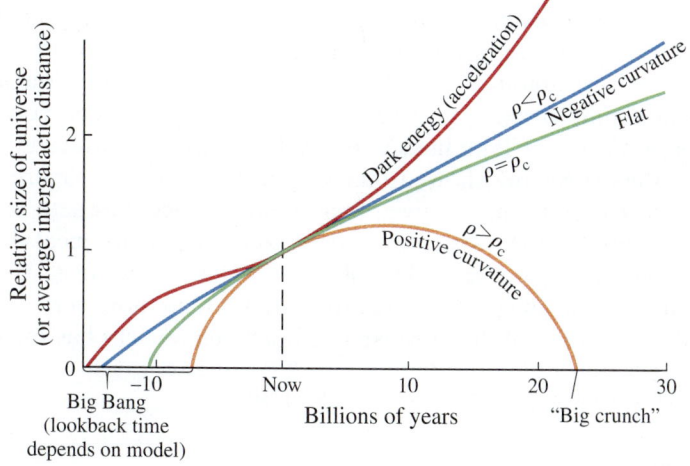

FIGURE 33–31 Three future possibilities for the universe, depending on the density ρ of ordinary matter, plus a fourth possibility that includes dark energy. Note that all curves have been chosen to have the same slope ($= H_0$, the Hubble parameter) right now. Looking back in time, the Big Bang occurs where each curve touches the horizontal (time) axis.

> **EXERCISE E** Return to the Chapter-Opening Questions, page 947, and answer them again. Try to explain why you may have answered differently the first time.

Dark Matter

COBE, WMAP, Planck, and other experiments have convinced scientists that the universe is flat and $\rho = \rho_c$. But this ρ cannot be only normal baryonic matter (atoms are 99.9% baryons—protons and neutrons—by weight). These recent experiments put the amount of normal baryonic matter in the universe at only about 5% of the critical density. What is the other 95%? There is strong evidence for a significant amount of nonluminous matter in the universe referred to as **dark matter**, which acts normally under gravity, but does not absorb or radiate light sufficiently to be visible. For example, observations of the rotation of galaxies suggest that they rotate as if they had considerably more mass than we can see. Recall from Chapter 5, Eq. 5–6, that for a satellite of mass m revolving around Earth (mass M)

$$m\frac{v^2}{r} = G\frac{mM}{r^2}$$

and hence $v = \sqrt{GM/r}$. If we apply this equation to stars in a galaxy, we see that their speed depends on galactic mass. Observations show that stars farther from the galactic center revolve much faster than expected if there is only the pull of visible matter, suggesting a great deal of invisible matter. Similarly, observations of the motion of galaxies within clusters also suggest that they have considerably more mass than can be seen. In fact, dark matter was first proposed in the 1930s, based on such observations.

Furthermore, theory suggests that without dark matter, galaxies and stars probably would not have formed and would not exist. Dark matter seems to hold the universe together.

What might this nonluminous matter in the universe be? We don't know yet. But we hope to find out soon. It cannot be made of ordinary baryonic matter (protons and neutrons), so it must consist of some other sort of elementary particle, perhaps created at a very early time. Perhaps it is made up of previously undetected *weakly interacting massive particles* (**WIMP**s), possibly supersymmetric particles (Section 32–12) such as neutralinos. We are anxiously awaiting the results of intense searches for such particles, looking both at what arrives from far out in the cosmos with underground detectors[†] and satellites, and by producing them in particle colliders (the LHC, Section 32–1).

Dark matter makes up roughly 25% of the mass–energy of the universe, according to the latest observations and models. Thus the total mass–energy is 25% dark matter plus 5% baryons for a total of about 30%, which does not bring ρ up to ρ_c. What is the other 70%? We are not sure about that either, but we have given it a name: "dark energy."

FIGURE 33–32 Saul Perlmutter, center, flanked by Adam G. Riess (left) and Brian P. Schmidt, at the Nobel Prize celebrations, December 2011.

Dark Energy—Cosmic Acceleration

In 1998, just before the turn of the millennium, two groups, one led by Saul Perlmutter and the other by Brian Schmidt and Adam Riess (Fig. 33–32), reported a huge surprise. Gravity was assumed to be the predominant force on a large scale in the universe, and it was thought that the expansion of the universe ought to be slowing down in time because gravity acts as an attractive force between objects. But measurements of Type Ia supernovae (our best standard candles—see Section 33–3) unexpectedly showed that very distant (high z) supernovae were dimmer than expected. That is, given their great distance d as determined from their low brightness, their speed v as determined from the measured z was less than expected according to Hubble's law. This result suggests that nearer galaxies are moving away from us relatively faster than those very distant ones, meaning the expansion of the universe in more recent epochs has sped up.

[†]In deep mines and under mountains to block out most other particles.

This **acceleration** in the expansion of the universe (in place of the expected deceleration due to gravitational attraction between masses) seems to have begun roughly 7 billion years ago (7 Gyr, which would be about halfway back to what we call the Big Bang).

What could be causing the universe to accelerate in its expansion, against the attractive force of gravity? Does our understanding of gravity need to be revised? We don't yet know the answers to these questions. There are several speculations. Somehow there seems to be a long-range *repulsive* effect on space, like a negative gravity, causing objects to speed away from each other ever faster. Whatever it is, it has been given the name **dark energy**. Many scientists say dark energy is the biggest mystery facing physical science today.

One idea is a sort of quantum field given the name **quintessence**. Another possibility suggests an energy latent in space itself (**vacuum energy**) and relates to an aspect of General Relativity known as the **cosmological constant** (symbol Λ). When Einstein developed his equations, he found that they offered no solutions for a static universe. In those days (1917) it was thought the universe was static—unchanging and everlasting. Einstein added an arbitrary constant (Λ) to his equations to provide solutions for a static universe.[†] A decade later, when Hubble showed us an expanding universe, Einstein discarded his cosmological constant as no longer needed ($\Lambda = 0$). But today, measurements are consistent with dark energy being due to a nonzero cosmological constant, although further measurements are needed to see subtle differences among theories.

There is increasing evidence that the effects of some form of dark energy are very real. Observations of the CMB, supernovae, and large-scale structure (Section 33–10) agree well with theories and computer models when they input dark energy as providing about 70% of the mass–energy in the universe, and when the total mass–energy density equals the critical density ρ_c.

Today's best estimate of how the mass–energy in the universe is distributed is approximately (see also Fig. 33–33):

70% dark energy

30% matter, subject to the known gravitational force.

Of this 30%, about

25% is dark matter

5% is baryons (what atoms are made of); of this 5% only $\frac{1}{10}$ is readily visible matter—stars and galaxies (that is, 0.5% of the total); the other $\frac{9}{10}$ of ordinary matter, which is not visible, is mainly gaseous plasma.

It is remarkable that only 0.5% of all the mass–energy in the universe is visible as stars and galaxies.

The idea that the universe is dominated by completely unknown forms of matter and energy seems bizarre. Nonetheless, the ability of our present model to precisely explain observations of the CMB anisotropy, cosmic expansion, and large-scale structure (next Section) presents a compelling case.

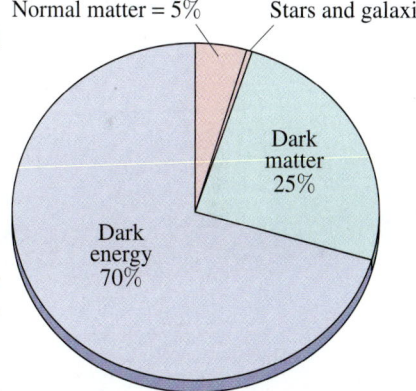

FIGURE 33–33 Portions of total mass–energy in the universe (approximate).

33–10 Large-Scale Structure of the Universe

The beautiful WMAP pictures of the sky (Fig. 33–26) show small but significant inhomogeneities in the temperature of the cosmic microwave background (CMB). These anisotropies reflect compressions and expansions in the primordial plasma just before decoupling (Fig. 33–29), from which galaxies and clusters of galaxies formed. Analyses of the irregularities in the CMB using mammoth computer

---

[†]It seems strange that Einstein and other scientists believed in a static universe. The ancients, including the Roman Lucretius argued against it: *If there was no birth-time of earth and heaven and they have been from everlasting, why before the [Trojan] war .... have not other poets as well sung other themes?* [The reference is to Homer being the oldest known writings.] The ancient Hebrews also argued for a beginning (like our Big Bang): see Genesis.

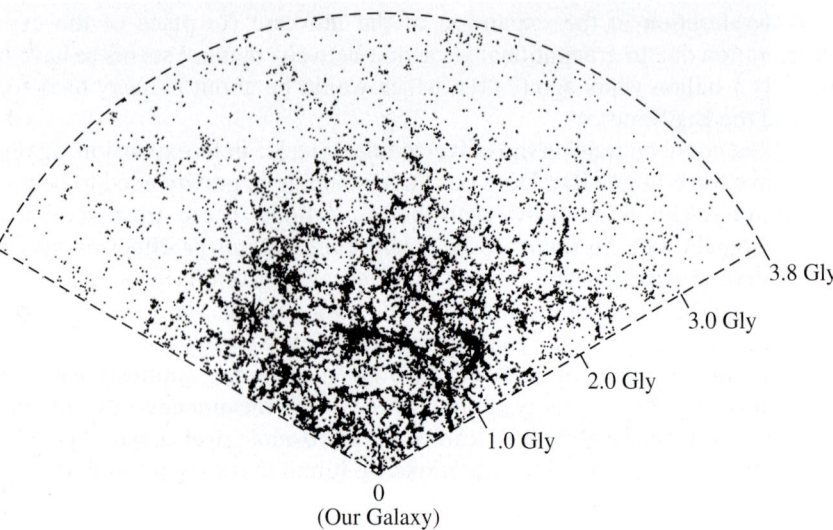

FIGURE 33–34 Distribution of some 50,000 galaxies in a 2.5° slice through almost half of the sky above the equator, as measured by the Sloan Digital Sky Survey (SDSS). Each dot represents a galaxy. The distance from us is obtained from the redshift and Hubble's law, and is given in units of 10^9 light-years (Gly). The point 0 represents us, our observation point. This diagram may seem to put us at the center, but remember that at greater distances, fewer galaxies are bright enough to be detected, thus resulting in an apparent thinning out of galaxies. Note the "walls" and "voids" of galaxies.

simulations predict a large-scale distribution of galaxies very similar to what is seen today (Fig. 33–34). These simulations are very successful if they contain dark energy and dark matter; and the dark matter needs to be *cold* (slow speed—think of Eq. 13–8, $\frac{1}{2}m\overline{v^2} = \frac{3}{2}kT$ where T is temperature), rather than "hot" dark matter such as neutrinos which move at or very near the speed of light. Indeed, the modern **cosmological model** is called the **ΛCDM** model, where lambda (Λ) stands for the cosmological constant, and CDM is **cold dark matter**.

Cosmologists have gained substantial confidence in this cosmological model from such a precise fit between observations and theory. They can also extract very precise values for cosmological parameters which previously were only known with low accuracy. The CMB is such an important cosmological observable that every effort is being made to extract all of the information it contains. A new generation of ground, balloon, and satellite experiments is observing the CMB with greater resolution and sensitivity. They may detect interaction of **gravity waves** (produced in the inflationary epoch) with the CMB and thereby provide direct evidence for cosmic inflation, and also provide information about elementary particle physics at energies far beyond the reach of man-made accelerators and colliders.

33–11 Finally . . .

When we look up into the night sky, we see stars; and with the best telescopes, we see galaxies and the exotic objects we discussed earlier, including rare supernovae. But even with our best instruments we do not see the processes going on inside stars and supernovae that we hypothesized (and believe). We are dependent on brilliant theorists who come up with viable theories and verifiable models. We depend on complicated computer models whose parameters are varied until the outputs compare favorably with our observations and analyses of WMAP and other experiments. And we now have a surprisingly precise idea about some aspects of our universe: it is flat, it is about 14 billion years old, it contains only 5% "normal" baryonic matter (for atoms), and so on.

The questions raised by cosmology are difficult and profound, and may seem removed from everyday "reality." We can always say, "the Sun is shining, it's going to shine on for an unimaginably long time, all is well." Nonetheless, the questions of cosmology are deep ones that fascinate the human intellect. One aspect that is especially intriguing is this: calculations on the formation and evolution of the universe have been performed that deliberately varied the values—just slightly—of certain fundamental physical constants. The result?

A universe in which life as we know it could not exist. [For example, if the difference in mass between a proton and a neutron were zero, or less than the mass of the electron, $0.511\ \text{MeV}/c^2$, there would be no atoms: electrons would be captured by protons to make neutrons.] Such results have contributed to a philosophical idea called the **anthropic principle**, which says that if the universe were even a little different than it is, we could not be here. We physicists are trying to find out if there are some undiscovered fundamental laws that determined those conditions that allowed us to exist. A poet might say that the universe is exquisitely tuned, almost as if to accommodate us.

Summary

The night sky contains myriads of stars including those in the Milky Way, which is a "side view" of our **Galaxy** looking along the plane of the disk. Our Galaxy includes over 10^{11} stars. Beyond our Galaxy are billions of other galaxies.

Astronomical distances are measured in **light-years** ($1\ \text{ly} \approx 10^{13}\ \text{km}$). The nearest star is about 4 ly away and the nearest large galaxy is 2 million ly away. Our Galactic disk has a diameter of about 100,000 ly. [Distances are sometimes specified in **parsecs**, where 1 parsec = 3.26 ly.]

Stars are believed to begin life as collapsing masses of gas (protostars), largely hydrogen. As they contract, they heat up (potential energy is transformed to kinetic energy). When the temperature reaches about 10 million degrees, nuclear fusion begins and forms heavier elements (**nucleosynthesis**), mainly helium at first. The energy released during these reactions heats the gas so its outward pressure balances the inward gravitational force, and the young star stabilizes as a **main-sequence** star. The tremendous luminosity of stars comes from the energy released during these thermonuclear reactions. After billions of years, as helium is collected in the core and hydrogen is used up, the core contracts and heats further. The outer envelope expands and cools, and the star becomes a **red giant** (larger diameter, redder color).

The next stage of stellar evolution depends on the mass of the star, which may have lost much of its original mass as its outer envelope escaped into space. Stars of residual mass less than about 1.4 solar masses cool further and become **white dwarfs**, eventually fading and going out altogether. Heavier stars contract further due to their greater gravity: the density approaches nuclear density, the huge pressure forces electrons to combine with protons to form neutrons, and the star becomes essentially a huge nucleus of neutrons. This is a **neutron star**, and the energy released during its final core collapse is believed to produce **supernova** explosions. If the star is very massive, it may contract even further and form a **black hole**, which is so dense that no matter or light can escape from it.

In the **general theory of relativity**, the **equivalence principle** states that an observer cannot distinguish acceleration from a gravitational field. Said another way, gravitational and inertial masses are the same. The theory predicts gravitational bending of light rays to a degree consistent with experiment. Gravity is treated as a curvature in space and time, the curvature being greater near massive objects. The universe as a whole may be curved. With sufficient mass, the curvature of the universe would be positive, and the universe is *closed* and *finite*; otherwise, it would be *open* and *infinite*. Today we believe the universe is **flat**.

Distant galaxies display a **redshift** in their spectral lines, originally interpreted as a Doppler shift. The universe is observed to be **expanding**, its galaxies racing away from each other at speeds (v) proportional to the distance (d) between them:

$$v = H_0 d, \qquad (33\text{--}4)$$

which is known as **Hubble's law** (H_0 is the **Hubble parameter**). This expansion of the universe suggests an explosive origin, the **Big Bang**, which occurred about 13.8 billion years ago. It is not like an ordinary explosion, but rather an expansion of space itself.

The **cosmological principle** assumes that the universe, on a large scale, is homogeneous and isotropic.

Important evidence for the Big Bang model of the universe was the discovery of the **cosmic microwave background** radiation (CMB), which conforms to a blackbody radiation curve at a temperature of 2.725 K.

The **Standard Model** of the Big Bang provides a possible scenario as to how the universe developed as it expanded and cooled after the Big Bang. Starting at 10^{-43} seconds after the Big Bang, according to this model, the universe underwent a brief but rapid exponential expansion, referred to as **inflation**. Shortly thereafter, quarks were **confined** into hadrons (the **hadron era**). About 10^{-4} s after the Big Bang, the majority of hadrons disappeared, having combined with anti-hadrons, producing photons, leptons, and energy, leaving mainly photons and leptons to freely move, thus introducing the **lepton era**. By the time the universe was about 10 s old, the electrons too had mostly disappeared, having combined with their antiparticles; the universe was **radiation-dominated**. A couple of minutes later, nucleosynthesis began, but lasted only a few minutes. It then took almost four hundred thousand years before the universe was cool enough for electrons to combine with nuclei to form atoms (**recombination**). Photons, up to then continually being scattered off of free electrons, could now move freely—they were **decoupled** from matter and the universe became transparent. The background radiation had expanded and cooled so much that its total energy became less than the energy in matter, and **matter dominated** increasingly over radiation. Then stars and galaxies formed, producing a universe not much different than it is today—some 14 billion years later.

Recent observations indicate that the universe is essentially flat, that it contains an as-yet unknown type of **dark matter**, and that it is dominated by a mysterious **dark energy** which exerts a sort of negative gravity causing the expansion of the universe to accelerate. The total contributions of baryonic (normal) matter, dark matter, and dark energy sum up to the **critical density**.

Questions

1. The Milky Way was once thought to be "murky" or "milky" but is now considered to be made up of point sources. Explain.
2. A star is in equilibrium when it radiates at its surface all the energy generated in its core. What happens when it begins to generate more energy than it radiates? Less energy? Explain.
3. Describe a red giant star. List some of its properties.
4. Does the H–R diagram directly reveal anything about the core of a star?
5. Why do some stars end up as white dwarfs, and others as neutron stars or black holes?
6. If you were measuring star parallaxes from the Moon instead of Earth, what corrections would you have to make? What changes would occur if you were measuring parallaxes from Mars?
7. *Cepheid variable* stars change in luminosity with a typical period of several days. The period has been found to have a definite relationship with the average intrinsic luminosity of the star. How could these stars be used to measure the distance to galaxies?
8. What is a geodesic? What is its role in General Relativity?
9. If it were discovered that the redshift of spectral lines of galaxies was due to something other than expansion, how might our view of the universe change? Would there be conflicting evidence? Discuss.
10. Almost all galaxies appear to be moving away from us. Are we therefore at the center of the universe? Explain.
11. If you were located in a galaxy near the boundary of our observable universe, would galaxies in the direction of the Milky Way appear to be approaching you or receding from you? Explain.
12. Compare an explosion on Earth to the Big Bang. Consider such questions as: Would the debris spread at a higher speed for more distant particles, as in the Big Bang? Would the debris come to rest? What type of universe would this correspond to, open or closed?
13. If nothing, not even light, escapes from a black hole, then how can we tell if one is there?
14. The Earth's age is often given as about 4.6 billion years. Find that time on Fig. 33–29. Modern humans have lived on Earth on the order of 200,000 years. Where is that on Fig. 33–29?
15. Why were atoms, as opposed to bare nuclei, unable to exist until hundreds of thousands of years after the Big Bang?
16. (*a*) Why are Type Ia supernovae so useful for determining the distances of galaxies? (*b*) How are their distances actually measured?
17. Under what circumstances would the universe eventually collapse in on itself?
18. (*a*) Why did astronomers expect that the expansion rate of the universe would be decreasing (decelerating) with time? (*b*) How, in principle, could astronomers hope to determine whether the universe used to expand faster than it does now?

MisConceptual Questions

1. Which one of the following is *not* expected to occur on an H–R diagram during the lifetime of a single star?
   (*a*) The star will move off the main sequence toward the upper right of the diagram.
   (*b*) Low-mass stars will become white dwarfs and end up toward the lower left of the diagram.
   (*c*) The star will move along the main sequence from one place to another.
   (*d*) All of the above.
2. When can parallax be used to determine the approximate distance from the Earth to a star?
   (*a*) Only during January and July.
   (*b*) Only when the star's distance is relatively small.
   (*c*) Only when the star's distance is relatively large.
   (*d*) Only when the star appears to move directly toward or away from the Earth.
   (*e*) Only when the star is the Sun.
   (*f*) Always.
   (*g*) Never.
3. Observations show that all galaxies tend to move away from Earth, and that more distant galaxies move away from Earth at faster velocities than do galaxies closer to the Earth. These observations imply that
   (*a*) the Earth is the center of the universe.
   (*b*) the universe is expanding.
   (*c*) the expansion of the universe will eventually stop.
   (*d*) All of the above.
4. Which process results in a tremendous amount of energy being emitted by the Sun?
   (*a*) Hydrogen atoms burn in the presence of oxygen—that is, hydrogen atoms oxidize.
   (*b*) The Sun contracts, decreasing its gravitational potential energy.
   (*c*) Protons in hydrogen atoms fuse, forming helium nuclei.
   (*d*) Radioactive atoms such as uranium, plutonium, and cesium emit gamma rays with high energy.
   (*e*) None of the above.
5. Which of the following methods can be used to find the distance from us to a star outside our galaxy? Choose all that apply.
   (*a*) Parallax.
   (*b*) Using luminosity and temperature from the H–R diagram and measuring the apparent brightness.
   (*c*) Using supernova explosions as a "standard candle."
   (*d*) Redshift in the line spectra of elements and compounds.
6. The history of the universe can be determined by observing astronomical objects at various (large) distances from the Earth. This method of discovery works because
   (*a*) time proceeds at different rates in different regions of the universe.
   (*b*) light travels at a finite speed.
   (*c*) matter warps space.
   (*d*) older galaxies are farther from the Earth than are younger galaxies.

7. Where did the Big Bang occur?
   (a) Near the Earth.
   (b) Near the center of the Milky Way Galaxy.
   (c) Several billion light-years away.
   (d) Throughout all space.
   (e) Near the Andromeda Galaxy.
8. When and how were virtually all of the elements of the Periodic Table formed?
   (a) In the very early universe a few seconds after the Big Bang.
   (b) At the centers of stars during their main-sequence phases.
   (c) At the centers of stars during novae.
   (d) At the centers of stars during supernovae.
   (e) On the surfaces of planets as they cooled and hardened.
9. We know that there must be dark matter in the universe because
   (a) we see dark dust clouds.
   (b) we see that the universe is expanding.
   (c) we see that stars far from the galactic center are moving faster than can be explained by visible matter.
   (d) we see that the expansion of the universe is accelerating.
10. Acceleration of the universe's expansion rate is due to
    (a) the repulsive effect of dark energy.
    (b) the attractive effect of dark matter.
    (c) the attractive effect of gravity.
    (d) the thermal expansion of stellar cores.

For assigned homework and other learning materials, go to the MasteringPhysics website.

Problems

33–1 to 33–3 Stars, Galaxies, Stellar Evolution, Distances

1. (I) The parallax angle of a star is 0.00029°. How far away is the star?
2. (I) A star exhibits a parallax of 0.27 seconds of arc. How far away is it?
3. (I) If one star is twice as far away from us as a second star, will the parallax angle of the farther star be greater or less than that of the nearer star? By what factor?
4. (II) What is the relative brightness of the Sun as seen from Jupiter, as compared to its brightness from Earth? (Jupiter is 5.2 times farther from the Sun than the Earth is.)
5. (II) When our Sun becomes a red giant, what will be its average density if it expands out to the orbit of Mercury (6×10^{10} m from the Sun)?
6. (II) We saw earlier (Chapter 14) that the rate energy reaches the Earth from the Sun (the "solar constant") is about $1.3 \times 10^3 \text{ W/m}^2$. What is (a) the apparent brightness b of the Sun, and (b) the intrinsic luminosity L of the Sun?
7. (II) Estimate the angular width that our Galaxy would subtend if observed from the nearest galaxy to us (Table 33–1). Compare to the angular width of the Moon from Earth.
8. (II) Assuming our Galaxy represents a good average for all other galaxies, how many stars are in the observable universe?
9. (II) Calculate the density of a white dwarf whose mass is equal to the Sun's and whose radius is equal to the Earth's. How many times larger than Earth's density is this?
10. (II) A neutron star whose mass is 1.5 solar masses has a radius of about 11 km. Calculate its average density and compare to that for a white dwarf (Problem 9) and to that of nuclear matter.
*11. (II) A star is 56 pc away. What is its parallax angle? State (a) in seconds of arc, and (b) in degrees.
*12. (II) What is the parallax angle for a star that is 65 ly away? How many parsecs is this?
*13. (II) A star is 85 pc away. How long does it take for its light to reach us?
14. (III) Suppose two stars of the same apparent brightness b are also believed to be the same size. The spectrum of one star peaks at 750 nm whereas that of the other peaks at 450 nm. Use Wien's law and the Stefan-Boltzmann equation (Eq. 14–6) to estimate their relative distances from us. [Hint: See Examples 33–4 and 33–5.]
15. (III) Stars located in a certain cluster are assumed to be about the same distance from us. Two such stars have spectra that peak at $\lambda_1 = 470$ nm and $\lambda_2 = 720$ nm, and the ratio of their apparent brightness is $b_1/b_2 = 0.091$. Estimate their relative sizes (give ratio of their diameters) using Wien's law and the Stefan-Boltzmann equation, Eq. 14–6.

33–4 General Relativity, Gravity and Curved Space

16. (I) Show that the Schwarzschild radius for Earth is 8.9 mm.
17. (II) What is the Schwarzschild radius for a typical galaxy (like ours)?
18. (II) What mass will give a Schwarzschild radius equal to that of the hydrogen atom in its ground state?
19. (II) What is the maximum sum-of-the-angles for a triangle on a sphere?
20. (II) Describe a triangle, drawn on the surface of a sphere, for which the sum of the angles is (a) 359°, and (b) 179°.
21. (III) What is the apparent deflection of a light beam in an elevator (Fig. 33–13) which is 2.4 m wide if the elevator is accelerating downward at 9.8 m/s^2?

33–5 Redshift, Hubble's Law

22. (I) The redshift of a galaxy indicates a recession velocity of 1850 km/s. How far away is it?
23. (I) If a galaxy is traveling away from us at 1.5% of the speed of light, roughly how far away is it?
24. (II) A galaxy is moving away from Earth. The "blue" hydrogen line at 434 nm emitted from the galaxy is measured on Earth to be 455 nm. (a) How fast is the galaxy moving? (b) How far is it from Earth based on Hubble's law?
25. (II) Estimate the wavelength shift for the 656.3-nm line in the Balmer series of hydrogen emitted from a galaxy whose distance from us is (a) 7.0×10^6 ly, (b) 7.0×10^7 ly.

26. (II) If an absorption line of calcium is normally found at a wavelength of 393.4 nm in a laboratory gas, and you measure it to be at 423.4 nm in the spectrum of a galaxy, what is the approximate distance to the galaxy?

27. (II) What is the speed of a galaxy with $z = 0.060$?

28. (II) What would be the redshift parameter z for a galaxy traveling away from us at $v = 0.075c$?

29. (II) Estimate the distance d from the Earth to a galaxy whose redshift parameter $z = 1$.

30. (II) Estimate the speed of a galaxy, and its distance from us, if the wavelength for the hydrogen line at 434 nm is measured on Earth as being 610 nm.

31. (II) Radiotelescopes are designed to observe 21-cm waves emitted by atomic hydrogen gas. A signal from a distant radio-emitting galaxy is found to have a wavelength that is 0.10 cm longer than the normal 21-cm wavelength. Estimate the distance to this galaxy.

32. (III) Starting from Eq. 33–3, show that the Doppler shift in wavelength is $\Delta\lambda/\lambda_{rest} \approx v/c$ (Eq. 33–6) for $v \ll c$. [*Hint*: Use the binomial expansion.]

33–6 to 33–8 The Big Bang, CMB, Universe Expansion

33. (I) Calculate the wavelength at the peak of the blackbody radiation distribution at 2.7 K using Wien's law.

34. (II) Calculate the peak wavelength of the CMB at 1.0 s after the birth of the universe. In what part of the EM spectrum is this radiation?

35. (II) The critical density for closure of the universe is $\rho_c \approx 10^{-26}$ kg/m^3. State ρ_c in terms of the average number of nucleons per cubic meter.

36. (II) The scale factor of the universe (average distance between galaxies) at any given time is believed to have been inversely proportional to the absolute temperature. Estimate the size of the universe, compared to today, at (*a*) $t = 10^6$ yr, (*b*) $t = 1$ s, (*c*) $t = 10^{-6}$ s, and (*d*) $t = 10^{-35}$ s.

37. (II) At approximately what time had the universe cooled below the threshold temperature for producing (*a*) kaons ($M \approx 500$ MeV/c^2), (*b*) Υ ($M \approx 9500$ MeV/c^2), and (*c*) muons ($M \approx 100$ MeV/c^2)?

33–9 Dark Matter, Dark Energy

38. (II) Only about 5% of the energy in the universe is composed of baryonic matter. (*a*) Estimate the average density of baryonic matter in the observable universe with a radius of 14 billion light-years that contains 10^{11} galaxies, each with about 10^{11} stars like our Sun. (*b*) Estimate the density of dark matter in the universe.

General Problems

39. Use conservation of angular momentum to estimate the angular velocity of a neutron star which has collapsed to a diameter of 16 km, from a star whose core radius was equal to that of Earth (6×10^6 m). Assume its mass is 1.5 times that of the Sun, and that it rotated (like our Sun) about once a month.

40. By what factor does the rotational kinetic energy change when the star in Problem 39 collapses to a neutron star?

41. Suppose that three main-sequence stars could undergo the three changes represented by the three arrows, A, B, and C, in the H–R diagram of Fig. 33–35. For each case, describe the changes in temperature, intrinsic luminosity, and size.

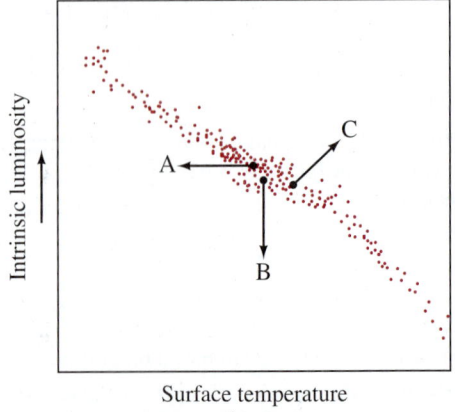

FIGURE 33–35 Problem 41.

42. Assume that the nearest stars to us have an intrinsic luminosity about the same as the Sun's. Their apparent brightness, however, is about 10^{11} times fainter than the Sun. From this, estimate the distance to the nearest stars.

43. A certain pulsar, believed to be a neutron star of mass 1.5 times that of the Sun, with diameter 16 km, is observed to have a rotation speed of 1.0 rev/s. If it loses rotational kinetic energy at the rate of 1 part in 10^9 per day, which is all transformed into radiation, what is the power output of the star?

44. The nearest large galaxy to our Galaxy is about 2×10^6 ly away. If both galaxies have a mass of 4×10^{41} kg, with what gravitational force does each galaxy attract the other? Ignore dark matter.

45. How large would the Sun be if its density equaled the critical density of the universe, $\rho_c \approx 10^{-26}$ kg/m^3? Express your answer in light-years and compare with the Earth–Sun distance and the diameter of our Galaxy.

46. Two stars, whose spectra peak at 660 nm and 480 nm, respectively, both lie on the main sequence. Use Wien's law, the Stefan-Boltzmann equation, and the H–R diagram (Fig. 33–6) to estimate the ratio of their diameters.

47. (*a*) In order to measure distances with parallax at 100 ly, what minimum angular resolution (in degrees) is needed? (*b*) What diameter mirror or lens would be needed?

48. (a) What temperature would correspond to 14-TeV collisions at the LHC? (b) To what era in cosmological history does this correspond? [*Hint*: See Fig. 33–29.]

49. In the later stages of stellar evolution, a star (if massive enough) will begin fusing carbon nuclei to form, for example, magnesium:

$$^{12}_{6}C + ^{12}_{6}C \rightarrow ^{24}_{12}Mg + \gamma.$$

(a) How much energy is released in this reaction (see Appendix B)? (b) How much kinetic energy must each carbon nucleus have (assume equal) in a head-on collision if they are just to "touch" (use Eq. 30–1) so that the strong force can come into play? (c) What temperature does this kinetic energy correspond to?

50. Consider the reaction

$$^{16}_{8}O + ^{16}_{8}O \rightarrow ^{28}_{14}Si + ^{4}_{2}He,$$

and answer the same questions as in Problem 49.

51. Use *dimensional analysis* with the fundamental constants c, G, and $\hbar$ to estimate the value of the so-called *Planck time*. It is thought that physics as we know it can say nothing about the universe before this time.

52. Estimate the mass of our observable universe using the following assumptions: Our universe is spherical in shape, it has been expanding at the speed of light since the Big Bang, and its density is the critical density.

Search and Learn

1. Estimate what neutrino mass (in eV/c^2) would provide the critical density to close the universe. Assume the neutrino density is, like photons, about 10^9 times that of nucleons, and that nucleons make up only (a) 2% of the mass needed, or (b) 5% of the mass needed.

2. Describe how we can estimate the distance from us to other stars. Which methods can we use for nearby stars, and which can we use for very distant stars? Which method gives the most accurate distance measurements for the most distant stars?

3. The evolution of stars, as discussed in Section 33–2, can lead to a white dwarf, a neutron star, or even a black hole, depending on the mass. (a) Referring to Sections 33–2 and 33–4, give the radius of (i) a white dwarf of 1 solar mass, (ii) a neutron star of 1.5 solar masses, and (iii) a black hole of 3 solar masses. (b) Express these three radii as ratios $(r_i : r_{ii} : r_{iii})$.

4. (a) Describe some of the evidence that the universe began with a "Big Bang." (b) How does the curvature of the universe affect its future destiny? (c) How does dark energy affect the possible future of the universe?

5. When stable nuclei first formed, about 3 minutes after the Big Bang, there were about 7 times more protons than neutrons. Explain how this leads to a ratio of the mass of hydrogen to the mass of helium of 3:1. This is about the actual ratio observed in the universe.

6. Explain what the 2.7-K cosmic microwave background radiation is. Where does it come from? Why is its temperature now so low?

7. We cannot use Hubble's law to measure the distances to nearby galaxies, because their random motions are larger than the overall expansion. Indeed, the closest galaxy to us, the Andromeda Galaxy, 2.5 million light-years away, is approaching us at a speed of about 130 km/s. (a) What is the shift in wavelength of the 656-nm line of hydrogen emitted from the Andromeda Galaxy, as seen by us? (b) Is this a redshift or a blueshift? (c) Ignoring the expansion, how soon will it and the Milky Way Galaxy collide?

ANSWERS TO EXERCISES

A: Our Earth and ourselves, 2 years ago.
B: 600 ly (estimating L from Fig. 33–6 as $L \approx 8 \times 10^{26}$ W; note that on a log scale, 6000 K is closer to 7000 K than it is to 5000 K).
C: 30 km.
D: (a); not the usual R^3, but R: see formula for the Schwarzschild radius.
E: (c); (d).

Mathematical Review

A–1 Relationships, Proportionality, and Equations

One of the important aspects of physics is the search for relationships between different quantities—that is, determining how one quantity affects another.

As a simple example, the ancients found that if one circle has twice the diameter of a second circle, the first also has twice the circumference. If the diameter is three times as large, the circumference is also three times as large. In other words, an increase in the diameter results in a proportional increase in the circumference. We say that the circumference is **directly proportional** to the diameter. This can be written in symbols as $C \propto D$, where "$\propto$" means "is proportional to," and C and D refer to the circumference and diameter of a circle, respectively. The next step is to change this proportionality to an equation, which will make it possible to link the two quantities numerically. This means inserting a proportionality constant, which in many cases is determined by measurement. The ancients found that the ratio of the circumference to the diameter of any circle was 3.1416 (to keep only the first few decimal places). This number is designated by the Greek letter π. It is the constant of proportionality for the relationship $C \propto D$. To obtain an equation, we insert π into the proportion and change the $\propto$ to $=$. Thus,

$$C = \pi D.$$

Other kinds of proportionality occur as well. For example, the area of a circle is proportional to the *square* of its radius. That is, if the radius is doubled, the area becomes four times as large; and so on. In this case we can write $A \propto r^2$, where A stands for the area and r for the radius of the circle. The constant of proportionality is found to be π again: $A = \pi r^2$.

Sometimes two quantities are related in such a way that an increase in one leads to a proportional *decrease* in the other. This is called **inverse proportion**. For example, the time required to travel a given distance is inversely proportional to the speed of travel. The greater the speed, the less time it takes. We can write this inverse proportion as

$$\text{time} \propto 1/\text{speed}.$$

The larger the denominator of a fraction, the lower the value of the fraction is as a whole. For example, $\frac{1}{4}$ is less than $\frac{1}{2}$. Thus, if the speed is doubled, the time is halved, which is what we want to express by this inverse proportionality relationship.

If you suspect that a relationship exists between two or more quantities, you can try to determine the precise nature of this relationship by varying one of the quantities and measuring how the other varies as a result. Sometimes a given quantity is affected by two or more quantities; for instance, the acceleration of an object is related to both its mass and the applied force. In such a case, only one quantity is varied at a time, while the others are held constant.

When one quantity affects another, we often use the expression **is a function of** to indicate this dependence; for example, we say that the pressure in a tire is a function of the temperature.

Whatever kind of proportion is found to hold, it can be changed to an equality by finding the proper proportionality constant. Quantitative statements or predictions about the physical world can then be made with the equation.

A-2 Exponents

When we write 10^4, we mean that you multiply 10 by itself four times: $10^4 = 10 \times 10 \times 10 \times 10 = 10{,}000$. The superscript 4 is called an **exponent**, and 10 is said to be raised to the fourth power. Any number or symbol can be raised to a power. Special names are used when the exponent is 2 (a^2 is "*a* squared") or 3 (a^3 is "*a* cubed"). For any other power, we say a^n is "*a* to the *n*th power." If the exponent is 1, it is usually dropped: $a^1 = a$, since no multiplication is involved.

The rules for multiplying numbers expressed as powers are as follows: first,

$$(a^n)(a^m) = a^{n+m}. \tag{A-1}$$

That is, the exponents are added. To see why, consider the result of the multiplication of 3^3 by 3^4:

$$(3^3)(3^4) = (3)(3)(3) \times (3)(3)(3)(3) = (3)^7.$$

Here the sum of the exponents is $3 + 4 = 7$, so rule A–1 works. Notice that this rule works only if the base numbers (*a* in Eq. A–1) are the same. Thus we *cannot* use the rule of summing exponents for $(6^3)(5^2)$; these numbers would have to be written out. However, if the base numbers are different but the exponents are the same, we can write a second rule:

$$(a^n)(b^n) = (ab)^n. \tag{A-2}$$

For example, $(5^3)(6^3) = (30)^3$ since

$$(5)(5)(5)(6)(6)(6) = (30)(30)(30).$$

The third rule involves a power raised to another power: $(a^3)^2$ means $(a^3)(a^3)$, which is equal to $a^{3+3} = a^6$. The general rule is then

$$(a^n)^m = a^{nm}. \tag{A-3}$$

In this case, the exponents are multiplied.

Negative exponents are used for reciprocals. Thus,

$$\frac{1}{a} = a^{-1}, \quad \frac{1}{a^3} = a^{-3},$$

and so on. The reason for using negative exponents is to allow us to use the multiplication rules given above. For example, $(a^5)(a^{-3})$ means

$$\frac{(a)(a)(a)(a)(a)}{(a)(a)(a)} = a^2,$$

after canceling 3 of the *a*'s. Rule A–1 gives us the same result:

$$(a^5)(a^{-3}) = a^{5-3} = a^2.$$

What does an exponent of zero mean? That is, what is a^0? Any number raised to the zeroth power is defined as being equal to 1:

$$a^0 = 1.$$

This definition is used because it follows from the rules for adding exponents. For example,

$$a^3 a^{-3} = a^{3-3} = a^0 = 1.$$

But *does* $a^3 a^{-3}$ actually equal 1? Yes, because

$$a^3 a^{-3} = \frac{a^3}{a^3} = 1.$$

Fractional exponents are used to represent *roots*. For example, $a^{\frac{1}{2}}$ means the square root of *a*; that is, $a^{\frac{1}{2}} = \sqrt{a}$. Similarly, $a^{\frac{1}{3}}$ means the cube root of *a*, and so on. The fourth root of *a* means that if you multiply the fourth root of *a* by itself four times, you again get *a*:

$$\left(a^{\frac{1}{4}}\right)^4 = a.$$

This is consistent with rule A–3 since $\left(a^{\frac{1}{4}}\right)^4 = a^{\frac{4}{4}} = a^1 = a$.

A–3 Powers of 10, or Exponential Notation

Writing out very large and very small numbers such as the distance of Neptune from the Sun, 4,500,000,000 km, or the diameter of a typical atom, 0.00000001 cm, is inconvenient and prone to error. It also leaves in question (see Section 1–4) the number of significant figures. (How many of the zeros are significant in the number 4,500,000,000 km?)

For these reasons we make use of the "powers of 10," or exponential notation. The distance from Neptune to the Sun is then expressed as 4.50×10^9 km (assuming that the value is significant to three digits), and the diameter of an atom 1.0×10^{-8} cm. This way of writing numbers is based on the use of exponents, where a^n signifies a multiplied by itself n times. For example, $10^4 = 10 \times 10 \times 10 \times 10 = 10,000$. Thus, $4.50 \times 10^9 = 4.50 \times 1,000,000,000 = 4,500,000,000$. Notice that the exponent (9 in this case) is just the number of places the decimal point is moved to the right to obtain the fully written-out number (4.500,000,000.)

When two numbers are multiplied (or divided), you first multiply (or divide) the simple parts and then the powers of 10. Thus, 2.0×10^3 multiplied by 5.5×10^4 equals $(2.0 \times 5.5) \times (10^3 \times 10^4) = 11 \times 10^7$, where we have used the rule for adding exponents (Appendix A–2). Similarly, 8.2×10^5 divided by 2.0×10^2 equals

$$\frac{8.2 \times 10^5}{2.0 \times 10^2} = \frac{8.2}{2.0} \times \frac{10^5}{10^2} = 4.1 \times 10^3.$$

For numbers less than 1, say 0.01, the exponent power of 10 is written with a negative sign (see previous page): $0.01 = 1/100 = 1/10^2 = 1 \times 10^{-2}$. Similarly, $0.002 = 2 \times 10^{-3}$. The decimal point has again been moved the number of places expressed in the exponent. For example, $0.020 \times 3600 = 72$, or in exponential notation $(2.0 \times 10^{-2}) \times (3.6 \times 10^3) = 7.2 \times 10^1 = 72$.

Notice also that $10^1 \times 10^{-1} = 10 \times 0.1 = 1$, and by the law of exponents, $10^1 \times 10^{-1} = 10^0$. Therefore, $10^0 = 1$.

When writing a number in exponential notation, it is usual to make the simple number be between 1 and 10. Thus it is conventional to write 4.5×10^9 rather than 45×10^8, although they are the same number.[†] This notation also allows the number of *significant figures* to be clearly expressed. We write 4.50×10^9 if this value is accurate to three significant figures, but 4.5×10^9 if it is accurate to only two.

A–4 Algebra

Physical relationships between quantities can be represented as equations involving symbols (usually letters of the Greek or Roman alphabet) that represent the quantities. The manipulation of such equations is the field of algebra, and it is used a great deal in physics. An equation involves an equals sign, which tells us that the quantities on either side of the equals sign have the same value. Examples of equations are

$$3 + 8 = 11$$
$$2x + 7 = 15$$
$$a^2b + c = 6.$$

The first equation involves only numbers, so is called an arithmetic equation. The other two equations are algebraic since they involve symbols. In the third equation, the quantity a^2b means the product of a times a times b: $a^2b = a \times a \times b$.

[†]Another convention used, particularly with computers, is that the simple number be between 0.1 and 1. Thus we could write 4,500,000,000 as 0.450×10^{10}. This is slightly less compact.

Solving for an Unknown

Often we wish to solve for one (or more) symbols, and we treat it as an *unknown*. For example, in the equation $2x + 7 = 15$, x is the unknown; this equation is true, however, only when $x = 4$. Determining what value (or values) the unknown can have to satisfy the equation is called *solving the equation*. To solve an equation, the following rule can be used:

> An equation will remain true if any operation performed on one side is also performed on the other side. For example: (*a*) addition or subtraction of a number or symbol; (*b*) multiplication or division by a number or symbol; (*c*) raising each side of the equation to the same power, or taking the same root (such as square root).

EXAMPLE A–1 Solve for x in the equation
$$2x + 7 = 15.$$

APPROACH We perform the same operations on both sides of the equation to isolate x as the only variable on the left side of the equals sign.

SOLUTION We first subtract 7 from both sides:
$$2x + 7 - 7 = 15 - 7$$
or
$$2x = 8.$$
Then we divide both sides by 2 to get
$$\frac{2x}{2} = \frac{8}{2},$$
or, carrying out the divisions,
$$x = 4,$$
and this solves the equation.

EXAMPLE A–2 (*a*) Solve the equation
$$a^2 b + c = 24$$
for the unknown a in terms of b and c. (*b*) Solve for a assuming that $b = 2$ and $c = 6$.

APPROACH We perform operations to isolate a as the only variable on the left side of the equals sign.

SOLUTION (*a*) We are trying to solve for a, so we first subtract c from both sides:
$$a^2 b = 24 - c,$$
then divide by b:
$$a^2 = \frac{24 - c}{b},$$
and finally take square roots:
$$a = \sqrt{\frac{24 - c}{b}}.$$

(*b*) If we are given that $b = 2$ and $c = 6$, then
$$a = \sqrt{\frac{24 - 6}{2}} = 3.$$

But this is not the only answer. Whenever we take a square root, the number can be either positive or negative. Thus $a = -3$ is also a solution. Why? Because $(-3)^2 = 9$, just as $(+3)^2 = 9$. So we actually get two solutions: $a = +3$ and $a = -3$.

NOTE When an unknown appears squared in an equation, there are generally two solutions for that unknown.

To check a solution, we put it back into the original equation (this is really a check that we did all the manipulations correctly). In the equation

$$a^2b + c = 24,$$

we put in $a = 3$, $b = 2$, $c = 6$ and find

$$(3)^2(2) + (6) \stackrel{?}{=} 24$$
$$24 = 24,$$

which checks.

EXERCISE A Put $a = -3$ into the equation of Example A–2 and show that it works too.

Two or More Unknowns

If we have two or more unknowns, one equation is not sufficient to find them. In general, if there are n unknowns, n independent equations are needed. For example, if there are two unknowns, we need two equations. If the unknowns are called x and y, a typical procedure is to solve one equation for x in terms of y, and substitute this into the second equation.

EXAMPLE A–3 Solve the following pair of equations for x and y:

$$3x - 2y = 19$$
$$x + 4y = -3.$$

APPROACH We have two unknowns and two equations; we can start by solving the second equation for x in terms of y. Then we substitute this result for x into the first equation.

SOLUTION We subtract $4y$ from both sides of the second equation:

$$x = -3 - 4y.$$

We substitute this expression for x into the first equation, and simplify:

$$3(-3 - 4y) - 2y = 19$$
$$-9 - 12y - 2y = 19 \quad \text{(carried out the multiplication by 3)}$$
$$-14y = 28 \quad \text{(added 9 to both sides)}$$
$$y = -2. \quad \text{(divided both sides by } -14)$$

Now that we know $y = -2$, we substitute this into the expression for x:

$$x = -3 - 4y$$
$$= -3 - 4(-2) = -3 + 8 = 5.$$

Our solution is $x = 5$, $y = -2$. We check this solution by putting these values back into the original equations:

$$3x - 2y \stackrel{?}{=} 19$$
$$3(5) - 2(-2) \stackrel{?}{=} 19$$
$$15 + 4 \stackrel{?}{=} 19$$
$$19 = 19 \quad \text{(it checks)}$$

and

$$x + 4y \stackrel{?}{=} -3$$
$$5 + 4(-2) \stackrel{?}{=} -3$$
$$-3 = -3. \quad \text{(it checks)}$$

Other methods for solving two or more equations, such as the method of determinants, can be found in an algebra textbook.

The Quadratic Formula

We sometimes encounter equations that involve an unknown, say x, that appears not only to the first power, but squared as well. Such a **quadratic equation** can be written in the general form

$$ax^2 + bx + c = 0.$$

The quantities a, b, and c are typically numbers or constants that are given.[†] The general solutions to such an equation are given by the **quadratic formula**:

$$x = \frac{-b \pm \sqrt{b^2 - 4ac}}{2a}. \qquad (A\text{–}4)$$

The $\pm$ sign indicates that there are two solutions for x: one where the plus sign is used, the other where the minus sign is used.

EXAMPLE A–4 Find the solutions for x in the equation

$$3x^2 - 5x = 2.$$

APPROACH Here x appears both to the first power and squared, so we use the quadratic equation.

SOLUTION First we write this equation in the standard form

$$ax^2 + bx + c = 0$$

by subtracting 2 from both sides:

$$3x^2 - 5x - 2 = 0.$$

In this case, a, b, and c in the standard formula take the values $a = 3$, $b = -5$, and $c = -2$. The two solutions for x are, using Eq. A–4,

$$x = \frac{+5 + \sqrt{25 - (4)(3)(-2)}}{(2)(3)} = \frac{5 + 7}{6} = 2$$

and

$$x = \frac{+5 - \sqrt{25 - (4)(3)(-2)}}{(2)(3)} = \frac{5 - 7}{6} = -\frac{1}{3}.$$

In this Example, the two solutions are $x = 2$ and $x = -\frac{1}{3}$. In physics problems, it sometimes happens that only one of the solutions corresponds to a real-life situation; in this case, the other solution is discarded. In other cases, both solutions may correspond to physical reality.

Notice, incidentally, that b^2 must be greater than $4ac$, so that $\sqrt{b^2 - 4ac}$ yields a real number. If $(b^2 - 4ac)$ is less than zero (negative), there is no real solution. The square root of a negative number is called **imaginary**.

A second-order equation—one in which the highest power of x is 2—has two solutions; a third-order equation—involving x^3—has three solutions; and so on.

A–5 The Binomial Expansion

Sometimes we end up with a quantity of the form $(1 + x)^n$. That is, the quantity $(1 + x)$ is raised to the nth power. This can be written as an infinite sum of terms known as the **binomial expansion**:

$$(1 + x)^n = 1 + nx + \frac{n(n-1)}{2}x^2 + \cdots. \qquad (A\text{–}5)$$

This formula is useful for us mainly when x is very small compared to one $(x \ll 1)$. In this case, each successive term is much smaller than the preceding

[†]Or one or more of them could be variables, in which case additional equations are needed.

term. For example, let $x = 0.01$, and $n = 2$. Then whereas the first term equals 1, the second term is $nx = (2)(0.01) = 0.02$, and the third term is $[(2)(1)/2](0.01)^2 = 0.0001$, and so on. Thus, when x is small, we can ignore all but the first two (or three) terms and can write

$$(1 + x)^n \approx 1 + nx. \quad\quad\quad (\text{A--6})$$

This approximation often allows us to solve an equation easily that otherwise might be very difficult. Some examples of the binomial expansion are

$$(1 + x)^2 \approx 1 + 2x,$$
$$\frac{1}{1 + x} = (1 + x)^{-1} \approx 1 - x,$$
$$\sqrt{1 + x} = (1 + x)^{\frac{1}{2}} \approx 1 + \tfrac{1}{2}x,$$
$$\frac{1}{\sqrt{1 + x}} = (1 + x)^{-\frac{1}{2}} \approx 1 - \tfrac{1}{2}x,$$

where $x \ll 1$.

As a numerical example, let us evaluate $\sqrt{1.02}$ using the binomial expansion since $x = 0.02$ is much smaller than 1:

$$\sqrt{1.02} = (1.02)^{\frac{1}{2}} = (1 + 0.02)^{\frac{1}{2}} \approx 1 + \tfrac{1}{2}(0.02) = 1.01.$$

You can check with a calculator (and maybe not even more quickly) that $\sqrt{1.02} \approx 1.01$.

A–6 Plane Geometry

We review here a number of theorems involving angles and triangles that are useful in physics.

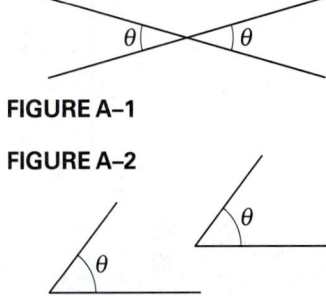

FIGURE A–1

FIGURE A–2

1. **Equal angles.** Two angles are equal if any of the following conditions are true:
   (a) They are vertical angles (Fig. A–1); or
   (b) the left side of one is parallel to the left side of the other, and the right side of one is parallel to the right side of the other (Fig. A–2; the left and right sides are as seen from the vertex, where the two sides meet); or
   (c) the left side of one is perpendicular to the left side of the other, and the right sides are likewise perpendicular (Fig. A–3).

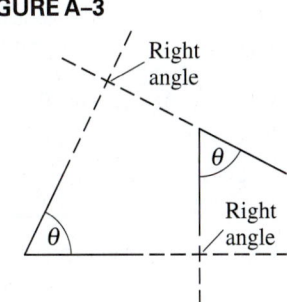

FIGURE A–3

2. *The sum of the angles* in any plane triangle is 180°.

3. **Similar triangles.** Two triangles are said to be similar if all three of their angles are equal (in Fig. A–4, $\theta_1 = \phi_1$, $\theta_2 = \phi_2$, and $\theta_3 = \phi_3$). Similar triangles thus have the same basic shape but may be different sizes and have different orientations. Two useful theorems about similar triangles are:
   (a) Two triangles are similar if any two of their angles are equal. (This follows because the third angles must also be equal since the sum of the angles of a triangle is 180°.)
   (b) The ratios of corresponding sides of two similar triangles are equal. That is (Fig. A–4),

$$\frac{a_1}{b_1} = \frac{a_2}{b_2} = \frac{a_3}{b_3}.$$

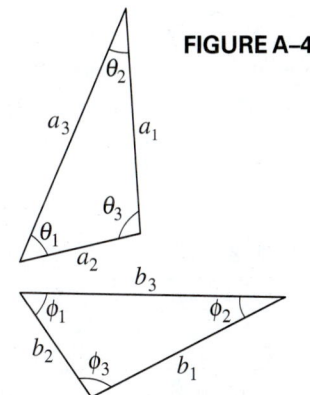

FIGURE A–4

4. **Congruent triangles.** Two triangles are congruent if one can be placed precisely on top of the other. That is, they are similar triangles and they have the same size. Two triangles are congruent if any of the following holds:
   (a) The three corresponding sides are equal.
   (b) Two sides and the enclosed angle are equal ("side-angle-side").
   (c) Two angles and the enclosed side are equal ("angle-side-angle").

FIGURE A–5

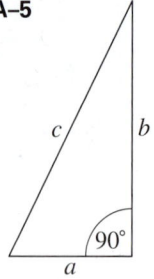

5. **Right triangles.** A right triangle has one angle that is 90° (a **right angle**); that is, the two sides that meet at the right angle are perpendicular (Fig. A–5). The two other (acute) angles in the right triangle add up to 90°.

6. **Pythagorean theorem.** In any right triangle, the square of the length of the hypotenuse (the side opposite the right angle) is equal to the sum of the squares of the lengths of the other two sides. In Fig. A–5,

$$c^2 = a^2 + b^2.$$

A–7 Trigonometric Functions and Identities

FIGURE A–6

Trigonometric functions for any angle θ are defined by constructing a right triangle about that angle as shown in Fig. A–6; opp and adj are the lengths of the sides opposite and adjacent to the angle θ, and hyp is the length of the hypotenuse:

$$\sin\theta = \frac{\text{opp}}{\text{hyp}} \qquad \csc\theta = \frac{1}{\sin\theta} = \frac{\text{hyp}}{\text{opp}}$$

$$\cos\theta = \frac{\text{adj}}{\text{hyp}} \qquad \sec\theta = \frac{1}{\cos\theta} = \frac{\text{hyp}}{\text{adj}}$$

$$\tan\theta = \frac{\text{opp}}{\text{adj}} = \frac{\sin\theta}{\cos\theta} \qquad \cot\theta = \frac{1}{\tan\theta} = \frac{\text{adj}}{\text{opp}}$$

$$\text{adj}^2 + \text{opp}^2 = \text{hyp}^2 \qquad \text{(Pythagorean theorem)}.$$

FIGURE A–7

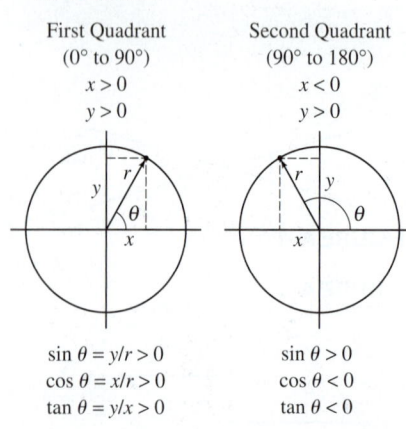

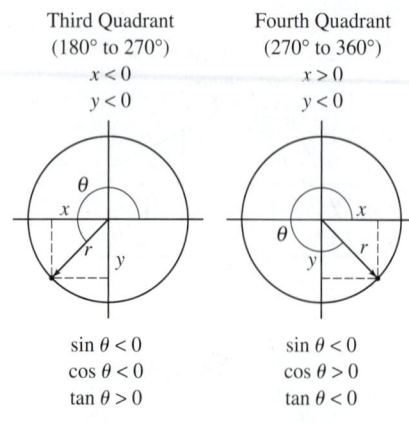

Figure A–7 shows the signs ($+$ or $-$) that cosine, sine, and tangent take on for angles θ in the four quadrants (0° to 360°). Note that angles are measured counterclockwise from the x axis as shown; negative angles are measured from *below* the x axis, clockwise: for example, $-30° = +330°$, and so on.

The following are some useful identities among the trigonometric functions:

$$\sin^2\theta + \cos^2\theta = 1$$

$$\sin 2\theta = 2\sin\theta\cos\theta$$

$$\cos 2\theta = \cos^2\theta - \sin^2\theta = 2\cos^2\theta - 1 = 1 - 2\sin^2\theta$$

$$\tan 2\theta = \frac{2\tan\theta}{1 - \tan^2\theta}$$

$$\sin(A \pm B) = \sin A \cos B \pm \cos A \sin B$$

$$\cos(A \pm B) = \cos A \cos B \mp \sin A \sin B$$

$$\tan(A \pm B) = \frac{\tan A \pm \tan B}{1 \mp \tan A \tan B}$$

$$\sin(180° - \theta) = \sin\theta$$

$$\cos(180° - \theta) = -\cos\theta$$

$$\sin(90° - \theta) = \cos\theta$$

$$\cos(90° - \theta) = \sin\theta$$

$$\sin\tfrac{1}{2}\theta = \sqrt{\frac{1 - \cos\theta}{2}}$$

$$\cos\tfrac{1}{2}\theta = \sqrt{\frac{1 + \cos\theta}{2}}$$

$$\tan\tfrac{1}{2}\theta = \sqrt{\frac{1 - \cos\theta}{1 + \cos\theta}}$$

$$\sin A \pm \sin B = 2\sin\left(\frac{A \pm B}{2}\right)\cos\left(\frac{A \mp B}{2}\right).$$

FIGURE A–8

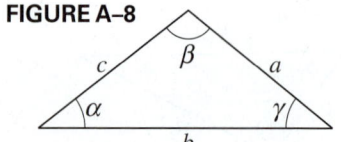

For any triangle (see Fig. A–8):

$$\frac{\sin\alpha}{a} = \frac{\sin\beta}{b} = \frac{\sin\gamma}{c} \qquad \text{(law of sines)}$$

$$c^2 = a^2 + b^2 - 2ab\cos\gamma. \qquad \text{(law of cosines)}$$

Trigonometric Table: Numerical Values of Sin, Cos, Tan

Angle in Degrees	Angle in Radians	Sine	Cosine	Tangent	Angle in Degrees	Angle in Radians	Sine	Cosine	Tangent
0°	0.000	0.000	1.000	0.000					
1°	0.017	0.017	1.000	0.017	46°	0.803	0.719	0.695	1.036
2°	0.035	0.035	0.999	0.035	47°	0.820	0.731	0.682	1.072
3°	0.052	0.052	0.999	0.052	48°	0.838	0.743	0.669	1.111
4°	0.070	0.070	0.998	0.070	49°	0.855	0.755	0.656	1.150
5°	0.087	0.087	0.996	0.087	50°	0.873	0.766	0.643	1.192
6°	0.105	0.105	0.995	0.105	51°	0.890	0.777	0.629	1.235
7°	0.122	0.122	0.993	0.123	52°	0.908	0.788	0.616	1.280
8°	0.140	0.139	0.990	0.141	53°	0.925	0.799	0.602	1.327
9°	0.157	0.156	0.988	0.158	54°	0.942	0.809	0.588	1.376
10°	0.175	0.174	0.985	0.176	55°	0.960	0.819	0.574	1.428
11°	0.192	0.191	0.982	0.194	56°	0.977	0.829	0.559	1.483
12°	0.209	0.208	0.978	0.213	57°	0.995	0.839	0.545	1.540
13°	0.227	0.225	0.974	0.231	58°	1.012	0.848	0.530	1.600
14°	0.244	0.242	0.970	0.249	59°	1.030	0.857	0.515	1.664
15°	0.262	0.259	0.966	0.268	60°	1.047	0.866	0.500	1.732
16°	0.279	0.276	0.961	0.287	61°	1.065	0.875	0.485	1.804
17°	0.297	0.292	0.956	0.306	62°	1.082	0.883	0.469	1.881
18°	0.314	0.309	0.951	0.325	63°	1.100	0.891	0.454	1.963
19°	0.332	0.326	0.946	0.344	64°	1.117	0.899	0.438	2.050
20°	0.349	0.342	0.940	0.364	65°	1.134	0.906	0.423	2.145
21°	0.367	0.358	0.934	0.384	66°	1.152	0.914	0.407	2.246
22°	0.384	0.375	0.927	0.404	67°	1.169	0.921	0.391	2.356
23°	0.401	0.391	0.921	0.424	68°	1.187	0.927	0.375	2.475
24°	0.419	0.407	0.914	0.445	69°	1.204	0.934	0.358	2.605
25°	0.436	0.423	0.906	0.466	70°	1.222	0.940	0.342	2.747
26°	0.454	0.438	0.899	0.488	71°	1.239	0.946	0.326	2.904
27°	0.471	0.454	0.891	0.510	72°	1.257	0.951	0.309	3.078
28°	0.489	0.469	0.883	0.532	73°	1.274	0.956	0.292	3.271
29°	0.506	0.485	0.875	0.554	74°	1.292	0.961	0.276	3.487
30°	0.524	0.500	0.866	0.577	75°	1.309	0.966	0.259	3.732
31°	0.541	0.515	0.857	0.601	76°	1.326	0.970	0.242	4.011
32°	0.559	0.530	0.848	0.625	77°	1.344	0.974	0.225	4.331
33°	0.576	0.545	0.839	0.649	78°	1.361	0.978	0.208	4.705
34°	0.593	0.559	0.829	0.675	79°	1.379	0.982	0.191	5.145
35°	0.611	0.574	0.819	0.700	80°	1.396	0.985	0.174	5.671
36°	0.628	0.588	0.809	0.727	81°	1.414	0.988	0.156	6.314
37°	0.646	0.602	0.799	0.754	82°	1.431	0.990	0.139	7.115
38°	0.663	0.616	0.788	0.781	83°	1.449	0.993	0.122	8.144
39°	0.681	0.629	0.777	0.810	84°	1.466	0.995	0.105	9.514
40°	0.698	0.643	0.766	0.839	85°	1.484	0.996	0.087	11.43
41°	0.716	0.656	0.755	0.869	86°	1.501	0.998	0.070	14.301
42°	0.733	0.669	0.743	0.900	87°	1.518	0.999	0.052	19.081
43°	0.750	0.682	0.731	0.933	88°	1.536	0.999	0.035	28.636
44°	0.768	0.695	0.719	0.966	89°	1.553	1.000	0.017	57.290
45°	0.785	0.707	0.707	1.000	90°	1.571	1.000	0.000	∞

A–8 Logarithms

Logarithms are defined in the following way:

$$\text{if } y = A^x, \quad \text{then } x = \log_A y.$$

That is, the logarithm of a number y to the base A is that number which, as the exponent of A, gives back the number y. For **common logarithms**, the base is 10, so

$$\text{if } y = 10^x, \quad \text{then } x = \log y.$$

The subscript 10 on $\log_{10}$ is usually omitted when dealing with common logs. Another base sometimes used is the exponential base $e = 2.718\cdots$, a natural number.† Such logarithms are called **natural logarithms** and are written "ln". Thus,

$$\text{if } y = e^x, \quad \text{then } x = \ln y.$$

For any number y, the two types of logarithm are related by

$$\ln y = 2.3026 \log y.$$

Some simple rules for logarithms include:

$$\log(ab) = \log a + \log b. \tag{A–7}$$

This is true because if $a = 10^n$ and $b = 10^m$, then $ab = 10^{n+m}$. From the definition of logarithm, $\log a = n$, $\log b = m$, and $\log(ab) = n + m$; hence, $\log(ab) = n + m = \log a + \log b$. In a similar way, we can show the rules

$$\log\left(\frac{a}{b}\right) = \log a - \log b \tag{A–8}$$

and

$$\log a^n = n \log a. \tag{A–9}$$

These three rules apply not only to common logs but to natural or any other kind of logarithm.

Logs were once used as a technique for simplifying certain types of calculation. Because of the advent of electronic calculators and computers, they are not often used any more for that purpose. However, logs do appear in certain physical equations, so it is helpful to know how to deal with them. If you do not have a calculator that calculates logs, you can use a **log table**, such as the small one shown here (Table A–1). The number N is given to two digits (some tables give N to three or more digits); the first digit is in the vertical column to the left, the second digit is in the horizontal row across the top. For example, the Table tells us that $\log 1.0 = 0.000$, $\log 1.1 = 0.041$, and $\log 4.1 = 0.613$. The Table gives logs for numbers between 1.0 and 9.9; for larger or smaller numbers, we use rule A–7:

$$\log(ab) = \log a + \log b.$$

For example,

$$\log(380) = \log(3.8 \times 10^2) = \log(3.8) + \log(10^2).$$

From the Table, $\log 3.8 = 0.580$; and from rule A–9,

$$\log(10^2) = 2\log(10) = 2,$$

since $\log(10) = 1$. [This follows from the definition of the logarithm: if

---

†The exponential base e can be written as an infinite series:

$$e = 1 + \frac{1}{1} + \frac{1}{1\cdot 2} + \frac{1}{1\cdot 2\cdot 3} + \frac{1}{1\cdot 2\cdot 3\cdot 4} + \cdots.$$

TABLE A–1 Short Table of Common Logarithms

N	0.0	0.1	0.2	0.3	0.4	0.5	0.6	0.7	0.8	0.9
1	.000	.041	.079	.114	.146	.176	.204	.230	.255	.279
2	.301	.322	.342	.362	.380	.398	.415	.431	.447	.462
3	.477	.491	.505	.519	.531	.544	.556	.568	.580	.591
4	.602	.613	.623	.633	.643	.653	.663	.672	.681	.690
5	.699	.708	.716	.724	.732	.740	.748	.756	.763	.771
6	.778	.785	.792	.799	.806	.813	.820	.826	.833	.839
7	.845	.851	.857	.863	.869	.875	.881	.886	.892	.898
8	.903	.908	.914	.919	.924	.929	.935	.940	.944	.949
9	.954	.959	.964	.968	.973	.978	.982	.987	.991	.996

$10 = 10^1$, then $1 = \log(10)$.] Thus,

$$\log(380) = \log(3.8) + \log(10^2)$$
$$= 0.580 + 2$$
$$= 2.580.$$

Similarly,

$$\log(0.081) = \log(8.1) + \log(10^{-2})$$
$$= 0.908 - 2 = -1.092.$$

Sometimes we need to do the reverse process: find the number N whose log is, say, 2.670. This is called "taking the **antilogarithm**." To do so, we separate our number 2.670 into two parts, making the separation at the decimal point:

$$\log N = 2.670 = 2 + 0.670$$
$$= \log 10^2 + 0.670.$$

We now look at Table A–1 to see what number has its log equal to 0.670; none does, so we must **interpolate**: we see that $\log 4.6 = 0.663$ and $\log 4.7 = 0.672$. So the number we want is between 4.6 and 4.7, and closer to the latter by $\frac{7}{9}$. Approximately we can say that $\log 4.68 = 0.670$. Thus

$$\log N = 2 + 0.670$$
$$= \log(10^2) + \log(4.68) = \log(4.68 \times 10^2),$$

so $N = 4.68 \times 10^2 = 468$.

If the given logarithm is negative, say, -2.180, we proceed as follows:

$$\log N = -2.180 = -3 + 0.820$$
$$= \log 10^{-3} + \log 6.6 = \log 6.6 \times 10^{-3},$$

so $N = 6.6 \times 10^{-3}$. Notice that we added to our given logarithm the next largest integer (3 in this case) so that we have an integer, plus a decimal number between 0 and 1.0 whose antilogarithm can be looked up in the Table.

Appendix B

Selected Isotopes

(1) Atomic Number Z	(2) Element	(3) Symbol	(4) Mass Number A	(5) Atomic Mass†	(6) % Abundance (or Radioactive Decay‡ Mode)	(7) Half-life (if radioactive)
0	(Neutron)	n	1	1.008665	β^-	10.183 min
1	Hydrogen	H	1	1.007825	99.9885%	
	[proton	p	1	1.007276]		
	Deuterium	$^2_1$H	2	2.014102	0.0115%	
	[deuteron	d or D	2	2.013553]		
	Tritium	$^3_1$H	3	3.016049	β^-	12.32 yr
	[triton	t or T	3	3.015500]		
2	Helium	He	3	3.016029	0.000137%	
			4	4.002603	99.999863%	
3	Lithium	Li	6	6.015123	7.59%	
			7	7.016003	92.41%	
4	Beryllium	Be	7	7.016929	EC, γ	53.24 days
			9	9.012183	100%	
5	Boron	B	10	10.012937	19.9%	
			11	11.009305	80.1%	
6	Carbon	C	11	11.011434	β^+, EC	20.334 min
			12	12.000000	98.93%	
			13	13.003355	1.07%	
			14	14.003242	β^-	5730 yr
7	Nitrogen	N	13	13.005739	β^+, EC	9.965 min
			14	14.003074	99.632%	
			15	15.000109	0.368%	
8	Oxygen	O	15	15.003066	β^+, EC	122.24 s
			16	15.994915	99.757%	
			18	17.999160	0.205%	
9	Fluorine	F	19	18.998403	100%	
10	Neon	Ne	20	19.992440	90.48%	
			22	21.991385	9.25%	
11	Sodium	Na	22	21.994437	β^+, EC, γ	2.6027 yr
			23	22.989769	100%	
			24	23.990963	β^-, γ	14.997 h
12	Magnesium	Mg	24	23.985042	78.99%	
13	Aluminum	Al	27	26.981539	100%	
14	Silicon	Si	28	27.976927	92.223%	
			31	30.975363	β^-, γ	157.3 min
15	Phosphorus	P	31	30.973762	100%	
			32	31.973908	β^-	14.262 days

†The masses (atomic mass units) given in column (5) are those for the neutral atom, including the Z electrons (except for the proton, deuteron, triton).
‡Chapter 30; EC = electron capture.

(1) Atomic Number Z	(2) Element	(3) Symbol	(4) Mass Number A	(5) Atomic Mass	(6) % Abundance (or Radioactive Decay Mode)	(7) Half-life (if radioactive)
16	Sulfur	S	32	31.972071	94.99%	
			35	34.969032	β^-	87.37 days
17	Chlorine	Cl	35	34.968853	75.76%	
			37	36.965903	24.24%	
18	Argon	Ar	40	39.962383	99.6035%	
19	Potassium	K	39	38.963706	93.2581%	
			40	39.963998	0.0117% β^-, EC, γ, β^+	1.248×10^9 yr
20	Calcium	Ca	40	39.962591	96.94%	
21	Scandium	Sc	45	44.955908	100%	
22	Titanium	Ti	48	47.947942	73.72%	
23	Vanadium	V	51	50.943957	99.750%	
24	Chromium	Cr	52	51.940506	83.789%	
25	Manganese	Mn	55	54.938044	100%	
26	Iron	Fe	56	55.934936	91.754%	
27	Cobalt	Co	59	58.933194	100%	
			60	59.933816	β^-, γ	5.2713 yr
28	Nickel	Ni	58	57.935342	68.077%	
			60	59.930786	26.223%	
29	Copper	Cu	63	62.929598	69.15%	
			65	64.927790	30.85%	
30	Zinc	Zn	64	63.929142	49.17%	
			66	65.926034	27.73%	
31	Gallium	Ga	69	68.925574	60.108%	
32	Germanium	Ge	72	71.922076	27.45%	
			74	73.921178	36.50%	
33	Arsenic	As	75	74.921595	100%	
34	Selenium	Se	80	79.916522	49.61%	
35	Bromine	Br	79	78.918338	50.69%	
36	Krypton	Kr	84	83.911498	56.987%	
37	Rubidium	Rb	85	84.911790	72.17%	
38	Strontium	Sr	86	85.909261	9.86%	
			88	87.905612	82.58%	
			90	89.907730	β^-	28.90 yr
39	Yttrium	Y	89	88.905840	100%	
40	Zirconium	Zr	90	89.904698	51.45%	
41	Niobium	Nb	93	92.906373	100%	
42	Molybdenum	Mo	98	97.905405	24.39%	
43	Technetium	Tc	98	97.907212	β^-, γ	4.2×10^6 yr
44	Ruthenium	Ru	102	101.904344	31.55%	
45	Rhodium	Rh	103	102.905498	100%	
46	Palladium	Pd	106	105.903480	27.33%	
47	Silver	Ag	107	106.905092	51.839%	
			109	108.904755	48.161%	
48	Cadmium	Cd	114	113.903365	28.73%	
49	Indium	In	115	114.903879	95.71%; β^-	4.41×10^{14} yr
50	Tin	Sn	120	119.902202	32.58%	
51	Antimony	Sb	121	120.903812	57.21%	

(1) Atomic Number Z	(2) Element	(3) Symbol	(4) Mass Number A	(5) Atomic Mass	(6) % Abundance (or Radioactive Decay Mode)	(7) Half-life (if radioactive)
52	Tellurium	Te	130	129.906223	34.08%; $\beta^-\beta^-$	$>3.0 \times 10^{24}$ yr
53	Iodine	I	127	126.904472	100%	
			131	130.906126	β^-, γ	8.0252 days
54	Xenon	Xe	132	131.904155	26.9086%	
			136	135.907214	8.8573%; $\beta^-\beta^-$	$>2.4 \times 10^{21}$ yr
55	Cesium	Cs	133	132.905452	100%	
56	Barium	Ba	137	136.905827	11.232%	
			138	137.905247	71.698%	
57	Lanthanum	La	139	138.906356	99.9119%	
58	Cerium	Ce	140	139.905443	88.450%	
59	Praseodymium	Pr	141	140.907658	100%	
60	Neodymium	Nd	142	141.907729	27.152%	
61	Promethium	Pm	145	144.912756	EC, α	17.7 yr
62	Samarium	Sm	152	151.919740	26.75%	
63	Europium	Eu	153	152.921238	52.19%	
64	Gadolinium	Gd	158	157.924112	24.84%	
65	Terbium	Tb	159	158.925355	100%	
66	Dysprosium	Dy	164	163.929182	28.260%	
67	Holmium	Ho	165	164.930329	100%	
68	Erbium	Er	166	165.930300	33.503%	
69	Thulium	Tm	169	168.934218	100%	
70	Ytterbium	Yb	174	173.938866	31.026%	
71	Lutetium	Lu	175	174.940775	97.401%	
72	Hafnium	Hf	180	179.946557	35.08%	
73	Tantalum	Ta	181	180.947996	99.98799%	
74	Tungsten (wolfram)	W	184	183.950931	30.64%; α	$>8.9 \times 10^{21}$ yr
75	Rhenium	Re	187	186.955750	62.60%; β^-	4.33×10^{10} yr
76	Osmium	Os	191	190.960926	β^-, γ	15.4 days
			192	191.961477	40.78%	
77	Iridium	Ir	191	190.960589	37.3%	
			193	192.962922	62.7%	
78	Platinum	Pt	195	194.964792	33.78%	
79	Gold	Au	197	196.966569	100%	
80	Mercury	Hg	199	198.968281	16.87%	
			202	201.970643	29.86%	
81	Thallium	Tl	205	204.974428	70.48%	
82	Lead	Pb	206	205.974466	24.1%	
			207	206.975897	22.1%	
			208	207.976652	52.4%	
			210	209.984189	β^-, γ, α	22.20 yr
			211	210.988737	β^-, γ	36.1 min
			212	211.991898	β^-, γ	10.64 h
			214	213.999806	β^-, γ	26.8 min
83	Bismuth	Bi	209	208.980399	100%	
			211	210.987270	α, γ, β^-	2.14 min
84	Polonium	Po	210	209.982874	α, γ, EC	138.376 days
			214	213.995202	α, γ	164.3 μs
85	Astatine	At	218	218.008695	α, β^-	1.5 s

(1) Atomic Number Z	(2) Element	(3) Symbol	(4) Mass Number A	(5) Atomic Mass	(6) % Abundance (or Radioactive Decay Mode)	(7) Half-life (if radioactive)
86	Radon	Rn	222	222.017578	α, γ	3.8235 days
87	Francium	Fr	223	223.019736	β^-, γ, α	22.00 min
88	Radium	Ra	226	226.025410	α, γ	1600 yr
89	Actinium	Ac	227	227.027752	β^-, γ, α	21.772 yr
90	Thorium	Th	228	228.028741	α, γ	1.9116 yr
			232	232.038056	100%; α, γ	1.40×10^{10} yr
91	Protactinium	Pa	231	231.035884	α, γ	3.276×10^4 yr
92	Uranium	U	232	232.037156	α, γ	68.9 yr
			233	233.039636	α, γ	1.592×10^5 yr
			235	235.043930	0.7204%; α, γ	7.04×10^8 yr
			236	236.045568	α, γ	2.342×10^7 yr
			238	238.050788	99.2742%; α, γ	4.468×10^9 yr
			239	239.054294	β^-, γ	23.45 min
93	Neptunium	Np	237	237.048174	α, γ	2.144×10^6 yr
			239	239.052939	β^-, γ	2.356 days
94	Plutonium	Pu	239	239.052164	α, γ	24,110 yr
			244	244.064205	α	8.00×10^7 yr
95	Americium	Am	243	243.061381	α, γ	7370 yr
96	Curium	Cm	247	247.070354	α, γ	1.56×10^7 yr
97	Berkelium	Bk	247	247.070307	α, γ	1380 yr
98	Californium	Cf	251	251.079589	α, γ	898 yr
99	Einsteinium	Es	252	252.082980	α, EC, γ	471.7 days
100	Fermium	Fm	257	257.095106	α, γ	100.5 days
101	Mendelevium	Md	258	258.098431	α, γ	51.5 days
102	Nobelium	No	259	259.101030	α, EC	58 min
103	Lawrencium	Lr	262	262.109610	α, EC, fission	$\approx$ 4 h
104	Rutherfordium	Rf	263	263.112500	fission	10 min
105	Dubnium	Db	268	268.125670	fission	32 h
106	Seaborgium	Sg	271	271.133930	α, fission	2.4 min
107	Bohrium	Bh	274	274.143550	α, fission	0.9 min
108	Hassium	Hs	270	270.134290	α	22 s
109	Meitnerium	Mt	278	278.156310	α, fission	8 s
110	Darmstadtium	Ds	281	281.164510	α, fission	20 s
111	Roentgenium	Rg	281	281.166360	α, fission	26 s
112	Copernicium	Cn	285	285.177120	α	30 s
113[†]			286	286.18210	α, fission	20 s
114	Flerovium	Fl	289	289.190420	α	2.7 s
115[†]			289	289.193630	α, fission	0.22 s
116	Livermorium	Lv	293	293.204490	α	53 ms
117[†]			294	294.210460	α	0.08 s
118[†]			294	294.213920	α, fission	0.9 ms

[†] Preliminary evidence (unconfirmed) has been reported for elements 113, 115, 117, and 118.

APPENDIX C

Rotating Frames of Reference; Inertial Forces; Coriolis Effect

Inertial and Noninertial Reference Frames

In Chapters 5 and 8 we examined the motion of objects, including circular and rotational motion, from the outside, as observers fixed on the Earth. Sometimes it is convenient to place ourselves (in theory, if not physically) into a reference frame that is rotating. Let us examine the motion of objects from the point of view, or frame of reference, of persons seated on a rotating platform such as a merry-go-round. It looks to them as if the rest of the world is going around *them*. But let us focus on what they observe when they place a tennis ball on the floor of the rotating platform, which we assume is frictionless. If they put the ball down gently, without giving it any push, they will observe that it accelerates from rest and moves outward as shown in Fig. C–1a. According to Newton's first law, an object initially at rest should stay at rest if no net force acts on it. But, according to the observers on the rotating platform, the ball starts moving even though there is no net force acting on it. To observers on the ground, this is all very clear: the ball has an initial velocity when it is released (because the platform is moving), and it simply continues moving in a straight-line path as shown in Fig. C–1b, in accordance with Newton's first law.

But what shall we do about the frame of reference of the observers on the rotating platform? Newton's first law, the law of inertia, does not hold in this rotating frame of reference since the ball starts moving with no net force on it. For this reason, such a frame is called a **noninertial reference frame**. An **inertial reference frame** (as discussed in Chapter 4) is one in which the law of inertia—Newton's first law—does hold, and so do Newton's second and third laws. In a noninertial reference frame, such as our rotating platform, Newton's second law also does not hold. For instance in the situation described above, there is no net force on the ball; yet, with respect to the rotating platform, the ball accelerates.

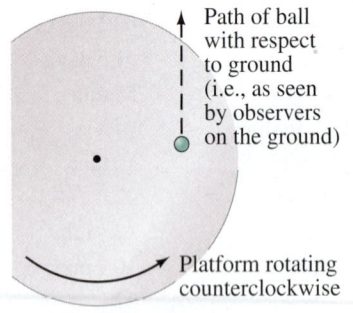

FIGURE C–1 Path of a ball released on a rotating merry-go-round (a) in the reference frame of the merry-go-round, and (b) in a reference frame fixed on the ground.

Fictitious (Inertial) Forces

Because Newton's laws do not hold when observations are made with respect to a rotating frame of reference, calculation of motion can be complicated. However, we can still apply Newton's laws in such a reference frame if we make use of a trick. The ball on the rotating platform of Fig. C–1a flies outward when released (even though no force is actually acting on it). So the trick we use is to write down the equation $\Sigma F = ma$ as if a force equal to mv^2/r (or $m\omega^2 r$) were acting radially outward on the object in addition to any other forces that may be acting. This extra force, which might be designated as "centrifugal force" since it *seems* to act outward, is called a **fictitious force** or **pseudoforce**. It is a pseudoforce ("pseudo" means "false") because there is no object that exerts this force. Furthermore, when viewed from an inertial reference frame, the effect doesn't exist at all. We have made up this pseudoforce so that we can make calculations in a noninertial frame using Newton's second law, $\Sigma F = ma$. Thus the observer in the noninertial frame of Fig. C–1a uses Newton's second law for the ball's outward motion by assuming that a force equal to mv^2/r acts on it. Such pseudoforces are also called **inertial forces** since they arise only because the reference frame is not an inertial one.

In Section 5–3 we discussed the forces on a person in a car going around a curve (Fig. 5–11) from the point of view of an inertial frame. The car, on the other hand, is not an inertial frame. Passengers in such a car could interpret this being pressed outward as the effect of a "centrifugal" force. But they need to recognize that it is a pseudoforce because there is no identifiable object exerting it. It is an effect of being in a noninertial frame of reference.

The Earth itself is rotating on its axis. Thus, strictly speaking, Newton's laws are not valid on the Earth. However, the effect of the Earth's rotation is usually so small that it can be ignored, although it does influence the movement of large air masses and ocean currents. Because of the Earth's rotation, the material of the Earth is concentrated slightly more at the equator. The Earth is thus not a perfect sphere but is slightly fatter at the equator than at the poles.

Coriolis Effect

In a reference frame that rotates at a constant angular speed ω (relative to an inertial frame), there exists another pseudoforce known as the *Coriolis force*. It appears to act on an object in a rotating reference frame only if the object is moving relative to that rotating reference frame, and it acts to deflect the object sideways. It, too, is an effect of the rotating reference frame being noninertial and hence is referred to as an *inertial force*. It also affects the weather.

To see how the Coriolis force arises, consider two people, A and B, at rest on a platform rotating with angular speed ω, as shown in Fig. C–2a. They are situated at distances r_A and r_B from the axis of rotation (at O). The woman at A throws a ball with a horizontal velocity $\vec{v}$ (in her reference frame) radially outward toward the man at B on the outer edge of the platform. In Fig. C–2a, we view the situation from an inertial reference frame. The ball initially has not only the velocity $\vec{v}$ radially outward, but also a tangential velocity $\vec{v}_A$ due to the rotation of the platform. Now Eq. 8–4 tells us that $v_A = r_A \omega$, where r_A is the woman's radial distance from the axis of rotation at O. If the man at B had this same velocity v_A, the ball would reach him perfectly. But his speed is $v_B = r_B \omega$, which is greater than v_A because $r_B > r_A$. Thus, when the ball reaches the outer edge of the platform, it passes a point that the man at B has already gone by because his speed in that direction is greater than the ball's. So the ball passes behind him.

Figure C–2b shows the situation as seen from the rotating platform as frame of reference. Both A and B are at rest, and the ball is thrown with velocity $\vec{v}$ toward B, but the ball deflects to the right as shown and passes behind B as previously described. This is not a centrifugal-force effect, because that would act radially outward. Instead, this effect acts sideways, perpendicular to $\vec{v}$, and is called a **Coriolis acceleration**; it is said to be due to the Coriolis force, which is a fictitious inertial force. Its explanation as seen from an inertial system was given above: it is an effect of being in a rotating system, for which a point farther from the rotation axis has a higher linear speed. On the other hand, when viewed from the rotating system, the motion can be described using Newton's second law, $\Sigma \vec{F} = m\vec{a}$, if we add a "pseudoforce" term corresponding to this Coriolis effect.

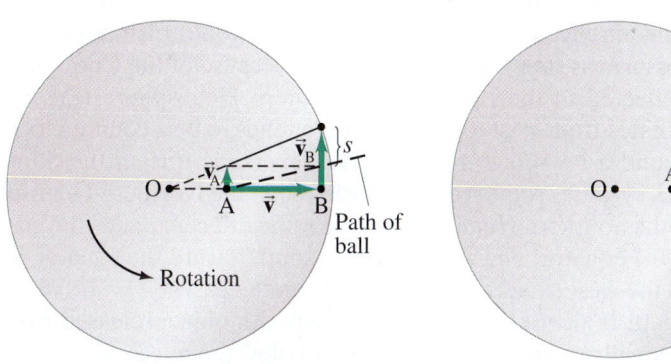

FIGURE C–2 The origin of the Coriolis effect. Looking down on a rotating platform, (a) as seen from a nonrotating inertial reference frame, and (b) as seen from the rotating platform as frame of reference.

(a) Inertial reference frame

(b) Rotating reference frame

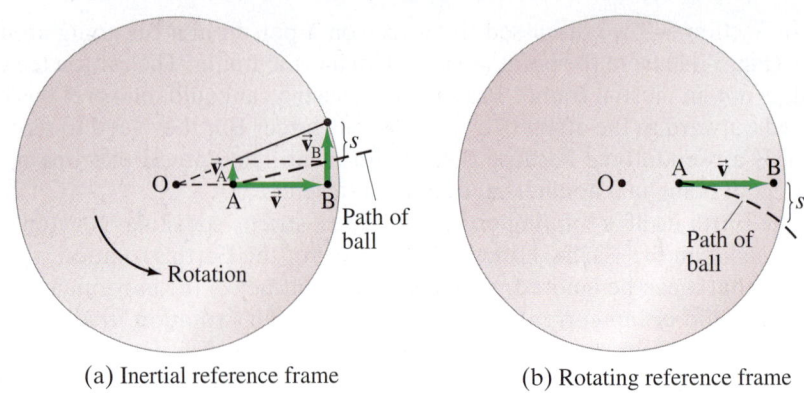

FIGURE C–2 (Repeated.) The origin of the Coriolis effect. Looking down on a rotating platform, (a) as seen from a nonrotating inertial reference frame, and (b) as seen from the rotating platform as frame of reference.

(a) Inertial reference frame   (b) Rotating reference frame

Let us determine the magnitude of the Coriolis acceleration for the simple case described above. (We assume v is large and distances short, so we can ignore gravity.) We do the calculation from the inertial reference frame (Fig. C–2a). The ball moves radially outward a distance $r_B - r_A$ at speed v in a short time t given by

$$r_B - r_A = vt.$$

During this time, the ball moves to the side a distance s_A given by

$$s_A = v_A t.$$

The man at B, in this time t, moves a distance

$$s_B = v_B t.$$

The ball therefore passes behind him a distance s (Fig. C–2a) given by

$$s = s_B - s_A = (v_B - v_A)t.$$

We saw earlier that $v_A = r_A \omega$ and $v_B = r_B \omega$, so

$$s = (r_B - r_A)\omega t.$$

We substitute $r_B - r_A = vt$ (see above) and get

$$s = \omega v t^2. \tag{C–1}$$

This same s equals the sideways displacement as seen from the noninertial rotating system (Fig. C–2b).

Equation C–1 corresponds to motion at constant acceleration, because as we saw in Chapter 2 (Eq. 2–11b), $y = \tfrac{1}{2}at^2$ for a constant acceleration (with zero initial velocity in the y direction). Thus, if we write Eq. C–1 in the form $s = \tfrac{1}{2}a_{\text{Cor}}t^2$, we see that the Coriolis acceleration a_{Cor} is

$$a_{\text{Cor}} = 2\omega v. \tag{C–2}$$

This relation is valid for any velocity in the plane of rotation perpendicular to the axis of rotation (in Fig. C–2, the axis through point O perpendicular to the page).

Because the Earth rotates, the Coriolis effect has some interesting manifestations on the Earth. It affects the movement of air masses and thus has an influence on weather. In the absence of the Coriolis effect, air would rush directly into a region of low pressure, as shown in Fig. C–3a. But because of the Coriolis effect, the winds are deflected to the right in the Northern Hemisphere (Fig. C–3b), since the Earth rotates from west to east. So there tends to be a counterclockwise wind pattern around a low-pressure area. The reverse is true in the Southern Hemisphere. Thus cyclones rotate counterclockwise in the Northern Hemisphere and clockwise in the Southern Hemisphere. The same effect explains the easterly trade winds near the equator: any winds heading south toward the equator will be deflected toward the west (that is, as if coming from the east).

The Coriolis effect also acts on a falling object. An object released from the top of a high tower will not hit the ground directly below the release point, but will be deflected slightly to the east. Viewed from an inertial frame, this happens because the top of the tower revolves with the Earth at a slightly higher speed than the bottom of the tower.

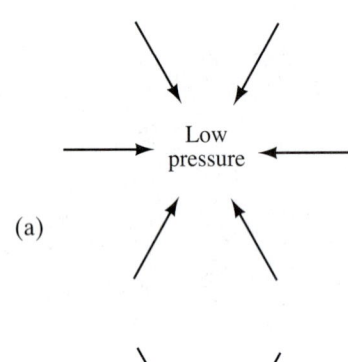

FIGURE C–3 (a) Winds (moving air masses) would flow directly toward a low-pressure area if the Earth did not rotate; (b) and (c): because of the Earth's rotation, the winds are deflected to the right in the Northern Hemisphere (as in Fig. C–2) as if a fictitious (Coriolis) force were acting.

Molar Specific Heats for Gases, and the Equipartition of Energy

Molar Specific Heats for Gases

The values of the specific heats for gases depend on how the thermodynamic process is carried out. Two important processes are those in which either the volume or the pressure is kept constant, and Table D–1 shows how different they can be.

The difference in specific heats for gases is nicely explained in terms of the first law of thermodynamics and kinetic theory. For gases we usually use **molar specific heats**, C_V and C_P, which are defined as the heat required to raise 1 mol of a gas by 1 K (or 1 C°) at constant volume and at constant pressure, respectively. In analogy to Eq. 14–2, the heat Q needed to raise the temperature of n moles of gas by ΔT is

$$Q = nC_V \Delta T \qquad \text{[volume constant]} \quad \textbf{(D–1a)}$$

$$Q = nC_P \Delta T. \qquad \text{[pressure constant]} \quad \textbf{(D–1b)}$$

We can see from the definition of molar specific heat (compare Eqs. 14–2 and D–1) that

$$C_V = Mc_V$$

$$C_P = Mc_P,$$

where M is the molecular mass of the gas ($M = m/n$ in grams/mol).[†] The values for molar specific heats are included in Table D–1. These values are nearly the same for different gases that have the same number of atoms per molecule.

Now we use kinetic theory and imagine that an ideal gas is slowly heated via two different processes—first at constant volume, and then at constant pressure. In both processes, we let the temperature increase by the same amount, ΔT.

[†]For example, $M = 2\,\text{g/mol}$ for He, and $M = 32\,\text{g/mol}$ for O_2.

TABLE D–1 Specific Heats of Gases at 15°C

Gas	Specific Heats (kcal/kg · K)		Molar Specific Heats (cal/mol · K)		$C_P - C_V$ (cal/mol · K)
	c_V	c_P	C_V	C_P	
Monatomic					
He	0.75	1.15	2.98	4.97	1.99
Ne	0.148	0.246	2.98	4.97	1.99
Diatomic					
N_2	0.177	0.248	4.96	6.95	1.99
O_2	0.155	0.218	5.03	7.03	2.00
Triatomic					
CO_2	0.153	0.199	6.80	8.83	2.03
H_2O (100°C)	0.350	0.482	6.20	8.20	2.00
Polyatomic					
C_2H_6	0.343	0.412	10.30	12.35	2.05

In the constant-volume process, no work is done since $\Delta V = 0$. Thus, according to the first law of thermodynamics ($Q = \Delta U + W$, Section 15–1), the heat added (which we denote by Q_V) all goes into increasing the internal energy of the gas:

$$Q_V = \Delta U.$$

In the constant-pressure process, work *is* done. Hence the heat added, Q_P, must not only increase the internal energy but also is used to do work $W = P\,\Delta V$. Thus, to increase the temperature by the same ΔT, more heat must be added in the process at constant pressure than in the process at constant volume. For the process at constant pressure, the first law of thermodynamics gives

$$Q_P = \Delta U + P\,\Delta V.$$

Since ΔU is the same in the two processes (we chose ΔT to be the same), we can combine the two above equations:

$$Q_P - Q_V = P\,\Delta V.$$

From the ideal gas law, $V = nRT/P$, so for a process at constant pressure $\Delta V = nR\,\Delta T/P$. Putting this into the last equation and using Eqs. D–1, we find

$$nC_P\,\Delta T - nC_V\,\Delta T = P\left(\frac{nR\,\Delta T}{P}\right)$$

or, after cancellations,

$$C_P - C_V = R. \tag{D–2}$$

Since the gas constant $R = 8.314\,\text{J/mol}\cdot\text{K} = 1.99\,\text{cal/mol}\cdot\text{K}$, our prediction is that C_P will be larger than C_V by about $1.99\,\text{cal/mol}\cdot\text{K}$. Indeed, this is very close to what is obtained experimentally, as shown in the last column in Table D–1.

Now we calculate the molar specific heat of a monatomic gas using kinetic theory. For a process carried out at constant volume, no work is done, so the first law of thermodynamics tells us that

$$\Delta U = Q_V.$$

For an ideal monatomic gas, the internal energy, U, is the total kinetic energy of all the molecules,

$$U = N(\tfrac{1}{2}m\overline{v^2}) = \tfrac{3}{2}nRT$$

as we saw in Section 14–2. Then, using Eq. D–1a, we write $\Delta U = Q_V$ as

$$\Delta U = \tfrac{3}{2}nR\,\Delta T = nC_V\,\Delta T \tag{D–3}$$

or

$$C_V = \tfrac{3}{2}R. \tag{D–4}$$

Since $R = 8.314\,\text{J/mol}\cdot\text{K} = 1.99\,\text{cal/mol}\cdot\text{K}$, kinetic theory predicts that $C_V = 2.98\,\text{cal/mol}\cdot\text{K}$ for an ideal monatomic gas. This is very close to the experimental values for monatomic gases such as helium and neon (Table D–1). From Eq. D–2, C_P is predicted to be $R + C_V = (1.99 + 2.98)\,\text{cal/mol}\cdot\text{K} = 4.97\,\text{cal/mol}\cdot\text{K}$, also in agreement with experiment (Table D–1).

Equipartition of Energy

The measured molar specific heats for more complex gases (Table D–1), such as diatomic (two atoms) and triatomic (three atoms) gases, increase with the increased number of atoms per molecule. We can explain this by assuming that the internal energy includes not only translational kinetic energy but other forms of energy as well. For example, in a diatomic gas (Fig. D–1), the two atoms can rotate about two different axes (but rotation about a third axis passing through the two atoms would give rise to very little energy since the moment of inertia is so small). The molecules can have rotational as well as translational kinetic energy.

It is useful to introduce the idea of **degrees of freedom**, by which we mean the number of independent ways molecules can possess energy. For example, a monatomic gas has three degrees of freedom, because an atom can have velocity along the x, y, and z axes. These are considered to be three independent motions because a change in any one of the components would not affect the others. A diatomic molecule has the same three degrees of freedom associated with translational kinetic energy plus two more degrees of freedom associated with rotational kinetic energy (Fig. D–1), for a total of five degrees of freedom.

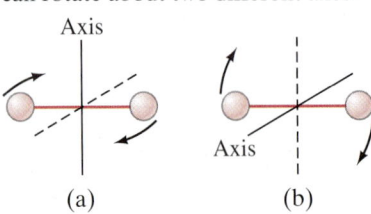

FIGURE D–1 A diatomic molecule can rotate about two different axes.

Table D–1 indicates that the C_V for diatomic gases is about $\frac{5}{3}$ times as great as for a monatomic gas—that is, in the same ratio as their degrees of freedom. This result led nineteenth-century physicists to the **principle of equipartition of energy**. This principle states that energy is shared equally among the active degrees of freedom, and each active degree of freedom of a molecule has on average an energy equal to $\frac{1}{2}kT$. Thus, the average energy for a molecule of a monatomic gas would be $\frac{3}{2}kT$ (which we already knew) and of a diatomic gas $\frac{5}{2}kT$. Hence the internal energy of a diatomic gas would be $U = N(\frac{5}{2}kT) = \frac{5}{2}nRT$, where n is the number of moles. Using the same argument we did for monatomic gases, we see that for diatomic gases the molar specific heat at constant volume would be $\frac{5}{2}R = 4.97 \text{ cal/mol} \cdot \text{K}$, close to measured values (Table D–1). More complex molecules have even more degrees of freedom and thus greater molar specific heats.

However, measurements showed that for diatomic gases at very low temperatures, C_V has a value of only $\frac{3}{2}R$, as if it had only three degrees of freedom. And at very high temperatures, C_V was about $\frac{7}{2}R$, as if there were seven degrees of freedom. The explanation is that at low temperatures, nearly all molecules have only translational kinetic energy; that is, no energy goes into rotational energy and only three degrees of freedom are "active." At very high temperatures, all five degrees of freedom are active plus two additional ones. We interpret the two new degrees of freedom as being associated with the two atoms vibrating, as if they were connected by a spring (Fig. D–2). One degree of freedom comes from the kinetic energy of the vibrational motion, and the second comes from the potential energy of vibrational motion $(\frac{1}{2}kx^2)$. At room temperature, these two degrees of freedom are apparently not active (Fig. D–3). Why fewer degrees of freedom are "active" at lower temperatures was eventually explained by Einstein using quantum theory.

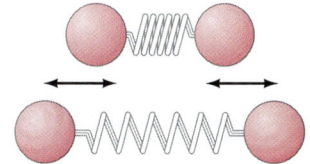

FIGURE D–2 A diatomic molecule can vibrate, as if the two atoms were connected by a spring. They aren't, but rather they exert forces on each other that are electrical in nature—of a form that resembles a spring force.

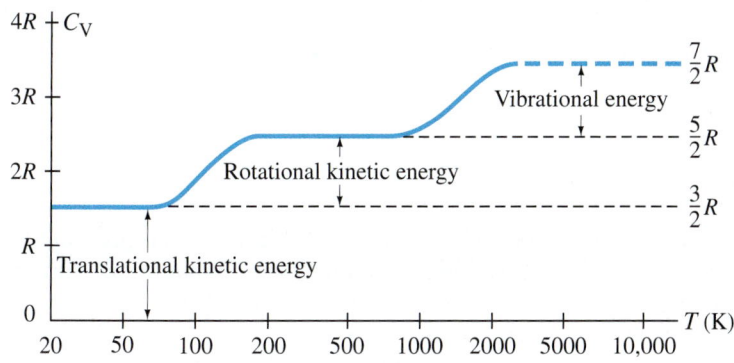

FIGURE D–3 Molar specific heat C_V as a function of temperature for hydrogen molecules (H_2). As the temperature is increased, some of the translational kinetic energy can be transferred in collisions into rotational kinetic energy and, at still higher temperature, into vibrational kinetic and potential energy. [Note: H_2 dissociates into two atoms at about 3200 K, so the last part of the curve is shown dashed.]

Solids

The principle of equipartition of energy can be applied to solids as well. The molar specific heat of any solid at high temperature is close to $3R$ $(6.0 \text{ cal/mol} \cdot \text{K})$, Fig. D–4. This is called the *Dulong and Petit value* after the scientists who first measured it in 1819. (Note that Table 14–1 gave the specific heats per kilogram, not per mole.) At high temperatures, each atom apparently has six degrees of freedom, although some are not active at low temperatures. Each atom in a crystalline solid can vibrate about its equilibrium position as if it were connected by springs to each of its neighbors (Fig. D–5). Thus it can have three degrees of freedom for kinetic energy and three more associated with potential energy of vibration in each of the x, y, and z directions, which is in accord with measured values.

FIGURE D–4 Molar specific heats of solids as a function of temperature.

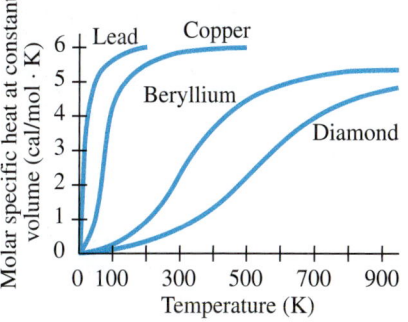

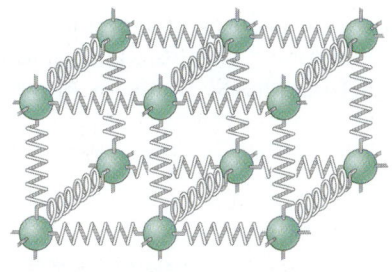

FIGURE D–5 The atoms in a crystalline solid can vibrate about their equilibrium positions as if they were connected to their neighbors by springs. (The forces between atoms are actually electrical in nature.)

APPENDIX E

Galilean and Lorentz Transformations

We examine in detail the mathematics of relating quantities in one inertial reference frame to the equivalent quantities in another. In particular, we will see how positions and velocities *transform* (that is, change) from one frame of reference to the other.

We begin with the classical, or Galilean, viewpoint. Consider two inertial reference frames S and S′ which are each characterized by a set of coordinate axes, Fig. E–1. The axes x and y (z is not shown) refer to S, and x' and y' refer to S′. The x' and x axes overlap one another, and we assume that frame S′ moves to the right (in the x direction) at speed v with respect to S. For simplicity let us assume the origins 0 and 0′ of the two reference frames are superimposed at time $t = 0$.

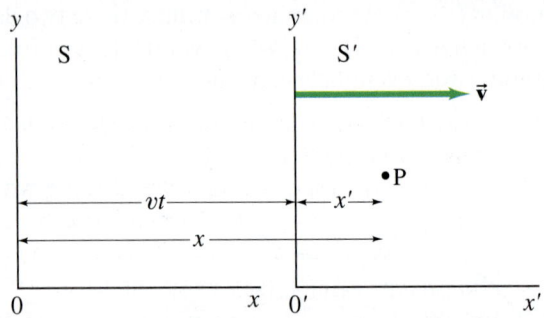

FIGURE E–1 Inertial reference frame S′ moves to the right at speed v with respect to inertial frame S.

Now consider an event that occurs at some point P (Fig. E–1) represented by the coordinates x', y', z' in reference frame S′ at the time t'. What will be the coordinates of P in S? Since S and S′ overlap precisely initially, after a time t', S′ will have moved a distance vt'. Therefore, at time t', $x = x' + vt'$. The y and z coordinates, on the other hand, are not altered by motion along the x axis; thus $y = y'$ and $z = z'$. Finally, since time is assumed to be absolute in Galilean–Newtonian physics, clocks in the two frames will agree with each other; so $t = t'$. We summarize these in the following **Galilean transformation equations**:

$$\begin{aligned} x &= x' + vt' \\ y &= y' \\ z &= z' \\ t &= t'. \end{aligned} \qquad \text{[Galilean]} \quad \text{(E–1)}$$

These equations give the coordinates of an event in the S frame when those in the S′ frame are known. If those in the S frame are known, then the S′ coordinates are obtained from

$$x' = x - vt, \quad y' = y, \quad z' = z, \quad t' = t. \qquad \text{[Galilean]}$$

These four equations are the "inverse" transformation, and are obtained from Eqs. E–1 by exchanging primed and unprimed quantities and replacing v by $-v$. This makes sense because, as seen from the S′ frame, S moves to the left (negative x direction) with speed v.

Now suppose that the point P in Fig. E–1 represents an object that is moving. Let the components of its velocity vector in S' be u'_x, u'_y, and u'_z. (We use u to distinguish it from the relative velocity of the two frames, v.) Now $u'_x = \Delta x'/\Delta t'$, $u'_y = \Delta y'/\Delta t'$, and $u'_z = \Delta z'/\Delta t'$, where all quantities are as measured in the S' frame. For example, if at time t'_1 the particle is at x'_1 and a short time later, t'_2, it is at x'_2, then

$$u'_x = \frac{x'_2 - x'_1}{t'_2 - t'_1} = \frac{\Delta x'}{\Delta t'}.$$

The velocity of P as seen from S will have components u_x, u_y, and u_z. We can show how these are related to the velocity components in S' by using Eqs. E–1. For example,

$$u_x = \frac{\Delta x}{\Delta t} = \frac{x_2 - x_1}{t_2 - t_1} = \frac{(x'_2 + vt'_2) - (x'_1 + vt'_1)}{t'_2 - t'_1}$$

$$= \frac{(x'_2 - x'_1) + v(t'_2 - t'_1)}{t'_2 - t'_1}$$

$$= \frac{\Delta x'}{\Delta t'} + v = u'_x + v.$$

For the other components, $u'_y = u_y$ and $u'_z = u_z$, so we have

$$u_x = u'_x + v$$
$$u_y = u'_y \qquad\qquad \text{[Galilean]} \quad \text{(E–2)}$$
$$u_z = u'_z.$$

These are known as the **Galilean velocity transformation equations**. We see that the y and z components of velocity are unchanged, but the x components differ by v. This is just what we have used before when dealing with relative velocity (Section 3–8). For example, if S' is a train and S the Earth, and the train moves with speed v with respect to Earth, a person walking toward the front of the train with speed u'_x will have a speed with respect to the Earth of $u_x = u'_x + v$.

The Galilean transformations, Eqs. E–1 and E–2, are accurate only when the velocities involved are not relativistic (Chapter 26)—that is, much less than the speed of light, c. We can see, for example, that the first of Eqs. E–2 will not work for the speed of light, c, which is the same in all inertial reference frames (a basic postulate in the theory of relativity). That is, light traveling in S' with speed $u'_x = c$ will have speed $c + v$ in S, according to Eq. E–2, whereas the theory of relativity insists it must be c in S. Clearly, then, a new set of transformation equations is needed to deal with relativistic velocities.

We will derive the required equations, again looking at Fig. E–1. We assume the transformation is linear and for x is of the form

$$x = \gamma(x' + vt'). \qquad\qquad \text{(i)}$$

That is, we modify the first of Eqs. E–1 by multiplying by a factor γ which is yet to be determined.[†] We assume the y and z equations are unchanged

$$y = y', \quad z = z'$$

because there is no length contraction in these directions. We will not assume a form for t, but will derive it. The inverse equations must have the same form with v replaced by $-v$. (The principle of relativity demands it, since S' moving to the right with respect to S is equivalent to S moving to the left with respect to S'.) Therefore

$$x' = \gamma(x - vt). \qquad\qquad \text{(ii)}$$

Suppose a light pulse leaves the common origin of S and S' at time $t = t' = 0$.

[†] We are NOT assuming γ is $1/\sqrt{1 - v^2/c^2}$, as in Chapter 26. Our minds are open. Let's see.

Then after a time t it will have traveled along the x axis a distance $x = ct$ (in S), or $x' = ct'$ (in S'). Therefore, from Eqs. (i) and (ii) above,

$$ct = x = \gamma(ct' + vt') = \gamma(c + v)t', \qquad \text{(iii)}$$
$$ct' = x' = \gamma(ct - vt) = \gamma(c - v)t. \qquad \text{(iv)}$$

From Eq. (iv), $t' = \gamma(c - v)(t/c)$, and we substitute this into Eq. (iii) and find $ct = \gamma(c + v)\gamma(c - v)(t/c) = \gamma^2(c^2 - v^2)t/c$. We cancel out the t on each side and solve for γ to find

$$\gamma = \frac{1}{\sqrt{1 - v^2/c^2}}.$$

We have found that γ is, in fact, the value for γ we used in Chapter 26, Eq. 26–2.

Now that we have found γ, we need only find the relation between t and t'. To do so, we combine $x' = \gamma(x - vt)$ with $x = \gamma(x' + vt')$:

$$x' = \gamma(x - vt) = \gamma[\gamma(x' + vt') - vt].$$

We solve for t, doing some algebra, and find $t = \gamma(t' + vx'/c^2)$. In summary,

$$x = \frac{1}{\sqrt{1 - v^2/c^2}}(x' + vt')$$
$$y = y'$$
$$z = z' \qquad \text{(E-3)}$$
$$t = \frac{1}{\sqrt{1 - v^2/c^2}}\left(t' + \frac{vx'}{c^2}\right).$$

LORENTZ TRANSFORMATIONS

These are called the **Lorentz transformation equations**. They were first proposed, in a slightly different form, by Lorentz in 1904 to explain the null result of the Michelson–Morley experiment and to make Maxwell's equations take the same form in all inertial reference frames. A year later, Einstein derived them independently based on his theory of relativity. Notice that not only is the x equation modified as compared to the Galilean transformation, but so is the t equation. Indeed, we see directly in this last equation how the space and time coordinates mix.

The relativistically correct velocity equations are obtained using Eqs. E–3 (we let $\gamma = 1/\sqrt{1 - v^2/c^2}$) and $u_x = \Delta x/\Delta t$, $u'_x = \Delta x'/\Delta t'$:

$$u_x = \frac{\Delta x}{\Delta t} = \frac{\gamma(\Delta x' + v\,\Delta t')}{\gamma(\Delta t' + v\,\Delta x'/c^2)} = \frac{(\Delta x'/\Delta t') + v}{1 + (v/c^2)(\Delta x'/\Delta t')}$$
$$= \frac{u'_x + v}{1 + vu'_x/c^2}.$$

The others are obtained in the same way, and we collect them here:

$$u_x = \frac{u'_x + v}{1 + vu'_x/c^2}$$
$$u_y = \frac{u'_y\sqrt{1 - v^2/c^2}}{1 + vu'_x/c^2} \qquad \text{(E-4)}$$
$$u_z = \frac{u'_z\sqrt{1 - v^2/c^2}}{1 + vu'_x/c^2}.$$

RELATIVISTIC VELOCITY TRANSFORMATIONS

The first of these equations is Eq. 26–11, which we used in Section 26–10 where we discussed how velocities do not add in our commonsense (Galilean) way, because of the denominator $(1 + vu'_x/c^2)$. We can now also see that the y and z components of velocity are also altered and that they depend on the x' component of velocity.

EXAMPLE E–1 Length contraction. Derive the length contraction formula, Eq. 26–3, from the Lorentz transformation equations.

APPROACH We consider measurements in two reference frames, S and S′, that move with speed v relative to each other, as in Fig. E–1.

SOLUTION Let an object of length ℓ_0 be at rest on the x axis in S. The coordinates of its two end points are x_1 and x_2, so that $x_2 - x_1 = \ell_0$. At any instant in S′, the end points will be at x_1' and x_2' as given by the Lorentz transformation equations. The length measured in S′ is $\ell = x_2' - x_1'$. An observer in S′ measures this length by measuring x_2' and x_1' at the same time (in the S′ frame), so $t_2' = t_1'$. Then, from the first of Eqs. E–3,

$$\ell_0 = x_2 - x_1 = \frac{1}{\sqrt{1 - v^2/c^2}} (x_2' + vt_2' - x_1' - vt_1').$$

Since $t_2' = t_1'$, we have

$$\ell_0 = \frac{1}{\sqrt{1 - v^2/c^2}} (x_2' - x_1') = \frac{\ell}{\sqrt{1 - v^2/c^2}},$$

or

$$\ell = \ell_0 \sqrt{1 - v^2/c^2},$$

which is Eq. 26–3a: the length contraction formula.

EXAMPLE E–2 Time dilation. Derive the time dilation formula, Eq. 26–1, from the Lorentz transformation equations.

APPROACH Again we compare measurements in two reference frames, S and S′, that move with speed v relative to each other, Fig. E–1.

SOLUTION The time Δt_0 between two events that occur at the same place $(x_2' = x_1')$ in S′ is measured to be $\Delta t_0 = t_2' - t_1'$. Since $x_2' = x_1'$, then from the last of Eqs. E–3, the time Δt between the events as measured in S is

$$\Delta t = t_2 - t_1 = \frac{1}{\sqrt{1 - v^2/c^2}} \left(t_2' + \frac{vx_2'}{c^2} - t_1' - \frac{vx_1'}{c^2} \right)$$

$$= \frac{1}{\sqrt{1 - v^2/c^2}} (t_2' - t_1')$$

$$= \frac{\Delta t_0}{\sqrt{1 - v^2/c^2}},$$

which is the time dilation formula, Eq. 26–1. Notice that we chose S′ to be the frame in which the two events occur at the same place, so that $x_1' = x_2'$, and then the terms containing x_1' and x_2' cancel out.

Answers to Odd-Numbered Problems

Chapter 1
1. (*a*) 3;
   (*b*) 4;
   (*c*) 3;
   (*d*) 1;
   (*e*) 2;
   (*f*) 4;
   (*g*) 2.
3. (*a*) 86,900;
   (*b*) 9100;
   (*c*) 0.88;
   (*d*) 476;
   (*e*) 0.0000362.
5. 4.6%.
7. 1.00×10^5 s.
9. 1%.
11. $(3.0 \pm 0.2) \times 10^9$ cm².
13. (*a*) 1 megavolt;
    (*b*) 2 micrometers;
    (*c*) 6 kilodays;
    (*d*) 18 hectobucks;
    (*e*) 700 nanoseconds.
15. (*a*) 1.5×10^{11} m;
    (*b*) 1.5×10^8 km.
17. (*a*) 3.9×10^{-9} in.;
    (*b*) 1.0×10^8 atoms.
19. (*a*) 9.46×10^{15} m;
    (*b*) 6.31×10^4 AU.
21. Soccer; 9.4 yd, 8.6 m, 9.4%.
23. (*a*) 10^{12} protons or neutrons;
    (*b*) 10^{40} protons or neutrons;
    (*c*) 10^{29} protons or neutrons;
    (*d*) 10^{68} protons or neutrons.
25. (*a*) 10^3;
    (*b*) 10^5;
    (*c*) 10^{-2};
    (*d*) 10^9.
27. 500 hr.
29. 2.5 hr.
31. (*a*) 700;
    (*b*) answers vary.
33. Second method.
35. 8.8 s.
37. (*a*) L/T^4, L/T^2;
    (*b*) m/s⁴, m/s².
39. 10^{-35} m.
41. 3.3×10^5 chips/cylinder.
43. 46,000 years.
45. 400 jelly beans.
47. 75 minutes.
49. 5×10^5 metric tons, 1×10^8 gal.
51. 3000 m.
53. (*a*) 0.10 nm;
    (*b*) 1.0×10^5 fm;
    (*c*) 1.0×10^{10} angstroms;
    (*d*) 9.5×10^{25} angstroms.
55. (*a*) 3%, 3%;
    (*b*) 0.7%, 0.2%.
57. 8×10^{-2} m³.
59. 1.18×10^9 atoms/m².
61. 4×10^{51} kg.

Chapter 2
1. 53 m.
3. 0.57 cm/s, no, we need the distance traveled.
5. 0.14 h.
7. (*a*) 350 km;
   (*b*) 78 km/h.
9. (*a*) 3.68 m/s;
   (*b*) 0.
11. 38 s.
13. 1.6 min.
15. 6.00 m/s.
17. 6.1 m/s².
19. 6.0 m/s², 0.61*g*'s.
21. (*a*) 21.2 m/s;
    (*b*) 2.00 m/s².
23. 1 m/s², 110 m.
25. 260 m/s².
27. 112 m.
29. 44*g*'s.
31. (*a*) 130 m;
    (*b*) 69 m.
33. 21 m/s.
35. 6.3 s.
37. 0.70 m/s².
39. 61.8 m.
41. 17 m/s, 14 m.
43. 1.09 s.

45. (*a*)

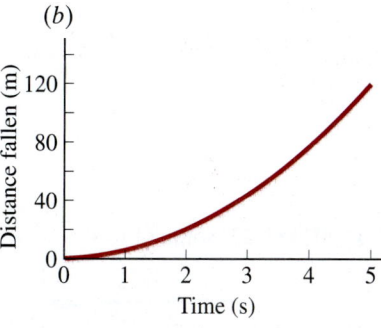

(*b*)

49. 5.21 s.
51. 12 m/s.
53. 1.6 m.
55. (*a*) 48 s;
    (*b*) 90 s to 108 s;
    (*c*) 0 s to 42 s, 65 s to 83 s, 90 s to 108 s;
    (*d*) 65 s to 83 s.
57. (*a*) 0.3 m/s;
    (*b*) 1.2 m/s;
    (*c*) 0.30 m/s;
    (*d*) 1.4 m/s;
    (*e*) −0.95 m/s.
59.

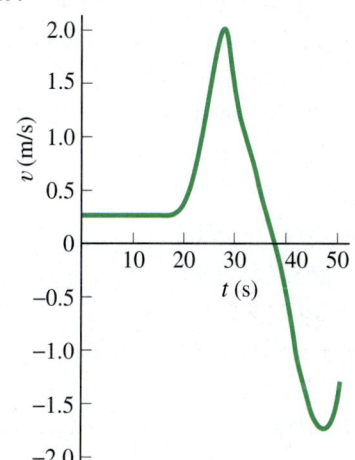

61. 1.2 m.

A-27

63. 3.1 m.
65. (a) 14.4 s, no;
(b) no, 4.6 s.
67. (b) 4.8 m;
(c) 36 m.
69. -20 m/s^2.
71. (a) 5.80 s;
(b) 41.4 m/s;
(c) 99.5 m.
73. She should try to stop the car.
75. 1.5 poles.
77. 30%.
79. 245.0 km/h.
81. 23.7 s, 840 km/h.
83. (a) 4.3×10^6 bits;
(b) 67%.

Chapter 3

1. 302 km, 13° south of west.

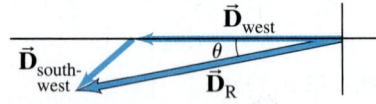

3. 11.70 units, $-33.1°$.
5. (a)

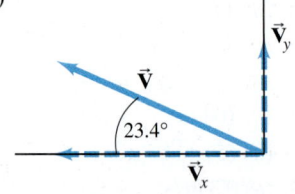

(b) -22.8 units, 9.85 units;
(c) 24.8 units, 23.4° above the $-x$ axis.
7. (a) 1.3 units, positive x direction;
(b) 12.3 units, positive x direction;
(c) 12.3 units, negative x direction.
9. (a) x component 24.0, y component 11.6;
(b) 26.7 units, 25.8°.
11. 64.6, 53.1°.
13. (a) 62.6, $-31.0°$;
(b) 77.5, 71.9°;
(c) 77.5, 251.9°.
15. $(-2845$ m, 3589 m, 2450 m), 5190 m.
17. 3.7 m.
19. 6 times farther.
21. 14.5 m.

23. 18°, 72°.

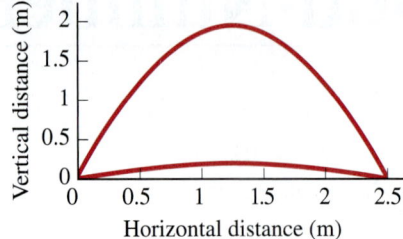

25. 2.0 m/s.
27. (a) 30.8 m;
(b) 5.02 s;
(c) 136 m;
(d) 28.9 m/s.
29. 22.3 m.
31. 481 m.
33. (a) 4.0 m.s, 55° above the horizontal;
(b) 4.6 m;
(c) 9.7 m/s, 76° below the horizontal.
35. No, 0.81 m too low; 4.6 m to 34.7 m.
37. (a) 4.0×10^1 m/s;
(b) 24 m/s.
39. 1.66 m/s, 25°.
41. 23.1 s.
43. (a) 10.4 m/s, 17° above the horizontal;
(b) 10.4 m/s, 17° below the horizontal.
45. 5.31° west of south.
47. (a) 56°;
(b) 140 s.
49. 23 s, 23 m.
51. 65 km/h, 58° west of north; 65 km/h, 32° south of east.
53. Horizontal: 3.9 m/s^2 leftward; vertical: 1.9 m/s^2 downward.
55. 0.88 s, 0.95 m.
57. 1.7 m/s^2.
59. (a) 9.96 s;
(b) 531 m;
(c) 53.2 m/s, -60.4 m/s;
(d) 80.5 m/s;
(e) 48.6° below the horizon;
(f) 70.9 m.
61. $v_T/\tan\theta$.
63. (a) 13.3 m;
(b) 22.1°.
65. 33 m/s.
67. 54°.

69. (a) 2.51 m/s, 61.4°;
(b) 3.60 m downstream, 6.60 m across the river.
71. (a) 13 m;
(b) 31° below the horizontal.
73. (a) 68 m;
(b) 7.3 m/s.
75. 0.51%.

Chapter 4

1. 77 N.
3. 1450 N.
5. -1.3×10^6 N, 39%, 1.3×10^6 N.
7. -3100 N.
9. 780 N, backward.
11. (a) 196 N, 196 N;
(b) 98.0 N, 294 N.
13. Descend with downward $a \geq 2.2$ m/s^2.
15. -2800 m/s^2, $280g's$, 1.9×10^5 N.
17. (a) -7.35 m/s^2;
(b) 1290 N.
19. (a) 7.4 m/s, downward;
(b) 2100 N, upward.
21. (a)   (b)

23. 1410 N.
25. (a) 31 N (lower cord), 63 N (upper cord);
(b) 35 N (lower cord), 71 N (upper cord).
27. $F_{T1}/F_{T2} = 2$.
29. (a) 1.0×10^1 m/s^2;
(b) 3.3 m/s.
31. (a) 23°;
(b) toward the windshield.
33. (a) 2.7 m/s^2;
(b) 0.96 s;
(c) 99 kg.
35. 34 N.

37. (a) 0.60;
    (b) 0.53.
39. 42°.
41. 0.46.
43. 1200 N.
45. 1.4.
47. (a) 5.0 kg;
    (b) 1.0×10^1 kg.
49. (a) 1.7 m/s^2;
    (b) 430 N;
    (c) 1.7 m/s^2, 220 N.
51. 1.20×10^2 N, in the direction opposite to the child's velocity.
53. 4.8 s.
55. 4.0×10^2 m.
57. (a) 3.67 m/s^2;
    (b) 9.39 m/s.
59. (a) 2.5 m/s^2;
    (b) 6.3 m/s.
61. -5.3 m/s^2.
63. 4.0×10^1 N.
65. -2.2 m/s^2.
67. (a) 1.6 m/s^2;
    (b) 0.53.
69. 0.86 m/s^2.
71. 9.9°.
73. 73 m/s.
75. (a) 0.67;
    (b) 6.1 m/s;
    (c) 15 m/s.
77. 4.2°.
79. (a) 8.76×10^4 N, upward;
    (b) 1.14×10^4 N;
    (c) 1.14×10^4 N, downward.
81. (a) 45 N (10 lb);
    (b) 37 N (8.4 lb);
    (c) no, the fish cannot be lifted vertically by a 10-lb force.
83. 380 N (between second and last climbers), 760 N (between first and second climbers).
85. (a) 3.0 times her weight;
    (b) 7.7 times his weight; Jim.
87. 23 m/s (85 km/h).
89. 4.90×10^2 N.
91. (a) 0.9 m/s^2;
    (b) 0.98 m/s^2.
93. (b) Yes.
95. (a) 16 m/s;
    (b) 13 m/s.
97. 1100 N, opposite to the velocity.

Chapter 5

1. (a) 1.01 m/s^2;
   (b) 22.7 N.
3. 12 m/s.
5. 13 m/s.
7. 34 m/s.
9. 24 m/s, yes.
11. 8.5 rpm.
13. 1700 rev/day.
15. 0.210.
17. $F_A = 4\pi^2 f^2 (m_A r_A + m_B r_B)$,
    $F_B = 4\pi^2 f^2 m_B r_B$.
19. (a) 5970 N;
    (b) 379 N;
    (c) 29.4 m/s.
21. 930 m.
23. 59 km/h to 110 km/h.
25. $a_{\text{tan}} = 4.1 \text{ m/s}^2$, $a_{\text{rad}} = 13 \text{ m/s}^2$; 1.4.
27. (a) 0.930 m/s;
    (b) 2.83 m/s.
29. (a) 24.0 kg on both;
    (b) $w_{\text{Earth}} = 235$ N,
        $w_{\text{Planet}} = 288$ N.
31. 3.94 kg, 0.06 kg.
33. 1.62 m/s^2.
35. 6.5×10^{23} kg.
37. 27.4 m/s^2.
39. (a) 9.78 m/s^2;
    (b) 2.44 m/s^2.
41. 9.6×10^{17} N; 2.7×10^{-5}.
43. 2.02×10^7 m.
45. 7460 m/s.
47. 2.4 m/s^2 upward.
49. 7.05×10^3 s.
51. (a) 568 N;
    (b) 568 N;
    (c) 699 N;
    (d) 440 N;
    (e) 0.
53. (a) 59 N, away from the Moon;
    (b) 76 N, toward the Moon.
55. 9.6 s.
57. 160 yr.
59. 84.5 min.
61. 2×10^8 yr.
63. Europa: 671×10^3 km,
    Ganymede: 1070×10^3 km,
    Callisto: 1880×10^3 km.
65. 5.4×10^{12} m; yes; Pluto.
67. $5.97 \times 10^{-3} \text{ m/s}^2$, 3.56×10^{22} N; the Sun.
69. 28.3 m/s, 0.410 rev/s.
71. 0.18; no; the wall pushes against the riders, so by Newton's third law, they push against the wall.
73. 9.2 m/s.
75. (a) In circular motion, they accelerate toward each other without moving toward each other.
    (b) 9.6×10^{29} kg.
77. Yes; $\sqrt{rg}$, where r is the radius of the vertical circle.
79. $T_{\text{inner}} = 2.0 \times 10^4$ s,
    $T_{\text{outer}} = 7.1 \times 10^4$ s.
81. (a) 3900 m/s;
    (b) 4.4×10^4 s.
83. 25.0 m/s.
85. (a) 7.6×10^6 m;
    (b) 3.8×10^4 N;
    (c) 1.2×10^6 m.
87. 1.21×10^6 m.
89. $0.44r$.

Chapter 6

1. 2.06×10^4 J.
3. 2300 J.
5. 1.0×10^6 J.
7. 1960 J.
9. 390 J.
11. 2 m.
13. (a) 2800 J;
    (b) 2100 J.
15. 484 m/s.
17. -5.51×10^{-19} J.
19. The lighter one, $\sqrt{2}$; both the same.
21. 43 m/s.
23. 21 m/s.
25. (a) 3010 N;
    (b) 7480 J;
    (c) 5.42×10^4 J;
    (d) -4.67×10^4 J;
    (e) 7.51 m/s.
27. 1.01 m.
29. (a) 9.06×10^5 J;
    (b) 9.06×10^5 J;
    (c) yes.
31. 45.4 m/s.
33. 4.89 m/s.
35. 74 cm.
37. 1.4×10^5 N/m.
39. (a) 7.47 m/s;
    (b) 3.01 m.

41. 52 m.
43. 12 Mg/h.
45. (a) 9.19×10^4 J;
    (b) 433 N.
47. 332 J.
49. (a) 15.3 m/s;
    (b) 1.03 N, upward.
51. 0.091.
53. 1.4×10^5 J.
55. (a) 2.8 m;
    (b) 1.5 m;
    (c) 1.5 m.
57. 22.0 s.
59. (a) 1100 J;
    (b) 1100 W.
61. 2700 N.
63. 2.9×10^4 W, 38 hp.
65. 5.3×10^4 W.
67. 15.4 W.
69. 33 hp.
71. 610 W.
73. 14.9 m/s.
75. (a) $\sqrt{Fx/m}$;
    (b) $\sqrt{3Fx/4m}$.
77. (a) 0.014 J;
    (b) 0.039 J.
79. (a) -9.0×10^4 J;
    (b) 8.2×10^4 N;
    (c) -2.3×10^5 J.
81. 340 W.
83. (a) 1.0×10^4 J;
    (b) 16 m/s.
85. (a) 42 m/s;
    (b) 3.2×10^5 W.
87. (b) 420 kWh;
    (c) 1.5×10^9 J;
    (d) $50, no.
89. (a) 8.9×10^5 J;
    (b) 54 W, 0.072 hp;
    (c) 360 W, 0.48 hp.
91. (a) 0.39 m;
    (b) $\mu_s < 0.53$;
    (c) 1.4 m/s.
93. 1.7×10^5 m^3.

Chapter 7

1. 0.24 kg·m/s.
3. 10.2 m/s.
5. 5.9×10^7 N, opposite the gas velocity.
7. -0.898 m/s.
9. 2500 m/s.
11. 0.99 m/s.
13. 4.9×10^6 N.
15. 2230 N, toward the pitcher.
17. (a) 9.0×10^1 kg·m/s;
    (b) 1.1×10^4 N.
19. (a) -0.16 m/s;
    (b) 521 N;
    (c) astronaut: 391 J; capsule: 26 J.
21. (a) 290 kg·m/s eastward;
    (b) 290 kg·m/s westward;
    (c) 290 kg·m/s eastward;
    (d) 340 N eastward.
23. (a) 5 N·s;
    (b) 80 m/s.
25. 0.440-kg ball: 1.27 m/s, east; 0.220-kg ball: 5.07 m/s, east.
27. Tennis ball: 2.50 m/s; other ball: 5.00 m/s; both in direction of tennis ball's initial motion.
29. (a) 0.840 kg;
    (b) 0.75.
31. (a) 1.7 m/s, in direction of initial incoming velocity;
    (b) 1.2 kg.
33. $\sqrt{2}$.
35. Vertical: 0.15 m; horizontal: 0.90 m.
37. 21 m/s.
39. 0.42.
41. (a) 12.1 m/s;
    (b) 56.4 J before, 13.7 J after.
43. (a) 920 m/s;
    (b) 0.999.
45. 1.14×10^{-22} kg·m/s, 147° from the electron's momentum, 123° from the neutrino's momentum.
47. (a) 30°;
    (b) $v_{\text{nucleus}} = v_{\text{target}} = v/\sqrt{3}$;
    (c) 2/3.
49. 6.5×10^{-11} m.
51. 2.62 m.
53. $(1.2\ell, 0.9\ell)$ relative to back left corner.
55. $0.27R$ to the left of C.
57. 19% of the person's height along the line from shoulder to hand.
59. 4.3% of their height; no.
61. (a) 4.66×10^6 m from center of Earth.
63. (a) 5.8 m;
    (b) 4.0 m;
    (c) 4.2 m.
65. 0.45 m toward initial position of 85-kg person.
67. 2.0×10^1 m.
69. 8.
71. (a) $v'_A = 3.65$ m/s, $v'_B = 4.45$ m/s;
    (b) $\Delta p_A = -370$ kg·m/s, $\Delta p_B = 370$ kg·m/s.
73. 110 km/h $\approx$ 70 mi/h.
75. 340 m/s.
77. (a) 8.6 m;
    (b) 38 m.
79. (a) $v'_m = 3.98$ m/s, $v'_M = 4.42$ m/s;
    (b) 1.62 m.
81. (a) 1.5×10^{21} J;
    (b) 38,000.
83. (a) No;
    (b) m_B/m_A;
    (c) m_B/m_A;
    (d) stays at rest.
85. $v_m = 3D\sqrt{\dfrac{k}{12m}}$, $v_{3m} = D\sqrt{\dfrac{k}{12m}}$.

Chapter 8

1. (a) 0.785 rad, $\pi/4$ rad;
    (b) 1.05 rad, $\pi/3$ rad;
    (c) 1.57 rad, $\pi/2$ rad;
    (d) 6.28 rad, 2π rad;
    (e) 7.77 rad, $89\pi/36$ rad.
3. 5.3×10^3 m.
5. (a) 750 rad/s;
    (b) 23 m/s;
    (c) 4.5×10^7 bit/s.
7. (a) 230 rad/s;
    (b) 4.0×10^1 m/s, 9.3×10^3 m/s^2.
9. (a) 0.105 rad/s;
    (b) 1.75×10^{-3} rad/s;
    (c) 1.45×10^{-4} rad/s;
    (d) 0.
11. (a) 464 m/s;
    (b) 185 m/s;
    (c) 345 m/s.
13. 3.3×10^4 rpm.
15. (a) 1.5×10^{-4} rad/s^2;
    (b) $a_{\text{rad}} = 1.2 \times 10^{-2}$ m/s^2, $a_{\text{tan}} = 6.2 \times 10^{-4}$ m/s^2.
17. (a) -96 rad/s^2;
    (b) 98 rev.
19. (a) 46 rev/min^2;
    (b) 46 rpm.
21. 33 m.
23. (a) 0.53 rad/s^2;
    (b) 13 s.
25. 1.2 m·N, clockwise.
27. $mg(\ell_2 - \ell_1)$, clockwise.

29. (a) 14 m·N;
    (b) −13 m·N.
31. 0.12 kg·m².
33. 1.2 × 10⁻¹⁰ m.
35. (a) 7.8 m·N;
    (b) 310 N.
37. 22 m·N.
39. (a) 7.0 kg·m²;
    (b) 0.70 kg·m²;
    (c) y axis.
41. 320 m·N; 130 N.
43. (a) 1.90 × 10³ kg·m²;
    (b) 8.9 × 10³ m·N.
45. 31 N.
47. (a) a_A = 0.69 m/s², upward;
        a_B = 0.69 m/s², downward;
    (b) 2%.
49. 125 hp.
51. 9.70 m/s.
53. (a) 2.6 × 10²⁹ J;
    (b) 2.7 × 10³³ J.
55. 1.63 × 10⁴ J.
57. $\sqrt{\tfrac{10}{7}g(R_0 - r_0)}$.
59. 7.27 m/s.
61. (a) 15 kg·m²/s;
    (b) −2.5 m·N.
63. $\tfrac{1}{2}\omega$.
65. 1.2 kg·m²; by pulling her arms in toward the center of her body.
67. (a) 0.52 rad/s;
    (b) KE_{before} = 370 J,
        KE_{after} = 2.0 × 10² J.
69. (a) 0.43 rad/s;
    (b) 0.80 rad/s.
71. (a) 5 × 10⁻² rad/s;
    (b) KE_f = 2 × 10⁴ KE_i.
73. (3.2 × 10⁻¹⁶)%.
75. 52 kg.
77. f_{R_1} = 480 rpm, f_{R_2} = 210 rpm.
79. 4.50 m/s.
81. (a) $\omega_R/\omega_F = N_F/N_R$;
    (b) 4.0;
    (c) 1.5.
83. (a) 3.5 m;
    (b) 4.7 s.
85. $\dfrac{Mg\sqrt{2Rh - h^2}}{R - h}$.
87. (a) 4.84 J;
    (b) $F_{4\,kg}$ = 26.3 N, $F_{3\,kg}$ = 19.8 N.
89. (a) $3g/2\ell$;
    (b) $\tfrac{3}{2}g$.
91. $\ell/2$; $\ell/2$.
93. 27 h.
95. (a) 820 kg·m²/s²;
    (b) 820 m·N;
    (c) 930 W.

Chapter 9

1. 528 N, 120° clockwise from $\vec{F}_A$.
3. (a) 2.3 m from vertical support;
   (b) 4200 kg.
5. (a) F_A = 1.5 × 10³ N, down;
       F_B = 2.0 × 10³ N, up;
   (b) F_A = 1.8 × 10³ N, down;
       F_B = 2.6 × 10³ N, up.
7. 1200 N.
9. F_{closer} = 2900 N, down;
   $F_{farther}$ = 1300 N, down.
11. (a) 2.3 m from the adult;
    (b) 2.5 m from the adult.
13. F_{left} = 260 N, F_{right} = 190 N.
15. 20 N to 50 N.
17. 0.64 m to right of fulcrum rock.
19. (a) 410 N;
    (b) horizontal: 410 N;
        vertical: 328 N.
21. F_A = 1.7 × 10⁴ N,
    F_B = 7.7 × 10³ N.
23. 6.0 × 10¹ N; the angle is small.
25. (a) 230 N;
    (b) 1.0 × 10² N.
27. (a)

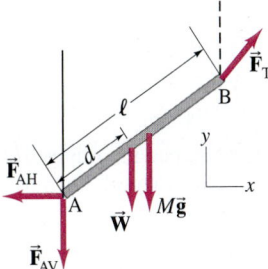

    (b) $F_{A_{vertical}}$ = 9 N,
        $F_{A_{horizontal}}$ = 51 N;
    (c) 2.4 m.
29. $F_{top_{horizontal}}$ = 55.2 N, right;
    $F_{top_{vertical}}$ = 63.7 N, up;
    $F_{bottom_{horizontal}}$ = 55.2 N, left;
    $F_{bottom_{vertical}}$ = 63.7 N, up;
31. 7.0 kg.
33. $2.4w$.
35. 1600 N.
37. 1800 N.
39. (b) Yes, by $\tfrac{1}{24}$ of a brick length;
    (c) $D = \sum_{i=1}^{n} \dfrac{\ell}{2i}$;
    (d) 35 bricks;
41. (a) 1.8 × 10⁵ N/m²;
    (b) 3.5 × 10⁻⁶.
43. (a) 1.4 × 10⁶ N/m²;
    (b) 6.9 × 10⁻⁶;
    (c) 6.6 × 10⁻⁵ m.
45. 9.0 × 10⁷ N/m², 9.0 × 10² atm.
47. 25 kg.
49. 1.7 × 10⁻² J.
51. (a) 393 N;
    (b) thicker.
53. (a) 3.7 × 10⁻⁵ m²;
    (b) 2.7 × 10⁻³ m.
55. 1.3 cm.
57. 12 m.
59. (a) F_{left} = 310 N, up;
        F_{right} = 210 N, down;
    (b) 0.65 m from right hand;
    (c) 1.2 m from right hand.
61. 2.9 × 10⁹ m·N, clockwise; no.
63. (a) 0.78 N;
    (b) 0.98 N.
65. 3.5 × 10⁻⁴ m.
67. A: 230 N; B: 110 N.
69. 2.51 m.
71. 45°.
73. (a) 2100 N;
    (b) 1.3.
75. 2500 N, no.
77. 2.6 × 10⁻⁴ m².
79. (a) 1.6 m;
    (b) 1.2 × 10⁴ N, no.
81. (4.0 × 10¹)°.

Chapter 10

1. 3 × 10¹¹ kg.
3. 710 kg.
5. 0.8547.
7. (a) 5501 kg/m³;
   (b) 5497 kg/m³, −0.07%.
9. (a) 6.1 × 10⁶ N/m²;
   (b) 1.7 × 10⁵ N/m².
11. (a) 4.5 × 10⁵ N;
    (b) 4.5 × 10⁵ N.
13. 1.2 m.
15. 1900 kg.
17. (a) 7.0 × 10⁵ N/m²;
    (b) 72 m.
19. 1.60 × 10⁴ m.
21. 4.0 × 10⁷ N/m².
23. 0.57.
25. (a) 1.5 × 10⁵ N;
    (b) 1.8 × 10⁵ N.
27. Iron or steel.
29. 9.9 × 10⁻³ m³.

31. 10.5%.
33. 32 bottles.
35. 0.88.
37. (a) 6.68×10^{-2} m^3;
    (b) 1.07;
    (c) 12%.
39. 9 N, down, 21 N, up.
41. 4.4 m/s.
43. 9.6 m/s.
45. 4.12×10^{-3} m^3/s.
47. 1.6×10^5 N/m^2.
49. 1.2×10^5 N.
51. 9.7×10^4 Pa.
53. 2.5 m/s, 2.2 atm.
57. 1100 N.
59. 8.2×10^3 Pa.
61. 0.094 m.
63. (a) Laminar;
    (b) 2940, turbulent.
65. 0.89 Pa/cm.
67. 2.4×10^{-2} N/m.
69. (a) $\gamma = F/4\pi r$;
    (b) 1.7×10^{-2} N/m.
71. 1.5 mm.
73. (a) 7.6×10^{-4} N;
    (b) 1.3 N.
75. $F_{T_{string}} = 0.71$ N, 984.2 g.
77. (a) 1.0×10^{-3} m^2;
    (b) 4.0×10^3 J;
    (c) 5.3×10^{-3} m;
    (d) 80 strokes.
79. 0.6 atm.
81. 1.0 m.
83. 2×10^7 Pa.
85. 1.89×10^4 m^3.
87. 5.29×10^{18} kg.
89. (a) 7.9 m/s;
    (b) 0.22 L/s;
    (c) 0.79 m/s.
91. 130 N.
93. 1.2×10^4 N/m^2.
95. 3.5×10^{-3} Pa·s.
97. 0.27 kg.
99. 68%.

Chapter 11

1. 0.84 m.
3. 560 N/m.
5. (a) 650 N/m;
   (b) 2.1 cm, 2.6 Hz.
7. 0.85 kg.

9. (a) $A_A = 2.5$ m, $A_B = 3.5$ m;
   (b) $f_A = 0.25$ Hz, $f_B = 0.50$ Hz;
   (c) $T_A = 4.0$ s, $T_B = 2.0$ s.
11. $\pm 70.7\%$ of the amplitude.
13. 0.233 s.
15. (a) 2.1 m/s;
    (b) 1.5 m/s;
    (c) 0.54 J;
    (d) $x = (0.15$ m$)\cos(4.4\pi t)$.
17. $\sqrt{3} : 1$.
19. (a) 430 N/m;
    (b) 4.6 kg.
21. (a) 0.436 s, 2.29 Hz;
    (b) 0.157 m;
    (c) 32.6 m/s^2;
    (d) 2.26 J;
    (e) 1.90 J.
23. 68.0 N/m, 15.6 m.
25. (a) $y = (0.16$ m$)\cos(14\,t)$;
    (b) 0.11 s;
    (c) 2.2 m/s;
    (d) 31 m/s^2, at the release point.
27. 3.0 s.
29. (a) 1.8 s;
    (b) 0.56 Hz.
31. Shorten it by 0.5 mm.
33. $\frac{1}{3}$.
35. 2.3 m/s.
37. (a) 1400 m/s;
    (b) 4100 m/s;
    (c) 5100 m/s.
39. (a) 1400 km;
    (b) No; need readings from at least two other stations.
41. 4.8 N.
43. 21 m.
45. (a) 8.7×10^9 J/m^2·s;
    (b) 1.7×10^{10} W.
47. $\sqrt{5} : 1$.
49. 440 Hz, 880 Hz, 1320 Hz, 1760 Hz.
51. 60 Hz, fundamental or first harmonic; 120 Hz, first overtone or second harmonic; 180 Hz, second overtone or third harmonic.
53. 70 Hz.
55. (a) 1.2 kg;
    (b) 0.29 kg;
    (c) 4.6×10^{-2} kg.
57. 1.3 m/s.
59. 24°.
61. 460 Hz; $f < 460$ Hz.
63. 0.11 m.

65. Mg/k.
67. (a) $1.16\,f$;
    (b) $0.81\,f$.
69. 2.6×10^{13} Hz.
71. (a) 1.2 Hz;
    (b) 12 J.
73. (a) 3.7×10^4 N/m;
    (b) 0.50 s.
75. 8.40×10^2 N/m.
77. 0.13 m/s, 0.12 m/s^2; 1.2%.
79. (a) 0.06 m;
    (b) 7.1.
81. (a) G: 784 Hz, 1180 Hz; B: 988 Hz, 1480 Hz;
    (b) 1.59 : 1;
    (c) 1.26 : 1;
    (d) 0.630 : 1.
83. 2.30×10^2 Hz.
85. 18 W.
87. (a) $\theta_{iM} = \sin^{-1}(v_{air}/v_{water})$
        $= \sin^{-1}(v_i/v_r)$;
    (b) 0.44 m.

Chapter 12

1. 430 m.
3. (a) 1.7 cm to 17 m;
   (b) 1.9×10^{-5} m.
5. (a) 0.994 s;
   (b) 4.52 s.
7. 33 m.
9. 62 dB.
11. 82 dB.
13. 82-dB player: 1.6×10^8; 98-dB player: 6.3×10^9.
15. (a) 790 W;
    (b) 440 m.
17. (a) 12;
    (b) 11 dB.
19. 130 dB.
21. (a) 220-W: 122 dB; 45-W: 115 dB;
    (b) no.
23. 80 Hz, 15,000 Hz.
25. 10 octaves.
27. 87 N.
29. (a) 360 Hz;
    (b) 540 Hz.
31. 260 Hz.
33. (a) 17 cm;
    (b) 1.02 m;
    (c) $f = 440$ Hz, $\lambda = 0.78$ m.
35. 1.9%.

37. (a) 0.585 m;
    (b) 858 Hz.
39. (a) 55 Hz;
    (b) 190 m/s.
41. (a) 253 overtones;
    (b) 253 overtones.
43. 4.2 cm, 8.2 cm, 11.9 cm, 15.5 cm, 18.8 cm, 22.0 cm.
45. $I_2/I_1 = 0.64$; $I_3/I_1 = 0.20$; $\beta_{2-1} = -2$ dB; $\beta_{3-1} = -7$ dB.
47. 28.5 kHz.
49. 347 Hz.
51. (a) 0.562 m;
    (b) 0.
53. (a) 343 Hz;
    (b) 1000 Hz, 1700 Hz.
55. (a) 8.9 beats per second;
    (b) 38 m.
57. 4.27×10^4 Hz.
59. 3.11×10^4 Hz.
61. (a) Every 1.4 s;
    (b) every 11 s.
63. 0.0821 m/s.
65. 11 km/h.
67. (a) 99;
    (b) 0.58°.
69. (a) 36°;
    (b) 560 m/s, 1.7.
71. 0.12 s.
73. 88 dB.
75. 14 W.
77. (a) 51 dB;
    (b) 5×10^{-9} W.
79. 635 Hz.
81. $\mu_{2nd} = 0.44\mu_{lowest}$,
    $\mu_{3rd} = 0.20\mu_{lowest}$,
    $\mu_{4th} = 0.088\mu_{lowest}$.
83. 150 Hz, 460 Hz, 770 Hz, 1100 Hz.
85. 2.35 m/s.
87. 2.62 m.
89. 11.5 m.
91. $\frac{1}{1000}$.
93. 36 Hz, 48 Hz, 61 Hz.
95. 10^6.

Chapter 13

1. $N_{gold} = 0.548 N_{silver}$.
3. (a) 20°C;
    (b) 3500°F.
5. 102.0°F.
7. $-40°C = -40°F$.
9. 0.08 m.
11. 2.2×10^{-6} m; $\frac{1}{60}$ of the change for steel.
13. $-70°C$.
15. 0.98%.
17. $-210°C$.
19. 4.0×10^7 N/m².
21. $-459.67°F$.
23. 1.25 m³.
25. (a) 0.2754 m³;
    (b) $-63°C$.
27. (a) 22.8 m³;
    (b) 2.16×10^5 Pa.
29. 1.69×10^8 Pa.
31. 7.4%.
33. 33%.
35. Actual: 0.598 kg/m³, ideal: 0.588 kg/m³; near a phase change.
37. 1.07 cm.
39. 55.51 mol, 3.343×10^{25} molecules.
41. 300 molecules/cm³.
43. (a) 5.65×10^{-21} J;
    (b) 3700 J.
45. 1.22.
47. 3.5×10^{-9} m/s.
49. $\sqrt{3}$.
53. $\dfrac{(v_{rms})_{^{235}UF_6}}{(v_{rms})_{^{238}UF_6}} = 1.004$.
55. Vapor.
57. (a) Vapor;
    (b) solid.
59. 3200 Pa.
61. 18°C.
63. 0.91 kg.
65. 2.5 kg.
67. (a) Greater than;
    (b) (-2.0×10^{-4})%;
    (c) 0.603%.
69. (b) 4×10^{-11} mol/s;
    (c) 0.6 s.
71. (a) Low;
    (b) (2.8×10^{-2})%.
73. 18%.
75. (a) 1500 kg;
    (b) 200 kg enters.
77. (a) Lower;
    (b) 0.36%.
79. 910 min.
81. (a) 0.66×10^3 kg/m³;
    (b) -3.0%.
83. 2300 m.
85. (a) 290 m/s;
    (b) 9.5 m/s.
87. PE/KE = 8.50×10^{-5}, yes.
89. 0.30 kg.
91. 2.4 kg.

Chapter 14

1. 10.7°C.
3. 0.04 candy bars.
7. 250 kg/h.
9. 6.0×10^6 J.
11. (a) 3.3×10^5 J;
    (b) 5600 s.
13. 4.0×10^2 s.
15. 42.6°C.
17. 2.3×10^3 J/kg·C°.
19. 43 C°.
21. 0.39 C°.
23. 473 kcal.
25. 7.1×10^6 J.
27. 0.18 kg.
29. (a) 5.2×10^5 J;
    (b) 1.5×10^5 J.
31. 11.2 kJ/kg.
33. 2.7 g.
35. 5.2 g.
37. 93 J/s = 93 W.
39. 7.5×10^4 s.
41. 20 bulbs.
43. 3.1×10^4 s.
45. 350 Btu/h.
47. (a) 3.2×10^{26} W;
    (b) 1.1×10^3 W/m².
49. A mixture of $\frac{1}{3}$ steam and $\frac{2}{3}$ liquid water at 100°C.
51. 2 C°.
53. 6.6×10^3 kcal.
55. 4.0×10^2 m/s.
57. 450°C.
59. 0.14 C°.
61. 1.43×10^3 m/s, toward the Earth.
63. 19 min.
65. (a) 3.4 W;
    (b) 2.3 C°/s;
    (c) no, $T > 8000°C$ in less than an hour;
    (d) 86°C;
    (e) conduction, convection, evaporation.
67. (a) 3.6×10^7 J;
    (b) 63 min.

Chapter 15

1. (a) 0;
   (b) 4.30×10^3 J.
3.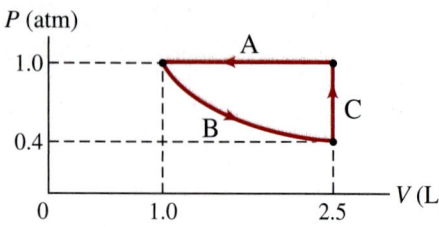
5.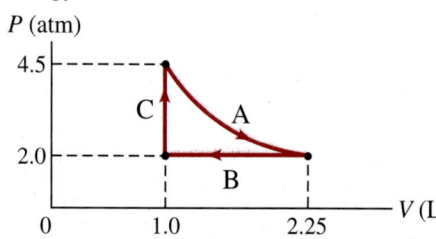
7. (a) 0;
   (b) 2630 J;
   (c) rise.
9. -78 K.
13. (a) -196 J;
    (b) 28 J;
    (c) 168 J;
    (d) 140 J;
    (e) 98 J.
15. 170 W.
17. (a) 1.4×10^7 J;
    (b) 3500 Cal.
19. 25.8%.
21. 8.8%.
23. 10°C decrease in the low-temperature reservoir.
25. 1.7×10^{13} J/h.
27. 1800 W.
29. 420°C.
31. 0.15.
33. 6.5.
35. (a) 1.0×10^3 J;
    (b) 1.0×10^3 J;
    (c) 230 J at 0°C, 390 J at -15°C.
37. 78 L.
39. -1.9×10^3 J/K.
41. -1.22×10^6 J/K.
43. 4×10^4 J/K.
45. 0.64 J/K.
47. 1.1 J/K.
49. (a) $\frac{1}{12}$;
    (b) $\frac{1}{12}$.

51. (a)

(r = red, o = orange, g = green)

Macrostate	Microstates			Number of microstates
3 r, 0 o, 0 g	r r r			1
2 r, 1 o, 0 g	r r o	r o r	o r r	3
2 r, 0 o, 1 g	r r g	r g r	g r r	3
1 r, 2 o, 0 g	r o o	o r o	o o r	3
1 r, 0 o, 2 g	r g g	g r g	g g r	3
1 r, 1 o, 1 g	r o g r g o o r g o g r g r o g o r			6
0 r, 3 o, 0 g	o o o			1
0 r, 2 o, 1 g	g o o	o g o	o o g	3
0 r, 1 o, 2 g	o g g	g o g	g g o	3
0 r, 0 o, 3 g	g g g			1

(b) $\frac{1}{27}$;
(c) $\frac{1}{9}$.
53. 70 m², yes.
55. 1.5×10^7 W.
57. (a) 2.2×10^5 J;
    (b) 3.6×10^5 J;
    (c)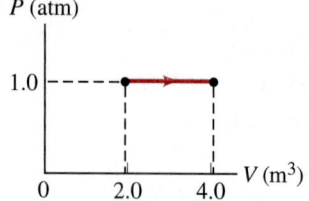
59. 86°C.
61. (a) 7.7%;
    (b) the large volume of "fuel" (ocean water) available.
63. $0.43/h.
65. (a) 44°C;
    (b) 4.3×10^{-2} J/K.
67. 60 K.
69. (a) 13 km³/day, possibly;
    (b) 73 km².
71. (a) 0.281;
    (b) 1.01×10^5 W, 2.1×10^9 J, 4.9×10^5 kcal.
73. (a) 0.22 kg;
    (b) 4.5 days.
75. 4.6×10^6 J.
77. (a) -4°C;
    (b) 29%.

Chapter 16

1. 2.7×10^{-3} N.
3. 2.2×10^4 N.
5. (1.9×10^{-13})%.
7. 3.76 cm.
9. -4.6×10^8 C, 0.
11. $F_{\text{left}} = 120$ N, to the left;
    $F_{\text{center}} = 560$ N, to the right;
    $F_{\text{right}} = 450$ N, to the left.
13. 2.1×10^{12} electrons.
15. $10.1 \frac{kQ^2}{\ell^2}$, at 61°.
17. (a) 88.8×10^{-6} C, 1.2×10^{-6} C;
    (b) 91.1×10^{-6} C, -1.1×10^{-6} C.
19. 3.94×10^{-16} N, west.
21. 6.30×10^6 N/C, upward.
23. 1.33×10^{14} m/s², opposite to the field.
25.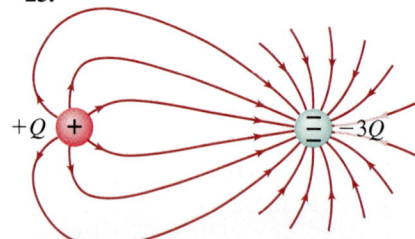
27. 5.97×10^{-10} N/C, south.
29. Upper right corner, $E = 3.76 \times 10^4$ N/C, at 45.0°.
31. $\frac{4kQxa}{(x^2-a^2)^2}$, to the left.
33. 3.7×10^7 N/C, 330°.
35. $E_A = 3.0 \times 10^6$ N/C, at 90°;
    $E_B = 7.8 \times 10^7$ N/C, at 56°; yes.
37. (a) 5×10^{-10} N;
    (b) 7×10^{-10} N;
    (c) 6×10^{-5} N.
39. (a) -1.1×10^5 N·m²/C;
    (b) 0.
41. 8.3×10^{-10} C.
43. (a) $k\frac{Q}{r^2}$;
    (b) 0;
    (c) $k\frac{Q}{r^2}$;
    (d) The shell causes the field to be 0 in the shell material. The charge polarizes the shell.
45. 4.0×10^9 C.
47. 6.8×10^5 C, negative.
49. 1.0×10^7 electron charges.
51. 5.2×10^{-11} m.
53. 4.3 m.
55. 0.14 N, rightward.
57. 8.2×10^{-7} C, positive.
59. (a) 4×10^{10} particles,
    (b) 4×10^{-5} kg.
61. 9.90×10^6 N/C, downward.

63. $x = d(\sqrt{2} + 1) \approx 2.41d$.
65. $QE\ell$, counterclockwise.
67. 8.94×10^{-19}.

Chapter 17

1. 5.0×10^{-4} J.
3. -1.0 V.
5. 4030 V, plate B.
7. 5.78 V.
9. -4.25×10^4 V.
11. -157 V.
13. 9.0×10^5 m/s.
15. 3000 V; only a small amount of charge was transferred.
17.

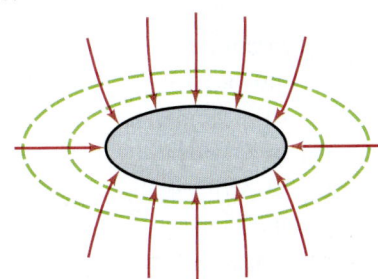

19. 2.8×10^{-9} C.
21. (a) 5.8×10^5 V;
    (b) 9.2×10^{-14} J.
23. 9.15×10^6 m/s.
25. (a) 18 cm from $-$ charge, on opposite side from $+$ charge;
    (b) 1.6 cm from $-$ charge, toward $+$ charge, and 8.0 cm from $-$ charge, away from $+$ charge.
27. (a) 1.6×10^4 V;
    (b) 9.9×10^4 V/m, 64°.
29. 4.2×10^6 V.
31. (a) 27 V;
    (b) 2.2×10^{-18} J, or 14 eV;
    (c) -2.2×10^{-18} J, or -14 eV;
    (d) 2.2×10^{-18} J, or 14 eV.
33. (a) 6.6×10^{-3} V;
    (b) 4.6×10^{-3} V;
    (c) -4.6×10^{-3} V.
35. 2.6×10^{-6} F.
37. 6.00×10^{-5} C.
39. 6.3×10^{-7} F.
41. 0.24 m².
43. 9×10^{-16} m, no.
45. $V_{2.50\,\mu F} = V_{6.80\,\mu F} = 611$ V, $Q_{2.50\,\mu F} = 1.53 \times 10^{-3}$ C, $Q_{6.80\,\mu F} = 4.16 \times 10^{-3}$ C.
47. 4.7×10^{-11} F.
49. 9.5 V.
51. 4.20×10^{-9} F, 0.247 m².
53. 9.6×10^{-5} F.
55. (a) 9×10^{-12} F;
    (b) 8×10^{-11} C;
    (c) 200 V/m;
    (d) 4×10^{-10} J;
    (e) capacitance, charge, work done.
57. 1.0×10^{-7} J/m³.
59. 1110100.
61. 43,690.
63. (a) 65,536;
    (b) 16,777,216;
    (c) 16,777,216.
65. (b) 56 Hz.
67. $+2.0 \times 10^5$ V/m to -2.0×10^5 V/m.
69. Yes, 1.3×10^{-12} V.
71. (a) Multiplied by 2;
    (b) multiplied by 2.
73. Alpha particle, 2.
75. Left: $-6.85kQ/\ell$, top: $-3.46kQ/\ell$, right: $-5.15kQ/\ell$.
77. (a) 17 cm from $-$ charge, on opposite side from $+$ charge;
    (b) 1.1 cm from $-$ charge, toward $+$ charge, and 8.1 cm from $-$ charge, away from $+$ charge.
79. (a) 31 J;
    (b) 5.9×10^5 W.
81. 1.8 J.
83. 3.7×10^{-10} C.
85. (a) 6.4×10^{-11} C;
    (b) 6.4×10^{-11} C;
    (c) 18 V;
    (d) 2×10^{-10} J.
87. (a) 3.6×10^3 m/s;
    (b) 2.8×10^3 m/s.
89. 1.7×10^6 V.
91. 1.3×10^{-6} C.
93. 16°.
95. (a) 0.32 μm²;
    (b) 59 megabytes.

Chapter 18

1. 1.00×10^{19} electrons/s.
3. 6.2×10^{-11} A.
5. 1200 V.
7. (a) 28 A;
   (b) 8.4×10^4 C.
9. (a) 8.9 Ω;
   (b) 1.2×10^4 C.
11. (a) 4.8 A;
    (b) 6.6 A.
13. 5.1×10^{-2} Ω.
15. Yes, for length 4.0 mm.
17. 2.0 V.
19. (a) 3.8×10^{-4} Ω;
    (b) 1.5×10^{-3} Ω;
    (c) 6.0×10^{-3} Ω.
21. 18C°.
23. 2400°C.
25. $R_{carbon} = 1.42$ kΩ, $R_{Nichrome} = 1.78$ kΩ.
27. 0.72 W.
29. 31 V.
31. 1.7×10^5 C.
33. (a) 950 W;
    (b) 15 Ω;
    (c) 9.9 Ω.
35. (a) 1.1 A;
    (b) 110 Ω.
37. 0.046 kWh; 6.6 cents per month.
39. 2.8×10^6 J.
41. 24 bulbs.
43. 1.5 m; power increases 36× and could start a fire.
45. (a) 7.2 A;
    (b) 1.7 Ω.
47. 0.12 A.
49. (a) Infinite resistance;
    (b) 96 Ω.
51. (a) 930 V;
    (b) 3.9 A.
53. (a) 3300 W;
    (b) 9.7 A.
55. 6.0×10^{-10} m/s.
57. 2.2 A/m², north.
59. 32 m/s (possible delay between nerve stimulation and generation of action potential).
61. 9.8 h.
63. 6.22 A.
65. 2.4×10^{-4} m.
67. $3200 per hour per meter.
69. 4.2×10^{-3} m.
71. (a) 33 Hz;
    (b) 0.990 A;
    (c) $V = (33.6 \sin 210t)$ V.
73. 2.25 Ω.
75. (b) As large as possible.
77. (a) 7.4 hp;
    (b) 220 km.
79. 1.7×10^{-4} m.
81. 32% increase.

83. (a) $I_A = 0.33$ A, $I_B = 3.3$ A;
(b) $R_A = 360\ \Omega$, $R_B = 3.6\ \Omega$;
(c) $Q_A = 1.2 \times 10^3$ C,
$Q_B = 1.2 \times 10^4$ C;
(d) $E_A = E_B = 1.4 \times 10^5$ J;
(e) Bulb B.

85. (a) 4×10^6 J;
(b) 2×10^4 m.

87. (a) 12 W;
(b) 4.6 W.

89. $1.34 \times 10^{-4}\ \Omega$.

91. $f = 1 - \dfrac{V}{V_0}$.

Chapter 19

1. (a) 5.92 V;
(b) 5.99 V.

3. $0.034\ \Omega$; $0.093\ \Omega$.

5. (a) $330\ \Omega$;
(b) $8.9\ \Omega$.

7. 2.

9. Connect 18 resistors in series; then measure voltage across 7 consecutive series resistors.

11. $0.3\ \Omega$.

13. $560\ \Omega$, 0.020.

15. $32\ \Omega$.

17. $140\ \Omega$.

19. $\tfrac{13}{8} R$.

21. $4.8\ \text{k}\Omega$.

23. 55 V.

25. 0.35 A.

27. 0.

29. (a) 34 V;
(b) 85-V battery: 82 V;
45-V battery: 43 V.

31. $I_1 = 0.68$ A, left; $I_2 = 0.33$ A, left.

33. (a) $\mathcal{E}/R$;
(b) R.

35. 0.56 A.

37. 3 parallel sets, each with 100 cells in series.

39. 3.71×10^{-6} F.

41. 2.0×10^{-9} F, yes.

43. 1.90×10^{-8} F in parallel,
1.7×10^{-9} F in series.

45. $2:1$.

47. In parallel, 750 pF.

49. $29.3\ \mu\text{F}$, $5.7\ \mu\text{F}$.

51. (a) $\tfrac{3}{5} C$;
(b) $Q_1 = Q_2 = \tfrac{1}{5} CV$, $Q_3 = \tfrac{2}{5} CV$,
$Q_4 = \tfrac{3}{5} CV$; $V_1 = V_2 = \tfrac{1}{5} V$,
$V_3 = \tfrac{2}{5} V$, $V_4 = \tfrac{3}{5} V$.

53. $1.0 \times 10^6\ \Omega$.

55. 7.4×10^{-3} s.

57. (a) $I_1 = \dfrac{2\mathcal{E}}{3R}$, $I_2 = I_3 = \dfrac{\mathcal{E}}{3R}$

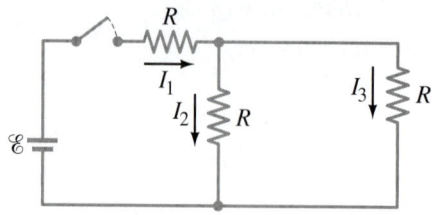

(b) $I_1 = I_2 = \dfrac{\mathcal{E}}{2R}$, $I_3 = 0$;

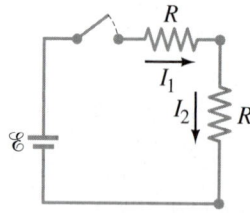

(c) $\tfrac{1}{2}\mathcal{E}$.

59. (a) 2.9×10^{-5} A;
(b) $8.8 \times 10^6\ \Omega$.

61. Add $710\ \Omega$ in series with ammeter, $29\ \Omega/\text{V}$.

63. 9.60×10^{-4} A, 4.8 V;
current: $+20\%$, voltage: -20%.

65. 9.8 V.

67. Put $9.0\ \text{k}\Omega$ in series with the body.

69. $\tfrac{1}{4} C$, $\tfrac{2}{5} C$, $\tfrac{3}{5} C$, $\tfrac{3}{4} C$, C, $\tfrac{4}{3} C$, $\tfrac{5}{3} C$, $\tfrac{5}{2} C$, $4C$.

71. $9.2 \times 10^4\ \Omega$.

73. (a) $3.6\ \Omega$;
(b) 14 W.

77. 600 cells; $0.54\ \text{m}^2$, 4 banks in parallel, each containing 150 cells in series.

79. (a) $6.0\ \Omega$;
(b) 2.2 V.

81. 11 V.

83. $100\ \Omega$.

87. $9.0\ \Omega$.

89. $Q_{12\,\mu\text{F}} = 1.0 \times 10^{-4}$ C,
$Q_{48\,\mu\text{F}} = 4.1 \times 10^{-4}$ C.

91. (a) 1.9×10^{-4} J;
(b) 4.0×10^{-5} J;
(c) $Q_a = 16\ \mu\text{C}$; $Q_b = 3.3\ \mu\text{C}$.

93. $Q_1 = \dfrac{C_1 C_2}{C_2 + C_1} V_0$,
$Q_2 = \dfrac{C_2^2}{C_2 + C_1} V$.

95. (a) In parallel;
(b) 7.7 pF to 35 pF.

97. $Q_1 = 11\ \mu\text{C}$, $V_1 = 11$ V;
$Q_2 = 13\ \mu\text{C}$, $V_2 = 6.3$ V;
$Q_3 = 13\ \mu\text{C}$, $V_3 = 5.2$ V.

Chapter 20

1. (a) 5.8 N/m;
(b) 3.3 N/m.

3. 1.3 N.

5. 27°.

7. (a) South pole;
(b) 3.86 A;
(c) 8.50×10^{-2} N.

9. 5.6×10^{-14} N, north.

11. 0.24 T.

13. (a) To the right;
(b) downward;
(c) into the page.

15. (a) 6.0×10^5 m/s;
(b) 3.6×10^{-2} m;
(c) 3.8×10^{-7} s.

17. 0.59 m.

19. $r_{\text{proton}}/r_{\text{electron}} = 42.8$.

21. 1.97×10^{-6} m.

23. (a) Sign determines polarity but not magnitude of Hall emf.
(b) 0.56 m/s.

25. 2.9×10^{-4} T, about 5.8 times larger.

27. 7.8×10^{-2} N, toward other wire.

29.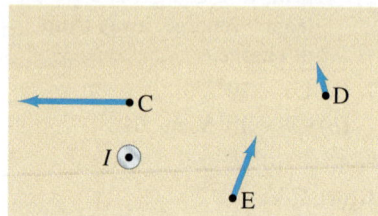

31. 5.1×10^{-6} N, toward wire.

33. 3.8×10^{-5} T, 17° below horizontal.

35. (a) $(2.0 \times 10^{-5}\ \text{T/A})(I - 25\ \text{A})$;
(b) $(2.0 \times 10^{-5}\ \text{T/A})(I + 25\ \text{A})$.

37.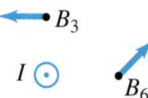

39. 15 A, downward.

A-36 Answers to Odd-Numbered Problems

41. Closer wire: 4.4×10^{-2} N/m, attract; farther wire: 2.2×10^{-2} N/m, repel.
43. 4.66×10^{-5} T.
45. 1.19 A.
47. 0.12 N, south.
49. (c) No; inversely as distance from center of toroid: $B \propto 1/R$.
51. 1.18 T.
53. 69.7 μA.
55. 1.87×10^6 V/m; perpendicular to velocity and magnetic field, and in opposite direction to magnetic force on protons.
57. 1.3×10^{-3} m; 6.5×10^{-4} m.
59. $^2_1$H nucleus or $^4_2$He nucleus.
61. 0.5 T.
63.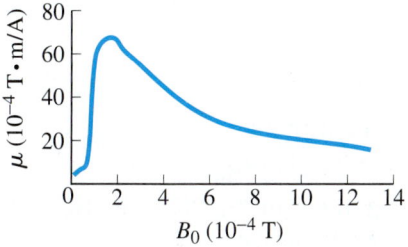
65. 2.7×10^{-2} T, upward.
67. 0.30 N, northerly, 68° above horizontal.
69. 7.7×10^{-6} N.
71. 1.1×10^{-6} m/s, west.
73. $\dfrac{2\mu_0 I}{\pi \ell}$, to the left.
75. They will exit above or below second tube; 52°.
77. -2.1×10^{-20} J.
79. $r = 5.3 \times 10^{-5}$ m, $p = 3.3 \times 10^{-4}$ m.
81. 1.9×10^{-3} T.
83. $\dfrac{5.0 mg}{\ell B}$, to the left.
85. (a) Negative;
(b) $\dfrac{qB_0 (d^2 + \ell^2)}{2d}$.
87. 1.2 A; downward.
89. (a) M: 5.8×10^{-4} N/m, upward; N: 3.4×10^{-4} N/m, at 300°; P: 3.4×10^{-4} N/m, at 240°;
(b) 1.75×10^{-4} T, at $-14°$.
91. 2.9×10^{-6} T, $B_{\text{wire}} \approx 0.06 B_{\text{Earth}}$.

Chapter 21

1. 560 V.
3. Counterclockwise.
5. 0.20 V.
7. (a) 1.0×10^{-2} Wb;
(b) 48°;
(c) 6.7×10^{-3} Wb.
9. (a) 1.73×10^{-2} V;
(b) 0.114 V/m, downward.
11. (a) 0;
(b) clockwise;
(c) counterclockwise;
(d) clockwise.
13. 0.65 mV, east or west.
15. (a) Magnetic force on current in moving bar, $B^2 \ell^2 v / R$;
(b) $B^2 \ell^2 v^2 / R$.
17. (a) 0.17 V;
(b) 7.1×10^{-3} A;
(c) 7.5×10^{-4} N, to the right.
19. 5.23 C.
21. (a) 810 V;
(b) double the rotation frequency.
23. 17 rotations per second.
25. 92 V.
27. 1.71×10^4 turns.
29. $I_S = 0.21 I_P$.
31. (a) 6.2 V;
(b) step-down.
33. 450 V, 56 A.
35. 6×10^9 m.
37. 55 MW.
39. 6.9 V.
41. 0.10 H.
43. (a) 1.5×10^{-2} H;
(b) 75 turns.
45. 46 m, 21 km; 0.70 kΩ.
47. 23 J.
49. 5×10^{15} J.
51. 3.7.
53. (a) 2.3;
(b) 4.6;
(c) 6.9.
55. 3300 Hz.
57. 1.6×10^4 Ω, 1.47×10^{-2} A.
59. (a) 7400 Ω;
(b) 0.38 A.
61. (a) 3.6×10^4 Ω;
(b) 3.7×10^4 Ω.
63. 205 Ω.
65. 270 Hz; the voltages are out of phase.
67. (a) 1.77×10^{-2} A;
(b) $-12.6°$;
(c) $V_R = 117$ V, $V_C = 26.1$ V.
69. 3.6×10^5 Hz.
71. (a) 1.3×10^{-7} F;
(b) 37 A.
73. (a) 0.032 H;
(b) 0.032 A;
(c) 16 μJ.
75. 6.01×10^{-3} J.
77. Coil radius = 1.5 cm, 10,000 turns.
79. 200 kV.
81. (a) 41 kV;
(b) 31 MW;
(c) 1.0 MW;
(d) 30 MW.
85. Put a 98-mH inductor in series with it.
87. 82 V.
89. 93 mH.
91. 2.

Chapter 22

1. 1.1×10^5 V/m/s.
3. 1.7×10^{15} V/m/s.
5. 2.4×10^{-13} T.
7. 90.0 kHz, 2.33 V/m, along the horizontal north–south line.
9. 1.25 s.
11. 4.20×10^{-7} m, violet visible light.
13. 2.00×10^{10} Hz.
15. (a) 1.319×10^{-2} m;
(b) 2.5×10^{18} Hz.
17. 4.0×10^{16} m.
19. 9600 wavelengths; 3.54×10^{-15} s.
21. 1.6×10^6 revolutions/s.
23. 3.02×10^7 s.
25. 5.7 W/m^2, 46 V/m.
27. 4.51×10^{-6} J.
29. 3.80×10^{26} W.
31. 7.3×10^{-7} N/m^2; 7.3×10^{-11} N away from bulb.
33. 400 m^2.
35. 0.16 m.
37. Channel 2: 5.56 m, Channel 51: 0.434 m.
39. 1.5×10^{-12} F.
41. (a) 1.3×10^{-6} H;
(b) 9.9×10^{-11} F.
43. 6.25×10^{-4} V/m; 1.04×10^{-9} W/m^2.
45. 10 ns.
47. 1.36 s; inside.
49. 5.00×10^2 s.
51. 13 V/m, 4.4×10^{-8} T.
53. (a) 1.2×10^{-10} J;
(b) 8.7×10^{-6} V/m, 2.9×10^{-14} T.

55. 61 km.
57. (a) 2.8×10^{-3} J/s;
    (b) 1.0 V/m;
    (c) 1.0 V;
    (d) 2.0×10^{-2} V.
59. 6×10^{10} W.
61. 35 kW.

Chapter 23

1.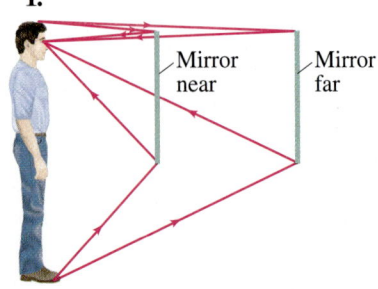
3. 11°.
5.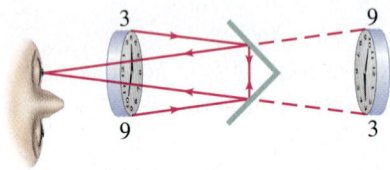
7. 4.0×10^{-6} m².
9. 10.5 cm.
11. (a) $d_i \approx -5$ cm;

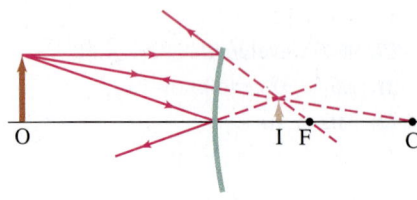

(b) $d_i = -5.3$ cm;
(c) 1.0 mm.
13. −6.8 m.
15. 5.7 m.
17. 2.0 cm behind ball's front surface; virtual; upright.
19. 1.0 m.
21. (a) Concave;
    (b) upright, virtual, and magnified;
    (c) 1.40 m.
23.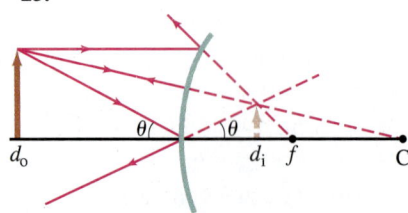

25. 1.31.
27. 1.62.
29. 50.1°.
31. 38.6°.
33. 81.9°.
35. 61.0°, crown glass.
37. At least 93.5 cm away.
39. (a) 1.4;
    (b) no;
    (c) 1.9.
41. (a) ∼500 mm;

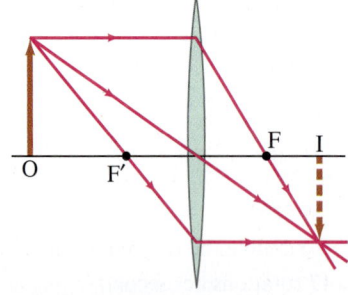

(b) 478 mm.
43. (a) 3.08 D, converging;
    (b) −0.148 m, diverging.
45. (a) 106 mm;
    (b) 109 mm;
    (c) 117 mm;
    (d) 513 mm.
47. (a) 37 cm behind lens;
    (b) +2.3×.
49. $d_i = -6.67$ cm behind lens, virtual and upright, $h_i = 0.534$ mm.
51. (a) 70.0 mm;
    (b) 30.0 mm.
53. 64 cm.
55. 21.3 cm or 64.7 cm from object.
57. 0.105 m; 5.8 m.
59. 18.5 cm beyond second lens; −0.651× (inverted).
61. (a) 10 cm beyond second lens;
    (b) −1.0×;
    (c)

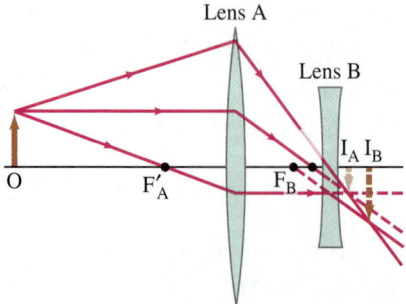

63. −29.7 cm.
65. 9.0 cm.
67. 0.34 m.
69. 1.25 s.
71. 6.04 m.
73. 1.58, light flint glass.
75. (a) Convex;
    (b) 25 cm behind mirror;
    (c) −110 cm;
    (d) −220 cm.
77. 67°.
79. 9 cm, 12 cm.
81. $n \geq 1.60$.
83. (a) −0.33 mm;
    (b) −0.47 mm;
    (c) −0.98 mm.
85. Left: converging; right: diverging.
89. 5.7 cm from object, between it and lens.

Chapter 24

1. 5.4×10^{-7} m.
3. 6.3×10^{-7} m, 4.8×10^{14} Hz.
5. 17° and 64°.
7. 1.5×10^{-4} m.
9. 3.1 cm.
11.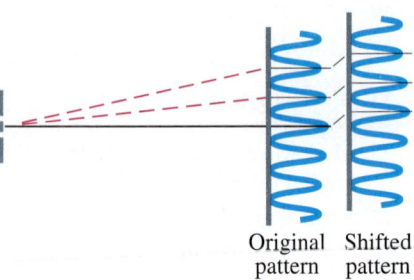

13. 2.2×10^{-3} m.
15. 14 mm.
17. 570 nm.
19. 0.23°.
21. 1.8°.
23. 1.0×10^{-6} m.
25. 7.38×10^{-2} m.
27. (a) λ;
    (b) 400 nm.
29. 3.4×10^{-6} m.
31. Entire pattern is shifted, with central maximum at 28.0° to the normal.
33. 330 nm.
35. 556 nm.
37. 490 nm, 610 nm, 640 nm, 670 nm.

39. $(3.0 \times 10^1)°$.
41. Second order.
43. 7140 slits/cm.
45. 0.878 cm.
47. 230 nm.
49. 9 lenses.
51. 33 dark bands.
53. 110 nm; 230 nm.
55. 482 nm.
57. 691 nm.
59. 1.004328.
61. 57.3°.
63. 48.1%.
65. No; for diamond, $\theta_p = 59.4°$.
67. 61.2°.
69. 36.9°; 48.8°; 53.1°.
71. 28° relative to first polarizer.
73. (a) 1.3×10^{-4} m;
   (b) 3.9×10^{-7} m.
75. H: 13.7°; Ne: 13.5°; Ar: 14.5°.
77. 480 nm.
79. $\lambda_2 > 600$ nm overlaps with $\lambda_3 < 467$ nm.
81. (a) 82 nm;
   (b) 130 nm.
83. 4.8×10^4 m.
85. 3.19×10^{-5} m.
87. 580 nm.
89. 0.6 m.
91. 658 nm; 782 slits/cm.
93. 400 nm, 600 nm.
95. (a) 0;
   (b) 0.11;
   (c) 0.

Chapter 25

1. $f/8$.
3. 3.00 mm to 46 mm.
5. 110 mm; 220 mm.
7. 5.1 m to infinity.
9. 16 mm.
11. $f/2.5$.
13. 1.1 D.
15. (a) -1.2 D;
   (b) 35 cm.
17. -8.3 D; -7.1 D.
19. (a) 2.0 cm;
   (b) 1.9 cm.
21. 18.4 cm, 1.00 m.
23. $1.6\times$.
25. 6.2 cm from lens, $4.0\times$.

27. (a) $3.6\times$;
   (b) 14 mm;
   (c) 6.9 cm.
29. 4 cm toward contract.
31. $-29\times$; 85 cm.
33. 25 cm.
35. $-22\times$.
37. $-110\times$.
39. Objective: 1.09 m; eyepiece: 9.09 mm.
41. 3.0 m.
43. $(4.0 \times 10^2)\times$.
45. $520\times$.
47. (a) $(9.00 \times 10^2)\times$;
   (b) eyepiece: 1.8 cm; objective, 0.300 cm;
   (c) 0.306 cm from objective.
49. (a) 14 cm;
   (b) $130\times$.
51. (a) Converging;
   (b) 281 cm.
53. $(1.54 \times 10^{-5})°$.
55. 1.7×10^{11} m.
57. 1.0×10^4 m.
59. 8.5 cm, 6.2×10^{-6} rad (distance of 2.4 km).
61. (a) 53.8°;
   (b) 0.19 nm.
63. (a) $1\times$;
   (b) $1\times$ (at back of body) to $2.7\times$ (at front of body).
65. $\frac{1}{6}$ s.
67. (a) 1.8×10^4 m;
   (b) $23''$; atmospheric effects and aberrations in the eye.
69. 0.82 mm.
71. 8.2.
73. -0.75 D (upper part), $+2.0$ D (lower part).
75. (a) $-2.8\times$;
   (b) $+5.5$-D lens.
77. $+3.9$ D.
79. 4.8×10^3 C°/min.
81. $-15\times$.
83. 0.4 m.
85. 110 m.
87. 2 m.
89. -8.4 cm.

Chapter 26

1. 83.9 m.
3. 35 ly.

5. $0.70c$.
7. (a) $0.141c$;
   (b) $0.140c$.
9. (a) 6.92 m, 1.35 m;
   (b) 13.9 s;
   (c) $0.720c$;
   (d) 13.9 s.
11. (a) 3.6 yr;
   (b) 7.0 yr.
13. $(6.97 \times 10^{-8})\%$.
15. $0.9716c$.
17. (a) -1.1%;
   (b) -34%.
19. $0.95c$.
21. 8.209×10^{-14} J, 0.512 MeV.
23. 1000 kg.
25. 5.36×10^{-13} kg.
27. 1.6 GeV/c.
29. (a) 4.5 GeV;
   (b) 5.4 GeV/c.
31. $0.866c$, 0.886 MeV/c.
33. 9.0×10^{13} J; 9.2×10^9 kg.
35. 1.30 MeV/c^2.
37. $0.333c$.
39. (a) $0.866c$;
   (b) $0.745c$.
41. (a) 1.8×10^{19} J;
   (b) -1.7%.
43. $\dfrac{2m}{\sqrt{1 - v^2/c^2}}$; 0; $\left(\dfrac{1}{\sqrt{1 - v^2/c^2}} - 1\right)2mc^2$.
47. $0.69c$.
49. (a) $0.959c$;
   (b) $0.36c$.
51. $0.962c$.
53. $0.3c$.
55. (a) $0.66c$;
   (b) 6.5 yr.
57. $0.91c$.
59. 1.022 MeV.
61. 0.79 MeV.
63. (a) 4×10^9 kg/s;
   (b) 4×10^7 yr;
   (c) 1×10^{13} yr.
67. 5.38×10^{-12} kg; $(1.5 \times 10^{-8})\%$.

69. (a)

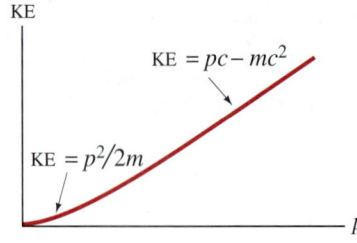

(b)

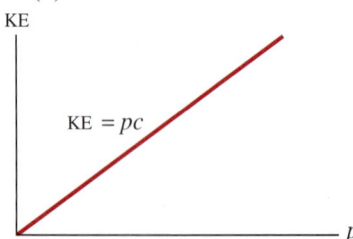

71. $0.981c$.
73. 1.7×10^{21} J; $\sim 20\times$ greater.
75. (a) 15 m;
    (b) 42 min.
77. (a) $0.986c$;
    (b) $(1 - 5 \times 10^{-7})c$.
79. Yes, in barn's reference frame, if his speed is $\geq 0.639c$; no, in boy's reference frame.
81. 8.0×10^{-8} s.

Chapter 27

1. 6.2×10^4 C/kg.
3. 5 electrons.
5. (a) 10.6 μm, far infrared;
   (b) 940 nm, near infrared;
   (c) 0.7 mm, microwave.
7. 5.4×10^{-20} J, 0.34 eV.
9. 9.35×10^{-6} m.
11. 2.7×10^{-19} J to 5.0×10^{-19} J, 1.7 eV to 3.1 eV.
13. 1.14×10^{-27} kg·m/s.
17. 7.2×10^{14} Hz.
19. 429 nm.
21. (a) 2.3 eV;
    (b) 0.85 V.
23. 0.92 eV; 5.7×10^5 m/s.
25. (a) 0.78 eV;
    (b) no ejected electrons.
27. 3.32 eV.
31. (a) 2.43×10^{-12} m;
    (b) 1.32×10^{-15} m.
33. 2.62 MeV.
35. 212 MeV; 5.86×10^{-15} m.
37. 1.592 MeV, 7.81×10^{-13} m.
39. 4.7×10^{-12} m.

41. 1840.
43. (a) 4×10^{-10} m;
    (b) 1×10^{-10} m;
    (c) 3.9×10^{-11} m.
45. 6.4×10^{-12} m; yes; much less than 5 cm; no.
47. 22 V.
49. (a) Absorption; final state; largest energy photon;
    (b) emission, initial state;
    (c) absorption, final state.
51. $n = 5$ to $n' = 3$.
53. (a) 486 nm;
    (b) 103 nm;
    (c) 434 nm.
55. 91.2 nm.
57. Yes; from $n = 1$ to $n = 3$.
59.

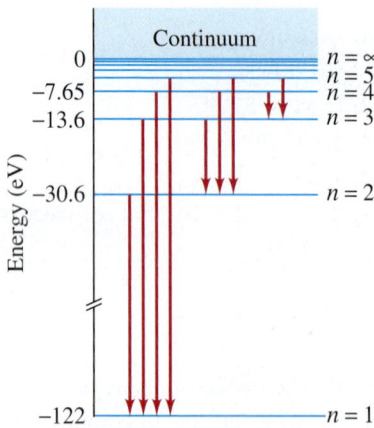

61. -0.544 eV.
63. Yes; $(7.30 \times 10^{-3})c$, $1/\gamma = 0.99997$.
65. 1.20×10^{29} m, -4.22×10^{-97} J.
67. 3.28×10^{15} Hz.
69. 4.4×10^{26} photons/s.
71. 5.5×10^{18} photons/s.
73. (a) 1.7 eV;
    (b) 3.0 eV.
75. 0.34 MeV for both.
77. 4.0×10^{-14} m.
79. 4.40×10^{-40}; yes.
81. 0.32 V.
83. 4.0×10^{-7} W/m²; 1.7×10^{-2} V/m.
85. 8.2×10^{-10} m.
87. 3.5×10^{-12} m.
89. 1200 m/s.
91. Paschen series, level 4.
93. 1.8×10^{11} C/kg.

95. $E_n = -\dfrac{2.84 \times 10^{165} \text{ J}}{n^2}$,
    $r_n = n^2(5.16 \times 10^{-129} \text{ m})$;
    not apparent.

Chapter 28

1. 4.5×10^{-7} m.
3. 5.3×10^{-11} m.
5. 7×10^{-8} eV.
7. 3.00×10^{-10} eV/c^2.
9. Electron: 1.5×10^{-3} m; baseball: 9.7×10^{-33} m; $\Delta x_{\text{electron}}/\Delta x_{\text{baseball}} = 1.5 \times 10^{29}$.
11. 7700 m/s; 1.7×10^{-4} eV.
13. 5.53×10^{-10} m.
15. $\ell = 0, 1, 2, 3, 4, 5$.
17. 14 electrons.
19. (a) $(1, 0, 0, -\tfrac{1}{2}), (1, 0, 0, +\tfrac{1}{2})$,
    $(2, 0, 0, -\tfrac{1}{2}), (2, 0, 0, +\tfrac{1}{2})$,
    $(2, 1, -1, -\tfrac{1}{2}), (2, 1, -1, +\tfrac{1}{2})$;
    (b) $(1, 0, 0, -\tfrac{1}{2}), (1, 0, 0, +\tfrac{1}{2})$,
    $(2, 0, 0, -\tfrac{1}{2}), (2, 0, 0, +\tfrac{1}{2})$,
    $(2, 1, -1, -\tfrac{1}{2}), (2, 1, -1, +\tfrac{1}{2})$,
    $(2, 1, 0, -\tfrac{1}{2}), (2, 1, 0, +\tfrac{1}{2})$,
    $(2, 1, 1, -\tfrac{1}{2}), (2, 1, 1, +\tfrac{1}{2})$,
    $(3, 0, 0, -\tfrac{1}{2}), (3, 0, 0, +\tfrac{1}{2})$,
    $(3, 1, -1, -\tfrac{1}{2})$.
21. $\sqrt{12}\hbar$, or 3.65×10^{-34} J·s.
23. $n \geq 4$; $3 \leq \ell \leq n - 1$; $m_s = -\tfrac{1}{2}, +\tfrac{1}{2}$.
25. (a) $1s^2 2s^2 2p^6 3s^2 3p^6 3d^6 4s^2$;
    (b) $1s^2 2s^2 2p^6 3s^2 3p^6 3d^{10} 4s^2 4p^4$;
    (c) $1s^2 2s^2 2p^6 3s^2 3p^6 3d^{10} 4s^2 4p^6 5s^2$.
27. (a) 5;
    (b) -0.544 eV;
    (c) $\sqrt{6}\hbar$, 2;
    (d) $-2, -1, 0, 1, 2$.
31. $n = 3$, $\ell = 2$.
33. 4.36×10^{-2} nm; no long wavelength cutoff—intensity just gets smaller and smaller.
37. 1.798×10^{-10} m.
39. 6.12×10^{-11} m; partial shielding by $n = 2$ shell.
41. 0.017 J, 5.5×10^{16} photons.
43. 634 nm.
45. $n \geq 7$, $\ell = 6$, $m_\ell = 2$.
47. $L_{\max} = \sqrt{30}\,\hbar$, $L_{\min} = 0$.
49. 5.3×10^{-27} kg·m/s, 1.3×10^{-7} m.
51. 5.75×10^{-13} m, 115 keV.
53. 22 electrons.
57. $L_{\text{Bohr}} = 2\hbar$; $L_{\text{qm}} = 0$ or $\sqrt{2}\,\hbar$.
59. 3.7×10^{-37} m.
61. $\Delta p_{\text{electron}}/\Delta p_{\text{proton}} = 0.0234$.

63. 2.0×10^{-35} m/s; yes; 10^{34} s.
65. Copper.

Chapter 29

1. 5.1 eV.
3. HN: 110 pm, CN: 150 pm, NO: 133 pm.
5. 4.6 eV.
9. 1.10×10^{-10} m.
11. 5.22×10^{-4} m.
13. (a) 680 N/m;
    (b) 2.1×10^{-6} m.
15. 0.315 nm.
17. 2.0 eV.
19. $\lambda \leq 1.7\ \mu$m.
21. 1.2×10^6 electrons.
23. 3×10^6.
25. 1.7 eV.
27. 14 V.
29. 7.3 mA.
31. (a) 5.1 mA;
    (b) 3.6 mA.
33.

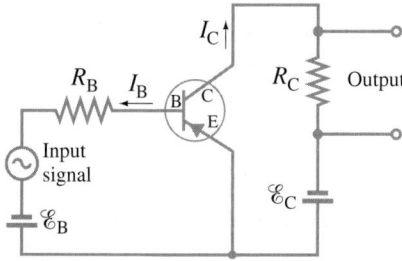

35. 0.65 V.
37. (a) 130;
    (b) 8700.
39. (a) 3.1×10^4 K;
    (b) 930 K.
41. (a) -5.3 eV;
    (b) 4.4 eV.
43. 1.94×10^{-46} kg·m².
45. 6.03×10^{-4} eV.
47. $\lambda \leq 3.5 \times 10^{-7}$ m.
49. Yes; 1.11×10^{-6} m.
51. 980 J/mol.

Chapter 30

1. 0.149 u.
3. 0.855%.
5. 3727 MeV/c^2.
7. (a) 180 m;
   (b) 3.5×10^4.
9. 30 MeV.
11. 6.0×10^{26} nucleons; no; all nucleons have about the same mass.
13. 550 MeV.
15. 7.699 MeV/nucleon.
17. 12.42 MeV.
19. $^{23}_{11}$Na: 8.113 MeV/nucleon; $^{24}_{11}$Na: 8.063 MeV/nucleon.
21. (b) Yes, binding energy is positive.
25. 0.782 MeV.
27. (a) β^- emitter;
    (b) $^{24}_{11}$Na $\rightarrow$ $^{24}_{12}$Mg + β^- + $\bar{\nu}$, 5.515 MeV.
29. 2.822 MeV.
31. α: 6.114 MeV; β^-: 0.259 MeV.
33. 10.8 MeV.
35. For both: $\text{KE}_{max} = 0.9612$ MeV, $\text{KE}_{min} = 0$.
37. $\text{KE}_{recoil} = 0.0718$ MeV, $Q = 4.27$ MeV.
39. 1.2 h.
41. 2.5×10^9 decays/s.
43. (a) 3.60×10^{12} decays/s;
    (b) 3.58×10^{12} decays/s;
    (c) 1.34×10^9 decays/s.
45. 7 α particles; 4 β^- particles.
47. 0.91 g.
49. (a) 1.38×10^{-13} s^{-1};
    (b) 3.73×10^7 decays/min.
51. 86 decays/s.
53. 1.78×10^9 yr.
55. 7900 yr.
57. 15.8 d.
59. $N_D = N_0\left(1 - e^{-\lambda t}\right)$.
61. 2.6×10^4 yr.
63. 41 yr.
65. 3.0×10^{-14} m, $4.2\times$ nuclear radius.
67. $^{40}_{19}$K: 0.16 mg; $^{39}_{19}$K: 1.2 g.
69. 2.71×10^{-11} g.
71. (a) 4.6×10^{15} decays/s;
    (b) 1.2×10^4 decays/s.
73. (a) 0.002603 u, 2.425 MeV/c^2;
    (b) 0, 0;
    (c) -0.090739 u, -84.52 MeV/c^2;
    (d) 0.043930 u, 40.92 MeV/c^2;
    (e) $\Delta \geq 0$ for $0 \leq Z \leq 8$ and $Z \geq 85$;
    $\Delta < 0$ for $9 \leq Z \leq 84$;
    $\Delta \geq 0$ for $0 \leq A \leq 15$ and $A \geq 218$;
    $\Delta < 0$ for $16 \leq A < 218$.
75. 0.2 decays/s.
77. (a) 180 m;
    (b) 13 km.
79. (a) There would be no atoms—just neutrons;
    (b) 0.083%.
81. 1.1×10^{11} yr.
83. 13 decays/s.
85. (b) 98.2%.

Chapter 31

1. $^{28}_{13}$Al; β^-; $^{28}_{14}$Si.
3. Yes, $Q > 0$.
5. 18.000937 u.
7. (a) Yes;
   (b) 20.4 MeV.
9. 4.730 MeV.
11. n + $^{14}_{7}$N $\rightarrow$ $^{14}_{6}$C + p, 0.626 MeV.
13. (a) He picks up a neutron from C;
    (b) $^{11}_{6}$C;
    (c) 1.856 MeV; exothermic.
15. 5.702 MeV released.
17. 126.5 MeV.
19. 1/1100.
21. (a) 5 neutrons;
    (b) 171.1 MeV.
23. 1100 kg.
25. 260 MeV; about 30% > fission energy released.
27. 3000 eV.
31. (a) a: 6.03×10^{23} MeV/g;
    b: 4.89×10^{23} MeV/g;
    c: 2.11×10^{24} MeV/g;
    (b) 5.13×10^{23} MeV/g;
    a: 0.851; b: 1.05; c: 0.243.
33. 0.39 g.
35. 5.6×10^3 kg/h.
37. 2.46×10^9 J; $50\times$ > gasoline.
39. 7000 rads.
41. 144 rads.
43. (a) 1.0 rad, or 0.010 Gy;
    (b) 1.0×10^{10} protons.
47. 5.61×10^{-10} kg.
49. 9×10^{16} e$^-$.
51. (a) 2×10^{-7} Sv/yr, 2×10^{-5} rem/yr; $(2 \times 10^{-4}) \times$ allowed dose;
    (b) 2×10^{-6} Sv/yr, 2×10^{-4} rem/yr; $(2 \times 10^{-3}) \times$ allowed dose.
53. 7.041 m; radio wave.
55. (a) $^{12}_{6}$C;
    (b) 5.702 MeV.
57. 1.0043 : 1.
59. 6.6×10^{-2} rem/yr.

61. (a) $^{226}_{88}\text{Ra} \rightarrow {}^{4}_{2}\text{He} + {}^{222}_{86}\text{Rn}$;
    (b) 4.871 MeV;
    (c) 190.6 MeV/c for both;
    (d) 8.78×10^{-2} MeV.
63. (a) 3.7×10^{26} W;
    (b) 3.5×10^{38} protons/s;
    (c) 1.1×10^{11} yr.
65. 8×10^{12} J.
67. (a) 3700 decays/s;
    (b) 4.0×10^{-4} Sv/yr; 11% of background.
69. 7.274 MeV.
71. 990 kg.
73. (a) $^{141}_{54}\text{Xe}$;
    (b) 2 neutrons escape or are absorbed, 1 causes another fission;
    (c) 176.0 MeV.
75. Most to least dangerous: C > B > A.

Chapter 32

1. 5.59 GeV.
3. 1.3×10^{7} Hz.
5. 2.0 T.
7. Alpha particles; $\lambda_\alpha \approx d_{\text{nucleon}}$, $\lambda_p \approx 2d_{\text{nucleon}}$.
9. 0.9 MeV/rev.
13. 33.9 MeV.
15. 1879.2 MeV.
17. 67.5 MeV.
19. 1.32×10^{-15} m.
21. 1.3×10^{-12} m.
23. First, second, and fourth will not happen.
25. (a) 37.8 MeV;
    (b) $KE_p = 5.4$ MeV, $KE_{\pi^-} = 32.4$ MeV.
27. 9×10^{-3} MeV.
29. 7.5×10^{-21} s.
31. (a) 1300 eV;
    (b) 150 MeV.
33. 8×10^{-27} s.
35. (a) udd;
    (b) $\bar{u}\bar{d}\bar{d}$;
    (c) uds;
    (d) $\bar{u}\bar{d}\bar{s}$.
37. $c\bar{u}$.
39.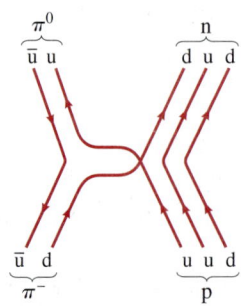
41. 16 GeV; 7.8×10^{-17} m.
43. (a) 1.022 MeV;
    (b) 1876.6 MeV.
45. 1.2×10^{10} m, 4.0×10^{1} s.
47. (a) Possible, strong interaction;
    (b) possible, strong interaction;
    (c) possible, strong interaction;
    (d) not possible; charge is not conserved;
    (e) possible, weak interaction.
49. 10^{-18} m corresponds to 200 GeV.
51. 64 fundamental fermions.
53. (b) 10^{29} K.
55. 798.7 MeV.
57. 9.3×10^{10} eV; 1.3×10^{-17} m, 0.99995c.
59. (a) π^0;
    (b) $\bar{\nu}_\mu$.
63. 10^{-25} s.

Chapter 33

1. 3.1 ly.
3. Less than; by a factor of 2.
5. 2×10^{-3} kg/m^3.
7. 4.2×10^{-2} rad, or 2.4°; about 4.5× Moon's width.
9. 1.83×10^{9} kg/m^3; 3.33×10^{5} times larger.
11. (a) 0.018″;
    (b) (5.0×10^{-6})°.
13. 280 yr.
15. $D_1/D_2 = 0.13$.
17. 3×10^{14} m.
19. 540°.
21. 3.1×10^{-16} m.
23. 2.1×10^{8} ly.
25. (a) 0.3 nm;
    (b) 3.2 nm.
27. 0.058c.
29. 9×10^{9} ly.
31. 6.8×10^{7} ly.
33. 1.1×10^{-3} m.
35. 6 nucleons/m^3.
37. (a) 10^{-5} s;
    (b) 10^{-7} s;
    (c) 10^{-4} s.
39. 0.2 rev/s.
41. A: T increases, L doesn't change, size decreases.
    B: T unchanged, L decreases, size decreases.
    C: T decreases, L increases, size increases.
43. 1.7×10^{25} W.
45. 400 ly; $r_{\text{Sun}}/d_{\text{Earth-Sun}} = 2 \times 10^{7}$, $r_{\text{Sun}}/d_{\text{Galaxy}} = 4 \times 10^{-3}$.
47. (a) (9×10^{-6})°;
    (b) 4 m.
49. (a) 13.93 MeV;
    (b) 4.71 MeV;
    (c) 5.46×10^{10} K.
51. $t_P = 5.38 \times 10^{-44}$ s.

Index

Note: The abbreviation *defn* means the page cited gives the definition of the term; *fn* means the reference is in a footnote; *pr* means it is found in a Problem or Question; *ff* means "also the following pages."

A (atomic mass number), 858
Aberration:
   chromatic, 725 *fn*, 728, 732
   of lenses, 727–28, 729, 731
   spherical, 650, 727, 728, 732
Absolute pressure, 264
Absolute space, 746, 748
Absolute temperature scale, 362, 368
Absolute time, 746
Absolute zero, 368, 424
   kinetic energy near, 376
Absorbed dose, 900
Absorption lines, 692–93, 787, 838
Absorption spectrum, 692–93, 787, 838
Absorption wavelength, 793
Abundances, natural, 858
Ac (*defn*), 514
Ac circuits, 514–15, 526 *fn*, 611–16
Ac generator, 597, 599
Ac motor, 577
Accademia del Cimento, 361
Accelerating reference frames, 77, 80, A-16–A-18
Acceleration, 26–38, 40, 58–63
   angular, 201–4, 208–12
   average, 26–27, 40
   centripetal, 110 *ff*
   constant, 28–38
   constant angular, 203–4
   Coriolis, A-17–A-18
   cosmic, 976–77
   of expansion of the universe, 976–77
   as a function of time (SHM), 301
   in g's, 37
   due to gravity, 33–38, 58–63, 79 *fn*, 84, 121–22
   instantaneous (*defn*), 26, 40
   of the Moon, 112, 119
   motion at constant, 28–38, 58–64
   radial, 110 *ff*, 118
   related to force, 78–80
   of simple harmonic oscillator, 301
   and slope, 40
   tangential, 118, 201–3
   uniform, 28–38, 58–64
   velocity vs., 27
Accelerators, particle, 916–21

Accelerometer, 92
Acceptor level, 845
Accommodation of eye, 719
Accuracy, 8
   precision vs., 8
Achromatic doublet, 728
Achromatic lens, 728
Actinides, 817
Action at a distance, 119
Action potential, 518–19
Action–reaction (Newton's third law), 81–83
Activation, on an LCD screen, 491
Activation energy, 377, 833, 834
Active galactic nuclei (AGN), 951
Active matrix, 492
Active solar heating, 435
Activity, 870
   and half-life, 872
   source, 900
ADC, 488–89
Addition of vectors, 50–57, 87, 450
Addition of velocities:
   classical, 65–66
   relativistic, 764
Addressing pixels, 491–92
Adenine, 460
Adhesion (*defn*), 281–82
Adiabatic processes, 415–16
ADP, 833
AF signal, 637
AFM, 786
AGN, 951
Air circulation, underground, 278
Air columns, vibrations of, 337–40
Air conditioners, 425–27
Air gap, 694
Air pollution, 434–35
Air resistance, 33
Airplane:
   noise, 333
   wing, 277
Airy disk, 729
Algebra, review of, A-3–A-6
Alkali metals, 817
Allowed transitions, 814, 838–39
Alpenhorn, 358 *pr*

Alpha decay, 864–66, 869
   and tunneling, 876
Alpha particle (or ray), 786–87, 864–66
Alternating current (ac), 514–15, 526 *fn*, 611–16
Alternators, 598
AM, 737
AM radio, 637
Amino acids, 836–37
Ammeter, 546–48, 576
   digital, 546, 548
   connecting, 547
   resistance, effect of, 547–48
AMOLED, 849
Amorphous solids, 840
Ampère, André, 504, 573
Ampere (A) (unit), 504, 572
   operational definition of, 572
Ampere-hour (A·h) (unit), 505
Ampère's law, 573–75, 626–27
Amplifiers, 850–51
Amplitude, 294, 306, 319
   intensity related to, 333
   of vibration, 294
   of wave, 294, 306, 310, 319, 333, 804–6
Amplitude modulation (AM), 637
Analog meters, 546–48, 576
Analog signals, 488–89, 604
Analog-to-digital converter (ADC), 488–89
Analyzer (of polarized light), 700
Anderson, Carl, 924
Andromeda, 950, 983 *pr*
Aneroid barometer, 266
Aneroid gauge, 266
Angle, 11 *fn*, 199
   attack, 277
   Brewster's, 702, 710 *pr*
   critical, 659
   of dip, 562
   of incidence (*defn*), 313, 317, 645, 657
   phase, 615
   polarizing, 702
   radian measure of, 199–200
   of reflection (*defn*), 313, 645
   of refraction, 317, 657
   solid, 11 *fn*
Angstrom (Å) (unit), 20 *pr*, 685 *fn*

Angular acceleration, 201–4, 208–12
   average, 201
   constant, 203–4
   instantaneous, 201
Angular displacement, 200, 302
Angular magnification, 722
Angular momentum, 215–18, 789, 795
   in atoms, 789, 812–14
   conservation, law of, 215–17, 869
   quantized in atoms, 812–13
   quantized in molecules, 837–38
   vector, 218
Angular position, 199
Angular quantities, 199 *ff*
   vector nature, 217–18
Angular velocity, 200–3
   average, 200
   instantaneous, 200
Animals, and sound waves, 309
Anisotropy of CMB, 968, 977
Annihilation of particles, 781, 925, 971–72
Anode, 490
Antenna, 627–28, 631, 638
Anthropic principle, 979
Antiatoms, 925
Anticodon, 836
Antilock brakes, 116
Antilogarithm, A-3, A-11
Antimatter, 925, 941, 943 *pr*
   (*see also* Antiparticle)
Antineutrino, 867–68, 930
Antineutron, 925
Antinodes, 315, 337, 338, 339
Antiparticle, 868, 924–26, 930–31 (*see also* Antimatter)
Antiproton, 924–25, 934
Antiquark, 930–31, 934–35, 936
Apparent brightness, 951–52, 958
Apparent weight, 124–25, 270
Apparent weightlessness, 124–25
Approximations, 8, 13–15
Arago, F., 687
Archeological dating, 875
Arches, 246–49
Archimedes, 268–70
Archimedes' principle, 268–72
   and geology, 272

A-43

Area, 12, A-1, inside back cover
  under a curve or graph, 142, 176–77
Arecibo radiotelescope, 643 pr, 731
Aristotle, 2, 76
Armature, 577, 597
Arteriosclerosis, 280
Artificial radioactivity, 863
ASA number, 715 fn
Asteroids, 135 pr, 136 pr, 196 pr, 228 pr
Astigmatism, 720, 727
Astronomical telescope, 650, 724–25, 743 pr
Astrophysics, 916, 947 ff
Asymptotic freedom, 937
ATLAS, 921, 939
Atmosphere (atm) (unit), 264
Atmosphere, scattering of light by, 704
Atmospheric pressure, 264, 266–67
  decrease with altitude, 264
Atom, model of, 445
Atom trap, 802 pr
Atomic bomb, 893–94, 897
Atomic density, 859
Atomic emission spectra, 692–93, 787–89
Atomic force microscope (AFM), 786
Atomic mass, 360
Atomic mass number (A), 858
Atomic mass unit, unified, 10, 360, 860
Atomic number (Z), 815, 817–19, 858
Atomic spectra, 787–89, 792–94
Atomic structure:
  Bohr model of, 789–95, 803–4, 811–12
  of complex atoms, 816–17
  early models of, 786–87
  of hydrogen atoms, 812–14
  of multielectron atoms, 816–17
  nuclear model of, 787
  planetary model of, 787
  quantum mechanics of, 803–24
  shells and subshells in, 816–17
Atomic theory of matter, 359–61, 443
Atomic weight, 360
Atomizer, 277
Atoms, 359–61, 372–77, 786–96, 916 fn
  angular momentum in, 789, 812–14
  binding energy in, 791
  Bohr model of, 789–95
  as cloud, 811
  complex, 815–17
  crystal lattice of, 840

and de Broglie's hypothesis, 795–96
  distance between, 361
  electric charge in, 445
  energy levels in, 789–95, 812–13, 815–16, 818
  hydrogen, 787–96, 812–14
  ionization energy in, 791–94
  multielectron, 815–17
  neutral, 860
  probability distributions in, 805, 811
  quantum mechanics of, 803–24
  shells and subshells in, 816–17
  (see also Atomic structure; Kinetic theory)
ATP, 833
Attack angle, 277
Attractive forces, 832–33, 922
Atwood machine, 91, 225 pr
Audible range, 329
Audible sound, frequency of, 329, 334–35
Audiofrequency (AF) signal, 637
Aurora borealis, 569
Autofocusing camera, 330
Autoradiography, 904
Average acceleration (defn), 26–27, 40
Average angular acceleration, 201
Average angular velocity, 200
Average speed (defn), 23–24, 376
Average velocity (defn), 23–25, 28, 39
Avogadro, Amedeo, 372
Avogadro's hypothesis, 372
Avogadro's number, 372–73
Axis of lens, 661
Axis of rotation (defn), 199
  forces tilting, 208
Axon, 517–19

Back, forces in, 238–39
Back emf, 599–600
Background radiation, cosmic microwave, 967–70, 973, 974, 977–78
Backlight, 491
Bacterium, 785
Bainbridge-type mass spectrometer, 578
Balance, human, 240
Ballistic galvanometer, 624 pr
Ballistic pendulum, 181
Balloons:
  helium, 272, 371
  hot-air, 43 pr, 359, 389 pr
Balmer, J. J., 788
Balmer formula, 789, 792
Balmer series, 788, 792–93
Band gap, 842–43
Band spectra, 837

Band theory of solids, 842–43
  and doped semiconductors, 845
Bandwidth, 489
Banking of curves, 115–17
Bar (unit), 264, 267
Bar codes, 822
Barometer, 266–67
Barrel distortion, 728
Barrier, Coulomb, 876, 954–55
Barrier tunneling, 876–77
Baryon, 930–31, 935, 936–38, 976–78
  decay, 931
  and quark theory, 934–35
Baryon number, 926–27, 930–31, 933–35, 940, 971
  conservation of, 927
Base, of transistor, 850
Base bias voltage, 850
Base quantities, 10
Base semiconductor, 850
Base units (defn), 10
Baseball, 68 pr, 70 pr, 74 pr, 102 pr, 138, 220 pr, 278
  position uncertainty, 809
Baseball curve, and Bernoulli's principle, 278
Bases, nucleotide, 460–61, 834–36
Basketball, 73 pr, 102 pr
Battery, 476, 502–3, 505, 507, 527
  automobile, charging, 536–37, 604
  chargers, inductive, 604
  electric, 502–3
  jump starting, 536–37
  lithium-ion, 504
  rechargeable, 604
  symbol, 504, 526
  voltaic, 502
Bayer mosaic, 714, 717
Beam splitter, 698
Beams, 235–36, 245–47
Bear sling, 105, 252
Beat frequency, 343
Beats, 342–43
Becquerel, Henri, 863
Becquerel (Bq) (unit), 900
Bel (unit), 331
Bell, Alexander Graham, 331
Benzene ring, 849
Bernoulli, Daniel, 274–75
Bernoulli's equation, 274–79
Bernoulli's principle, 274–78
Beta decay, 863–64, 866–68, 869, 873, 938
  inverse, 956
Beta particle (or ray), 864, 866
  (see also Electron)
Bethe, Hans, 896
Biasing and bias voltage, 845–46, 850–51
Biceps, and torque, 238, 255 pr
Bicycle, 205, 218, 227 pr, 229 pr
Bicycle gears, 227 pr

Big Bang theory, 801 pr, 941, 947, 964, 966–79
Big crunch, 975
Bimetallic strip, 362
Bimetallic-strip thermometer, 362
Binary code, 488, 604
Binary numbers, 488
Binary system, 956–57, 963
Binding energy:
  in atoms, 791
  for iron, 862
  in molecules, 830, 832–33, 834–35
  of nuclei, 860–62
  in solids, 840
  total, 769 pr, 861
Binding energy per nucleon (defn), 861, 894
Binoculars, 660, 725
Binomial expansion, 753, 755, 763, A-1, A-6–A-7, inside back cover
Biochemical analysis by spectroscopy, 693
Biological cells, 460–62
Biological damage by radiation, 899
Biological development, and entropy, 430–31
Bipolar junction transistor, 850–51
Birth and death of stars, 954–57
Bismuth-strontium-calcium-copper oxide (BSCCO), 517
Bit depth, 488–89
Bit-line, 605
Bits, 488, 604
Black holes, 136 pr, 951, 956, 962–63, 975
Blackbody, 774
Blackbody radiation, 774, 952, 968
Blinking flashers, 542
Blood flow, 274, 280, 282–83, 288 pr
  convection by, 402
  rate, 584 pr
  TIAs and, 278
Blood-flow measurement, electromagnetic, 596
Blood-flow meter, Doppler, 347, 358 pr
Blood pressure, measuring, 280, 283, 288 pr, 289 pr, 290 pr
Blood transfusion, 288 pr, 289 pr
Blue sky, 704
Blueshift, 965
Body, human:
  balance, 240
  energy in, 418–19
  heat loss from, 402, 404
  metabolism, 418–19

A-44 Index

Body, human (*continued*)
(*see also* Muscles and joints;
specific topics)
Body fat, 287 *pr*
Body parts, CM of, 186–87
Body temperature, 363, 400
Bohr, Niels, 782, 789, 795, 803,
 811, 867
Bohr model of atom, 789–95,
 803–4, 811, 812
Bohr radius, 790, 811
Bohr theory, 803, 811, 812
Boiling, 380 (*see also* Phase,
 changes of)
Boiling point, 362, 380,
 397 table
Boltzmann, Ludwig, 432
Boltzmann constant, 372
Bomb:
 atomic, 893–94, 897
 fission, 893–94
 fusion, 897
 hydrogen, 897
Bomb calorimeter, 396, 409 *pr*
Bond (*defn*), 829–31
 covalent, 830, 831–32, 840
 dipole–dipole, 834
 dipole–induced dipole, 834
 hydrogen, 460, 461, 834–37
 ionic, 831, 833, 840
 metallic, 840
 molecular, 829–32
 partially ionic and covalent,
 831–32
 in solids, 840–41
 strong, 829–32, 834–35, 840
 van der Waals, 834–37
 weak, 460, 461, 834–37, 840
Bond energy, 830, 834–35
Bond length, 854 *pr*
Bonding:
 in molecules, 829–32
 in solids, 840–41
Bone density, measuring, 780
Born, Max, 805
Bose, Satyendranath, 816
Bosons, 816, 930–31, 935–36
Bottom quark, 931 *fn*, 934–35
Boundary layer, 278
Bow wave, 348–49
Boyle, Robert, 368
Boyle's law, 368, 374
Braces, and forces on teeth,
 231
Bragg, W. H., 734
Bragg, W. L., 734
Bragg equation, 734
Bragg peak, 904
Bragg scattering of X-rays, 824
Brahe, Tycho, 125
Brakes:
 antilock, 116
 hydraulic, 265
Braking a car, 32, 116, 145
 LED lights to signal, 848
Branes, 942

Break-even (fusion), 898
Breakdown voltage, 477
Breaking point, 241
Breaking the sound barrier, 349
Breath, molecules in, 373
Breeder reactor, 892
Bremsstrahlung, 819
Brewster, D., 702
Brewster's angle and law, 702,
 710 *pr*
Bridge circuit, 556 *pr*
Bridge collapse, 304
Bridge-type full-wave rectifier,
 853 *pr*
Brightness, apparent, 951–53,
 958
British engineering system of
 units, 10
Broglie, Louis de, 782, 795–96
Bronchoscope, 660
Brown, Robert, 360
Brownian motion, 360
Brunelleschi, Filippo, 248
Brushes, 577, 597
BSCCO, 517
Btu (unit), 391
Bubble chamber, 878, 925
Building dampers, 303
Building materials, *R*-values
 for, 402, 410 *pr*
Bulk modulus, 242, 244–45, 309
Buoyancy, 268–72
 center of, 291 *pr*
Buoyant force, 268–72
Burglar alarms, 778
Burning (= fusion), 954 *fn*
Burning out, motor, 600
Burns, 487
Bytes, 488

Cable television, 639
Calculator errors, 7
Caloric, 391
Calories (cal) (unit), 391
 measuring, 396
 relation to joule, 391
 working off, 392
Calorimeter, 396, 409 *pr*,
 877–78
Calorimetry, 394–400
Camera, digital and film,
 713–18
 adjustments, 715
 autofocusing, 330
 gamma, 905
 pinhole, 742 *pr*
Camera flash unit, 486–87
Cancer, 899, 903–4, 917
Candles, standard, 957, 958
Cantilever, 235
Capacitance, 482–87
 derivation of, 485
Capacitive reactance, 613
Capacitor charging, 539–41
Capacitor discharging, 541–42,
 627

Capacitors, 482–87, 612–13,
 847, 851
 in circuits, 538–43, 613
 energy stored in, 486–87
 as filters, 613
 as power backup, 484
 reactance of, 613
 with R or L, 539–43, 610 *ff*
 in series and in parallel,
 538–39
 symbol, 526
 uses of, 482, 613
Capacity, 482–87
Capillaries, 274, 282
Capillarity, 281–82
Capture, electron, 868
Car (*see also specific parts*):
 battery charging, 536–37
 brake lights, 848
 efficiency, 422
 electric, 504, 524 *pr*, 604
 flywheel, 229 *pr*
 ignition system, 602
 jump starting, 536–37
 power needs, 160
 skidding, 116
 starter, 573
 stopping of, 32, 145
 windshield wipers,
 intermittent, 543
Carbon (CNO) cycle, 896,
 912 *pr*, 914 *pr*
Carbon dating, 874–76
Carbon dioxide emissions,
 offset of, 442 *pr*
Carbon footprint, 434
Carbon isotopes, 858
Carnot, N. L. Sadi, 422
Carnot cycle, 423
Carnot efficiency, 422–24
 and second law of
 thermodynamics, 422–24
Carnot engine, 422–24
Carousel, 198, 201, 202
Carrier frequency, 637
Carrier of force, 916, 922–24,
 936
Carry, forces, 922
Caruso, Enrico, 304
Cassegrainian focus, 725
CAT scan, 735–36, 905–6
Catalysts, 834
Cathedrals, 247–48
Cathode, 490
Cathode ray tube (CRT), 490
Cathode rays, 490, 772–73 (*see
 also* Electron)
Causal laws, 128
Causality, 128
Cavendish, Henry, 120–21
CCD, 714
CD player, 822
CDs, 20 *pr*, 43 *pr*, 48 *pr*, 226 *pr*,
 489, 605, 822
CDF, 878

CDM model of universe,
 977–78
CDMA cell phone, 639
Celestial sphere, 125
Cell:
 electric, 505, 527
 solar (photovoltaic), 435,
 556 *pr*, 847
Cell (biological), 460–62
 ATP and energy in, 833–34
 pressure on, 264
 radiation taken up by, 900
Cell phone, 318, 602, 604, 631,
 639
Celsius temperature scale, 362
Center of buoyancy, 291 *pr*
Center of gravity (CG), 186,
 240
Center of mass (CM), 184–89
 for human body, 186
 and moment of inertia, 209,
 213–14
 and sport, 153
 and statics, 233
 and translational motion,
 187–89
Centi- (prefix), 10
Centigrade temperature scale,
 362
Centiliter (cL) (unit), 10
Centimeter (cm) (unit), 10
Centipoise (cP) (unit), 279
Centrifugal (pseudo) force,
 113, A-16–A-17
Centrifugal pump, 282
Centrifuge, 204
Centripetal acceleration, 110 *ff*
Centripetal force, 112
Cepheid variables, 958, 980 *pr*
CERN, 919, 920, 924, 939, 942
CG (center of gravity), 186, 240
Cgs system of units, 10
Chadwick, James, 858, 913 *pr*
Chain reaction, 890–91, 893–94
Chamberlain, Owen, 925
Chandrasekhar limit, 955
Change of phase (or state),
 377–81, 397–400
Change in a quantity, 23
Characteristic expansion time,
 967
Characteristic X-rays, 818
Charge-coupled device (CCD),
 714
Charge density, 465
Charging, capacitor, 539–41
Charging a battery, 536–37
Charging by conduction,
 446–47
Charging by induction, 446–47,
 604
Charles, Jacques, 368
Charles's law, 368
Charm, 931 *fn*, 933–35
Charm quark, 934
Chart of the Nuclides, 857

Index **A-45**

Chemical analysis by spectroscopy, 693
Chemical bonds, 829–37
Chemical lasers, 822
Chemical reactions, rate of, 377
Chemical shift, 907
Chernobyl, 892, 912 *pr*, 913 *pr*
Chimney, and Bernoulli effect, 278
Chip, computer, 19 *pr*, 829, 845, 851
Cholesterol, 280
Chord, 39, 200 *fn*
Chromatic aberration, 725 *fn*, 728, 732
Chromodynamics, quantum (QCD), 923, 937–39
Chromosomes, 460
Circle of confusion, 716
Circle of least confusion, 727
Circuit, digital, 851
Circuit, electric (*see* Electric circuits)
Circuit breaker, 512–13, 544–45, 573, 607
Circular apertures, 728–30
Circular motion, 110–18
 nonuniform, 118
 and simple harmonic motion, 299–300
 uniform, 110–15
Circular standing wave, as electron wave, 795–96
Circulating pump, 282
Classical physics (*defn*), 2, 745, 804
Clausius, R. J. E., 420, 428
Clausius statement of second law of thermodynamics, 420, 423 *pr*, 425
Clock, pendulum, 302
Cloning, 462
Closed system (*defn*), 394
Closed tube, 338
Clothing:
 dark vs. light, 403
 insulating properties of, 401
Cloud, electron, 811, 814, 830–32
Cloud chamber, 182, 878
Cloud color, 704
Clusters, of galaxies, 950, 974, 977
 of stars, 950
CM, 184–89 (*see also* Center of mass)
CMB, 967–70, 973, 974, 977–78
CMB anisotropy, 968, 977
CMB uniformity, 974
CMOS, 714, 921
CNO cycle, 896, 912 *pr*, 914 *pr*
CO molecule, 838–39
Coal, energy in, vs. uranium, 893
Coating of lenses, optical, 697–98

Coaxial cable, 586 *pr*, 631
COBE, 968, 970
Codon, 836
Coefficient:
 of kinetic friction, 93–94
 of linear expansion, 364–65
 of performance (COP), 426, 427
 of static friction, 93–94
 of thermal expansion, 364
 of viscosity, 279
 of volume expansion, 364, 366
Coherence, 685
Coherent light, 685, 820–23
Cohesion, 281–82
Coil (*see* Inductor)
Cold dark matter (CDM) model of universe, 977–78
Collector (of transistor), 850
Collider Detector at Fermilab (CDF), 878
Colliding beams, 920–21
Collimated beam, 735, 905 *fn*
Collimated gamma-ray detector, 905
Collision:
 completely inelastic, 180
 conservation of energy and momentum in, 173–75, 177–84
 elastic, 178–80, 183
 and impulse, 176–77
 inelastic, 178, 180–82
 nuclear, 180, 182–83
 in two dimensions, 182–83
Colloids, 261
Colonoscope, 660
Color:
 in digital camera, 714
 of light related to frequency and wavelength, 682, 685–87, 696–97
 of quarks, 936–37
 of star, 774, 952–53
 in thin soap film, 696–97
 in visible spectrum, 686
Color charge, 936–37
Color-corrected lens, 728
Color force, 937, 939–40
Color screens, 490–91
Coma, 727
Comets, 135 *pr*
Common logarithms, A-10–A-11
Commutative property, 51
Commutator, 577, 597–98
Compact disc (or disk) (CD), 20 *pr*, 43 *pr*, 48 *pr*, 226 *pr*, 489, 822
Compact disc (CD) player, 822
Compass, magnetic, 560–63, 570
Complementarity, principle of, 782

Complementary metal oxide semiconductor (CMOS), 714, 921
Complete circuit, 504
Completely inelastic collisions, 180
Complex atoms, 815–17
Complex wave, 341
Components of vector, 53–57
Composite particles, 930–31, 937 *fn*
Composite wave, 340
Composition resistor, 506
Compound lenses, 728
Compound microscope, 726–27
Compound nucleus, 889
Compound semiconductors, 847
Compounds, 360
 organic, 849
Compressed, digital data, 489, 717
Compression (longitudinal wave), 307, 309
Compressive strength, 245
Compressive stress, 243–44
Compton, A. H., 780, 890
Compton effect, 780, 781, 899
Compton shift, 780
Compton wavelength, 780, 800 *pr*
Computed tomography (CT), 735–36, 905–6
Computer:
 and digital information, 604
 disks, 604–5
 hard drive, 19 *pr*, 222 *pr*
 keyboard, 484
 memory, 500 *pr*
 monitor, 490–92, 703–4
 printers, 462–63
Computer chips, 19 *pr*, 829, 845, 851
Computer-assisted tomography (CAT), 735–36, 905–6
Computerized axial tomography (CAT), 735–36, 905–6
Concave mirror, 649, 653–54, 725
Concentration gradient, 382, 400 *fn*
Concrete, prestressed and reinforced, 246
Condensation, 379
Condensed-matter physics, 840
Condenser, 482
Condenser microphone, 484
Conductance, 523 *pr*
Conduction:
 charging by, 446–47
 electrical, 445, 501–48
 of heat, 400–2
 in nervous system, 517–19
 to skin, 410 *pr*
Conduction band, 842–43

Conduction current (*defn*), 627
Conduction electrons, 445
Conductive layer, 849
Conductivity:
 electrical, 508, 517
 thermal, 400–1
Conductors:
 charge of, 845
 electric, 445, 459, 501–2, 504 *ff*
 heat, 401
 quantum theory of, 842–43
Cones, 719
Configuration, electron, 816–17
 in fusion, 897–98
 of quarks, 937, 971
Conservation of energy, 150–58, 394–96, 413–19, 776, 865, 867, 869, 926–28
 in collisions, 173–75
 in an isolated system, 394–96
Conservation laws, 138, 150–51
 of angular momentum, 215–17, 869
 apparent violation of, in beta decay, 867
 of baryon number, 927, 940, 971
 and collisions, 173–75, 177–84
 of electric charge, 444, 869, 926
 in elementary particle interactions, 923, 926–28
 of energy, 150–58, 394–96, 413–19, 776, 865, 867, 869, 926–28
 in isolated systems, 394–96
 of lepton number, 927, 940, 971
 of linear momentum, 173–84, 869
 of mechanical energy, 150–55
 of momentum, 173–84, 926–28
 in nuclear and particle physics, 869, 926–27
 in nuclear processes, 867
 of nucleon number, 869, 926–28, 931
 of strangeness, 933
Conservative forces, 149–51
Conserved quantity, 138, 150
Constant acceleration, 28–38, 58–64
Constant angular acceleration, 203–4
Constant-volume gas thermometer, 363
Constants, values of: inside front cover
Constructive interference, 313–14, 341–43, 683 *ff*, 697, 830
Contact, thermal, 363
Contact force, 76, 84, 87

A-46 Index

Contact lenses, 721
Continent, 272, 289 *pr*
Continental drift, 272, 289 *pr*
Continuity, equation of, 273
Continuous laser, 822
Continuous spectrum, 692, 774
Continuous wave, 306
Contrast, 733
Control rods, 891
Convection, 402
Conventional current (*defn*), 505
Conventions, sign (geometric optics), 653, 655, 665
Converging lens, 661 *ff*
Conversion factors, 11, inside front cover
Converting units, 11–12
Convex mirror, 649, 655–56
Cooksey, Donald, 917
Coordinate axes, 22
COP, 426, 427
Copenhagen interpretation of quantum mechanics, 811
Copernicus, Nicolaus, 3, 125
Copier, electrostatic, 454, 462–63
Cord, tension in, 89
Cordless phone, 604
Core, of reactor, 892
Coriolis acceleration, A-17–A-18
Coriolis effect, A-17–A-18
Coriolis force, A-17, A-18
Cornea, 719
Corrective lenses, 719–21
Correspondence principle, 765, 795, 804
Cosine, 54, A-8
Cosmic acceleration, 976–77
Cosmic Background Explorer (COBE), 949, 968
Cosmic microwave background radiation (CMB), 967–70, 973, 974, 977–78
 anisotropy of, 968, 977
 uniformity of, 968, 974
Cosmic rays, 916
Cosmological constant, 977–78
Cosmological model, 970–78
Cosmological principle, 966
 perfect, 967
Cosmological redshift, 965
Cosmology, 941, 947–79
Coulomb, Charles, 447
Coulomb (C) (unit), 448, 572
 operational definition of, 572
Coulomb barrier, 876, 954–55
Coulomb force, 450, 461
Coulomb potential (*defn*), 479
Coulomb's law, 447–53, 463–64, 626, 628, 790, 864
 vector form of, 450–53
Counter emf, 599–600
Counter torque, 600
Counters, 877–78

Counterweight, 91
Covalent bond, 830, 831–32, 840
Cowan, Clyde, 928 *fn*
Creativity in science, 2–3
Credit card reader, 606
Crest, wave, 306, 313–14
Crick, F., 735
Critical angle, 659
Critical damping, 303
Critical density, of universe, 975
Critical mass, 891–94
Critical point, 377
Critical reaction, 891–94
Critical temperature, 377, 517
Cross section, 888–89
Crossed Polaroids, 700–1
CRT, 490
Crystal, liquid, 261, 378
Crystal lattice, 360, 840
Crystalline solids, 840
Crystallography, 734
CT scan, 735–36, 905–6
Curie, Marie, 863
Curie, Pierre, 863
Curie (Ci) (unit), 900
Curie temperature, 579
Current, electric (*see* Electric current)
Current, induced, 590 *ff*
Current gain, 851
Current sensitivity, 546
Curvature of field, 727
Curvature of space, 961–63, 974–75
Curvature of universe (space–time), 961–63, 974–75
Curveball, 278
Curves, banking of, 115–17
Cutoff wavelength, 818–19
Cycle (*defn*), 294
Cyclotron, 588 *pr*, 917–19
Cyclotron frequency, 568, 918
Cygnus X-1, 963
Cytosine, 460

da Vinci, Leonardo, 4
DAC, 489, 559 *pr*
Damage, done by radiation, 899
Dampers, building, 303
Damping and damped harmonic motion, 303
Dante, 321 *pr*
Dark energy, 916, 926, 973, 975–77
Dark matter, 916, 942, 973, 975–78
 hot and cold, 977–78
Data lines, 492
Data stream, 491
Dating:
 archeological, 875
 geological, 876

 radioactive, 874–76
Daughter nucleus (*defn*), 864
Davisson, C. J., 783
dB (unit), 331–33
DC, or dc (*defn*), 514
DC circuits, 526–48
DC generator, 597, 599
DC motor, 577
DC power transmission, 603 *fn*
de Broglie, Louis, 782, 795, 804
de Broglie wavelength, 782–83, 795–96, 805, 917
 applied to atoms, 795–96
Debye (unit), 482
Decay, 863
 alpha, 864–66, 869
 beta, 863–64, 866–68, 869, 873, 938, 956
 of elementary particles, 927–42
 exponential, 540–41, 610, 870
 gamma, 863–64, 868–69
 proton, 930, 940–41
 radioactive, 863–78
 rate of, 869–73
 types of radioactive, 863–64, 869
Decay constant, 869–70
Decay rates, 872–73
Decay series, 873–74, 884 *pr*
Deceleration, 27
Decibels (dB) (unit), 331–33
Declination, magnetic, 562
Decommissioning nuclear power plant, 892
Decoupled photons, 969, 973
Dee, 917–18
Defects of the eye, 719–21, 728
Defibrillator, heart, 487, 498 *pr*, 543 *fn*
Definite proportions, law of, 360 *fn*
Degeneracy:
 electron, 955
 neutron, 956
Degradation of energy, 431
Degrees of freedom, A-20
Dehumidifier, 389 *pr*, 442 *pr*
Delayed neutrons, 891
Delta particle, 932
Democritus, 359
Demodulator, 638
Dendrites, 517
Density, 261–62
 charge, 465
 and floating, 271
 nuclear and atomic, 859
 probability, 805, 811, 814, 830
Density of occupied states, 841
Density of universe, 975
Deoxyribonucleic acid, 460–62
Depletion layer, 845
Depth of field, 716
Depth finding, 349
Derived quantities, 11

Descriptive laws, 5
Destructive interference, 313–14, 341–43, 683, 697, 698, 830
Detection of radiation, 877–78, 901
Detectors, of particles and radiation, 877–78
Detergents and surface tension, 281
Determinism, 128, 810–11
Deuterium, 858, 883 *pr*, 886, 891, 895–98, 914 *pr*, A-12
Deuterium–tritium fusion (d–t), 897
Deuteron, 858, 886, A-12
Dew point, 381
Diagrams:
 energy-level, 792, 815
 Feynman, 922, 938
 force, 87
 free-body, 87–88
 H–R, 952–55, 958
 phase, 378
 phasor, 614
 potential, energy, 832–34
 PT, 378
 PV, 377–78, 414–16
 ray, 651, 655, 666
 for solving problems, 30, 57, 60, 88, 115, 141, 158, 184, 211, 234, 456, 655, 666
Diamagnetism, 580 *fn*
Diamond, 686
Diastolic pressure, 283
Diatomic molecules, 838–39, A-20–A-21
Dielectric constant, 485
Dielectric strength, 485
Dielectrics, 485–86
 molecular description of, 486
Diesel engine, 415
Difference in potential, electric, 474 *ff* (*see also* Electric potential; Voltage)
Diffraction, 680, 687–93, 821
 by circular opening, 728–30
 in double-slit experiment, 690
 of electrons, 783–84
 Fresnel, 687
 of light, 680, 687–93
 as limit to resolution, 728–30
 by single slit, 687–89
 of water waves, 318
 X-ray, 733–35
Diffraction grating, 690–91
 resolving power of, 937–38
Diffraction limit of lens resolution, 728–30
Diffraction patterns, 687
 of disk, 687–91
 of circular opening, 729
 of electrons, 783
 of single slit, 687–91
 X-ray, 733–35

Index **A-47**

Diffraction spot or disk, 729
Diffuse reflection, 646
Diffusion, 381–83
   Fick's law of, 382
Diffusion constant, 382
Diffusion equation, 382
Diffusion time, 382
Digital, 559
Digital ammeter, 546, 548
Digital artifact, 714
Digital camera, 714–18
Digital circuits, 851
Digital information, 48 pr, 604
Digital signals, 488–89
Digital TV, 491–92
Digital video disk (DVD) players, 822–23
Digital voltmeter, 546, 548
Digital zoom, 718
Digital-to-analog converter (DAC), 489, 559 pr
Dilation, time, 750–55
Dimensional analysis, 16, 19 pr, 136 pr, 983 pr
Dimensions, 16
Diodes, 845–50, 878
   forward-biased, 845
   junction, 850
   lasers, semiconductor, 822, 848
   light-emitting (LED), 491, 847–50
   photo-, 778, 847–48
   reverse-biased, 845
   semiconductor, 845–50
   zener, 846
Diopter (D) (unit), 662
Dip, angle of, 562
Dipole layer, 518
Dipole moment, 482
Dipole–dipole bonds, 834
Dipole–induced dipole bonds, 834
Dipoles and dipole moments:
   electric, 458, 478, 482, 493
   magnetic, 575–76, 580 fn
Dirac, P. A. M., 812, 841 fn, 925–26
Dirac equation, 925
Direct current (dc), 514 (see also Electric current)
Direct proportion, A-1
Direction of magnetic field, 563
Direction of vector, 23, 50
Discharge tube, 772, 787
Discharging, capacitor, 541–42
Discovery in science, 773
Disintegration, 863
Disintegration energy (defn), 865
Disorder and order, 430–31
Displacement, 23–24, 294, 302, 319
   angular, 200, 302
   net, 50–52
   resultant, 50–52

vector, 23, 50–52
   in vibrational motion, 294
   of wave, 319, 804–5
Displacement current, 627
Dissipative forces, 156–57
   energy conservation with, 156–58
Dissociation energy, 830
Distance:
   astronomical, 948, 951, 953, 957–59
   image, 647, 652, 664–65
   object, 647, 652, 664–65
   relativity of, 756–59
   table of, typical, 9
   traveled, 23
Distant stars and galaxies, 958–59
Distortion, by lenses, 728
Distribution, probability:
   in atoms, 805, 811, 814
   in molecules, 830
Diver, 216
Diverging lens, 661 ff
DNA, 460–62, 693, 735, 834–37, 899, 904
   structure and replication, 460–62, 834, 836
Domains, magnetic, 579
Domes, 248–49
Donor level, 845
Door opener, automatic, 778
Doorbell, 573
Doping of semiconductors, 844–45
Doppler, J. C., 344 fn
Doppler blood-flow meter, 347, 358 pr
Doppler effect:
   for light, 348, 964
   medical use of, 347
   for sound, 344–47
   in weather forecasting, 348
Doppler redshift, 965
Dose, 899–903
   effective, 901
Dosimetry, 899–903
Double helix (DNA), 460, 461
Double-slit experiment (electrons), 805–6
Double-slit experiment (light), 682–85, 690
   intensity in pattern, 690
Down quark, 934
Drag force, 277
Drain (in MOSFET), 851
Drain terminal, 605
DRAM, 500 pr, 605–6
Drift velocity, 516, 569
Drinking fountain, 290 pr
Dry cell, 503
Dry ice, 378
d–t (deuterium–tritium) fusion, 897
Duality, wave–particle, 782–84, 795–96

Dulong and Petit value, A-21
Dust, interstellar, 950
DVD player, 822–23
DVDs, 605
Dwarfs, white, 951, 953, 955–57
Dye lasers, 822
Dynamic lift, 277
Dynamic random access memory (DRAM), 500 pr, 605–6
Dynamic rope, 107 pr
Dynamics, 21, 76 ff
   fluid, 272–83
   hydro-, 272
   of rotational motion, 208 ff
   of uniform circular motion, 112–15
Dynamo, 597–98
Dyne (unit), 79
Dynodes, 877

Ear, 334
   discomfort, altitude, 289 pr
   range of hearing, 331
   response of, 334–35
   sensitivity of, 333
Ear popping, 289 pr
Earth, 1, 3, 9, 15, 33–34, 77, 109, 119, 121, 122, 125, 126, 128, 134 pr, 137 pr
   density of, 285 pr
   electric field of, 471 pr
   gravitational field of, 458
   gravity due to, 33 ff, 119, 121–22
   magnetic field and magnetic poles of, 562
   mass, radius, etc.: inside front cover
   mass determination, 121
   not inertial frame, 77, 137 pr
   radius, circumference of, 9, 15, 19 pr
   rocks and earliest life, 876
   varying density within, core, mantle, 285 pr
Earth soundings, 349
Earthquake waves, 309, 310–11, 318
ECG, 476, 492, 493
Echo, 312
Echolocation, 309
Eclipse, 129, 229 pr
Eddy currents (electric), 600–1
Eddy currents (fluids), 273
Edison, Thomas, 490
Effective dose, 901
Effective values, 515
Efficiency (defn), 161, 422
   Carnot, 422–24
Einstein, Albert, 4, 360, 745, 747, 748–49, 752, 760–61, 764, 775–76, 893, 938, 959–63, 977
Einstein cross, 961
Einstein ring, 961

EKG, 476, 492, 493
El Capitan, 69 pr, 285 pr
EL (electroluminescent) device, 849
Elapsed time, 23–24
Elastic collisions, 178–80, 183
Elastic limit, 241
Elastic moduli, 241
   and speed of sound waves, 308–9
Elastic potential energy, 148, 154–55, 295
Elastic region, 241
Elasticity, 241–45
Electric battery, 476, 502–3, 507, 527
Electric car, 504, 524 pr, 604
Electric cell, 505, 527
Electric charge, 444 ff
   in atom, 445
   conservation of, 444, 869, 926
   and Coulomb's law, 447–53
   of electron, 448, 579
   elementary, 448
   induced, 446–47, 486
   motion of, in magnetic field, 566–69
   negative, 444, 475, 503, 505
   point (defn), 449
   positive, 444, 475, 503, 505
   quantization of, 448
   test, 453
   types of, 444
Electric circuits, 504–5, 512–15, 526–48, 610–16
   ac, 514–15, 526 fn, 611–16
   complete, 504
   containing capacitors, 538–43, 612 ff
   dc, 514, 526–48
   digital, 851
   induced, 590 ff
   integrated, 851
   and Kirchhoff's rules, 532–35
   LC, 616
   LR, 610
   LRC, 614–16
   open, 504
   parallel, 513, 528
   RC, 538–43
   rectifier, 846–47
   resonant, 616
   series, 503, 528
   time constants of, 540, 610
Electric conductivity, 508, 517
   in nervous system, 517–19
Electric current, 501, 504–8, 512–16, 532 ff
   alternating (ac) (defn), 514–15, 526 fn, 611–16
   conduction (defn), 627
   conventional, 505
   direct (dc) (defn), 514
   displacement, 627
   eddy, 600–1
   hazards of, 543–45

A-48   Index

Electric current (*continued*)
 induced, 591
 leakage, 545
 magnetic force on, 564–76
 microscopic view of, 516
 and Ohm's law, 505–8
 peak, 514
 produced by magnetic field, 591–95
 produces magnetic field, 563–65, 579–80
 rms, 514–15
 (*see also* Electric circuits)
Electric dipole, 458, 478, 482, 493, 834
Electric energy, 474–76, 478–79, 486–87, 510–12
 stored in capacitor, 486–87
 stored in electric field, 487
Electric energy resources, 434–35
Electric field, 453–59, 463–66, 477, 478–80, 516, 597
 calculation of, 453–58, 477
 and conductors, 459, 502
 in dielectric, 485–86
 of Earth, 471 *pr*
 in EM wave, 627–29
 energy stored in, 487
 and Gauss's law, 463–66
 produced by changing magnetic field, 591–95, 597
 produces magnetic field, 626–27
 relation to electric potential, 477
 work done by, 474
Electric field lines, 457–58, 478
Electric flux, 463–65, 627
Electric force (*defn*), 443, 447–53
 adding, 450–52
 Coulomb's law for, 447–53
 and ionization, 899
 in molecular biology, 460–62, 482, 834–37
Electric generator, 597–98
Electric hazards, 543–45
Electric heater, 511
Electric motor, 577
 counter emf in, 599–600
Electric plug, 544–45
Electric potential, 474–77
 of dipole, 482
 due to point charges, 479–81
 equipotential surfaces, 478
 relation to electric field, 477, 482
 (*see also* Potential difference)
Electric potential energy, 474–76, 481, 486–87
Electric power, 510–13
 in ac circuits, 514–15, 610–16
 generation, 597–98
 in household circuits, 512–13
 transmission of, 601–4

Electric shock, 543–45
Electric stove burner, 510
Electric toothbrush, 604
Electric vehicle, 524 *pr*
Electrical conduction, 445, 501–48
Electrical grounding, 446, 505
Electrical shielding, 459
Electrical wiring, 545
Electricity, 443–639, 847 *fn*
 hazards of, 512–13, 543–45
Electricity, static, 444 *ff*
Electrocardiogram (ECG, EKG), 476, 492, 493
Electrochemical series, 502
Electrode, 503
Electroluminescent (EL) device, 849
Electrolyte, 503
Electromagnet, 572
Electromagnetic energy, 919
Electromagnetic force, 129, 922–24, 930, 938–41, 959
Electromagnetic induction, 590 *ff*
Electromagnetic oscillation, 616
Electromagnetic pumping, 589 *pr*
Electromagnetic spectrum, 630, 685–87
Electromagnetic (EM) waves, 625–39
 Doppler effect for, 348
 intensity for, 634
 production of, 627–29
 (*see also* Light)
Electrometer, 447
Electromotive force (emf), 527–28, 590–97, 599–600
 (*see also* Emf)
Electron, 445
 as beta particle, 864, 866
 as cathode rays, 490, 773
 charge on, 448, 579, 772–73
 cloud, 811, 814, 830–32
 conduction, 445
 defined, 784
 discovery of, 772–73
 in double-slit experiment, 805–6
 as elementary particle, 916, 926–27
 free, 445, 840
 mass of, 772–73, 860
 measurement of charge on, 772–73
 measurement of *e/m*, 772–73
 momentum of, 759
 in pair production, 781
 path in magnetic field, 567
 photoelectron, 778
 position uncertainty, 809
 properties of, 772–73
 sharing, 460, 830
 speed of, 516
 spin, 579

 wave nature, 806
 wavelength of, 783
 what is it?, 784
Electron capture, 868
Electron cloud, 811, 814, 830–32
Electron configuration, 816–17
Electron degeneracy, 955
Electron diffraction, 783–84
Electron flow, 505
Electron lepton number, 927, 930
Electron micrographs, 785
Electron microscope (EM), 771, 785–86, 807
Electron neutrino, 930–31
Electron–positron annihilation, 781
Electron sharing, 460, 830
Electron spin, 579, 812–13, 830
Electron transport layer (ETL), 850
Electron volt (eV) (unit), 478–79, 860
Electronic circuits, 846–51
Electronic devices, 844–51
Electronic pacemakers, 543, 608
Electrons, sea of, 925–26
Electroscope, 446–47, 502 *fn*
Electrostatic copier, 454, 462–63
Electrostatic force, 447–53, 460–62, 834
 defined, 449
 potential energy for, 474
Electrostatic unit (esu), 448 *fn*
Electrostatics, 444–493
Electroweak era, 971
Electroweak force, 129, 155, 443 *fn*, 938–41
Electroweak theory, 938–41
Elementary charge, 448
Elementary particle physics, 915–42
Elementary particles, 752, 915–42
Elements, 360, 816–17
 in compound lenses, 728
 origin of in universe, 955–56
 Periodic Table of, 816–17, inside back cover
 production of, 955–56
 transmutation of, 864, 885–89
 transuranic, 888
Elevator and counterweight, 91
Ellipse, 126
EM (electron microscope), 771, 785–86, 807
EM (electromagnetic) waves, 625–39 (*see also* Light)
Emf, 527–28, 590–97, 599–600
 back, 599–600
 counter, 599–600
 of generator, 597–99
 Hall, 569
 induced, 590–97

 motional, 596
 and photons, 922
 RC circuit with, 541
 in series and in parallel, 536–37
 sources of, 527, 590–97
Emission spectra, 774, 787–96
 atomic, 692–93, 787–89
Emission tomography, 905–6
Emissive layer (EML), 849–50
Emissivity, 403
Emitter (transistor), 850
EML, 849–50
Emulsion, photographic, 714, 878
Endoergic reaction (*defn*), 886
Endoscopes, 660
Endothermic reaction (*defn*), 886
Energy, 138, 142–48, 150–58, 177–84, 212–14, 412–19, 473 *ff*
 activation, 377, 833, 834
 and ATP, 833
 binding, 769 *pr*, 791, 830, 832–33, 834–35, 860–62
 bond, 830, 834–35
 conservation of, 150–58, 394–96, 413–19, 776, 865, 867, 869, 926–28
 dark, 916, 926, 973, 975–77
 degradation of, 431
 disintegration, 865
 dissociation, 830
 elastic potential, 148, 154–55, 295
 electric, 474–76, 478–79, 486–87, 510–12
 in EM waves, 627–28, 633–34, 919
 equipartition of, A-20–A-21
 Fermi, 841
 and first law of thermodynamics, 413–19
 geothermal, 435
 gravitational potential, 145–47, 152–53, 154–55
 in human body, 418–19
 internal, 392–93
 ionization, 791–94
 kinetic, 142–45, 212–14, 460, 760–64
 and mass, 760–64
 mechanical, 150–55, 295–96
 molecular kinetic, 374–75
 nuclear, 421 *fn*, 435, 885–908
 nucleotide, 835
 photon, 775–79
 potential, 145–48, 474–76, 481, 486–87 (*see also* Electric potential; Potential energy)
 power vs., 159
 quantization of, 775, 789–95
 reaction (*defn*), 886
 relation to mass, 760–63

Index **A-49**

Energy (continued)
　relation to work, 142–47, 155, 157–61, 212–14, 760
　relativistic, 760–64
　rest, 760–64, 809
　rotational, 212–14 and ff, 393, 838–39
　in simple harmonic motion, 295–97
　solar, 405, 434–35, 643 pr
　thermal, 156–57, 392
　threshold, 887
　total binding, 769 pr
　total mechanical (defn), 150
　transformation of, 155–56, 159
　translational kinetic (defn), 142–45
　transported by waves, 310–11
　unavailability of, 431
　and uncertainty principle, 808–9
　units of, 139, 144, 207 fn
　vacuum, 977
　vibrational, 295–97, 393, 839
　zero-point, 839
Energy bands, 842–43
Energy conservation, law of, 150–58, 394–96, 413, 865, 867, 869, 926–28
Energy density:
　in electric field, 487
　in magnetic field, 610, 633
Energy gap, 842–43
Energy-level diagram, 792, 815
Energy levels:
　in atoms, 789–95, 812–14
　for fluorescence, 820
　for lasers, 820–23
　in molecules, 837–39
　nuclear, 868–69
　in solids, 842–43
Energy resources, 434–35
Energy states, in atoms, 789–95
Energy transfer, heat as, 391–92
Engine:
　Carnot, 422–24
　diesel, 415
　internal combustion, 420–22
　power, 159–61
　steam, 420–21, 424
Enriched uranium, 891
Entire universe, 970
Entropy, 428–33
　and biological development, 430–31
　as order to disorder, 430–31
　and probability, 432–33
　and second law of thermodynamics, 428–33
　as a state variable, 428 fn
　statistical interpretation, 432–33
　and time's arrow, 431
Enzymes, 461, 834, 836

Equally tempered chromatic scale, 335
Equation of continuity, 273
Equation of state, 367
　ideal gas, 370
Equations, A-3–A-6
Equilibrium (defn), 231–33, 240
　first condition for, 232
　force in, 231–33
　neutral, 240
　second condition for, 232–33
　stable, 240
　static, 230–49
　thermal, 363, 394–95
　unstable, 240
Equilibrium distance, 839, 854 pr
Equilibrium position (vibrational motion), 293
Equilibrium state, 367
Equipartition of energy, A-20–A-21
Equipotential lines, 478
Equipotential surfaces, 478
Equivalence, principle of, 959–60
Erg (unit), 139
Escape velocity, 122, 384 pr
Escher drawing, 162 pr
Escherichia coli, 785
Estimated uncertainty, 6
Estimating (introduction), 13–15
esu (unit), 448 fn
Eta (particle), 931
Ether, 747
ETL, 850
Euclidean space, 961–62
European Center for Nuclear Research (CERN), 919, 920, 924, 939, 942
Evaporation, 379
　and latent heat, 399–400
Event, 749 ff
Event horizon, 963
Everest, Mt., 9, 11, 121, 137 pr, 380
Evolution:
　and entropy, 430–31
　stellar, 954–57
Exchange particles (carriers of force), 922–24
Excited state:
　of atom, 781, 791 ff
　of nucleon, 932
　of nucleus, 868–69
Exciton, 850
Exclusion principle, 815–16, 830, 841, 867, 936, 955, 956
Exoergic reaction (defn), 886
Exothermic reaction (defn), 886
Expansion:
　binomial, A-6–A-7, inside back cover

　linear and volume, 241–45, 364–66
　thermal, 364–67
　of universe, 964–67, 975–77
　of water, 366
Expansion joints, 361, 365, 367
Expansions, in waves, 307
Exponential curves, 540–41, 610, 869
Exponential decay, 540–41, 610, 869
Exponential notation, A-3
Exponents, A-2–A-3, inside back cover
Exposure time, 715
Extension cord, 513
Extensor muscle, 238
External force, 174, 188
Extragalactic (defn), 950
Extraterrestrials, possible communication with, 834 pr
Eye:
　aberrations of, 728
　accommodation, 719
　defects of, 719–21, 728
　far and near points of, 719
　lens of, 719
　normal (defn), 719
　resolution of, 730, 732
　structure and function of, 719–21
Eyeglass lenses, 719–21
Eyepiece, 724

Fahrenheit temperature scale, 362–63
Falling objects, 33–38
Fallout, radioactive, 894
False-color image, 736
Fan-beam scanner, 736
Far point of eye, 719
Farad (F) (unit of capacitance), 483
Faraday, Michael, 453, 590–92
Faraday cage, 459
Faraday's law of induction, 590, 592–93, 626
Farsightedness, 720
Fat, 287 pr
Femtometer (fm) (unit), 859
Fermi, Enrico, 14, 782, 804, 816, 841 fn, 867, 888, 890, 932
Fermi (fm) (unit), 859
Fermi–Dirac statistics, 841
Fermi energy, 841
Fermi gas, 841
Fermi level, 841
Fermilab, 762, 919, 920, 934
Fermions, 816, 841, 936
Ferris wheel, 114, 198
Ferromagnetism and ferromagnetic materials, 561, 579–80
　sources of, 579
FET, 851

Feynman, R., 922
Feynman diagram, 922, 938
Fiber optics, 660–61
Fick's law of diffusion, 382
Fictitious (inertial) force, A-16–A-17
Field:
　electric, 453–59, 463–66, 477, 478–80, 516 (see also Electric field)
　in elementary particles, 922
　gravitational, 458, 959–63
　Higgs, 939
　magnetic, 560–75 (see also Magnetic field)
　vector, 457
Field-effect transistor (FET), 851
Film badge, 878, 901
Film speed, 715 fn
Filter, 613, 704
Filter circuit, 613
Fine structure, 803, 813
Finnegans Wake, 934 fn
First harmonic, 316
First law of motion, 76–78
First law of thermodynamics, 413–19
　applications, 414–18
　extended, 414
　human metabolism and, 418–19
First overtone, 316, 338
Fission, nuclear, 435, 889–94
Fission bomb, 893–94
Fission fragments, 889–92
Fixed stars, 3, 125
Flash memory, 606
Flasher unit, 542
Flashlight, 505, 506, 522 pr, 848
Flat screens, 491–92, 703–4
Flatness, 974
Flavor (of elementary particles), 928, 936
Flavor oscillation, 928
Flexible cord, tension in, 89
Flexor muscle, 238
Flip coil, 624 pr
Floating, 271
Floating gate, 606
Floor vibrations, 299
Florence, 248, 361
Flow:
　of fluids, 272–83
　laminar, 272–73
　meter, Doppler, 337, 358 pr
　streamline, 272–73
　in tubes, 273–76, 278, 279–80
　turbulent, 273, 277
　volume rate of, 273
Flow rate (defn), 273
Fluid dynamics, 272–83
Fluids, 260–83 (see also Flow of fluids; Gases; Liquids; Pressure)

A-50　Index

Fluorescence, 820
Fluorescence analysis, 820
Fluorescent lightbulb, 820
Flux:
   electric, 463–65, 627
   magnetic, 592 *ff*, 597
Flying buttresses, 247
Flywheel, 229 *pr*
FM radio, 637, 638 *fn*
f-number, 715
Focal length:
   of lens, 662, 669, 670, 718, 719
   of spherical mirror, 649–50, 655
Focal plane, 662
Focal point, 649–50, 655, 661, 719
Focus, 650
Focusing, of camera, 716
Foot (ft) (unit), 9
Foot-pounds (unit), 139
Football kicks, 62, 64
Forbidden energy gap, 842
Forbidden transitions, 814, 821 *fn*, 839 *fn*
Force, 75–98, 129, 149–50, 171, 187–89, 924, 941
   addition of vectors, 87
   attractive, 832–33, 922
   buoyant, 268–72
   centrifugal (pseudo), 113, A-16–A-17
   centripetal, 112–14
   color, 937, 939–40
   conservative, 149–51
   contact, 76, 84, 87
   Coriolis, A-17, A-18
   Coulomb, 450, 461
   definition of, 79
   diagram, 87
   dissipative, 156–57
   drag, 277
   electric, 443, 447–53
   electromagnetic, 129, 922–24, 930–31, 938–41, 959
   electrostatic, 447–53, 460–62, 834
   electroweak, 129, 443 *fn*, 941
   in equilibrium, 231–33
   exerted by inanimate object, 82
   external, 174, 188
   fictitious, A-16–A-17
   of friction, 77–78, 93–96
   of gravity, 76, 84–86, 119–29, 465, 924, 941, 942, 948, 955–56, 959–63, 975–77
   and impulse, 177
   inertial, A-16–A-17
   long-range, 863, 959
   magnetic, 560–61, 564–76
   measurement of, 76
   on Moon, 119
   in muscles and joints, 207, 223 *pr*, 238–39, 255 *pr*
   net, 77–80, 87 *ff*
   in Newton's laws, 75–98, 171, 174, 187–89
   nonconservative, 149–50
   normal, 84–86
   nuclear, 129, 862–63, 867, 922–42, 959
   pseudoforce, A-16–A-17
   relation of acceleration to, 78–80
   relation of momentum to, 171–72, 174, 176–77, 760
   repulsive, 832–33, 922
   restoring, 148, 293
   short-range, 862–63, 959
   strong nuclear, 129, 862, 888 *fn*, 922–42, 959
   types of, in nature, 129, 443 *fn*, 924, 941
   units of, 79
   van der Waals, 834–37
   varying, 142
   viscous, 279–80
   weak nuclear, 129, 863, 867, 924–42, 959
   (*see also* Electric force; Magnetic force)
Force diagrams, 87
Force pumps, 267, 282
Forced convection, 402
Forced oscillations, 304
Forward-biased diode, 845
Fossil-fuel power plants, 434, 435
Foucault, J., 681
Four-dimensional space–time, 758–59, 961
Fourier analysis, 341
Fovea, 719
Foveon, 714, 717
Fractional exponents, A-2–A-3
Fracture, 241, 245–46
Frame of reference, 22, 59, 65, 77, 218, 745 *ff*, A-16–A-17
   accelerating, 77, 80, A-16–A-17
   Earth's, 128
   inertial, 77, 80, 137 *pr*, 745 *ff*, A-16
   noninertial, 78, 80, 745, A-16
   rotating, 218, A-16–A-18
   Sun's, 128
Franklin, Benjamin, 444
Franklin, Rosalind, 735
Free-body diagrams, 87–88
Free-electron theory of metals, 841
Free electrons, 445, 841
Free fall, 33–38, 124
Freedom, degrees of, A-20
Freezing (*see* Phase, changes of)
Freezing point, 362, 397
French Academy of Sciences, 9
Frequency, 111, 203, 294, 306
   of audible sound, 329, 334–35
   beat, 342–43
   of circular motion, 111
   cyclotron, 568, 918
   fundamental, 316, 336, 337–40
   infrasonic, 330
   of light, 630, 686–87
   natural, 304, 315–16
   resonant, 304, 315–16, 335–40
   of rotation, 203
   ultrasonic, 329, 350
   of vibration, 294, 303, 315–17
   of wave, 306
Frequency modulation (FM), 637, 638 *fn*
Fresnel, A., 687
Friction, 77–78, 93–96
   coefficients of, 93–94
   force of, 77–78, 93–96
   helping us to walk, 82
   kinetic, 93 *ff*
   reducing, 95
   rolling, 93, 213–14
   static, 93–94, 204, 214
Fringes, interference, 683–85
Frisch, Otto, 889
f-stop (*defn*) 715
Fukushima, 892
Fulcrum, 233
Full moon, 129, 137 *pr*
Full-scale current sensitivity, 546
Full-wave rectifier, 846–47, 853 *pr*
Fundamental constants: inside front cover
Fundamental frequency, 316, 336, 337–40
Fundamental particles, 915, 916, 930–31, 935, 938
Fuse, 512–13
Fusion, heat of, latent, 397–398, 400
Fusion, nuclear, 435, 894–98
   in stars, 894–96, 954–55
Fusion bomb, 897
Fusion reactor, 896–98

Gain (amplifier), 357 *pr*
Galaxies, 947, 948–51, 964–67, 973, 974, 976–78
   black hole at center of, 951, 963
   clusters of, 950, 974, 977
   mass of, 949
   origin of, 974, 977–78
   redshift of, 964–65
   seeds, 968, 974
   superclusters of, 950
   walls and voids, 978
Galilean–Newtonian relativity, 745–47, A-23–A-24
Galilean telescope, 723, 723 *fn*, 725
Galilean transformation, A-22–A-25
Galilean velocity transformations, A-23–A-24
Galileo, 2, 21, 33–34, 45 *pr*, 49, 58–59, 76–77, 125, 267, 302, 329, 361, 632, 646, 723, 723 *fn*, 745, 948
Galvani, Luigi, 4, 502
Galvanometer, 546–48, 576, 624 *pr*
Gamma camera, 905
Gamma decay, 863–64, 868–69
Gamma particle, 863–64, 868–69, 898–99, 916, 922
Gamma ray, 863–64, 868–69, 898–99, 922
Gamow, George, 744, 968
Garden hose, 287 *pr*, 288 *pr*, 290 *pr*
Gas constant, 370
Gas lasers, 822
Gas laws, 367–69
Gas vs. vapor, 378
Gas-discharge tube, 787
Gases, 261, 360–61, 367–83, 414–18
   atoms and molecules of, 374–75, 392–93
   Fermi, 841
   ideal, 369 *ff*
   kinetic theory of, 373–83
   molar specific heats for, A-19–A-20
   real, 377–78
   specific heats for, 394
   work done by, 414–18
Gate, transistor, 492, 605, 606, 851
Gate electrode, in a TFT, 492
Gate terminal, 605
Gauge bosons, 916, 930–31, 935
Gauge pressure, 264
Gauge theory, 938
Gauges, pressure, 266–67
Gauss, K. F., 463
Gauss (G) (unit), 565
Gauss's law, 463–66, 626
Gay-Lussac, Joseph, 368
Gay-Lussac's law, 368, 369, 372
Geiger counter, 877
Gell-Mann, M., 934
General motion, 184
General theory of relativity, 948, 959–63
Generator, 434–35
   ac, 597
   dc, 597, 599
   electric, 597–98
   emf of, 597–99
   Van de Graaff, 459
Generator equation, deriving, 598–99
Genes, 460
Genetic code, 836

Genetic information, 461, 836
Geocentric, 3, 125, 128
Geodesic, 961
Geographic poles, 562
Geological dating, 876
Geology, and Archimedes' principle, 272
Geometric optics, 645–70
Geometry (review), A-7–A-8
Geometry, plane, A-7–A-8, inside back cover
Geosynchronous satellite, 123
Geothermal energy, 435
Germanium, 844
Germer, L. H., 783
GFCI, 545, 607
Gianotti, Fabiola, 939
Giants, red, 951, 953–55
Giraffe, 289 *pr*
Glaser, D. A., 878
Glashow, S., 938
Glass, testing for flatness, 695
Glasses, eye, 719–21
Global positioning satellite (GPS), 19 *pr*, 136 *pr*, 642 *pr*, 755
Global System for Mobile Communication (GSM), 639
Global warming, 434
Glueballs, 937 *fn*
Gluino, 942
Gluons, 923, 930–31, 935, 936, 938
Golf putt, 47 *pr*
GPS, 19 *pr*, 136 *pr*, 642 *pr*, 755
Gradient:
  concentration, 382, 400 *fn*
  pressure, 280, 400 *fn*
  temperature, 400–1
  velocity, 279
Gram (g) (unit), 10, 79
Grand unified era, 971
Grand unified theories (GUT), 129, 939–41
Graphical analysis:
  of linear motion, 39–40
  for work, 142
Graphite, 891
Grating, 690–93
Gravitation, universal law of, 119–21, 448, 959
Gravitational collapse, 963
Gravitational constant (G), 120
Gravitational field (*defn*), 458, 959–63
Gravitational force, 76, 84–86, 119–29, 465, 924, 941, 942, 948, 955–56, 959–63, 975–77
Gravitational lensing, 961
Gravitational mass, 960
Gravitational potential, 476, 478
Gravitational potential energy, 145–47, 151–55

Gravitational redshift, 965
Gravitational slingshot effect, 197 *pr*
Gravitino, 942
Graviton, 924, 935, 942
Gravity, 33–38, 76, 84–86, 119–29, 465, 924, 941, 942, 948, 955–56, 959–63, 975–77
  acceleration of, 33–38, 79 *fn*, 84, 121–22
  center of, 186, 240
  and curvature of space, 959–63
  on Earth, 33 *ff*, 119, 121–22
  effect on light, 960–61, 963
  force of, 76, 84–86, 119 *ff*, 924, 941, 942, 948, 955–56, 959–63, 975, 975–77
  free fall under, 33–38, 124
  specific, 262, 271
Gravity waves, 978
Gray (Gy) (unit), 900
Greek alphabet: inside front cover
Grid, in cathode ray tube, 490
Grimaldi, F., 680, 685
Ground fault, 607
Ground fault circuit interrupter (GFCI), 545, 607
Ground state, of atom, 791
Ground wire, 544–45
Grounding, electrical, 446, 505
  and shocks, 544
  symbol, 526
Groves, Leslie, 893
g's, acceleration in, 37
GSM, 639
Guanine, 460
Guest compound, 850
GUT, 129, 939–41
Guth, Alan, 971, 973

h-bar ($\hbar$), 808, 814
Hadron era, 970–71
Hadrons, 930–31, 933–37, 971
Hahn, Otto, 889
Hair dryer, 515
Hale-Bopp comet, 135 *pr*
Hale telescope, 725
Half-life, 870–73
  calculations involving, 872–73
  formula, derivation of, 871
Half-wave rectification, 846
Hall, E. H., 569
Hall effect, Hall emf, Hall field, Hall probe, 569, 584 *pr*, 844 *fn*
Halley's comet, 135 *pr*
Halogen bulb, 501, 503
Halogens, 817
Hard drive, 19 *pr*, 222 *pr*, 604
Harmonic motion:
  damped, 303
  simple (SHM), 295–303
Harmonic oscillator, 295–303

Harmonics, 316, 336–40
Hazards of electricity, 512–13, 543–45
HD television, 491–92
Headlights, 476, 511, 526
Heads, magnetic, 604
Headsets, 577
Hearing, 328–49 (*see* Sound)
  range of human, 331
  threshold of, 335
Heart, 282–83, 289 *pr*, 290 *pr*
  defibrillator, 487, 498 *pr*, 543 *fn*
  pacemaker, 543, 608
Heart disease, and blood flow, 280
Heartbeat, 290 *pr*
Heat, 155 *fn*, 156, 390 *ff*, 412–19
  calorimetry, 394–400
  compared to work, 412
  conduction, 400–2
  convection, 402
  distinguished from internal energy and temperature, 392
  as energy transfer, 391–92
  in first law of thermodynamics, 413–19
  of fusion, 397–98
  and human metabolism, 418
  latent, 397–98, 400
  lost by human body, 402
  mechanical equivalent of, 391
  radiation, 403–6
  of vaporization, 397–98
Heat capacity, 409 *pr* (*see also* Specific heat)
Heat conduction to skin, 410 *pr*
Heat death, 431
Heat death not for stars, 431 *fn*
Heat engine, 420–25, 434, 891
  Carnot, 422–24
  efficiency of, 422
  internal combustion, 420–21, 22
  operating temperatures, 420
  steam, 420–21
  temperature difference, 421
  and thermal pollution, 434
Heat of fusion, 397–98
Heat of vaporization, 397–98
Heat pump, 425–27
Heat reservoir, 414, 423
Heat transfer, 400–6
  conduction, 400–2
  convection, 402
  radiation, 403–6
Heating duct, 274
Heating element, 510, 515
Heavy elements, 955–56
Heavy water, 891
Heisenberg, W., 771, 804
Heisenberg uncertainty principle, 806–9, 830
  and particle resonance, 932
  and tunneling, 877

Helicopter drop, 49
Heliocentric, 3, 125, 128
Helium, 815, 860–61, 864, 886–87, 894
  I and II, 378
  balloons, 272, 371
  ionization energy, 794
  primordial production of, 972, 973 *fn*
  spectrum of, 787
  and stellar evolution, 954–56
Helium–neon laser, 822
Helix (DNA), 460–61
Henry, Joseph, 590, 608
Henry (H) (unit), 608
Hertz, Heinrich, 629–30
Hertz (Hz) (unit of frequency), 203, 294
Hertzsprung–Russell diagram, 952–55, 958
Higgs, Peter, 939
Higgs boson, 915, 916, 935, 939
Higgs field, 939
High definition (HD) television, 491, 492
High-energy accelerators, 916–21
High-energy particles, 916–21
High-energy physics, 916–42
High mass stars, 955–56
High-pass filter, 613
High heels, 286 *pr*
High jump, 165 *pr*, 187
Highest occupied molecular orbitals (HOMO), 849–50
Highway:
  curves, banked and unbanked, 115–17
  mirages, 682
Hiroshima, 894
Hole transport layer (HTL), 850
Holes (in semiconductors), 843, 844–45, 849–50
Holes expand, 365
Hologram and holography, 823–24
HOMO, 849–50
Homogeneous (universe), 966
Hooke, Robert, 241, 694 *fn*
Hooke's law, 148, 241, 293
Horizon, 970
  event, 963
Horizontal (*defn*), 84 *fn*
Horizontal range (*defn*), 63
Horsepower (hp) (unit), 159
Hose, 287 *pr*, 288 *pr*, 290 *pr*
Host material, 850
Hot-air balloons, 43 *pr*, 359, 389 *pr*
Hot dark matter, 977–78
Hot wire, 544–45
Household circuits, 512–13
H–R diagram, 952–55, 958
HST (*see* Hubble Space Telescope)

A-52 Index

HTL, 850
Hubble, Edwin, 950, 964
Hubble age, 967
Hubble eXtreme Deep Field (XDF), 947, 961
Hubble parameter, 964, 967
Hubble Space Telescope (HST), 136 pr, 730, 743 pr, 961
Hubble's constant, 964
Hubble's law, 964–67, 976
Human body, 287 pr (see also Muscles and joints and specific topics):
　balance and, 240
　center of mass for, 186–87
　energy, metabolism of, 418–19
　nervous system, and electrical conduction, 517–19
　radiative heat loss of, 404
　temperature, 363, 400
Human ear (see Ear)
Human eye (see Eye)
Human radiation exposure, 901
Humidity, 380–81
　and comfort, 380
　relative (defn), 380
Hurricane, 287 pr, 290 pr
Huygens, C., 680
Huygens' principle, 680–82
Hydraulic brake, 265
Hydraulic lift, 265
Hydraulic press, 286 pr
Hydraulic pressure, 265
Hydrodynamics, 272
Hydroelectric power, 435
Hydrogen atom:
　Bohr theory of, 789–95
　quantum mechanics of, 812–14
　spectrum of, 692–93, 787–88
Hydrogen bomb, 893–94, 897
Hydrogen bond, 460, 461, 834–37
Hydrogen isotopes, 858
Hydrogen molecule, 830–33, 837, 839
Hydrogenlike atoms, 790 fn, 794, 795–96
Hydrometer, 271
Hyperopia, 720
Hypodermic needle, 289 pr
Hysteresis, 580
　hysteresis loop, 580

Icarus, asteroid, 135 pr
Ice, life under, 367
Ice skater, 81, 108 pr, 216
ICF, 898
Ideal gas, 369 ff, 841
　internal energy of, 392–93
　kinetic theory of, 373–77, 841
Ideal gas law, 369–73, 377
　in terms of molecules, 372–73
Identical (electrons), 816

Identities, trigonometric, A-7, inside back cover
Ignition:
　car, 476, 602
　fusion, 898
ILC, 921
Image:
　CAT scan, 735–36, 905–6
　false-color, 736
　formed by lens, 661 ff
　formed by plane mirror, 645–49
　formed by spherical mirror, 649–56, 725
　MRI, 907–8
　NMR, 906–8
　PET and SPECT, 905–6
　real, 647, 651, 663
　seeing, 654–55, 663
　as tiny diffraction pattern, 729
　tomography, 735–36
　ultrasound, 350–51, 445–46
　virtual, 647, 664
Image artifact, 714
Image distance, 647, 652, 664–65
Image formation, 651, 736
Image point, 647
Image reconstruction, 736–37
Imaginary number, A-6
Imaging, medical, 350–51, 905–8
Impedance (defn), 614–16
Impulse, 176–77
In-phase waves, 314, 611
Inanimate object, force exerted by, 82
Inch (in.) (unit), 9
Incidence, angle of, 313, 317, 645, 657
Incident waves, 313, 317
Inclines, motion on, 97–98, 213–14
Incoherent source of light, 685
Indeterminacy principle, 807 (see Uncertainty principle)
Index of refraction, 656
　dependence on wavelength (dispersion), 686
　in Snell's law, 657–58
Indium–tin oxide (ITO), 849
Induced current, 590 ff
Induced electric charge, 446–47, 486
Induced emf, 590–97
　counter, 599–600
　in electric generator, 597–99
　in transformer, 601–3
Inductance, 608–9
　in ac circuits, 610–16
　mutual, 608
　self-, 608–9
Induction:
　applications, 606–7

charging by, 446–47
electromagnetic, 590 ff
Faraday's law of, 590, 592–93, 626
Induction stove, 594
Inductive charging, 604
Inductive reactance, 612
Inductor, 608, 851
　in circuits, 610–16
　energy stored in, 610
　reactance of, 612
　symbol, 608
Inelastic collisions, 178, 180–82
Inertia, 77
　moment of (rotational), 208–10
Inertial confinement, 897–98
Inertial confinement fusion (ICF), 898
Inertial forces, A-16–A-17
Inertial mass, 960
Inertial reference frame, 77, 80, 137 pr, 745 ff, A-16
　Earth as, 77, 137 pr
　equivalence of all, 746–47, 748
Inflationary scenario, 971, 973–75
Information storage:
　magnetic, 604–5
　semiconductor, 605–6
Infrared (IR) radiation, 403, 405 fn, 630, 686, 693
Infrasonic waves, 330
Initial conditions, 300
Inkjet printer, 463
In-phase waves, 683, 690, 694–98
Insertions (muscle attachment points), 238
Instantaneous acceleration (defn), 26, 40
Instantaneous angular acceleration, 201
Instantaneous angular velocity, 200
Instantaneous speed, 25
Instantaneous velocity (defn), 25, 39–40
Instruments, musical, 335–40
Instruments, optical, 713–37
Insulators:
　electrical, 445, 508, 842–43
　thermal, 401–2, 842–43
Integrated circuits, 851
Integration by parts, A-6, A-7
Intensity, 310–11, 331 ff
　for EM waves, 634
　in interference and diffraction patterns, 688–90
　of light, 685, 700–1, 804–5
　of sound, 331–33
　of waves, 310–11, 331–33

Interference, 313–14, 341–43
　beats, 342–43
　constructive, 313–14, 341–42, 683, 697, 698, 830
　destructive, 313–14, 341–42, 683, 697, 698, 830
　of electrons, 805–6, 830
　of light waves, 682–85
　of sound waves, 341–43
　by thin films, 693–98
　of water waves, 314
　wave-phenomenon, 682
　of waves on a cord, 313
Interference fringes, 683–85
Interference microscope, 733
Interference pattern:
　double-slit, 682–85, 690, 805–6
　including diffraction, 690
　multiple slit, 690–93
Interferometers, 698, 747
Interlaced, 490
Intermittent windshield wipers, 543
Internal combustion engine, 420–22
Internal conversion, 869
Internal energy, 392–93, 413–14
　distinguished from heat and temperature, 392
　of an ideal gas, 392–93
Internal reflection, total, 327 pr, 659–60
Internal resistance, 527–28
International Linear Collider (ILC), 921
International Thermonuclear Experimental Reactor (ITER), 898
Interpolation, 6, A-11
Interstellar dust, 950
Intervertebral disc, 239
Intravenous transfusion, 288 pr, 289 pr
Intrinsic luminosity, 951–53, 958
Intrinsic semiconductor, 843, 844
Invariant energy–momentum, 763
Invariant quantity, 759, 763
Inverse proportion, A-1
Inverse square law, 120, 310, 332, 448, 628
Inverted population, 821–22, 848
Ion (defn), 445
Ionic bonds, 831, 833, 840
Ionization energy, 791–94
Ionizing radiation (defn), 898
IR (infrared) radiation, 403, 405 fn, 630, 686, 693
Iris, 719
Iron, binding energy for, 862
Irreversible process, 423

Index A-53

ISO number, 715 *fn*
Isobaric processes, 415
Isochoric processes, 415
Isolated system, 174, 394–96
Isomer, 869
Isotherm, 414
Isothermal processes, 414–15
Isotopes, 578, 858–59, 863–64
  half-life of, 870–71
  mean life of, 871
  in medicine, 904
  table of, A-14–A-17
Isotropic, 680, 966
Isovolumetric (isochoric)
  process, 415
ITER, 898
Iterative technique, 737
ITO, 849

J/ψ particle, 809, 935
Jars and lids, 365, 369
JET, 898
Jet plane noise, 333
Jets (particle), 934, 935, 1164
Jeweler's loupe, 723
Joint European Torus (JET), 898
Joints:
  expansion, 361
  human, forces in muscles
    and, 207, 223 *pr*, 238–39,
    255 *pr*, 256 *pr*, 259 *pr*
Joule, James Prescott, 391
Joule (J) (unit), 139, 144,
  207 *fn*, 478–79, 511
  relation to calorie, 391
Joyce, James, 934 *fn*
jpeg, 489, 717
Jump starting a car, 536–37
Junction diode, 850
Junction rule, Kirchhoff's, 533 *ff*
Jupiter, 3, 126, 128, 134 *pr*,
  135 *pr*, 137 *pr*
  moons of, 125, 128, 134 *pr*,
    135 *pr*, 632, 723

K-capture, 868
K lines, 818–19
K particle (kaon), 931, 932–33
K2, 11
Kant, Immanuel, 950
Kaon, 931, 932–33
Karate blow, 177
Keck telescope, 725
Kelvin (K) (unit), 368
Kelvin-Planck statement of
  second law of
  thermodynamics, 424
Kelvin temperature scale, 362,
  368
Kepler, Johannes, 126, 723 *fn*
Keplerian telescope, 723 *fn*, 724
Kepler's laws, 125–28
Keyboard, computer, 484
Kilby, Jack, 851
Kilo- (prefix), 10
Kilocalorie (kcal) (unit), 391

Kilogram (kg) (unit), 10, 78, 79
Kilometer (km) (unit), 10
Kilowatt-hour (kWh) (unit),
  168 *pr*, 511
Kinematic equations, 29
Kinematics, 21–40, 49–66,
  198–205
  for rotational motion,
    198–205
  translational motion, 21–40,
    49–66
  for uniform circular motion,
    110–12
Kinetic energy, 142–45, 150 *ff*,
  212–14, 460, 760–64
  in collisions, 178–81
  and electric potential energy,
    474
  of gas atoms and molecules,
    374–75, 392–93, A-20–A-21
  molecular, relation to
    temperature, 374–75,
    392–93, A-20–A-21
  of photon, 779
  relativistic, 760–64
  rotational, 212–14
  translational, 142–45
Kinetic friction, 93 *ff*
  coefficient of, 93–94
Kinetic theory, 359, 373–83
  basic postulates, 373
  boiling, 380
  diffusion, 381–83
  evaporation, 379
  ideal gas, 373–77
  in cells, 460
  kinetic energy near absolute
    zero, 376
  of latent heat, 400
  molecular speeds,
    distribution of, 376–77
  and probability, 373–77
  of real gases, 377–78
Kirchhoff, G. R., 532
Kirchhoff's rules, 532–35
  junction rule, 533 *ff*
  loop rule, 533 *ff*

Ladder, forces on, 237, 259 *pr*
Lag time, 610
Lambda (particle), 931, 932–33
Laminar flow, 272–73
Land, Edwin, 699
Lanthanides, 817
Large-diameter objective lens,
  731
Large Hadron Collider (LHC),
  915, 919–21, 939, 942
Large-scale structure of
  universe, 977–78
Laser printer, 463
Lasers, 707 *pr*, 820–23
  applications, 822–23
  chemical, 822
  diode, 848
  gas, 822

  helium–neon, 822
  medical uses, 823
  surgery, 823
Latent heats, 397–398, 400
Lattice structure, 360, 840, 844
Laue, Max von, 734
Law (*defn*), 5 (*see proper name*)
  causal, 128
Law of inertia, 77
Lawrence, E. O., 917
Lawson, J. D., 898
Lawson criterion, 898
LC circuit, 616
LC oscillation, 616
LCD, 491–92, 703–4, 714
Leakage current, 545
LED, 491, 847–50
  applications, 847–48
  lighting, 847–48
  white-light, 848
Lemaître, Georges, 964
Length:
  focal, 649–50, 655, 662, 669,
    670, 718, 719
  Planck, 19 *pr*, 970
  proper, 756
  relativity of, 756–59
  standard of, 9, 698
Length contraction, 756–59,
  A-25
Lens, 661–70
  achromatic, 728
  axis of, 661
  coating of, 697–98
  color-corrected, 728
  combinations of, 668–69
  compound, 728
  contact, 721
  converging, 661 *ff*
  corrective, 719–21
  cylindrical, 720
  diverging, 661 *ff*
  of eye, 719
  eyeglass, 719–21
  eyepiece, 724
  focal length of, 662, 669, 670
  magnetic, 785
  magnification of, 665
  negative, 665
  normal, 718, 743 *pr*
  objective, 724, 725, 726
  ocular, 726
  positive, 665
  power of (diopters), 662
  resolution of, 717, 728–32
  sign conventions, 665
  telephoto, 718
  thin (*defn*), 661
  wide-angle, 718, 728
  zoom, 718
Lens aberrations, 727–28, 729,
  731
Lens elements, 728
Lens opening, 715
Lensing, gravitational, 961
Lensmaker's equation, 670

Lenz's law, 593–95
Leonardo da Vinci, 4
Lepton era, 970, 972
Lepton number, 927, 930–31,
  933, 940
Leptons, 916, 922, 927, 930–31,
  933–35, 937–39, 942, 970–71
Level:
  acceptor, 845
  donor, 845
  energy (*see* Energy levels)
  Fermi, 841
  loudness, 334–35
  sound, 331–33
Level horizontal range, 63
Level range formula, 63–64
Lever, 164 *pr*, 233
Lever arm, 206, 238
LHC, 915, 919–21, 939, 942,
  983 *pr*
Lids and jars, 365, 369
Life under ice, 367
Lifetime, 930 (*see also* Mean
  life)
Lift, dynamic, 277
Lift, hydraulic, 265
Light, 629–31, 632–33, 644–704
  coherent sources of, 685
  color of, and wavelength,
    682, 685–87, 696–97
  dispersion of, 686
  Doppler shift for, 348, 964
  as electromagnetic wave,
    629–31
  frequencies of, 630, 686
  gravitational deflection of,
    960–61, 963
  incoherent sources of, 685
  infrared (IR), 630, 686, 693,
    711 *pr*
  intensity of, 685, 700–1, 804–5
  monochromatic (*defn*), 682
  as particles, 681, 775–82
  photon (particle) theory of,
    775–82
  polarized, 699–702, 711 *pr*
  ray model of, 645 *ff*, 661 *ff*
  scattering, 704
  from sky, 704
  spectrometer, 692–93
  speed of, 9, 629, 632–33, 656,
    681, 747, 748
  ultraviolet (UV), 630, 686,
    693
  unpolarized (*defn*), 699
  velocity of, 6, 629, 632–33,
    656, 681, 747, 748
  visible, 630, 685–87
  wave–particle duality of, 782
  wave theory of, 679–704
  wavelengths of, 630, 682,
    685–87, 696
  white, 686
  (*see also* Diffraction;
    Intensity; Interference;
    Reflection; Refraction)

Light-emitting diode (LED), 491, 847–50
 applications, 847–48
Light-gathering power, 725
Light-hour, 948 fn
Light meter (photographic), 778
Light-minute, 948
Light pipe, 660
Light rays, 645 ff, 661 ff
Light-second, 948
Light sources, 919
Light-year (ly) (unit), 18 pr, 948
Lightbulb, 501, 503, 504, 505, 506, 511, 556 pr, 777
 fluorescent, 820
 in RC circuit, 542
Lightning, 329, 512
Linac, 920
Linde, Andreí, 971, 973
Line spectrum, 692–93, 787 ff, 803
Line voltage, 515
Linear accelerator, 920
Linear expansion (thermal), 364–65
 coefficient of, 364
Linear momentum, 170–89
 conservation of, 173–84, 869
Linear motion, 21–40
Linear waves, 310
Linearly polarized light, 699 ff
Lines of force, 457–58, 561
Liquefaction, 367–70, 373, 377
Liquid crystal, 261, 378, 703–4
Liquid crystal display (LCD), 491–92, 703–4, 714
Liquid-drop model, 889
Liquid-in-glass thermometer, 362
Liquid scintillators, 878
Liquids, 261 ff, 360 (see also Phase, changes of)
Lithium, 815
Log table, A-10–A-11
Logarithms, A-10–A-11, inside back cover
Long jump, 70 pr, 72 pr
Long-range force, 863, 959
Longitudinal waves, 307 ff
 and earthquakes, 309
 speed of, 308–9
 (see also Sound waves)
Lookback time, 951, 969
Loop rule, Kirchhoff's, 533 ff
Lorentz transformation, A-24–A-25
Los Alamos laboratory, 893
Loudness, 329, 331, 332 (see also Intensity)
Loudness control, 335
Loudness level, 334–35
Loudspeaker cross-over, 613
Loudspeakers, 332, 341–42, 356 pr, 577, 613
Loupe, jeweler's, 723

Low mass stars, 955
Low-pass filter, 613
Lowest unoccupied molecular orbitals (LUMO), 849–50
LR circuit, 610
LRC circuit, 614–16
Lucretius, 977 fn
Luminosity (stars and galaxies), 951–53, 958
LUMO, 849–50
Lunar eclipse, 129
Lyman series, 788, 792, 793, 817

M-theory, 942
Mach, E., 348 fn
Mach number, 348
Macroscopic properties, 360
Macrostate of system, 432–33
Magellanic clouds, 950 fn
Magnet, 560–62, 579–80
 domains of, 579
 electro-, 572
 permanent, 579–80
 superconducting, 572
Magnetic bottle, 897–98
Magnetic circuit breakers, 573
Magnetic confinement, 897–98
Magnetic damping, 618 pr
Magnetic declination, 562
Magnetic deflection coils, 490
Magnetic dipoles and magnetic dipole moments, 575–76, 580 fn
Magnetic domains, 579
Magnetic field, 560–75
 between two currents, 570
 of circular loop, 563
 definition of, 564–65
 determination of, 565, 570–71, 573–75
 direction of, 561, 563, 568
 of Earth, 562
 energy stored in, 610
 hysteresis, 580
 induces emf, 591–97
 of long straight wire, 570–71
 motion of charged particle in, 566–69
 produced by changing electric field, 626–27
 produced by electric current, 563–65, 579–80 (see also Ampère's law)
 produces electric field and current, 597
 inside solenoid, 574–75
 of solenoid, 572–73
 of straight wire, 564–65, 574
 of toroid, 580
 uniform, 562
Magnetic field lines, 561
Magnetic flux, 592 ff, 597
Magnetic force, 560–61, 564–76
 on current-carrying wire, 564–65
 on electric current, 564–76

 on moving electric charges, 566–69
 on proton, 566
 between two parallel wires, 571–72
Magnetic information storage, 604–5
Magnetic lens, 785
Magnetic moment, 575–76
Magnetic monopole, 561, 975
Magnetic permeability, 570, 579–80
Magnetic poles, 560–62
 of Earth, 562
Magnetic quantum number, 812–13
Magnetic resonance imaging (MRI), 907–8
Magnetic tape and disks, 604
Magnetism, 560–610
Magnetoresistive random access memory (MRAM), 606
Magnification:
 angular, 722
 of lens, 665
 of lens combination, 668–69
 of magnifying glass, 722–23
 of microscope, 726–27, 731–32, 785
 of mirror, 652–53
 sign conventions for, 653, 655, 665
 of telescope, 724, 730–32
 useful, 732, 785
Magnifier, simple, 722–23
Magnifying glass, 661, 722–23
Magnifying mirror, 649, 654
Magnifying power, 722 (see also Magnification)
 total, 724
Magnitude of vector, 23, 50
Main-sequence (stars), 953–55, 982 pr
Majorana, Ettore, 929 fn
Majorana particles, 929
Malus' law, 700
Manhattan Project, 893
Manometer, 266
Marconi, Guglielmo, 636
Mars, 3, 126, 128
Mass, 6, 78–80, 261
 of air in a room, 371
 atomic, 360
 center of, 184–89
 critical, 891–94
 of Galaxy, 949
 gravitational vs. inertial, 960
 and luminosity, 952
 as measure of inertia, 78
 molecular, 360, 369
 of neutrinos, 929–30
 nuclear, 859–60
 of photon, 779
 relation to energy, 760–63
 in relativity theory, 760
 rest, 760

 standard of, 10
 table of, 10
 units of, 6–7, 79
 vs. weight, 78, 84
Mass excess (defn), 883 pr
Mass number, 858
Mass spectrometer (spectrograph), 578
Mass–energy, distribution in universe, 975–77
Mass–energy transformation, 760–64
Mathematical review, A-1–A-11
Mathematical signs and symbols: inside front cover
Mather, John, 968
Matrix, on an LCD screen, 491, 492
Matter:
 anti-, 925, 941, 943 pr
 atomic theory of, 359–61
 dark, 916, 942, 973, 975–78
 passage of radiation through, 898–99
 phases (states) of, 261, 360–61
 wave nature of, 782–84, 795–96
Matter waves, 782–84, 795–96, 804 ff
Matter–antimatter problem, 941
Matter-dominated universe, 972, 973
Maxwell, James Clerk, 376, 625, 627, 629–30, 635, 747
Maxwell distribution of molecular speeds, 376–77, 433, 897
Maxwell's equations, 626–27, 744, 747, 747 fn, 748
Maxwell's preferred reference frame, 747
Mean life, 871, 930
 of proton, 941
Measurements, 3, 5–8
 of astronomical distances, 948, 953, 957–59
 of blood flow, 358 pr
 of Calorie content, 396
 of charge on electron, 772–73
 electromagnetic, of blood flow, 596
 of e/m, 772–73
 of force, 76
 precision of, 5–8, 806–8
 of pressure, 266–67
 of radiation, 899–903
 of speed of light, 632–33
 uncertainty in, 5–8, 806–9
 units of, 8–10
Mechanical advantage, 92, 233, 265

Index A-55

Mechanical energy, 150–55
  total (*defn*), 150, 295–96
Mechanical equivalent of heat, 391
Mechanical oscillations, 293
Mechanical waves, 305–19
Mechanics, 21–445 (*see also* Motion)
  definition, 21
Mediate, forces, 922
Medical imaging, 350–51, 405, 905–8
  techniques, 908
Medical instruments, 660–61
Medical uses of lasers, 823
Meitner, Lise, 804, 889
Melting point, 397–400 (*see also* Phase, changes of)
Memory:
  dynamic random access (DRAM), 500 *pr*, 605–6
  flash, 606
  magnetoresistive random access (MRAM), 606
  random access (RAM), 482, 500 *pr*, 605–6
  volatile and nonvolatile, 606
Mendeleev, Dmitri, 816
Mercury (planet), 3, 126, 128
Mercury, surface tension of, 281
Mercury barometer, 267
Mercury-in-glass thermometer, 361–62
Merry-go-round, 198, 201, 202
Meson exchange, 923
Meson lifetime, 809
Mesons, 816, 916, 923, 926–27, 930–31, 935, 936–38
Messenger RNA (m-RNA), 836–37
Metabolic pathway, 820
Metabolism, human, 418–19
Metal detector, 601
Metal-oxide semiconductor field-effect transistor (MOSFET), 605, 851
Metallic bond, 840
Metals:
  alkali, 817
  free-electron theory of, 841
Metastable state, 820, 821–22, 869
Meter (m) (unit), 9
Meters, electric, 546–48, 576
  correction for resistance of, 547–48
Metric (SI) multipliers: inside front cover
Metric (SI) system, 10
Mho (unit), 523 *pr*
Michelson, A. A., 632–33, 698, 747, 748
Michelson interferometer, 698, 747

Michelson–Morley experiment, 747, 748, 748 *fn*
Microampere (μA) (unit), 504
Micrometer (measuring device), 14
Micron (μm) (unit), 10, 484
Microphones, magnetic, 606
Microscope, 726–27, 730–32
  atomic force, 786
  compound, 726–27
  electron, 771, 785–86, 807
  interference, 733
  magnification of, 726–27, 731–32, 785
  phase-contrast, 733
  resolving power of, 731
  scanning tunneling electron (STM), 786
  specialty, 733
  useful magnification, 732, 785
Microscopic description, 360, 373 *ff*
Microscopic properties, 360, 373 *ff*
Microstate of a system, 432–33
Microwave background radiation, cosmic, 967–70, 973, 974, 977–78
Microwaves, 630, 639, 967–68
Milky Way, 136 *pr*, 948–50
Milliampere (mA) (unit), 504
Millikan, R. A., 773, 777
Millikan oil-drop experiment, 773
Millimeter (mm) (unit), 10
Mirage, 682
Mirror equation, 652–56
Mirrors, 645–56
  aberrations of, 725 *fn*, 727–28
  concave, 649, 653–54, 725
  convex, 649, 655–56
  focal length of, 649–50, 655
  magnifying, 649, 654
  plane, 645–49
  sign conventions, 653
  used in telescope, 725
MKS (meter-kilogram-second) system (*defn*), 10
mm-Hg (unit), 266
Model of solar system:
  geocentric, 3, 125
  heliocentric, 3, 125
Models, 5, 22, 460
Moderator, 891–92
Modern physics (*defn*), 2, 745
Modulation:
  amplitude, 637
  frequency, 637, 638 *fn*
Moduli of elasticity, 241–42, 308–9
Molar specific heats, A-19–A-21
Mole (mol) (unit), 369
  volume of, for ideal gas, 369

Molecular biology, electric force in, 460–62, 482, 834–37
Molecular kinetic energy, 374–75, 392–93, A-20–A-21
Molecular mass, 360, 369
Molecular oscillation, 326 *pr*, 393, 839
Molecular rotation, 838–39
Molecular spectra, 837–39
Molecular speeds, 376–77, 433
Molecular vibration, 326 *pr*, 393, 839
Molecular weight, 360
Molecules, 360, 372–77, 829–39
  bonding in, 829–32
  in a breath, 373
  diatomic, A-20–A-21
  polar, 445, 461, 482, 486, 831–32
  potential-energy diagrams for, 832–34
  spectra of, 837–39
  weak bonds between, 834–37
Molybdenum, 818
Moment arm, 206
Moment of a force about an axis, 206
Moment of inertia, 208–10
Momentum, 138, 170–89
  angular, 215–18, 789, 795
  center of mass (CM), 184–86
  in collisions, 173–75
  conservation of angular, 215–17
  conservation of linear, 173–76, 177–84, 188, 926–28
  linear, 170–89
  of photon, 779
  relation of force to, 171–72, 174, 176–77, 760
  relativistic, 759–60, 763
  uncertainty in measurement of, 807
  units of, 171
Momentum transfer, 635
Monochromatic aberration, 728
Monochromatic light (*defn*), 682
Monitors, TV and computer, 490–92, 703–4
Moon, 3, 129, 948
  acceleration due to gravity on, 46 *pr*, 134 *pr*
  acceleration toward the Earth, 119
  centripetal acceleration of, 112, 119
  force on, 119
  Full moon, 129, 137 *pr*
  New moon, 129
  orbit of, 129
  phases of, 129

  rising of, 129, 137 *pr*
  sidereal period, 129, 137 *pr*
  synodic period, 129, 137 *pr*
  wireless communication from, 639
  work on, 142
Moonrise, 129
Moons of Jupiter, 125, 128, 134 *pr*, 135 *pr*
Morley, E. W., 747, 748
Moseley, H. G. J., 818
Moseley plot, 818
MOSFET, 605, 851
Most probable speed, 376
Motion, 21–229, 744–65
  Brownian, 360
  circular, 110–18
  at constant acceleration, 28–38, 58–64
  damped, 303
  description of (kinematics), 21–40, 49–66, 110–12
  of electric charge in magnetic field, 566–69
  and frames of reference, 22
  in free fall, 33–38, 124
  general, 184
  graphical analysis of linear, 39–40
  harmonic, 295–303
  on inclines, 97–98
  Kepler's laws of planetary, 125–28
  linear, 21–40
  Newton's laws of, 75–83, 87–98, 138, 141, 157, 170, 171, 174, 188, 208–12, 804, 810
  nonuniform circular, 118
  oscillatory, 292 *ff*
  parabolic, 58, 64
  periodic (*defn*), 293 *ff*
  planetary, 125–28
  projectile, 49, 58–64
  rectilinear, 21–40
  and reference frames, 22
  relative, 65–66, 744–65
  rolling, 204–5
  rotational, 198–218
  of satellites, 122–23
  simple harmonic (SHM), 295–303
  translational, 21–189
  uniform circular, 110–15
  uniformly accelerated, 28–38
  vibrational, 292 *ff*
  of waves, 305–19
Motional emf, 596
Motor:
  ac, 577
  back emf in, 599–600
  burning out, 600
  dc, 577
  electric, 577
  overload, 600

A-56 Index

Mountaineering, 105 *pr*, 106 *pr*, 107 *pr*, 258 *pr*
Mr Tompkins in Wonderland (Gamow), 744, 766 *pr*
MRAM (magnetoresistive random access memory), 606
MRI, 907–8
m-RNA, 836–37
Mt. Everest, 9, 11, 121, 137 *pr*, 380
Mt. Wilson, 950
Muller, Richard, 968
Multielectron atoms, 815–17
Multimeter, 548
Multiplication factor, 891–92
Multiplication of vectors, 52–53
Multiwire chamber, 878, 934
Muon, 927, 930–31, 934
  decay, 927
  lifetime, 752, 770 *pr*
Muon lepton number, 927
Muon neutrino, 930–31
Muscle insertion, 238
Muscles and joints, forces in, 207, 223 *pr*, 238–39, 255 *pr*, 256 *pr*, 259 *pr*
Musical instruments, 335–40
Musical scale, 335
Mutation, 899
Mutual inductance, 608
Myopia, 719–20

n-type semiconductor, 844–46
Nagasaki, 894
Nanometer, 845 *fn*
National Ignition Facility (NIF), 898
Natural abundances, 858
Natural convection, 402
Natural frequency, 304, 315–16 (*see also* Resonant frequency)
Natural logarithms, A-10
Natural radioactive background, 866, 901
Natural radioactivity, 863
Nautical mile, 20 *pr*
Navstar Global Positioning System (GPS), 136 *pr*
Nd:YAG laser, 822
Near Earth Asteroid Rendezvous (NEAR), 136 *pr*
Near-Earth orbit, 134 *pr*
Near field, 628
Near point, of eye, 719
Nearsightedness, 719–20, 721
Nebulae, 950
Negative, photographic, 714 *fn*
Negative, of a vector (*defn*), 52
Negative curvature, 962, 975
Negative electric charge (*defn*), 444, 475, 503, 505

Negative exponents, A-2
Negative lens, 665
Negative sea of electrons, 925–26
Negative work, 140
Neon tubes, 803
Neptune, 126, 127
Neptunium, 888
Nerve pulse, 518, 567
Nervi, Pier Luigi, 249
Nervous system, electrical conduction in, 517–19
Net displacement, 50–52
Net force, 77–80, 87 *ff*
Net resistance, 528
Neuron, 517–18
Neutral atom, 445, 860
Neutral equilibrium, 240
Neutral wire, 545
Neutrino, 190 *pr*, 866–68, 916, 927, 928–30, 972
  mass of, 929–30, 931
  types of, 927–29
Neutrino flavor oscillation, 928
Neutrinoless double beta decay, 946 *pr*
Neutron, 445, 858, 916, 930–31
  delayed, 891
  in nuclear reactions, 889–94
  role in fission, 889 *ff*
Neutron capture, 956
Neutron degeneracy, 956
Neutron multiplication factor, 891
Neutron number, 858
Neutron physics, 888
Neutron star, 217, 951, 956–57
New moon, 129
Newton, Isaac, 21, 77–78, 81, 119–20, 129, 453, 681, 694 *fn*, 725 *fn*, 745–47, 959, 963 *fn*
Newton (N) (unit), 79
Newton-meter (unit), 139
Newtonian focus, 725
Newtonian mechanics, 75–129, 744
Newton's first law of motion, 76–78, 745
Newton's law of universal gravitation, 109, 119–21, 448, 959
Newton's laws of motion, 75–83, 87–98, 138, 141, 157, 171, 174, 187–89, 208–12, 804, 810
Newton's rings, 694–95
Newton's second law, 78–80, 82, 87–89, 171, 174, 187–89, 747, 760
  for rotation, 208–12, 215–17
  for a system of particles, 187–89
Newton's synthesis, 128
Newton's third law of motion, 81–83

NIF (National Ignition Facility), 898
NMR, 906–8
Noble gases, 817, 840
Nodes, 315, 337, 338, 339, 340
Noise, 333, 340–41, 489
Nonconductors, 445
Nonconservative forces, 149–50
Non-Euclidean space, 961–62
Noninertial reference frames, 78, 80, A-16
Nonlinear device, 846
Nonohmic device, 506
Nonreflecting glass, 697–98
Nonuniform circular motion, 118
Nonvolatile memory, 606
Nonvolatile memory (NVM) cell, 606
Normal eye (*defn*), 719
Normal force, 84–86
Normal lens, 718, 743 *pr*
Normal X-ray image, 735
North pole, Earth, 562
North pole, of magnet, 561
Notre Dame, Paris, 247
Nova, 951, 956–57
Noyce, Robert, 851
npn transistors, 850
n-type semiconductor, 844–46
Nuclear binding energy, 860–62
Nuclear collision, 180, 182–83
Nuclear decay, 762
Nuclear density, 859
Nuclear energy, 421 *fn*, 435, 885–908
Nuclear fission, 435, 889–94
Nuclear forces, 129, 862–63, 867, 922–42, 959
Nuclear fusion, 435, 894–98, 954–55
Nuclear magnetic resonance (NMR), 906–8
Nuclear masses, 858 and *ff*
Nuclear medicine, 903–8
Nuclear physics, 857–908
Nuclear power, 891–94
Nuclear power plants, 435, 597, 891–93
Nuclear radius, 859
Nuclear reactions, 885–89
Nuclear reactors, 890–93, 896–98
Nuclear spin, 860
Nuclear structure, 858–60
Nuclear weapons testing, 894
Nucleon, 858, 916, 938, 971–72
Nucleon number, conservation of, 869, 926–28
Nucleosynthesis, 954–55, 972
Nucleotide bases, 460–61, 834–36
Nucleus, 858 *ff*
  compound, 889
  daughter and parent (*defn*), 864

  half-lives of, 869–71
  liquid-drop model of, 889
  masses of, 858–60
  radioactive decay of unstable, 863–77
  size of, 859
  structure and properties of, 858–60
Nuclide (*defn*), 858
Null result, 747, 748
Null technique, 559 *pr*
NVM (nonvolatile memory) cell, 606

Object distance, 647, 652, 664–65
Objective lens, 724, 725, 726, 731–32
  large diameter, 731
Observable universe, 967, 969–70, 974
Observations, 2, 745
  and uncertainty, 806–7
Occhialini, G., 923
Occupied states, density of, 841
Ocean currents, and convection, 402
Octave, 335
Oersted, H. C., 563
Off-axis astigmatism, 727
Ohm, G. S., 505
Ohm (Ω) (unit), 506
Ohmmeter, 548, 576
Ohm's law, 505–8, 528, 534
Oil film, 697, 709 *pr*
Oil-drop experiment, 773
OLED, 849–50
  applications, 849
Omega (particle), 931
One-dimensional waves, 310
Onnes, H. K., 517
Open circuit, 504
Open system, 394
Open tube, 338
Open-tube manometer, 266
Operating temperatures, heat engines, 420
Operational definitions, 11, 572
Oppenheimer, J. Robert, 893
Optical coating, 697–98
Optical drive, 605
Optical illusion, 657, 682
Optical instruments, 697–98, 713–37
Optical pumping, 821
Optical sound track, 778
Optical tweezers, 636
Optical zoom, 718
Optics:
  fiber, 660–61
  geometric, 645–70
  physical, 679–704
Orbit, of the Moon, 129, 137 *pr*
Orbit, near-Earth, 134 *pr*

Index **A-57**

Orbital angular momentum, in atoms, 812–13
Orbital quantum number, 812
Orbitals, atomic, 814
Order and disorder, 430–31
Order of interference or diffraction pattern, 683–85, 690–91, 692, 711 *pr*, 734
Order-of-magnitude estimate, 13–15
Organ pipe, 339
Organic compounds, 849
Organic light-emitting diode (OLED), 849–50
Origin, of coordinate axes, 22
Orion, 950
Oscillations, 292 *ff*
   of air columns, 337–40
   damped harmonic motion, 303
   displacement, 294
   electromagnetic, 616
   forced, 304
   LC, 616
   mechanical, 293
   molecular, 326 *pr*, 393, 839
   resonant, 304
   simple harmonic motion (SHM), 295–303
   as source of waves, 306
   of a spring, 293–95
   on strings, 315–16, 335–37
Oscillator, simple harmonic, 295–303
Oscilloscope, 492
Osteoporosis, diagnosis of, 780
Out-of-phase waves, 314, 611, 683, 690, 698
Overdamped system, 303
Overexposure, 715
Overtones, 316, 336, 337, 338, 340

p-type semiconductor, 844–46
P waves, 309, 311
Pacemaker, heart, 543, 604, 608
Packet, wave, 327 *pr*
Packing of atoms, 840
Page thickness, 14
Pain, threshold of, 335
Pair production, 781, 926
Pantheon, dome of, 248
Parabola, 49, 58, 64
Parabolic reflector, 650
Parallax, 957–58
Parallel capacitors, 538–39
Parallel electric circuits, 513, 528
Parallel emf, 536–37
Parallel resistors, 528–32
Parallelogram method of adding vectors, 52
Paramagnetism, 580 *fn*
Paraxial rays (*defn*), 650
Parent nucleus (*defn*), 864
Paris, 247, 386
Parsec (pc) (unit), 958

Partial ionic character, 831–32
Partial pressure, 380–81
Partially conserved quantities, 933
Partially polarized, 700, 704
Particle (*defn*), 22
Particle accelerators, 916–21
Particle classification, 930–31
Particle detectors, 847, 877–78, 921, 934
Particle exchange, 922–24, 937
Particle interactions, 924 *ff*
Particle physics, 915–42
Particle resonance, 932
Particle–antiparticle pair, 926
Particles, elementary, 752, 915–42
Particles vs. waves, 680
Pascal, Blaise, 262, 265, 291 *pr*
Pascal (Pa) (unit of pressure), 262
Pascal's principle, 265
Paschen series, 788, 792
Passive solar heating, 435
Path difference, 683, 694
Pauli, Wolfgang, 804, 815, 867
Pauli exclusion principle, 815–16, 830, 841, 867, 936, 955, 956
Peak current, 514
Peak voltage, 514
Peaks, tallest, 11
Pendulum:
   ballistic, 181
   simple, 16, 301–3
Pendulum clock, 302
Penetration, barrier, 876–77
Penzias, Arno, 967–68
Percent uncertainty, 6, 8
   and significant figures, 8
Perfect cosmological principle, 967
Performance, coefficient of (COP), 426, 427
Perfume atomizer, 277
Period, 111, 203, 294, 298–99, 306
   of circular motion, 111
   of pendulums, 16, 302
   of planets, 126–28
   of rotation, 203
   sidereal, 129, 137 *pr*
   of simple harmonic motion, 298–99
   synodic, 129, 137 *pr*
   of vibration, 294
   of wave, 306
Periodic motion, 293 *ff*
Periodic Table, 816–17, 858 *fn*, inside back cover
Periodic wave, 306
Perlmutter, Saul, 976
Permeability, magnetic, 570, 579–80
Permeability of free space, 570
Permittivity, 448, 485

Perturbations, 127
PET, 905–6
Phase:
   in ac circuit, 611–16
   changes of, 377–79, 397–400
   of matter, 261, 360–61
   of waves, 314, 683, 690, 694–98
Phase angle, 615
Phase-contrast microscope, 733
Phase diagram, 378
Phase plate, 733
Phase shift, 695, 697, 698
Phase transitions, 377–81, 397–400
Phases of the Moon, 129
Phasor diagram, 614
Phon (unit), 334
Phone:
   cell, 318, 602, 604, 631, 639
   cordless, 604
Phonon, 847 *fn*
Phosphor, 877
Phosphorescence, 820
Photino, 942
Photocathode, 877
Photocell, 499 *pr*, 776
Photocell circuit, 776, 778
Photoconductivity, 462
Photocopier, 454, 462–63
Photodiode, 778, 847–48
Photoelectric effect, 776–78, 781, 899
   applications, 778
Photographic emulsion, 878
Photographic film, 714, 715
Photomultiplier (PM) tube, 801 *pr*, 877–78
Photon, 775–82, 804–6, 816, 916, 922–24, 926, 930–31, 971–73
   absorption of, 820
   decoupled (early universe), 969, 973
   and emf, 922
   energy of, 777, 779
   mass of, 779
   mediation of (force), 922
   momentum of, 779
   virtual, 922
Photon exchange, 922–24
Photon interactions, 781
Photon theory of light, 775–82
   predictions, 777
Photosynthesis, 779
Photovoltaic (solar) cells, 435, 556 *pr*, 847
Photovoltaic panel, 847
Physics:
   classical (*defn*), 2, 847
   modern (*defn*), 2, 745
   relation to other fields, 4–5
Pi meson, 816, 923, 930–31, 935
Piano tuners, number of, 14–15
Piano tuning, 343
Pick-up nuclear reaction, 911 *pr*

Picture sharpness, 717–18
Pin, structural, 246
Pincushion distortion, 728
Pinhole camera, 742 *pr*
Pion, 923, 930–31, 932
   kinetic energy of, 761
Pipe, light, 660
Pipe, vibrating air columns in, 335 *ff*
Pipe organ, 339
Pisa, tower, 34, 302
Pitch of a sound, 329
Pixels, 10, 488–89, 491–92, 703–4, 714, 717, 736
Pixels, addressing, 491–92
Planck, Max, 775
Planck length, 19 *pr*, 970
Planck time, 941, 970
Planck's constant, 775, 808
Planck's quantum hypothesis, 771, 775
Plane:
   focal, 662
   mirror, 645–49
   polarization of light by, 699–702
Plane geometry, A-7–A-8
Plane waves, 312–13, 628, 682
Plane-polarized light, 700
Planetary motion, 125–28
Planets, 3, 125–28, 134 *pr*, 137 *pr*, 189, 197 *pr*, 228 *pr*
   period of, 126–28
Plasma, 261, 897
Plastic region, 241
Plate tectonics, 272, 289 *pr*
Plum-pudding model of atom, 786
Plumb bob, 137 *pr*
Pluto, 126, 127, 948
Plutonium, 888, 892, 894
PM tube, 801 *pr*, 877–78
pn diode laser, 848
pn junction, 845–50
pn junction diode, 845–50, 878
pn junction diode laser, 822, 848
pnp transistor, 850
Point:
   boiling, 362, 380, 397
   breaking, 241
   critical, 377
   dew, 381
   far, 719
   focal, 649–50, 655, 661, 719
   freezing, 362, 397
   image, 647
   mathematical, 22
   melting, 397–400
   near, 719
   sublimation, 378
   triple, 378
Point charge (*defn*), 449
   potential, 479–81
Point particle, 22, 88
Poise (P) (unit), 279

Poiseuille, J. L., 280
Poiseuille's equation, 280
Poisson, Siméon, 687
Polar, 460
Polar ice caps, 405
Polar molecules, 445, 461, 482, 486, 831–32
Polarization of light, 699–702, 711 pr
  by absorption, 699–701
  plane, 699–702
  by reflection, 701–2
  of skylight, 704
Polarizer, 700–4
Polarizing angle, 702
Polaroid, 699–701
Polaroid sheet, 699
Pole vault, 153
Poles, geographic, 562
Poles, magnetic, 560–62
  of Earth, 562
Pollution:
  air, 434–35
  thermal, 434–35
Poloidal field, 898
Polonium, 863
Polymers, 849
Pontecorvo, Bruno, 928 fn
Pool depth, apparent, 658
Pope, Alexander, 963 fn
Population, inverted, 821–22, 848
Position, 22
  angular, 199
  equilibrium (vibrational motion), 293
  as a function of time (SHM), 299–300
  uncertainty in, 806–9
Positive curvature, 962, 975
Positive electric charge (defn), 444, 475, 503, 505
Positive holes, 844
Positive lens, 665
Positron, 781, 868, 905–6, 916, 924–26
Positron emission tomography (PET), 905–6
Post-and-beam construction, 243
Potassium-40, 883 pr, 912, 913
Potential (see Electric potential)
Potential difference, electric, 474 ff, 527, 845 (see also Electric potential; Voltage)
Potential drop, 507
Potential energy (defn), 145–48 and ff
  diagrams, 832–34
  elastic, 148, 154–55, 295–97
  electric, 474–76, 481, 486–87
  gravitational, 145–47, 154–55
  for molecules, 832–34, 839, 840
  for nucleus, 876
  as stored energy, 148

Potentiometer, 559 pr
Pound (lb) (unit), 79
Powell, C. F., 923
Power (defn), 159–61, 510–15, 615
  energy vs., 159
  rating of an engine, 159–61
  wind, 435
Power, magnifying, 722
  total, 724
  (see also Electric power)
Power backup, 484
Power factor (ac circuit), 615
Power generation, 434–35, 597
Power of a lens, 662
Power plants:
  fossil-fuel, 434–35
  nuclear, 435, 597, 891–93
Power reactor, 891–92
Power transmission, 601–4
Powers of ten, 7, A-3
Poynting vector, 634 fn
Precision, 5–8
  accuracy vs., 8
Prefixes, metric, 10
Presbyopia, 720
Prescriptive laws, 5
Pressure (defn), 244, 262–64
  absolute, 264
  atmospheric, 264, 266–67
  blood, 283
  in fluids, 262–63
  in a gas, 264, 367–72, 374, 377–81
  gauge, 264
  head, 263
  hydraulic, 265
  on living cells, 264
  measurement of, 266–67
  partial, 380–81
  and Pascal's principle, 265
  tire, 372
  units for and conversions, 262, 264, 266–67
  vapor, 379
  volume changes under, 244, 415–17
Pressure amplitude, 330, 333
Pressure cooker, 380, 388 pr
Pressure gauges, 266–67
Pressure gradient, 280, 400 fn
Pressure head, 263
Pressure waves, 309, 330 ff
Prestressed concrete, 246
Primary coil, 600
Principal axis (defn), 650
Principal quantum number, 789 ff, 812–14
Principia (Newton), 77
Principle, 5 (see proper name)
Principle of complementarity, 782
Principle of correspondence, 765, 795, 804

Principle of equipartition of energy, A-21
Principle of equivalence, 959–60
Principle of superposition, 313–14, 340–41, 450–53, 455
Printers, inkjet and laser, 463
Prism, 660, 686
Prism binoculars, 660, 725
Probability:
  and entropy, 432–33
  in kinetic theory, 373–77
  in nuclear decay, 869
  in quantum mechanics, 805, 806, 810–11, 830–32
Probability density (probability distribution):
  in atoms, 805, 811, 814
  in molecules, 830–32
Problem-solving strategies, 30, 57, 60, 88, 115, 141, 158, 184, 211, 234, 399, 436, 456, 534, 568, 594, 655, 666, 697
Processes:
  adiabatic, 415–16
  isobaric, 415
  isochoric, 415
  isothermal, 414–15
  isovolumetric, 415
  reversible and irreversible (defn), 423
Projectile, horizontal range of, 63–64
Projectile motion, 49, 58–64
  kinematic equations for (table), 60
  parabolic, 49, 58, 64
Proper length, 756
Proper time, 753, 945 pr
Proportion, direct and inverse, A-1–A-2
Proportional limit, 241
Proteins, 460, 836–37
  shape of, 837
  synthesis of, 836–37
Proton, 445, 762, 858 ff, 885, 894–96, 904, 916, 930–31, 934
  decay, 930, 940–41
  magnetic force on, 566
  mean life of, 941
Proton–antiproton collision, 934, 1164
Proton centers, 904
Proton decay, 930, 940–41
Proton–proton chain, 895, 954
Proton therapy, 904
Protostar, 954
Protractor (for angles), 7
Proxima Centauri, 948
Pseudoforce, A-16–A-17
Psi (pound per square inch), 262
p-type semiconductor, 844–46

PT diagram, 378
Ptolemy, 2, 3, 125
Pulley, 91–92, 211, 212, 225 pr
Pulse, wave, 306
Pulse-echo technique, 349–51, 908
Pulsed laser, 822
Pulsed oximeter, 848
Pumps, 267, 282–83
  centrifugal, 282
  circulating, 282
  force, 267, 282
  heat, 425–27
  vacuum, 267, 282
Pupil, 719
Putting, in golf, 47 pr
PV diagrams, 377–78, 414
Pythagorean theorem, 51, 55 fn, A-8

QCD, 923, 937–39
QED, 922
QSOs, 951, 961
Quadratic equation, 36, 38
Quadratic formula, 38, A-6, inside back cover
Quality factor (Q-value) of a resonant system, 624 pr
Quality of sound, 340–41
Quantities, base and derived, 10–11
Quantization:
  of angular momentum, 789, 812–13
  of electric charge, 448
  of energy, 775, 789–95
Quantization error, 488
Quantum chromodynamics (QCD), 923, 937–39
Quantum condition, Bohr's, 789, 795–96
Quantum electrodynamics (QED), 922
Quantum (quanta) of energy, 775
Quantum fluctuations, 974
Quantum hypothesis, Planck's, 771, 775
Quantum mechanics, 796, 804–51
  of atoms, 796, 803–24
  Copenhagen interpretation of, 811
  of molecules and solids, 829–51
Quantum numbers, 775, 789–91, 812–14, 815–16, 837–39
  ground state, 815
  principal, 789 ff
Quantum theory, 745, 771–96, 804–51
  of atoms, 789–96, 803–24
  of blackbody radiation, 774
  of light, 774–82
  of specific heat, A-21

Quarks, 448 *fn*, 916, 922–24, 930–31, 933–38, 971–72
  combinations, 936
  confinement, 937, 971
Quasars (quasistellar objects, QSOs), 951, 961
Quintessence, 977
Q-value (disintegration energy), 865
Q-value (quality factor) of a resonant system, 624 *pr*
Q-value (reaction energy), 886

Rad (unit), 900
Rad equivalent man (rem), 901
Radar, 348, 349, 630, 639
Radial acceleration, 110 *ff*, 118
Radian (rad), measure for angles, 199–200
Radiation, electromagnetic:
  blackbody, 774, 952, 968
  cosmic microwave background, 967–70, 973, 974, 977–78
  emissivity of, 403
  gamma, 863–64, 868–69, 898–99
  from human body, 404
  infrared (IR), 630, 686, 693
  microwave, 630, 639
  seasons and, 405
  solar constant and, 405
  synchrotron, 919
  thermal, 403–6
  ultraviolet (UV), 630, 686, 693
  X-ray, 630, 733–35, 819 (*see also* X-rays)
Radiation, nuclear:
  activity of, 870, 872, 900
  alpha, 863–66, 869
  beta, 863–64, 866–68, 869, 956
  damage by, 899
  detection of, 877–78, 901
  dosimetry for, 899–903
  gamma, 863–64, 868–69, 898–99
  human exposure to, 901–3
  ionizing (*defn*), 898
  measurement of, 899–903
  medical uses of, 903–8
  types of, 863–64, 869
Radiation biology, 903–8
Radiation damage, 899
Radiation-dominated universe, 972–73
Radiation dosimetry, 899–903
Radiation era, 972–73
Radiation field, 628
Radiation film badge, 878, 901
Radiation pressure, 635–36
Radiation sickness, 901
Radiation therapy, 903–4

Radiation worker exposure, 901
Radio, 636–39
Radio receiver, 638
Radio transmitter, 637
Radio waves, 630, 639, 731
Radioactive background, natural, 866, 901
Radioactive dating, 874–76
Radioactive decay, 863–78
Radioactive decay constant, 869–70
Radioactive decay law, 870
Radioactive decay series, 873–74
Radioactive fallout, 894
Radioactive tracers, 904–5
Radioactive waste, 892–94
Radioactivity, 857–78
  artificial (*defn*), 863
  natural (*defn*), 863
Radiofrequency (RF) signal, 637, 907–8
Radioisotope (*defn*), 863
Radionuclide (*defn*), 863, 900
Radiotelescope, 731
Radium, 863, 864, 873
Radius, of nuclei, 859
Radius of curvature (*defn*), 117
Radius of Earth, estimate of, 15, 19 *pr*
Radon, 864, 901, 902–3, 912 *pr*
  exposure, 902–3
Rae Lakes, 13
Rainbow, 686
RAM (random access memory), 482, 500 *pr*, 605–6
Ramp vs. stair analogy, 775
Random access memory (RAM), 482, 500 *pr*, 605–6
Range of projectile, 63–64
Raphael, 2
Rapid estimating, 13–15
Rapid transit system, 47 *pr*
Rare-earth solid-state lasers, 822
Rarefactions, in waves, 307
Raster, 490
Rate of nuclear decay, 869–73
Ray, 312, 645 *ff*, 661 *ff*
  paraxial (*defn*), 650
Ray diagram, 651, 655, 666
Ray model of light, 645 *ff*, 661 *ff*
Ray tracing, 645 *ff*, 661 *ff*
Rayleigh, Lord, 729
Rayleigh criterion, 729
RBE, 901
RC circuits, 539–43
  applications, 542–43
Reactance, 612, 613
  capacitive, 613
  inductive, 612
  (*see also* Impedance)
Reaction energy, 886
Reaction time, 32, 610

Reactions:
  chain, 890–91, 893–94
  chemical, rate of, 377
  endoergic, 886
  endothermic, 886
  nuclear, 885–89
  slow-neutron, 886–87
  subcritical, 891, 894
  supercritical, 891, 894
Reactors, nuclear, 890–93, 896–98
Reading data, 605
Read/Write head, 604
Real gases, 377–78
Real image, 647, 651, 663
Rearview mirror, convex, 656
Receivers, radio and television, 638
Recoil, 176
Recombination epoch, 973
Rectangular coordinates, 22
Rectifiers, 846–47, 853 *pr*
Rectilinear motion, 21–40
Recurrent novae, 956
Red giants, 951, 953–55
Redshift, 348, 959, 964–65, 969
  Doppler, 965
  origins, 965
Redshift parameter, 947, 965
Reducing friction, 95
Reference frames, 22, 65, 77–78, 218, 745 *ff*, A-16–A-18
  accelerating, 77, 80
  Earth's, 128
  inertial, 77, 80, 745 *ff*, A-16
  noninertial, 78, 80, 137 *pr*, 745, A-16
  rotating, 218, A-16–A-18
  Sun's, 128
Reflecting telescope, 725
Reflection:
  angle of (*defn*), 313, 645
  diffuse, 646
  law of, 313, 645
  and lens coating, 697–98
  of light, 644–49
  phase changes during, 693–98
  polarization by, 701–2
  specular, 646
  from thin films, 693–98
  total internal, 327 *pr*, 659–61
  of waves on a cord, 312–13
Reflection grating, 690
Refracting telescope, 724
Refraction, 317–18, 656–70, 681–82
  angle of, 317, 657
  of earthquake waves, 318
  index of, 656
  law of, 317, 657, 681–82
  of light, 656–58, 681–82
  and Snell's law, 657–58
  by thin lenses, 661–64
  of water waves, 317
Refresh rate, 492

Refrigerators, 425–27
  coefficient of performance (COP) of, 426
Regelation, 384 *pr*
Reheating, 971
Reines, Frederick, 928 *fn*
Reinforced concrete, 246
Relative biological effectiveness (RBE), 901
Relative humidity (*defn*), 380
Relative motion, 65–66, 591, 744–65
Relative velocity, 65–66, 178–79, 749 *ff*, 764
Relativistic addition of velocities, 764
Relativistic energy, 760–64
Relativistic formulas, when to use, 763–64
Relativistic mass, 760
Relativistic momentum, 759–60, 763
Relativistic velocity transformations, A-24
Relativity, Einstein's theory of, 4
Relativity, Galilean–Newtonian, 745–47, A-22–A-23
Relativity, general theory of, 948, 959–63
Relativity, special theory of, 744–65, 959
  constancy of speed of light, 748
  four-dimensional space–time, 758–59
  impact of, 765
  and length, 756–59
  and mass, 760
  mass–energy relation in, 760–64
  postulates of, 748–49
  simultaneity in, 749–50
  and time, 749–55, 758–59
Relativity principle, 746–47, 748 *ff*
Relay, 582 *pr*
Rem (unit), 901
Remote controls, 639
Replication, DNA, 460–62
Repulsive forces, 832–33, 922
Research chamber, undersea, 287 *pr*
Research reactor, 891
Reservoir, heat, 414, 423
Resistance:
  air, 33
  thermal, 402
Resistance and resistors, 506–7, 511, 611
  in ac circuit, 611 *ff*
  with capacitor, 539–43, 611–16
  color code, 507
  and electric currents, 501 *ff*
  with inductor, 610, 611–16

Resistance and resistors (*continued*)
  internal, in battery, 527–28
  in *LRC* circuit, 611–16
  of meters, 547–48
  net, 528
  in series and parallel, 528–32
  shunt, 546, 548
  and superconductivity, 517
Resistance thermometer, 510
Resistivity, 508–10
  temperature coefficient of, 508, 509–10
  temperature dependence of, 509–10
Resistor, 506
  composition, 506
  shunt, 546, 548
  symbol, 507, 526
  wire-wound, 506
Resistors, in series and in parallel, 528–32
Resolution:
  of electron microscope, 785–86
  of eye, 730, 732
  of high-energy accelerators, 917
  of lens, 717, 728–32
  of light microscope, 731–32
  limits of, 728–32
  and pixels, 717
  of telescope, 730–32
  of voltage, 488–89
Resolution, of vectors, 53–57
Resolving power (RP), 731
Resonance, 304, 315–17
  in ac circuit, 616
  elementary particle, 932
  nuclear magnetic, 906–8
Resonant collapse, 304
Resonant frequency, 304, 315–16, 335–40, 616
Resonant oscillation, 304
Rest energy, 760–64, 809
Rest mass, 760
Resting potential, 518
Restoring force, 148, 293
Resultant displacement, 50–52
Resultant vector, 50–52, 56–57
Retina, 719, 785
Reverse-biased diode, 845
Reversible process, 423
Revolutions per second (rev/s), 203
Reynold's number, 288 *pr*
RF signal, 637, 907–8
Rho (particle), 931
Ribosome, 836–37
Richards, P., 968
Riess, Adam G., 976
Rifle:
  muzzle velocity, 323 *pr*
  recoil, 176
Right angle, A–8
Right triangle, A–8

Right-hand rule, 217, 219 *fn*, 563, 564, 566, 568, 570, 594
Rigid object (*defn*), 198
  rotational motion of, 198–218
  translational motion of, 187–89, 213–14
Ripple voltage, 847
Rms (root-mean-square):
  current, 514–15
  voltage, 514–15
Rms (root-mean-square) speed, 376–77
RNA, 836–37
Rock climbing, 105 *pr*, 106 *pr*, 107 *pr*, 258 *pr*
Rocket propulsion, 75, 82, 175
Rocks, dating oldest Earth, 876
Rods, 719
Roemer, Ole, 632
Roentgen (R) (unit), 900
Roentgen, W. C., 733
Roller coaster, 147, 152, 158
Rolling friction, 93, 213–14
Rolling motion, 204–5, 212–14
  without slipping, 204–5
  total kinetic energy, 213
Rome, 4, 247, 248, 249
Roosevelt, Franklin, 893
Root-mean-square (rms) current, 514–15
Root-mean-square (rms) speed, 376–77
Root-mean-square (rms) voltage, 514–15
Roots, A-2–A-3
Rotating reference frames, 218, A-16–A-18
Rotation, 198–218
  axis of (*defn*), 199, 208
  frequency of (*defn*), 203
  and Newton's second law, 208–12, 215–17
  period of, 203
  of rigid body, 198–218
Rotational angular momentum quantum number, 838–39
Rotational dynamics, 208 *ff*
Rotational inertia, 208–10 (see also Moment of inertia)
Rotational kinetic energy, 212–14
  molecular, 393, A-20–A-21
Rotational motion, 198–218
  kinematics for, 198–205
  uniformly accelerated, 203–4
Rotational transitions, 838–39
Rotor, 577, 598
Rotor-ride, 135 *pr*
Rough calculations, 13–15
RP (resolving power), 731
Rubidium–strontium dating, 882 *pr*
Ruby laser, 821
Rulers, 6

Runway, 29
Russell, Bertrand, 784
Rutherford, Ernest, 786, 859, 863, 885
Rutherford's model of the atom, 786–89
R-value, 402, 410 *pr*
Rydberg constant, 788, 792

S wave, 309
SAE, viscosity numbers, 279 *fn*
Safety:
  in electrical wiring and circuits, 512–13, 545
  in jump starting a car, 537
Safety factor, 245
Sailboats, and Bernoulli's principle, 277
Salam, A., 938
Sampling rate, 488–89
Satellite dish, 638, 650
Satellite television and radio, 639
Satellites, 109, 122–25
  geosynchronous, 123
  global positioning (GPS), 19 *pr*, 136 *pr*, 755
Saturated vapor pressure, 379
Saturation (magnetic), 580
Saturn, 3, 126, 127, 136 *pr*, 197 *pr*
Sawtooth oscillator, 542
Sawtooth voltage, 542
Scalar (*defn*), 50
Scalar components, 53
Scalar quantities, 50
Scale, musical, 335
Scale, spring, 76
Scale factor of universe, 965
Scanner, fan-beam, 736
Scanning electron microscope (SEM), 771, 785–86
Scanning tunneling electron microscope (STM), 786
Scattering:
  of light, 704
  of X-rays, Bragg, 824
Schmidt, Brian P., 976
Schrödinger, Erwin, 771, 804, 812
Schwarzschild radius, 963
Scientific notation, 7
Scintigram, 905
Scintillation counter, 877
Scintillator, 877–78, 905
Scuba tank, 287 *pr*, 386 *pr*, 389 *pr*
SDSS, 978
Sea of electrons, 925–26
Search coil, 624 *pr*
Seasonal Energy Efficiency Ratio (SEER) rating, 427
Seasons, 405
Second (s) (unit), 9
Second harmonic, 316, 338

Second law of motion, 78–80, 82, 87–89, 171, 174, 187–89, 747
  for rotation, 208–12, 215–17
  for a system of particles, 187–89
Second law of thermodynamics, 419–33
  and Carnot efficiency, 422–24
  Clausius statement of, 420, 423 *pr*, 425
  and efficiency, 422–24
  and entropy, 428–33, 436
  general statement of, 429, 430
  heat engine, 420–25
  and irreversible processes, 423
  Kelvin-Planck statement of, 424
  refrigerators, air conditioners, and heat pumps, 425–27
  reversible processes, 423
  and statistical interpretation of entropy, 432–33
  and time's arrow, 431
Secondary coil, 600
Seeds, of galaxies, 968, 974
SEER (Seasonal Energy Efficiency Ratio) rating, 427
Seesaw, 234–35
Segrè, Emilio, 925
Seismograph, 607
Selection rules, 814, 838, 839
Self-inductance, 608–9
Self-sustaining chain reaction, 890–94
SEM, 771, 785–86
Semiconductor detector, 878
Semiconductor diode lasers, 822
Semiconductor diodes, 845–50
Semiconductor doping, 844–45
Semiconductor information storage, 605–6
Semiconductors, 445, 508, 843–50
  compound, 847
  intrinsic, 843, 844
  n and *p* types, 844–46
  resistivity of, 508
  silicon wafer, 878
Sensitivity, full-scale current, 546
Sensitivity of meters, 547–48
Series capacitors, 538–39
Series electric circuits, 503, 528
Series emf, 536–37
Series resistors, 528–32
Sharing of electrons, 460, 830
Sharpness, picture, 717–18
Shear modulus, 242, 244
Shear strength, 245
Shear stress, 244

Index   A-61

Shear wave, 309
Shells, atomic, 816–17
Shielded cable, 617 pr
Shielding, electrical, 459
SHM, see Simple harmonic motion
SHO, see Simple harmonic oscillator
Shock absorbers, 292, 295, 303
Shock waves, 348–49
Shocks, 487
 and grounding, 544
Short circuit, 512–13
Short-range forces, 862–63, 959
Shunt resistor, 546, 548
Shutter speed, 715, 717
SI (Système International) units, 10
SI derived units: inside front cover
Sicily, 243
Sidereal period, 129, 137 pr
Siemens (S) (unit), 523 pr
Sievert (Sv) (unit), 901
Sigma (particle), 931
Sign conventions (geometric optics), 653, 655, 665
Signal voltage, 488–89
Signals, analog and digital, 488–89
Significant figures, 6–7, A-3
 percent uncertainty vs., 8
Silicon, 843 ff
Silicon wafer semiconductor, 878
Simple harmonic motion (SHM), 295–303
 applied to pendulums, 301–3
 energy in, 295–97
 period of, 298–99
 related to uniform circular motion, 299–300
 sinusoidal nature of, 300–1
Simple harmonic oscillator (SHO), 295–303
 acceleration of, 301
 energy in, 295–97
 molecular vibration as, 839
 velocity and acceleration of, 301
Simple machines:
 lever, 164 pr, 233
 pulley, 91–92, 211, 212
Simple magnifier, 722–23
Simple pendulum, 16, 301–3
Simultaneity, 749–50
Sine, 54, A-8
Single-lens reflex (SLR) camera, 718
Single photon emission computed tomography (SPECT), 905–6
Single photon emission tomography (SPET), 905–6
Single-slit diffraction, 687–89

Singularity, 963
Sinusoidal curve, 300 ff
Sinusoidal traveling wave, 319
Siphon, 284 pr, 290 pr
Size of star, 952
Skater, 81, 108 pr, 216
Skidding car, 116
Skier, 97, 125, 138, 168 pr
Sky color, 704
Sky diver, 69 pr, 101 pr
SLAC, 920
Slepton, 942
Slingshot effect, gravitational, 197 pr
Sloan Digital Sky Survey (SDSS), 978
Slope, of a curve, 39–40
Slow-neutron reaction, 886–87
SLR camera, 718
Slug (unit), 79
SM (Standard Model):
 cosmological, 970–73
 elementary particles, 448 fn, 915, 916, 930, 935–39
Small Sports Palace, dome of, 249
Smoke, and Bernoulli effect, 278
Smoke detector, 866
Smoot, George, 968
Snell, W., 657
Snell's law, 657–58, 681
SNIa (type Ia) supernovae, 956, 957, 958, 976
SN1987a, 929, 956
Snowboarder, 49, 107 pr
Soap bubble, 679, 693, 696–97
Soaps, 281
Society of Automotive Engineers (SAE), 279 fn
Sodium, 815
Sodium chloride, bonding in, 831, 833, 840
Sodium pump, 518 fn
Solar absorption spectrum, 692–93, 787
Solar (photovoltaic) cell, 435, 556 pr, 847
Solar constant, 405
Solar eclipse, 129, 229 pr
Solar energy, 405, 434–35, 643 pr
Solar neutrino problem, 928
Solar pressure, 635
Solar sail, 636
Solar system, models, 3, 125
Solenoid, 572–73, 579–80, 609
Solid angle, 11 fn
Solid-state lighting, 848
Solid-state physics, 840
Solids, 241 ff, 261, 360, 840–43, A-21 (see also Phase, changes of)
 amorphous, 840
 band theory of, 842–43
 bonding in, 840–41

 energy levels in, 842–43
 equipartition of energy for, A-21
 molar specific heats of, A-21
Solving for unknowns, A-4–A-5
Sonar, 349–50
Sonic boom, 349
Sonogram, 350
Sound, 328–51
 audible range of, 329, 334–35
 and beats, 342–43
 dBs of, 331–33
 Doppler effect of, 344–47
 ear's response to, 334–35
 infrasonic, 330
 intensity of, 331–33
 interference of, 341–43
 level, 331–33
 loudness of, 329, 331, 332
 loudness level of, 334
 pitch of, 329
 pressure amplitude of, 330, 333
 quality of, 340–41
 shock waves of, 348–49
 and sonic boom, 349
 sources of, 335–40
 spectrum, 341
 speed of, 329
 supersonic, 329, 348–49
 timbre of, 340
 tone color of, 340
 ultrasonic, 329, 350
Sound barrier, 349
Sound level, 331–33
Sound spectrum, 341
Sound track, optical, 778
Sound waves, 307, 309, 328–51 (see also Sound)
Sounding board, 337
Sounding box, 337
Soundings, 349
Source (in MOSFET), 851
Source activity, 900
Source terminal, 605
Sources of emf, 527, 590–97
Sources of ferromagnetism, 579
South pole, of Earth, 562
South pole, of magnet, 561
Space:
 absolute, 746, 748
 curvature of, 961–63, 974–75
 Euclidean and non-Euclidean, 961–62
 relativity of, 756–59
 telescope, 730, 961
Space perception, using sound waves, 309
Space quantization, 812
Space shuttle, 21, 75, 109
Space station, 122, 125, 130 pr
Space–time (4-D), 758–59
 curvature of, 961–63, 974–75
Space–time interval, 759
Space travel, 754

Spatial interference, of sound, 341–42
Speaker wires, 509
Special theory of relativity, 744–65, 959 (see also Relativity, special theory of)
Specialty microscopes, 733
Specific gravity (defn), 262, 271
Specific heat, 393–94, A-19–A-21
 for gases, 394
 molar, A-19–A-21
 for solids, A-21
 for water, 393
SPECT, 905–6
Spectrometer:
 light, 692–93
 mass, 578
Spectroscope and spectroscopy, 692–93
Spectrum, 690–91
 absorption, 692–93, 787
 atomic emission, 692–93, 787–89
 band, 837
 continuous, 692, 774
 electromagnetic, 630, 685–87
 emitted by hot object, 774
 line, 692–93, 787 ff, 803
 molecular, 837–39
 sound, 341
 visible light, 685–87
 X-ray, 817–19
Specular reflection, 646
Speed, 23
 average (defn), 23–24, 376
 of EM waves, 629, 631
 instantaneous, 25
 of light (see separate entry below)
 molecular, 376–77, 433
 most probable, 376
 relative, 178–79
 rms (root-mean-square), 376–77
 of sound (see separate entry below)
 of waves, 306, 308–9
 (see also Velocity)
Speed of light, 9, 629–33, 656, 681, 747, 748
 constancy of, 748
 measurement of, 632–33
 as ultimate speed, 760
Speed of light principle, 748 ff
Speed of sound, 329
 infrasonic, 330
 supersonic, 329, 348–49
SPET, 905–6
Spherical aberration, 650, 727, 728. 729
Spherical mirrors, image formed by, 649–56, 725, 728

Spherical wave, 310, 312
Spiderman, 167 *pr*
Spin:
  boson, 936
  down, 813, 907
  electron, 579, 812–13, 830
  fermion, 936
  nuclear, 860
  up, 813, 907
Spin angular momentum, 813
Spin quantum number, 812–13
Spin-echo technique, 908
Spin–orbit interaction, 813 *fn*
Spine, forces on, 238–39
Spiral galaxy, 950
Splitting of atomic energy levels, 842, 907
Spring:
  oscillation of, 293 *ff*
  potential energy of, 148, 154–55, 295–97
Spring constant, 148, 293
Spring equation, 148, 293
Spring force, 148
Spring scale, 76
Spring stiffness constant, 148, 293
Springs, car, 295, 303
SPS, 920
Spyglass, 725
Squark, 942
Stability, of particles, 932
  and tunneling, 876–77
Stable equilibrium, 240
Stable nucleus, 863
Standard candle, 957, 958
Standard conditions (STP), 370
Standard length, 9, 698
Standard mass, 10
Standard Model (SM):
  cosmological, 970–73
  elementary particles, 448 *fn*, 915, 916, 930, 935–39
Standard temperature and pressure (STP), 370
Standard temperature scale, 363
Standard of time, 9
Standards and units, 8–11
Standing waves, 315–17
  circular, 795–96
  fundamental frequency of, 316, 336, 337, 338
  natural frequencies of, 315–16
  resonant frequencies of, 315–16, 335, 336
  and sources of sound, 335–40
Stanford Linear Accelerator Center (SLAC), 920
Star clusters, 950
Stars: 894–96, 948–59 and *ff*
  birth and death of, 954–57
  black holes, 136 *pr*, 951, 956, 962–63, 975
  clusters of, 950

color of, 774, 952–53
distance to, 957–59
evolution of, 954–57
fixed, 3, 125
H–R diagram, 952–55, 958
heat death not from, 431
high mass, 955–56
low mass, 955
main-sequence, 953–55, 982 *pr*
neutron, 217, 951, 956–57
numbers of, 949
quasars, 951, 961
red giants, 951, 953–55
size of, 406, 952
source of energy of, 894–96, 954–56
Sun (*see* Sun)
supernovae, 929–30, 951, 955–58
temperature of, 952
types of, 951 and *ff*
variable, 958
white dwarfs, 951, 953, 955–57
Starter, car, 573
Statcoulomb, 448 *fn*
State:
  changes of, 377–81, 397–400
  energy, in atoms, 789–95
  equation of, for an ideal gas, 367, 370, 372
  of matter, 261, 360–61
  metastable, 820, 821–22, 869
  as physical condition of system, 367
State variable, 367, 413, 428 *fn*
Static electricity, 444 *ff*
Static equilibrium, 230–49
Static friction, 94, 204, 214
  coefficient of, 93, 94
Static rope, 107 *pr*
Statics, 230–49
  and center of mass (CM), 233
Stationary states, in atom, 789–96
Statistics:
  and entropy, 432–33
  Fermi–Dirac, 841
Stator, 598
Steady-state model of universe, 967
Steam engine, 420–21
  efficiency, 424
Steam power plants, 434, 435, 892–93
Stefan-Boltzmann constant, 403
Stefan-Boltzmann law (or equation), 403, 952
Stellar evolution, 954–57
Stellar fusion, 894–96
Step-down transformer, 601–2
Step-up transformer, 601–2
Stereo, 638 *fn*
Sterilization, 904

Stimulated emission, 820–23
STM, 786
Stopping a car, 32, 145
Stopping potential, 776
Stopping voltage, 776
Storage, information, 604–6
Storage rings, 920
Stored energy, as potential energy, 148
Stove, induction, 594
STP, 370
Strain, 243–44
Strain gauge, 525 *pr*
Strange particles, 932–33
Strange quark, 934
Strangeness, 931 *fn*, 932–33
  conservation of, 933
Strassman, Fritz, 889
Straw, 264, 267, 289 *pr*
Streamline (*defn*), 272–73
Streamline flow, 272–73
Strength of materials, 242, 245
Stress, 243–44
  compressive, 243–44
  shear, 244
  tensile, 243–44
  thermal, 367
String, 942
String theory, 19 *pr*, 942
Stringed instruments, 317, 336–37
Strings, vibrating, 315–16, 335–37
Stripping nuclear reaction, 911 *pr*
Stromboli, 64
Strong bonds, 829–32, 834–35, 840
Strong nuclear force, 129, 862, 888 *fn*, 922–42, 959
  and elementary particles, 922–42
Strongly interacting particles (*defn*), 930
Strontium-90, 883 *pr*, 913, 914
Structure:
  fine, 803, 813
  of universe, 973–75, 977–78
Subcritical reactions, 891, 894
Sublimation, 378
Sublimation point, 378
Subpixels, 491–92
Subshells, atomic, 816–17
Subtraction of vectors, 52–53
Suction, 267, 289 *pr*
Sun, 3, 126–29, 134 *pr*, 135 *pr*, 136 *pr*, 894–96, 948–49, 951–55
  energy source of, 894–96, 954
  mass of, 126–27
  reference frame of, 3, 125, 128
  surface temperature of, 403, 774
Sunglasses, polarized, 701

Sunsets, color, 704
Super Invar, 386 *pr*
Super Proton Synchrotron (SPS), 920
Supercluster, 950
Superconducting magnets, 572
Superconductivity, 517
Supercritical reactions, 891, 894
Superdome (New Orleans, LA), 248
Superfluidity, 378
Supernovae, 929–30, 951, 955–58
  as source of elements on Earth, 956
  type Ia, 956, 957, 958, 976
Superposition, principle of, 313–14, 340–41, 450–53, 455
Supersaturated air, 381
Supersonic speed, 329, 348–49
Superstring theory, 942
Supersymmetry, 939, 942
Supply voltage, 488
Surface area formulas, inside back cover
Surface of last scattering, 969
Surface tension, 280–82
Surface waves, 309–10
Surfactants, 281
Surgery, laser, 823
SUSY, 942
Swing, children's, 304
Symmetry, 14, 37, 62, 119, 183, 186, 218, 233, 447 *fn*, 449, 456, 457, 464, 465, 539, 574, 626, 627, 653, 670, 782, 934, 940, 942, 971
Symmetry breaking, 940, 971
Synapse, 517
Synchrocyclotron, 919
Synchrotron, 919
Synchrotron radiation, 919
Synodic period, 129, 137 *pr*
Système International (SI), 10, inside front cover
Systems, 90, 174, 394–96, 413
  closed, 394
  isolated, 174, 394–96
  open, 394
  overdamped, 303
  of particles, 187–89
  as set of objects, 90, 413
  underdamped, 303
  of units, 10
Systolic pressure, 283

Tacoma Narrows Bridge, 304
Tail-to-tip method of adding vectors, 51–52
Tangent, 39, 54, A-8
Tangential acceleration, 118, 201–3
Tape recorder, 604
Tau lepton, 927, 930–31, 935
Tau lepton number, 927, 930–31

Index **A-63**

Tau neutrino, 930–31
Technetium-99, 905
Technology generation, 851
Teeth, braces and forces on, 231
Telephone, cell, 318, 602, 604, 631, 639
Telephoto lens, 718
Telescope(s), 723–25, 730–32
   Arecibo, 731
   astronomical, 650, 724–25, 743 pr
   Galilean, 723, 723 fn, 725
   Hale, 725
   Hubble Space (HST), 136 pr, 730, 743 pr, 961
   Keck, 725
   Keplerian, 723 fn, 724
   magnification of, 724
   reflecting, 725
   refracting, 724
   resolution of, 730–32
   space, 730, 950, 961
   terrestrial, 725
Television, 490–92, 636–39, 703–4
   digital, 491–92
   high definition, 491–92
Television receiver, 638
TEM, 785
Temperature, 361–63, 368
   absolute, 362, 368
   Celsius (or centigrade), 362
   critical, 377, 517
   Curie, 579
   distinguished from heat and internal energy, 392
   Fahrenheit, 362
   gradient, 400–1
   human body, 363, 400
   Kelvin, 362, 368
   molecular interpretation of, 373–76
   operating (of heat engine), 420
   relation to chemical reactions, 377
   relation to molecular kinetic energy, 374–75, 392–93
   relation to molecular velocities, 373–77
   scales of, 362–63, 368
   standard scale, 363
   of star, 952
   transition, 517
   of the universe, 941
Temperature coefficient of resistivity, 508, 509–10
Temperature dependence of resistivity, 509–10
Tennis serve, 73 pr, 172, 176, 193 pr
Tensile strength, 245
Tensile stress, 243–44
Tension (stress), 243–44
Tension, surface, 280–82
Tension in flexible cord, 89
Terminal, of battery, 503, 505

Terminal velocity, 34 fn
Terminal voltage, 527–28
Terrestrial telescope, 725
Tesla (T) (unit), 565
Test charge, 453
Testing, of ideas/theories, 3
Tevatron, 762, 919, 920
TFT, 492
TFTR, 898
Theories (general), 3–5
Theories of everything, 942
Therm (unit), 391
Thermal conductivity, 400–1
Thermal conductor, 401
Thermal contact, 363
Thermal energy, 156–57, 392
   distinguished from heat and temperature, 392
   transformation of electric to, 510
   (see also Internal energy)
Thermal equilibrium, 363, 394–95
Thermal expansion, 364–67
   anomalous behavior of water below 4°C, 366–67
   coefficients of, 364
   linear expansion, 364–65
   volume expansion, 366
Thermal insulator, 401
Thermal pollution, 434–35
Thermal radiation, 403–6
Thermal resistance, 402
Thermal stress, 367
Thermal windows, 401
Thermionic emission, 490
Thermistor, 510
Thermodynamic processes, 414–18
   adiabatic, 415–16
   isobaric, 415
   isothermal, 414–15
   isovolumetric, 415
   work done in volume changes, 415–17
Thermodynamics, 363, 412–36
   first law of, 413–19
   second law of, 419–33
   third law of, 424
   zeroth law of, 363
Thermography, 405
Thermoluminescent dosimeter (TLD), 901
Thermometers, 361–63
   bimetallic-strip, 362
   constant-volume gas, 363
   liquid-in-glass, 362
   mercury-in-glass, 361–62
   resistance, 510
Thermonuclear devices, 897
Thermonuclear runaway, 957
Thermostat, 384 pr, 411 pr
Thin lens equation, 664–67
Thin lenses, 661–70
Thin-film interference, 693–98
Thin-film transistor (TFT), 492

Third law of motion, 81–83
Third law of thermodynamics, 424
Thomson, G. P., 783
Thomson, J. J., 772–73, 783, 784
Thought experiment, 749 and ff, 807
   definition, 749
Three Mile Island, 892
Three-dimensional waves, 310
Three-way lightbulb, 556 pr
Threshold energy, 887
Threshold of hearing, 335
Threshold of pain, 335
Thymine, 460
TIA, 278
Tidal wave, 306
Timbre, 340
Time:
   absolute, 746
   characteristic expansion, 967
   lookback, 951, 969
   Planck, 941, 970
   proper, 753, 945 pr
   relativity of, 749–55, 758–59
   standard of, 9
Time constant, 540, 610, 871
Time dilation, 750–55, A-25
Time-of-flight, 906
Time intervals, 9, 24
Time's arrow, 431
Tire pressure, 372
Tire pressure gauge, 266
TLD, 901
Tokamak, 898
Tokamak Fusion Test Reactor (TFTR), 898
Tomography, 735–37
   image formation, 736
   image reconstruction, 736–37
Tone color, 340
Toner, 462
Toothbrush, electric, 604
Top quark, 934
Topographic map, 478
Toroid, 580, 586 pr, 898
Toroidal field, 898
Torque, 206–8 ff
   counter, 600
   on current loop, 575–76
   work done by, 214
Torr (unit), 266
Torricelli, Evangelista, 266–67, 276
Torricelli's theorem, 276
Torsion balance, 447
Torus, 580, 586, 898
Total binding energy, 769 pr, 861
Total internal reflection, 327 pr, 659–61
Total magnifying power, 724
Total mechanical energy (defn), 150, 295–97
Townsend, J. S., 773
Tracers, 904–5
Traffic light, LED, 848

Transfer-RNA (t-RNA), 836–37
Transformation of energy, 155–56, 159
Transformations:
   Galilean, A-22–A-25
   Galilean velocity, A-23–A-24
   Lorentz, A-24–A-25
Transformer, 601–3, 608
Transformer equation, 601
Transfusion, blood, 288 pr, 289 pr
Transient ischemic attack (TIA), 278
Transistors, 492, 605–6, 845, 850–51
   bipolar junction, 850–51
   field-effect, 851
   floating gate, 606
   metal-oxide semiconductor field-effect (MOSFET), 605, 851
   thin-film, 492
Transition elements, 817
Transition temperature, 517
Transitions, in atoms, 789–95, 812, 817–19, 820–23
   in molecules, 837–39
   forbidden and allowed, 814, 821 fn, 838–39
   for lasers, 820–23
   in semiconductors, 843–45, 847–50
   in nuclei, 868–69
Translational kinetic energy (defn), 142–45
Translational motion, 21–189
   and center of mass (CM), 187–89
   kinematics for, 21–40, 49–66
Transmission axis, of Polaroid, 699–701
Transmission electron microscope (TEM), 785
Transmission grating, 690 ff
Transmission lines, 603, 631
Transmission of electricity, 603
Transmission of power, wireless, 604
Transmission of waves, 312–13
Transmutation of elements, 864, 885–89
Transparency, 843
Transuranic elements, 888
Transverse waves, 307 ff
   and earthquakes, 309
   EM waves, 629
   speed of, 308
Traveling sinusoidal wave, 319
Trees, offsetting CO_2 emissions, 442 pr
Triangle, on a curved surface, 961
Triangulation, 14, 957
Trigonometric functions and identities, 54, 56, 63, A-8–A-9, inside back cover
Trigonometric table, A-9

Triple point, 378
Tritium, 858, 883 *pr*, 897, A-12
Tritium dating, 883 *pr*
Triton, 858, A-12
t-RNA, 836–37
Trough, wave, 306, 313–14
True north, 562
Tsunami, 306, 327 *pr*
Tubes:
   discharge, 772, 787
   flow in, 273–76, 278, 279–80
   open and closed, 338
   Venturi, 278
   vibrating column of air in, 335 *ff*
Tunneling:
   through a barrier, 876–77
   in a microscope, 786
   quantum mechanical, 606, 876–77
   and stability, 876–77
Turbine, 434–35, 597
Turbulent flow, 273, 277
TV, 490–92, 636–39, 703–4
22-nm technology, 851
Twin paradox, 754–55
Two-dimensional collisions, 182–83
Two-dimensional waves, 310
Tycho Brahe, 125
Type Ia supernovae (SNIa), 956, 957, 958, 976
Tyrolean traverse, 258 *pr*

UA1 detector, 924
Ultimate speed, 760
Ultimate strength, 241, 245
Ultracapacitors, 499 *pr*
Ultrasonic frequencies, 329, 350
Ultrasonic waves, 329, 347, 350
Ultrasound, 350
Ultrasound imaging, 350–51
Ultraviolet (UV) light, 630, 686, 693
Unavailability of energy, 431
Unbanked curves, 115–17
Uncertainty (in measurements), 5–8, 806–9
   estimated, 6
   percent, 6, 8
Uncertainty principle, 806–9, 830
   and particle resonance, 932
   and tunneling, 877
Underdamped system, 303
Underexposure, 715
Underground air circulation, 278
Undersea research chamber, 287 *pr*
Underwater vision, 721
Unification distance, 946 *pr*

Unification scale, 940
Unified (basis of forces), 938
Unified atomic mass units (u), 10, 360, 860
Unified theories, grand (GUT), 129, 939–41
Uniform circular motion, 110–15
   dynamics of, 112–15
   kinematics of, 110–12
   related to simple harmonic motion, 299–300
Uniform magnetic field, 562
Uniformly accelerated motion, 28 *ff*, 58 *ff*
Uniformly accelerated rotational motion, 203–4
Unit conversion, 11–12, inside front cover
Units and standards, 8–11
Units of measurement, 8–10
   converting, 11–12, inside front cover
   prefixes, 10
   in problem solving, 12, 30
Universal gas constant, 370
Universal law of gravitation, 119–21, 448, 959
Universe:
   age of, 940 *fn*, 967
   Big Bang theory of, 941, 966 *ff*
   CDM model of, 977–78
   critical density of, 975
   curvature of, 961–63, 974–75
   entire, 970
   expanding, 964–67, 975–77
   finite or infinite, 948, 962–63, 967, 975
   future of, 975–77
   history of, 970–73
   homogeneous, 966
   inflationary scenario of, 971, 973–75
   isotropic, 966
   large-scale structure, 977–78
   matter-dominated, 972, 973
   observable, 967, 969–70, 974
   origin of elements in, 955–56
   radiation-dominated, 972–73
   Standard Model of, 970–73
   steady-state model of, 967
   temperature of, 941
Unknowns, solving for, A-4–A-5
Unphysical solution, 31, 36
Unpolarized light (*defn*), 699
Unstable equilibrium, 240
Unstable nucleus, 863 *ff*
Up quark, 934
Updating monitor screen, 492
Uranium:
   in dating, 873–77
   decay, 865
   enriched, 891

   fission of, 889–94
   in reactors, 889–94
Uranus, 126, 127
Useful magnification, 732, 785
UV light, 630, 686, 693

Vacuum energy, 977
Vacuum pump, 267, 282
Vacuum state, 925–26, 974
Valence, 817
Valence bands, 842–43
Van de Graaff generator, 459
van der Waals bonds and forces, 834–37
Vapor (*defn*), 378 (*see also* Gases)
Vapor pressure, 379
Vaporization, latent heat of, 397–98, 399, 400
Variable stars, 958
Varying force, 142
Vector, escape, 384 *pr*
Vector displacement, 23, 50–52
Vector field, 457
Vector form of Coulomb's law, 450–53
Vector quantities, 50
Vector sum, 50–57, 87, 173
Vectors, 23, 50–57
   addition of, 50–57, 87, 450
   components of, 53–57
   direction of, 23, 50
   displacement, 23, 50–52
   force, addition of, 87
   magnitude of, 23, 50
   multiplication, by a scalar, 52–53
   multiplication of, 52–53
   negative of (*defn*), 52
   parallelogram method of adding, 52
   resolution of, 53–57
   resultant, 50–52, 56–57
   subtraction of, 52–53
   sum, 50–57, 87
   tail-to-tip method of adding, 51–52
Velocity, 23–25, 50
   acceleration vs., 27
   addition of, 65–66
   angular, 200–3
   average (*defn*), 23–25, 28, 39
   drift, 516, 569
   of EM waves, 629
   as a function of time (SHM), 301
   gradient, 279
   instantaneous (*defn*), 25, 39–40
   of light, 9, 629–33, 656, 681, 747, 748
   molecular:
      and relation to temperature, 376–77
      and probability, 433
   relative, 65–66, 179

   relativistic addition of, 764
   relativistic transformations, A-24
   rms (root-mean-square), 376
   and slope, 39–40
   of sound, 329
   supersonic, 329, 348–49
   terminal, 34 *fn*
   of waves, 306, 308–9
Velocity selector, 578
Veneziano, Gabriele, 942
Ventricular fibrillation, 487, 543
Venturi meter, 278, 288 *pr*
Venturi tube, 278
Venus, 3, 125, 126, 128, 134 *pr*, 723
Vertical (*defn*), 84 *fn*
Vibrating strings, 315–16, 335–37
Vibration, 292 *ff*
   of air columns, 337–40
   amplitude of, 294
   floor, 299
   forced, 304
   frequency of, 294, 303, 315–17
   molecular, 326 *pr*, 393, 839
   period of, 294
   as source of waves, 306
   of spring, 293 *ff*
   on strings, 315–16, 335–37
   (*see also* Oscillations)
Vibrational energy, 295–97
   molecular, 393, 839
Vibrational quantum number, 839
Vibrational transition, 839
Virtual image, 647, 664
Virtual particles, 922
Virtual photon, 922
Virus, 9, 785
Viscosity, 273, 279–80
   coefficient of, 279
Viscous force, 279–80
Visible light, wavelengths of, 630, 685–87
Visible spectrum, 685–87
Vitreous humor, 719
Volatile memory, 606
Volt (V) (unit), 475
Volt-Ohm-Meter/Volt-Ohm-Milliammeter (VOM), 548
Volta, Alessandro, 475, 483, 502
Voltage, 473, 475 *ff*, 503 *ff*, 527 *ff*
   base bias, 850
   bias, 845
   breakdown, 477
   electric field related to, 477
   hazards of, 543–45
   measuring, 546–48
   peak, 514
   ripple, 847
   rms, 514–15
   signal, 488–89
   supply, 488

Voltage (continued)
  terminal, 527–28
  (see also Electric potential)
Voltage divider, 530
Voltage drop, 507, 533 (see Voltage)
Voltage gain (defn), 851
Voltaic battery, 502
Voltmeter, 546–48, 576
  connecting, 547
  digital, 546, 548
  resistance, effect of, 547–48
  sensitivity, 547
Volume change under pressure, 244, 415–17
Volume expansion (thermal), 364, 366
  coefficient of, 364, 366
Volume formulas, inside back cover
Volume holograms, 824
Volume rate of flow, 273
VOM, 548
von Laue, Max, 734
Vonn, Lindsey, 97

W$^\pm$ particles, 924, 930–32, 937
Walking, 82
Walls and voids of galaxies, 978
Watch-face LCD display, 704
Water:
  anomalous behavior below 4°C, 366–67
  cohesion of, 281
  density of, 261–62, 271
  dipole moment of, 482
  and electric shock, 544
  expansion of, 366
  heavy, 891
  latent heats of, 397
  molecule, 497 pr, 832, 833
  polar nature of, 445, 482, 832
  properties of: inside front cover
  saturated vapor pressure, 379
  specific gravity of, 262, 271
  specific heat of, 393
  supply, 263
  thermal expansion of, 366
  triple point of, 378
  waves, 305 ff, 314, 317, 318
Water barometer, 267
Water strider, 281
Watson, J., 735
Watt, James, 159 fn
Watt (W) (unit), 159, 511
Wave(s), 305–19, 627 ff, 679–704
  amplitude, 294, 306, 310, 319, 333, 804–6
  bow, 348–49
  complex, 341
  composite, 340
  compression, 307, 309
  continuous (defn), 306

crest, 306, 313–14
diffraction of, 318, 680, 687–93
dispersion, 686
displacement of, 319
earthquake, 309, 310–11, 318
electromagnetic, 625–39 (see also Light)
energy in, 310–11
expansions in, 307
frequency, 306
front, 312, 680
function, 804–6
gravity, 978
in-phase, 314
incident, 313, 317
infrasonic, 330
intensity, 310–11, 331–33, 634–35, 688–90
interference of, 313–14, 341–43, 682–85
light, 629–33, 679–704 (see also Light)
linear, 310
longitudinal (defn), 307 ff
mathematical representation of, 319
of matter, 782–84, 795–96, 804 ff
mechanical, 305–19
motion of, 305–19
one-dimensional, 310
out-of-phase, 314
P, 309, 311
packet, 327 pr
particles vs., 680
period of, 306
periodic (defn), 306
phase of, 314
plane, 312–13, 628, 682
power, 310
pressure, 309, 330 ff
pulse, 306
radio, 630, 639, 731
rarefactions in, 307
reflection of, 312–13
refraction of, 317–18
S, 309
shear, 309
shock, 348–49
sinusoidal traveling, 319
sound, 307, 309, 328–51, 631
source of, oscillations as, 306
speed of, 306, 308–9 (see also Speed of light; Speed of sound)
spherical, 310, 312
standing, 315–17, 335–40
  on a string, 315–16, 335–37
surface, 309
three-dimensional, 310
tidal, 306
transmission of, 312–13
transverse, 307 ff, 308, 309, 629, 699
traveling, 319

trough, 306, 313–14
two-dimensional, 310
types of, 307–10 (see also Light)
ultrasonic, 329, 347, 350
velocity of, 306, 308–9, 629
water, 305 ff (see also Light)
Wave displacement, 319, 804–5
Wave front, 312, 680
Wave function, 804–6
  for H atom, 830
Wave intensity, 310–11, 331–33, 634–35, 688–90
Wave-interference phenomenon, 682
Wave motion (see Wave(s); Light; Sound)
Wave nature of electron, 806
Wave nature of matter, 782–84, 795–96
Wave packet, 327 pr
Wave theory of light, 679–704, 776
Wave velocity, 306, 308–9, 629 (see also Light; Sound)
Wave–particle duality:
  of light, 782
  of matter, 782–84, 795–96, 804–9
Waveform, 340–41
Wavelength (defn), 306, 314 fn
  absorption, 793
  Compton, 780
  cutoff, 818–19
  de Broglie, 782–83, 795–96, 805, 917
  depending on index of refraction, 681, 686
  as limit to resolution, 732, 917
  of material particles, 782–83, 795–96
  of spectral lines, 792–93
Weak bonds, 460, 461, 834–37, 840
Weak charge, 937
Weak nuclear force, 129, 863, 867, 924–42, 959
  range of, 938
Weakly interacting massive particles (WIMPS), 976
Weather, 381
  and Coriolis effect, A-18
  forecasting, and Doppler effect, 348
Weber (Wb) (unit), 592
Weight, 76, 78, 84–86, 121–22
  apparent, 124–25, 270
  atomic, 360
  as a force, 78, 84
  force of gravity, 76, 84–86, 121–22
  mass compared to, 78, 84
  molecular, 360
Weightlessness, 124–25

Weinberg, S., 938
Wess, J., 942
Whales, echolocation in, 309
Wheatstone bridge, 556 pr
Whirlpool galaxy, 950
White dwarfs, 951, 953, 955–57
White light, 686
White-light holograms, 824
White-light LED, 848
Whole-body dose, 902
Wide-angle lens, 718, 728
Width, of resonance, 932
Wien's (displacement) law, 774, 952, 953
Wilkinson, D., 968
Wilkinson Microwave Anisotropy Probe (WMAP), 931 fn, 968
Wilson, Robert, 919 fn, 967–68
WIMPS, 976
Wind:
  as convection, 402
  and Coriolis effect, A-18
  noise, 340
  power, 435
Wind instruments, 317, 337–40
Windings, 577
Windows:
  heat loss through, 401
  thermal, 401
Windshield wipers, intermittent, 543
Wing of an airplane, lift on, 277
Wire, ground, 544–45
Wire drift chamber, 878 fn
Wire proportional chamber, 878 fn
Wire-wound resistor, 506
Wireless communication, 625, 636–39
Wireless transmission of power, 604
Wiring, electrical, 545
Witten, Edward, 942
WMAP, 931 fn, 968
Word-line, 605
Work, 138–45, 155, 391, 412–19
  to bring positive charges together, 480
  compared to heat, 412
  defined, 139, 412 ff
  done by a constant force (defn), 139–42
  done by an electric field, 474
  done by a gas, 414 ff
  done by torque, 214
  done by a varying force, 142
  done in volume changes, 415–17
  in first law of thermodynamics, 413–19
  graphical analysis for, 142
  from heat engines, 420 ff
  on the Moon, 142
  negative, 140
  and power, 159–61

Work (*continued*)
   relation to energy, 142–47, 155, 157–61
   units of, 139
Work function, 776–77
Work-energy principle, 142–45, 150, 760, 921
   energy conservation vs., 157
   as reformulation of Newton's laws, 144
Working off calories, 392
Working substance (*defn*), 421
Wrench, 223 *pr*
Wright, Thomas, 948
Writing data, 605

XDF (Hubble eXtreme Deep Field), 947, 961
Xerox (*see* Photocopier)
Xi (particle), 931
Xi—anti-Xi pair, 925
X-rays, 630, 733–36, 817–19, 869
   and atomic number, 817–19
   characteristic, 818
   in electromagnetic spectrum, 630
   spectra, 817–19
X-ray crystallography, 734
X-ray diffraction, 733–35
X-ray image, normal, 735
X-ray scattering, 780

YBCO superconductor, 517
Yerkes Observatory, 724
Yosemite Falls, 155
Young, Thomas, 682, 685
Young's double-slit experiment, 682–85, 690, 805–6
Young's modulus, 241–42
Yo-yo, 227 *pr*
Yttrium, barium, copper, oxygen superconductor (YBCO), 517
Yukawa, Hideki, 922–23
Yukawa particle, 922–23

Z (atomic number), 815, 817–19, 858

Z^0 particle, 826 *pr*, 915, 924, 930–32, 937
Z-particle decay, 924
Zeeman effect, 812
Zener diode, 846
Zero, absolute, temperature of, 368, 424
Zero-point energy, 839
Zeroth law of thermodynamics, 363
Zoom, digital, 718
Zoom lens, 718
Zumino, B., 942
Zweig, G., 934

Photo Credits

Front cover D. Giancoli **Cover inset** Science Photo Library/Alamy **Back cover** D. Giancoli

p. iii Reuters/NASA **p. iv** Scott Boehm/BCI **p. v** Pixland/age fotostock **p. vi** AFP/Getty Images/Newscom **p. vii** Giuseppe Molesini, Istituto Nazionale di Ottica, Florence **p. viii** Professors Pietro M. Motta & Silvia Correr/Photo Researchers, Inc. **p. ix** NASA, ESA, R. Ellis (Caltech), and the UDF 2012 Team **p. xvii** D. Giancoli

CO–1 Reuters/NASA **1–1** Erich Lessing/Art Resource **1–2a/b, 1–3** Franca Principe/Istituto e Museo di Storia della Scienza, Florence **1–4a** fotoVoyager/iStockphoto.com **1–4b** AP Photo/The Minnesota Daily, Stacy Bengs **1–5** Mary Teresa Giancoli **1–6a/b** Pearson Education/Travis Amos **1–7** Paul Silverman/Fundamental Photographs **1–8a** Oliver Meckes/Ottawa/Photo Researchers, Inc. **1–8b** D. Giancoli **1–9** Art Wolfe/Getty Images **1–10a** D. Giancoli **1–11** Larry Voight/Photo Researchers, Inc. **1–16** David Parker/Science Photo Library/Photo Researchers, Inc. **1–17** Evan Sklar/Food Pix/Getty Images

CO–2 NASA/Bill Ingalls **2–8** BlueMoon Stock/Superstock **2–13** Yamaha Motor Racing **2–14** Staten Island Academy **2–18** Scala/Art Resource, NY **2–19** Harold E. Edgerton/Palm Press, Inc. **2–22** Fred de Noyelle/Godong/Corbis **2–24** jsaunders84/iStockphoto

CO–3 Lucas Kane/Lucas Kane Photography **3–17a** Berenice Abbott/Commerce Graphics Ltd., Inc. **3–17b** American Association of Physics Teachers/Greg Gentile **3–19** Richard Megna/Fundamental Photographs **3–27 top** American Association of Physics Teachers/Joey Moro **3–27 bottom** Westend61 GmbH/Alamy

CO–4 NASA/John F. Kennedy Space Center **4–1** Daly & Newton/Stone Allstock/Getty Images **4–4** Bettmann/Corbis **4–5** Gerard Vandystadt/Agence Vandystadt/Photo Researchers, Inc. **4–7** David Jones/Photo Researchers, Inc. **4–10** NASA/John F. Kennedy Space Center **4–33** AFP PHOTO/DIMITAR DILKOFF/Getty Images/Newscom **4–37** Lars Ternblad/Amana Japan **4–39** Prof. Melissa Vigil **4–41** Prof. Nicholas Murgo **4–42** Prof. Martin Hackworth **4–44** Kathleen Schiaparelli **4–46** Brian Bahr/Allsport Photography/Getty Images **4–71** Tyler Stableford/The Image Bank/Getty Images

CO–5 Earth Imaging/Stone Allstock/Getty Images **5–6c** PeterFactors/Fotolia **5–12** Guido Alberto Rossi/TIPS North America **5–20** D. Giancoli **5–21, 5–25** NASA **5–26a** AP Wide World Photos **5–26b** Mickey Pfleger **5–26c** Dave Cannon/Getty Images **5–27** David Nunuk/Photo Researchers, Inc. **5–40** Photofest **5–47** Daniel L. Feicht/Cedar Fair Entertainment Company **5–48** Tierbild Okapia/Photo Researchers, Inc.

CO–6 Scott Boehm/BCI **6–20** Harold E. Edgerton/Palm Press **6–21** AP Photo/John Marshall Mantel **6–24** Inga Spence/Getty Images **6–25** AP Photo/Rob Griffith **6–28** Nick Rowe/Getty Images **6–32** M.C. Escher's "Waterfall" © 2013 The M.C. Escher Company—The Netherlands. All rights reserved. www.mcescher.com <http://www.mcescher.com> **6–33** Prof. Walter H. G. Lewin **6–39** Airman Krisopher Wilson/US Department of Defense **6–44** Columbia Pictures/Photofest **6–46** age fotostock **6–47** Bettmann/Corbis

CO–7 Richard Megna/Fundamental Photographs **7–1** ADRIAN DENNIS/Staff/AFP/Getty Images **7–8** Ted Kinsman/Photo Researchers, Inc. **7–11** Edward Kinsman/Photo Researchers, Inc. **7–14** D. J. Johnson **7–17** Science Photo Library/Photo Researchers, Inc. **7–21** Berenice Abbott/Photo Researchers, Inc. **7–28** Kathy Ferguson/PhotoEdit

CO–8 David R. Frazier/The Image Works **8–7** Mary Teresa Giancoli **8–11a** Richard Megna/Fundamental Photographs **8–11b** Photoquest **8–33** Tim Davis/Lynn Images/Corbis **8–34** AP Photo/The Canadian Press/Clement Allard **8–35** AP Photo **8–41** Karl Weatherly/Getty Images **8–42** Tom Stewart/Corbis **8–68a** Friedemann Vogel/Staff/Bongarts/Getty Images

CO–9 Fraser Hall/Getty Images **9–1** AP Photo **9–21** D. Giancoli **9–23a/b** Mary Teresa Giancoli **9–26** Tarek El Sombati/iStockphoto **9–28a** D. Giancoli **9–28b** Christopher Talbot Frank/Ambient Images, Inc./Alamy Images **9–30** D. Giancoli **9–32** Richard Carafelli/National Gallery of Art **9–33** D. Giancoli **9–34** Italian Government Tourist Board **9–48a** James Lemass/Photolibrary

CO–10 top 1929 Massimo Terzano - © Ass. Ardito Desio/Maria Emanuela Desio **CO–10 bottom** 2009 Fabiano Ventura - © Archive F. Ventura **10–9** Corbis/Bettmann **10–18a/b** Gary S. Settles and Jason Listak/Photo Researchers, Inc. **10–26** Mike Brake/Shutterstock **10–32a** Corbis/Bettmann **10–32b** Biophoto Associates/Photo Researchers, Inc. **10–33** Larry West/Stockbyte/Getty Images **10–35a** Alan Blank/Photoshot Holdings Ltd./Bruce Coleman **10–35b** American Association of Physics Teachers/Shilpa Hampole **10–44** D. Giancoli **10–47** Adam Jones/Photo Researchers, Inc. **10–48** American Association of Physics Teachers/Anna Russell **10–52** NASA Goddard Space Flight Center/Science Source/Photo Researchers, Inc.

CO–11 left Ford Motor Company **CO–11 right** Jonathan Nourok/PhotoEdit **11–4** Ford Motor Company **11–6** Paul Springett A/Alamy **11–11** Berenice Abbott/Photo Researchers, Inc. **11–13** D. Giancoli **11–17** Taylor Devices **11–19** Martin Bough/Fundamental Photographs **11–20a** AP Wide World Photos **11–20b** Paul X. Scott **11–28** Art Wolfe/Getty Images **11–21a/b/c/d, 11–38** D. Giancoli **11–43** Jacynthroode/iStockphoto **11–45a/b** Richard Megna/Fundamental Photographs **11–49** Andre Gallant/Getty Images **11–59** Richard Megna/Fundamental Photographs

CO–12 Scala/Art Resource **12–4** Willie Maldonado/Getty Images **12–9a** Ben Clark/Getty Images **12–9b** Tony Cenicola/The New York Times **12–10** Bob Daemmrich/The Image Works **12–24** Prof. Vickie Frohne **12–25** Gary S. Settles/Photo Researchers, Inc. **12–26** GE Medical Systems/Photo Researchers, Inc. **12–28a** Visions of America/SuperStock **12–35** Nation Wong/Corbis **12–42** D. Giancoli

CO–13 left Niall Edwards/Alamy **CO–13 right** Richard Price/Getty **13–3** Ingolfson **13–4b** Franca Principe/Istituto e Museo di Storia della Scienza, Florence **13–6** Leonard Lessin/Photolibrary **13–12** Pearson Education/Eric Schrader **13–16** D-BASE/Getty Images **13–17** Michael Newman/PhotoEdit **13–25** Paul Silverman/Fundamental Photographs **13–26a** Norbert Rosing/National Geographic Stock **13–26b** American Association of Physics Teachers/Mrinalini Modak **13–26c** Prof. Gary Wysin **13–27a/b/c** Mary Teresa Giancoli **13–30** Kennan Harvey/Getty Images **13–31** Reed Kaestner/Corbis

CO–14 Pixland/age fotostock **14–10** Ernst Haas/Hulton Archive/Getty Images **14–13a/b** Science Photo Library/Photo Researchers, Inc. **14–14** Prof. Vickie Frohne **14–16** Stacey Bates/Shutterstock **14–21** Tobias Titz/Getty Images

CO–15 Taxi/Getty Images **15–9** Will Hart **15–10 left, middle, right** Leonard Lessin/Photolibrary **15–20a** Carsten Koall/Getty Images **15–20b** Kevin Burke/Corbis **15–20c** David Woodfall **Table 15–4 top left** Michael Collier **Table 15–4 bottom left** Michel de Nijs/iStockphoto **Table 15–4 mid-top** Larry Lee Photography/Corbis **Table 15–4 mid-bottom** Richard Schmidt-Zuper/iStockphoto **15–25** Patrick Landmann/Photo Researchers, Inc. **15–26** Richard Schmidt-Zuper/iStockphoto **15–27** Michael Collier

CO–16 Mike Dunning/Dorling Kindersley **16–37** Peter Menzel/Photo Researchers, Inc. **16–38** Dr. Gopal Murti/Science Source/Photo Researchers, Inc. **16–48** American Association of Physics Teachers/Matthew Claspill

CO–17 Emily Michot/Miami Herald/MCT/Newscom **17–8** D. Giancoli **17–13c, 17–18, 17–20** Eric Schrader/Pearson Education **17–21** tunart/iStockphoto **17–31 left** Eric Schrader/Pearson Education **17–31 right** Robnil **17–36** beerkoff/Shutterstock **17–44** Andrea Sordini

CO–18 left Mahaux Photography **CO–18 right** Eric Schrader/Pearson Education **18–1** Jean-Loup Charmet/Science Photo Library/Photo Researchers, Inc. **18–2** The Burndy Library Collection/Huntington Library **18–6a** Richard Megna/Fundamental Photographs **18–11** T. J. Florian/Rainbow Image Library **18–15** Richard Megna/Fundamental Photographs **18–16** Tony Freeman/PhotoEdit **18–18** Clint Spencer/iStockphoto **18–32** Alexandra Truitt & Jerry Marshall **18–34** Scott T. Smith/Corbis **18–36** Jim Wehtje/Getty Images

CO–19 Patrik Stoffarz/AFP/Getty Images/Newscom **19–15** David R. Frazier/Photolibrary, Inc./Alamy **19–24** Apogee/Photo Researchers, Inc. **19–27a** Photodisc/Getty Images **19–27b, 19–28, 19–30a** Eric Schrader/Pearson Education **19–30b** Olaf Doring/Imagebroker/AGE Fotostock **19–67** Raymond Forbes/AGE fotostock **19–74** Eric Schrader/Pearson Education

CO–20 Richard Megna/Fundamental Photographs **20–1** Dorling Kindersley **20–4a** Stephen Oliver/Dorling Kindersley **20–6** Mary Teresa Giancoli **20–8a/b, 20–18** Richard Megna/Fundamental Photographs **20–20b** Jack Finch/Science Photo Library/Photo Researchers, Inc. **20–28b, 20–43** Richard Megna/Fundamental Photographs **20–50** Clive Streeter/Dorling Kindersley, Courtesy of The Science Museum, London

CO–21 Richard Megna/Fundamental Photographs **21–7** Photo Courtesy of Diva de Provence, Toronto, ON, Canada **21–12** Jeff Hunter/Getty Images **21–21** Associated Press Photo/Robert F. Bukaty **21–22** Photograph by Robert Fenton Houser **21–27a** Terence Kearey **21–34a** 4kodiak/iStockphoto **21–34b** Eric Schrader/Pearson Education

CO–22 NASA **22–1** Original photograph in the possession of Sir Henry Roscoe, courtesy AIP Emilio Segrè Visual Archives **22–9** The Image Works Archives **22–12** Time Life Pictures/Getty Images **22–19** David J. Green/Alamy Images **22–20** Don Baida **22–22** NASA

CO–23, 23–6 D. Giancoli **23–11a** Mary Teresa Giancoli and Suzanne Saylor **23–11b** Paul Silverman/Fundamental Photographs **23–20** John Lawrence/Travel Pix Ltd. **23–22a** Shannon Fagan/age fotostock **23–23** Giuseppe Molesini, Istituto Nazionale di Ottica, Florence **23–30b** Garo/Phanie/Photo Researchers, Inc. **23–31c/d** D. Giancoli **23–32** D. Giancoli and Howard Shugat **23–34** Kari Erik Marttila Photography **23–38a/b, 23–47** D. Giancoli **23–50** Mary Teresa Giancoli **23–65** American Association of Physics Teachers/Annacy Wilson **23–66** American Association of Physics Teachers/Matt Buck **23–69** American Association of Physics Teachers/Sarah Lampen **23–70a/b** Scott Dudley

CO–24 Giuseppe Molesini, Istituto Nazionale di Ottica, Florence **24–4a** Kent Wood/Photo Researchers, Inc. **24–9a** Bausch & Lomb Incorporated **24–13** David Parker/Science Photo Library/Photo Researchers, Inc. **24–16b** Lewis Kemper Photography **24–17** George Diebold **24–19a** P. M. Rinard/American Journal of Physics **24–19b** Richard Megna/Fundamental Photographs **24–19c** Ken Kay/Fundamental Photographs **24–28** Wabash Instrument Corp./Fundamental Photographs **24–29a** Giuseppe Molesini, Istituto Nazionale di Ottica, Florence **24–29b** Richard Megna/Fundamental Photographs **24–29c** Paul Silverman/Fundamental Photographs **24–31b** Ken Kay/Fundamental Photographs **24–33b/c** Bausch & Lomb Incorporated **24–35** D. Hurst/Alamy **24–44** Diane Schiumo/Fundamental Photographs **24–47a/b** JiarenLau Photography, http://creativecommons.org/licenses/by/2.0/deed.en **24–52** Suunto **24–53** Daniel Rutter/Dan's Data

CO–25, 25–5, 25–6a/b Mary Teresa Giancoli **25–08** Leonard Lessin/Photolibrary **25–19a/b** Museo Galileo - Istituto e Museo di Storia della Scienza **25–21** Yerkes Observatory, University of Chicago **25–22c** Sandy Huffaker/Stringer/Getty Images **25–22d** Inter-University Centre for Astronomy and Astrophysics/Laurie Hatch **25–24b** Leica Microsystems **25–29a/b** Reproduced by permission from M. Cagnet, M. Francon, and J. Thrier, The Atlas of Optical Phenomena. Berlin: Springer-Verlag, 1962. **25–32** Space Telescope Science Institute **25–33** David Parker/Photo Researchers, Inc. **25–37** The Burndy Library Collection/Huntington Library **25–40** Rosalind Franklin/Photo Researchers, Inc. **25–44a** Martin M. Rotker **25–44b** Scott Camazine/Alamy **25–47** Ron Chapple/Ron Chapple Photography **25–48** NOAA Space Environment Center

CO–26 "The City Blocks Became Still Shorter" from page 4 of the book "Mr Tompkins in Paperback" by George Gamow. Reprinted with the permission of Cambridge University Press. **26–1** Bettmann/Corbis **26–7** D. Giancoli **26–12** "Unbelievably Shortened" from page 3 of "Mr Tompkins in Paperback" by George Gamow. Reprinted with the permission of Cambridge University Press.

CO–27 left Paul Gunning/Photo Researchers, Inc. **CO–27 right** Professors Pietro M. Motta & Silvia Correr/Photo Researchers, Inc. **27–11** Samuel Goudsmit/AIP Emilio Segrè Visual Archives, Goudsmit Collection **27–12** Education Development Center, Inc. **27–16a** Lee D. Simon/Science Source/Photo Researchers, Inc. **27–16b** Oliver Meckes/Max-Planck-Institut-Tubingen/Photo Researchers, Inc. **27–16c** Science Photo Library/Alamy **27–22b** Richard Megna/Fundamental Photographs **27–23a/b/c** Wabash Instrument Corp./Fundamental Photographs

CO–28 Richard Cummins/Corbis **28–1** Niels Bohr Institute, courtesy AIP Emilio Segrè Visual Archives **28–2** F. D. Rosetti/American Institute of Physics/Emilio Segrè Visual Archives **28–4** Hitachi, Ltd., Advanced Research Laboratory **28–15** Paul Silverman/Fundamental Photographs **28–22** NIH/Photo Researchers, Inc. **28–23** AP Photo/Dusan Vranic

CO–29 Intel **29–35** Jeff J. Daly/Fundamental Photographs **29–36** D. Giancoli **29–39** General Motors

CO–30 D. Giancoli **30–3** French Government Tourist Office **30–7** Enrico Fermi Stamp Design/2001 United States Postal Service. All Rights Reserved. Used with Permission from the US Postal Service and Rachel Fermi. **30–16** Fermilab Visual Media Services

CO–31 Peter Beck/Corbis **31–6** Archival Photofiles, [apf2-00502], Special Collections Research Center, University of Chicago Library **31–9** Tokyo Electric Power Company/Jana Press/ZUMAPRESS.com/Newscom **31–10** LeRoy N. Sanchez/Los Alamos National Laboratory **31–11** Corbis/Bettmann **31–15a** UPPA/Photoshot/Newscom **31–15b** National Ignition Facility/Lawrence Livermore National Laboratory **31–19a** Robert Turgeon, Cornell University **31–19b** Jack Van't Hof/Brookhaven National Laboratory **31–20b** Needell M.D./Custom Medical Stock Photography **31–22a/b** National Institute of Health **31–25b** Slaven/Custom Medical Stock Photo **31–27** Scott Camazine/Photo Researchers, Inc.

CO–32 CERN **32–1** Lawrence Berkeley National Laboratory **32–3** Reidar Hahn/Fermilab Visual Media Services **32–5a/b, 32–6** CERN **32–10a** David Parker/Science Photo Library/Photo Researchers, Inc. **32–10b** Science Photo Library/Photo Researchers, Inc. **32–11** Brookhaven National Laboratory **32–12** Lawrence Berkeley National Laboratory **32–15** Fermilab/Science Photo Library/Photo Researchers, Inc. **32–20** Denis Balibouse/AFP/Getty Images

CO–33 NASA, ESA, R. Ellis (Caltech), and the UDF 2012 Team **33–1a** NASA **33–1b** Allan Morton/Dennis Milon/Science Photo Library/Photo Researchers, Inc. **33–2c** NASA **33–3** U.S. Naval Observatory/NASA **33–4** National Optical Astronomy Observatories **33–5a** Reginald J. Dofour, Rice University **33–5b** U.S. Naval Observatory/NASA **33–5c** National Optical Astronomy Observatories **33–9a/b** Australian Astronomical Observatory/David Main Images **33–15a** NASA **33–23** Alcatel-Lucent **33–25** AP Photo/Fredrik Persson **33–26** WMAP Science Team/NASA **33–32** Bertil Ericson/SCANPIX/AP/Corbis

Useful Geometry Formulas — Areas, Volumes

Circumference of circle $C = \pi d = 2\pi r$

Area of circle $A = \pi r^2 = \dfrac{\pi d^2}{4}$

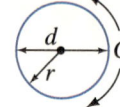

Area of rectangle $A = \ell w$

Area of parallelogram $A = bh$

Area of triangle $A = \tfrac{1}{2} hb$

Right triangle (Pythagoras) $c^2 = a^2 + b^2$

Sphere: surface area $A = 4\pi r^2$
        volume $V = \tfrac{4}{3} \pi r^3$

Rectangular solid: volume $V = \ell w h$

Cylinder (right):
   surface area $A = 2\pi r \ell + 2\pi r^2$
   volume $V = \pi r^2 \ell$

Right circular cone:
   surface area $A = \pi r^2 + \pi r \sqrt{r^2 + h^2}$
   volume $V = \tfrac{1}{3} \pi r^2 h$

Exponents [See Appendix A–2 for details]

$(a^n)(a^m) = a^{n+m}$ [Example: $(a^3)(a^2) = a^5$]
$(a^n)(b^n) = (ab)^n$ [Example: $(a^3)(b^3) = (ab)^3$]
$(a^n)^m = a^{nm}$ [Example: $(a^3)^2 = a^6$]
             [Example: $(a^{\frac{1}{4}})^4 = a$]

$a^{-1} = \dfrac{1}{a}$ $a^{-n} = \dfrac{1}{a^n}$ $a^0 = 1$

$a^{\frac{1}{2}} = \sqrt{a}$ $a^{\frac{1}{4}} = \sqrt{\sqrt{a}}$

$(a^n)(a^{-m}) = \dfrac{a^n}{a^m} = a^{n-m}$ [Ex.: $(a^5)(a^{-2}) = a^3$]

$\dfrac{a^n}{b^n} = \left(\dfrac{a}{b}\right)^n$

Quadratic Formula [Appendix A–4]

Equation with unknown x, in the form
$$ax^2 + bx + c = 0,$$
has solutions
$$x = \dfrac{-b \pm \sqrt{b^2 - 4ac}}{2a}.$$

Logarithms [Appendix A–8; Table p. A-11]

If $y = 10^x$,   then $x = \log_{10} y = \log y$.
If $y = e^x$,   then $x = \log_e y = \ln y$.
$\log(ab) = \log a + \log b$
$\log\left(\dfrac{a}{b}\right) = \log a - \log b$
$\log a^n = n \log a$

Binomial Expansion [Appendix A–5]

$(1 + x)^n = 1 + nx + \dfrac{n(n-1)}{2 \cdot 1} x^2 + \dfrac{n(n-1)(n-2)}{3 \cdot 2 \cdot 1} x^3 + \cdots$ [for $x^2 < 1$]

$\approx 1 + nx$ if $x \ll 1$

[Example: $(1 + 0.01)^3 \approx 1.03$]

[Example: $\dfrac{1}{\sqrt{0.99}} = \dfrac{1}{\sqrt{1 - 0.01}} = (1 - 0.01)^{-\frac{1}{2}} \approx 1 - (-\tfrac{1}{2})(0.01) \approx 1.005$]

Fractions

$\dfrac{a}{b} = \dfrac{c}{d}$ is the same as $ad = bc$

$\dfrac{\left(\dfrac{a}{b}\right)}{\left(\dfrac{c}{d}\right)} = \dfrac{ad}{bc}$

Trigonometric Formulas [Appendix A–7]

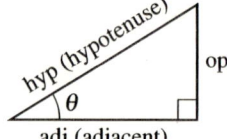

$\sin \theta = \dfrac{\text{opp}}{\text{hyp}}$

$\cos \theta = \dfrac{\text{adj}}{\text{hyp}}$

$\tan \theta = \dfrac{\text{opp}}{\text{adj}}$

$\text{adj}^2 + \text{opp}^2 = \text{hyp}^2$ (Pythagorean theorem)

$\tan \theta = \dfrac{\sin \theta}{\cos \theta}$

$\sin^2 \theta + \cos^2 \theta = 1$

$\sin 2\theta = 2 \sin \theta \cos \theta$

$\cos 2\theta = (\cos^2 \theta - \sin^2 \theta) = (1 - 2\sin^2 \theta) = (2\cos^2 \theta - 1)$

$\sin(180° - \theta) = \sin \theta$ $\cos(180° - \theta) = -\cos \theta$
$\sin(90° - \theta) = \cos \theta$
$\cos(90° - \theta) = \sin \theta$
$\sin \tfrac{1}{2} \theta = \sqrt{(1 - \cos \theta)/2}$ $\cos \tfrac{1}{2} \theta = \sqrt{(1 + \cos \theta)/2}$
$\sin \theta \approx \theta$ [for small $\theta \lesssim 0.2$ rad]
$\cos \theta \approx 1 - \dfrac{\theta^2}{2}$ [for small $\theta \lesssim 0.2$ rad]
$\sin(A \pm B) = \sin A \cos B \pm \cos A \sin B$
$\cos(A \pm B) = \cos A \cos B \mp \sin A \sin B$

For any triangle:
$c^2 = a^2 + b^2 - 2ab \cos \gamma$ (law of cosines)

$\dfrac{\sin \alpha}{a} = \dfrac{\sin \beta}{b} = \dfrac{\sin \gamma}{c}$ (law of sines)

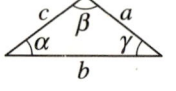